P. Fellows

D1742145

JOHNSTON-MACCO
Schlumberger

THE TECHNOLOGY OF ARTIFICIAL LIFT METHODS
Volume 3a

THE TECHNOLOGY OF ARTIFICIAL LIFT METHODS

Volume 3a
Pressure Gradient Curves

Vertical Multiphase Tubing Flow
Horizontal Multiphase Pipe Flow

Kermit E. Brown

Eduardo Proaño
James P. Brill
Joe Mach
Dana Turner
Eni P. Kanu

PennWell Books

Division of
PennWell Publishing Company

Tulsa, Oklahoma

To the many fine individuals
of our great Petroleum Industry who
will find need for this book

Copyright © 1980 by
PennWell Publishing Company
1421 South Sheridan Road/P. O. Box 1260
Tulsa, Oklahoma 74101

Library of Congress cataloging in publication data

All rights reserved. No part of this book may be
reproduced, stored in a retrieval system, or
transcribed in any form or by any means, electronic
or mechanical, including photocopying and recording,
without the prior written permission of the publisher.

Printed in the United States of America

1 2 3 4 5   84 83 82 81 80

Library of Congress Cataloging in Publication Data (Revised)

Brown, Kermit E
  The technology of artificial lift methods.

  Includes bibliographical references.
  CONTENTS: v. 1. Inflow performance, multiphase
flow in pipes, the flowing well.—[etc.]—3a.
Pressure gradient curves, vertical multiphase
tubing flow, horizontal multiphase pipe flow.
  1. Oil wells—Artificial lift. I. Title.
TN871.B819   622'.33'8   76-53201
ISBN 0-87814-031-X

Preface

Volume 3 consists of many gradient curves to assist in production systems analysis and the design of artificial lift installations. Because of the large number of curves, the book is in two volumes, 3a and 3b. The following groups of curves are presented and their corresponding volume is noted:

Curves	Section	Volume
Vertical Multiphase Tubing Flow	A	3a
Horizontal Multiphase Pipe Flow	B	3a
Vertical Multiphase Annular Flow	C	3b
Vertical Tubing Gas Production	D	3b
Vertical Gas Injection	E	3b
Vertical Tubing Water Injection	F	3b

Discussions concerning the correlations used and example problems can be found for those curves located in the respective texts. Also, refer to volumes 1 and 2 of this series for detailed specific examples in using these curves.

We want to express our sincere appreciation to Johnston-Macco Schlumberger for furnishing all of the curves which were computer-plotted in their Houston office.

Acknowledgement is extended to all the individuals that made this book possible and their respective companies.

Eduardo Proaño	— Johnston-Macco Schlumberger
James P. Brill	— University of Tulsa
Joe Mach	— Johnston-Macco Schlumberger
Dana Turner	— Johnston-Macco Schlumberger
Eni P. Kanu	— Johnston-Macco Schlumberger

Vertical Multiphase Tubing Flow

The vertical multiphase tubing flow curves were prepared using the Hagedorn and Brown correlation. This decision was made after observing results from hundreds of field cases; this correlation more consistently gave the best results. Tubing sizes from 1.049 in. through 12 in. have been included.

Note that curves for vertical flow are available for 10%, 50%, and 100% oil. For other oil percentages, interpolation is permissible.

Refer to Volume I (pp. 160–167) for numerous example problems that can be worked using vertical multiphase flow curves. Two very basic examples are presented here.

Example Problem A-1
How to Find the Flowing Bottom Hole Pressure

Given Data: 2-⅞-in. OD tubing (2.441-in. ID)
                 Liquid flow rate = 1,000 b/d
                 (50% water) Depth = 12,000 ft
                 Producing GOR = 800 scf/bbl
                 Producing GLR = 800/2 = 400 scf/bbl
                 Wellhead pressure = 160 psi

Find the flowing bottom hole pressure.

Solution:

1. Find the equivalent depth corresponding to 160 psi wellhead pressure. To do this, proceed vertically downward from 160 psi at zero depth until intersecting the 400 scf/bbl GLR line. This is at a depth of 1400 ft. Note the pressure scale is in 80-psi increments and the depth scale is in 200-ft increments.
2. Add the equivalent depth of 1,400 ft to the well depth of 12,000 ft and obtain 13,400 ft.
3. From 13,400 ft on the vertical scale, proceed horizontally to the 400 scf/bbl line and read a flowing pressure of 3,360 psi.

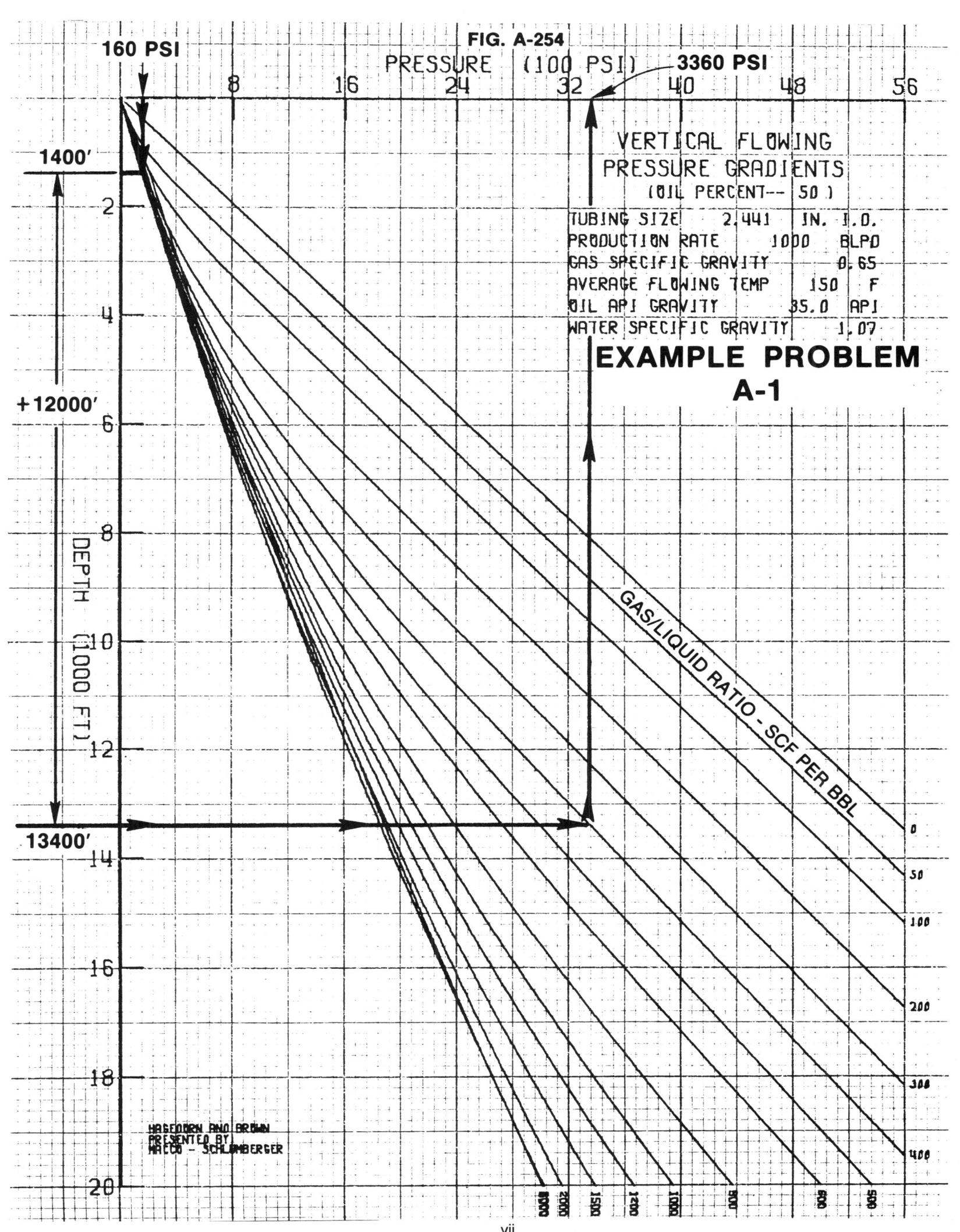

FIG. A-254

VERTICAL FLOWING
PRESSURE GRADIENTS
(OIL PERCENT— 50)

TUBING SIZE	2.441	IN. I.D.
PRODUCTION RATE	1000	BLPD
GAS SPECIFIC GRAVITY	0.65	
AVERAGE FLOWING TEMP	150	F
OIL API GRAVITY	35.0	API
WATER SPECIFIC GRAVITY	1.07	

EXAMPLE PROBLEM
A-1

GAS/LIQUID RATIO - SCF PER BBL

HAGEDORN AND BROWN
PRESENTED BY
MAECO - SCHLUMBERGER

Example Problem A-2
How to Find the Flowing Wellhead Pressure

Given Data: Tubing ID $= 2.441$
Liquid flow rate $= 1{,}000$ b/d (50% water)
Depth $= 12{,}000$ ft
GOR $= 800$ scf/bbl
GLR $= 400$ scf/bbl
Static pressure $= 4{,}000$ psi
Productivity index $= 5$ b/d psi (assume constant)

Find the flowing wellhead pressure.

Solution:

1. Calculate the flowing bottom hole pressure or obtain it from an inflow performance curve.
   For a constant PI

   $$P_{wf} = P_r - \frac{q}{J} = 4{,}000 - \frac{1{,}000}{5} = 3{,}800 \text{ psi}$$

2. Locate 3,800 psi on the 400 scf/bbl line and note a depth of 14,600 ft.
3. Subtract the well depth of 12,000 ft from the depth of 14,600 ft to obtain 2600 ft.
4. From 2600 ft on the graph, proceed horizontally to the 400 scf/bbl line and read 320 psi. This is the permissible wellhead pressure.

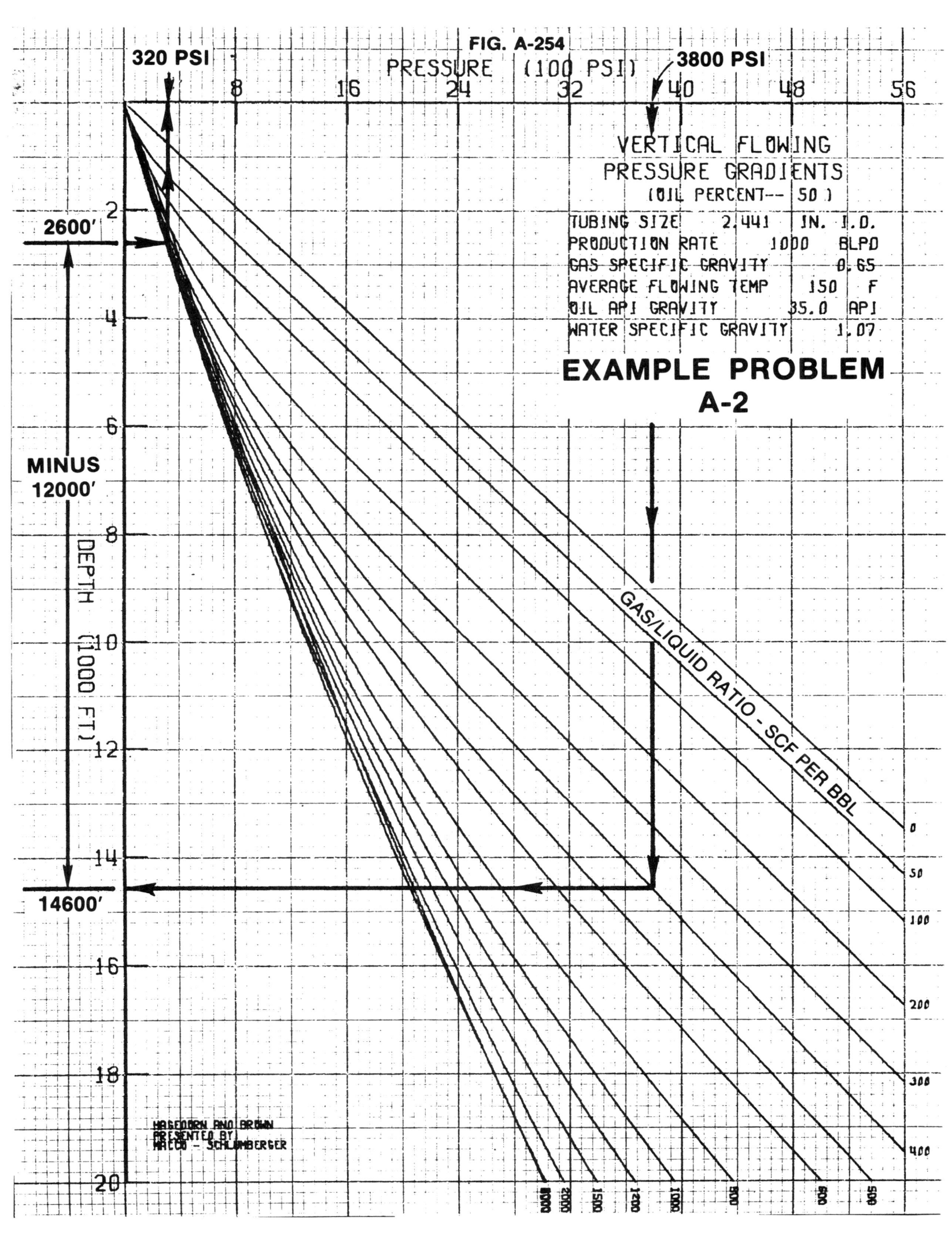

FIG. A-254

PRESSURE (100 PSI)

VERTICAL FLOWING
PRESSURE GRADIENTS
(OIL PERCENT-- 50)

TUBING SIZE   2.441   IN. I.D.
PRODUCTION RATE   1000   BLPD
GAS SPECIFIC GRAVITY   0.65
AVERAGE FLOWING TEMP   150   F
OIL API GRAVITY   35.0   API
WATER SPECIFIC GRAVITY   1.07

**EXAMPLE PROBLEM
A-2**

320 PSI

3800 PSI

2600'

MINUS
12000'

14600'

DEPTH (1000 FT)

GAS/LIQUID RATIO - SCF PER BBL

0
50
100
200
300
400

HAGEDORN AND BROWN
PRESENTED BY
MAECO - SCHLUMBERGER

Vertical Multiphase Tubing Flow

Pipe Size, in.	Percent Oil	Rate, b/d	Figure Number	Page Number	Pipe Size, in.	Percent Oil	Rate, b/d	Figure Number	Page Number
1.049	10	25	A-1	1	1.380	10	200	A-49	49
1.049	50	25	A-2	2	1.380	50	200	A-50	50
1.049	100	25	A-3	3	1.380	100	200	A-51	51
1.049	10	50	A-4	4	1.380	10	250	A-52	52
1.049	50	50	A-5	5	1.380	50	250	A-53	53
1.049	100	50	A-6	6	1.380	100	250	A-54	54
1.049	10	100	A-7	7	1.380	10	300	A-55	55
1.049	50	100	A-8	8	1.380	50	300	A-56	56
1.049	100	100	A-9	9	1.380	100	300	A-57	57
1.049	10	150	A-10	10	1.380	10	400	A-58	58
1.049	50	150	A-11	11	1.380	50	400	A-59	59
1.049	100	150	A-12	12	1.380	100	400	A-60	60
1.049	10	200	A-13	13	1.380	10	500	A-61	61
1.049	50	200	A-14	14	1.380	50	500	A-62	62
1.049	100	200	A-15	15	1.380	100	500	A-63	63
1.049	10	250	A-16	16	1.380	10	600	A-64	64
1.049	50	250	A-17	17	1.380	50	600	A-65	65
1.049	100	250	A-18	18	1.380	100	600	A-66	66
1.049	10	300	A-19	19	1.380	10	700	A-67	67
1.049	50	300	A-20	20	1.380	50	700	A-68	68
1.049	100	300	A-21	21	1.380	100	700	A-69	69
1.049	10	400	A-22	22	1.380	10	800	A-70	70
1.049	50	400	A-23	23	1.380	50	800	A-71	71
1.049	100	400	A-24	24	1.380	100	800	A-72	72
1.049	10	500	A-25	25	1.380	10	1,000	A-73	73
1.049	50	500	A-26	26	1.380	50	1,000	A-74	74
1.049	100	500	A-27	27	1.380	100	1,000	A-75	75
1.049	10	600	A-28	28	1.380	10	1,200	A-76	76
1.049	50	600	A-29	29	1.380	50	1,200	A-77	77
1.049	100	600	A-30	30	1.380	100	1,200	A-78	78
1.049	10	700	A-31	31	1.610	10	25	A-79	79
1.049	50	700	A-32	32	1.610	50	25	A-80	80
1.049	100	700	A-33	33	1.610	100	25	A-81	81
1.049	10	800	A-34	34	1.610	10	50	A-82	82
1.049	50	800	A-35	35	1.610	50	50	A-83	83
1.049	100	800	A-36	36	1.610	100	50	A-84	84
1.380	10	25	A-37	37	1.610	10	100	A-85	85
1.380	50	25	A-38	38	1.610	50	100	A-86	86
1.380	100	25	A-39	39	1.610	100	100	A-87	87
1.380	10	50	A-40	40	1.610	10	150	A-88	88
1.380	50	50	A-41	41	1.610	50	150	A-89	89
1.380	100	50	A-42	42	1.610	100	150	A-90	90
1.380	10	100	A-43	43	1.610	10	200	A-91	91
1.380	50	100	A-44	44	1.610	50	200	A-92	92
1.380	100	100	A-45	45	1.610	100	200	A-93	93
1.380	10	150	A-46	46	1.610	10	250	A-94	94
1.380	50	150	A-47	47	1.610	50	250	A-95	95
1.380	100	150	A-48	48	1.610	100	250	A-96	96

Pipe Size, in.	Percent Oil	Rate, b/d	Figure Number	Page Number	Pipe Size, in.	Percent Oil	Rate, b/d	Figure Number	Page Number
1.610	10	300	A-97	97	1.751	10	600	A-145	145
1.610	50	300	A-98	98	1.751	50	600	A-146	146
1.610	100	300	A-99	99	1.751	100	600	A-147	147
1.610	10	400	A-100	100	1.751	10	700	A-148	148
1.610	50	400	A-101	101	1.751	50	700	A-149	149
1.610	100	400	A-102	102	1.751	100	700	A-150	150
1.610	10	500	A-103	103	1.751	10	800	A-151	151
1.610	50	500	A-104	104	1.751	50	800	A-152	152
1.610	100	500	A-105	105	1.751	100	800	A-153	153
1.610	10	600	A-106	106	1.751	10	900	A-154	154
1.610	50	600	A-107	107	1.751	50	900	A-155	155
1.610	100	600	A-108	108	1.751	100	900	A-156	156
1.610	10	700	A-109	109	1.751	10	1,000	A-157	157
1.610	50	700	A-110	110	1.751	50	1,000	A-158	158
1.610	100	700	A-111	111	1.751	100	1,000	A-159	159
1.610	10	800	A-112	112	1.751	10	1,200	A-160	160
1.610	50	800	A-113	113	1.751	50	1,200	A-161	161
1.610	100	800	A-114	114	1.751	100	1,200	A-162	162
1.610	10	1,000	A-115	115	1.751	10	1,500	A-163	163
1.610	50	1,000	A-116	116	1.751	50	1,500	A-164	164
1.610	100	1,000	A-117	117	1.751	100	1,500	A-165	165
1.610	10	1,200	A-118	118	1.751	10	2,000	A-166	166
1.610	50	1,200	A-119	119	1.751	50	2,000	A-167	167
1.610	100	1,200	A-120	120	1.751	100	2,000	A-168	168
1.610	10	1,500	A-121	121	1.751	10	2,500	A-169	169
1.610	50	1,500	A-122	122	1.751	50	2,500	A-170	170
1.610	100	1,500	A-123	123	1.751	100	2,500	A-171	171
1.610	10	2,000	A-124	124	1.751	10	3,000	A-172	172
1.610	50	2,000	A-125	125	1.751	50	3,000	A-173	173
1.610	100	2,000	A-126	126	1.751	100	3,000	A-174	174
1.751	10	50	A-127	127	1.995	10	50	A-175	175
1.751	50	50	A-128	128	1.995	50	50	A-176	176
1.751	100	50	A-129	129	1.995	100	50	A-177	177
1.751	10	100	A-130	130	1.995	10	100	A-178	178
1.751	50	100	A-131	131	1.995	50	100	A-179	179
1.751	100	100	A-132	132	1.995	100	100	A-180	180
1.751	10	200	A-133	133	1.995	10	200	A-181	181
1.751	50	200	A-134	134	1.995	50	200	A-182	182
1.751	100	200	A-135	135	1.995	100	200	A-183	183
1.751	10	300	A-136	136	1.995	10	300	A-184	184
1.751	50	300	A-137	137	1.995	50	300	A-185	185
1.751	100	300	A-138	138	1.995	100	300	A-186	186
1.751	10	400	A-139	139	1.995	10	400	A-187	187
1.751	50	400	A-140	140	1.995	50	400	A-188	188
1.751	100	400	A-141	141	1.995	100	400	A-189	189
1.751	10	500	A-142	142	1.995	10	500	A-190	190
1.751	50	500	A-143	143	1.995	50	500	A-191	191
1.751	100	500	A-144	144	1.995	100	500	A-192	192

Pipe Size, in.	Percent Oil	Rate, b/d	Figure Number	Page Number	Pipe Size, in.	Percent Oil	Rate, b/d	Figure Number	Page Number
1.995	10	600	A-193	193	2.441	10	600	A-241	241
1.995	50	600	A-194	194	2.441	50	600	A-242	242
1.995	100	600	A-195	195	2.441	100	600	A-243	243
1.995	10	700	A-196	196	2.441	10	700	A-244	244
1.995	50	700	A-197	197	2.441	50	700	A-245	245
1.995	100	700	A-198	198	2.441	100	700	A-246	246
1.995	10	800	A-199	199	2.441	10	800	A-247	247
1.995	50	800	A-200	200	2.441	50	800	A-248	248
1.995	100	800	A-201	201	2.441	100	800	A-249	249
1.995	10	900	A-202	202	2.441	10	900	A-250	250
1.995	50	900	A-203	203	2.441	50	900	A-251	251
1.995	100	900	A-204	204	2.441	100	900	A-252	252
1.995	10	1,000	A-205	205	2.441	10	1,000	A-253	253
1.995	50	1,000	A-206	206	2.441	50	1,000	A-254	254
1.995	100	1,000	A-207	207	2.441	100	1,000	A-255	255
1.995	10	1,200	A-208	208	2.441	10	1,200	A-256	256
1.995	50	1,200	A-209	209	2.441	50	1,200	A-257	257
1.995	100	1,200	A-210	210	2.441	100	1,200	A-258	258
1.995	10	1,500	A-211	211	2.441	10	1,500	A-259	259
1.995	50	1,500	A-212	212	2.441	50	1,500	A-260	260
1.995	100	1,500	A-213	213	2.441	100	1,500	A-261	261
1.995	10	2,000	A-214	214	2.441	10	2,000	A-262	262
1.995	50	2,000	A-215	215	2.441	50	2,000	A-263	263
1.995	100	2,000	A-216	216	2.441	100	2,000	A-264	264
1.995	10	2,500	A-217	217	2.441	10	2,500	A-265	265
1.995	50	2,500	A-218	218	2.441	50	2,500	A-266	266
1.995	100	2,500	A-219	219	2.441	100	2,500	A-267	267
1.995	10	3,000	A-220	220	2.441	10	3,000	A-268	268
1.995	50	3,000	A-221	221	2.441	50	3,000	A-269	269
1.995	100	3,000	A-222	222	2.441	100	3,000	A-270	270
2.441	10	50	A-223	223	2.441	10	4,000	A-271	271
2.441	50	50	A-224	224	2.441	50	4,000	A-272	272
2.441	100	50	A-225	225	2.441	100	4,000	A-273	273
2.441	10	100	A-226	226	2.441	10	5,000	A-274	274
2.441	50	100	A-227	227	2.441	50	5,000	A-275	275
2.441	100	100	A-228	228	2.441	100	5,000	A-276	276
2.441	10	200	A-229	229	2.992	10	100	A-277	277
2.441	50	200	A-230	230	2.992	50	100	A-278	278
2.441	100	200	A-231	231	2.992	100	100	A-279	279
2.441	10	300	A-232	232	2.992	10	200	A-280	280
2.441	50	300	A-233	233	2.992	50	200	A-281	281
2.441	100	300	A-234	234	2.992	100	200	A-282	282
2.441	10	400	A-235	235	2.992	10	300	A-283	283
2.441	50	400	A-236	236	2.992	50	300	A-284	284
2.441	100	400	A-237	237	2.992	100	300	A-285	285
2.441	10	500	A-238	238	2.992	10	400	A-286	286
2.441	50	500	A-239	239	2.992	50	400	A-287	287
2.441	100	500	A-240	240	2.992	100	400	A-288	288

Pipe Size, in.	Percent Oil	Rate, b/d	Figure Number	Page Number	Pipe Size, in.	Percent Oil	Rate, b/d	Figure Number	Page Number
2.992	10	500	A-289	289	3.476	10	100	A-337	337
2.992	50	500	A-290	290	3.476	50	100	A-338	338
2.992	100	500	A-291	291	3.476	100	100	A-339	339
2.992	10	600	A-292	292	3.476	10	200	A-340	340
2.992	50	600	A-293	293	3.476	50	200	A-341	341
2.992	100	600	A-294	294	3.476	100	200	A-342	342
2.992	10	700	A-295	295	3.476	10	300	A-343	343
2.992	50	700	A-296	296	3.476	50	300	A-344	344
2.992	100	700	A-297	297	3.476	100	300	A-345	345
2.992	10	800	A-298	298	3.476	10	400	A-346	346
2.992	50	800	A-299	299	3.476	50	400	A-347	347
2.992	100	800	A-300	300	3.476	100	400	A-348	348
2.992	10	900	A-301	301	3.476	10	500	A-349	349
2.992	50	900	A-302	302	3.476	50	500	A-350	350
2.992	100	900	A-303	303	3.476	100	500	A-351	351
2.992	10	1,000	A-304	304	3.476	10	600	A-352	352
2.992	50	1,000	A-305	305	3.476	50	600	A-353	353
2.992	100	1,000	A-306	306	3.476	100	600	A-354	354
2.992	10	1,200	A-307	307	3.476	10	700	A-355	355
2.992	50	1,200	A-308	308	3.476	50	700	A-356	356
2.992	100	1,200	A-309	309	3.476	100	700	A-357	357
2.992	10	1,500	A-310	310	3.476	10	800	A-358	358
2.992	50	1,500	A-311	311	3.476	50	800	A-359	359
2.992	100	1,500	A-312	312	3.476	100	800	A-360	360
2.992	10	2,000	A-313	313	3.476	10	900	A-361	361
2.992	50	2,000	A-314	314	3.476	50	900	A-362	362
2.992	100	2,000	A-315	315	3.476	100	900	A-363	363
2.992	10	2,500	A-316	316	3.476	10	1,000	A-364	364
2.992	50	2,500	A-317	317	3.476	50	1,000	A-365	365
2.992	100	2,500	A-318	318	3.476	100	1,000	A-366	366
2.992	10	3,000	A-319	319	3.476	10	1,200	A-367	367
2.992	50	3,000	A-320	320	3.476	50	1,200	A-368	368
2.992	100	3,000	A-321	321	3.476	100	1,200	A-369	369
2.992	10	4,000	A-322	322	3.476	10	1,500	A-370	370
2.992	50	4,000	A-323	323	3.476	50	1,500	A-371	371
2.992	100	4,000	A-324	324	3.476	100	1,500	A-372	372
2.992	10	5,000	A-325	325	3.476	10	2,000	A-373	373
2.992	50	5,000	A-326	326	3.476	50	2,000	A-374	374
2.992	100	5,000	A-327	327	3.476	100	2,000	A-375	375
2.992	10	6,000	A-328	328	3.476	10	2,500	A-376	376
2.992	50	6,000	A-329	329	3.476	50	2,500	A-377	377
2.992	100	6,000	A-330	330	3.476	100	2,500	A-378	378
2.992	10	8,000	A-331	331	3.476	10	3,000	A-379	379
2.992	50	8,000	A-332	332	3.476	50	3,000	A-380	380
2.992	100	8,000	A-333	333	3.476	100	3,000	A-381	381
2.992	10	10,000	A-334	334	3.476	10	4,000	A-382	382
2.992	50	10,000	A-335	335	3.476	50	4,000	A-383	383
2.992	100	10,000	A-336	336	3.476	100	4,000	A-384	384

Pipe Size, in.	Percent Oil	Rate, b/d	Figure Number	Page Number	Pipe Size, in.	Percent Oil	Rate, b/d	Figure Number	Page Number
3.476	10	5,000	A-385	385	3.958	10	1,200	A-433	433
3.476	50	5,000	A-386	386	3.958	50	1,200	A-434	434
3.476	100	5,000	A-387	387	3.958	100	1,200	A-435	435
3.476	10	6,000	A-388	388	3.958	10	1,500	A-436	436
3.476	50	6,000	A-389	389	3.958	50	1,500	A-437	437
3.476	100	6,000	A-390	390	3.958	100	1,500	A-438	438
3.476	10	8,000	A-391	391	3.958	10	2,000	A-439	439
3.476	50	8,000	A-392	392	3.958	50	2,000	A-440	440
3.476	100	8,000	A-393	393	3.958	100	2,000	A-441	441
3.476	10	10,000	A-394	394	3.958	10	2,500	A-442	442
3.476	50	10,000	A-395	395	3.958	50	2,500	A-443	443
3.476	100	10,000	A-396	396	3.958	100	2,500	A-444	444
3.476	10	15,000	A-397	397	3.958	10	3,000	A-445	445
3.476	50	15,000	A-398	398	3.958	50	3,000	A-446	446
3.476	100	15,000	A-399	399	3.958	100	3,000	A-447	447
3.476	10	20,000	A-400	400	3.958	10	4,000	A-448	448
3.476	50	20,000	A-401	401	3.958	50	4,000	A-449	449
3.476	100	20,000	A-402	402	3.958	100	4,000	A-450	450
3.958	10	100	A-403	403	3.958	10	5,000	A-451	451
3.958	50	100	A-404	404	3.958	50	5,000	A-452	452
3.958	100	100	A-405	405	3.958	100	5,000	A-453	453
3.958	10	200	A-406	406	3.958	10	6,000	A-454	454
3.958	50	200	A-407	407	3.958	50	6,000	A-455	455
3.958	100	200	A-408	408	3.958	100	6,000	A-456	456
3.958	10	300	A-409	409	3.958	10	8,000	A-457	457
3.958	50	300	A-410	410	3.958	50	8,000	A-458	458
3.958	100	300	A-411	411	3.958	100	8,000	A-459	459
3.958	10	400	A-412	412	3.958	10	10,000	A-460	460
3.958	50	400	A-413	413	3.958	50	10,000	A-461	461
3.958	100	400	A-414	414	3.958	100	10,000	A-462	462
3.958	10	500	A-415	415	3.958	10	15,000	A-463	463
3.958	50	500	A-416	416	3.958	50	15,000	A-464	464
3.958	100	500	A-417	417	3.958	100	15,000	A-465	465
3.958	10	600	A-418	418	3.958	10	20,000	A-466	466
3.958	50	600	A-419	419	3.958	50	20,000	A-467	467
3.958	100	600	A-420	420	3.958	100	20,000	A-468	468
3.958	10	700	A-421	421	3.958	10	30,000	A-469	469
3.958	50	700	A-422	422	3.958	50	30,000	A-470	470
3.958	100	700	A-423	423	3.958	100	30,000	A-471	471
3.958	10	800	A-424	424	4.494	10	1,000	A-472	472
3.958	50	800	A-425	425	4.494	50	1,000	A-473	473
3.958	100	800	A-426	426	4.494	100	1,000	A-474	474
3.958	10	900	A-427	427	4.494	10	2,000	A-475	475
3.958	50	900	A-428	428	4.494	50	2,000	A-476	476
3.958	100	900	A-429	429	4.494	100	2,000	A-477	477
3.958	10	1,000	A-430	430	4.494	10	3,000	A-478	478
3.958	50	1,000	A-431	431	4.494	50	3,000	A-479	479
3.958	100	1,000	A-432	432	4.494	100	3,000	A-480	480

Pipe Size, in.	Percent Oil	Rate, b/d	Figure Number	Page Number	Pipe Size, in.	Percent Oil	Rate, b/d	Figure Number	Page Number
5.921	10	50,000	A-577	577	6.366	10	60,000	A-625	625
5.921	50	50,000	A-578	578	6.366	50	60,000	A-626	626
5.921	100	50,000	A-579	579	6.366	100	60,000	A-627	627
5.921	10	60,000	A-580	580	6.366	10	70,000	A-628	628
5.921	50	60,000	A-581	581	6.366	50	70,000	A-629	629
5.921	100	60,000	A-582	582	6.366	100	70,000	A-630	630
5.921	10	70,000	A-583	583	6.366	10	80,000	A-631	631
5.921	50	70,000	A-584	584	6.366	50	80,000	A-632	632
5.921	100	70,000	A-585	585	6.366	100	80,000	A-633	633
5.921	10	80,000	A-586	586	8.921	10	1,000	A-634	634
5.921	50	80,000	A-587	587	8.921	50	1,000	A-635	635
5.921	100	80,000	A-588	588	8.921	100	1,000	A-636	636
6.366	10	1,000	A-589	589	8.921	10	2,000	A-637	637
6.366	50	1,000	A-590	590	8.921	50	2,000	A-638	638
6.366	100	1,000	A-591	591	8.921	100	2,000	A-639	639
6.366	10	2,000	A-592	592	8.921	10	3,000	A-640	640
6.366	50	2,000	A-593	593	8.921	50	3,000	A-641	641
6.366	100	2,000	A-594	594	8.921	100	3,000	A-642	642
6.366	10	3,000	A-595	595	8.921	10	4,000	A-643	643
6.366	50	3,000	A-596	596	8.921	50	4,000	A-644	644
6.366	100	3,000	A-597	597	8.921	100	4,000	A-645	645
6.366	10	4,000	A-598	598	8.921	10	5,000	A-646	646
6.366	50	4,000	A-599	599	8.921	50	5,000	A-647	647
6.366	100	4,000	A-600	600	8.921	100	5,000	A-648	648
6.366	10	5,000	A-601	601	8.921	10	6,000	A-649	649
6.366	50	5,000	A-602	602	8.921	50	6,000	A-650	650
6.366	100	5,000	A-603	603	8.921	100	6,000	A-651	651
6.366	10	6,000	A-604	604	8.921	10	8,000	A-652	652
6.366	50	6,000	A-605	605	8.921	50	8,000	A-653	653
6.366	100	6,000	A-606	606	8.921	100	8,000	A-654	654
6.366	10	8,000	A-607	607	8.921	10	10,000	A-655	655
6.366	50	8,000	A-608	608	8.921	50	10,000	A-656	656
6.366	100	8,000	A-609	609	8.921	100	10,000	A-657	657
6.366	10	10,000	A-610	610	8.921	10	15,000	A-658	658
6.366	50	10,000	A-611	611	8.921	50	15,000	A-659	659
6.366	100	10,000	A-612	612	8.921	100	15,000	A-660	660
6.366	10	15,000	A-613	613	8.921	10	20,000	A-661	661
6.366	50	15,000	A-614	614	8.921	50	20,000	A-662	662
6.366	100	15,000	A-615	615	8.921	100	20,000	A-663	663
6.366	10	20,000	A-616	616	8.921	10	30,000	A-664	664
6.366	50	20,000	A-617	617	8.921	50	30,000	A-665	665
6.366	100	20,000	A-618	618	8.921	100	30,000	A-666	666
6.366	10	30,000	A-619	619	8.921	10	50,000	A-667	667
6.366	50	30,000	A-620	620	8.921	50	50,000	A-668	668
6.366	100	30,000	A-621	621	8.921	100	50,000	A-669	669
6.366	10	50,000	A-622	622	8.921	10	60,000	A-670	670
6.366	50	50,000	A-623	623	8.921	50	60,000	A-671	671
6.366	100	50,000	A-624	624	8.921	100	60,000	A-672	672

Section B

Horizontal Multiphase Pipe Flow

The horizontal multiphase flow curves were prepared by Johnston–Macco Schlumberger using the correlation of Beggs and Brill. The Beggs and Brill correlation for horizontal flow is the most accurate, as well as predicting very consistent results for a wide range of conditions. Pipe sizes from 2 in. through 12 in. are included and curves were prepared for 100% oil and 100% water flow. Interpolation is permissible for various water percentages although no allowance has been made for the formation of emulsions.

Refer to Volume I (pp. 191–196) for numerous example problems that can be worked using horizontal multiphase flow curves. One very basic example problem is worked here.

Example Problem B-1

How to Find the Flowing Wellhead Pressure

Given Data: 2½-in. ID flowline
Length = 6,000 ft
Rate = 1,500 b/d (100% oil)
GOR = 800 scf/bbl
Separator pressure = 100 psi

Find the required wellhead pressure.

Solution:

1. Find the equivalent length corresponding to 100 psi separator pressure. Locate 100 psi at zero length and proceed vertically downwards until intersecting the 800 scf/bbl gas-oil ratio line at a length of 600 ft.
2. Add the equivalent length of 600 ft to the line length of 6,000 ft and obtain 6,600 ft.
3. From 6,600 ft on Fig. B-1, proceed horizontally to the 800 scf/bbl line and read a flowing wellhead pressure of 490 psi.

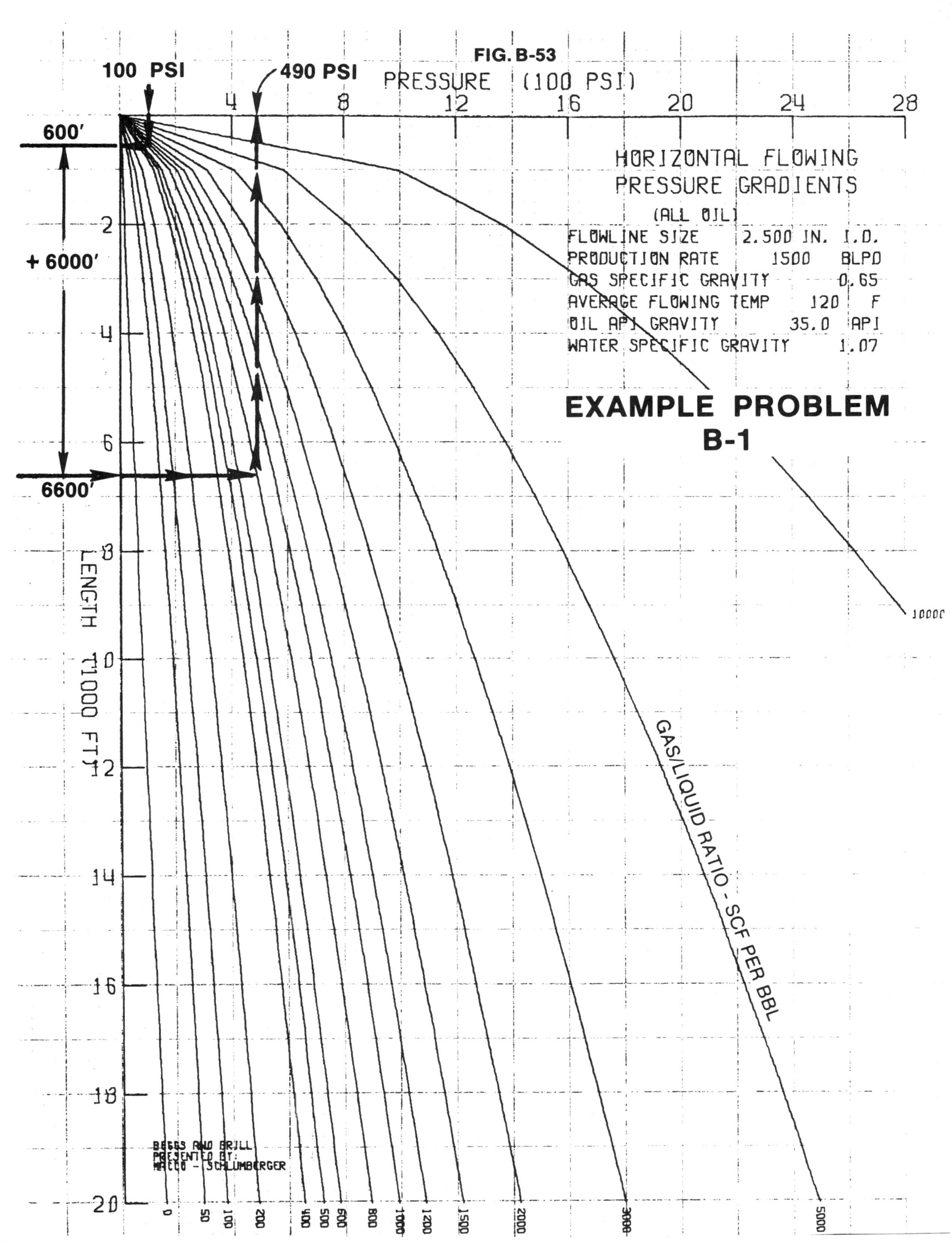

FIG. B-53

HORIZONTAL FLOWING
PRESSURE GRADIENTS
(ALL OIL)

FLOWLINE SIZE	2.500 IN. I.D.
PRODUCTION RATE	1500 BLPD
GAS SPECIFIC GRAVITY	0.65
AVERAGE FLOWING TEMP	120 F
OIL API GRAVITY	35.0 API
WATER SPECIFIC GRAVITY	1.07

EXAMPLE PROBLEM
B-1

GAS/LIQUID RATIO - SCF PER BBL

BEGGS AND BRILL
PRESENTED BY:
HALCO - SCHLUMBERGER

Horizontal Multiphase Pipe Flow

Pipe Size, in.	Percent Oil	Rate, b/d	Figure Number	Page Number	Pipe Size, in.	Percent Oil	Rate, b/d	Figure Number	Page Number
2.000	100	100	B-1	732	2.500	100	1,000	B-49	780
2.000	0	100	B-2	733	2.500	0	1,000	B-50	781
2.000	100	200	B-3	734	2.500	100	1,200	B-51	782
2.000	0	200	B-4	735	2.500	0	1,200	B-52	783
2.000	100	300	B-5	736	2.500	100	1,500	B-53	784
2.000	0	300	B-6	737	2.500	0	1,500	B-54	785
2.000	100	400	B-7	738	2.500	100	2,000	B-55	786
2.000	0	400	B-8	739	2.500	0	2,000	B-56	787
2.000	100	500	B-9	740	2.500	100	2,500	B-57	788
2.000	0	500	B-10	741	2.500	0	2,500	B-58	789
2.000	100	600	B-11	742	2.500	100	3,000	B-59	790
2.000	0	600	B-12	743	2.500	0	3,000	B-60	791
2.000	100	700	B-13	744	2.500	100	4,000	B-61	792
2.000	0	700	B-14	745	2.500	0	4,000	B-62	793
2.000	100	800	B-15	746	2.500	100	5,000	B-63	794
2.000	0	800	B-16	747	2.500	0	5,000	B-64	795
2.000	100	900	B-17	748	3.000	100	200	B-65	796
2.000	0	900	B-18	749	3.000	0	200	B-66	797
2.000	100	1,000	B-19	750	3.000	100	400	B-67	798
2.000	0	1,000	B-20	751	3.000	0	400	B-68	799
2.000	100	1,200	B-21	752	3.000	100	600	B-69	800
2.000	0	1,200	B-22	753	3.000	0	600	B-70	801
2.000	100	1,500	B-23	754	3.000	100	800	B-71	802
2.000	0	1,500	B-24	755	3.000	0	800	B-72	803
2.000	100	2,000	B-25	756	3.000	100	1,000	B-73	804
2.000	0	2,000	B-26	757	3.000	0	1,000	B-74	805
2.000	100	2,500	B-27	758	3.000	100	1,500	B-75	806
2.000	0	2,500	B-28	759	3.000	0	1,500	B-76	807
2.000	100	3,000	B-29	760	3.000	100	2,000	B-77	808
2.000	0	3,000	B-30	761	3.000	0	2,000	B-78	809
2.500	100	100	B-31	762	3.000	100	3,000	B-79	810
2.500	0	100	B-32	763	3.000	0	3,000	B-80	811
2.500	100	200	B-33	764	3.000	100	4,000	B-81	812
2.500	0	200	B-34	765	3.000	0	4,000	B-82	813
2.500	100	300	B-35	766	3.000	100	5,000	B-83	814
2.500	0	300	B-36	767	3.000	0	5,000	B-84	815
2.500	100	400	B-37	768	3.000	100	6,000	B-85	816
2.500	0	400	B-38	769	3.000	0	6,000	B-86	817
2.500	100	500	B-39	770	3.000	100	8,000	B-87	818
2.500	0	500	B-40	771	3.000	0	8,000	B-88	819
2.500	100	600	B-41	772	3.000	100	10,000	B-89	820
2.500	0	600	B-42	773	3.000	0	10,000	B-90	821
2.500	100	700	B-43	774	3.500	100	200	B-91	822
2.500	0	700	B-44	775	3.500	0	200	B-92	823
2.500	100	800	B-45	776	3.500	100	400	B-93	824
2.500	0	800	B-46	777	3.500	0	400	B-94	825
2.500	100	900	B-47	778	3.500	100	600	B-95	826
2.500	0	900	B-48	779	3.500	0	600	B-96	827

Pipe Size, in.	Percent Oil	Rate, b/d	Figure Number	Page Number	Pipe Size, in.	Percent Oil	Rate, b/d	Figure Number	Page Number
3.500	100	800	B-97	828	4.000	100	15,000	B-145	876
3.500	0	800	B-98	829	4.000	0	15,000	B-146	877
3.500	100	1,000	B-99	830	4.000	100	20,000	B-147	878
3.500	0	1,000	B-100	831	4.000	0	20,000	B-148	879
3.500	100	1,500	B-101	832	4.000	100	30,000	B-149	880
3.500	0	1,500	B-102	833	4.000	0	30,000	B-150	881
3.500	100	2,000	B-103	834	5.000	100	400	B-151	882
3.500	0	2,000	B-104	835	5.000	0	400	B-152	883
3.500	100	3,000	B-105	836	5.000	100	600	B-153	884
3.500	0	3,000	B-106	837	5.000	0	600	B-154	885
3.500	100	4,000	B-107	838	5.000	100	800	B-155	886
3.500	0	4,000	B-108	839	5.000	0	800	B-156	887
3.500	100	5,000	B-109	840	5.000	100	1,000	B-157	888
3.500	0	5,000	B-110	841	5.000	0	1,000	B-158	889
3.500	100	6,000	B-111	842	5.000	100	1,500	B-159	890
3.500	0	6,000	B-112	843	5.000	0	1,500	B-160	891
3.500	100	8,000	B-113	844	5.000	100	2,000	B-161	892
3.500	0	8,000	B-114	845	5.000	0	2,000	B-162	893
3.500	100	10,000	B-115	846	5.000	100	3,000	B-163	894
3.500	0	10,000	B-116	847	5.000	0	3,000	B-164	895
3.500	100	15,000	B-117	848	5.000	100	4,000	B-165	896
3.500	0	15,000	B-118	849	5.000	0	4,000	B-166	897
3.500	100	20,000	B-119	850	5.000	100	5,000	B-167	898
3.500	0	20,000	B-120	851	5.000	0	5,000	B-168	899
4.000	100	400	B-121	852	5.000	100	6,000	B-169	900
4.000	0	400	B-122	853	5.000	0	6,000	B-170	901
4.000	100	600	B-123	854	5.000	100	8,000	B-171	902
4.000	0	600	B-124	855	5.000	0	8,000	B-172	903
4.000	100	800	B-125	856	5.000	100	10,000	B-173	904
4.000	0	800	B-126	857	5.000	0	10,000	B-174	905
4.000	100	1,000	B-127	858	5.000	100	15,000	B-175	906
4.000	0	1,000	B-128	859	5.000	0	15,000	B-176	907
4.000	100	1,500	B-129	860	5.000	100	20,000	B-177	908
4.000	0	1,500	B-130	861	5.000	0	20,000	B-178	909
4.000	100	2,000	B-131	862	5.000	100	30,000	B-179	910
4.000	0	2,000	B-132	863	5.000	0	30,000	B-180	911
4.000	100	3,000	B-133	864	5.000	100	50,000	B-181	912
4.000	0	3,000	B-134	865	5.000	0	50,000	B-182	913
4.000	100	4,000	B-135	866	6.000	100	400	B-183	914
4.000	0	4,000	B-136	867	6.000	0	400	B-184	915
4.000	100	5,000	B-137	868	6.000	100	600	B-185	916
4.000	0	5,000	B-138	869	6.000	0	600	B-186	917
4.000	100	6,000	B-139	870	6.000	100	800	B-187	918
4.000	0	6,000	B-140	871	6.000	0	800	B-188	919
4.000	100	8,000	B-141	872	6.000	100	1,000	B-189	920
4.000	0	8,000	B-142	873	6.000	0	1,000	B-190	921
4.000	100	10,000	B-143	874	6.000	100	1,500	B-191	922
4.000	0	10,000	B-144	875	6.000	0	1,500	B-192	923

Pipe Size, in.	Percent Oil	Rate, b/d	Figure Number	Page Number	Pipe Size, in.	Percent Oil	Rate, b/d	Figure Number	Page Number
6.000	100	2,000	B-193	924	8.000	100	20,000	B-241	972
6.000	0	2,000	B-194	925	8.000	0	20,000	B-242	973
6.000	100	3,000	B-195	926	8.000	100	30,000	B-243	974
6.000	0	3,000	B-196	927	8.000	0	30,000	B-244	975
6.000	100	4,000	B-197	928	8.000	100	50,000	B-245	976
6.000	0	4,000	B-198	929	8.000	0	50,000	B-246	977
6.000	100	5,000	B-199	930	8.000	100	60,000	B-247	978
6.000	0	5,000	B-200	931	8.000	0	60,000	B-248	979
6.000	100	6,000	B-201	932	8.000	100	70,000	B-249	980
6.000	0	6,000	B-202	933	8.000	0	70,000	B-250	981
6.000	100	8,000	B-203	934	8.000	100	80,000	B-251	982
6.000	0	8,000	B-204	935	8.000	0	80,000	B-252	983
6.000	100	10,000	B-205	936	10.000	100	400	B-253	984
6.000	0	10,000	B-206	937	10.000	0	400	B-254	985
6.000	100	15,000	B-207	938	10.000	100	600	B-255	986
6.000	0	15,000	B-208	939	10.000	0	600	B-256	987
6.000	100	20,000	B-209	940	10.000	100	800	B-257	988
6.000	0	20,000	B-210	941	10.000	0	800	B-258	989
6.000	100	30,000	B-211	942	10.000	100	1,000	B-259	990
6.000	0	30,000	B-212	943	10.000	0	1,000	B-260	991
6.000	100	50,000	B-213	944	10.000	100	1,500	B-261	992
6.000	0	50,000	B-214	945	10.000	0	1,500	B-262	993
8.000	100	400	B-215	946	10.000	100	2,000	B-263	994
8.000	0	400	B-216	947	10.000	0	2,000	B-264	995
8.000	100	600	B-217	948	10.000	100	3,000	B-265	996
8.000	0	600	B-218	949	10.000	0	3,000	B-266	997
8.000	100	800	B-219	950	10.000	100	4,000	B-267	998
8.000	0	800	B-220	951	10.000	0	4,000	B-268	999
8.000	100	1,000	B-221	952	10.000	100	5,000	B-269	1000
8.000	0	1,000	B-222	953	10.000	0	5,000	B-270	1001
8.000	100	1,500	B-223	954	10.000	100	6,000	B-271	1002
8.000	0	1,500	B-224	955	10.000	0	6,000	B-272	1003
8.000	100	2,000	B-225	956	10.000	100	8,000	B-273	1004
8.000	0	2,000	B-226	957	10.000	0	8,000	B-274	1005
8.000	100	3,000	B-227	958	10.000	100	10,000	B-275	1006
8.000	0	3,000	B-228	959	10.000	0	10,000	B-276	1007
8.000	100	4,000	B-229	960	10.000	100	15,000	B-277	1008
8.000	0	4,000	B-230	961	10.000	0	15,000	B-278	1009
8.000	100	5,000	B-231	962	10.000	100	20,000	B-279	1010
8.000	0	5,000	B-232	963	10.000	0	20,000	B-280	1011
8.000	100	6,000	B-233	964	10.000	100	30,000	B-281	1012
8.000	0	6,000	B-234	965	10.000	0	30,000	B-282	1013
8.000	100	8,000	B-235	966	10.000	100	50,000	B-283	1014
8.000	0	8,000	B-236	967	10.000	0	50,000	B-284	1015
8.000	100	10,000	B-237	968	10.000	100	60,000	B-285	1016
8.000	0	10,000	B-238	969	10.000	0	60,000	B-286	1017
8.000	100	15,000	B-239	970	10.000	100	70,000	B-287	1018
8.000	0	15,000	B-240	971	10.000	0	70,000	B-288	1019

Pipe Size, in.	Percent Oil	Rate, b/d	Figure Number	Page Number
10.000	100	80,000	B-289	1020
10.000	0	80,000	B-290	1021
10.000	100	10,000	B-291	1022
10.000	0	10,000	B-292	1023
12.000	100	600	B-293	1024
12.000	0	600	B-294	1025
12.000	100	800	B-295	1026
12.000	0	800	B-296	1027
12.000	100	1,000	B-297	1028
12.000	0	1,000	B-298	1029
12.000	100	1,500	B-299	1030
12.000	0	1,500	B-300	1031
12.000	100	2,000	B-301	1032
12.000	0	2,000	B-302	1033
12.000	100	3,000	B-303	1034
12.000	0	3,000	B-304	1035
12.000	100	4,000	B-305	1036
12.000	0	4,000	B-306	1037
12.000	100	5,000	B-307	1038
12.000	0	5,000	B-308	1039
12.000	100	6,000	B-309	1040
12.000	0	6,000	B-310	1041
12.000	100	8,000	B-311	1042
12.000	0	8,000	B-312	1043
12.000	100	10,000	B-313	1044
12.000	0	10,000	B-314	1045
12.000	100	15,000	B-315	1046
12.000	0	15,000	B-316	1047
12.000	100	20,000	B-317	1048
12.000	0	20,000	B-318	1049
12.000	100	30,000	B-319	1050
12.000	0	30,000	B-320	1051
12.000	100	50,000	B-321	1052
12.000	0	50,000	B-322	1053
12.000	100	60,000	B-323	1054
12.000	0	60,000	B-324	1055
12.000	100	70,000	B-325	1056
12.000	0	70,000	B-326	1057
12.000	100	80,000	B-327	1058
12.000	0	80,000	B-328	1059
12.000	100	100,000	B-329	1060
12.000	0	100,000	B-330	1061

SECTION A

Vertical Multiphase Tubing Flow

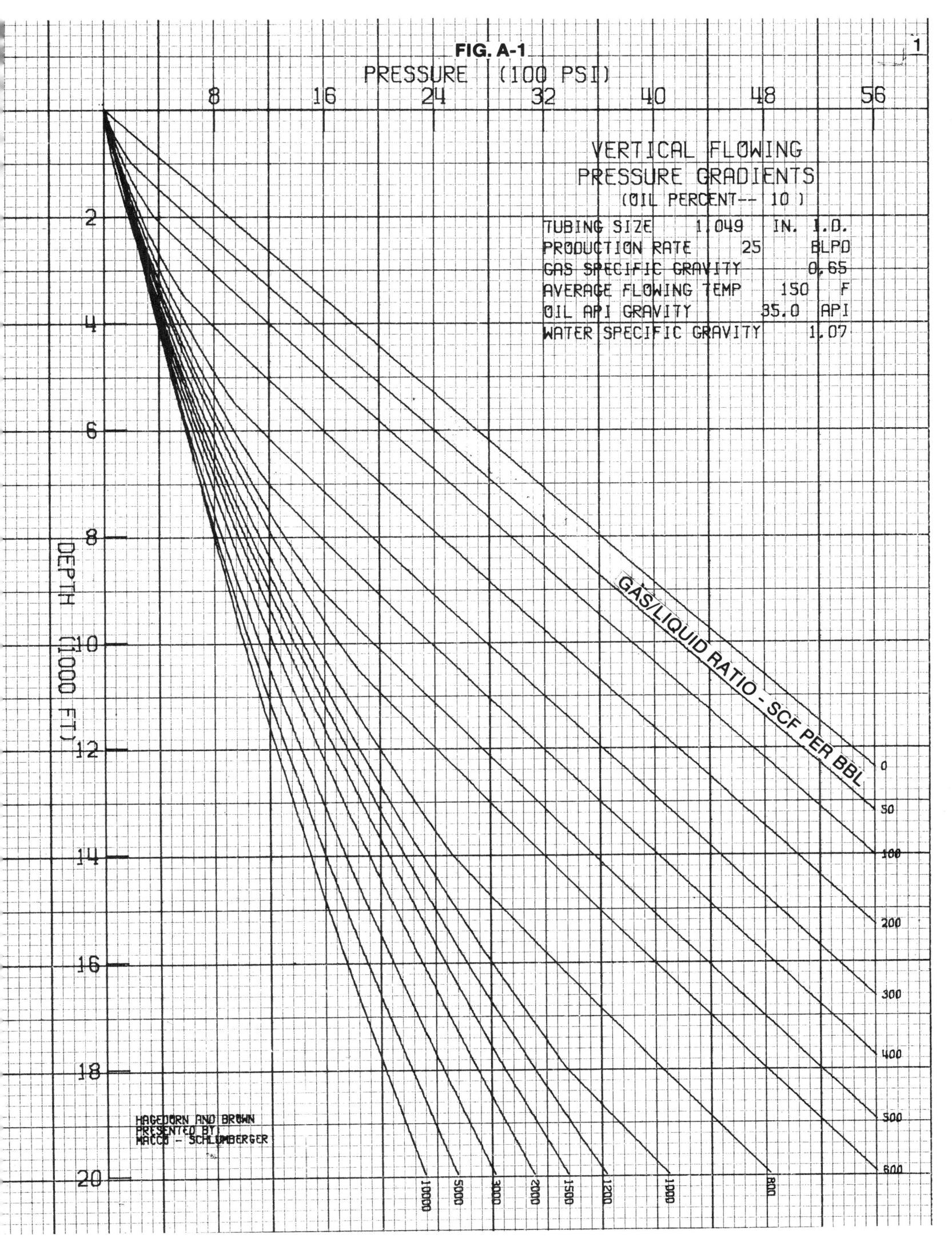

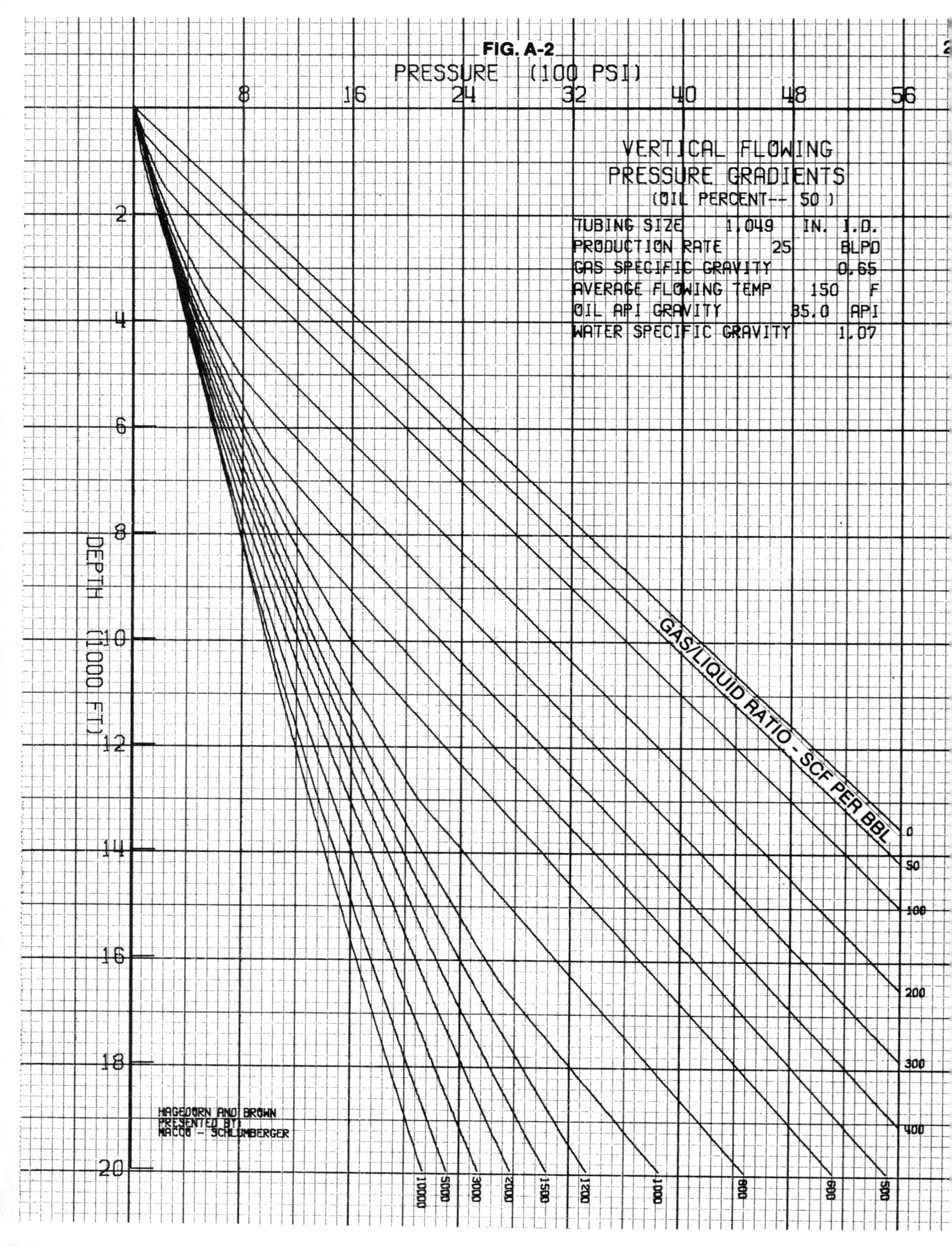

FIG. A-2

VERTICAL FLOWING
PRESSURE GRADIENTS
(OIL PERCENT-- 50)

TUBING SIZE	1.049	IN. I.D.
PRODUCTION RATE	25	BLPD
GAS SPECIFIC GRAVITY	0.65	
AVERAGE FLOWING TEMP	150	F
OIL API GRAVITY	35.0	API
WATER SPECIFIC GRAVITY	1.07	

PRESSURE (100 PSI)

DEPTH (1000 FT)

GAS/LIQUID RATIO - SCF PER BBL

HAGEDORN AND BROWN
PRESENTED BY:
NACCO - SCHLUMBERGER

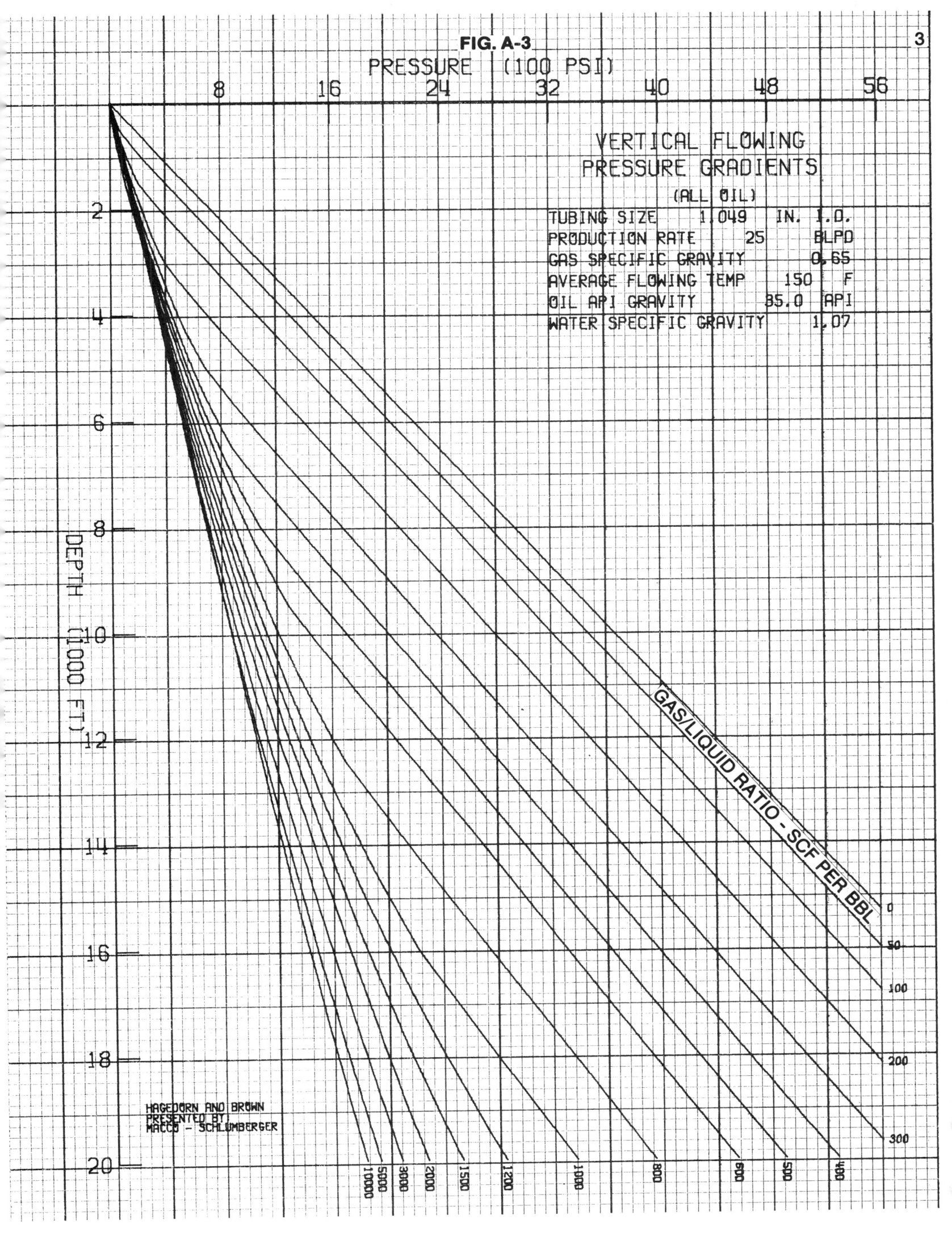

FIG. A-3

VERTICAL FLOWING
PRESSURE GRADIENTS
(ALL OIL)

TUBING SIZE	1.049	IN. I.D.
PRODUCTION RATE	25	BLPD
GAS SPECIFIC GRAVITY	0.65	
AVERAGE FLOWING TEMP	150	F
OIL API GRAVITY	35.0	API
WATER SPECIFIC GRAVITY	1.07	

PRESSURE (100 PSI)

DEPTH (1,000 FT)

GAS/LIQUID RATIO - SCF PER BBL

HAGEDORN AND BROWN
PRESENTED BY
MACCO - SCHLUMBERGER

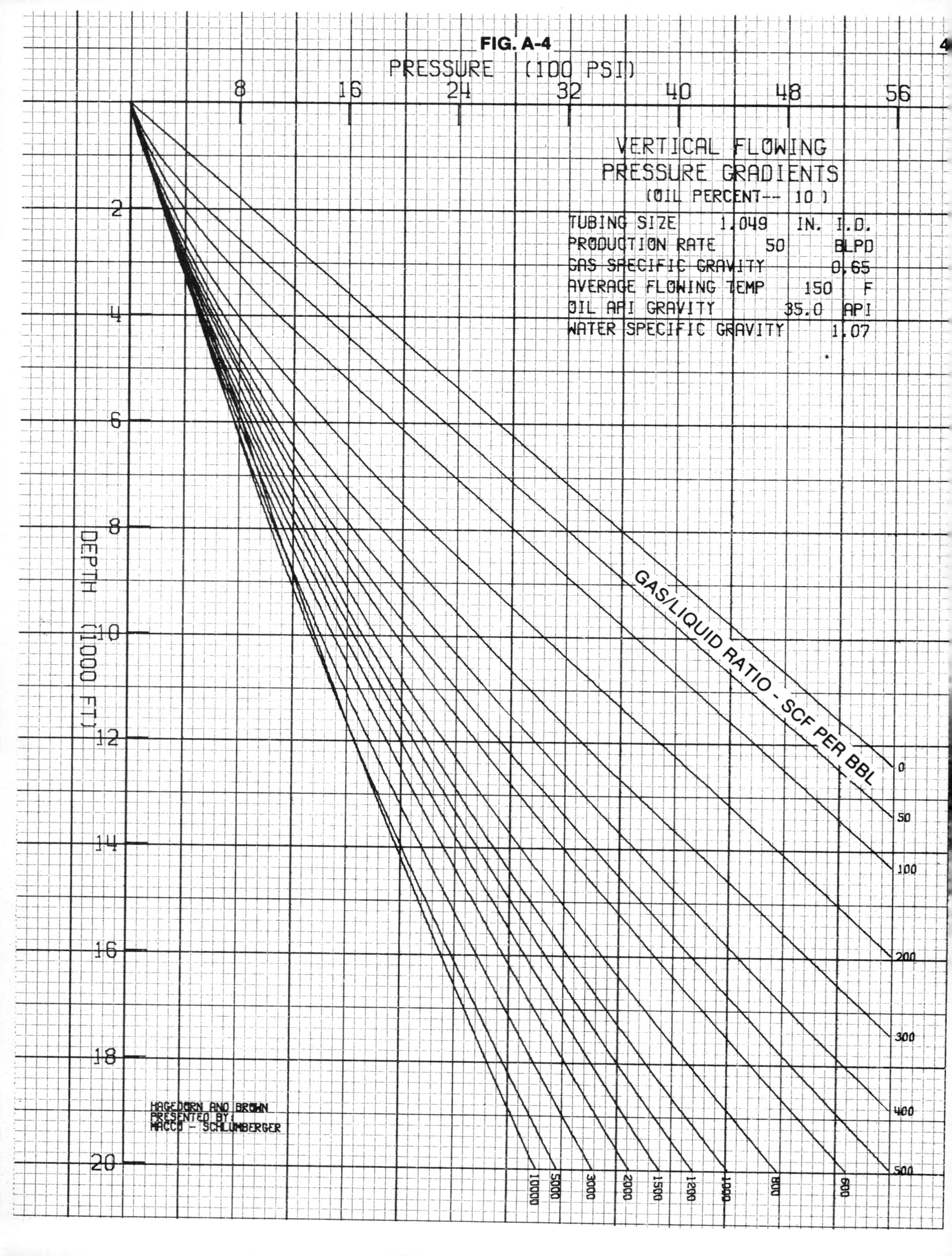

VERTICAL FLOWING
PRESSURE GRADIENTS
(OIL PERCENT-- 10)

TUBING SIZE	1.049	IN. I.D.
PRODUCTION RATE	50	BLPD
GAS SPECIFIC GRAVITY	0.65	
AVERAGE FLOWING TEMP	150	F
OIL API GRAVITY	35.0	API
WATER SPECIFIC GRAVITY	1.07	

PRESSURE (100 PSI)

DEPTH (1000 FT)

GAS/LIQUID RATIO - SCF PER BBL

HAGEDORN AND BROWN
PRESENTED BY:
MACCO - SCHLUMBERGER

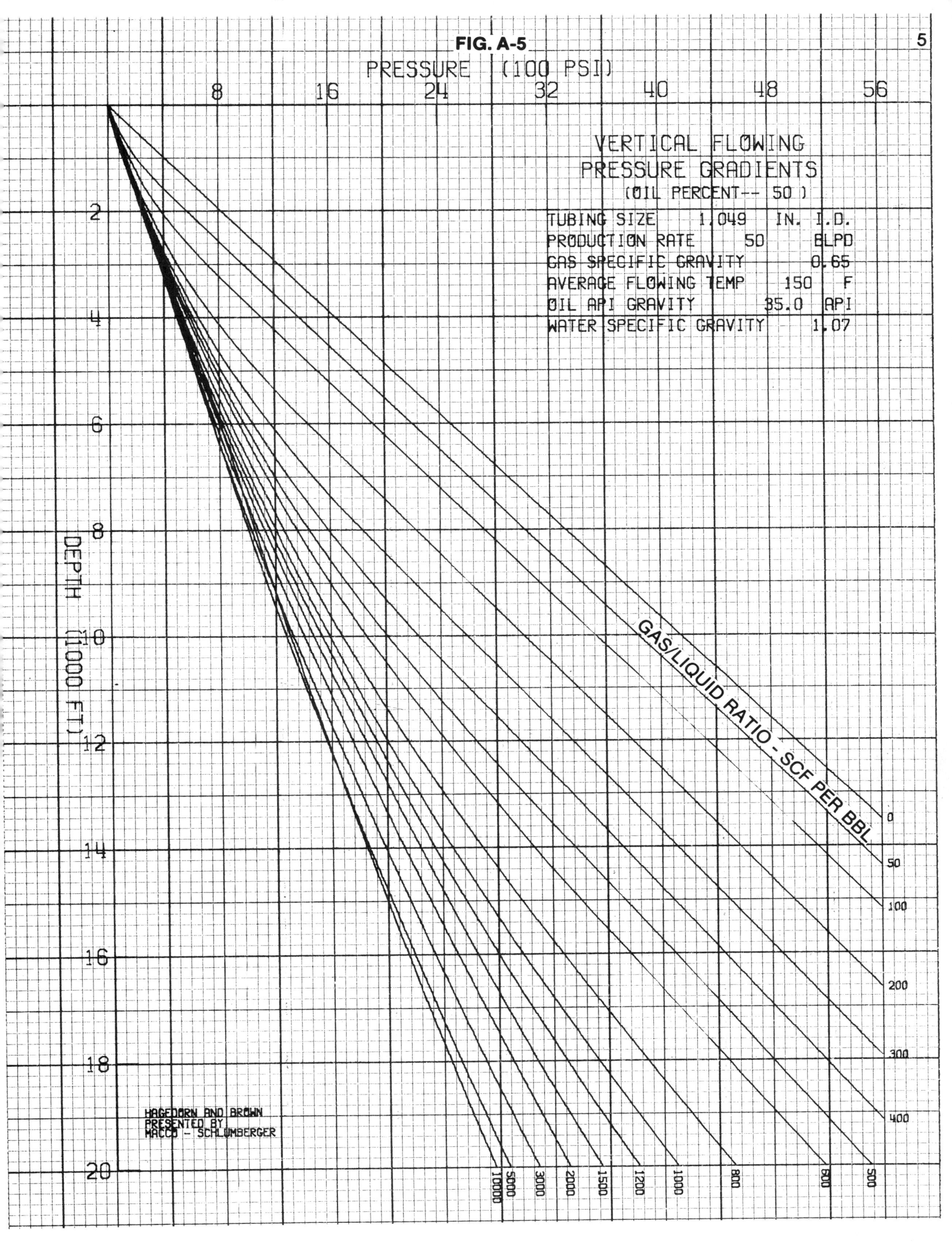

FIG. A-5

PRESSURE (100 PSI)

VERTICAL FLOWING
PRESSURE GRADIENTS
(OIL PERCENT-- 50)

TUBING SIZE   1.049   IN. I.D.
PRODUCTION RATE   50   BLPD
GAS SPECIFIC GRAVITY   0.65
AVERAGE FLOWING TEMP   150   F
OIL API GRAVITY   35.0   API
WATER SPECIFIC GRAVITY   1.07

DEPTH (1000 FT)

GAS/LIQUID RATIO - SCF PER BBL

HAGEDORN AND BROWN
PRESENTED BY
MACCO - SCHLUMBERGER

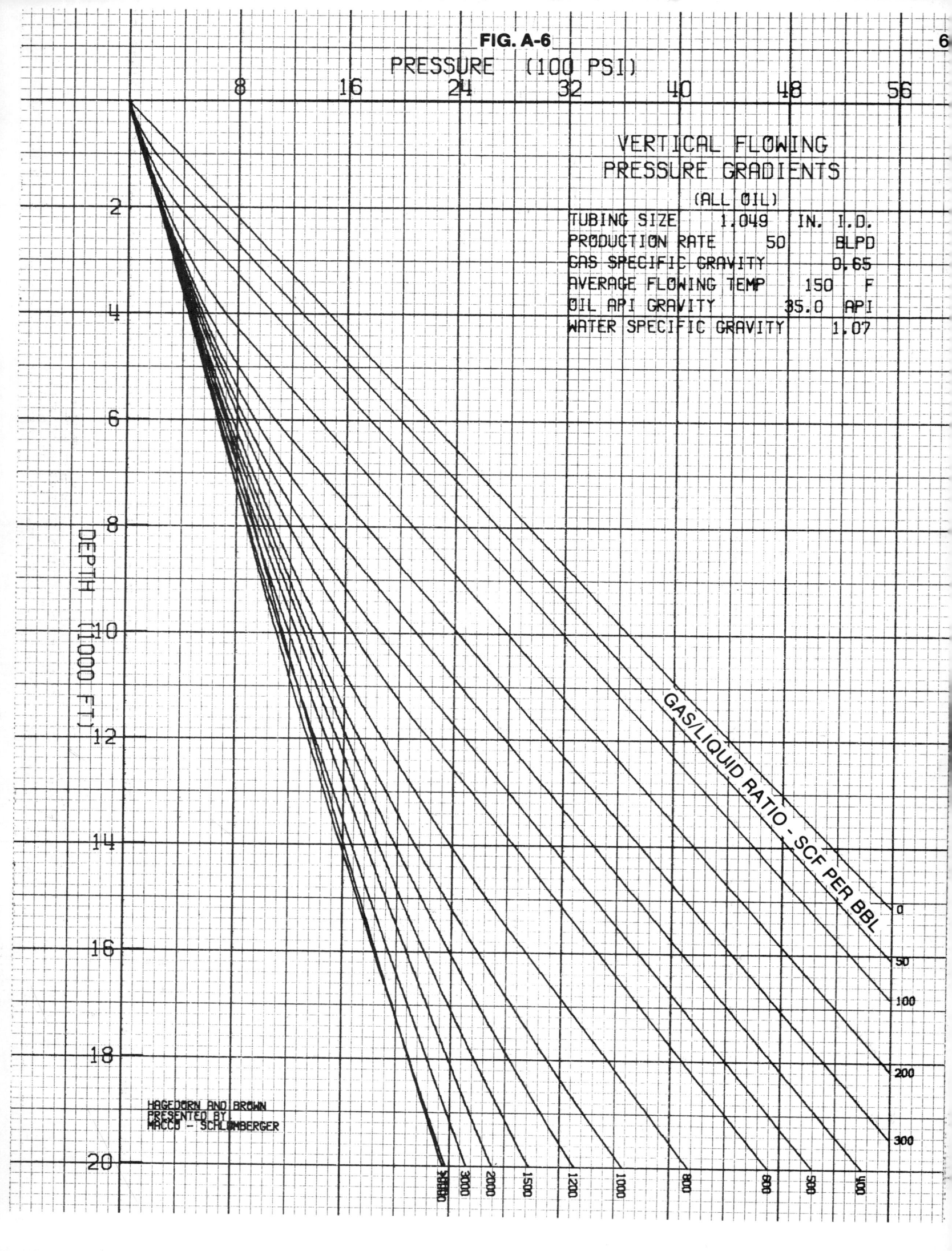

PRESSURE (100 PSI)

VERTICAL FLOWING
PRESSURE GRADIENTS
(ALL OIL)

TUBING SIZE	1.049	IN. I.D.
PRODUCTION RATE	50	BLPD
GAS SPECIFIC GRAVITY	0.65	
AVERAGE FLOWING TEMP	150	F
OIL API GRAVITY	35.0	API
WATER SPECIFIC GRAVITY	1.07	

DEPTH (1000 FT)

GAS/LIQUID RATIO - SCF PER BBL

HAGEDORN AND BROWN
PRESENTED BY
MACCO - SCHLUMBERGER

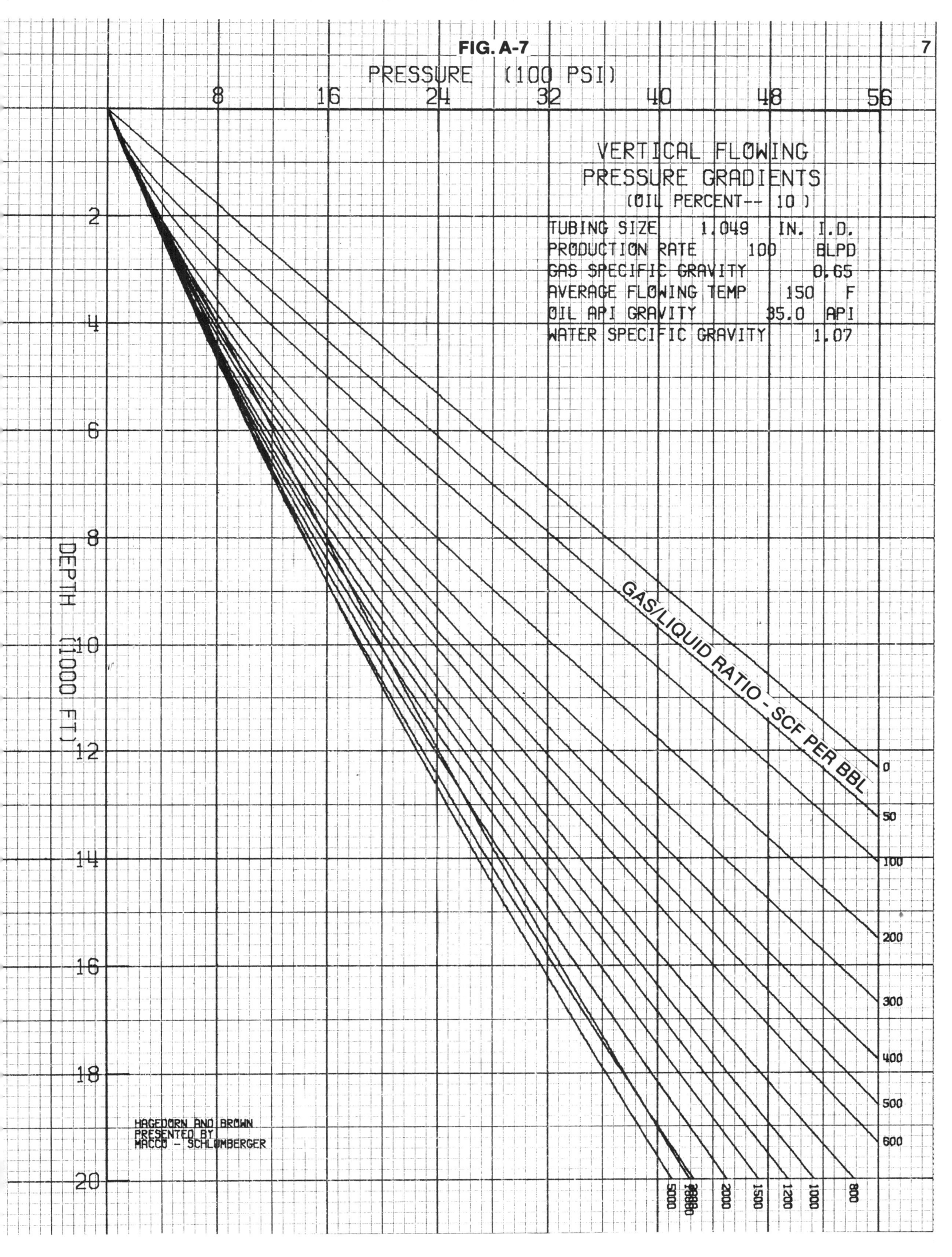

PRESSURE (100 PSI)

DEPTH (1000 FT)

VERTICAL FLOWING
PRESSURE GRADIENTS
(OIL PERCENT-- 10)

TUBING SIZE	1.049	IN. I.D.
PRODUCTION RATE	100	BLPD
GAS SPECIFIC GRAVITY		0.65
AVERAGE FLOWING TEMP	150	F
OIL API GRAVITY	35.0	API
WATER SPECIFIC GRAVITY		1.07

GAS/LIQUID RATIO - SCF PER BBL

HAGEDORN AND BROWN
PRESENTED BY
MACCO - SCHLUMBERGER

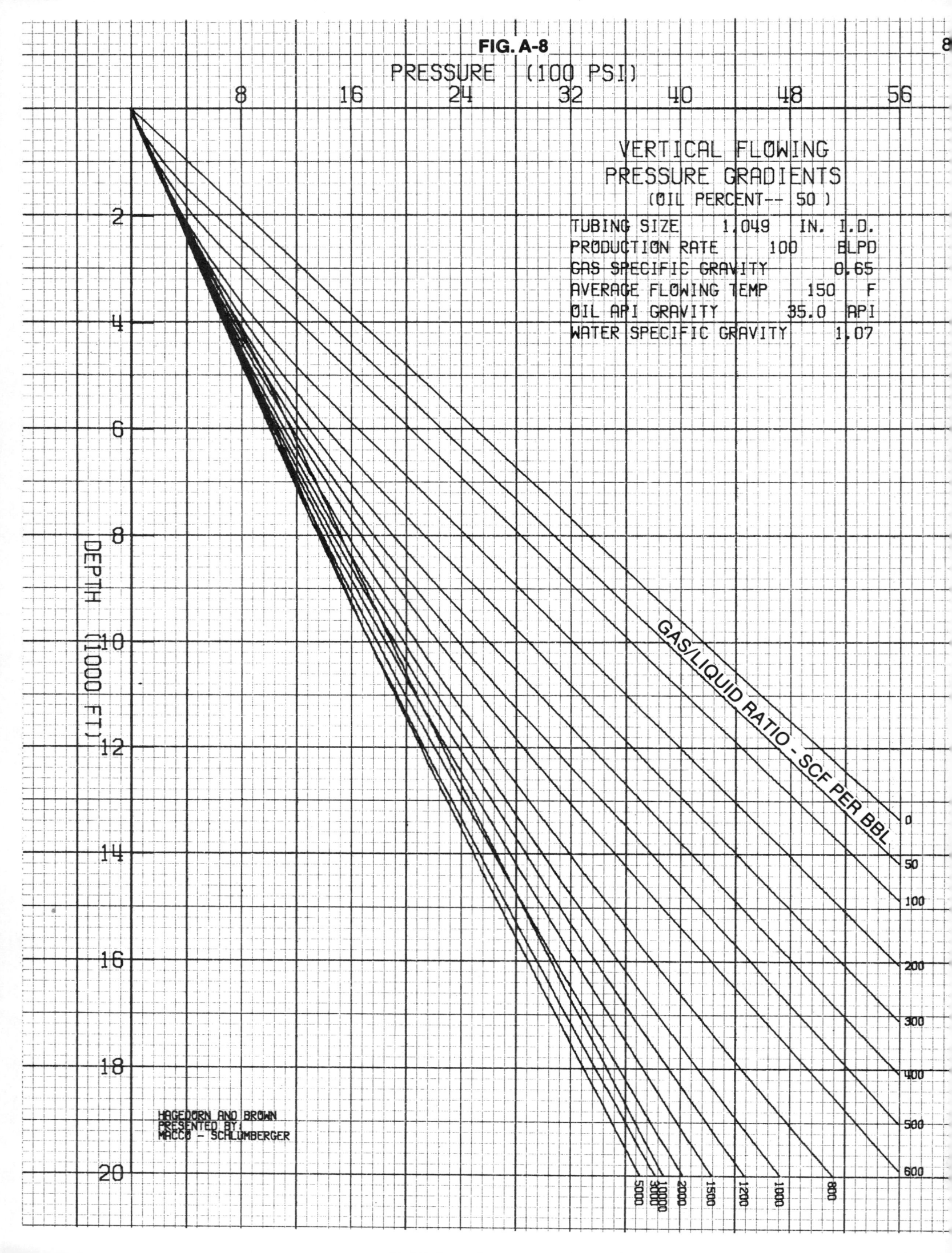

FIG. A-8

VERTICAL FLOWING
PRESSURE GRADIENTS
(OIL PERCENT-- 50)

TUBING SIZE	1.049	IN. I.D.
PRODUCTION RATE	100	BLPD
GAS SPECIFIC GRAVITY		0.65
AVERAGE FLOWING TEMP	150	F
OIL API GRAVITY	35.0	API
WATER SPECIFIC GRAVITY		1.07

GAS/LIQUID RATIO - SCF PER BBL

HAGEDORN AND BROWN
PRESENTED BY /
MACCO - SCHLUMBERGER

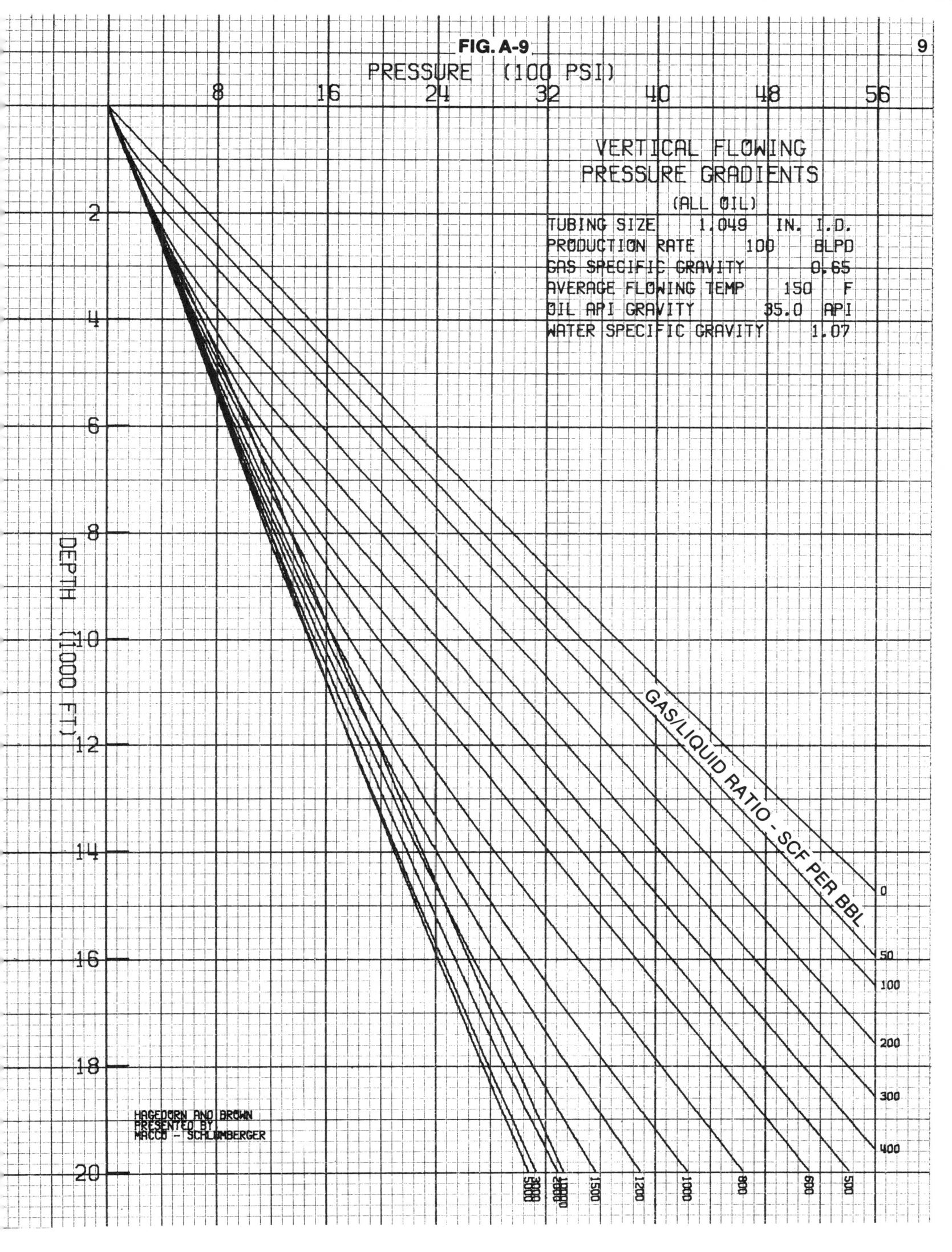

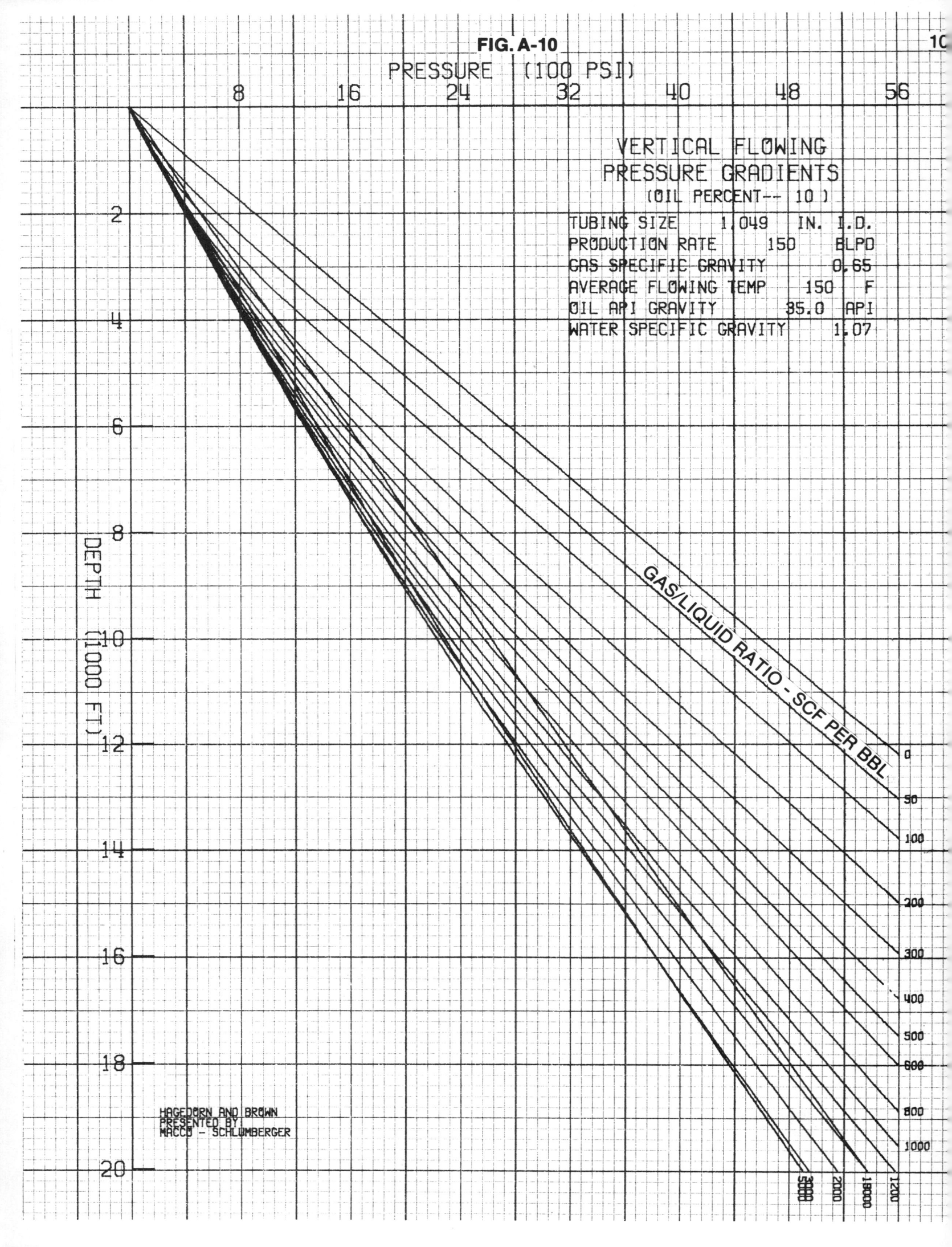

FIG. A-10

VERTICAL FLOWING
PRESSURE GRADIENTS
(OIL PERCENT-- 10)

TUBING SIZE	1.049	IN. I.D.
PRODUCTION RATE	150	BLPD
GAS SPECIFIC GRAVITY	0.65	
AVERAGE FLOWING TEMP	150	F
OIL API GRAVITY	35.0	API
WATER SPECIFIC GRAVITY	1.07	

PRESSURE (100 PSI)

DEPTH (1000 FT)

GAS/LIQUID RATIO - SCF PER BBL

HAGEDORN AND BROWN
PRESENTED BY
MACCO - SCHLUMBERGER

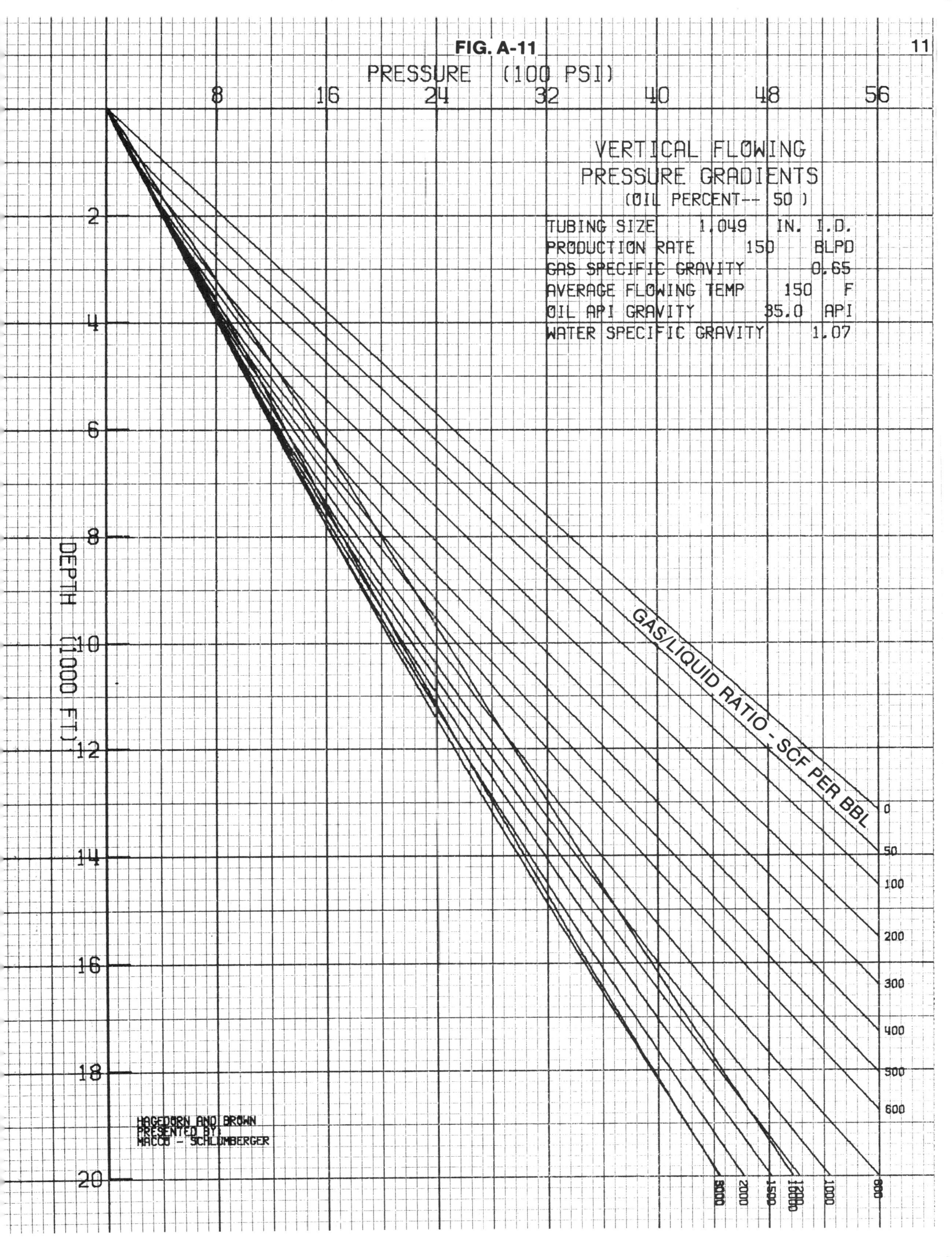

FIG. A-11

PRESSURE (100 PSI)

VERTICAL FLOWING
PRESSURE GRADIENTS
(OIL PERCENT-- 50)

TUBING SIZE	1.049	IN. I.D.
PRODUCTION RATE	150	BLPD
GAS SPECIFIC GRAVITY	0.65	
AVERAGE FLOWING TEMP	150	F
OIL API GRAVITY	35.0	API
WATER SPECIFIC GRAVITY	1.07	

DEPTH (1,000 FT)

GAS/LIQUID RATIO - SCF PER BBL

HAGEDORN AND BROWN
PRESENTED BY:
MACCO - SCHLUMBERGER

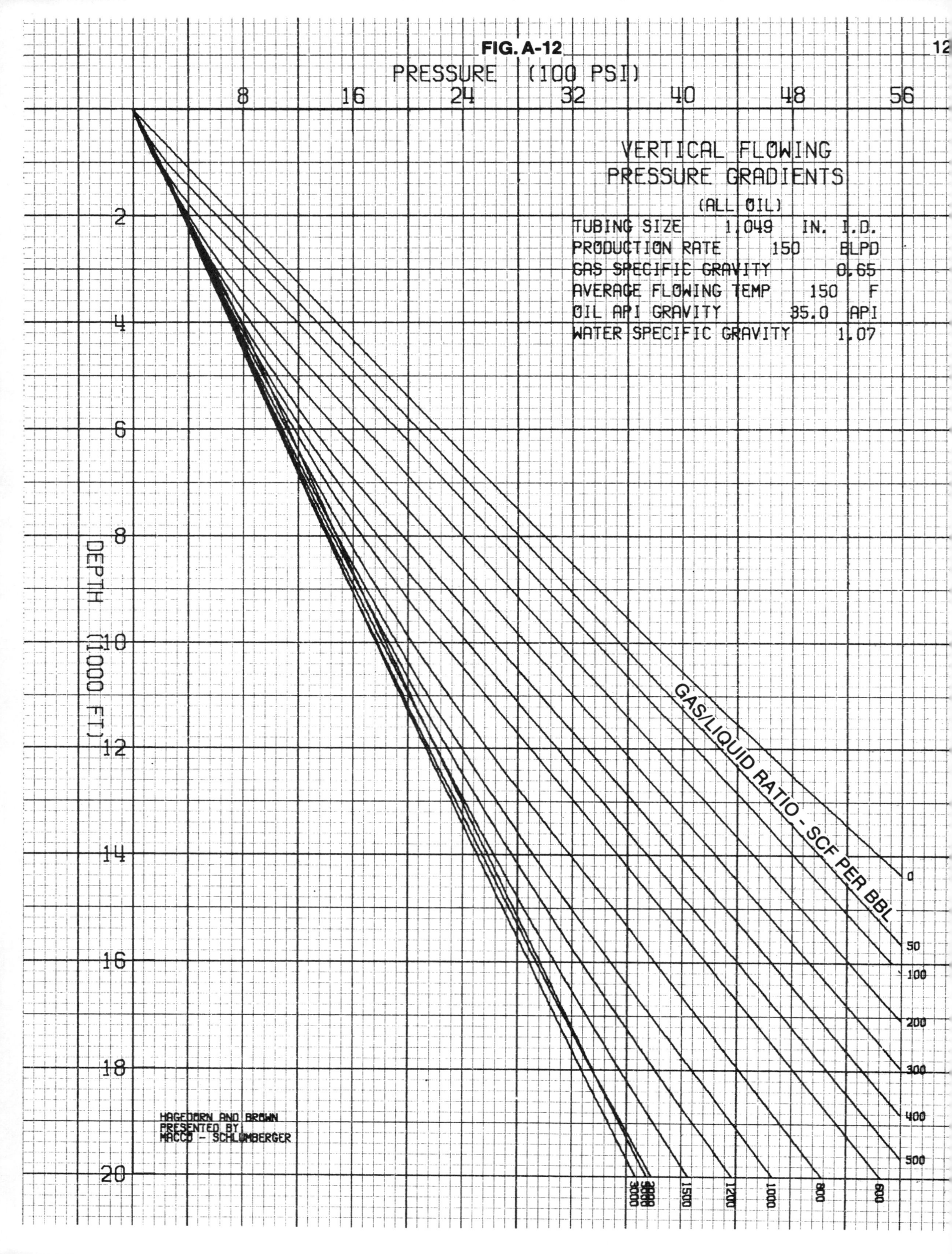

FIG. A-12

12

VERTICAL FLOWING
PRESSURE GRADIENTS
(ALL OIL)

TUBING SIZE	1.049	IN. I.D.
PRODUCTION RATE	150	BLPD
GAS SPECIFIC GRAVITY	0.65	
AVERAGE FLOWING TEMP	150	F
OIL API GRAVITY	35.0	API
WATER SPECIFIC GRAVITY	1.07	

PRESSURE (100 PSI)

DEPTH (1000 FT)

GAS/LIQUID RATIO - SCF PER BBL

HAGEDORN AND BROWN
PRESENTED BY
MACCO - SCHLUMBERGER

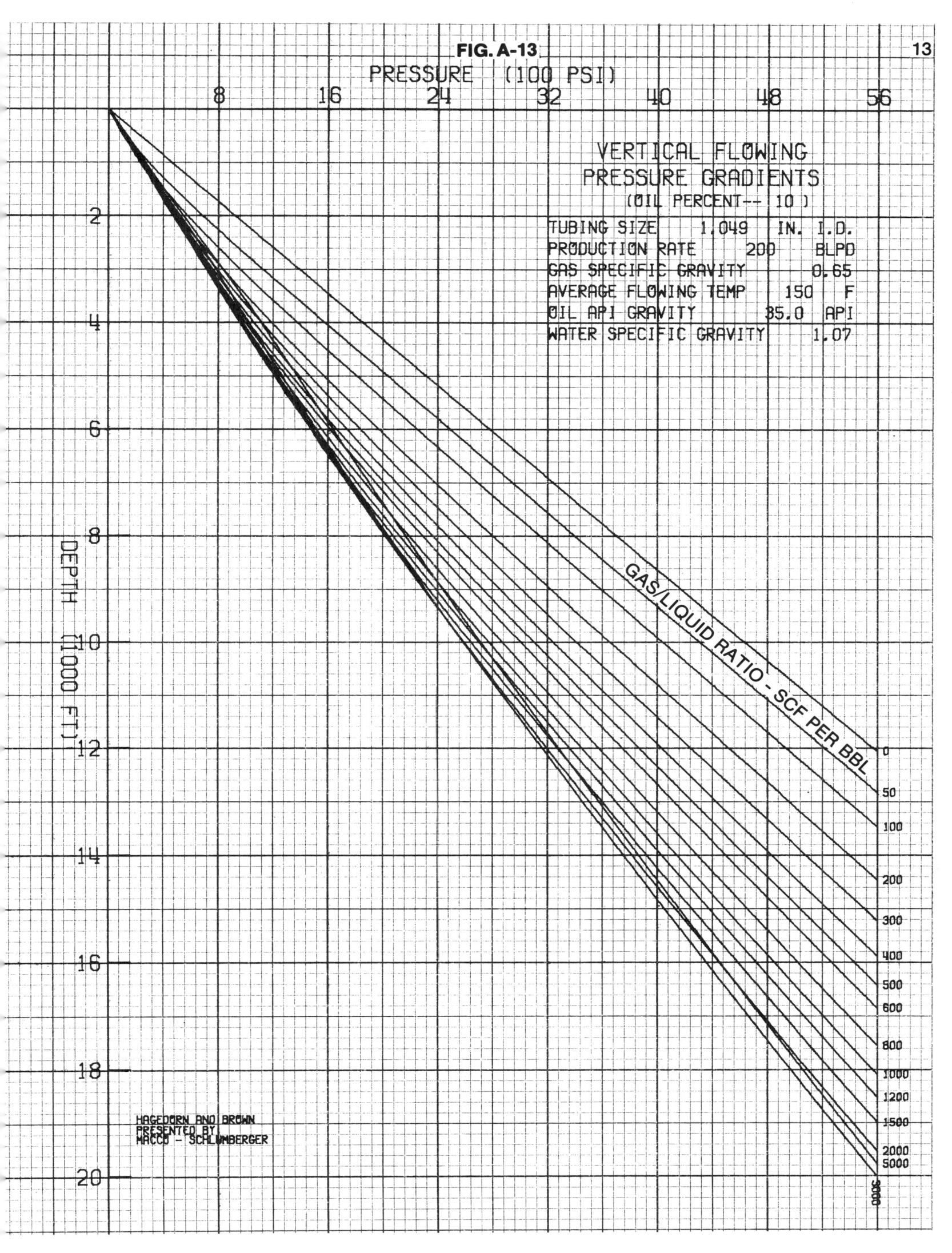

FIG. A-13   13

PRESSURE (100 PSI)

VERTICAL FLOWING
PRESSURE GRADIENTS
(OIL PERCENT-- 10)

TUBING SIZE	1.049	IN. I.D.
PRODUCTION RATE	200	BLPD
GAS SPECIFIC GRAVITY		0.65
AVERAGE FLOWING TEMP	150	F
OIL API GRAVITY	35.0	API
WATER SPECIFIC GRAVITY		1.07

GAS/LIQUID RATIO - SCF PER BBL

DEPTH (1000 FT)

HAGEDORN AND BROWN
PRESENTED BY
MACCO - SCHLUMBERGER

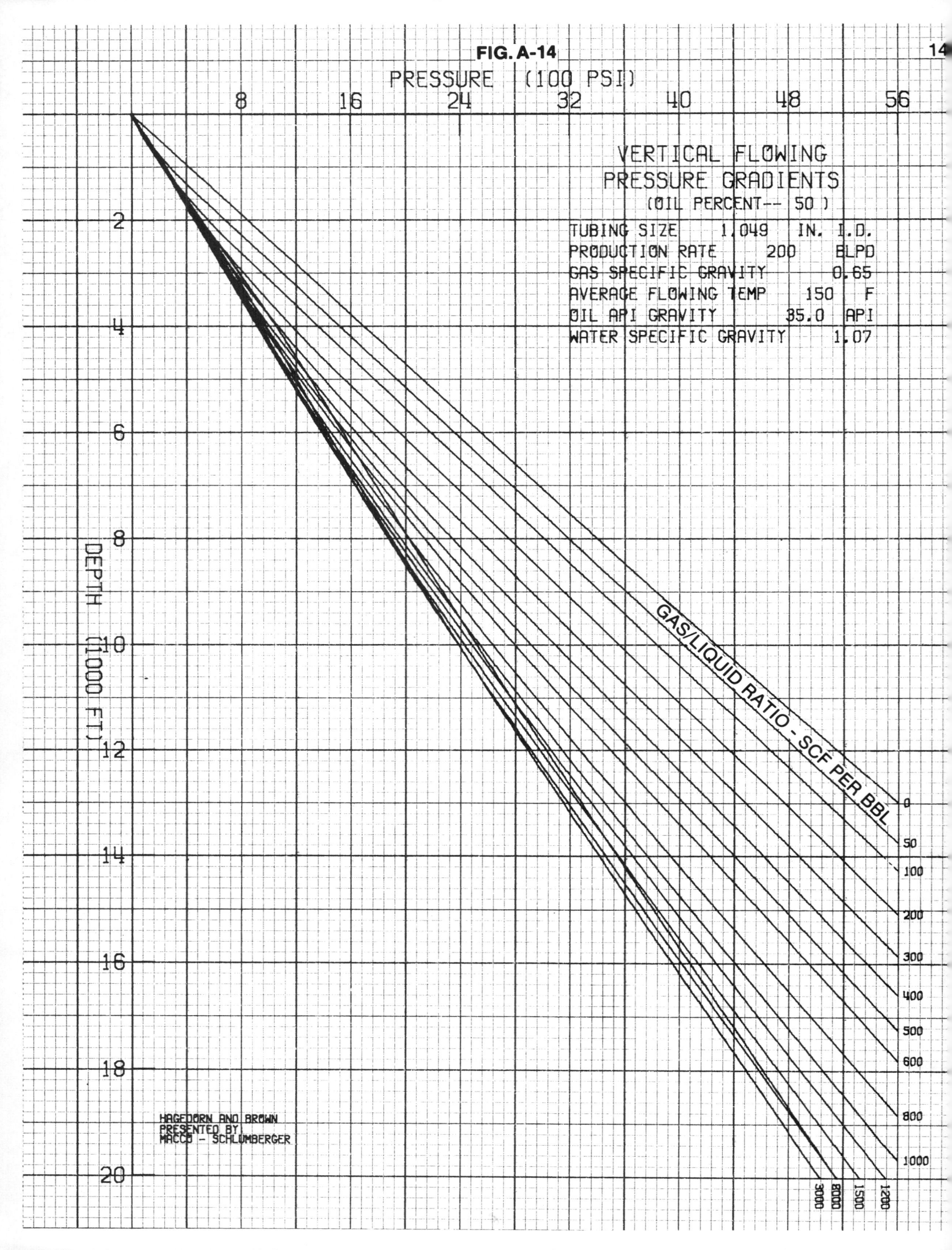

FIG. A-14

PRESSURE (100 PSI)

VERTICAL FLOWING
PRESSURE GRADIENTS
(OIL PERCENT-- 50)

TUBING SIZE	1.049	IN. I.D.
PRODUCTION RATE	200	BLPD
GAS SPECIFIC GRAVITY	0.65	
AVERAGE FLOWING TEMP	150	F
OIL API GRAVITY	35.0	API
WATER SPECIFIC GRAVITY	1.07	

DEPTH (1000 FT)

GAS/LIQUID RATIO - SCF PER BBL

HAGEDORN AND BROWN
PRESENTED BY
MACCO - SCHLUMBERGER

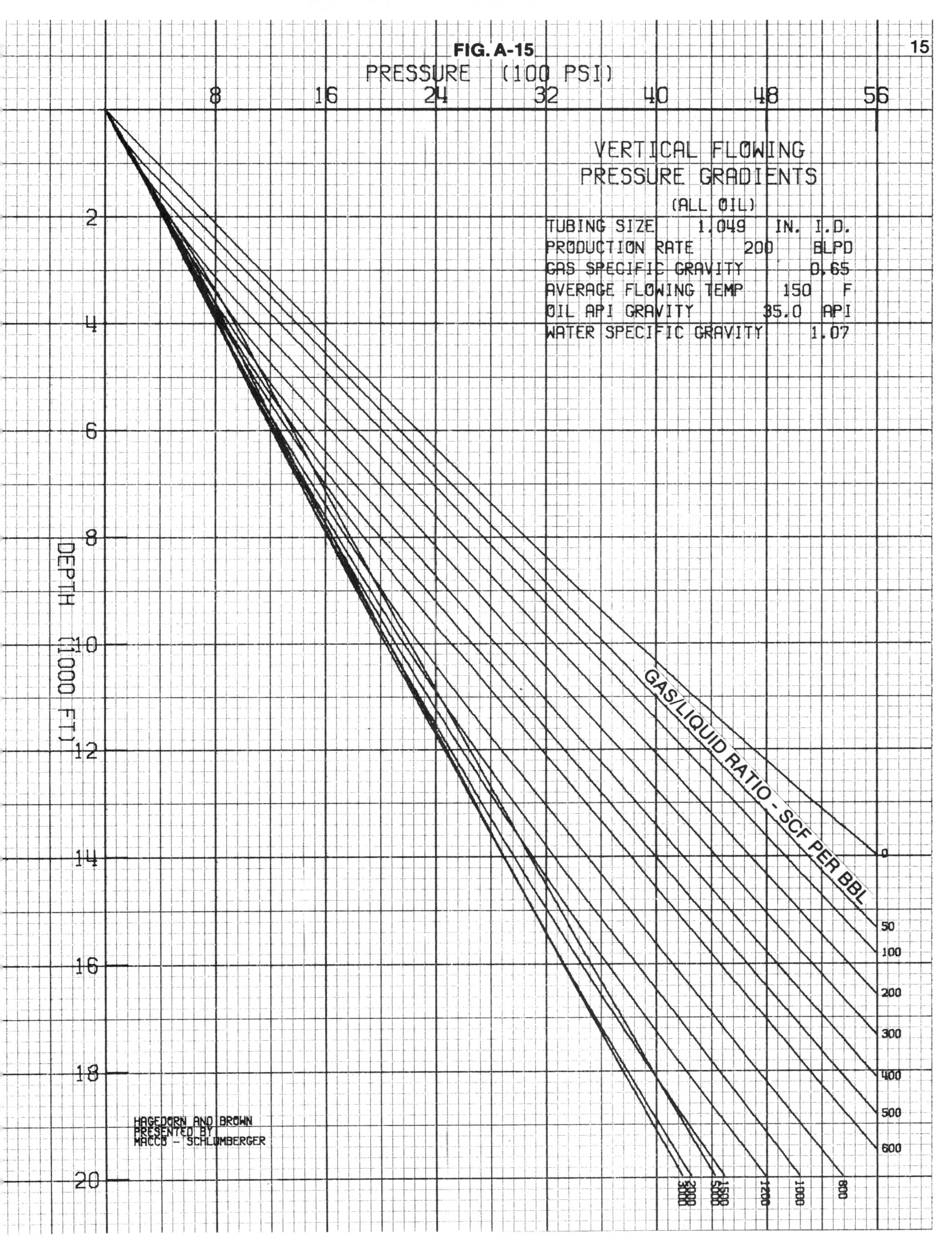

PRESSURE (100 PSI)

DEPTH (1000 FT)

VERTICAL FLOWING
PRESSURE GRADIENTS
(ALL OIL)

TUBING SIZE	1.049	IN. I.D.
PRODUCTION RATE	200	BLPD
GAS SPECIFIC GRAVITY		0.65
AVERAGE FLOWING TEMP	150	F
OIL API GRAVITY	35.0	API
WATER SPECIFIC GRAVITY		1.07

GAS/LIQUID RATIO - SCF PER BBL

HAGEDORN AND BROWN
PRESENTED BY
MACCO - SCHLUMBERGER

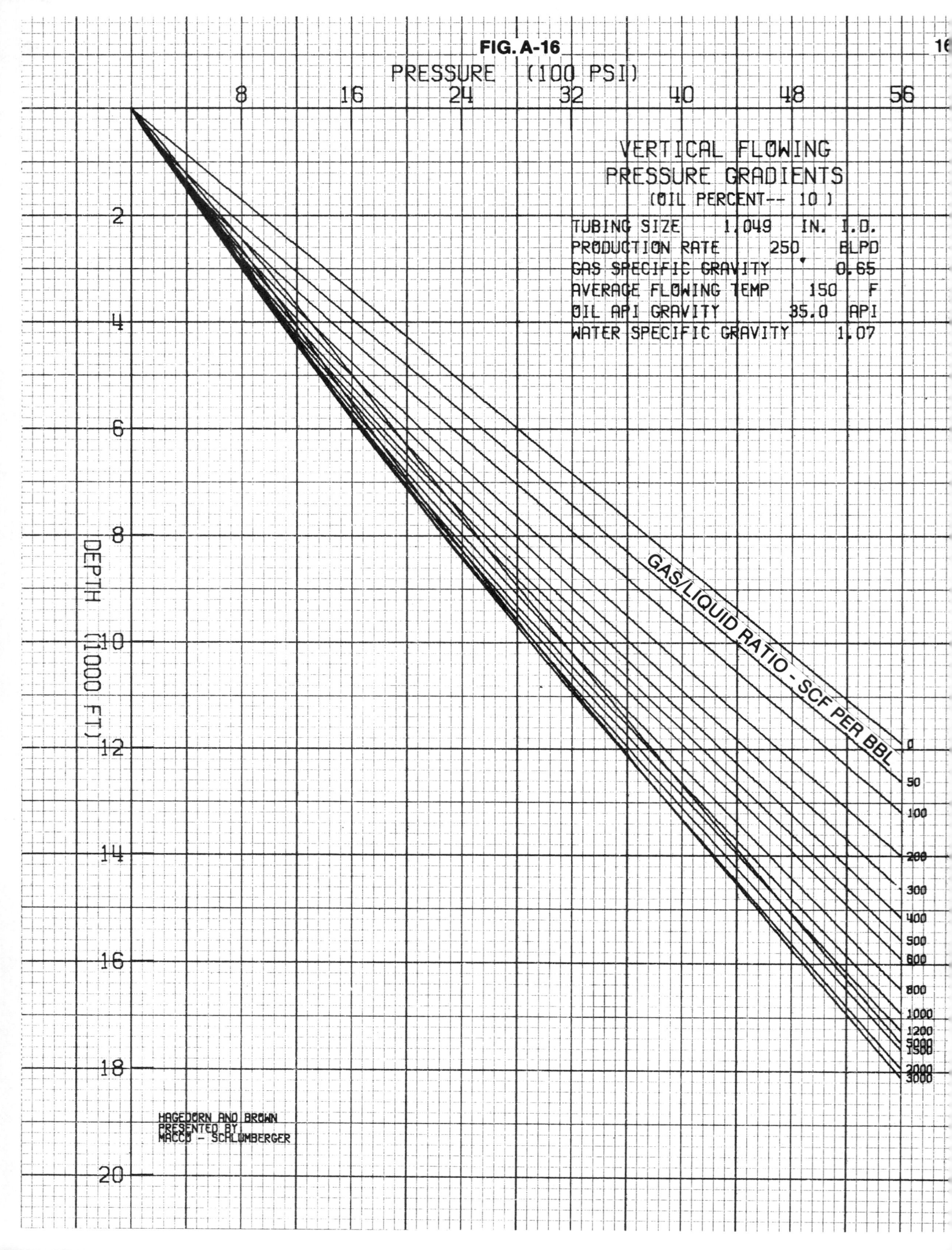

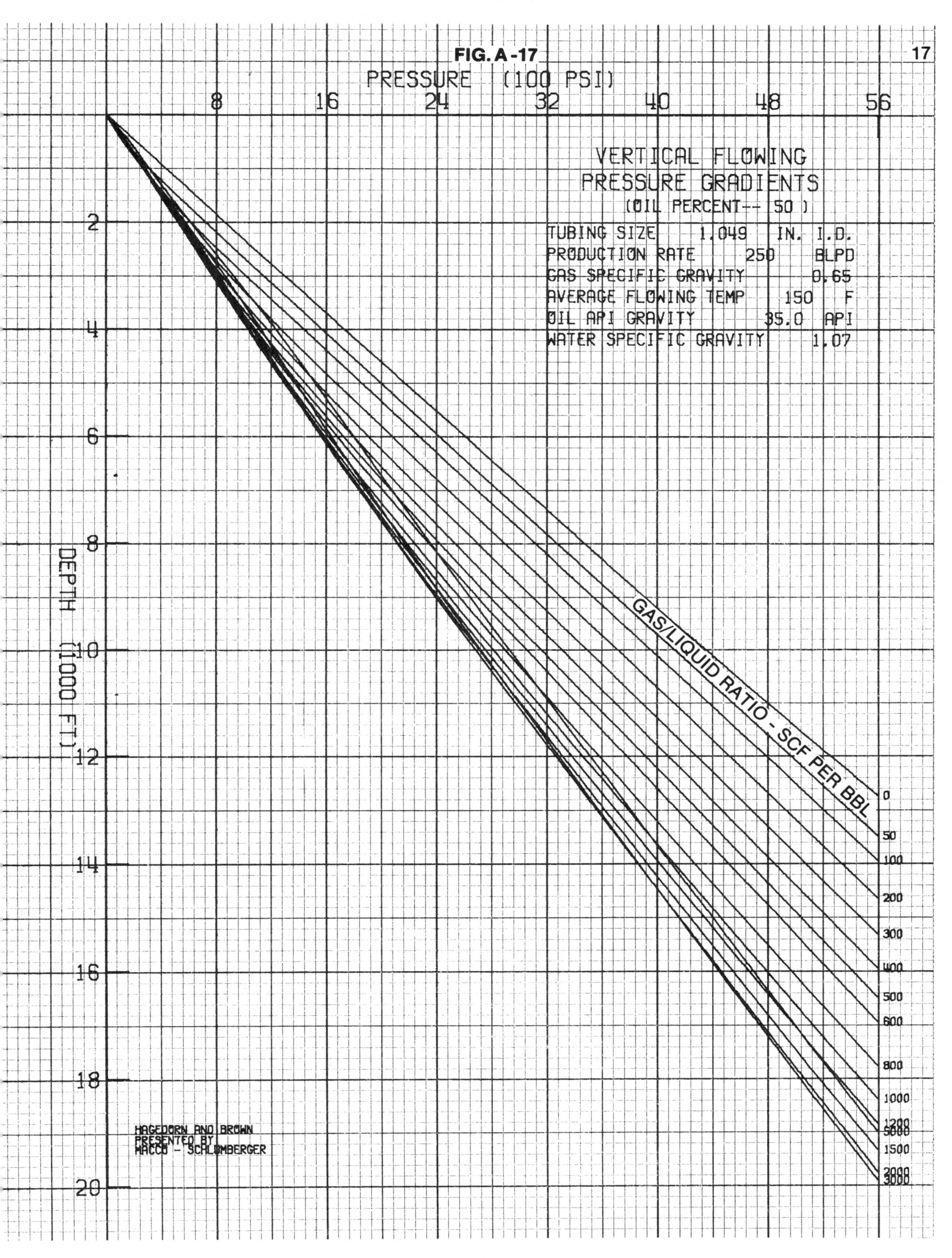

FIG. A-17

17

PRESSURE (100 PSI)

DEPTH (1000 FT)

VERTICAL FLOWING
PRESSURE GRADIENTS
(OIL PERCENT-- 50)

TUBING SIZE   1.049   IN. I.D.
PRODUCTION RATE   250   BLPD
GAS SPECIFIC GRAVITY   0.65
AVERAGE FLOWING TEMP   150   F
OIL API GRAVITY   35.0 API
WATER SPECIFIC GRAVITY   1.07

GAS/LIQUID RATIO - SCF PER BBL

0
50
100
200
300
400
500
600
800
1000
1200
5000
1500
2000
3000

HAGEDORN AND BROWN
PRESENTED BY
MACCO - SCHLUMBERGER

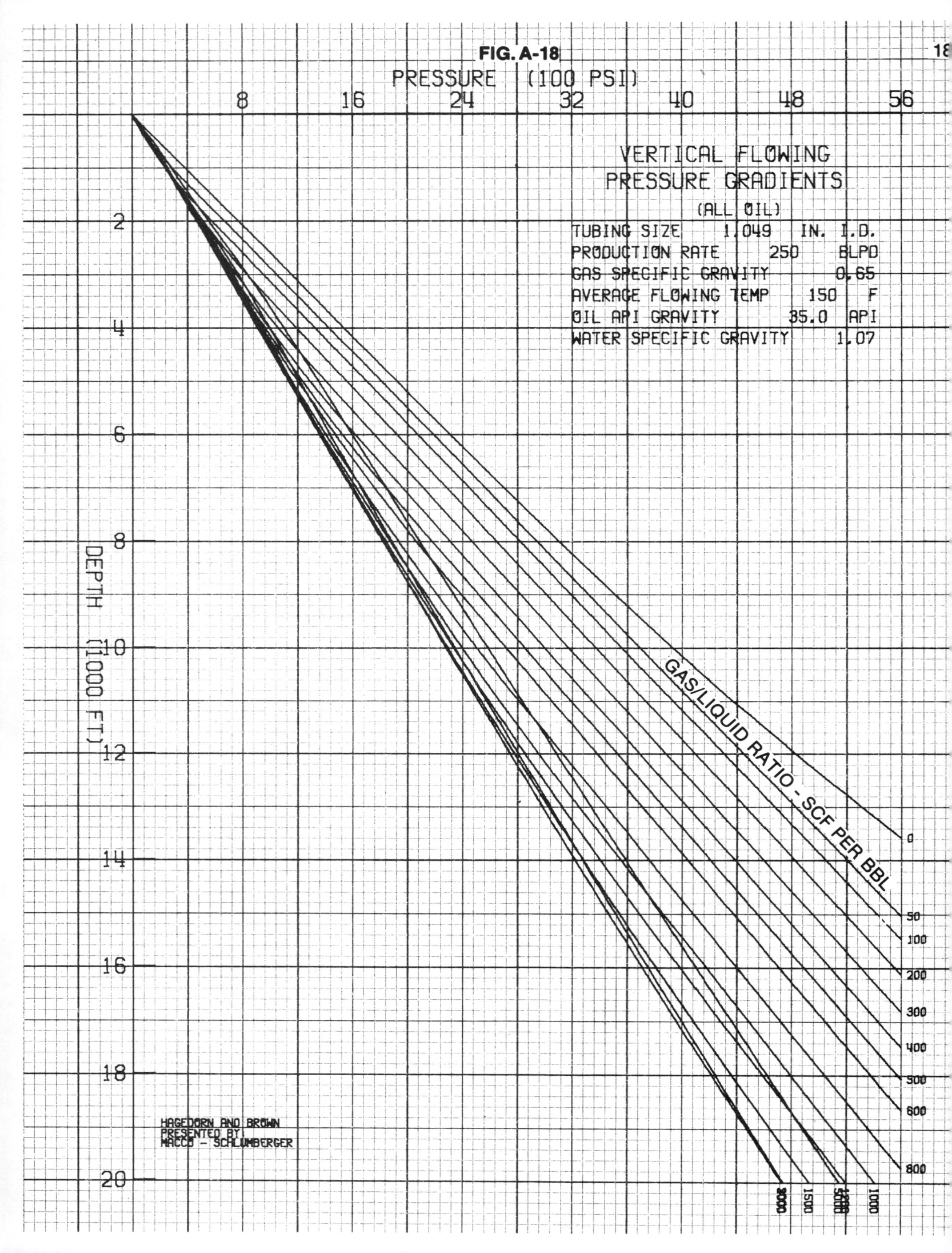

FIG. A-18

PRESSURE (100 PSI)

VERTICAL FLOWING
PRESSURE GRADIENTS
(ALL OIL)

TUBING SIZE	1.049	IN. I.D.
PRODUCTION RATE	250	BLPD
GAS SPECIFIC GRAVITY	0.65	
AVERAGE FLOWING TEMP	150	F
OIL API GRAVITY	35.0	API
WATER SPECIFIC GRAVITY	1.07	

DEPTH (1000 FT.)

GAS/LIQUID RATIO - SCF PER BBL

HAGEDORN AND BROWN
PRESENTED BY
MACCO - SCHLUMBERGER

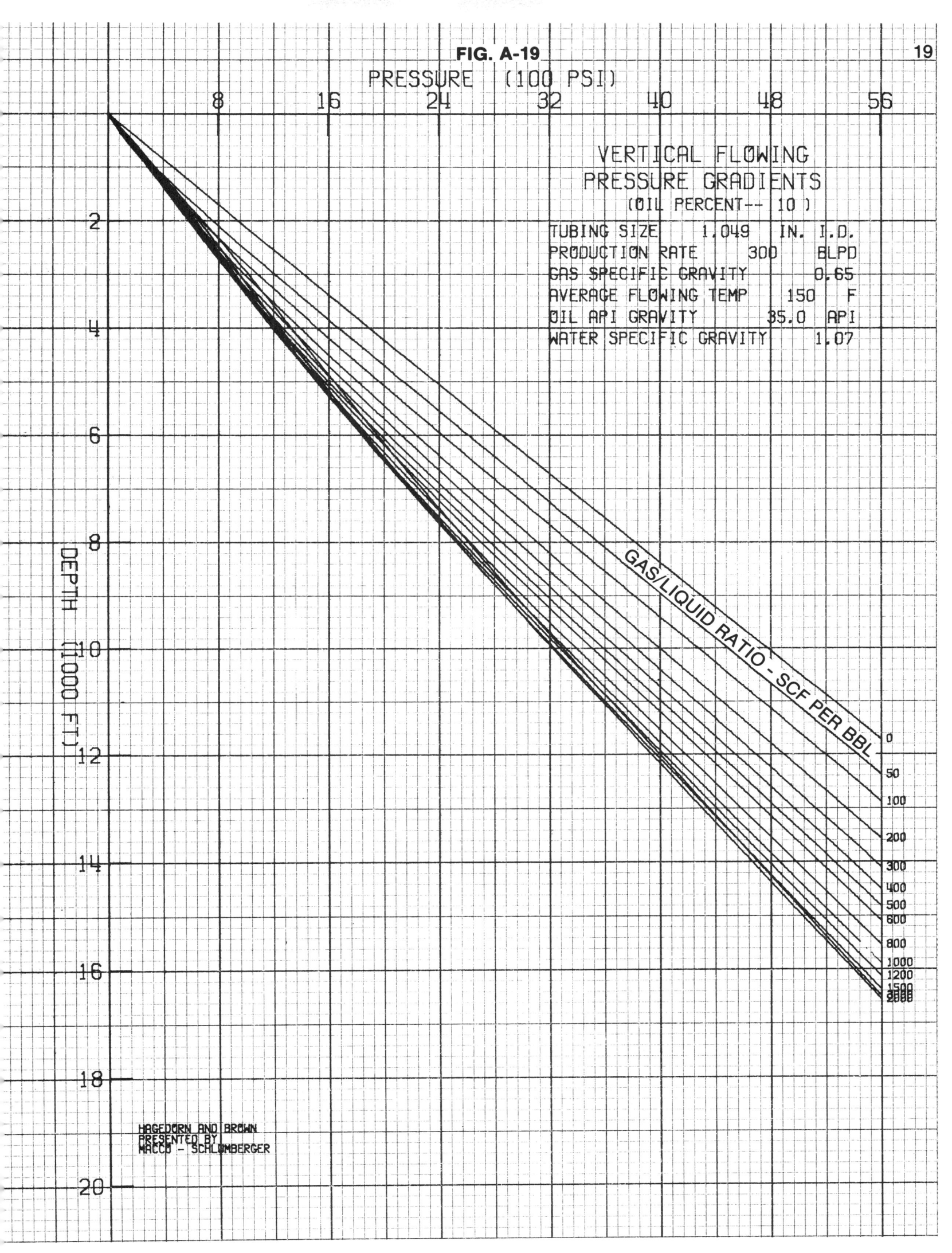

FIG. A-19

PRESSURE (100 PSI)

VERTICAL FLOWING
PRESSURE GRADIENTS
(OIL PERCENT-- 10)

TUBING SIZE	1.049	IN. I.D.
PRODUCTION RATE	300	BLPD
GAS SPECIFIC GRAVITY	0.65	
AVERAGE FLOWING TEMP	150	F
OIL API GRAVITY	35.0	API
WATER SPECIFIC GRAVITY	1.07	

DEPTH (1000 FT)

GAS/LIQUID RATIO - SCF PER BBL

0
50
100
200
300
400
500
600
800
1000
1200
1500
2000

HAGEDORN AND BROWN
PRESENTED BY
MACCO - SCHLUMBERGER

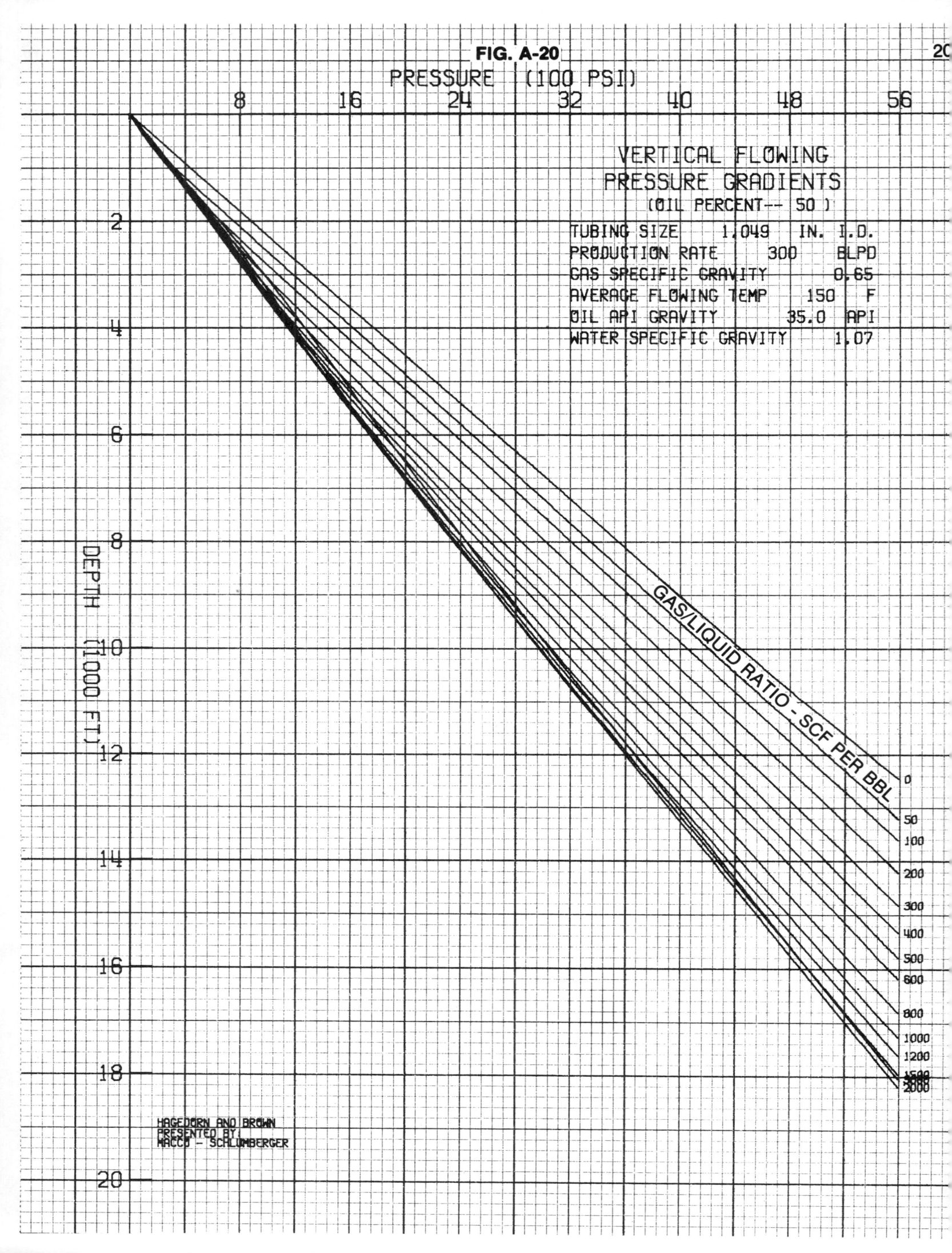

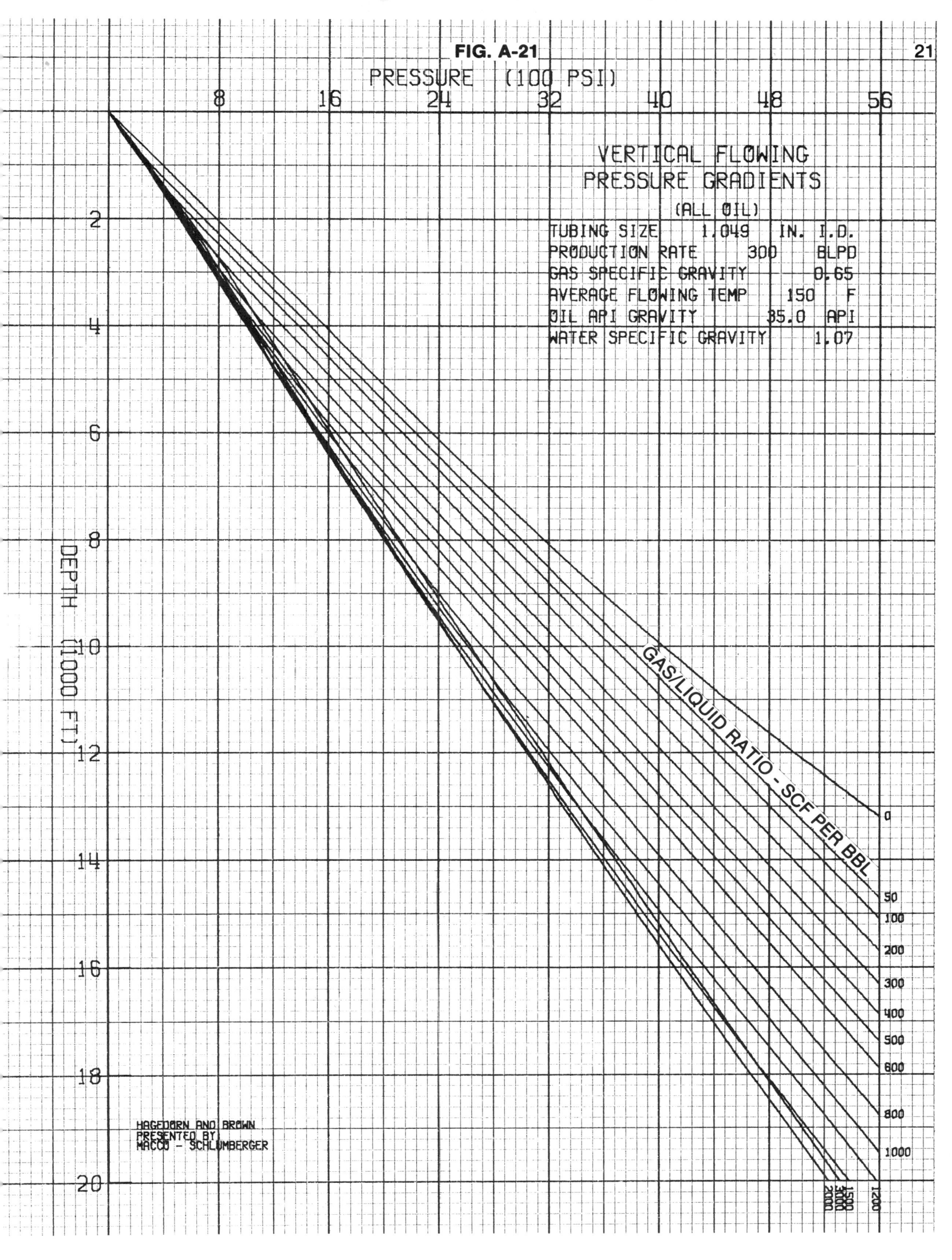

PRESSURE (100 PSI)

DEPTH (1000 FT)

VERTICAL FLOWING
PRESSURE GRADIENTS
(ALL OIL)

TUBING SIZE	1.049	IN. I.D.
PRODUCTION RATE	300	BLPD
GAS SPECIFIC GRAVITY	0.65	
AVERAGE FLOWING TEMP	150	F
OIL API GRAVITY	35.0	API
WATER SPECIFIC GRAVITY	1.07	

GAS/LIQUID RATIO - SCF PER BBL

HAGEDORN AND BROWN
PRESENTED BY
MACCO - SCHLUMBERGER

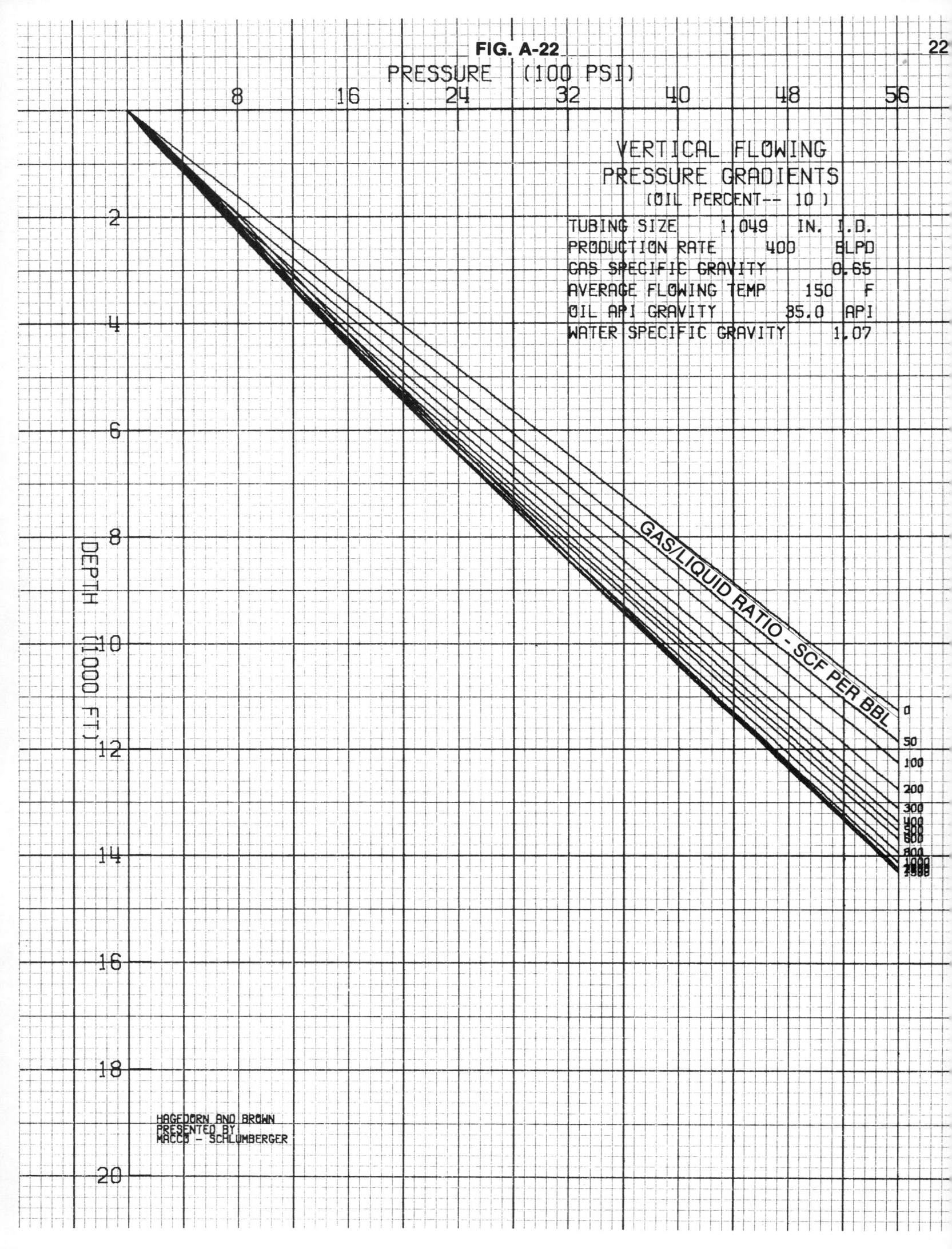

PRESSURE (100 PSI)

DEPTH (1000 FT)

VERTICAL FLOWING
PRESSURE GRADIENTS
(OIL PERCENT-- 10)

TUBING SIZE   1.049   IN. I.D.
PRODUCTION RATE   400   BLPD
GAS SPECIFIC GRAVITY   0.65
AVERAGE FLOWING TEMP   150   F
OIL API GRAVITY   35.0   API
WATER SPECIFIC GRAVITY   1.07

GAS/LIQUID RATIO - SCF PER BBL

0
50
100
200
300
400
500
600
800
1000

HAGEDORN AND BROWN
PRESENTED BY
MACCO - SCHLUMBERGER

FIG. A-23

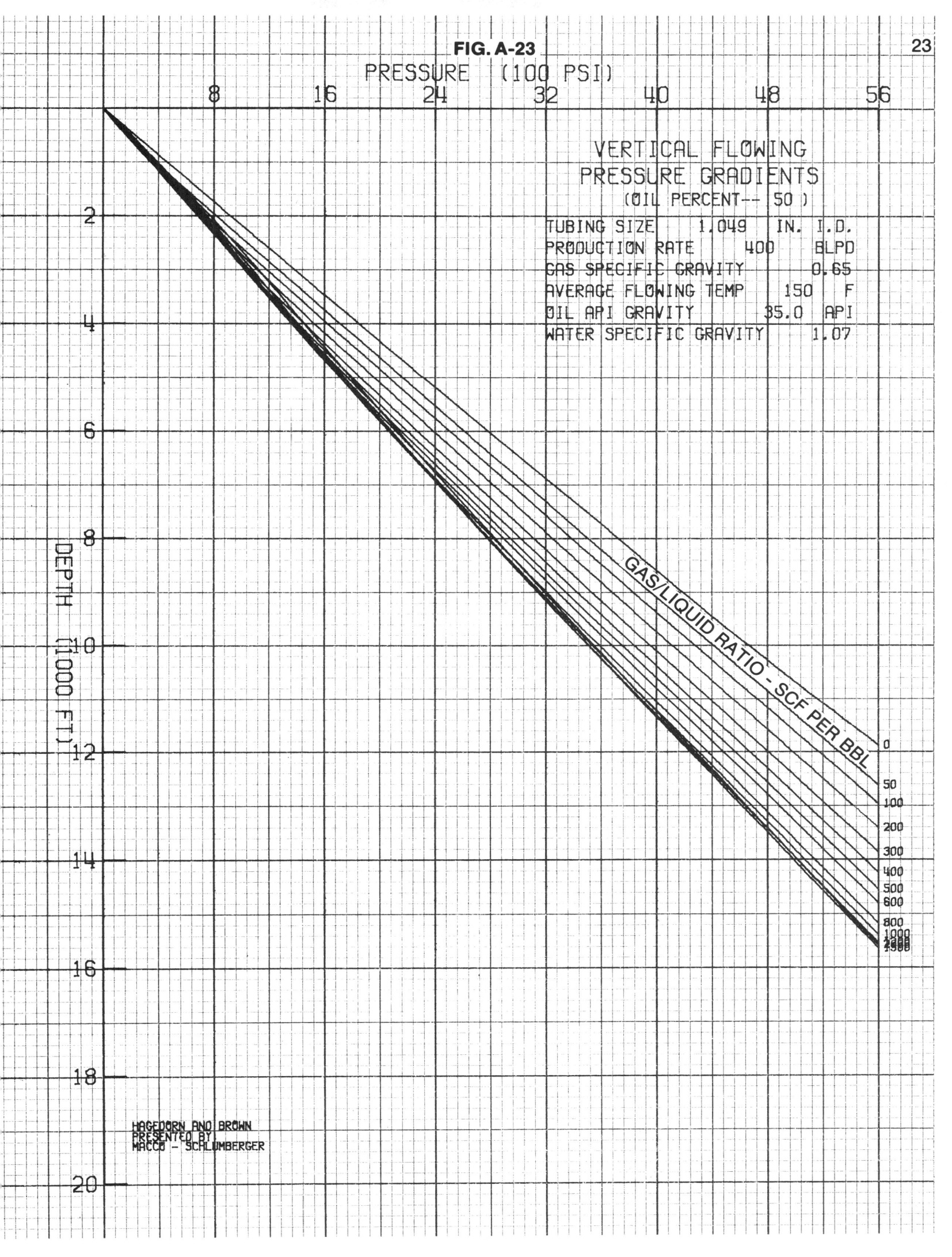

PRESSURE (100 PSI)

VERTICAL FLOWING
PRESSURE GRADIENTS
(OIL PERCENT-- 50)

TUBING SIZE	1.049	IN. I.D.
PRODUCTION RATE	400	BLPD
GAS SPECIFIC GRAVITY		0.65
AVERAGE FLOWING TEMP	150	F
OIL API GRAVITY	35.0	API
WATER SPECIFIC GRAVITY		1.07

GAS/LIQUID RATIO - SCF PER BBL

DEPTH (1000 FT)

HAGEDORN AND BROWN
PRESENTED BY
MACCO - SCHLUMBERGER

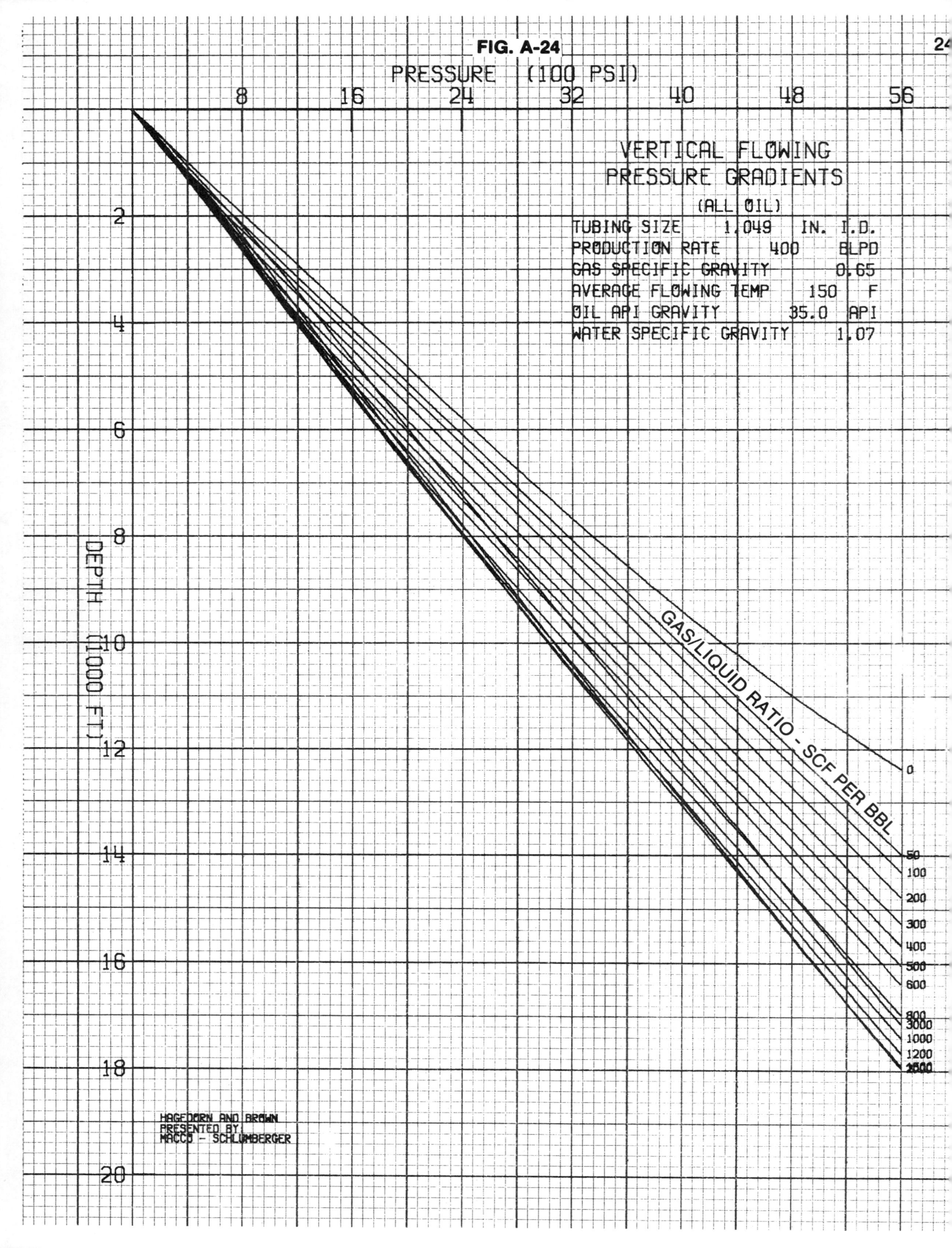

FIG. A-24

VERTICAL FLOWING
PRESSURE GRADIENTS
(ALL OIL)

TUBING SIZE	1.049	IN. I.D.
PRODUCTION RATE	400	BLPD
GAS SPECIFIC GRAVITY	0.65	
AVERAGE FLOWING TEMP	150	F
OIL API GRAVITY	35.0	API
WATER SPECIFIC GRAVITY	1.07	

PRESSURE (100 PSI)

DEPTH (1000 FT)

GAS/LIQUID RATIO - SCF PER BBL

0
50
100
200
300
400
500
600
800
3000
1000
1200
1500

HAGEDORN AND BROWN
PRESENTED BY
MACCO - SCHLUMBERGER

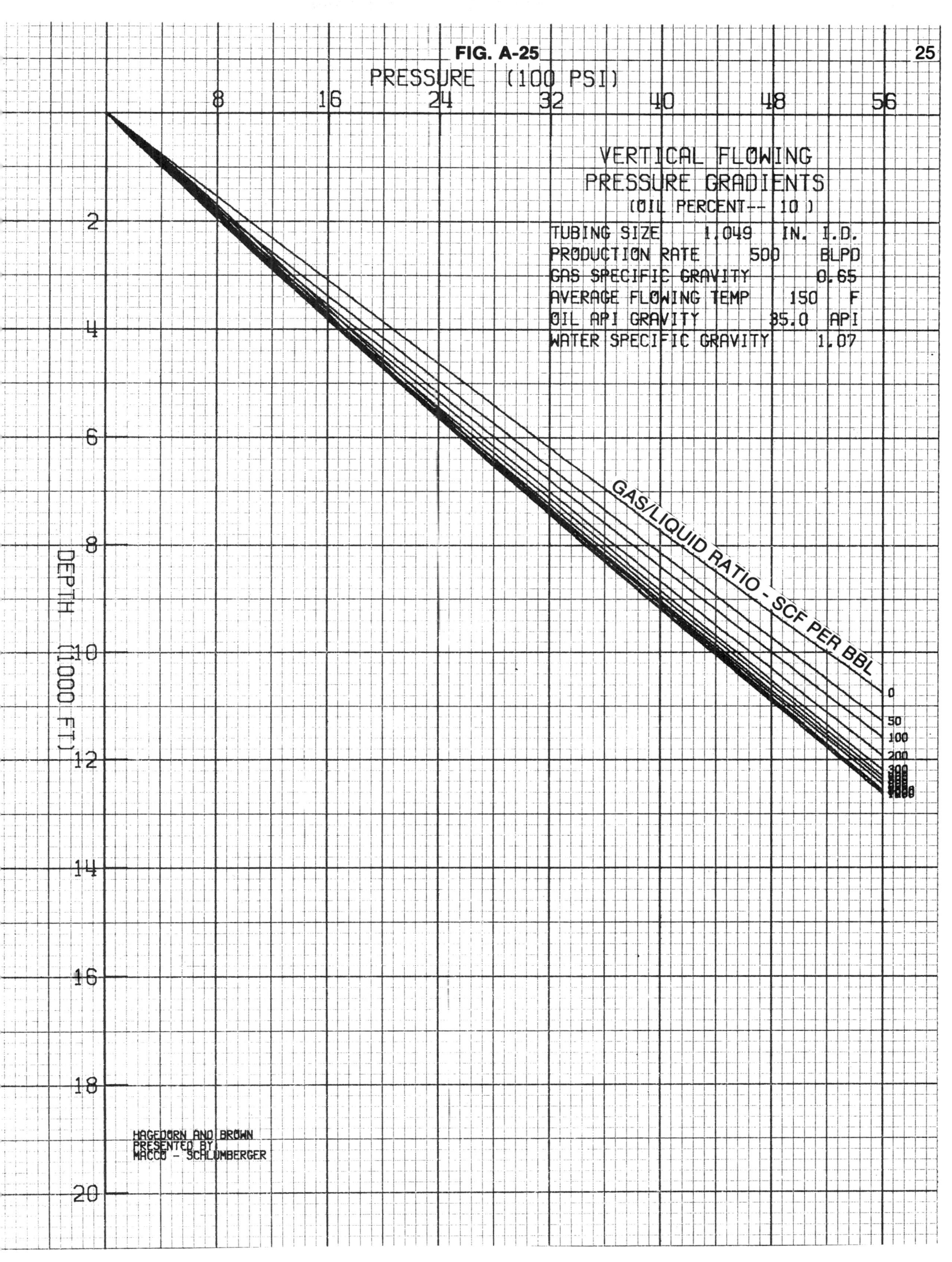

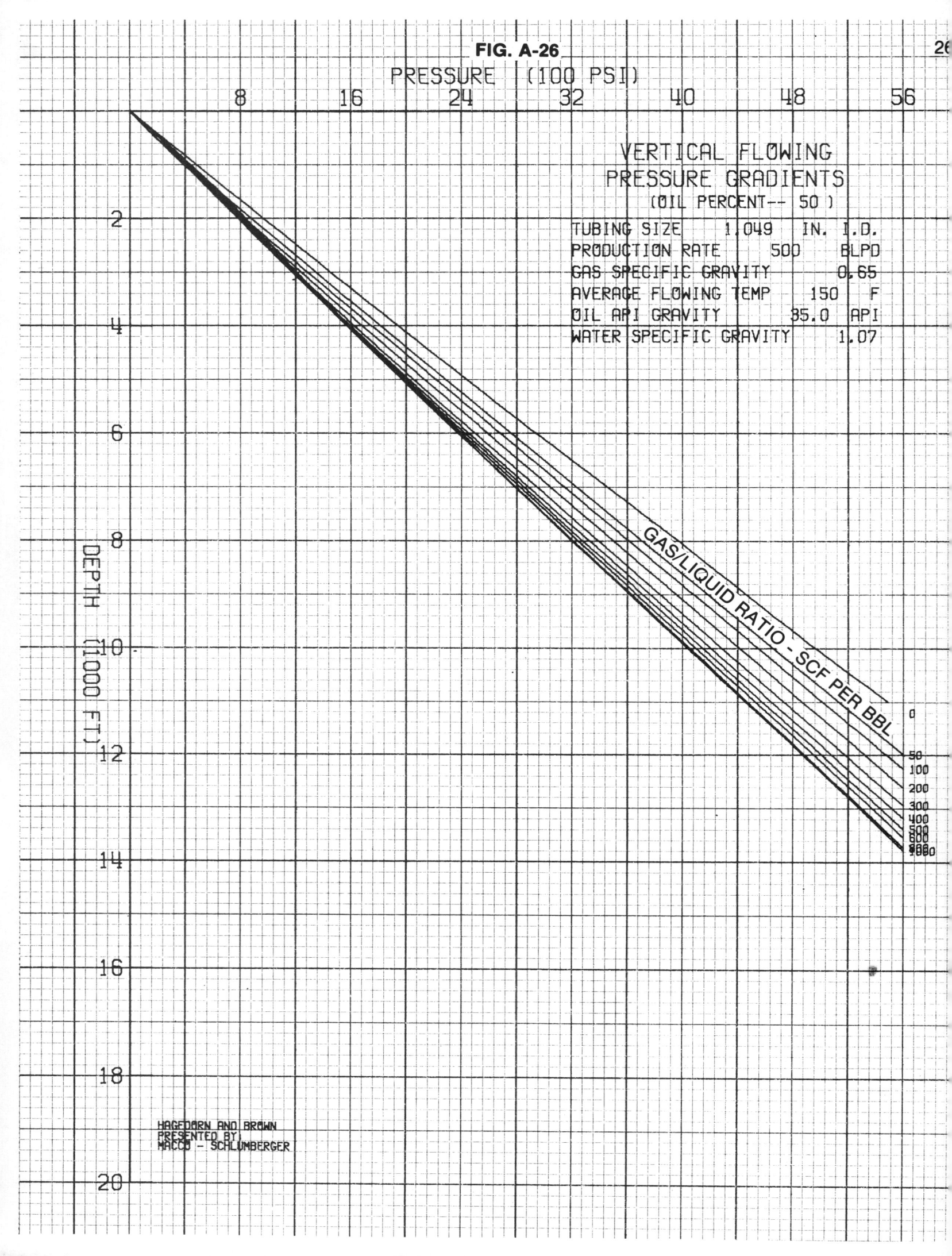

FIG. A-26

VERTICAL FLOWING PRESSURE GRADIENTS
(OIL PERCENT-- 50)

TUBING SIZE	1.049	IN. I.D.
PRODUCTION RATE	500	BLPD
GAS SPECIFIC GRAVITY	0.65	
AVERAGE FLOWING TEMP	150	F
OIL API GRAVITY	35.0	API
WATER SPECIFIC GRAVITY	1.07	

GAS/LIQUID RATIO - SCF PER BBL

HAGEDORN AND BROWN
PRESENTED BY
MACCO — SCHLUMBERGER

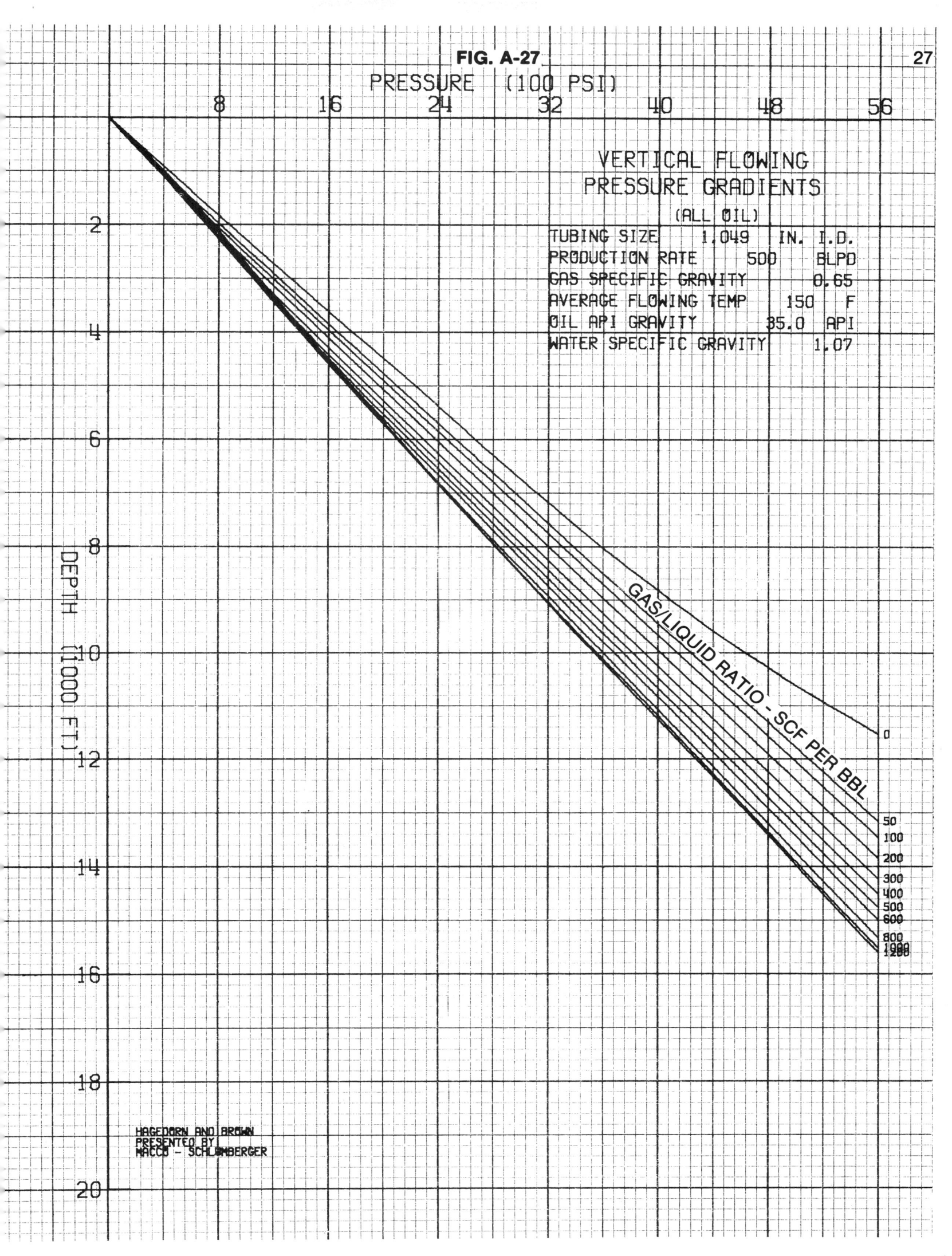

VERTICAL FLOWING
PRESSURE GRADIENTS
(ALL OIL)

TUBING SIZE	1.049	IN.	I.D.
PRODUCTION RATE	500		BLPD
GAS SPECIFIC GRAVITY	0.65		
AVERAGE FLOWING TEMP	150		F
OIL API GRAVITY	35.0		API
WATER SPECIFIC GRAVITY	1.07		

PRESSURE (100 PSI)

DEPTH (1000 FT)

GAS/LIQUID RATIO - SCF PER BBL

HAGEDORN AND BROWN
PRESENTED BY
NACCO - SCHLUMBERGER

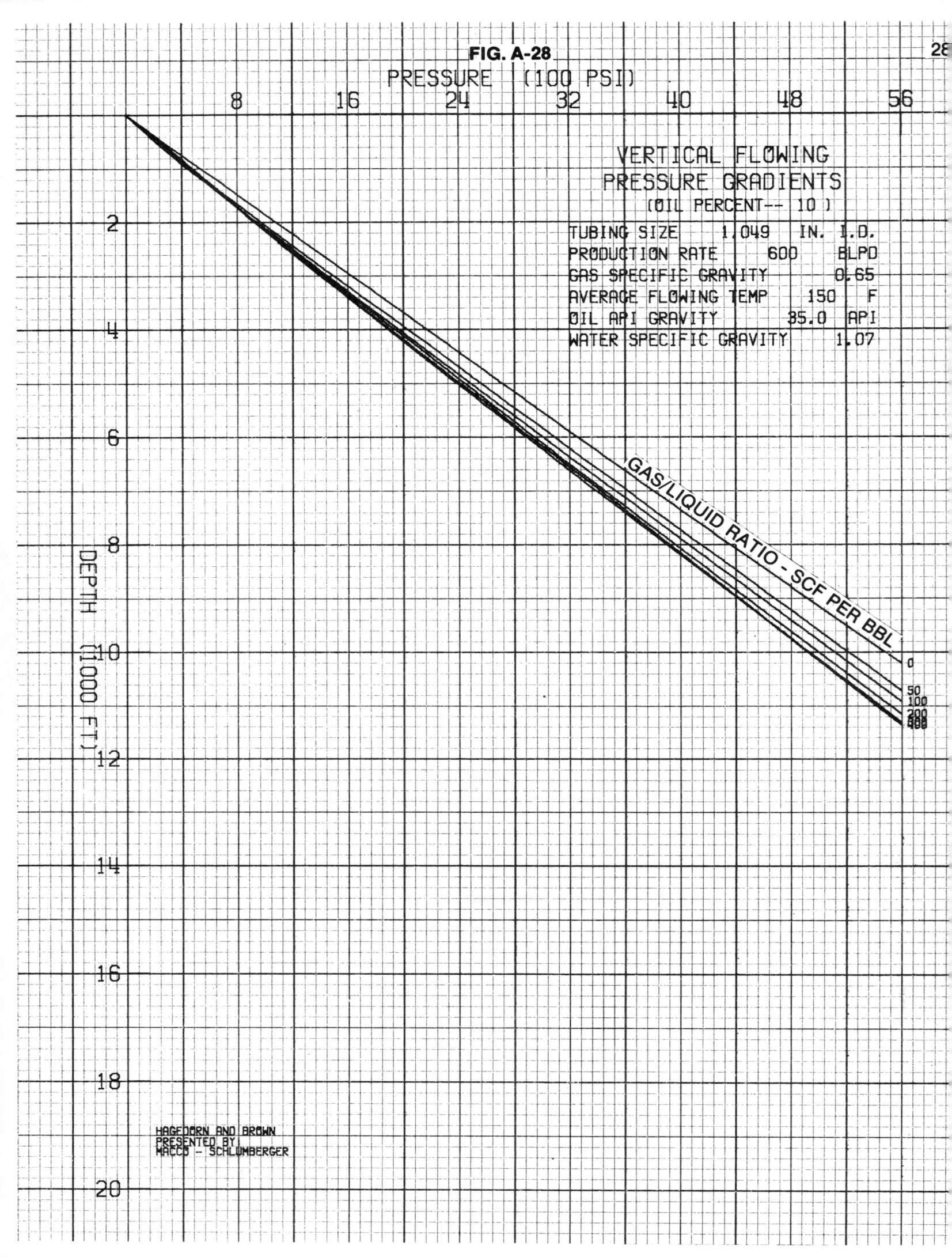

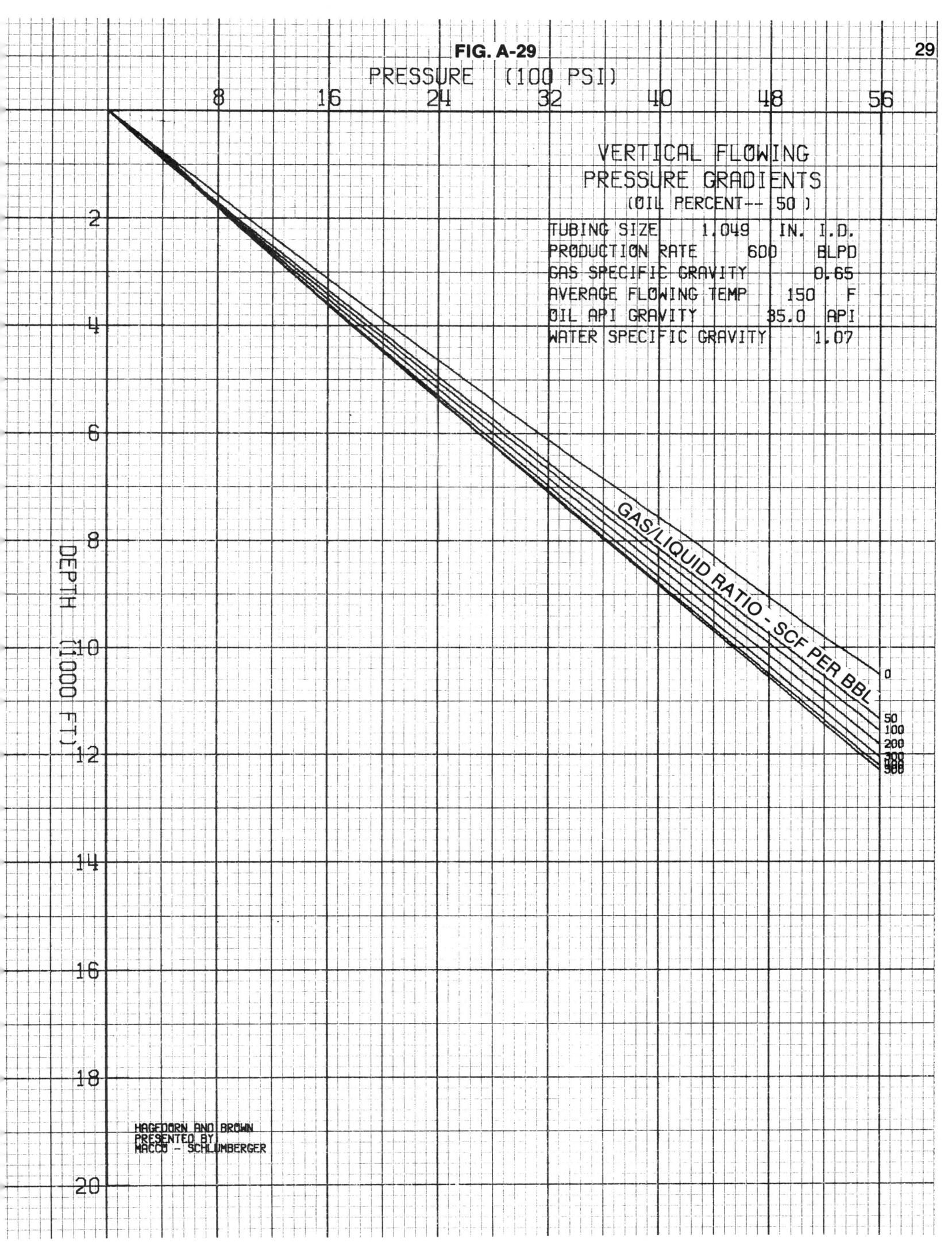

PRESSURE (100 PSI)

DEPTH (1000 FT)

VERTICAL FLOWING
PRESSURE GRADIENTS
(OIL PERCENT-- 50)

TUBING SIZE	1.049	IN. I.D.
PRODUCTION RATE	600	BLPD
GAS SPECIFIC GRAVITY		0.65
AVERAGE FLOWING TEMP	150	F
OIL API GRAVITY	35.0	API
WATER SPECIFIC GRAVITY		1.07

GAS/LIQUID RATIO - SCF PER BBL

0
50
100
200
300
600
900

HAGEDORN AND BROWN
PRESENTED BY
MACCO - SCHLUMBERGER

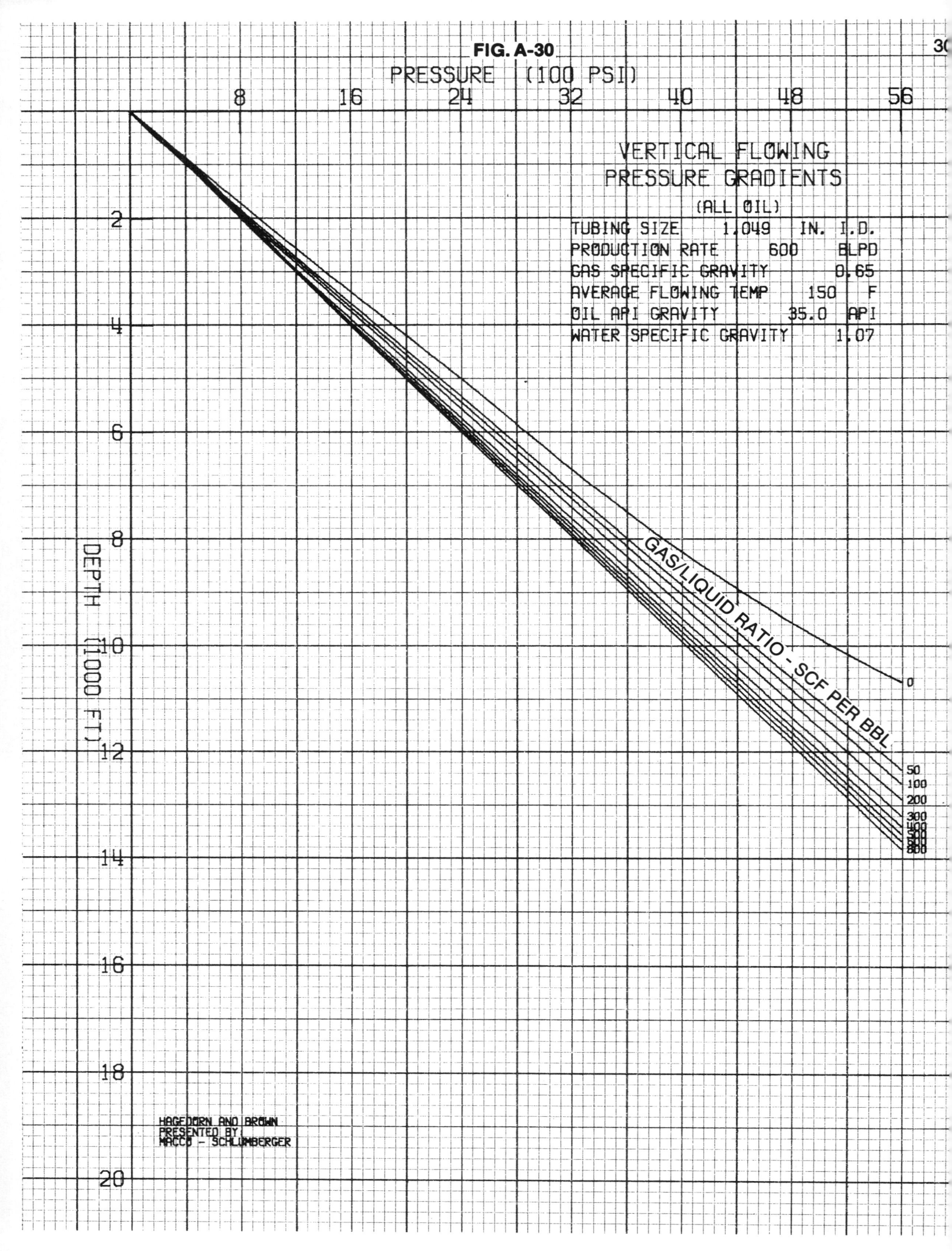

VERTICAL FLOWING
PRESSURE GRADIENTS
(ALL OIL)

TUBING SIZE	1.049	IN. I.D.
PRODUCTION RATE	600	BLPD
GAS SPECIFIC GRAVITY	0.65	
AVERAGE FLOWING TEMP	150	F
OIL API GRAVITY	35.0	API
WATER SPECIFIC GRAVITY	1.07	

PRESSURE (100 PSI)

DEPTH (1000 FT)

GAS/LIQUID RATIO - SCF PER BBL

0
50
100
200
300
400
500
600
800

HAGEDORN AND BROWN
PRESENTED BY
MACCO - SCHLUMBERGER

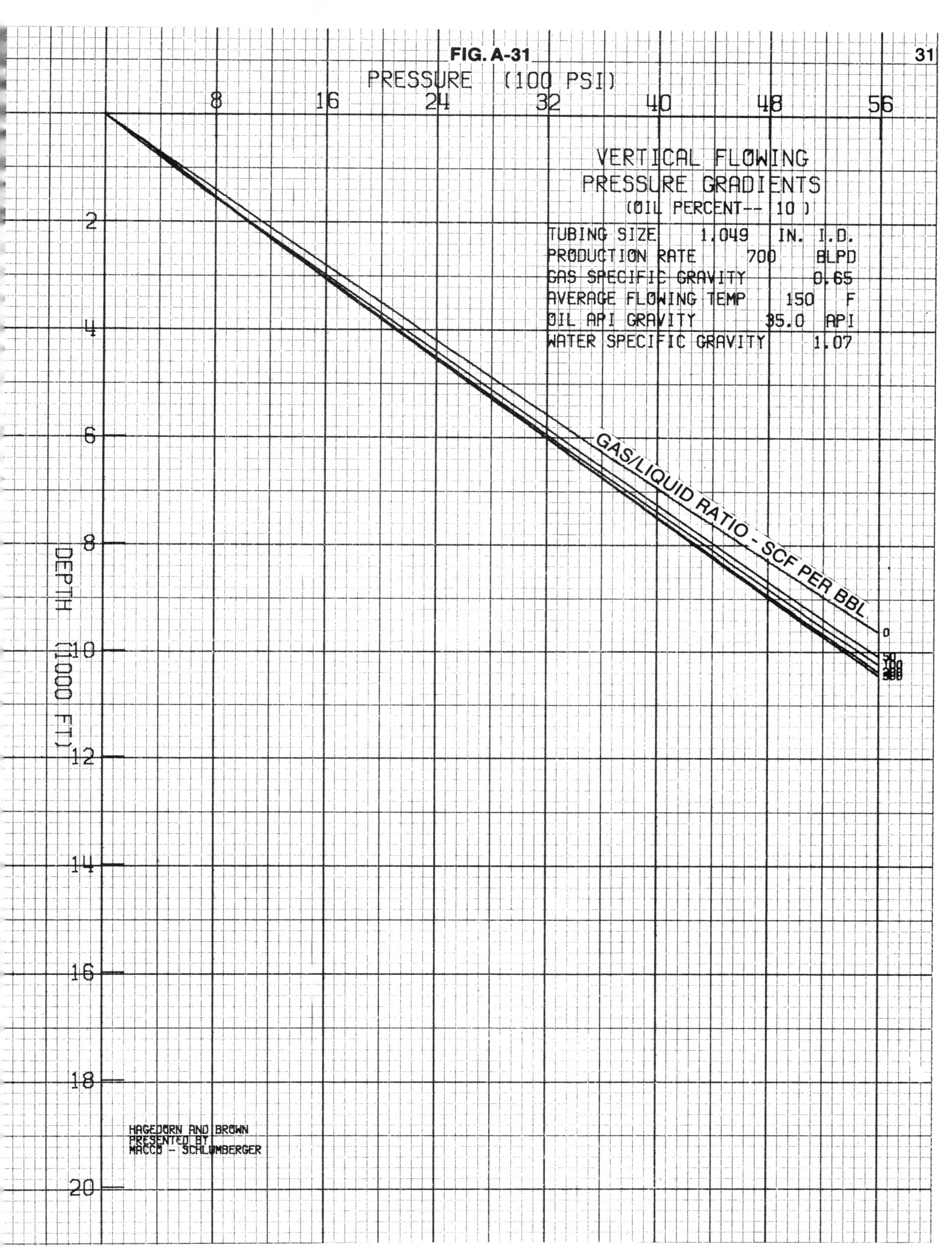

PRESSURE (100 PSI)

VERTICAL FLOWING
PRESSURE GRADIENTS
(OIL PERCENT-- 10)

TUBING SIZE	1.049	IN. I.D.
PRODUCTION RATE	700	BLPD
GAS SPECIFIC GRAVITY	0.65	
AVERAGE FLOWING TEMP	150	F
OIL API GRAVITY	35.0	API
WATER SPECIFIC GRAVITY	1.07	

GAS/LIQUID RATIO - SCF PER BBL

DEPTH (1000 FT)

HAGEDORN AND BROWN
PRESENTED BY
MACCO - SCHLUMBERGER

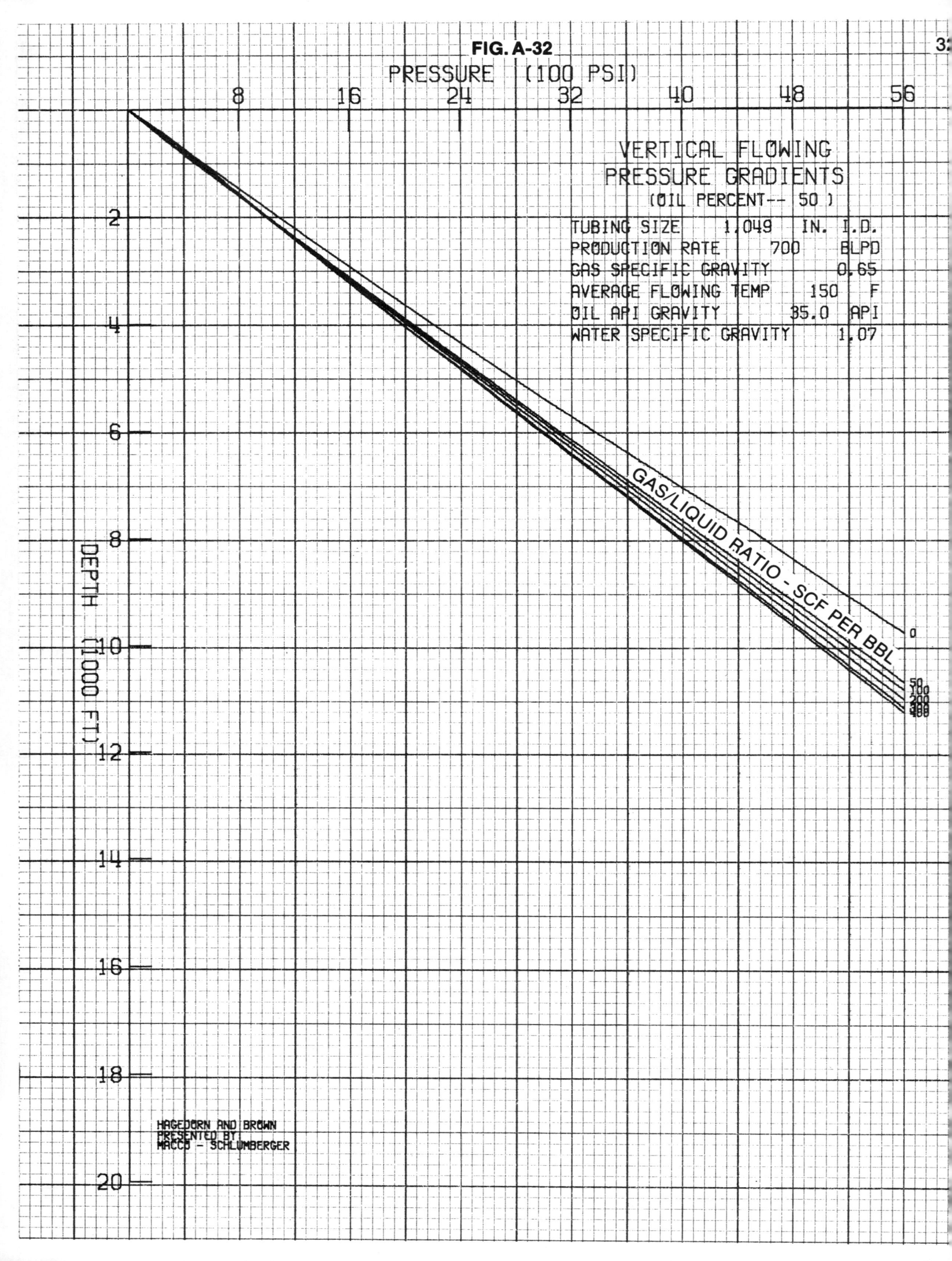

VERTICAL FLOWING
PRESSURE GRADIENTS
(OIL PERCENT-- 50)

TUBING SIZE	1.049	IN. I.D.
PRODUCTION RATE	700	BLPD
GAS SPECIFIC GRAVITY		0.65
AVERAGE FLOWING TEMP	150	F
OIL API GRAVITY	35.0	API
WATER SPECIFIC GRAVITY		1.07

PRESSURE (100 PSI)

DEPTH (1000 FT)

GAS/LIQUID RATIO - SCF PER BBL

0
50
100
200
400

HAGEDORN AND BROWN
PRESENTED BY
MACCO -- SCHLUMBERGER

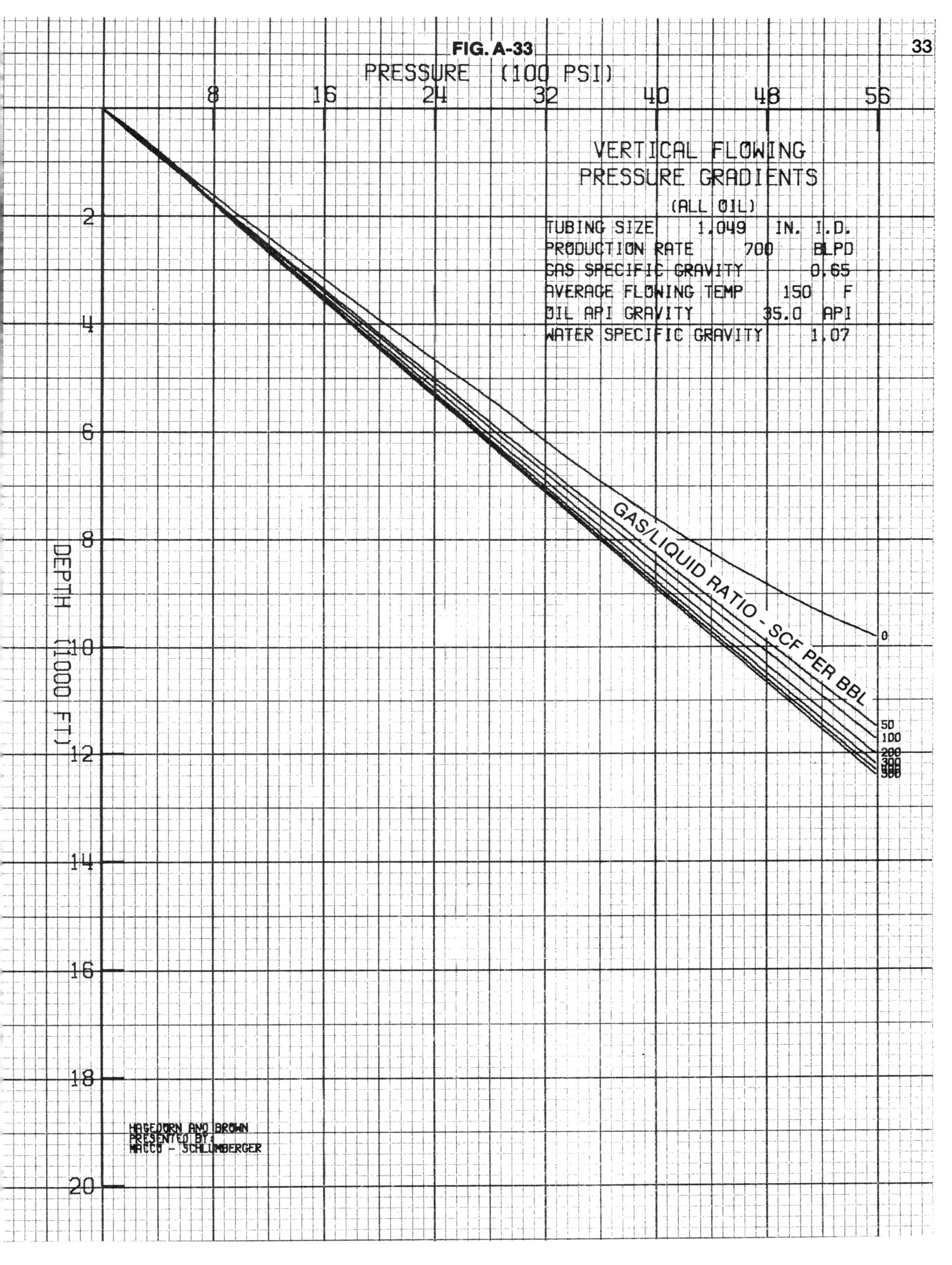

VERTICAL FLOWING
PRESSURE GRADIENTS
(ALL OIL)

TUBING SIZE	1.049	IN. I.D.
PRODUCTION RATE	700	BLPD
GAS SPECIFIC GRAVITY	0.65	
AVERAGE FLOWING TEMP	150	F
OIL API GRAVITY	35.0	API
WATER SPECIFIC GRAVITY	1.07	

PRESSURE (100 PSI)

DEPTH (1000 FT)

GAS/LIQUID RATIO - SCF PER BBL

0
50
100
200
300
500

HAGEDORN AND BROWN
PRESENTED BY:
MACCO - SCHLUMBERGER

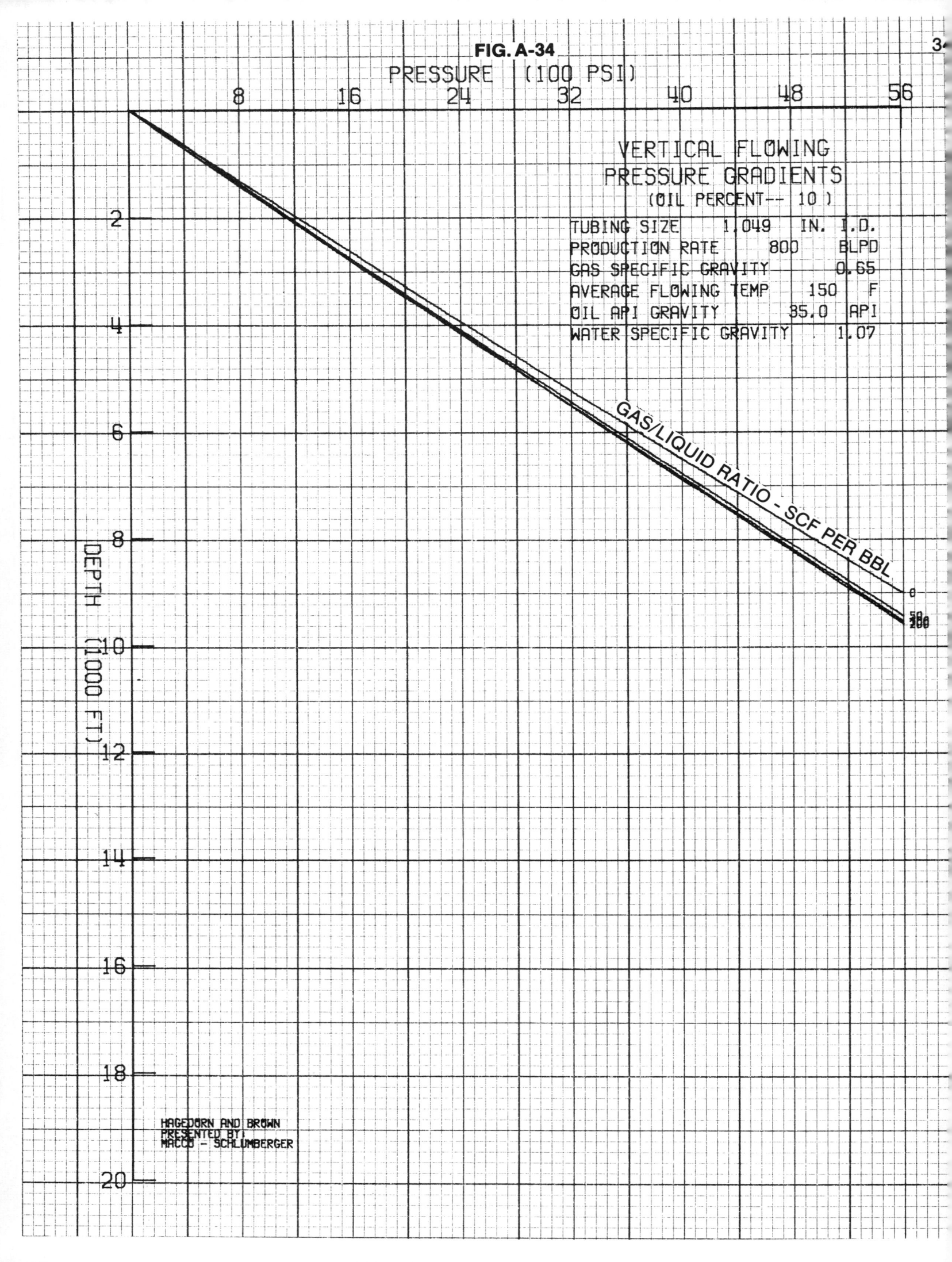

VERTICAL FLOWING
PRESSURE GRADIENTS
(OIL PERCENT-- 10)

TUBING SIZE	1.049	IN. I.D.
PRODUCTION RATE	800	BLPD
GAS SPECIFIC GRAVITY	0.65	
AVERAGE FLOWING TEMP	150	F
OIL API GRAVITY	35.0	API
WATER SPECIFIC GRAVITY	1.07	

PRESSURE (100 PSI)

DEPTH (1000 FT)

GAS/LIQUID RATIO - SCF PER BBL

HAGEDORN AND BROWN
PRESENTED BY:
MAECO - SCHLUMBERGER

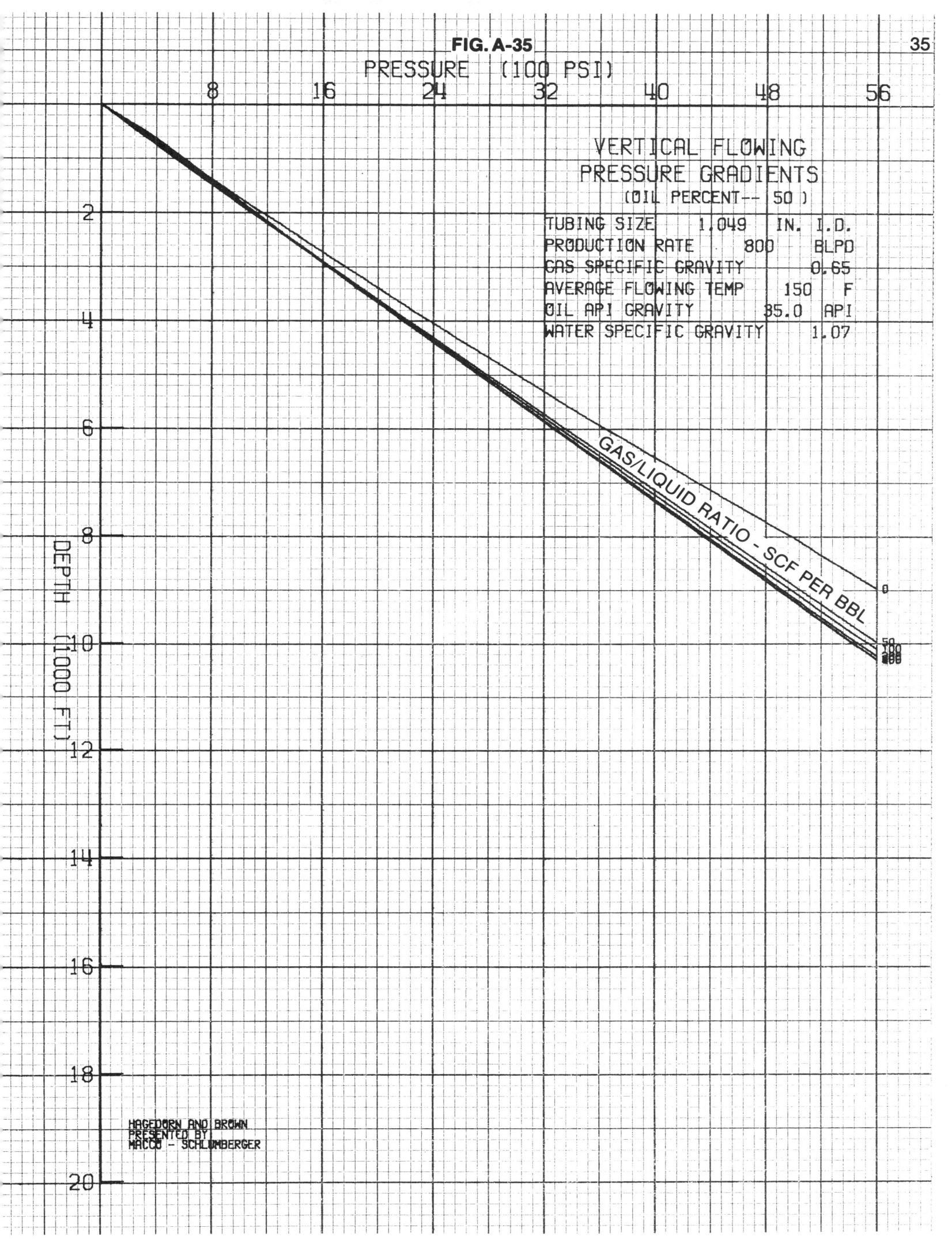

HAGEDORN AND BROWN
PRESENTED BY
MACCO — SCHLUMBERGER

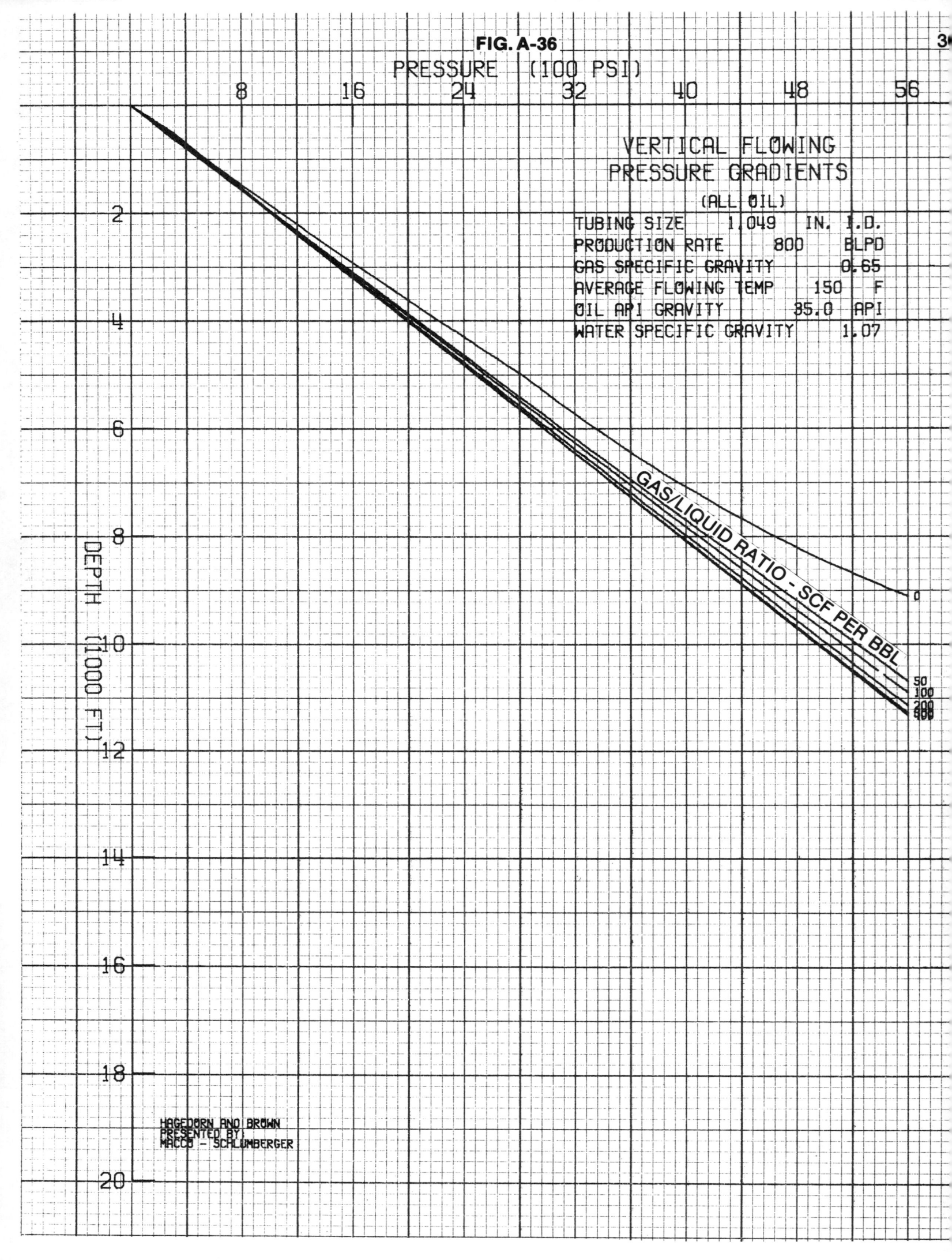

PRESSURE (100 PSI)

VERTICAL FLOWING
PRESSURE GRADIENTS

(ALL OIL)

TUBING SIZE	1.049	IN. I.D.
PRODUCTION RATE	800	BLPD
GAS SPECIFIC GRAVITY	0.65	
AVERAGE FLOWING TEMP	150	F
OIL API GRAVITY	35.0	API
WATER SPECIFIC GRAVITY	1.07	

DEPTH (1000 FT)

GAS/LIQUID RATIO - SCF PER BBL

0
50
100
200
400
600

HAGEDORN AND BROWN
PRESENTED BY:
MACCO - SCHLUMBERGER

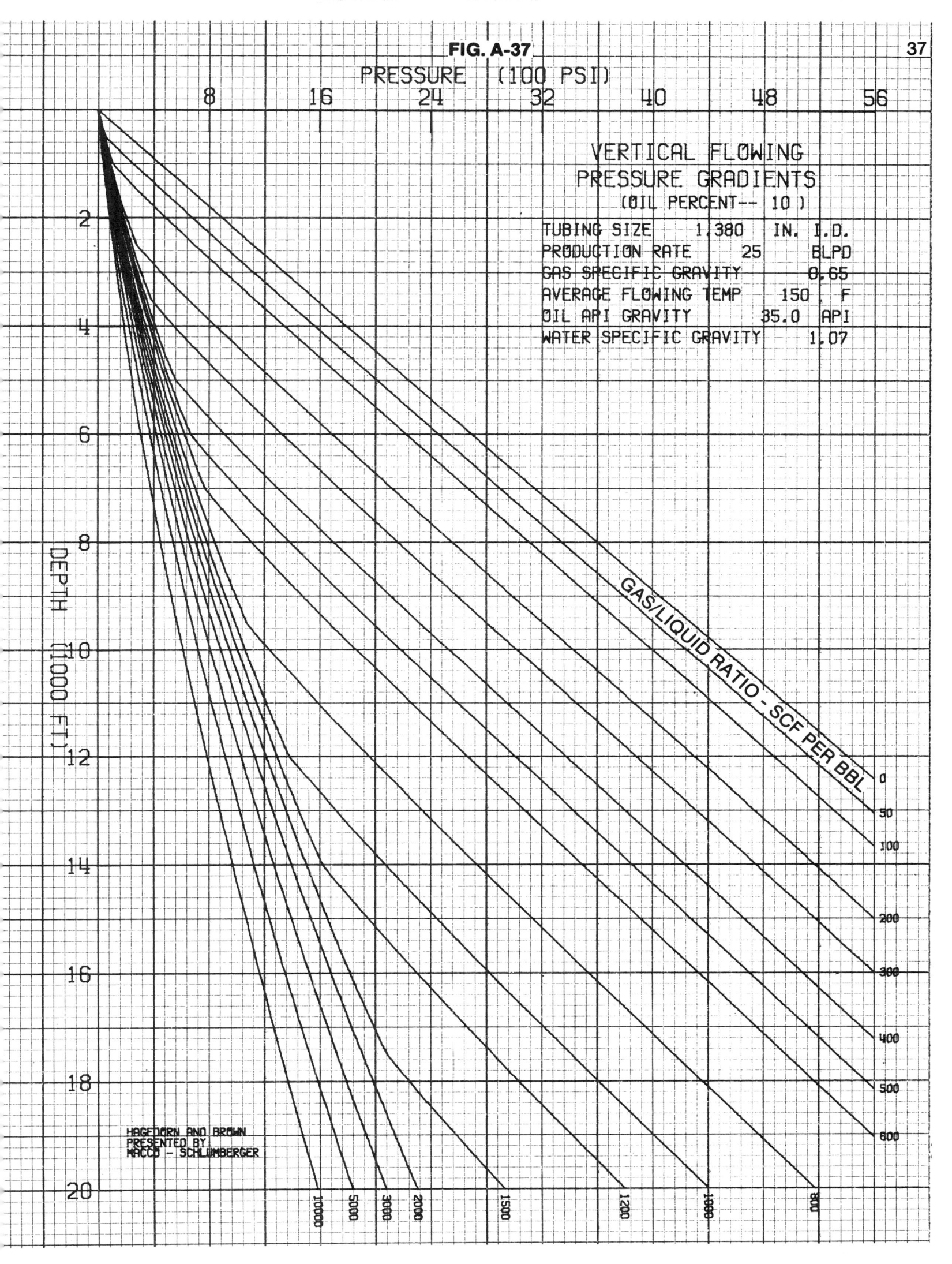

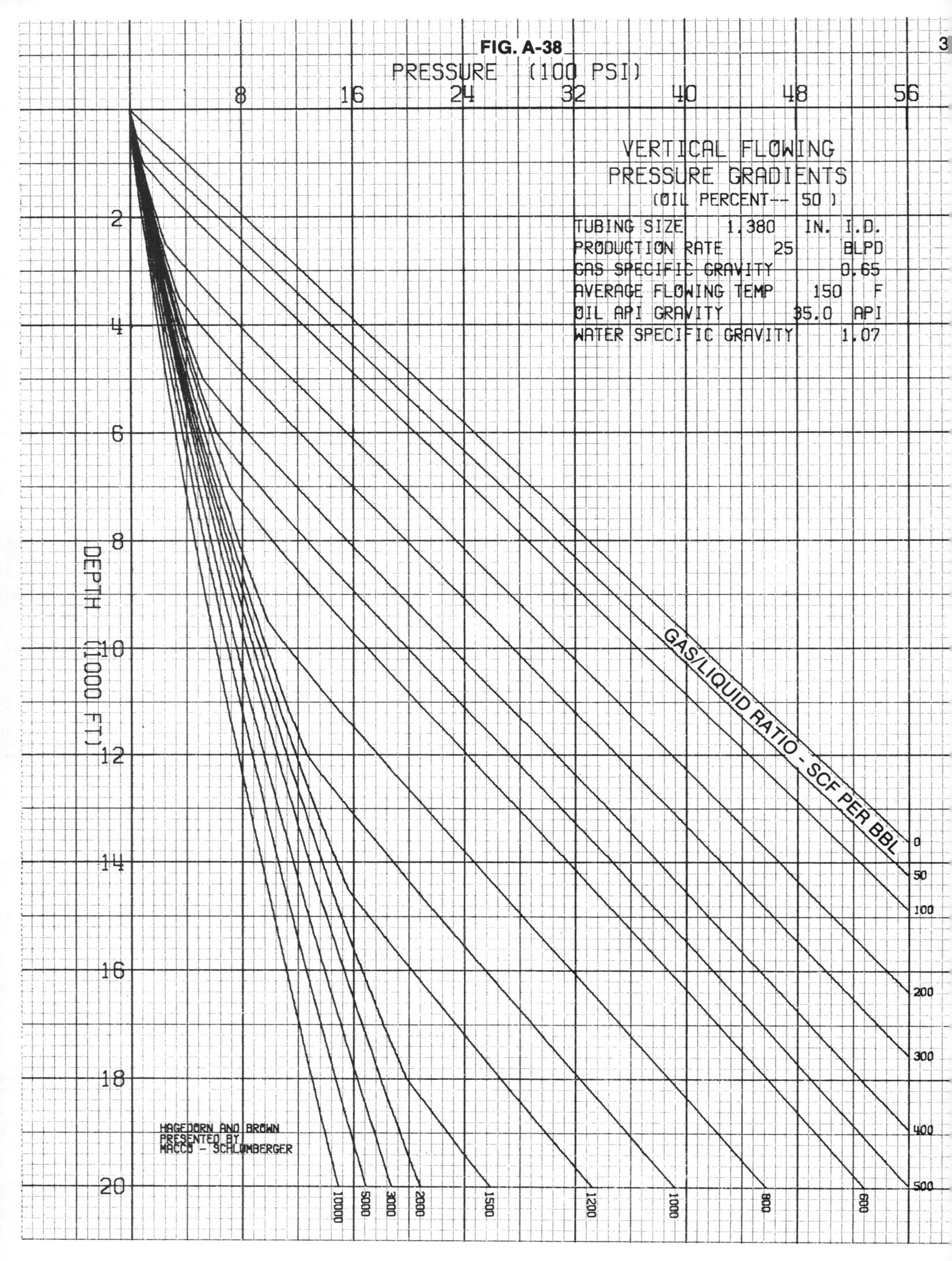

PRESSURE (100 PSI)

VERTICAL FLOWING
PRESSURE GRADIENTS
(OIL PERCENT--- 50)

TUBING SIZE	1.380	IN. I.D.
PRODUCTION RATE	25	BLPD
GAS SPECIFIC GRAVITY		0.65
AVERAGE FLOWING TEMP	150	F
OIL API GRAVITY	35.0	API
WATER SPECIFIC GRAVITY		1.07

DEPTH (1000 FT)

GAS/LIQUID RATIO - SCF PER BBL

HAGEDORN AND BROWN
PRESENTED BY
MACCO - SCHLUMBERGER

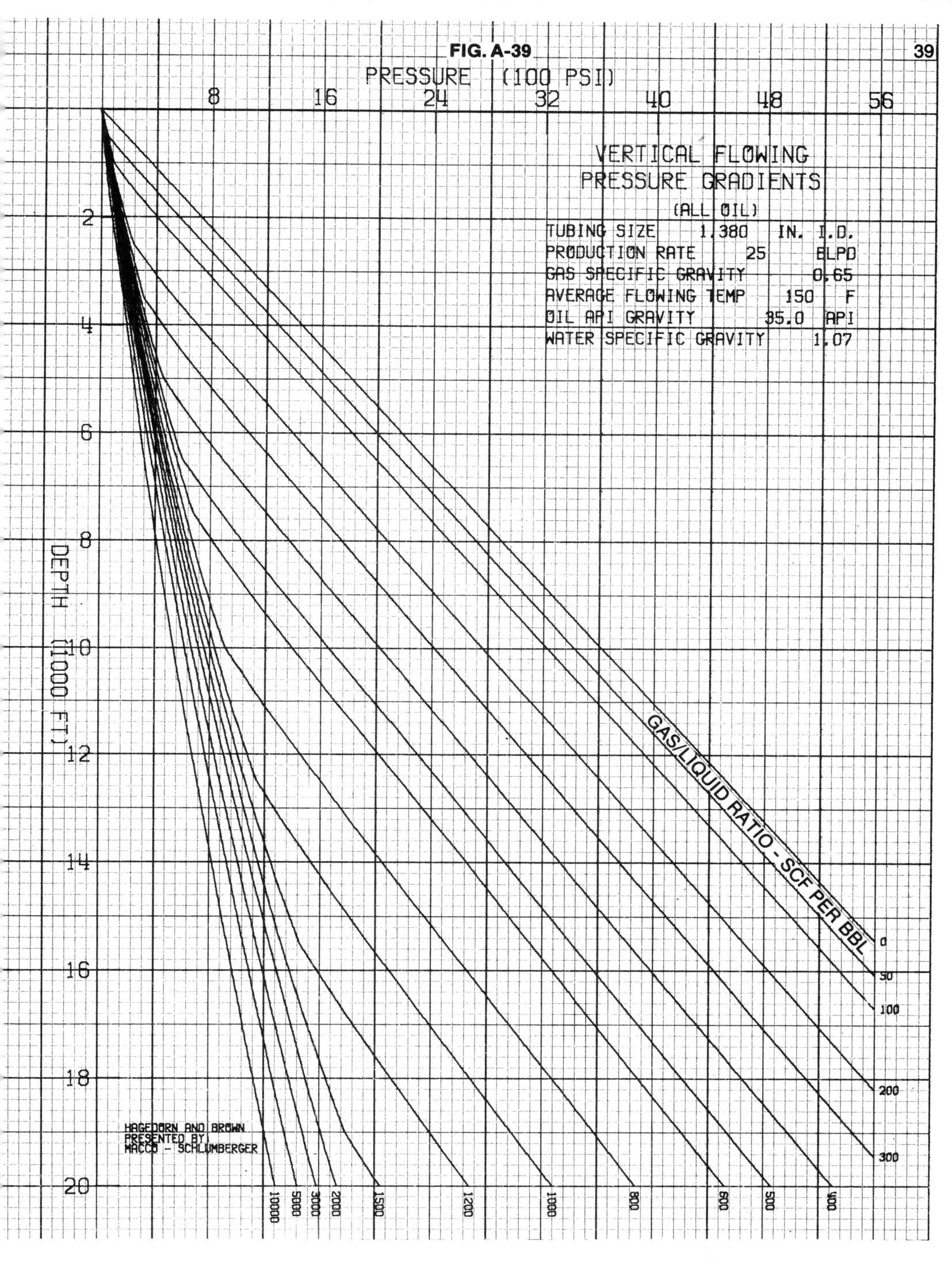

PRESSURE (100 PSI)

VERTICAL FLOWING
PRESSURE GRADIENTS
(ALL OIL)

TUBING SIZE	1.380	IN. I.D.
PRODUCTION RATE	25	BLPD
GAS SPECIFIC GRAVITY	0.65	
AVERAGE FLOWING TEMP	150	F
OIL API GRAVITY	35.0	API
WATER SPECIFIC GRAVITY	1.07	

DEPTH (1000 FT)

GAS/LIQUID RATIO - SCF PER BBL

HAGEDORN AND BROWN
PRESENTED BY
MACCO - SCHLUMBERGER

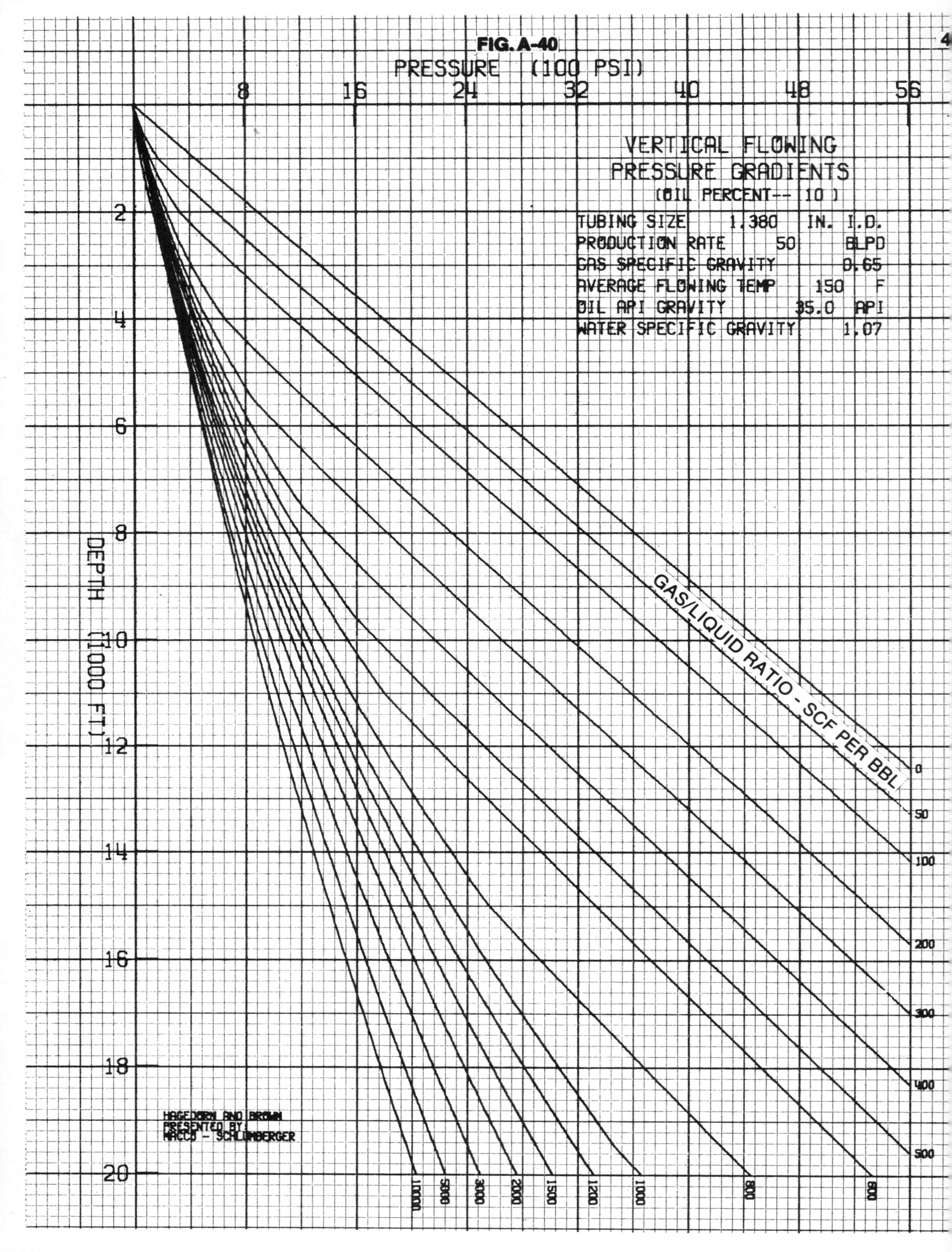

FIG. A-40

PRESSURE (100 PSI)

VERTICAL FLOWING
PRESSURE GRADIENTS
(OIL PERCENT -- 10)

TUBING SIZE	1.380	IN. I.D.
PRODUCTION RATE	50	BLPD
GAS SPECIFIC GRAVITY	0.65	
AVERAGE FLOWING TEMP	150	F
OIL API GRAVITY	35.0	API
WATER SPECIFIC GRAVITY	1.07	

GAS/LIQUID RATIO - SCF PER BBL

DEPTH (1000 FT)

HAGEDORN AND BROWN
PRESENTED BY
KALCO - SCHLUMBERGER

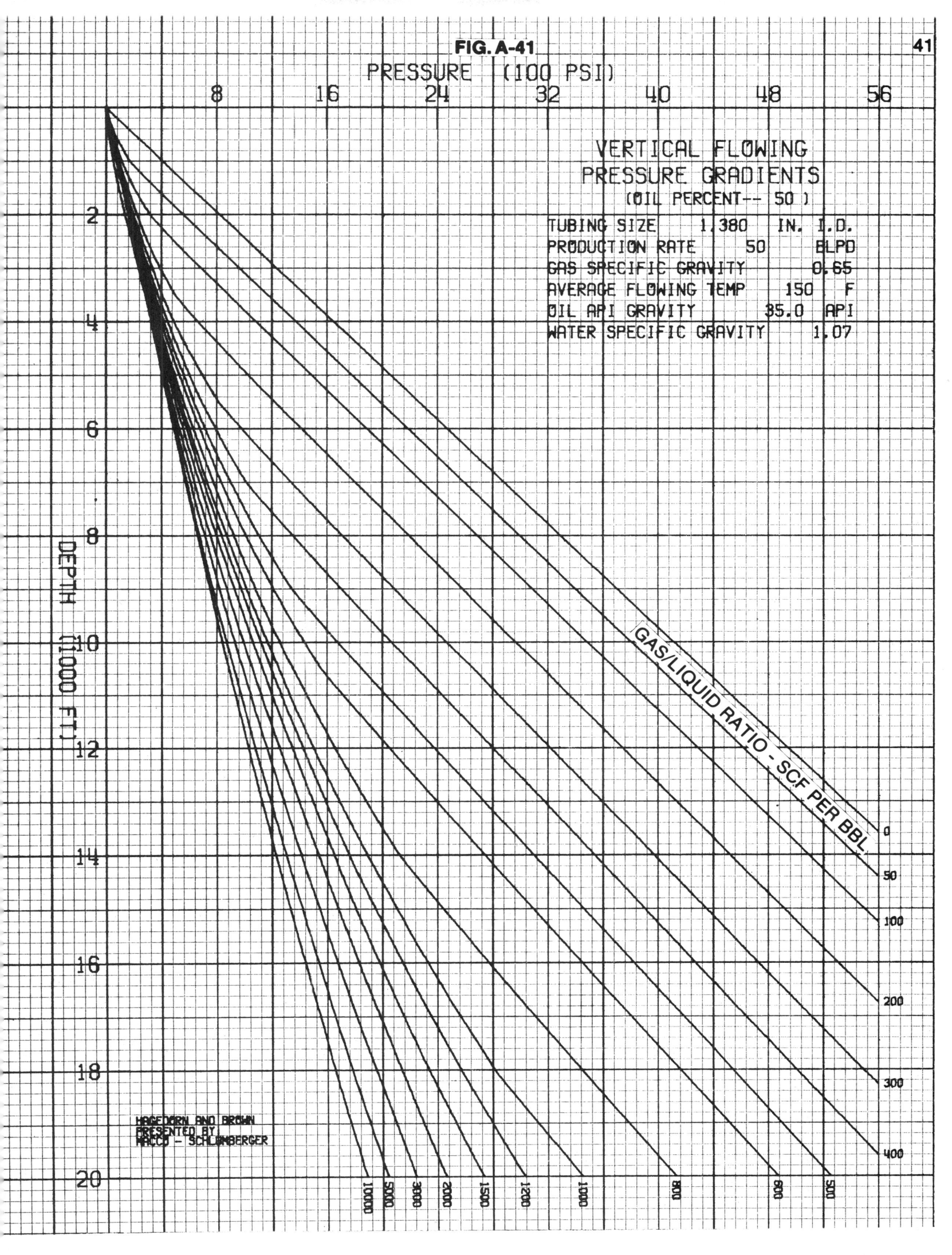

VERTICAL FLOWING
PRESSURE GRADIENTS
(OIL PERCENT-- 50)

TUBING SIZE	1.380 IN. I.D.
PRODUCTION RATE	50 BLPD
GAS SPECIFIC GRAVITY	0.65
AVERAGE FLOWING TEMP	150 F
OIL API GRAVITY	35.0 API
WATER SPECIFIC GRAVITY	1.07

GAS/LIQUID RATIO - SCF PER BBL

HAGEDORN AND BROWN
PRESENTED BY
HALCO - SCHLUMBERGER

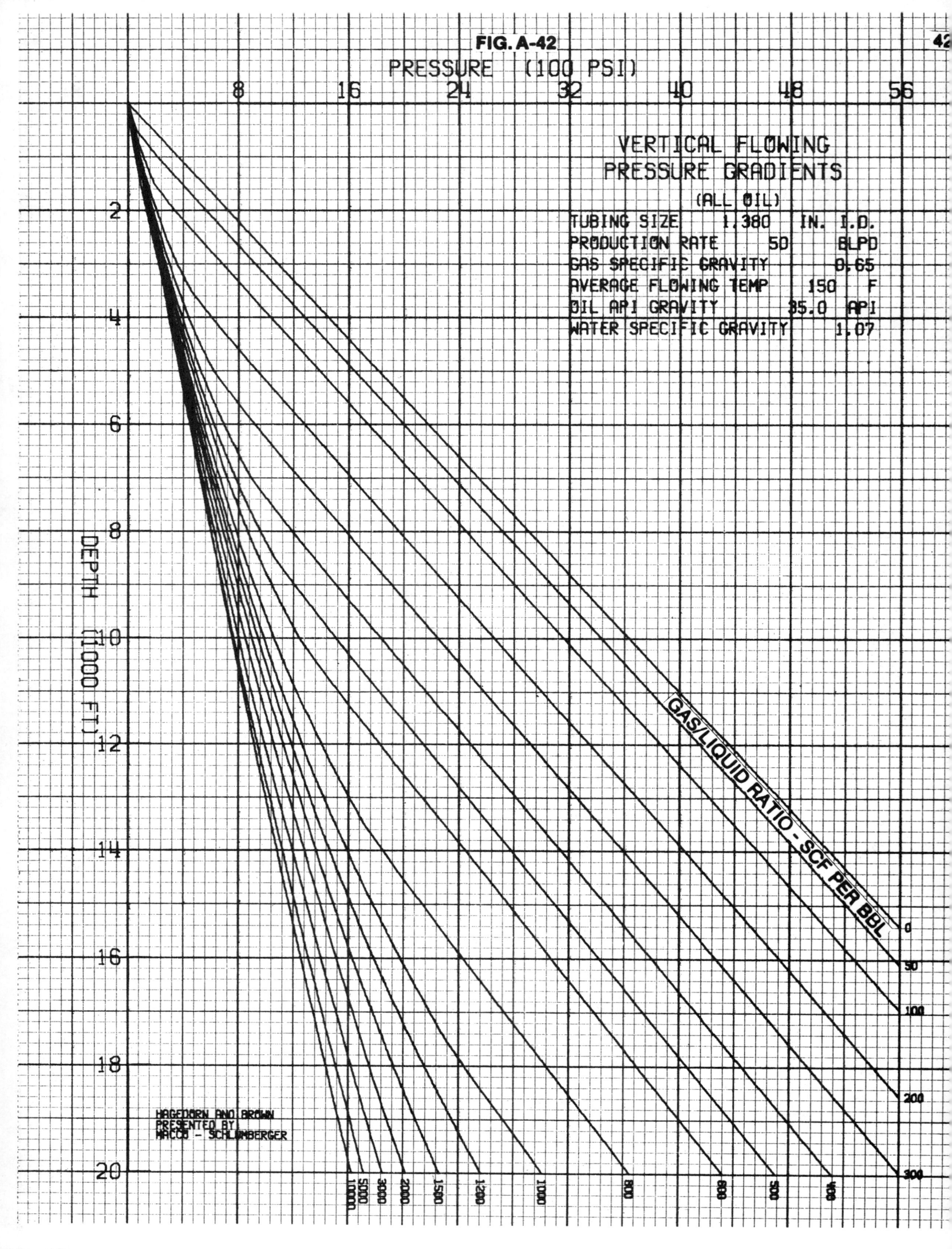

FIG. A-42

42

PRESSURE (100 PSI)

VERTICAL FLOWING
PRESSURE GRADIENTS
(ALL OIL)

TUBING SIZE	1.380	IN.	I.D.
PRODUCTION RATE	50		BLPD
GAS SPECIFIC GRAVITY		0.65	
AVERAGE FLOWING TEMP	150		F
OIL API GRAVITY		35.0	API
WATER SPECIFIC GRAVITY		1.07	

DEPTH (1,000 FT)

GAS/LIQUID RATIO - SCF PER BBL

HAGEDORN AND BROWN
PRESENTED BY
MACCO - SCHLUMBERGER

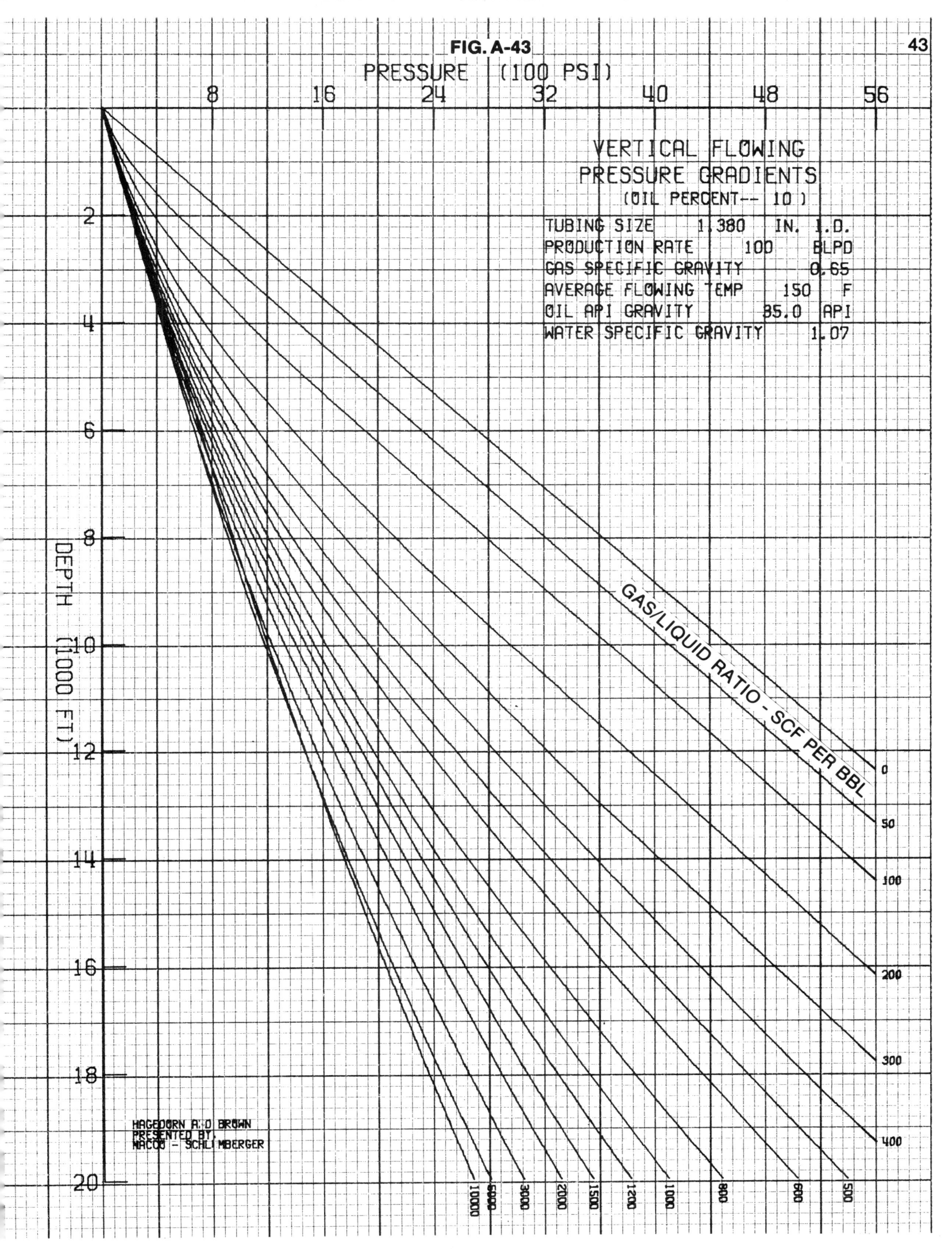

VERTICAL FLOWING
PRESSURE GRADIENTS
(OIL PERCENT-- 10)

TUBING SIZE	1.380	IN. I.D.
PRODUCTION RATE	100	BLPD
GAS SPECIFIC GRAVITY	0.65	
AVERAGE FLOWING TEMP	150	F
OIL API GRAVITY	85.0	API
WATER SPECIFIC GRAVITY	1.07	

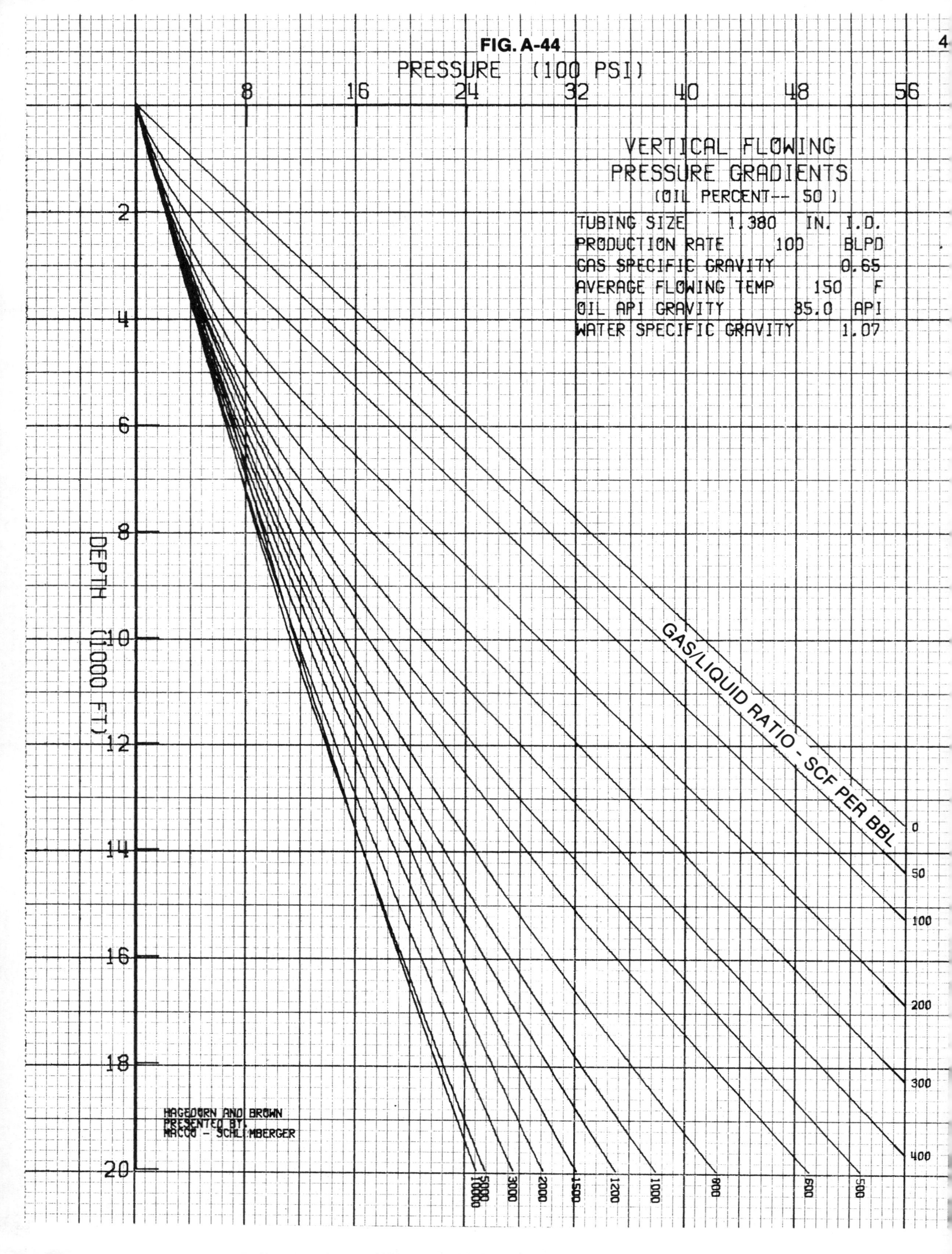

PRESSURE (100 PSI)

VERTICAL FLOWING
PRESSURE GRADIENTS
(OIL PERCENT-- 50)

TUBING SIZE	1.380	IN. I.D.
PRODUCTION RATE	100	BLPD
GAS SPECIFIC GRAVITY	0.65	
AVERAGE FLOWING TEMP	150	F
OIL API GRAVITY	85.0	API
WATER SPECIFIC GRAVITY	1.07	

DEPTH (1000 FT)

GAS/LIQUID RATIO - SCF PER BBL

HAGEDORN AND BROWN
PRESENTED BY:
MACCO - SCHLUMBERGER

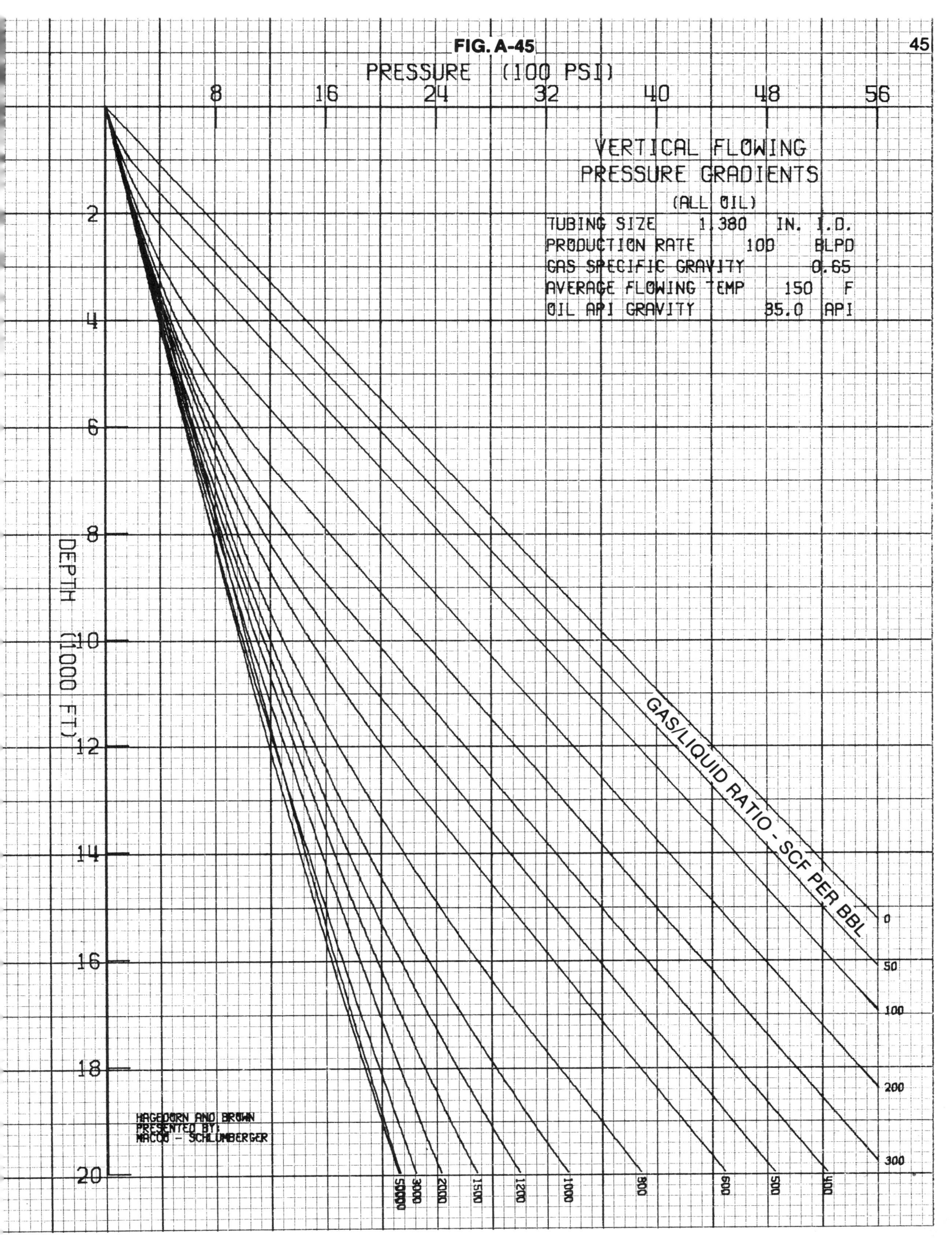

VERTICAL FLOWING
PRESSURE GRADIENTS
(ALL OIL)

TUBING SIZE	1.380	IN. I.D.
PRODUCTION RATE	100	BLPD
GAS SPECIFIC GRAVITY		0.65
AVERAGE FLOWING TEMP	150	F
OIL API GRAVITY	35.0	API

PRESSURE (100 PSI)

DEPTH (1000 FT)

GAS/LIQUID RATIO - SCF PER BBL

HAGEDORN AND BROWN
PRESENTED BY:
NACOO - SCHLUMBERGER

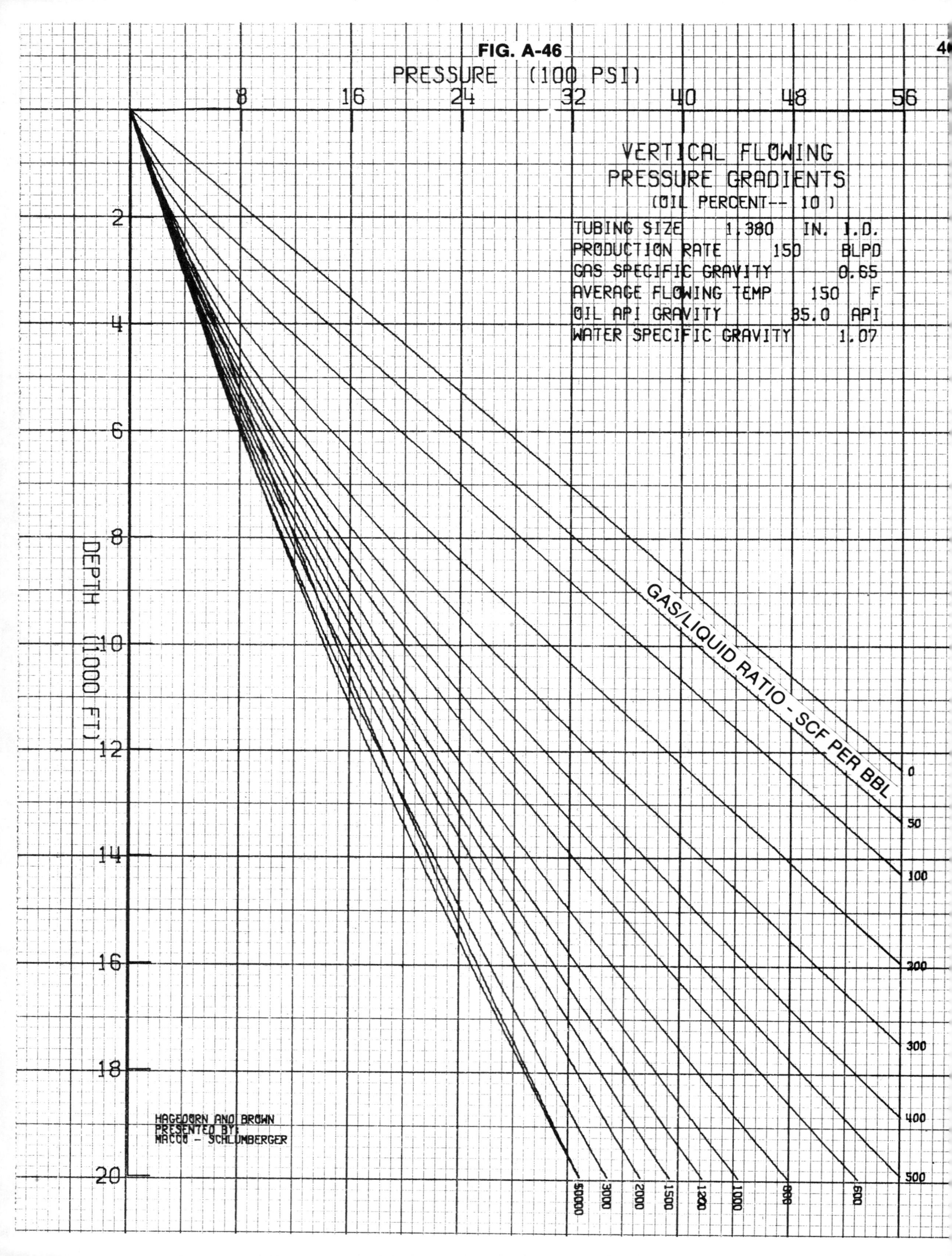

VERTICAL FLOWING
PRESSURE GRADIENTS
(OIL PERCENT-- 10)

TUBING SIZE	1.380	IN. I.D.
PRODUCTION RATE	150	BLPD
GAS SPECIFIC GRAVITY		0.65
AVERAGE FLOWING TEMP	150	F
OIL API GRAVITY	35.0	API
WATER SPECIFIC GRAVITY		1.07

PRESSURE (100 PSI)

DEPTH (1000 FT)

GAS/LIQUID RATIO - SCF PER BBL

HAGEDORN AND BROWN
PRESENTED BY:
MACCO - SCHLUMBERGER

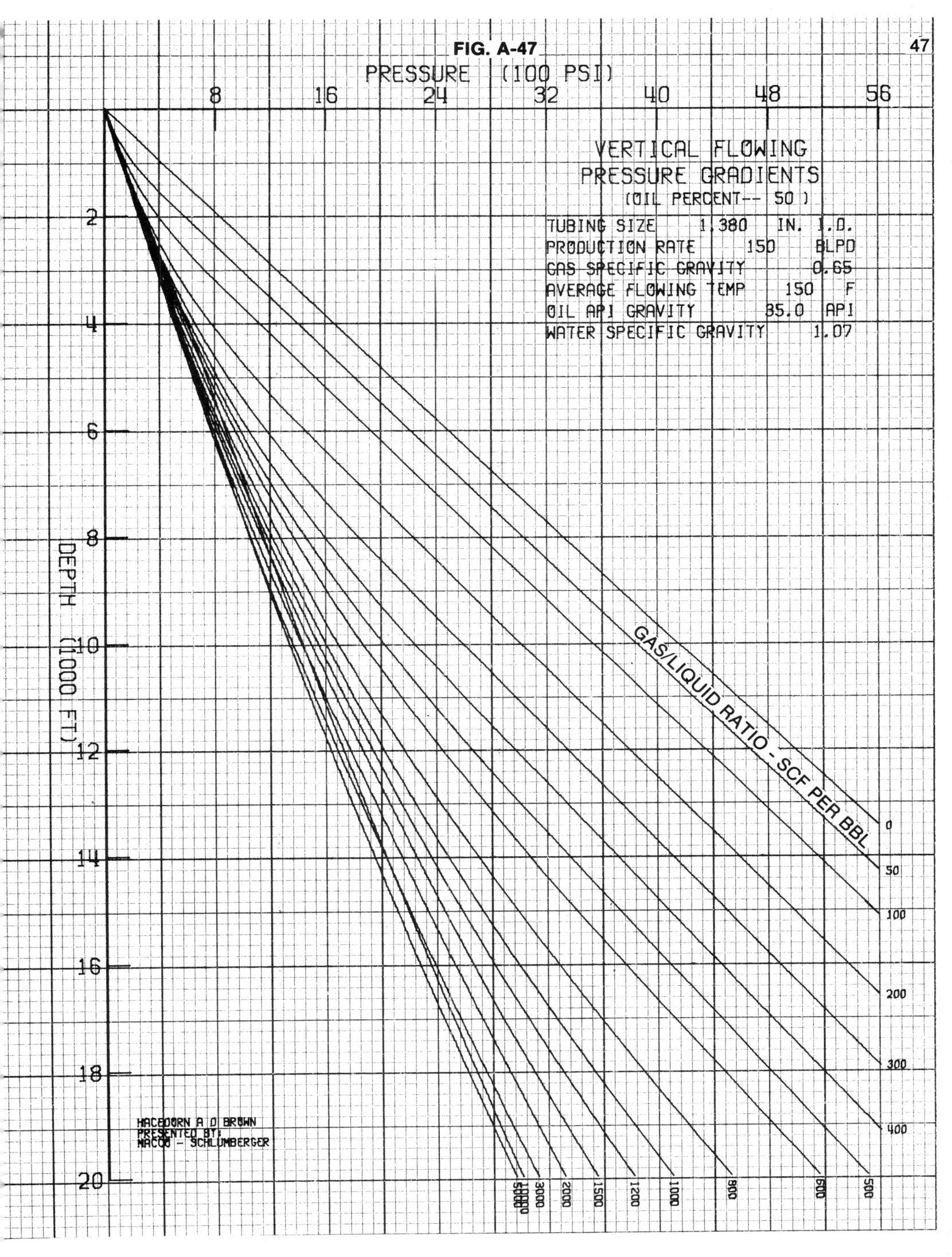

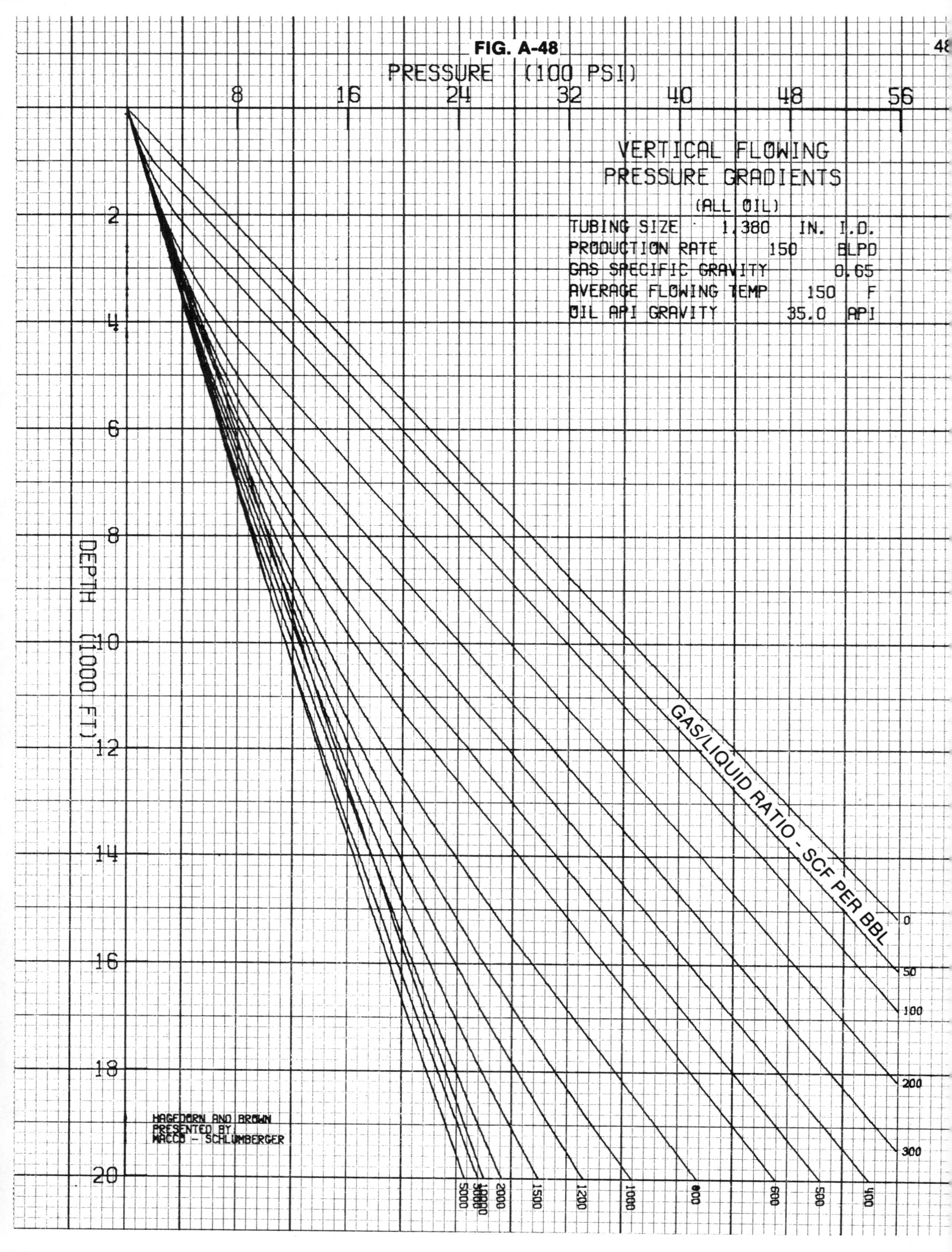

FIG. A-48

PRESSURE (100 PSI)

VERTICAL FLOWING
PRESSURE GRADIENTS
(ALL OIL)

TUBING SIZE   1.380   IN. I.D.
PRODUCTION RATE   150   BLPD
GAS SPECIFIC GRAVITY   0.65
AVERAGE FLOWING TEMP   150   F
OIL API GRAVITY   35.0   API

DEPTH (1000 FT)

GAS/LIQUID RATIO - SCF PER BBL

HAGEDORN AND BROWN
PRESENTED BY
WACCO - SCHLUMBERGER

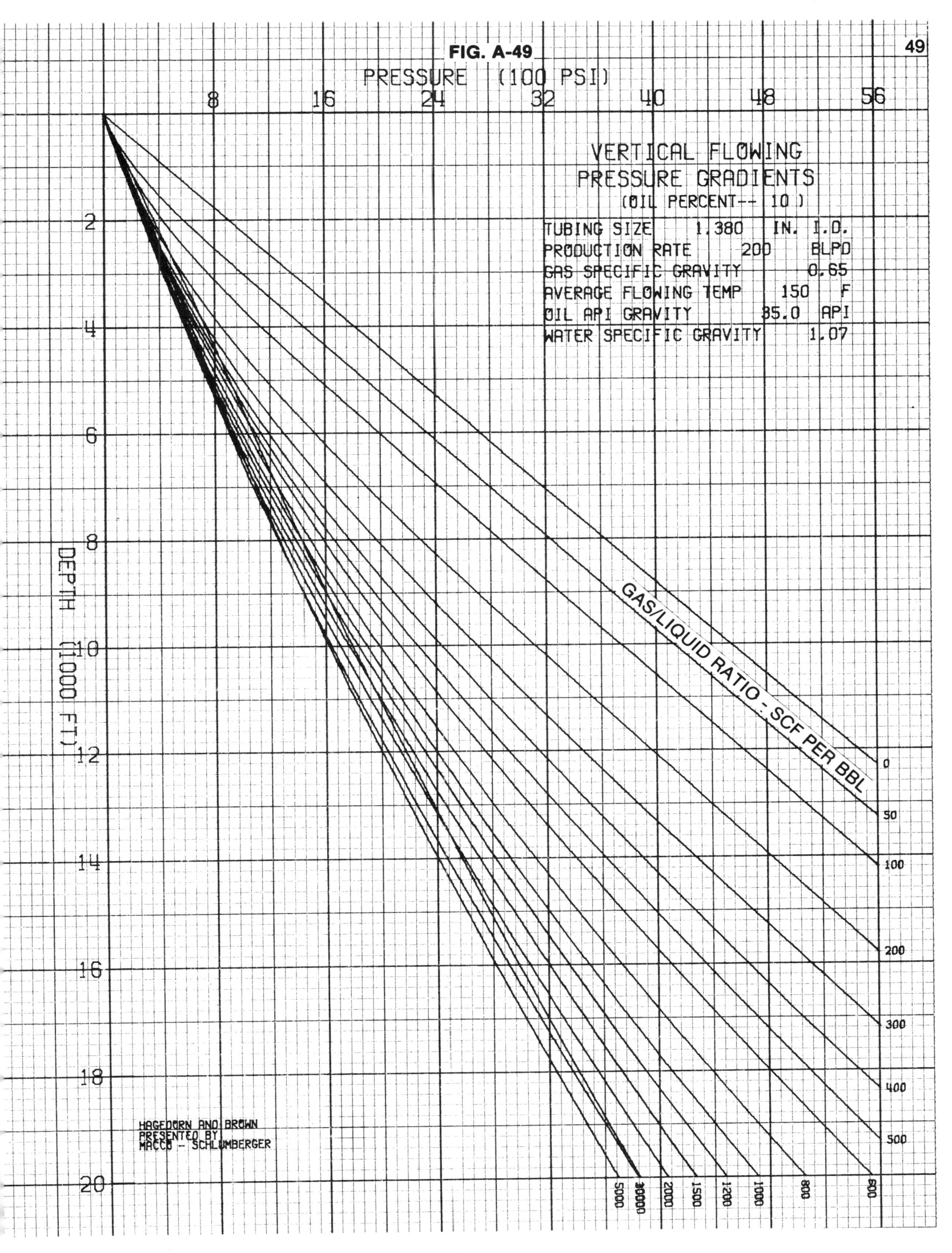

VERTICAL FLOWING
PRESSURE GRADIENTS
(OIL PERCENT-- 10)

TUBING SIZE	1.380	IN. I.D.
PRODUCTION RATE	200	BLPD
GAS SPECIFIC GRAVITY	0.65	
AVERAGE FLOWING TEMP	150	F
OIL API GRAVITY	35.0	API
WATER SPECIFIC GRAVITY	1.07	

PRESSURE (100 PSI)

DEPTH (1000 FT)

GAS/LIQUID RATIO : SCF PER BBL

HAGEDORN AND BROWN
PRESENTED BY
MACCO - SCHLUMBERGER

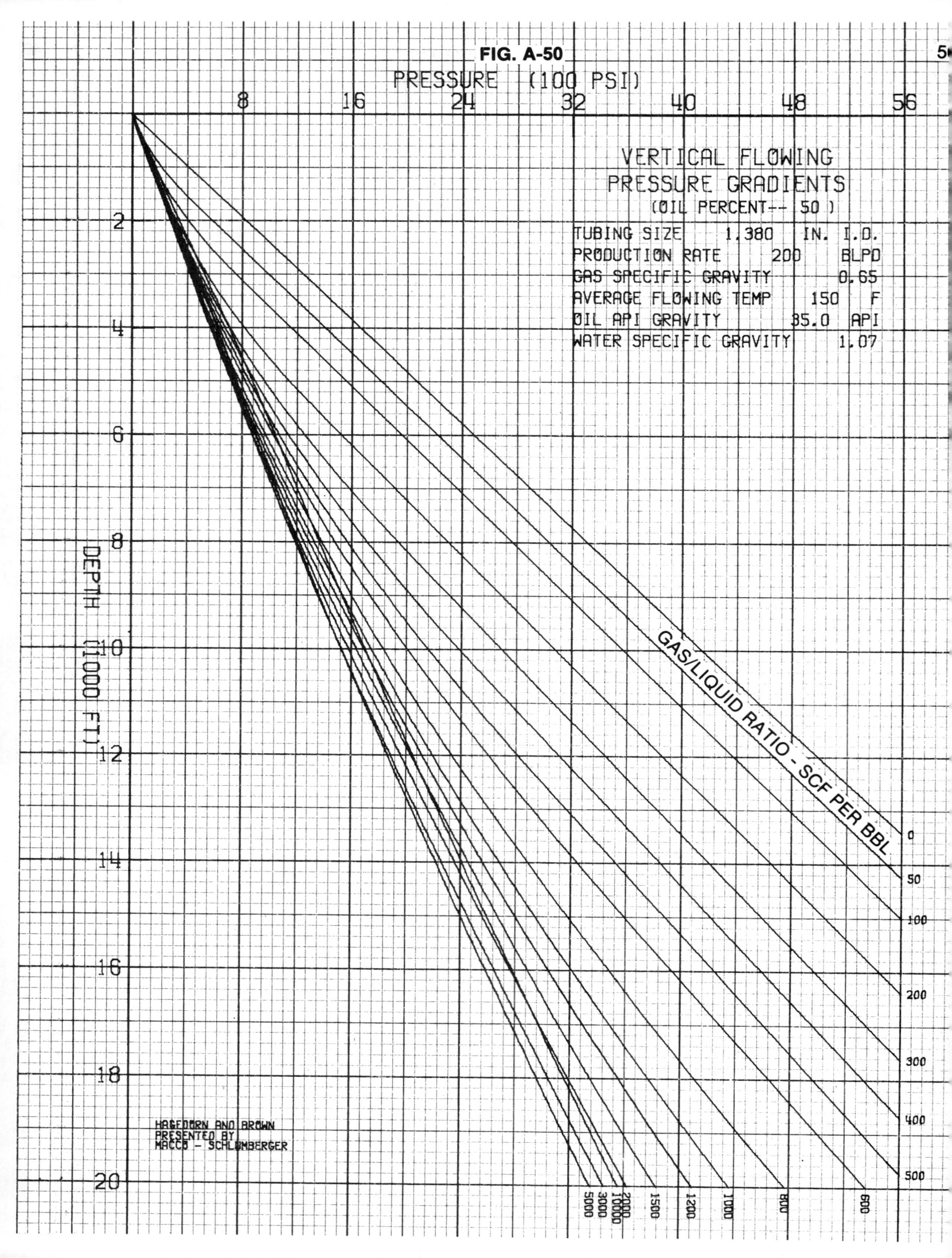

FIG. A-50

VERTICAL FLOWING
PRESSURE GRADIENTS
(OIL PERCENT-- 50)

TUBING SIZE	1.380	IN. I.D.
PRODUCTION RATE	200	BLPD
GAS SPECIFIC GRAVITY	0.65	
AVERAGE FLOWING TEMP	150	F
OIL API GRAVITY	85.0	API
WATER SPECIFIC GRAVITY	1.07	

PRESSURE (100 PSI)

DEPTH (1000 FT)

GAS/LIQUID RATIO - SCF PER BBL

HAGEDORN AND BROWN
PRESENTED BY
MACCO - SCHLUMBERGER

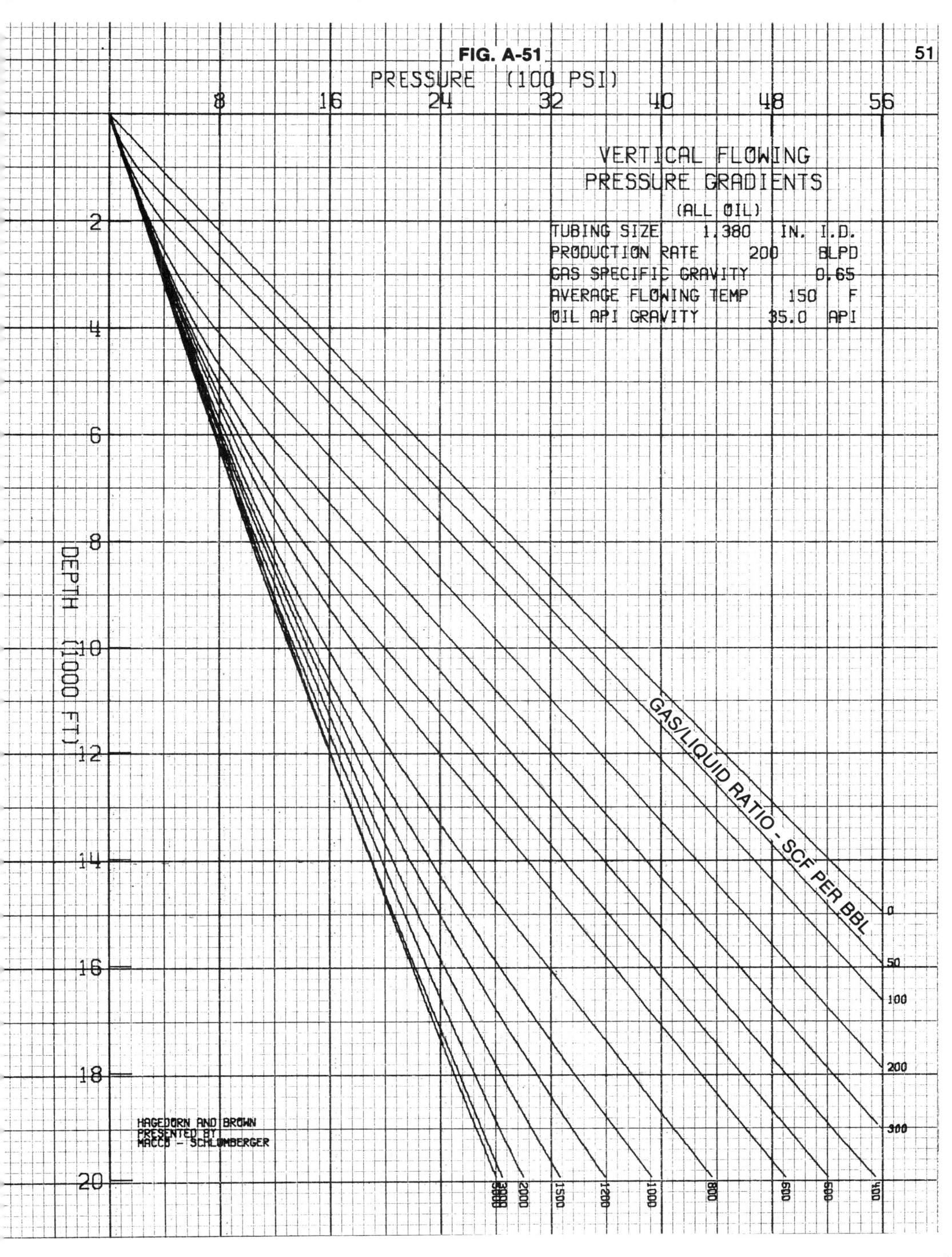

PRESSURE (100 PSI)

VERTICAL FLOWING
PRESSURE GRADIENTS
(ALL OIL)

TUBING SIZE	1.380	IN. I.D.
PRODUCTION RATE	200	BLPD
GAS SPECIFIC GRAVITY	0.65	
AVERAGE FLOWING TEMP	150	F
OIL API GRAVITY	35.0	API

DEPTH (1000 FT)

GAS/LIQUID RATIO - SCF PER BBL

0
50
100
200
300

400
600
800
1000
1200
1500
2000
3000

HAGEDORN AND BROWN
PRESENTED BY
MAECO - SCHLUMBERGER

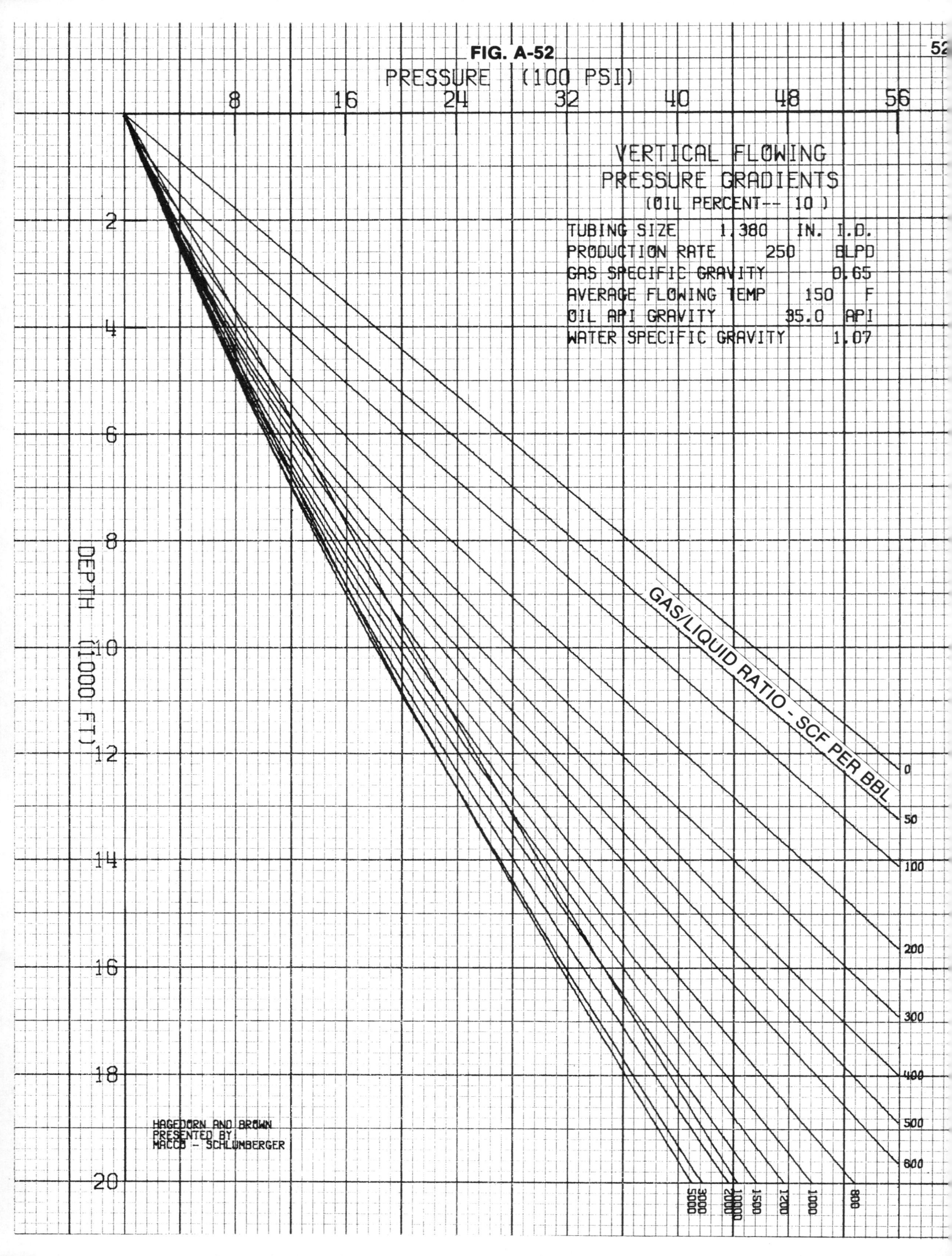

PRESSURE (100 PSI)

VERTICAL FLOWING
PRESSURE GRADIENTS
(OIL PERCENT-- 10)

TUBING SIZE   1.380   IN. I.D.
PRODUCTION RATE   250   BLPD
GAS SPECIFIC GRAVITY   0.65
AVERAGE FLOWING TEMP   150   F
OIL API GRAVITY   35.0   API
WATER SPECIFIC GRAVITY   1.07

GAS/LIQUID RATIO - SCF PER BBL

DEPTH (1000 FT)

HAGEDORN AND BROWN
PRESENTED BY
MACCO - SCHLUMBERGER

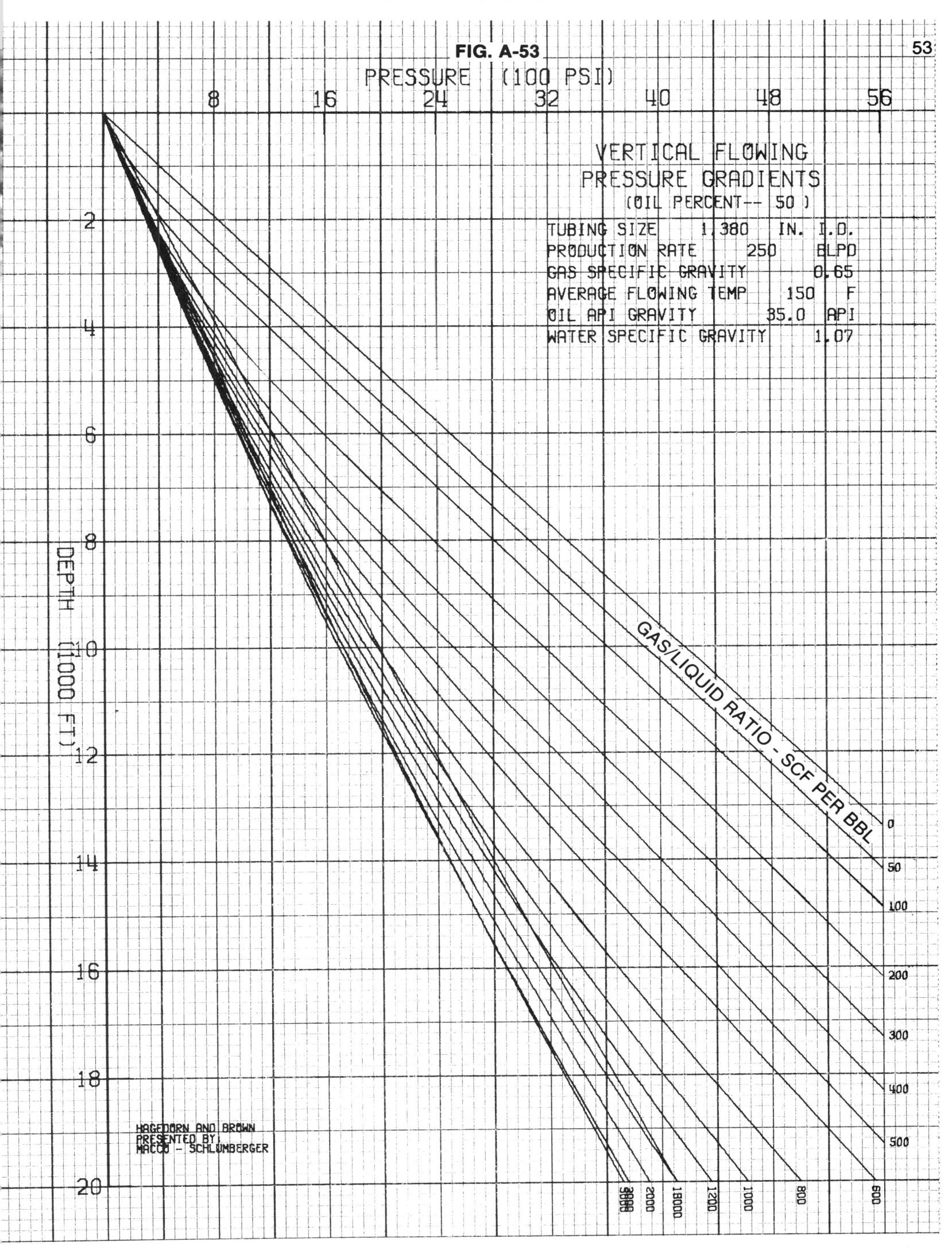

FIG. A-53

53

PRESSURE (100 PSI)

VERTICAL FLOWING
PRESSURE GRADIENTS
(OIL PERCENT-- 50)

TUBING SIZE	1.380	IN. I.D.
PRODUCTION RATE	250	BLPD
GAS SPECIFIC GRAVITY	0.65	
AVERAGE FLOWING TEMP	150	F
OIL API GRAVITY	35.0	API
WATER SPECIFIC GRAVITY	1.07	

DEPTH (1000 FT)

GAS/LIQUID RATIO - SCF PER BBL

HAGEDORN AND BROWN
PRESENTED BY,
MACCO - SCHLUMBERGER

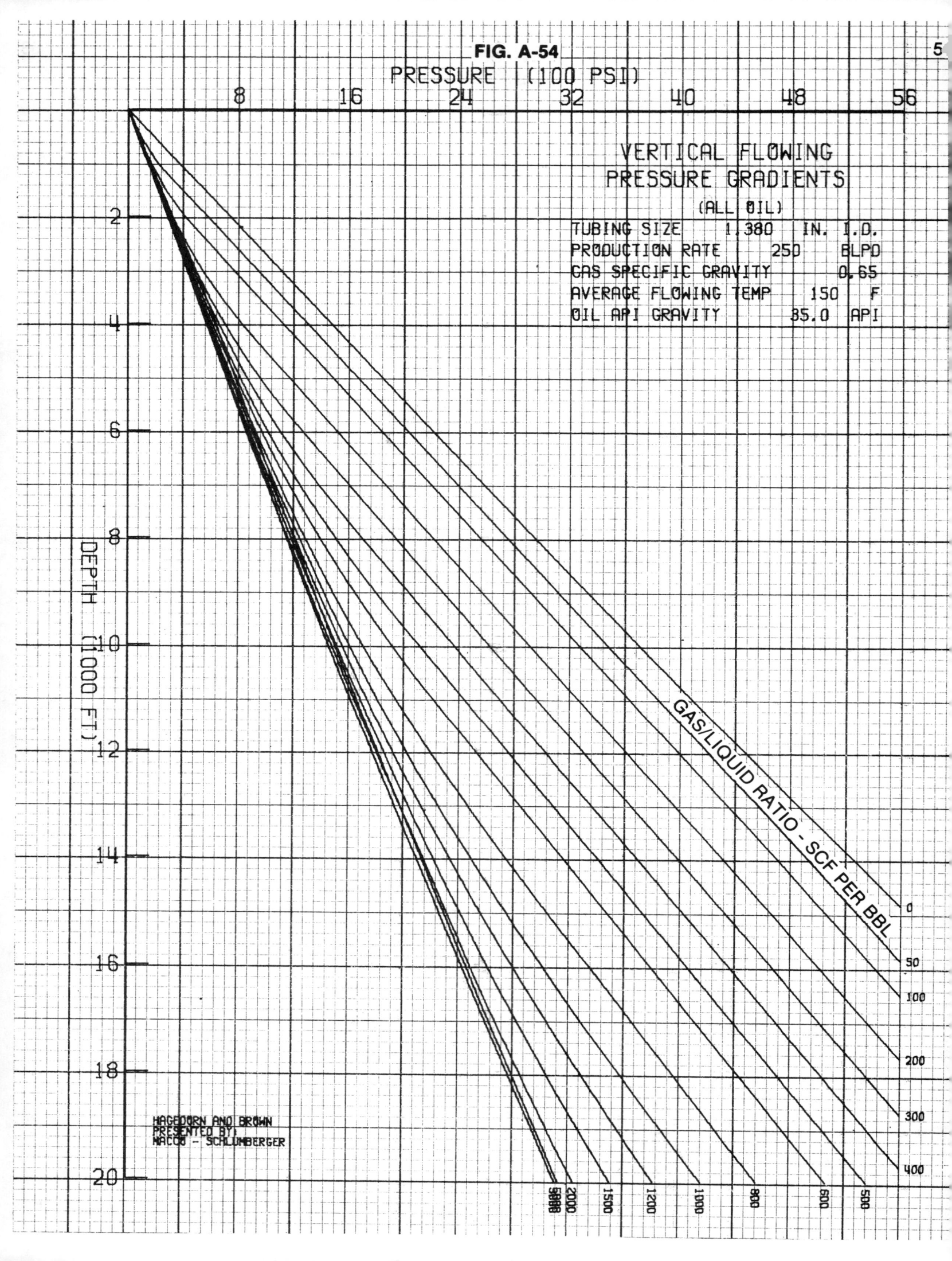

FIG. A-54

5

VERTICAL FLOWING
PRESSURE GRADIENTS
(ALL OIL)

TUBING SIZE	1.380	IN. I.D.
PRODUCTION RATE	250	BLPD
GAS SPECIFIC GRAVITY	0.65	
AVERAGE FLOWING TEMP	150	F
OIL API GRAVITY	35.0	API

PRESSURE (100 PSI)

DEPTH (1 000 FT)

GAS/LIQUID RATIO - SCF PER BBL

HAGEDORN AND BROWN
PRESENTED BY:
NACCO - SCHLUMBERGER

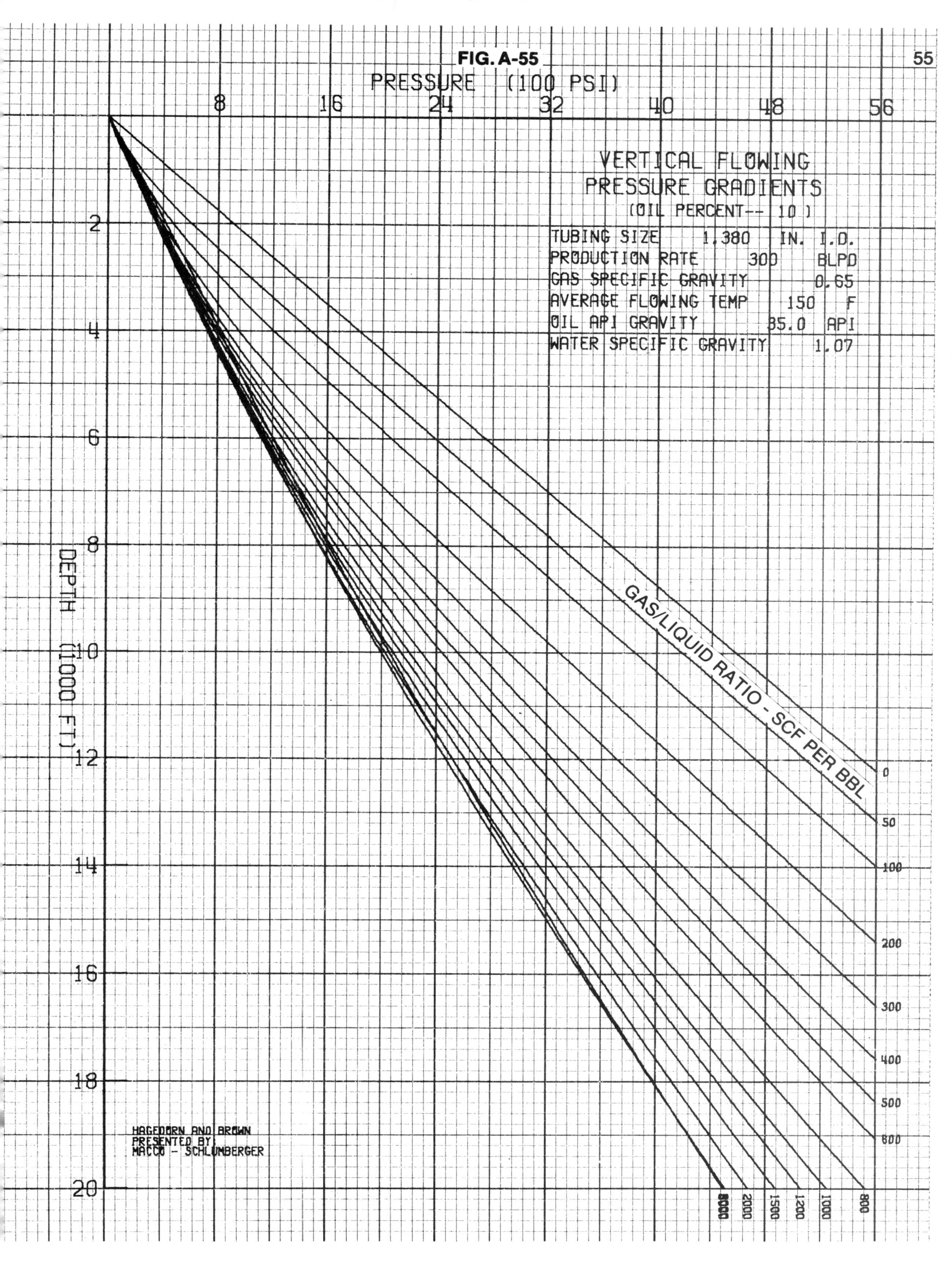

FIG. A-55

PRESSURE (100 PSI)

VERTICAL FLOWING
PRESSURE GRADIENTS
(OIL PERCENT-- 10)

TUBING SIZE   1.380   IN. I.D.
PRODUCTION RATE   300   BLPD
GAS SPECIFIC GRAVITY   0.65
AVERAGE FLOWING TEMP   150   F
OIL API GRAVITY   35.0 API
WATER SPECIFIC GRAVITY   1.07

GAS/LIQUID RATIO - SCF PER BBL

DEPTH (1000 FT)

HAGEDORN AND BROWN
PRESENTED BY
MACCO - SCHLUMBERGER

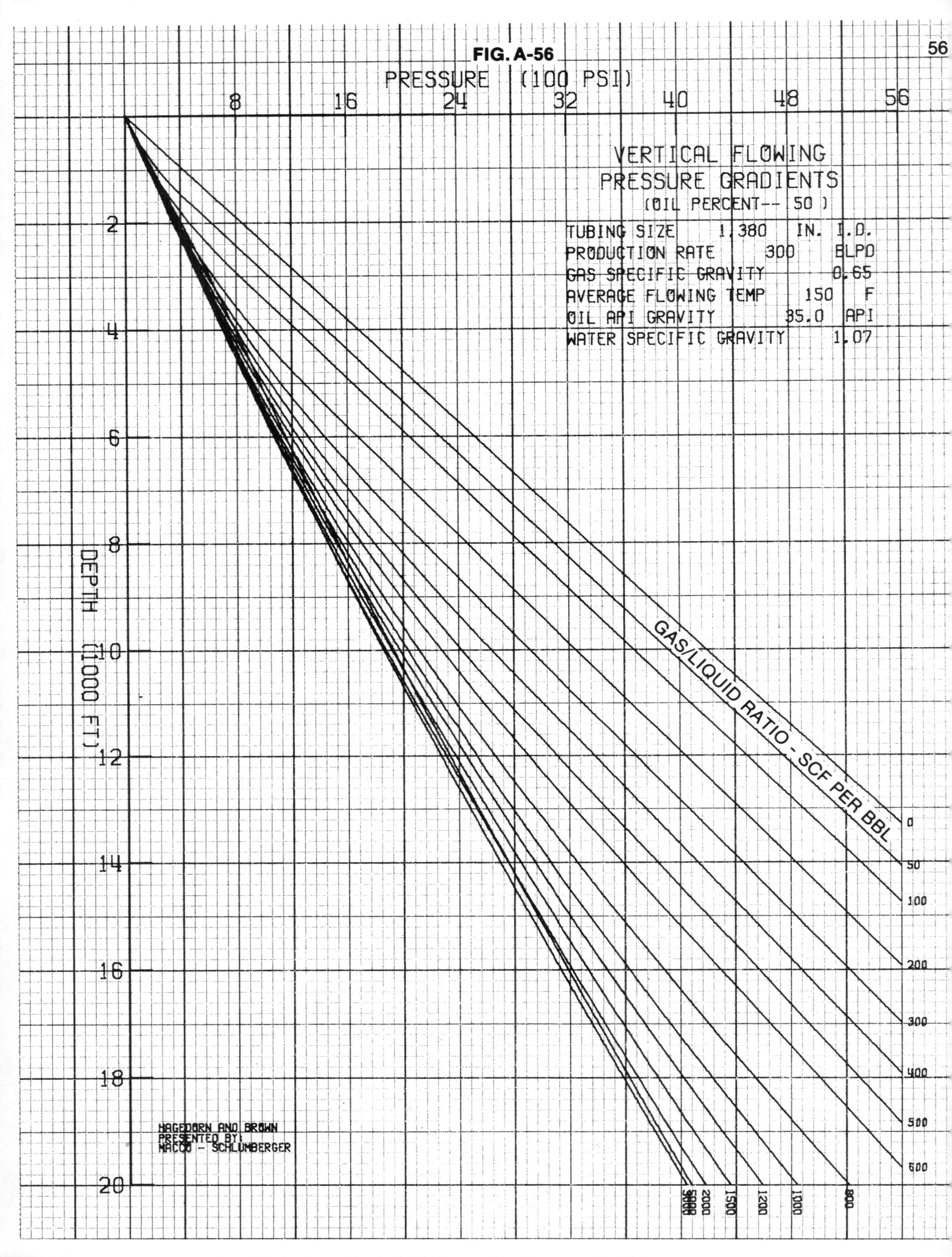

PRESSURE (100 PSI)

DEPTH (1000 FT)

VERTICAL FLOWING
PRESSURE GRADIENTS
(OIL PERCENT-- 50)

TUBING SIZE   1.380   IN. I.D.
PRODUCTION RATE   300   BLPD
GAS SPECIFIC GRAVITY   0.65
AVERAGE FLOWING TEMP   150   F
OIL API GRAVITY   35.0   API
WATER SPECIFIC GRAVITY   1.07

GAS/LIQUID RATIO - SCF PER BBL

HAGEDORN AND BROWN
PRESENTED BY:
NACOO - SCHLUMBERGER

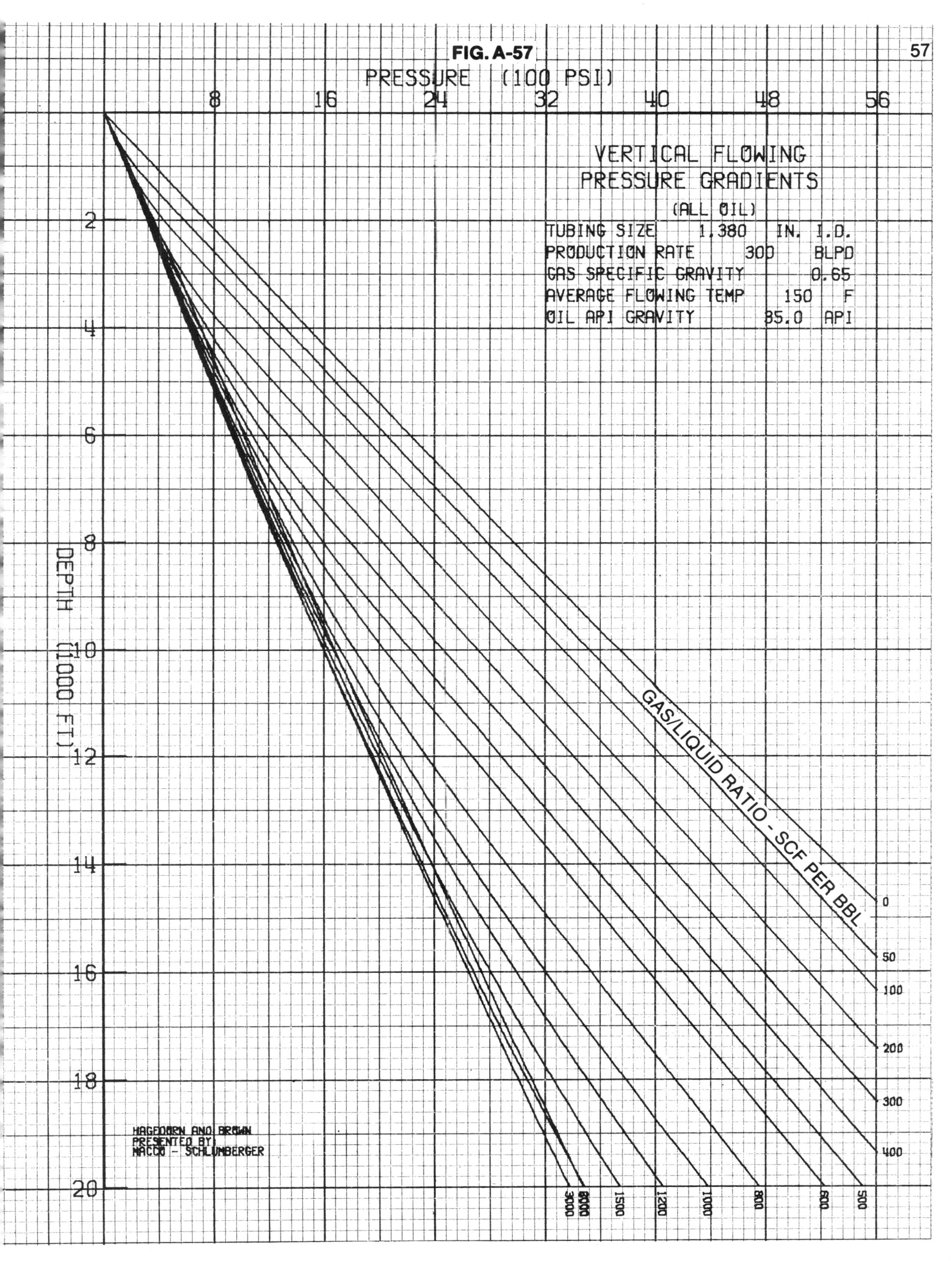

PRESSURE (100 PSI)

VERTICAL FLOWING
PRESSURE GRADIENTS
(ALL OIL)

TUBING SIZE	1.380	IN.	I.D.
PRODUCTION RATE	300		BLPD
GAS SPECIFIC GRAVITY		0.65	
AVERAGE FLOWING TEMP	150		F
OIL API GRAVITY	85.0		API

DEPTH (1000 FT)

GAS/LIQUID RATIO - SCF PER BBL

HAGEDORN AND BROWN
PRESENTED BY
NACCO - SCHLUMBERGER

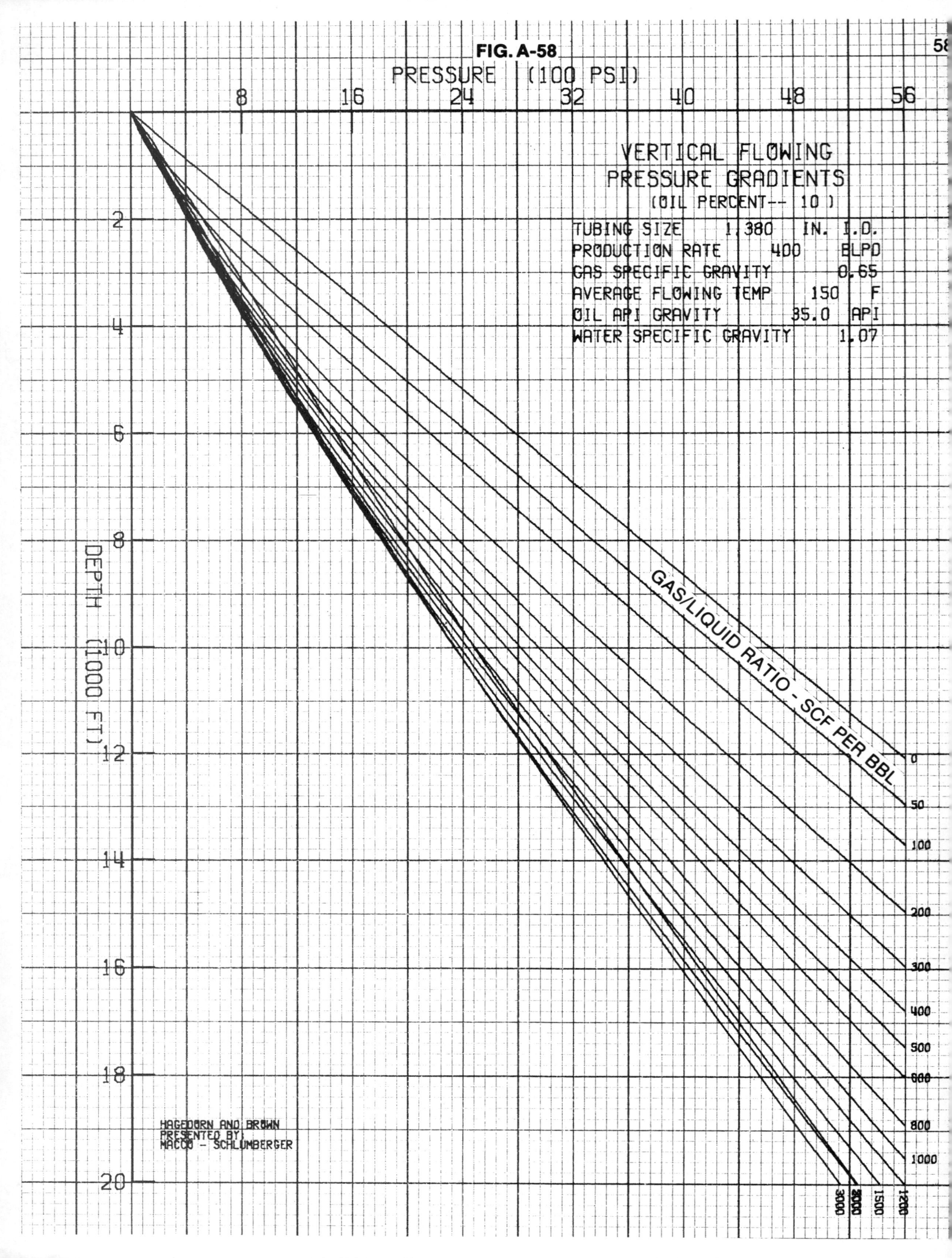

FIG. A-58

VERTICAL FLOWING
PRESSURE GRADIENTS
(OIL PERCENT-- 10)

TUBING SIZE   1.380   IN. I.D.
PRODUCTION RATE   400   BLPD
GAS SPECIFIC GRAVITY   0.65
AVERAGE FLOWING TEMP   150   F
OIL API GRAVITY   35.0   API
WATER SPECIFIC GRAVITY   1.07

GAS/LIQUID RATIO - SCF PER BBL

HAGEDORN AND BROWN
PRESENTED BY:
MACCO - SCHLUMBERGER

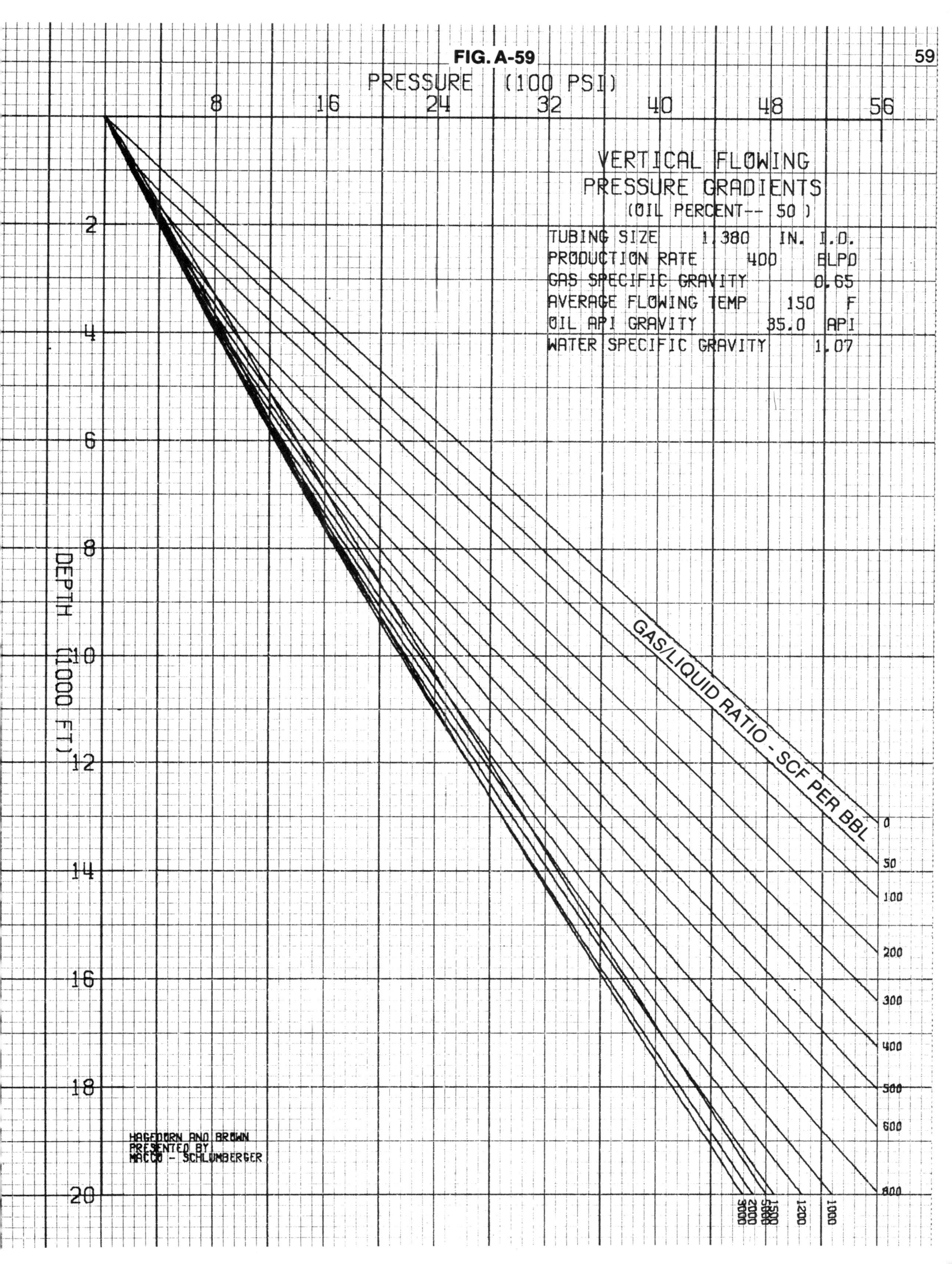

FIG. A-59

VERTICAL FLOWING
PRESSURE GRADIENTS
(OIL PERCENT-- 50)

TUBING SIZE	1.380	IN. I.D.
PRODUCTION RATE	400	BLPD
GAS SPECIFIC GRAVITY	0.65	
AVERAGE FLOWING TEMP	150	F
OIL API GRAVITY	35.0	API
WATER SPECIFIC GRAVITY	1.07	

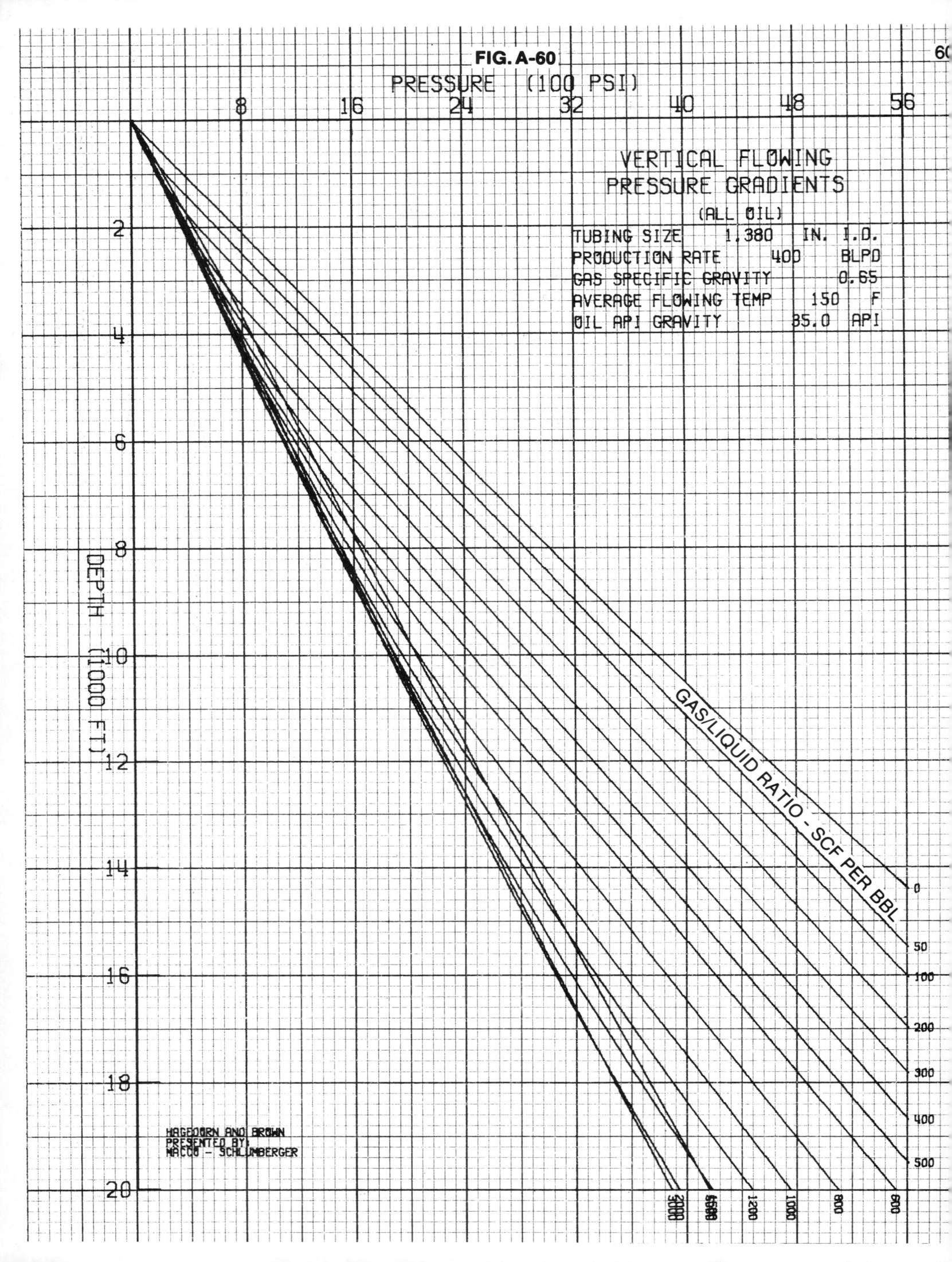

FIG. A-60

60

PRESSURE (100 PSI)

VERTICAL FLOWING
PRESSURE GRADIENTS
(ALL OIL)

TUBING SIZE	1.380	IN. I.D.
PRODUCTION RATE	400	BLPD
GAS SPECIFIC GRAVITY	0.65	
AVERAGE FLOWING TEMP	150	F
OIL API GRAVITY	35.0	API

DEPTH (1000 FT)

GAS/LIQUID RATIO - SCF PER BBL

0

50

100

200

300

400

500

HAGEDORN AND BROWN
PRESENTED BY:
MACCO – SCHLUMBERGER

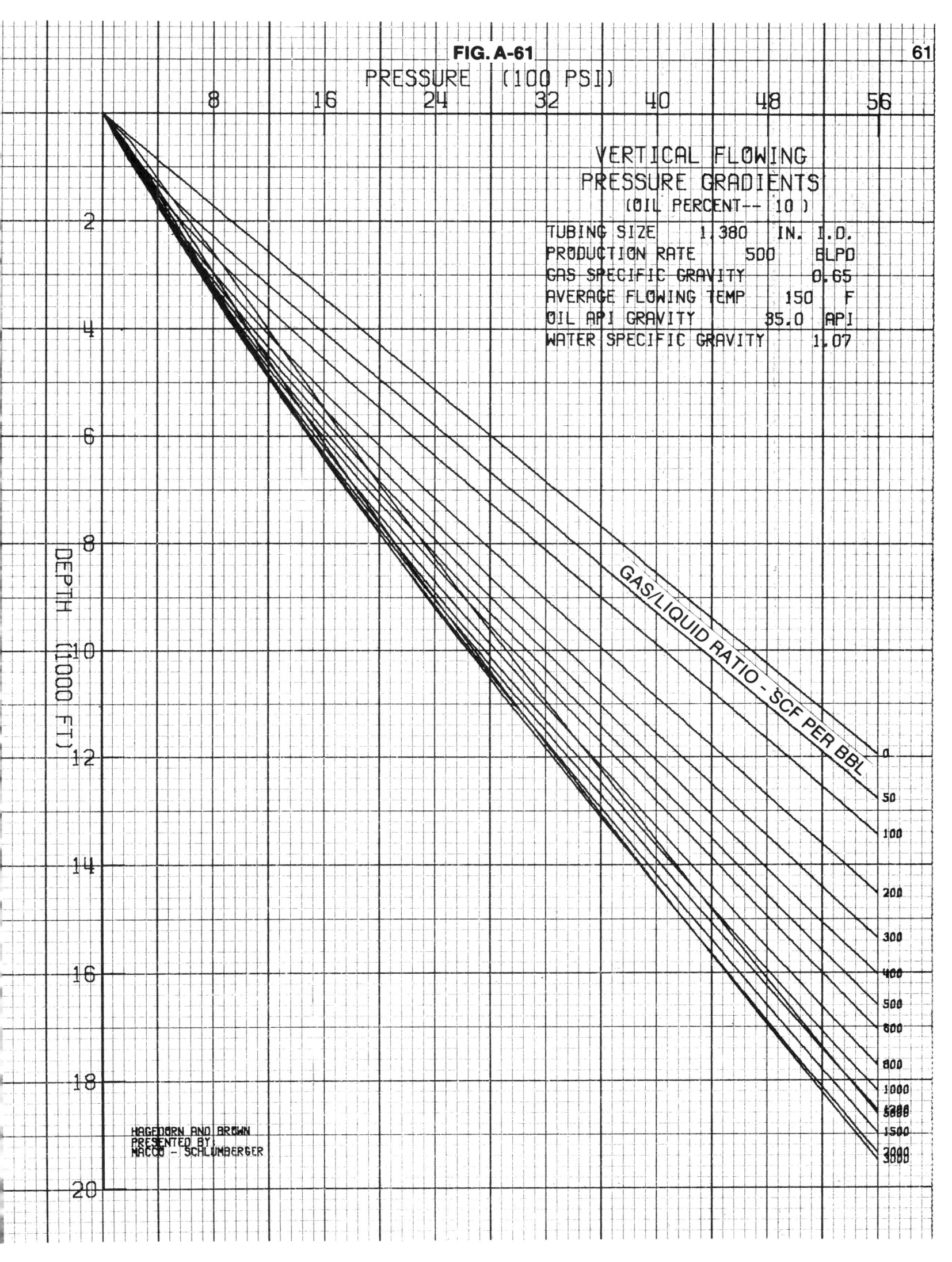

VERTICAL FLOWING
PRESSURE GRADIENTS
(OIL PERCENT-- 10)

TUBING SIZE	1.380	IN. I.D.
PRODUCTION RATE	500	BLPD
GAS SPECIFIC GRAVITY	0.65	
AVERAGE FLOWING TEMP	150	F
OIL API GRAVITY	35.0	API
WATER SPECIFIC GRAVITY	1.07	

PRESSURE (100 PSI)

DEPTH (1000 FT)

GAS/LIQUID RATIO - SCF PER BBL

HAGEDORN AND BROWN
PRESENTED BY
NACCO - SCHLUMBERGER

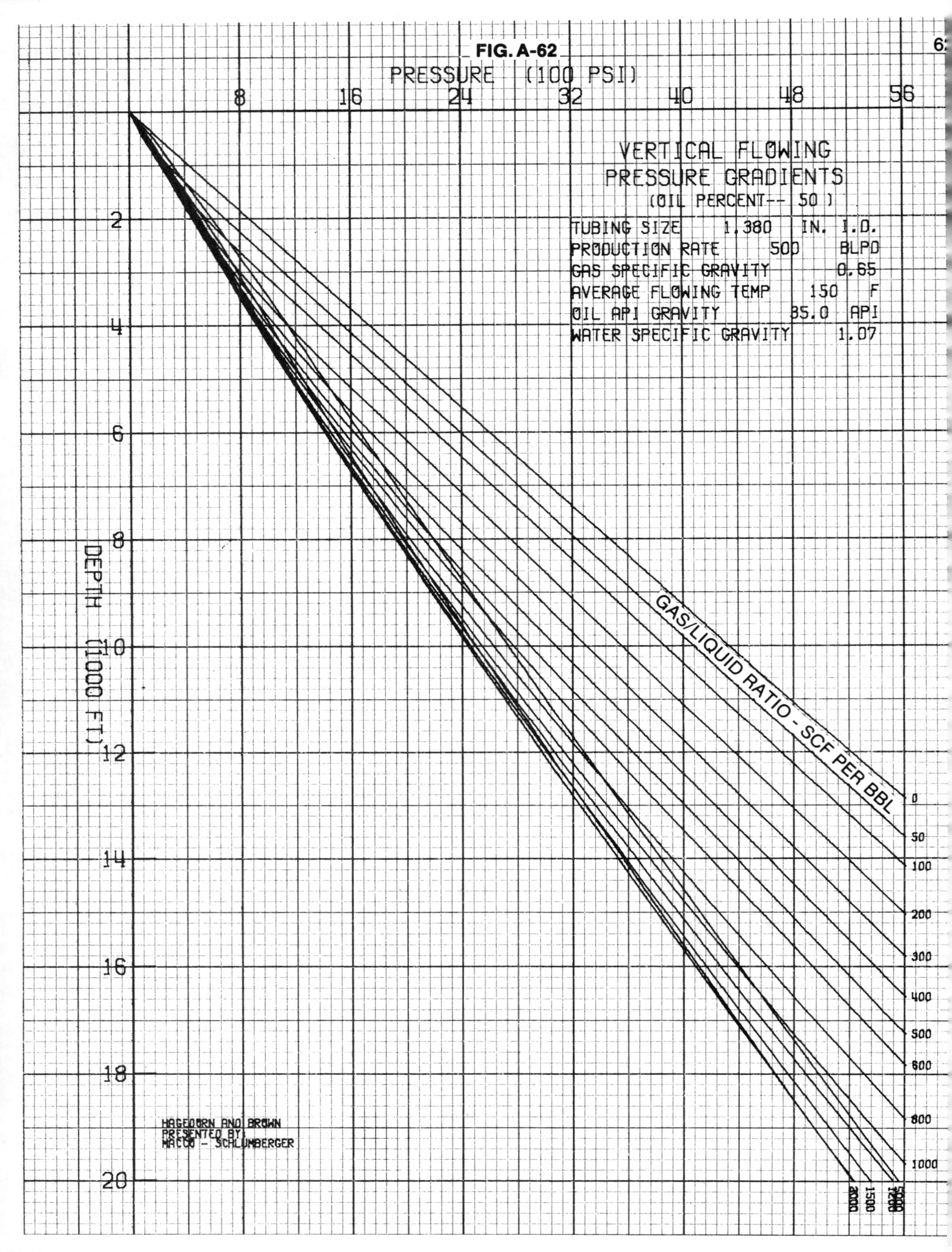

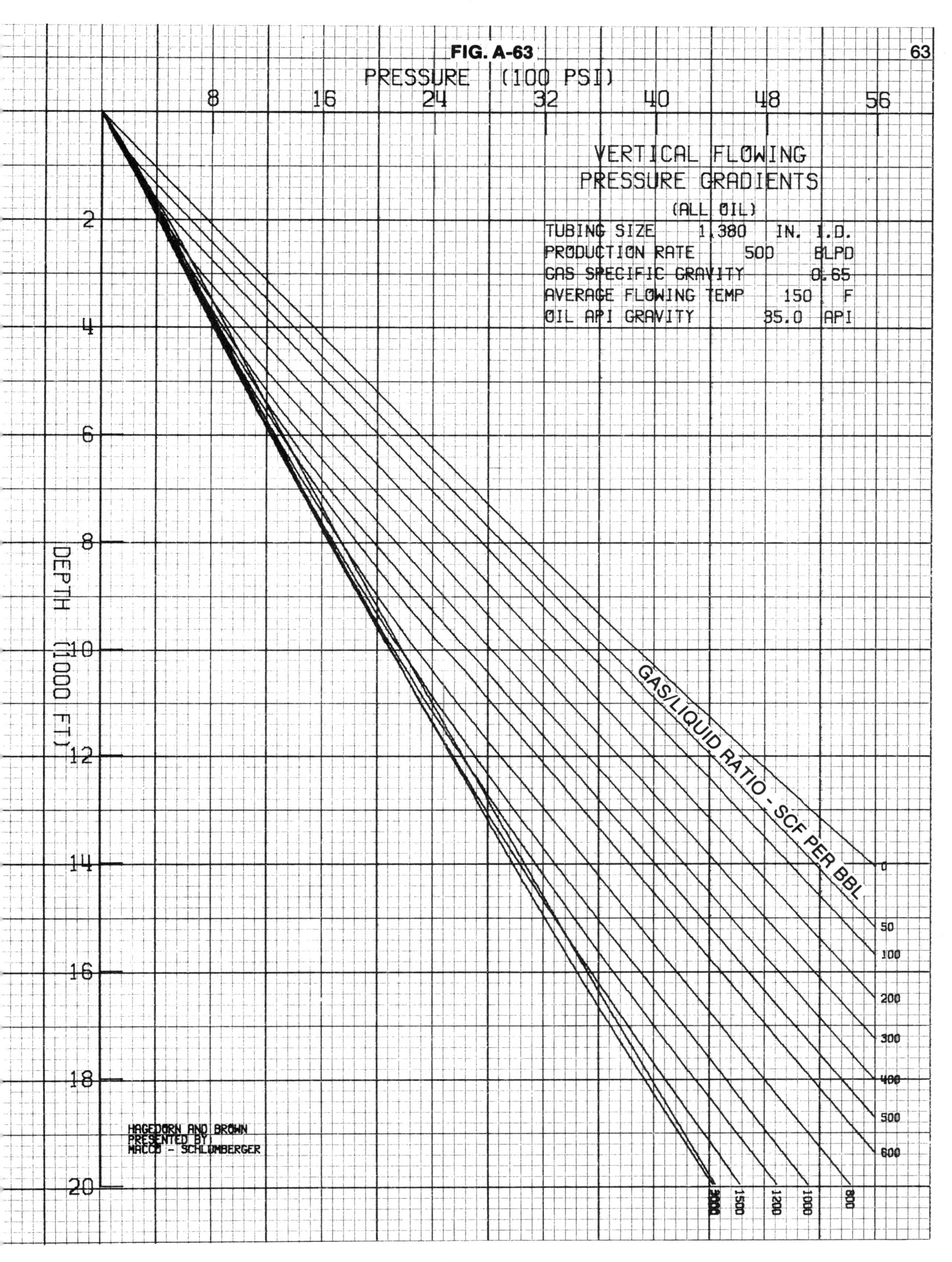

FIG. A-63

63

PRESSURE (100 PSI)

VERTICAL FLOWING
PRESSURE GRADIENTS
(ALL OIL)

TUBING SIZE	1.380	IN. I.D.
PRODUCTION RATE	500	BLPD
GAS SPECIFIC GRAVITY		0.65
AVERAGE FLOWING TEMP	150	F
OIL API GRAVITY	35.0	API

DEPTH (1000 FT)

GAS/LIQUID RATIO - SCF PER BBL

HAGEDORN AND BROWN
PRESENTED BY
MACCO - SCHLUMBERGER

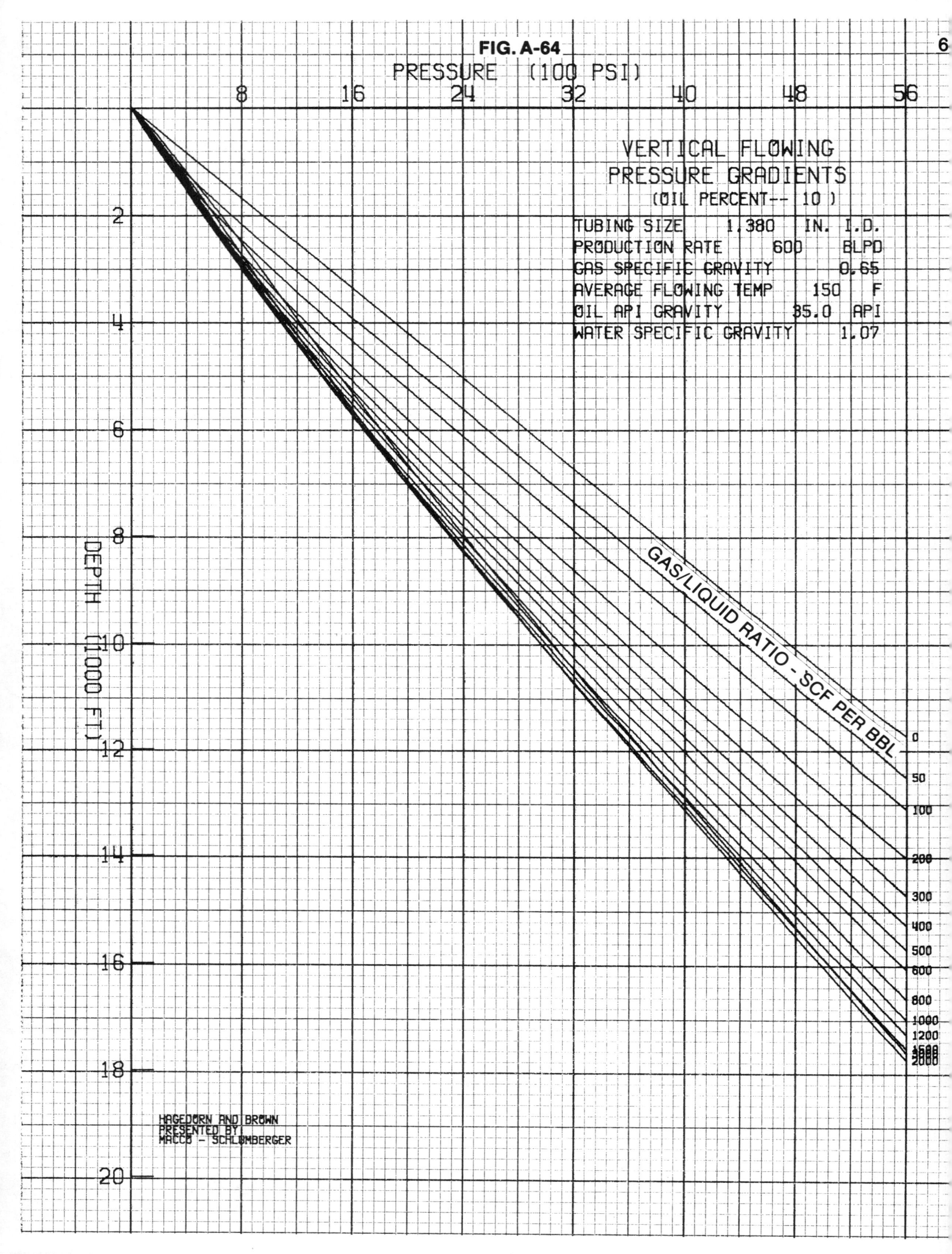

FIG. A-64

VERTICAL FLOWING
PRESSURE GRADIENTS
(OIL PERCENT-- 10)

TUBING SIZE	1.380	IN. I.D.
PRODUCTION RATE	600	BLPD
GAS SPECIFIC GRAVITY		0.65
AVERAGE FLOWING TEMP	150	F
OIL API GRAVITY	35.0	API
WATER SPECIFIC GRAVITY		1.07

PRESSURE (100 PSI)

DEPTH (1000 FT)

GAS/LIQUID RATIO - SCF PER BBL

HAGEDORN AND BROWN
PRESENTED BY
MACCO - SCHLUMBERGER

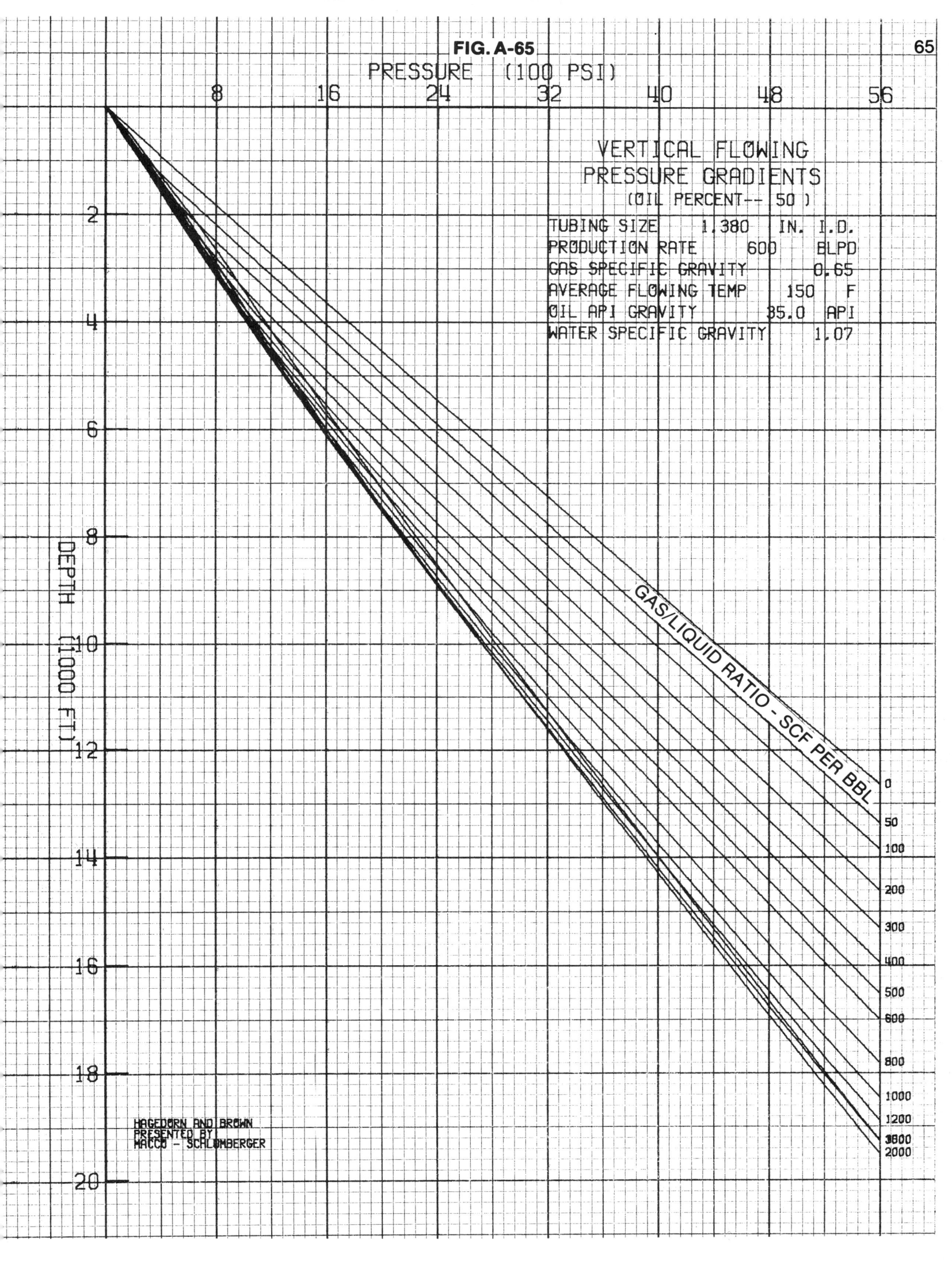

VERTICAL FLOWING
PRESSURE GRADIENTS
(OIL PERCENT-- 50)

TUBING SIZE	1.380	IN. I.D.
PRODUCTION RATE	600	BLPD
GAS SPECIFIC GRAVITY	0.65	
AVERAGE FLOWING TEMP	150	F
OIL API GRAVITY	35.0	API
WATER SPECIFIC GRAVITY	1.07	

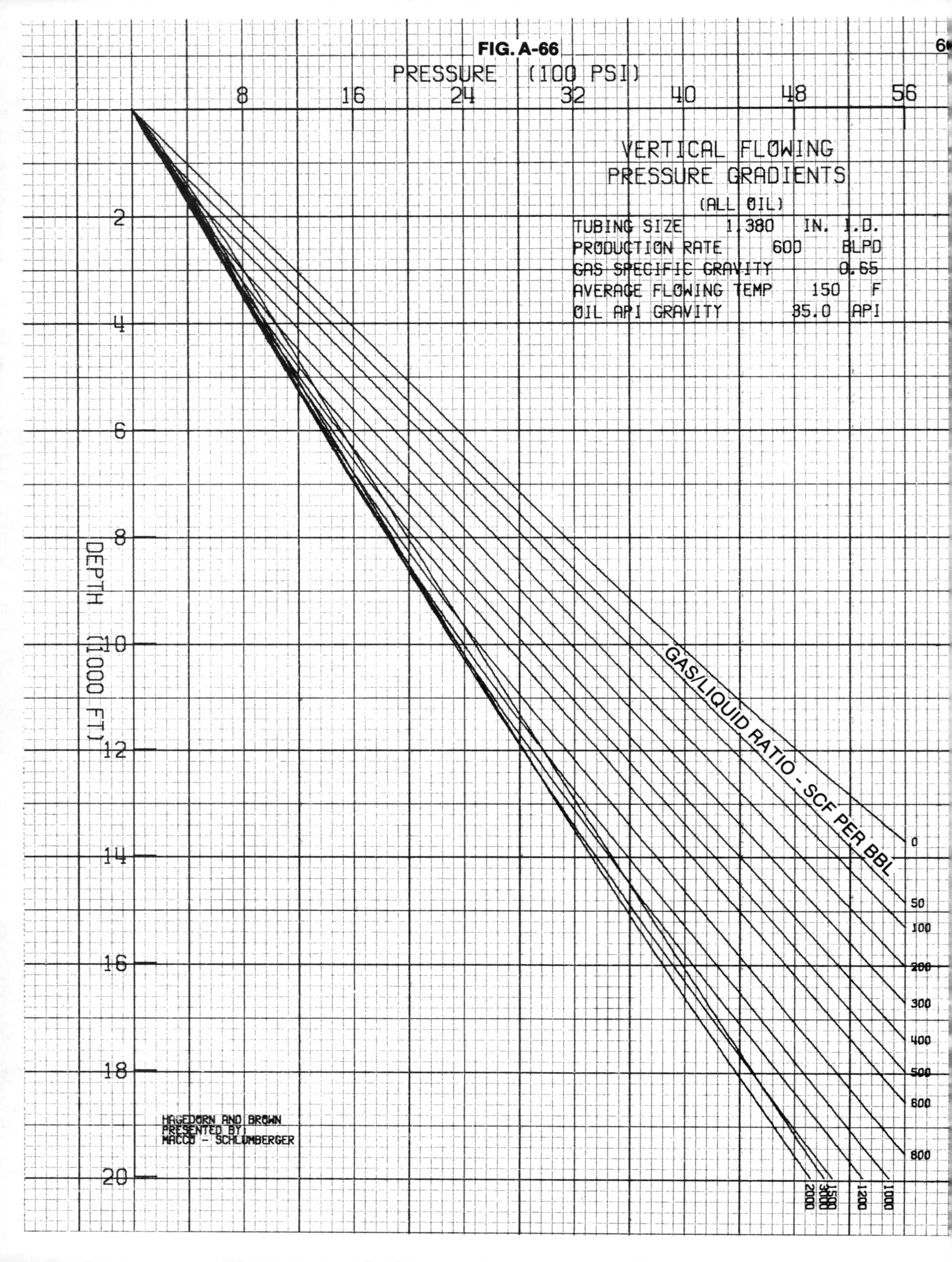

VERTICAL FLOWING
PRESSURE GRADIENTS
(ALL OIL)

TUBING SIZE	1.380	IN.	I.D.
PRODUCTION RATE	600	BLPD	
GAS SPECIFIC GRAVITY	0.65		
AVERAGE FLOWING TEMP	150	F	
OIL API GRAVITY	35.0	API	

PRESSURE (100 PSI)

DEPTH (1000 FT)

GAS/LIQUID RATIO - SCF PER BBL

HAGEDORN AND BROWN
PRESENTED BY:
MACCO – SCHLUMBERGER

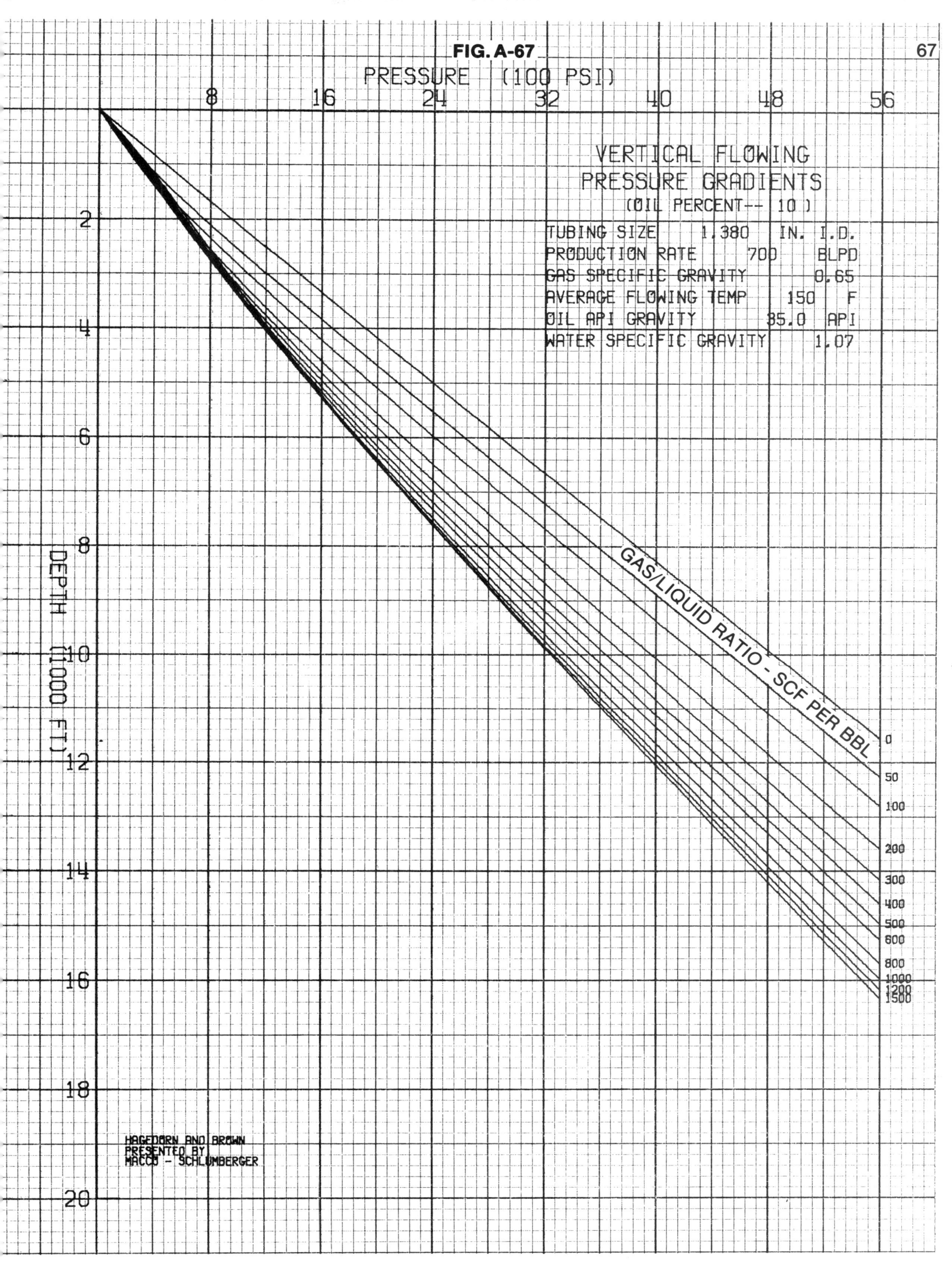

PRESSURE (100 PSI)

DEPTH (1000 FT)

VERTICAL FLOWING
PRESSURE GRADIENTS
(OIL PERCENT-- 10)

TUBING SIZE	1.380	IN. I.D.
PRODUCTION RATE	700	BLPD
GAS SPECIFIC GRAVITY	0.65	
AVERAGE FLOWING TEMP	150	F
OIL API GRAVITY	35.0	API
WATER SPECIFIC GRAVITY	1.07	

GAS/LIQUID RATIO - SCF PER BBL

0
50
100
200
300
400
500
600
800
1000
1200
1500

HAGEDORN AND BROWN
PRESENTED BY
MACCO - SCHLUMBERGER

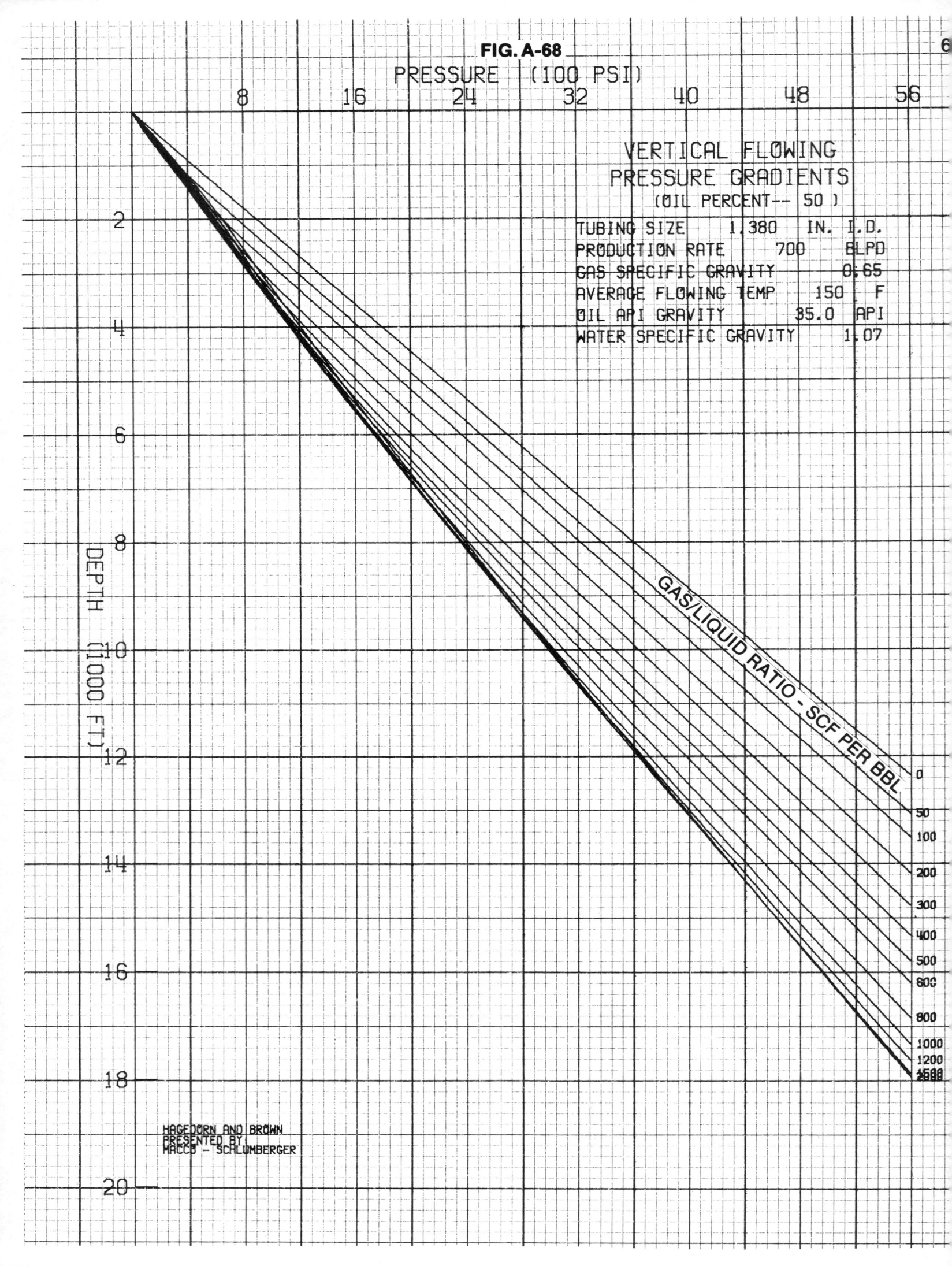

FIG. A-68

VERTICAL FLOWING
PRESSURE GRADIENTS
(OIL PERCENT-- 50)

TUBING SIZE	1.380	IN. I.D.
PRODUCTION RATE	700	BLPD
GAS SPECIFIC GRAVITY	0.65	
AVERAGE FLOWING TEMP	150	F
OIL API GRAVITY	35.0	API
WATER SPECIFIC GRAVITY	1.07	

PRESSURE (100 PSI)

DEPTH (1000 FT)

GAS/LIQUID RATIO - SCF PER BBL

0
50
100
200
300
400
500
600
800
1000
1200
1500

HAGEDORN AND BROWN
PRESENTED BY
MACCO - SCHLUMBERGER

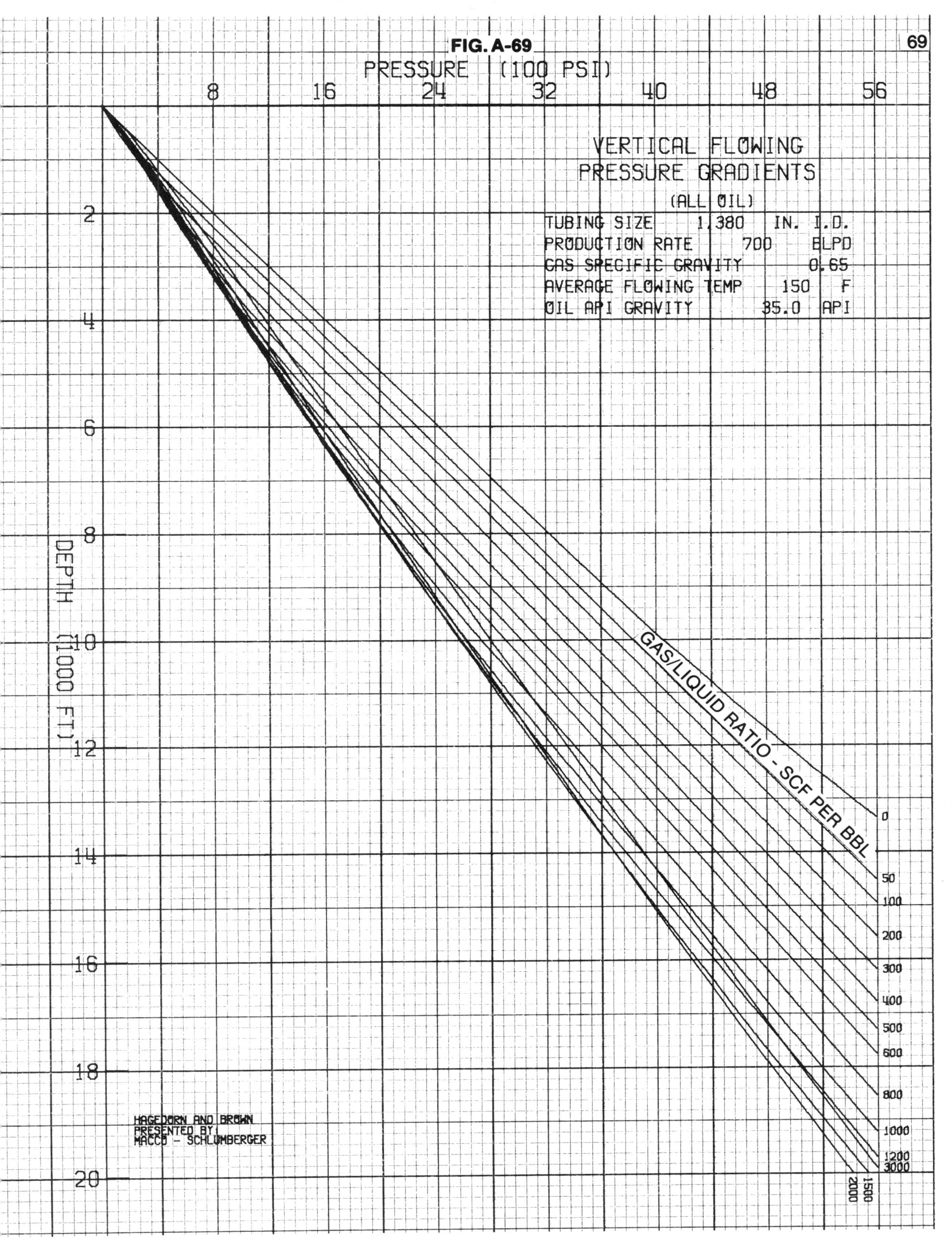

VERTICAL FLOWING
PRESSURE GRADIENTS
(ALL OIL)

TUBING SIZE	1.380	IN. I.D.
PRODUCTION RATE	700	BLPD
GAS SPECIFIC GRAVITY	0.65	
AVERAGE FLOWING TEMP	150	F
OIL API GRAVITY	35.0	API

PRESSURE (100 PSI)

DEPTH (1000 FT)

GAS/LIQUID RATIO - SCF PER BBL

HAGEDORN AND BROWN
PRESENTED BY
MACCO - SCHLUMBERGER

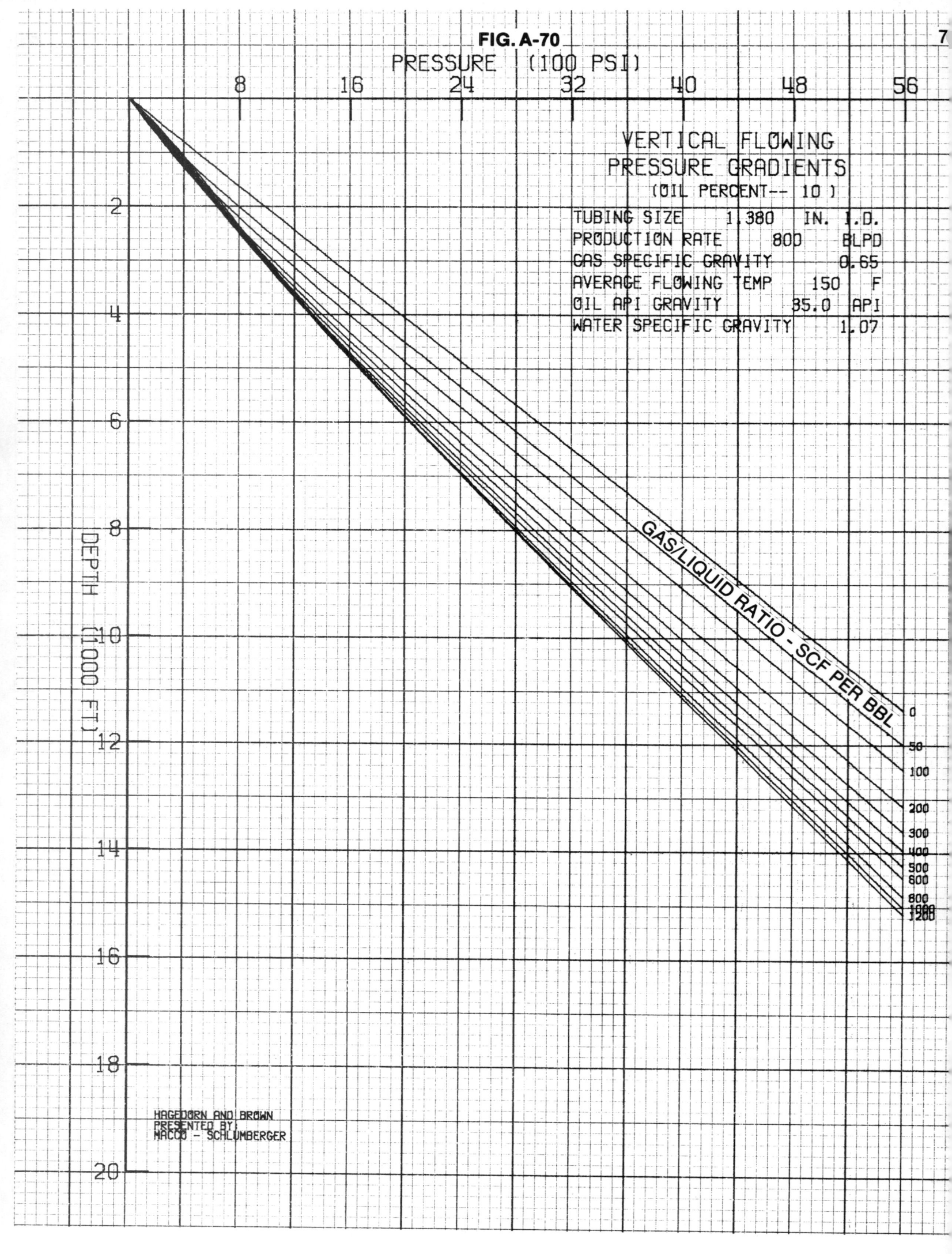

PRESSURE (100 PSI)

VERTICAL FLOWING
PRESSURE GRADIENTS
(OIL PERCENT-- 10)

TUBING SIZE	1.380	IN. I.D.
PRODUCTION RATE	800	BLPD
GAS SPECIFIC GRAVITY	0.65	
AVERAGE FLOWING TEMP	150	F
OIL API GRAVITY	35.0	API
WATER SPECIFIC GRAVITY	1.07	

GAS/LIQUID RATIO - SCF PER BBL

DEPTH (1,000 FT)

HAGEDORN AND BROWN
PRESENTED BY
NACCO - SCHLUMBERGER

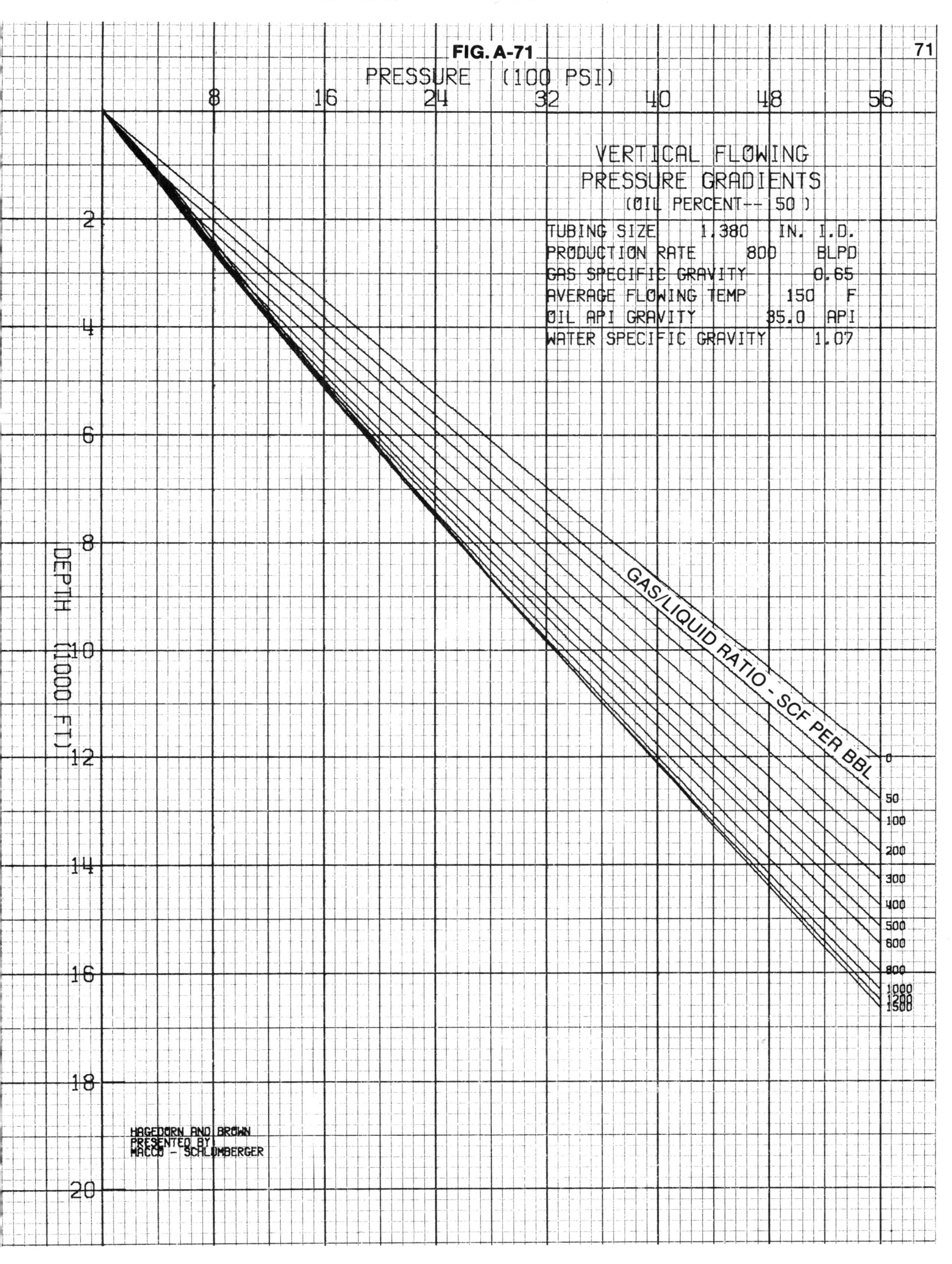

FIG. A-71

71

PRESSURE (100 PSI)

VERTICAL FLOWING
PRESSURE GRADIENTS
(OIL PERCENT-- 50)

TUBING SIZE	1.380	IN. I.D.
PRODUCTION RATE	800	BLPD
GAS SPECIFIC GRAVITY		0.65
AVERAGE FLOWING TEMP	150	F
OIL API GRAVITY	35.0	API
WATER SPECIFIC GRAVITY		1.07

DEPTH (1000 FT)

GAS/LIQUID RATIO - SCF PER BBL

HAGEDORN AND BROWN
PRESENTED BY
MACCO - SCHLUMBERGER

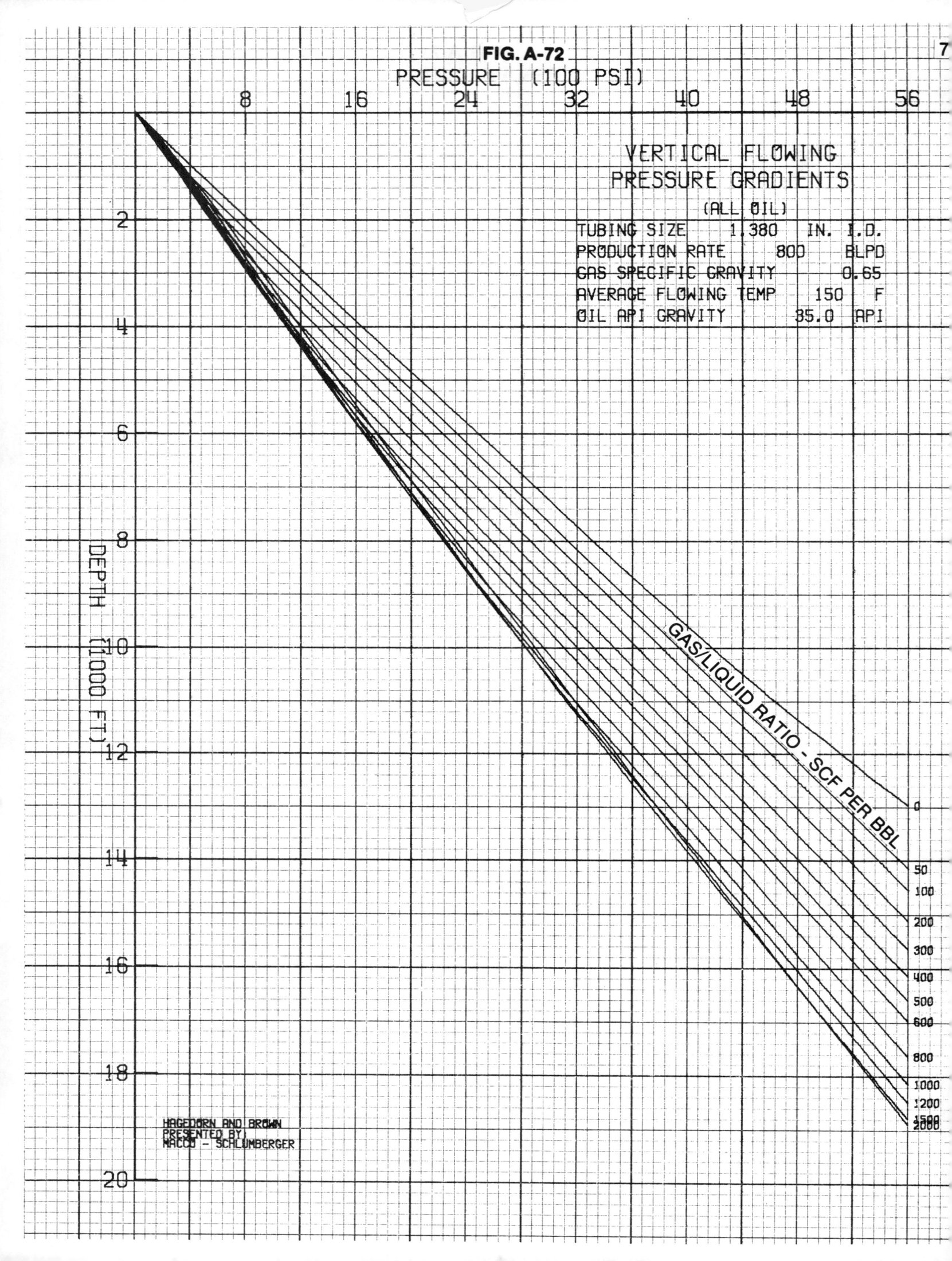

FIG. A-72

PRESSURE (100 PSI)

VERTICAL FLOWING
PRESSURE GRADIENTS
(ALL OIL)

TUBING SIZE	1.380	IN. I.D.
PRODUCTION RATE	800	BLPD
GAS SPECIFIC GRAVITY		0.65
AVERAGE FLOWING TEMP	150	F
OIL API GRAVITY	35.0	API

DEPTH (1000 FT)

GAS/LIQUID RATIO - SCF PER BBL

HAGEDORN AND BROWN
PRESENTED BY
NACCO - SCHLUMBERGER

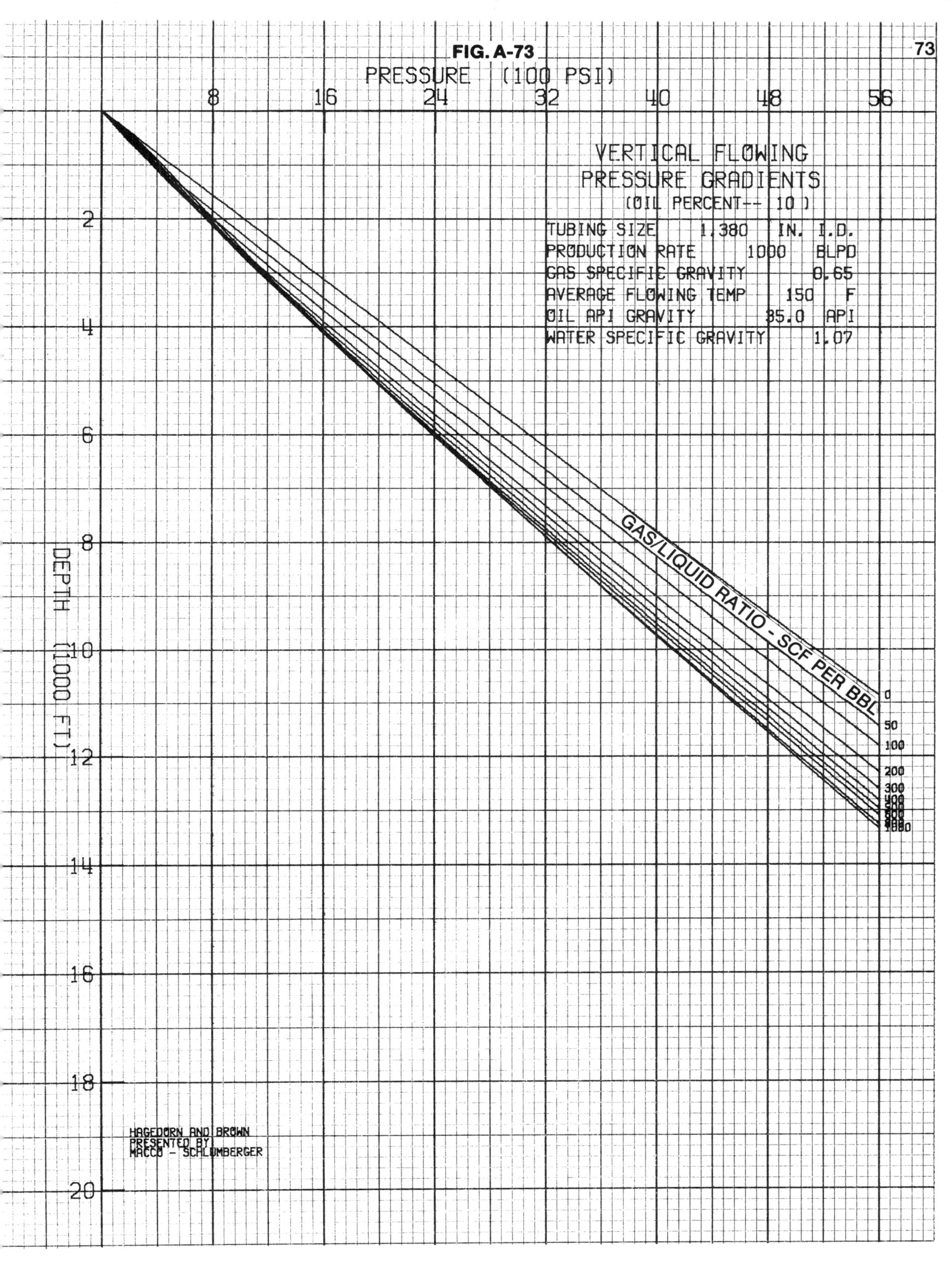

VERTICAL FLOWING
PRESSURE GRADIENTS
(OIL PERCENT-- 10)

TUBING SIZE	1.380	IN. I.D.
PRODUCTION RATE	1000	BLPD
GAS SPECIFIC GRAVITY	0.65	
AVERAGE FLOWING TEMP	150	F
OIL API GRAVITY	35.0	API
WATER SPECIFIC GRAVITY	1.07	

HAGEDORN AND BROWN
PRESENTED BY
MACCO - SCHLUMBERGER

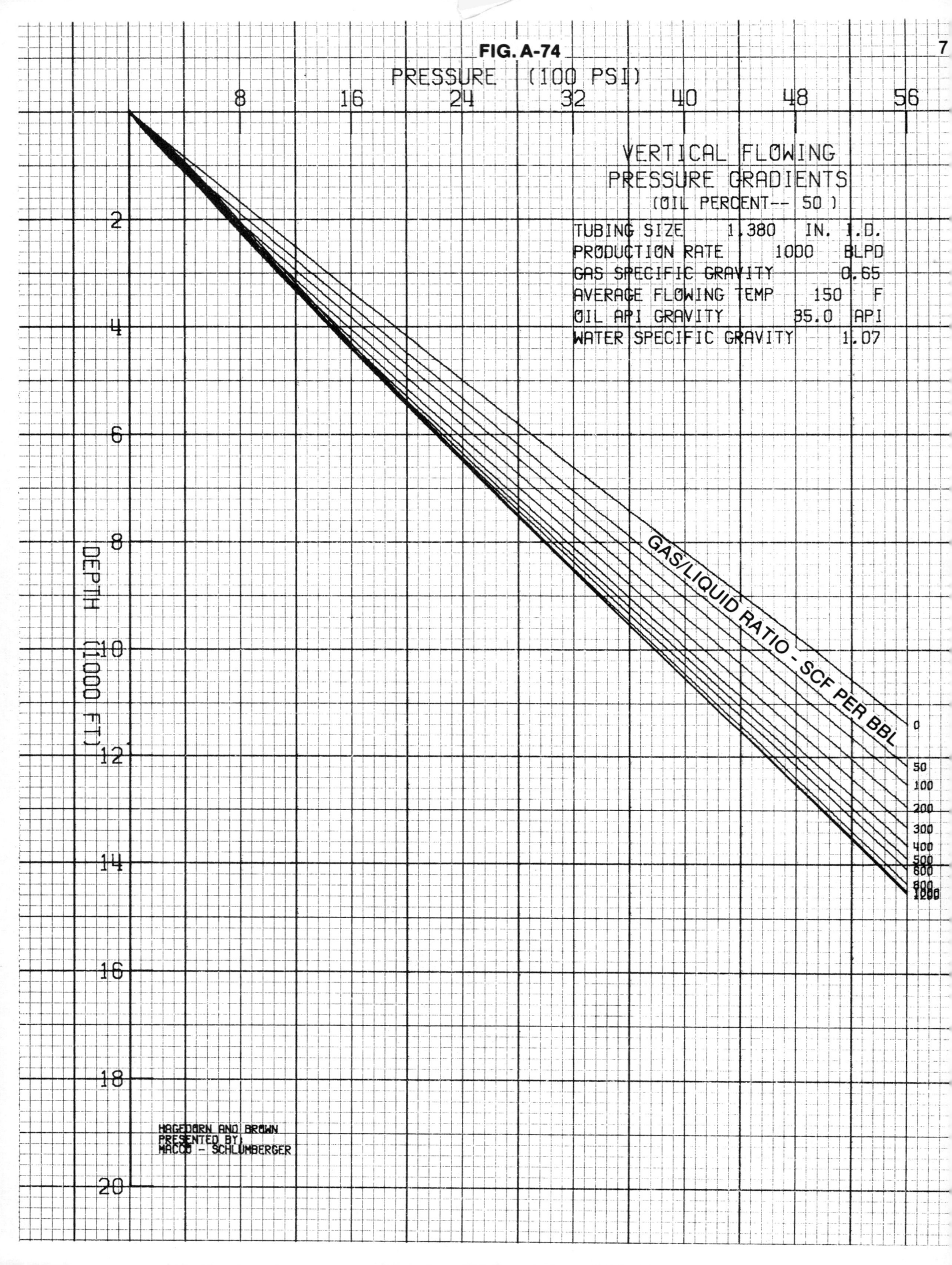

VERTICAL FLOWING
PRESSURE GRADIENTS
(OIL PERCENT-- 50)

TUBING SIZE	1.380	IN.	I.D.
PRODUCTION RATE	1000		BLPD
GAS SPECIFIC GRAVITY		0.65	
AVERAGE FLOWING TEMP	150		F
OIL API GRAVITY		35.0	API
WATER SPECIFIC GRAVITY		1.07	

PRESSURE (100 PSI)

DEPTH (1000 FT)

GAS/LIQUID RATIO - SCF PER BBL

0
50
100
200
300
400
500
600
800
1200

HAGEDORN AND BROWN
PRESENTED BY:
MACCO - SCHLUMBERGER

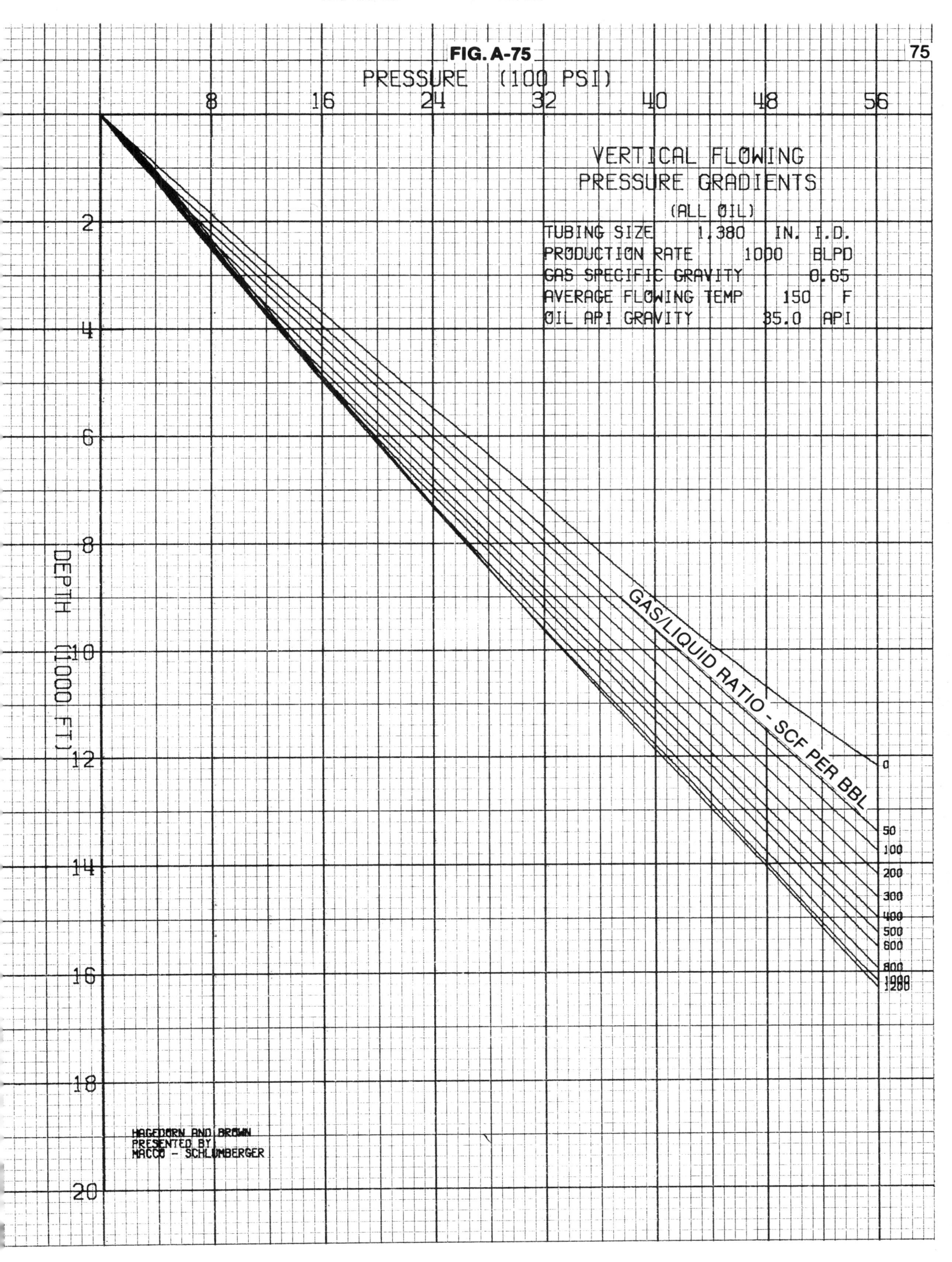

PRESSURE (100 PSI)

DEPTH (1000 FT)

VERTICAL FLOWING
PRESSURE GRADIENTS
(ALL OIL)

TUBING SIZE	1.380	IN. I.D.
PRODUCTION RATE	1000	BLPD
GAS SPECIFIC GRAVITY		0.65
AVERAGE FLOWING TEMP	150	F
OIL API GRAVITY	35.0	API

GAS/LIQUID RATIO - SCF PER BBL

0
50
100
200
300
400
500
600
800
1000

HAGEDORN AND BROWN
PRESENTED BY
MACCO - SCHLUMBERGER

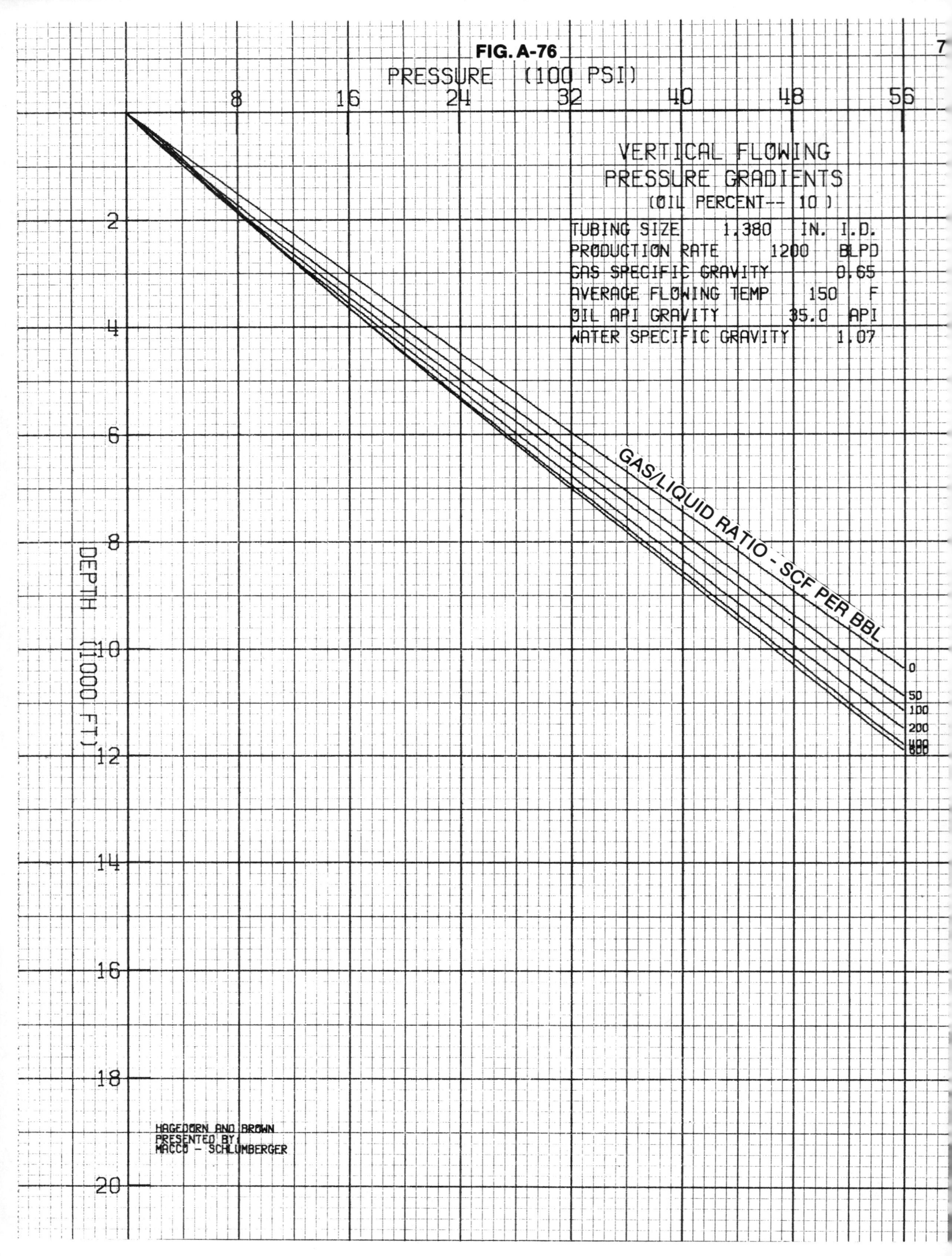

VERTICAL FLOWING
PRESSURE GRADIENTS
(OIL PERCENT-- 10)

TUBING SIZE	1.380	IN. I.D.
PRODUCTION RATE	1200	BLPD
GAS SPECIFIC GRAVITY	0.65	
AVERAGE FLOWING TEMP	150	F
OIL API GRAVITY	35.0	API
WATER SPECIFIC GRAVITY	1.07	

PRESSURE (100 PSI)

DEPTH (1000 FT)

GAS/LIQUID RATIO - SCF PER BBL

0
50
100
200
400
600

HAGEDORN AND BROWN
PRESENTED BY:
MACCO - SCHLUMBERGER

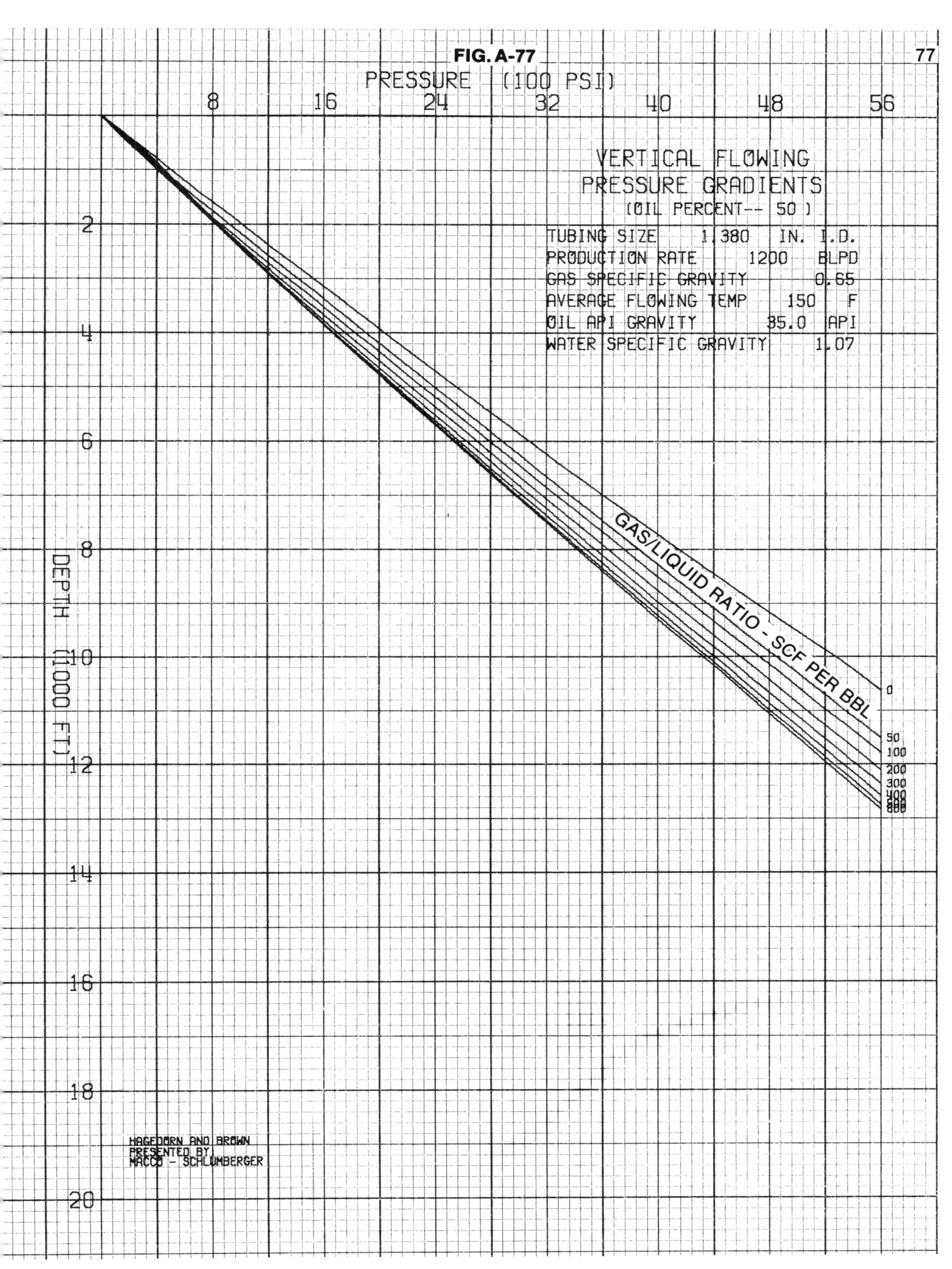

PRESSURE (100 PSI)

VERTICAL FLOWING
PRESSURE GRADIENTS
(OIL PERCENT-- 50)

TUBING SIZE	1.380	IN. I.D.
PRODUCTION RATE	1200	BLPD
GAS SPECIFIC GRAVITY	0.65	
AVERAGE FLOWING TEMP	150	F
OIL API GRAVITY	35.0	API
WATER SPECIFIC GRAVITY	1.07	

DEPTH (1000 FT)

GAS/LIQUID RATIO - SCF PER BBL

0
50
100
200
300
400
600
800

HAGEDORN AND BROWN
PRESENTED BY
MACCO - SCHLUMBERGER

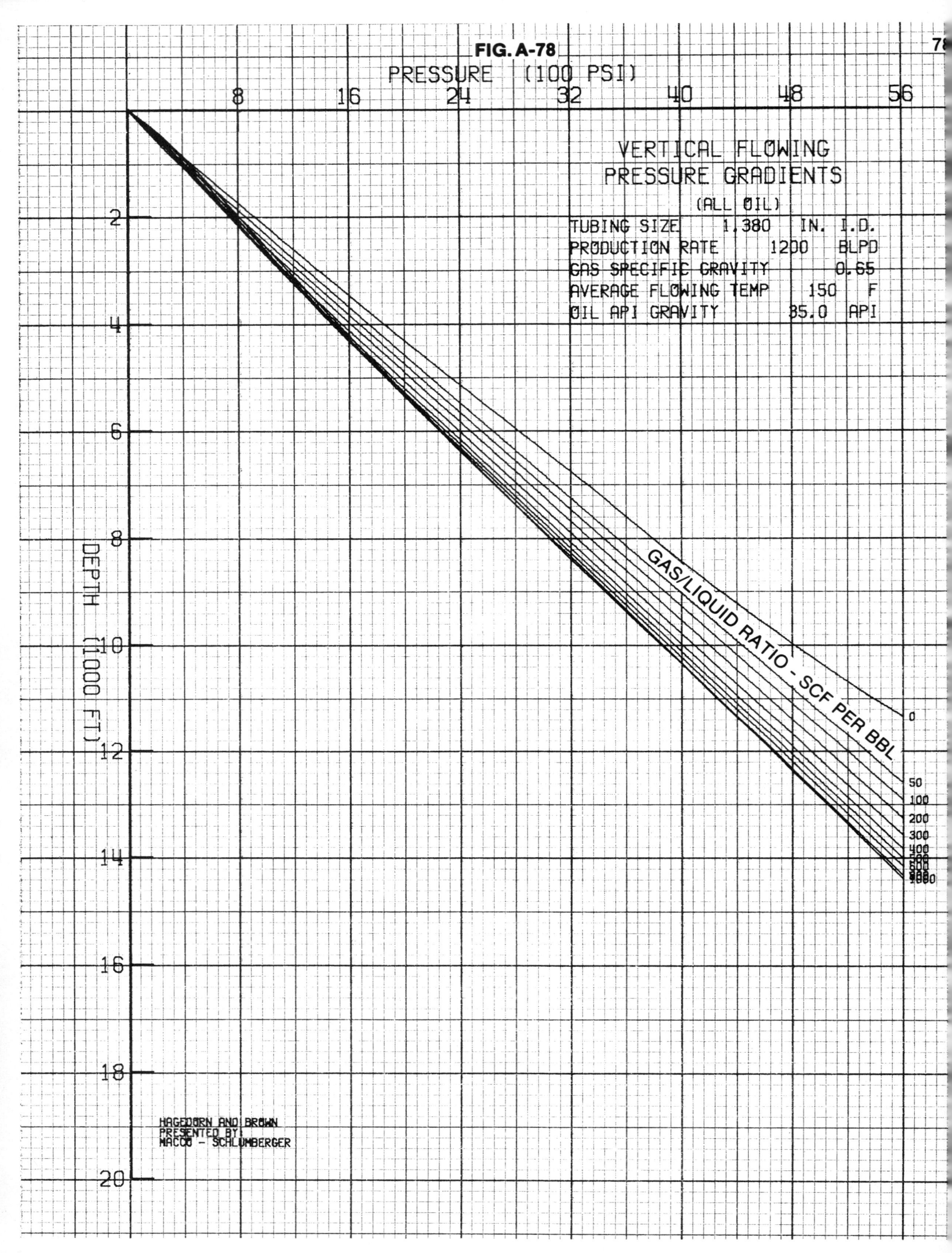

FIG. A-78

VERTICAL FLOWING
PRESSURE GRADIENTS
(ALL OIL)

TUBING SIZE	1.380	IN. I.D.
PRODUCTION RATE	1200	BLPD
GAS SPECIFIC GRAVITY	0.65	
AVERAGE FLOWING TEMP	150	F
OIL API GRAVITY	35.0	API

PRESSURE (100 PSI)

DEPTH (1000 FT)

GAS/LIQUID RATIO - SCF PER BBL

0
50
100
200
300
400
500
600
800
1000

HAGEDORN AND BROWN
PRESENTED BY
MACCO - SCHLUMBERGER

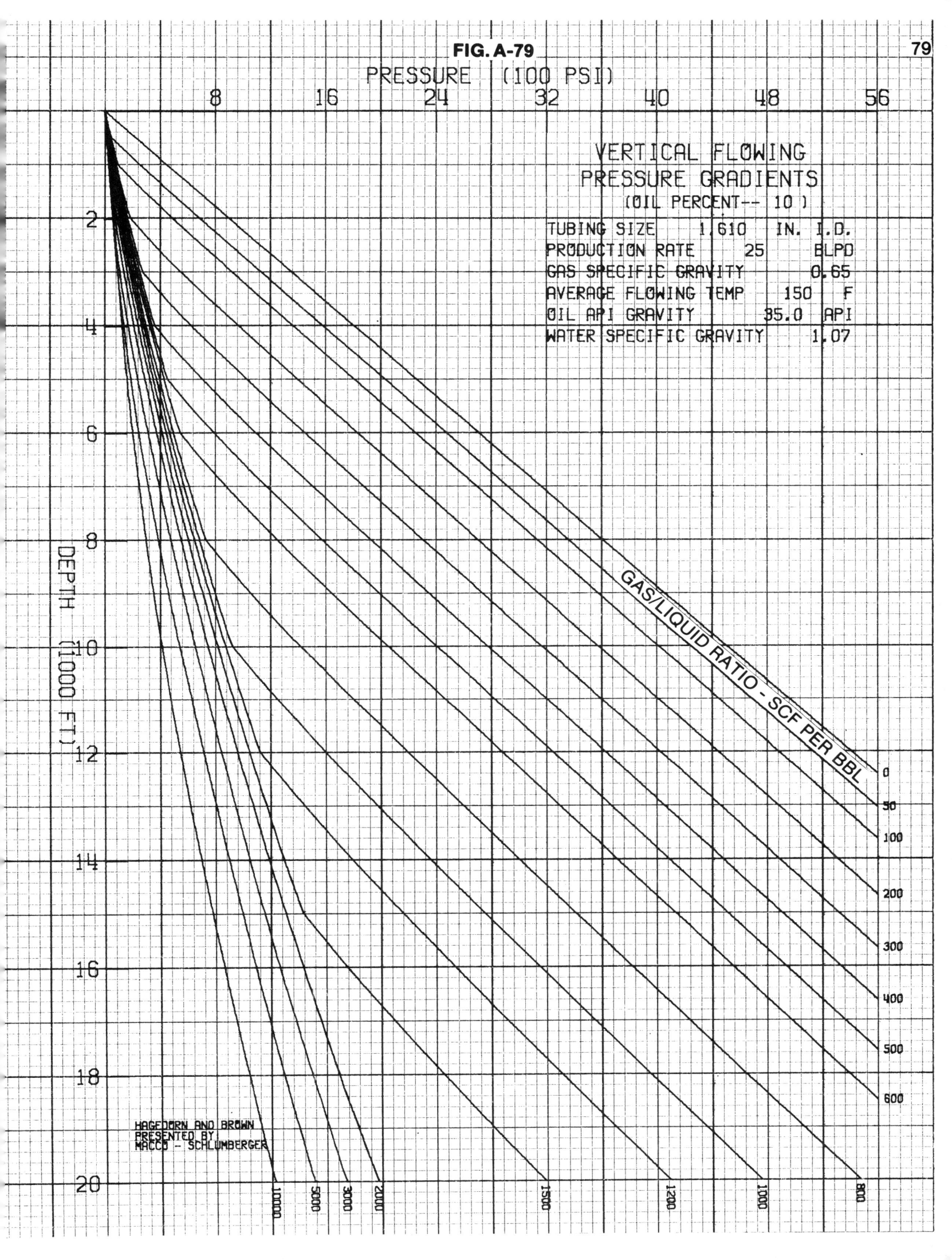

PRESSURE (100 PSI)

DEPTH (1000 FT.)

VERTICAL FLOWING
PRESSURE GRADIENTS
(OIL PERCENT-- 10)

TUBING SIZE	1.610	IN. I.D.
PRODUCTION RATE	25	BLPD
GAS SPECIFIC GRAVITY	0.65	
AVERAGE FLOWING TEMP	150	F
OIL API GRAVITY	35.0	API
WATER SPECIFIC GRAVITY	1.07	

GAS/LIQUID RATIO - SCF PER BBL

HAGEDORN AND BROWN
PRESENTED BY
MACCO - SCHLUMBERGER

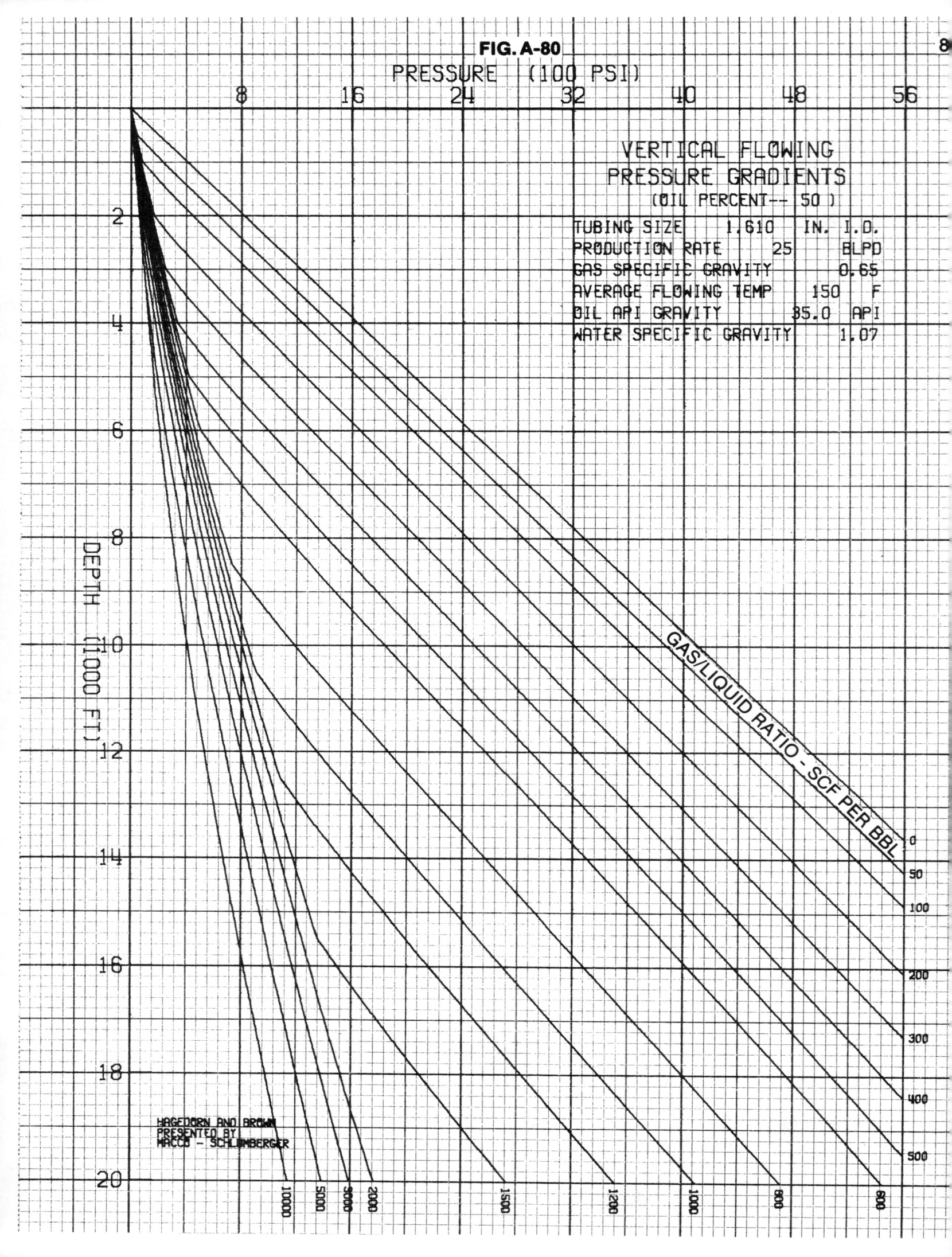

FIG. A-80

PRESSURE (100 PSI)

DEPTH (1000 FT)

VERTICAL FLOWING
PRESSURE GRADIENTS
(OIL PERCENT-- 50)

TUBING SIZE	1.610	IN. I.D.
PRODUCTION RATE	25	BLPD
GAS SPECIFIC GRAVITY		0.65
AVERAGE FLOWING TEMP	150	F
OIL API GRAVITY	35.0	API
WATER SPECIFIC GRAVITY		1.07

GAS/LIQUID RATIO - SCF PER BBL

HAGEDORN AND BROWN
PRESENTED BY
MACCO - SCHLUMBERGER

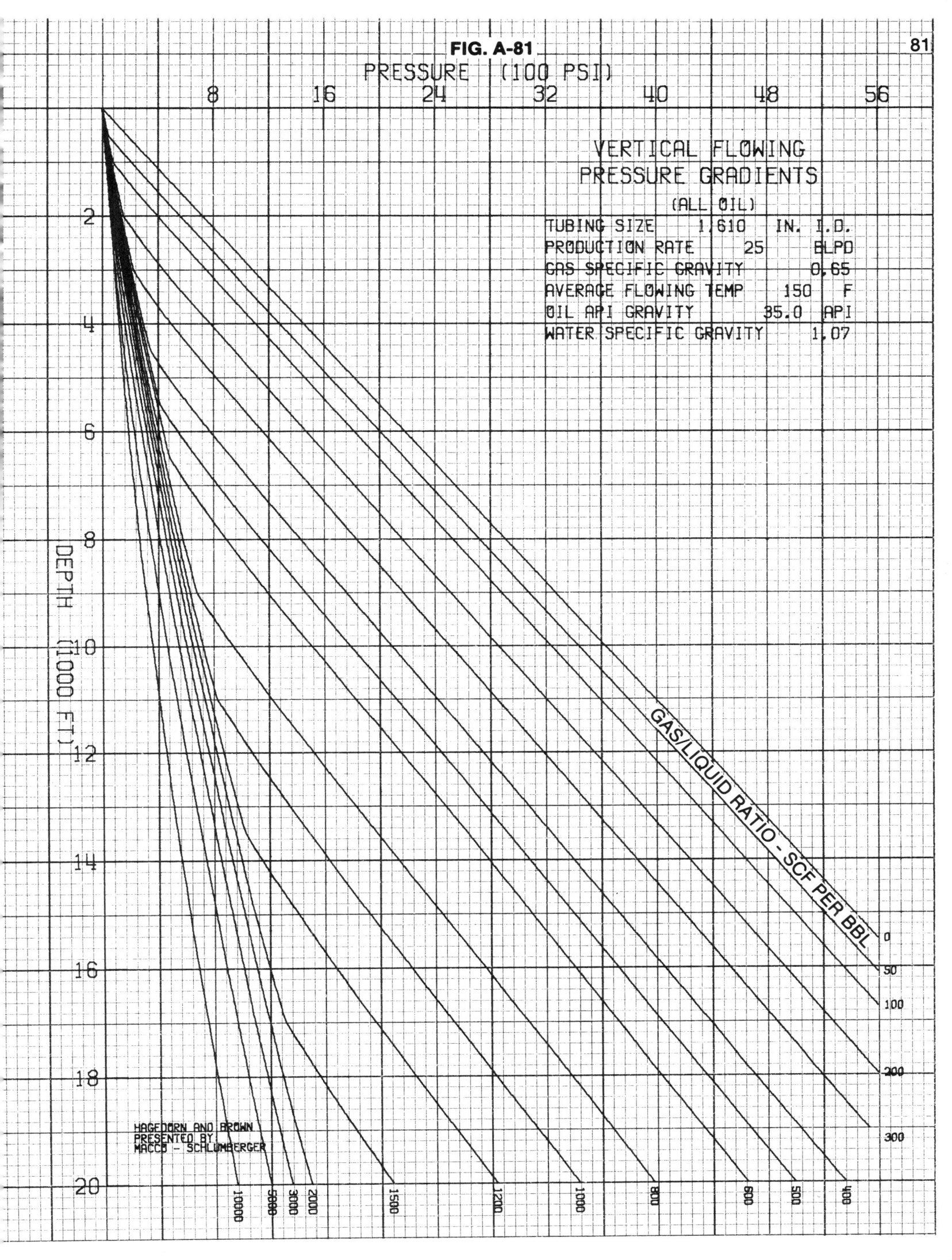

PRESSURE (100 PSI)

VERTICAL FLOWING
PRESSURE GRADIENTS
(ALL OIL)

TUBING SIZE	1.610	IN. I.D.
PRODUCTION RATE	25	BLPD
GAS SPECIFIC GRAVITY		0.65
AVERAGE FLOWING TEMP	150	F
OIL API GRAVITY	35.0	API
WATER SPECIFIC GRAVITY		1.07

DEPTH (1000 FT)

GAS/LIQUID RATIO - SCF PER BBL

HAGEDORN AND BROWN
PRESENTED BY
MACCO - SCHLUMBERGER

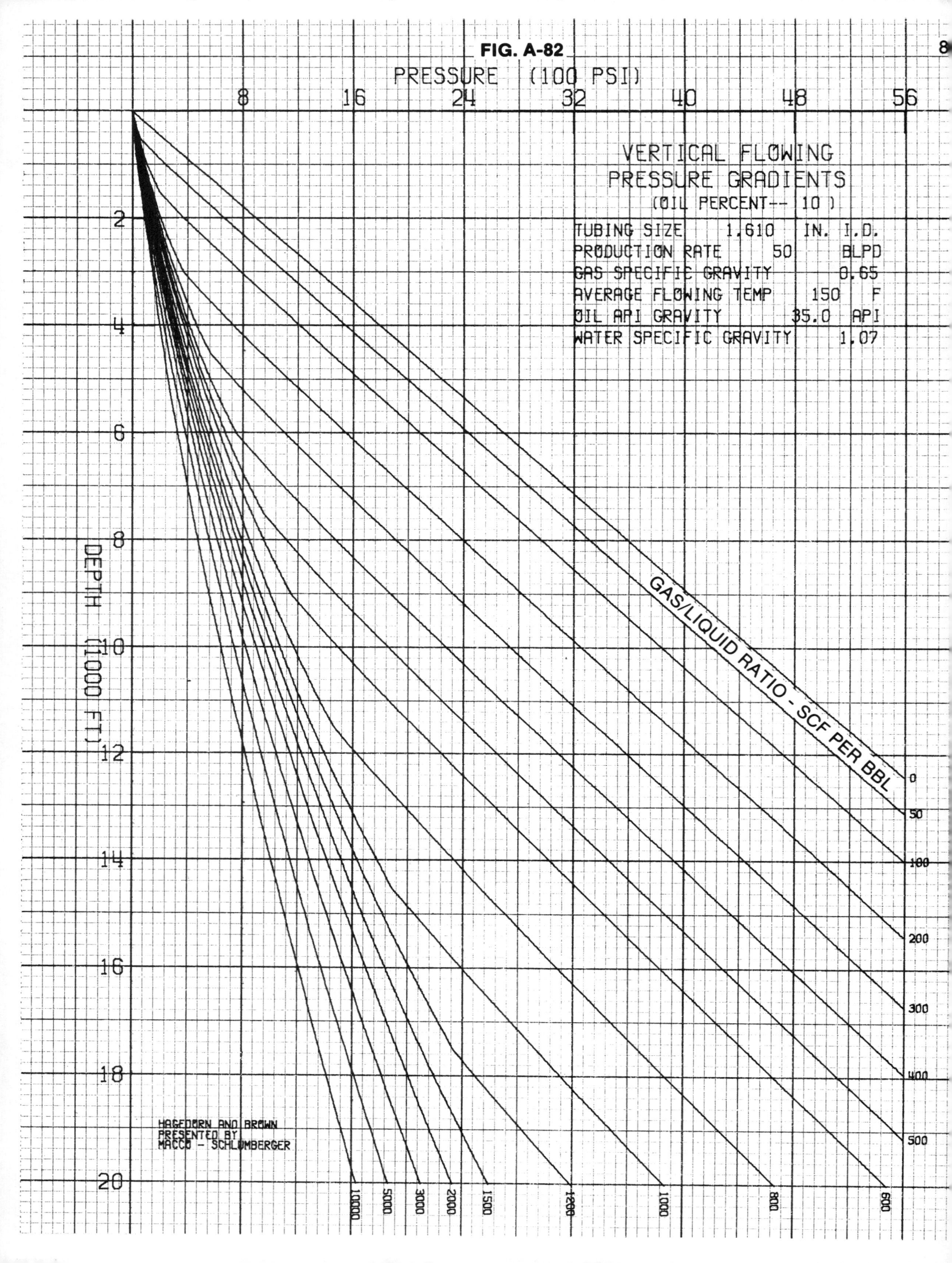

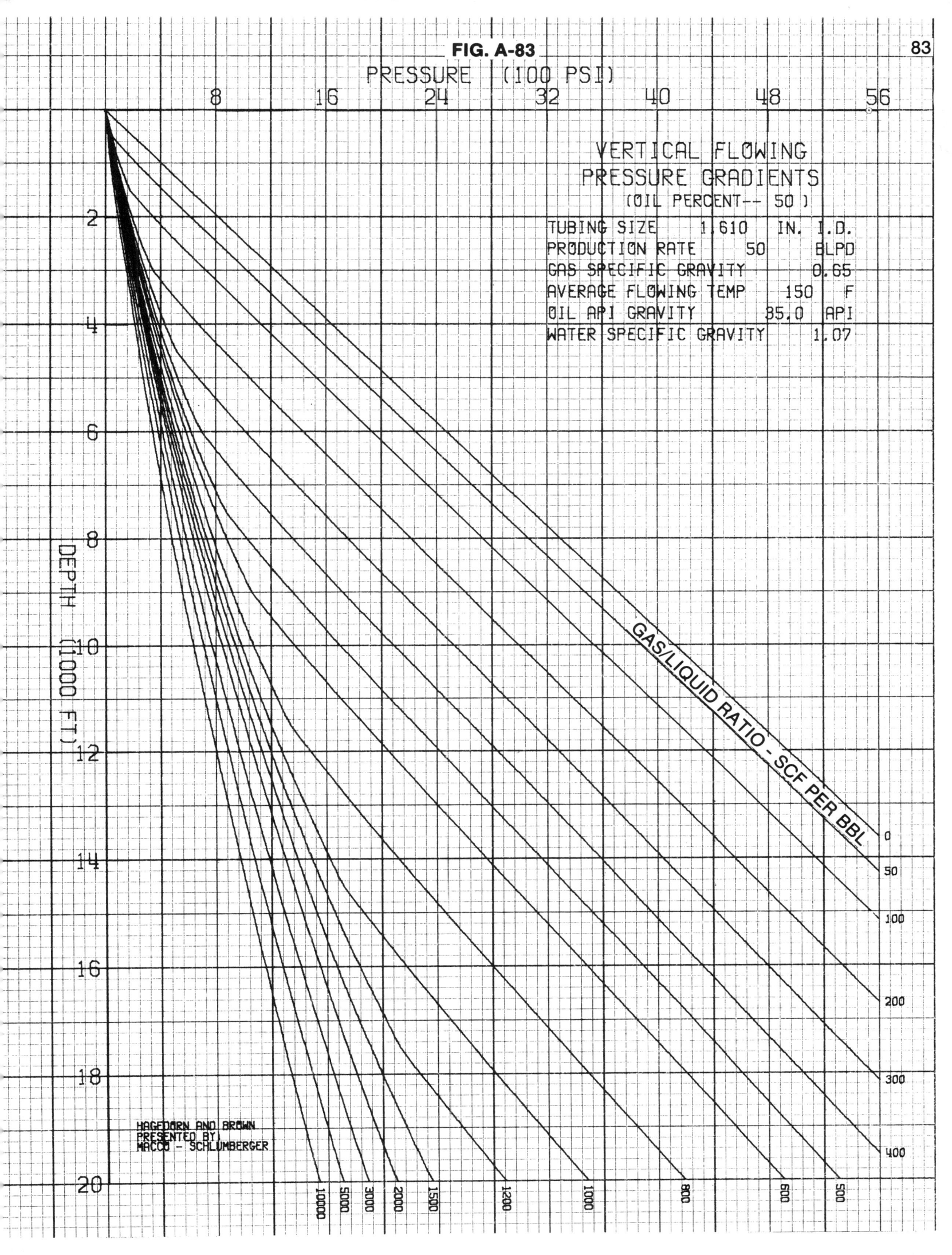

PRESSURE (100 PSI)

VERTICAL FLOWING
PRESSURE GRADIENTS
(OIL PERCENT-- 50)

TUBING SIZE   1.610   IN. I.D.
PRODUCTION RATE   50   BLPD
GAS SPECIFIC GRAVITY   0.65
AVERAGE FLOWING TEMP   150   F
OIL API GRAVITY   85.0 API
WATER SPECIFIC GRAVITY   1.07

DEPTH (1000 FT)

GAS/LIQUID RATIO - SCF PER BBL

HAGEDORN AND BROWN
PRESENTED BY
MACCO - SCHLUMBERGER

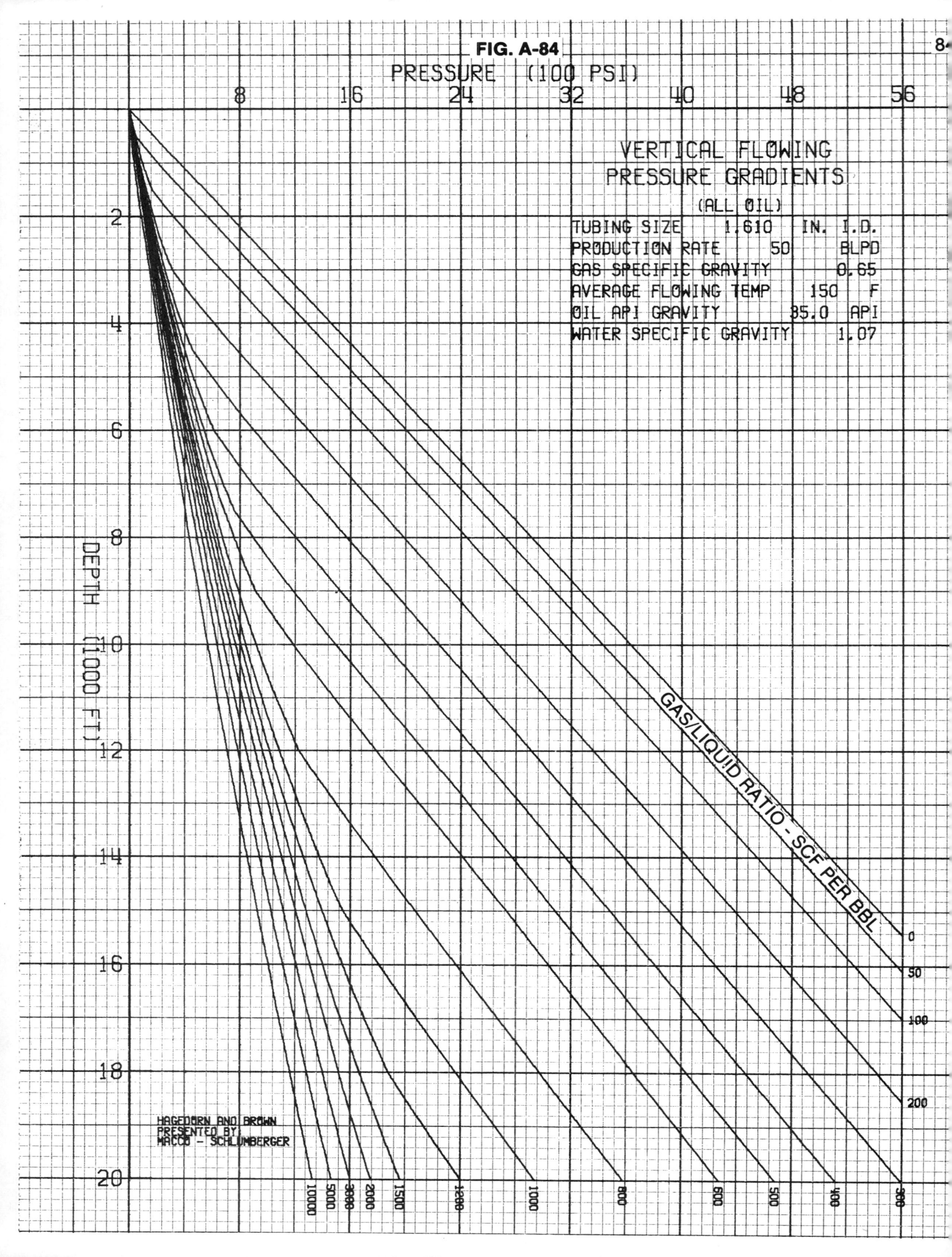

FIG. A-84

VERTICAL FLOWING
PRESSURE GRADIENTS
(ALL OIL)

TUBING SIZE	1.610	IN. I.D.
PRODUCTION RATE	50	BLPD
GAS SPECIFIC GRAVITY	0.65	
AVERAGE FLOWING TEMP	150	F
OIL API GRAVITY	85.0	API
WATER SPECIFIC GRAVITY	1.07	

PRESSURE (100 PSI)

DEPTH (1000 FT)

GAS/LIQUID RATIO - SCF PER BBL

HAGEDORN AND BROWN
PRESENTED BY
MACCO - SCHLUMBERGER

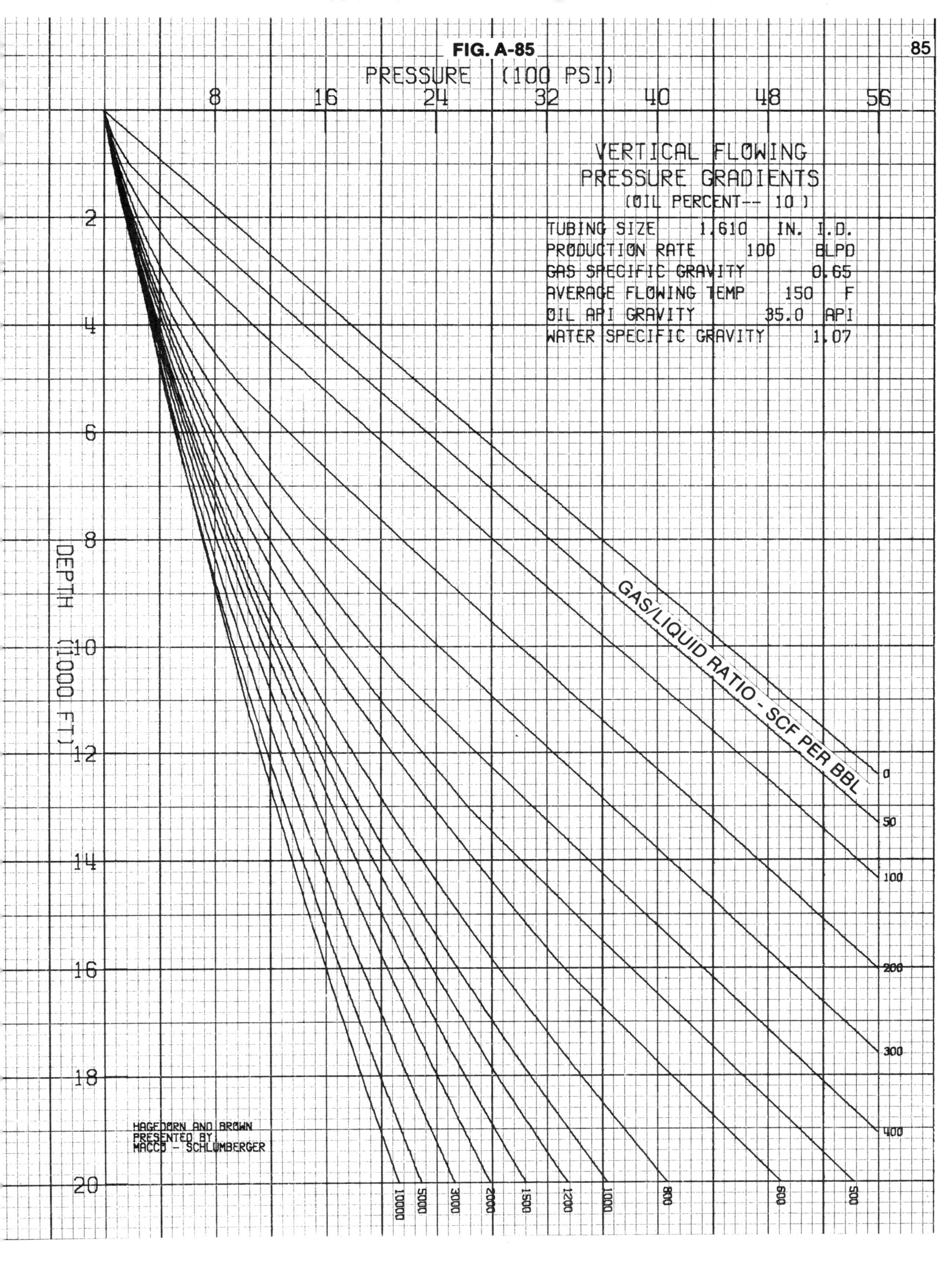

PRESSURE (100 PSI)

VERTICAL FLOWING
PRESSURE GRADIENTS
(OIL PERCENT-- 10)

TUBING SIZE	1.610	IN. I.D.
PRODUCTION RATE	100	BLPD
GAS SPECIFIC GRAVITY	0.65	
AVERAGE FLOWING TEMP	150	F
OIL API GRAVITY	35.0	API
WATER SPECIFIC GRAVITY	1.07	

DEPTH (1000 FT)

GAS/LIQUID RATIO - SCF PER BBL

HAGEDORN AND BROWN
PRESENTED BY
MACCO - SCHLUMBERGER

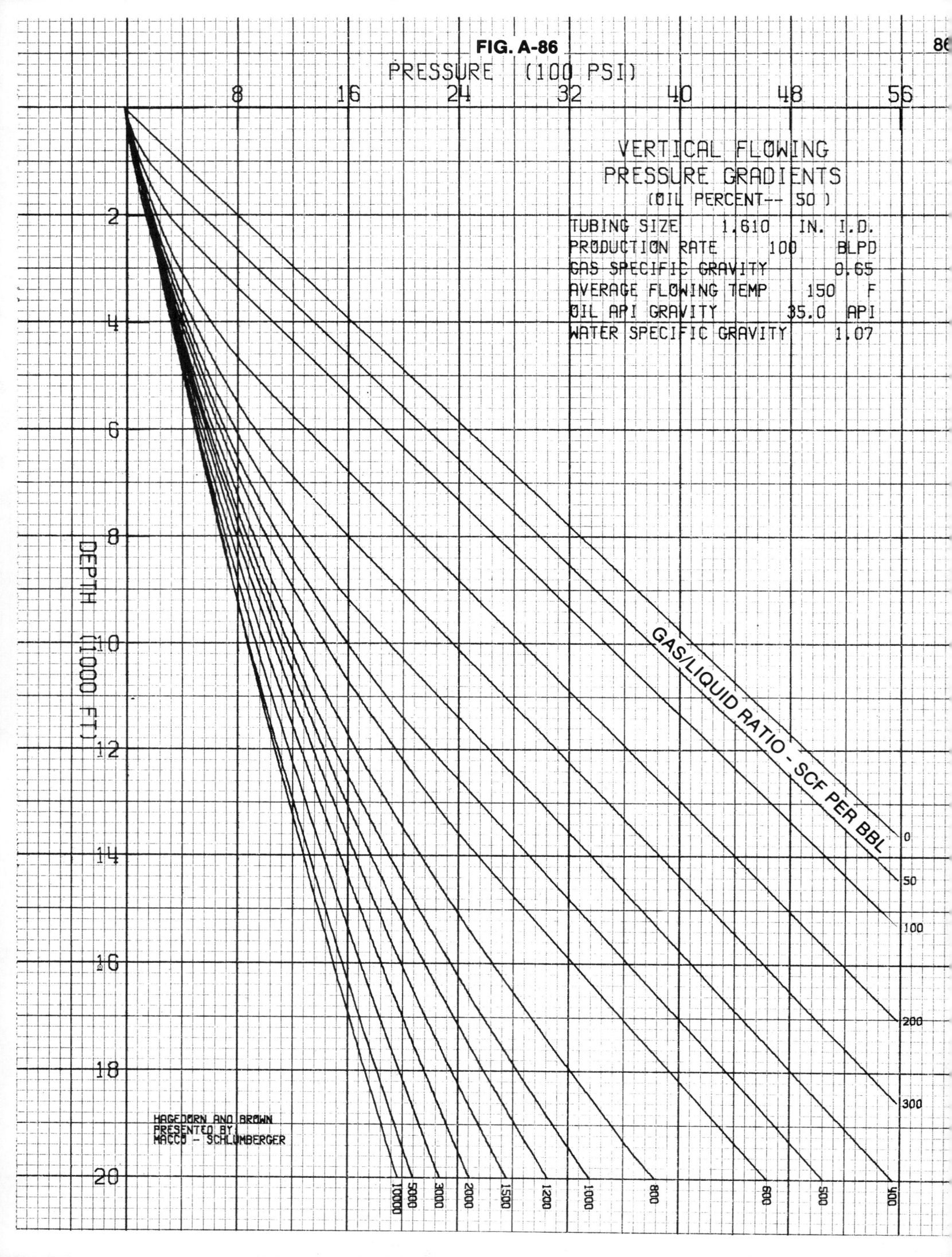

VERTICAL FLOWING
PRESSURE GRADIENTS
(OIL PERCENT-- 50)

TUBING SIZE	1.610	IN. I.D.
PRODUCTION RATE	100	BLPD
GAS SPECIFIC GRAVITY		0.65
AVERAGE FLOWING TEMP	150	F
OIL API GRAVITY	35.0	API
WATER SPECIFIC GRAVITY		1.07

PRESSURE (100 PSI)

DEPTH (1000 FT)

GAS/LIQUID RATIO - SCF PER BBL

HAGEDORN AND BROWN
PRESENTED BY
MACCO - SCHLUMBERGER

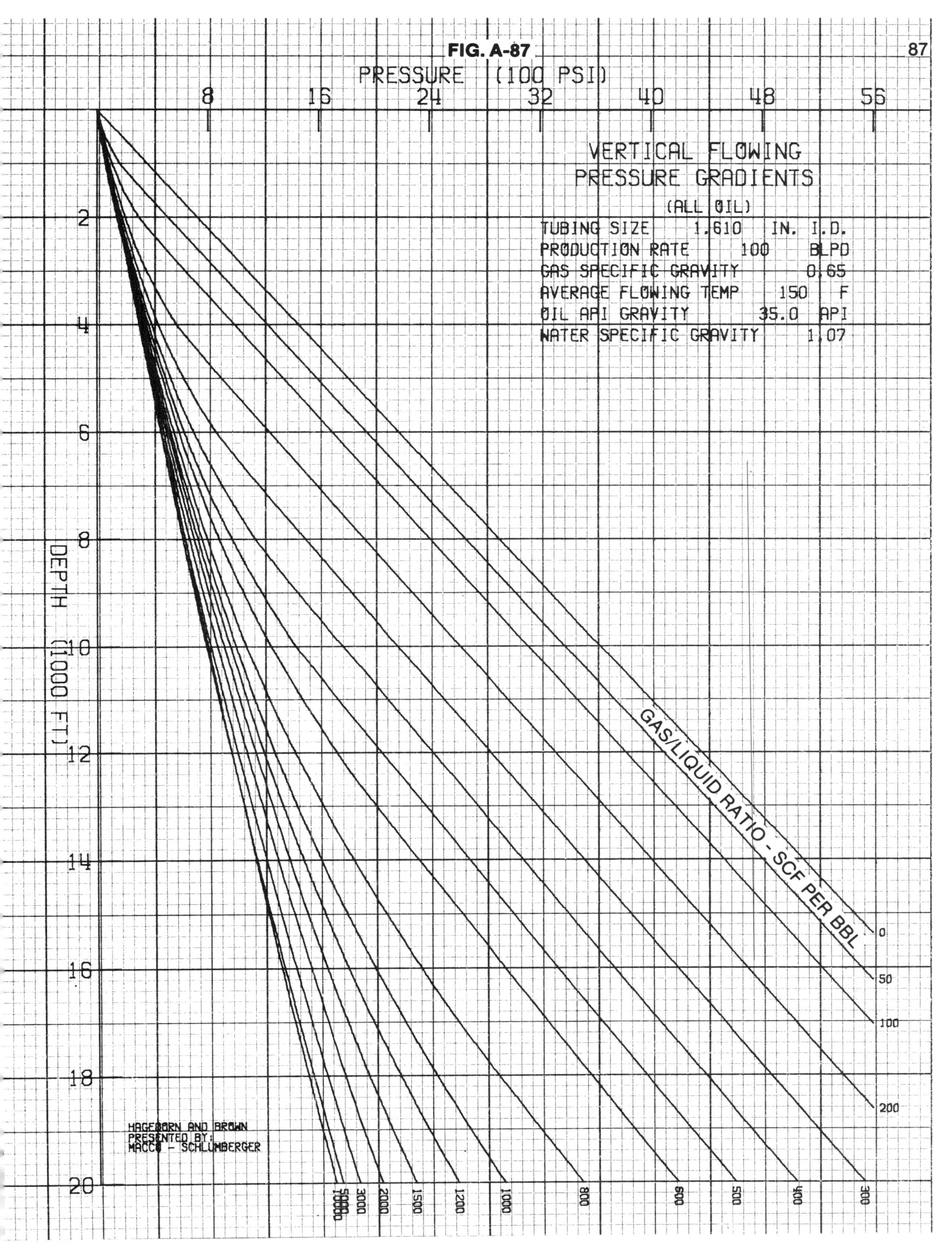

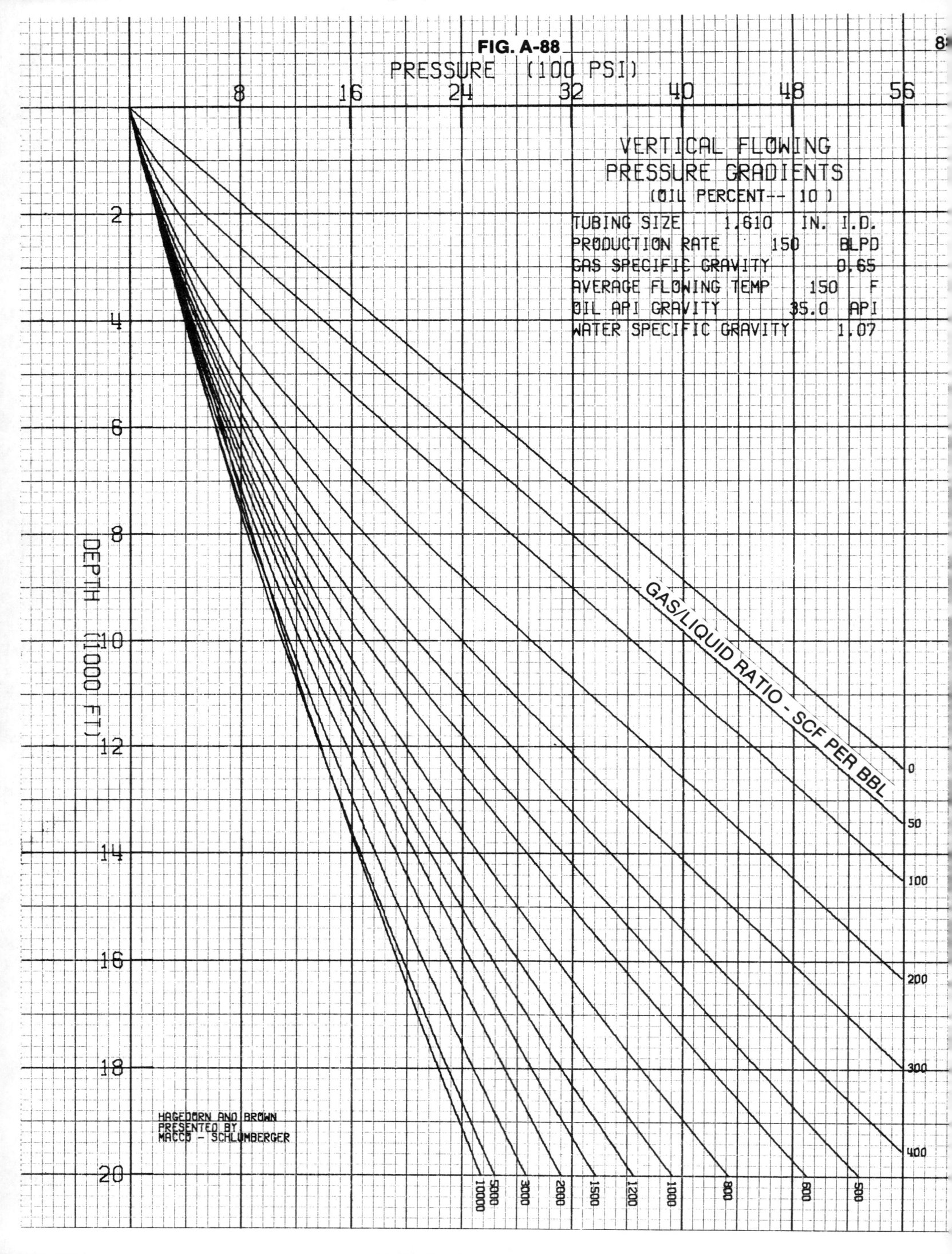

FIG. A-88

PRESSURE (100 PSI)

VERTICAL FLOWING
PRESSURE GRADIENTS
(OIL PERCENT-- 10)

TUBING SIZE	1.610	IN. I.D.
PRODUCTION RATE	150	BLPD
GAS SPECIFIC GRAVITY	0.65	
AVERAGE FLOWING TEMP	150	F
OIL API GRAVITY	35.0	API
WATER SPECIFIC GRAVITY	1.07	

DEPTH (1000 FT)

GAS/LIQUID RATIO - SCF PER BBL

HAGEDORN AND BROWN
PRESENTED BY
MACCO - SCHLUMBERGER

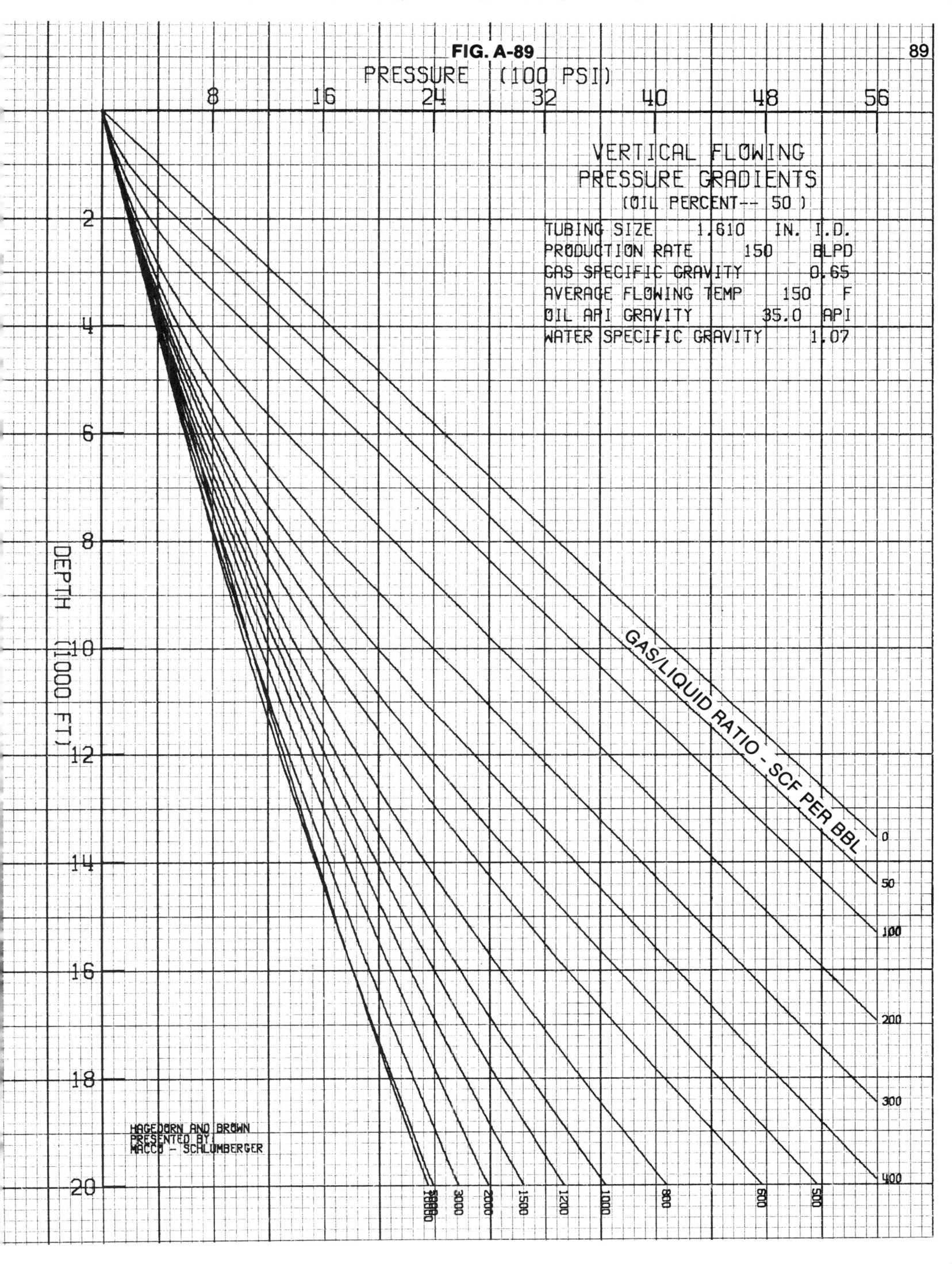

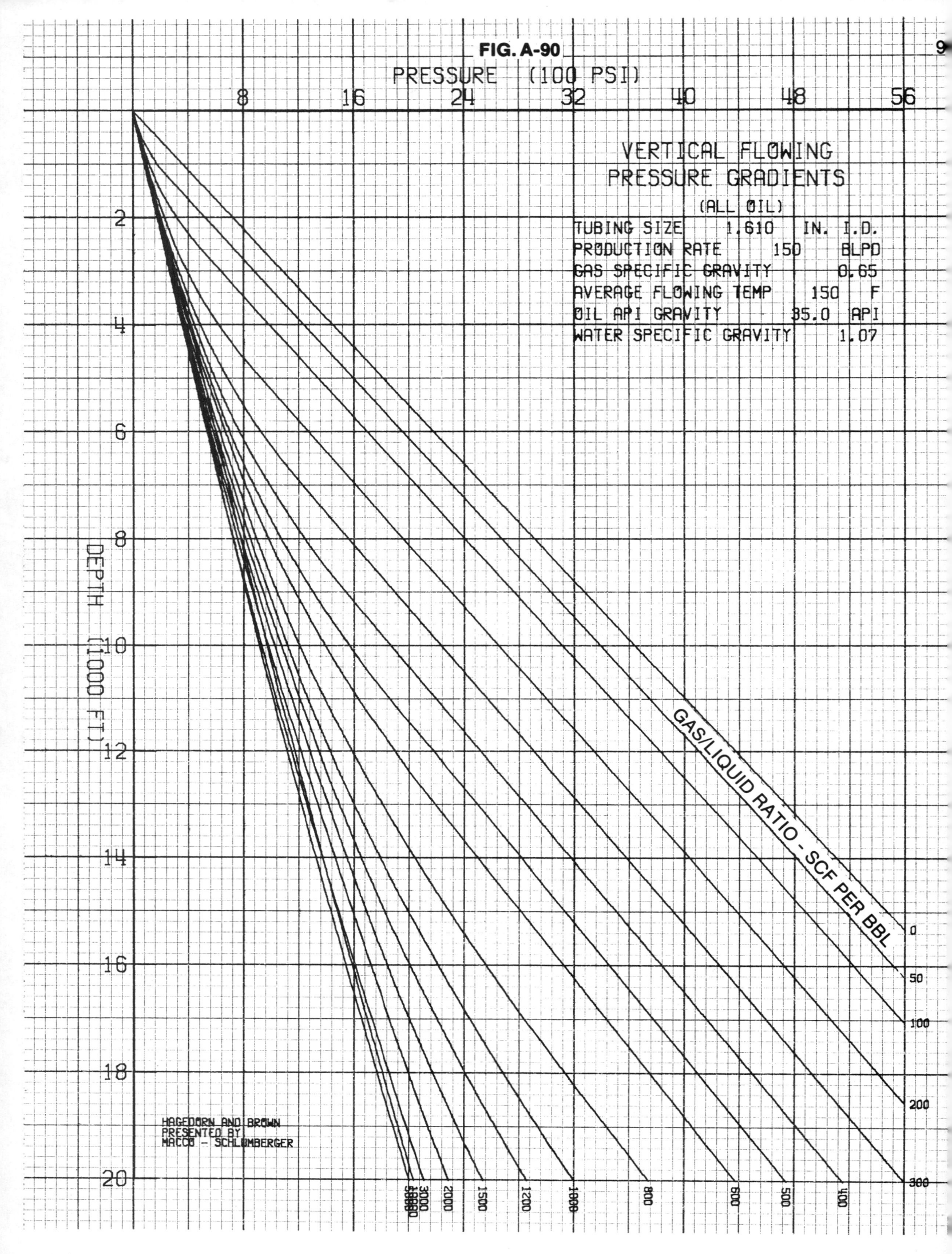

FIG. A-90

PRESSURE (100 PSI)

VERTICAL FLOWING
PRESSURE GRADIENTS
(ALL OIL)

TUBING SIZE	1.610	IN. I.D.
PRODUCTION RATE	150	BLPD
GAS SPECIFIC GRAVITY		0.65
AVERAGE FLOWING TEMP	150	F
OIL API GRAVITY	35.0	API
WATER SPECIFIC GRAVITY		1.07

DEPTH (1000 FT)

GAS/LIQUID RATIO - SCF PER BBL

HAGEDORN AND BROWN
PRESENTED BY
MACCO – SCHLUMBERGER

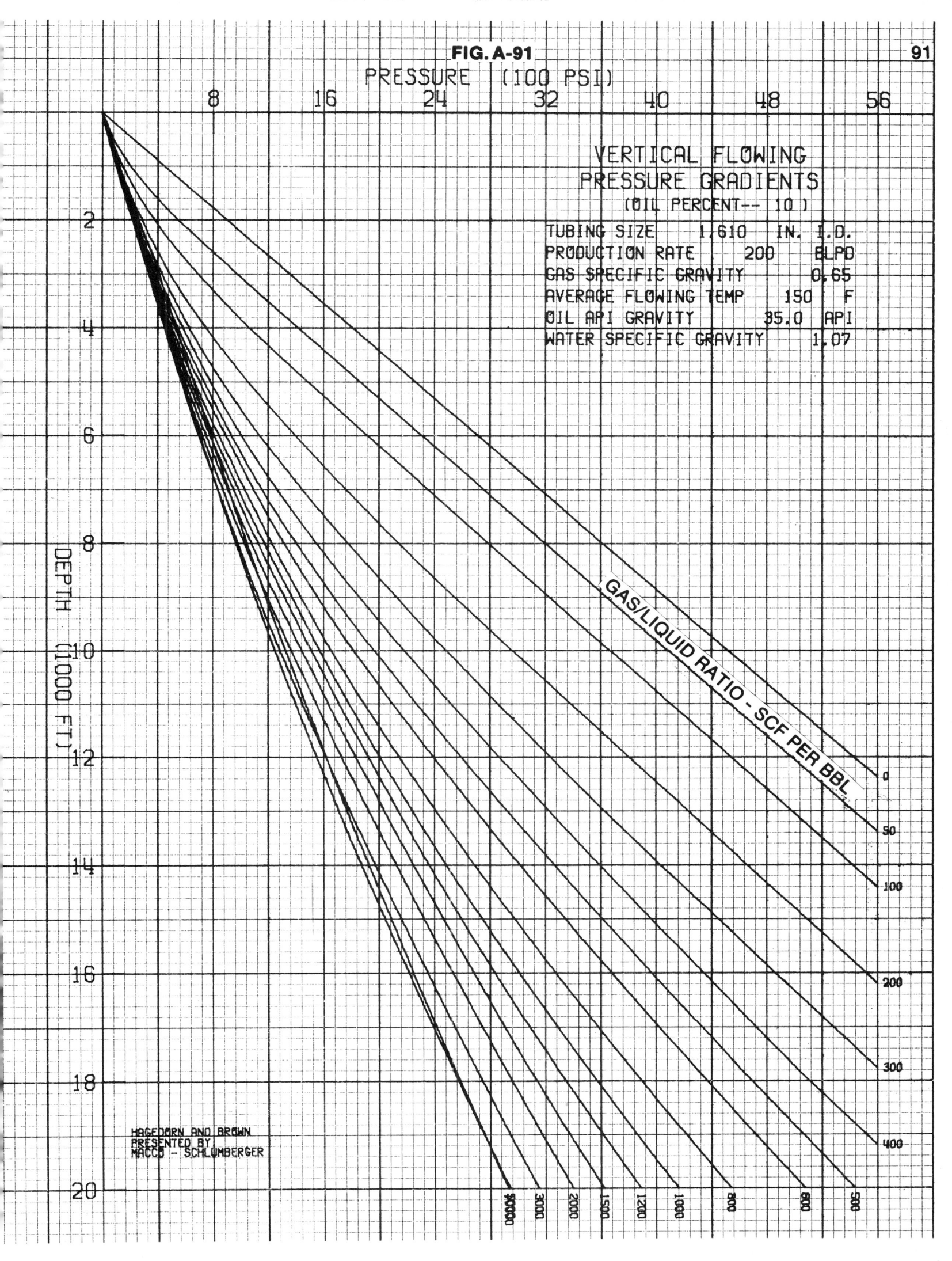

VERTICAL FLOWING
PRESSURE GRADIENTS
(OIL PERCENT-- 10)

TUBING SIZE	1.610	IN. I.D.
PRODUCTION RATE	200	BLPD
GAS SPECIFIC GRAVITY		0.65
AVERAGE FLOWING TEMP	150	F
OIL API GRAVITY	35.0	API
WATER SPECIFIC GRAVITY		1.07

PRESSURE (100 PSI)

DEPTH (1000 FT)

GAS/LIQUID RATIO - SCF PER BBL

HAGEDORN AND BROWN
PRESENTED BY
MACCO - SCHLUMBERGER

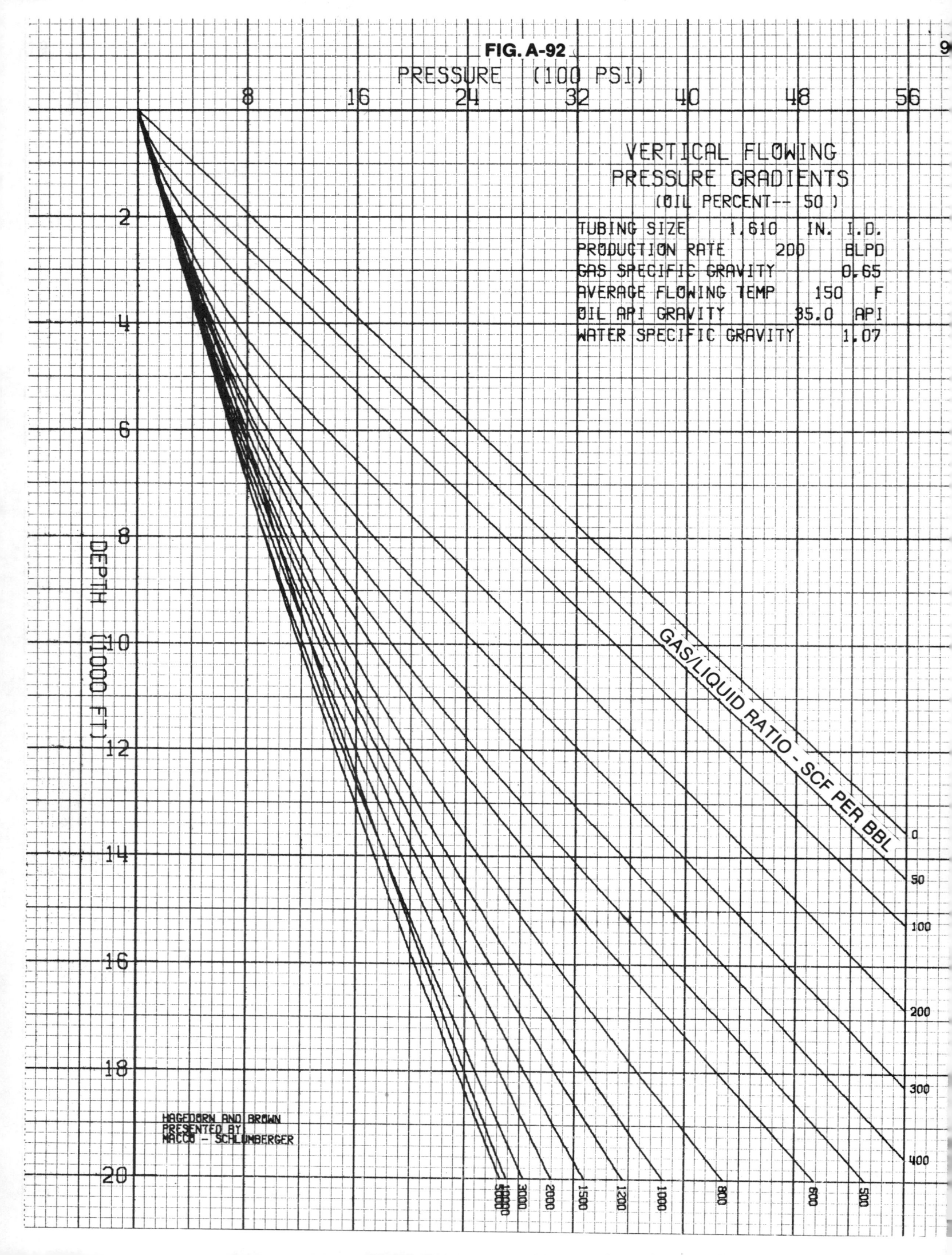

FIG. A-92

PRESSURE (100 PSI)

VERTICAL FLOWING
PRESSURE GRADIENTS
(OIL PERCENT-- 50)

TUBING SIZE	1.610	IN. I.D.
PRODUCTION RATE	200	BLPD
GAS SPECIFIC GRAVITY	0.65	
AVERAGE FLOWING TEMP	150	F
OIL API GRAVITY	35.0	API
WATER SPECIFIC GRAVITY	1.07	

DEPTH (1000 FT)

GAS/LIQUID RATIO - SCF PER BBL

HAGEDORN AND BROWN
PRESENTED BY
MACCO - SCHLUMBERGER

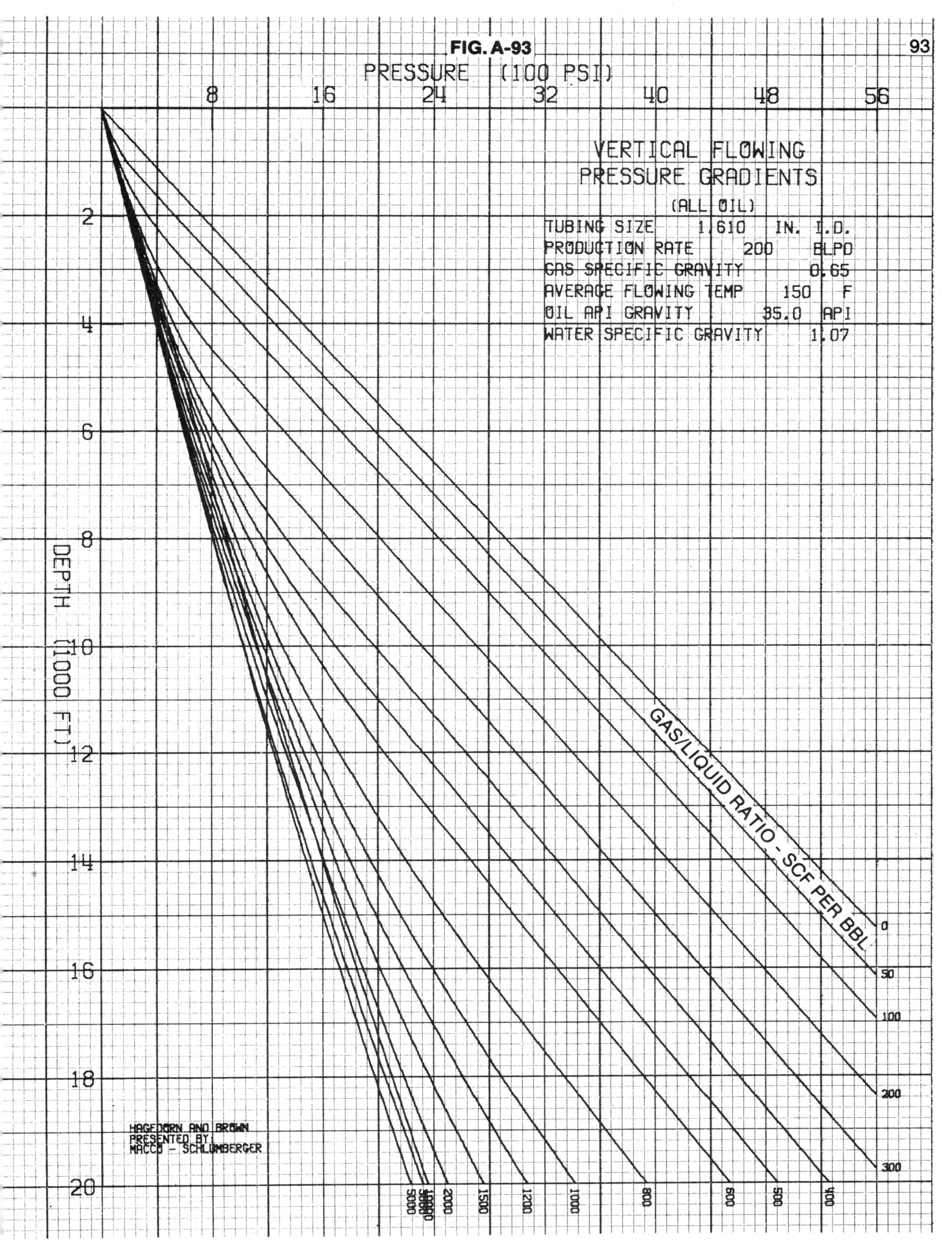

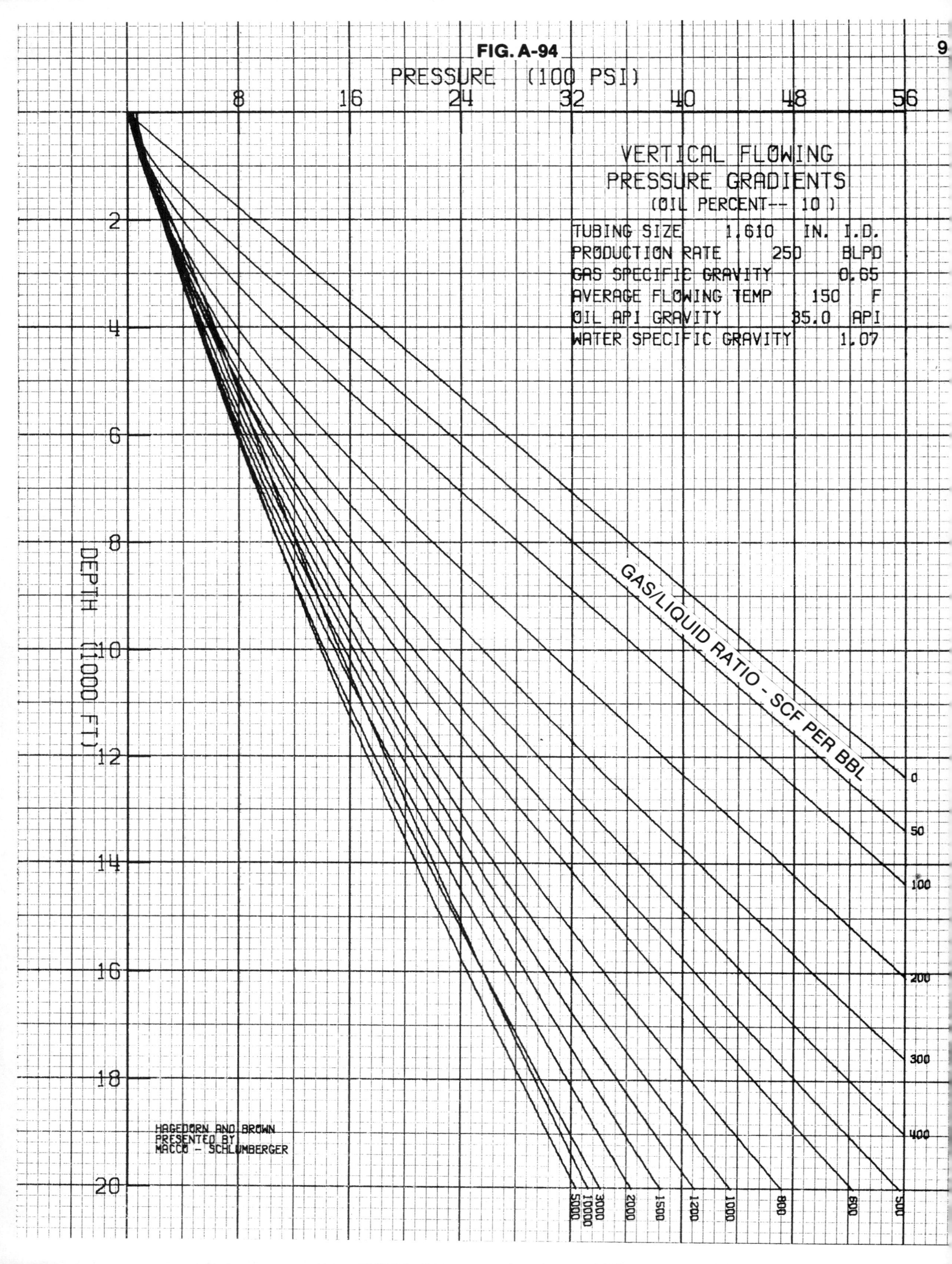

FIG. A-94

VERTICAL FLOWING
PRESSURE GRADIENTS
(OIL PERCENT-- 10)

TUBING SIZE	1.610	IN. I.D.
PRODUCTION RATE	250	BLPD
GAS SPECIFIC GRAVITY		0.65
AVERAGE FLOWING TEMP	150	F
OIL API GRAVITY	85.0	API
WATER SPECIFIC GRAVITY		1.07

PRESSURE (100 PSI)

DEPTH (1000 FT)

GAS/LIQUID RATIO - SCF PER BBL

HAGEDORN AND BROWN
PRESENTED BY
MACCO - SCHLUMBERGER

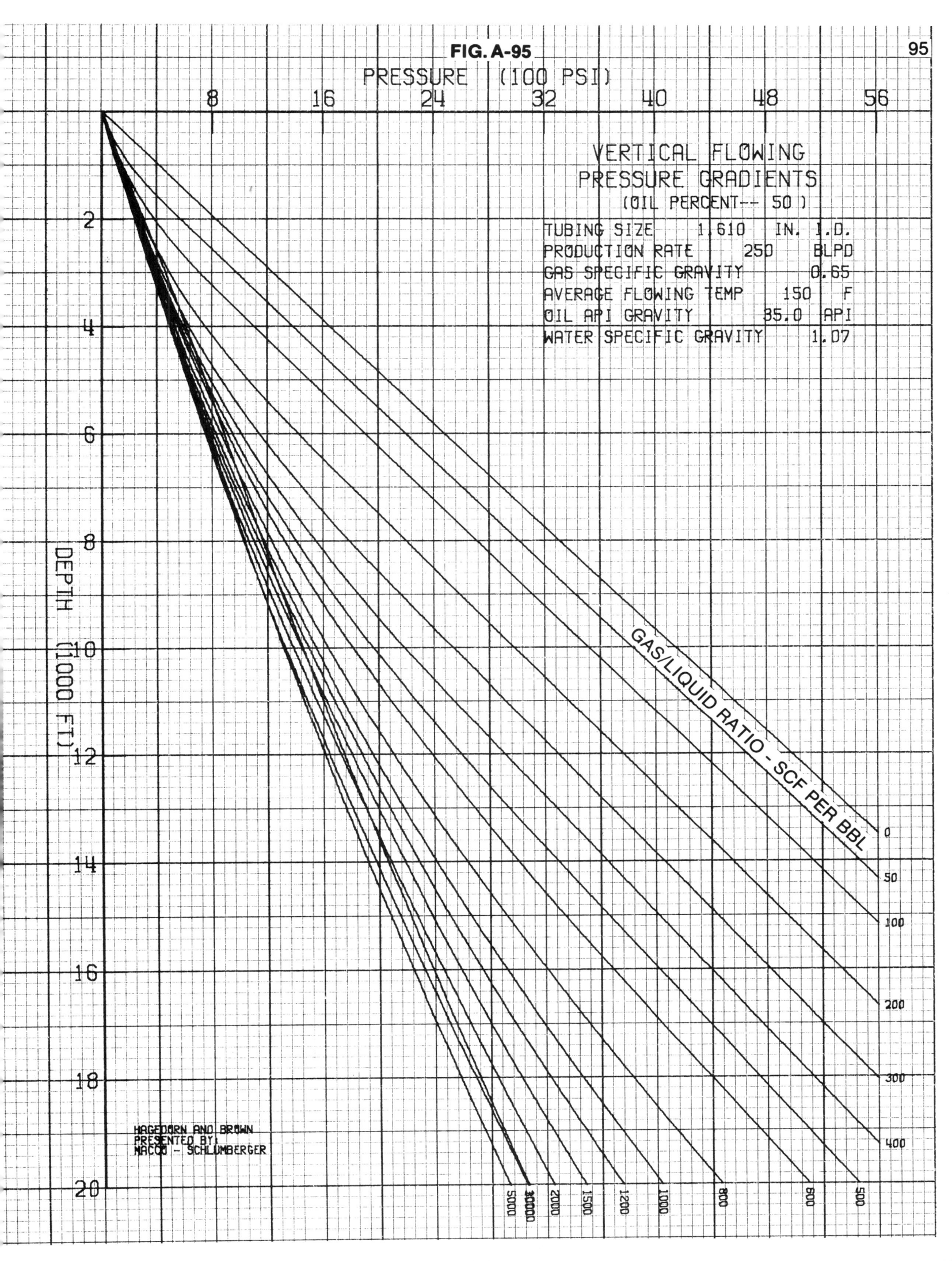

PRESSURE (100 PSI)

VERTICAL FLOWING
PRESSURE GRADIENTS
(OIL PERCENT-- 50)

TUBING SIZE	1.610	IN. I.D.
PRODUCTION RATE	250	BLPD
GAS SPECIFIC GRAVITY	0.65	
AVERAGE FLOWING TEMP	150	F
OIL API GRAVITY	35.0	API
WATER SPECIFIC GRAVITY	1.07	

DEPTH (1000 FT)

GAS/LIQUID RATIO - SCF PER BBL

HAGEDORN AND BROWN
PRESENTED BY
MACCO - SCHLUMBERGER

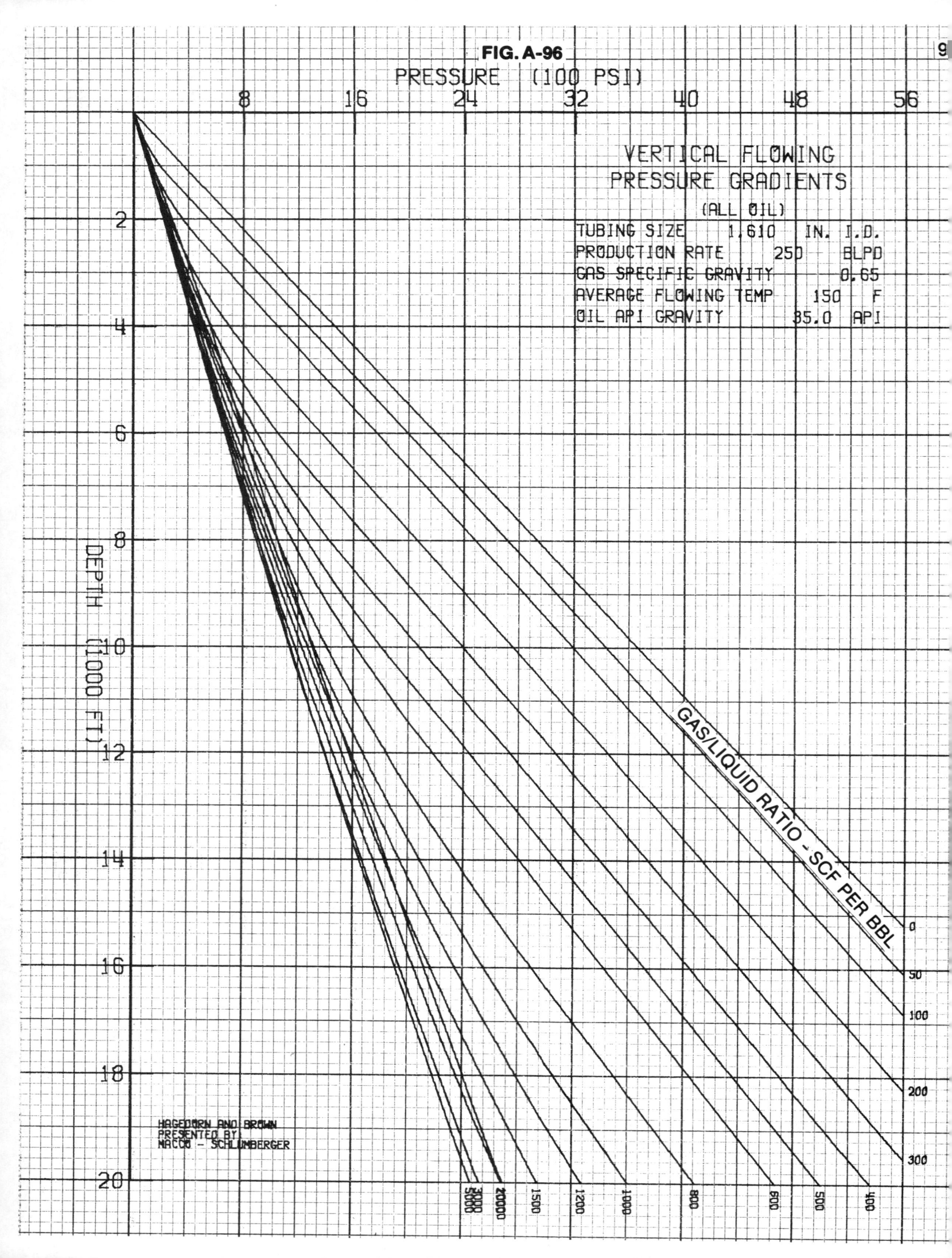

FIG. A-96

PRESSURE (100 PSI)

VERTICAL FLOWING
PRESSURE GRADIENTS
(ALL OIL)

TUBING SIZE	1.610	IN. I.D.
PRODUCTION RATE	250	BLPD
GAS SPECIFIC GRAVITY		0.65
AVERAGE FLOWING TEMP	150	F
OIL API GRAVITY	35.0	API

DEPTH (1000 FT)

GAS/LIQUID RATIO - SCF PER BBL

HAGEDORN AND BROWN
PRESENTED BY
NACCO - SCHLUMBERGER

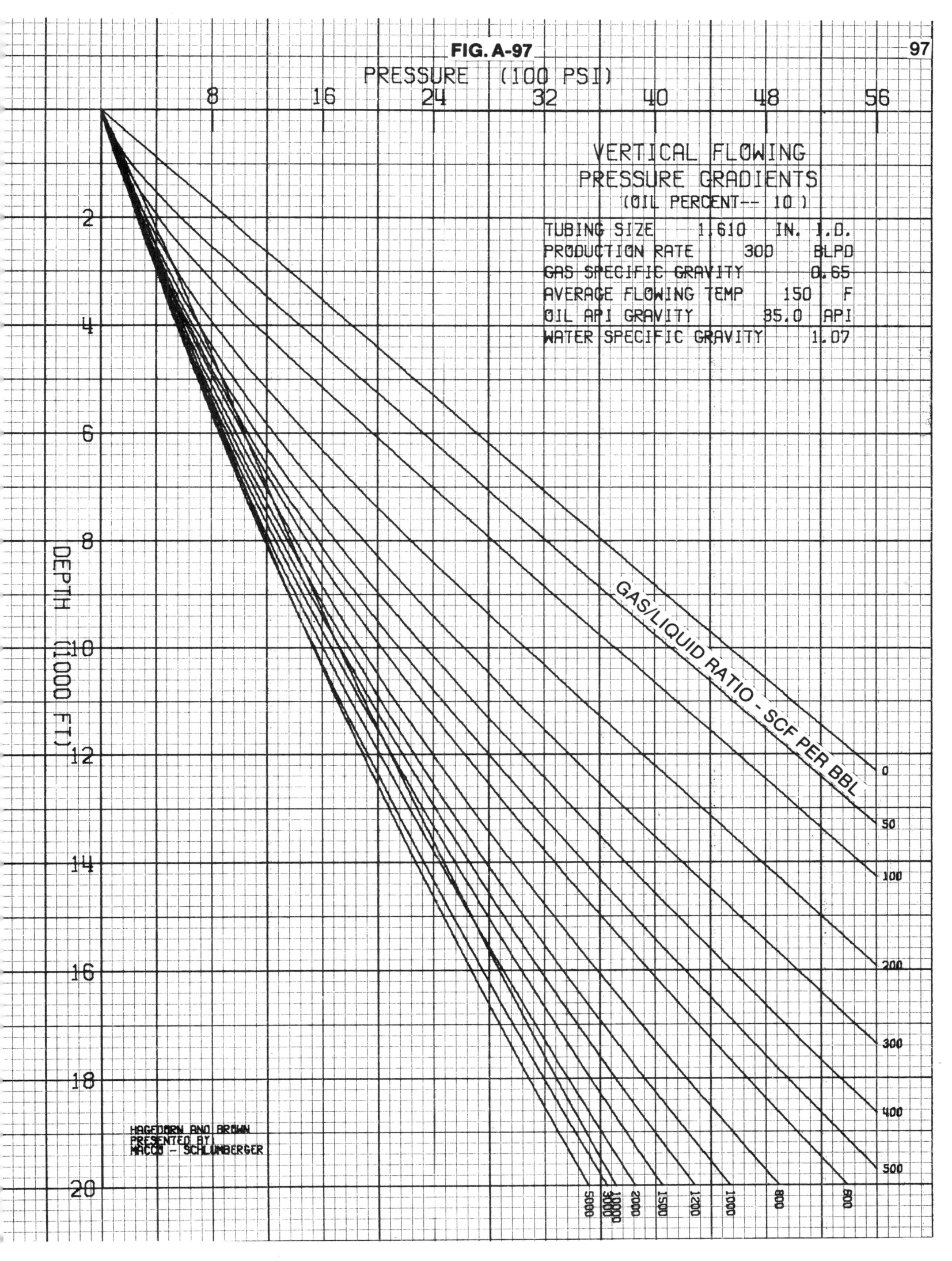

PRESSURE (100 PSI)

VERTICAL FLOWING
PRESSURE GRADIENTS
(OIL PERCENT-- 10)

TUBING SIZE	1.610	IN. I.D.
PRODUCTION RATE	300	BLPD
GAS SPECIFIC GRAVITY		0.65
AVERAGE FLOWING TEMP	150	F
OIL API GRAVITY	35.0	API
WATER SPECIFIC GRAVITY		1.07

DEPTH (1000 FT)

GAS/LIQUID RATIO - SCF PER BBL

HAGEDORN AND BROWN
PRESENTED BY
MACCO - SCHLUMBERGER

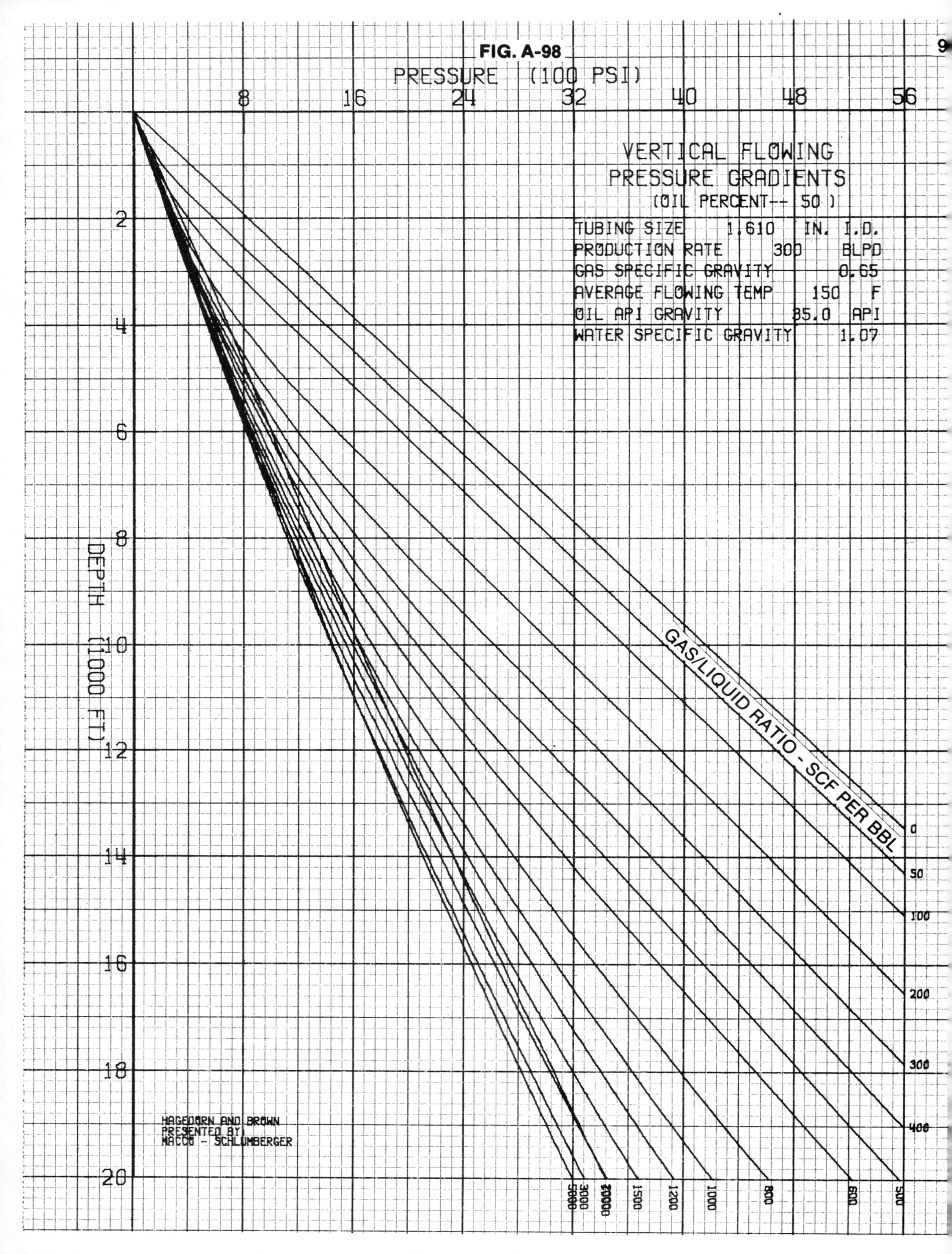

FIG. A-98

PRESSURE (100 PSI)

VERTICAL FLOWING
PRESSURE GRADIENTS
(OIL PERCENT-- 50)

TUBING SIZE	1.610	IN. I.D.
PRODUCTION RATE	300	BLPD
GAS SPECIFIC GRAVITY		0.65
AVERAGE FLOWING TEMP	150	F
OIL API GRAVITY	85.0	API
WATER SPECIFIC GRAVITY		1.07

DEPTH (1000 FT)

GAS/LIQUID RATIO - SCF PER BBL

HAGEDORN AND BROWN
PRESENTED BY
NACCO - SCHLUMBERGER

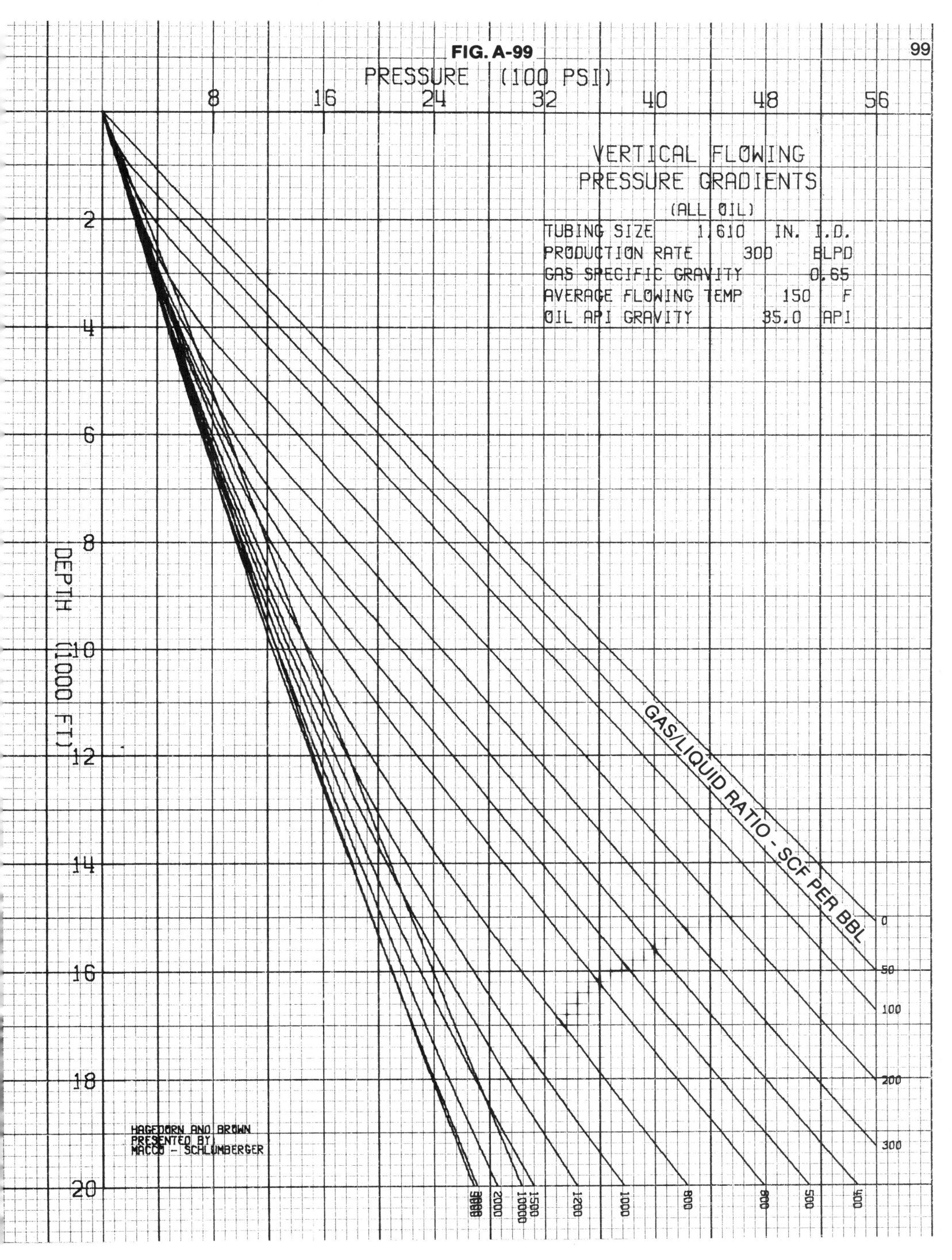

PRESSURE (100 PSI)

VERTICAL FLOWING
PRESSURE GRADIENTS
(ALL OIL)

TUBING SIZE	1.610	IN. I.D.
PRODUCTION RATE	300	BLPD
GAS SPECIFIC GRAVITY	0.65	
AVERAGE FLOWING TEMP	150	F
OIL API GRAVITY	35.0	API

DEPTH (1000 FT)

GAS/LIQUID RATIO - SCF PER BBL

HAGEDORN AND BROWN
PRESENTED BY
MACCO - SCHLUMBERGER

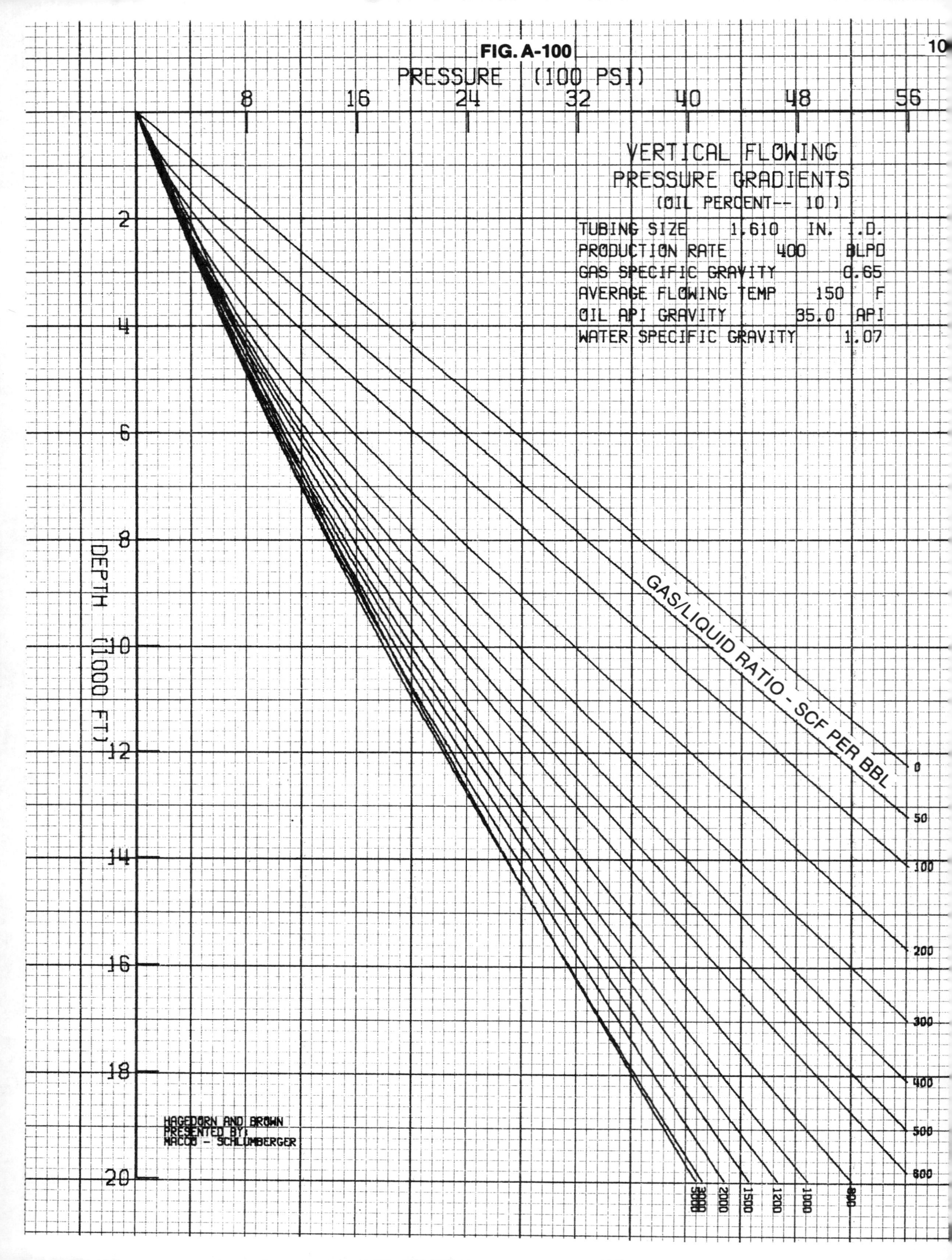

FIG. A-100

PRESSURE (100 PSI)

VERTICAL FLOWING PRESSURE GRADIENTS
(OIL PERCENT-- 10)

TUBING SIZE	1.610 IN. I.D.
PRODUCTION RATE	400 BLPD
GAS SPECIFIC GRAVITY	0.65
AVERAGE FLOWING TEMP	150 F
OIL API GRAVITY	35.0 API
WATER SPECIFIC GRAVITY	1.07

GAS/LIQUID RATIO - SCF PER BBL

DEPTH (1000 FT)

HAGEDORN AND BROWN
PRESENTED BY:
MACCO - SCHLUMBERGER

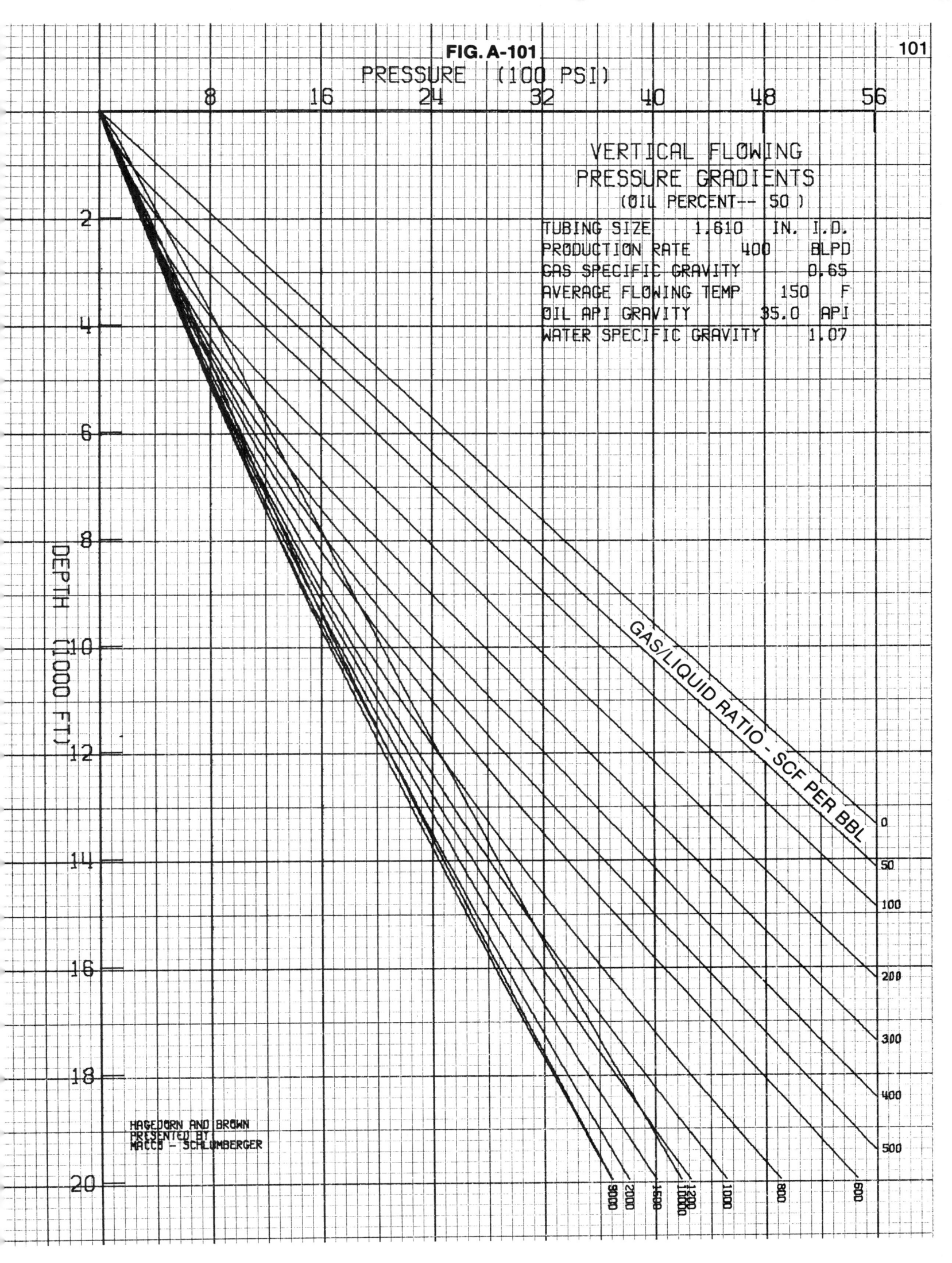

PRESSURE (100 PSI)

VERTICAL FLOWING
PRESSURE GRADIENTS
(OIL PERCENT-- 50)

TUBING SIZE	1.610	IN. I.D.
PRODUCTION RATE	400	BLPD
GAS SPECIFIC GRAVITY	0.65	
AVERAGE FLOWING TEMP	150	F
OIL API GRAVITY	35.0	API
WATER SPECIFIC GRAVITY	1.07	

GAS/LIQUID RATIO - SCF PER BBL

DEPTH (1000 FT)

HAGEDORN AND BROWN
PRESENTED BY
MACCO - SCHLUMBERGER

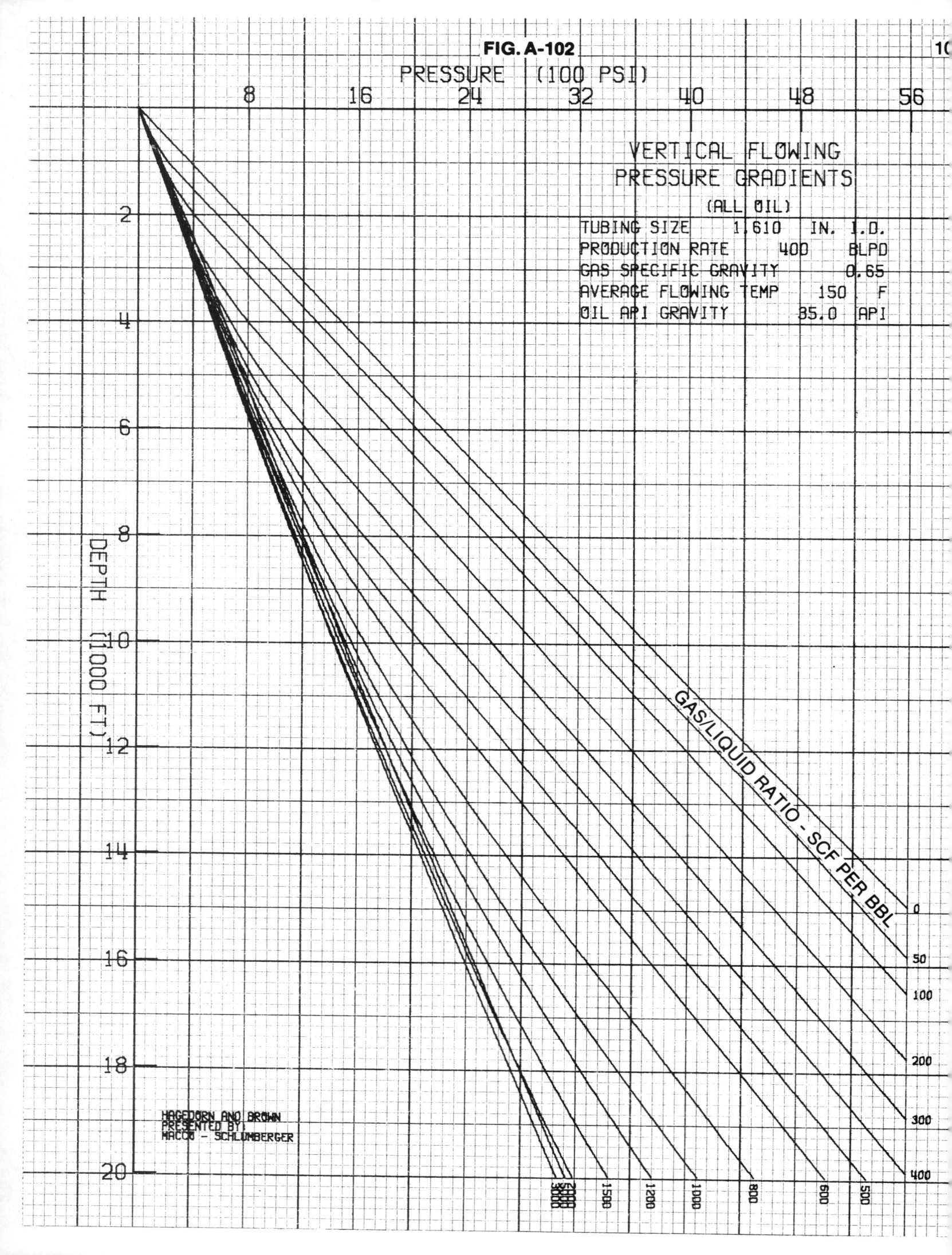

PRESSURE (100 PSI)

8 16 24 32 40 48 56

VERTICAL FLOWING
PRESSURE GRADIENTS
(ALL OIL)

TUBING SIZE	1.610	IN. I.D.
PRODUCTION RATE	400	BLPD
GAS SPECIFIC GRAVITY	0.65	
AVERAGE FLOWING TEMP	150	F
OIL API GRAVITY	35.0	API

DEPTH (1000 FT)

2
4
6
8
10
12
14
16
18
20

GAS/LIQUID RATIO - SCF PER BBL

0
50
100
200
300
400

HAGEDORN AND BROWN
PRESENTED BY:
MACCO - SCHLUMBERGER

3000 1500 1200 1000 800 600 500

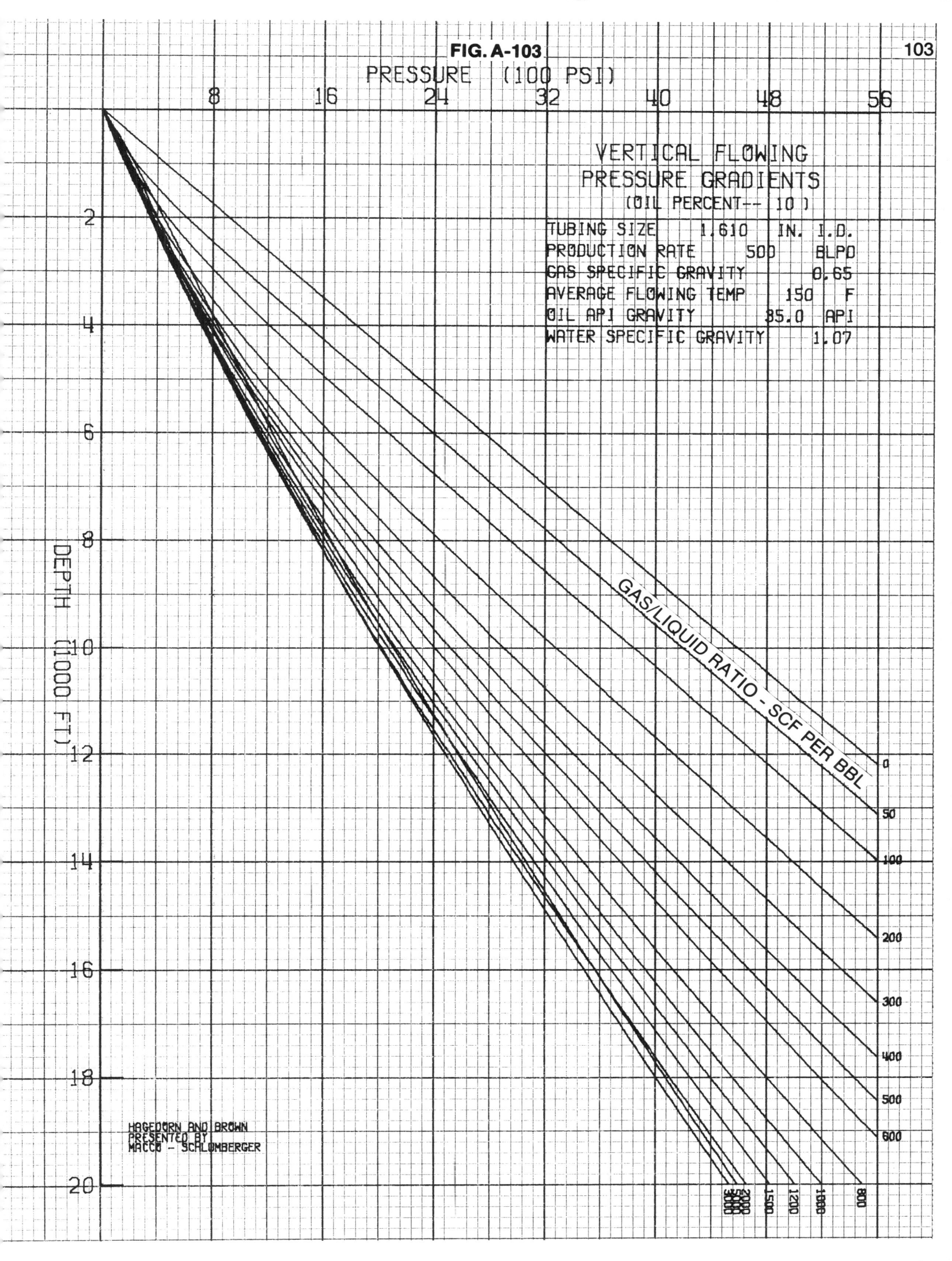

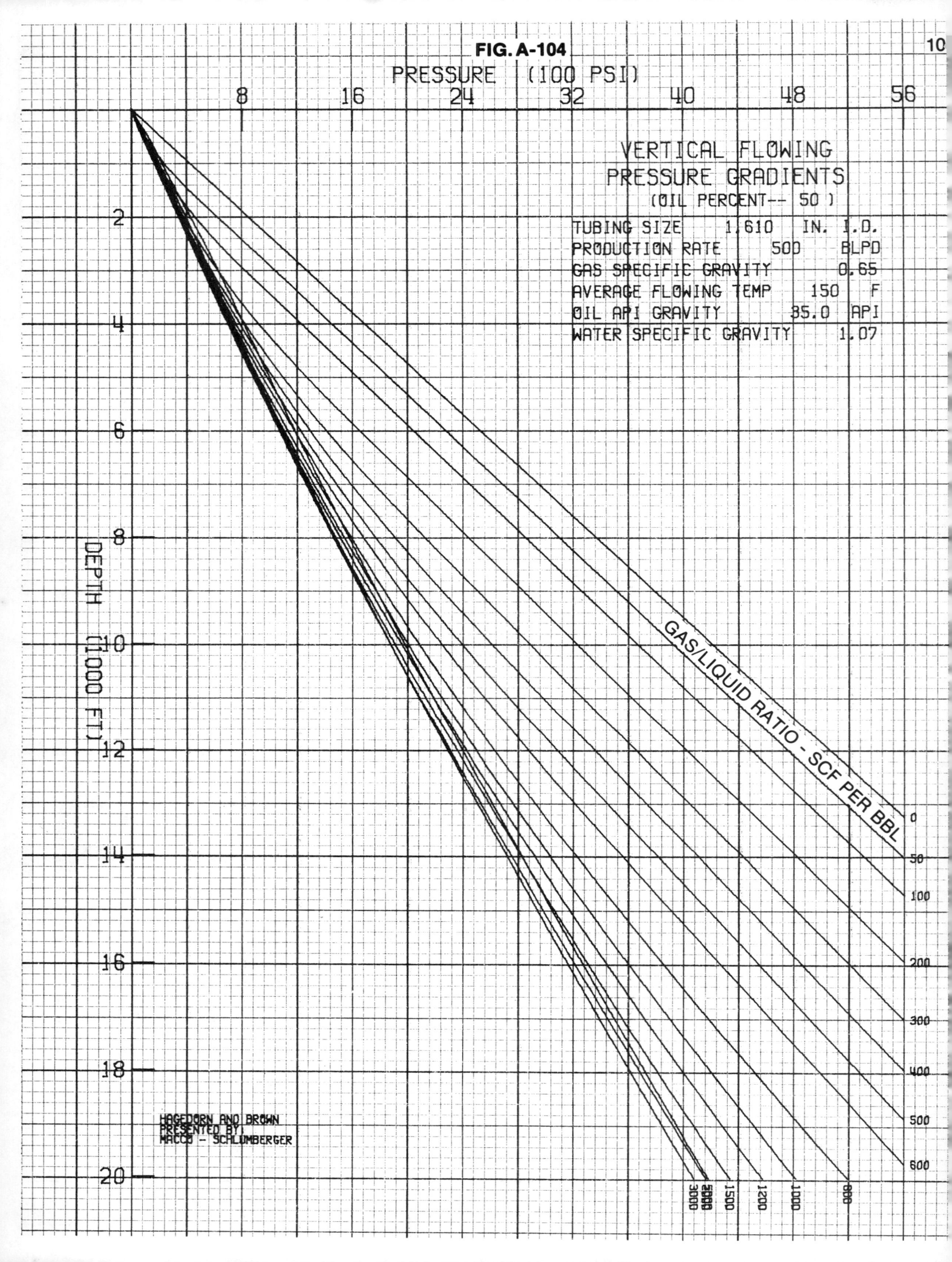

FIG. A-104

VERTICAL FLOWING
PRESSURE GRADIENTS
(OIL PERCENT-- 50)

TUBING SIZE	1.610	IN. I.D.
PRODUCTION RATE	500	BLPD
GAS SPECIFIC GRAVITY	0.65	
AVERAGE FLOWING TEMP	150	F
OIL API GRAVITY	35.0	API
WATER SPECIFIC GRAVITY	1.07	

PRESSURE (100 PSI)

DEPTH (1,000 FT)

GAS/LIQUID RATIO - SCF PER BBL

HAGEDORN AND BROWN
PRESENTED BY:
MACCO - SCHLUMBERGER

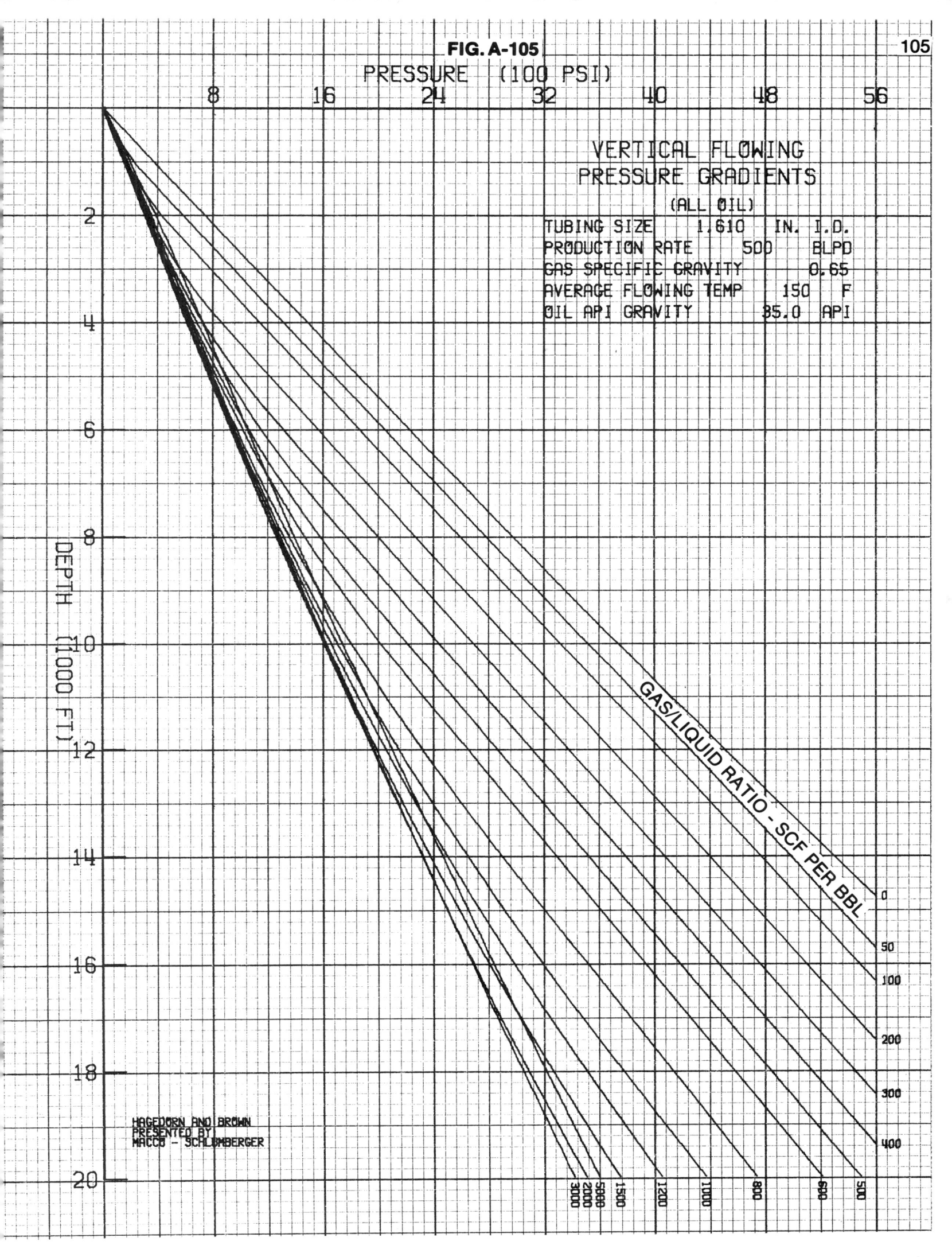

PRESSURE (100 PSI)

VERTICAL FLOWING
PRESSURE GRADIENTS
(ALL OIL)

TUBING SIZE	1.610	IN.	I.D.
PRODUCTION RATE	500	BLPD	
GAS SPECIFIC GRAVITY	0.65		
AVERAGE FLOWING TEMP	150	F	
OIL API GRAVITY	35.0	API	

DEPTH (1000 FT)

GAS/LIQUID RATIO - SCF PER BBL

HAGEDORN AND BROWN
PRESENTED BY
MACCO - SCHLUMBERGER

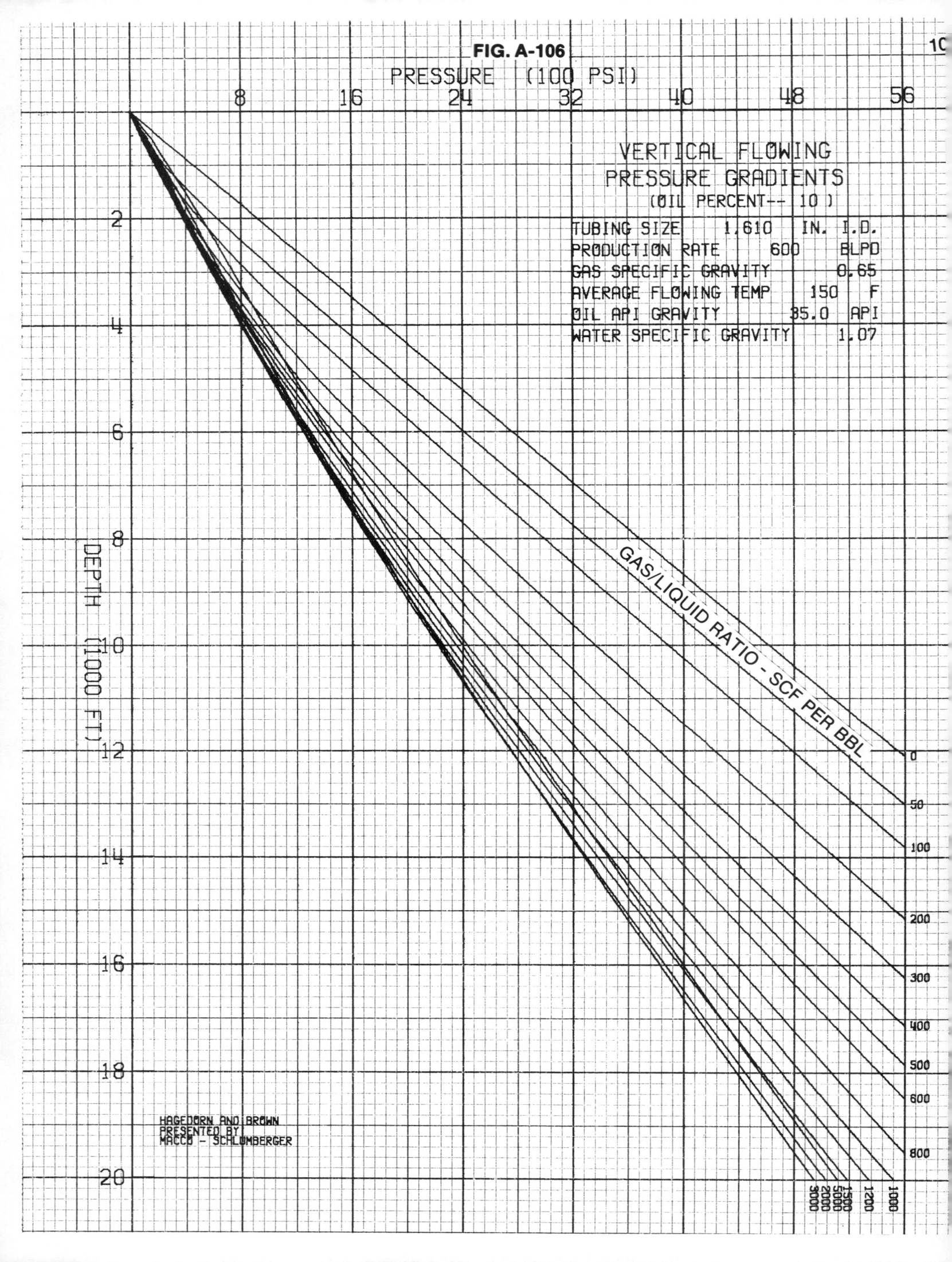

FIG. A-106

VERTICAL FLOWING
PRESSURE GRADIENTS
(OIL PERCENT-- 10)

TUBING SIZE	1.610	IN. I.D.
PRODUCTION RATE	600	BLPD
GAS SPECIFIC GRAVITY	0.65	
AVERAGE FLOWING TEMP	150	F
OIL API GRAVITY	35.0	API
WATER SPECIFIC GRAVITY	1.07	

PRESSURE (100 PSI)

DEPTH (1000 FT)

GAS/LIQUID RATIO - SCF PER BBL

HAGEDORN AND BROWN
PRESENTED BY
MACCO - SCHLUMBERGER

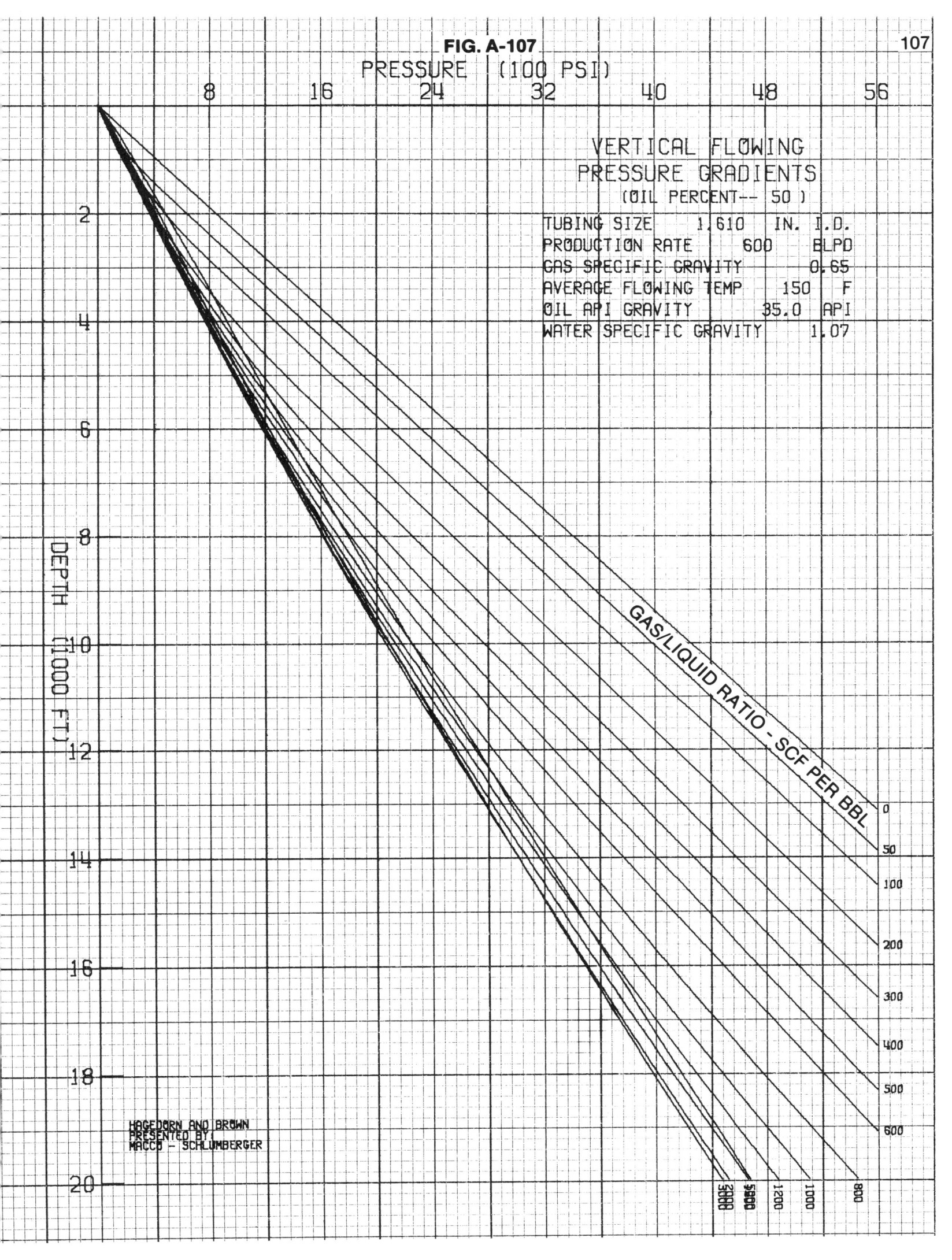

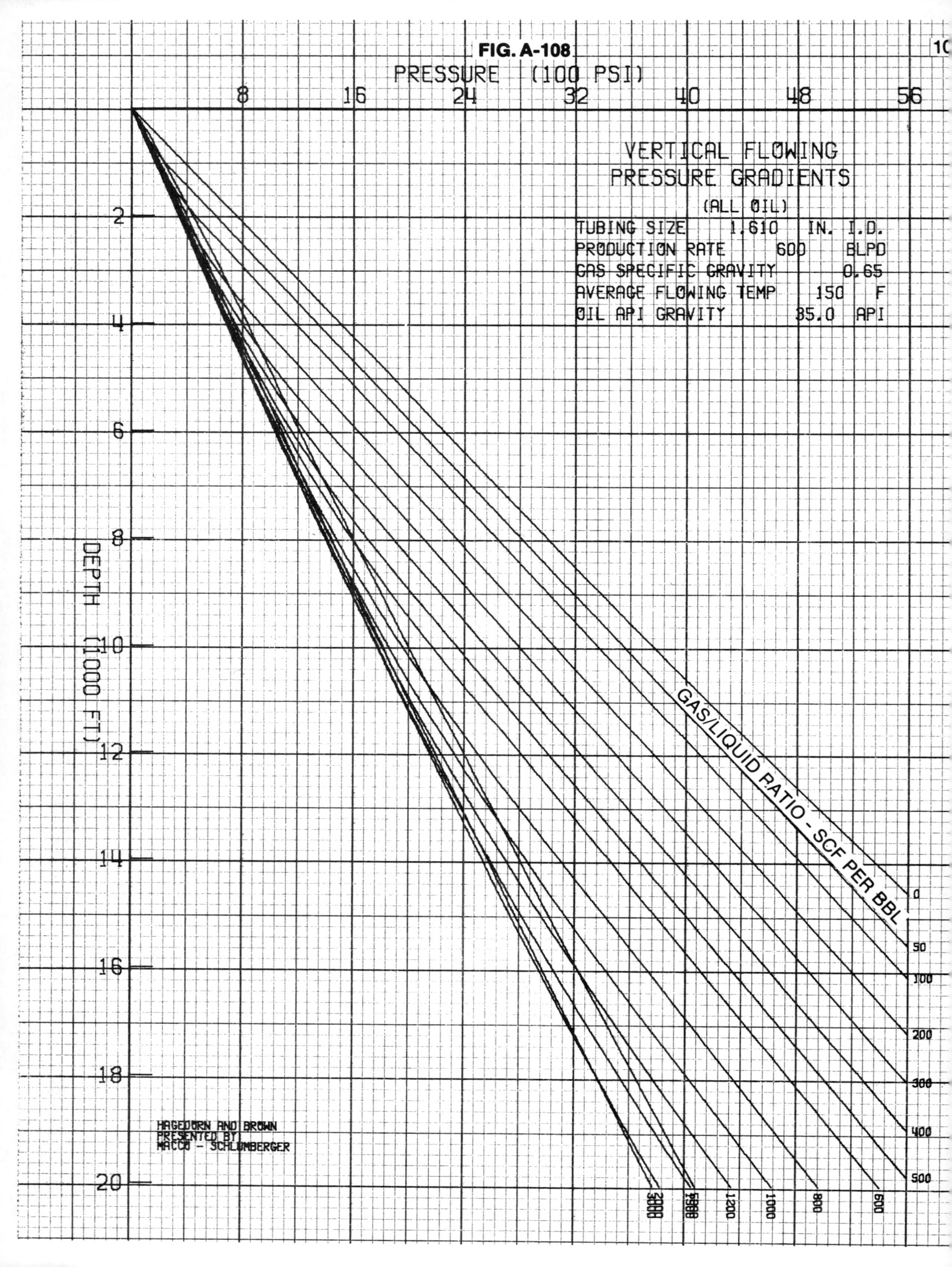

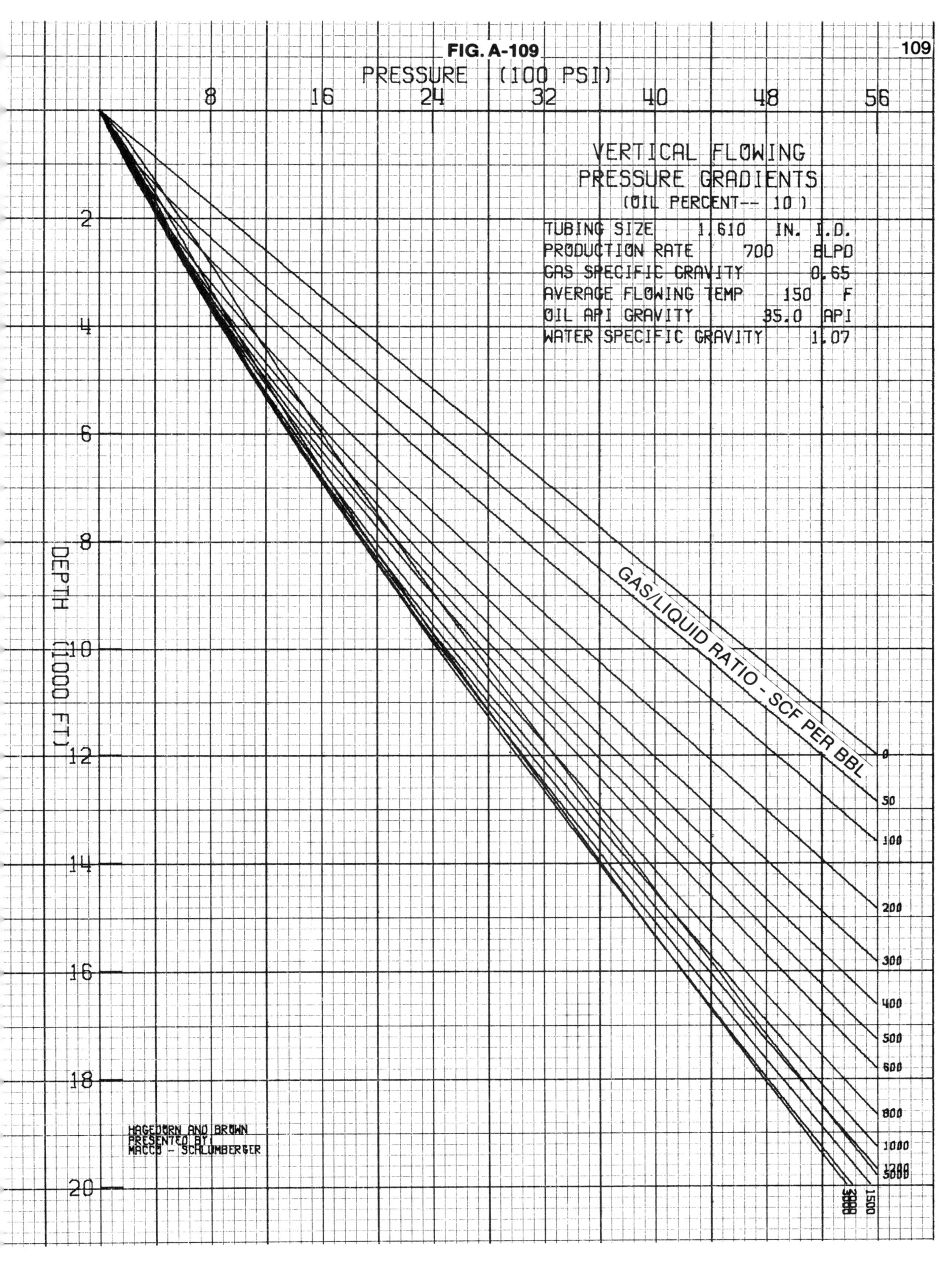

PRESSURE (100 PSI)

VERTICAL FLOWING
PRESSURE GRADIENTS
(OIL PERCENT-- 10)

TUBING SIZE	1.610	IN. I.D.
PRODUCTION RATE	700	BLPD
GAS SPECIFIC GRAVITY	0.65	
AVERAGE FLOWING TEMP	150	F
OIL API GRAVITY	35.0	API
WATER SPECIFIC GRAVITY	1.07	

DEPTH (1000 FT)

GAS/LIQUID RATIO - SCF PER BBL

HAGEDORN AND BROWN
PRESENTED BY:
MACCO - SCHLUMBERGER

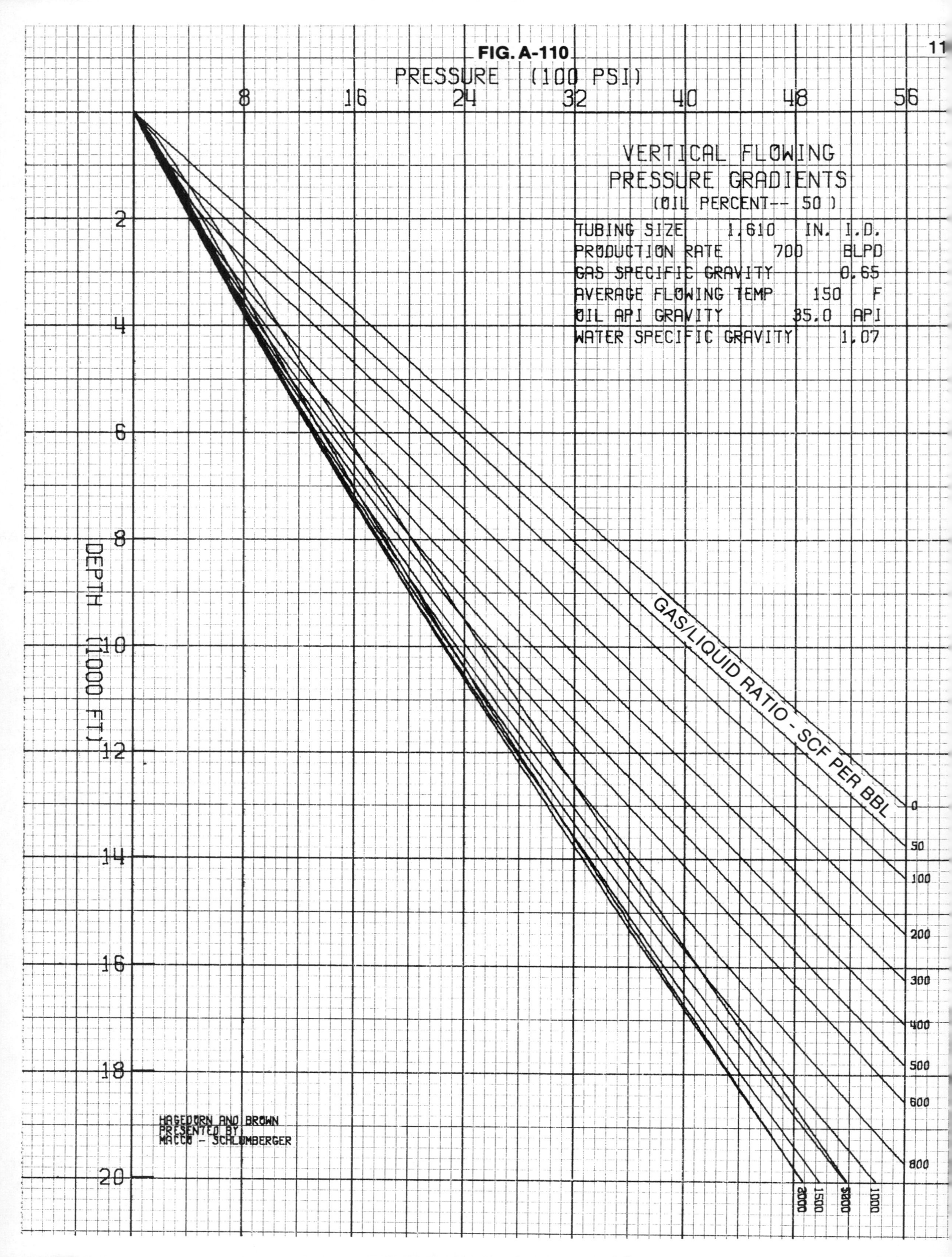

FIG. A-110

VERTICAL FLOWING
PRESSURE GRADIENTS
(OIL PERCENT-- 50)

TUBING SIZE	1.610	IN. I.D.
PRODUCTION RATE	700	BLPD
GAS SPECIFIC GRAVITY		0.65
AVERAGE FLOWING TEMP	150	F
OIL API GRAVITY	35.0	API
WATER SPECIFIC GRAVITY		1.07

PRESSURE (100 PSI)

DEPTH (1000 FT)

GAS/LIQUID RATIO - SCF PER BBL

HAGEDORN AND BROWN
PRESENTED BY
MACCO - SCHLUMBERGER

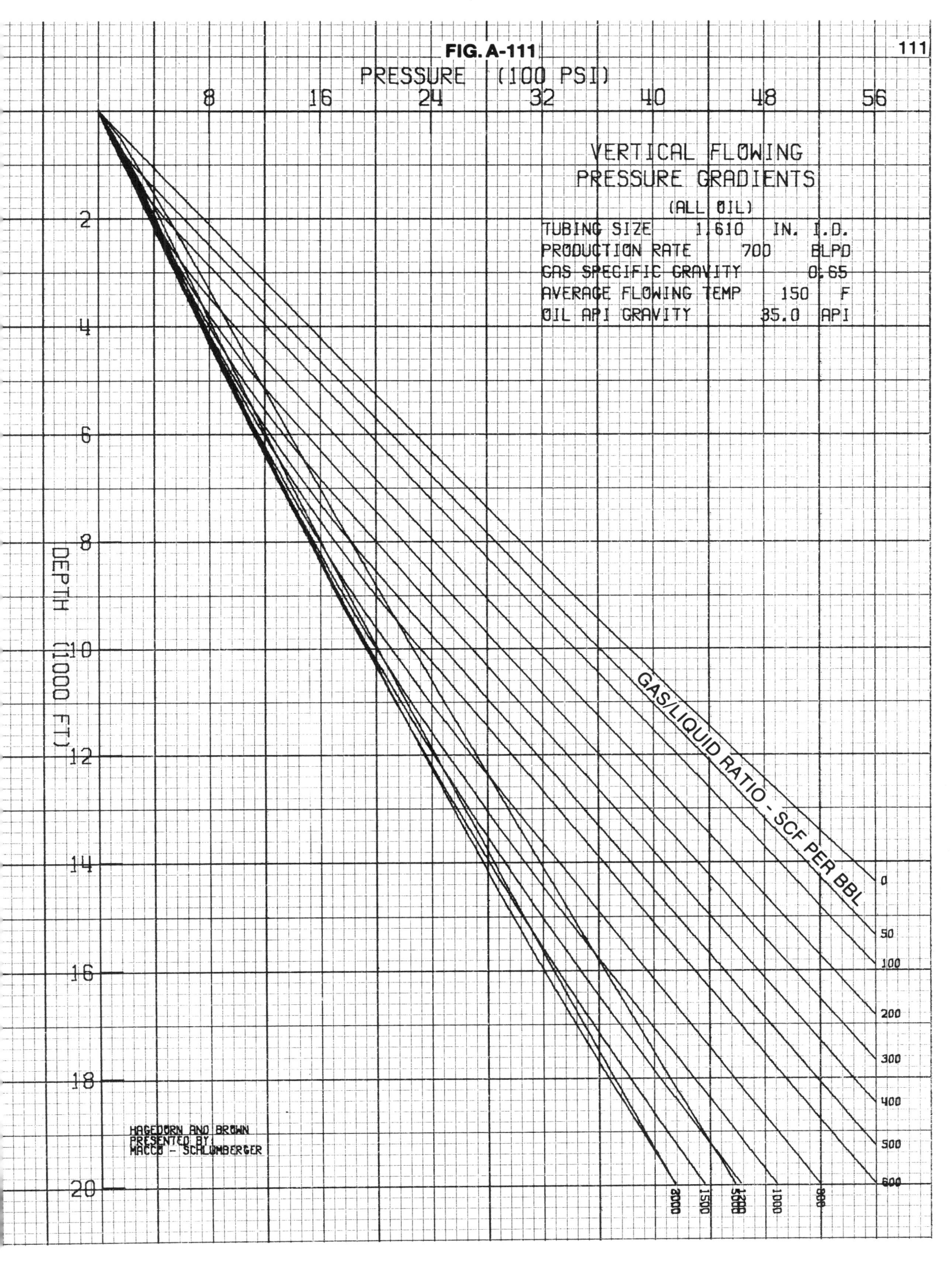

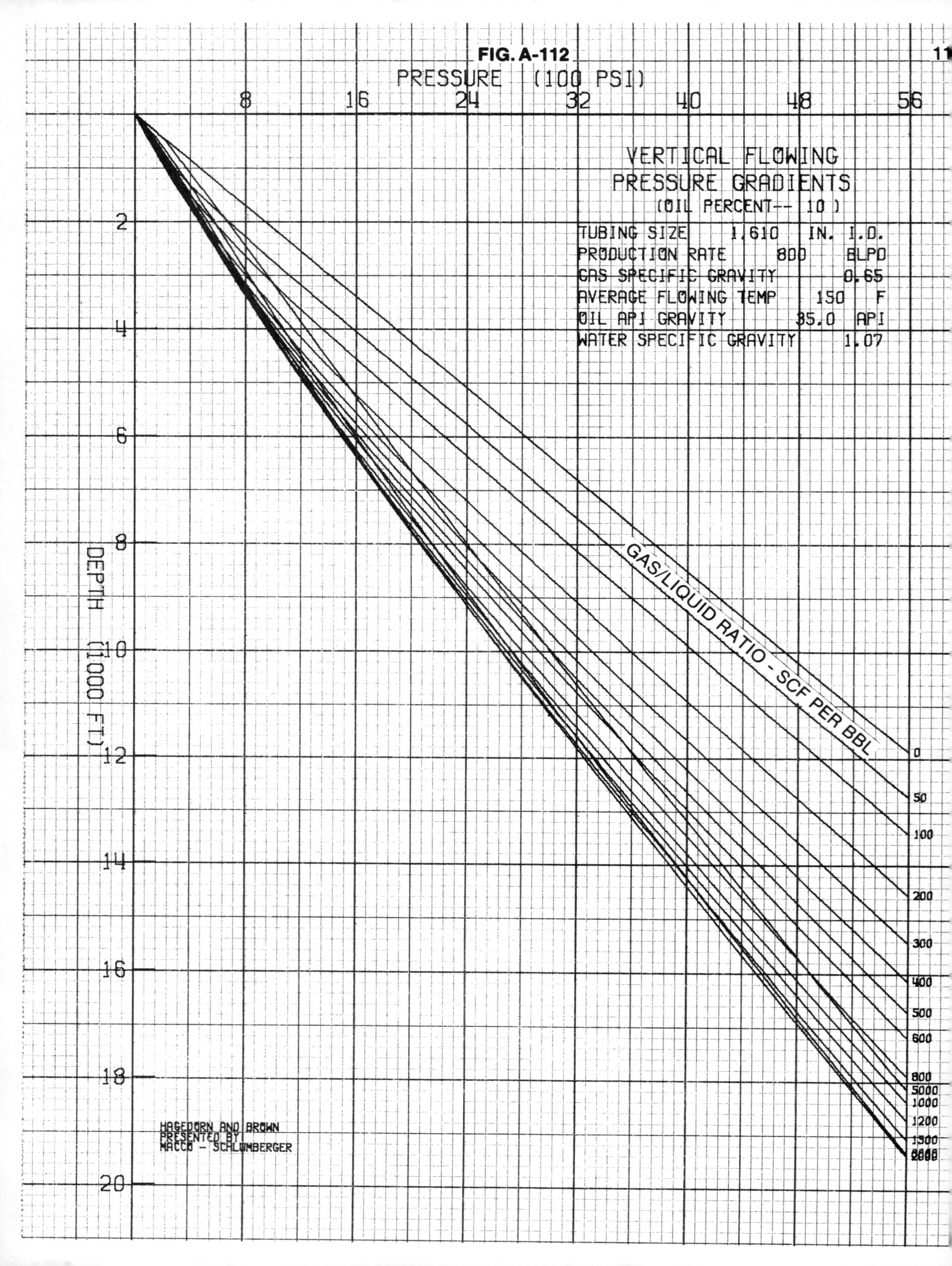

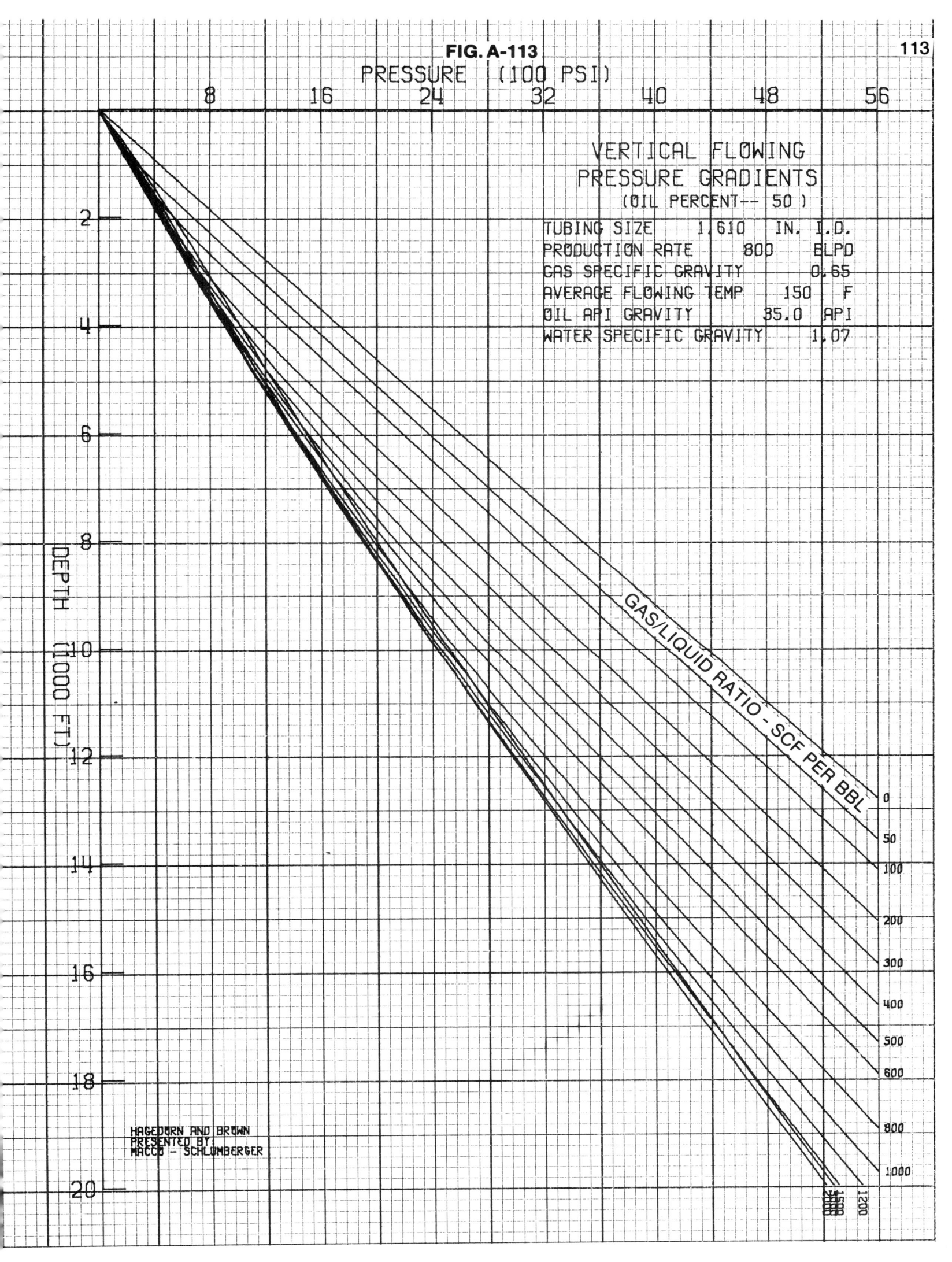

PRESSURE (100 PSI)

VERTICAL FLOWING
PRESSURE GRADIENTS
(OIL PERCENT-- 50)

TUBING SIZE	1.610	IN. I.D.
PRODUCTION RATE	800	BLPD
GAS SPECIFIC GRAVITY	0.65	
AVERAGE FLOWING TEMP	150	F
OIL API GRAVITY	35.0	API
WATER SPECIFIC GRAVITY	1.07	

DEPTH (1000 FT)

GAS/LIQUID RATIO - SCF PER BBL

HAGEDORN AND BROWN
PRESENTED BY
MACCO - SCHLUMBERGER

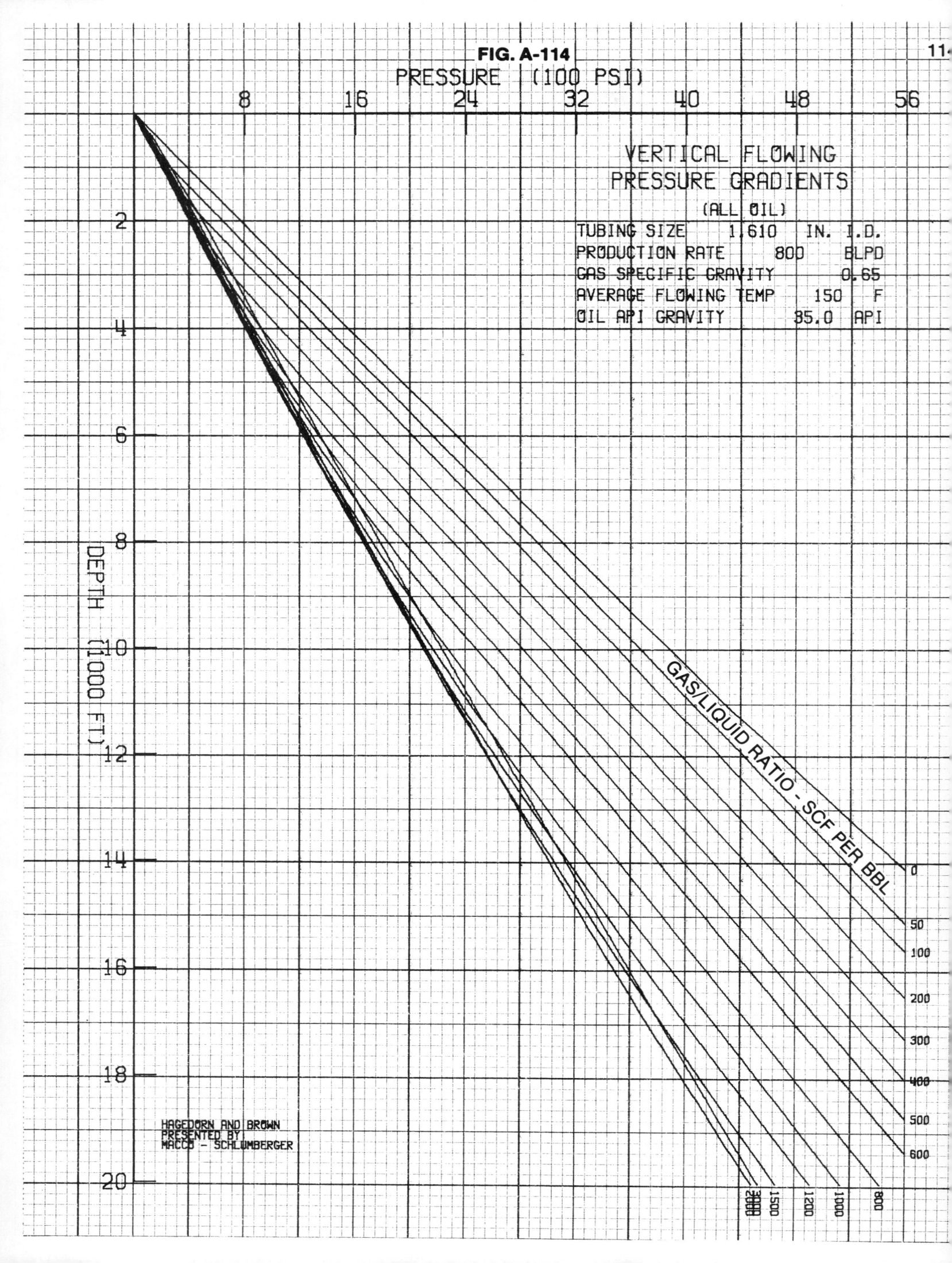

PRESSURE (100 PSI)

VERTICAL FLOWING
PRESSURE GRADIENTS
(ALL OIL)

TUBING SIZE	1.610	IN. I.D.
PRODUCTION RATE	800	BLPD
GAS SPECIFIC GRAVITY	0.65	
AVERAGE FLOWING TEMP	150	F
OIL API GRAVITY	35.0	API

DEPTH (1000 FT)

GAS/LIQUID RATIO - SCF PER BBL

HAGEDORN AND BROWN
PRESENTED BY
MACCO - SCHLUMBERGER

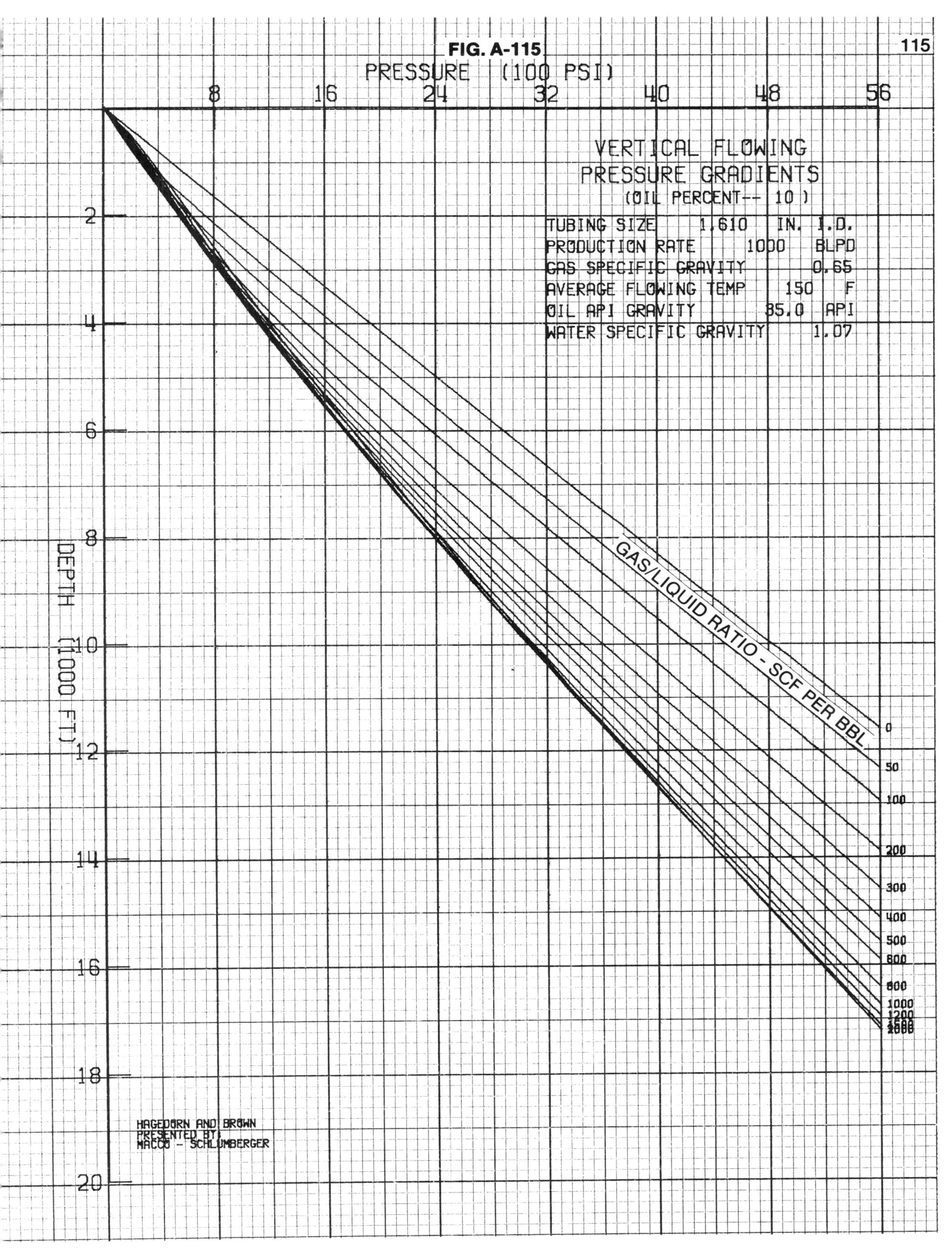

FIG. A-115

115

PRESSURE (100 PSI)

VERTICAL FLOWING
PRESSURE GRADIENTS
(OIL PERCENT-- 10)

TUBING SIZE	1.610	IN. I.D.
PRODUCTION RATE	1000	BLPD
GAS SPECIFIC GRAVITY		0.65
AVERAGE FLOWING TEMP	150	F
OIL API GRAVITY	35.0	API
WATER SPECIFIC GRAVITY		1.07

GAS/LIQUID RATIO - SCF PER BBL

DEPTH (1000 FT)

0
50
100
200
300
400
500
600
800
1000
1200
1500
2000

HAGEDORN AND BROWN
PRESENTED BY:
MAECO - SCHLUMBERGER

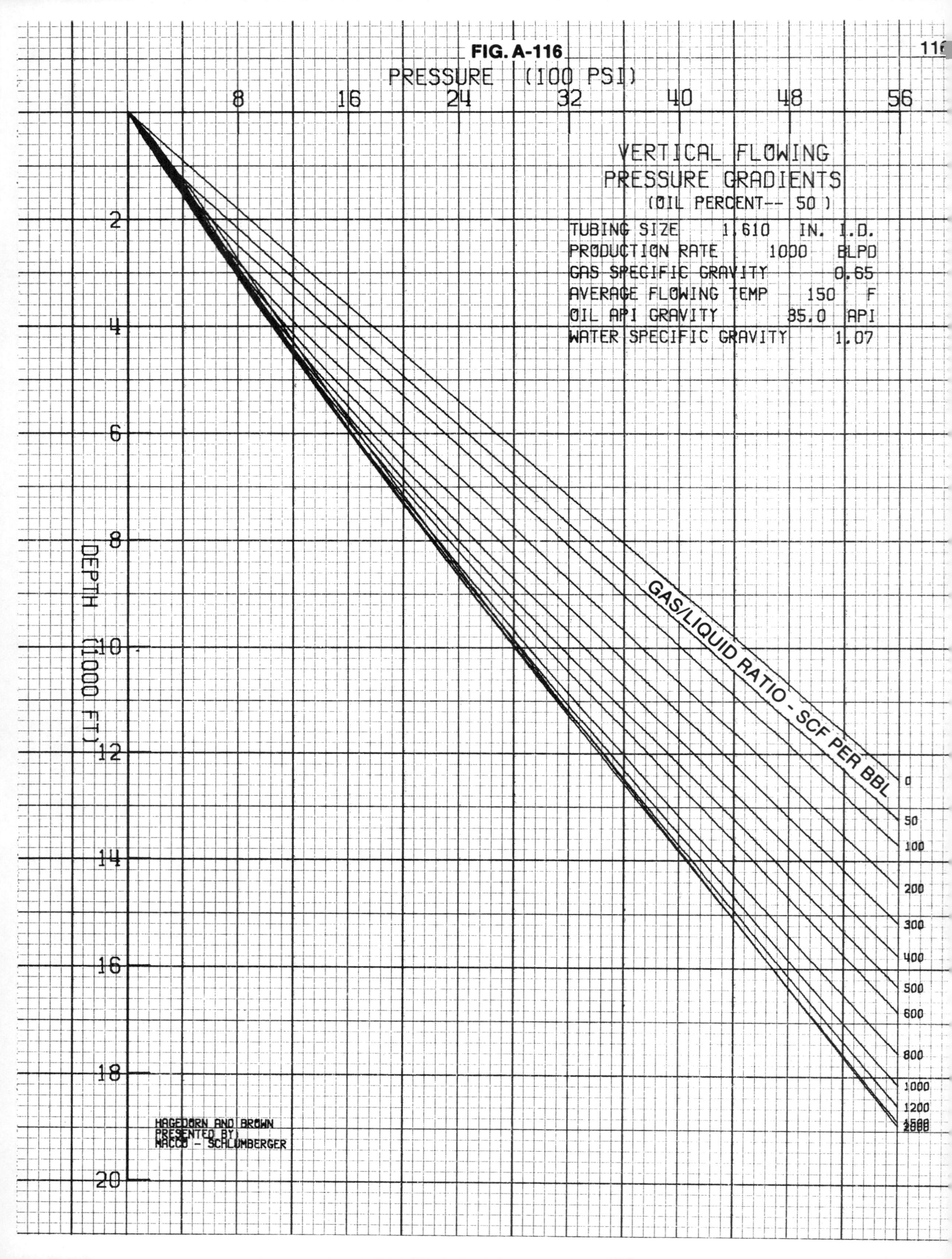

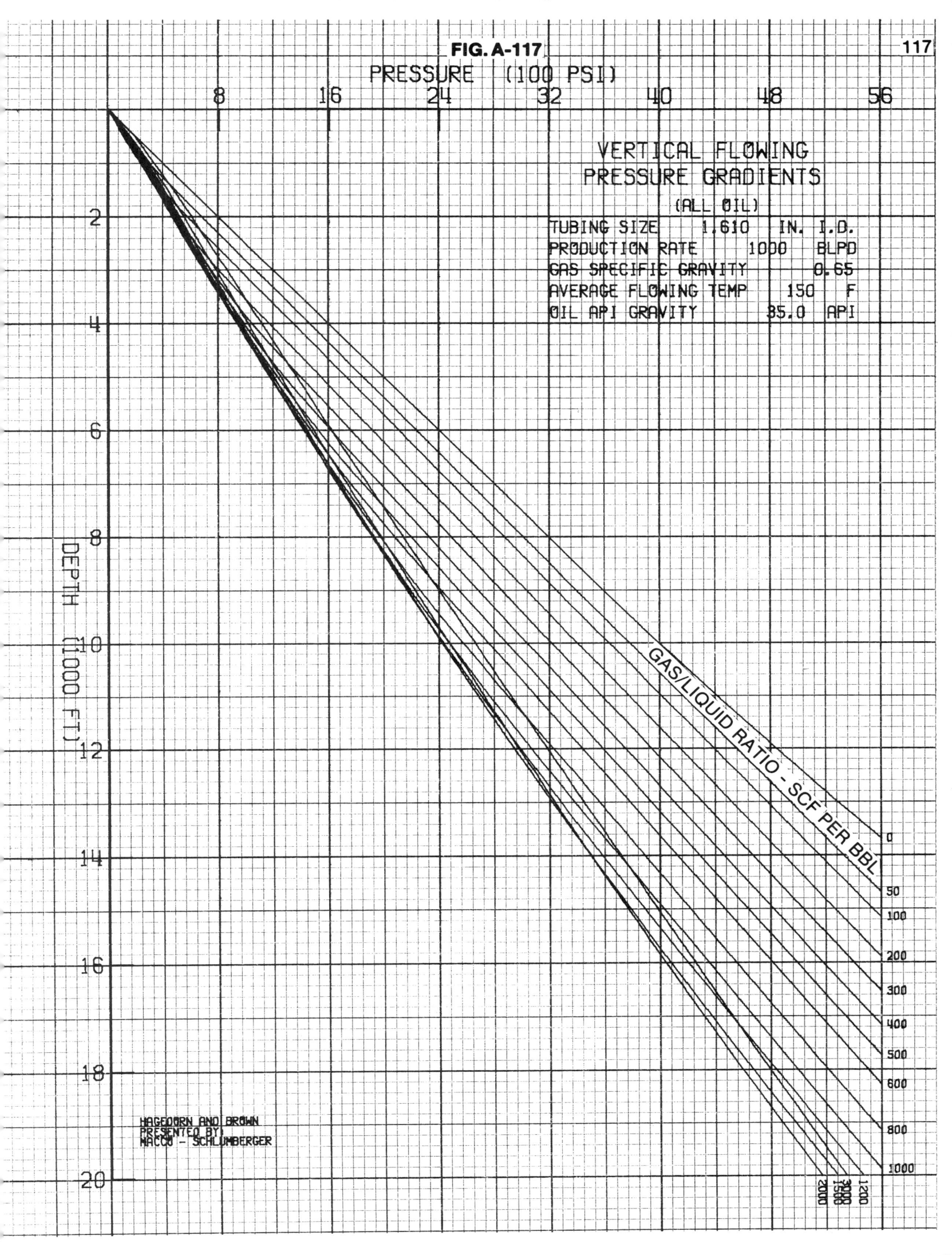

PRESSURE (100 PSI)

VERTICAL FLOWING
PRESSURE GRADIENTS
(ALL OIL)

TUBING SIZE	1.610	IN. I.D.
PRODUCTION RATE	1000	BLPD
GAS SPECIFIC GRAVITY	0.65	
AVERAGE FLOWING TEMP	150	F
OIL API GRAVITY	35.0	API

DEPTH (1000 FT)

GAS/LIQUID RATIO - SCF PER BBL

HAGEDORN AND BROWN
PRESENTED BY:
MACCO - SCHLUMBERGER

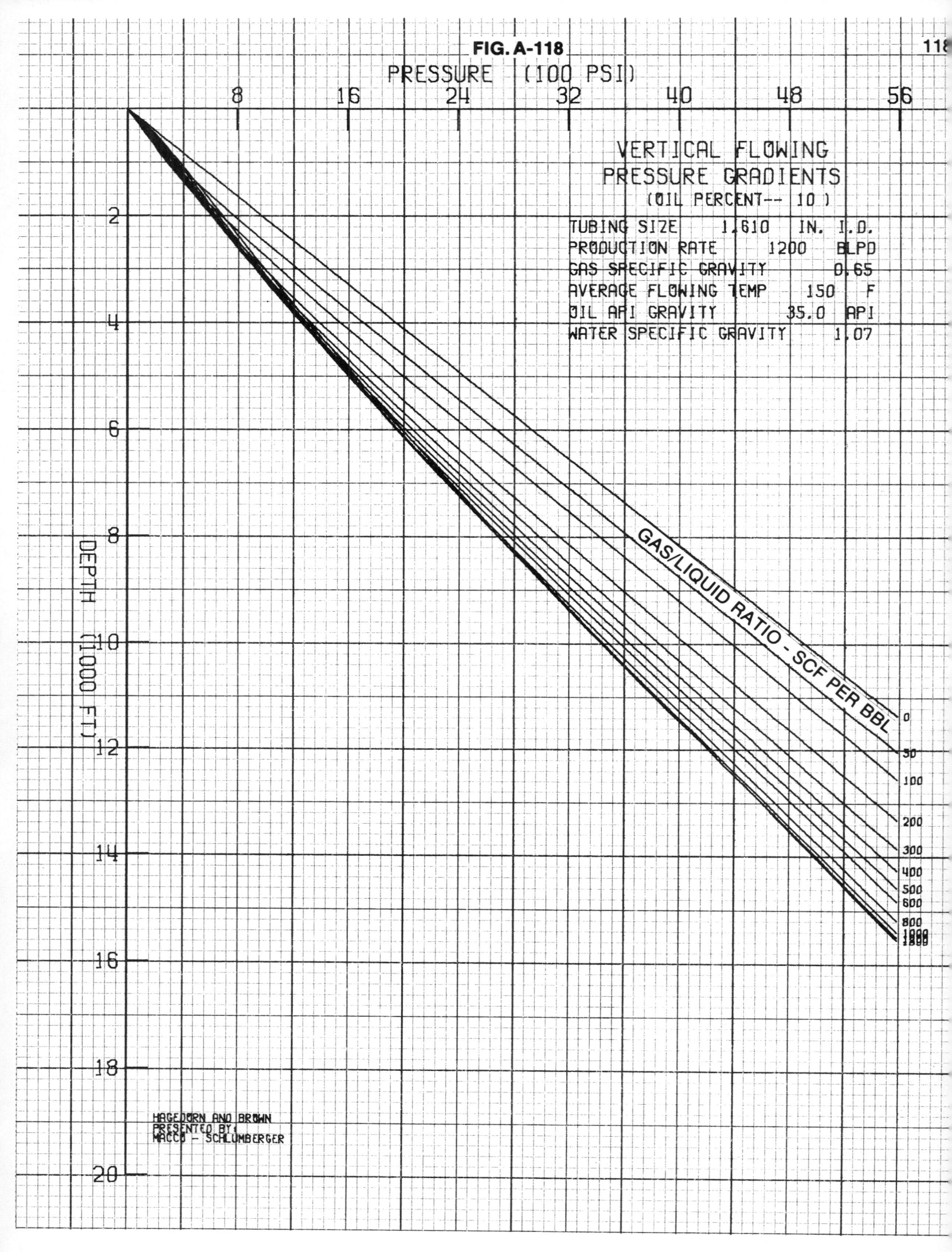

FIG. A-118

118

PRESSURE (100 PSI)

VERTICAL FLOWING
PRESSURE GRADIENTS
(OIL PERCENT-- 10)

TUBING SIZE	1.610	IN. I.D.
PRODUCTION RATE	1200	BLPD
GAS SPECIFIC GRAVITY	0.65	
AVERAGE FLOWING TEMP	150	F
OIL API GRAVITY	35.0	API
WATER SPECIFIC GRAVITY	1.07	

GAS/LIQUID RATIO - SCF PER BBL

DEPTH (1000 FT)

HAGEDORN AND BROWN
PRESENTED BY:
MACCO - SCHLUMBERGER

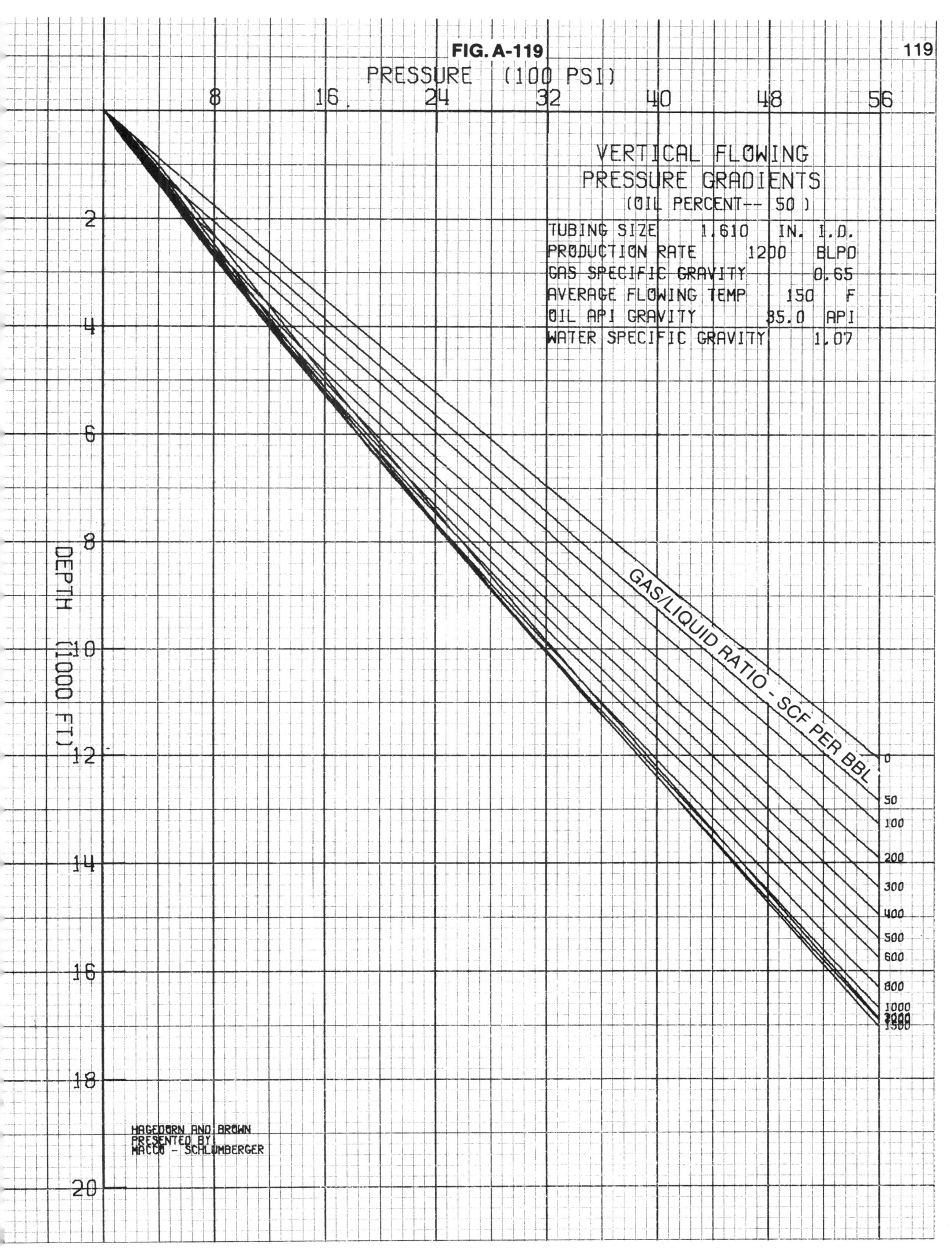

PRESSURE (100 PSI)

VERTICAL FLOWING
PRESSURE GRADIENTS
(OIL PERCENT-- 50)

TUBING SIZE	1.610	IN. I.D.
PRODUCTION RATE	1200	BLPD
GAS SPECIFIC GRAVITY		0.65
AVERAGE FLOWING TEMP	150	F
OIL API GRAVITY	35.0	API
WATER SPECIFIC GRAVITY		1.07

DEPTH (1000 FT)

GAS/LIQUID RATIO - SCF PER BBL

0
50
100
200
300
400
500
600
800
1000
2000
1500

HAGEDORN AND BROWN
PRESENTED BY
MACCO - SCHLUMBERGER

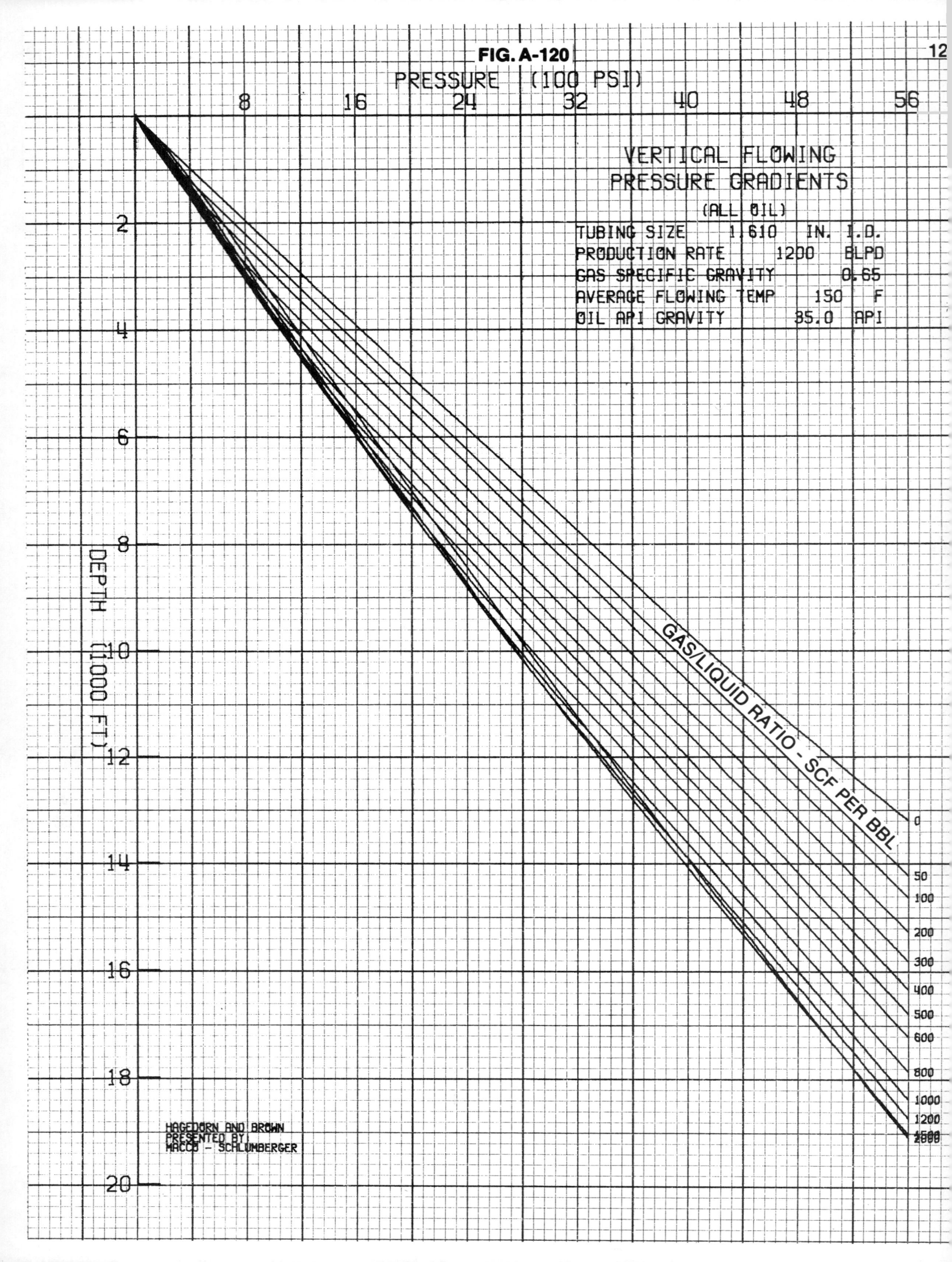

FIG. A-120

VERTICAL FLOWING
PRESSURE GRADIENTS
(ALL OIL)

TUBING SIZE	1.610	IN. I.D.
PRODUCTION RATE	1200	BLPD
GAS SPECIFIC GRAVITY	0.65	
AVERAGE FLOWING TEMP	150	F
OIL API GRAVITY	85.0	API

PRESSURE (100 PSI)

DEPTH (1000 FT)

GAS/LIQUID RATIO - SCF PER BBL

HAGEDORN AND BROWN
PRESENTED BY
MACCO - SCHLUMBERGER

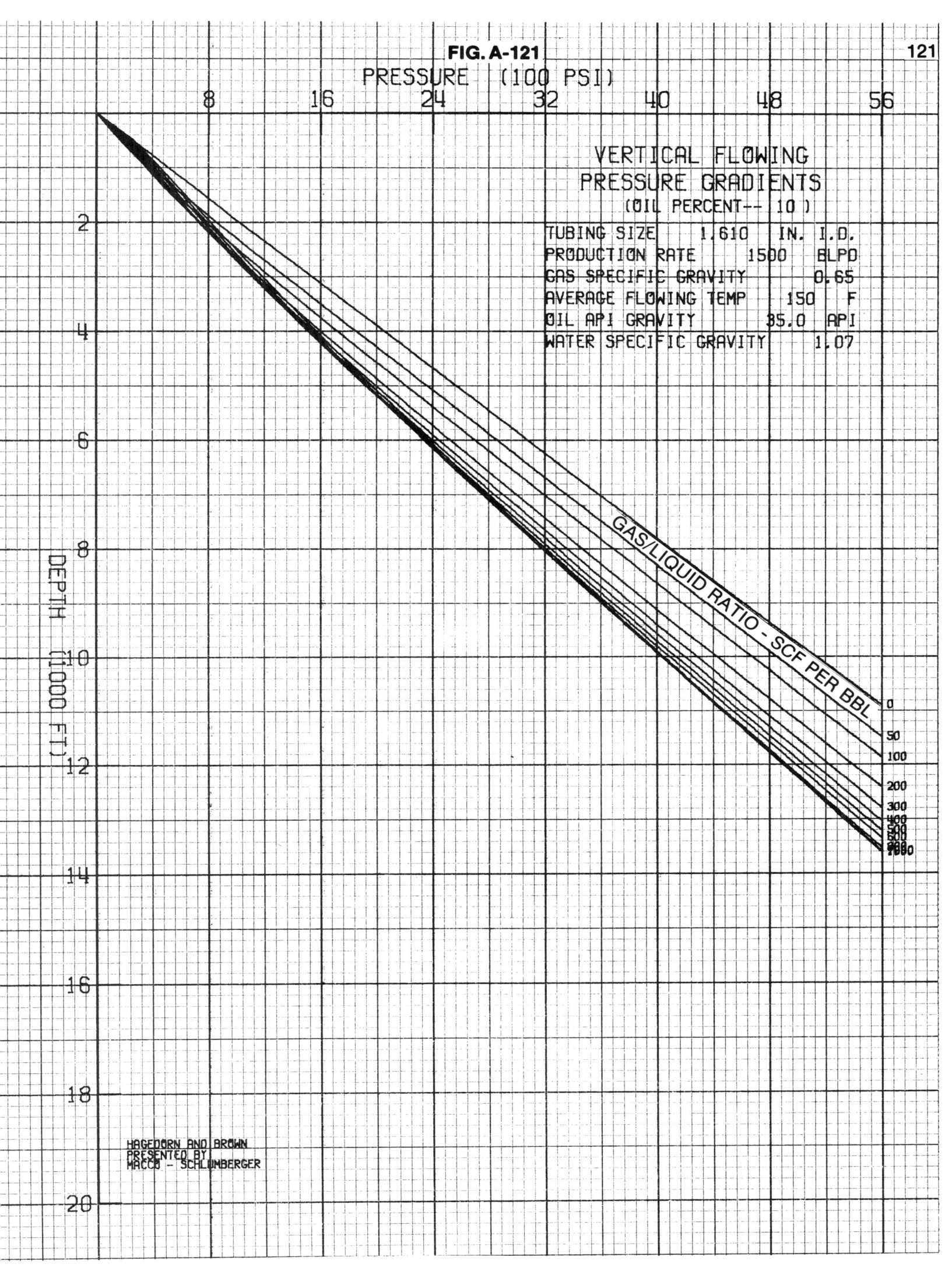

PRESSURE (100 PSI)

VERTICAL FLOWING
PRESSURE GRADIENTS
(OIL PERCENT-- 10)

TUBING SIZE	1.610	IN. I.D.
PRODUCTION RATE	1500	BLPD
GAS SPECIFIC GRAVITY	0.65	
AVERAGE FLOWING TEMP	150	F
OIL API GRAVITY	35.0	API
WATER SPECIFIC GRAVITY	1.07	

GAS/LIQUID RATIO - SCF PER BBL

DEPTH (1000 FT)

0
50
100
200
300
400
500
600
800
1000

HAGEDORN AND BROWN
PRESENTED BY
MACCO - SCHLUMBERGER

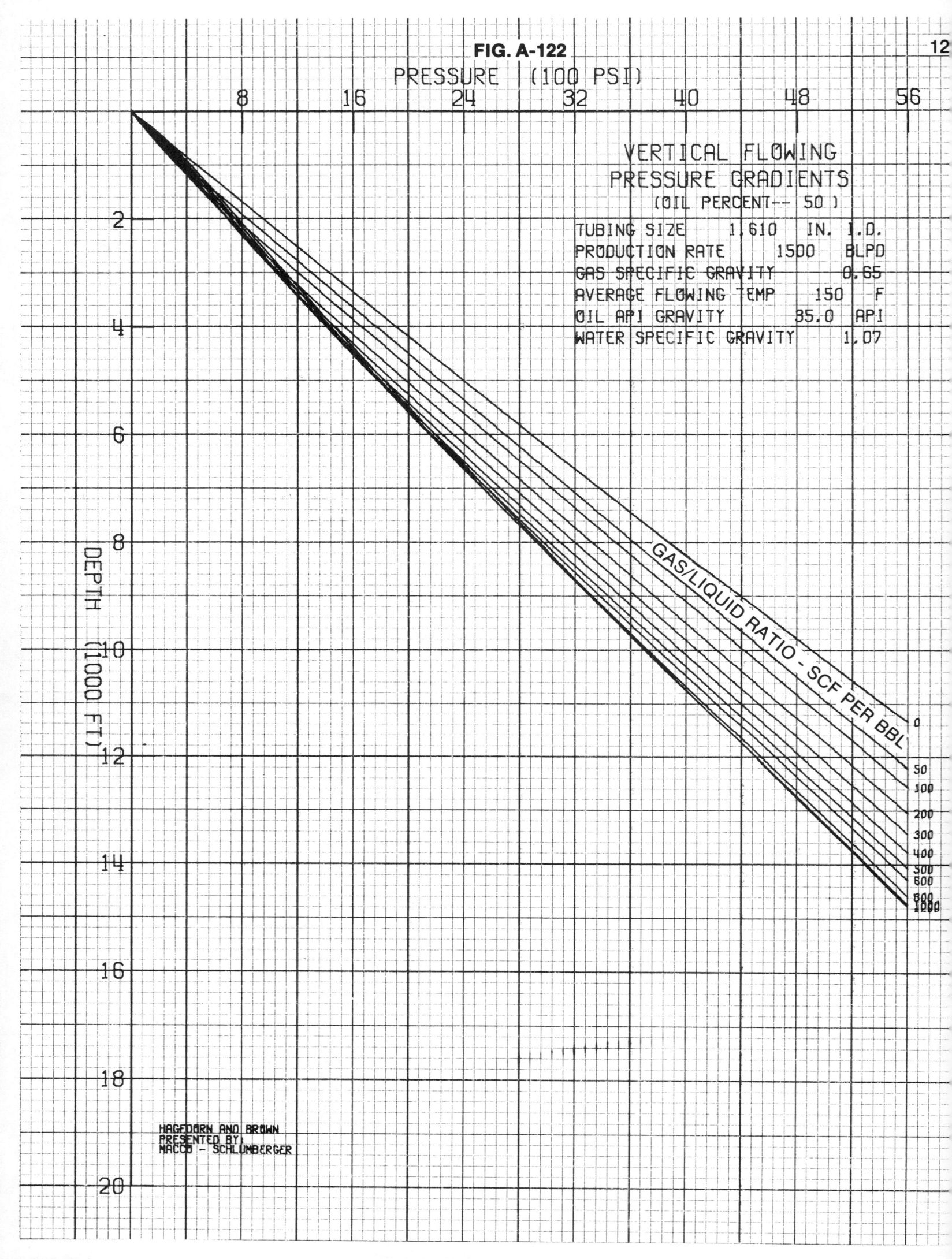

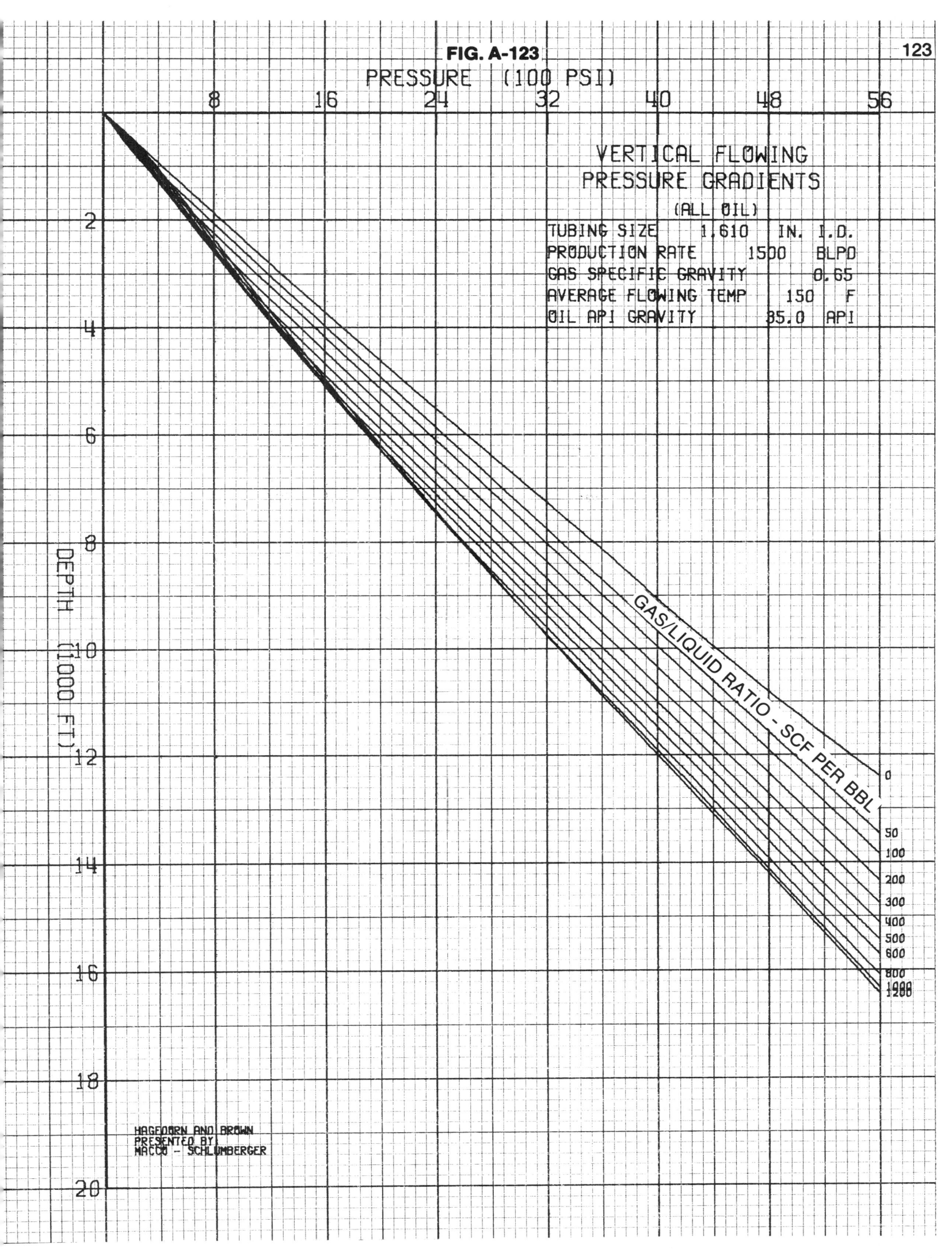

VERTICAL FLOWING
PRESSURE GRADIENTS
(ALL OIL)

TUBING SIZE	1.610	IN. I.D.
PRODUCTION RATE	1500	BLPD
GAS SPECIFIC GRAVITY	0.65	
AVERAGE FLOWING TEMP	150	F
OIL API GRAVITY	35.0	API

PRESSURE (100 PSI)

DEPTH (1,000 FT)

GAS/LIQUID RATIO - SCF PER BBL

0
50
100
200
300
400
500
600
800
1000

HAGEDORN AND BROWN
PRESENTED BY
MACCO - SCHLUMBERGER

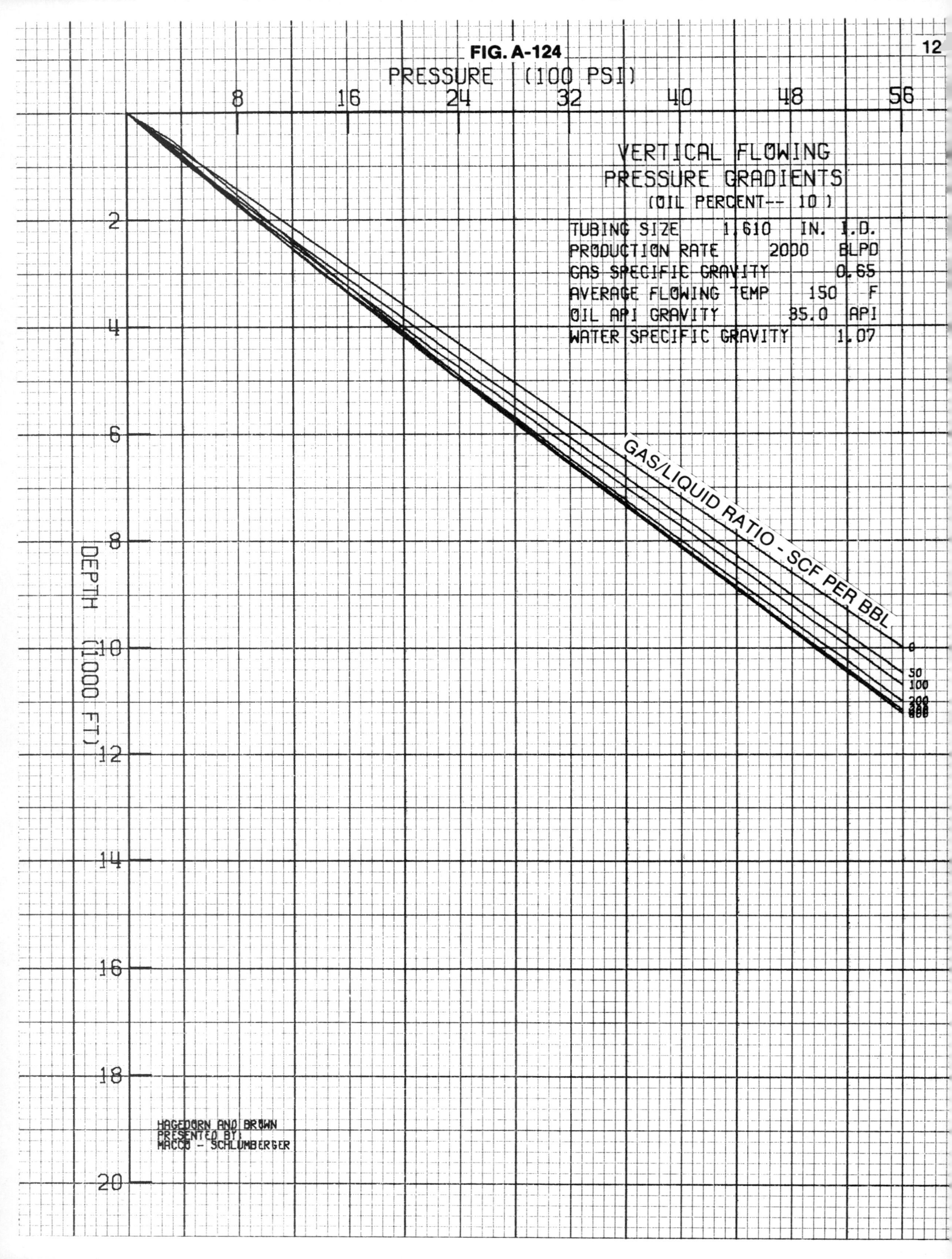

FIG. A-124

PRESSURE (100 PSI)

VERTICAL FLOWING
PRESSURE GRADIENTS
(OIL PERCENT-- 10)

TUBING SIZE	1.610	IN. I.D.
PRODUCTION RATE	2000	BLPD
GAS SPECIFIC GRAVITY	0.65	
AVERAGE FLOWING TEMP	150	F
OIL API GRAVITY	35.0	API
WATER SPECIFIC GRAVITY	1.07	

GAS/LIQUID RATIO - SCF PER BBL

DEPTH (1000 FT)

HAGEDORN AND BROWN
PRESENTED BY:
MACCO - SCHLUMBERGER

12

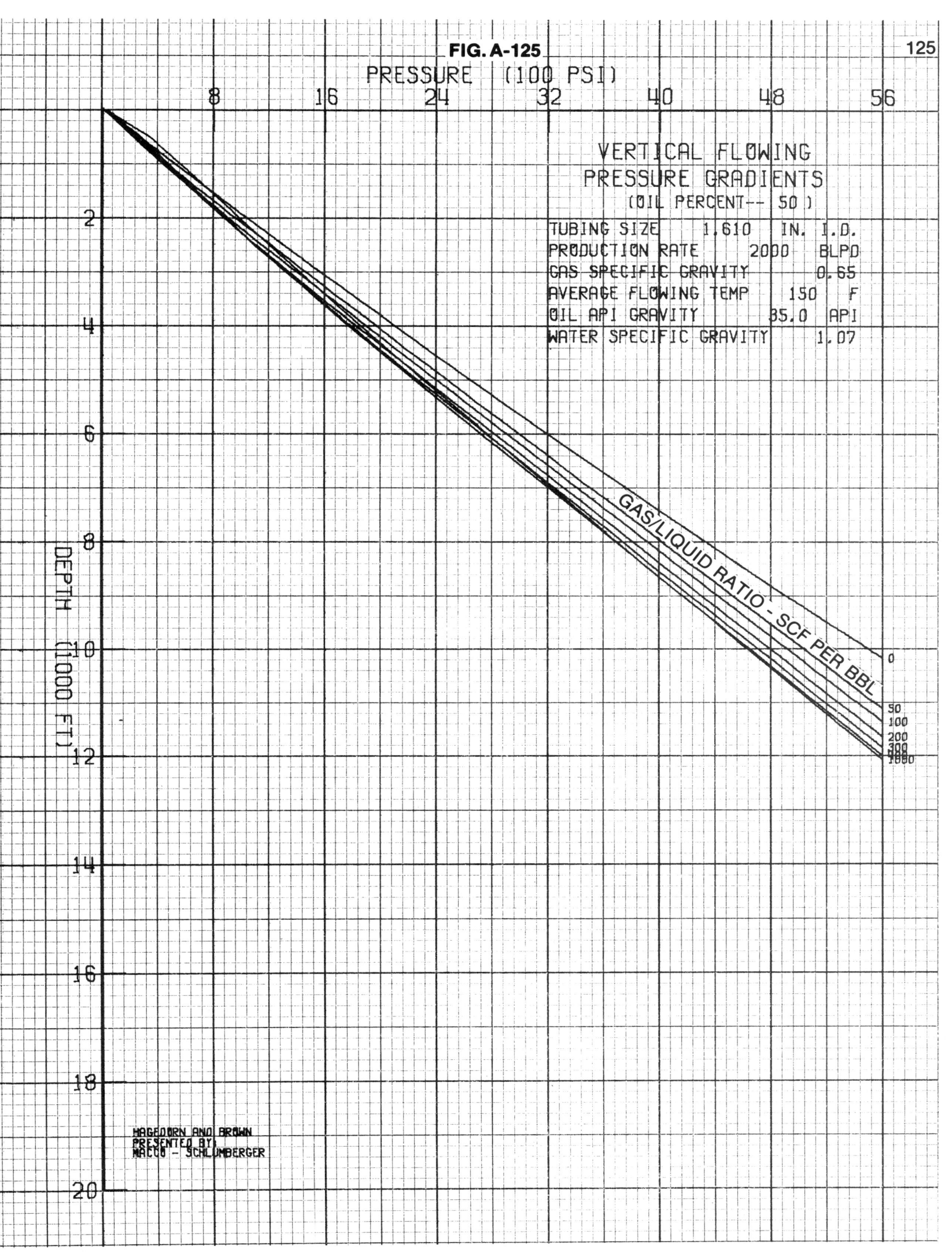

PRESSURE (100 PSI)

DEPTH (1000 FT)

VERTICAL FLOWING
PRESSURE GRADIENTS
(OIL PERCENT-- 50)

TUBING SIZE	1.610	IN. I.D.
PRODUCTION RATE	2000	BLPD
GAS SPECIFIC GRAVITY	0.65	
AVERAGE FLOWING TEMP	150	F
OIL API GRAVITY	35.0	API
WATER SPECIFIC GRAVITY	1.07	

GAS/LIQUID RATIO - SCF PER BBL

0
50
100
200
300
1000

HAGEDORN AND BROWN
PRESENTED BY
MACCO - SCHLUMBERGER

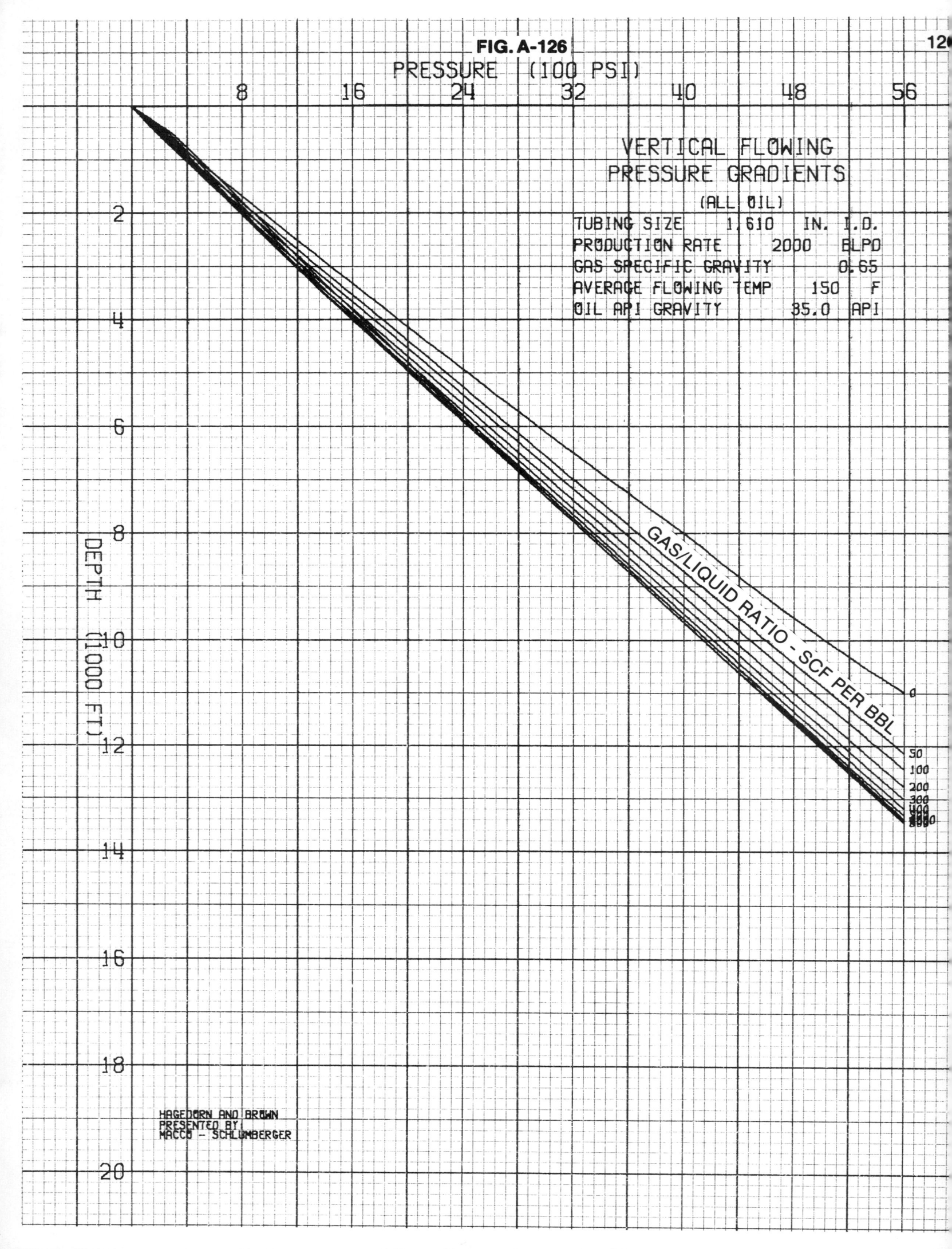

PRESSURE (100 PSI)

VERTICAL FLOWING
PRESSURE GRADIENTS
(ALL OIL)

TUBING SIZE	1.610	IN. I.D.
PRODUCTION RATE	2000	BLPD
GAS SPECIFIC GRAVITY	0.65	
AVERAGE FLOWING TEMP	150	F
OIL API GRAVITY	35.0	API

DEPTH (1000 FT)

GAS/LIQUID RATIO - SCF PER BBL

0
50
100
200
300
400
1000

HAGEDORN AND BROWN
PRESENTED BY
MACCO - SCHLUMBERGER

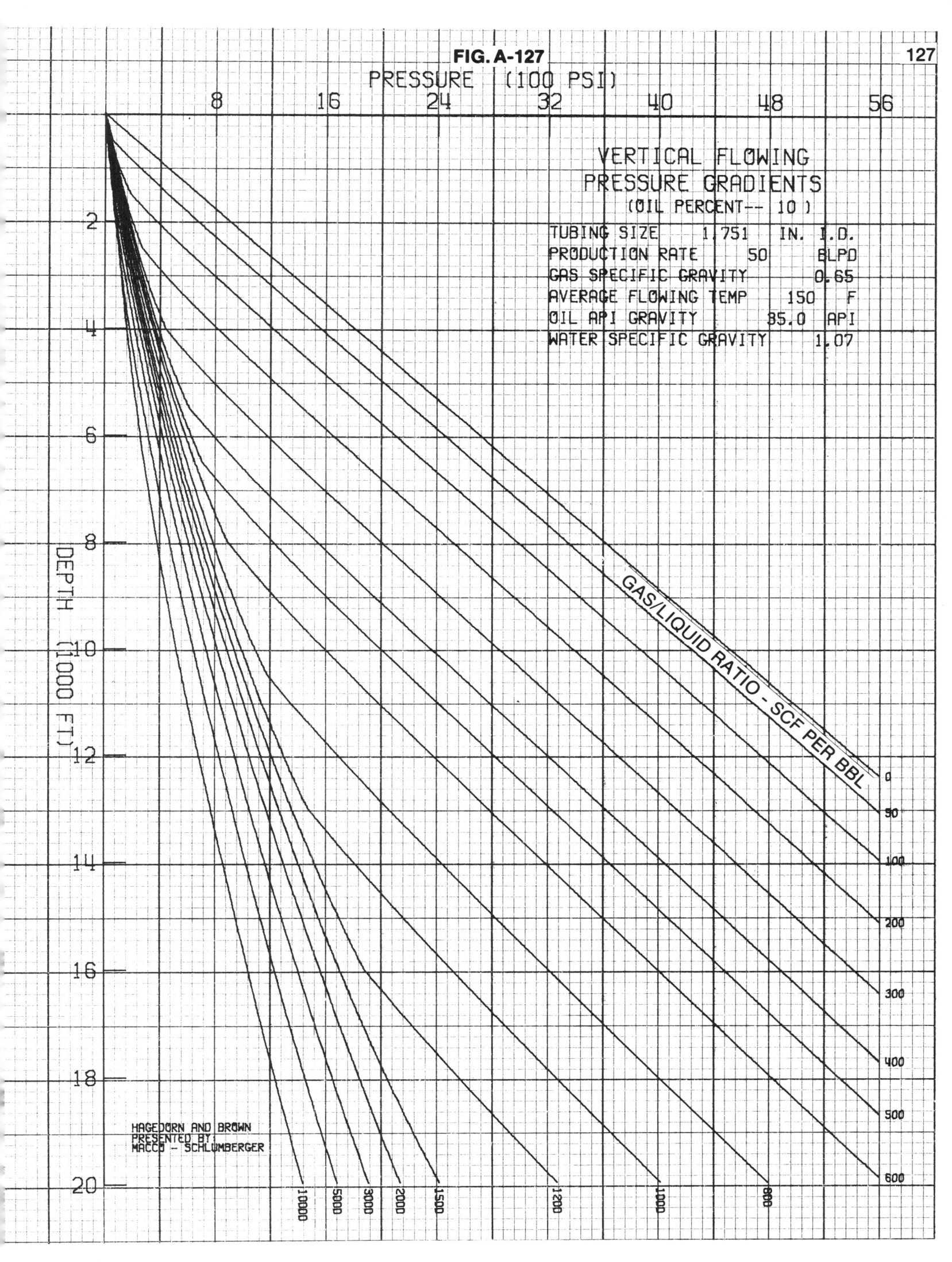

FIG. A-127

127

VERTICAL FLOWING
PRESSURE GRADIENTS
(OIL PERCENT-- 10)

TUBING SIZE	1.751	IN. I.D.
PRODUCTION RATE	50	BLPD
GAS SPECIFIC GRAVITY	0.65	
AVERAGE FLOWING TEMP	150	F
OIL API GRAVITY	35.0	API
WATER SPECIFIC GRAVITY	1.07	

PRESSURE (100 PSI)

DEPTH (1000 FT)

GAS/LIQUID RATIO - SCF PER BBL

HAGEDORN AND BROWN
PRESENTED BY:
MACCO - SCHLUMBERGER

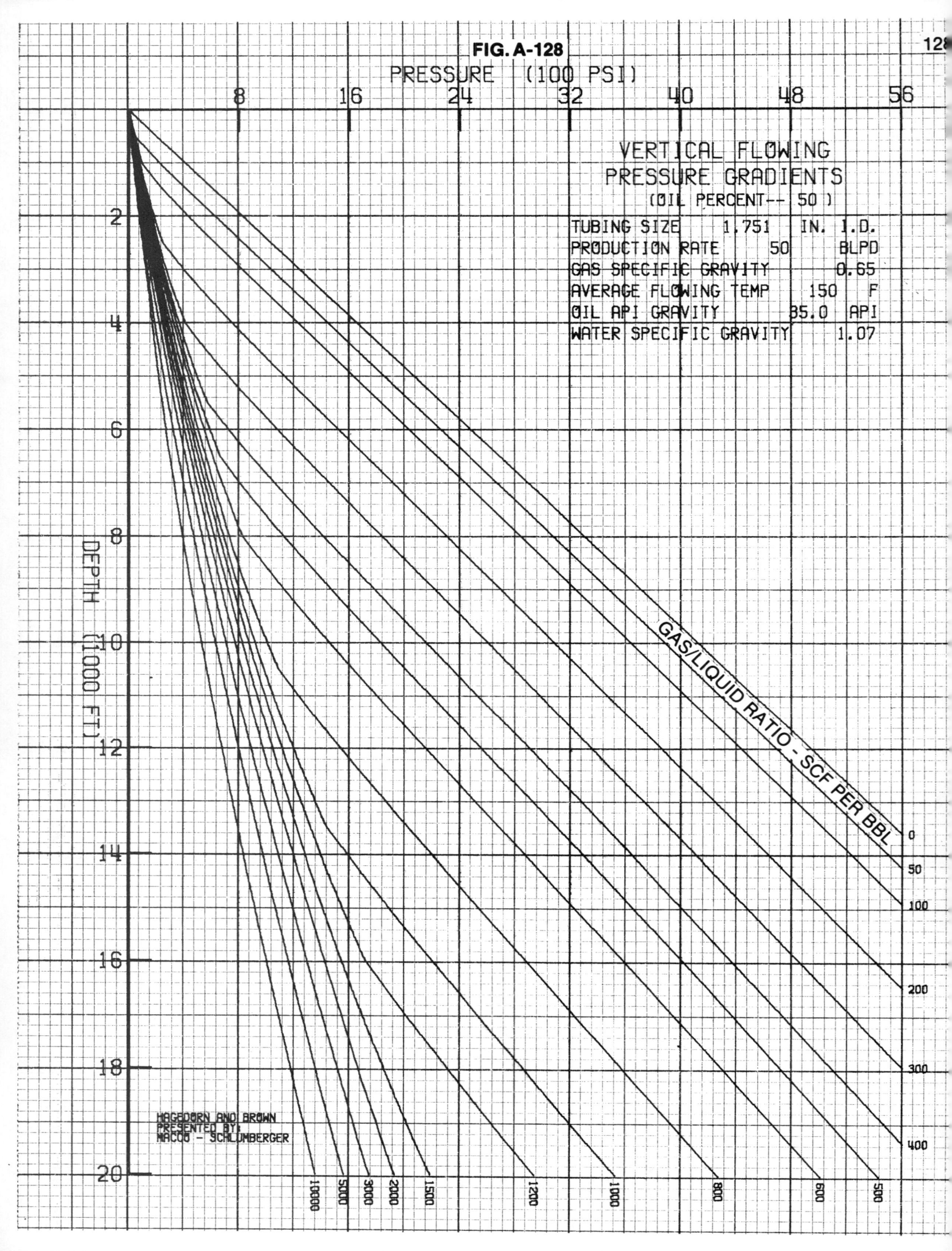

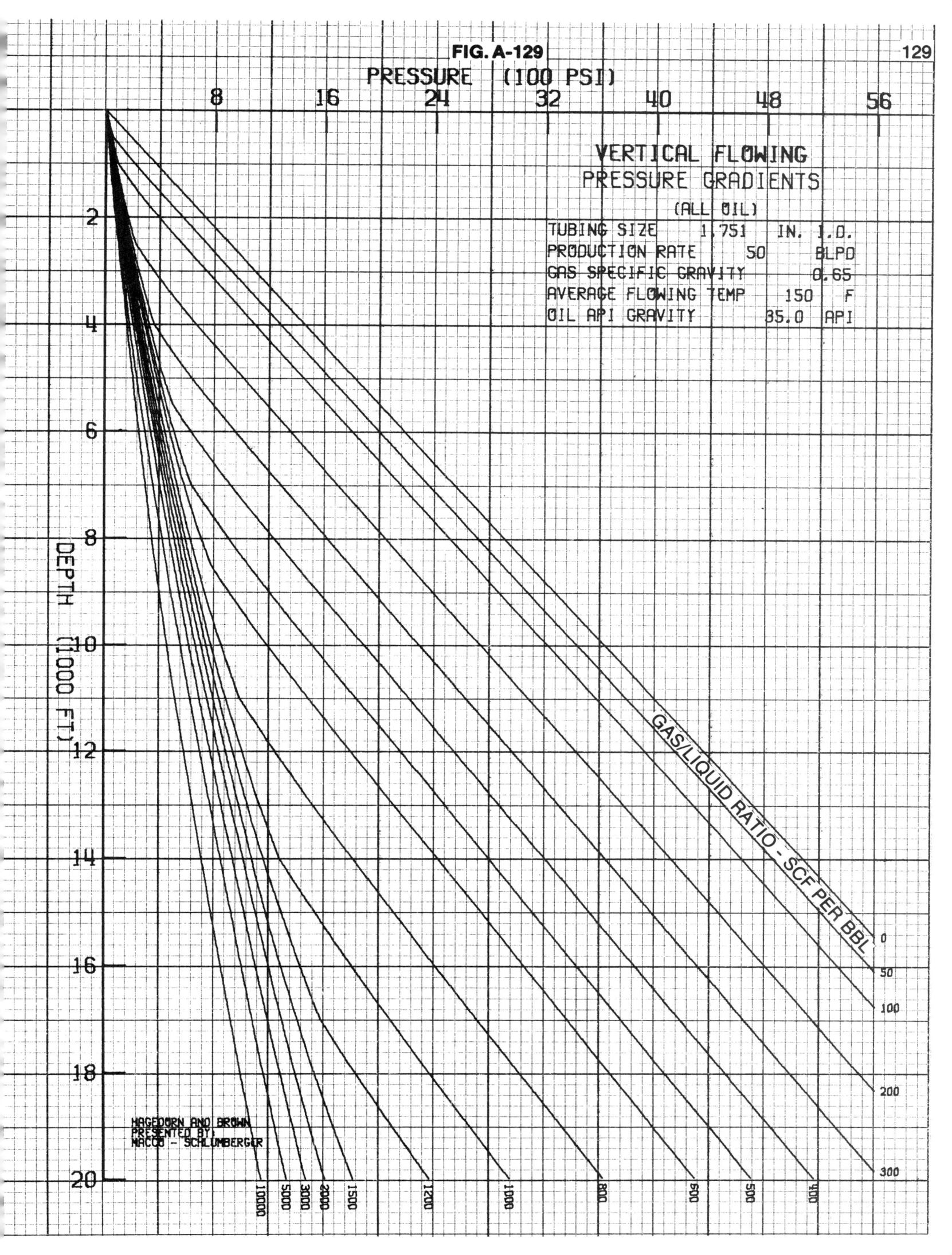

FIG. A-129

PRESSURE (100 PSI)

VERTICAL FLOWING
PRESSURE GRADIENTS
(ALL OIL)

TUBING SIZE	1.751	IN. I.D.
PRODUCTION RATE	50	BLPD
GAS SPECIFIC GRAVITY	0.65	
AVERAGE FLOWING TEMP	150	F
OIL API GRAVITY	35.0	API

GAS/LIQUID RATIO - SCF PER BBL

DEPTH (1000 FT)

HAGEDORN AND BROWN
PRESENTED BY:
NACCO - SCHLUMBERGER

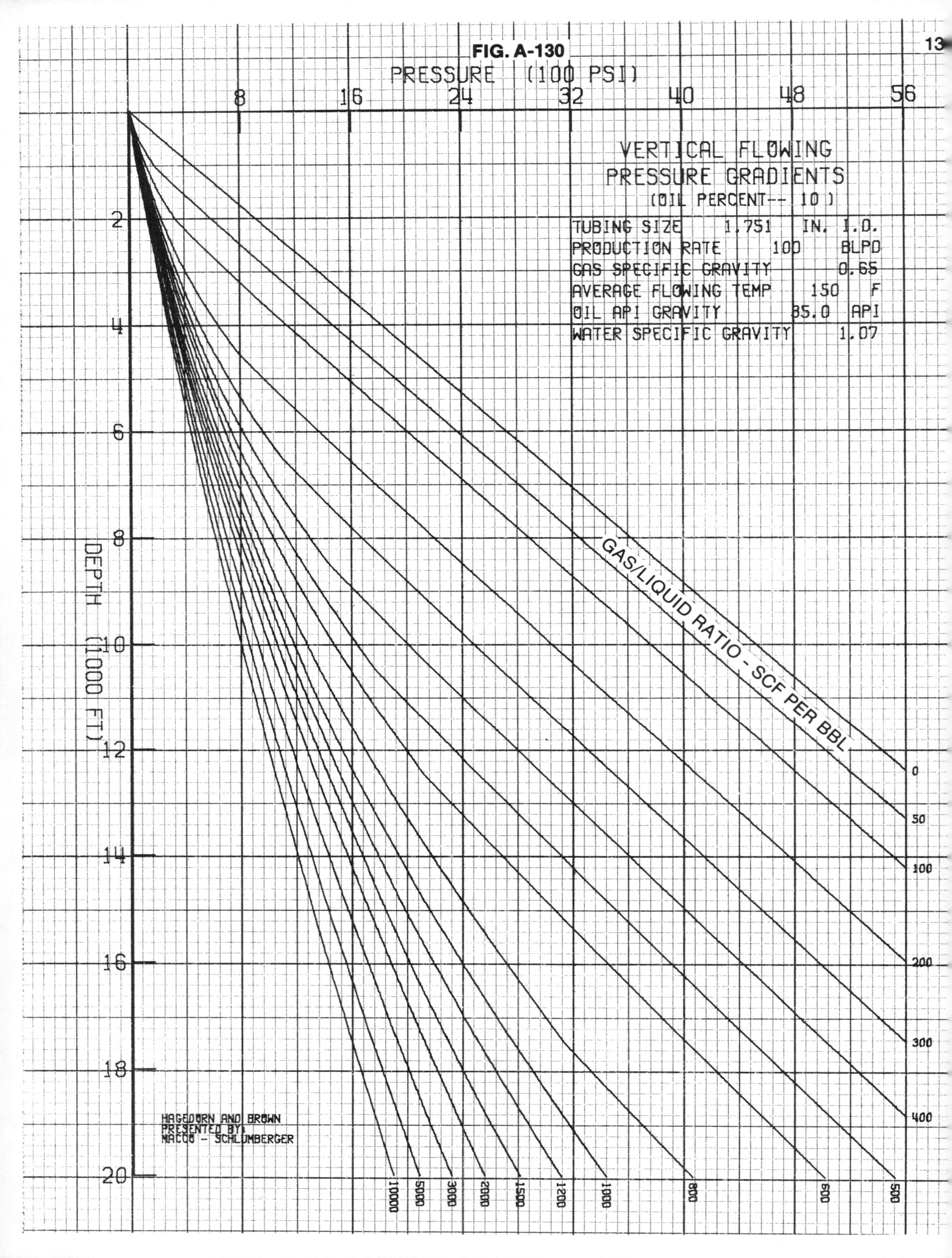

FIG. A-130

13

PRESSURE (100 PSI)

VERTICAL FLOWING
PRESSURE GRADIENTS
(OIL PERCENT-- 10)

TUBING SIZE	1.751	IN. I.D.
PRODUCTION RATE	100	BLPD
GAS SPECIFIC GRAVITY	0.65	
AVERAGE FLOWING TEMP	150	F
OIL API GRAVITY	85.0	API
WATER SPECIFIC GRAVITY	1.07	

DEPTH (1000 FT)

GAS/LIQUID RATIO - SCF PER BBL

HAGEDORN AND BROWN
PRESENTED BY:
NACOO - SCHLUMBERGER

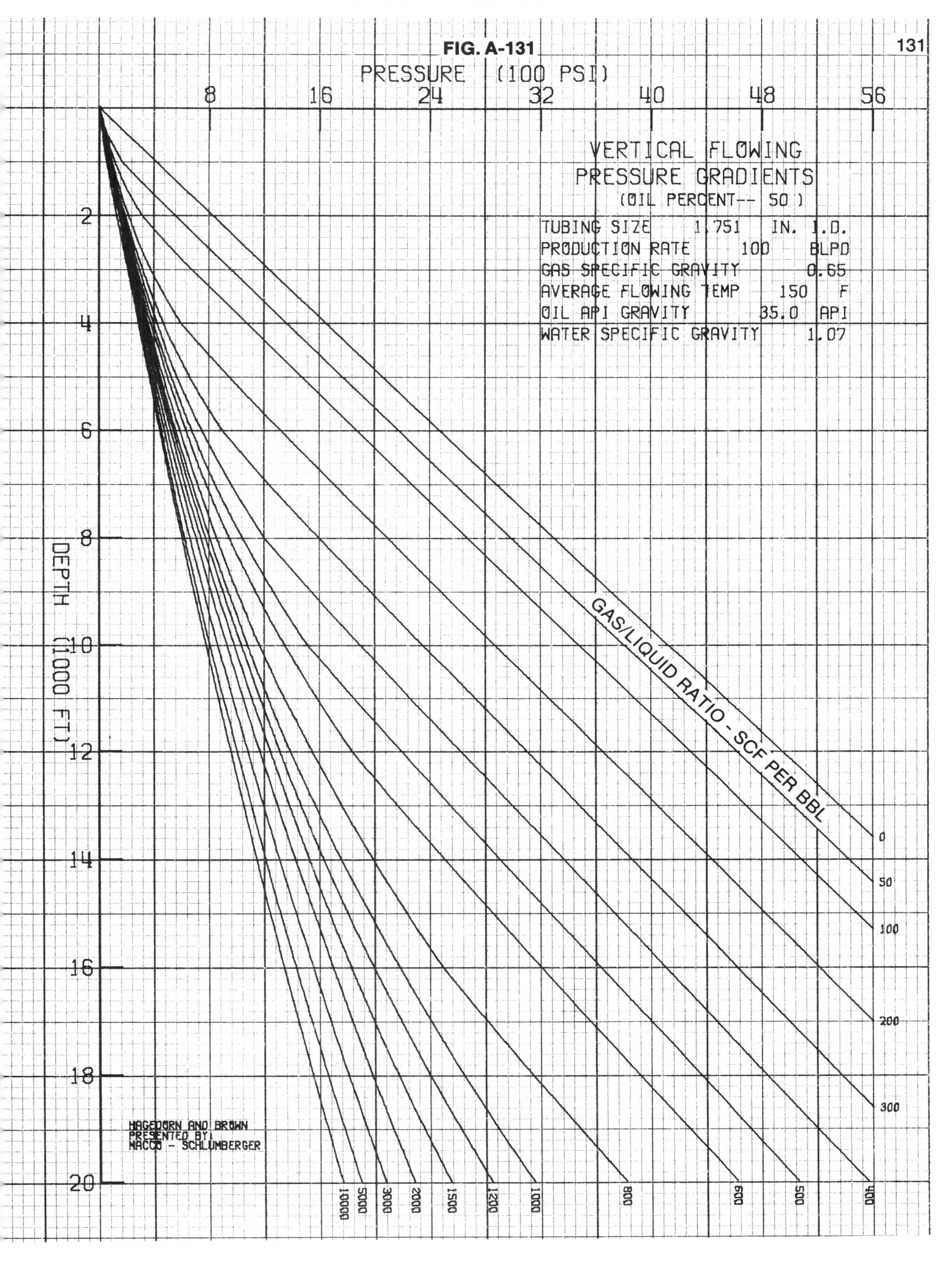

PRESSURE (100 PSI)

VERTICAL FLOWING
PRESSURE GRADIENTS
(OIL PERCENT-- 50)

TUBING SIZE	1.751	IN. I.D.
PRODUCTION RATE	100	BLPD
GAS SPECIFIC GRAVITY	0.65	
AVERAGE FLOWING TEMP	150	F
OIL API GRAVITY	35.0	API
WATER SPECIFIC GRAVITY	1.07	

DEPTH (1000 FT)

GAS/LIQUID RATIO - SCF PER BBL

0
50
100
200
300

HAGEDORN AND BROWN
PRESENTED BY:
MACCO - SCHLUMBERGER

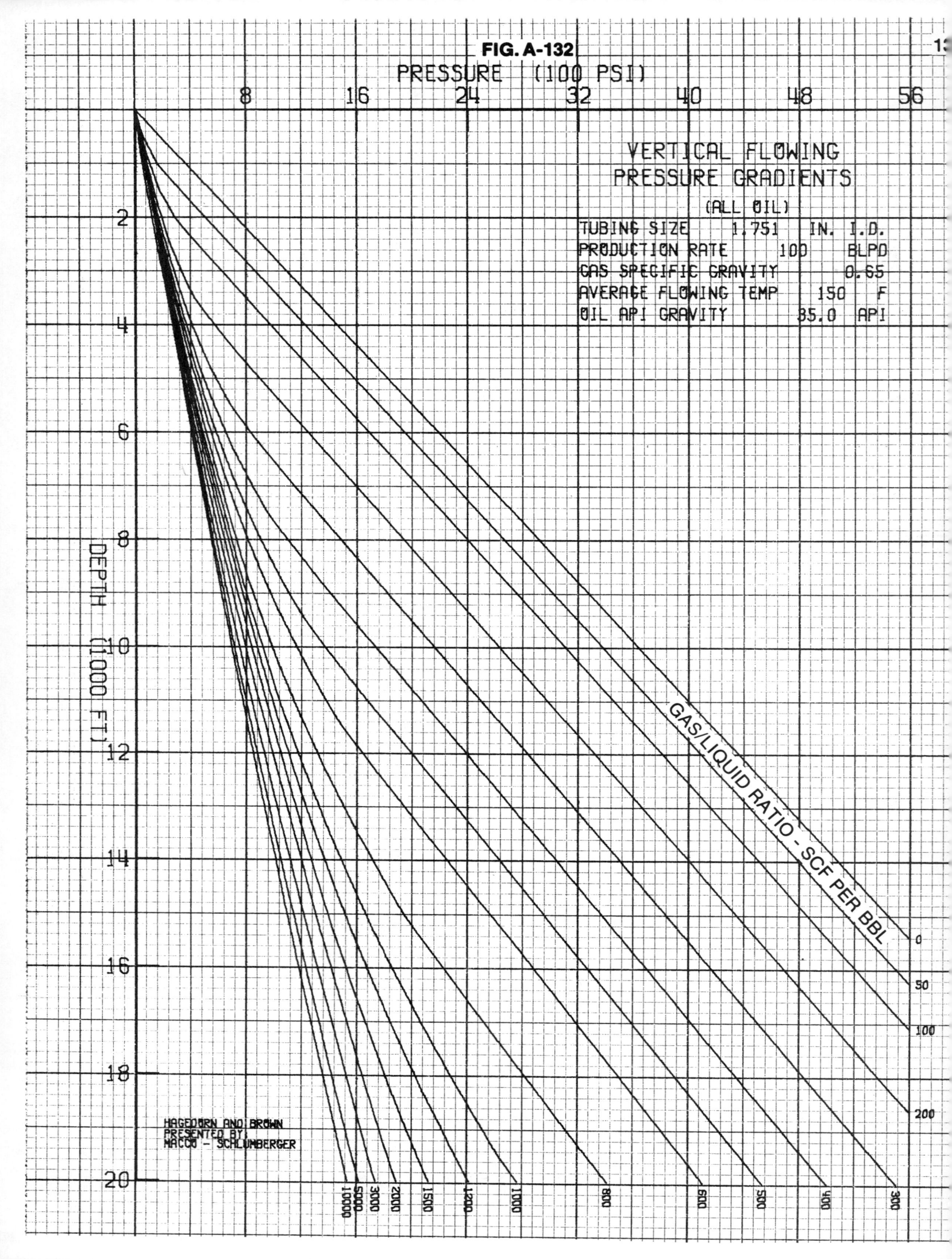

FIG. A-132

VERTICAL FLOWING
PRESSURE GRADIENTS
(ALL OIL)

TUBING SIZE	1.751	IN. I.D.
PRODUCTION RATE	100	BLPD
GAS SPECIFIC GRAVITY		0.65
AVERAGE FLOWING TEMP	150	F
OIL API GRAVITY	85.0	API

PRESSURE (100 PSI)

DEPTH (1000 FT)

GAS/LIQUID RATIO - SCF PER BBL

HAGEDORN AND BROWN
PRESENTED BY
MACCO - SCHLUMBERGER

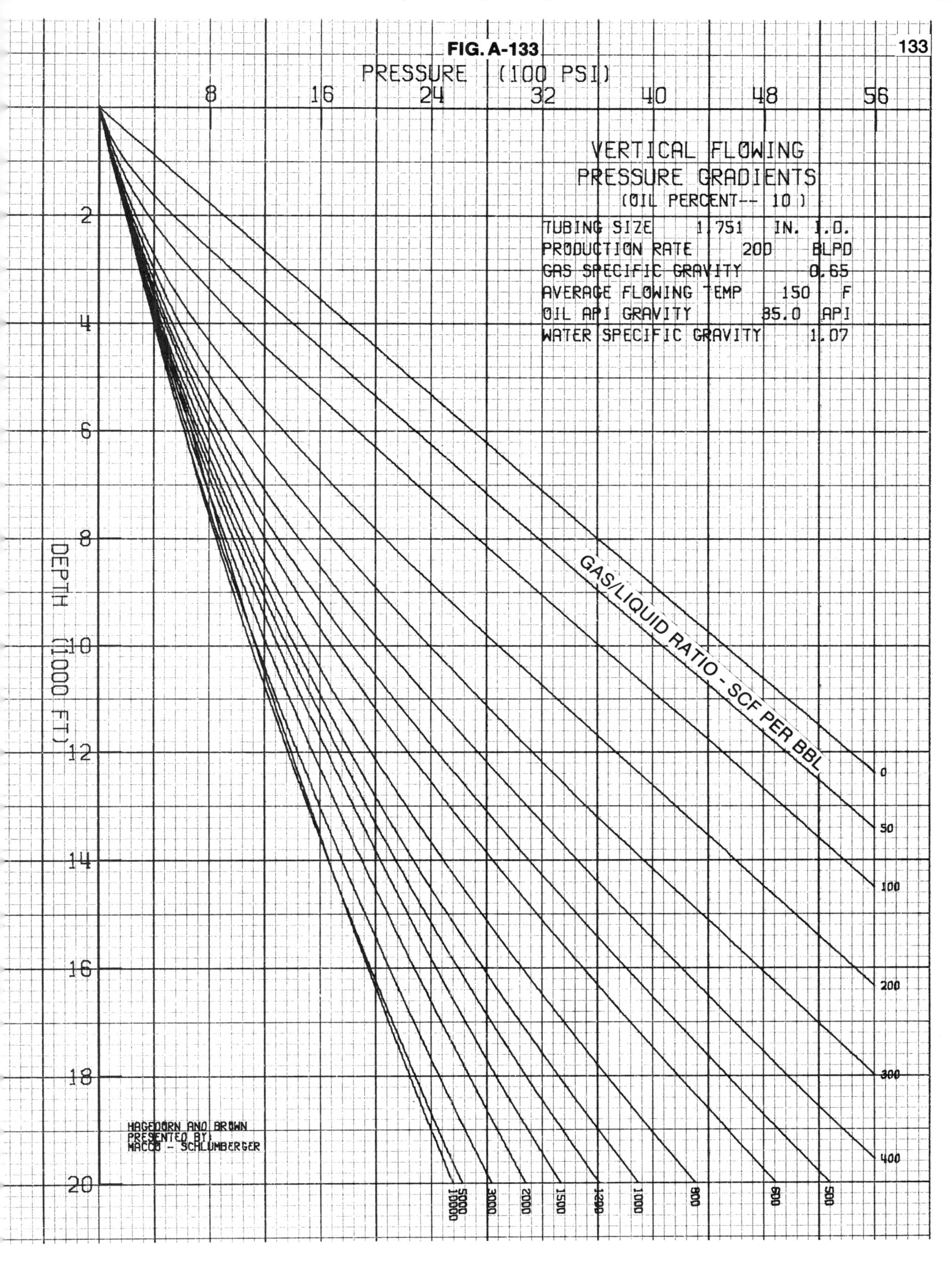

PRESSURE (100 PSI)

VERTICAL FLOWING
PRESSURE GRADIENTS
(OIL PERCENT-- 10)

TUBING SIZE	1.751	IN. I.D.
PRODUCTION RATE	200	BLPD
GAS SPECIFIC GRAVITY		0.65
AVERAGE FLOWING TEMP	150	F
OIL API GRAVITY	85.0	API
WATER SPECIFIC GRAVITY		1.07

DEPTH (1000 FT)

GAS/LIQUID RATIO - SCF PER BBL

HAGEDORN AND BROWN
PRESENTED BY
MACCO - SCHLUMBERGER

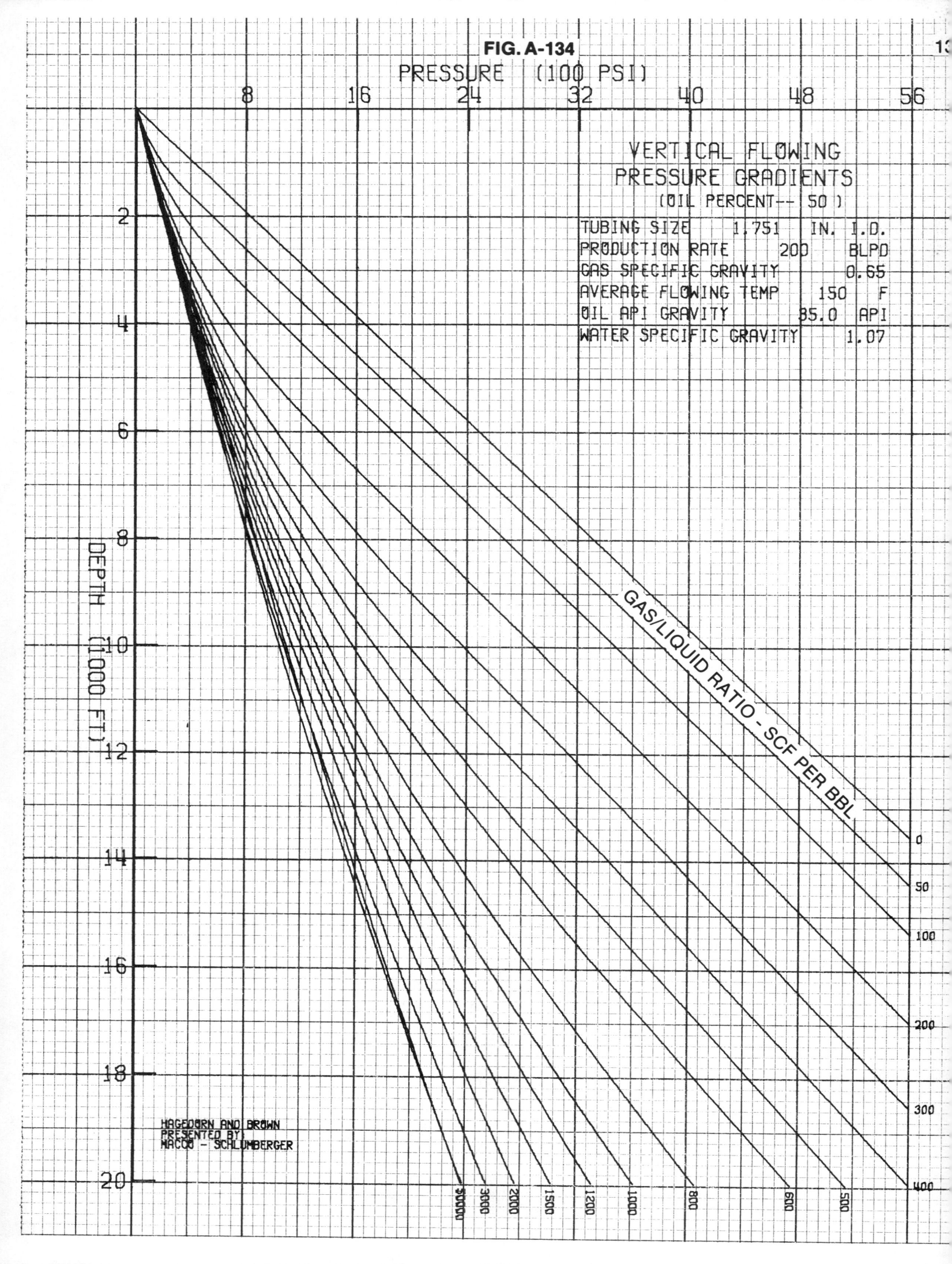

FIG. A-134

VERTICAL FLOWING
PRESSURE GRADIENTS
(OIL PERCENT-- 50)

TUBING SIZE	1.751	IN. I.D.
PRODUCTION RATE	200	BLPD
GAS SPECIFIC GRAVITY	0.65	
AVERAGE FLOWING TEMP	150	F
OIL API GRAVITY	85.0	API
WATER SPECIFIC GRAVITY	1.07	

PRESSURE (100 PSI)

DEPTH (1000 FT)

GAS/LIQUID RATIO - SCF PER BBL

HAGEDORN AND BROWN
PRESENTED BY
MACCO - SCHLUMBERGER

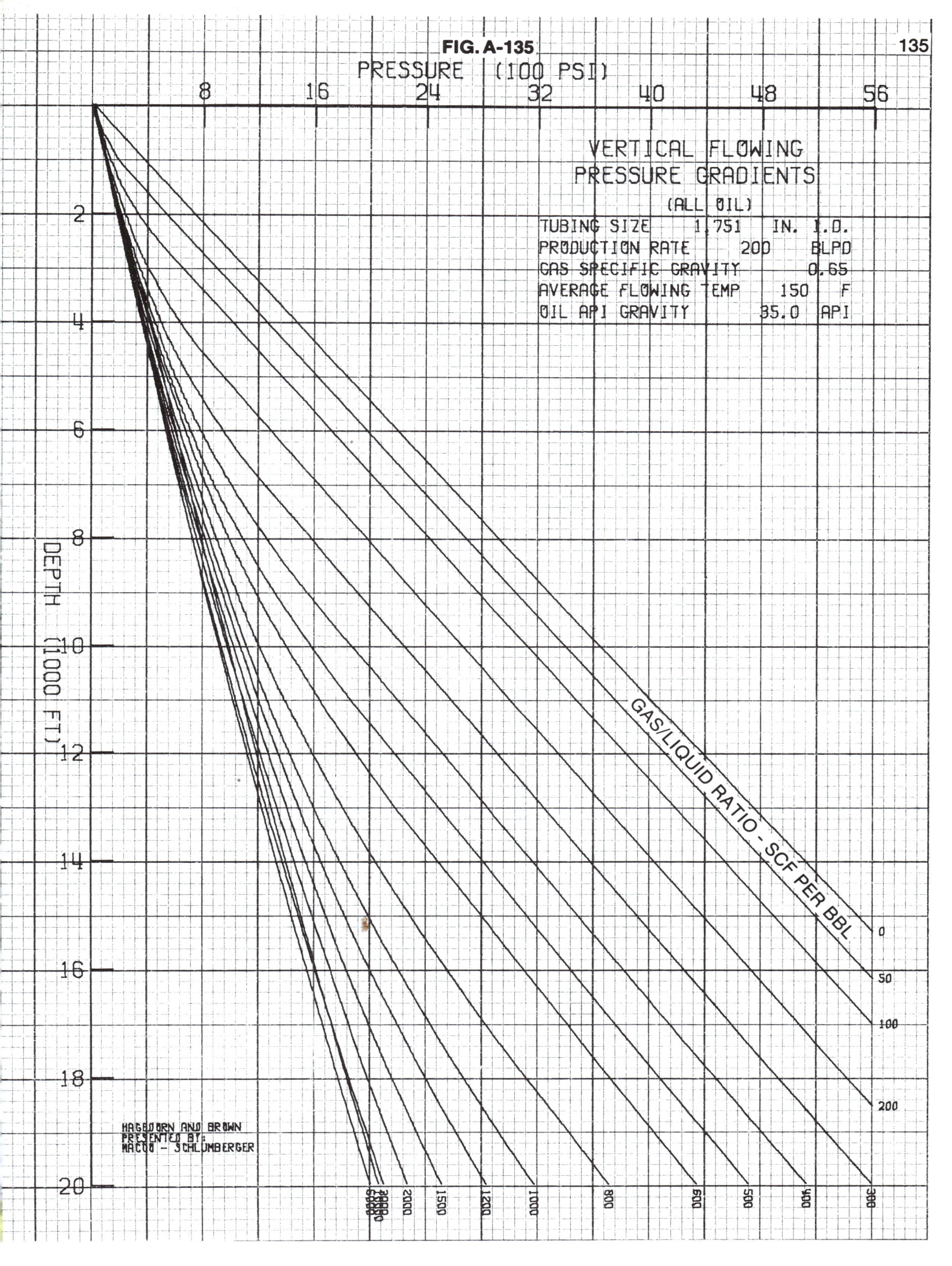

PRESSURE (100 PSI)

VERTICAL FLOWING
PRESSURE GRADIENTS
(ALL OIL)

TUBING SIZE	1,751	IN. I.D.
PRODUCTION RATE	200	BLPD
GAS SPECIFIC GRAVITY	0.65	
AVERAGE FLOWING TEMP	150	F
OIL API GRAVITY	35.0	API

DEPTH (1000 FT)

GAS/LIQUID RATIO - SCF PER BBL

HAGEDORN AND BROWN
PRESENTED BY:
MACCO - SCHLUMBERGER

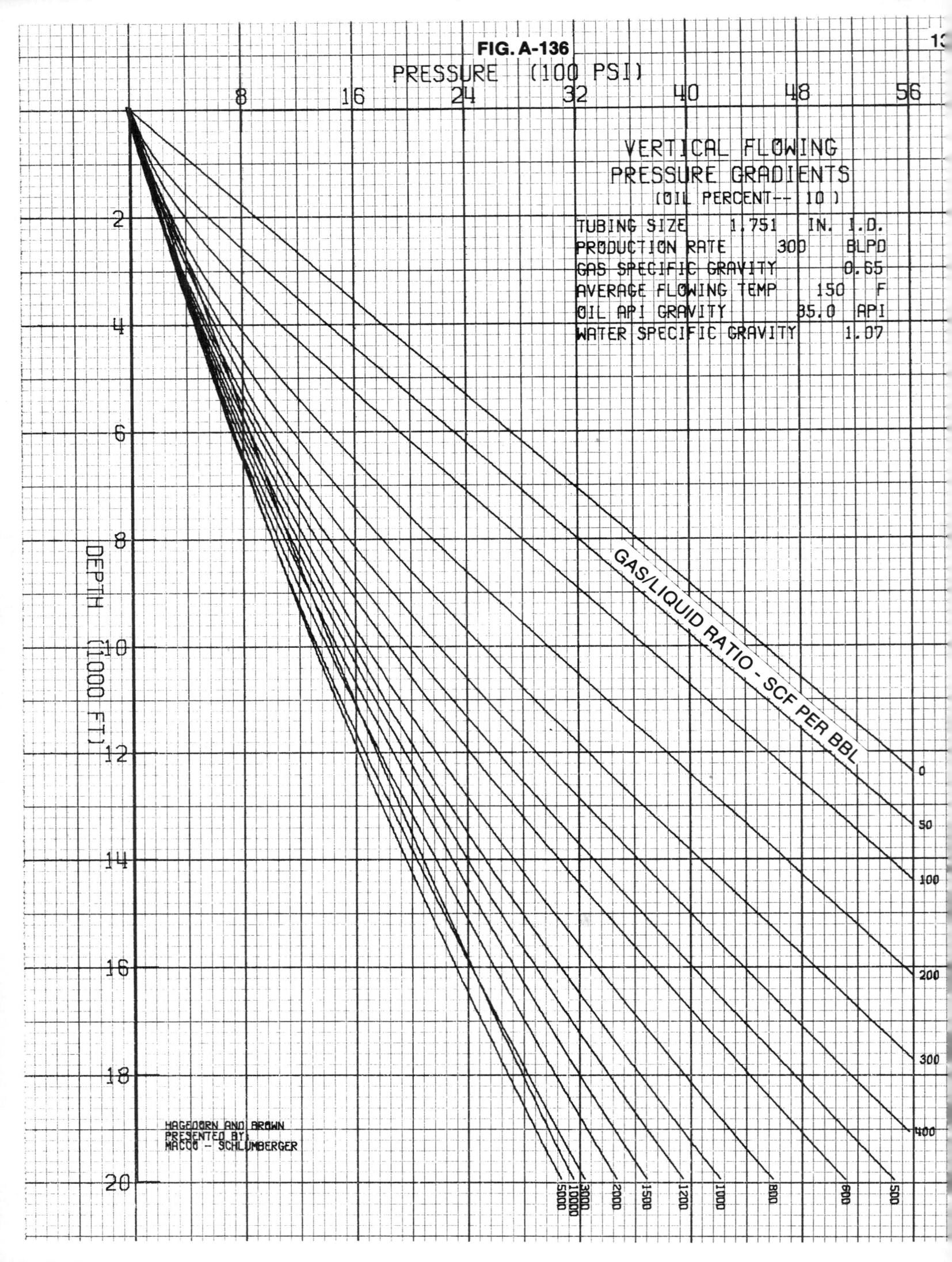

FIG. A-136

PRESSURE (100 PSI)

VERTICAL FLOWING
PRESSURE GRADIENTS
(OIL PERCENT-- 10)

TUBING SIZE	1.751	IN. 1.D.
PRODUCTION RATE	300	BLPD
GAS SPECIFIC GRAVITY	0.65	
AVERAGE FLOWING TEMP	150	F
OIL API GRAVITY	35.0	API
WATER SPECIFIC GRAVITY	1.07	

DEPTH (1000 FT)

GAS/LIQUID RATIO - SCF PER BBL

HAGEDORN AND BROWN
PRESENTED BY:
MACCO - SCHLUMBERGER

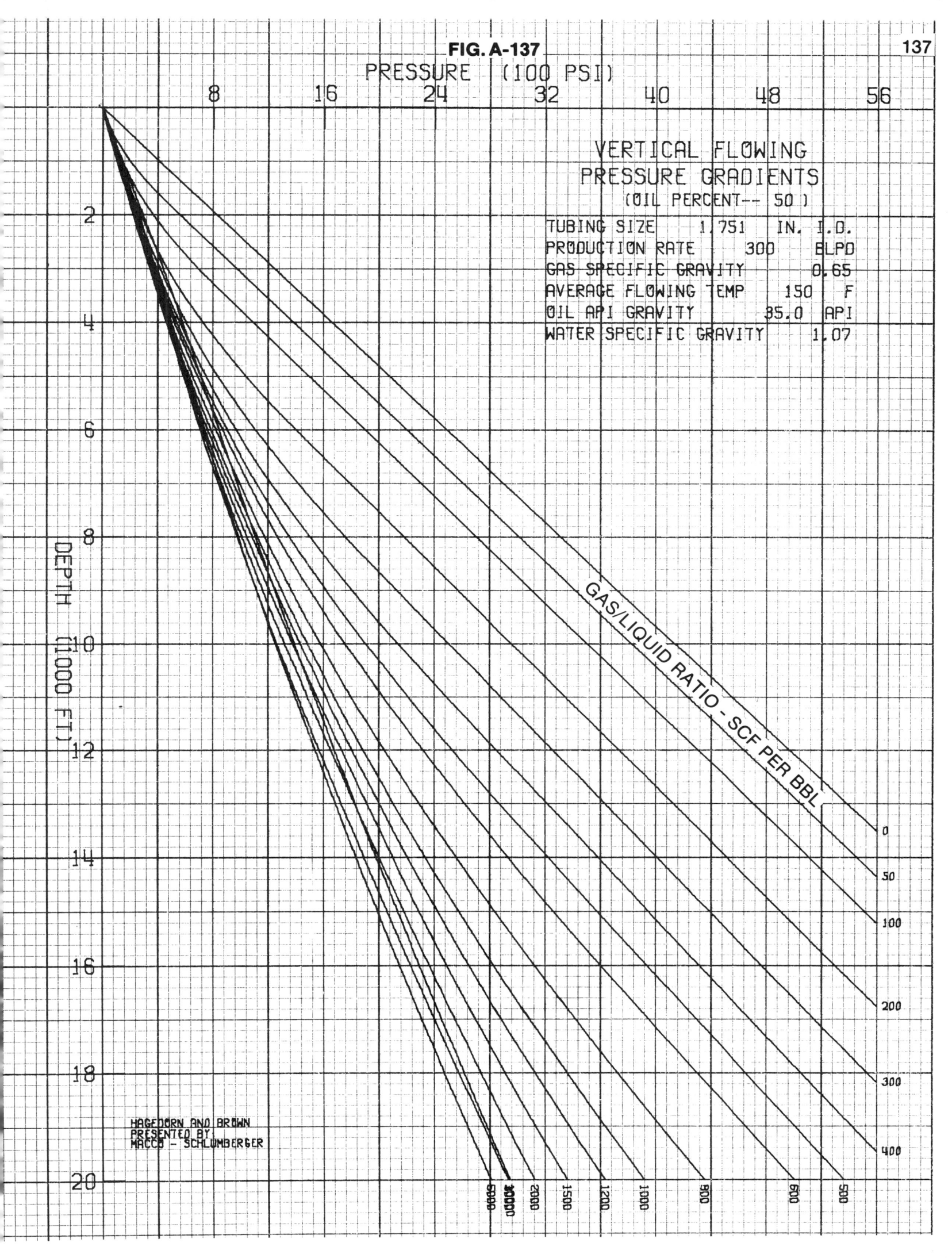

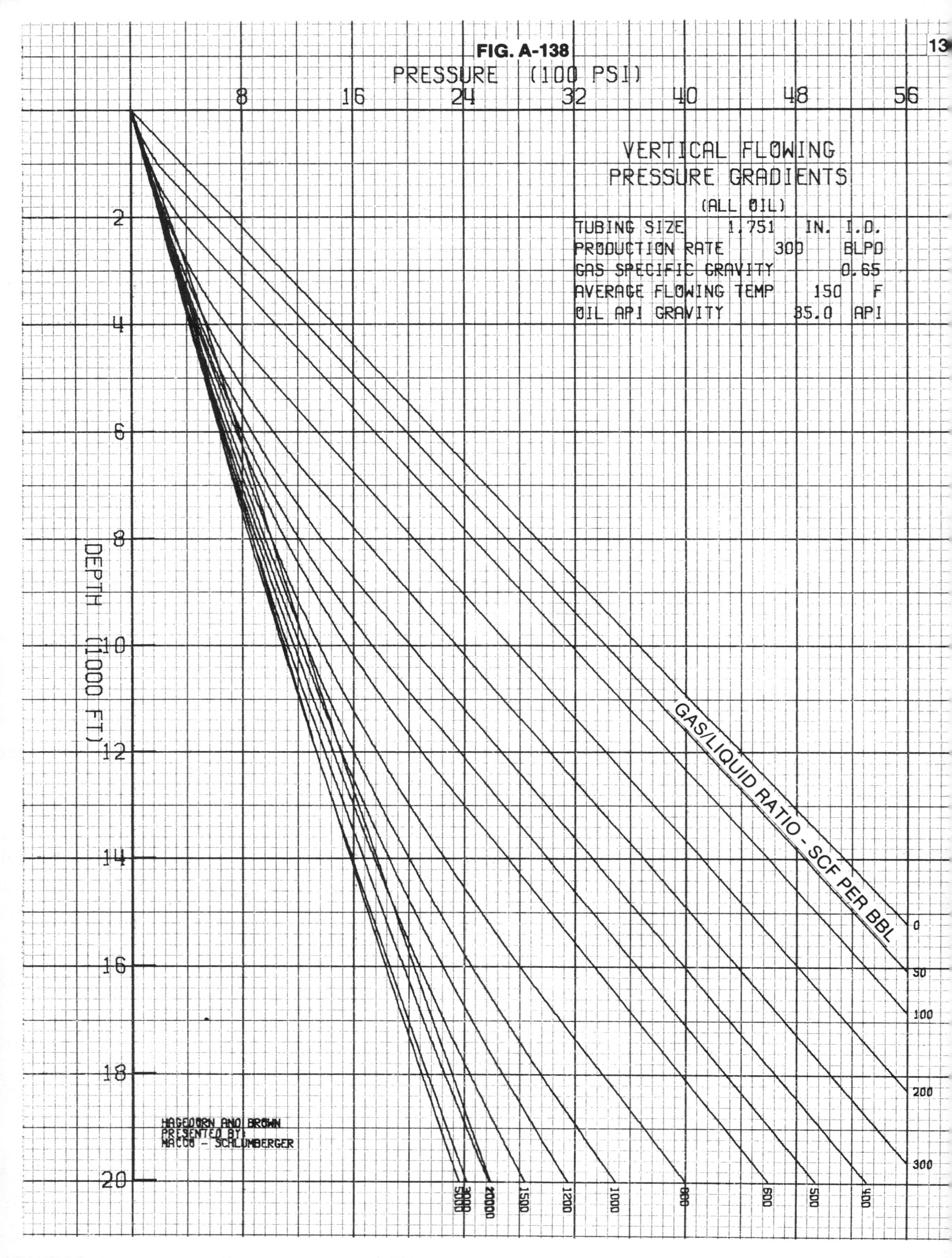

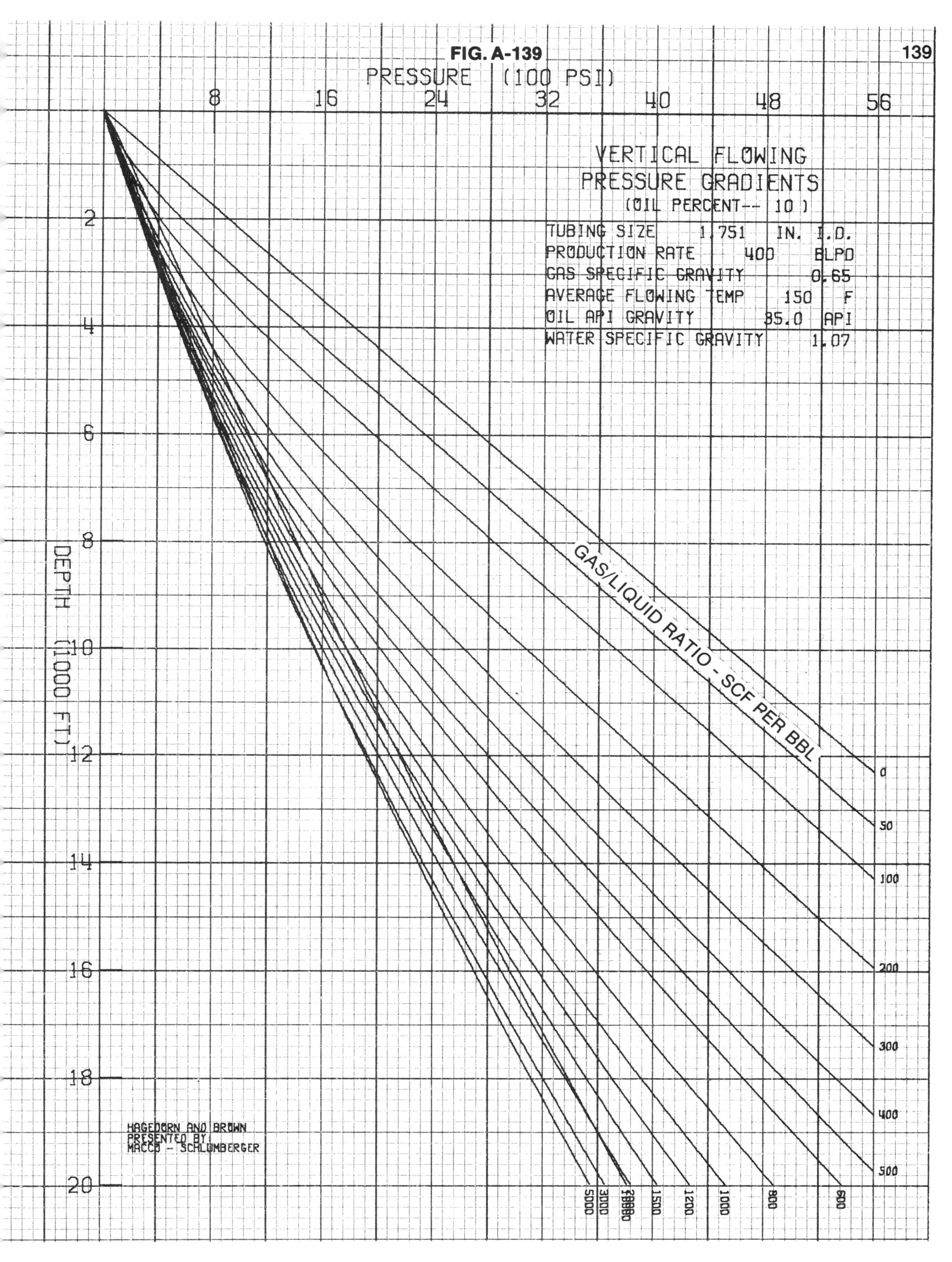

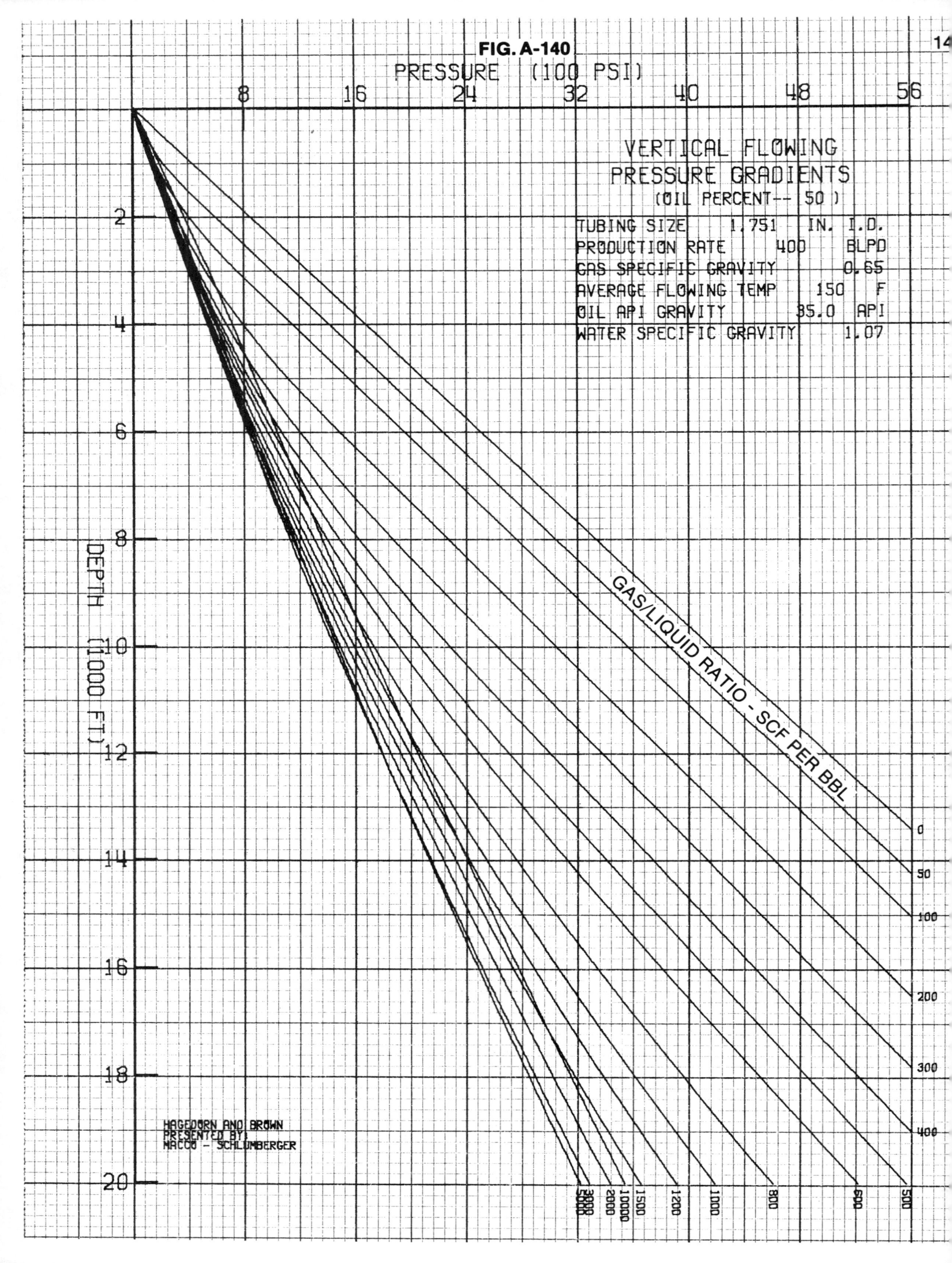

VERTICAL FLOWING
PRESSURE GRADIENTS
(OIL PERCENT-- 50)

TUBING SIZE	1.751	IN. I.D.
PRODUCTION RATE	400	BLPD
GAS SPECIFIC GRAVITY	0.65	
AVERAGE FLOWING TEMP	150	F
OIL API GRAVITY	35.0	API
WATER SPECIFIC GRAVITY	1.07	

PRESSURE (100 PSI)

DEPTH (1,000 FT)

GAS/LIQUID RATIO - SCF PER BBL

HAGEDORN AND BROWN
PRESENTED BY
MACCO - SCHLUMBERGER

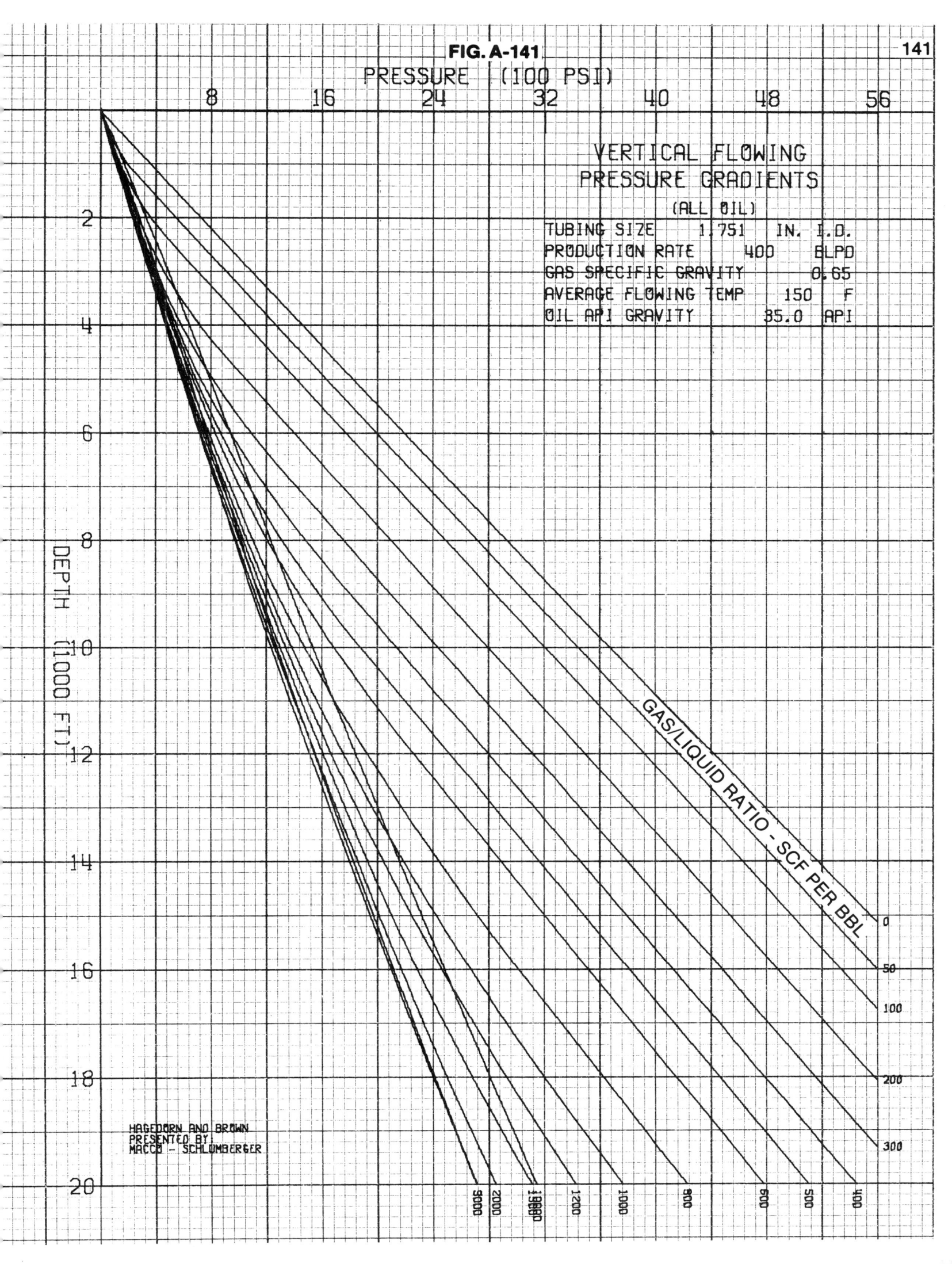

FIG. A-141

141

VERTICAL FLOWING
PRESSURE GRADIENTS
(ALL OIL)

TUBING SIZE	1.751	IN. I.D.
PRODUCTION RATE	400	BLPD
GAS SPECIFIC GRAVITY		0.65
AVERAGE FLOWING TEMP	150	F
OIL API GRAVITY	35.0	API

PRESSURE (100 PSI)

DEPTH (1000 FT)

GAS/LIQUID RATIO - SCF PER BBL

HAGEDORN AND BROWN
PRESENTED BY
MACCO - SCHLUMBERGER

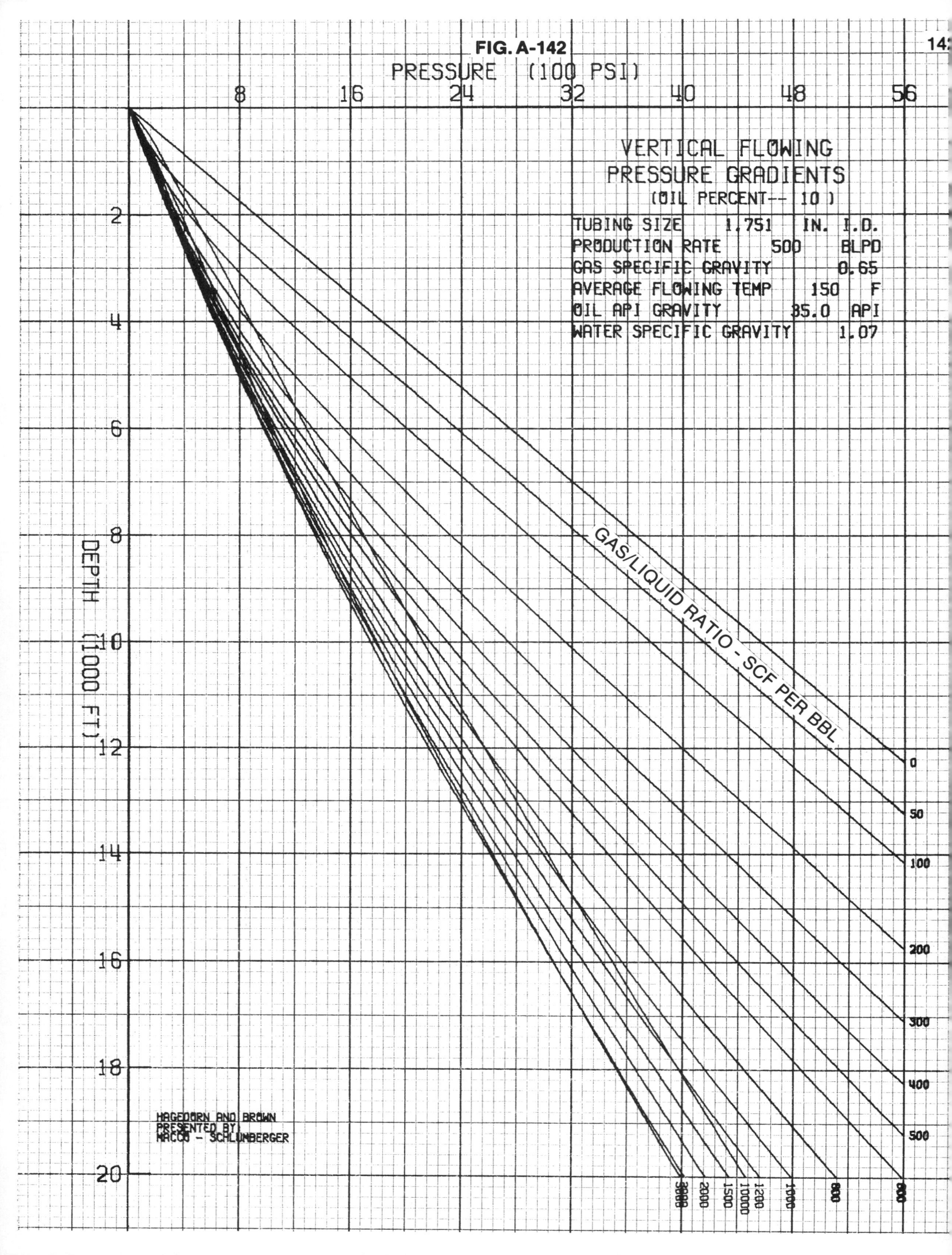

PRESSURE (100 PSI)

VERTICAL FLOWING
PRESSURE GRADIENTS
(OIL PERCENT-- 10)

TUBING SIZE	1.751	IN. I.D.
PRODUCTION RATE	500	BLPD
GAS SPECIFIC GRAVITY	0.65	
AVERAGE FLOWING TEMP	150	F
OIL API GRAVITY	35.0	API
WATER SPECIFIC GRAVITY	1.07	

GAS/LIQUID RATIO - SCF PER BBL.

DEPTH (1000 FT)

HAGEDORN AND BROWN
PRESENTED BY
NACCO - SCHLUMBERGER

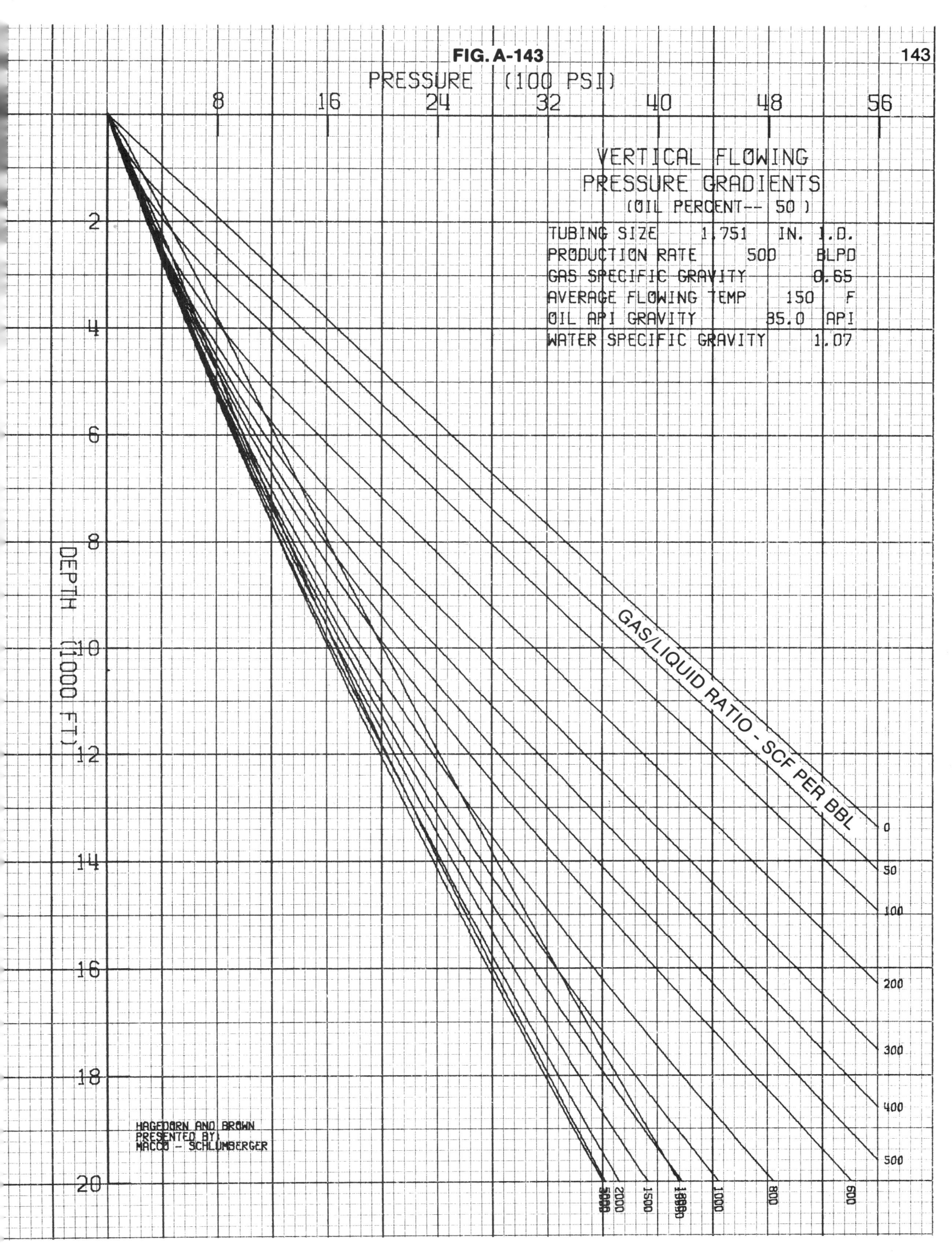

FIG. A-143

143

PRESSURE (100 PSI)

VERTICAL FLOWING
PRESSURE GRADIENTS
(OIL PERCENT-- 50)

TUBING SIZE	1.751	IN. I.D.
PRODUCTION RATE	500	BLPD
GAS SPECIFIC GRAVITY	0.65	
AVERAGE FLOWING TEMP	150	F
OIL API GRAVITY	85.0	API
WATER SPECIFIC GRAVITY	1.07	

DEPTH (1000 FT)

GAS/LIQUID RATIO - SCF PER BBL

HAGEDORN AND BROWN
PRESENTED BY:
MACCO - SCHLUMBERGER

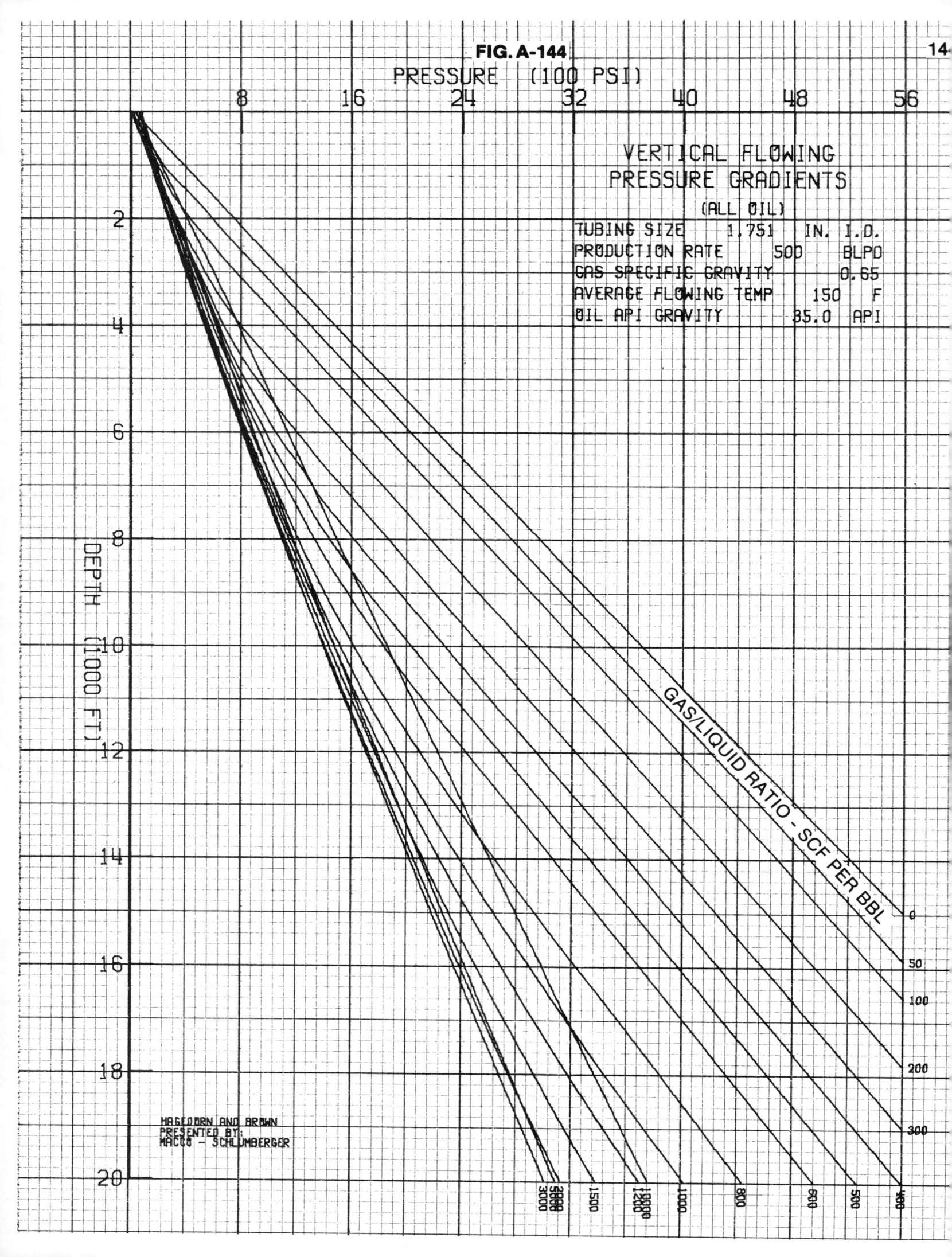

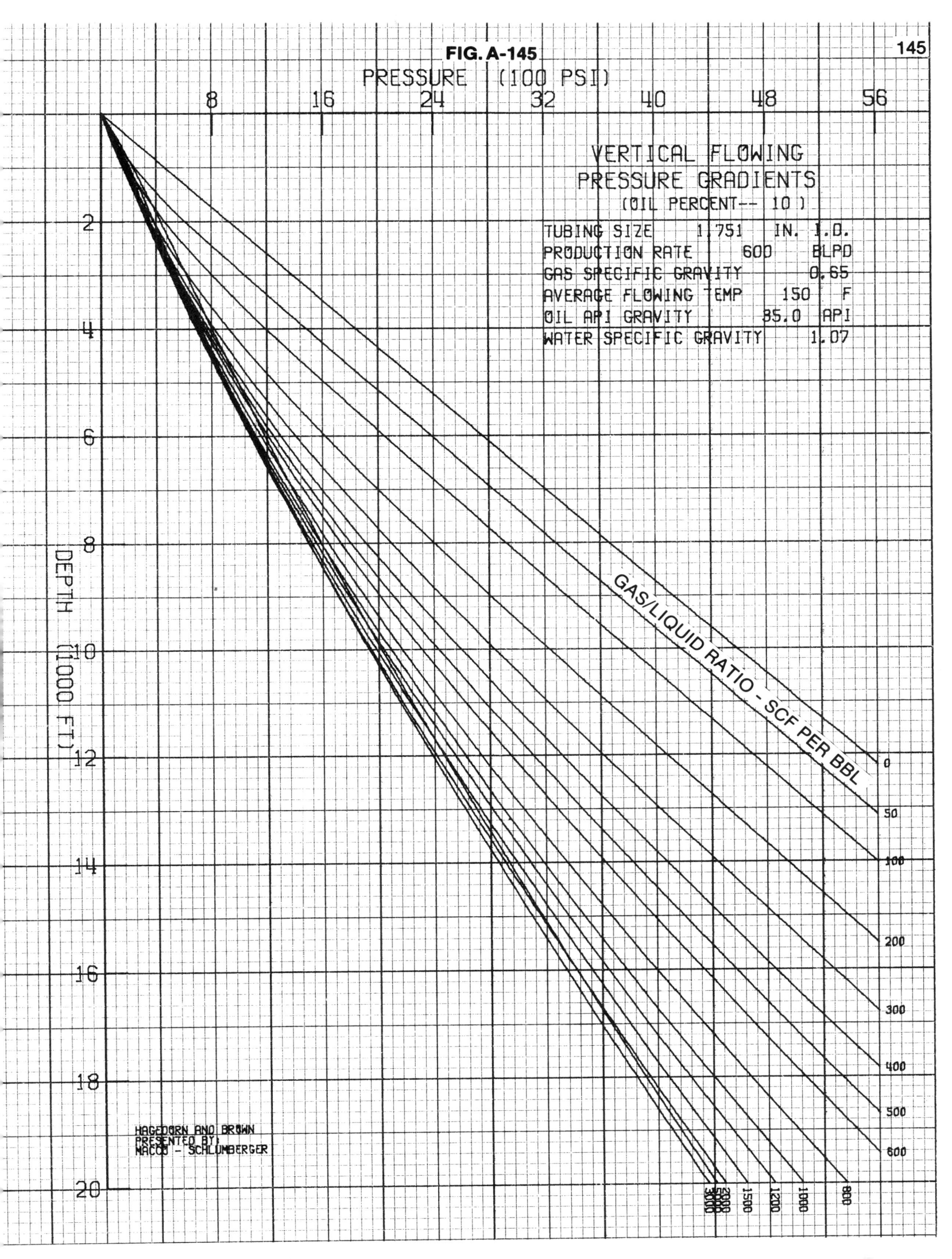

FIG. A-145

145

PRESSURE (100 PSI)

VERTICAL FLOWING
PRESSURE GRADIENTS
(OIL PERCENT-- 10)

TUBING SIZE   1.751   IN. I.D.
PRODUCTION RATE   600   BLPD
GAS SPECIFIC GRAVITY   0.65
AVERAGE FLOWING TEMP   150   F
OIL API GRAVITY   85.0   API
WATER SPECIFIC GRAVITY   1.07

DEPTH (1000 FT)

GAS/LIQUID RATIO - SCF PER BBL

0
50
100
200
300
400
500
600

3000
2500
2000
1500
1200
1000
800

HAGEDORN AND BROWN
PRESENTED BY
NACO - SCHLUMBERGER

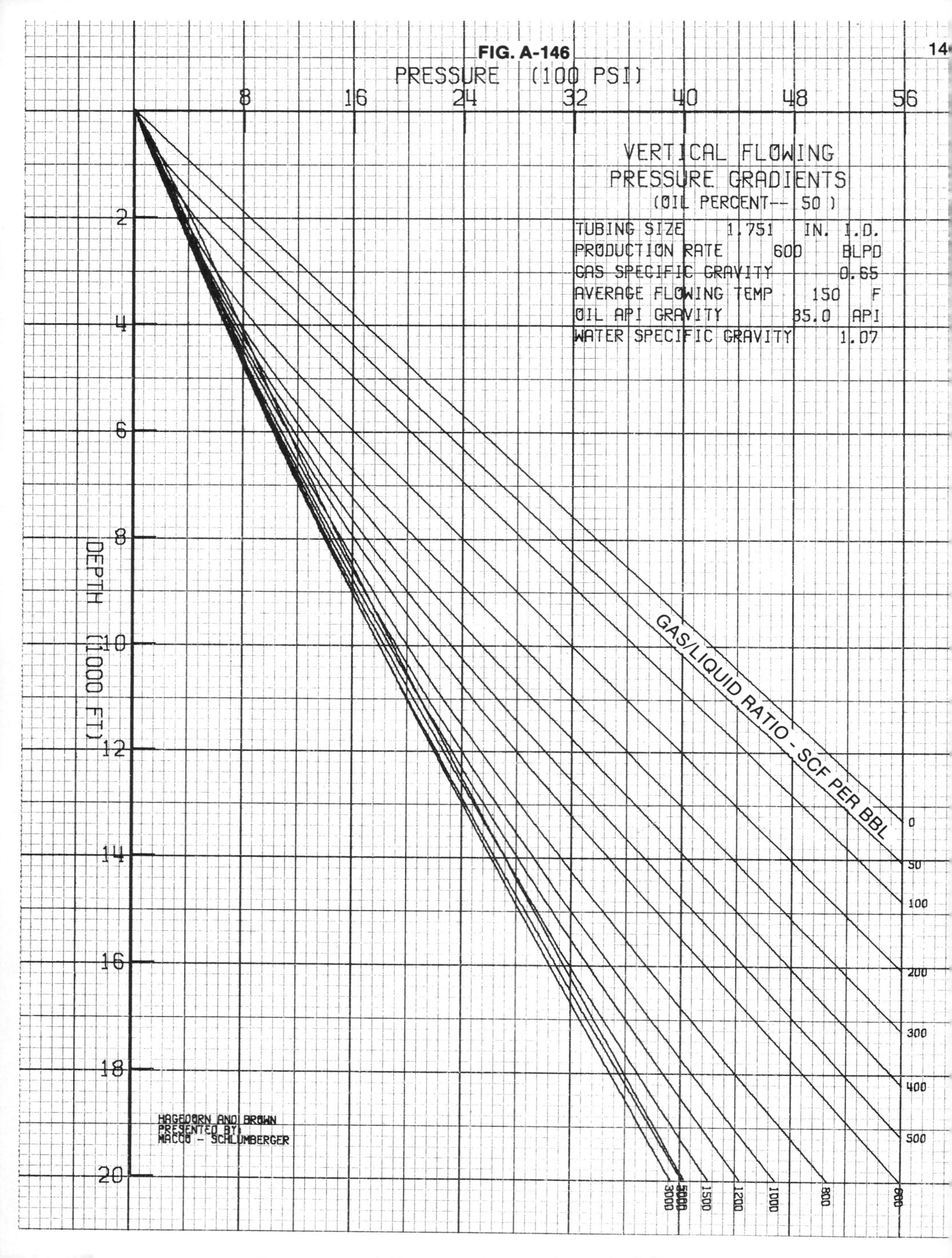

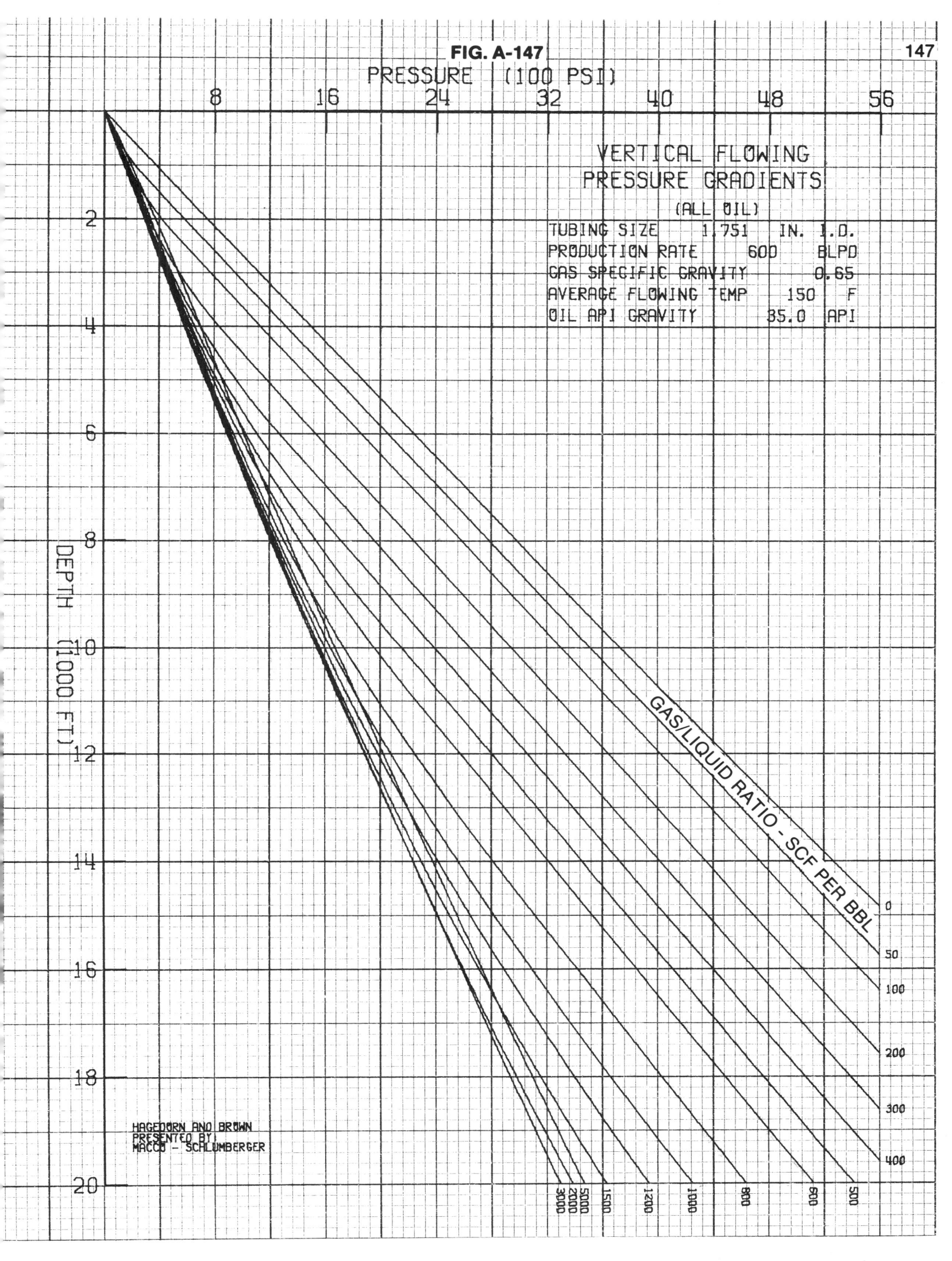

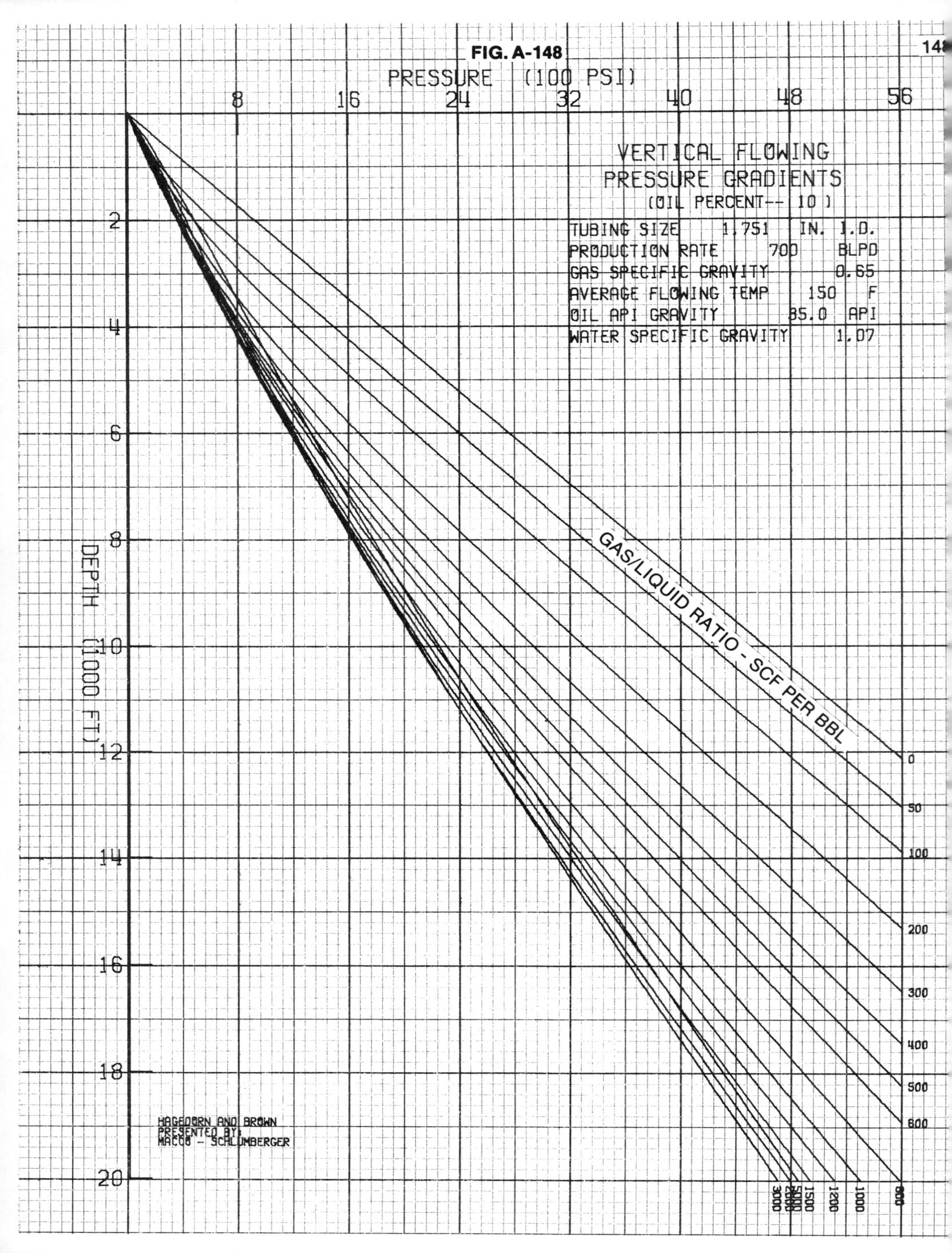

PRESSURE (100 PSI)

VERTICAL FLOWING
PRESSURE GRADIENTS
(OIL PERCENT-- 10)

TUBING SIZE	1.751	IN. I.D.
PRODUCTION RATE	700	BLPD
GAS SPECIFIC GRAVITY		0.65
AVERAGE FLOWING TEMP	150	F
OIL API GRAVITY	35.0	API
WATER SPECIFIC GRAVITY		1.07

DEPTH (1,000 FT)

GAS/LIQUID RATIO - SCF PER BBL

HAGEDORN AND BROWN
PRESENTED BY:
MACCO - SCHLUMBERGER

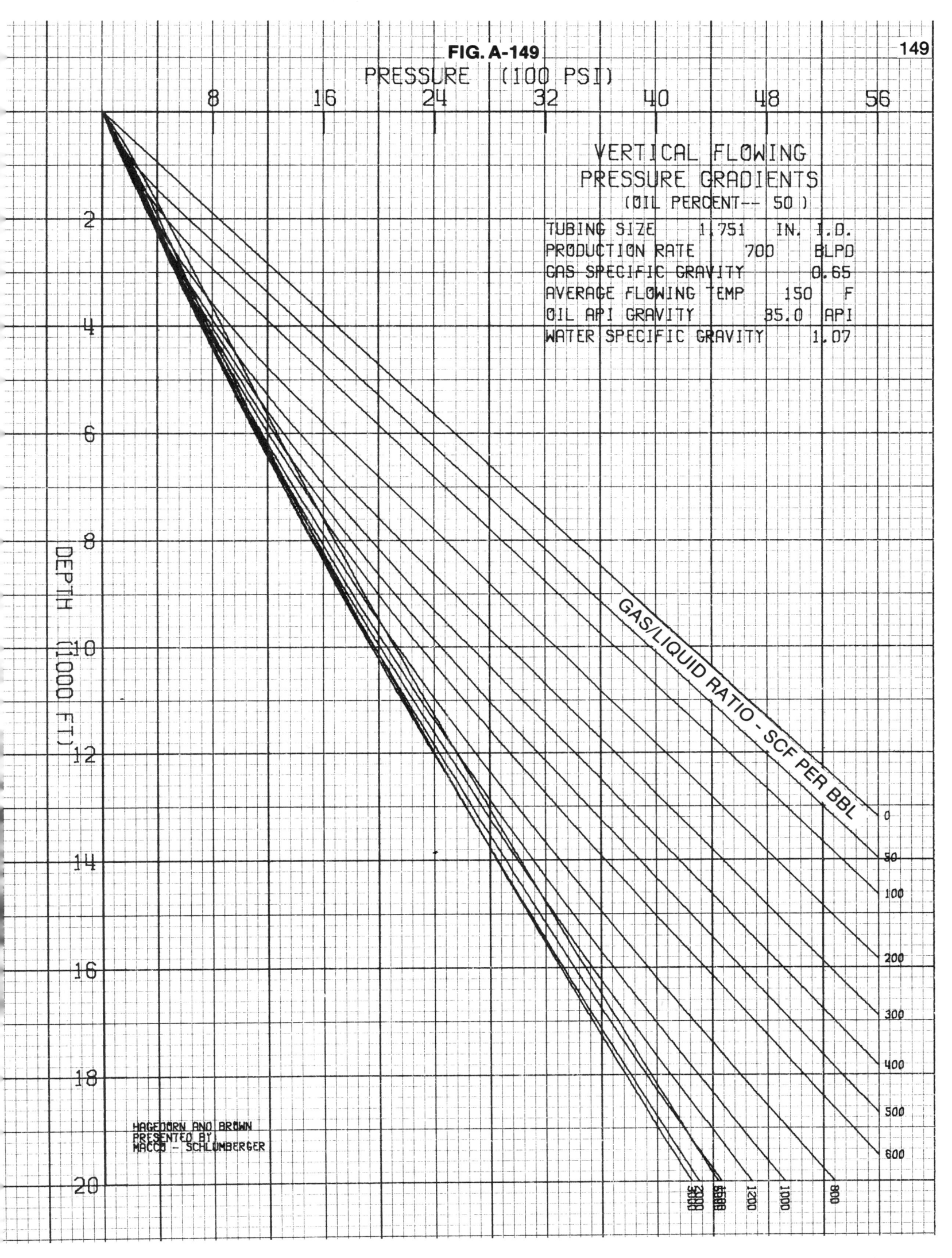

PRESSURE (100 PSI)

DEPTH (1000 FT)

VERTICAL FLOWING
PRESSURE GRADIENTS
(OIL PERCENT-- 50)

TUBING SIZE	1.751	IN. I.D.
PRODUCTION RATE	700	BLPD
GAS SPECIFIC GRAVITY		0.65
AVERAGE FLOWING TEMP	150	F
OIL API GRAVITY	35.0	API
WATER SPECIFIC GRAVITY		1.07

GAS/LIQUID RATIO - SCF PER BBL

HAGEDORN AND BROWN
PRESENTED BY
MACCO - SCHLUMBERGER

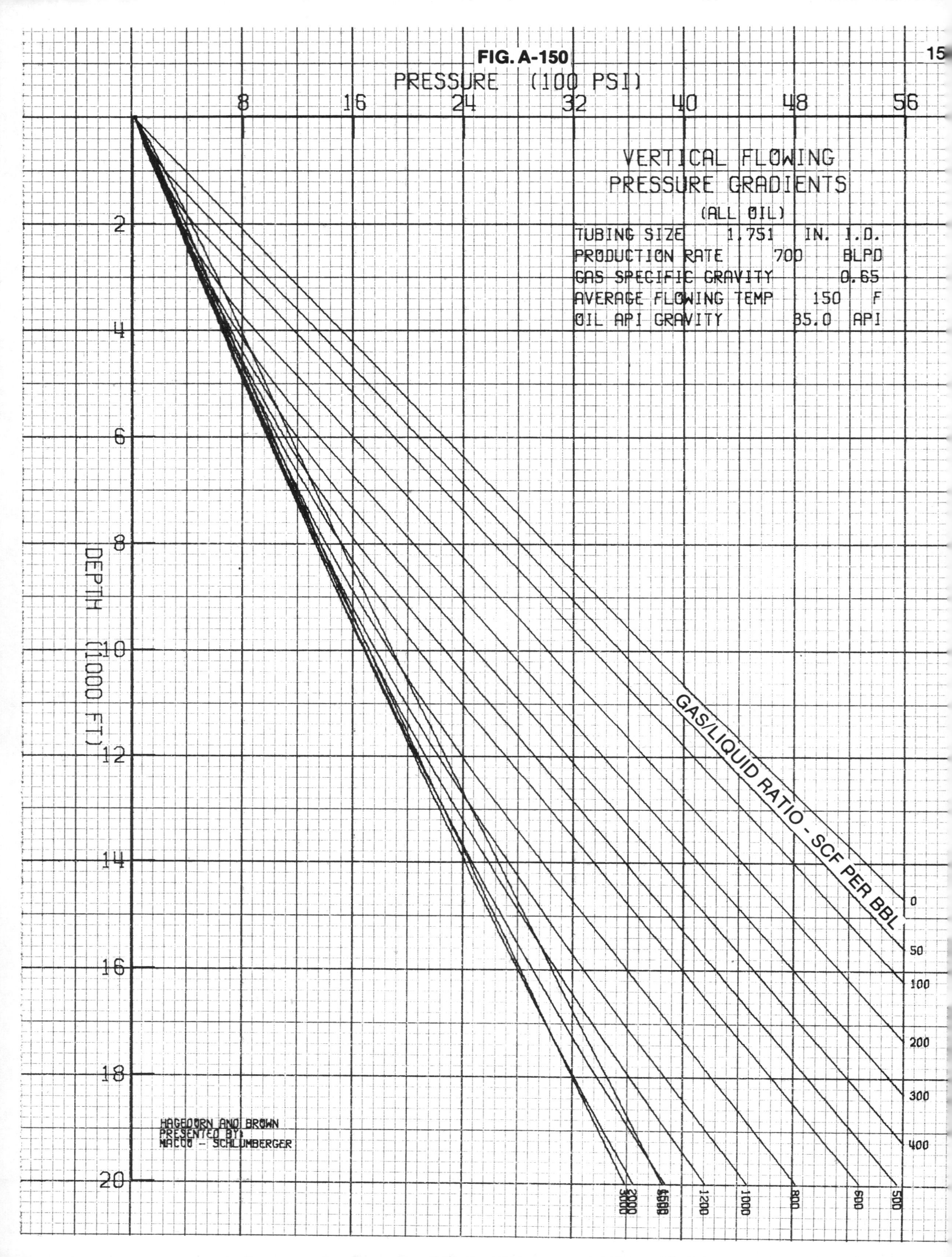

PRESSURE (100 PSI)

VERTICAL FLOWING
PRESSURE GRADIENTS
(ALL OIL)

TUBING SIZE	1.751	IN. I.D.
PRODUCTION RATE	700	BLPD
GAS SPECIFIC GRAVITY	0.65	
AVERAGE FLOWING TEMP	150	F
OIL API GRAVITY	35.0	API

DEPTH (1,000 FT)

GAS/LIQUID RATIO - SCF PER BBL

HAGEDORN AND BROWN
PRESENTED BY:
NACOO - SCHLUMBERGER

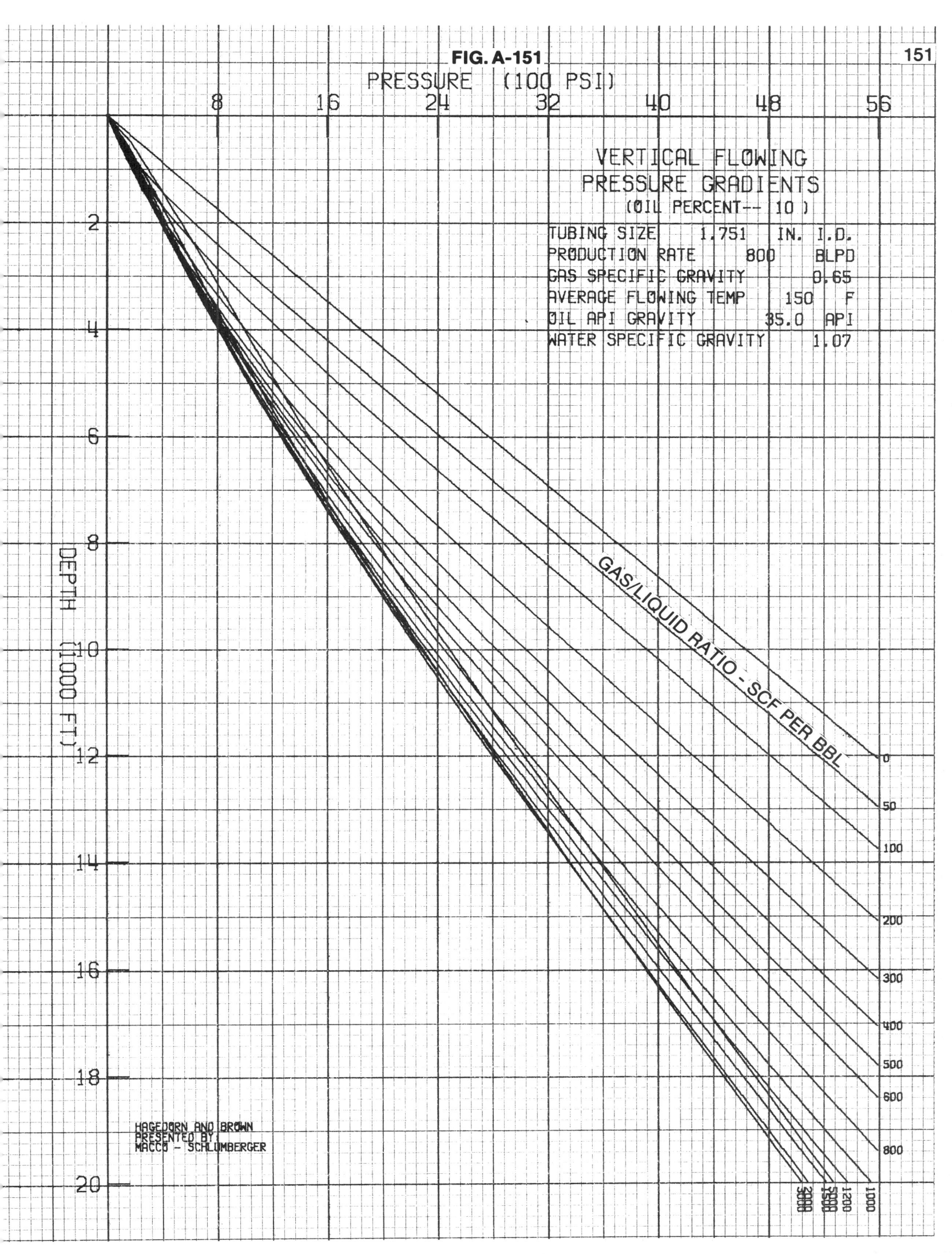

FIG. A-151

151

PRESSURE (100 PSI)

VERTICAL FLOWING
PRESSURE GRADIENTS
(OIL PERCENT-- 10)

TUBING SIZE	1.751	IN. I.D.
PRODUCTION RATE	800	BLPD
GAS SPECIFIC GRAVITY	0.65	
AVERAGE FLOWING TEMP	150	F
OIL API GRAVITY	35.0	API
WATER SPECIFIC GRAVITY	1.07	

GAS/LIQUID RATIO - SCF PER BBL

DEPTH (1000 FT)

HAGEDORN AND BROWN
PRESENTED BY
MACCO - SCHLUMBERGER

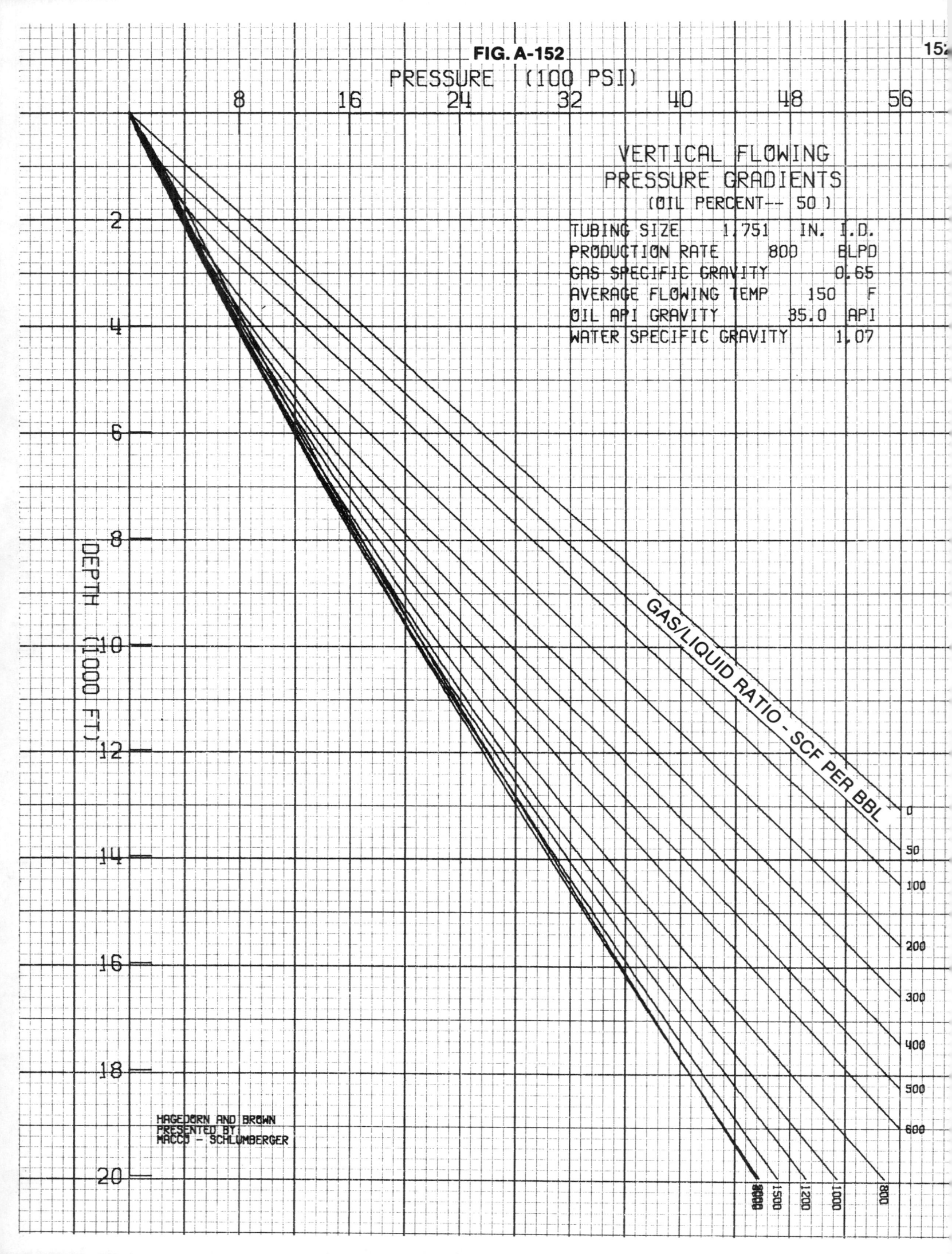

FIG. A-152

VERTICAL FLOWING
PRESSURE GRADIENTS
(OIL PERCENT—— 50)

TUBING SIZE	1.751	IN. I.D.
PRODUCTION RATE	800	BLPD
GAS SPECIFIC GRAVITY	0.65	
AVERAGE FLOWING TEMP	150	F
OIL API GRAVITY	35.0	API
WATER SPECIFIC GRAVITY	1.07	

PRESSURE (100 PSI)

DEPTH (1000 FT)

GAS/LIQUID RATIO - SCF PER BBL

HAGEDORN AND BROWN
PRESENTED BY
MACCO - SCHLUMBERGER

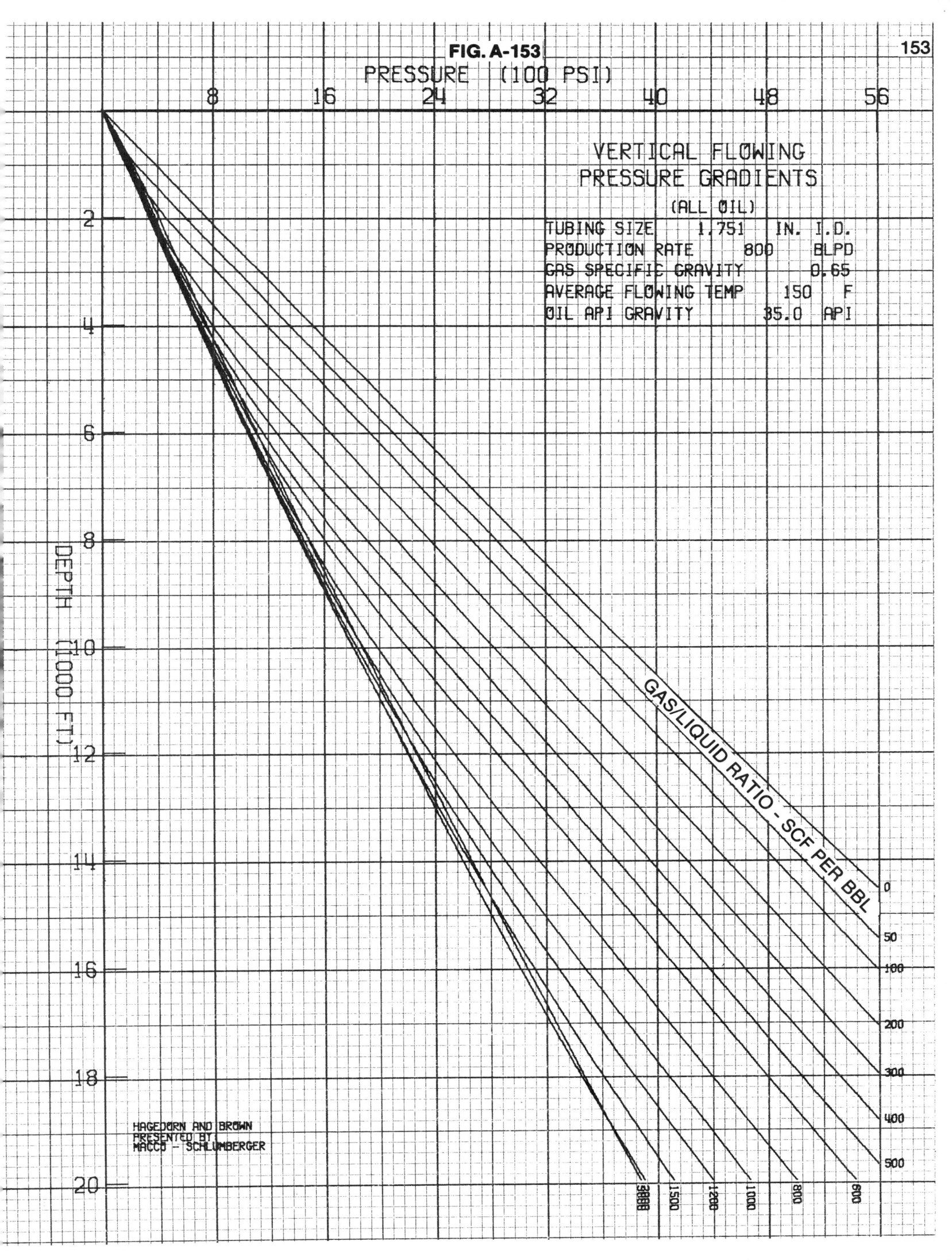

PRESSURE (100 PSI)

VERTICAL FLOWING
PRESSURE GRADIENTS
(ALL OIL)

TUBING SIZE	1.751	IN. I.D.
PRODUCTION RATE	800	BLPD
GAS SPECIFIC GRAVITY	0.65	
AVERAGE FLOWING TEMP	150	F
OIL API GRAVITY	35.0	API

DEPTH (1000 FT)

GAS/LIQUID RATIO - SCF PER BBL

0
50
100
200
300
400
500

3000
1500
1200
1000
800
600

HAGEDORN AND BROWN
PRESENTED BY
MACCO - SCHLUMBERGER

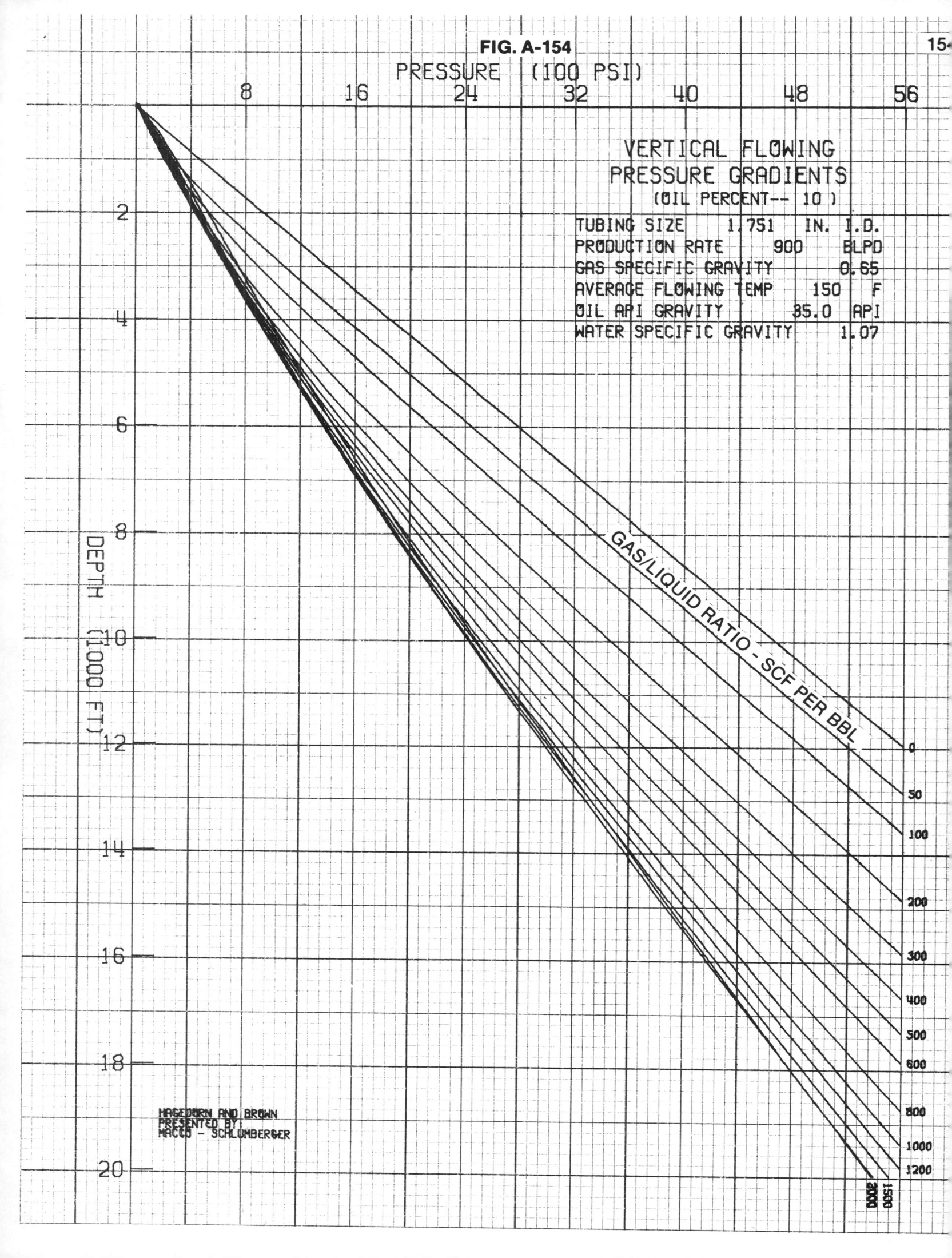

FIG. A-154

PRESSURE (100 PSI)

VERTICAL FLOWING
PRESSURE GRADIENTS
(OIL PERCENT-- 10)

TUBING SIZE	1.751	IN. I.D.
PRODUCTION RATE	900	BLPD
GAS SPECIFIC GRAVITY	0.65	
AVERAGE FLOWING TEMP	150	F
OIL API GRAVITY	35.0	API
WATER SPECIFIC GRAVITY	1.07	

DEPTH (1,000 FT)

GAS/LIQUID RATIO - SCF PER BBL

HAGEDORN AND BROWN
PRESENTED BY
MACCO - SCHLUMBERGER

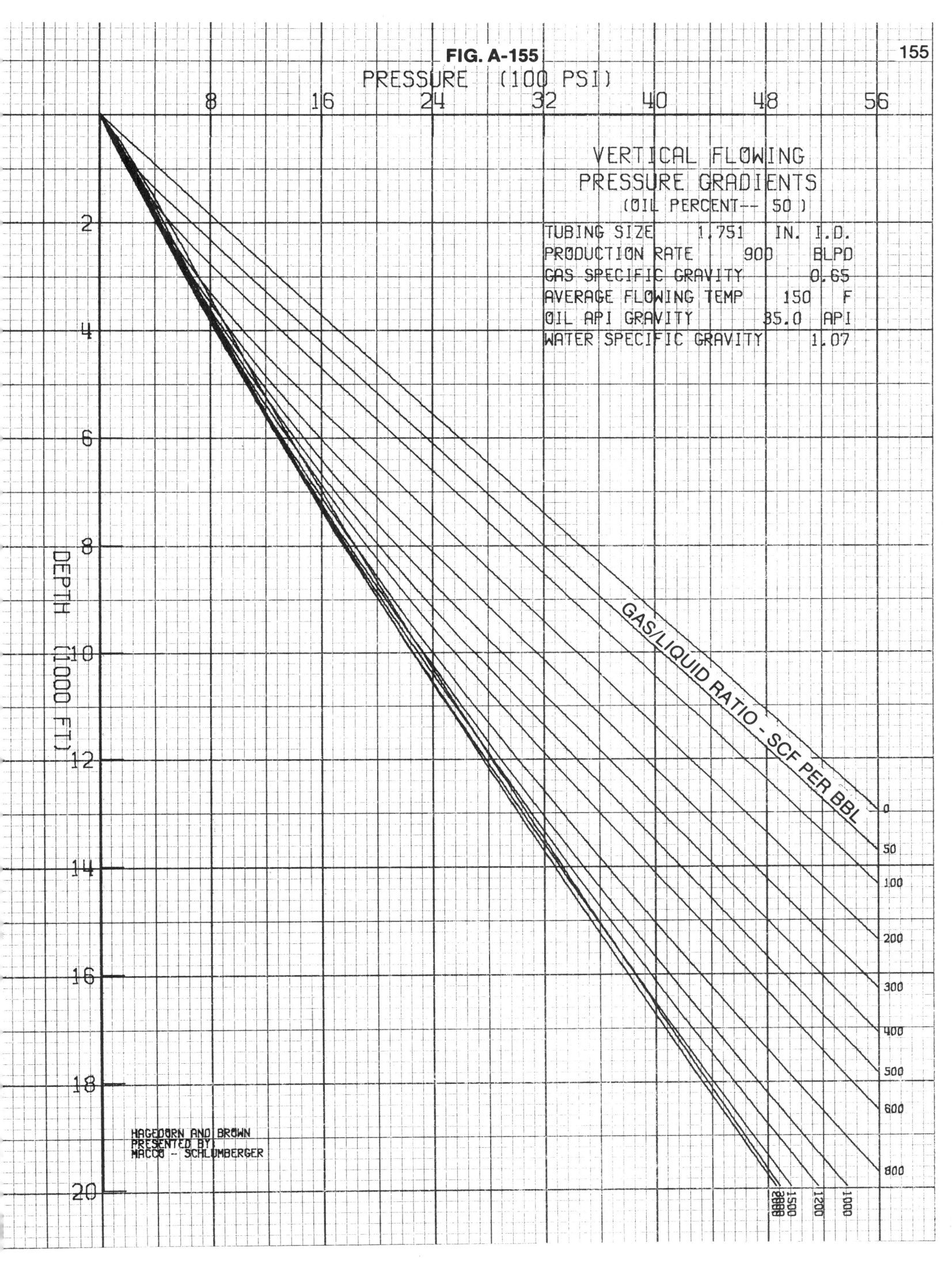

FIG. A-155

155

PRESSURE (100 PSI)

VERTICAL FLOWING
PRESSURE GRADIENTS
(OIL PERCENT-- 50)

TUBING SIZE	1.751	IN. I.D.
PRODUCTION RATE	900	BLPD
GAS SPECIFIC GRAVITY	0.65	
AVERAGE FLOWING TEMP	150	F
OIL API GRAVITY	35.0	API
WATER SPECIFIC GRAVITY	1.07	

DEPTH (1000 FT)

GAS/LIQUID RATIO - SCF PER BBL

HAGEDORN AND BROWN
PRESENTED BY
MACCO - SCHLUMBERGER

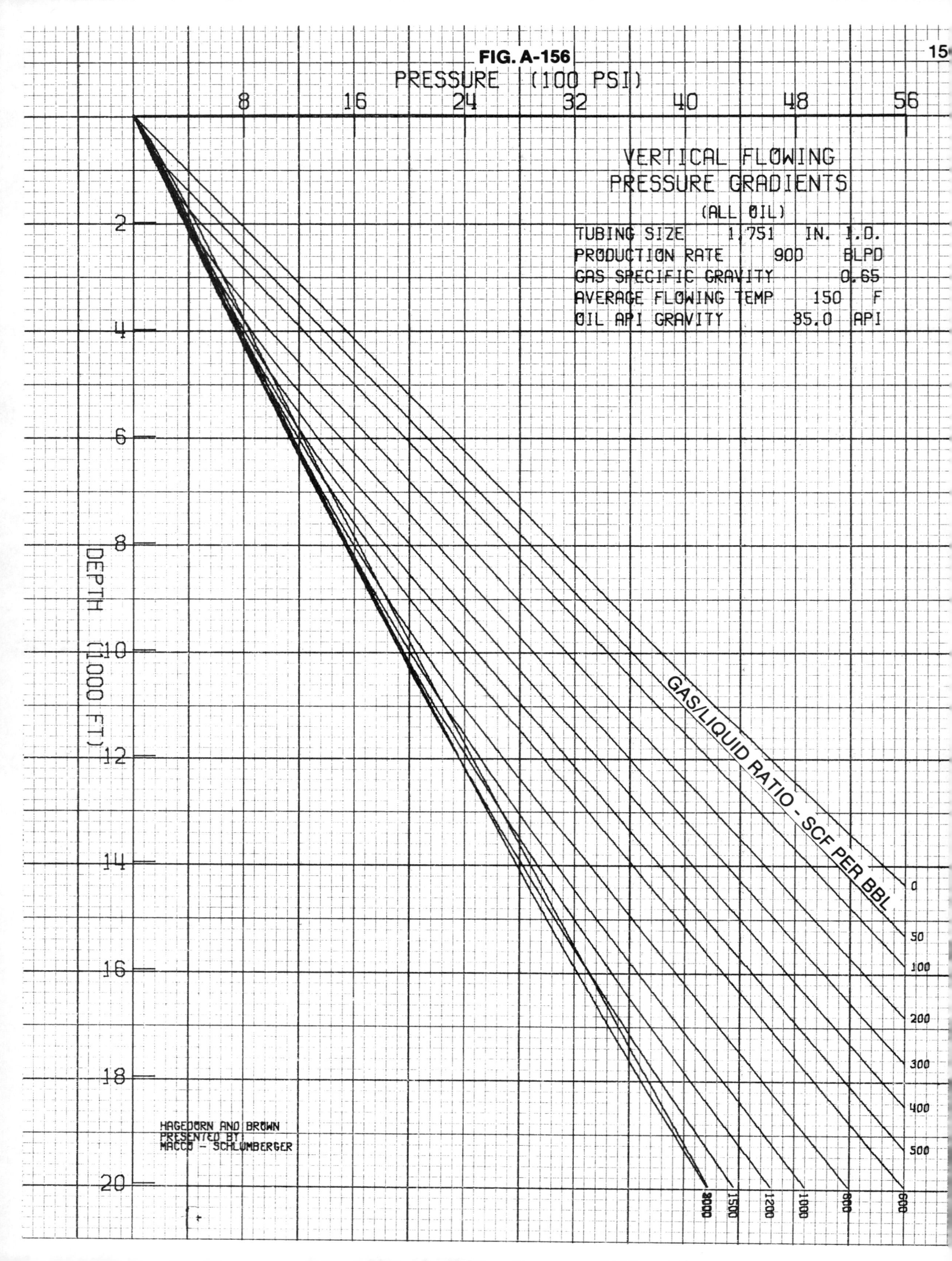

FIG. A-156

VERTICAL FLOWING
PRESSURE GRADIENTS
(ALL OIL)

TUBING SIZE	1.751	IN. I.D.
PRODUCTION RATE	900	BLPD
GAS SPECIFIC GRAVITY	0.65	
AVERAGE FLOWING TEMP	150	F
OIL API GRAVITY	35.0	API

PRESSURE (100 PSI)

DEPTH (1,000 FT)

GAS/LIQUID RATIO - SCF PER BBL

HAGEDORN AND BROWN
PRESENTED BY
MACCO - SCHLUMBERGER

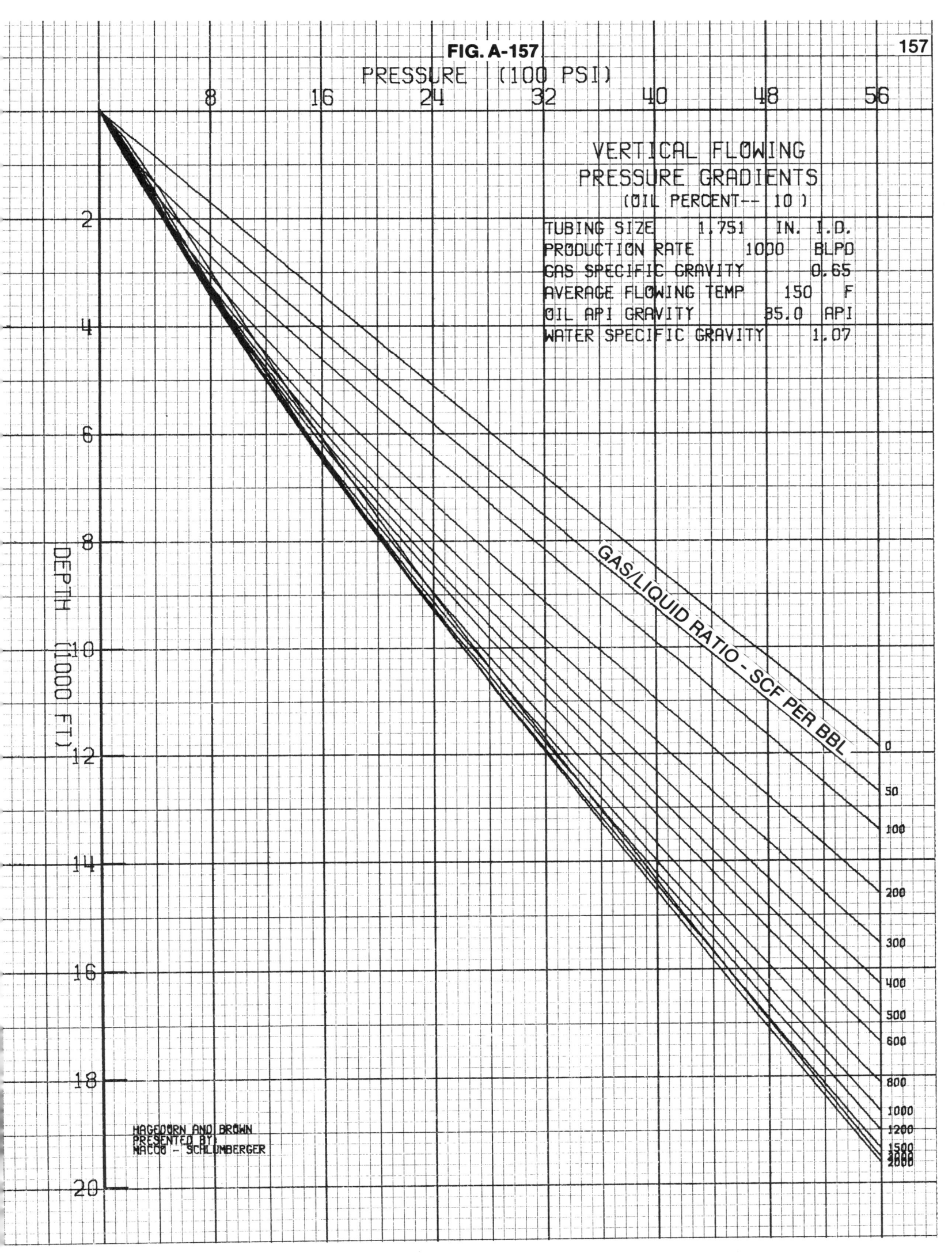

PRESSURE (100 PSI)

DEPTH (1000 FT)

VERTICAL FLOWING
PRESSURE GRADIENTS
(OIL PERCENT-- 10)

TUBING SIZE	1.751	IN. I.D.
PRODUCTION RATE	1000	BLPD
GAS SPECIFIC GRAVITY		0.65
AVERAGE FLOWING TEMP	150	F
OIL API GRAVITY	35.0	API
WATER SPECIFIC GRAVITY		1.07

GAS/LIQUID RATIO - SCF PER BBL

HAGEDORN AND BROWN
PRESENTED BY
NACOO - SCHLUMBERGER

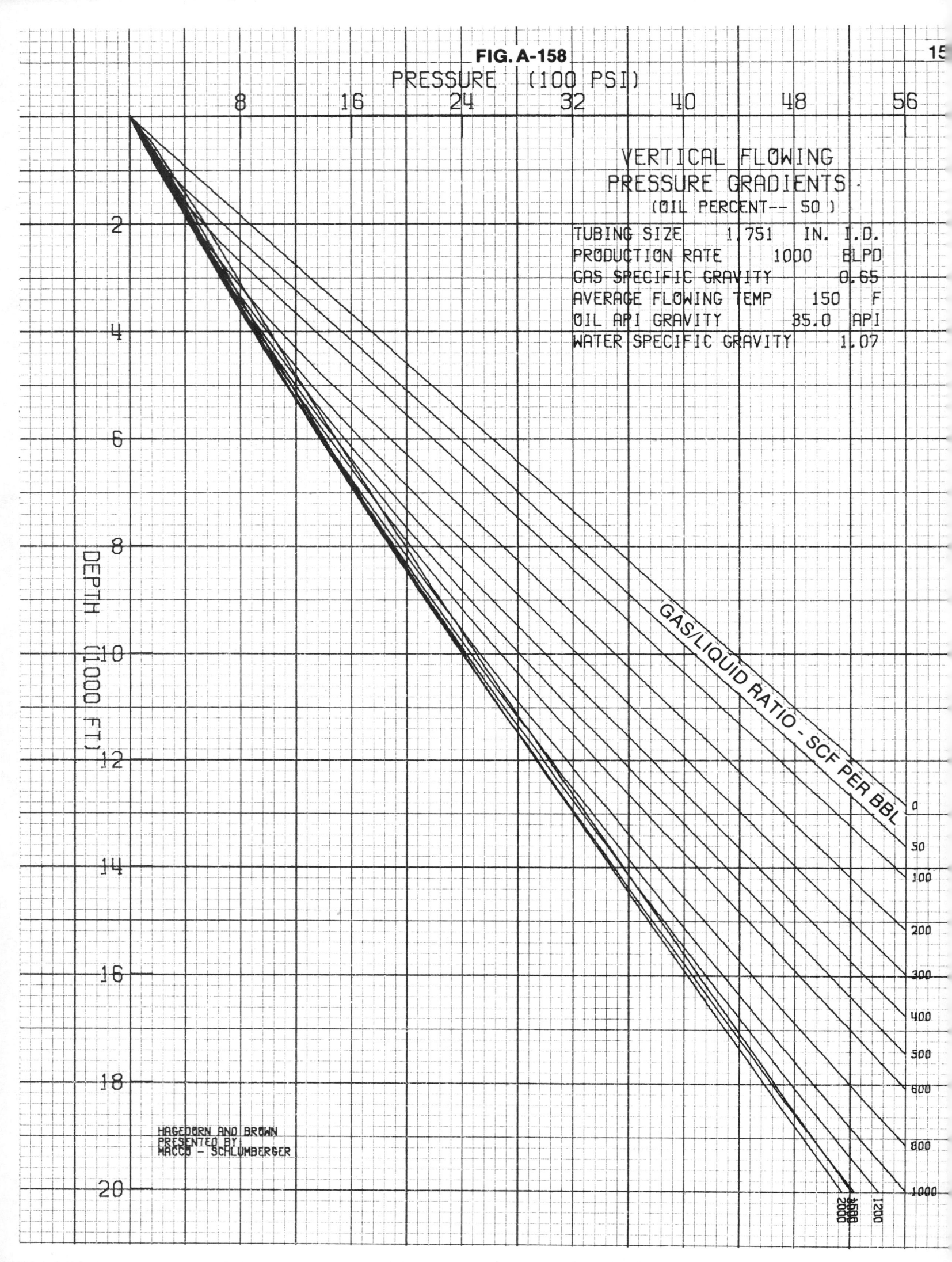

FIG. A-158

VERTICAL FLOWING
PRESSURE GRADIENTS
(OIL PERCENT -- 50)

TUBING SIZE	1.751	IN. I.D.
PRODUCTION RATE	1000	BLPD
GAS SPECIFIC GRAVITY		0.65
AVERAGE FLOWING TEMP	150	F
OIL API GRAVITY	35.0	API
WATER SPECIFIC GRAVITY		1.07

PRESSURE (100 PSI)

DEPTH (1000 FT)

GAS/LIQUID RATIO - SCF PER BBL

HAGEDORN AND BROWN
PRESENTED BY
MACCO - SCHLUMBERGER

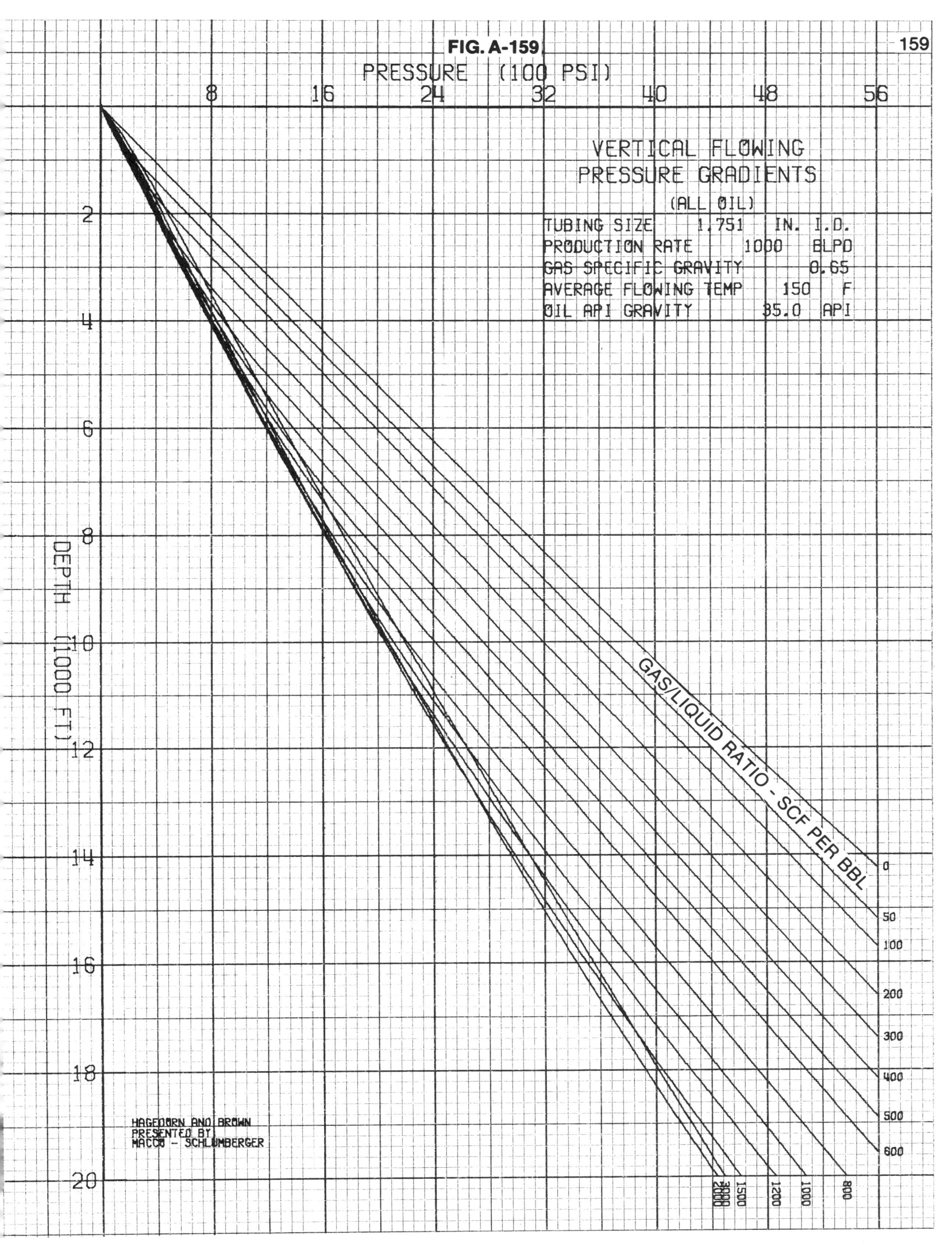

PRESSURE (100 PSI)

VERTICAL FLOWING
PRESSURE GRADIENTS
(ALL OIL)

TUBING SIZE	1.751	IN. I.D.
PRODUCTION RATE	1000	BLPD
GAS SPECIFIC GRAVITY	0.65	
AVERAGE FLOWING TEMP	150	F
OIL API GRAVITY	35.0	API

DEPTH (1000 FT)

GAS/LIQUID RATIO - SCF PER BBL

0

50

100

200

300

400

500

600

HAGEDORN AND BROWN
PRESENTED BY
MACCO — SCHLUMBERGER

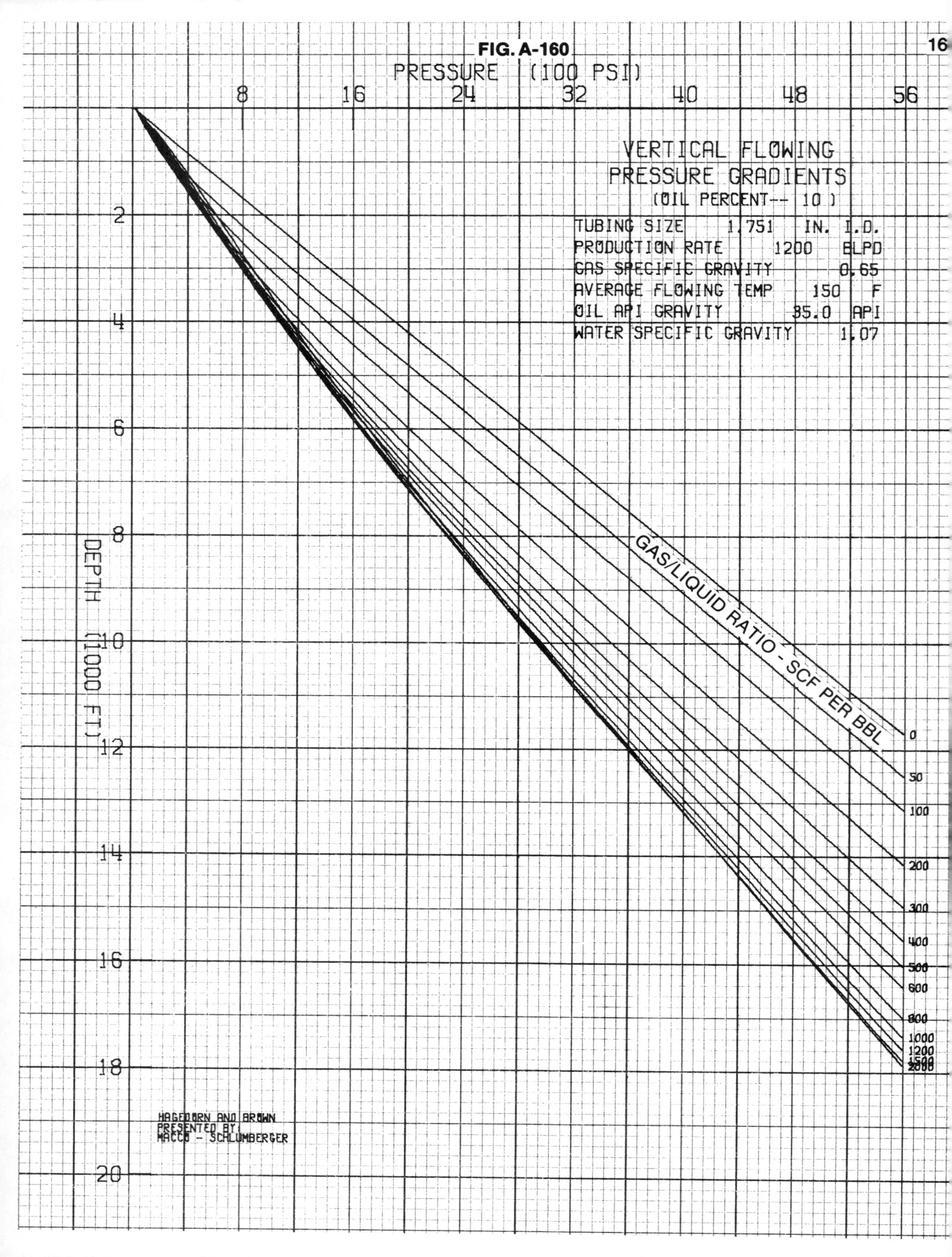

PRESSURE (100 PSI)

VERTICAL FLOWING
PRESSURE GRADIENTS
(OIL PERCENT-- 10)

TUBING SIZE	1.751	IN. I.D.
PRODUCTION RATE	1200	BLPD
GAS SPECIFIC GRAVITY	0.65	
AVERAGE FLOWING TEMP	150	F
OIL API GRAVITY	35.0	API
WATER SPECIFIC GRAVITY	1.07	

GAS/LIQUID RATIO - SCF PER BBL

DEPTH (1000 FT)

HAGEDORN AND BROWN
PRESENTED BY
MACCO - SCHLUMBERGER

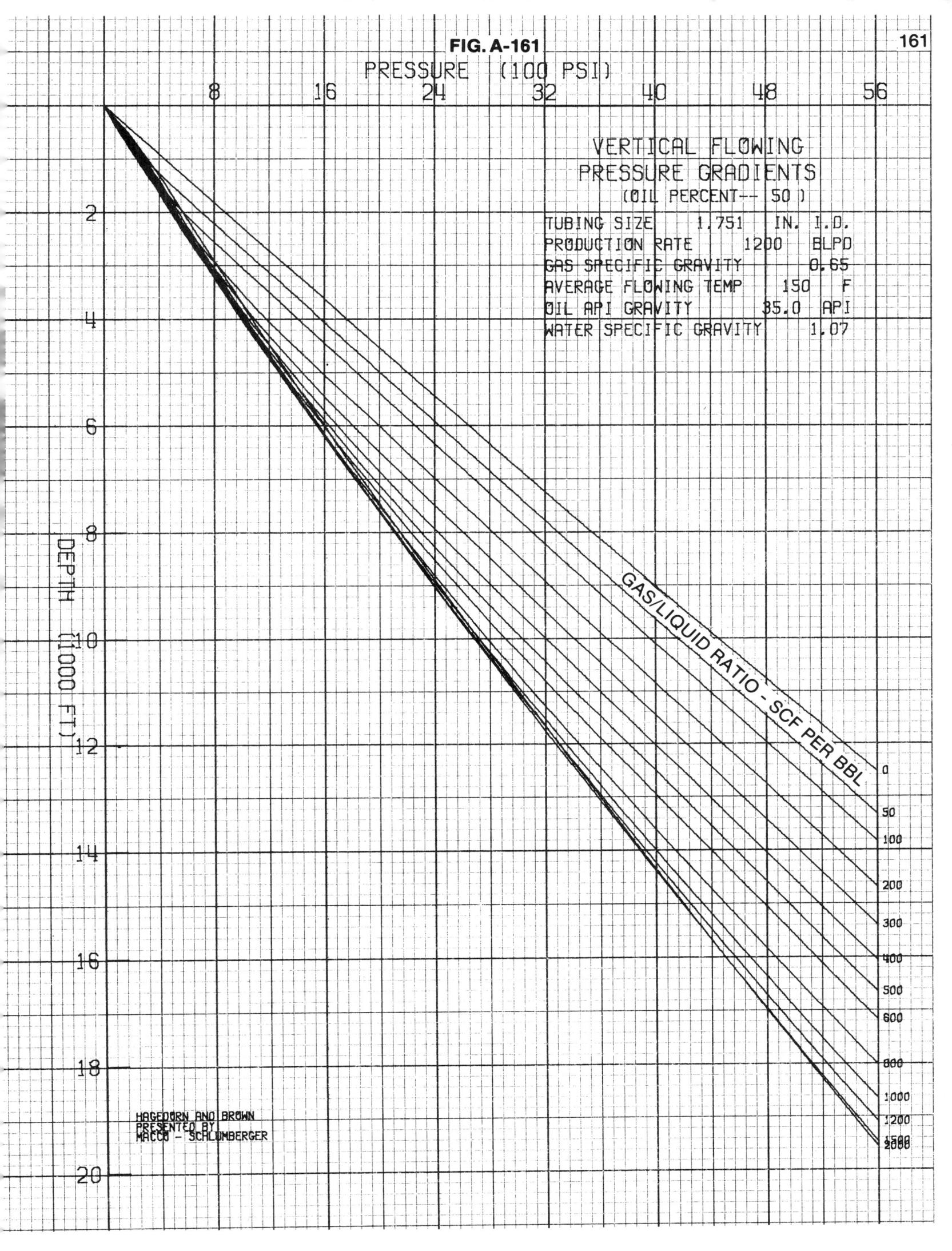

PRESSURE (100 PSI)

DEPTH (1000 FT)

VERTICAL FLOWING
PRESSURE GRADIENTS
(OIL PERCENT--- 50)

TUBING SIZE	1.751	IN. I.D.
PRODUCTION RATE	1200	BLPD
GAS SPECIFIC GRAVITY	0.65	
AVERAGE FLOWING TEMP	150	F
OIL API GRAVITY	35.0	API
WATER SPECIFIC GRAVITY	1.07	

GAS/LIQUID RATIO - SCF PER BBL

HAGEDORN AND BROWN
PRESENTED BY
MACCO - SCHLUMBERGER

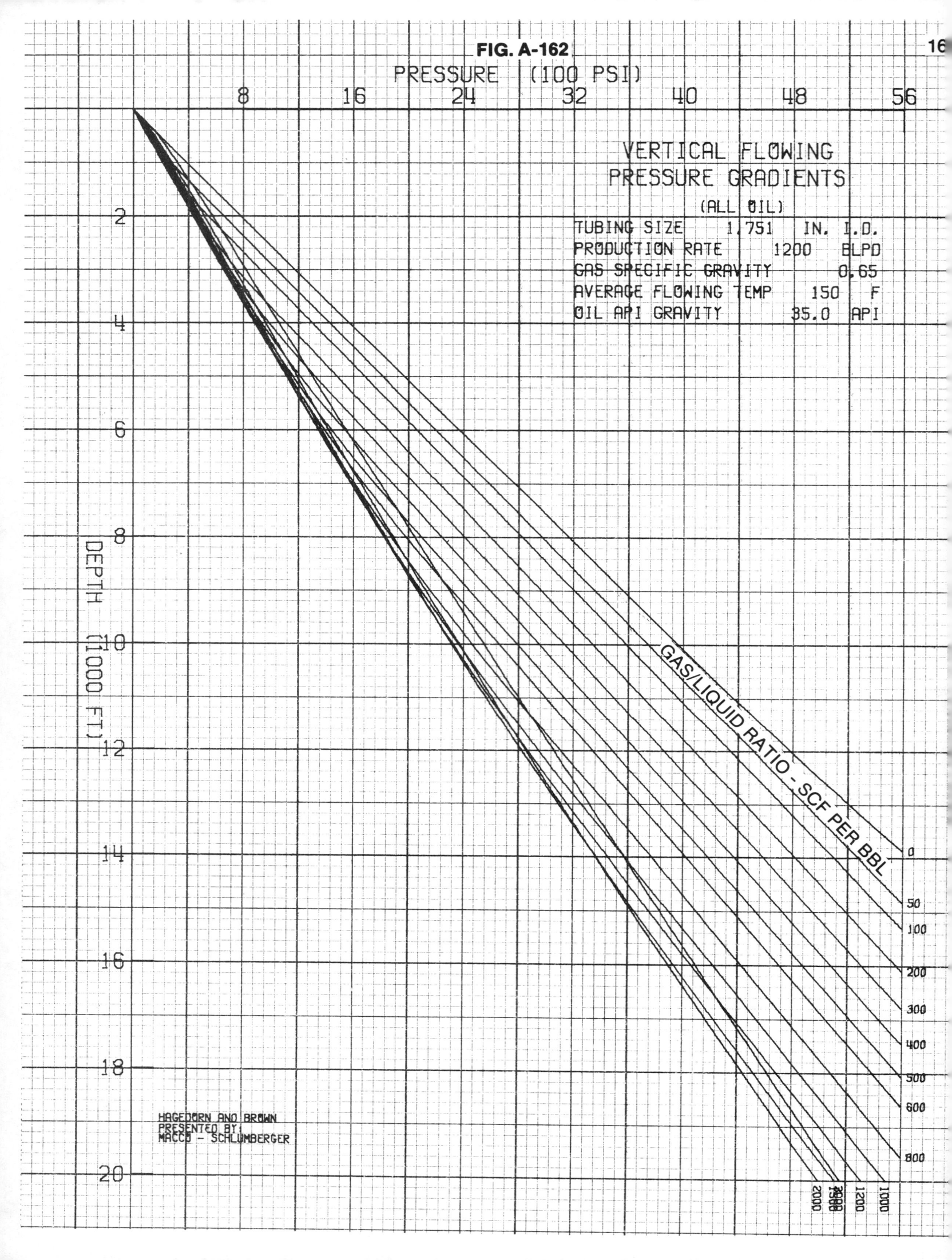

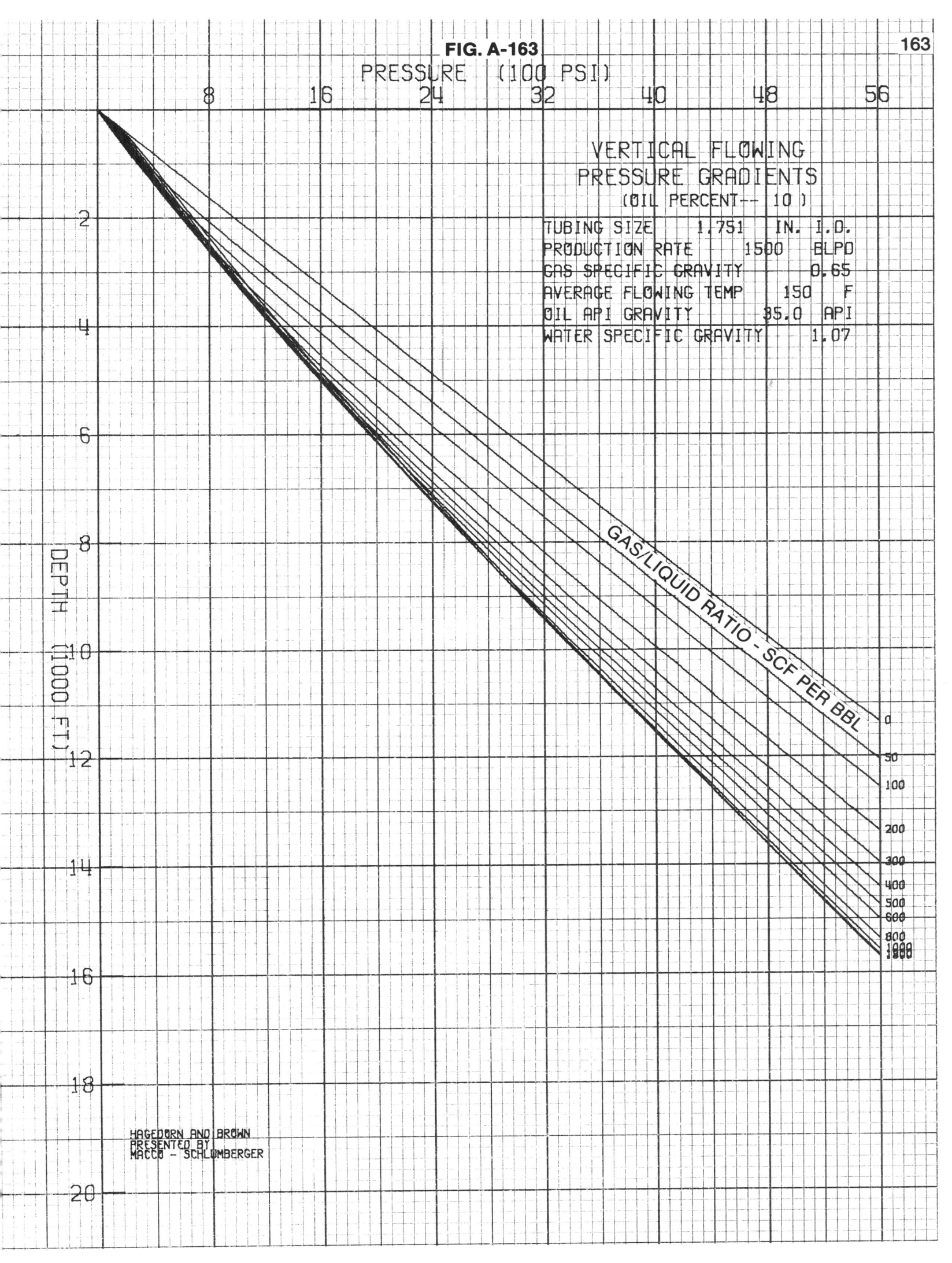

FIG. A-163

163

PRESSURE (100 PSI)

VERTICAL FLOWING
PRESSURE GRADIENTS
(OIL PERCENT-- 10)

TUBING SIZE	1.751	IN. I.D.
PRODUCTION RATE	1500	BLPD
GAS SPECIFIC GRAVITY	0.65	
AVERAGE FLOWING TEMP	150	F
OIL API GRAVITY	35.0	API
WATER SPECIFIC GRAVITY	1.07	

GAS/LIQUID RATIO - SCF PER BBL

DEPTH (1000 FT)

HAGEDORN AND BROWN
PRESENTED BY
MAECO - SCHLUMBERGER

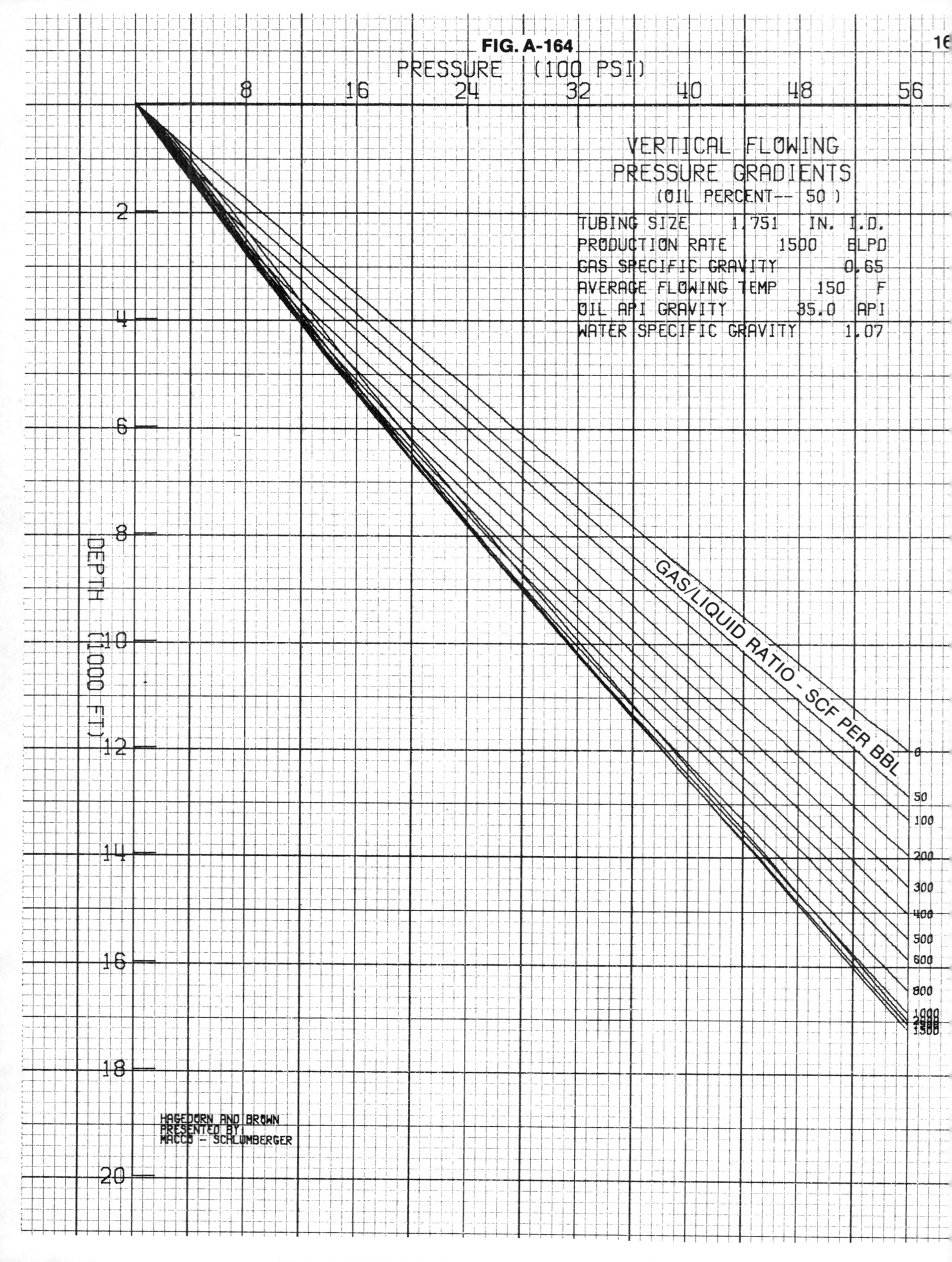

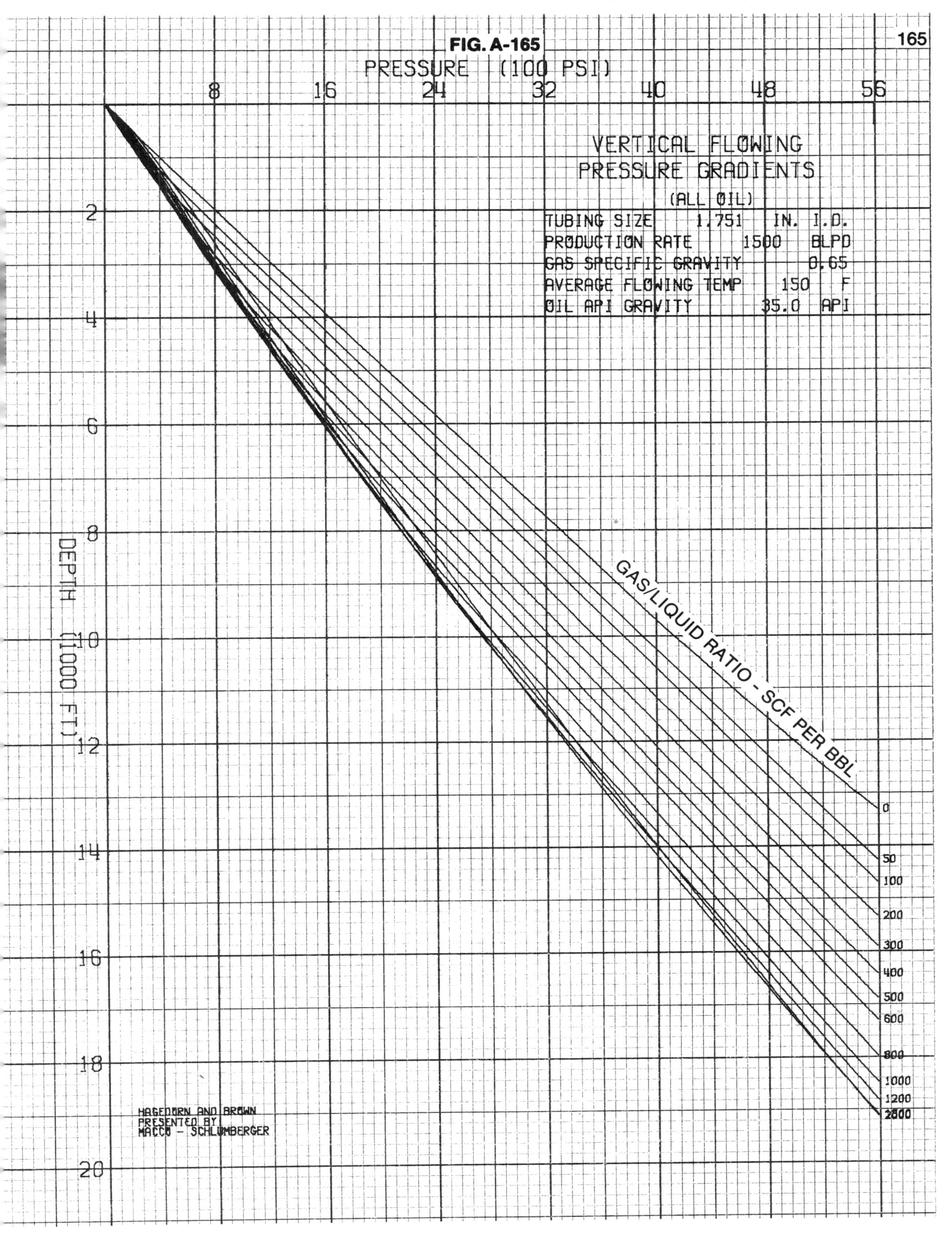

FIG. A-165

165

PRESSURE (100 PSI)

VERTICAL FLOWING
PRESSURE GRADIENTS
(ALL OIL)

TUBING SIZE	1.751	IN. I.D.
PRODUCTION RATE	1500	BLPD
GAS SPECIFIC GRAVITY	0.65	
AVERAGE FLOWING TEMP	150	F
OIL API GRAVITY	35.0	API

DEPTH (1000 FT)

GAS/LIQUID RATIO - SCF PER BBL

0
50
100
200
300
400
500
600
800
1000
1200
2500

HAGEDORN AND BROWN
PRESENTED BY
MACCO - SCHLUMBERGER

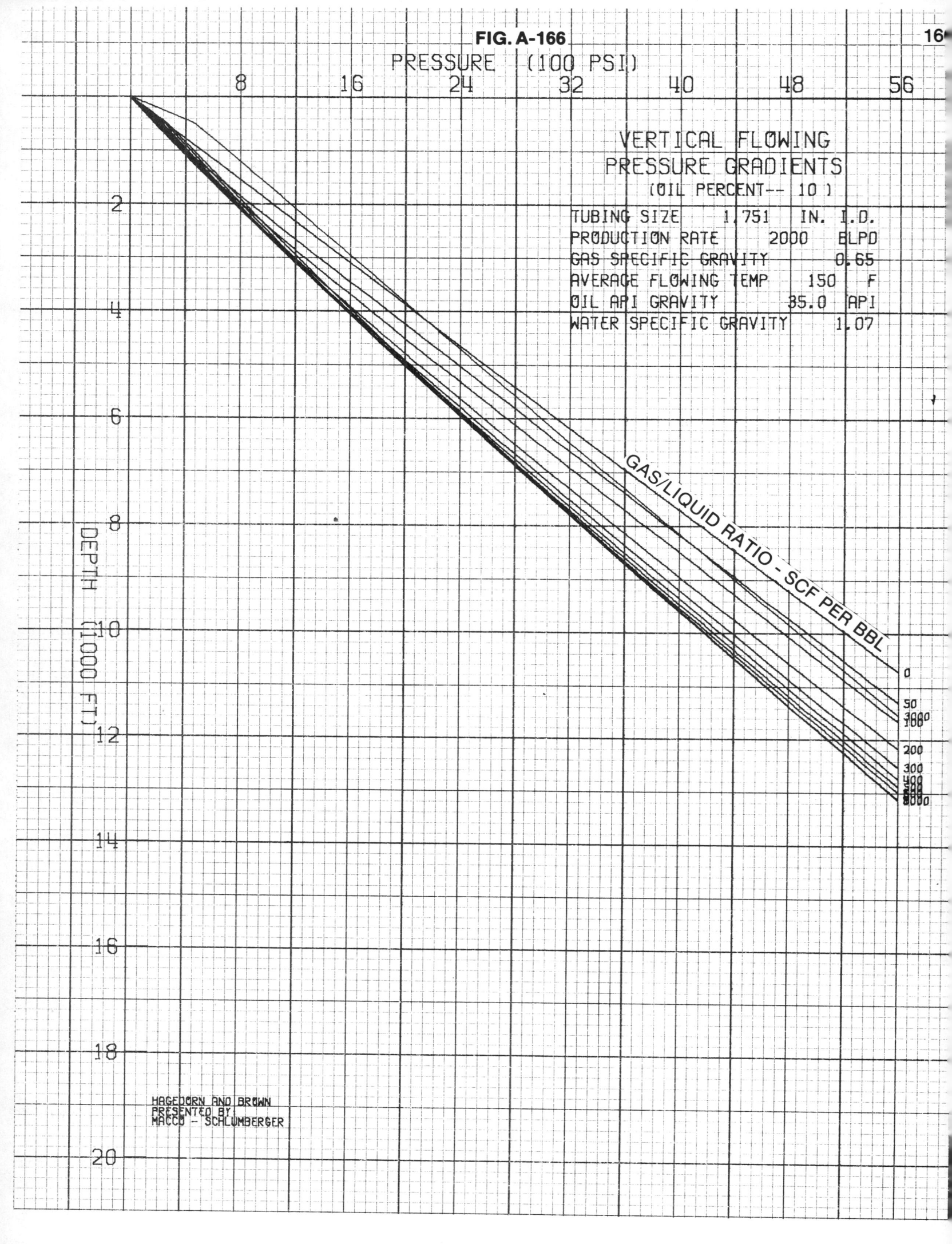

FIG. A-166

VERTICAL FLOWING
PRESSURE GRADIENTS
(OIL PERCENT-- 10)

TUBING SIZE	1.751	IN. I.D.
PRODUCTION RATE	2000	BLPD
GAS SPECIFIC GRAVITY	0.65	
AVERAGE FLOWING TEMP	150	F
OIL API GRAVITY	35.0	API
WATER SPECIFIC GRAVITY	1.07	

PRESSURE (100 PSI)

DEPTH (1000 FT)

GAS/LIQUID RATIO - SCF PER BBL

0
50
100
200
300
400

HAGEDORN AND BROWN
PRESENTED BY
MACCO - SCHLUMBERGER

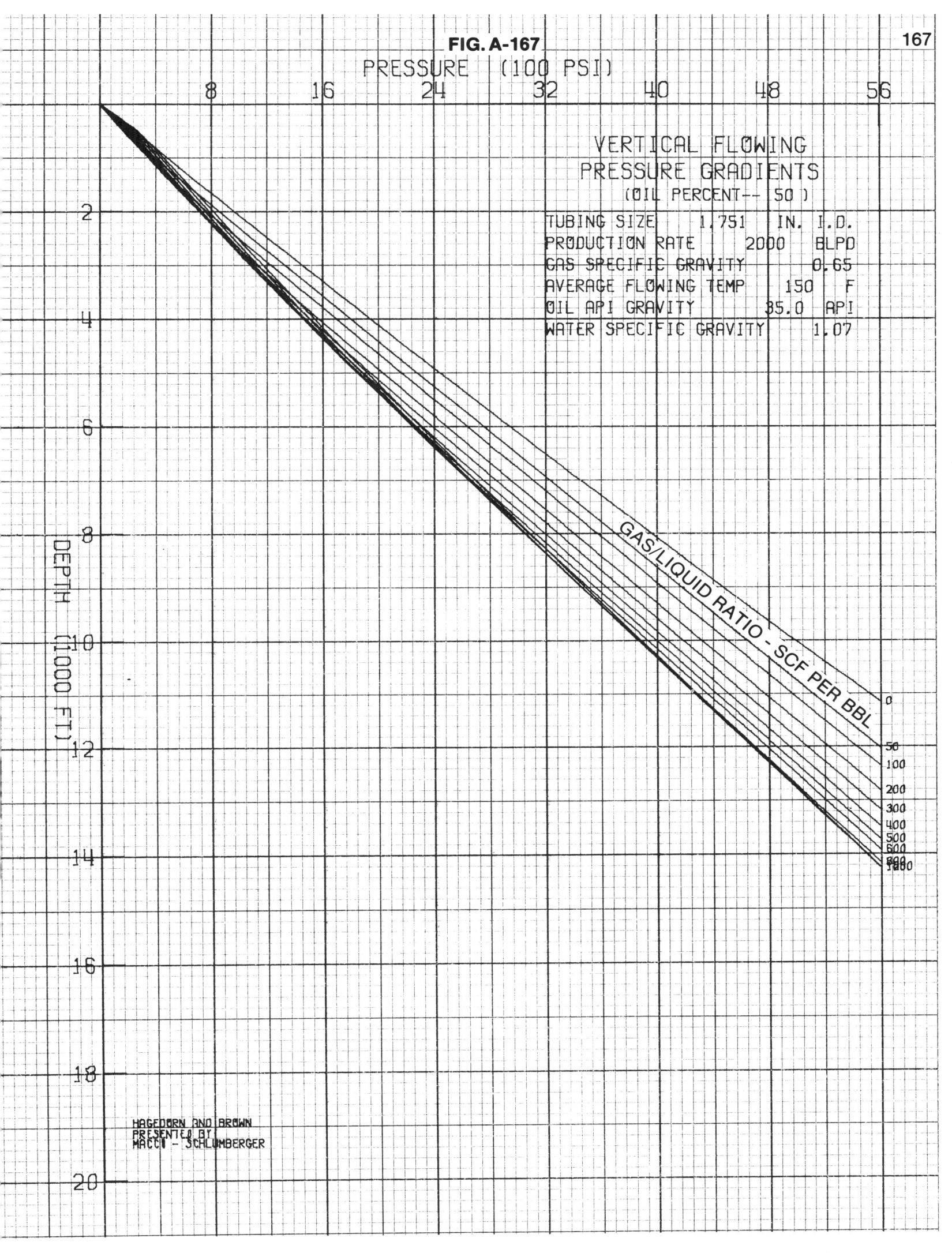

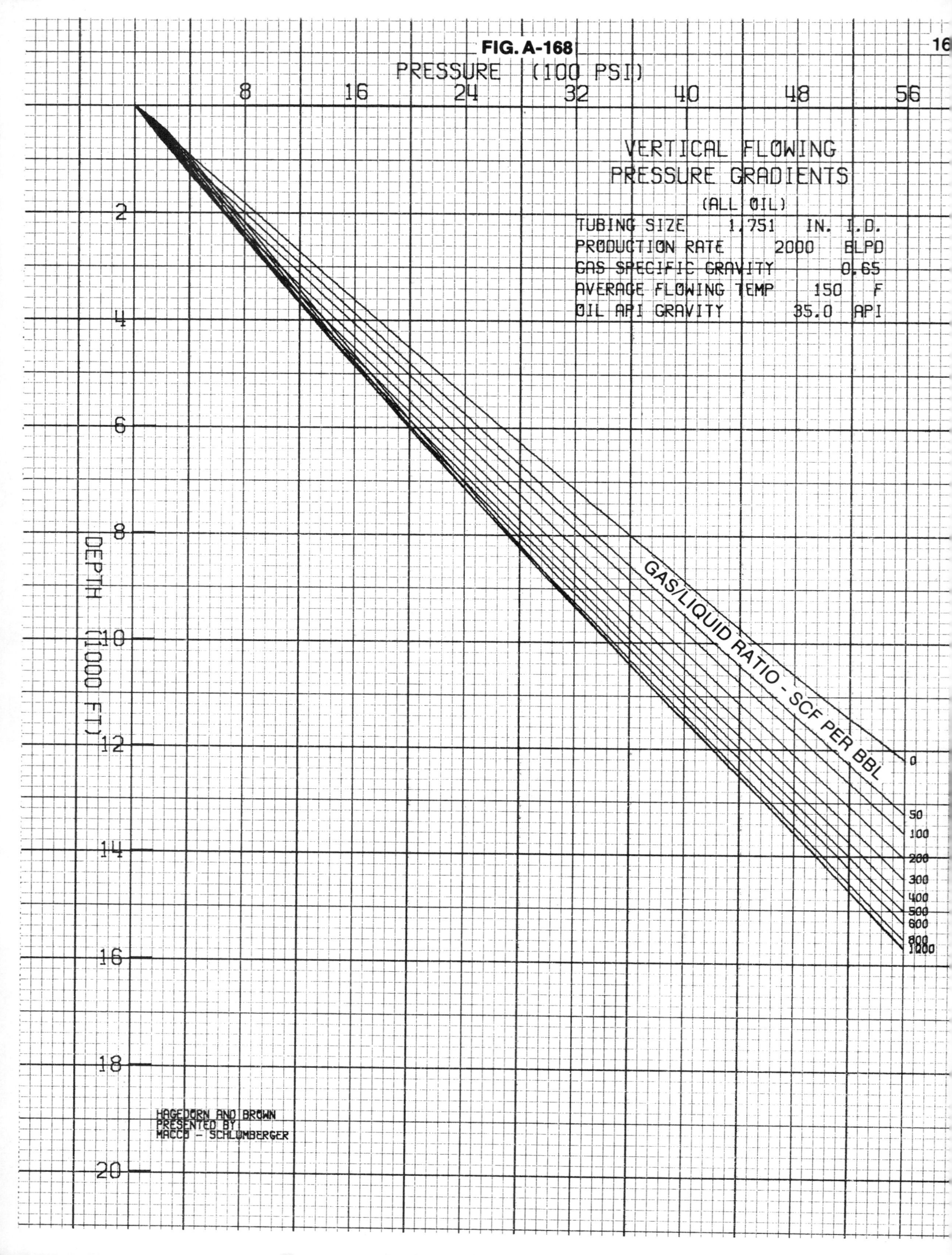

PRESSURE (100 PSI)

DEPTH (1000 FT)

VERTICAL FLOWING
PRESSURE GRADIENTS
(ALL OIL)

TUBING SIZE	1.751	IN. I.D.
PRODUCTION RATE	2000	BLPD
GAS SPECIFIC GRAVITY		0.65
AVERAGE FLOWING TEMP	150	F
OIL API GRAVITY	35.0	API

GAS/LIQUID RATIO - SCF PER BBL

HAGEDORN AND BROWN
PRESENTED BY
MACCO - SCHLUMBERGER

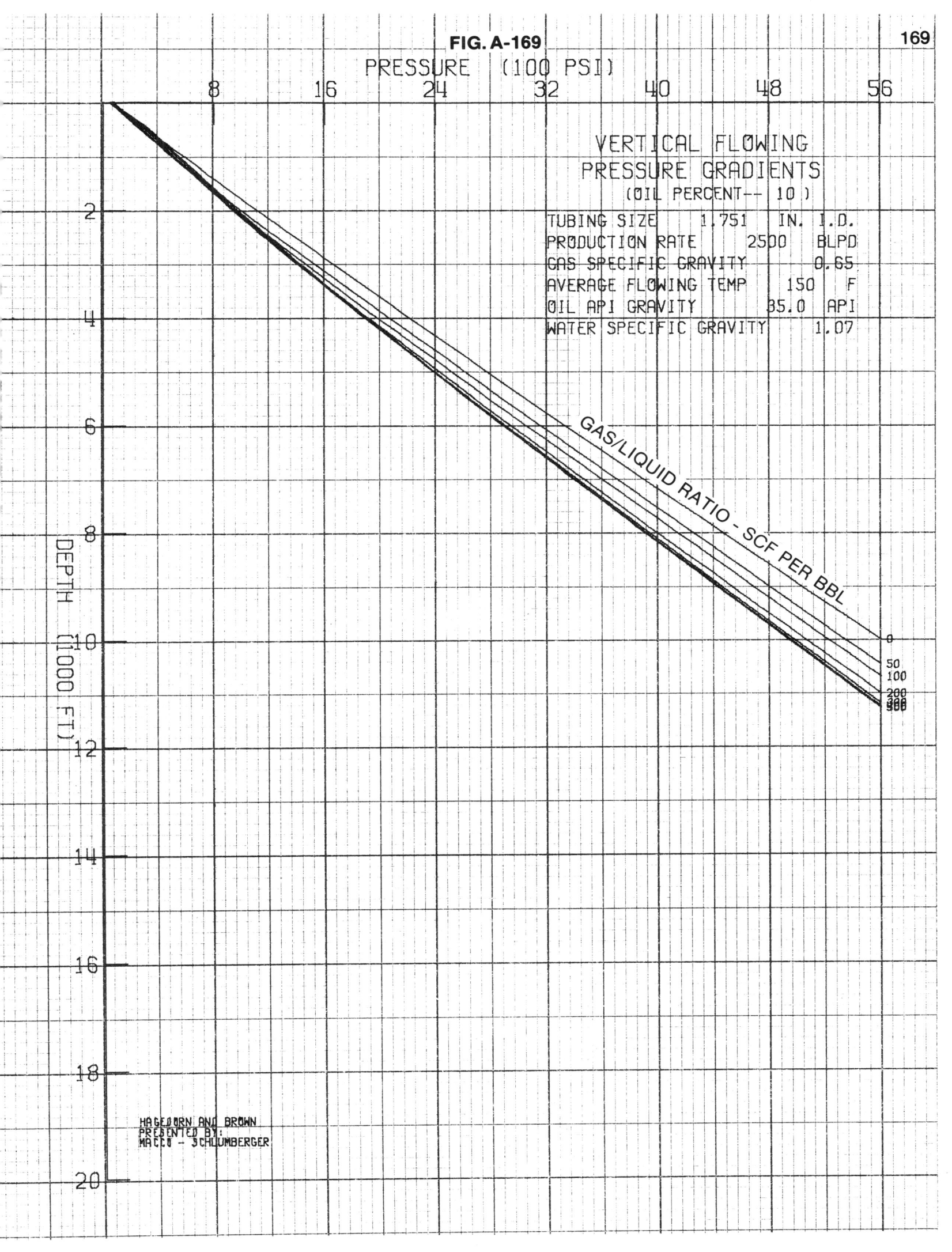

FIG. A-169

PRESSURE (100 PSI)

VERTICAL FLOWING
PRESSURE GRADIENTS
(OIL PERCENT-- 10)

TUBING SIZE	1.751	IN. I.D.
PRODUCTION RATE	2500	BLPD
GAS SPECIFIC GRAVITY	0.65	
AVERAGE FLOWING TEMP	150	F
OIL API GRAVITY	35.0	API
WATER SPECIFIC GRAVITY	1.07	

DEPTH (1000 FT)

GAS/LIQUID RATIO - SCF PER BBL

0
50
100
200
300

HAGEDORN AND BROWN
PRESENTED BY:
MACCO - SCHLUMBERGER

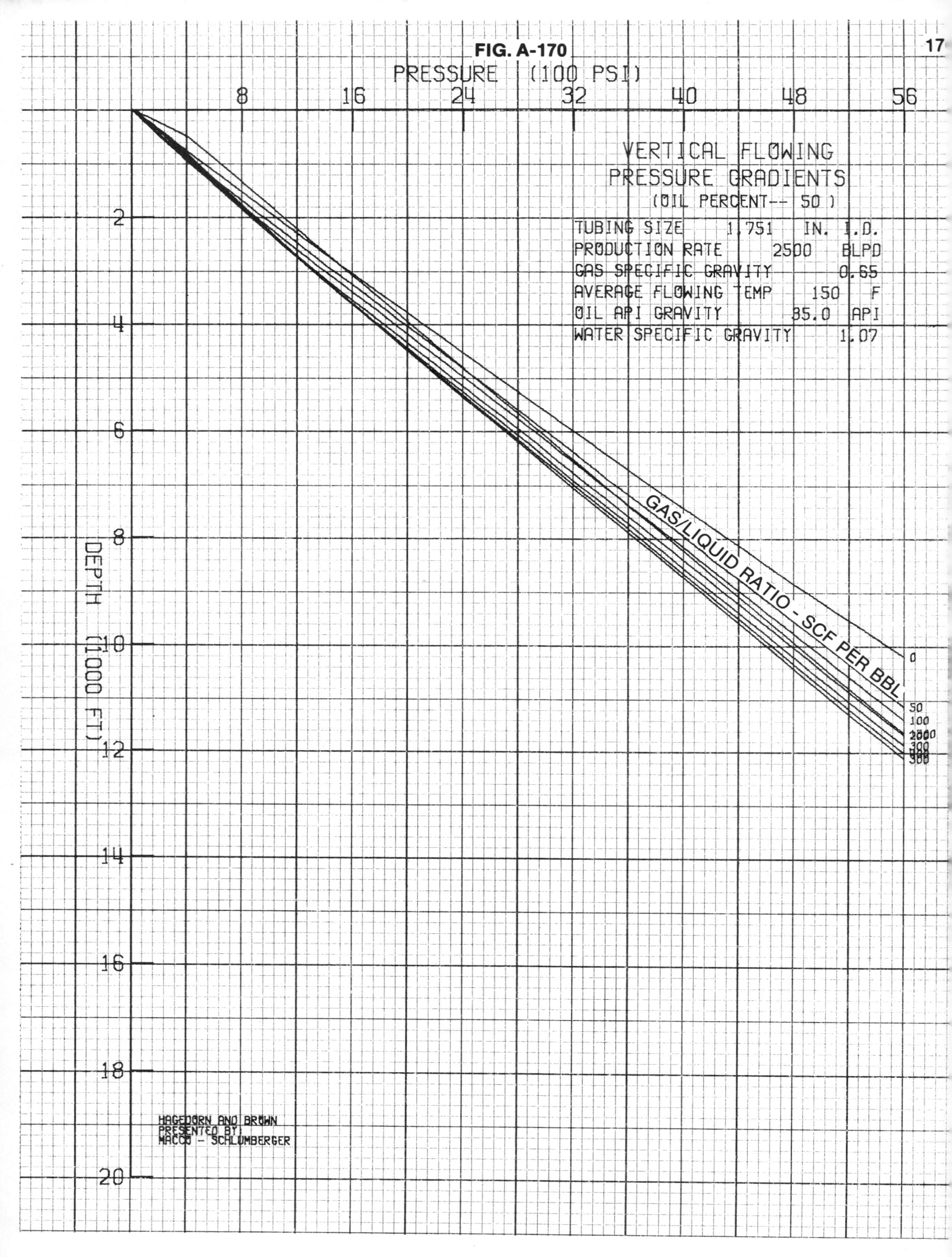

PRESSURE (100 PSI)

VERTICAL FLOWING
PRESSURE GRADIENTS
(OIL PERCENT-- 50)

TUBING SIZE	1.751	IN. I.D.
PRODUCTION RATE	2500	BLPD
GAS SPECIFIC GRAVITY	0.65	
AVERAGE FLOWING TEMP	150	F
OIL API GRAVITY	35.0	API
WATER SPECIFIC GRAVITY	1.07	

DEPTH (1,000 FT)

GAS/LIQUID RATIO - SCF PER BBL

0
50
100
200
150
300
500

HAGEDORN AND BROWN
PRESENTED BY
MACCO - SCHLUMBERGER

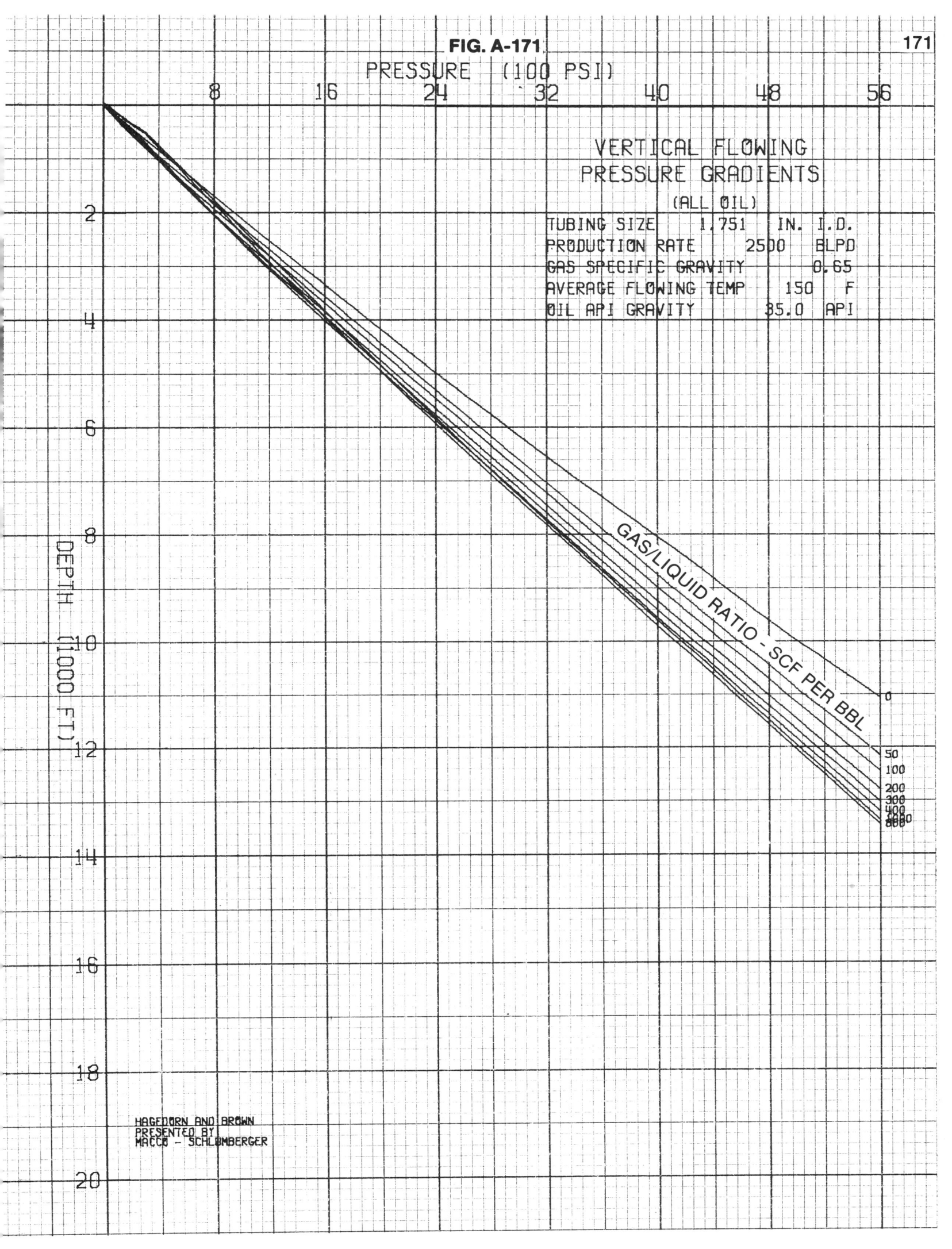

VERTICAL FLOWING
PRESSURE GRADIENTS
(ALL OIL)

TUBING SIZE	1.751	IN. I.D.
PRODUCTION RATE	2500	BLPD
GAS SPECIFIC GRAVITY	0.65	
AVERAGE FLOWING TEMP	150	F
OIL API GRAVITY	35.0	API

PRESSURE (100 PSI)

DEPTH (1000 FT)

GAS/LIQUID RATIO - SCF PER BBL

0
50
100
200
300
400
600
800

HAGEDORN AND BROWN
PRESENTED BY
MACCO - SCHLUMBERGER

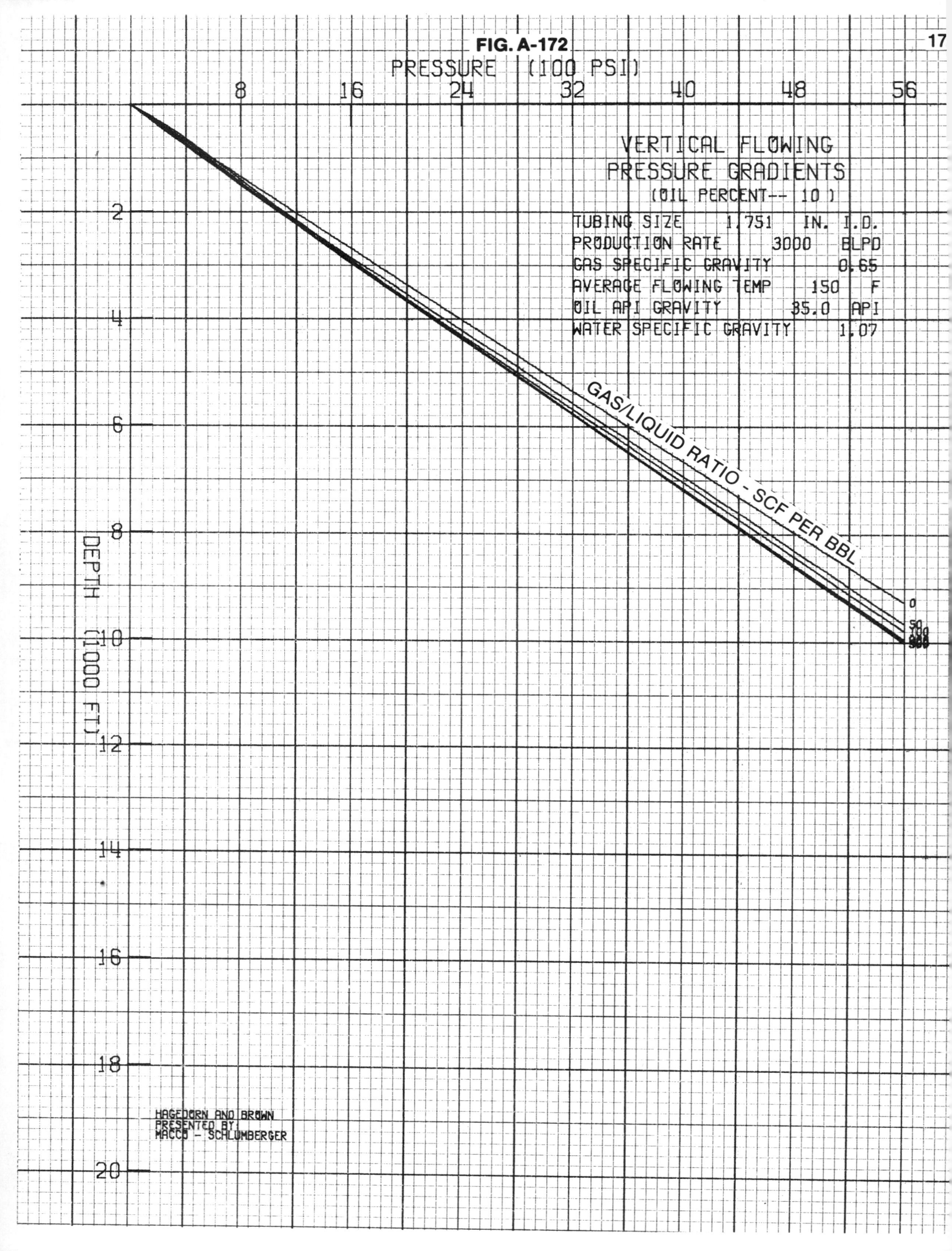

PRESSURE (100 PSI)

VERTICAL FLOWING
PRESSURE GRADIENTS
(OIL PERCENT-- 10)

TUBING SIZE	1.751	IN. I.D.
PRODUCTION RATE	3000	BLPD
GAS SPECIFIC GRAVITY	0.65	
AVERAGE FLOWING TEMP	150	F
OIL API GRAVITY	35.0	API
WATER SPECIFIC GRAVITY	1.07	

GAS/LIQUID RATIO - SCF PER BBL

DEPTH (1000 FT)

HAGEDORN AND BROWN
PRESENTED BY
MACCO - SCHLUMBERGER

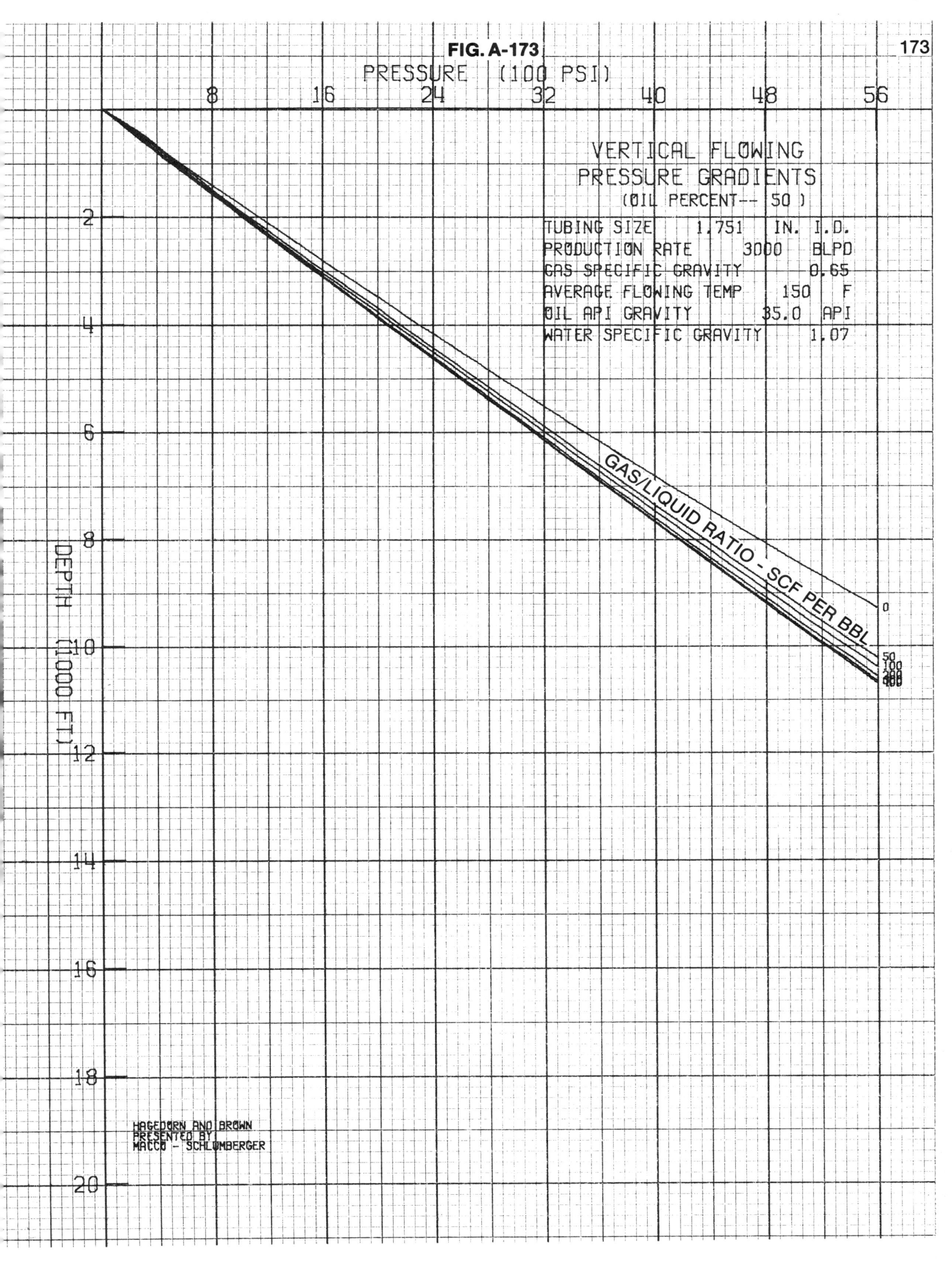

PRESSURE (100 PSI)

VERTICAL FLOWING
PRESSURE GRADIENTS
(OIL PERCENT-- 50)

TUBING SIZE	1.751	IN. I.D.
PRODUCTION RATE	3000	BLPD
GAS SPECIFIC GRAVITY	0.65	
AVERAGE FLOWING TEMP	150	F
OIL API GRAVITY	35.0	API
WATER SPECIFIC GRAVITY	1.07	

GAS/LIQUID RATIO - SCF PER BBL

DEPTH (1000 FT)

HAGEDORN AND BROWN
PRESENTED BY
MACCO - SCHLUMBERGER

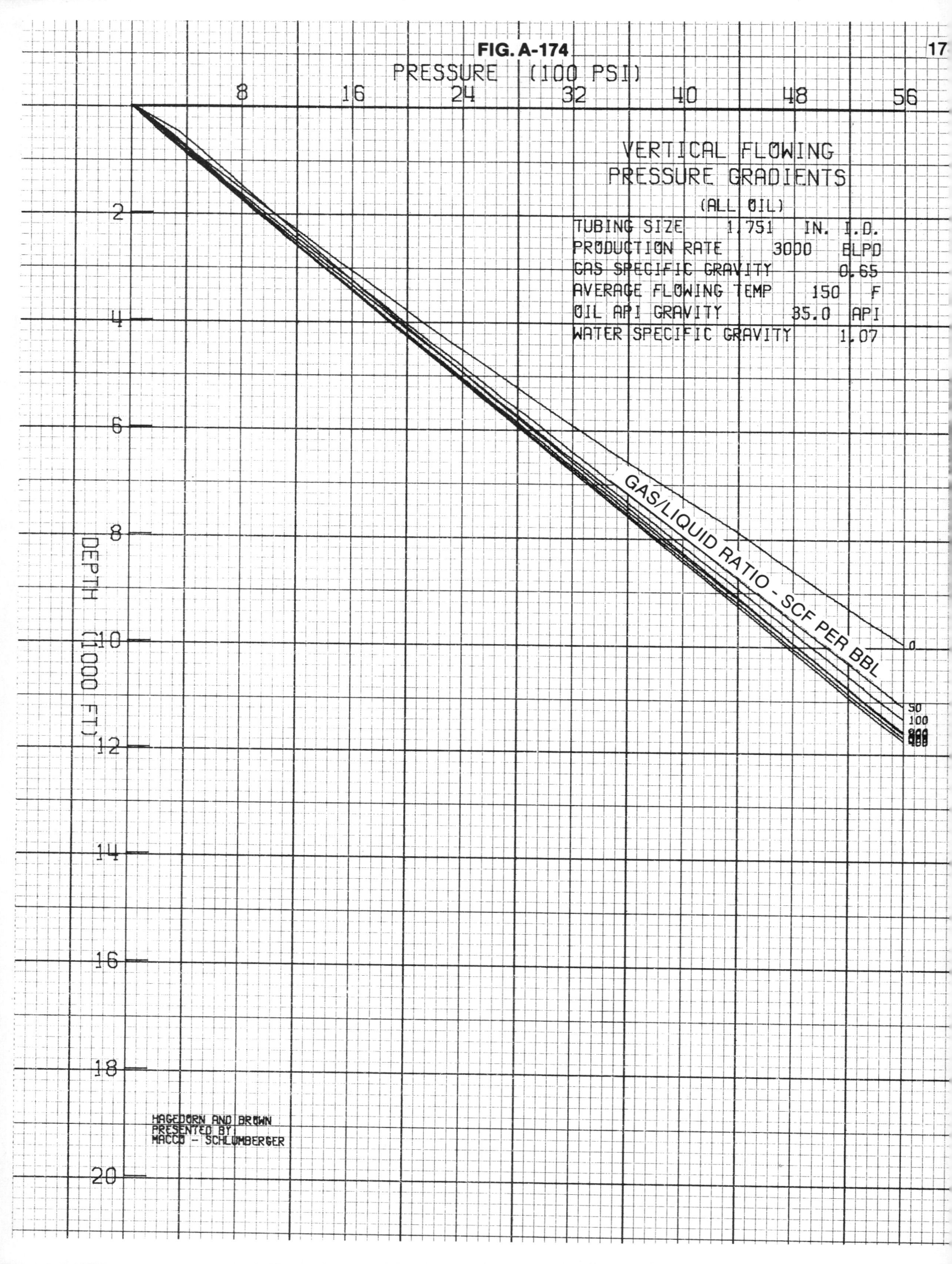

VERTICAL FLOWING
PRESSURE GRADIENTS
(ALL OIL)

TUBING SIZE	1.751	IN. I.D.
PRODUCTION RATE	3000	BLPD
GAS SPECIFIC GRAVITY	0.65	
AVERAGE FLOWING TEMP	150	F
OIL API GRAVITY	35.0	API
WATER SPECIFIC GRAVITY	1.07	

PRESSURE (100 PSI)

DEPTH (1000 FT)

GAS/LIQUID RATIO - SCF PER BBL

HAGEDORN AND BROWN
PRESENTED BY
MACCO - SCHLUMBERGER

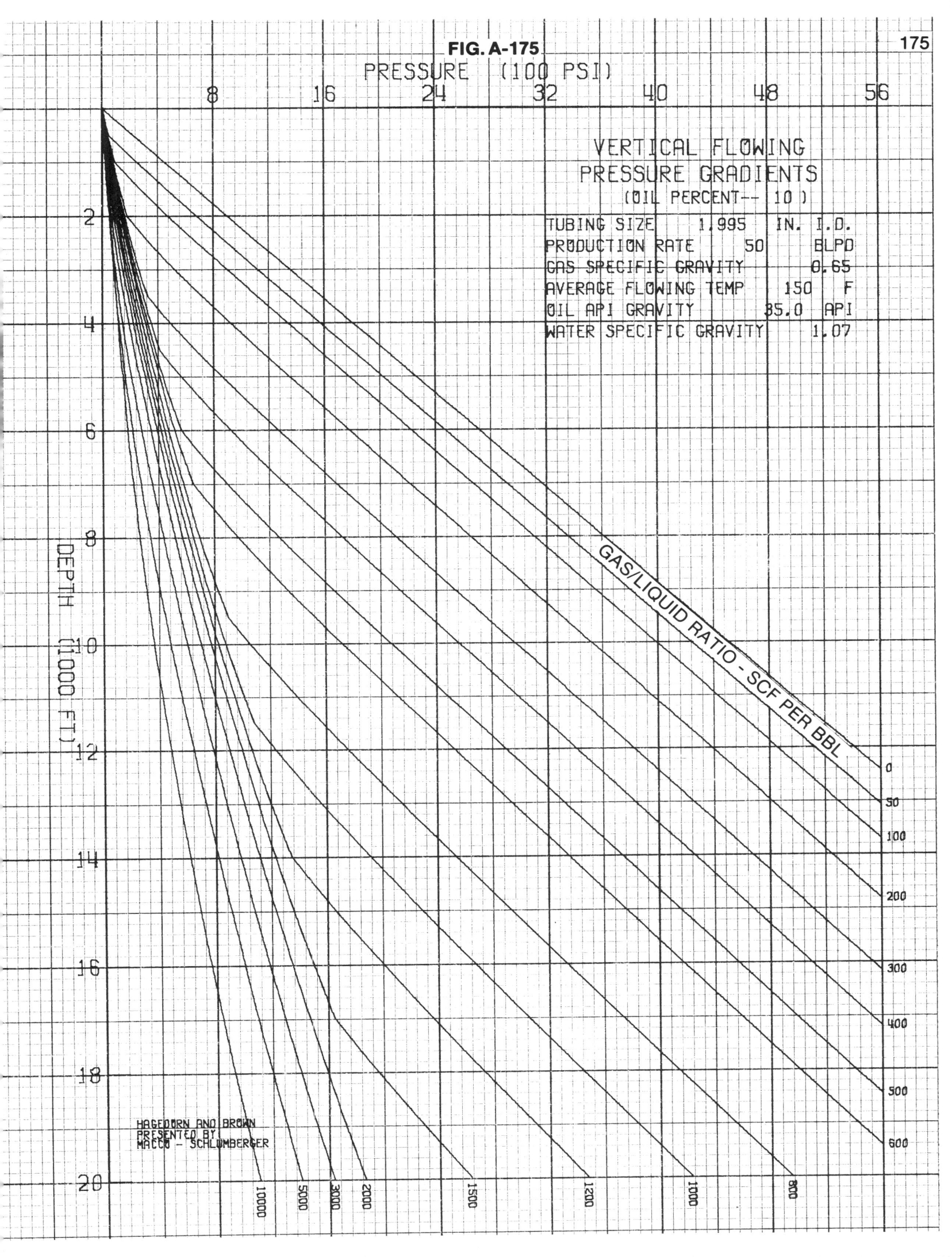

PRESSURE (100 PSI)

DEPTH (1000 FT)

VERTICAL FLOWING
PRESSURE GRADIENTS
(OIL PERCENT-- 10)

TUBING SIZE   1.995   IN. I.D.
PRODUCTION RATE   50   BLPD
GAS SPECIFIC GRAVITY   0.65
AVERAGE FLOWING TEMP   150   F
OIL API GRAVITY   35.0   API
WATER SPECIFIC GRAVITY   1.07

GAS/LIQUID RATIO - SCF PER BBL

HAGEDORN AND BROWN
PRESENTED BY
MACCO - SCHLUMBERGER

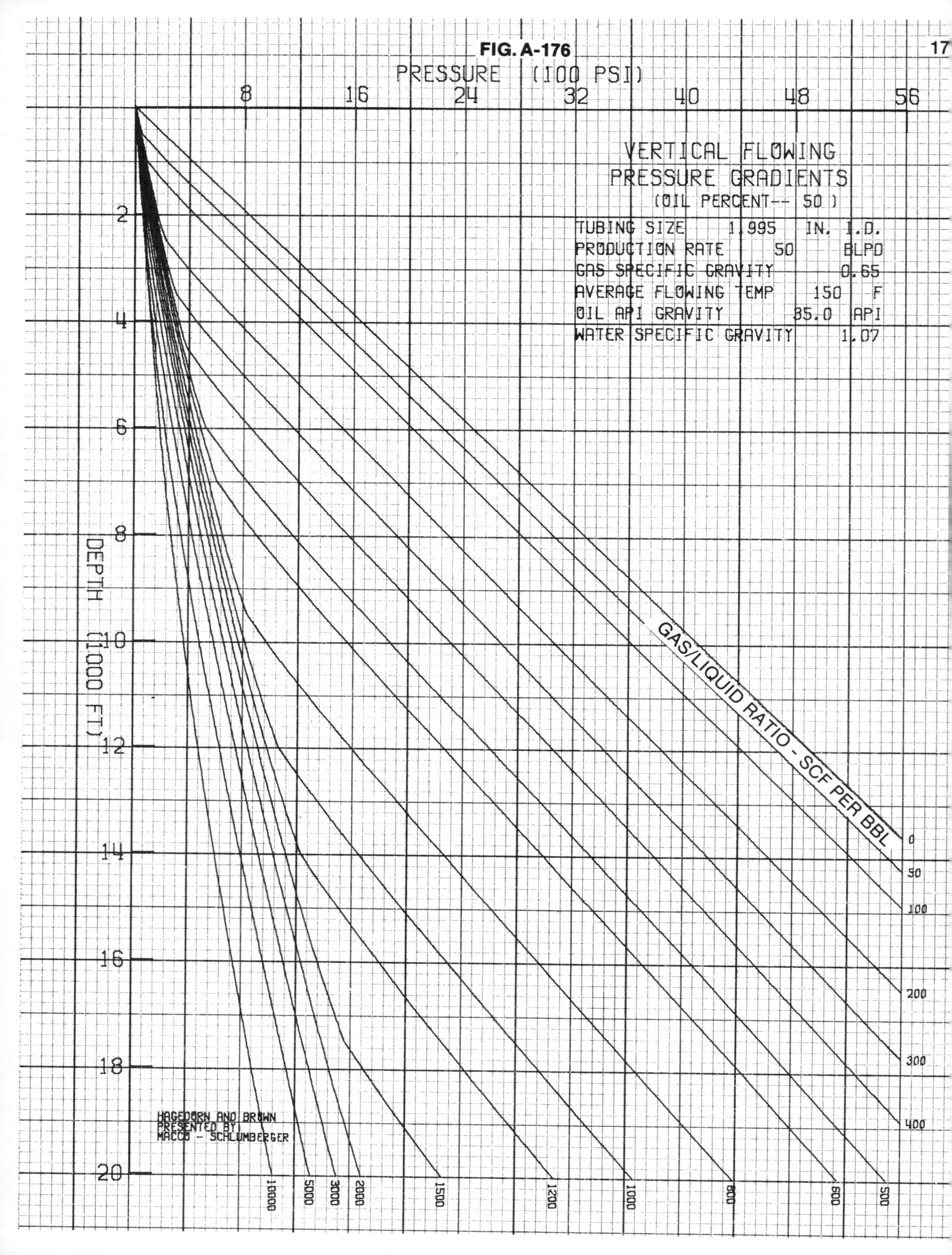

PRESSURE (100 PSI)

VERTICAL FLOWING
PRESSURE GRADIENTS
(OIL PERCENT-- 50)

TUBING SIZE	1.995	IN. I.D.
PRODUCTION RATE	50	BLPD
GAS SPECIFIC GRAVITY		0.65
AVERAGE FLOWING TEMP	150	F
OIL API GRAVITY	35.0	API
WATER SPECIFIC GRAVITY		1.07

DEPTH (1,000 FT)

GAS/LIQUID RATIO - SCF PER BBL

HAGEDORN AND BROWN
PRESENTED BY
MACCO - SCHLUMBERGER

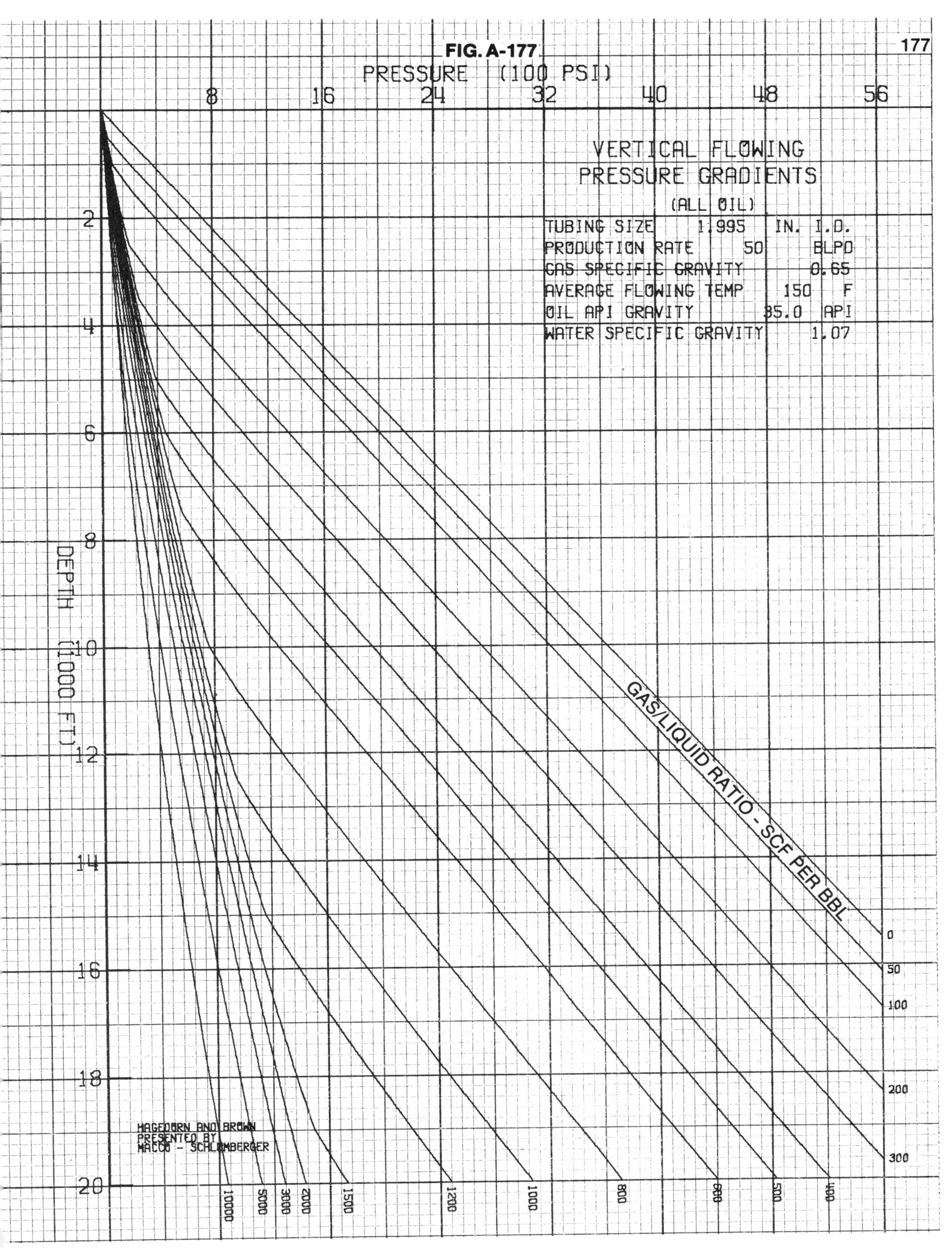

PRESSURE (100 PSI)

VERTICAL FLOWING
PRESSURE GRADIENTS
(ALL OIL)

TUBING SIZE	1.995	IN. I.D.
PRODUCTION RATE	50	BLPD
GAS SPECIFIC GRAVITY	0.65	
AVERAGE FLOWING TEMP	150	F
OIL API GRAVITY	85.0	API
WATER SPECIFIC GRAVITY	1.07	

GAS/LIQUID RATIO - SCF PER BBL

DEPTH (1000 FT)

HAGEDOORN AND BROWN
PRESENTED BY
MACCO - SCHLUMBERGER

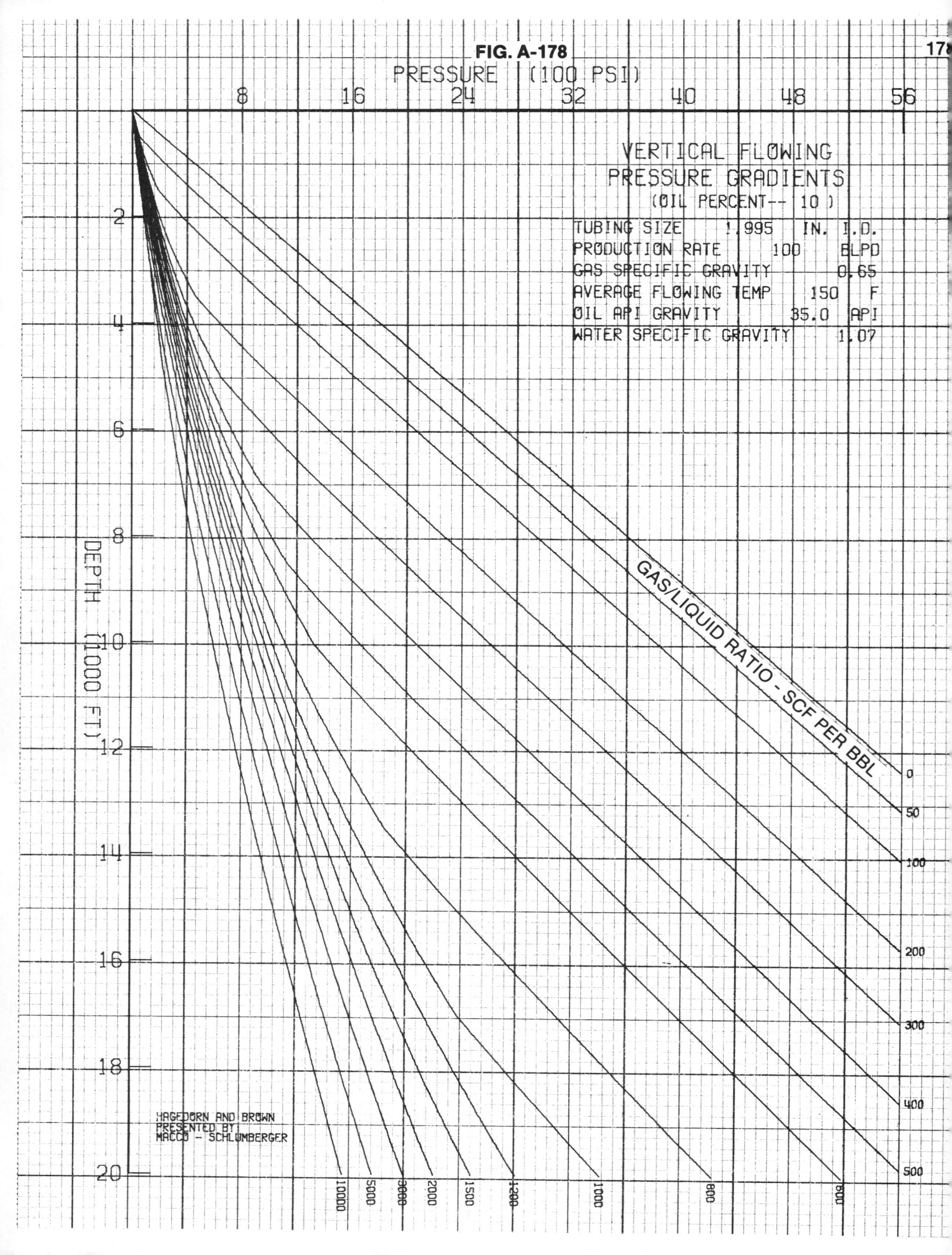

FIG. A-178

VERTICAL FLOWING
PRESSURE GRADIENTS
(OIL PERCENT-- 10)

TUBING SIZE	1.995	IN. I.D.
PRODUCTION RATE	100	BLPD
GAS SPECIFIC GRAVITY	0.65	
AVERAGE FLOWING TEMP	150	F
OIL API GRAVITY	35.0	API
WATER SPECIFIC GRAVITY	1.07	

GAS/LIQUID RATIO - SCF PER BBL

HAGEDORN AND BROWN
PRESENTED BY
MACCO - SCHLUMBERGER

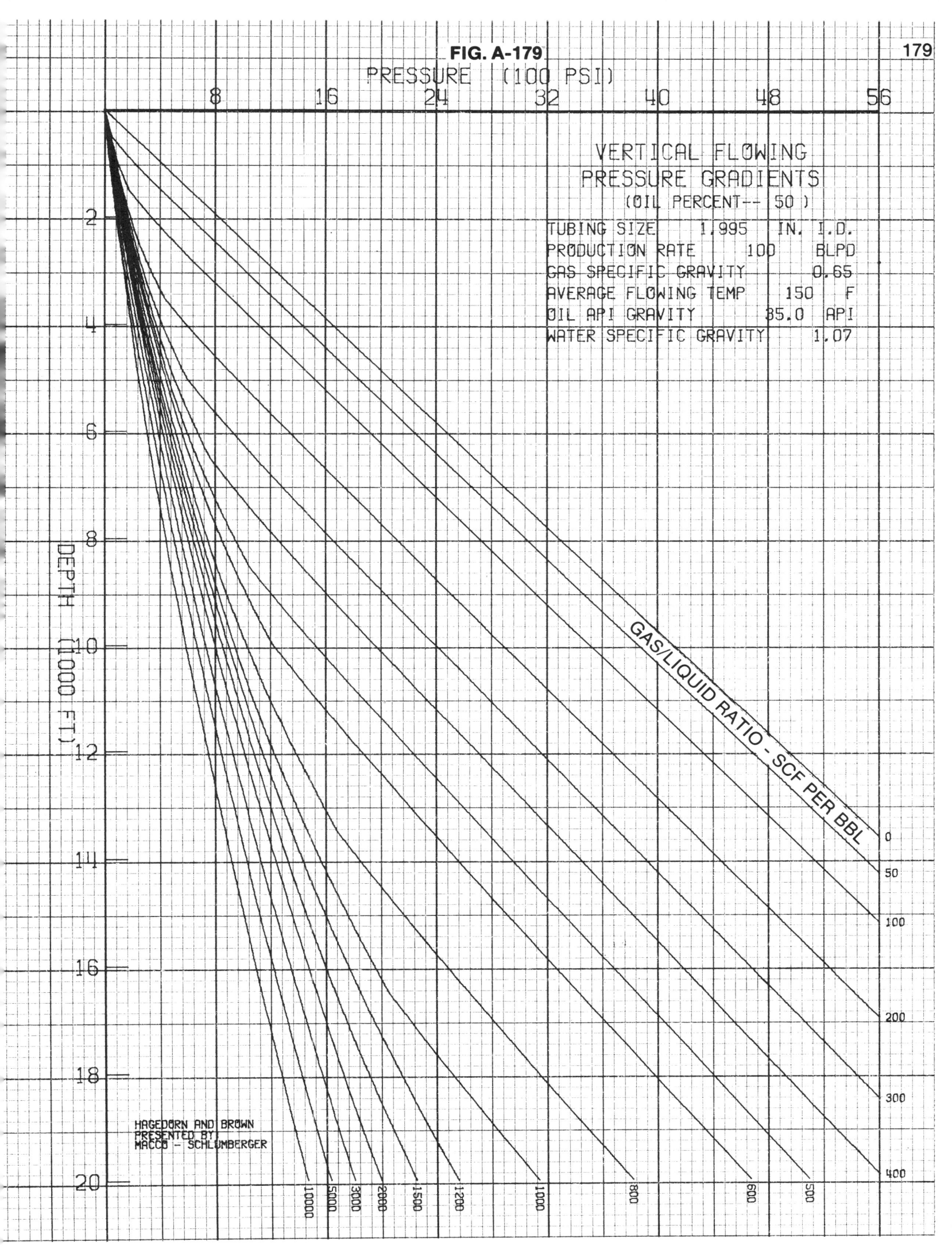

FIG. A-179

179

VERTICAL FLOWING
PRESSURE GRADIENTS
(OIL PERCENT-- 50)

TUBING SIZE	1.995	IN. I.D.
PRODUCTION RATE	100	BLPD
GAS SPECIFIC GRAVITY	0.65	
AVERAGE FLOWING TEMP	150	F
OIL API GRAVITY	85.0	API
WATER SPECIFIC GRAVITY	1.07	

GAS/LIQUID RATIO - SCF PER BBL

HAGEDORN AND BROWN
PRESENTED BY
MACCO — SCHLUMBERGER

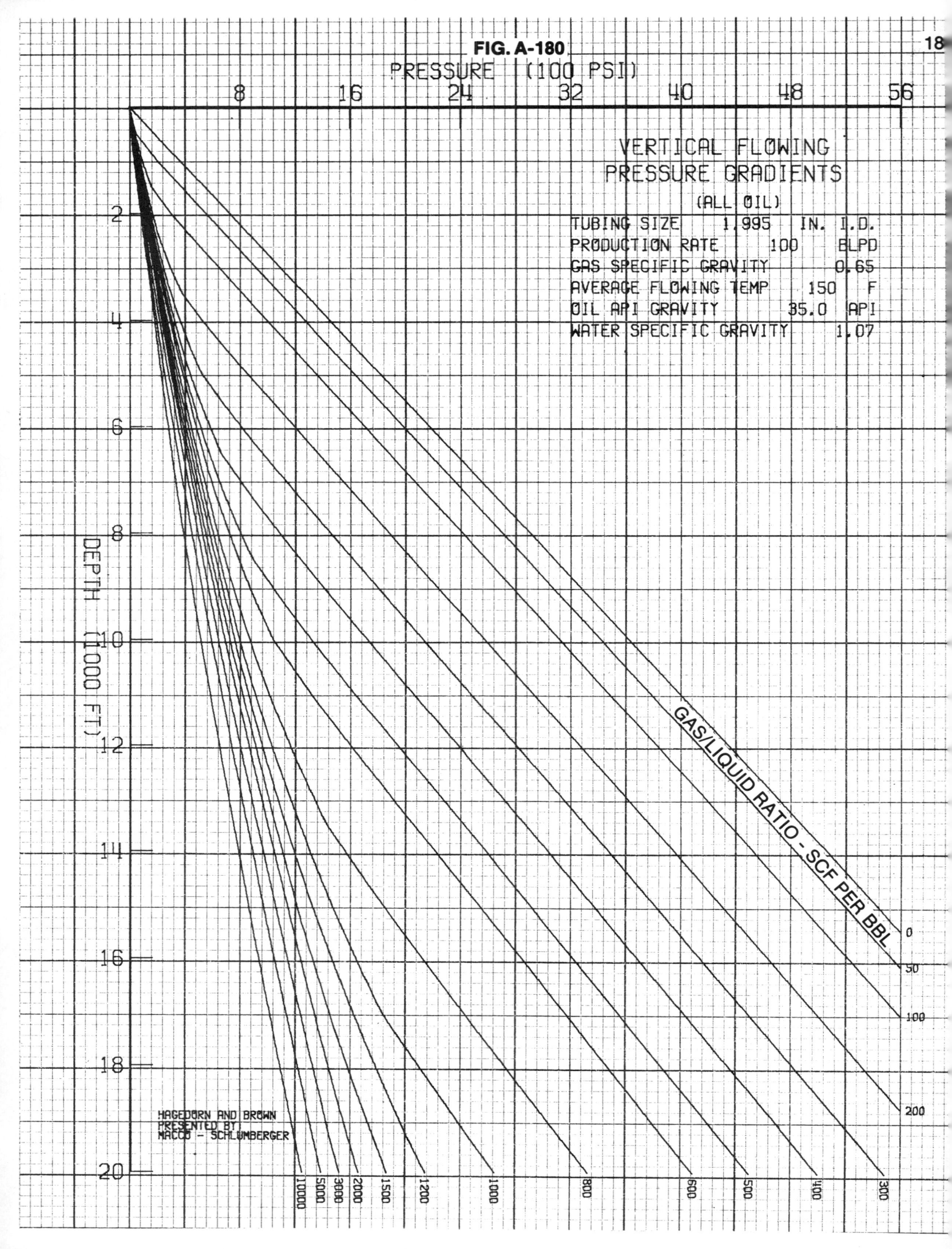

PRESSURE (100 PSI)

VERTICAL FLOWING
PRESSURE GRADIENTS
(ALL OIL)

TUBING SIZE	1.995	IN. I.D.
PRODUCTION RATE	100	BLPD
GAS SPECIFIC GRAVITY	0.65	
AVERAGE FLOWING TEMP	150	F
OIL API GRAVITY	35.0	API
WATER SPECIFIC GRAVITY	1.07	

DEPTH (1000 FT)

GAS/LIQUID RATIO - SCF PER BBL

HAGEDORN AND BROWN
PRESENTED BY
NACCO - SCHLUMBERGER

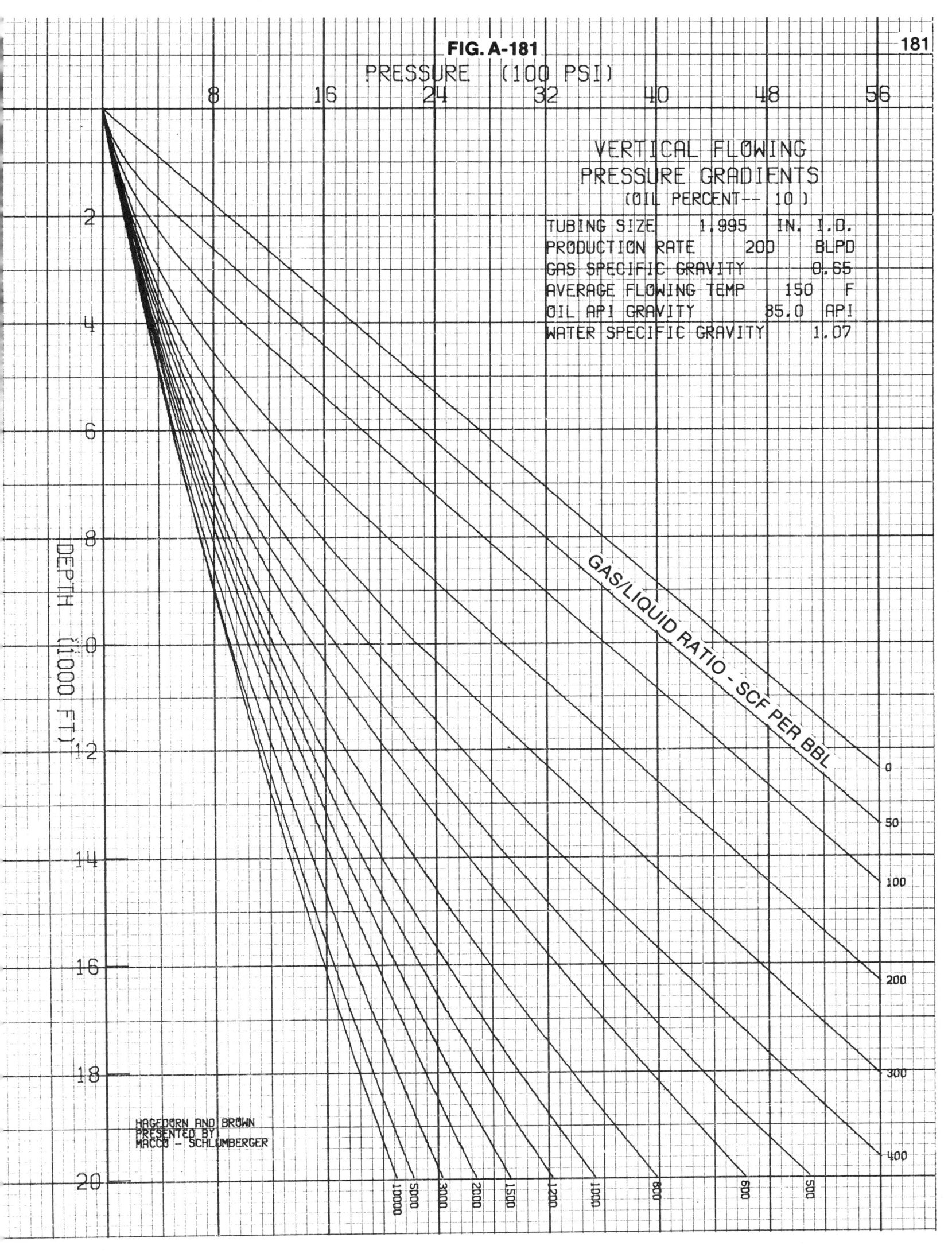

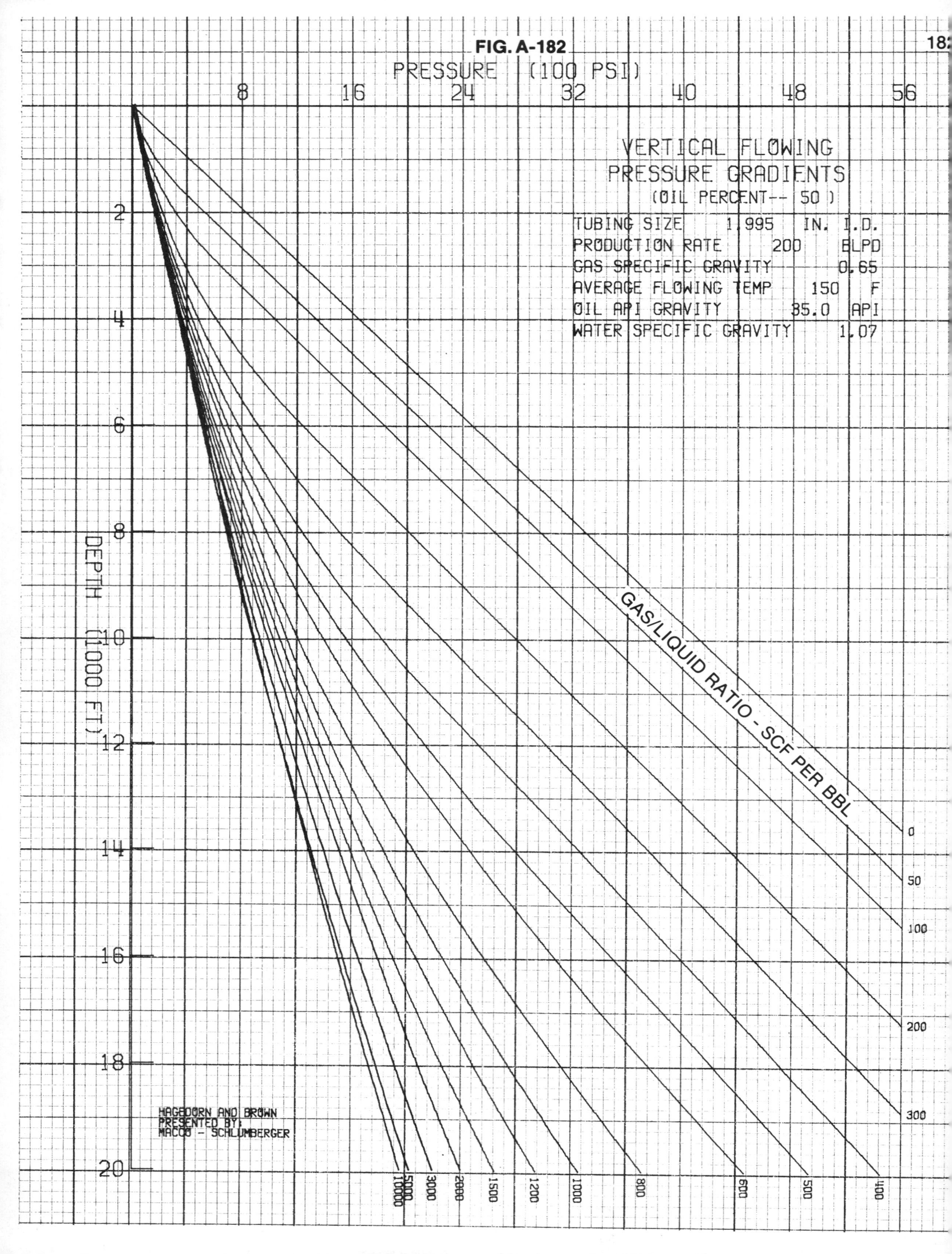

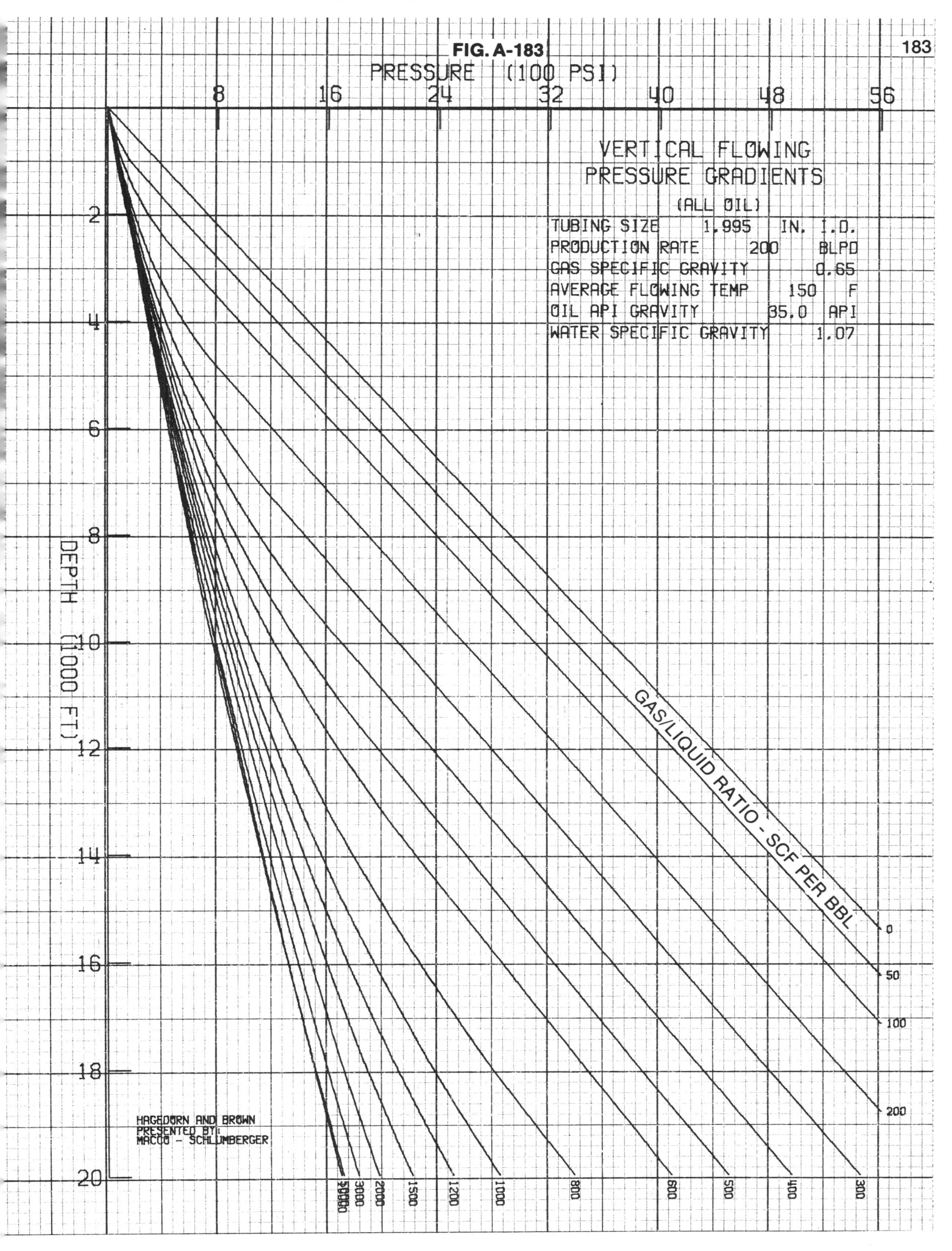

FIG. A-183

183

PRESSURE (100 PSI)

VERTICAL FLOWING
PRESSURE GRADIENTS
(ALL OIL)

TUBING SIZE	1.995	IN. I.D.
PRODUCTION RATE	200	BLPD
GAS SPECIFIC GRAVITY		0.65
AVERAGE FLOWING TEMP	150	F
OIL API GRAVITY	35.0	API
WATER SPECIFIC GRAVITY		1.07

DEPTH (1000 FT)

GAS/LIQUID RATIO - SCF PER BBL

HAGEDORN AND BROWN
PRESENTED BY:
MACCO - SCHLUMBERGER

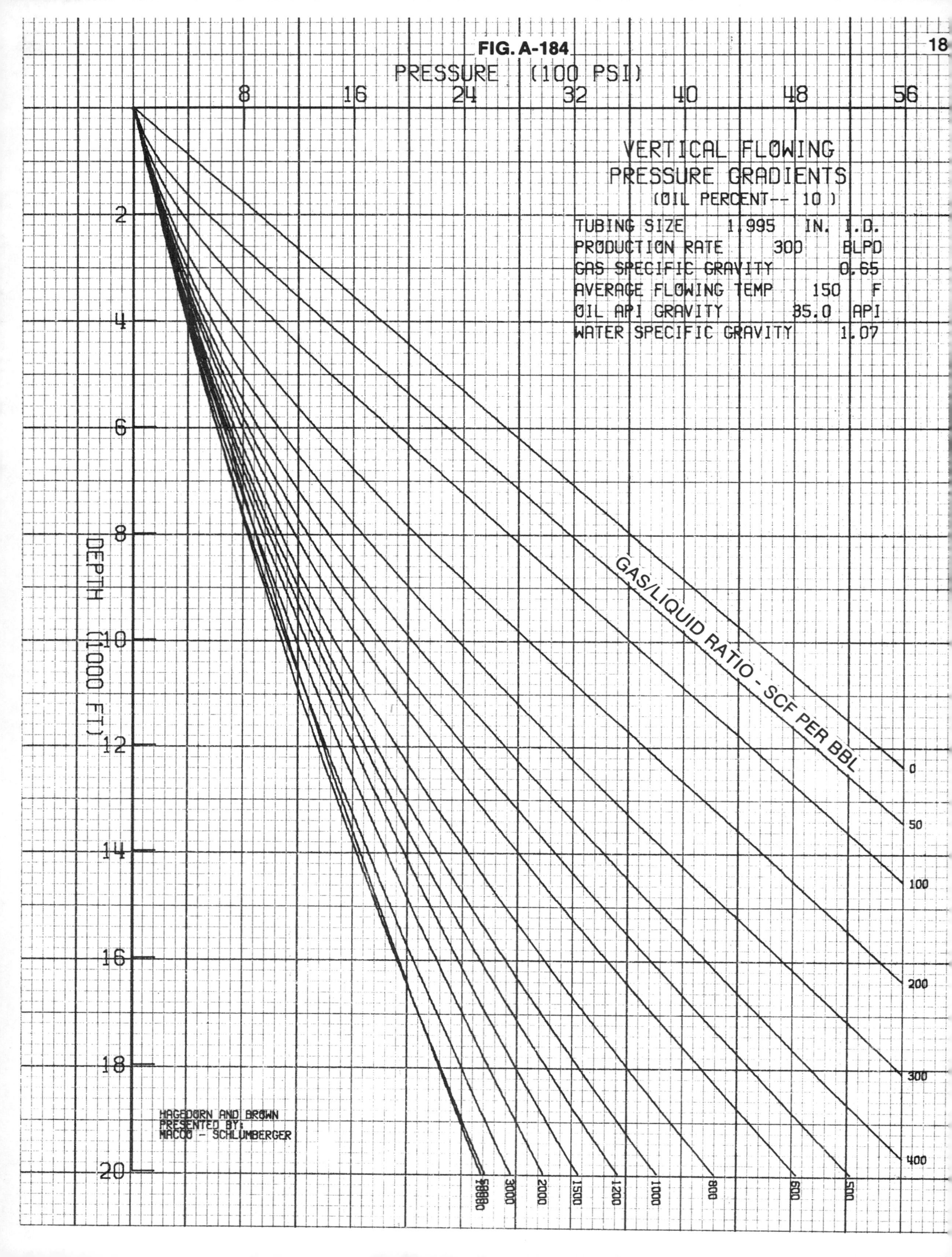

PRESSURE (100 PSI)

DEPTH (1000 FT)

VERTICAL FLOWING
PRESSURE GRADIENTS
(OIL PERCENT-- 10)

TUBING SIZE	1.995	IN. I.D.
PRODUCTION RATE	300	BLPD
GAS SPECIFIC GRAVITY		0.65
AVERAGE FLOWING TEMP	150	F
OIL API GRAVITY	35.0	API
WATER SPECIFIC GRAVITY		1.07

GAS/LIQUID RATIO - SCF PER BBL

HAGEDORN AND BROWN
PRESENTED BY:
MACCO - SCHLUMBERGER

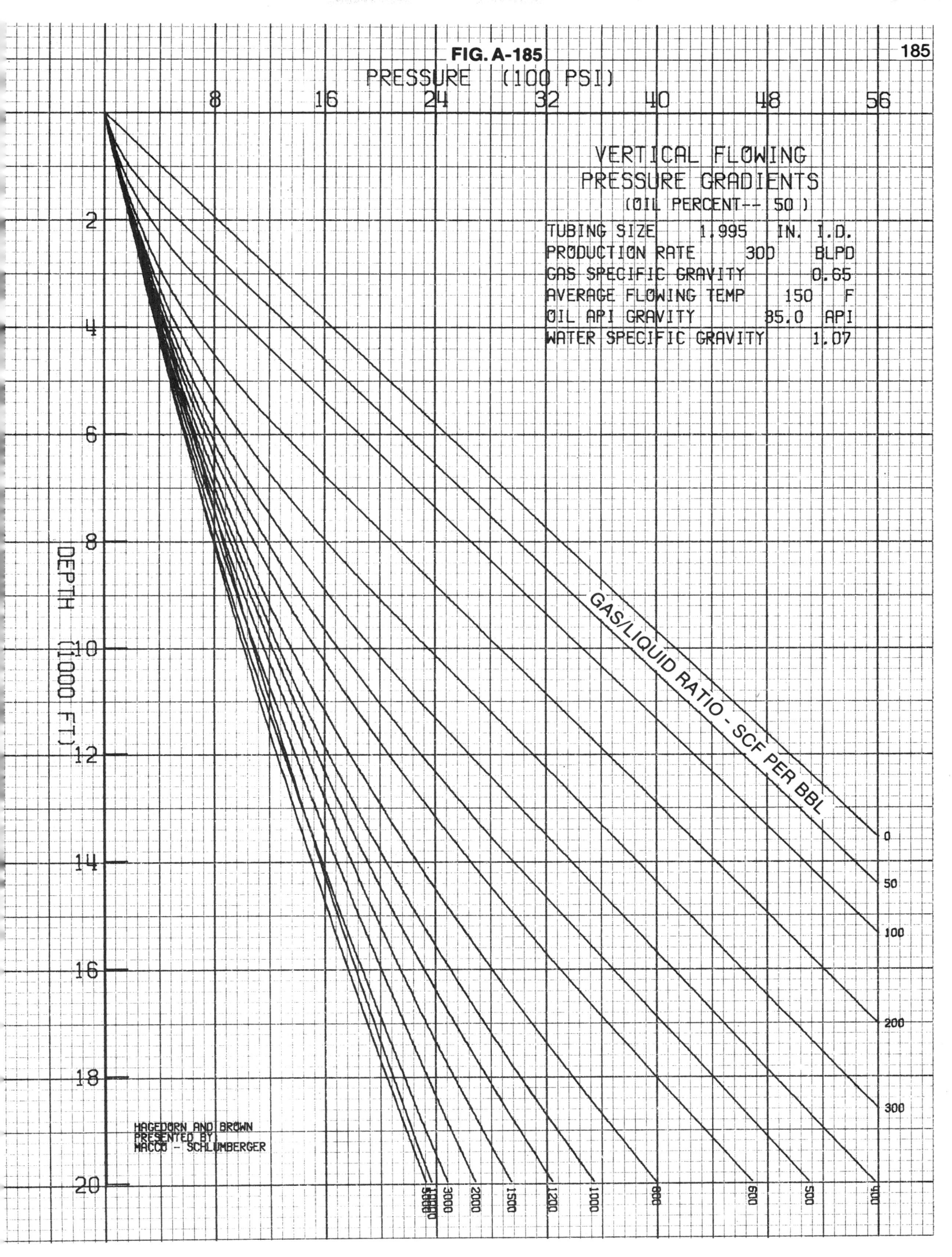

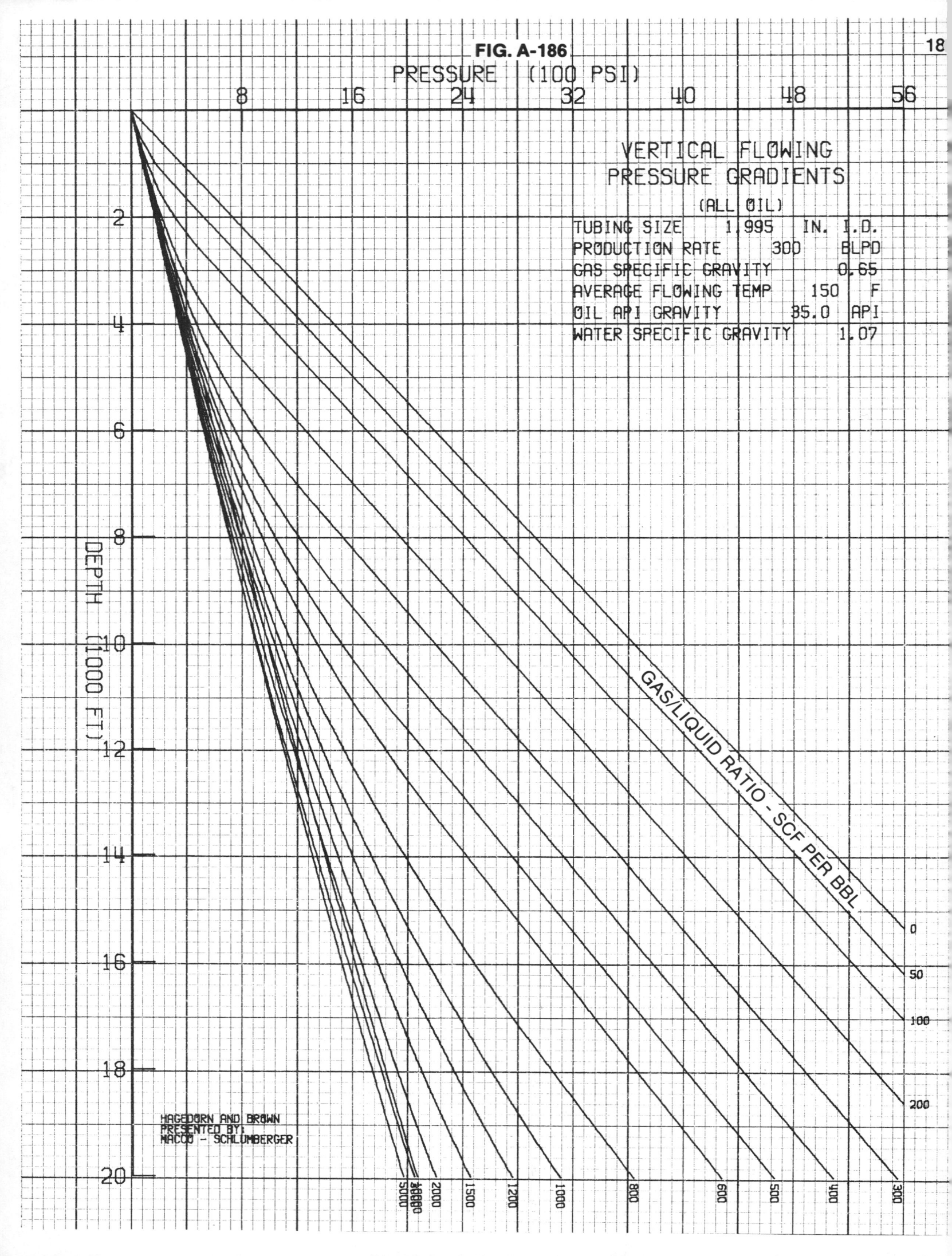

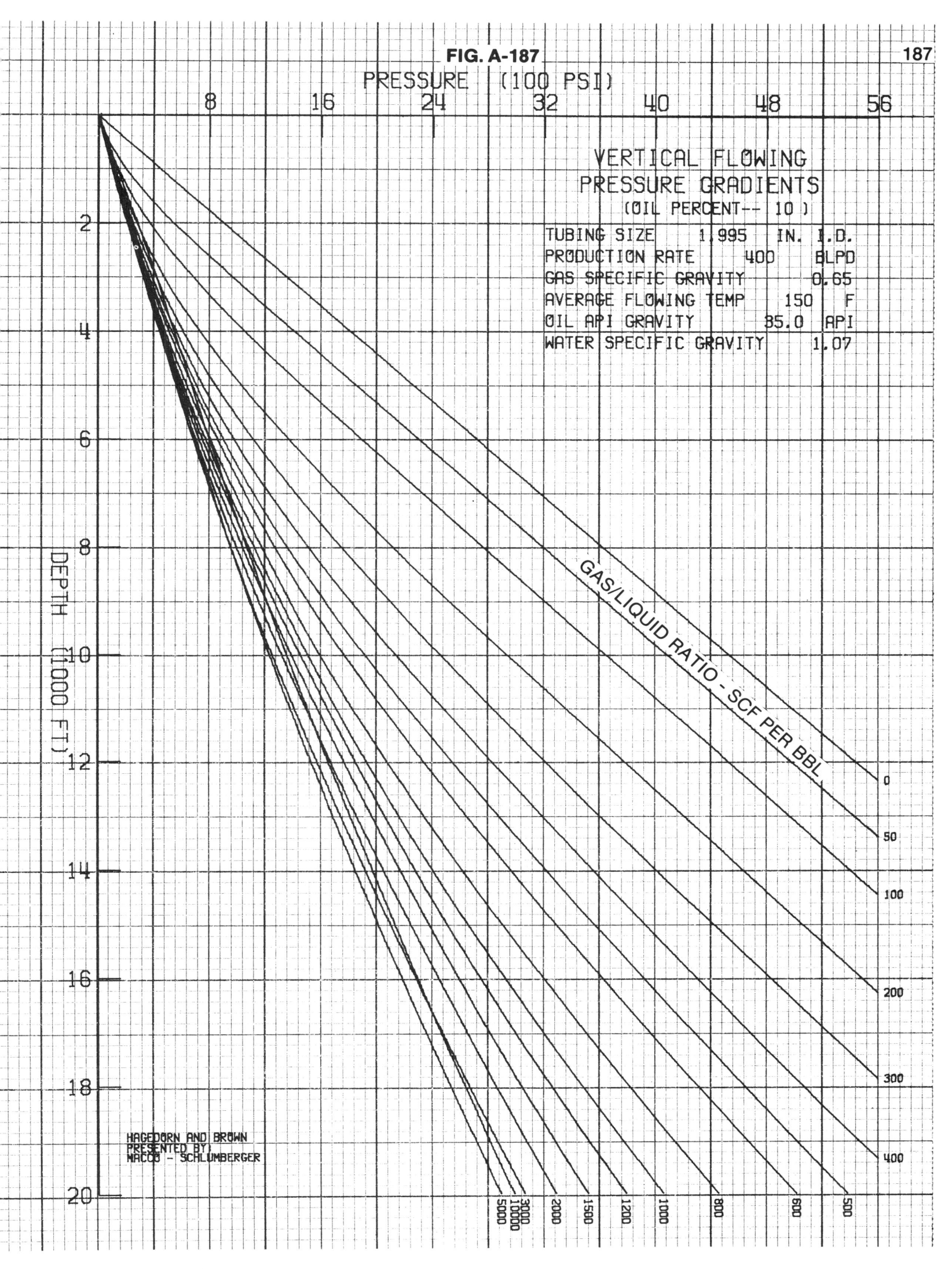

VERTICAL FLOWING
PRESSURE GRADIENTS
(OIL PERCENT-- 10)

TUBING SIZE	1.995	IN. I.D.
PRODUCTION RATE	400	BLPD
GAS SPECIFIC GRAVITY	0.65	
AVERAGE FLOWING TEMP	150	F
OIL API GRAVITY	35.0	API
WATER SPECIFIC GRAVITY	1.07	

PRESSURE (100 PSI)

DEPTH (1000 FT)

GAS/LIQUID RATIO - SCF PER BBL

HAGEDORN AND BROWN
PRESENTED BY
MACCO - SCHLUMBERGER

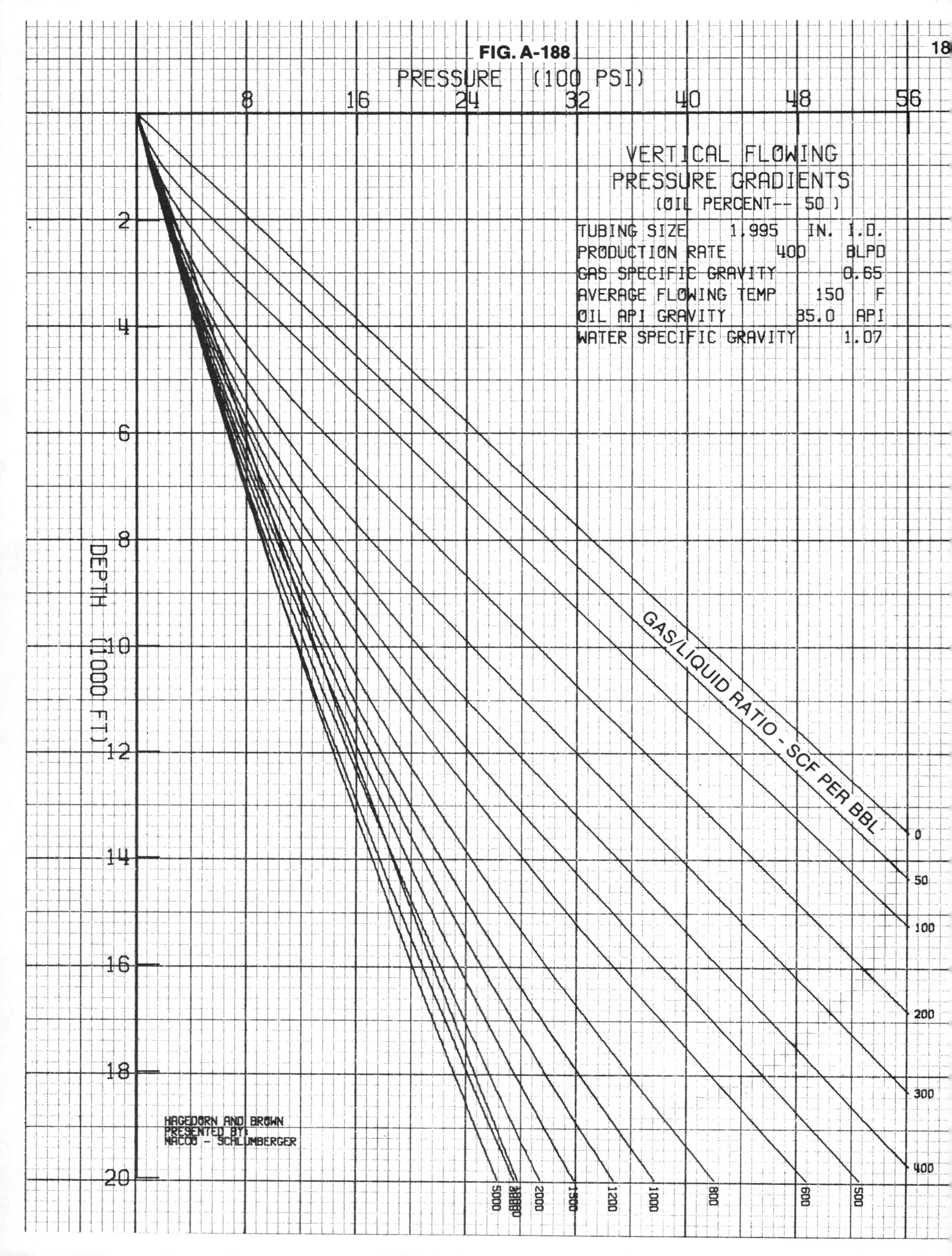

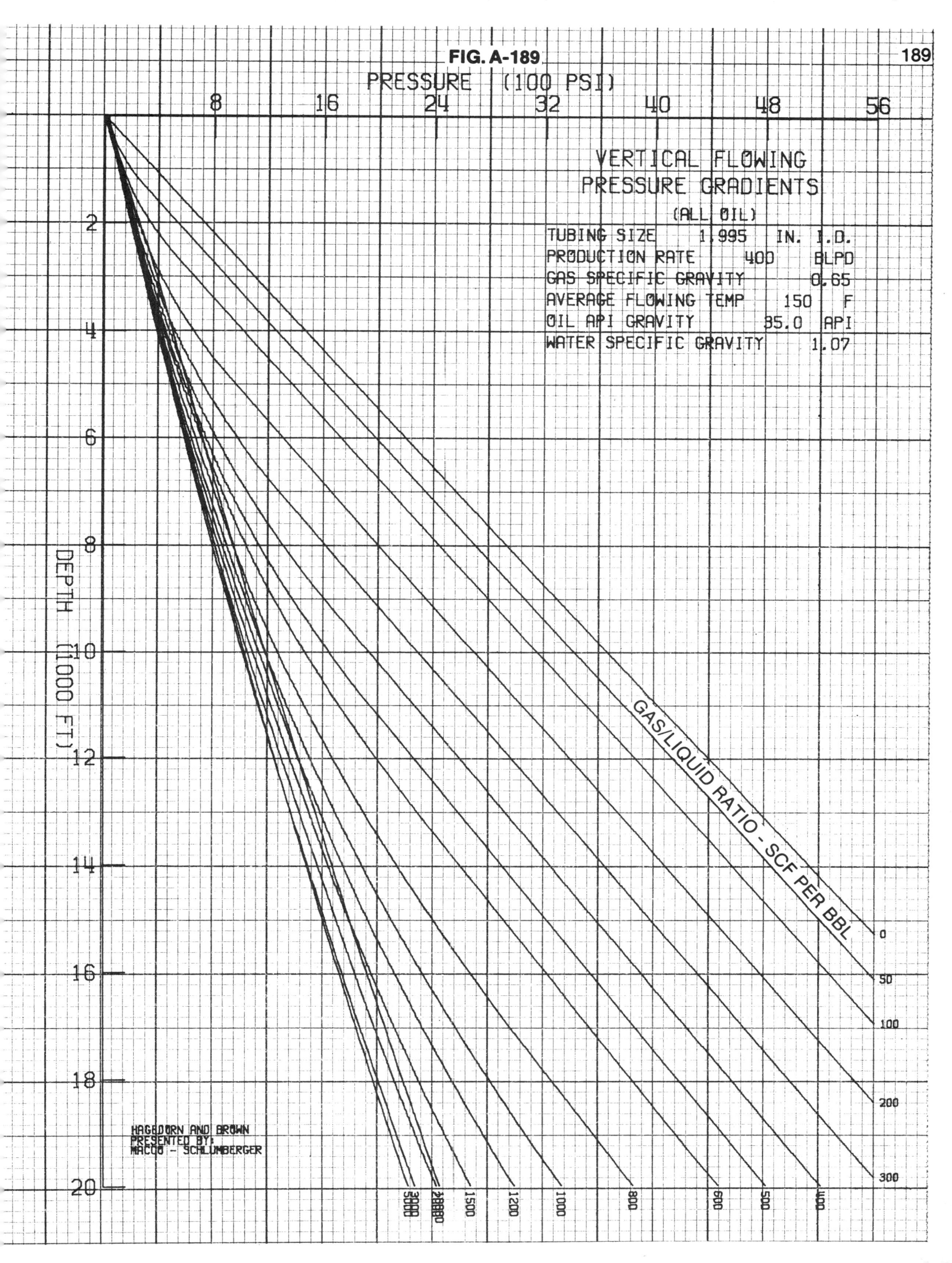

FIG. A-189

189

PRESSURE (100 PSI)

VERTICAL FLOWING
PRESSURE GRADIENTS
(ALL OIL)

TUBING SIZE	1.995	IN. I.D.
PRODUCTION RATE	400	BLPD
GAS SPECIFIC GRAVITY	0.65	
AVERAGE FLOWING TEMP	150	F
OIL API GRAVITY	35.0	API
WATER SPECIFIC GRAVITY	1.07	

DEPTH (1000 FT)

GAS/LIQUID RATIO - SCF PER BBL

HAGEDORN AND BROWN
PRESENTED BY:
MACCO - SCHLUMBERGER

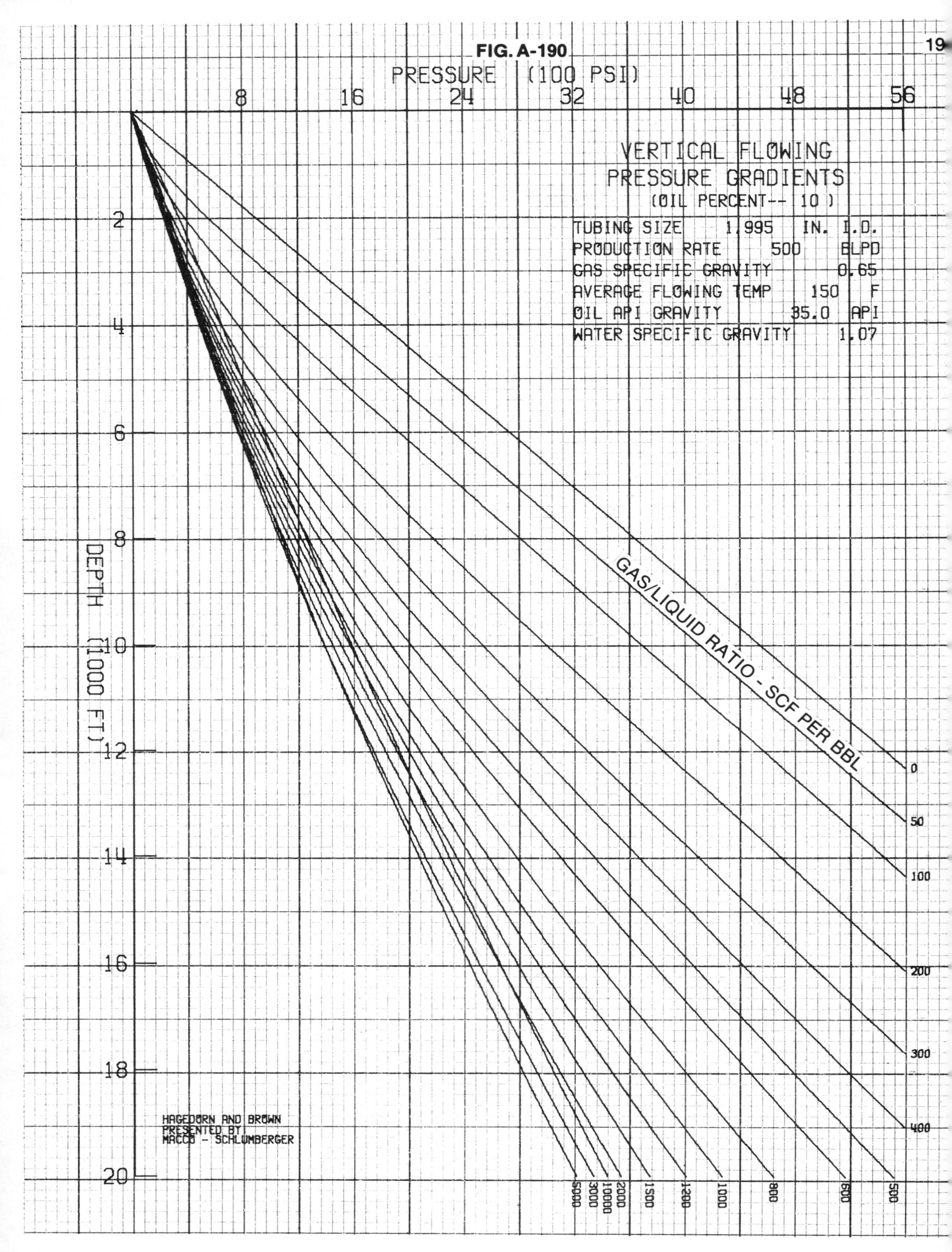

PRESSURE (100 PSI)

VERTICAL FLOWING
PRESSURE GRADIENTS
(OIL PERCENT-- 10)

TUBING SIZE	1.995	IN. I.D.
PRODUCTION RATE	500	BLPD
GAS SPECIFIC GRAVITY	0.65	
AVERAGE FLOWING TEMP	150	F
OIL API GRAVITY	35.0	API
WATER SPECIFIC GRAVITY	1.07	

DEPTH (1000 FT)

GAS/LIQUID RATIO - SCF PER BBL

HAGEDORN AND BROWN
PRESENTED BY
MACCO - SCHLUMBERGER

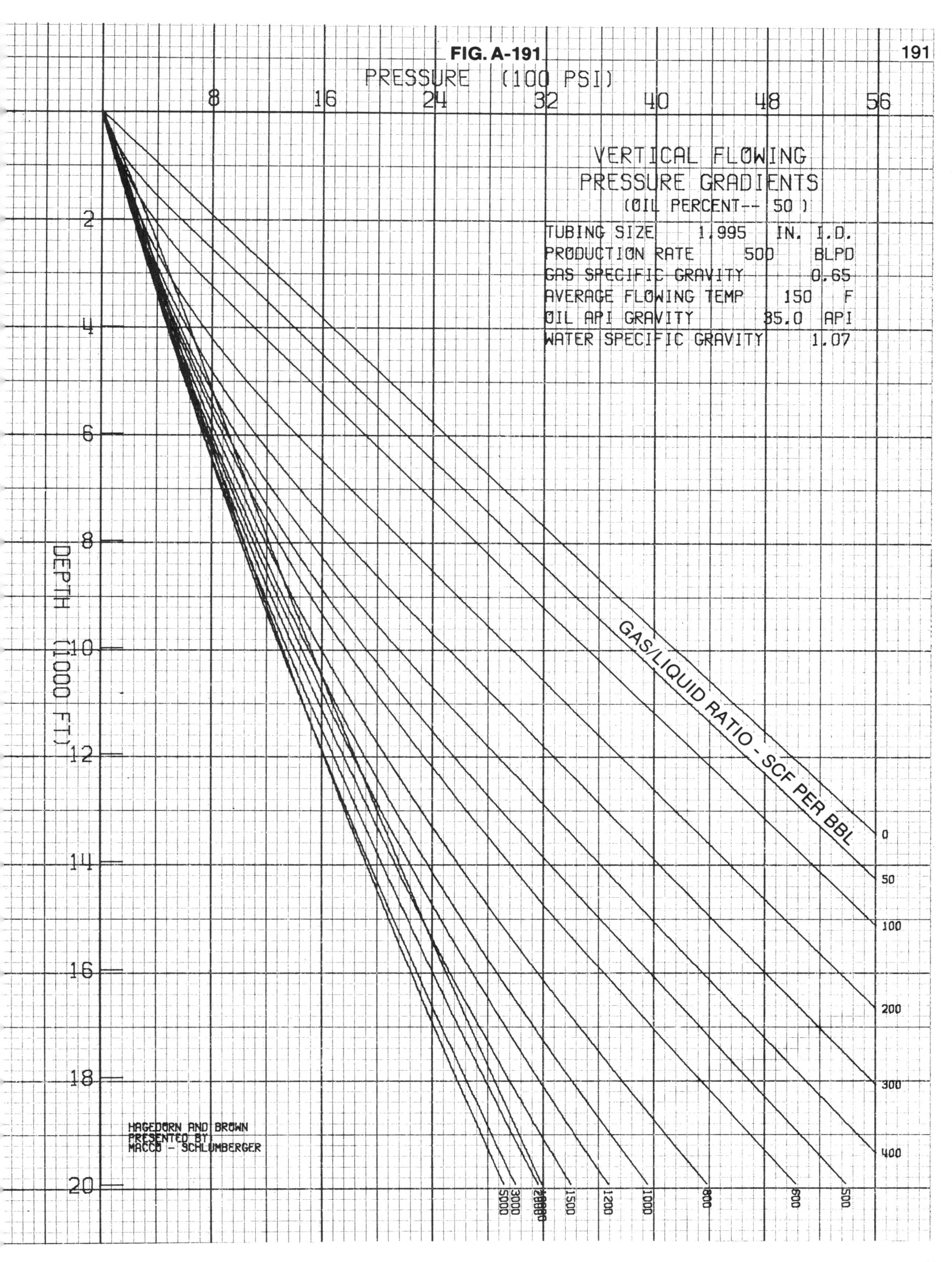

VERTICAL FLOWING
PRESSURE GRADIENTS
(OIL PERCENT-- 50)

TUBING SIZE   1.995   IN. I.D.
PRODUCTION RATE   500   BLPD
GAS SPECIFIC GRAVITY   0.65
AVERAGE FLOWING TEMP   150   F
OIL API GRAVITY   35.0   API
WATER SPECIFIC GRAVITY   1.07

PRESSURE (100 PSI)

DEPTH (1000 FT)

GAS/LIQUID RATIO - SCF PER BBL

HAGEDORN AND BROWN
PRESENTED BY
MACCO - SCHLUMBERGER

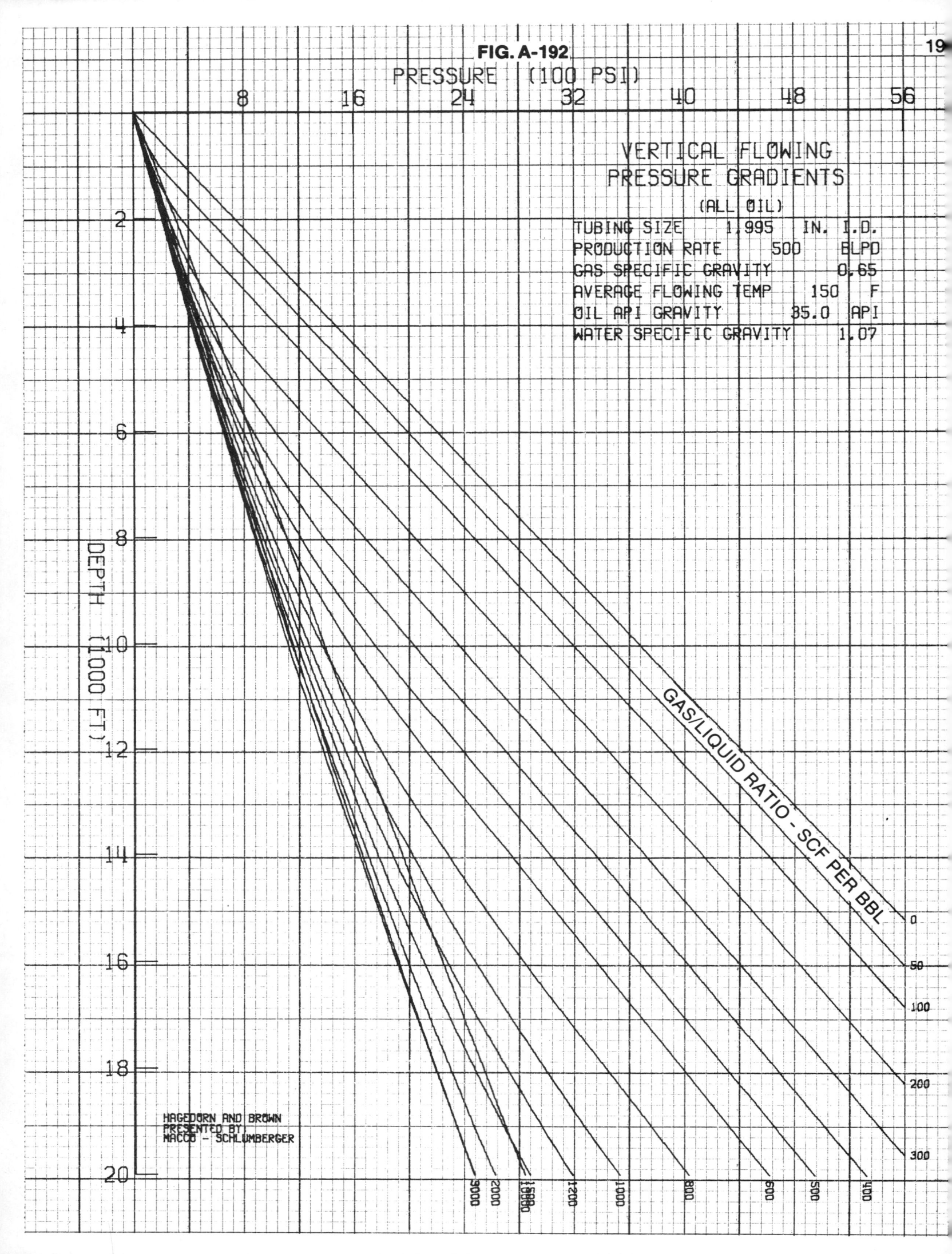

PRESSURE (100 PSI)

VERTICAL FLOWING
PRESSURE GRADIENTS
(ALL OIL)

TUBING SIZE	1,995	IN. I.D.
PRODUCTION RATE	500	BLPD
GAS SPECIFIC GRAVITY		0.65
AVERAGE FLOWING TEMP	150	F
OIL API GRAVITY	35.0	API
WATER SPECIFIC GRAVITY		1.07

DEPTH (1000 FT)

GAS/LIQUID RATIO - SCF PER BBL

HAGEDORN AND BROWN
PRESENTED BY
MACCO - SCHLUMBERGER

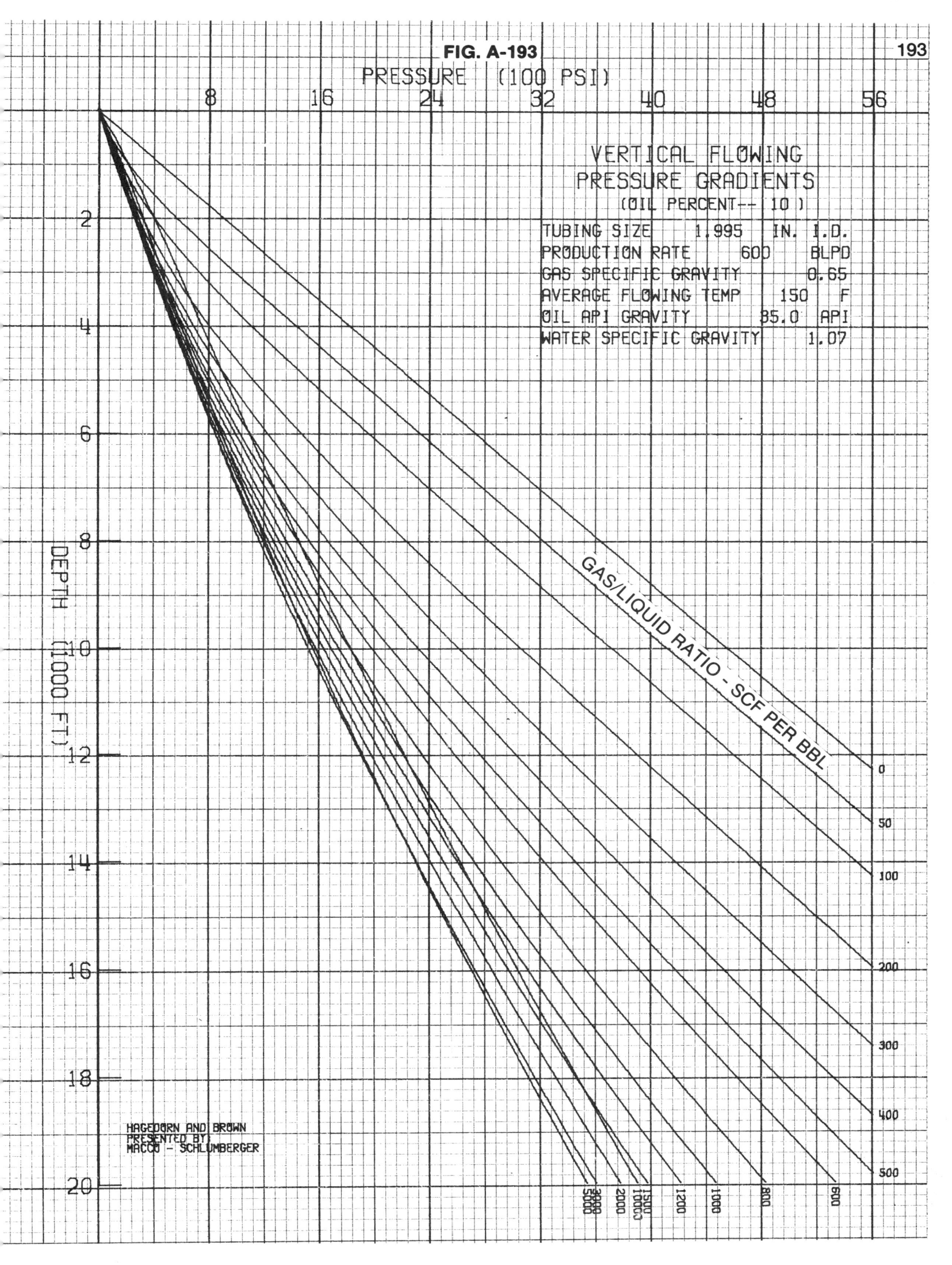

PRESSURE (100 PSI)

VERTICAL FLOWING
PRESSURE GRADIENTS
(OIL PERCENT-- 10)

TUBING SIZE	1.995	IN. I.D.
PRODUCTION RATE	600	BLPD
GAS SPECIFIC GRAVITY		0.65
AVERAGE FLOWING TEMP	150	F
OIL API GRAVITY	35.0	API
WATER SPECIFIC GRAVITY		1.07

GAS/LIQUID RATIO - SCF PER BBL

DEPTH (1000 FT)

HAGEDORN AND BROWN
PRESENTED BY
MACCO - SCHLUMBERGER

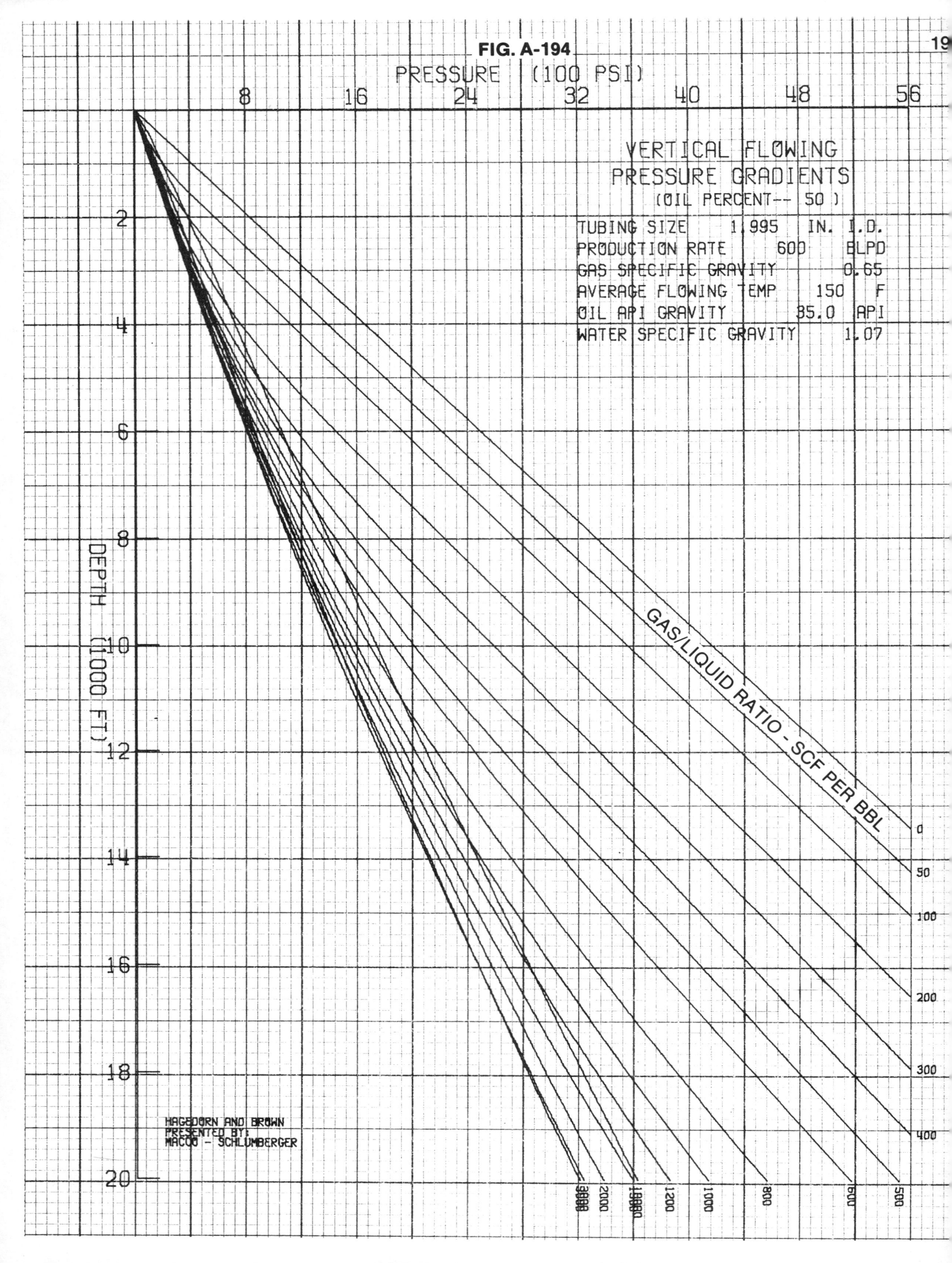

PRESSURE (100 PSI)

VERTICAL FLOWING
PRESSURE GRADIENTS
(OIL PERCENT-- 50)

TUBING SIZE	1.995	IN. I.D.
PRODUCTION RATE	600	BLPD
GAS SPECIFIC GRAVITY	0.65	
AVERAGE FLOWING TEMP	150	F
OIL API GRAVITY	85.0	API
WATER SPECIFIC GRAVITY	1.07	

DEPTH (1000 FT)

GAS/LIQUID RATIO - SCF PER BBL

HAGEDORN AND BROWN
PRESENTED BY:
MACCO - SCHLUMBERGER

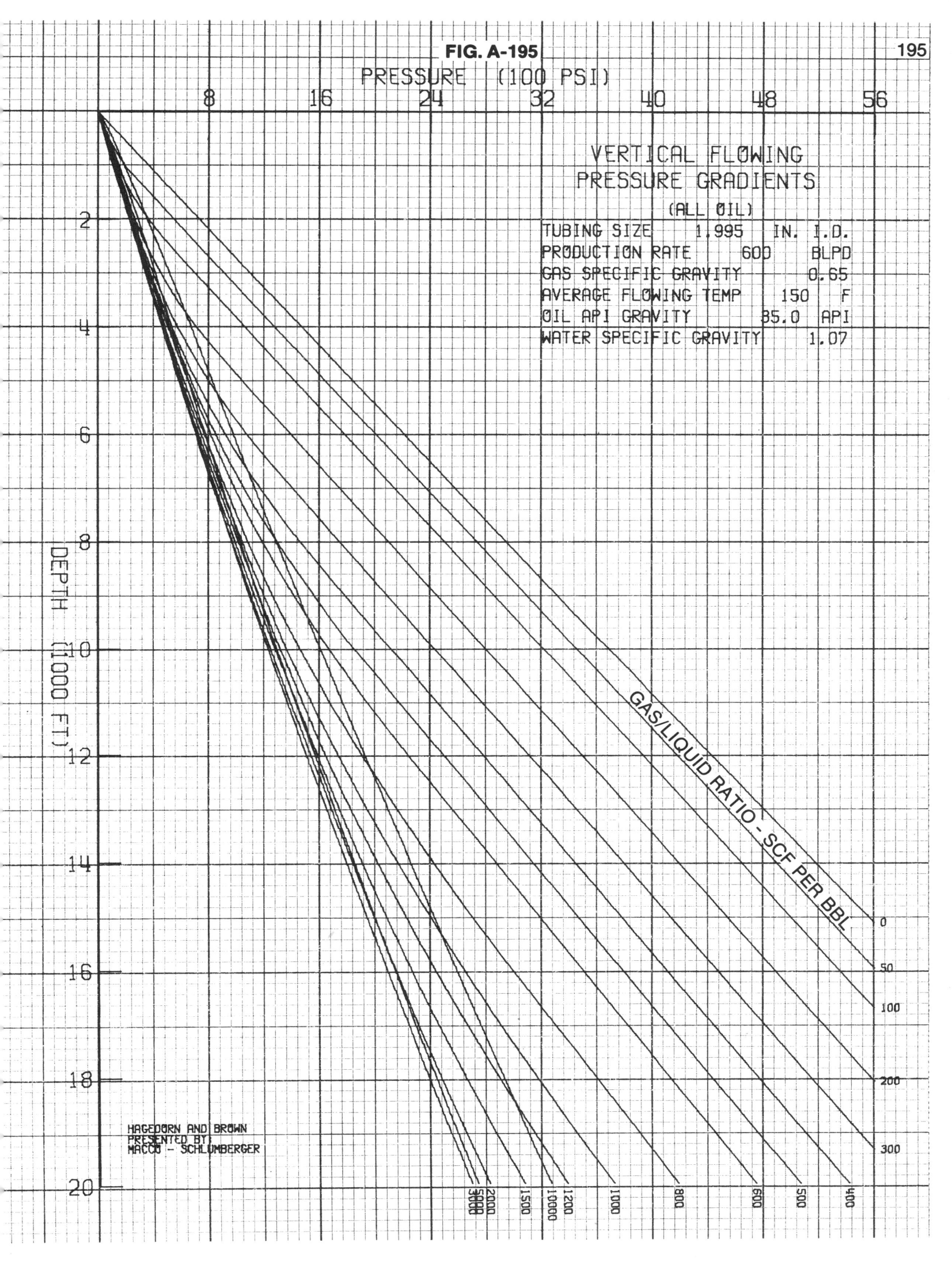

PRESSURE (100 PSI)

VERTICAL FLOWING
PRESSURE GRADIENTS
(ALL OIL)

TUBING SIZE	1.995	IN. I.D.
PRODUCTION RATE	600	BLPD
GAS SPECIFIC GRAVITY	0.65	
AVERAGE FLOWING TEMP	150	F
OIL API GRAVITY	35.0	API
WATER SPECIFIC GRAVITY	1.07	

DEPTH (1000 FT)

GAS/LIQUID RATIO - SCF PER BBL

HAGEDORN AND BROWN
PRESENTED BY
MACCO – SCHLUMBERGER

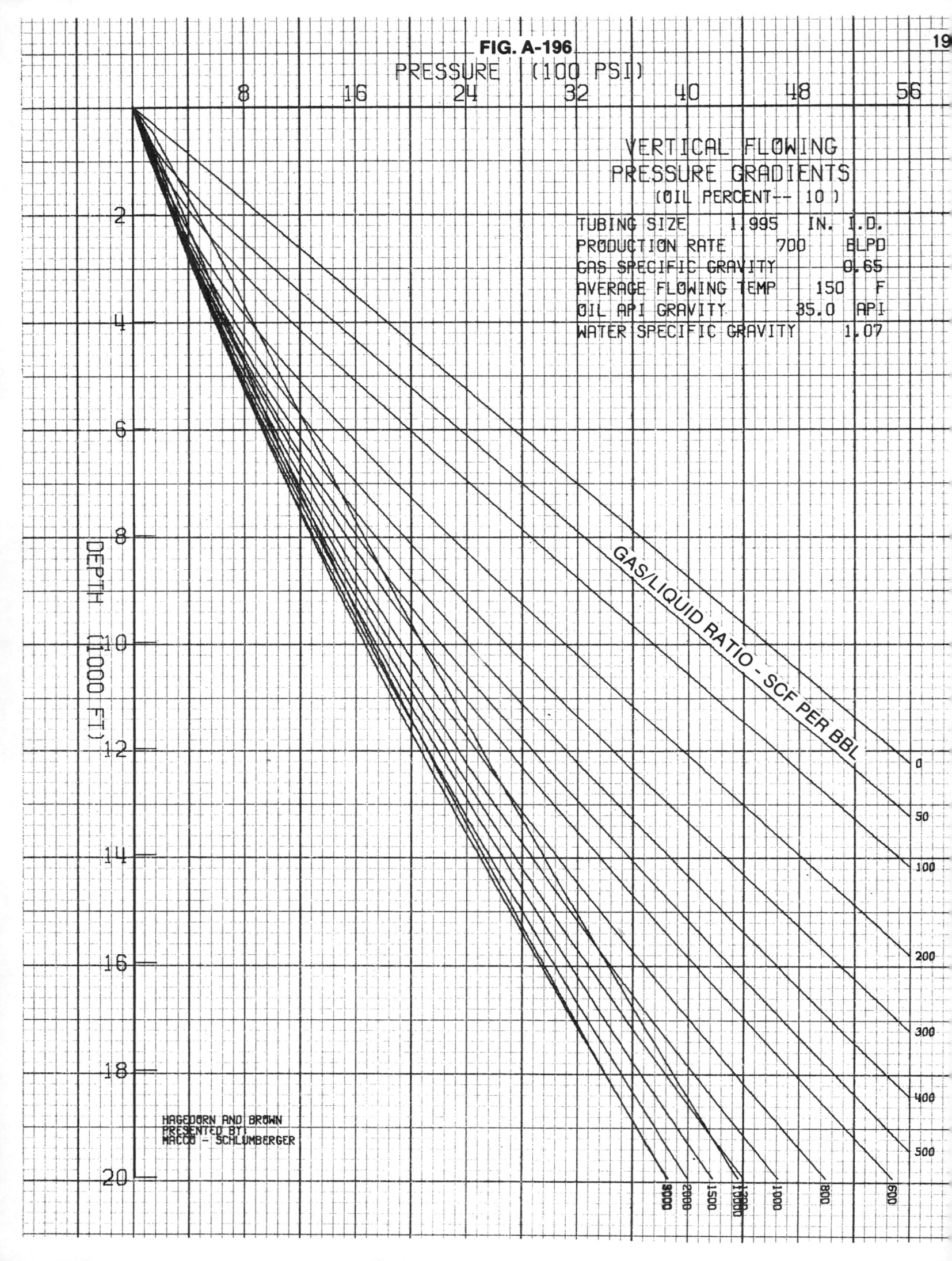

FIG. A-196

VERTICAL FLOWING
PRESSURE GRADIENTS
(OIL PERCENT-- 10)

TUBING SIZE	1.995	IN. I.D.
PRODUCTION RATE	700	BLPD
GAS SPECIFIC GRAVITY	0.65	
AVERAGE FLOWING TEMP	150	F
OIL API GRAVITY	35.0	API
WATER SPECIFIC GRAVITY	1.07	

GAS/LIQUID RATIO - SCF PER BBL

HAGEDORN AND BROWN
PRESENTED BY:
MACCO - SCHLUMBERGER

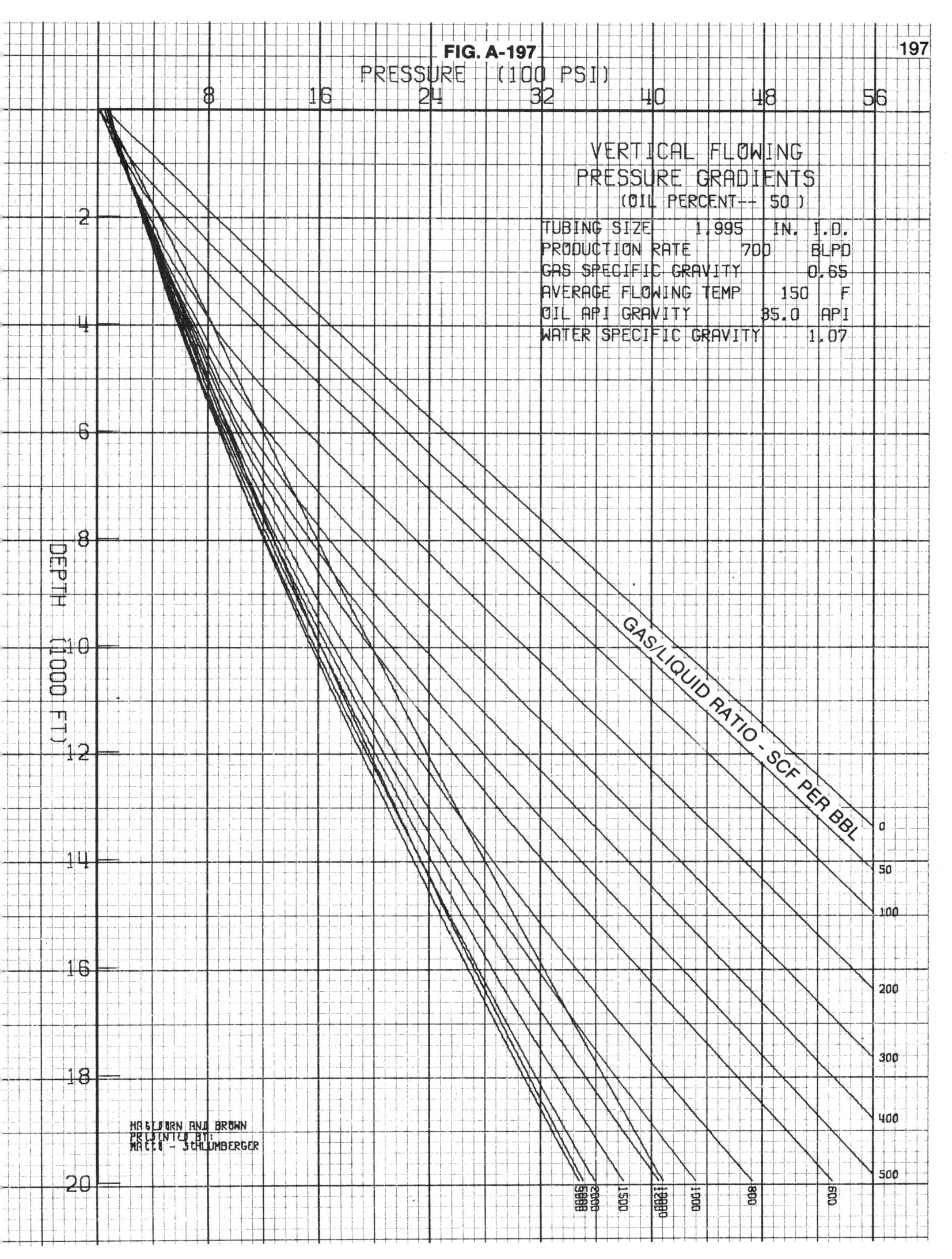

VERTICAL FLOWING
PRESSURE GRADIENTS
(OIL PERCENT-- 50)

TUBING SIZE	1.995	IN. I.D.
PRODUCTION RATE	700	BLPD
GAS SPECIFIC GRAVITY		0.65
AVERAGE FLOWING TEMP	150	F
OIL API GRAVITY	35.0	API
WATER SPECIFIC GRAVITY		1.07

PRESSURE (100 PSI)

DEPTH (1000 FT)

GAS/LIQUID RATIO - SCF PER BBL

HAGEDORN AND BROWN
PRESENTED BY:
MACCO - SCHLUMBERGER

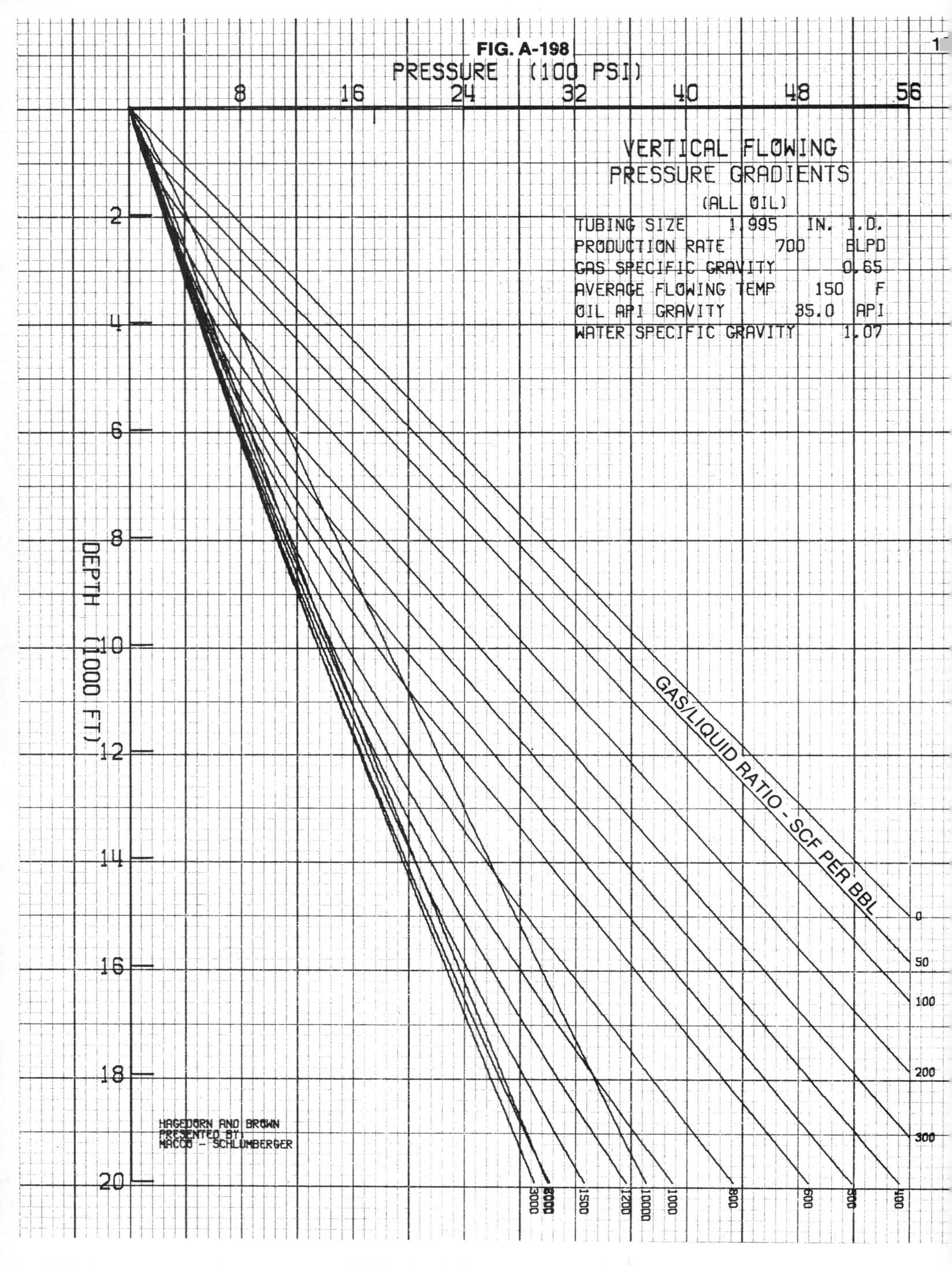

VERTICAL FLOWING
PRESSURE GRADIENTS
(ALL OIL)

TUBING SIZE	1.995	IN. I.D.
PRODUCTION RATE	700	BLPD
GAS SPECIFIC GRAVITY	0.65	
AVERAGE FLOWING TEMP	150	F
OIL API GRAVITY	35.0	API
WATER SPECIFIC GRAVITY	1.07	

PRESSURE (100 PSI)

DEPTH (1000 FT)

GAS/LIQUID RATIO - SCF PER BBL

HAGEDORN AND BROWN
PRESENTED BY
MACCO - SCHLUMBERGER

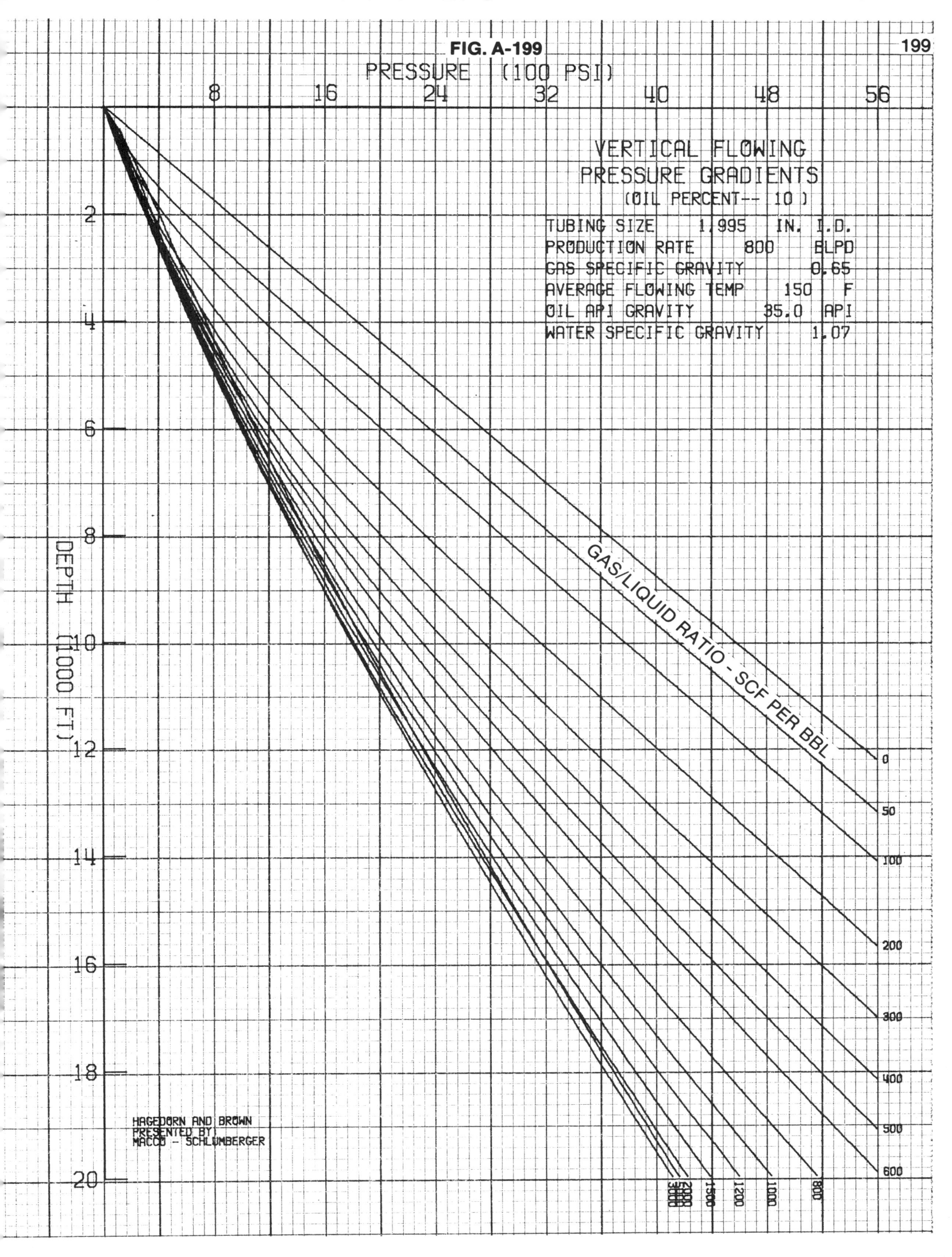

FIG. A-199

VERTICAL FLOWING
PRESSURE GRADIENTS
(OIL PERCENT-- 10)

TUBING SIZE   1.995   IN. I.D.
PRODUCTION RATE   800   BLPD
GAS SPECIFIC GRAVITY   0.65
AVERAGE FLOWING TEMP   150   F
OIL API GRAVITY   35.0 API
WATER SPECIFIC GRAVITY   1.07

GAS/LIQUID RATIO - SCF PER BBL

PRESSURE (100 PSI)

DEPTH (1000 FT)

HAGEDORN AND BROWN
PRESENTED BY
MACCO - SCHLUMBERGER

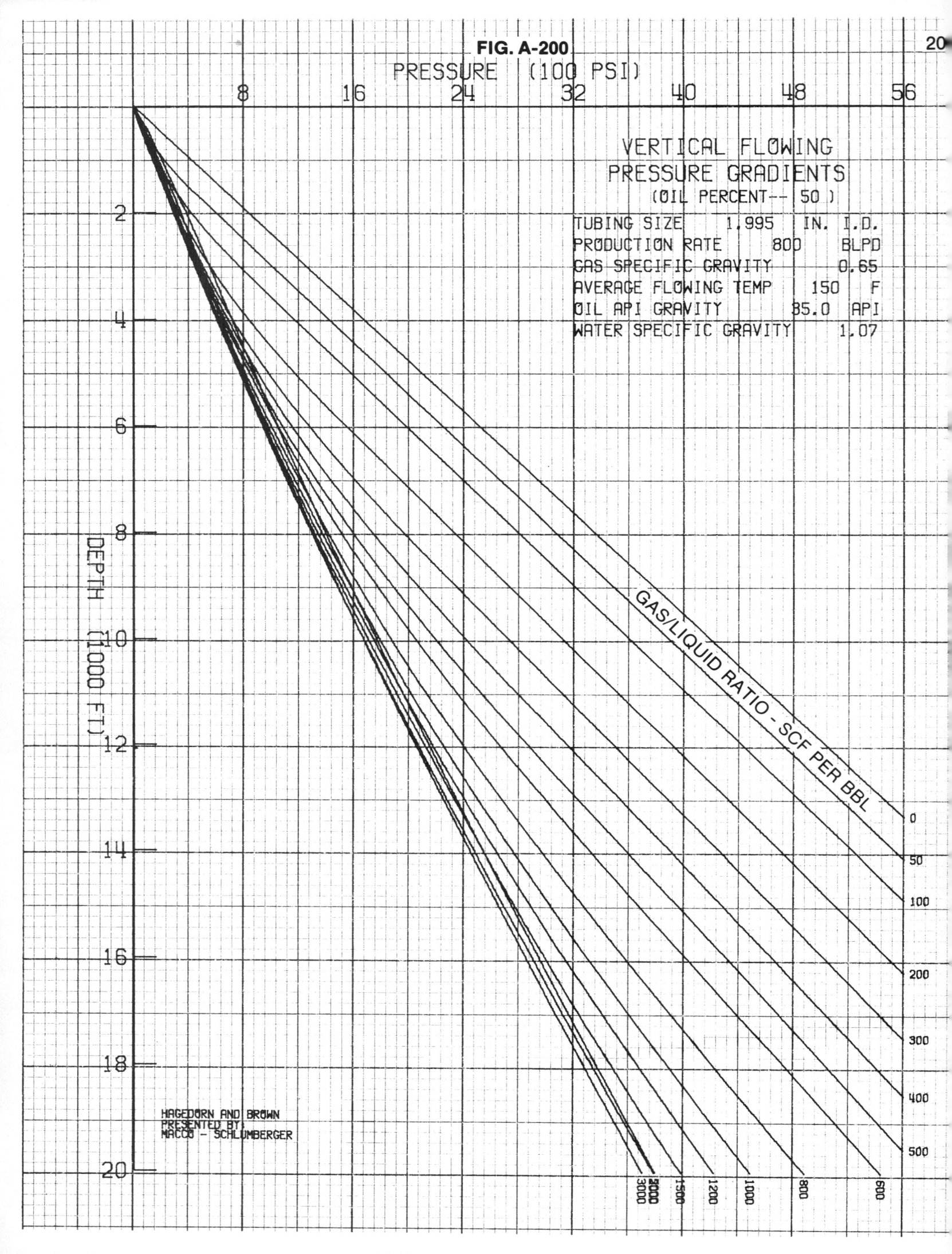

PRESSURE (100 PSI)

8   16   24   32   40   48   56

VERTICAL FLOWING
PRESSURE GRADIENTS
(OIL PERCENT-- 50)

TUBING SIZE	1.995	IN. I.D.
PRODUCTION RATE	800	BLPD
GAS SPECIFIC GRAVITY	0.65	
AVERAGE FLOWING TEMP	150	F
OIL API GRAVITY	35.0	API
WATER SPECIFIC GRAVITY	1.07	

DEPTH (1000 FT)

GAS/LIQUID RATIO - SCF PER BBL

HAGEDORN AND BROWN
PRESENTED BY
MACCO - SCHLUMBERGER

0
50
100
200
300
400
500

3000 2000 1500 1200 1000 800 600

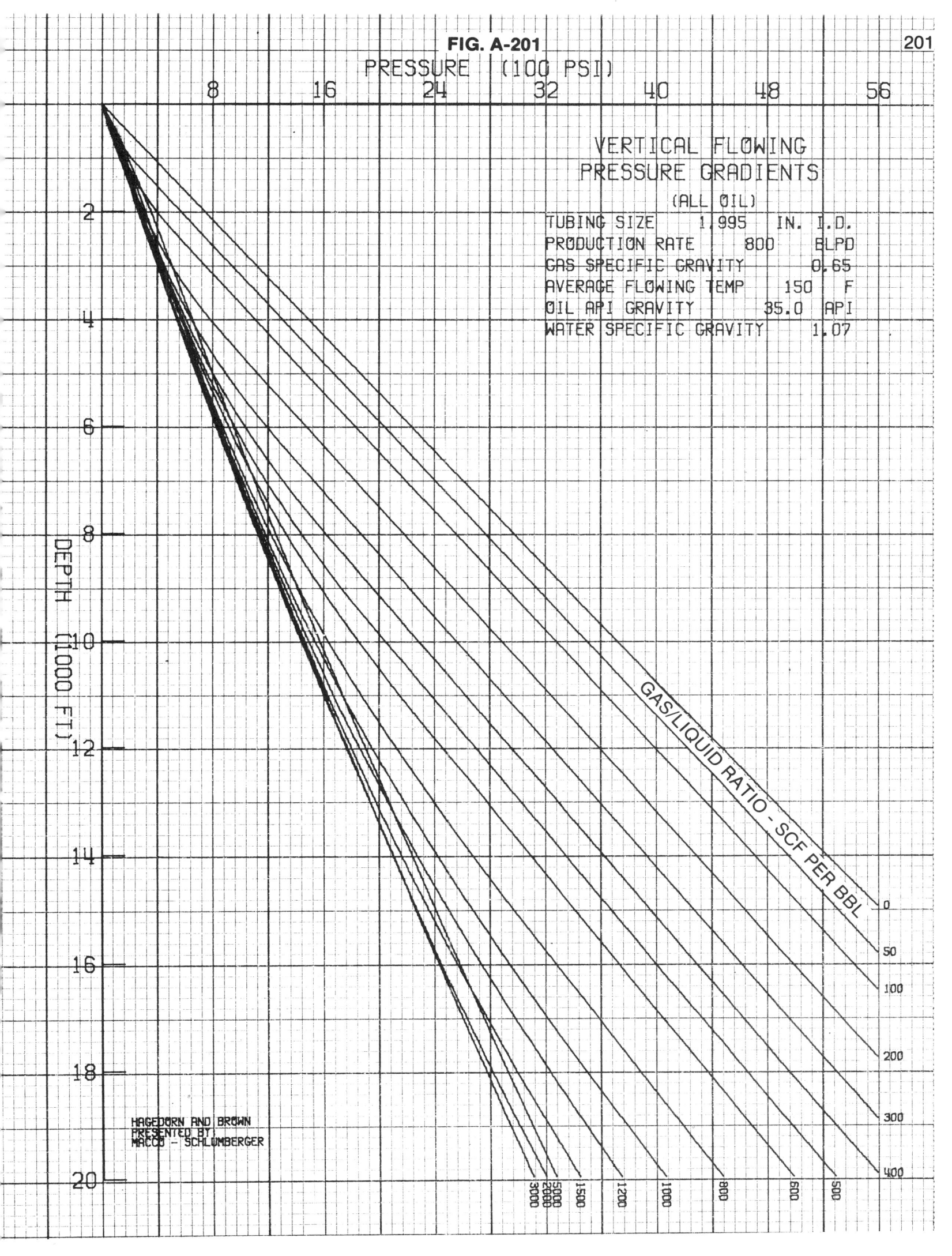

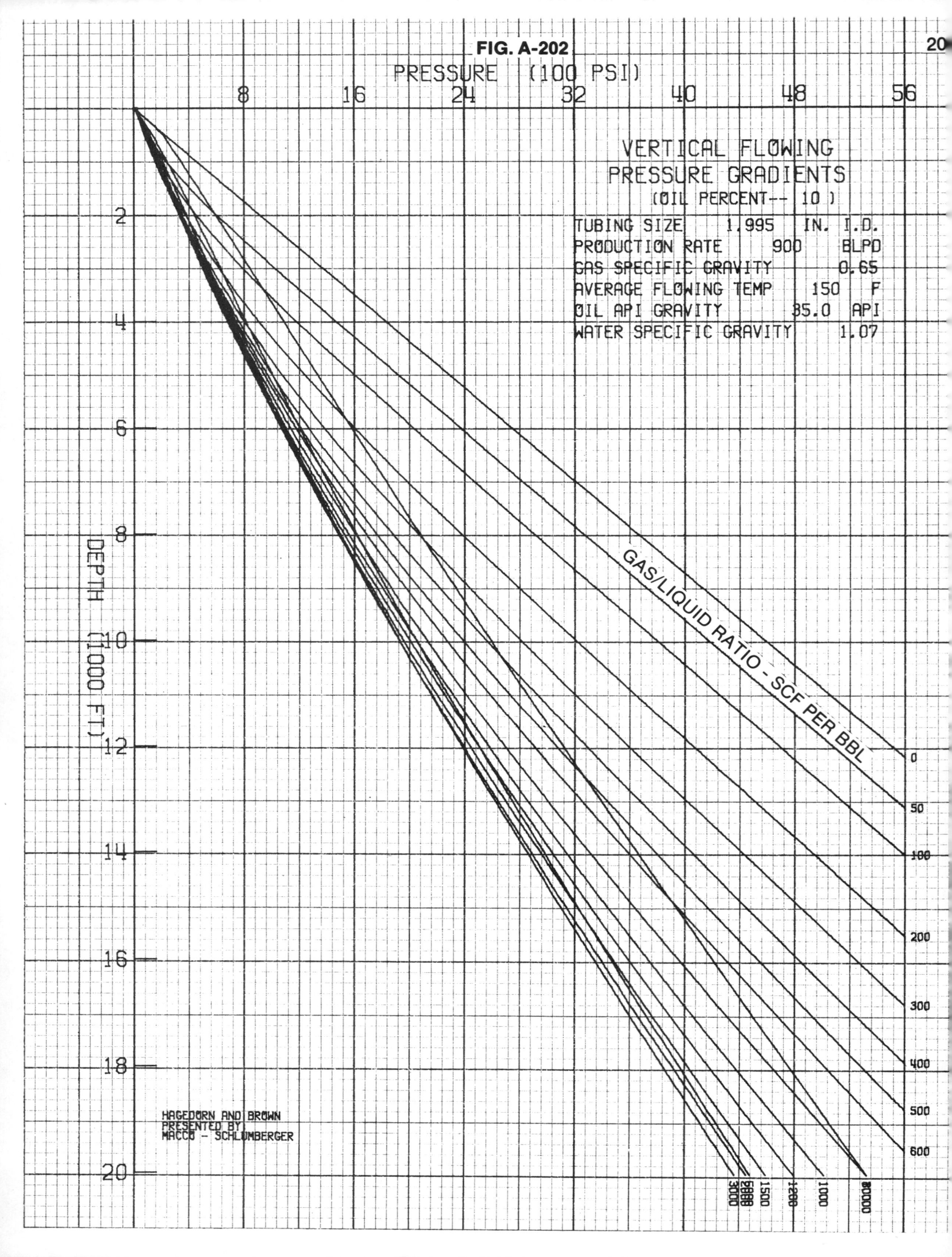

PRESSURE (100 PSI)

VERTICAL FLOWING
PRESSURE GRADIENTS
(OIL PERCENT-- 10)

TUBING SIZE	1.995	IN. I.D.
PRODUCTION RATE	900	BLPD
GAS SPECIFIC GRAVITY	0.65	
AVERAGE FLOWING TEMP	150	F
OIL API GRAVITY	35.0	API
WATER SPECIFIC GRAVITY	1.07	

GAS/LIQUID RATIO - SCF PER BBL

DEPTH (1000 FT)

HAGEDORN AND BROWN
PRESENTED BY
MACCO - SCHLUMBERGER

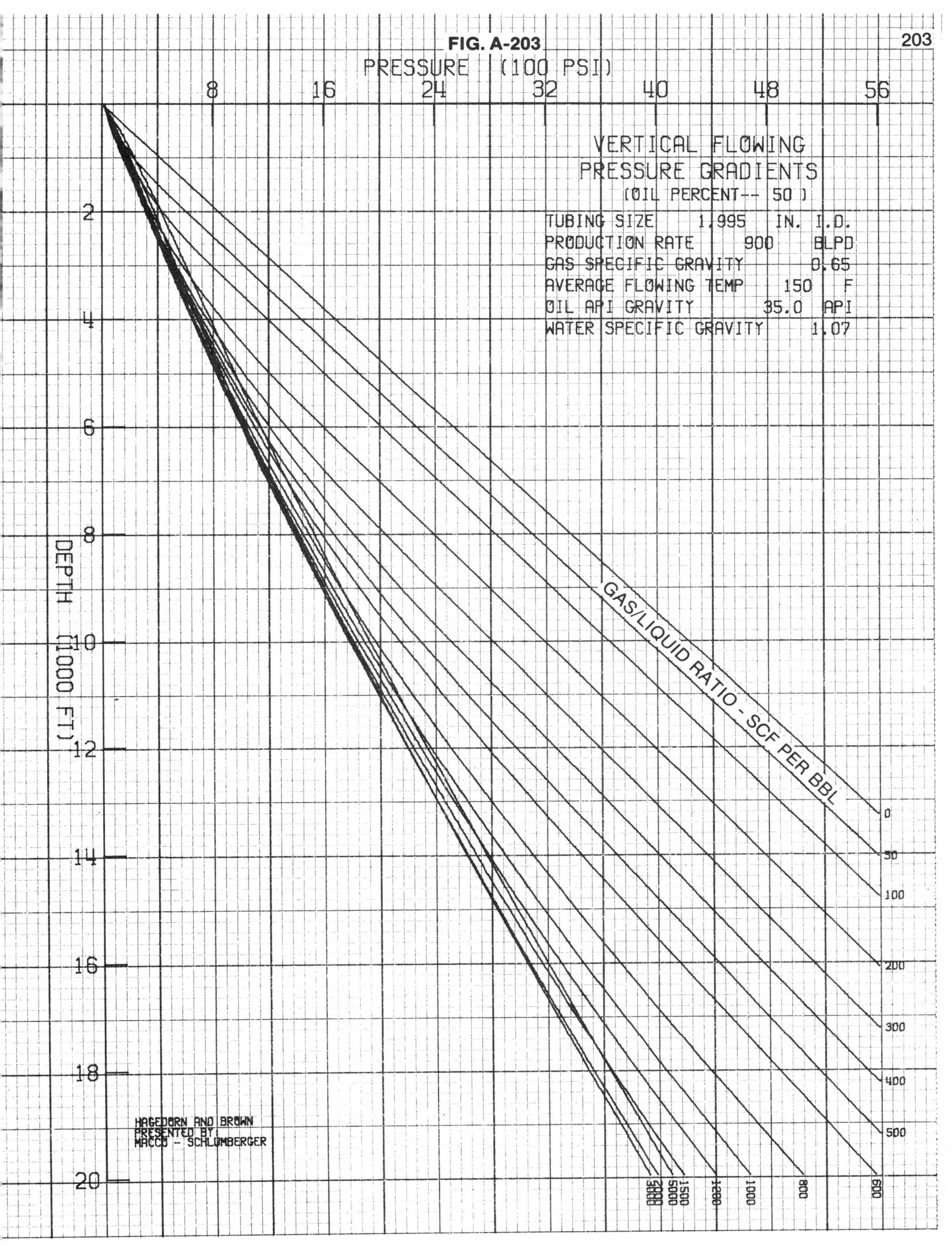

PRESSURE (100 PSI)

VERTICAL FLOWING
PRESSURE GRADIENTS
(OIL PERCENT-- 50)

TUBING SIZE	1.995	IN. I.D.
PRODUCTION RATE	900	BLPD
GAS SPECIFIC GRAVITY	0.65	
AVERAGE FLOWING TEMP	150	F
OIL API GRAVITY	35.0	API
WATER SPECIFIC GRAVITY	1.07	

DEPTH (1000 FT)

GAS/LIQUID RATIO - SCF PER BBL

HAGEDORN AND BROWN
PRESENTED BY
MACCO - SCHLUMBERGER

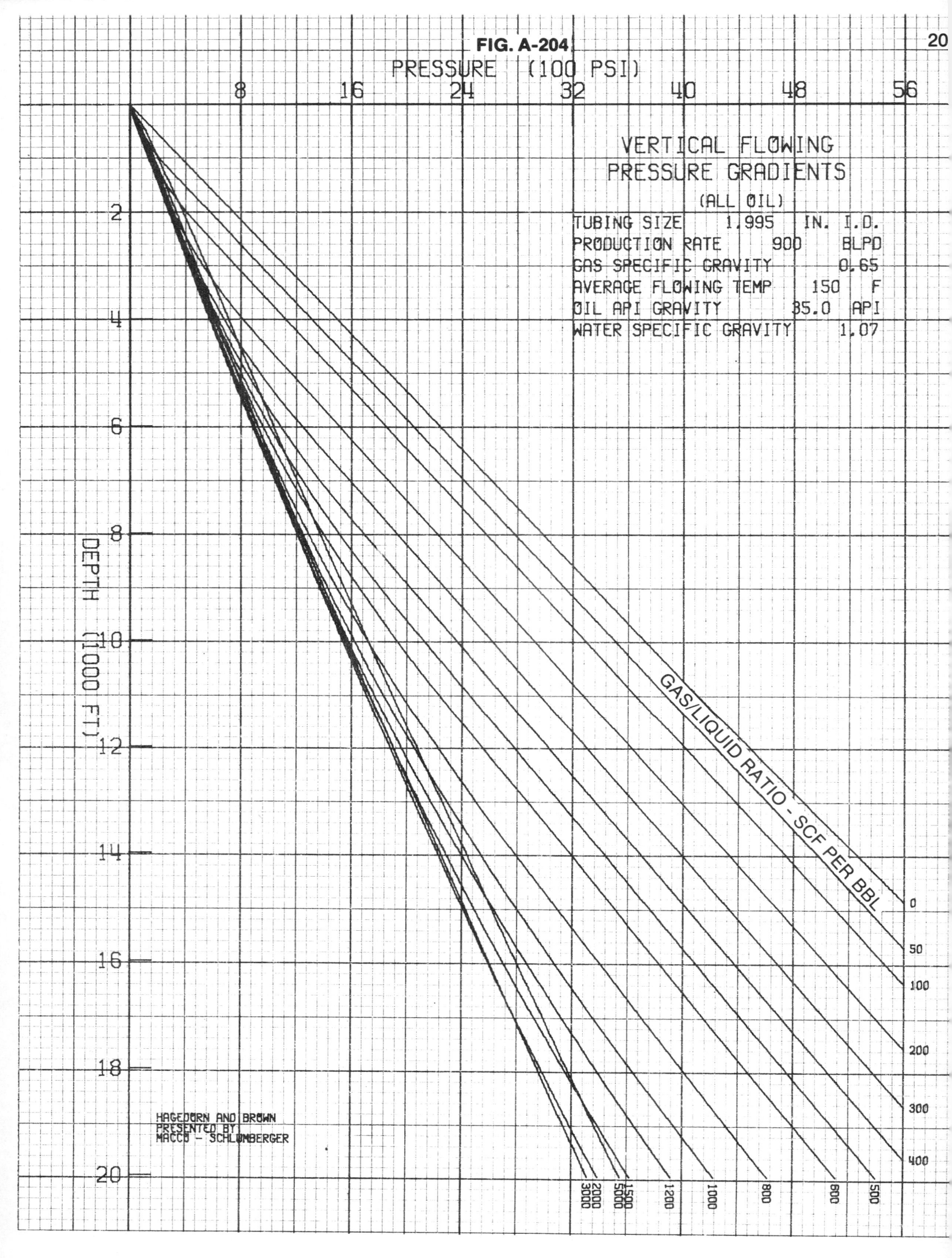

FIG. A-204

PRESSURE (100 PSI)

VERTICAL FLOWING
PRESSURE GRADIENTS
(ALL OIL)

TUBING SIZE	1.995	IN. I.D.
PRODUCTION RATE	900	BLPD
GAS SPECIFIC GRAVITY	0.65	
AVERAGE FLOWING TEMP	150	F
OIL API GRAVITY	35.0	API
WATER SPECIFIC GRAVITY	1.07	

DEPTH (1000 FT)

GAS/LIQUID RATIO - SCF PER BBL

HAGEDORN AND BROWN
PRESENTED BY
MACCO — SCHLUMBERGER

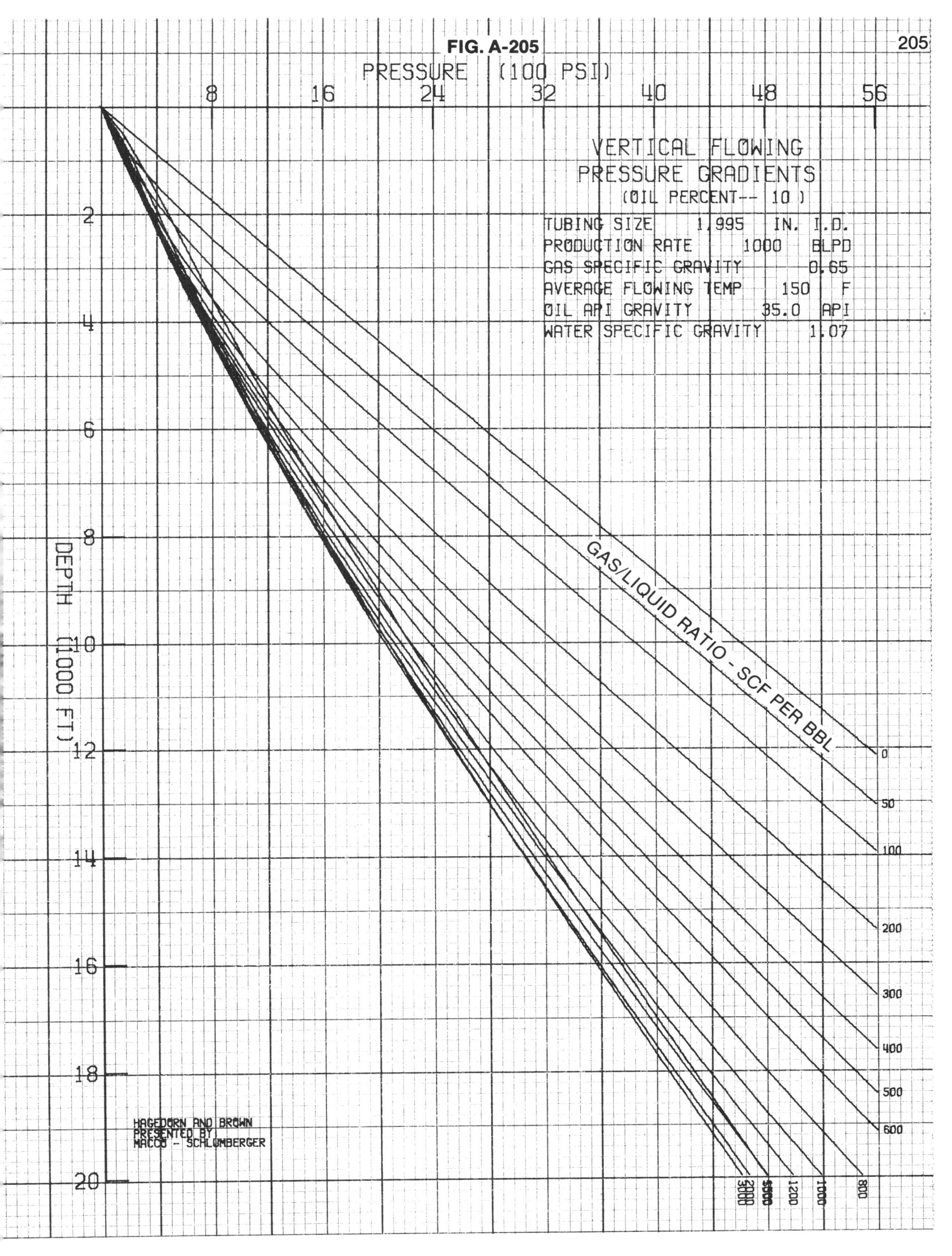

PRESSURE (100 PSI)

VERTICAL FLOWING
PRESSURE GRADIENTS
(OIL PERCENT-- 10)

TUBING SIZE	1.995	IN. I.D.
PRODUCTION RATE	1000	BLPD
GAS SPECIFIC GRAVITY		0.65
AVERAGE FLOWING TEMP	150	F
OIL API GRAVITY	35.0	API
WATER SPECIFIC GRAVITY		1.07

DEPTH (1000 FT)

GAS/LIQUID RATIO - SCF PER BBL

HAGEDORN AND BROWN
PRESENTED BY
MACCO - SCHLUMBERGER

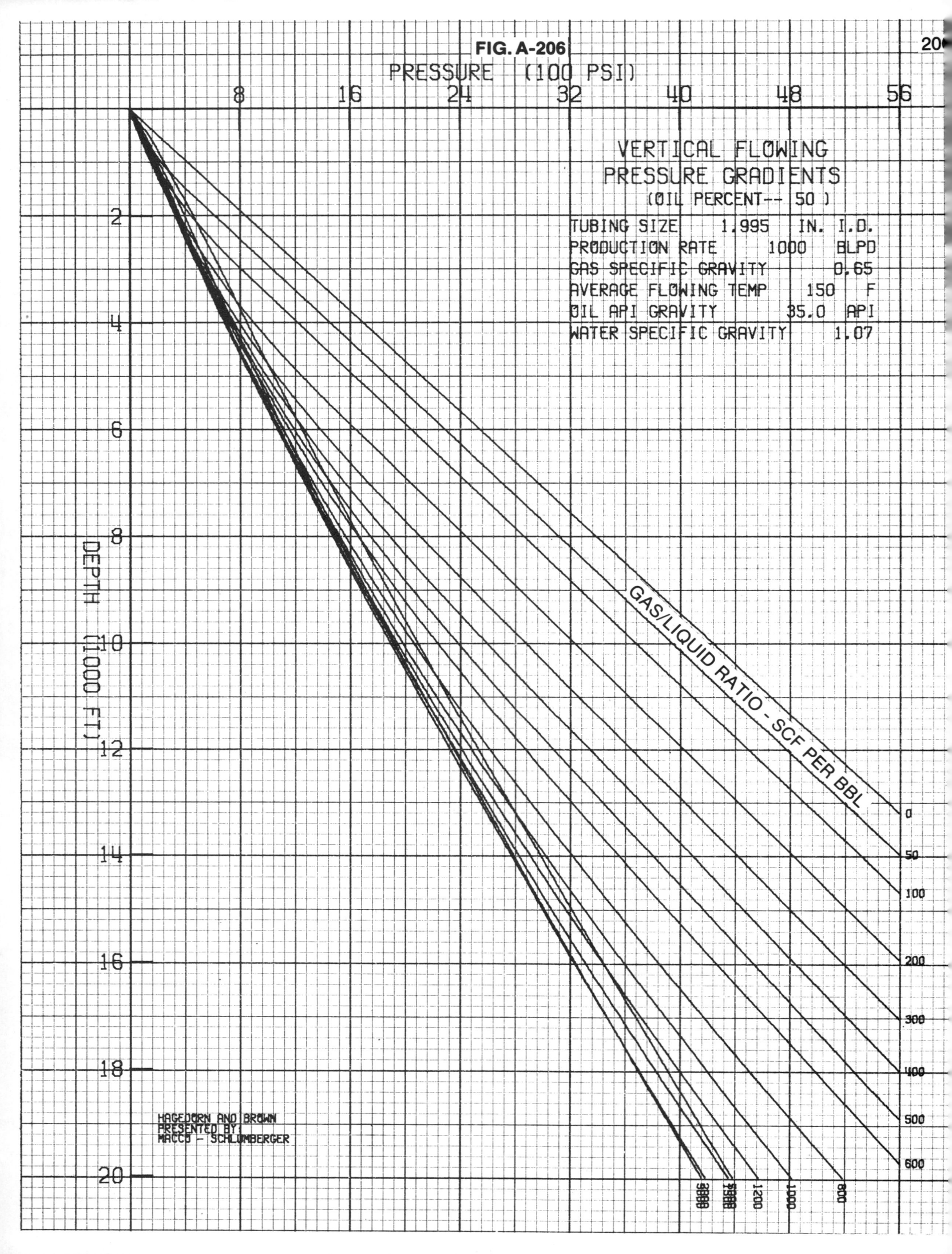

FIG. A-206

VERTICAL FLOWING
PRESSURE GRADIENTS
(OIL PERCENT-- 50)

TUBING SIZE   1.995   IN. I.D.
PRODUCTION RATE   1000   BLPD
GAS SPECIFIC GRAVITY   0.65
AVERAGE FLOWING TEMP   150   F
OIL API GRAVITY   35.0   API
WATER SPECIFIC GRAVITY   1.07

GAS/LIQUID RATIO - SCF PER BBL

HAGEDORN AND BROWN
PRESENTED BY:
MACCO - SCHLUMBERGER

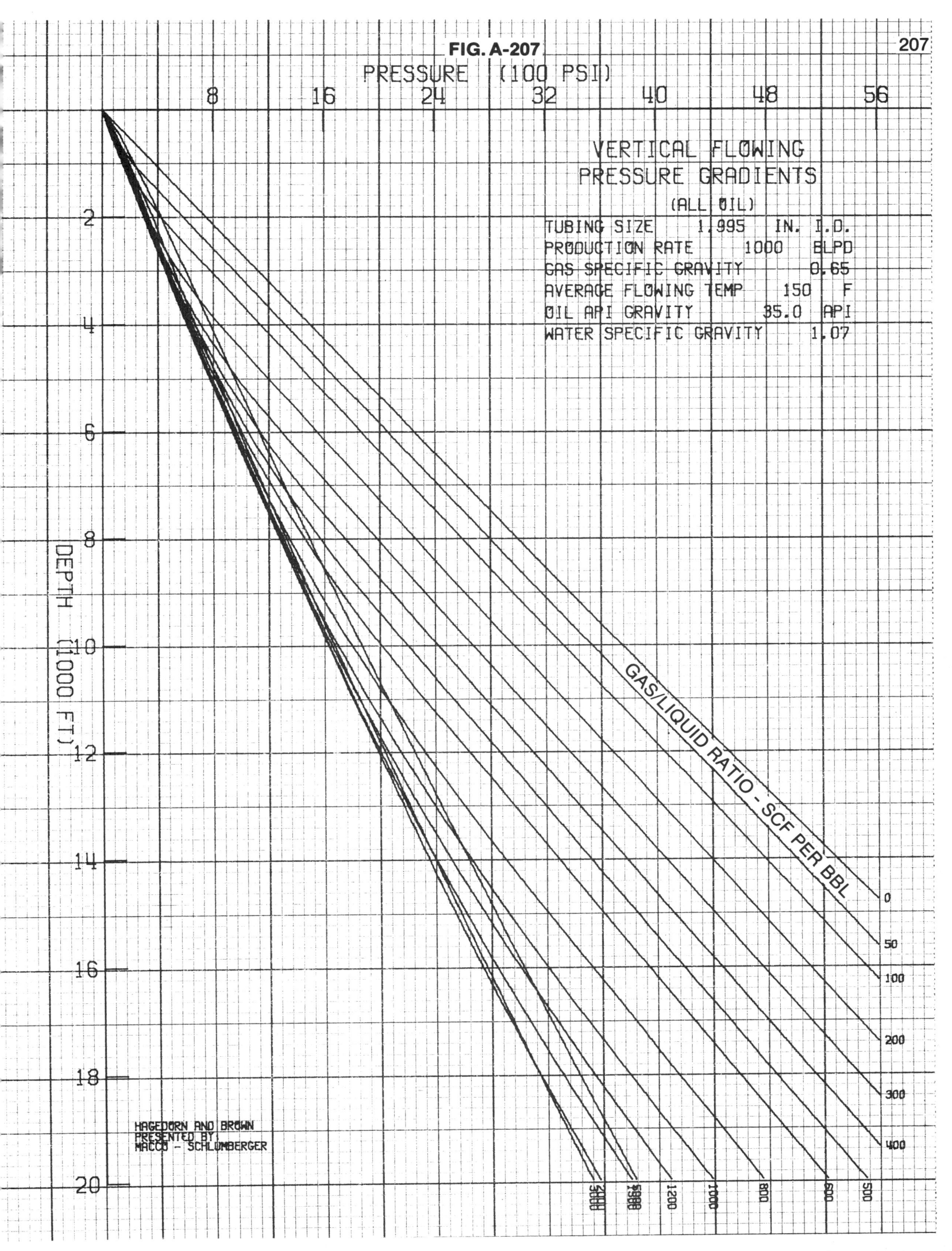

FIG. A-207

207

VERTICAL FLOWING
PRESSURE GRADIENTS
(ALL OIL)

TUBING SIZE	1.995	IN. I.D.
PRODUCTION RATE	1000	BLPD
GAS SPECIFIC GRAVITY	0.65	
AVERAGE FLOWING TEMP	150	F
OIL API GRAVITY	35.0	API
WATER SPECIFIC GRAVITY	1.07	

PRESSURE (100 PSI)

DEPTH (1000 FT)

GAS/LIQUID RATIO - SCF PER BBL

HAGEDORN AND BROWN
PRESENTED BY
MACCO - SCHLUMBERGER

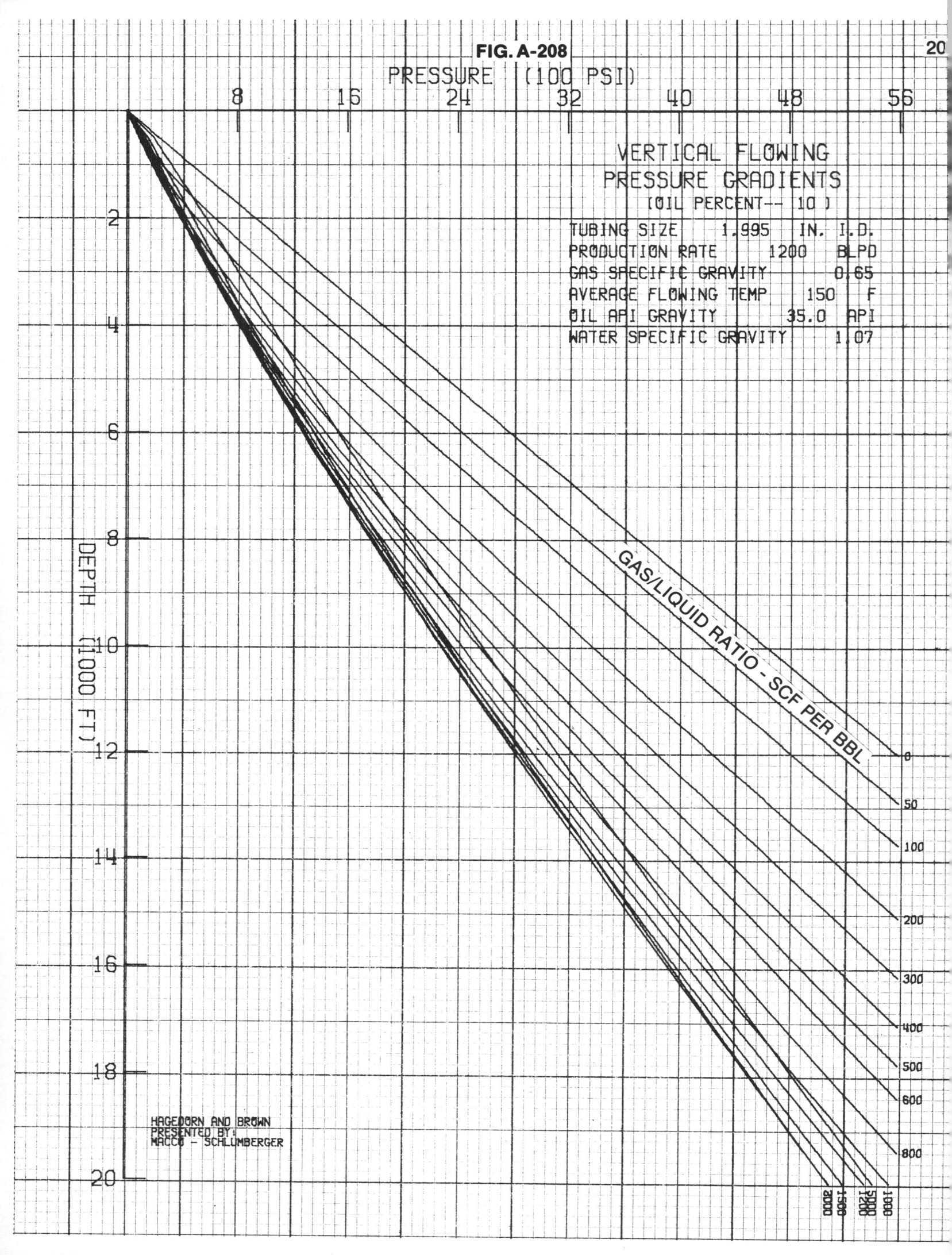

FIG. A-208

20

PRESSURE (100 PSI)

VERTICAL FLOWING
PRESSURE GRADIENTS
(OIL PERCENT-- 10)

TUBING SIZE	1.995	IN. I.D.
PRODUCTION RATE	1200	BLPD
GAS SPECIFIC GRAVITY	0.65	
AVERAGE FLOWING TEMP	150	F
OIL API GRAVITY	35.0	API
WATER SPECIFIC GRAVITY	1.07	

GAS/LIQUID RATIO - SCF PER BBL

DEPTH (1000 FT)

HAGEDORN AND BROWN
PRESENTED BY:
MACCO - SCHLUMBERGER

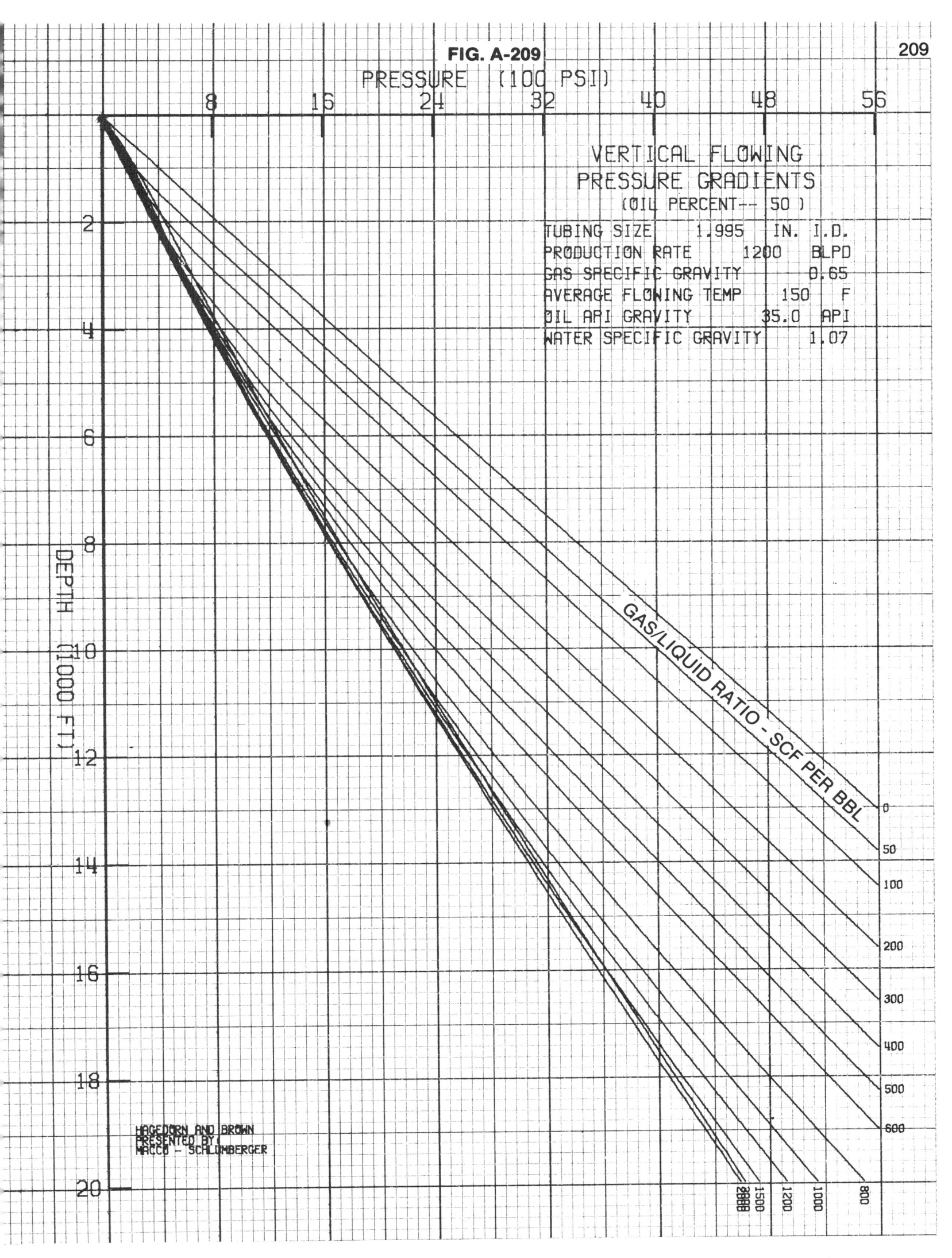

PRESSURE (100 PSI)

VERTICAL FLOWING
PRESSURE GRADIENTS
(OIL PERCENT-- 50)

TUBING SIZE	1.995	IN. I.D.
PRODUCTION RATE	1200	BLPD
GAS SPECIFIC GRAVITY	0.65	
AVERAGE FLOWING TEMP	150	F
OIL API GRAVITY	35.0	API
WATER SPECIFIC GRAVITY	1.07	

GAS/LIQUID RATIO - SCF PER BBL

DEPTH (1000 FT)

HAGEDORN AND BROWN
PRESENTED BY:
MACCO - SCHLUMBERGER

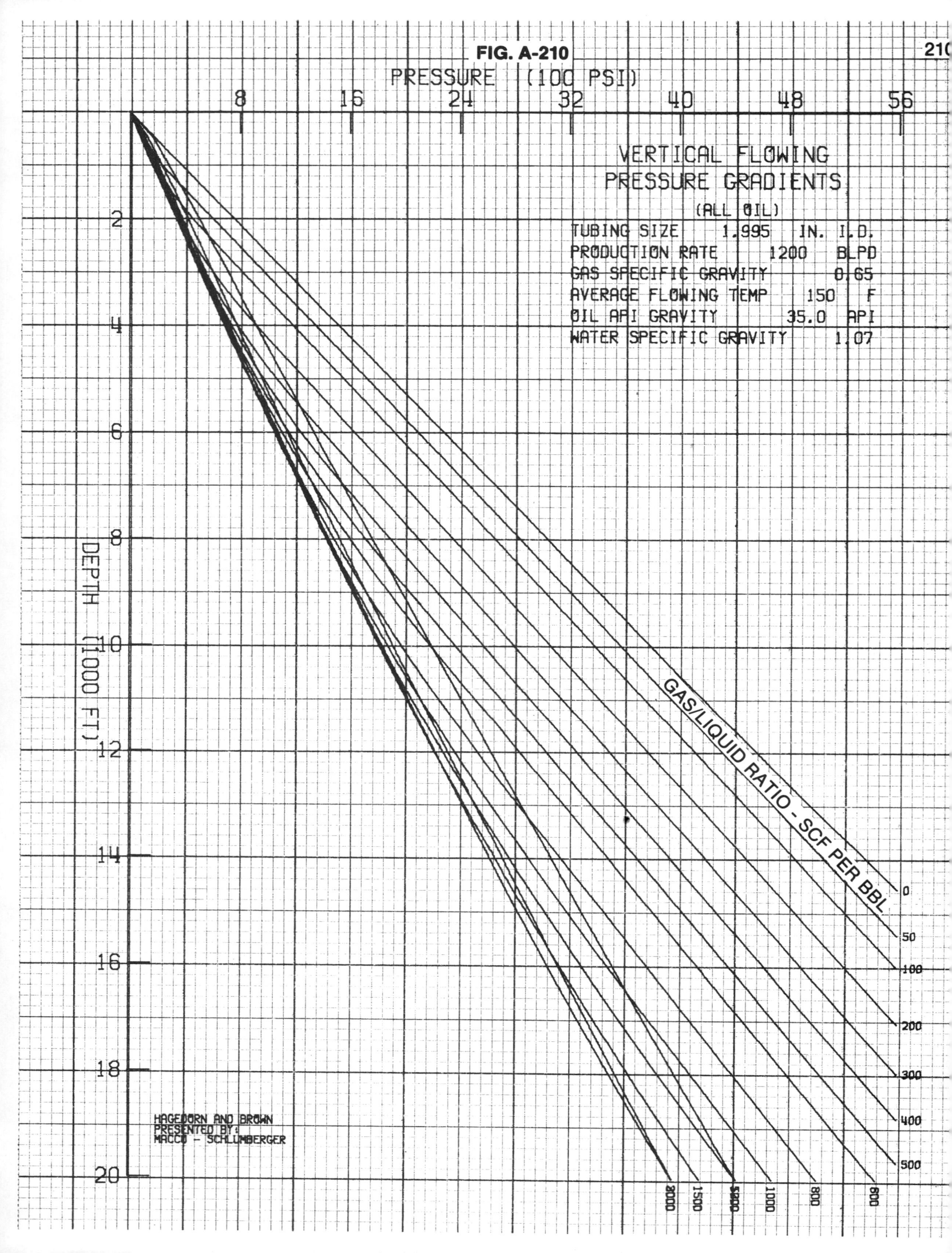

FIG. A-210

210

PRESSURE (100 PSI)

VERTICAL FLOWING
PRESSURE GRADIENTS
(ALL OIL)

TUBING SIZE	1.995	IN. I.D.
PRODUCTION RATE	1200	BLPD
GAS SPECIFIC GRAVITY	0.65	
AVERAGE FLOWING TEMP	150	F
OIL API GRAVITY	35.0	API
WATER SPECIFIC GRAVITY	1.07	

DEPTH (1000 FT)

GAS/LIQUID RATIO - SCF PER BBL

0
50
100
200
300
400
500

2000 1500 1200 1000 800 600

HAGEDORN AND BROWN
PRESENTED BY:
MACCO – SCHLUMBERGER

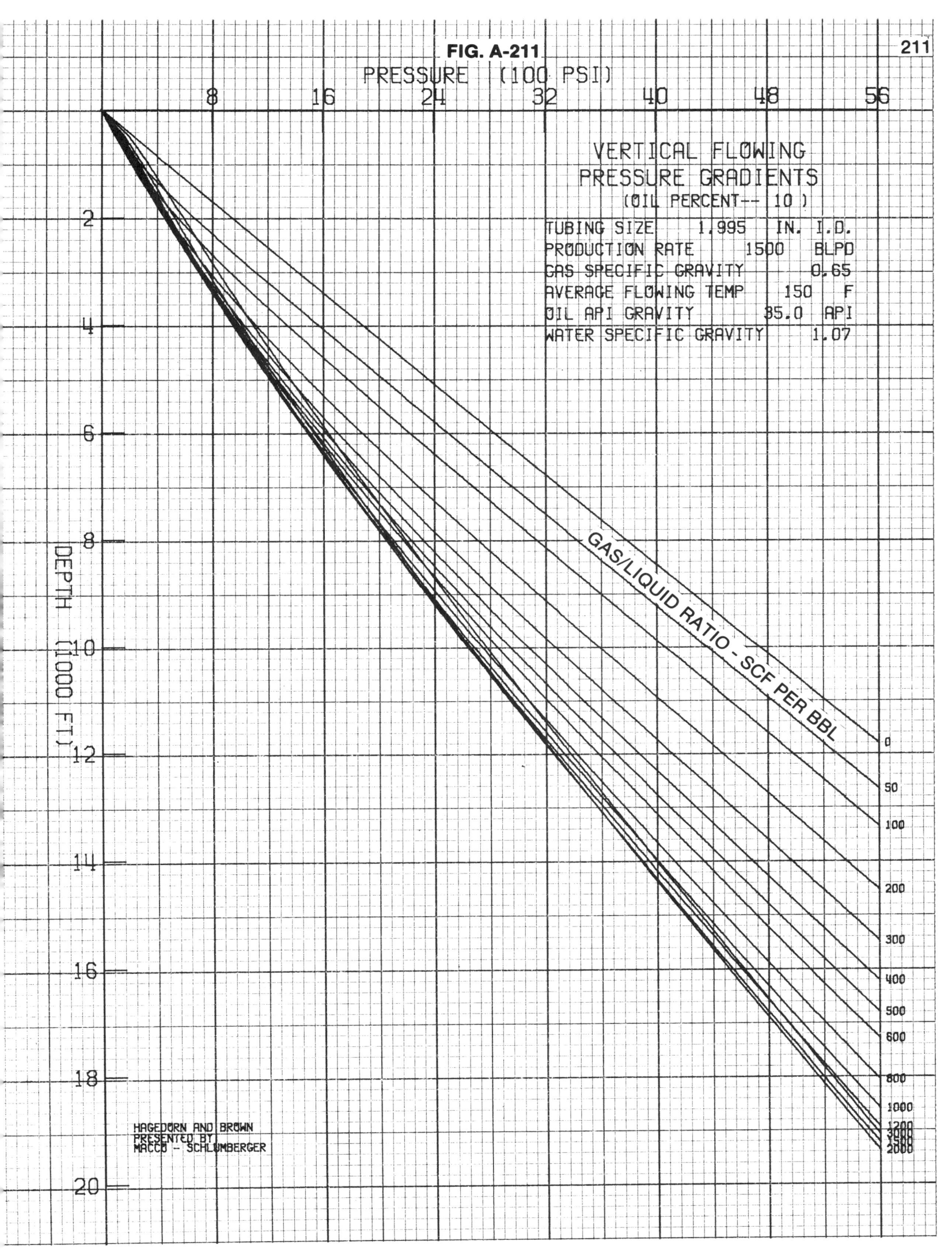

PRESSURE (100 PSI)

DEPTH (1000 FT)

VERTICAL FLOWING
PRESSURE GRADIENTS
(OIL PERCENT-- 10)

TUBING SIZE   1.995   IN. I.D.
PRODUCTION RATE   1500   BLPD
GAS SPECIFIC GRAVITY   0.65
AVERAGE FLOWING TEMP   150   F
OIL API GRAVITY   85.0 API
WATER SPECIFIC GRAVITY   1.07

GAS/LIQUID RATIO - SCF PER BBL

0
50
100
200
300
400
500
600
800
1000
1200
3000
2000

HAGEDORN AND BROWN
PRESENTED BY
MACCO - SCHLUMBERGER

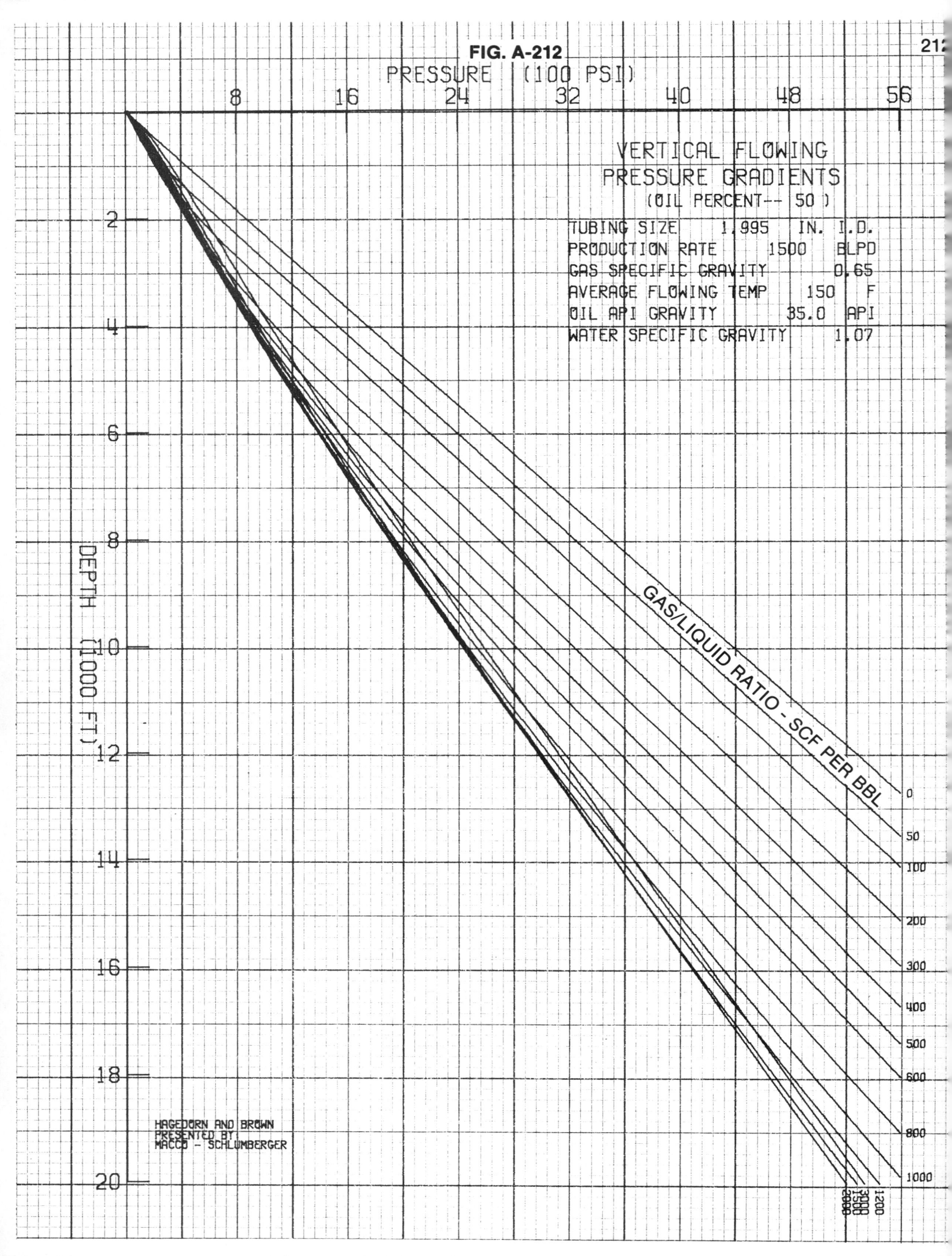

PRESSURE (100 PSI)

VERTICAL FLOWING
PRESSURE GRADIENTS
(OIL PERCENT-- 50)

TUBING SIZE   1.995   IN. I.D.
PRODUCTION RATE   1500   BLPD
GAS SPECIFIC GRAVITY   0.65
AVERAGE FLOWING TEMP   150   F
OIL API GRAVITY   35.0   API
WATER SPECIFIC GRAVITY   1.07

DEPTH (1000 FT)

GAS/LIQUID RATIO - SCF PER BBL

HAGEDORN AND BROWN
PRESENTED BY
MACCO - SCHLUMBERGER

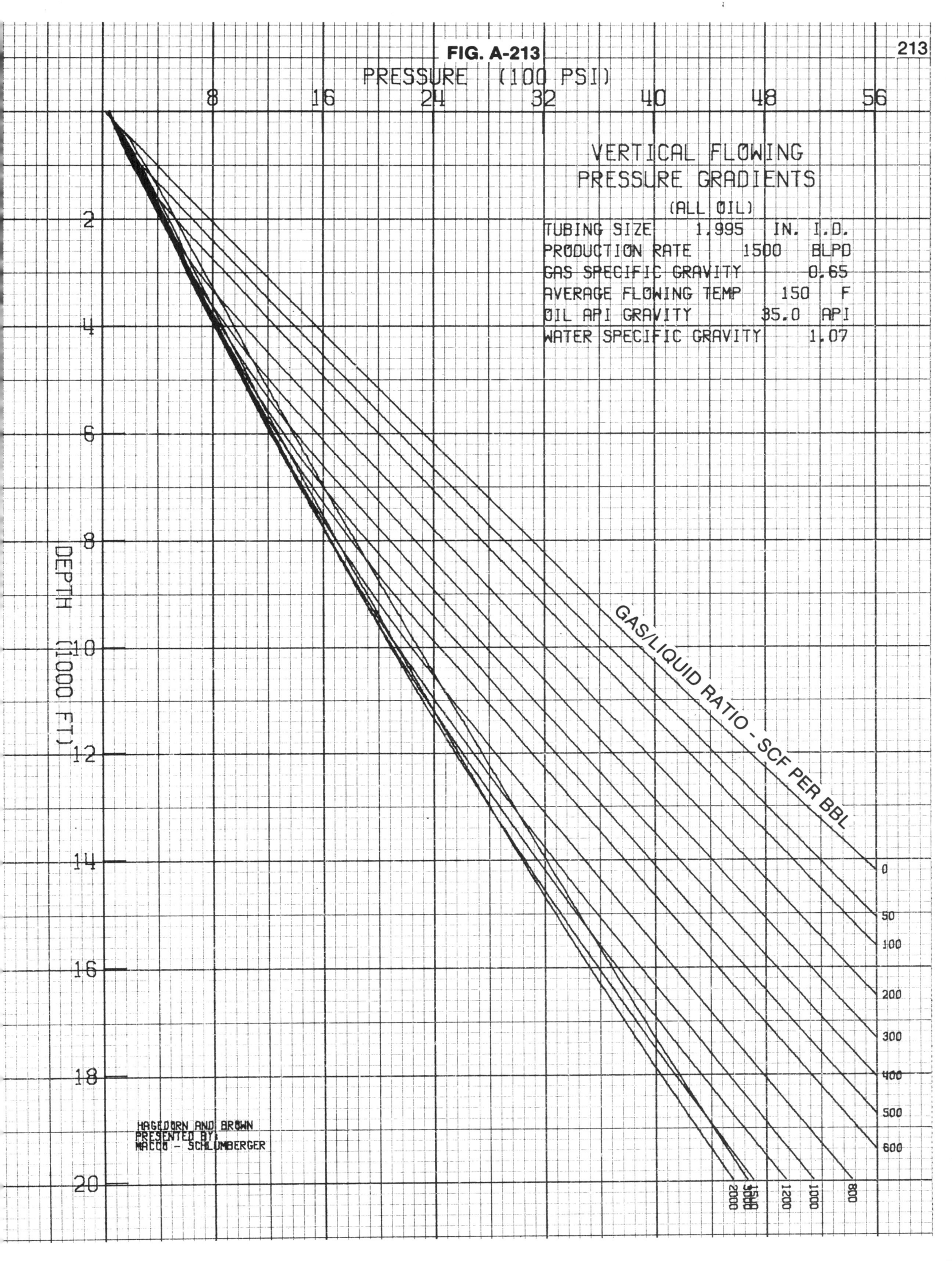

FIG. A-213

213

PRESSURE (100 PSI)

VERTICAL FLOWING
PRESSURE GRADIENTS
(ALL OIL)

TUBING SIZE	1.995	IN. I.D.
PRODUCTION RATE	1500	BLPD
GAS SPECIFIC GRAVITY		0.65
AVERAGE FLOWING TEMP	150	F
OIL API GRAVITY	35.0	API
WATER SPECIFIC GRAVITY		1.07

DEPTH (1000 FT)

GAS/LIQUID RATIO - SCF PER BBL

HAGEDORN AND BROWN
PRESENTED BY:
MACCO - SCHLUMBERGER

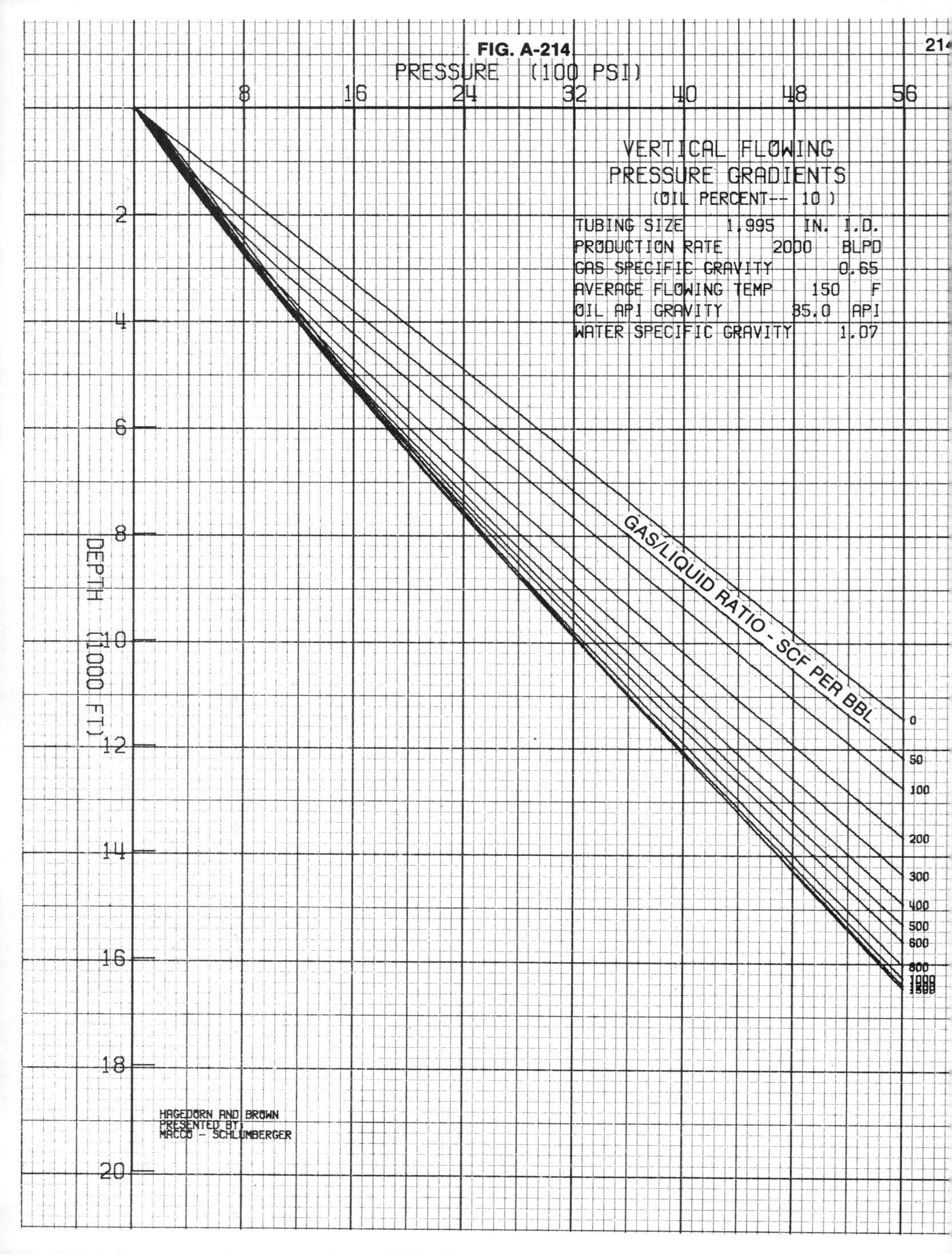

FIG. A-214

PRESSURE (100 PSI)

DEPTH (1000 FT)

VERTICAL FLOWING
PRESSURE GRADIENTS
(OIL PERCENT-- 10)

TUBING SIZE	1.995	IN. I.D.
PRODUCTION RATE	2000	BLPD
GAS SPECIFIC GRAVITY		0.65
AVERAGE FLOWING TEMP	150	F
OIL API GRAVITY	35.0	API
WATER SPECIFIC GRAVITY		1.07

GAS/LIQUID RATIO - SCF PER BBL

0
50
100
200
300
400
500
600
800
1000
1500

HAGEDORN AND BROWN
PRESENTED BY
MACCO - SCHLUMBERGER

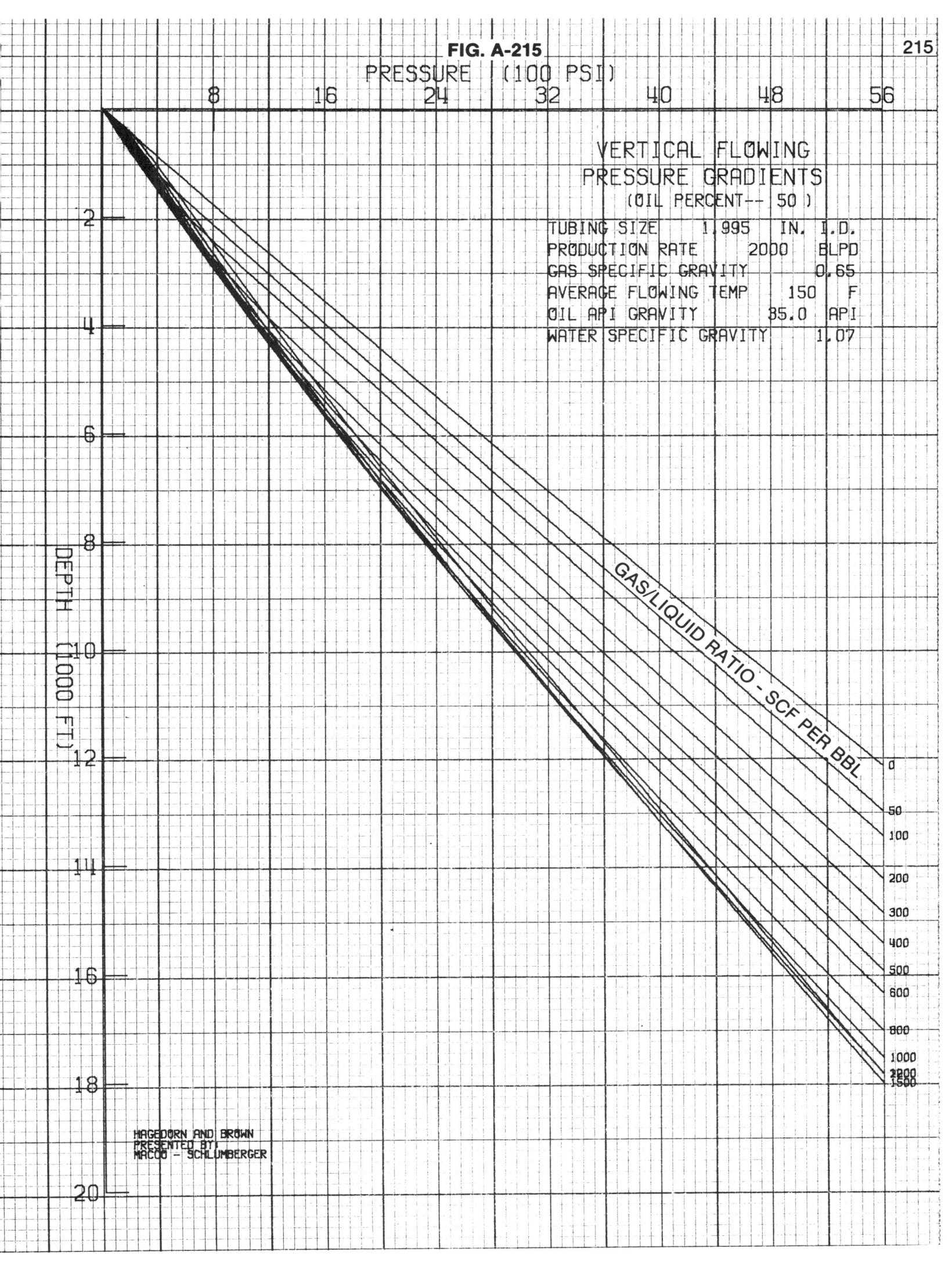

FIG. A-215

215

VERTICAL FLOWING
PRESSURE GRADIENTS
(OIL PERCENT-- 50)

TUBING SIZE	1.995	IN. I.D.
PRODUCTION RATE	2000	BLPD
GAS SPECIFIC GRAVITY	0.65	
AVERAGE FLOWING TEMP	150	F
OIL API GRAVITY	85.0	API
WATER SPECIFIC GRAVITY	1.07	

PRESSURE (100 PSI)

DEPTH (1000 FT)

GAS/LIQUID RATIO - SCF PER BBL

HAGEDORN AND BROWN
PRESENTED BY:
MACCO - SCHLUMBERGER

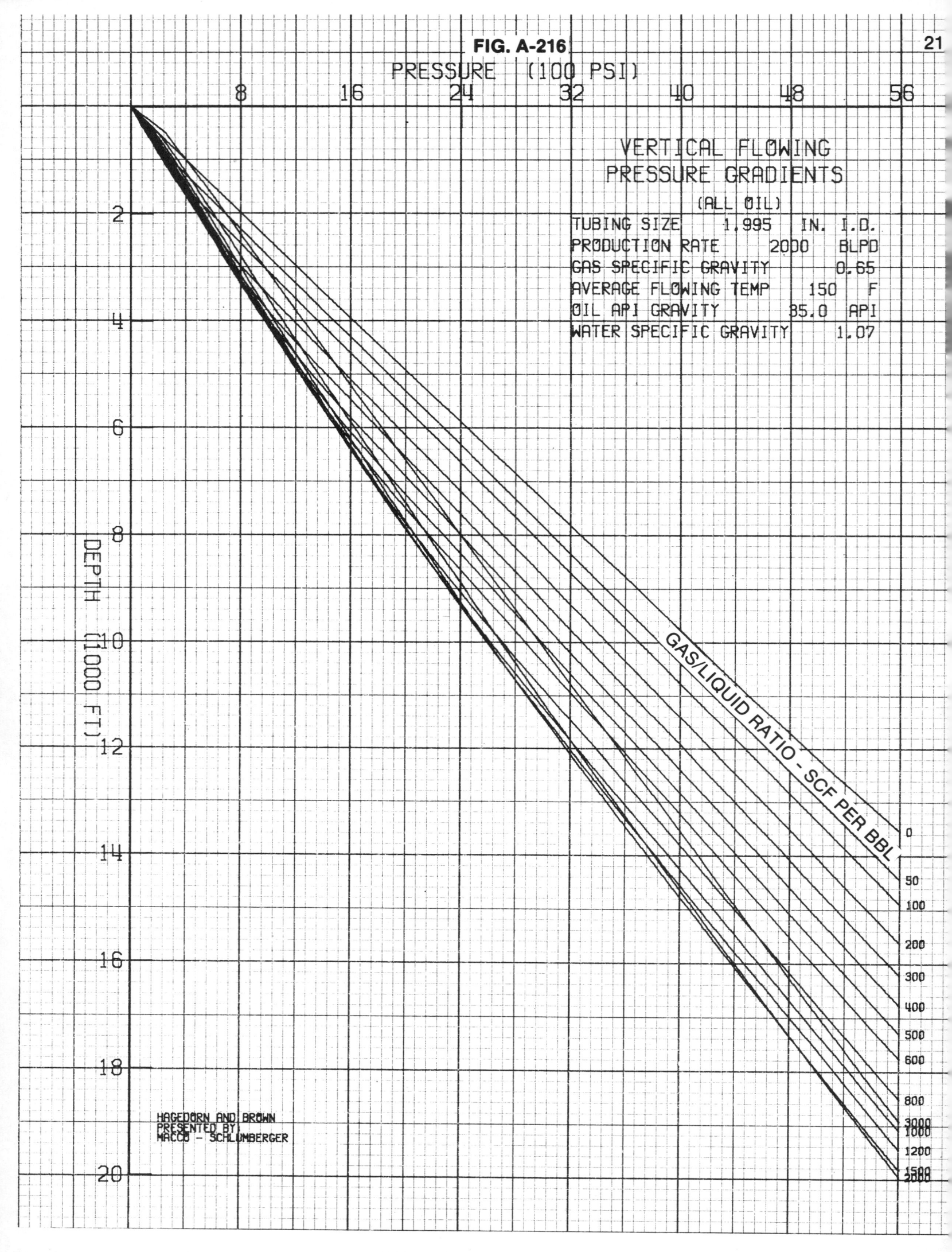

FIG. A-216

VERTICAL FLOWING
PRESSURE GRADIENTS
(ALL OIL)

TUBING SIZE	1.995 IN. I.D.
PRODUCTION RATE	2000 BLPD
GAS SPECIFIC GRAVITY	0.65
AVERAGE FLOWING TEMP	150 F
OIL API GRAVITY	35.0 API
WATER SPECIFIC GRAVITY	1.07

PRESSURE (100 PSI)

DEPTH (1000 FT)

GAS/LIQUID RATIO - SCF PER BBL

HAGEDORN AND BROWN
PRESENTED BY
MACCO - SCHLUMBERGER

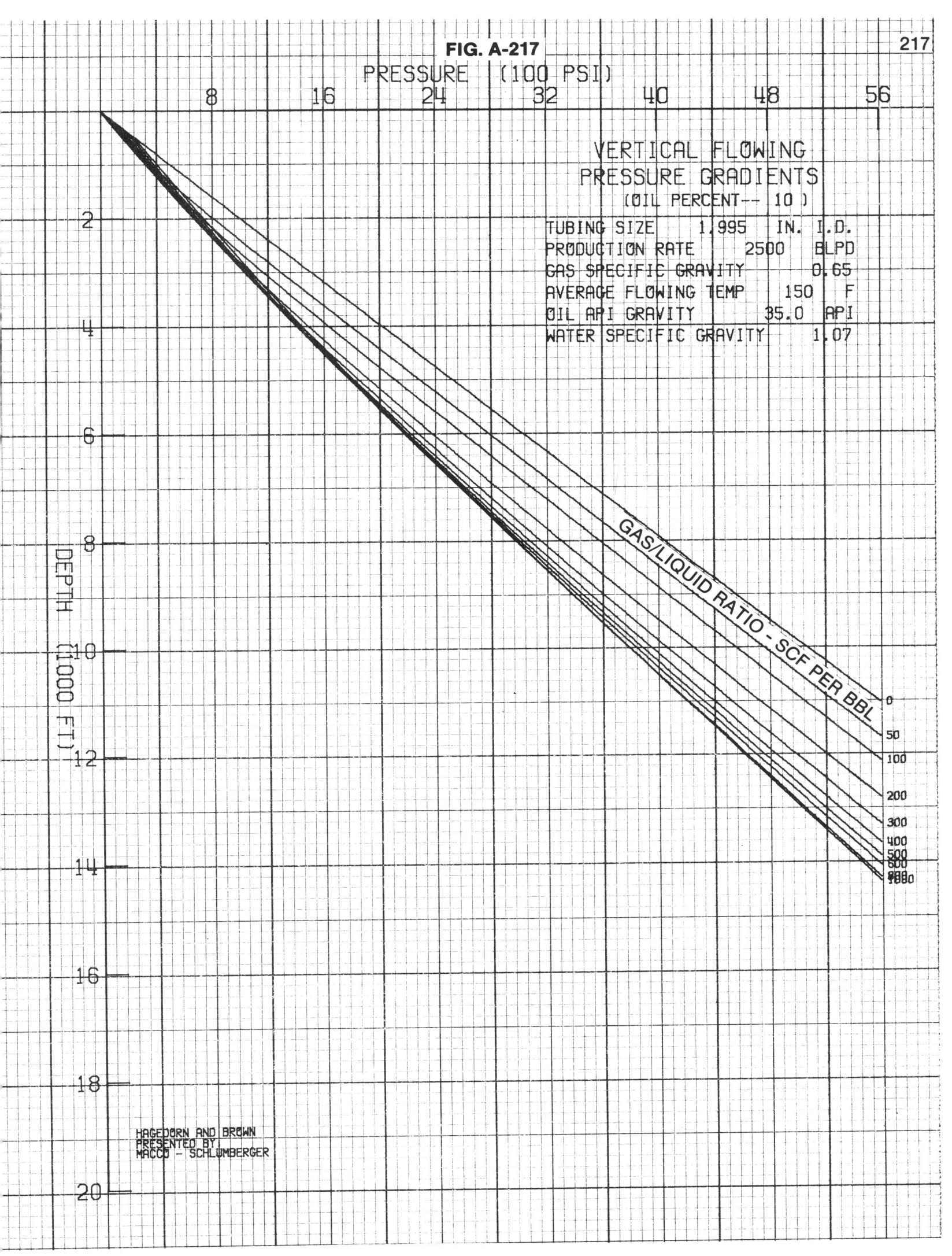

FIG. A-217

217

PRESSURE (100 PSI)

VERTICAL FLOWING
PRESSURE GRADIENTS
(OIL PERCENT-- 10)

TUBING SIZE	1.995	IN. I.D.
PRODUCTION RATE	2500	BLPD
GAS SPECIFIC GRAVITY		0.65
AVERAGE FLOWING TEMP	150	F
OIL API GRAVITY	35.0	API
WATER SPECIFIC GRAVITY		1.07

DEPTH (1000 FT)

GAS/LIQUID RATIO - SCF PER BBL

0
50
100
200
300
400
500
600
800

HAGEDORN AND BROWN
PRESENTED BY
MACCO - SCHLUMBERGER

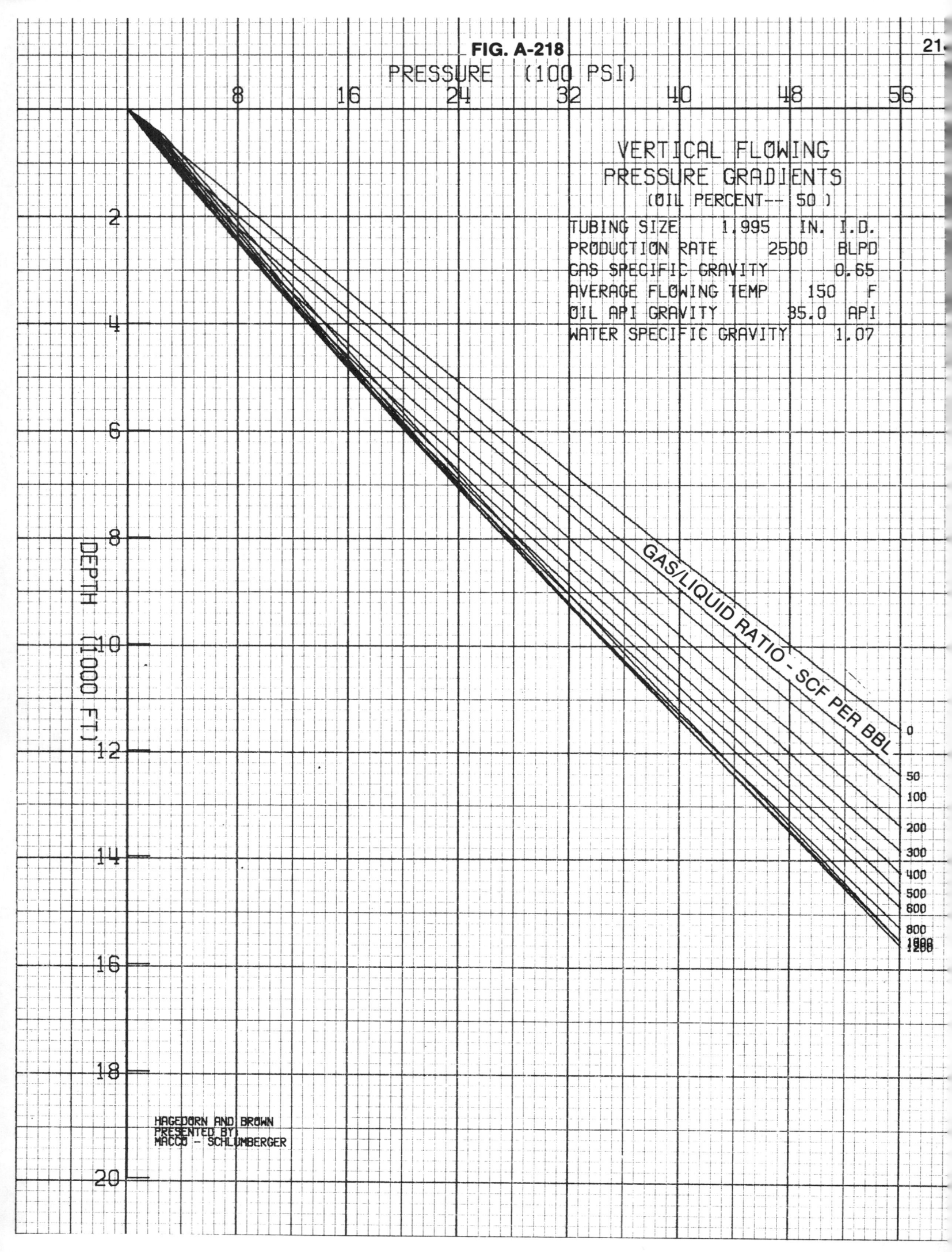

PRESSURE (100 PSI)

VERTICAL FLOWING
PRESSURE GRADIENTS
(OIL PERCENT-- 50)

TUBING SIZE	1.995	IN. I.D.
PRODUCTION RATE	2500	BLPD
GAS SPECIFIC GRAVITY		0.65
AVERAGE FLOWING TEMP	150	F
OIL API GRAVITY	35.0	API
WATER SPECIFIC GRAVITY		1.07

DEPTH (1000 FT)

GAS/LIQUID RATIO - SCF PER BBL

0
50
100
200
300
400
500
600
800
1000
1200

HAGEDORN AND BROWN
PRESENTED BY
MACCO - SCHLUMBERGER

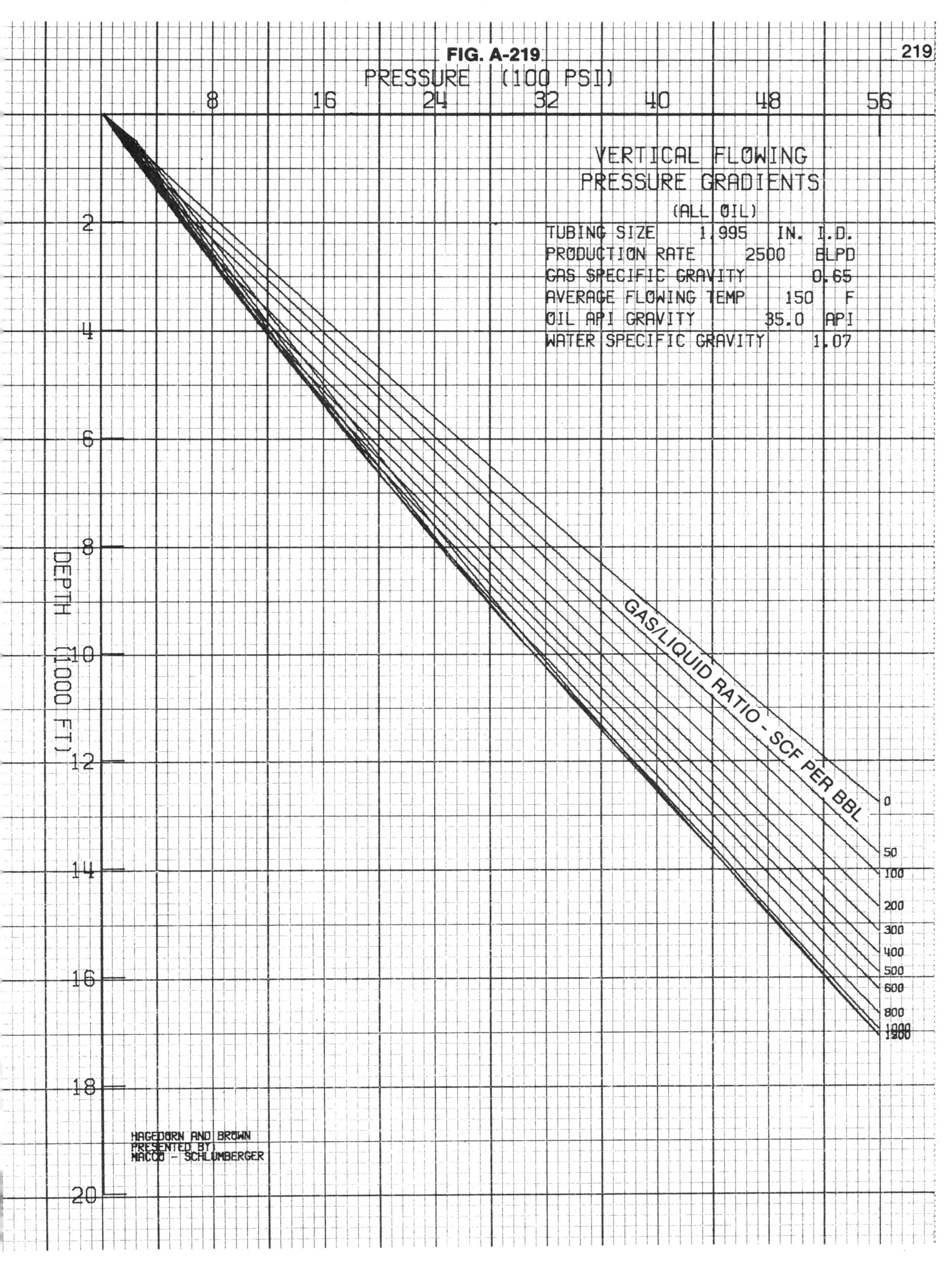

FIG. A-219

219

PRESSURE (100 PSI)

VERTICAL FLOWING
PRESSURE GRADIENTS
(ALL OIL)

TUBING SIZE	1.995	IN. I.D.
PRODUCTION RATE	2500	BLPD
GAS SPECIFIC GRAVITY	0.65	
AVERAGE FLOWING TEMP	150	F
OIL API GRAVITY	35.0	API
WATER SPECIFIC GRAVITY	1.07	

DEPTH (1000 FT)

GAS/LIQUID RATIO - SCF PER BBL

0
50
100
200
300
400
500
600
800
1000
1200

HAGEDORN AND BROWN
PRESENTED BY
NACCO - SCHLUMBERGER

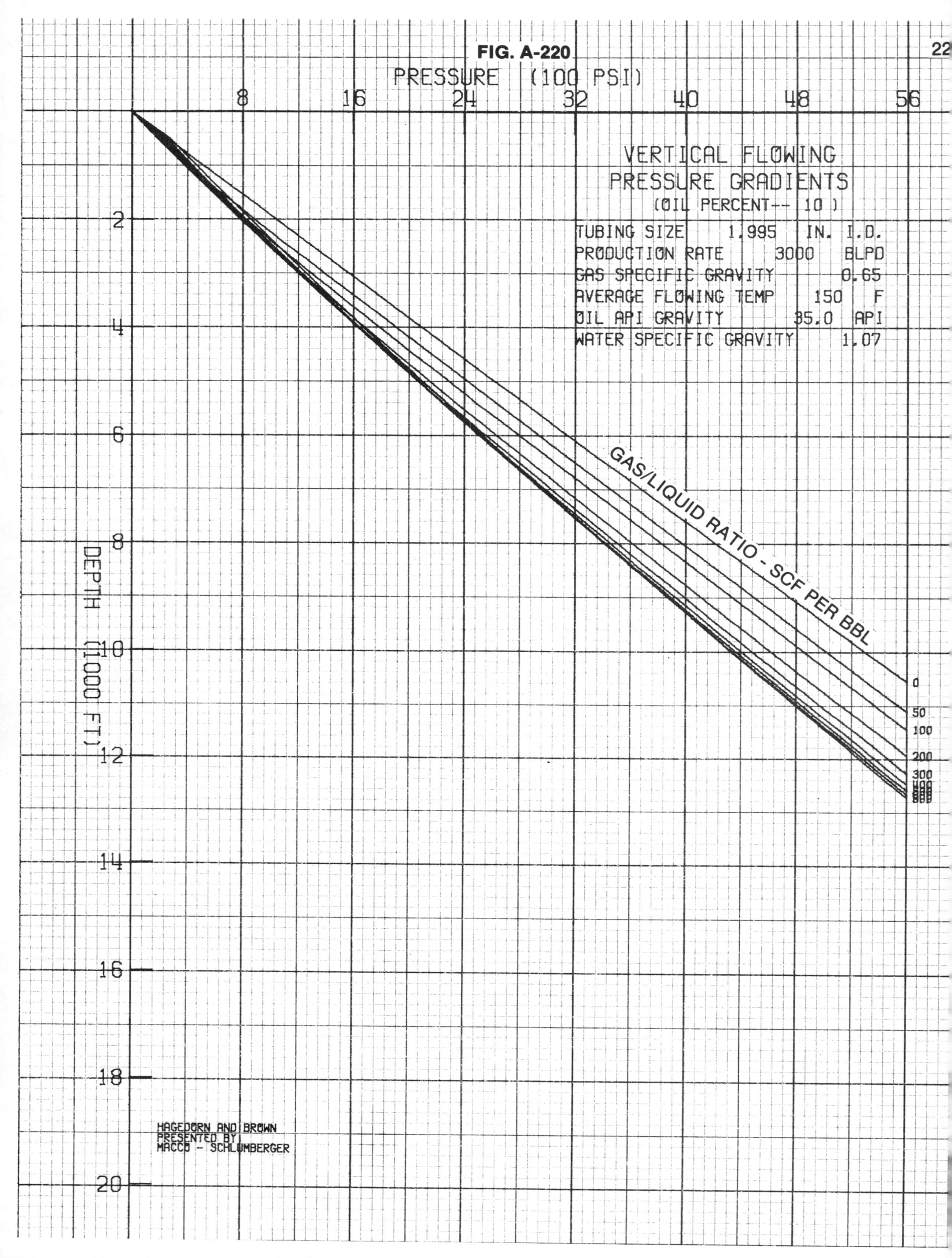

VERTICAL FLOWING
PRESSURE GRADIENTS
(OIL PERCENT--- 10)

TUBING SIZE	1.995	IN. I.D.
PRODUCTION RATE	3000	BLPD
GAS SPECIFIC GRAVITY	0.65	
AVERAGE FLOWING TEMP	150	F
OIL API GRAVITY	35.0	API
WATER SPECIFIC GRAVITY	1.07	

PRESSURE (100 PSI)

DEPTH (1000 FT)

GAS/LIQUID RATIO - SCF PER BBL

0
50
100
200
300

HAGEDORN AND BROWN
PRESENTED BY
MACCO - SCHLUMBERGER

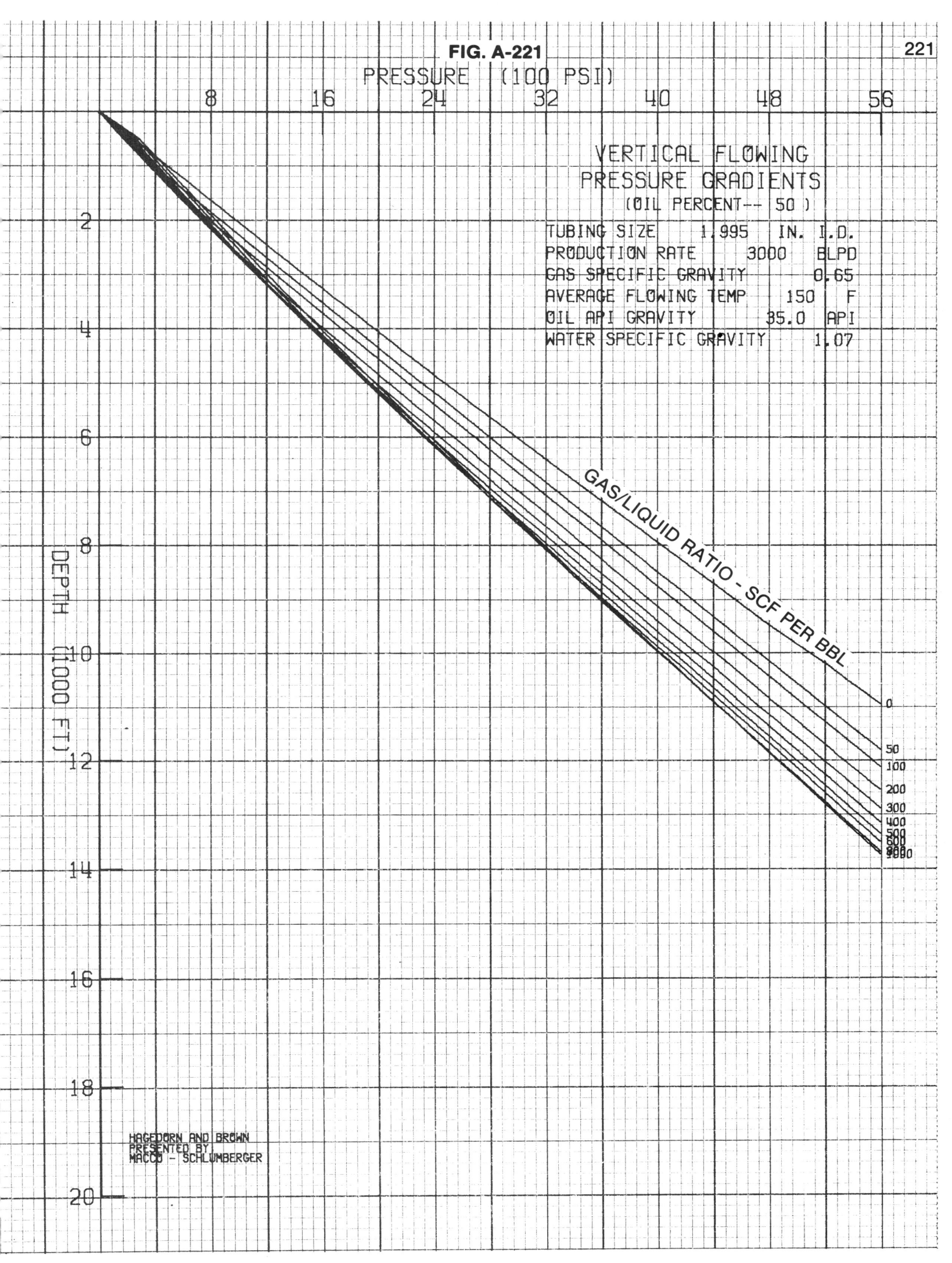

VERTICAL FLOWING
PRESSURE GRADIENTS
(OIL PERCENT-- 50)

TUBING SIZE	1.995	IN. I.D.
PRODUCTION RATE	3000	BLPD
GAS SPECIFIC GRAVITY	0.65	
AVERAGE FLOWING TEMP	150	F
OIL API GRAVITY	35.0	API
WATER SPECIFIC GRAVITY	1.07	

GAS/LIQUID RATIO - SCF PER BBL

PRESSURE (100 PSI)

DEPTH (1000 FT)

0
50
100
200
300
400
500
600
800
1000

HAGEDORN AND BROWN
PRESENTED BY
MACCO - SCHLUMBERGER

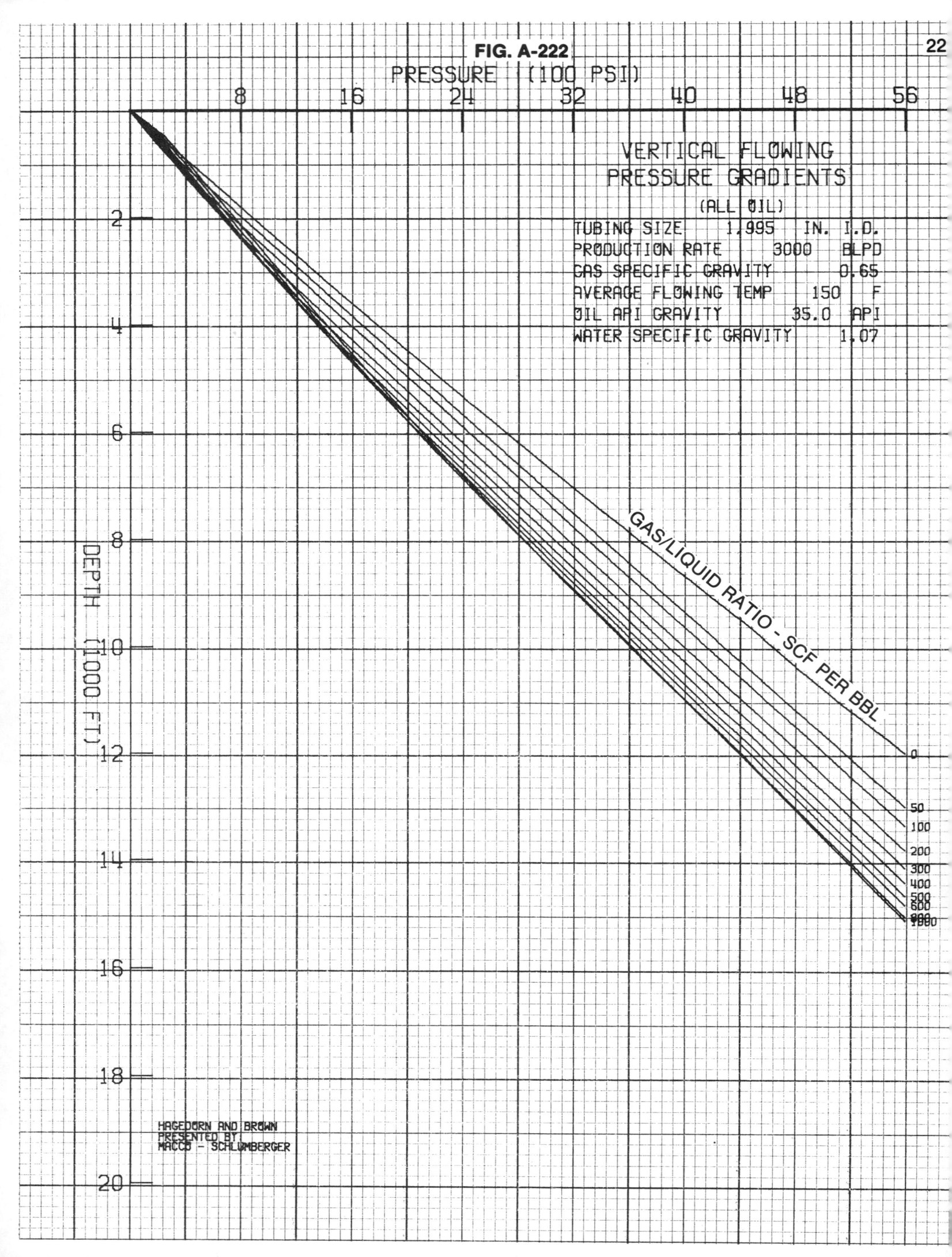

PRESSURE (100 PSI)

VERTICAL FLOWING
PRESSURE GRADIENTS
(ALL OIL)

TUBING SIZE	1.995	IN. I.D.
PRODUCTION RATE	3000	BLPD
GAS SPECIFIC GRAVITY	0.65	
AVERAGE FLOWING TEMP	150	F
OIL API GRAVITY	35.0	API
WATER SPECIFIC GRAVITY	1.07	

DEPTH (1000 FT)

GAS/LIQUID RATIO - SCF PER BBL

0
50
100
200
300
400
500
600
800
1000

HAGEDORN AND BROWN
PRESENTED BY
MACCO - SCHLUMBERGER

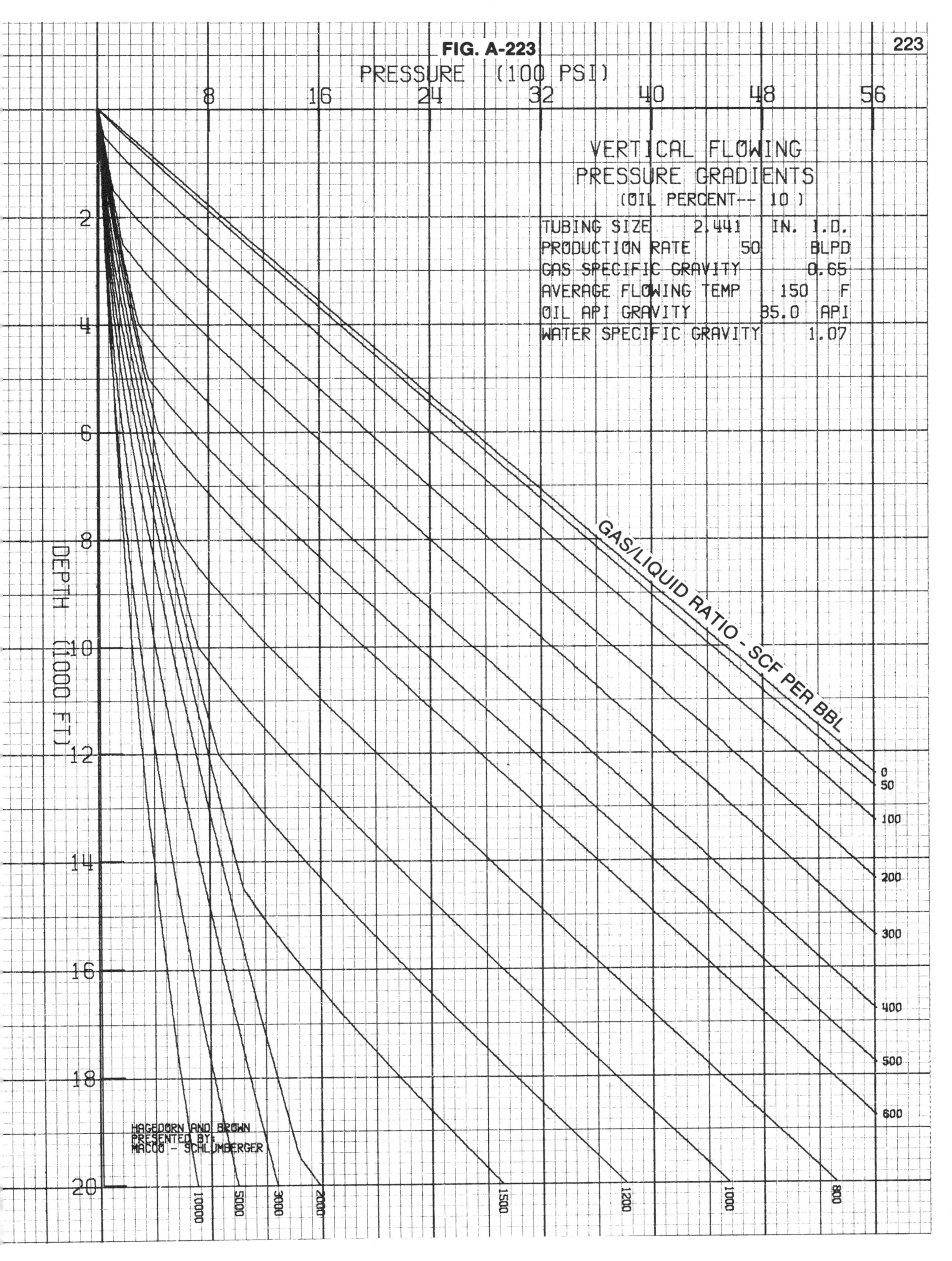

PRESSURE (100 PSI)

VERTICAL FLOWING
PRESSURE GRADIENTS
(OIL PERCENT-- 10)

TUBING SIZE	2.441	IN. I.D.
PRODUCTION RATE	50	BLPD
GAS SPECIFIC GRAVITY	0.65	
AVERAGE FLOWING TEMP	150	F
OIL API GRAVITY	35.0	API
WATER SPECIFIC GRAVITY	1.07	

GAS/LIQUID RATIO - SCF PER BBL

DEPTH (1000 FT)

HAGEDORN AND BROWN
PRESENTED BY:
MACCO - SCHLUMBERGER

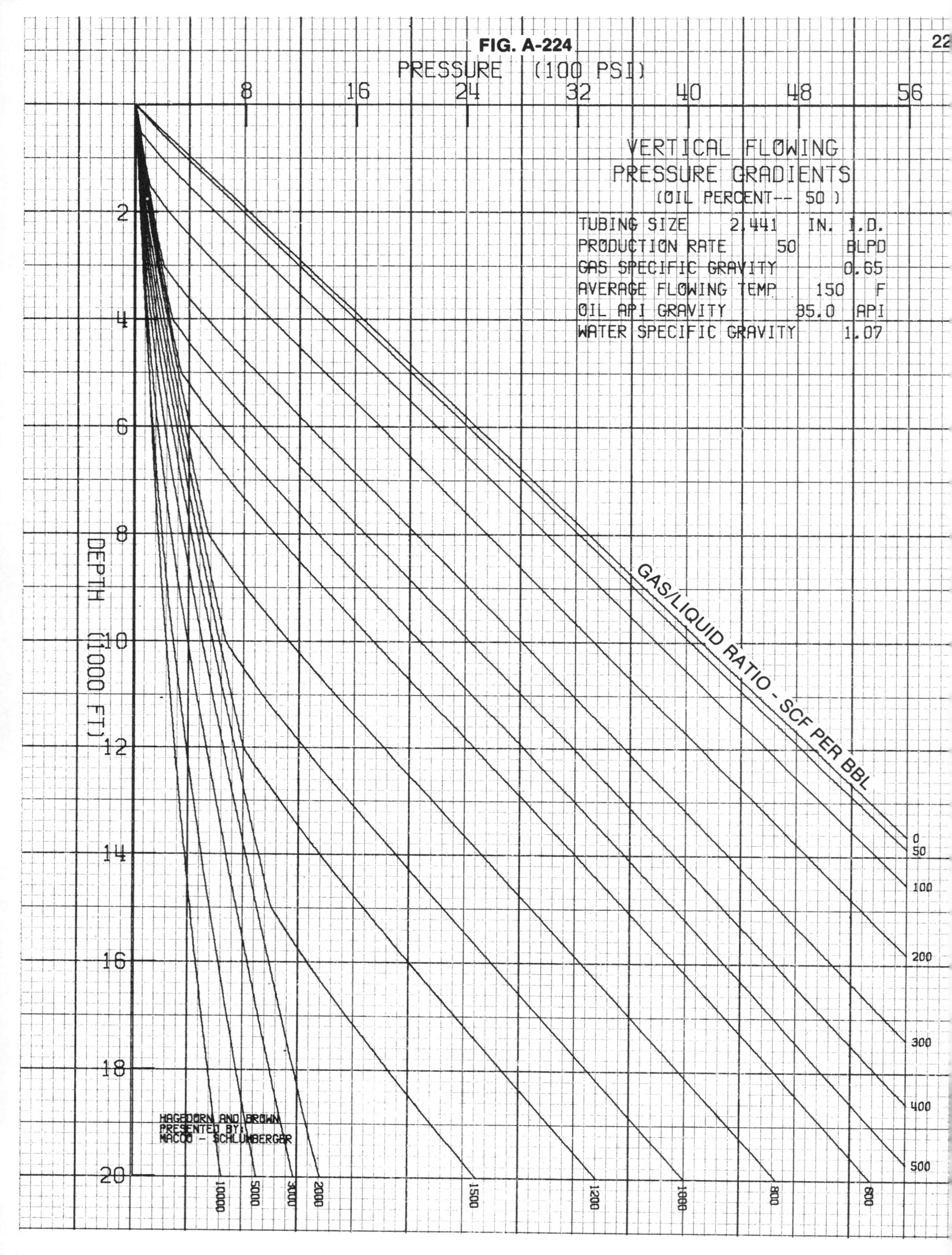

PRESSURE (100 PSI)

VERTICAL FLOWING
PRESSURE GRADIENTS
(OIL PERCENT-- 50)

TUBING SIZE	2.441	IN. I.D.
PRODUCTION RATE	50	BLPD
GAS SPECIFIC GRAVITY		0.65
AVERAGE FLOWING TEMP	150	F
OIL API GRAVITY	35.0	API
WATER SPECIFIC GRAVITY		1.07

GAS/LIQUID RATIO - SCF PER BBL

DEPTH (1000 FT)

HAGEDORN AND BROWN
PRESENTED BY
NACOO - SCHLUMBERGER

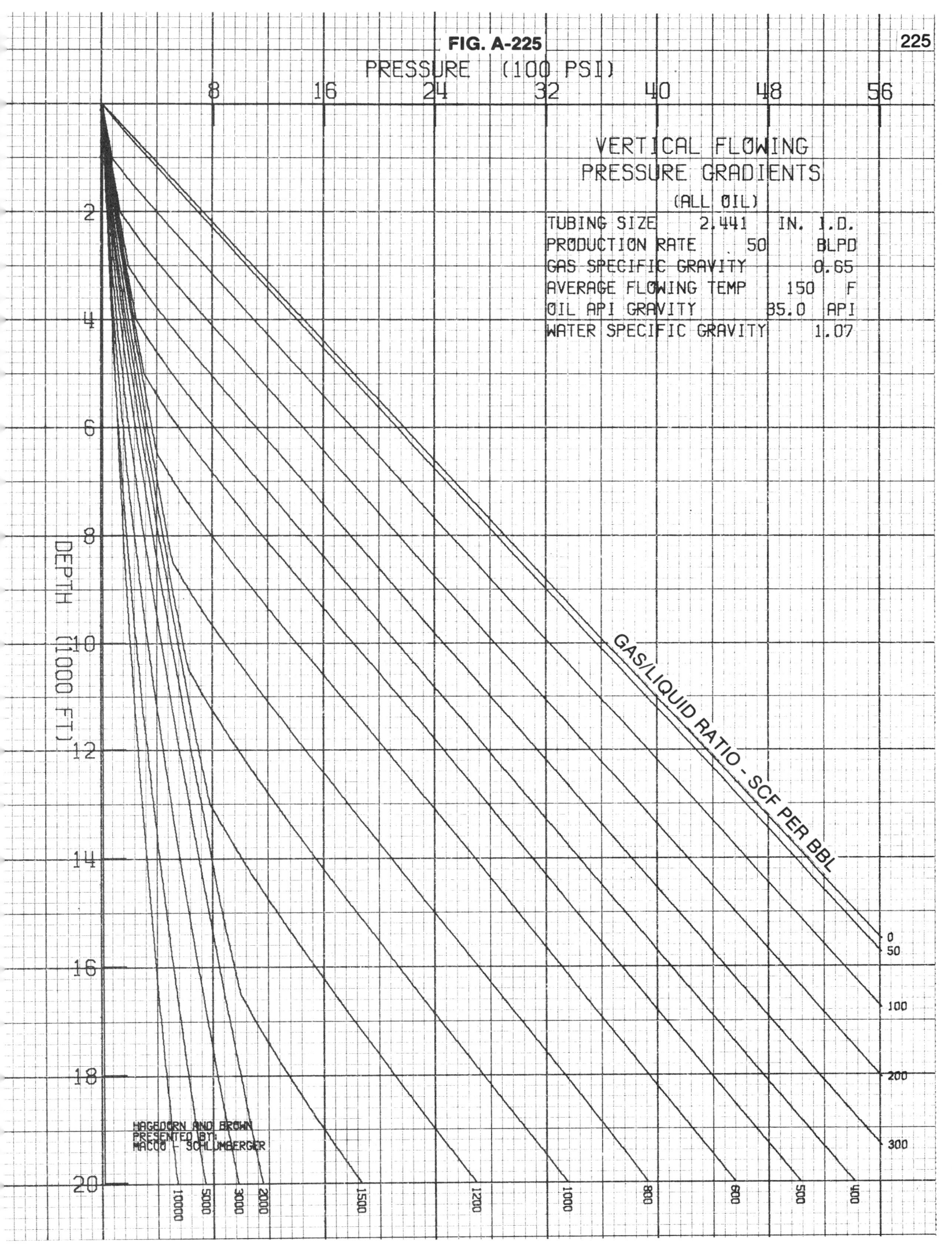

PRESSURE (100 PSI)

VERTICAL FLOWING
PRESSURE GRADIENTS
(ALL OIL)

TUBING SIZE	2.441	IN. I.D.
PRODUCTION RATE	50	BLPD
GAS SPECIFIC GRAVITY	0.65	
AVERAGE FLOWING TEMP	150	F
OIL API GRAVITY	35.0	API
WATER SPECIFIC GRAVITY	1.07	

DEPTH (1000 FT)

GAS/LIQUID RATIO - SCF PER BBL

HAGEDORN AND BROWN
PRESENTED BY:
MACCO — SCHLUMBERGER

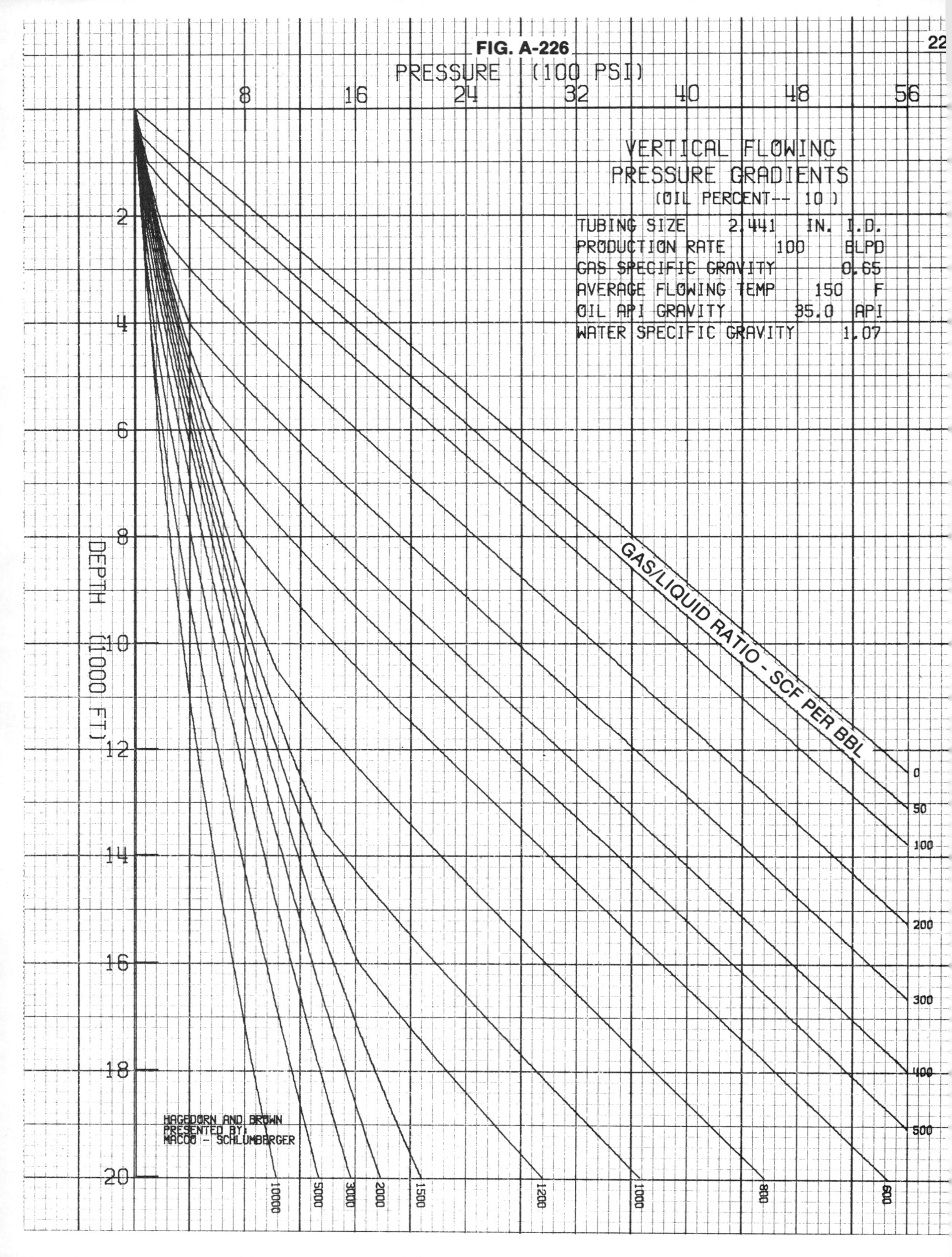

FIG. A-226

22

VERTICAL FLOWING
PRESSURE GRADIENTS
(OIL PERCENT-- 10)

TUBING SIZE	2.441	IN. I.D.
PRODUCTION RATE	100	BLPD
GAS SPECIFIC GRAVITY	0.65	
AVERAGE FLOWING TEMP	150	F
OIL API GRAVITY	35.0	API
WATER SPECIFIC GRAVITY	1.07	

PRESSURE (100 PSI)

DEPTH (1000 FT)

GAS/LIQUID RATIO - SCF PER BBL

HAGEDORN AND BROWN
PRESENTED BY:
NACOO - SCHLUMBERGER

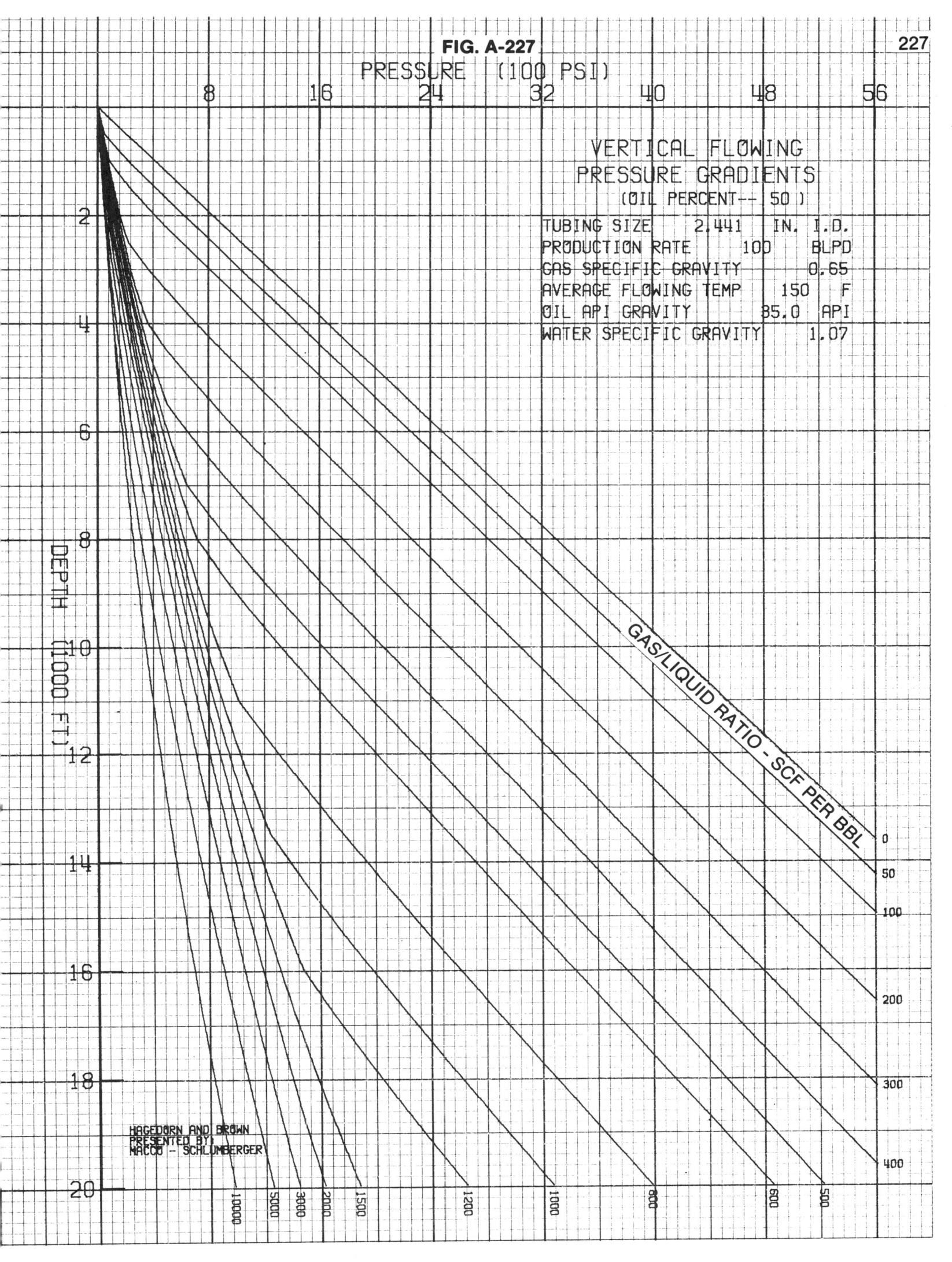

PRESSURE (100 PSI)

VERTICAL FLOWING
PRESSURE GRADIENTS
(OIL PERCENT-- 50)

TUBING SIZE	2.441	IN. I.D.
PRODUCTION RATE	100	BLPD
GAS SPECIFIC GRAVITY	0.65	
AVERAGE FLOWING TEMP	150	F
OIL API GRAVITY	35.0	API
WATER SPECIFIC GRAVITY	1.07	

DEPTH (1000 FT)

GAS/LIQUID RATIO - SCF PER BBL

HAGEDORN AND BROWN
PRESENTED BY:
NACCO - SCHLUMBERGER

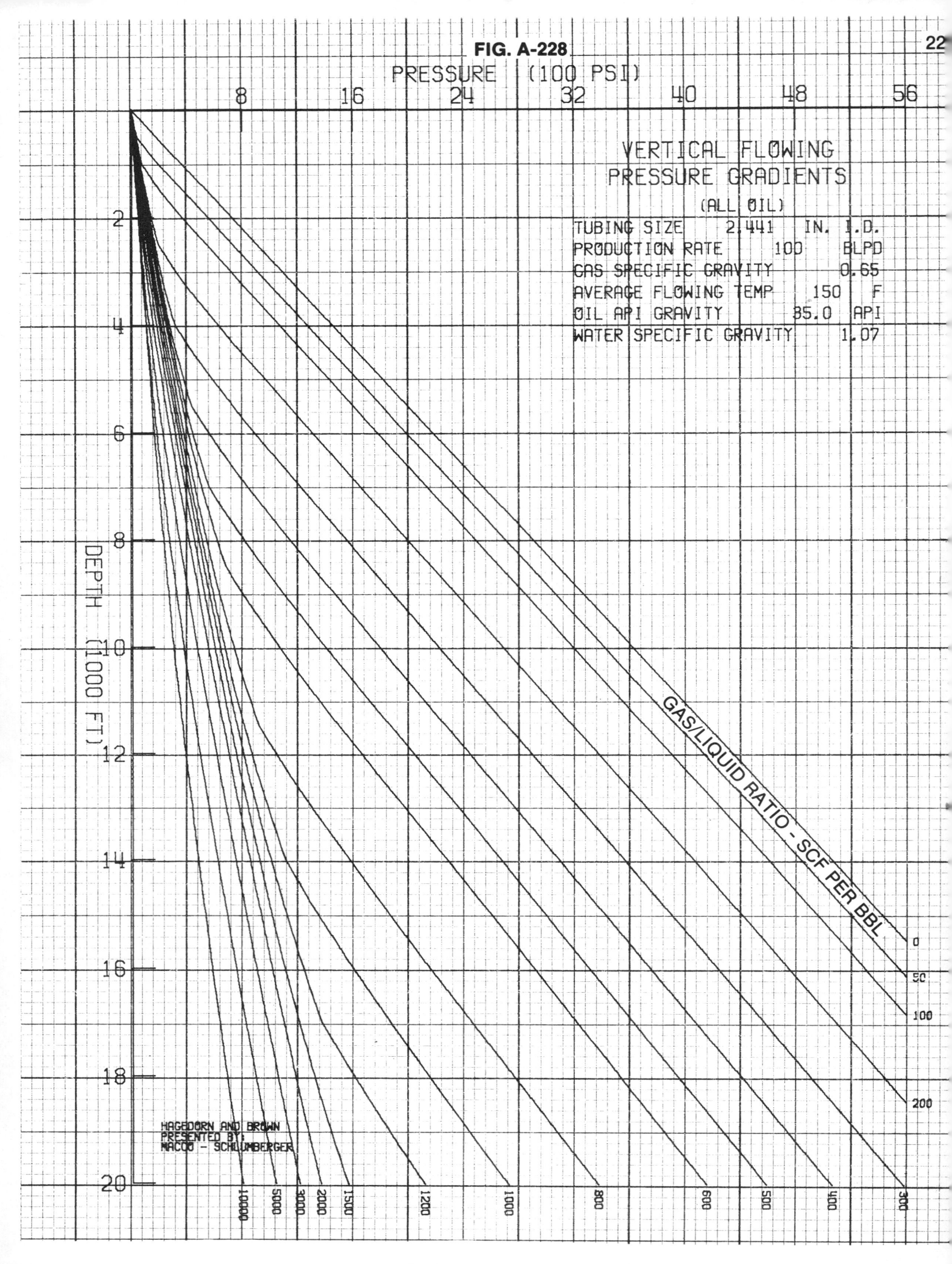

PRESSURE (100 PSI)

VERTICAL FLOWING
PRESSURE GRADIENTS
(ALL OIL)

TUBING SIZE	2.441	IN. I.D.
PRODUCTION RATE	100	BLPD
GAS SPECIFIC GRAVITY	0.65	
AVERAGE FLOWING TEMP	150	F
OIL API GRAVITY	35.0	API
WATER SPECIFIC GRAVITY	1.07	

DEPTH (1000 FT)

GAS/LIQUID RATIO - SCF PER BBL

HAGEDORN AND BROWN
PRESENTED BY:
MACCO - SCHLUMBERGER

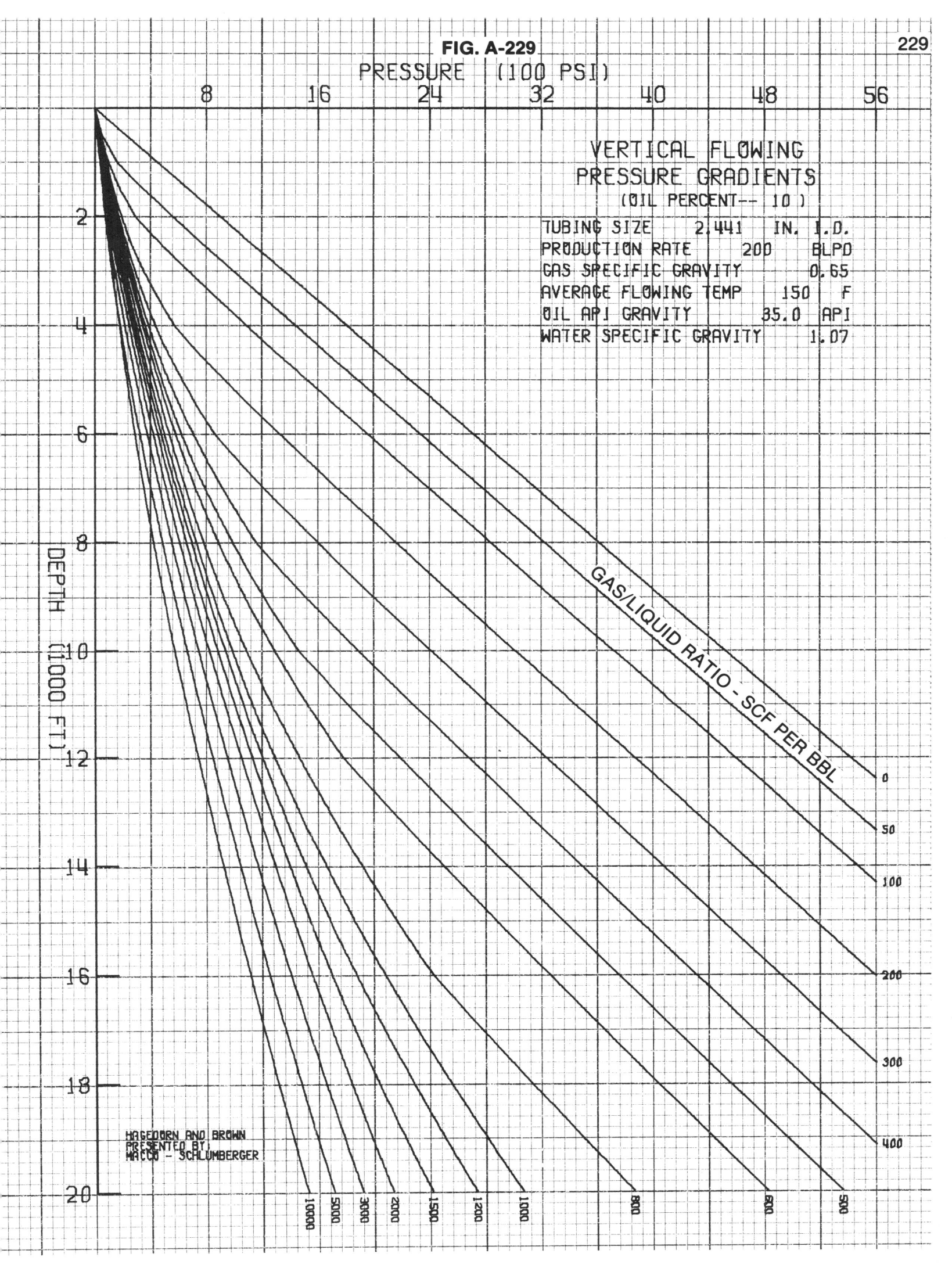

PRESSURE (100 PSI)

VERTICAL FLOWING
PRESSURE GRADIENTS
(OIL PERCENT-- 10)

TUBING SIZE	2.441	IN. I.D.
PRODUCTION RATE	200	BLPD
GAS SPECIFIC GRAVITY	0.65	
AVERAGE FLOWING TEMP	150	F
OIL API GRAVITY	35.0	API
WATER SPECIFIC GRAVITY	1.07	

DEPTH (1000 FT)

GAS/LIQUID RATIO - SCF PER BBL

HAGEDORN AND BROWN
PRESENTED BY
MACCO - SCHLUMBERGER

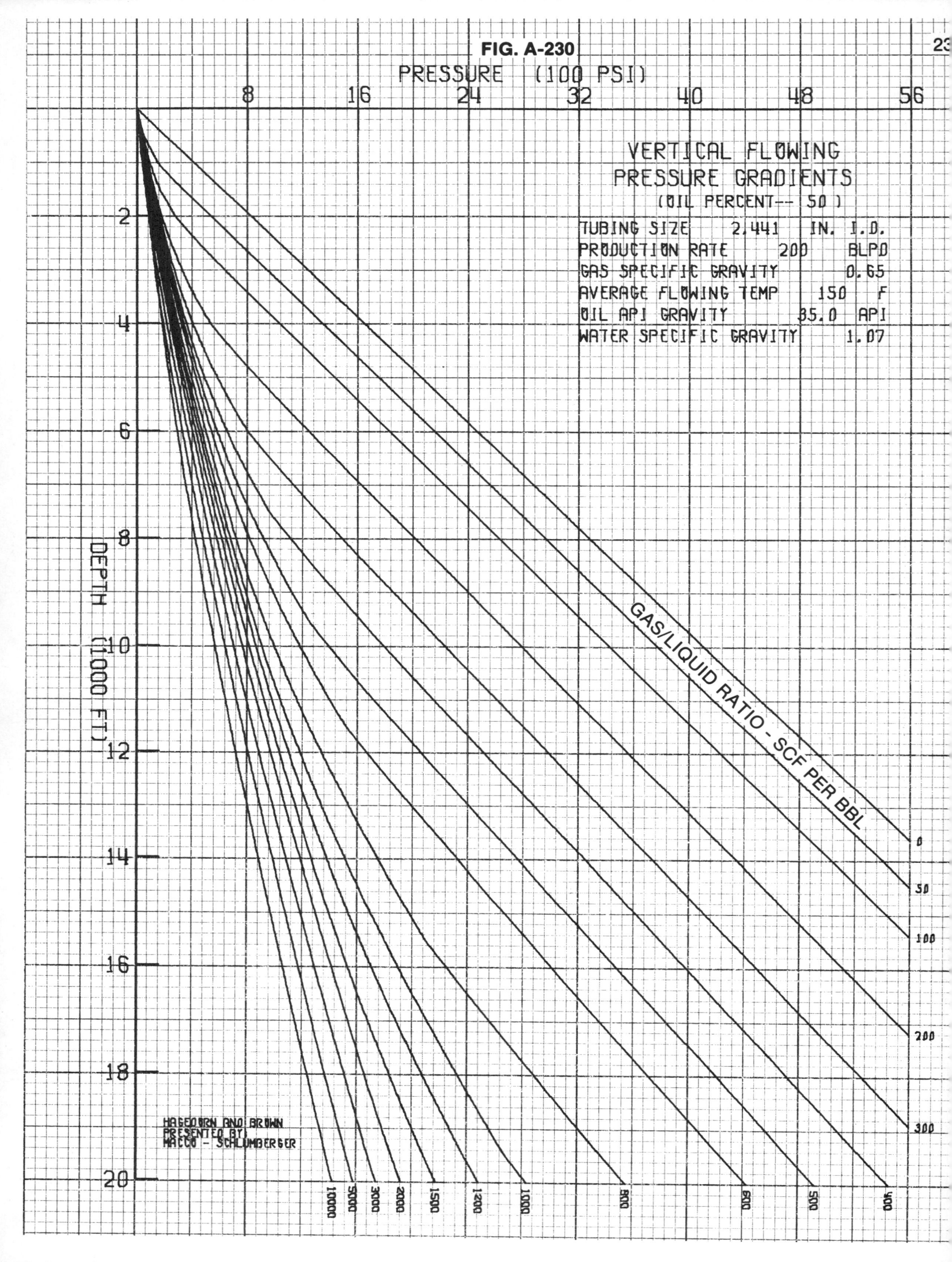

FIG. A-230

23

PRESSURE (100 PSI)

VERTICAL FLOWING
PRESSURE GRADIENTS
(OIL PERCENT— 50)

TUBING SIZE   2.441   IN. I.D.
PRODUCTION RATE   200   BLPD
GAS SPECIFIC GRAVITY   0.65
AVERAGE FLOWING TEMP   150   F
OIL API GRAVITY   35.0 API
WATER SPECIFIC GRAVITY   1.07

GAS/LIQUID RATIO - SCF PER BBL

DEPTH (1,000 FT)

HAGEDORN AND BROWN
PRESENTED BY
MAECO - SCHLUMBERGER

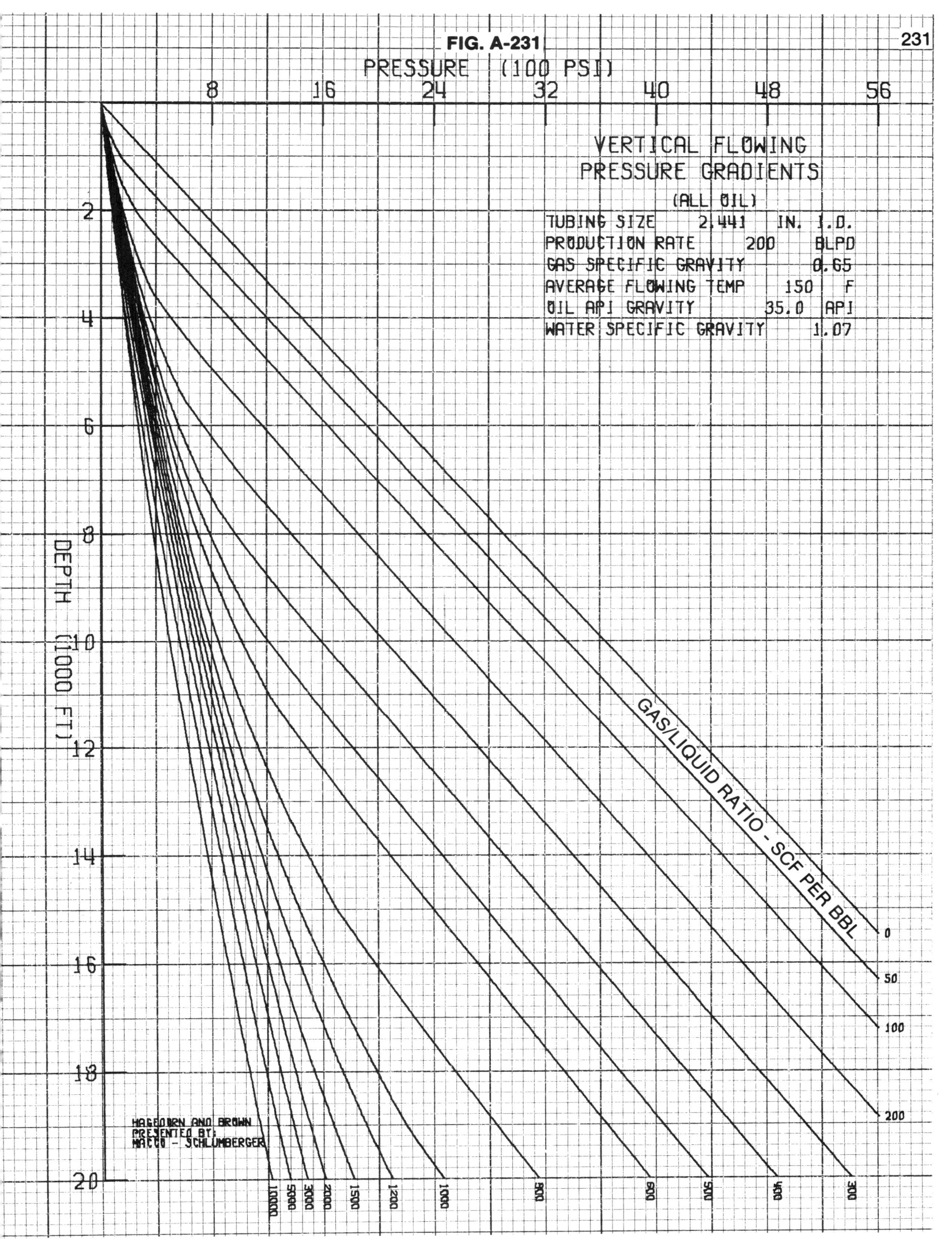

PRESSURE (100 PSI)

VERTICAL FLOWING
PRESSURE GRADIENTS
(ALL OIL)

TUBING SIZE	2.441	IN. I.D.
PRODUCTION RATE	200	BLPD
GAS SPECIFIC GRAVITY	0.65	
AVERAGE FLOWING TEMP	150	F
OIL API GRAVITY	35.0	API
WATER SPECIFIC GRAVITY	1.07	

DEPTH (1000 FT)

GAS/LIQUID RATIO - SCF PER BBL

0

50

100

200

HAGEDORN AND BROWN
PRESENTED BY:
MACCO - SCHLUMBERGER

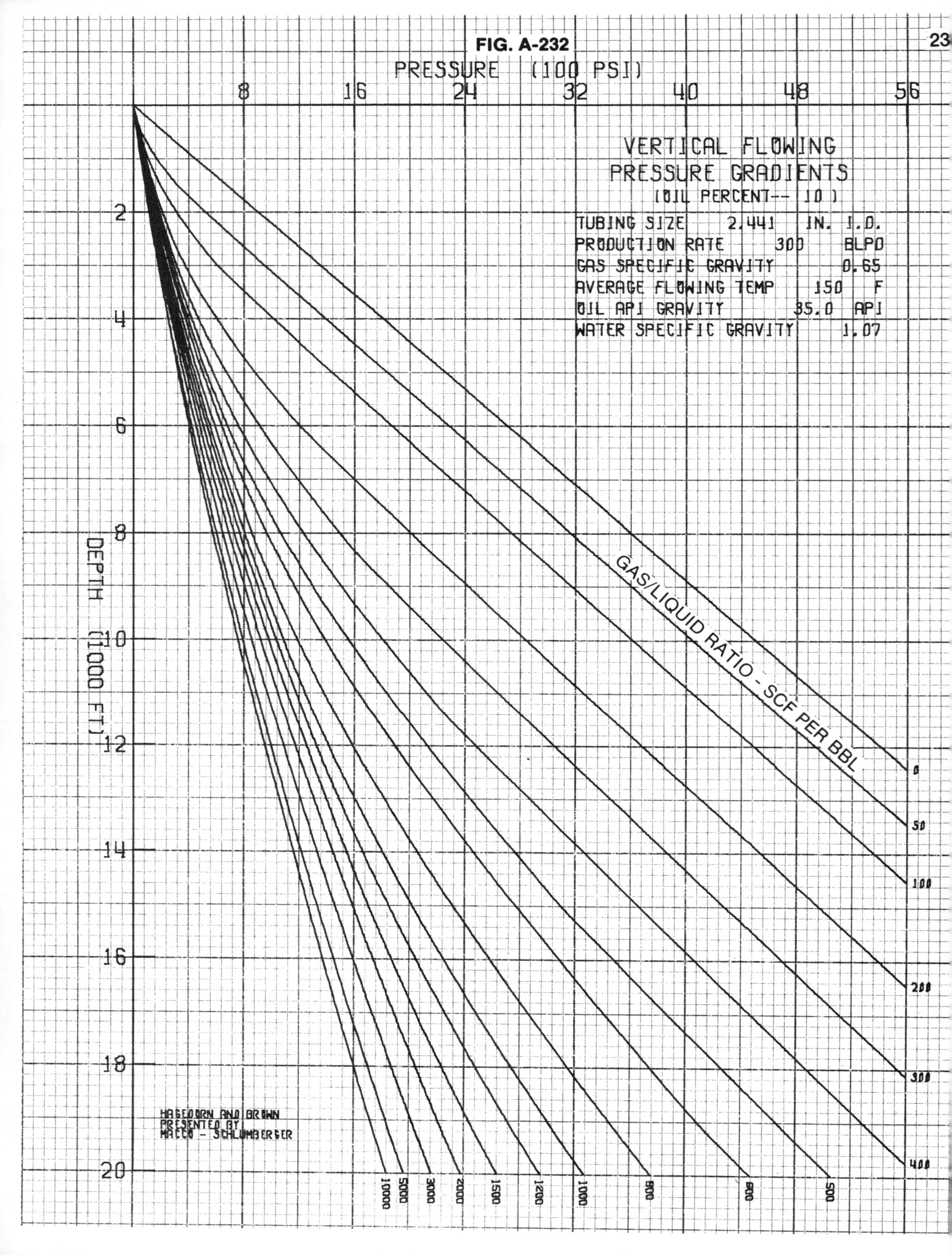

PRESSURE (100 PSI)

VERTICAL FLOWING
PRESSURE GRADIENTS
(OIL PERCENT-- 10)

TUBING SIZE	2.441	IN. I.D.
PRODUCTION RATE	300	BLPD
GAS SPECIFIC GRAVITY	0.65	
AVERAGE FLOWING TEMP	150	F
OIL API GRAVITY	35.0	API
WATER SPECIFIC GRAVITY	1.07	

DEPTH (1000 FT)

GAS/LIQUID RATIO - SCF PER BBL

HAGEDORN AND BROWN
PRESENTED BY
MACCO - SCHLUMBERGER

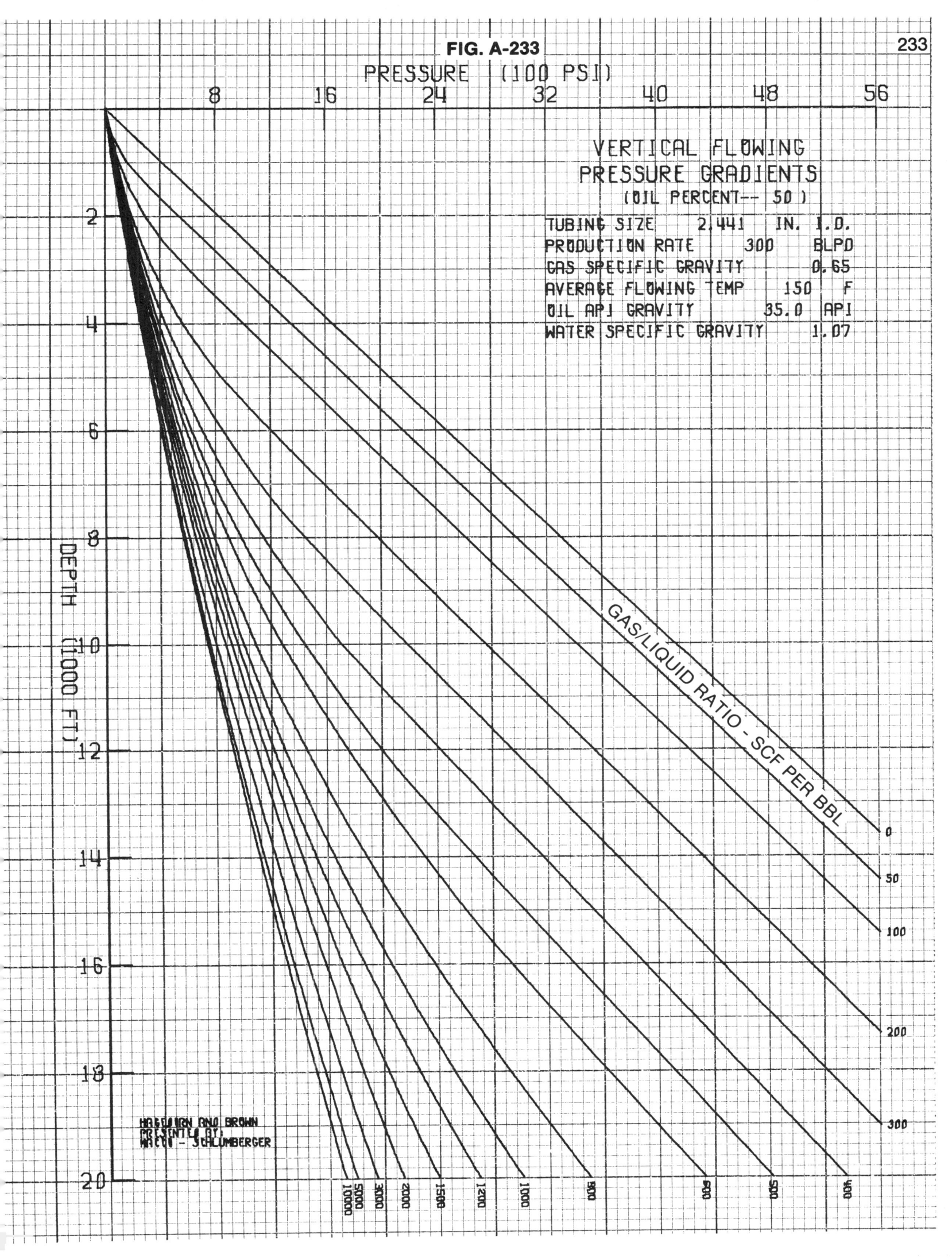

FIG. A-233

233

VERTICAL FLOWING
PRESSURE GRADIENTS
(OIL PERCENT-- 50)

TUBING SIZE	2.441	IN. I.D.
PRODUCTION RATE	300	BLPD
GAS SPECIFIC GRAVITY	0.65	
AVERAGE FLOWING TEMP	150	F
OIL API GRAVITY	35.0	API
WATER SPECIFIC GRAVITY	1.07	

PRESSURE (100 PSI)

DEPTH (1000 FT)

GAS/LIQUID RATIO - SCF PER BBL

HAGEDORN AND BROWN
PRESENTED BY:
WATCO - SCHLUMBERGER

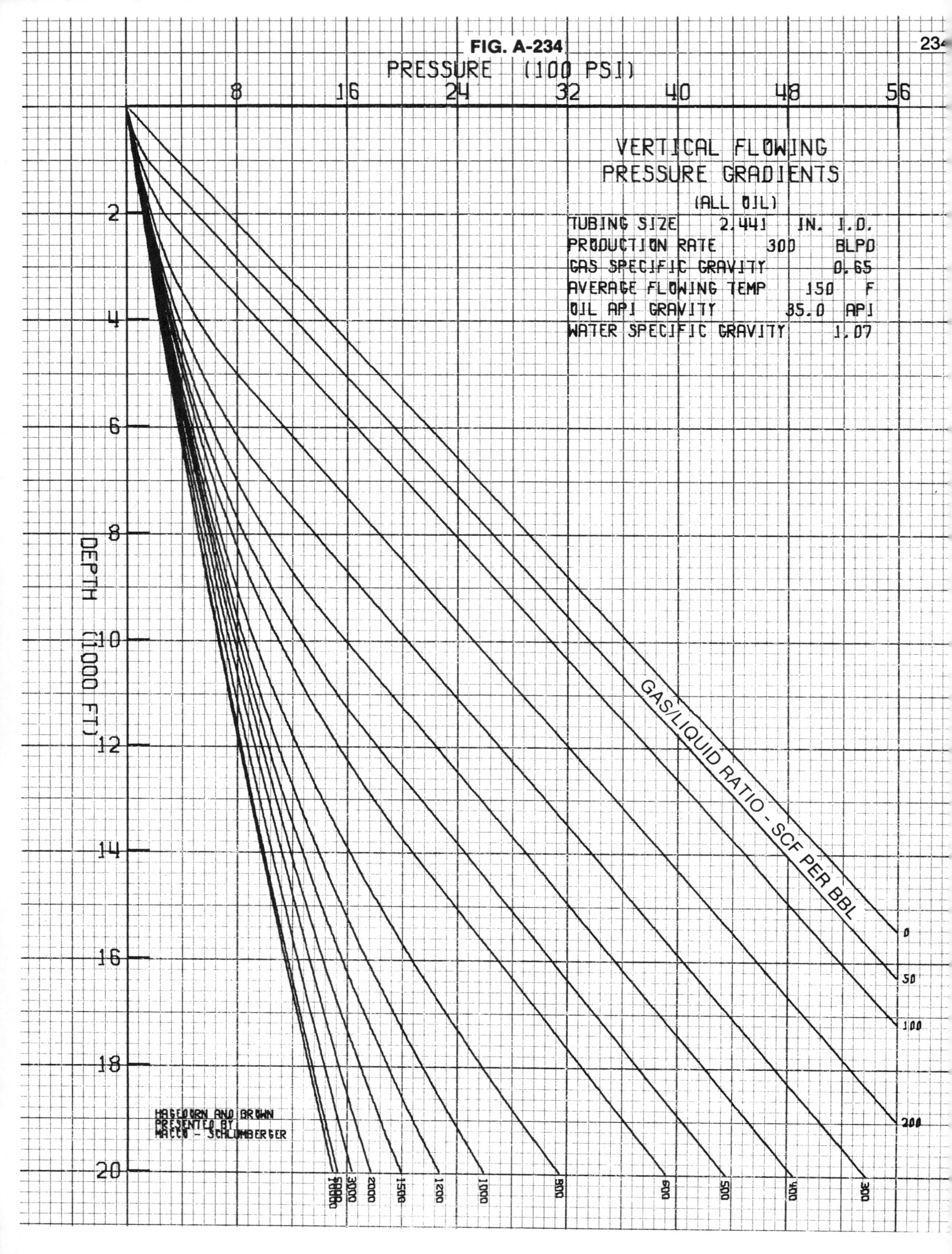

PRESSURE (100 PSI)

DEPTH (1000 FT)

VERTICAL FLOWING
PRESSURE GRADIENTS
(ALL OIL)

TUBING SIZE	2.441	IN. I.D.
PRODUCTION RATE	300	BLPD
GAS SPECIFIC GRAVITY	0.65	
AVERAGE FLOWING TEMP	150	F
OIL API GRAVITY	35.0	API
WATER SPECIFIC GRAVITY	1.07	

GAS/LIQUID RATIO - SCF PER BBL

HAGEDORN AND BROWN
PRESENTED BY
WILCO - SCHLUMBERGER

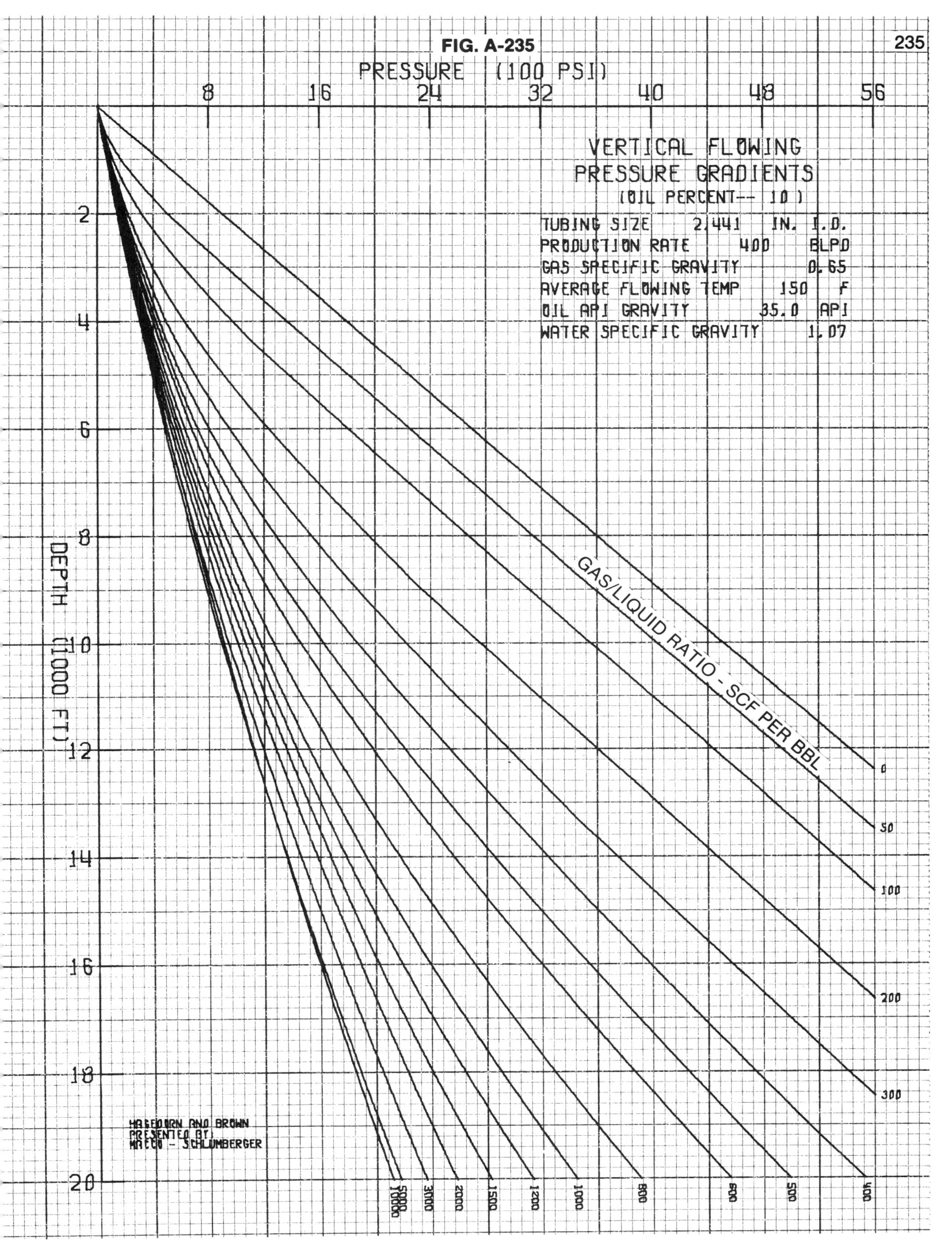

VERTICAL FLOWING
PRESSURE GRADIENTS
(OIL PERCENT-- 10)

TUBING SIZE	2.441	IN. I.D.
PRODUCTION RATE	400	BLPD
GAS SPECIFIC GRAVITY	0.65	
AVERAGE FLOWING TEMP	150	F
OIL API GRAVITY	35.0	API
WATER SPECIFIC GRAVITY	1.07	

PRESSURE (100 PSI)

DEPTH (1000 FT)

GAS/LIQUID RATIO - SCF PER BBL

HAGEDORN AND BROWN
PRESENTED BY:
MATCO - SCHLUMBERGER

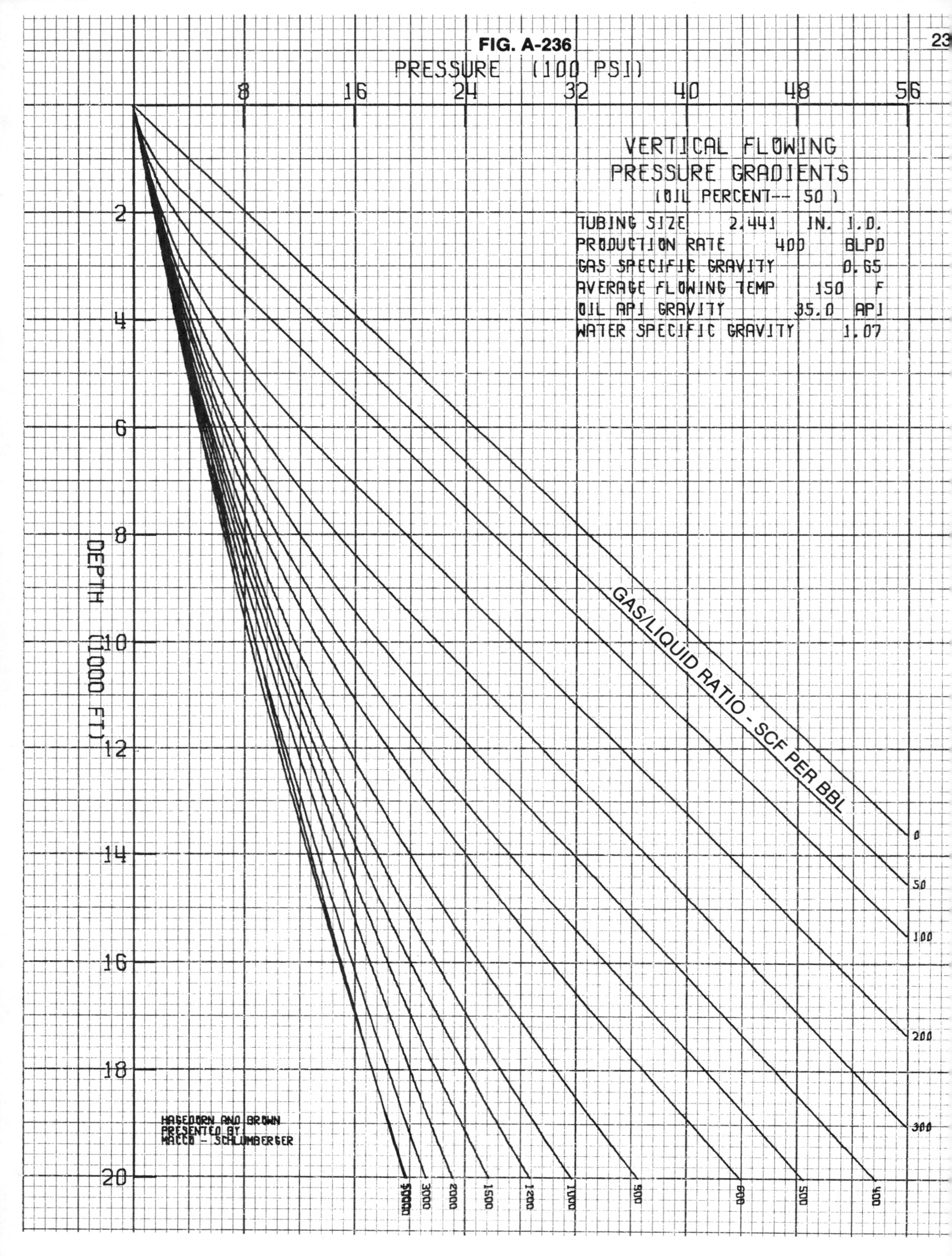

FIG. A-236

23

VERTICAL FLOWING
PRESSURE GRADIENTS
(OIL PERCENT--- 50)

TUBING SIZE   2.441   IN. I.D.
PRODUCTION RATE   400   BLPD
GAS SPECIFIC GRAVITY   0.65
AVERAGE FLOWING TEMP   150   F
OIL API GRAVITY   35.0 API
WATER SPECIFIC GRAVITY   1.07

PRESSURE (100 PSI)

DEPTH (1000 FT)

GAS/LIQUID RATIO - SCF PER BBL

HAGEDORN AND BROWN
PRESENTED BY
MAECO - SCHLUMBERGER

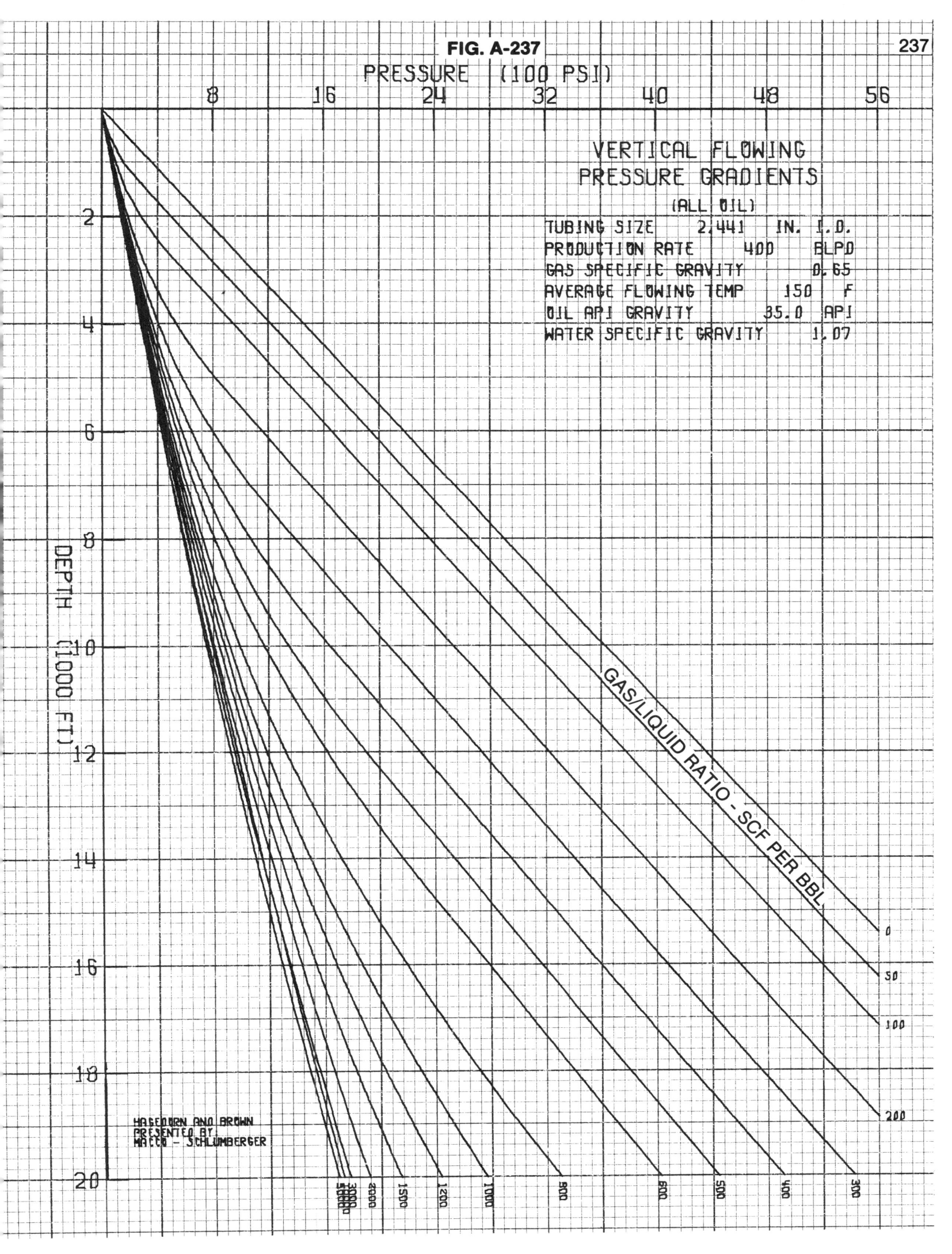

PRESSURE (100 PSI)

VERTICAL FLOWING
PRESSURE GRADIENTS
(ALL OIL)

TUBING SIZE   2.441   IN. I.D.
PRODUCTION RATE   400   BLPD
GAS SPECIFIC GRAVITY   0.65
AVERAGE FLOWING TEMP   150   F
OIL API GRAVITY   35.0   API
WATER SPECIFIC GRAVITY   1.07

DEPTH (1000 FT)

GAS/LIQUID RATIO - SCF PER BBL

HAGEDORN AND BROWN
PRESENTED BY
MACCO - SCHLUMBERGER

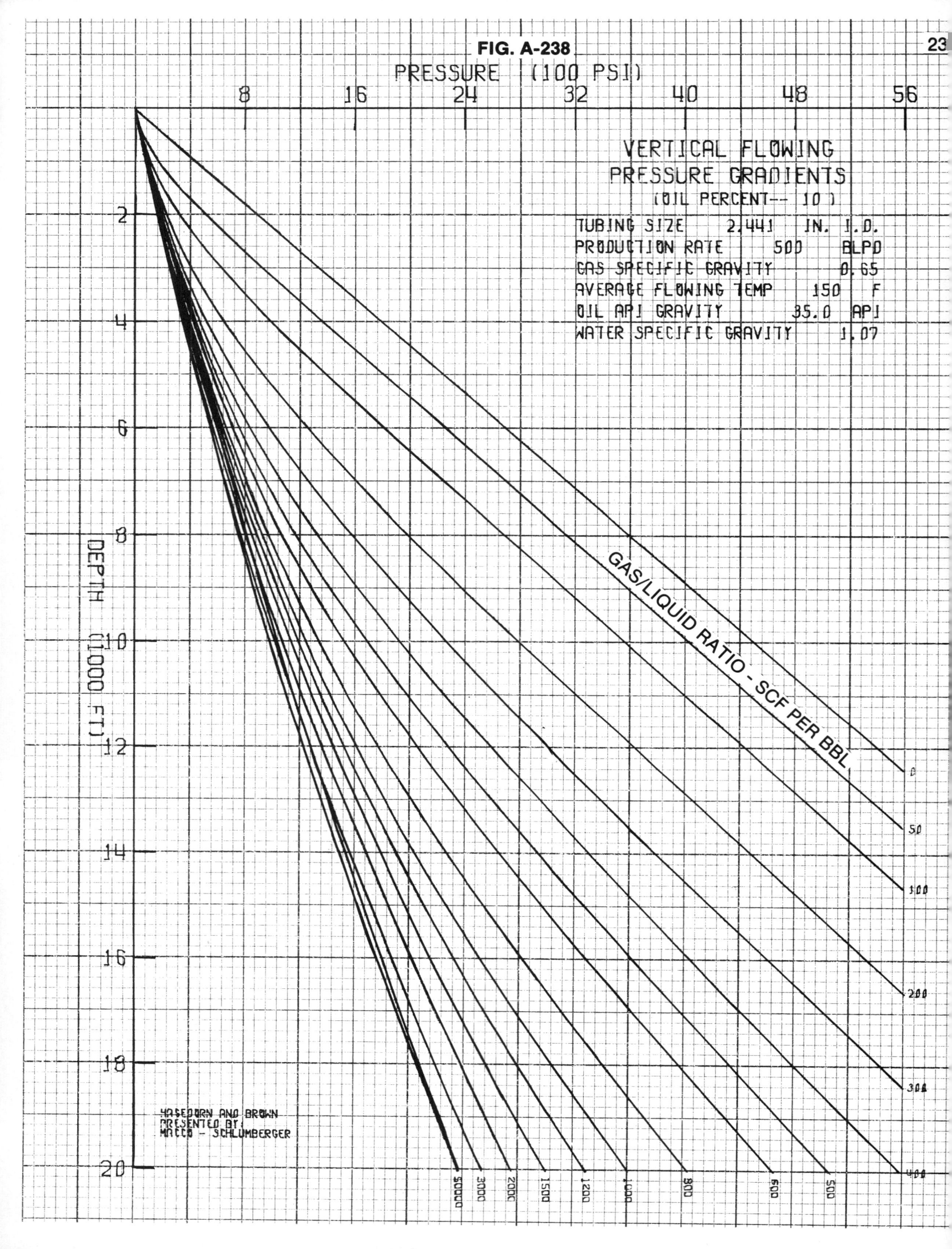

PRESSURE (100 PSI)

DEPTH (1000 FT)

VERTICAL FLOWING
PRESSURE GRADIENTS
(OIL PERCENT— 100)

TUBING SIZE	2.441	IN. I.D.
PRODUCTION RATE	500	BLPD
GAS SPECIFIC GRAVITY		0.65
AVERAGE FLOWING TEMP	150	F
OIL API GRAVITY	35.0	API
WATER SPECIFIC GRAVITY		1.07

GAS/LIQUID RATIO - SCF PER BBL

HAGEDORN AND BROWN
PRESENTED BY
MACCO - SCHLUMBERGER

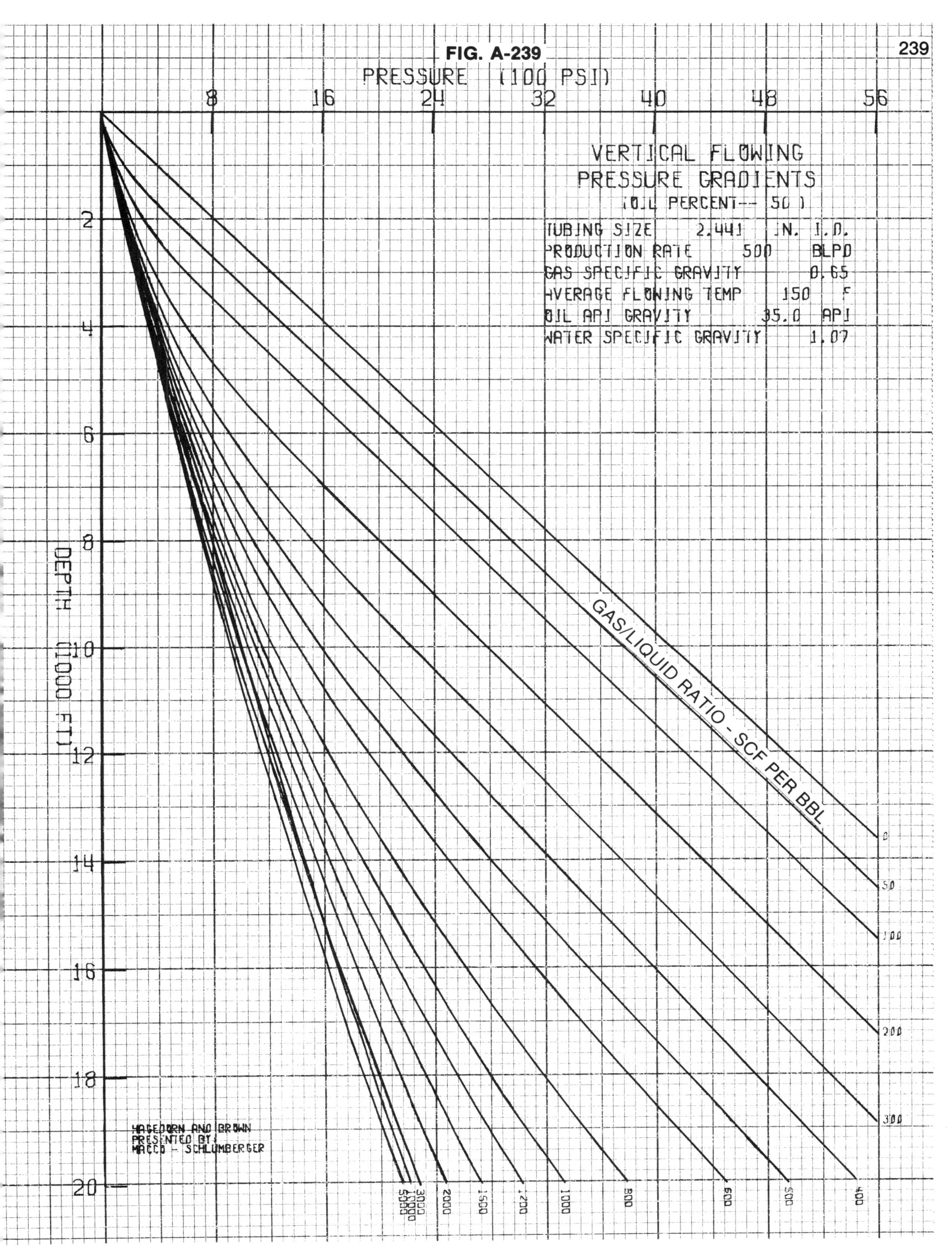

PRESSURE (100 PSI)

VERTICAL FLOWING
PRESSURE GRADIENTS
(OIL PERCENT-- 50)

TUBING SIZE	2.441	IN. I.D.
PRODUCTION RATE	500	BLPD
GAS SPECIFIC GRAVITY	0.65	
AVERAGE FLOWING TEMP	150	F
OIL API GRAVITY	35.0	API
WATER SPECIFIC GRAVITY	1.07	

DEPTH (1000 FT)

GAS/LIQUID RATIO - SCF PER BBL

HAGEDORN AND BROWN
PRESENTED BY:
MACCO - SCHLUMBERGER

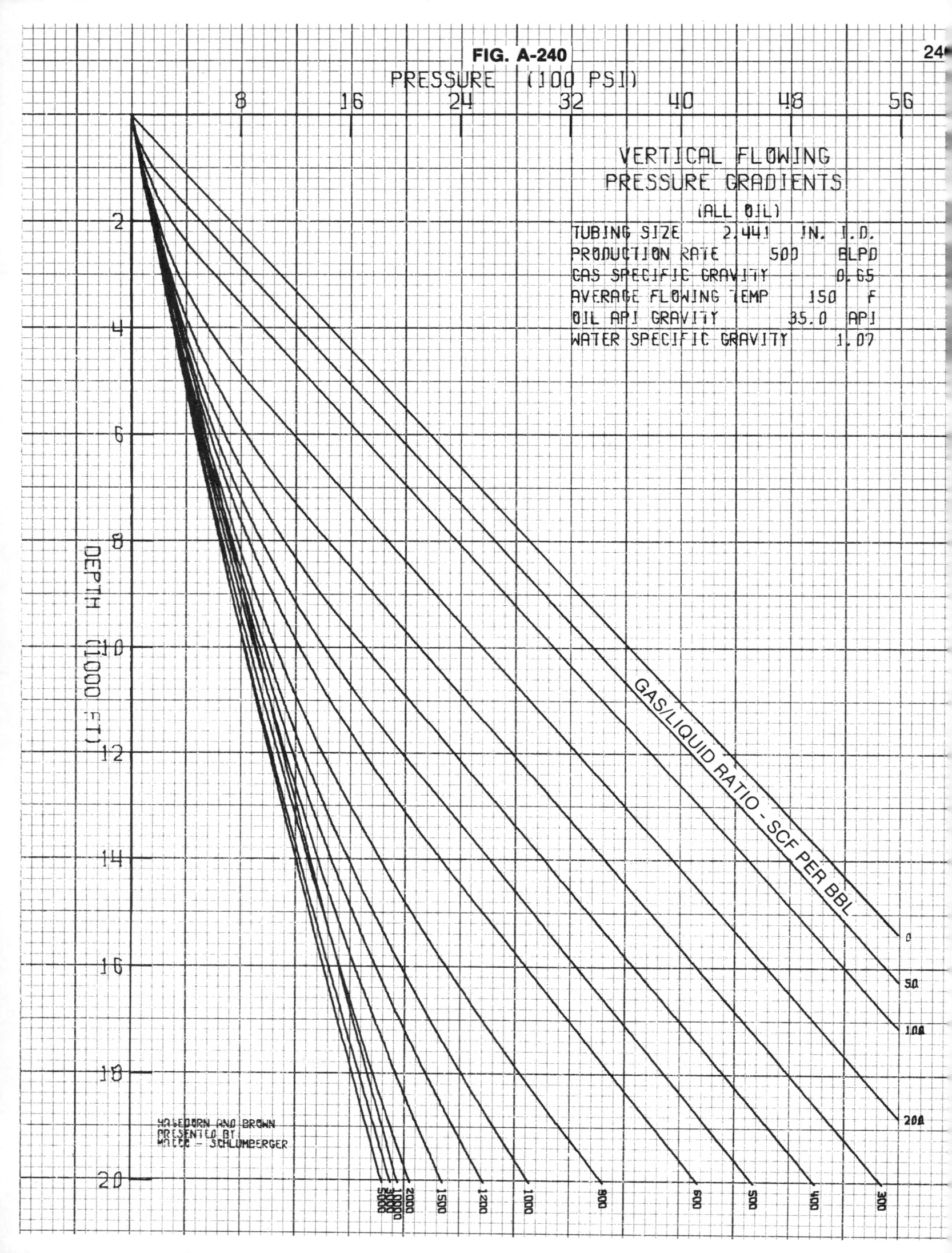

VERTICAL FLOWING
PRESSURE GRADIENTS
(ALL OIL)

TUBING SIZE	2.441	IN. I.D.
PRODUCTION RATE	500	BLPD
GAS SPECIFIC GRAVITY	0.65	
AVERAGE FLOWING TEMP	150	F
OIL API GRAVITY	35.0	API
WATER SPECIFIC GRAVITY	1.07	

PRESSURE (100 PSI)

DEPTH (1000 FT)

GAS/LIQUID RATIO - SCF PER BBL

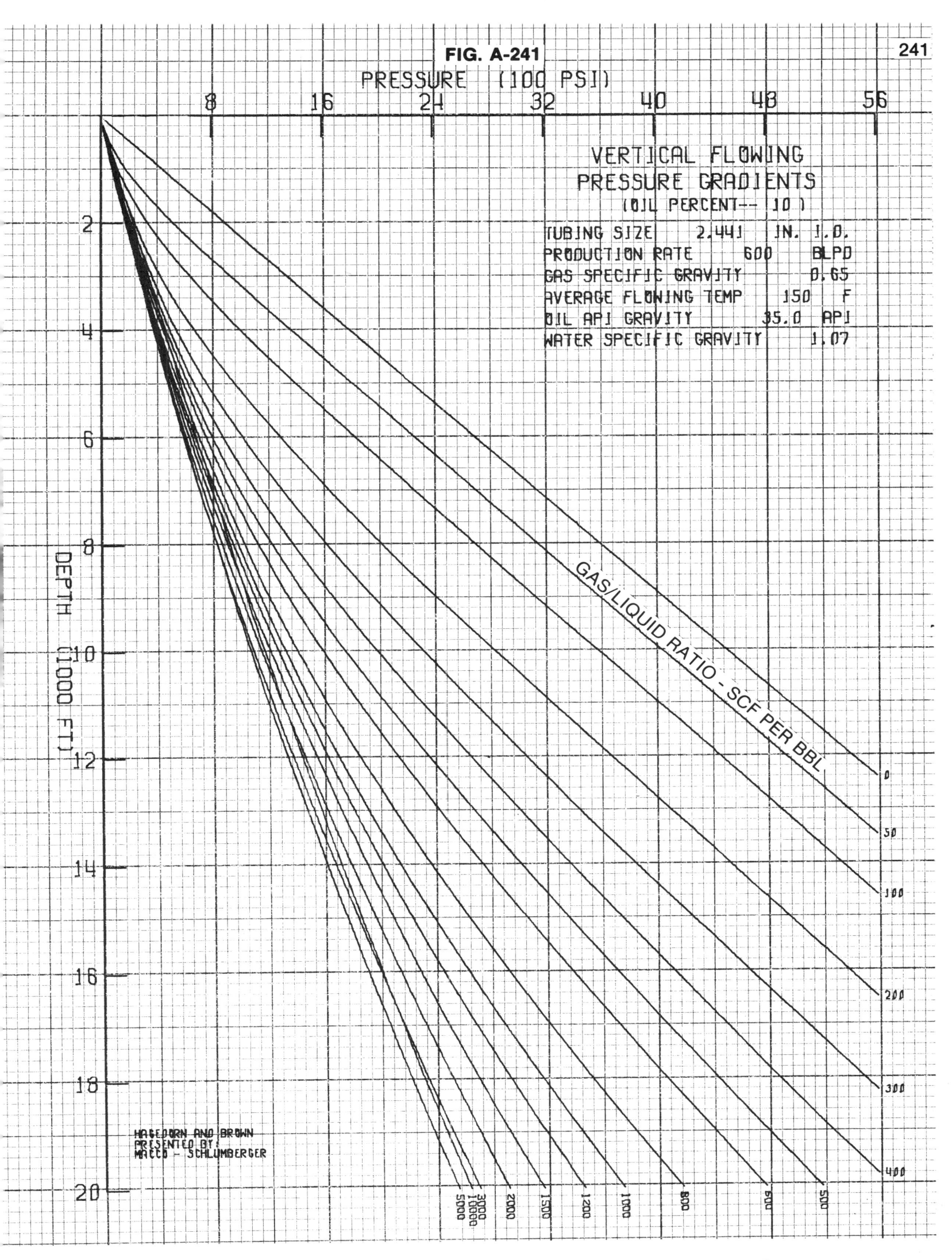

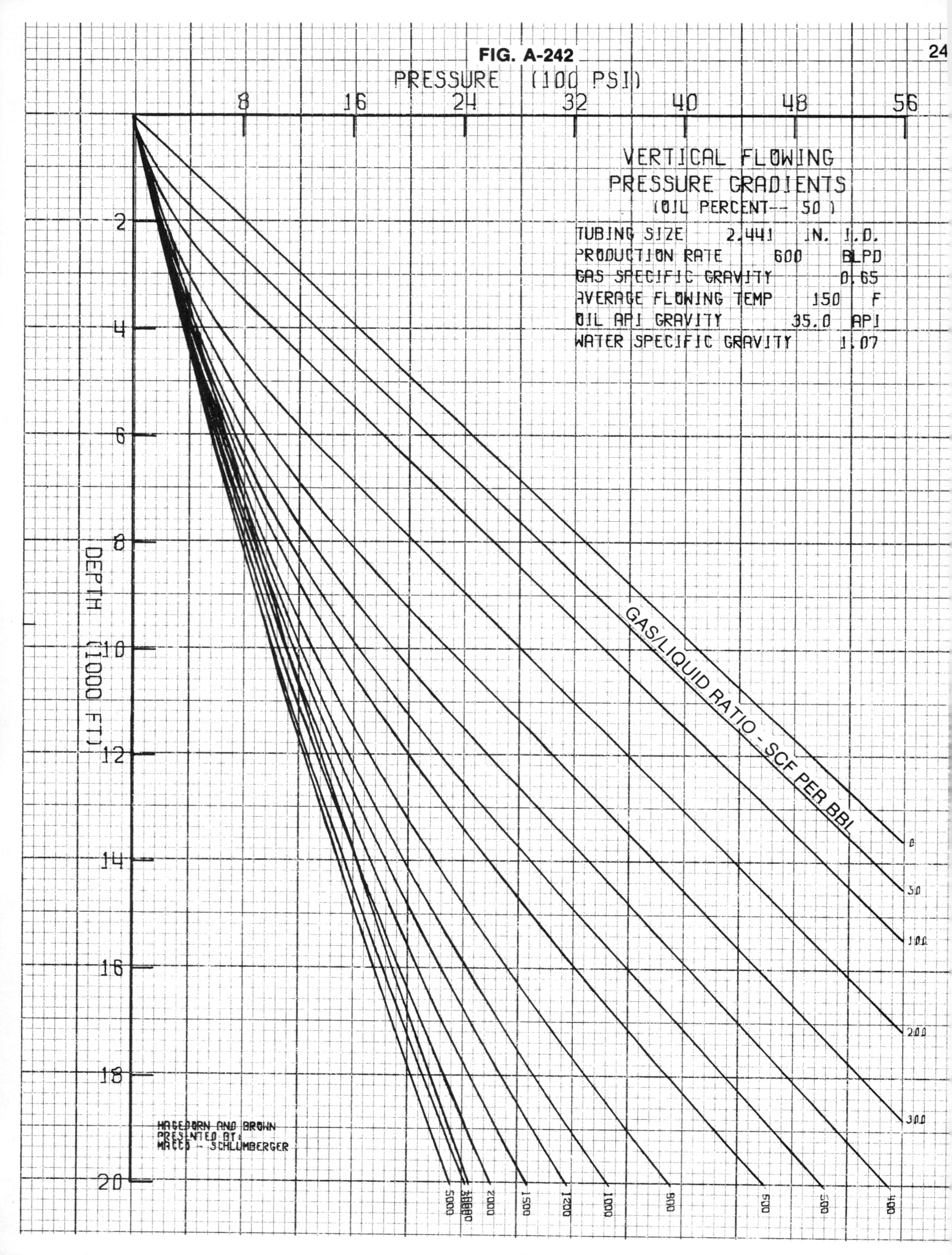

FIG. A-242

24

VERTICAL FLOWING
PRESSURE GRADIENTS
(OIL PERCENT-- 50)

TUBING SIZE	2.441	IN. I.D.
PRODUCTION RATE	600	BLPD
GAS SPECIFIC GRAVITY	0.65	
AVERAGE FLOWING TEMP	150	F
OIL API GRAVITY	35.0	API
WATER SPECIFIC GRAVITY	1.07	

PRESSURE (100 PSI)

DEPTH (1000 FT)

GAS/LIQUID RATIO - SCF PER BBL

HAGEDORN AND BROWN
PRESENTED BY
MACCO - SCHLUMBERGER

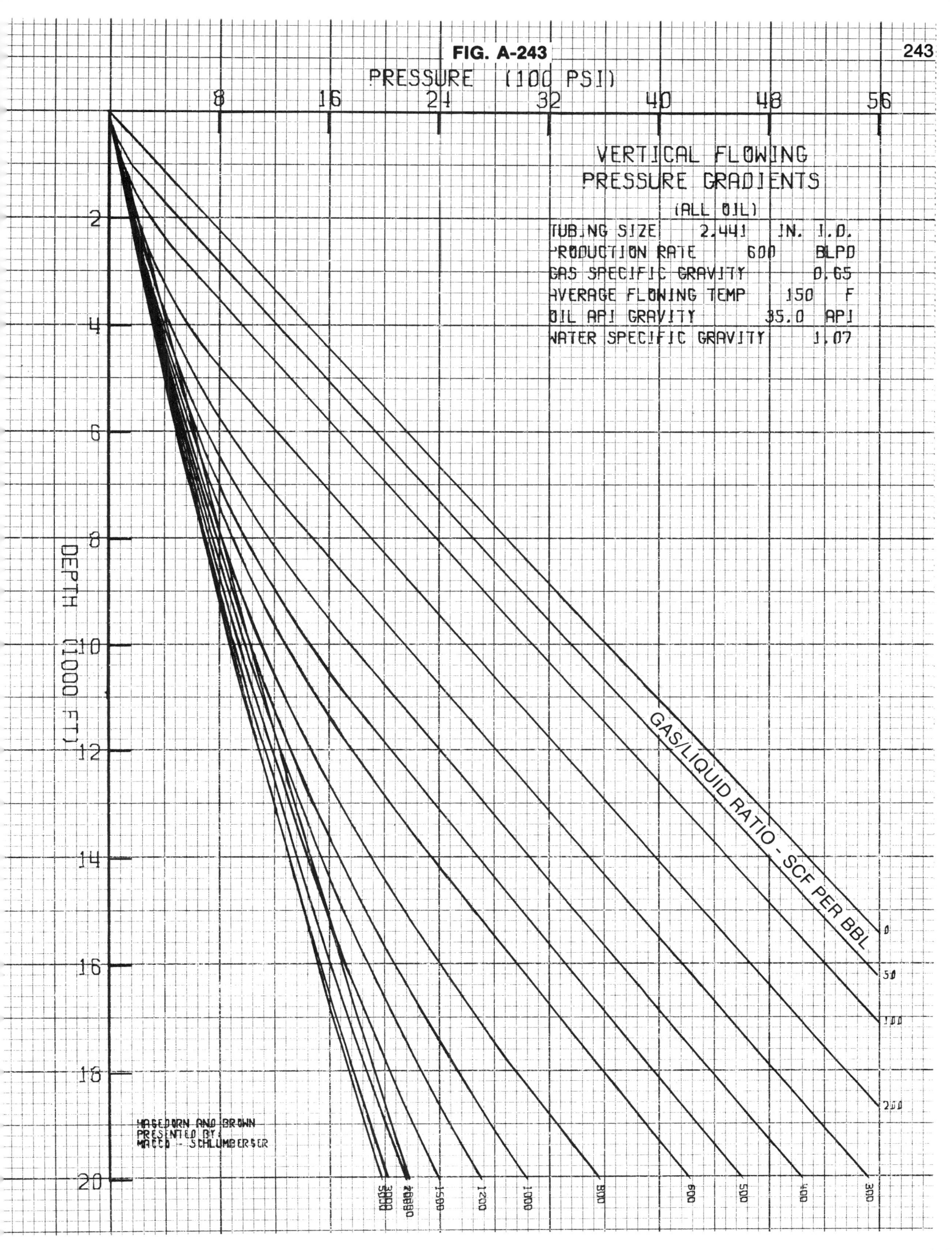

PRESSURE (100 PSI)

VERTICAL FLOWING
PRESSURE GRADIENTS
(ALL OIL)

TUBING SIZE	2.441	IN. I.D.
PRODUCTION RATE	600	BLPD
GAS SPECIFIC GRAVITY	0.65	
AVERAGE FLOWING TEMP	150	F
OIL API GRAVITY	35.0	API
WATER SPECIFIC GRAVITY	1.07	

DEPTH (1000 FT)

GAS/LIQUID RATIO - SCF PER BBL

HAGEDORN AND BROWN
PRESENTED BY:
MACCO - SCHLUMBERGER

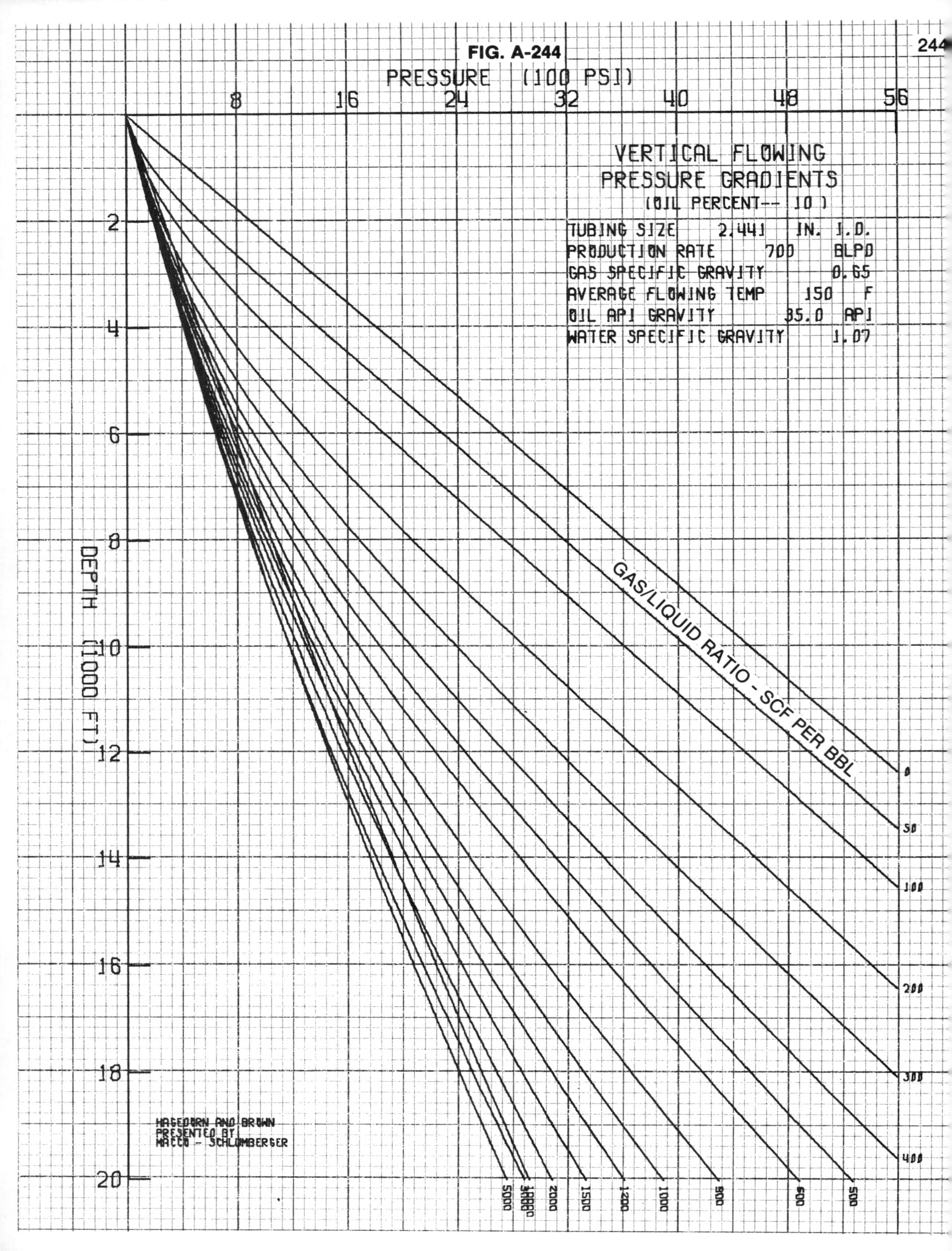

FIG. A-244

244

PRESSURE (100 PSI)

VERTICAL FLOWING
PRESSURE GRADIENTS
(OIL PERCENT-- 10)

TUBING SIZE	2.441	IN. I.D.
PRODUCTION RATE	700	BLPD
GAS SPECIFIC GRAVITY		0.65
AVERAGE FLOWING TEMP	150	F
OIL API GRAVITY	35.0	API
WATER SPECIFIC GRAVITY		1.07

DEPTH (1000 FT)

GAS/LIQUID RATIO - SCF PER BBL

HAGEDORN AND BROWN
PRESENTED BY
MACCO - SCHLUMBERGER

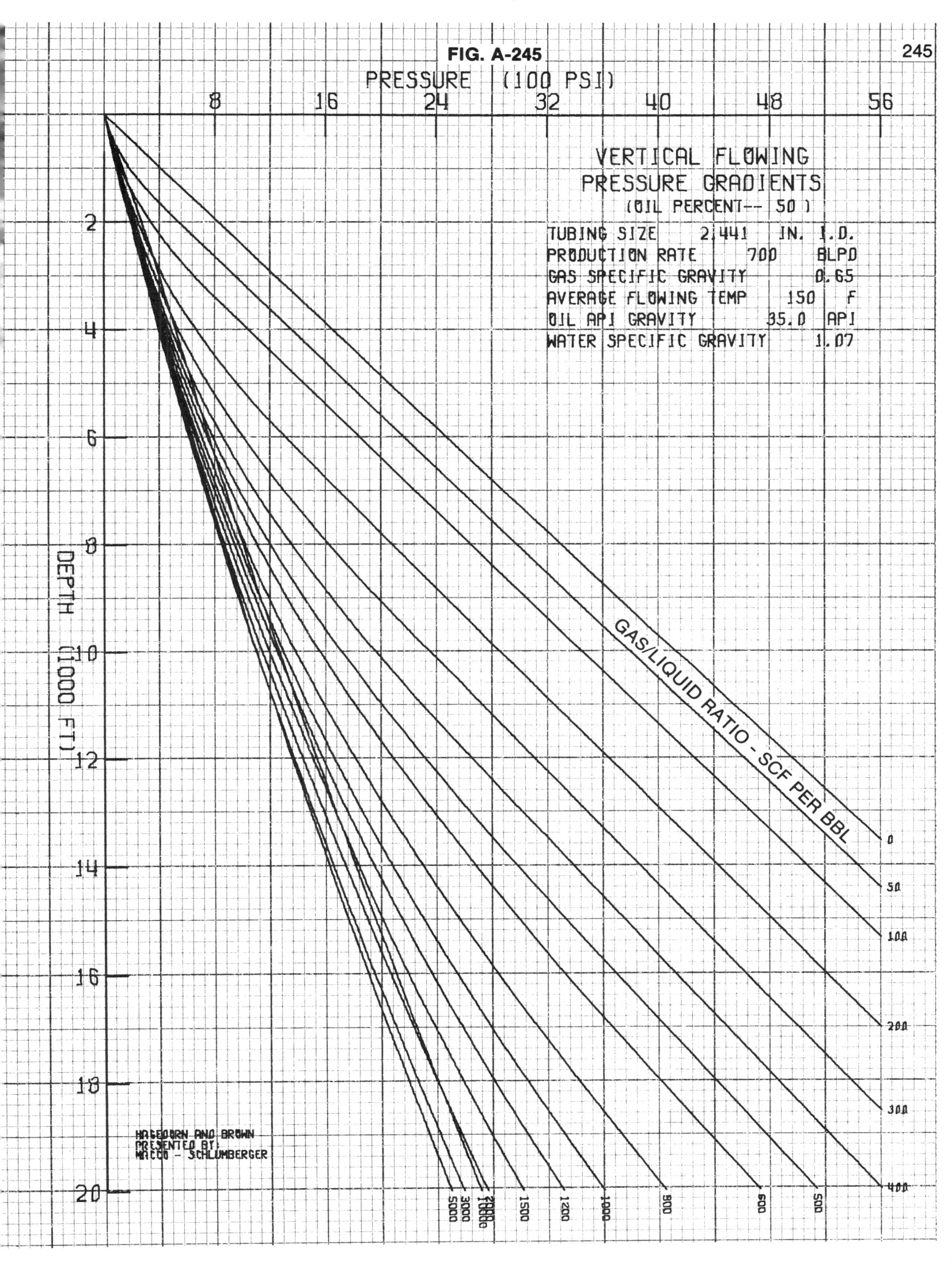

FIG. A-245

245

VERTICAL FLOWING
PRESSURE GRADIENTS
(OIL PERCENT-- 50)

TUBING SIZE   2.441   IN. I.D.
PRODUCTION RATE   700   BLPD
GAS SPECIFIC GRAVITY   0.65
AVERAGE FLOWING TEMP   150   F
OIL API GRAVITY   35.0   API
WATER SPECIFIC GRAVITY   1.07

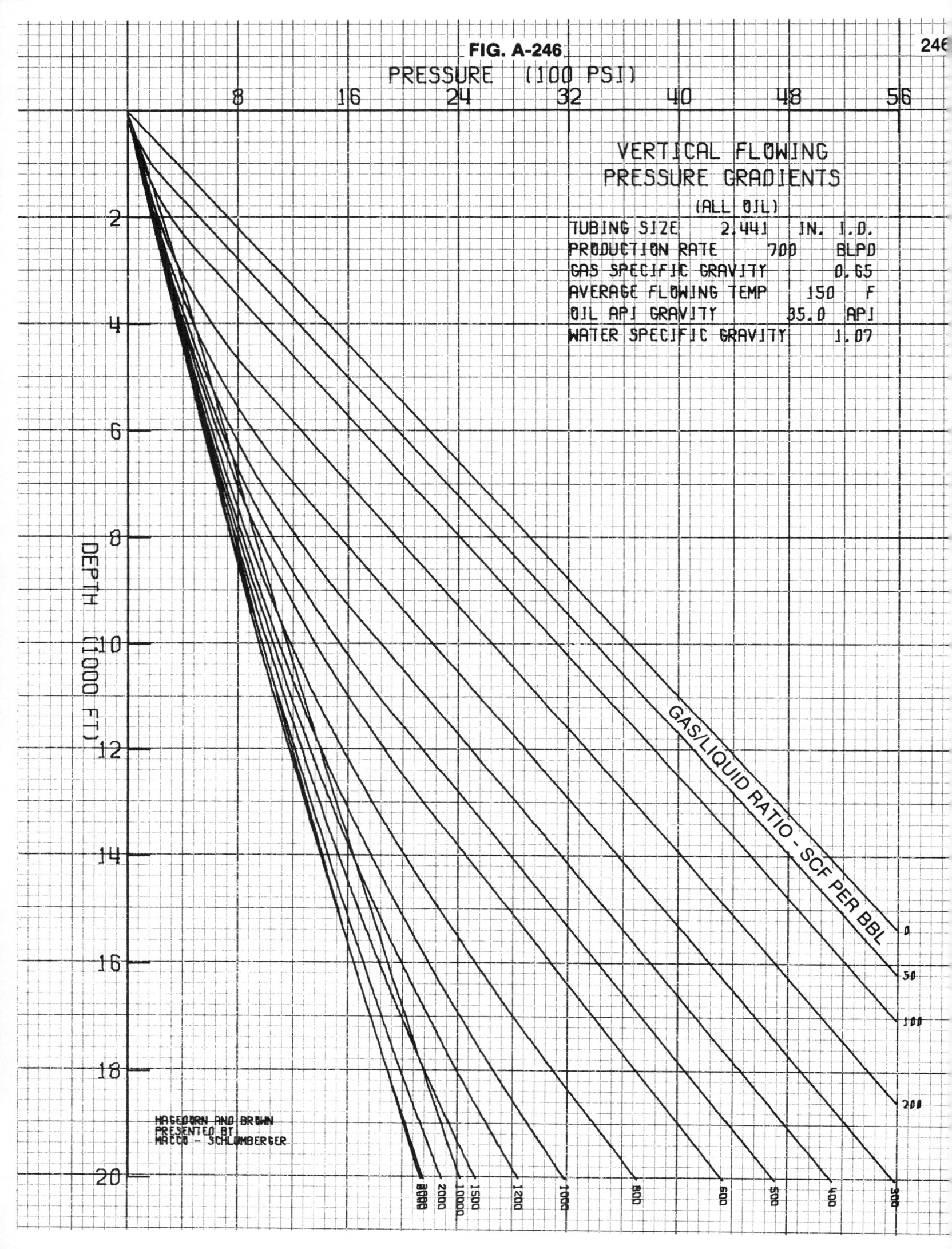

FIG. A-246

VERTICAL FLOWING
PRESSURE GRADIENTS
(ALL OIL)

TUBING SIZE	2.441	IN. I.D.
PRODUCTION RATE	700	BLPD
GAS SPECIFIC GRAVITY		0.65
AVERAGE FLOWING TEMP	150	F
OIL API GRAVITY	35.0	API
WATER SPECIFIC GRAVITY		1.07

PRESSURE (100 PSI)

DEPTH (1000 FT)

GAS/LIQUID RATIO - SCF PER BBL

HAGEDORN AND BROWN
PRESENTED BY
MATCO - SCHLUMBERGER

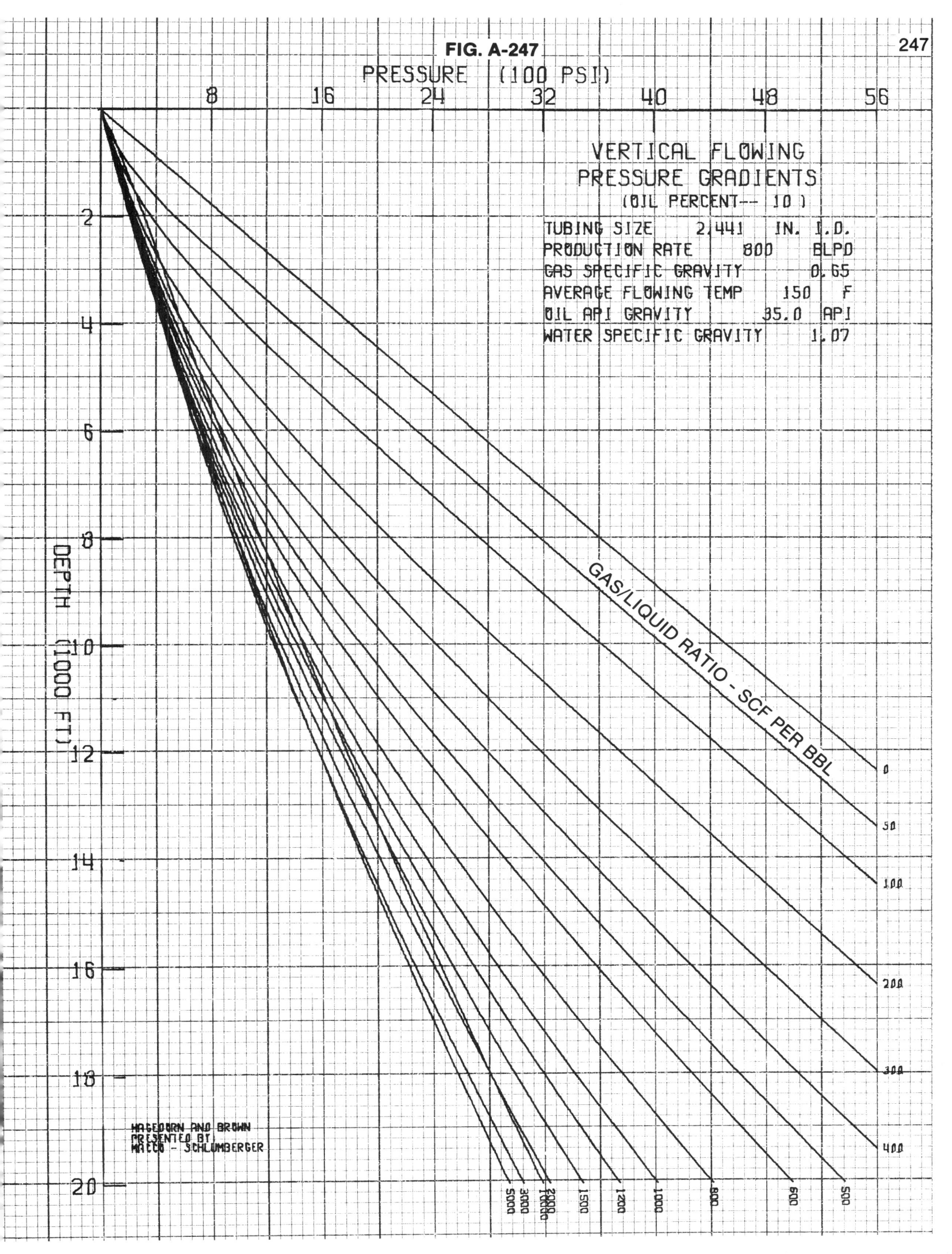

FIG. A-247

247

PRESSURE (100 PSI)

VERTICAL FLOWING
PRESSURE GRADIENTS
(OIL PERCENT--- 10)

TUBING SIZE   2.441   IN. I.D.
PRODUCTION RATE   800   BLPD
GAS SPECIFIC GRAVITY   0.65
AVERAGE FLOWING TEMP   150   F
OIL API GRAVITY   35.0   API
WATER SPECIFIC GRAVITY   1.07

GAS/LIQUID RATIO - SCF PER BBL

DEPTH (1000 FT)

HAGEDORN AND BROWN
PRESENTED BY
MAECO - SCHLUMBERGER

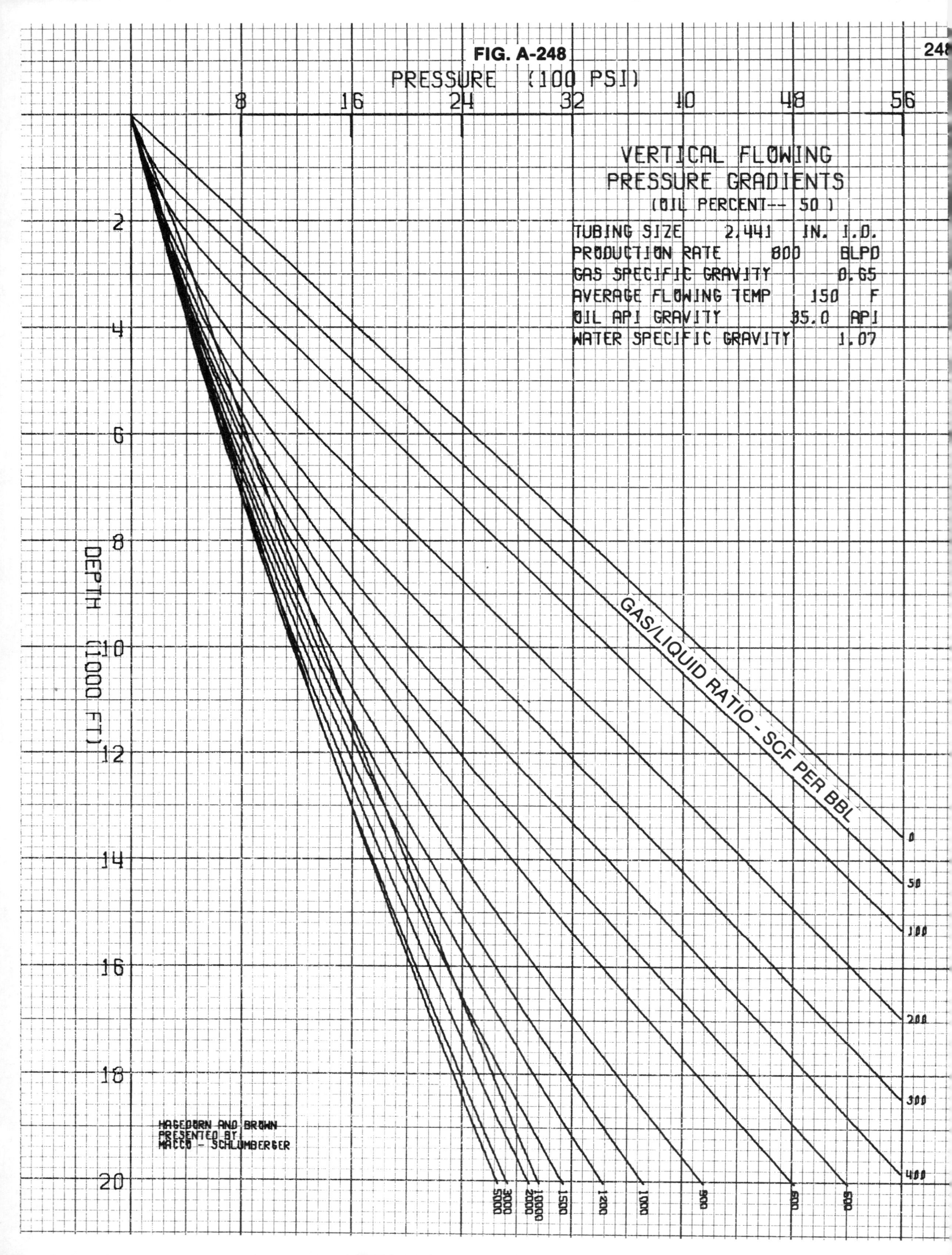

FIG. A-248

248

PRESSURE (100 PSI)

VERTICAL FLOWING
PRESSURE GRADIENTS
(OIL PERCENT-- 50)

TUBING SIZE	2.441	IN. I.D.
PRODUCTION RATE	800	BLPD
GAS SPECIFIC GRAVITY	0.65	
AVERAGE FLOWING TEMP	150	F
OIL API GRAVITY	35.0	API
WATER SPECIFIC GRAVITY	1.07	

DEPTH (1000 FT)

GAS/LIQUID RATIO - SCF PER BBL

HAGEDORN AND BROWN
PRESENTED BY
MACCO - SCHLUMBERGER

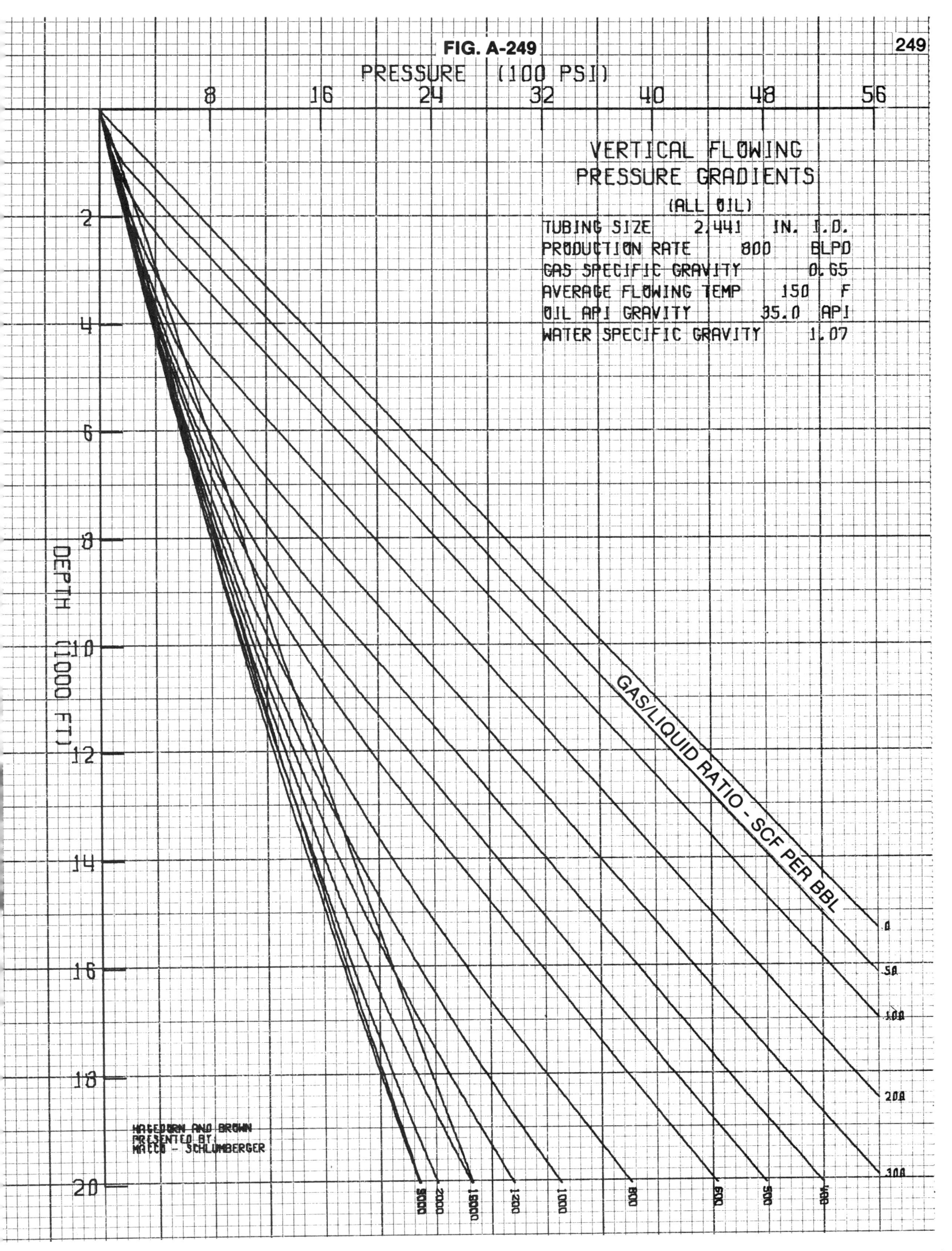

PRESSURE (100 PSI)

VERTICAL FLOWING
PRESSURE GRADIENTS
(ALL OIL)

TUBING SIZE	2.441	IN. I.D.
PRODUCTION RATE	800	BLPD
GAS SPECIFIC GRAVITY	0.65	
AVERAGE FLOWING TEMP	150	F
OIL API GRAVITY	35.0	API
WATER SPECIFIC GRAVITY	1.07	

DEPTH (1000 FT)

GAS/LIQUID RATIO - SCF PER BBL

HAGEDORN AND BROWN
PRESENTED BY
MAECO - SCHLUMBERGER

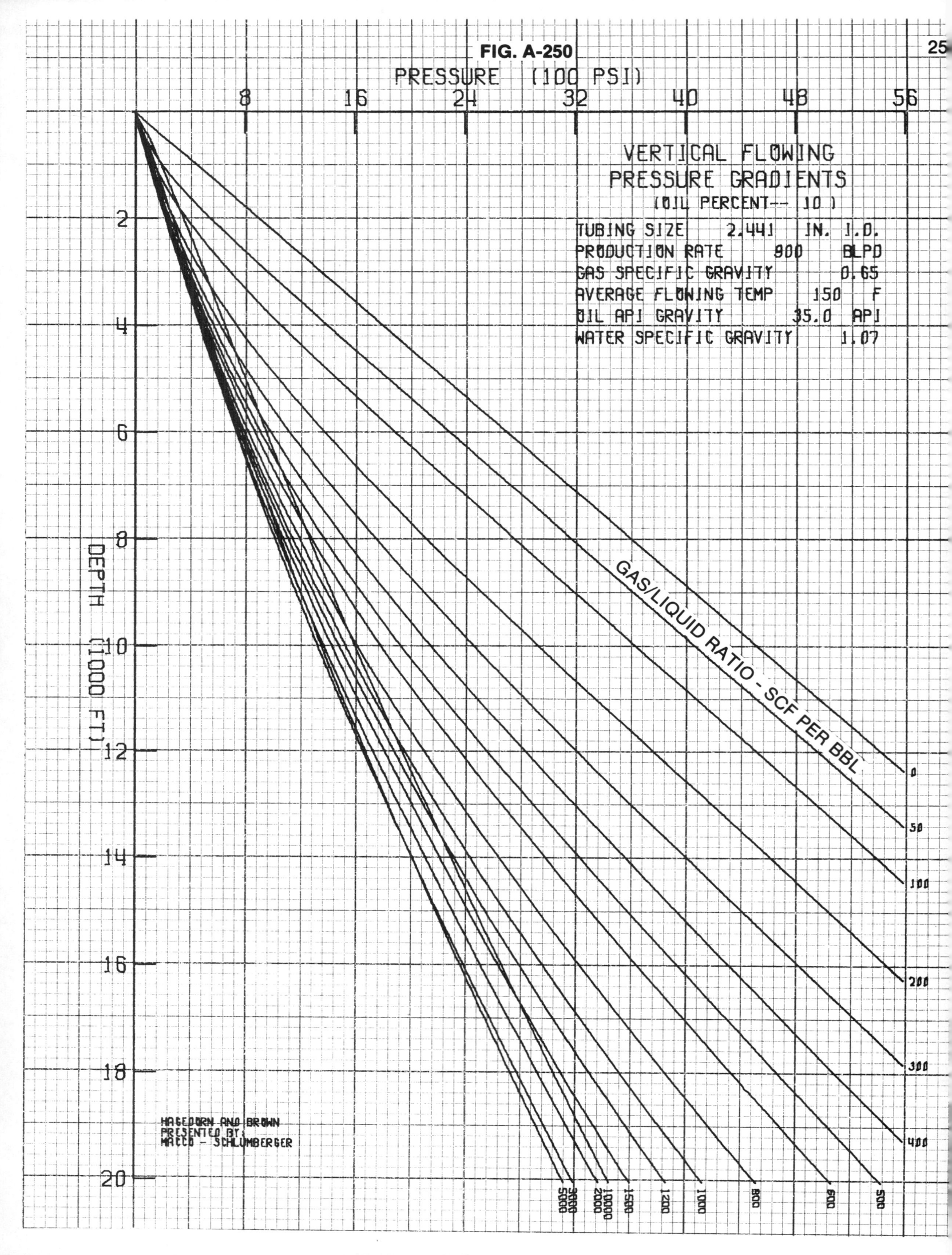

PRESSURE (100 PSI)

VERTICAL FLOWING
PRESSURE GRADIENTS
(OIL PERCENT--- 10)

TUBING SIZE	2.441	IN. I.D.
PRODUCTION RATE	900	BLPD
GAS SPECIFIC GRAVITY		0.65
AVERAGE FLOWING TEMP	150	F
OIL API GRAVITY	35.0	API
WATER SPECIFIC GRAVITY		1.07

DEPTH (1000 FT.)

GAS/LIQUID RATIO - SCF PER BBL

HAGEDORN AND BROWN
PRESENTED BY:
MACCO - SCHLUMBERGER

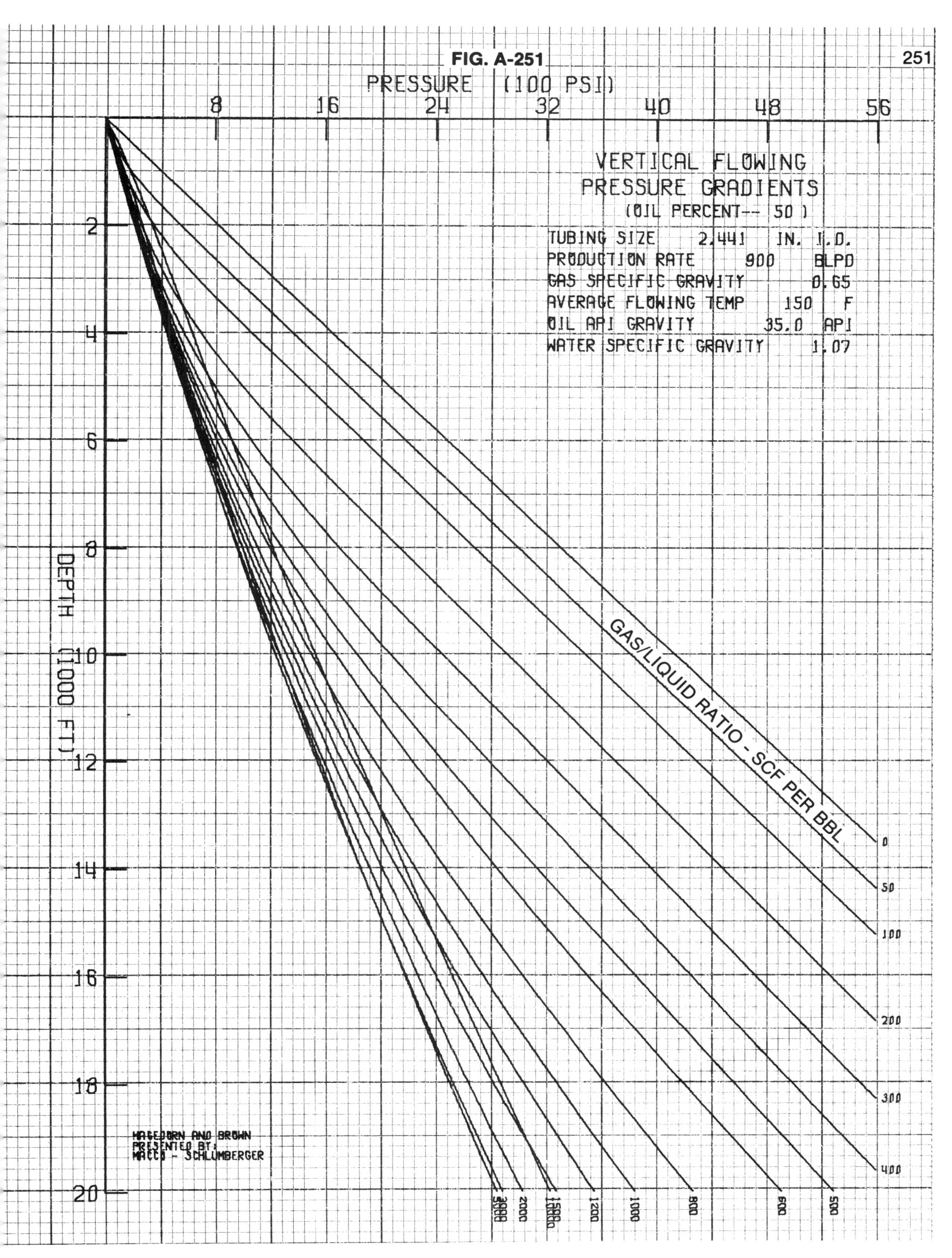

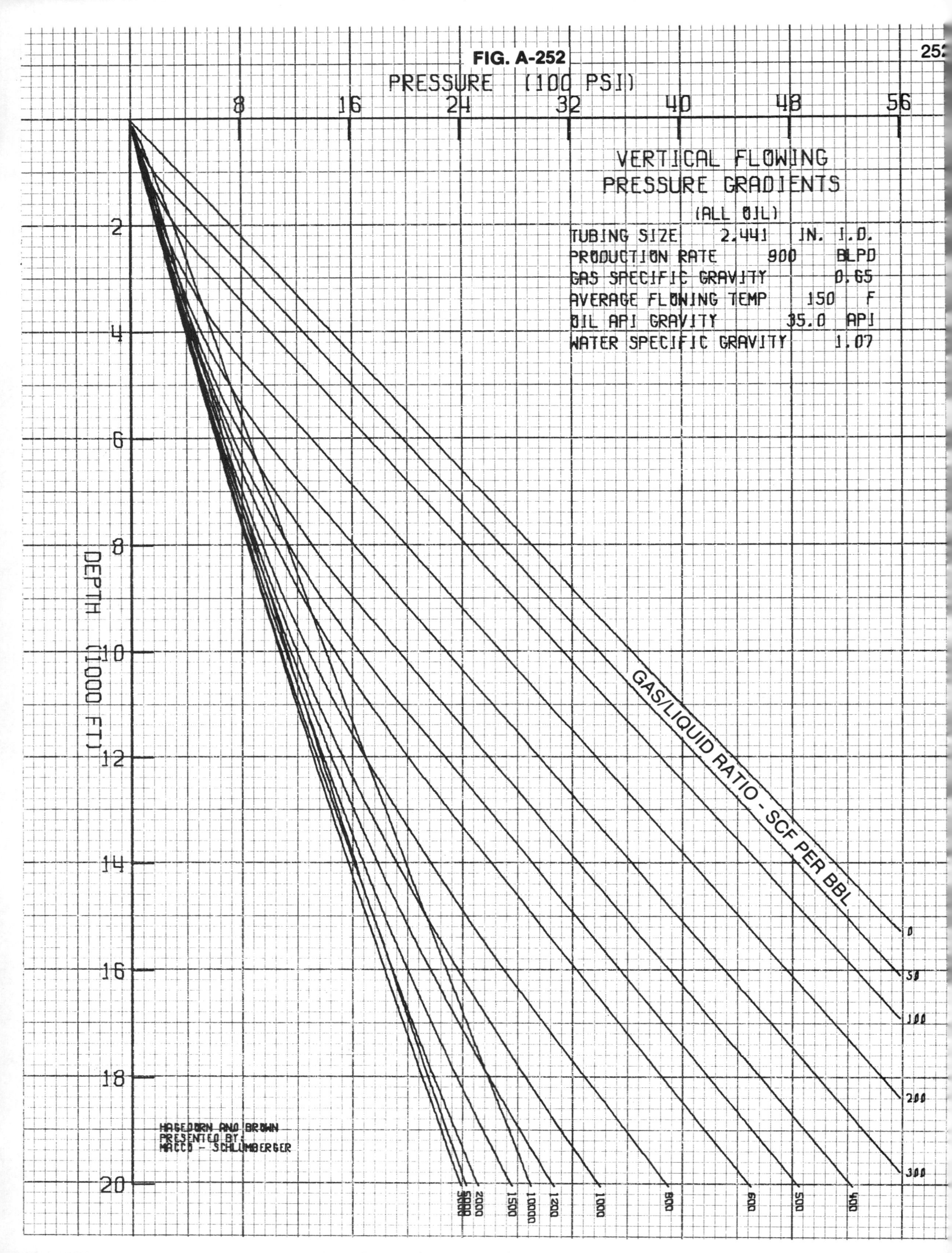

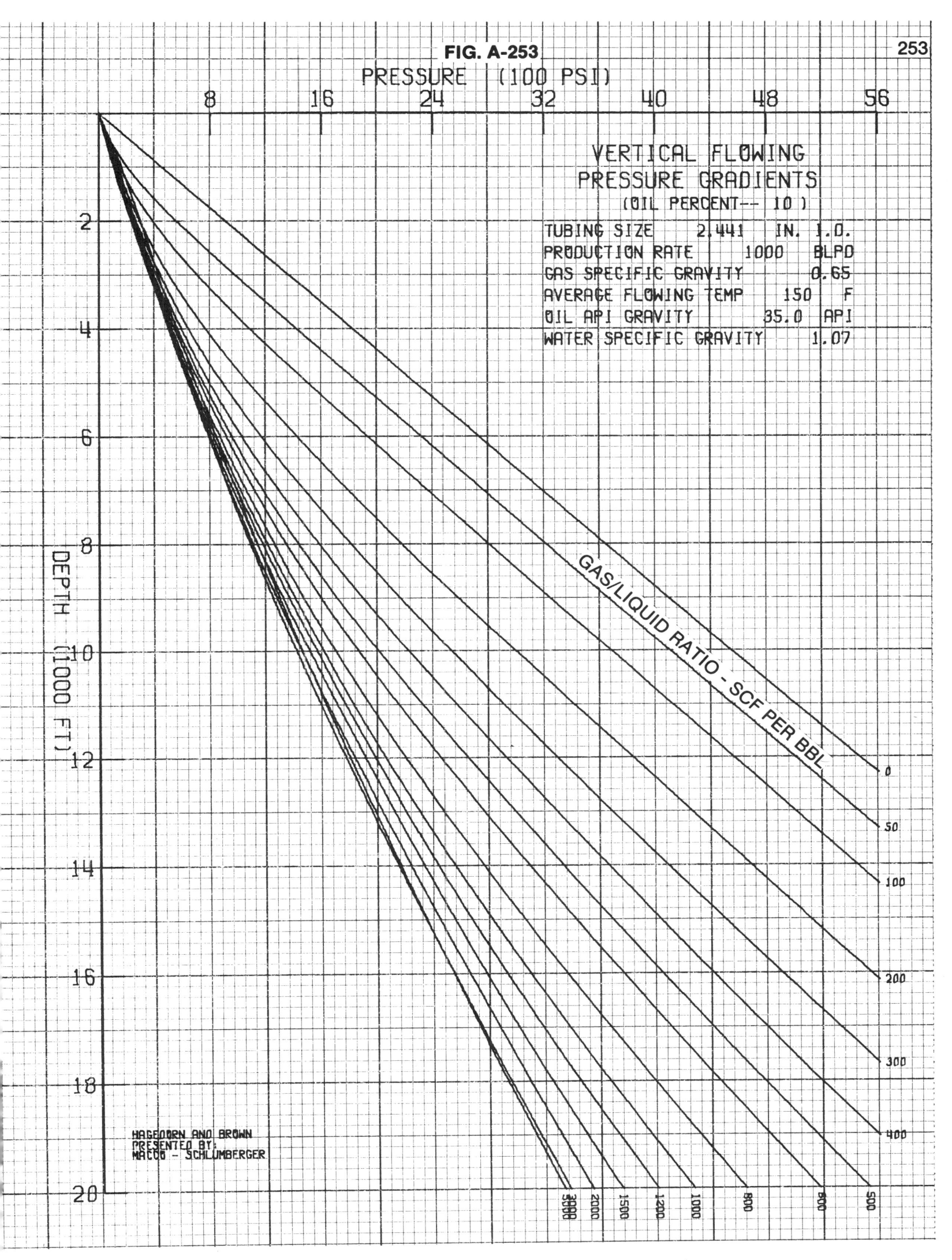

FIG. A-253

253

PRESSURE (100 PSI)

VERTICAL FLOWING
PRESSURE GRADIENTS
(OIL PERCENT— 10)

TUBING SIZE   2.441   IN. I.D.
PRODUCTION RATE   1000   BLPD
GAS SPECIFIC GRAVITY   0.65
AVERAGE FLOWING TEMP   150 F
OIL API GRAVITY   35.0 API
WATER SPECIFIC GRAVITY   1.07

GAS/LIQUID RATIO - SCF PER BBL

DEPTH (1000 FT)

HAGEDORN AND BROWN
PRESENTED BY:
MAECO - SCHLUMBERGER

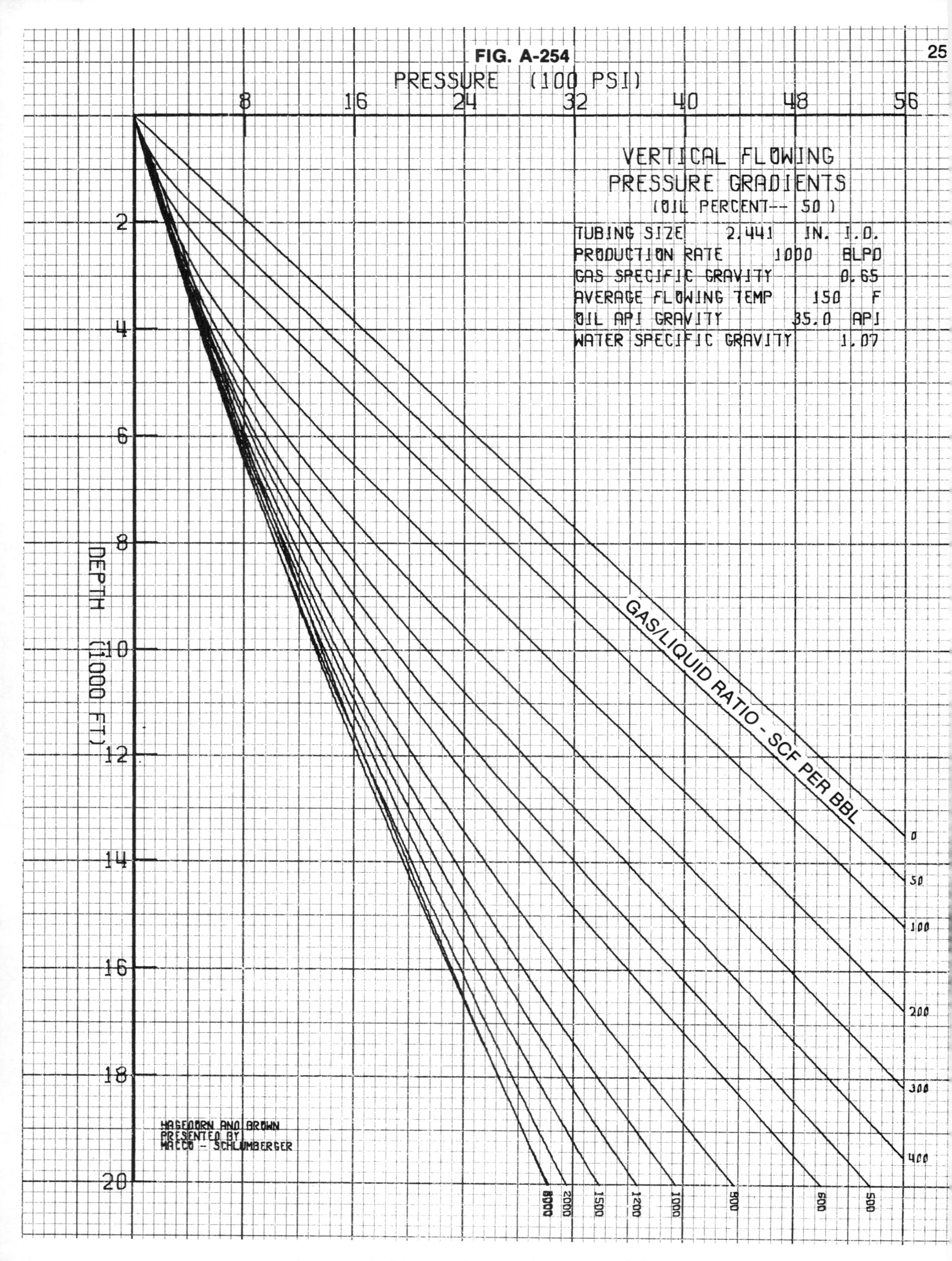

FIG. A-254

25

VERTICAL FLOWING
PRESSURE GRADIENTS
(OIL PERCENT-- 50)

TUBING SIZE   2.441   IN. I.D.
PRODUCTION RATE   1000   BLPD
GAS SPECIFIC GRAVITY   0.65
AVERAGE FLOWING TEMP   150   F
OIL API GRAVITY   35.0   API
WATER SPECIFIC GRAVITY   1.07

GAS/LIQUID RATIO - SCF PER BBL

HAGEDORN AND BROWN
PRESENTED BY
MAECO - SCHLUMBERGER

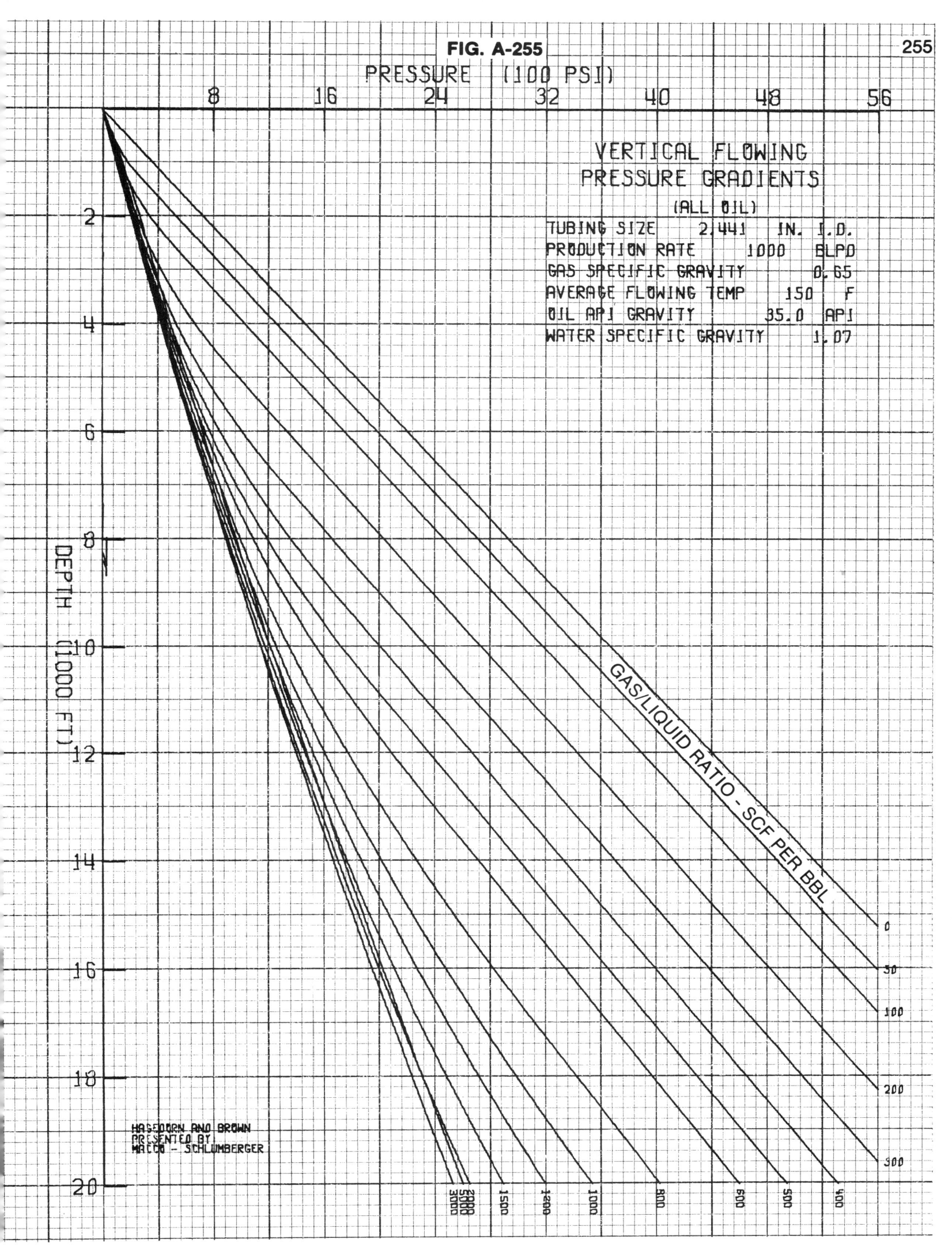

FIG. A-255

255

PRESSURE (100 PSI)

VERTICAL FLOWING
PRESSURE GRADIENTS
(ALL OIL)

TUBING SIZE	2.441	IN. I.D.
PRODUCTION RATE	1000	BLPD
GAS SPECIFIC GRAVITY	0.65	
AVERAGE FLOWING TEMP	150	F
OIL API GRAVITY	35.0	API
WATER SPECIFIC GRAVITY	1.07	

GAS/LIQUID RATIO - SCF PER BBL

DEPTH (1000 FT)

HASEDORN AND BROWN
PRESENTED BY
MACCO - SCHLUMBERGER

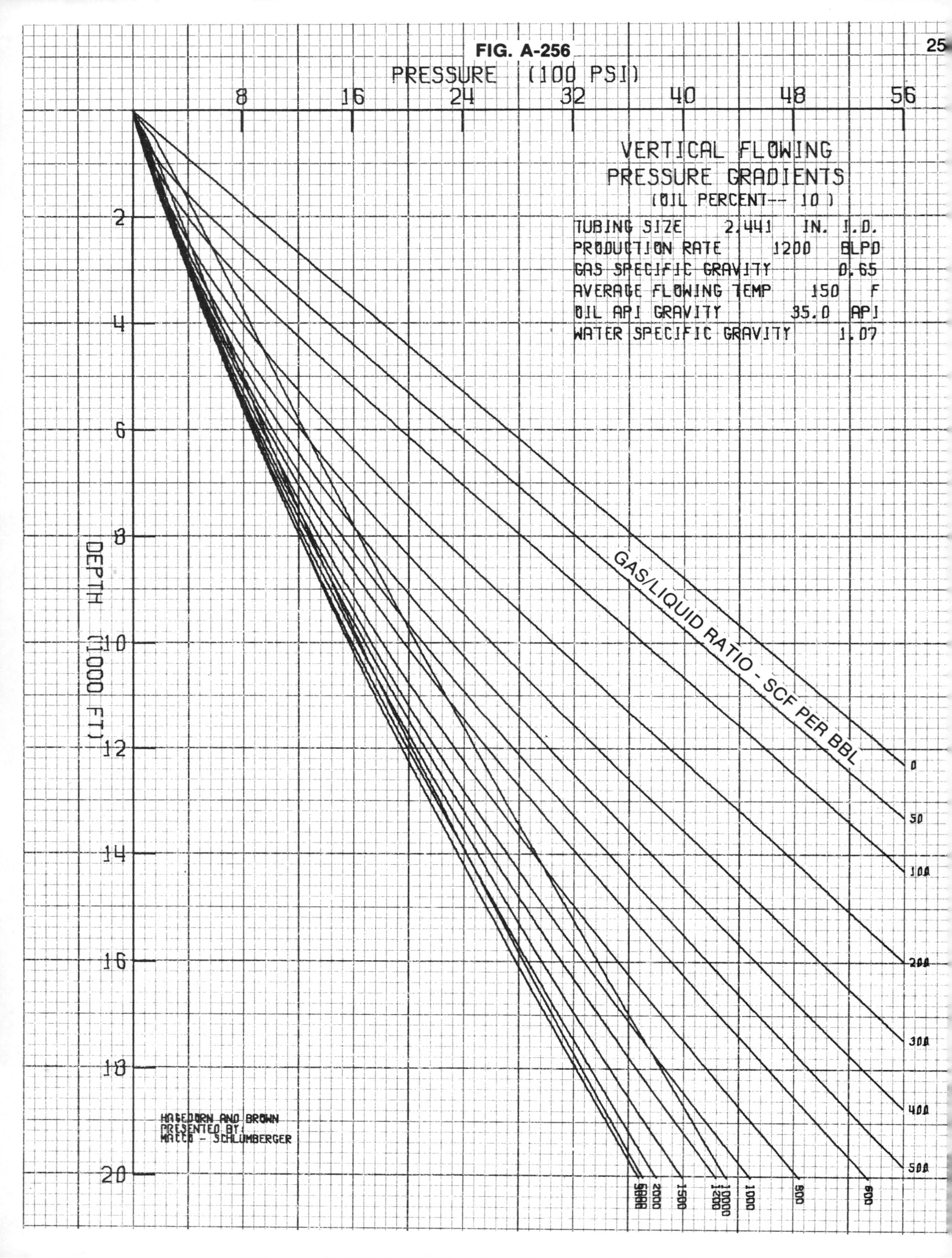

VERTICAL FLOWING
PRESSURE GRADIENTS
(OIL PERCENT-- 10)

TUBING SIZE	2.441	IN. I.D.
PRODUCTION RATE	1200	BLPD
GAS SPECIFIC GRAVITY	0.65	
AVERAGE FLOWING TEMP	150	F
OIL API GRAVITY	35.0	API
WATER SPECIFIC GRAVITY	1.07	

PRESSURE (100 PSI)

DEPTH (1000 FT)

GAS/LIQUID RATIO - SCF PER BBL

HAGEDORN AND BROWN
PRESENTED BY
MACCO - SCHLUMBERGER

FIG. A-257

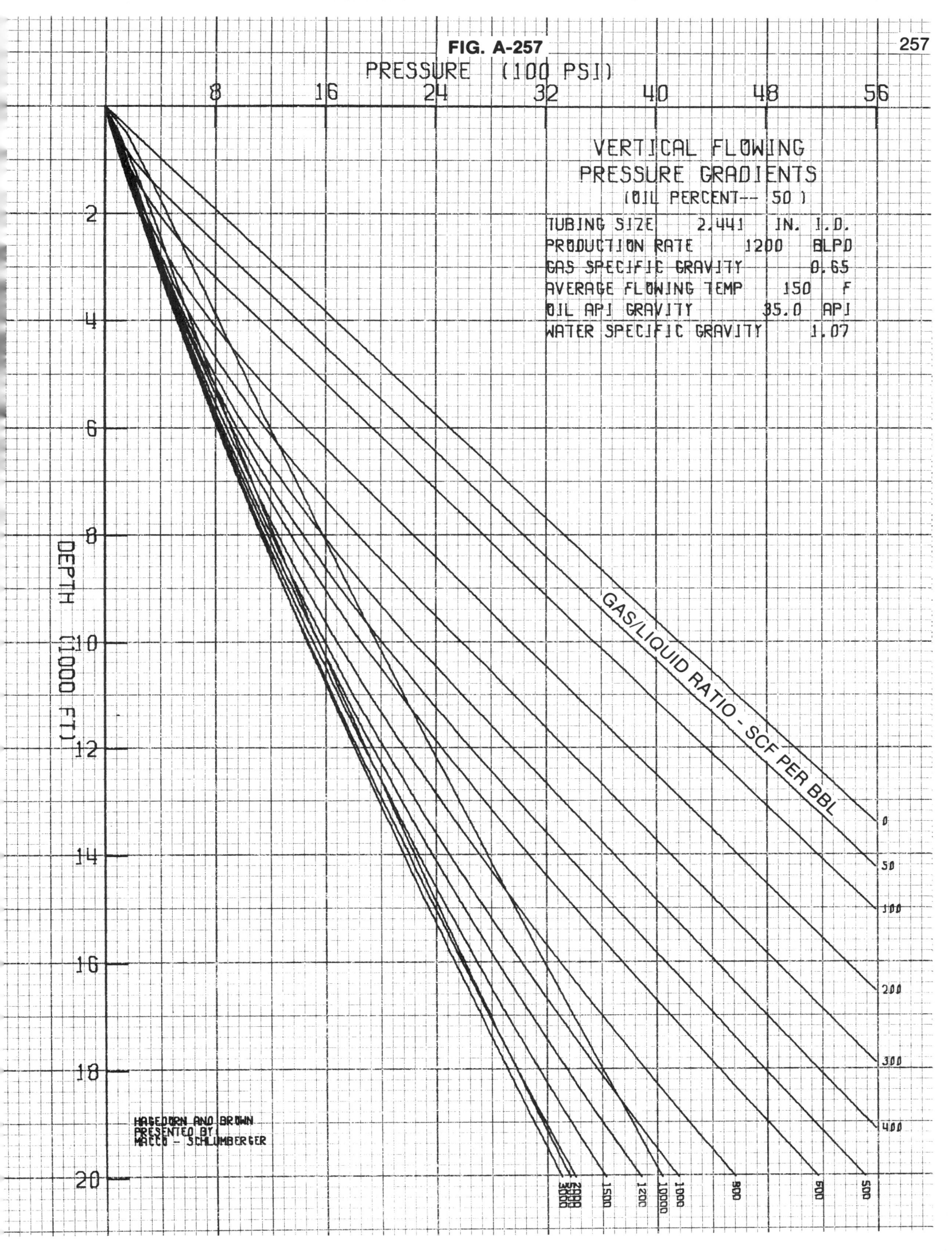

PRESSURE (100 PSI)

VERTICAL FLOWING
PRESSURE GRADIENTS
(OIL PERCENT-- 50)

TUBING SIZE	2.441	IN. I.D.
PRODUCTION RATE	1200	BLPD
GAS SPECIFIC GRAVITY		0.65
AVERAGE FLOWING TEMP	150	F
OIL API GRAVITY	35.0	API
WATER SPECIFIC GRAVITY		1.07

DEPTH (1000 FT)

GAS/LIQUID RATIO - SCF PER BBL

HAGEDORN AND BROWN
PRESENTED BY
WALCO - SCHLUMBERGER

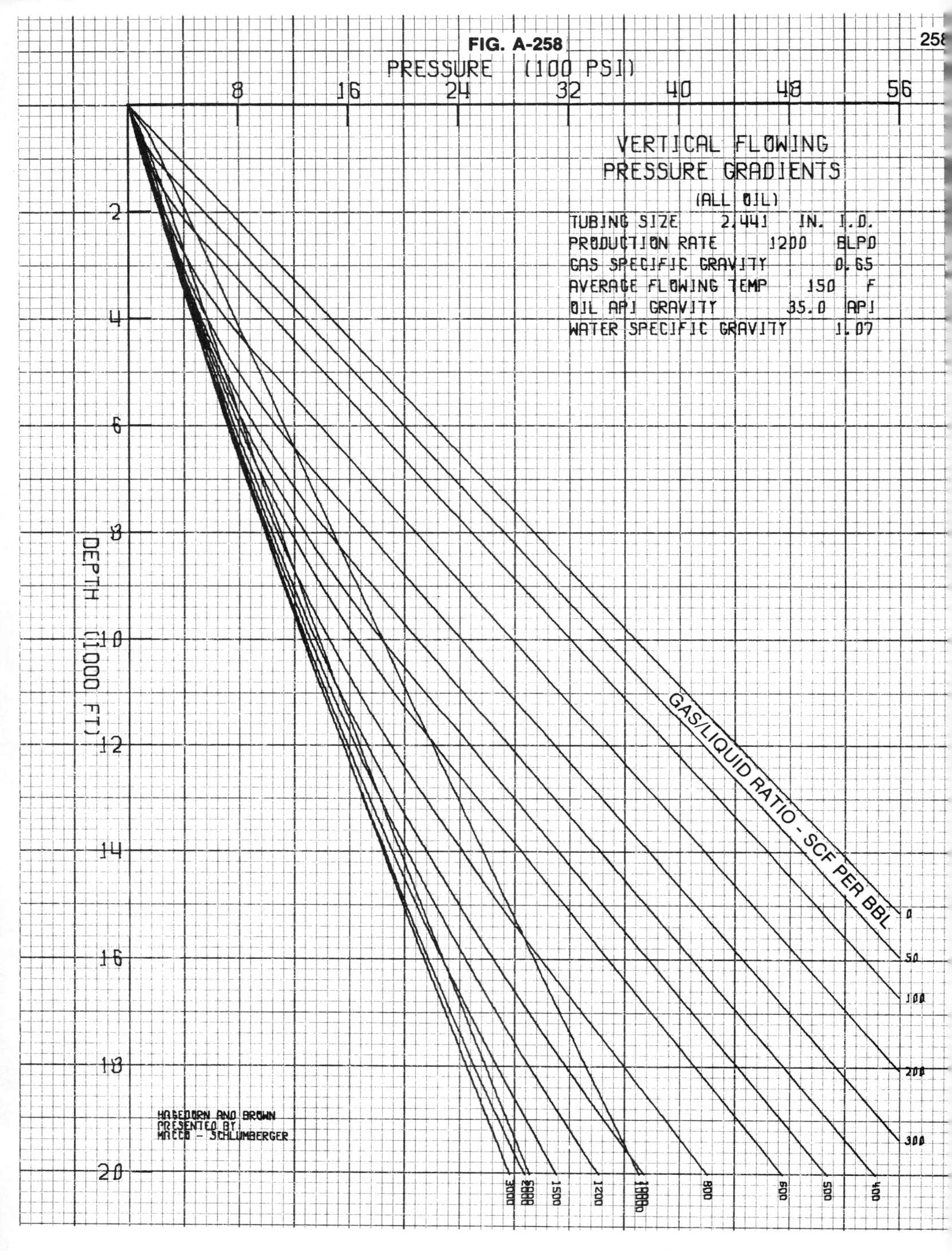

FIG. A-258

258

PRESSURE (100 PSI)

VERTICAL FLOWING
PRESSURE GRADIENTS
(ALL OIL)

TUBING SIZE   2.441   IN. I.D.
PRODUCTION RATE   1200   BLPD
GAS SPECIFIC GRAVITY   0.65
AVERAGE FLOWING TEMP   150   F
OIL API GRAVITY   35.0 API
WATER SPECIFIC GRAVITY   1.07

DEPTH (1000 FT)

GAS/LIQUID RATIO - SCF PER BBL

HAGEDORN AND BROWN
PRESENTED BY
MACCO - SCHLUMBERGER

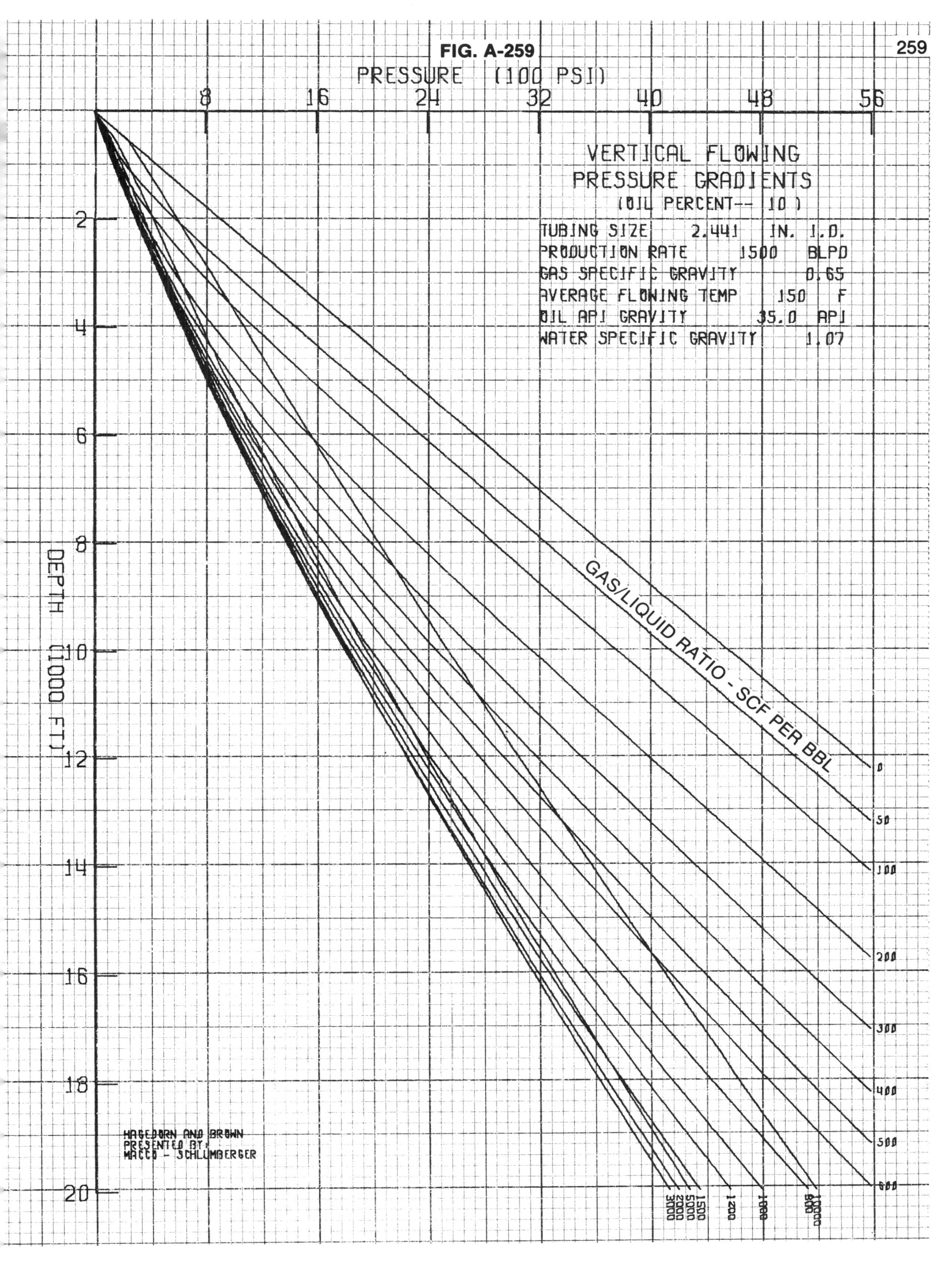

FIG. A-259

PRESSURE (100 PSI)

VERTICAL FLOWING
PRESSURE GRADIENTS
(OIL PERCENT-- 10)

TUBING SIZE	2.441	IN. I.D.
PRODUCTION RATE	1500	BLPD
GAS SPECIFIC GRAVITY	0.65	
AVERAGE FLOWING TEMP	150	F
OIL API GRAVITY	35.0	API
WATER SPECIFIC GRAVITY	1.07	

DEPTH (1000 FT)

GAS/LIQUID RATIO - SCF PER BBL

HAGEDORN AND BROWN
PRESENTED BY:
MACCO - SCHLUMBERGER

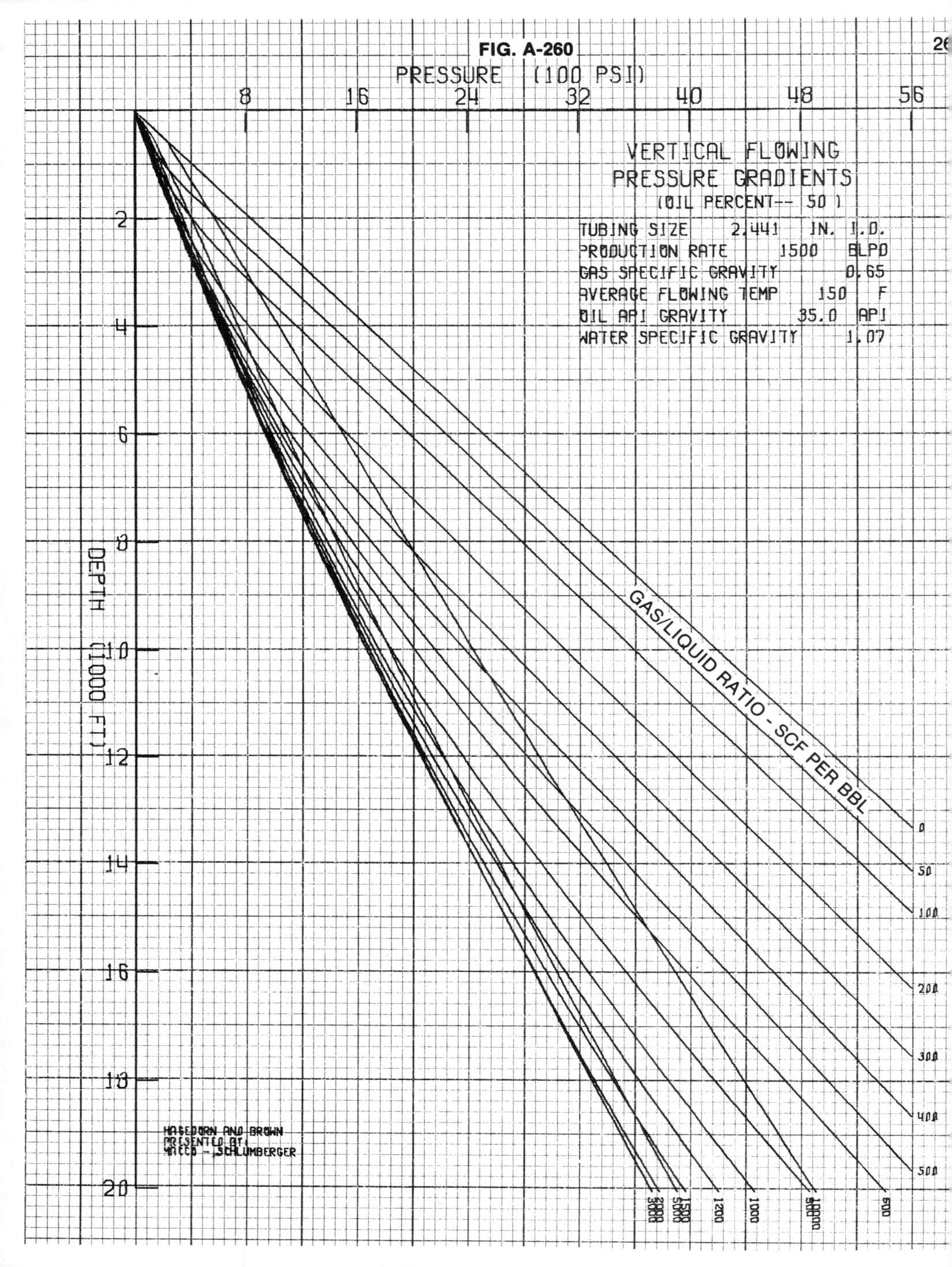

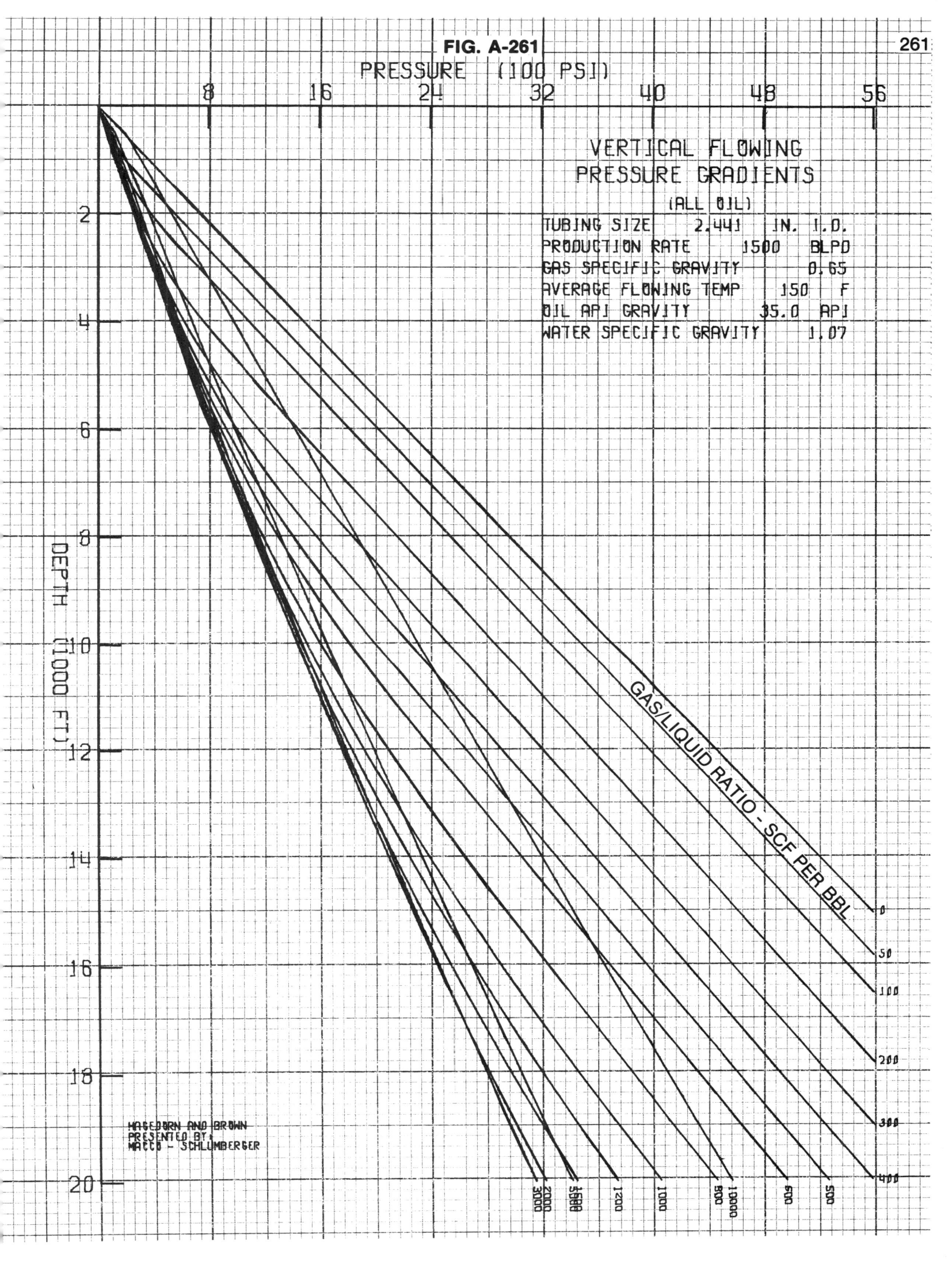

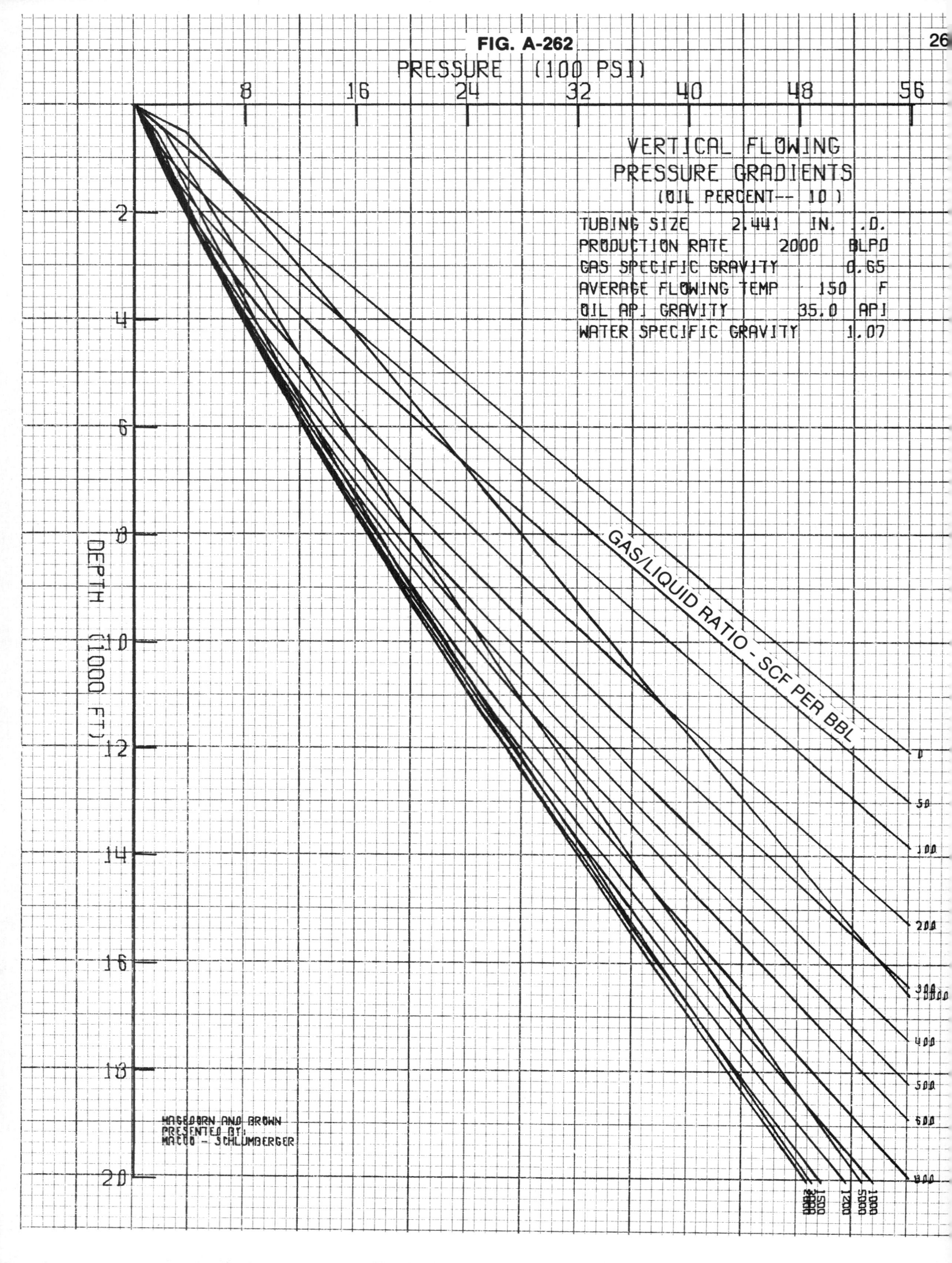

PRESSURE (100 PSI)

VERTICAL FLOWING
PRESSURE GRADIENTS
(OIL PERCENT-- 10)

TUBING SIZE	2.441	IN. I.D.
PRODUCTION RATE	2000	BLPD
GAS SPECIFIC GRAVITY		0.65
AVERAGE FLOWING TEMP	150	F
OIL API GRAVITY	35.0	API
WATER SPECIFIC GRAVITY		1.07

GAS/LIQUID RATIO - SCF PER BBL

DEPTH (1000 FT)

HASBORN AND BROWN
PRESENTED BY:
MACCO - SCHLUMBERGER

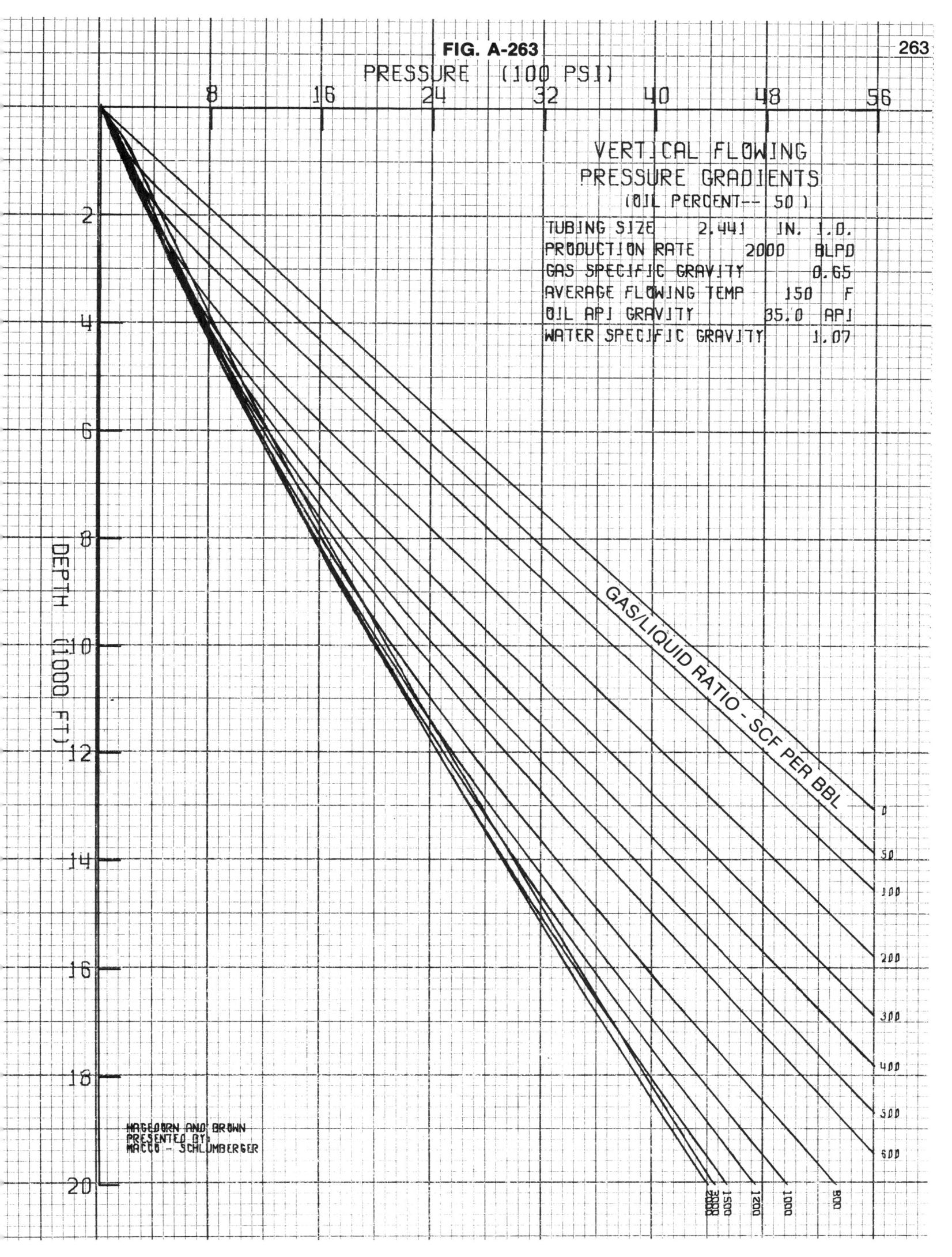

PRESSURE (100 PSI)

VERTICAL FLOWING
PRESSURE GRADIENTS
(OIL PERCENT-- 50)

TUBING SIZE 2.441 IN. I.D.
PRODUCTION RATE 2000 BLPD
GAS SPECIFIC GRAVITY 0.65
AVERAGE FLOWING TEMP 150 F
OIL API GRAVITY 35.0 API
WATER SPECIFIC GRAVITY 1.07

GAS/LIQUID RATIO - SCF PER BBL

DEPTH (1000 FT)

HAGEDORN AND BROWN
PRESENTED BY
MACCO - SCHLUMBERGER

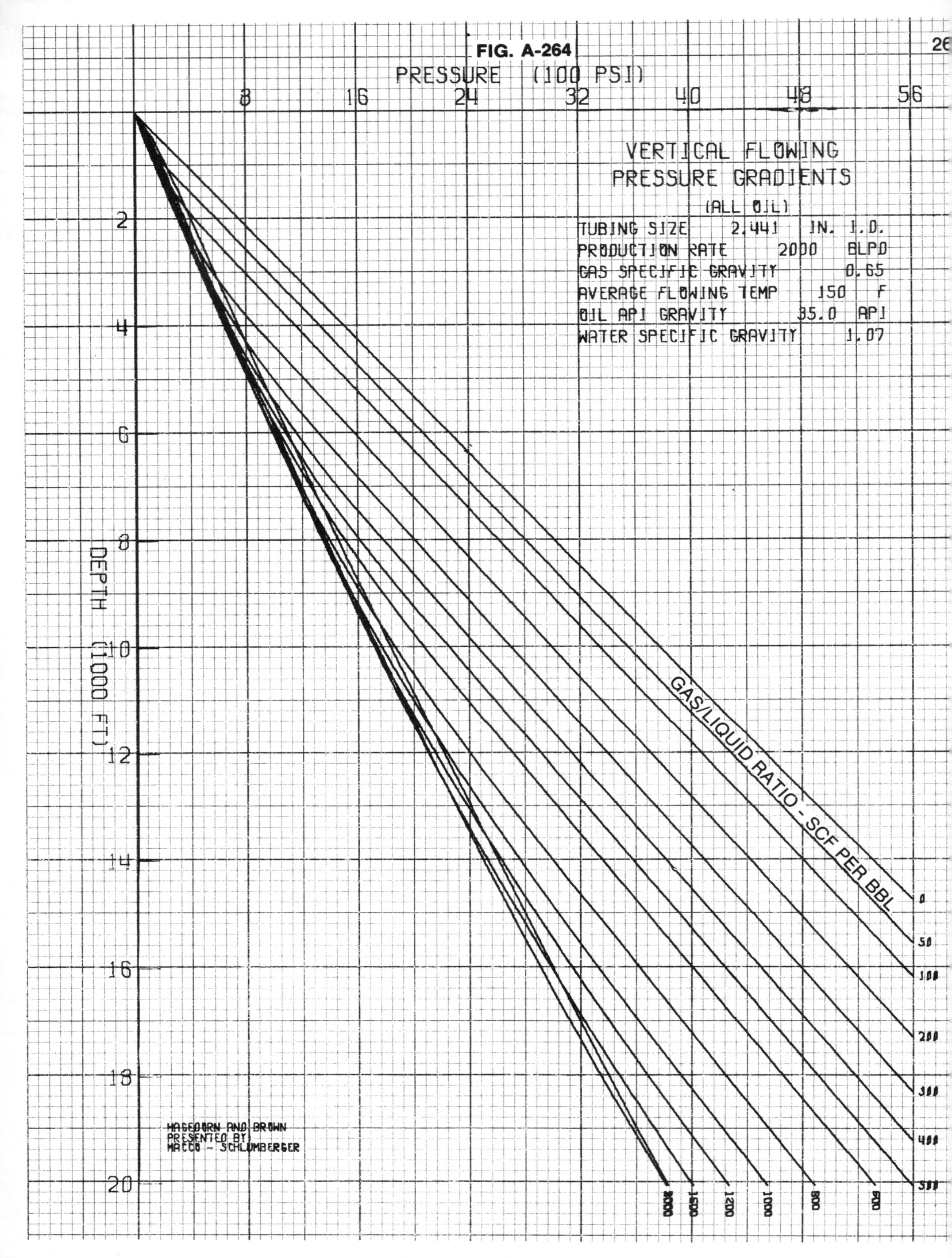

PRESSURE (100 PSI)

VERTICAL FLOWING
PRESSURE GRADIENTS
(ALL OIL)

TUBING SIZE	2.441	IN. I.D.
PRODUCTION RATE	2000	BLPD
GAS SPECIFIC GRAVITY	0.65	
AVERAGE FLOWING TEMP	150	F
OIL API GRAVITY	35.0	API
WATER SPECIFIC GRAVITY	1.07	

DEPTH (1000 FT)

GAS/LIQUID RATIO - SCF PER BBL

HAGEDORN AND BROWN
PRESENTED BY
MACCO - SCHLUMBERGER

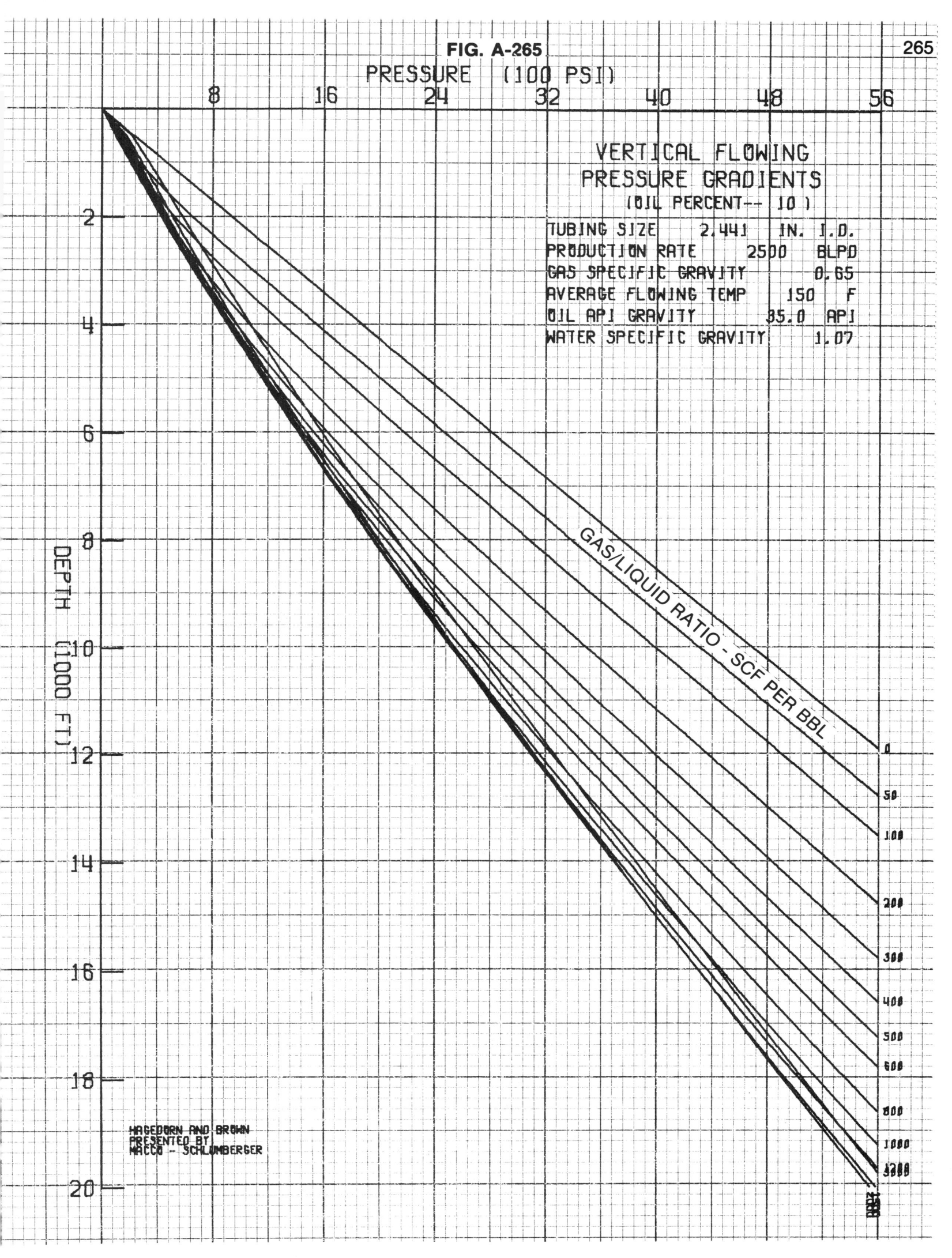

FIG. A-265

265

PRESSURE (100 PSI)

VERTICAL FLOWING
PRESSURE GRADIENTS
(OIL PERCENT-- 10)

TUBING SIZE	2.441	IN. I.D.
PRODUCTION RATE	2500	BLPD
GAS SPECIFIC GRAVITY		0.65
AVERAGE FLOWING TEMP	150	F
OIL API GRAVITY	35.0	API
WATER SPECIFIC GRAVITY		1.07

DEPTH (1000 FT)

GAS/LIQUID RATIO - SCF PER BBL

HAGEDORN AND BROWN
PRESENTED BY
MAECO - SCHLUMBERGER

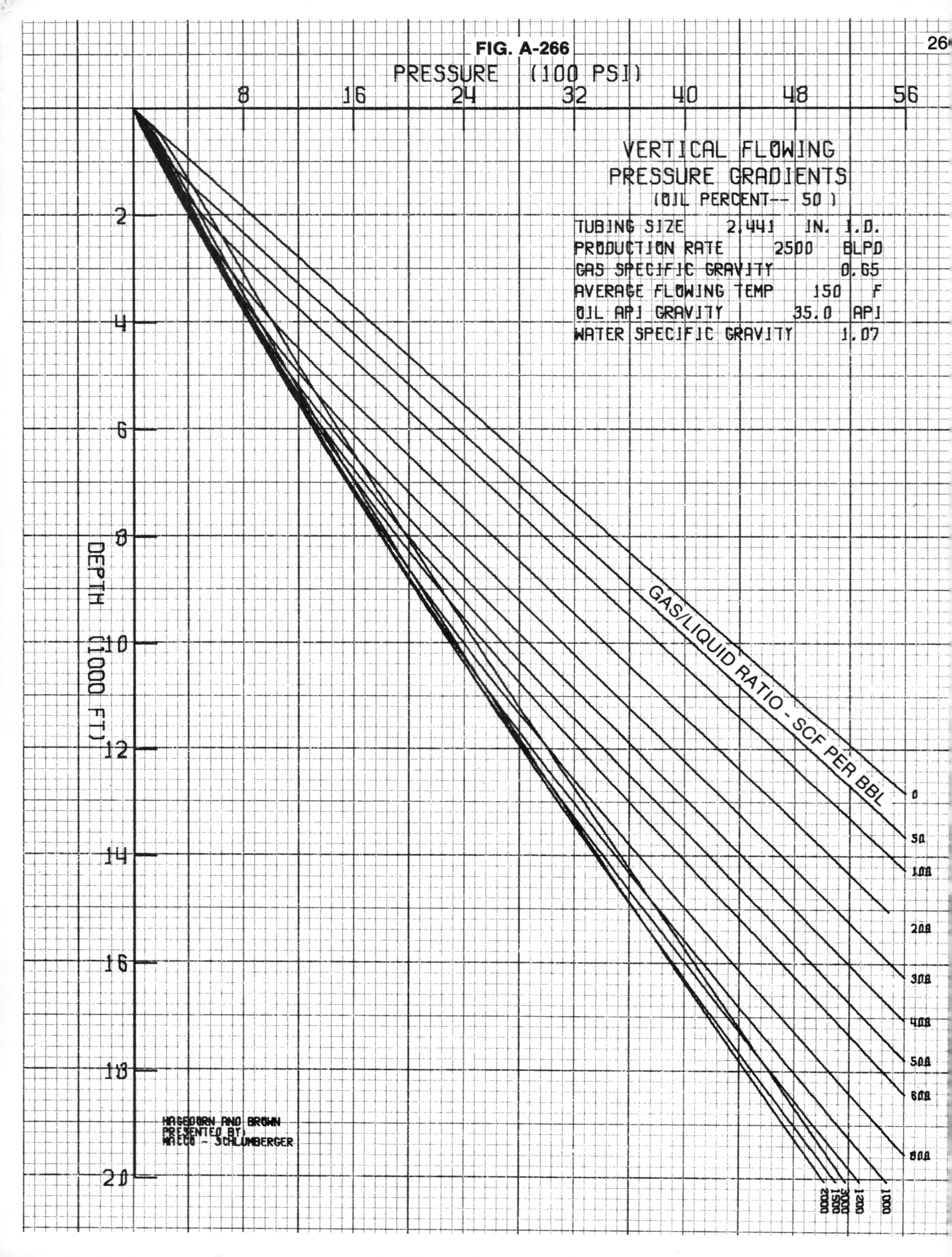

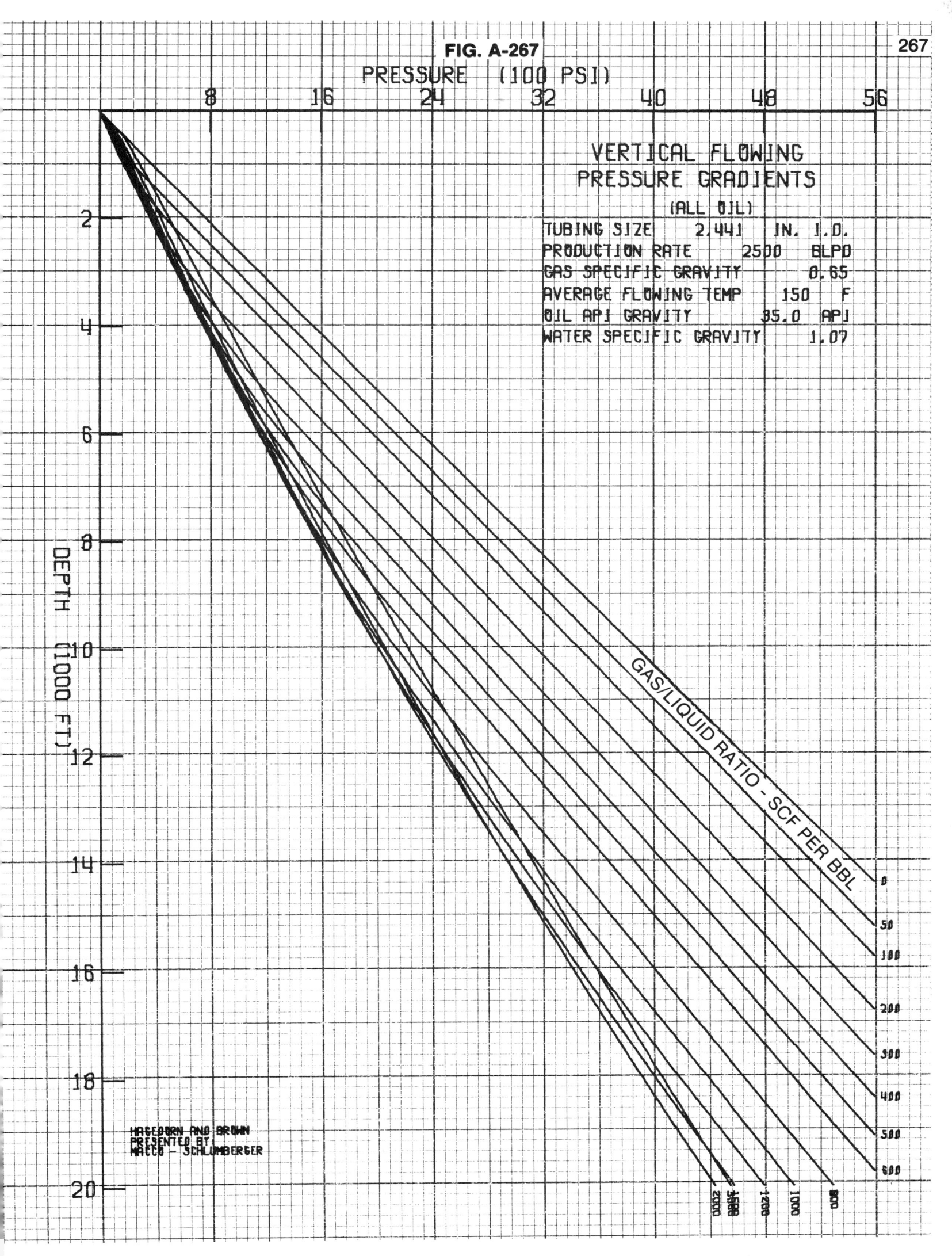

VERTICAL FLOWING
PRESSURE GRADIENTS
(ALL OIL)

TUBING SIZE	2.441	IN. I.D.
PRODUCTION RATE	2500	BLPD
GAS SPECIFIC GRAVITY	0.65	
AVERAGE FLOWING TEMP	150	F
OIL API GRAVITY	35.0	API
WATER SPECIFIC GRAVITY	1.07	

PRESSURE (100 PSI)

DEPTH (1000 FT)

GAS/LIQUID RATIO - SCF PER BBL

HAGEDORN AND BROWN
PRESENTED BY
MAECO - SCHLUMBERGER

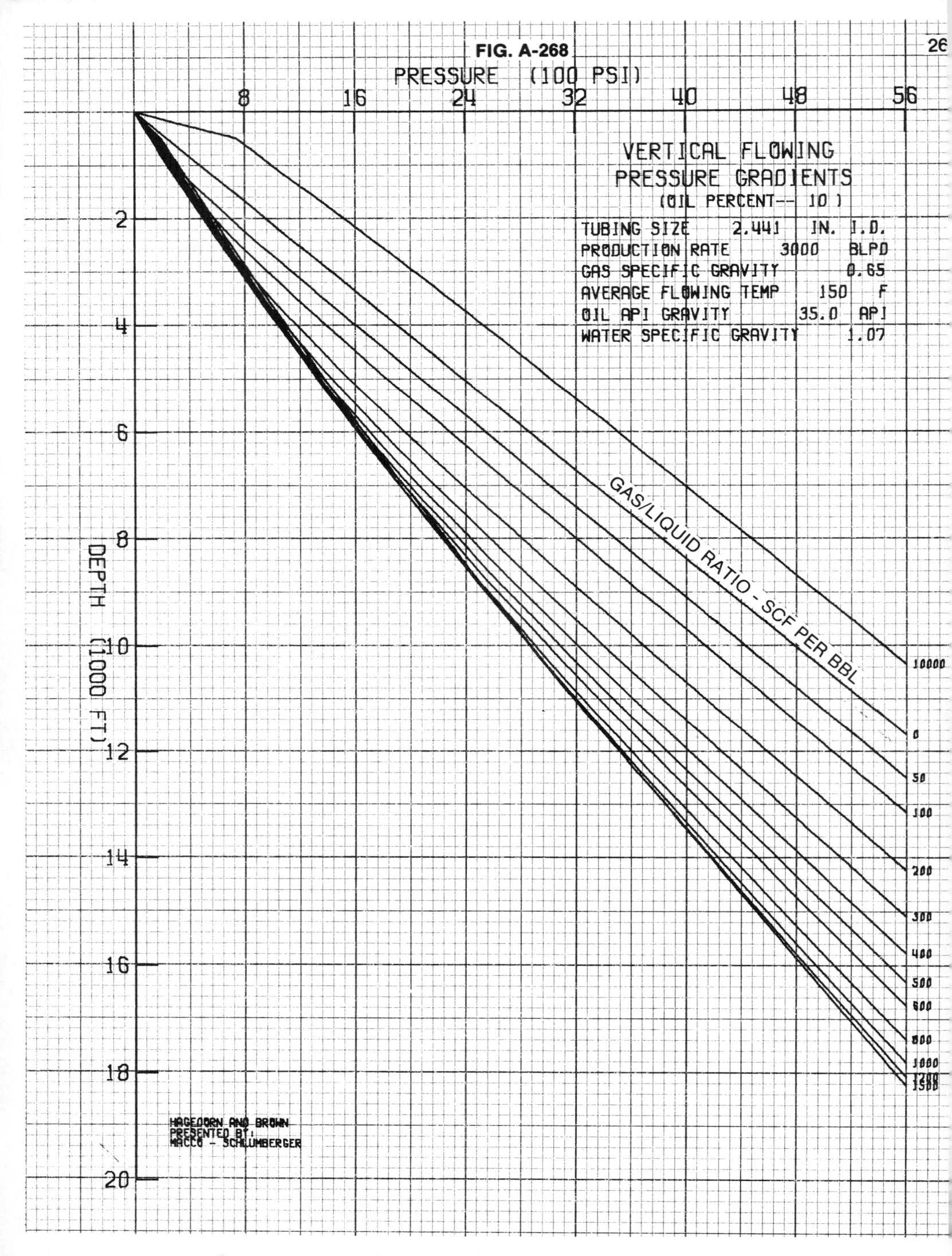

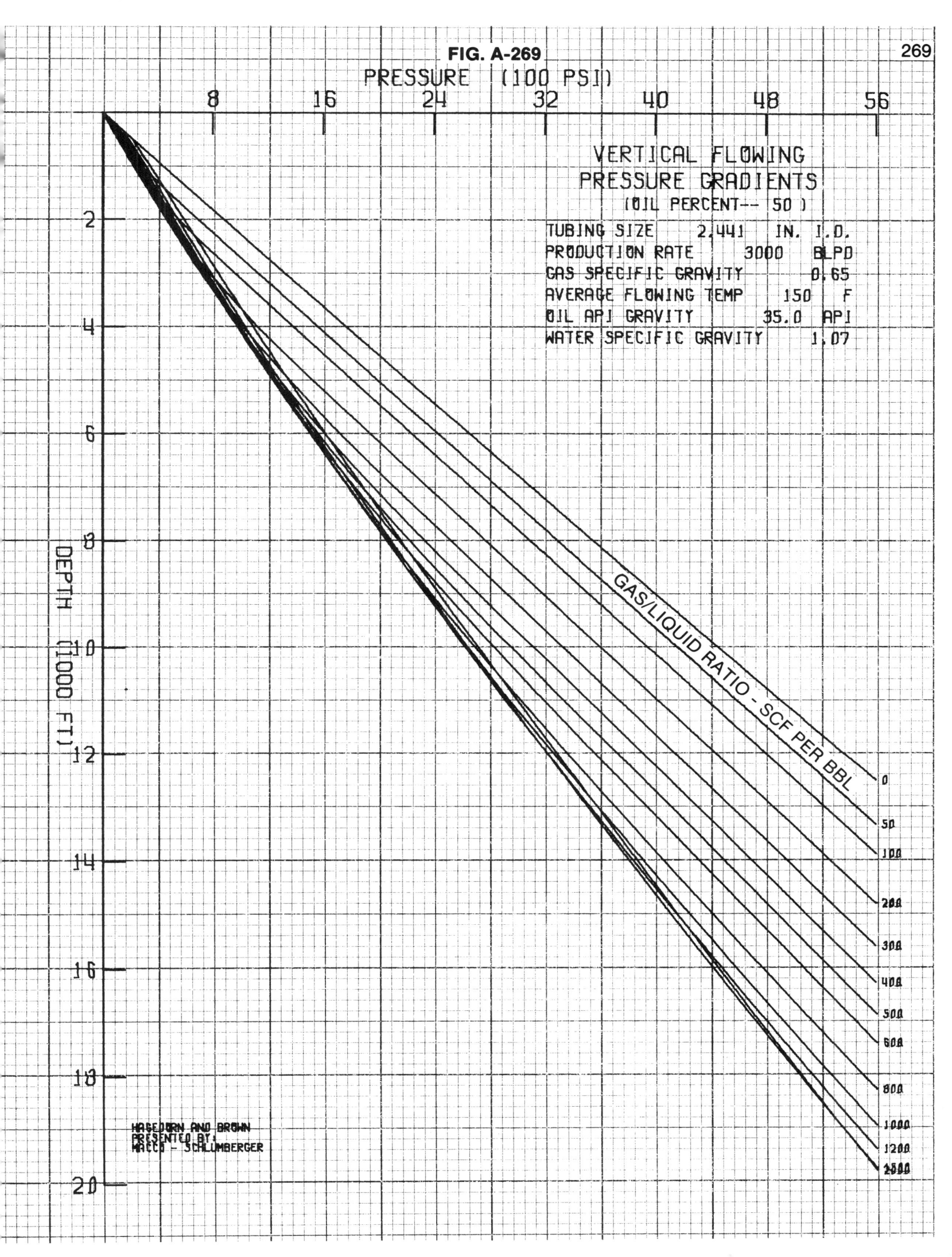

FIG. A-269

269

PRESSURE (100 PSI)

VERTICAL FLOWING
PRESSURE GRADIENTS
(OIL PERCENT-- 50)

TUBING SIZE	2.441	IN. I.D.
PRODUCTION RATE	3000	BLPD
GAS SPECIFIC GRAVITY	0.65	
AVERAGE FLOWING TEMP	150	F
OIL API GRAVITY	35.0	API
WATER SPECIFIC GRAVITY	1.07	

DEPTH (1000 FT)

GAS/LIQUID RATIO - SCF PER BBL

0
50
100
200
300
400
500
600
800
1000
1200
1500

HAGEDORN AND BROWN
PRESENTED BY:
MAECO - SCHLUMBERGER

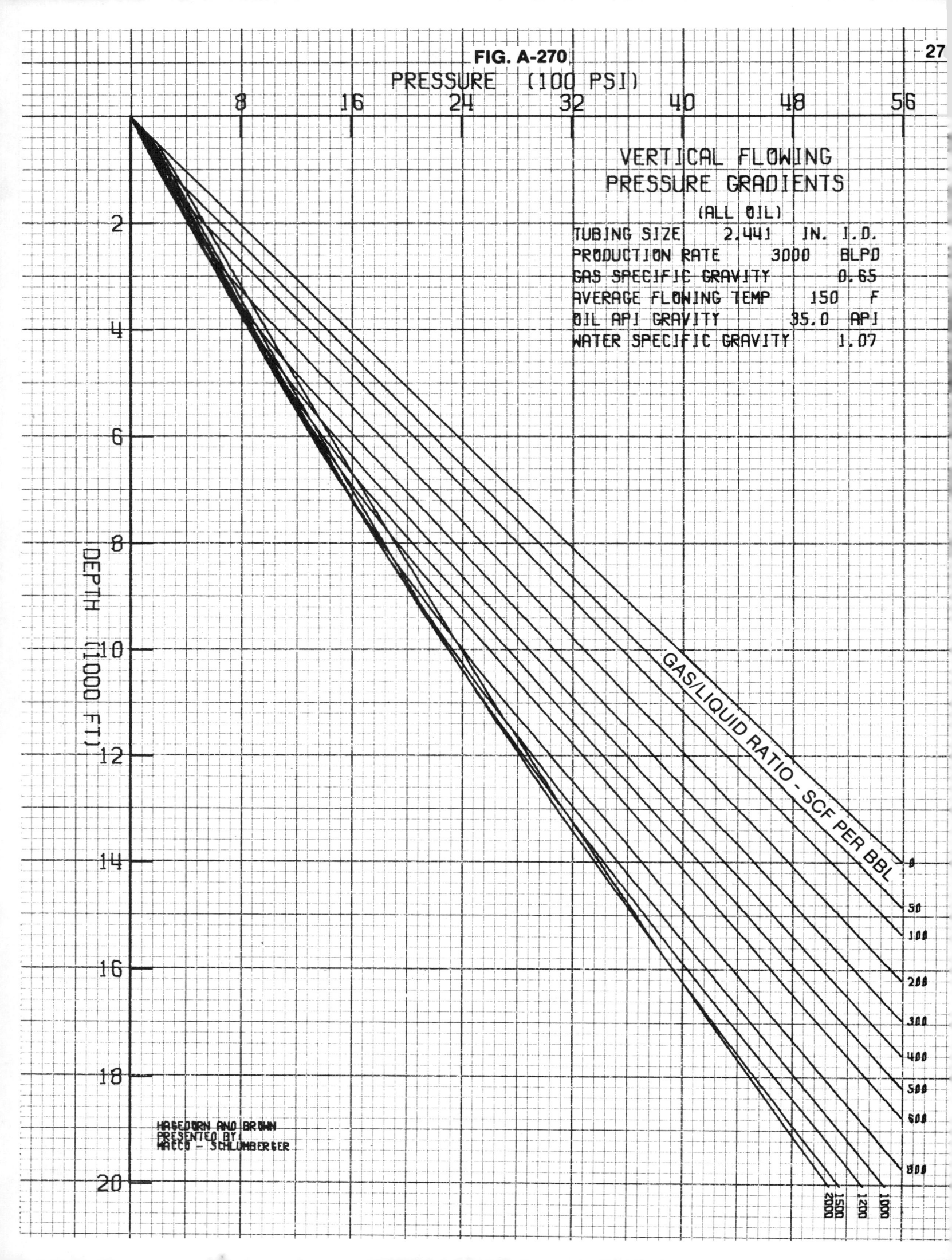

PRESSURE (100 PSI)

VERTICAL FLOWING
PRESSURE GRADIENTS
(ALL OIL)

TUBING SIZE	2.441	IN. I.D.
PRODUCTION RATE	3000	BLPD
GAS SPECIFIC GRAVITY	0.65	
AVERAGE FLOWING TEMP	150	F
OIL API GRAVITY	35.0	API
WATER SPECIFIC GRAVITY	1.07	

DEPTH (1000 FT)

GAS/LIQUID RATIO - SCF PER BBL

HAGEDORN AND BROWN
PRESENTED BY:
MALCO - SCHLUMBERGER

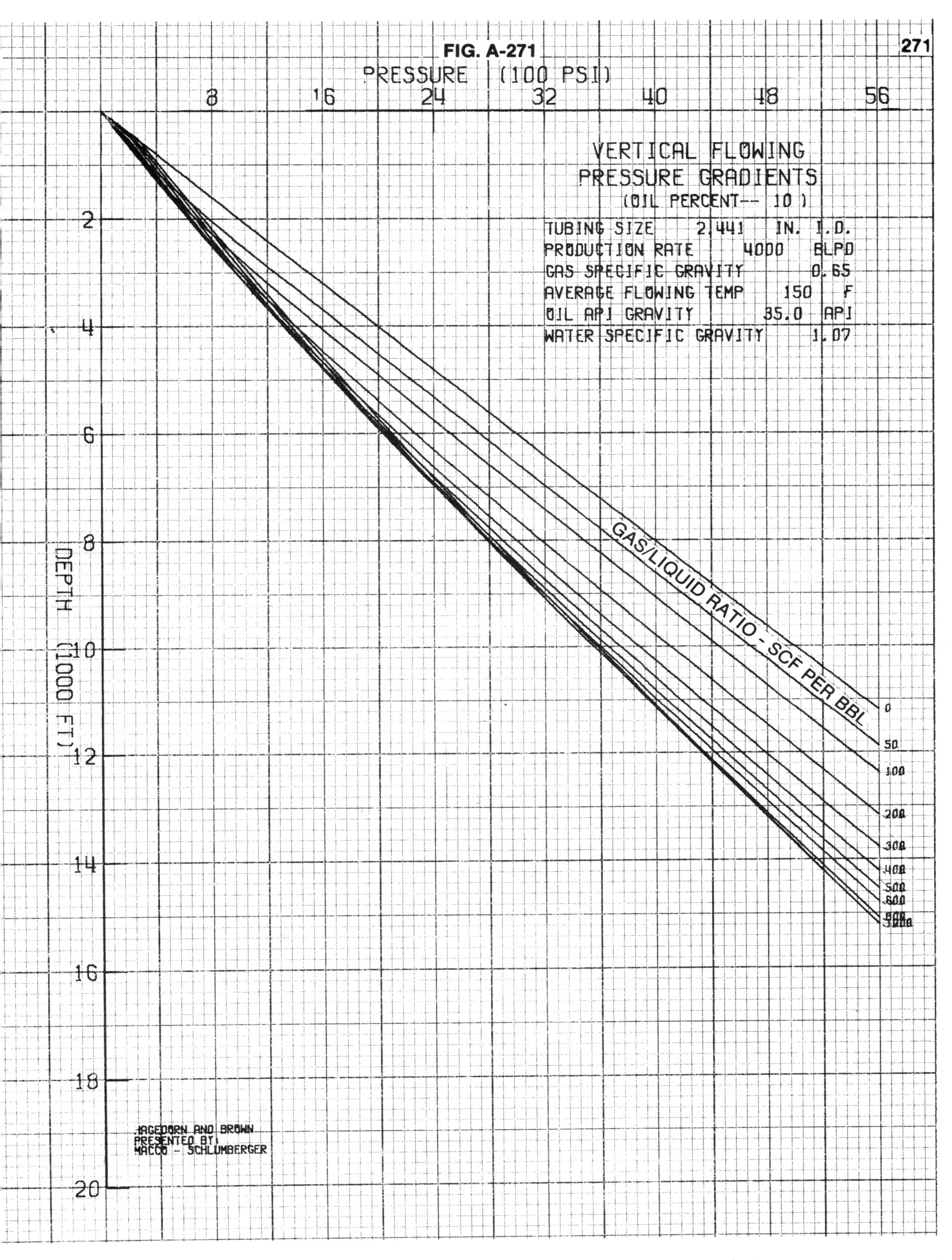

FIG. A-271

271

VERTICAL FLOWING
PRESSURE GRADIENTS
(OIL PERCENT-- 10)

TUBING SIZE	2.441	IN. I.D.
PRODUCTION RATE	4000	BLPD
GAS SPECIFIC GRAVITY	0.65	
AVERAGE FLOWING TEMP	150	F
OIL API GRAVITY	35.0	API
WATER SPECIFIC GRAVITY	1.07	

PRESSURE (100 PSI)

DEPTH (1000 FT)

GAS/LIQUID RATIO - SCF PER BBL

0
50
100
200
300
400
500
600
800
1000

HAGEDORN AND BROWN
PRESENTED BY
MACCO - SCHLUMBERGER

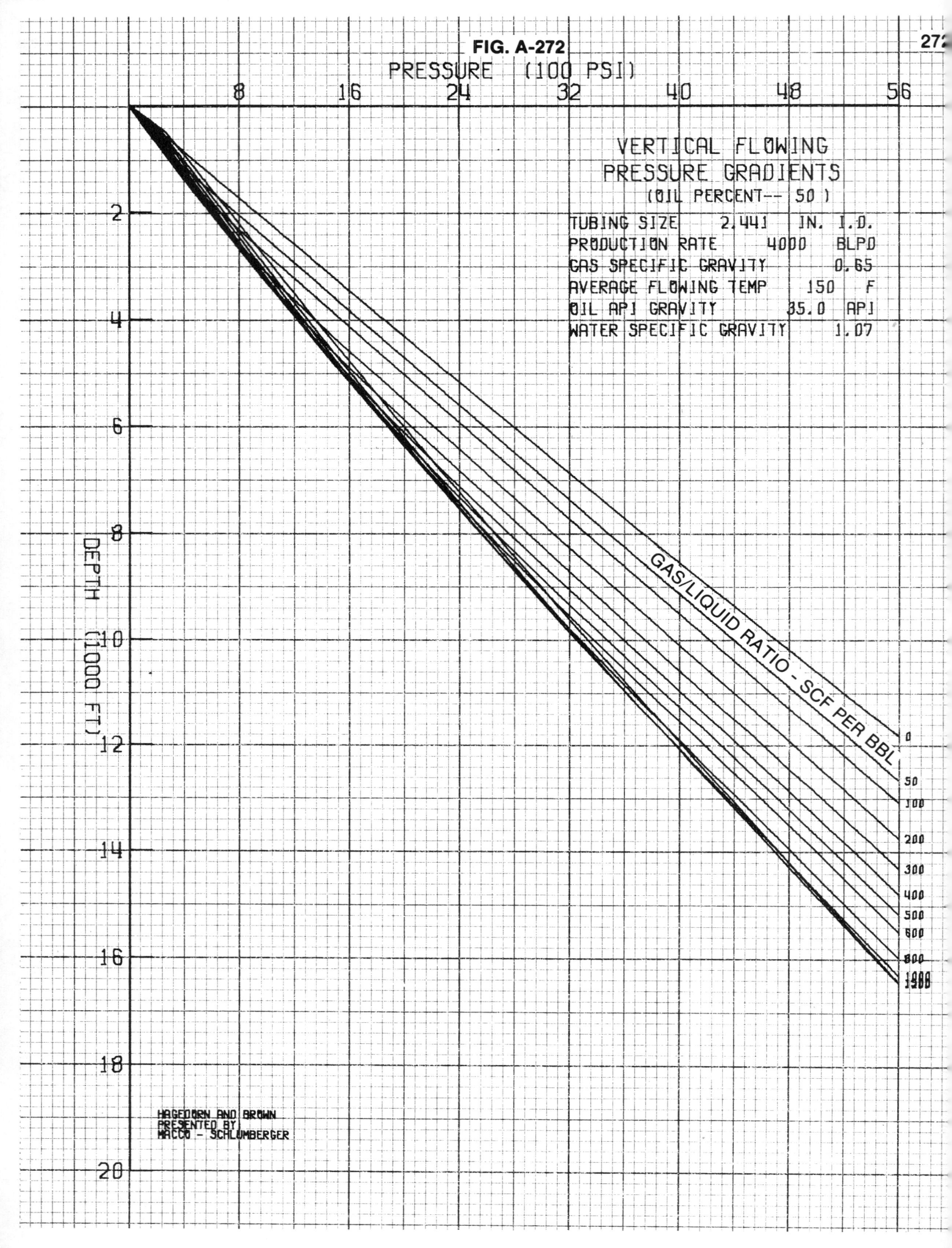

PRESSURE (100 PSI)

VERTICAL FLOWING
PRESSURE GRADIENTS
(OIL PERCENT-- 50)

TUBING SIZE	2.441	IN. I.D.
PRODUCTION RATE	4000	BLPD
GAS SPECIFIC GRAVITY	0.65	
AVERAGE FLOWING TEMP	150	F
OIL API GRAVITY	35.0	API
WATER SPECIFIC GRAVITY	1.07	

GAS/LIQUID RATIO - SCF PER BBL

DEPTH (1000 FT)

0
50
100
200
300
400
500
600
800
1000

HAGEDORN AND BROWN
PRESENTED BY
MACCO - SCHLUMBERGER

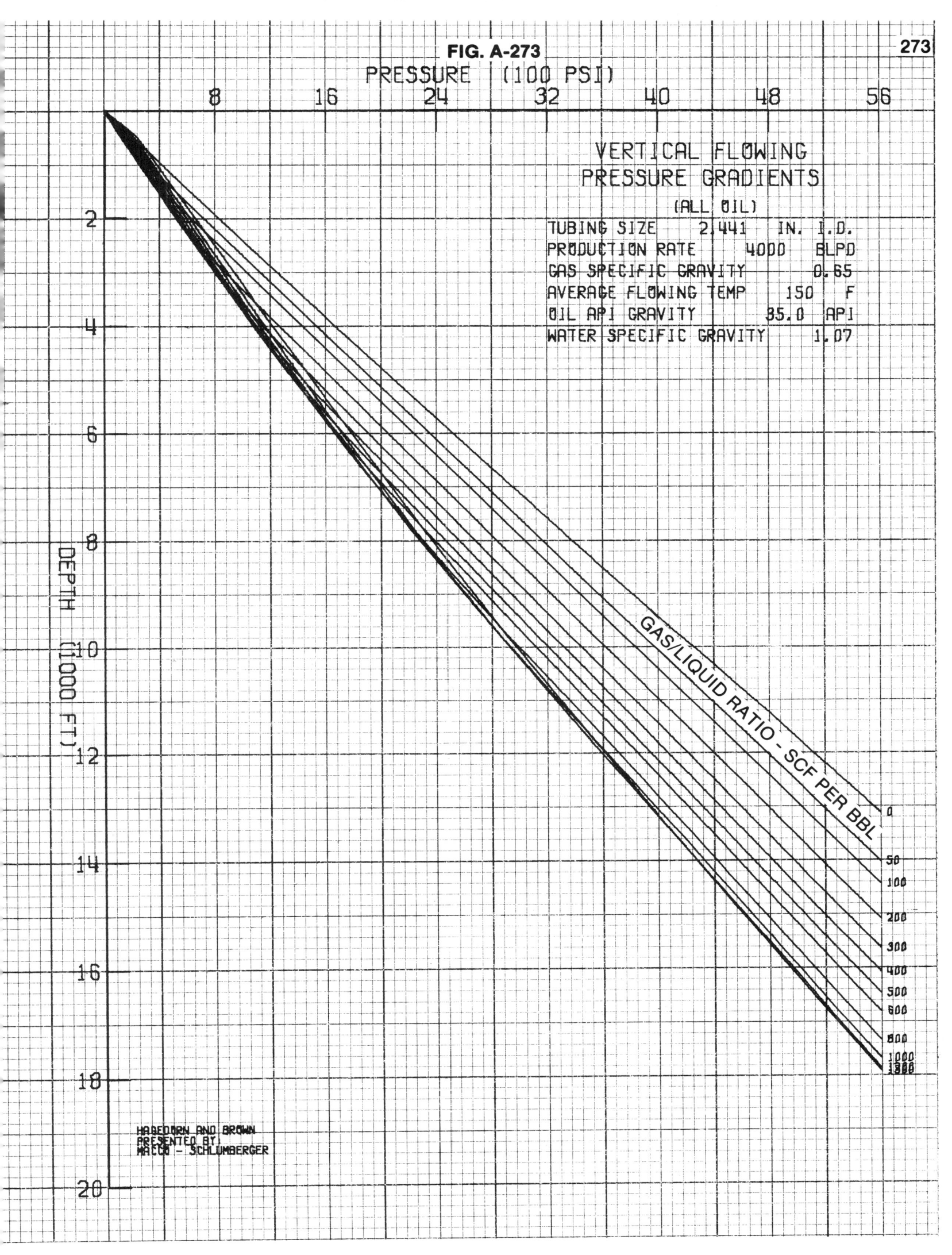

FIG. A-273

273

PRESSURE (100 PSI)

DEPTH (1000 FT)

VERTICAL FLOWING
PRESSURE GRADIENTS
(ALL OIL)

TUBING SIZE	2.441	IN. I.D.
PRODUCTION RATE	4000	BLPD
GAS SPECIFIC GRAVITY	0.65	
AVERAGE FLOWING TEMP	150	F
OIL API GRAVITY	35.0	API
WATER SPECIFIC GRAVITY	1.07	

GAS/LIQUID RATIO - SCF PER BBL

0
50
100
200
300
400
500
600
800
1000
1500

HAGEDORN AND BROWN
PRESENTED BY
MACCO - SCHLUMBERGER

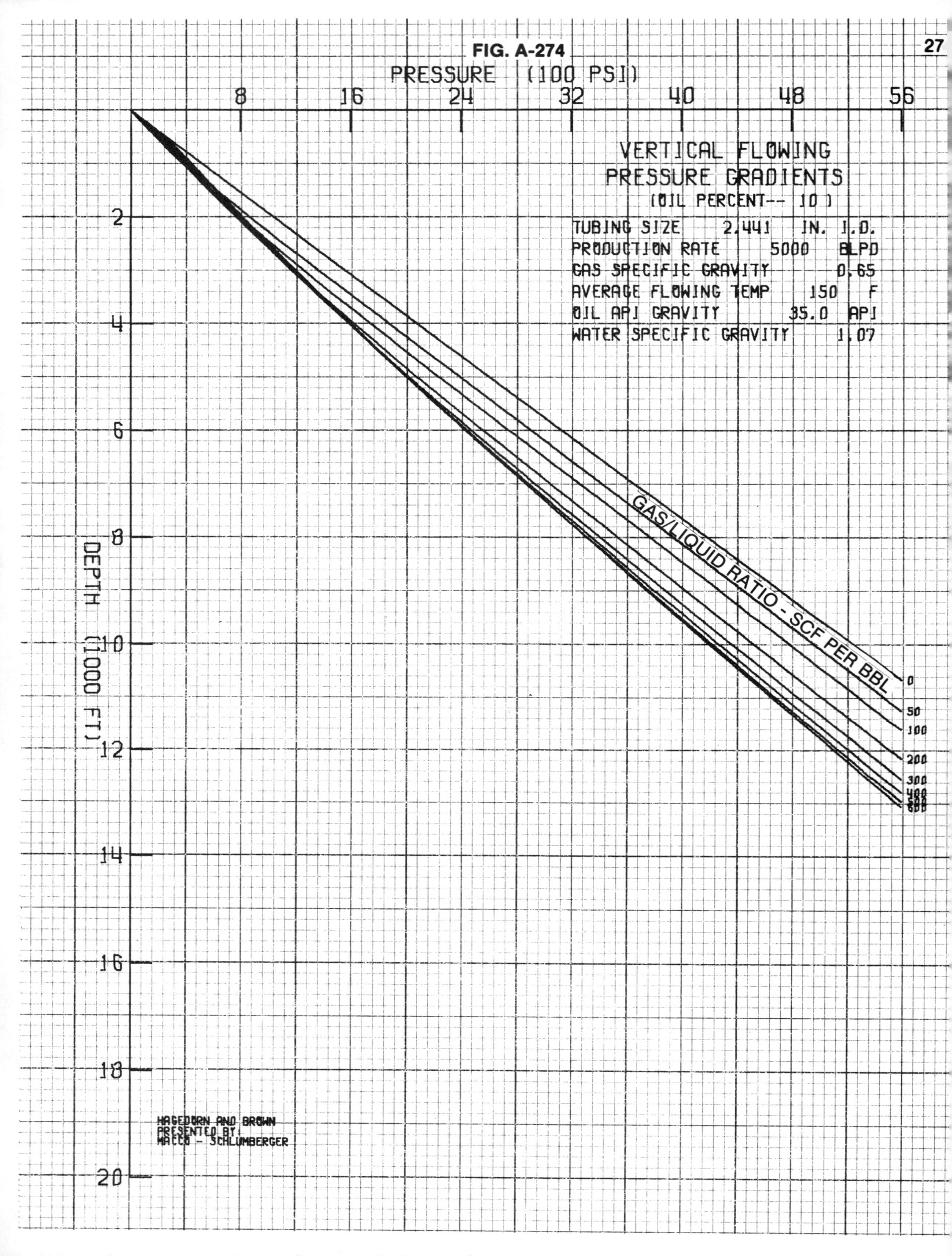

FIG. A-274

27

PRESSURE (100 PSI)

VERTICAL FLOWING
PRESSURE GRADIENTS
(OIL PERCENT-- 10)

TUBING SIZE	2.441	IN. I.D.
PRODUCTION RATE	5000	BLPD
GAS SPECIFIC GRAVITY		0.65
AVERAGE FLOWING TEMP	150	F
OIL API GRAVITY	35.0	API
WATER SPECIFIC GRAVITY		1.07

DEPTH (1000 FT)

GAS/LIQUID RATIO - SCF PER BBL

0
50
100
200
300
400
500
600

HAGEDORN AND BROWN
PRESENTED BY
MACCO - SCHLUMBERGER

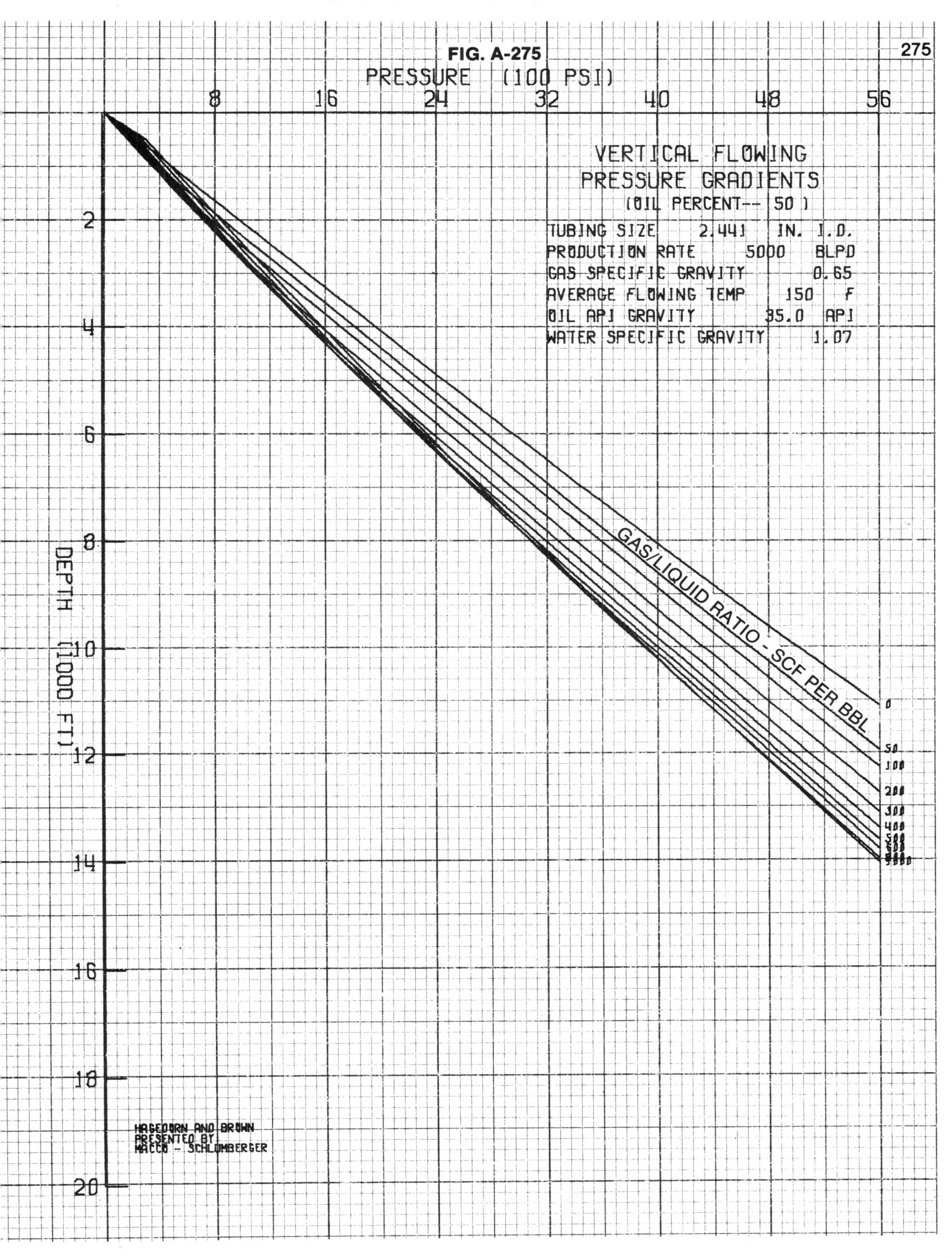

VERTICAL FLOWING
PRESSURE GRADIENTS
(OIL PERCENT-- 50)

TUBING SIZE	2.441	IN. I.D.
PRODUCTION RATE	5000	BLPD
GAS SPECIFIC GRAVITY	0.65	
AVERAGE FLOWING TEMP	150	F
OIL API GRAVITY	35.0	API
WATER SPECIFIC GRAVITY	1.07	

PRESSURE (100 PSI)

DEPTH (1000 FT)

GAS/LIQUID RATIO - SCF PER BBL

HAGEDORN AND BROWN
PRESENTED BY
MACCO - SCHLUMBERGER

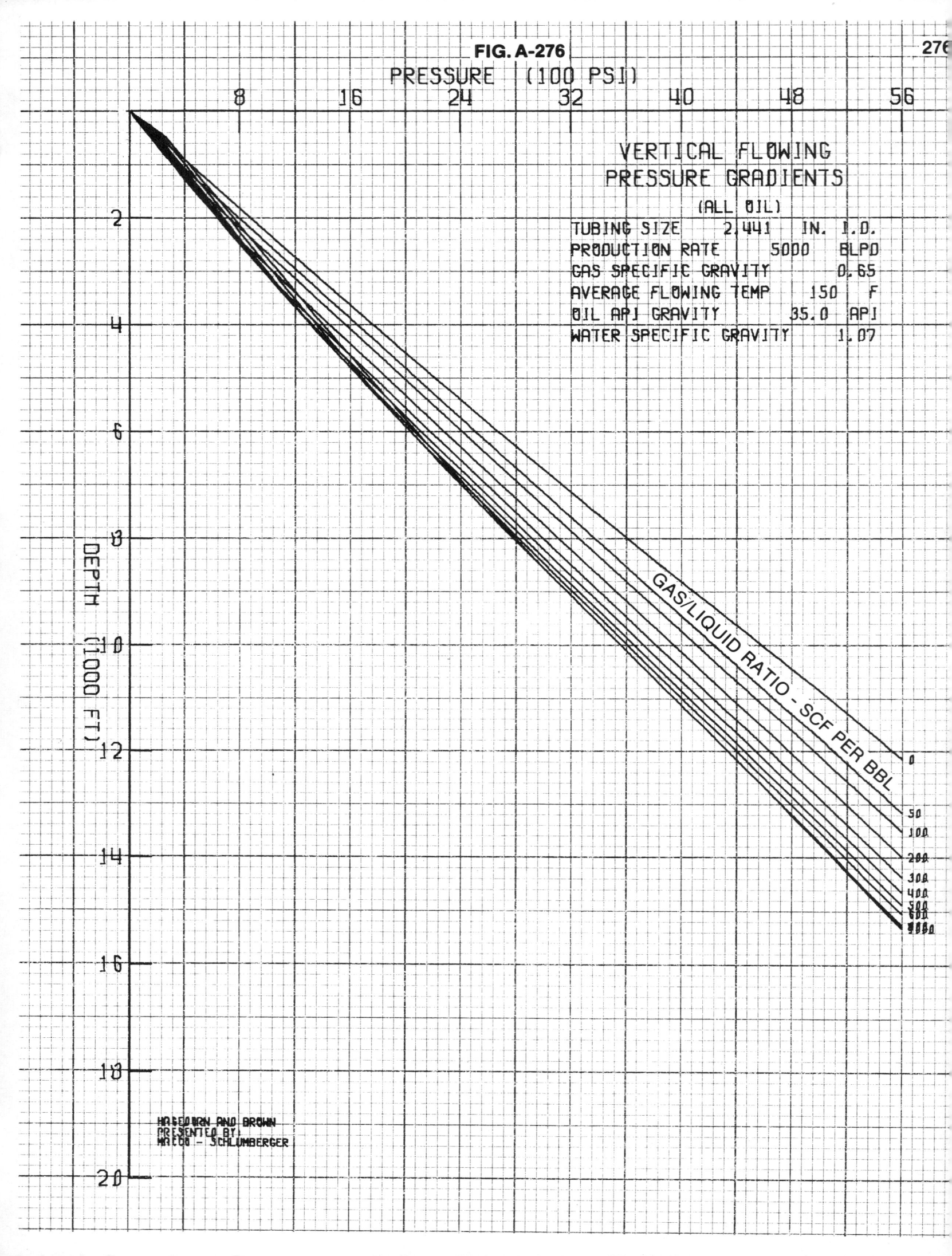

VERTICAL FLOWING
PRESSURE GRADIENTS
(ALL OIL)

TUBING SIZE	2.441	IN. I.D.
PRODUCTION RATE	5000	BLPD
GAS SPECIFIC GRAVITY	0.65	
AVERAGE FLOWING TEMP	150	F
OIL API GRAVITY	35.0	API
WATER SPECIFIC GRAVITY	1.07	

PRESSURE (100 PSI)

DEPTH (1000 FT)

GAS/LIQUID RATIO - SCF PER BBL

0
50
100
200
300
400
500
600
1000

HAGEDORN AND BROWN
PRESENTED BY
WILCOX - SCHLUMBERGER

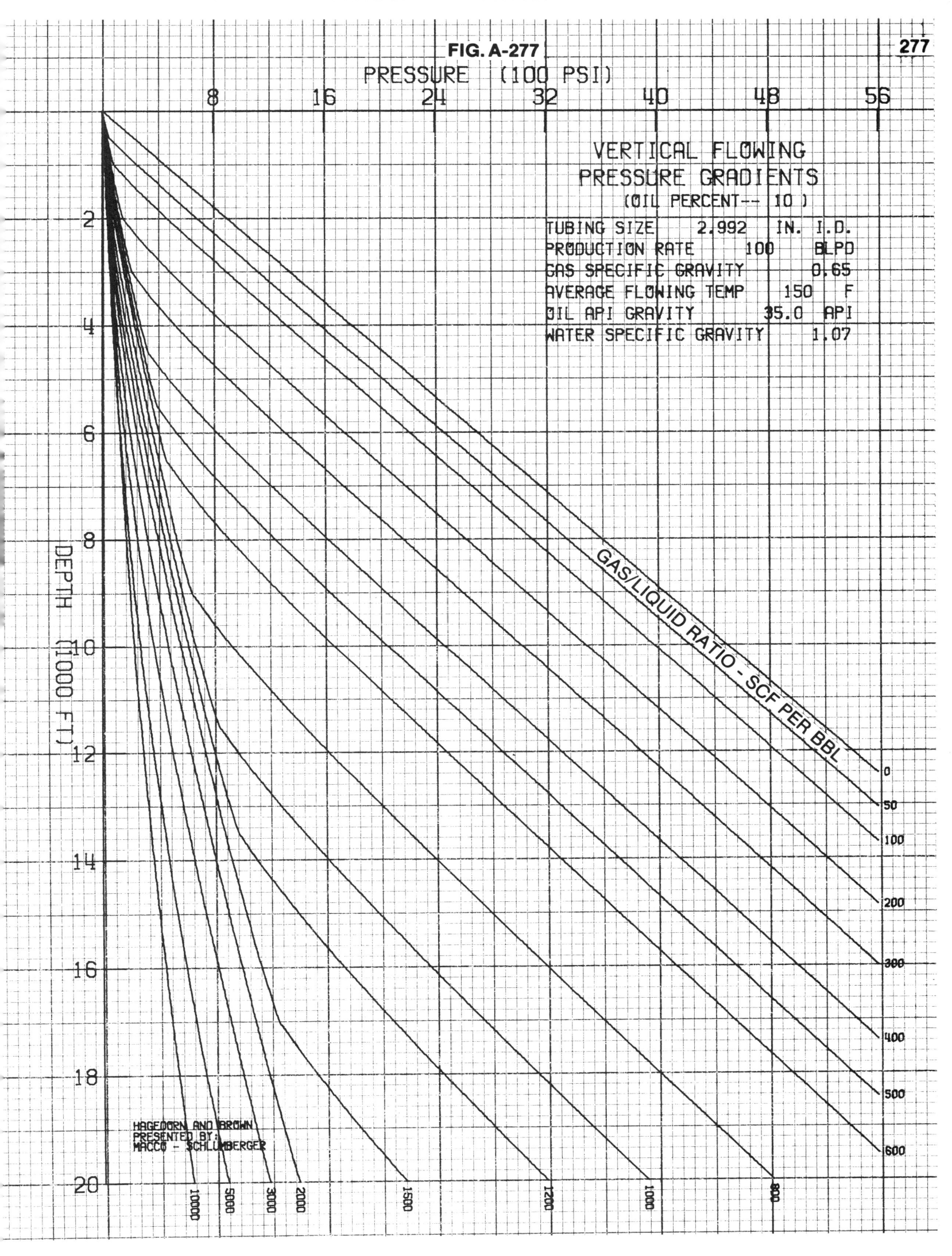

PRESSURE (100 PSI)

DEPTH (1000 FT)

VERTICAL FLOWING
PRESSURE GRADIENTS
(OIL PERCENT-- 10)

TUBING SIZE	2.992	IN. I.D.
PRODUCTION RATE	100	BLPD
GAS SPECIFIC GRAVITY	0.65	
AVERAGE FLOWING TEMP	150	F
OIL API GRAVITY	35.0	API
WATER SPECIFIC GRAVITY	1.07	

GAS/LIQUID RATIO - SCF PER BBL

HAGEDORN AND BROWN
PRESENTED BY:
MACCO - SCHLUMBERGER

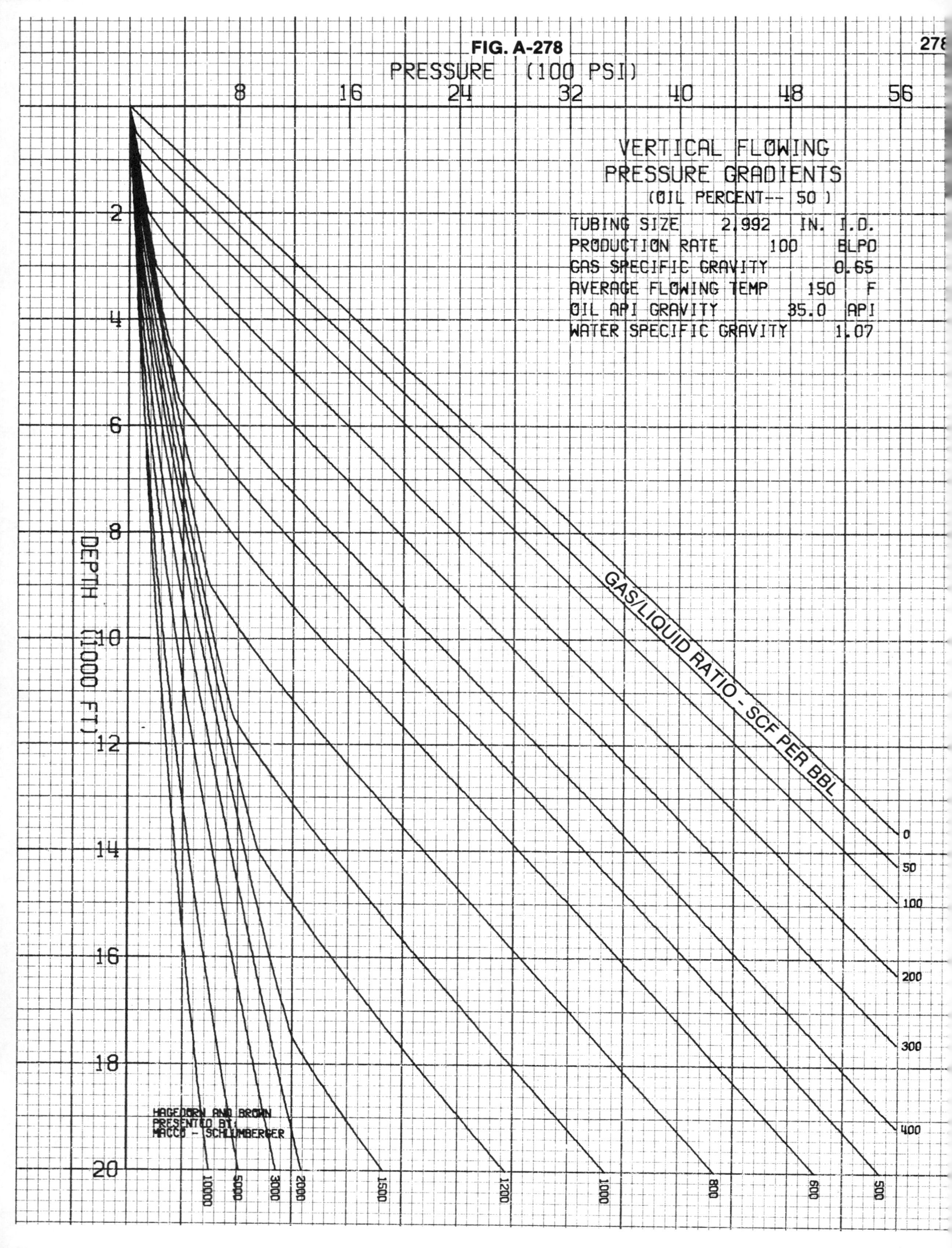

FIG. A-278

278

VERTICAL FLOWING
PRESSURE GRADIENTS
(OIL PERCENT-- 50)

TUBING SIZE	2.992	IN. I.D.
PRODUCTION RATE	100	BLPD
GAS SPECIFIC GRAVITY	0.65	
AVERAGE FLOWING TEMP	150	F
OIL API GRAVITY	35.0	API
WATER SPECIFIC GRAVITY	1.07	

PRESSURE (100 PSI)

DEPTH (1000 FT)

GAS/LIQUID RATIO - SCF PER BBL

HAGEDORN AND BROWN
PRESENTED BY
MACCO - SCHLUMBERGER

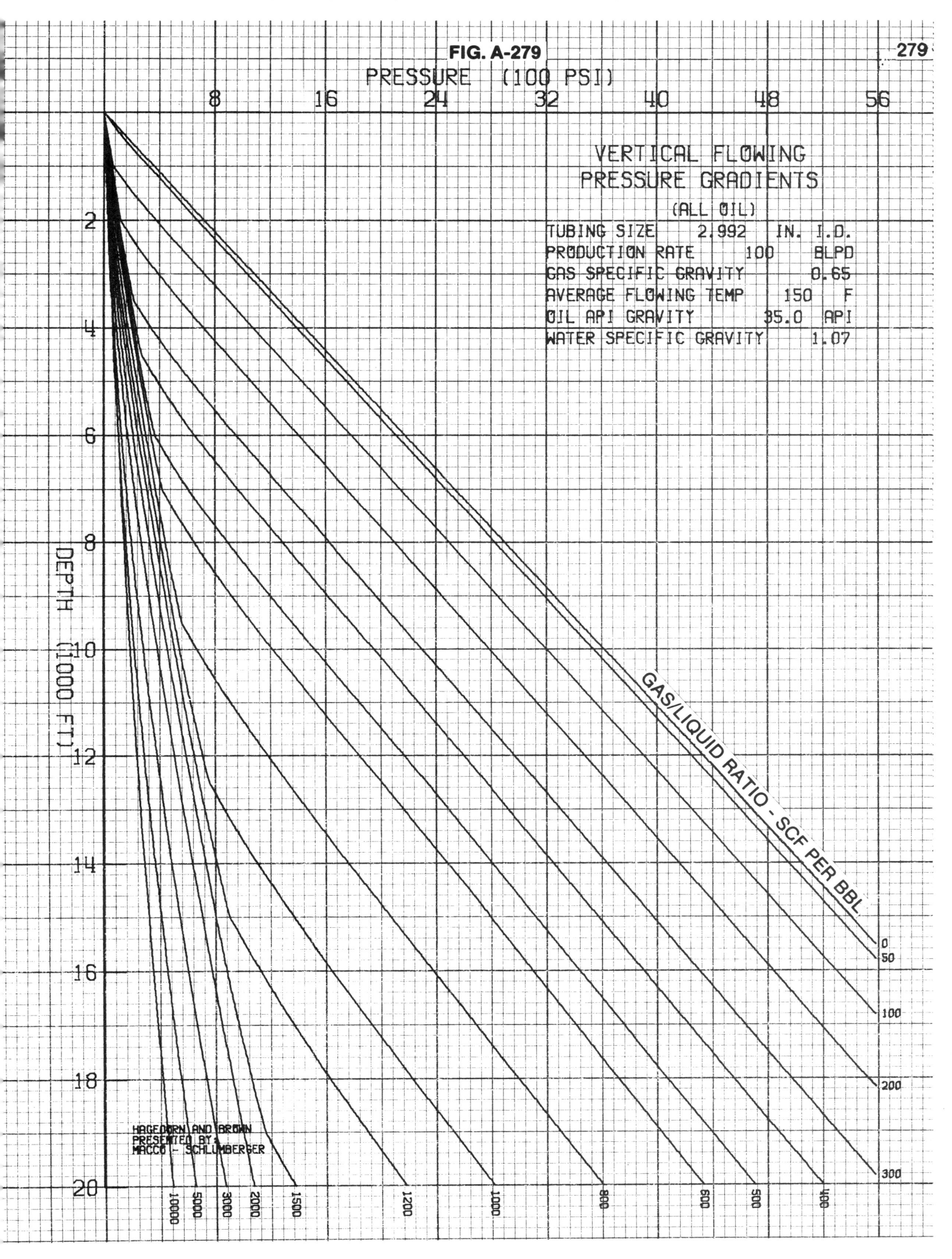

PRESSURE (100 PSI)

VERTICAL FLOWING
PRESSURE GRADIENTS
(ALL OIL)

TUBING SIZE	2.992	IN. I.D.
PRODUCTION RATE	100	BLPD
GAS SPECIFIC GRAVITY	0.65	
AVERAGE FLOWING TEMP	150	F
OIL API GRAVITY	35.0	API
WATER SPECIFIC GRAVITY	1.07	

GAS/LIQUID RATIO - SCF PER BBL

DEPTH (1000 FT)

HAGEDORN AND BROWN
PRESENTED BY
MACCO - SCHLUMBERGER

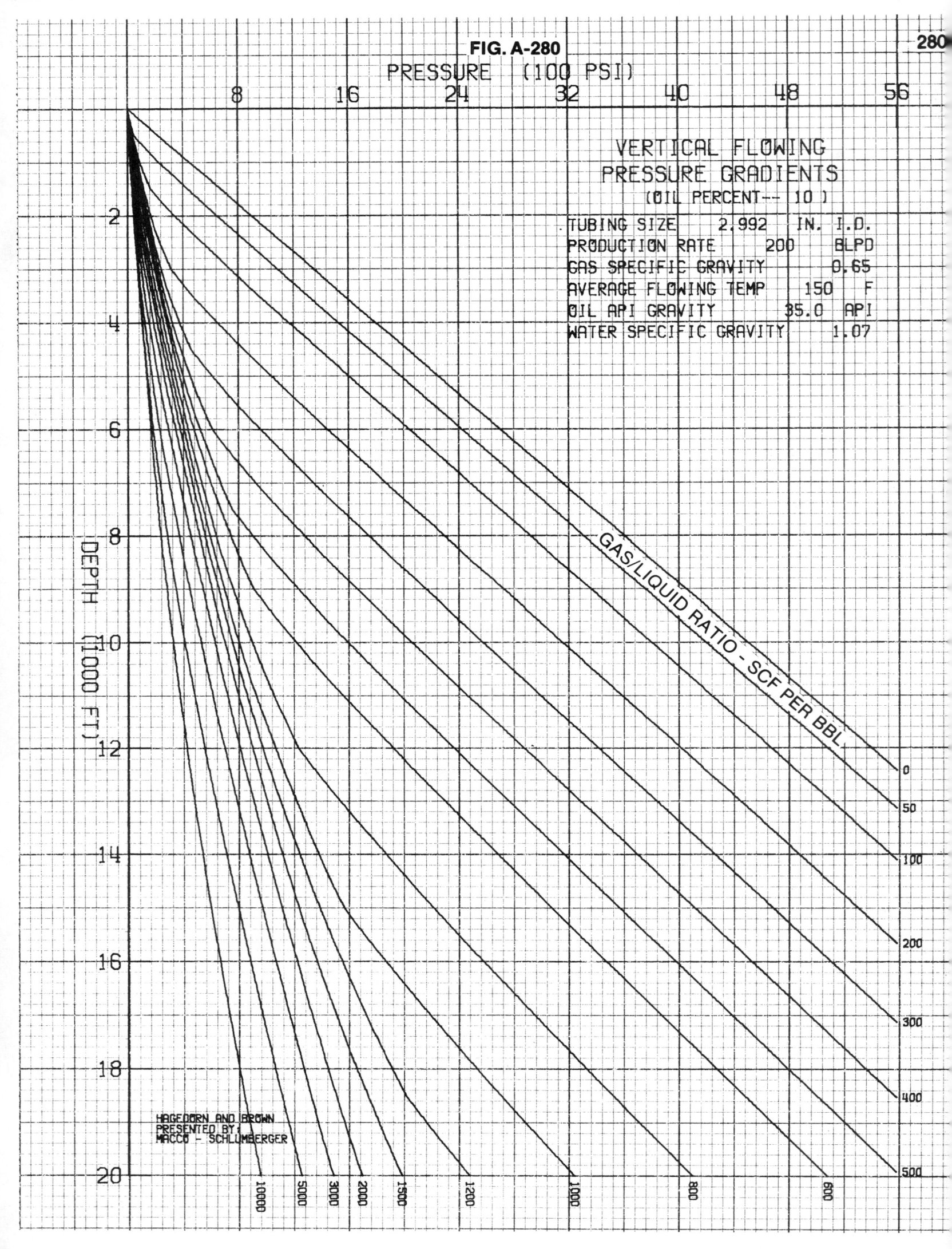

PRESSURE (100 PSI)

VERTICAL FLOWING
PRESSURE GRADIENTS
(OIL PERCENT-- 10)

TUBING SIZE	2.992	IN. I.D.
PRODUCTION RATE	200	BLPD
GAS SPECIFIC GRAVITY		0.65
AVERAGE FLOWING TEMP	150	F
OIL API GRAVITY	35.0	API
WATER SPECIFIC GRAVITY		1.07

DEPTH (1000 FT)

GAS/LIQUID RATIO - SCF PER BBL

HAGEDORN AND BROWN
PRESENTED BY:
MACCO - SCHLUMBERGER

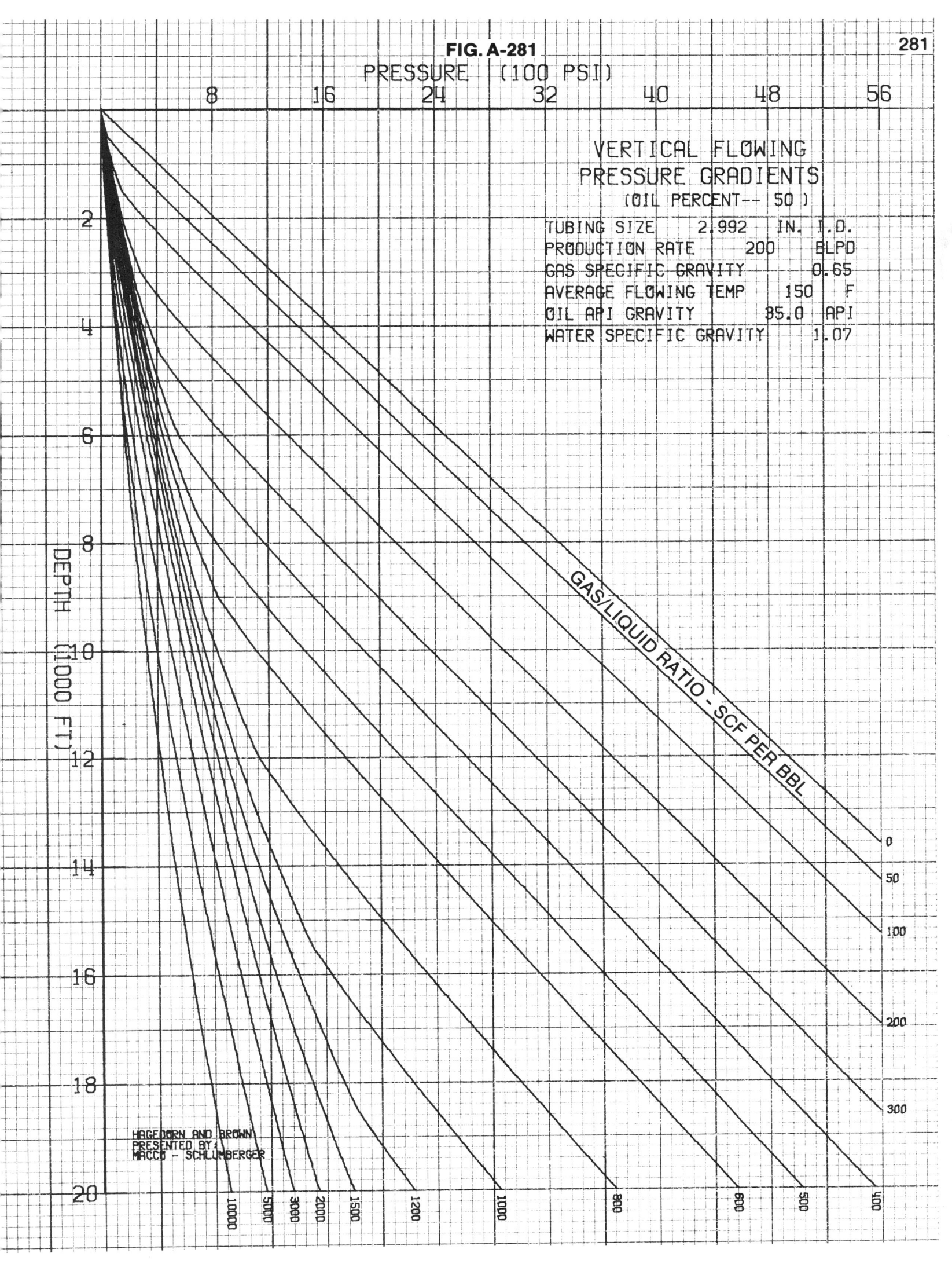

FIG. A-281

281

PRESSURE (100 PSI)

VERTICAL FLOWING
PRESSURE GRADIENTS
(OIL PERCENT--- 50)

TUBING SIZE   2.992   IN. I.D.
PRODUCTION RATE   200   BLPD
GAS SPECIFIC GRAVITY   0.65
AVERAGE FLOWING TEMP   150   F
OIL API GRAVITY   85.0   API
WATER SPECIFIC GRAVITY   1.07

GAS/LIQUID RATIO - SCF PER BBL

DEPTH (1000 FT)

HAGEDORN AND BROWN
PRESENTED BY:
MACCO - SCHLUMBERGER

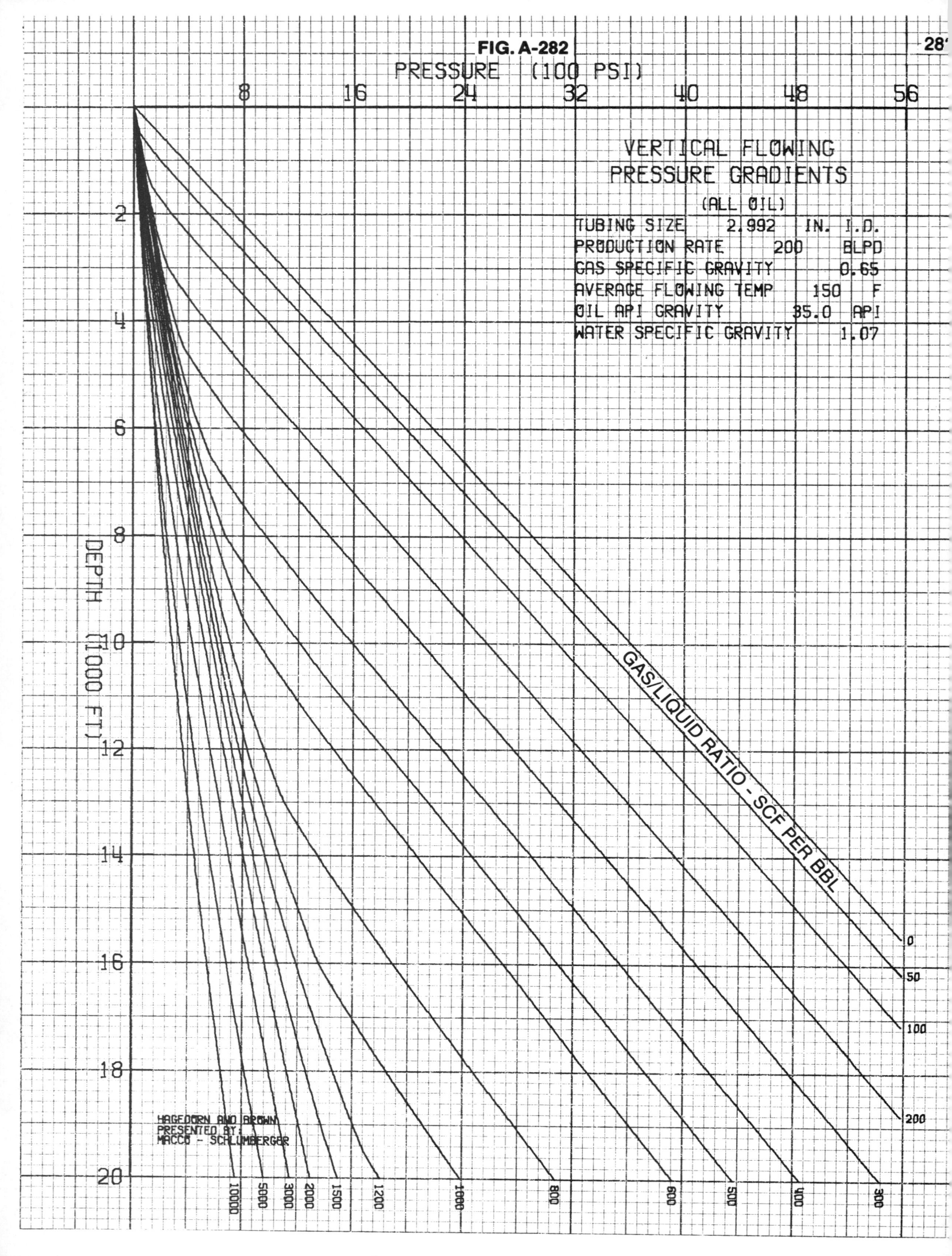

VERTICAL FLOWING
PRESSURE GRADIENTS
(ALL OIL)

TUBING SIZE	2.992	IN. I.D.
PRODUCTION RATE	200	BLPD
GAS SPECIFIC GRAVITY	0.65	
AVERAGE FLOWING TEMP	150	F
OIL API GRAVITY	35.0	API
WATER SPECIFIC GRAVITY	1.07	

PRESSURE (100 PSI)

DEPTH (1000 FT)

GAS/LIQUID RATIO - SCF PER BBL

HAGEDORN AND BROWN
PRESENTED BY:
MACCO - SCHLUMBERGER

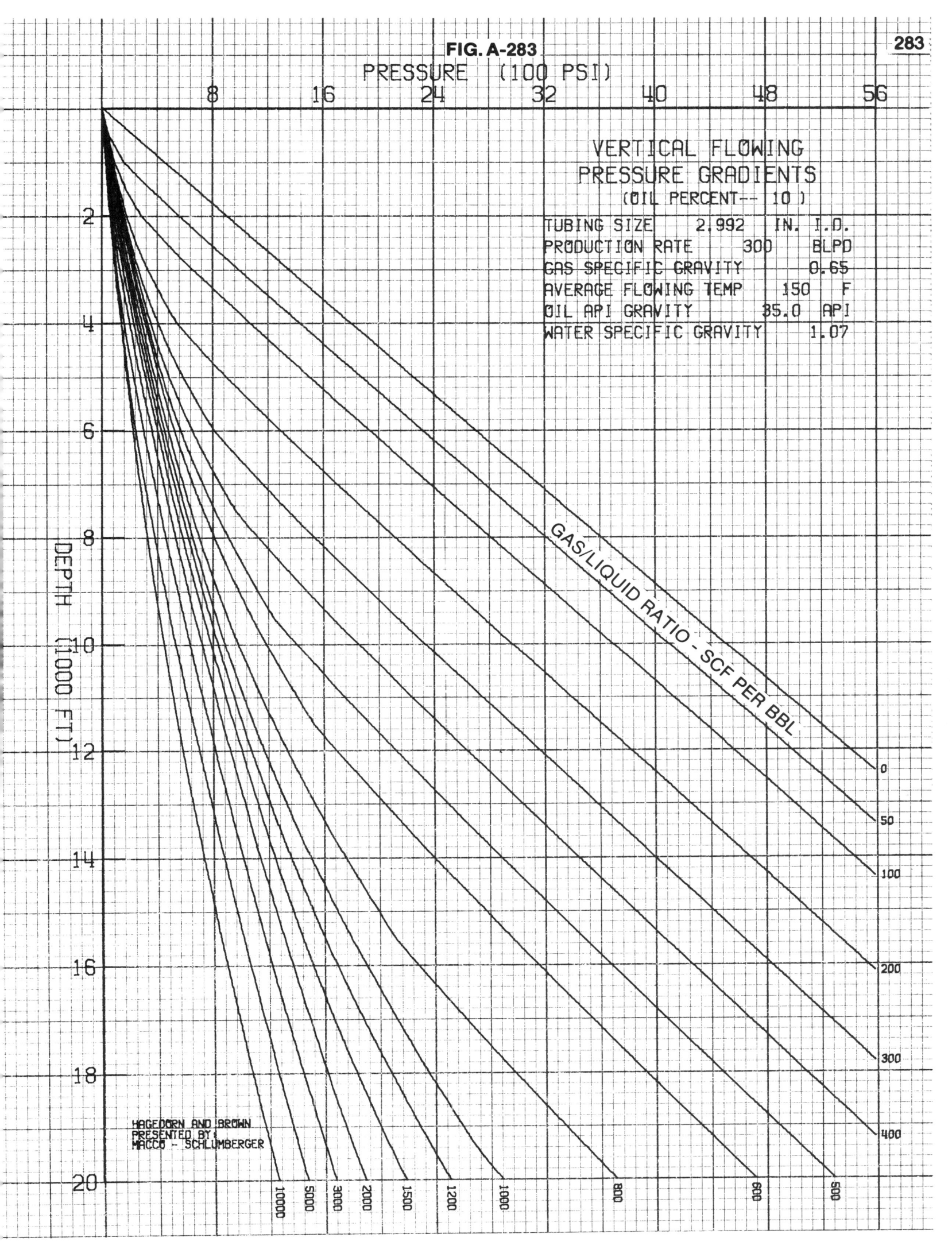

PRESSURE (100 PSI)

VERTICAL FLOWING
PRESSURE GRADIENTS
(OIL PERCENT— 10)

TUBING SIZE	2.992	IN. I.D.
PRODUCTION RATE	300	BLPD
GAS SPECIFIC GRAVITY	0.65	
AVERAGE FLOWING TEMP	150	F
OIL API GRAVITY	35.0	API
WATER SPECIFIC GRAVITY	1.07	

GAS/LIQUID RATIO - SCF PER BBL

DEPTH (1000 FT)

HAGEDORN AND BROWN
PRESENTED BY:
MACCO - SCHLUMBERGER

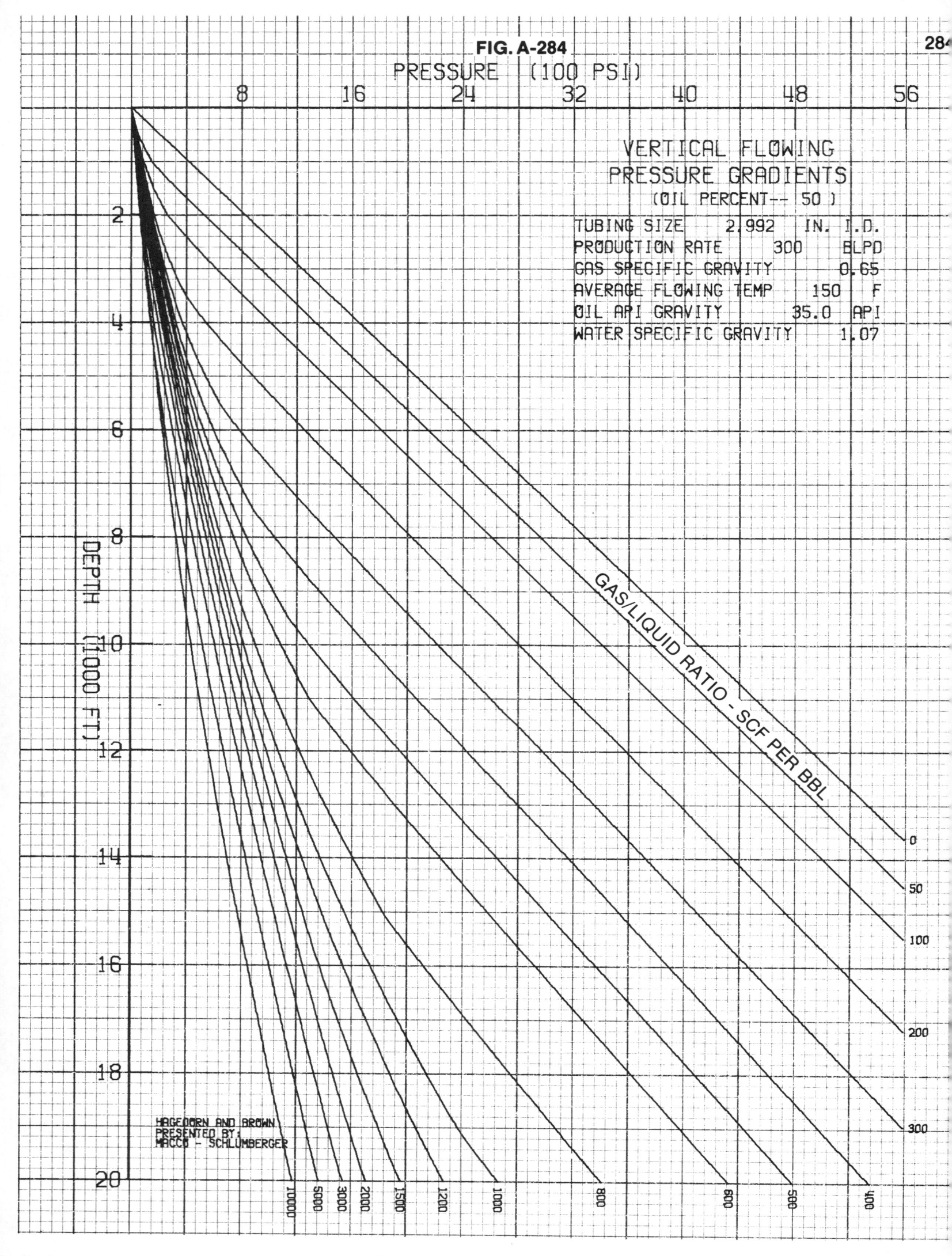

PRESSURE (100 PSI)

VERTICAL FLOWING
PRESSURE GRADIENTS
(OIL PERCENT-- 50)

TUBING SIZE	2.992	IN. I.D.
PRODUCTION RATE	300	BLPD
GAS SPECIFIC GRAVITY		0.65
AVERAGE FLOWING TEMP	150	F
OIL API GRAVITY	35.0	API
WATER SPECIFIC GRAVITY		1.07

DEPTH (1000 FT)

GAS/LIQUID RATIO - SCF PER BBL

HAGEDORN AND BROWN
PRESENTED BY,
MACCO - SCHLUMBERGER

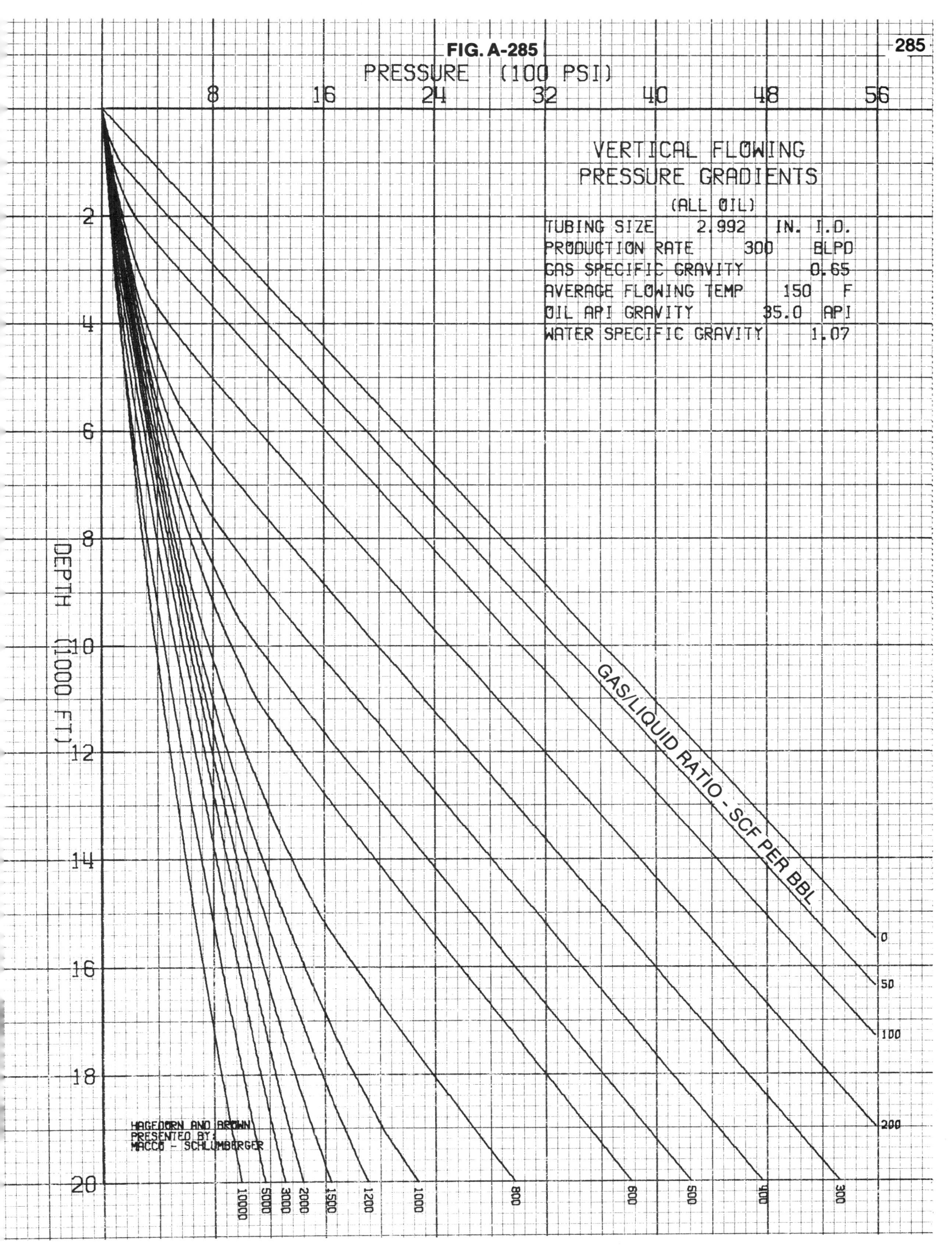

PRESSURE (100 PSI)

DEPTH (1000 FT)

VERTICAL FLOWING
PRESSURE GRADIENTS
(ALL OIL)

TUBING SIZE	2.992	IN. I.D.
PRODUCTION RATE	300	BLPD
GAS SPECIFIC GRAVITY	0.65	
AVERAGE FLOWING TEMP	150	F
OIL API GRAVITY	35.0	API
WATER SPECIFIC GRAVITY	1.07	

GAS/LIQUID RATIO - SCF PER BBL

HAGEDORN AND BROWN
PRESENTED BY:
MACCO - SCHLUMBERGER

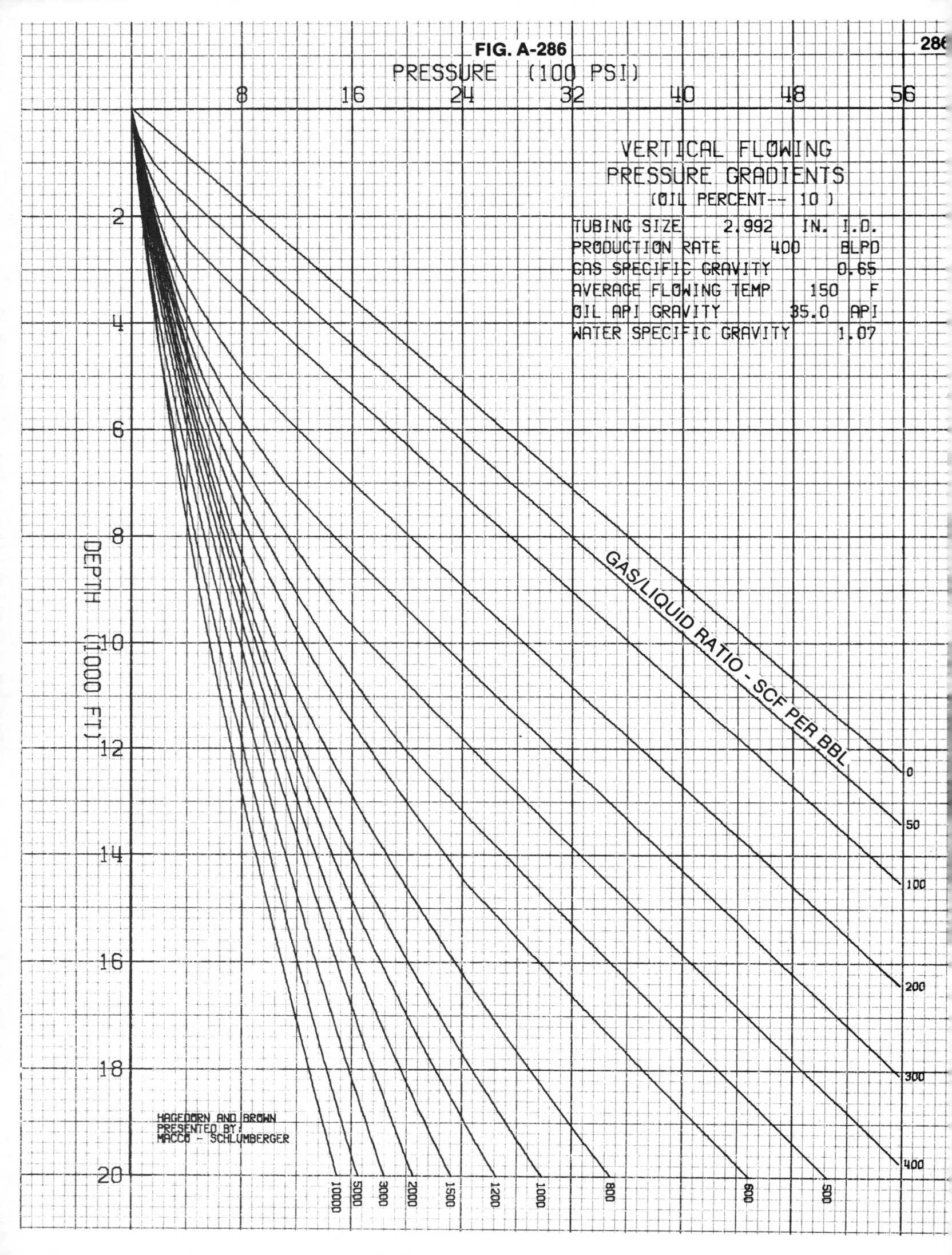

FIG. A-286

286

PRESSURE (100 PSI)

VERTICAL FLOWING
PRESSURE GRADIENTS
(OIL PERCENT-- 10)

TUBING SIZE	2.992	IN. I.D.
PRODUCTION RATE	400	BLPD
GAS SPECIFIC GRAVITY		0.65
AVERAGE FLOWING TEMP	150	F
OIL API GRAVITY	35.0	API
WATER SPECIFIC GRAVITY		1.07

GAS/LIQUID RATIO - SCF PER BBL

DEPTH (1000 FT)

HAGEDORN AND BROWN
PRESENTED BY:
MACCO - SCHLUMBERGER

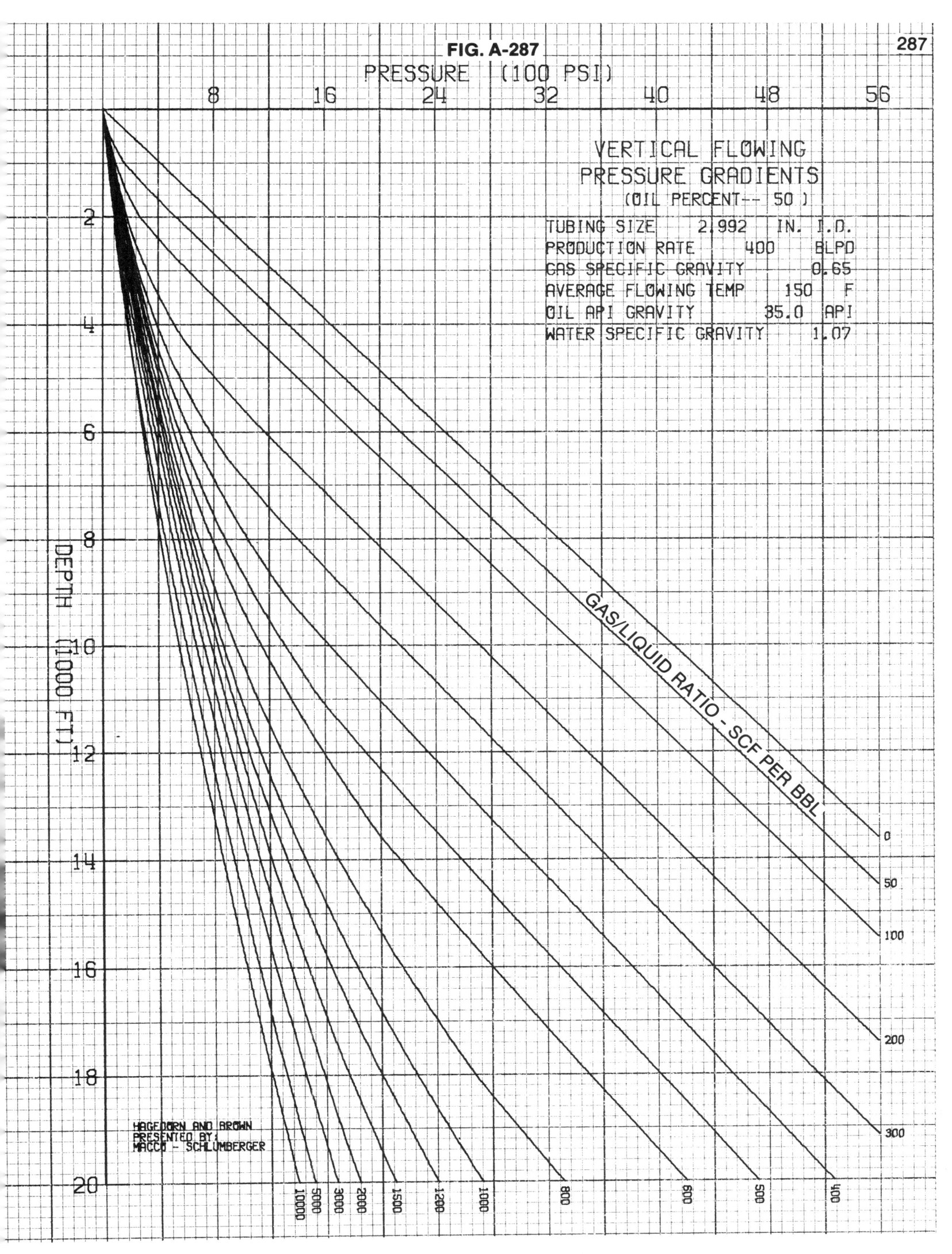

FIG. A-287

287

VERTICAL FLOWING
PRESSURE GRADIENTS
(OIL PERCENT-- 50)

TUBING SIZE	2.992	IN. I.D.
PRODUCTION RATE	400	BLPD
GAS SPECIFIC GRAVITY	0.65	
AVERAGE FLOWING TEMP	150	F
OIL API GRAVITY	35.0	API
WATER SPECIFIC GRAVITY	1.07	

PRESSURE (100 PSI)

DEPTH (1000 FT)

GAS/LIQUID RATIO - SCF PER BBL

HAGEDORN AND BROWN
PRESENTED BY:
MACCO - SCHLUMBERGER

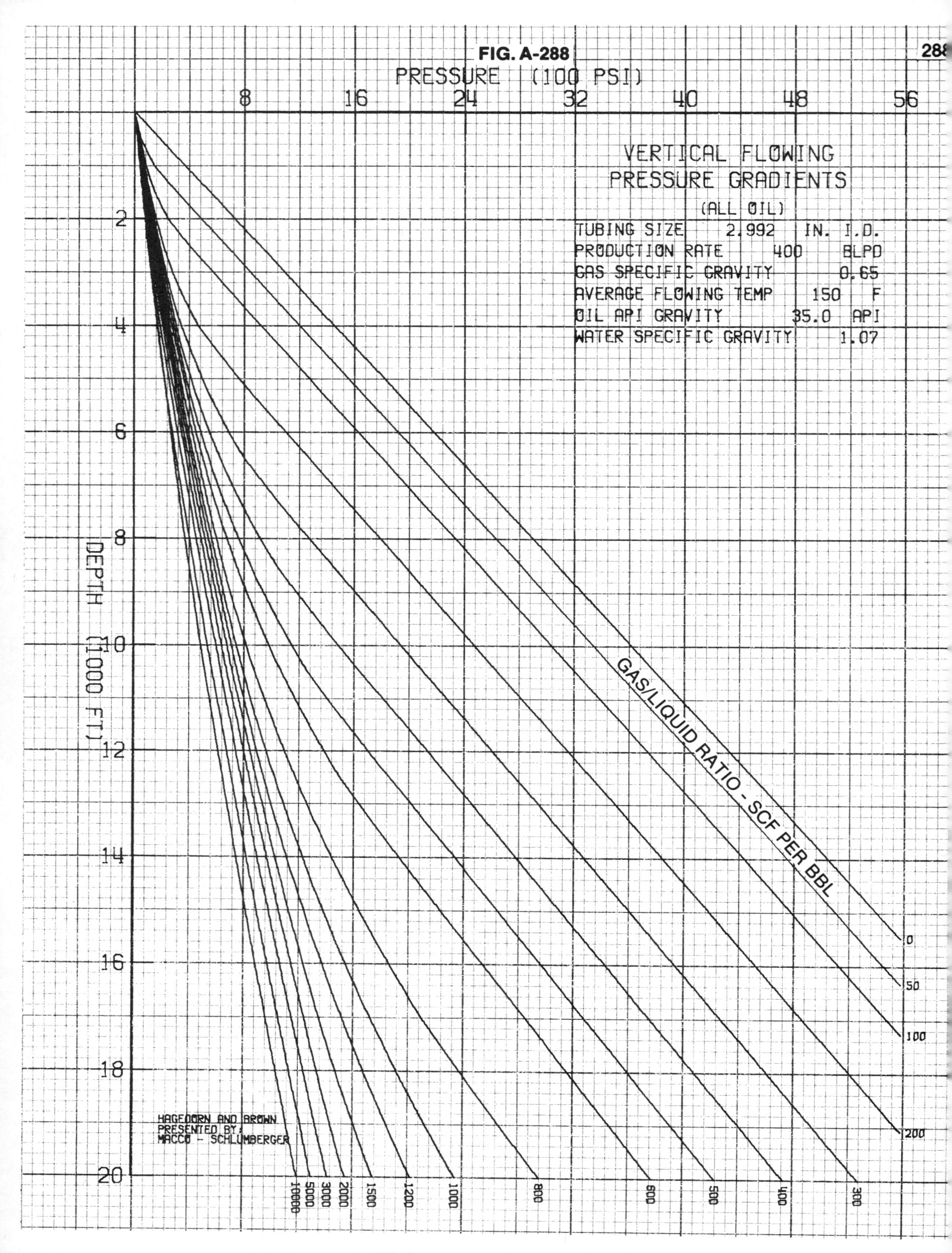

VERTICAL FLOWING
PRESSURE GRADIENTS
(ALL OIL)

TUBING SIZE	2.992	IN. I.D.
PRODUCTION RATE	400	BLPD
GAS SPECIFIC GRAVITY	0.65	
AVERAGE FLOWING TEMP	150	F
OIL API GRAVITY	35.0	API
WATER SPECIFIC GRAVITY	1.07	

PRESSURE (100 PSI)

DEPTH (1000 FT)

GAS/LIQUID RATIO - SCF PER BBL

HAGEDORN AND BROWN
PRESENTED BY:
MACCO - SCHLUMBERGER

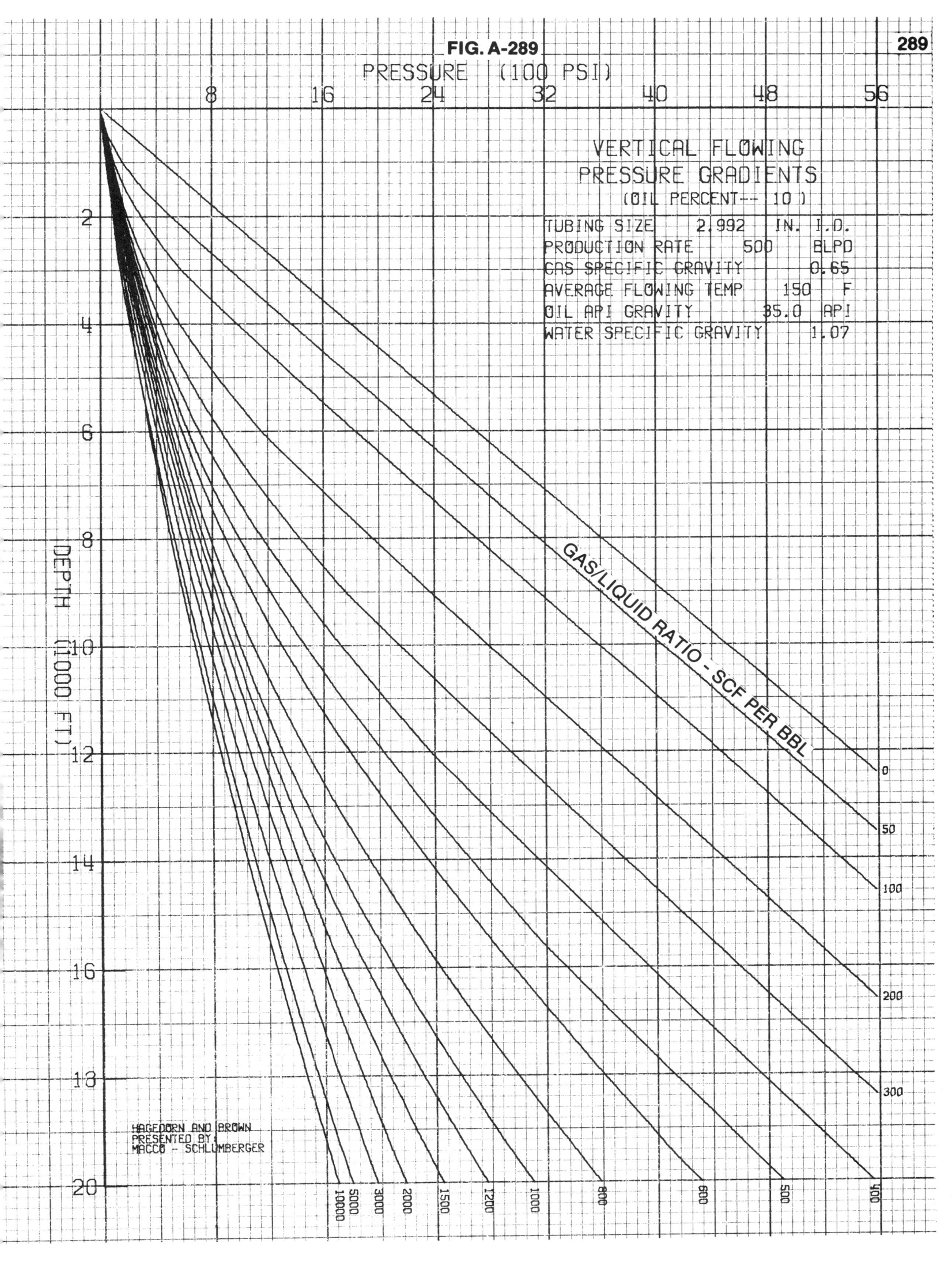

PRESSURE (100 PSI)

VERTICAL FLOWING
PRESSURE GRADIENTS
(OIL PERCENT-- 10)

TUBING SIZE	2.992	IN. I.D.
PRODUCTION RATE	500	BLPD
GAS SPECIFIC GRAVITY	0.65	
AVERAGE FLOWING TEMP	150	F
OIL API GRAVITY	35.0	API
WATER SPECIFIC GRAVITY	1.07	

DEPTH (1000 FT)

GAS/LIQUID RATIO - SCF PER BBL

HAGEDORN AND BROWN
PRESENTED BY:
MACCO - SCHLUMBERGER

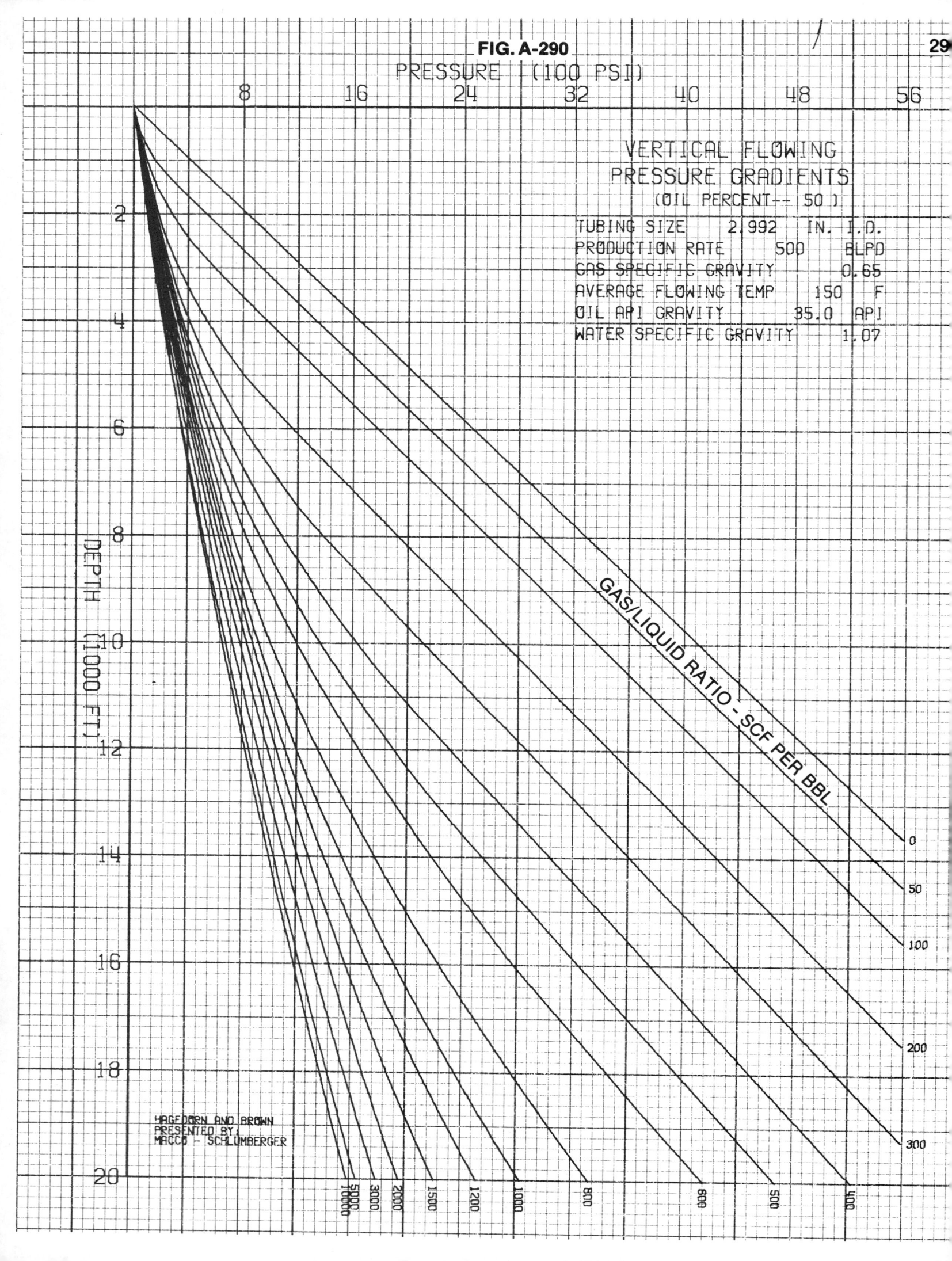

VERTICAL FLOWING
PRESSURE GRADIENTS
(OIL PERCENT -- 50)

TUBING SIZE	2.992	IN. I.D.
PRODUCTION RATE	500	BLPD
GAS SPECIFIC GRAVITY		0.65
AVERAGE FLOWING TEMP	150	F
OIL API GRAVITY	35.0	API
WATER SPECIFIC GRAVITY		1.07

PRESSURE (100 PSI)

DEPTH (1000 FT)

GAS/LIQUID RATIO - SCF PER BBL

HAGEDORN AND BROWN
PRESENTED BY
MACCO - SCHLUMBERGER

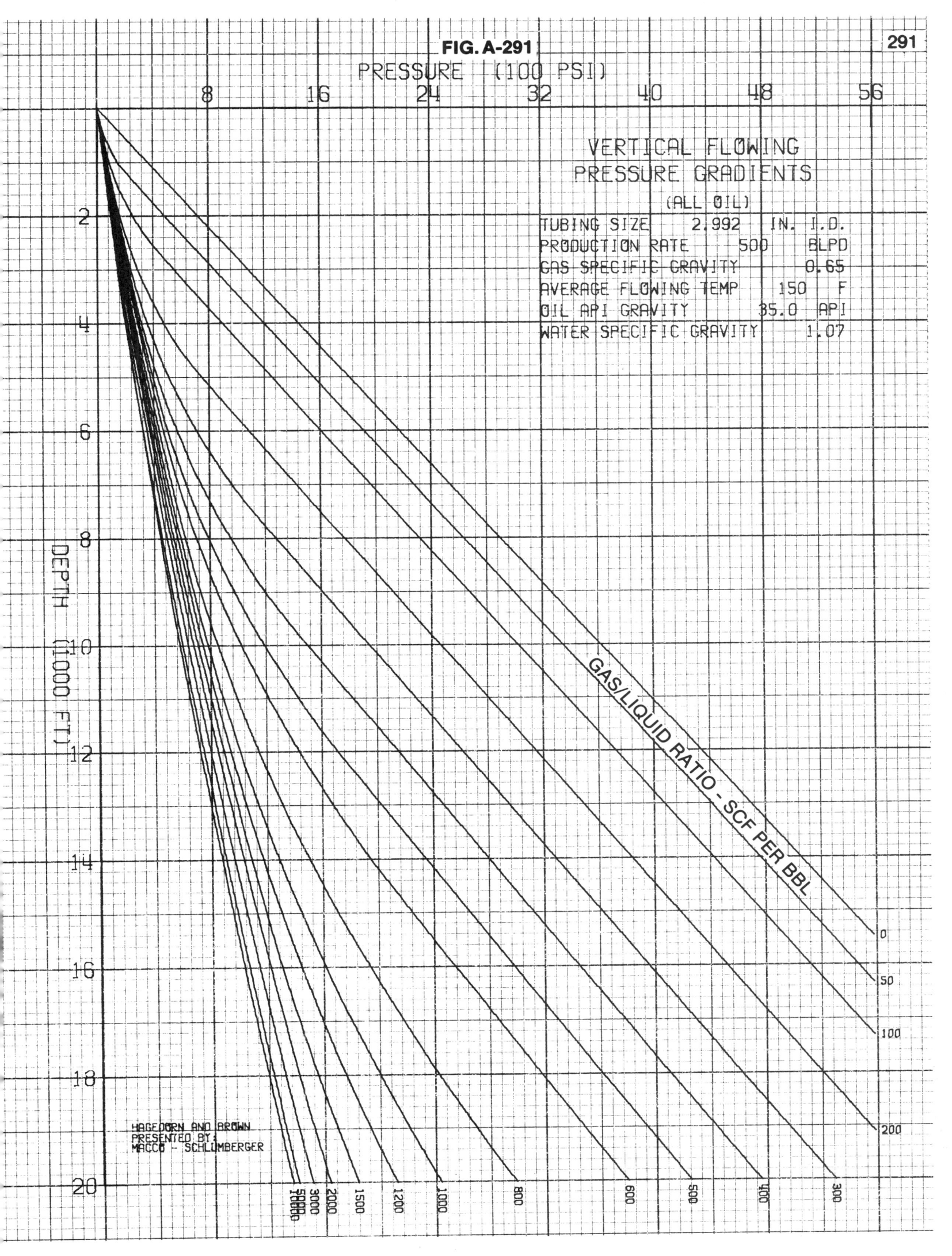

VERTICAL FLOWING
PRESSURE GRADIENTS
(ALL OIL)

TUBING SIZE	2.992	IN. I.D.
PRODUCTION RATE	500	BLPD
GAS SPECIFIC GRAVITY	0.65	
AVERAGE FLOWING TEMP	150	F
OIL API GRAVITY	35.0	API
WATER SPECIFIC GRAVITY	1.07	

GAS/LIQUID RATIO - SCF PER BBL

HAGEDORN AND BROWN
PRESENTED BY:
MACCO - SCHLUMBERGER

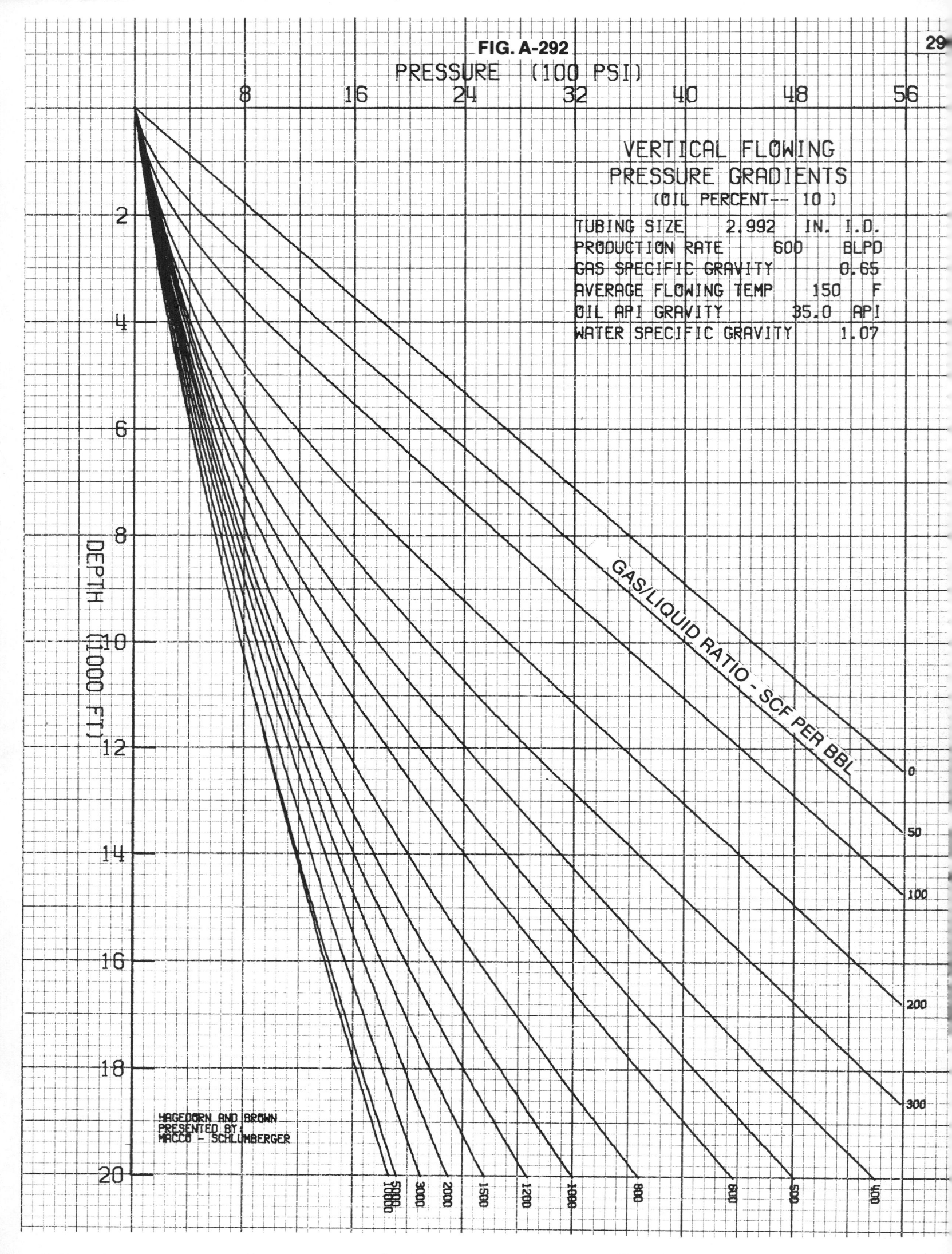

PRESSURE (100 PSI)

VERTICAL FLOWING
PRESSURE GRADIENTS
(OIL PERCENT-- 10)

TUBING SIZE	2.992	IN. I.D.
PRODUCTION RATE	600	BLPD
GAS SPECIFIC GRAVITY	0.65	
AVERAGE FLOWING TEMP	150	F
OIL API GRAVITY	35.0	API
WATER SPECIFIC GRAVITY	1.07	

DEPTH (1000 FT)

GAS/LIQUID RATIO - SCF PER BBL

HAGEDORN AND BROWN
PRESENTED BY
MACCO - SCHLUMBERGER

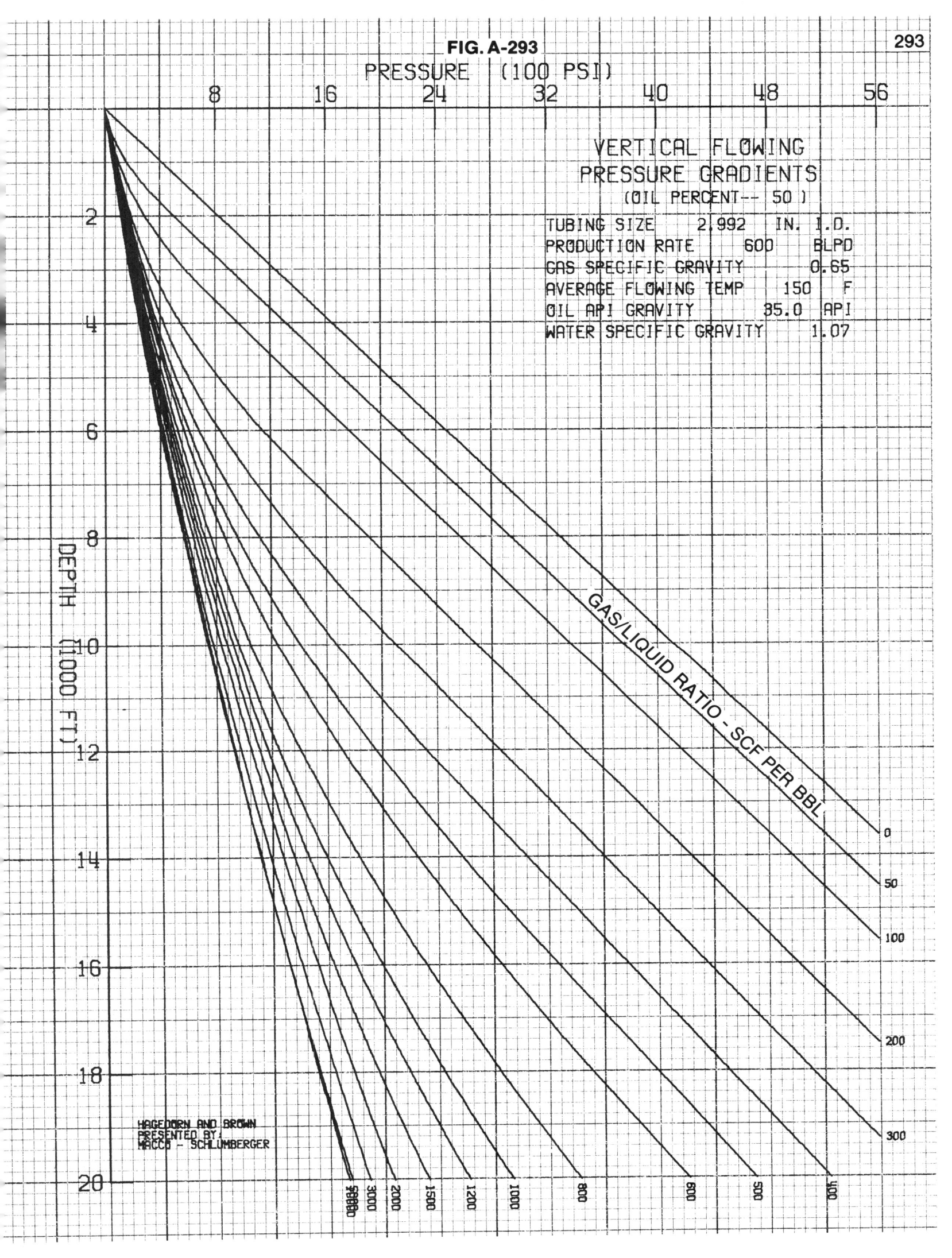

FIG. A-293

293

PRESSURE (100 PSI)

VERTICAL FLOWING
PRESSURE GRADIENTS
(OIL PERCENT-- 50)

TUBING SIZE	2.992	IN. I.D.
PRODUCTION RATE	600	BLPD
GAS SPECIFIC GRAVITY	0.65	
AVERAGE FLOWING TEMP	150	F
OIL API GRAVITY	35.0	API
WATER SPECIFIC GRAVITY	1.07	

GAS/LIQUID RATIO - SCF PER BBL

DEPTH (1000 FT)

HAGEDORN AND BROWN
PRESENTED BY:
MACCO - SCHLUMBERGER

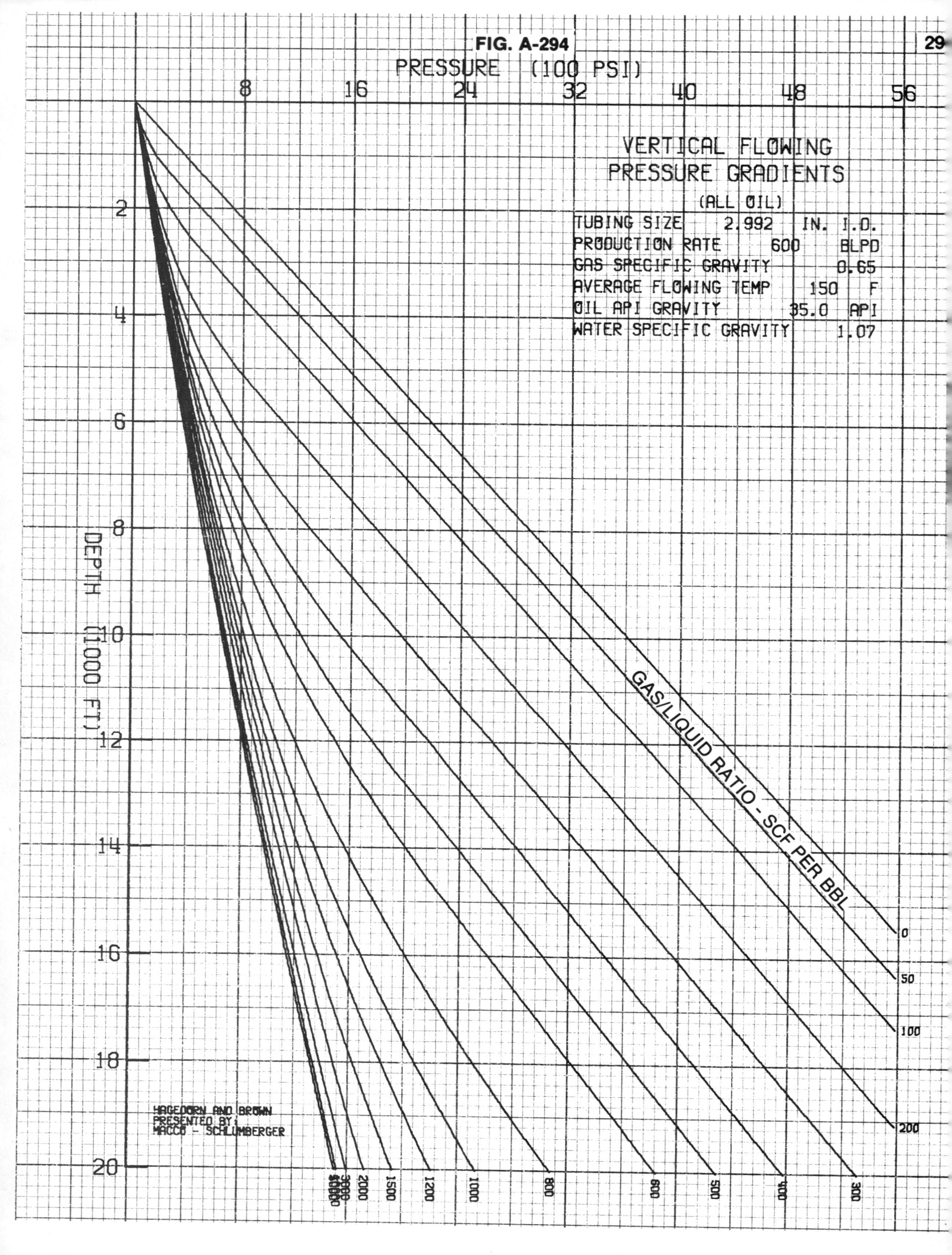

PRESSURE (100 PSI)

VERTICAL FLOWING
PRESSURE GRADIENTS
(ALL OIL)

TUBING SIZE	2.992	IN. I.D.
PRODUCTION RATE	600	BLPD
GAS SPECIFIC GRAVITY	0.65	
AVERAGE FLOWING TEMP	150	F
OIL API GRAVITY	35.0	API
WATER SPECIFIC GRAVITY	1.07	

DEPTH (1000 FT)

GAS/LIQUID RATIO - SCF PER BBL

HAGEDORN AND BROWN
PRESENTED BY
MACCO - SCHLUMBERGER

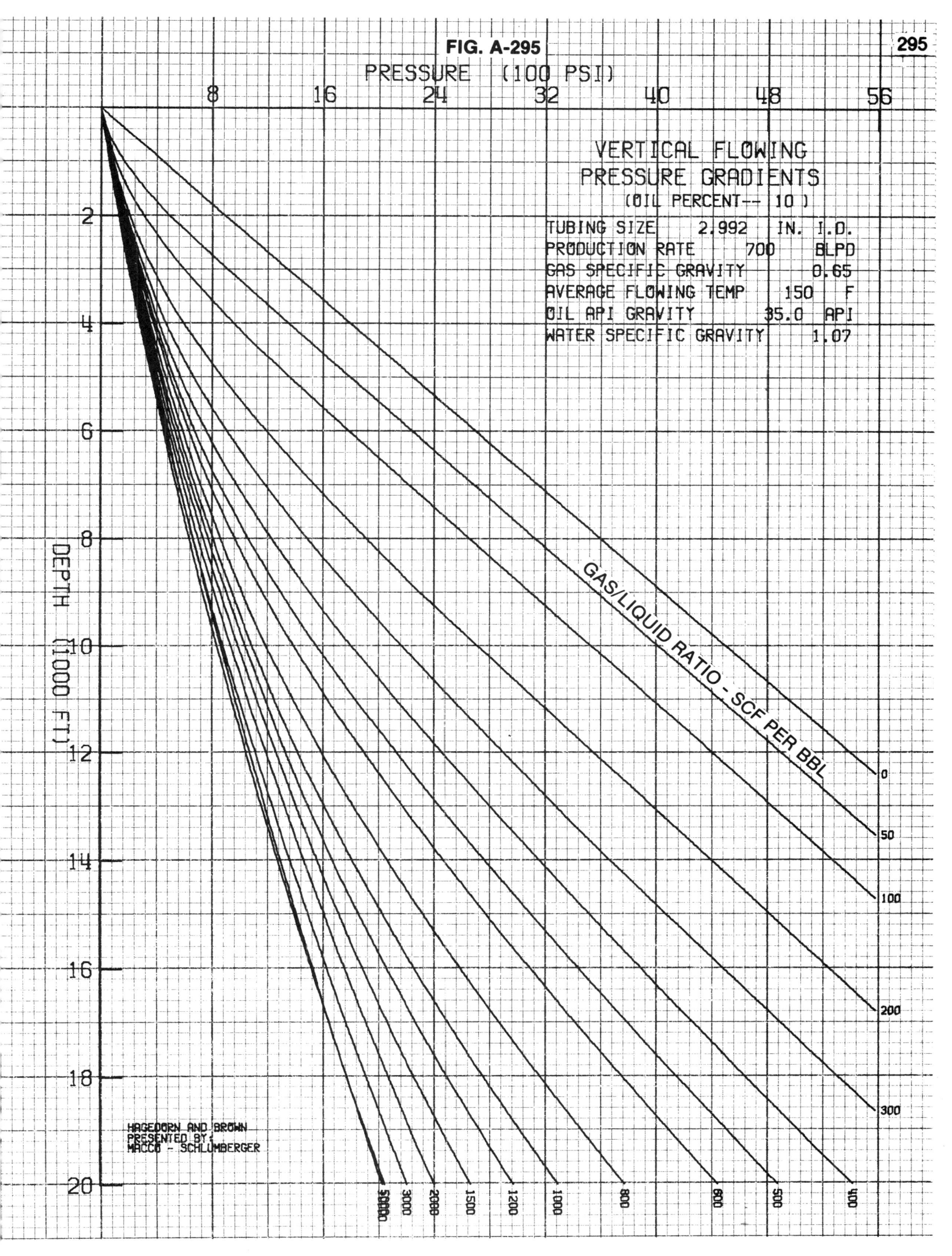

FIG. A-295

VERTICAL FLOWING
PRESSURE GRADIENTS
(OIL PERCENT-- 10)

TUBING SIZE	2.992	IN. I.D.
PRODUCTION RATE	700	BLPD
GAS SPECIFIC GRAVITY		0.65
AVERAGE FLOWING TEMP	150	F
OIL API GRAVITY	35.0	API
WATER SPECIFIC GRAVITY		1.07

GAS/LIQUID RATIO - SCF PER BBL

PRESSURE (100 PSI)

DEPTH (1000 FT)

HAGEDORN AND BROWN
PRESENTED BY:
MACCO - SCHLUMBERGER

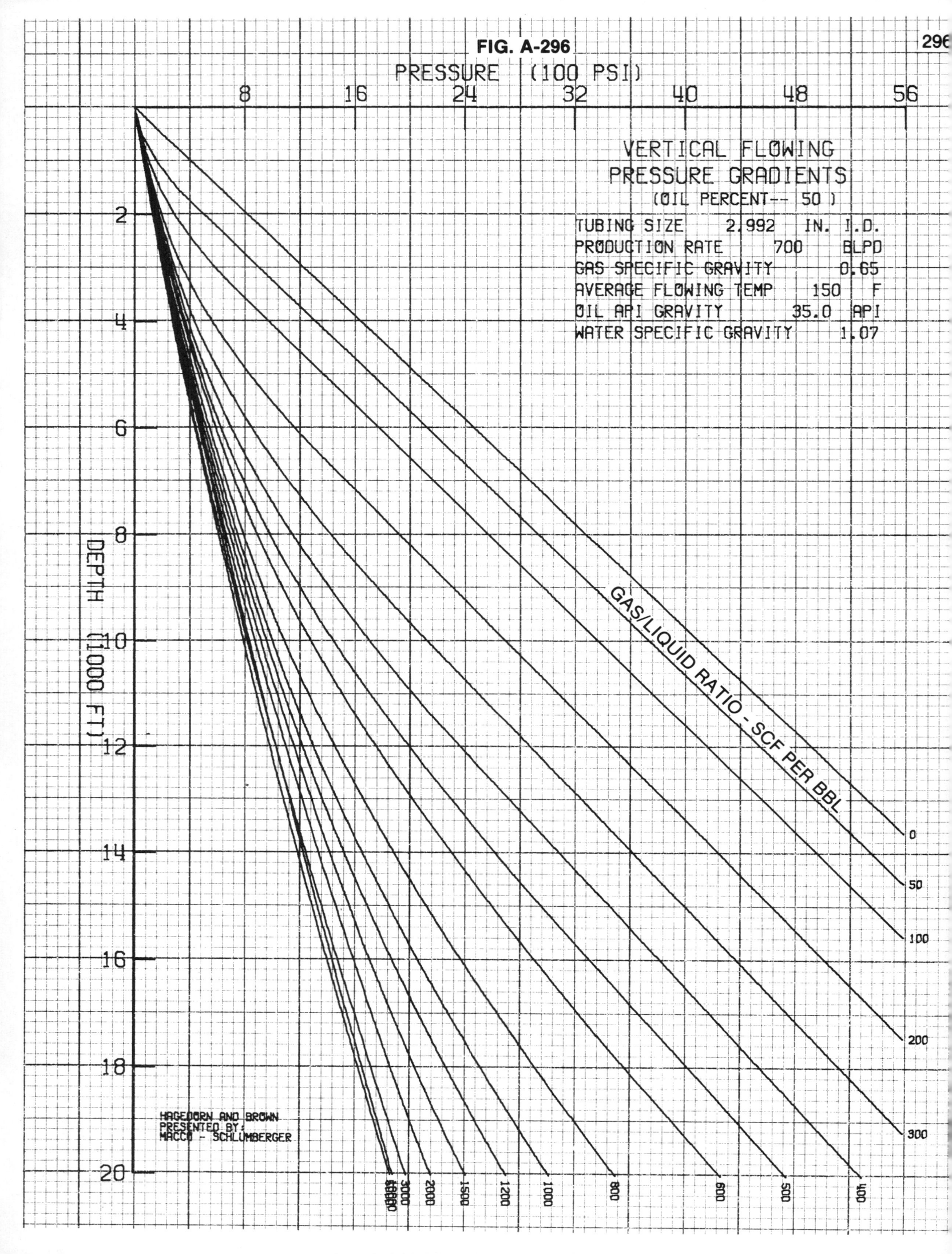

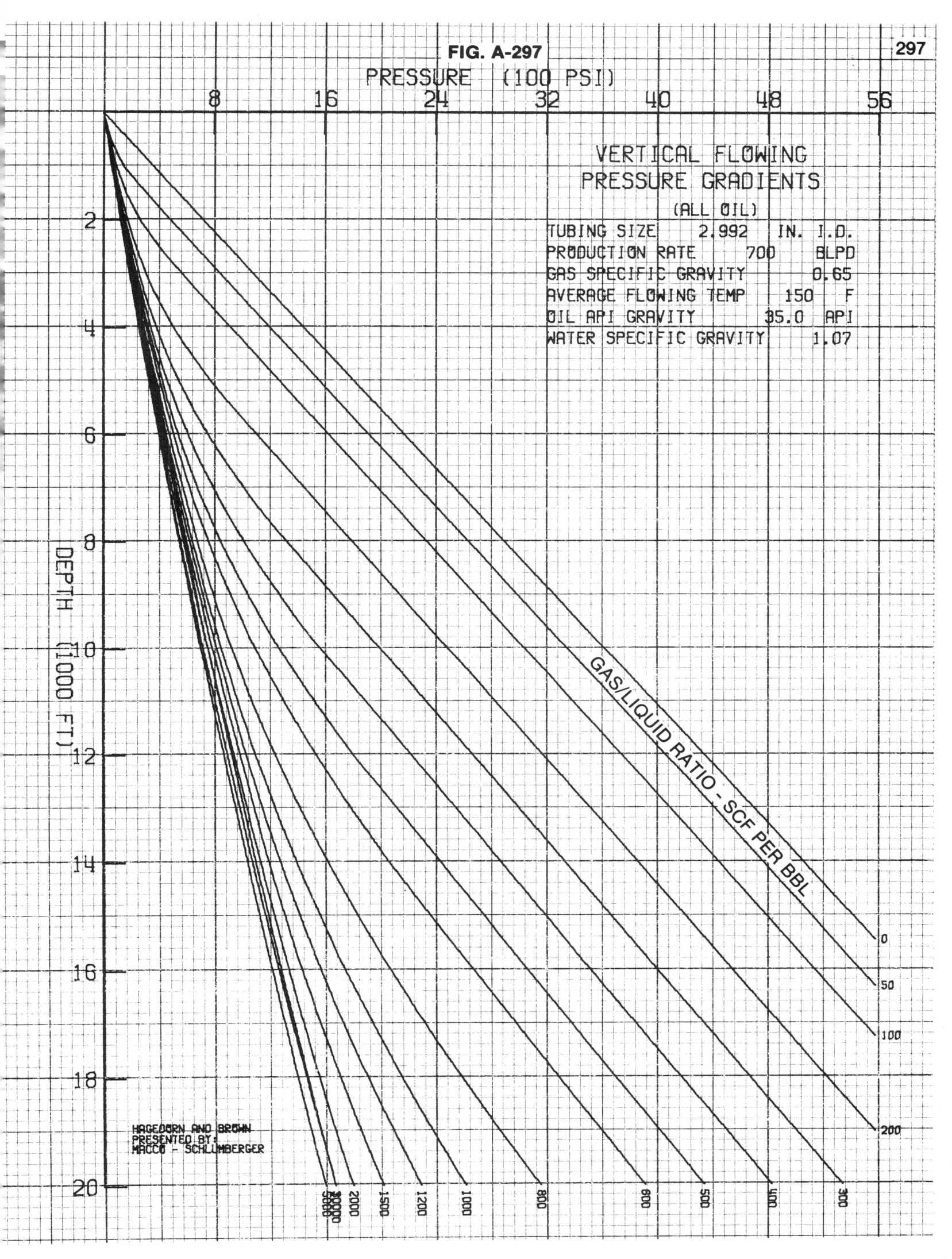

VERTICAL FLOWING
PRESSURE GRADIENTS
(ALL OIL)

TUBING SIZE	2.992	IN. I.D.
PRODUCTION RATE	700	BLPD
GAS SPECIFIC GRAVITY	0.65	
AVERAGE FLOWING TEMP	150	F
OIL API GRAVITY	35.0	API
WATER SPECIFIC GRAVITY	1.07	

PRESSURE (100 PSI)

DEPTH (1000 FT)

GAS/LIQUID RATIO - SCF PER BBL

HAGEDORN AND BROWN
PRESENTED BY:
MACCO - SCHLUMBERGER

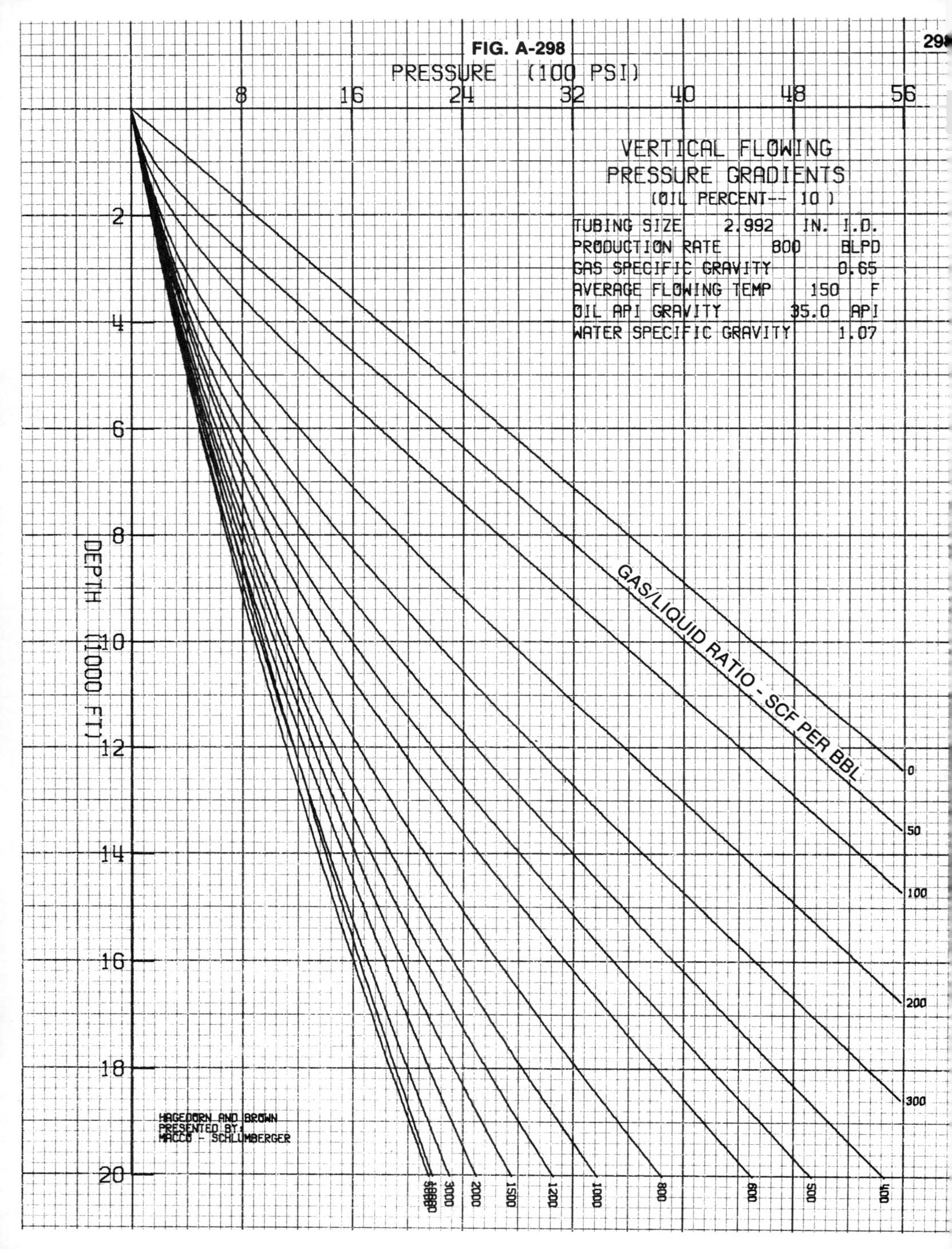

FIG. A-298

VERTICAL FLOWING
PRESSURE GRADIENTS
(OIL PERCENT-- 10)

TUBING SIZE	2.992	IN. I.D.
PRODUCTION RATE	800	BLPD
GAS SPECIFIC GRAVITY		0.65
AVERAGE FLOWING TEMP	150	F
OIL API GRAVITY	35.0	API
WATER SPECIFIC GRAVITY		1.07

PRESSURE (100 PSI)

DEPTH (1000 FT)

GAS/LIQUID RATIO - SCF PER BBL

HAGEDORN AND BROWN
PRESENTED BY:
MACCO - SCHLUMBERGER

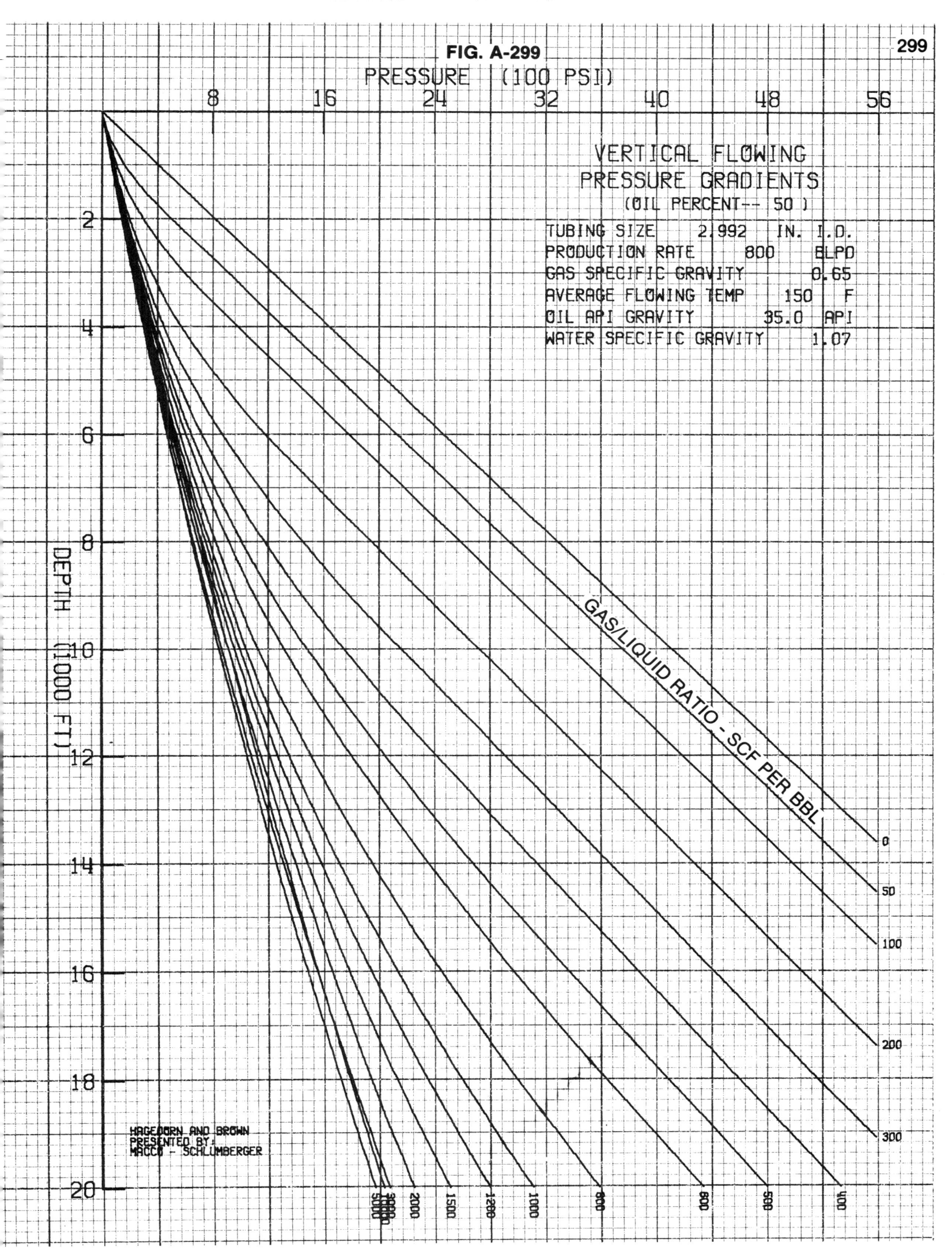

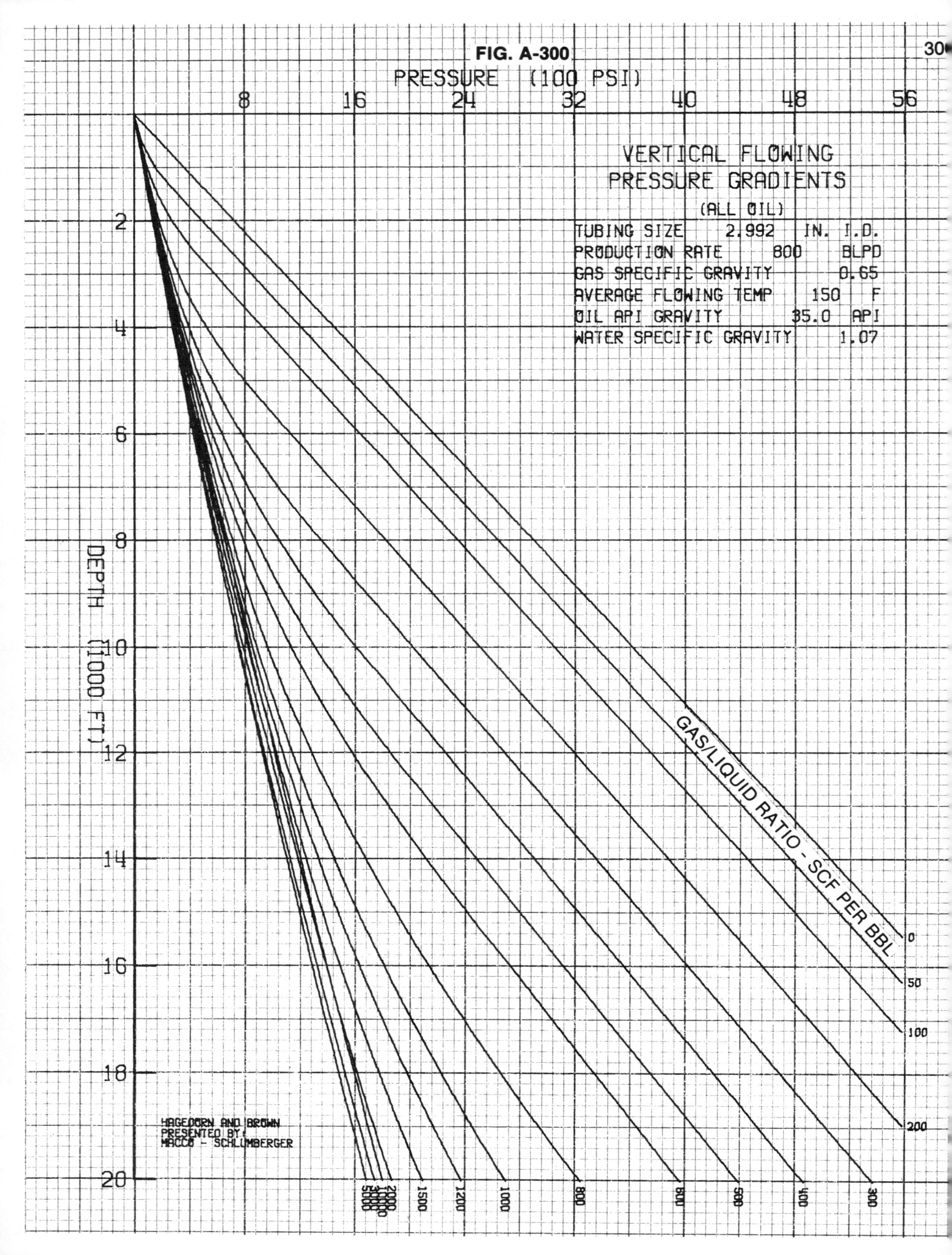

FIG. A-300

VERTICAL FLOWING
PRESSURE GRADIENTS
(ALL OIL)

TUBING SIZE	2.992	IN. I.D.
PRODUCTION RATE	800	BLPD
GAS SPECIFIC GRAVITY	0.65	
AVERAGE FLOWING TEMP	150	F
OIL API GRAVITY	35.0	API
WATER SPECIFIC GRAVITY	1.07	

PRESSURE (100 PSI)

DEPTH (1000 FT)

GAS/LIQUID RATIO - SCF PER BBL

HAGEDORN AND BROWN
PRESENTED BY:
MACCO - SCHLUMBERGER

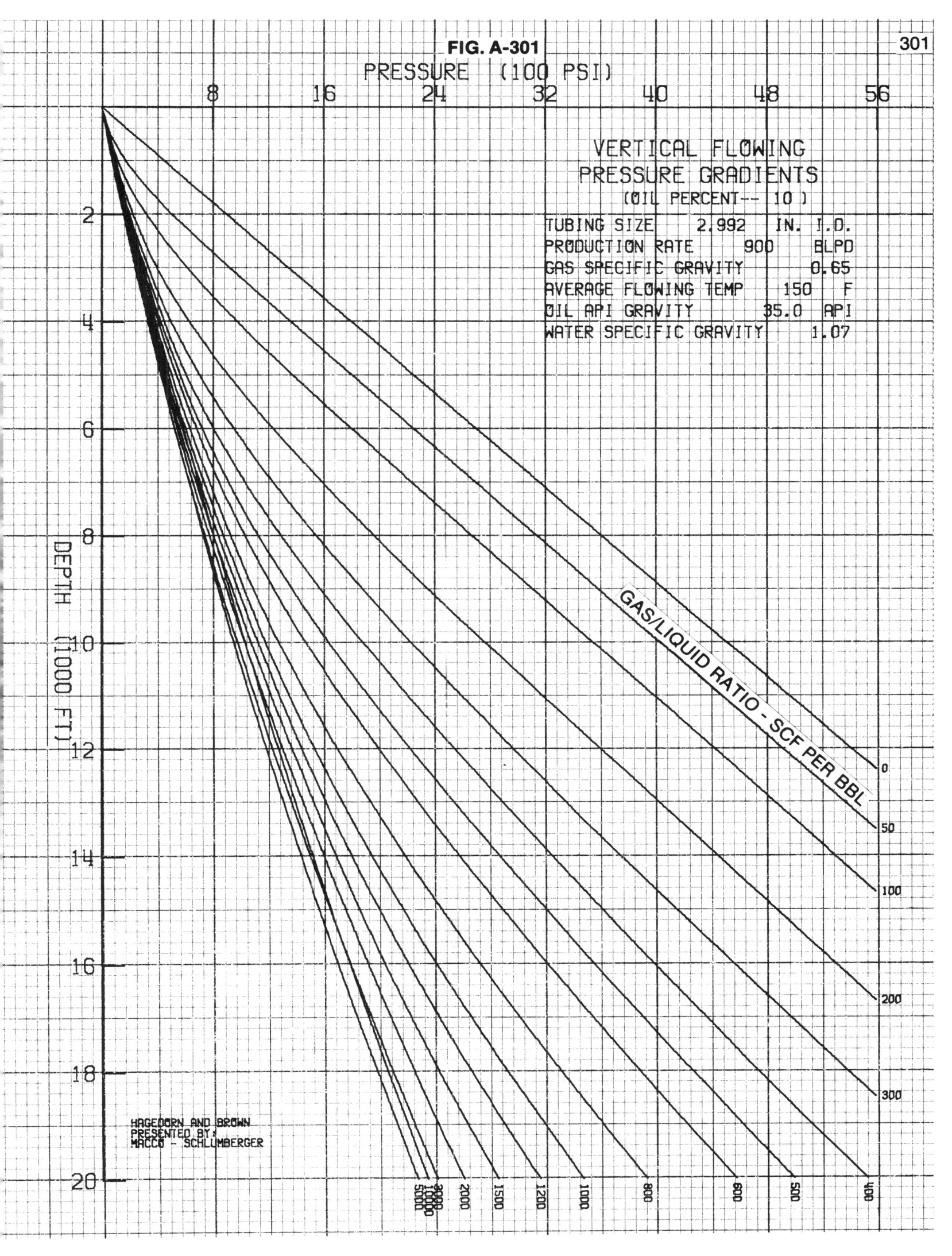

PRESSURE (100 PSI)

VERTICAL FLOWING
PRESSURE GRADIENTS
(OIL PERCENT--- 10)

TUBING SIZE	2.992	IN. I.D.
PRODUCTION RATE	900	BLPD
GAS SPECIFIC GRAVITY	0.65	
AVERAGE FLOWING TEMP	150	F
OIL API GRAVITY	35.0	API
WATER SPECIFIC GRAVITY	1.07	

DEPTH (1000 FT)

GAS/LIQUID RATIO - SCF PER BBL

HAGEDORN AND BROWN
PRESENTED BY :
MACCO - SCHLUMBERGER

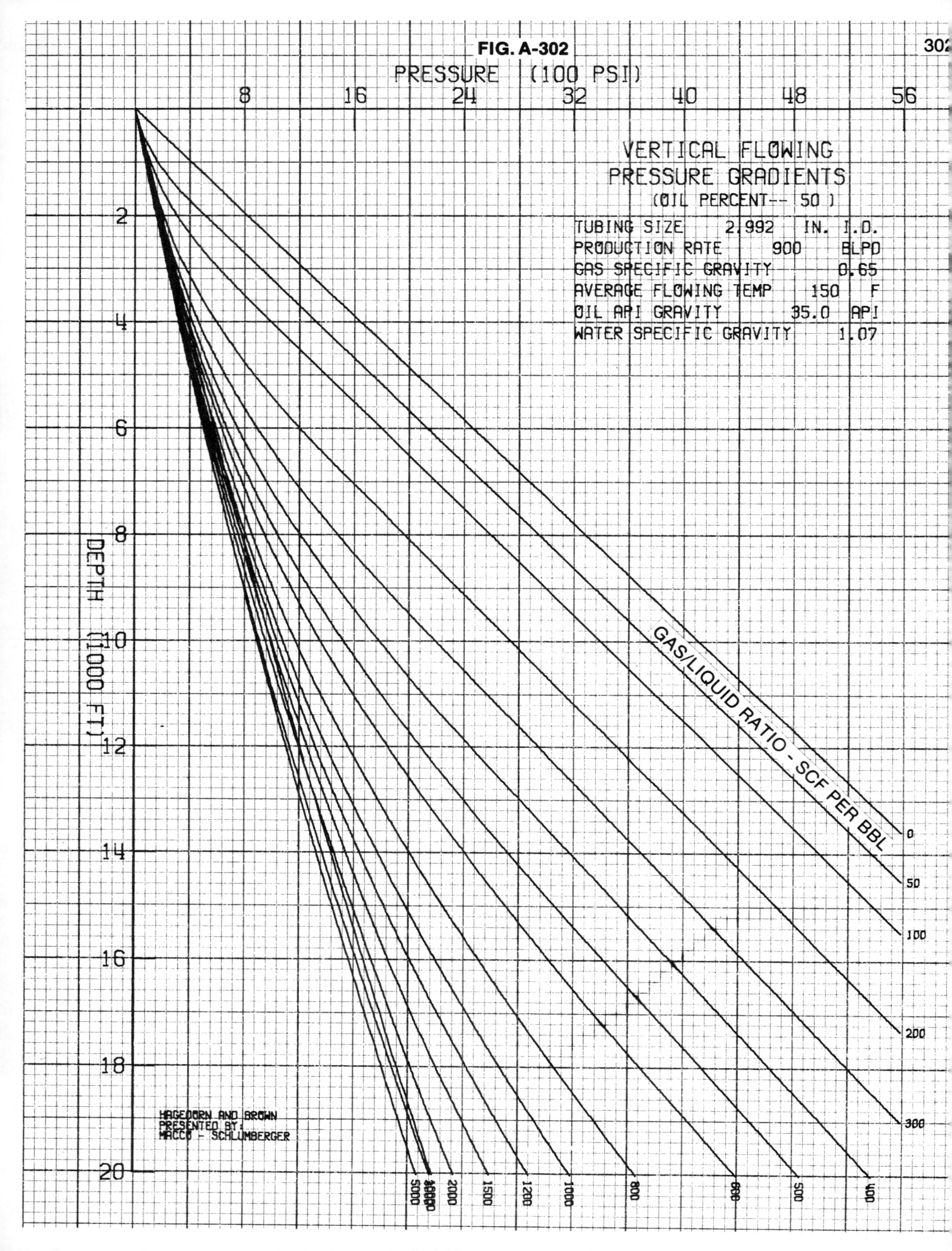

PRESSURE (100 PSI)

VERTICAL FLOWING
PRESSURE GRADIENTS
(OIL PERCENT-- 50)

TUBING SIZE	2.992	IN. I.D.
PRODUCTION RATE	900	BLPD
GAS SPECIFIC GRAVITY	0.65	
AVERAGE FLOWING TEMP	150	F
OIL API GRAVITY	35.0	API
WATER SPECIFIC GRAVITY	1.07	

DEPTH (1000 FT)

GAS/LIQUID RATIO - SCF PER BBL

HAGEDORN AND BROWN
PRESENTED BY:
MACCO - SCHLUMBERGER

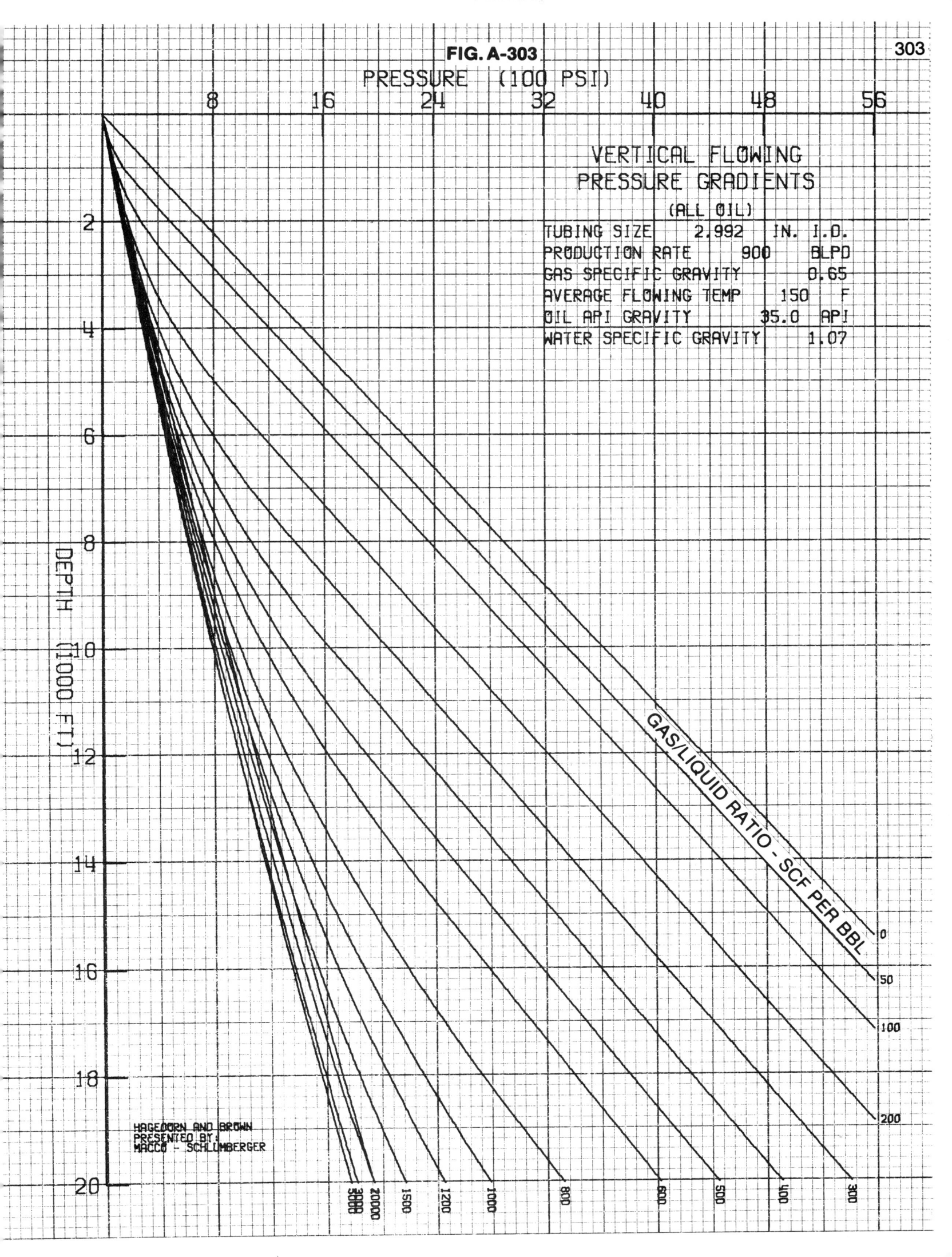

VERTICAL FLOWING
PRESSURE GRADIENTS
(ALL OIL)

TUBING SIZE	2.992	IN. I.D.
PRODUCTION RATE	900	BLPD
GAS SPECIFIC GRAVITY	0.65	
AVERAGE FLOWING TEMP	150	F
OIL API GRAVITY	35.0	API
WATER SPECIFIC GRAVITY	1.07	

PRESSURE (100 PSI)

DEPTH (1000 FT)

GAS/LIQUID RATIO - SCF PER BBL

HAGEDORN AND BROWN
PRESENTED BY:
MACCO - SCHLUMBERGER

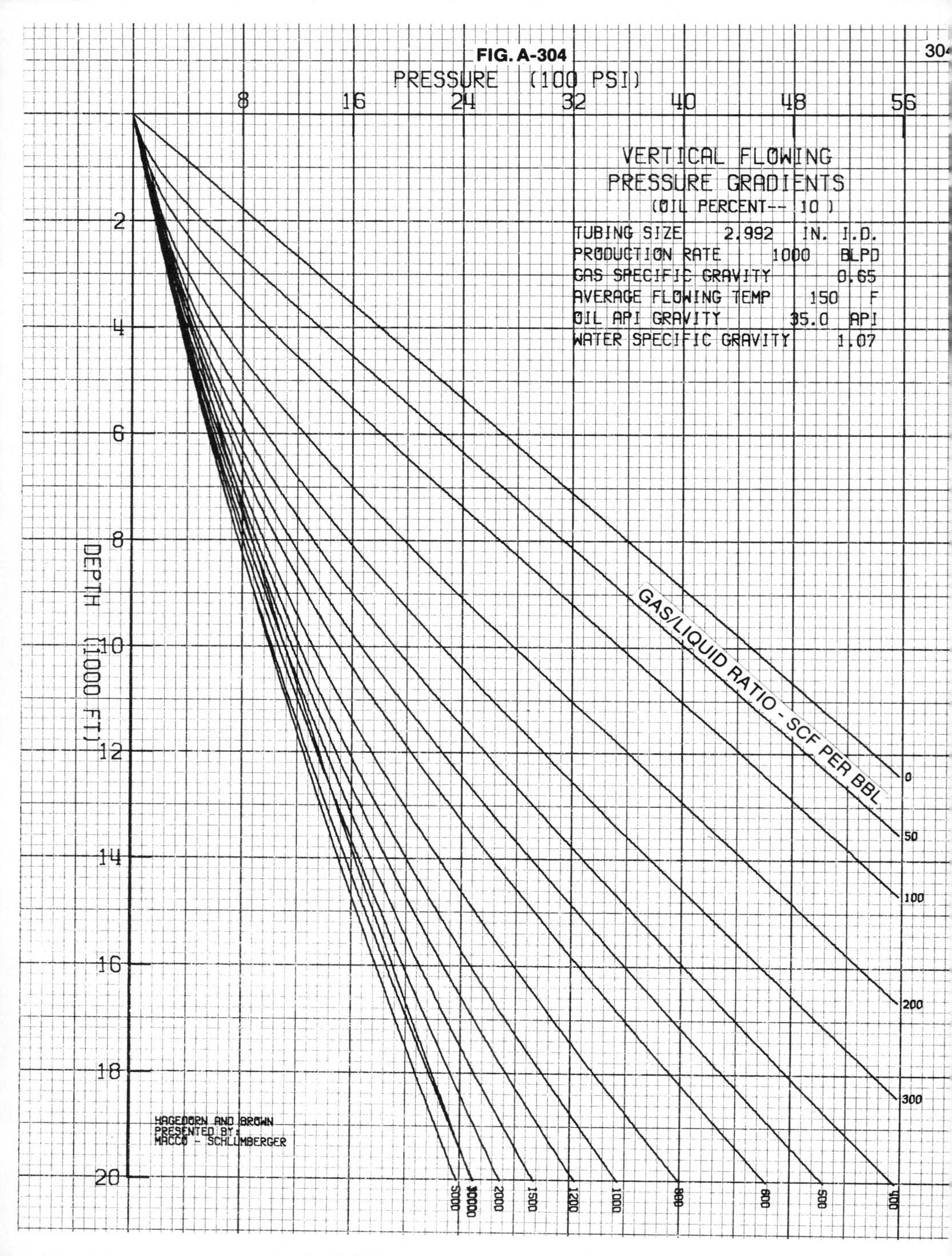

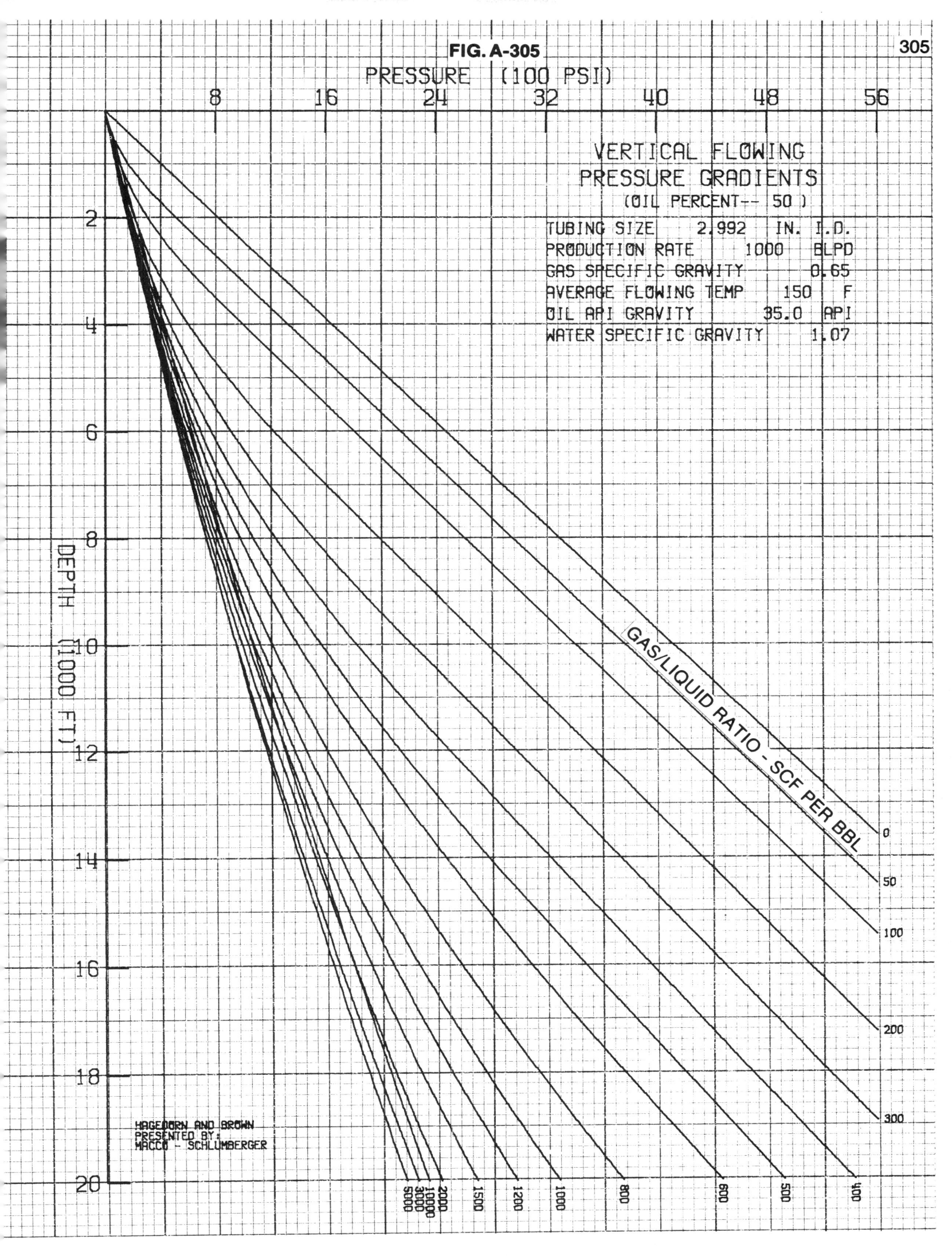

FIG. A-305

305

PRESSURE (100 PSI)

VERTICAL FLOWING
PRESSURE GRADIENTS
(OIL PERCENT— 50)

TUBING SIZE   2.992   IN. I.D.
PRODUCTION RATE   1000   BLPD
GAS SPECIFIC GRAVITY   0.65
AVERAGE FLOWING TEMP   150   F
OIL API GRAVITY   35.0   API
WATER SPECIFIC GRAVITY   1.07

DEPTH (1000 FT)

GAS/LIQUID RATIO - SCF PER BBL

0
50
100
200
300

HAGEDORN AND BROWN
PRESENTED BY:
MACCO - SCHLUMBERGER

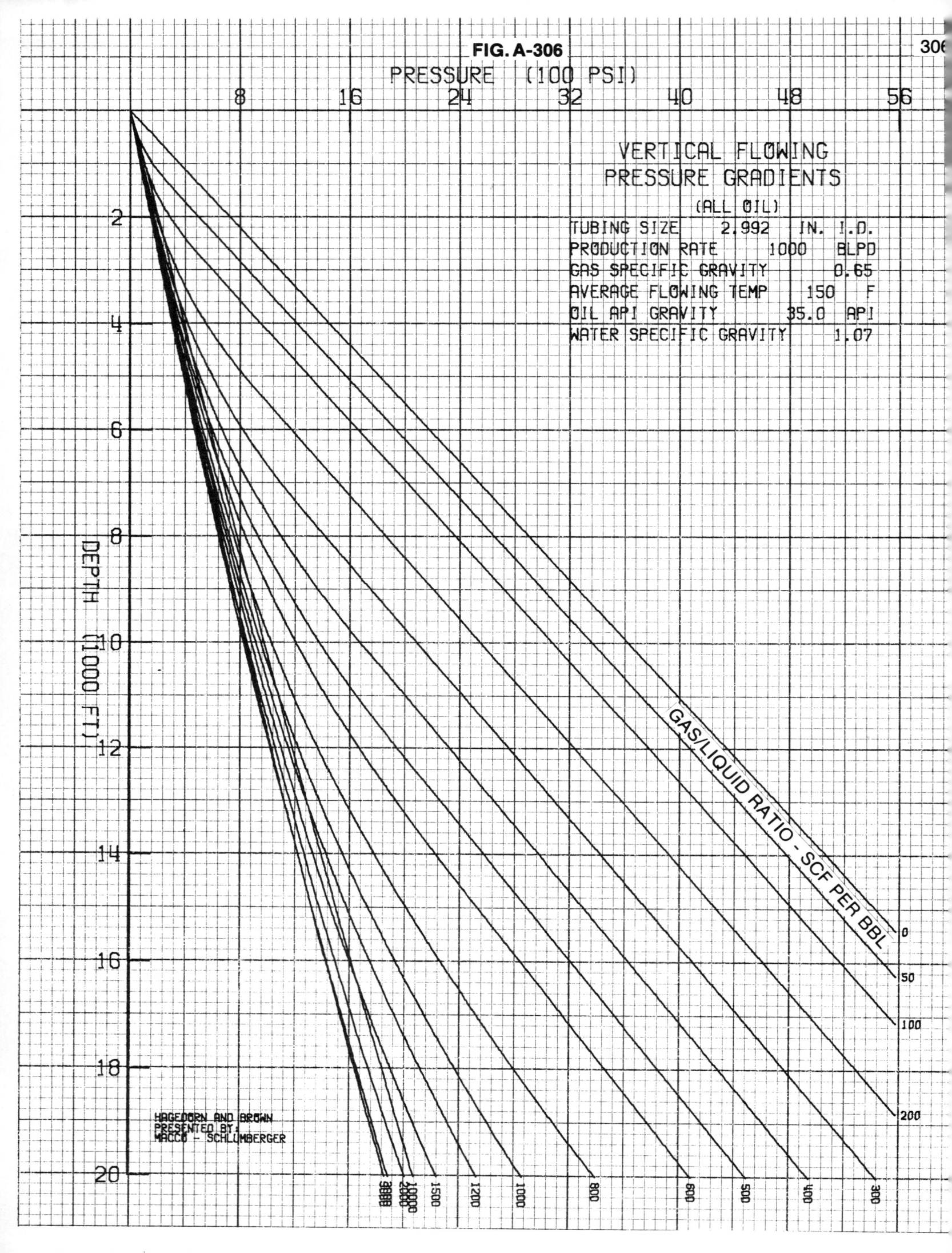

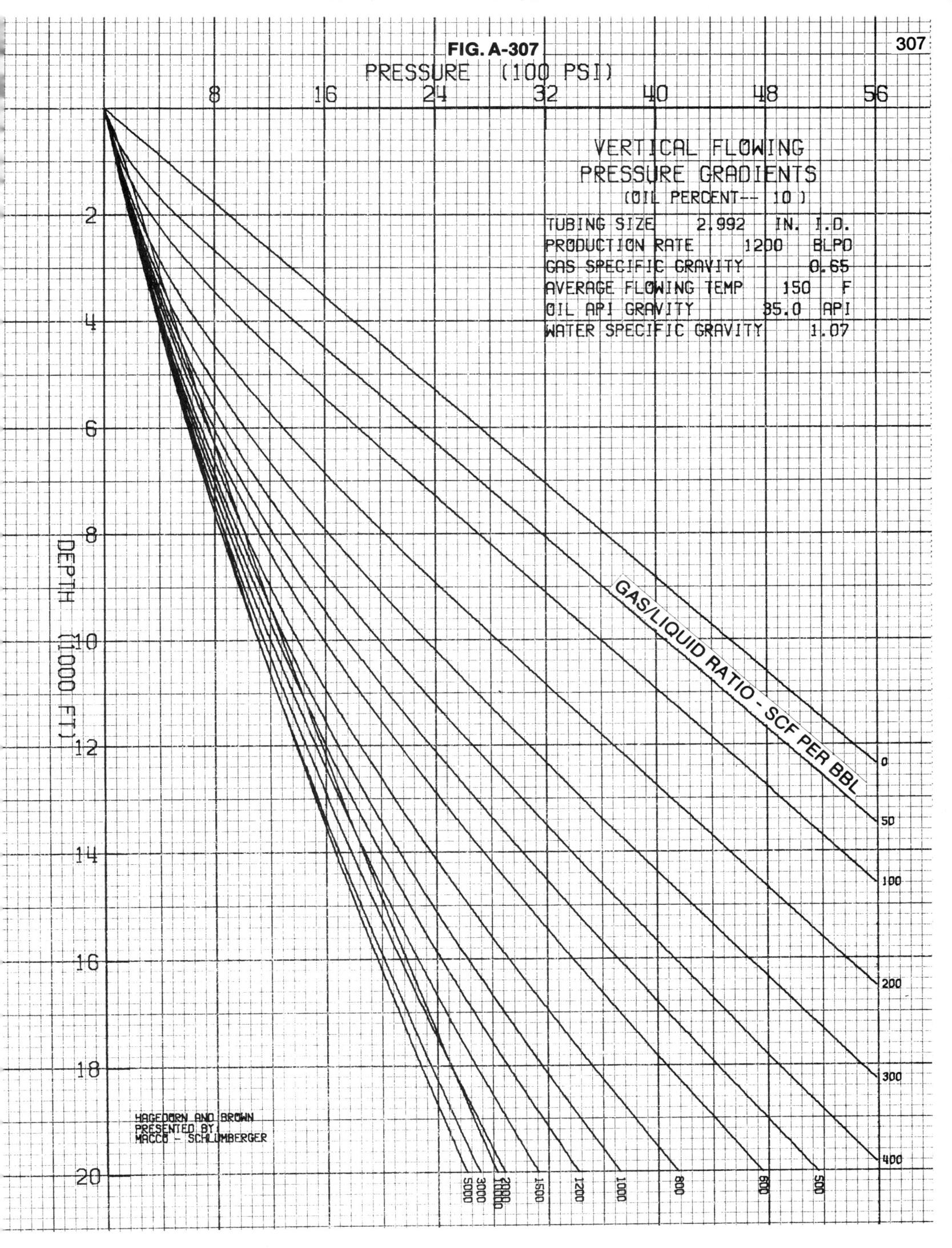

PRESSURE (100 PSI)

VERTICAL FLOWING
PRESSURE GRADIENTS
(OIL PERCENT-- 10)

TUBING SIZE	2.992	IN. I.D.
PRODUCTION RATE	1200	BLPD
GAS SPECIFIC GRAVITY	0.65	
AVERAGE FLOWING TEMP	150	F
OIL API GRAVITY	35.0	API
WATER SPECIFIC GRAVITY	1.07	

GAS/LIQUID RATIO - SCF PER BBL

DEPTH (1000 FT)

HAGEDORN AND BROWN
PRESENTED BY :
MACCO - SCHLUMBERGER

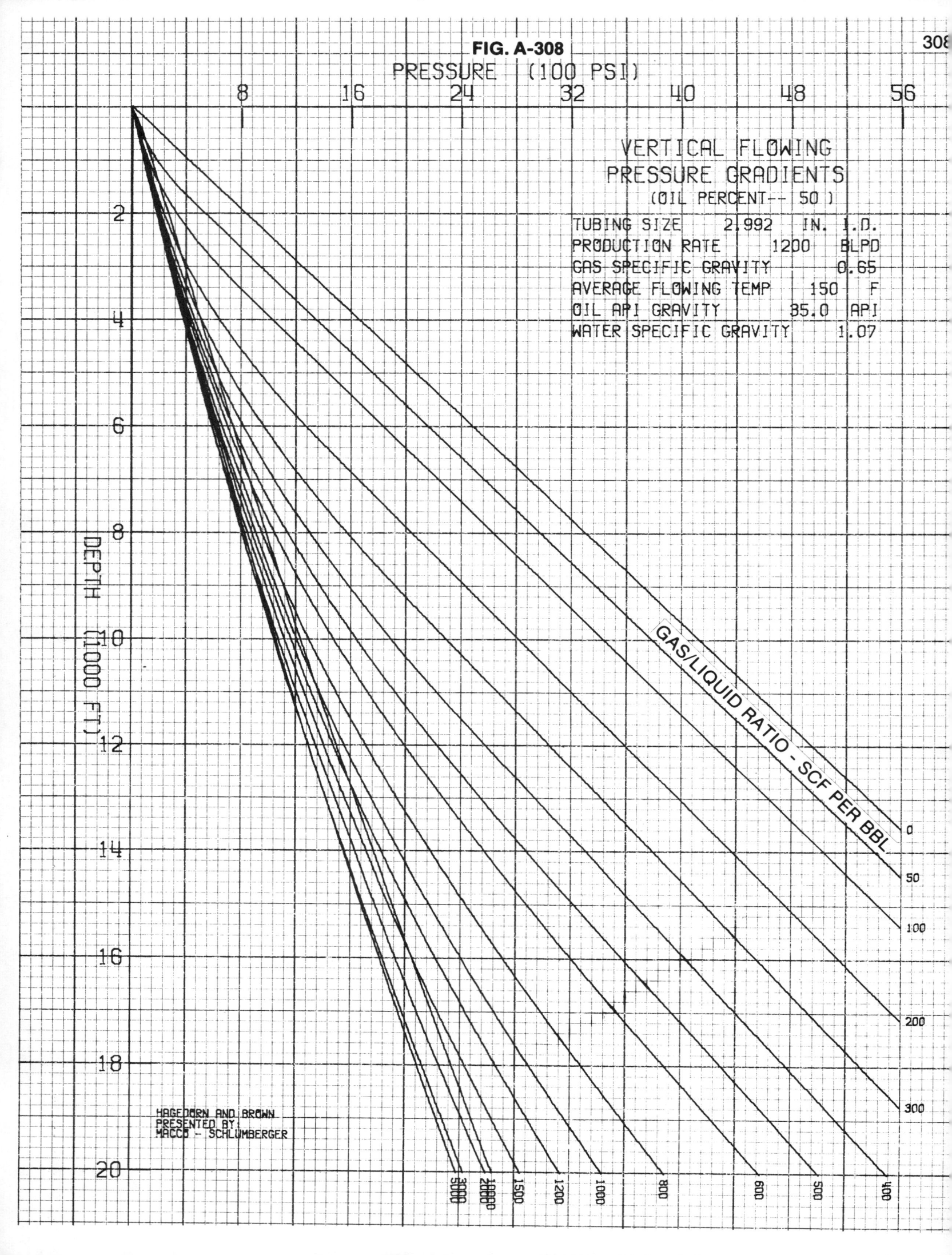

VERTICAL FLOWING
PRESSURE GRADIENTS
(OIL PERCENT-- 50)

TUBING SIZE	2.992	IN. I.D.
PRODUCTION RATE	1200	BLPD
GAS SPECIFIC GRAVITY	0.65	
AVERAGE FLOWING TEMP	150	F
OIL API GRAVITY	35.0	API
WATER SPECIFIC GRAVITY	1.07	

PRESSURE (100 PSI)

DEPTH (1000 FT)

GAS/LIQUID RATIO - SCF PER BBL

HAGEDORN AND BROWN
PRESENTED BY:
MACCO - SCHLUMBERGER

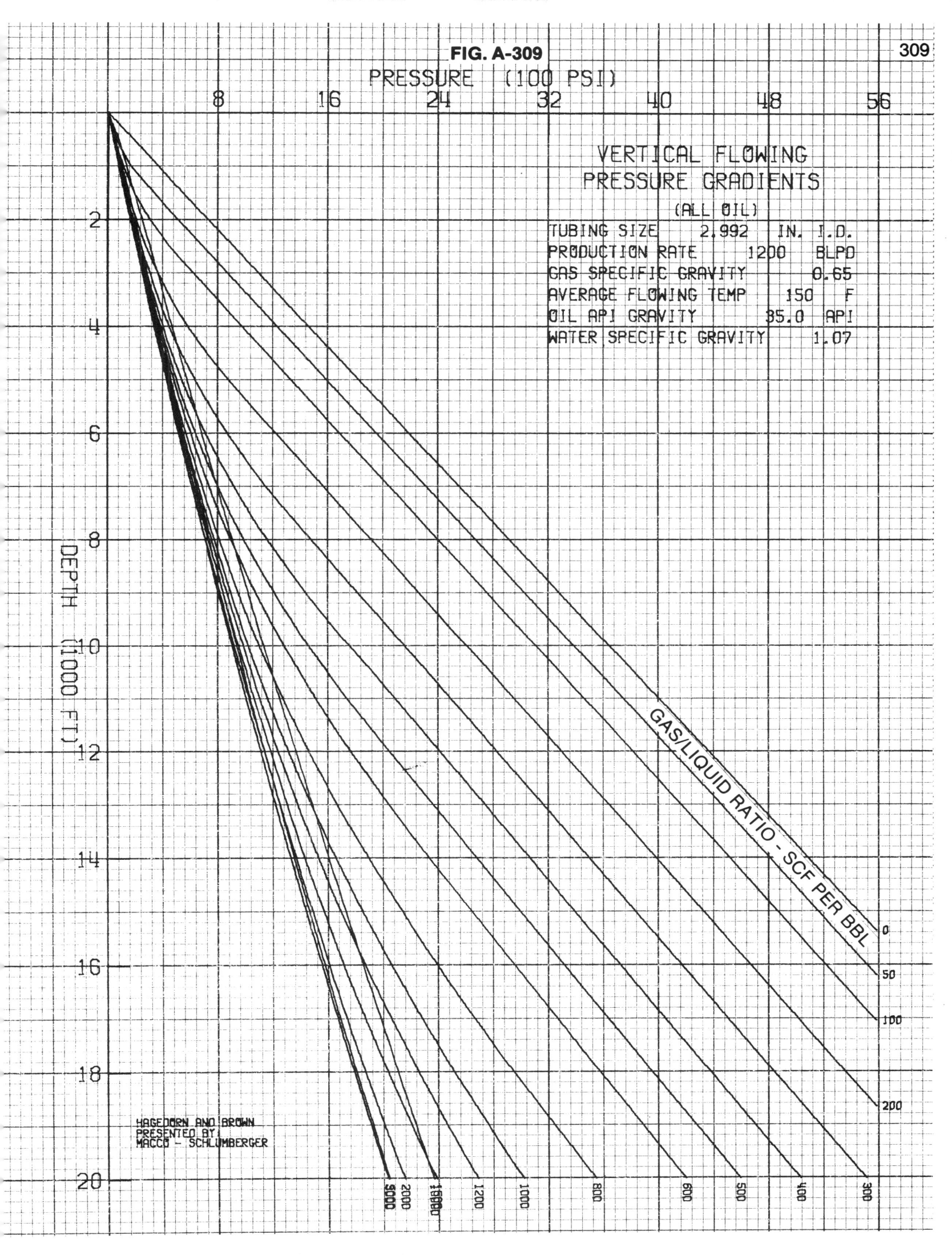

PRESSURE (100 PSI)

VERTICAL FLOWING
PRESSURE GRADIENTS
(ALL OIL)

TUBING SIZE	2.992	IN. I.D.
PRODUCTION RATE	1200	BLPD
GAS SPECIFIC GRAVITY	0.65	
AVERAGE FLOWING TEMP	150	F
OIL API GRAVITY	35.0	API
WATER SPECIFIC GRAVITY	1.07	

DEPTH (1000 FT)

GAS/LIQUID RATIO - SCF PER BBL

HAGEDORN AND BROWN
PRESENTED BY
MACCO - SCHLUMBERGER

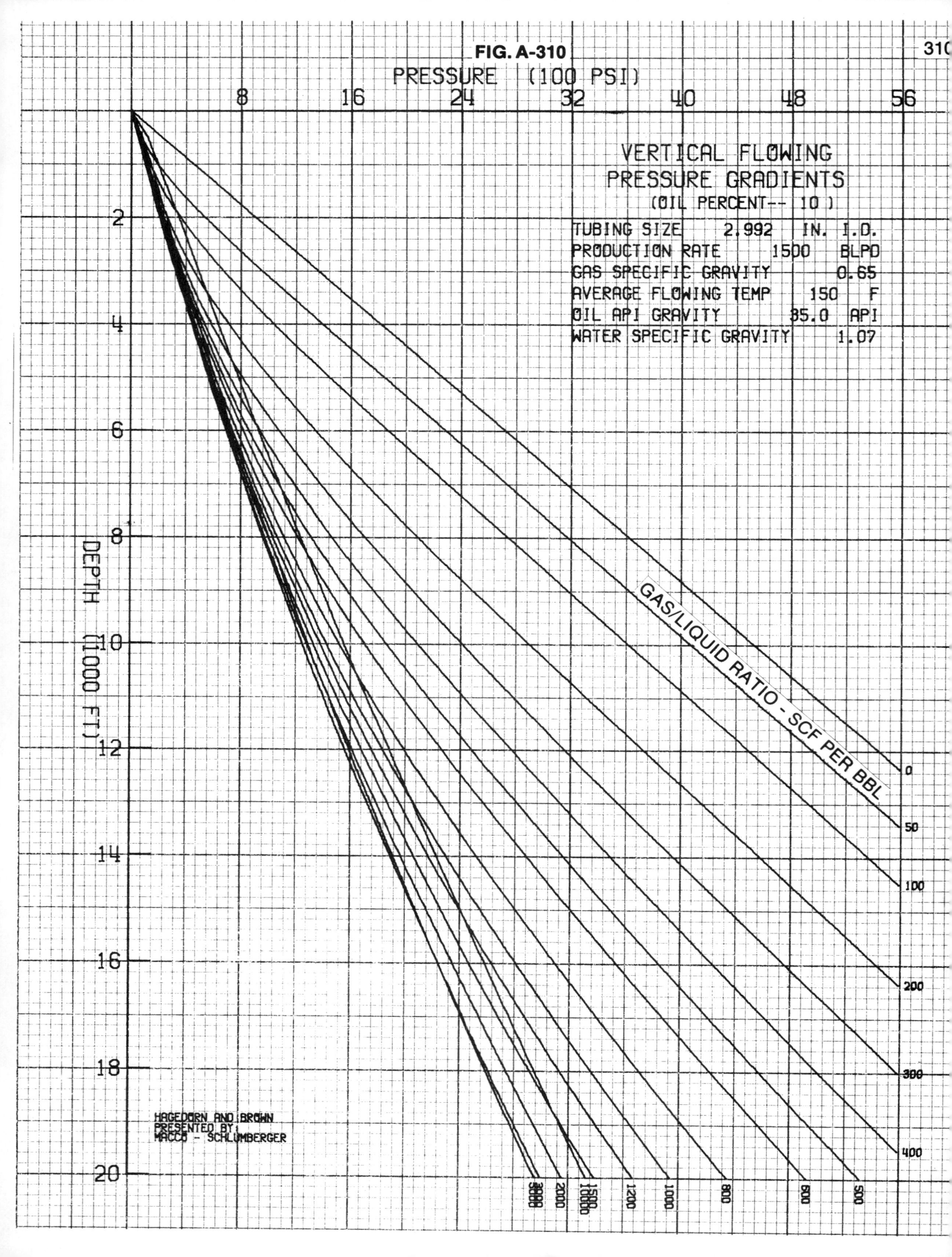

VERTICAL FLOWING
PRESSURE GRADIENTS
(OIL PERCENT-- 10)

TUBING SIZE	2.992	IN. I.D.
PRODUCTION RATE	1500	BLPD
GAS SPECIFIC GRAVITY		0.65
AVERAGE FLOWING TEMP	150	F
OIL API GRAVITY	35.0	API
WATER SPECIFIC GRAVITY		1.07

PRESSURE (100 PSI)

DEPTH (1000 FT)

GAS/LIQUID RATIO - SCF PER BBL

HAGEDORN AND BROWN
PRESENTED BY
MACCO - SCHLUMBERGER

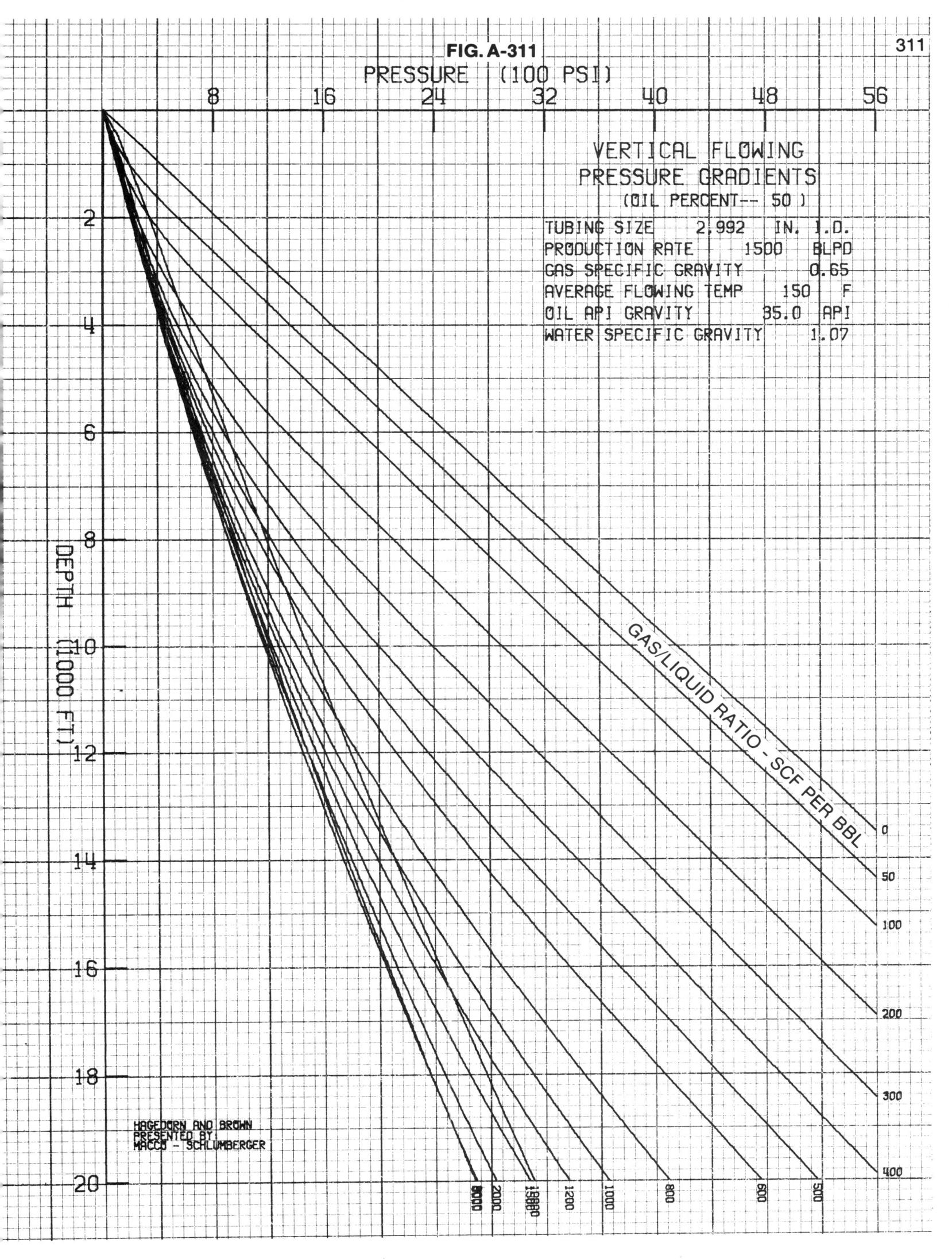

FIG. A-311

PRESSURE (100 PSI)

VERTICAL FLOWING
PRESSURE GRADIENTS
(OIL PERCENT-- 50)

TUBING SIZE	2.992	IN. I.D.
PRODUCTION RATE	1500	BLPD
GAS SPECIFIC GRAVITY	0.65	
AVERAGE FLOWING TEMP	150	F
OIL API GRAVITY	35.0	API
WATER SPECIFIC GRAVITY	1.07	

DEPTH (1000 FT)

GAS/LIQUID RATIO - SCF PER BBL

HAGEDORN AND BROWN
PRESENTED BY:
MACCO - SCHLUMBERGER

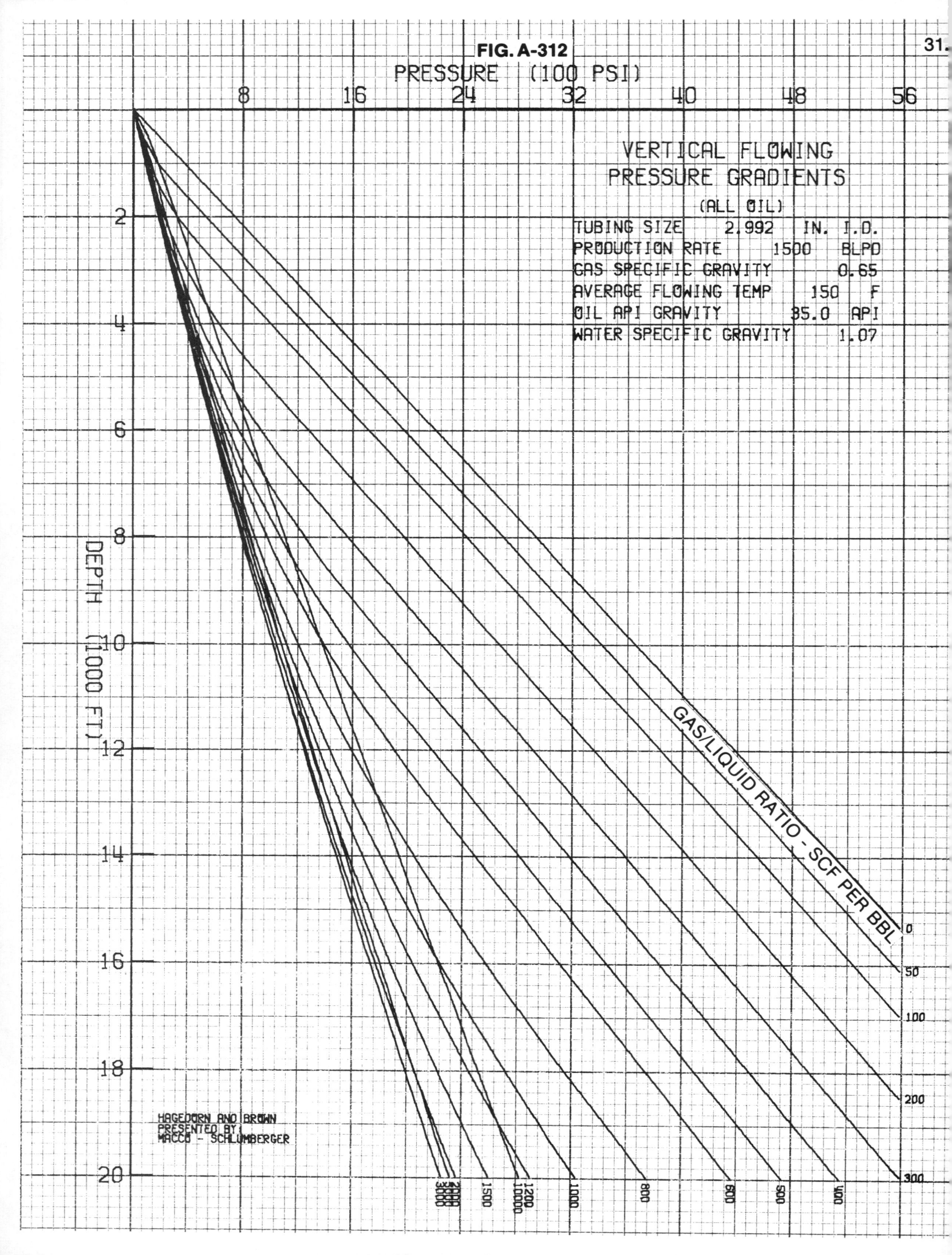

FIG. A-312

31.

VERTICAL FLOWING
PRESSURE GRADIENTS
(ALL OIL)

TUBING SIZE   2.992   IN. I.D.
PRODUCTION RATE   1500   BLPD
GAS SPECIFIC GRAVITY   0.65
AVERAGE FLOWING TEMP 150   F
OIL API GRAVITY   35.0   API
WATER SPECIFIC GRAVITY   1.07

PRESSURE   (100 PSI)

DEPTH   (1000 FT)

GAS/LIQUID RATIO - SCF PER BBL

0
50
100
200
300

HAGEDORN AND BROWN
PRESENTED BY
MACCO - SCHLUMBERGER

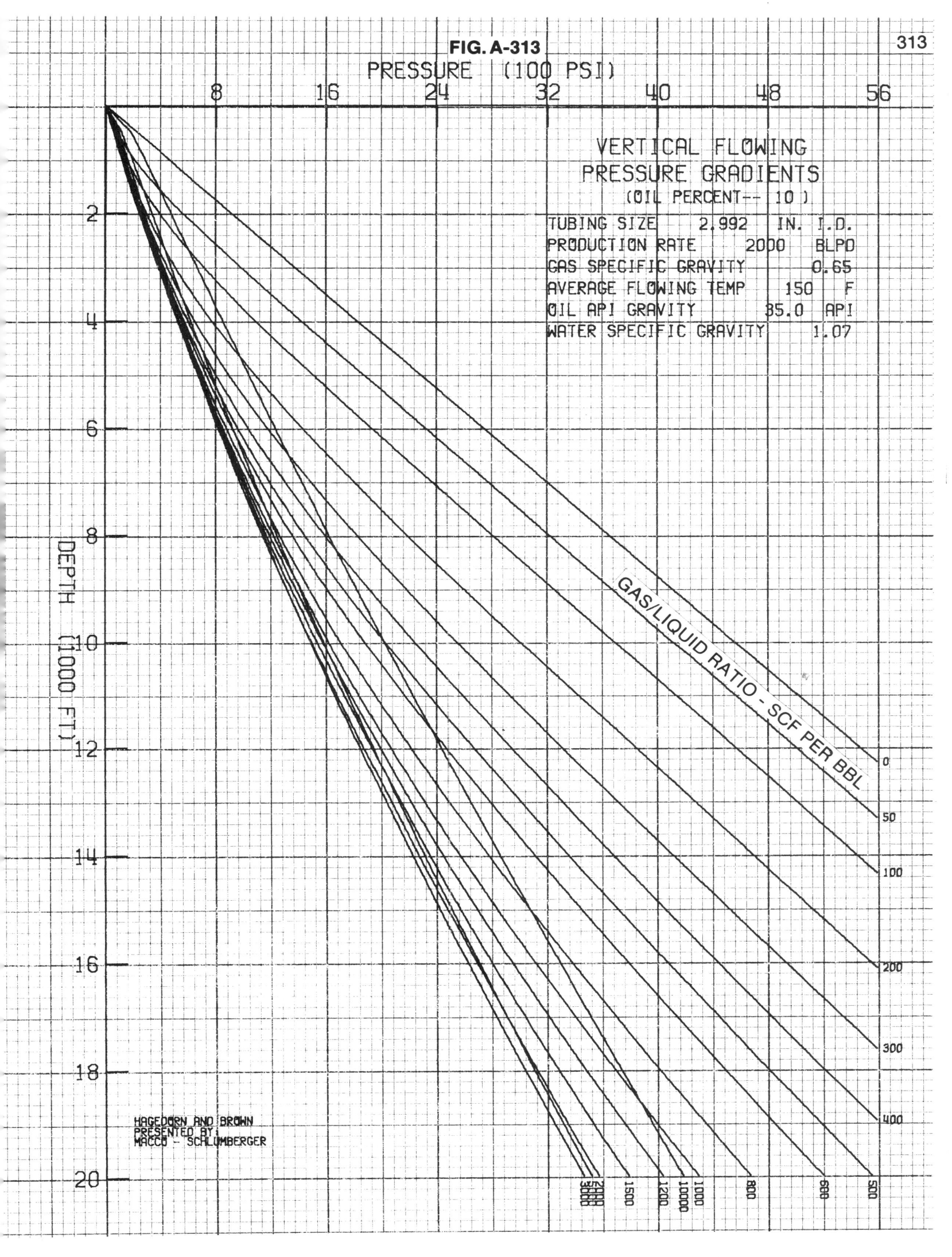

PRESSURE (100 PSI)

VERTICAL FLOWING
PRESSURE GRADIENTS
(OIL PERCENT-- 10)

TUBING SIZE	2.992	IN. I.D.
PRODUCTION RATE	2000	BLPD
GAS SPECIFIC GRAVITY	0.65	
AVERAGE FLOWING TEMP	150	F
OIL API GRAVITY	35.0	API
WATER SPECIFIC GRAVITY	1.07	

DEPTH (1000 FT)

GAS/LIQUID RATIO - SCF PER BBL

HAGEDORN AND BROWN
PRESENTED BY
MACCO -- SCHLUMBERGER

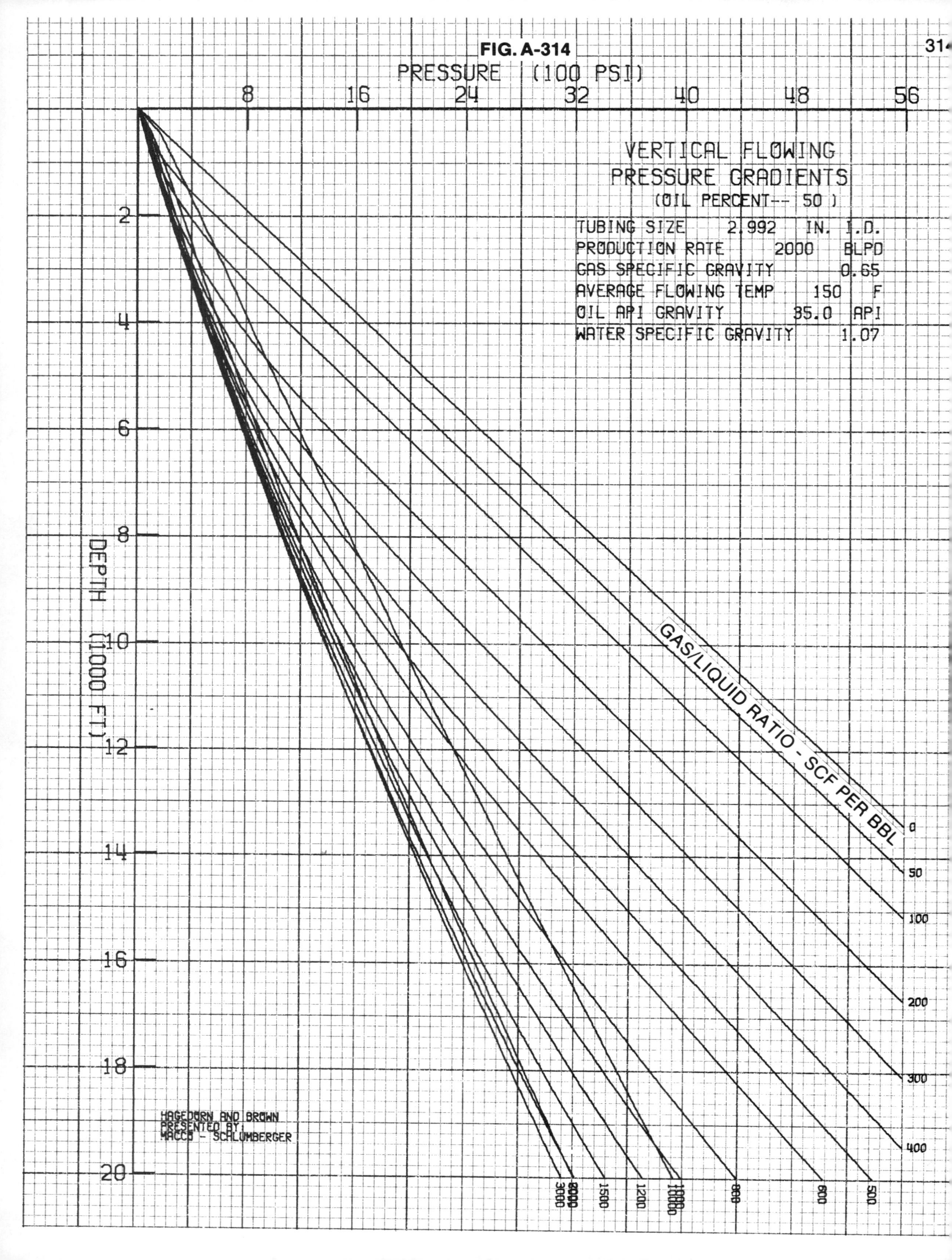

VERTICAL FLOWING
PRESSURE GRADIENTS
(OIL PERCENT-- 50)

TUBING SIZE	2.992	IN. I.D.
PRODUCTION RATE	2000	BLPD
GAS SPECIFIC GRAVITY	0.65	
AVERAGE FLOWING TEMP	150	F
OIL API GRAVITY	35.0	API
WATER SPECIFIC GRAVITY	1.07	

PRESSURE (100 PSI)

DEPTH (1000 FT.)

GAS/LIQUID RATIO - SCF PER BBL

HAGEDORN AND BROWN
PRESENTED BY
MACCO - SCHLUMBERGER

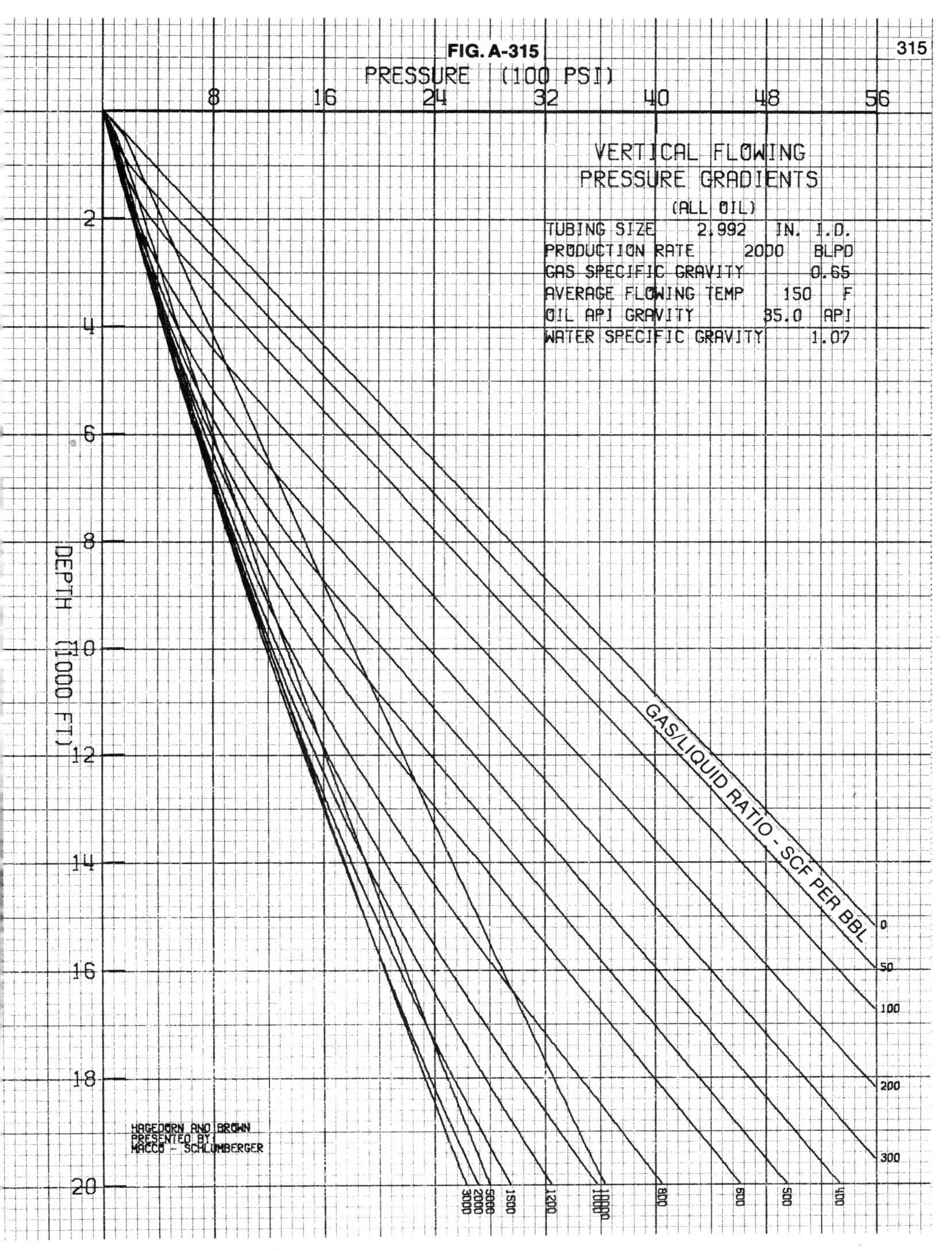

FIG. A-315

315

VERTICAL FLOWING
PRESSURE GRADIENTS
(ALL OIL)

TUBING SIZE	2.992	IN. I.D.
PRODUCTION RATE	2000	BLPD
GAS SPECIFIC GRAVITY	0.65	
AVERAGE FLOWING TEMP	150	F
OIL API GRAVITY	35.0	API
WATER SPECIFIC GRAVITY	1.07	

GAS/LIQUID RATIO - SCF PER BBL

HAGEDORN AND BROWN
PRESENTED BY
MACCO - SCHLUMBERGER

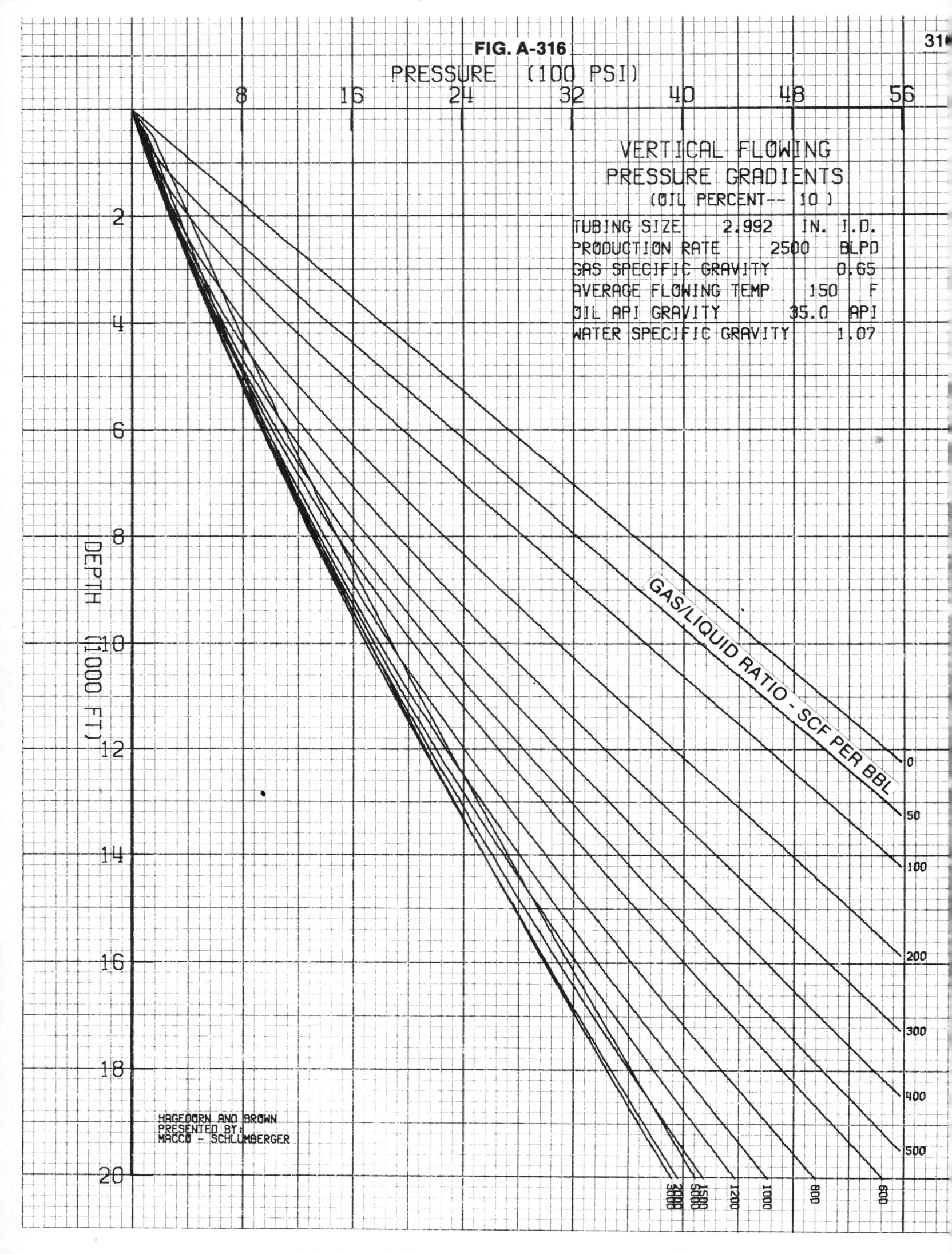

PRESSURE (100 PSI)

DEPTH (1000 FT)

VERTICAL FLOWING
PRESSURE GRADIENTS
(OIL PERCENT-- 10)

TUBING SIZE	2.992	IN. I.D.
PRODUCTION RATE	2500	BLPD
GAS SPECIFIC GRAVITY	0.65	
AVERAGE FLOWING TEMP	150	F
OIL API GRAVITY	35.0	API
WATER SPECIFIC GRAVITY	1.07	

GAS/LIQUID RATIO - SCF PER BBL

0
50
100
200
300
400
500

600
800
1000
1200
1500
1800
2000

HAGEDORN AND BROWN
PRESENTED BY:
MACCO - SCHLUMBERGER

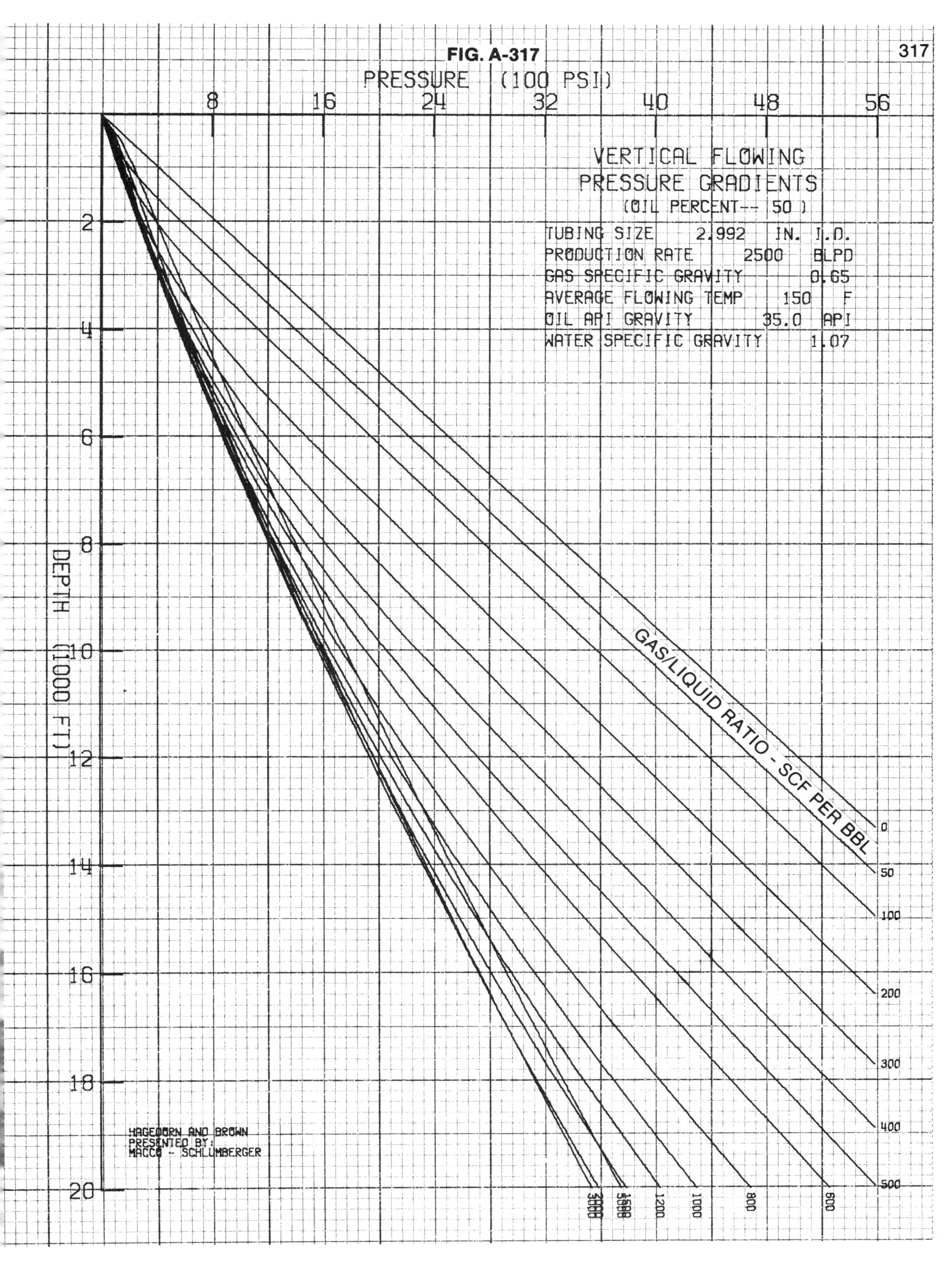

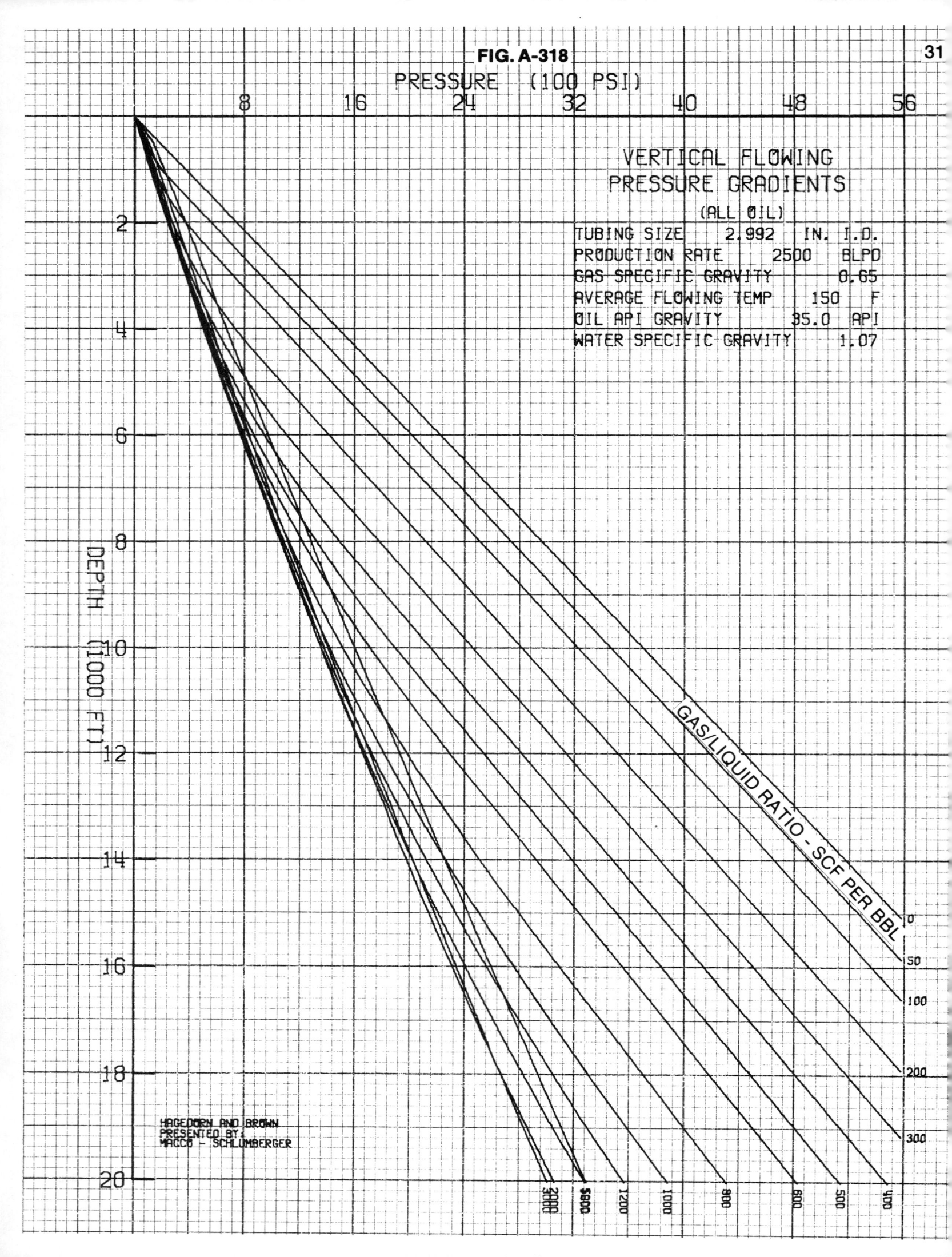

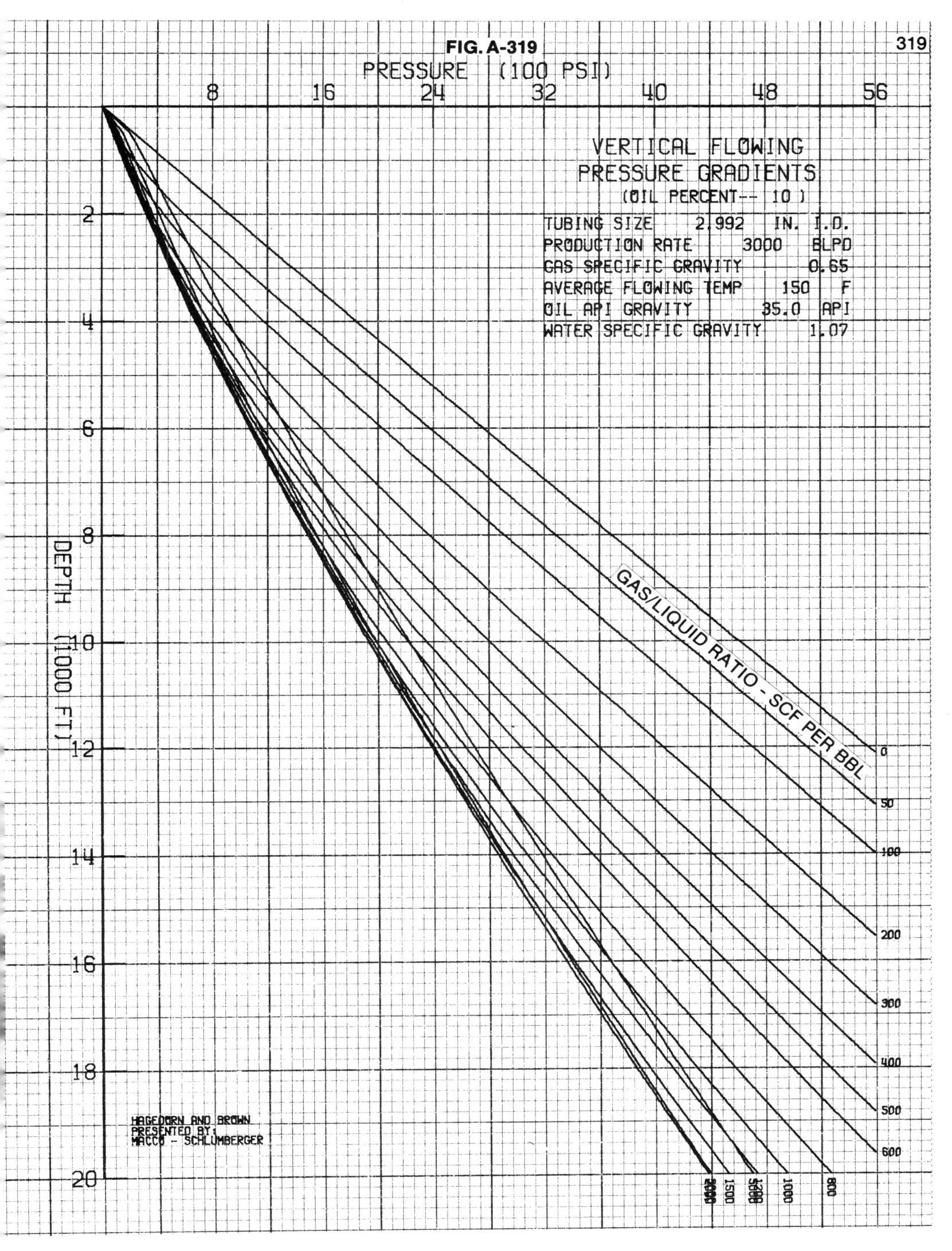

FIG. A-320

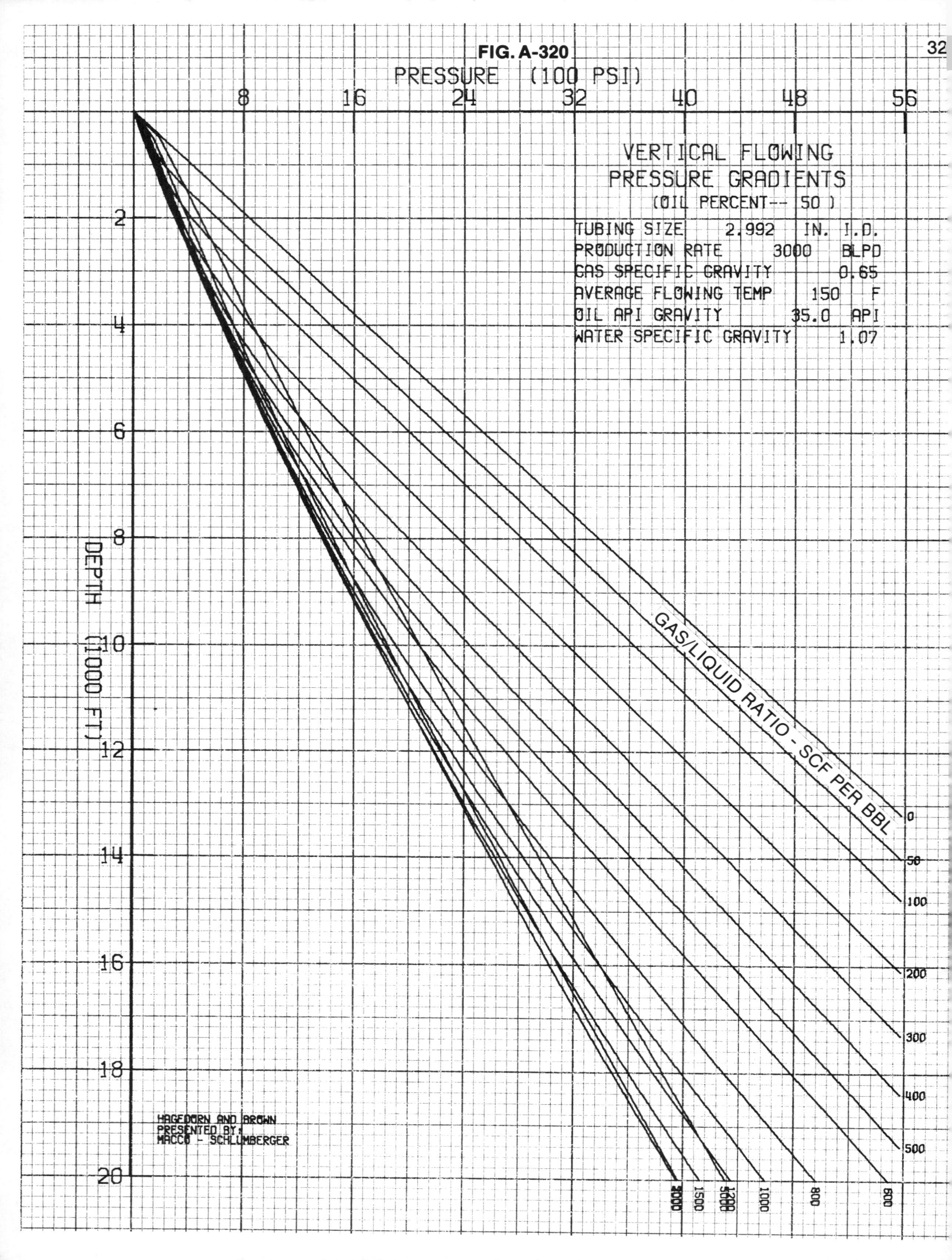

PRESSURE (100 PSI)

VERTICAL FLOWING
PRESSURE GRADIENTS
(OIL PERCENT-- 50)

TUBING SIZE	2.992	IN. I.D.
PRODUCTION RATE	3000	BLPD
GAS SPECIFIC GRAVITY		0.65
AVERAGE FLOWING TEMP	150	F
OIL API GRAVITY	35.0	API
WATER SPECIFIC GRAVITY		1.07

DEPTH (1000 FT)

GAS/LIQUID RATIO - SCF PER BBL

HAGEDORN AND BROWN
PRESENTED BY:
MACCO - SCHLUMBERGER

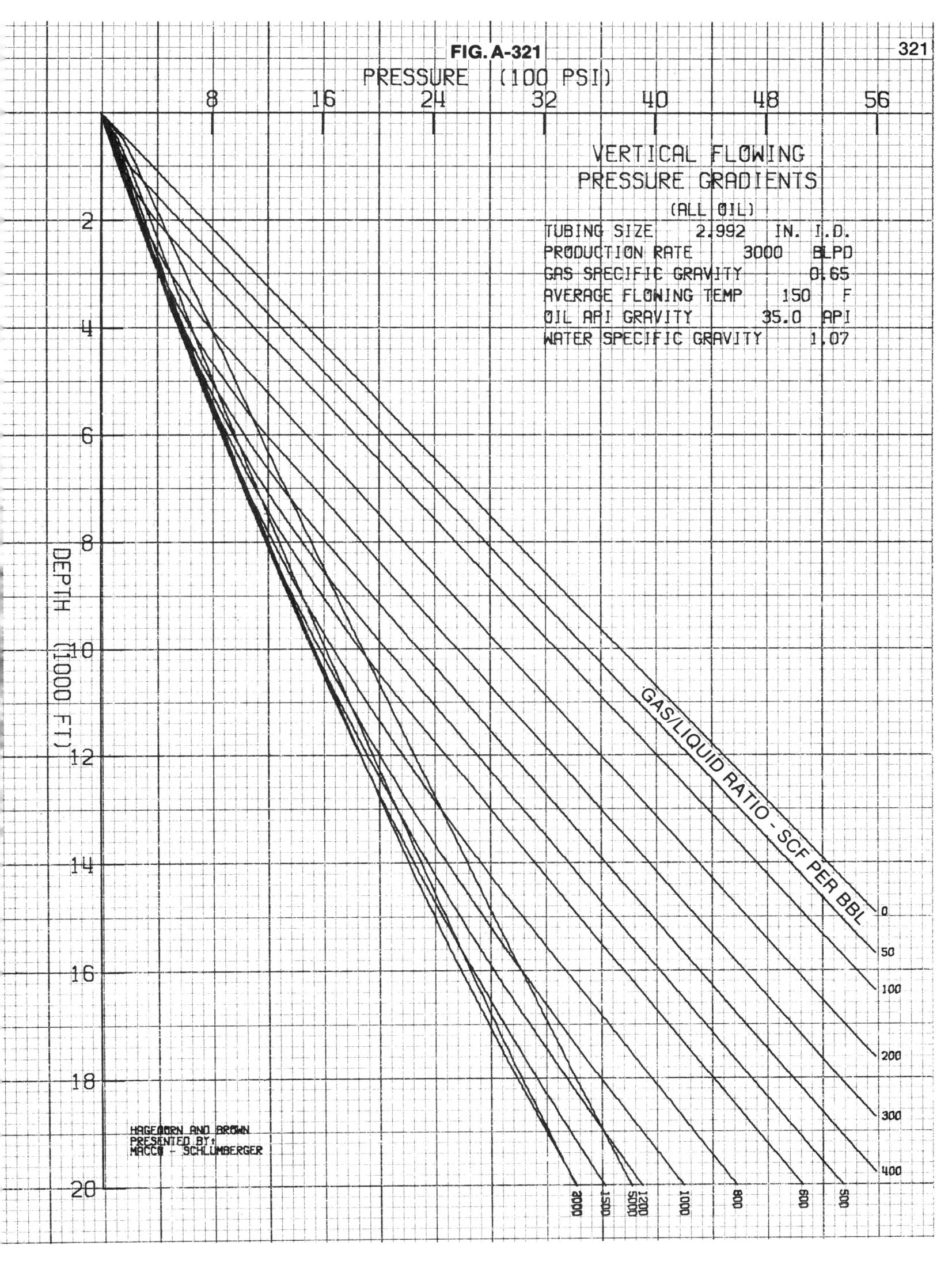

PRESSURE (100 PSI)

VERTICAL FLOWING
PRESSURE GRADIENTS
(ALL OIL)

TUBING SIZE	2.992	IN. I.D.
PRODUCTION RATE	3000	BLPD
GAS SPECIFIC GRAVITY	0.65	
AVERAGE FLOWING TEMP	150	F
OIL API GRAVITY	35.0	API
WATER SPECIFIC GRAVITY	1.07	

DEPTH (1000 FT)

GAS/LIQUID RATIO - SCF PER BBL

HAGEDORN AND BROWN
PRESENTED BY:
MACCO - SCHLUMBERGER

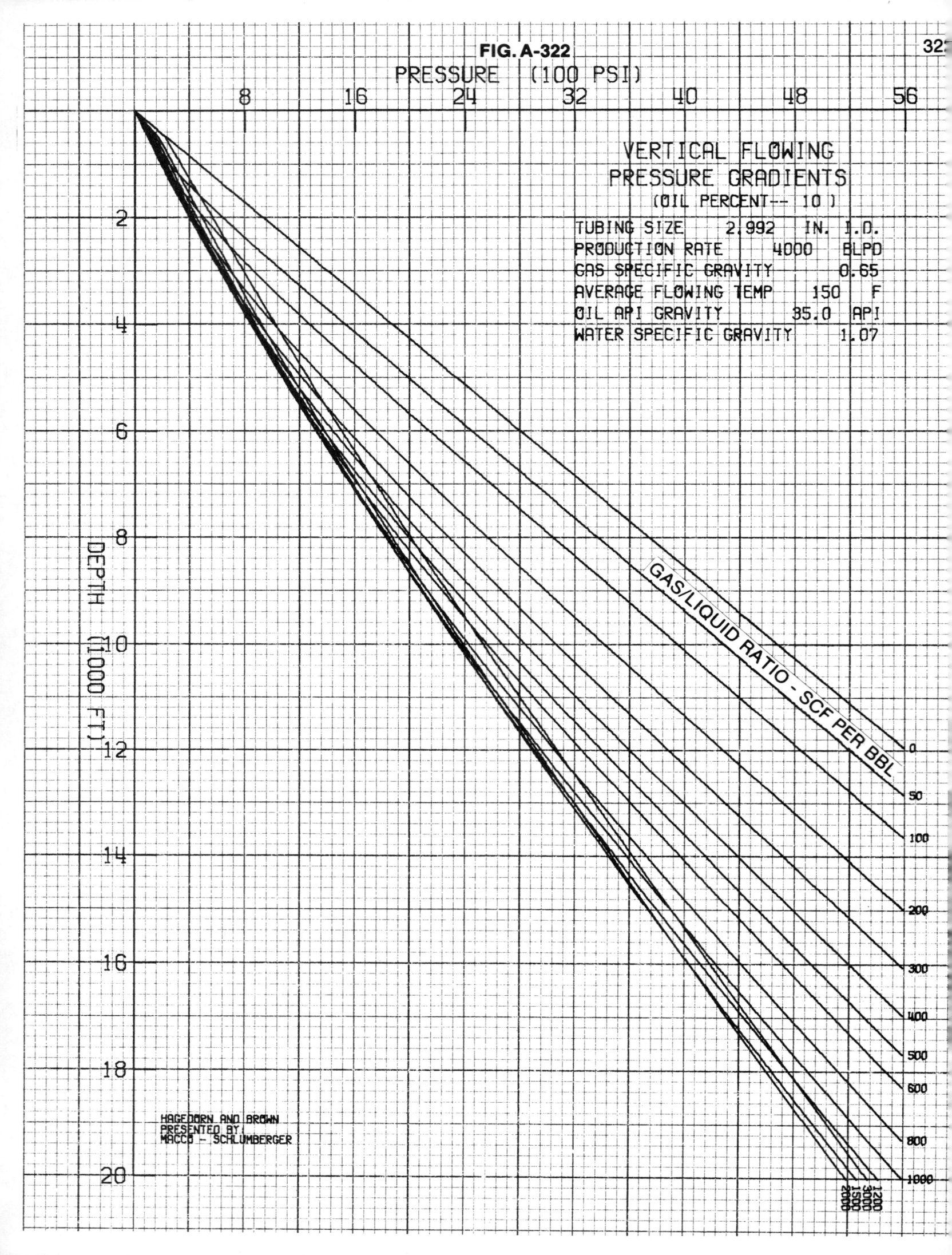

PRESSURE (100 PSI)

DEPTH (1000 FT)

VERTICAL FLOWING
PRESSURE GRADIENTS
(OIL PERCENT-- 10)

TUBING SIZE	2.992 IN. I.D.
PRODUCTION RATE	4000 BLPD
GAS SPECIFIC GRAVITY	0.65
AVERAGE FLOWING TEMP	150 F
OIL API GRAVITY	35.0 API
WATER SPECIFIC GRAVITY	1.07

GAS/LIQUID RATIO - SCF PER BBL

HAGEDORN AND BROWN
PRESENTED BY
MACCO - SCHLUMBERGER

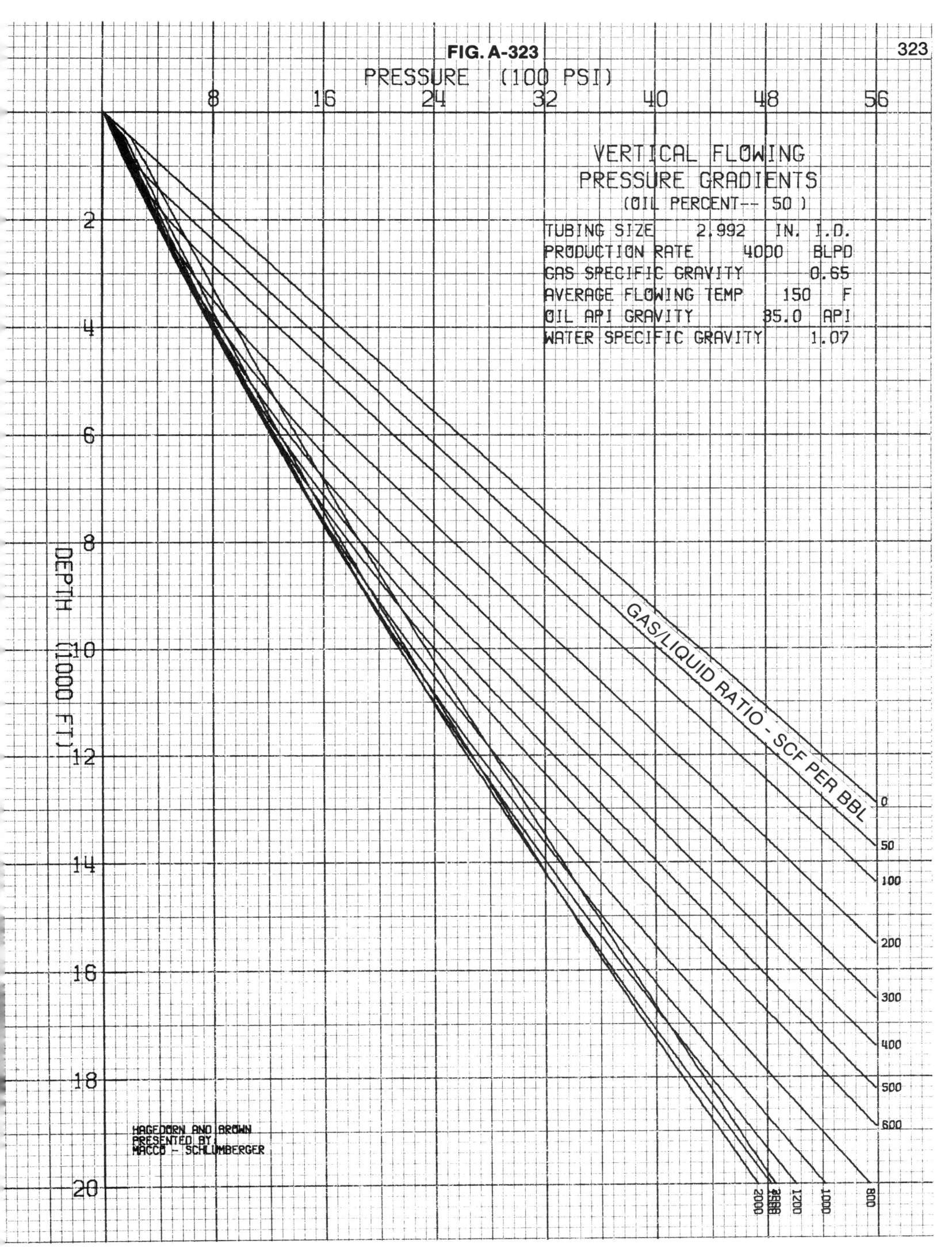

PRESSURE (100 PSI)

VERTICAL FLOWING
PRESSURE GRADIENTS
(OIL PERCENT-- 50)

TUBING SIZE	2.992	IN. I.D.
PRODUCTION RATE	4000	BLPD
GAS SPECIFIC GRAVITY		0.65
AVERAGE FLOWING TEMP	150	F
OIL API GRAVITY	35.0	API
WATER SPECIFIC GRAVITY		1.07

DEPTH (1000 FT)

GAS/LIQUID RATIO - SCF PER BBL

HAGEDORN AND BROWN
PRESENTED BY
MACCO - SCHLUMBERGER

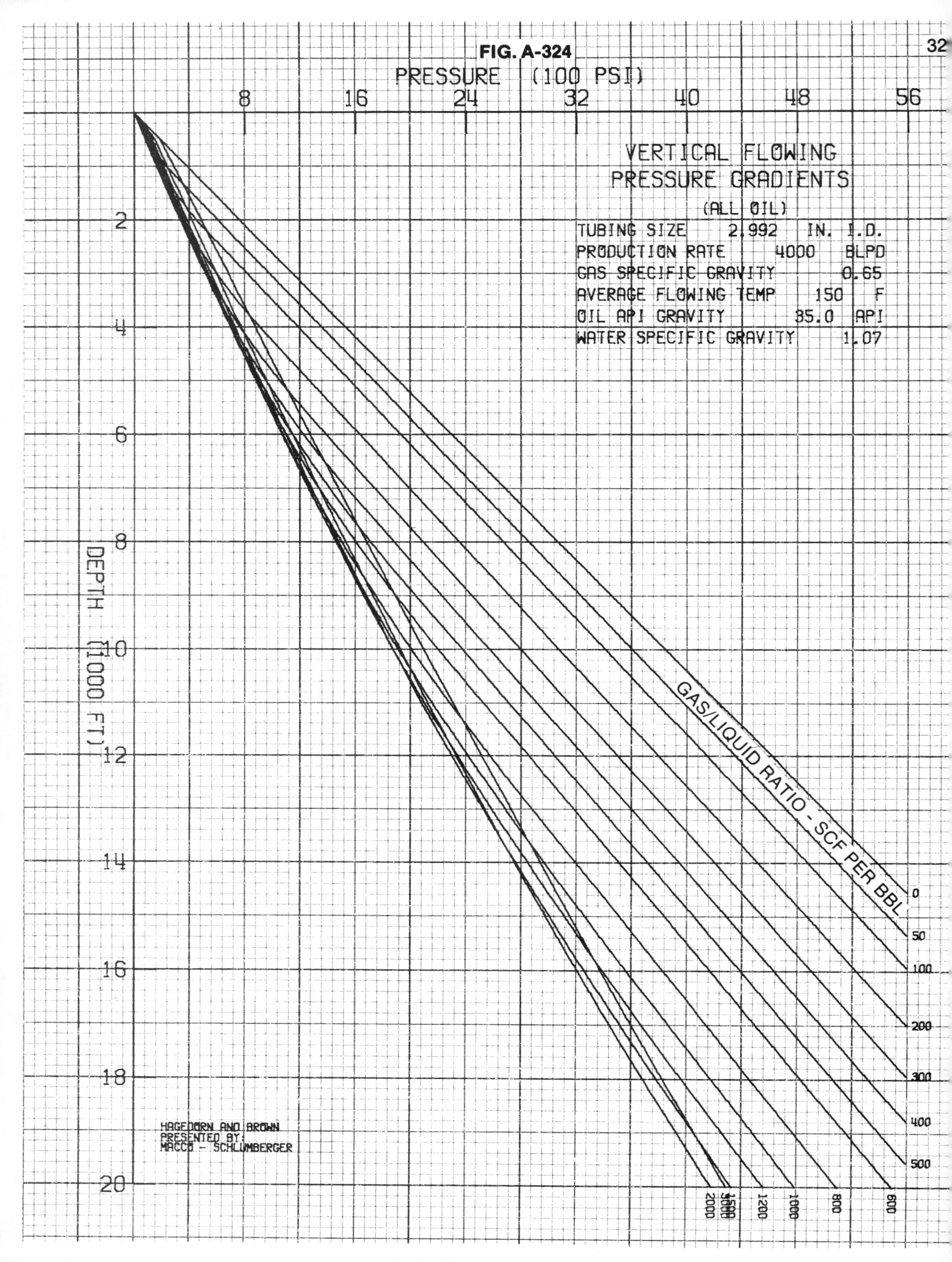

VERTICAL FLOWING
PRESSURE GRADIENTS
(ALL OIL)

TUBING SIZE	2.992	IN. I.D.
PRODUCTION RATE	4000	BLPD
GAS SPECIFIC GRAVITY	0.65	
AVERAGE FLOWING TEMP	150	F
OIL API GRAVITY	35.0	API
WATER SPECIFIC GRAVITY	1.07	

PRESSURE (100 PSI)

DEPTH (1000 FT)

GAS/LIQUID RATIO - SCF PER BBL

HAGEDORN AND BROWN
PRESENTED BY:
MACCO - SCHLUMBERGER

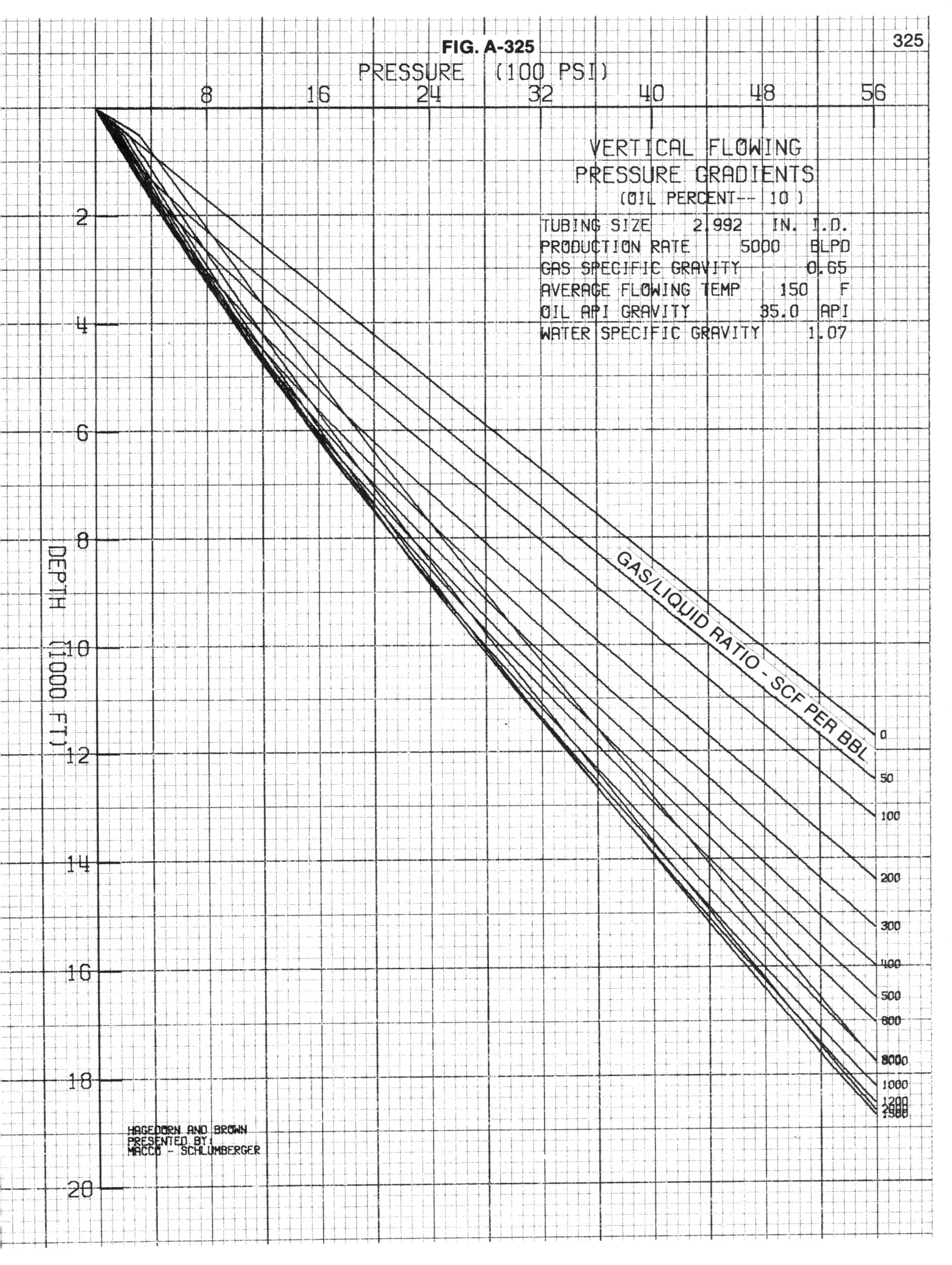

FIG. A-325

325

PRESSURE (100 PSI)

VERTICAL FLOWING
PRESSURE GRADIENTS
(OIL PERCENT-- 10)

TUBING SIZE   2.992   IN. I.D.
PRODUCTION RATE   5000   BLPD
GAS SPECIFIC GRAVITY   0.65
AVERAGE FLOWING TEMP   150   F
OIL API GRAVITY   35.0   API
WATER SPECIFIC GRAVITY   1.07

GAS/LIQUID RATIO - SCF PER BBL

DEPTH (1000 FT)

HAGEDORN AND BROWN
PRESENTED BY:
MACCO - SCHLUMBERGER

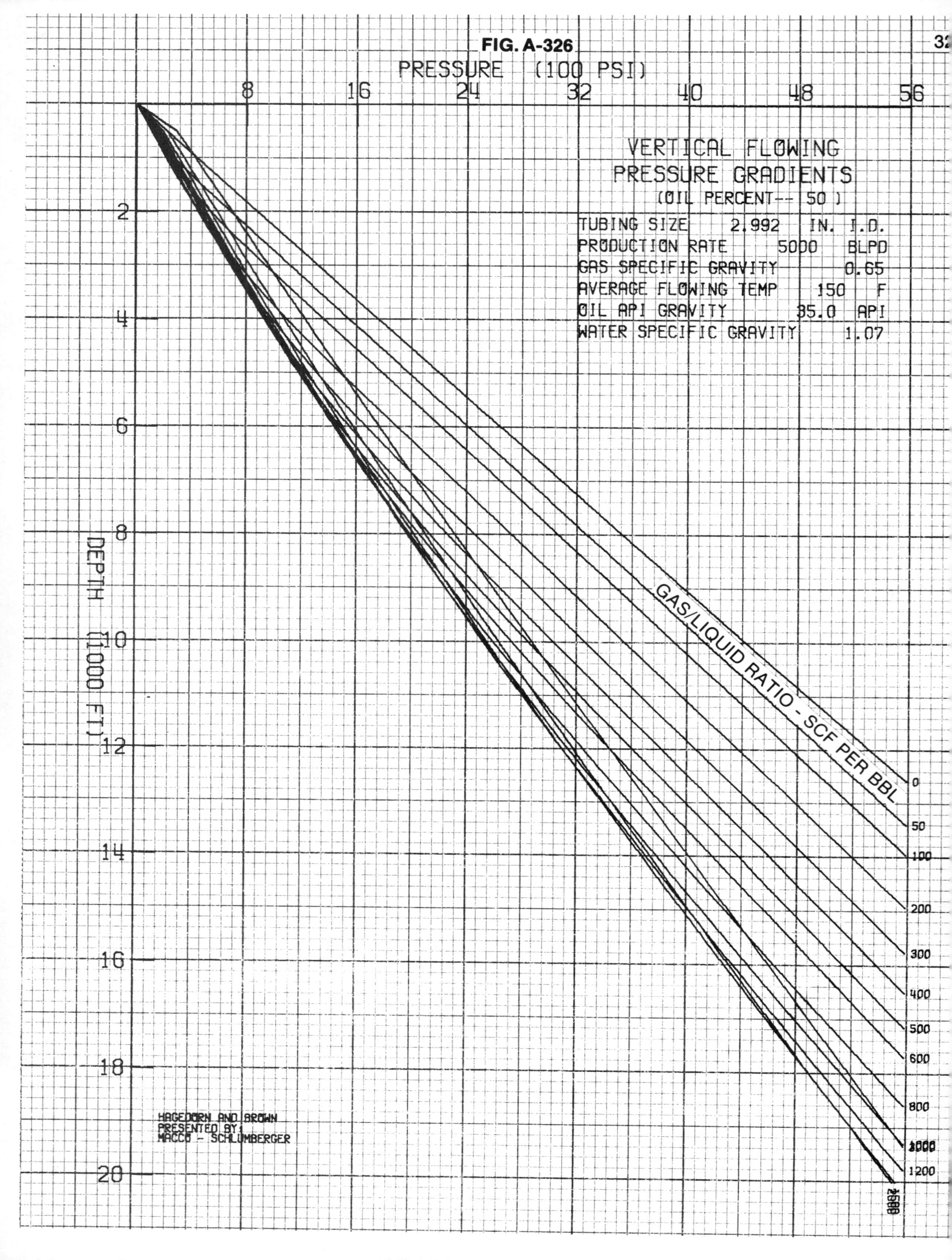

FIG. A-326

VERTICAL FLOWING
PRESSURE GRADIENTS
(OIL PERCENT-- 50)

TUBING SIZE	2.992	IN. I.D.
PRODUCTION RATE	5000	BLPD
GAS SPECIFIC GRAVITY	0.65	
AVERAGE FLOWING TEMP	150	F
OIL API GRAVITY	35.0	API
WATER SPECIFIC GRAVITY	1.07	

GAS/LIQUID RATIO - SCF PER BBL

PRESSURE (100 PSI)

DEPTH (1000 FT)

HAGEDORN AND BROWN
PRESENTED BY:
MACCO - SCHLUMBERGER

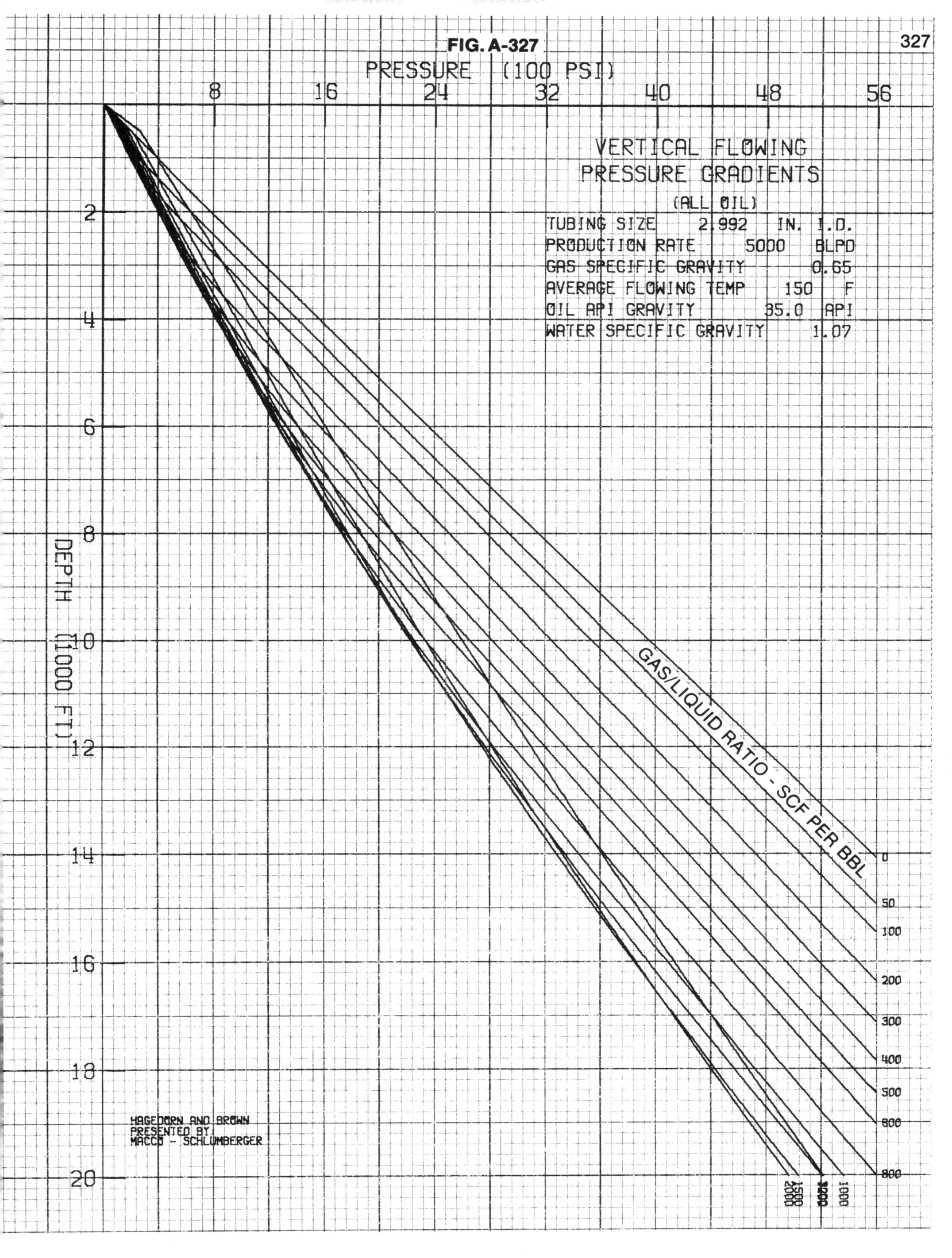

PRESSURE (100 PSI)

VERTICAL FLOWING
PRESSURE GRADIENTS
(ALL OIL)

TUBING SIZE	2.992	IN. I.D.
PRODUCTION RATE	5000	BLPD
GAS SPECIFIC GRAVITY	0.65	
AVERAGE FLOWING TEMP	150	F
OIL API GRAVITY	35.0	API
WATER SPECIFIC GRAVITY	1.07	

DEPTH (1000 FT)

GAS/LIQUID RATIO - SCF PER BBL

HAGEDORN AND BROWN
PRESENTED BY
MACCO - SCHLUMBERGER

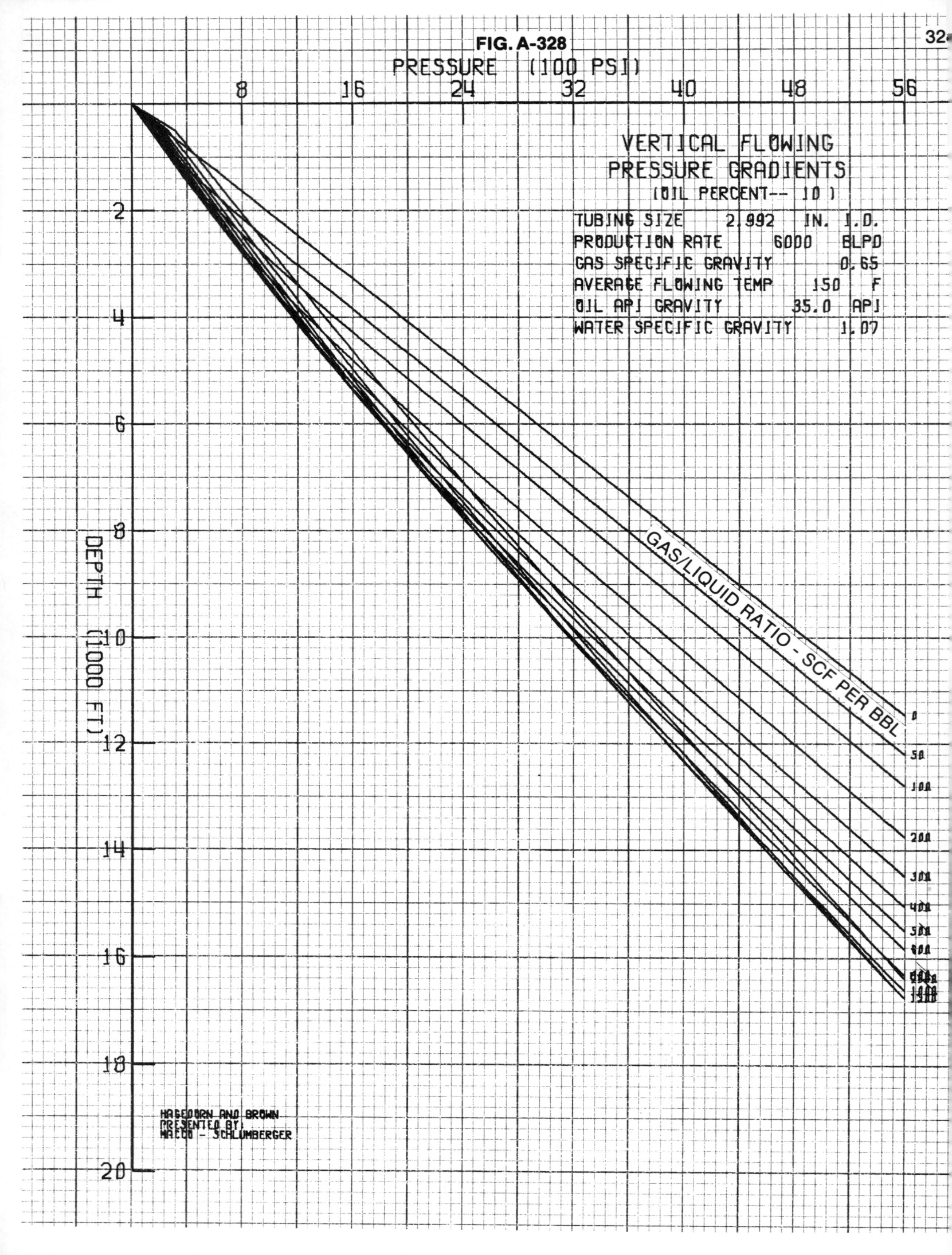

PRESSURE (100 PSI)

VERTICAL FLOWING
PRESSURE GRADIENTS
(OIL PERCENT-- 10)

TUBING SIZE   2.992   IN. I.D.
PRODUCTION RATE   6000   BLPD
GAS SPECIFIC GRAVITY   0.65
AVERAGE FLOWING TEMP   150   F
OIL API GRAVITY   35.0 API
WATER SPECIFIC GRAVITY   1.07

GAS/LIQUID RATIO - SCF PER BBL

DEPTH (1000 FT)

HAGEDORN AND BROWN
PRESENTED BY,
HAECO - SCHLUMBERGER

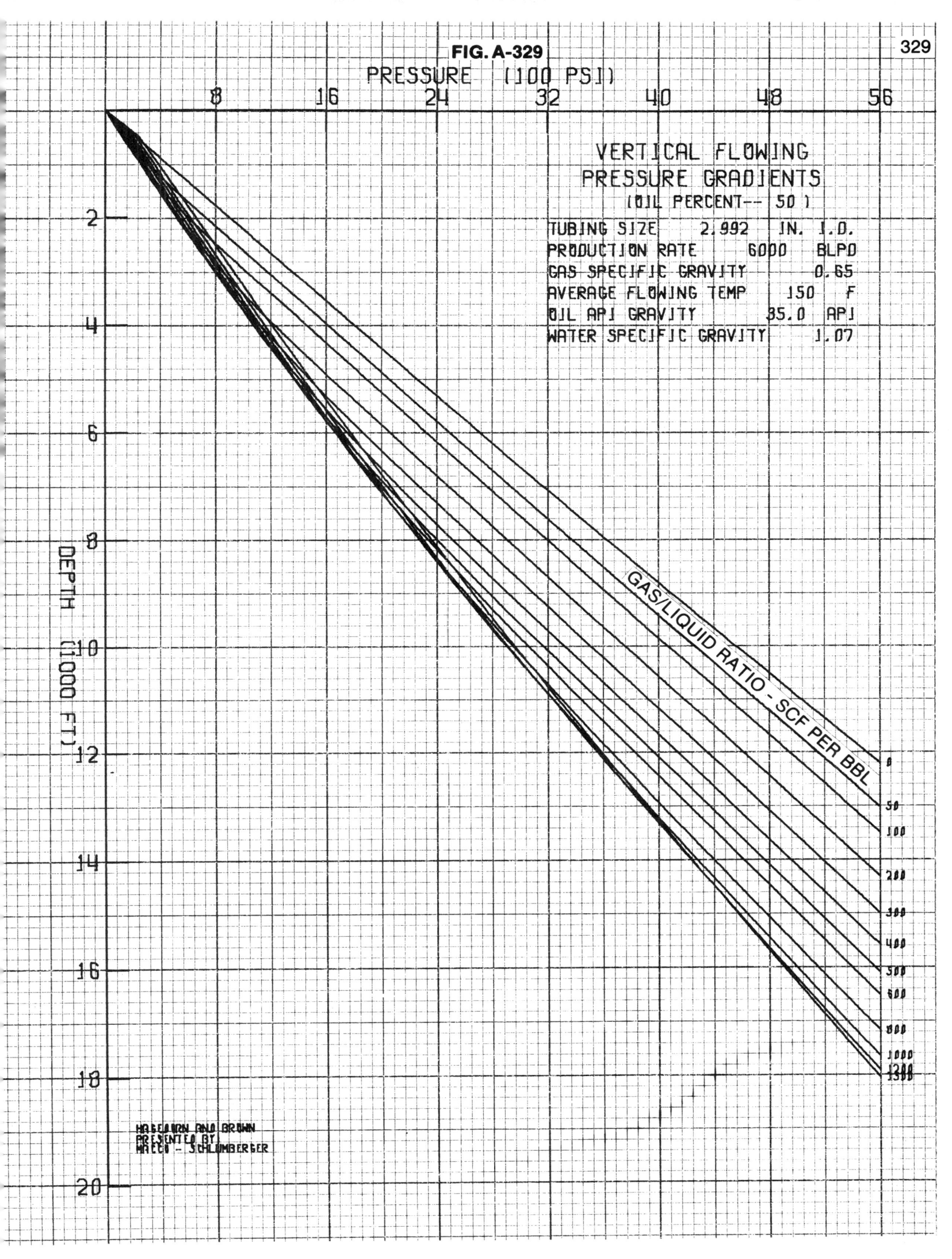

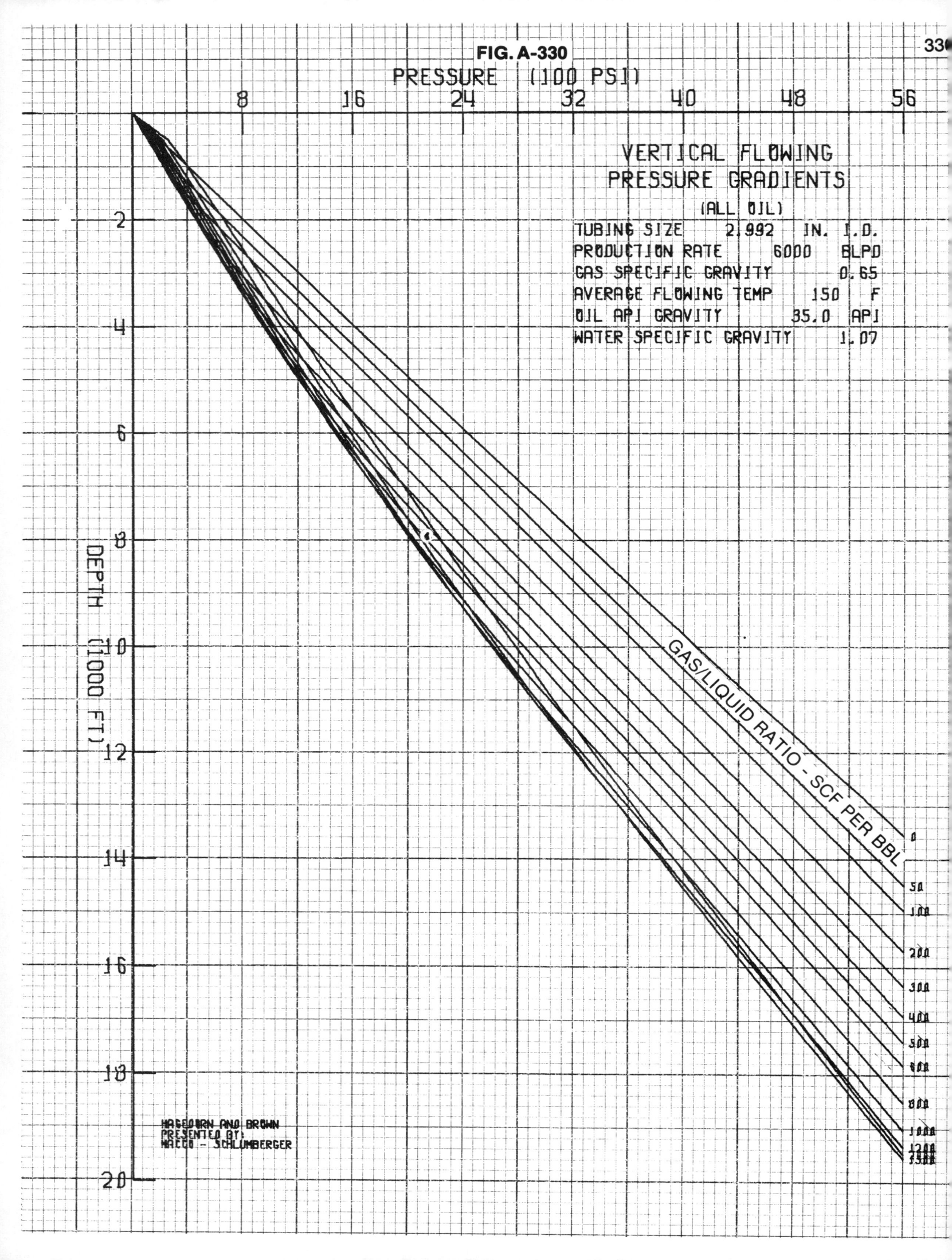

PRESSURE (100 PSI)

VERTICAL FLOWING
PRESSURE GRADIENTS
(ALL OIL)

TUBING SIZE	2.992	IN. I.D.
PRODUCTION RATE	6000	BLPD
GAS SPECIFIC GRAVITY		0.65
AVERAGE FLOWING TEMP	150	F
OIL API GRAVITY	35.0	API
WATER SPECIFIC GRAVITY		1.07

DEPTH (1000 FT)

GAS/LIQUID RATIO - SCF PER BBL

0
50
100
200
300
400
500
600
800
1000
1200
1500

HAGEDORN AND BROWN
PRESENTED BY:
NACCO - SCHLUMBERGER

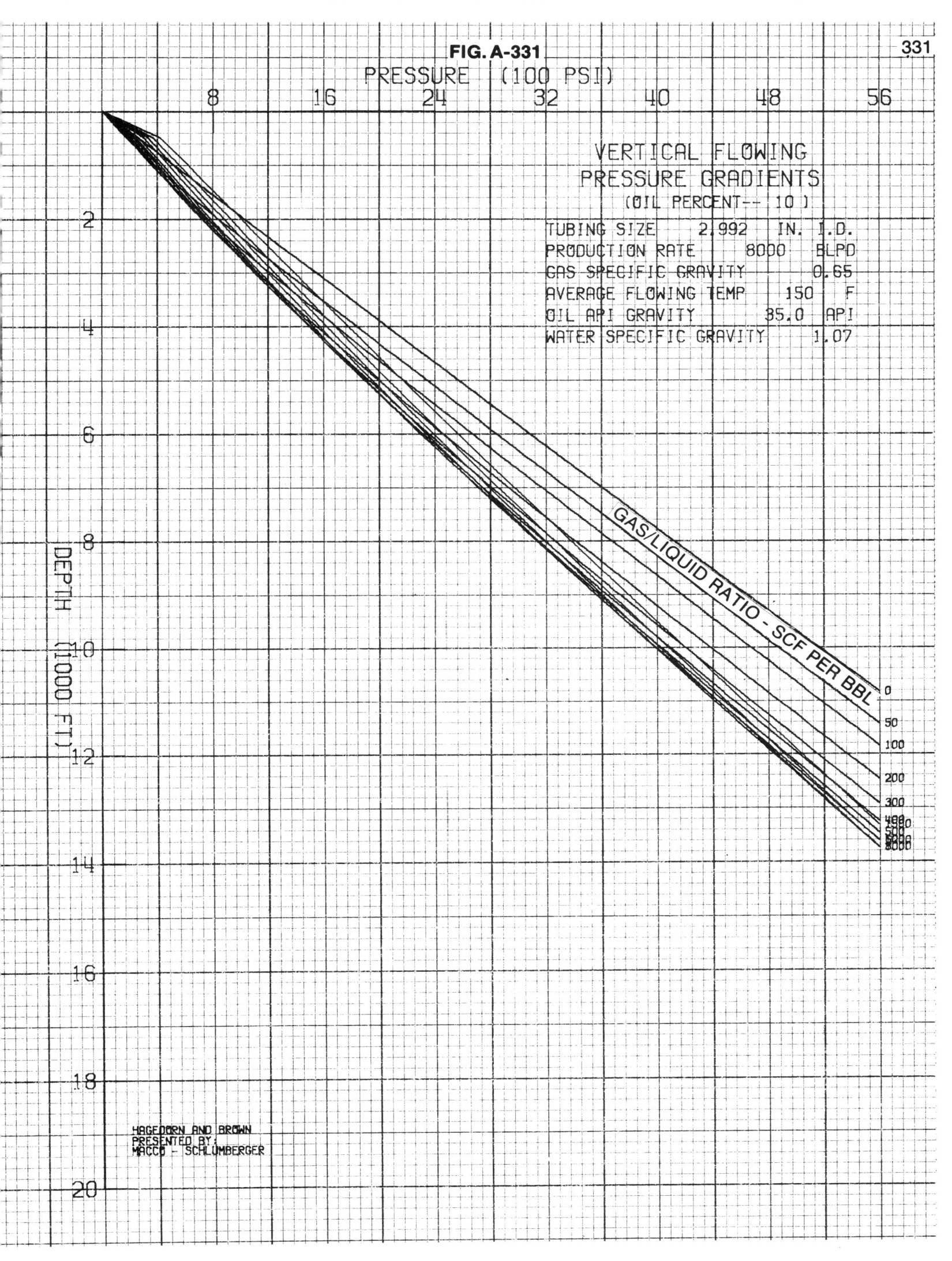

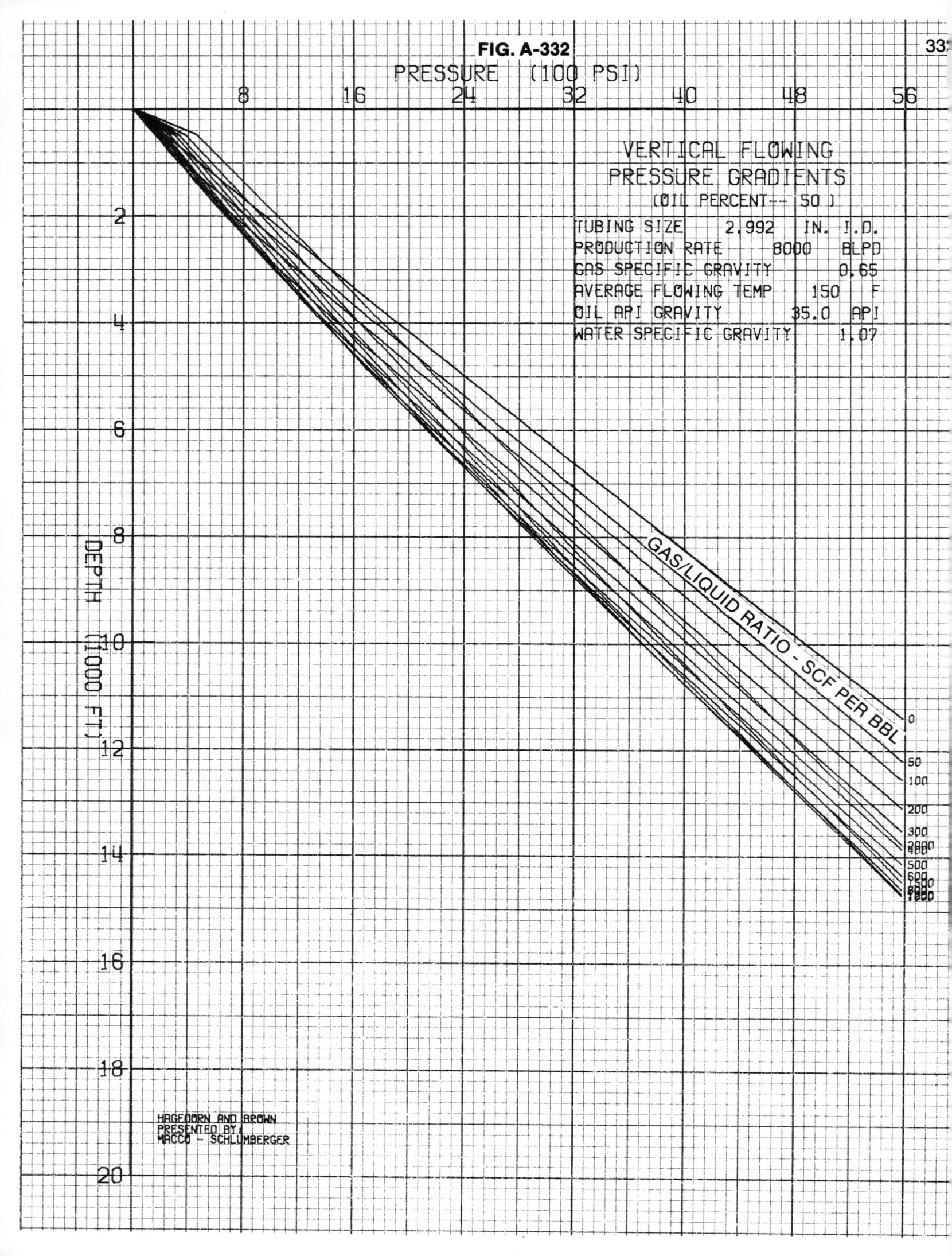

FIG. A-332

33?

VERTICAL FLOWING
PRESSURE GRADIENTS
(OIL PERCENT-- 50)

TUBING SIZE	2.992	IN. I.D.
PRODUCTION RATE	8000	BLPD
GAS SPECIFIC GRAVITY	0.65	
AVERAGE FLOWING TEMP	150	F
OIL API GRAVITY	35.0	API
WATER SPECIFIC GRAVITY	1.07	

PRESSURE (100 PSI)

DEPTH (1000 FT)

GAS/LIQUID RATIO - SCF PER BBL

0
50
100
200
300
400
500
600
800
1000

HAGEDORN AND BROWN
PRESENTED BY:
MACCO - SCHLUMBERGER

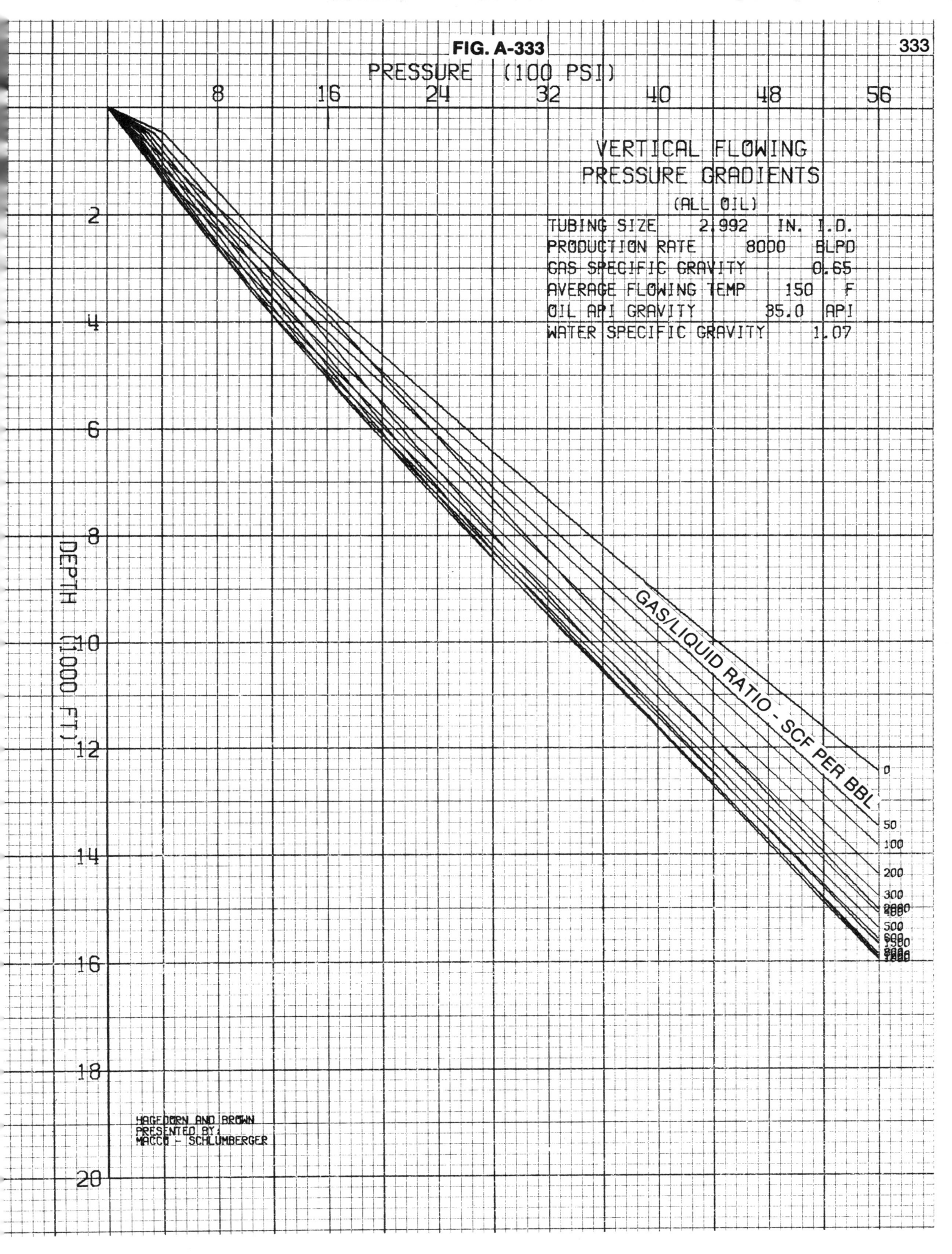

FIG. A-333

333

PRESSURE (100 PSI)

VERTICAL FLOWING
PRESSURE GRADIENTS
(ALL OIL)

TUBING SIZE	2.992	IN. I.D.
PRODUCTION RATE	8000	BLPD
GAS SPECIFIC GRAVITY	0.65	
AVERAGE FLOWING TEMP	150	F
OIL API GRAVITY	35.0	API
WATER SPECIFIC GRAVITY	1.07	

DEPTH (1000 FT)

GAS/LIQUID RATIO - SCF PER BBL

0
50
100
200
300

500
600

HAGEDORN AND BROWN
PRESENTED BY
MACCO - SCHLUMBERGER

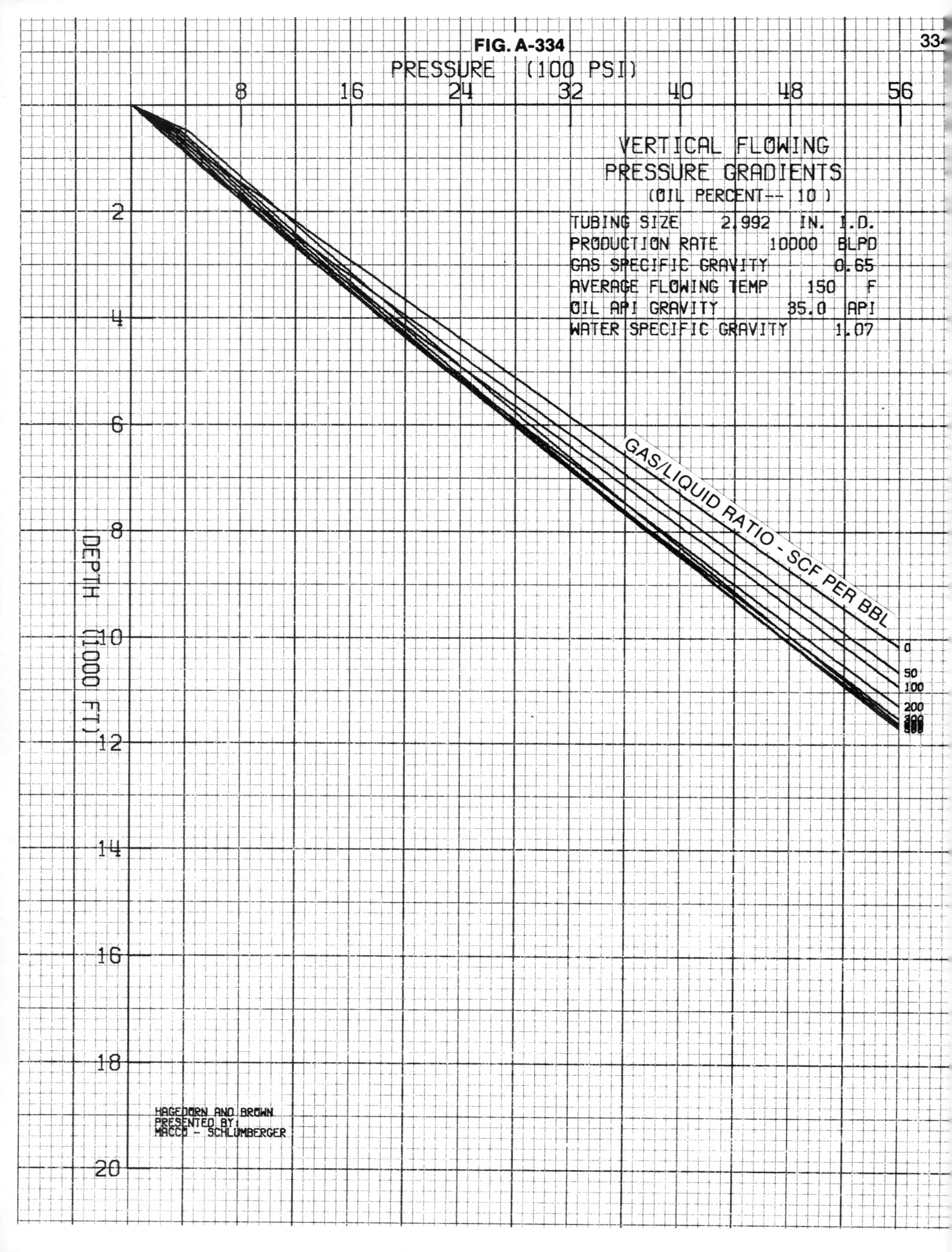

VERTICAL FLOWING
PRESSURE GRADIENTS
(OIL PERCENT-- 10)

TUBING SIZE	2.992	IN. I.D.
PRODUCTION RATE	10000	BLPD
GAS SPECIFIC GRAVITY		0.65
AVERAGE FLOWING TEMP	150	F
OIL API GRAVITY	35.0	API
WATER SPECIFIC GRAVITY		1.07

PRESSURE (100 PSI)

DEPTH (1000 FT)

GAS/LIQUID RATIO - SCF PER BBL

0
50
100
200

HAGEDORN AND BROWN
PRESENTED BY:
MACCO - SCHLUMBERGER

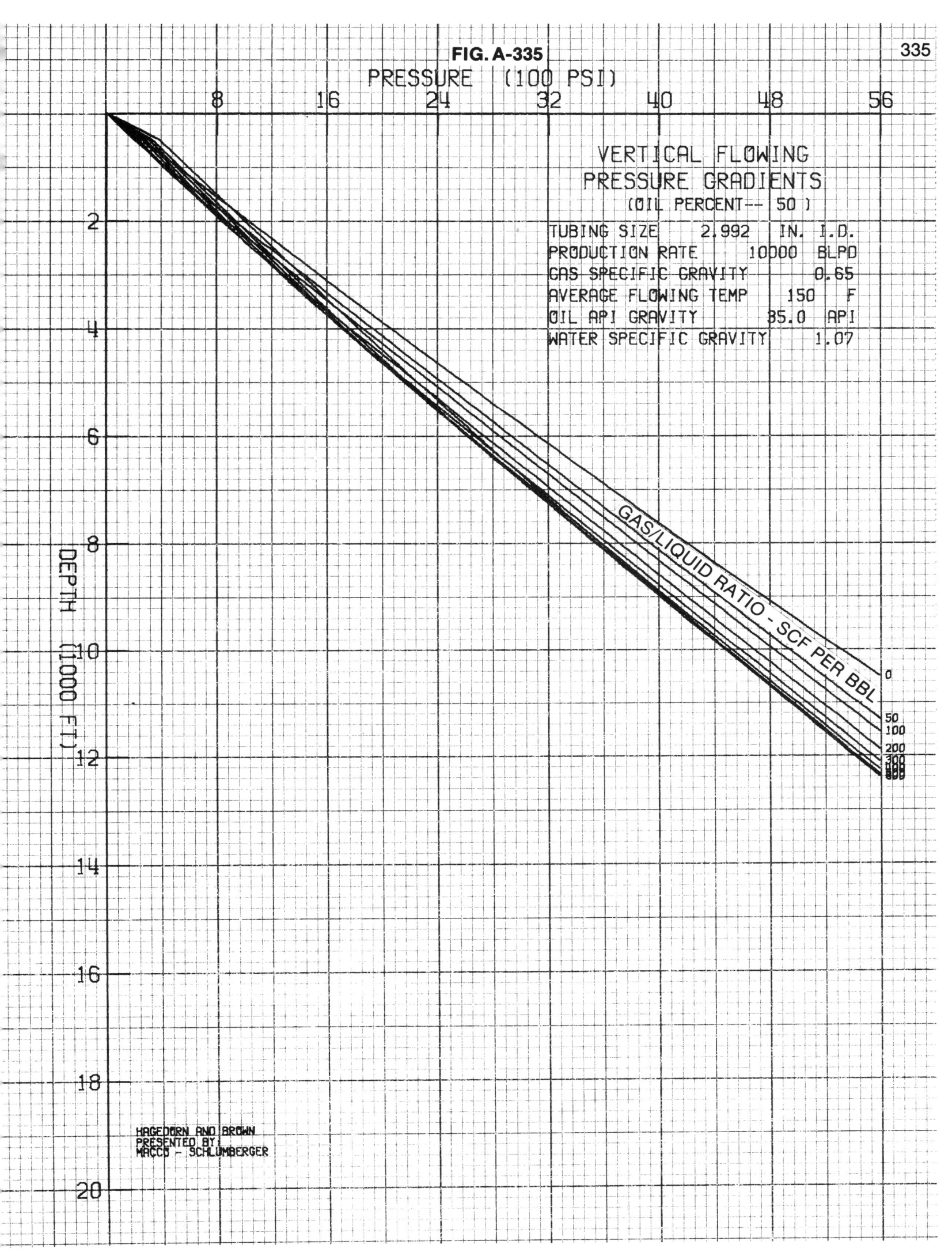

FIG. A-335

PRESSURE (100 PSI)

VERTICAL FLOWING
PRESSURE GRADIENTS
(OIL PERCENT-- 50)

TUBING SIZE	2.992	IN. I.D.
PRODUCTION RATE	10000	BLPD
GAS SPECIFIC GRAVITY	0.65	
AVERAGE FLOWING TEMP	150	F
OIL API GRAVITY	35.0	API
WATER SPECIFIC GRAVITY	1.07	

DEPTH (1000 FT)

GAS/LIQUID RATIO - SCF PER BBL

0
50
100
200
300

HAGEDORN AND BROWN
PRESENTED BY
MACCO - SCHLUMBERGER

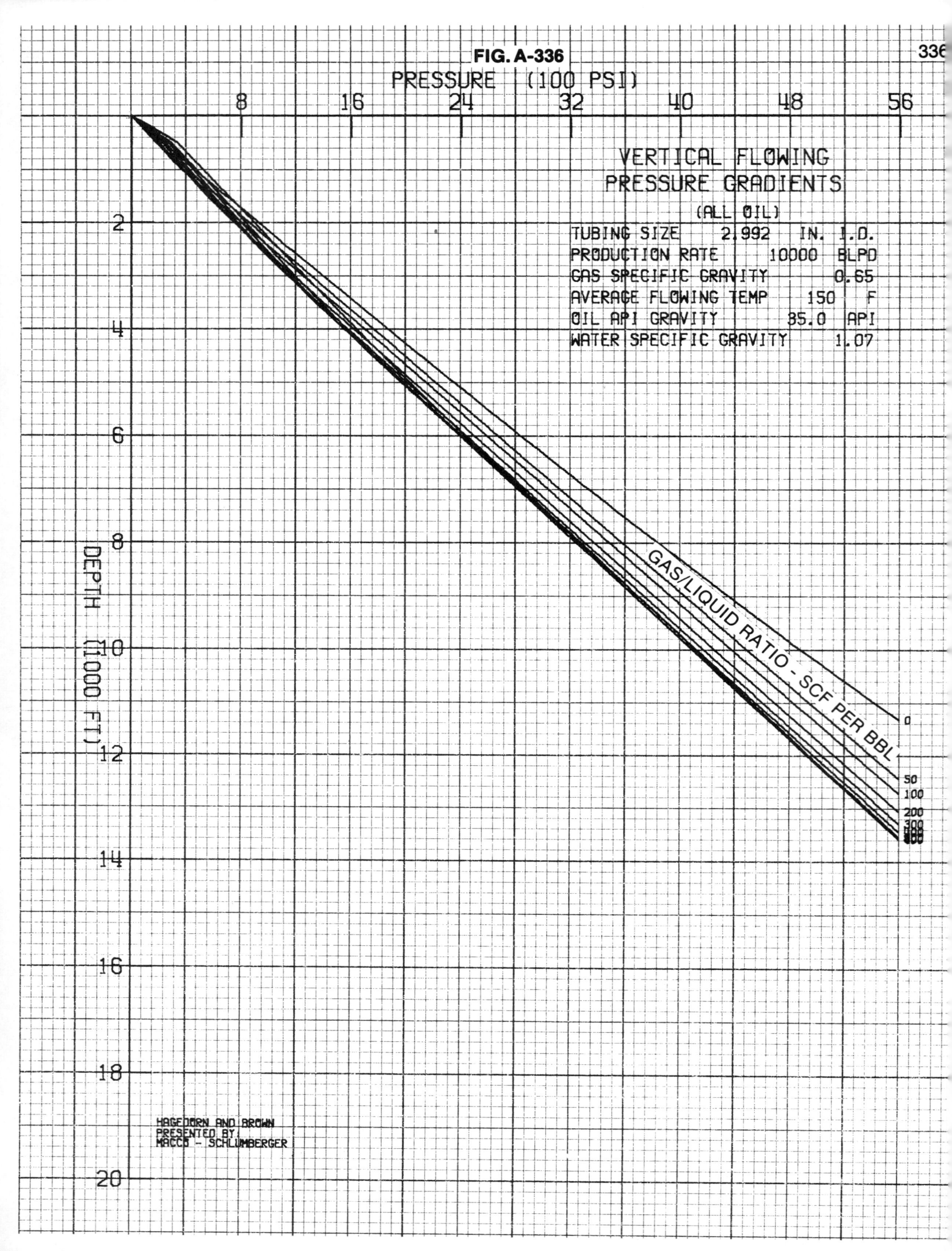

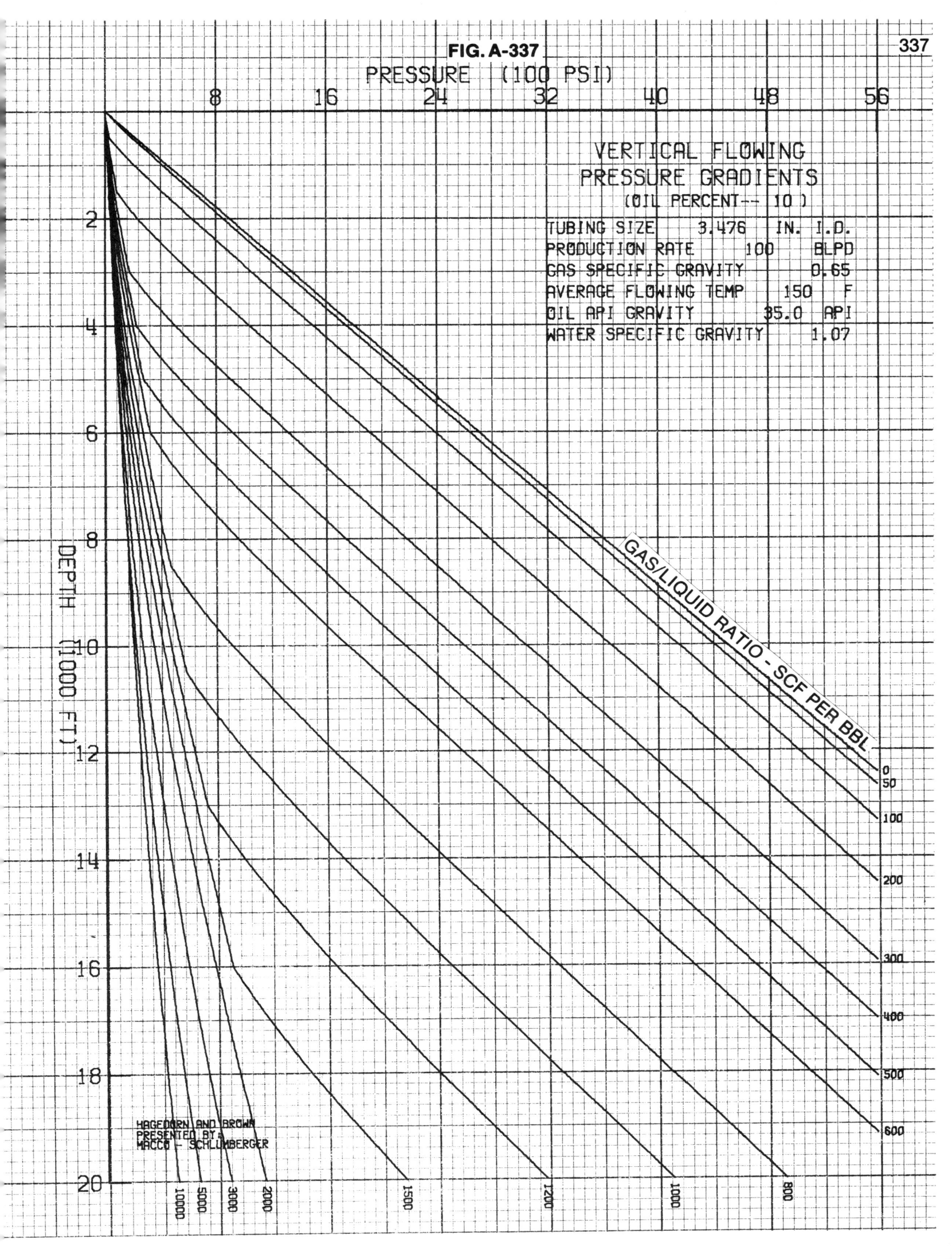

PRESSURE (100 PSI)

DEPTH (1000 FT)

VERTICAL FLOWING
PRESSURE GRADIENTS
(OIL PERCENT-- 10)

TUBING SIZE   3.476   IN. I.D.
PRODUCTION RATE   100   BLPD
GAS SPECIFIC GRAVITY   0.65
AVERAGE FLOWING TEMP   150   F
OIL API GRAVITY   35.0 API
WATER SPECIFIC GRAVITY   1.07

GAS/LIQUID RATIO - SCF PER BBL

0
50
100
200
300
400
500
600

HAGEDORN AND BROWN
PRESENTED BY
MACCO - SCHLUMBERGER

10000
5000
3000
2000
1500
1200
1000
800

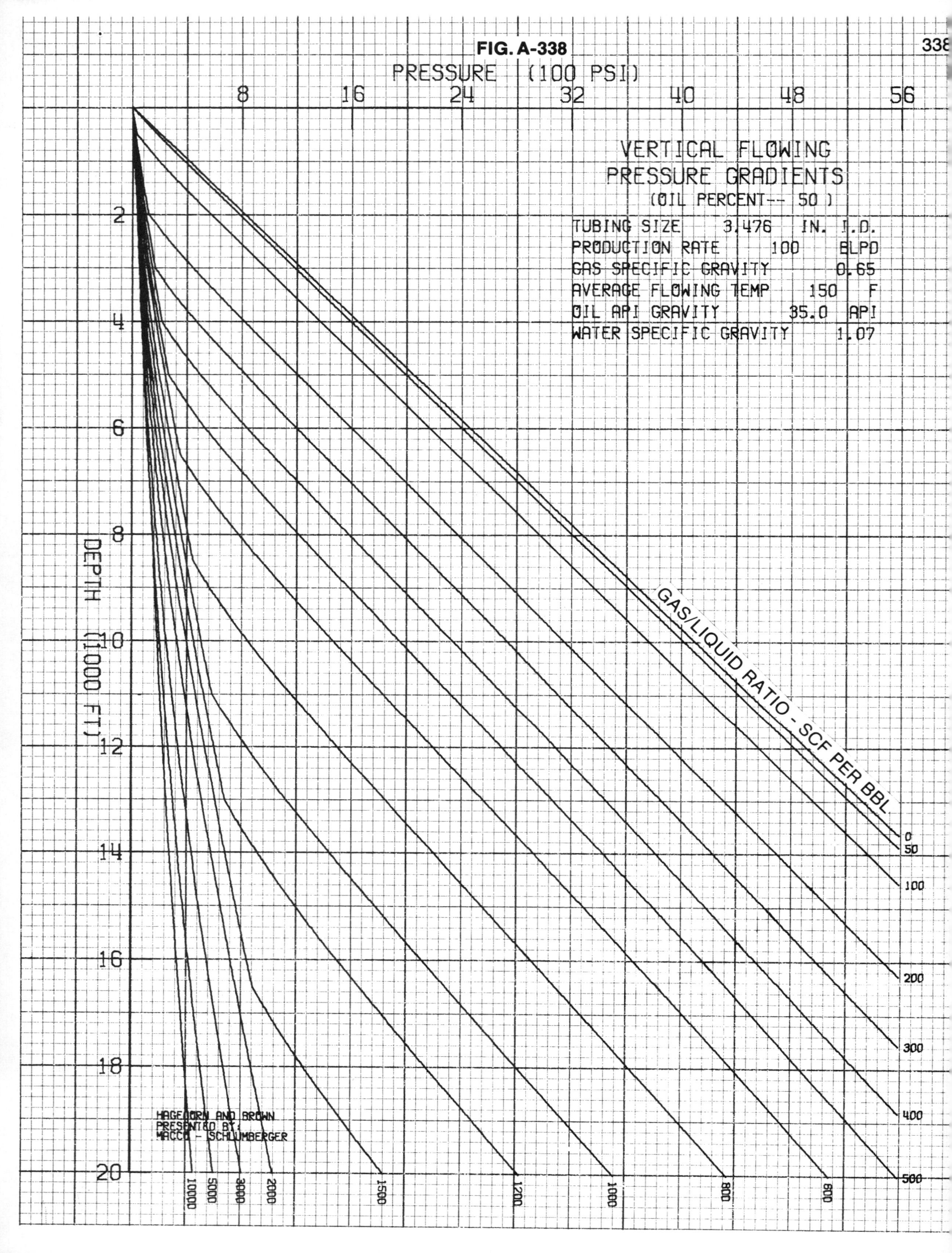

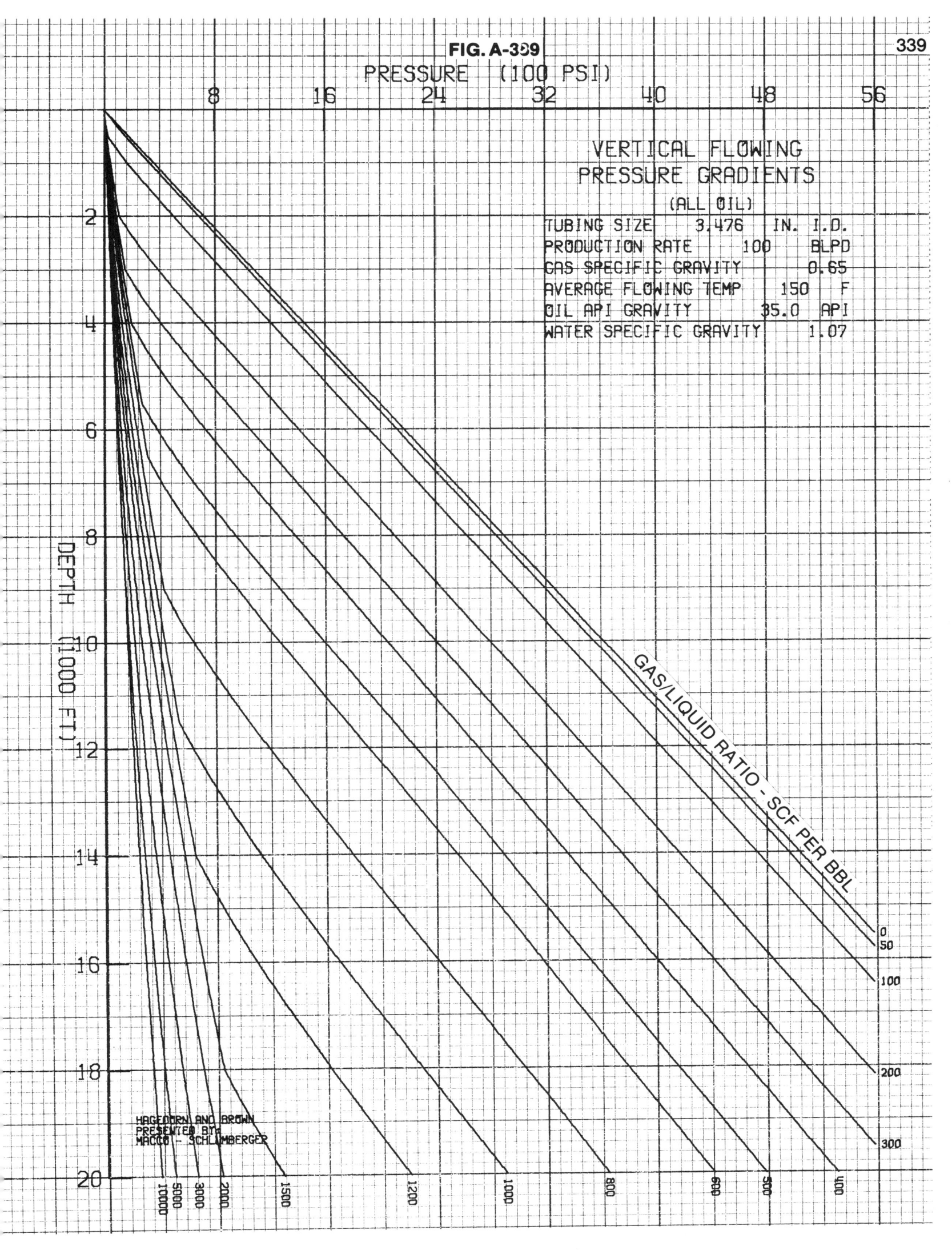

PRESSURE (100 PSI)

VERTICAL FLOWING
PRESSURE GRADIENTS
(ALL OIL)

TUBING SIZE	3.476	IN. I.D.
PRODUCTION RATE	100	BLPD
GAS SPECIFIC GRAVITY	0.65	
AVERAGE FLOWING TEMP	150	F
OIL API GRAVITY	35.0	API
WATER SPECIFIC GRAVITY	1.07	

DEPTH (1000 FT)

GAS/LIQUID RATIO - SCF PER BBL

HAGEDORN AND BROWN
PRESENTED BY:
MACCO - SCHLUMBERGER

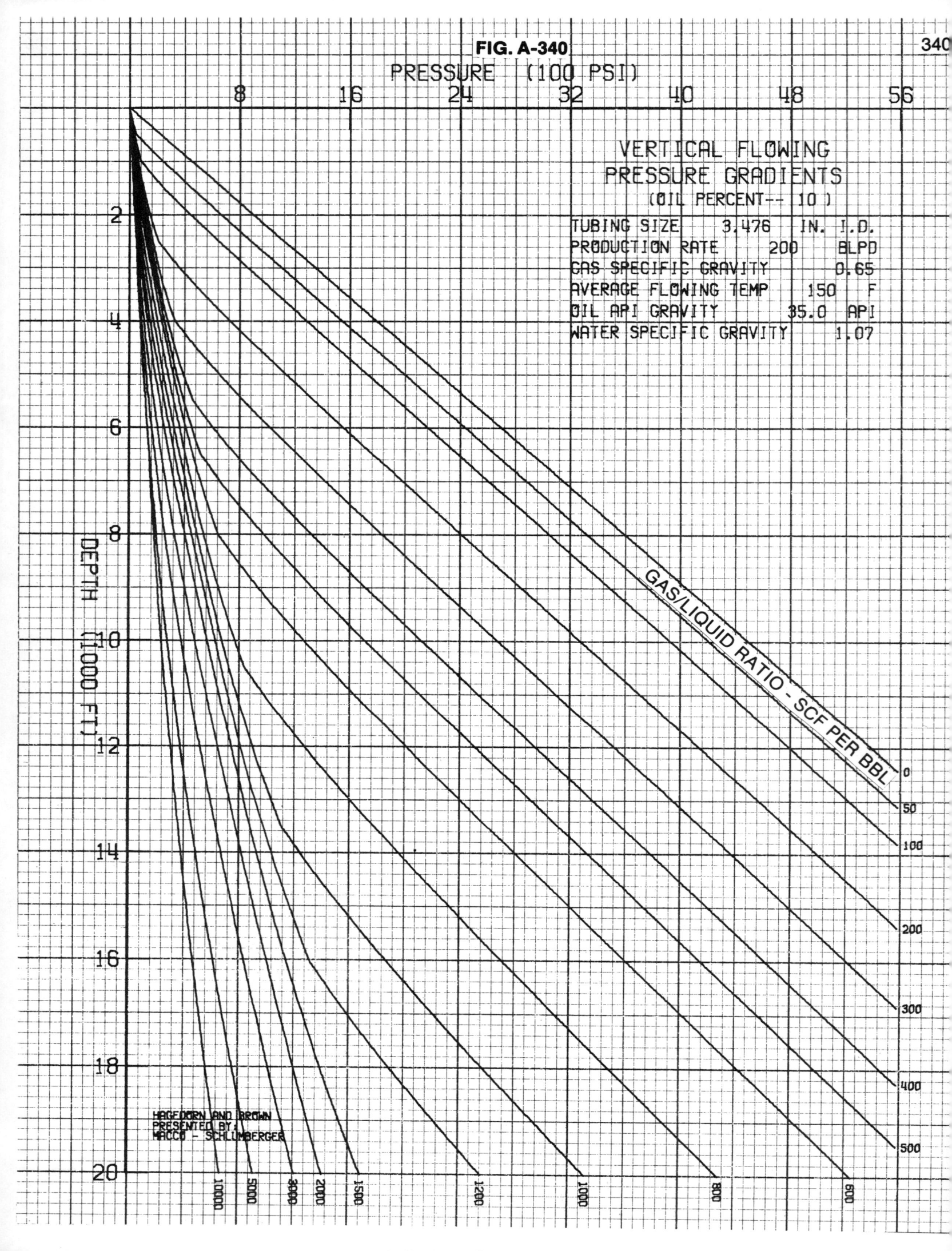

FIG. A-340

340

VERTICAL FLOWING
PRESSURE GRADIENTS
(OIL PERCENT-- 10)

TUBING SIZE	3.476	IN. I.D.
PRODUCTION RATE	200	BLPD
GAS SPECIFIC GRAVITY		0.65
AVERAGE FLOWING TEMP	150	F
OIL API GRAVITY	35.0	API
WATER SPECIFIC GRAVITY		1.07

PRESSURE (100 PSI)

DEPTH (1000 FT)

GAS/LIQUID RATIO - SCF PER BBL

HAGEDORN AND BROWN
PRESENTED BY
MACCO -- SCHLUMBERGER

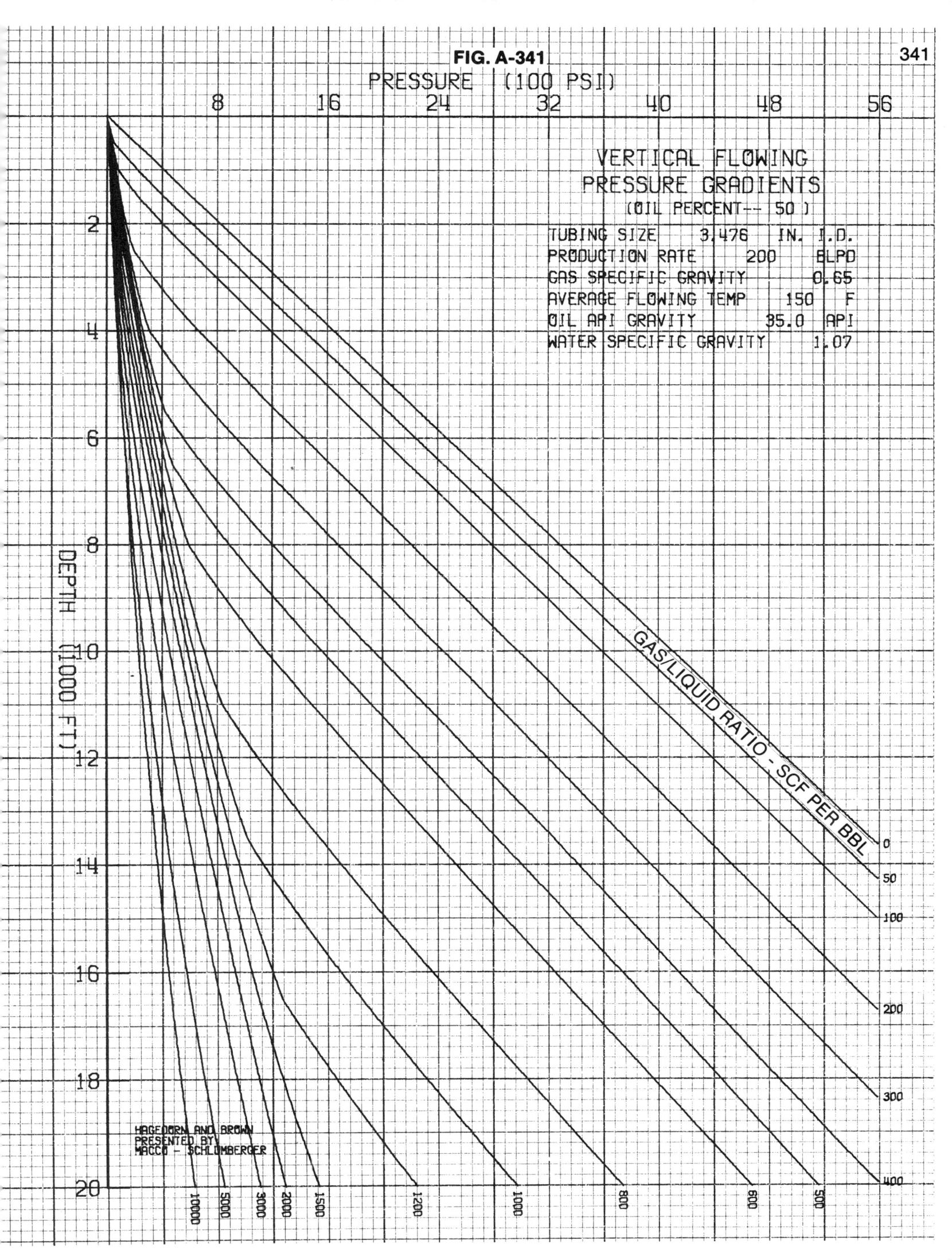

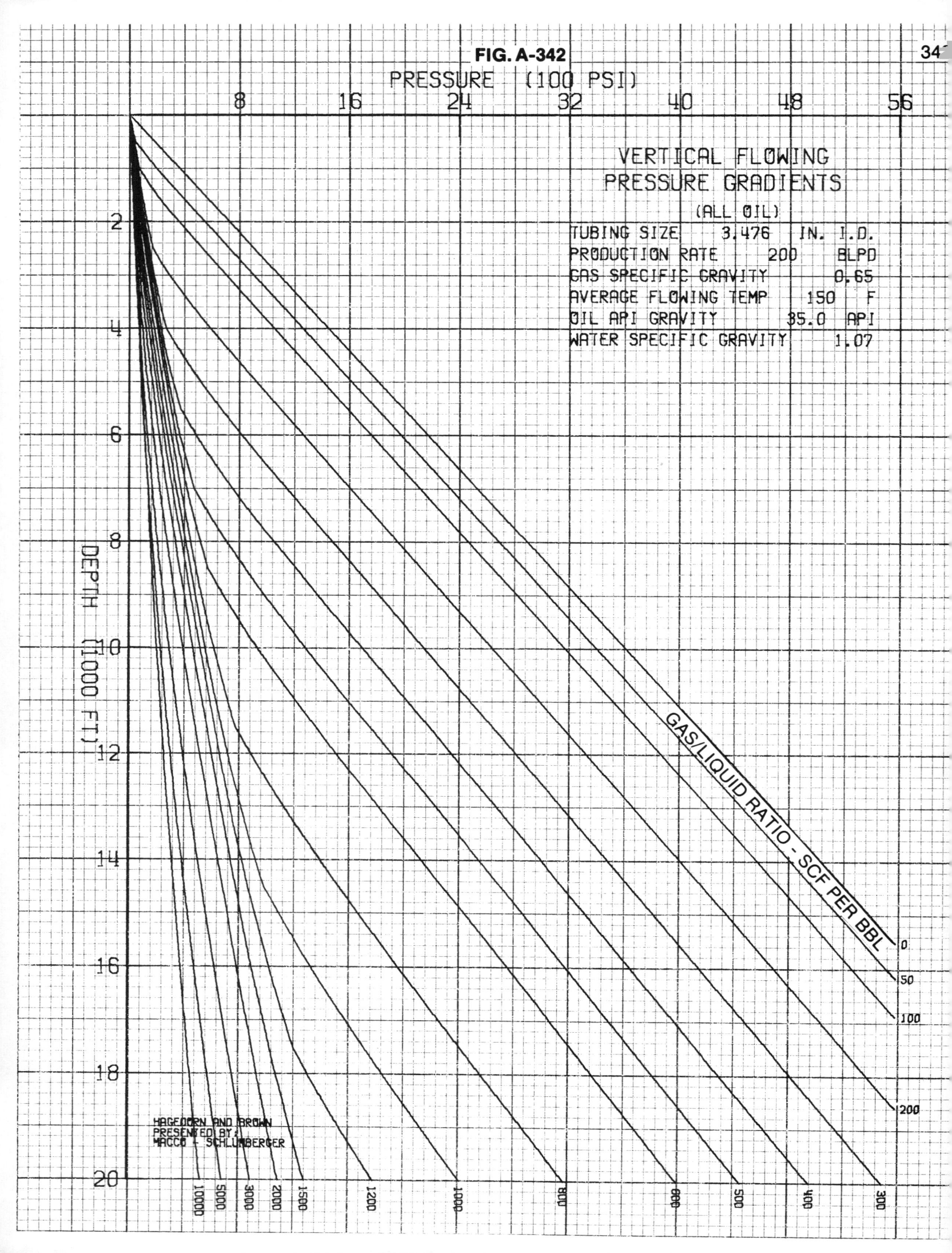

FIG. A-342

VERTICAL FLOWING
PRESSURE GRADIENTS
(ALL OIL)

TUBING SIZE	3.476	IN. I.D.
PRODUCTION RATE	200	BLPD
GAS SPECIFIC GRAVITY	0.65	
AVERAGE FLOWING TEMP	150	F
OIL API GRAVITY	35.0	API
WATER SPECIFIC GRAVITY	1.07	

PRESSURE (100 PSI)

DEPTH (1000 FT)

GAS/LIQUID RATIO - SCF PER BBL

HAGEDORN AND BROWN
PRESENTED BY
MACCO I. SCHLUMBERGER

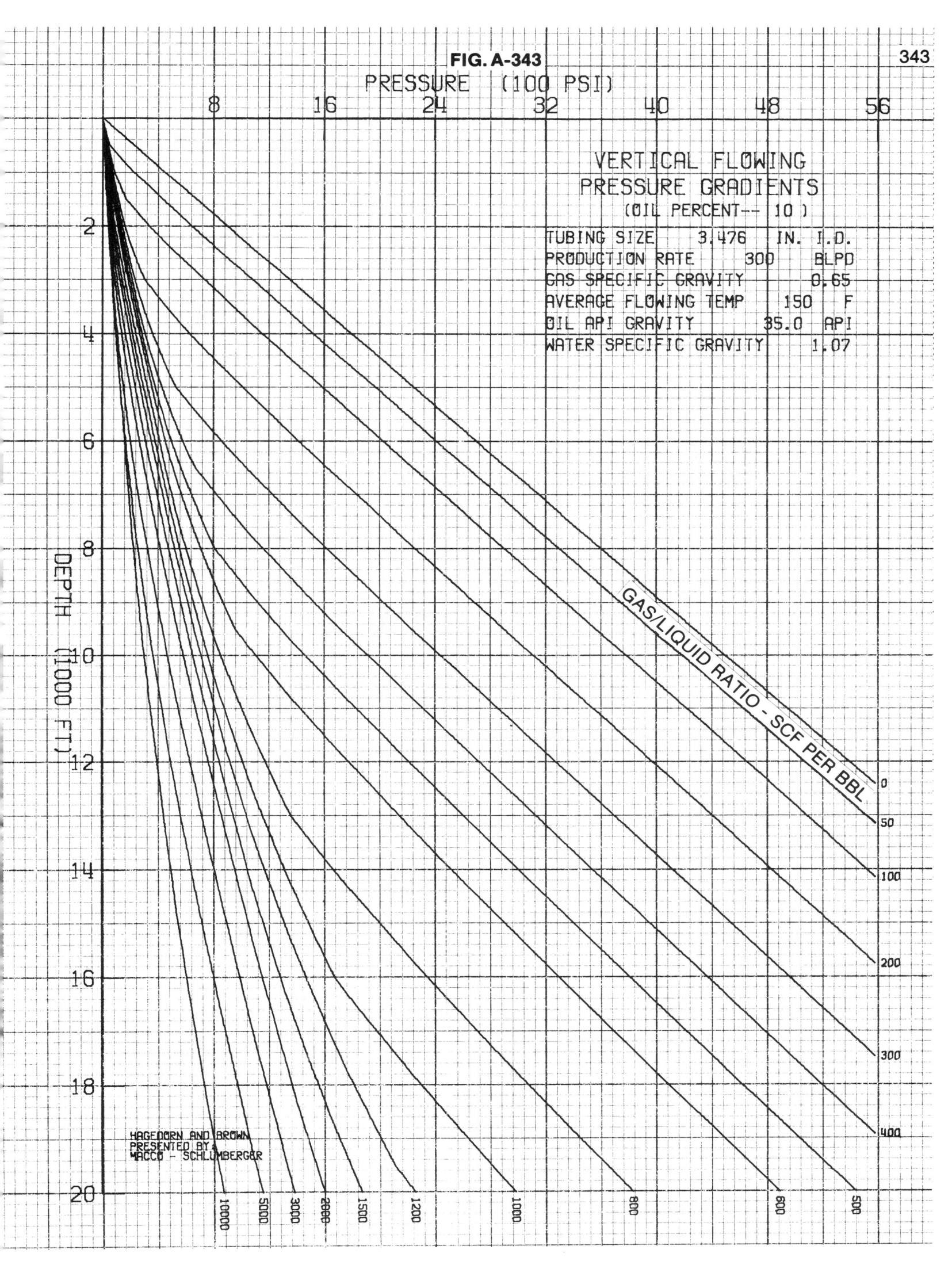

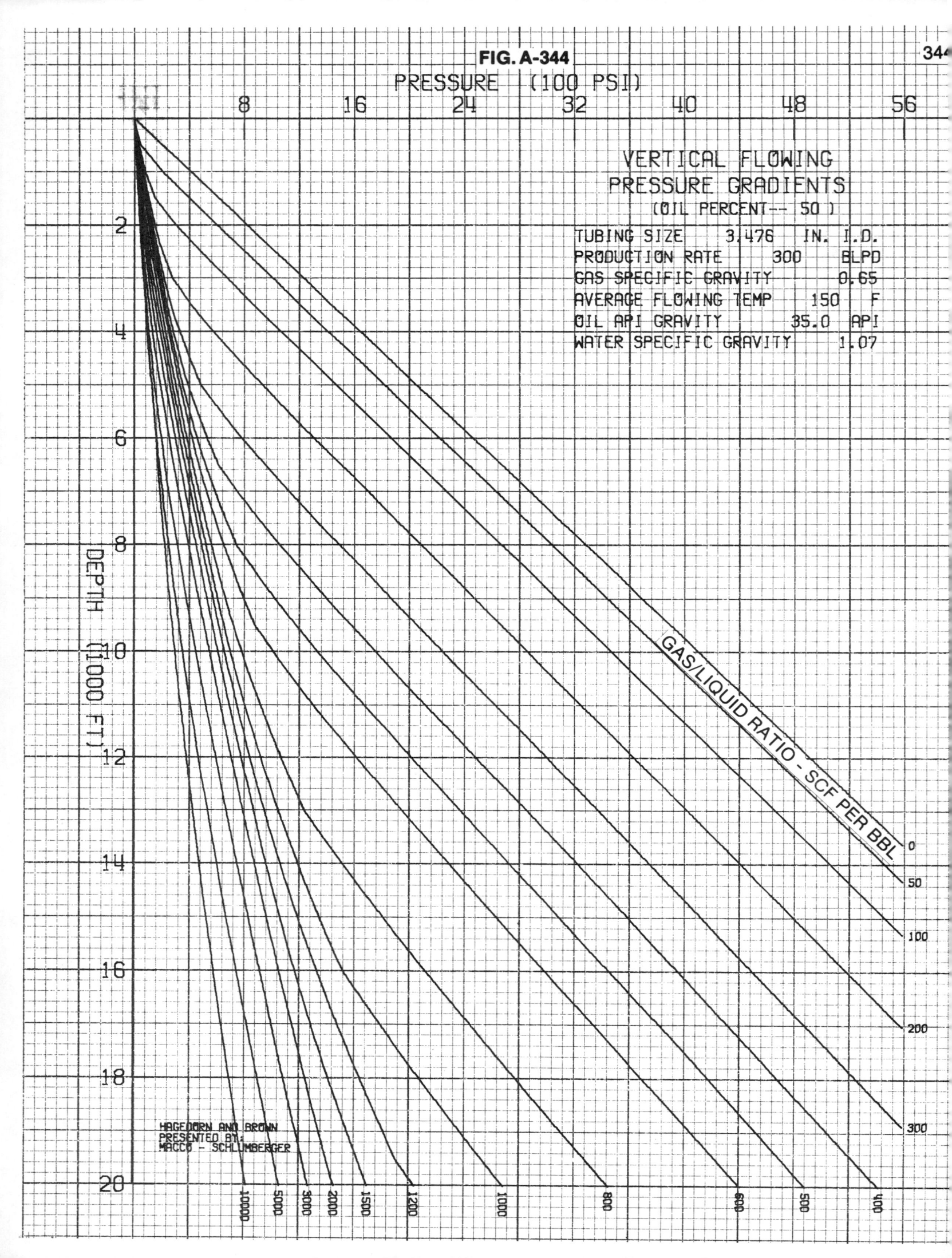

FIG. A-344

344

PRESSURE (100 PSI)

VERTICAL FLOWING
PRESSURE GRADIENTS
(OIL PERCENT-- 50)

TUBING SIZE   3.476   IN. I.D.
PRODUCTION RATE   300   BLPD
GAS SPECIFIC GRAVITY   0.65
AVERAGE FLOWING TEMP   150   F
OIL API GRAVITY   35.0   API
WATER SPECIFIC GRAVITY   1.07

DEPTH (1000 FT)

GAS/LIQUID RATIO - SCF PER BBL

HAGEDORN AND BROWN
PRESENTED BY:
MACCO - SCHLUMBERGER

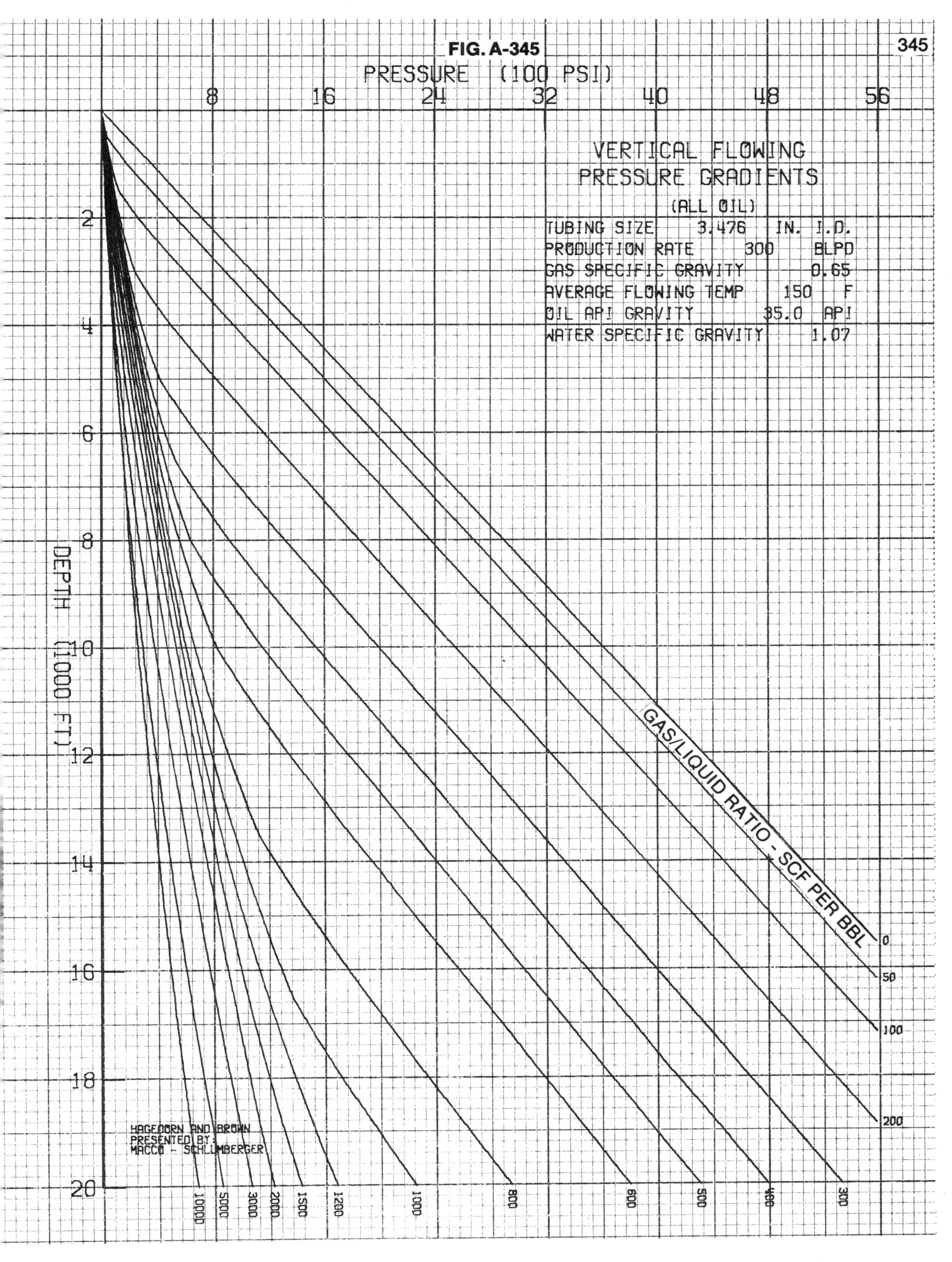

VERTICAL FLOWING
PRESSURE GRADIENTS
(ALL OIL)

TUBING SIZE	3.476	IN. I.D.
PRODUCTION RATE	300	BLPD
GAS SPECIFIC GRAVITY	0.65	
AVERAGE FLOWING TEMP	150	F
OIL API GRAVITY	35.0	API
WATER SPECIFIC GRAVITY	1.07	

PRESSURE (100 PSI)

DEPTH (1000 FT)

GAS/LIQUID RATIO - SCF PER BBL

HAGEDORN AND BROWN
PRESENTED BY:
MACCO - SCHLUMBERGER

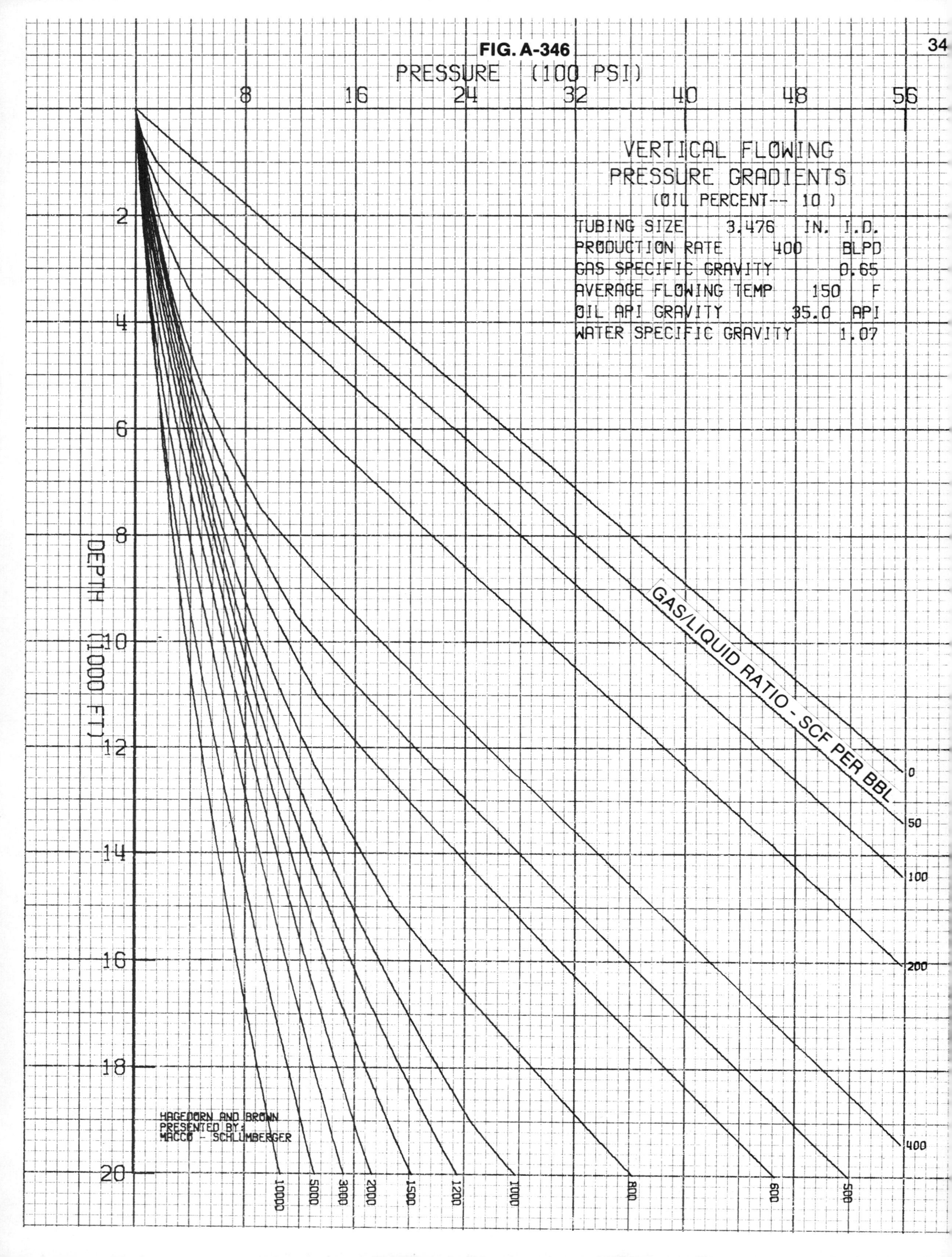

PRESSURE (100 PSI)

DEPTH (1000 FT)

VERTICAL FLOWING
PRESSURE GRADIENTS
(OIL PERCENT-- 10)

TUBING SIZE	3.476	IN. I.D.
PRODUCTION RATE	400	BLPD
GAS SPECIFIC GRAVITY		0.65
AVERAGE FLOWING TEMP	150	F
OIL API GRAVITY	35.0	API
WATER SPECIFIC GRAVITY		1.07

GAS/LIQUID RATIO - SCF PER BBL

HAGEDORN AND BROWN
PRESENTED BY:
MACCO - SCHLUMBERGER

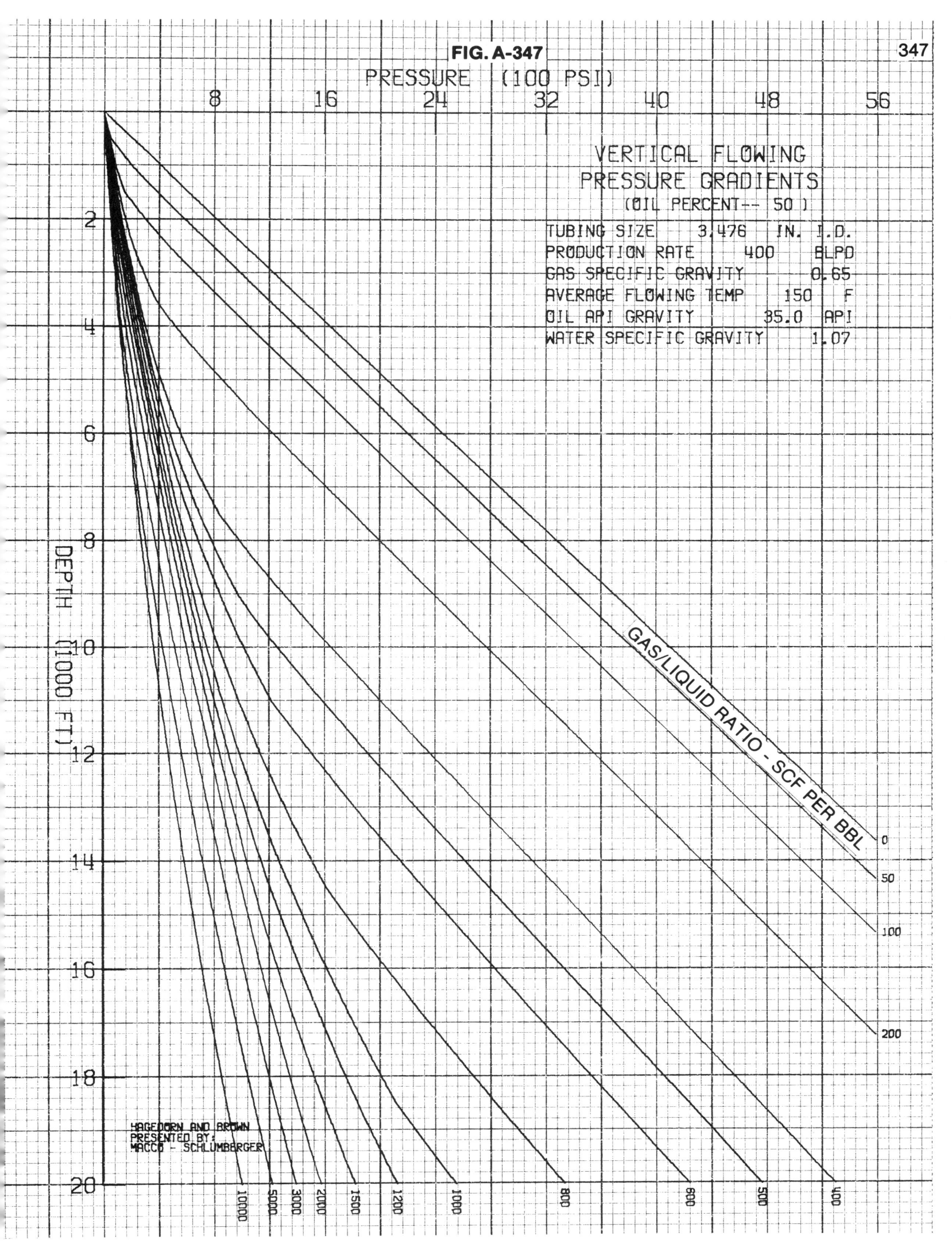

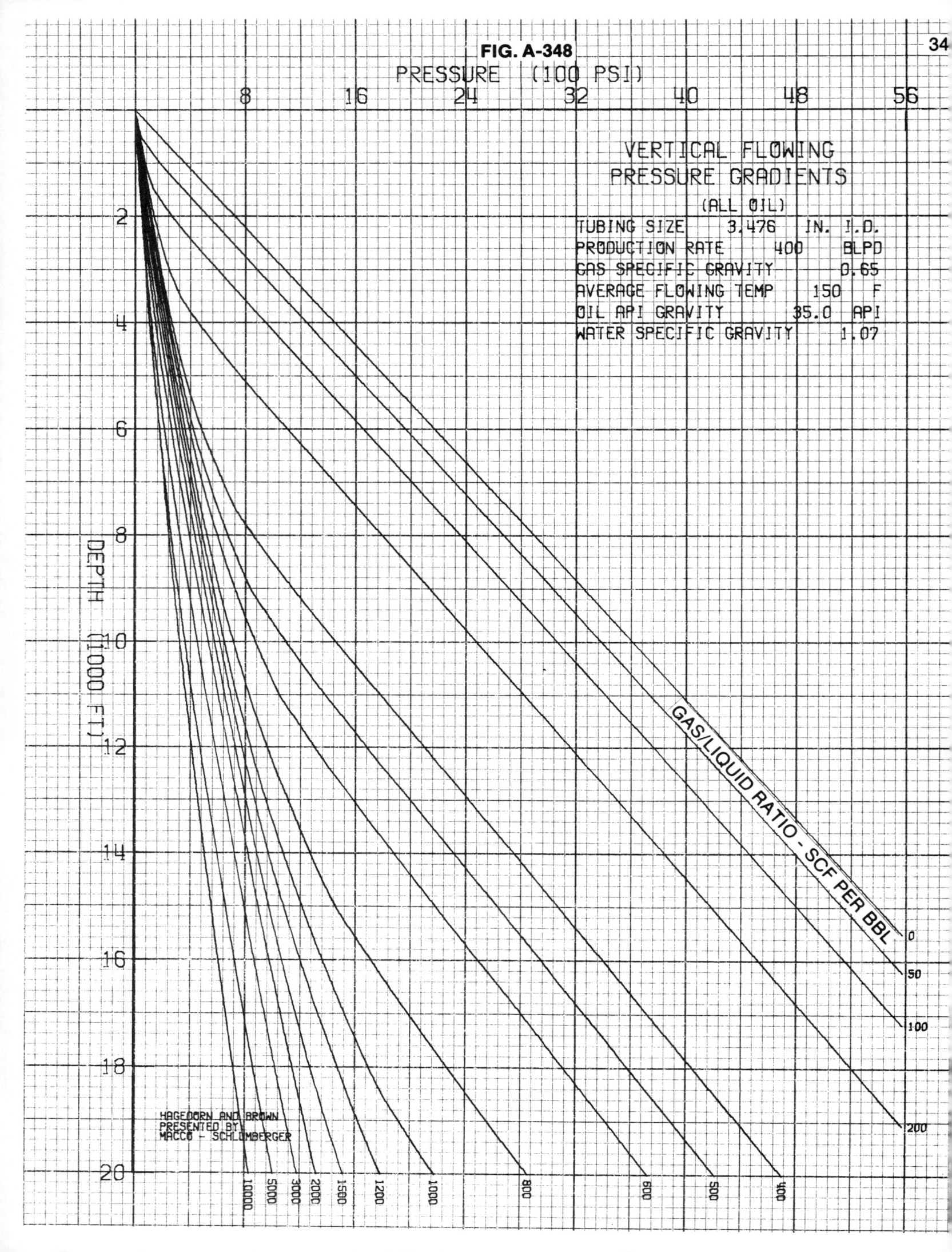

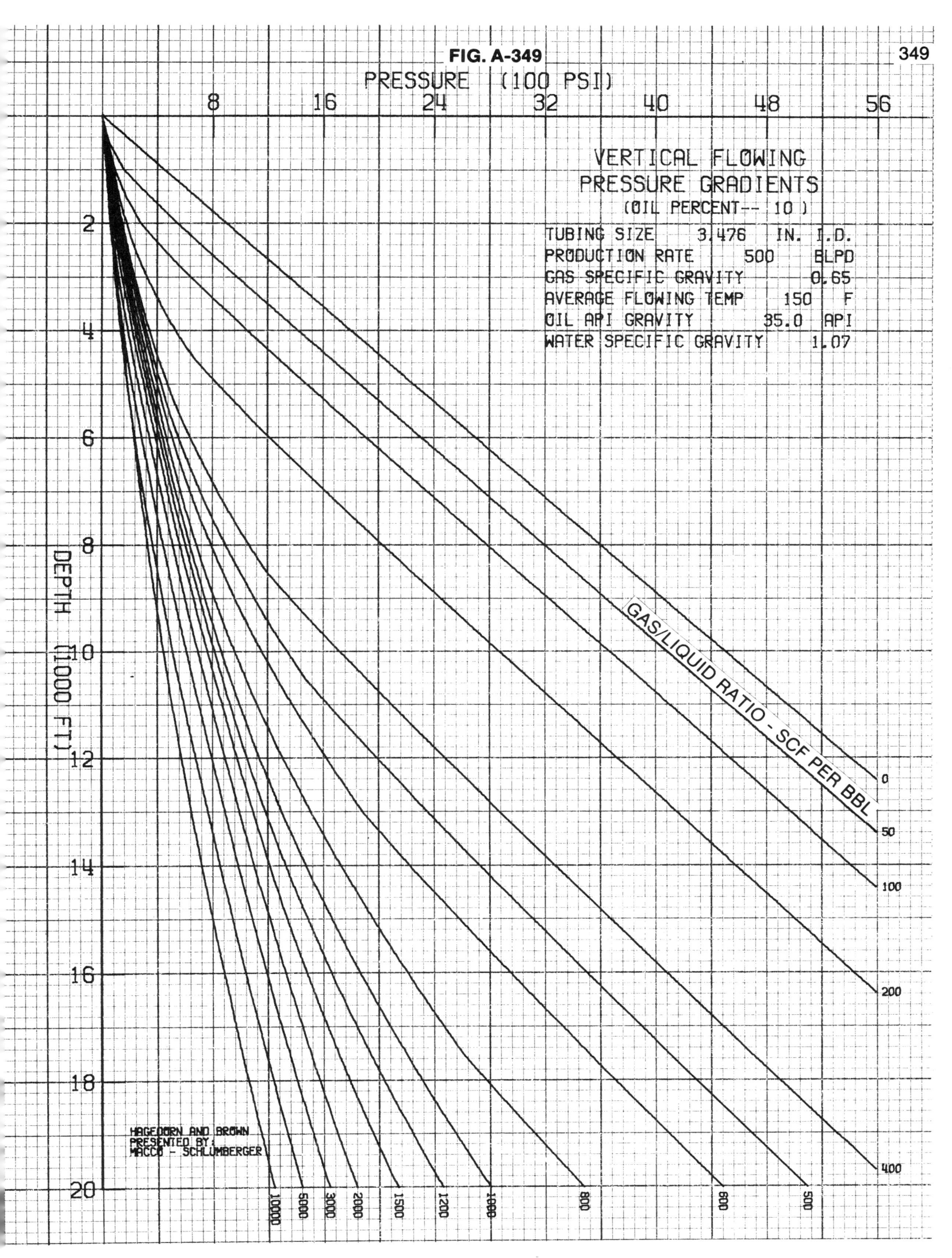

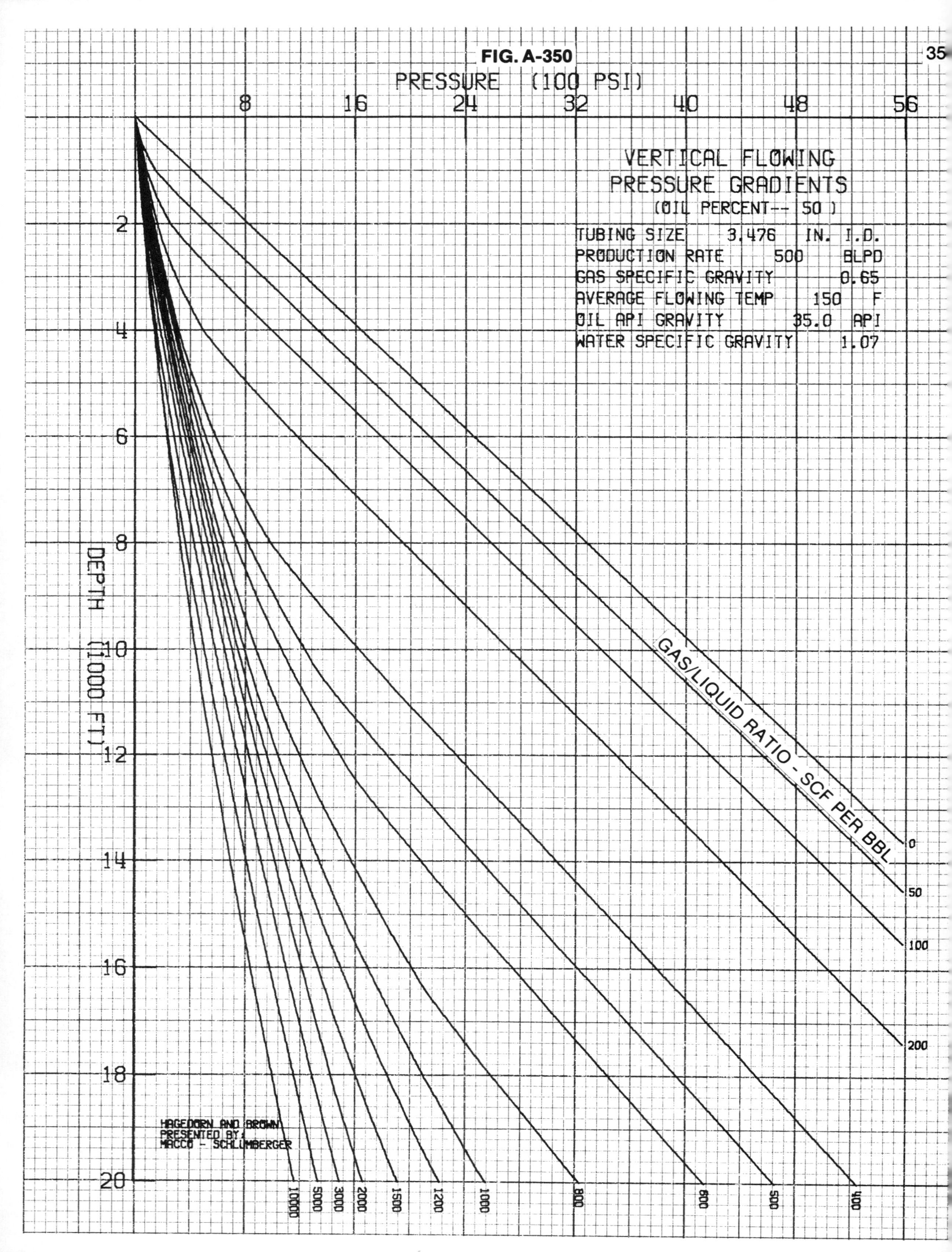

FIG. A-350

35

PRESSURE (100 PSI)

VERTICAL FLOWING
PRESSURE GRADIENTS
(OIL PERCENT--- 50)

TUBING SIZE	3.476	IN. I.D.
PRODUCTION RATE	500	BLPD
GAS SPECIFIC GRAVITY	0.65	
AVERAGE FLOWING TEMP	150	F
OIL API GRAVITY	35.0	API
WATER SPECIFIC GRAVITY	1.07	

DEPTH (1000 FT)

GAS/LIQUID RATIO - SCF PER BBL

HAGEDORN AND BROWN
PRESENTED BY
MACCO - SCHLUMBERGER

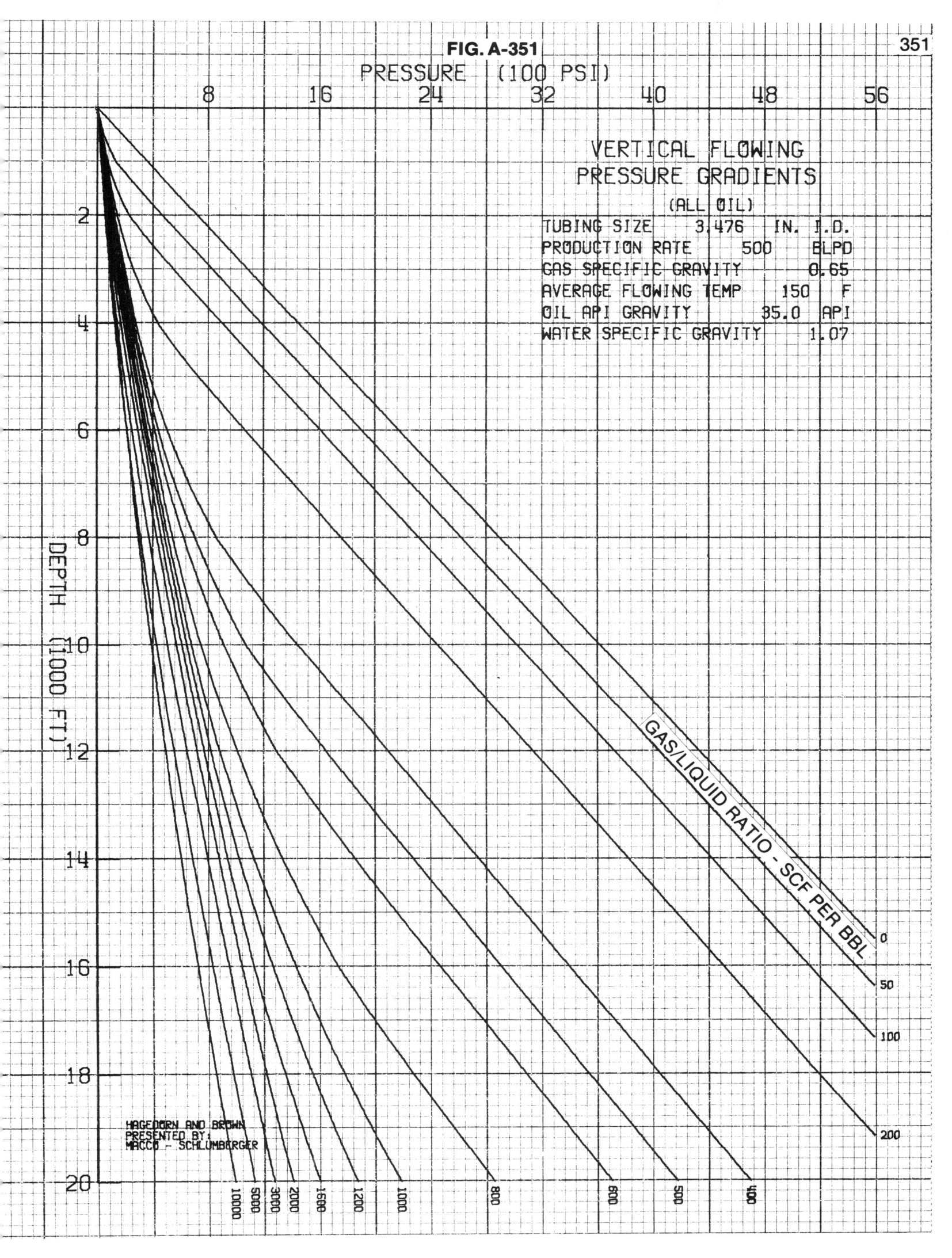

PRESSURE (100 PSI)

VERTICAL FLOWING
PRESSURE GRADIENTS
(ALL OIL)

TUBING SIZE	3.476	IN. I.D.
PRODUCTION RATE	500	BLPD
GAS SPECIFIC GRAVITY	0.65	
AVERAGE FLOWING TEMP	150	F
OIL API GRAVITY	35.0	API
WATER SPECIFIC GRAVITY	1.07	

DEPTH (1000 FT)

GAS/LIQUID RATIO - SCF PER BBL

HAGEDORN AND BROWN
PRESENTED BY
MACCO — SCHLUMBERGER

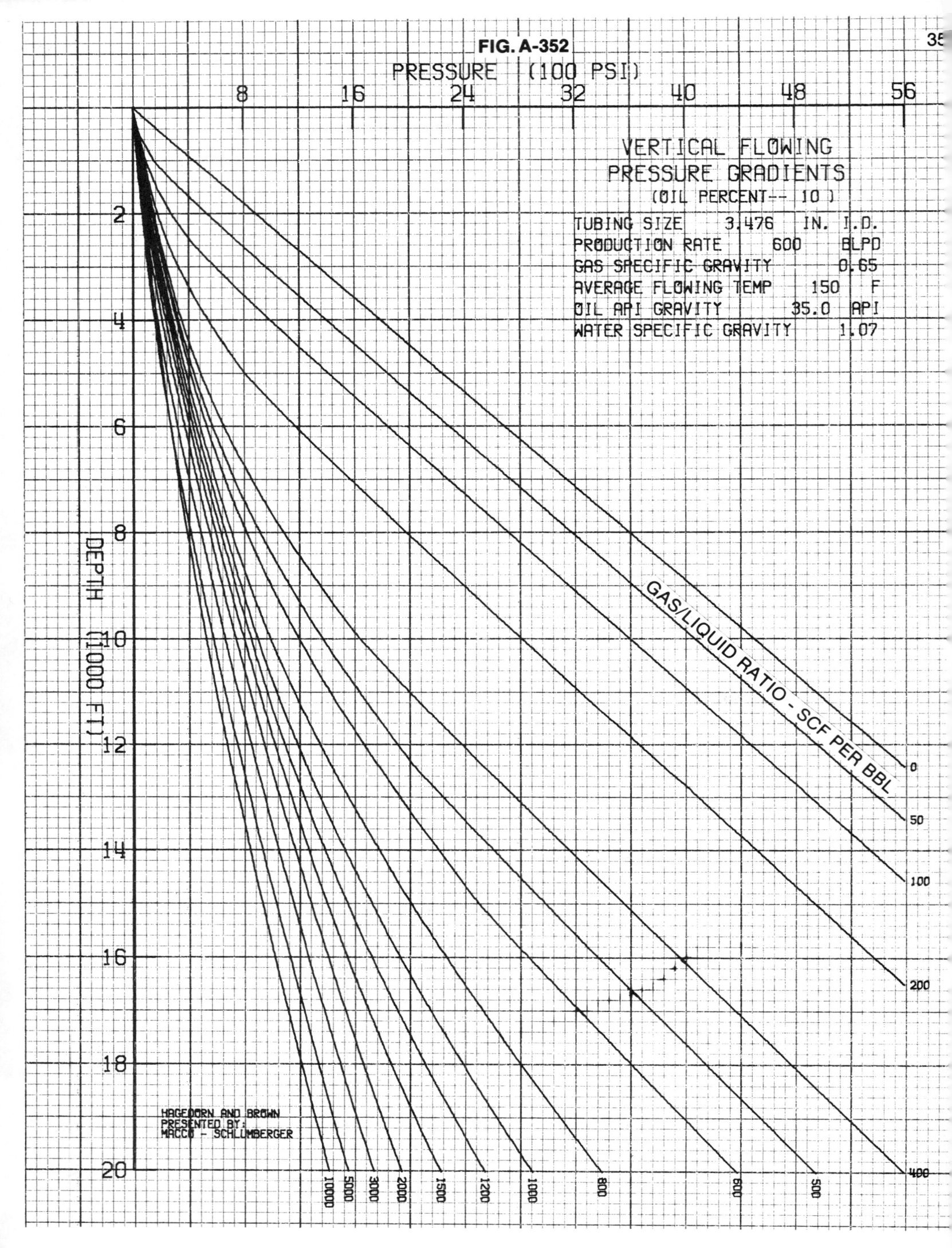

FIG. A-352

VERTICAL FLOWING PRESSURE GRADIENTS
(OIL PERCENT-- 10)

TUBING SIZE	3.476	IN. I.D.
PRODUCTION RATE	600	BLPD
GAS SPECIFIC GRAVITY		0.65
AVERAGE FLOWING TEMP	150	F
OIL API GRAVITY	35.0	API
WATER SPECIFIC GRAVITY		1.07

GAS/LIQUID RATIO - SCF PER BBL

HAGEDORN AND BROWN
PRESENTED BY:
MACCO - SCHLUMBERGER

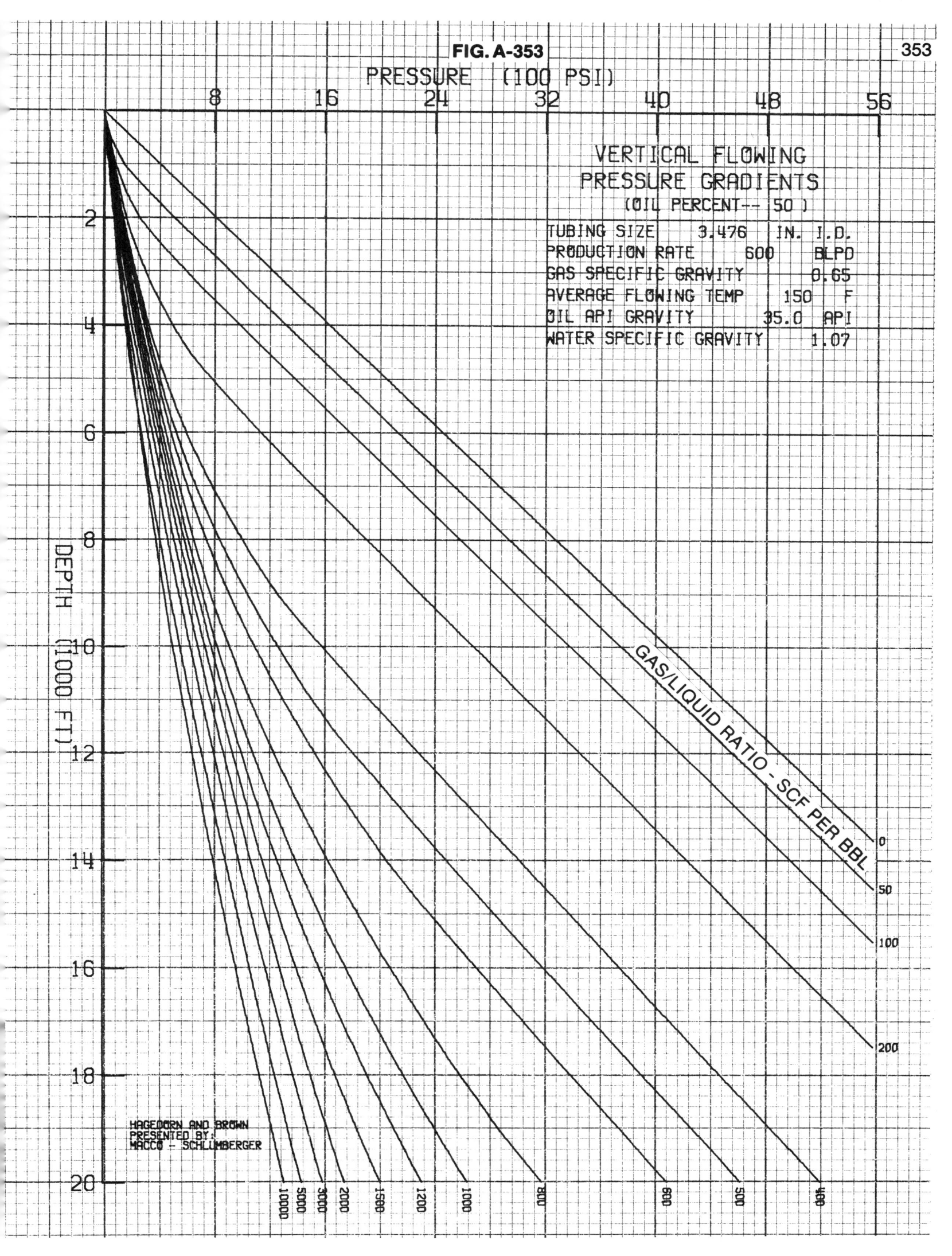

PRESSURE (100 PSI)

VERTICAL FLOWING
PRESSURE GRADIENTS
(OIL PERCENT-- 50)

TUBING SIZE	3.476	IN. I.D.
PRODUCTION RATE	600	BLPD
GAS SPECIFIC GRAVITY		0.65
AVERAGE FLOWING TEMP	150	F
OIL API GRAVITY	35.0	API
WATER SPECIFIC GRAVITY		1.07

DEPTH (1000 FT)

GAS/LIQUID RATIO - SCF PER BBL

HAGEDORN AND BROWN
PRESENTED BY:
MACCO - SCHLUMBERGER

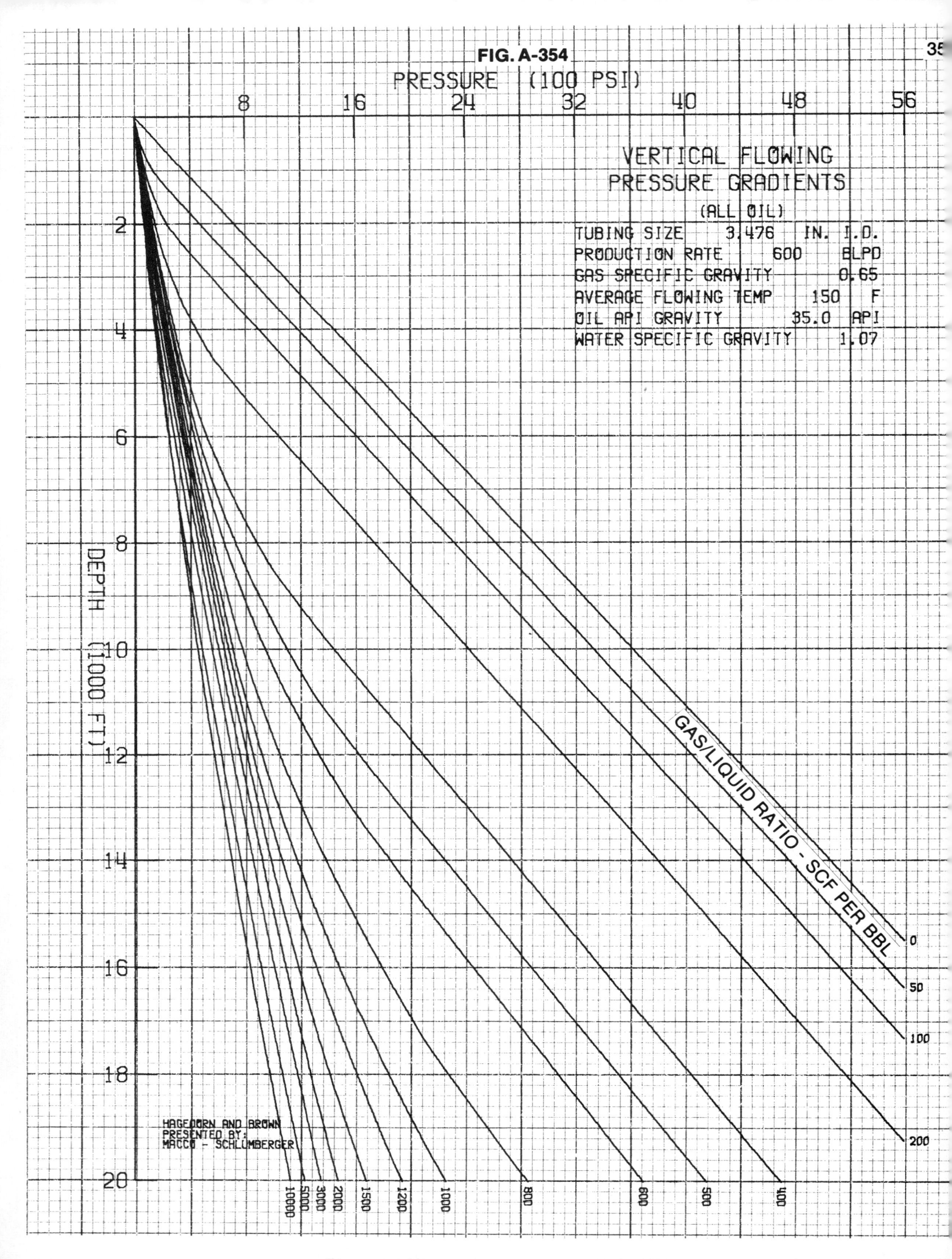

FIG. A-354

VERTICAL FLOWING
PRESSURE GRADIENTS
(ALL OIL)

TUBING SIZE	3.476	IN. I.D.
PRODUCTION RATE	600	BLPD
GAS SPECIFIC GRAVITY	0.65	
AVERAGE FLOWING TEMP	150	F
OIL API GRAVITY	35.0	API
WATER SPECIFIC GRAVITY	1.07	

PRESSURE (100 PSI)

DEPTH (1000 FT)

GAS/LIQUID RATIO - SCF PER BBL

HAGEDORN AND BROWN
PRESENTED BY:
MACCO - SCHLUMBERGER

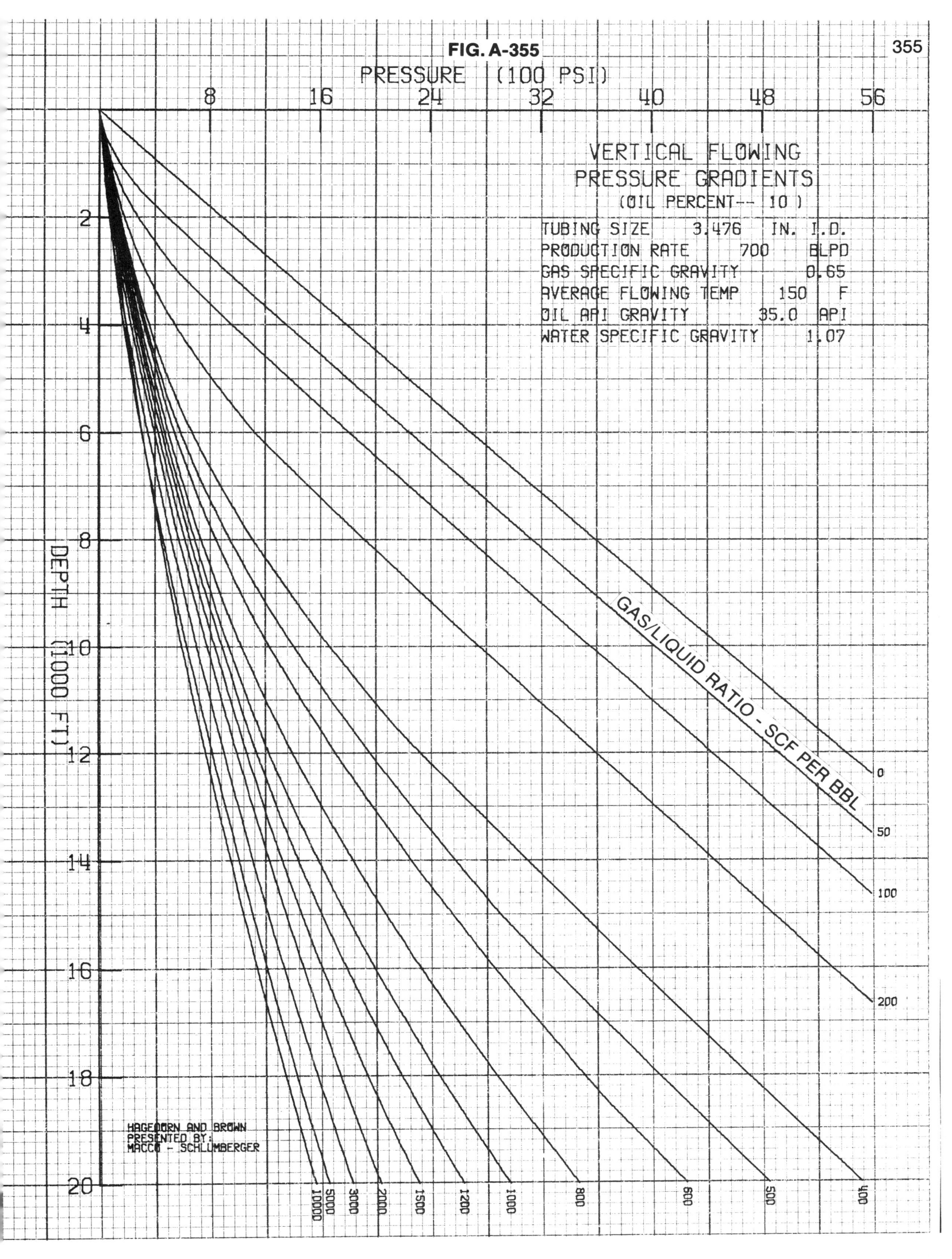

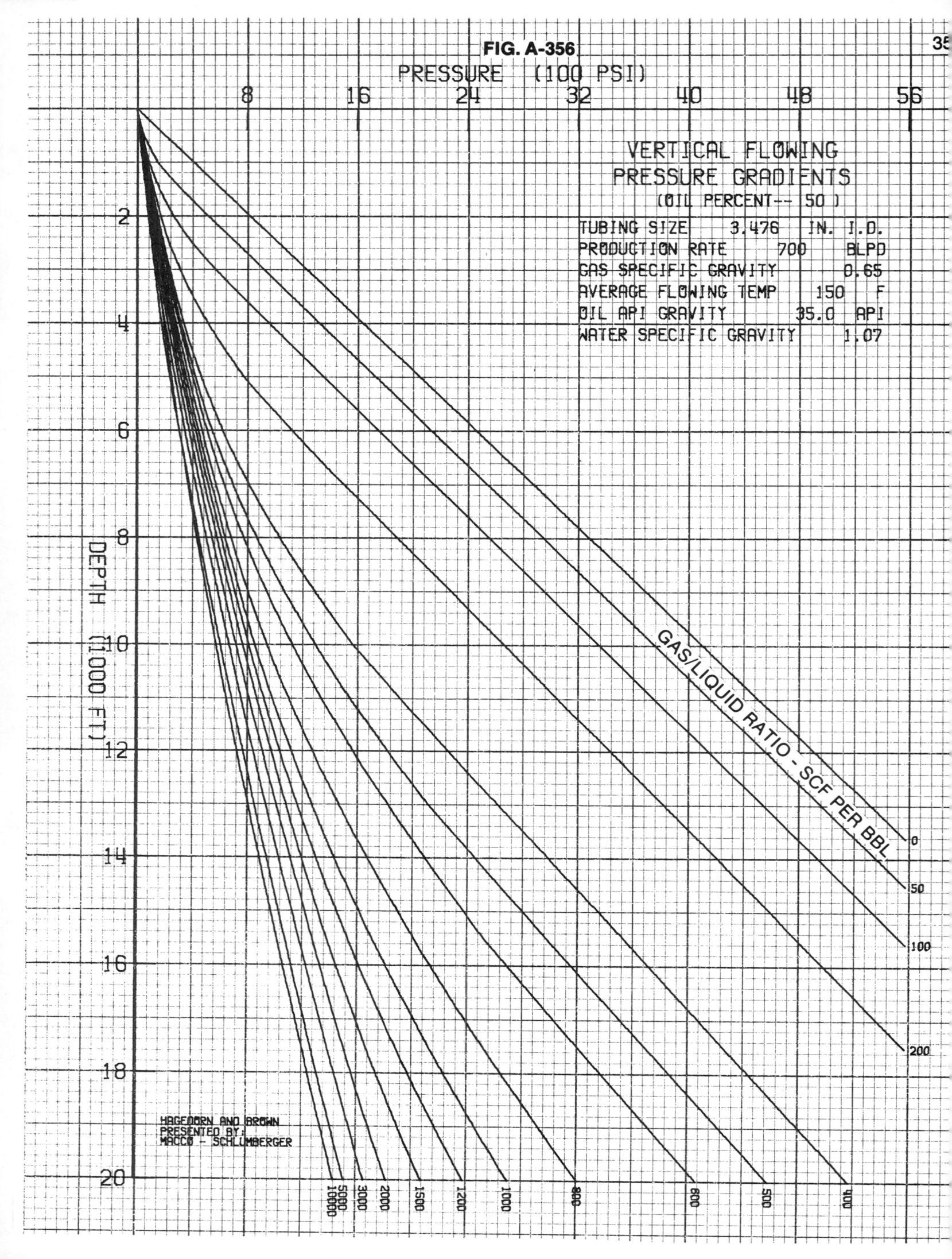

FIG. A-356

VERTICAL FLOWING
PRESSURE GRADIENTS
(OIL PERCENT--- 50)

TUBING SIZE	3.476	IN. I.D.
PRODUCTION RATE	700	BLPD
GAS SPECIFIC GRAVITY	0.65	
AVERAGE FLOWING TEMP	150	F
OIL API GRAVITY	35.0	API
WATER SPECIFIC GRAVITY	1.07	

PRESSURE (100 PSI)

DEPTH (1000 FT)

GAS/LIQUID RATIO - SCF PER BBL

HAGEDORN AND BROWN
PRESENTED BY:
MACCO - SCHLUMBERGER

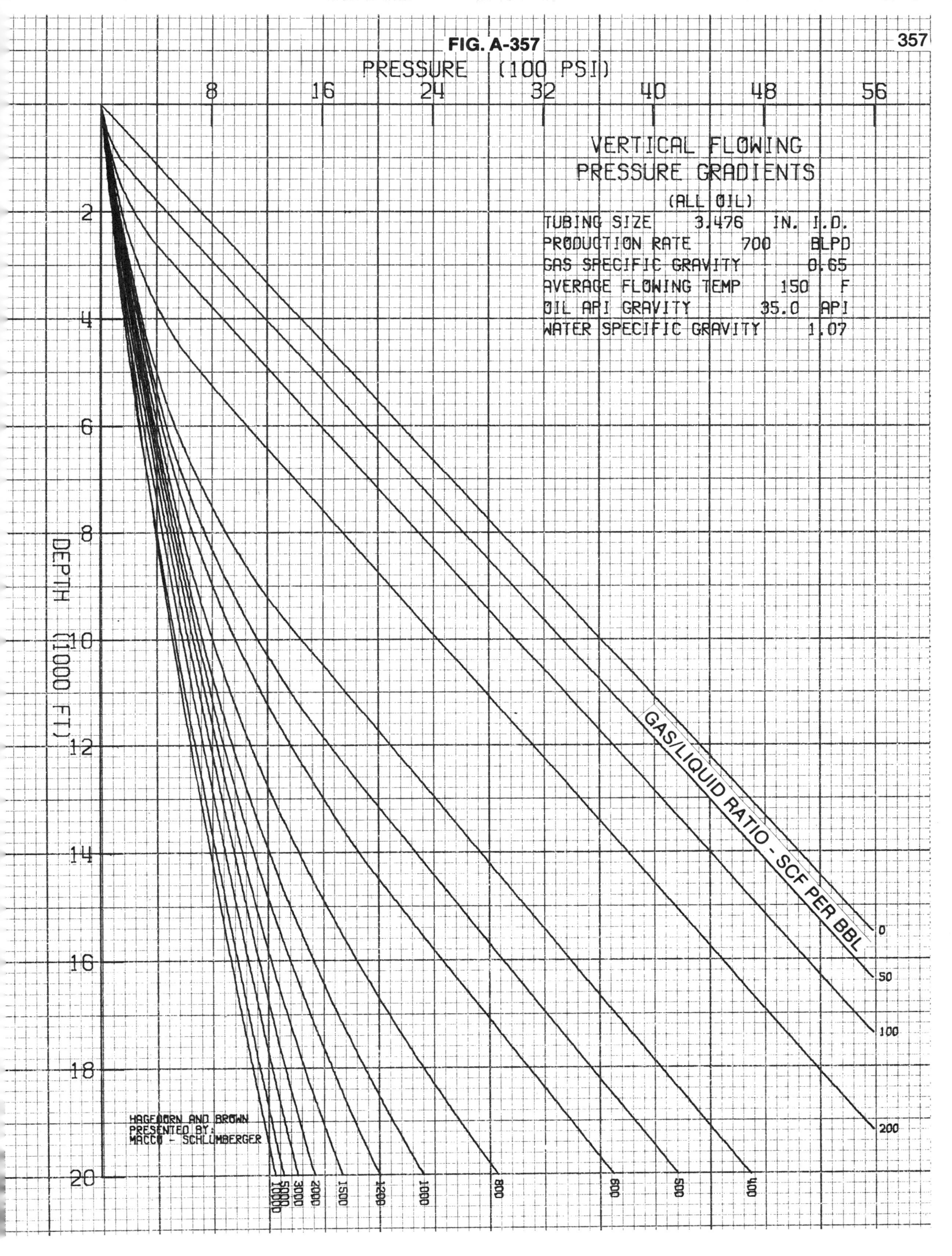

VERTICAL FLOWING
PRESSURE GRADIENTS
(ALL OIL)

TUBING SIZE	3.476	IN. I.D.
PRODUCTION RATE	700	BLPD
GAS SPECIFIC GRAVITY	0.65	
AVERAGE FLOWING TEMP	150	F
OIL API GRAVITY	35.0	API
WATER SPECIFIC GRAVITY	1.07	

PRESSURE (100 PSI)

DEPTH (1000 FT)

GAS/LIQUID RATIO - SCF PER BBL

HAGEDORN AND BROWN
PRESENTED BY:
MACCO - SCHLUMBERGER

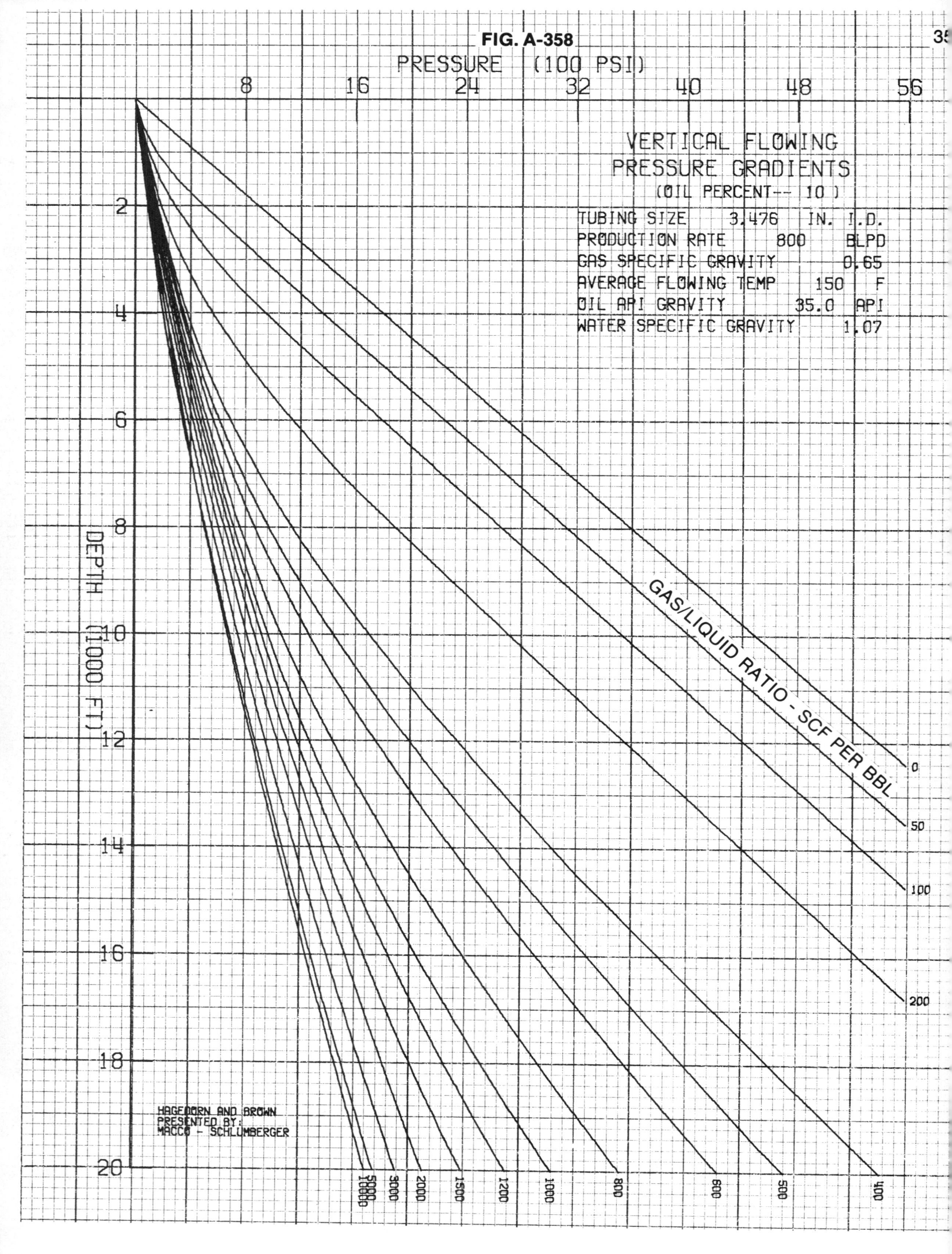

FIG. A-358

VERTICAL FLOWING
PRESSURE GRADIENTS
(OIL PERCENT-- 10)

| TUBING SIZE | 3.476 | IN. I.D. |
| AVERAGE FLOWING TEMP | 150 | F |

GAS/LIQUID RATIO - SCF PER BBL

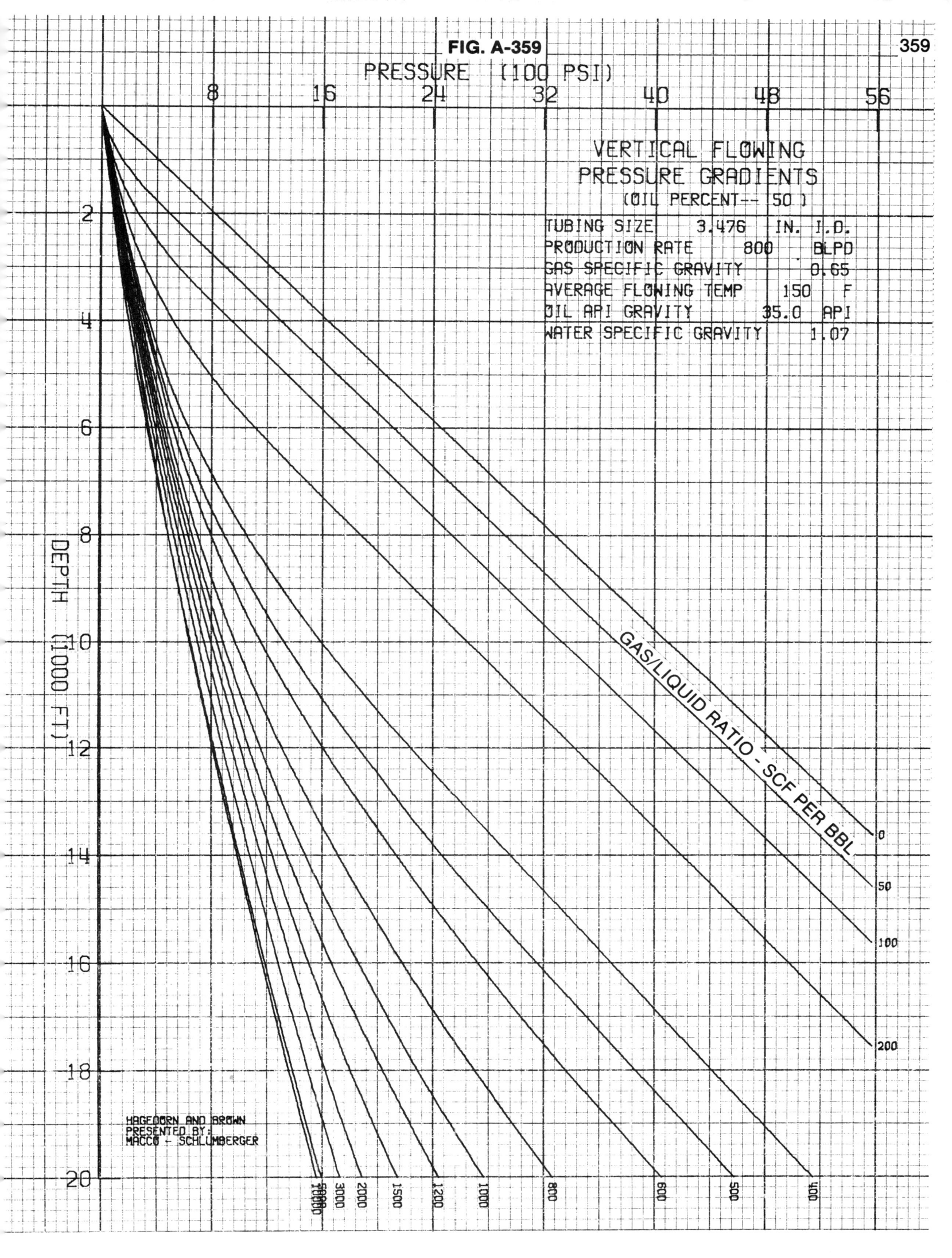

FIG. A-359

359

PRESSURE (100 PSI)

VERTICAL FLOWING
PRESSURE GRADIENTS
(OIL PERCENT-- 50)

TUBING SIZE	3.476	IN. I.D.
PRODUCTION RATE	800	BLPD
GAS SPECIFIC GRAVITY	0.65	
AVERAGE FLOWING TEMP	150	F
OIL API GRAVITY	35.0	API
WATER SPECIFIC GRAVITY	1.07	

DEPTH (1000 FT)

GAS/LIQUID RATIO - SCF PER BBL

HAGEDORN AND BROWN
PRESENTED BY:
MACCO - SCHLUMBERGER

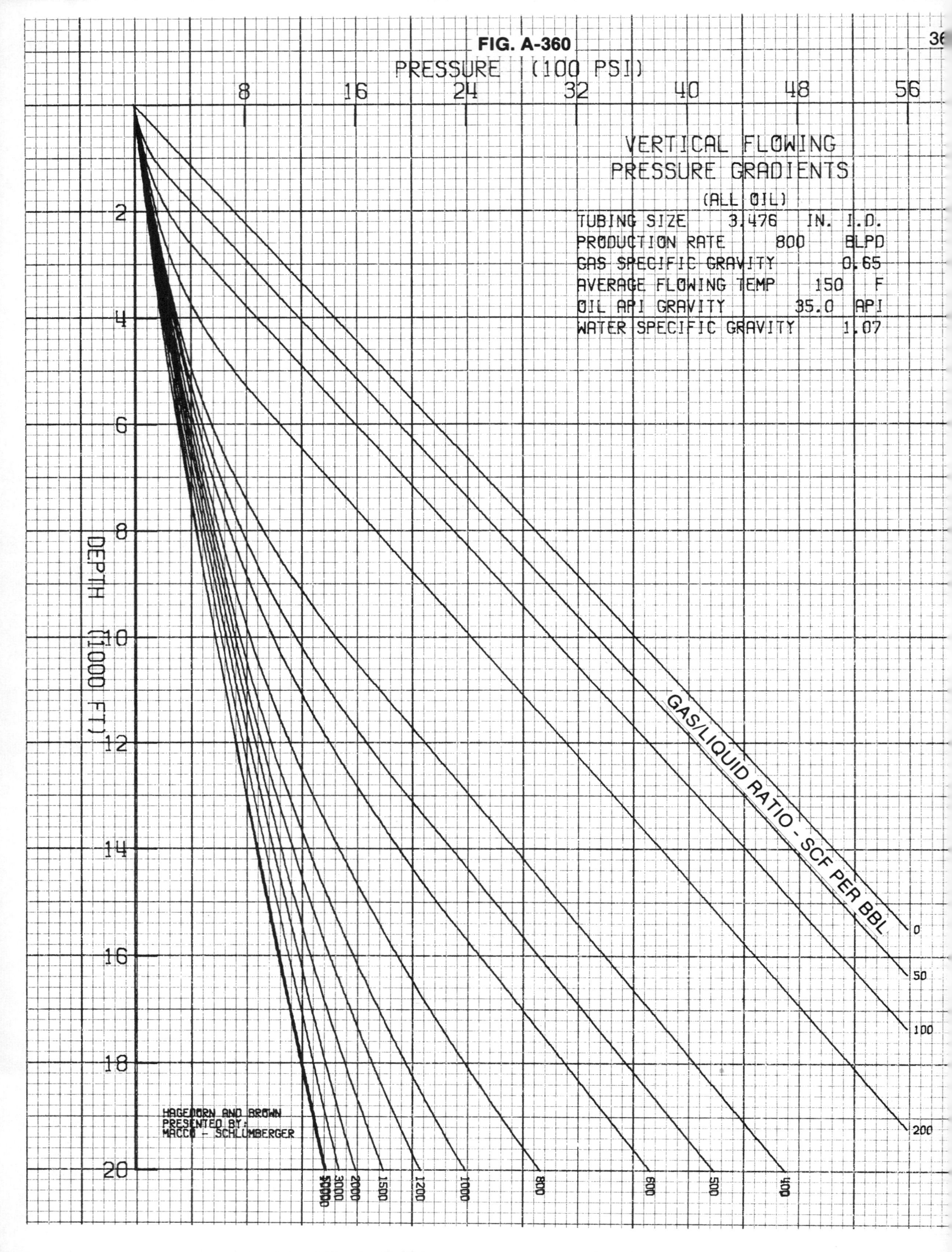

FIG. A-360

36

VERTICAL FLOWING PRESSURE GRADIENTS
(ALL OIL)

TUBING SIZE	3.476	IN. I.D.
PRODUCTION RATE	800	BLPD
GAS SPECIFIC GRAVITY	0.65	
AVERAGE FLOWING TEMP	150	F
OIL API GRAVITY	35.0	API
WATER SPECIFIC GRAVITY	1.07	

PRESSURE (100 PSI)

DEPTH (1000 FT)

GAS/LIQUID RATIO - SCF PER BBL

HAGEDORN AND BROWN
PRESENTED BY:
MACCO - SCHLUMBERGER

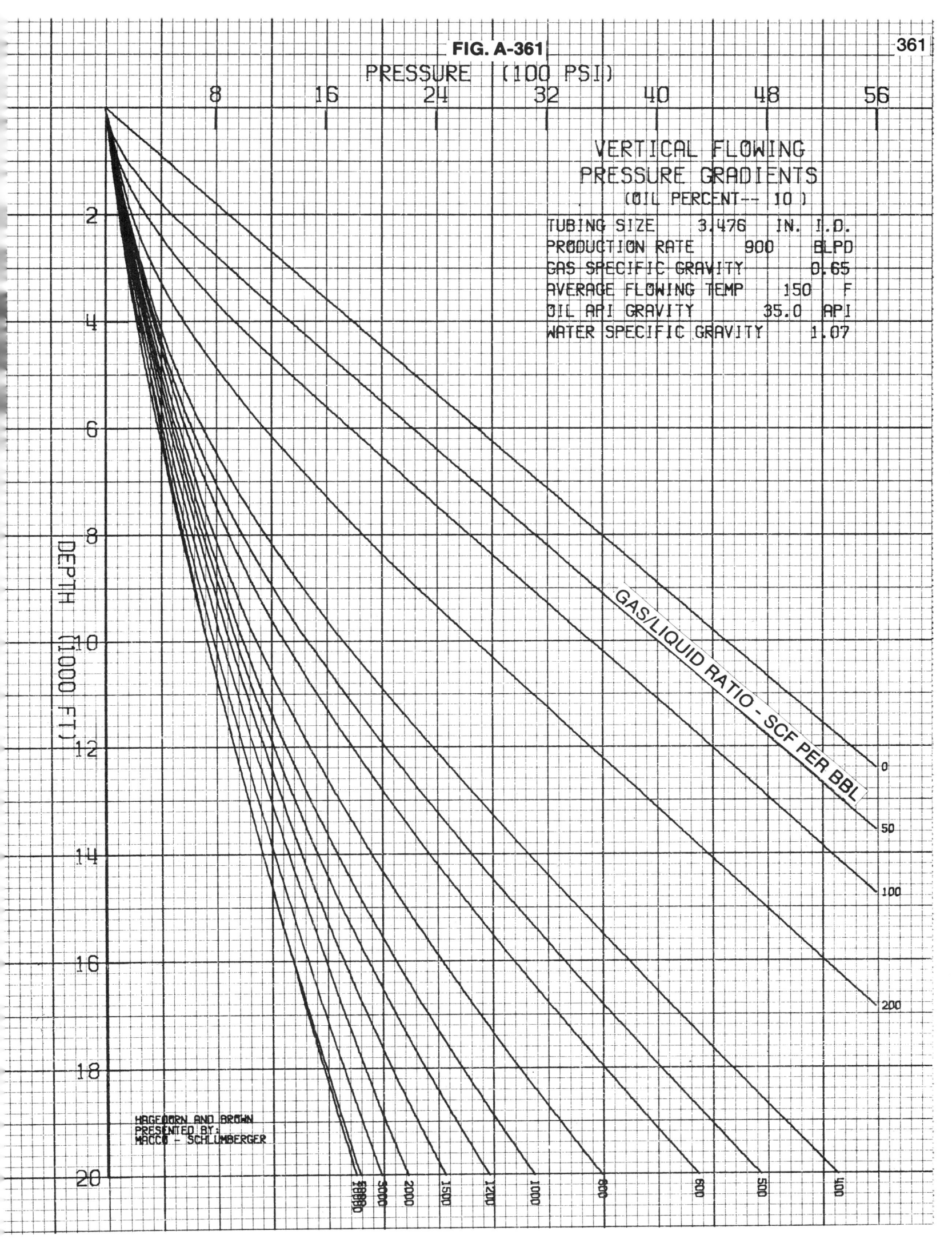

FIG. A-361

361

PRESSURE (100 PSI)

VERTICAL FLOWING
PRESSURE GRADIENTS
(OIL PERCENT-- 10)

TUBING SIZE	3.476	IN. I.D.
PRODUCTION RATE	900	BLPD
GAS SPECIFIC GRAVITY		0.65
AVERAGE FLOWING TEMP	150	F
OIL API GRAVITY	35.0	API
WATER SPECIFIC GRAVITY		1.07

GAS/LIQUID RATIO - SCF PER BBL

DEPTH (1000 FT)

HAGEDORN AND BROWN
PRESENTED BY:
MACCO - SCHLUMBERGER

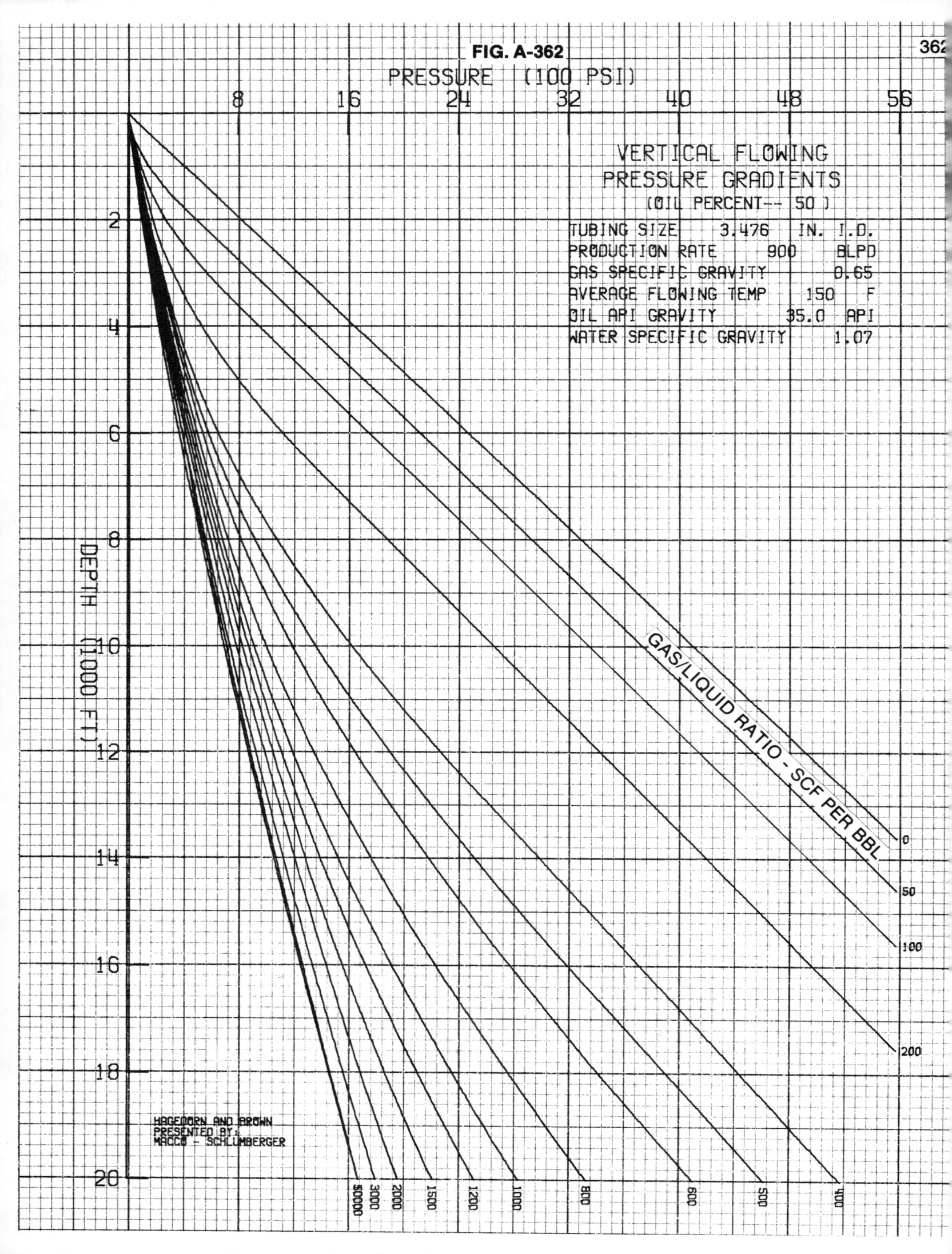

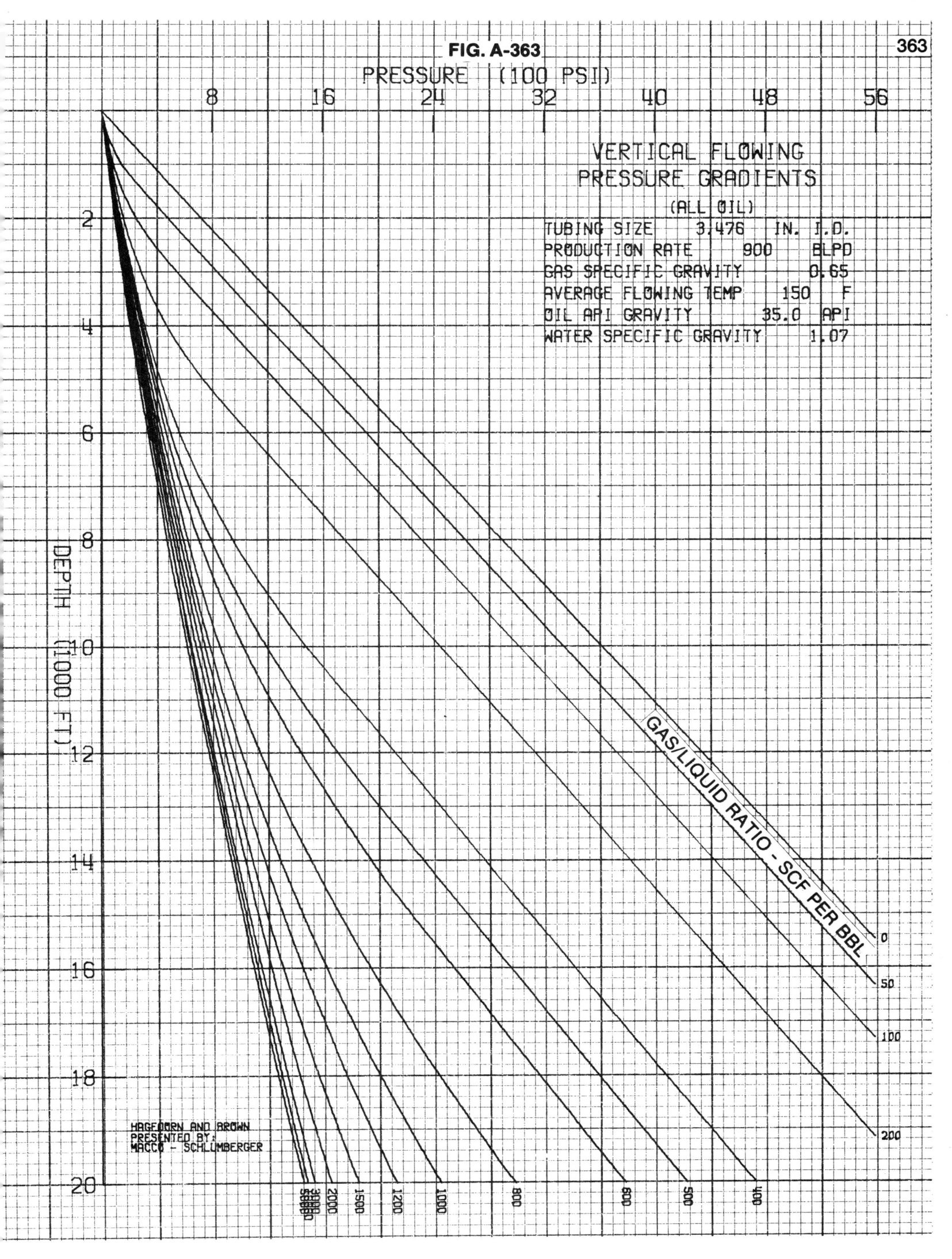

PRESSURE (100 PSI)

VERTICAL FLOWING
PRESSURE GRADIENTS
(ALL OIL)

TUBING SIZE	3.476	IN. I.D.
PRODUCTION RATE	900	BLPD
GAS SPECIFIC GRAVITY	0.65	
AVERAGE FLOWING TEMP	150	F
OIL API GRAVITY	35.0	API
WATER SPECIFIC GRAVITY	1.07	

DEPTH (1000 FT)

GAS/LIQUID RATIO - SCF PER BBL

HAGEDORN AND BROWN
PRESENTED BY:
MACCO - SCHLUMBERGER

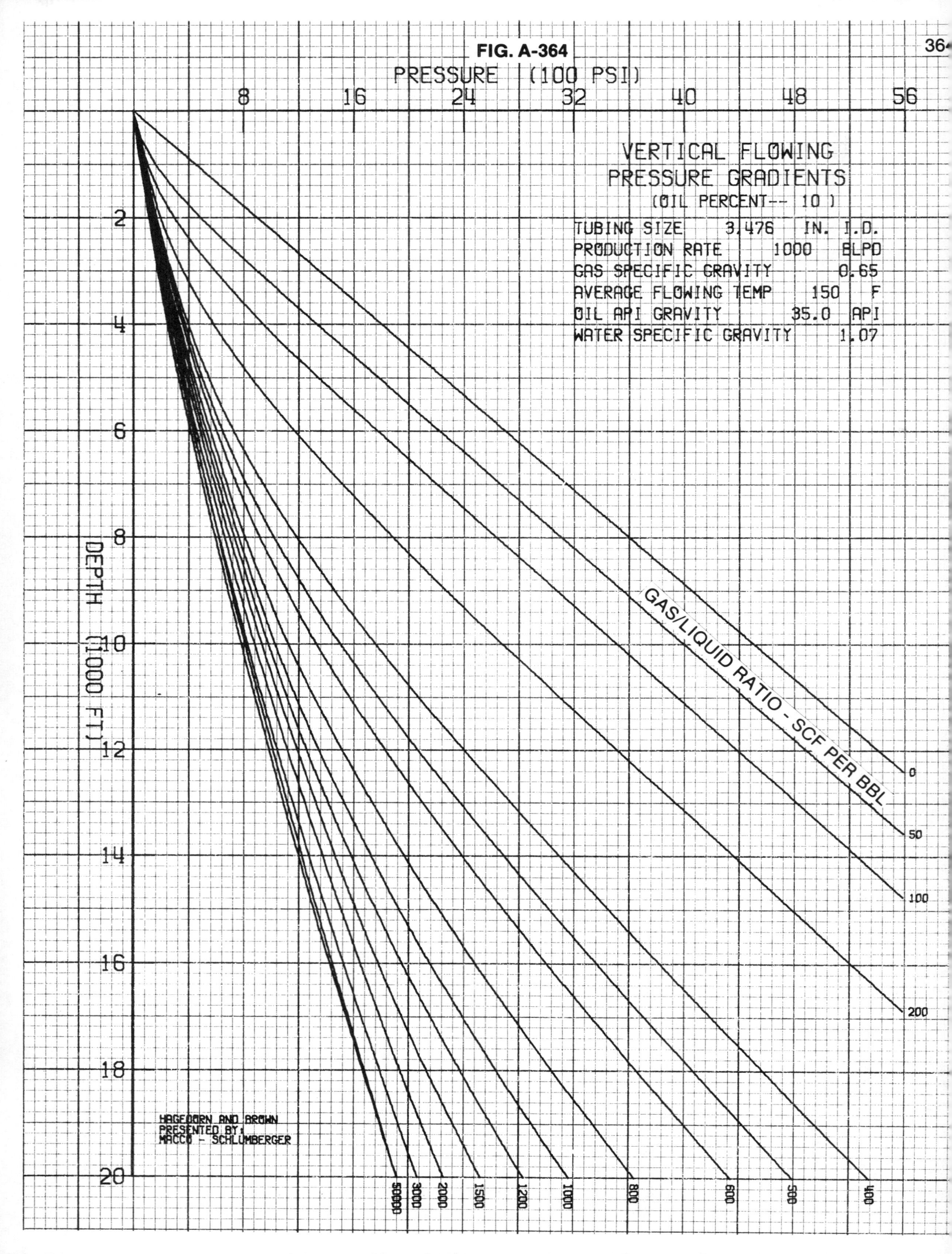

FIG. A-364

PRESSURE (100 PSI)

VERTICAL FLOWING
PRESSURE GRADIENTS
(OIL PERCENT-- 10)

TUBING SIZE   3.476   IN. I.D.
PRODUCTION RATE   1000   BLPD
GAS SPECIFIC GRAVITY   0.65
AVERAGE FLOWING TEMP   150   F
OIL API GRAVITY   35.0 API
WATER SPECIFIC GRAVITY   1.07

GAS/LIQUID RATIO - SCF PER BBL

DEPTH (1000 FT)

HAGEDORN AND BROWN
PRESENTED BY:
MACCO - SCHLUMBERGER

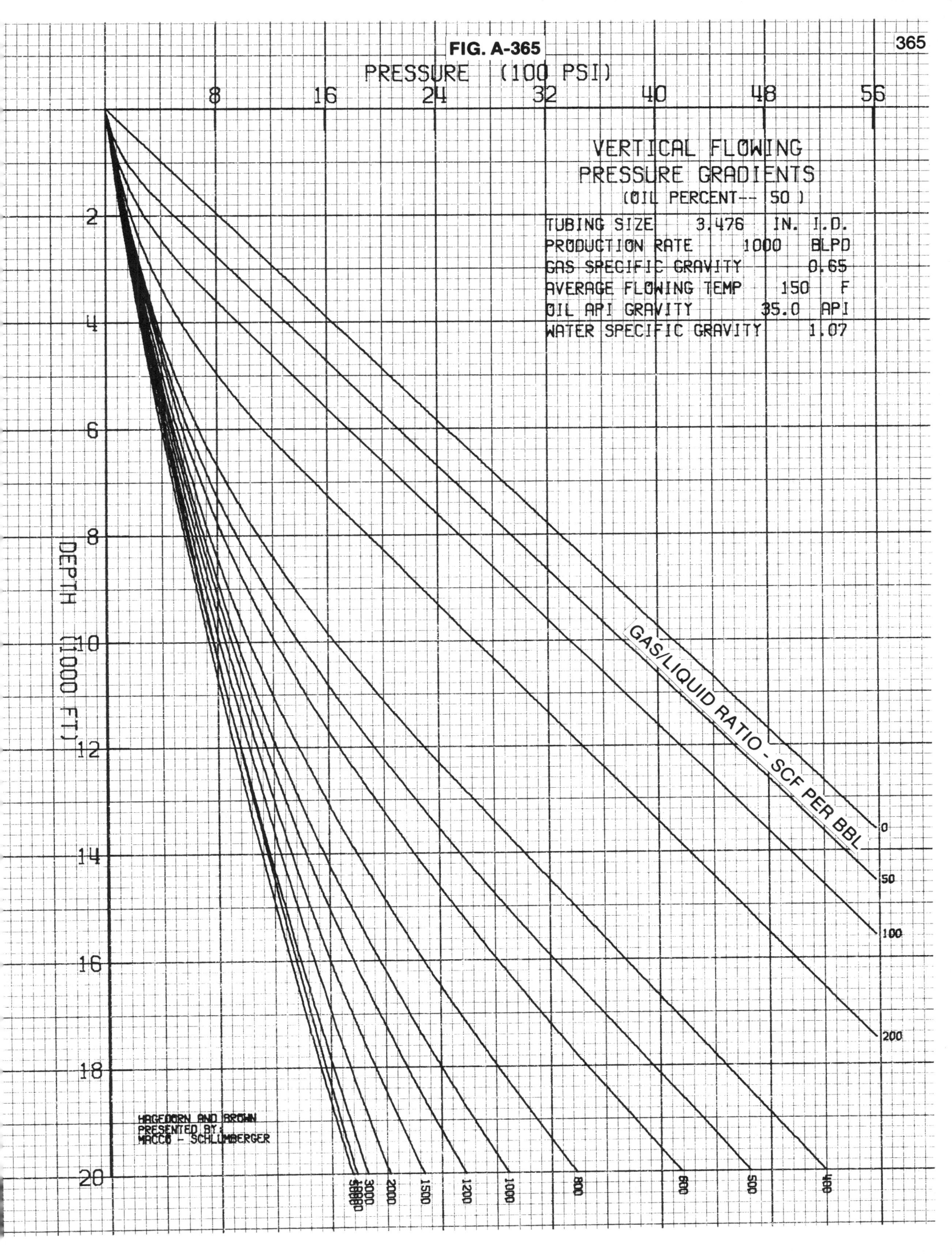

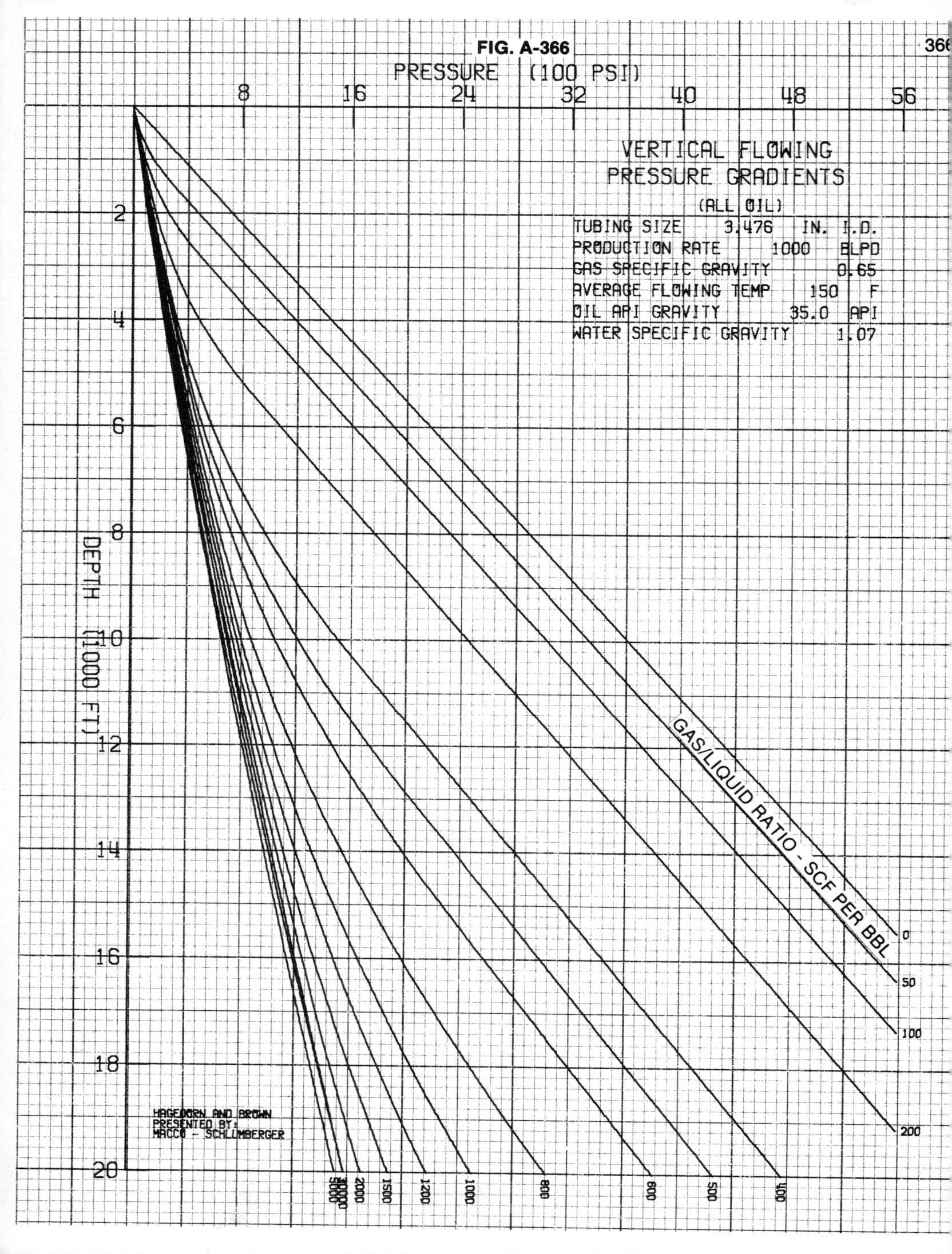

PRESSURE (100 PSI)

DEPTH (1000 FT)

VERTICAL FLOWING
PRESSURE GRADIENTS
(ALL OIL)

TUBING SIZE   3.476   IN. I.D.
PRODUCTION RATE   1000   BLPD
GAS SPECIFIC GRAVITY   0.65
AVERAGE FLOWING TEMP 150   F
OIL API GRAVITY   35.0   API
WATER SPECIFIC GRAVITY   1.07

GAS/LIQUID RATIO - SCF PER BBL

0
50
100
200

HAGEDORN AND BROWN
PRESENTED BY:
MACCO - SCHLUMBERGER

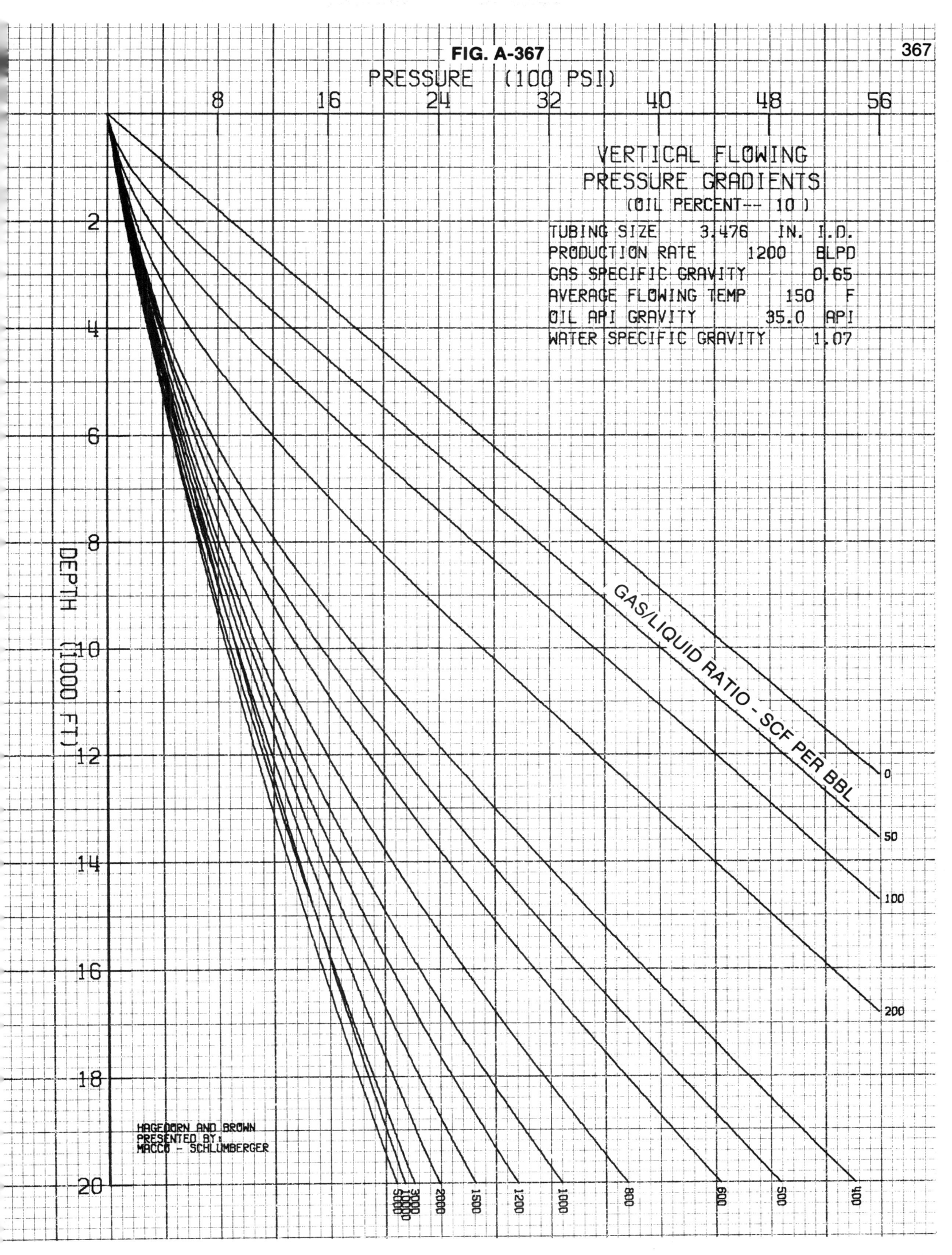

FIG. A-367

Page 367

VERTICAL FLOWING PRESSURE GRADIENTS
(OIL PERCENT-- 10)

TUBING SIZE	3.476	IN. I.D.
PRODUCTION RATE	1200	BLPD
GAS SPECIFIC GRAVITY	0.65	
AVERAGE FLOWING TEMP	150	F
OIL API GRAVITY	35.0	API
WATER SPECIFIC GRAVITY	1.07	

PRESSURE (100 PSI)

DEPTH (1000 FT)

GAS/LIQUID RATIO - SCF PER BBL

HAGEDORN AND BROWN
PRESENTED BY:
MACCO - SCHLUMBERGER

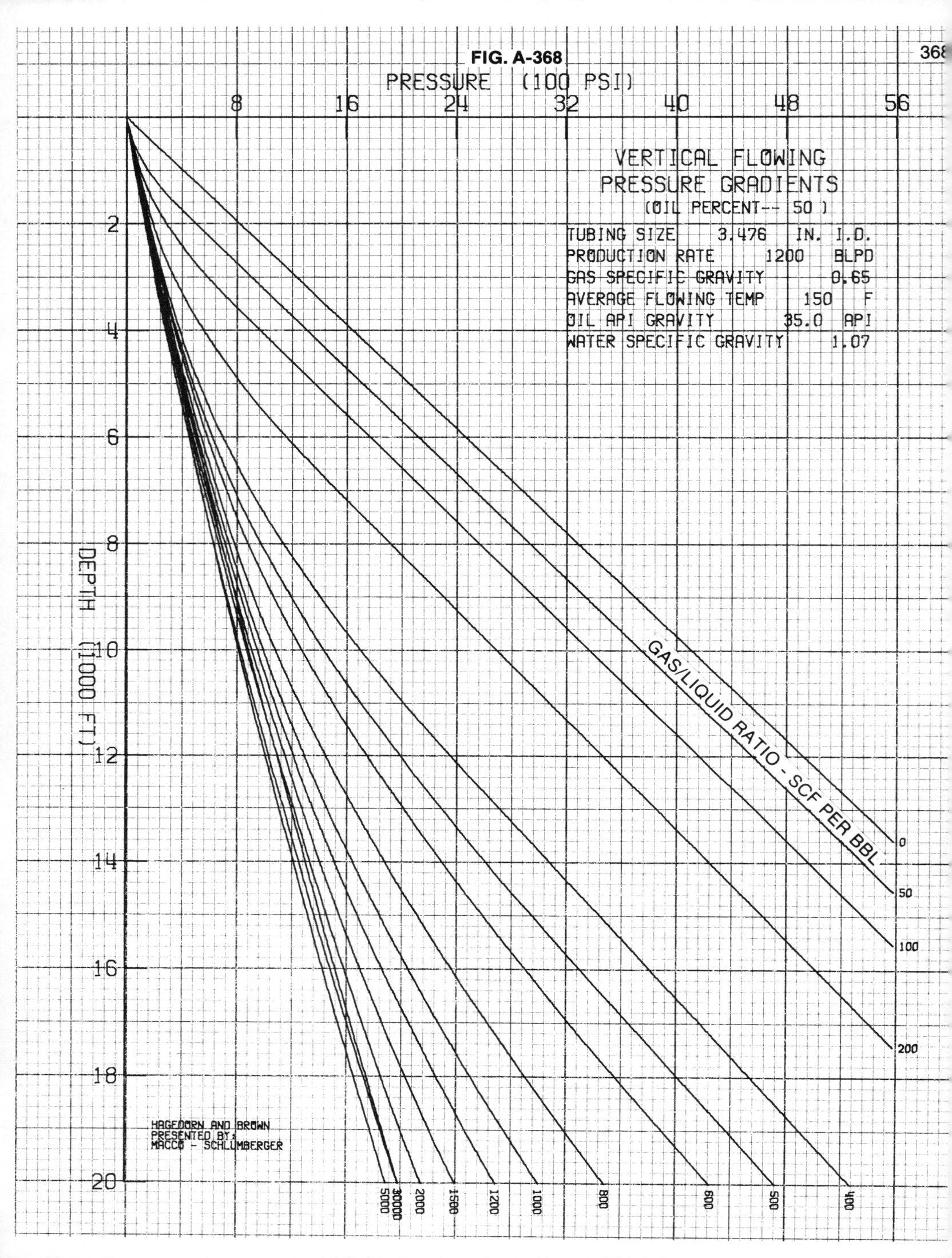

FIG. A-368

368

PRESSURE (100 PSI)

VERTICAL FLOWING
PRESSURE GRADIENTS
(OIL PERCENT-- 50)

TUBING SIZE	3.476	IN. I.D.
PRODUCTION RATE	1200	BLPD
GAS SPECIFIC GRAVITY	0.65	
AVERAGE FLOWING TEMP	150	F
OIL API GRAVITY	35.0	API
WATER SPECIFIC GRAVITY	1.07	

DEPTH (1000 FT)

GAS/LIQUID RATIO - SCF PER BBL

HAGEDORN AND BROWN
PRESENTED BY:
MACCO - SCHLUMBERGER

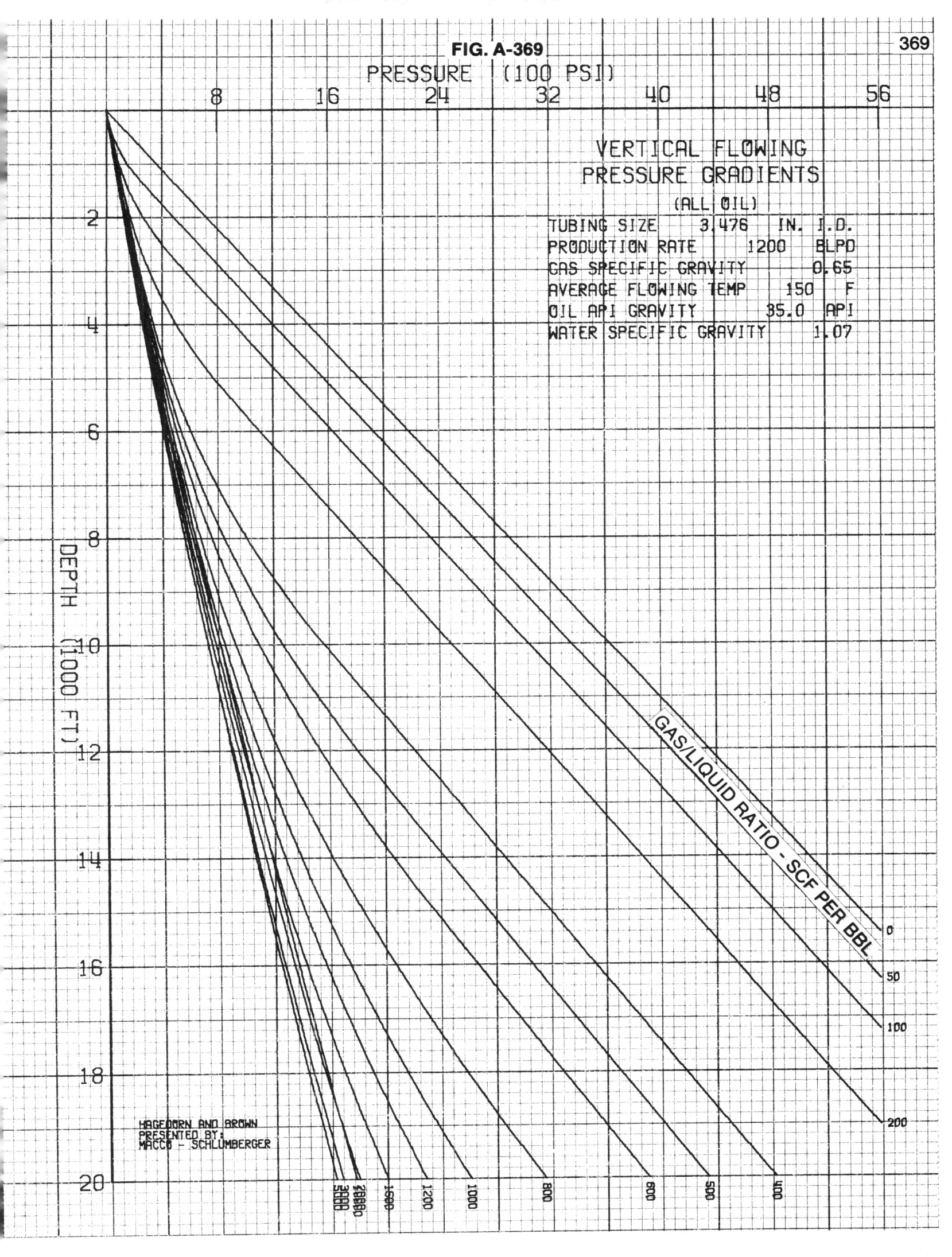

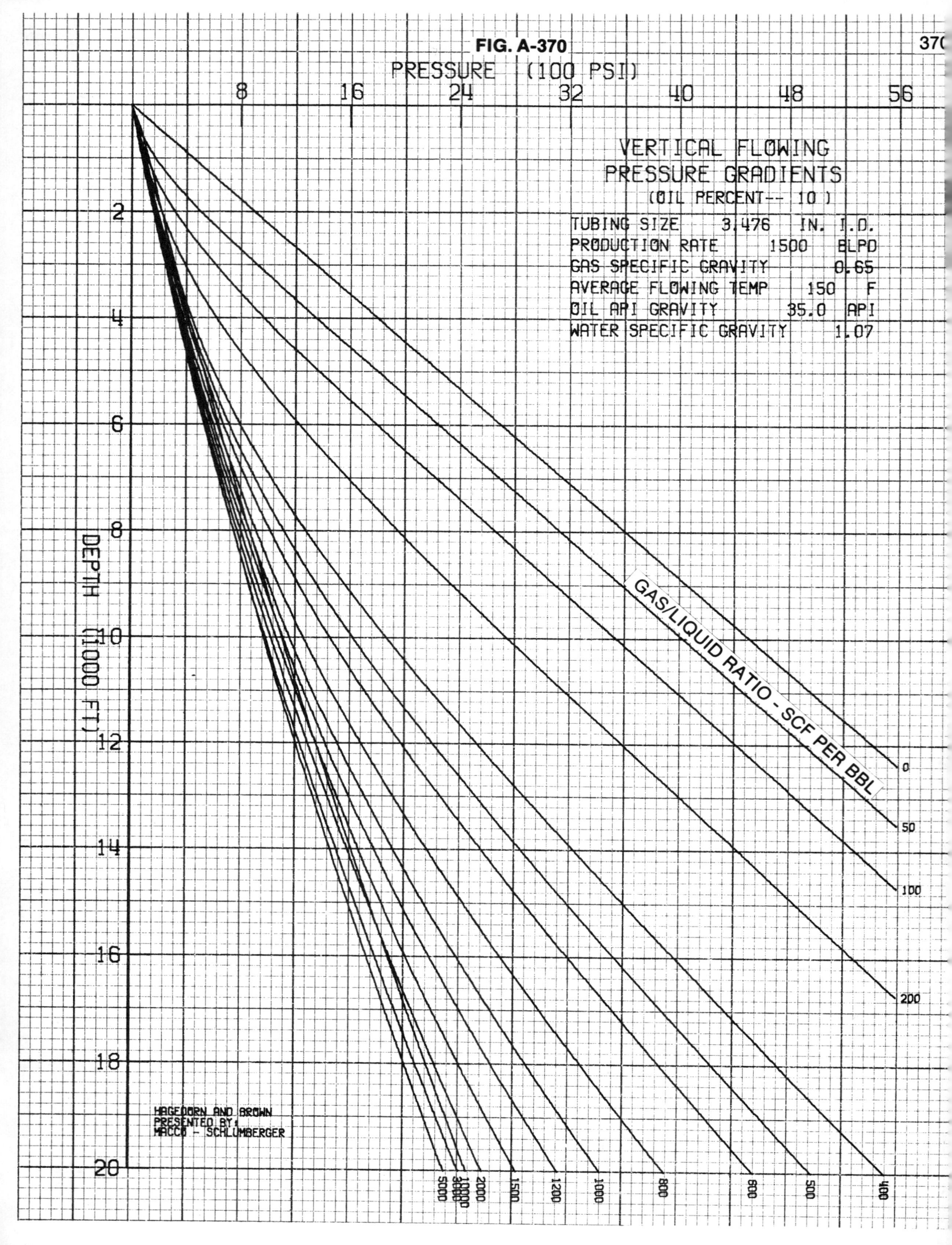

PRESSURE (100 PSI)

VERTICAL FLOWING
PRESSURE GRADIENTS
(OIL PERCENT-- 10)

TUBING SIZE	3.476	IN. I.D.
PRODUCTION RATE	1500	BLPD
GAS SPECIFIC GRAVITY	0.65	
AVERAGE FLOWING TEMP	150	F
OIL API GRAVITY	35.0	API
WATER SPECIFIC GRAVITY	1.07	

DEPTH (1000 FT)

GAS/LIQUID RATIO - SCF PER BBL

0
50
100
200

HAGEDORN AND BROWN
PRESENTED BY:
MACCO - SCHLUMBERGER

5000 3000 2000 1500 1200 1000 800 600 500 400

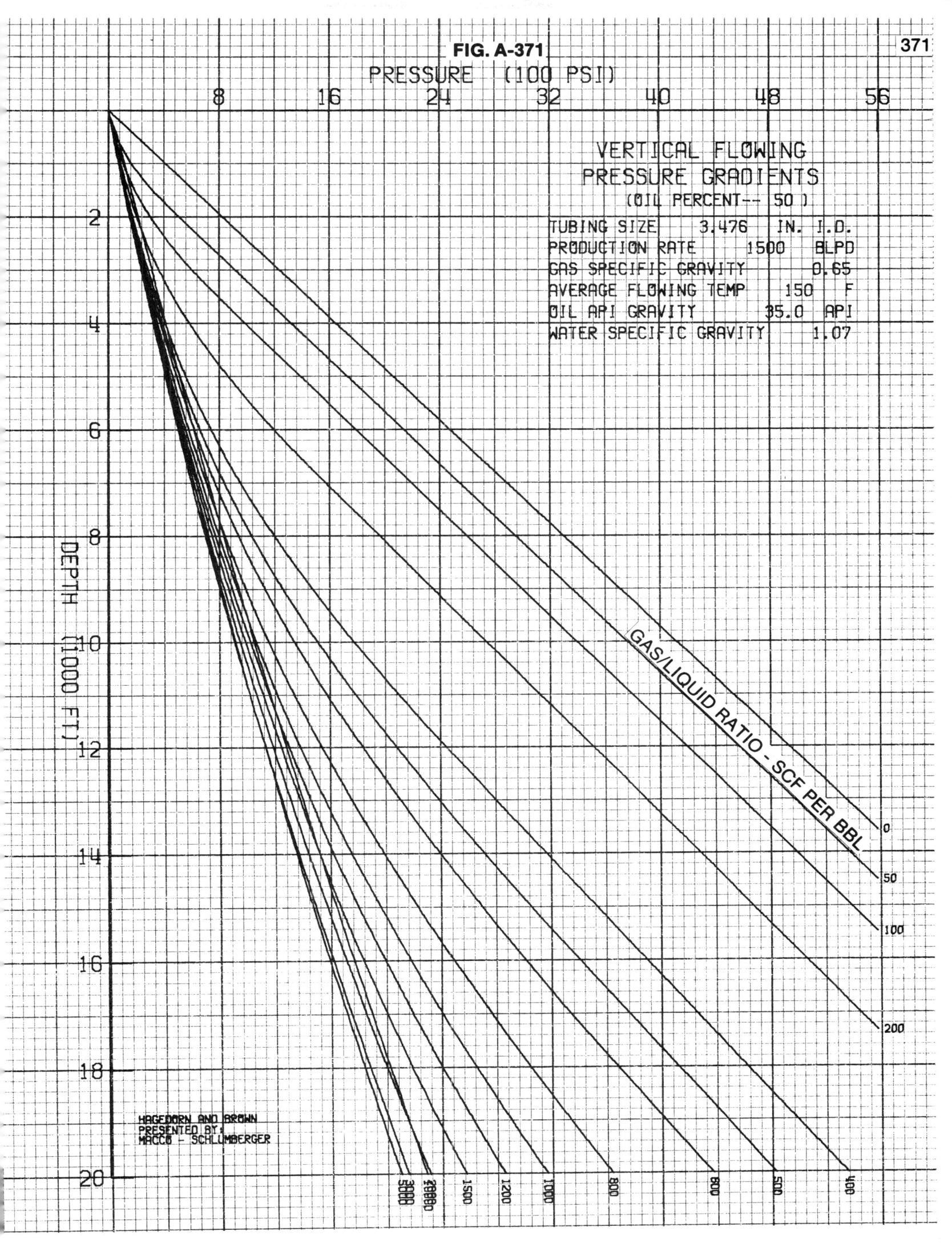

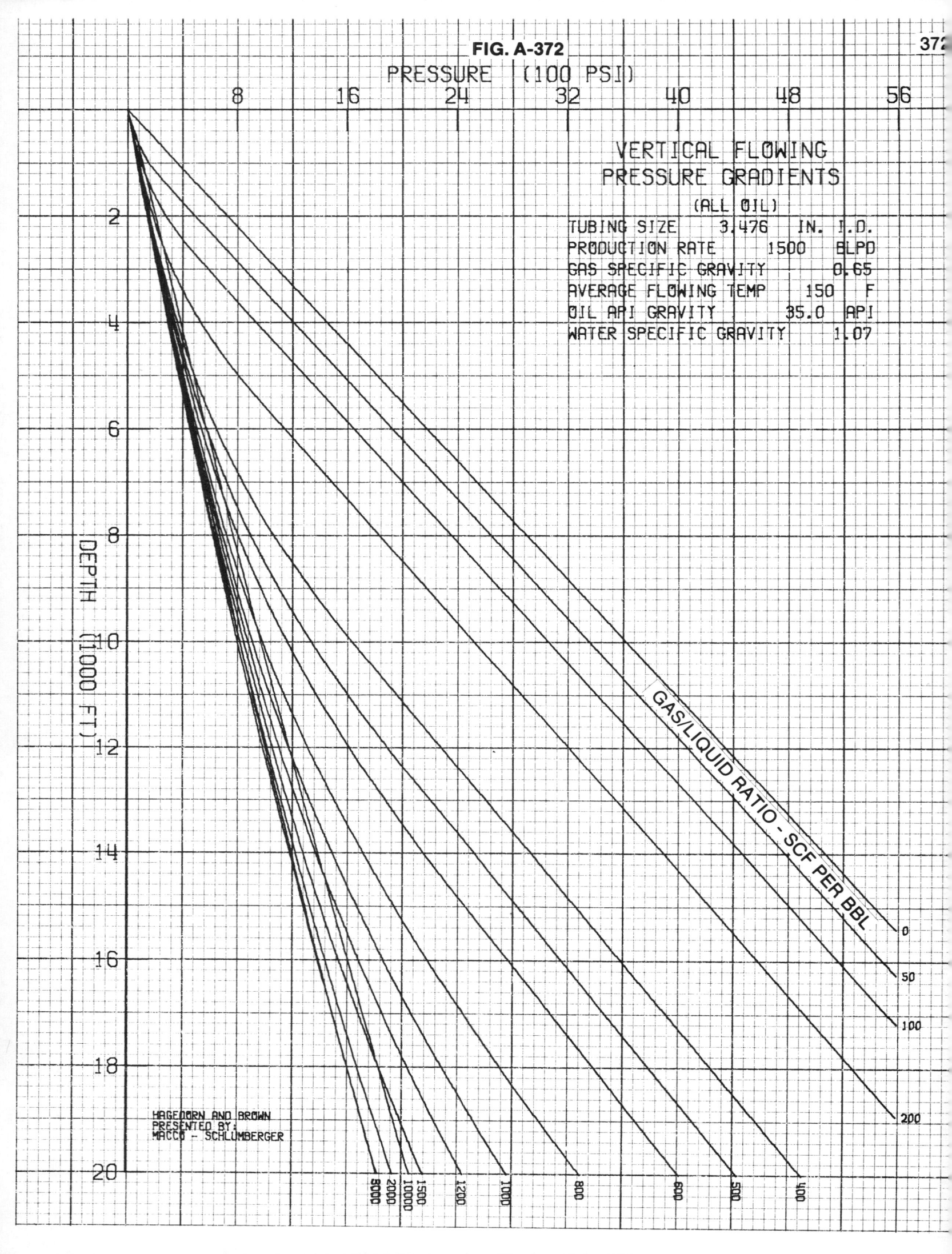

FIG. A-372

PRESSURE (100 PSI)

VERTICAL FLOWING
PRESSURE GRADIENTS
(ALL OIL)

TUBING SIZE	3.476	IN. I.D.
PRODUCTION RATE	1500	BLPD
GAS SPECIFIC GRAVITY	0.65	
AVERAGE FLOWING TEMP	150	F
OIL API GRAVITY	35.0	API
WATER SPECIFIC GRAVITY	1.07	

GAS/LIQUID RATIO - SCF PER BBL

DEPTH (1000 FT)

HAGEDORN AND BROWN
PRESENTED BY:
MACCO - SCHLUMBERGER

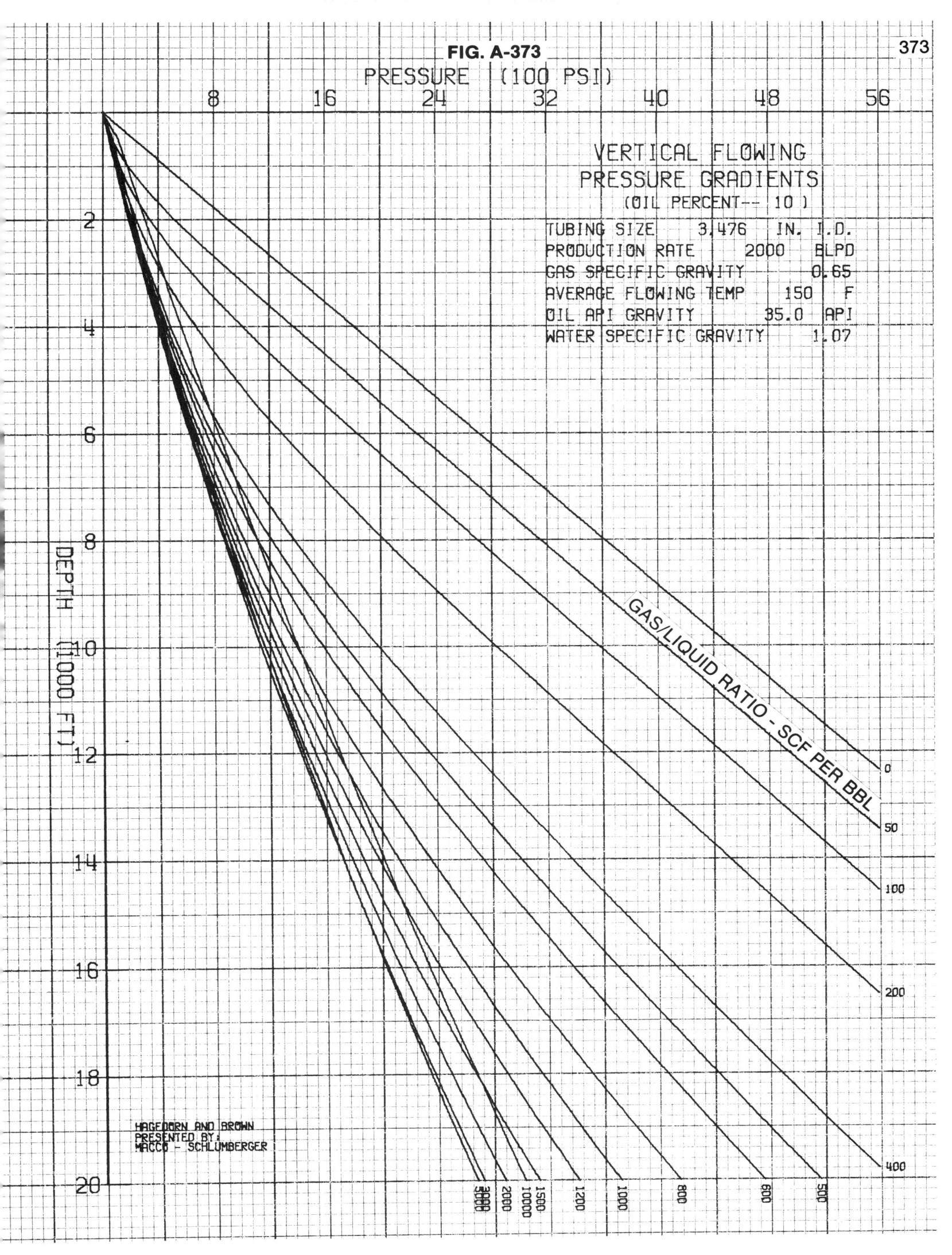

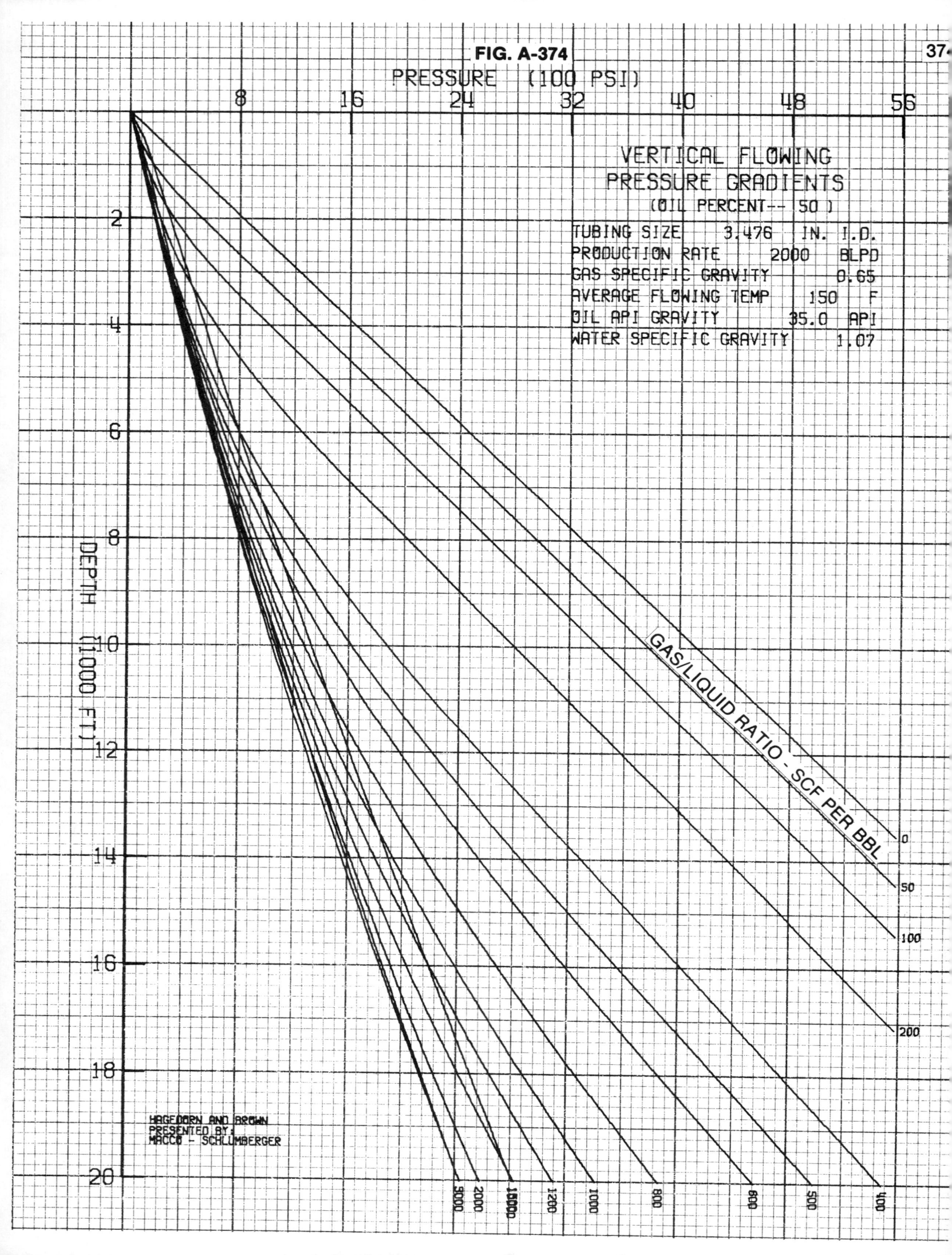

PRESSURE (100 PSI)

VERTICAL FLOWING
PRESSURE GRADIENTS
(OIL PERCENT-- 50)

TUBING SIZE	3.476	IN. I.D.
PRODUCTION RATE	2000	BLPD
GAS SPECIFIC GRAVITY		0.65
AVERAGE FLOWING TEMP	150	F
OIL API GRAVITY	35.0	API
WATER SPECIFIC GRAVITY		1.07

GAS/LIQUID RATIO - SCF PER BBL

DEPTH (1000 FT)

HAGEDORN AND BROWN
PRESENTED BY:
MACCO - SCHLUMBERGER

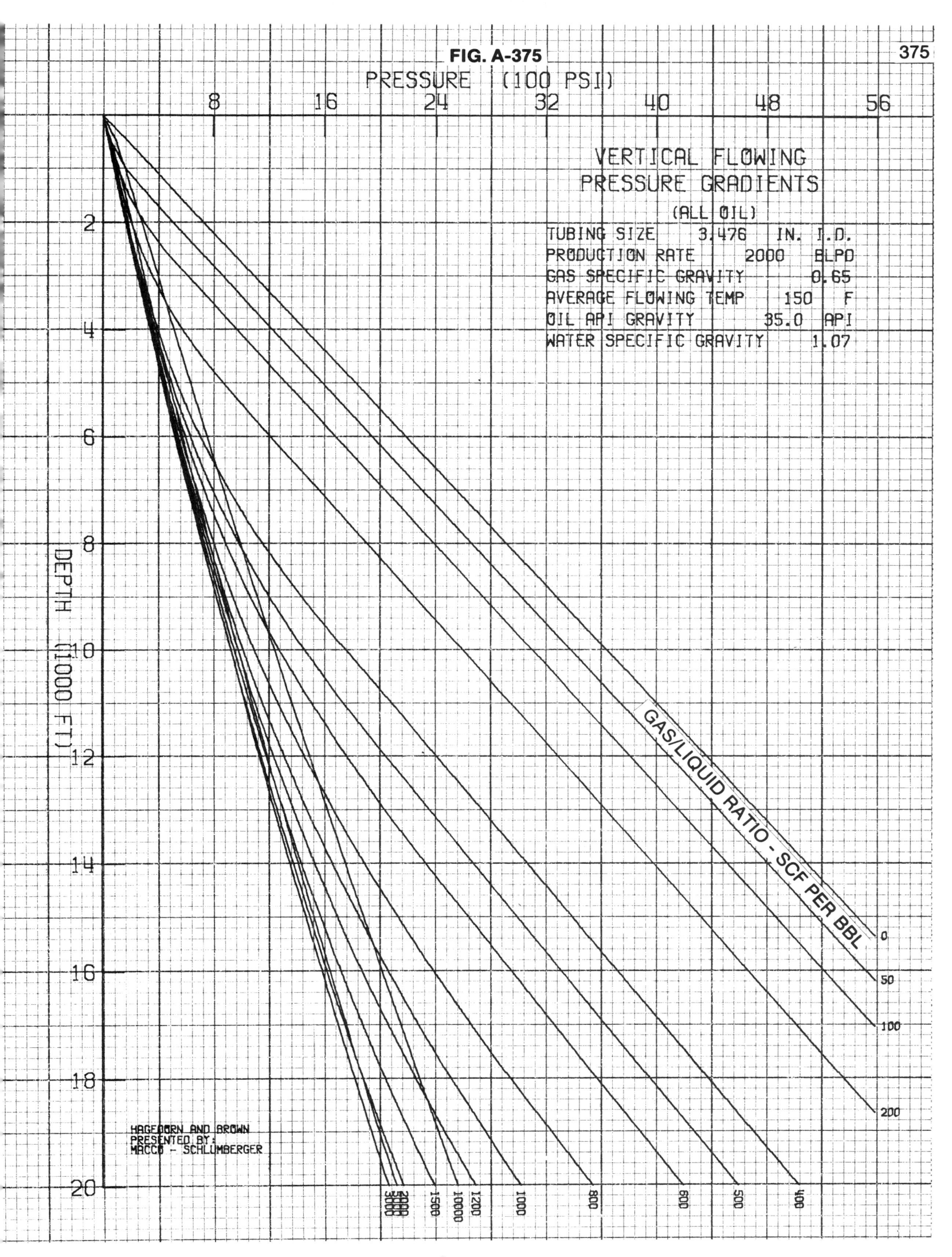

FIG. A-375

VERTICAL FLOWING
PRESSURE GRADIENTS
(ALL OIL)

TUBING SIZE	3.476	IN. I.D.
PRODUCTION RATE	2000	BLPD
GAS SPECIFIC GRAVITY	0.65	
AVERAGE FLOWING TEMP	150	F
OIL API GRAVITY	35.0	API
WATER SPECIFIC GRAVITY	1.07	

PRESSURE (100 PSI)

DEPTH (1000 FT)

GAS/LIQUID RATIO - SCF PER BBL

HAGEDORN AND BROWN
PRESENTED BY:
MACCO - SCHLUMBERGER

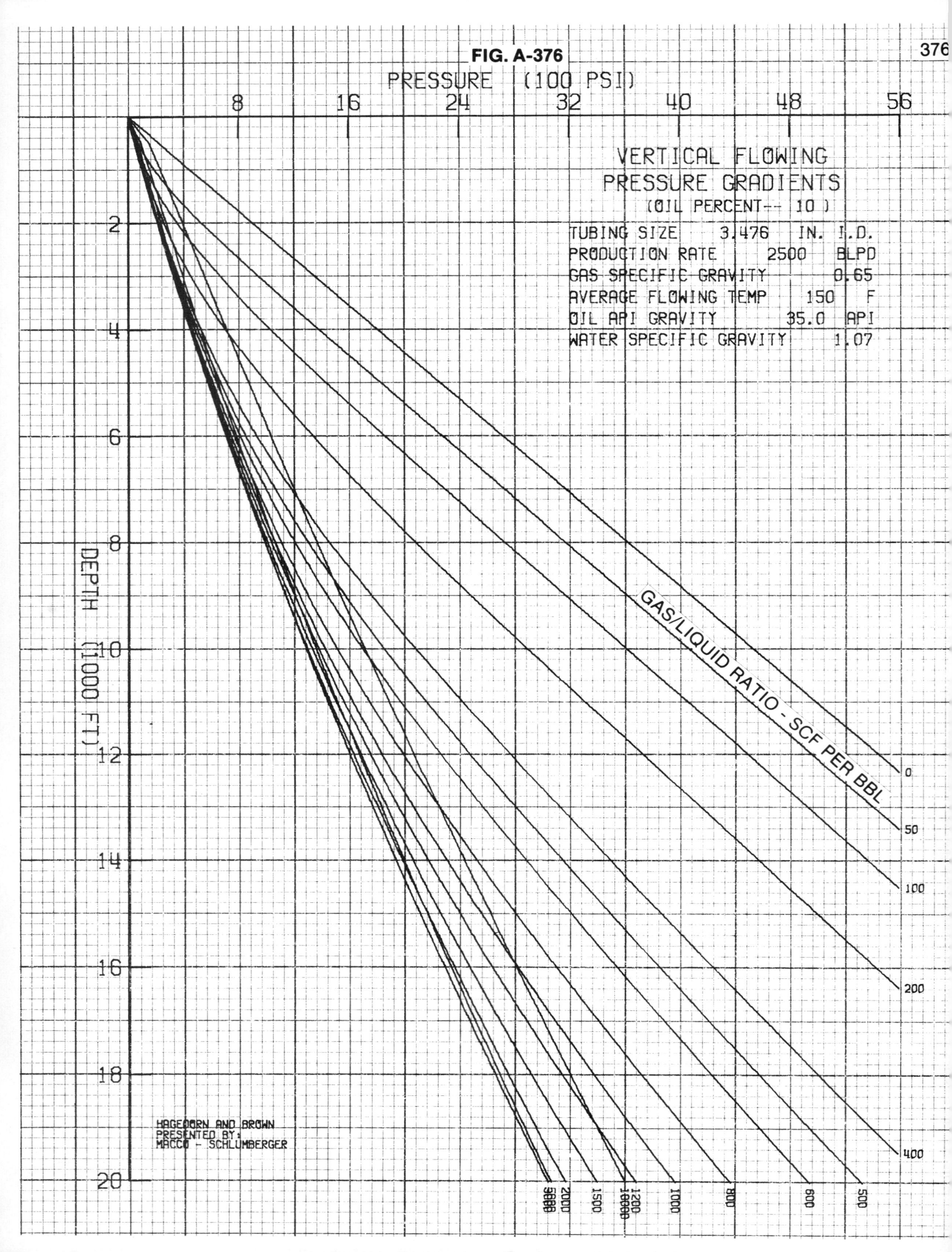

FIG. A-376

376

PRESSURE (100 PSI)

VERTICAL FLOWING
PRESSURE GRADIENTS
(OIL PERCENT-- 10)

TUBING SIZE	3.476	IN. I.D.
PRODUCTION RATE	2500	BLPD
GAS SPECIFIC GRAVITY	0.65	
AVERAGE FLOWING TEMP	150	F
OIL API GRAVITY	35.0	API
WATER SPECIFIC GRAVITY	1.07	

DEPTH (1000 FT)

GAS/LIQUID RATIO - SCF PER BBL

HAGEDORN AND BROWN
PRESENTED BY:
MACCO - SCHLUMBERGER

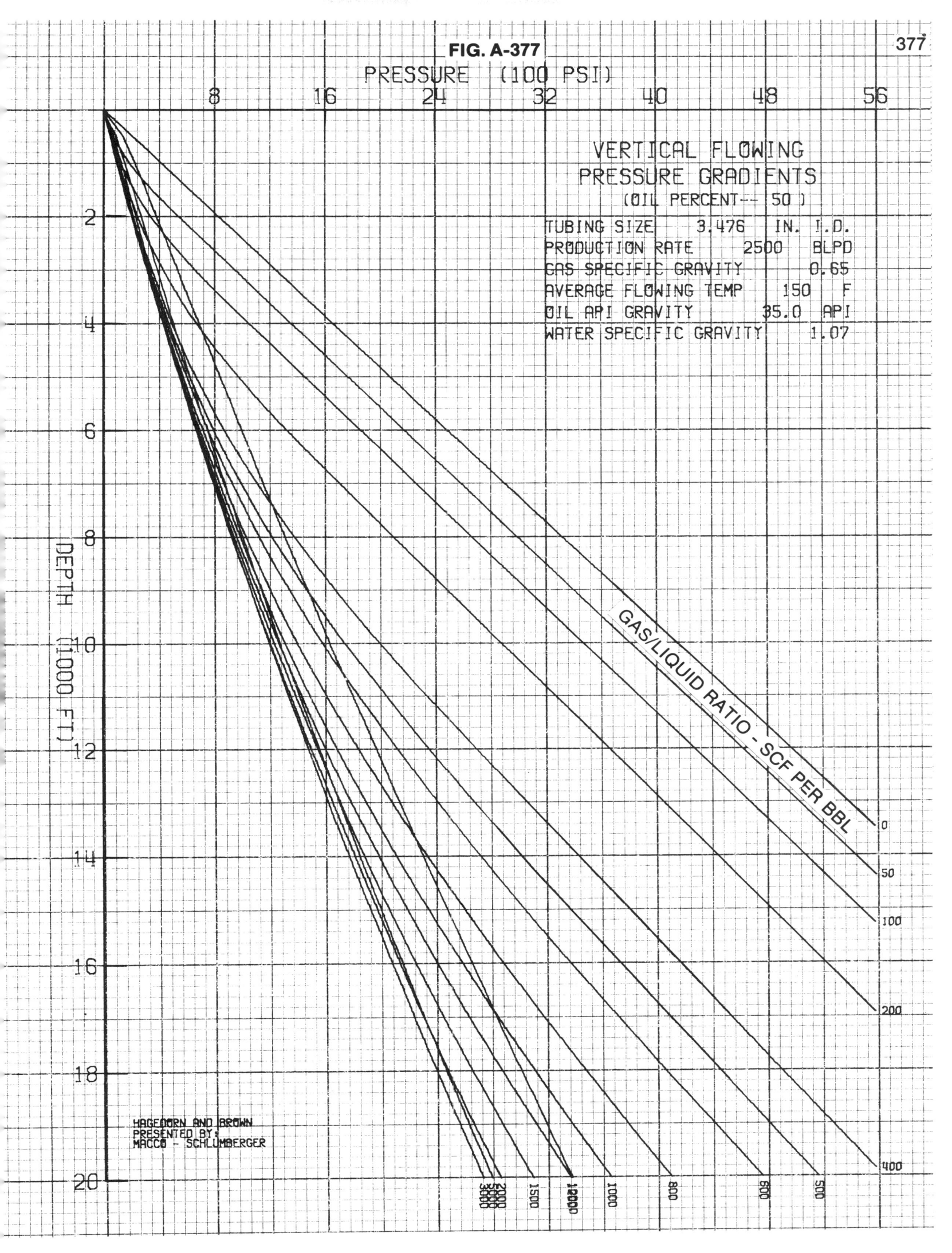

PRESSURE (100 PSI)

VERTICAL FLOWING
PRESSURE GRADIENTS
(OIL PERCENT--- 50)

TUBING SIZE	3.476	IN. I.D.
PRODUCTION RATE	2500	BLPD
GAS SPECIFIC GRAVITY	0.65	
AVERAGE FLOWING TEMP	150	F
OIL API GRAVITY	35.0	API
WATER SPECIFIC GRAVITY	1.07	

DEPTH (1000 FT)

GAS/LIQUID RATIO - SCF PER BBL

HAGEDORN AND BROWN
PRESENTED BY:
MACCO - SCHLUMBERGER

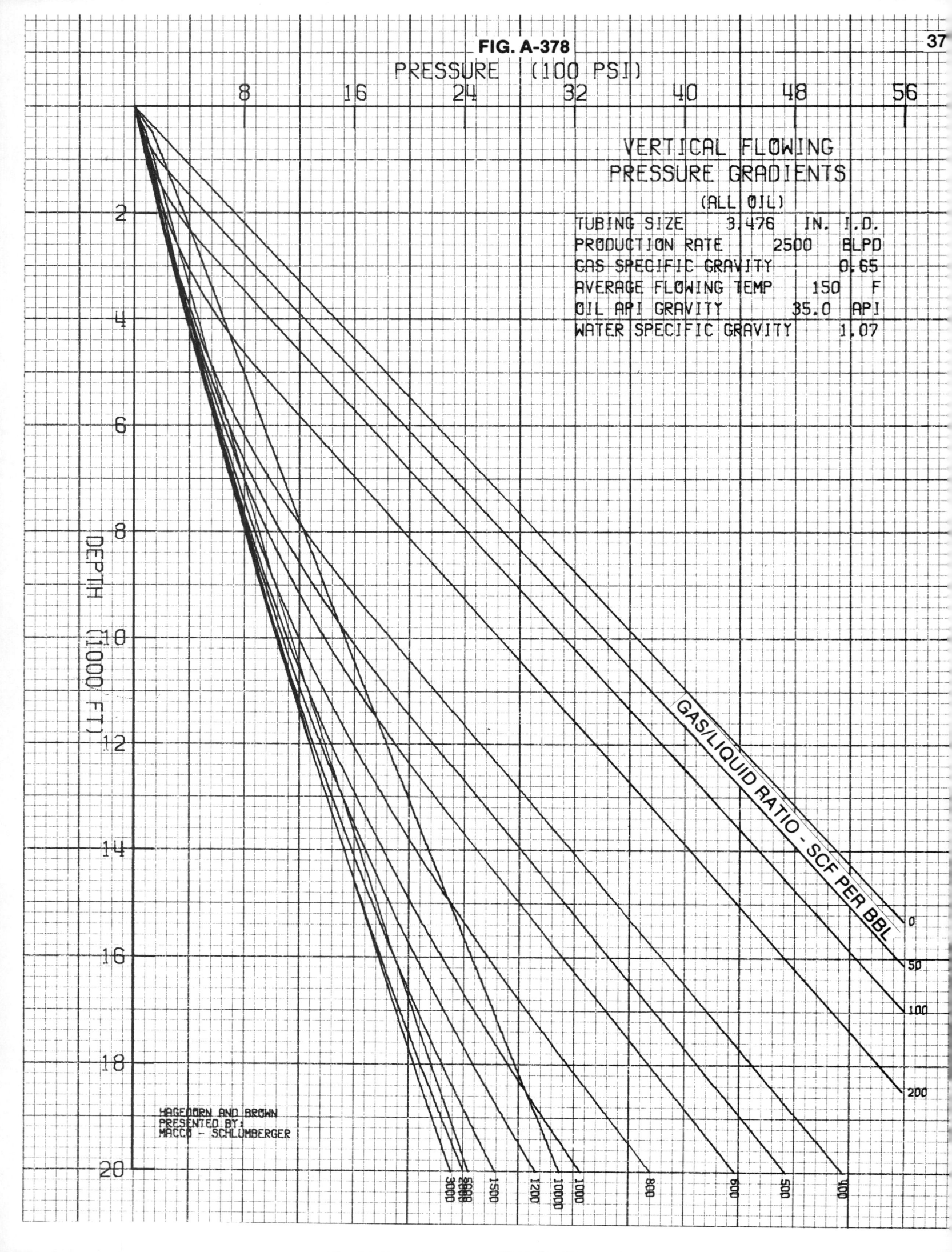

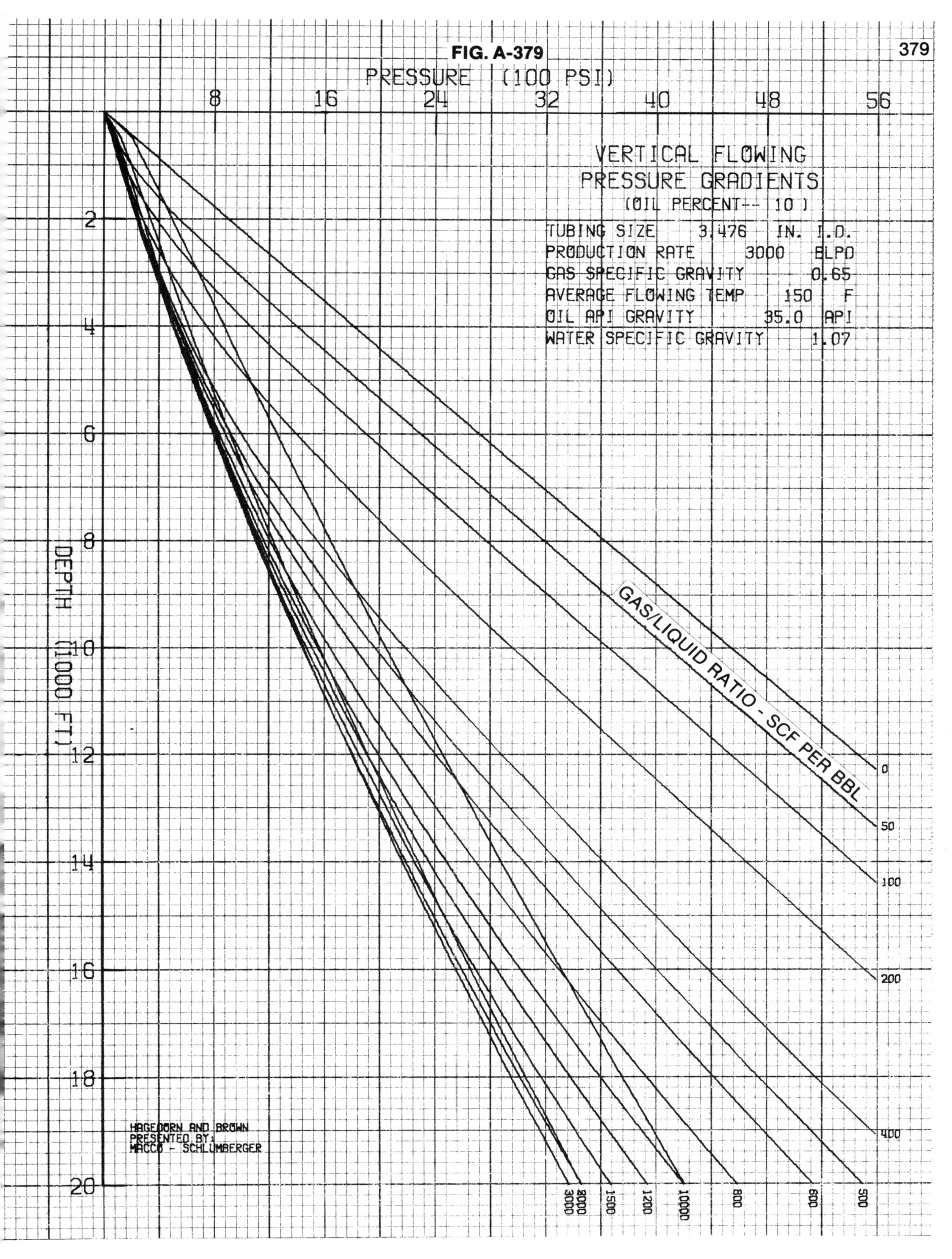

FIG. A-379

379

PRESSURE (100 PSI)

VERTICAL FLOWING
PRESSURE GRADIENTS
(OIL PERCENT-- 10)

TUBING SIZE   3.476   IN. I.D.
PRODUCTION RATE   3000   BLPD
GAS SPECIFIC GRAVITY   0.65
AVERAGE FLOWING TEMP   150   F
OIL API GRAVITY   35.0   API
WATER SPECIFIC GRAVITY   1.07

GAS/LIQUID RATIO - SCF PER BBL

DEPTH (1000 FT)

HAGEDORN AND BROWN
PRESENTED BY:
MACCO - SCHLUMBERGER

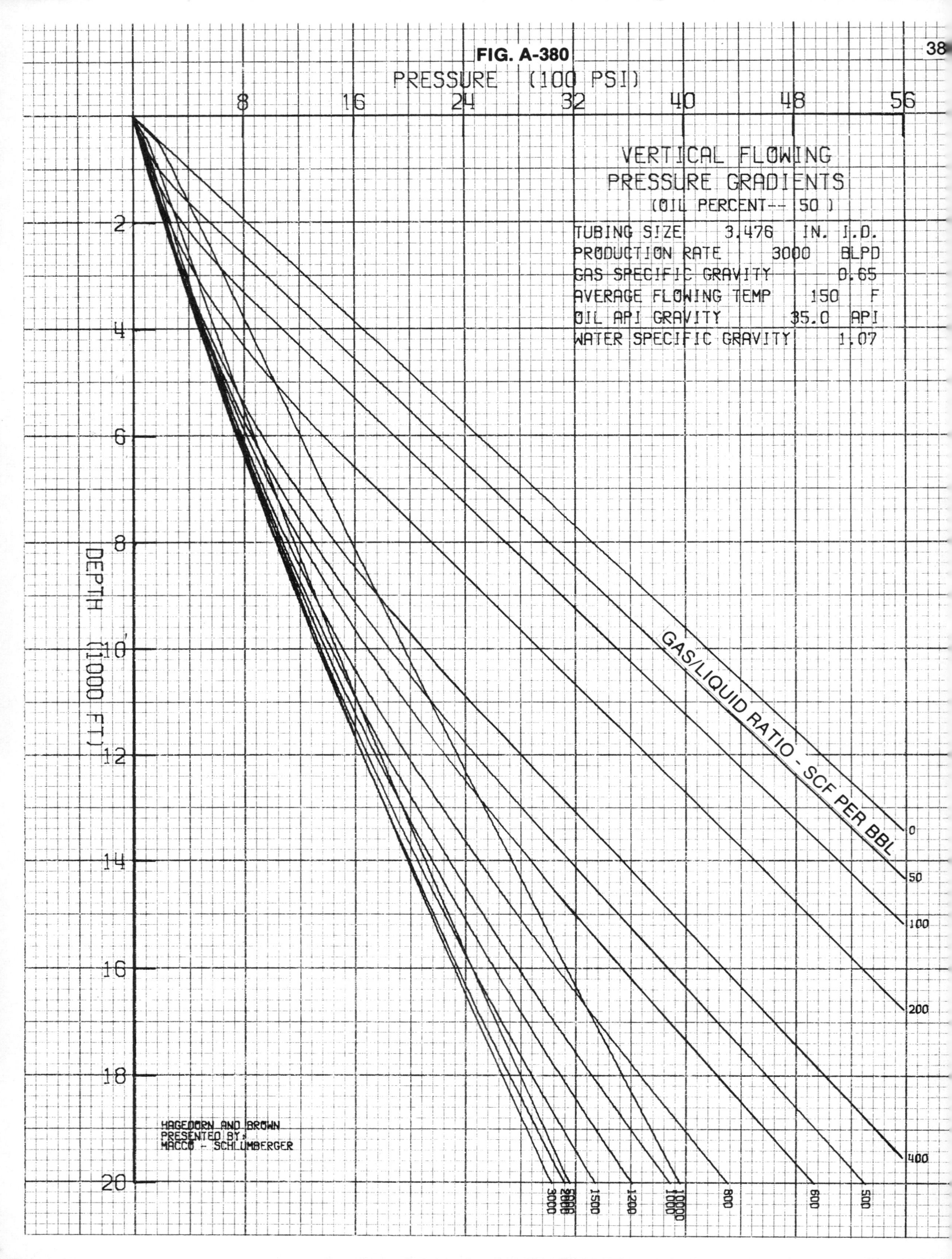

FIG. A-380

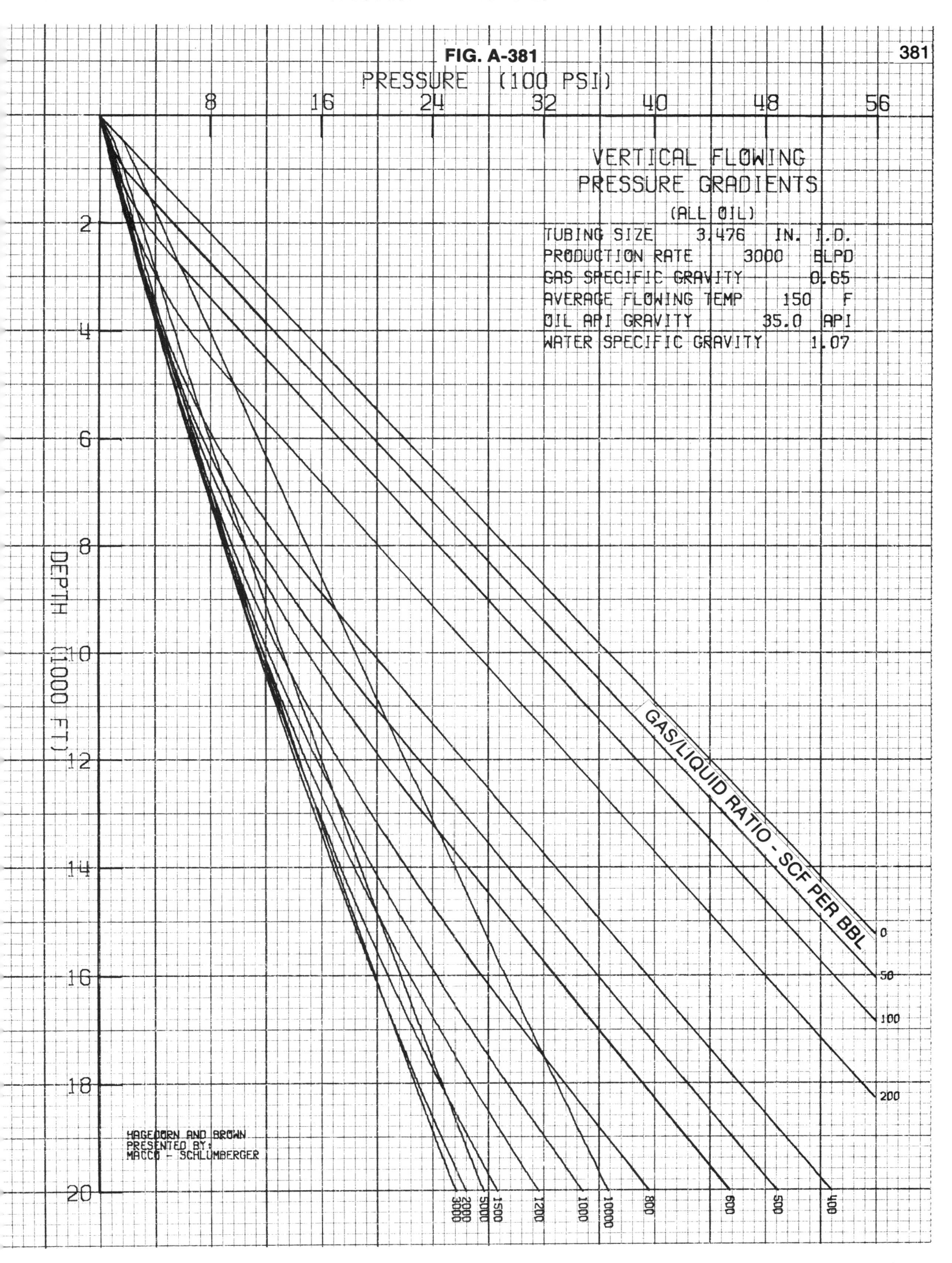

FIG. A-381

381

VERTICAL FLOWING
PRESSURE GRADIENTS
(ALL OIL)

TUBING SIZE	3.476	IN. I.D.
PRODUCTION RATE	3000	BLPD
GAS SPECIFIC GRAVITY	0.65	
AVERAGE FLOWING TEMP	150	F
OIL API GRAVITY	35.0	API
WATER SPECIFIC GRAVITY	1.07	

PRESSURE (100 PSI)

DEPTH (1000 FT)

GAS/LIQUID RATIO - SCF PER BBL

HAGEDORN AND BROWN
PRESENTED BY:
MACCO - SCHLUMBERGER

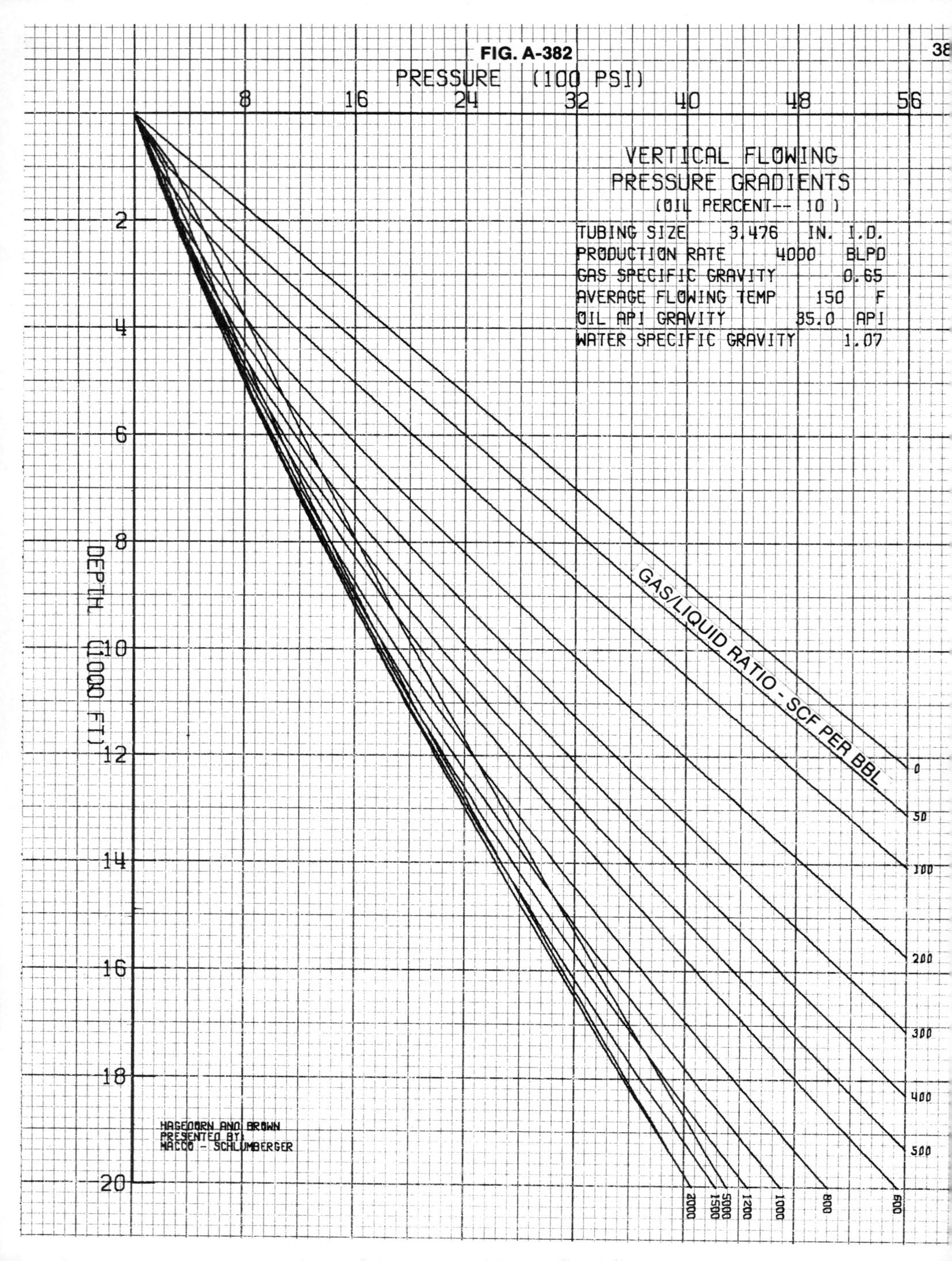

FIG. A-382

VERTICAL FLOWING
PRESSURE GRADIENTS
(OIL PERCENT-- 10)

TUBING SIZE	3.476	IN. I.D.
PRODUCTION RATE	4000	BLPD
GAS SPECIFIC GRAVITY	0.65	
AVERAGE FLOWING TEMP	150	F
OIL API GRAVITY	35.0	API
WATER SPECIFIC GRAVITY	1.07	

PRESSURE (100 PSI)

DEPTH (1000 FT)

GAS/LIQUID RATIO - SCF PER BBL

HAGEDORN AND BROWN
PRESENTED BY
MACCO - SCHLUMBERGER

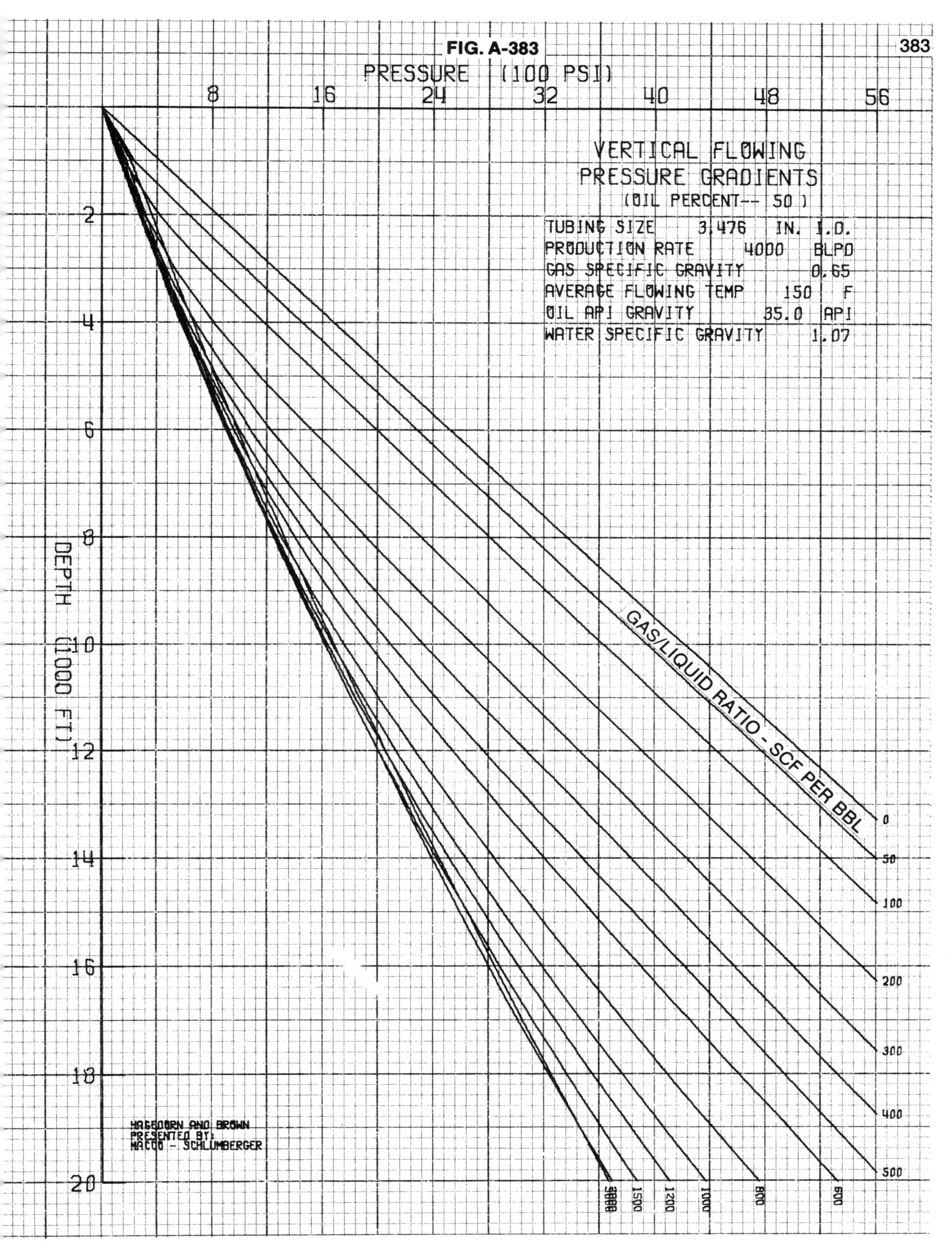

PRESSURE (100 PSI)

VERTICAL FLOWING
PRESSURE GRADIENTS
(OIL PERCENT-- 50)

TUBING SIZE	3.476	IN. I.D.
PRODUCTION RATE	4000	BLPD
GAS SPECIFIC GRAVITY		0.65
AVERAGE FLOWING TEMP	150	F
OIL API GRAVITY	35.0	API
WATER SPECIFIC GRAVITY		1.07

DEPTH (1000 FT)

GAS/LIQUID RATIO - SCF PER BBL

HAGEDORN AND BROWN
PRESENTED BY:
MAECO - SCHLUMBERGER

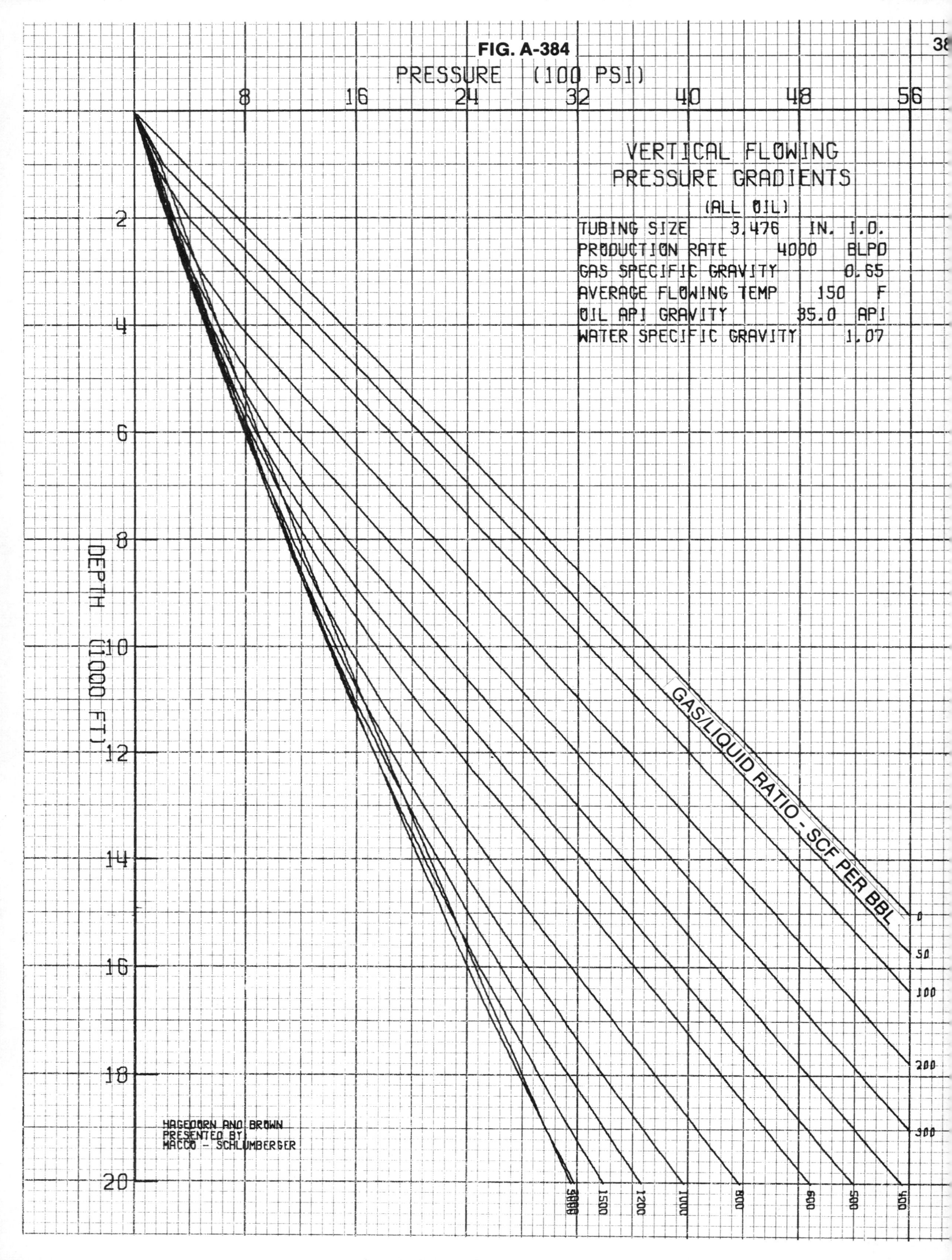

FIG. A-384

VERTICAL FLOWING
PRESSURE GRADIENTS
(ALL OIL)

TUBING SIZE	3.476	IN. I.D.
PRODUCTION RATE	4000	BLPD
GAS SPECIFIC GRAVITY	0.65	
AVERAGE FLOWING TEMP	150	F
OIL API GRAVITY	35.0	API
WATER SPECIFIC GRAVITY	1.07	

PRESSURE (100 PSI)

DEPTH (1000 FT)

GAS/LIQUID RATIO - SCF PER BBL

HAGEDORN AND BROWN
PRESENTED BY
MACCO - SCHLUMBERGER

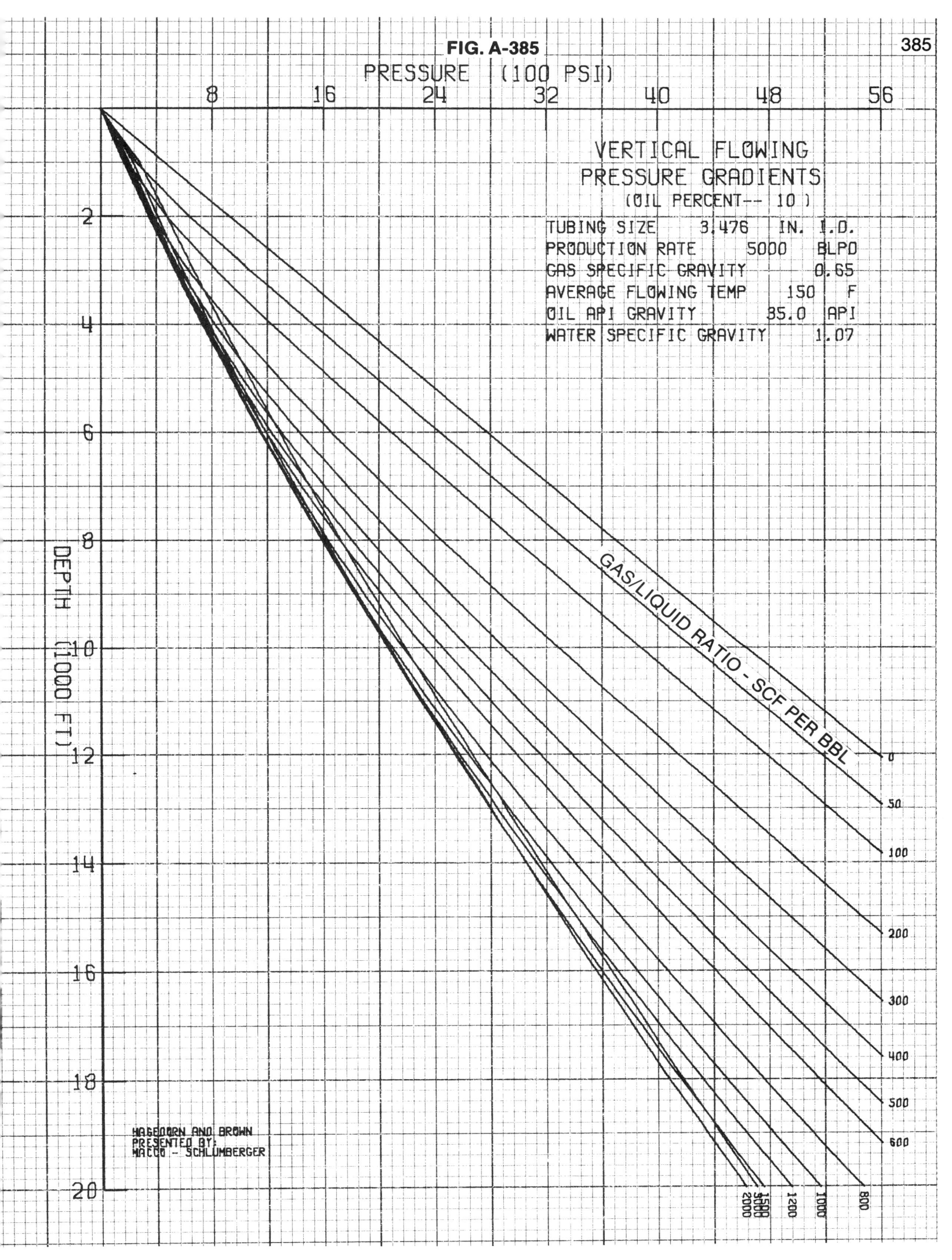

FIG. A-385

385

VERTICAL FLOWING
PRESSURE GRADIENTS
(OIL PERCENT-- 10)

TUBING SIZE	3.476	IN. I.D.
PRODUCTION RATE	5000	BLPD
GAS SPECIFIC GRAVITY		0.65
AVERAGE FLOWING TEMP	150	F
OIL API GRAVITY	35.0	API
WATER SPECIFIC GRAVITY		1.07

PRESSURE (100 PSI)

DEPTH (1000 FT)

GAS/LIQUID RATIO - SCF PER BBL

HAGEDORN AND BROWN
PRESENTED BY:
WATCO - SCHLUMBERGER

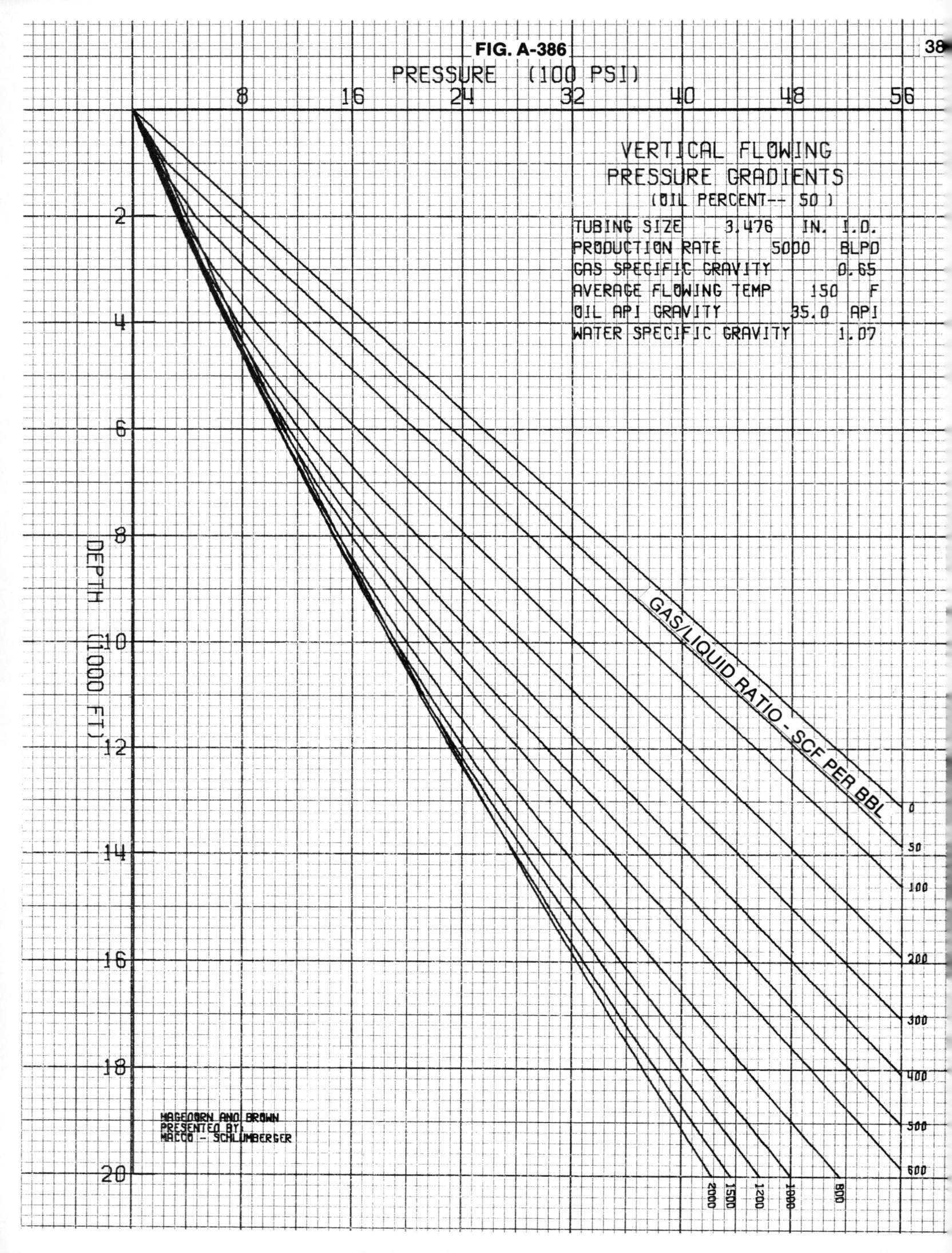

PRESSURE (100 PSI)

VERTICAL FLOWING
PRESSURE GRADIENTS
(OIL PERCENT-- 50)

TUBING SIZE	3.476	IN. I.D.
PRODUCTION RATE	5000	BLPD
GAS SPECIFIC GRAVITY		0.65
AVERAGE FLOWING TEMP	150	F
OIL API GRAVITY	35.0	API
WATER SPECIFIC GRAVITY		1.07

GAS/LIQUID RATIO - SCF PER BBL

DEPTH (1000 FT)

HAGEDORN AND BROWN
PRESENTED BY:
MACCO - SCHLUMBERGER

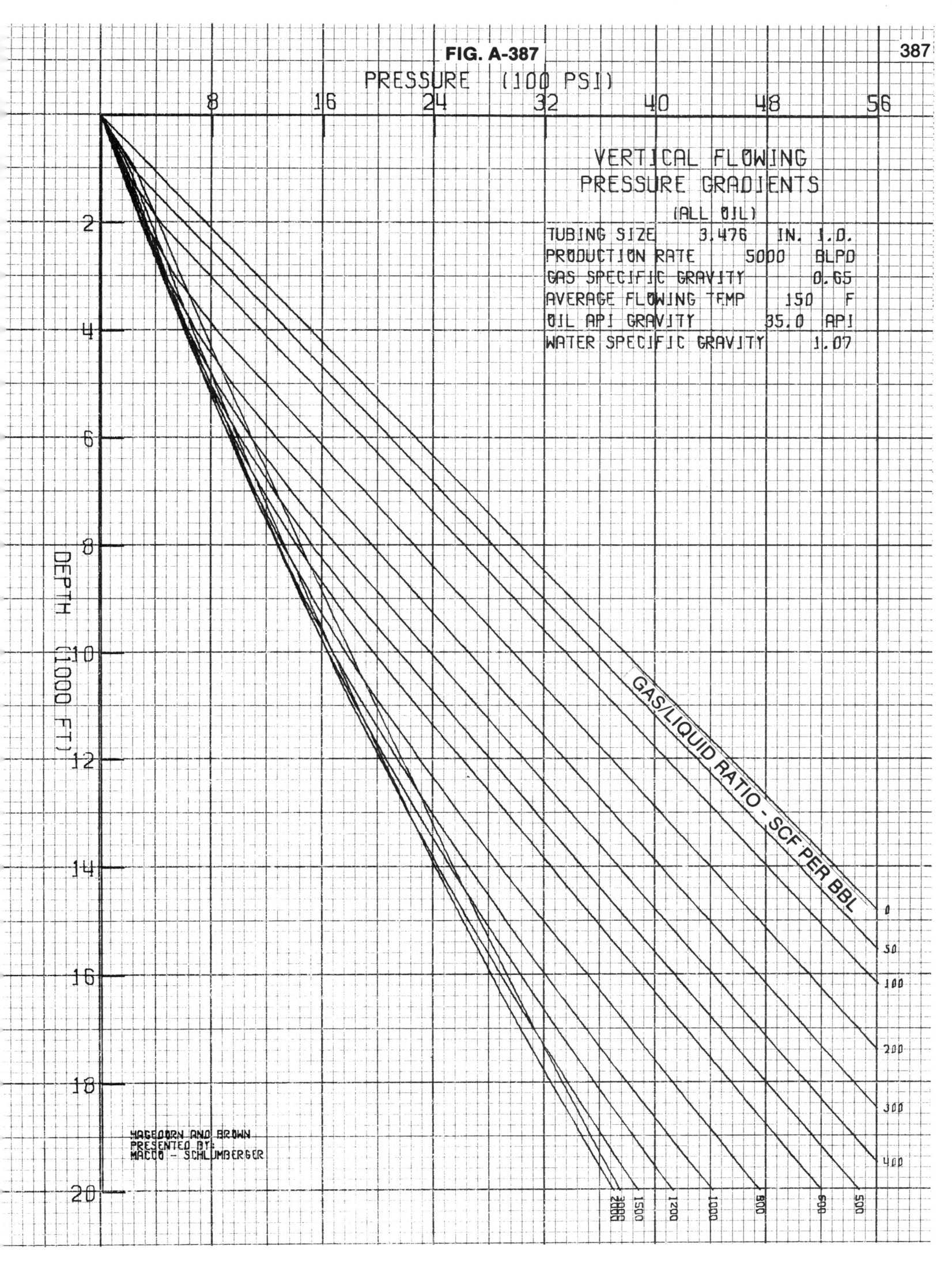

FIG. A-387

387

PRESSURE (100 PSI)

VERTICAL FLOWING
PRESSURE GRADIENTS
(ALL OIL)

TUBING SIZE	3.476	IN. I.D.
PRODUCTION RATE	5000	BLPD
GAS SPECIFIC GRAVITY	0.65	
AVERAGE FLOWING TEMP	150	F
OIL API GRAVITY	35.0	API
WATER SPECIFIC GRAVITY	1.07	

DEPTH (1000 FT)

GAS/LIQUID RATIO - SCF PER BBL

HAGEDORN AND BROWN
PRESENTED BY:
MAEDO - SCHLUMBERGER

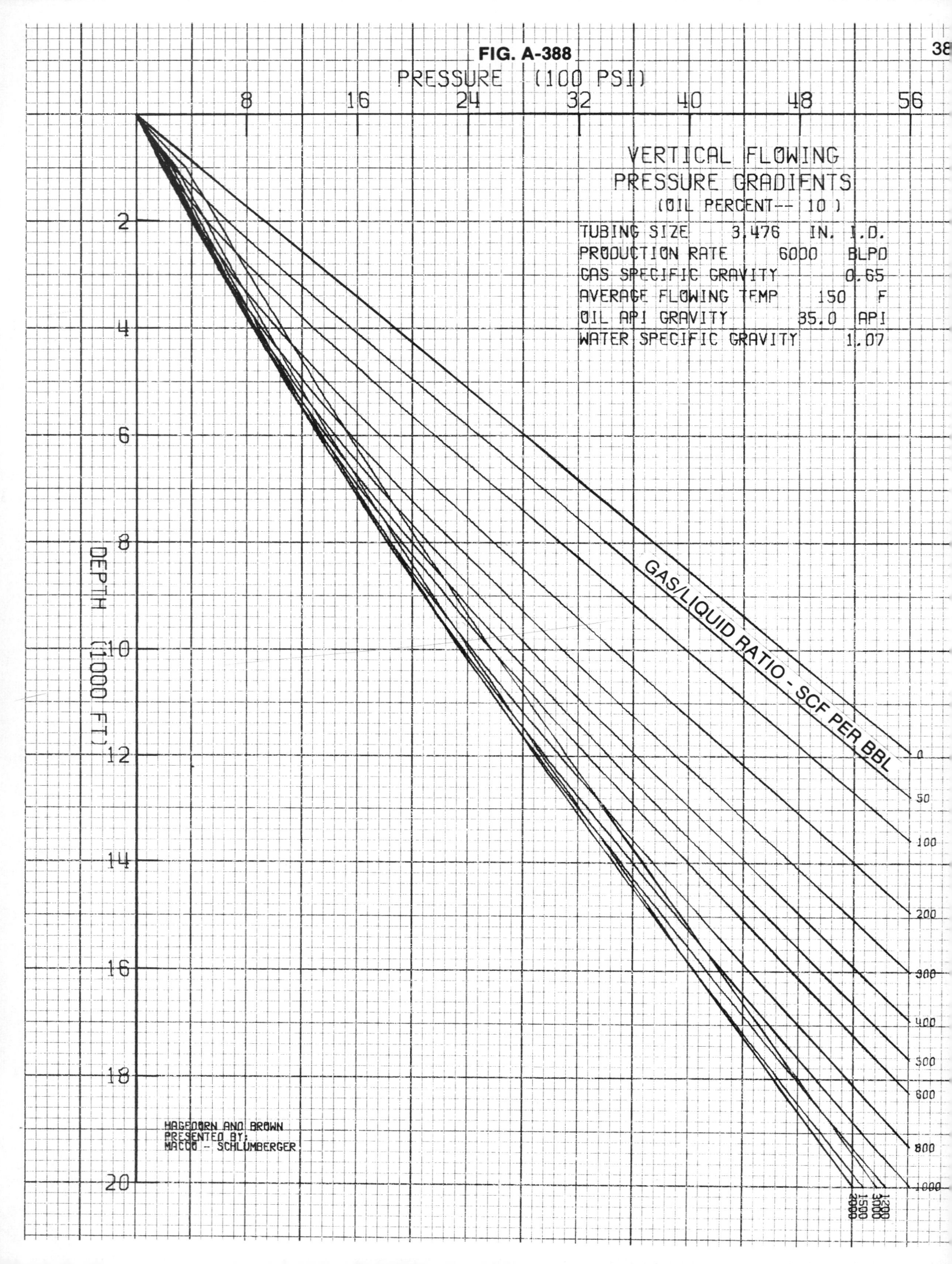

FIG. A-388

VERTICAL FLOWING
PRESSURE GRADIENTS
(OIL PERCENT— 10)

TUBING SIZE   3.476   IN. I.D.
PRODUCTION RATE   6000   BLPD
GAS SPECIFIC GRAVITY   0.65
AVERAGE FLOWING TEMP   150   F
OIL API GRAVITY   35.0 API
WATER SPECIFIC GRAVITY   1.07

GAS/LIQUID RATIO - SCF PER BBL

HAGEDORN AND BROWN
PRESENTED BY:
MACCO — SCHLUMBERGER

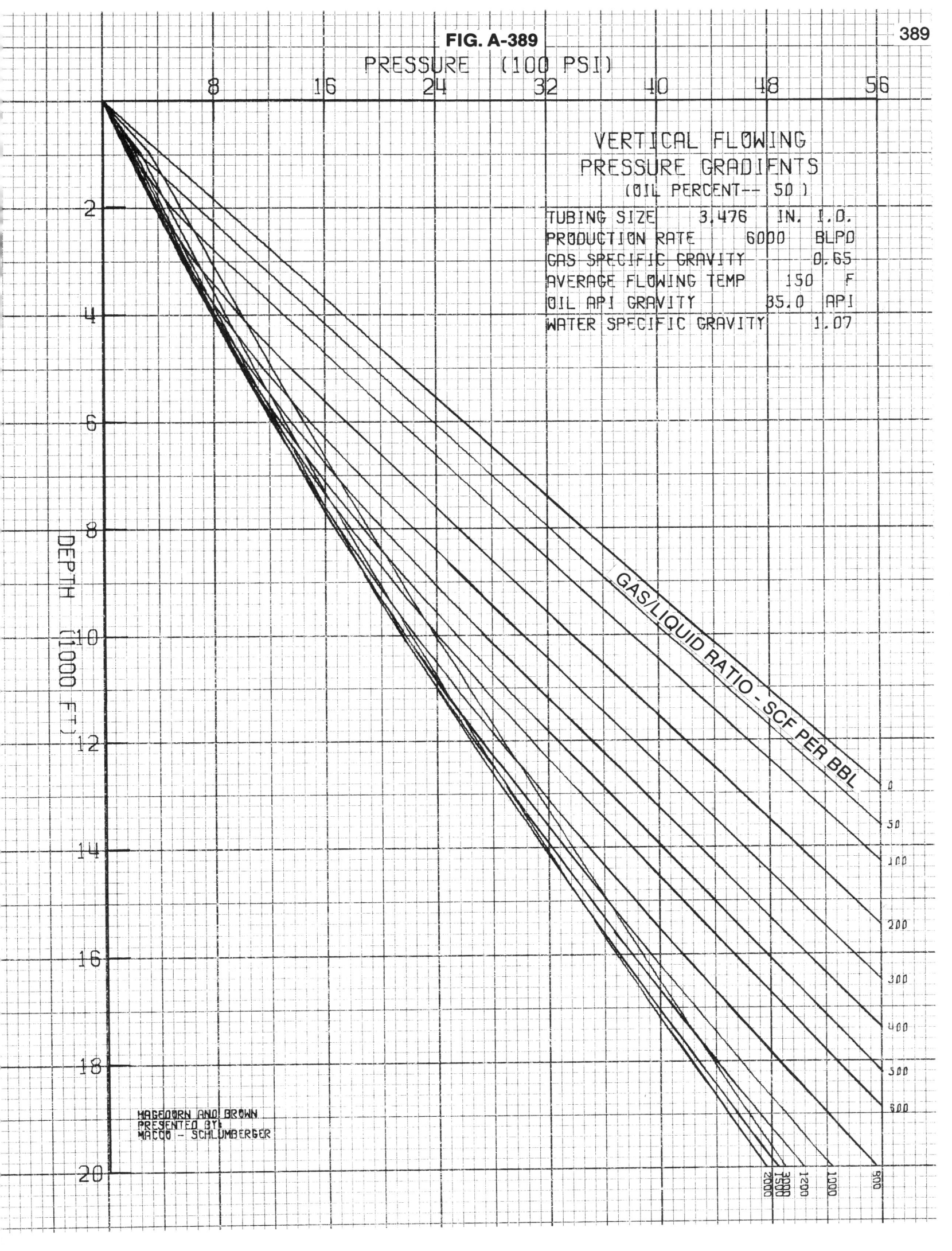

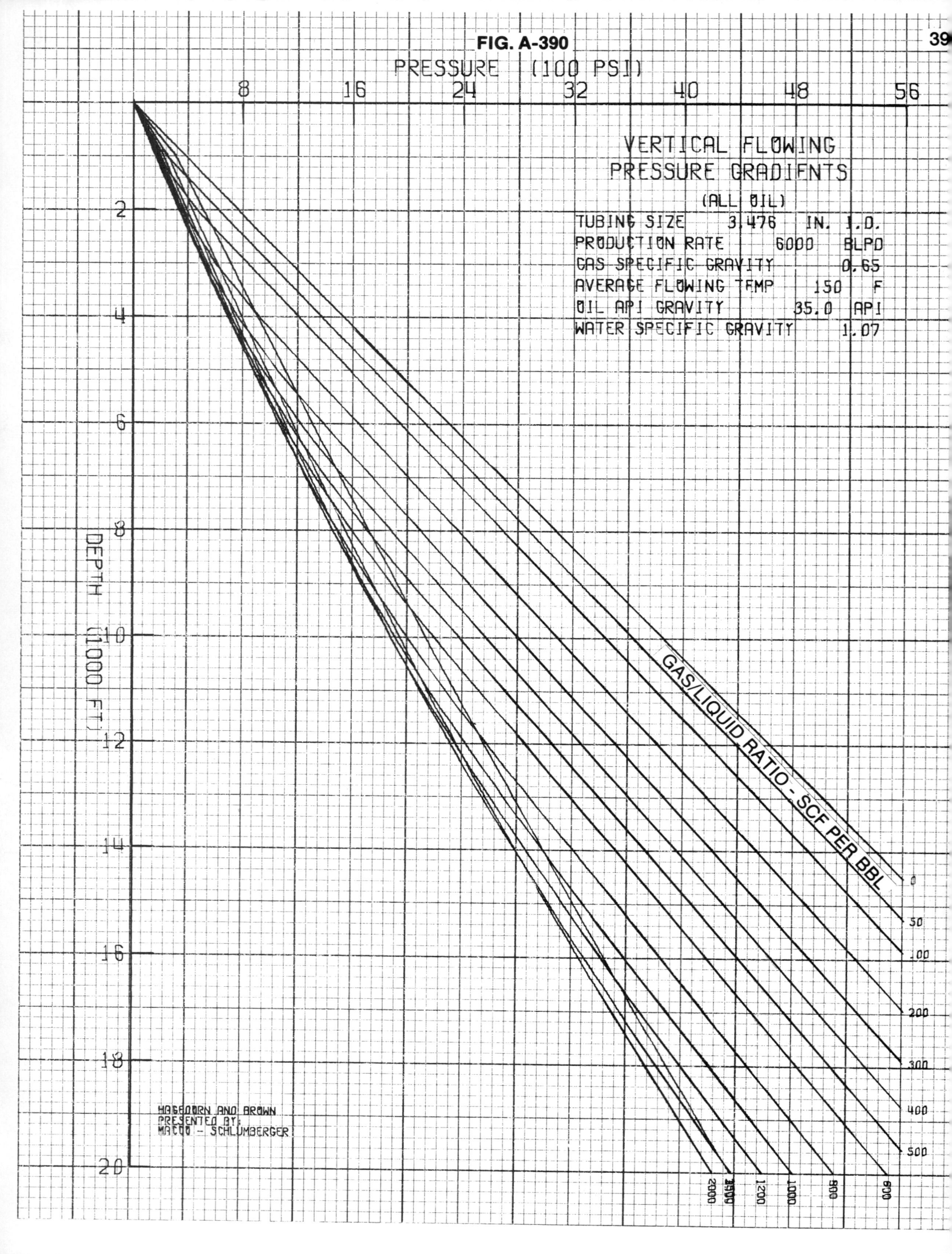

PRESSURE (100 PSI)

DEPTH (1000 FT)

VERTICAL FLOWING
PRESSURE GRADIENTS
(ALL OIL)

TUBING SIZE	3.476	IN. I.D.
PRODUCTION RATE	6000	BLPD
GAS SPECIFIC GRAVITY	0.65	
AVERAGE FLOWING TEMP	150	F
OIL API GRAVITY	35.0	API
WATER SPECIFIC GRAVITY	1.07	

GAS/LIQUID RATIO - SCF PER BBL

HAGEDORN AND BROWN
PRESENTED BY:
MACCO – SCHLUMBERGER

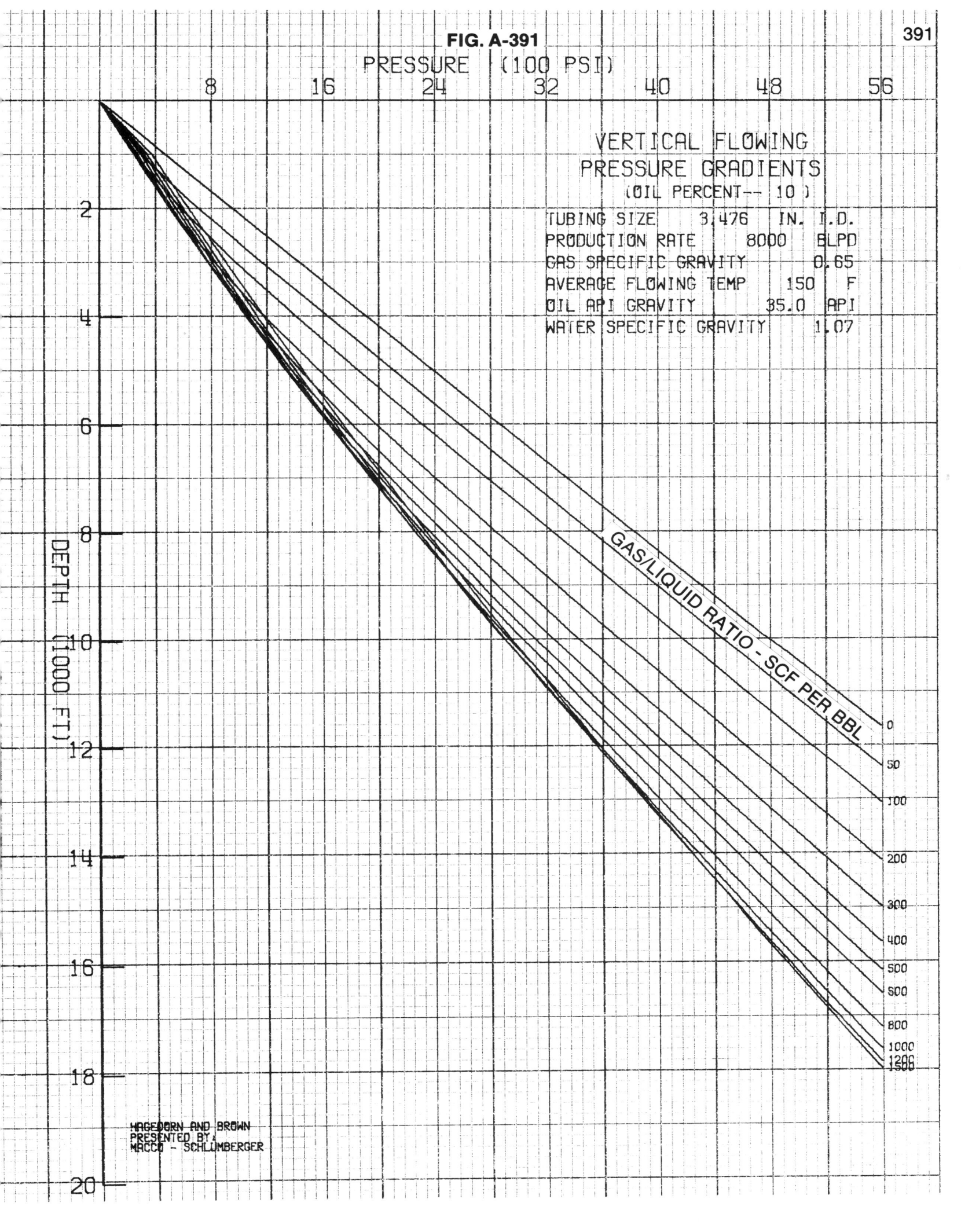

PRESSURE (100 PSI)

VERTICAL FLOWING
PRESSURE GRADIENTS
(OIL PERCENT-- 10)

TUBING SIZE	3.476	IN. I.D.
PRODUCTION RATE	8000	BLPD
GAS SPECIFIC GRAVITY		0.65
AVERAGE FLOWING TEMP	150	F
OIL API GRAVITY	35.0	API
WATER SPECIFIC GRAVITY		1.07

DEPTH (1000 FT)

GAS/LIQUID RATIO - SCF PER BBL

0
50
100
200
300
400
500
600
800
1000
1200
1500

HAGEDORN AND BROWN
PRESENTED BY
MACCO - SCHLUMBERGER

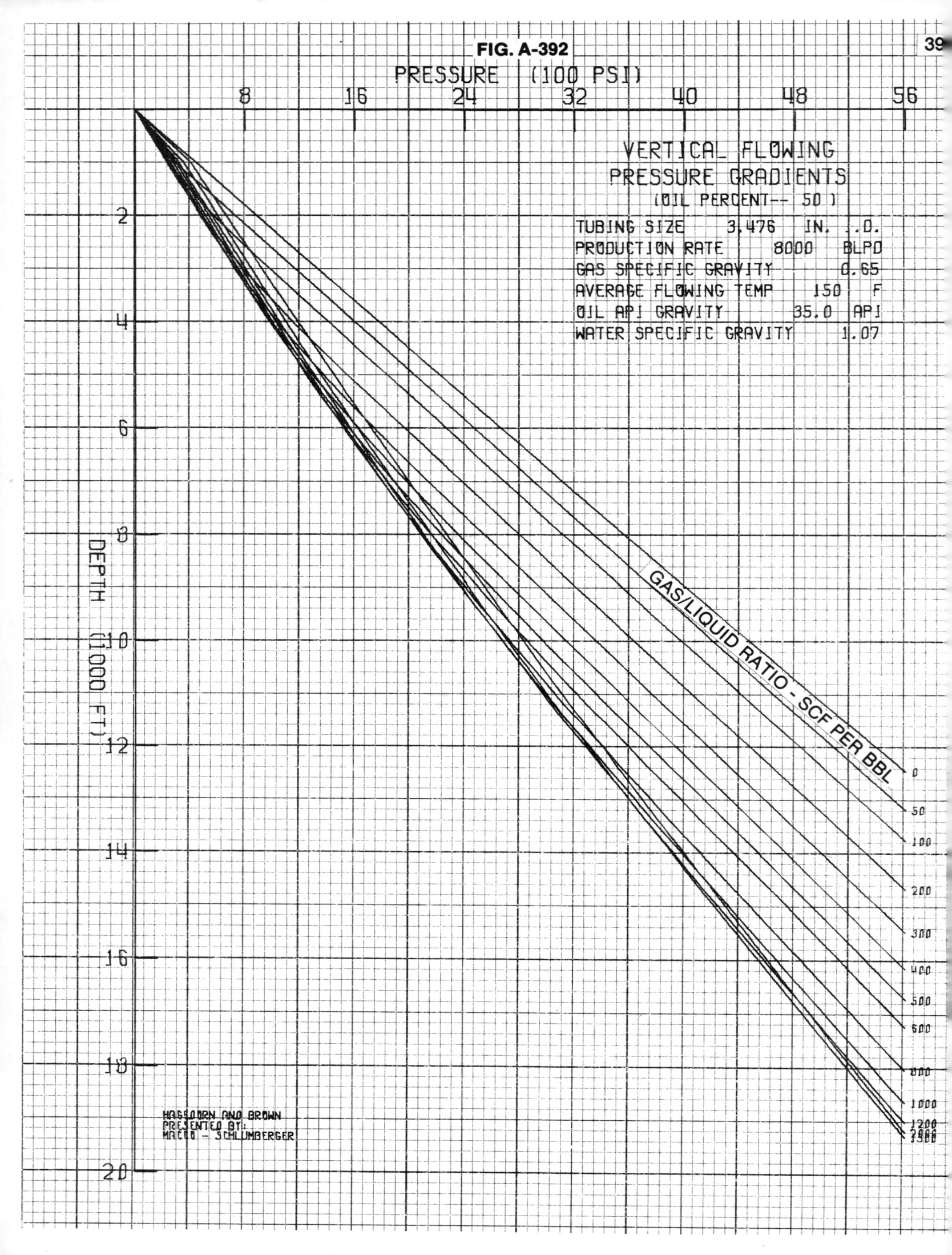

PRESSURE (100 PSI)

VERTICAL FLOWING
PRESSURE GRADIENTS
(OIL PERCENT-- 50)

TUBING SIZE	3.476	IN. I.D.
PRODUCTION RATE	8000	BLPD
GAS SPECIFIC GRAVITY	0.65	
AVERAGE FLOWING TEMP	150	F
OIL API GRAVITY	35.0	API
WATER SPECIFIC GRAVITY	1.07	

DEPTH (1000 FT)

GAS/LIQUID RATIO - SCF PER BBL

0
50
100
200
300
400
500
600
800
1000
1200
1900

HAGBOORN AND BROWN
PRESENTED BY:
MACCO - SCHLUMBERGER

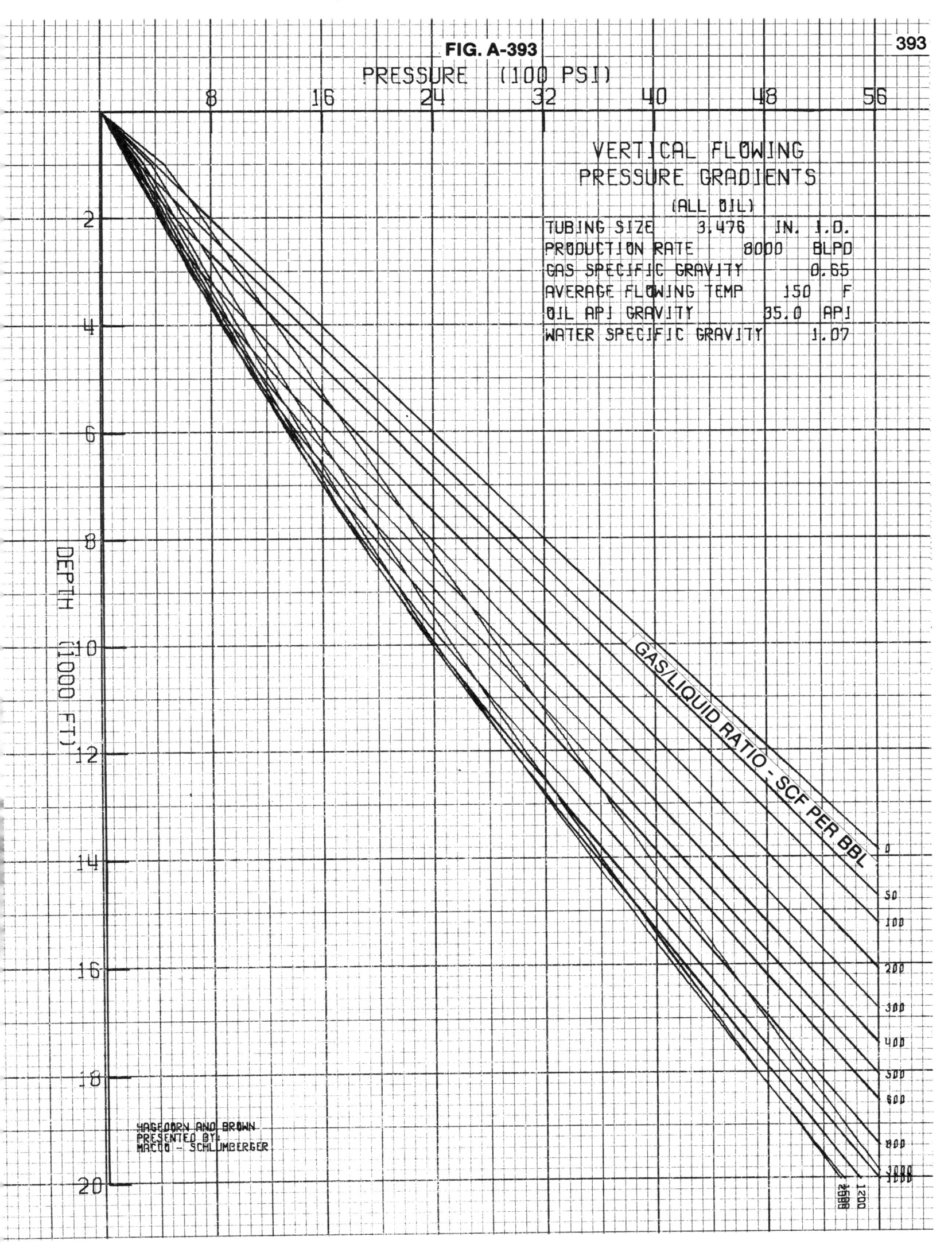

PRESSURE (100 PSI)

DEPTH (1000 FT)

VERTICAL FLOWING
PRESSURE GRADIENTS
(ALL OIL)

TUBING SIZE	3.476	IN. I.D.
PRODUCTION RATE	8000	BLPD
GAS SPECIFIC GRAVITY	0.65	
AVERAGE FLOWING TEMP	150	F
OIL API GRAVITY	35.0	API
WATER SPECIFIC GRAVITY	1.07	

GAS/LIQUID RATIO - SCF PER BBL

HAGEDORN AND BROWN
PRESENTED BY:
MACCO - SCHLUMBERGER

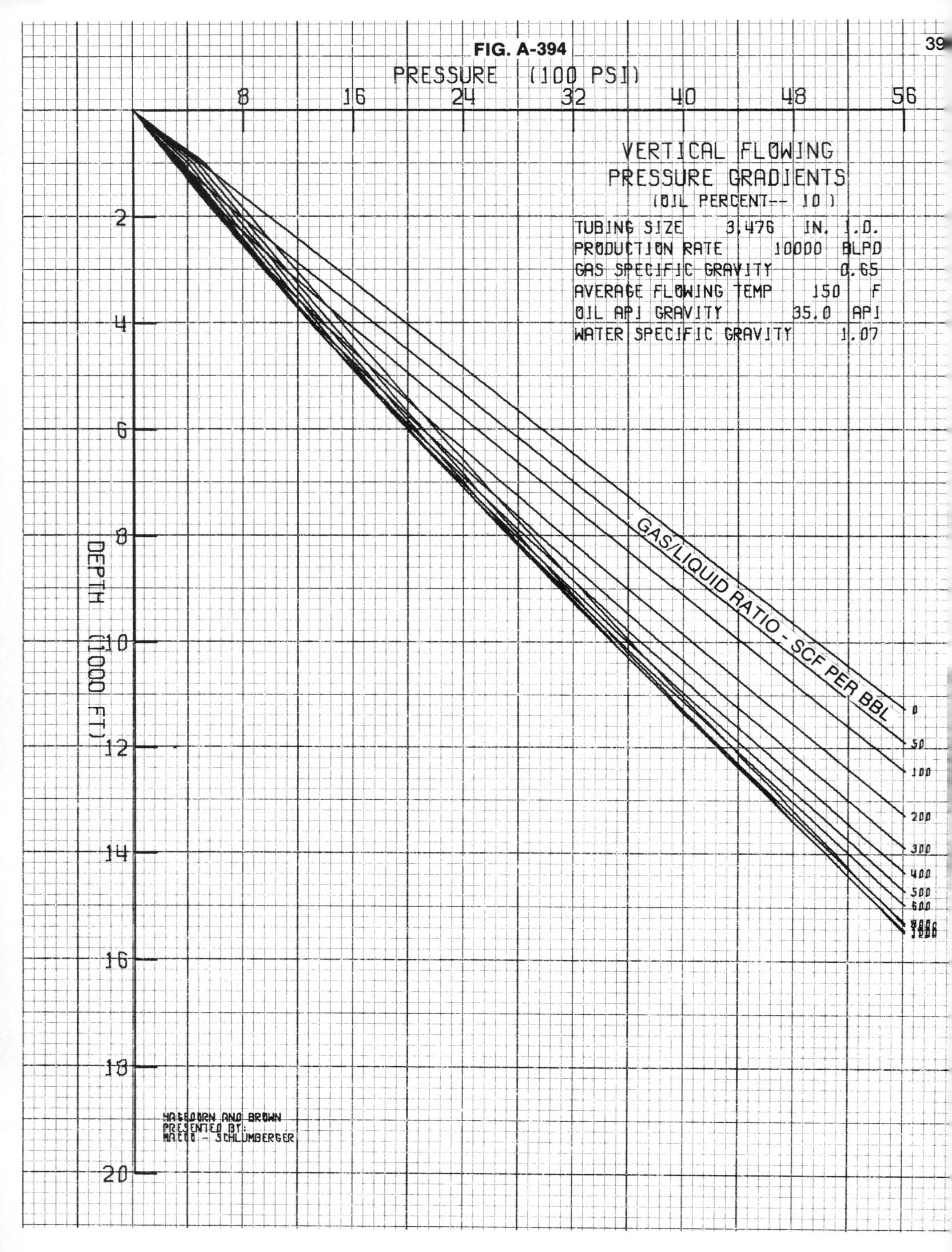

PRESSURE (100 PSI)

DEPTH (1000 FT)

VERTICAL FLOWING
PRESSURE GRADIENTS
(OIL PERCENT-- 10)

TUBING SIZE	3.476	IN. I.D.
PRODUCTION RATE	10000	BLPD
GAS SPECIFIC GRAVITY		0.65
AVERAGE FLOWING TEMP	150	F
OIL API GRAVITY	35.0	API
WATER SPECIFIC GRAVITY		1.07

GAS/LIQUID RATIO - SCF PER BBL

0
50
100
200
300
400
500
600
800
1000

HAGEDORN AND BROWN
PRESENTED BY:
MACCO - SCHLUMBERGER

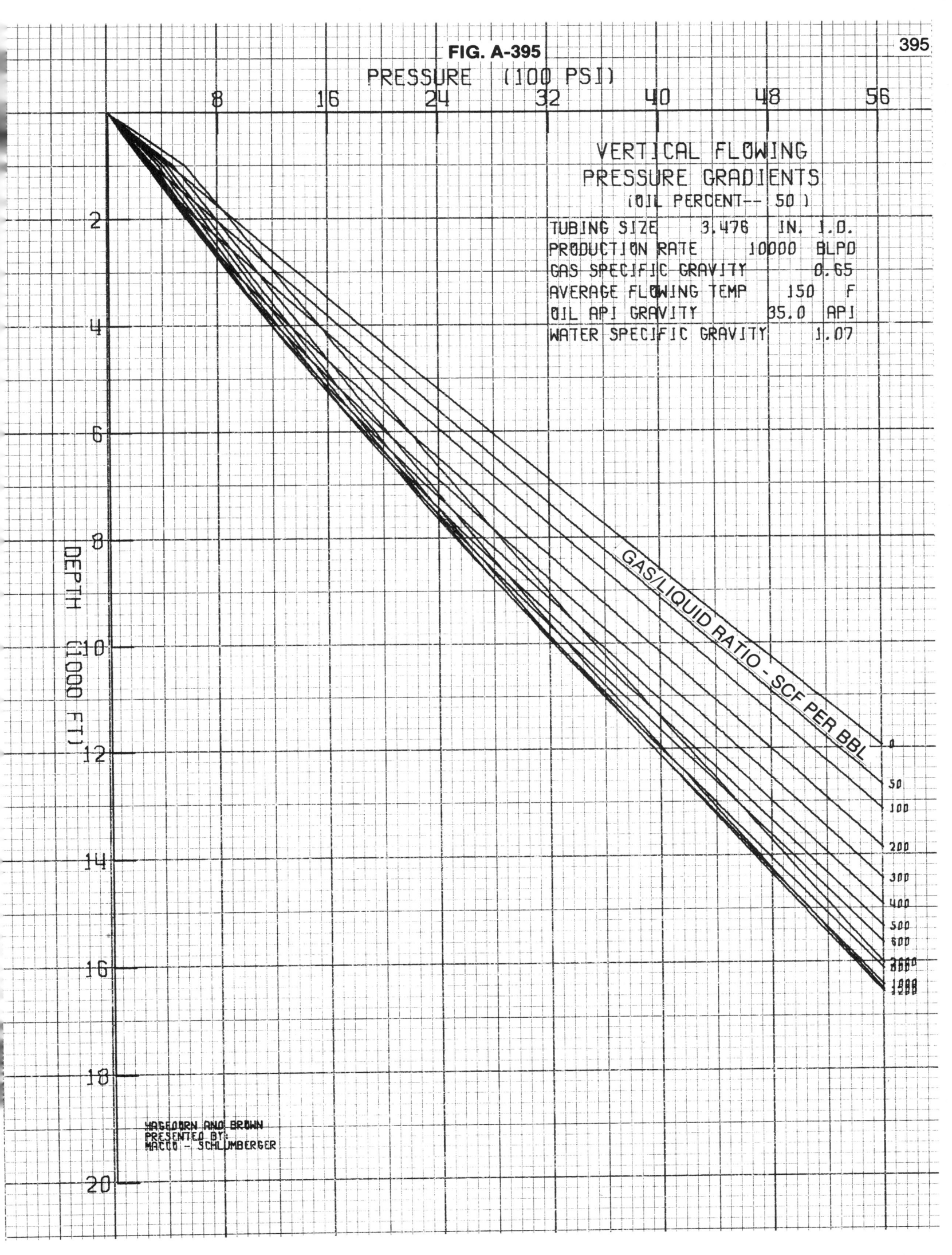

FIG. A-395

395

VERTICAL FLOWING
PRESSURE GRADIENTS
(OIL PERCENT-- 50)

TUBING SIZE	3.476	IN. I.D.
PRODUCTION RATE	10000	BLPD
GAS SPECIFIC GRAVITY	0.65	
AVERAGE FLOWING TEMP	150	F
OIL API GRAVITY	35.0	API
WATER SPECIFIC GRAVITY	1.07	

PRESSURE (100 PSI)

DEPTH (1000 FT)

GAS/LIQUID RATIO - SCF PER BBL

0
50
100
200
300
400
500
600
800
1000

HAGEDORN AND BROWN
PRESENTED BY:
MACCO - SCHLUMBERGER

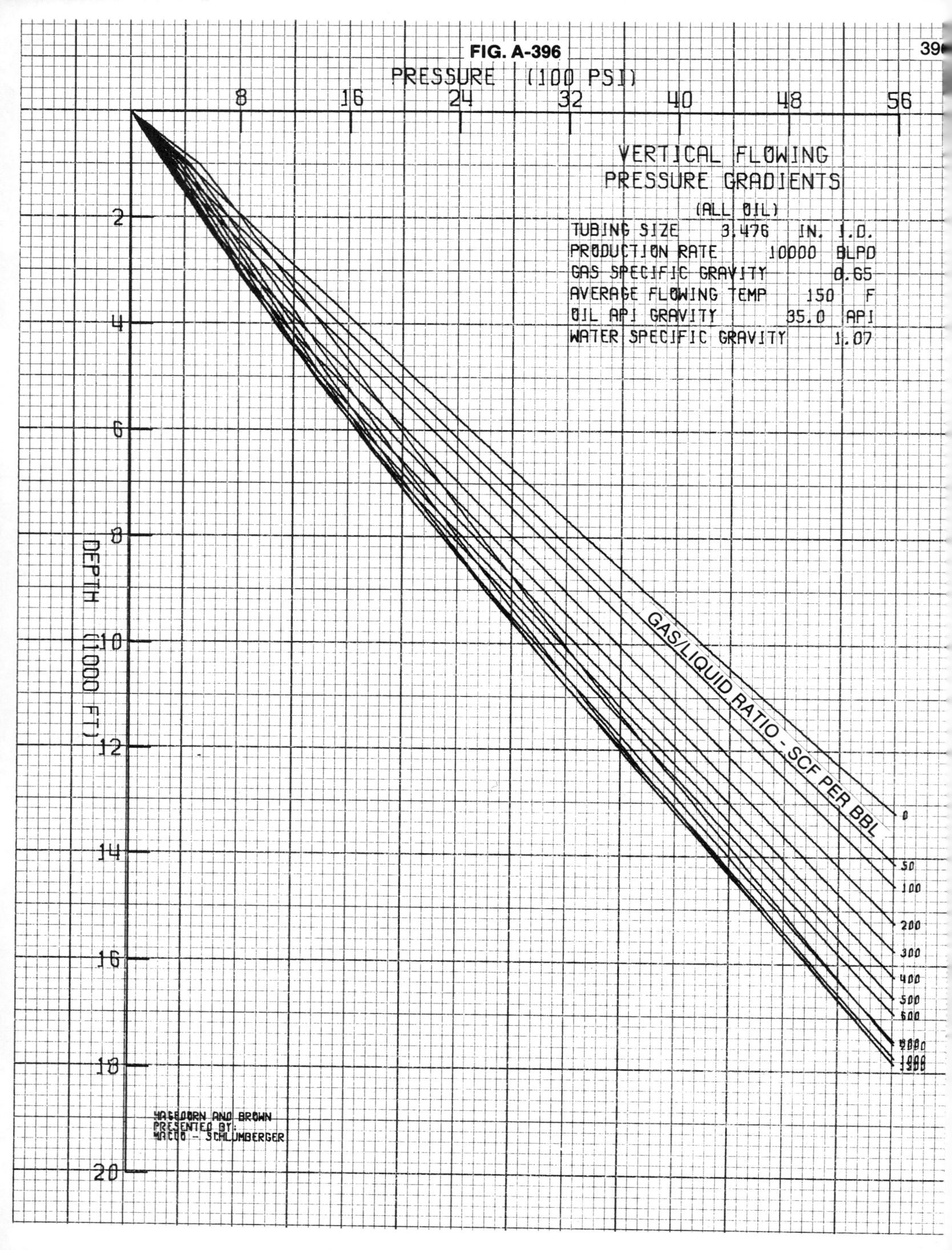

VERTICAL FLOWING
PRESSURE GRADIENTS
(ALL OIL)

TUBING SIZE	3.476	IN. I.D.
PRODUCTION RATE	10000	BLPD
GAS SPECIFIC GRAVITY	0.65	
AVERAGE FLOWING TEMP	150	F
OIL API GRAVITY	35.0	API
WATER SPECIFIC GRAVITY	1.07	

HAGEDORN AND BROWN
PRESENTED BY:
MACCO — SCHLUMBERGER

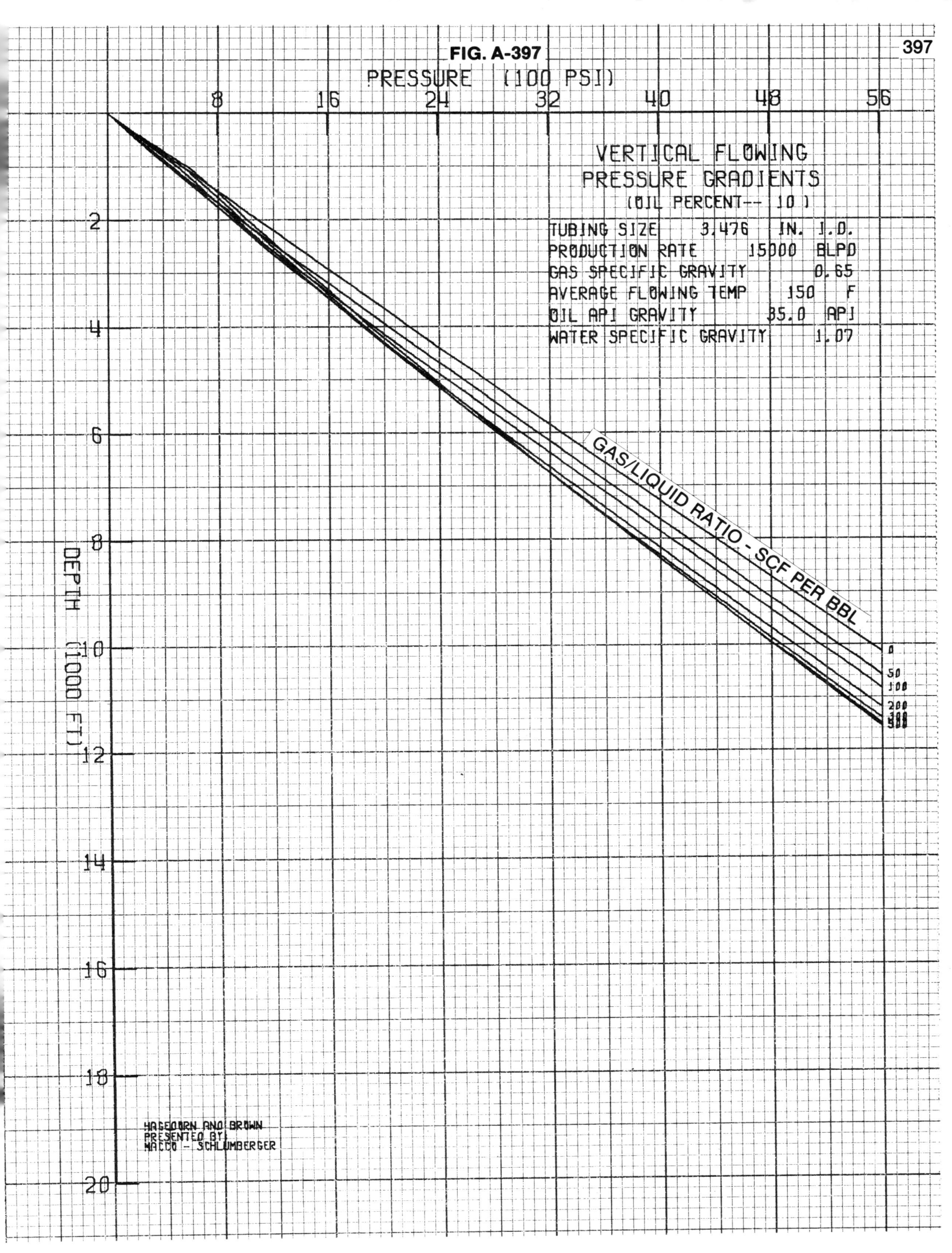

PRESSURE (100 PSI)

DEPTH (1000 FT)

VERTICAL FLOWING
PRESSURE GRADIENTS
(OIL PERCENT--- 10)

TUBING SIZE	3.476	IN. I.D.
PRODUCTION RATE	15000	BLPD
GAS SPECIFIC GRAVITY	0.65	
AVERAGE FLOWING TEMP	150	F
OIL API GRAVITY	35.0	API
WATER SPECIFIC GRAVITY	1.07	

GAS/LIQUID RATIO - SCF PER BBL

0
50
100
200
300
400

HAGEDORN AND BROWN
PRESENTED BY
NALCO - SCHLUMBERGER

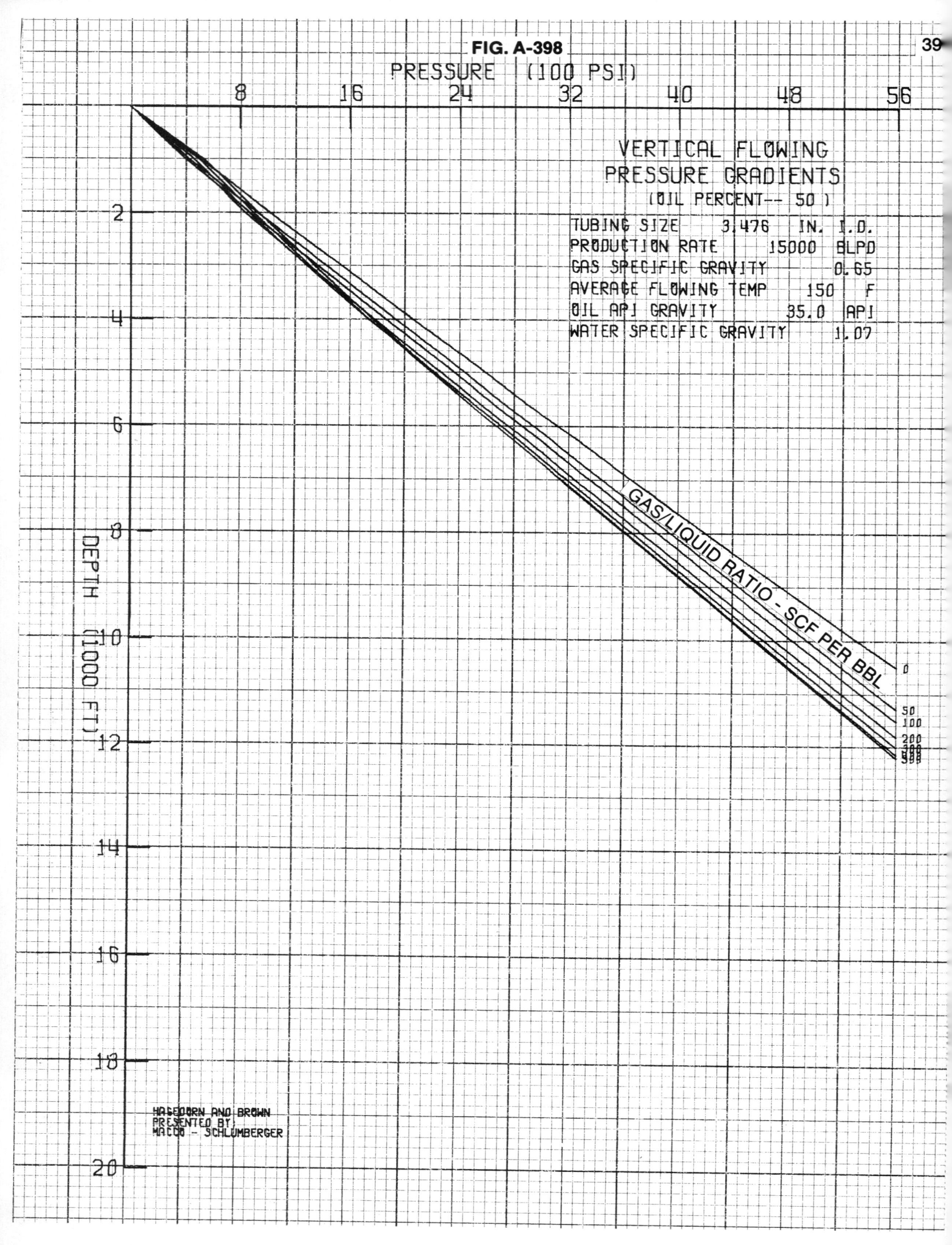

VERTICAL FLOWING
PRESSURE GRADIENTS
(OIL PERCENT-- 50)

TUBING SIZE   3.476   IN. I.D.
PRODUCTION RATE   15000 BLPD
GAS SPECIFIC GRAVITY   0.65
AVERAGE FLOWING TEMP   150   F
OIL API GRAVITY   35.0   API
WATER SPECIFIC GRAVITY   1.07

PRESSURE   (100 PSI)

DEPTH (1000 FT)

GAS/LIQUID RATIO - SCF PER BBL

0
50
100
200
300
400

HAGEDORN AND BROWN
PRESENTED BY
MACCO - SCHLUMBERGER

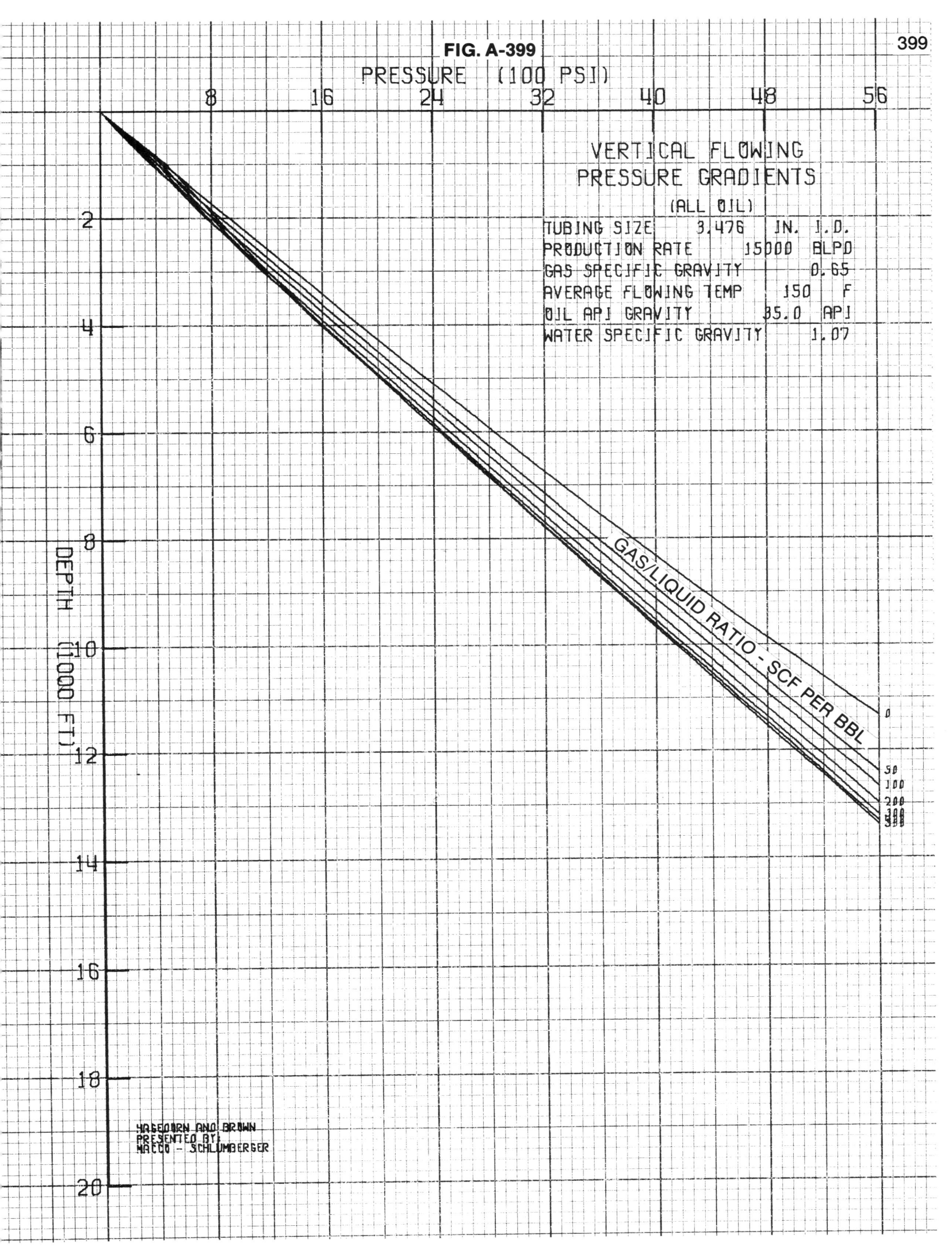

FIG. A-399

399

PRESSURE (100 PSI)

DEPTH (1000 FT)

VERTICAL FLOWING
PRESSURE GRADIENTS
(ALL OIL)

TUBING SIZE	3.476	IN. I.D.
PRODUCTION RATE	15000	BLPD
GAS SPECIFIC GRAVITY	0.65	
AVERAGE FLOWING TEMP	150	F
OIL API GRAVITY	35.0	API
WATER SPECIFIC GRAVITY	1.07	

GAS/LIQUID RATIO - SCF PER BBL

0

50
100
200
300
400

HAGEDORN AND BROWN
PRESENTED BY:
NACCO - SCHLUMBERGER

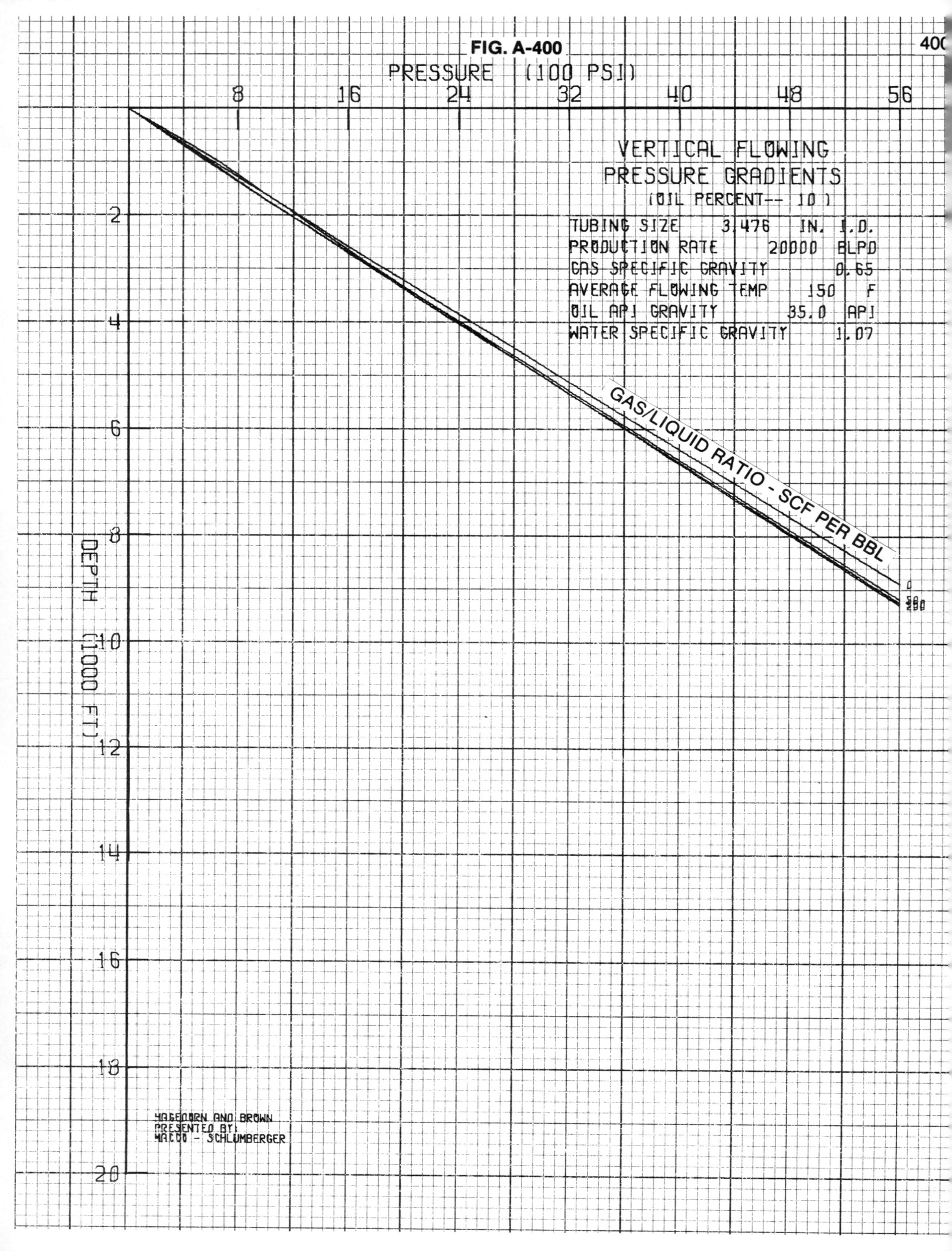

FIG. A-400

VERTICAL FLOWING
PRESSURE GRADIENTS
(OIL PERCENT-- 10)

TUBING SIZE	3.476	IN. I.D.
PRODUCTION RATE	20000	BLPD
GAS SPECIFIC GRAVITY	0.65	
AVERAGE FLOWING TEMP	150	F
OIL API GRAVITY	35.0	API
WATER SPECIFIC GRAVITY	1.07	

GAS/LIQUID RATIO - SCF PER BBL

HAGEDORN AND BROWN
PRESENTED BY:
MAECO - SCHLUMBERGER

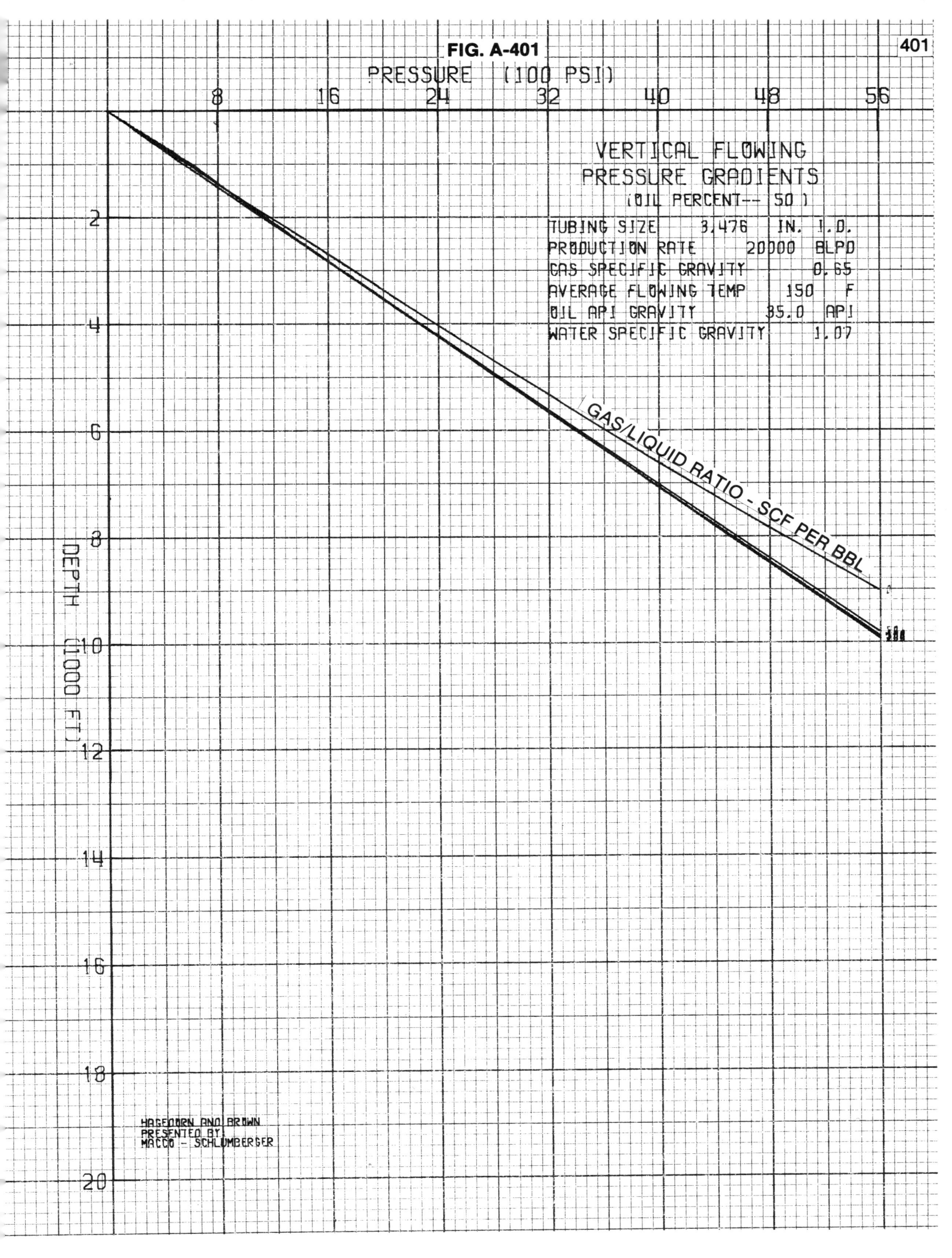

FIG. A-401

401

PRESSURE (100 PSI)

VERTICAL FLOWING
PRESSURE GRADIENTS
(OIL PERCENT--- 50)

TUBING SIZE	3.476	IN. I.D.
PRODUCTION RATE	20000	BLPD
GAS SPECIFIC GRAVITY	0.65	
AVERAGE FLOWING TEMP	150	F
OIL API GRAVITY	35.0	API
WATER SPECIFIC GRAVITY	1.07	

GAS/LIQUID RATIO - SCF PER BBL

DEPTH (1000 FT)

HAGEDORN AND BROWN
PRESENTED BY
MACCO - SCHLUMBERGER

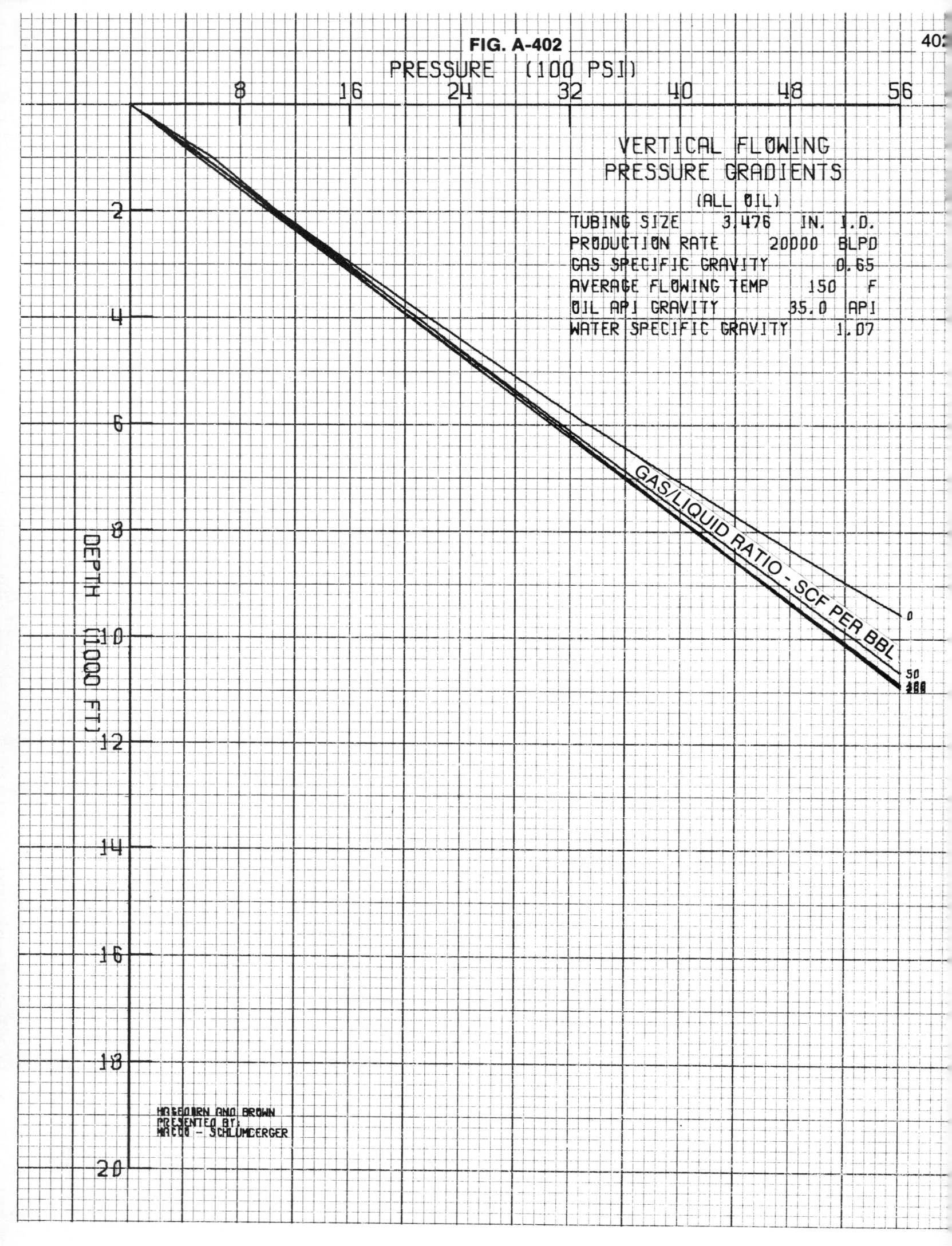

PRESSURE (100 PSI)

VERTICAL FLOWING
PRESSURE GRADIENTS
(ALL OIL)

TUBING SIZE   3.476   IN. I.D.
PRODUCTION RATE   20000 BLPD
GAS SPECIFIC GRAVITY   0.65
AVERAGE FLOWING TEMP   150   F
OIL API GRAVITY   35.0 API
WATER SPECIFIC GRAVITY   1.07

GAS/LIQUID RATIO - SCF PER BBL

DEPTH (1000 FT)

HAGEDORN AND BROWN
PRESENTED BY:
WACCO - SCHLUMBERGER

FIG. A-403

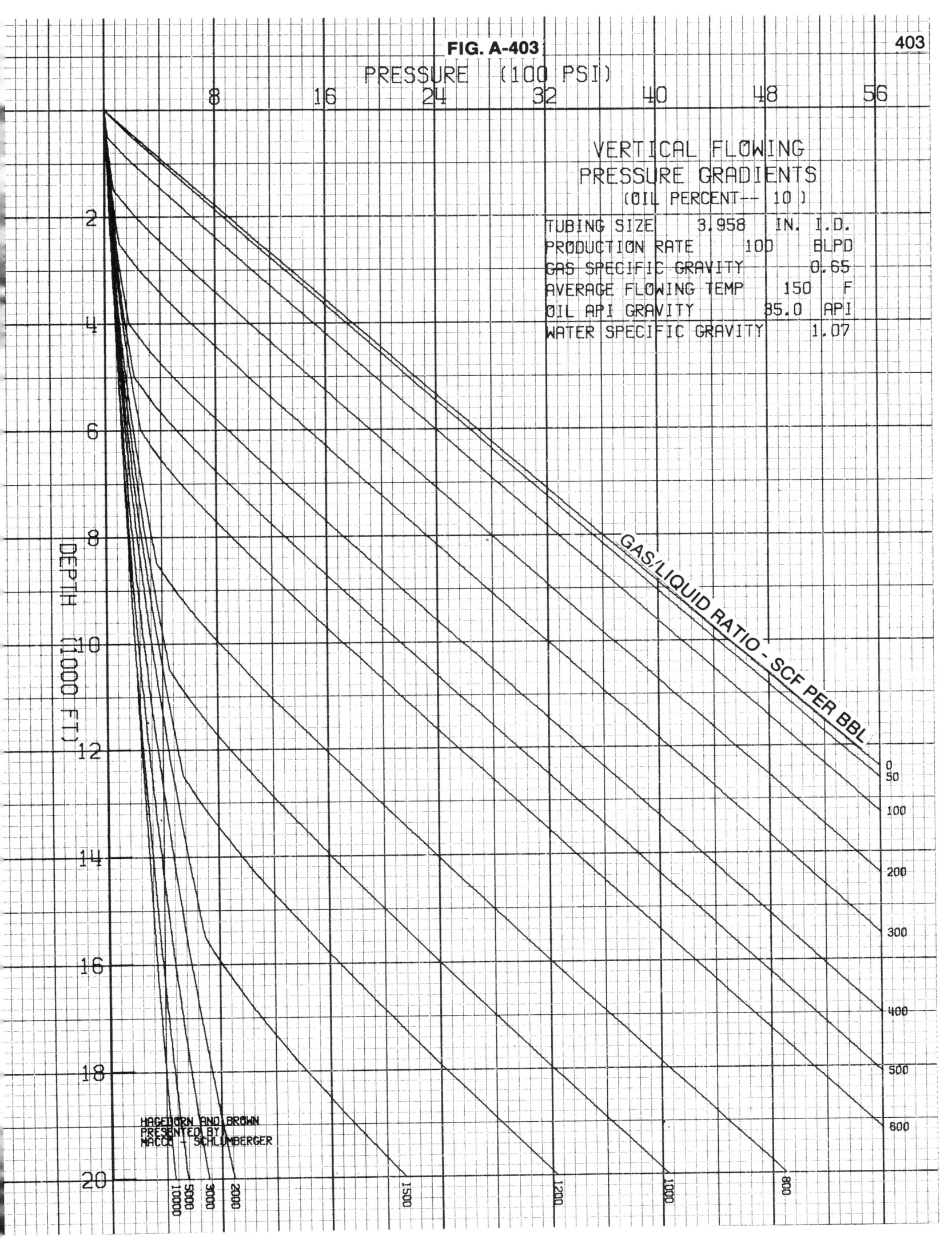

PRESSURE (100 PSI)

VERTICAL FLOWING
PRESSURE GRADIENTS
(OIL PERCENT-- 10)

TUBING SIZE 3.958 IN. I.D.
PRODUCTION RATE 100 BLPD
GAS SPECIFIC GRAVITY 0.65
AVERAGE FLOWING TEMP 150 F
OIL API GRAVITY 35.0 API
WATER SPECIFIC GRAVITY 1.07

GAS/LIQUID RATIO - SCF PER BBL

DEPTH (1000 FT)

HAGEDORN AND BROWN
PRESENTED BY
MACCO - SCHLUMBERGER

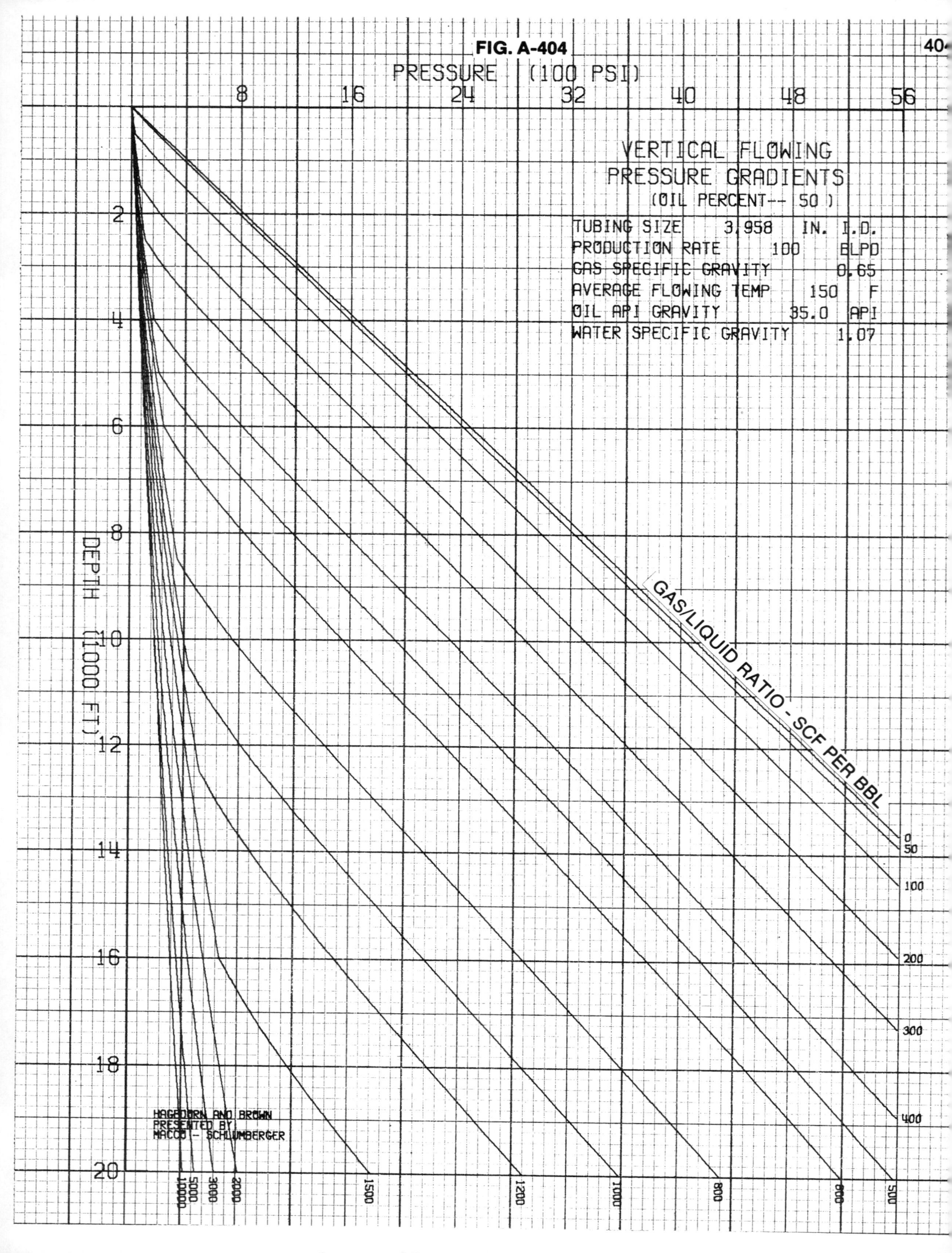

PRESSURE (100 PSI)

VERTICAL FLOWING
PRESSURE GRADIENTS
(OIL PERCENT-- 50)

TUBING SIZE	3.958	IN. I.D.
PRODUCTION RATE	100	BLPD
GAS SPECIFIC GRAVITY	0.65	
AVERAGE FLOWING TEMP	150	F
OIL API GRAVITY	35.0	API
WATER SPECIFIC GRAVITY	1.07	

DEPTH (1000 FT)

GAS/LIQUID RATIO - SCF PER BBL

HAGEDORN AND BROWN
PRESENTED BY
MACCO - SCHLUMBERGER

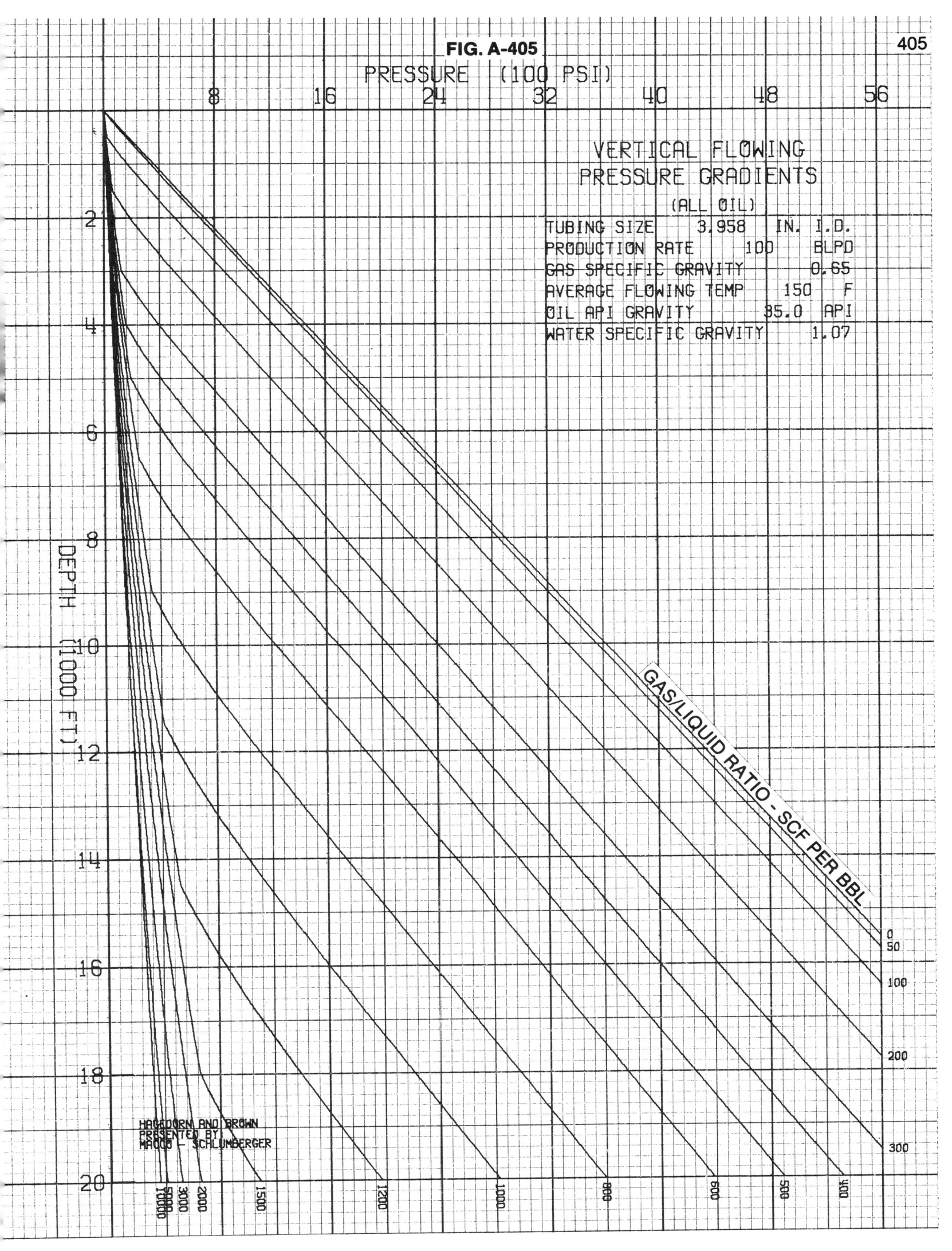

PRESSURE (100 PSI)

VERTICAL FLOWING
PRESSURE GRADIENTS
(ALL OIL)

TUBING SIZE	3.958	IN. I.D.
PRODUCTION RATE	100	BLPD
GAS SPECIFIC GRAVITY	0.65	
AVERAGE FLOWING TEMP	150	F
OIL API GRAVITY	35.0	API
WATER SPECIFIC GRAVITY	1.07	

DEPTH (1000 FT)

GAS/LIQUID RATIO - SCF PER BBL

HAGEDORN AND BROWN
PRESENTED BY
MAGCO - SCHLUMBERGER

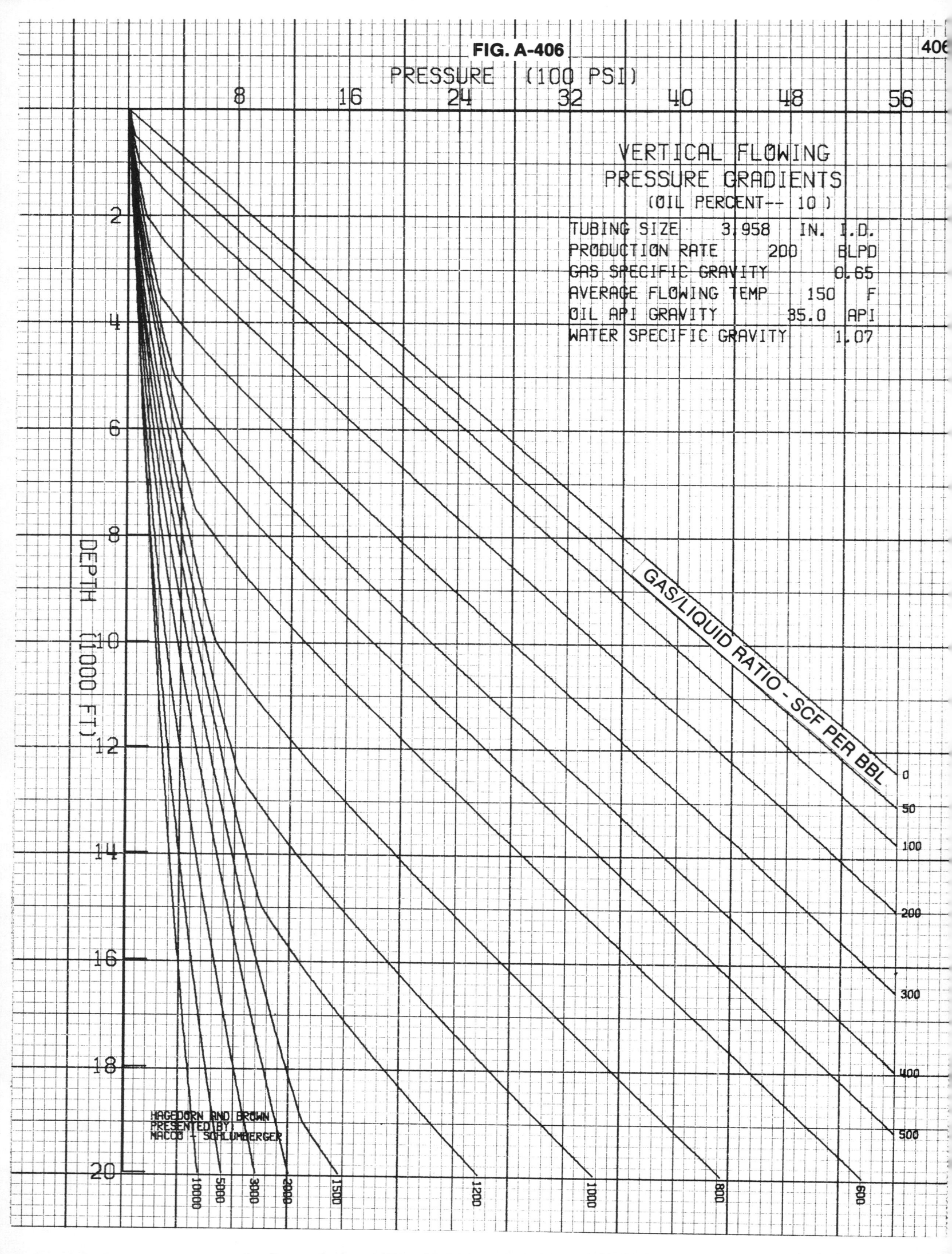

FIG. A-406

VERTICAL FLOWING
PRESSURE GRADIENTS
(OIL PERCENT-- 10)

TUBING SIZE	3.958	IN. I.D.
PRODUCTION RATE	200	BLPD
GAS SPECIFIC GRAVITY		0.65
AVERAGE FLOWING TEMP	150	F
OIL API GRAVITY	35.0	API
WATER SPECIFIC GRAVITY		1.07

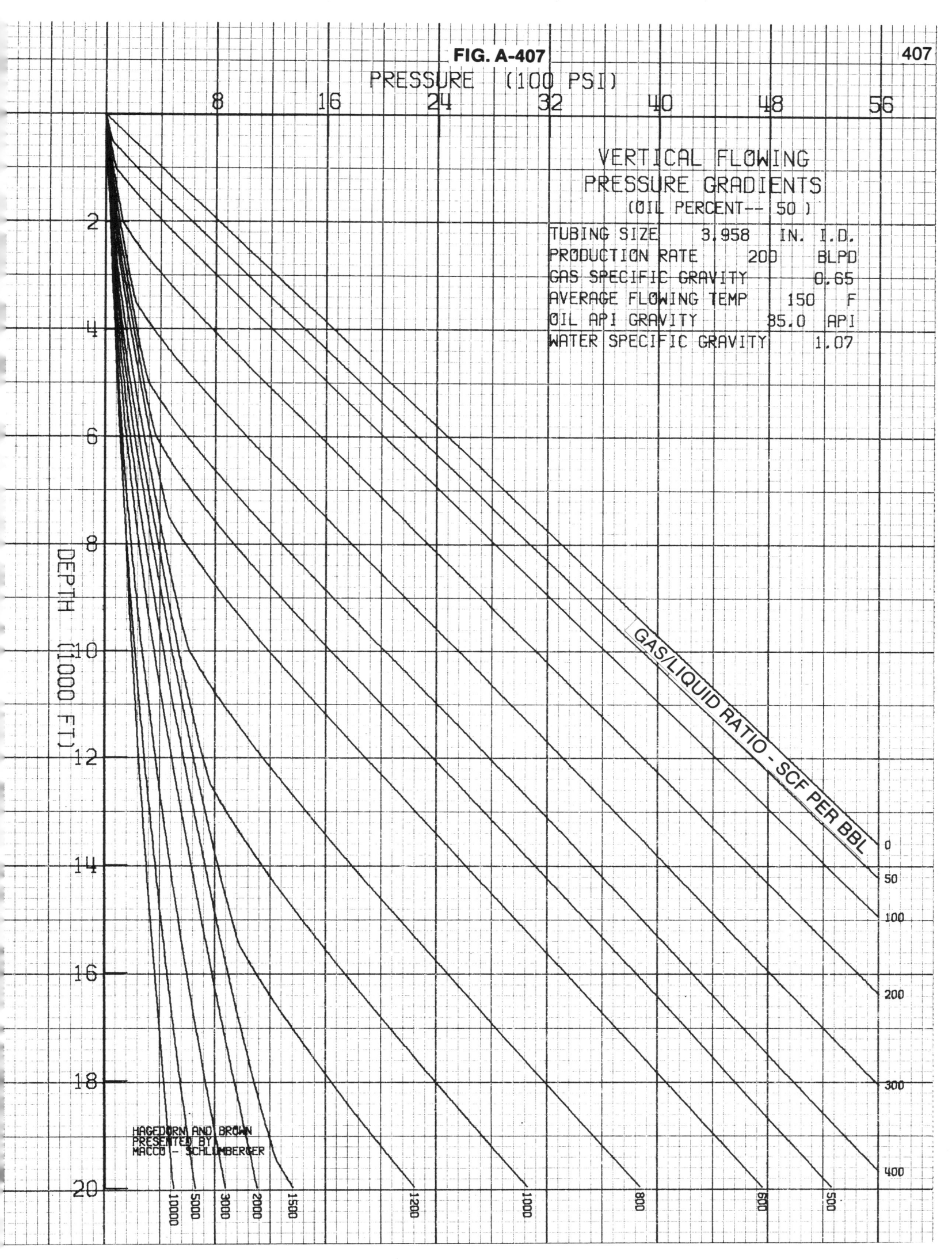

FIG. A-407

407

PRESSURE (100 PSI)

VERTICAL FLOWING
PRESSURE GRADIENTS
(OIL PERCENT-- 50)

TUBING SIZE	3.958	IN. I.D.
PRODUCTION RATE	200	BLPD
GAS SPECIFIC GRAVITY	0.65	
AVERAGE FLOWING TEMP	150	F
OIL API GRAVITY	85.0	API
WATER SPECIFIC GRAVITY	1.07	

DEPTH (1000 FT)

GAS/LIQUID RATIO - SCF PER BBL

HAGEDORN AND BROWN
PRESENTED BY
MACCO - SCHLUMBERGER

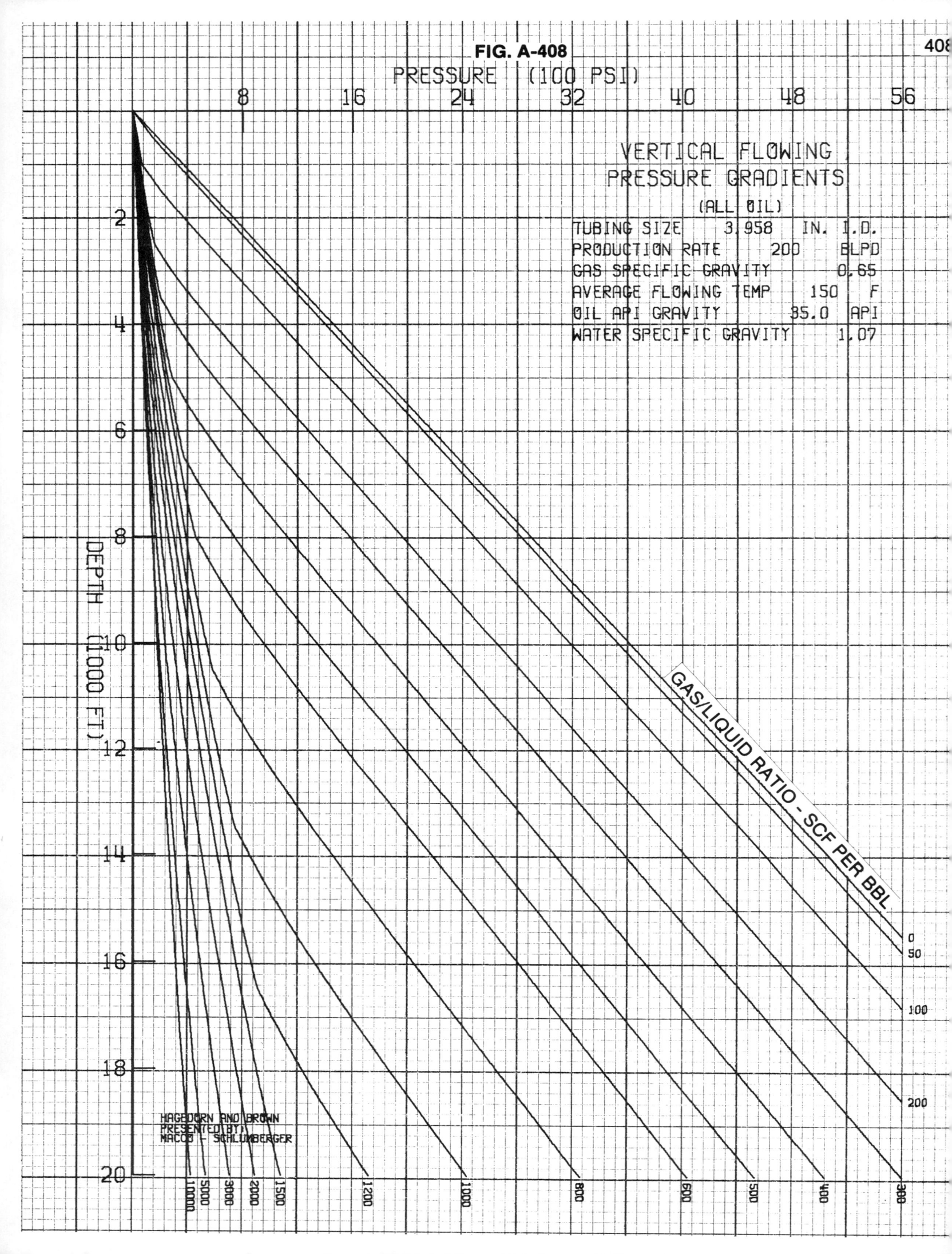

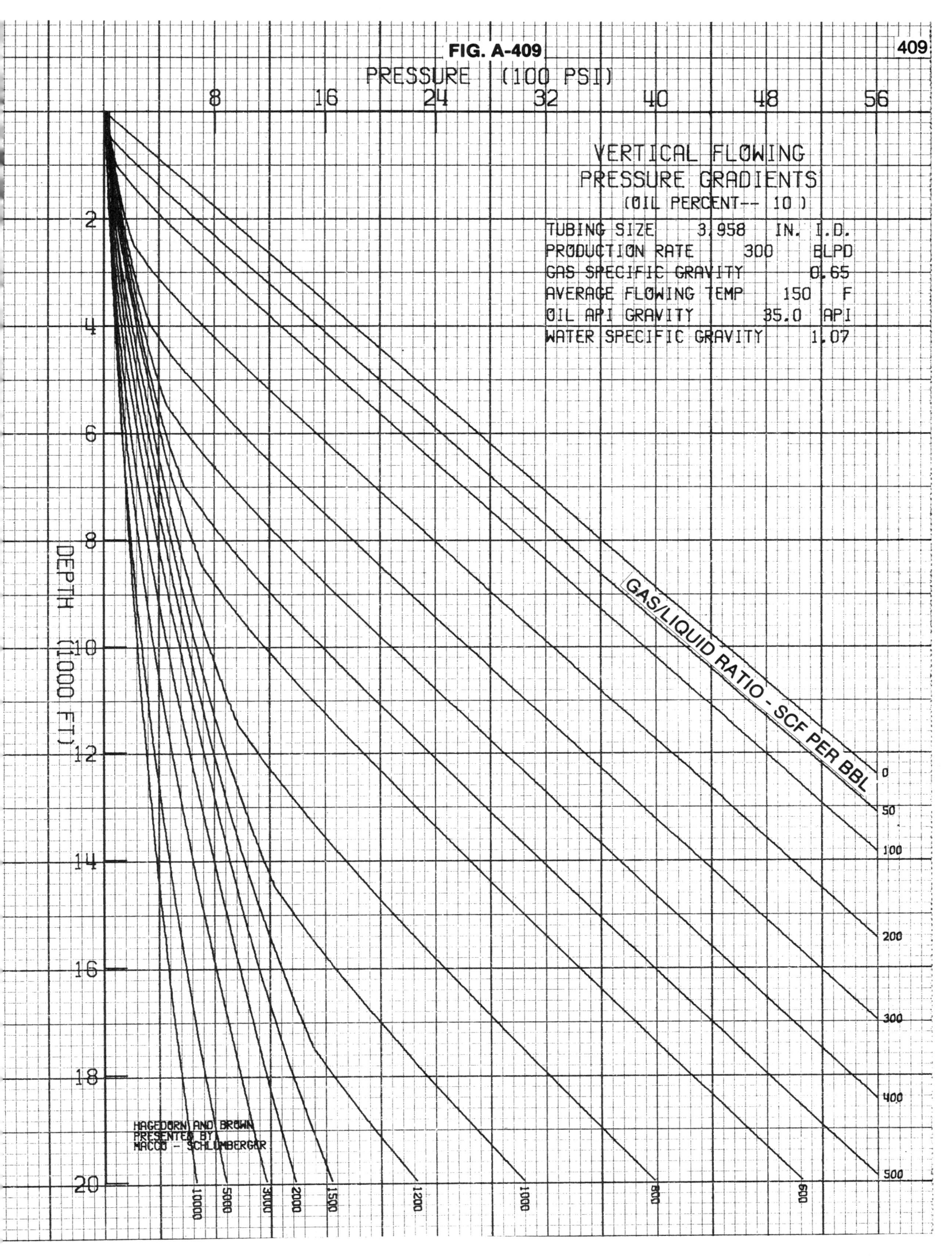

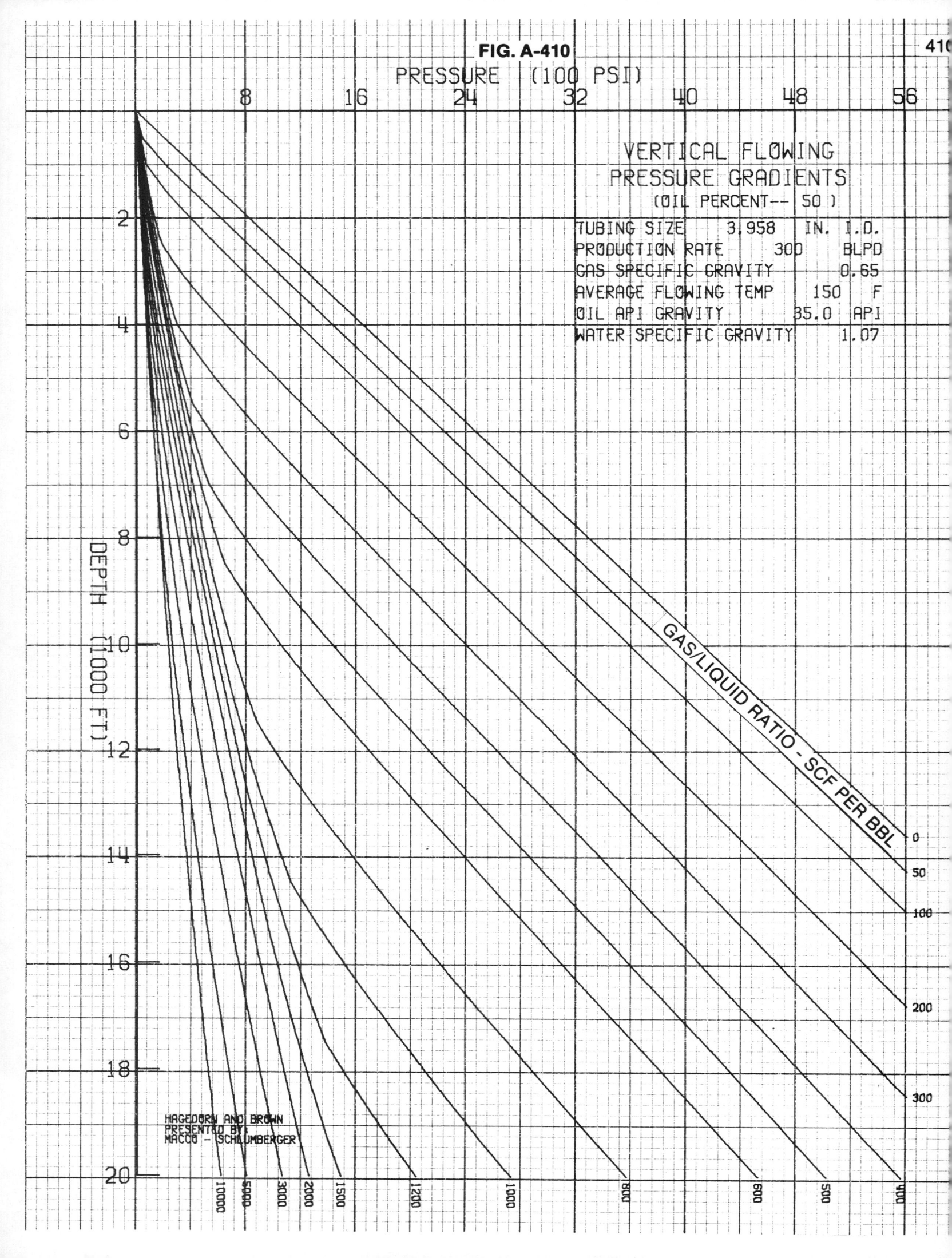

PRESSURE (100 PSI)

VERTICAL FLOWING
PRESSURE GRADIENTS
(OIL PERCENT-- 50)

TUBING SIZE	3.958	IN. I.D.
PRODUCTION RATE	300	BLPD
GAS SPECIFIC GRAVITY	0.65	
AVERAGE FLOWING TEMP	150	F
OIL API GRAVITY	35.0	API
WATER SPECIFIC GRAVITY	1.07	

DEPTH (1000 FT)

GAS/LIQUID RATIO - SCF PER BBL

HAGEDORN AND BROWN
PRESENTED BY:
MACCO - SCHLUMBERGER

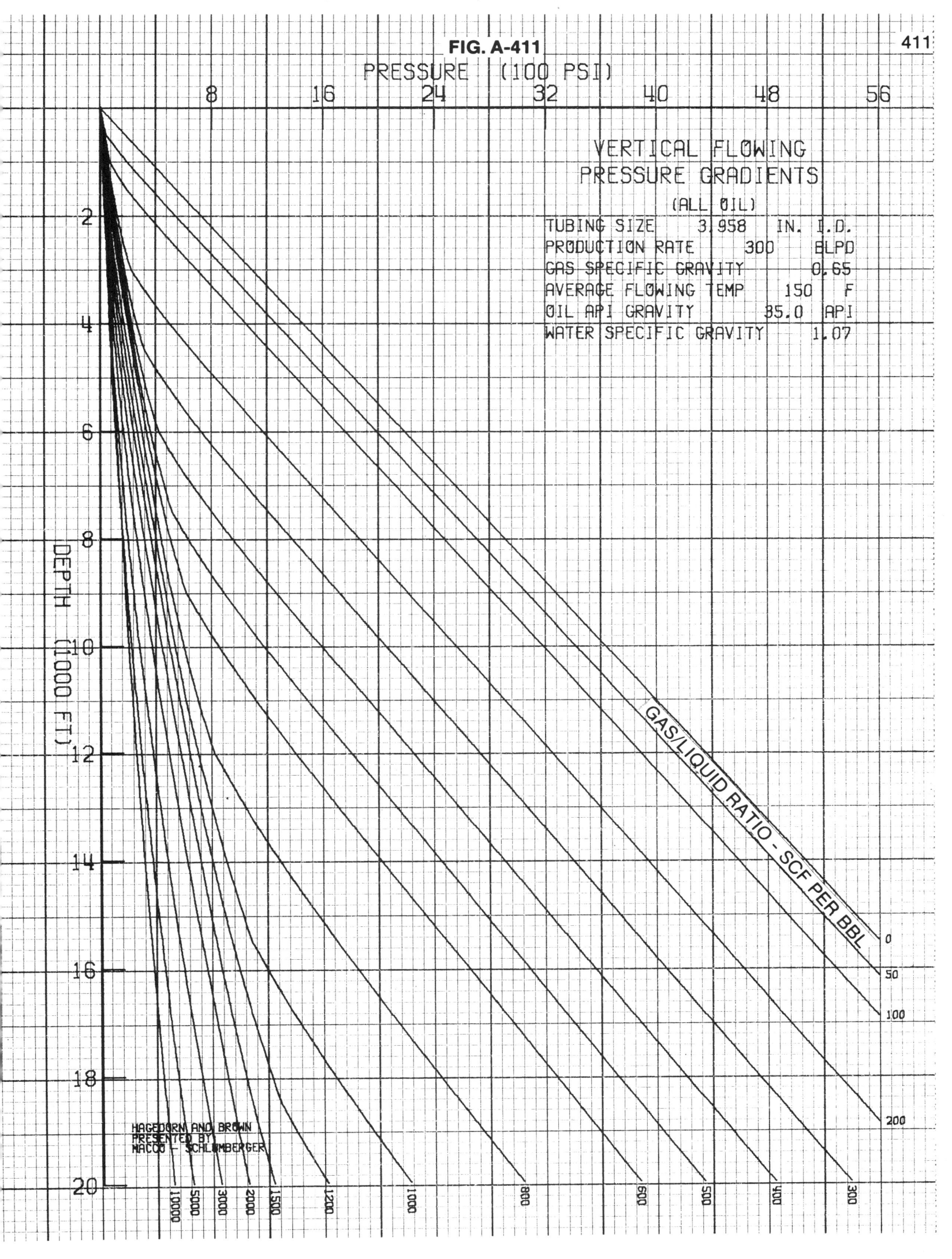

PRESSURE (100 PSI)

VERTICAL FLOWING
PRESSURE GRADIENTS
(ALL OIL)

TUBING SIZE	3.958	IN. I.D.
PRODUCTION RATE	300	BLPD
GAS SPECIFIC GRAVITY	0.65	
AVERAGE FLOWING TEMP	150	F
OIL API GRAVITY	35.0	API
WATER SPECIFIC GRAVITY	1.07	

DEPTH (1000 FT)

GAS/LIQUID RATIO - SCF PER BBL

HAGEDORN AND BROWN
PRESENTED BY
NACCO – SCHLUMBERGER

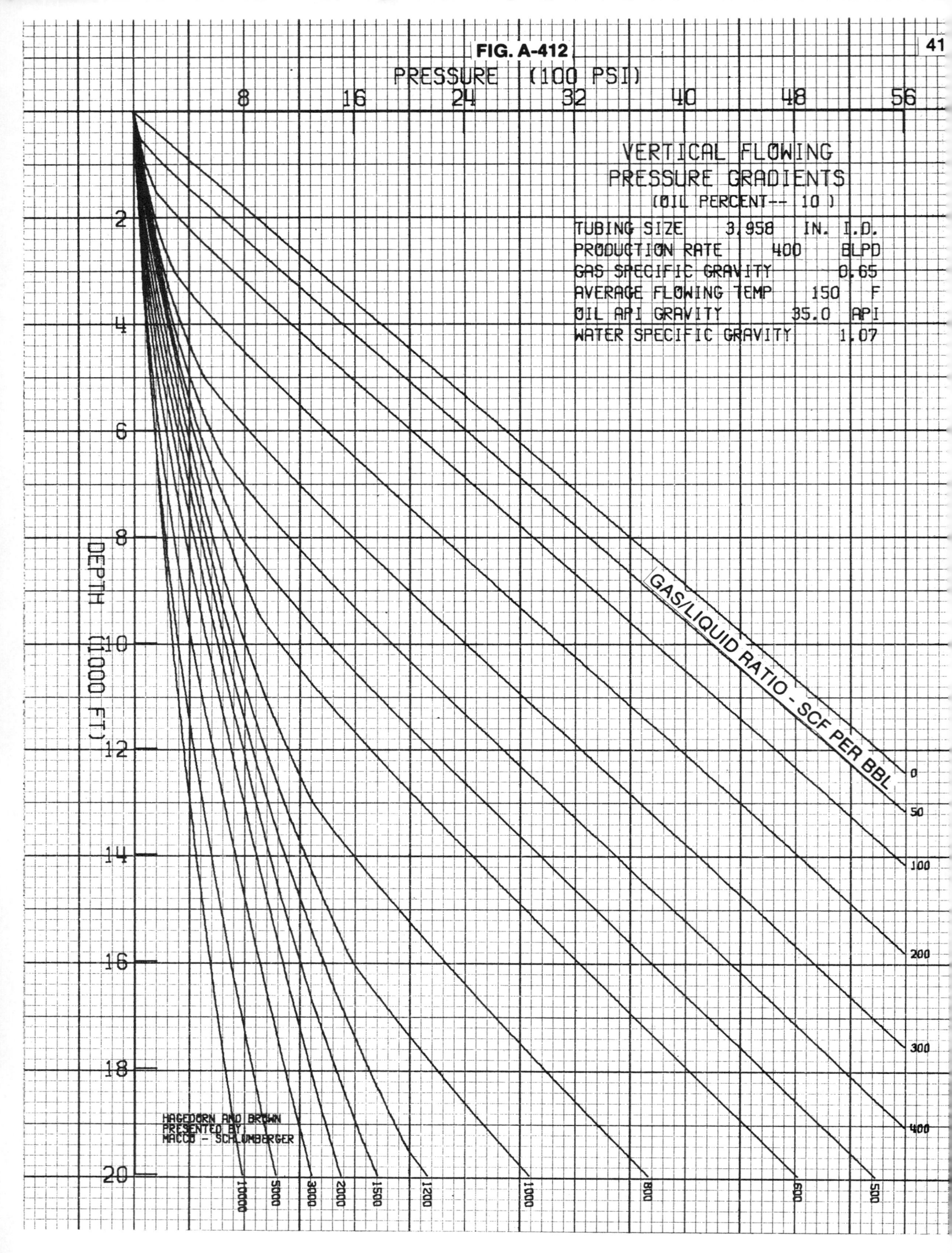

PRESSURE (100 PSI)

VERTICAL FLOWING
PRESSURE GRADIENTS
(OIL PERCENT--- 10)

TUBING SIZE	3.958	IN. I.D.
PRODUCTION RATE	400	BLPD
GAS SPECIFIC GRAVITY		0.65
AVERAGE FLOWING TEMP	150	F
OIL API GRAVITY	35.0	API
WATER SPECIFIC GRAVITY		1.07

DEPTH (1,000 FT)

GAS/LIQUID RATIO - SCF PER BBL

HAGEDORN AND BROWN
PRESENTED BY
MACCO - SCHLUMBERGER

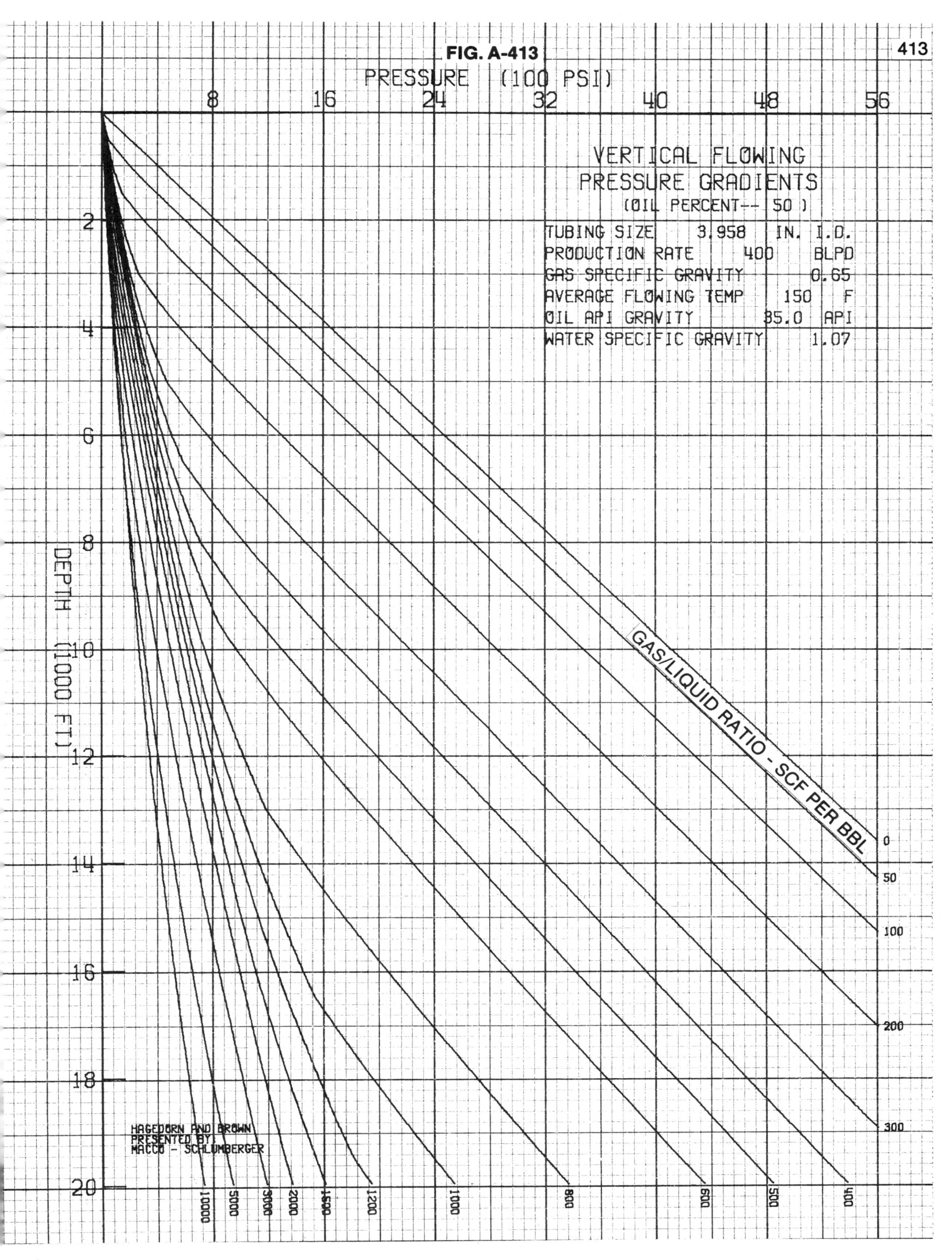

FIG. A-413

413

VERTICAL FLOWING
PRESSURE GRADIENTS
(OIL PERCENT-- 50)

TUBING SIZE	3.958	IN. I.D.
PRODUCTION RATE	400	BLPD
GAS SPECIFIC GRAVITY		0.65
AVERAGE FLOWING TEMP	150	F
OIL API GRAVITY	35.0	API
WATER SPECIFIC GRAVITY		1.07

PRESSURE (100 PSI)

DEPTH (1000 FT)

GAS/LIQUID RATIO - SCF PER BBL

HAGEDORN AND BROWN
PRESENTED BY
MACCO - SCHLUMBERGER

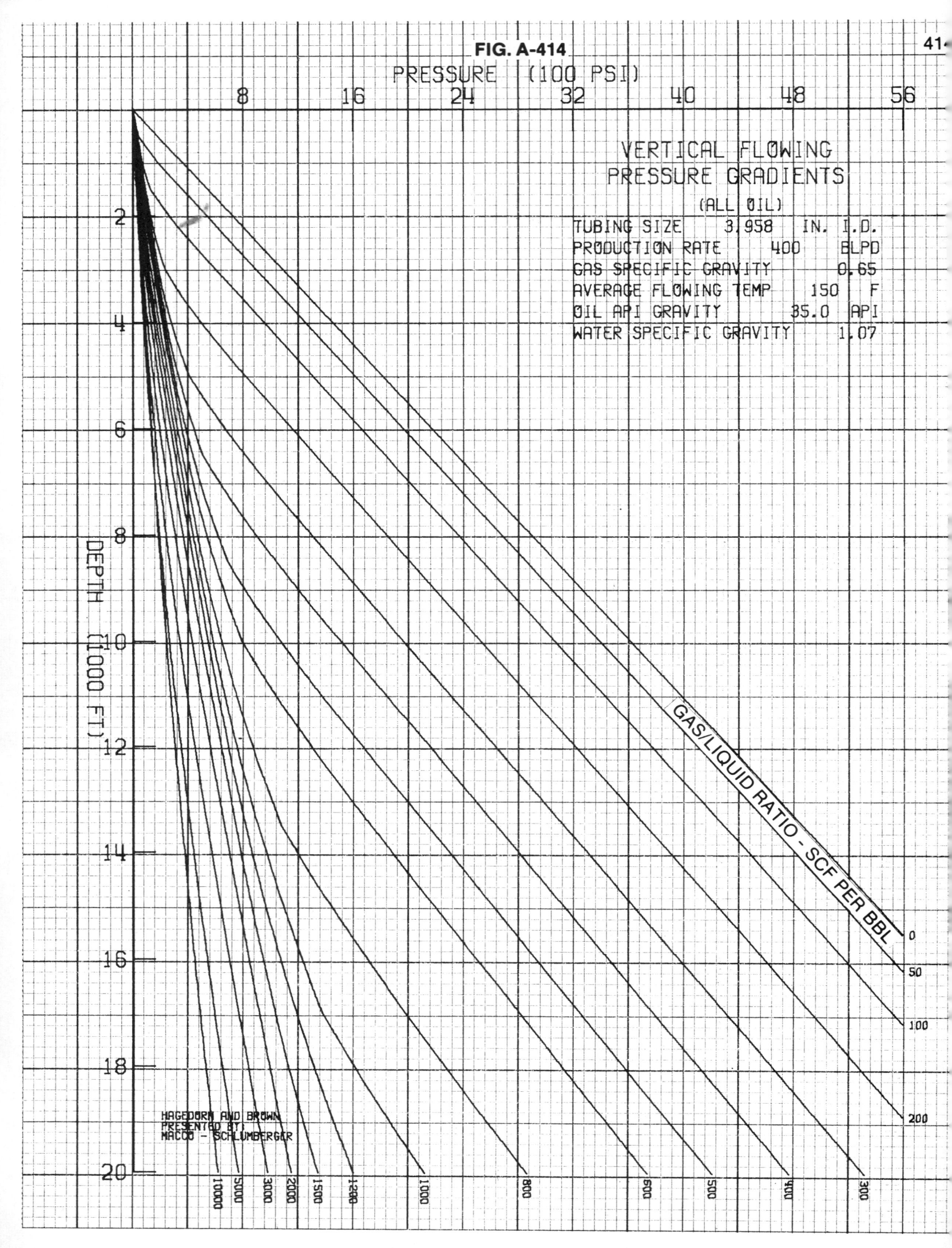

PRESSURE (100 PSI)

VERTICAL FLOWING
PRESSURE GRADIENTS
(ALL OIL)

TUBING SIZE	3.958	IN. I.D.
PRODUCTION RATE	400	BLPD
GAS SPECIFIC GRAVITY	0.65	
AVERAGE FLOWING TEMP	150	F
OIL API GRAVITY	35.0	API
WATER SPECIFIC GRAVITY	1.07	

DEPTH (1000 FT)

GAS/LIQUID RATIO - SCF PER BBL

HAGEDORN AND BROWN
PRESENTED BY
MACCO - SCHLUMBERGER

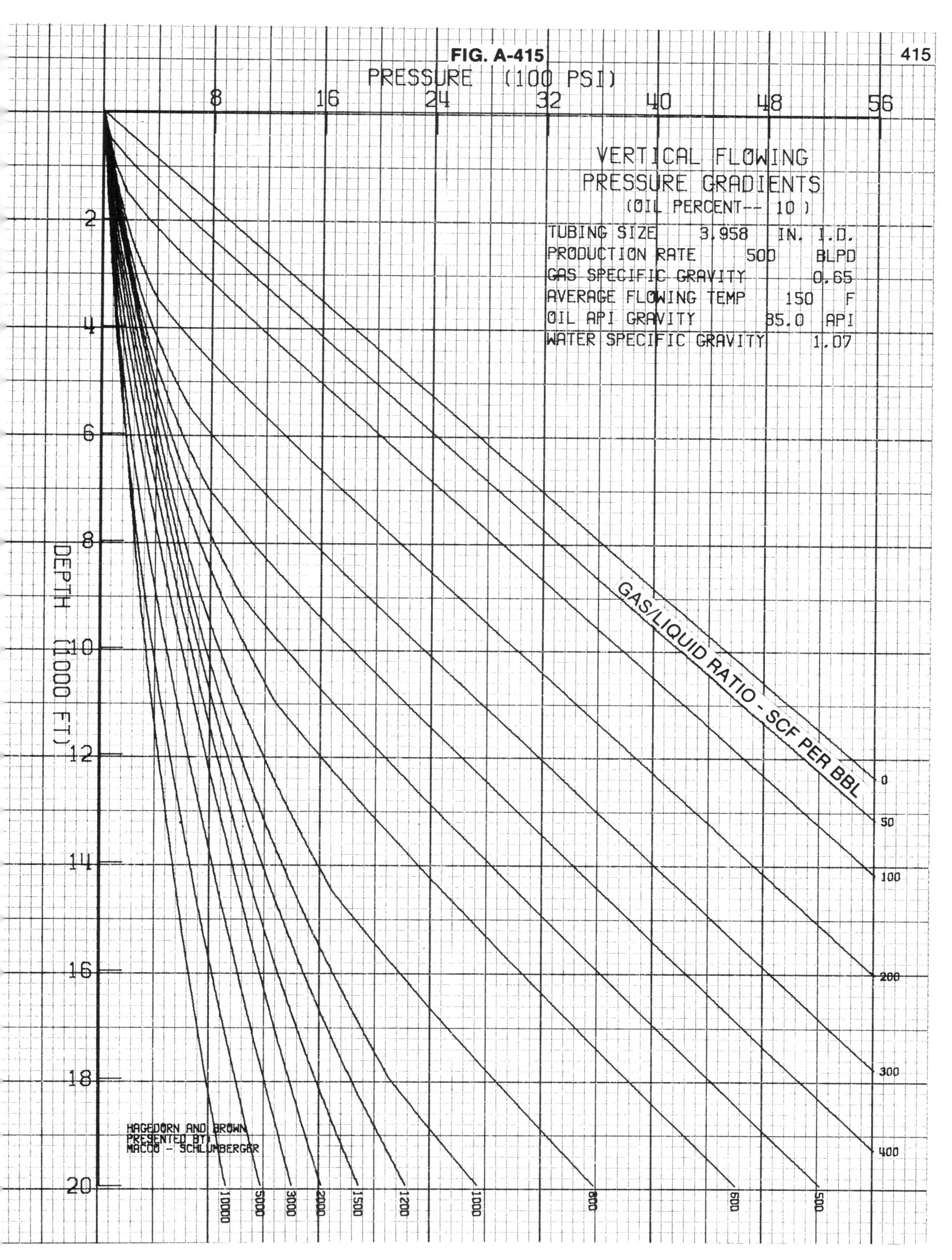

FIG. A-415

415

VERTICAL FLOWING
PRESSURE GRADIENTS
(OIL PERCENT-- 10)

TUBING SIZE	3.958	IN. I.D.
PRODUCTION RATE	500	BLPD
GAS SPECIFIC GRAVITY	0.65	
AVERAGE FLOWING TEMP	150	F
OIL API GRAVITY	35.0	API
WATER SPECIFIC GRAVITY	1.07	

PRESSURE (100 PSI)

DEPTH (1000 FT)

GAS/LIQUID RATIO - SCF PER BBL

HAGEDORN AND BROWN
PRESENTED BY
MACCO - SCHLUMBERGER

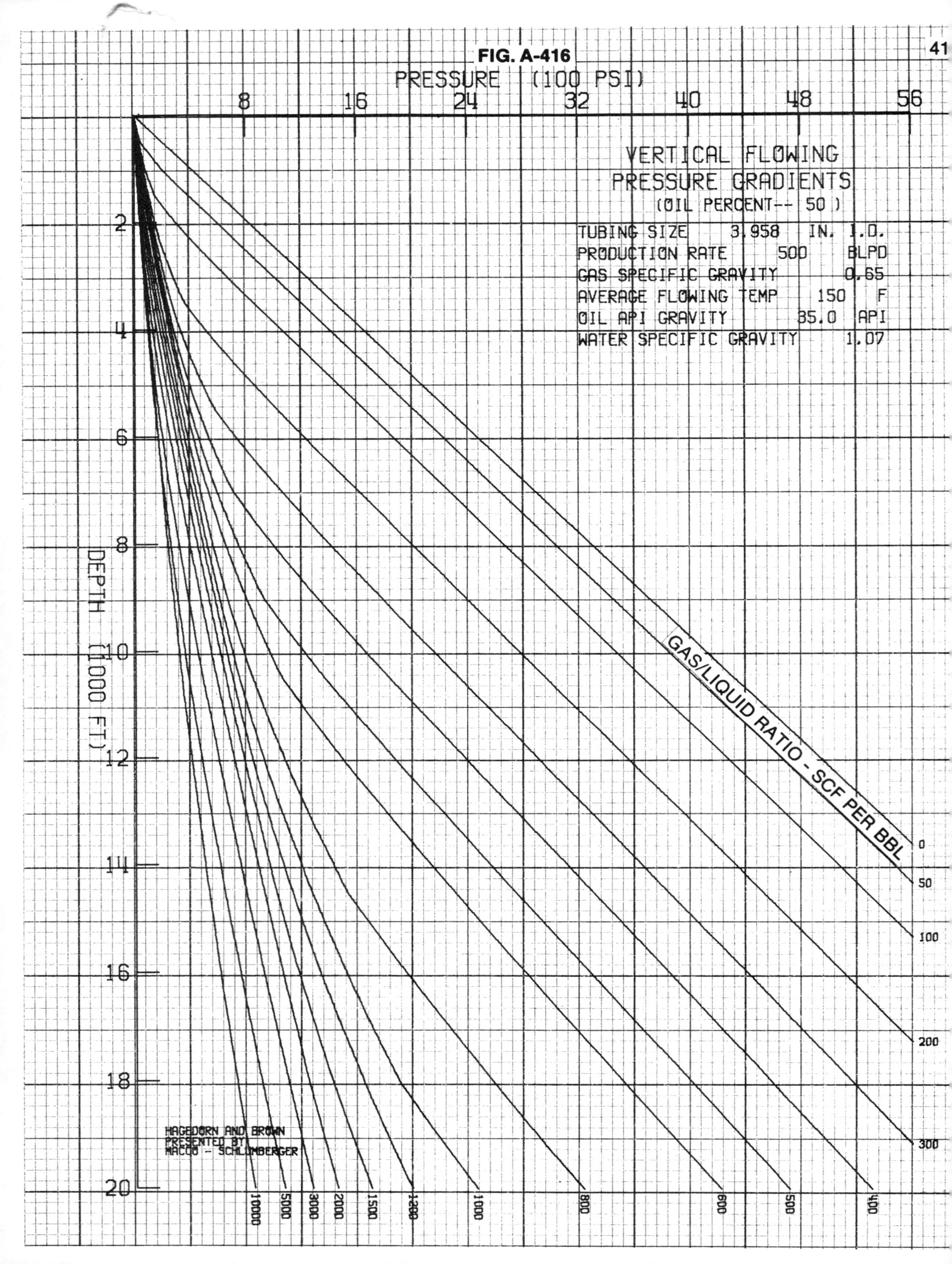

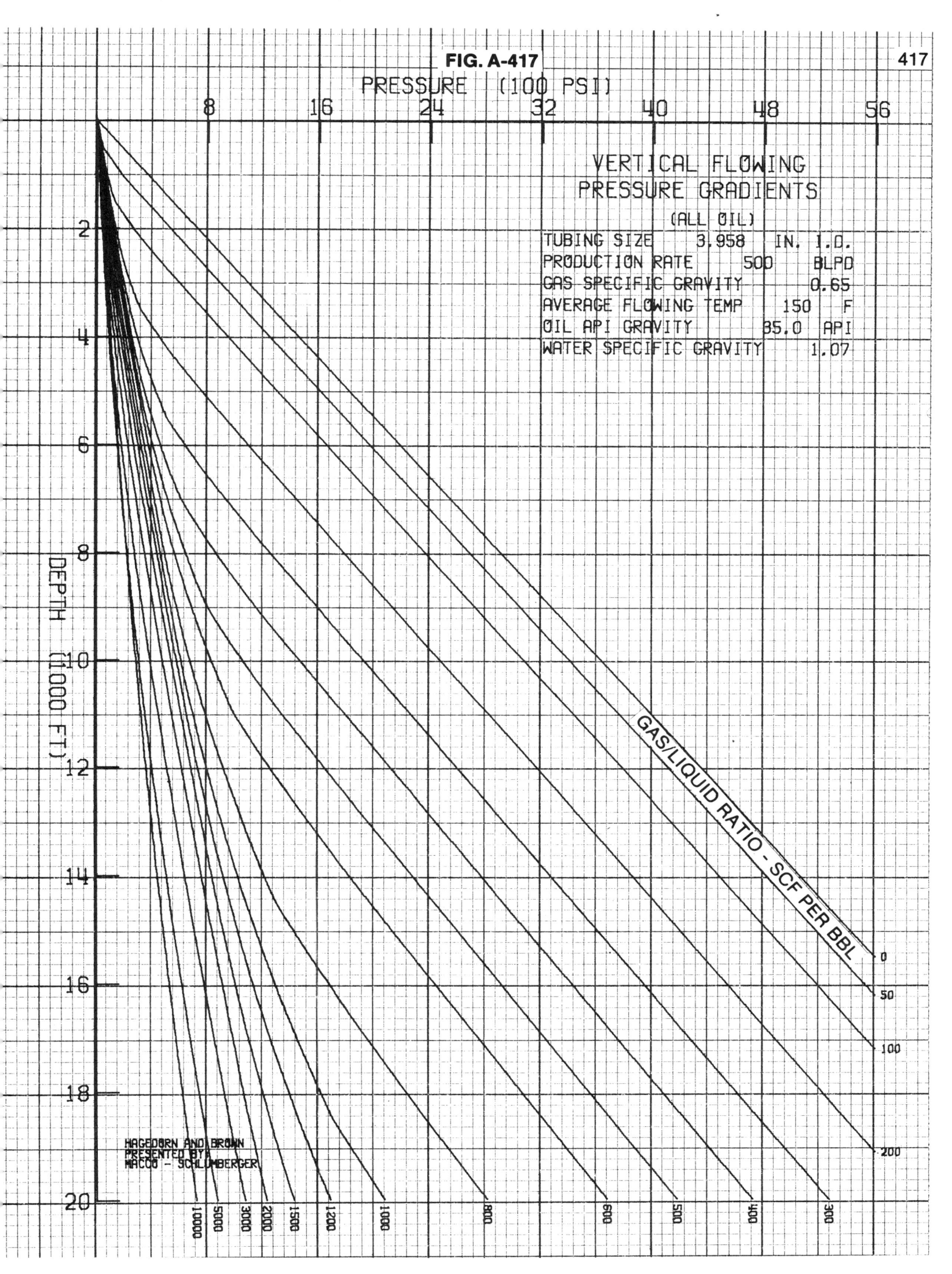

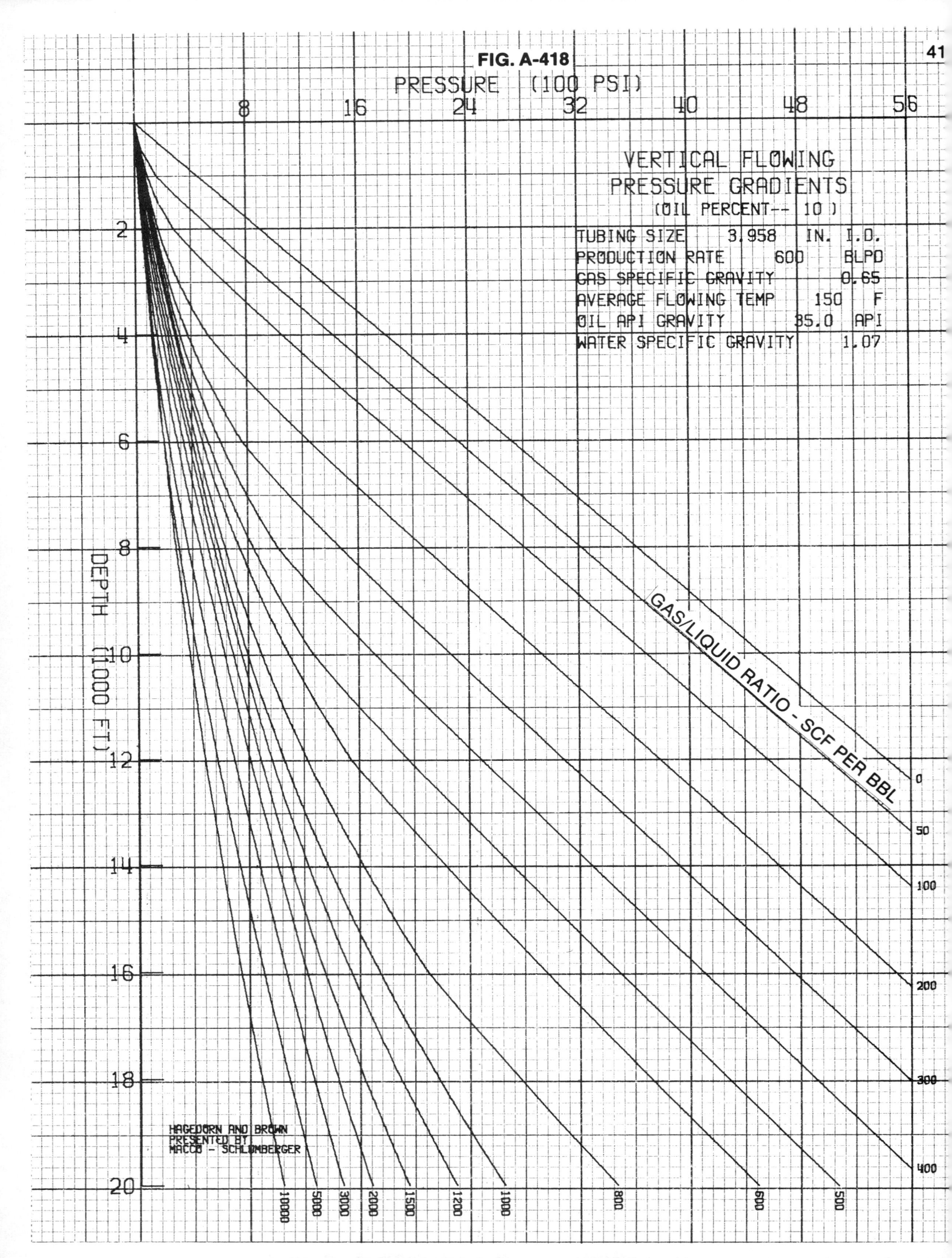

FIG. A-418

41

VERTICAL FLOWING
PRESSURE GRADIENTS
(OIL PERCENT-- 10)

| TUBING SIZE | 3.958 | IN. I.D. |
| AVERAGE FLOWING TEMP | 150 | F |

PRESSURE (100 PSI)

DEPTH (1000 FT)

GAS/LIQUID RATIO - SCF PER BBL

TUBING SIZE   3.958   IN. I.D.
PRODUCTION RATE   600   BLPD
GAS SPECIFIC GRAVITY   0.65
AVERAGE FLOWING TEMP   150   F
OIL API GRAVITY   35.0   API
WATER SPECIFIC GRAVITY   1.07

HAGEDORN AND BROWN
PRESENTED BY
MACCO - SCHLUMBERGER

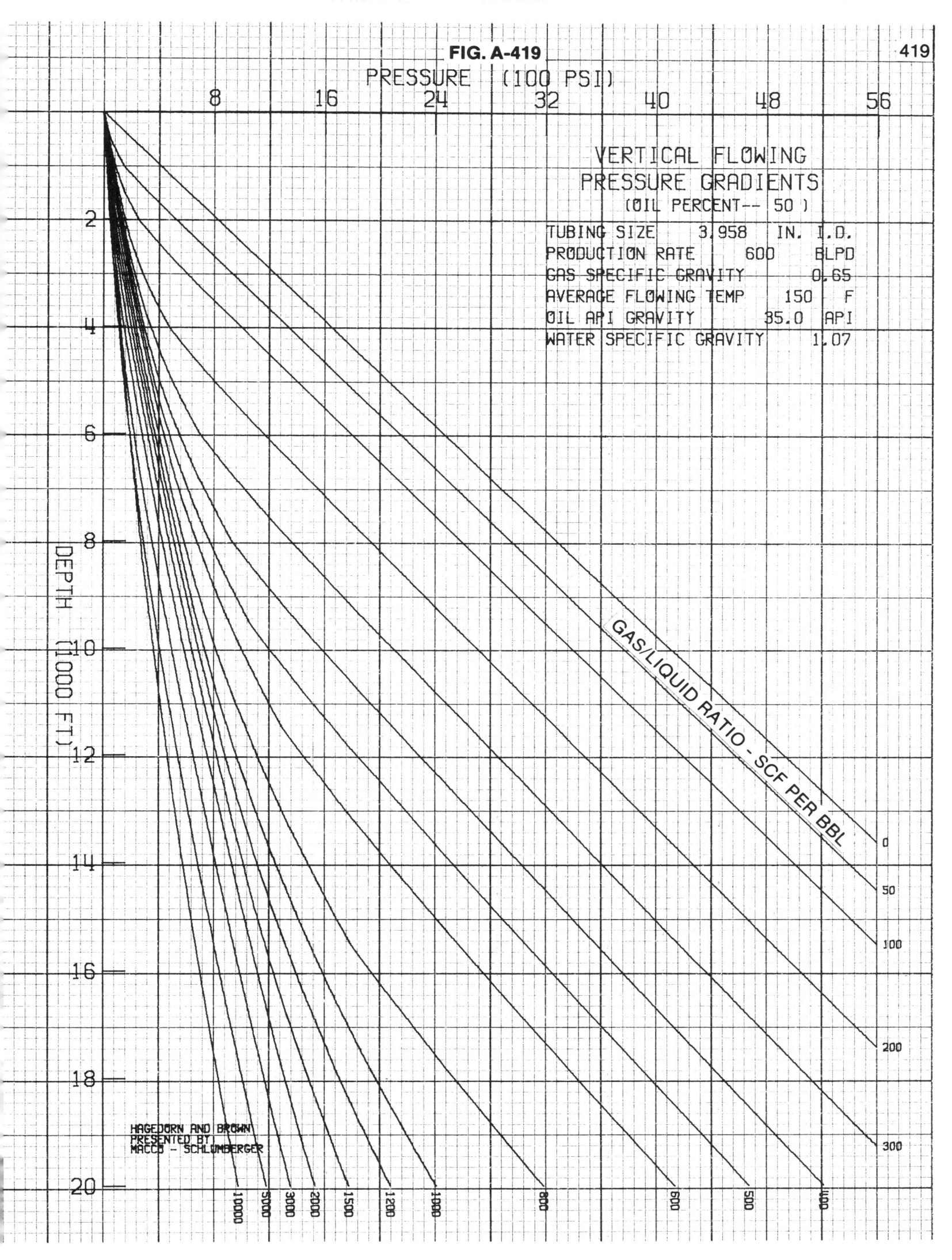

PRESSURE (100 PSI)

VERTICAL FLOWING
PRESSURE GRADIENTS
(OIL PERCENT-- 50)

TUBING SIZE	3.958	IN. I.D.
PRODUCTION RATE	600	BLPD
GAS SPECIFIC GRAVITY	0.65	
AVERAGE FLOWING TEMP	150	F
OIL API GRAVITY	35.0	API
WATER SPECIFIC GRAVITY	1.07	

GAS/LIQUID RATIO - SCF PER BBL

DEPTH (1000 FT)

HAGEDORN AND BROWN
PRESENTED BY
MACCO - SCHLUMBERGER

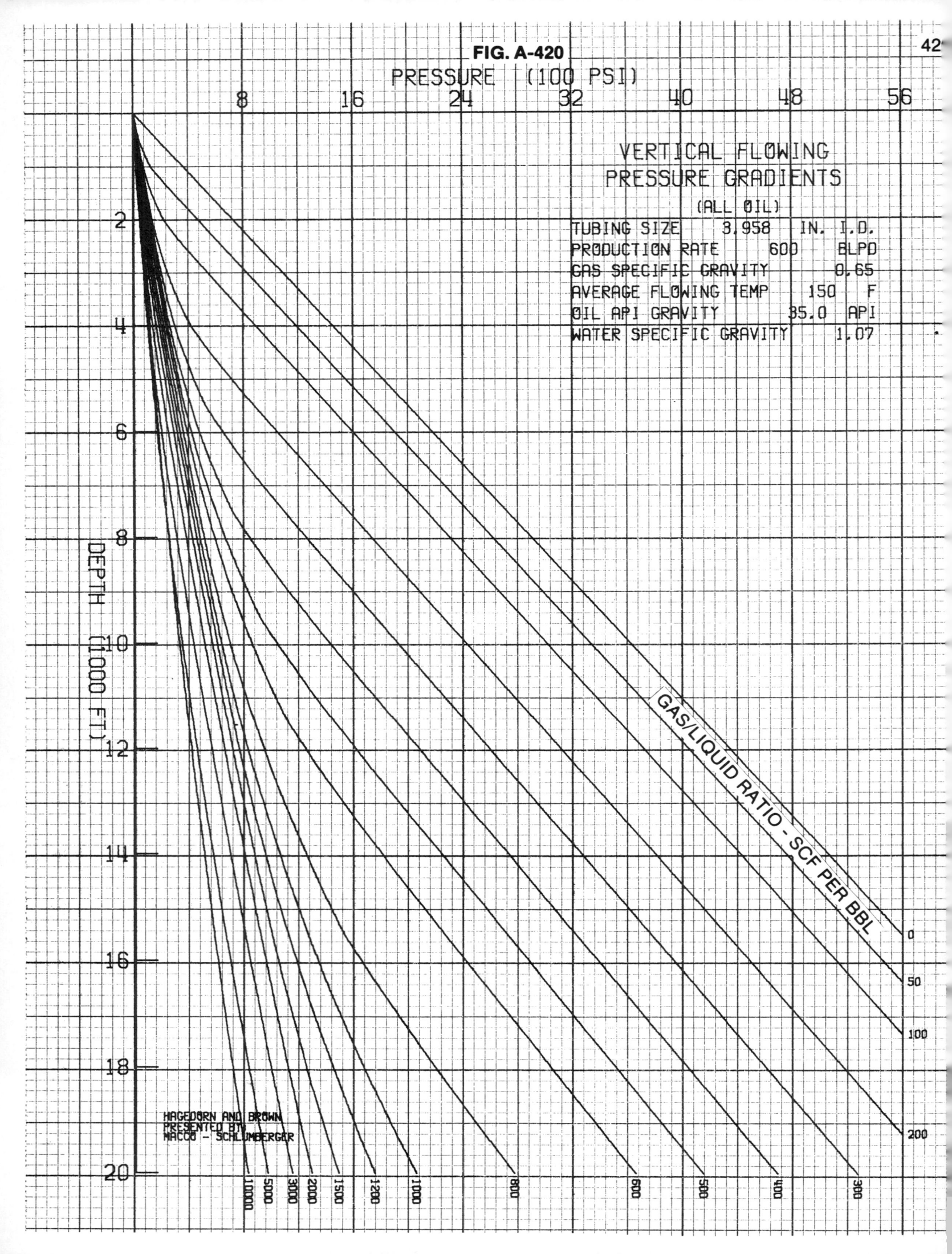

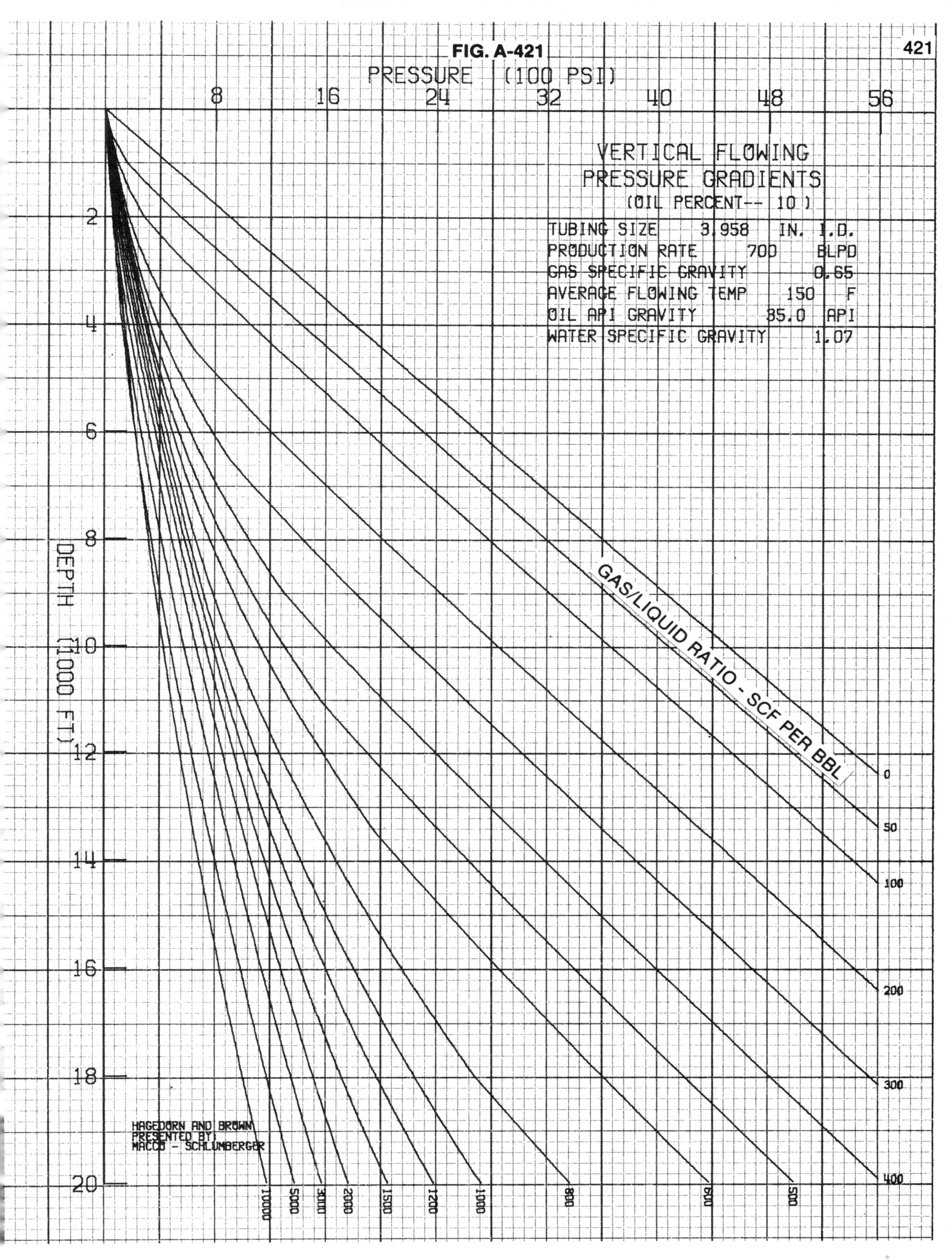

PRESSURE (100 PSI)

VERTICAL FLOWING
PRESSURE GRADIENTS
(OIL PERCENT-- 10)

TUBING SIZE	3.958	IN. I.D.
PRODUCTION RATE	700	BLPD
GAS SPECIFIC GRAVITY	0.65	
AVERAGE FLOWING TEMP	150	F
OIL API GRAVITY	35.0	API
WATER SPECIFIC GRAVITY	1.07	

GAS/LIQUID RATIO - SCF PER BBL

DEPTH (1000 FT)

HAGEDORN AND BROWN
PRESENTED BY
MACCO - SCHLUMBERGER

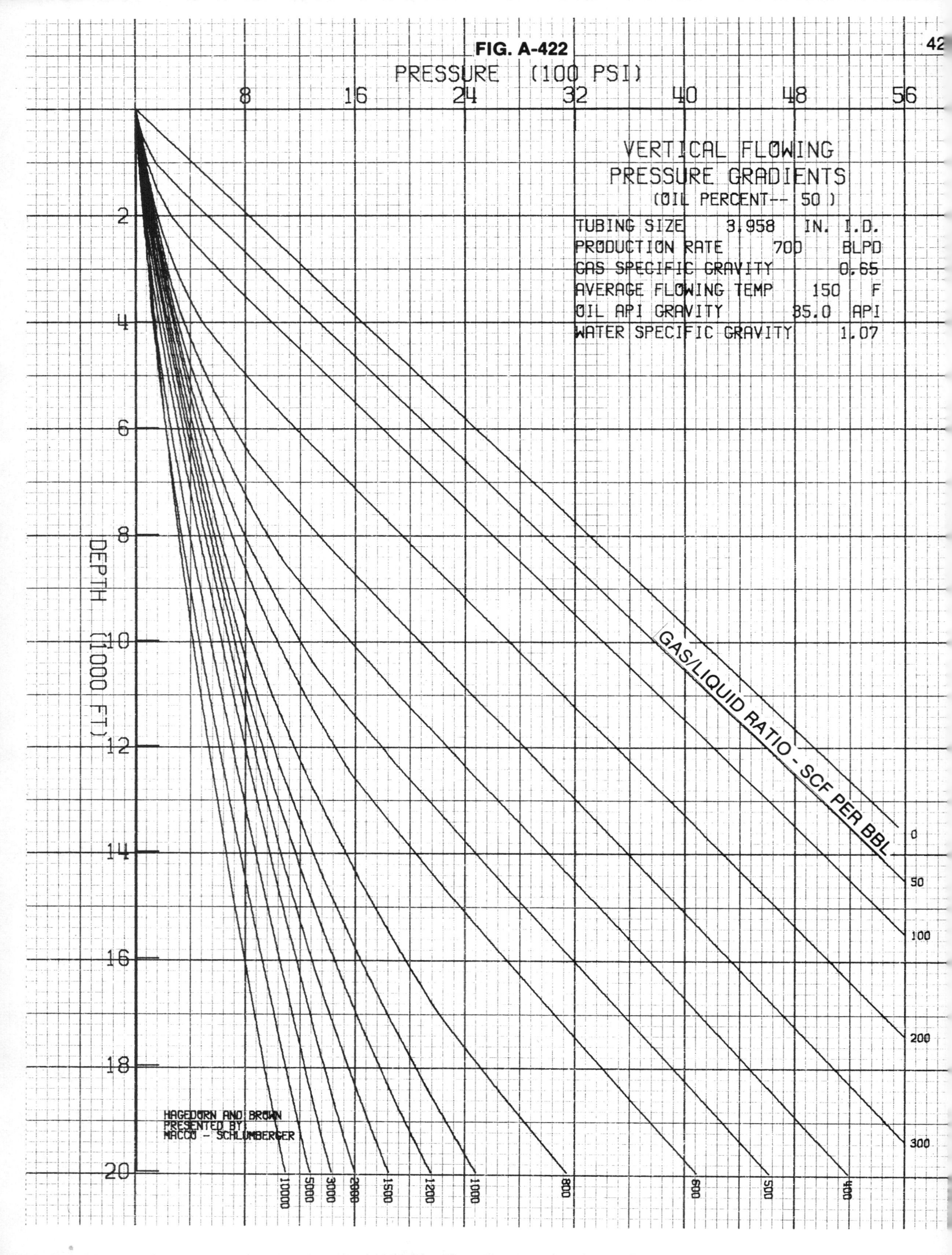

FIG. A-422

VERTICAL FLOWING
PRESSURE GRADIENTS
(OIL PERCENT-- 50)

TUBING SIZE	3.958	IN. I.D.
PRODUCTION RATE	700	BLPD
GAS SPECIFIC GRAVITY	0.65	
AVERAGE FLOWING TEMP	150	F
OIL API GRAVITY	35.0	API
WATER SPECIFIC GRAVITY	1.07	

PRESSURE (100 PSI)

DEPTH (1000 FT)

GAS/LIQUID RATIO - SCF PER BBL

HAGEDORN AND BROWN
PRESENTED BY
MACCO - SCHLUMBERGER

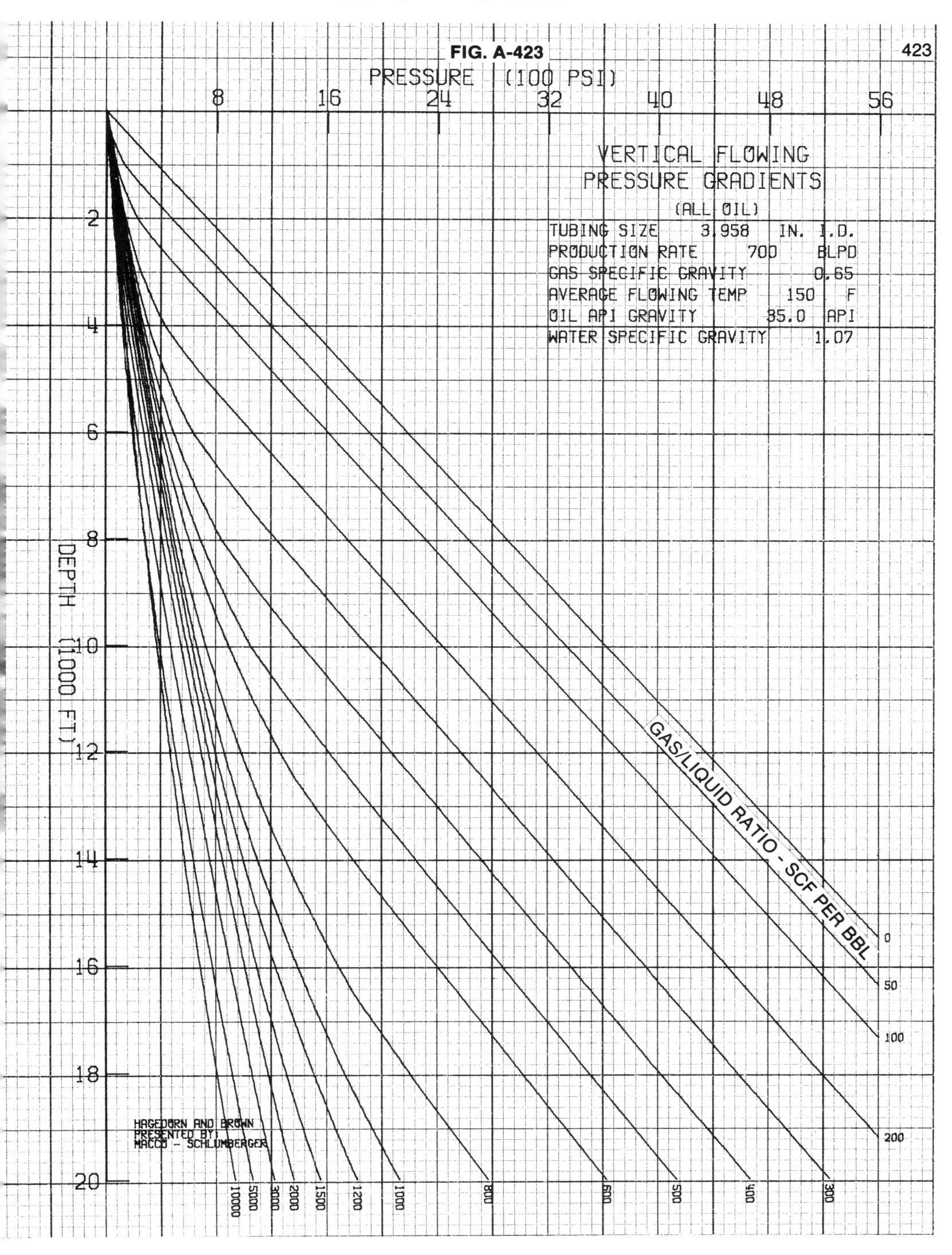

FIG. A-423

423

PRESSURE (100 PSI)

VERTICAL FLOWING
PRESSURE GRADIENTS
(ALL OIL)

TUBING SIZE	3.958	IN. I.D.
PRODUCTION RATE	700	BLPD
GAS SPECIFIC GRAVITY	0.65	
AVERAGE FLOWING TEMP	150	F
OIL API GRAVITY	35.0	API
WATER SPECIFIC GRAVITY	1.07	

DEPTH (1000 FT)

GAS/LIQUID RATIO - SCF PER BBL

HAGEDORN AND BROWN
PRESENTED BY:
MACCO — SCHLUMBERGER

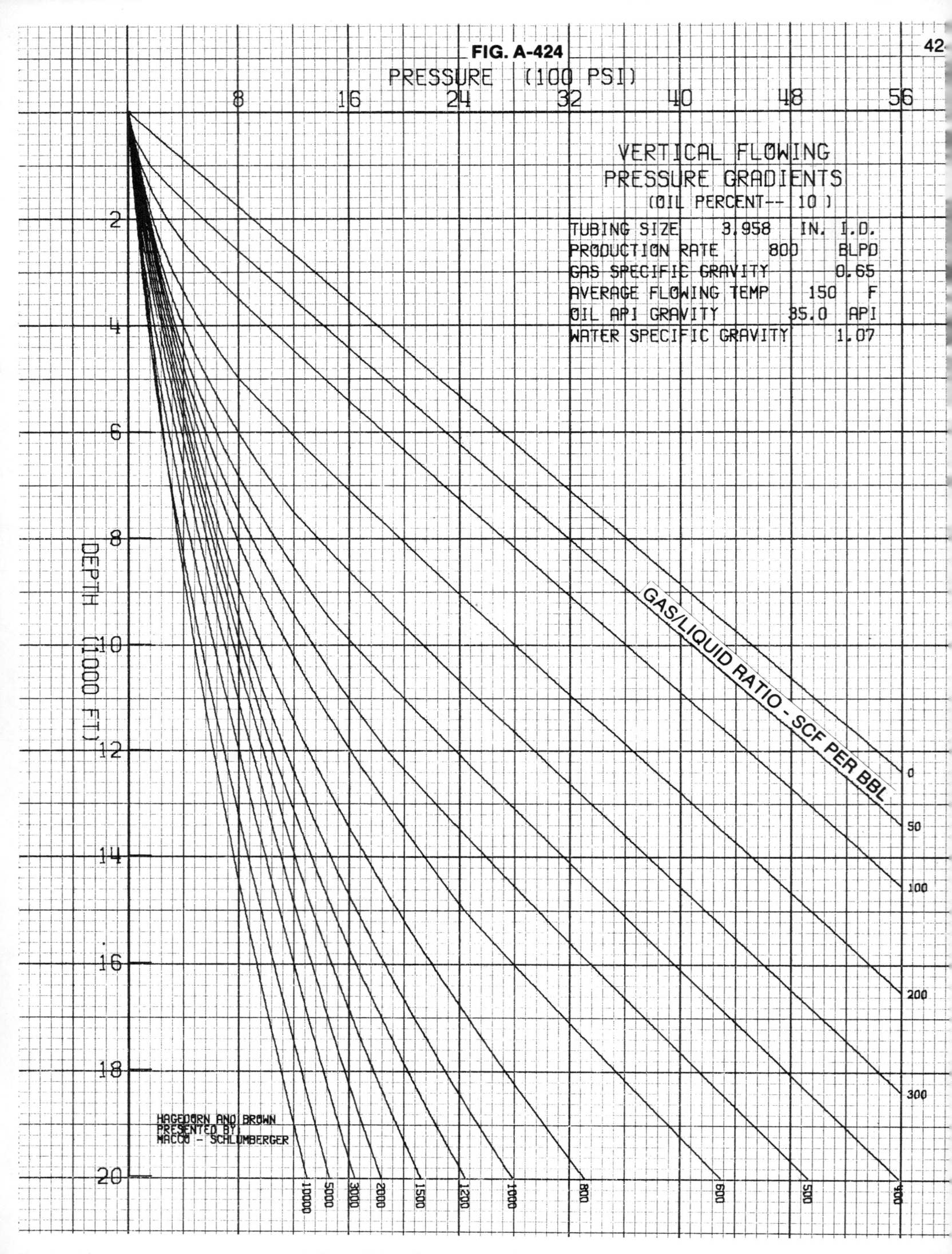

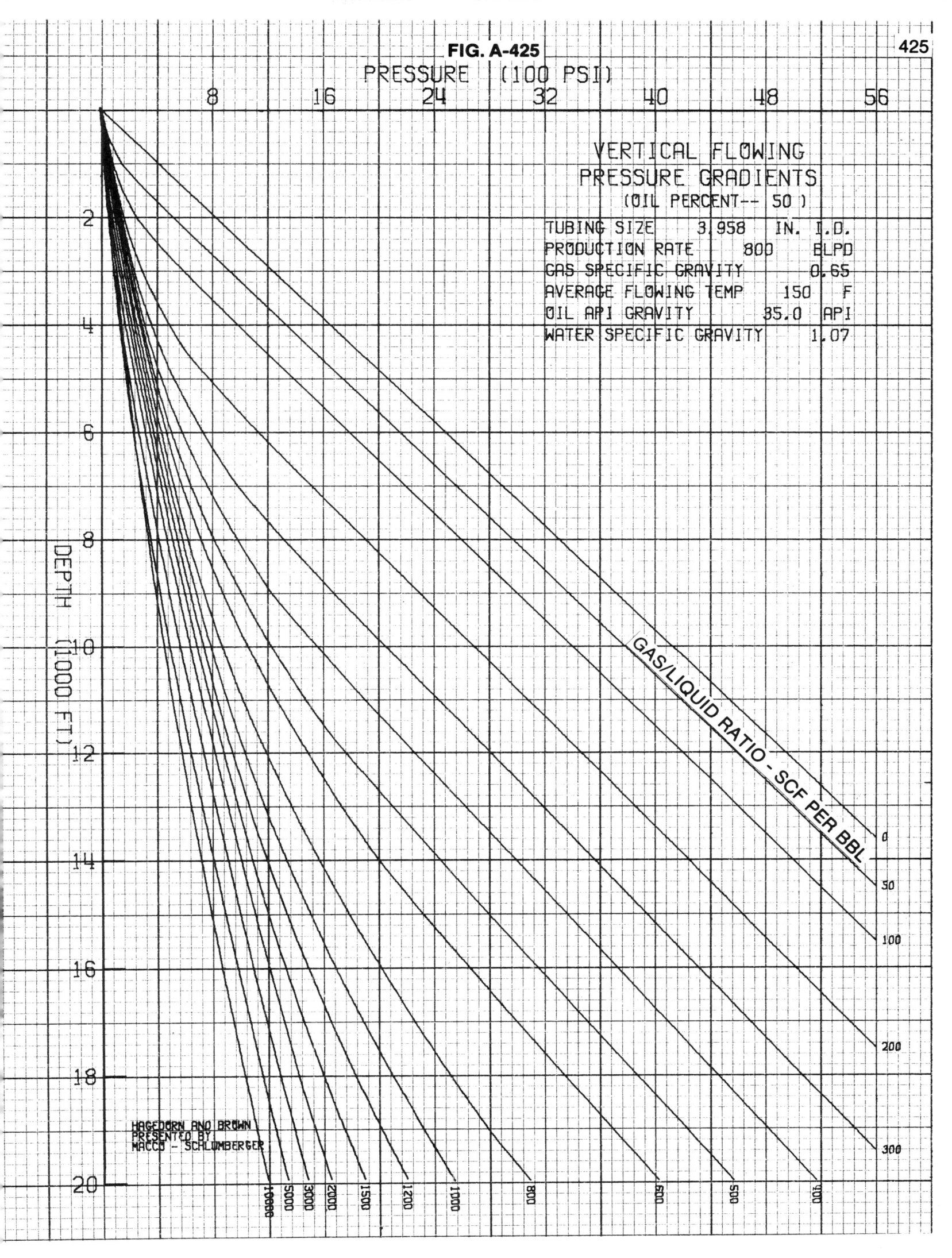

VERTICAL FLOWING
PRESSURE GRADIENTS
(OIL PERCENT-- 50)

TUBING SIZE	3.958	IN. I.D.
PRODUCTION RATE	800	BLPD
GAS SPECIFIC GRAVITY	0.65	
AVERAGE FLOWING TEMP	150	F
OIL API GRAVITY	35.0	API
WATER SPECIFIC GRAVITY	1.07	

PRESSURE (100 PSI)

DEPTH (1000 FT)

GAS/LIQUID RATIO - SCF PER BBL

HAGEDORN AND BROWN
PRESENTED BY
MACCO - SCHLUMBERGER

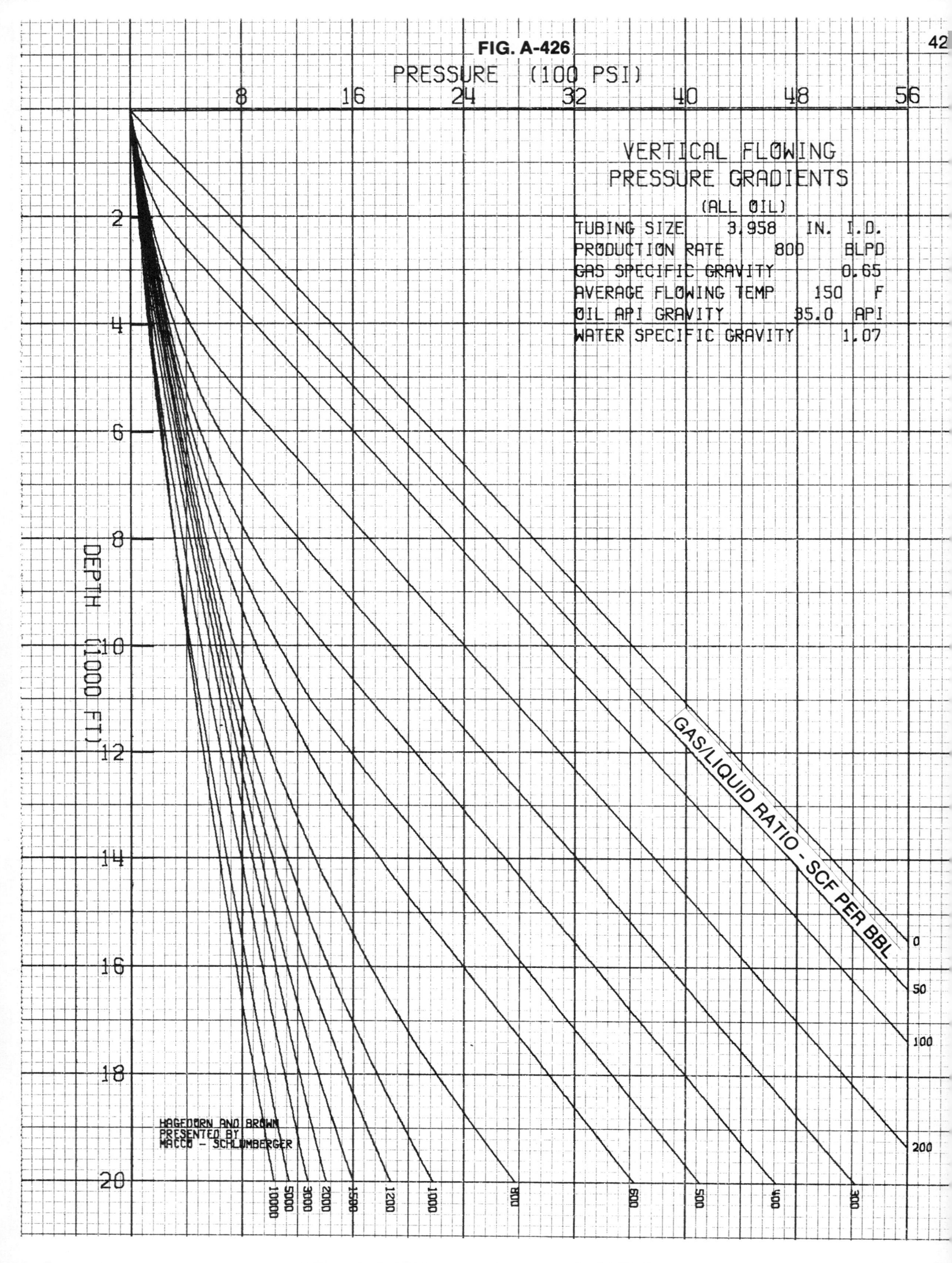

FIG. A-426

42

VERTICAL FLOWING
PRESSURE GRADIENTS
(ALL OIL)

TUBING SIZE	3.958	IN. I.D.
PRODUCTION RATE	800	BLPD
GAS SPECIFIC GRAVITY	0.65	
AVERAGE FLOWING TEMP	150	F
OIL API GRAVITY	35.0	API
WATER SPECIFIC GRAVITY	1.07	

PRESSURE (100 PSI)

DEPTH (1000 FT)

GAS/LIQUID RATIO - SCF PER BBL

HAGEDORN AND BROWN
PRESENTED BY
MACCO - SCHLUMBERGER

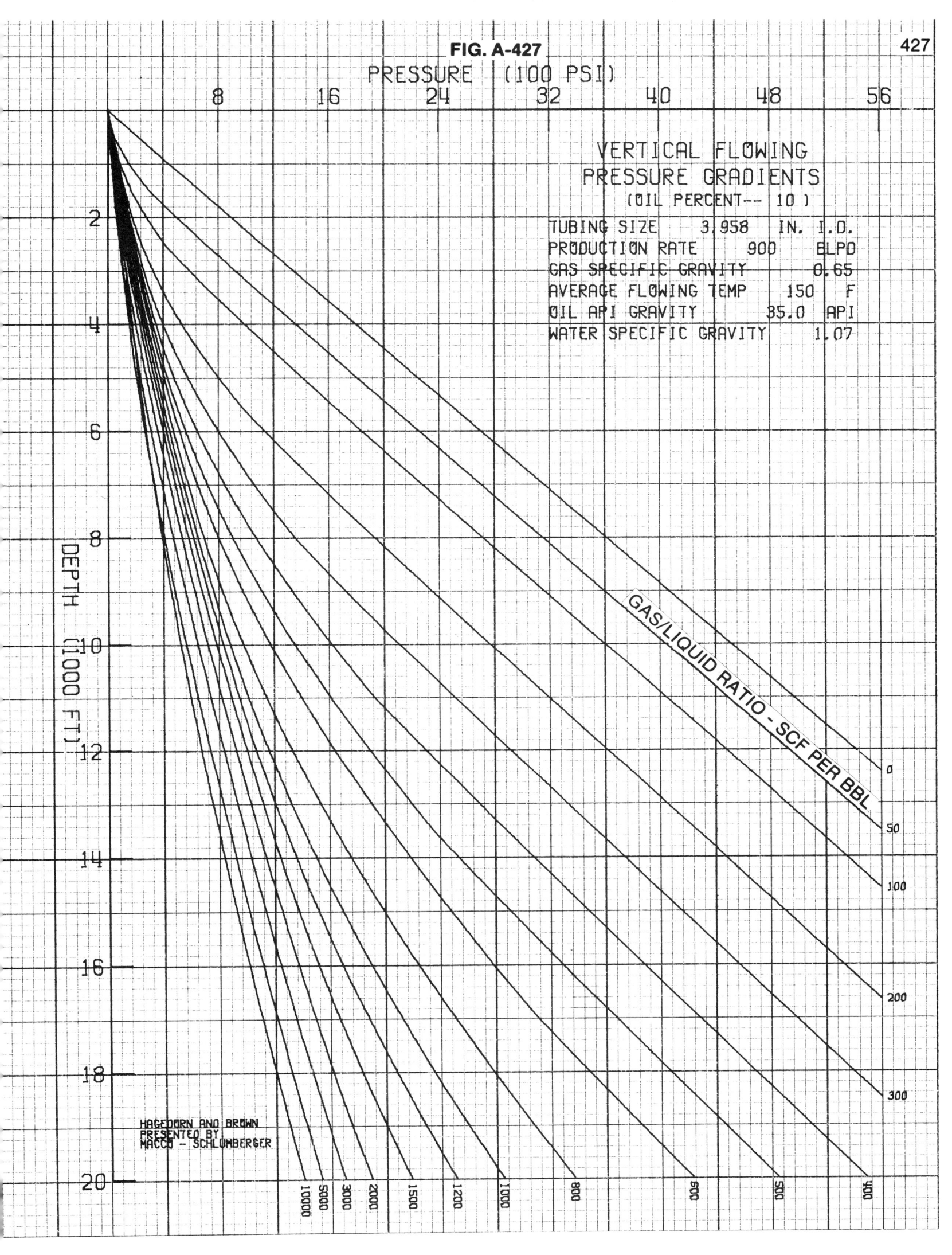

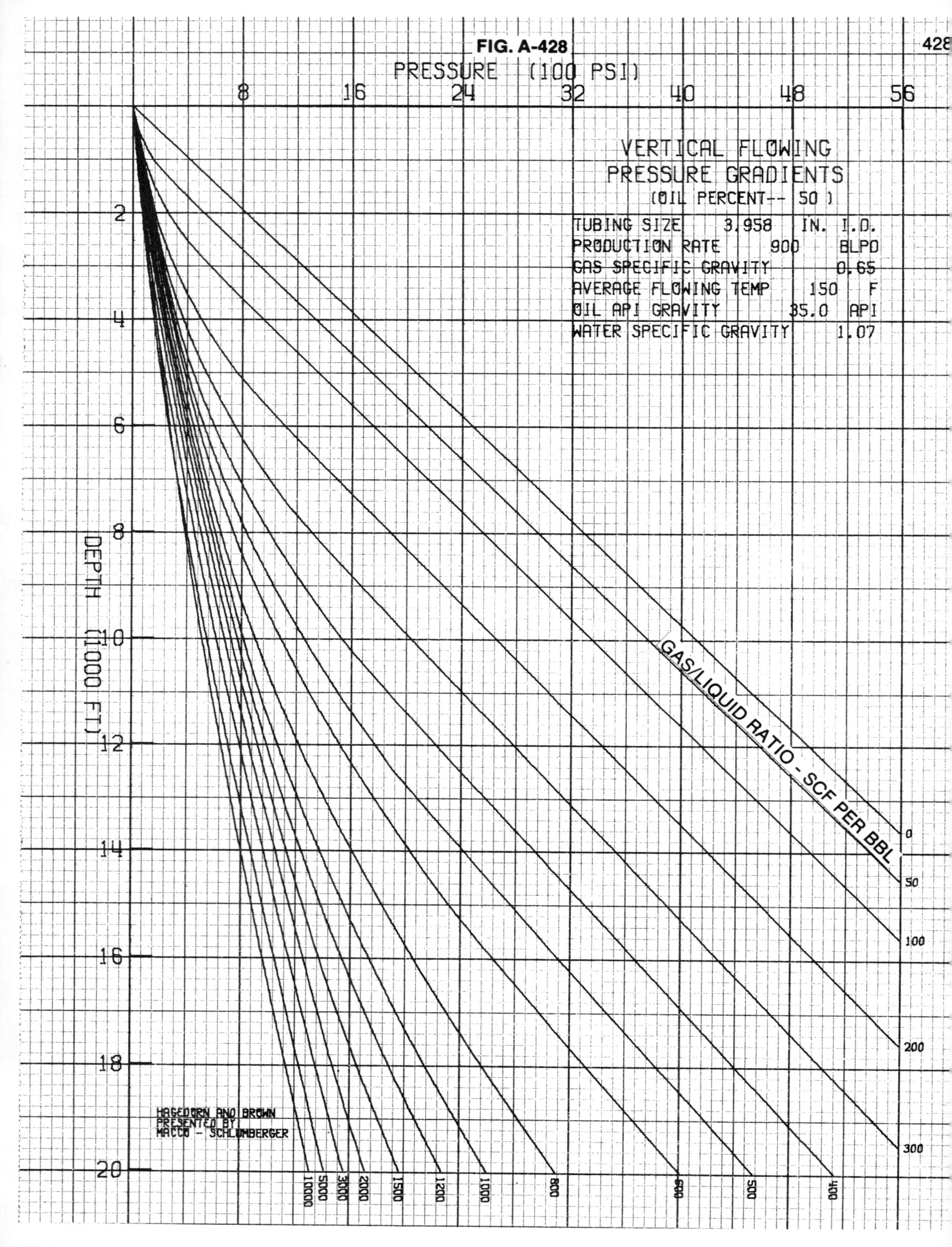

PRESSURE (100 PSI)

VERTICAL FLOWING
PRESSURE GRADIENTS
(OIL PERCENT-- 50)

TUBING SIZE	3.958	IN. I.D.
PRODUCTION RATE	900	BLPD
GAS SPECIFIC GRAVITY		0.65
AVERAGE FLOWING TEMP	150	F
OIL API GRAVITY	35.0	API
WATER SPECIFIC GRAVITY		1.07

DEPTH (1000 FT)

GAS/LIQUID RATIO - SCF PER BBL

0
50
100
200
300

HAGEDORN AND BROWN
PRESENTED BY
MACCO - SCHLUMBERGER

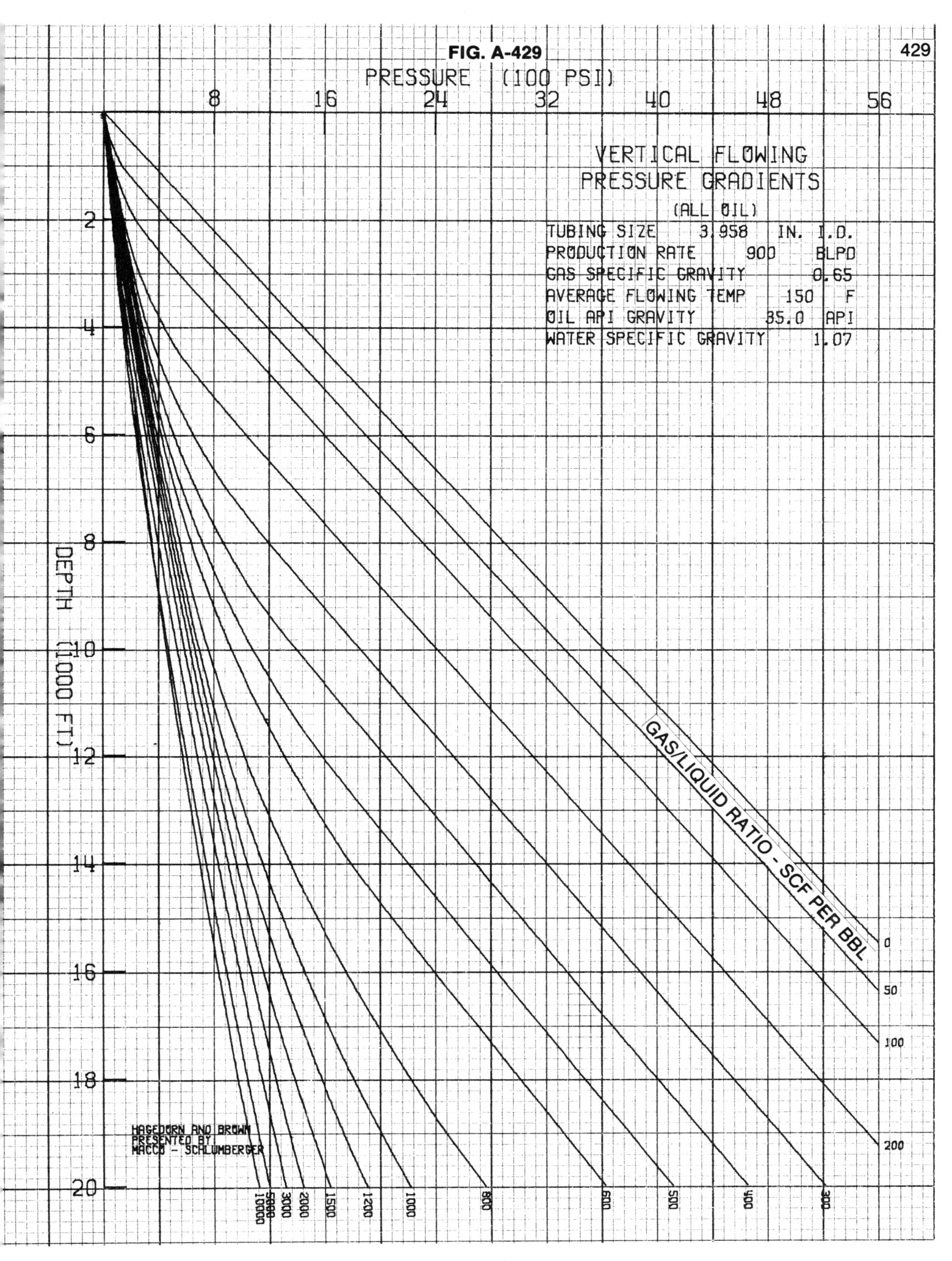

FIG. A-429

429

VERTICAL FLOWING
PRESSURE GRADIENTS
(ALL OIL)

TUBING SIZE	3.958	IN. I.D.
PRODUCTION RATE	900	BLPD
GAS SPECIFIC GRAVITY	0.65	
AVERAGE FLOWING TEMP	150	F
OIL API GRAVITY	35.0	API
WATER SPECIFIC GRAVITY	1.07	

PRESSURE (100 PSI)

DEPTH (1000 FT)

GAS/LIQUID RATIO - SCF PER BBL

HAGEDORN AND BROWN
PRESENTED BY
MACCO - SCHLUMBERGER

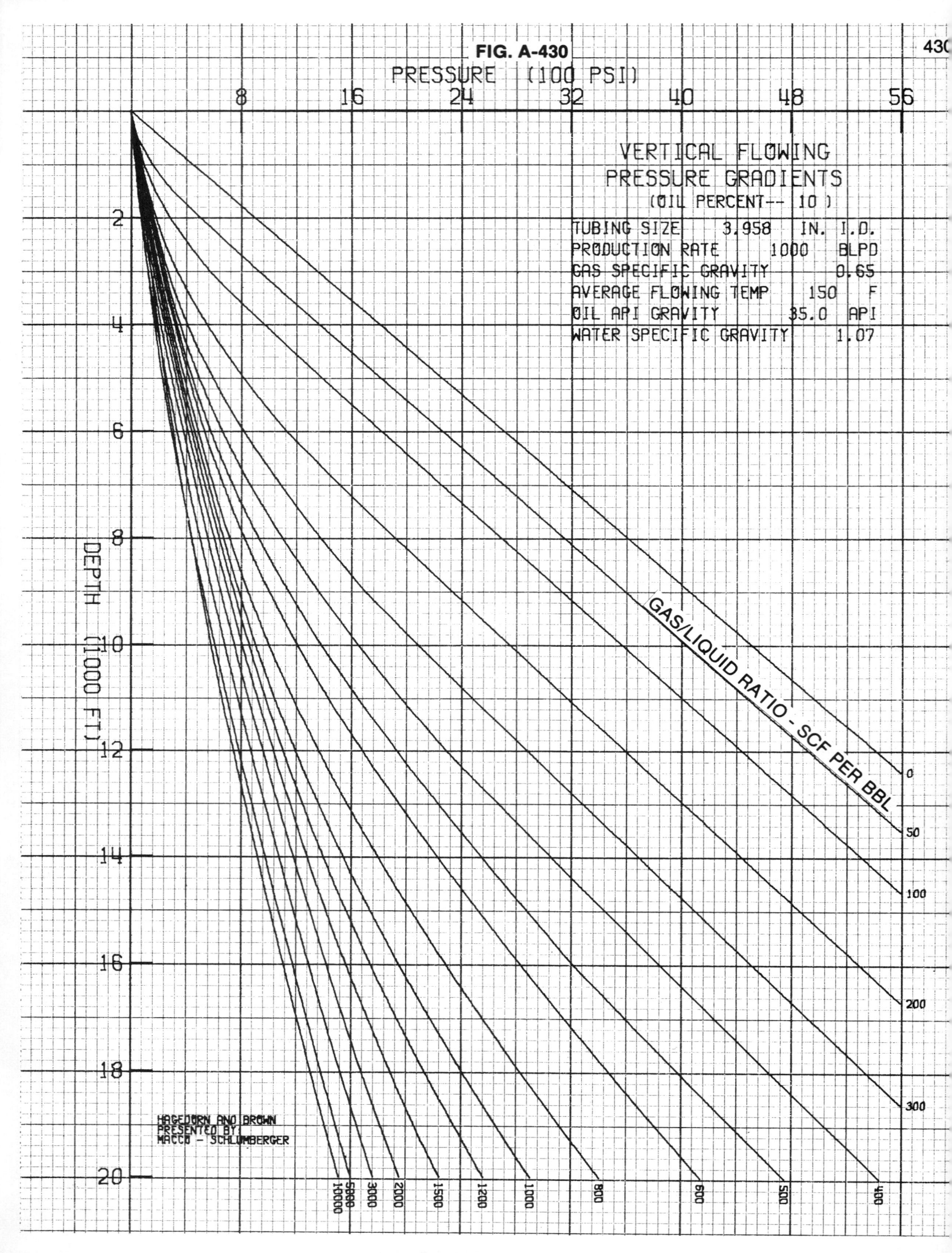

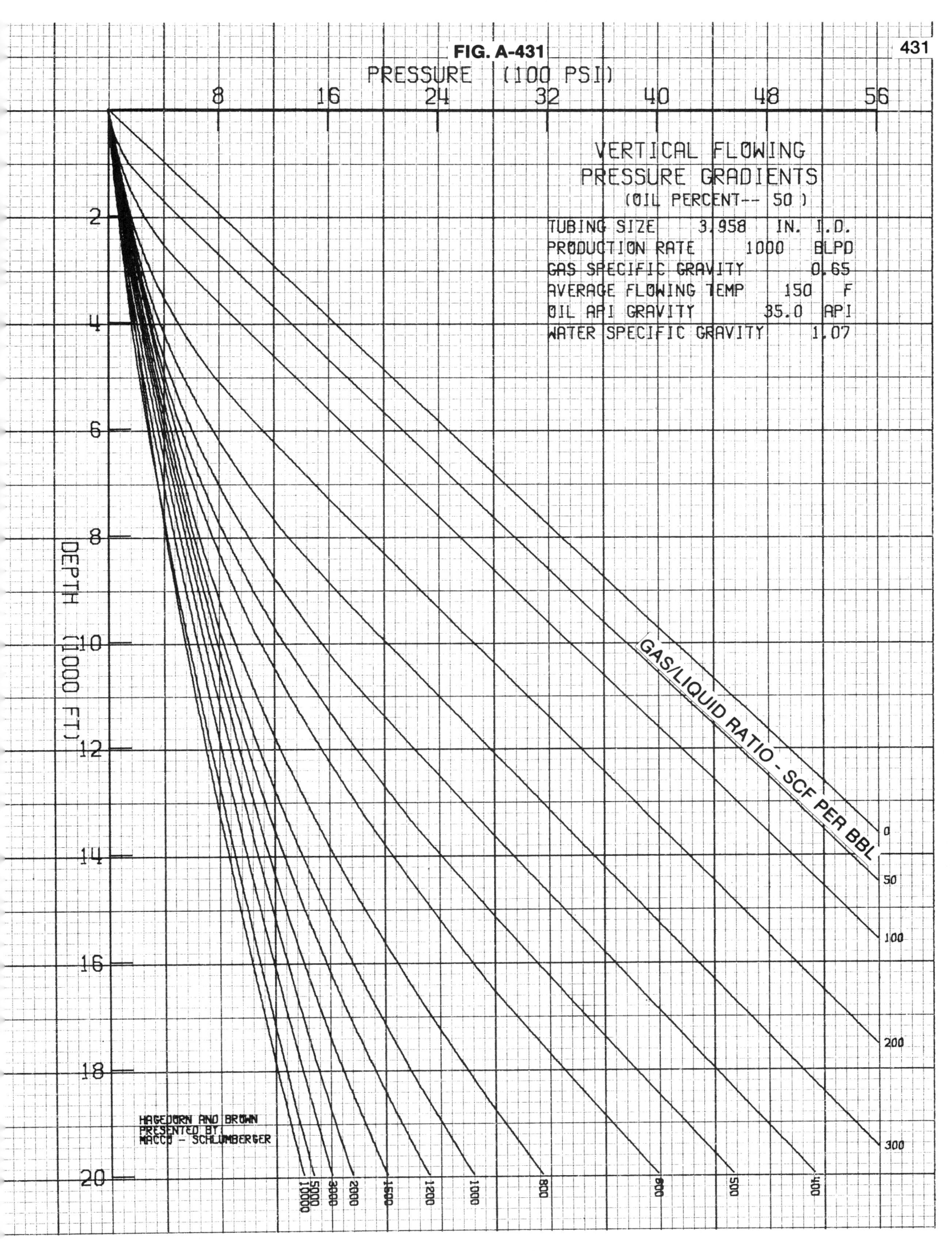

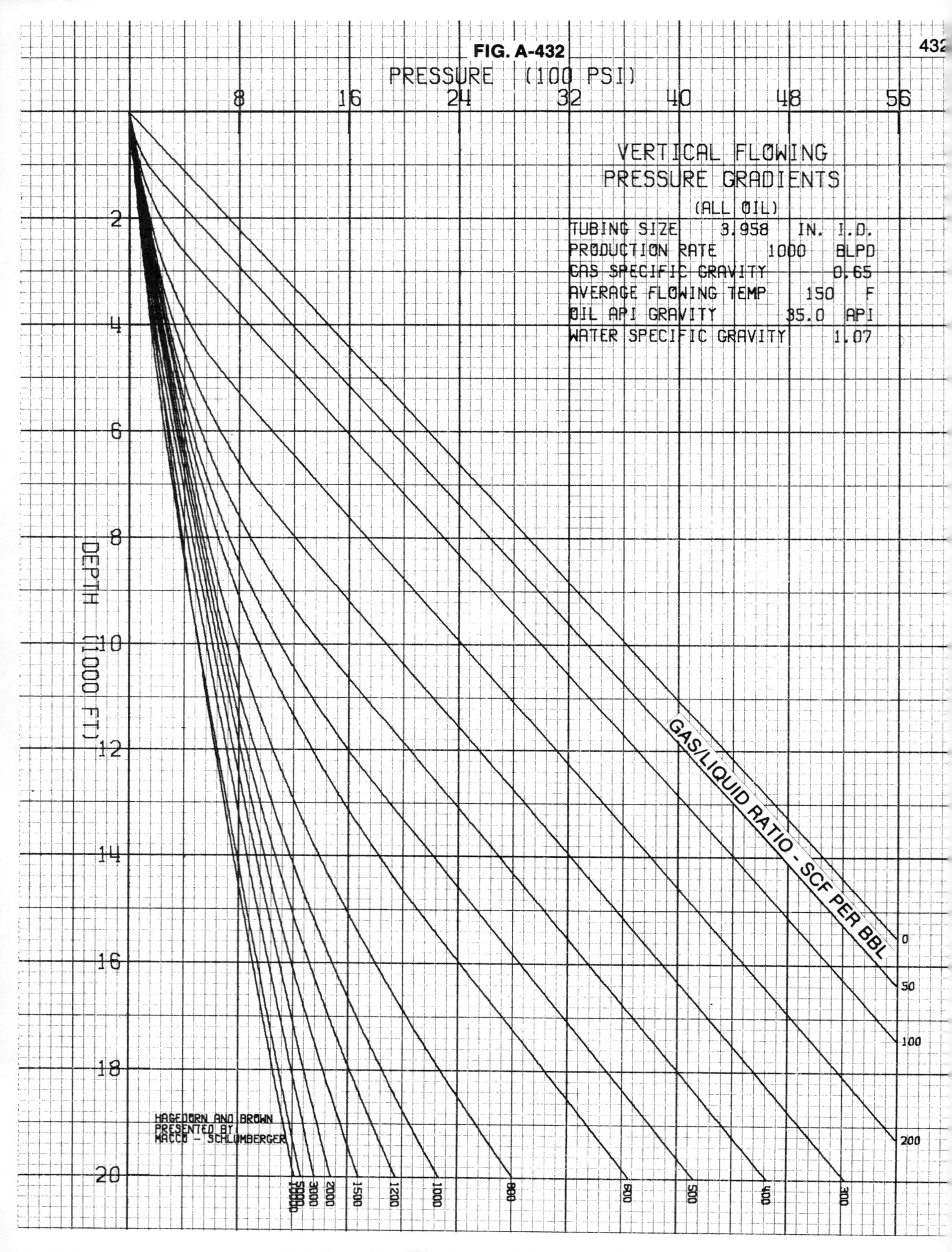

PRESSURE (100 PSI)

VERTICAL FLOWING
PRESSURE GRADIENTS
(ALL OIL)

TUBING SIZE	3.958	IN. I.D.
PRODUCTION RATE	1000	BLPD
GAS SPECIFIC GRAVITY	0.65	
AVERAGE FLOWING TEMP	150	F
OIL API GRAVITY	35.0	API
WATER SPECIFIC GRAVITY	1.07	

DEPTH (1000 FT)

GAS/LIQUID RATIO - SCF PER BBL

HAGEDORN AND BROWN
PRESENTED BY
MACCO - SCHLUMBERGER

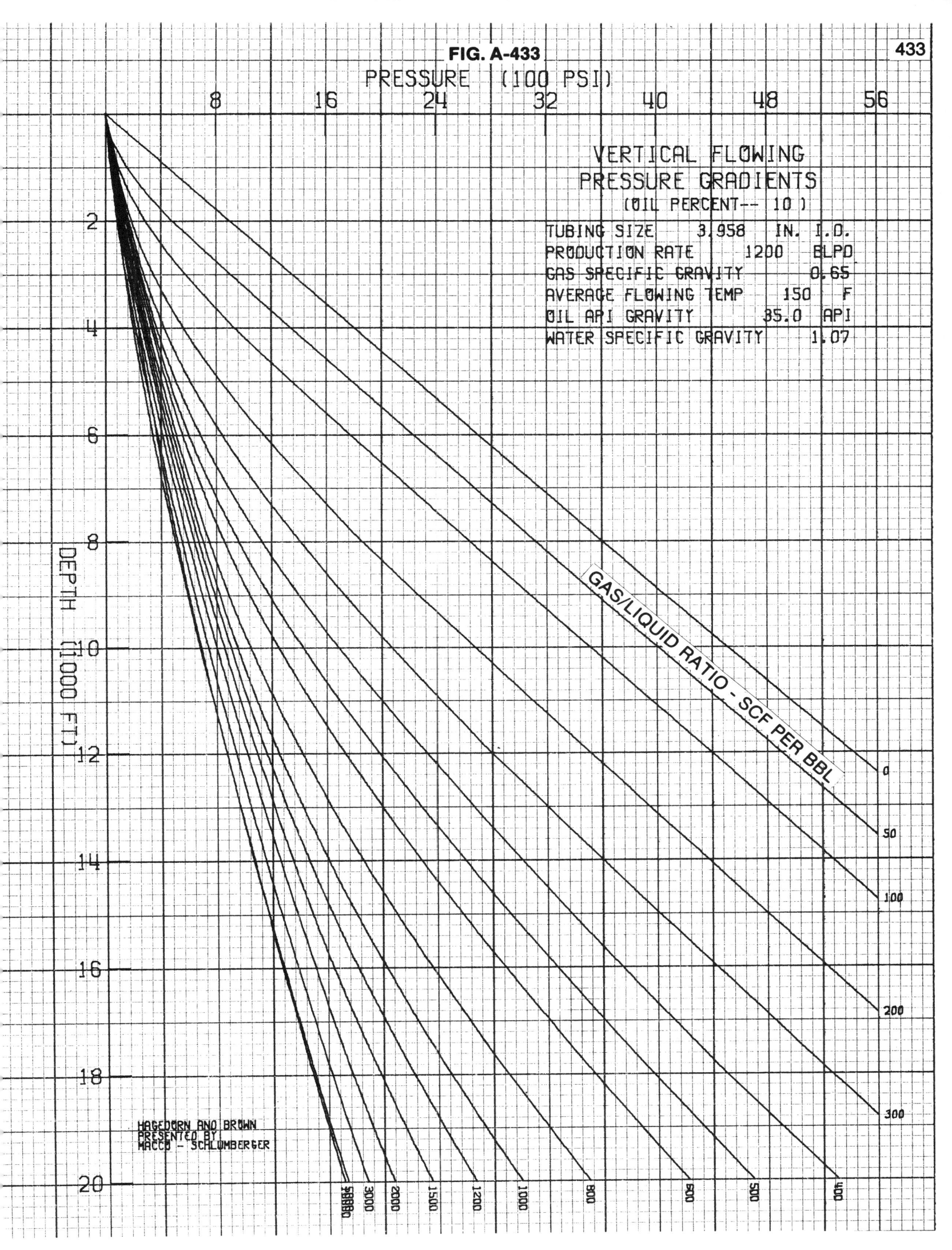

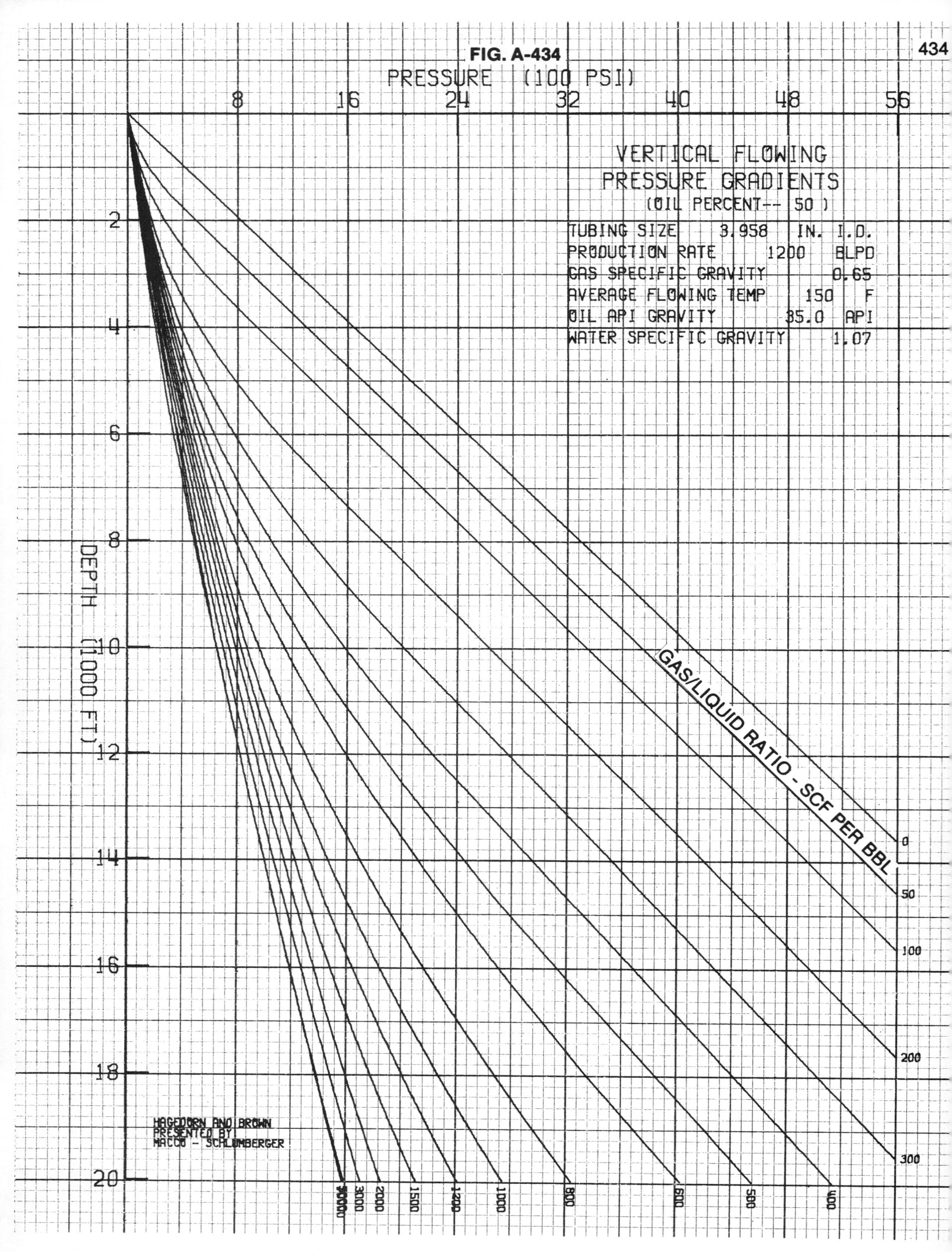

PRESSURE (100 PSI)

VERTICAL FLOWING
PRESSURE GRADIENTS
(OIL PERCENT-- 50)

TUBING SIZE	3.958	IN. I.D.
PRODUCTION RATE	1200	BLPD
GAS SPECIFIC GRAVITY		0.65
AVERAGE FLOWING TEMP	150	F
OIL API GRAVITY	35.0	API
WATER SPECIFIC GRAVITY		1.07

DEPTH (1000 FT)

GAS/LIQUID RATIO - SCF PER BBL

HAGEDORN AND BROWN
PRESENTED BY
MACCO - SCHLUMBERGER

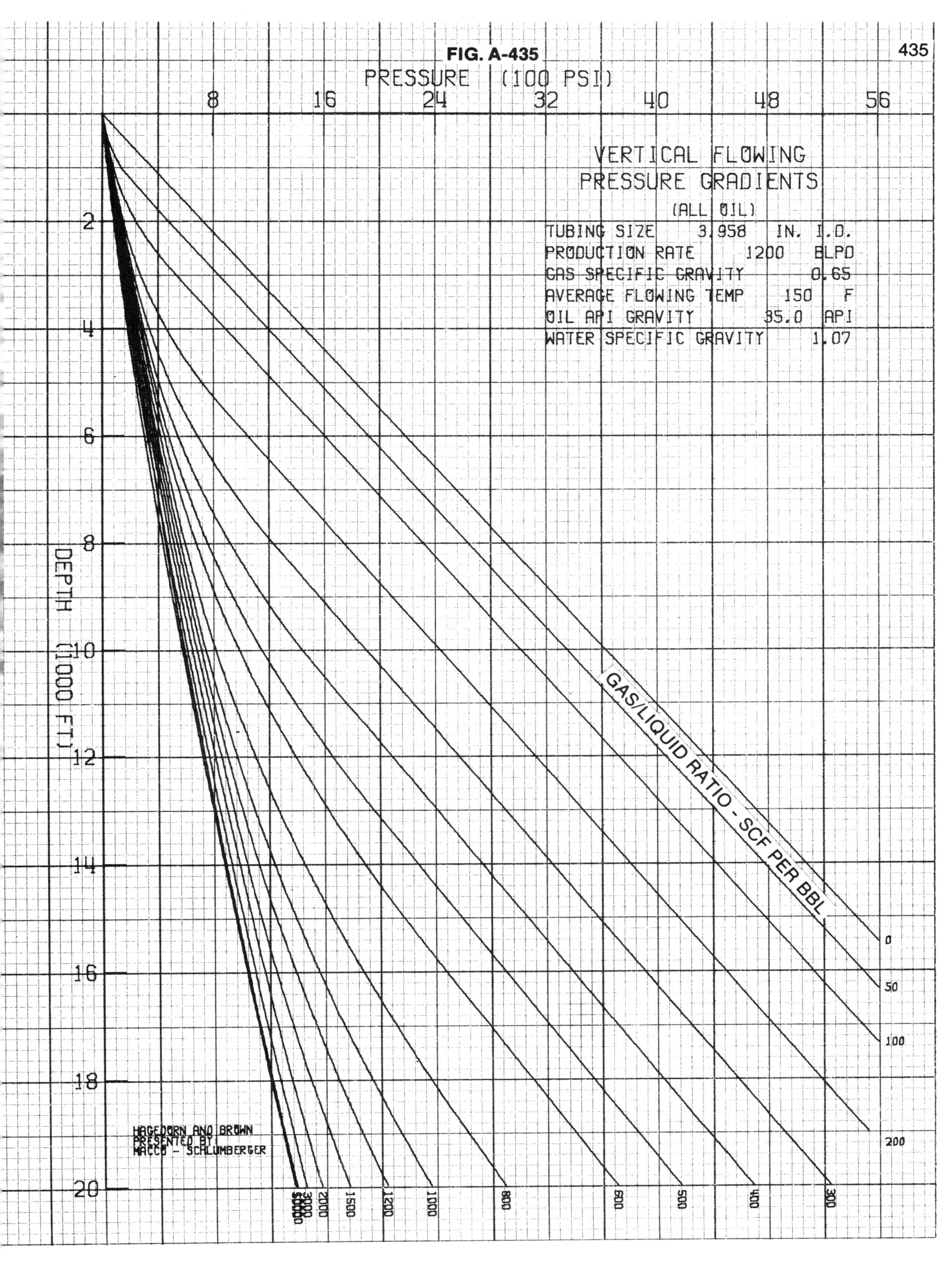

PRESSURE (100 PSI)

VERTICAL FLOWING
PRESSURE GRADIENTS
(ALL OIL)

TUBING SIZE   3.958   IN. I.D.
PRODUCTION RATE   1200   BLPD
GAS SPECIFIC GRAVITY   0.65
AVERAGE FLOWING TEMP   150   F
OIL API GRAVITY   35.0   API
WATER SPECIFIC GRAVITY   1.07

DEPTH (1000 FT)

GAS/LIQUID RATIO - SCF PER BBL

HAGEDORN AND BROWN
PRESENTED BY
MACCO - SCHLUMBERGER

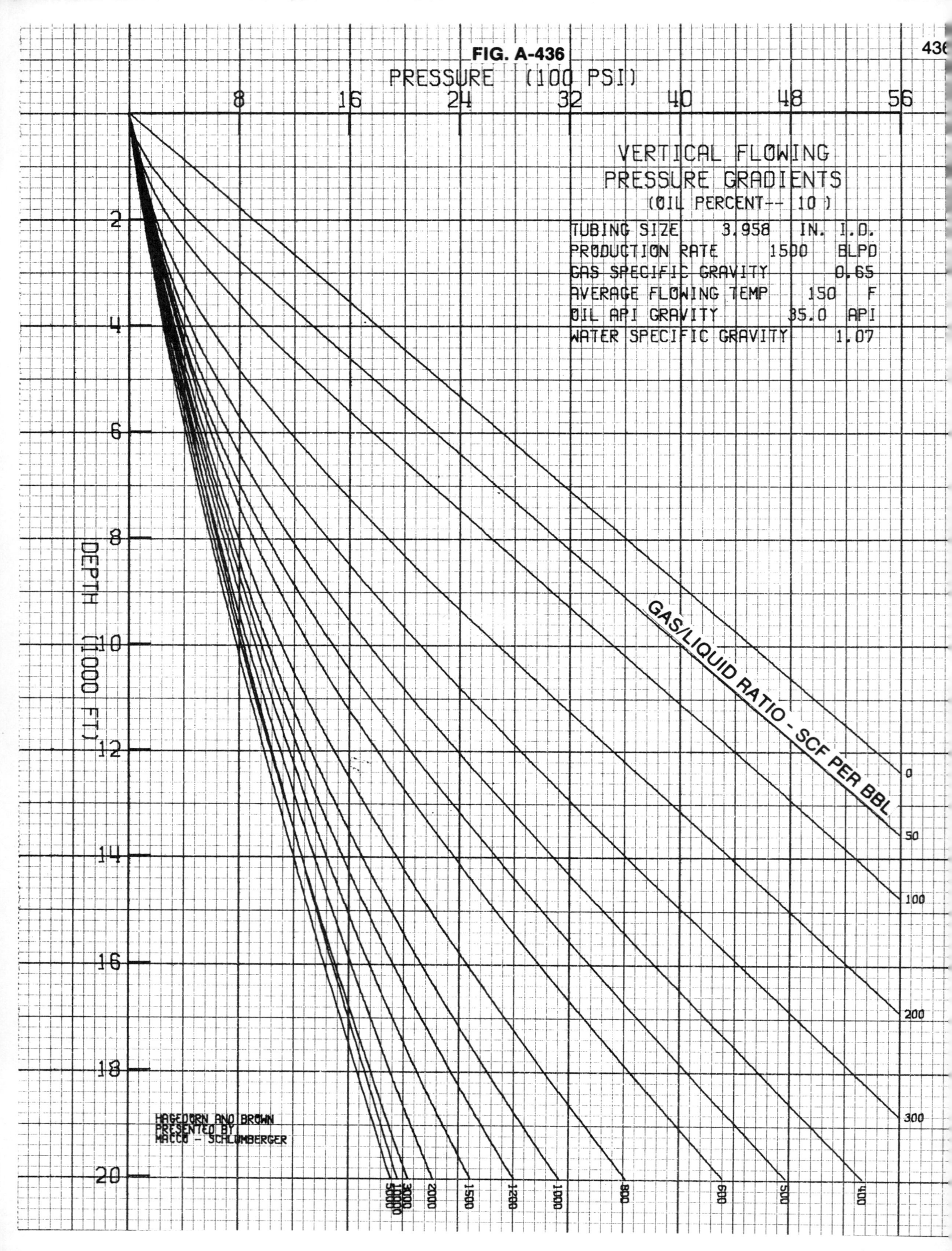

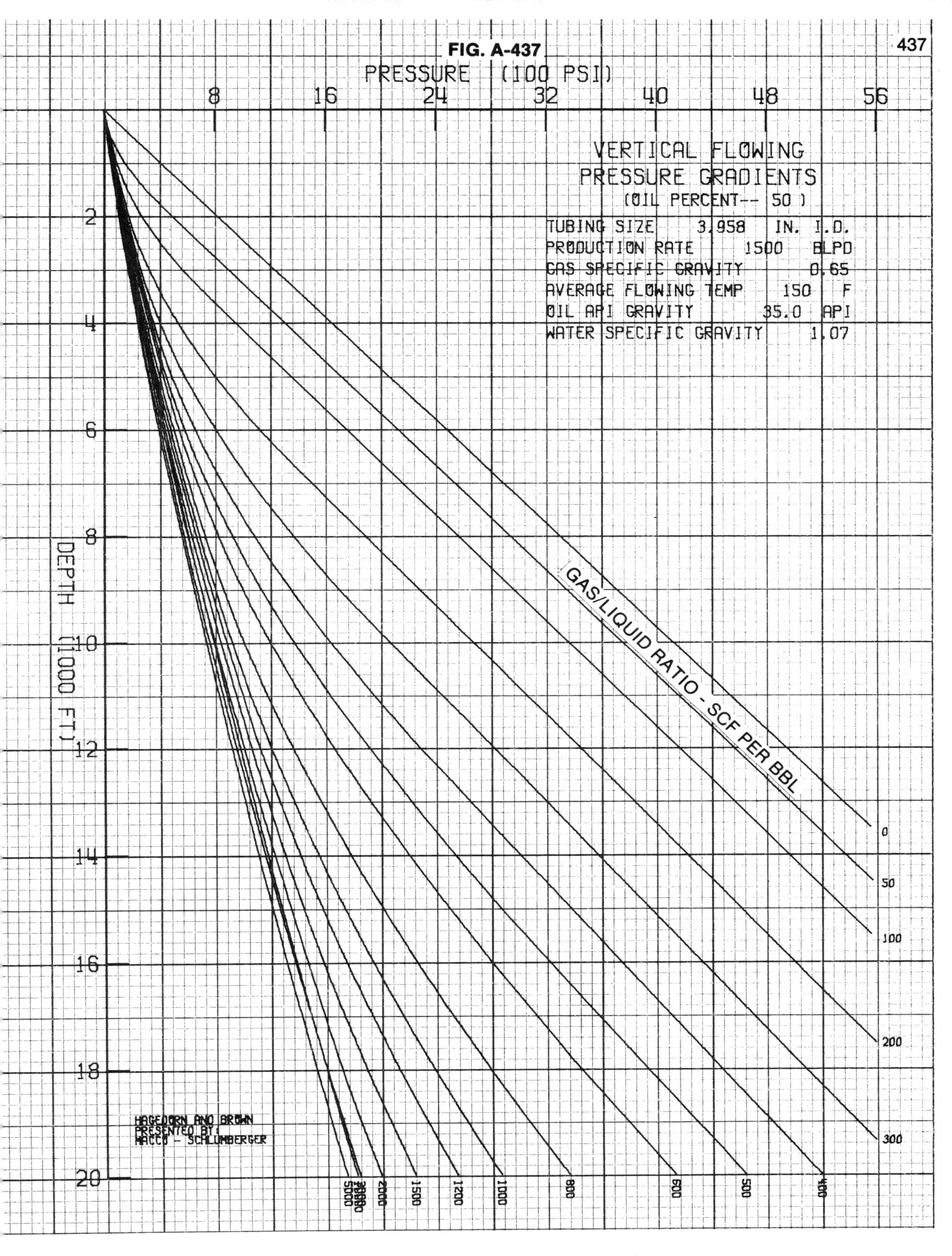

PRESSURE (100 PSI)

VERTICAL FLOWING
PRESSURE GRADIENTS
(OIL PERCENT--- 50)

TUBING SIZE	3.958	IN. I.D.
PRODUCTION RATE	1500	BLPD
GAS SPECIFIC GRAVITY	0.65	
AVERAGE FLOWING TEMP	150	F
OIL API GRAVITY	35.0	API
WATER SPECIFIC GRAVITY	1.07	

GAS/LIQUID RATIO - SCF PER BBL

DEPTH (1000 FT)

HAGEDORN AND BROWN
PRESENTED BY:
NACCO - SCHLUMBERGER

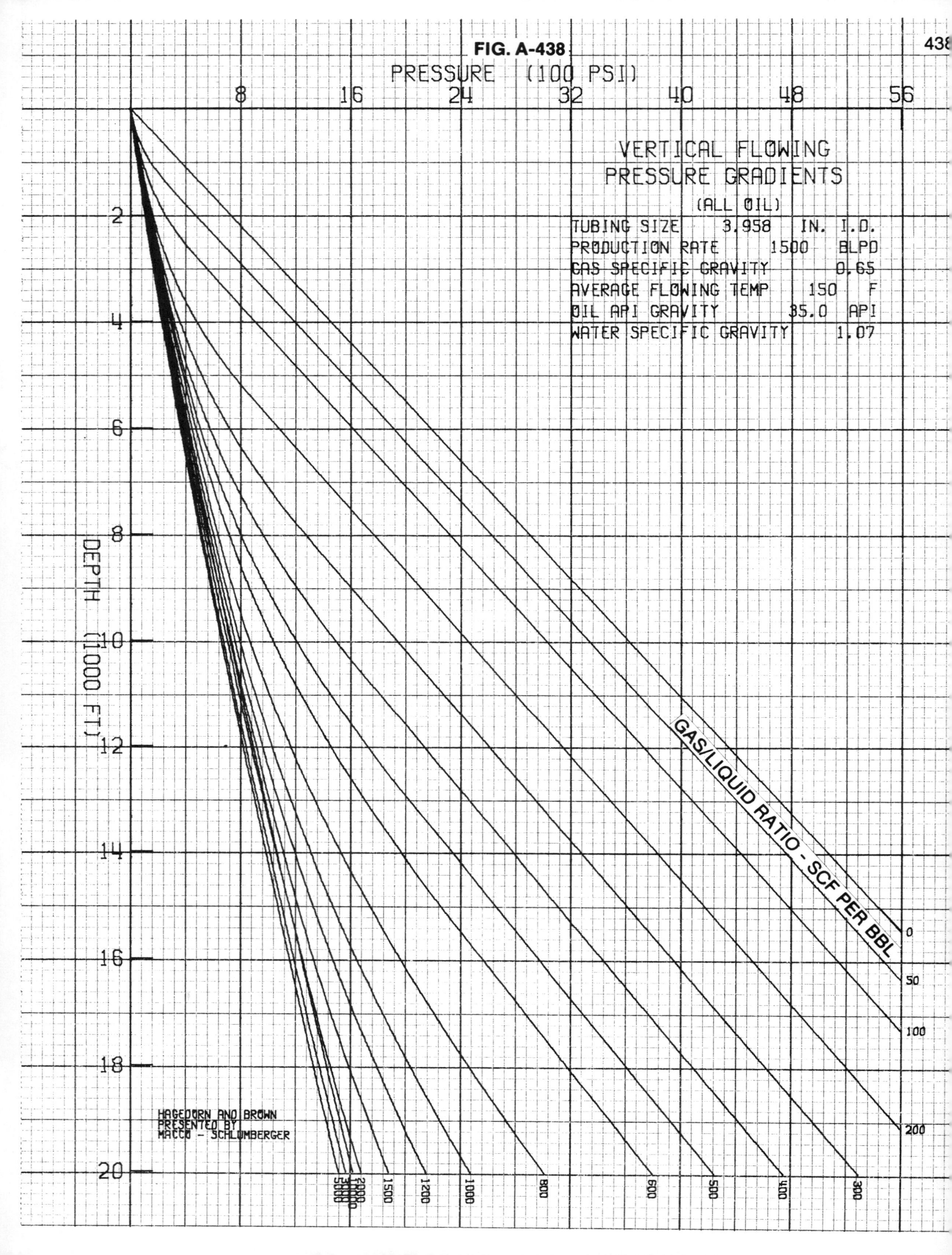

PRESSURE (100 PSI)

VERTICAL FLOWING
PRESSURE GRADIENTS
(ALL OIL)

TUBING SIZE	3.958	IN. I.D.
PRODUCTION RATE	1500	BLPD
GAS SPECIFIC GRAVITY	0.65	
AVERAGE FLOWING TEMP	150	F
OIL API GRAVITY	35.0	API
WATER SPECIFIC GRAVITY	1.07	

DEPTH (1,000 FT)

GAS/LIQUID RATIO - SCF PER BBL

HAGEDORN AND BROWN
PRESENTED BY
MACCO – SCHLUMBERGER

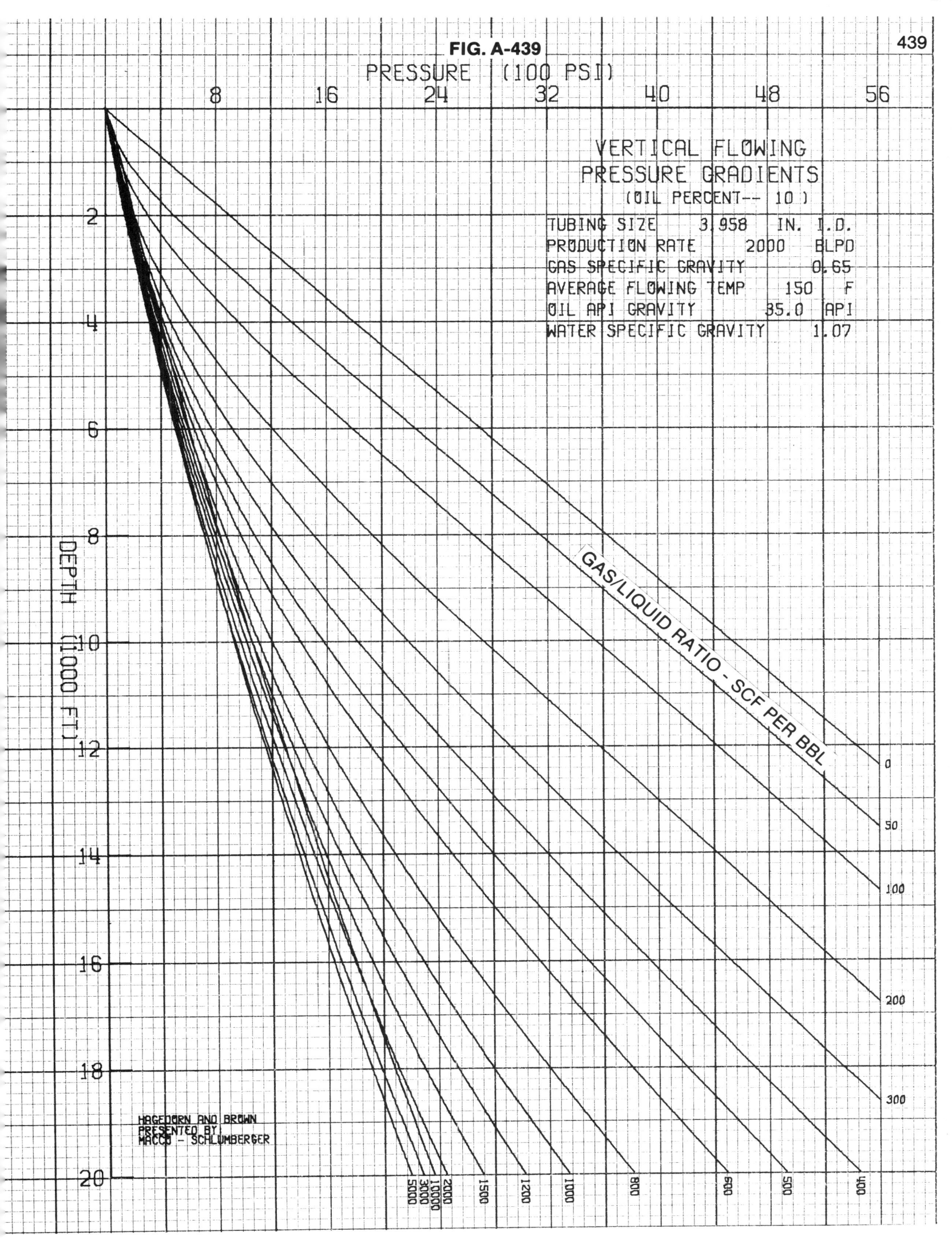

PRESSURE (100 PSI)

VERTICAL FLOWING
PRESSURE GRADIENTS
(OIL PERCENT-- 10)

TUBING SIZE	3.958	IN. I.D.
PRODUCTION RATE	2000	BLPD
GAS SPECIFIC GRAVITY	0.65	
AVERAGE FLOWING TEMP	150	F
OIL API GRAVITY	35.0	API
WATER SPECIFIC GRAVITY	1.07	

DEPTH (1,000 FT)

GAS/LIQUID RATIO - SCF PER BBL

HAGEDORN AND BROWN
PRESENTED BY
MACCO - SCHLUMBERGER

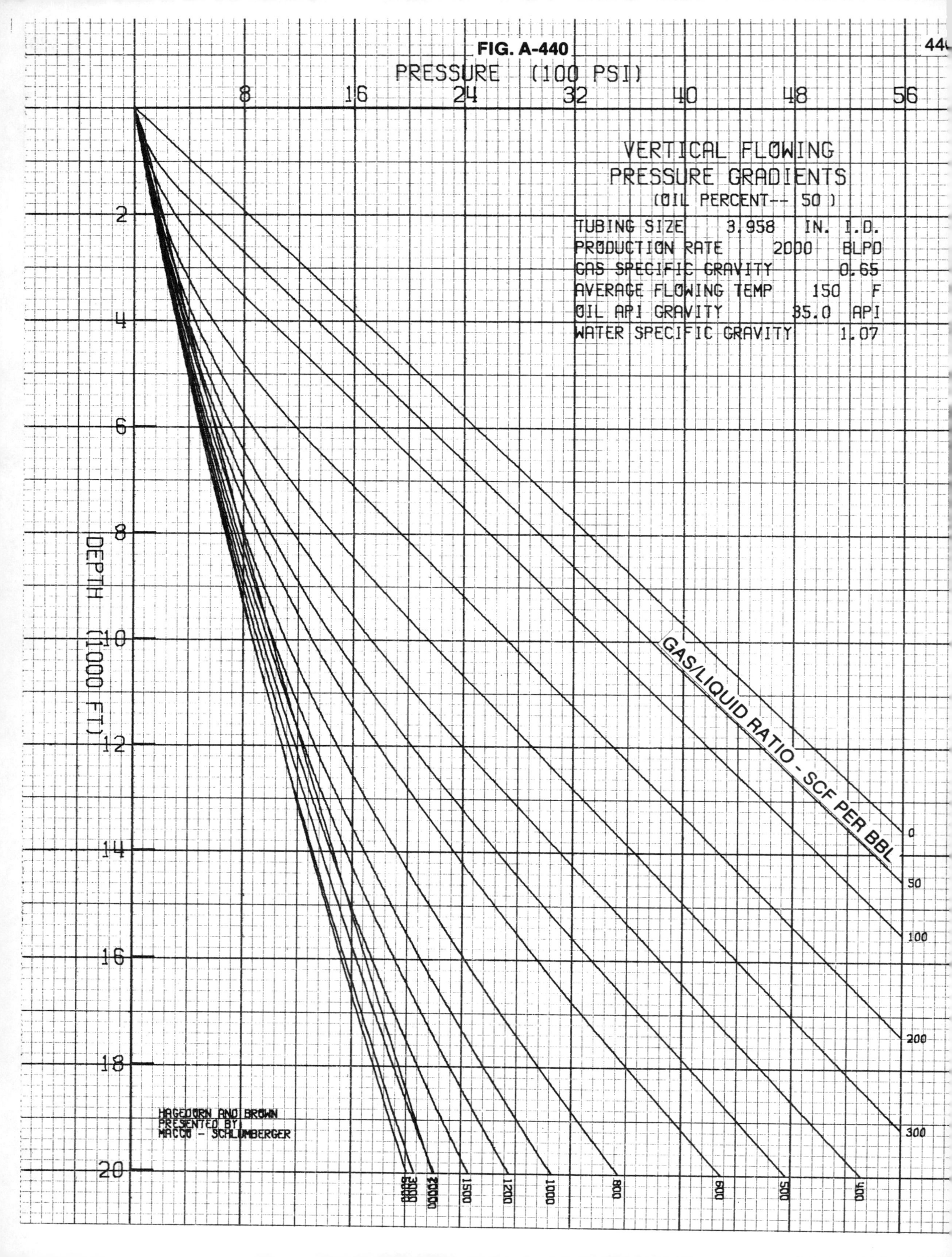

VERTICAL FLOWING
PRESSURE GRADIENTS
(OIL PERCENT-- 50)

TUBING SIZE	3.958	IN. I.D.
PRODUCTION RATE	2000	BLPD
GAS SPECIFIC GRAVITY		0.65
AVERAGE FLOWING TEMP	150	F
OIL API GRAVITY	85.0	API
WATER SPECIFIC GRAVITY		1.07

PRESSURE (100 PSI)

DEPTH (1000 FT)

GAS/LIQUID RATIO - SCF PER BBL

HAGEDORN AND BROWN
PRESENTED BY
MACCO - SCHLUMBERGER

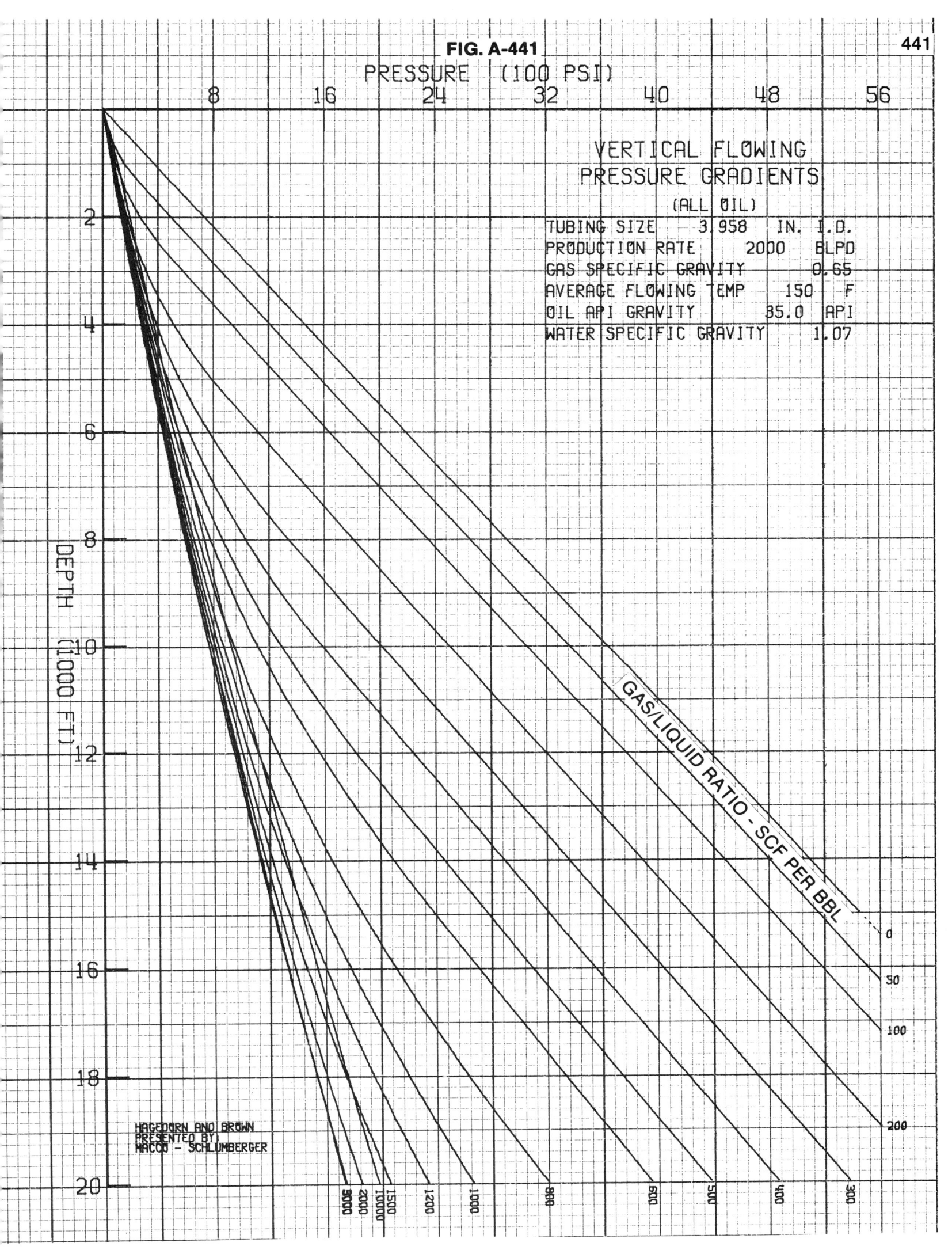

PRESSURE (100 PSI)

VERTICAL FLOWING
PRESSURE GRADIENTS
(ALL OIL)

TUBING SIZE	3.958	IN. I.D.
PRODUCTION RATE	2000	BLPD
GAS SPECIFIC GRAVITY	0.65	
AVERAGE FLOWING TEMP	150	F
OIL API GRAVITY	35.0	API
WATER SPECIFIC GRAVITY	1.07	

GAS/LIQUID RATIO - SCF PER BBL

DEPTH (1000 FT)

HAGEDORN AND BROWN
PRESENTED BY
MACCO - SCHLUMBERGER

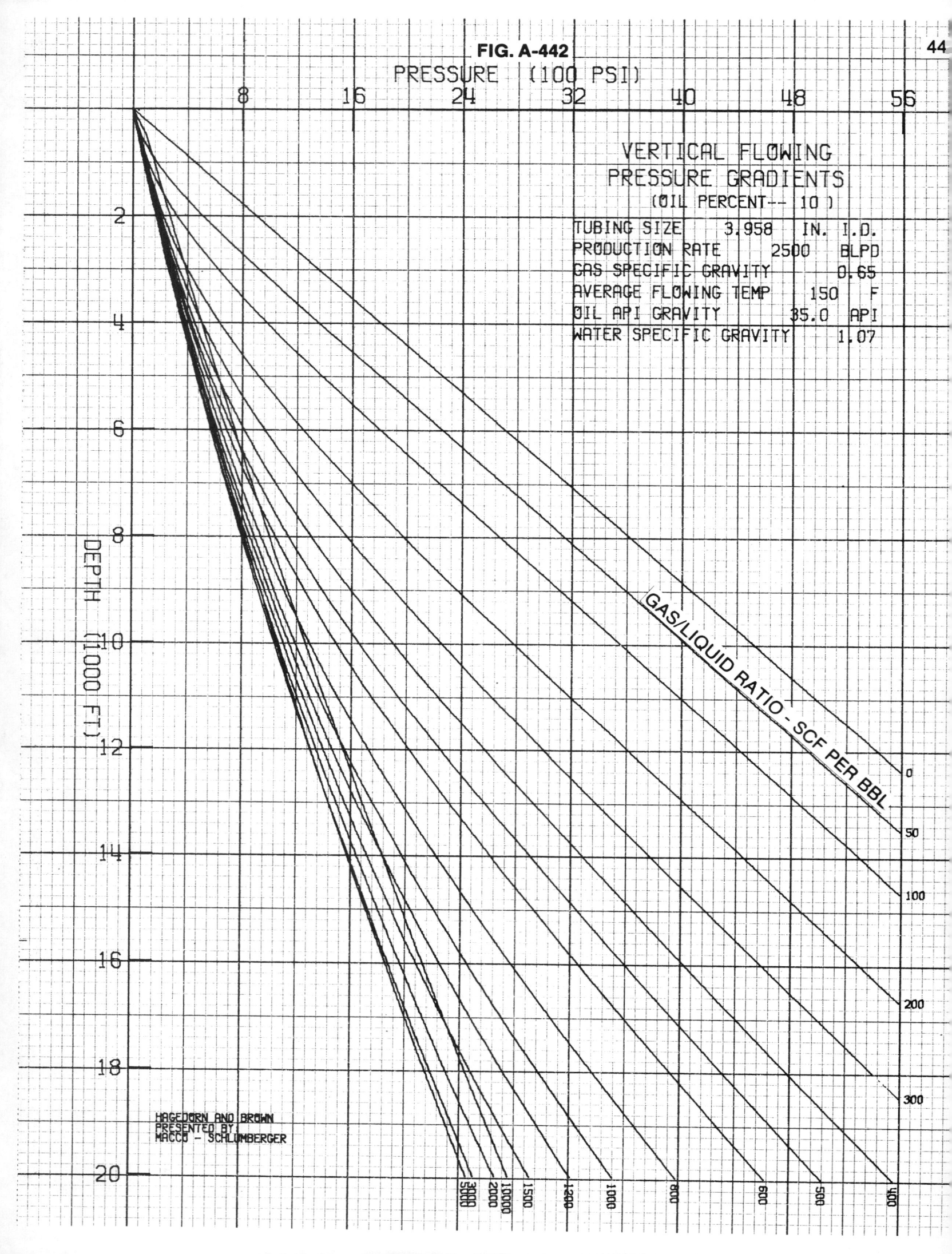

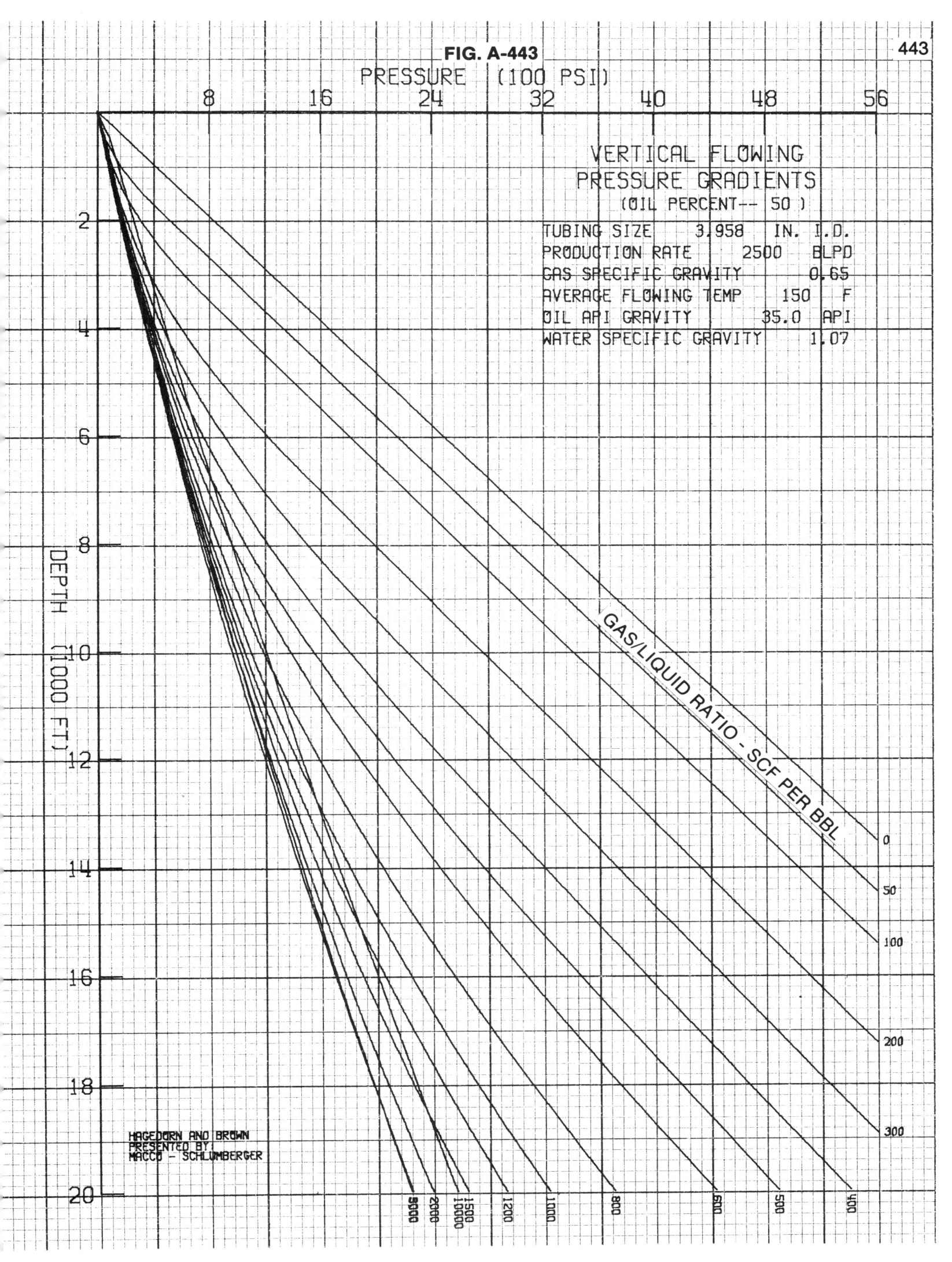

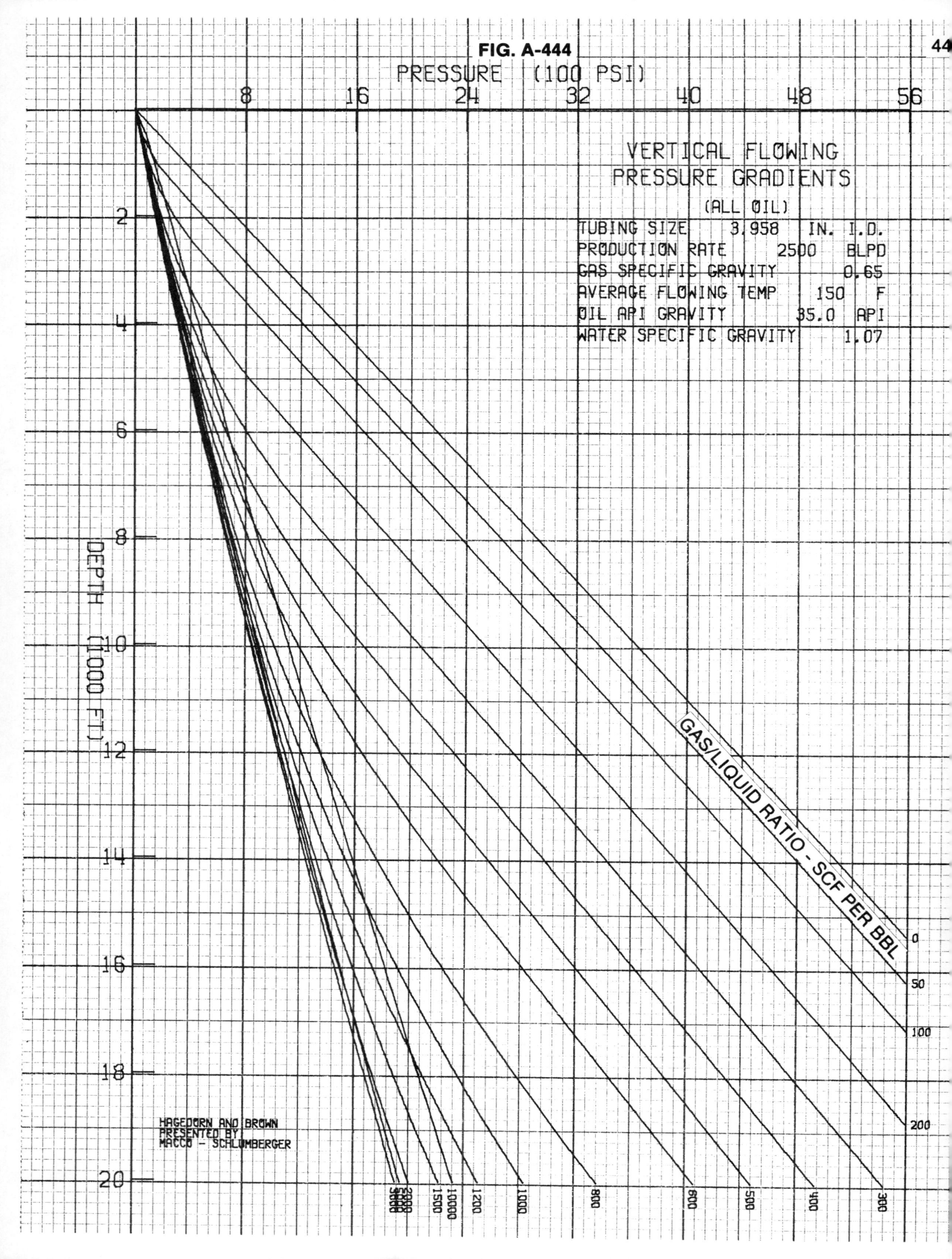

VERTICAL FLOWING
PRESSURE GRADIENTS
(ALL OIL)

TUBING SIZE	3.958	IN. I.D.
PRODUCTION RATE	2500	BLPD
GAS SPECIFIC GRAVITY	0.65	
AVERAGE FLOWING TEMP	150	F
OIL API GRAVITY	35.0	API
WATER SPECIFIC GRAVITY	1.07	

PRESSURE (100 PSI)

DEPTH (1000 FT)

GAS/LIQUID RATIO - SCF PER BBL

HAGEDORN AND BROWN
PRESENTED BY
MACCO - SCHLUMBERGER

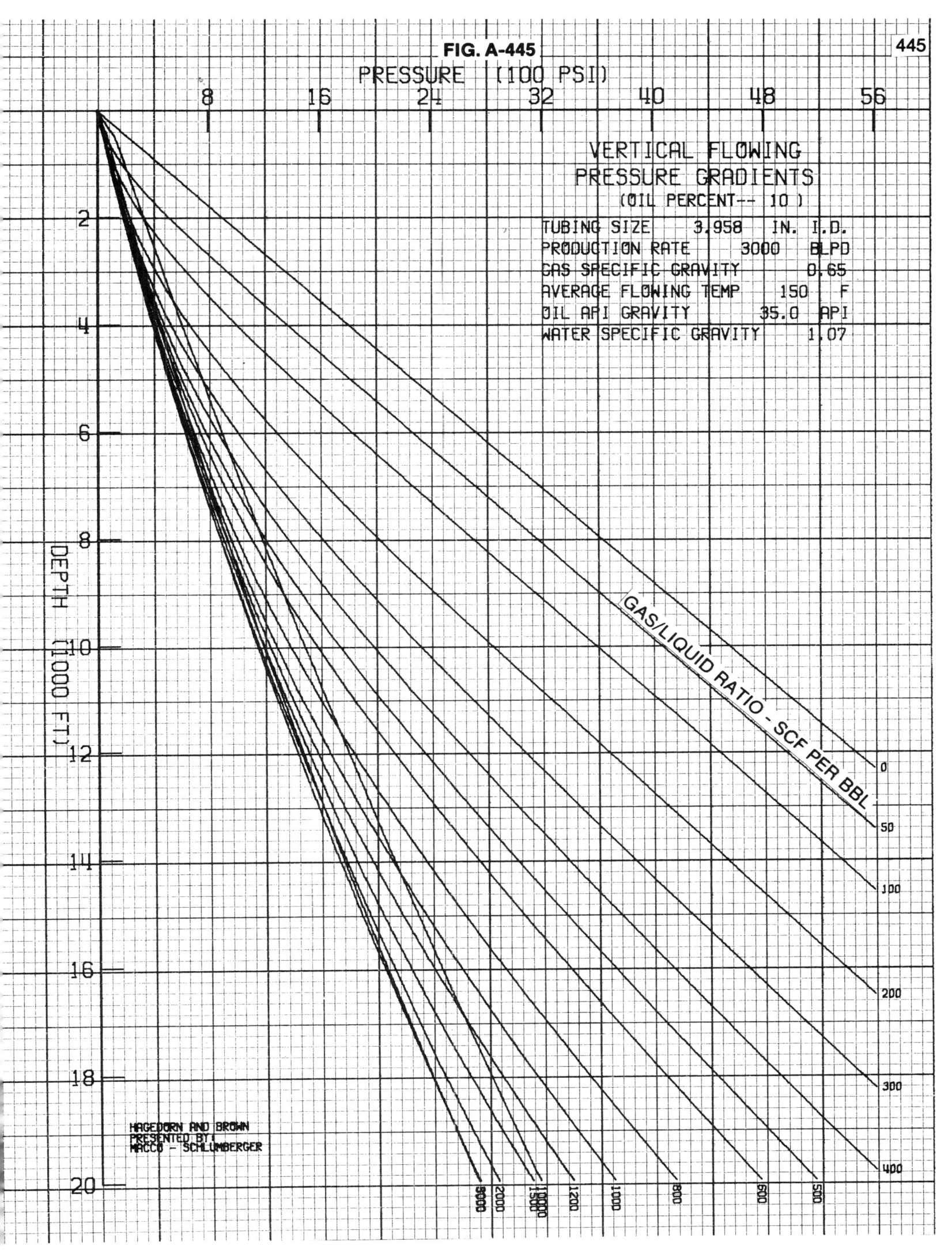

FIG. A-445

445

PRESSURE (100 PSI)

VERTICAL FLOWING
PRESSURE GRADIENTS
(OIL PERCENT-- 10)

TUBING SIZE	3.958	IN. I.D.
PRODUCTION RATE	3000	BLPD
GAS SPECIFIC GRAVITY	0.65	
AVERAGE FLOWING TEMP	150	F
OIL API GRAVITY	35.0	API
WATER SPECIFIC GRAVITY	1.07	

GAS/LIQUID RATIO - SCF PER BBL

DEPTH (1000 FT)

HAGEDORN AND BROWN
PRESENTED BY
MACCO - SCHLUMBERGER

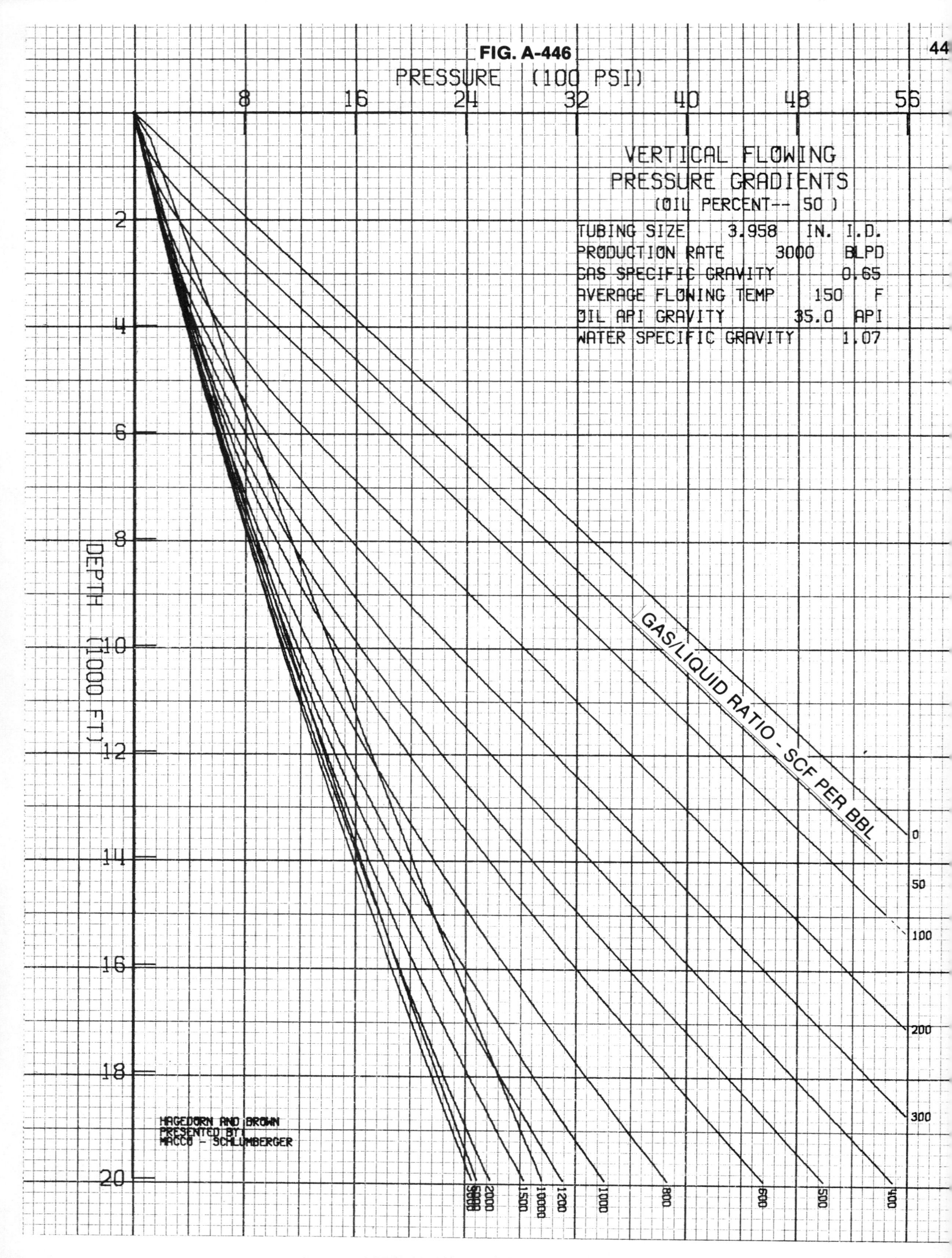

PRESSURE (100 PSI)

VERTICAL FLOWING
PRESSURE GRADIENTS
(OIL PERCENT-- 50)

TUBING SIZE	3.958	IN. I.D.
PRODUCTION RATE	3000	BLPD
GAS SPECIFIC GRAVITY	0.65	
AVERAGE FLOWING TEMP	150	F
OIL API GRAVITY	35.0	API
WATER SPECIFIC GRAVITY	1.07	

DEPTH (1000 FT)

GAS/LIQUID RATIO - SCF PER BBL

HAGEDORN AND BROWN
PRESENTED BY
MACCO - SCHLUMBERGER

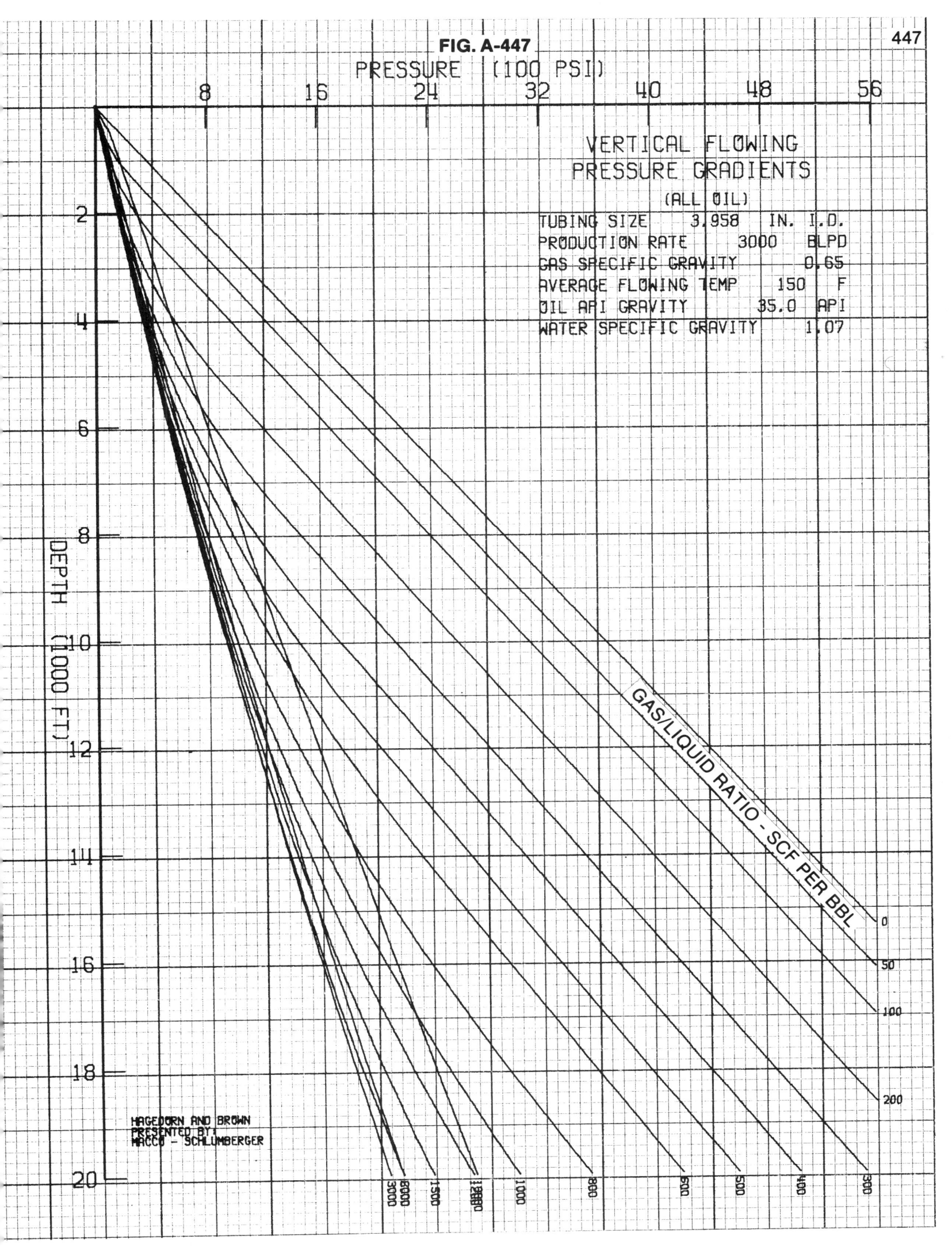

FIG. A-447

447

VERTICAL FLOWING
PRESSURE GRADIENTS
(ALL OIL)

TUBING SIZE	3.958	IN. I.D.
PRODUCTION RATE	3000	BLPD
GAS SPECIFIC GRAVITY	0.65	
AVERAGE FLOWING TEMP	150	F
OIL API GRAVITY	35.0	API
WATER SPECIFIC GRAVITY	1.07	

PRESSURE (100 PSI)

DEPTH (1000 FT)

GAS/LIQUID RATIO - SCF PER BBL

HAGEDORN AND BROWN
PRESENTED BY:
MACCO - SCHLUMBERGER

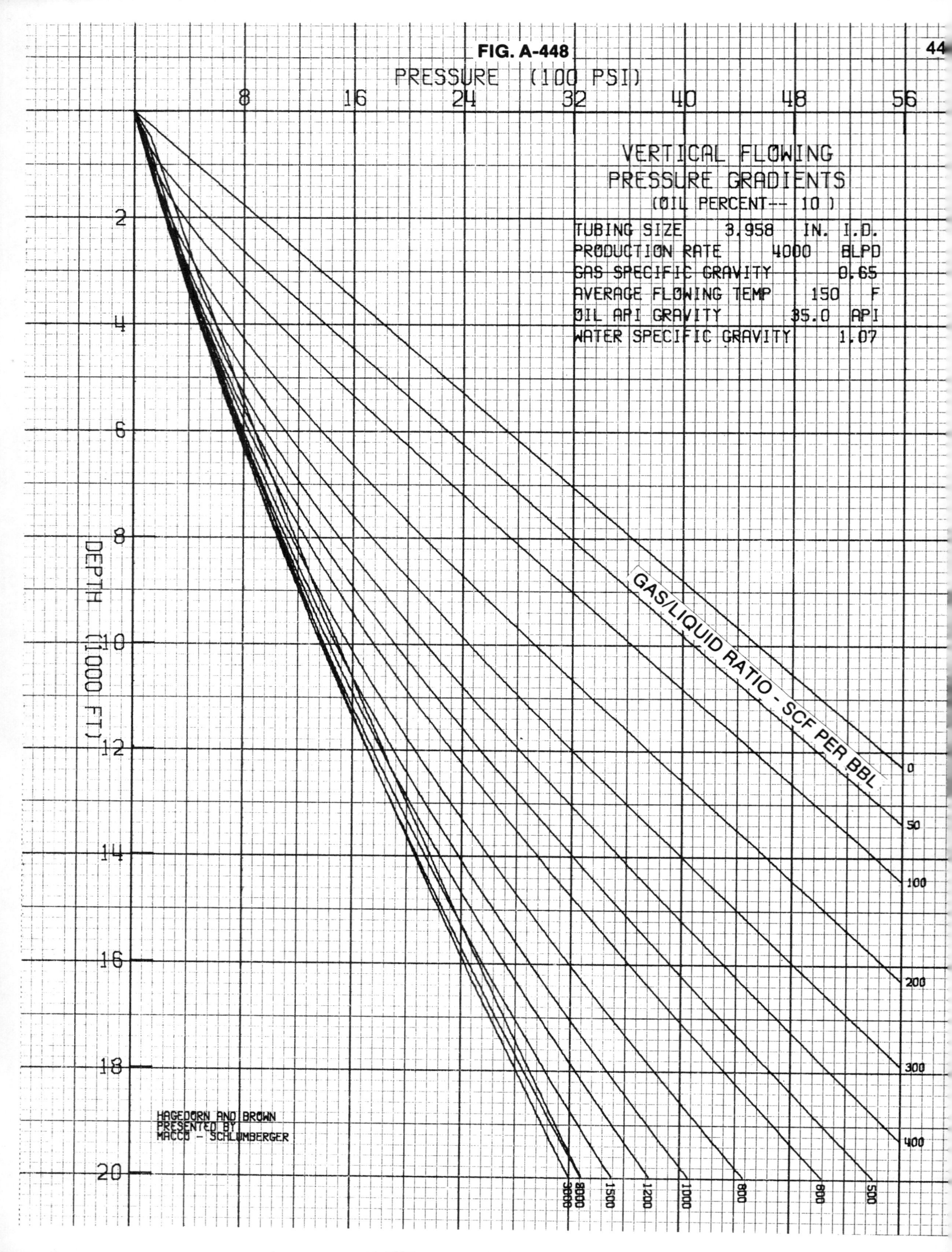

PRESSURE (100 PSI)

VERTICAL FLOWING
PRESSURE GRADIENTS
(OIL PERCENT-- 10)

TUBING SIZE	3.958	IN. I.D.
PRODUCTION RATE	4000	BLPD
GAS SPECIFIC GRAVITY		0.65
AVERAGE FLOWING TEMP	150	F
OIL API GRAVITY	35.0	API
WATER SPECIFIC GRAVITY		1.07

DEPTH (1000 FT)

GAS/LIQUID RATIO - SCF PER BBL

HAGEDORN AND BROWN
PRESENTED BY
MACCO - SCHLUMBERGER

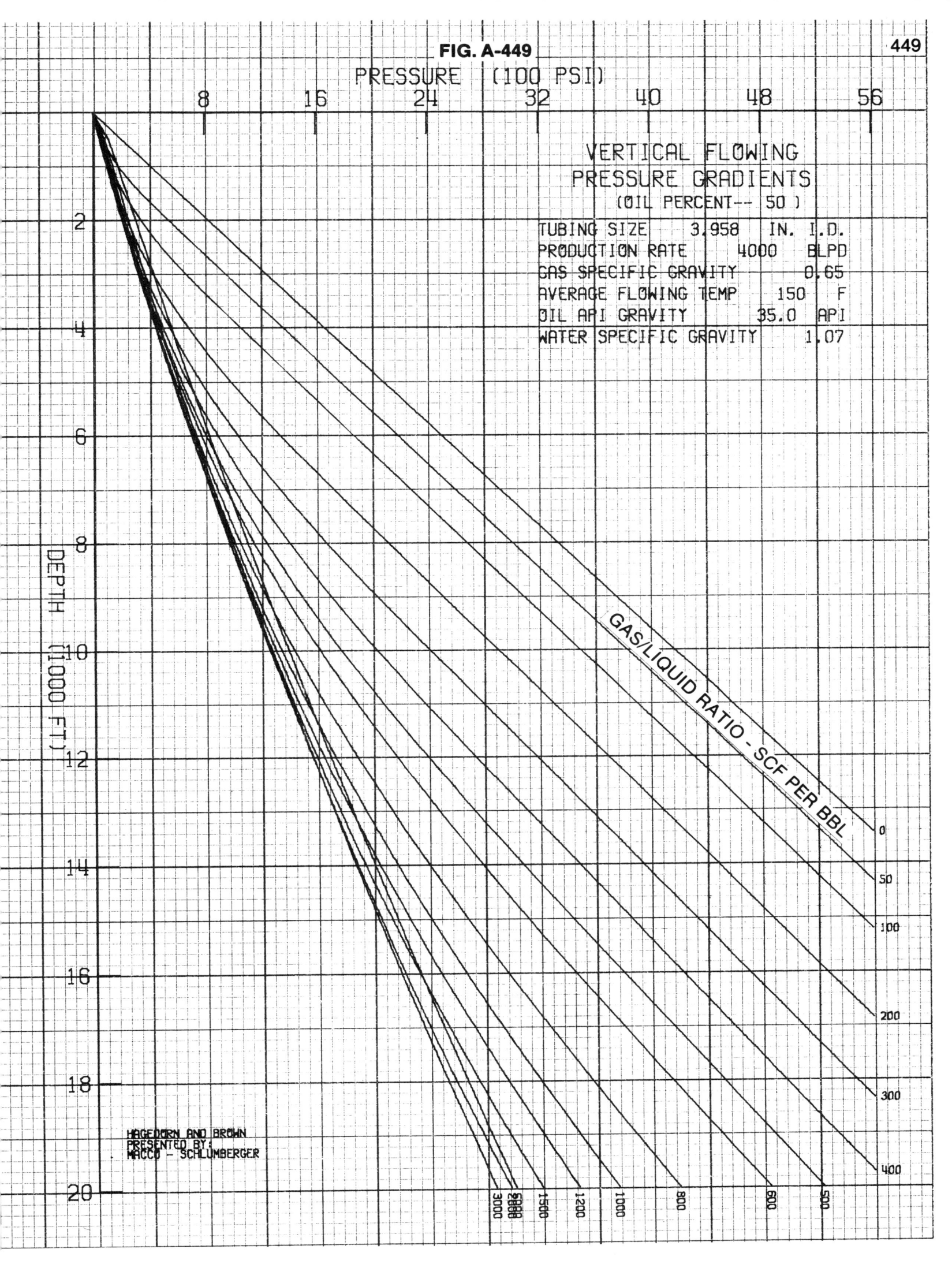

FIG. A-449

449

VERTICAL FLOWING
PRESSURE GRADIENTS
(OIL PERCENT— 50)

TUBING SIZE	3.958	IN. I.D.
PRODUCTION RATE	4000	BLPD
GAS SPECIFIC GRAVITY	0.65	
AVERAGE FLOWING TEMP	150	F
OIL API GRAVITY	35.0	API
WATER SPECIFIC GRAVITY	1.07	

PRESSURE (100 PSI)

DEPTH (1000 FT)

GAS/LIQUID RATIO - SCF PER BBL

HAGEDORN AND BROWN
PRESENTED BY:
MACCO - SCHLUMBERGER

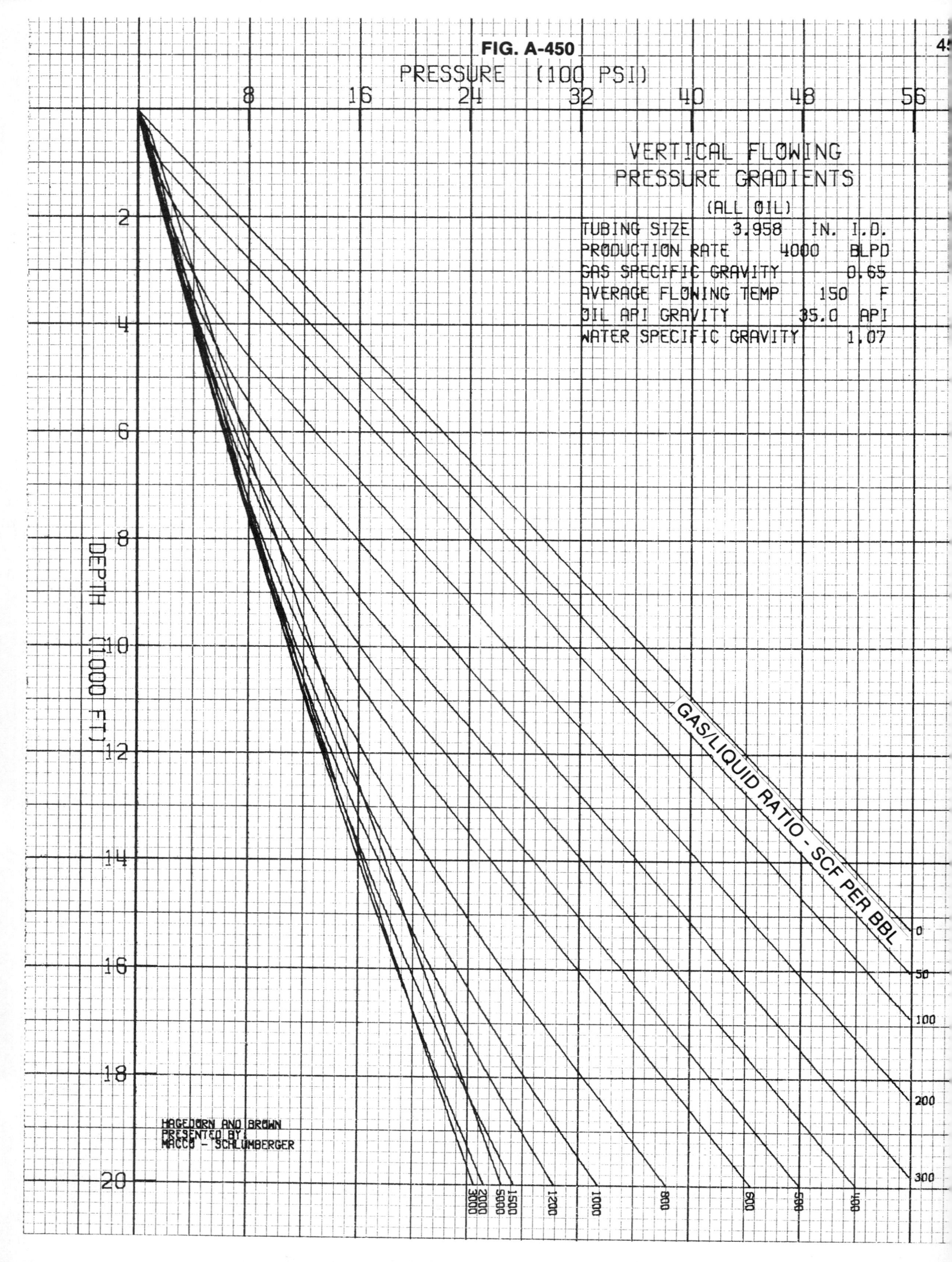

FIG. A-450

VERTICAL FLOWING
PRESSURE GRADIENTS
(ALL OIL)

TUBING SIZE	3.958	IN. I.D.
PRODUCTION RATE	4000	BLPD
GAS SPECIFIC GRAVITY	0.65	
AVERAGE FLOWING TEMP	150	F
OIL API GRAVITY	35.0	API
WATER SPECIFIC GRAVITY	1.07	

GAS/LIQUID RATIO - SCF PER BBL

HAGEDORN AND BROWN
PRESENTED BY
MACCO - SCHLUMBERGER

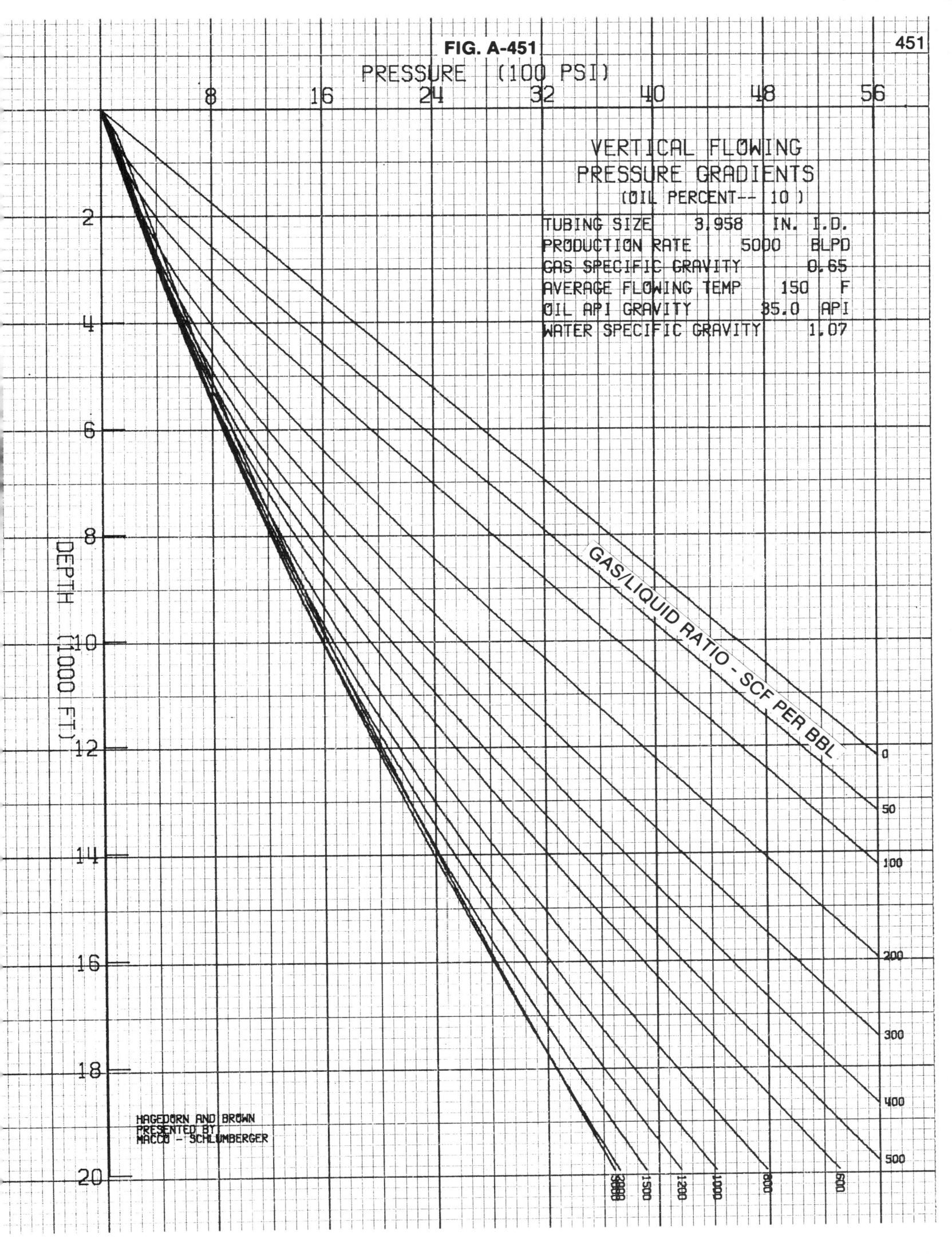

PRESSURE (100 PSI)

VERTICAL FLOWING
PRESSURE GRADIENTS
(OIL PERCENT-- 10)

TUBING SIZE   3.958   IN. I.D.
PRODUCTION RATE   5000   BLPD
GAS SPECIFIC GRAVITY   0.65
AVERAGE FLOWING TEMP   150   F
OIL API GRAVITY   35.0   API
WATER SPECIFIC GRAVITY   1.07

GAS/LIQUID RATIO - SCF PER BBL

DEPTH (1000 FT)

HAGEDORN AND BROWN
PRESENTED BY
MACCO - SCHLUMBERGER

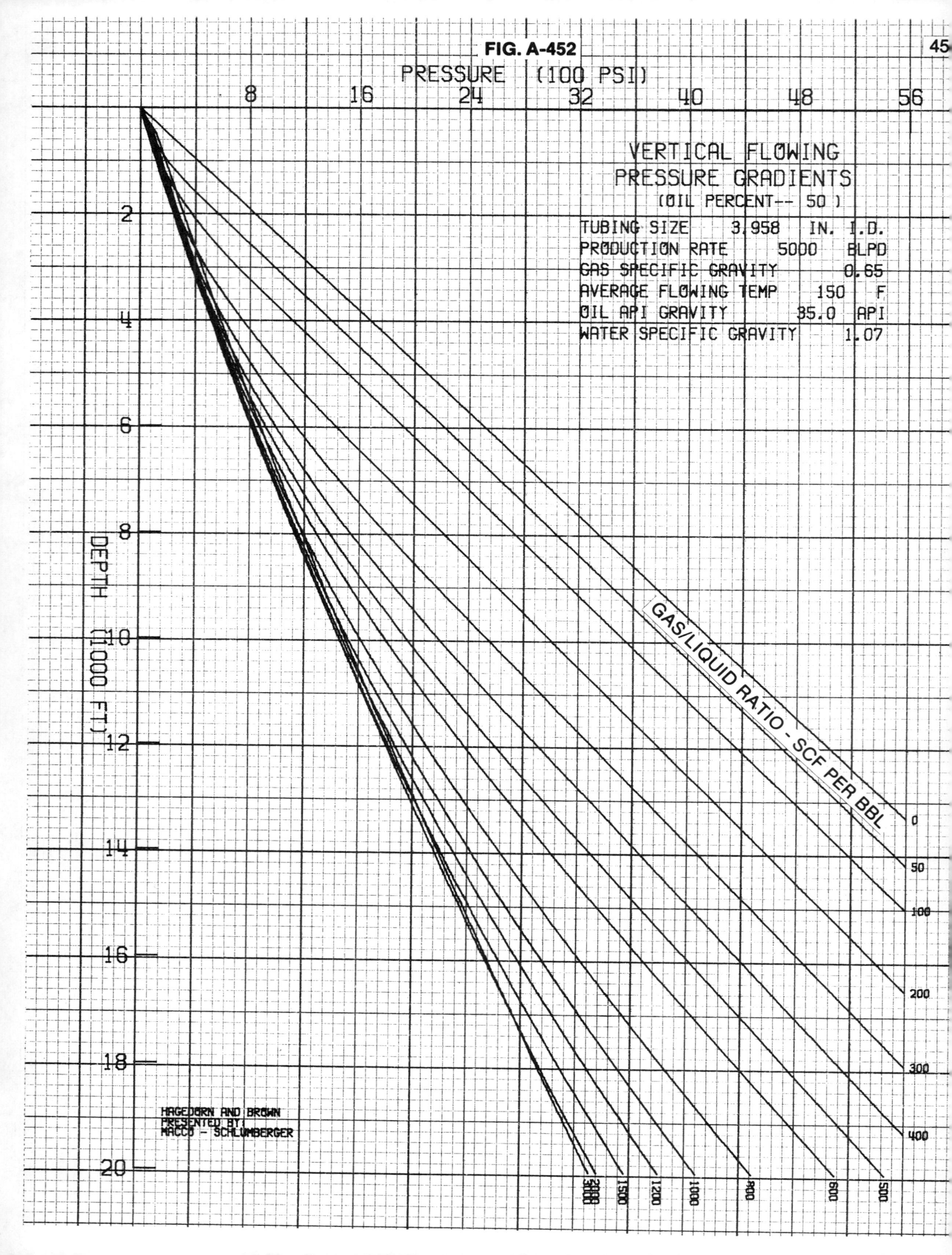

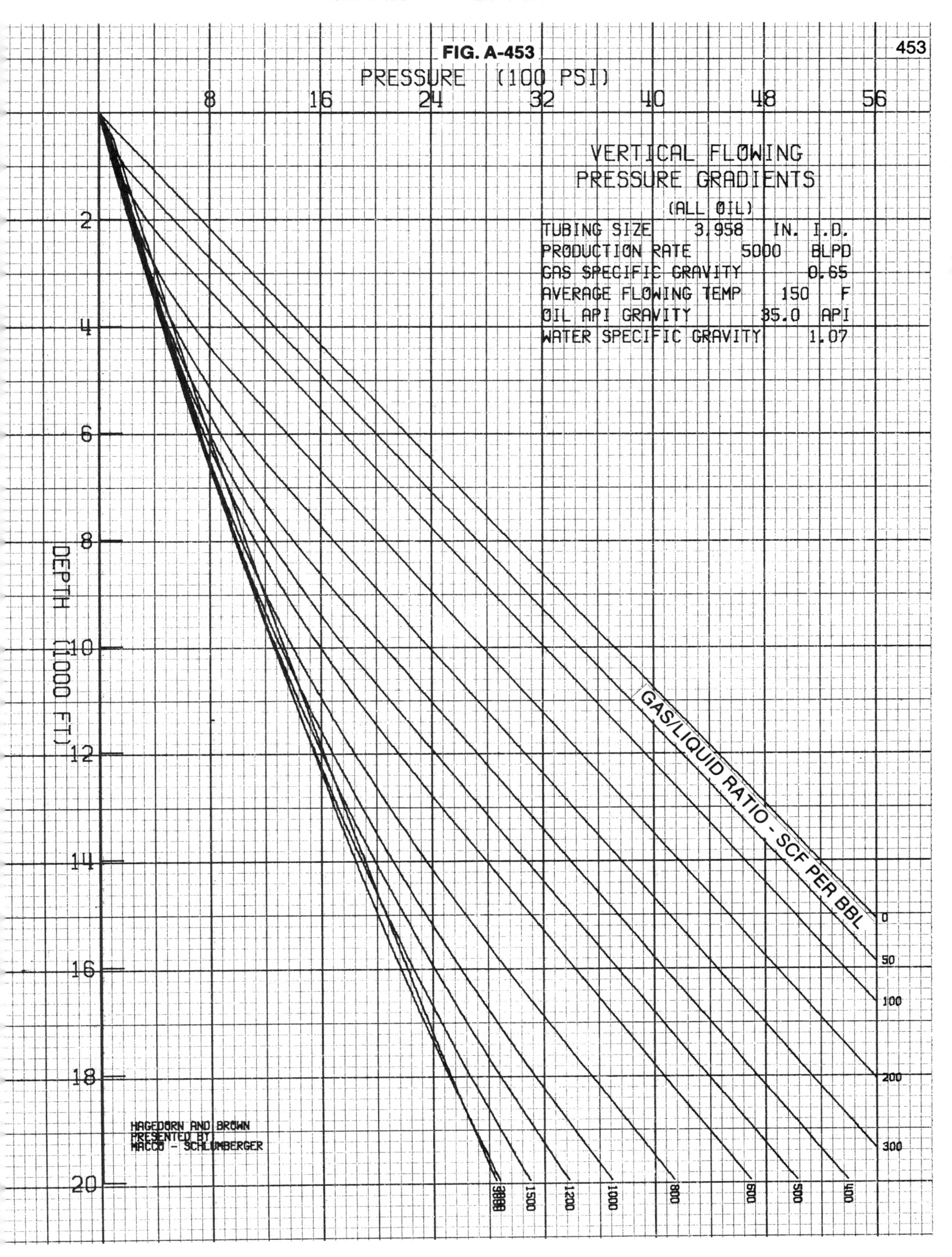

PRESSURE (100 PSI)

VERTICAL FLOWING
PRESSURE GRADIENTS
(ALL OIL)

TUBING SIZE	3.958	IN. I.D.
PRODUCTION RATE	5000	BLPD
GAS SPECIFIC GRAVITY		0.65
AVERAGE FLOWING TEMP	150	F
OIL API GRAVITY	35.0	API
WATER SPECIFIC GRAVITY		1.07

DEPTH (1000 FT)

GAS/LIQUID RATIO - SCF PER BBL

HAGEDORN AND BROWN
PRESENTED BY
KACCO - SCHLUMBERGER

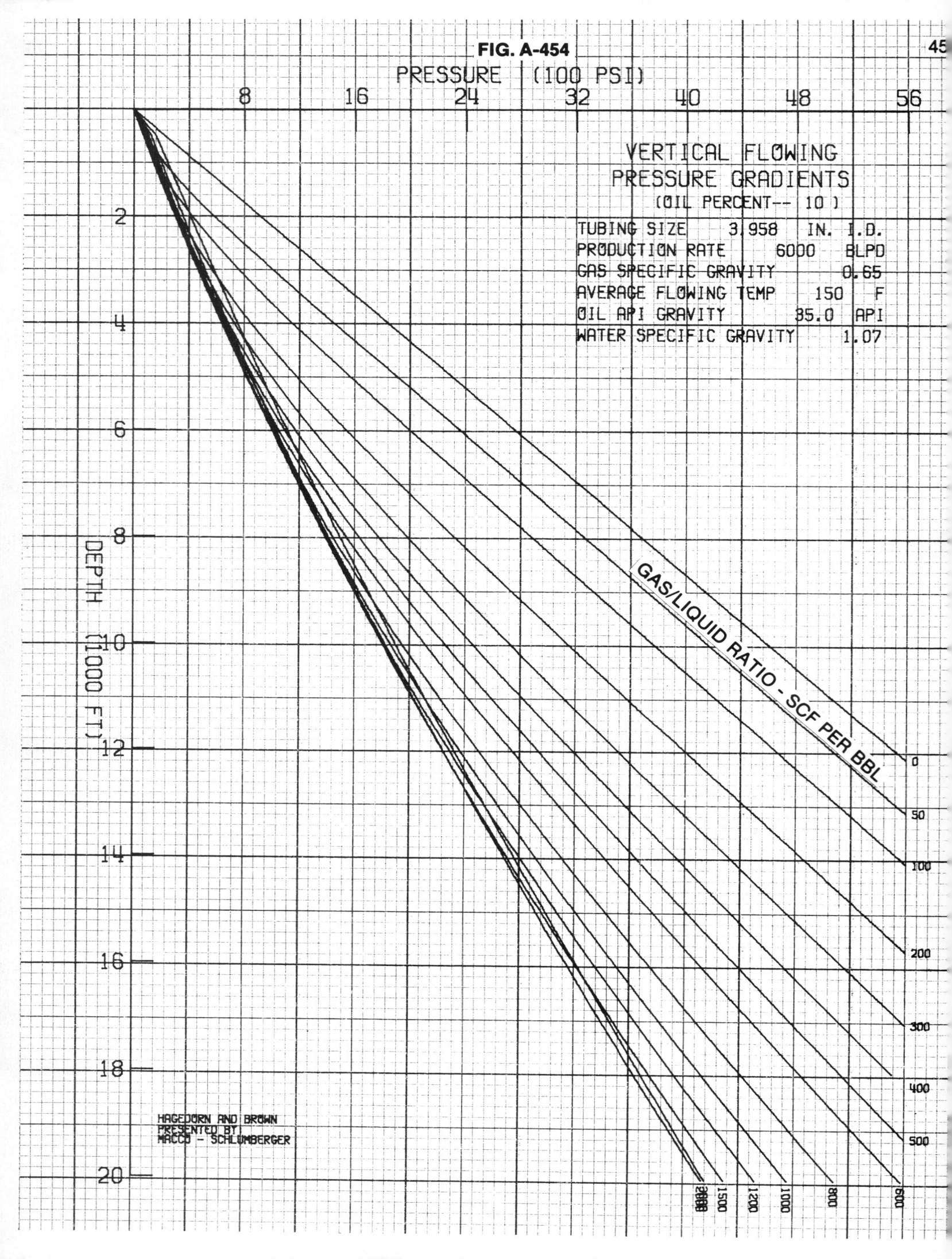

FIG. A-454

VERTICAL FLOWING
PRESSURE GRADIENTS
(OIL PERCENT-- 10)

TUBING SIZE	3.958	IN. I.D.
PRODUCTION RATE	6000	BLPD
GAS SPECIFIC GRAVITY		0.65
AVERAGE FLOWING TEMP	150	F
OIL API GRAVITY	35.0	API
WATER SPECIFIC GRAVITY		1.07

PRESSURE (100 PSI)

DEPTH (1000 FT)

GAS/LIQUID RATIO - SCF PER BBL

HAGEDORN AND BROWN
PRESENTED BY
MACCO - SCHLUMBERGER

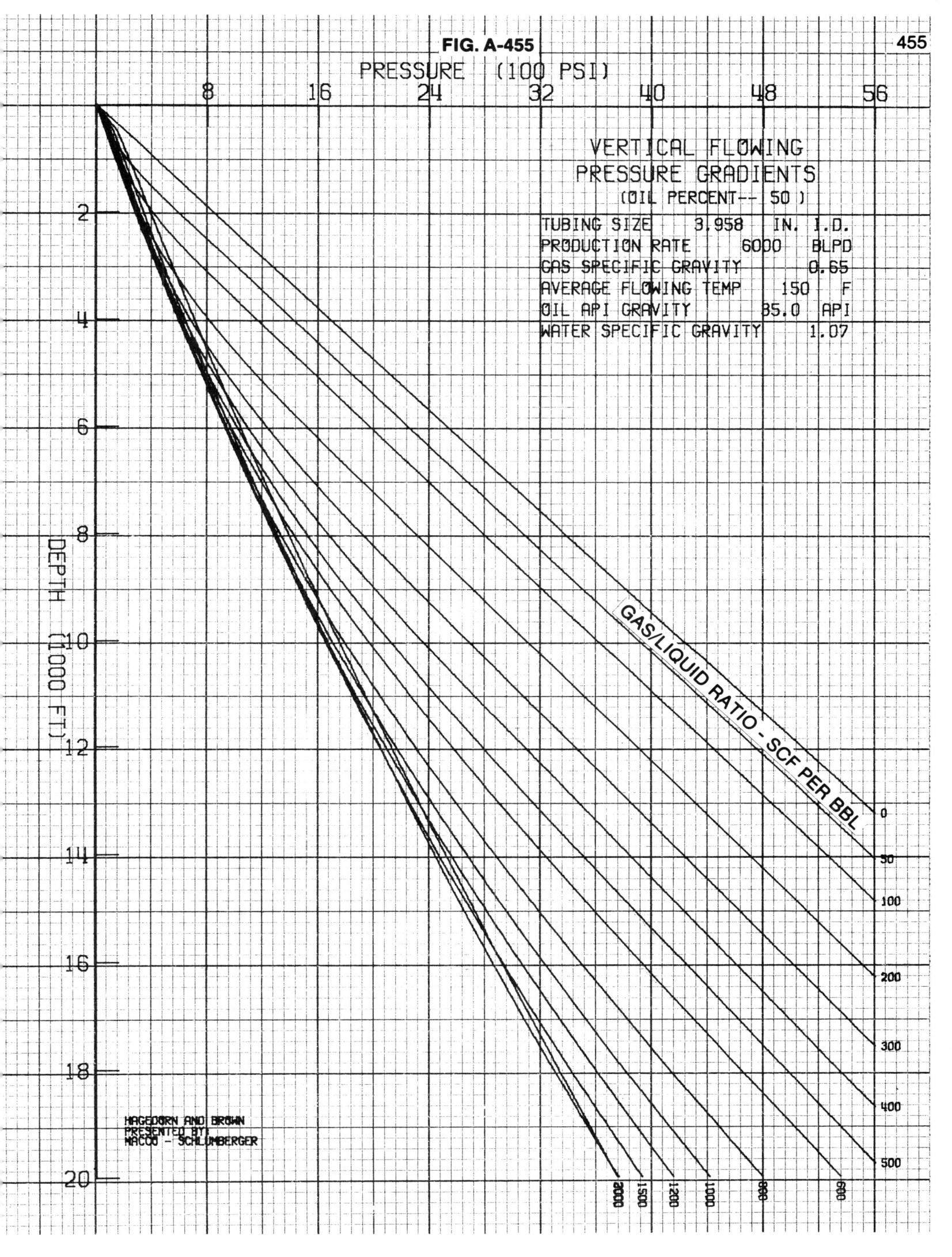

VERTICAL FLOWING
PRESSURE GRADIENTS
(OIL PERCENT-- 50)

TUBING SIZE	3.958	IN. I.D.
PRODUCTION RATE	6000	BLPD
GAS SPECIFIC GRAVITY	0.65	
AVERAGE FLOWING TEMP	150	F
OIL API GRAVITY	35.0	API
WATER SPECIFIC GRAVITY	1.07	

GAS/LIQUID RATIO - SCF PER BBL

HAGEDORN AND BROWN
PRESENTED BY:
NACCO - SCHLUMBERGER

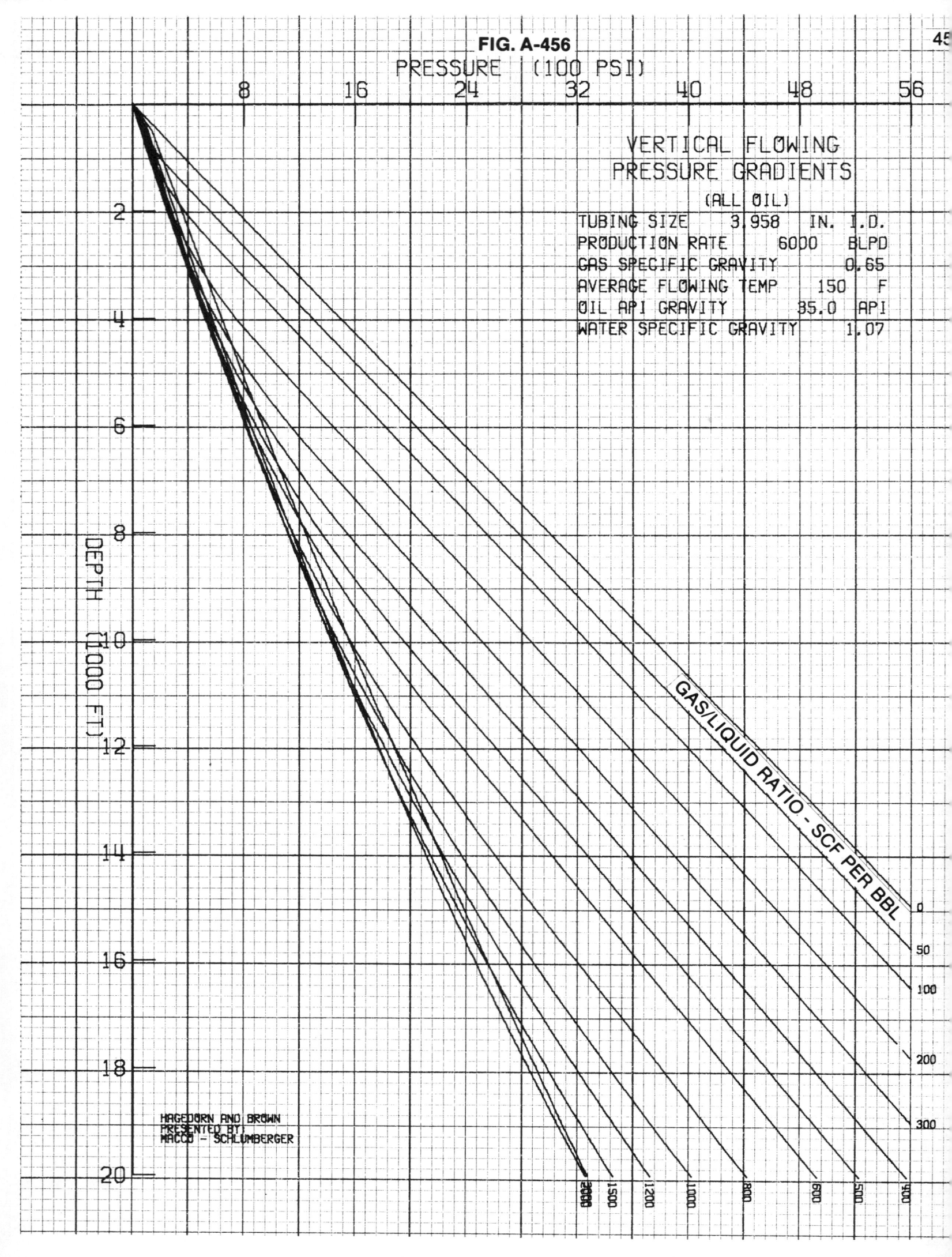

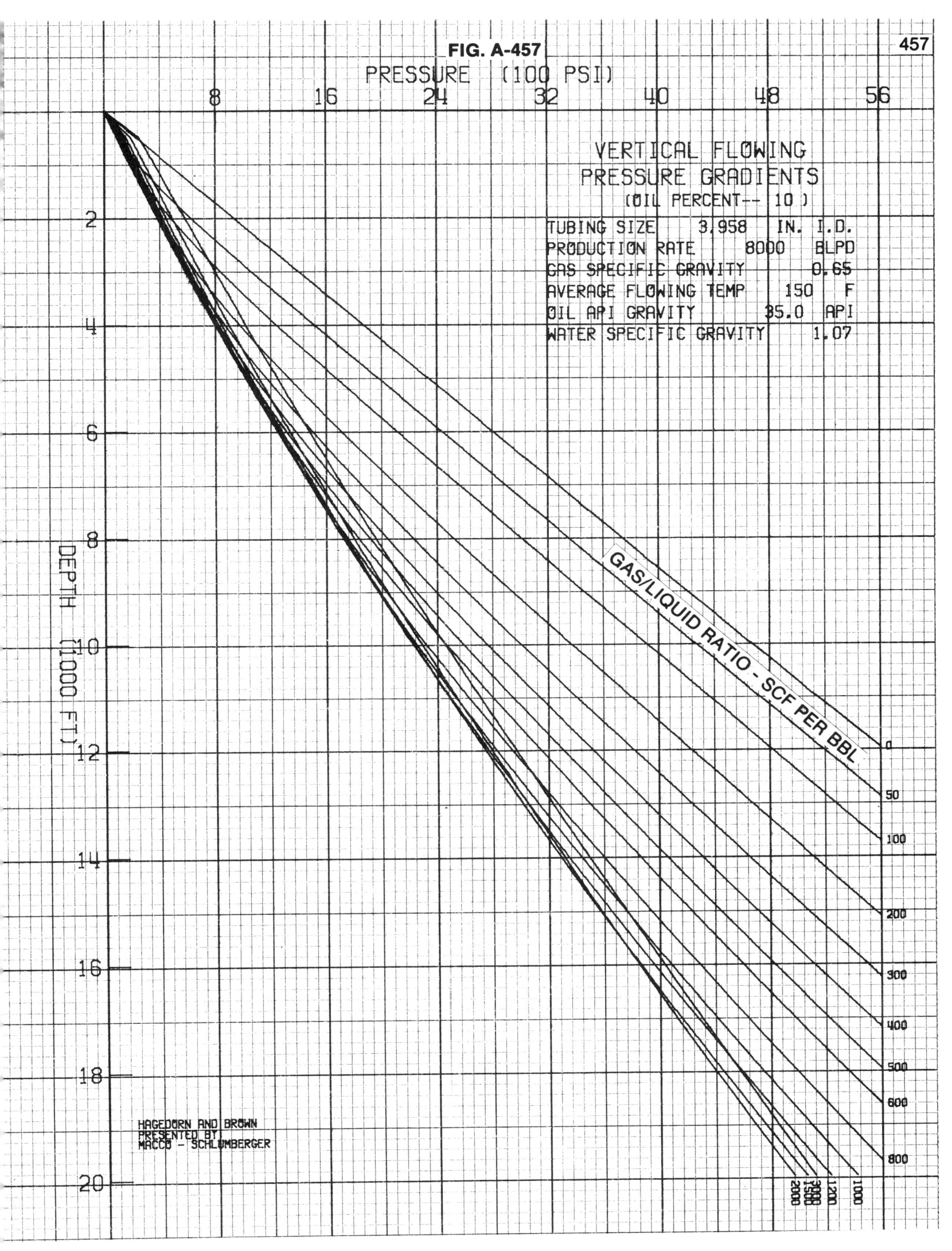

VERTICAL FLOWING
PRESSURE GRADIENTS
(OIL PERCENT-- 10)

TUBING SIZE	3.958	IN.	I.D.
PRODUCTION RATE	8000	BLPD	
GAS SPECIFIC GRAVITY		0.65	
AVERAGE FLOWING TEMP	150	F	
OIL API GRAVITY	35.0	API	
WATER SPECIFIC GRAVITY	1.07		

PRESSURE (100 PSI)

DEPTH (1000 FT)

GAS/LIQUID RATIO - SCF PER BBL

HAGEDORN AND BROWN
PRESENTED BY
MACCO - SCHLUMBERGER

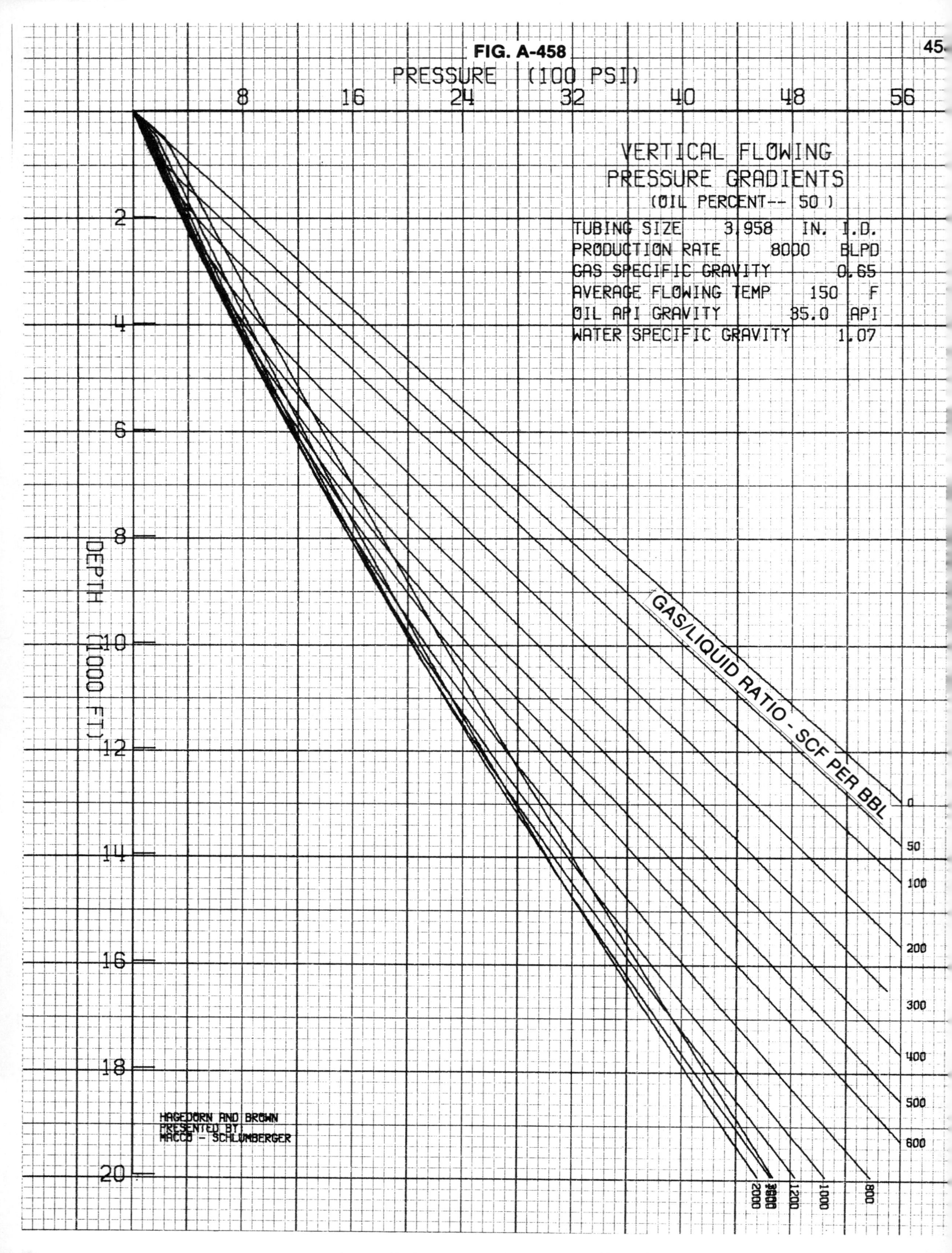

PRESSURE (100 PSI)

DEPTH (1000 FT)

VERTICAL FLOWING
PRESSURE GRADIENTS
(OIL PERCENT-- 50)

TUBING SIZE	3.958	IN. I.D.
PRODUCTION RATE	8000	BLPD
GAS SPECIFIC GRAVITY	0.65	
AVERAGE FLOWING TEMP	150	F
OIL API GRAVITY	35.0	API
WATER SPECIFIC GRAVITY	1.07	

GAS/LIQUID RATIO - SCF PER BBL

HAGEDORN AND BROWN
PRESENTED BY
MACCO - SCHLUMBERGER

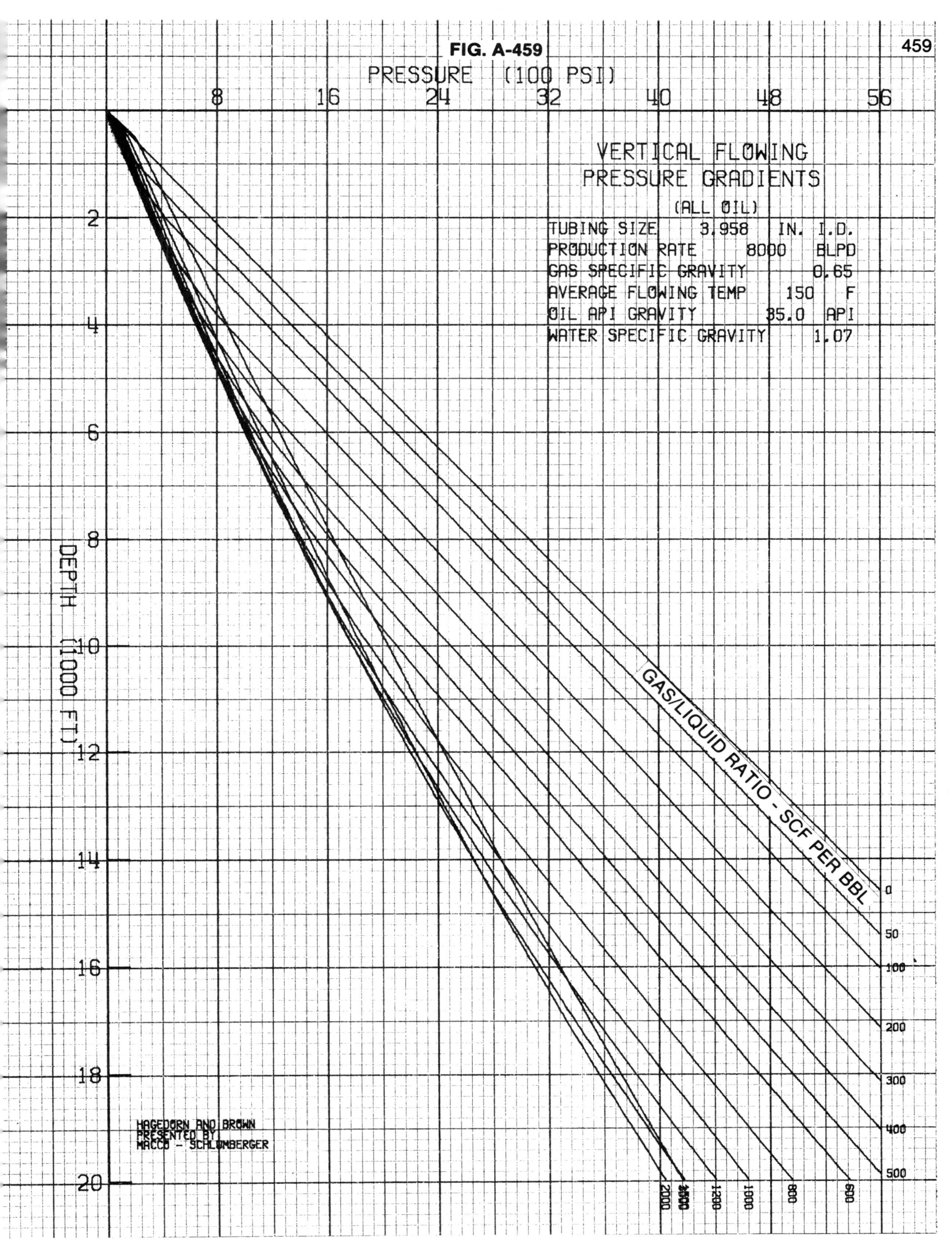

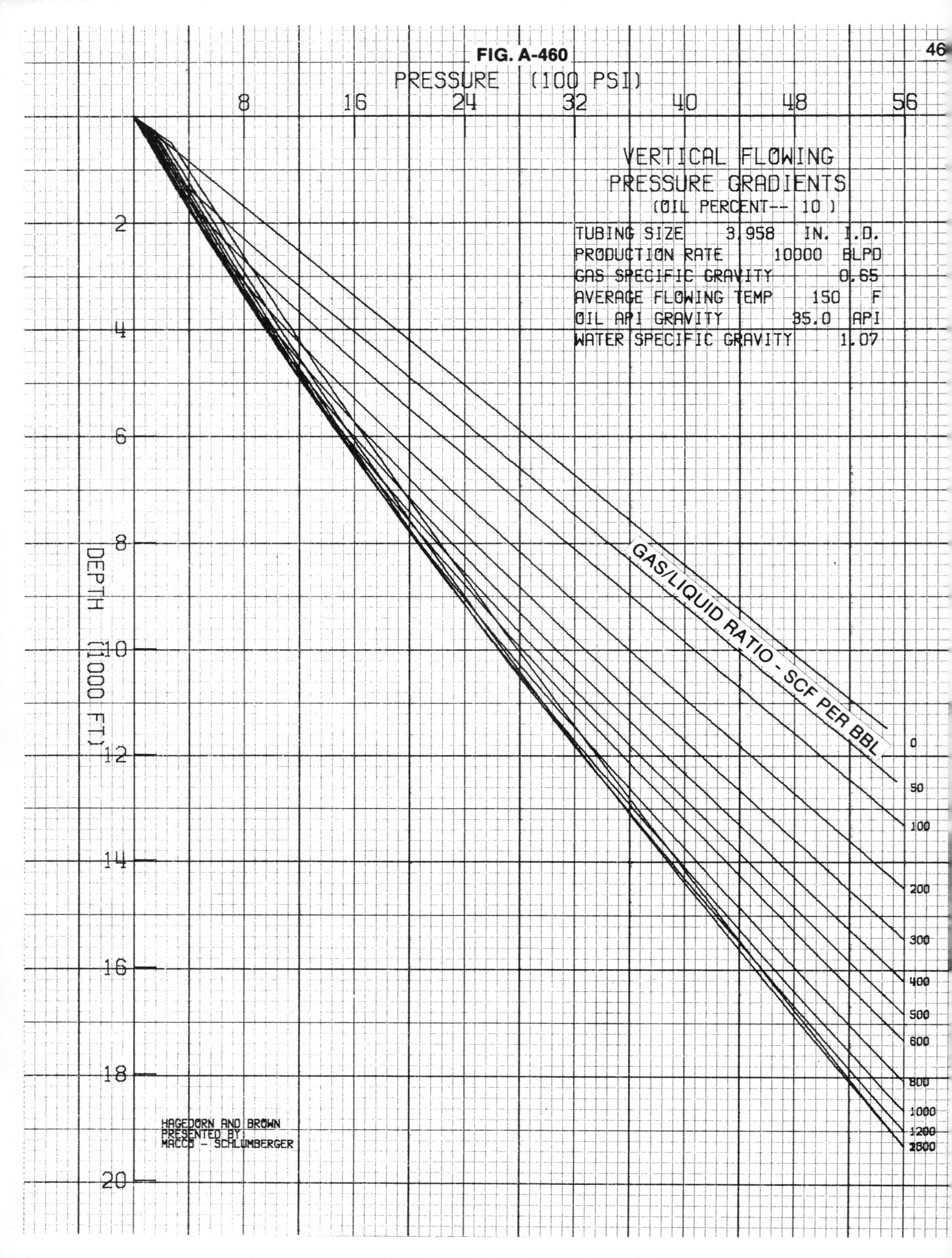

PRESSURE (100 PSI)

VERTICAL FLOWING
PRESSURE GRADIENTS
(OIL PERCENT-- 10)

TUBING SIZE	3.958	IN. I.D.
PRODUCTION RATE	10000	BLPD
GAS SPECIFIC GRAVITY		0.65
AVERAGE FLOWING TEMP	150	F
OIL API GRAVITY	35.0	API
WATER SPECIFIC GRAVITY		1.07

DEPTH (1000 FT.)

GAS/LIQUID RATIO - SCF PER BBL

0
50
100
200
300
400
500
600
800
1000
1200
2500

HAGEDORN AND BROWN
PRESENTED BY
MACCO – SCHLUMBERGER

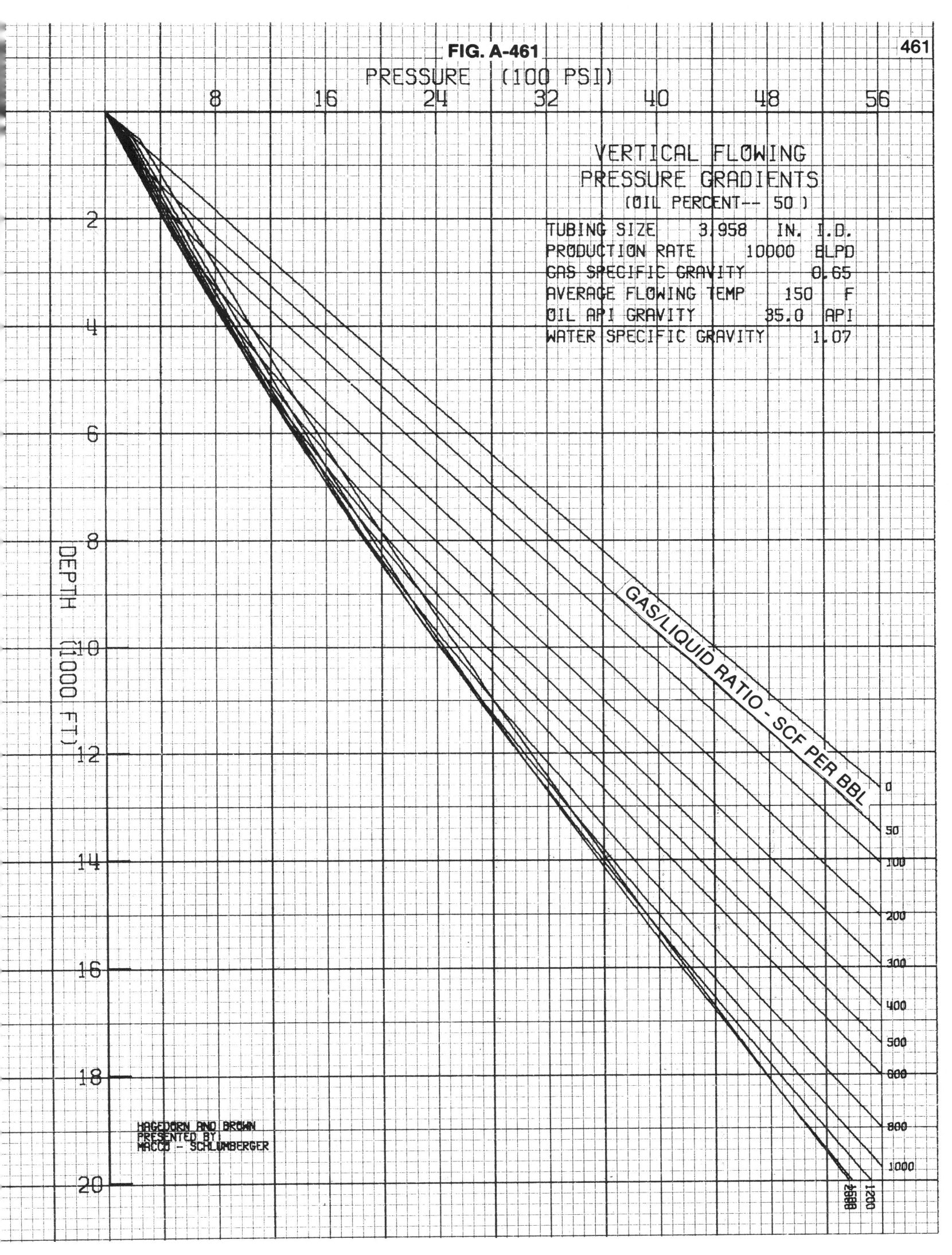

FIG. A-461

461

VERTICAL FLOWING
PRESSURE GRADIENTS
(OIL PERCENT-- 50)

TUBING SIZE   3.958   IN. I.D.
PRODUCTION RATE   10000   BLPD
GAS SPECIFIC GRAVITY   0.65
AVERAGE FLOWING TEMP   150   F
OIL API GRAVITY   35.0   API
WATER SPECIFIC GRAVITY   1.07

PRESSURE (100 PSI)

DEPTH (1000 FT)

GAS/LIQUID RATIO - SCF PER BBL

HAGEDORN AND BROWN
PRESENTED BY
MACCO - SCHLUMBERGER

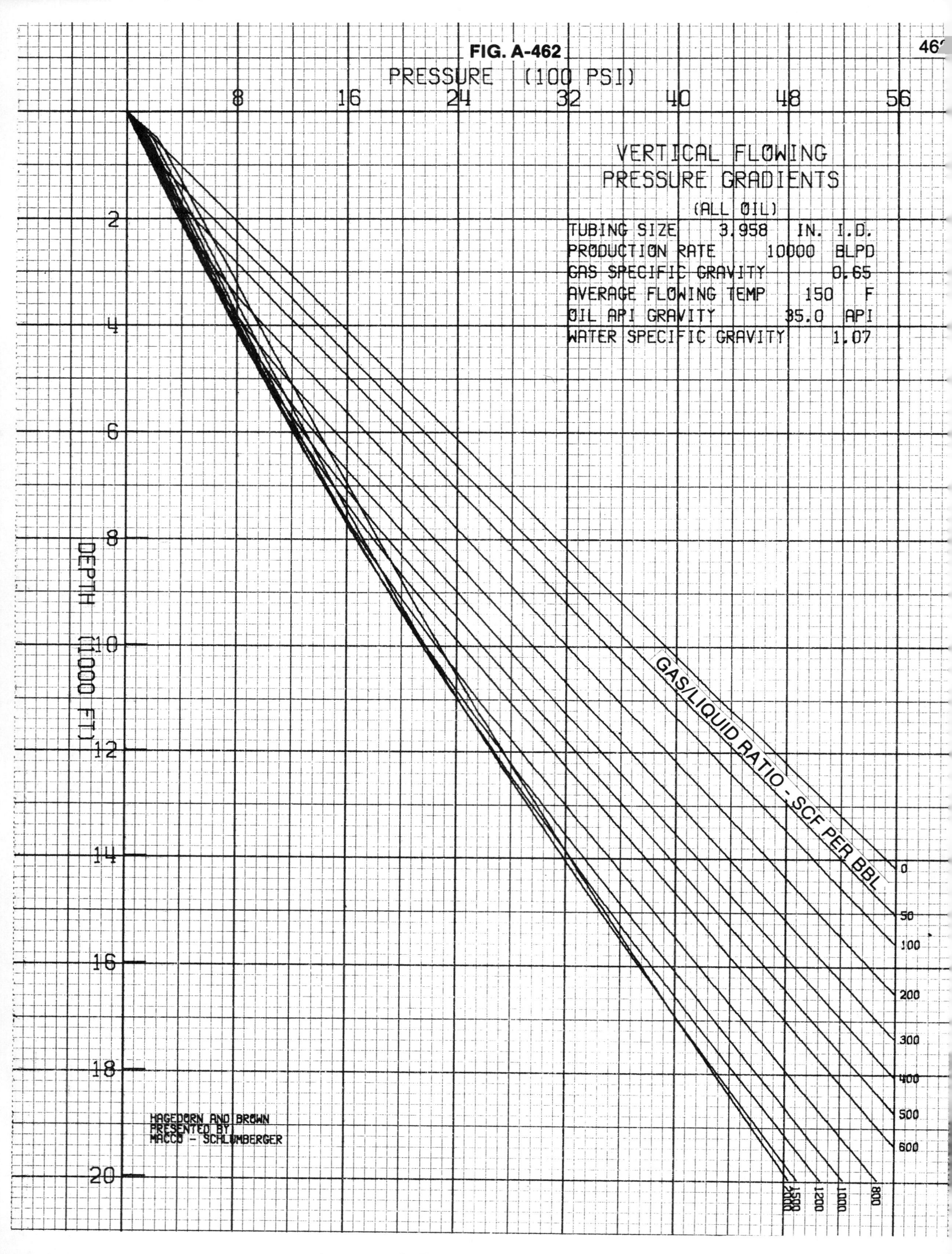

FIG. A-462

VERTICAL FLOWING
PRESSURE GRADIENTS
(ALL OIL)

TUBING SIZE	3.958	IN. I.D.
PRODUCTION RATE	10000	BLPD
GAS SPECIFIC GRAVITY		0.65
AVERAGE FLOWING TEMP	150	F
OIL API GRAVITY	35.0	API
WATER SPECIFIC GRAVITY		1.07

PRESSURE (100 PSI)

DEPTH (1000 FT)

GAS/LIQUID RATIO - SCF PER BBL

HAGEDORN AND BROWN
PRESENTED BY
MACCO - SCHLUMBERGER

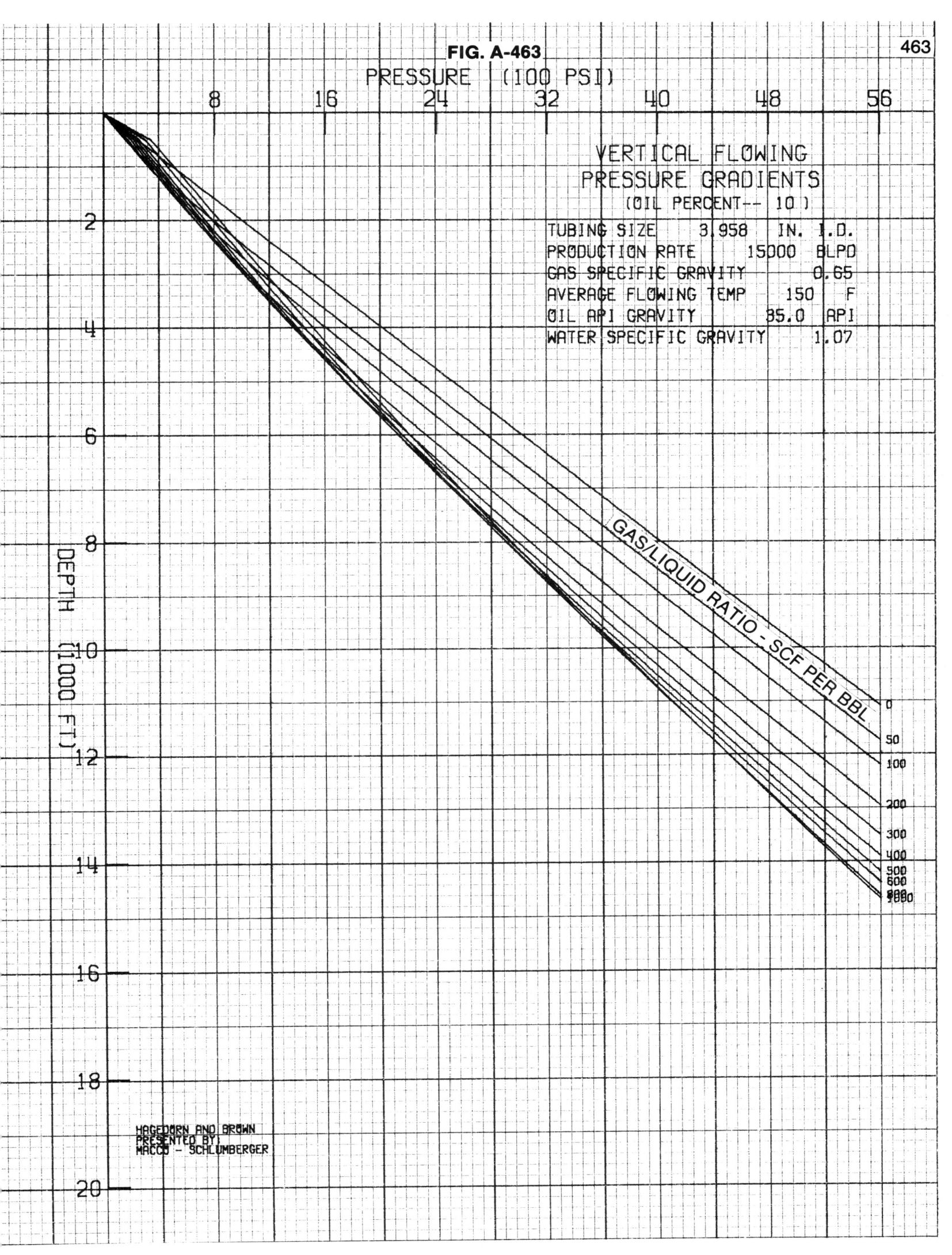

FIG. A-463

PRESSURE (100 PSI)

VERTICAL FLOWING
PRESSURE GRADIENTS
(OIL PERCENT-- 10)

TUBING SIZE	3.958	IN. I.D.
PRODUCTION RATE	15000	BLPD
GAS SPECIFIC GRAVITY	0.65	
AVERAGE FLOWING TEMP	150	F
OIL API GRAVITY	35.0	API
WATER SPECIFIC GRAVITY	1.07	

DEPTH (1,000 FT)

GAS/LIQUID RATIO - SCF PER BBL

0
50
100
200
300
400
500
600
800
1000

HAGEDORN AND BROWN
PRESENTED BY:
MACCO - SCHLUMBERGER

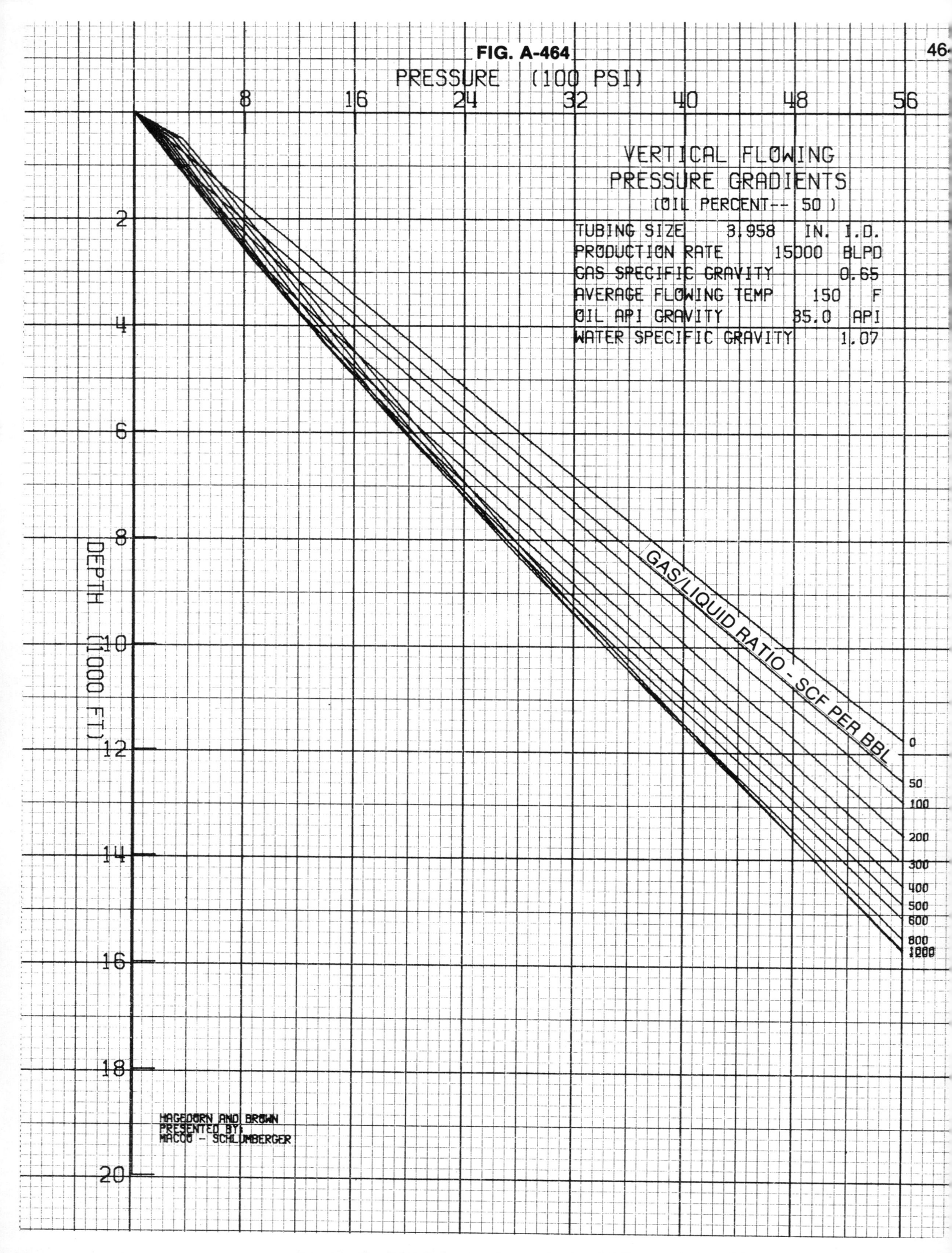

FIG. A-464

VERTICAL FLOWING
PRESSURE GRADIENTS
(OIL PERCENT-- 50)

TUBING SIZE	3.958	IN. I.D.
PRODUCTION RATE	15000	BLPD
GAS SPECIFIC GRAVITY	0.65	
AVERAGE FLOWING TEMP	150	F
OIL API GRAVITY	35.0	API
WATER SPECIFIC GRAVITY	1.07	

PRESSURE (100 PSI)

DEPTH (1000 FT)

GAS/LIQUID RATIO - SCF PER BBL

0
50
100
200
300
400
500
600
800
1000

HAGEDORN AND BROWN
PRESENTED BY:
MACCO - SCHLUMBERGER

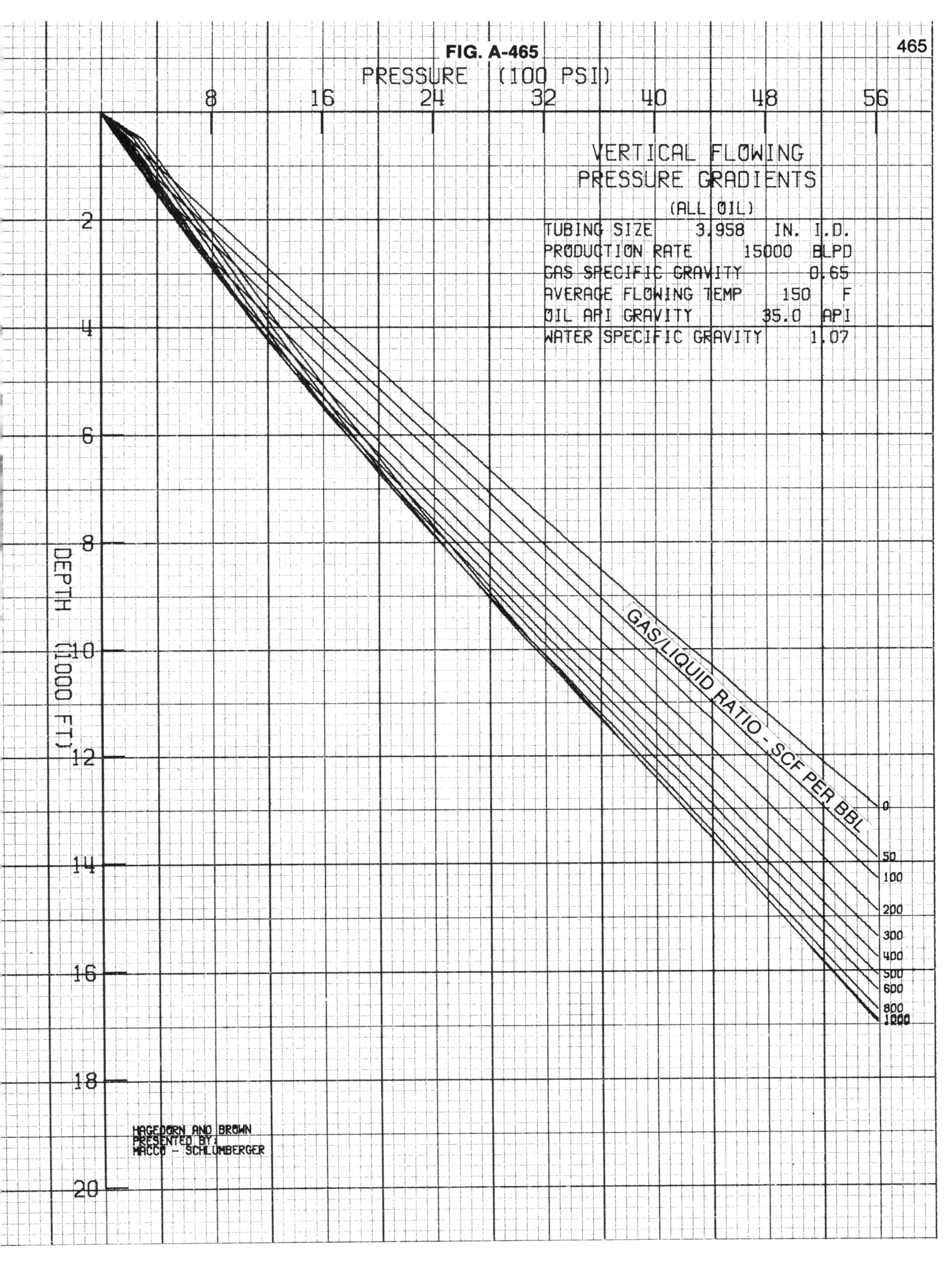

FIG. A-465

465

VERTICAL FLOWING
PRESSURE GRADIENTS
(ALL OIL)

TUBING SIZE	3.958	IN. I.D.
PRODUCTION RATE	15000	BLPD
GAS SPECIFIC GRAVITY	0.65	
AVERAGE FLOWING TEMP	150	F
OIL API GRAVITY	35.0	API
WATER SPECIFIC GRAVITY	1.07	

PRESSURE (100 PSI)

DEPTH (1000 FT.)

GAS/LIQUID RATIO - SCF PER BBL

HAGEDORN AND BROWN
PRESENTED BY:
MACCO - SCHLUMBERGER

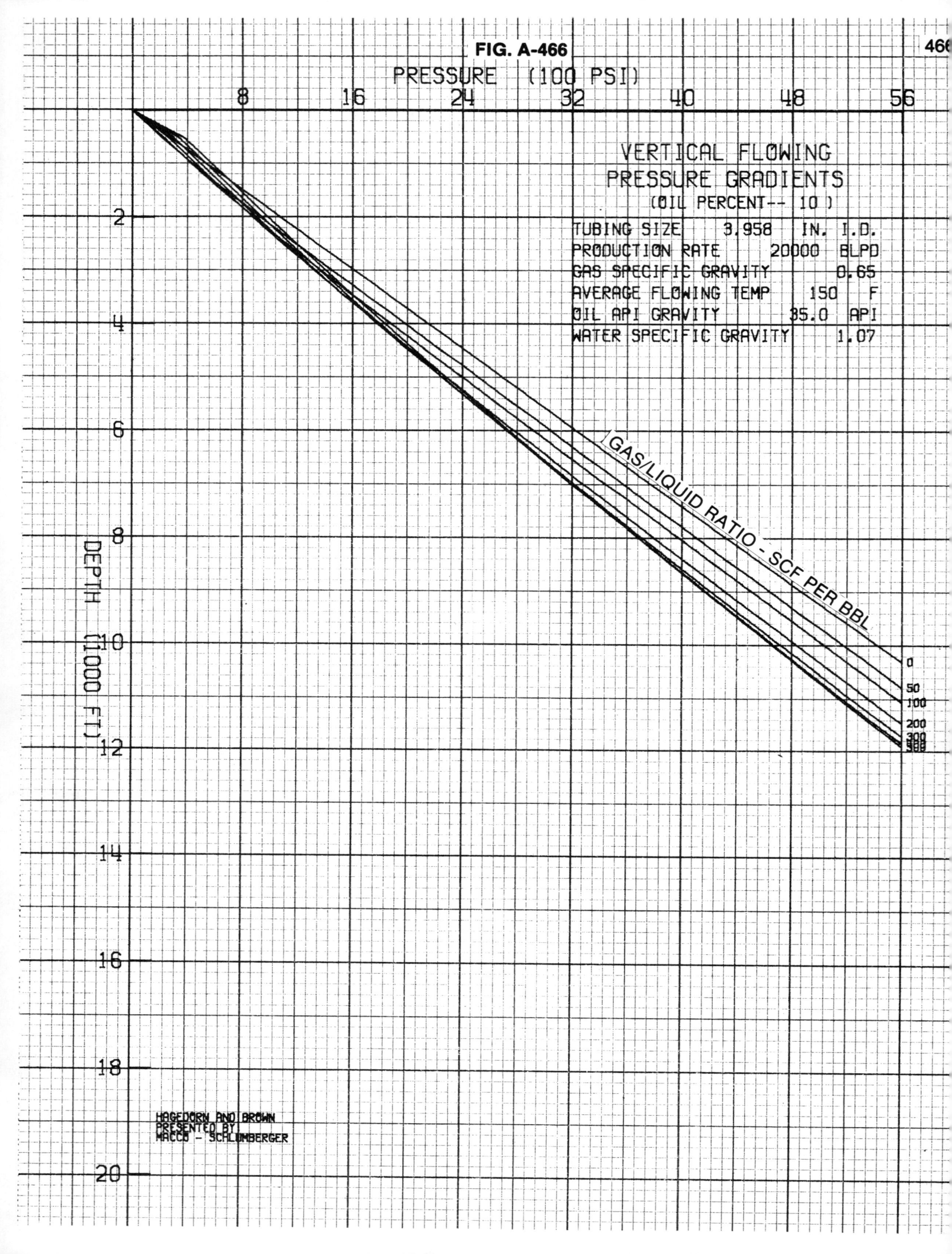

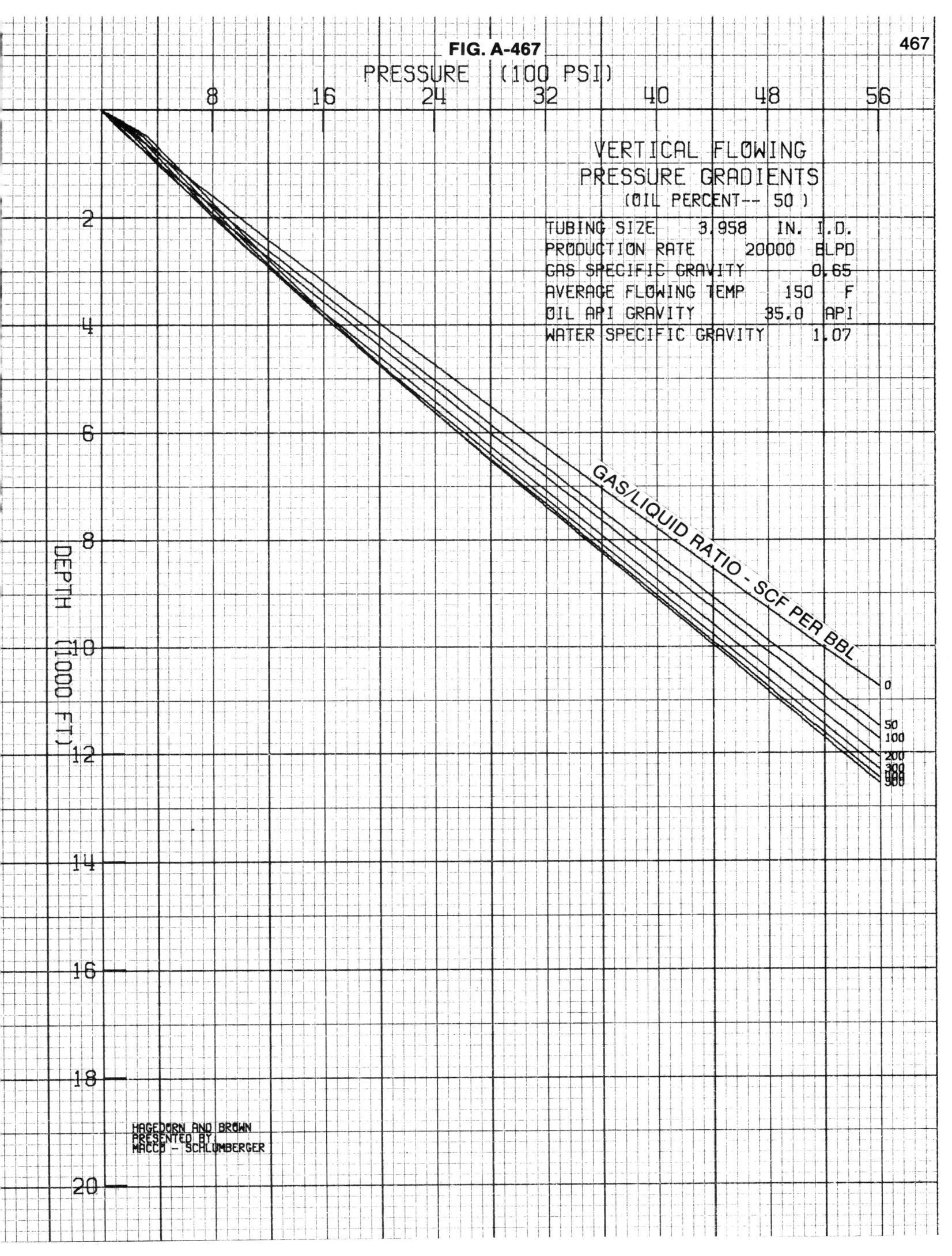

PRESSURE (100 PSI)

DEPTH (1000 FT)

VERTICAL FLOWING
PRESSURE GRADIENTS
(OIL PERCENT-- 50)

TUBING SIZE	3.958	IN. I.D.
PRODUCTION RATE	20000	BLPD
GAS SPECIFIC GRAVITY	0.65	
AVERAGE FLOWING TEMP	150	F
OIL API GRAVITY	35.0	API
WATER SPECIFIC GRAVITY	1.07	

GAS/LIQUID RATIO - SCF PER BBL

0

50
100

200
300
900

HAGEDORN AND BROWN
PRESENTED BY
MACCO - SCHLUMBERGER

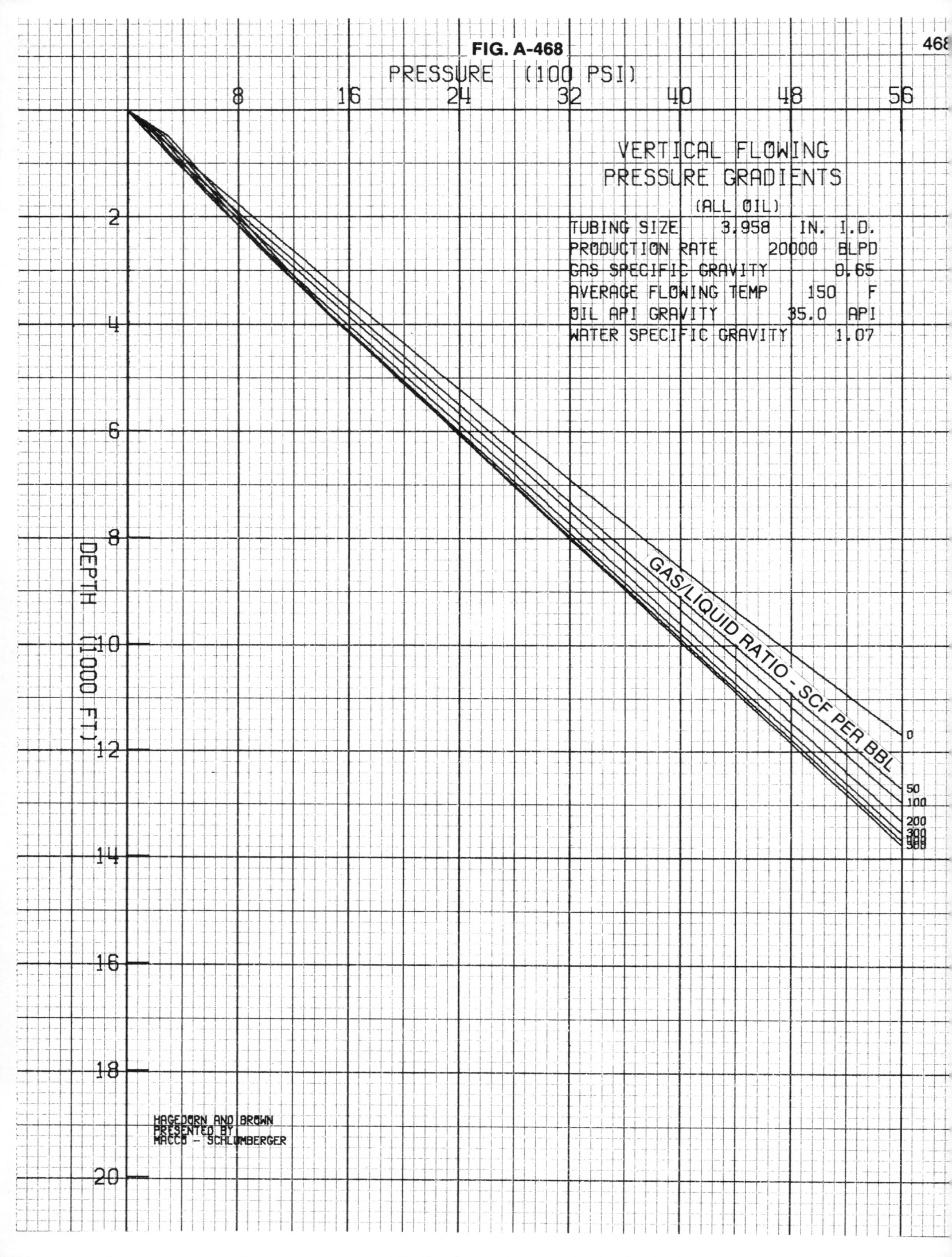

FIG. A-468

468

PRESSURE (100 PSI)

VERTICAL FLOWING
PRESSURE GRADIENTS
(ALL OIL)

TUBING SIZE	3.958	IN. I.D.
PRODUCTION RATE	20000	BLPD
GAS SPECIFIC GRAVITY	0.65	
AVERAGE FLOWING TEMP	150	F
OIL API GRAVITY	35.0	API
WATER SPECIFIC GRAVITY	1.07	

DEPTH (1000 FT)

GAS/LIQUID RATIO - SCF PER BBL

0
50
100
200
300
500

HAGEDORN AND BROWN
PRESENTED BY
MACCO - SCHLUMBERGER

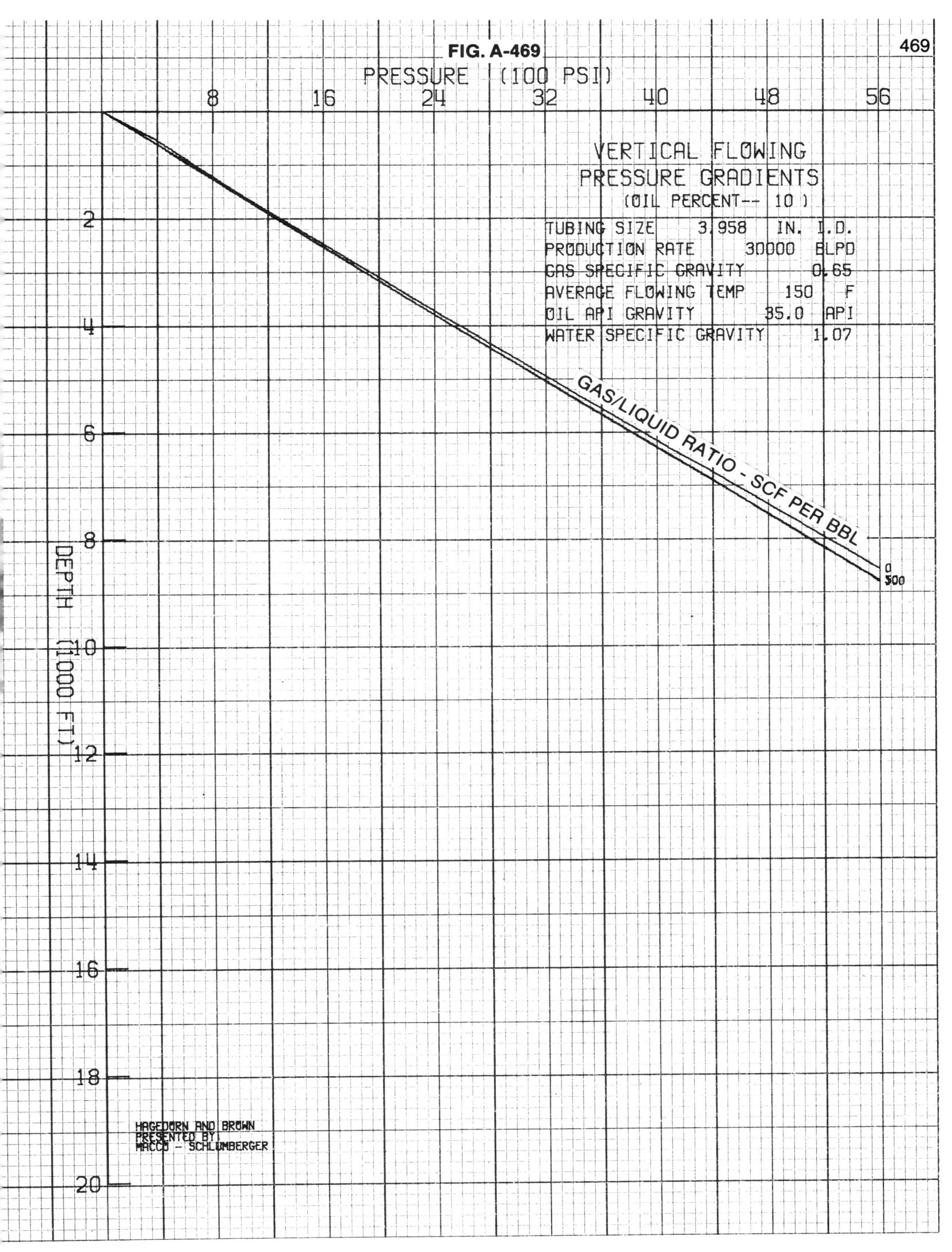

PRESSURE (100 PSI)

8   16   24   32   40   48   56

VERTICAL FLOWING
PRESSURE GRADIENTS
(OIL PERCENT-- 10)

TUBING SIZE	3.958	IN. I.D.
PRODUCTION RATE	30000	BLPD
GAS SPECIFIC GRAVITY	0.65	
AVERAGE FLOWING TEMP	150	F
OIL API GRAVITY	35.0	API
WATER SPECIFIC GRAVITY	1.07	

GAS/LIQUID RATIO - SCF PER BBL

0
500

DEPTH (1,000 FT)

HAGEDORN AND BROWN
PRESENTED BY,
MACCO - SCHLUMBERGER

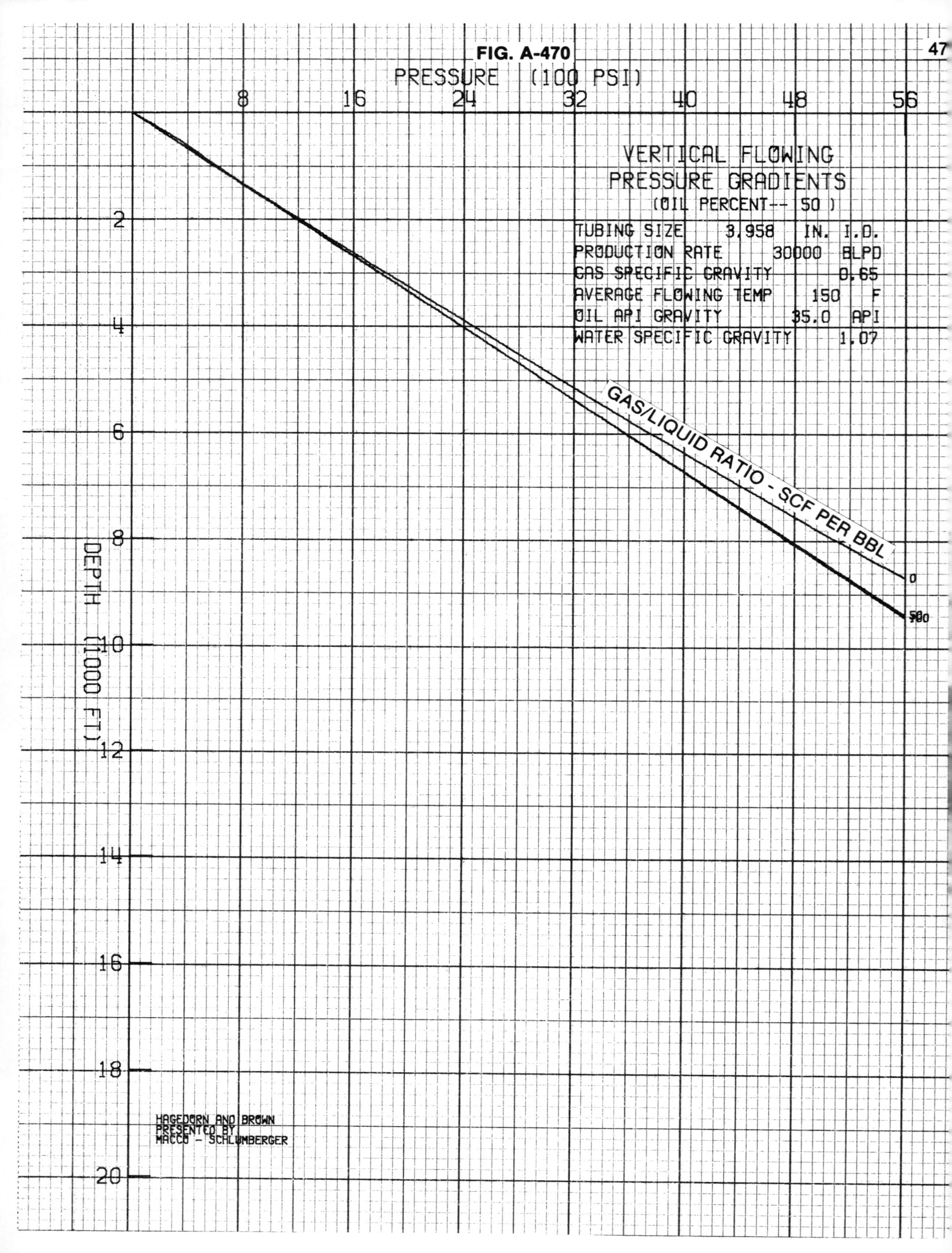

PRESSURE (100 PSI)

VERTICAL FLOWING
PRESSURE GRADIENTS
(OIL PERCENT-- 50)

TUBING SIZE	3.958	IN. I.D.
PRODUCTION RATE	30000	BLPD
GAS SPECIFIC GRAVITY	0.65	
AVERAGE FLOWING TEMP	150	F
OIL API GRAVITY	35.0	API
WATER SPECIFIC GRAVITY	1.07	

GAS/LIQUID RATIO - SCF PER BBL

DEPTH (1000 FT)

HAGEDORN AND BROWN
PRESENTED BY
MACCO - SCHLUMBERGER

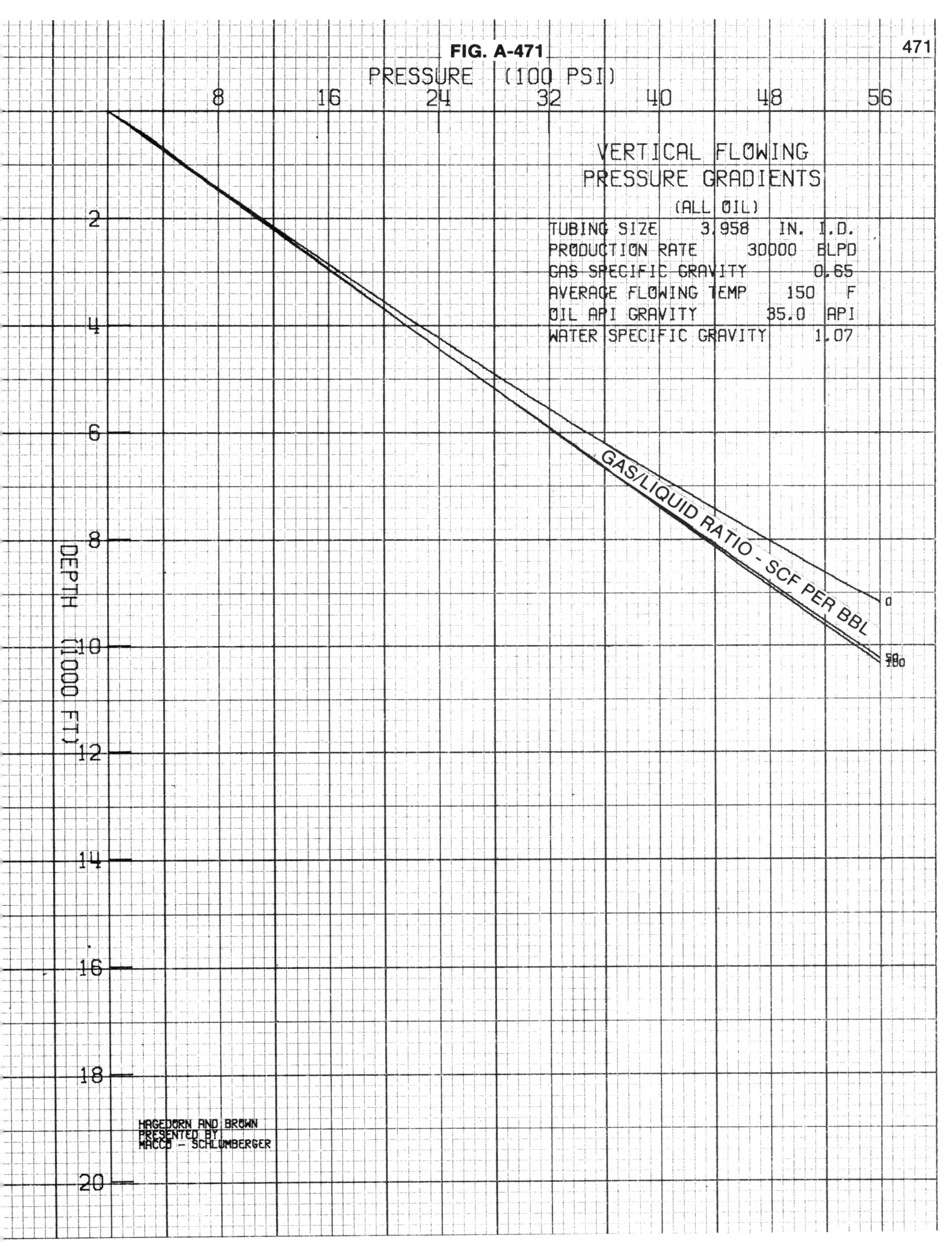

VERTICAL FLOWING
PRESSURE GRADIENTS
(ALL OIL)

TUBING SIZE	3.958	IN. I.D.
PRODUCTION RATE	30000	BLPD
GAS SPECIFIC GRAVITY	0.65	
AVERAGE FLOWING TEMP	150	F
OIL API GRAVITY	35.0	API
WATER SPECIFIC GRAVITY	1.07	

PRESSURE (100 PSI)

DEPTH (1000 FT.)

GAS/LIQUID RATIO - SCF PER BBL

HAGEDORN AND BROWN
PRESENTED BY
MACCO — SCHLUMBERGER

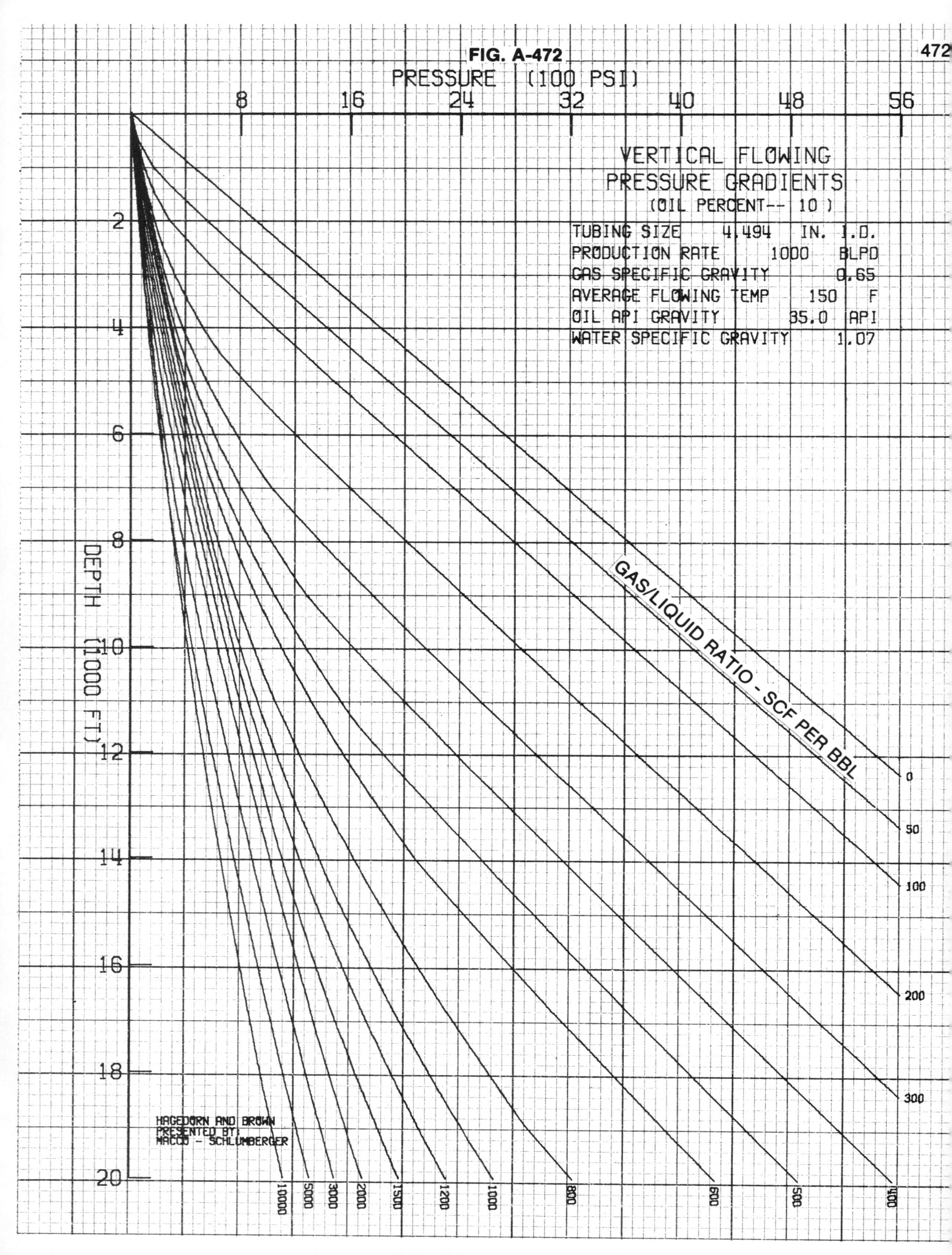

FIG. A-472

472

VERTICAL FLOWING
PRESSURE GRADIENTS
(OIL PERCENT-- 10)

TUBING SIZE	4.494	IN. I.D.
PRODUCTION RATE	1000	BLPD
GAS SPECIFIC GRAVITY	0.65	
AVERAGE FLOWING TEMP	150	F
OIL API GRAVITY	35.0	API
WATER SPECIFIC GRAVITY	1.07	

PRESSURE (100 PSI)

DEPTH (1000 FT)

GAS/LIQUID RATIO - SCF PER BBL

HAGEDORN AND BROWN
PRESENTED BY:
MACCO - SCHLUMBERGER

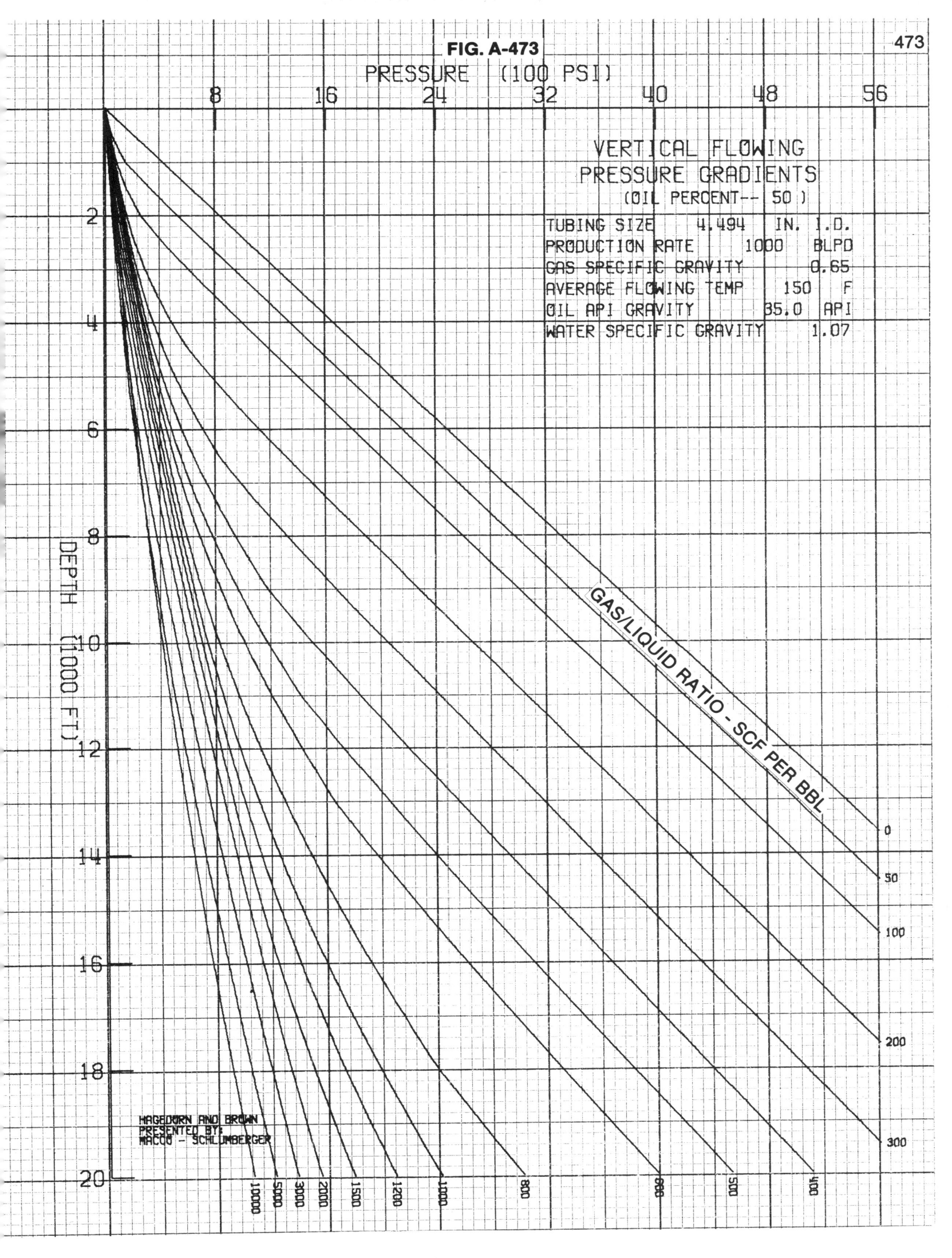

PRESSURE (100 PSI)

DEPTH (1000 FT)

VERTICAL FLOWING
PRESSURE GRADIENTS
(OIL PERCENT-- 50)

TUBING SIZE	4.494	IN. I.D.
PRODUCTION RATE	1000	BLPD
GAS SPECIFIC GRAVITY		0.65
AVERAGE FLOWING TEMP	150	F
OIL API GRAVITY	35.0	API
WATER SPECIFIC GRAVITY		1.07

GAS/LIQUID RATIO - SCF PER BBL

0
50
100
200
300

HAGEDORN AND BROWN
PRESENTED BY:
MACCO - SCHLUMBERGER

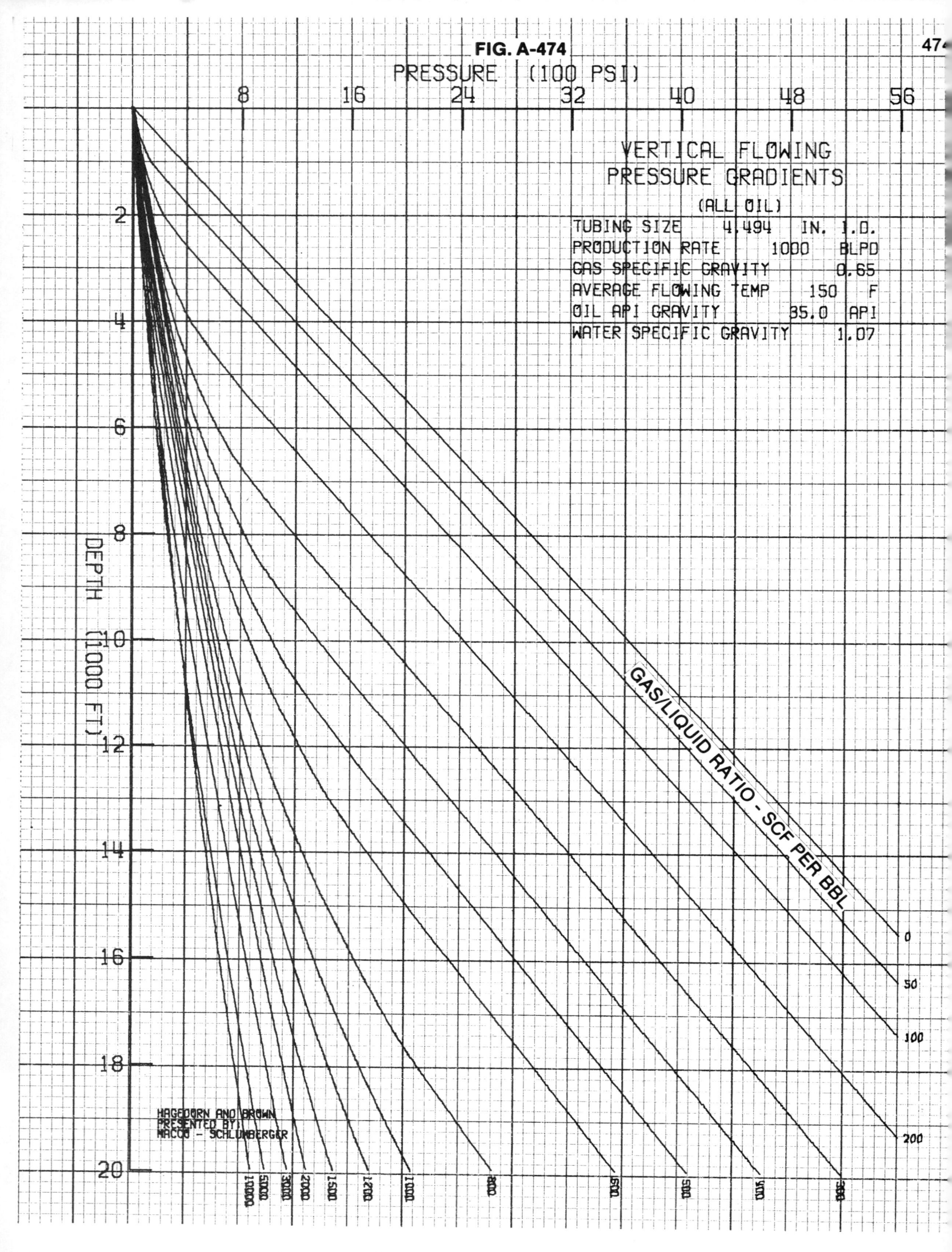

FIG. A-474

VERTICAL FLOWING PRESSURE GRADIENTS
(ALL OIL)

TUBING SIZE	4.494	IN. I.D.
PRODUCTION RATE	1000	BLPD
GAS SPECIFIC GRAVITY	0.65	
AVERAGE FLOWING TEMP	150	F
OIL API GRAVITY	35.0	API
WATER SPECIFIC GRAVITY	1.07	

GAS/LIQUID RATIO - SCF PER BBL

HAGEDORN AND BROWN
PRESENTED BY
WACOO - SCHLUMBERGER

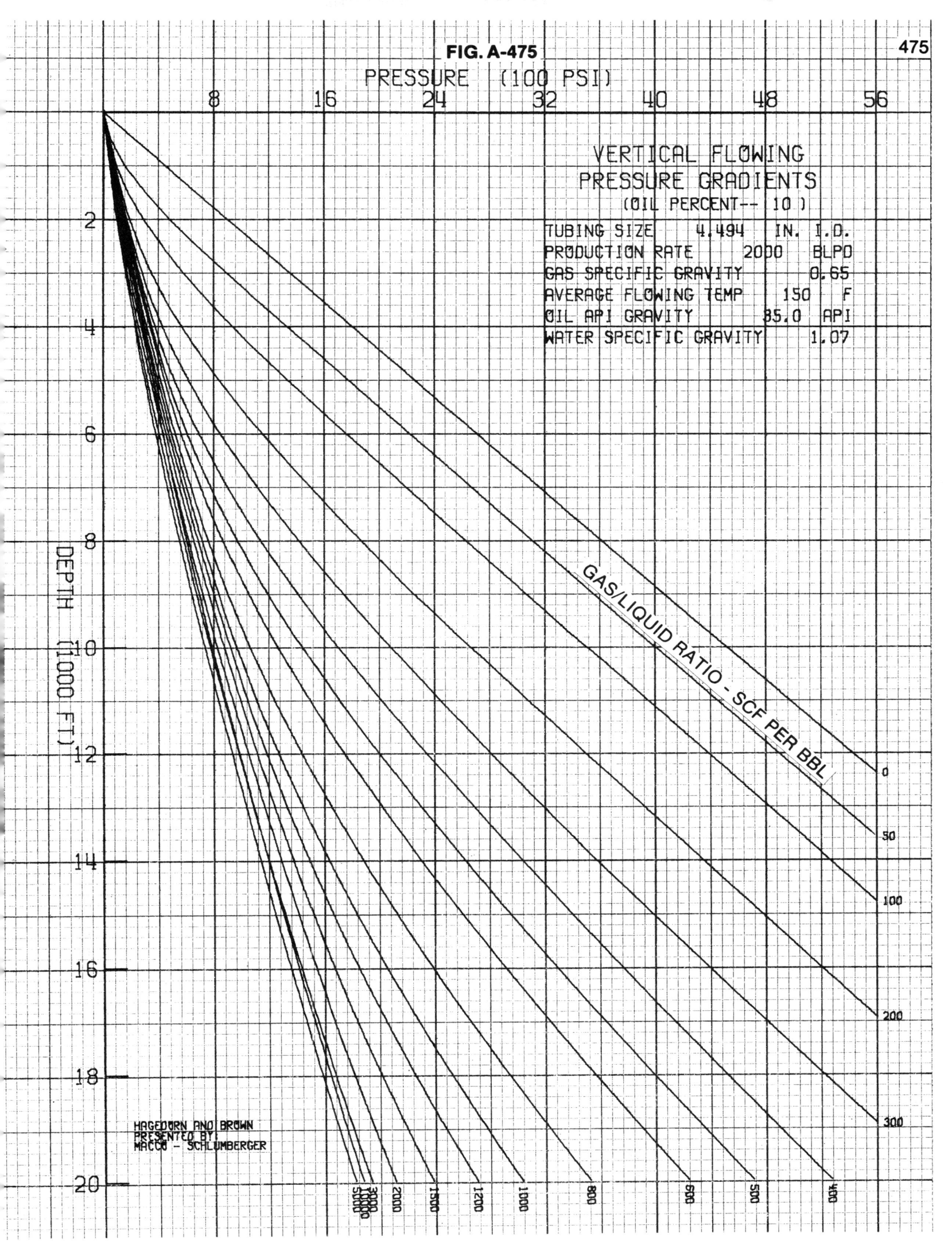

PRESSURE (100 PSI)

DEPTH (1000 FT)

VERTICAL FLOWING
PRESSURE GRADIENTS
(OIL PERCENT-- 10)

TUBING SIZE	4.494	IN. I.D.
PRODUCTION RATE	2000	BLPD
GAS SPECIFIC GRAVITY		0.65
AVERAGE FLOWING TEMP	150	F
OIL API GRAVITY	85.0	API
WATER SPECIFIC GRAVITY		1.07

GAS/LIQUID RATIO - SCF PER BBL

HAGEDORN AND BROWN
PRESENTED BY
MACCO - SCHLUMBERGER

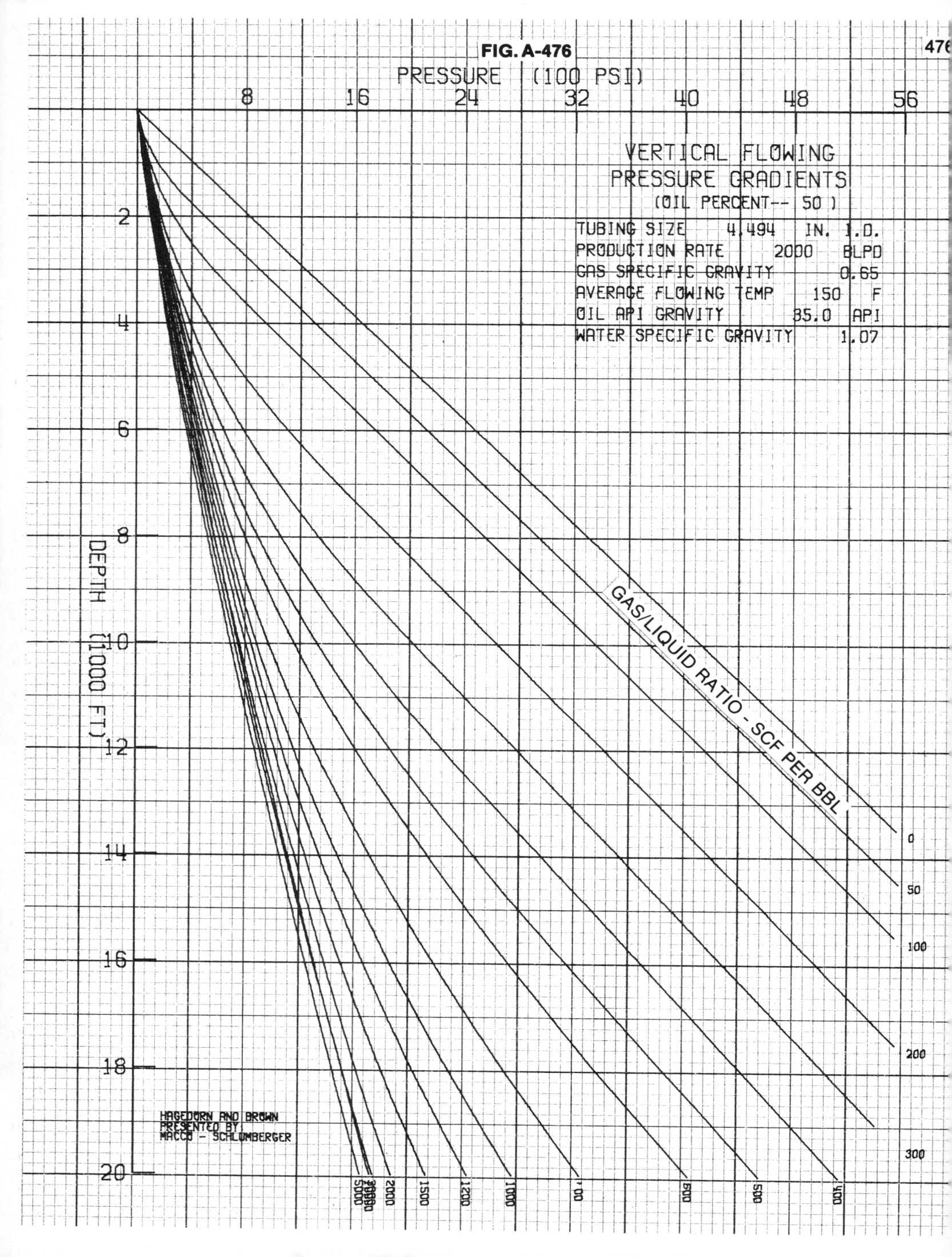

FIG. A-476

476

VERTICAL FLOWING
PRESSURE GRADIENTS
(OIL PERCENT-- 50)

TUBING SIZE	4.494	IN. I.D.
PRODUCTION RATE	2000	BLPD
GAS SPECIFIC GRAVITY	0.65	
AVERAGE FLOWING TEMP	150	F
OIL API GRAVITY	85.0	API
WATER SPECIFIC GRAVITY	1.07	

GAS/LIQUID RATIO - SCF PER BBL

HAGEDORN AND BROWN
PRESENTED BY
MACCO - SCHLUMBERGER

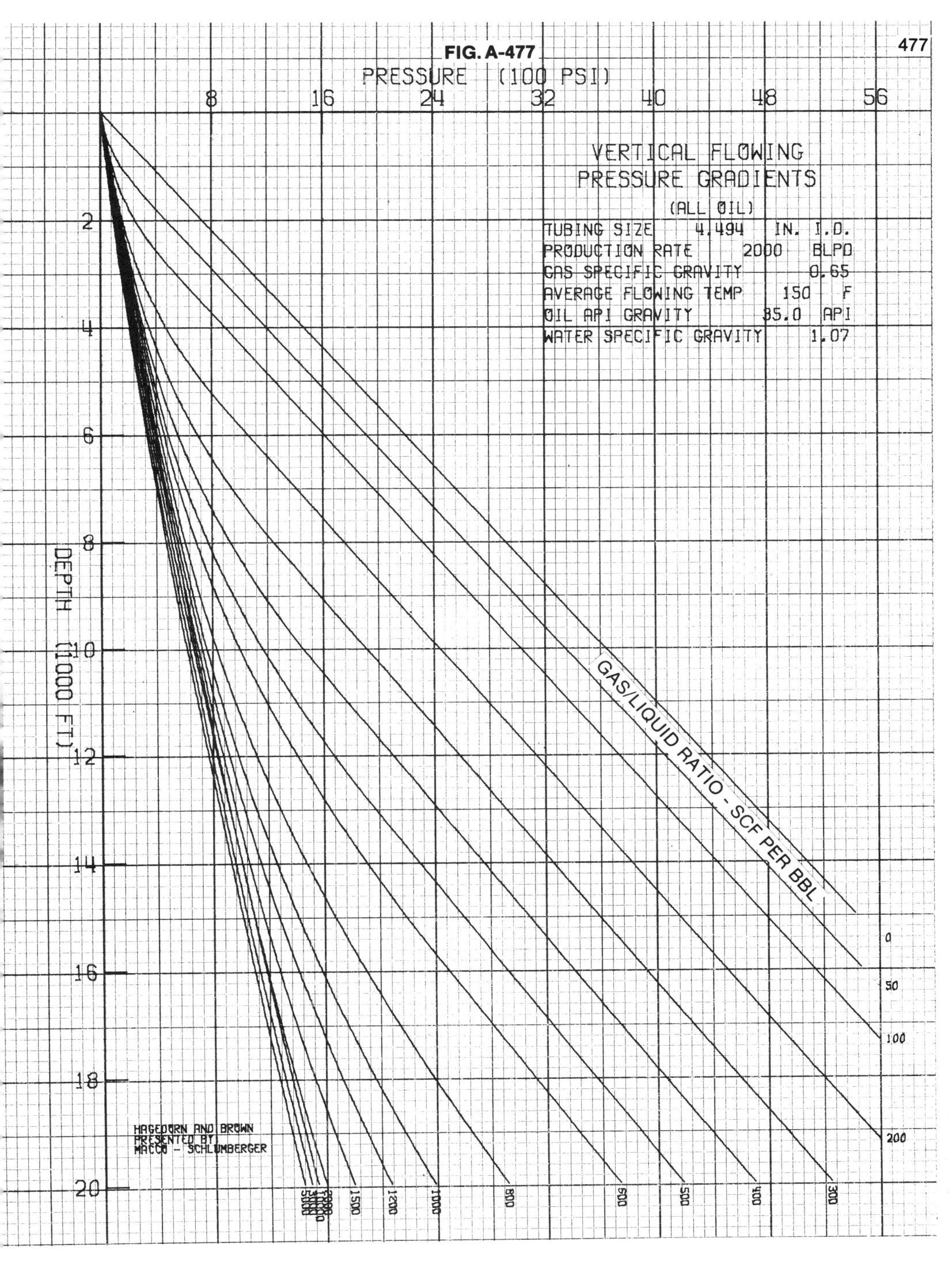

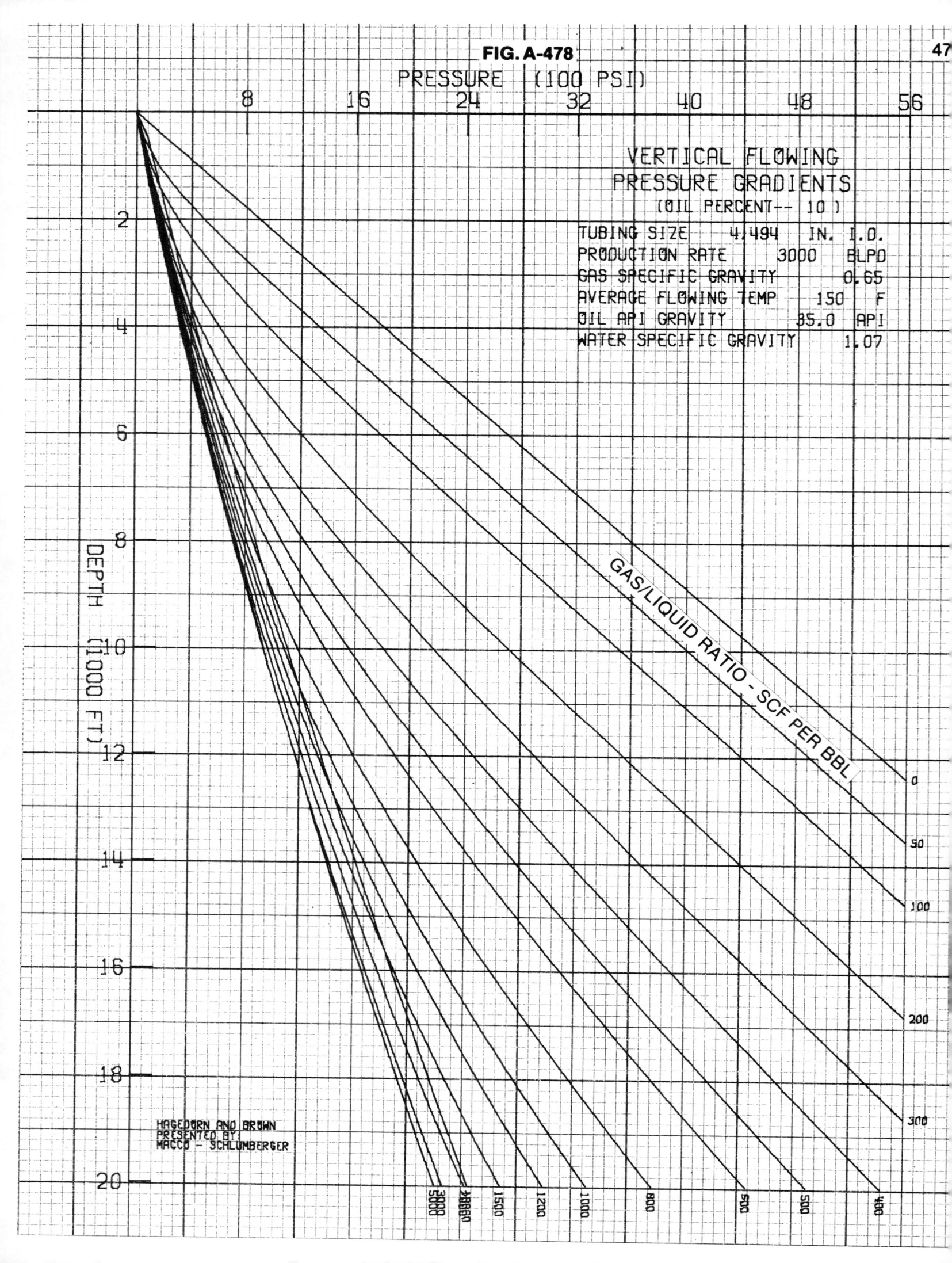

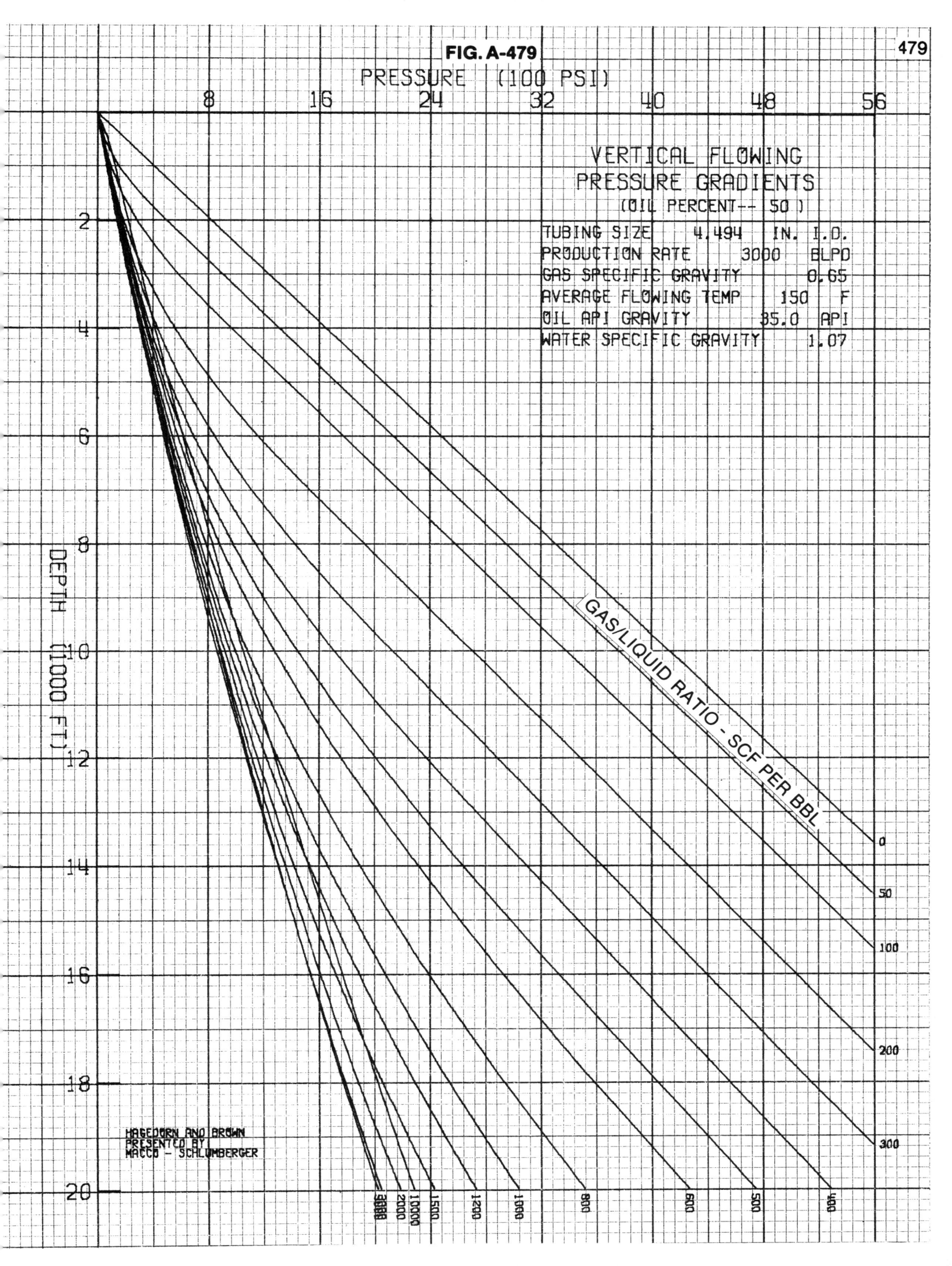

PRESSURE (100 PSI)

DEPTH (1000 FT)

VERTICAL FLOWING
PRESSURE GRADIENTS
(OIL PERCENT-- 50)

TUBING SIZE	4.494	IN. I.D.
PRODUCTION RATE	3000	BLPD
GAS SPECIFIC GRAVITY	0.65	
AVERAGE FLOWING TEMP	150	F
OIL API GRAVITY	35.0	API
WATER SPECIFIC GRAVITY	1.07	

GAS/LIQUID RATIO - SCF PER BBL

HAGEDORN AND BROWN
PRESENTED BY
MACCO - SCHLUMBERGER

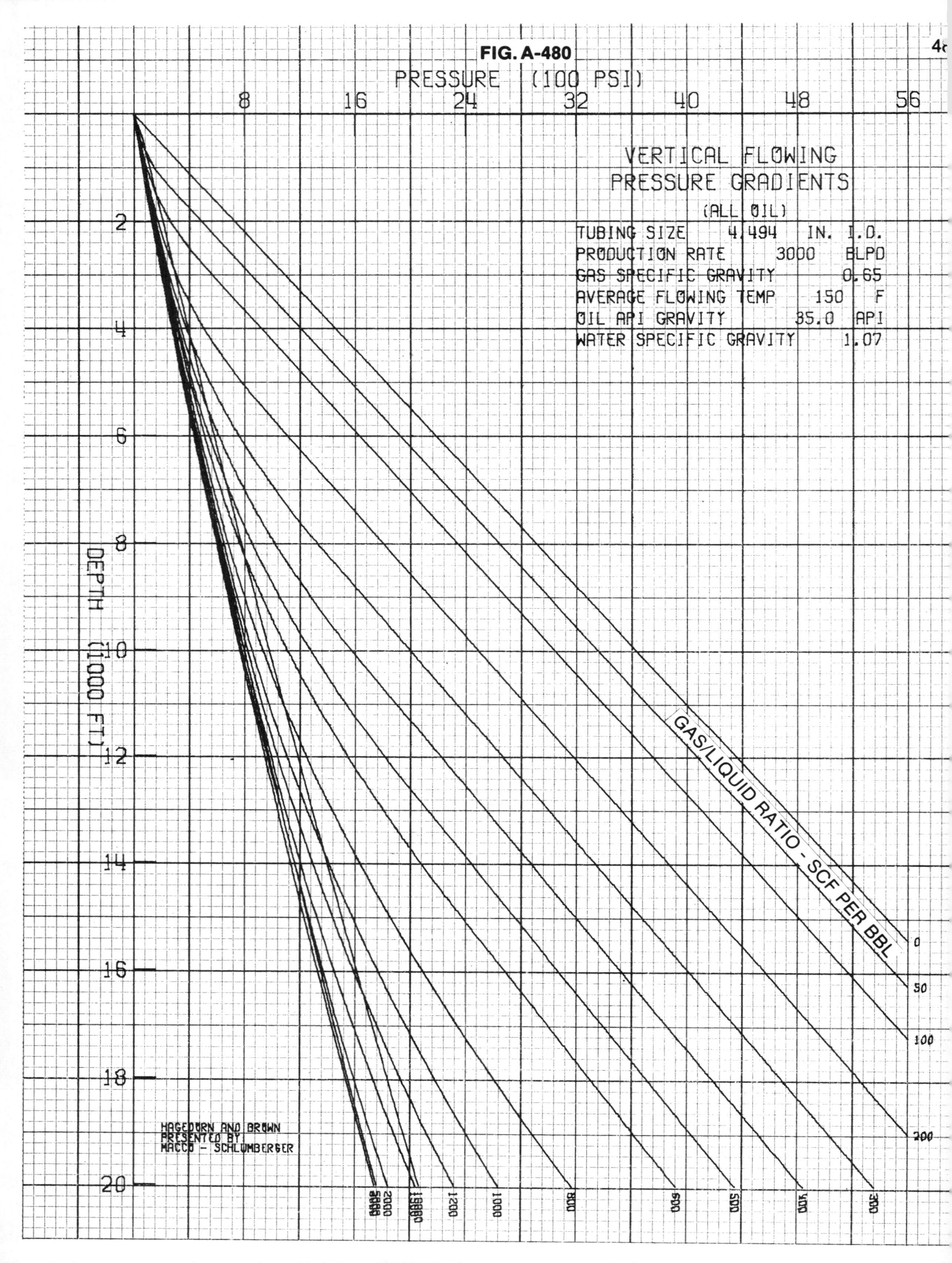

FIG. A-480

VERTICAL FLOWING
PRESSURE GRADIENTS
(ALL OIL)

TUBING SIZE   4.494 IN. I.D.
PRODUCTION RATE   3000   BLPD
GAS SPECIFIC GRAVITY   0.65
AVERAGE FLOWING TEMP   150   F
OIL API GRAVITY   35.0   API
WATER SPECIFIC GRAVITY   1.07

PRESSURE   (100 PSI)

DEPTH (1000 FT)

GAS/LIQUID RATIO - SCF PER BBL

HAGEDORN AND BROWN
PRESENTED BY
MACCO - SCHLUMBERGER

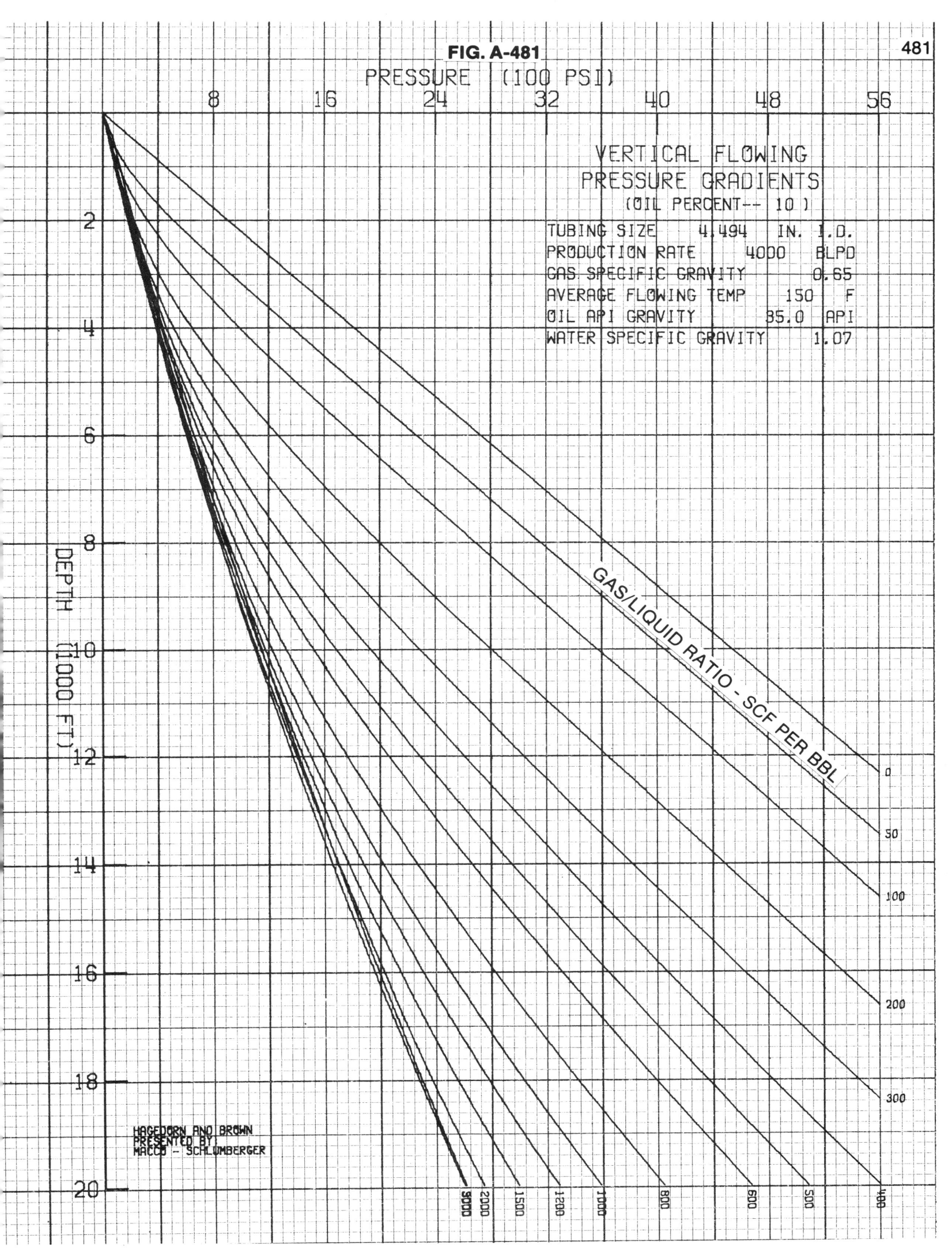

PRESSURE (100 PSI)

VERTICAL FLOWING
PRESSURE GRADIENTS
(OIL PERCENT-- 10)

TUBING SIZE	4.494	IN. I.D.
PRODUCTION RATE	4000	BLPD
GAS SPECIFIC GRAVITY		0.65
AVERAGE FLOWING TEMP	150	F
OIL API GRAVITY	35.0	API
WATER SPECIFIC GRAVITY		1.07

DEPTH (1000 FT)

GAS/LIQUID RATIO - SCF PER BBL

HAGEDORN AND BROWN
PRESENTED BY:
MACCO - SCHLUMBERGER

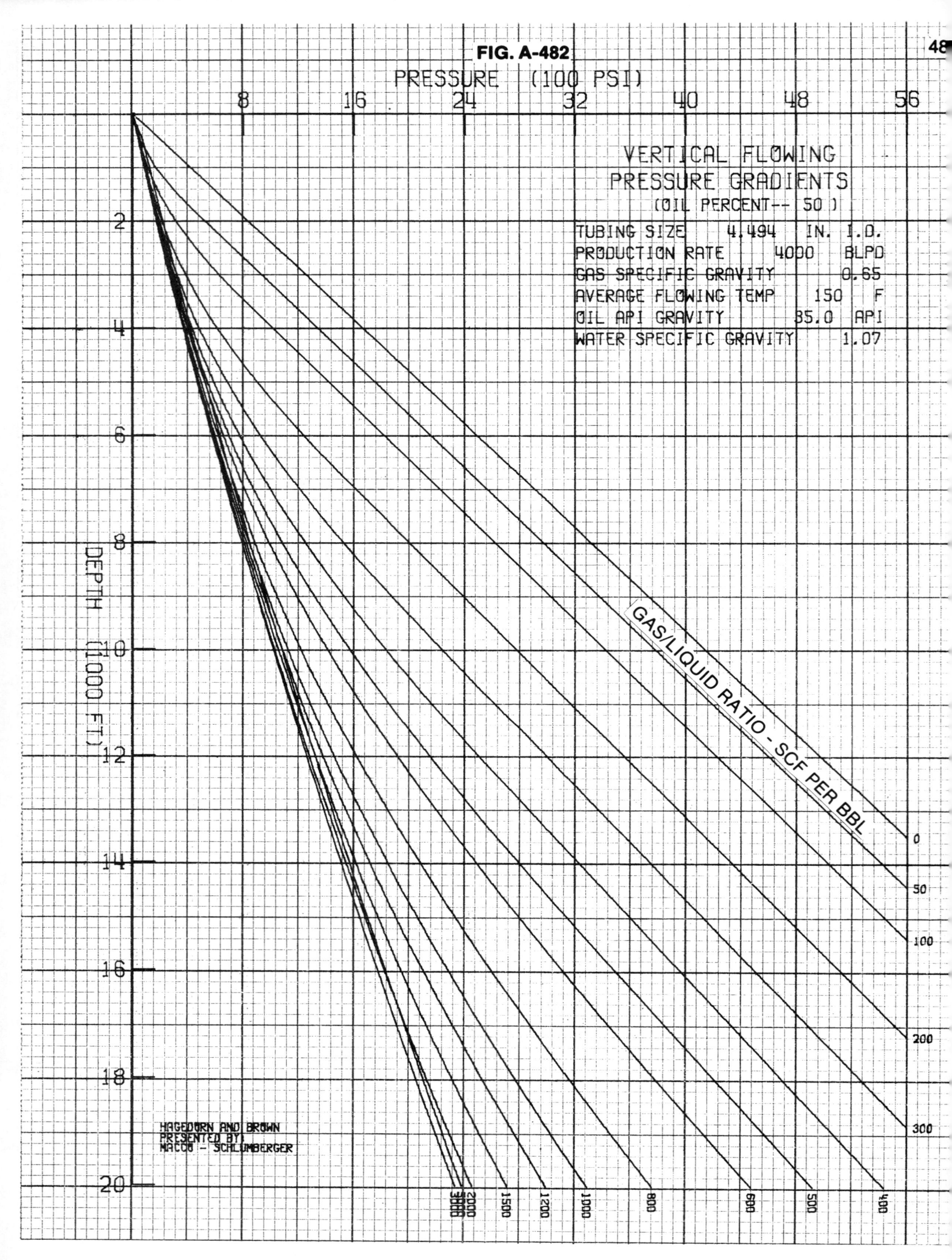

FIG. A-482

VERTICAL FLOWING
PRESSURE GRADIENTS
(OIL PERCENT-- 50)

TUBING SIZE	4.494	IN. I.D.
PRODUCTION RATE	4000	BLPD
GAS SPECIFIC GRAVITY	0.65	
AVERAGE FLOWING TEMP	150	F
OIL API GRAVITY	35.0	API
WATER SPECIFIC GRAVITY	1.07	

PRESSURE (100 PSI)

DEPTH (1000 FT)

GAS/LIQUID RATIO - SCF PER BBL

HAGEDORN AND BROWN
PRESENTED BY
MACCO - SCHLUMBERGER

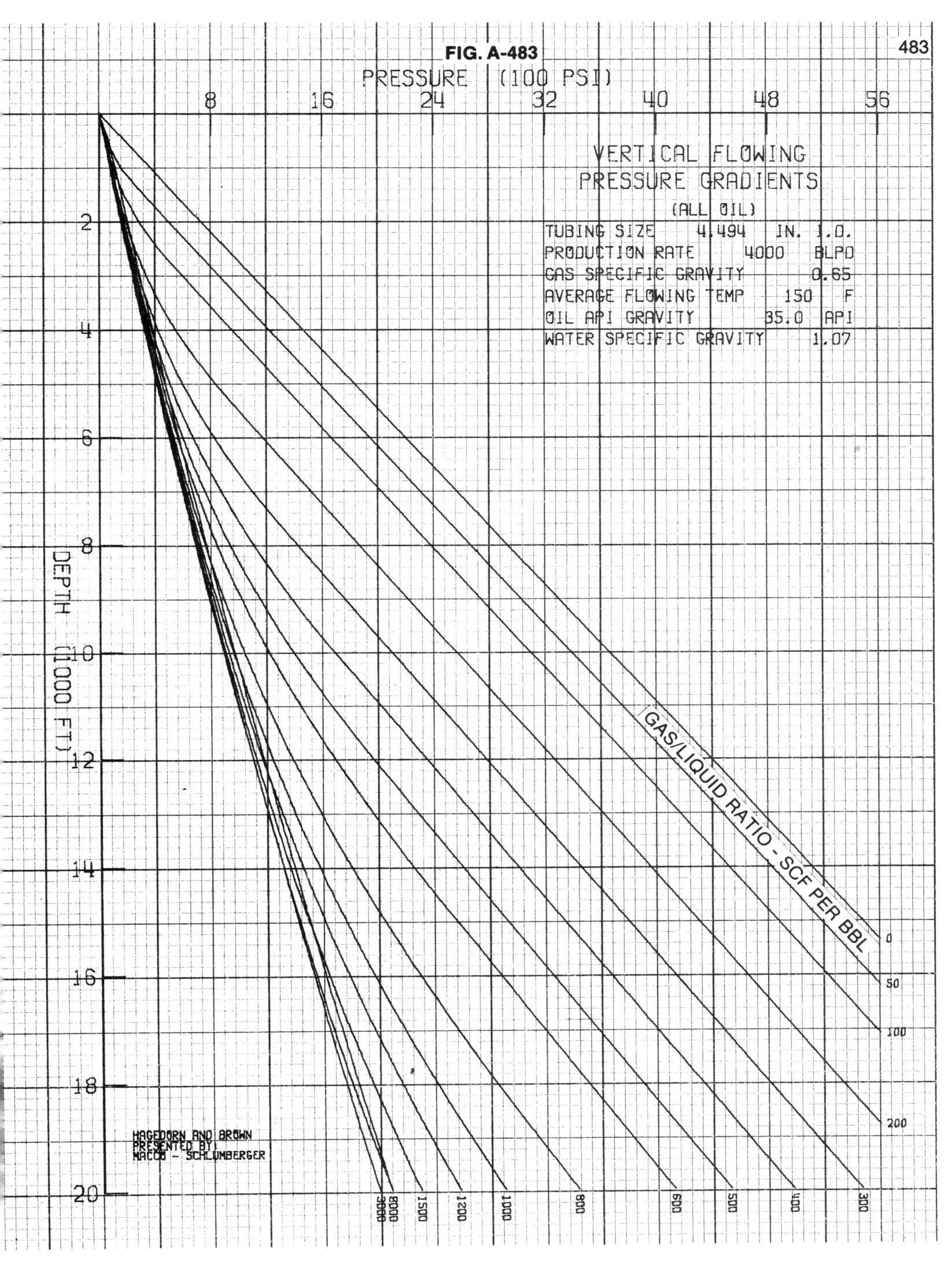

PRESSURE (100 PSI)

VERTICAL FLOWING
PRESSURE GRADIENTS
(ALL OIL)

TUBING SIZE	4.494	IN. I.D.
PRODUCTION RATE	4000	BLPD
GAS SPECIFIC GRAVITY	0.65	
AVERAGE FLOWING TEMP	150	F
OIL API GRAVITY	35.0	API
WATER SPECIFIC GRAVITY	1.07	

DEPTH (1,000 FT)

GAS/LIQUID RATIO - SCF PER BBL

HAGEDORN AND BROWN
PRESENTED BY
MACCO - SCHLUMBERGER

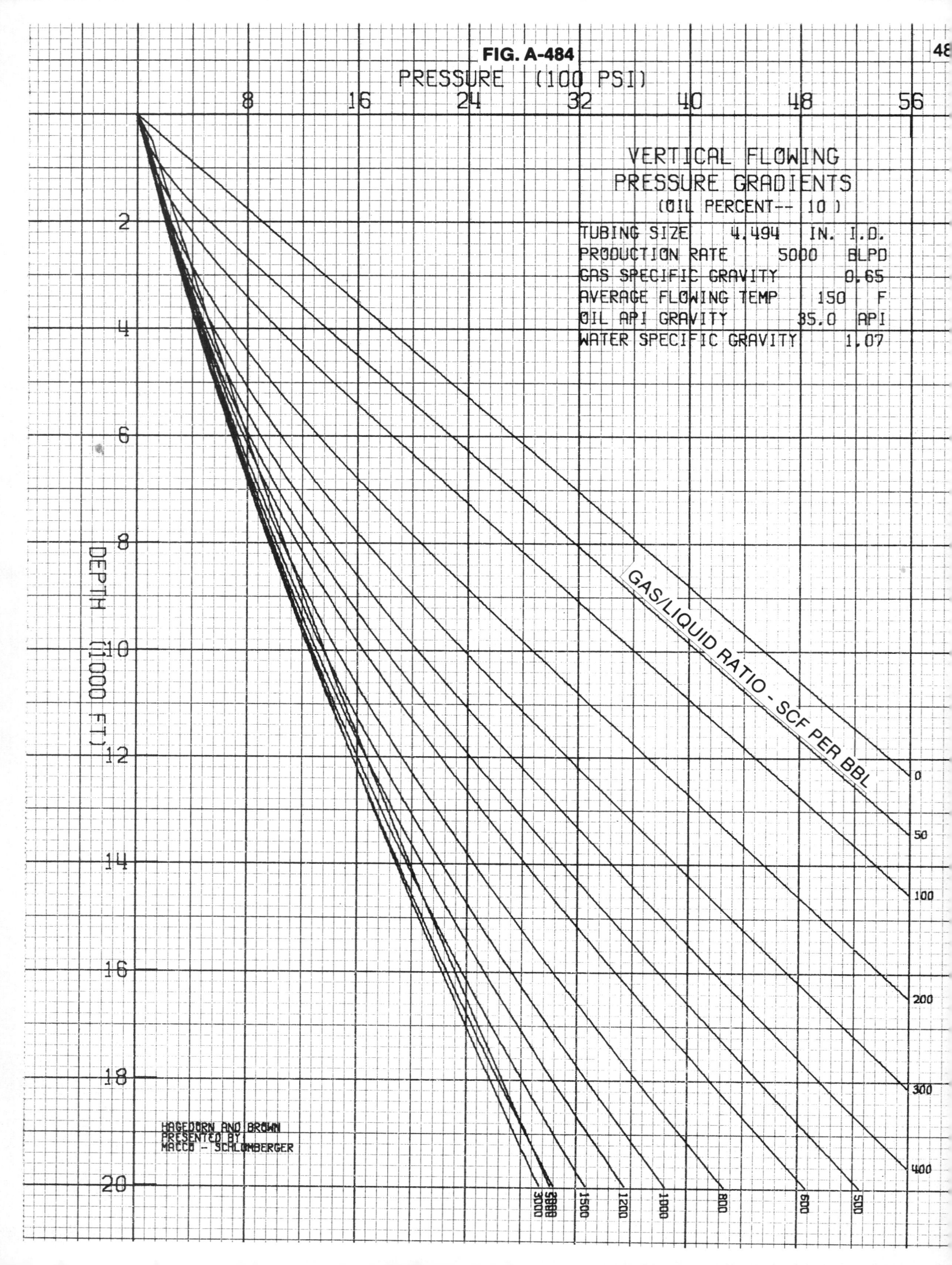

FIG. A-484

PRESSURE (100 PSI)

DEPTH (1000 FT)

VERTICAL FLOWING
PRESSURE GRADIENTS
(OIL PERCENT-- 10)

TUBING SIZE	4.494	IN. I.D.
PRODUCTION RATE	5000	BLPD
GAS SPECIFIC GRAVITY		0.65
AVERAGE FLOWING TEMP	150	F
OIL API GRAVITY	35.0	API
WATER SPECIFIC GRAVITY		1.07

GAS/LIQUID RATIO - SCF PER BBL

HAGEDORN AND BROWN
PRESENTED BY
MACCO - SCHLUMBERGER

FIG. A-485

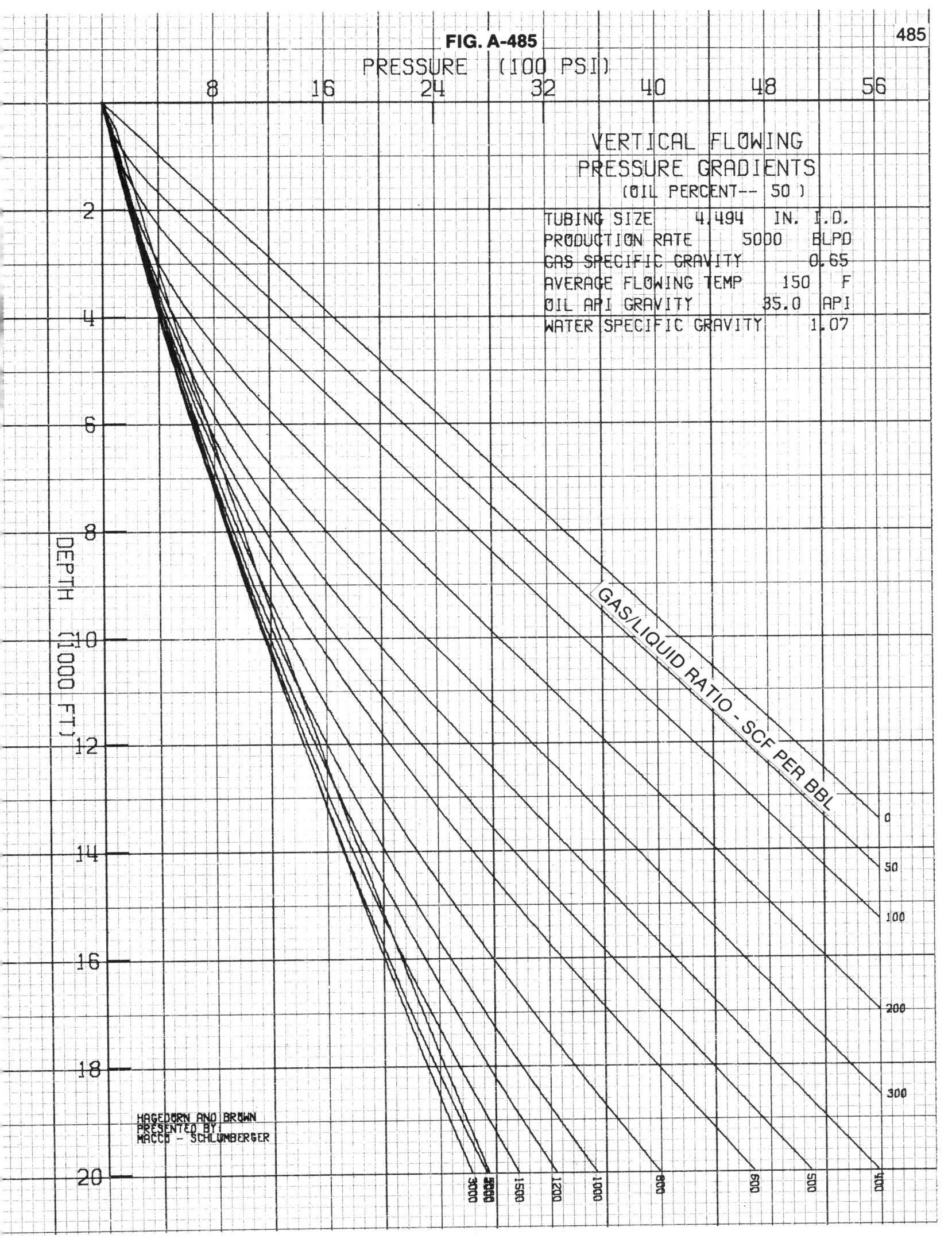

PRESSURE (100 PSI)

VERTICAL FLOWING
PRESSURE GRADIENTS
(OIL PERCENT-- 50)

TUBING SIZE	4.494	IN. I.D.
PRODUCTION RATE	5000	BLPD
GAS SPECIFIC GRAVITY	0.65	
AVERAGE FLOWING TEMP	150	F
OIL API GRAVITY	35.0	API
WATER SPECIFIC GRAVITY	1.07	

GAS/LIQUID RATIO - SCF PER BBL

DEPTH (1000 FT)

HAGEDORN AND BROWN
PRESENTED BY:
MACCO - SCHLUMBERGER

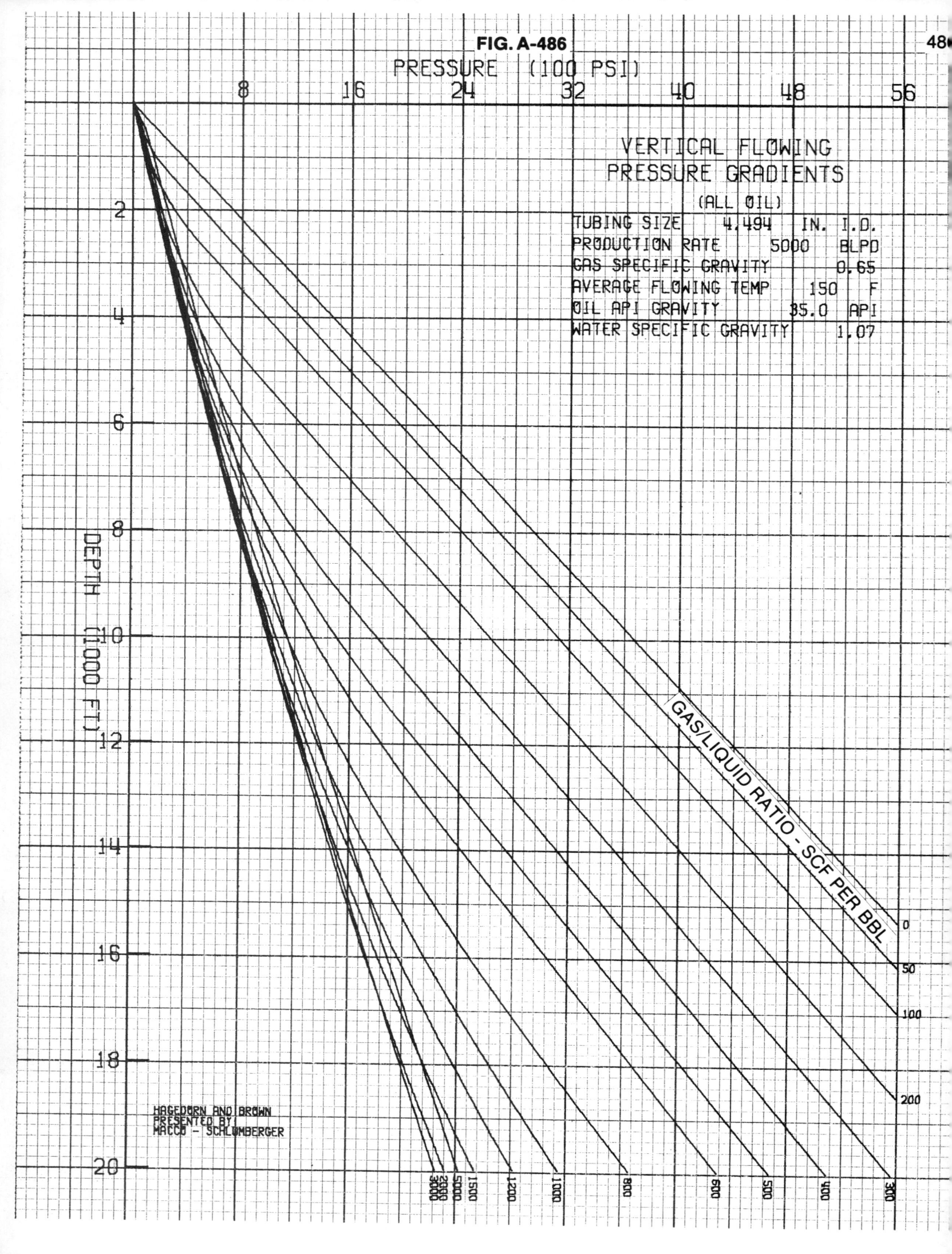

FIG. A-486

VERTICAL FLOWING
PRESSURE GRADIENTS
(ALL OIL)

TUBING SIZE	4.494	IN. I.D.
PRODUCTION RATE	5000	BLPD
GAS SPECIFIC GRAVITY	0.65	
AVERAGE FLOWING TEMP	150	F
OIL API GRAVITY	35.0	API
WATER SPECIFIC GRAVITY	1.07	

GAS/LIQUID RATIO - SCF PER BBL

HAGEDORN AND BROWN
PRESENTED BY
MACCO - SCHLUMBERGER

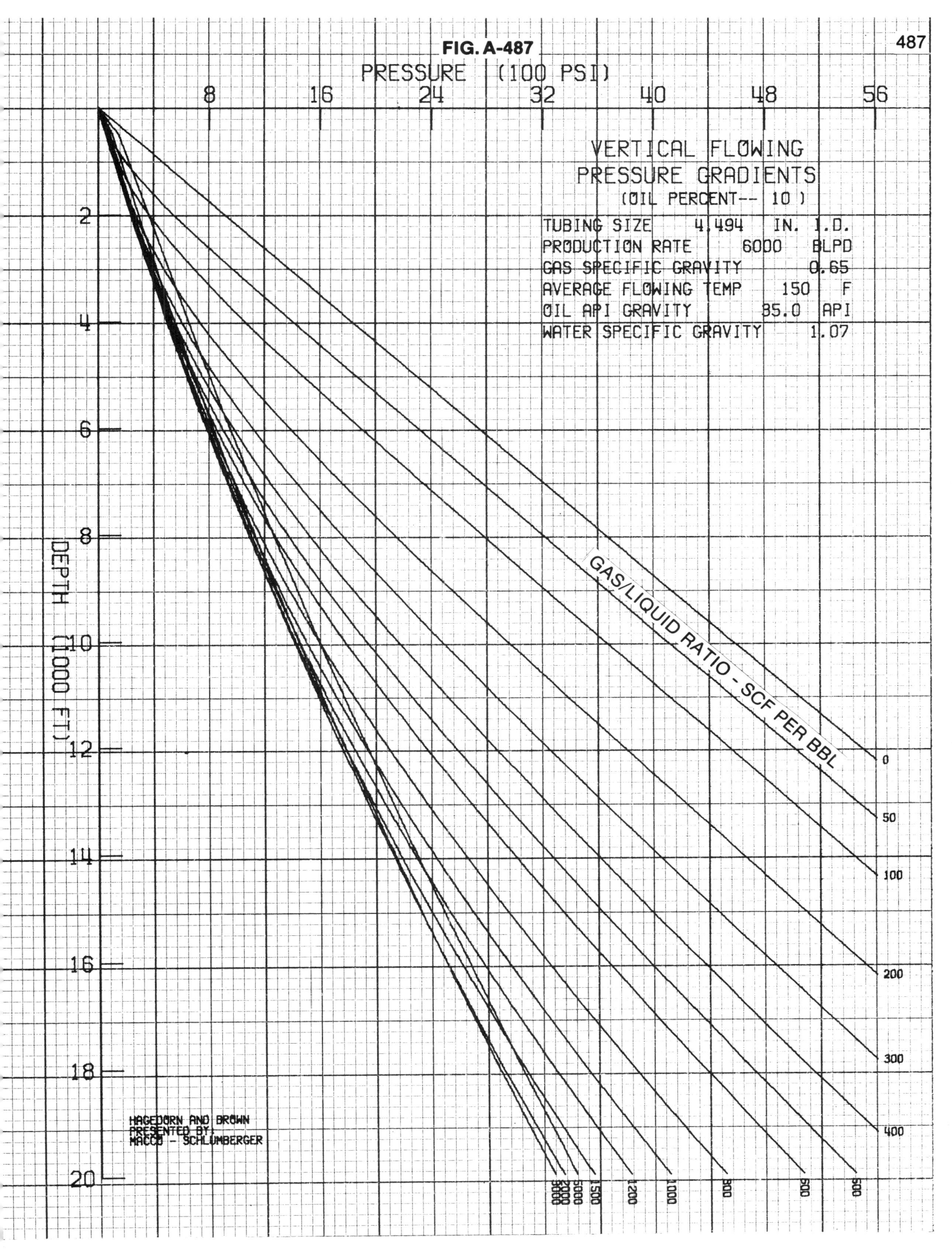

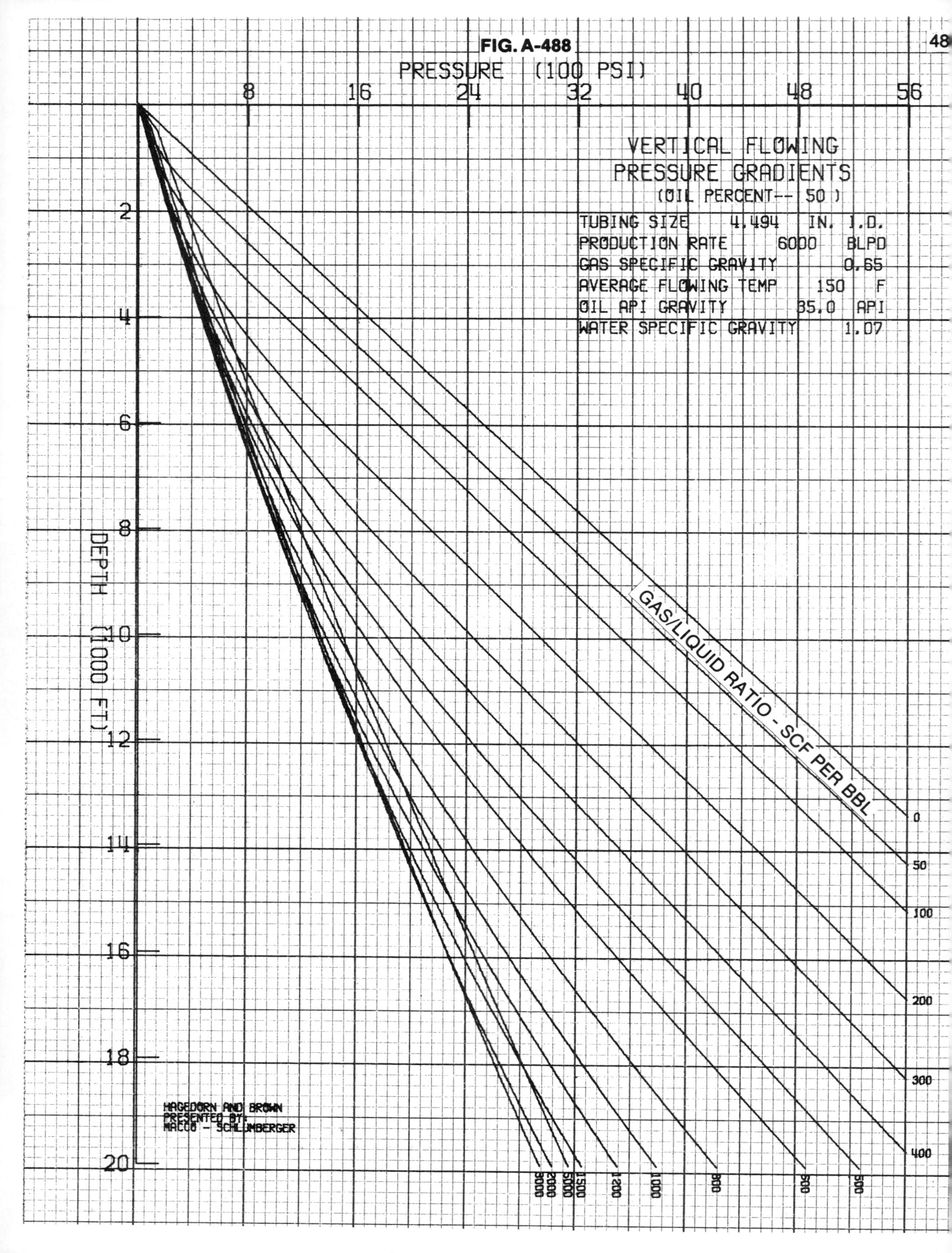

FIG. A-488

VERTICAL FLOWING
PRESSURE GRADIENTS
(OIL PERCENT-- 50)

TUBING SIZE	4.494	IN. I.D.
PRODUCTION RATE	6000	BLPD
GAS SPECIFIC GRAVITY	0.65	
AVERAGE FLOWING TEMP	150	F
OIL API GRAVITY	35.0	API
WATER SPECIFIC GRAVITY	1.07	

GAS/LIQUID RATIO - SCF PER BBL

PRESSURE (100 PSI)

DEPTH (1000 FT)

HAGEDORN AND BROWN
PRESENTED BY:
MACCO - SCHLUMBERGER

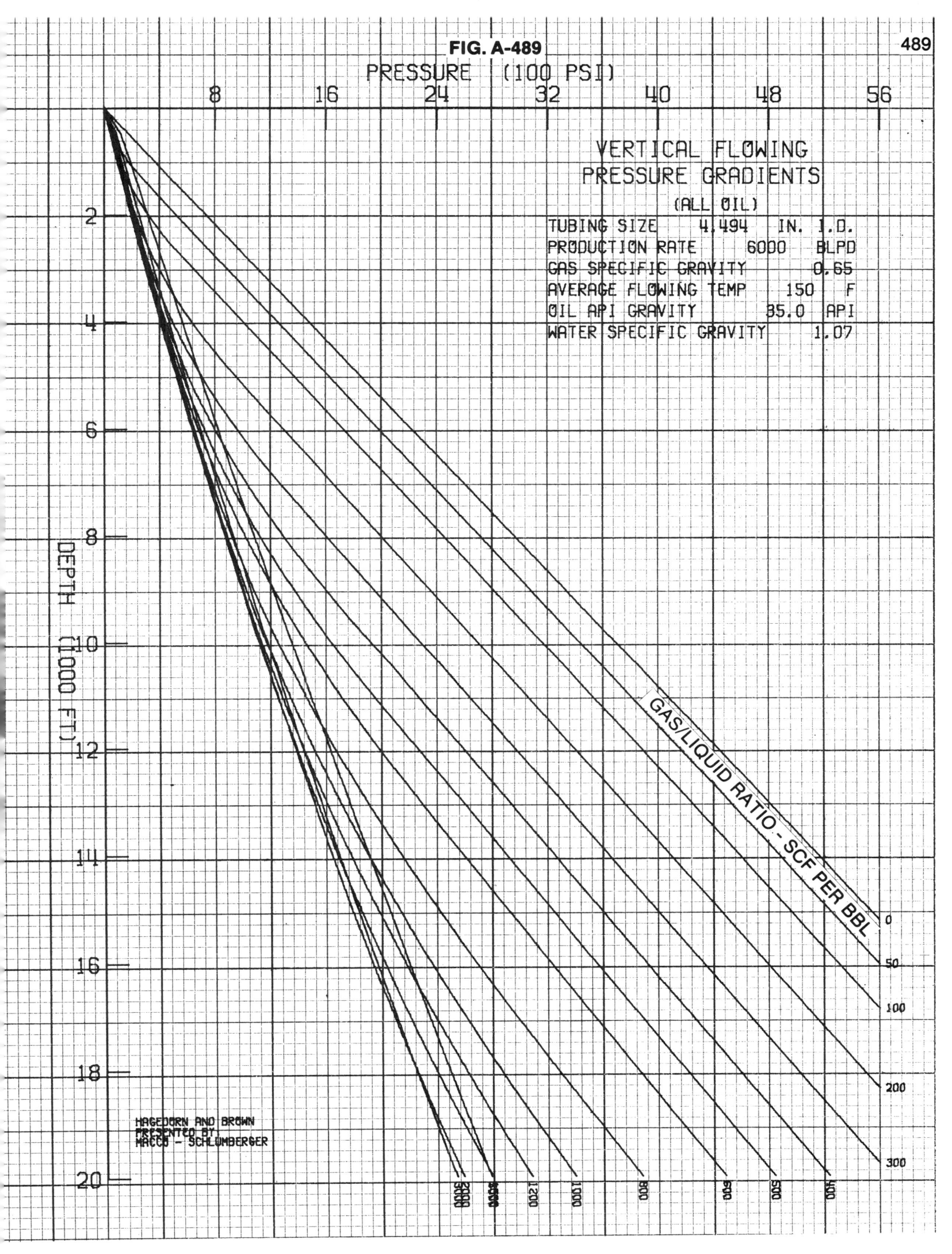

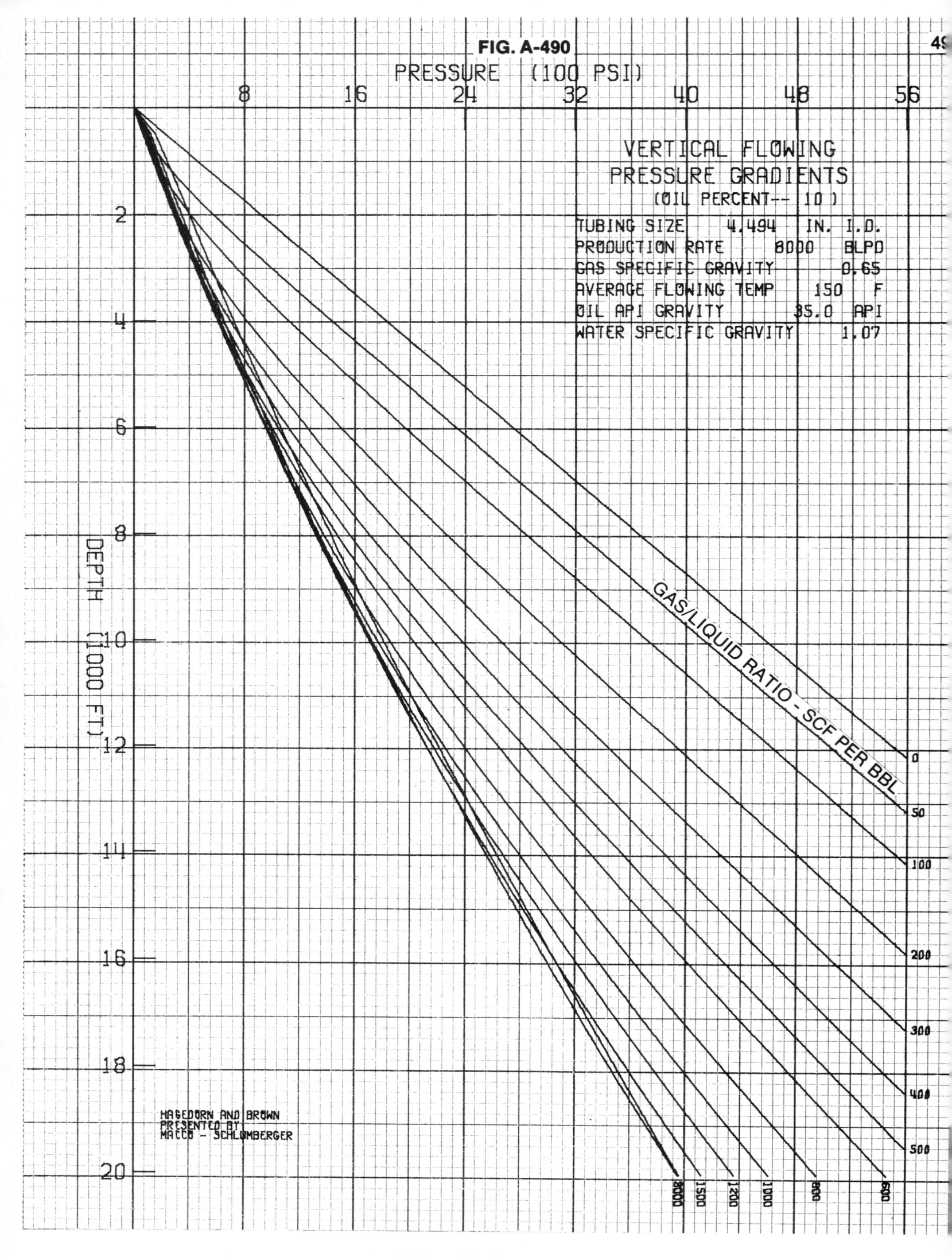

FIG. A-490

VERTICAL FLOWING
PRESSURE GRADIENTS
(OIL PERCENT-- 10)

TUBING SIZE	4.494	IN. I.D.
PRODUCTION RATE	8000	BLPD
GAS SPECIFIC GRAVITY	0.65	
AVERAGE FLOWING TEMP	150	F
OIL API GRAVITY	85.0	API
WATER SPECIFIC GRAVITY	1.07	

PRESSURE (100 PSI)

DEPTH (1000 FT)

GAS/LIQUID RATIO - SCF PER BBL

HAGEDORN AND BROWN
PRESENTED BY
MACCO - SCHLUMBERGER

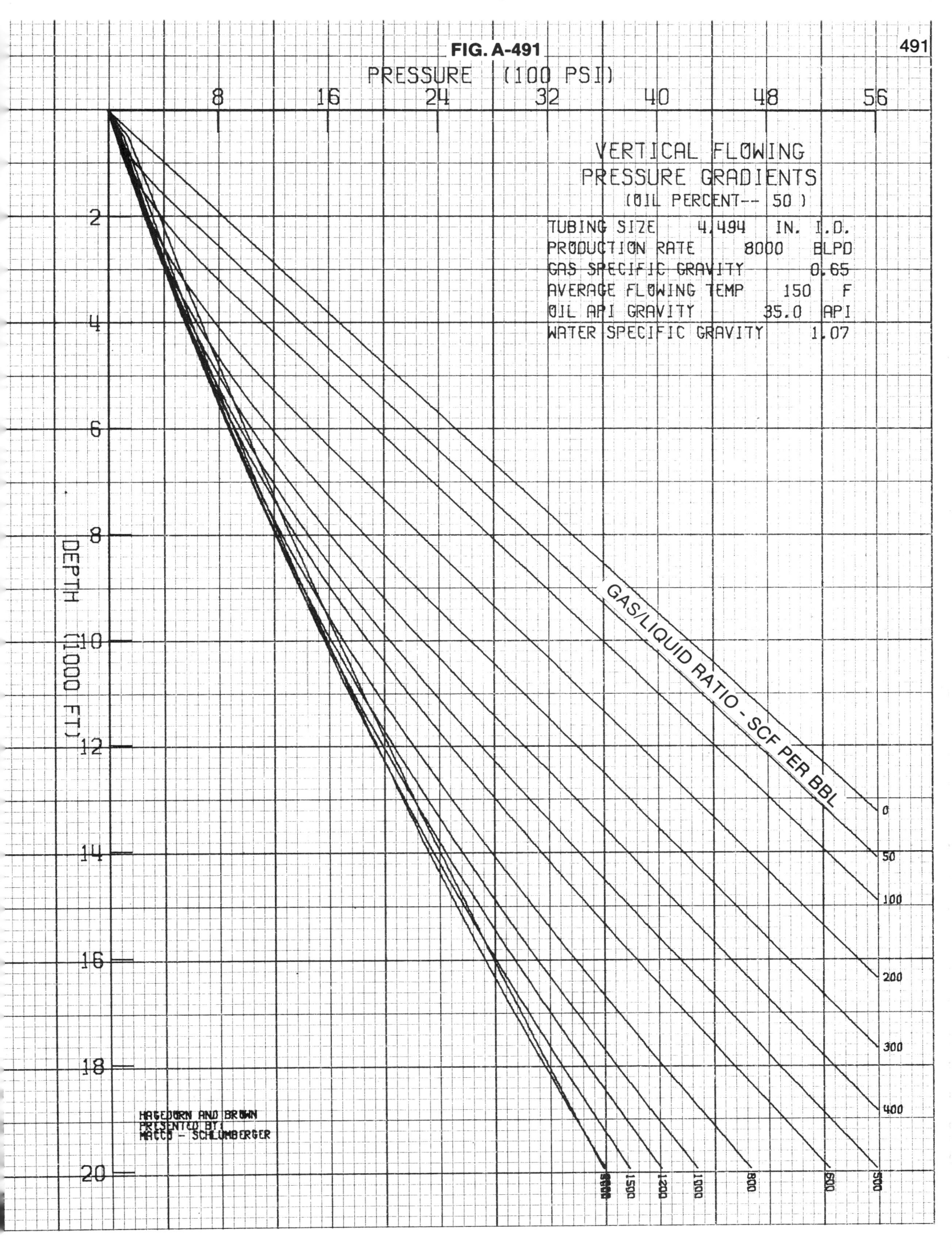

PRESSURE (100 PSI)

VERTICAL FLOWING
PRESSURE GRADIENTS
(OIL PERCENT-- 50)

TUBING SIZE   4.494   IN. I.D.
PRODUCTION RATE   8000   BLPD
GAS SPECIFIC GRAVITY   0.65
AVERAGE FLOWING TEMP 150   F
OIL API GRAVITY   35.0 API
WATER SPECIFIC GRAVITY   1.07

GAS/LIQUID RATIO - SCF PER BBL

DEPTH (1000 FT)

HAGEDORN AND BROWN
PRESENTED BY:
MACCO - SCHLUMBERGER

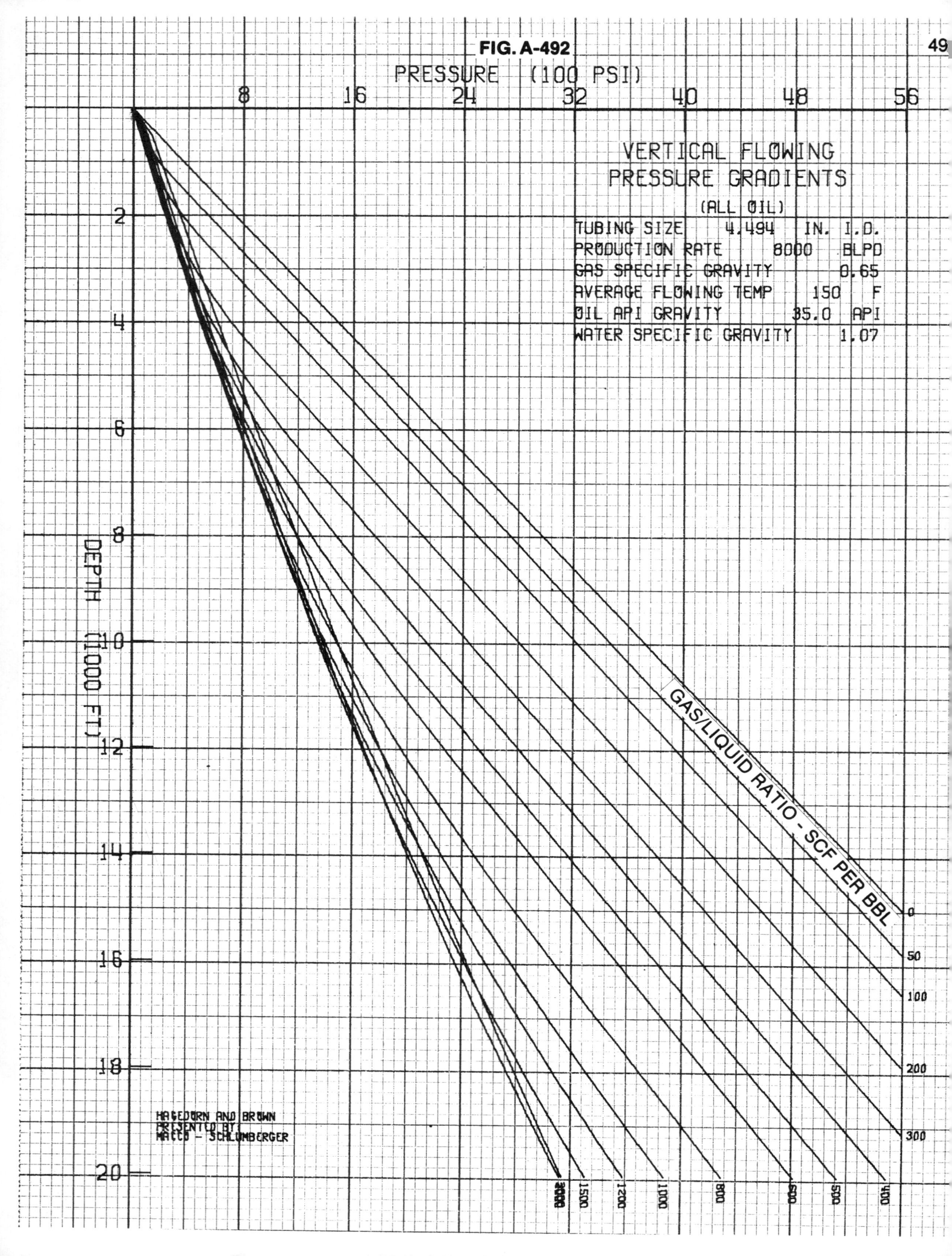

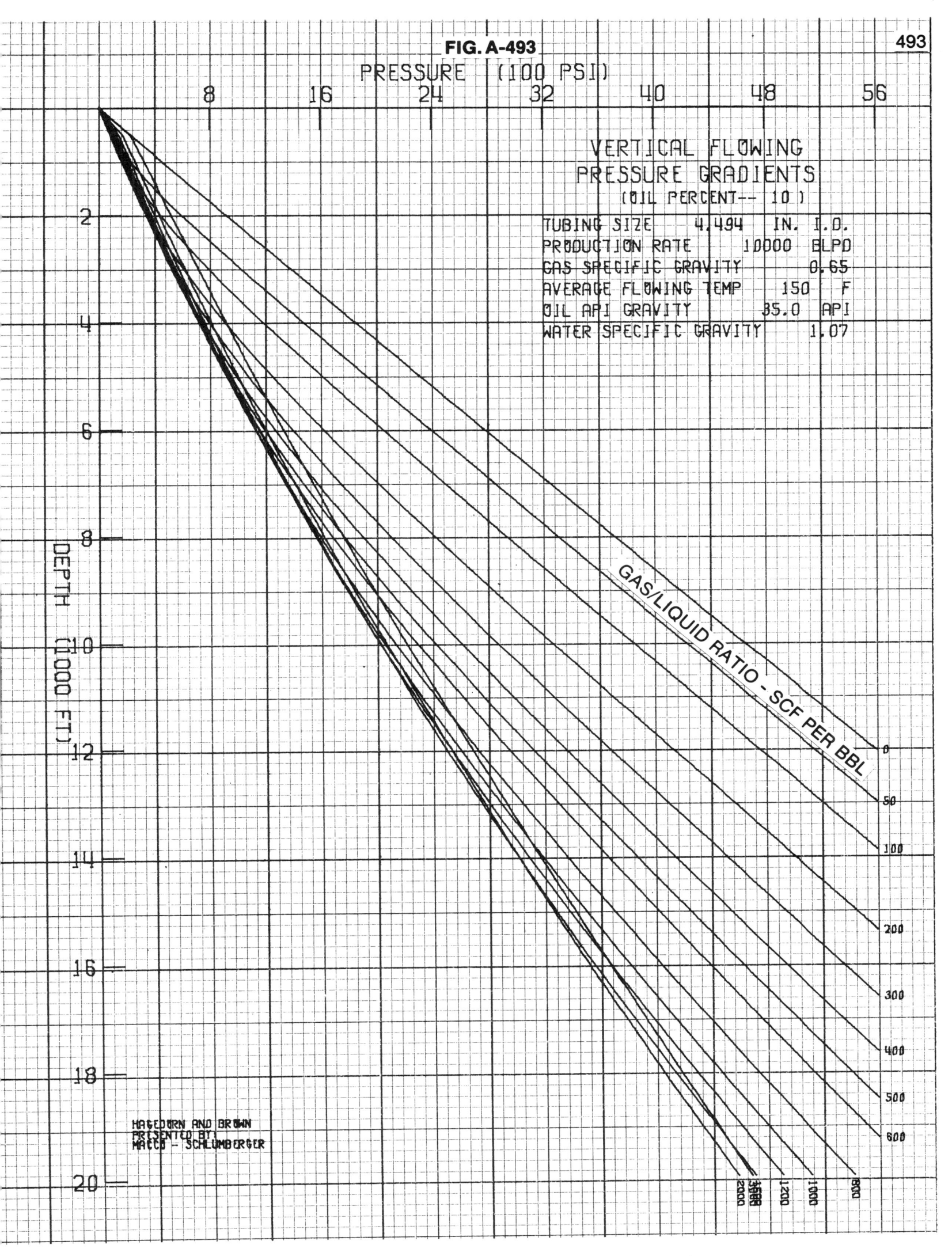

FIG. A-493

493

PRESSURE (100 PSI)

VERTICAL FLOWING
PRESSURE GRADIENTS
(OIL PERCENT-- 10)

TUBING SIZE	4.494	IN. I.D.
PRODUCTION RATE	10000	BLPD
GAS SPECIFIC GRAVITY	0.65	
AVERAGE FLOWING TEMP	150	F
OIL API GRAVITY	35.0	API
WATER SPECIFIC GRAVITY	1.07	

GAS/LIQUID RATIO - SCF PER BBL

DEPTH (1000 FT)

HAGEDORN AND BROWN
PRESENTED BY
MACCO - SCHLUMBERGER

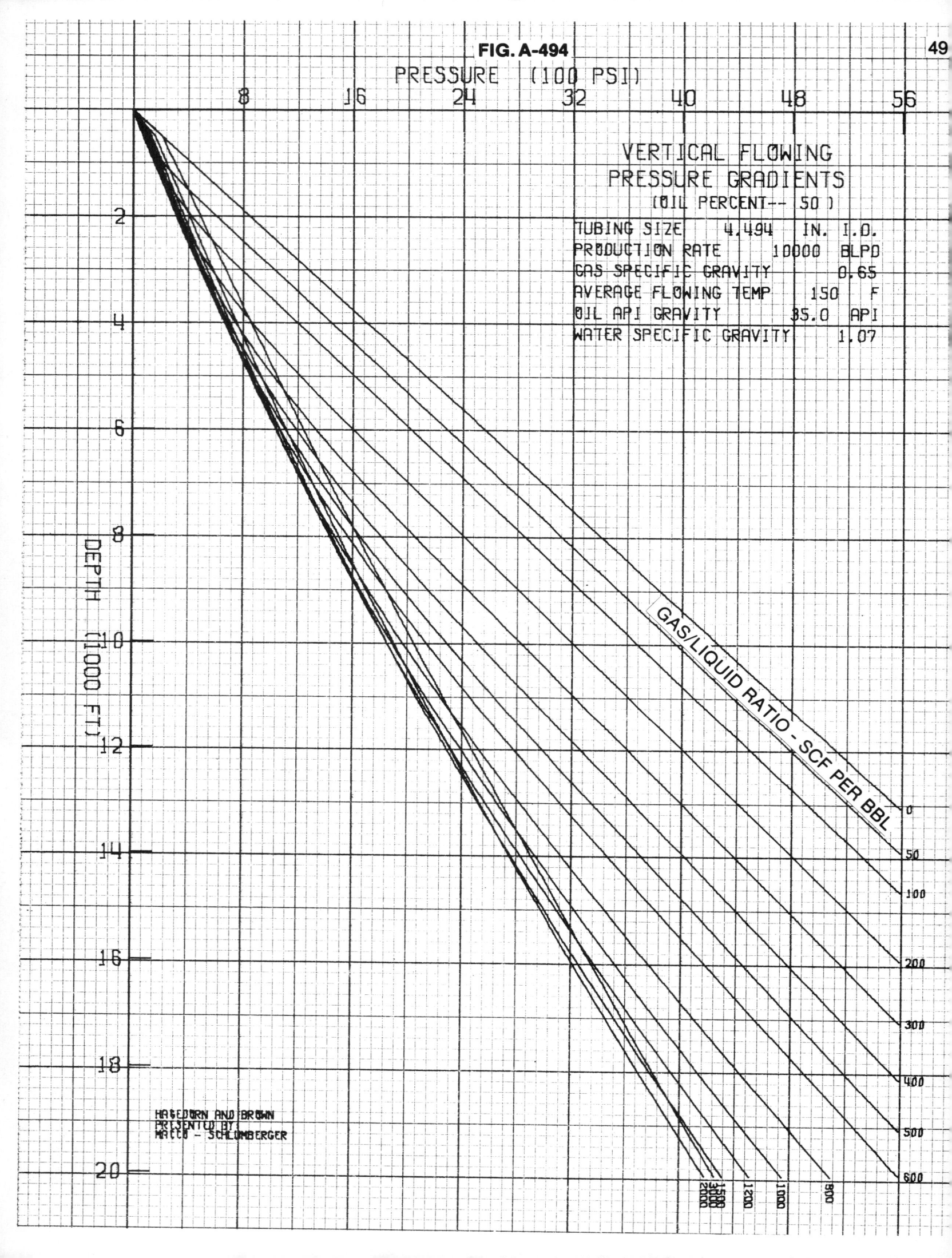

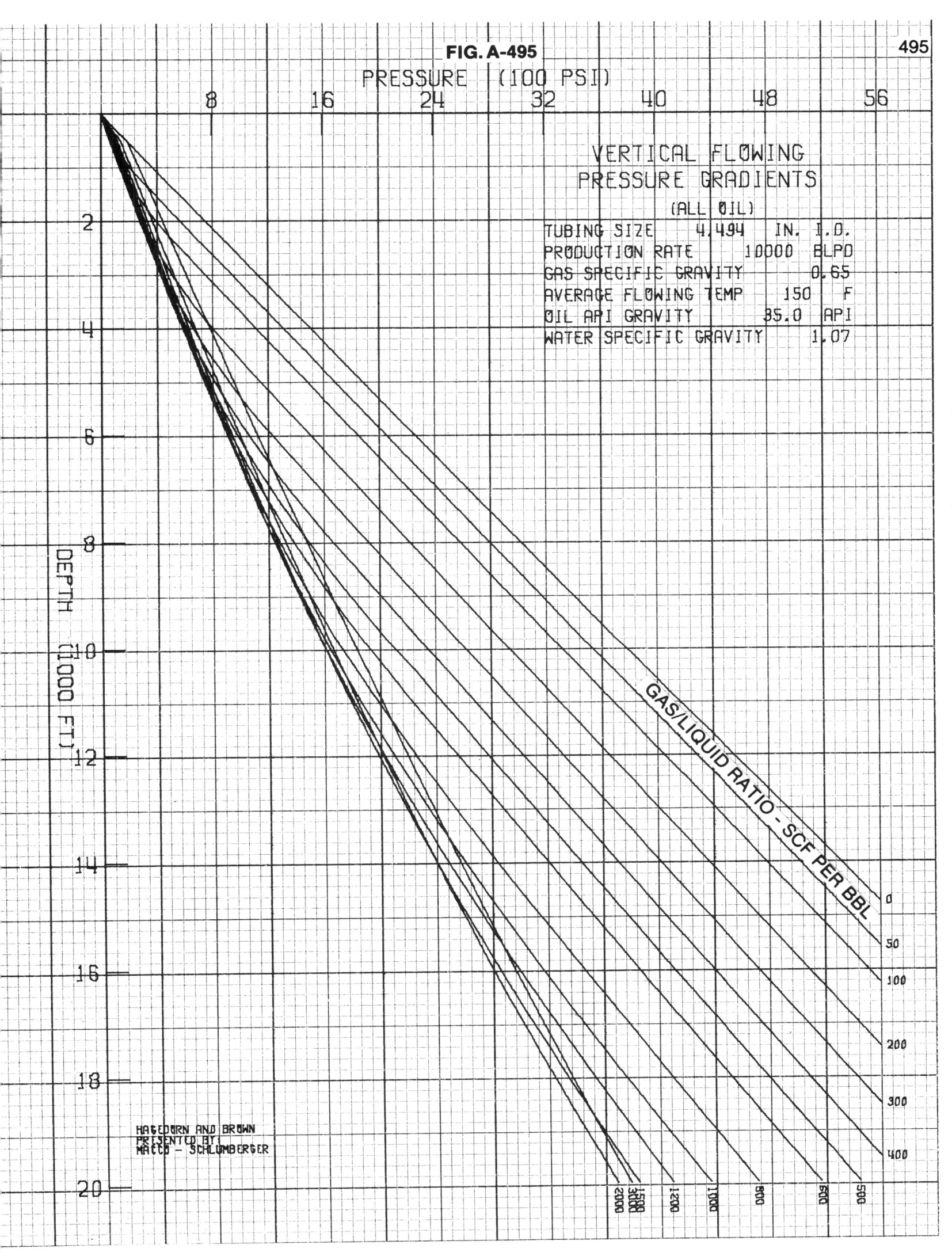

PRESSURE (100 PSI)

VERTICAL FLOWING
PRESSURE GRADIENTS
(ALL OIL)

TUBING SIZE	4.494	IN. I.D.
PRODUCTION RATE	10000	BLPD
GAS SPECIFIC GRAVITY	0.65	
AVERAGE FLOWING TEMP	150	F
OIL API GRAVITY	35.0	API
WATER SPECIFIC GRAVITY	1.07	

DEPTH (1000 FT)

GAS/LIQUID RATIO - SCF PER BBL

0
50
100
200
300
400

2000 1500 1200 1000 800 600 500

HAGEDORN AND BROWN
PRESENTED BY
MAECO - SCHLUMBERGER

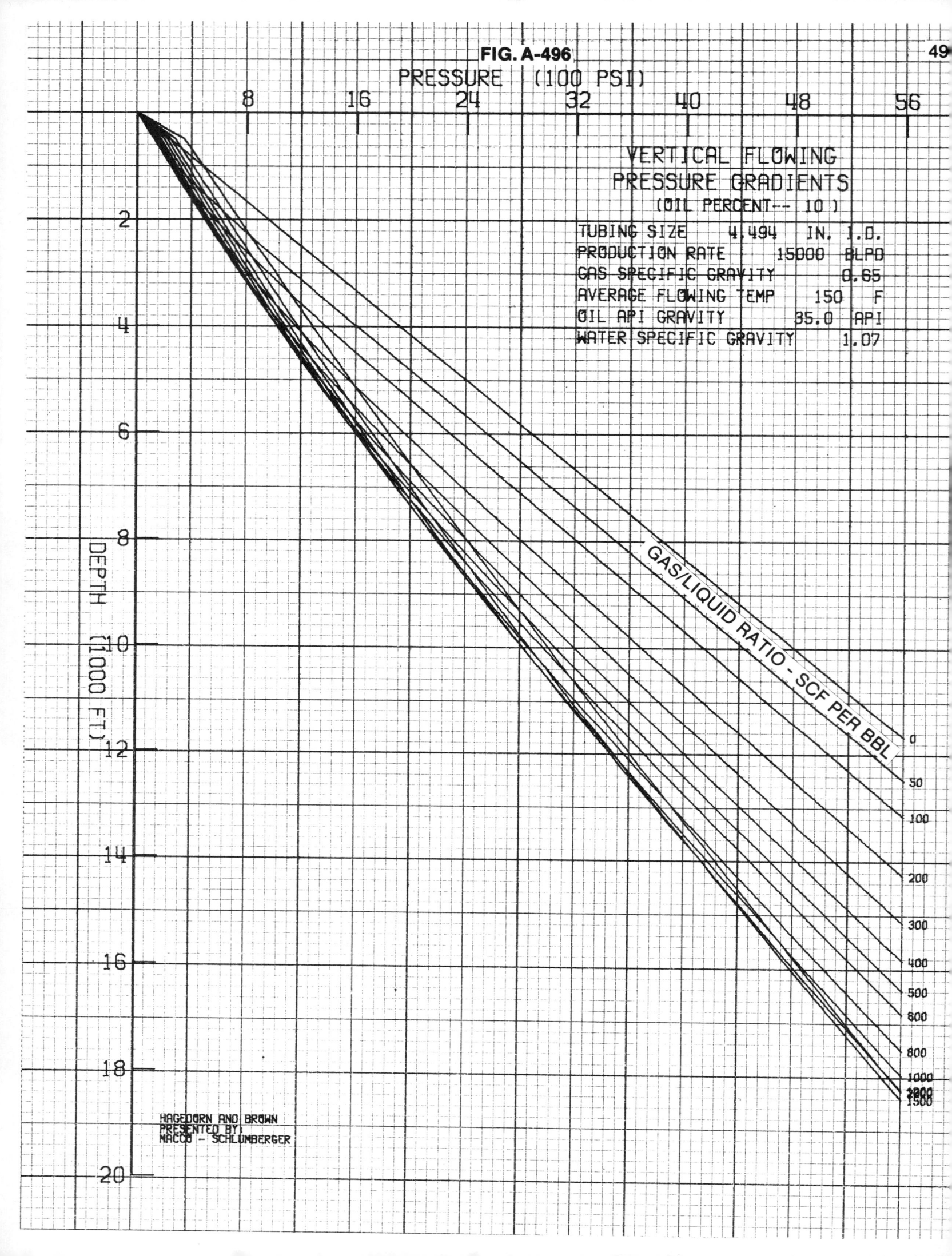

VERTICAL FLOWING
PRESSURE GRADIENTS
(OIL PERCENT-- 10)

TUBING SIZE	4.494	IN. I.D.
PRODUCTION RATE	15000	BLPD
GAS SPECIFIC GRAVITY	0.65	
AVERAGE FLOWING TEMP	150	F
OIL API GRAVITY	35.0	API
WATER SPECIFIC GRAVITY	1.07	

PRESSURE (100 PSI)

DEPTH (1000 FT)

GAS/LIQUID RATIO - SCF PER BBL

HAGEDORN AND BROWN
PRESENTED BY:
NACOO - SCHLUMBERGER

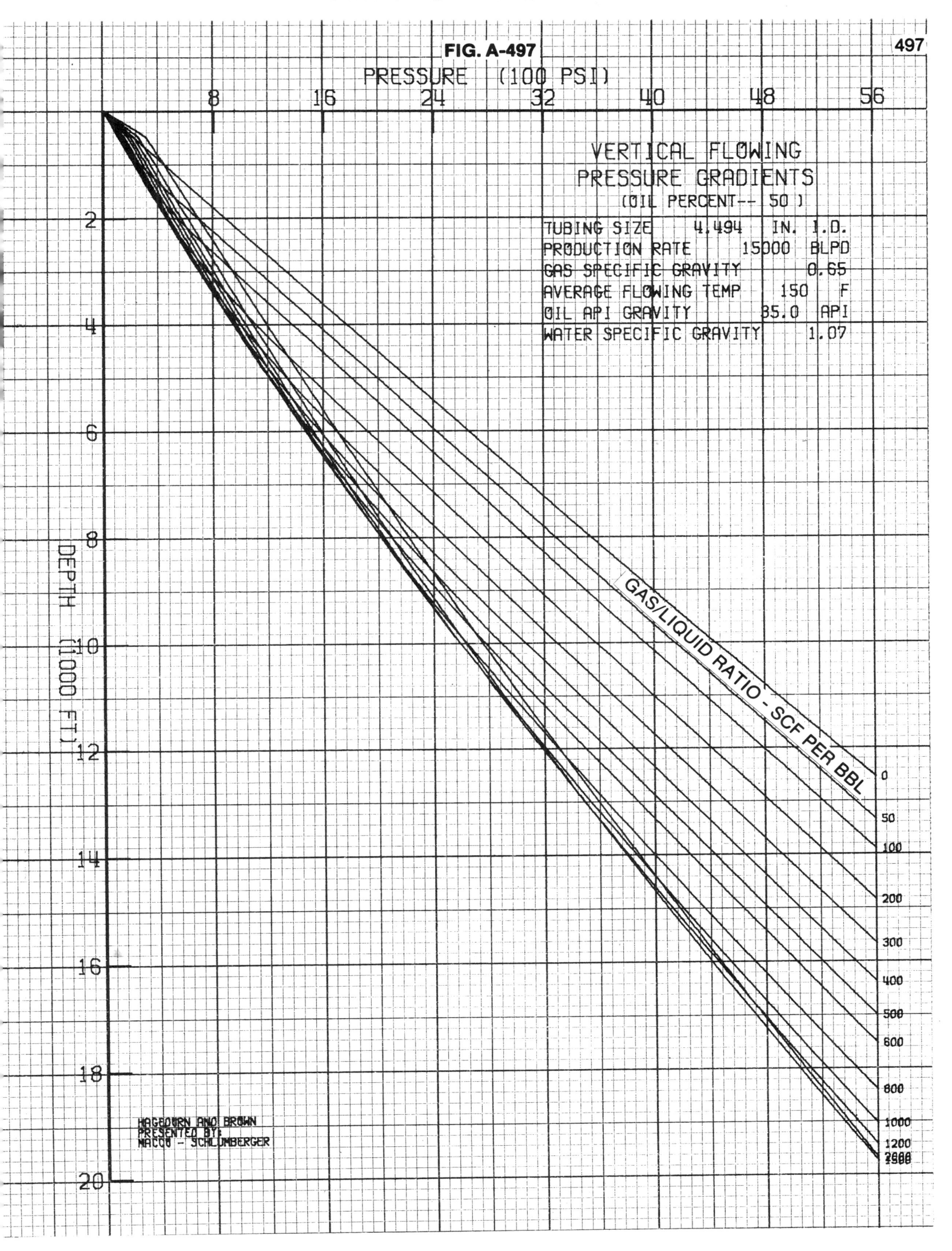

PRESSURE (100 PSI)

DEPTH (1000 FT)

VERTICAL FLOWING
PRESSURE GRADIENTS
(OIL PERCENT-- 50)

TUBING SIZE	4.494	IN. I.D.
PRODUCTION RATE	15000	BLPD
GAS SPECIFIC GRAVITY	0.65	
AVERAGE FLOWING TEMP	150	F
OIL API GRAVITY	35.0	API
WATER SPECIFIC GRAVITY	1.07	

GAS/LIQUID RATIO - SCF PER BBL

HAGEDORN AND BROWN
PRESENTED BY:
MACCO - SCHLUMBERGER

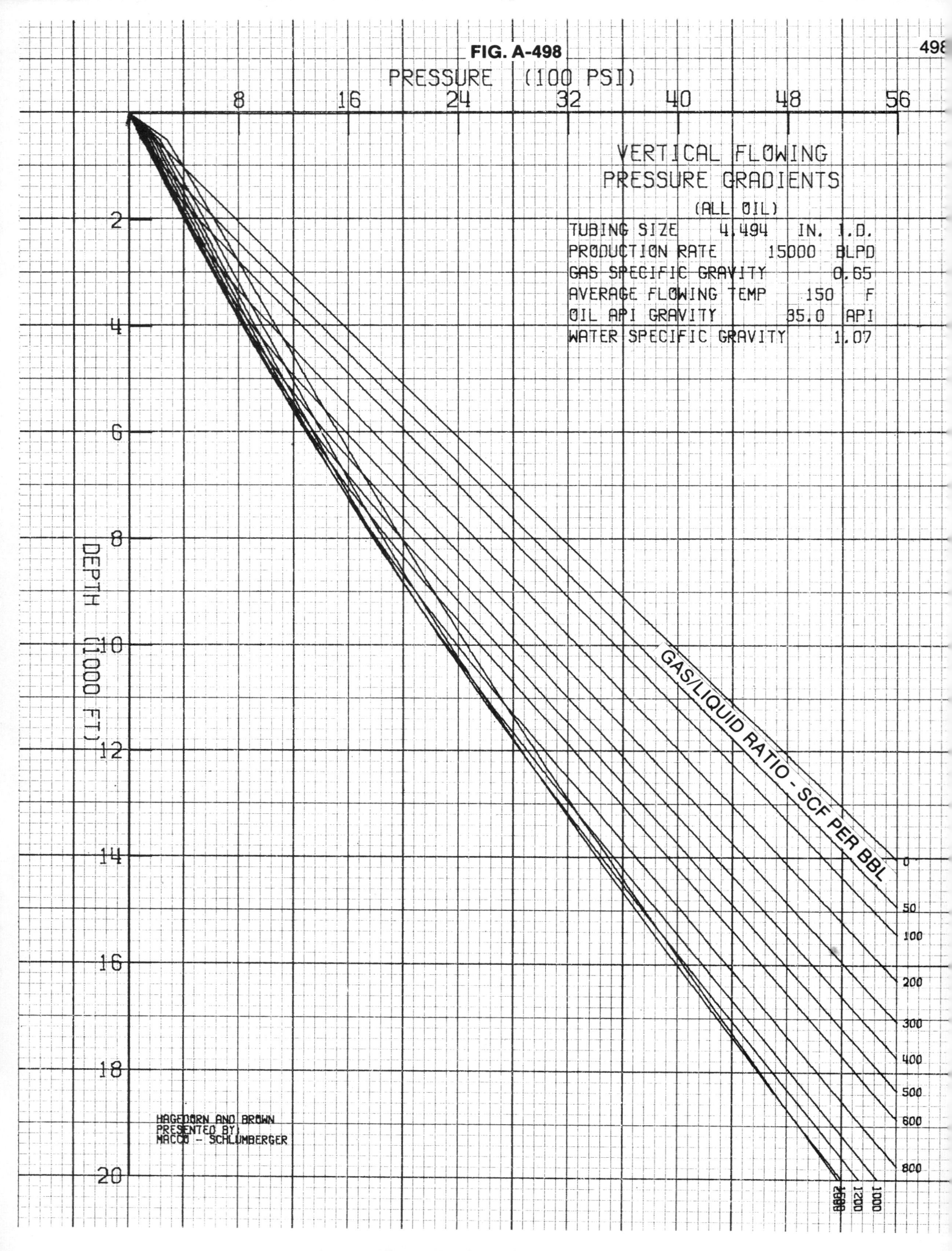

FIG. A-498

498

PRESSURE (100 PSI)

VERTICAL FLOWING
PRESSURE GRADIENTS
(ALL OIL)

TUBING SIZE	4.494	IN. I.D.
PRODUCTION RATE	15000	BLPD
GAS SPECIFIC GRAVITY	0.65	
AVERAGE FLOWING TEMP	150	F
OIL API GRAVITY	35.0	API
WATER SPECIFIC GRAVITY	1.07	

DEPTH (1000 FT)

GAS/LIQUID RATIO - SCF PER BBL

HAGEDORN AND BROWN
PRESENTED BY:
MACCO - SCHLUMBERGER

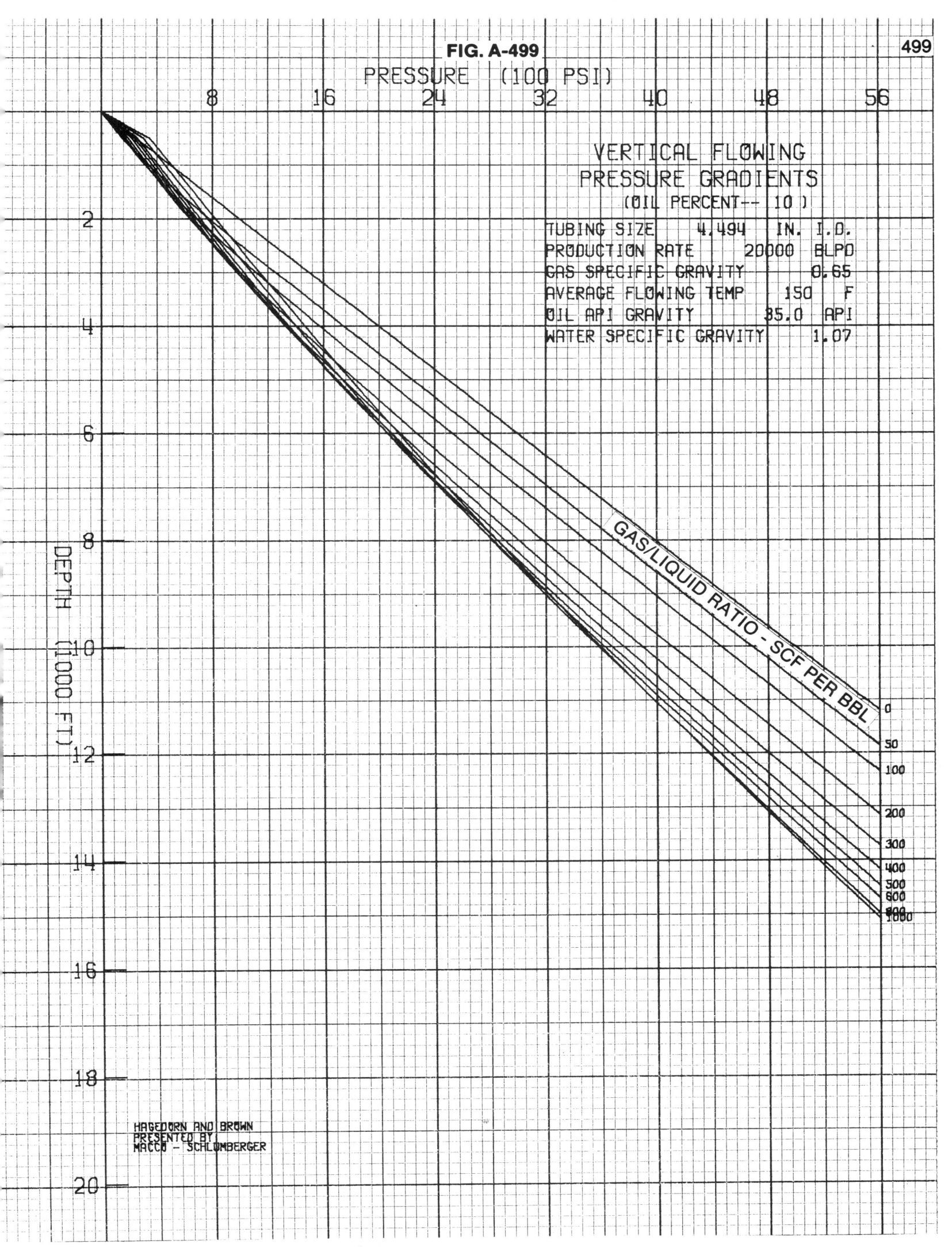

VERTICAL FLOWING
PRESSURE GRADIENTS
(OIL PERCENT-- 10)

TUBING SIZE	4.494	IN. I.D.
PRODUCTION RATE	20000	BLPD
GAS SPECIFIC GRAVITY		0.65
AVERAGE FLOWING TEMP	150	F
OIL API GRAVITY	35.0	API
WATER SPECIFIC GRAVITY		1.07

PRESSURE (100 PSI)

DEPTH (1000 FT)

GAS/LIQUID RATIO - SCF PER BBL

0
50
100
200
300
400
500
600
800
1000

HAGEDORN AND BROWN
PRESENTED BY
MACCO – SCHLUMBERGER

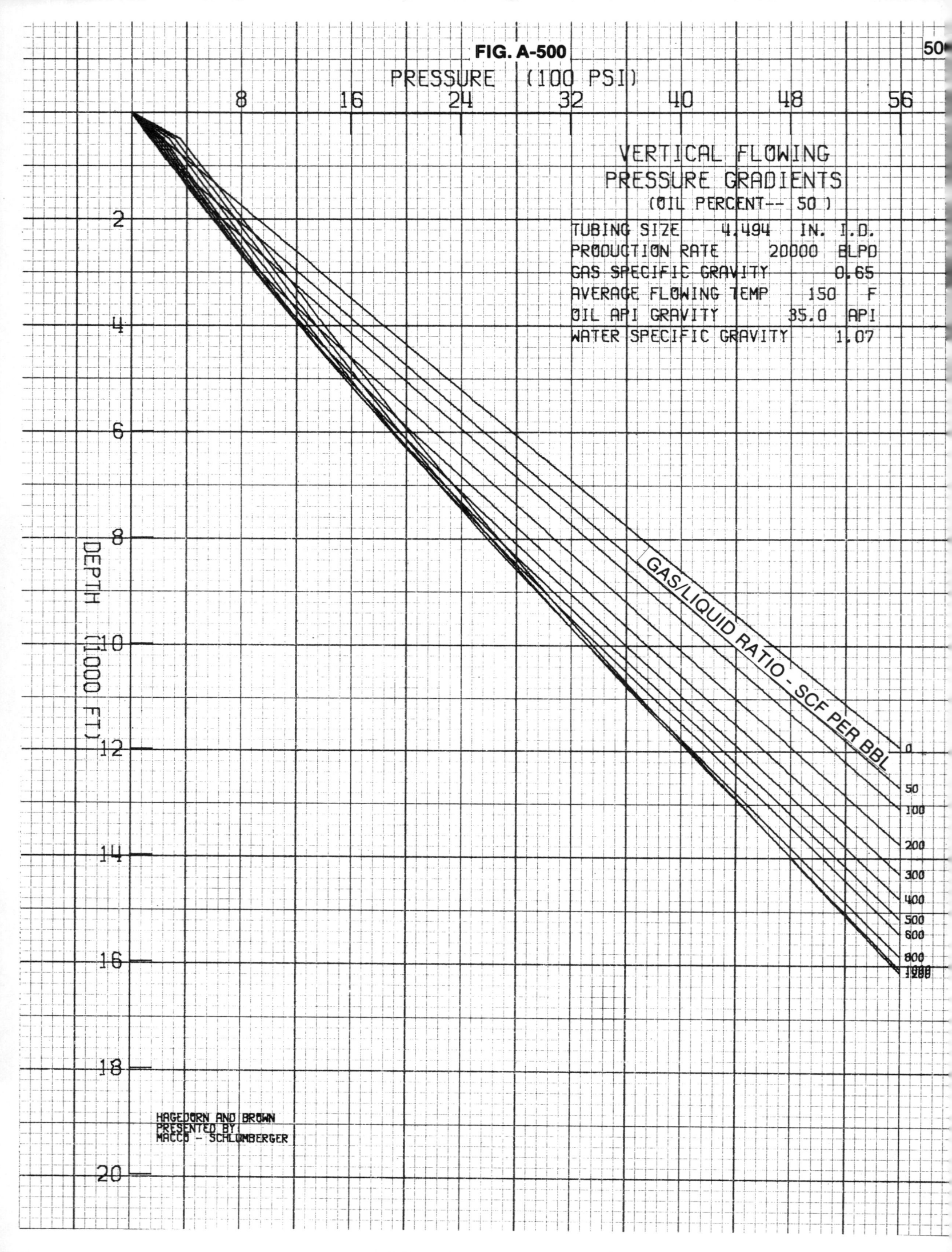

FIG. A-500

VERTICAL FLOWING
PRESSURE GRADIENTS
(OIL PERCENT-- 50)

TUBING SIZE	4.494	IN. I.D.
PRODUCTION RATE	20000	BLPD
GAS SPECIFIC GRAVITY	0.65	
AVERAGE FLOWING TEMP	150	F
OIL API GRAVITY	35.0	API
WATER SPECIFIC GRAVITY	1.07	

PRESSURE (100 PSI)

DEPTH (1000 FT)

GAS/LIQUID RATIO - SCF PER BBL

HAGEDORN AND BROWN
PRESENTED BY
MACCO - SCHLUMBERGER

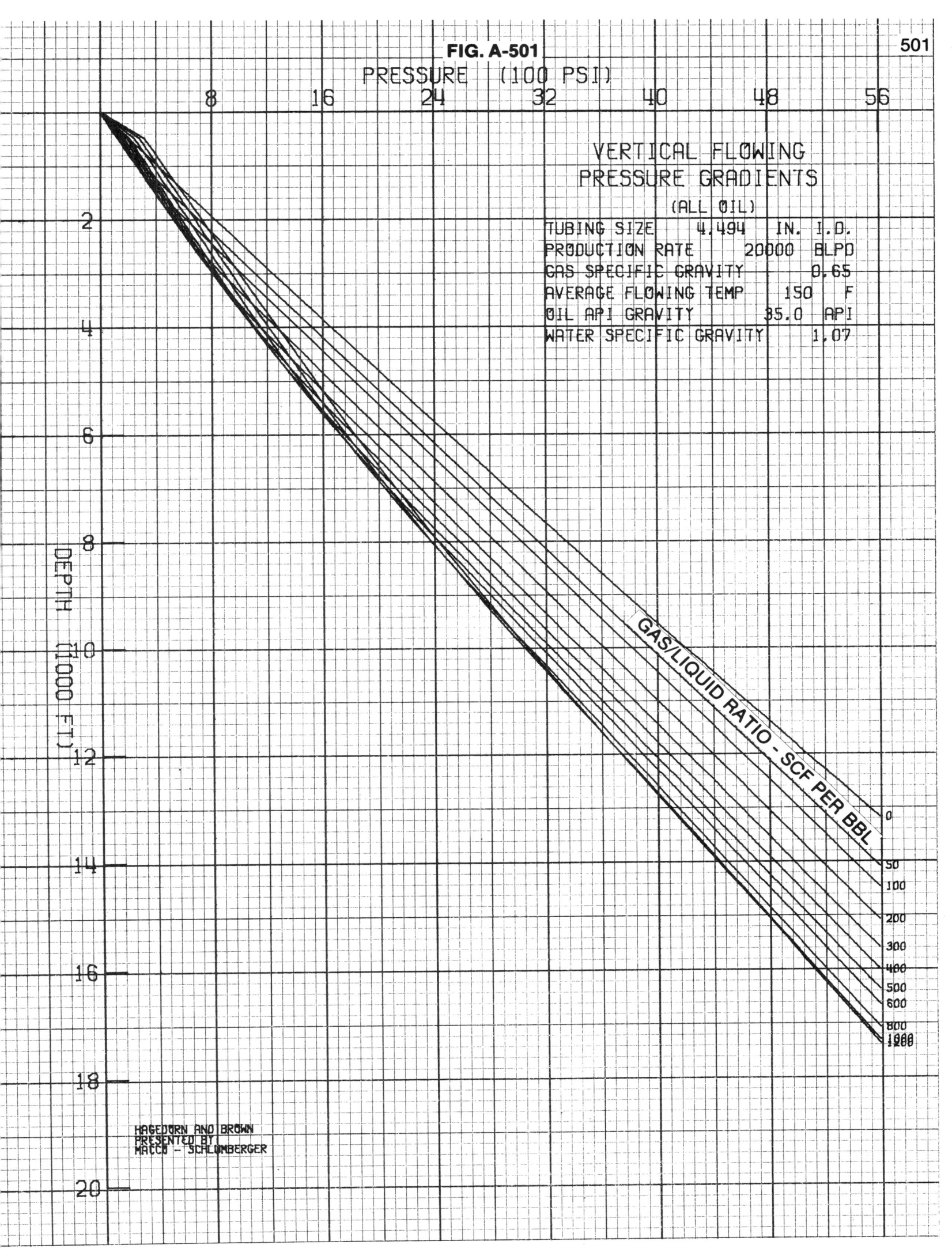

FIG. A-501

501

PRESSURE (100 PSI)

VERTICAL FLOWING
PRESSURE GRADIENTS
(ALL OIL)

TUBING SIZE	4.494	IN. I.D.
PRODUCTION RATE	20000	BLPD
GAS SPECIFIC GRAVITY		0.65
AVERAGE FLOWING TEMP	150	F
OIL API GRAVITY	35.0	API
WATER SPECIFIC GRAVITY		1.07

DEPTH (1000 FT)

GAS/LIQUID RATIO - SCF PER BBL

0
50
100
200
300
400
500
600
800
1000

HAGEDORN AND BROWN
PRESENTED BY
MACCO - SCHLUMBERGER

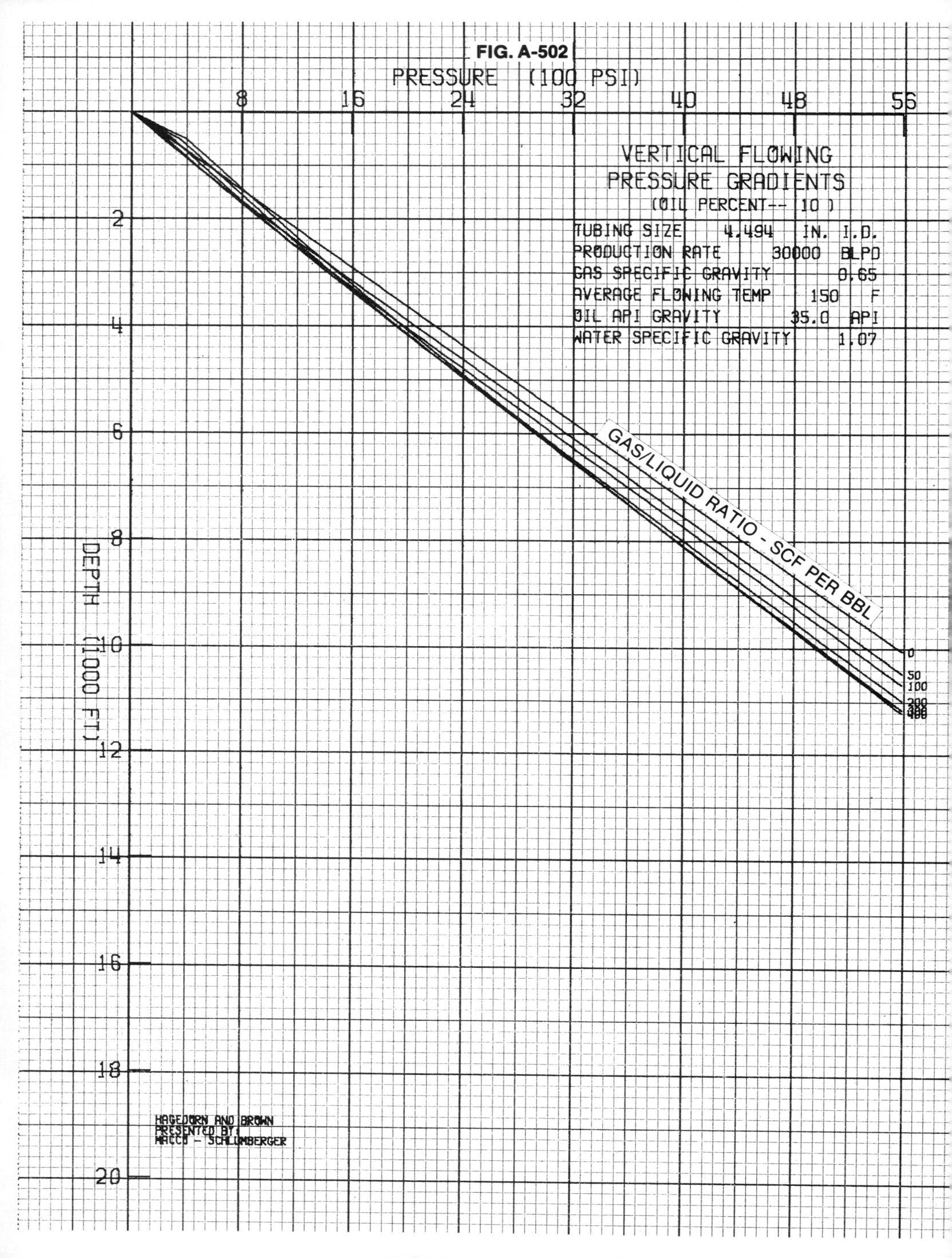

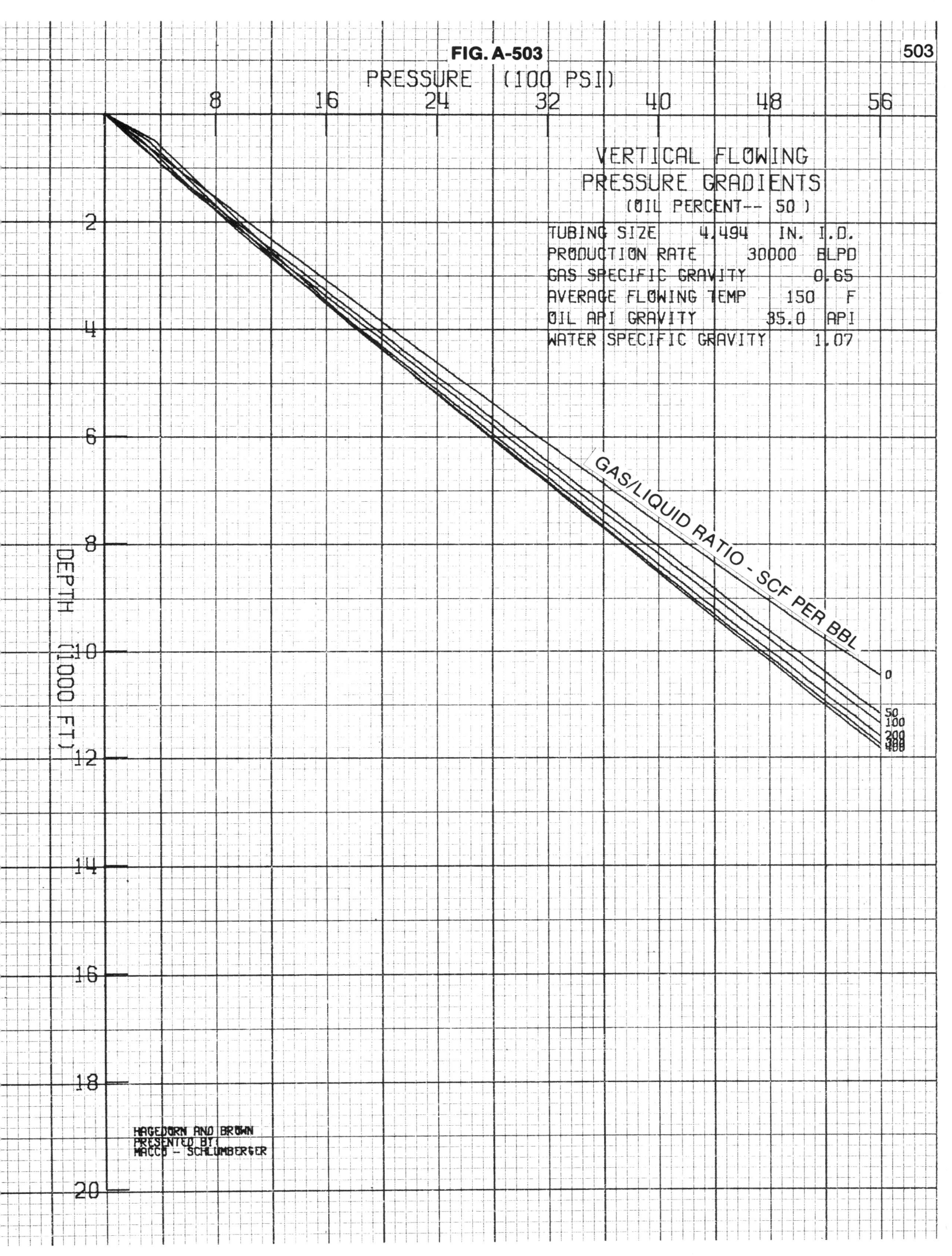

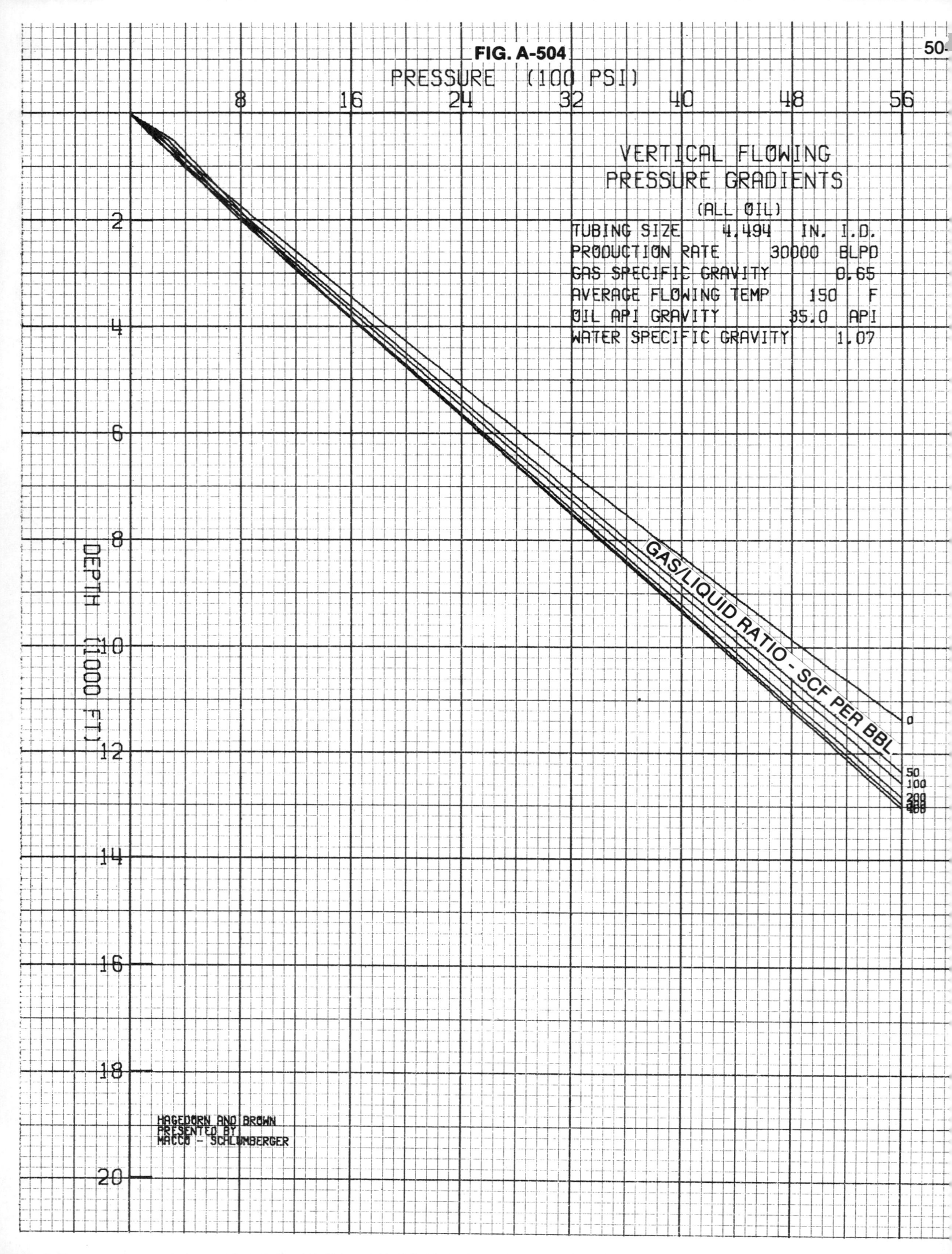

VERTICAL FLOWING PRESSURE GRADIENTS

(ALL OIL)

TUBING SIZE	4.494	IN. I.D.
PRODUCTION RATE	30000	BLPD
GAS SPECIFIC GRAVITY	0.65	
AVERAGE FLOWING TEMP	150	F
OIL API GRAVITY	35.0	API
WATER SPECIFIC GRAVITY	1.07	

PRESSURE (100 PSI)

DEPTH (1000 FT)

GAS/LIQUID RATIO - SCF PER BBL

0
50
100
200
400

HAGEDORN AND BROWN
PRESENTED BY
MACCO - SCHLUMBERGER

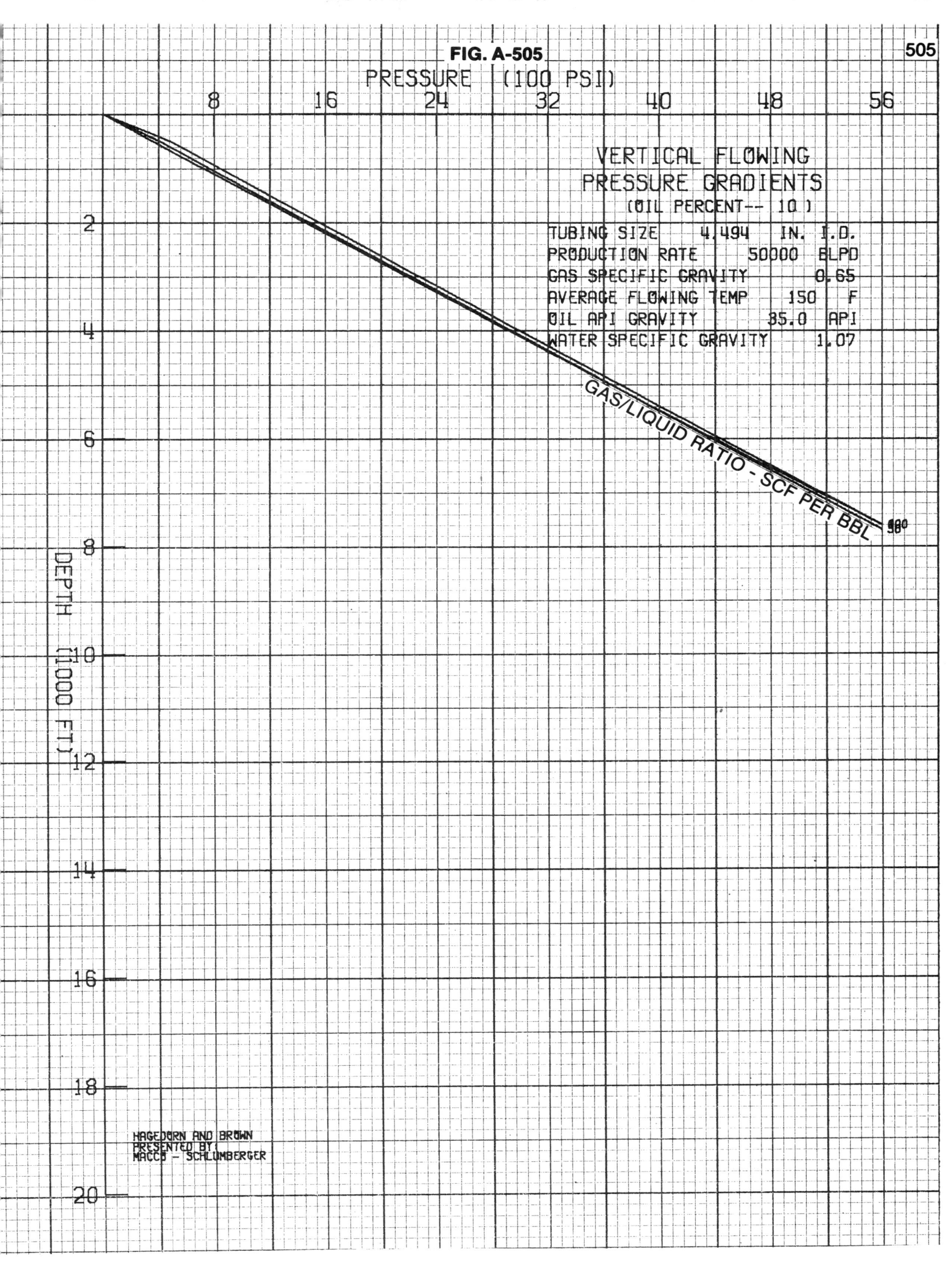

VERTICAL FLOWING
PRESSURE GRADIENTS
(OIL PERCENT-- 10)

TUBING SIZE	4.494	IN. I.D.
PRODUCTION RATE	50000	BLPD
GAS SPECIFIC GRAVITY		0.65
AVERAGE FLOWING TEMP	150	F
OIL API GRAVITY	35.0	API
WATER SPECIFIC GRAVITY		1.07

PRESSURE (100 PSI)

DEPTH (1000 FT)

GAS/LIQUID RATIO - SCF PER BBL

HAGEDORN AND BROWN
PRESENTED BY
MACCO - SCHLUMBERGER

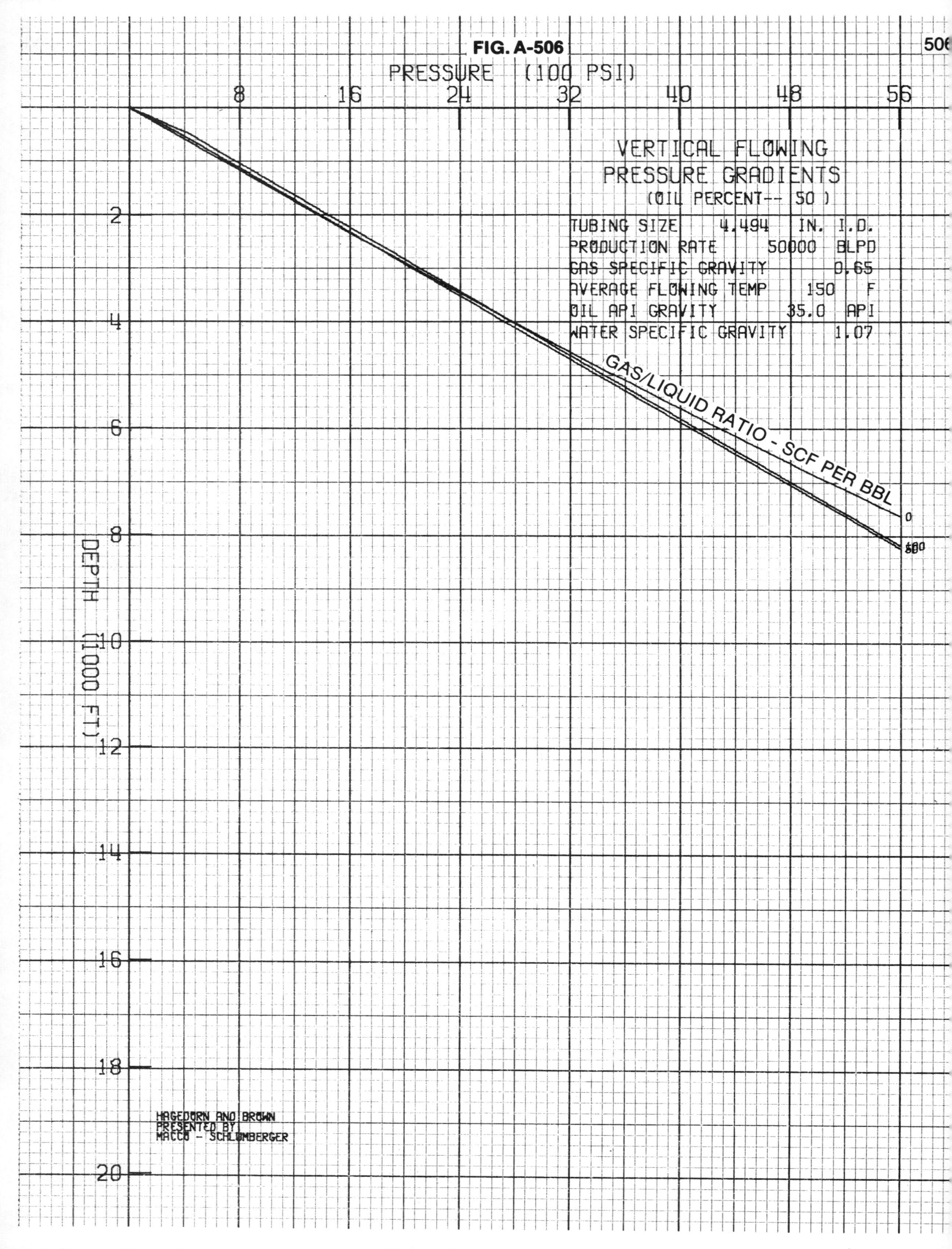

FIG. A-506

VERTICAL FLOWING
PRESSURE GRADIENTS
(OIL PERCENT-- 50)

TUBING SIZE	4.494	IN. I.D.
PRODUCTION RATE	50000	BLPD
GAS SPECIFIC GRAVITY	0.65	
AVERAGE FLOWING TEMP	150	F
OIL API GRAVITY	35.0	API
WATER SPECIFIC GRAVITY	1.07	

GAS/LIQUID RATIO - SCF PER BBL

HAGEDORN AND BROWN
PRESENTED BY
MACCO - SCHLUMBERGER

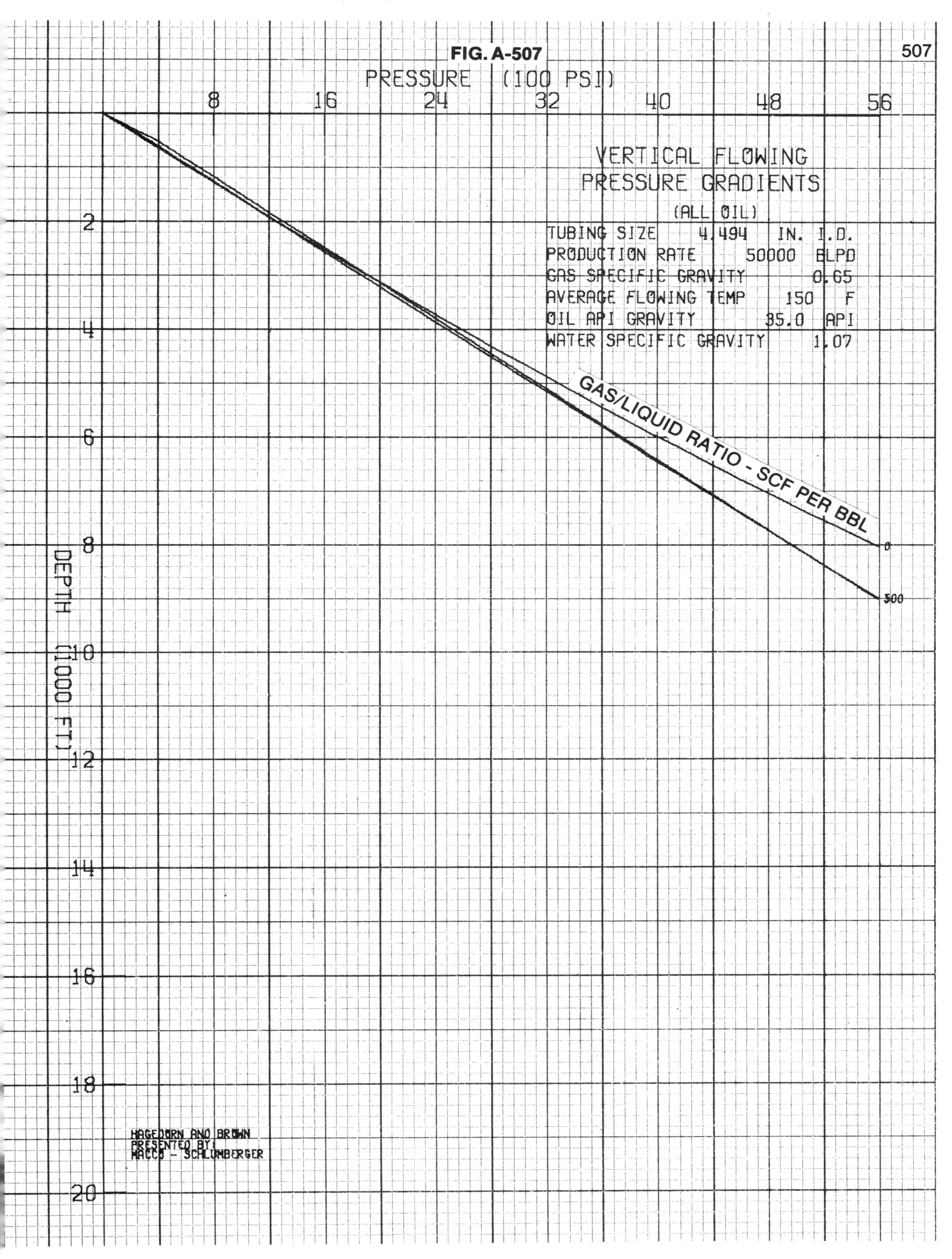

FIG. A-507

507

PRESSURE (100 PSI)

VERTICAL FLOWING
PRESSURE GRADIENTS
(ALL OIL)

TUBING SIZE	4.494	IN. I.D.
PRODUCTION RATE	50000	BLPD
GAS SPECIFIC GRAVITY	0.65	
AVERAGE FLOWING TEMP	150	F
OIL API GRAVITY	35.0	API
WATER SPECIFIC GRAVITY	1.07	

GAS/LIQUID RATIO - SCF PER BBL

DEPTH (1000 FT)

HAGEDORN AND BROWN
PRESENTED BY:
MACCO - SCHLUMBERGER

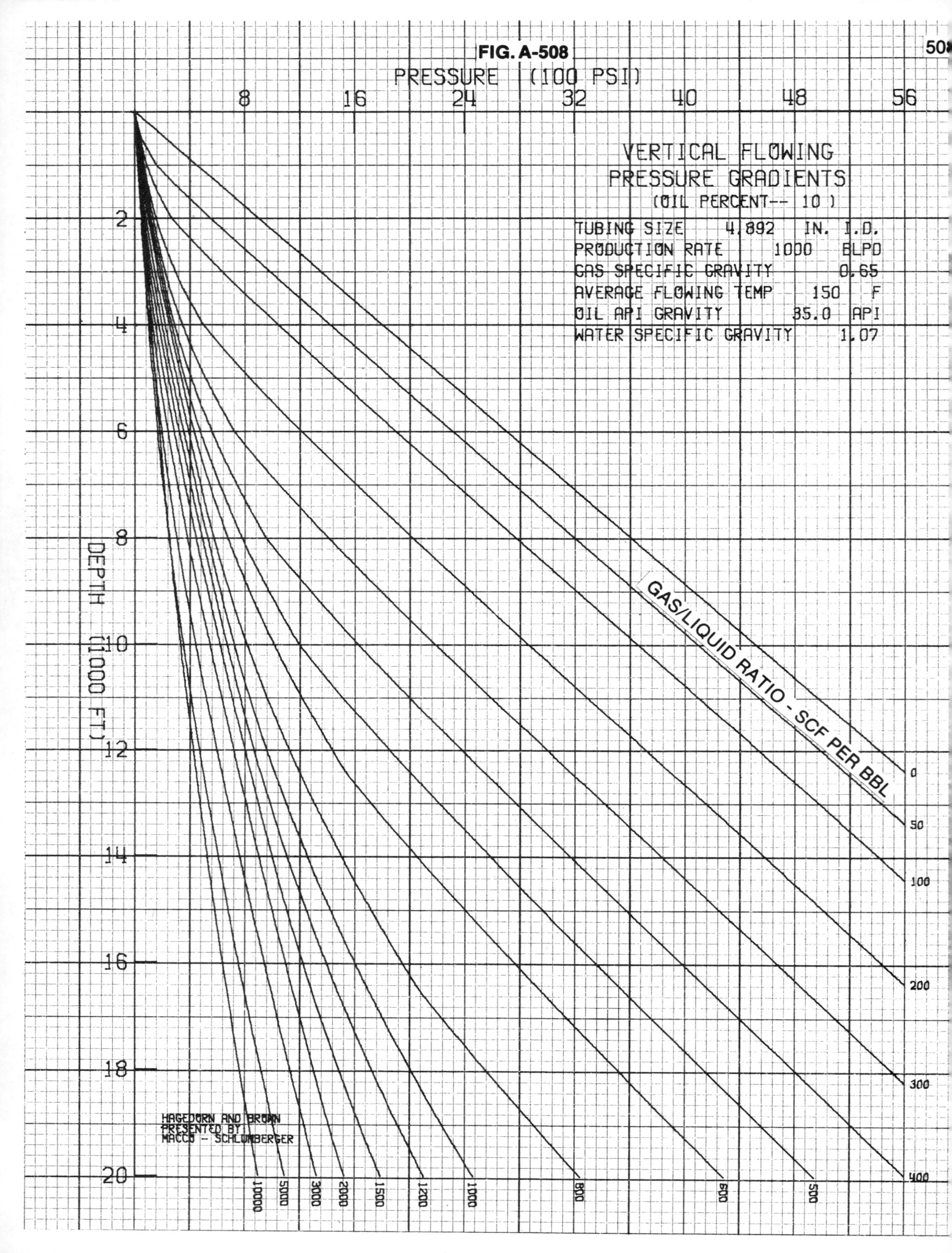

PRESSURE (100 PSI)

VERTICAL FLOWING
PRESSURE GRADIENTS
(OIL PERCENT-- 10)

TUBING SIZE   4.892   IN. I.D.
PRODUCTION RATE   1000   BLPD
GAS SPECIFIC GRAVITY   0.65
AVERAGE FLOWING TEMP   150   F
OIL API GRAVITY   35.0 API
WATER SPECIFIC GRAVITY   1.07

GAS/LIQUID RATIO - SCF PER BBL

DEPTH (1000 FT)

HAGEDORN AND BROWN
PRESENTED BY
MACCO - SCHLUMBERGER

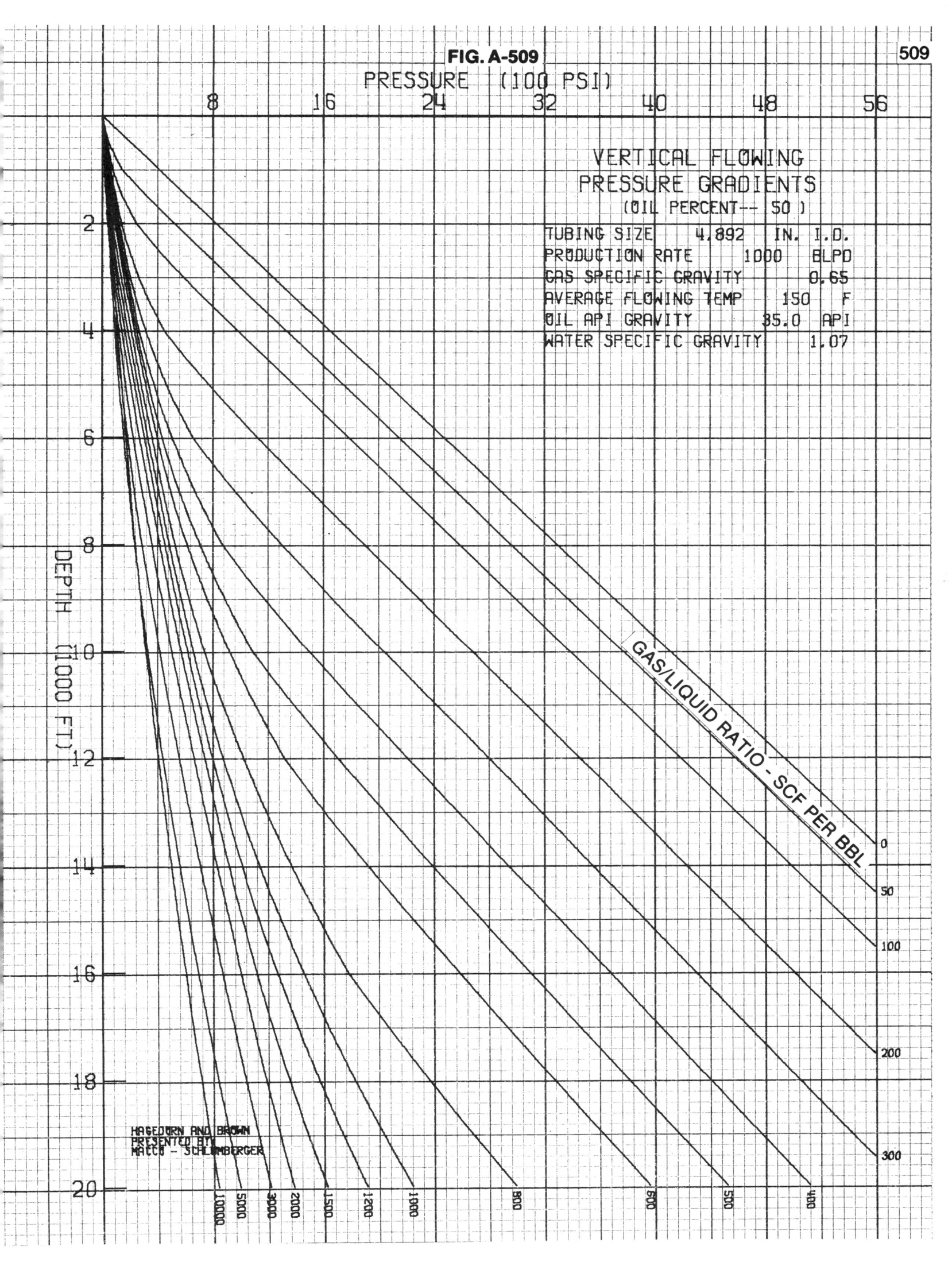

FIG. A-509

509

PRESSURE (100 PSI)

VERTICAL FLOWING
PRESSURE GRADIENTS
(OIL PERCENT -- 50)

TUBING SIZE	4.892	IN. I.D.
PRODUCTION RATE	1000	BLPD
GAS SPECIFIC GRAVITY	0.65	
AVERAGE FLOWING TEMP	150	F
OIL API GRAVITY	35.0	API
WATER SPECIFIC GRAVITY	1.07	

DEPTH (1000 FT)

GAS/LIQUID RATIO - SCF PER BBL

HAGEDORN AND BROWN
PRESENTED BY:
MAECO - SCHLUMBERGER

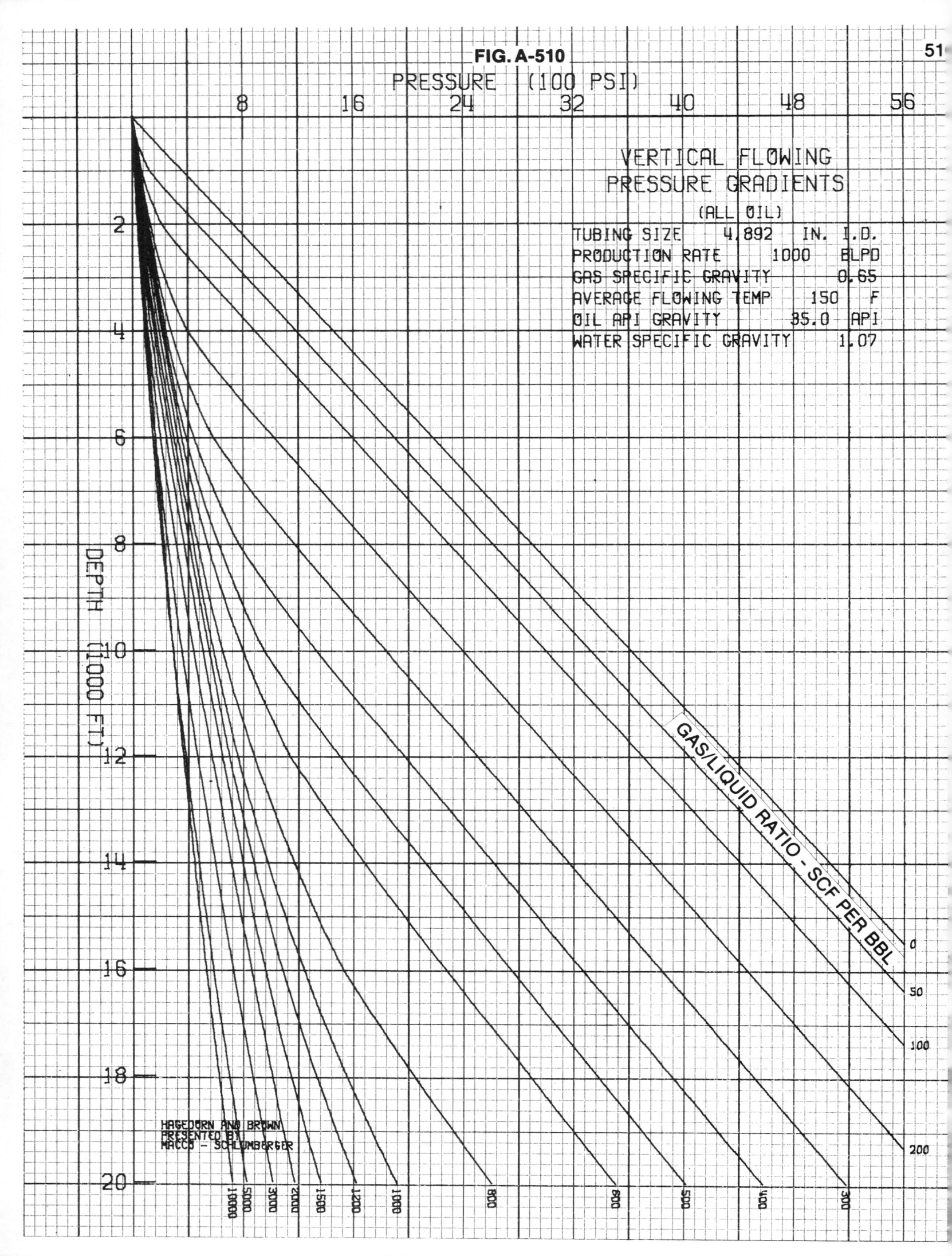

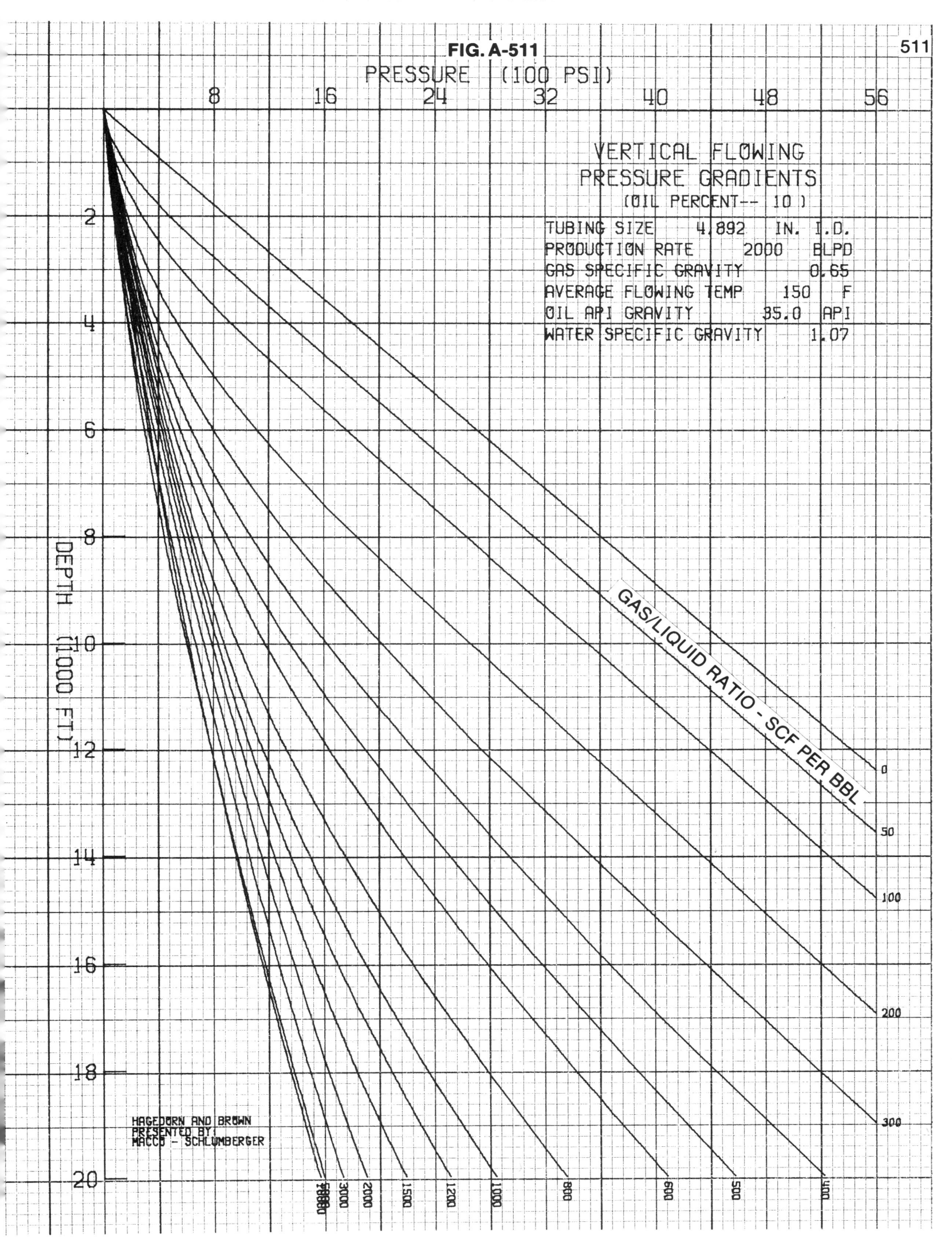

FIG. A-511

511

PRESSURE (100 PSI)

VERTICAL FLOWING
PRESSURE GRADIENTS
(OIL PERCENT-- 10)

TUBING SIZE	4.892	IN. I.D.
PRODUCTION RATE	2000	BLPD
GAS SPECIFIC GRAVITY		0.65
AVERAGE FLOWING TEMP	150	F
OIL API GRAVITY	35.0	API
WATER SPECIFIC GRAVITY		1.07

DEPTH (1000 FT)

GAS/LIQUID RATIO - SCF PER BBL

HAGEDORN AND BROWN
PRESENTED BY
MACCO - SCHLUMBERGER

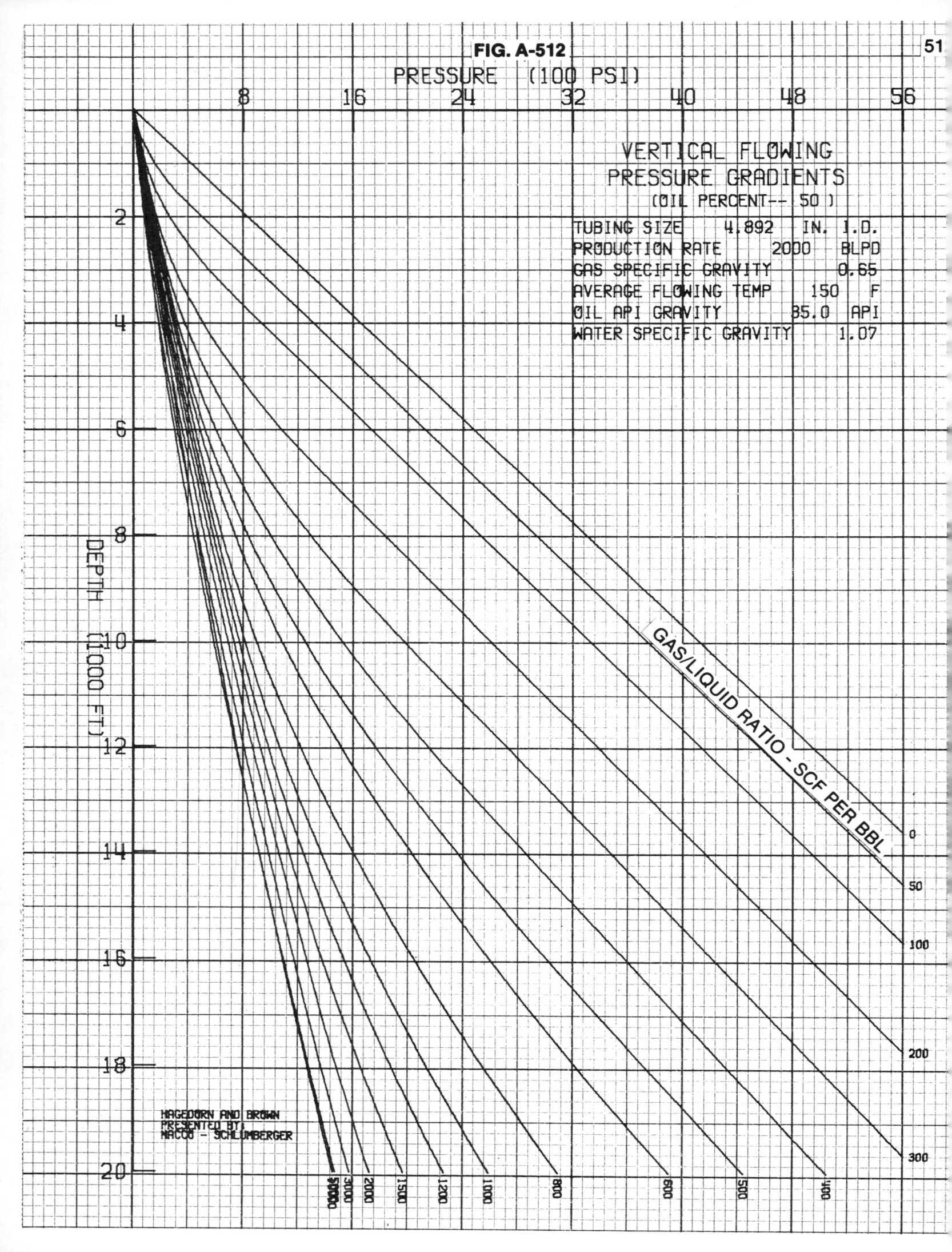

FIG. A-512

51

PRESSURE (100 PSI)

VERTICAL FLOWING
PRESSURE GRADIENTS
(OIL PERCENT-- 50)

TUBING SIZE	4.892	IN. I.D.
PRODUCTION RATE	2000	BLPD
GAS SPECIFIC GRAVITY		0.65
AVERAGE FLOWING TEMP	150	F
OIL API GRAVITY	35.0	API
WATER SPECIFIC GRAVITY		1.07

DEPTH (1000 FT)

GAS/LIQUID RATIO - SCF PER BBL

HAGEDORN AND BROWN
PRESENTED BY:
NACCO - SCHLUMBERGER

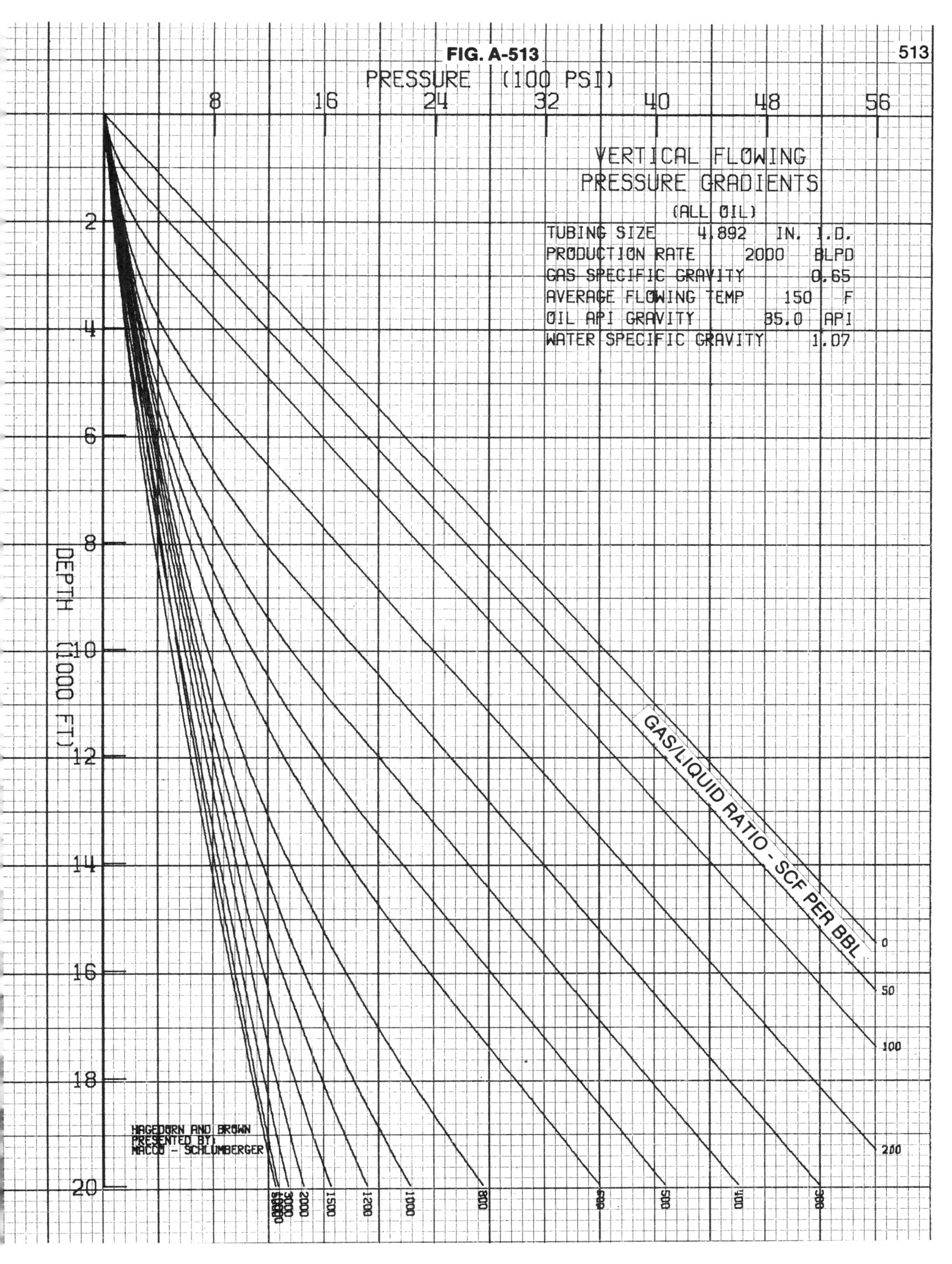

FIG. A-513

513

PRESSURE (100 PSI)

VERTICAL FLOWING
PRESSURE GRADIENTS
(ALL OIL)

TUBING SIZE	4.892	IN. I.D.
PRODUCTION RATE	2000	BLPD
GAS SPECIFIC GRAVITY	0.65	
AVERAGE FLOWING TEMP	150	F
OIL API GRAVITY	35.0	API
WATER SPECIFIC GRAVITY	1.07	

DEPTH (1000 FT)

GAS/LIQUID RATIO - SCF PER BBL

HAGEDORN AND BROWN
PRESENTED BY:
NACO - SCHLUMBERGER

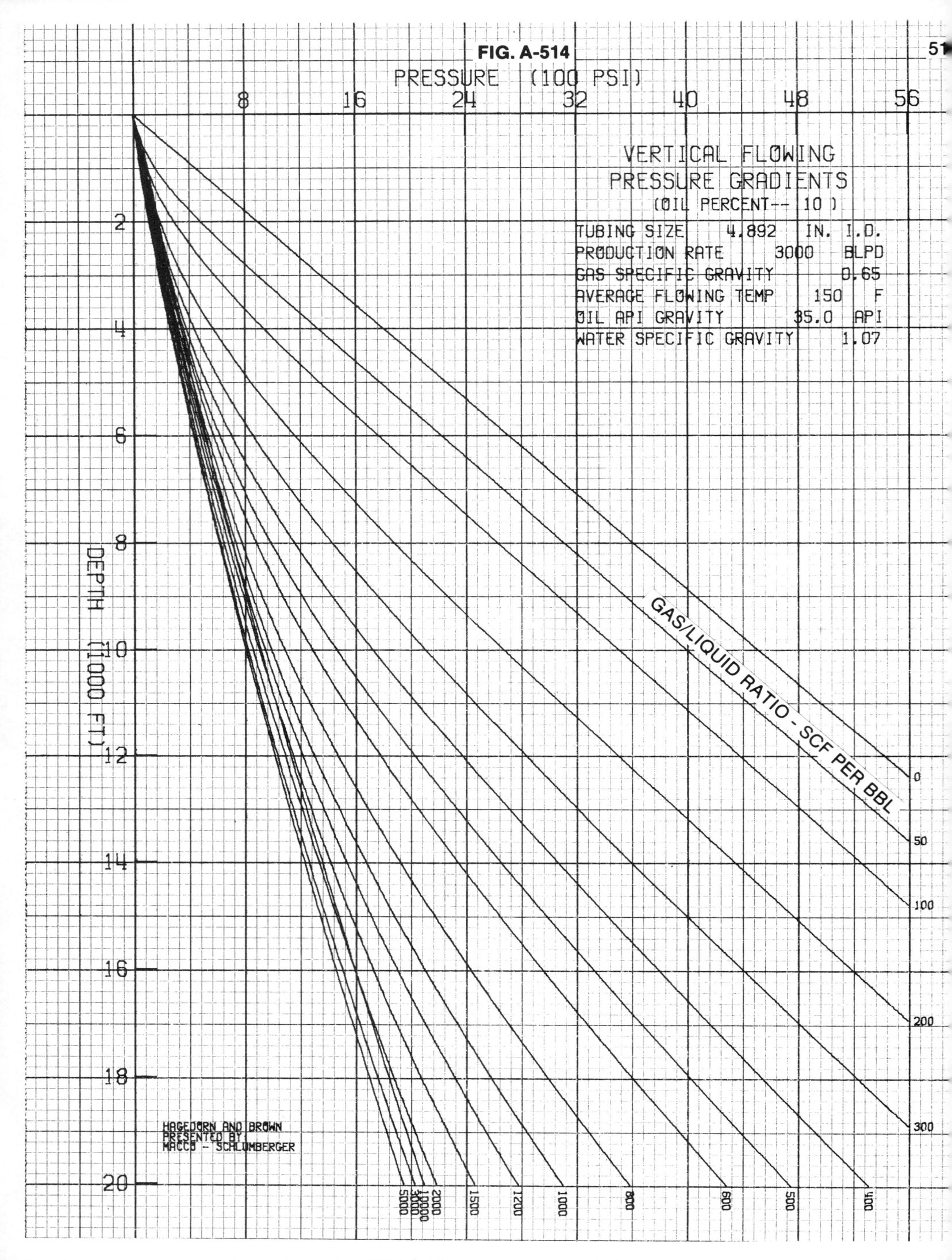

VERTICAL FLOWING
PRESSURE GRADIENTS
(OIL PERCENT-- 10)

TUBING SIZE	4.892	IN. I.D.
PRODUCTION RATE	3000	BLPD
GAS SPECIFIC GRAVITY	0.65	
AVERAGE FLOWING TEMP	150	F
OIL API GRAVITY	35.0	API
WATER SPECIFIC GRAVITY	1.07	

PRESSURE (100 PSI)

DEPTH (1000 FT)

GAS/LIQUID RATIO - SCF PER BBL

HAGEDORN AND BROWN
PRESENTED BY
MACCO - SCHLUMBERGER

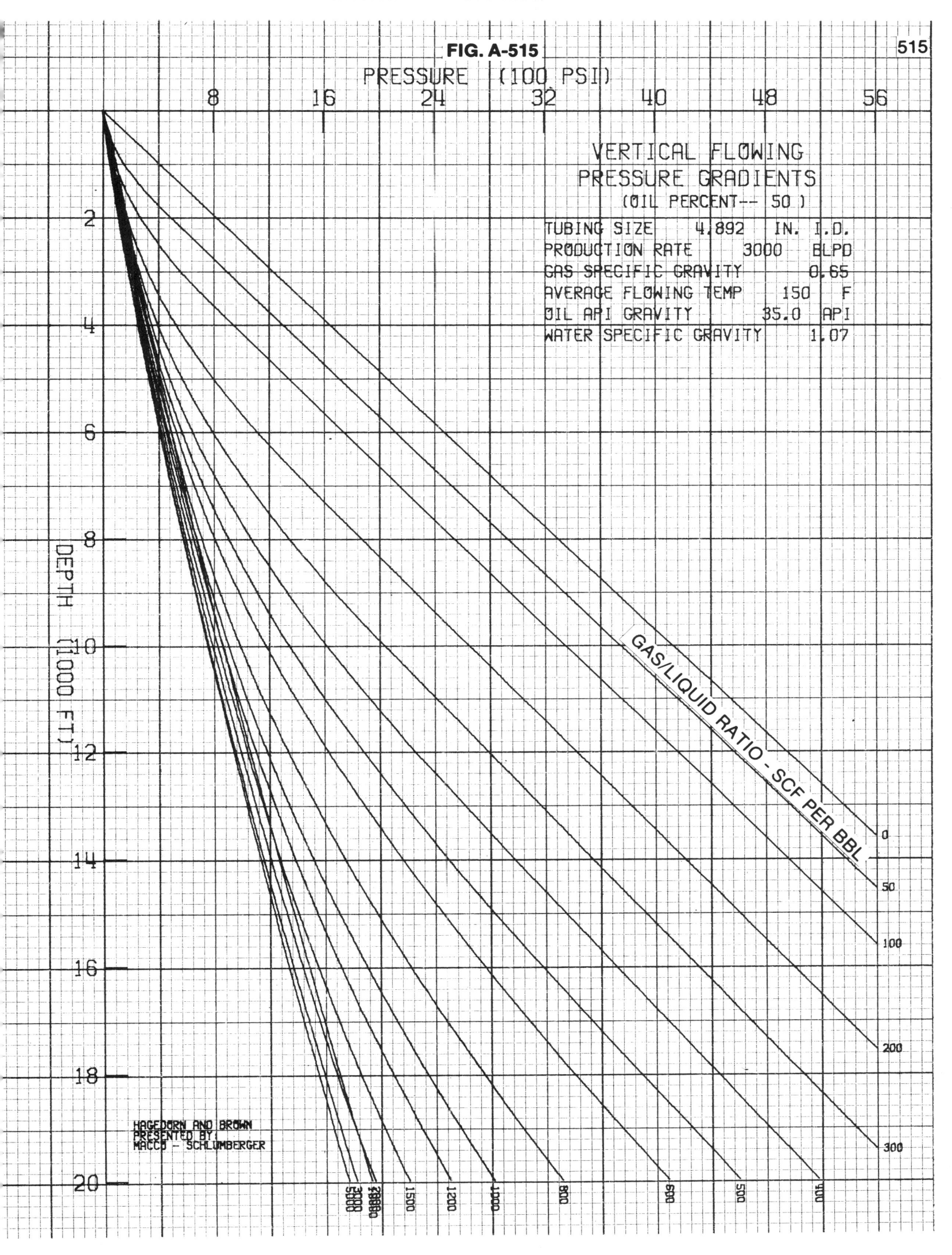

FIG. A-515

515

PRESSURE (100 PSI)

VERTICAL FLOWING
PRESSURE GRADIENTS
(OIL PERCENT-- 50)

TUBING SIZE	4.892	IN. I.D.
PRODUCTION RATE	3000	BLPD
GAS SPECIFIC GRAVITY	0.65	
AVERAGE FLOWING TEMP	150	F
OIL API GRAVITY	35.0	API
WATER SPECIFIC GRAVITY	1.07	

DEPTH (1000 FT)

GAS/LIQUID RATIO - SCF PER BBL

HAGEDORN AND BROWN
PRESENTED BY
MACCO - SCHLUMBERGER

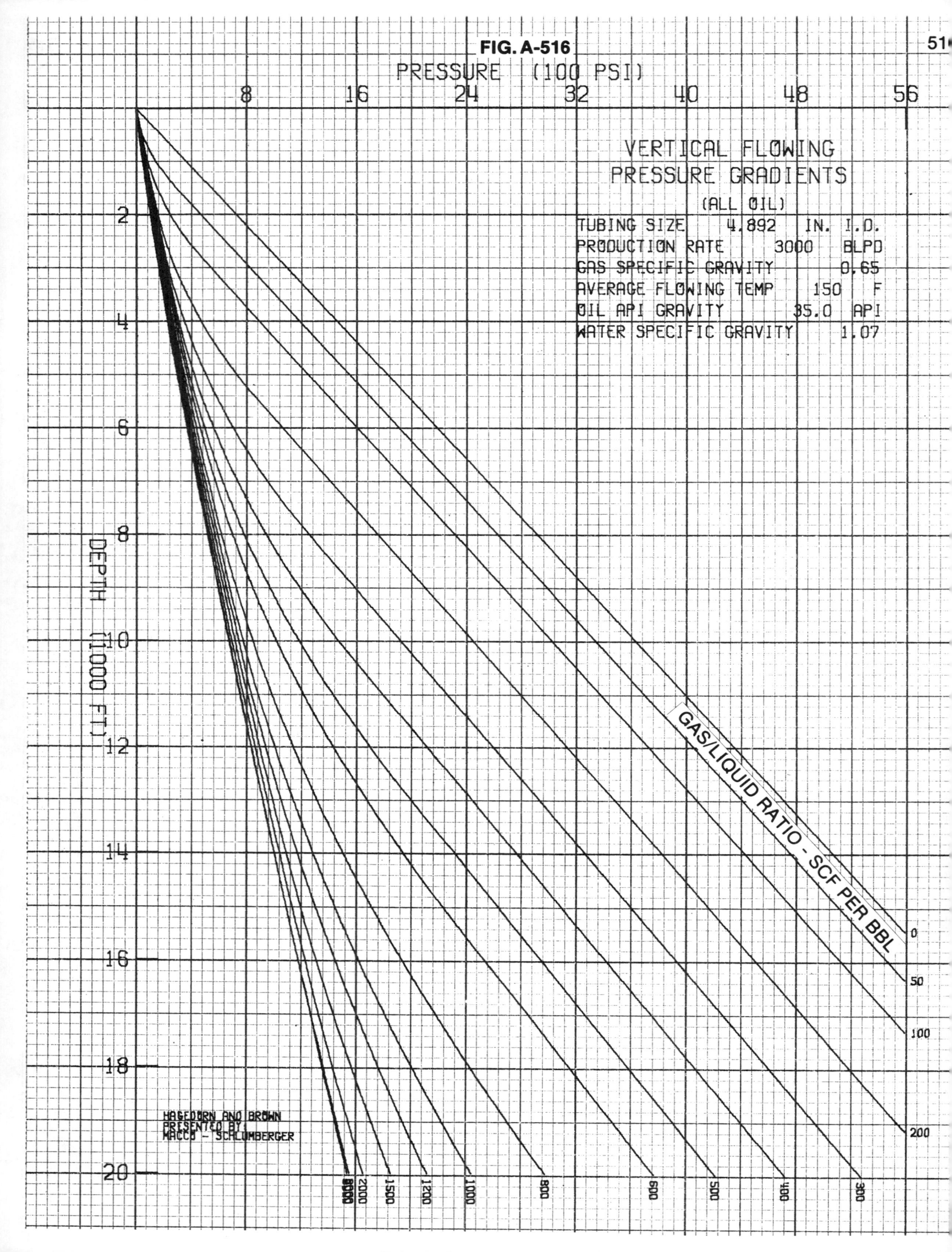

FIG. A-516

VERTICAL FLOWING
PRESSURE GRADIENTS
(ALL OIL)

TUBING SIZE	4.892	IN. I.D.
PRODUCTION RATE	3000	BLPD
GAS SPECIFIC GRAVITY	0.65	
AVERAGE FLOWING TEMP	150	F
OIL API GRAVITY	35.0	API
WATER SPECIFIC GRAVITY	1.07	

PRESSURE (100 PSI)

DEPTH (1000 FT)

GAS/LIQUID RATIO - SCF PER BBL

HAGEDORN AND BROWN
PRESENTED BY
MACCO - SCHLUMBERGER

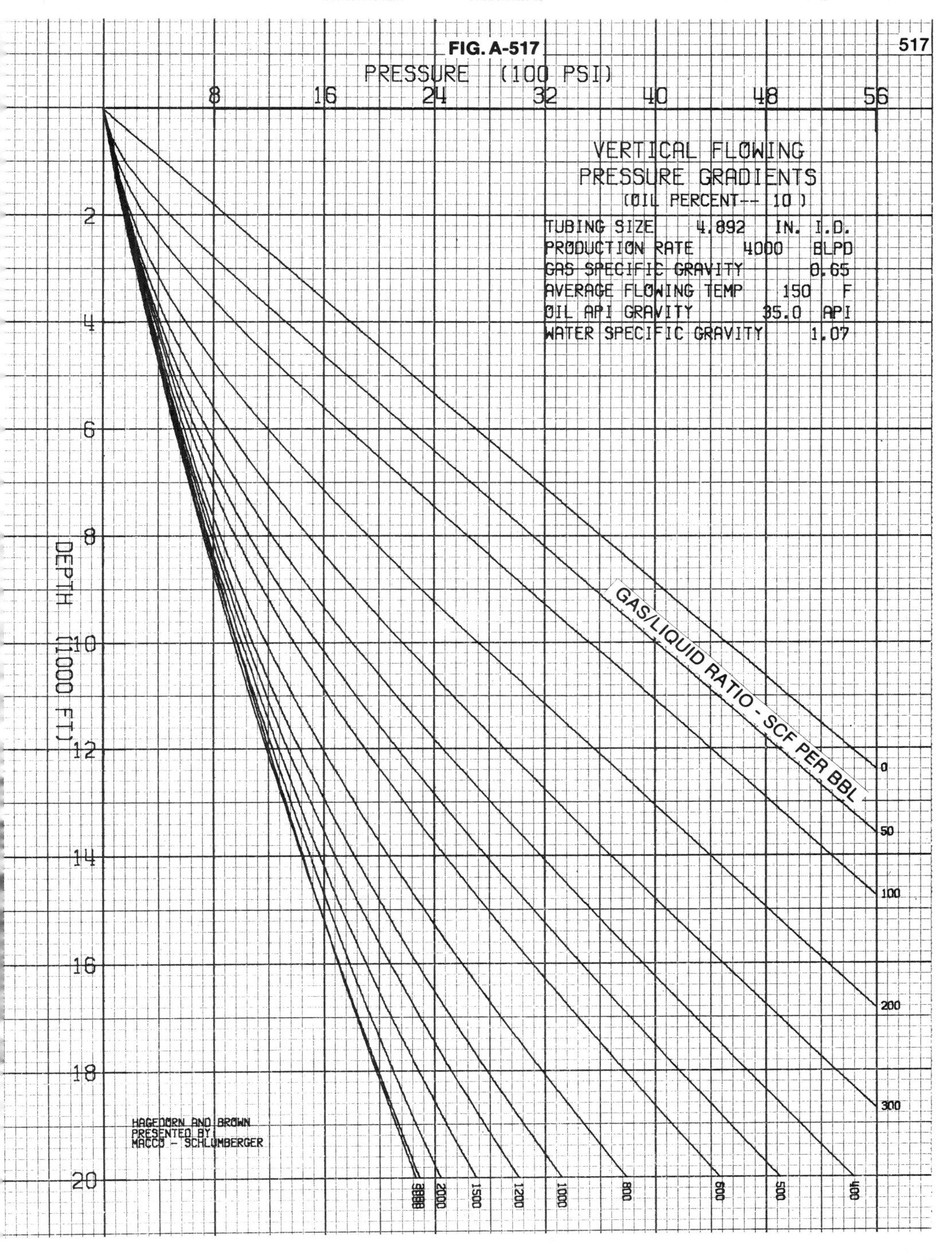

PRESSURE (100 PSI)

DEPTH (1000 FT)

VERTICAL FLOWING
PRESSURE GRADIENTS
(OIL PERCENT-- 10)

TUBING SIZE   4.892   IN. I.D.
PRODUCTION RATE   4000   BLPD
GAS SPECIFIC GRAVITY   0.65
AVERAGE FLOWING TEMP   150   F
OIL API GRAVITY   35.0   API
WATER SPECIFIC GRAVITY   1.07

GAS/LIQUID RATIO - SCF PER BBL

HAGEDORN AND BROWN
PRESENTED BY
MACCO - SCHLUMBERGER

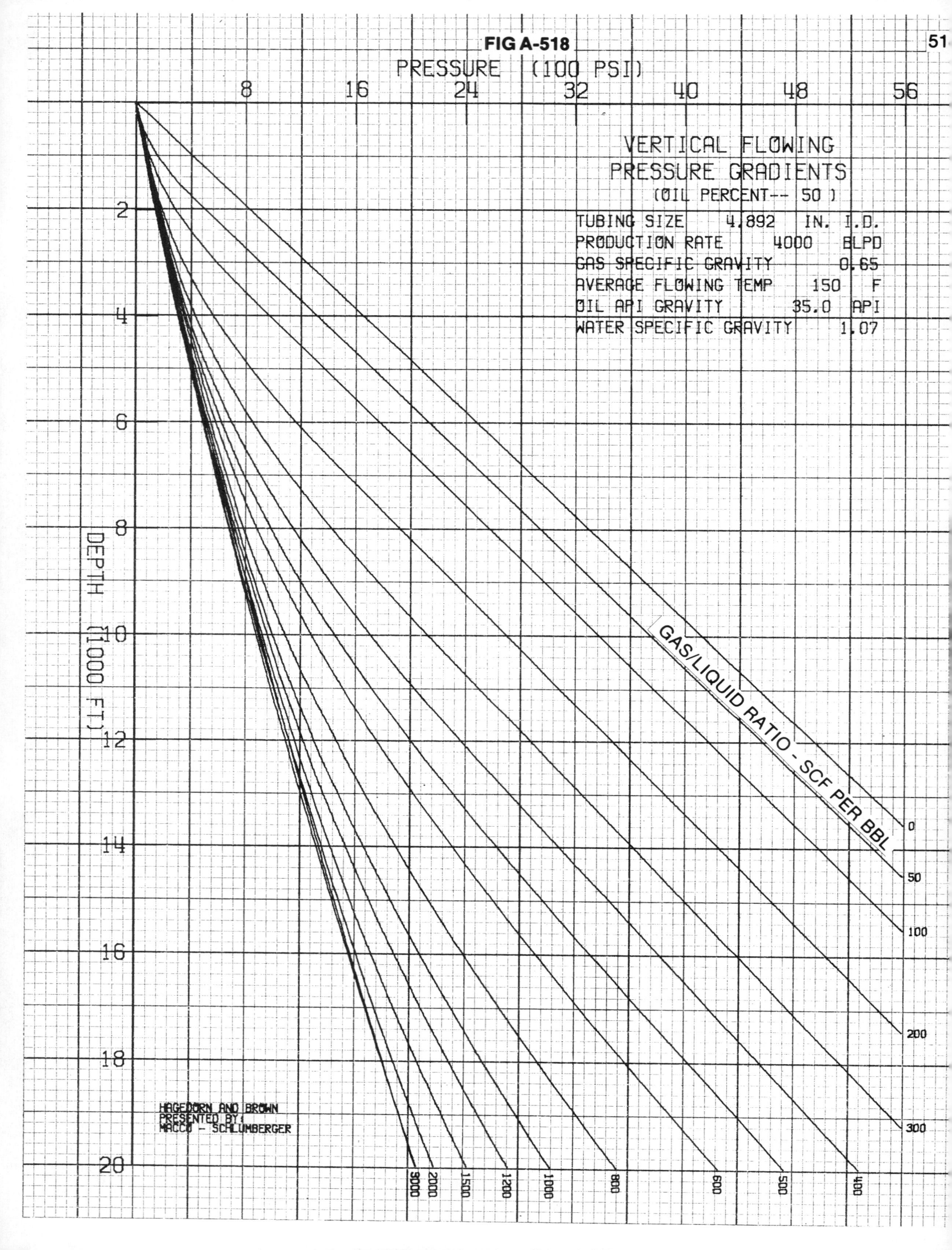

PRESSURE (100 PSI)

VERTICAL FLOWING
PRESSURE GRADIENTS
(OIL PERCENT--- 50)

TUBING SIZE	4.892	IN. I.D.
PRODUCTION RATE	4000	BLPD
GAS SPECIFIC GRAVITY	0.65	
AVERAGE FLOWING TEMP	150	F
OIL API GRAVITY	35.0	API
WATER SPECIFIC GRAVITY	1.07	

DEPTH (1000 FT)

GAS/LIQUID RATIO - SCF PER BBL

HAGEDORN AND BROWN
PRESENTED BY:
MACCO - SCHLUMBERGER

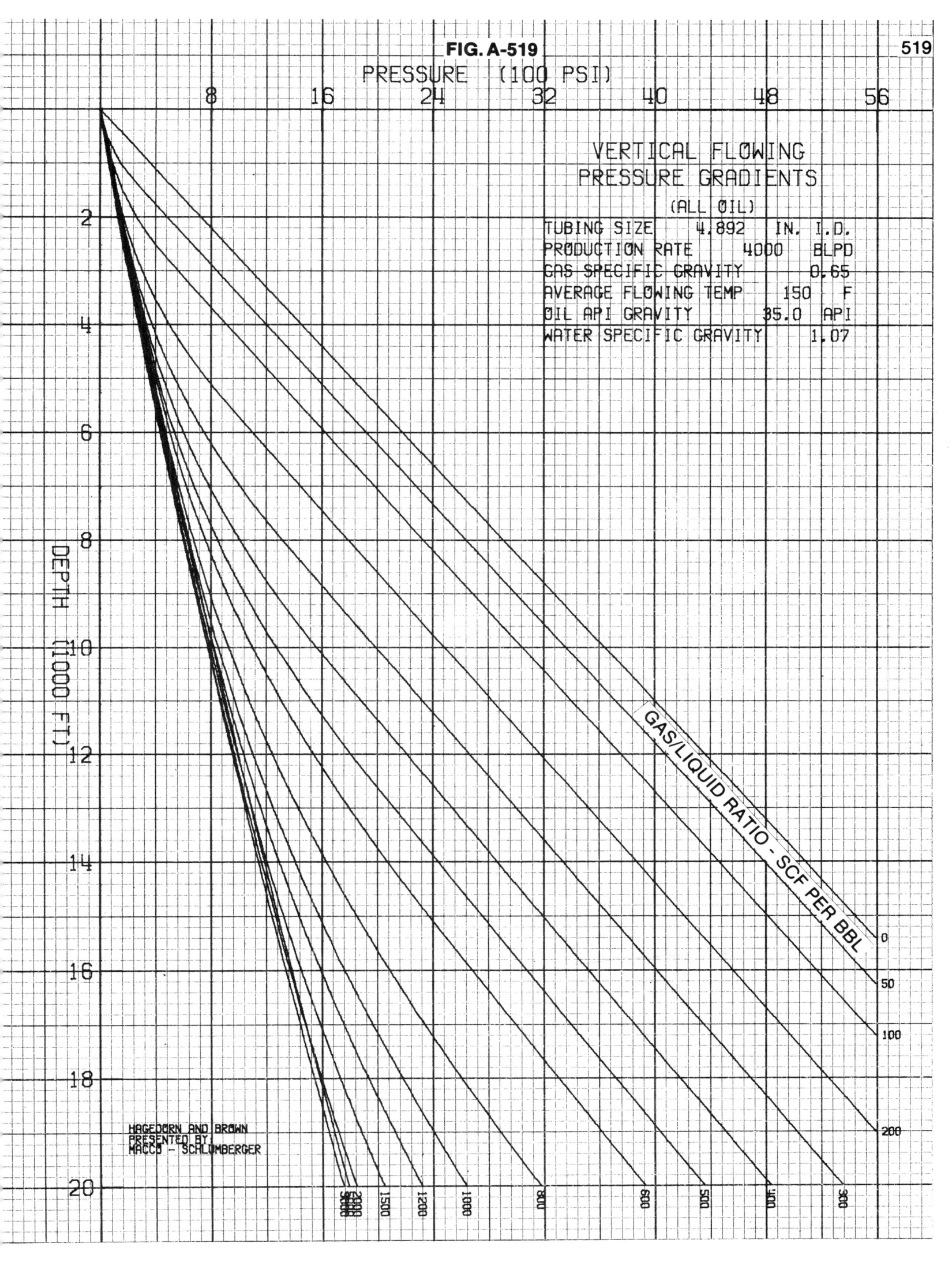

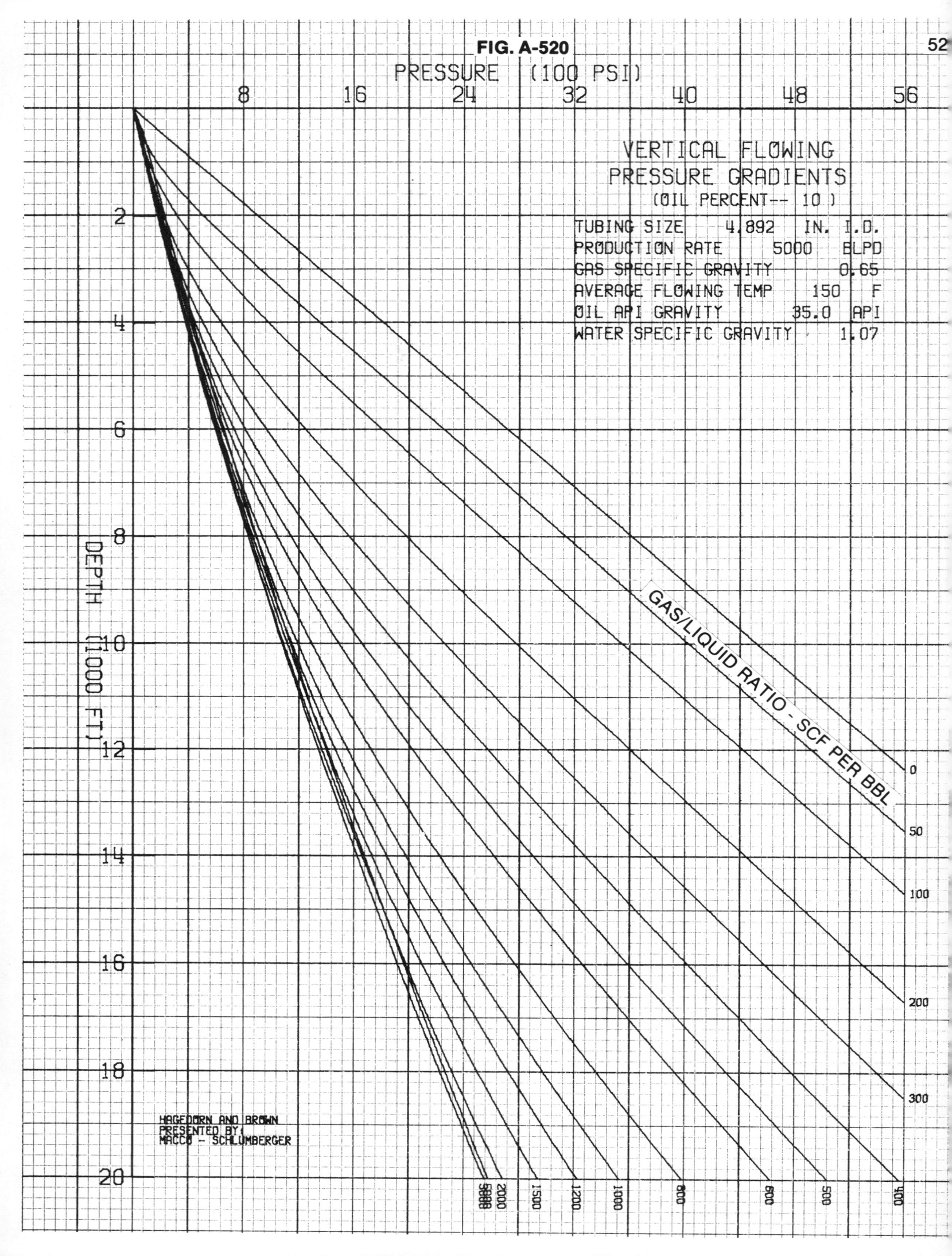

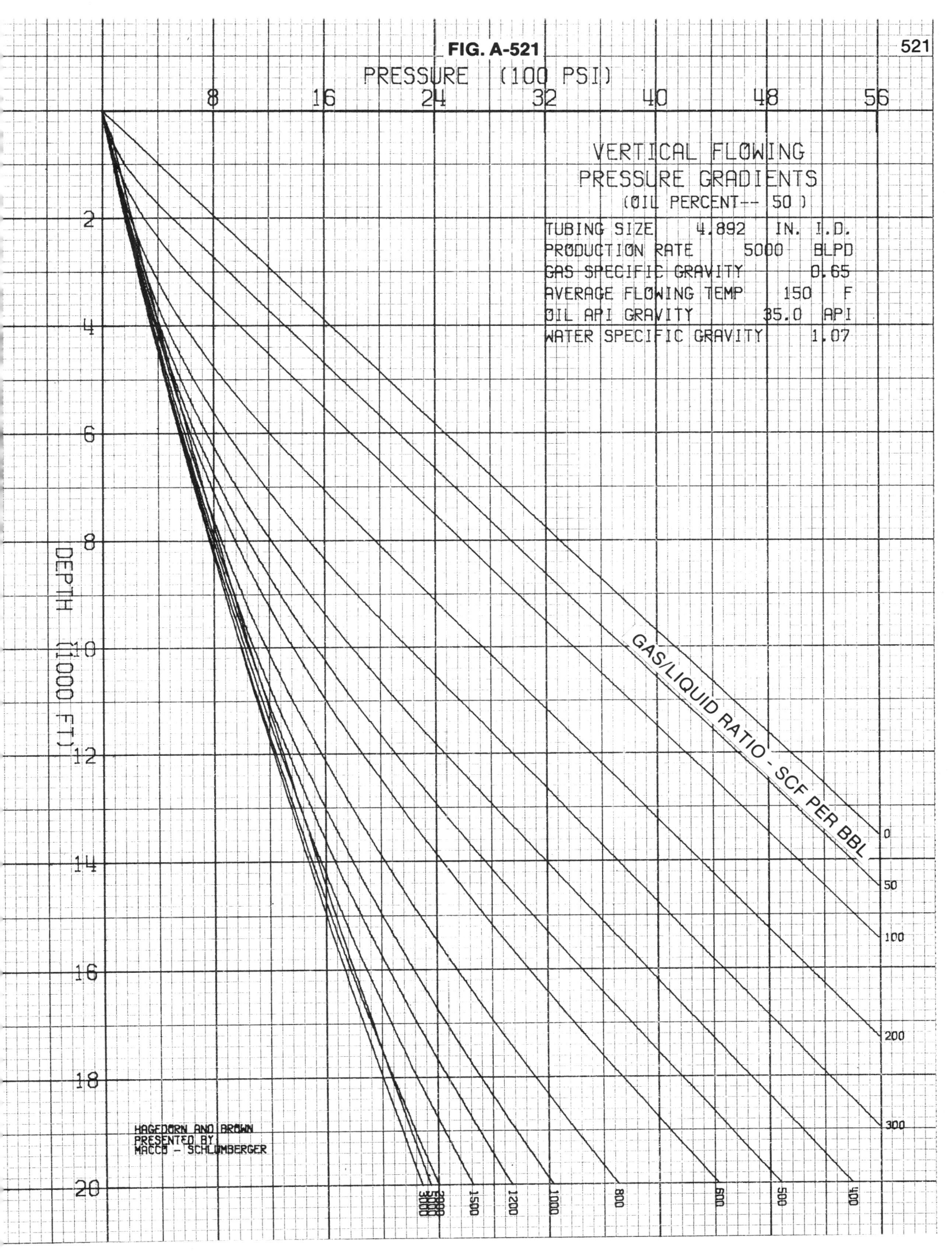

VERTICAL FLOWING
PRESSURE GRADIENTS
(OIL PERCENT — 50)

TUBING SIZE	4.892	IN. I.D.
PRODUCTION RATE	5000	BLPD
GAS SPECIFIC GRAVITY	0.65	
AVERAGE FLOWING TEMP	150	F
OIL API GRAVITY	35.0	API
WATER SPECIFIC GRAVITY	1.07	

PRESSURE (100 PSI)

DEPTH (1000 FT)

GAS/LIQUID RATIO - SCF PER BBL

HAGEDORN AND BROWN
PRESENTED BY
MACCO — SCHLUMBERGER

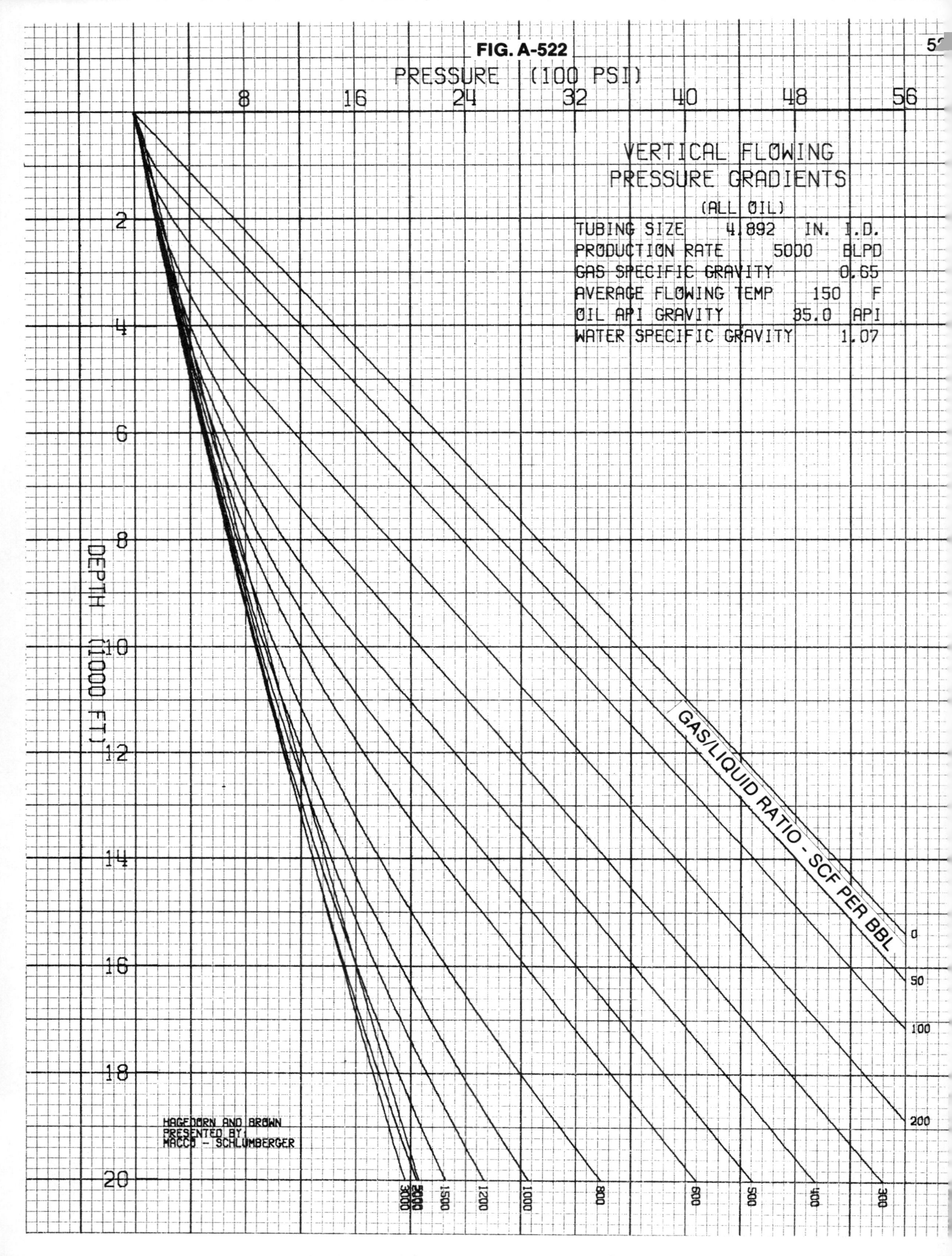

FIG. A-522

VERTICAL FLOWING
PRESSURE GRADIENTS
(ALL OIL)

TUBING SIZE	4.892	IN. I.D.
PRODUCTION RATE	5000	BLPD
GAS SPECIFIC GRAVITY	0.65	
AVERAGE FLOWING TEMP	150	F
OIL API GRAVITY	35.0	API
WATER SPECIFIC GRAVITY	1.07	

PRESSURE (100 PSI)

DEPTH (1000 FT)

GAS/LIQUID RATIO - SCF PER BBL

HAGEDORN AND BROWN
PRESENTED BY:
MACCO - SCHLUMBERGER

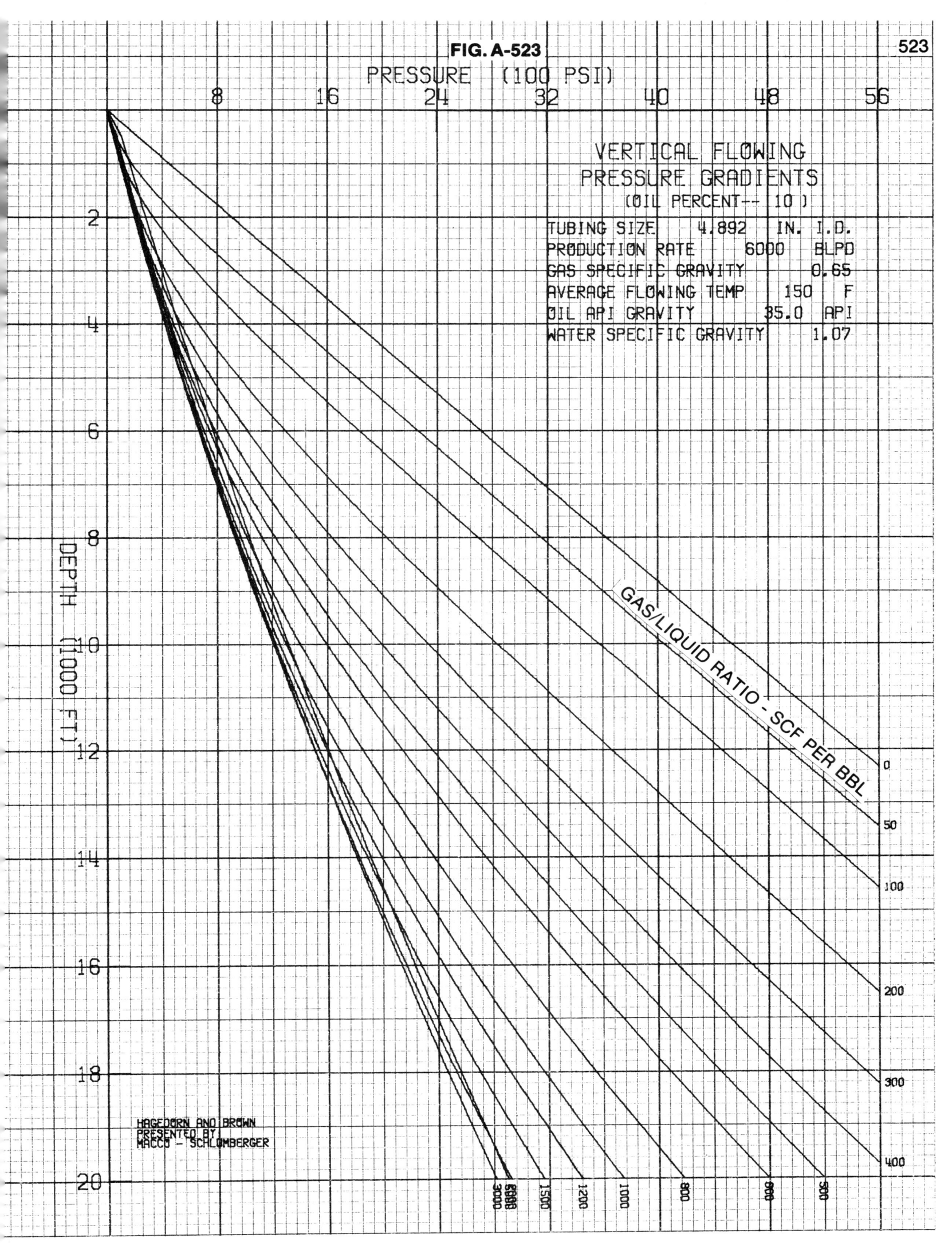

FIG. A-523

523

VERTICAL FLOWING
PRESSURE GRADIENTS
(OIL PERCENT-- 10)

TUBING SIZE	4.892	IN. I.D.
PRODUCTION RATE	6000	BLPD
GAS SPECIFIC GRAVITY		0.65
AVERAGE FLOWING TEMP	150	F
OIL API GRAVITY	35.0	API
WATER SPECIFIC GRAVITY		1.07

PRESSURE (100 PSI)

DEPTH (1000 FT)

GAS/LIQUID RATIO - SCF PER BBL

HAGEDORN AND BROWN
PRESENTED BY
MACCO - SCHLUMBERGER

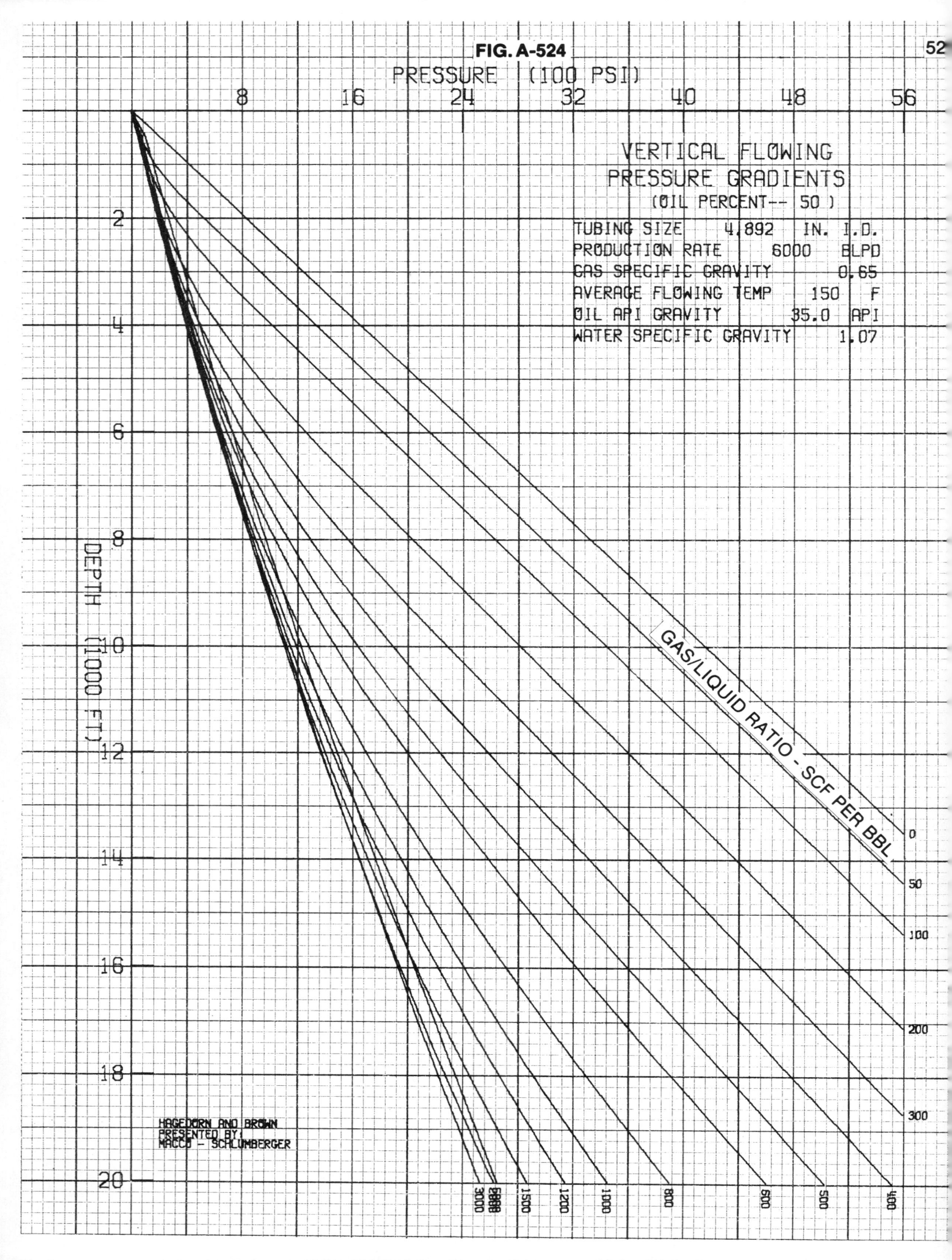

FIG. A-524

52

PRESSURE (100 PSI)

VERTICAL FLOWING
PRESSURE GRADIENTS
(OIL PERCENT-- 50)

TUBING SIZE	4.892	IN. I.D.
PRODUCTION RATE	6000	BLPD
GAS SPECIFIC GRAVITY		0.65
AVERAGE FLOWING TEMP	150	F
OIL API GRAVITY	35.0	API
WATER SPECIFIC GRAVITY		1.07

DEPTH (1000 FT)

GAS/LIQUID RATIO - SCF PER BBL

HAGEDORN AND BROWN
PRESENTED BY:
MACCO - SCHLUMBERGER

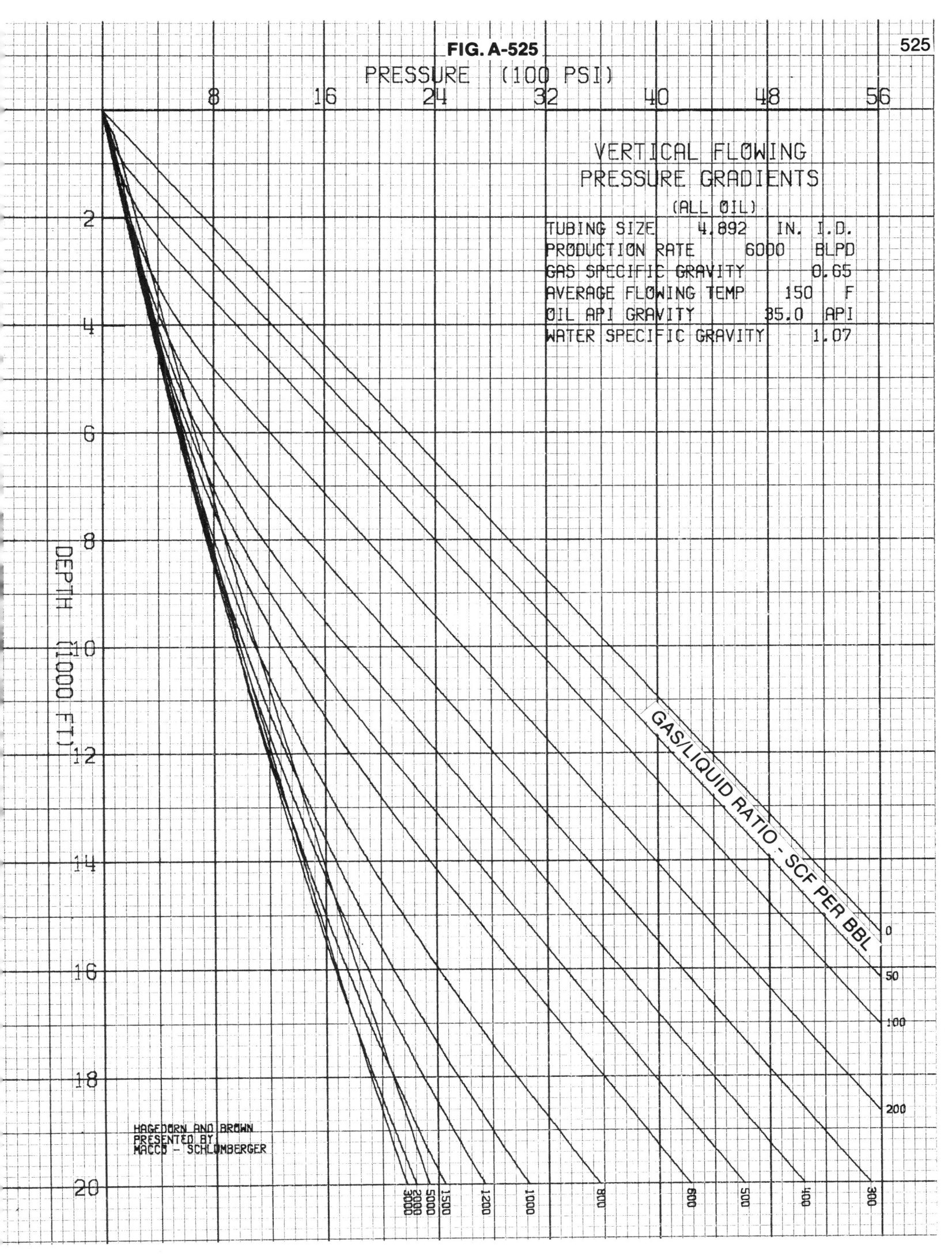

VERTICAL FLOWING
PRESSURE GRADIENTS
(ALL OIL)

TUBING SIZE	4.892	IN. I.D.
PRODUCTION RATE	6000	BLPD
GAS SPECIFIC GRAVITY	0.65	
AVERAGE FLOWING TEMP	150	F
OIL API GRAVITY	35.0	API
WATER SPECIFIC GRAVITY	1.07	

PRESSURE (100 PSI)

DEPTH (1000 FT)

GAS/LIQUID RATIO - SCF PER BBL

HAGEDORN AND BROWN
PRESENTED BY
MACCO - SCHLUMBERGER

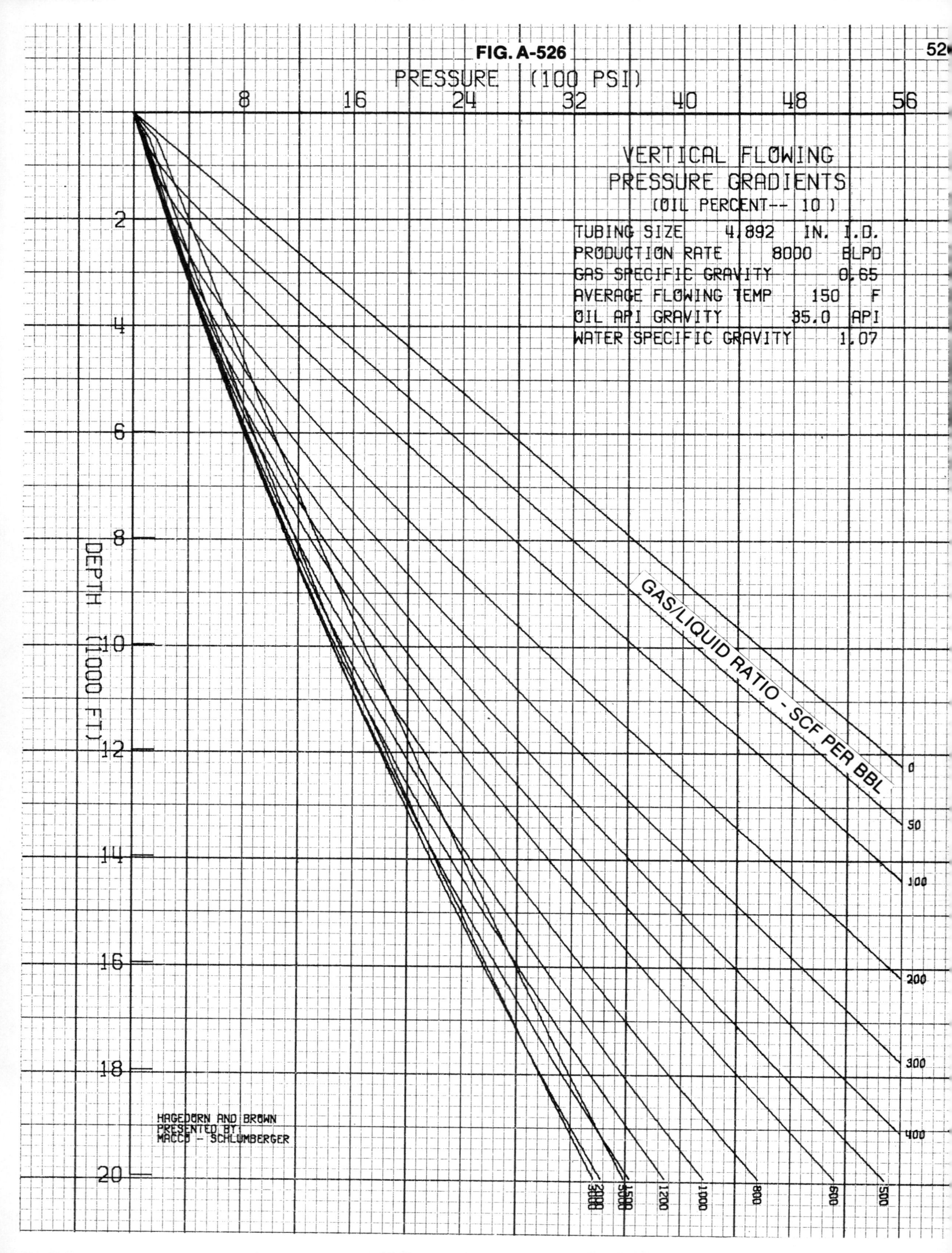

FIG. A-526

VERTICAL FLOWING
PRESSURE GRADIENTS
(OIL PERCENT-- 10)

TUBING SIZE	4.892	IN. I.D.
PRODUCTION RATE	8000	BLPD
GAS SPECIFIC GRAVITY	0.65	
AVERAGE FLOWING TEMP	150	F
OIL API GRAVITY	35.0	API
WATER SPECIFIC GRAVITY	1.07	

PRESSURE (100 PSI)

DEPTH (1000 FT)

GAS/LIQUID RATIO - SCF PER BBL

HAGEDORN AND BROWN
PRESENTED BY:
MACCO - SCHLUMBERGER

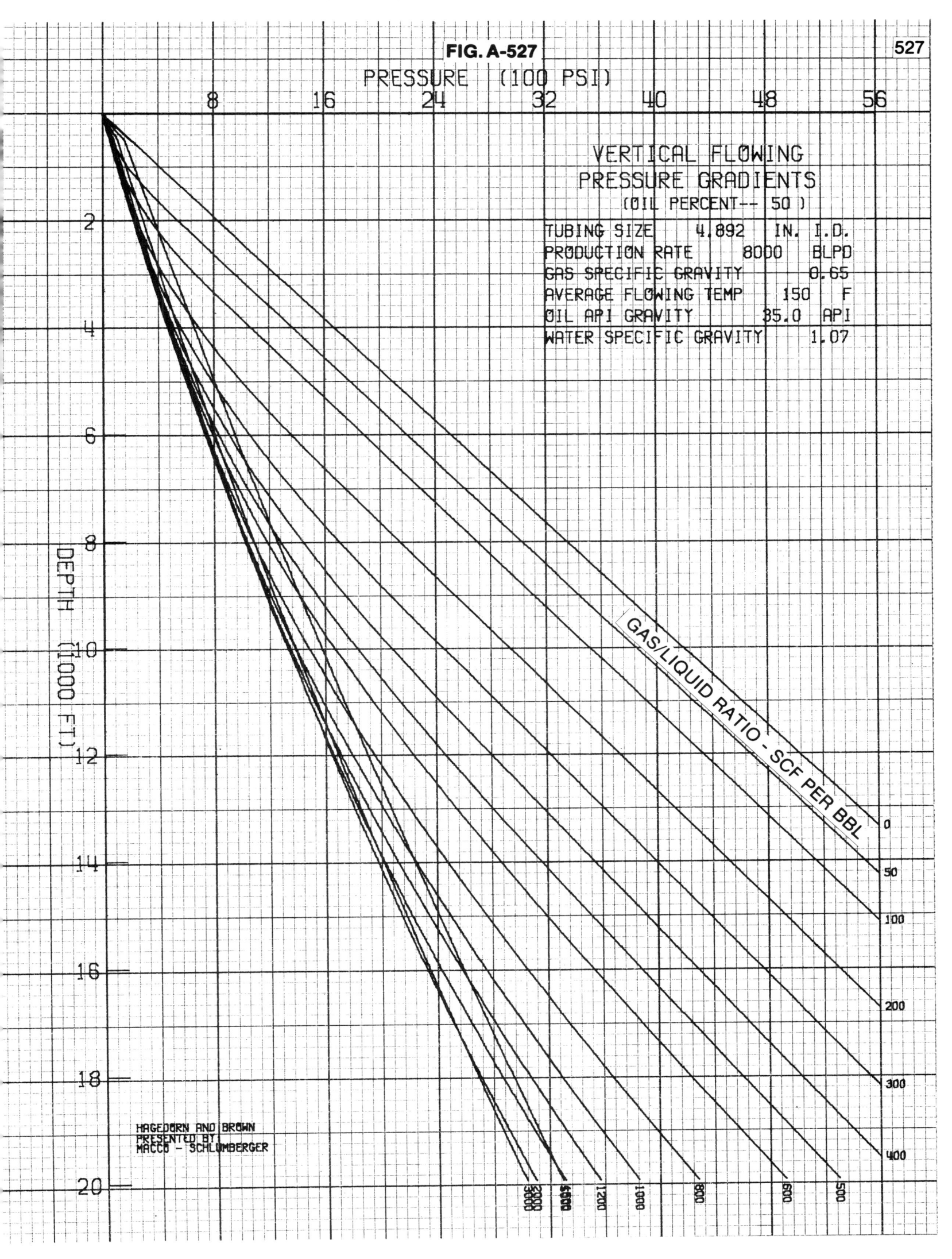

FIG. A-527

527

PRESSURE (100 PSI)

VERTICAL FLOWING
PRESSURE GRADIENTS
(OIL PERCENT-- 50)

TUBING SIZE	4.892	IN. I.D.
PRODUCTION RATE	8000	BLPD
GAS SPECIFIC GRAVITY	0.65	
AVERAGE FLOWING TEMP	150	F
OIL API GRAVITY	35.0	API
WATER SPECIFIC GRAVITY	1.07	

DEPTH (1000 FT)

GAS/LIQUID RATIO - SCF PER BBL

HAGEDORN AND BROWN
PRESENTED BY
MACCO - SCHLUMBERGER

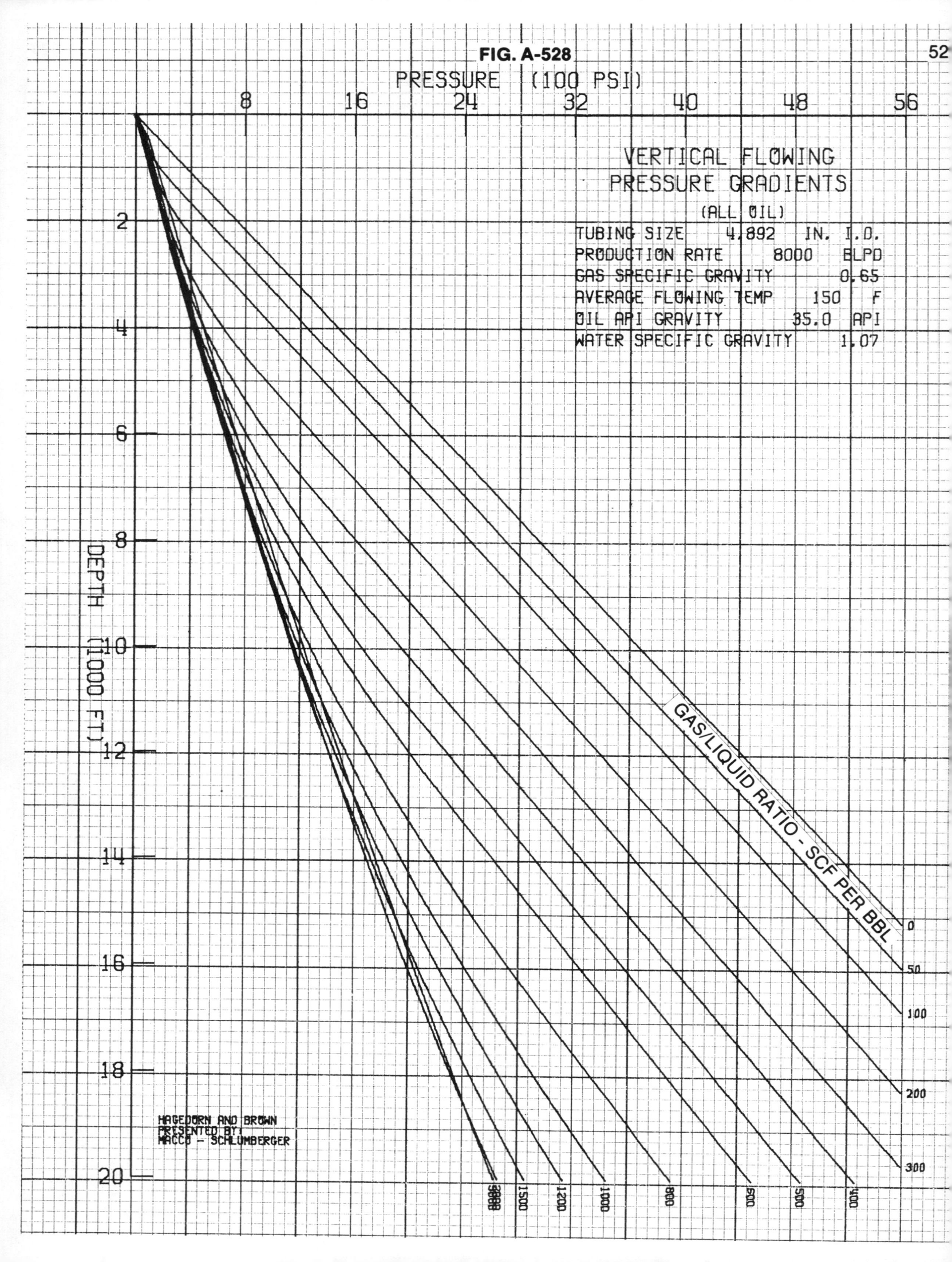

FIG. A-528

52

PRESSURE (100 PSI)

VERTICAL FLOWING
PRESSURE GRADIENTS
(ALL OIL)

TUBING SIZE	4.892	IN. I.D.
PRODUCTION RATE	8000	BLPD
GAS SPECIFIC GRAVITY	0.65	
AVERAGE FLOWING TEMP	150	F
OIL API GRAVITY	35.0	API
WATER SPECIFIC GRAVITY	1.07	

DEPTH (1000 FT)

GAS/LIQUID RATIO - SCF PER BBL

HAGEDORN AND BROWN
PRESENTED BY:
MACCO - SCHLUMBERGER

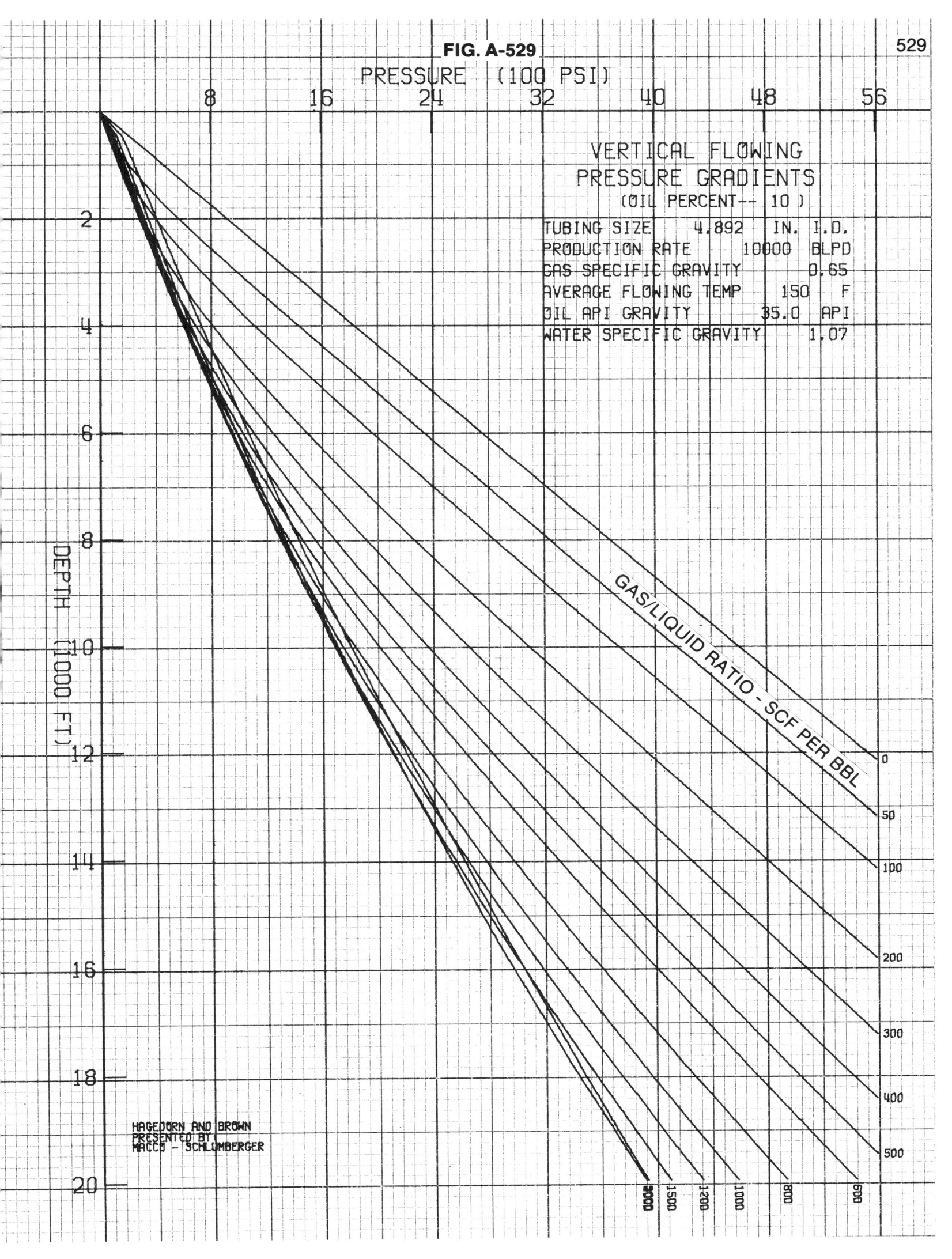

PRESSURE (100 PSI)

VERTICAL FLOWING
PRESSURE GRADIENTS
(OIL PERCENT--- 10)

TUBING SIZE	4.892	IN. I.D.
PRODUCTION RATE	10000	BLPD
GAS SPECIFIC GRAVITY		0.65
AVERAGE FLOWING TEMP	150	F
OIL API GRAVITY	35.0	API
WATER SPECIFIC GRAVITY		1.07

GAS/LIQUID RATIO - SCF PER BBL

DEPTH (1000 FT)

HAGEDORN AND BROWN
PRESENTED BY
MACCO - SCHLUMBERGER

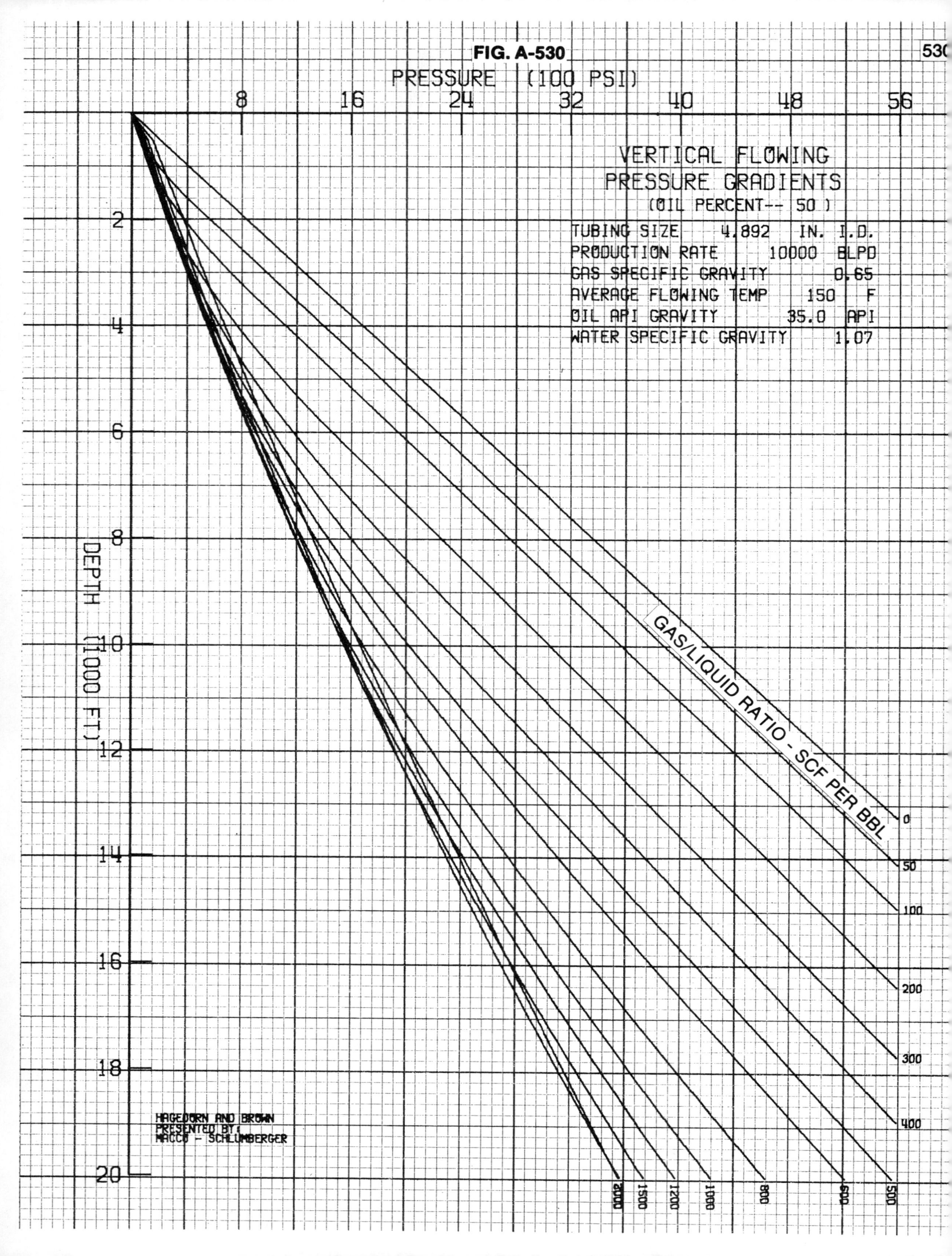

PRESSURE (100 PSI)

VERTICAL FLOWING
PRESSURE GRADIENTS
(OIL PERCENT-- 50)

TUBING SIZE	4.892	IN. I.D.
PRODUCTION RATE	10000	BLPD
GAS SPECIFIC GRAVITY	0.65	
AVERAGE FLOWING TEMP	150	F
OIL API GRAVITY	35.0	API
WATER SPECIFIC GRAVITY	1.07	

DEPTH (1000 FT)

GAS/LIQUID RATIO - SCF PER BBL

HAGEDORN AND BROWN
PRESENTED BY:
MACCO - SCHLUMBERGER

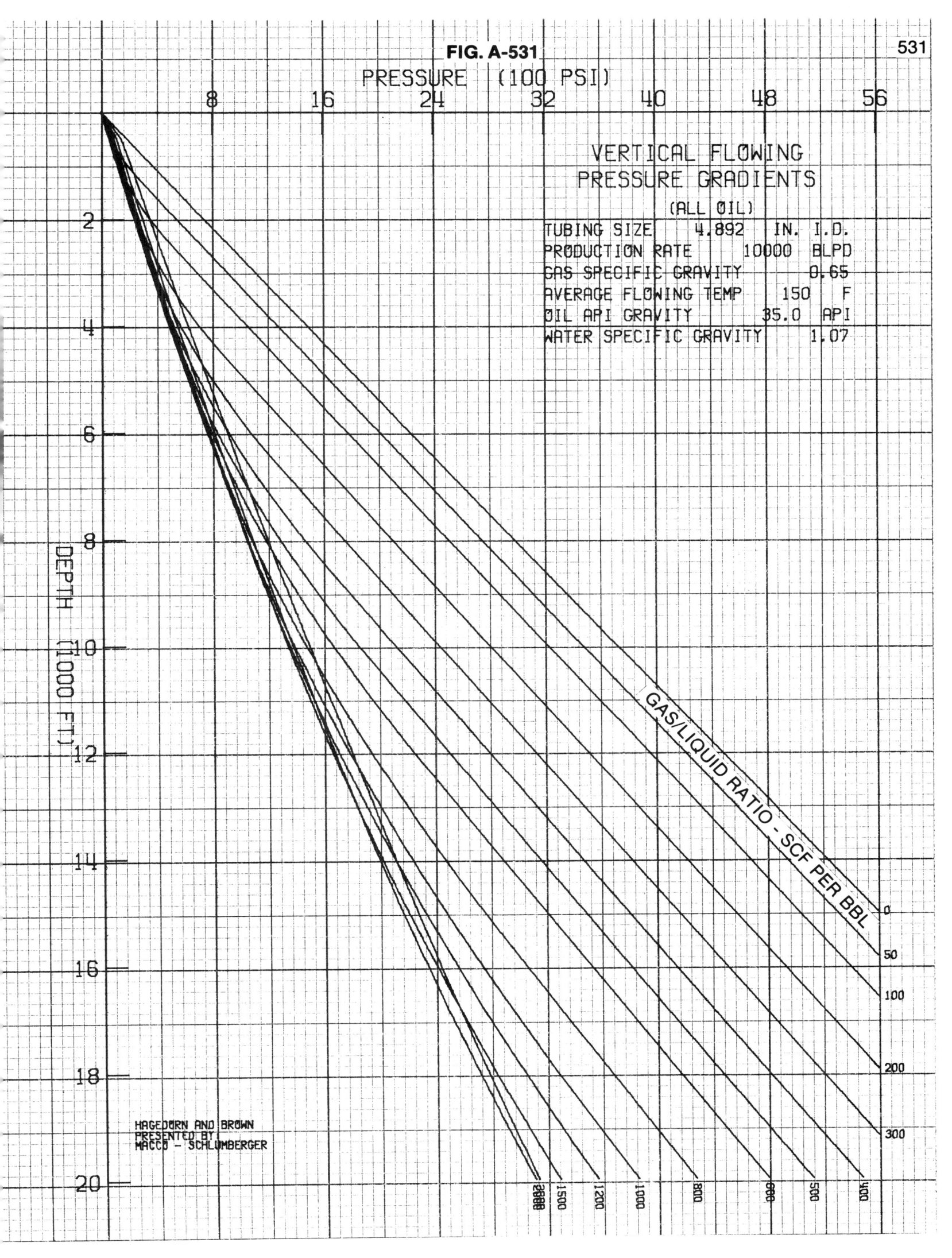

VERTICAL FLOWING
PRESSURE GRADIENTS
(ALL OIL)

TUBING SIZE	4.892	IN. I.D.
PRODUCTION RATE	10000	BLPD
GAS SPECIFIC GRAVITY	0.65	
AVERAGE FLOWING TEMP	150	F
OIL API GRAVITY	35.0	API
WATER SPECIFIC GRAVITY	1.07	

PRESSURE (100 PSI)

DEPTH (1000 FT)

GAS/LIQUID RATIO - SCF PER BBL

HAGEDORN AND BROWN
PRESENTED BY
MACCO - SCHLUMBERGER

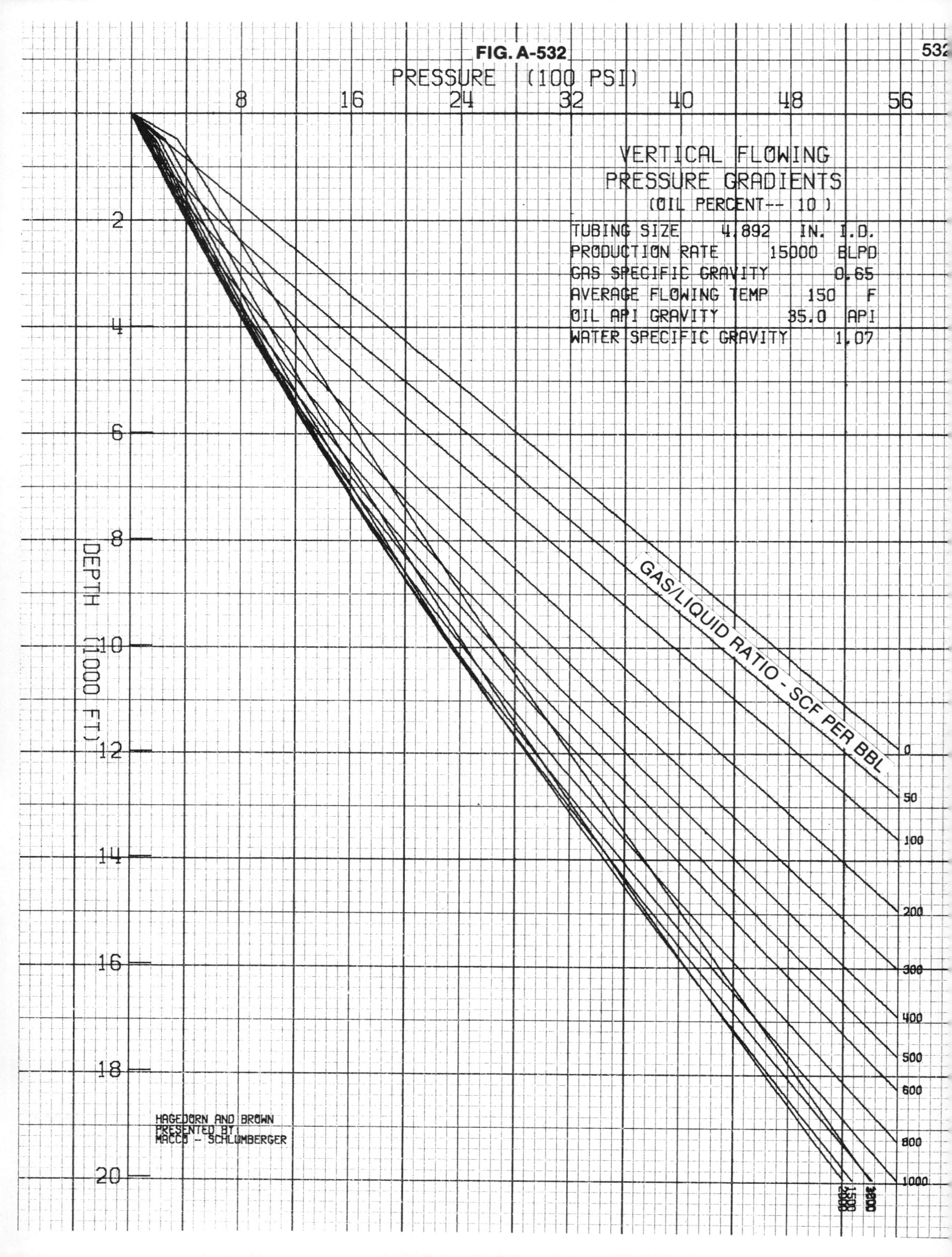

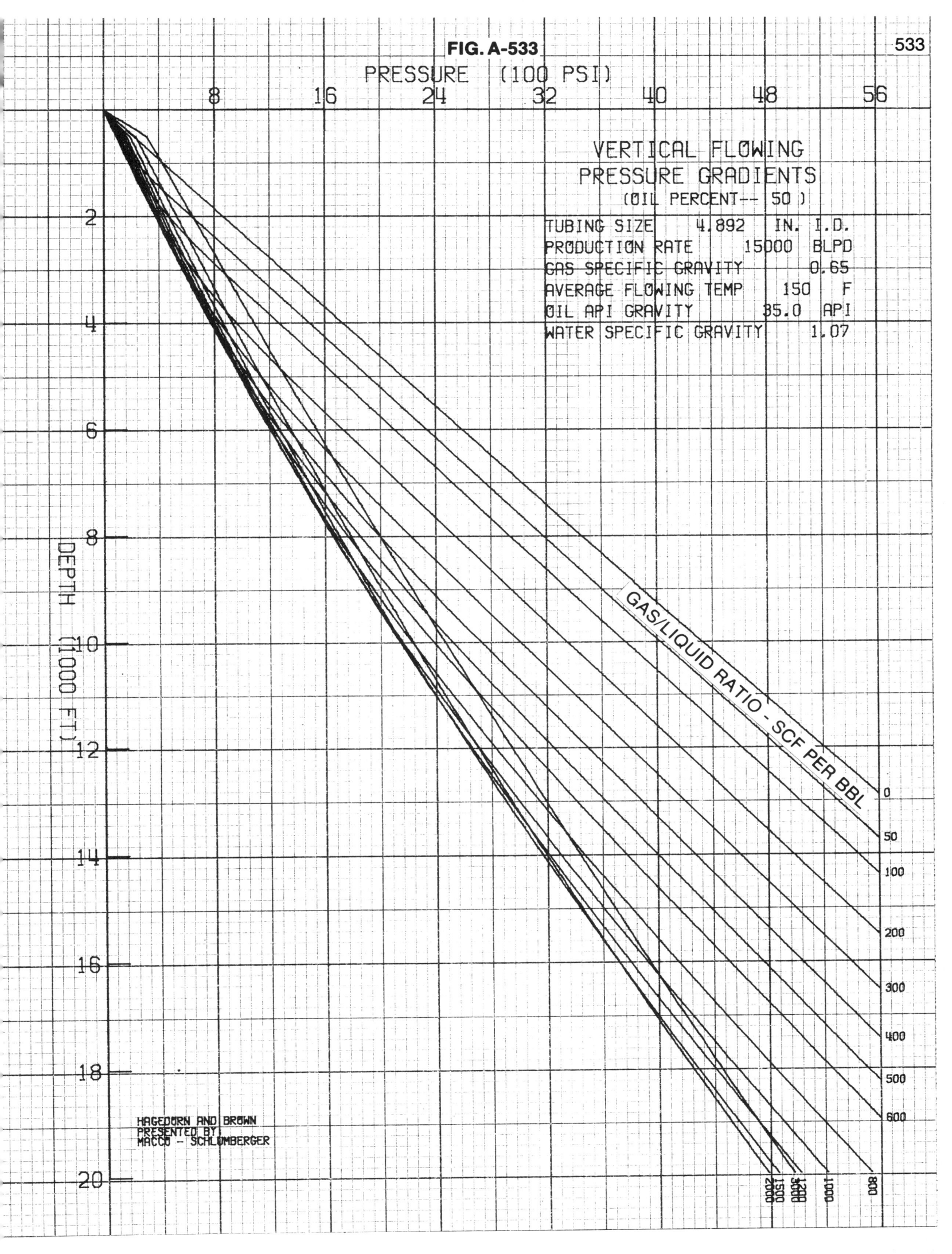

FIG. A-533

533

VERTICAL FLOWING
PRESSURE GRADIENTS
(OIL PERCENT-- 50)

TUBING SIZE	4.892	IN. I.D.
PRODUCTION RATE	15000	BLPD
GAS SPECIFIC GRAVITY	0.65	
AVERAGE FLOWING TEMP	150	F
OIL API GRAVITY	35.0	API
WATER SPECIFIC GRAVITY	1.07	

PRESSURE (100 PSI)

DEPTH (1000 FT)

GAS/LIQUID RATIO - SCF PER BBL

HAGEDORN AND BROWN
PRESENTED BY
MACCO - SCHLUMBERGER

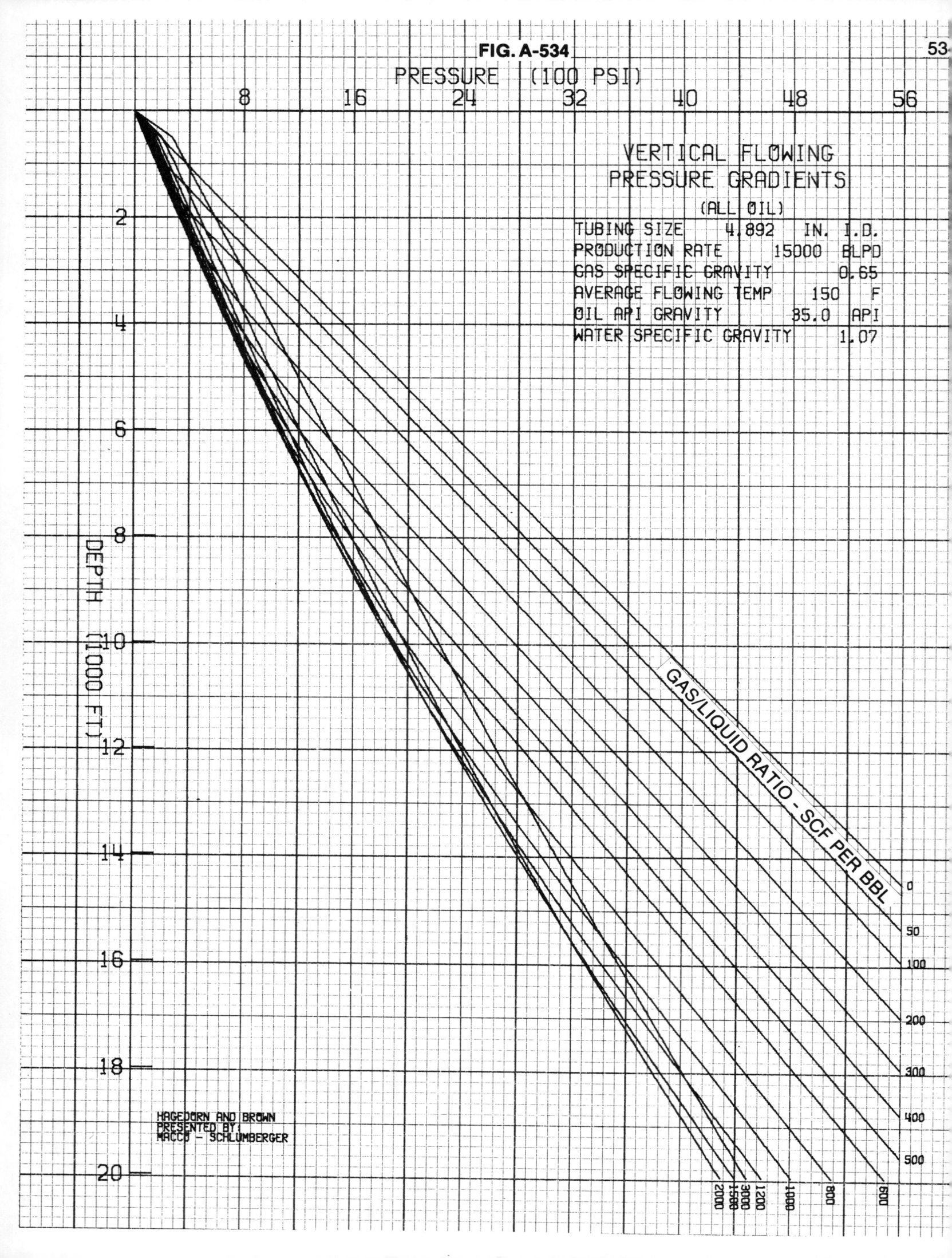

FIG. A-534

PRESSURE (100 PSI)

VERTICAL FLOWING
PRESSURE GRADIENTS
(ALL OIL)

TUBING SIZE	4.892	IN. I.D.
PRODUCTION RATE	15000	BLPD
GAS SPECIFIC GRAVITY		0.65
AVERAGE FLOWING TEMP	150	F
OIL API GRAVITY	35.0	API
WATER SPECIFIC GRAVITY		1.07

DEPTH (1000 FT)

GAS/LIQUID RATIO - SCF PER BBL

HAGEDORN AND BROWN
PRESENTED BY:
MACCO — SCHLUMBERGER

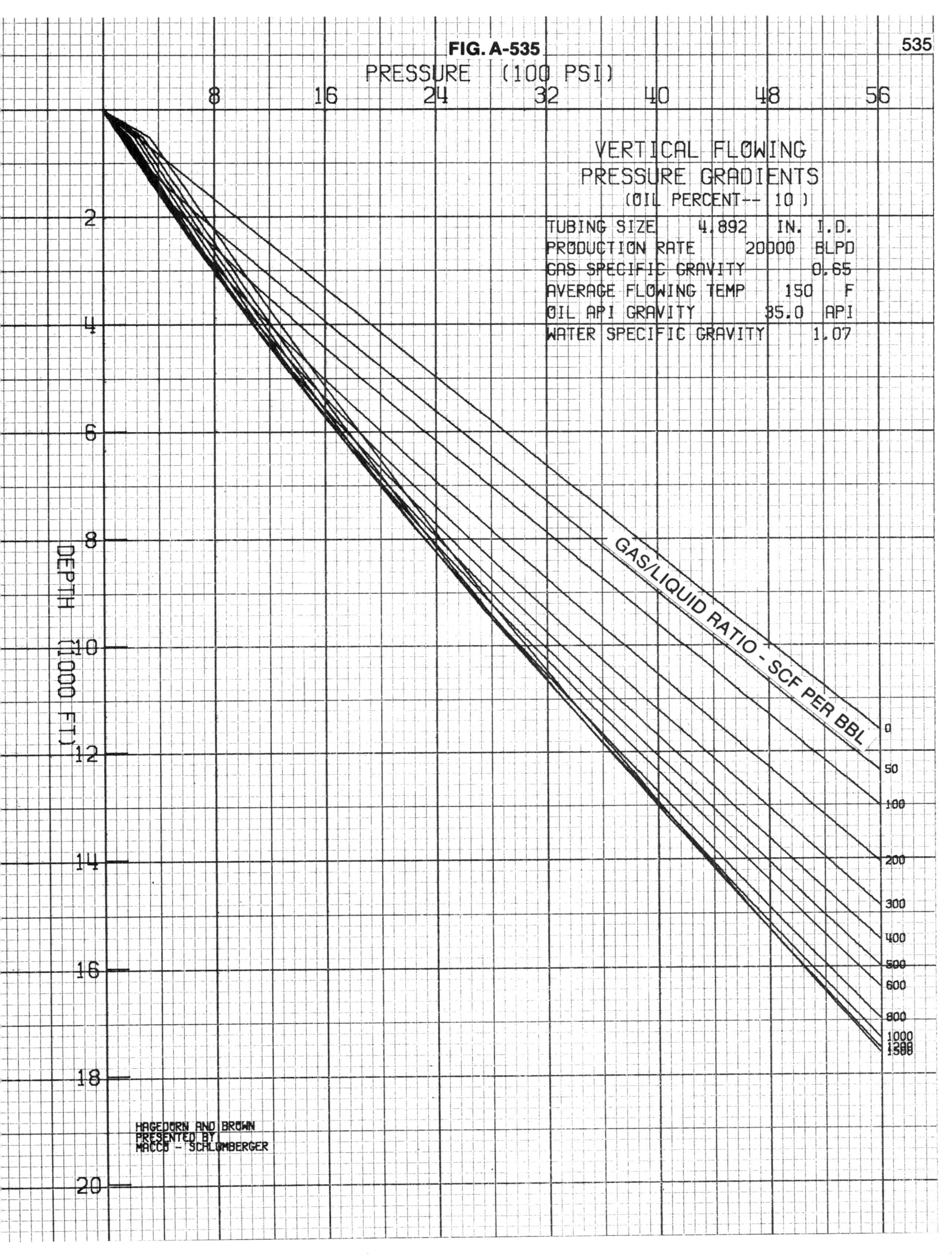

FIG. A-535

535

PRESSURE (100 PSI)

VERTICAL FLOWING
PRESSURE GRADIENTS
(OIL PERCENT-- 10)

TUBING SIZE	4.892	IN. I.D.
PRODUCTION RATE	20000	BLPD
GAS SPECIFIC GRAVITY	0.65	
AVERAGE FLOWING TEMP	150	F
OIL API GRAVITY	35.0	API
WATER SPECIFIC GRAVITY	1.07	

GAS/LIQUID RATIO - SCF PER BBL

DEPTH (1000 FT)

HAGEDORN AND BROWN
PRESENTED BY
MACCO - SCHLUMBERGER

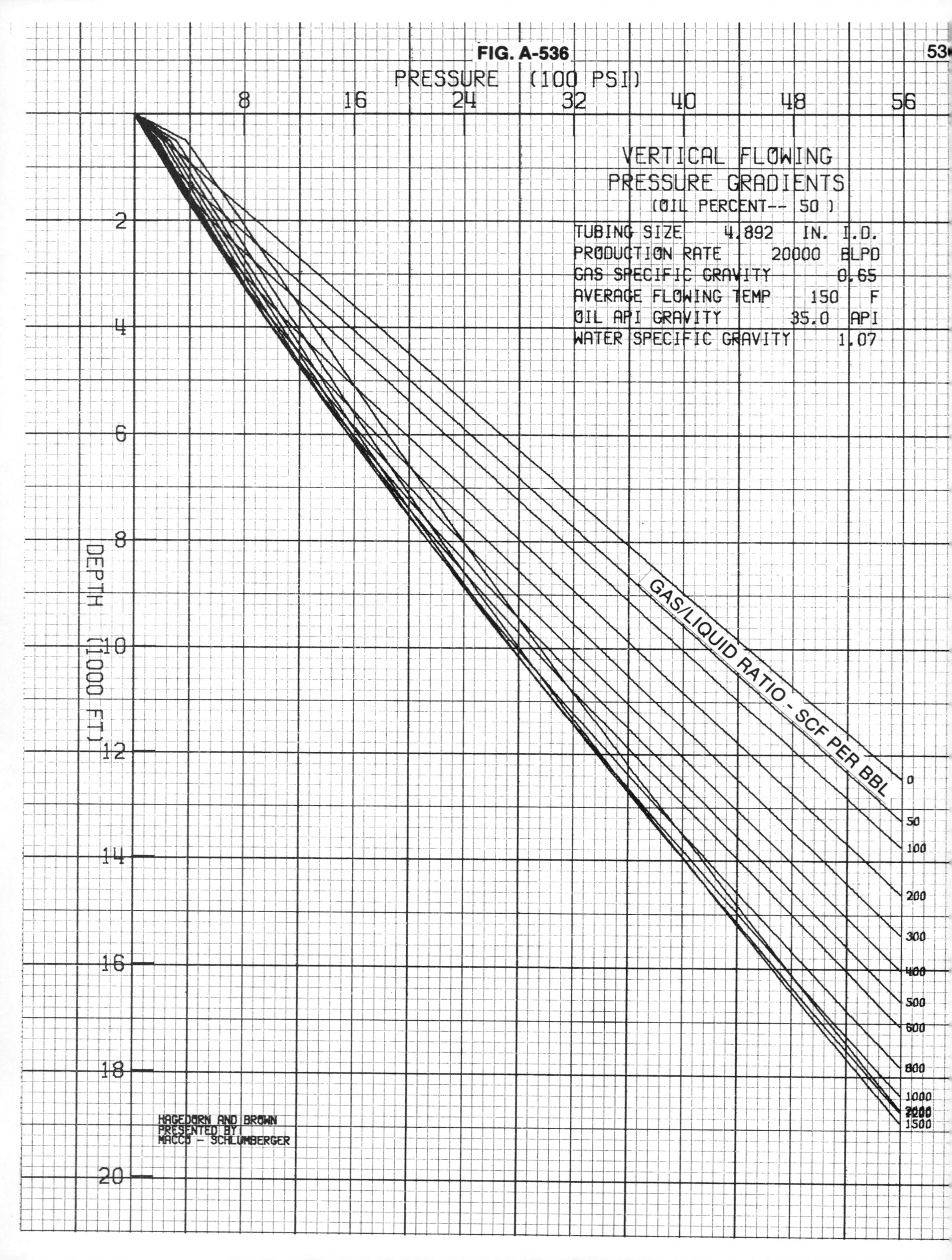

PRESSURE (100 PSI)

VERTICAL FLOWING
PRESSURE GRADIENTS
(OIL PERCENT-- 50)

TUBING SIZE	4.892	IN. I.D.
PRODUCTION RATE	20000	BLPD
GAS SPECIFIC GRAVITY	0.65	
AVERAGE FLOWING TEMP	150	F
OIL API GRAVITY	35.0	API
WATER SPECIFIC GRAVITY	1.07	

GAS/LIQUID RATIO - SCF PER BBL

DEPTH (1000 FT)

HAGEDORN AND BROWN
PRESENTED BY
MACCO – SCHLUMBERGER

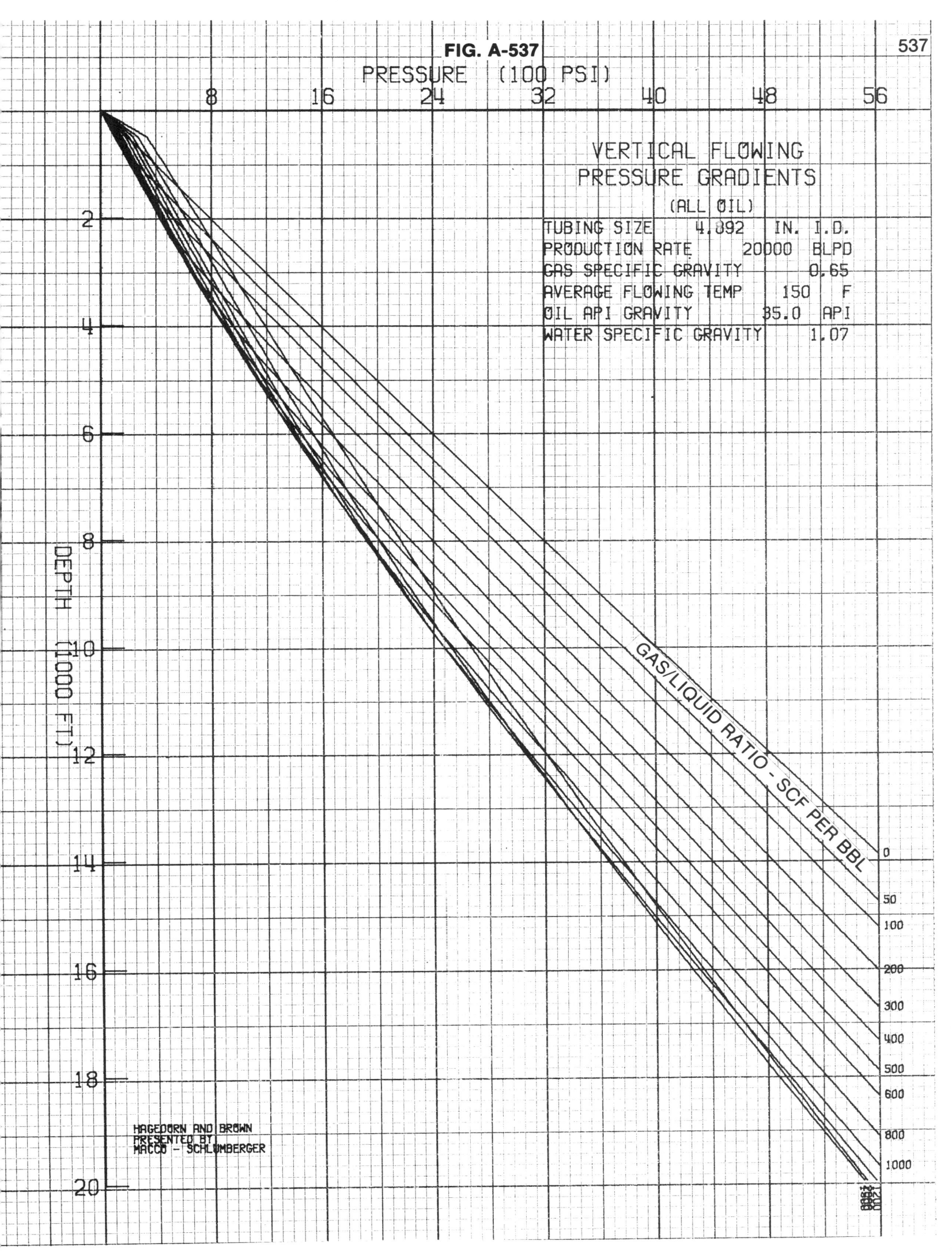

PRESSURE (100 PSI)

VERTICAL FLOWING
PRESSURE GRADIENTS
(ALL OIL)

TUBING SIZE	4.892 IN. I.D.
PRODUCTION RATE	20000 BLPD
GAS SPECIFIC GRAVITY	0.65
AVERAGE FLOWING TEMP	150 F
OIL API GRAVITY	35.0 API
WATER SPECIFIC GRAVITY	1.07

DEPTH (1000 FT.)

GAS/LIQUID RATIO - SCF PER BBL

HAGEDORN AND BROWN
PRESENTED BY
MACCO - SCHLUMBERGER

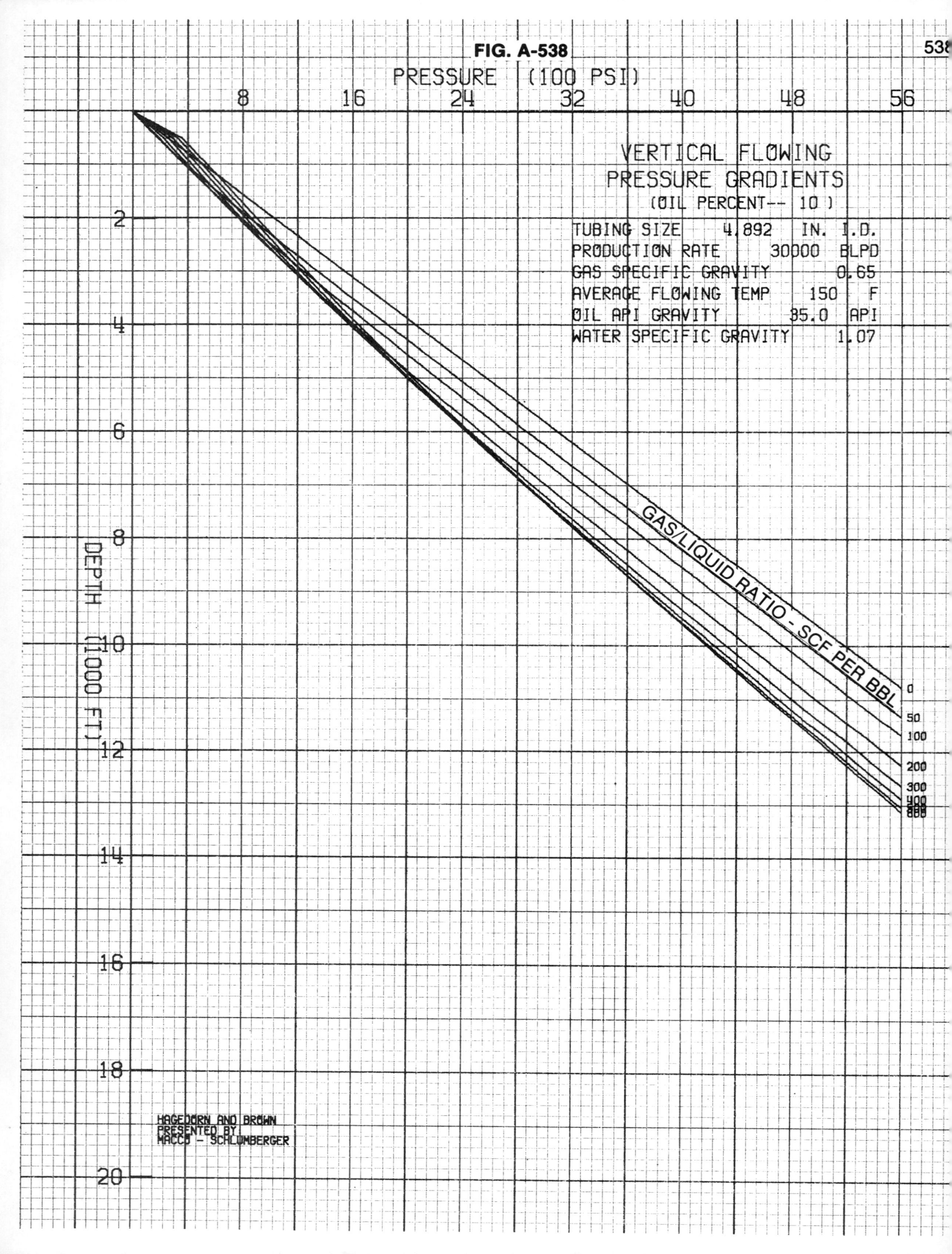

PRESSURE (100 PSI)

VERTICAL FLOWING
PRESSURE GRADIENTS
(OIL PERCENT-- 10)

TUBING SIZE	4.892	IN. I.D.
PRODUCTION RATE	30000	BLPD
GAS SPECIFIC GRAVITY	0.65	
AVERAGE FLOWING TEMP	150	F
OIL API GRAVITY	35.0	API
WATER SPECIFIC GRAVITY	1.07	

GAS/LIQUID RATIO - SCF PER BBL

DEPTH (1000 FT)

0
50
100
200
300
400
800

HAGEDORN AND BROWN
PRESENTED BY
MACCO - SCHLUMBERGER

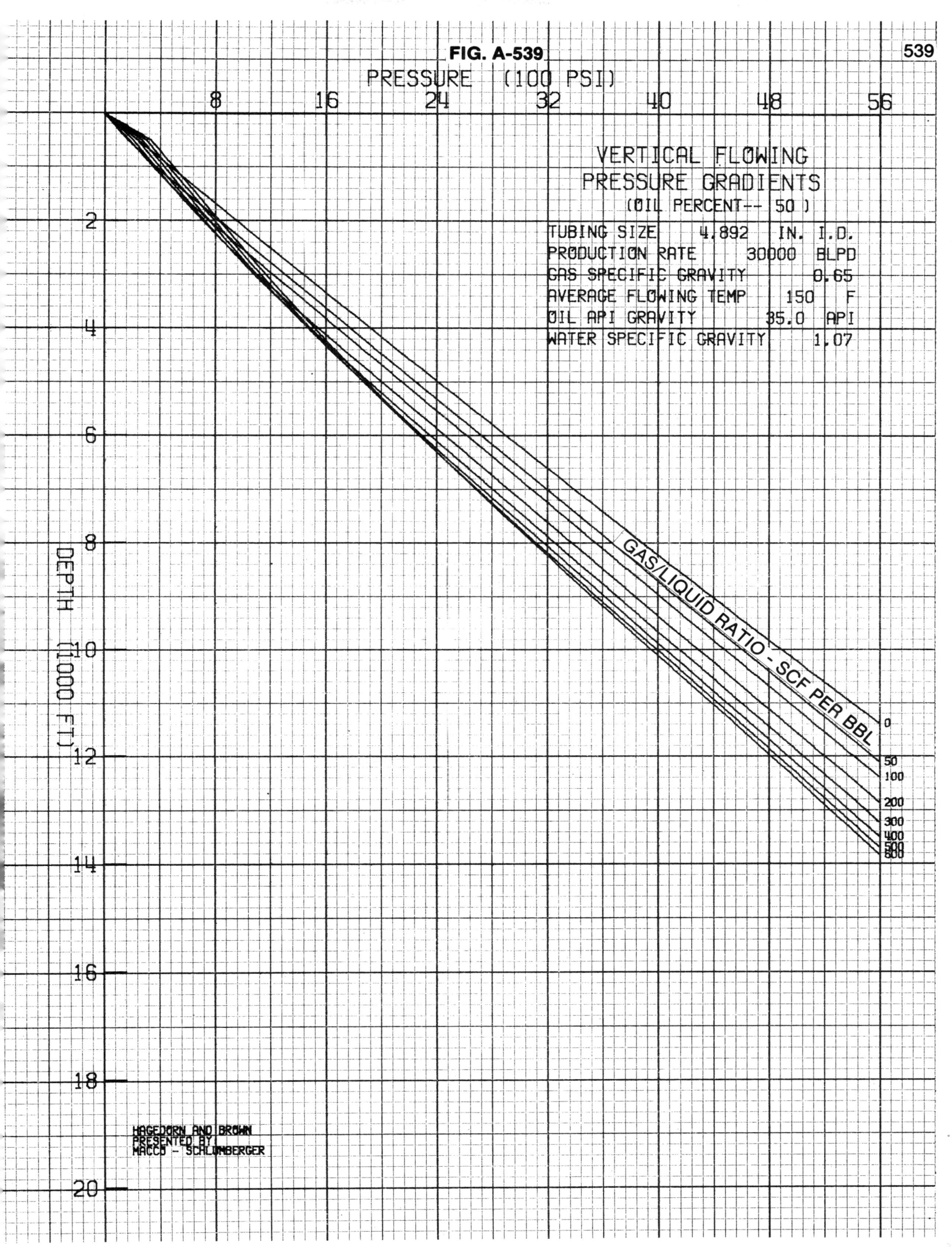

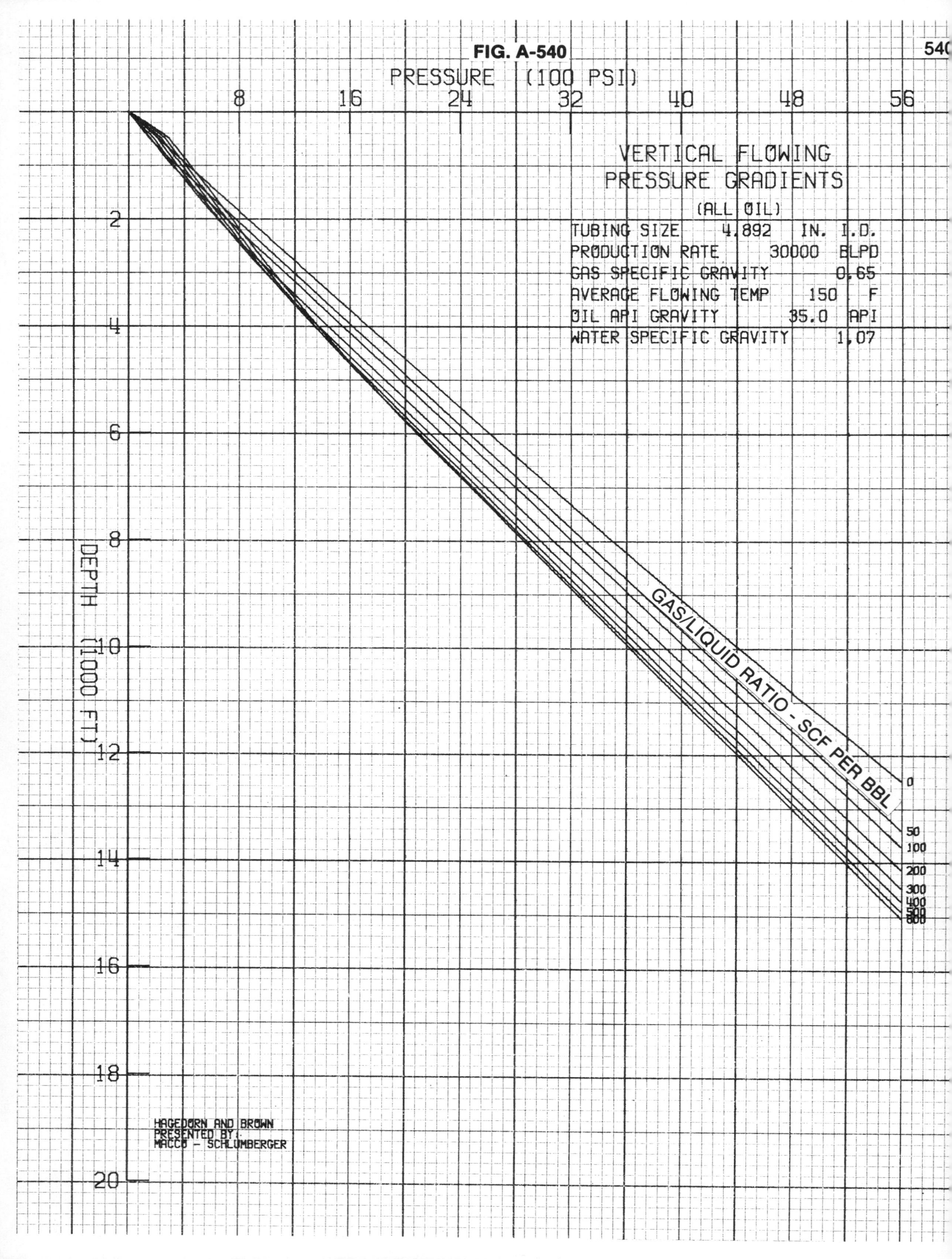

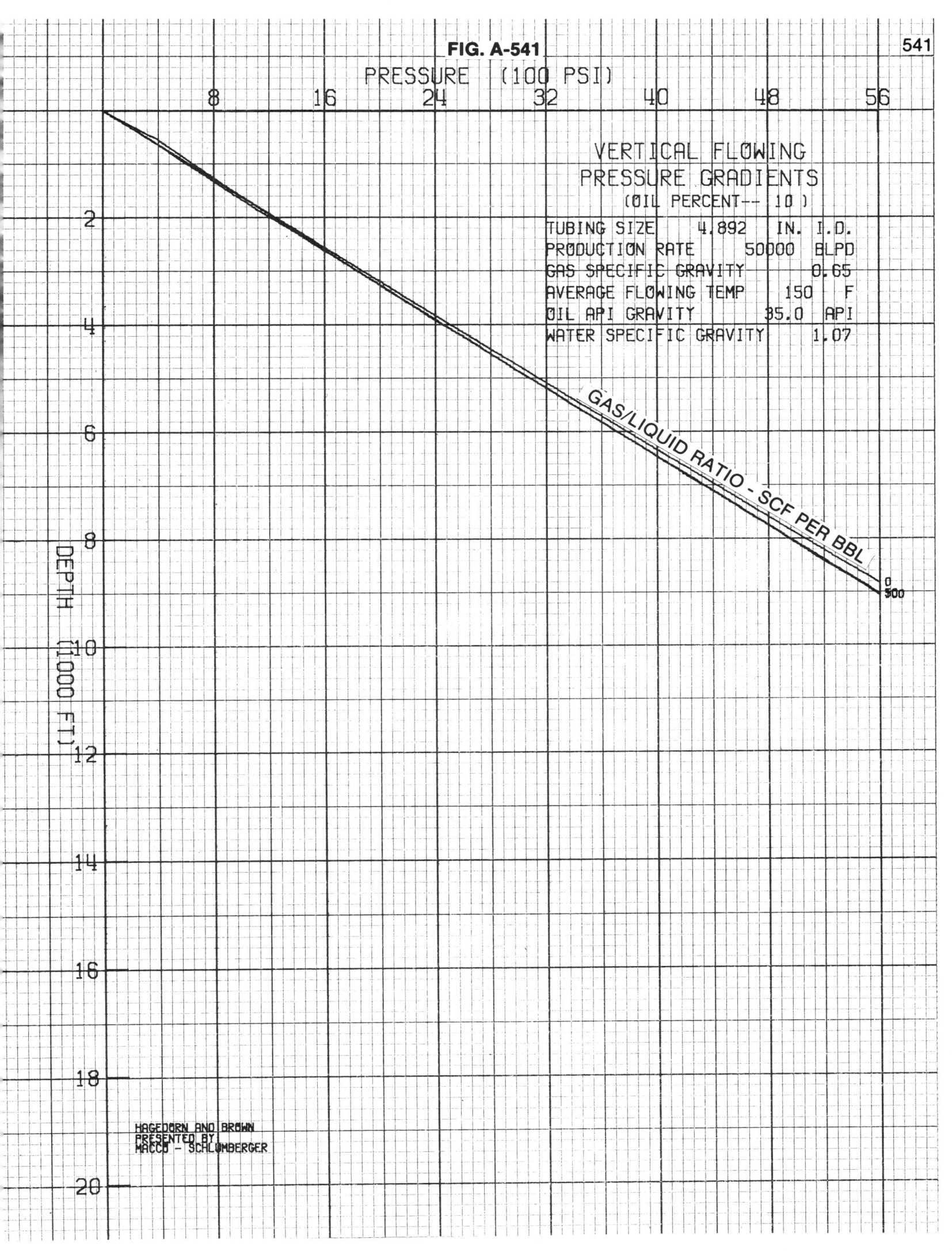

PRESSURE (100 PSI)

VERTICAL FLOWING
PRESSURE GRADIENTS
(OIL PERCENT-- 10)

TUBING SIZE	4.892	IN. I.D.
PRODUCTION RATE	50000	BLPD
GAS SPECIFIC GRAVITY	0.65	
AVERAGE FLOWING TEMP	150	F
OIL API GRAVITY	35.0	API
WATER SPECIFIC GRAVITY	1.07	

GAS/LIQUID RATIO - SCF PER BBL

DEPTH (1000 FT)

HAGEDORN AND BROWN
PRESENTED BY
MACCO - SCHLUMBERGER

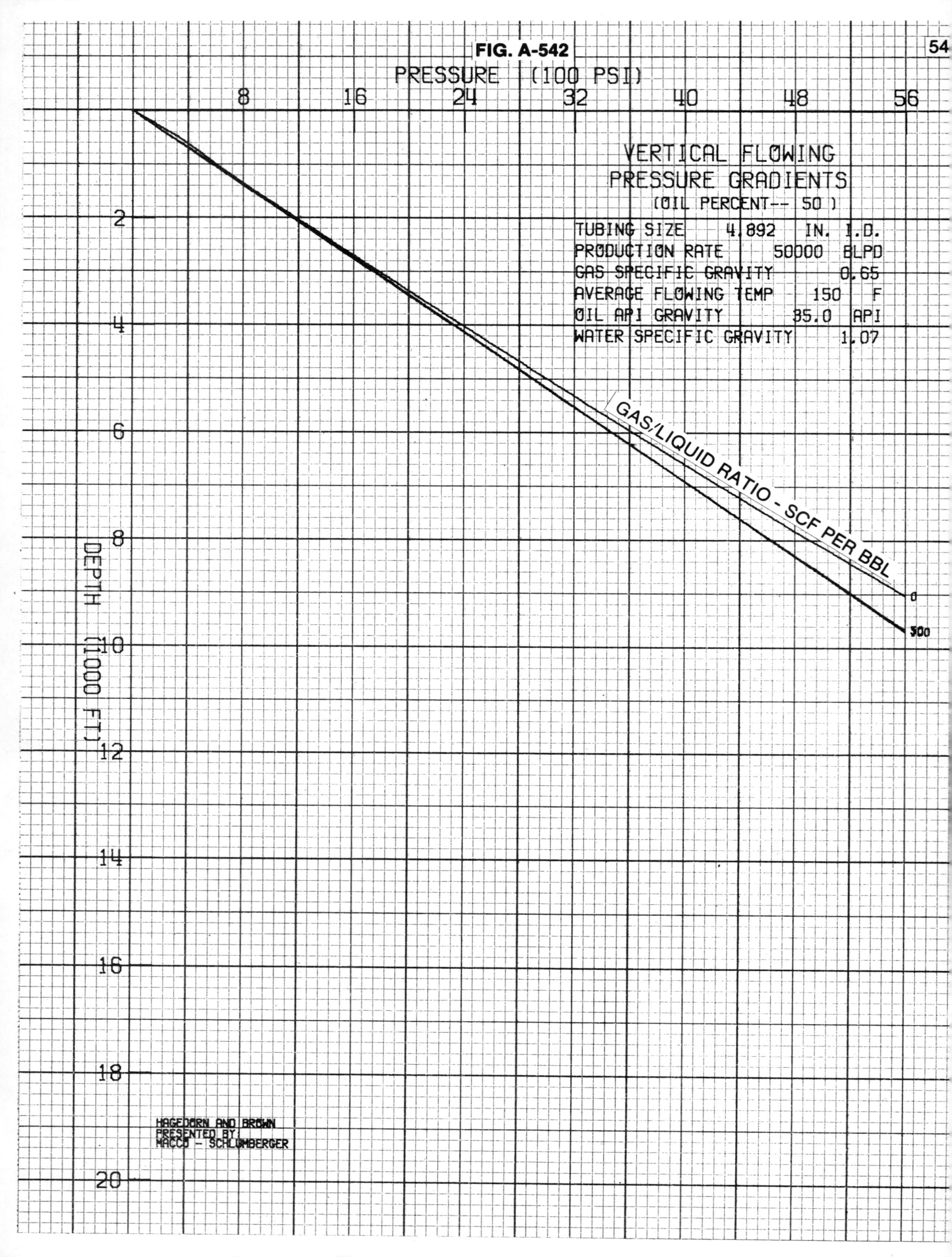

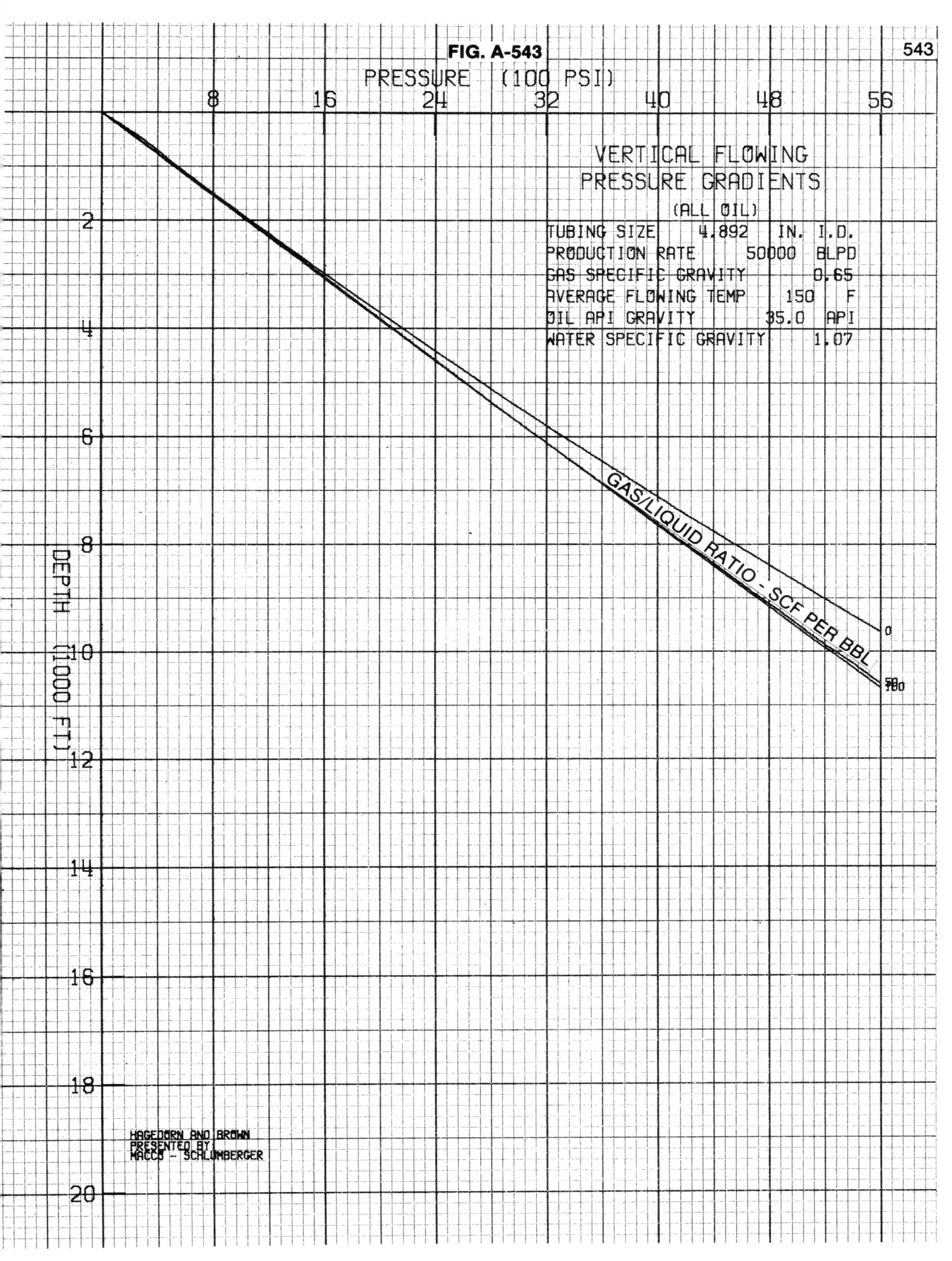

PRESSURE (100 PSI)

VERTICAL FLOWING
PRESSURE GRADIENTS
(ALL OIL)

TUBING SIZE	4.892	IN. I.D.
PRODUCTION RATE	50000	BLPD
GAS SPECIFIC GRAVITY	0.65	
AVERAGE FLOWING TEMP	150	F
OIL API GRAVITY	35.0	API
WATER SPECIFIC GRAVITY	1.07	

GAS/LIQUID RATIO - SCF PER BBL

DEPTH (1000 FT)

HAGEDORN AND BROWN
PRESENTED BY
MACCO - SCHLUMBERGER

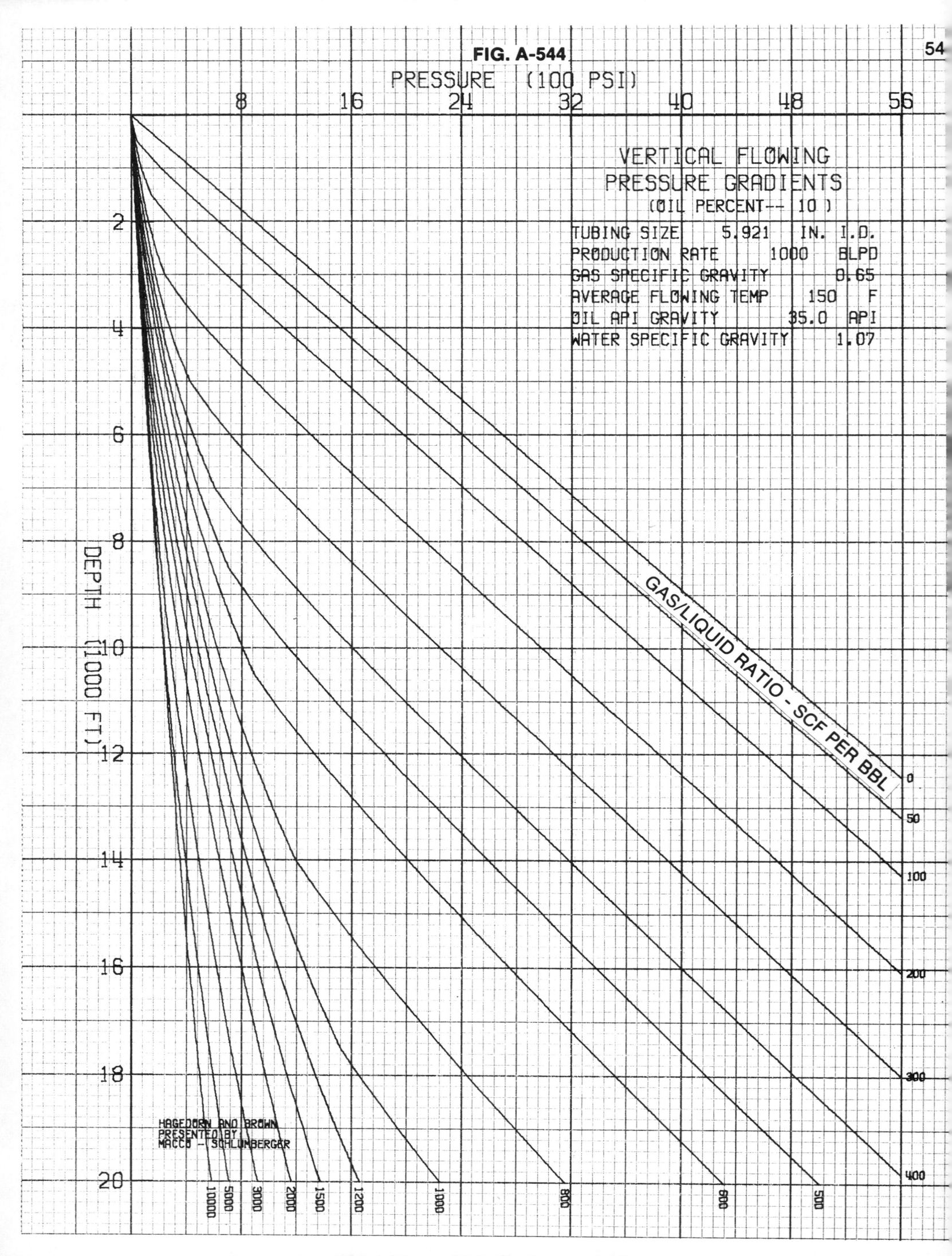

FIG. A-544

54

PRESSURE (100 PSI)

VERTICAL FLOWING
PRESSURE GRADIENTS
(OIL PERCENT-- 10)

TUBING SIZE	5.921	IN. I.D.
PRODUCTION RATE	1000	BLPD
GAS SPECIFIC GRAVITY	0.65	
AVERAGE FLOWING TEMP	150	F
OIL API GRAVITY	35.0	API
WATER SPECIFIC GRAVITY	1.07	

DEPTH (1000 FT)

GAS/LIQUID RATIO - SCF PER BBL

HAGEDORN AND BROWN
PRESENTED BY
MACCO - SCHLUMBERGER

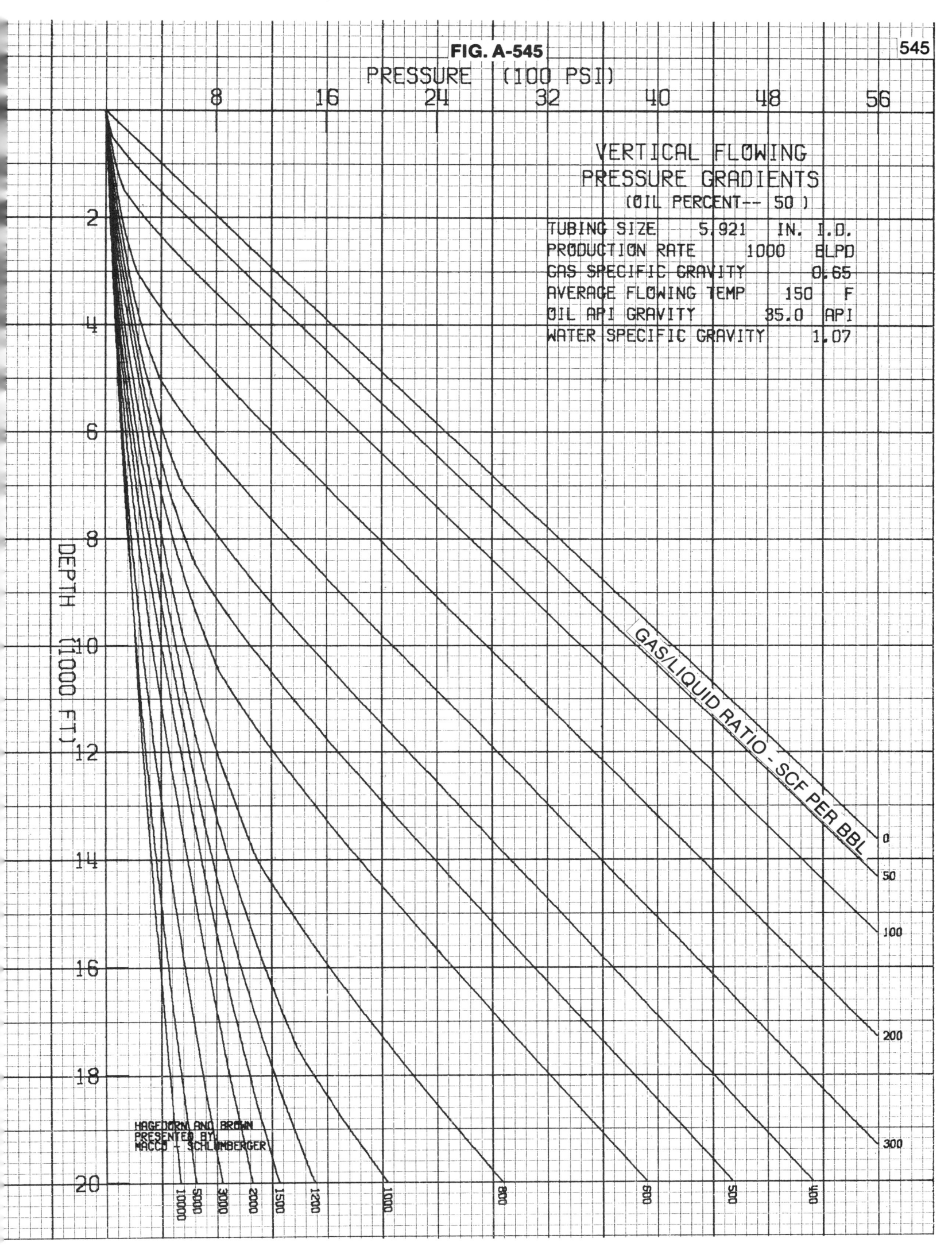

PRESSURE (100 PSI)

VERTICAL FLOWING
PRESSURE GRADIENTS
(OIL PERCENT-- 50)

TUBING SIZE	5.921	IN. I.D.
PRODUCTION RATE	1000	BLPD
GAS SPECIFIC GRAVITY		0.65
AVERAGE FLOWING TEMP	150	F
OIL API GRAVITY	35.0	API
WATER SPECIFIC GRAVITY		1.07

GAS/LIQUID RATIO - SCF PER BBL

DEPTH (1000 FT)

HAGEDORN AND BROWN
PRESENTED BY
MACCO - SCHLUMBERGER

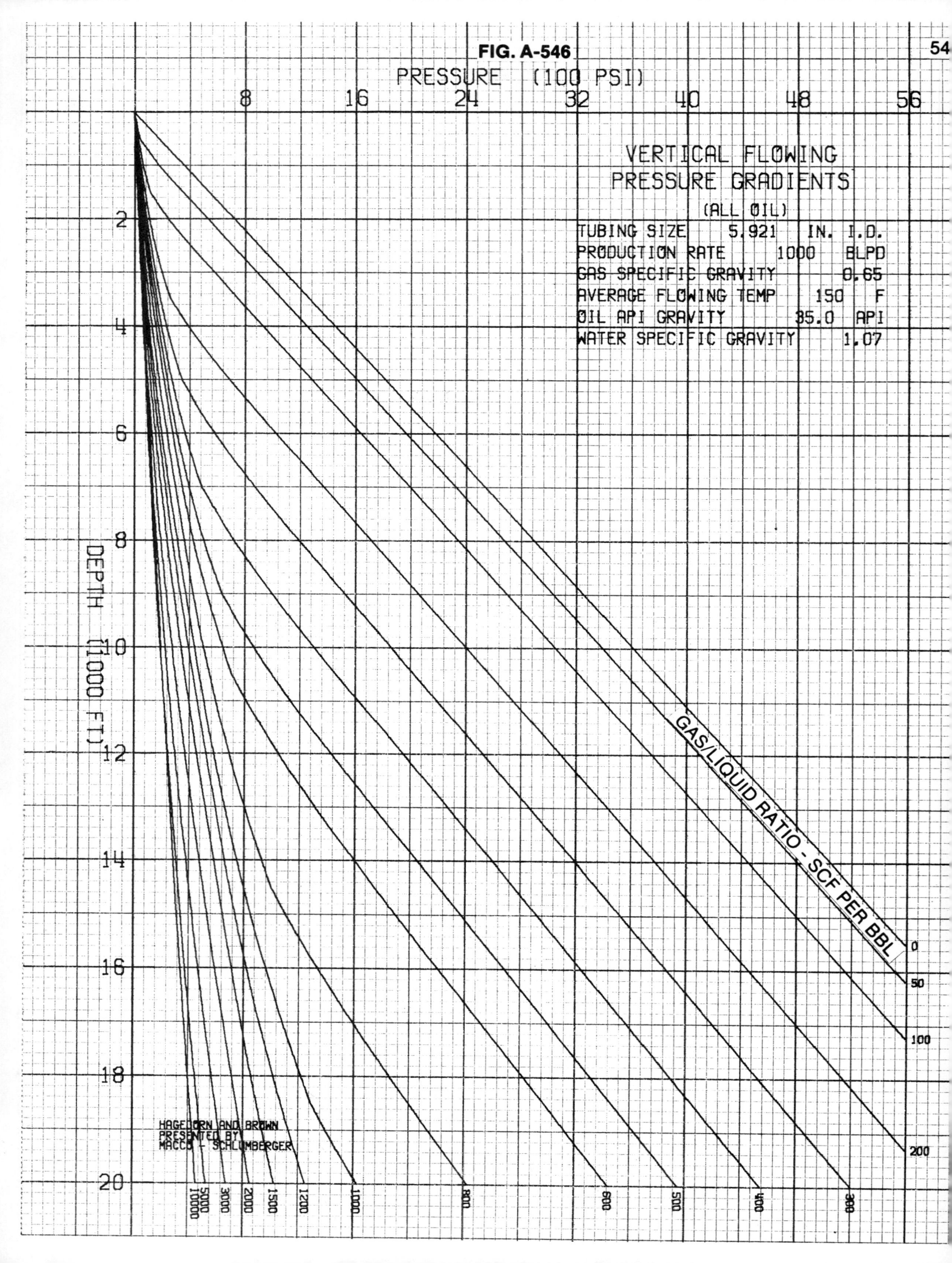

FIG. A-546

54

VERTICAL FLOWING
PRESSURE GRADIENTS
(ALL OIL)

TUBING SIZE	5.921	IN. I.D.
PRODUCTION RATE	1000	BLPD
GAS SPECIFIC GRAVITY	0.65	
AVERAGE FLOWING TEMP	150	F
OIL API GRAVITY	35.0	API
WATER SPECIFIC GRAVITY	1.07	

PRESSURE (100 PSI)

DEPTH (1000 FT)

GAS/LIQUID RATIO - SCF PER BBL

HAGEDORN AND BROWN
PRESENTED BY
MACCO - SCHLUMBERGER

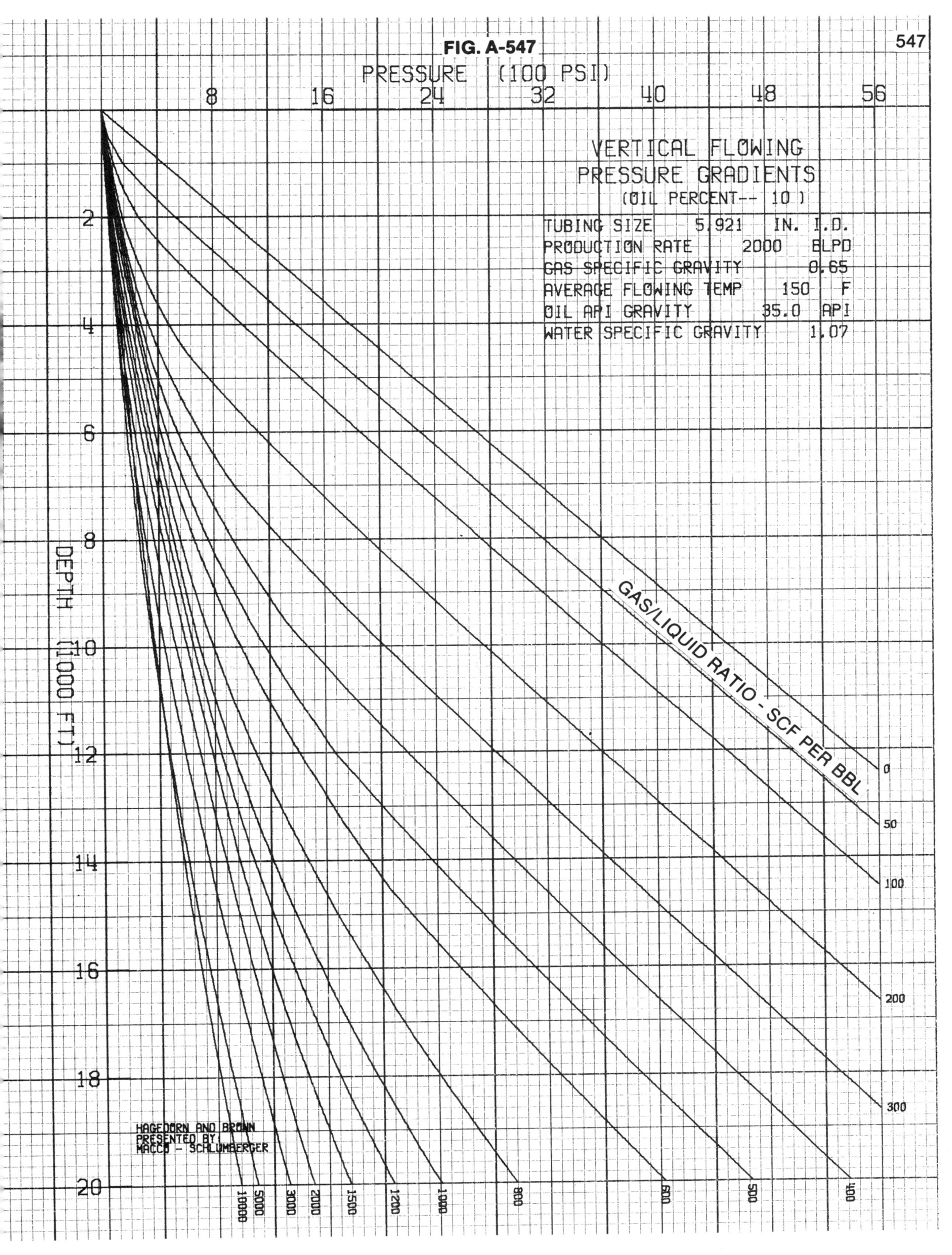

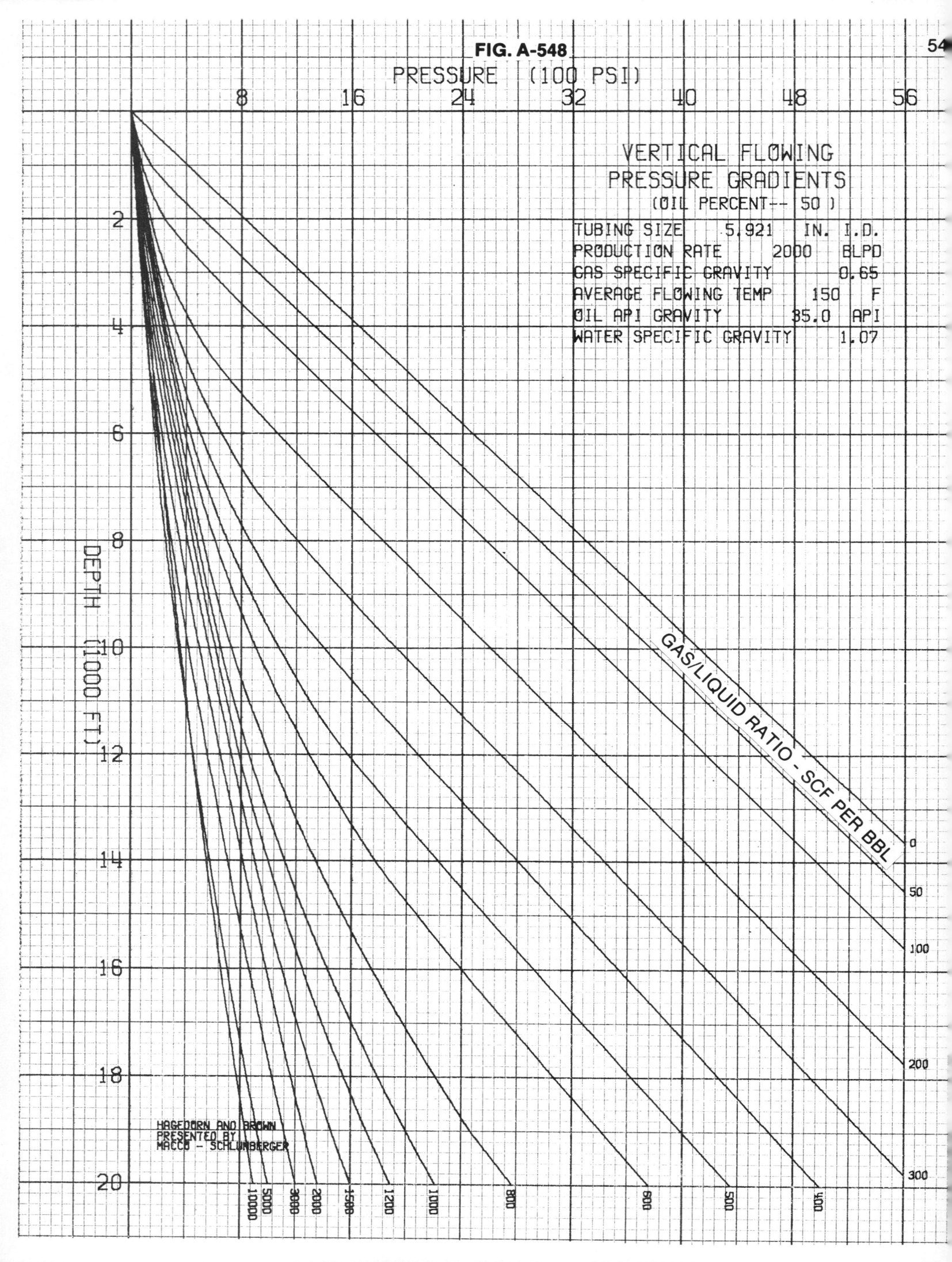

FIG. A-548

PRESSURE (100 PSI)

VERTICAL FLOWING
PRESSURE GRADIENTS
(OIL PERCENT-- 50)

TUBING SIZE   5.921   IN. I.D.
PRODUCTION RATE   2000   BLPD
GAS SPECIFIC GRAVITY   0.65
AVERAGE FLOWING TEMP   150   F
OIL API GRAVITY   35.0 API
WATER SPECIFIC GRAVITY   1.07

GAS/LIQUID RATIO - SCF PER BBL

DEPTH (1000 FT)

HAGEDORN AND BROWN
PRESENTED BY
MACCO -- SCHLUMBERGER

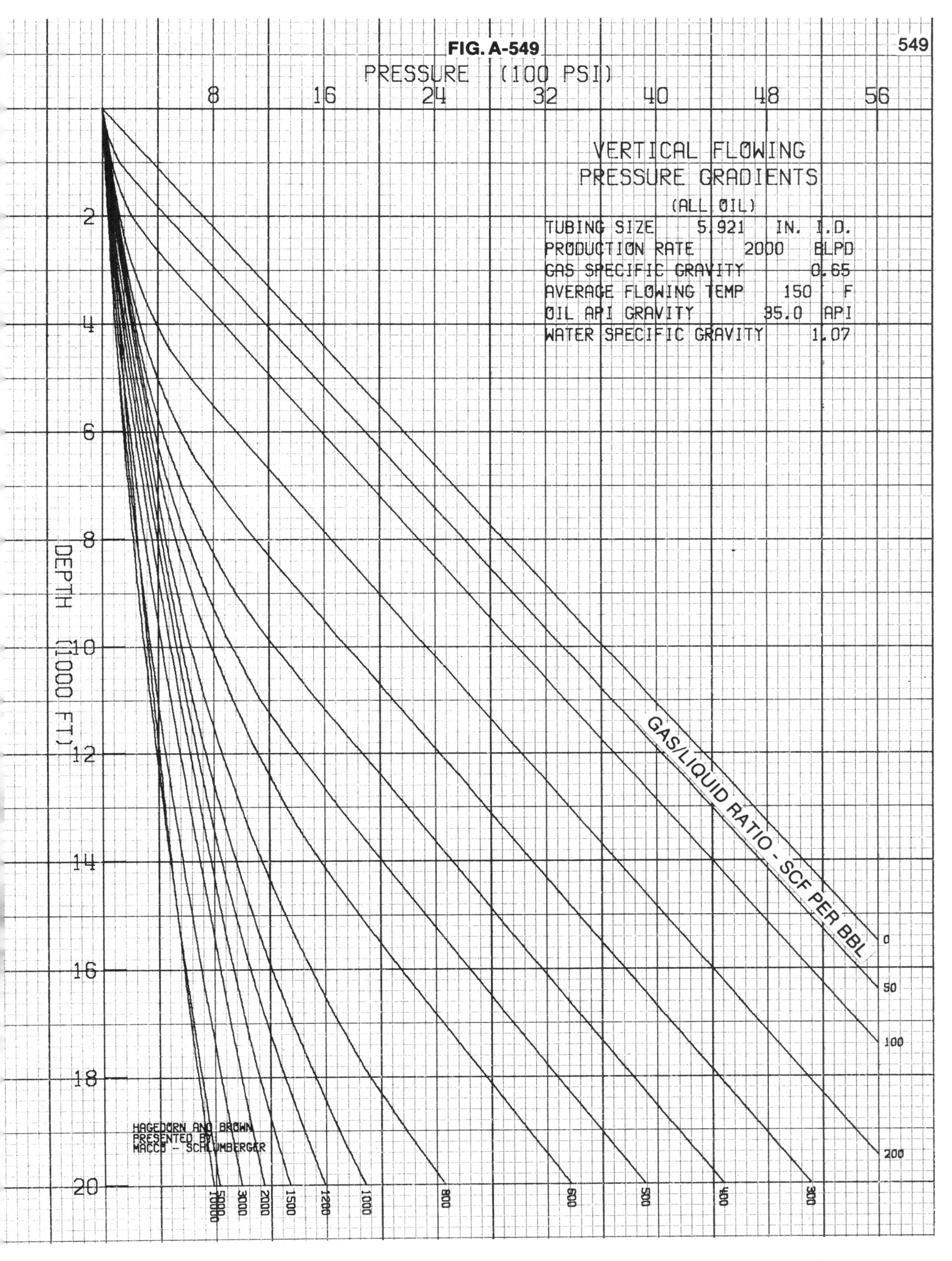

PRESSURE (100 PSI)

VERTICAL FLOWING
PRESSURE GRADIENTS
(ALL OIL)

TUBING SIZE	5.921	IN. I.D.
PRODUCTION RATE	2000	BLPD
GAS SPECIFIC GRAVITY	0.65	
AVERAGE FLOWING TEMP	150	F
OIL API GRAVITY	35.0	API
WATER SPECIFIC GRAVITY	1.07	

DEPTH (1000 FT)

GAS/LIQUID RATIO - SCF PER BBL

HAGEDORN AND BROWN
PRESENTED BY
MACCO - SCHLUMBERGER

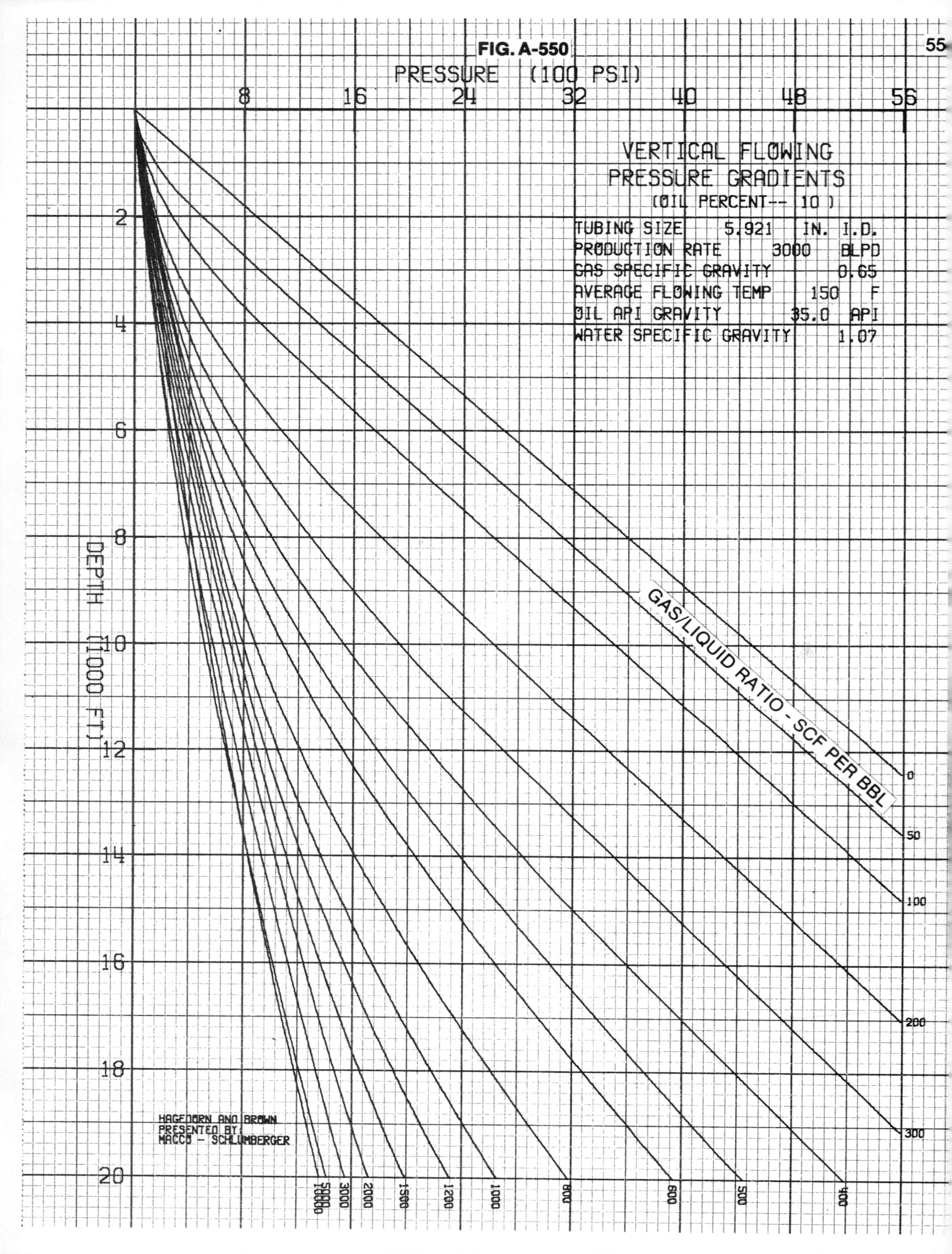

FIG. A-550

55

VERTICAL FLOWING
PRESSURE GRADIENTS
(OIL PERCENT-- 10)

TUBING SIZE	5.921	IN. I.D.
PRODUCTION RATE	3000	BLPD
GAS SPECIFIC GRAVITY	0.65	
AVERAGE FLOWING TEMP	150	F
OIL API GRAVITY	35.0	API
WATER SPECIFIC GRAVITY	1.07	

PRESSURE (100 PSI)

DEPTH (1000 FT)

GAS/LIQUID RATIO - SCF PER BBL

HAGEDORN AND BROWN
PRESENTED BY:
MACCO - SCHLUMBERGER

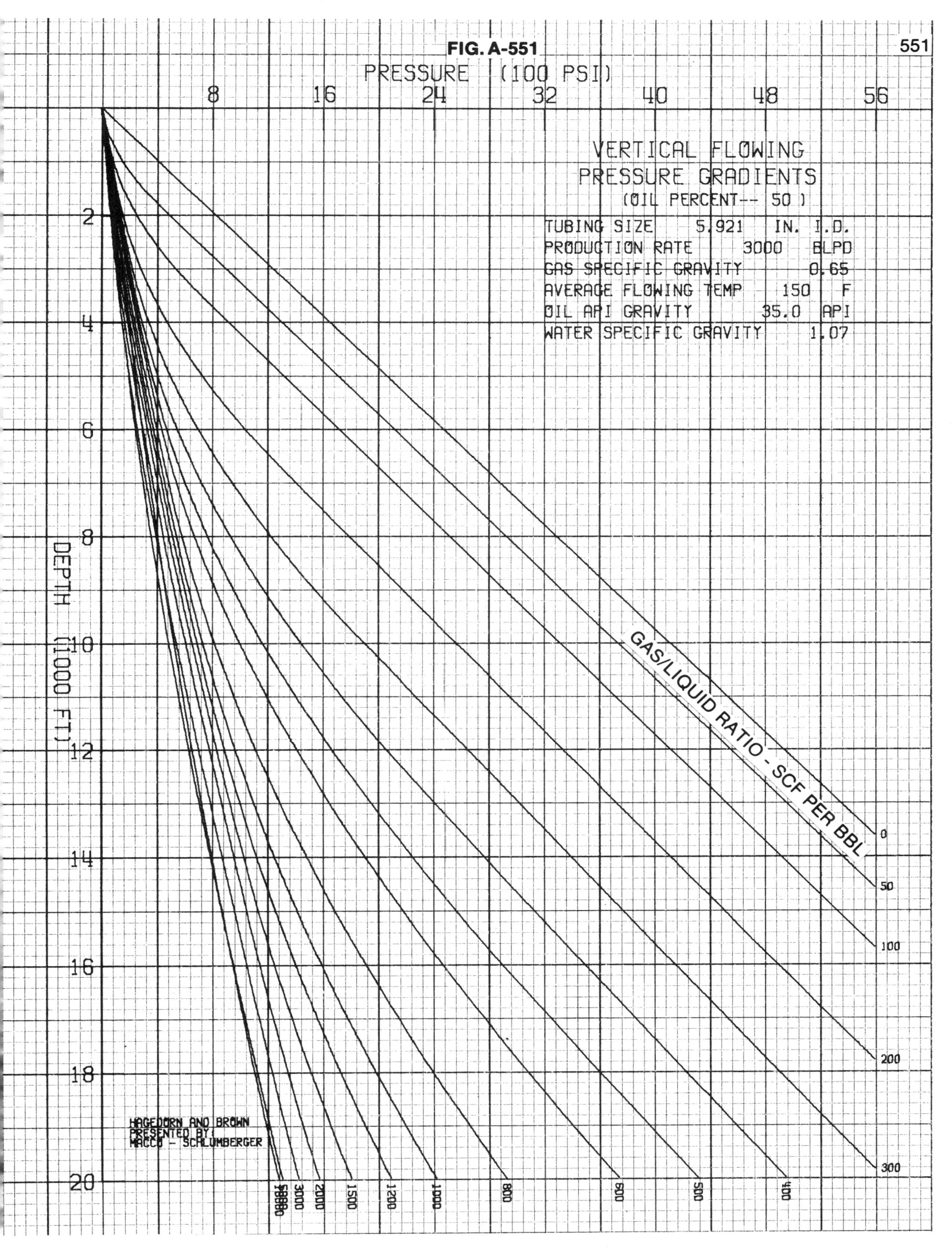

PRESSURE (100 PSI)

VERTICAL FLOWING
PRESSURE GRADIENTS
(OIL PERCENT-- 50)

TUBING SIZE	5.921	IN. I.D.
PRODUCTION RATE	3000	BLPD
GAS SPECIFIC GRAVITY	0.65	
AVERAGE FLOWING TEMP	150	F
OIL API GRAVITY	35.0	API
WATER SPECIFIC GRAVITY	1.07	

DEPTH (1000 FT)

GAS/LIQUID RATIO - SCF PER BBL

HAGEDORN AND BROWN
PRESENTED BY:
MACCO - SCHLUMBERGER

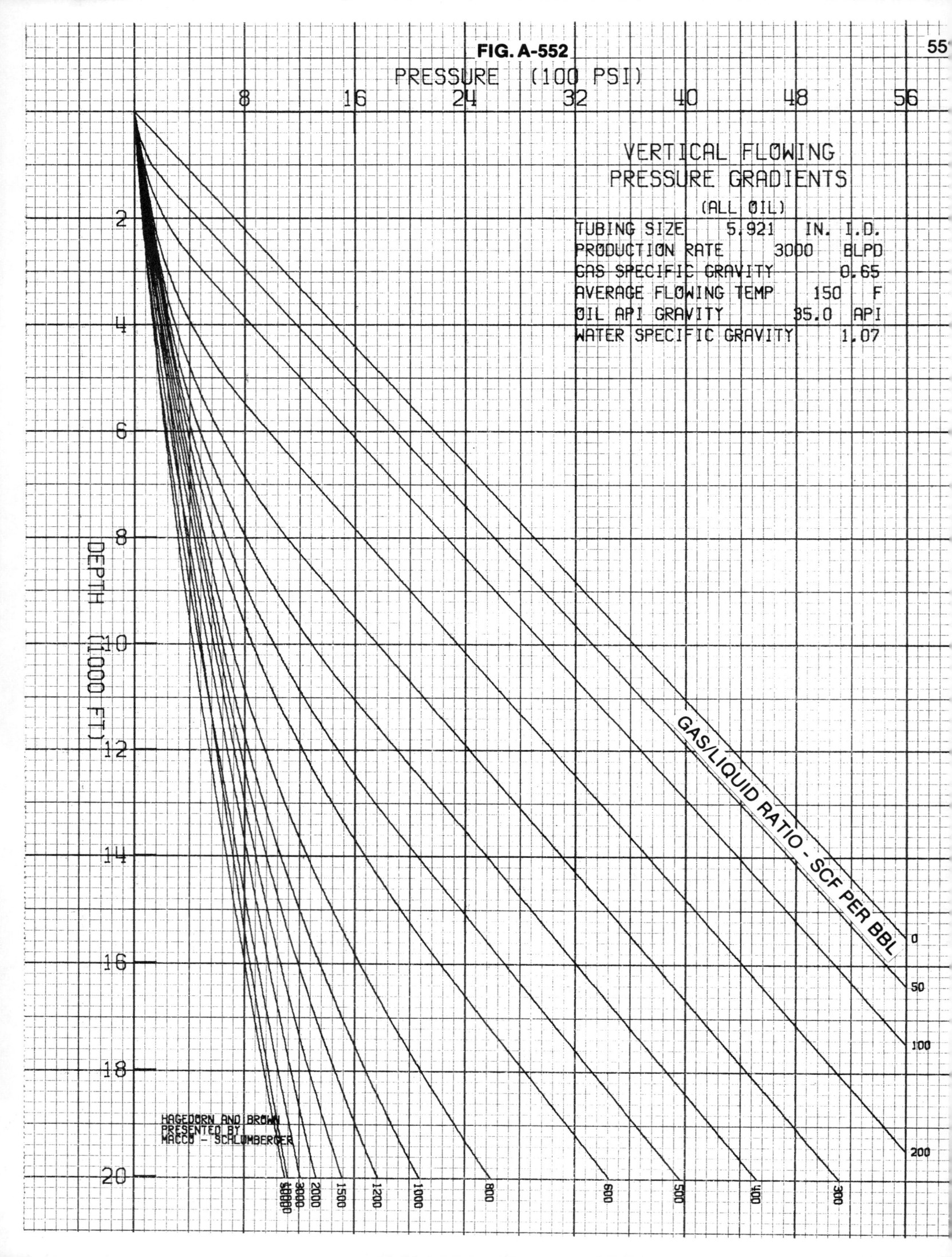

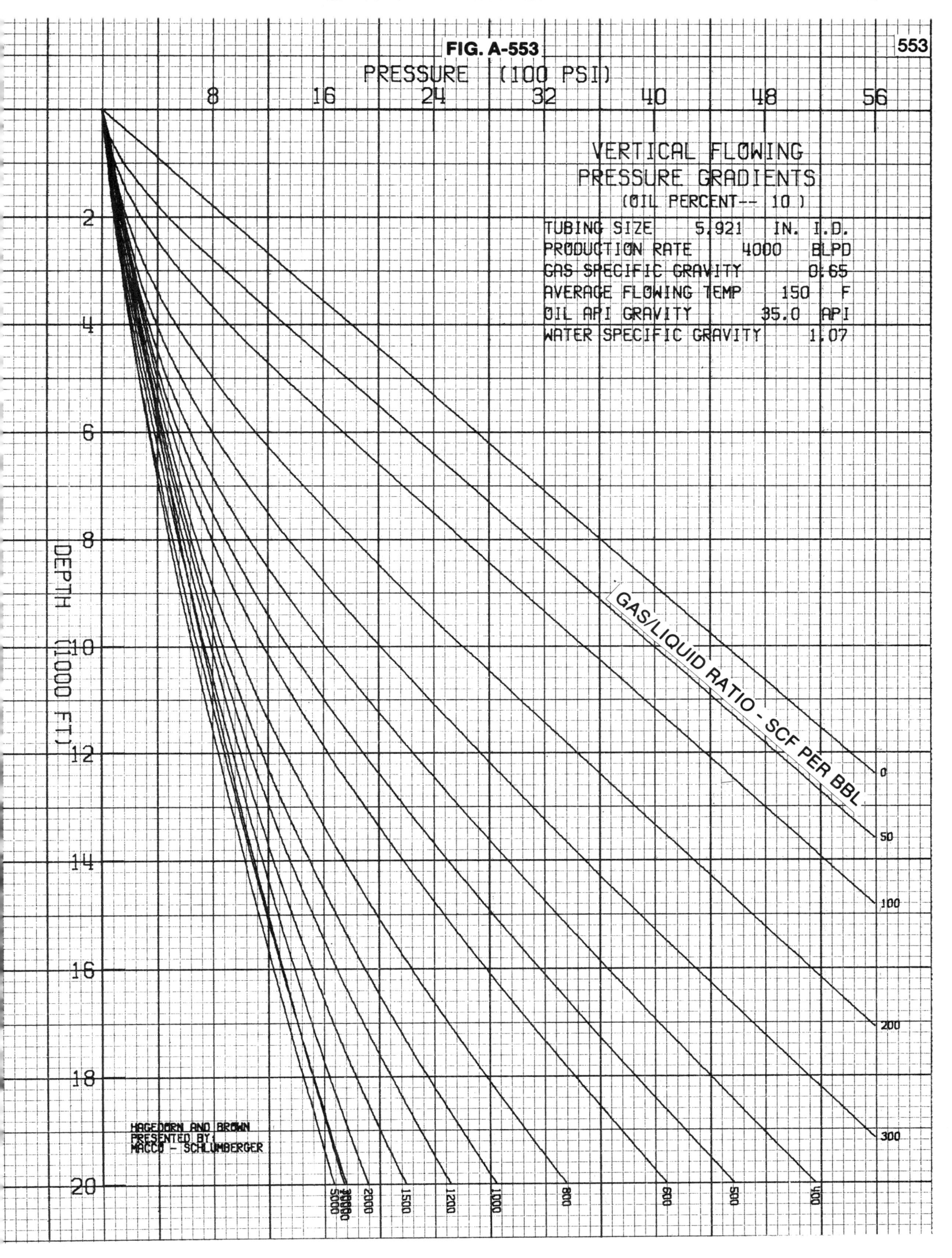

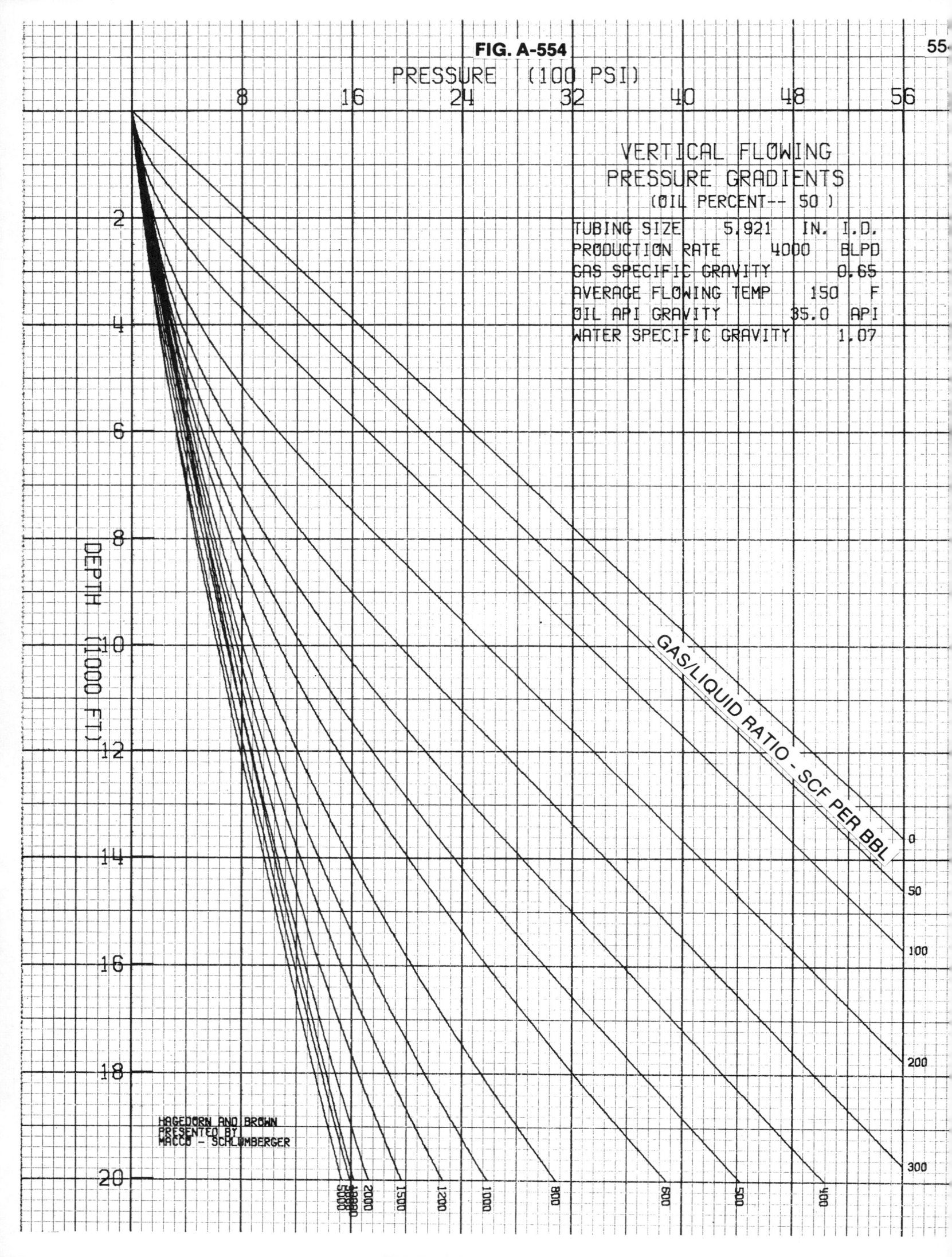

PRESSURE (100 PSI)

VERTICAL FLOWING
PRESSURE GRADIENTS
(OIL PERCENT-- 50)

TUBING SIZE	5.921	IN. I.D.
PRODUCTION RATE	4000	BLPD
GAS SPECIFIC GRAVITY		0.65
AVERAGE FLOWING TEMP	150	F
OIL API GRAVITY	35.0	API
WATER SPECIFIC GRAVITY		1.07

DEPTH (1000 FT)

GAS/LIQUID RATIO - SCF PER BBL

HAGEDORN AND BROWN
PRESENTED BY
MACCO - SCHLUMBERGER

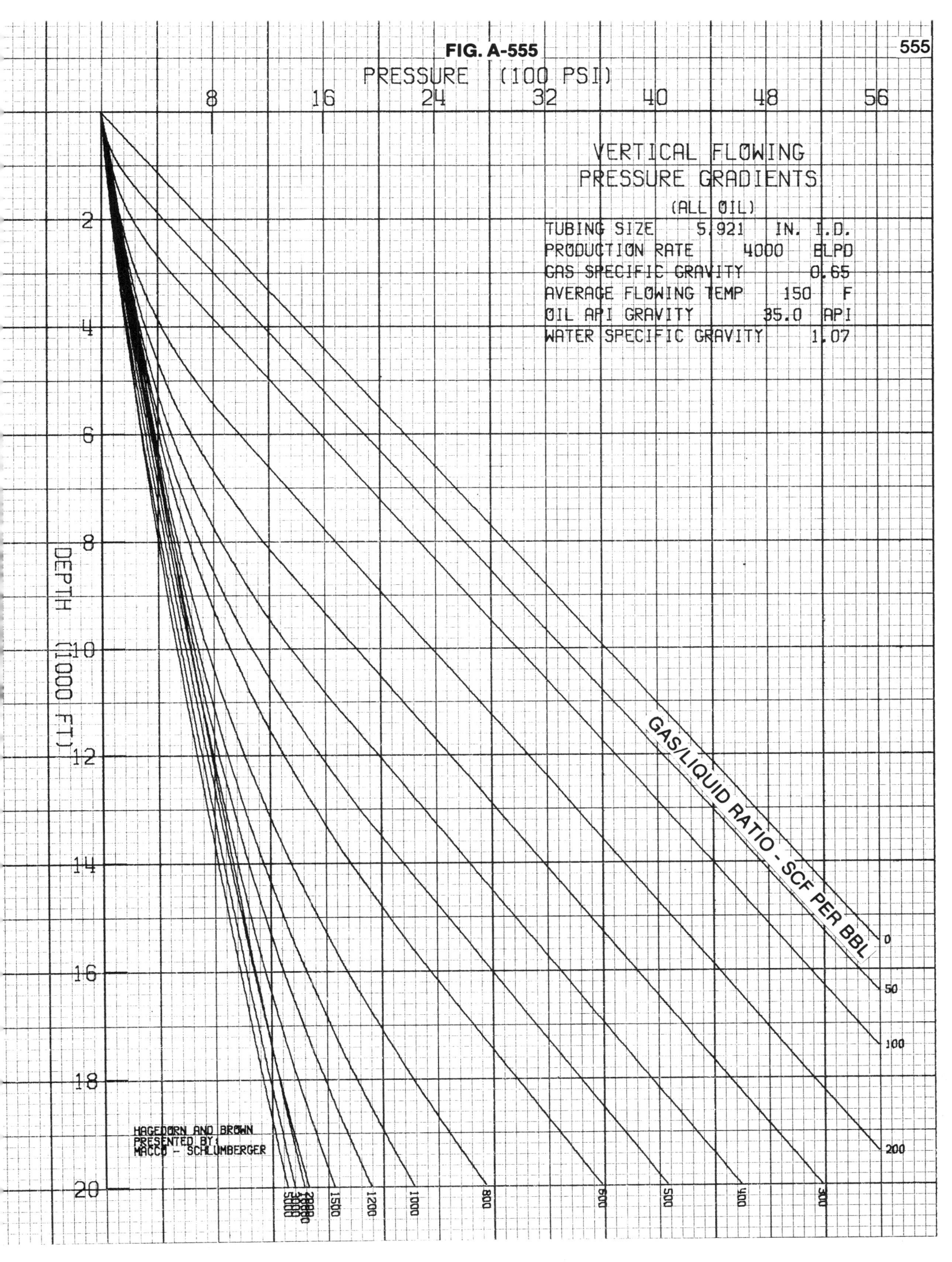

FIG. A-555

555

PRESSURE (100 PSI)

VERTICAL FLOWING
PRESSURE GRADIENTS
(ALL OIL)

TUBING SIZE	5.921	IN. I.D.
PRODUCTION RATE	4000	BLPD
GAS SPECIFIC GRAVITY	0.65	
AVERAGE FLOWING TEMP	150	F
OIL API GRAVITY	35.0	API
WATER SPECIFIC GRAVITY	1.07	

DEPTH (1000 FT)

GAS/LIQUID RATIO - SCF PER BBL

HAGEDORN AND BROWN
PRESENTED BY:
MACCO - SCHLUMBERGER

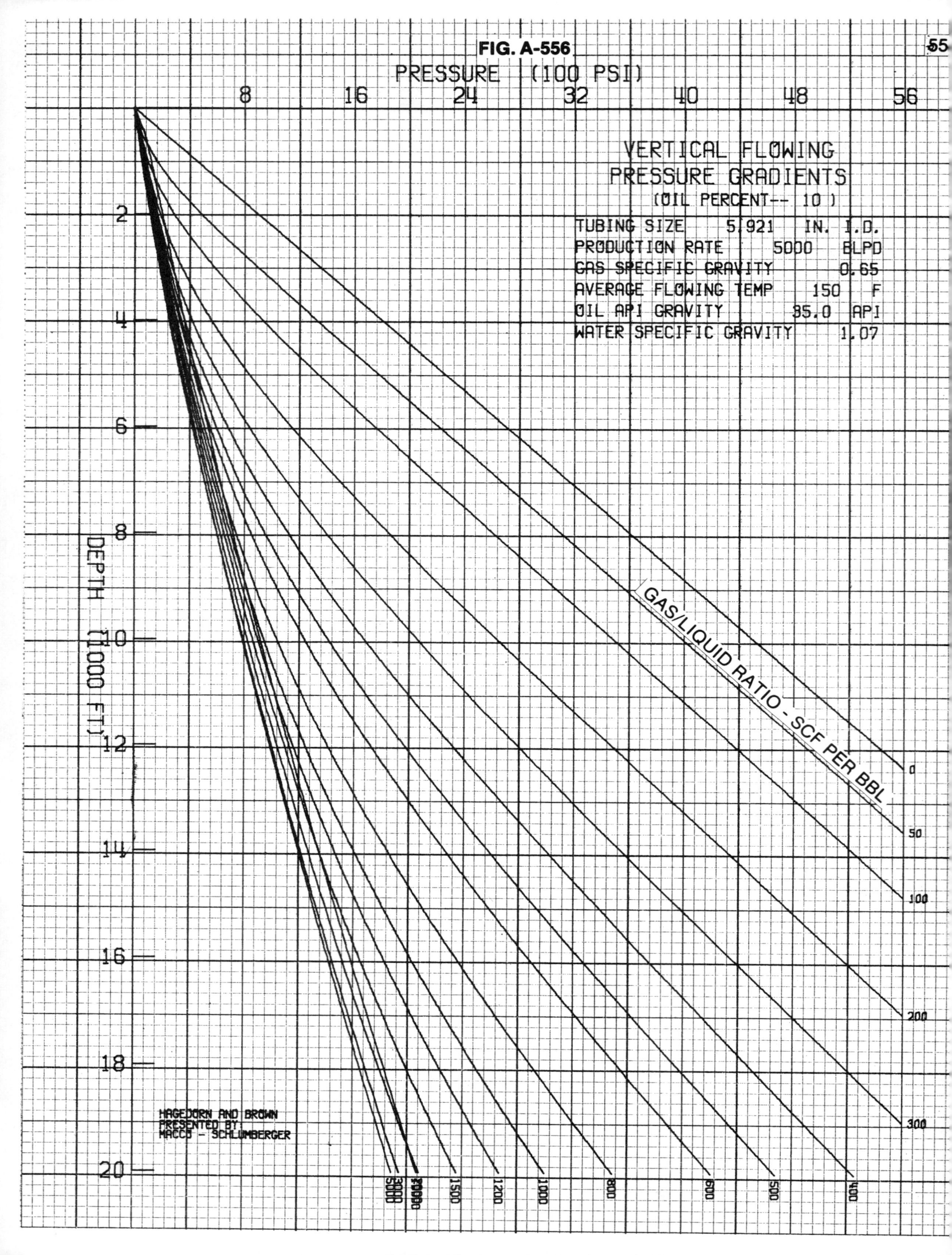

FIG. A-556

55

PRESSURE (100 PSI)

VERTICAL FLOWING
PRESSURE GRADIENTS
(OIL PERCENT-- 10)

TUBING SIZE	5.921	IN. I.D.
PRODUCTION RATE	5000	BLPD
GAS SPECIFIC GRAVITY		0.65
AVERAGE FLOWING TEMP	150	F
OIL API GRAVITY	35.0	API
WATER SPECIFIC GRAVITY		1.07

DEPTH (1000 FT)

GAS/LIQUID RATIO - SCF PER BBL

HAGEDORN AND BROWN
PRESENTED BY:
MACCO - SCHLUMBERGER

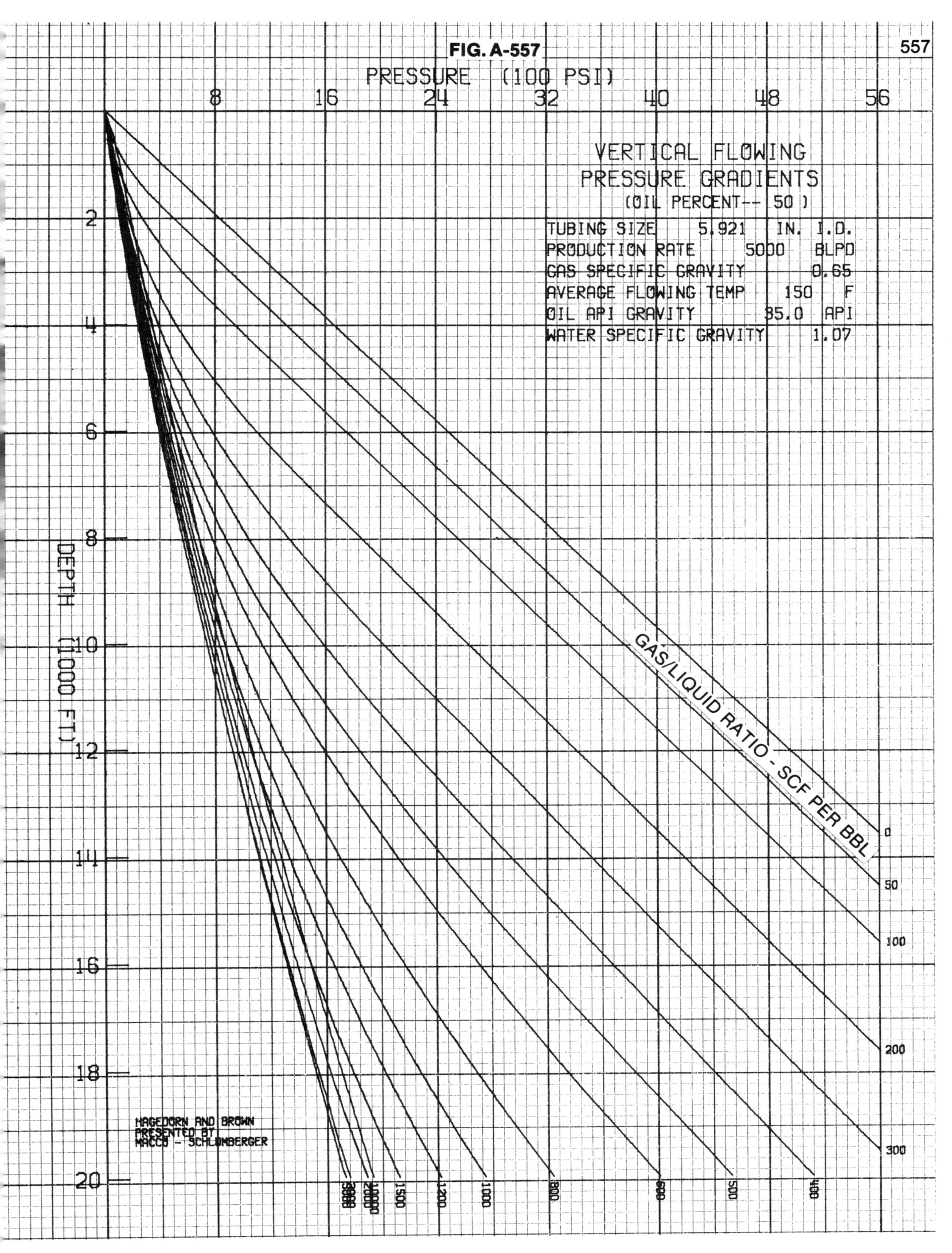

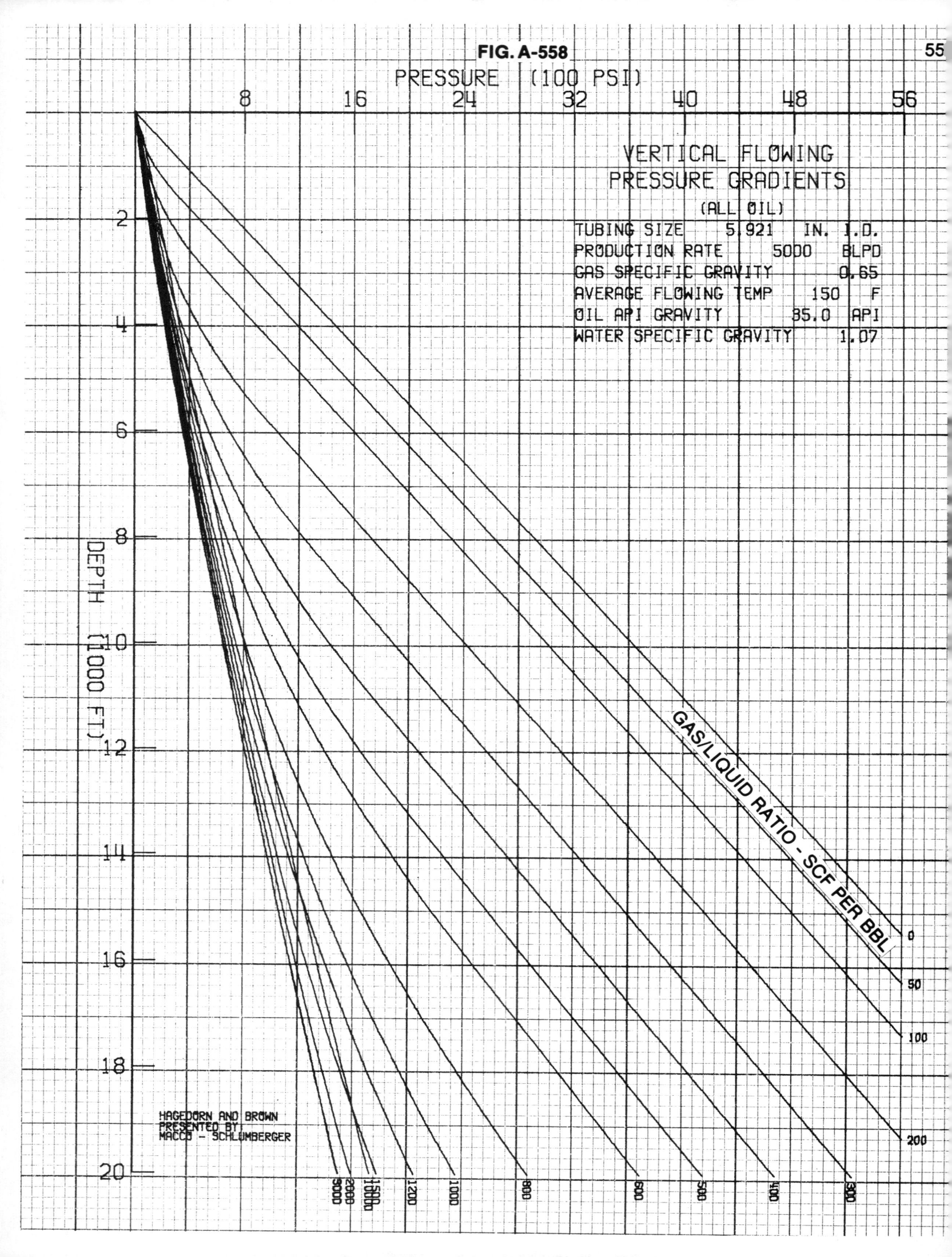

PRESSURE (100 PSI)

VERTICAL FLOWING
PRESSURE GRADIENTS
(ALL OIL)

TUBING SIZE	5.921	IN. I.D.
PRODUCTION RATE	5000	BLPD
GAS SPECIFIC GRAVITY		0.65
AVERAGE FLOWING TEMP	150	F
OIL API GRAVITY	35.0	API
WATER SPECIFIC GRAVITY		1.07

DEPTH (1000 FT)

GAS/LIQUID RATIO - SCF PER BBL

HAGEDORN AND BROWN
PRESENTED BY
MACCO - SCHLUMBERGER

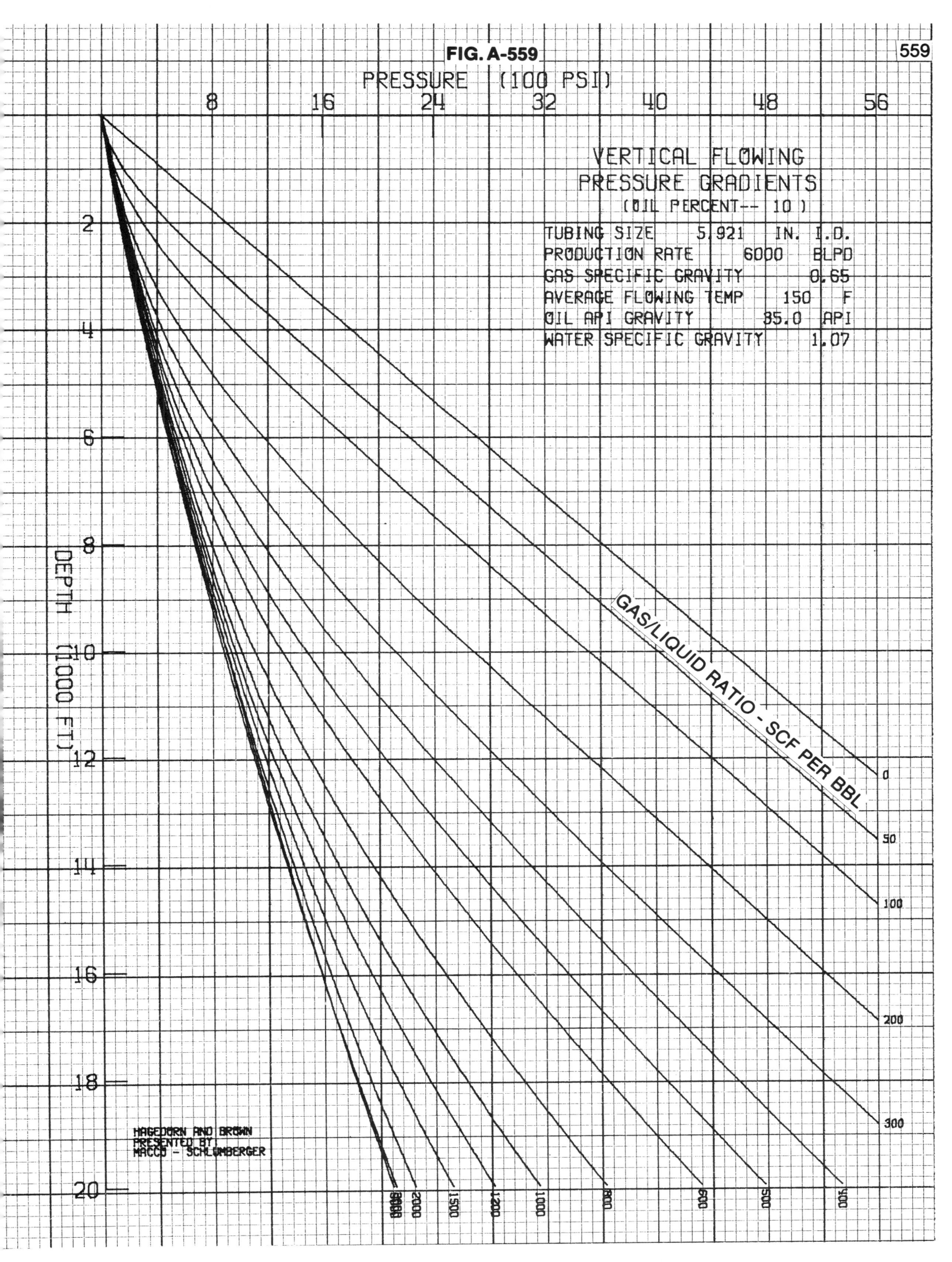

FIG. A-559

559

PRESSURE (100 PSI)

VERTICAL FLOWING
PRESSURE GRADIENTS
(OIL PERCENT-- 10)

TUBING SIZE   5.921   IN. I.D.
PRODUCTION RATE   6000   BLPD
GAS SPECIFIC GRAVITY   0.65
AVERAGE FLOWING TEMP   150   F
OIL API GRAVITY   85.0   API
WATER SPECIFIC GRAVITY   1.07

GAS/LIQUID RATIO - SCF PER BBL

DEPTH (1000 FT)

HAGEDORN AND BROWN
PRESENTED BY
MACCO - SCHLUMBERGER

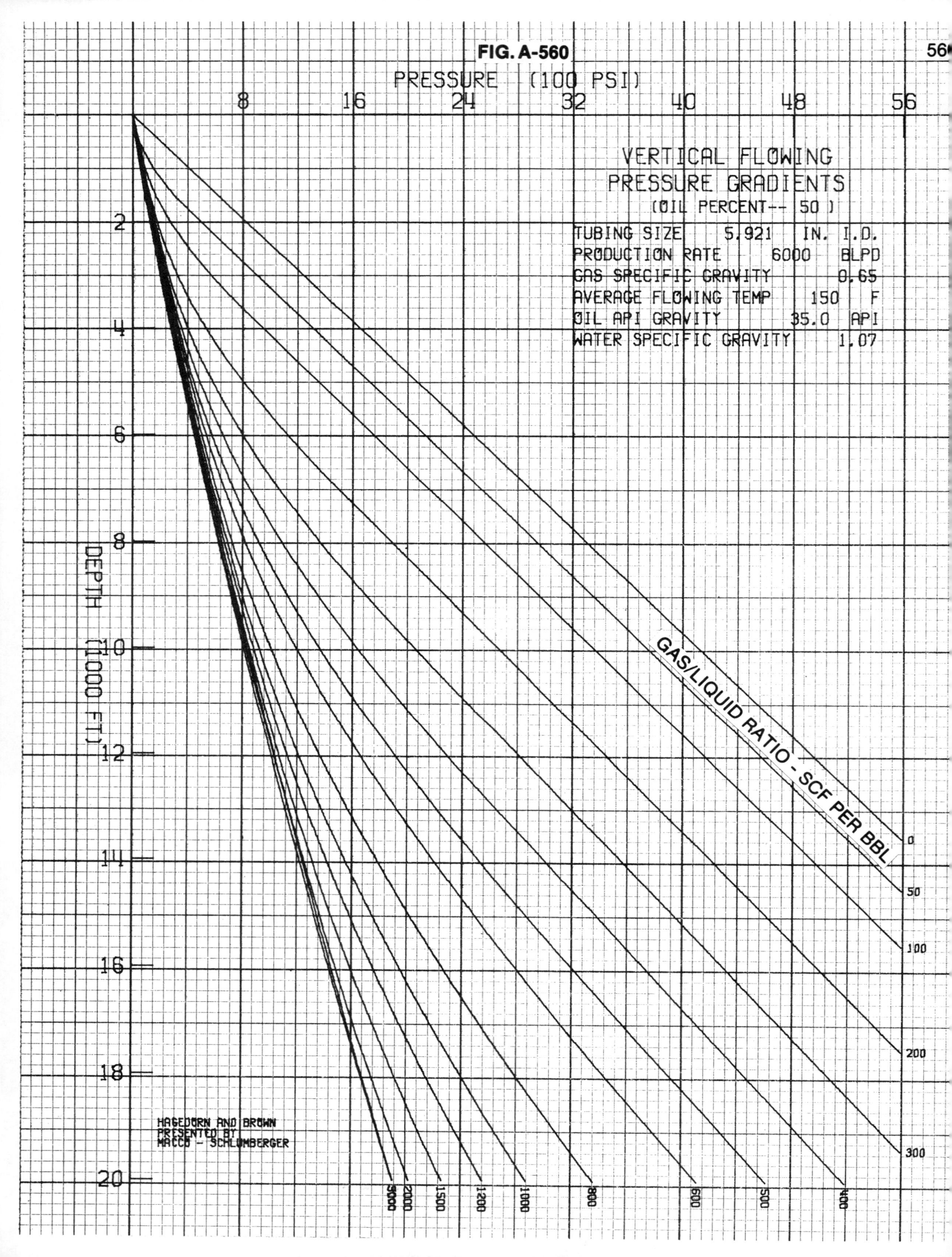

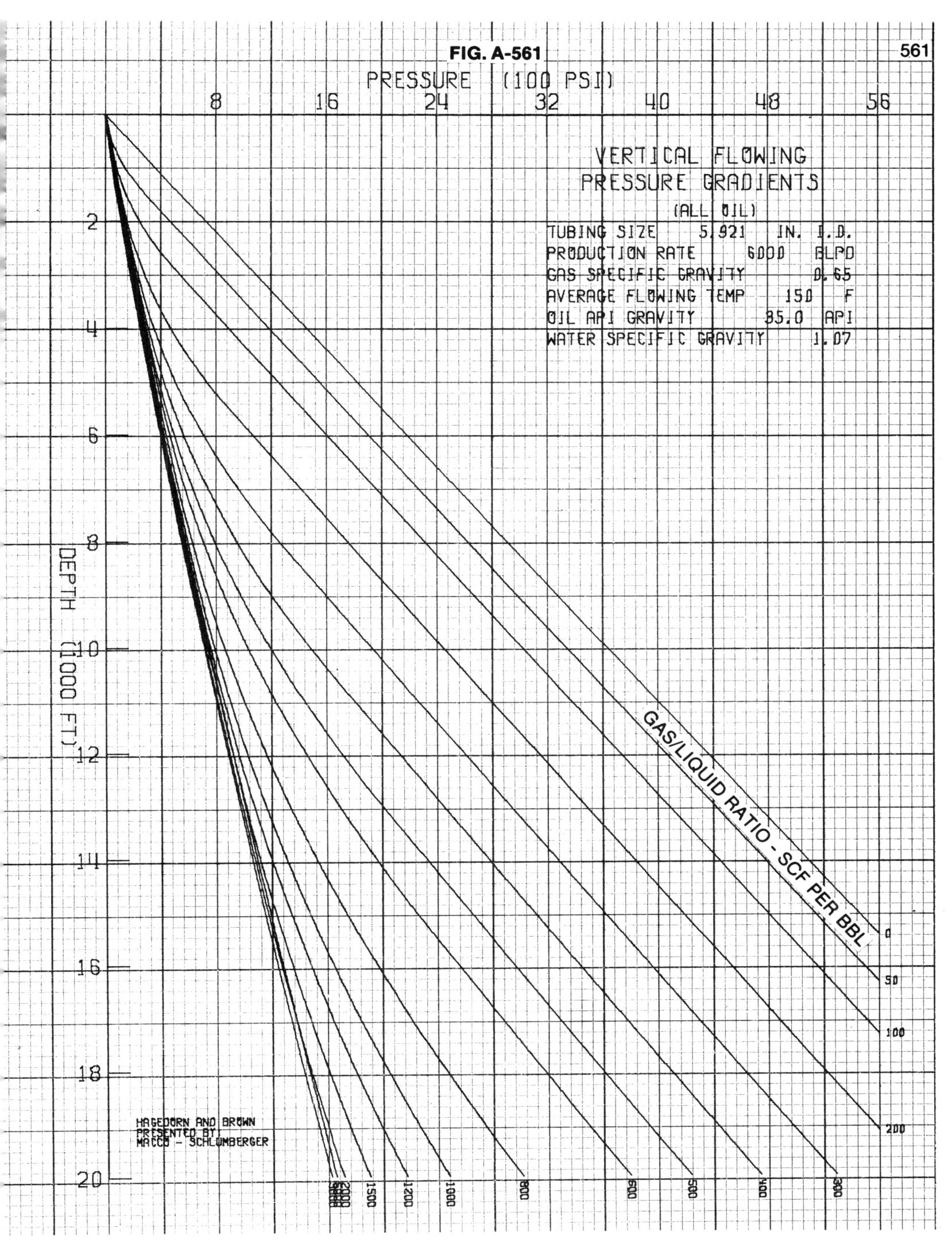

VERTICAL FLOWING
PRESSURE GRADIENTS
(ALL OIL)

TUBING SIZE	5.921	IN. I.D.
PRODUCTION RATE	6000	BLPD
GAS SPECIFIC GRAVITY	0.65	
AVERAGE FLOWING TEMP	150	F
OIL API GRAVITY	35.0	API
WATER SPECIFIC GRAVITY	1.07	

PRESSURE (100 PSI)

DEPTH (1000 FT)

GAS/LIQUID RATIO - SCF PER BBL

HAGEDORN AND BROWN
PRESENTED BY
MACCO - SCHLUMBERGER

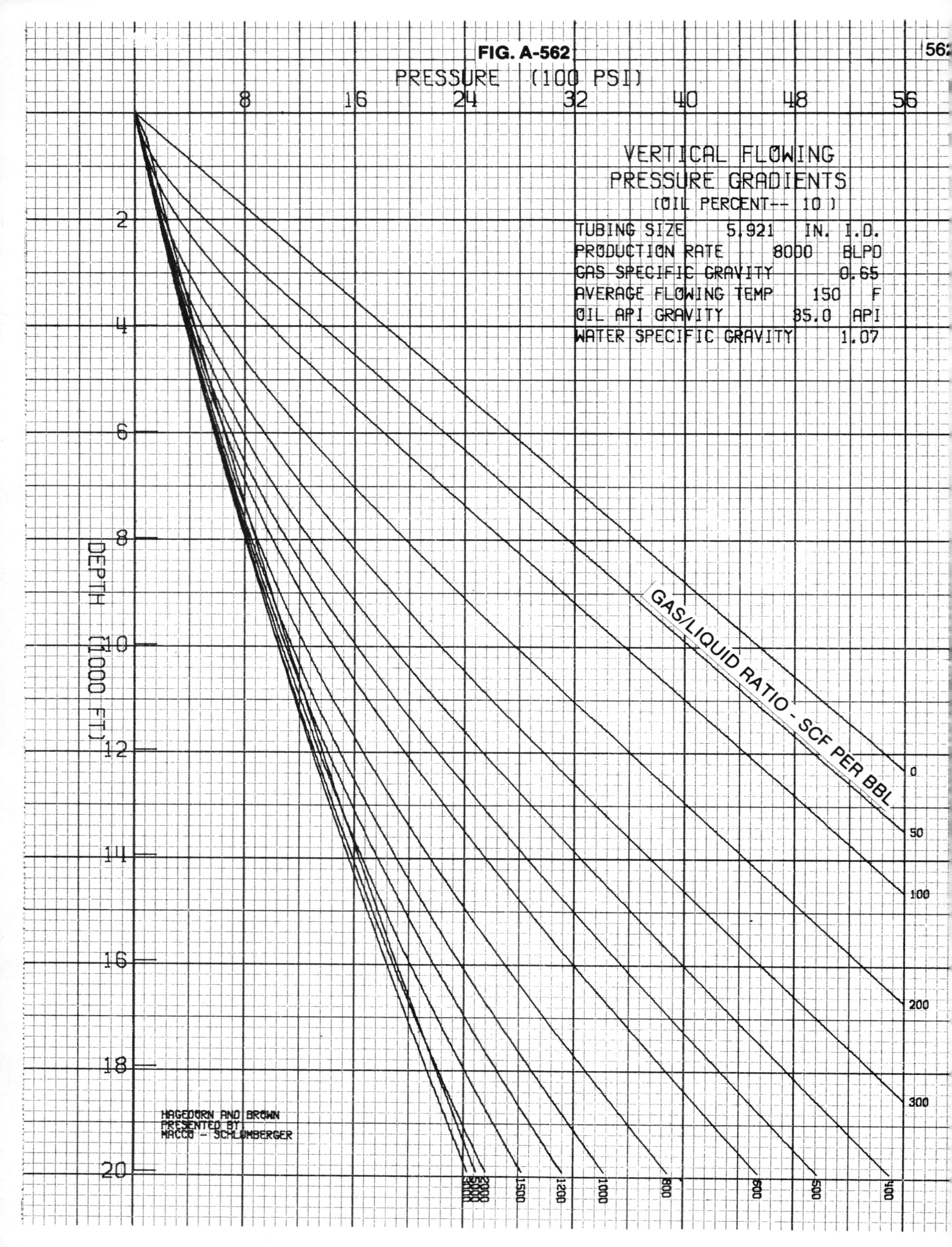

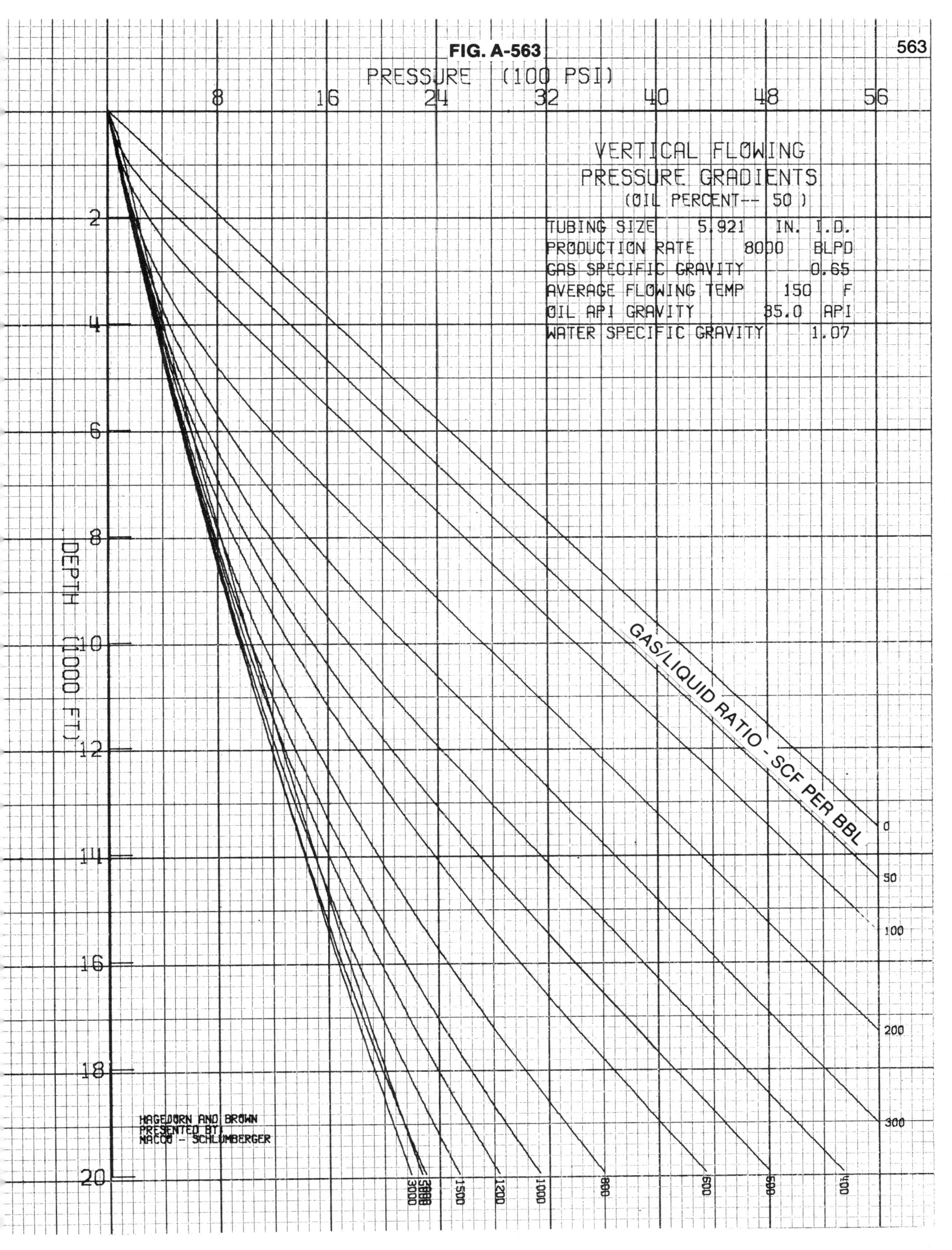

FIG. A-563

563

PRESSURE (100 PSI)

VERTICAL FLOWING
PRESSURE GRADIENTS
(OIL PERCENT-- 50)

TUBING SIZE	5.921	IN. I.D.
PRODUCTION RATE	8000	BLPD
GAS SPECIFIC GRAVITY		0.65
AVERAGE FLOWING TEMP	150	F
OIL API GRAVITY	35.0	API
WATER SPECIFIC GRAVITY		1.07

DEPTH (1000 FT)

GAS/LIQUID RATIO - SCF PER BBL

HAGEDORN AND BROWN
PRESENTED BY:
NACCO - SCHLUMBERGER

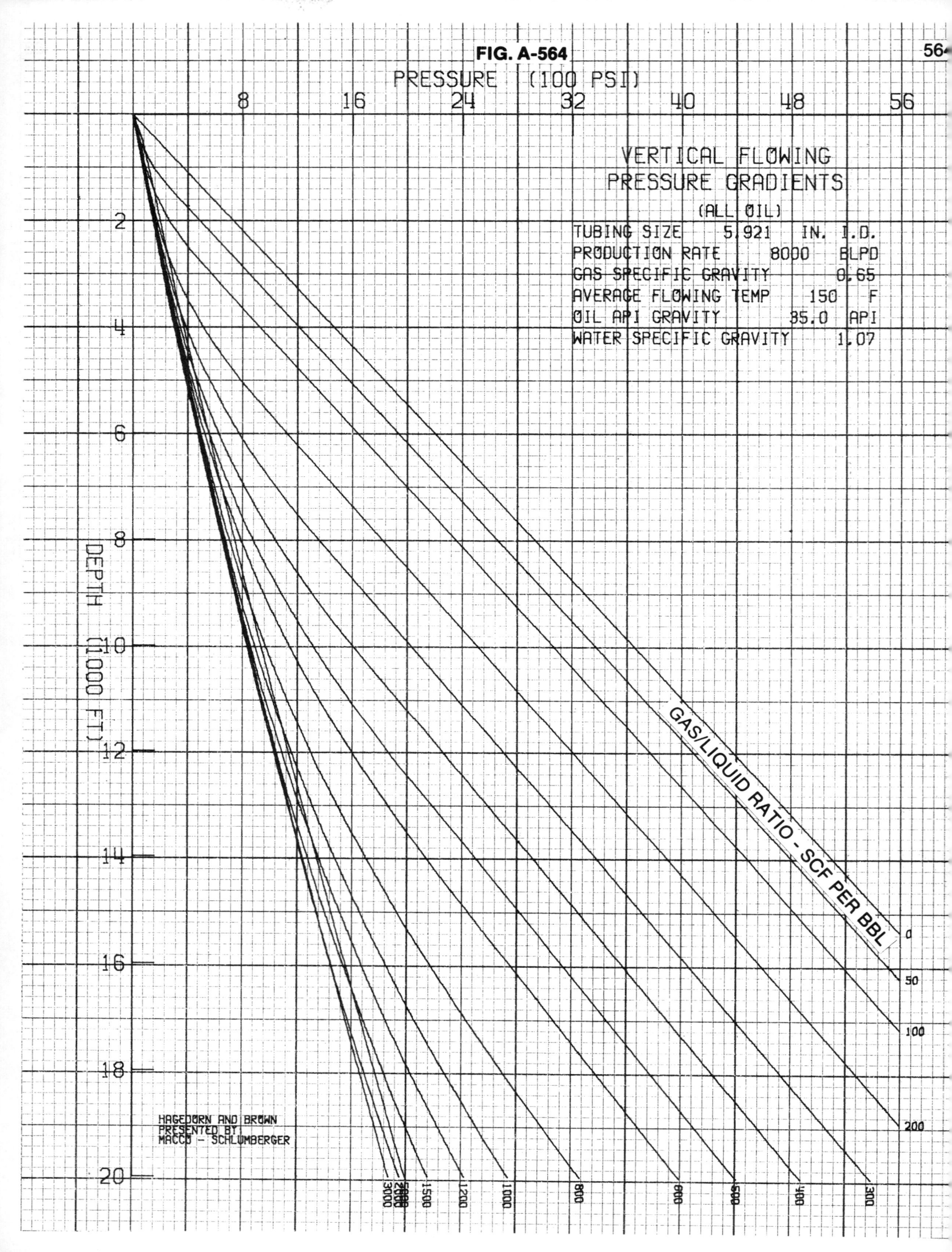

FIG. A-564

VERTICAL FLOWING
PRESSURE GRADIENTS
(ALL OIL)

TUBING SIZE	5.921	IN. I.D.
PRODUCTION RATE	8000	BLPD
GAS SPECIFIC GRAVITY	0.65	
AVERAGE FLOWING TEMP	150	F
OIL API GRAVITY	35.0	API
WATER SPECIFIC GRAVITY	1.07	

PRESSURE (100 PSI)

DEPTH (1000 FT)

GAS/LIQUID RATIO - SCF PER BBL

HAGEDORN AND BROWN
PRESENTED BY
MACCO - SCHLUMBERGER

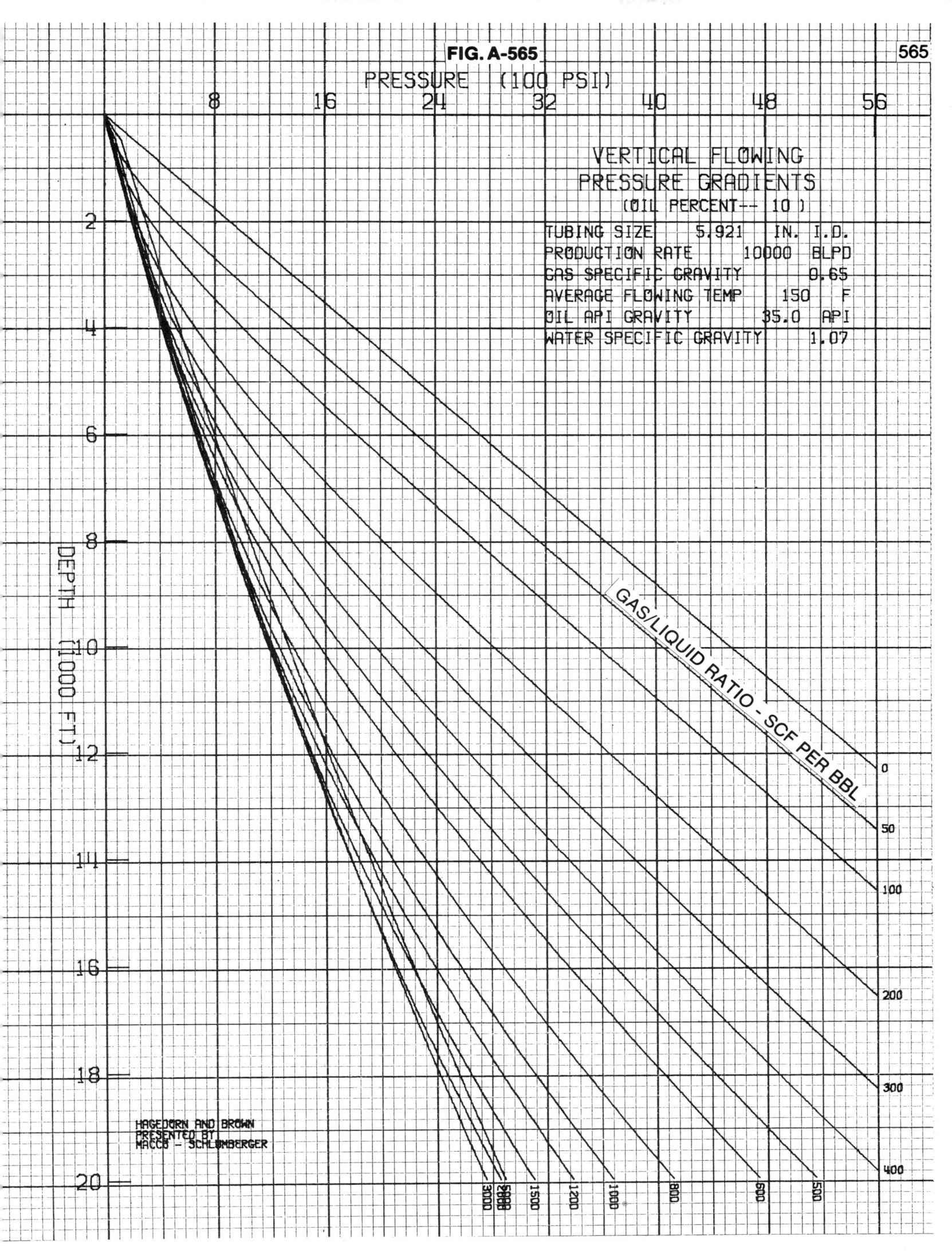

FIG. A-565

565

PRESSURE (100 PSI)

VERTICAL FLOWING
PRESSURE GRADIENTS
(OIL PERCENT-- 10)

TUBING SIZE	5.921	IN. I.D.
PRODUCTION RATE	10000	BLPD
GAS SPECIFIC GRAVITY	0.65	
AVERAGE FLOWING TEMP	150	F
OIL API GRAVITY	35.0	API
WATER SPECIFIC GRAVITY	1.07	

DEPTH (1000 FT)

GAS/LIQUID RATIO - SCF PER BBL

HAGEDORN AND BROWN
PRESENTED BY
MACCO - SCHLUMBERGER

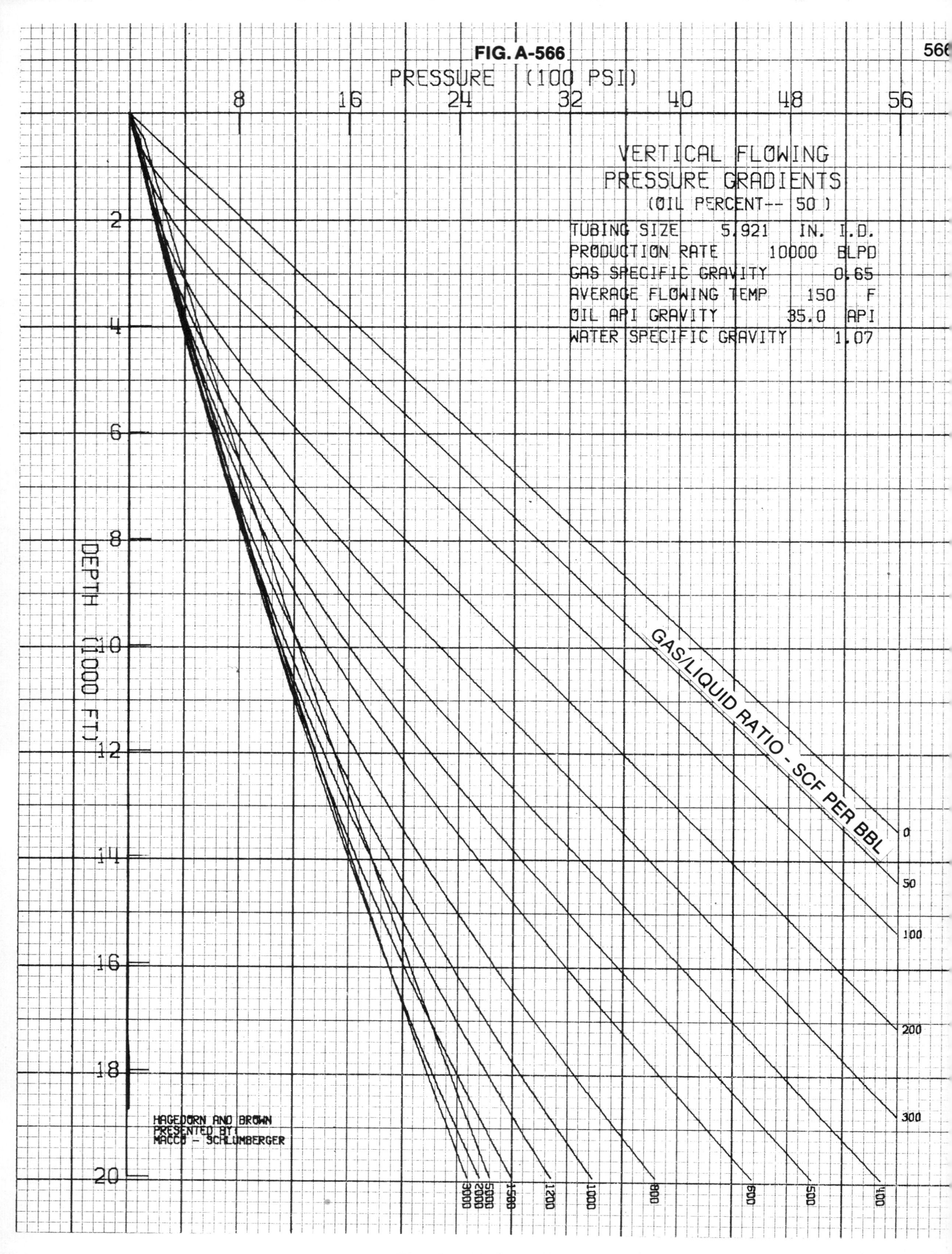

FIG. A-566

566

PRESSURE (100 PSI)

VERTICAL FLOWING
PRESSURE GRADIENTS
(OIL PERCENT-- 50)

TUBING SIZE	5.921	IN. I.D.
PRODUCTION RATE	10000	BLPD
GAS SPECIFIC GRAVITY	0.65	
AVERAGE FLOWING TEMP	150	F
OIL API GRAVITY	35.0	API
WATER SPECIFIC GRAVITY	1.07	

GAS/LIQUID RATIO - SCF PER BBL

DEPTH (1000 FT)

HAGEDORN AND BROWN
PRESENTED BY:
MACCO - SCHLUMBERGER

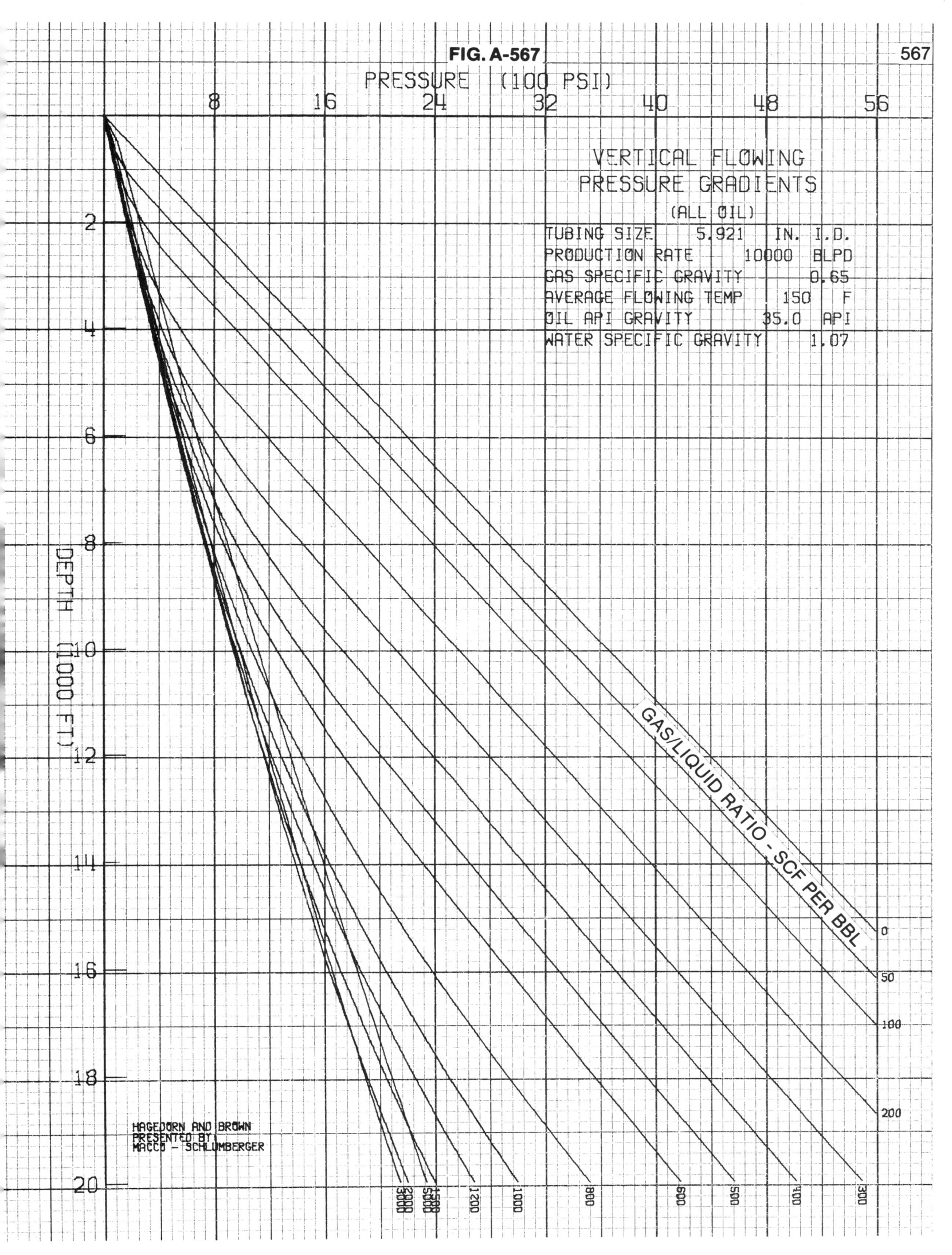

PRESSURE (100 PSI)

VERTICAL FLOWING
PRESSURE GRADIENTS
(ALL OIL)

TUBING SIZE	5.921	IN. I.D.
PRODUCTION RATE	10000	BLPD
GAS SPECIFIC GRAVITY	0.65	
AVERAGE FLOWING TEMP	150	F
OIL API GRAVITY	35.0	API
WATER SPECIFIC GRAVITY	1.07	

DEPTH (1000 FT)

GAS/LIQUID RATIO - SCF PER BBL

HAGEDORN AND BROWN
PRESENTED BY
MACCO - SCHLUMBERGER

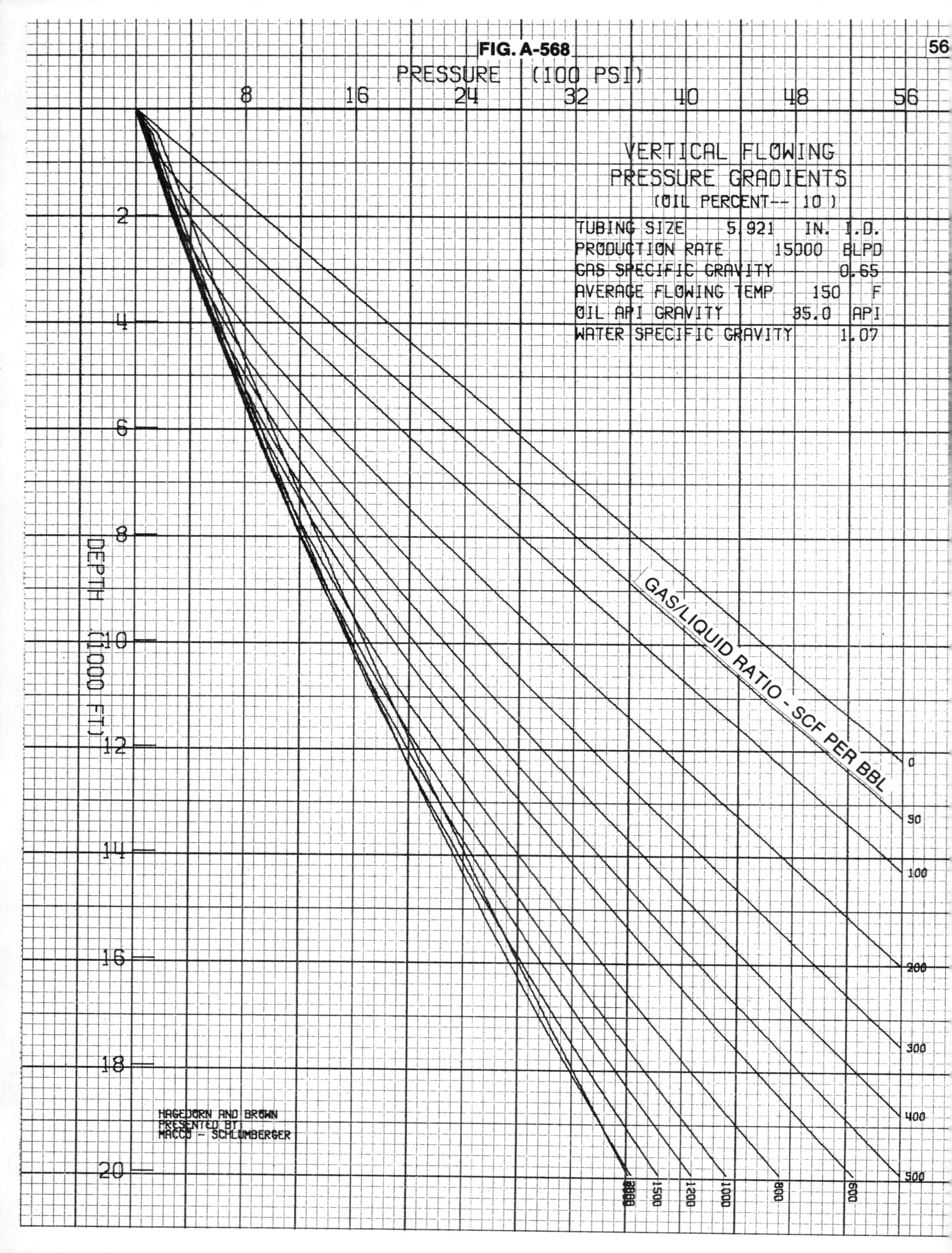

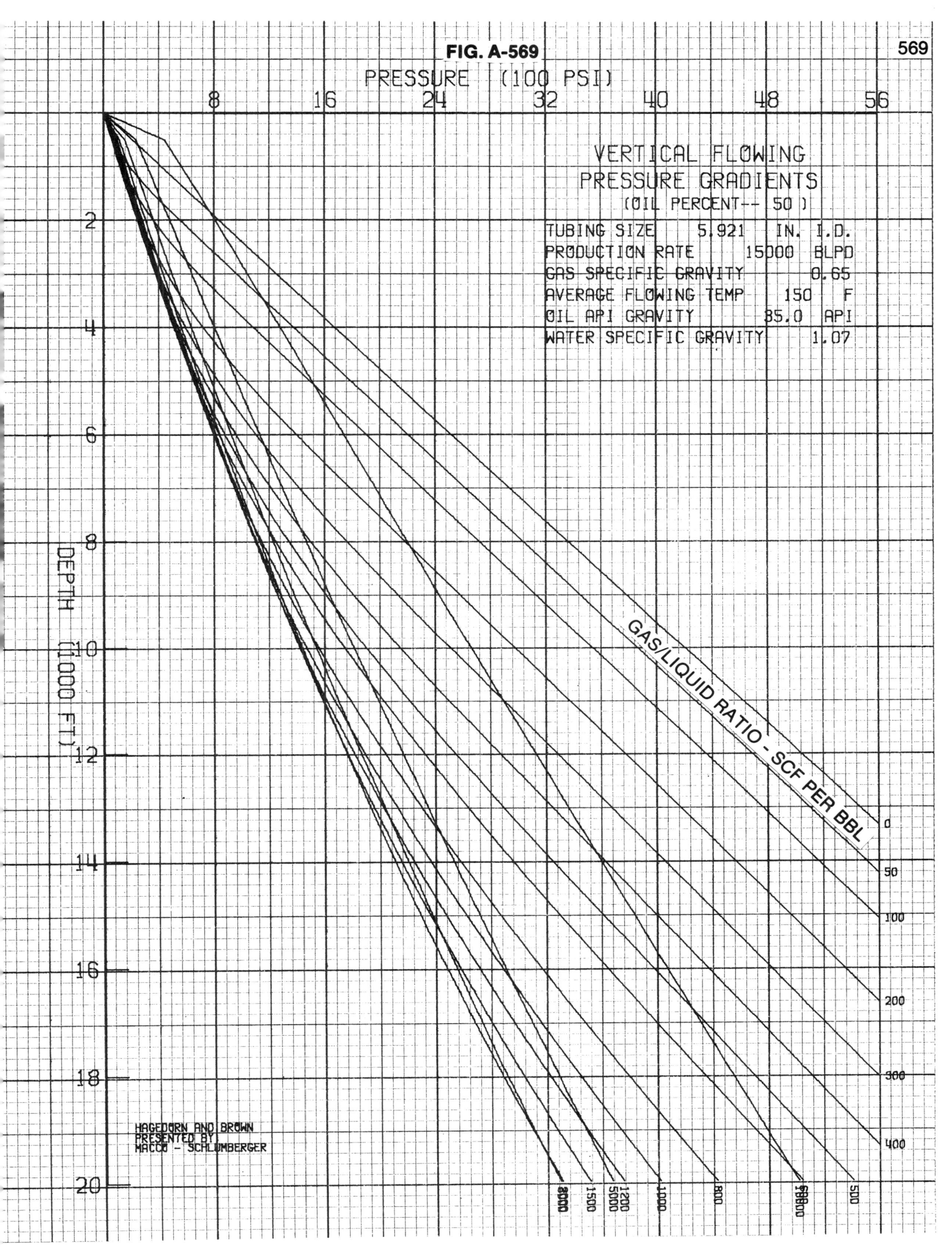

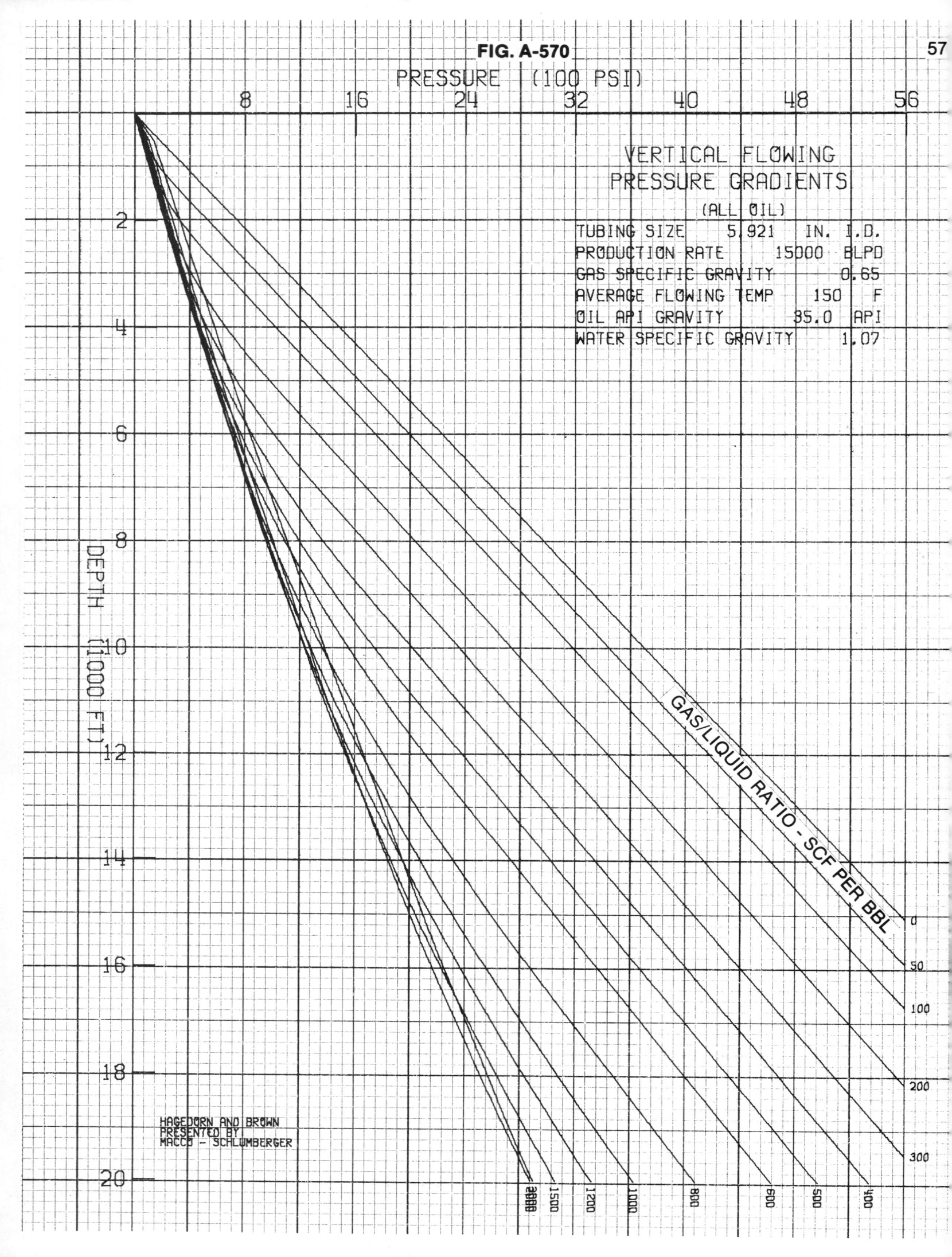

FIG. A-570

57

PRESSURE (100 PSI)

VERTICAL FLOWING
PRESSURE GRADIENTS
(ALL OIL)

TUBING SIZE   5.921   IN. I.D.
PRODUCTION RATE   15000   BLPD
GAS SPECIFIC GRAVITY   0.65
AVERAGE FLOWING TEMP   150   F
OIL API GRAVITY   35.0   API
WATER SPECIFIC GRAVITY   1.07

DEPTH (1000 FT)

GAS/LIQUID RATIO - SCF PER BBL

HAGEDORN AND BROWN
PRESENTED BY
MACCO - SCHLUMBERGER

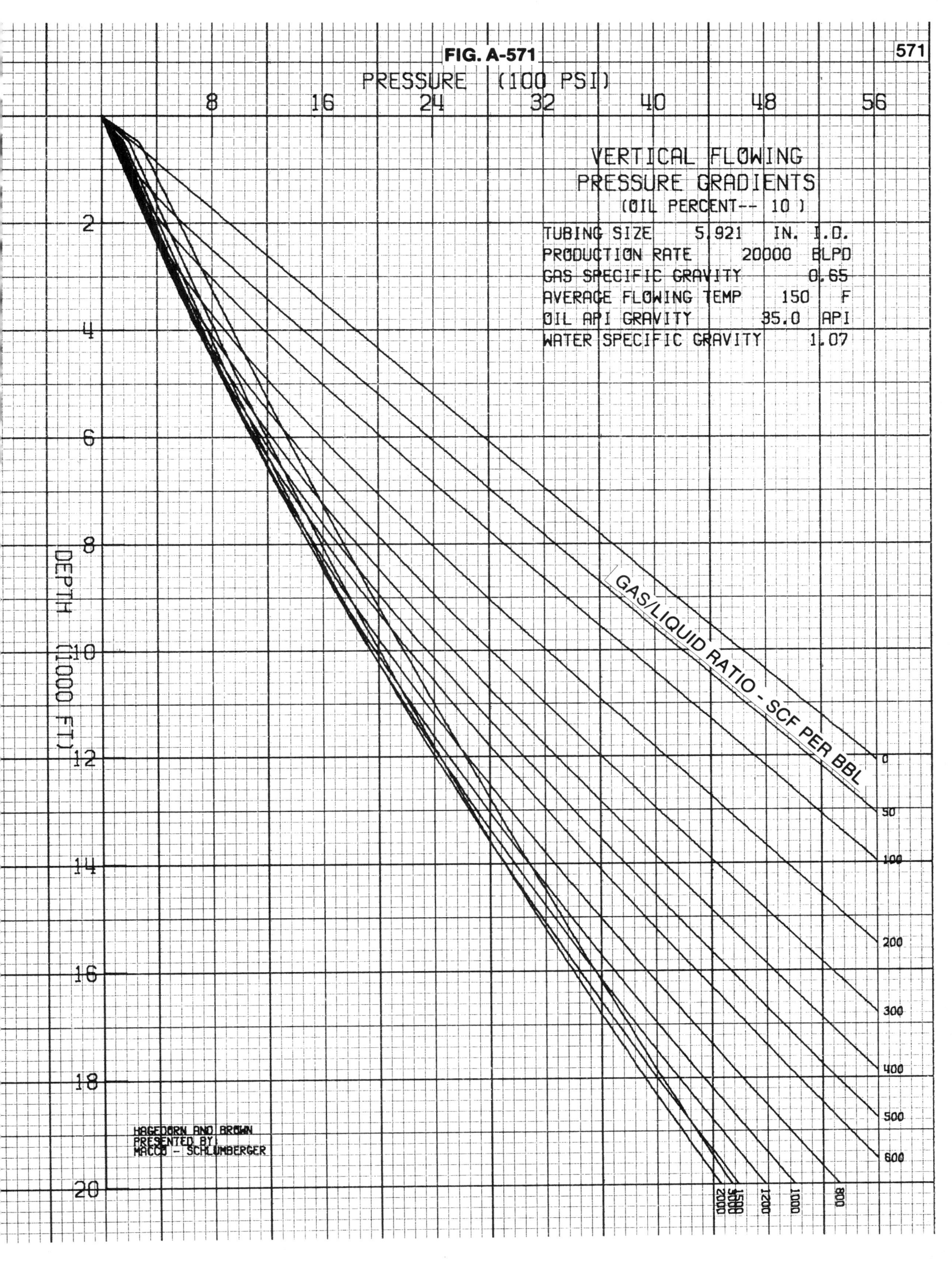

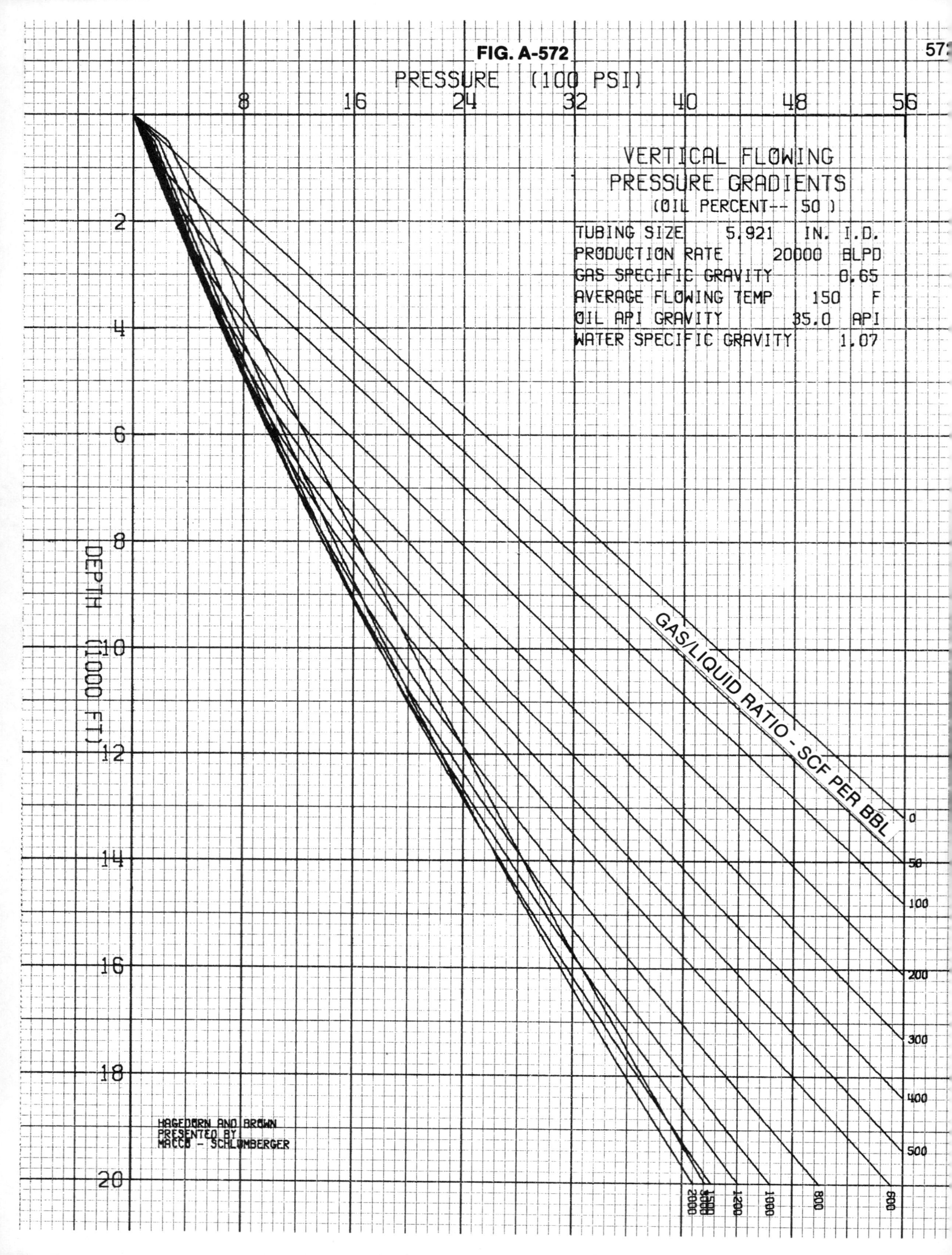

FIG. A-572

PRESSURE (100 PSI)

VERTICAL FLOWING
PRESSURE GRADIENTS
(OIL PERCENT-- 50)

TUBING SIZE	5.921	IN. I.D.
PRODUCTION RATE	20000	BLPD
GAS SPECIFIC GRAVITY	0.65	
AVERAGE FLOWING TEMP	150	F
OIL API GRAVITY	35.0	API
WATER SPECIFIC GRAVITY	1.07	

DEPTH (1000 FT)

GAS/LIQUID RATIO - SCF PER BBL

HAGEDORN AND BROWN
PRESENTED BY
MAECO - SCHLUMBERGER

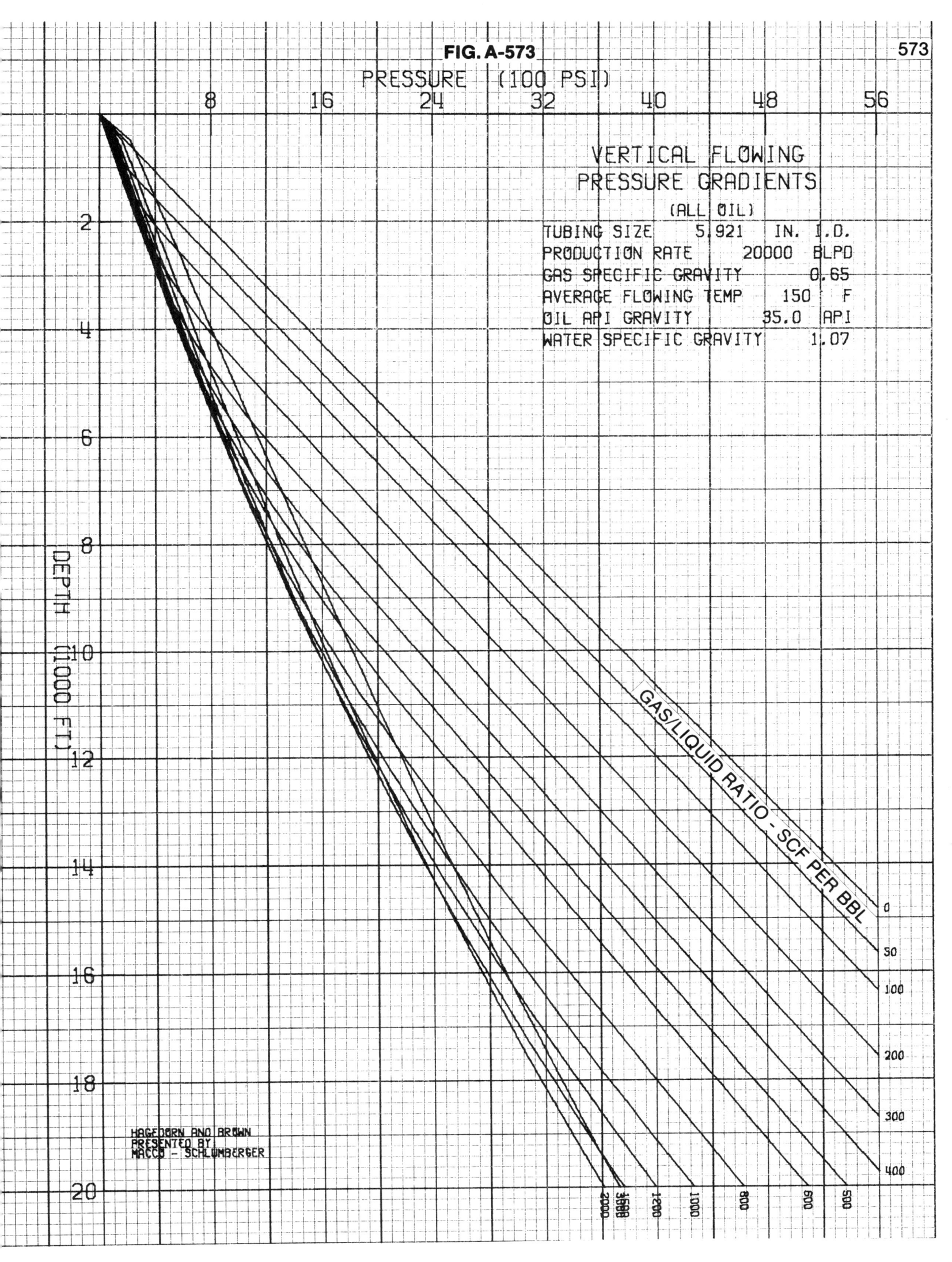

FIG. A-573

573

PRESSURE (100 PSI)

VERTICAL FLOWING
PRESSURE GRADIENTS
(ALL OIL)

TUBING SIZE	5.921	IN. I.D.
PRODUCTION RATE	20000	BLPD
GAS SPECIFIC GRAVITY	0.65	
AVERAGE FLOWING TEMP	150	F
OIL API GRAVITY	35.0	API
WATER SPECIFIC GRAVITY	1.07	

DEPTH (1000 FT)

GAS/LIQUID RATIO - SCF PER BBL

HAGEDORN AND BROWN
PRESENTED BY
MACCO - SCHLUMBERGER

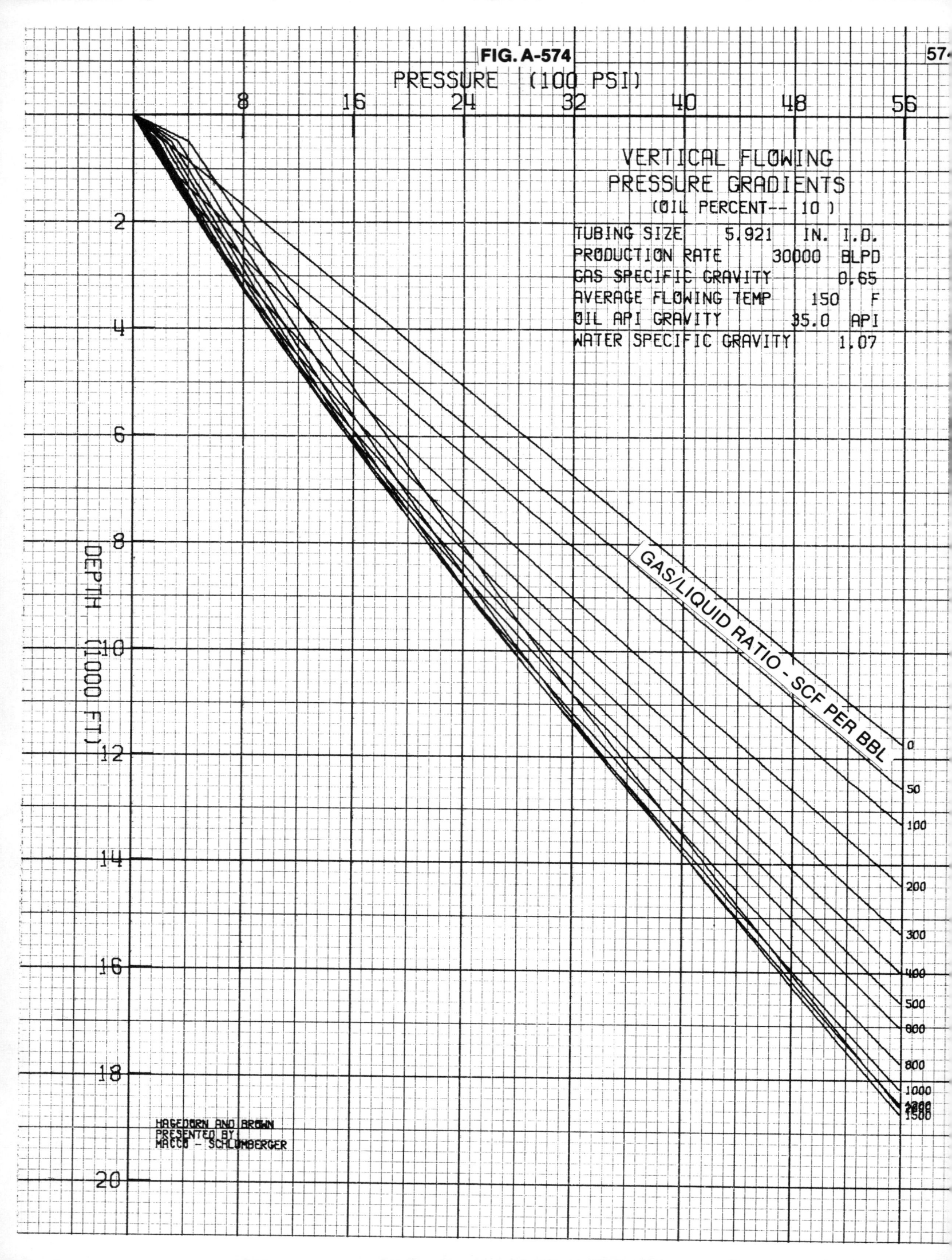

PRESSURE (100 PSI)

VERTICAL FLOWING
PRESSURE GRADIENTS
(OIL PERCENT-- 10)

TUBING SIZE	5.921	IN. I.D.
PRODUCTION RATE	30000	BLPD
GAS SPECIFIC GRAVITY	0.65	
AVERAGE FLOWING TEMP	150	F
OIL API GRAVITY	35.0	API
WATER SPECIFIC GRAVITY	1.07	

GAS/LIQUID RATIO - SCF PER BBL

DEPTH (1000 FT)

HAGEDORN AND BROWN
PRESENTED BY
MACCO - SCHLUMBERGER

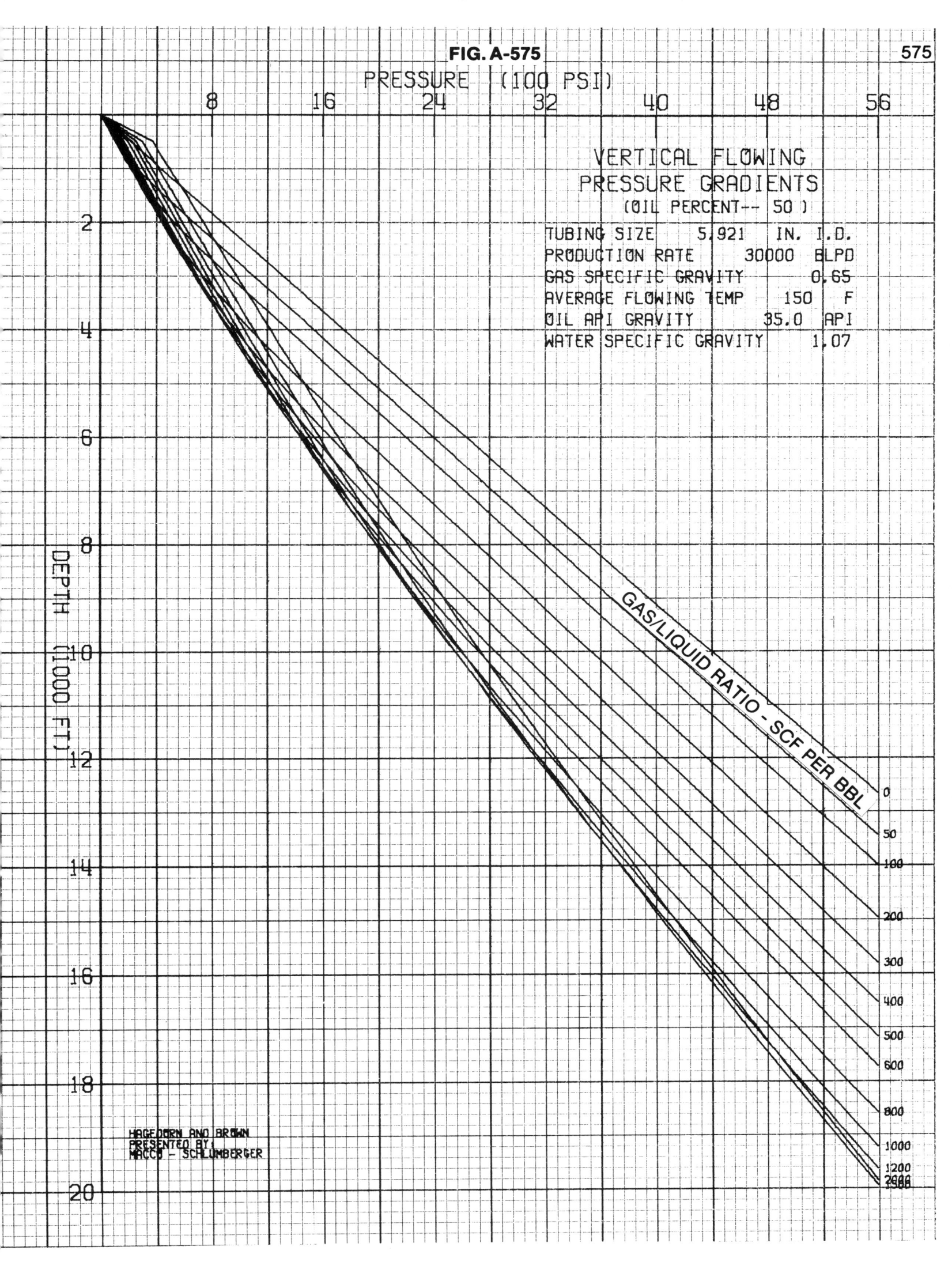

FIG. A-575

575

VERTICAL FLOWING
PRESSURE GRADIENTS
(OIL PERCENT-- 50)

TUBING SIZE	5.921	IN. I.D.
PRODUCTION RATE	30000	BLPD
GAS SPECIFIC GRAVITY	0.65	
AVERAGE FLOWING TEMP	150	F
OIL API GRAVITY	35.0	API
WATER SPECIFIC GRAVITY	1.07	

GAS/LIQUID RATIO - SCF PER BBL

HAGEDORN AND BROWN
PRESENTED BY
MACCO - SCHLUMBERGER

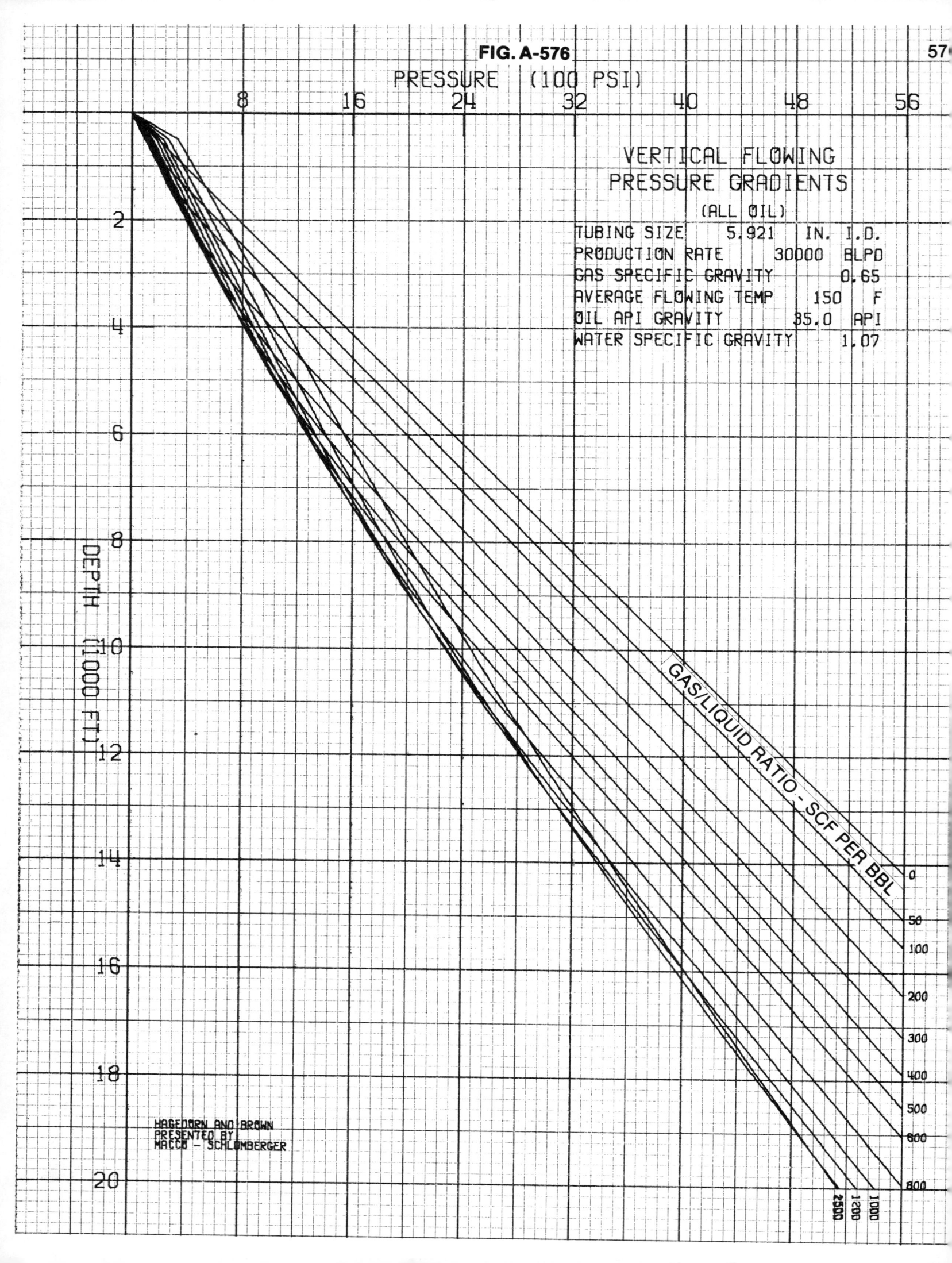

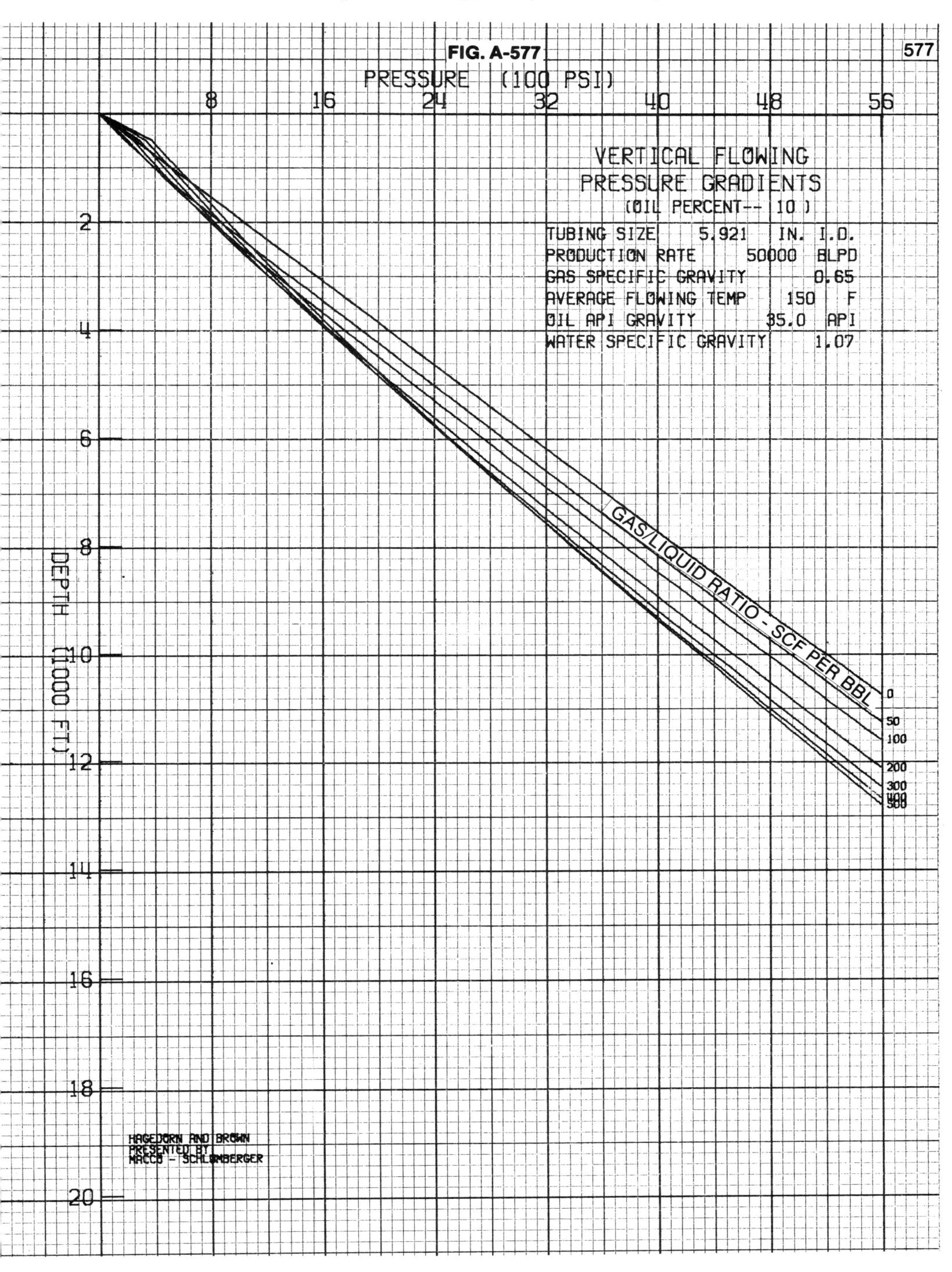

FIG. A-577

577

PRESSURE (100 PSI)

VERTICAL FLOWING
PRESSURE GRADIENTS
(OIL PERCENT-- 10)

TUBING SIZE	5.921	IN. I.D.
PRODUCTION RATE	50000	BLPD
GAS SPECIFIC GRAVITY	0.65	
AVERAGE FLOWING TEMP	150	F
OIL API GRAVITY	35.0	API
WATER SPECIFIC GRAVITY	1.07	

DEPTH (1000 FT)

GAS/LIQUID RATIO - SCF PER BBL

0
50
100
200
300
400
500

HAGEDORN AND BROWN
PRESENTED BY
MACCO - SCHLUMBERGER

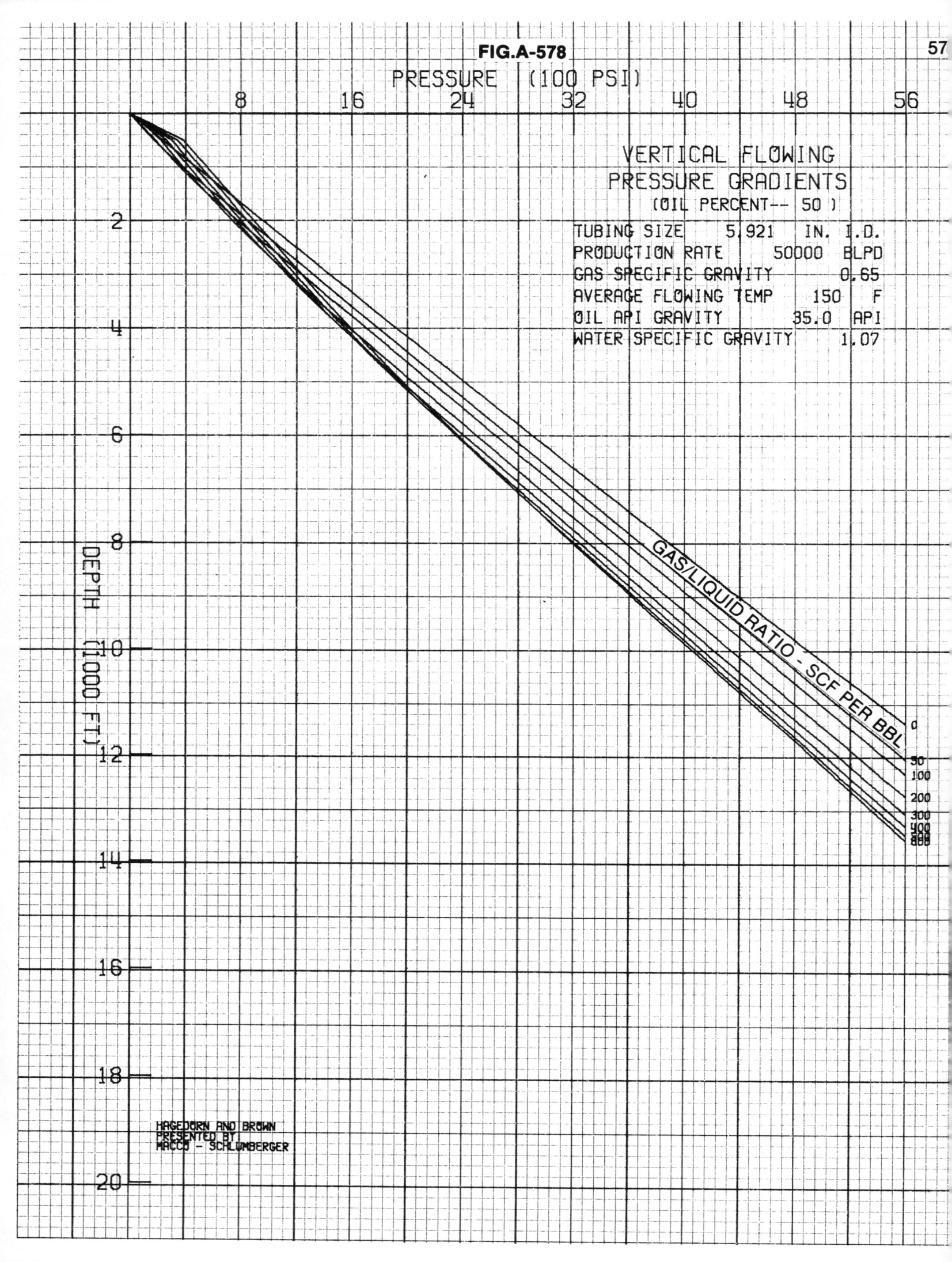

FIG.A-578

VERTICAL FLOWING
PRESSURE GRADIENTS
(OIL PERCENT-- 50)

TUBING SIZE	5.921	IN. I.D.
PRODUCTION RATE	50000	BLPD
GAS SPECIFIC GRAVITY	0.65	
AVERAGE FLOWING TEMP	150	F
OIL API GRAVITY	35.0	API
WATER SPECIFIC GRAVITY	1.07	

PRESSURE (100 PSI)

DEPTH (1000 FT)

GAS/LIQUID RATIO - SCF PER BBL

0
50
100
200
300
400
600
800

HAGEDORN AND BROWN
PRESENTED BY
MACCO - SCHLUMBERGER

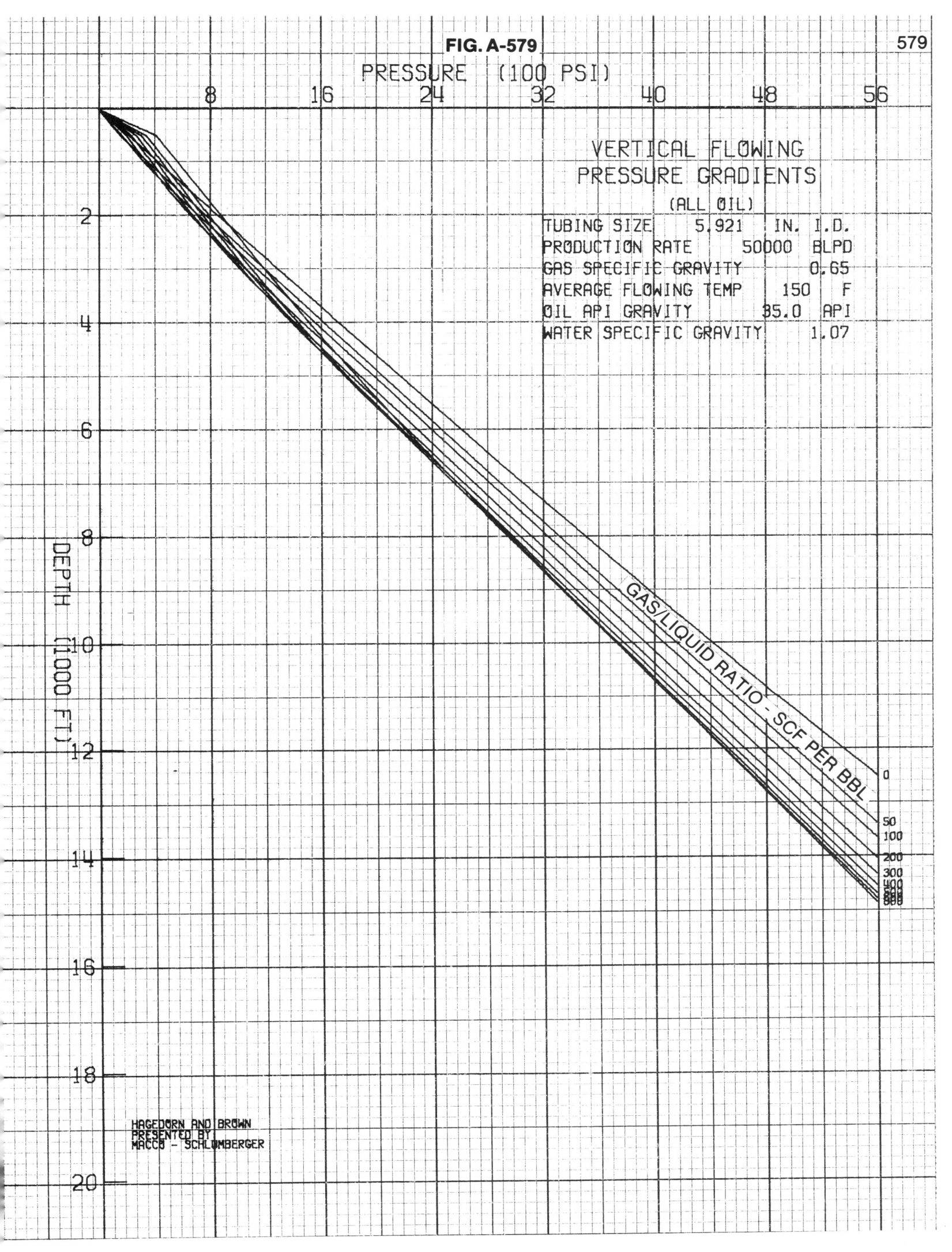

PRESSURE (100 PSI)

DEPTH (1000 FT)

VERTICAL FLOWING
PRESSURE GRADIENTS
(ALL OIL)

TUBING SIZE	5.921	IN. I.D.
PRODUCTION RATE	50000	BLPD
GAS SPECIFIC GRAVITY	0.65	
AVERAGE FLOWING TEMP	150	F
OIL API GRAVITY	35.0	API
WATER SPECIFIC GRAVITY	1.07	

GAS/LIQUID RATIO - SCF PER BBL

0

50
100

200
300
400
500
600
800

HAGEDORN AND BROWN
PRESENTED BY
MACCO - SCHLUMBERGER

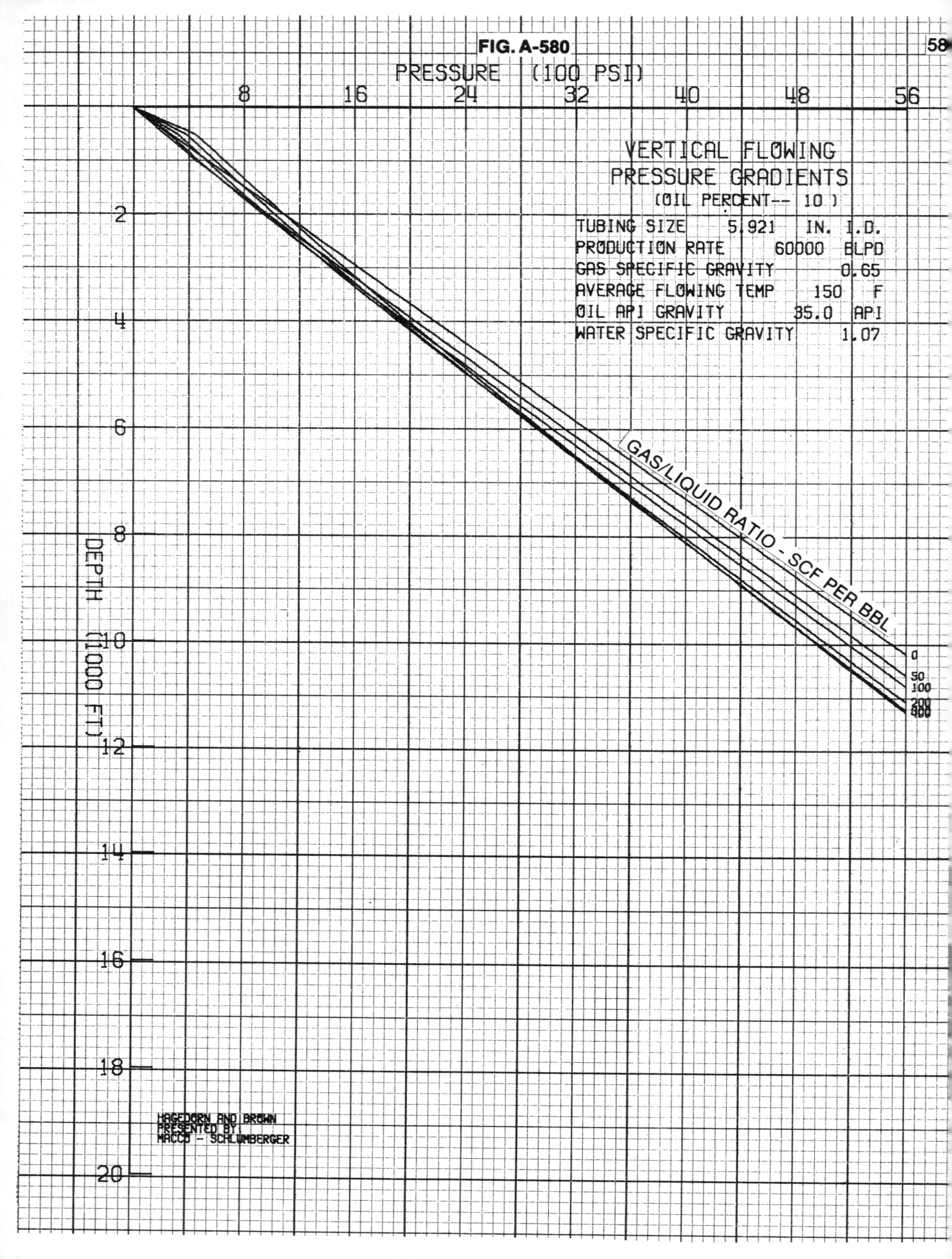

FIG. A-580

VERTICAL FLOWING
PRESSURE GRADIENTS
(OIL PERCENT-- 10)

TUBING SIZE	5.921	IN. I.D.
PRODUCTION RATE	60000	BLPD
GAS SPECIFIC GRAVITY	0.65	
AVERAGE FLOWING TEMP	150	F
OIL API GRAVITY	35.0	API
WATER SPECIFIC GRAVITY	1.07	

GAS/LIQUID RATIO - SCF PER BBL

0
50
100
200
400
600

HAGEDORN AND BROWN
PRESENTED BY
MACCO - SCHLUMBERGER

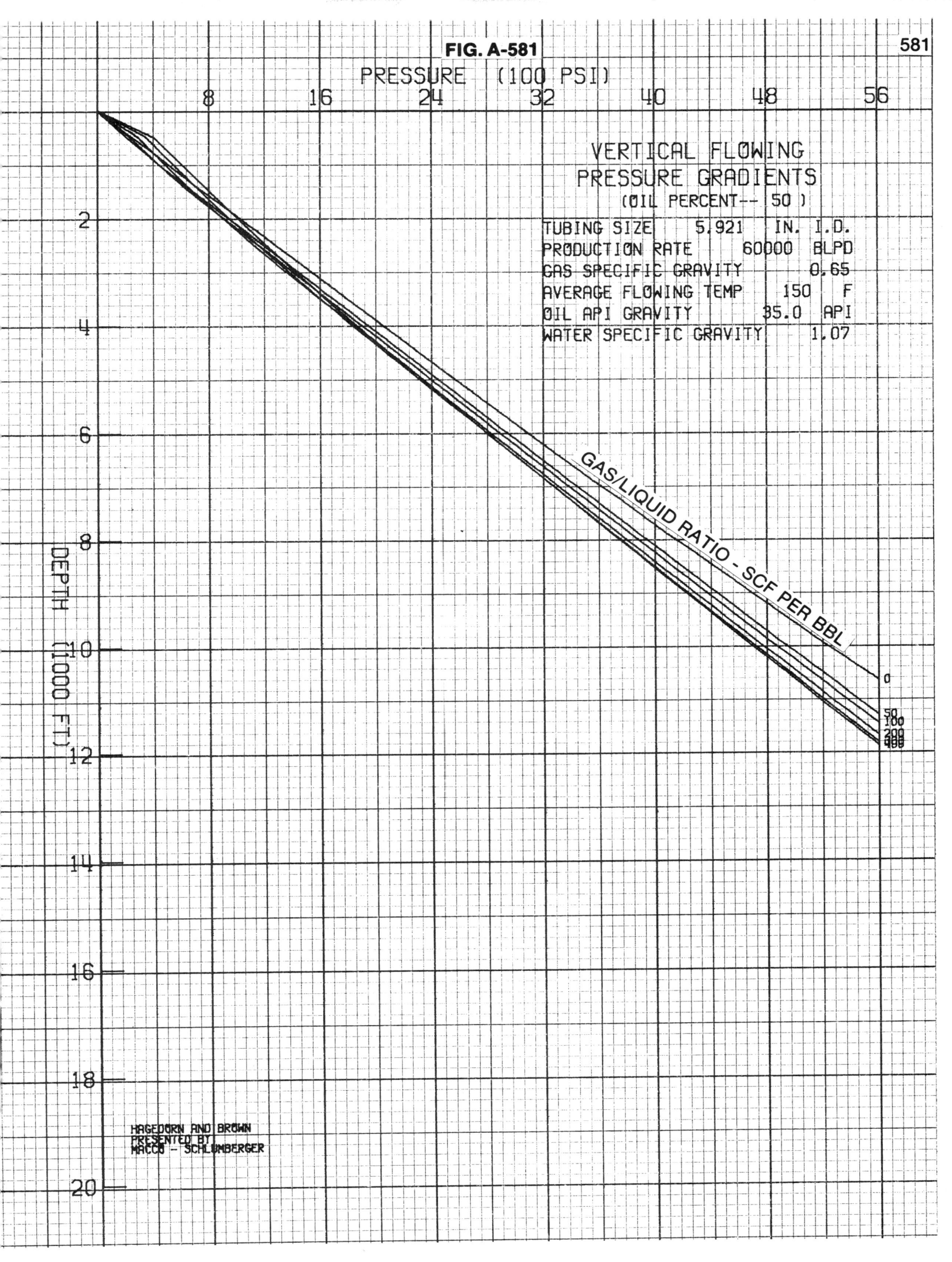

PRESSURE (100 PSI)

VERTICAL FLOWING
PRESSURE GRADIENTS
(OIL PERCENT-- 50)

TUBING SIZE   5.921   IN. I.D.
PRODUCTION RATE   60000   BLPD
GAS SPECIFIC GRAVITY   0.65
AVERAGE FLOWING TEMP   150   F
OIL API GRAVITY   35.0   API
WATER SPECIFIC GRAVITY   1.07

GAS/LIQUID RATIO - SCF PER BBL

DEPTH (1000 FT)

HAGEDORN AND BROWN
PRESENTED BY
MACCO - SCHLUMBERGER

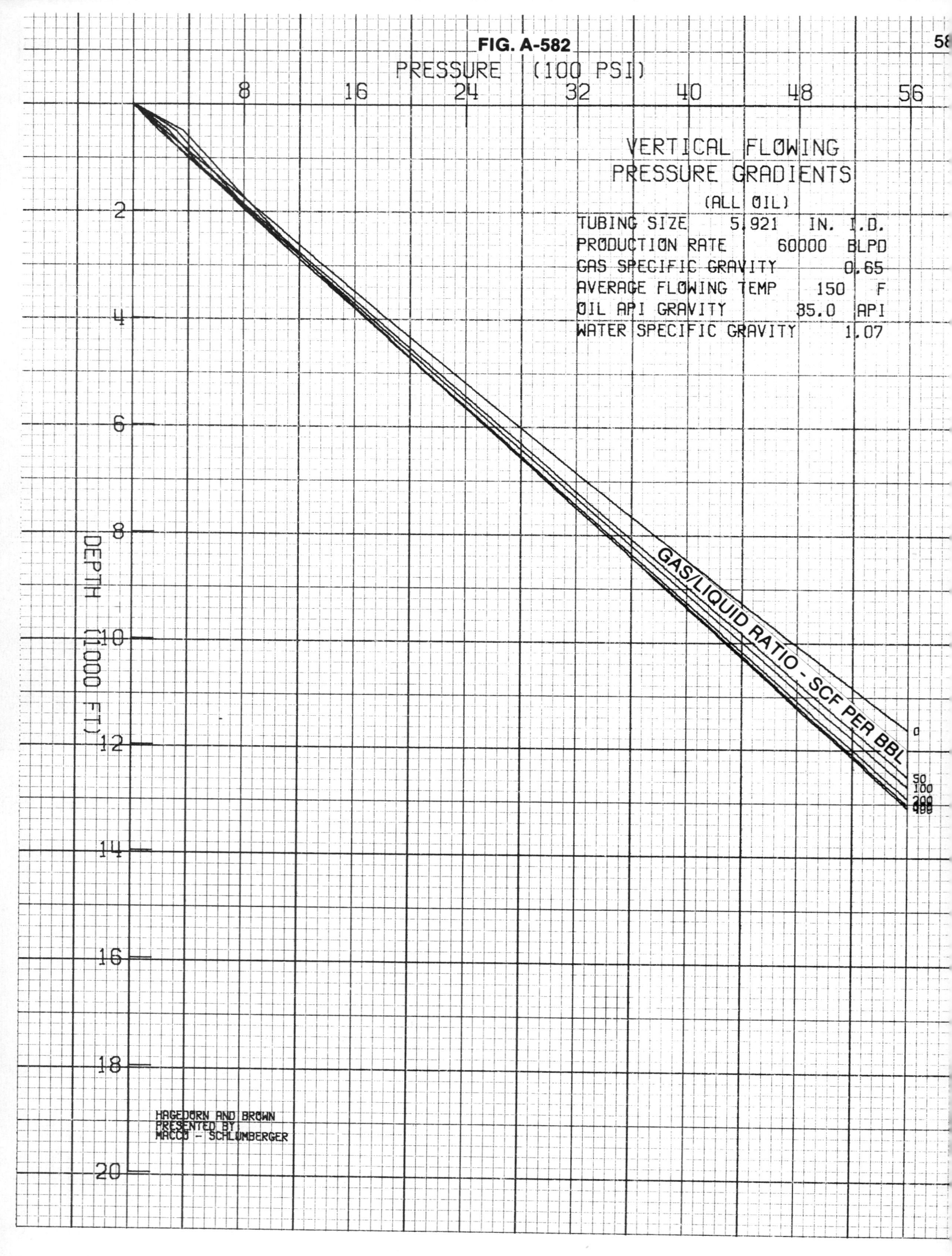

FIG. A-582

VERTICAL FLOWING
PRESSURE GRADIENTS
(ALL OIL)

TUBING SIZE	5.921	IN. I.D.
PRODUCTION RATE	60000	BLPD
GAS SPECIFIC GRAVITY	0.65	
AVERAGE FLOWING TEMP	150	F
OIL API GRAVITY	85.0	API
WATER SPECIFIC GRAVITY	1.07	

PRESSURE (100 PSI)

DEPTH (1000 FT)

GAS/LIQUID RATIO - SCF PER BBL

HAGEDORN AND BROWN
PRESENTED BY
MACCO - SCHLUMBERGER

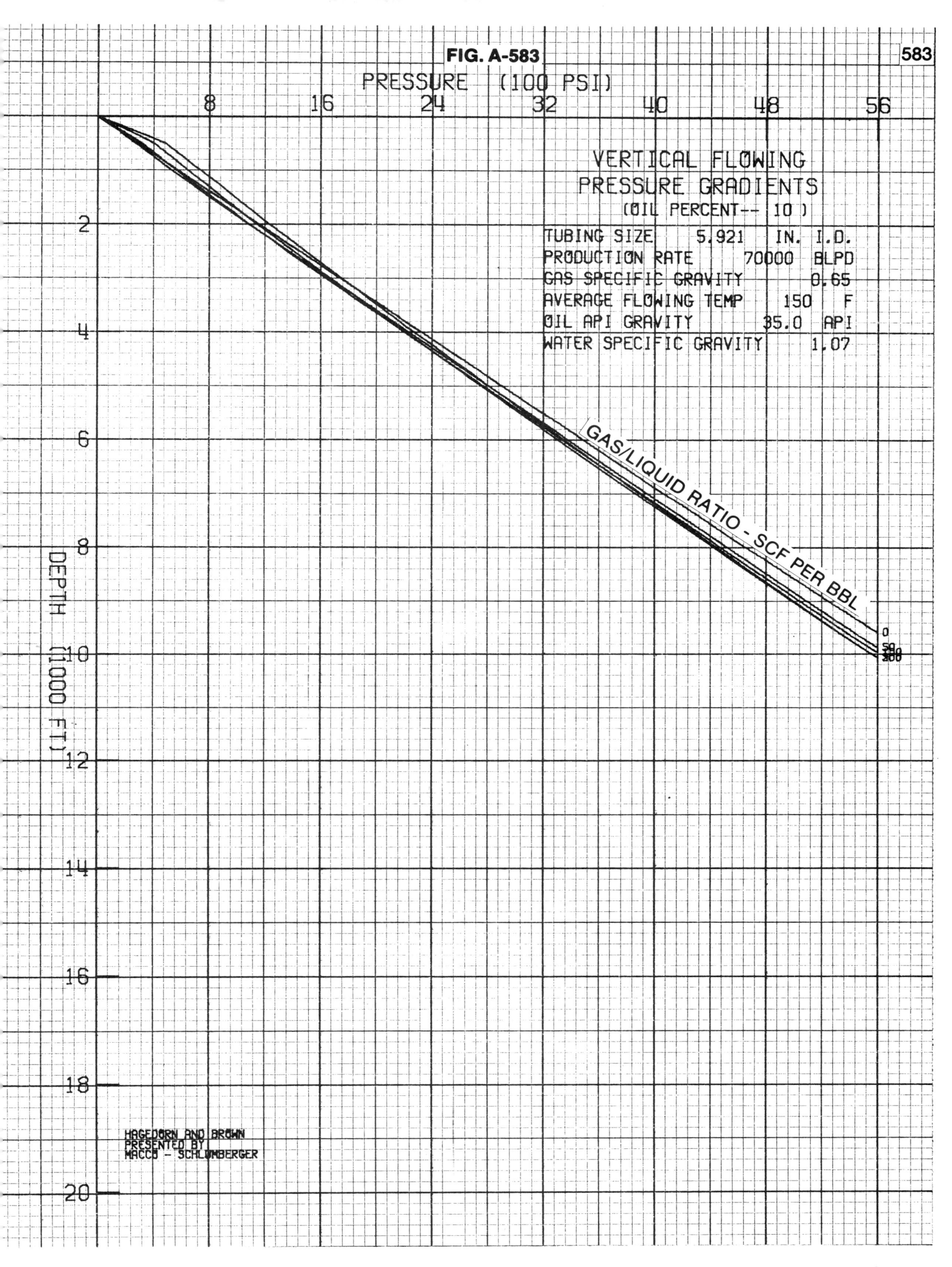

FIG. A-583

VERTICAL FLOWING
PRESSURE GRADIENTS
(OIL PERCENT-- 10)

TUBING SIZE   5.921   IN. I.D.
PRODUCTION RATE   70000   BLPD
GAS SPECIFIC GRAVITY   0.65
AVERAGE FLOWING TEMP   150   F
OIL API GRAVITY   35.0   API
WATER SPECIFIC GRAVITY   1.07

GAS/LIQUID RATIO - SCF PER BBL

HAGEDORN AND BROWN
PRESENTED BY
MACCO - SCHLUMBERGER

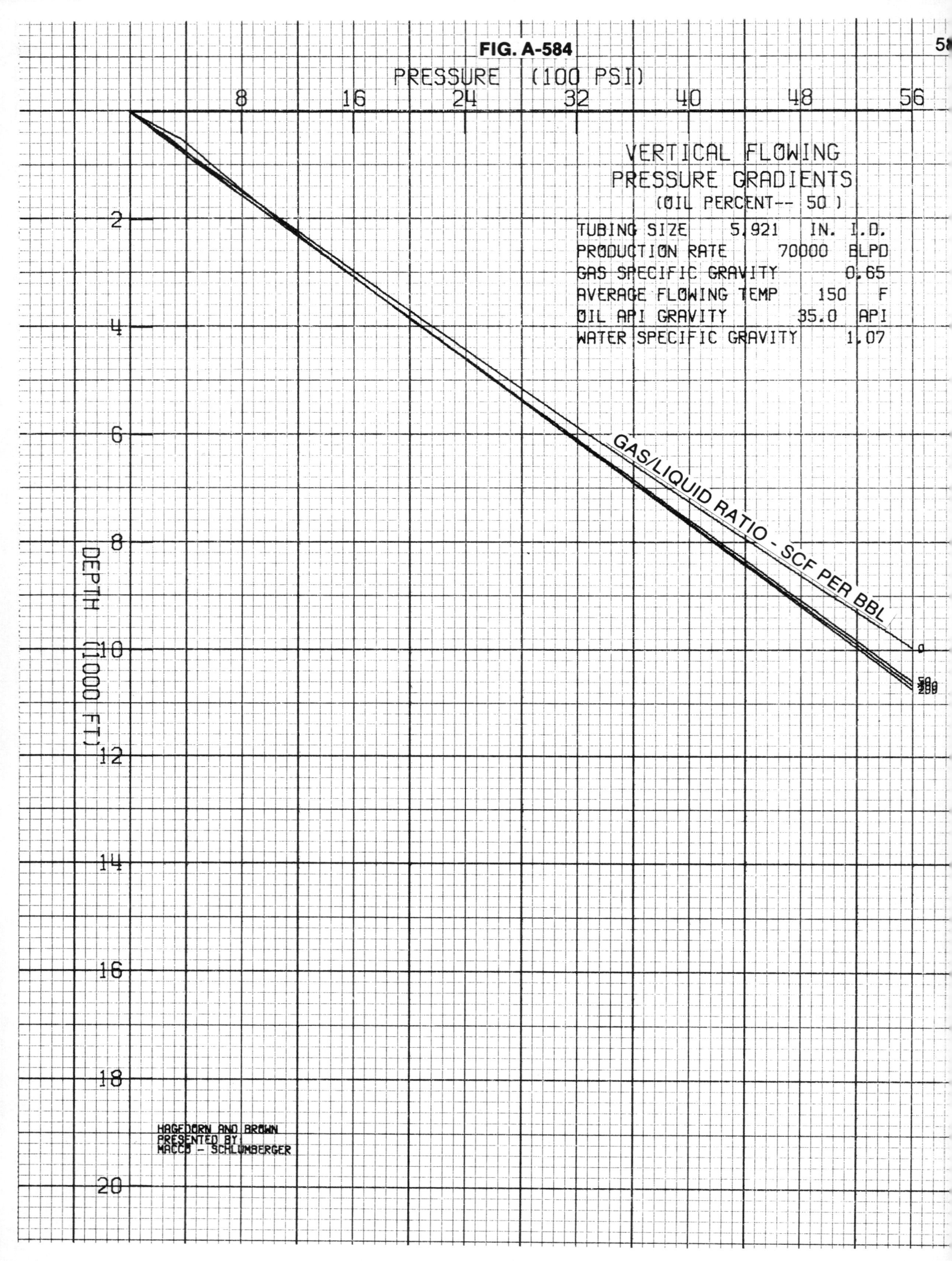

VERTICAL FLOWING
PRESSURE GRADIENTS
(OIL PERCENT-- 50)

TUBING SIZE	5.921	IN. I.D.
PRODUCTION RATE	70000	BLPD
GAS SPECIFIC GRAVITY	0.65	
AVERAGE FLOWING TEMP	150	F
OIL API GRAVITY	35.0	API
WATER SPECIFIC GRAVITY	1.07	

GAS/LIQUID RATIO - SCF PER BBL

PRESSURE (100 PSI)

DEPTH (1000 FT)

HAGEDORN AND BROWN
PRESENTED BY
MACCO - SCHLUMBERGER

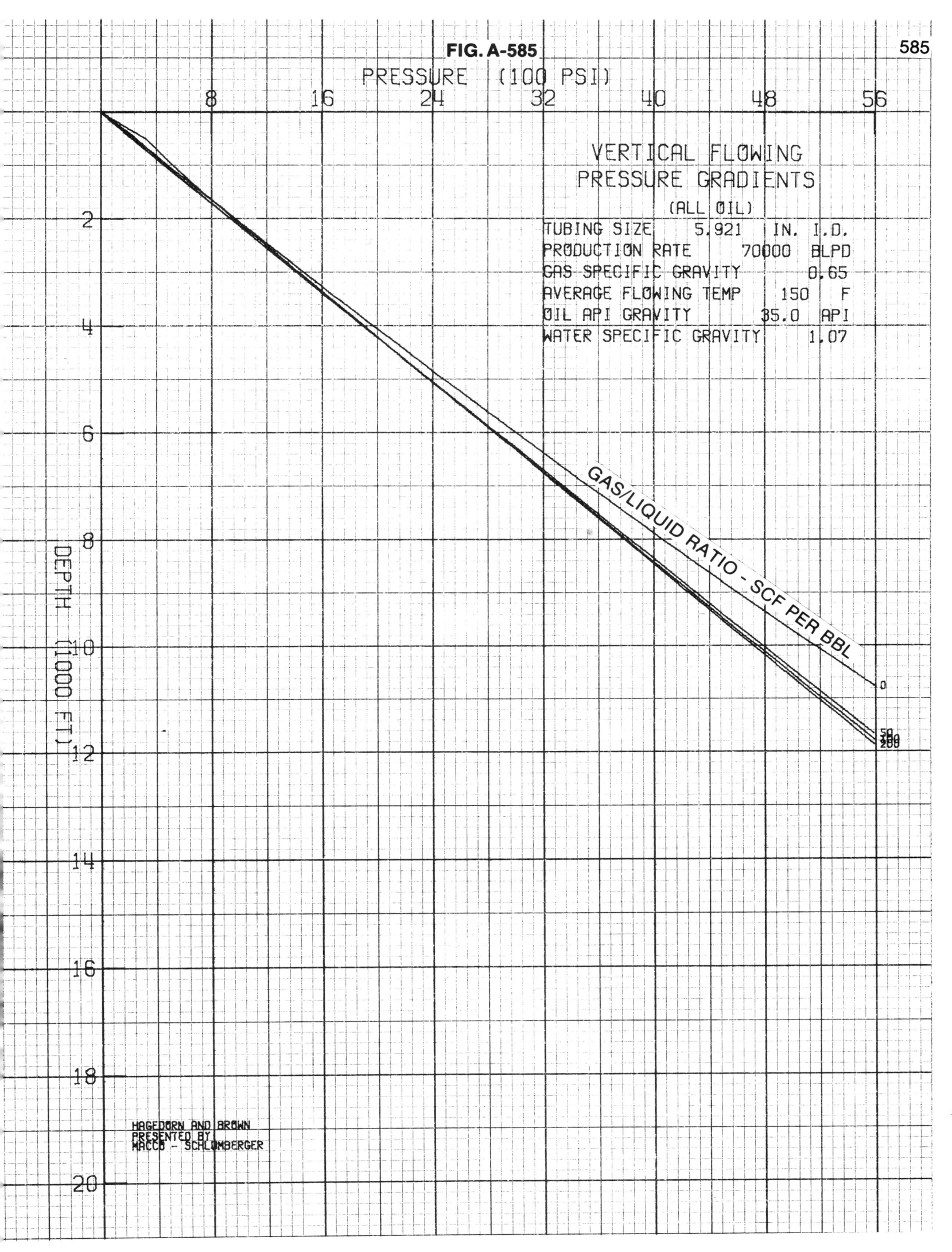

PRESSURE (100 PSI)

VERTICAL FLOWING
PRESSURE GRADIENTS
(ALL OIL)

TUBING SIZE	5.921	IN. I.D.
PRODUCTION RATE	70000	BLPD
GAS SPECIFIC GRAVITY	0.65	
AVERAGE FLOWING TEMP	150	F
OIL API GRAVITY	35.0	API
WATER SPECIFIC GRAVITY	1.07	

GAS/LIQUID RATIO - SCF PER BBL

DEPTH (1000 FT)

HAGEDORN AND BROWN
PRESENTED BY
MACCO - SCHLUMBERGER

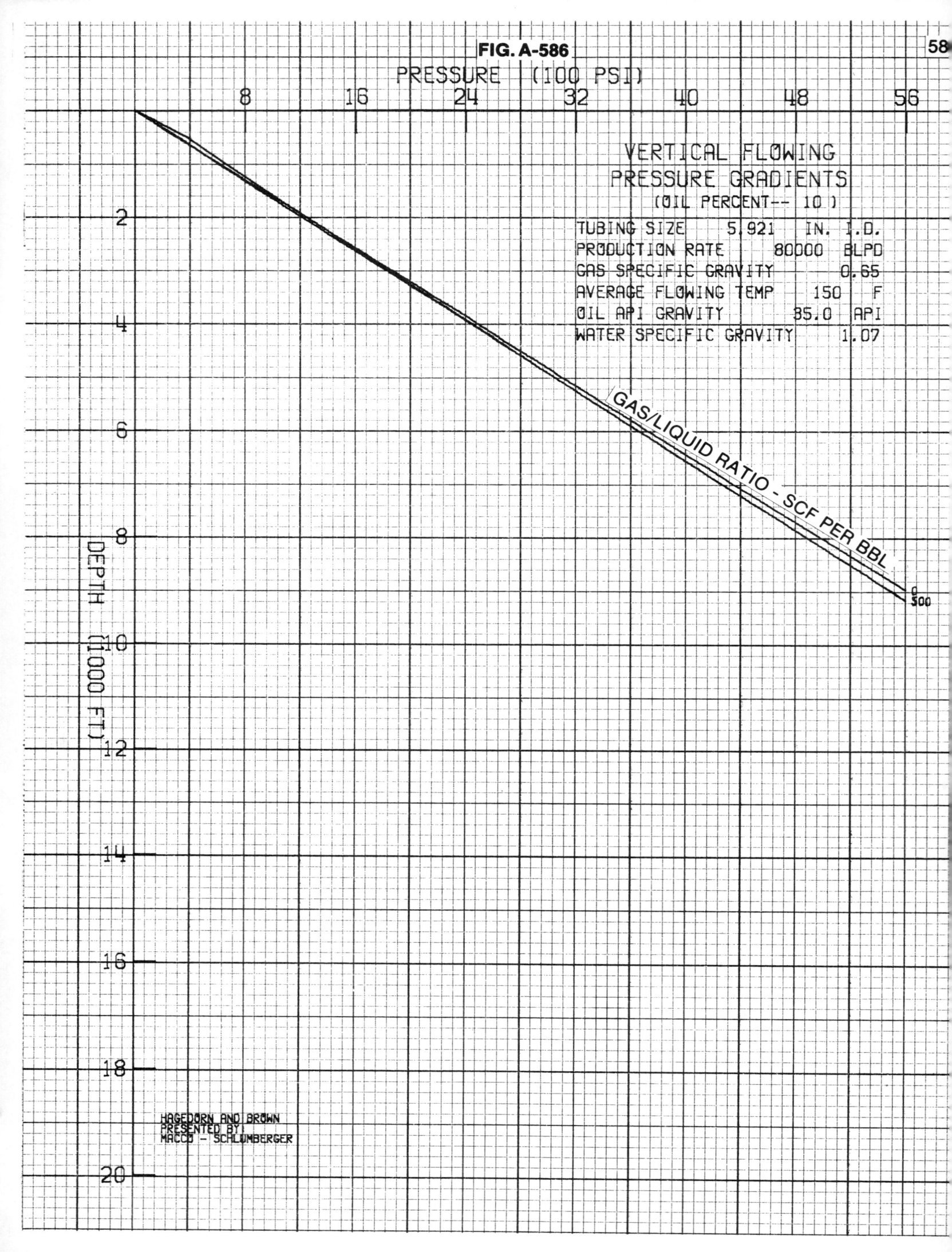

PRESSURE (100 PSI)

DEPTH (1000 FT)

VERTICAL FLOWING
PRESSURE GRADIENTS
(OIL PERCENT-- 10)

TUBING SIZE	5.921	IN. I.D.
PRODUCTION RATE	80000	BLPD
GAS SPECIFIC GRAVITY	0.65	
AVERAGE FLOWING TEMP	150	F
OIL API GRAVITY	85.0	API
WATER SPECIFIC GRAVITY	1.07	

GAS/LIQUID RATIO - SCF PER BBL

0
500

HAGEDORN AND BROWN
PRESENTED BY
MACCO - SCHLUMBERGER

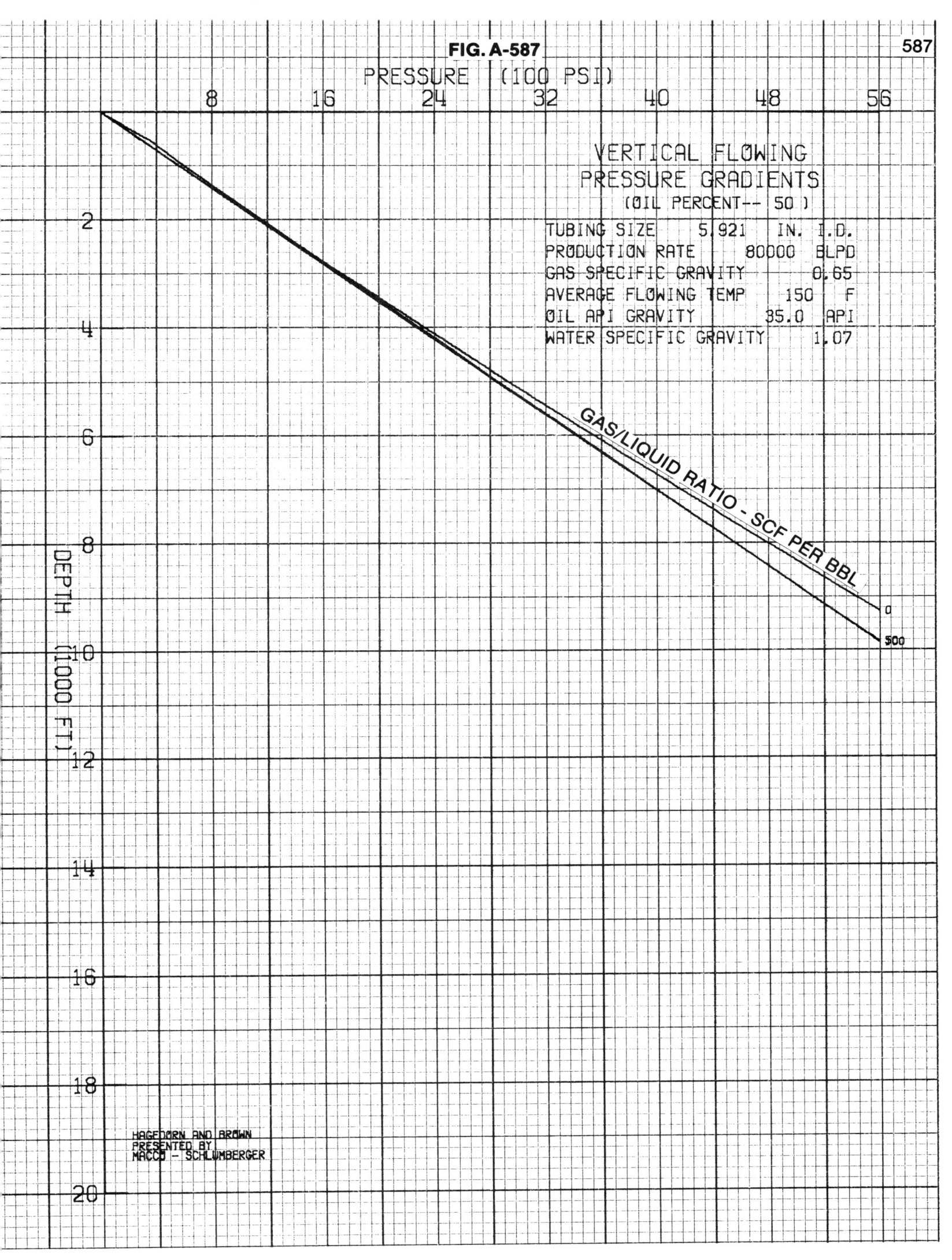

PRESSURE (100 PSI)

VERTICAL FLOWING
PRESSURE GRADIENTS
(OIL PERCENT-- 50)

TUBING SIZE	5.921	IN. I.D.
PRODUCTION RATE	80000	BLPD
GAS SPECIFIC GRAVITY	0.65	
AVERAGE FLOWING TEMP	150	F
OIL API GRAVITY	35.0	API
WATER SPECIFIC GRAVITY	1.07	

GAS/LIQUID RATIO - SCF PER BBL

DEPTH (1000 FT)

HAGEDORN AND BROWN
PRESENTED BY
MACCO - SCHLUMBERGER

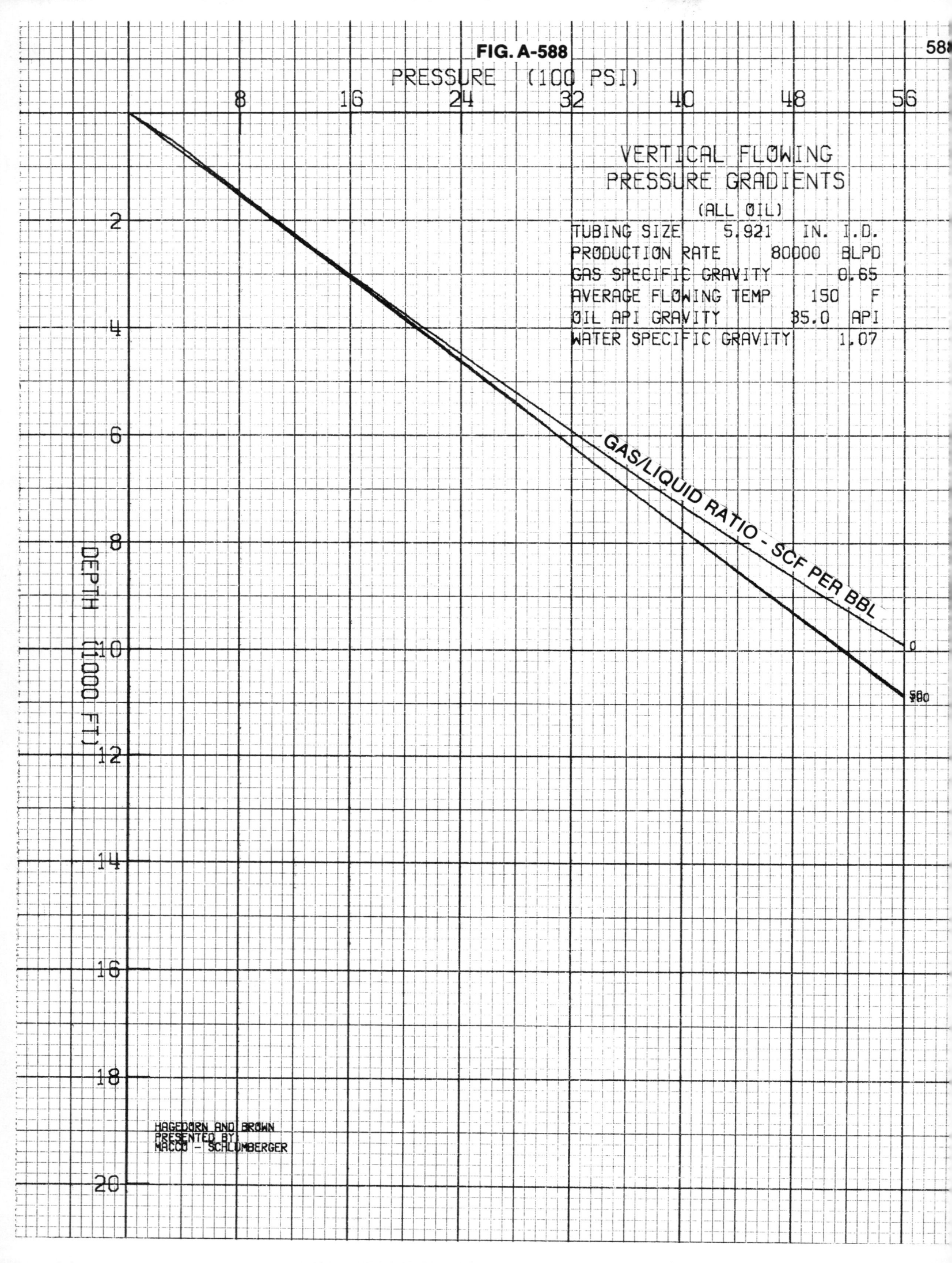

PRESSURE (100 PSI)

VERTICAL FLOWING
PRESSURE GRADIENTS
(ALL OIL)

TUBING SIZE	5.921	IN. I.D.
PRODUCTION RATE	80000	BLPD
GAS SPECIFIC GRAVITY		0.65
AVERAGE FLOWING TEMP	150	F
OIL API GRAVITY	35.0	API
WATER SPECIFIC GRAVITY		1.07

DEPTH (1000 FT)

GAS/LIQUID RATIO - SCF PER BBL

HAGEDORN AND BROWN
PRESENTED BY
MACCO - SCHLUMBERGER

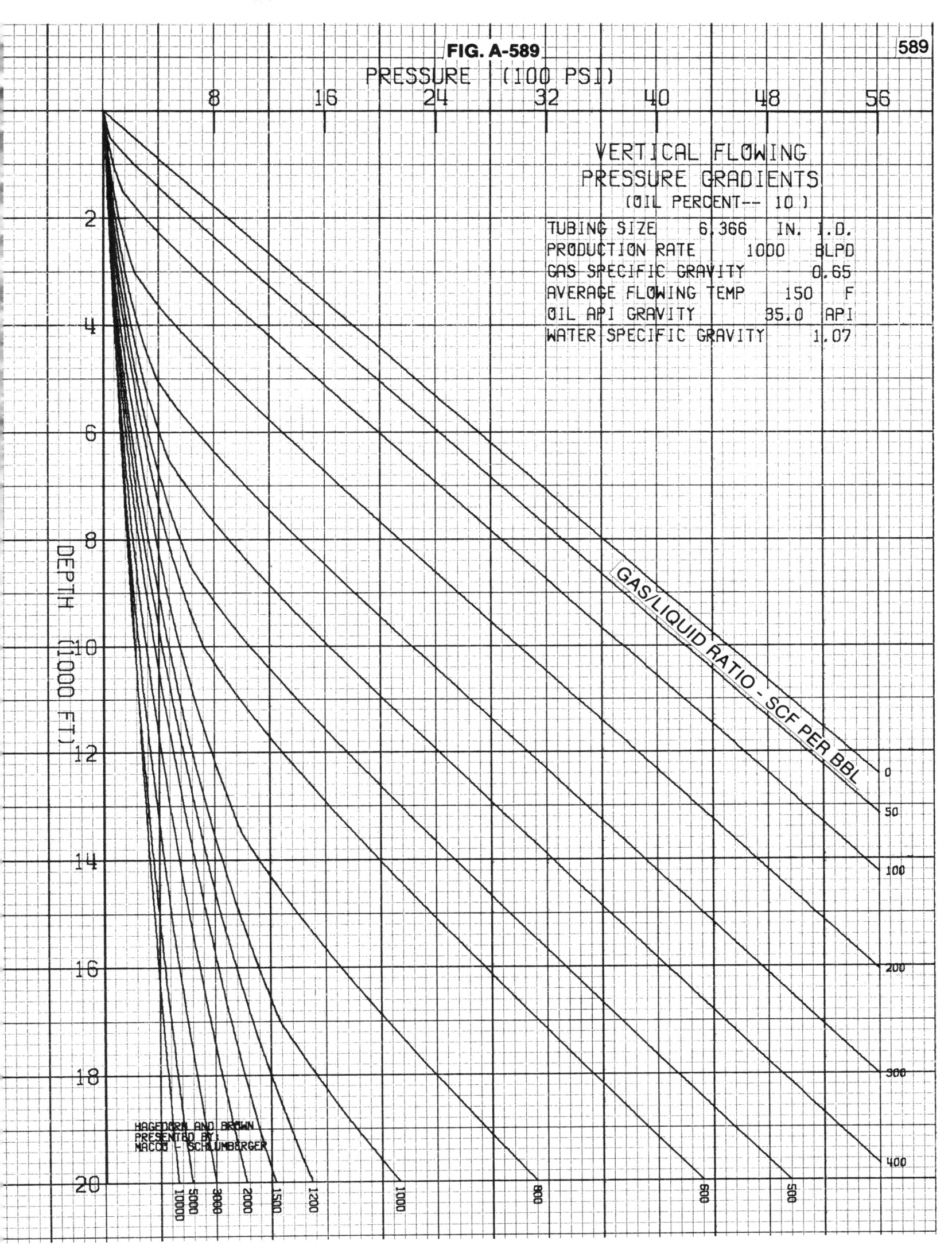

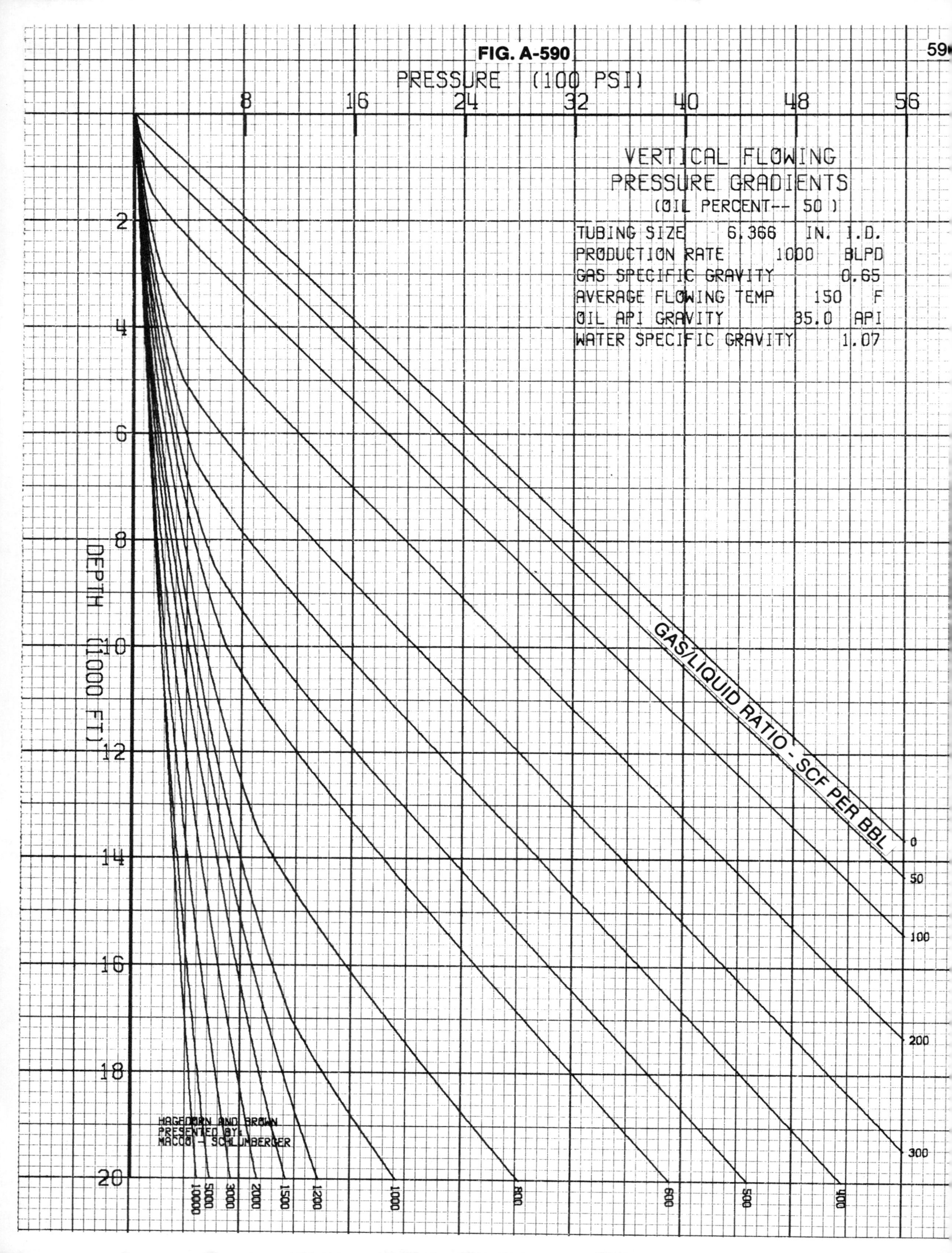

FIG. A-590

VERTICAL FLOWING
PRESSURE GRADIENTS
(OIL PERCENT-- 50)

TUBING SIZE	6.366	IN. I.D.
PRODUCTION RATE	1000	BLPD
GAS SPECIFIC GRAVITY	0.65	
AVERAGE FLOWING TEMP	150	F
OIL API GRAVITY	35.0	API
WATER SPECIFIC GRAVITY	1.07	

PRESSURE (100 PSI)

DEPTH (1000 FT)

GAS/LIQUID RATIO - SCF PER BBL

HAGEDORN AND BROWN
PRESENTED BY
MACCO - SCHLUMBERGER

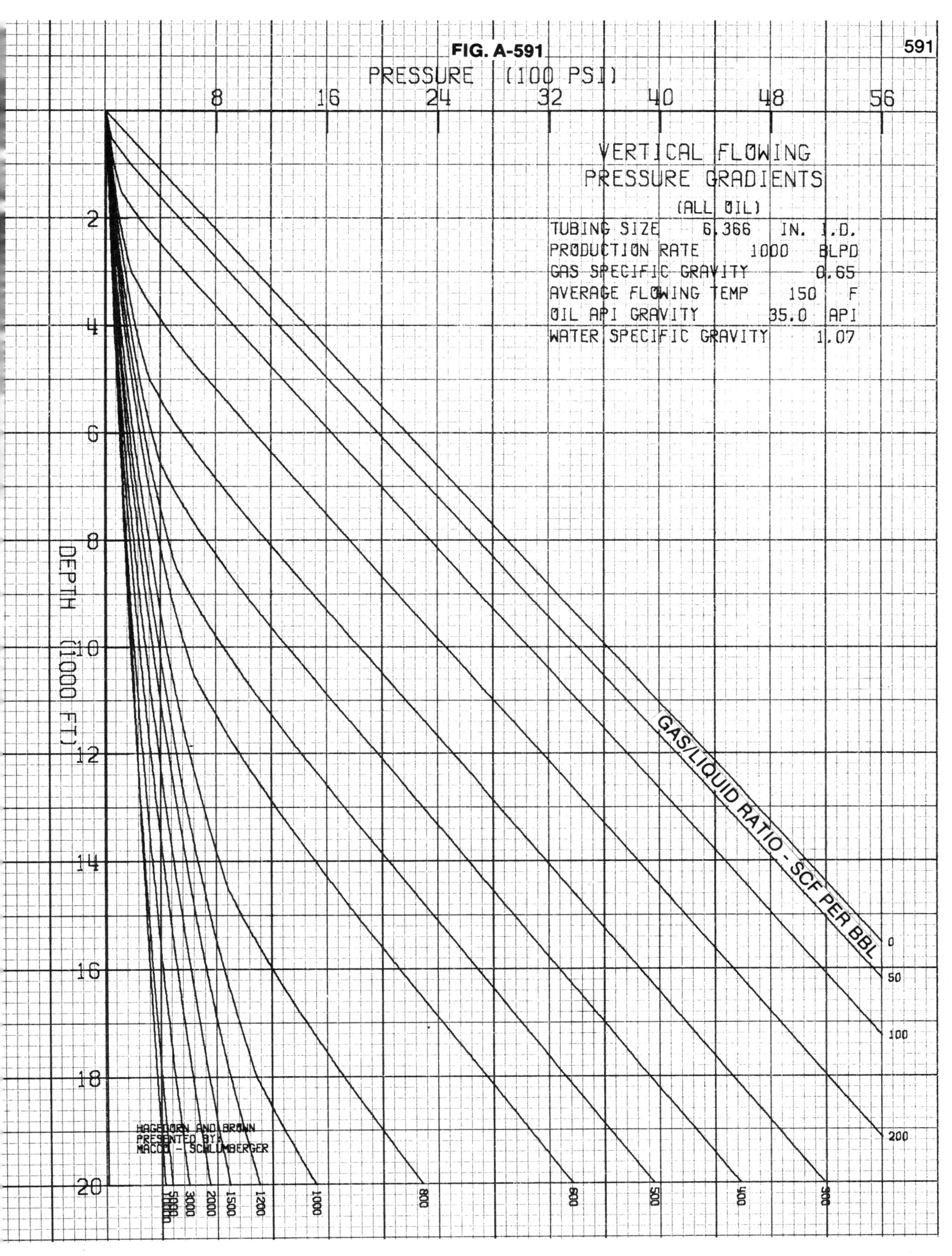

PRESSURE (100 PSI)

VERTICAL FLOWING
PRESSURE GRADIENTS
(ALL OIL)

TUBING SIZE	6.366	IN. I.D.
PRODUCTION RATE	1000	BLPD
GAS SPECIFIC GRAVITY	0.65	
AVERAGE FLOWING TEMP	150	F
OIL API GRAVITY	35.0	API
WATER SPECIFIC GRAVITY	1.07	

DEPTH (1000 FT)

GAS/LIQUID RATIO - SCF PER BBL

HAGEDORN AND BROWN
PRESENTED BY
MACCO - SCHLUMBERGER

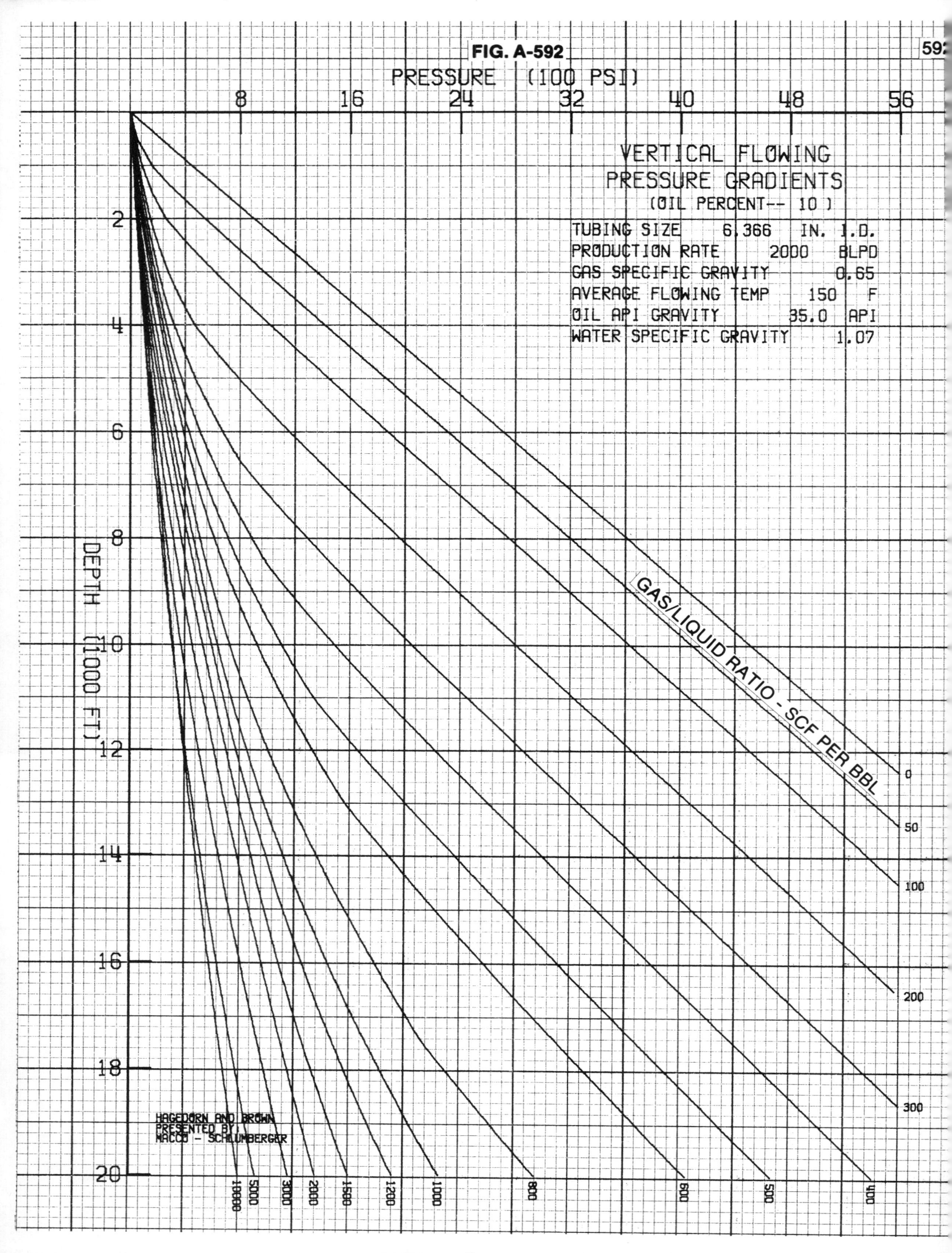

FIG. A-592

VERTICAL FLOWING
PRESSURE GRADIENTS
(OIL PERCENT-- 10)

TUBING SIZE	6.366	IN. I.D.
PRODUCTION RATE	2000	BLPD
GAS SPECIFIC GRAVITY	0.65	
AVERAGE FLOWING TEMP	150	F
OIL API GRAVITY	35.0	API
WATER SPECIFIC GRAVITY	1.07	

PRESSURE (100 PSI)

DEPTH (1000 FT)

GAS/LIQUID RATIO - SCF PER BBL

HAGEDORN AND BROWN
PRESENTED BY
MACCO -- SCHLUMBERGER

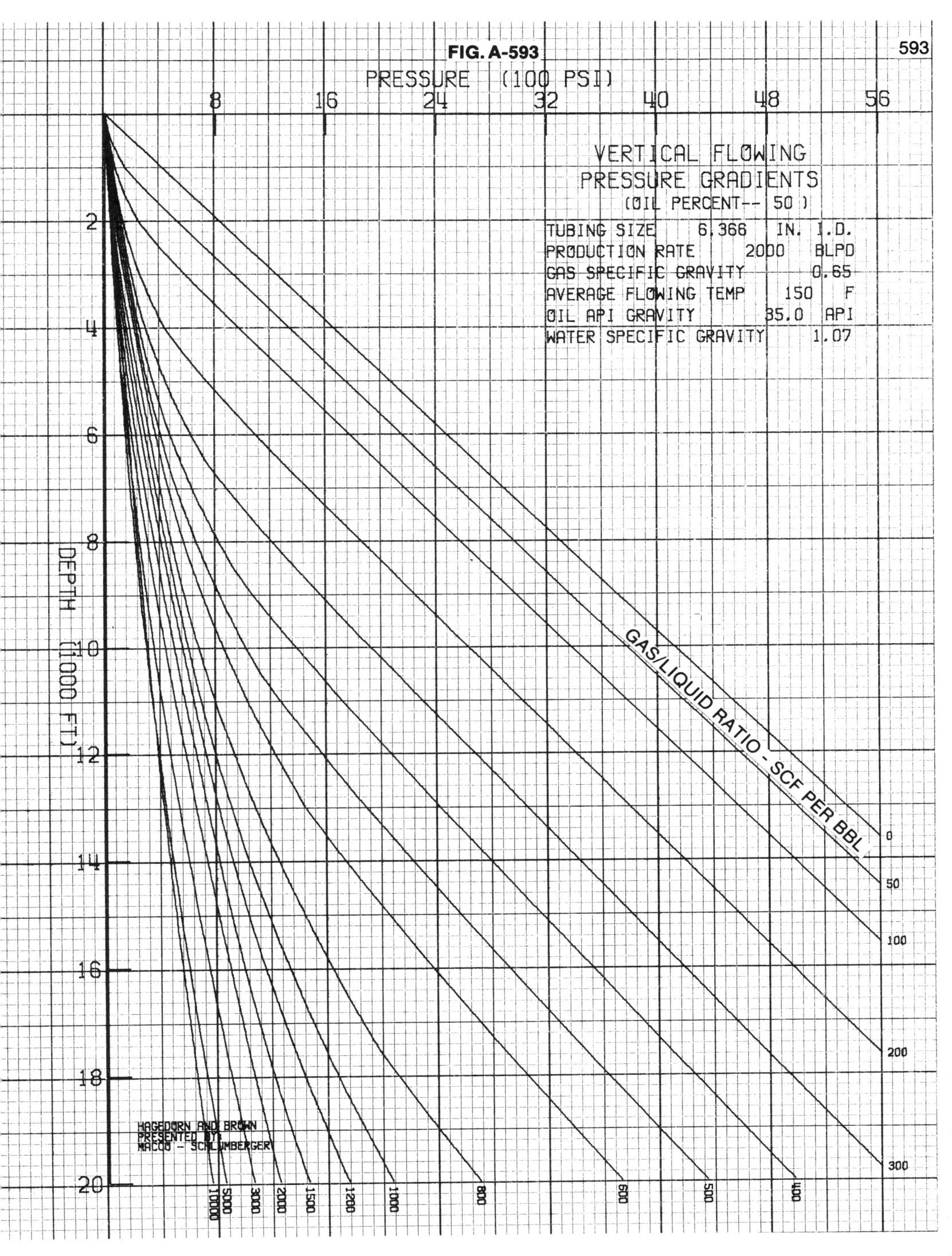

FIG. A-593

593

PRESSURE (100 PSI)

VERTICAL FLOWING
PRESSURE GRADIENTS
(OIL PERCENT-- 50)

TUBING SIZE	6.366	IN. I.D.
PRODUCTION RATE	2000	BLPD
GAS SPECIFIC GRAVITY		0.65
AVERAGE FLOWING TEMP	150	F
OIL API GRAVITY	35.0	API
WATER SPECIFIC GRAVITY		1.07

DEPTH (1,000 FT)

GAS/LIQUID RATIO - SCF PER BBL

HAGEDORN AND BROWN
PRESENTED BY:
MACCO - SCHLUMBERGER

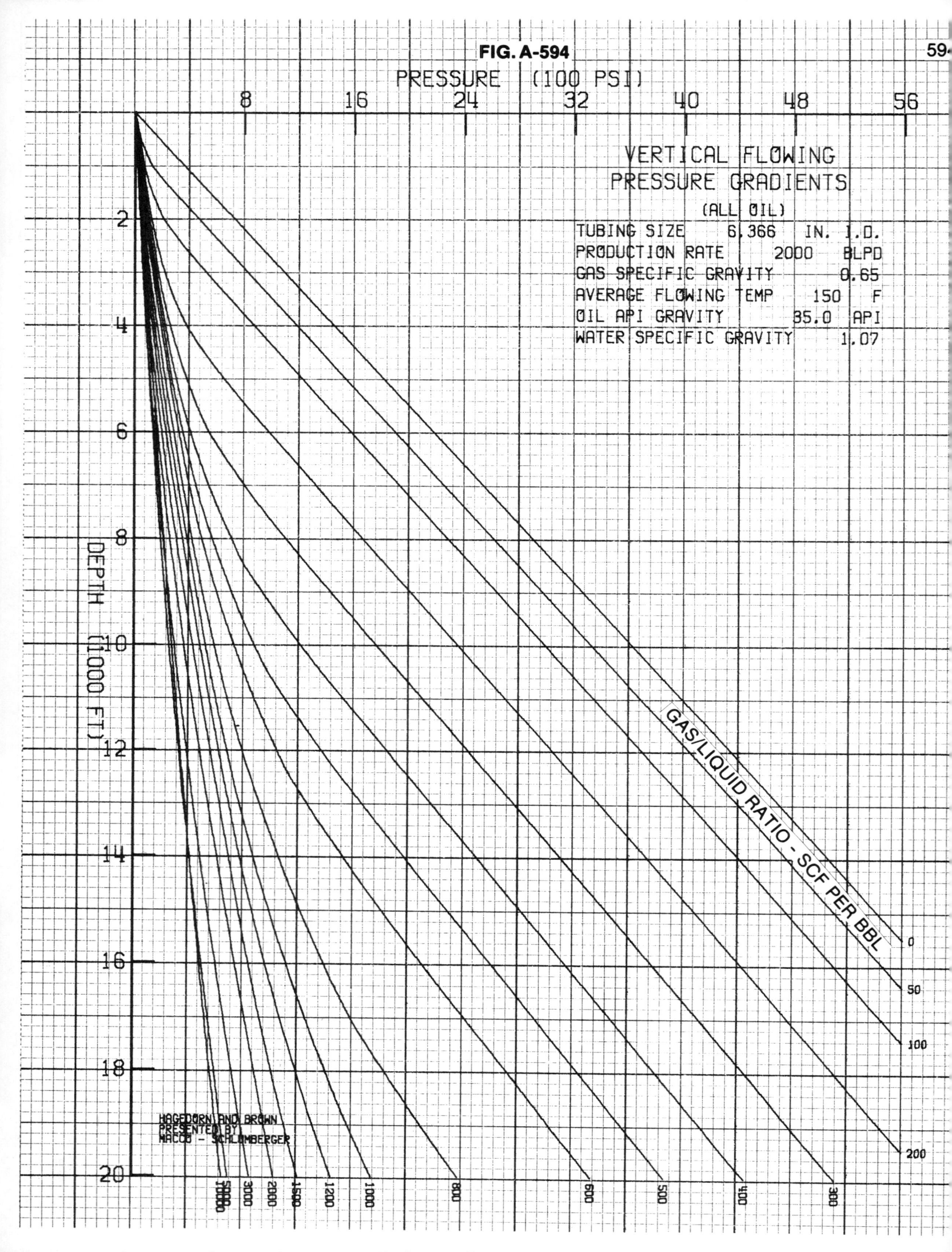

VERTICAL FLOWING
PRESSURE GRADIENTS
(ALL OIL)

TUBING SIZE	6.366	IN. I.D.
PRODUCTION RATE	2000	BLPD
GAS SPECIFIC GRAVITY	0.65	
AVERAGE FLOWING TEMP	150	F
OIL API GRAVITY	35.0	API
WATER SPECIFIC GRAVITY	1.07	

PRESSURE (100 PSI)

GAS/LIQUID RATIO - SCF PER BBL

DEPTH (1000 FT)

HAGEDORN AND BROWN
PRESENTED BY
MACCO - SCHLUMBERGER

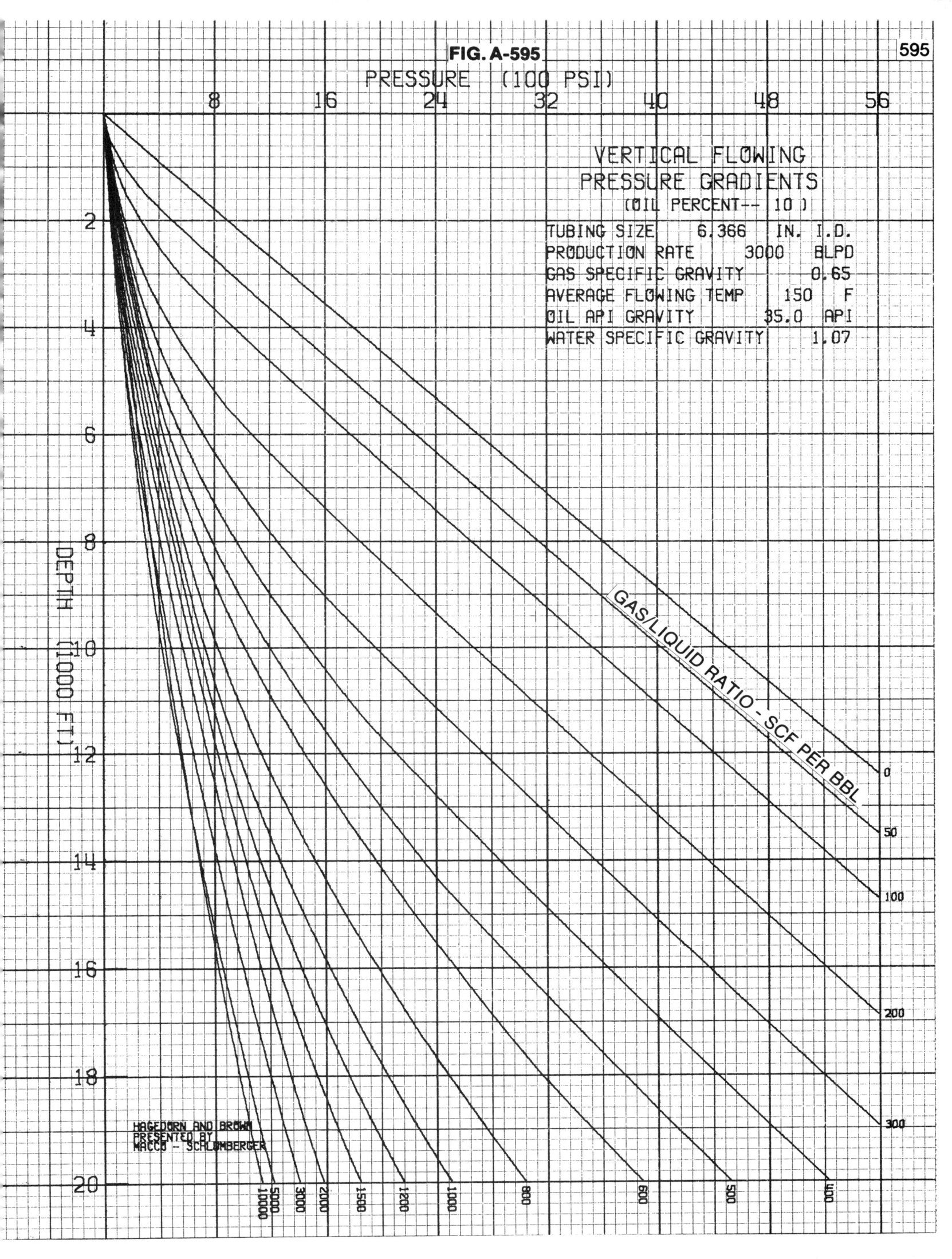

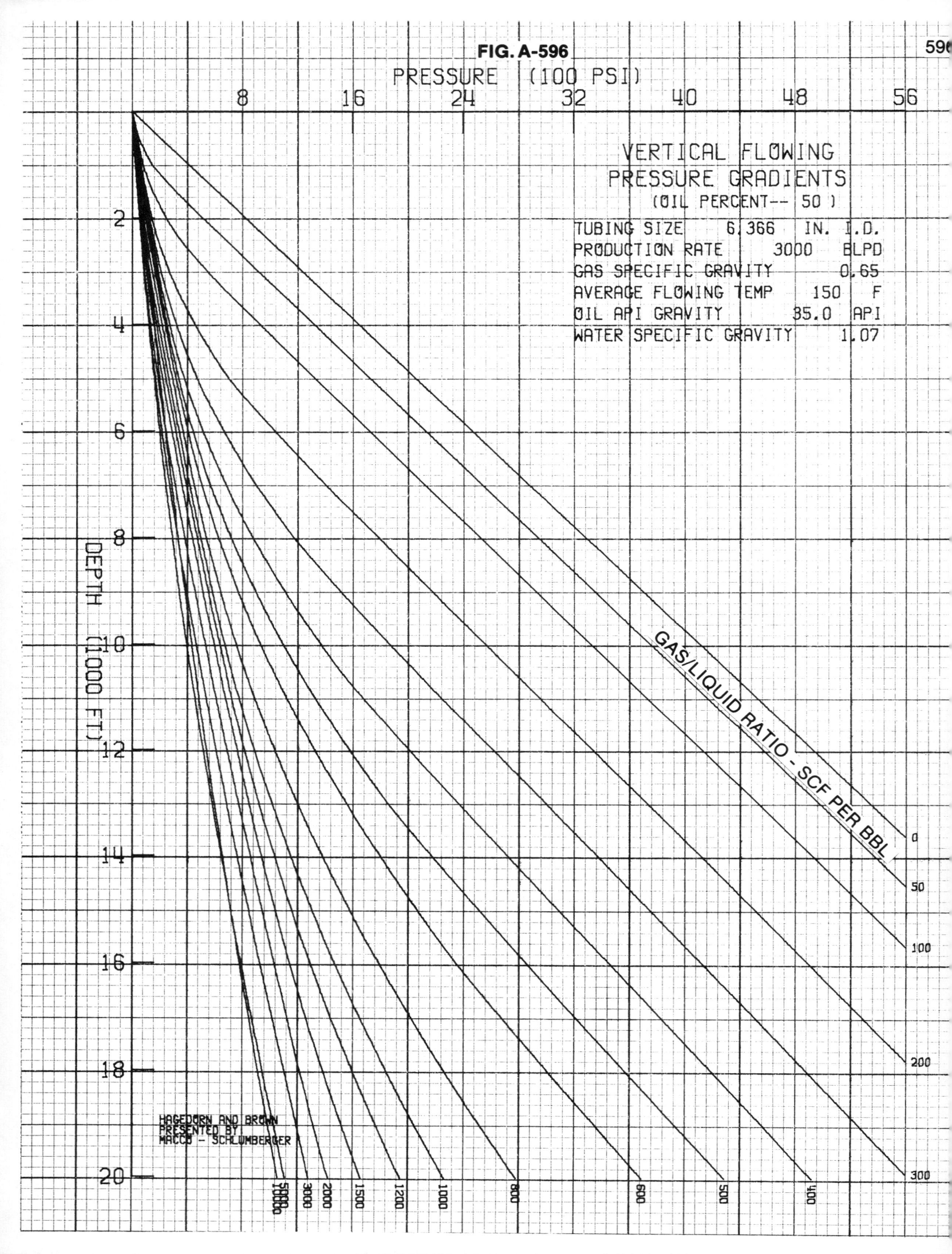

PRESSURE (100 PSI)

VERTICAL FLOWING
PRESSURE GRADIENTS
(OIL PERCENT-- 50)

TUBING SIZE	6.366	IN. I.D.
PRODUCTION RATE	3000	BLPD
GAS SPECIFIC GRAVITY	0.65	
AVERAGE FLOWING TEMP	150	F
OIL API GRAVITY	35.0	API
WATER SPECIFIC GRAVITY	1.07	

DEPTH (1000 FT)

GAS/LIQUID RATIO - SCF PER BBL

HAGEDORN AND BROWN
PRESENTED BY
MACCO - SCHLUMBERGER

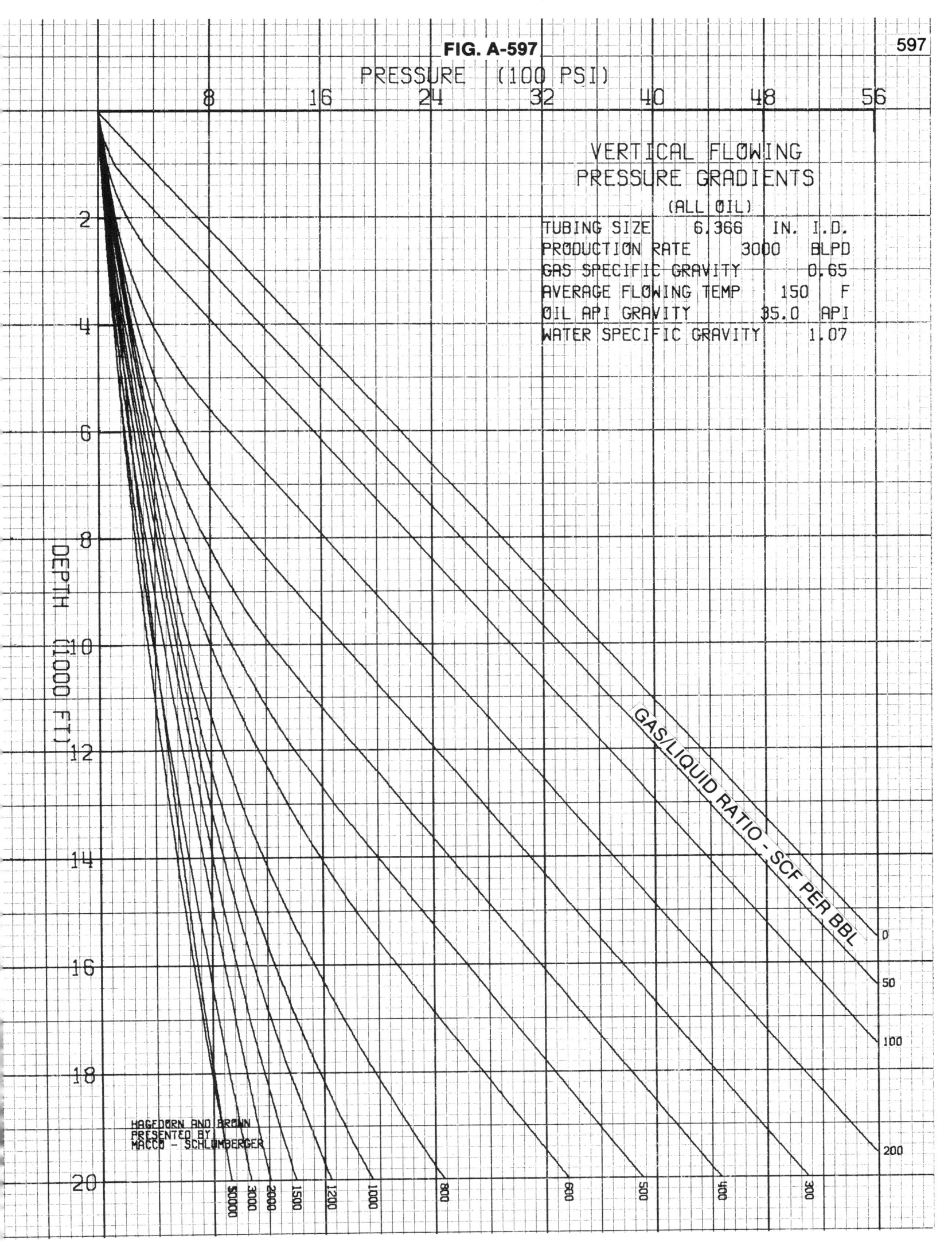

PRESSURE (100 PSI)

VERTICAL FLOWING
PRESSURE GRADIENTS
(ALL OIL)

TUBING SIZE 6.366 IN. I.D.
PRODUCTION RATE 3000 BLPD
GAS SPECIFIC GRAVITY 0.65
AVERAGE FLOWING TEMP 150 F
OIL API GRAVITY 35.0 API
WATER SPECIFIC GRAVITY 1.07

DEPTH (1000 FT)

GAS/LIQUID RATIO - SCF PER BBL

HAGEDORN AND BROWN
PRESENTED BY
MACCO - SCHLUMBERGER

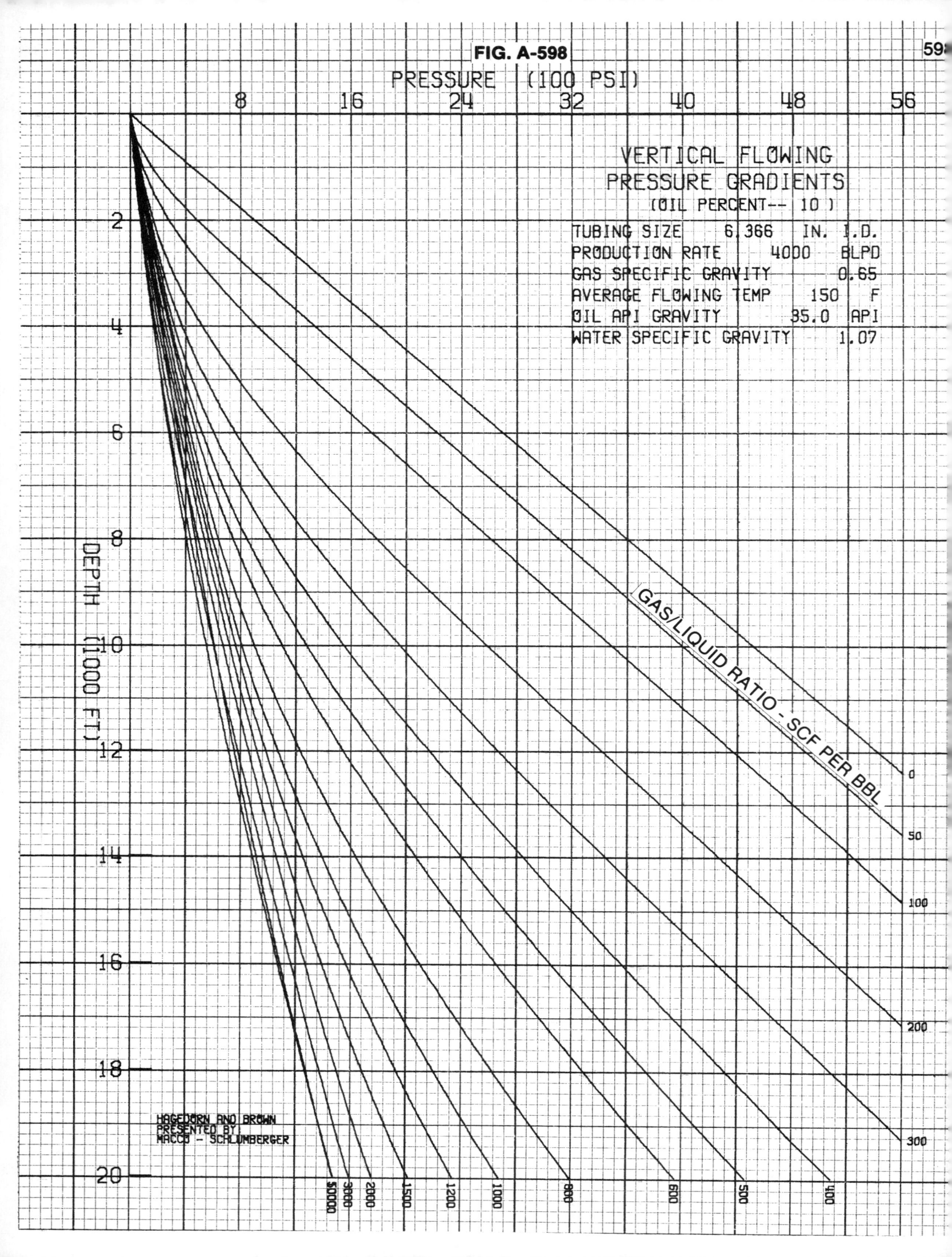

FIG. A-598

PRESSURE (100 PSI)

VERTICAL FLOWING
PRESSURE GRADIENTS
(OIL PERCENT-- 10)

TUBING SIZE	6.366	IN. I.D.
PRODUCTION RATE	4000	BLPD
GAS SPECIFIC GRAVITY	0.65	
AVERAGE FLOWING TEMP	150	F
OIL API GRAVITY	35.0	API
WATER SPECIFIC GRAVITY	1.07	

DEPTH (1000 FT)

GAS/LIQUID RATIO - SCF PER BBL

HAGEDORN AND BROWN
PRESENTED BY
MACCO - SCHLUMBERGER

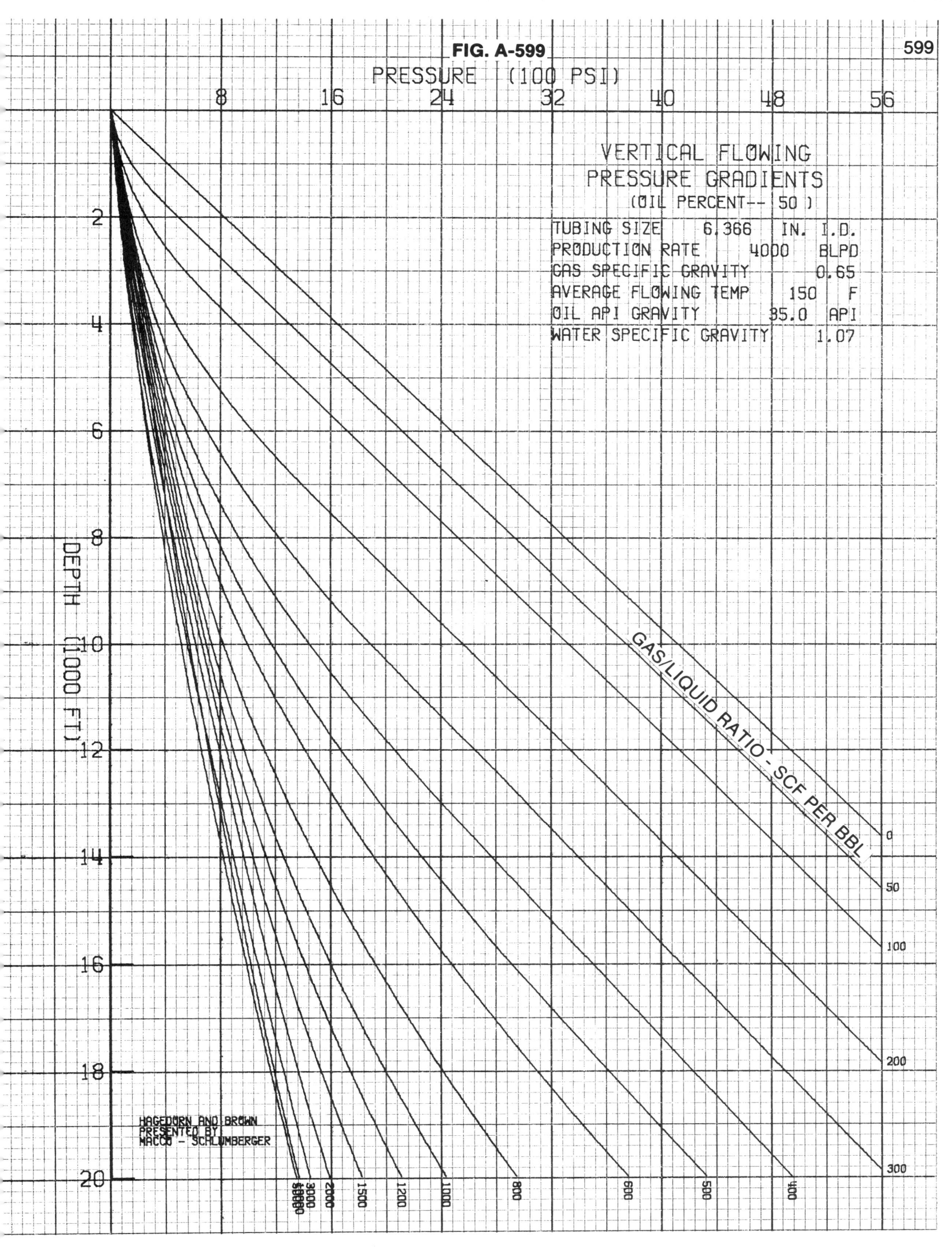

FIG. A-599

599

PRESSURE (100 PSI)

VERTICAL FLOWING
PRESSURE GRADIENTS
(OIL PERCENT-- 50)

TUBING SIZE	6.366	IN. I.D.
PRODUCTION RATE	4000	BLPD
GAS SPECIFIC GRAVITY	0.65	
AVERAGE FLOWING TEMP	150	F
OIL API GRAVITY	35.0	API
WATER SPECIFIC GRAVITY	1.07	

DEPTH (1000 FT)

GAS/LIQUID RATIO - SCF PER BBL

HAGEDORN AND BROWN
PRESENTED BY
MACCO - SCHLUMBERGER

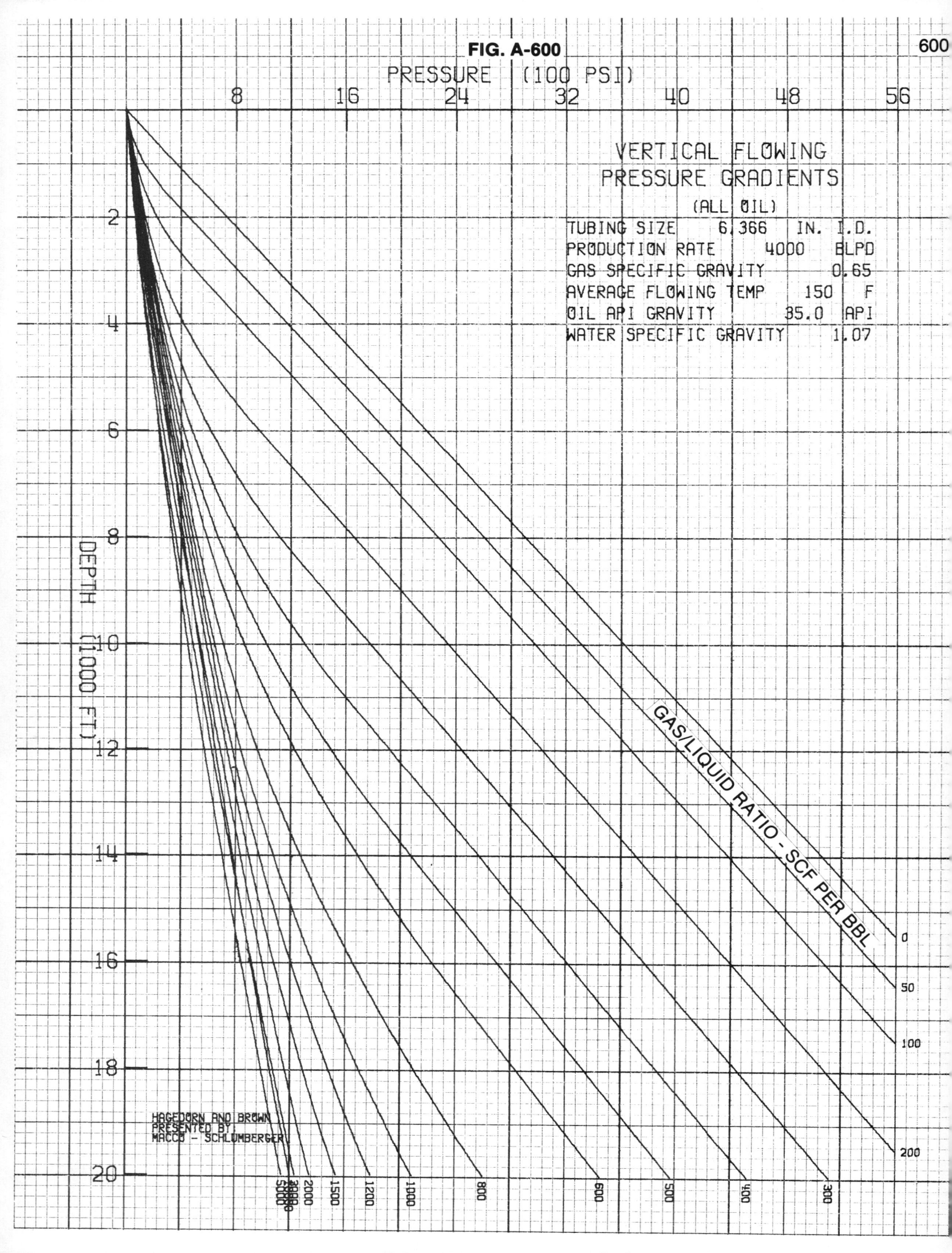

PRESSURE (100 PSI)

VERTICAL FLOWING
PRESSURE GRADIENTS
(ALL OIL)

TUBING SIZE	6.366	IN. I.D.
PRODUCTION RATE	4000	BLPD
GAS SPECIFIC GRAVITY	0.65	
AVERAGE FLOWING TEMP	150	F
OIL API GRAVITY	35.0	API
WATER SPECIFIC GRAVITY	1.07	

DEPTH (1000 FT)

GAS/LIQUID RATIO - SCF PER BBL

HAGEDORN AND BROWN
PRESENTED BY
MACCO - SCHLUMBERGER

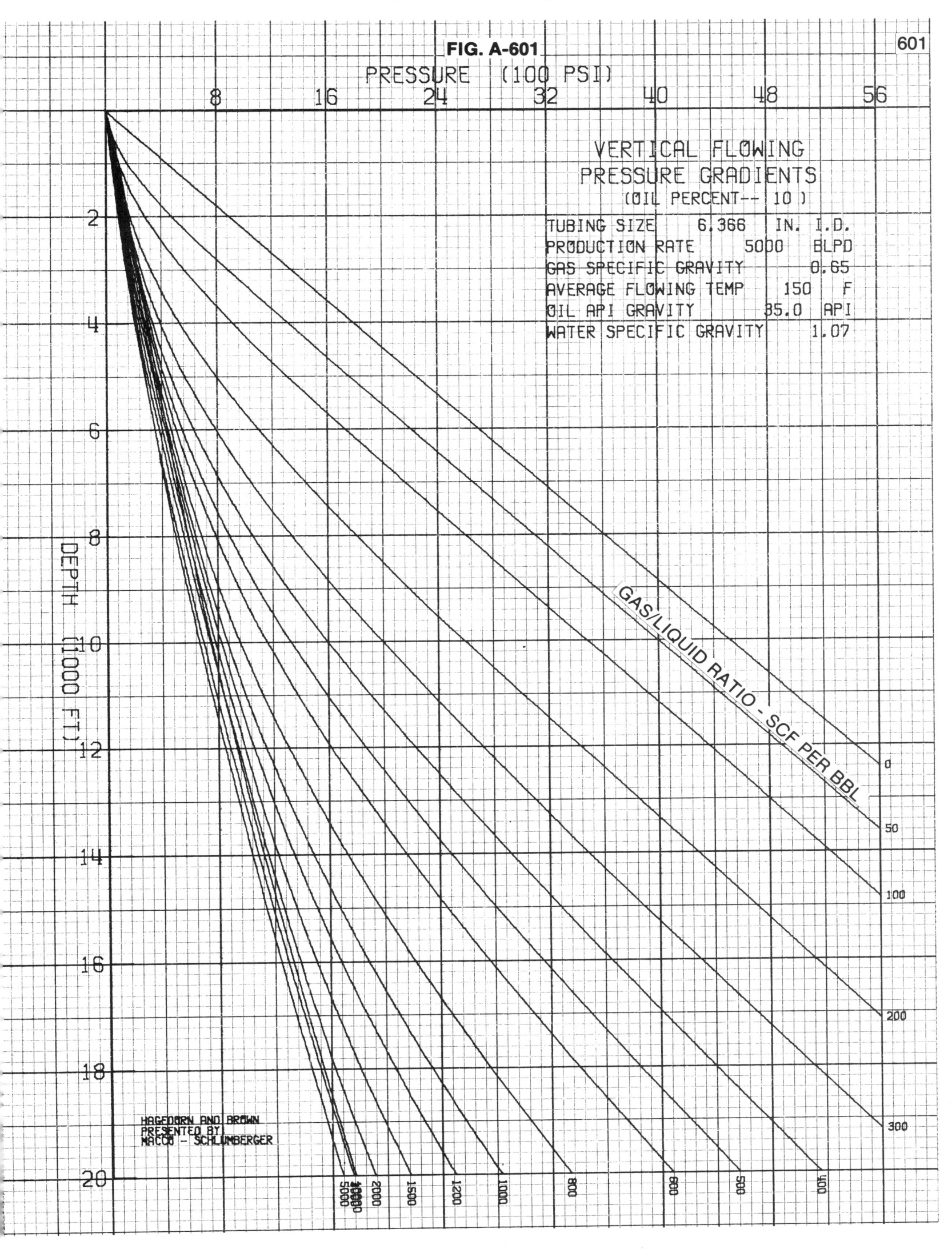

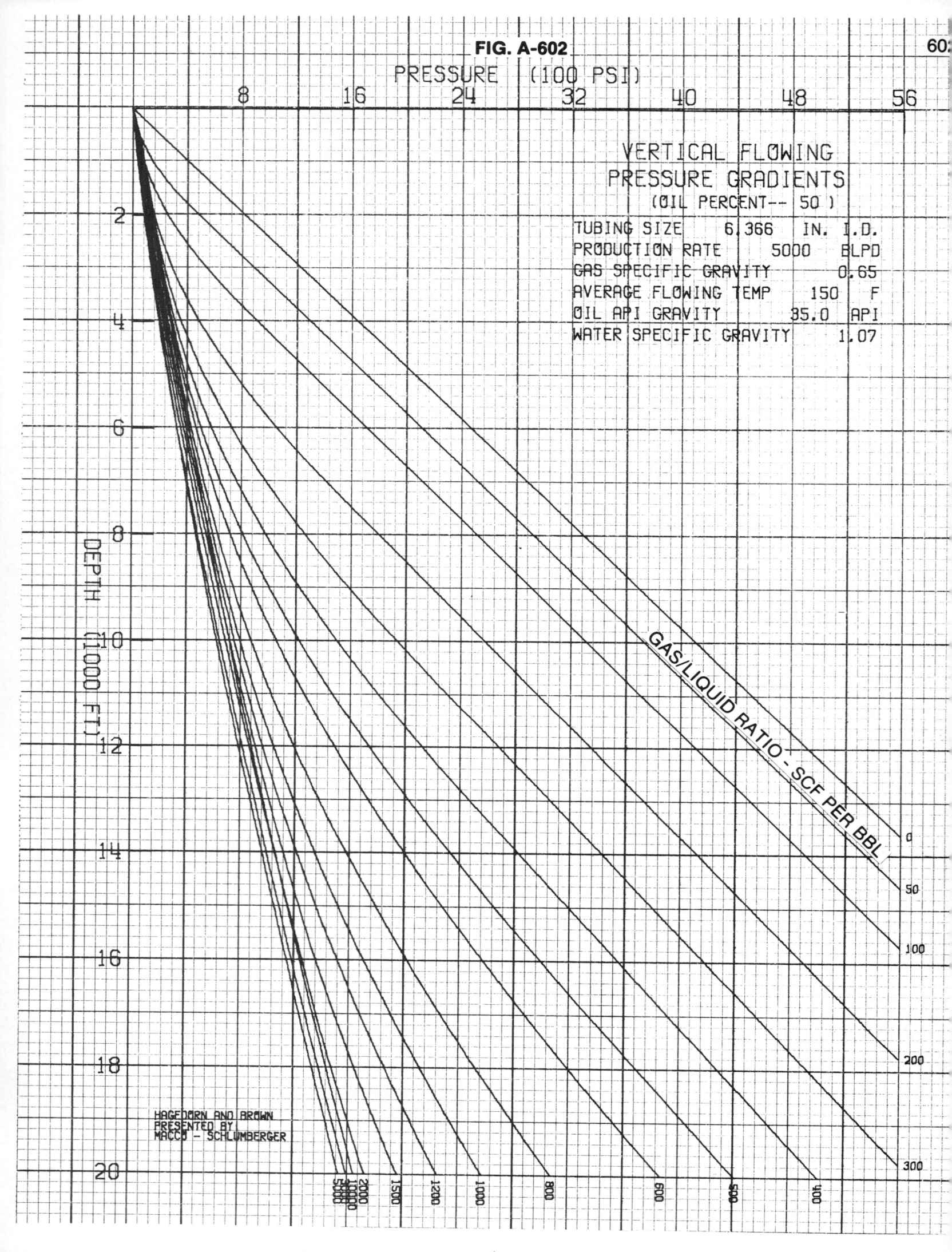

VERTICAL FLOWING
PRESSURE GRADIENTS
(OIL PERCENT-- 50)

TUBING SIZE   6.366   IN. I.D.
PRODUCTION RATE   5000   BLPD
GAS SPECIFIC GRAVITY   0.65
AVERAGE FLOWING TEMP   150   F
OIL API GRAVITY   35.0 API
WATER SPECIFIC GRAVITY   1.07

HAGEDORN AND BROWN
PRESENTED BY
MACCO - SCHLUMBERGER

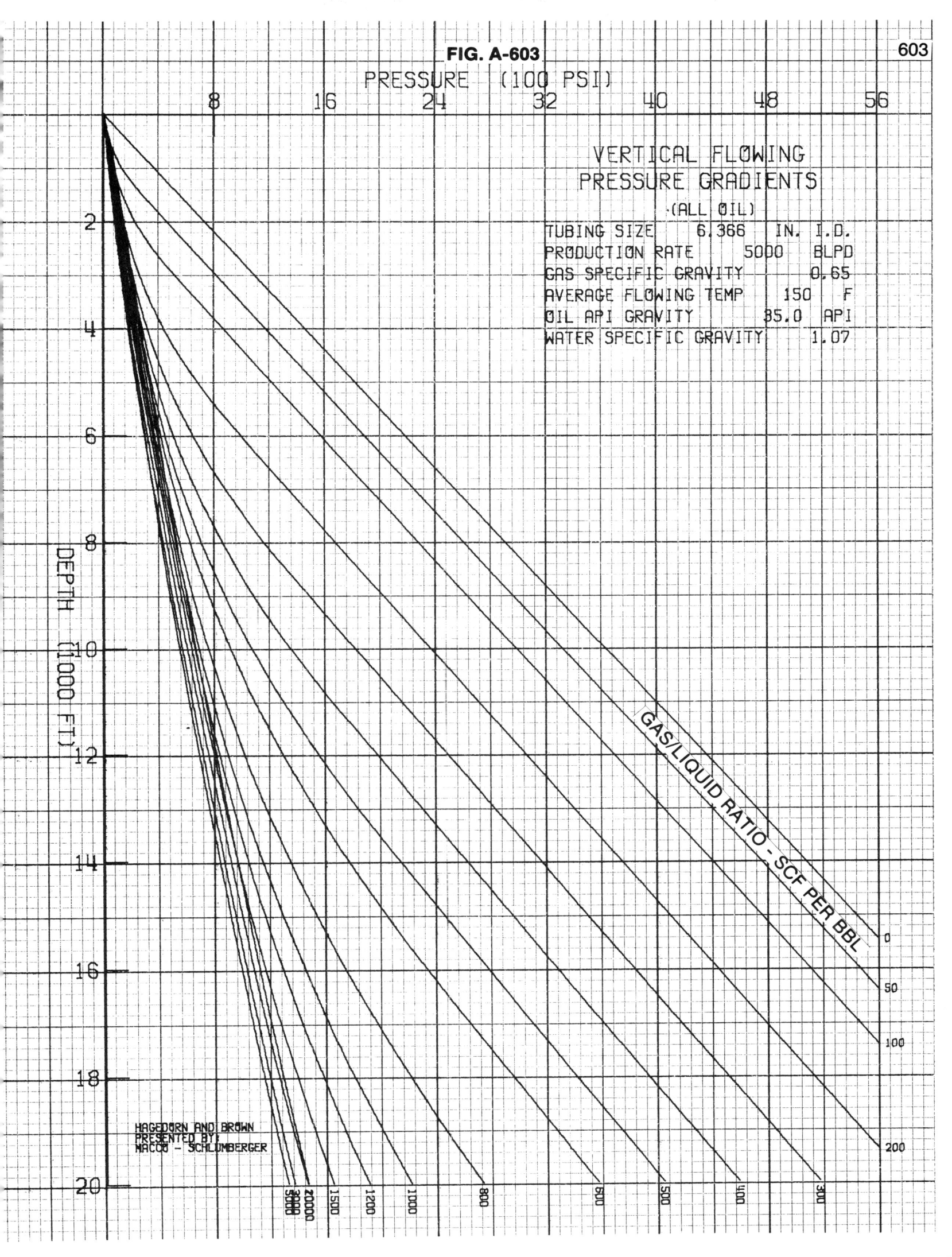

FIG. A-603

603

PRESSURE (100 PSI)

VERTICAL FLOWING
PRESSURE GRADIENTS
(ALL OIL)

TUBING SIZE	6.366	IN. I.D.
PRODUCTION RATE	5000	BLPD
GAS SPECIFIC GRAVITY	0.65	
AVERAGE FLOWING TEMP	150	F
OIL API GRAVITY	35.0	API
WATER SPECIFIC GRAVITY	1.07	

DEPTH (1000 FT)

GAS/LIQUID RATIO - SCF PER BBL

0

50

100

200

HAGEDORN AND BROWN
PRESENTED BY:
NACO - SCHLUMBERGER

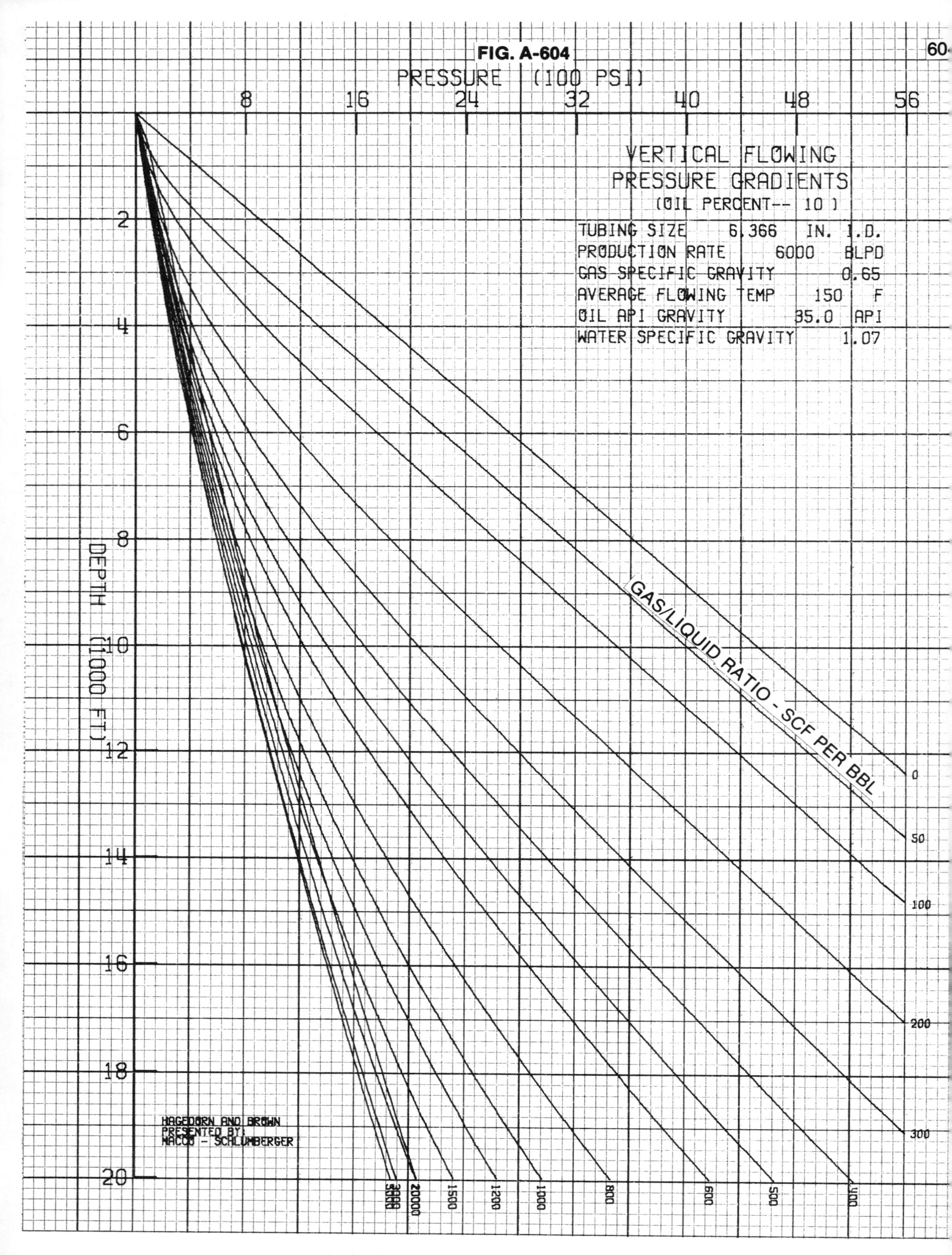

FIG. A-604

VERTICAL FLOWING
PRESSURE GRADIENTS
(OIL PERCENT-- 10)

TUBING SIZE	6.366	IN. I.D.
PRODUCTION RATE	6000	BLPD
GAS SPECIFIC GRAVITY	0.65	
AVERAGE FLOWING TEMP	150	F
OIL API GRAVITY	35.0	API
WATER SPECIFIC GRAVITY	1.07	

PRESSURE (100 PSI)

DEPTH (1000 FT)

GAS/LIQUID RATIO - SCF PER BBL

HAGEDORN AND BROWN
PRESENTED BY:
NACO - SCHLUMBERGER

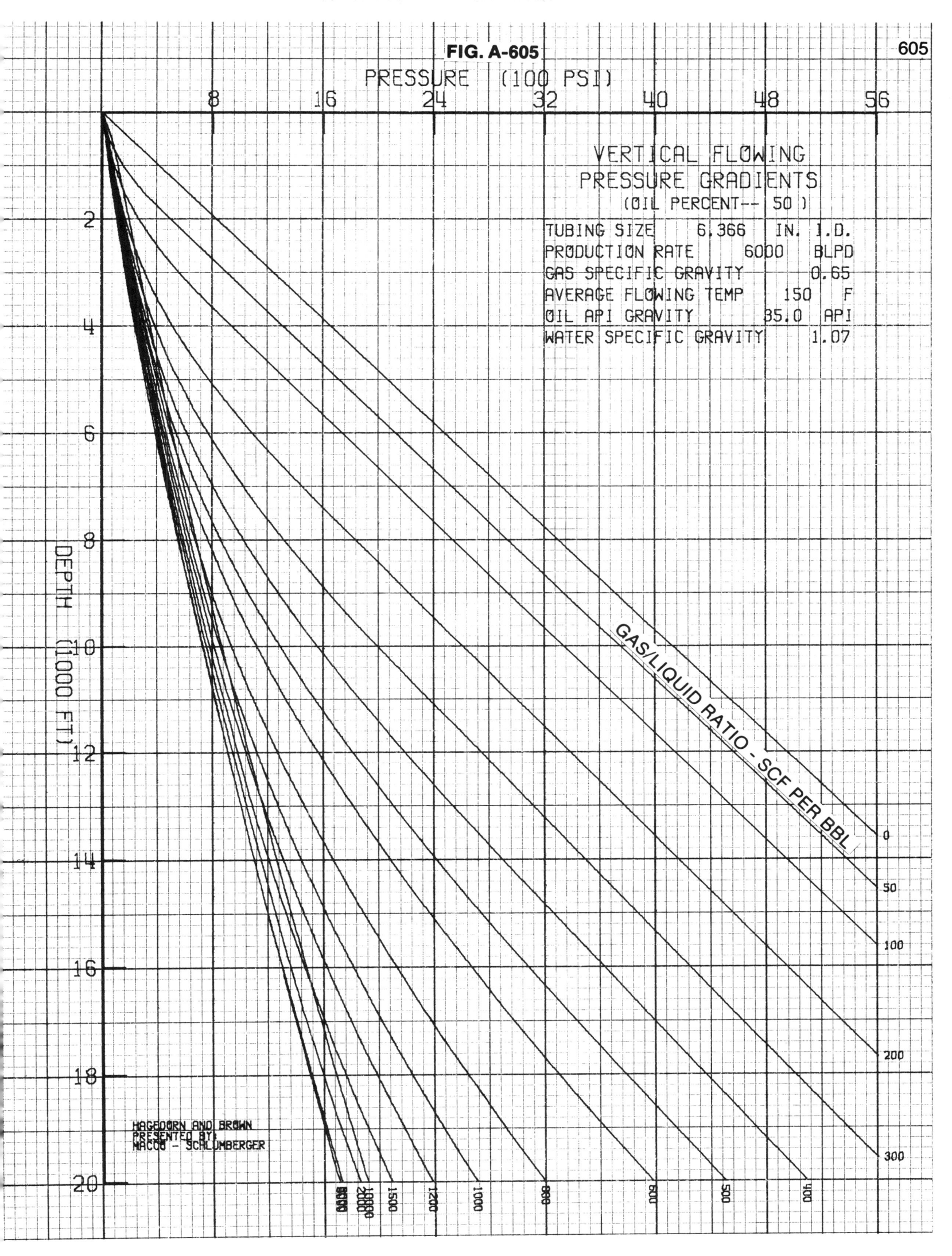

PRESSURE (100 PSI)

VERTICAL FLOWING
PRESSURE GRADIENTS
(OIL PERCENT-- 50)

TUBING SIZE	6.366	IN. I.D.
PRODUCTION RATE	6000	BLPD
GAS SPECIFIC GRAVITY	0.65	
AVERAGE FLOWING TEMP	150	F
OIL API GRAVITY	35.0	API
WATER SPECIFIC GRAVITY	1.07	

DEPTH (1000 FT)

GAS/LIQUID RATIO - SCF PER BBL

HAGEDORN AND BROWN
PRESENTED BY
NACCO - SCHLUMBERGER

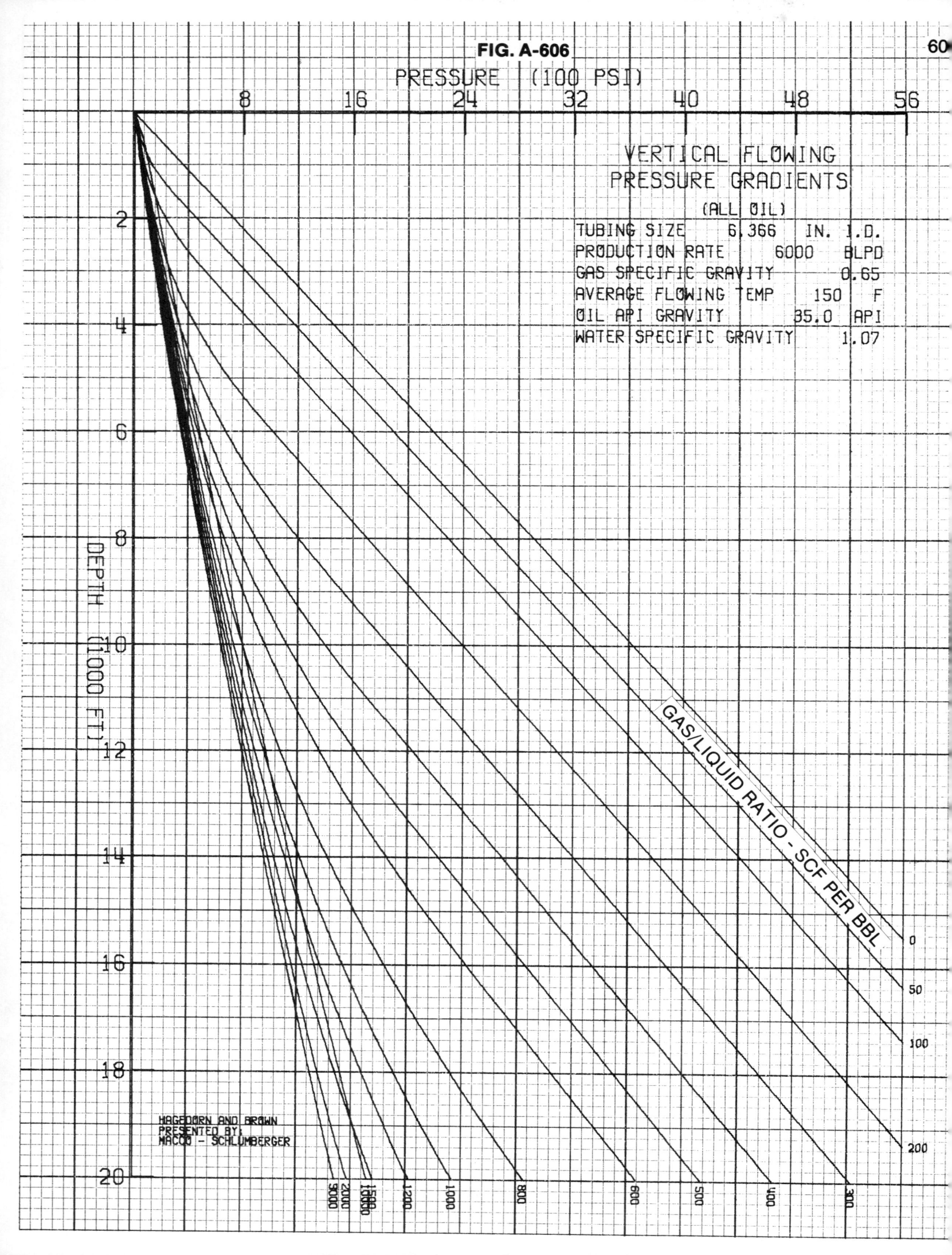

PRESSURE (100 PSI)

VERTICAL FLOWING
PRESSURE GRADIENTS
(ALL OIL)

TUBING SIZE	6.366	IN. I.D.
PRODUCTION RATE	6000	BLPD
GAS SPECIFIC GRAVITY	0.65	
AVERAGE FLOWING TEMP	150	F
OIL API GRAVITY	35.0	API
WATER SPECIFIC GRAVITY	1.07	

DEPTH (1000 FT)

GAS/LIQUID RATIO - SCF PER BBL

HAGEDORN AND BROWN
PRESENTED BY:
MACCO - SCHLUMBERGER

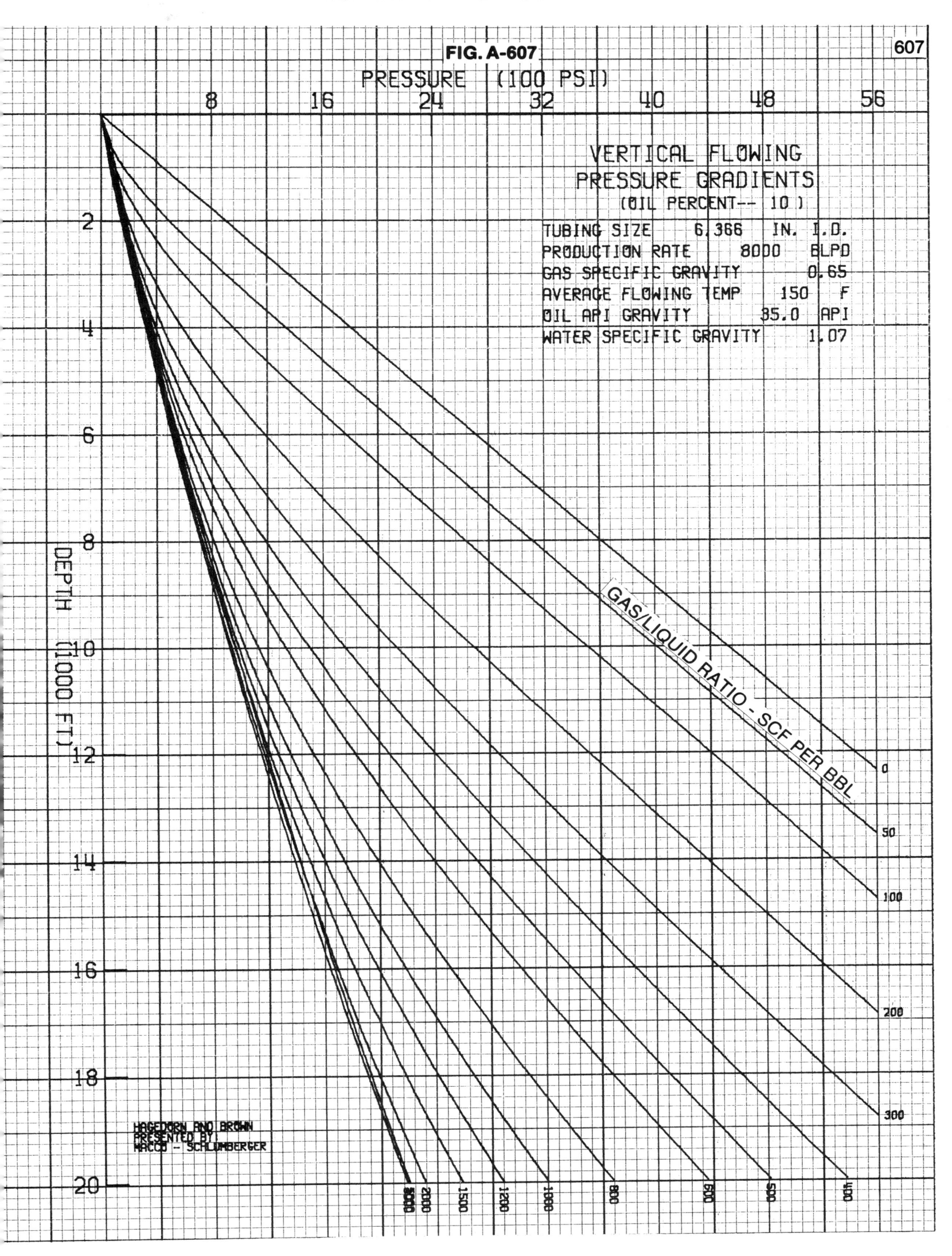

FIG. A-607

607

PRESSURE (100 PSI)

VERTICAL FLOWING
PRESSURE GRADIENTS
(OIL PERCENT-- 10)

TUBING SIZE   6.366   IN. I.D.
PRODUCTION RATE   8000   BLPD
GAS SPECIFIC GRAVITY   0.65
AVERAGE FLOWING TEMP   150   F
OIL API GRAVITY   35.0 API
WATER SPECIFIC GRAVITY   1.07

DEPTH (1000 FT)

GAS/LIQUID RATIO - SCF PER BBL

HAGEDORN AND BROWN
PRESENTED BY
MACCO - SCHLUMBERGER

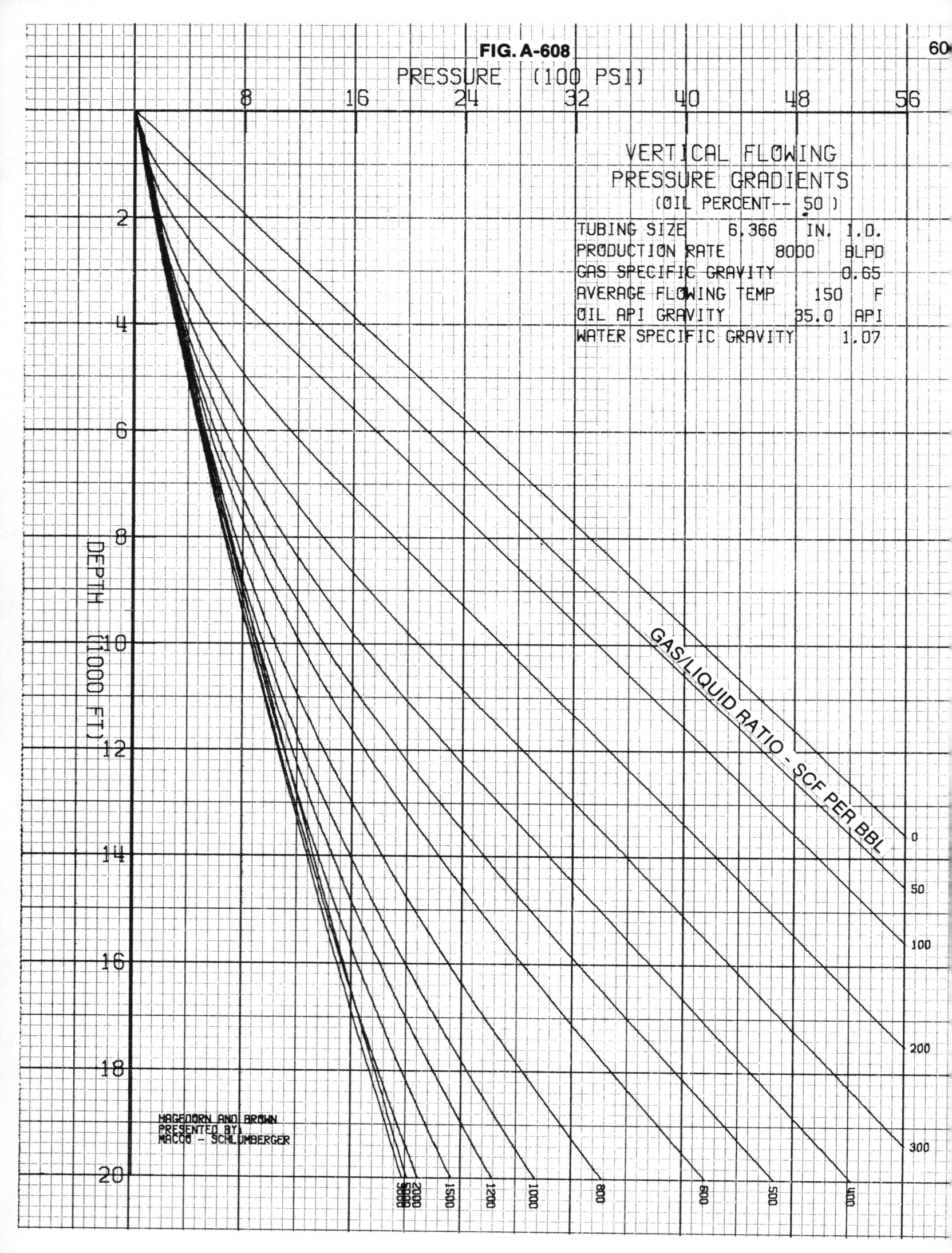

FIG. A-608

VERTICAL FLOWING
PRESSURE GRADIENTS
(OIL PERCENT-- 50)

TUBING SIZE	6.366	IN. I.D.
PRODUCTION RATE	8000	BLPD
GAS SPECIFIC GRAVITY	0.65	
AVERAGE FLOWING TEMP	150	F
OIL API GRAVITY	35.0	API
WATER SPECIFIC GRAVITY	1.07	

PRESSURE (100 PSI)

DEPTH (1000 FT)

GAS/LIQUID RATIO - SCF PER BBL

HAGEDORN AND BROWN
PRESENTED BY
MACCO - SCHLUMBERGER

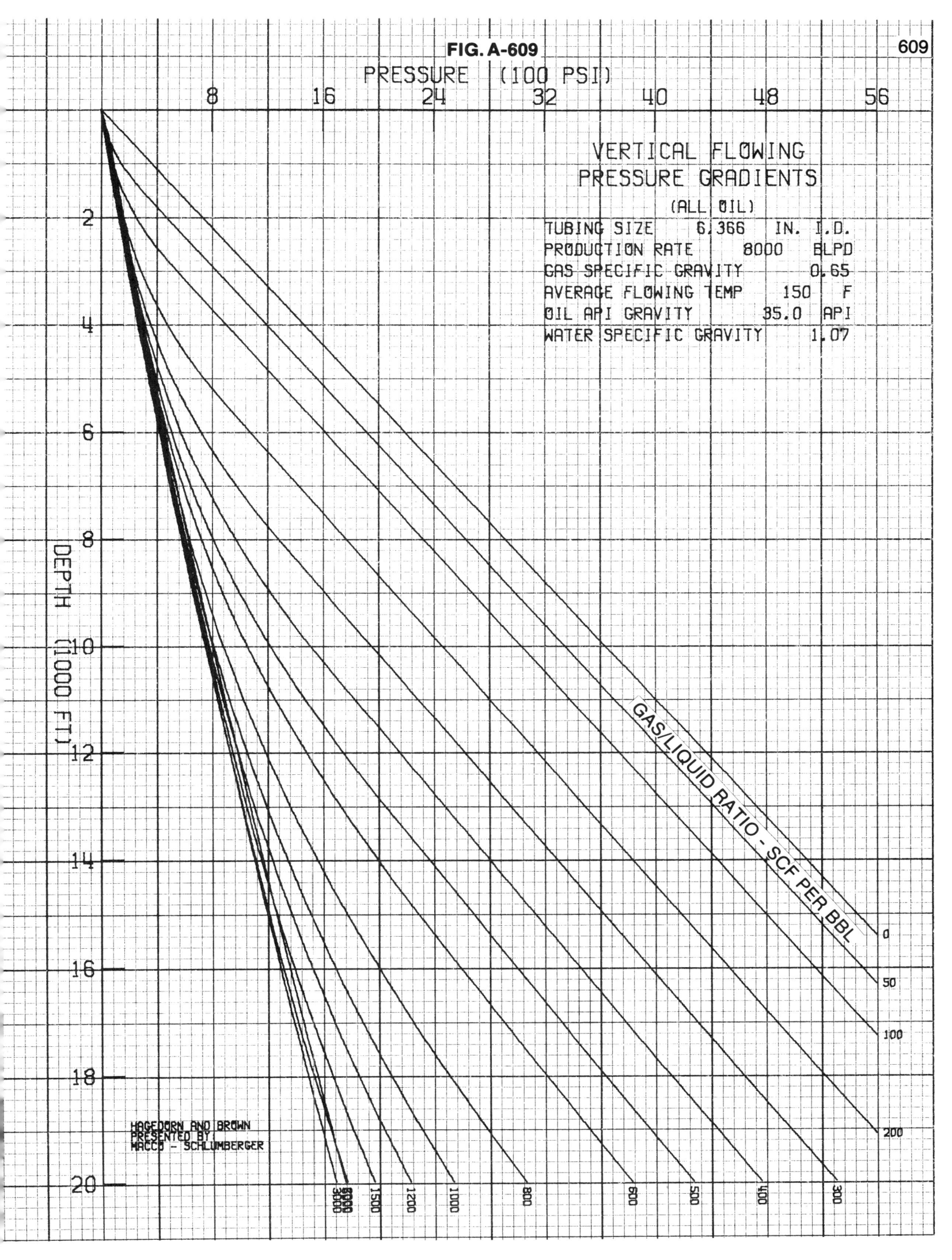

PRESSURE (100 PSI)

VERTICAL FLOWING
PRESSURE GRADIENTS
(ALL OIL)

TUBING SIZE	6.366	IN. I.D.
PRODUCTION RATE	8000	BLPD
GAS SPECIFIC GRAVITY	0.65	
AVERAGE FLOWING TEMP	150	F
OIL API GRAVITY	35.0	API
WATER SPECIFIC GRAVITY	1.07	

DEPTH (1000 FT)

GAS/LIQUID RATIO - SCF PER BBL

HAGEDORN AND BROWN
PRESENTED BY
MACCO - SCHLUMBERGER

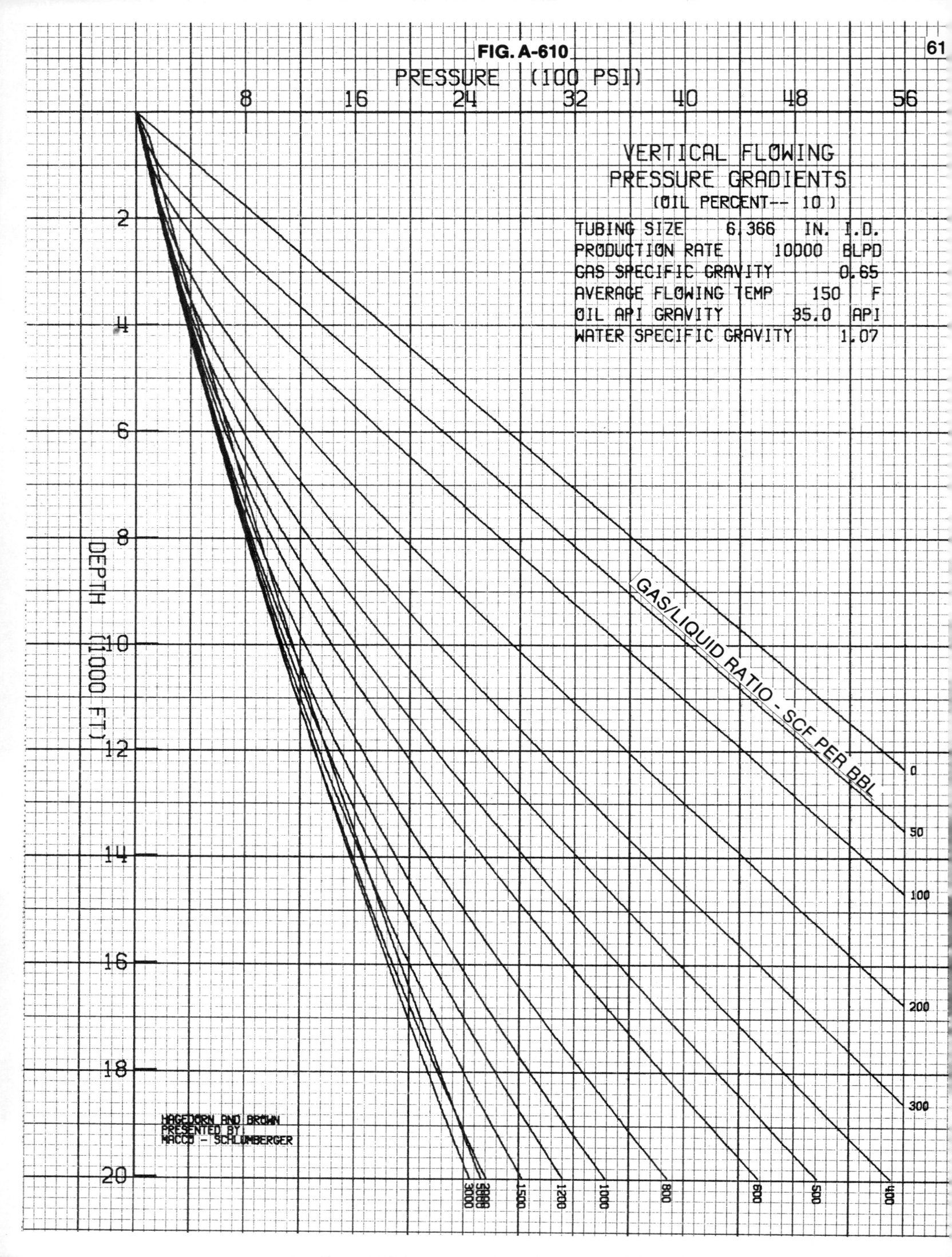

VERTICAL FLOWING
PRESSURE GRADIENTS
(OIL PERCENT–– 10)

TUBING SIZE	6.366	IN. I.D.
PRODUCTION RATE	10000	BLPD
GAS SPECIFIC GRAVITY	0.65	
AVERAGE FLOWING TEMP	150	F
OIL API GRAVITY	35.0	API
WATER SPECIFIC GRAVITY	1.07	

PRESSURE (100 PSI)

DEPTH (1000 FT)

GAS/LIQUID RATIO - SCF PER BBL

HAGEDORN AND BROWN
PRESENTED BY
MACCO – SCHLUMBERGER

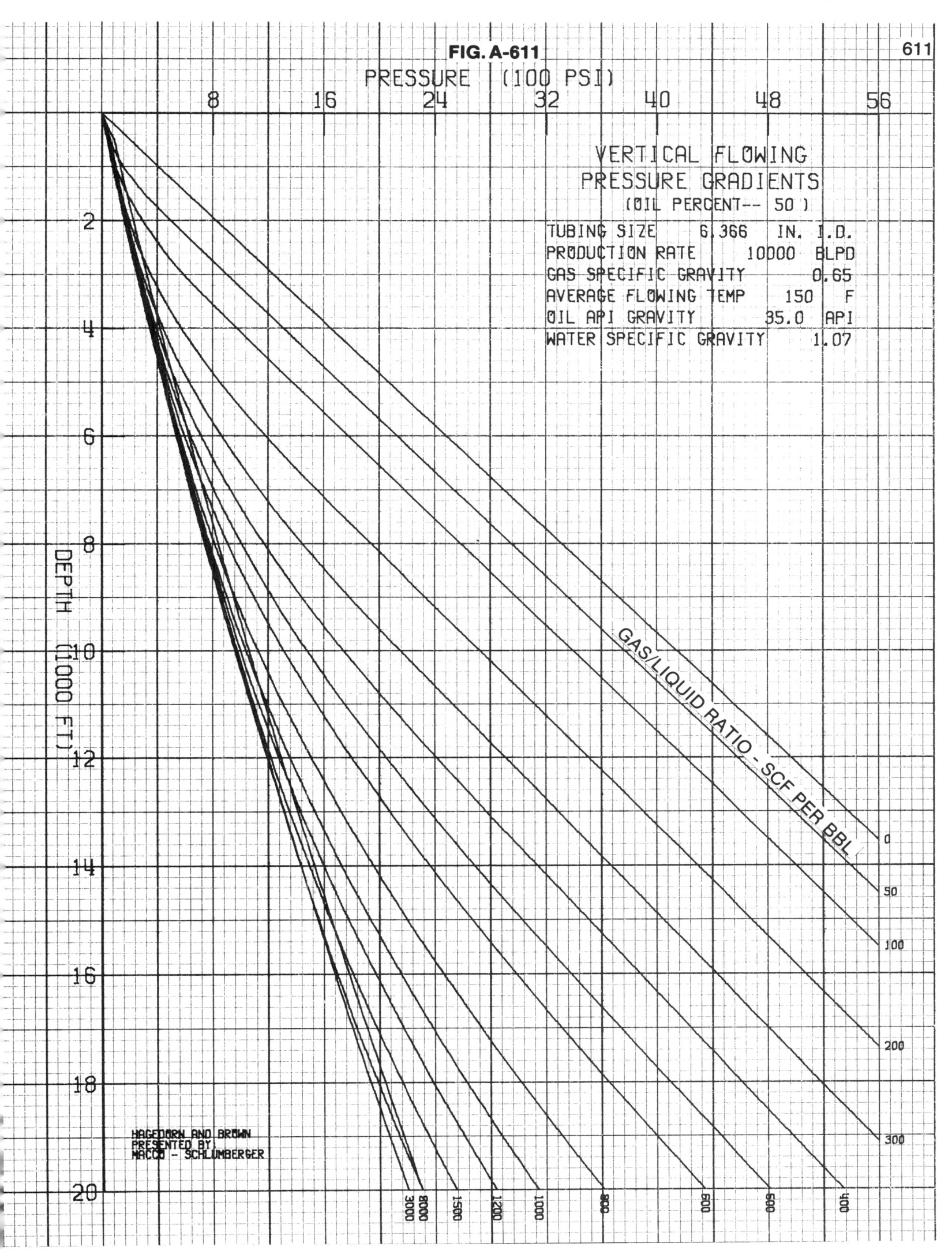

FIG. A-611

611

VERTICAL FLOWING
PRESSURE GRADIENTS
(OIL PERCENT-- 50)

TUBING SIZE	6.366	IN. I.D.
PRODUCTION RATE	10000	BLPD
GAS SPECIFIC GRAVITY	0.65	
AVERAGE FLOWING TEMP	150	F
OIL API GRAVITY	35.0	API
WATER SPECIFIC GRAVITY	1.07	

PRESSURE (100 PSI)

DEPTH (1000 FT)

GAS/LIQUID RATIO - SCF PER BBL

HAGEDORN AND BROWN
PRESENTED BY
MACCO - SCHLUMBERGER

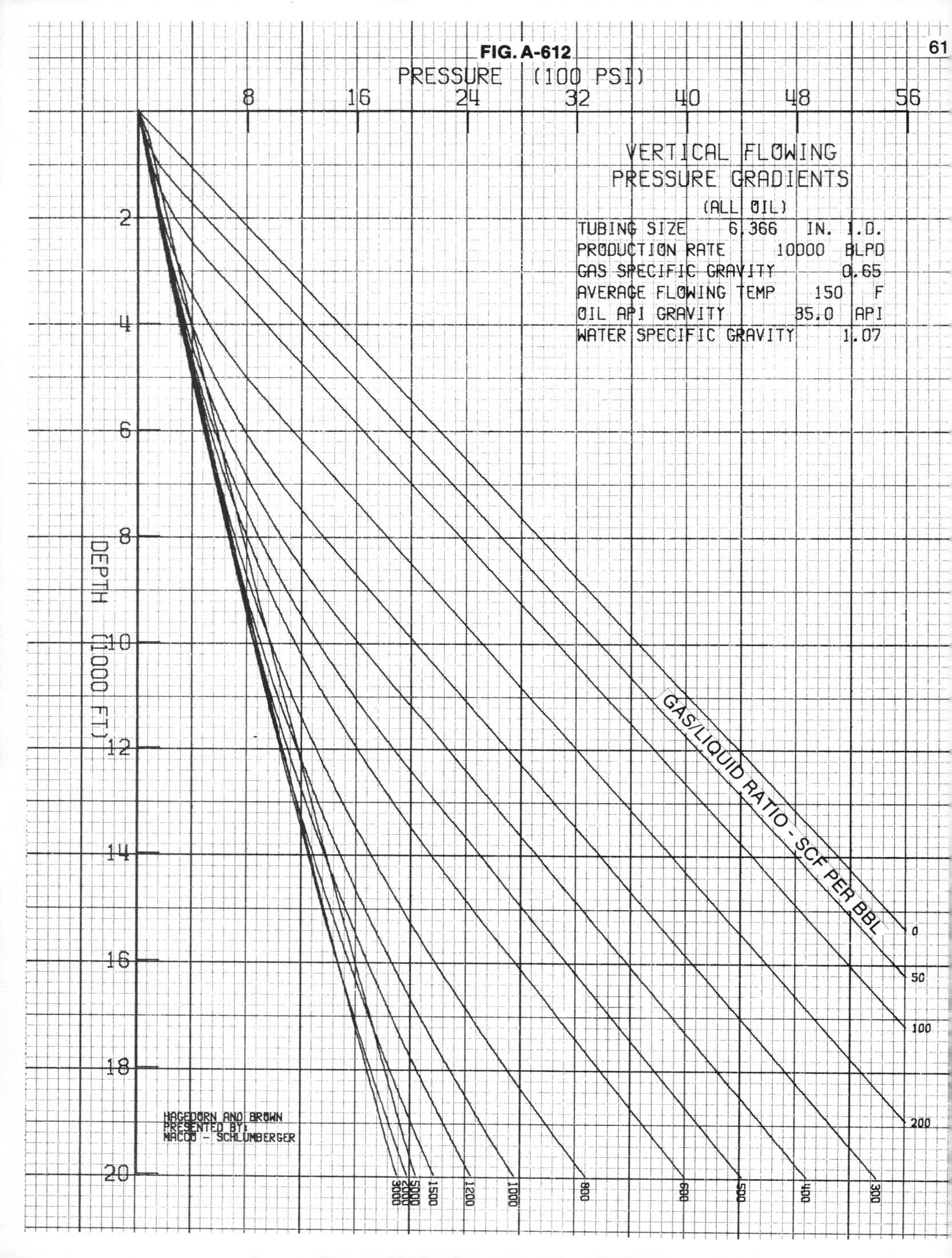

PRESSURE (100 PSI)

DEPTH (1000 FT)

VERTICAL FLOWING
PRESSURE GRADIENTS
(ALL OIL)

TUBING SIZE	6.366	IN. I.D.
PRODUCTION RATE	10000	BLPD
GAS SPECIFIC GRAVITY	0.65	
AVERAGE FLOWING TEMP	150	F
OIL API GRAVITY	35.0	API
WATER SPECIFIC GRAVITY	1.07	

GAS/LIQUID RATIO - SCF PER BBL

HAGEDORN AND BROWN
PRESENTED BY:
NACO - SCHLUMBERGER

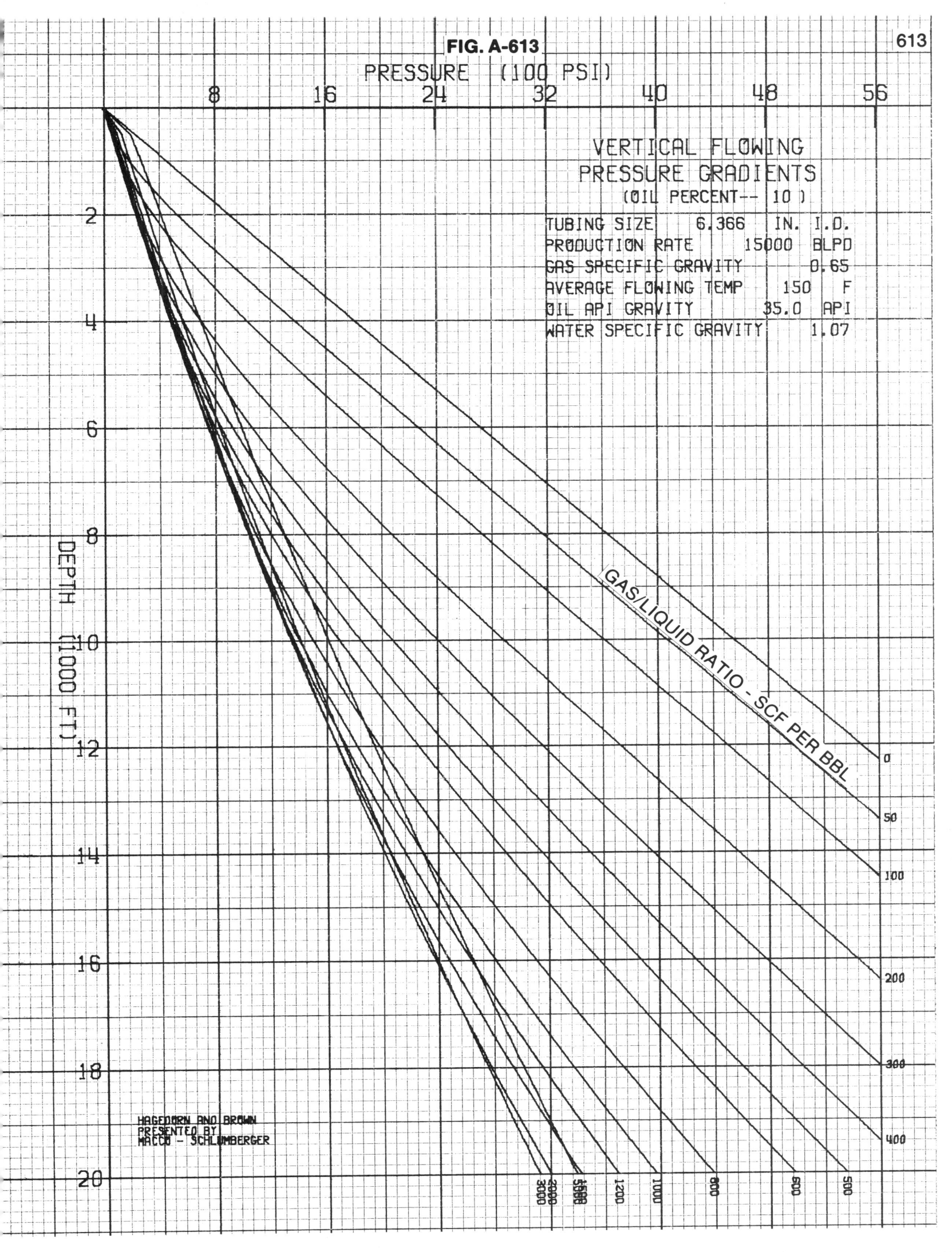

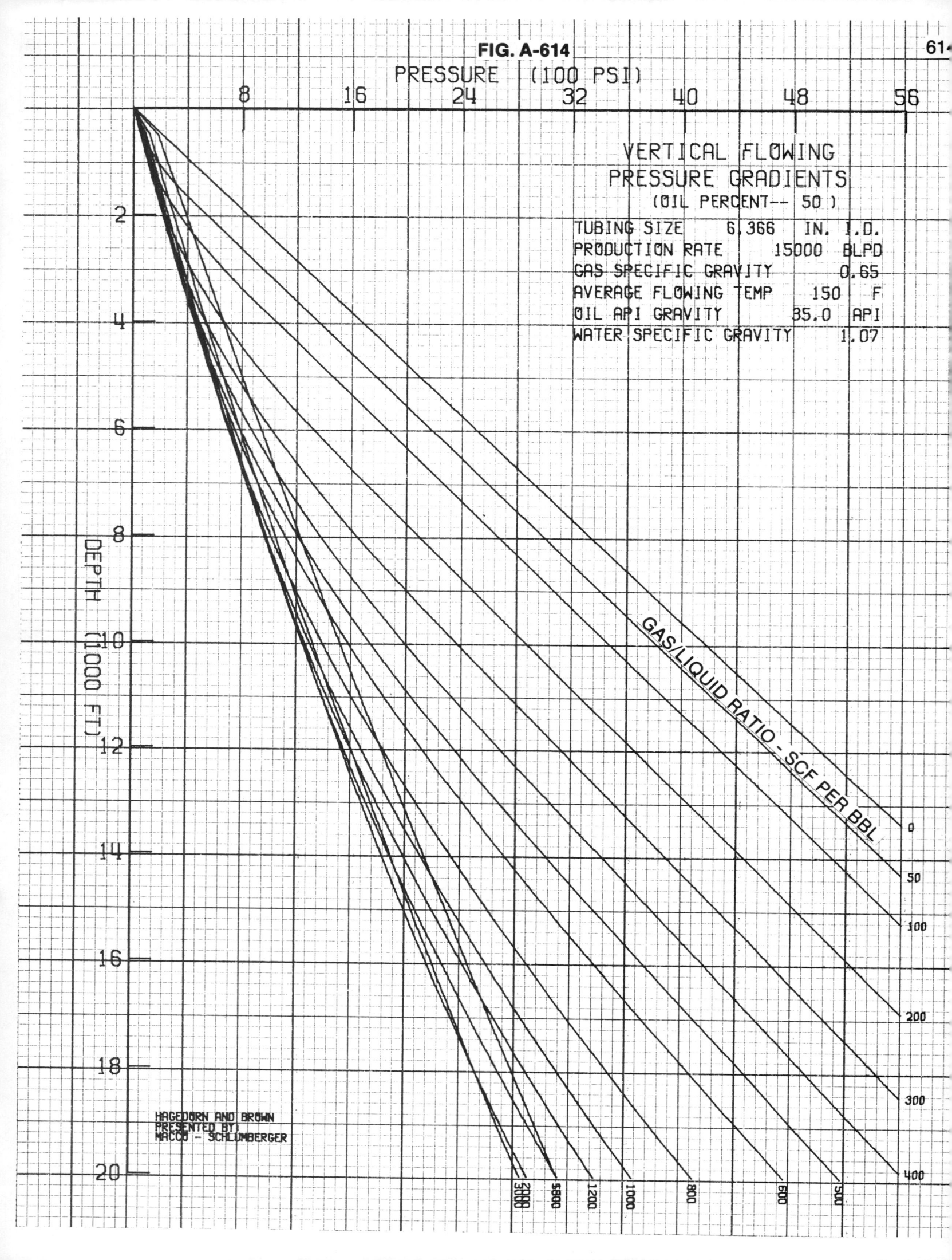

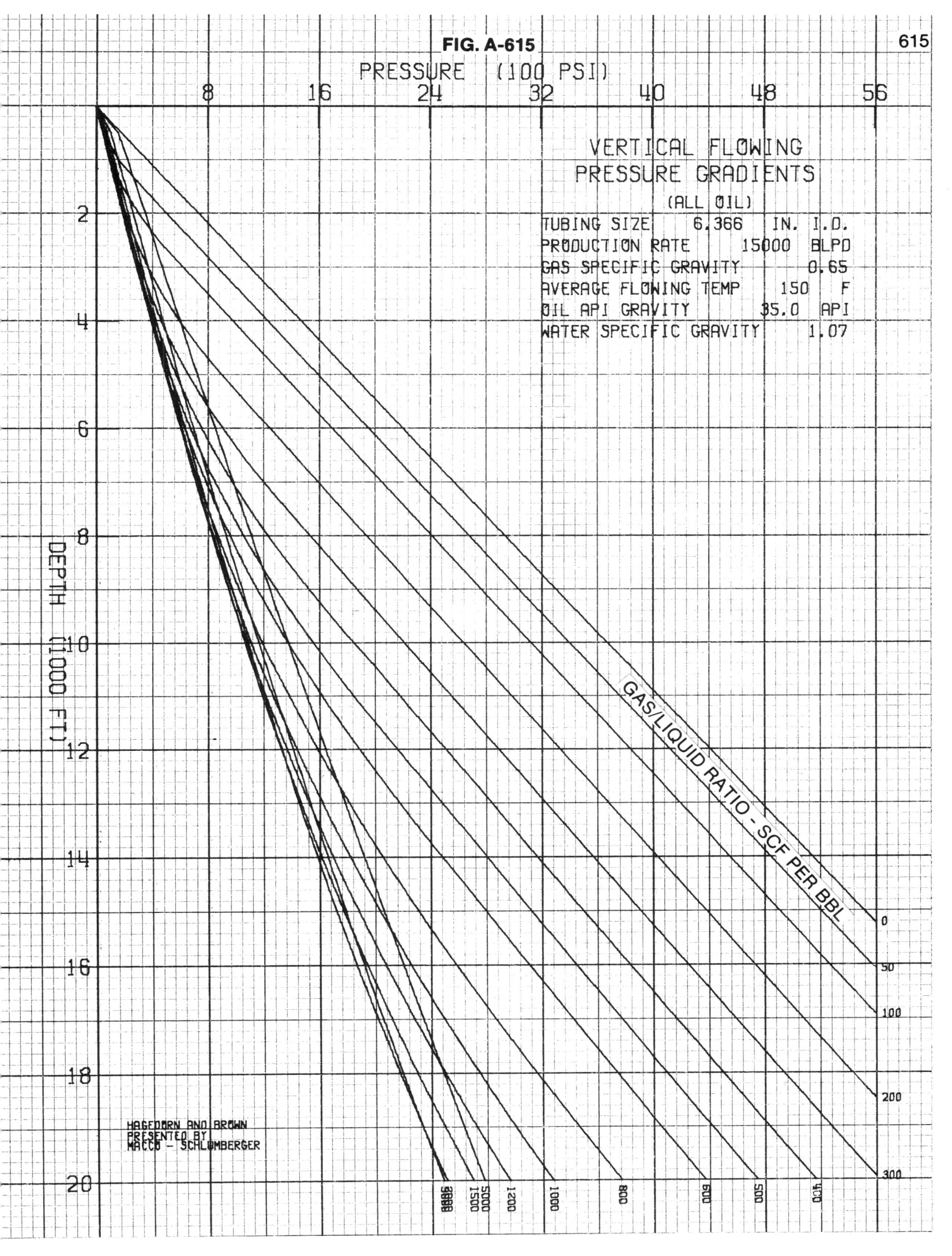

FIG. A-615

615

VERTICAL FLOWING
PRESSURE GRADIENTS
(ALL OIL)

TUBING SIZE	6.366	IN. I.D.
PRODUCTION RATE	15000	BLPD
GAS SPECIFIC GRAVITY	0.65	
AVERAGE FLOWING TEMP	150	F
OIL API GRAVITY	35.0	API
WATER SPECIFIC GRAVITY	1.07	

PRESSURE (100 PSI)

DEPTH (1,000 FT)

GAS/LIQUID RATIO - SCF PER BBL

HAGEDORN AND BROWN
PRESENTED BY
MAECO - SCHLUMBERGER

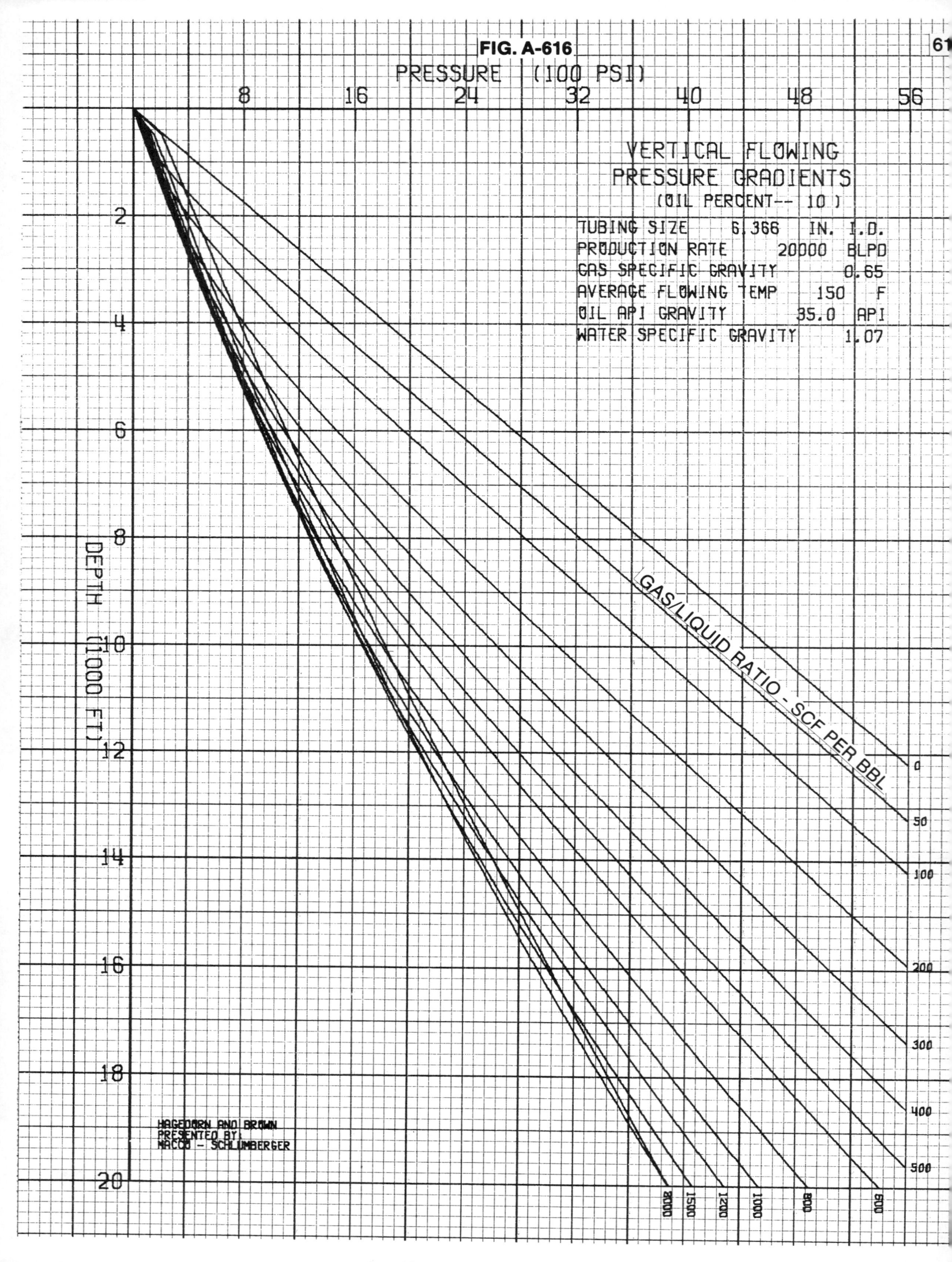

FIG. A-616

VERTICAL FLOWING
PRESSURE GRADIENTS
(OIL PERCENT-- 10)

TUBING SIZE	6.366 IN. I.D.
PRODUCTION RATE	20000 BLPD
GAS SPECIFIC GRAVITY	0.65
AVERAGE FLOWING TEMP	150 F
OIL API GRAVITY	35.0 API
WATER SPECIFIC GRAVITY	1.07

PRESSURE (100 PSI)

DEPTH (1000 FT)

GAS/LIQUID RATIO - SCF PER BBL

HAGEDORN AND BROWN
PRESENTED BY
NACCO - SCHLUMBERGER

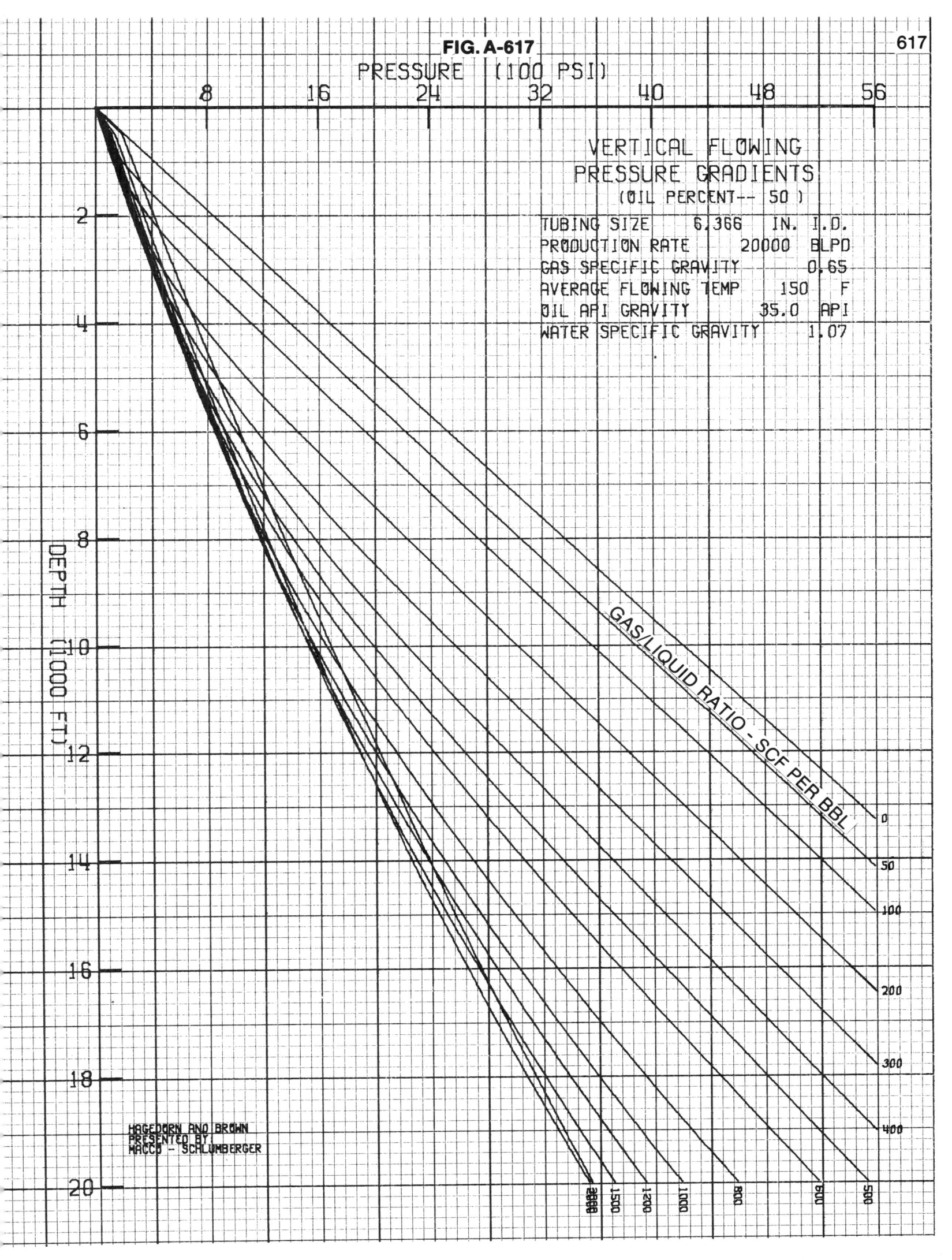

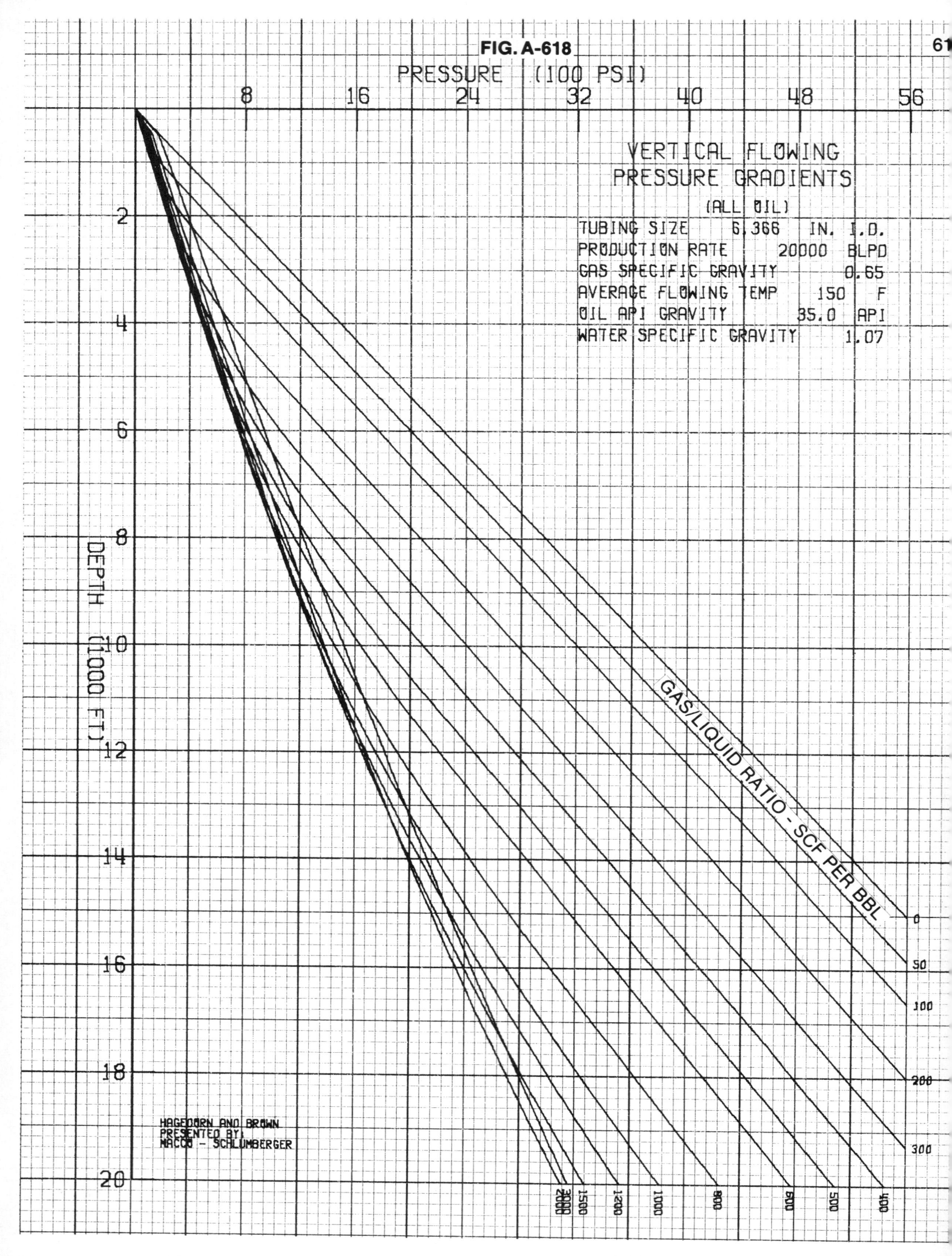

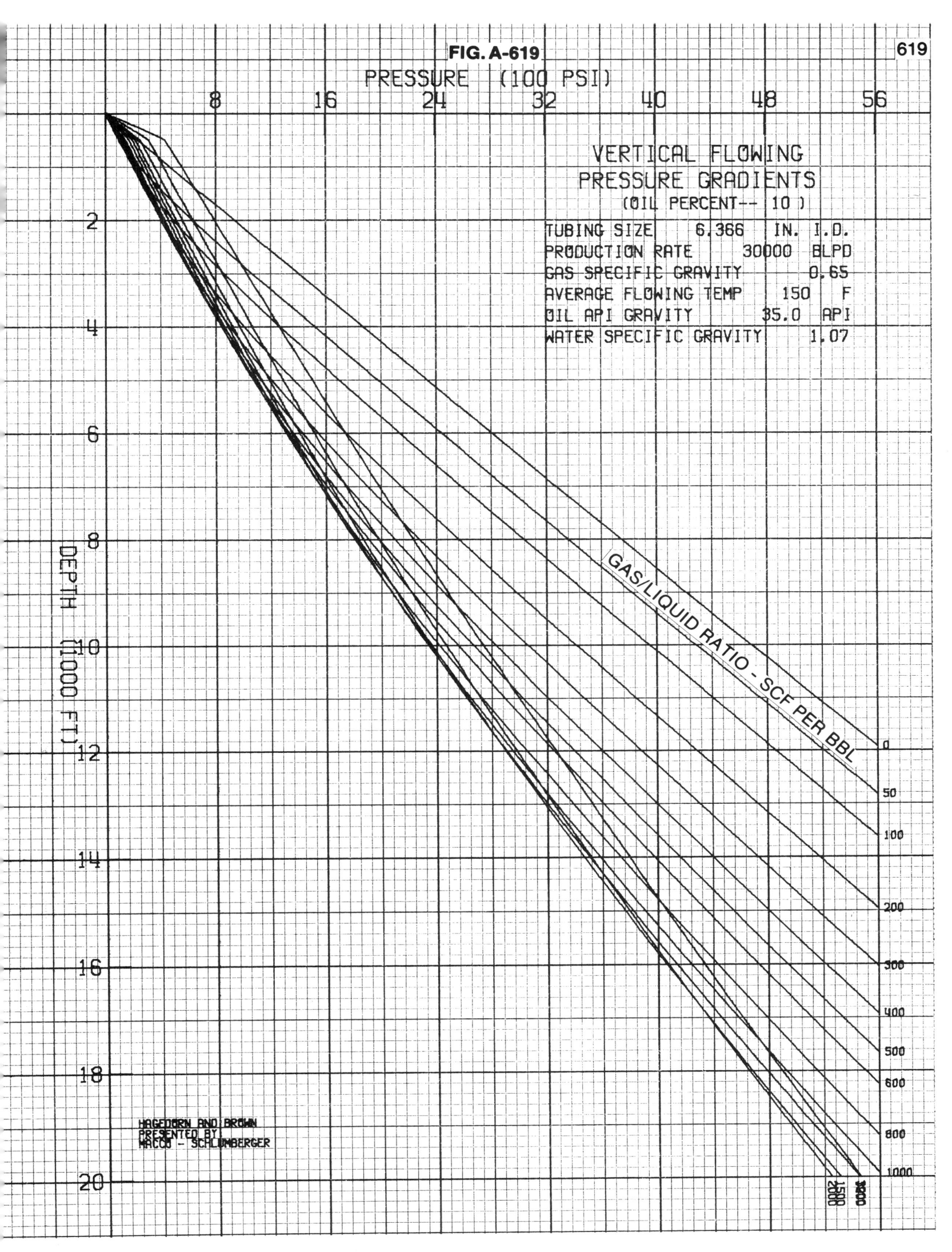

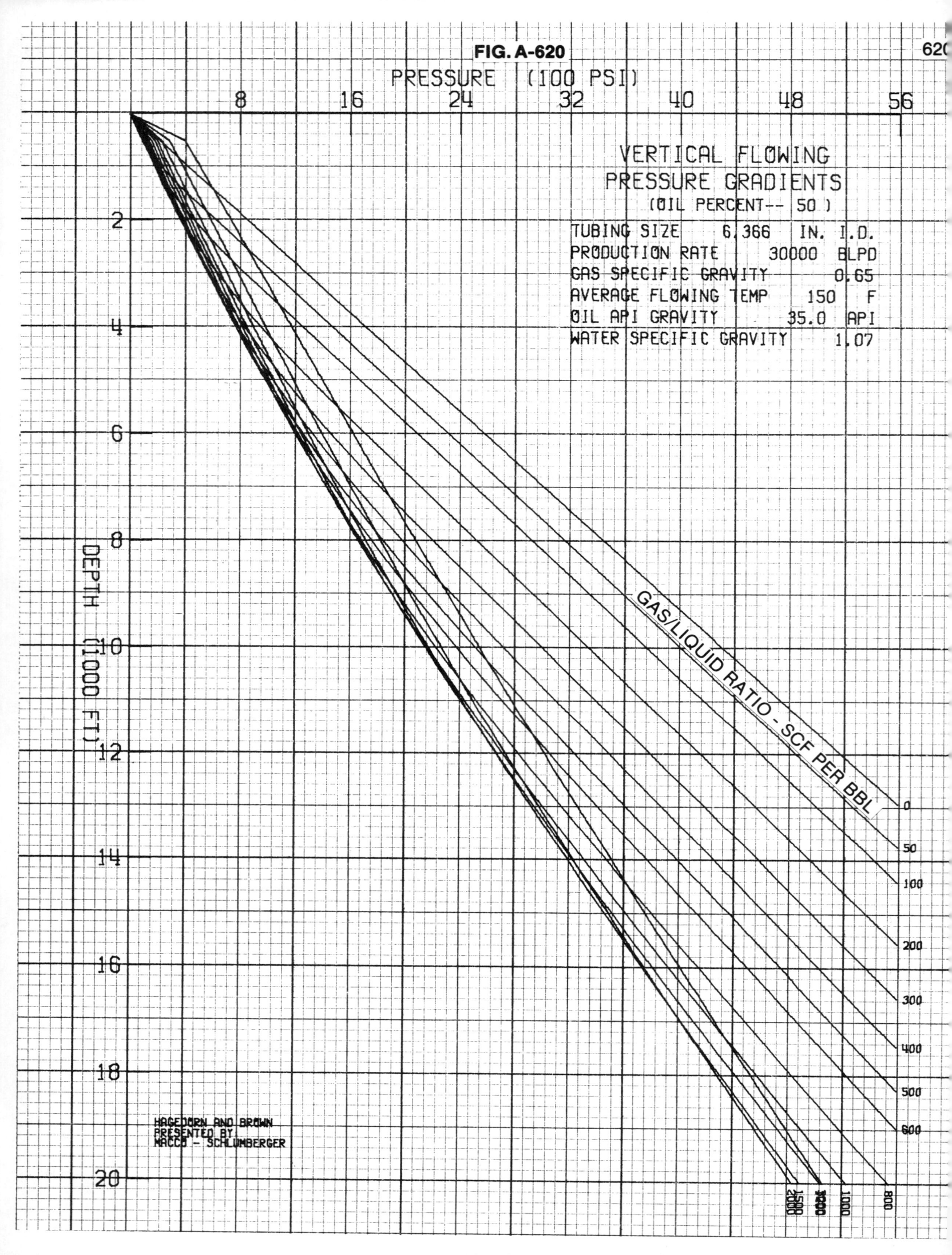

FIG. A-620

VERTICAL FLOWING
PRESSURE GRADIENTS
(OIL PERCENT-- 50)

TUBING SIZE   6.366   IN. I.D.
PRODUCTION RATE   30000   BLPD
GAS SPECIFIC GRAVITY   0.65
AVERAGE FLOWING TEMP   150   F
OIL API GRAVITY   35.0   API
WATER SPECIFIC GRAVITY   1.07

GAS/LIQUID RATIO - SCF PER BBL

PRESSURE (100 PSI)

DEPTH (1000 FT)

HAGEDORN AND BROWN
PRESENTED BY
NACCO - SCHLUMBERGER

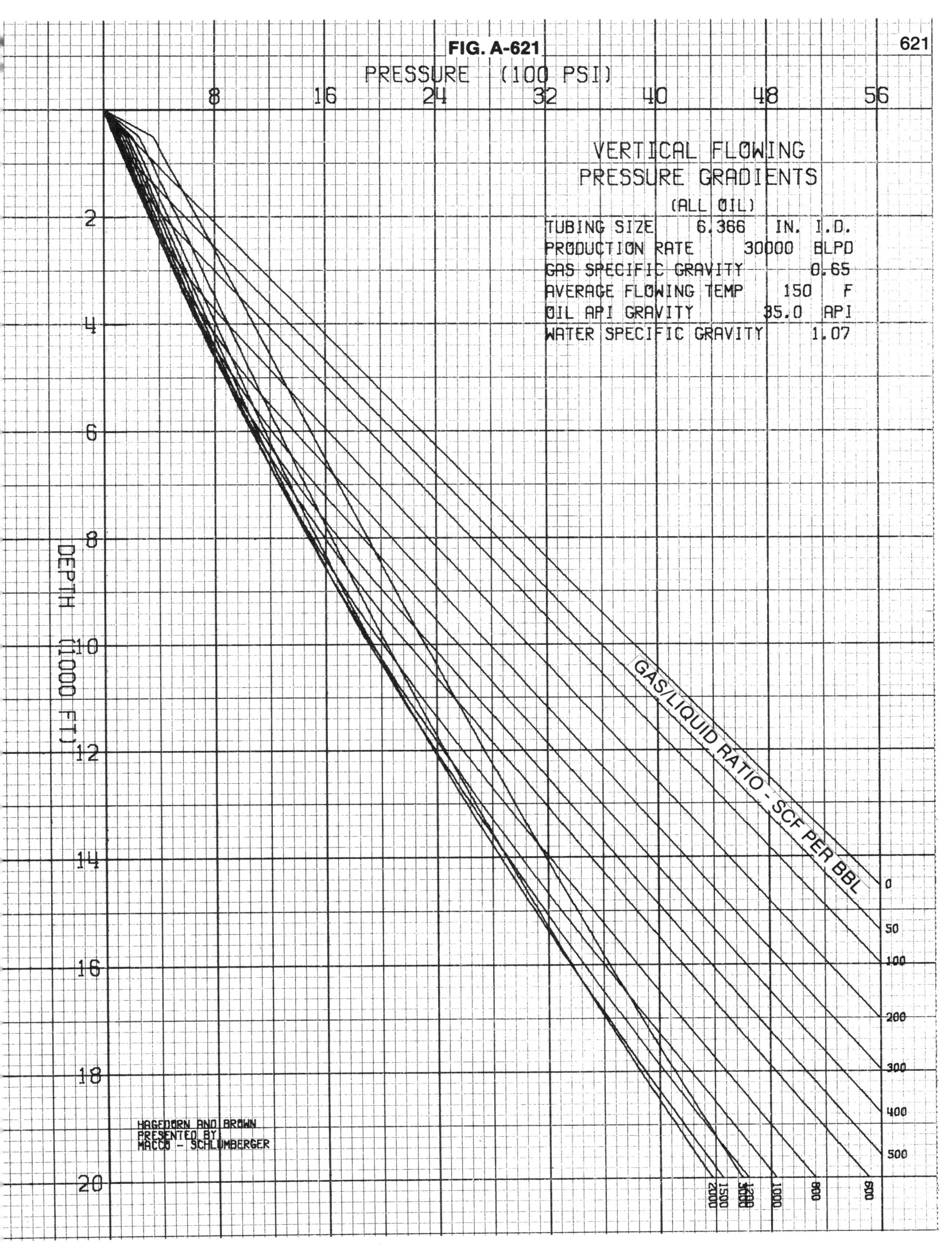

FIG. A-621

621

VERTICAL FLOWING
PRESSURE GRADIENTS
(ALL OIL)

TUBING SIZE	6.366 IN. I.D.
PRODUCTION RATE	30000 BLPD
GAS SPECIFIC GRAVITY	0.65
AVERAGE FLOWING TEMP	150 F
OIL API GRAVITY	35.0 API
WATER SPECIFIC GRAVITY	1.07

PRESSURE (100 PSI)

DEPTH (1000 FT)

GAS/LIQUID RATIO - SCF PER BBL

HAGEDORN AND BROWN
PRESENTED BY
MACCO - SCHLUMBERGER

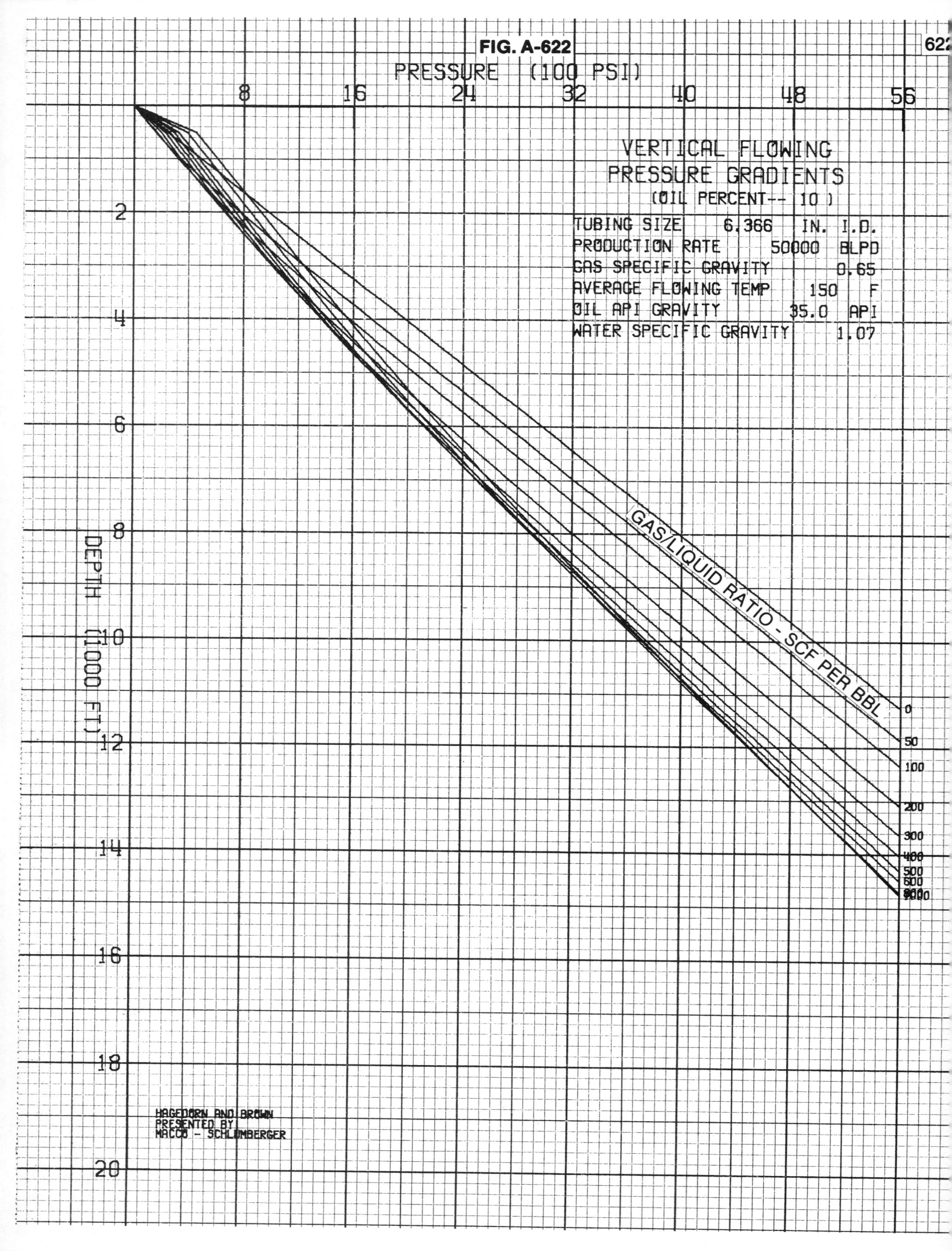

PRESSURE (100 PSI)

VERTICAL FLOWING
PRESSURE GRADIENTS
(OIL PERCENT-- 10)

TUBING SIZE	6.366	IN. I.D.
PRODUCTION RATE	50000	BLPD
GAS SPECIFIC GRAVITY		0.65
AVERAGE FLOWING TEMP	150	F
OIL API GRAVITY	35.0	API
WATER SPECIFIC GRAVITY		1.07

DEPTH (1000 FT)

GAS/LIQUID RATIO - SCF PER BBL

0
50
100
200
300
400
500
600
800
1000

HAGEDORN AND BROWN
PRESENTED BY
MACCO - SCHLUMBERGER

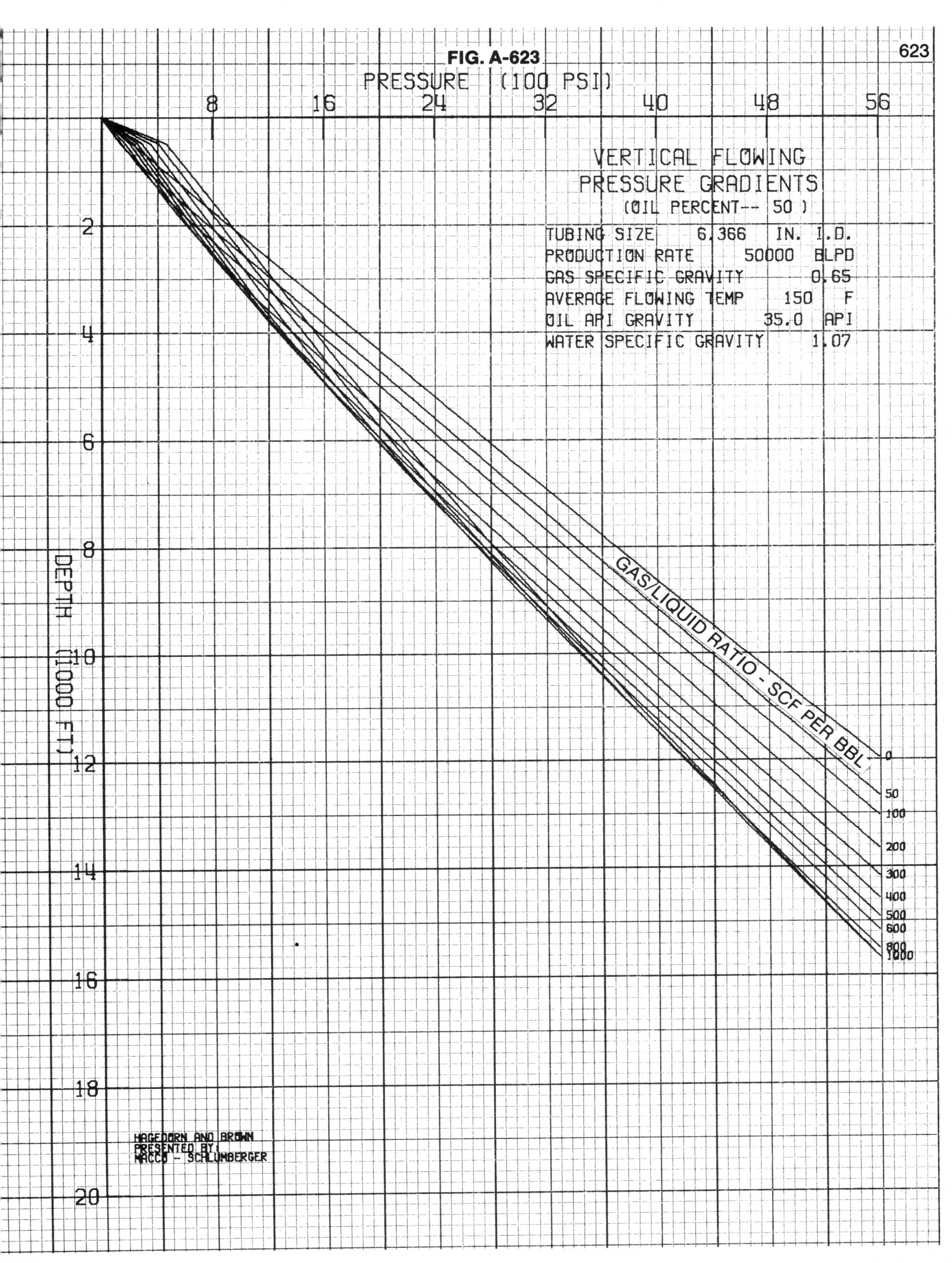

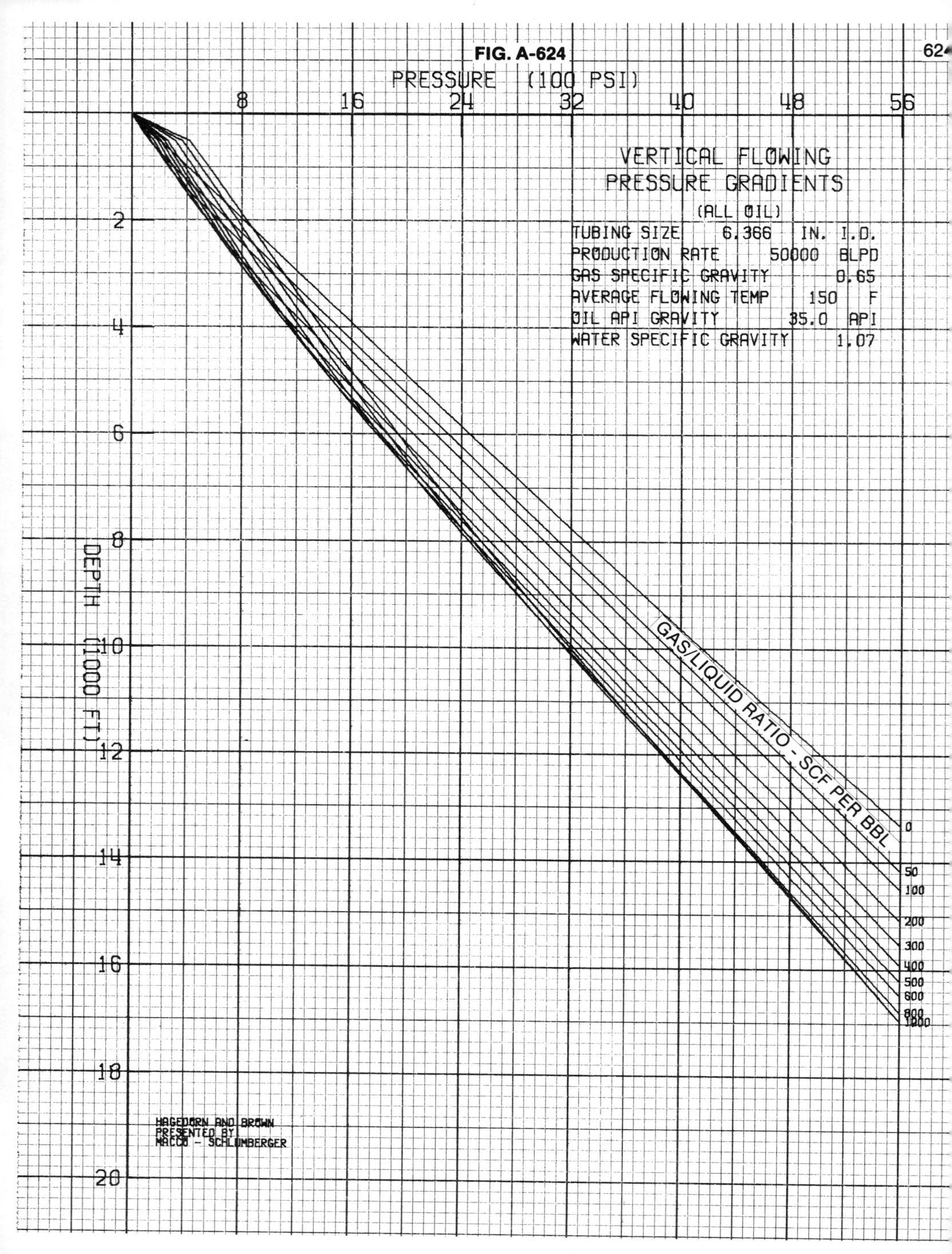

VERTICAL FLOWING
PRESSURE GRADIENTS
(ALL OIL)

TUBING SIZE	6.366	IN. I.D.
PRODUCTION RATE	50000	BLPD
GAS SPECIFIC GRAVITY	0.65	
AVERAGE FLOWING TEMP	150	F
OIL API GRAVITY	35.0	API
WATER SPECIFIC GRAVITY	1.07	

PRESSURE (100 PSI)

DEPTH (1000 FT)

GAS/LIQUID RATIO - SCF PER BBL

0
50
100
200
300
400
500
600
800
1000

HAGEDORN AND BROWN
PRESENTED BY
NACCO - SCHLUMBERGER

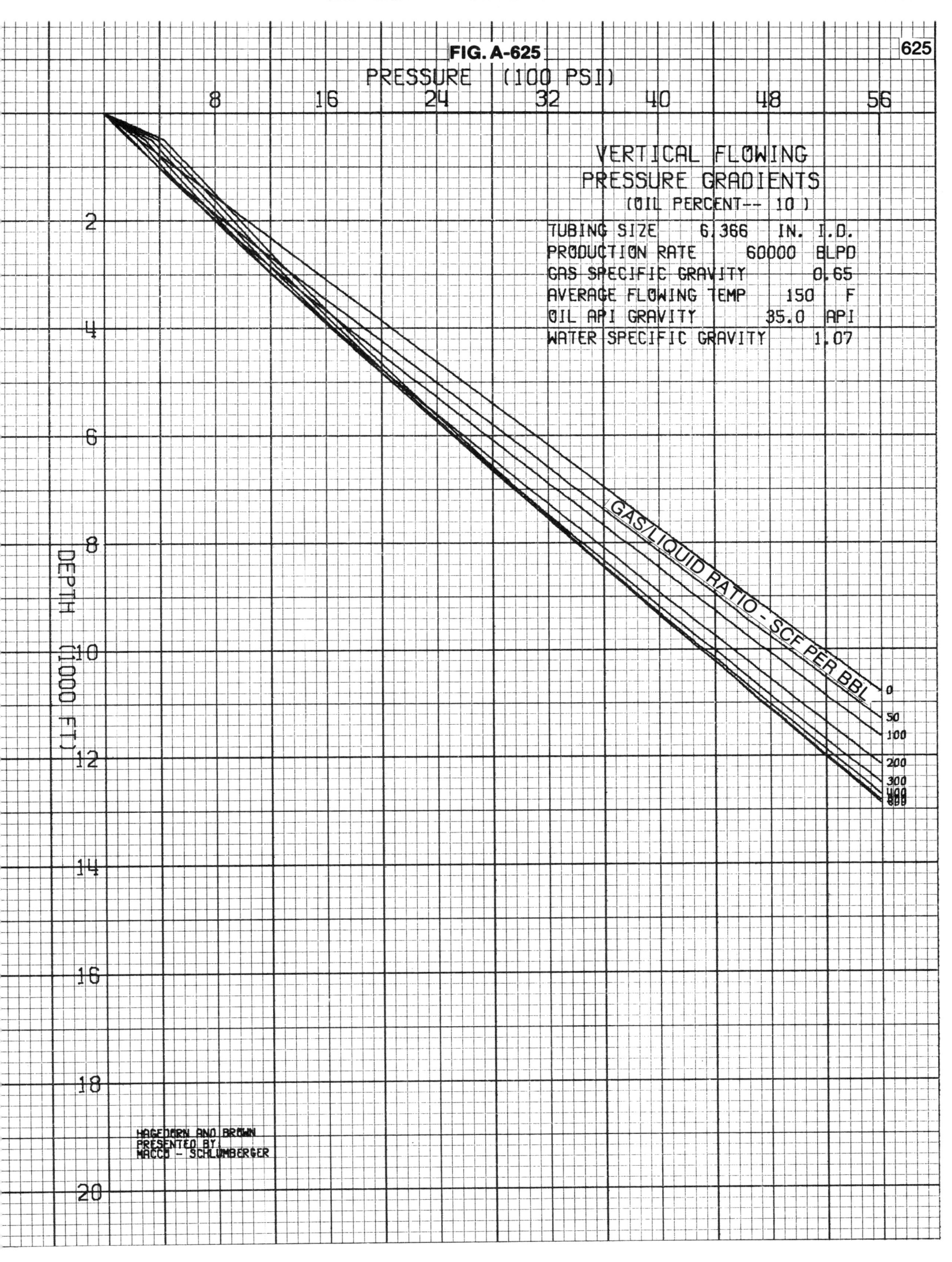

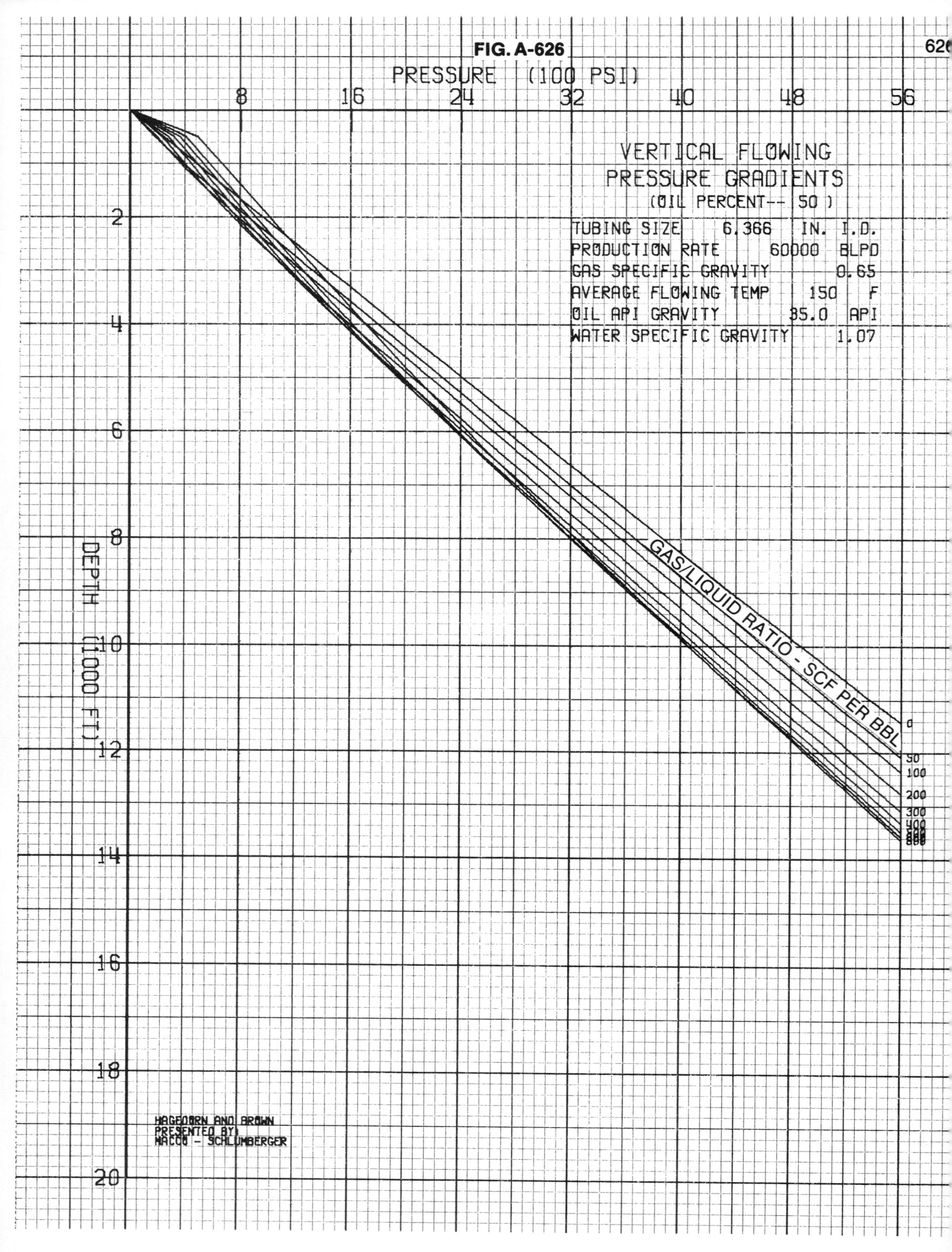

FIG. A-626

VERTICAL FLOWING
PRESSURE GRADIENTS
(OIL PERCENT-- 50)

TUBING SIZE	6.366	IN. I.D.
PRODUCTION RATE	60000	BLPD
GAS SPECIFIC GRAVITY		0.65
AVERAGE FLOWING TEMP	150	F
OIL API GRAVITY	35.0	API
WATER SPECIFIC GRAVITY		1.07

PRESSURE (100 PSI)

DEPTH (1000 FT)

GAS/LIQUID RATIO - SCF PER BBL

0
50
100
200
300
400

HAGEDORN AND BROWN
PRESENTED BY:
NACOG - SCHLUMBERGER

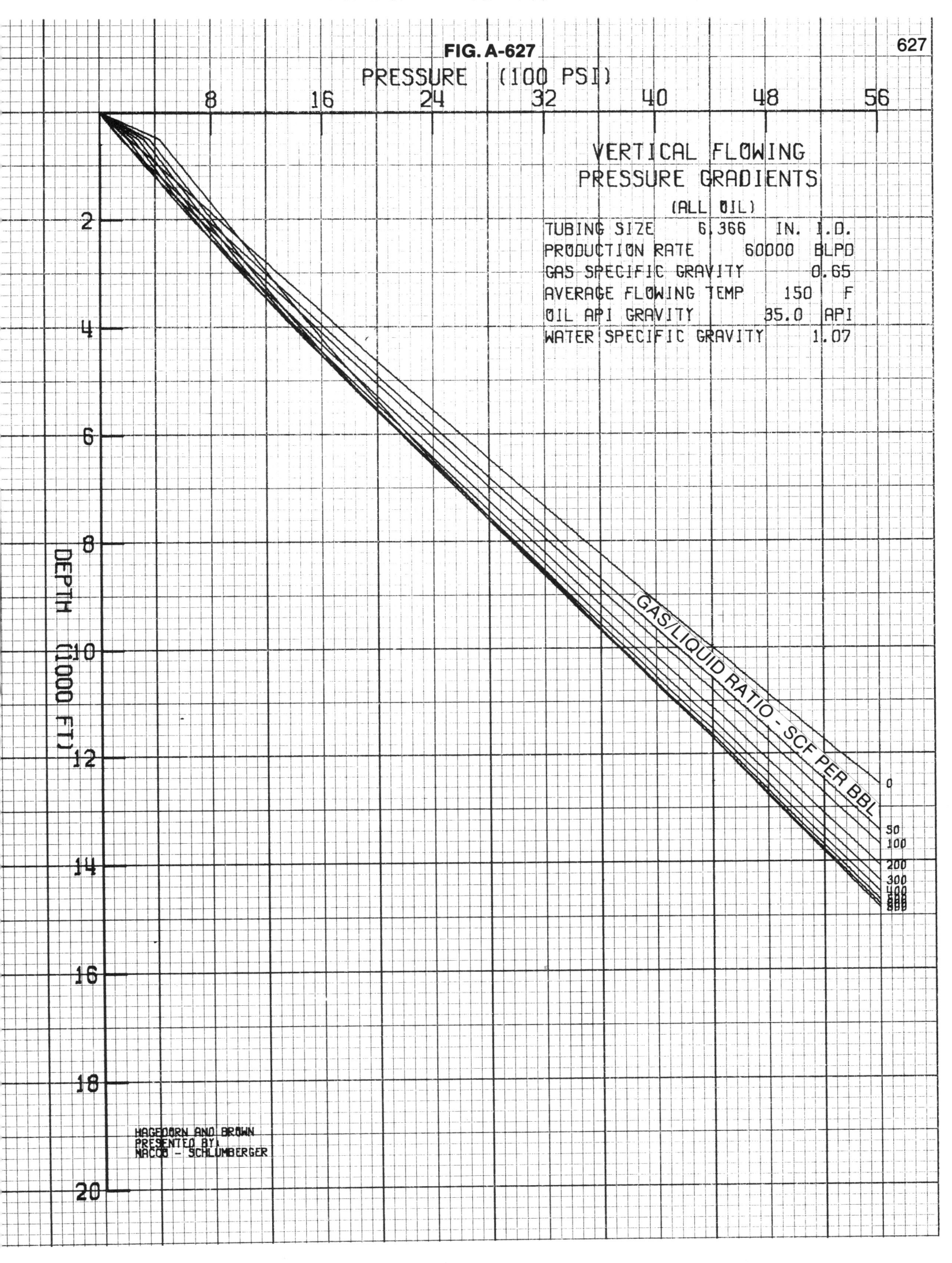

FIG. A-627

627

PRESSURE (100 PSI)

VERTICAL FLOWING
PRESSURE GRADIENTS
(ALL OIL)

TUBING SIZE	6.366	IN. I.D.
PRODUCTION RATE	60000	BLPD
GAS SPECIFIC GRAVITY	0.65	
AVERAGE FLOWING TEMP	150	F
OIL API GRAVITY	35.0	API
WATER SPECIFIC GRAVITY	1.07	

DEPTH (1000 FT)

GAS/LIQUID RATIO - SCF PER BBL

0

50
100

200
300
400
600
800

HAGEDORN AND BROWN
PRESENTED BY:
NACO - SCHLUMBERGER

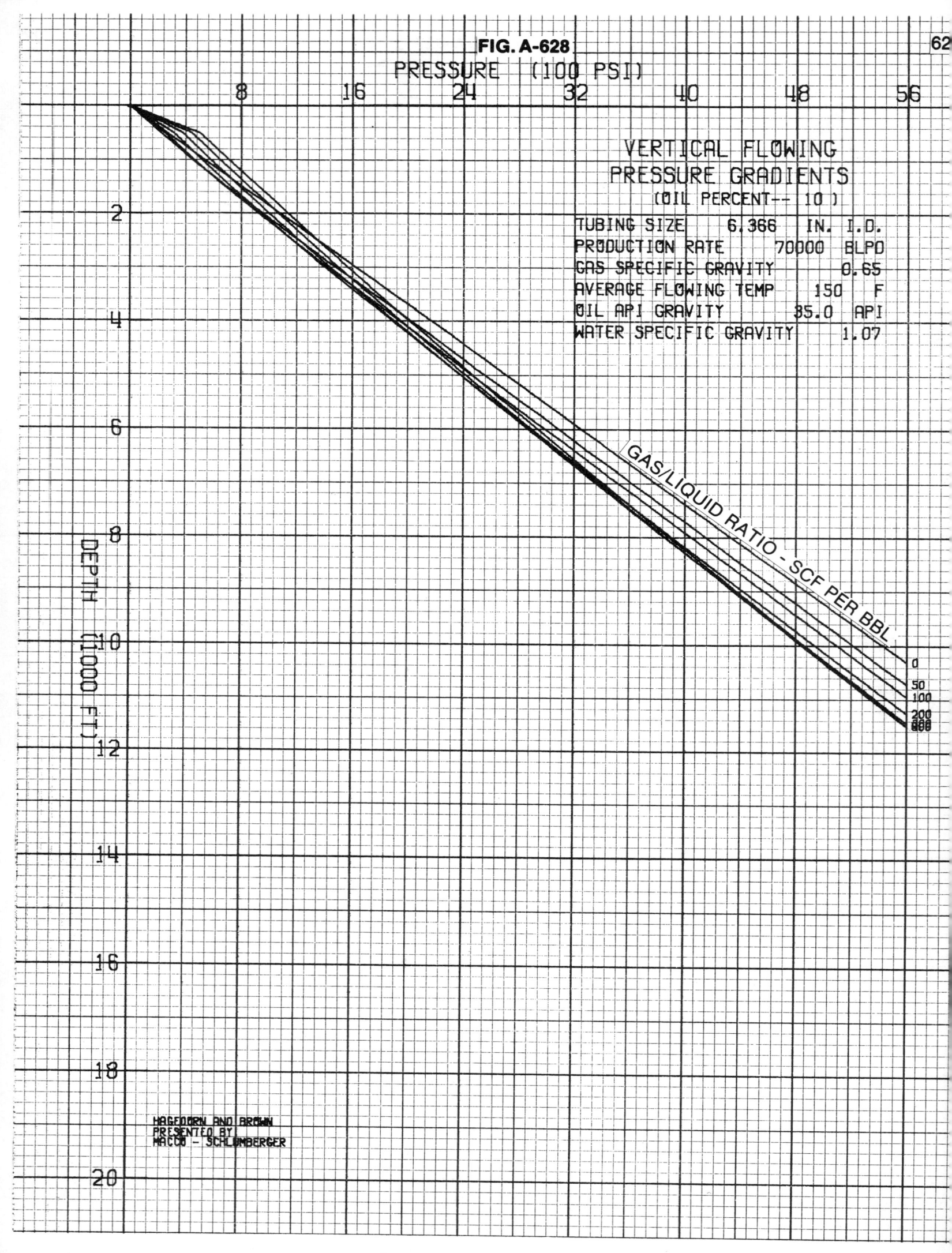

FIG. A-628

VERTICAL FLOWING
PRESSURE GRADIENTS
(OIL PERCENT-- 10)

TUBING SIZE	6.366	IN. I.D.
PRODUCTION RATE	70000	BLPD
GAS SPECIFIC GRAVITY	0.65	
AVERAGE FLOWING TEMP	150	F
OIL API GRAVITY	35.0	API
WATER SPECIFIC GRAVITY	1.07	

PRESSURE (100 PSI)

DEPTH (1000 FT)

GAS/LIQUID RATIO - SCF PER BBL

0
50
100
200

HAGEDORN AND BROWN
PRESENTED BY
MACCO - SCHLUMBERGER

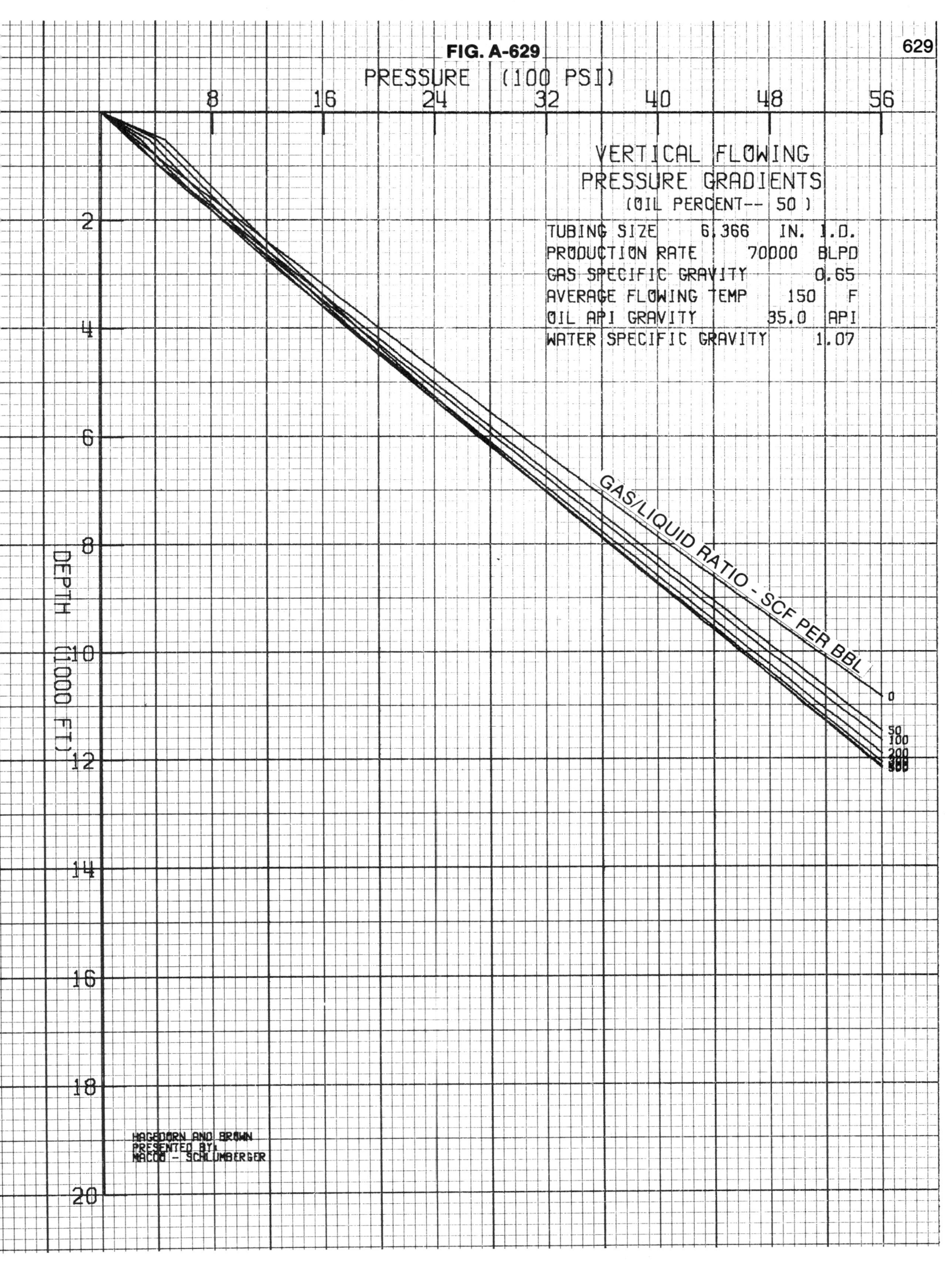

PRESSURE (100 PSI)

VERTICAL FLOWING
PRESSURE GRADIENTS
(OIL PERCENT-- 50)

TUBING SIZE	6.366	IN. I.D.
PRODUCTION RATE	70000	BLPD
GAS SPECIFIC GRAVITY		0.65
AVERAGE FLOWING TEMP	150	F
OIL API GRAVITY	35.0	API
WATER SPECIFIC GRAVITY		1.07

DEPTH (1000 FT)

GAS/LIQUID RATIO - SCF PER BBL

0
50
100
200
300

HAGEDORN AND BROWN
PRESENTED BY
NACOO - SCHLUMBERGER

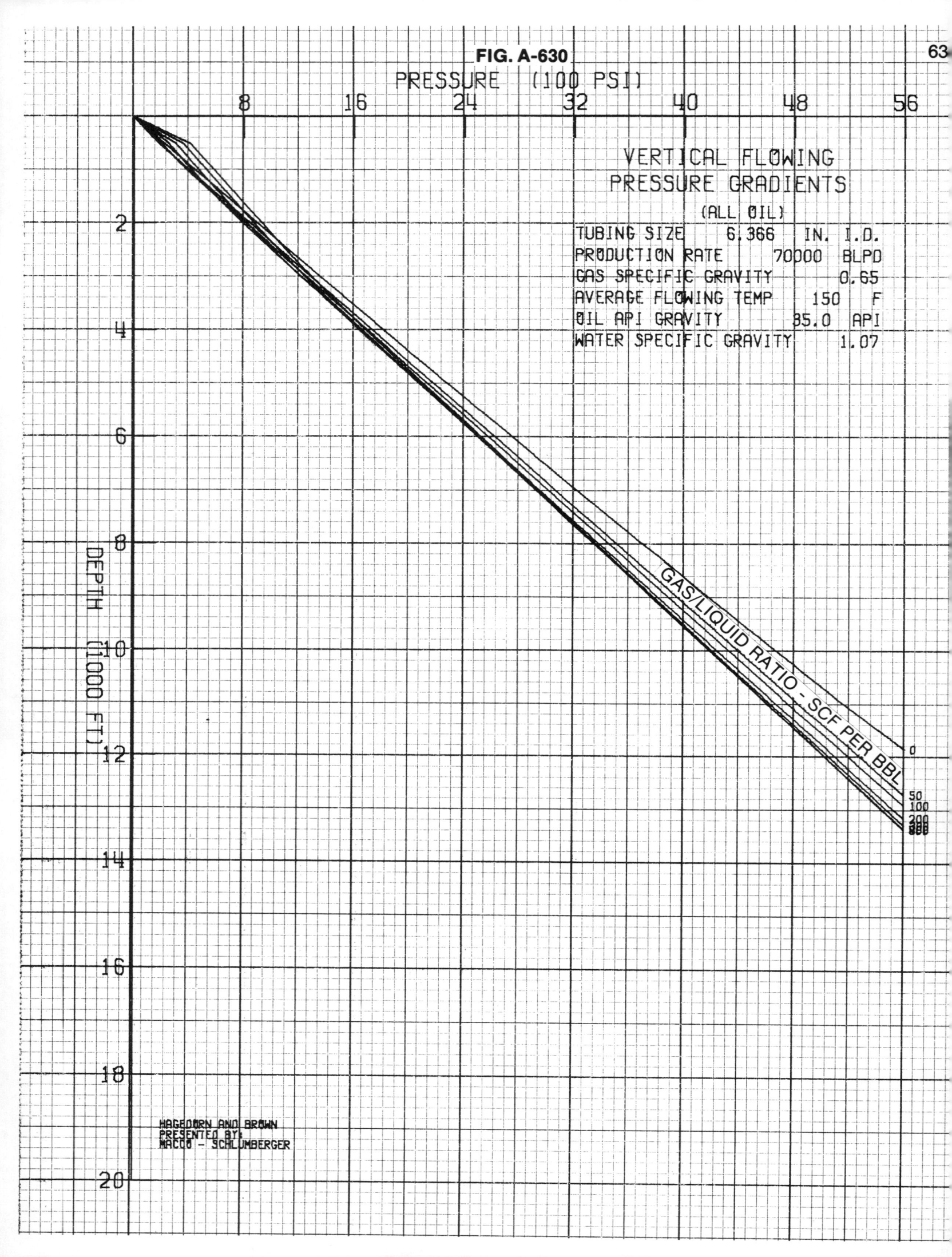

PRESSURE (100 PSI)

VERTICAL FLOWING
PRESSURE GRADIENTS
(ALL OIL)

TUBING SIZE	6.366	IN. I.D.
PRODUCTION RATE	70000	BLPD
GAS SPECIFIC GRAVITY	0.65	
AVERAGE FLOWING TEMP	150	F
OIL API GRAVITY	85.0	API
WATER SPECIFIC GRAVITY	1.07	

DEPTH (1000 FT)

GAS/LIQUID RATIO - SCF PER BBL

0

50
100
200
300

HAGEDORN AND BROWN
PRESENTED BY:
MACCO - SCHLUMBERGER

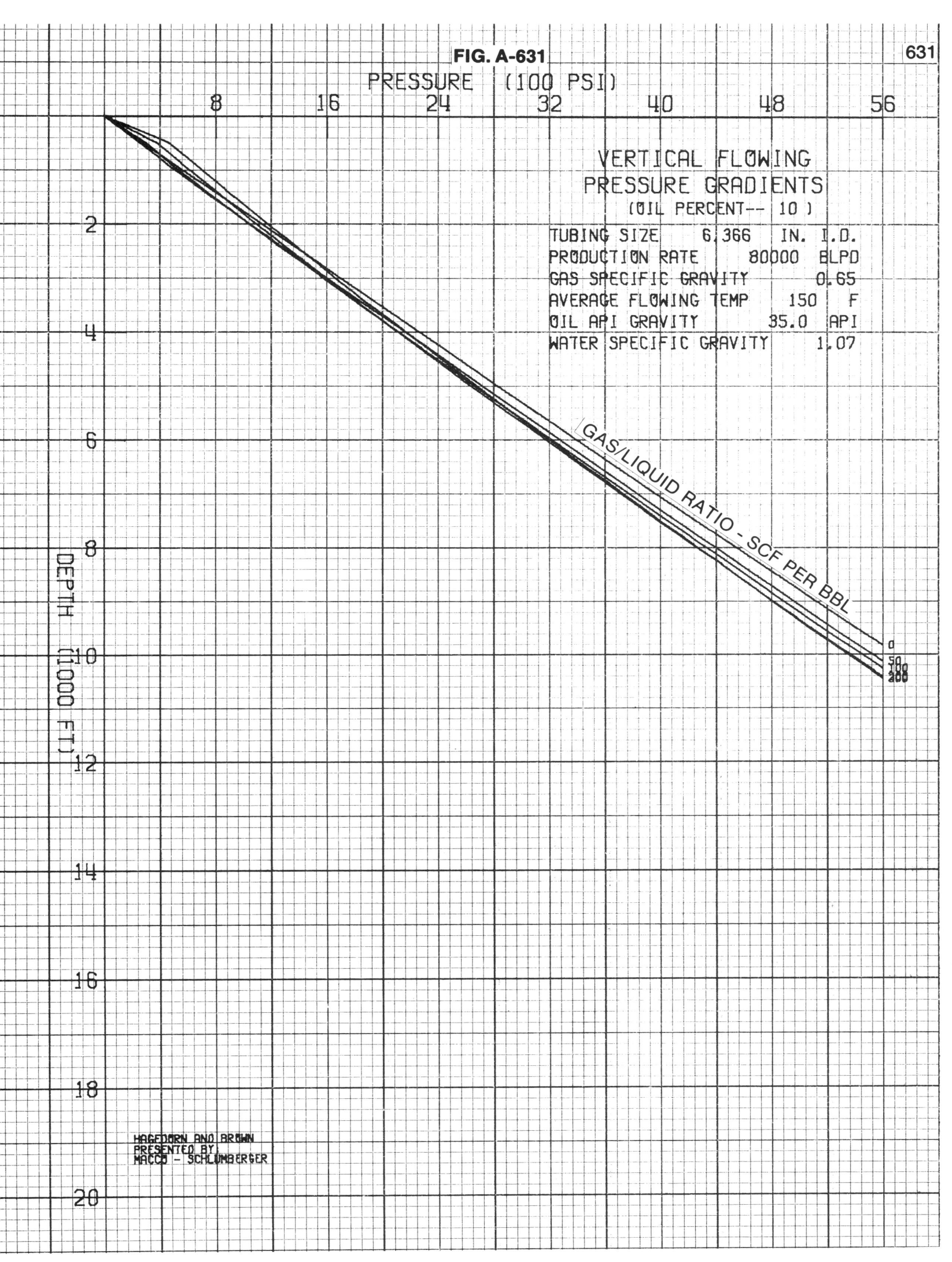

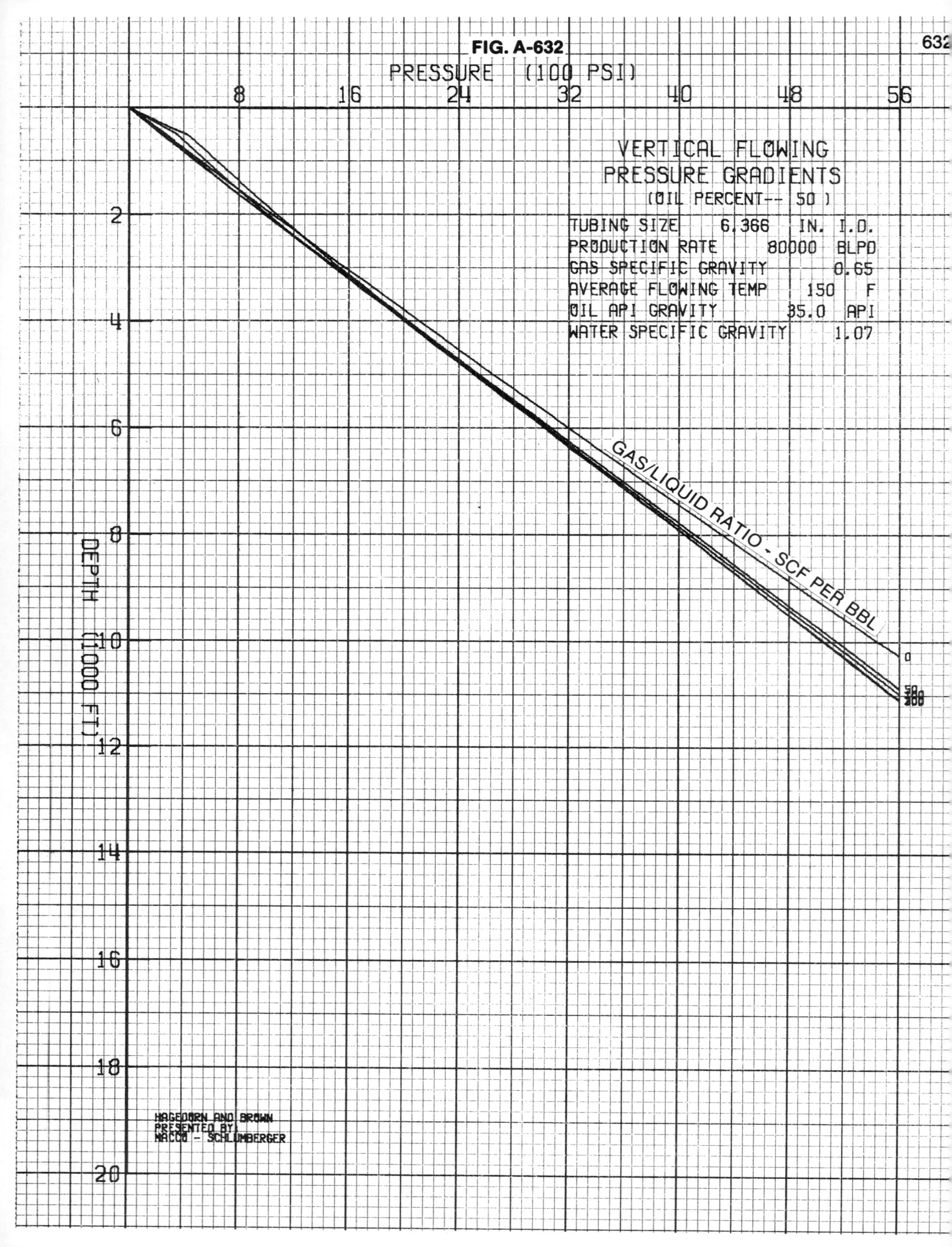

FIG. A-632

632

VERTICAL FLOWING
PRESSURE GRADIENTS
(OIL PERCENT-- 50)

TUBING SIZE   6.366   IN. I.D.
PRODUCTION RATE   80000   BLPD
GAS SPECIFIC GRAVITY   0.65
AVERAGE FLOWING TEMP   150   F
OIL API GRAVITY   35.0   API
WATER SPECIFIC GRAVITY   1.07

PRESSURE   (100 PSI)

DEPTH (1000 FT)

GAS/LIQUID RATIO - SCF PER BBL

HAGEDORN AND BROWN
PRESENTED BY
NACCO - SCHLUMBERGER

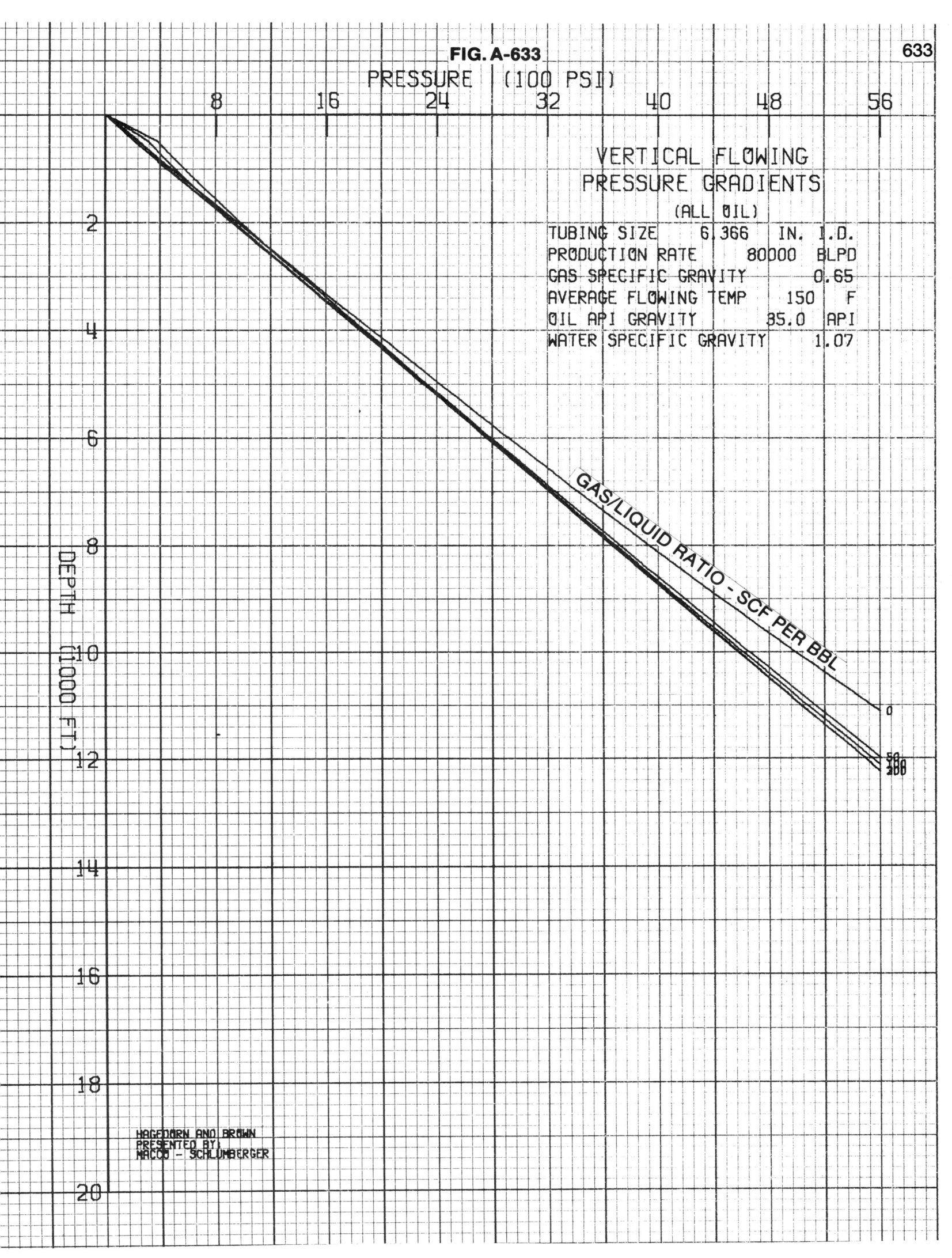

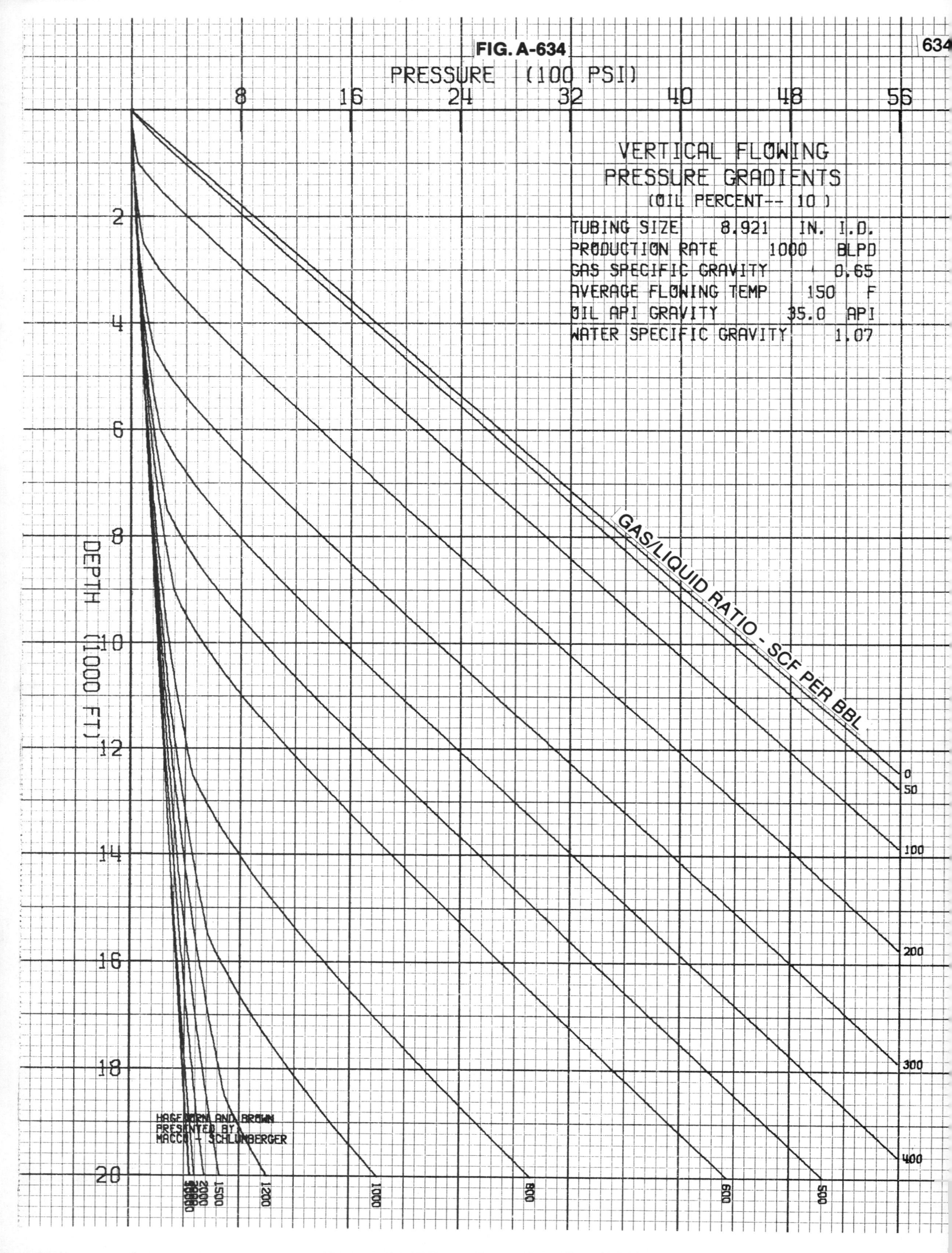

VERTICAL FLOWING
PRESSURE GRADIENTS
(OIL PERCENT-- 10)

TUBING SIZE	8.921	IN. I.D.
PRODUCTION RATE	1000	BLPD
GAS SPECIFIC GRAVITY		0.65
AVERAGE FLOWING TEMP	150	F
OIL API GRAVITY	35.0	API
WATER SPECIFIC GRAVITY		1.07

PRESSURE (100 PSI)

DEPTH (1000 FT)

GAS/LIQUID RATIO - SCF PER BBL

HAGEDORN AND BROWN
PRESENTED BY
MACCO - SCHLUMBERGER

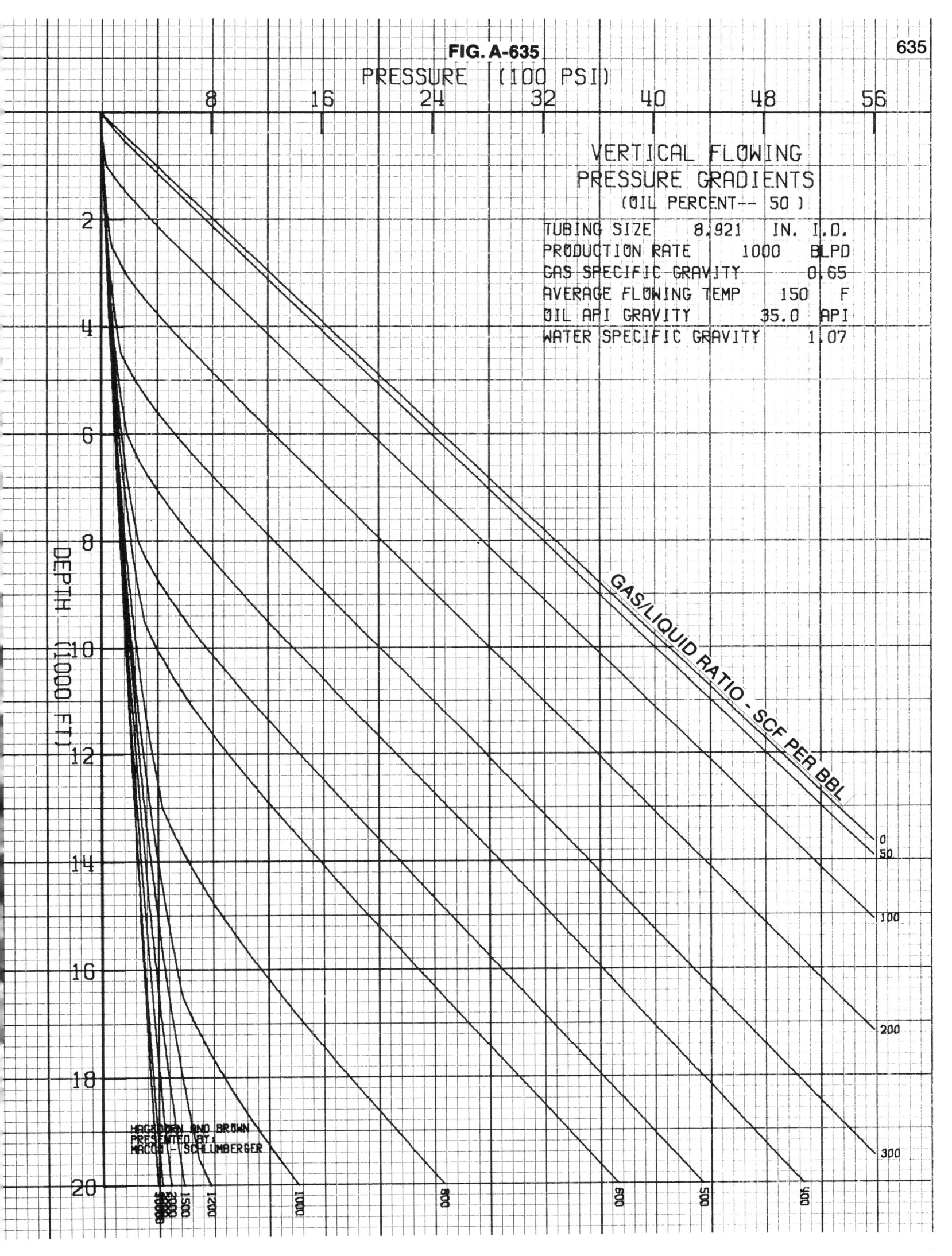

PRESSURE (100 PSI)

VERTICAL FLOWING
PRESSURE GRADIENTS
(OIL PERCENT-- 50)

TUBING SIZE	8.921	IN. I.D.
PRODUCTION RATE	1000	BLPD
GAS SPECIFIC GRAVITY	0.65	
AVERAGE FLOWING TEMP	150	F
OIL API GRAVITY	35.0	API
WATER SPECIFIC GRAVITY	1.07	

DEPTH (1000 FT)

GAS/LIQUID RATIO - SCF PER BBL

0
50
100
200
300

HAGEDORN AND BROWN
PRESENTED BY:
MACCO - SCHLUMBERGER

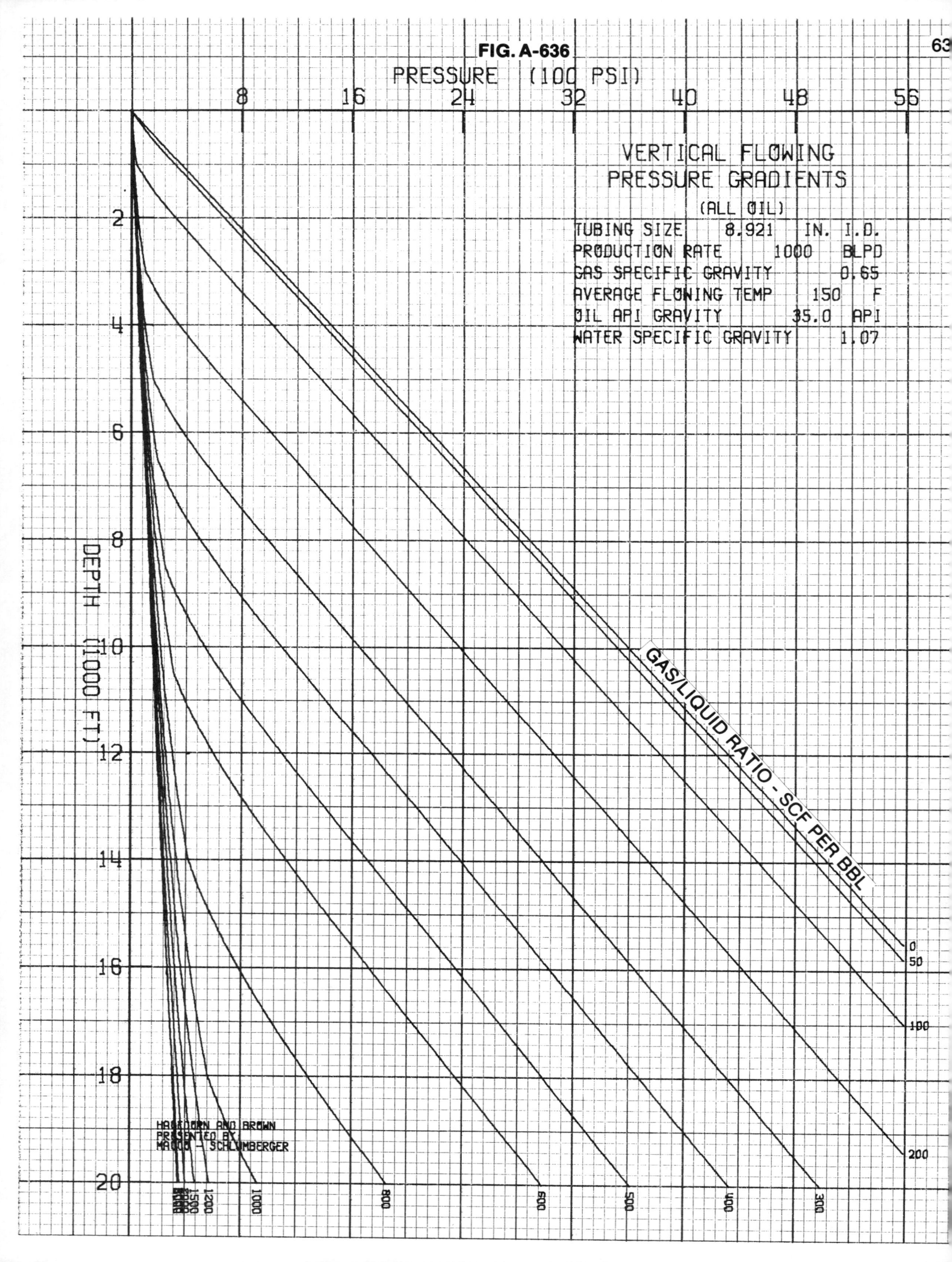

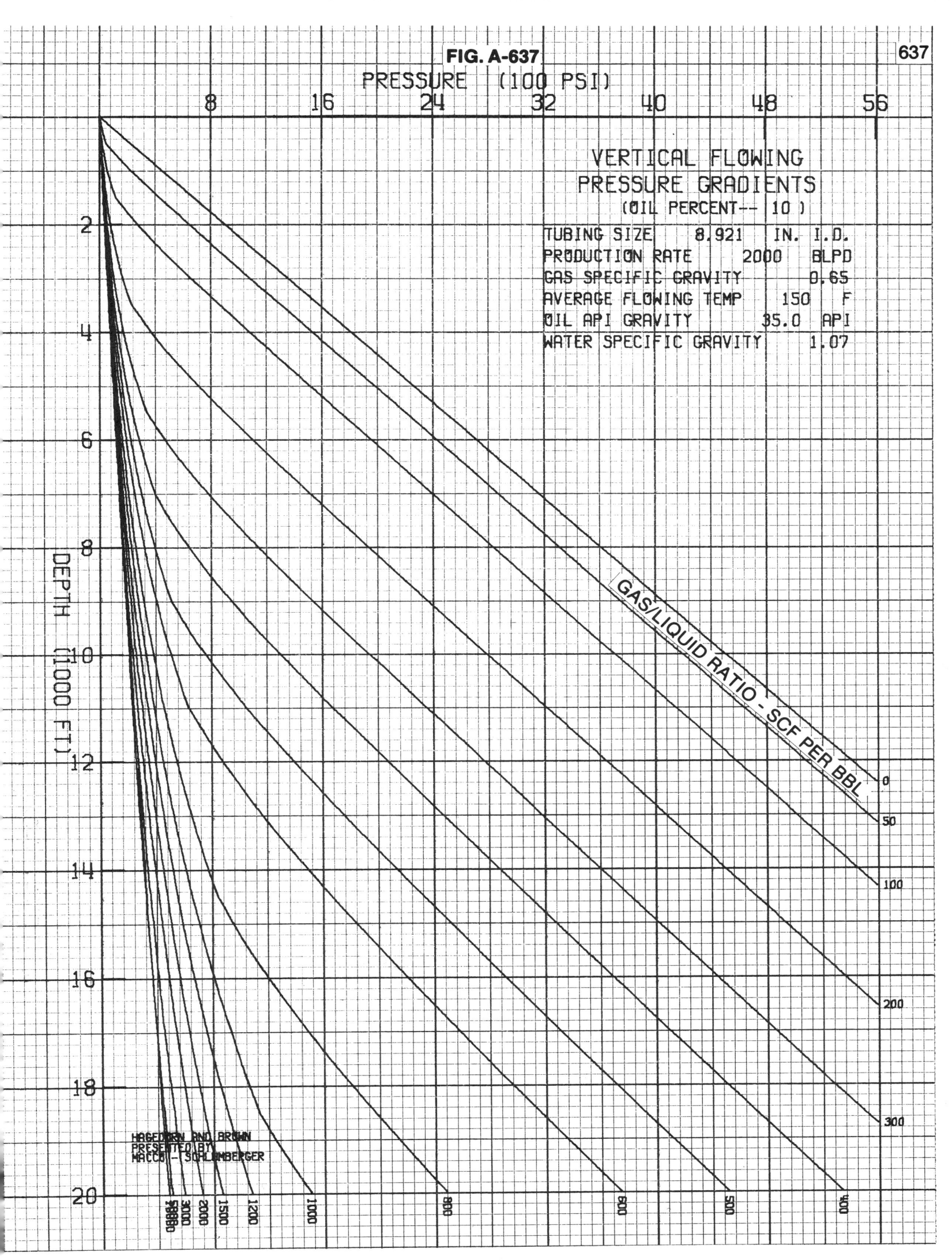

FIG. A-637

VERTICAL FLOWING
PRESSURE GRADIENTS
(OIL PERCENT-- 10)

TUBING SIZE	8.921	IN. I.D.
PRODUCTION RATE	2000	BLPD
GAS SPECIFIC GRAVITY	0.65	
AVERAGE FLOWING TEMP	150	F
OIL API GRAVITY	35.0	API
WATER SPECIFIC GRAVITY	1.07	

PRESSURE (100 PSI)

DEPTH (1000 FT)

GAS/LIQUID RATIO - SCF PER BBL

HAGEDORN AND BROWN
PRESENTED BY
MACCO - SCHLUMBERGER

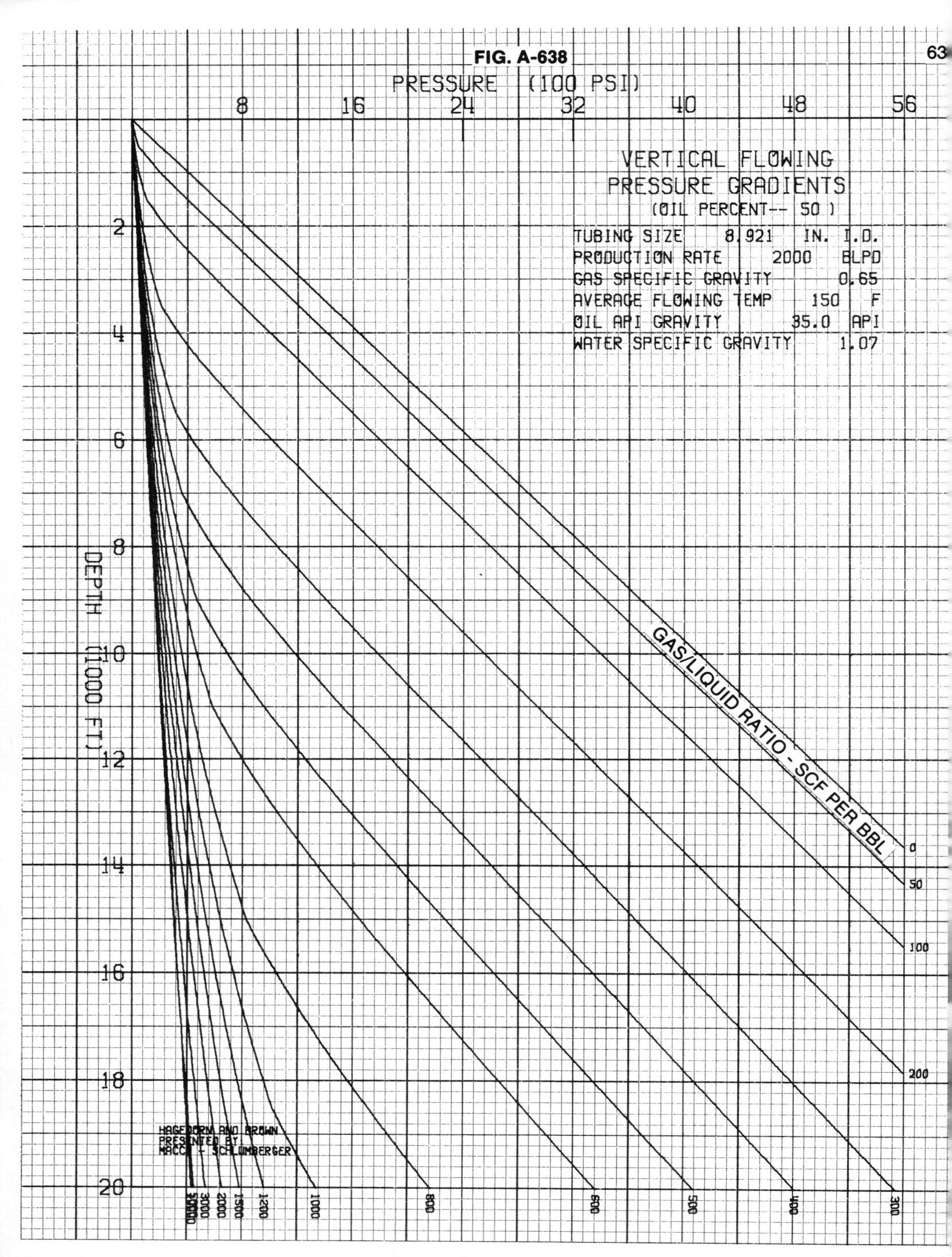

FIG. A-638

63

PRESSURE (100 PSI)

VERTICAL FLOWING
PRESSURE GRADIENTS
(OIL PERCENT-- 50)

TUBING SIZE   8.921   IN. I.D.
PRODUCTION RATE   2000   BLPD
GAS SPECIFIC GRAVITY   0.65
AVERAGE FLOWING TEMP   150   F
OIL API GRAVITY   35.0   API
WATER SPECIFIC GRAVITY   1.07

GAS/LIQUID RATIO - SCF PER BBL

DEPTH (1000 FT)

HAGEDORN AND BROWN
PRESENTED BY
NACCO - SCHLUMBERGER

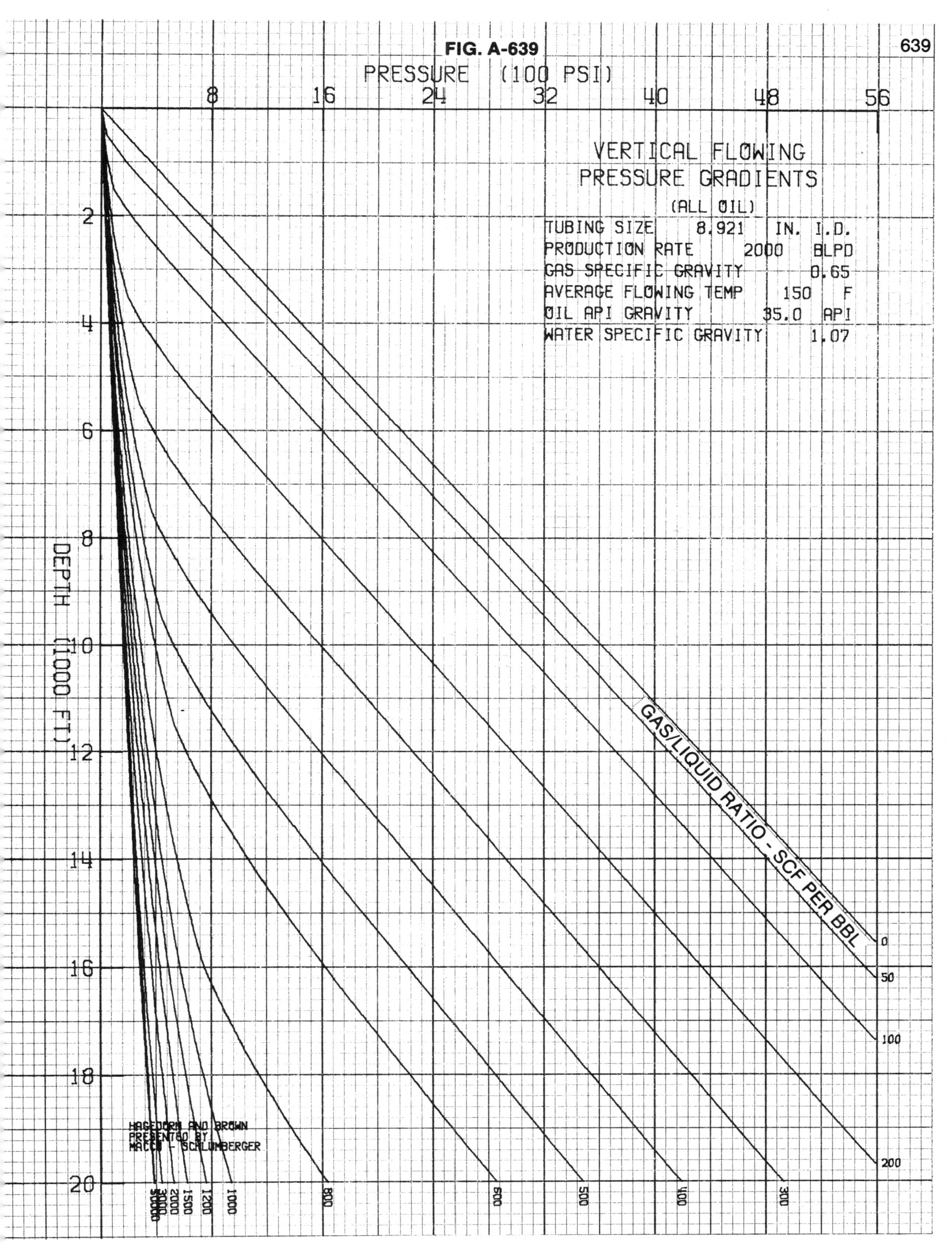

FIG. A-639

639

PRESSURE (100 PSI)

VERTICAL FLOWING
PRESSURE GRADIENTS
(ALL OIL)

TUBING SIZE	8.921	IN. I.D.
PRODUCTION RATE	2000	BLPD
GAS SPECIFIC GRAVITY	0.65	
AVERAGE FLOWING TEMP	150	F
OIL API GRAVITY	35.0	API
WATER SPECIFIC GRAVITY	1.07	

DEPTH (1000 FT)

GAS/LIQUID RATIO - SCF PER BBL

HAGEDORN AND BROWN
PRESENTED BY
MACCO - SCHLUMBERGER

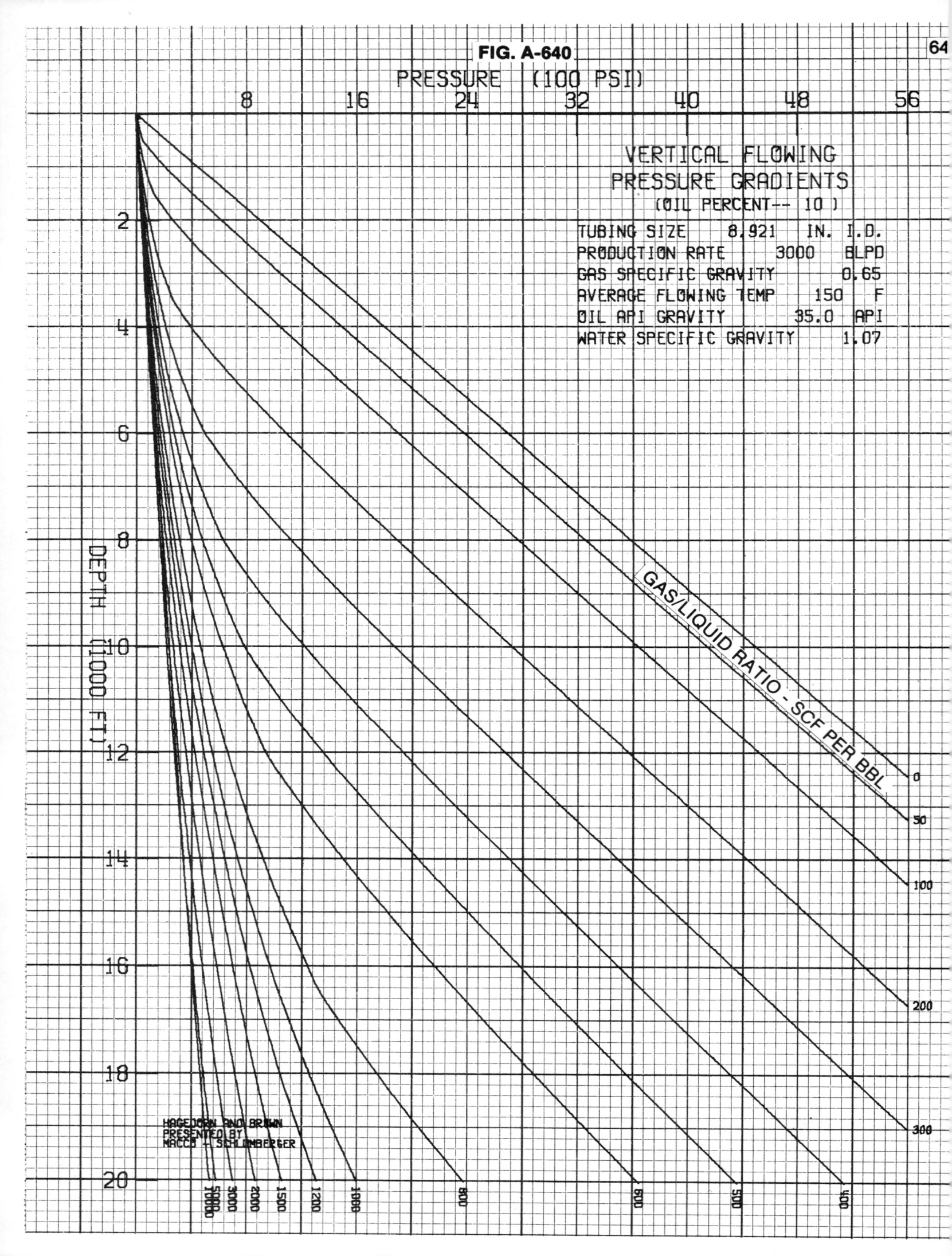

FIG. A-640

VERTICAL FLOWING
PRESSURE GRADIENTS
(OIL PERCENT-- 10)

TUBING SIZE	8.921	IN. I.D.
PRODUCTION RATE	3000	BLPD
GAS SPECIFIC GRAVITY	0.65	
AVERAGE FLOWING TEMP	150	F
OIL API GRAVITY	35.0	API
WATER SPECIFIC GRAVITY	1.07	

PRESSURE (100 PSI)

DEPTH (1000 FT)

GAS/LIQUID RATIO - SCF PER BBL

HAGEDORN AND BROWN
PRESENTED BY
NACCO - SCHLUMBERGER

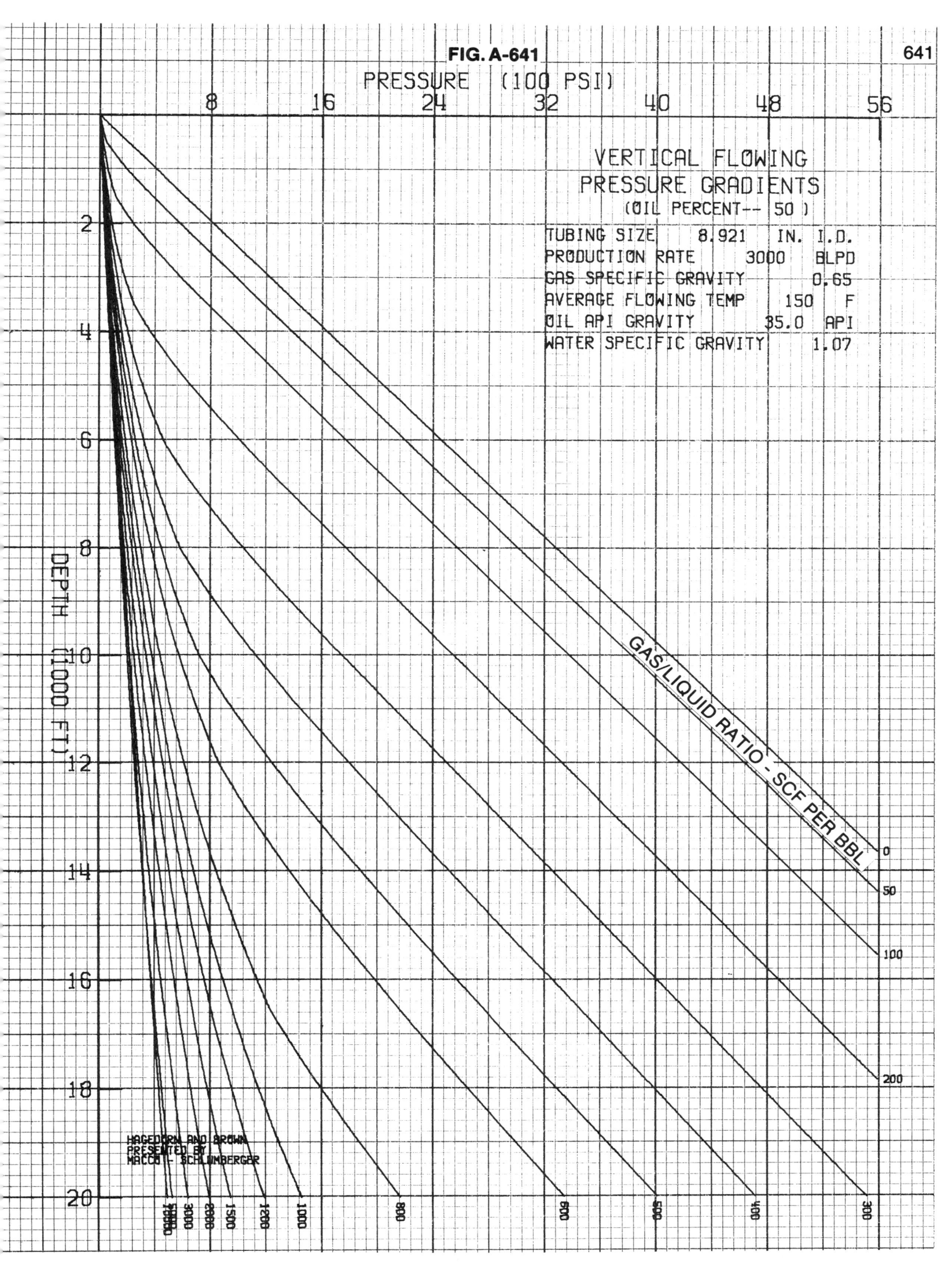

FIG. A-641

641

PRESSURE (100 PSI)

VERTICAL FLOWING
PRESSURE GRADIENTS
(OIL PERCENT-- 50)

TUBING SIZE   8.921   IN. I.D.
PRODUCTION RATE   3000   BLPD
GAS SPECIFIC GRAVITY   0.65
AVERAGE FLOWING TEMP   150   F
OIL API GRAVITY   35.0   API
WATER SPECIFIC GRAVITY   1.07

GAS/LIQUID RATIO - SCF PER BBL

DEPTH (1000 FT)

HAGEDORN AND BROWN
PRESENTED BY
MACCO - SCHLUMBERGER

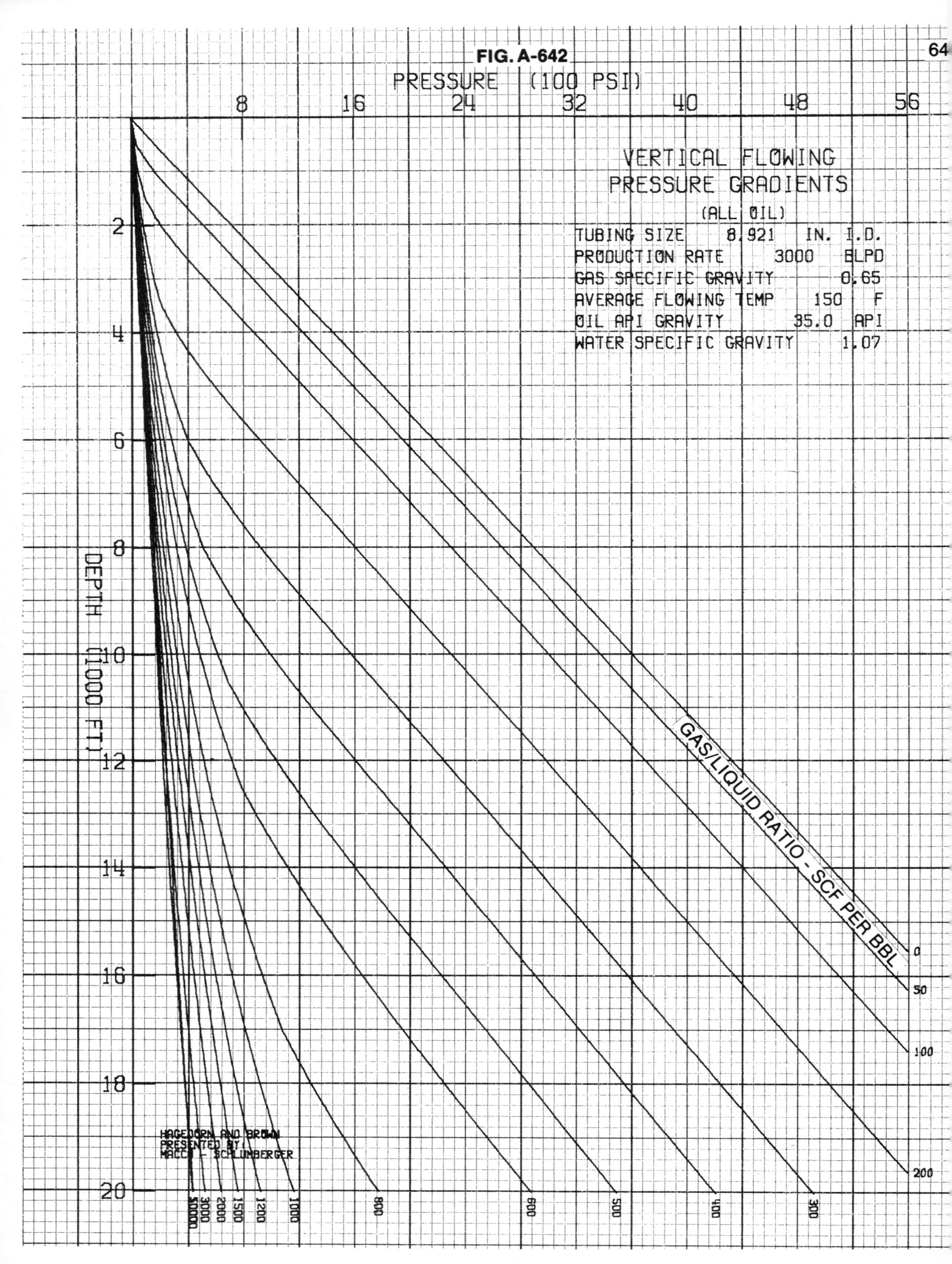

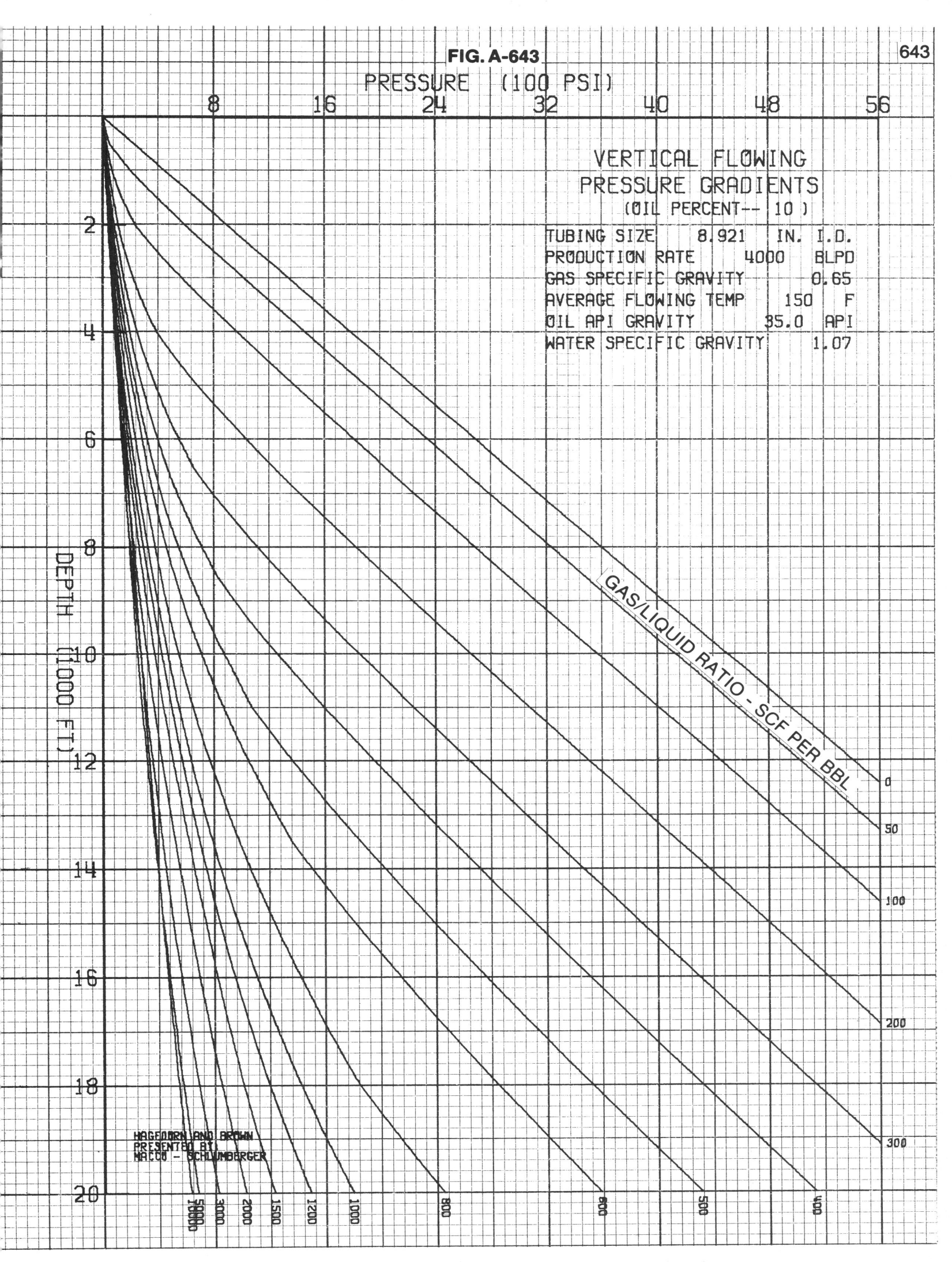

FIG. A-643

643

PRESSURE (100 PSI)

VERTICAL FLOWING
PRESSURE GRADIENTS
(OIL PERCENT-- 10)

TUBING SIZE	8.921	IN. I.D.
PRODUCTION RATE	4000	BLPD
GAS SPECIFIC GRAVITY	0.65	
AVERAGE FLOWING TEMP	150	F
OIL API GRAVITY	35.0	API
WATER SPECIFIC GRAVITY	1.07	

GAS/LIQUID RATIO - SCF PER BBL

DEPTH (1000 FT)

HAGEDORN AND BROWN
PRESENTED AT
NACOO - SCHLUMBERGER

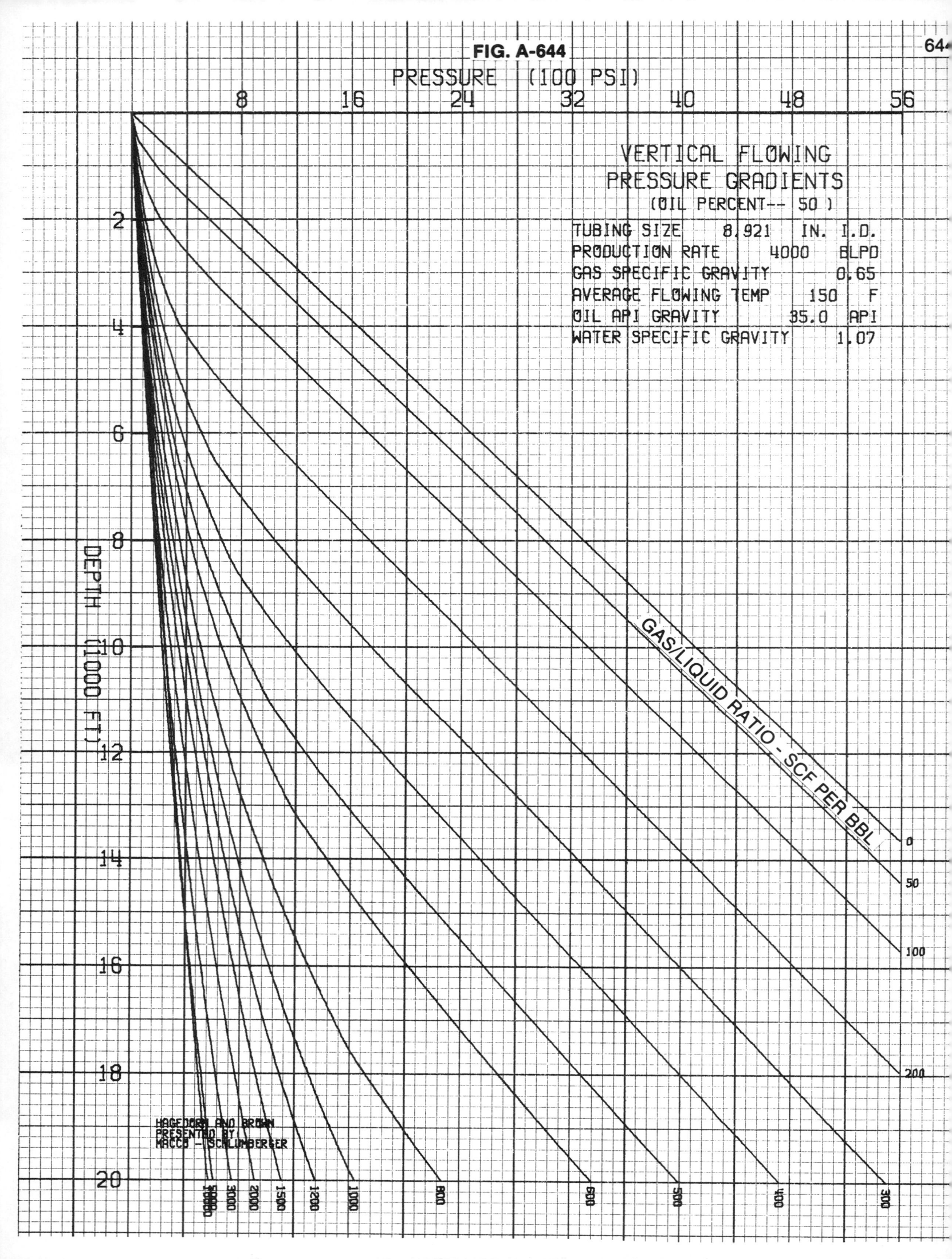

VERTICAL FLOWING
PRESSURE GRADIENTS
(OIL PERCENT-- 50)

TUBING SIZE   8.921   IN. I.D.
PRODUCTION RATE   4000   BLPD
GAS SPECIFIC GRAVITY   0.65
AVERAGE FLOWING TEMP   150   F
OIL API GRAVITY   35.0   API
WATER SPECIFIC GRAVITY   1.07

PRESSURE   (100 PSI)

DEPTH (1000 FT)

GAS/LIQUID RATIO - SCF PER BBL

HAGEDORN AND BROWN
PRESENTED BY
MACCO - SCHLUMBERGER

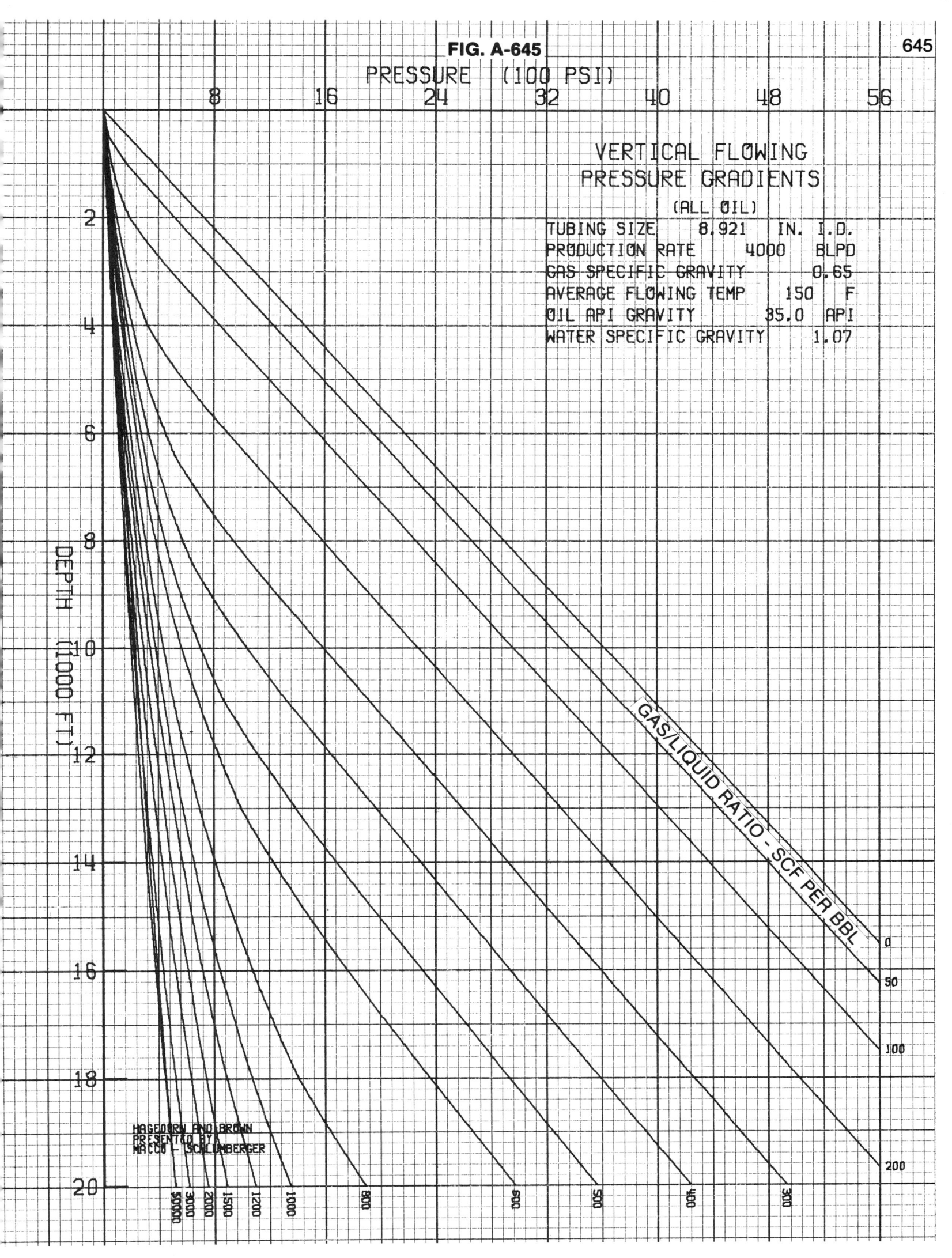

PRESSURE (100 PSI)

VERTICAL FLOWING
PRESSURE GRADIENTS
(ALL OIL)

TUBING SIZE	8.921	IN. I.D.
PRODUCTION RATE	4000	BLPD
GAS SPECIFIC GRAVITY		0.65
AVERAGE FLOWING TEMP	150	F
OIL API GRAVITY	35.0	API
WATER SPECIFIC GRAVITY		1.07

DEPTH (1000 FT)

GAS/LIQUID RATIO - SCF PER BBL

HAGEDORN AND BROWN
PRESENTED BY
MAECO - SCHLUMBERGER

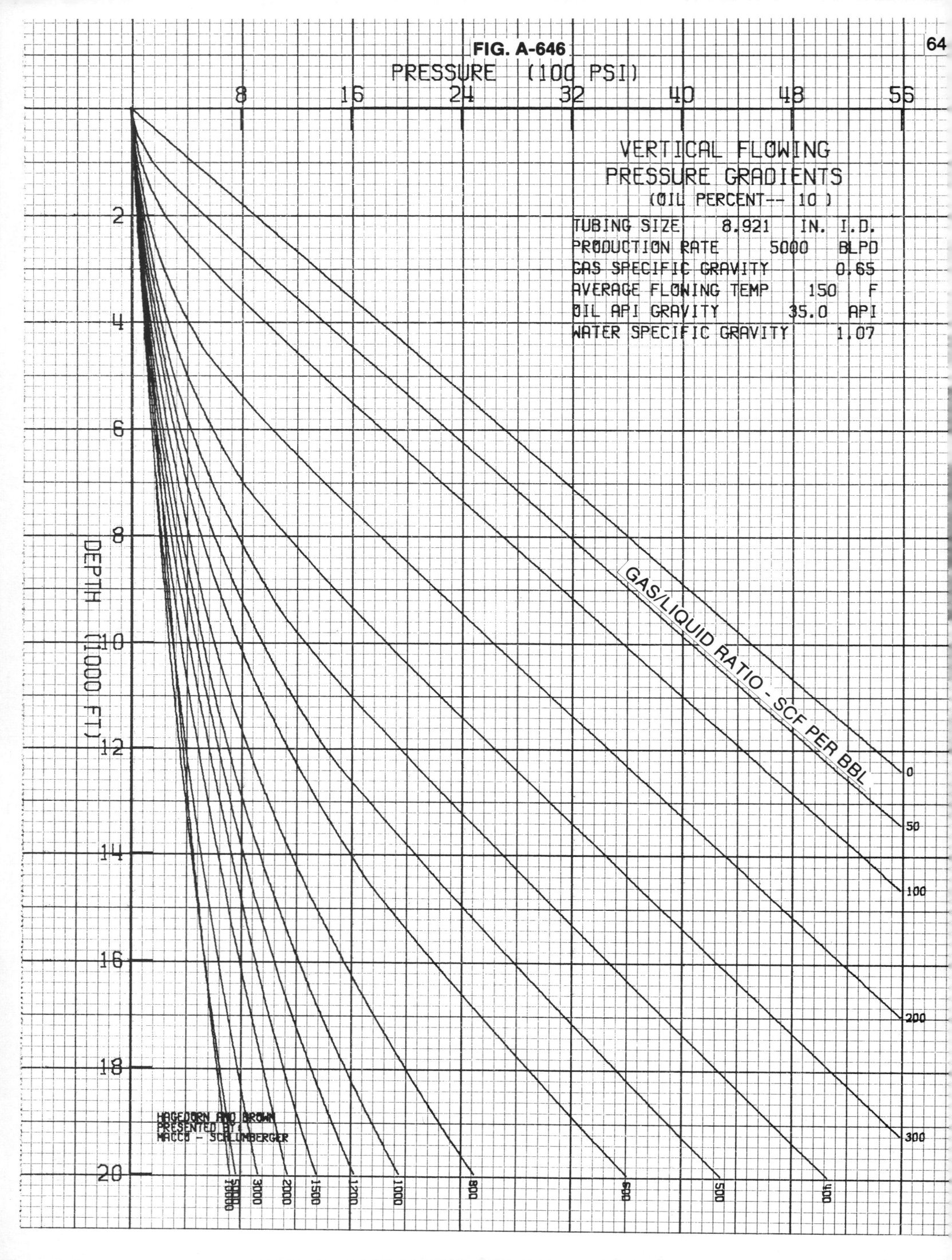

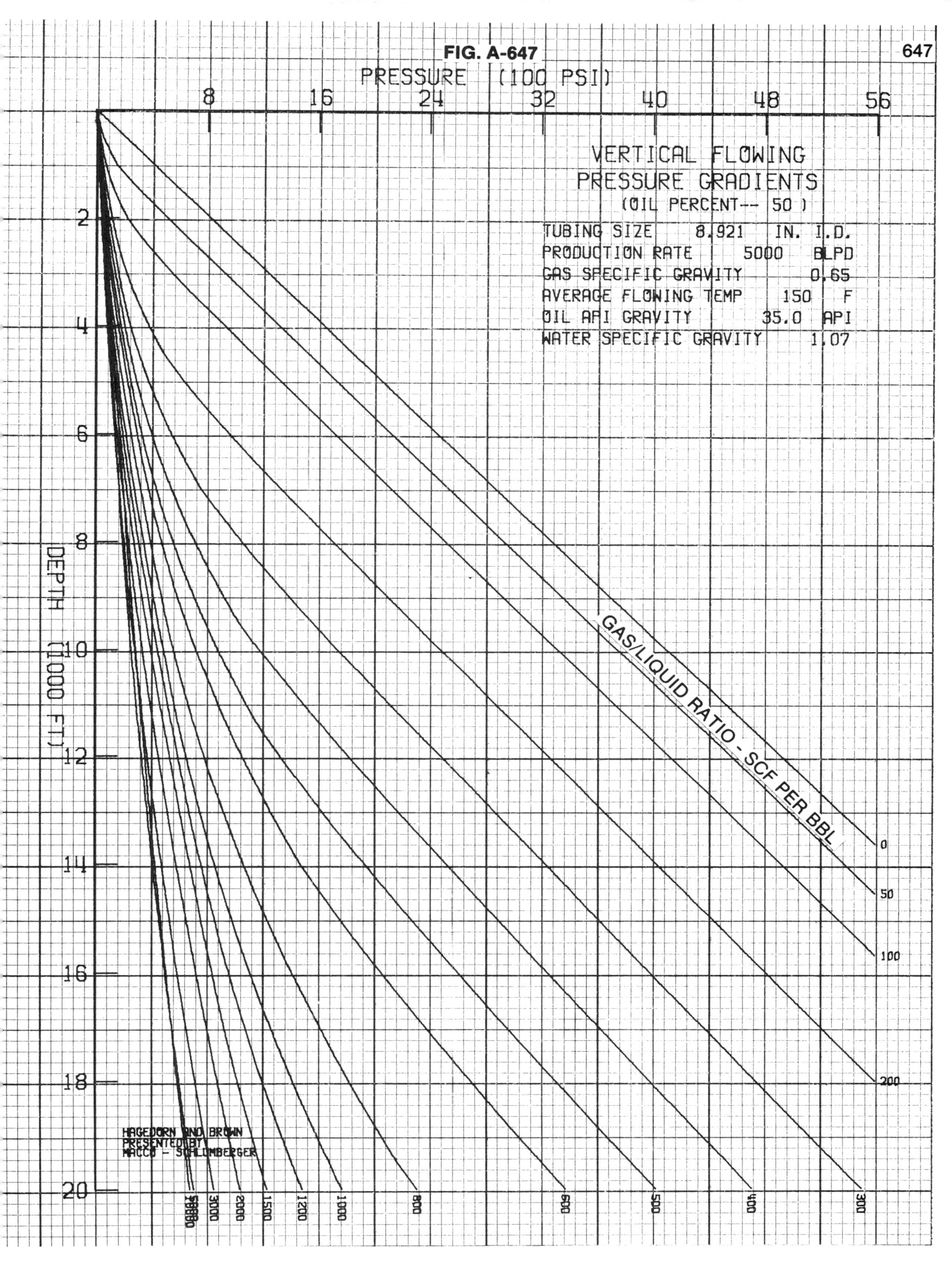

PRESSURE (100 PSI)

VERTICAL FLOWING
PRESSURE GRADIENTS
(OIL PERCENT--- 50)

TUBING SIZE	8.921	IN. I.D.
PRODUCTION RATE	5000	BLPD
GAS SPECIFIC GRAVITY	0.65	
AVERAGE FLOWING TEMP	150	F
OIL API GRAVITY	35.0	API
WATER SPECIFIC GRAVITY	1.07	

DEPTH (1000 FT)

GAS/LIQUID RATIO - SCF PER BBL

HAGEDORN AND BROWN
PRESENTED BY
MACCO - SCHLUMBERGER

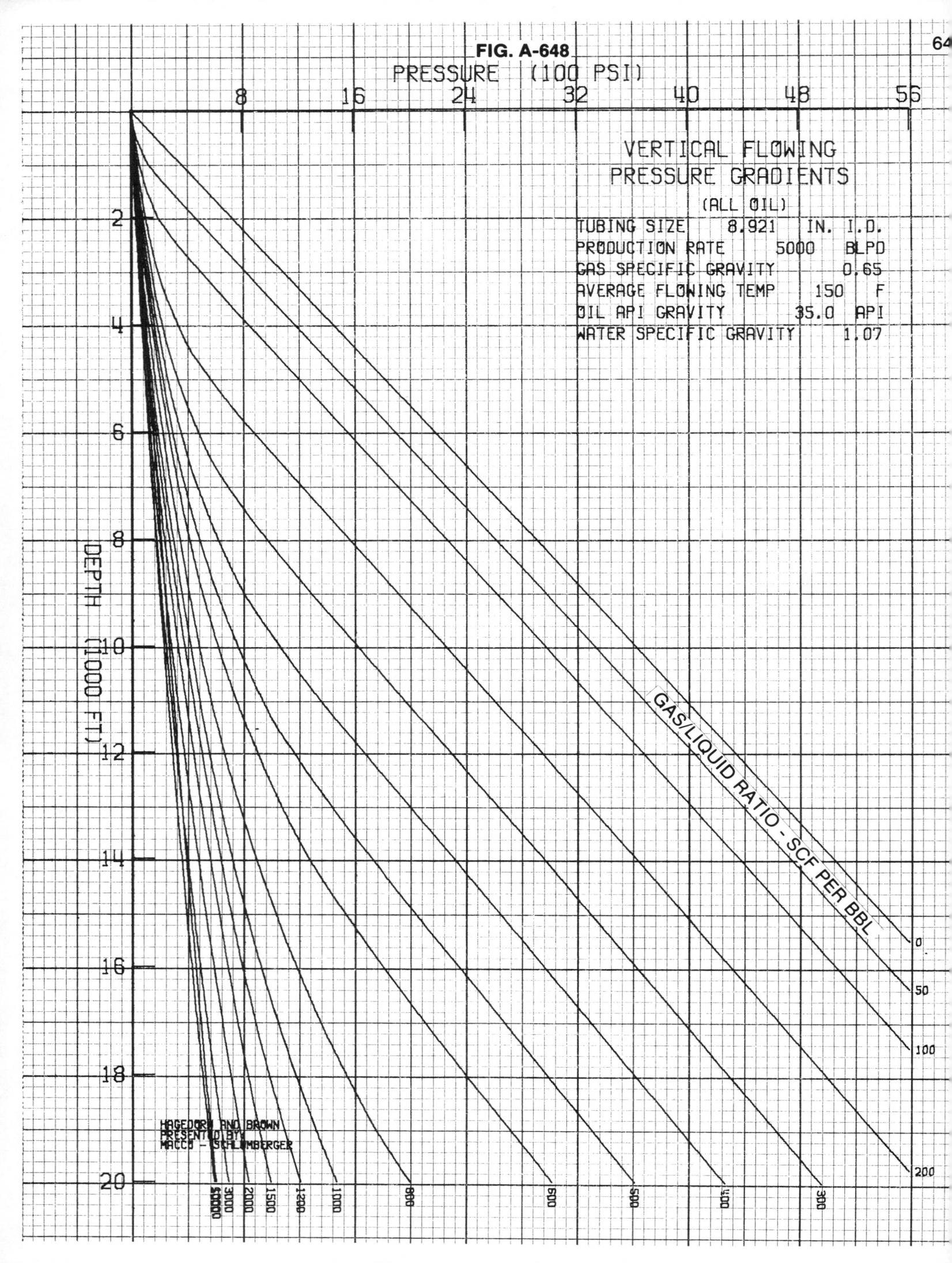

PRESSURE (100 PSI)

VERTICAL FLOWING
PRESSURE GRADIENTS
(ALL OIL)

TUBING SIZE	8.921	IN. I.D.
PRODUCTION RATE	5000	BLPD
GAS SPECIFIC GRAVITY	0.65	
AVERAGE FLOWING TEMP	150	F
OIL API GRAVITY	35.0	API
WATER SPECIFIC GRAVITY	1.07	

DEPTH (1000 FT)

GAS/LIQUID RATIO - SCF PER BBL

HAGEDORN AND BROWN
PRESENTED BY:
MACCO - SCHLUMBERGER

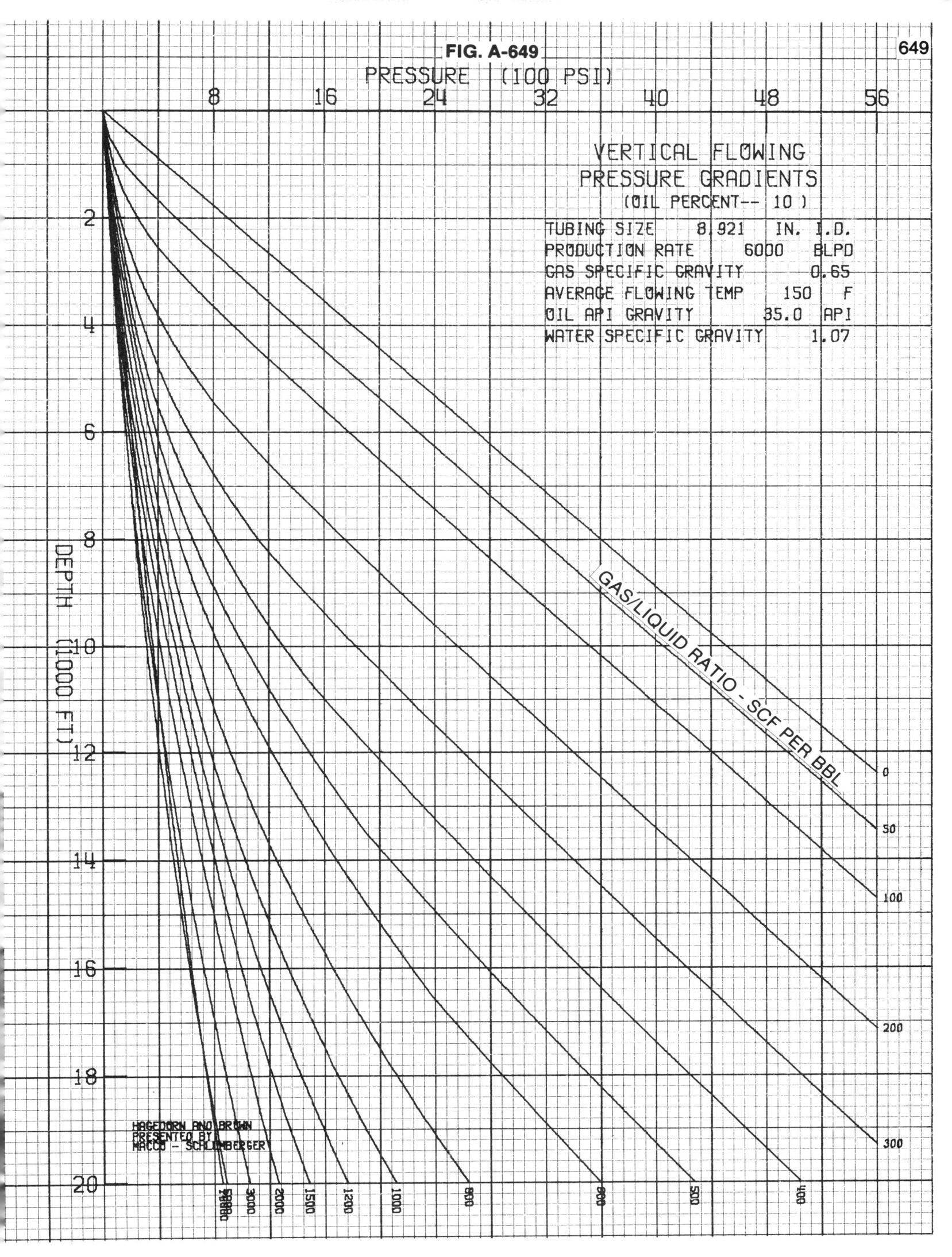

FIG. A-649

649

VERTICAL FLOWING
PRESSURE GRADIENTS
(OIL PERCENT-- 10)

TUBING SIZE   8.921   IN. I.D.
PRODUCTION RATE   6000   BLPD
GAS SPECIFIC GRAVITY   0.65
AVERAGE FLOWING TEMP   150   F
OIL API GRAVITY   35.0 API
WATER SPECIFIC GRAVITY   1.07

GAS/LIQUID RATIO - SCF PER BBL

HAGEDORN AND BROWN
PRESENTED BY
MACCO - SCHLUMBERGER

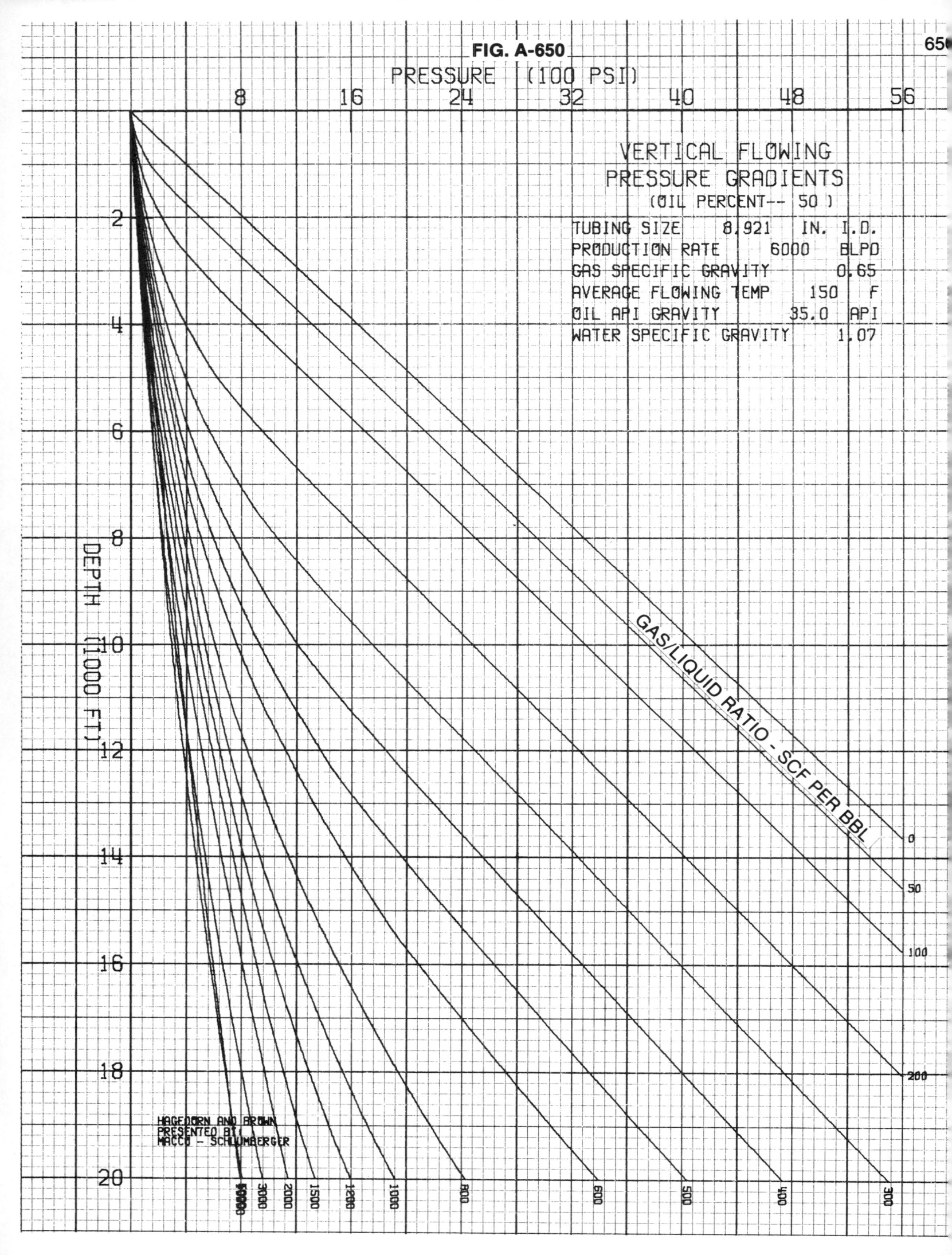

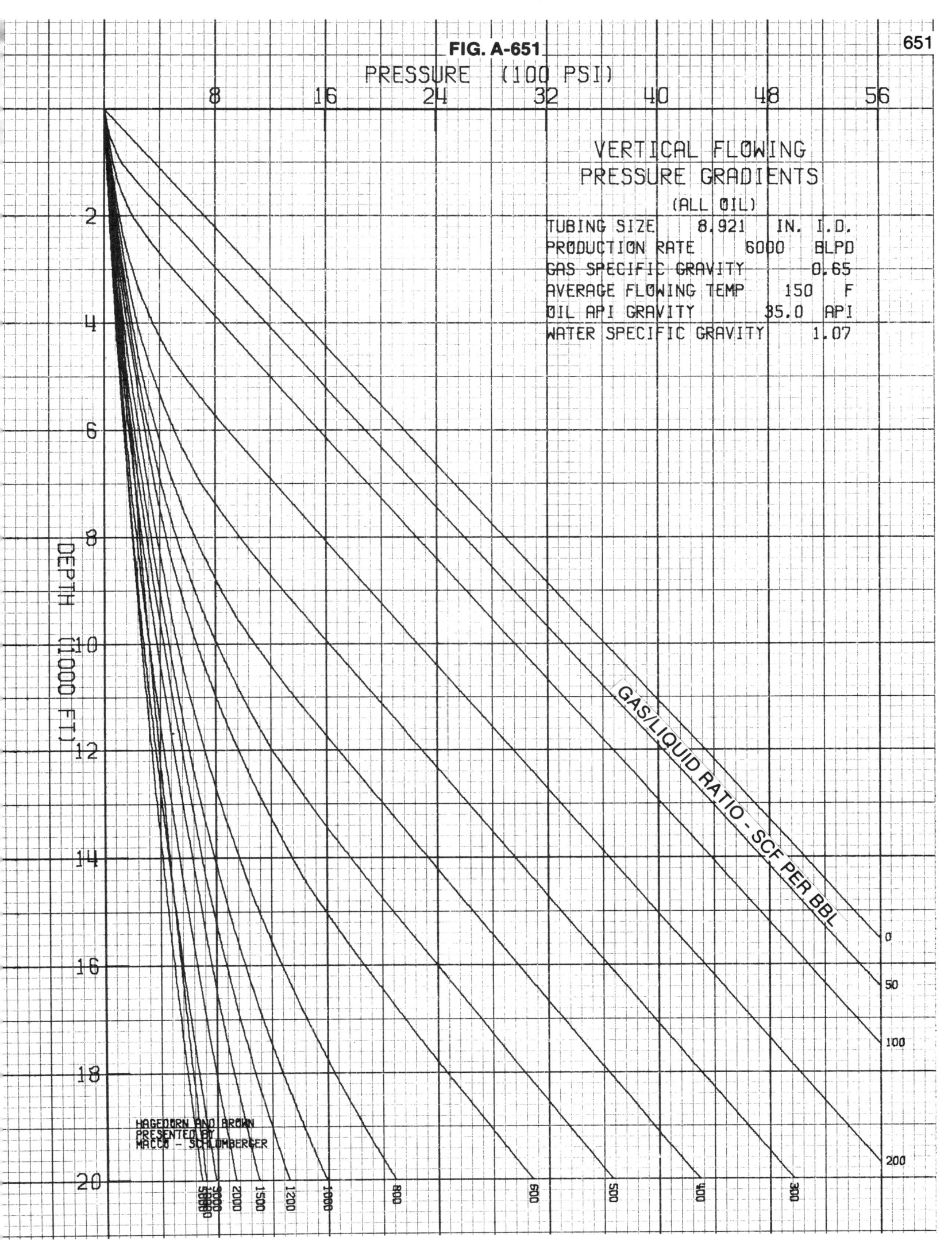

PRESSURE (100 PSI)

VERTICAL FLOWING
PRESSURE GRADIENTS
(ALL OIL)

TUBING SIZE	8.921	IN. I.D.
PRODUCTION RATE	6000	BLPD
GAS SPECIFIC GRAVITY		0.65
AVERAGE FLOWING TEMP	150	F
OIL API GRAVITY	35.0	API
WATER SPECIFIC GRAVITY		1.07

DEPTH (1000 FT)

GAS/LIQUID RATIO - SCF PER BBL

HAGEDORN AND BROWN
PRESENTED BY
MACCO - SCHLUMBERGER

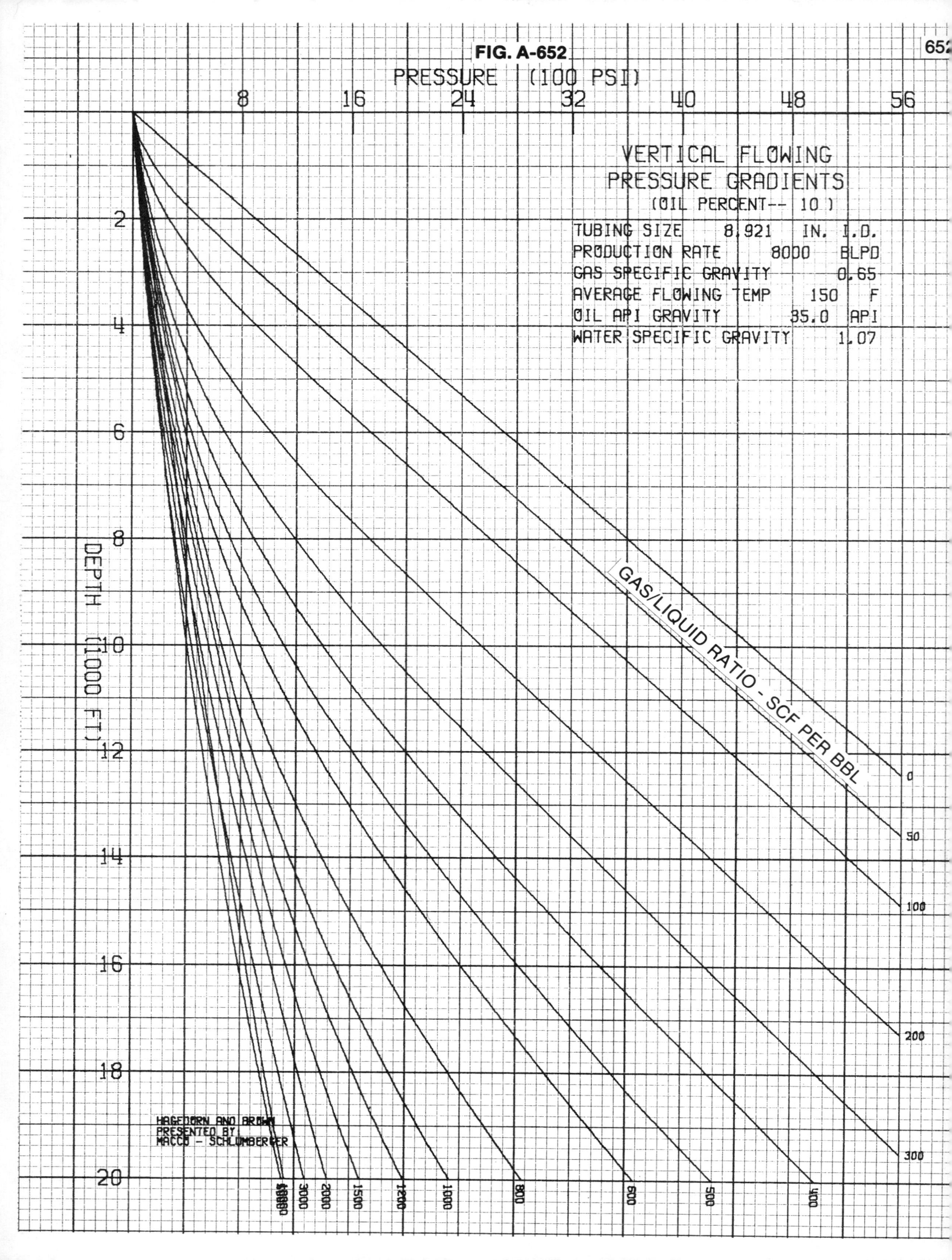

FIG. A-652

652

VERTICAL FLOWING
PRESSURE GRADIENTS
(OIL PERCENT-- 10)

TUBING SIZE   8.921   IN. I.D.
PRODUCTION RATE   8000   BLPD
GAS SPECIFIC GRAVITY   0.65
AVERAGE FLOWING TEMP   150   F
OIL API GRAVITY   35.0   API
WATER SPECIFIC GRAVITY   1.07

PRESSURE   (100 PSI)

DEPTH (1000 FT)

GAS/LIQUID RATIO - SCF PER BBL

HAGEDORN AND BROWN
PRESENTED BY
MACCO - SCHLUMBERGER

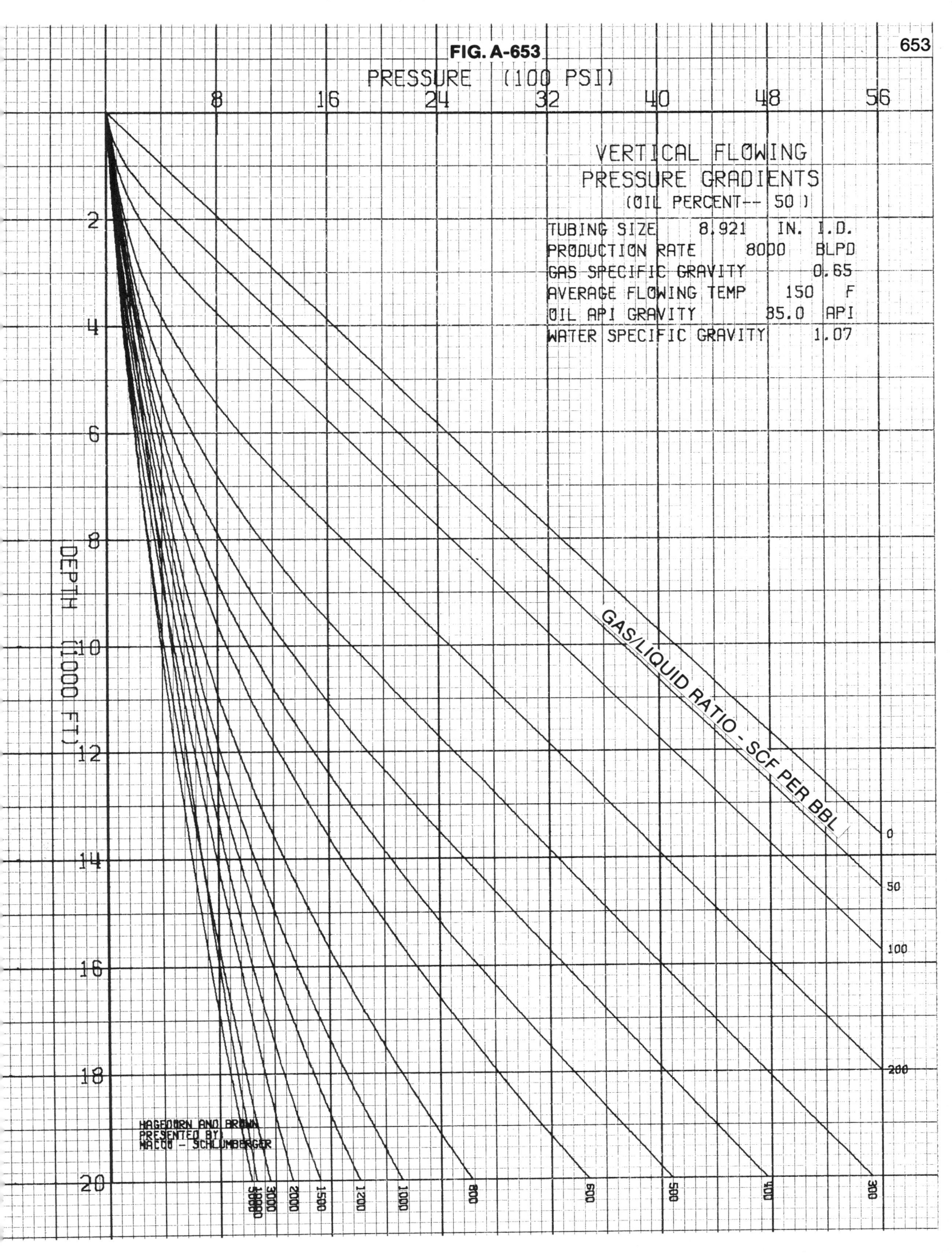

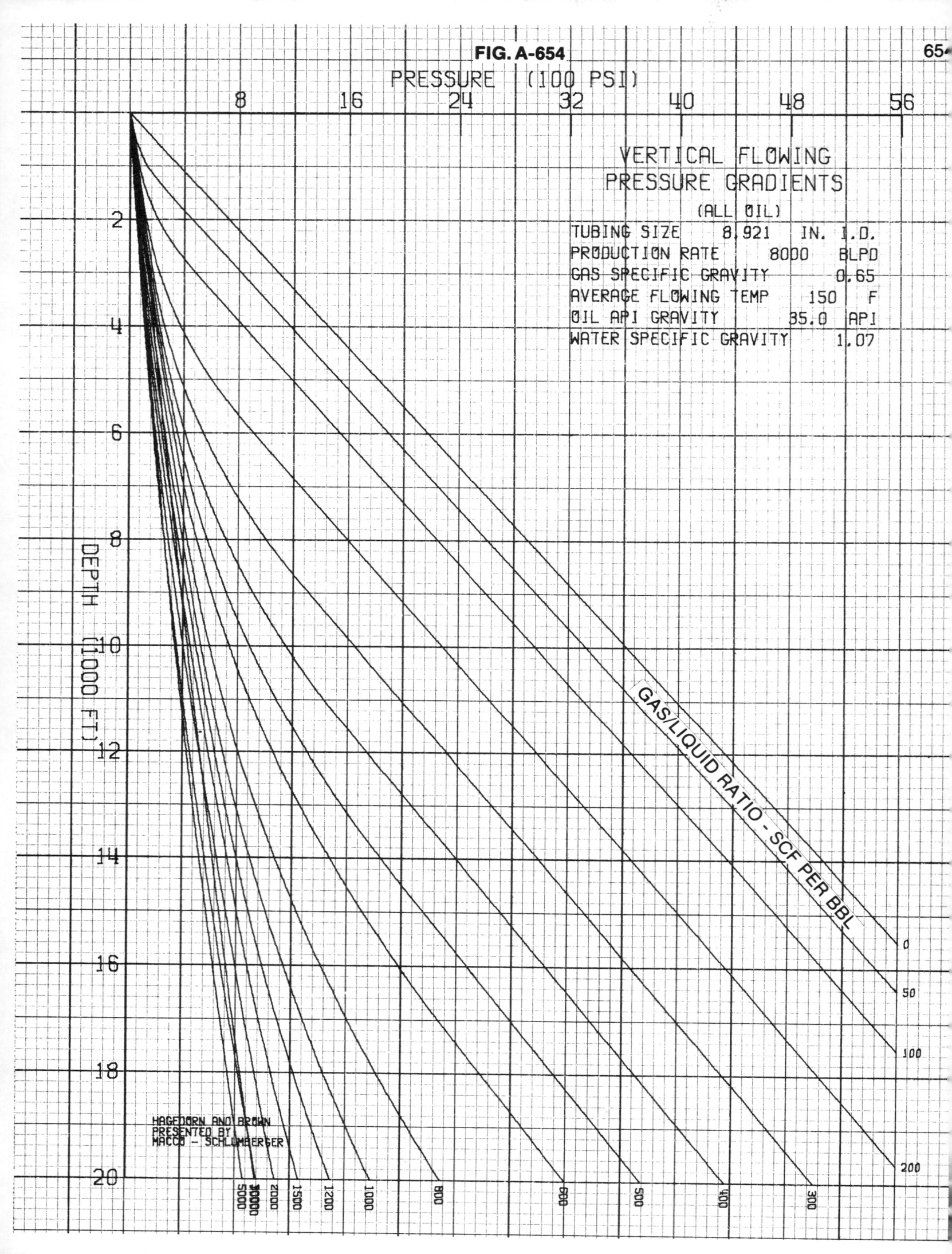

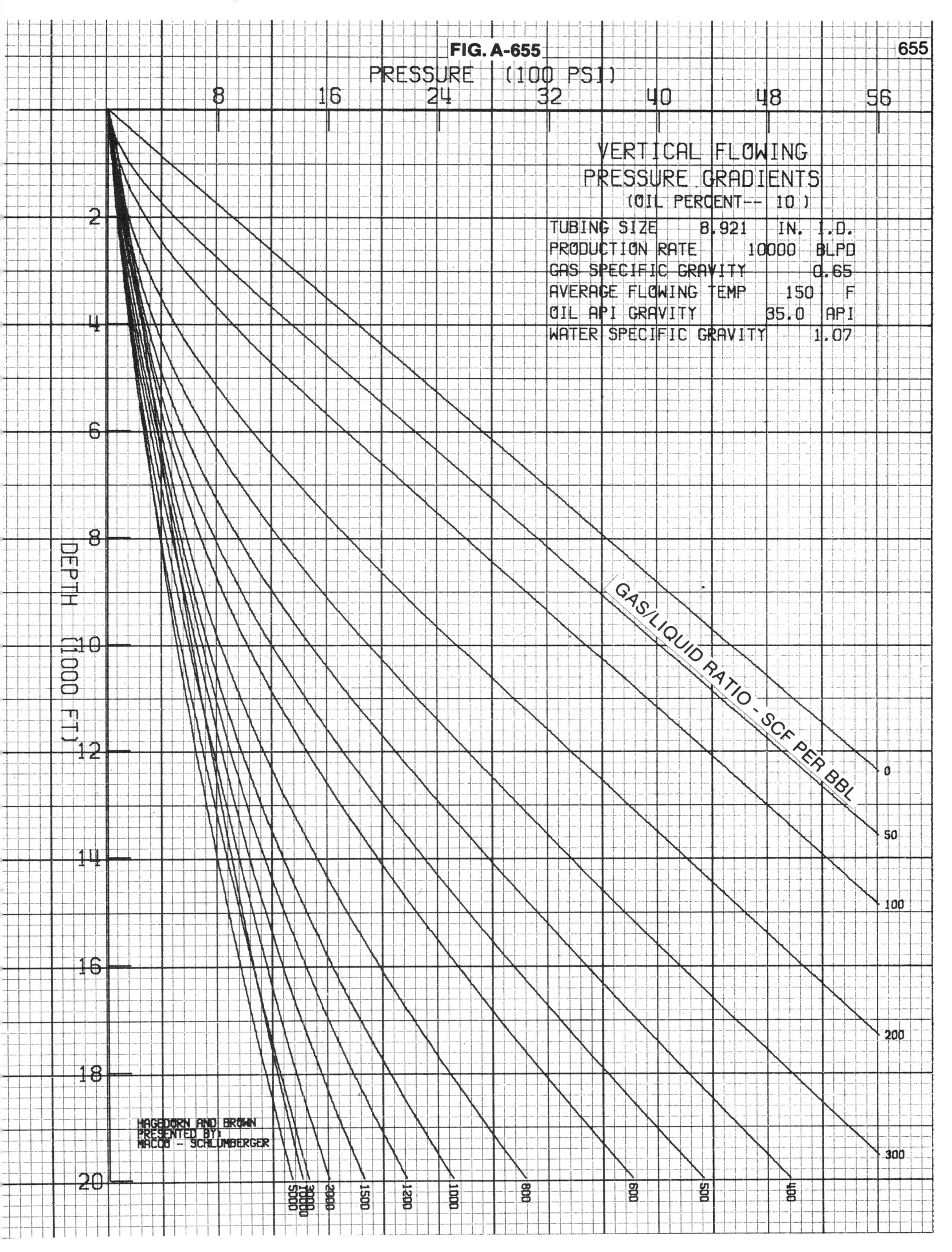

FIG. A-655

PRESSURE (100 PSI)

VERTICAL FLOWING
PRESSURE GRADIENTS
(OIL PERCENT-- 10)

TUBING SIZE	8.921	IN. I.D.
PRODUCTION RATE	10000	BLPD
GAS SPECIFIC GRAVITY		0.65
AVERAGE FLOWING TEMP	150	F
OIL API GRAVITY	35.0	API
WATER SPECIFIC GRAVITY		1.07

DEPTH (1000 FT)

GAS/LIQUID RATIO - SCF PER BBL

HAGEDORN AND BROWN
PRESENTED BY:
MACCO - SCHLUMBERGER

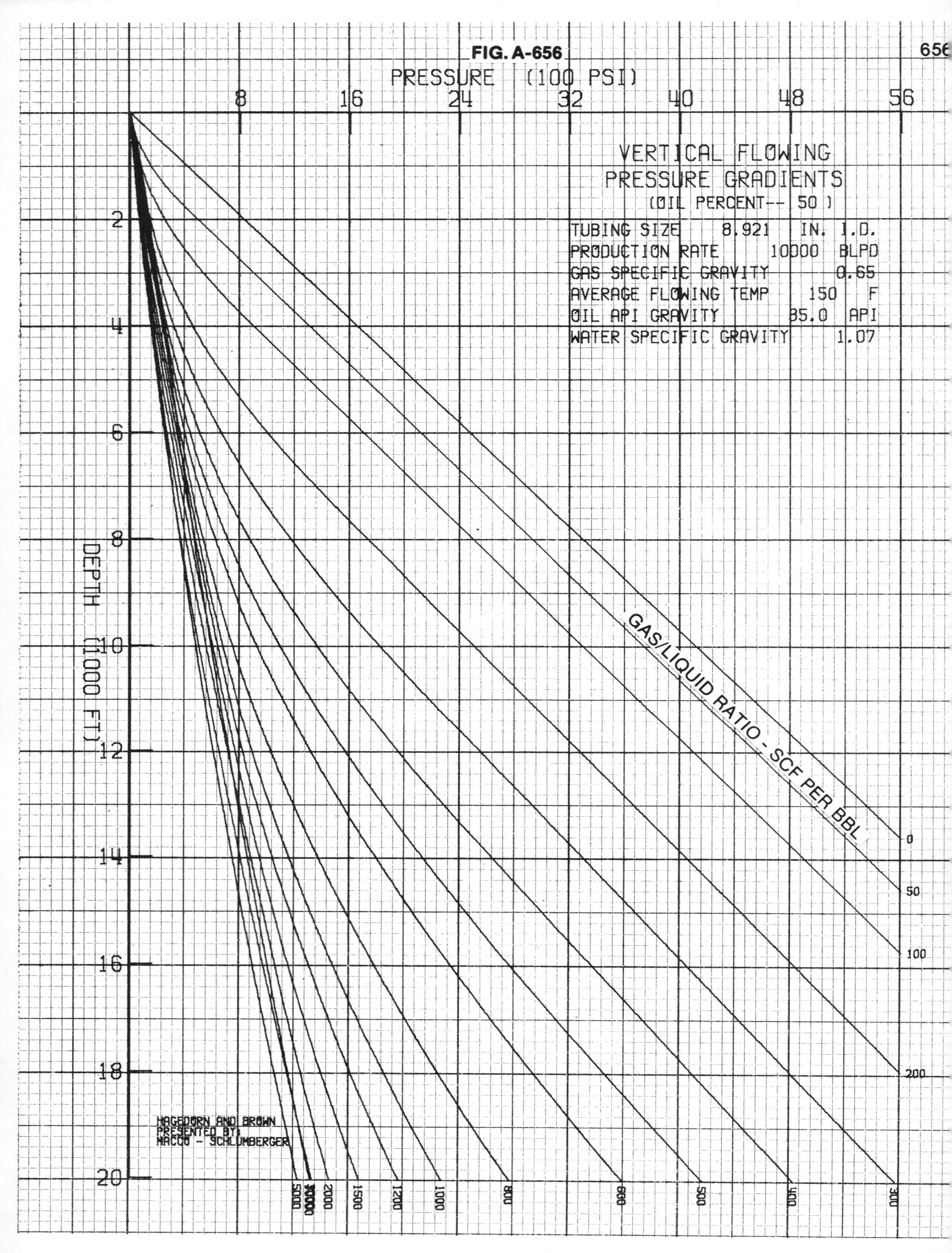

FIG. A-656

656

VERTICAL FLOWING
PRESSURE GRADIENTS
(OIL PERCENT-- 50)

TUBING SIZE	8.921	IN. I.D.
PRODUCTION RATE	10000	BLPD
GAS SPECIFIC GRAVITY	0.65	
AVERAGE FLOWING TEMP	150	F
OIL API GRAVITY	35.0	API
WATER SPECIFIC GRAVITY	1.07	

PRESSURE (100 PSI)

DEPTH (1000 FT)

GAS/LIQUID RATIO - SCF PER BBL

HAGEDORN AND BROWN
PRESENTED BY:
NACOO - SCHLUMBERGER

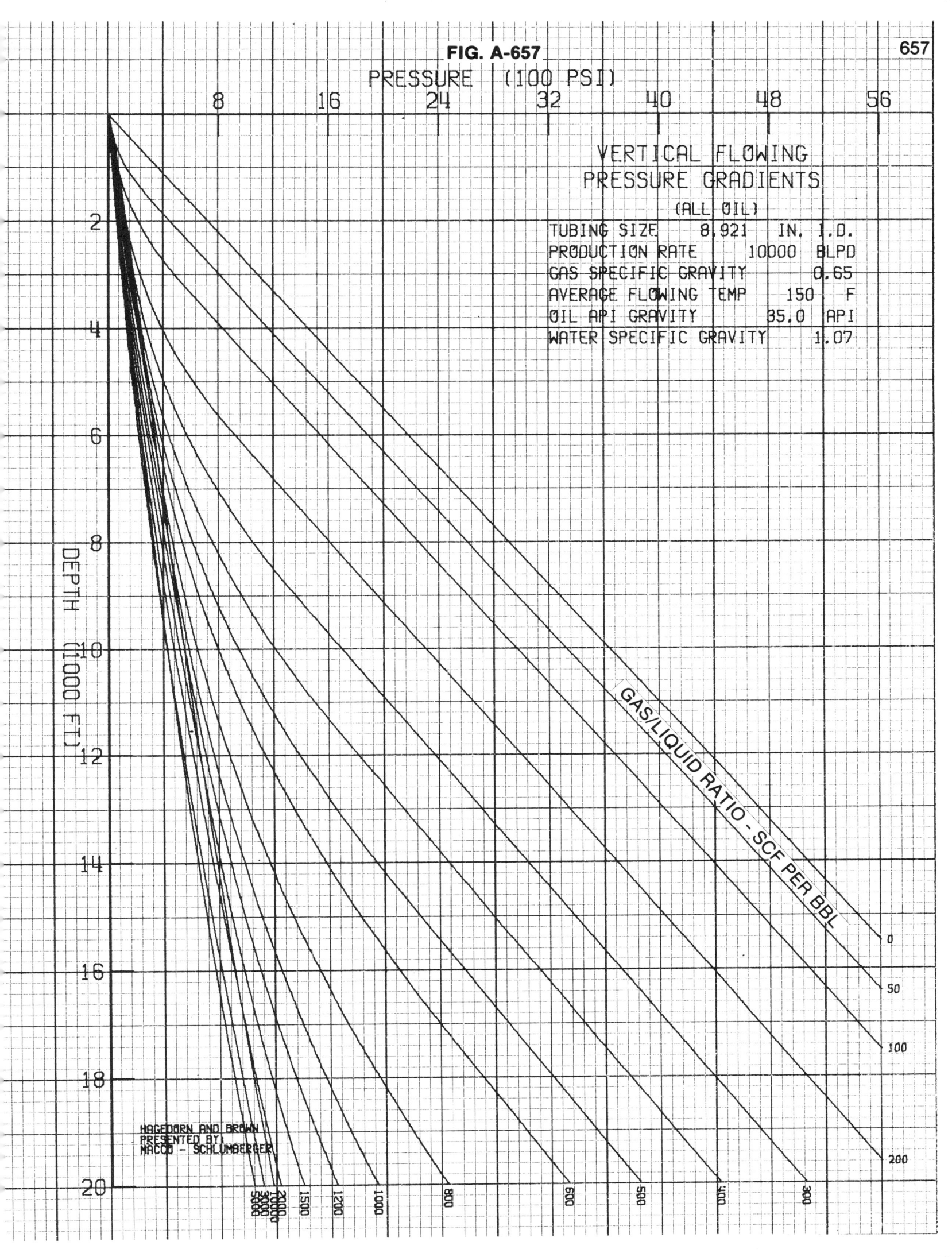

FIG. A-657

657

PRESSURE (100 PSI)

VERTICAL FLOWING
PRESSURE GRADIENTS
(ALL OIL)

TUBING SIZE	8.921	IN. I.D.
PRODUCTION RATE	10000	BLPD
GAS SPECIFIC GRAVITY	0.65	
AVERAGE FLOWING TEMP	150	F
OIL API GRAVITY	35.0	API
WATER SPECIFIC GRAVITY	1.07	

DEPTH (1000 FT)

GAS/LIQUID RATIO - SCF PER BBL

HAGEDORN AND BROWN
PRESENTED BY:
MACCO - SCHLUMBERGER

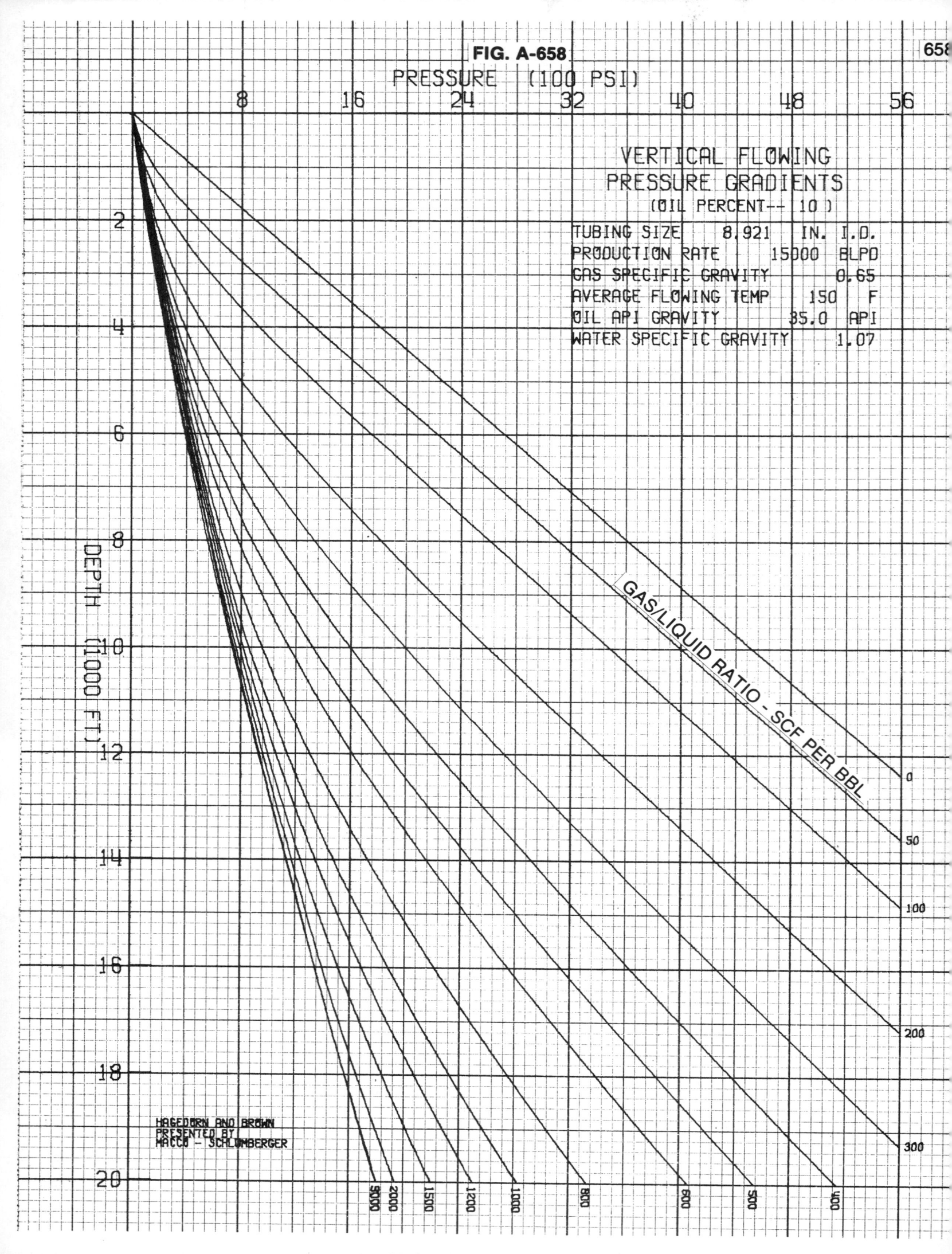

FIG. A-658

658

VERTICAL FLOWING
PRESSURE GRADIENTS
(OIL PERCENT-- 10)

TUBING SIZE	8.921	IN. I.D.
PRODUCTION RATE	15000	BLPD
GAS SPECIFIC GRAVITY	0.65	
AVERAGE FLOWING TEMP	150	F
OIL API GRAVITY	35.0	API
WATER SPECIFIC GRAVITY	1.07	

PRESSURE (100 PSI)

DEPTH (1000 FT)

GAS/LIQUID RATIO - SCF PER BBL

HAGEDORN AND BROWN
PRESENTED BY
MACCO - SCHLUMBERGER

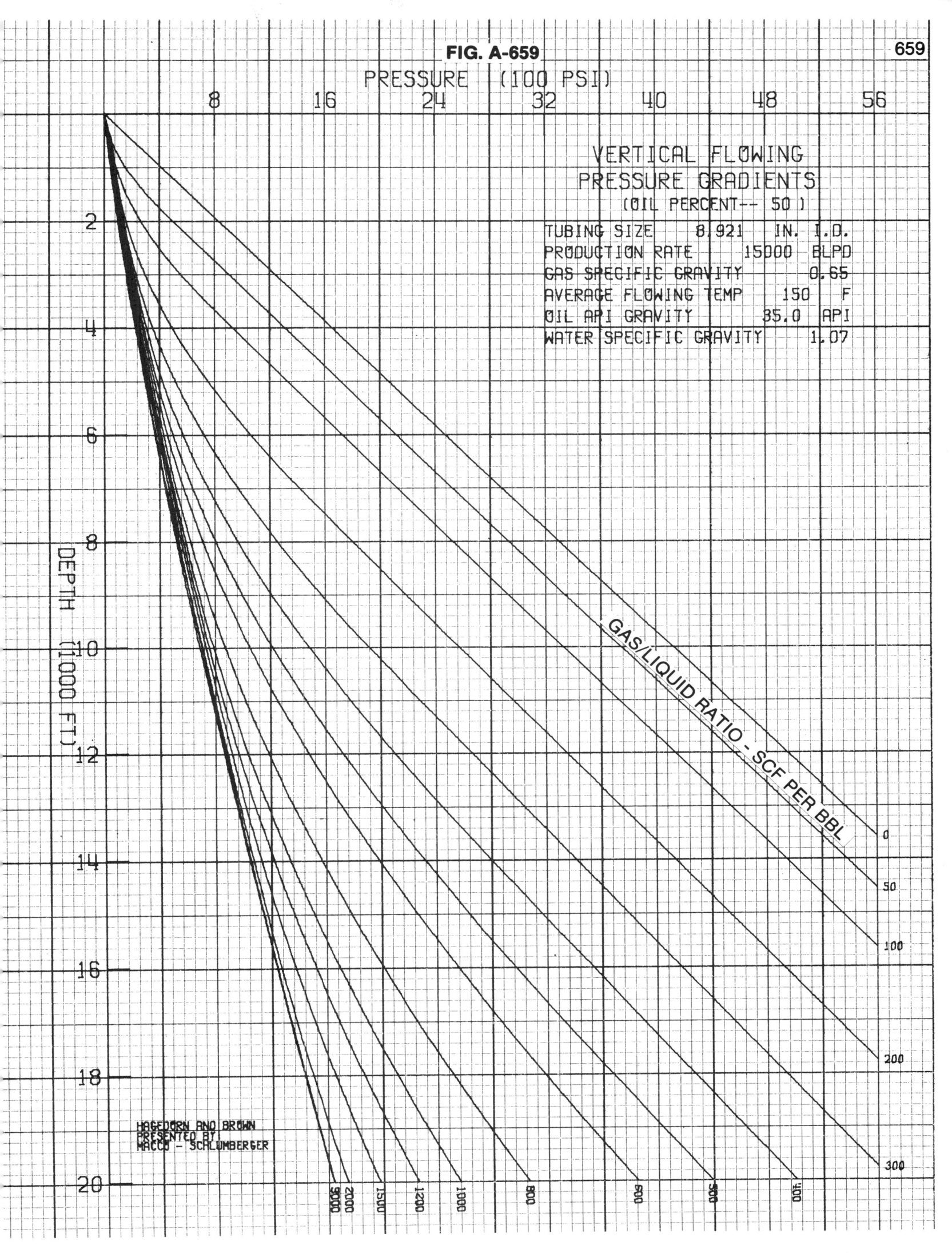

PRESSURE (100 PSI)

VERTICAL FLOWING
PRESSURE GRADIENTS
(OIL PERCENT-- 50)

TUBING SIZE	8.921	IN. I.D.
PRODUCTION RATE	15000	BLPD
GAS SPECIFIC GRAVITY	0.65	
AVERAGE FLOWING TEMP	150	F
OIL API GRAVITY	35.0	API
WATER SPECIFIC GRAVITY	1.07	

GAS/LIQUID RATIO - SCF PER BBL

DEPTH (1,000 FT)

HAGEDORN AND BROWN
PRESENTED BY
MACCO - SCHLUMBERGER

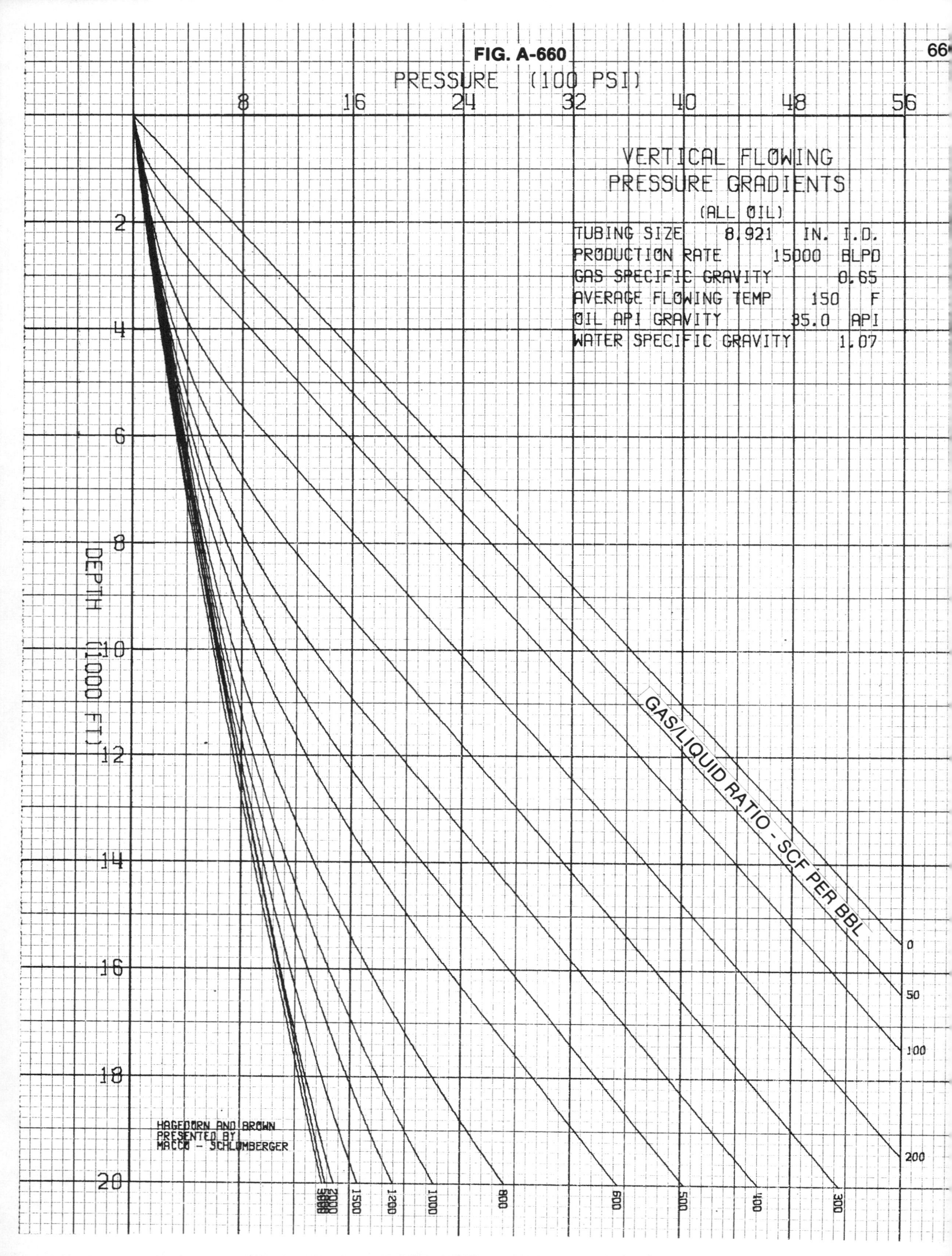

FIG. A-660

VERTICAL FLOWING
PRESSURE GRADIENTS
(ALL OIL)

TUBING SIZE   8.921   IN. I.D.
PRODUCTION RATE   15000   BLPD
GAS SPECIFIC GRAVITY   0.65
AVERAGE FLOWING TEMP   150   F
OIL API GRAVITY   35.0 API
WATER SPECIFIC GRAVITY   1.07

PRESSURE (100 PSI)

DEPTH (1000 FT)

GAS/LIQUID RATIO - SCF PER BBL

HAGEDORN AND BROWN
PRESENTED BY
MACCO - SCHLUMBERGER

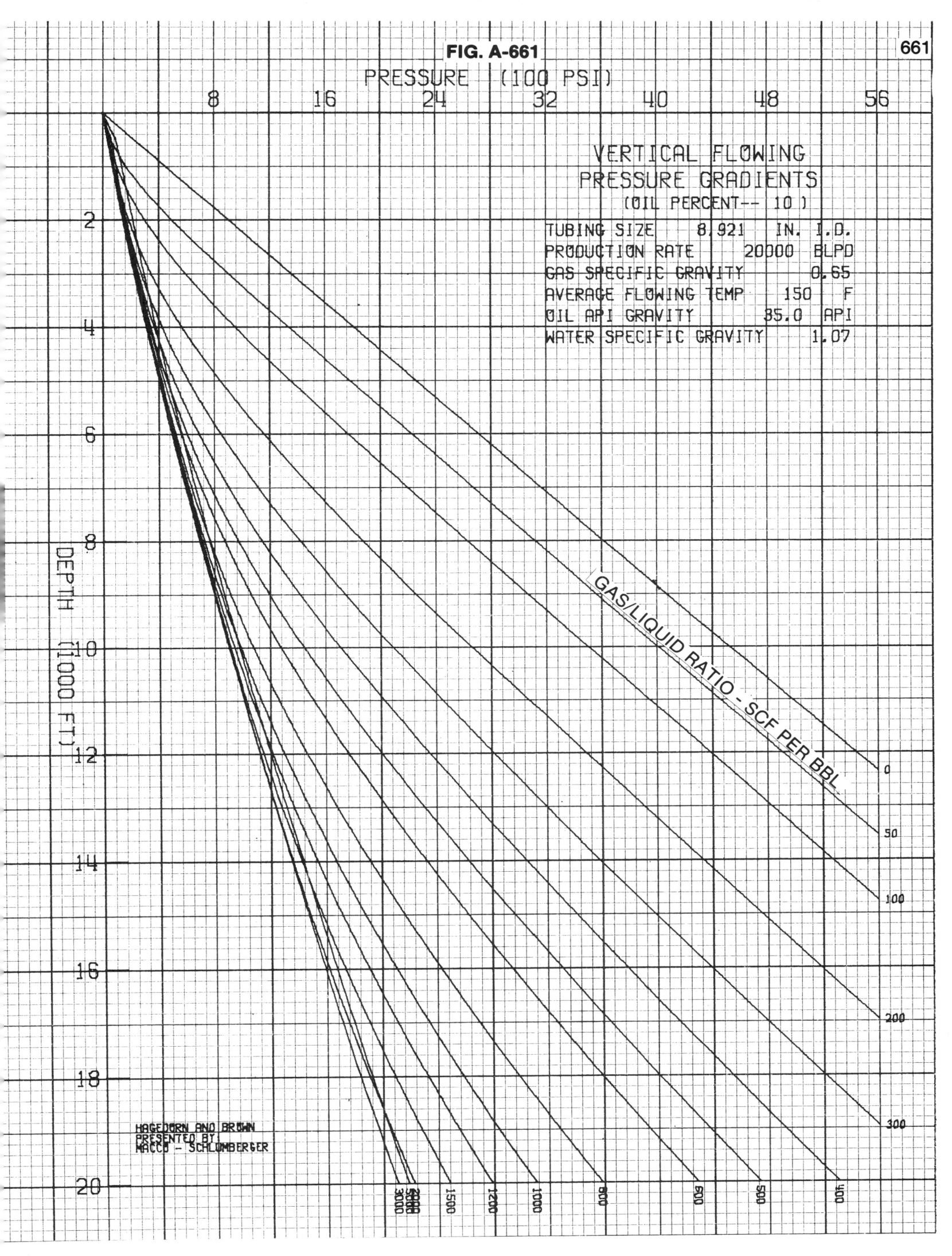

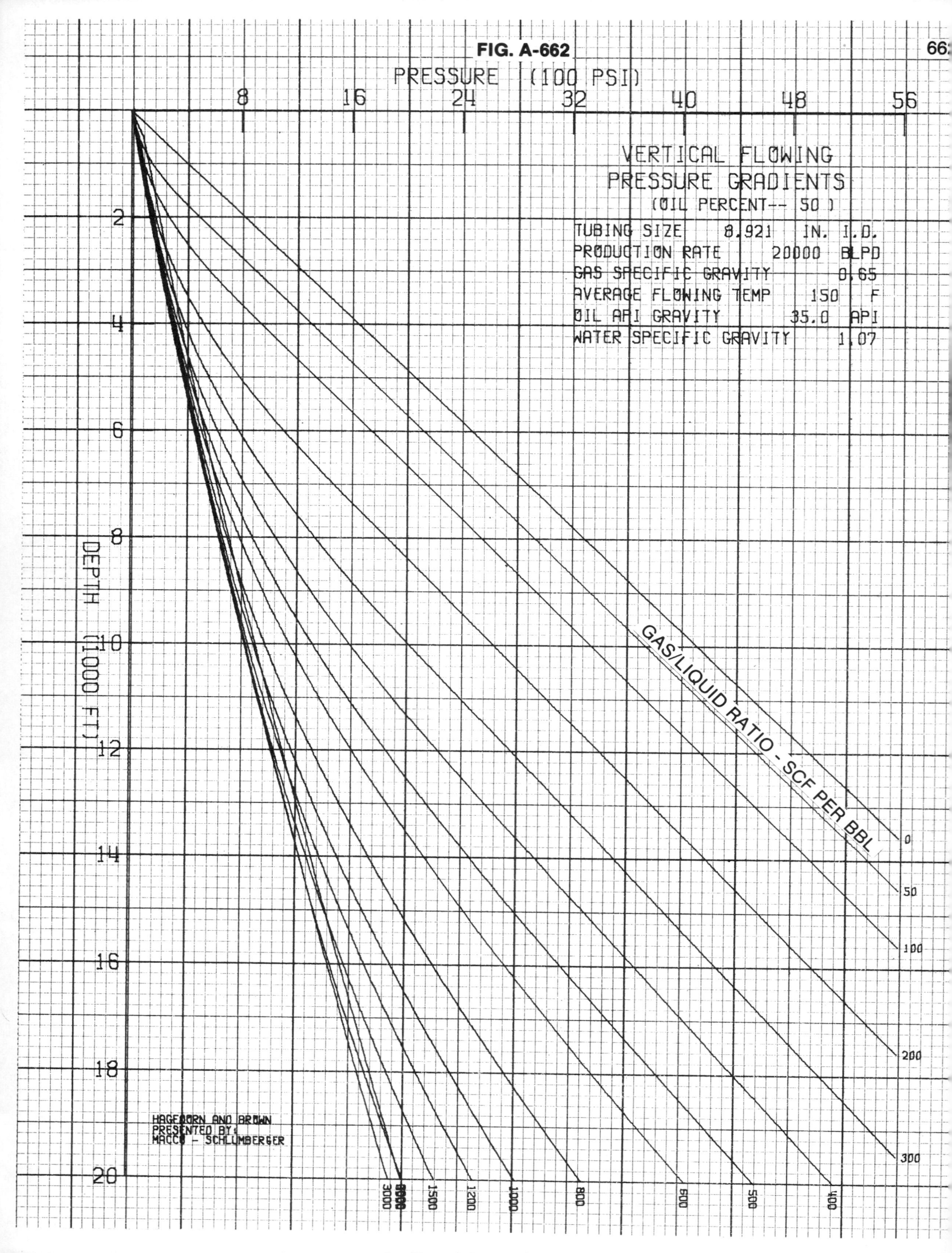

PRESSURE (100 PSI)

VERTICAL FLOWING
PRESSURE GRADIENTS
(OIL PERCENT-- 50)

TUBING SIZE	8.921	IN. I.D.
PRODUCTION RATE	20000	BLPD
GAS SPECIFIC GRAVITY	0.65	
AVERAGE FLOWING TEMP	150	F
OIL API GRAVITY	35.0	API
WATER SPECIFIC GRAVITY	1.07	

DEPTH (1000 FT)

GAS/LIQUID RATIO - SCF PER BBL

HAGEDORN AND BROWN
PRESENTED BY:
MACCO - SCHLUMBERGER

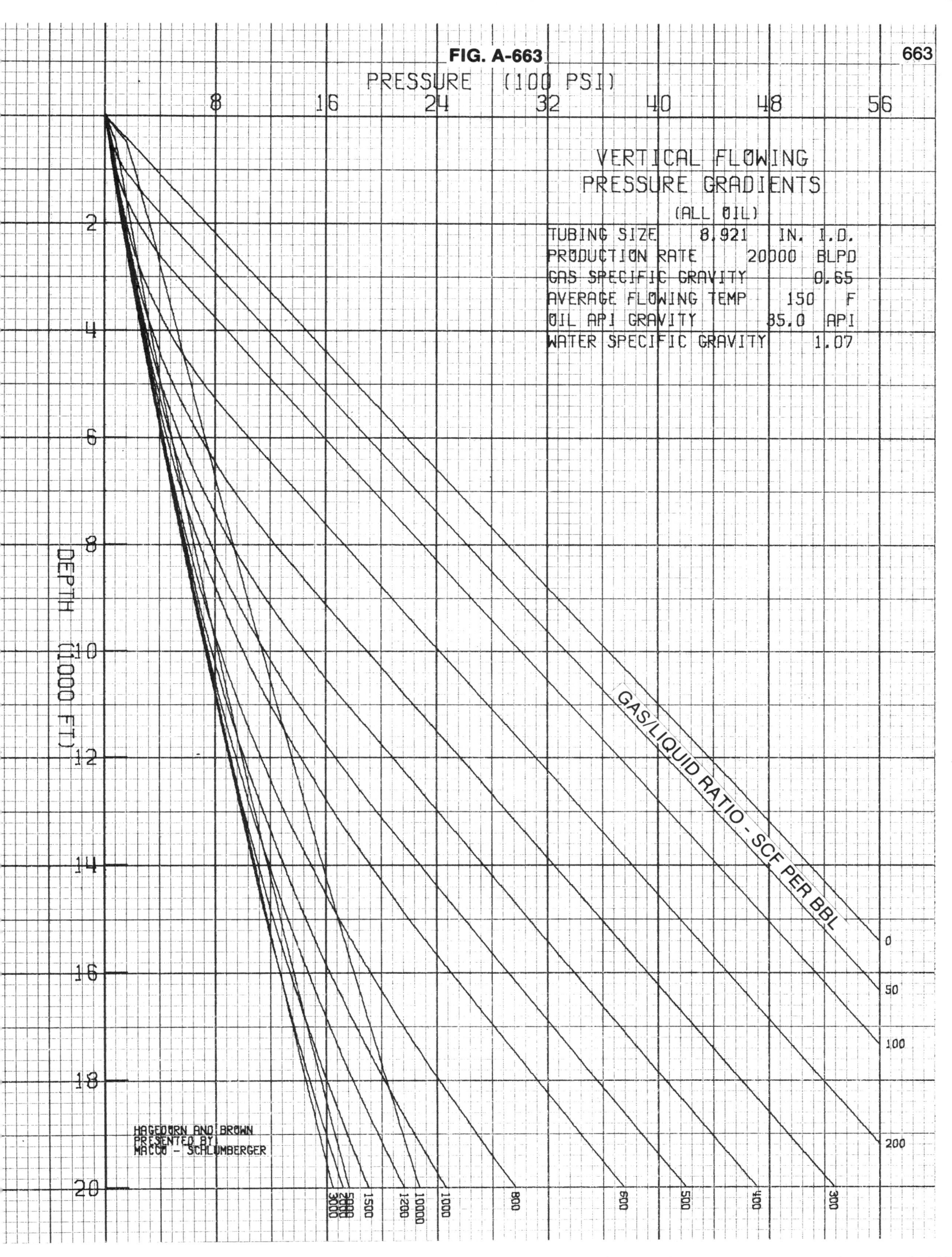

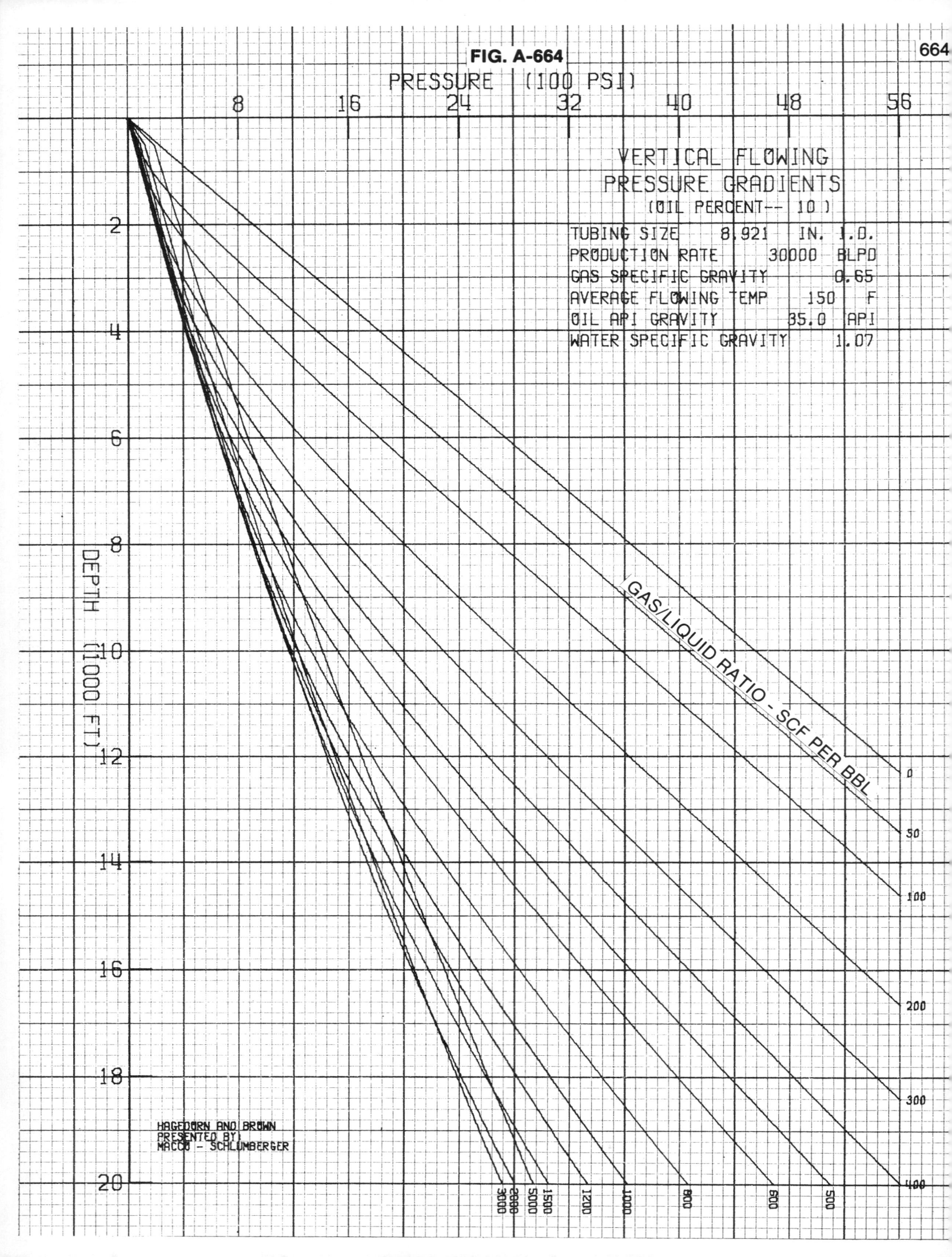

FIG. A-664

664

PRESSURE (100 PSI)

VERTICAL FLOWING
PRESSURE GRADIENTS
(OIL PERCENT-- 10)

TUBING SIZE	8.921	IN. I.D.
PRODUCTION RATE	30000	BLPD
GAS SPECIFIC GRAVITY	0.65	
AVERAGE FLOWING TEMP	150	F
OIL API GRAVITY	35.0	API
WATER SPECIFIC GRAVITY	1.07	

DEPTH (1000 FT)

GAS/LIQUID RATIO - SCF PER BBL

HAGEDORN AND BROWN
PRESENTED BY:
MACCO - SCHLUMBERGER

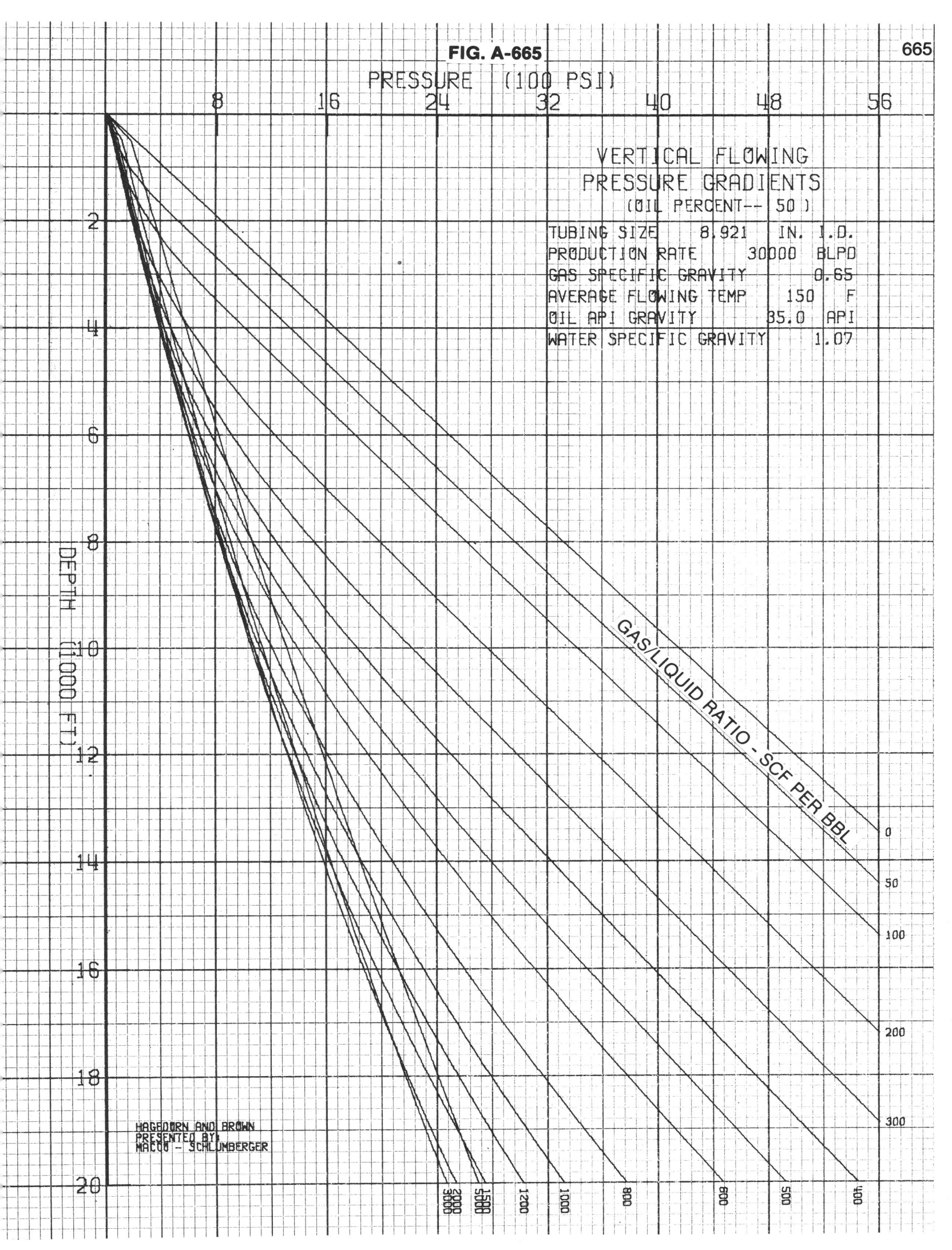

PRESSURE (100 PSI)

VERTICAL FLOWING
PRESSURE GRADIENTS
(OIL PERCENT-- 50)

TUBING SIZE	8.921	IN. I.D.
PRODUCTION RATE	30000	BLPD
GAS SPECIFIC GRAVITY		0.65
AVERAGE FLOWING TEMP	150	F
OIL API GRAVITY	35.0	API
WATER SPECIFIC GRAVITY		1.07

GAS/LIQUID RATIO - SCF PER BBL

DEPTH (1000 FT)

HAGEDORN AND BROWN
PRESENTED BY:
MAECO - SCHLUMBERGER

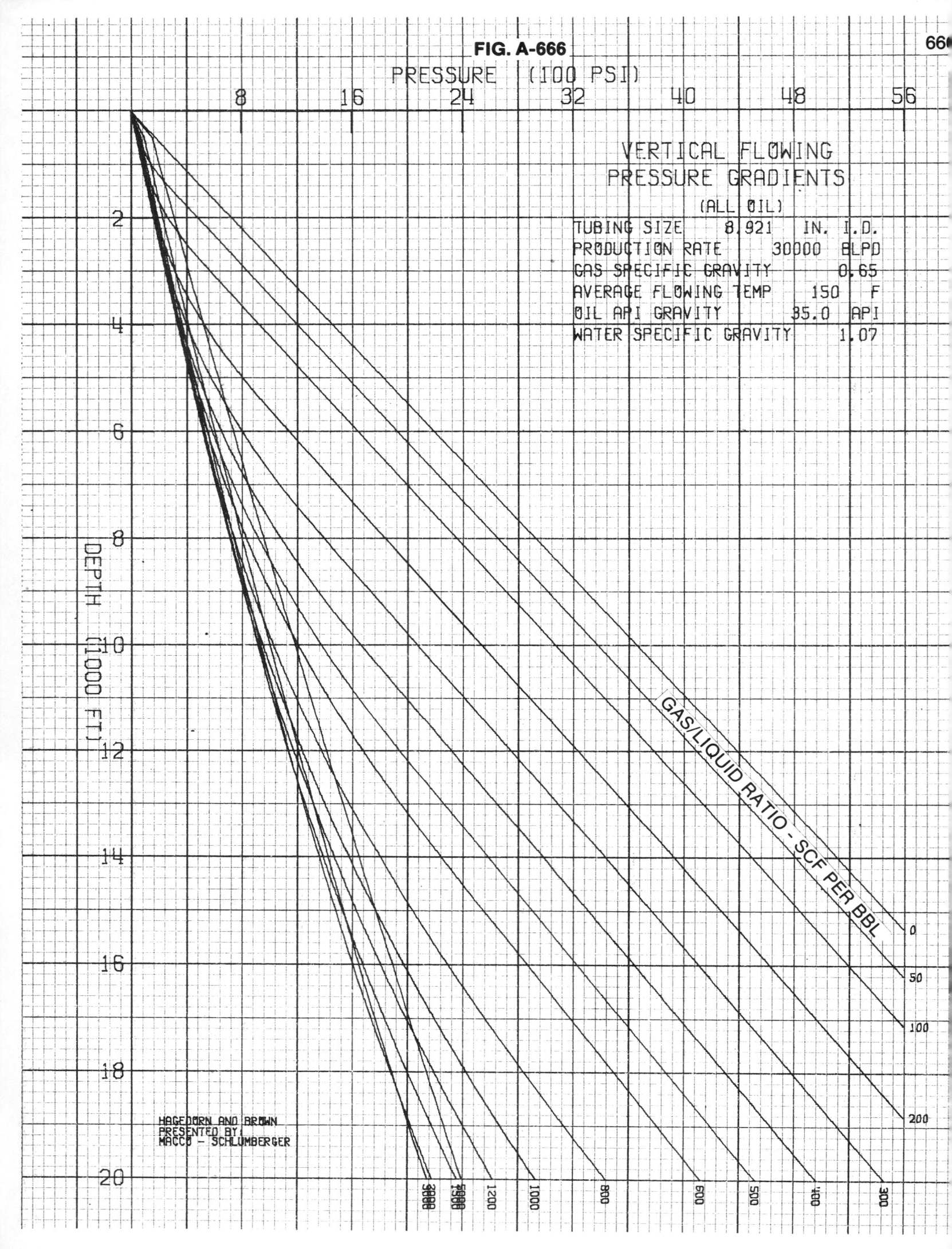

VERTICAL FLOWING
PRESSURE GRADIENTS
(ALL OIL)

TUBING SIZE	8.921	IN. I.D.
PRODUCTION RATE	30000	BLPD
GAS SPECIFIC GRAVITY	0.65	
AVERAGE FLOWING TEMP	150	F
OIL API GRAVITY	35.0	API
WATER SPECIFIC GRAVITY	1.07	

PRESSURE (100 PSI)

DEPTH (1000 FT)

GAS/LIQUID RATIO - SCF PER BBL

HAGEDORN AND BROWN
PRESENTED BY
MACCO - SCHLUMBERGER

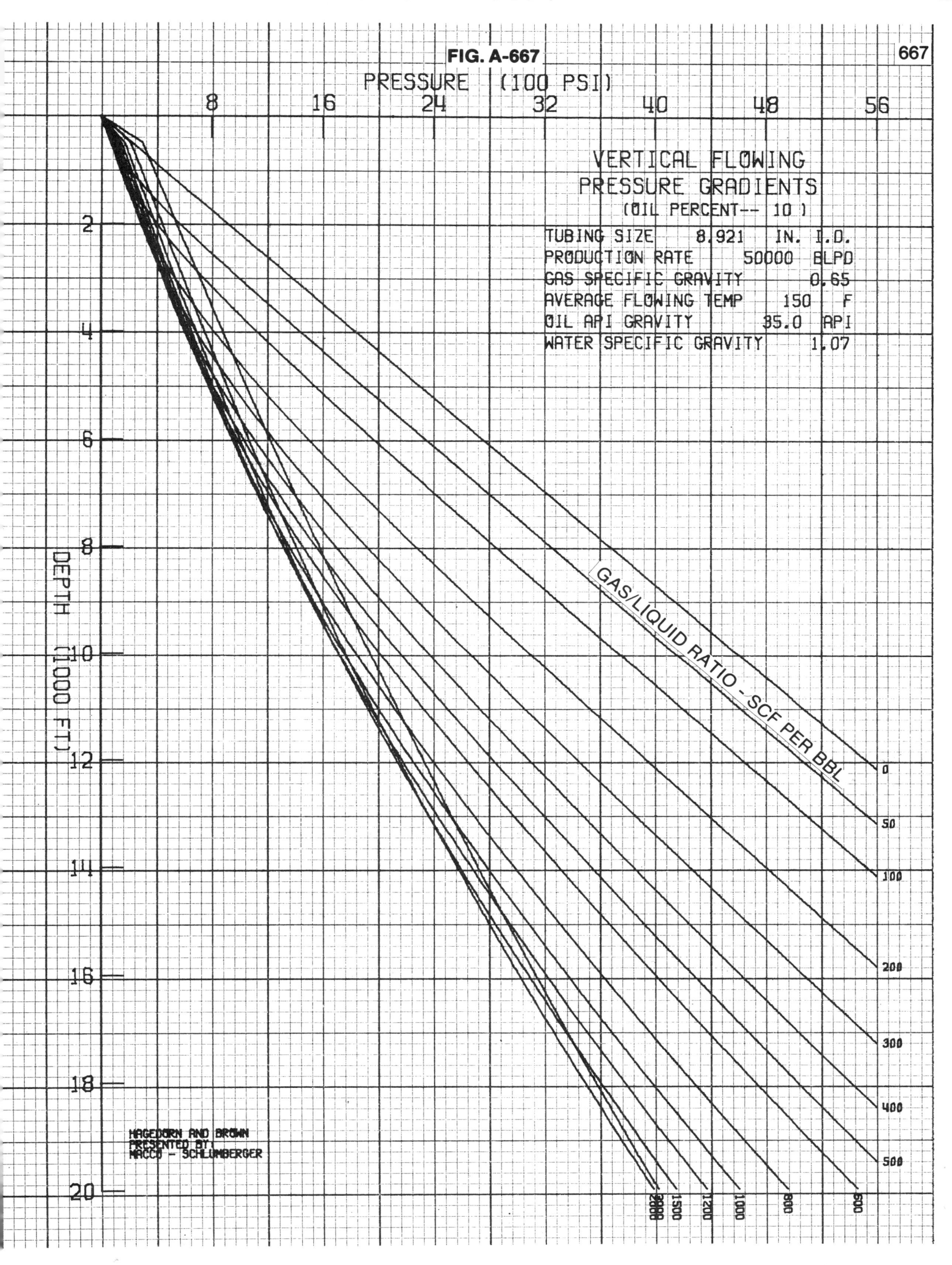

FIG. A-667

667

VERTICAL FLOWING
PRESSURE GRADIENTS
(OIL PERCENT-- 10)

TUBING SIZE	8.921	IN. I.D.
PRODUCTION RATE	50000	BLPD
GAS SPECIFIC GRAVITY	0.65	
AVERAGE FLOWING TEMP	150	F
OIL API GRAVITY	35.0	API
WATER SPECIFIC GRAVITY	1.07	

PRESSURE (100 PSI)

DEPTH (1000 FT)

GAS/LIQUID RATIO - SCF PER BBL

HAGEDORN AND BROWN
PRESENTED BY
MACCO - SCHLUMBERGER

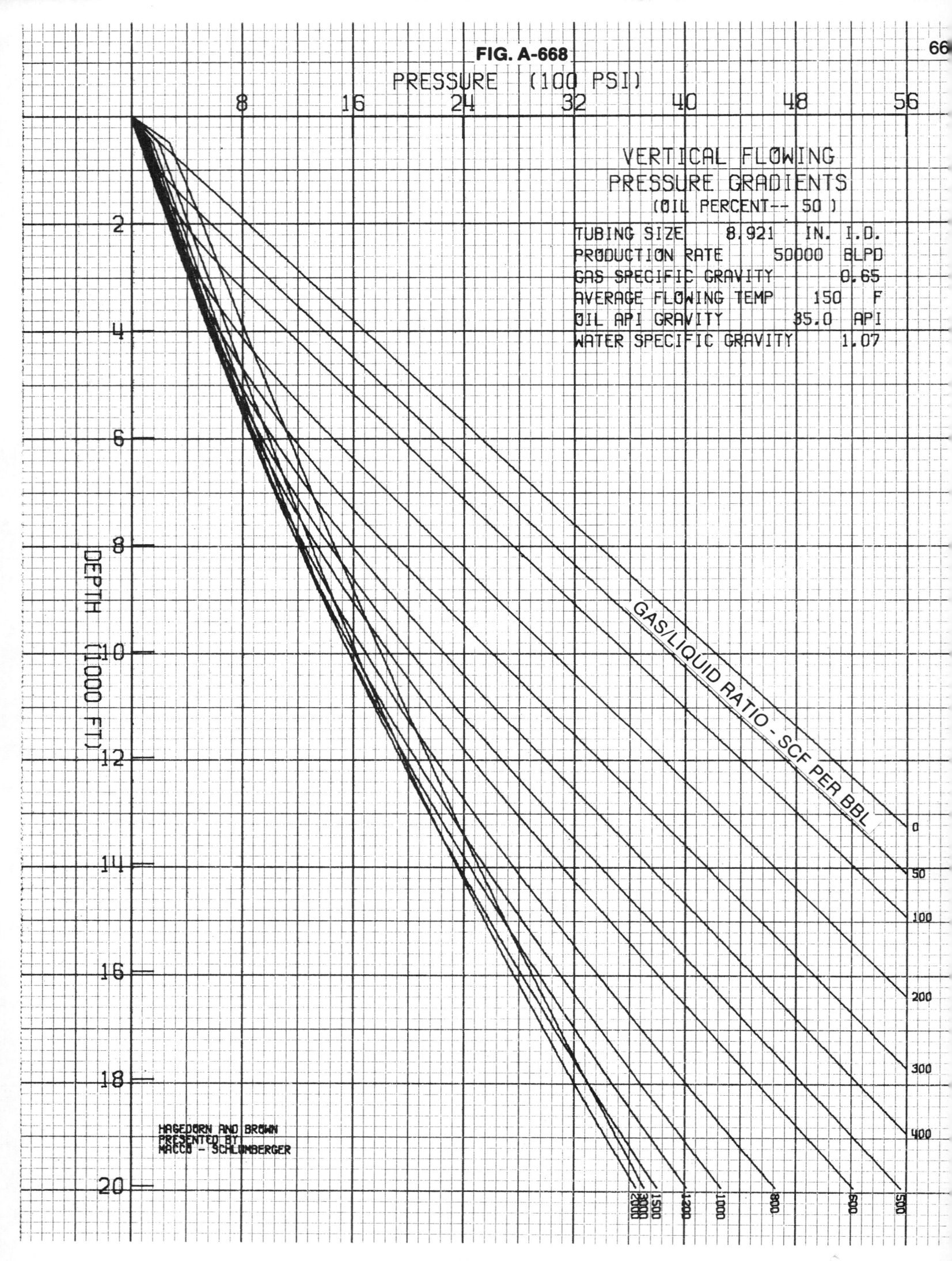

FIG. A-668

VERTICAL FLOWING
PRESSURE GRADIENTS
(OIL PERCENT-- 50)

TUBING SIZE	8.921	IN. I.D.
PRODUCTION RATE	50000	BLPD
GAS SPECIFIC GRAVITY	0.65	
AVERAGE FLOWING TEMP	150	F
OIL API GRAVITY	35.0	API
WATER SPECIFIC GRAVITY	1.07	

PRESSURE (100 PSI)

DEPTH (1000 FT)

GAS/LIQUID RATIO - SCF PER BBL

HAGEDORN AND BROWN
PRESENTED BY
MACCO - SCHLUMBERGER

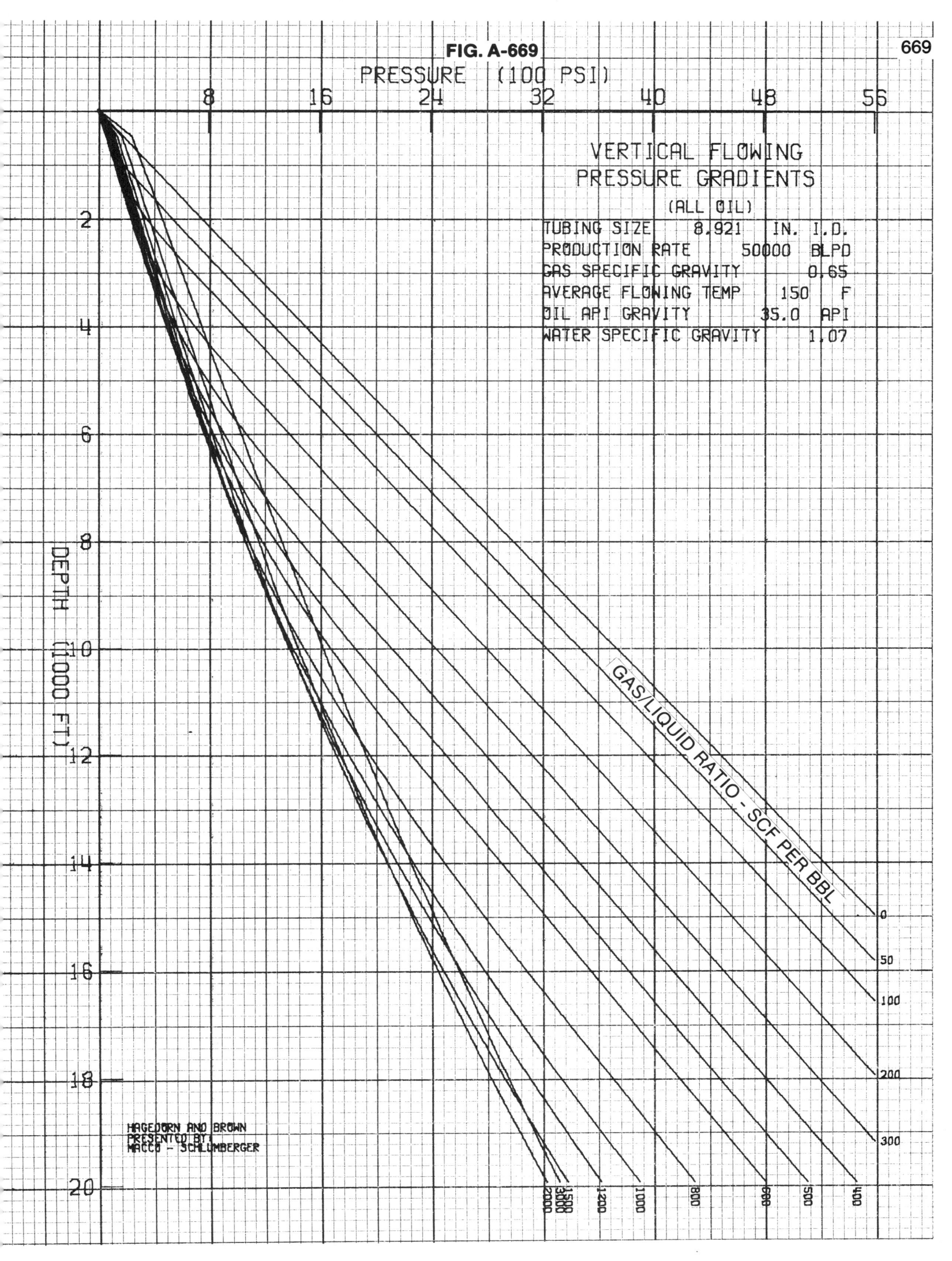

PRESSURE (100 PSI)

VERTICAL FLOWING
PRESSURE GRADIENTS
(ALL OIL)

TUBING SIZE	8.921	IN. I.D.
PRODUCTION RATE	50000	BLPD
GAS SPECIFIC GRAVITY	0.65	
AVERAGE FLOWING TEMP	150	F
OIL API GRAVITY	35.0	API
WATER SPECIFIC GRAVITY	1.07	

DEPTH (1000 FT)

GAS/LIQUID RATIO - SCF PER BBL

HAGEDORN AND BROWN
PRESENTED BY:
MACCO - SCHLUMBERGER

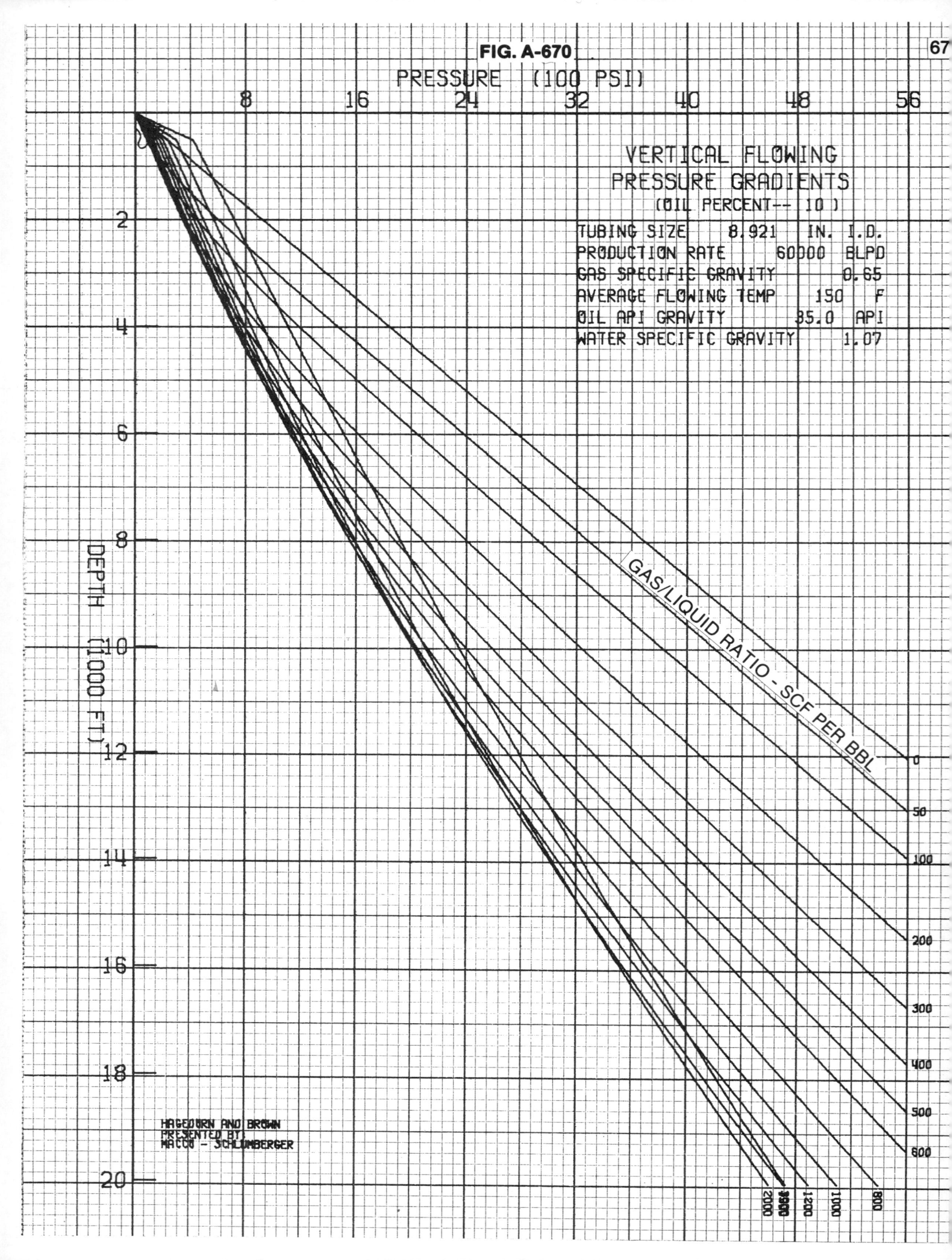

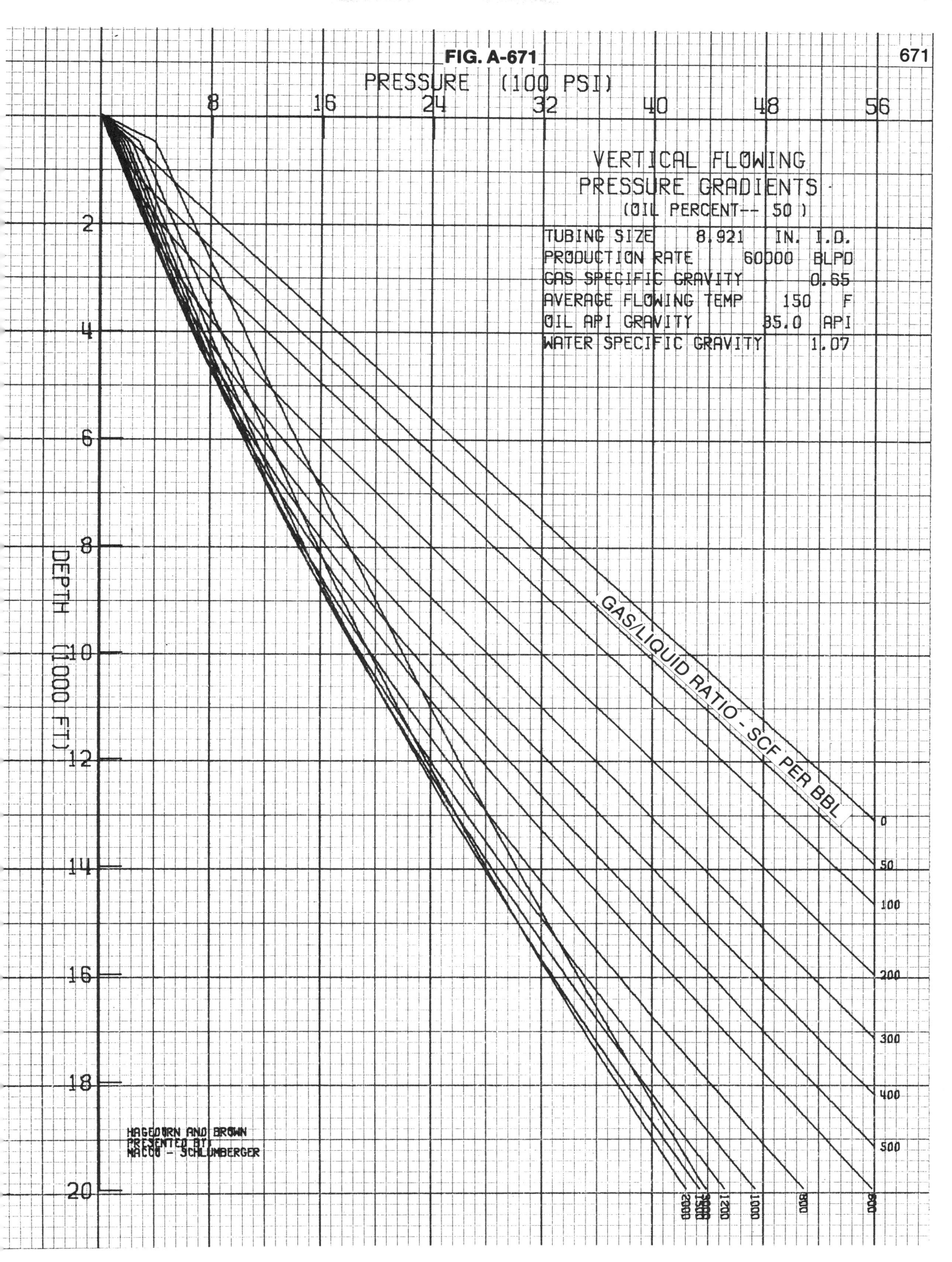

FIG. A-671

671

VERTICAL FLOWING
PRESSURE GRADIENTS
(OIL PERCENT-- 50)

TUBING SIZE	8.921	IN. I.D.
PRODUCTION RATE	60000	BLPD
GAS SPECIFIC GRAVITY	0.65	
AVERAGE FLOWING TEMP	150	F
OIL API GRAVITY	35.0	API
WATER SPECIFIC GRAVITY	1.07	

PRESSURE (100 PSI)

DEPTH (1,000 FT)

GAS/LIQUID RATIO - SCF PER BBL

HAGEDORN AND BROWN
PRESENTED BY
NACCO - SCHLUMBERGER

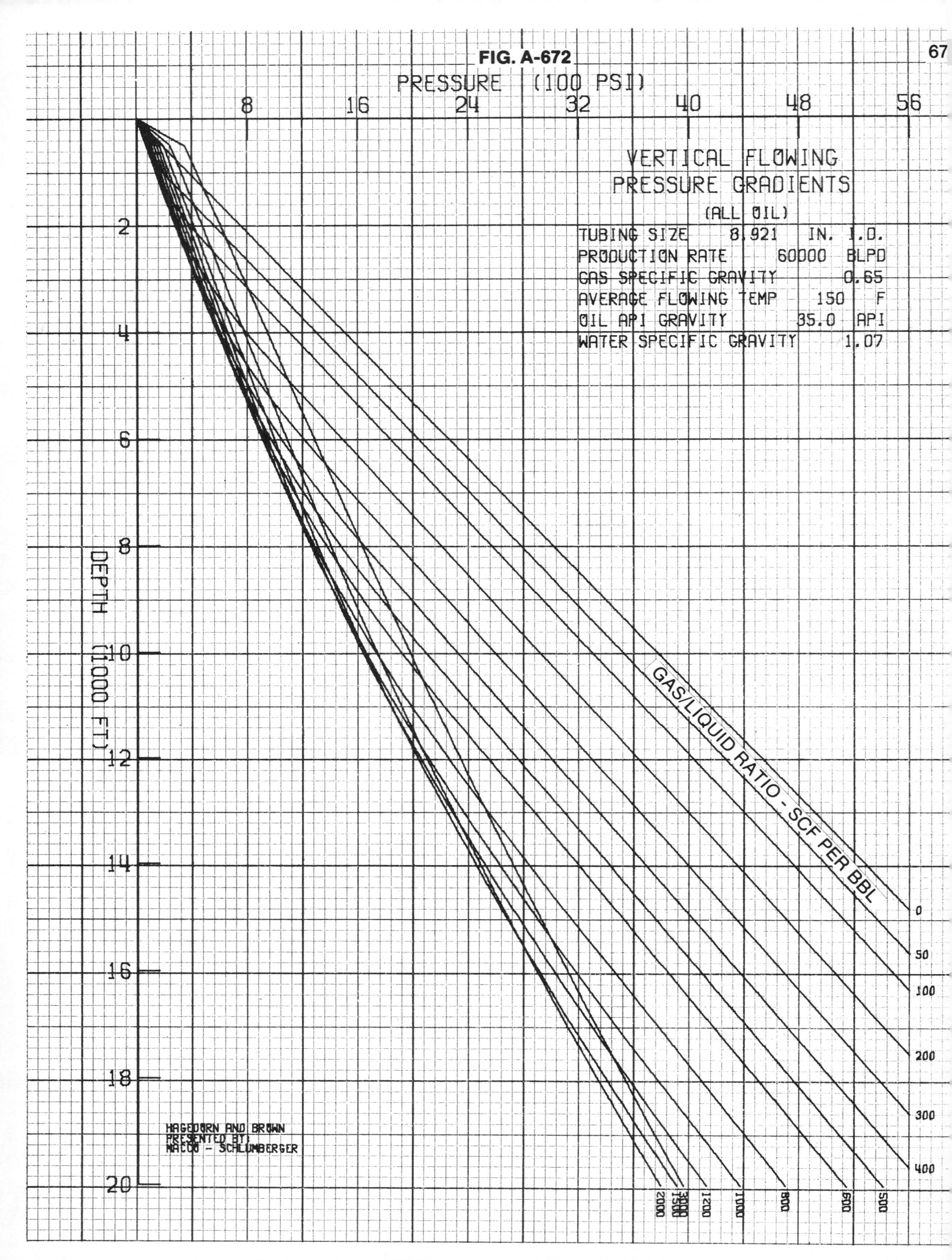

VERTICAL FLOWING
PRESSURE GRADIENTS
(ALL OIL)

TUBING SIZE	8.921	IN. I.D.
PRODUCTION RATE	60000	BLPD
GAS SPECIFIC GRAVITY	0.65	
AVERAGE FLOWING TEMP	150	F
OIL API GRAVITY	35.0	API
WATER SPECIFIC GRAVITY	1.07	

PRESSURE (100 PSI)

DEPTH (1000 FT)

GAS/LIQUID RATIO - SCF PER BBL

HAGEDORN AND BROWN
PRESENTED BY:
NACCO - SCHLUMBERGER

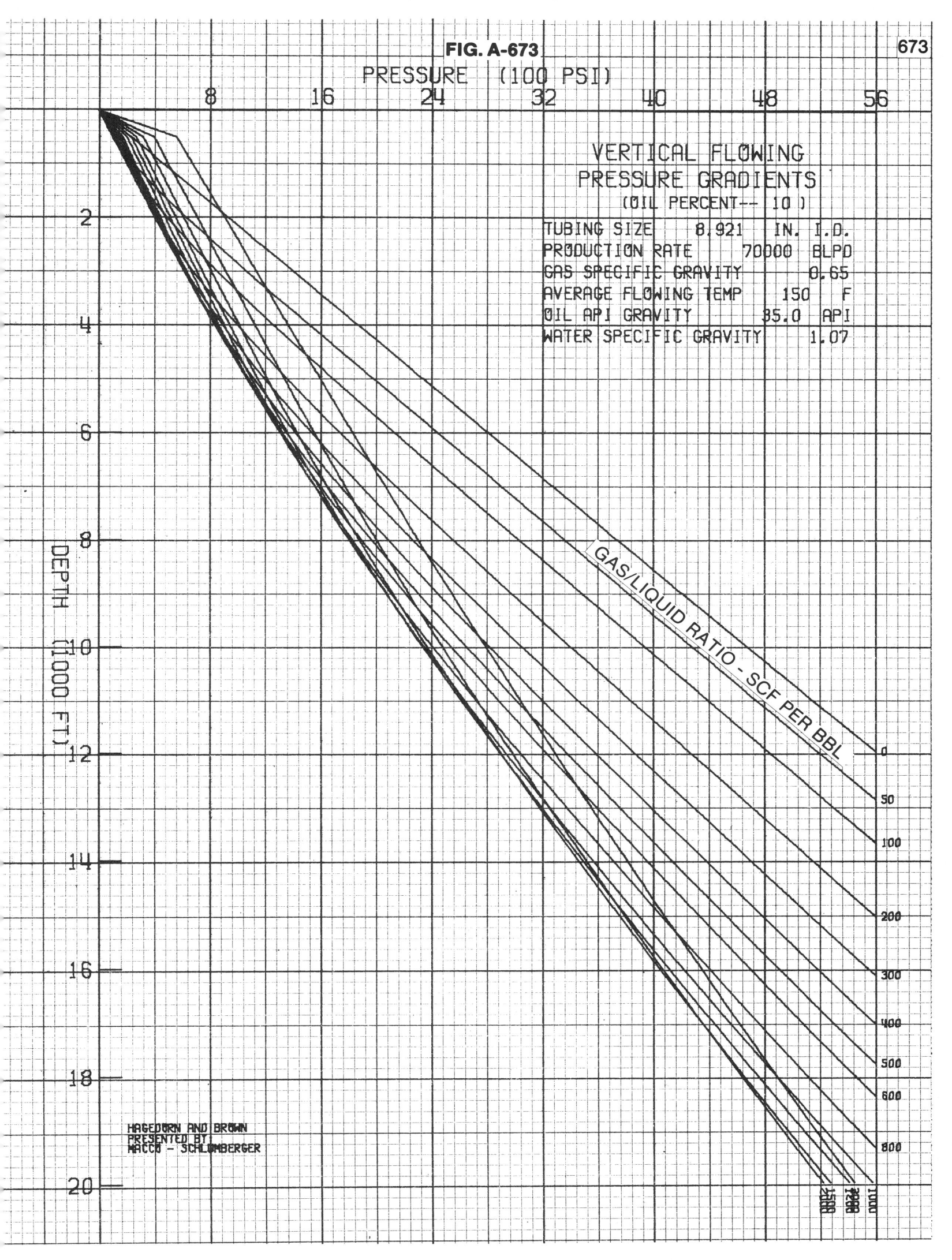

FIG. A-673

673

PRESSURE (100 PSI)

VERTICAL FLOWING
PRESSURE GRADIENTS
(OIL PERCENT-- 10)

TUBING SIZE   8.921   IN. I.D.
PRODUCTION RATE   70000   BLPD
GAS SPECIFIC GRAVITY   0.65
AVERAGE FLOWING TEMP   150   F
OIL API GRAVITY   35.0   API
WATER SPECIFIC GRAVITY   1.07

GAS/LIQUID RATIO - SCF PER BBL

DEPTH (1000 FT)

HAGEDORN AND BROWN
PRESENTED BY
MACCO - SCHLUMBERGER

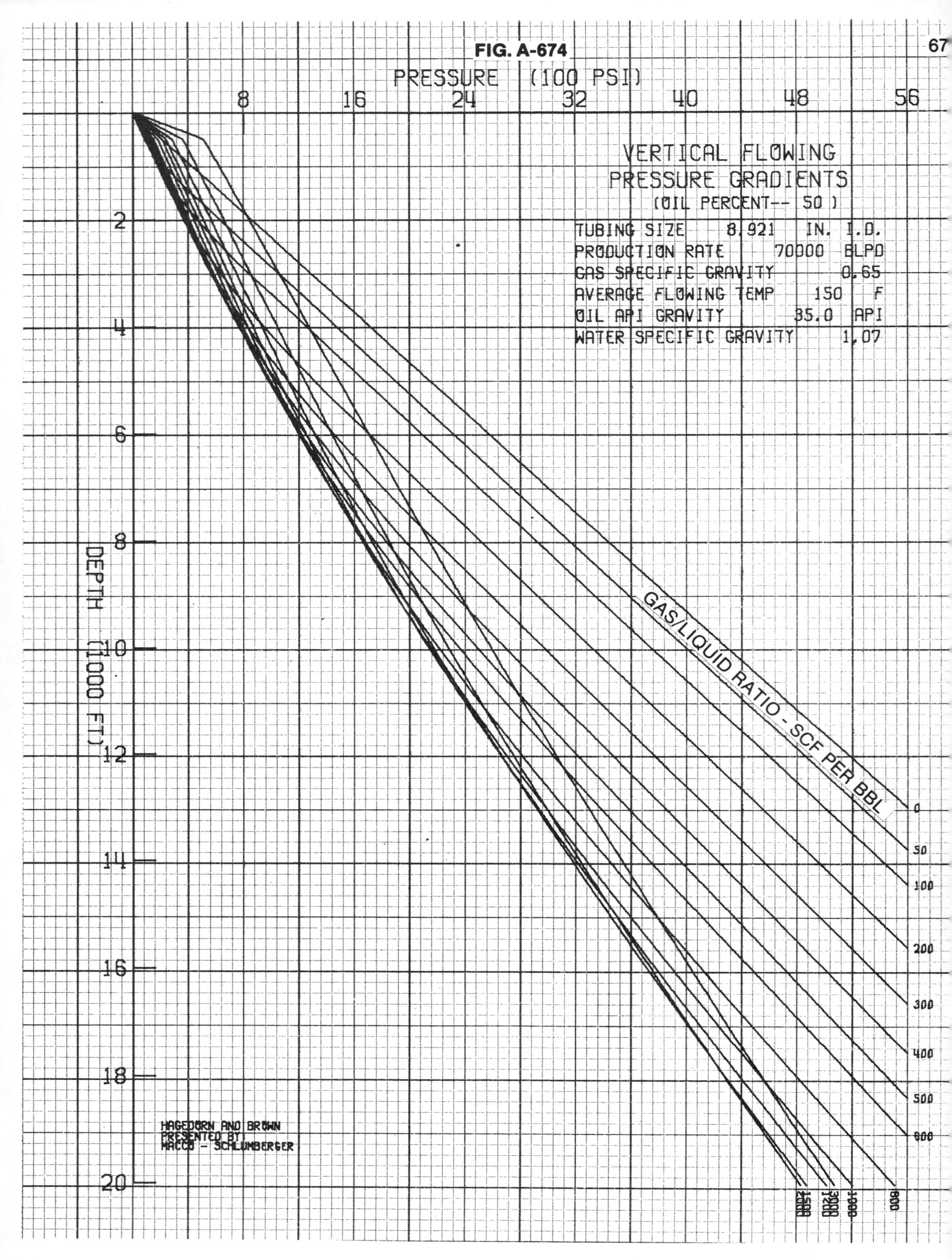

FIG. A-674

PRESSURE (100 PSI)

VERTICAL FLOWING
PRESSURE GRADIENTS
(OIL PERCENT-- 50)

TUBING SIZE	8.921	IN. I.D.
PRODUCTION RATE	70000	BLPD
GAS SPECIFIC GRAVITY	0.65	
AVERAGE FLOWING TEMP	150	F
OIL API GRAVITY	35.0	API
WATER SPECIFIC GRAVITY	1.07	

DEPTH (1000 FT)

GAS/LIQUID RATIO - SCF PER BBL

HAGEDORN AND BROWN
PRESENTED BY
MACCO — SCHLUMBERGER

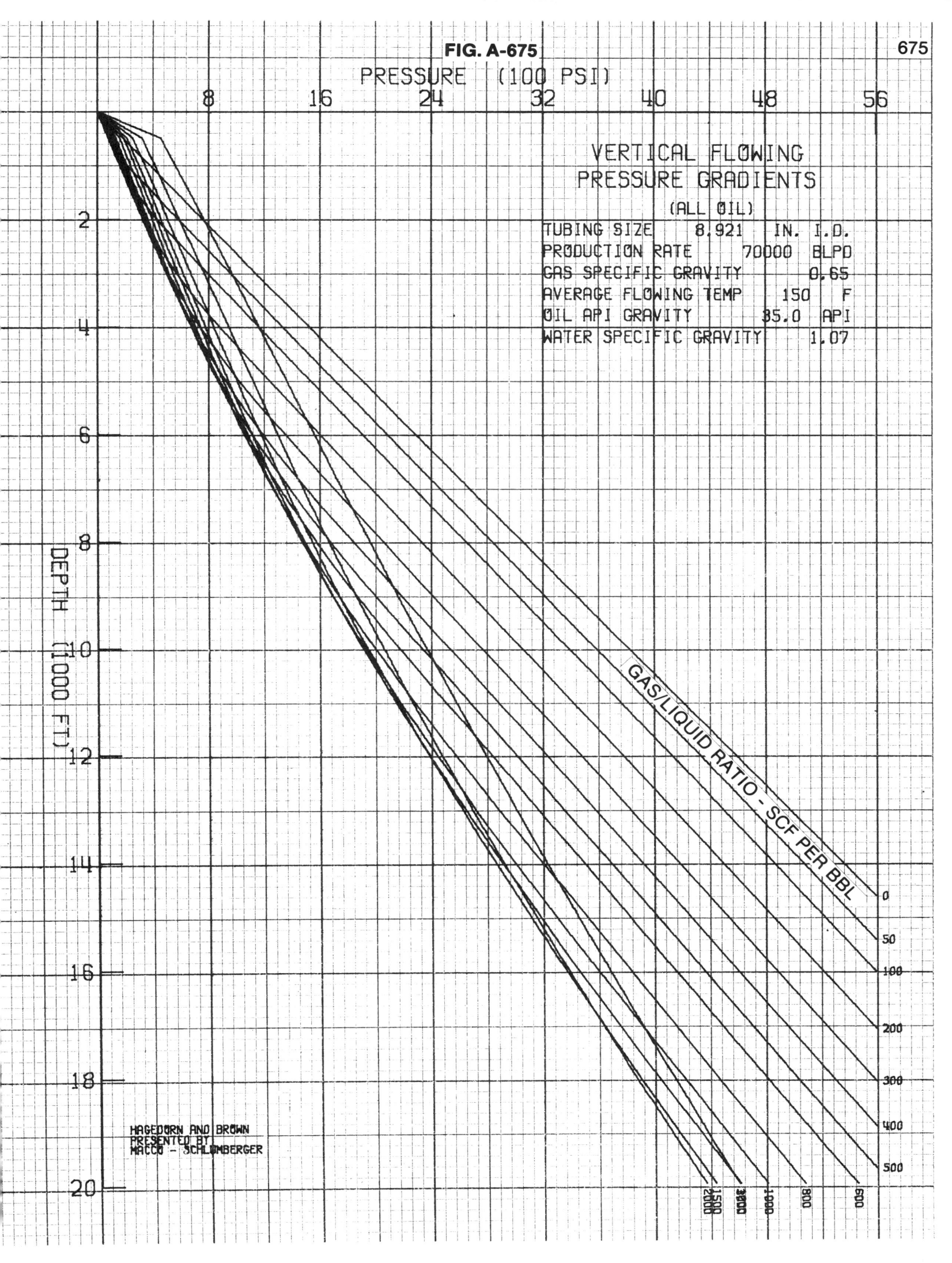

PRESSURE (100 PSI)

VERTICAL FLOWING
PRESSURE GRADIENTS
(ALL OIL)

TUBING SIZE	8.921	IN. I.D.
PRODUCTION RATE	70000	BLPD
GAS SPECIFIC GRAVITY		0.65
AVERAGE FLOWING TEMP	150	F
OIL API GRAVITY	35.0	API
WATER SPECIFIC GRAVITY		1.07

DEPTH (1000 FT)

GAS/LIQUID RATIO - SCF PER BBL

HAGEDORN AND BROWN
PRESENTED BY
MAECO - SCHLUMBERGER

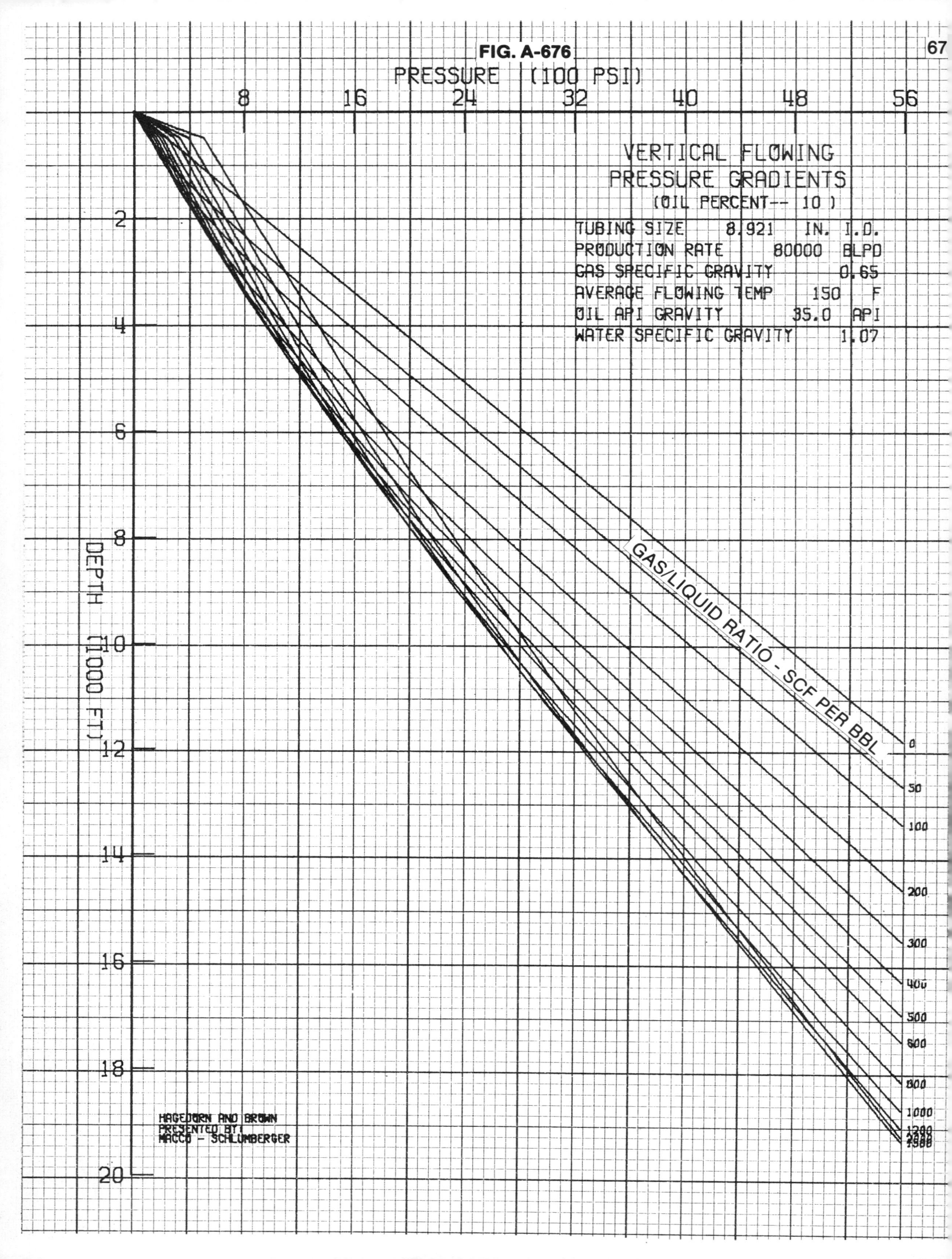

PRESSURE (100 PSI)

VERTICAL FLOWING
PRESSURE GRADIENTS
(OIL PERCENT-- 10)

TUBING SIZE	8.921	IN. I.D.
PRODUCTION RATE	80000	BLPD
GAS SPECIFIC GRAVITY		0.65
AVERAGE FLOWING TEMP	150	F
OIL API GRAVITY	35.0	API
WATER SPECIFIC GRAVITY		1.07

GAS/LIQUID RATIO - SCF PER BBL

DEPTH (1000 FT)

HAGEDORN AND BROWN
PRESENTED BY:
MACCO – SCHLUMBERGER

0
50
100
200
300
400
500
600
800
1000
1200
1500

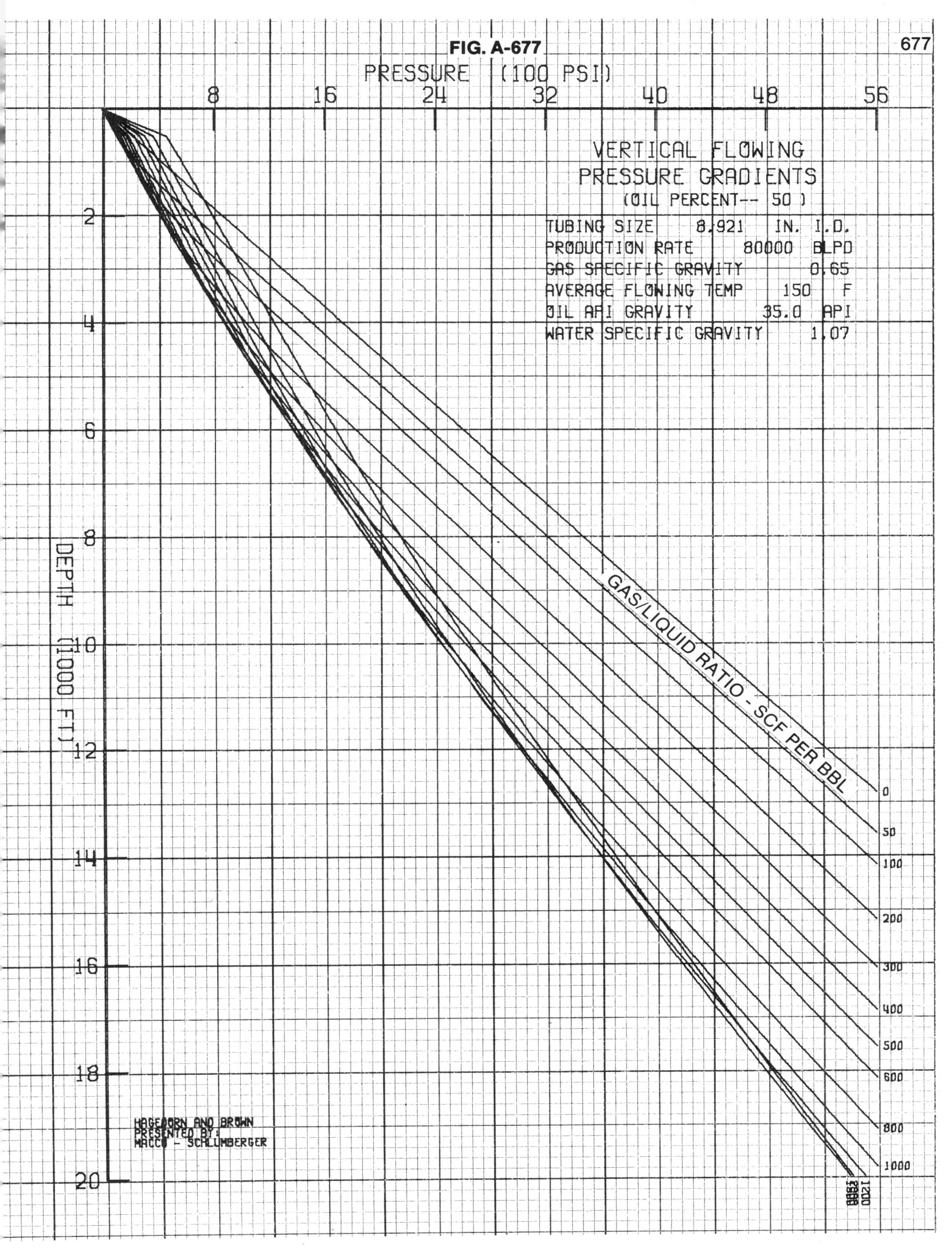

PRESSURE (100 PSI)

VERTICAL FLOWING
PRESSURE GRADIENTS
(OIL PERCENT-- 50)

TUBING SIZE	8.921	IN. I.D.
PRODUCTION RATE	80000	BLPD
GAS SPECIFIC GRAVITY	0.65	
AVERAGE FLOWING TEMP	150	F
OIL API GRAVITY	35.0	API
WATER SPECIFIC GRAVITY	1.07	

DEPTH (1000 FT)

GAS/LIQUID RATIO - SCF PER BBL

0
50
100
200
300
400
500
600
800
1000
1200

HAGEDORN AND BROWN
PRESENTED BY:
MACCO - SCHLUMBERGER

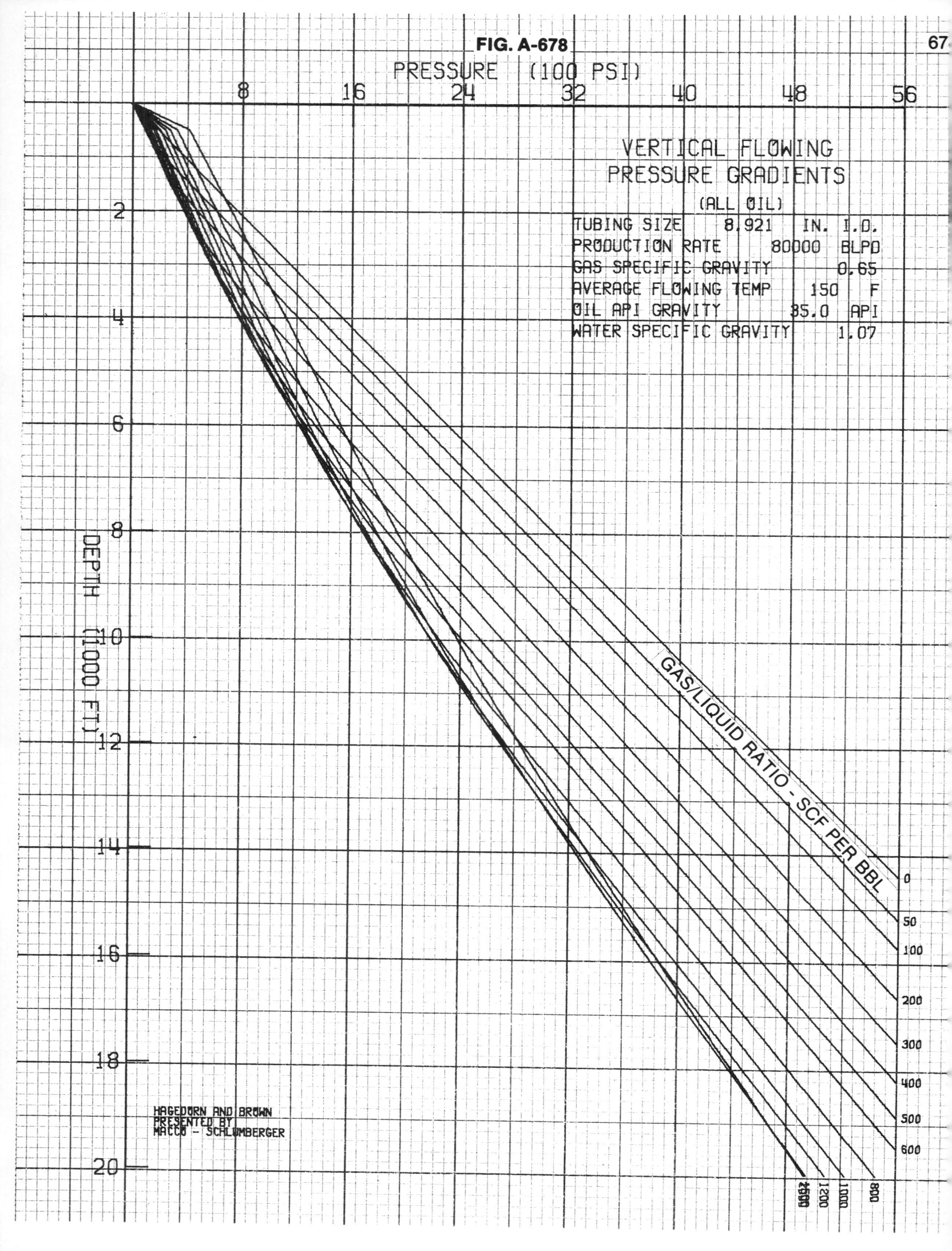

PRESSURE (100 PSI)

VERTICAL FLOWING
PRESSURE GRADIENTS
(ALL OIL)

TUBING SIZE	8.921	IN. I.D.
PRODUCTION RATE	80000	BLPD
GAS SPECIFIC GRAVITY	0.65	
AVERAGE FLOWING TEMP	150	F
OIL API GRAVITY	35.0	API
WATER SPECIFIC GRAVITY	1.07	

DEPTH (1000 FT)

GAS/LIQUID RATIO - SCF PER BBL

HAGEDORN AND BROWN
PRESENTED BY
MACCO - SCHLUMBERGER

0
50
100
200
300
400
500
600

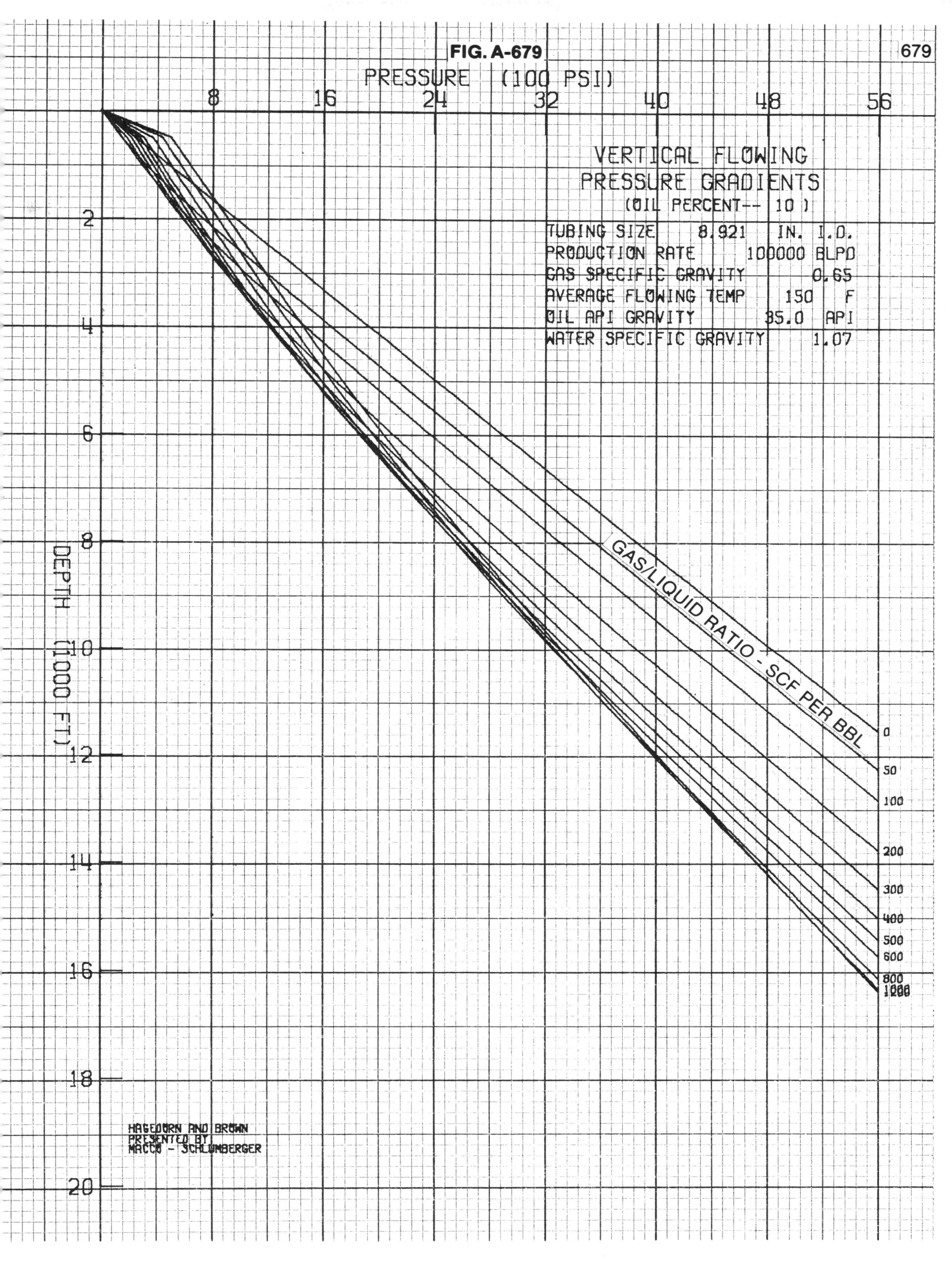

FIG. A-679

679

PRESSURE (100 PSI)

VERTICAL FLOWING
PRESSURE GRADIENTS
(OIL PERCENT-- 10)

TUBING SIZE	8.921	IN. I.D.
PRODUCTION RATE		100000 BLPD
GAS SPECIFIC GRAVITY		0.65
AVERAGE FLOWING TEMP	150	F
OIL API GRAVITY	35.0	API
WATER SPECIFIC GRAVITY		1.07

DEPTH (1000 FT)

GAS/LIQUID RATIO - SCF PER BBL

0
50
100
200
300
400
500
600
800
1000

HAGEDORN AND BROWN
PRESENTED BY
MACCO - SCHLUMBERGER

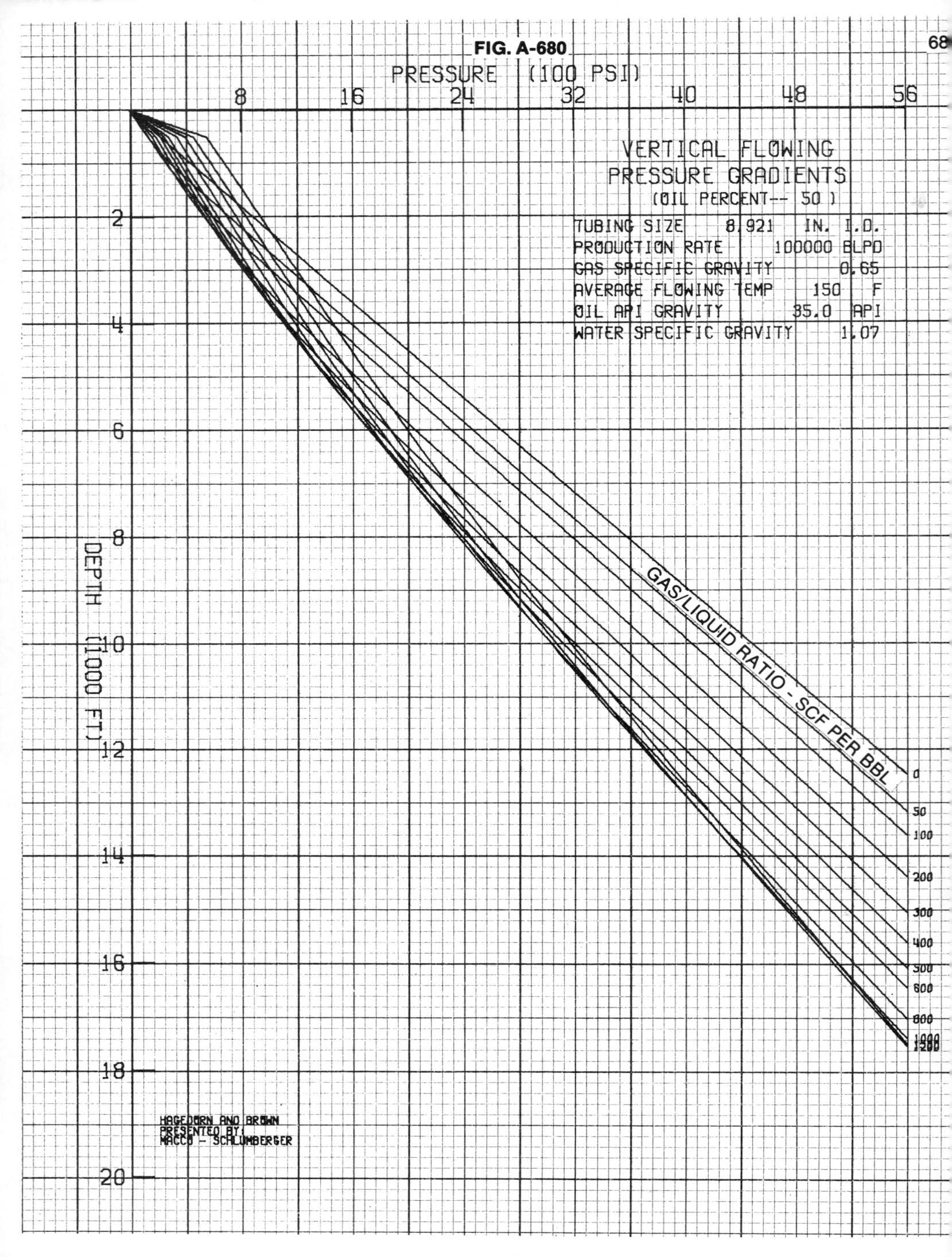

FIG. A-680

VERTICAL FLOWING
PRESSURE GRADIENTS
(OIL PERCENT-- 50)

TUBING SIZE	8.921	IN. I.D.
PRODUCTION RATE	100000	BLPD
GAS SPECIFIC GRAVITY	0.65	
AVERAGE FLOWING TEMP	150	F
OIL API GRAVITY	35.0	API
WATER SPECIFIC GRAVITY	1.07	

PRESSURE (100 PSI)

DEPTH (1000 FT)

GAS/LIQUID RATIO - SCF PER BBL

0
50
100
200
300
400
500
600
800
1000
1200

HAGEDORN AND BROWN
PRESENTED BY
KACCO — SCHLUMBERGER

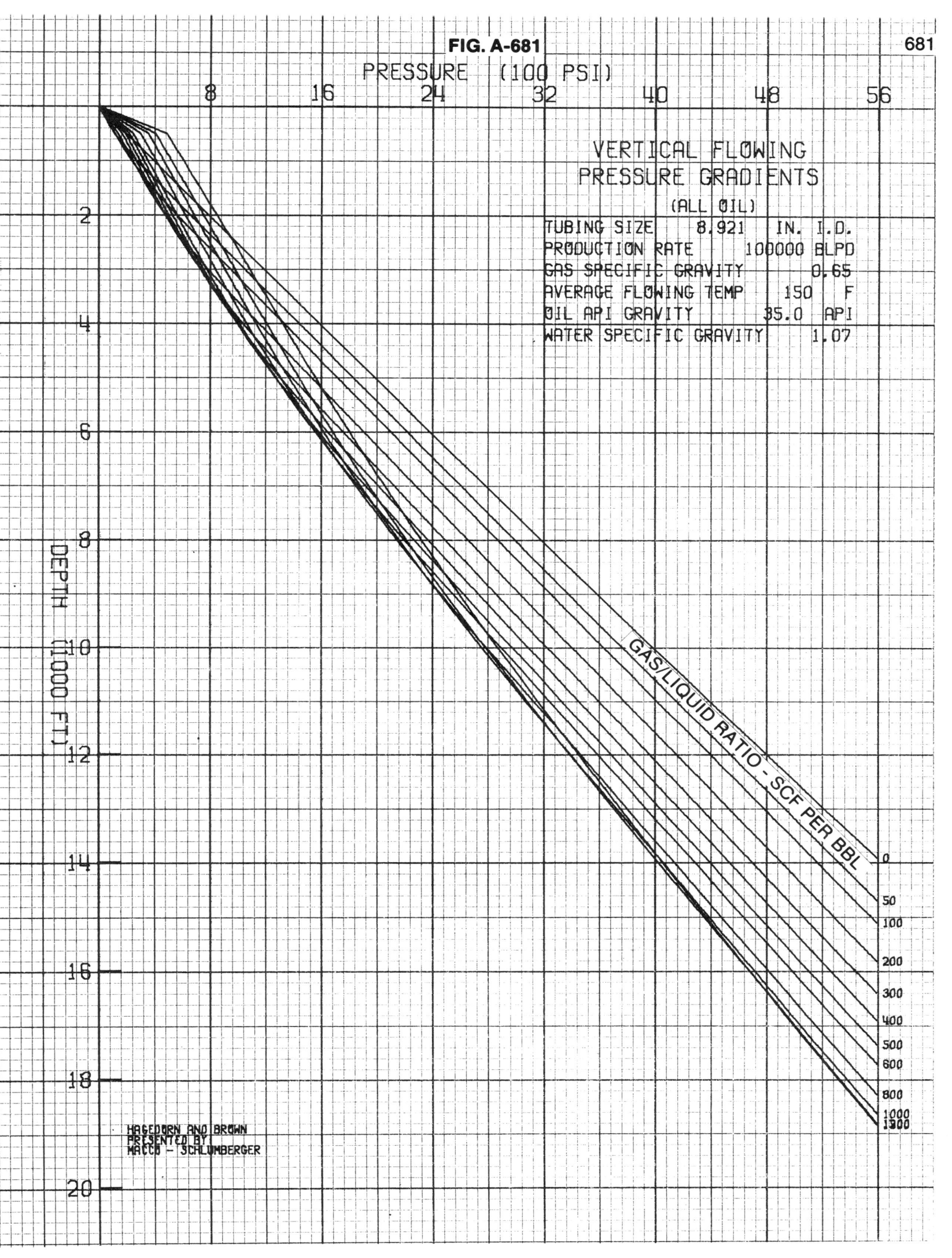

FIG. A-681

681

VERTICAL FLOWING
PRESSURE GRADIENTS
(ALL OIL)

TUBING SIZE	8.921	IN. I.D.
PRODUCTION RATE	100000	BLPD
GAS SPECIFIC GRAVITY	0.65	
AVERAGE FLOWING TEMP	150	F
OIL API GRAVITY	35.0	API
WATER SPECIFIC GRAVITY	1.07	

PRESSURE (100 PSI)

DEPTH (1000 FT)

GAS/LIQUID RATIO - SCF PER BBL

HAGEDORN AND BROWN
PRESENTED BY
MACCO - SCHLUMBERGER

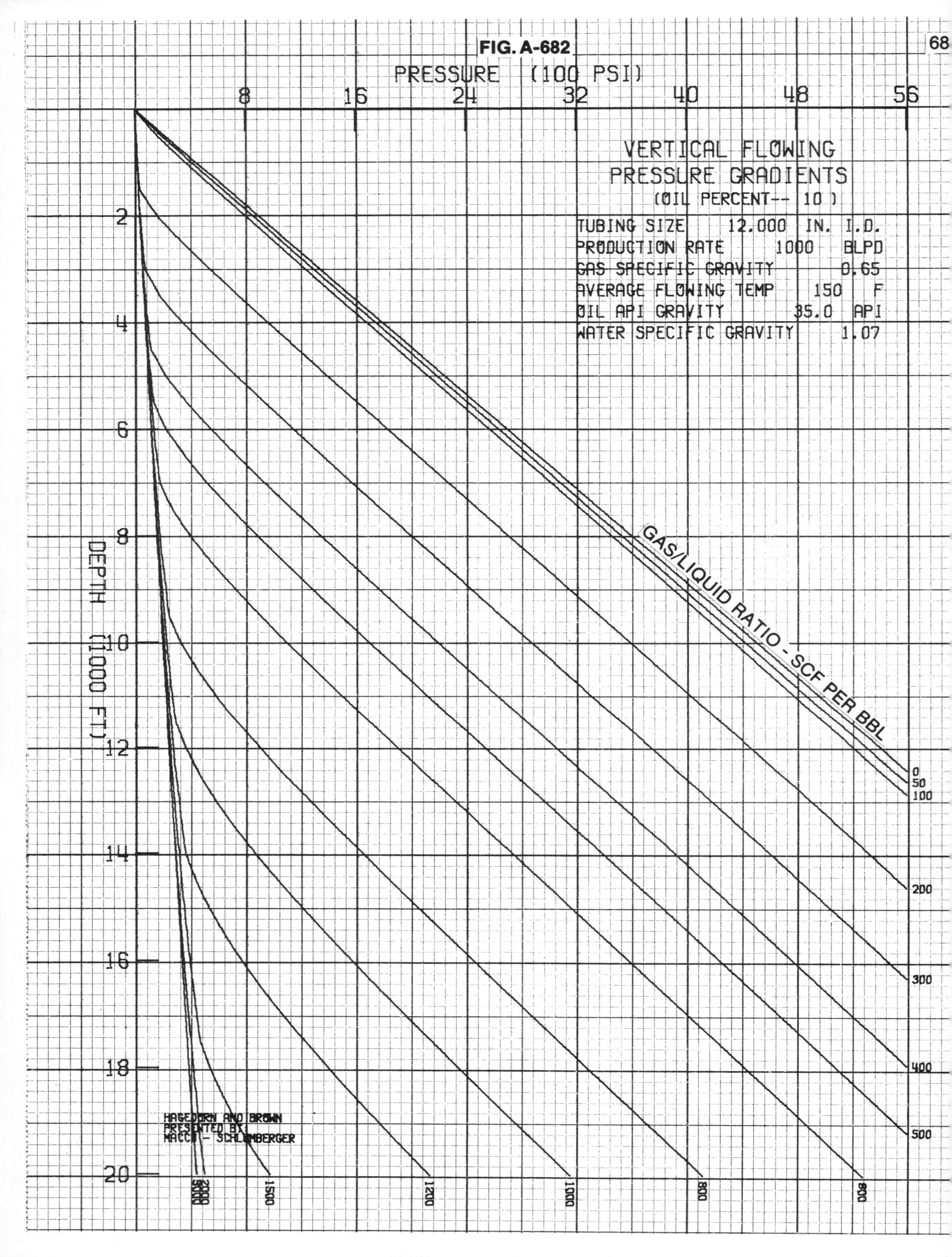

FIG. A-682

68

PRESSURE (100 PSI)

VERTICAL FLOWING
PRESSURE GRADIENTS
(OIL PERCENT-- 10)

TUBING SIZE	12.000	IN. I.D.
PRODUCTION RATE	1000	BLPD
GAS SPECIFIC GRAVITY	0.65	
AVERAGE FLOWING TEMP	150	F
OIL API GRAVITY	35.0	API
WATER SPECIFIC GRAVITY	1.07	

DEPTH (1000 FT)

GAS/LIQUID RATIO - SCF PER BBL

HAGEDORN AND BROWN
PRESENTED BY
MACCO - SCHLUMBERGER

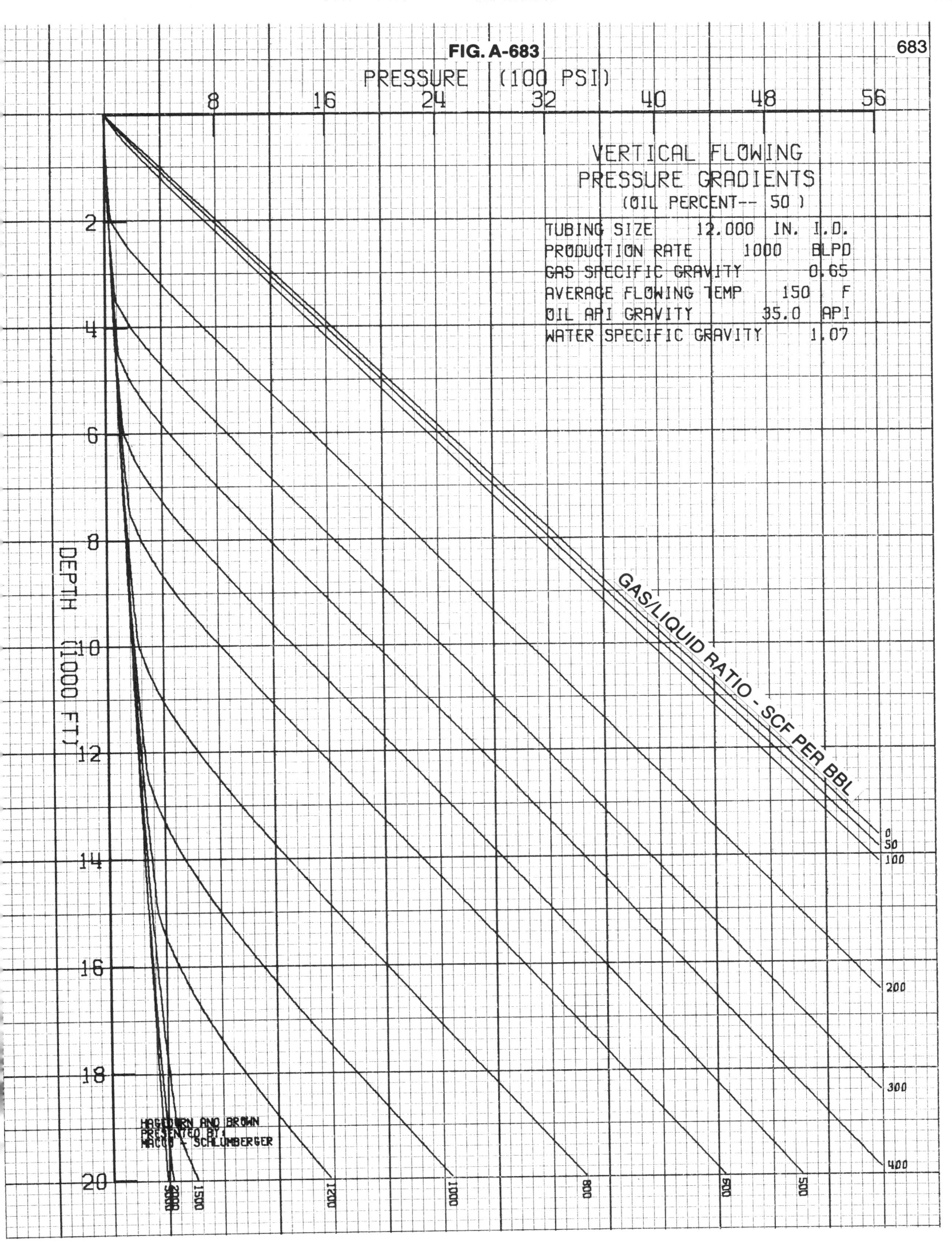

FIG. A-683

683

VERTICAL FLOWING
PRESSURE GRADIENTS
(OIL PERCENT-- 50)

TUBING SIZE	12.000	IN. I.D.
PRODUCTION RATE	1000	BLPD
GAS SPECIFIC GRAVITY	0.65	
AVERAGE FLOWING TEMP	150	F
OIL API GRAVITY	35.0	API
WATER SPECIFIC GRAVITY	1.07	

PRESSURE (100 PSI)

DEPTH (1000 FT)

GAS/LIQUID RATIO - SCF PER BBL

HAGEDORN AND BROWN
PRESENTED BY
MACCO - SCHLUMBERGER

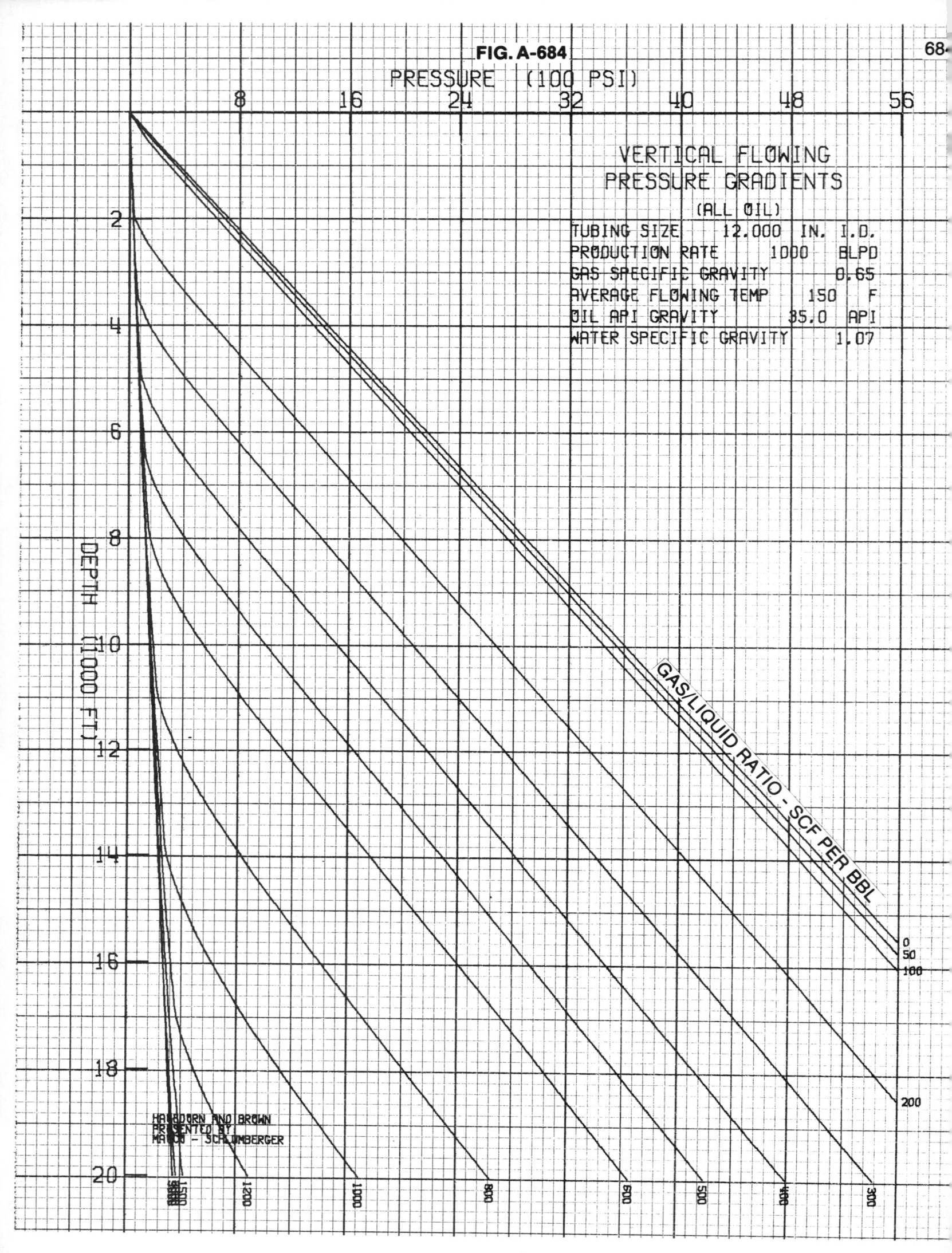

FIG. A-684

VERTICAL FLOWING
PRESSURE GRADIENTS
(ALL OIL)

TUBING SIZE	12.000	IN. I.D.
PRODUCTION RATE	1000	BLPD
GAS SPECIFIC GRAVITY	0.65	
AVERAGE FLOWING TEMP	150	F
OIL API GRAVITY	35.0	API
WATER SPECIFIC GRAVITY	1.07	

PRESSURE (100 PSI)

DEPTH (1000 FT)

GAS/LIQUID RATIO - SCF PER BBL

HAGEDORN AND BROWN
PRESENTED BY
MADCO - SCHLUMBERGER

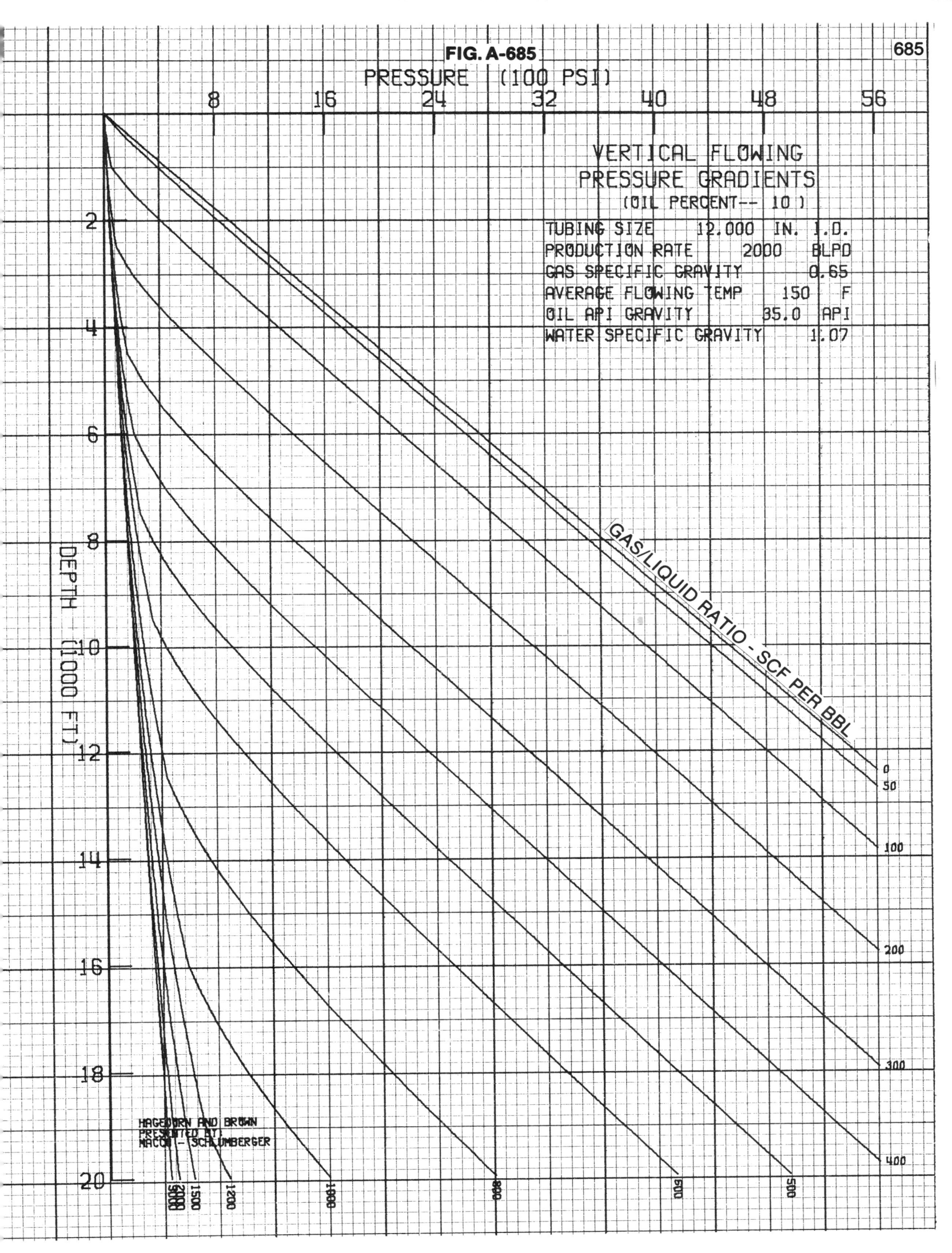

FIG. A-685

685

PRESSURE (100 PSI)

VERTICAL FLOWING
PRESSURE GRADIENTS
(OIL PERCENT-- 10)

TUBING SIZE	12.000	IN. I.D.
PRODUCTION RATE	2000	BLPD
GAS SPECIFIC GRAVITY	0.65	
AVERAGE FLOWING TEMP	150	F
OIL API GRAVITY	35.0	API
WATER SPECIFIC GRAVITY	1.07	

GAS/LIQUID RATIO - SCF PER BBL

DEPTH (1000 FT)

HAGEDORN AND BROWN
PRESENTED BY:
NACON - SCHLUMBERGER

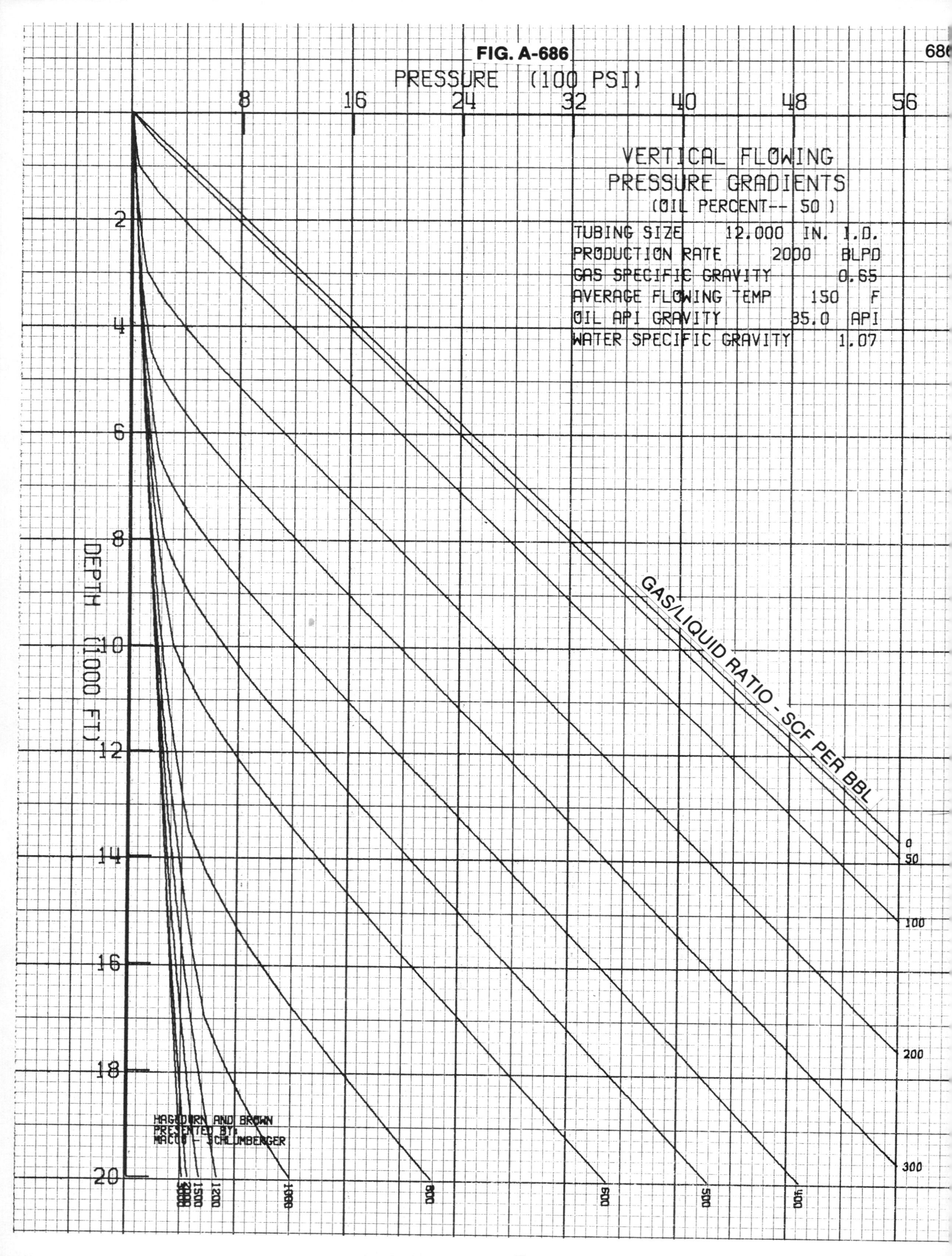

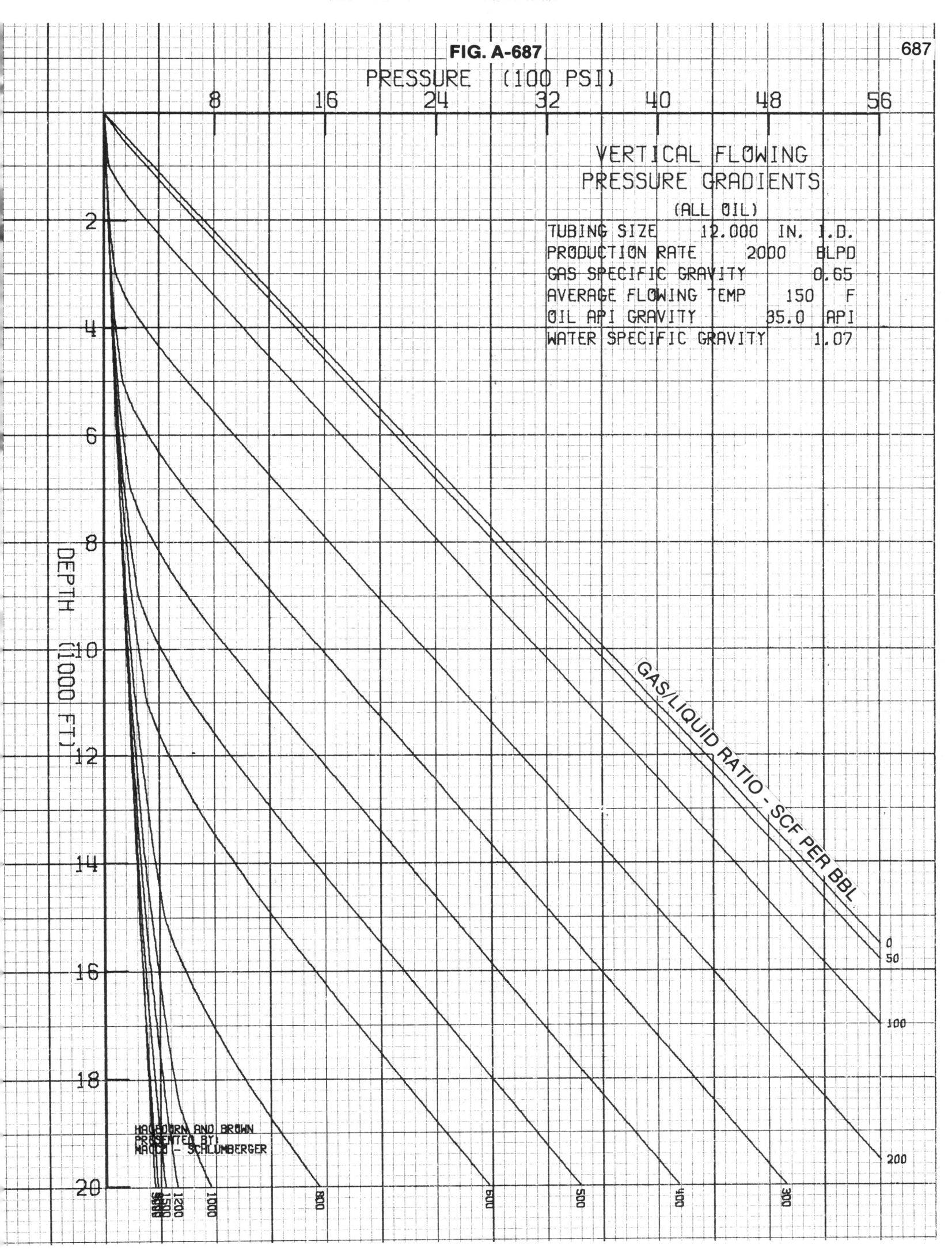

PRESSURE (100 PSI)

VERTICAL FLOWING
PRESSURE GRADIENTS
(ALL OIL)

TUBING SIZE   12.000   IN. I.D.
PRODUCTION RATE   2000   BLPD
GAS SPECIFIC GRAVITY   0.65
AVERAGE FLOWING TEMP   150   F
OIL API GRAVITY   35.0   API
WATER SPECIFIC GRAVITY   1.07

GAS/LIQUID RATIO - SCF PER BBL

DEPTH (1000 FT)

HAGEDORN AND BROWN
PRESENTED BY:
MACCO - SCHLUMBERGER

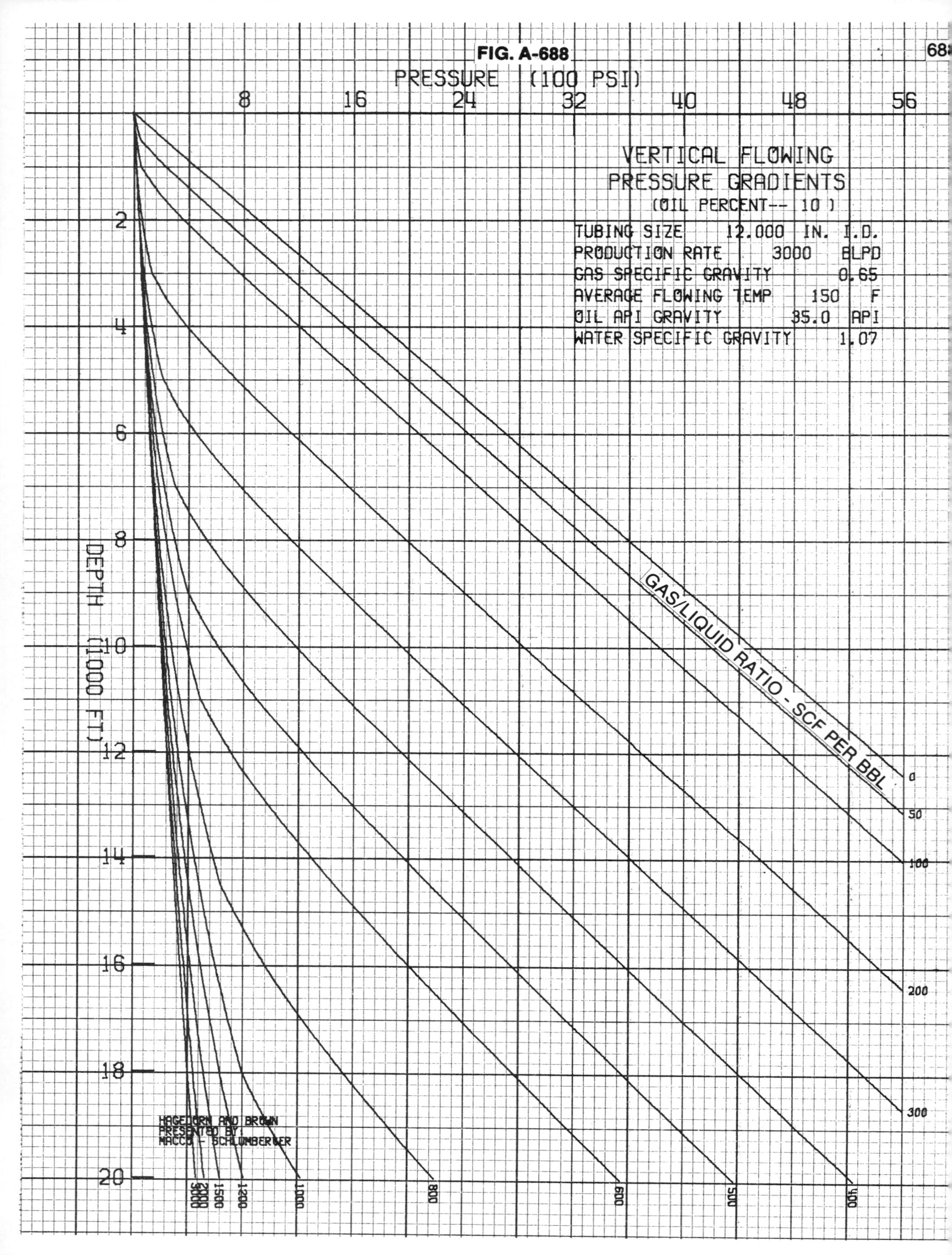

FIG. A-688

688

VERTICAL FLOWING PRESSURE GRADIENTS
(OIL PERCENT-- 10)

TUBING SIZE	12.000	IN. I.D.
PRODUCTION RATE	3000	BLPD
GAS SPECIFIC GRAVITY	0.65	
AVERAGE FLOWING TEMP	150	F
OIL API GRAVITY	35.0	API
WATER SPECIFIC GRAVITY	1.07	

PRESSURE (100 PSI)

DEPTH (1000 FT)

GAS/LIQUID RATIO - SCF PER BBL

HAGEDORN AND BROWN
PRESENTED BY
MACCO - SCHLUMBERGER

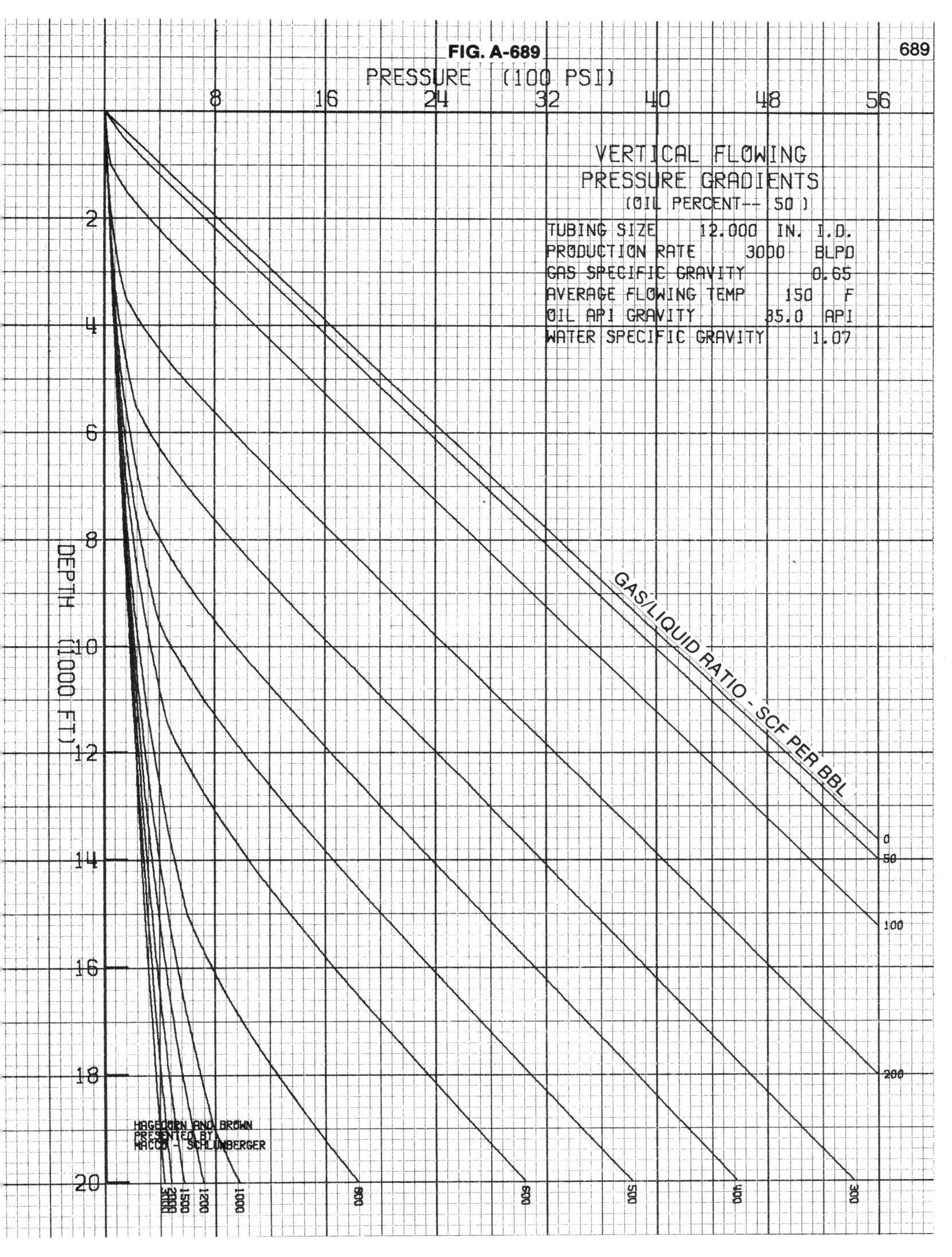

PRESSURE (100 PSI)

VERTICAL FLOWING
PRESSURE GRADIENTS
(OIL PERCENT-- 50)

TUBING SIZE	12.000	IN. I.D.
PRODUCTION RATE	3000	BLPD
GAS SPECIFIC GRAVITY		0.65
AVERAGE FLOWING TEMP	150	F
OIL API GRAVITY	35.0	API
WATER SPECIFIC GRAVITY		1.07

DEPTH (1000 FT)

GAS/LIQUID RATIO - SCF PER BBL

HAGEDORN AND BROWN
PRESENTED BY
MACCO - SCHLUMBERGER

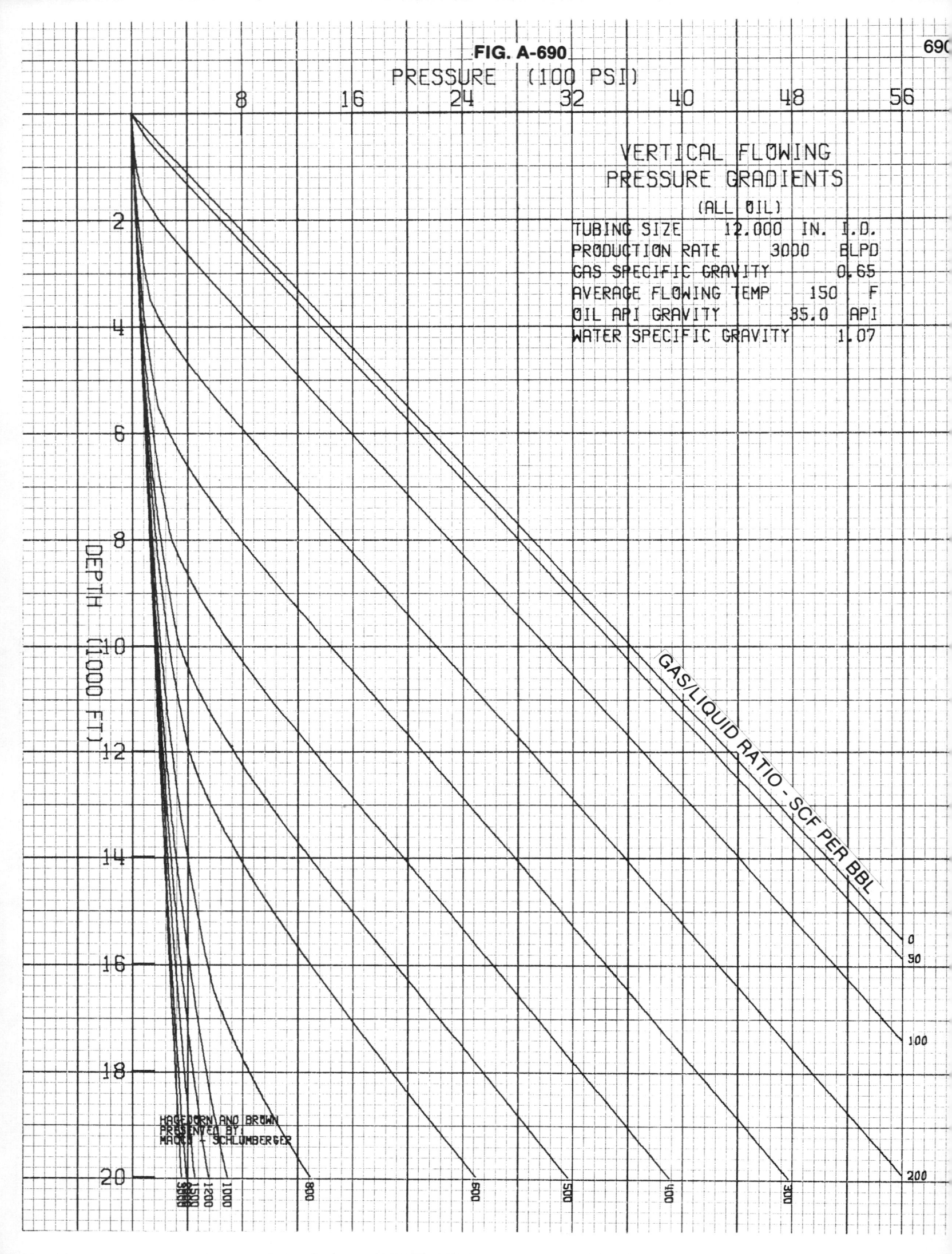

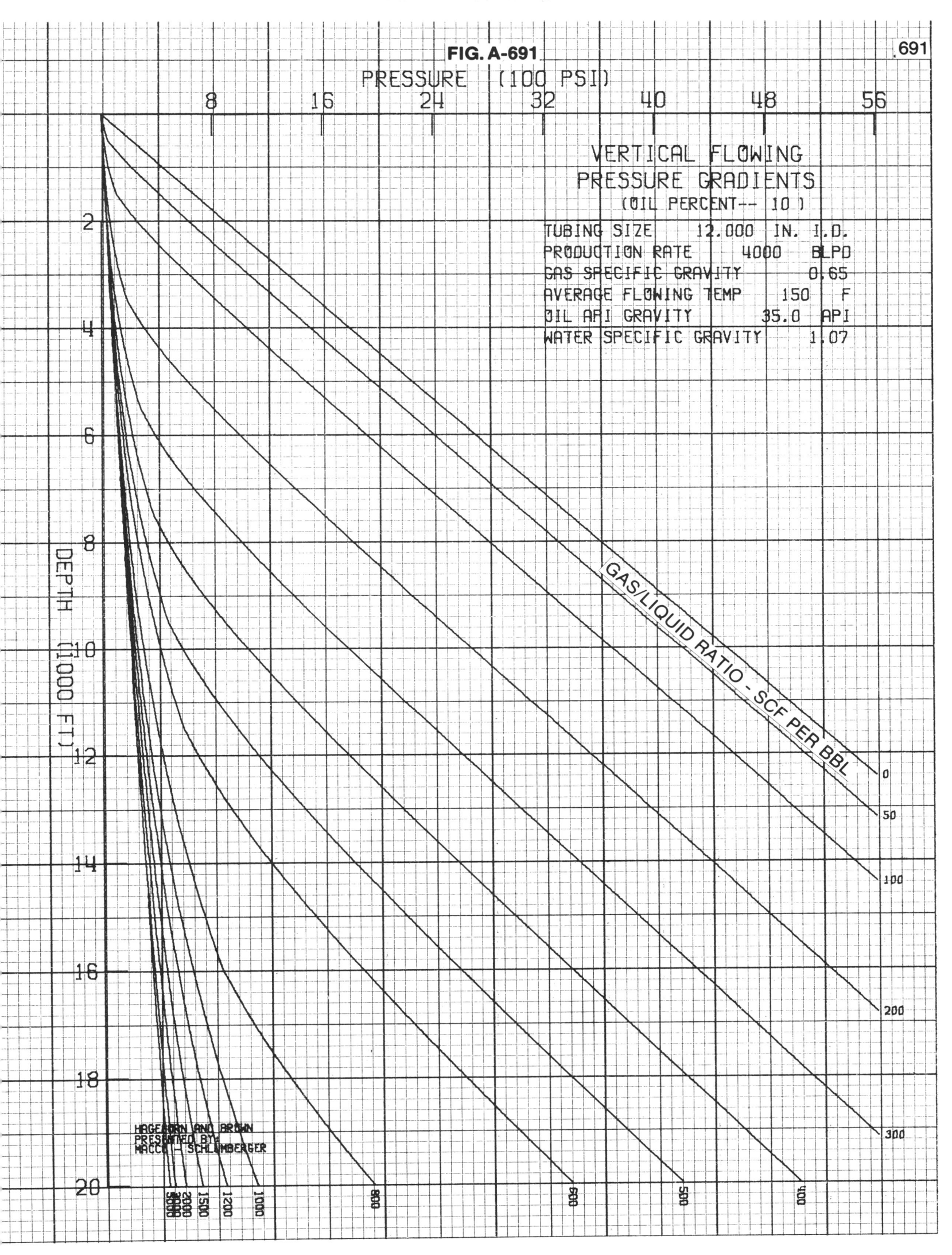

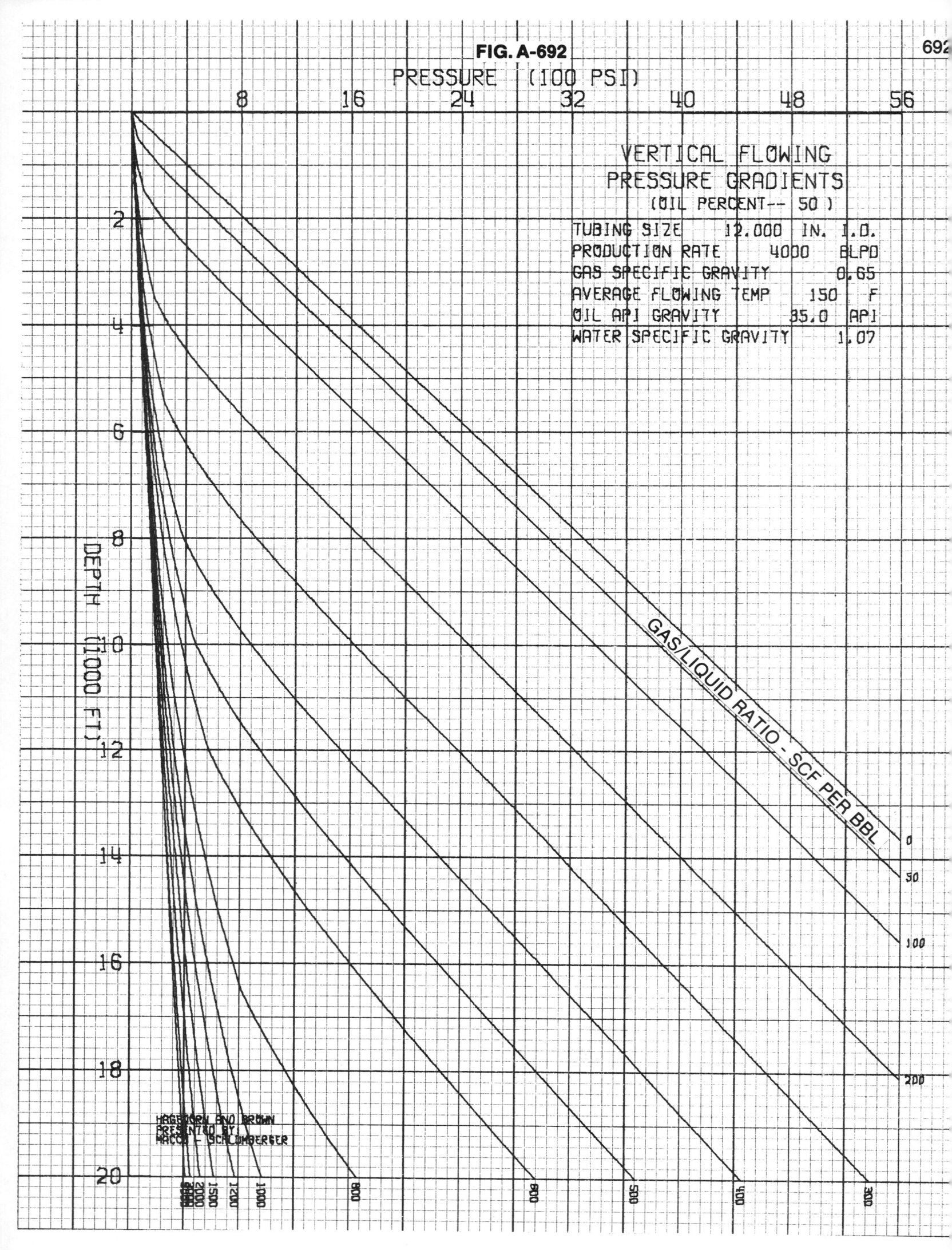

PRESSURE (100 PSI)

DEPTH (1000 FT)

VERTICAL FLOWING
PRESSURE GRADIENTS
(OIL PERCENT-- 50)

TUBING SIZE   12.000   IN. I.D.
PRODUCTION RATE   4000   BLPD
GAS SPECIFIC GRAVITY   0.65
AVERAGE FLOWING TEMP   150   F
OIL API GRAVITY   35.0   API
WATER SPECIFIC GRAVITY   1.07

GAS/LIQUID RATIO - SCF PER BBL

HAGEDORN AND BROWN
PRESENTED BY
MAICO - SCHLUMBERGER

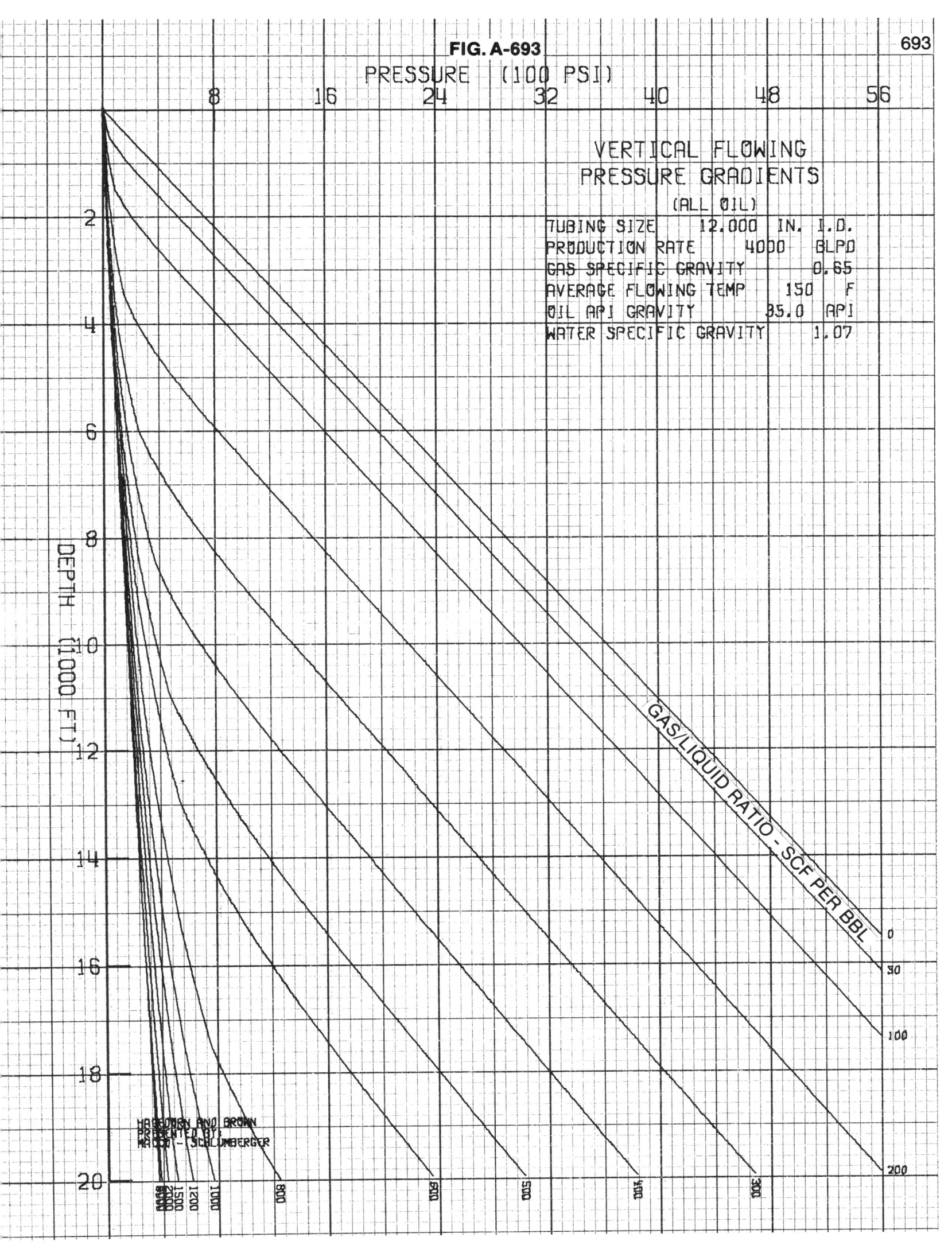

FIG. A-693

693

PRESSURE (100 PSI)

VERTICAL FLOWING
PRESSURE GRADIENTS
(ALL OIL)

TUBING SIZE   12.000 IN. I.D.
PRODUCTION RATE   4000   BLPD
GAS SPECIFIC GRAVITY   0.65
AVERAGE FLOWING TEMP   150   F
OIL API GRAVITY   35.0 API
WATER SPECIFIC GRAVITY   1.07

GAS/LIQUID RATIO - SCF PER BBL

DEPTH (1000 FT)

HAGEDORN AND BROWN
PRESENTED BY
HAGDDT - SCHLUMBERGER

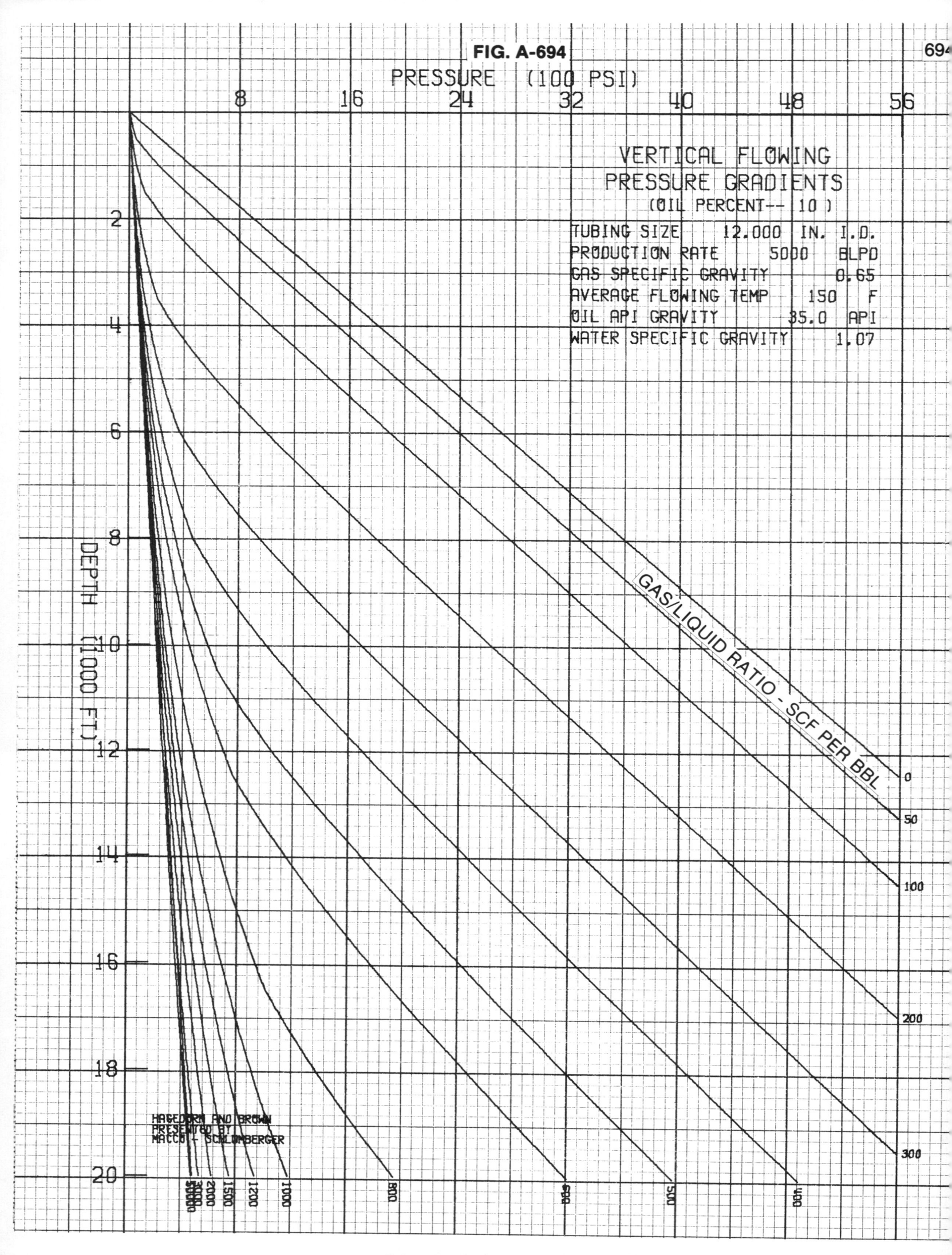

PRESSURE (100 PSI)

VERTICAL FLOWING
PRESSURE GRADIENTS
(OIL PERCENT-- 10)

TUBING SIZE	12.000	IN. I.D.
PRODUCTION RATE	5000	BLPD
GAS SPECIFIC GRAVITY	0.65	
AVERAGE FLOWING TEMP	150	F
OIL API GRAVITY	35.0	API
WATER SPECIFIC GRAVITY	1.07	

DEPTH (1000 FT)

GAS/LIQUID RATIO - SCF PER BBL

HAGEDORN AND BROWN
PRESENTED BY
MACCOLL SCHLUMBERGER

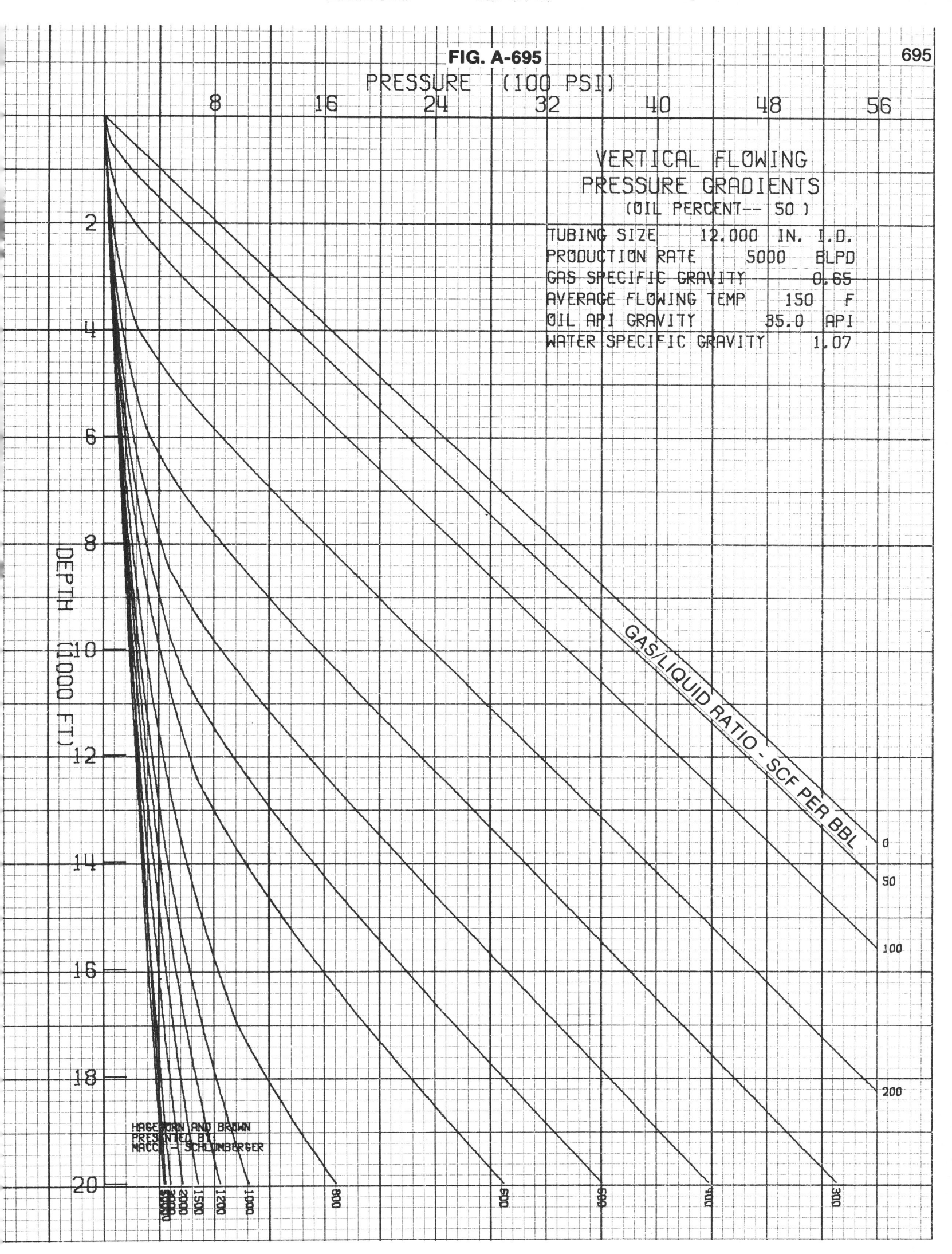

FIG. A-696

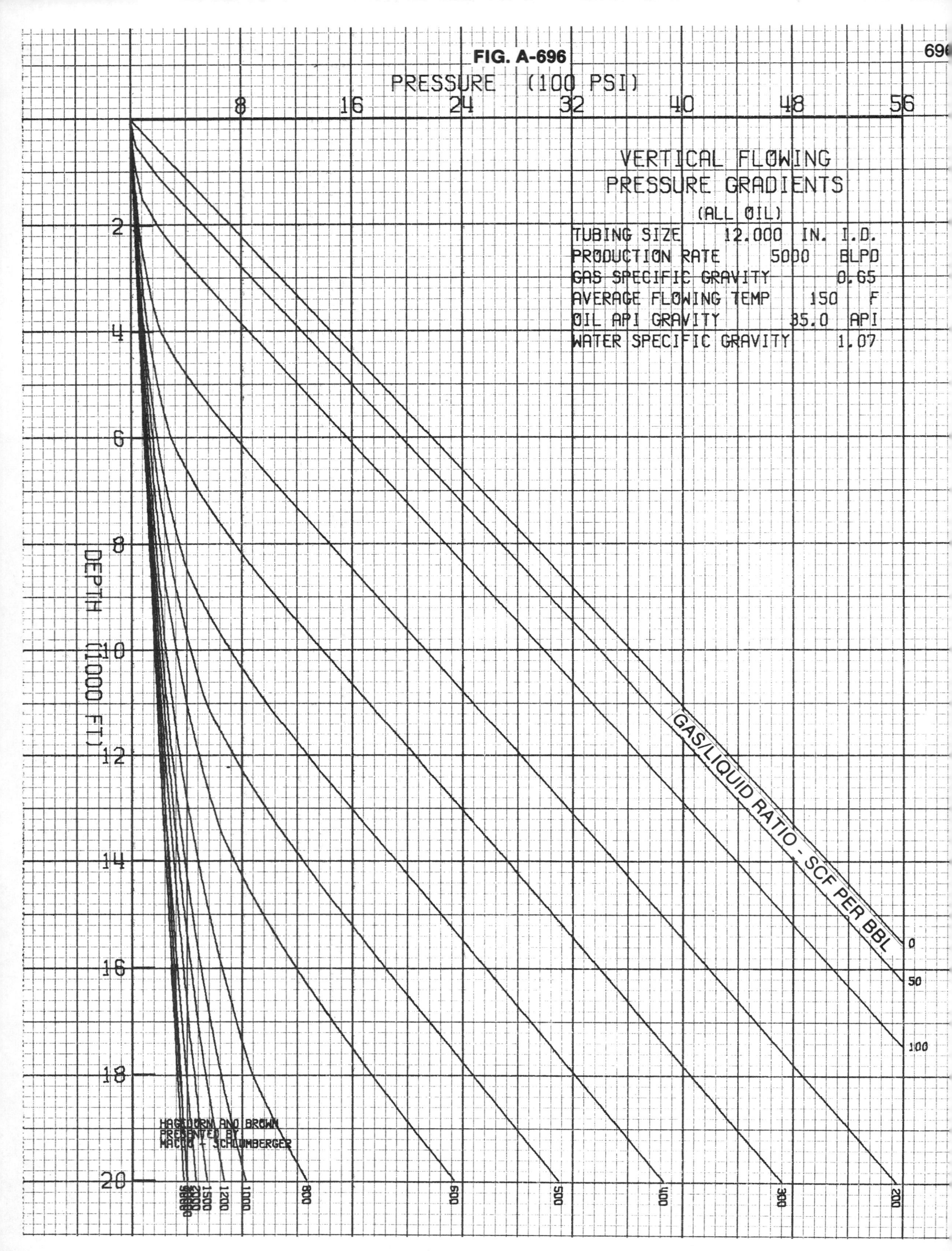

VERTICAL FLOWING
PRESSURE GRADIENTS
(ALL OIL)

TUBING SIZE	12.000	IN. I.D.
PRODUCTION RATE	5000	BLPD
GAS SPECIFIC GRAVITY	0.65	
AVERAGE FLOWING TEMP	150	F
OIL API GRAVITY	85.0	API
WATER SPECIFIC GRAVITY	1.07	

PRESSURE (100 PSI)

DEPTH (1000 FT)

GAS/LIQUID RATIO - SCF PER BBL

HAGEDORN AND BROWN
PRESENTED BY
MACCO + SCHLUMBERGER

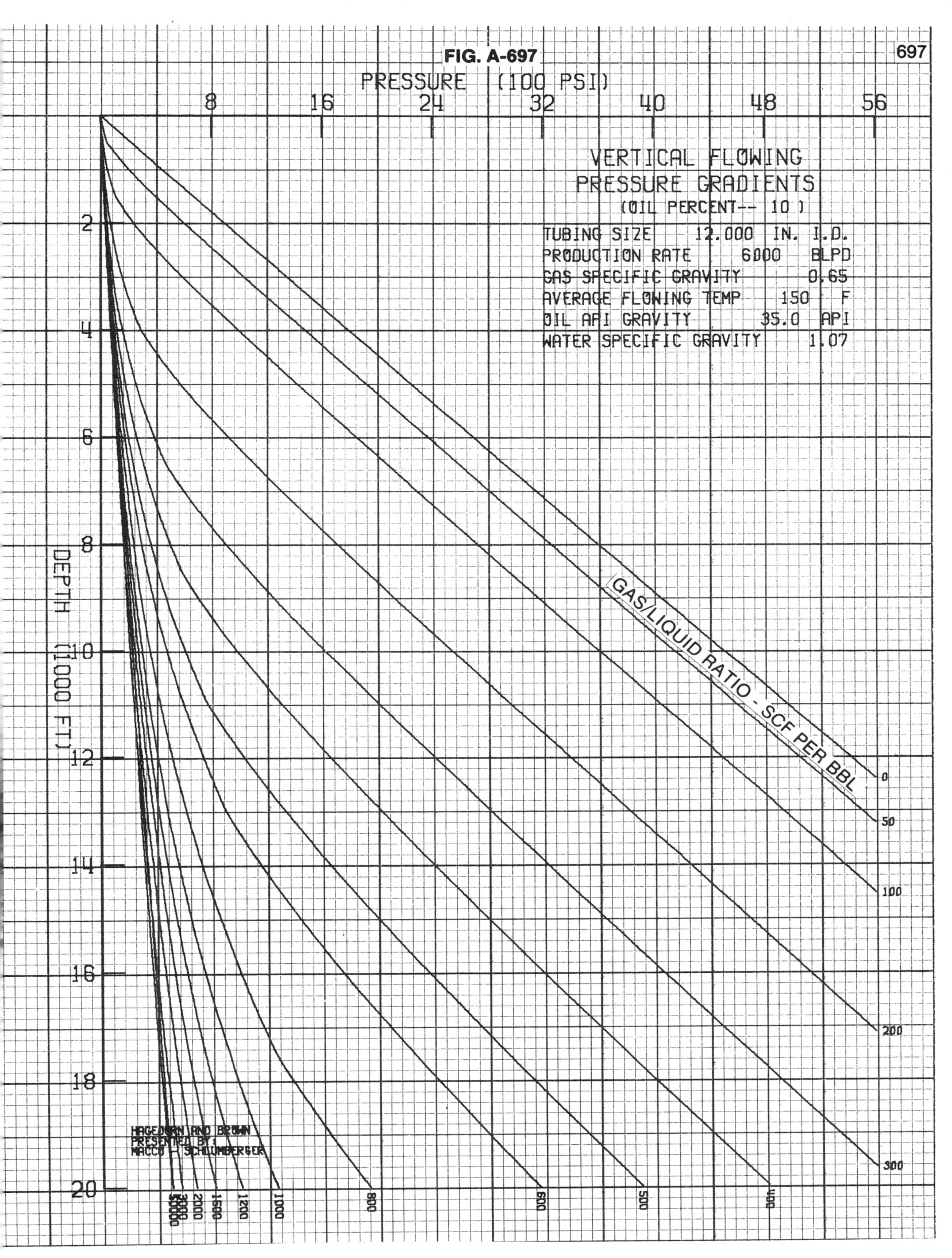

FIG. A-697

697

PRESSURE (100 PSI)

VERTICAL FLOWING
PRESSURE GRADIENTS
(OIL PERCENT--- 10)

TUBING SIZE   12.000 IN. I.D.
PRODUCTION RATE   6000   BLPD
GAS SPECIFIC GRAVITY   0.65
AVERAGE FLOWING TEMP   150   F
OIL API GRAVITY   35.0 API
WATER SPECIFIC GRAVITY   1.07

GAS/LIQUID RATIO - SCF PER BBL

DEPTH (1000 FT)

HAGEDORN AND BROWN
PRESENTED BY:
MACCO - SCHLUMBERGER

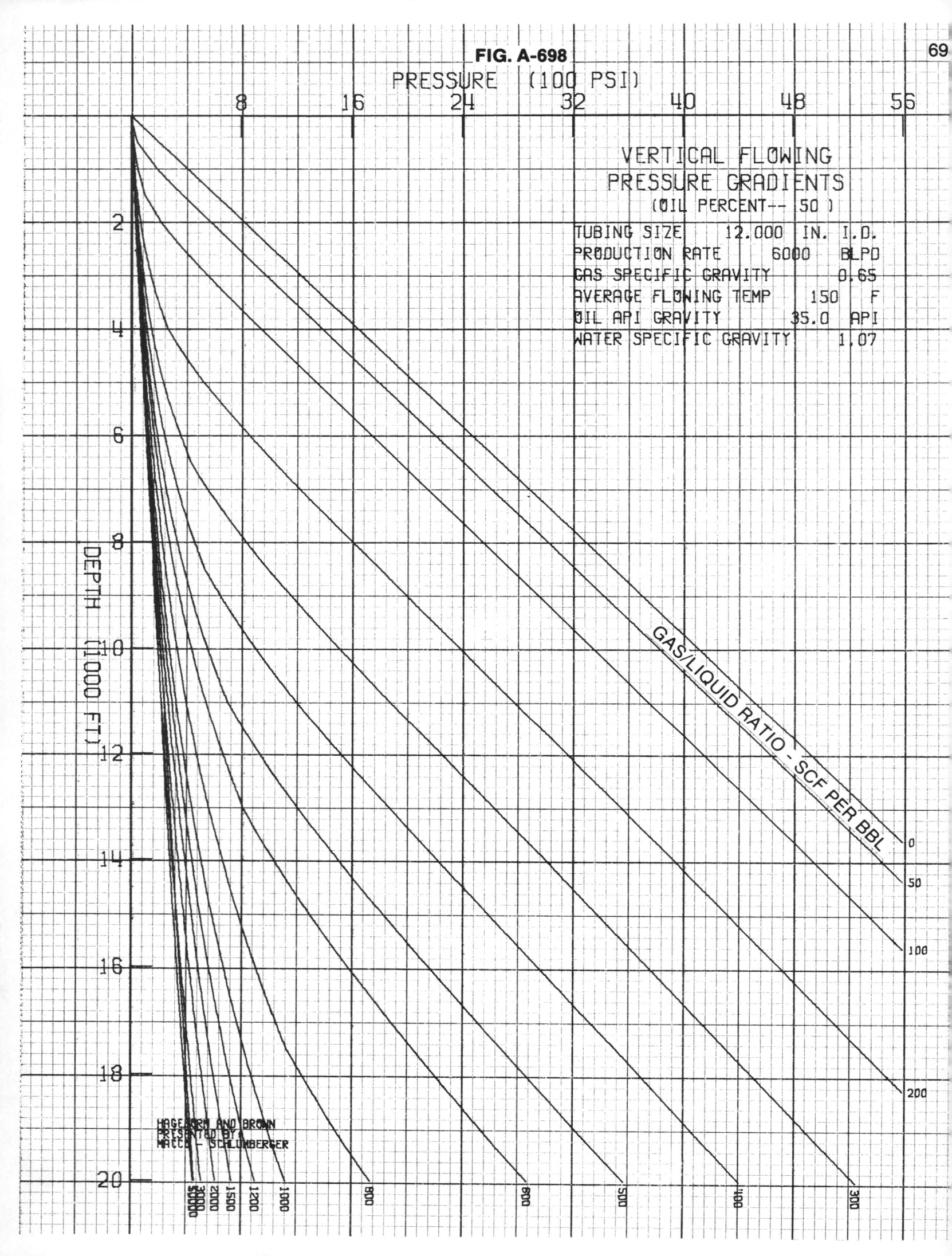

FIG. A-698

69

PRESSURE (100 PSI)

VERTICAL FLOWING
PRESSURE GRADIENTS
(OIL PERCENT-- 50)

TUBING SIZE	12.000	IN. I.D.
PRODUCTION RATE	6000	BLPD
GAS SPECIFIC GRAVITY	0.65	
AVERAGE FLOWING TEMP	150	F
OIL API GRAVITY	35.0	API
WATER SPECIFIC GRAVITY	1.07	

DEPTH (1000 FT)

GAS/LIQUID RATIO - SCF PER BBL

HAGEDORN AND BROWN
PRESENTED BY
MATCO - SCHLUMBERGER

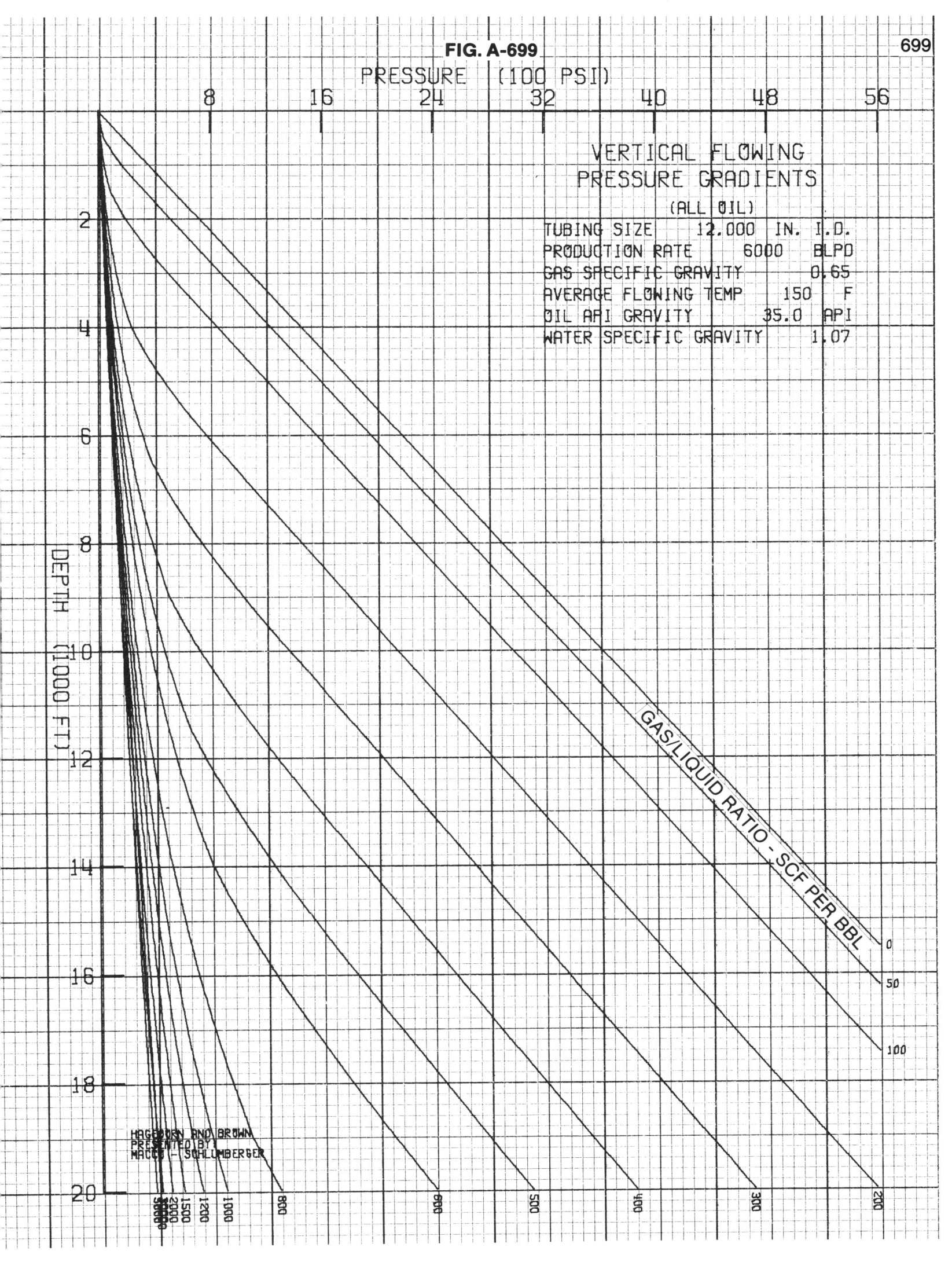

PRESSURE (100 PSI)

DEPTH (1000 FT)

VERTICAL FLOWING
PRESSURE GRADIENTS
(ALL OIL)

TUBING SIZE	12.000	IN.	I.D.
PRODUCTION RATE	6000	BLPD	
GAS SPECIFIC GRAVITY	0.65		
AVERAGE FLOWING TEMP	150	F	
OIL API GRAVITY	35.0	API	
WATER SPECIFIC GRAVITY	1.07		

GAS/LIQUID RATIO - SCF PER BBL

HAGEDORN AND BROWN
PRESENTED BY
MACCO - SCHLUMBERGER

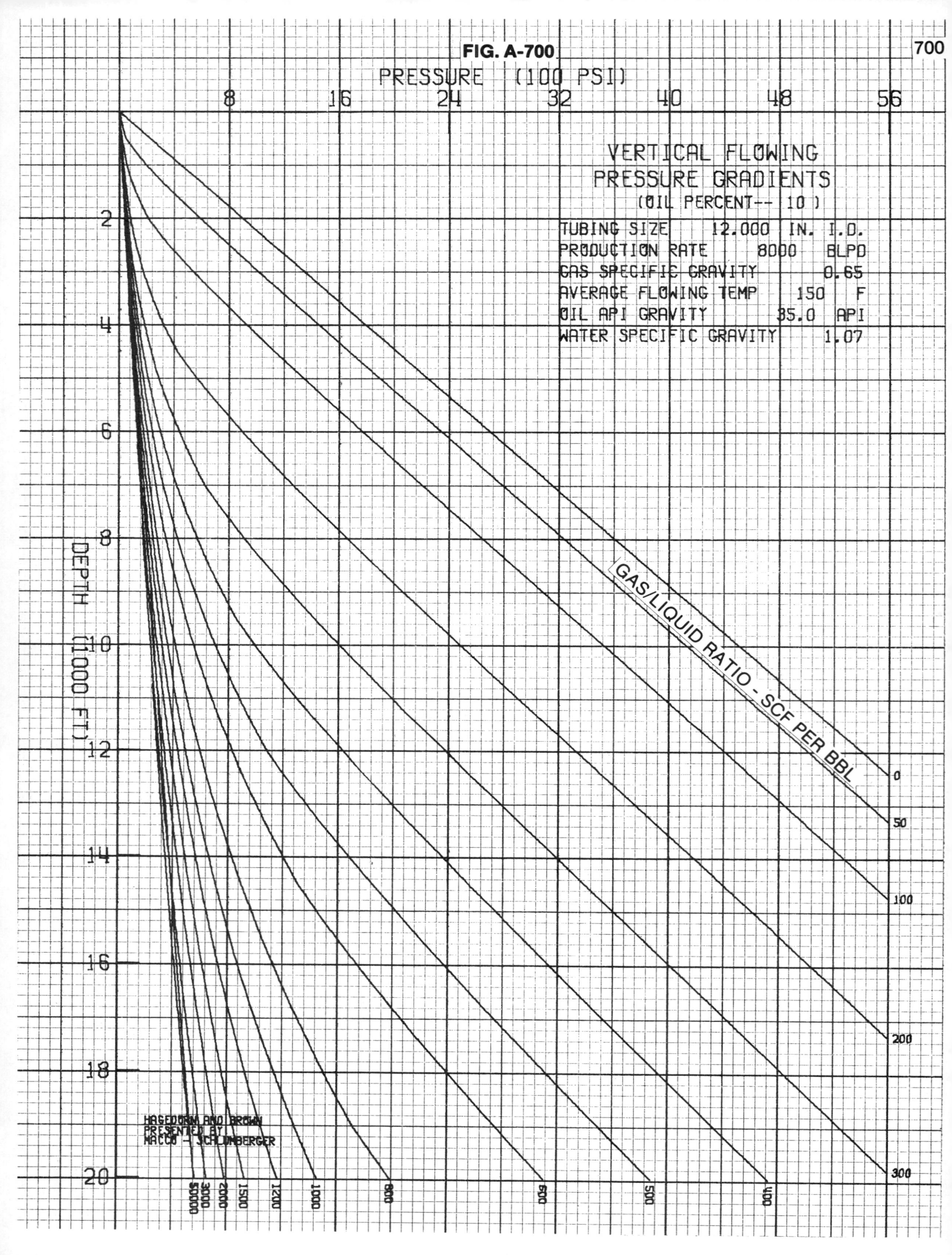

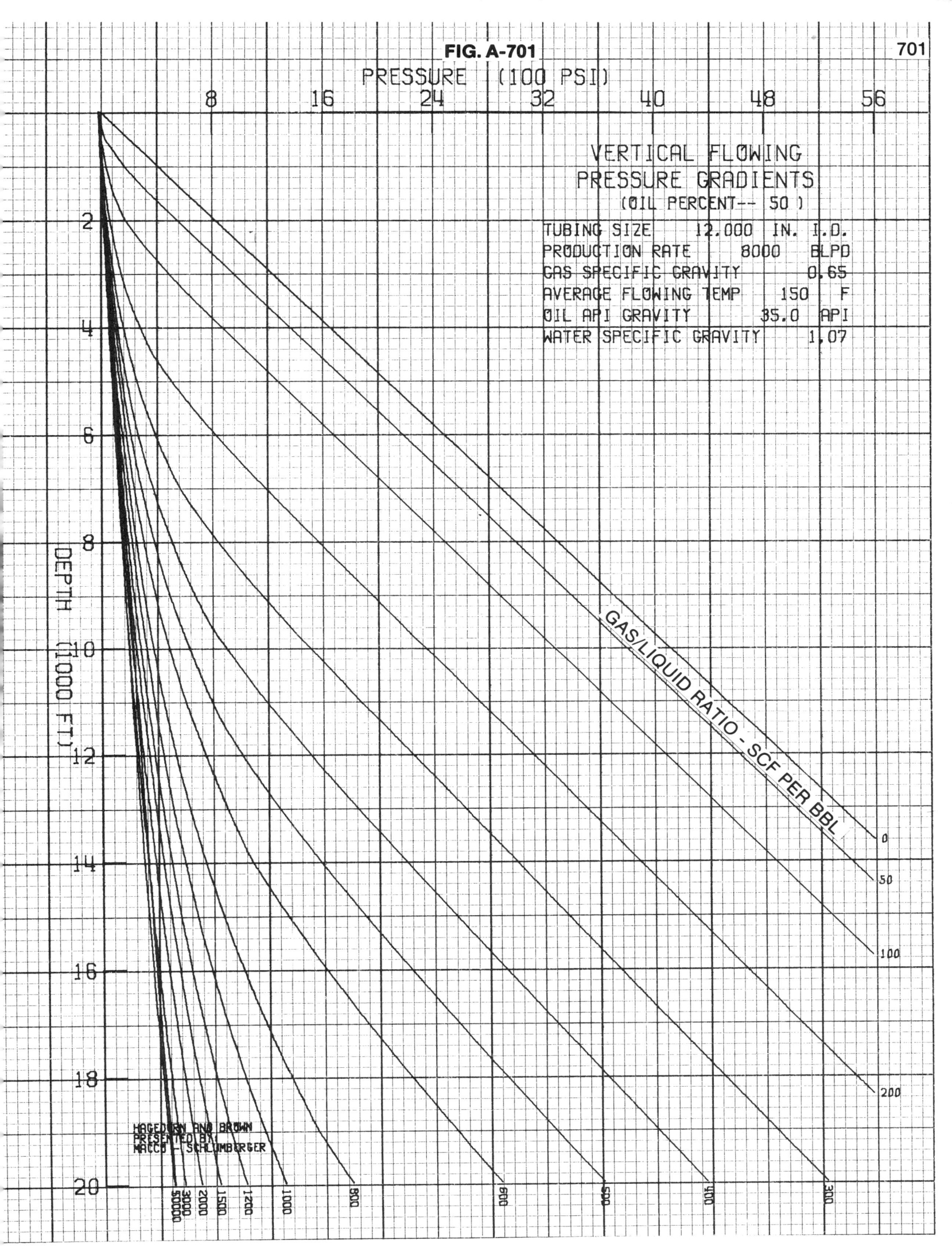

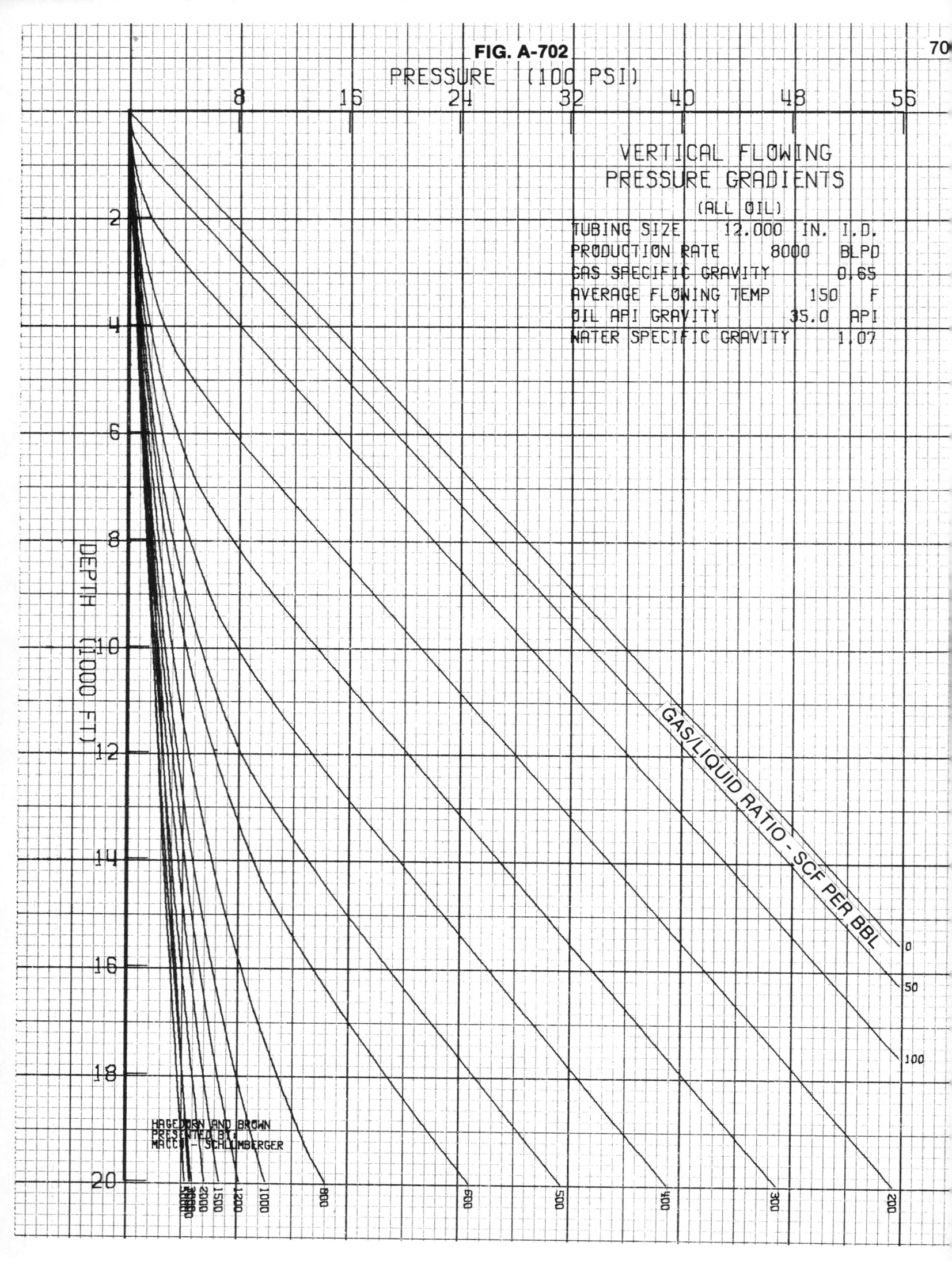

PRESSURE (100 PSI)

VERTICAL FLOWING
PRESSURE GRADIENTS
(ALL OIL)

TUBING SIZE	12.000 IN. I.D.
PRODUCTION RATE	8000 BLPD
GAS SPECIFIC GRAVITY	0.65
AVERAGE FLOWING TEMP	150 F
OIL API GRAVITY	35.0 API
WATER SPECIFIC GRAVITY	1.07

DEPTH (1000 FT)

GAS/LIQUID RATIO - SCF PER BBL

HAGEDORN AND BROWN
PRESENTED BY:
MACCO - SCHLUMBERGER

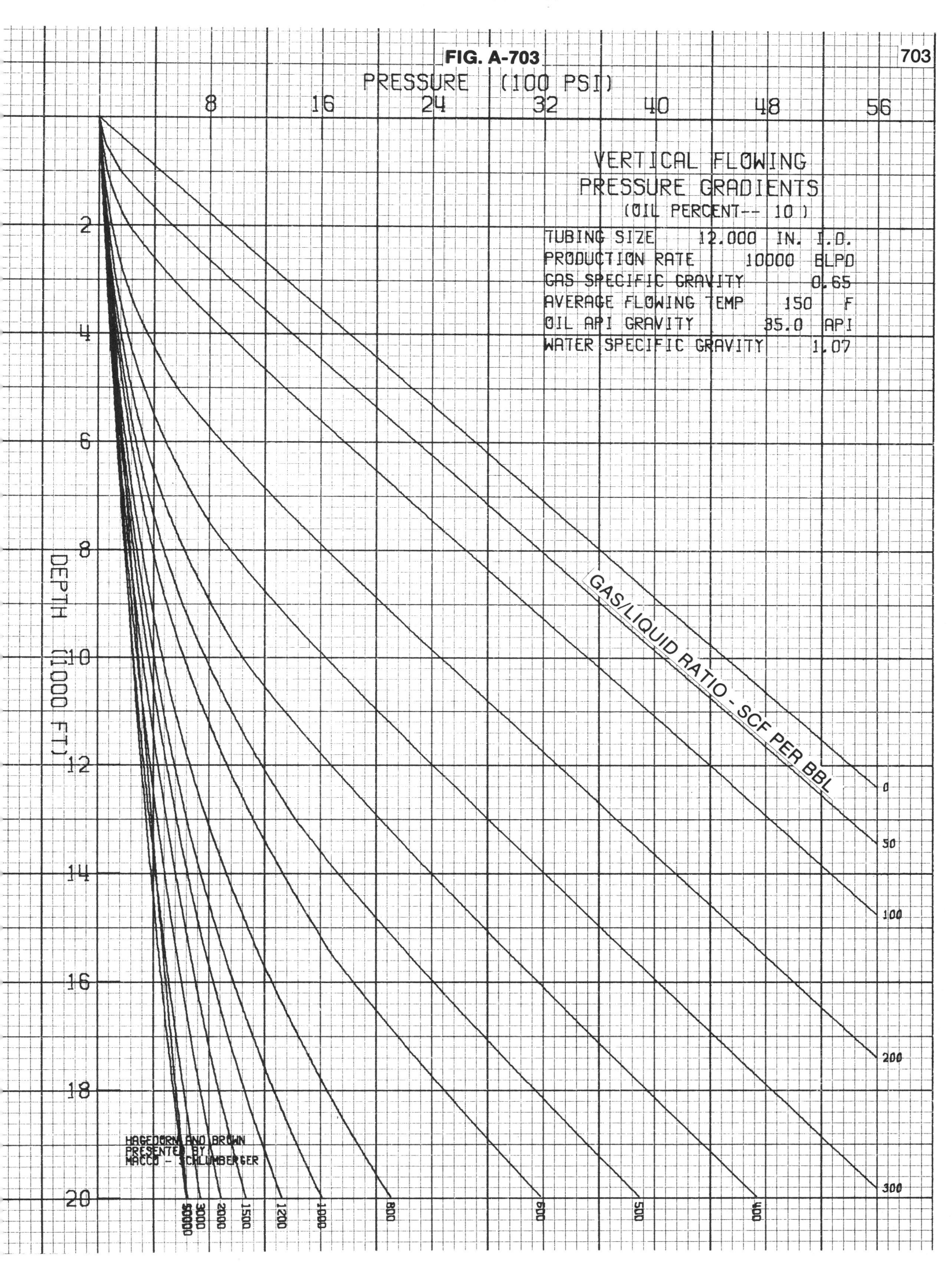

FIG. A-703

703

VERTICAL FLOWING
PRESSURE GRADIENTS
(OIL PERCENT-- 10)

TUBING SIZE   12.000   IN. I.D.
PRODUCTION RATE   10000   BLPD
GAS SPECIFIC GRAVITY   0.65
AVERAGE FLOWING TEMP   150   F
OIL API GRAVITY   35.0   API
WATER SPECIFIC GRAVITY   1.07

PRESSURE (100 PSI)

DEPTH (1000 FT)

GAS/LIQUID RATIO - SCF PER BBL

HAGEDORN AND BROWN
PRESENTED BY
MACCO - SCHLUMBERGER

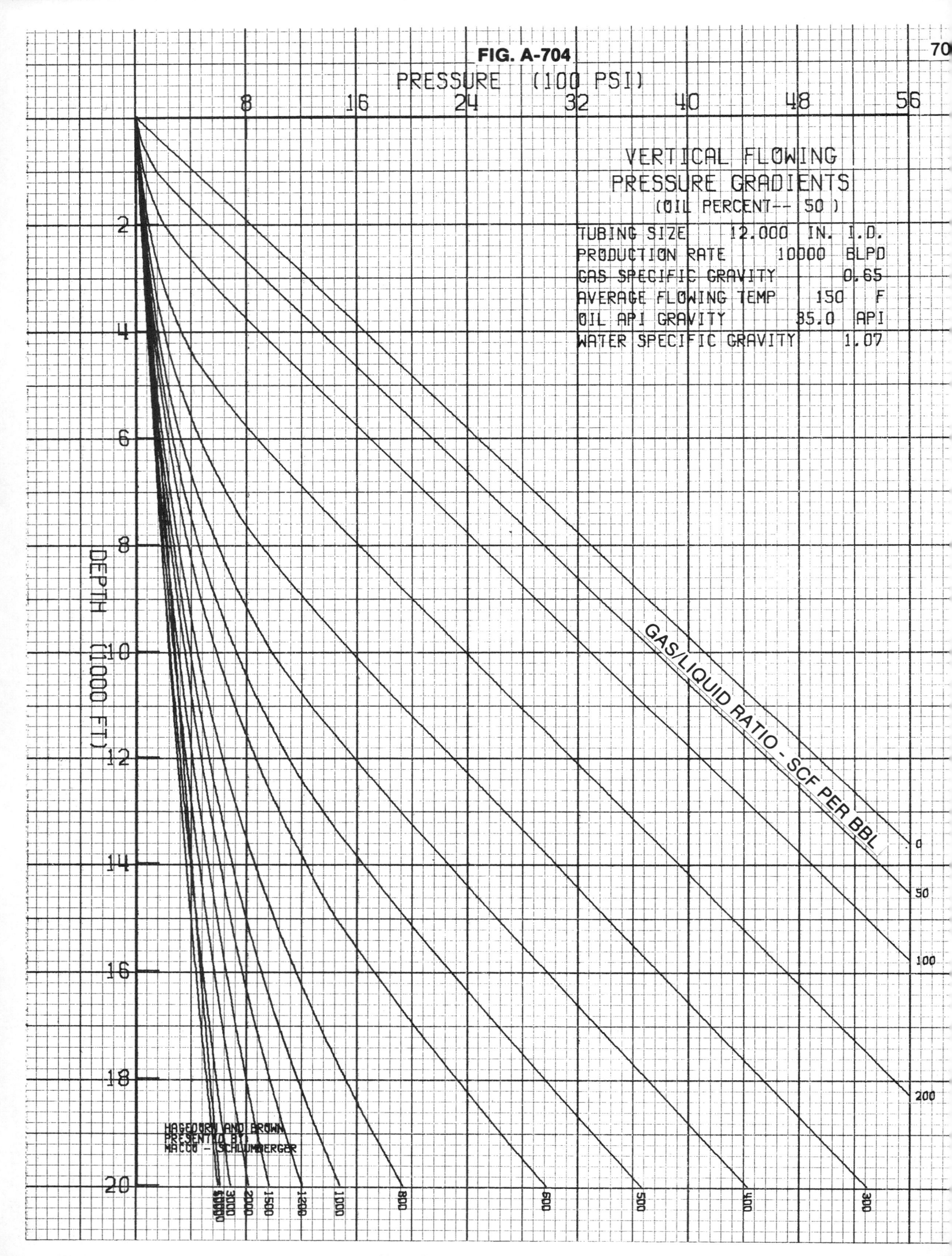

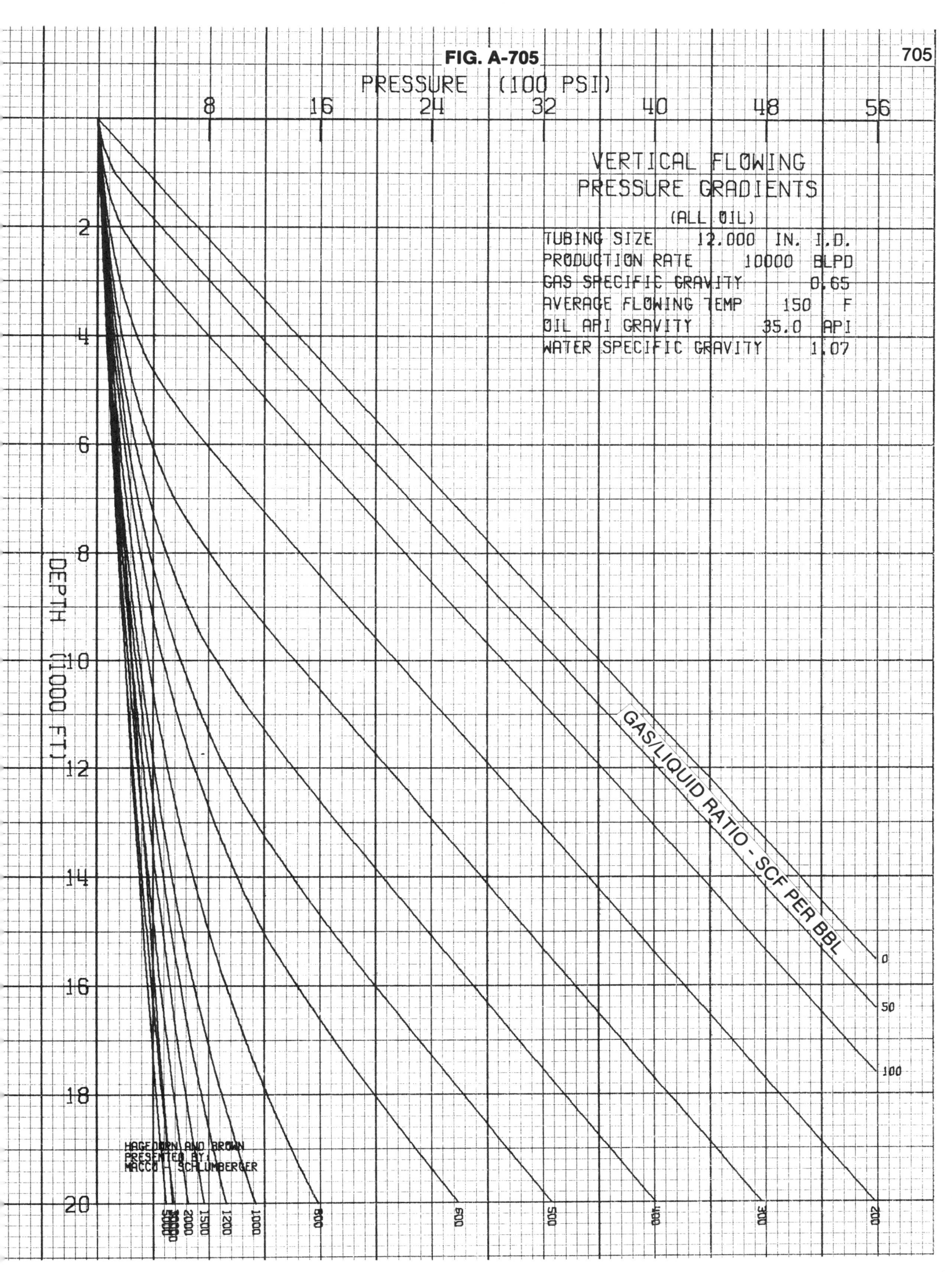

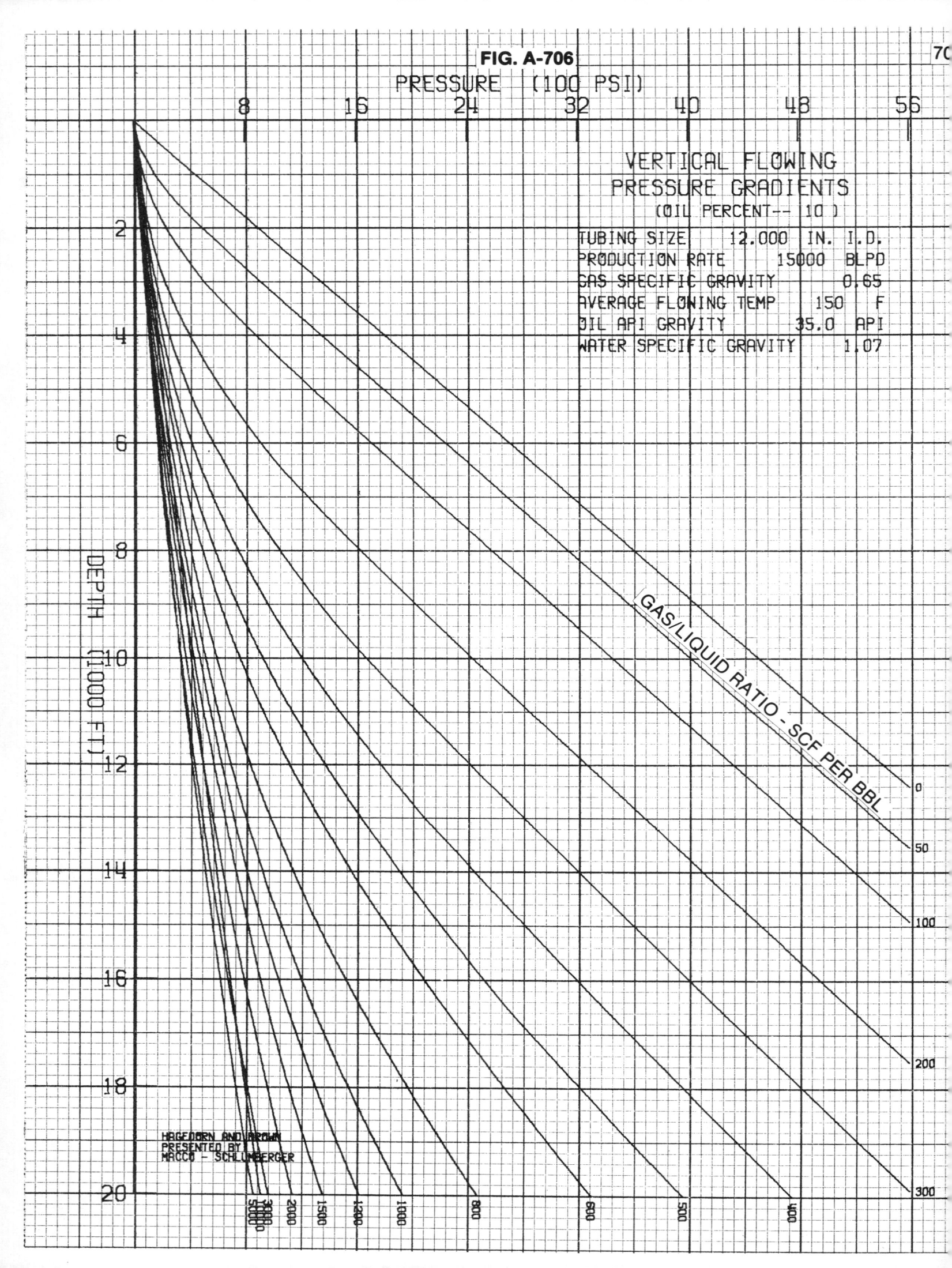

FIG. A-706

PRESSURE (100 PSI)

VERTICAL FLOWING
PRESSURE GRADIENTS
(OIL PERCENT-- 10)

TUBING SIZE   12.000 IN. I.D.
PRODUCTION RATE   15000 BLPD
GAS SPECIFIC GRAVITY   0.65
AVERAGE FLOWING TEMP   150   F
OIL API GRAVITY   35.0 API
WATER SPECIFIC GRAVITY   1.07

GAS/LIQUID RATIO - SCF PER BBL

DEPTH (1000 FT)

HAGEDORN AND BROWN
PRESENTED BY
MACCO - SCHLUMBERGER

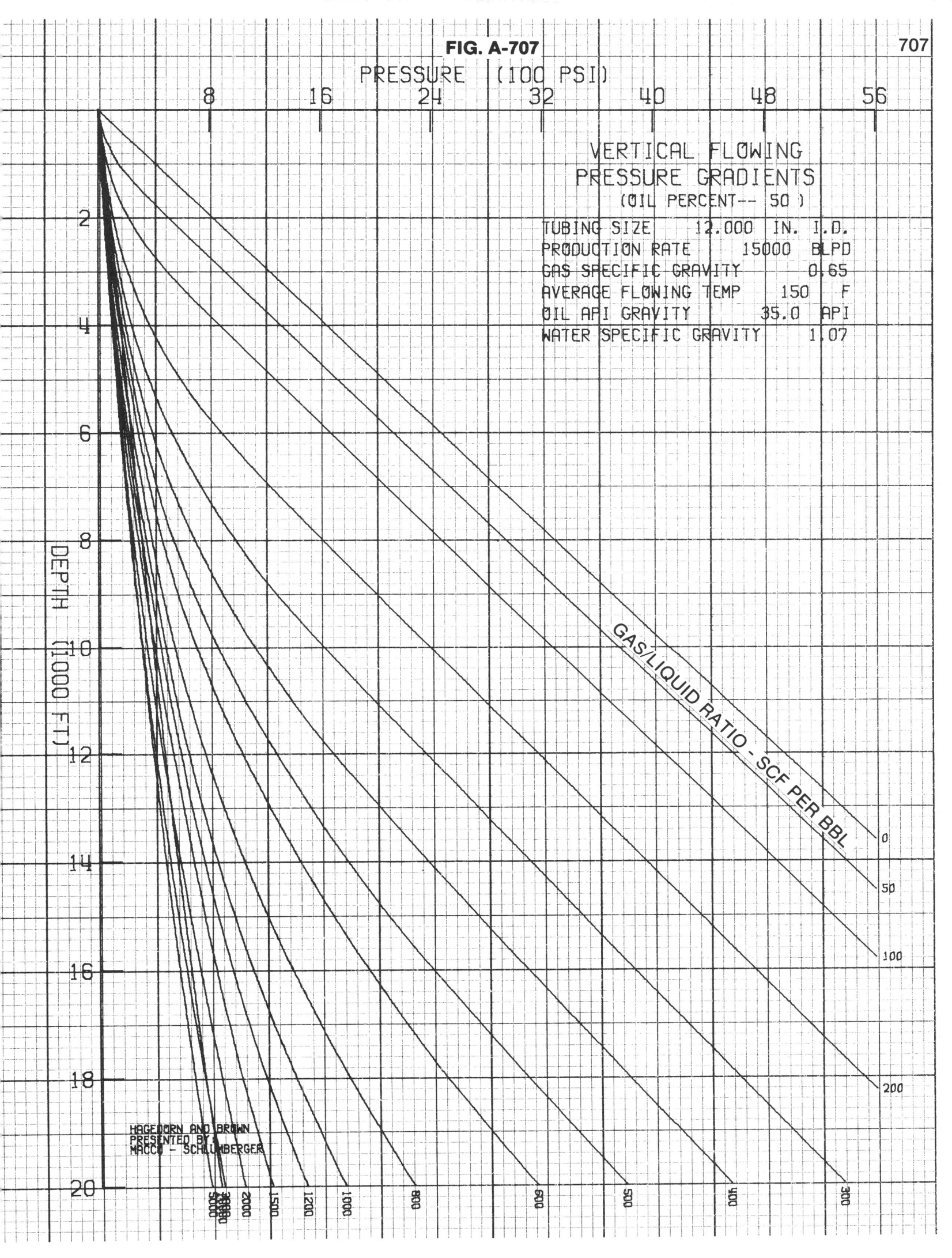

VERTICAL FLOWING
PRESSURE GRADIENTS
(OIL PERCENT— 50)

TUBING SIZE	12.000	IN. I.D.
PRODUCTION RATE	15000	BLPD
GAS SPECIFIC GRAVITY	0.65	
AVERAGE FLOWING TEMP	150	F
OIL API GRAVITY	35.0	API
WATER SPECIFIC GRAVITY	1.07	

PRESSURE (100 PSI)

DEPTH (1000 FT)

GAS/LIQUID RATIO - SCF PER BBL

HAGEDORN AND BROWN
PRESENTED BY:
MACCO - SCHLUMBERGER

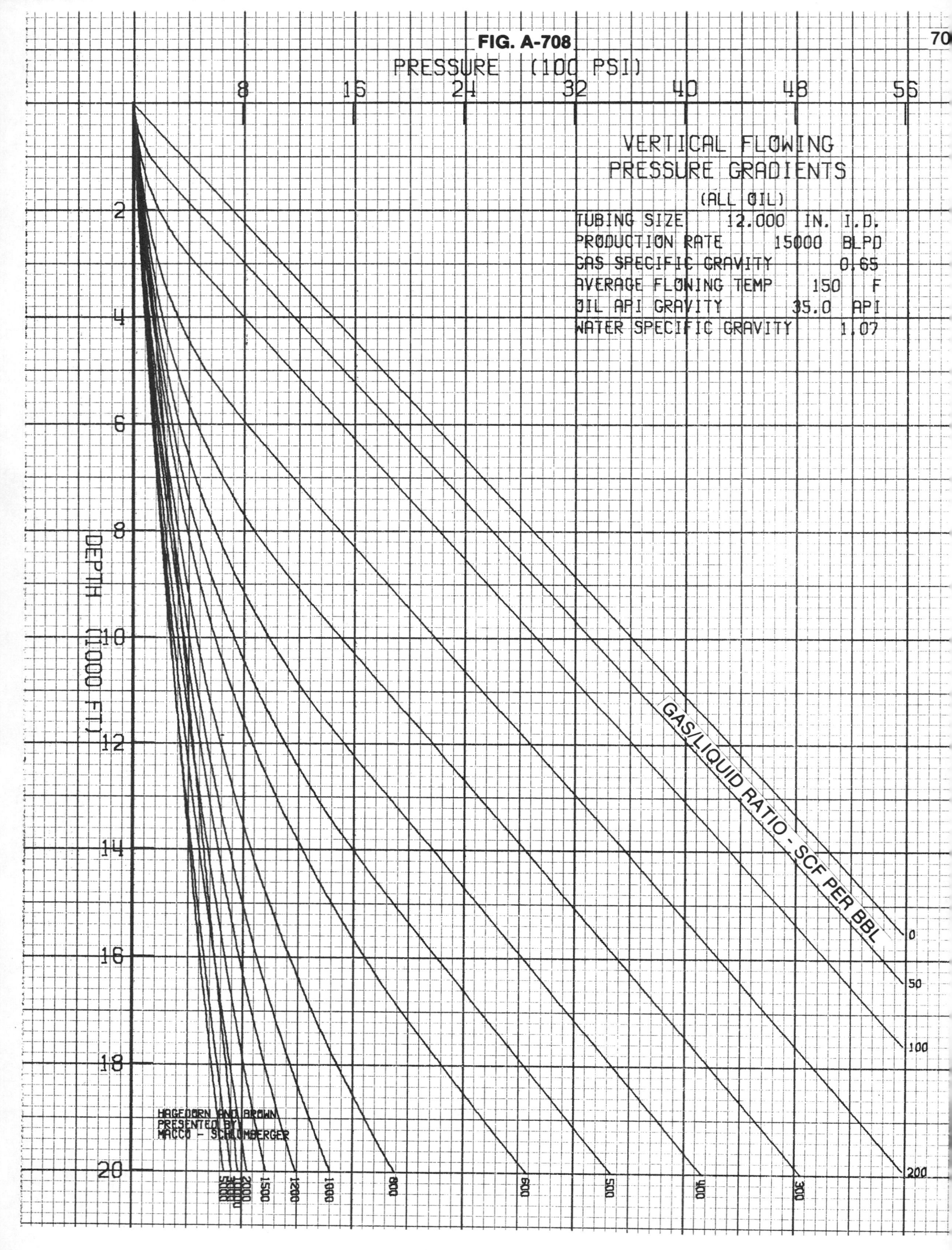

PRESSURE (100 PSI)

VERTICAL FLOWING
PRESSURE GRADIENTS
(ALL OIL)

TUBING SIZE	12.000 IN. I.D.
PRODUCTION RATE	15000 BLPD
GAS SPECIFIC GRAVITY	0.65
AVERAGE FLOWING TEMP	150 F
OIL API GRAVITY	35.0 API
WATER SPECIFIC GRAVITY	1.07

DEPTH (1000 FT)

GAS/LIQUID RATIO - SCF PER BBL

HAGEDORN AND BROWN
PRESENTED BY
MACCO - SCHLUMBERGER

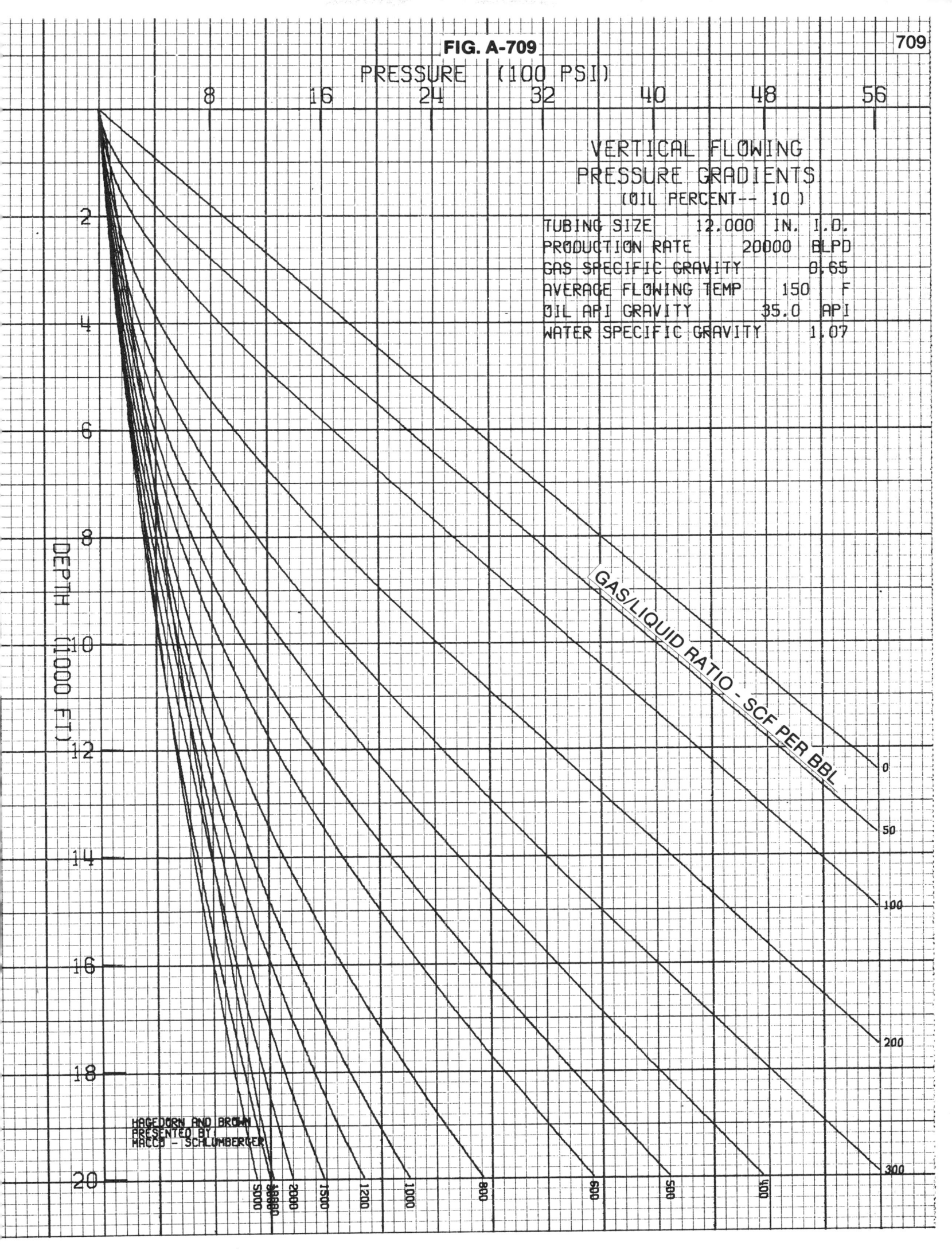

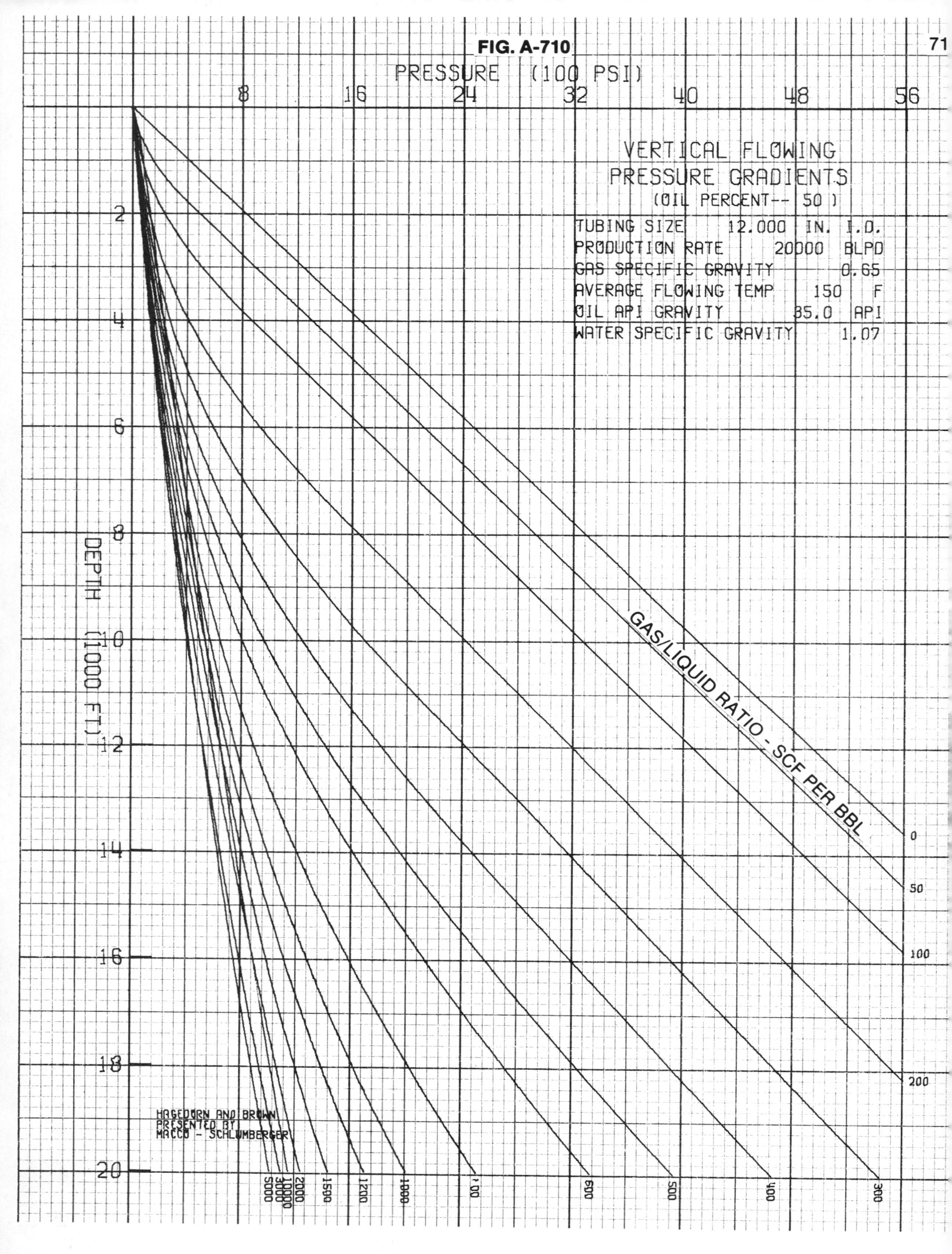

FIG. A-710

71

VERTICAL FLOWING
PRESSURE GRADIENTS
(OIL PERCENT-- 50)

TUBING SIZE	12.000	IN. I.D.
PRODUCTION RATE	20000	BLPD
GAS SPECIFIC GRAVITY	0.65	
AVERAGE FLOWING TEMP	150	F
OIL API GRAVITY	35.0	API
WATER SPECIFIC GRAVITY	1.07	

PRESSURE (100 PSI)

DEPTH (1000 FT)

GAS/LIQUID RATIO - SCF PER BBL

HAGEDORN AND BROWN
PRESENTED BY
MACCO - SCHLUMBERGER

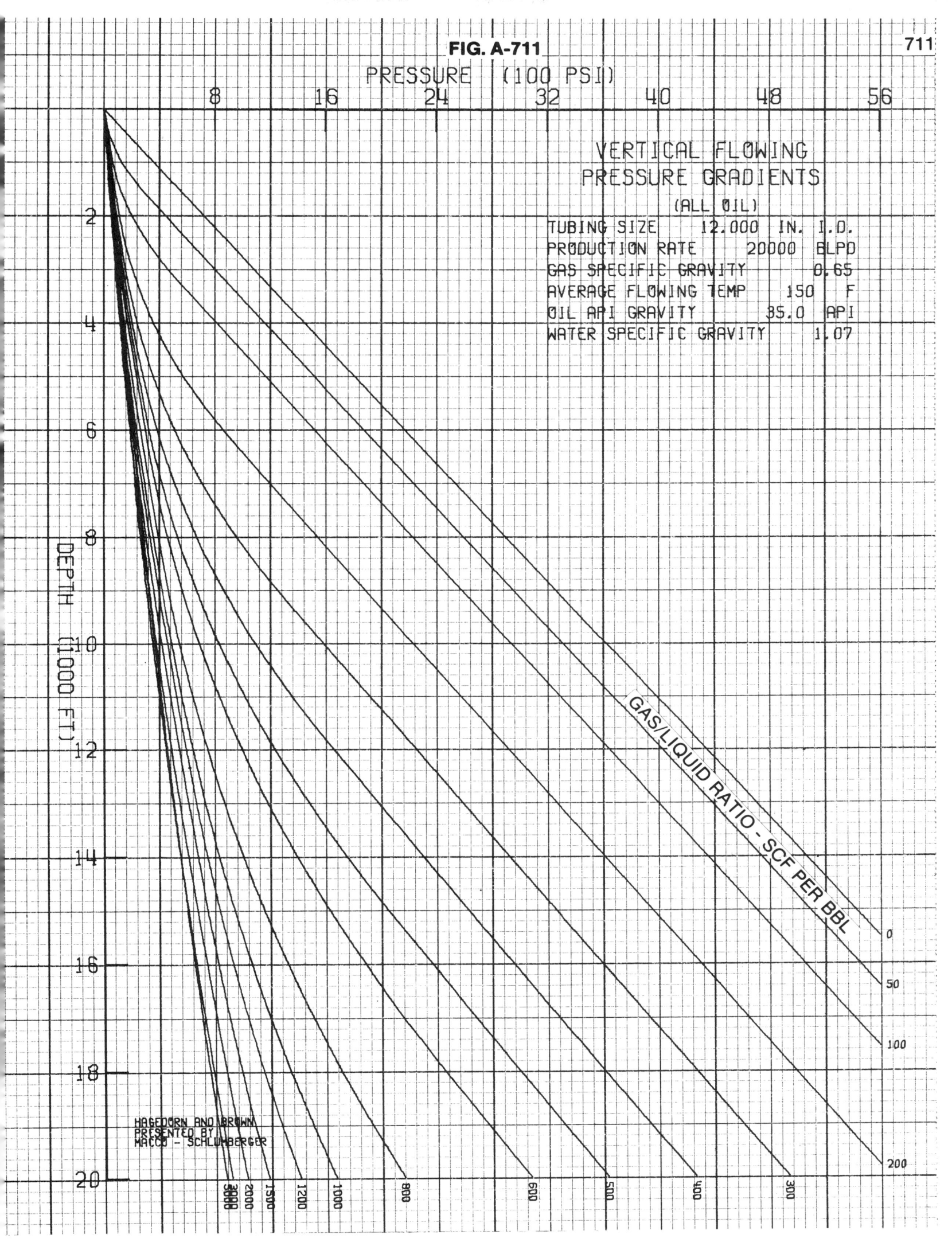

FIG. A-711

VERTICAL FLOWING
PRESSURE GRADIENTS
(ALL OIL)

TUBING SIZE	12.000	IN. I.D.
PRODUCTION RATE	20000	BLPD
GAS SPECIFIC GRAVITY	0.65	
AVERAGE FLOWING TEMP	150	F
OIL API GRAVITY	35.0	API
WATER SPECIFIC GRAVITY	1.07	

PRESSURE (100 PSI)

DEPTH (1000 FT)

GAS/LIQUID RATIO - SCF PER BBL

HAGEDORN AND BROWN
PRESENTED BY
MACCO - SCHLUMBERGER

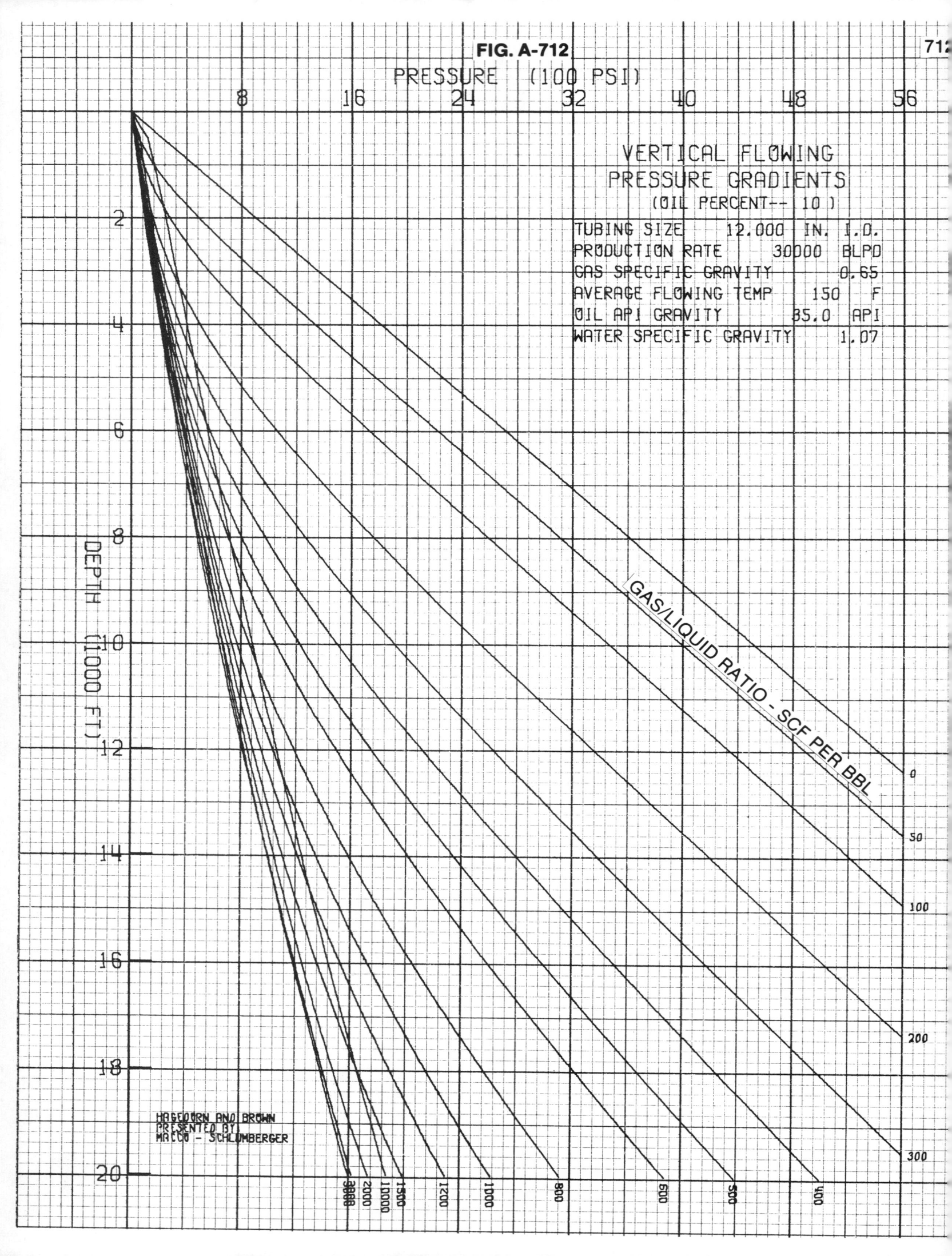

FIG. A-712

712

VERTICAL FLOWING
PRESSURE GRADIENTS
(OIL PERCENT-- 10)

TUBING SIZE	12.000 IN. I.D.
PRODUCTION RATE	30000 BLPD
GAS SPECIFIC GRAVITY	0.65
AVERAGE FLOWING TEMP	150 F
OIL API GRAVITY	35.0 API
WATER SPECIFIC GRAVITY	1.07

PRESSURE (100 PSI)

DEPTH (1000 FT)

GAS/LIQUID RATIO - SCF PER BBL

HAGEDORN AND BROWN
PRESENTED BY
MACCO - SCHLUMBERGER

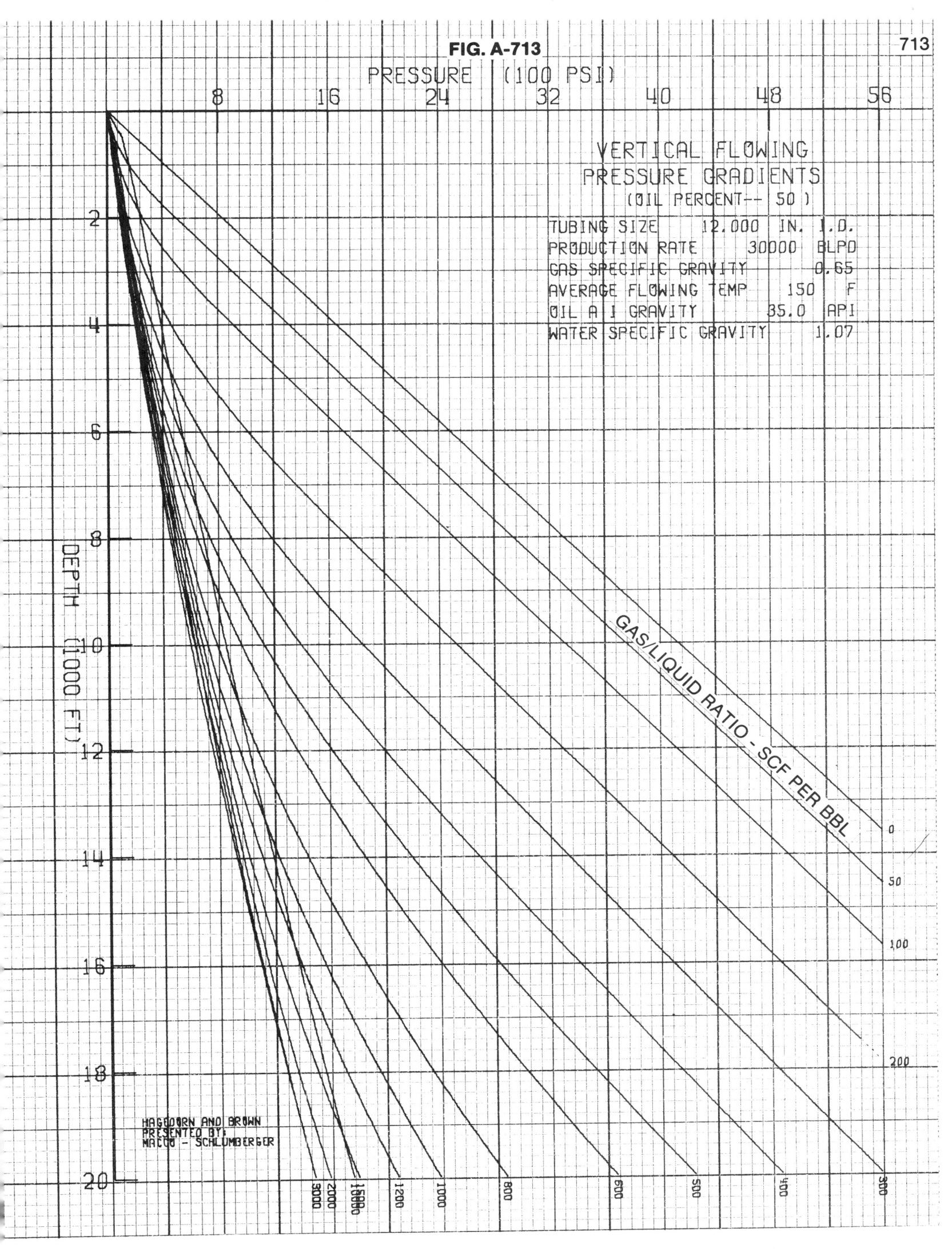

PRESSURE (100 PSI)

VERTICAL FLOWING
PRESSURE GRADIENTS
(OIL PERCENT-- 50)

TUBING SIZE	12.000	IN. I.D.
PRODUCTION RATE	30000	BLPD
GAS SPECIFIC GRAVITY	0.65	
AVERAGE FLOWING TEMP	150	F
OIL A I GRAVITY	35.0	API
WATER SPECIFIC GRAVITY	1.07	

GAS/LIQUID RATIO - SCF PER BBL

DEPTH (1000 FT)

HAGEDORN AND BROWN
PRESENTED BY:
MACCO - SCHLUMBERGER

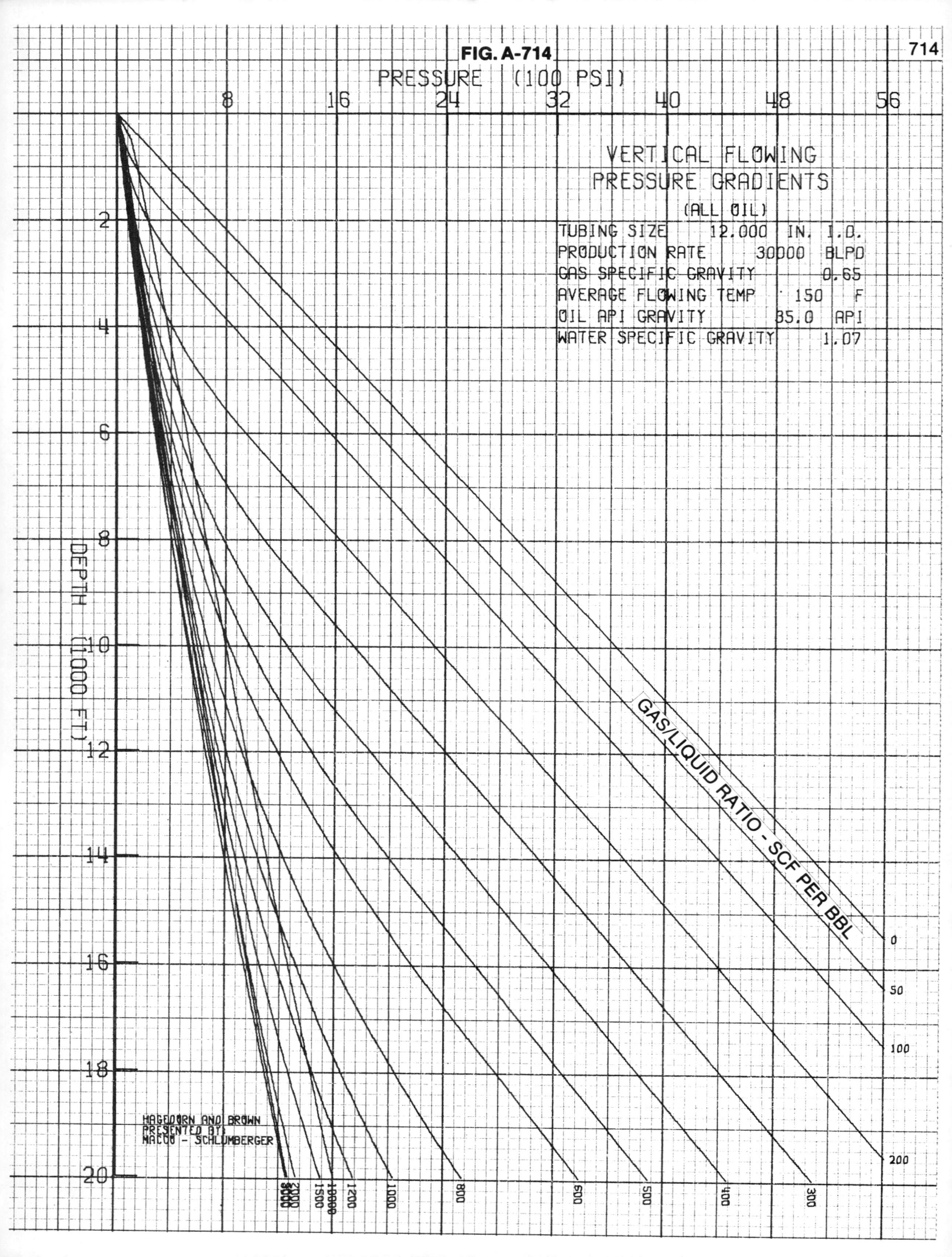

FIG. A-714

714

PRESSURE (100 PSI)

VERTICAL FLOWING
PRESSURE GRADIENTS
(ALL OIL)

TUBING SIZE	12.000	IN. I.D.
PRODUCTION RATE	30000	BLPD
GAS SPECIFIC GRAVITY	0.65	
AVERAGE FLOWING TEMP	150	F
OIL API GRAVITY	35.0	API
WATER SPECIFIC GRAVITY	1.07	

DEPTH (1000 FT)

GAS/LIQUID RATIO - SCF PER BBL

HAGEDOORN AND BROWN
PRESENTED BY:
MACCO - SCHLUMBERGER

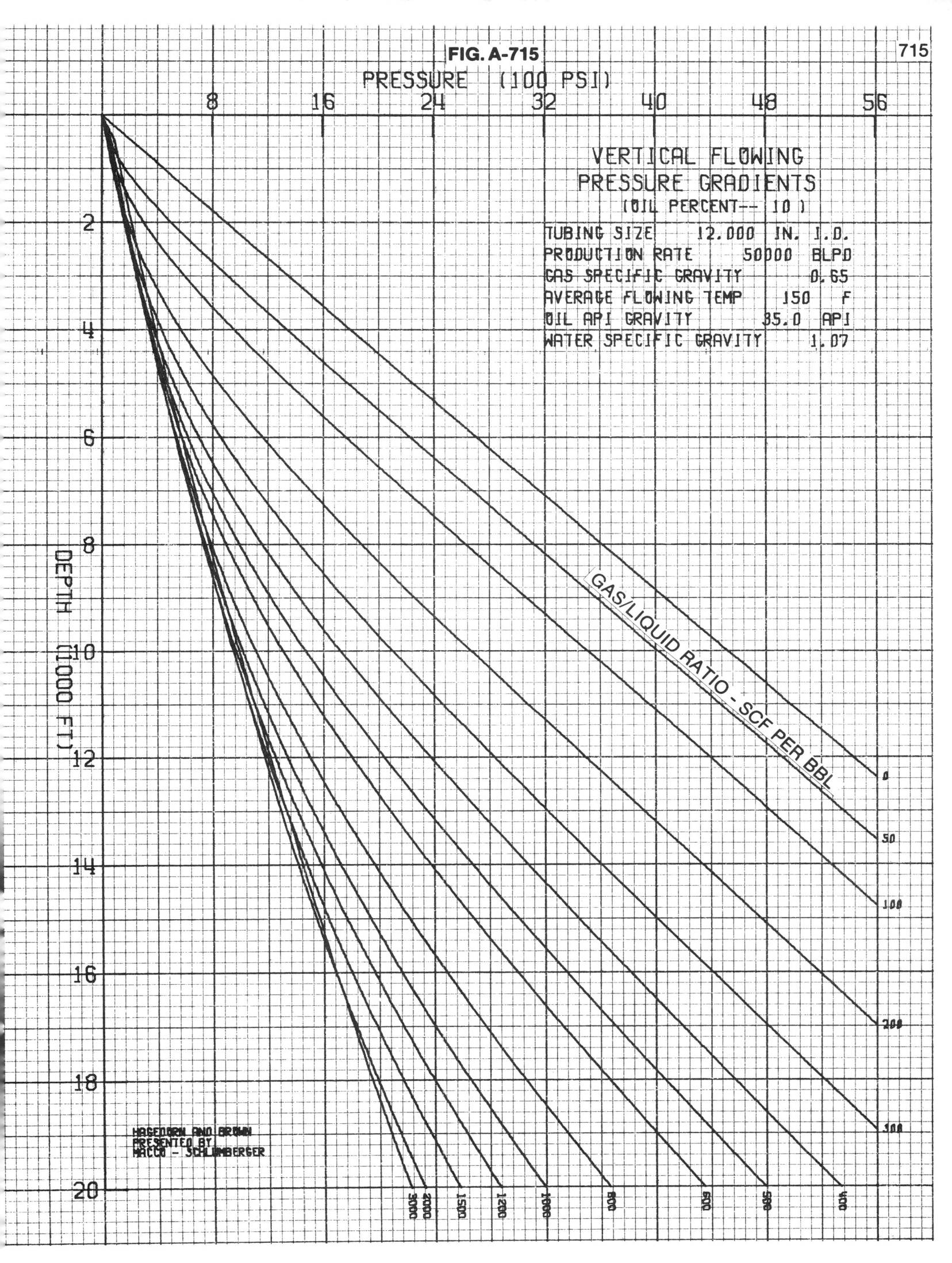

VERTICAL FLOWING
PRESSURE GRADIENTS
(OIL PERCENT-- 10)

TUBING SIZE	12.000	IN. I.D.
PRODUCTION RATE	50000	BLPD
GAS SPECIFIC GRAVITY	0.65	
AVERAGE FLOWING TEMP	150	F
OIL API GRAVITY	35.0	API
WATER SPECIFIC GRAVITY	1.07	

PRESSURE (100 PSI)

DEPTH (1000 FT)

GAS/LIQUID RATIO - SCF PER BBL

HAGEDORN AND BROWN
PRESENTED BY
MACCO - SCHLUMBERGER

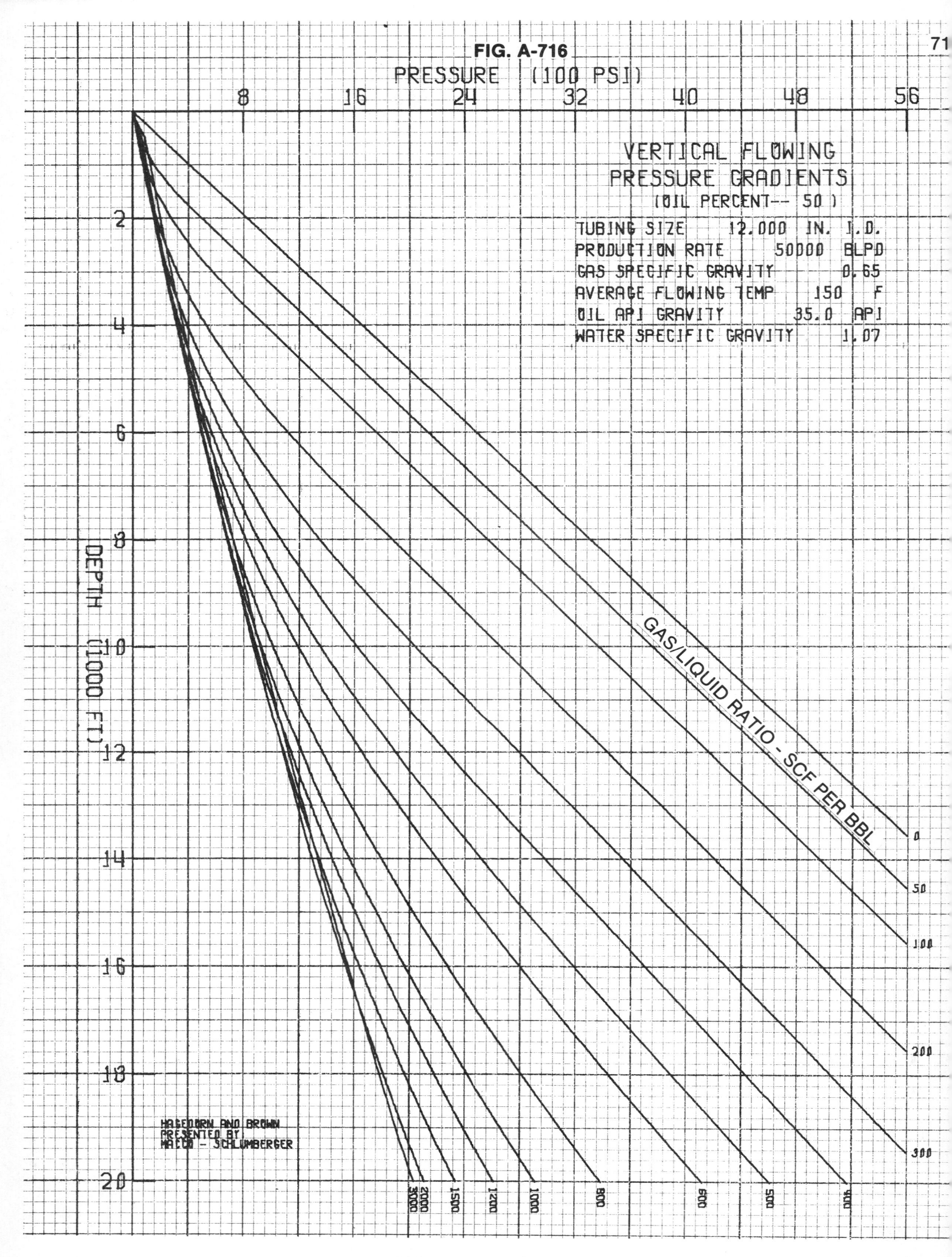

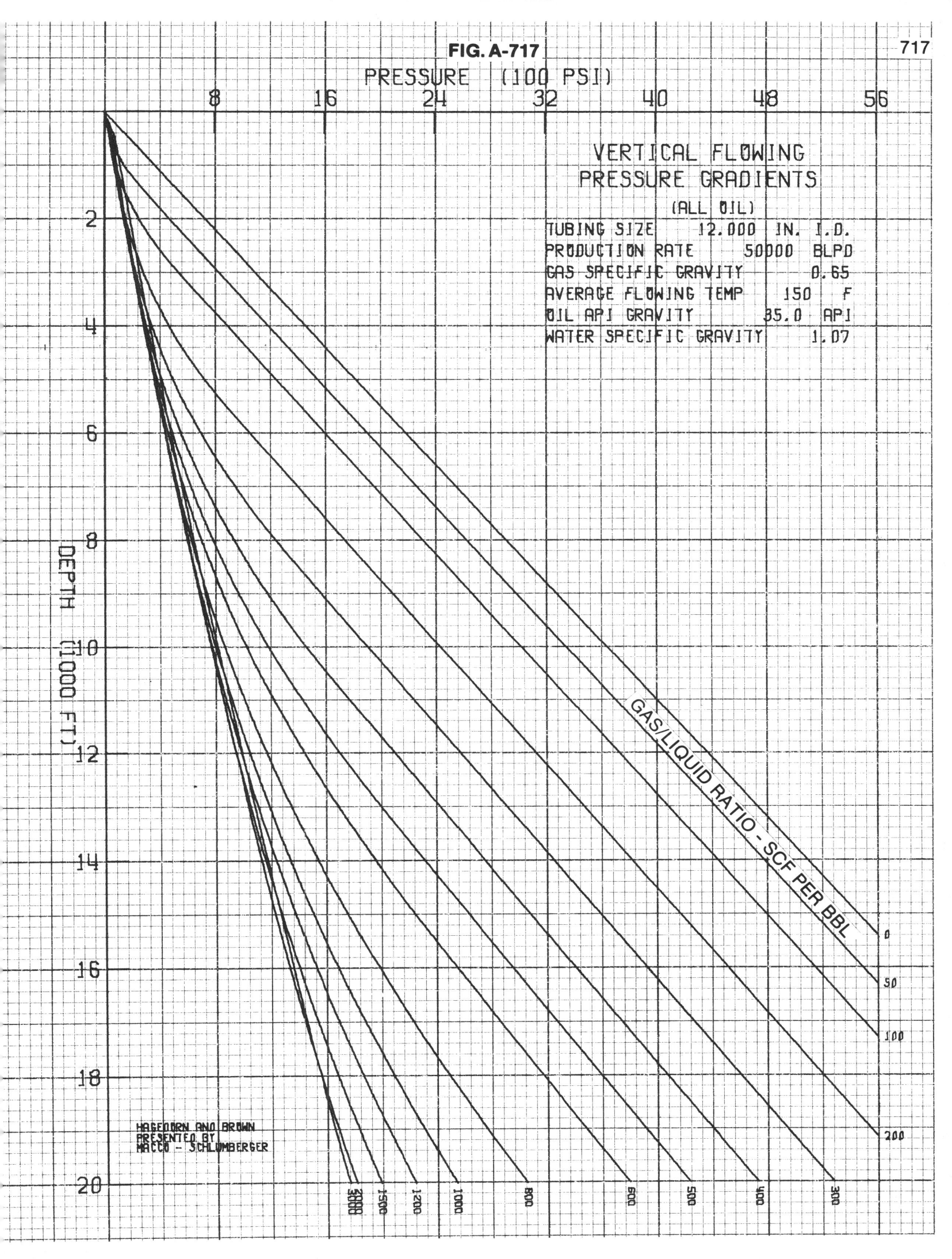

PRESSURE (100 PSI)

VERTICAL FLOWING
PRESSURE GRADIENTS
(ALL OIL)

TUBING SIZE	12.000	IN. I.D.
PRODUCTION RATE	50000	BLPD
GAS SPECIFIC GRAVITY	0.65	
AVERAGE FLOWING TEMP	150	F
OIL API GRAVITY	35.0	API
WATER SPECIFIC GRAVITY	1.07	

DEPTH (1000 FT)

GAS/LIQUID RATIO - SCF PER BBL

HAGEDORN AND BROWN
PRESENTED BY
MACCO - SCHLUMBERGER

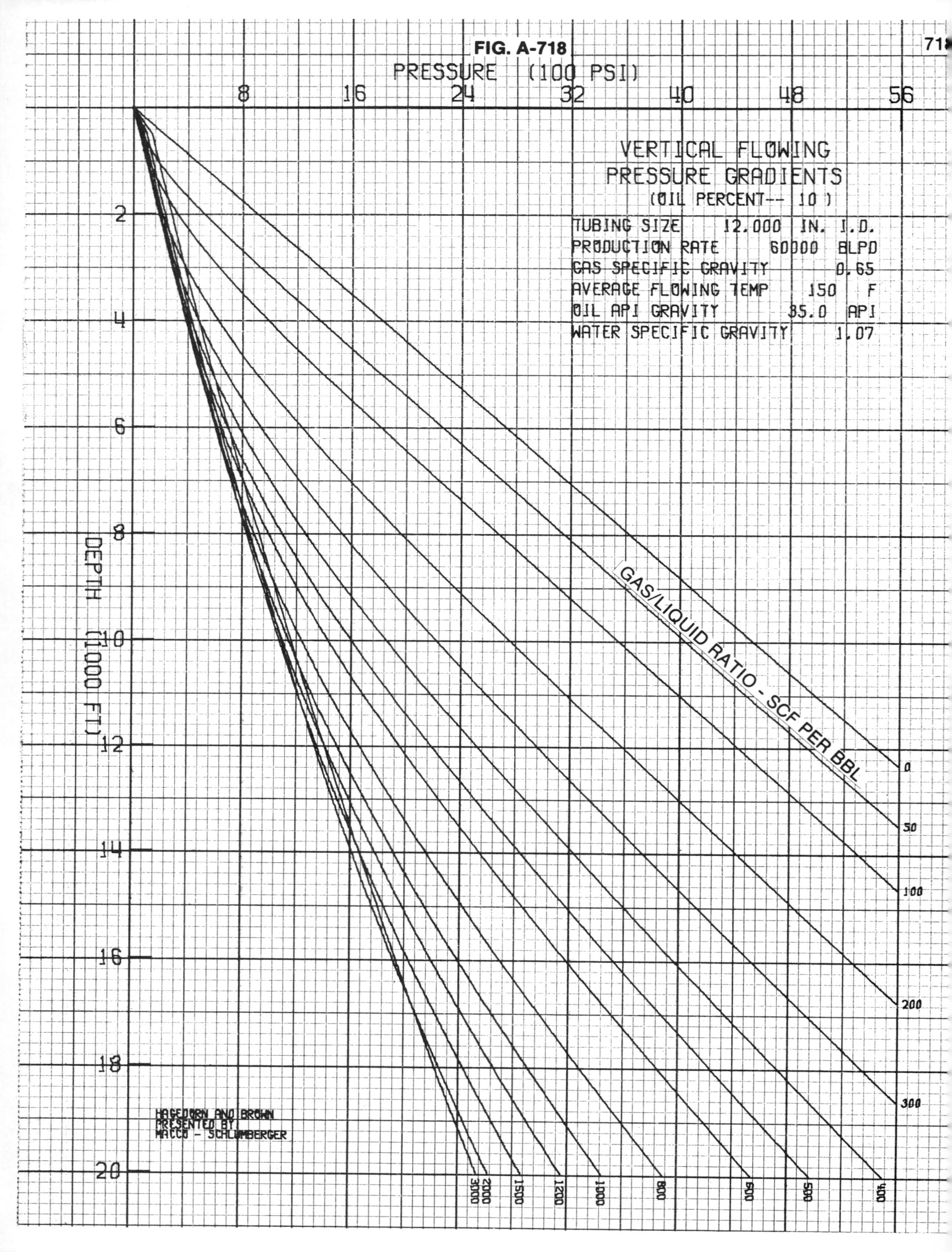

FIG. A-718

718

VERTICAL FLOWING
PRESSURE GRADIENTS
(OIL PERCENT-- 10)

TUBING SIZE	12.000	IN. I.D.
PRODUCTION RATE	60000	BLPD
GAS SPECIFIC GRAVITY	0.65	
AVERAGE FLOWING TEMP	150	F
OIL API GRAVITY	35.0	API
WATER SPECIFIC GRAVITY	1.07	

PRESSURE (100 PSI)

DEPTH (1000 FT)

GAS/LIQUID RATIO - SCF PER BBL

HAGEDORN AND BROWN
PRESENTED BY
WAECO - SCHLUMBERGER

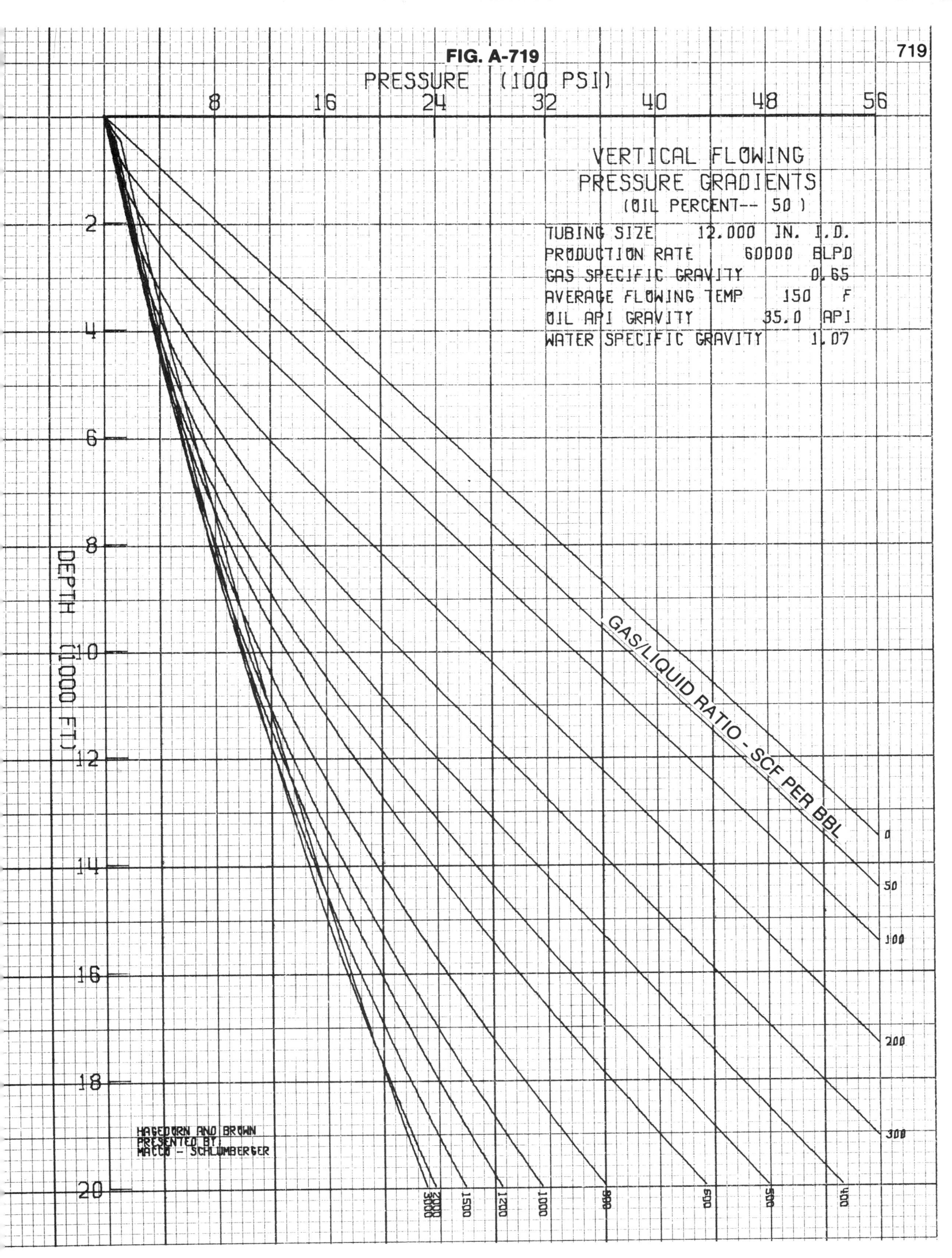

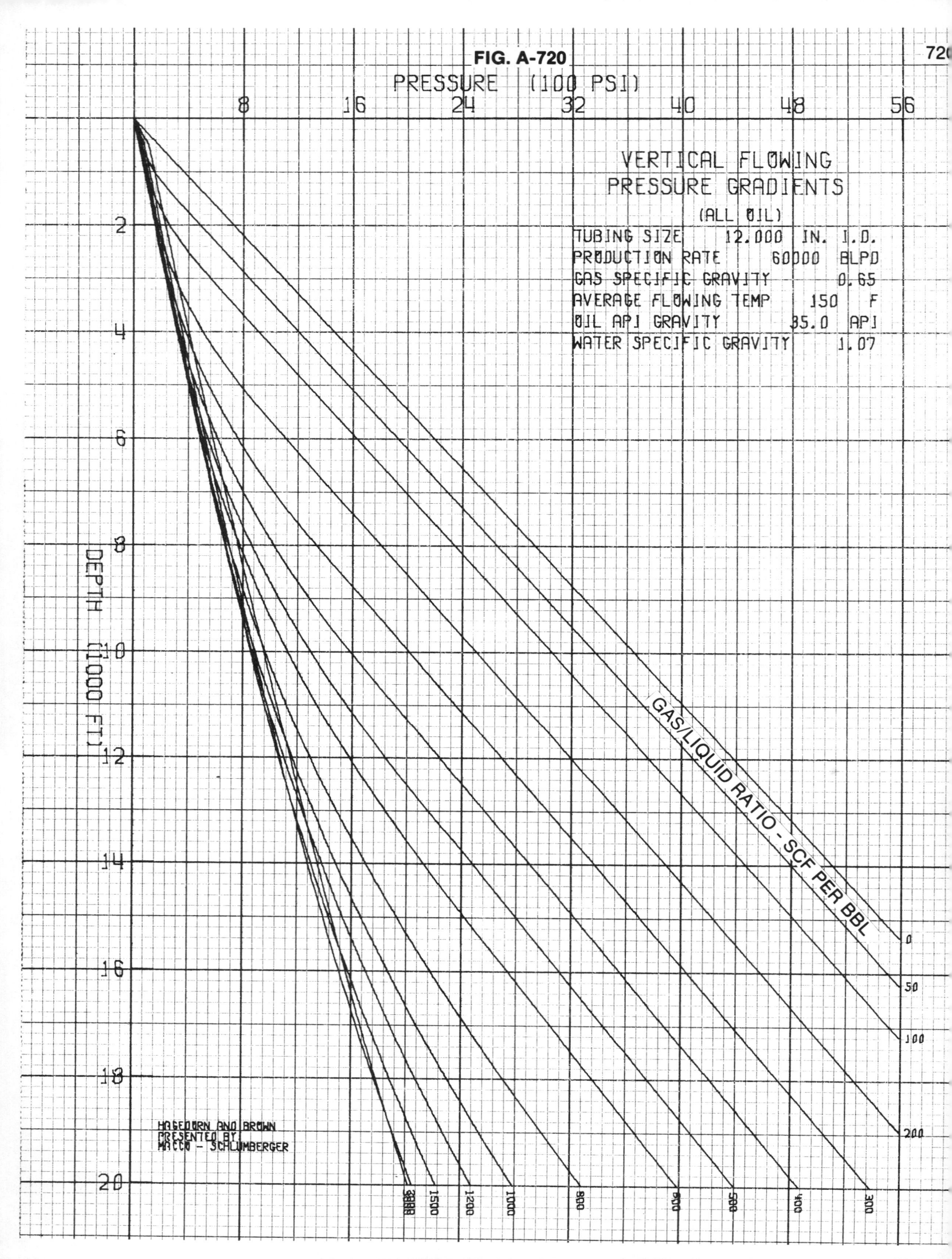

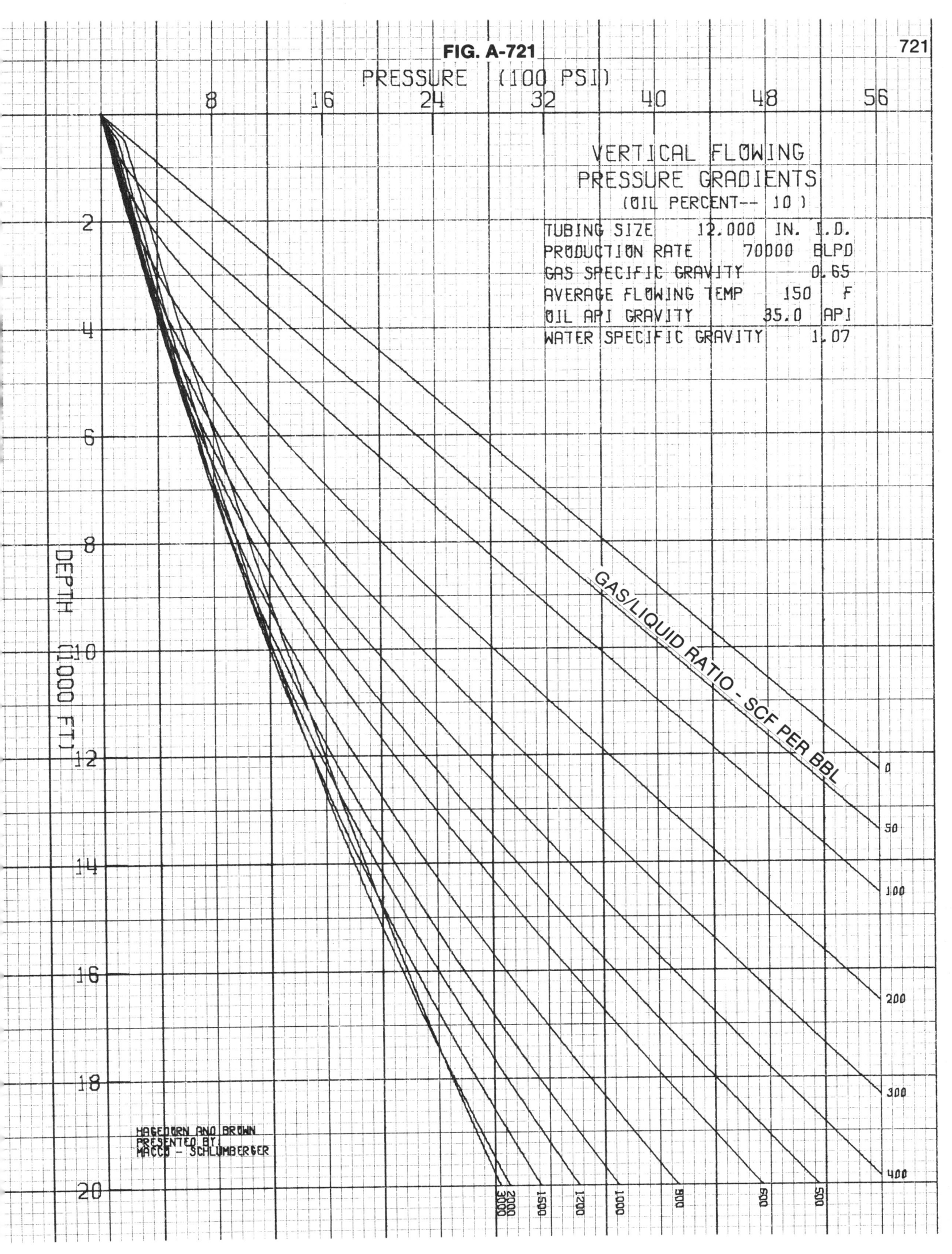

VERTICAL FLOWING
PRESSURE GRADIENTS
(OIL PERCENT-- 10)

TUBING SIZE	12.000	IN.	I.D.
PRODUCTION RATE	70000		BLPD
GAS SPECIFIC GRAVITY			0.65
AVERAGE FLOWING TEMP	150		F
OIL API GRAVITY	35.0		API
WATER SPECIFIC GRAVITY			1.07

PRESSURE (100 PSI)

DEPTH (1000 FT)

GAS/LIQUID RATIO - SCF PER BBL

HAGEDORN AND BROWN
PRESENTED BY:
MACCO - SCHLUMBERGER

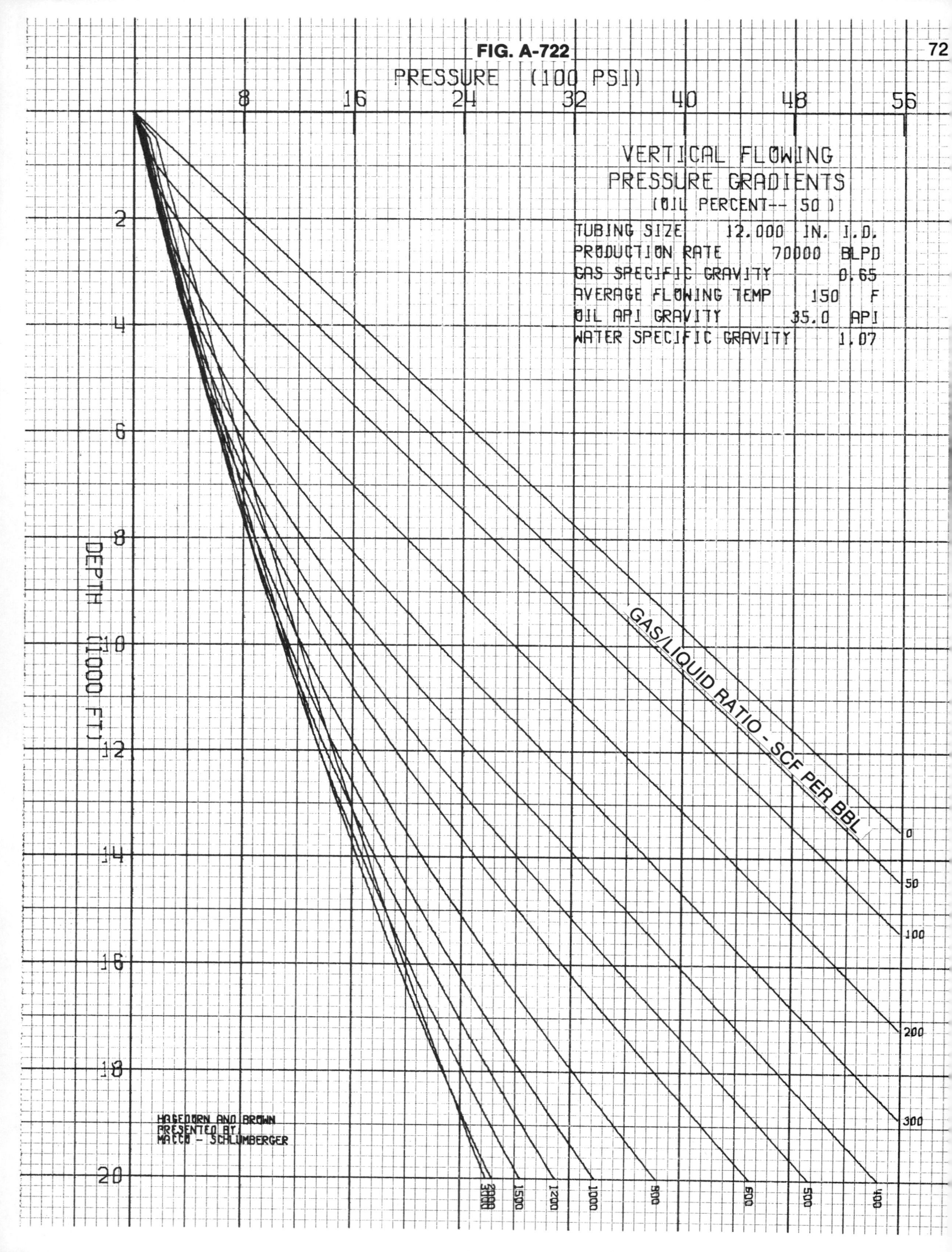

FIG. A-722

72

PRESSURE (100 PSI)

VERTICAL FLOWING
PRESSURE GRADIENTS
(OIL PERCENT-- 50)

TUBING SIZE   12.000 IN. I.D.
PRODUCTION RATE   70000 BLPD
GAS SPECIFIC GRAVITY   0.65
AVERAGE FLOWING TEMP   150   F
OIL API GRAVITY   35.0 API
WATER SPECIFIC GRAVITY   1.07

GAS/LIQUID RATIO - SCF PER BBL

DEPTH (1000 FT)

HAGEDORN AND BROWN
PRESENTED BY
MAECO - SCHLUMBERGER

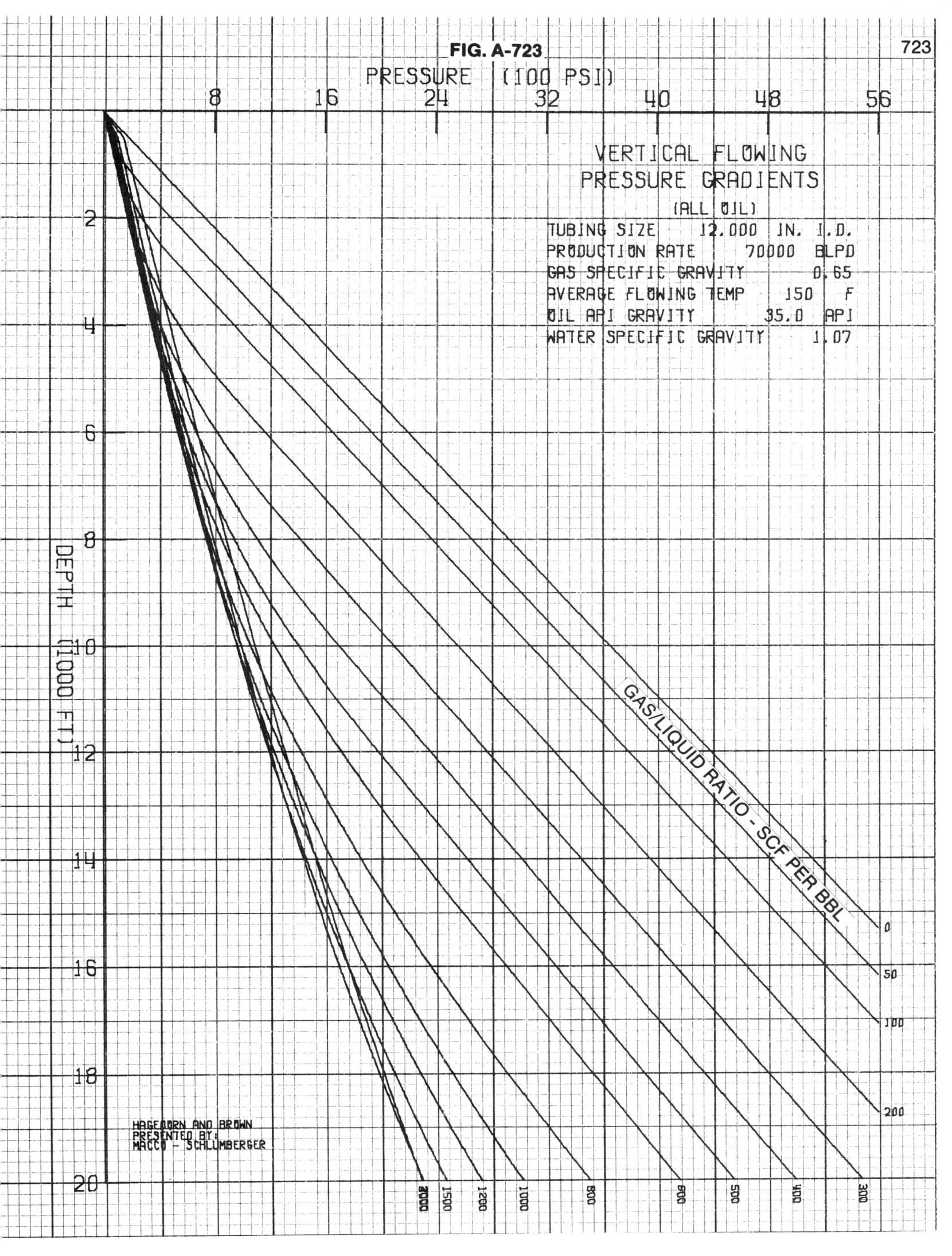

VERTICAL FLOWING
PRESSURE GRADIENTS
(ALL OIL)

TUBING SIZE	12.000	IN. I.D.
PRODUCTION RATE	70000	BLPD
GAS SPECIFIC GRAVITY	0.65	
AVERAGE FLOWING TEMP	150	F
OIL API GRAVITY	35.0	API
WATER SPECIFIC GRAVITY	1.07	

GAS/LIQUID RATIO - SCF PER BBL

HAGEDORN AND BROWN
PRESENTED BY:
MACCO - SCHLUMBERGER

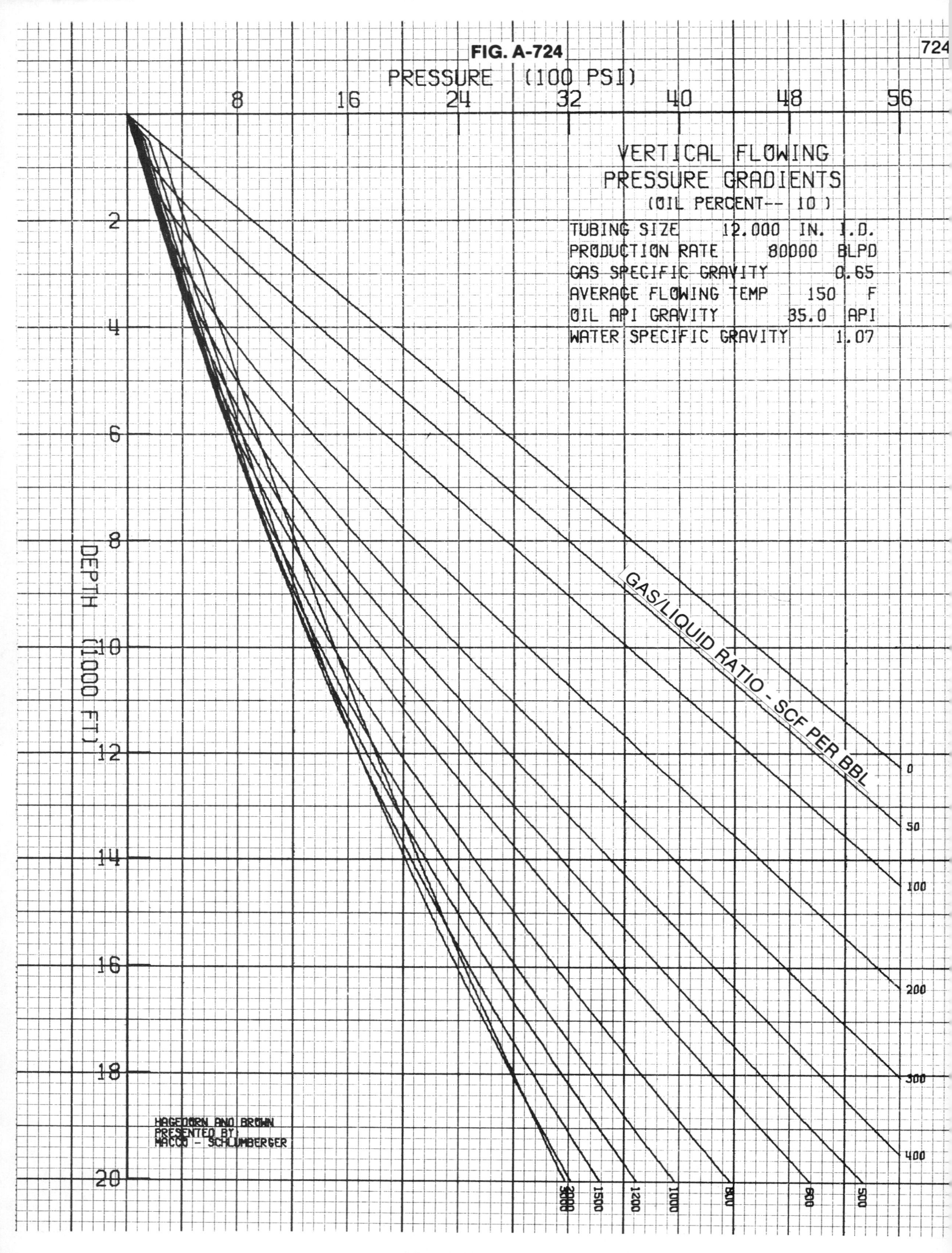

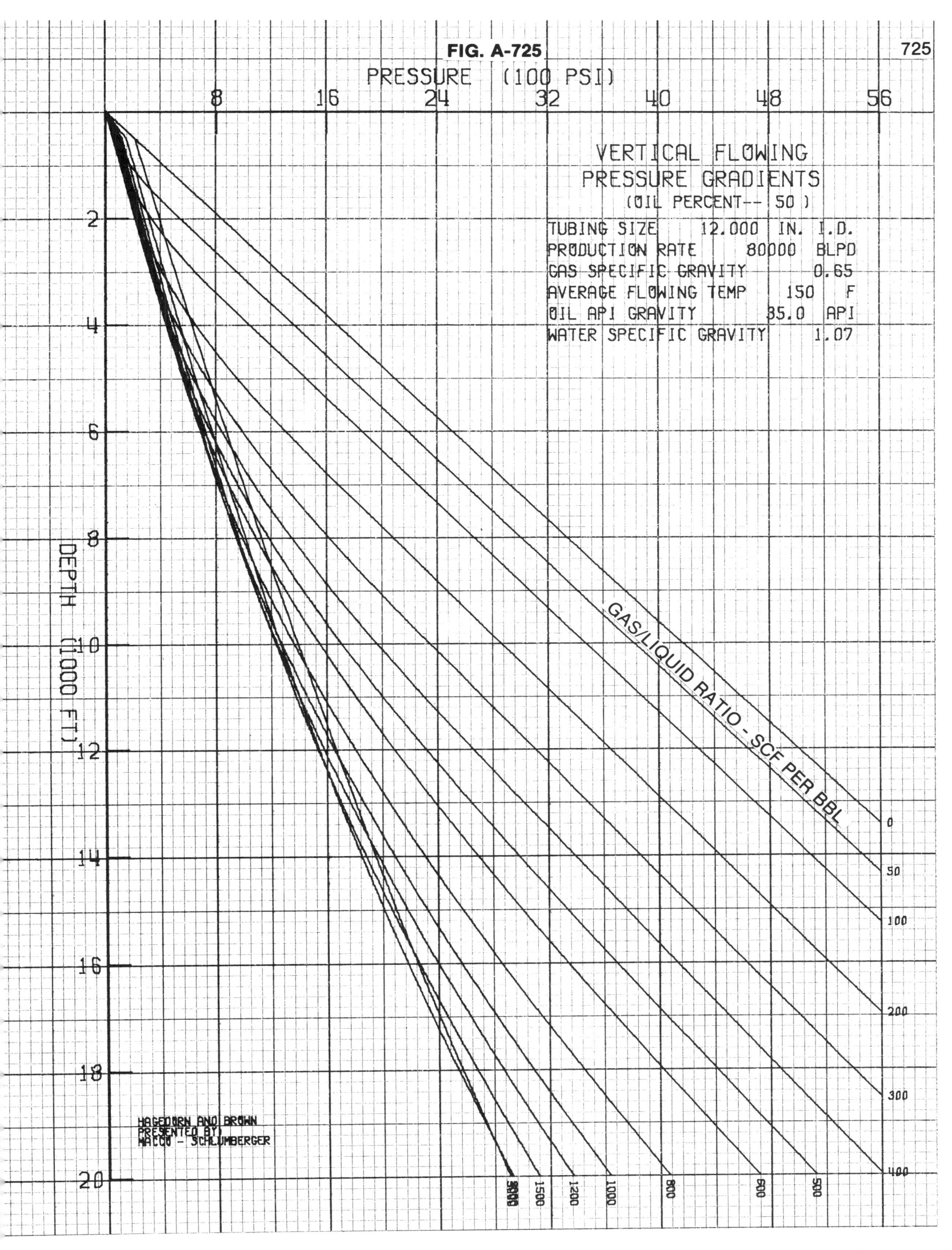

VERTICAL FLOWING
PRESSURE GRADIENTS
(OIL PERCENT-- 50)

TUBING SIZE	12.000	IN. I.D.
PRODUCTION RATE	80000	BLPD
GAS SPECIFIC GRAVITY	0.65	
AVERAGE FLOWING TEMP	150	F
OIL API GRAVITY	35.0	API
WATER SPECIFIC GRAVITY	1.07	

PRESSURE (100 PSI)

DEPTH (1000 FT)

GAS/LIQUID RATIO - SCF PER BBL

HAGEDORN AND BROWN
PRESENTED BY
HALCO - SCHLUMBERGER

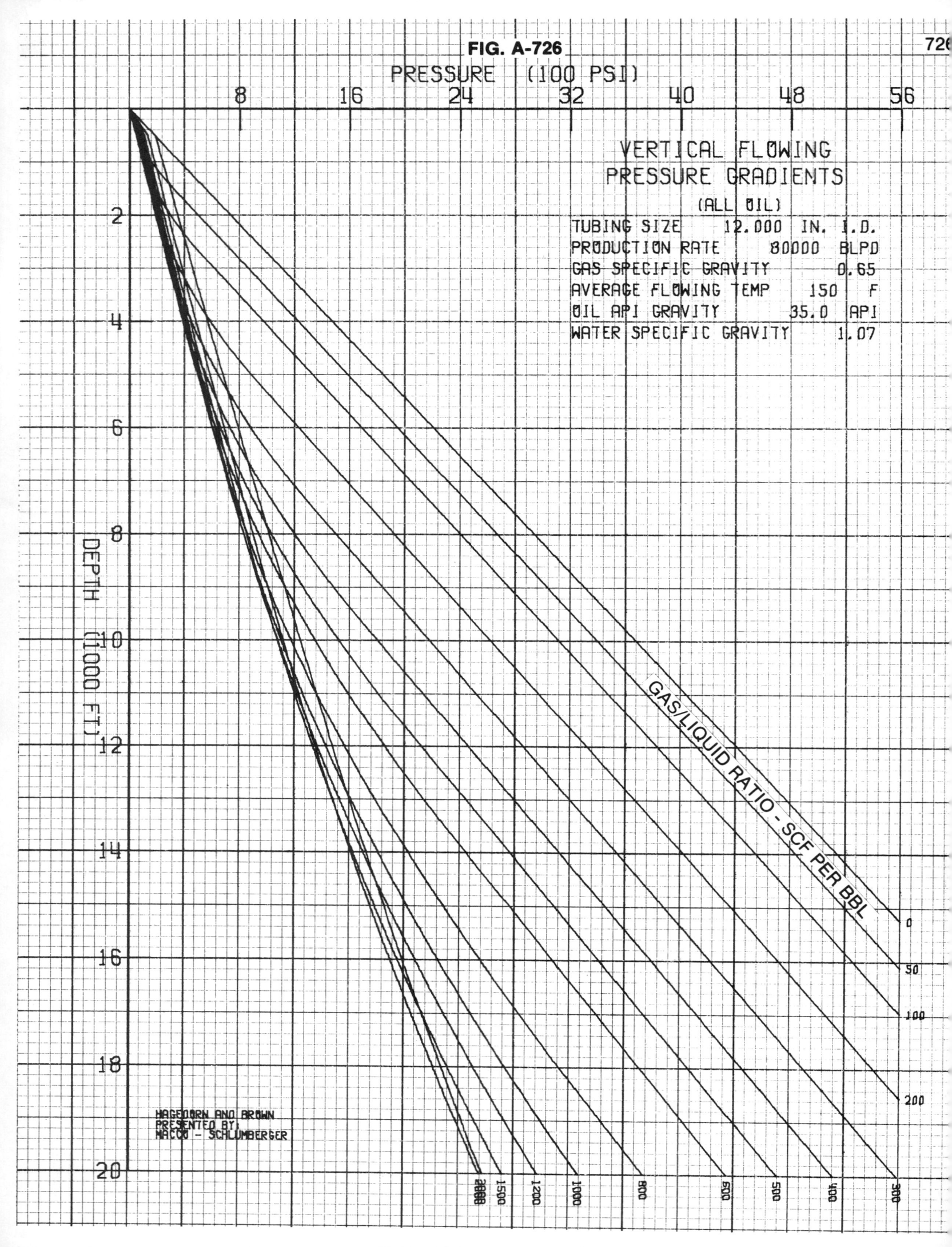

VERTICAL FLOWING
PRESSURE GRADIENTS
(ALL OIL)

PRESSURE (100 PSI)

TUBING SIZE	12.000	IN. I.D.
PRODUCTION RATE	80000	BLPD
GAS SPECIFIC GRAVITY	0.65	
AVERAGE FLOWING TEMP	150	F
OIL API GRAVITY	35.0	API
WATER SPECIFIC GRAVITY	1.07	

DEPTH (1000 FT)

GAS/LIQUID RATIO - SCF PER BBL

HAGEDORN AND BROWN
PRESENTED BY
MACCO — SCHLUMBERGER

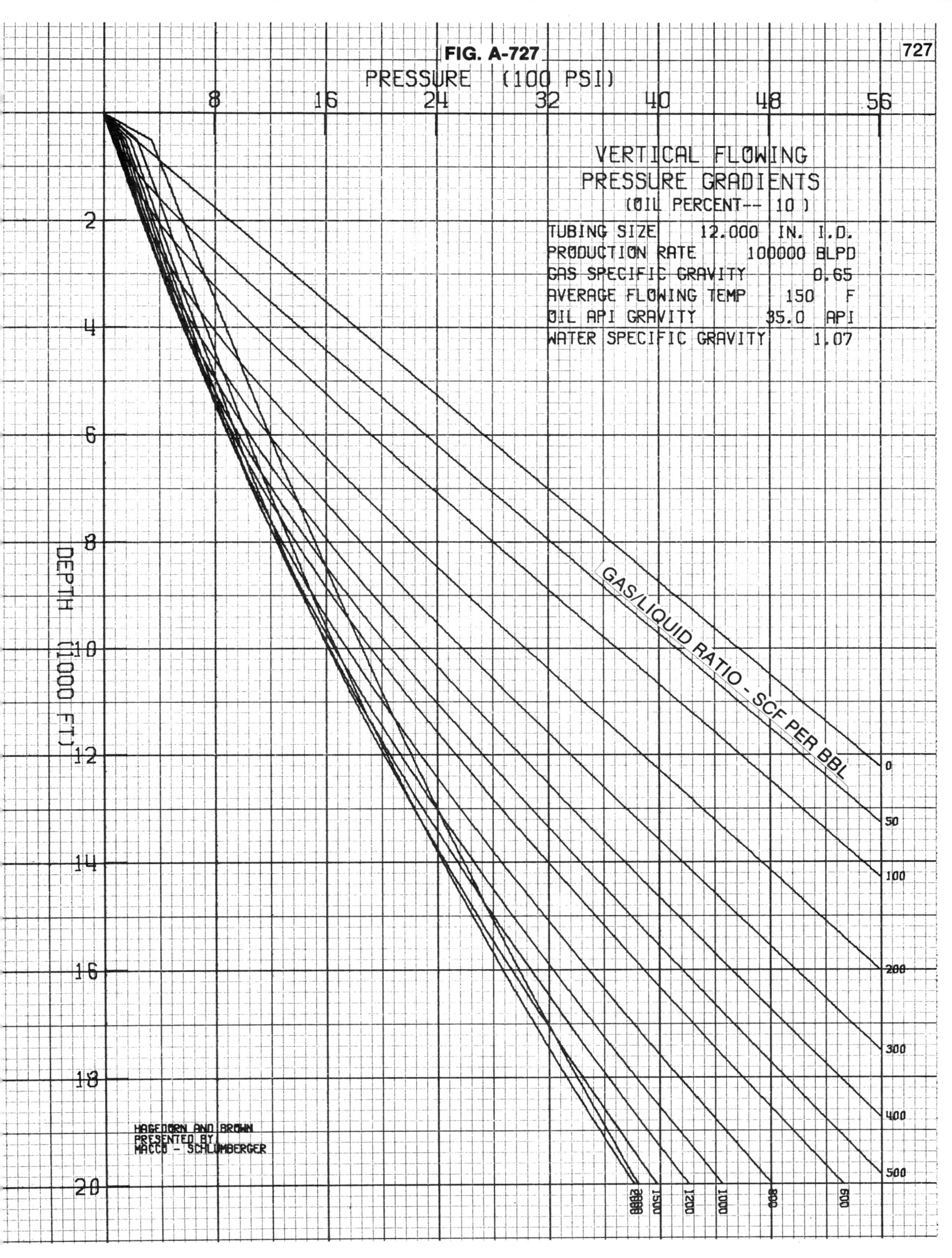

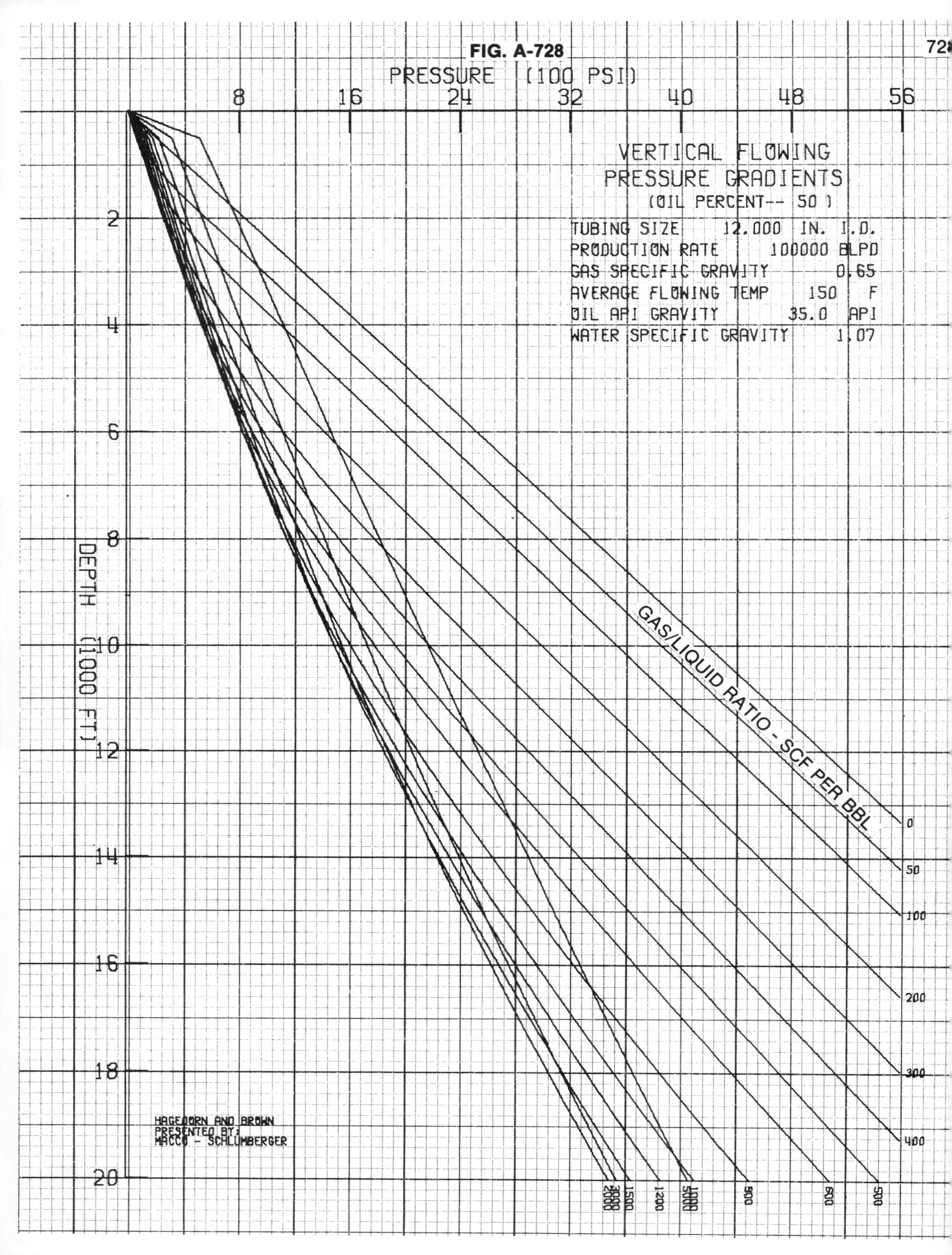

VERTICAL FLOWING
PRESSURE GRADIENTS
(OIL PERCENT-- 50)

TUBING SIZE   12.000 IN. I.D.
PRODUCTION RATE   100000 BLPD
GAS SPECIFIC GRAVITY   0.65
AVERAGE FLOWING TEMP   150   F
OIL API GRAVITY   35.0 API
WATER SPECIFIC GRAVITY   1.07

GAS/LIQUID RATIO - SCF PER BBL

HAGEDORN AND BROWN
PRESENTED BY:
MACCO - SCHLUMBERGER

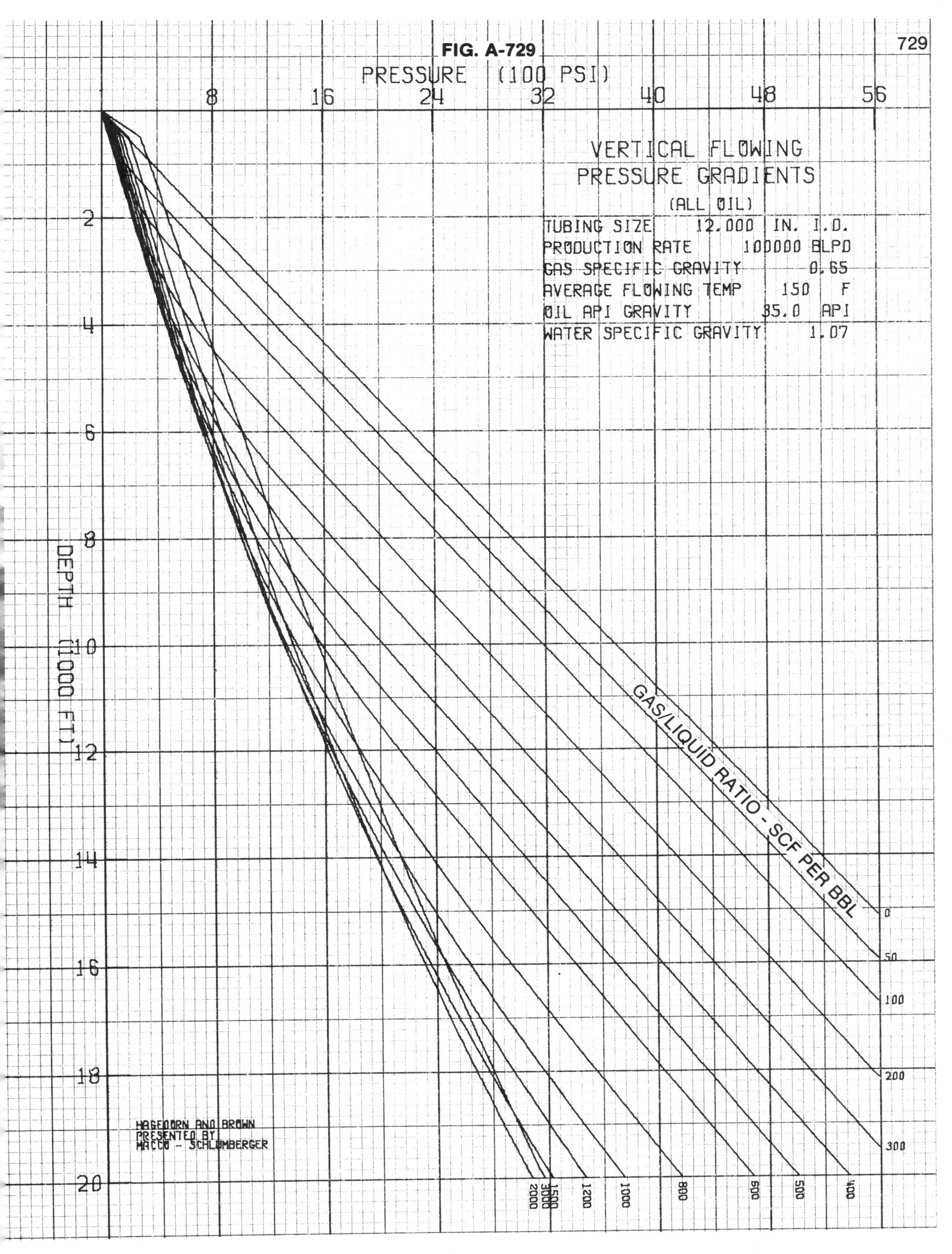

SECTION B

Horizontal Multiphase Pipe Flow

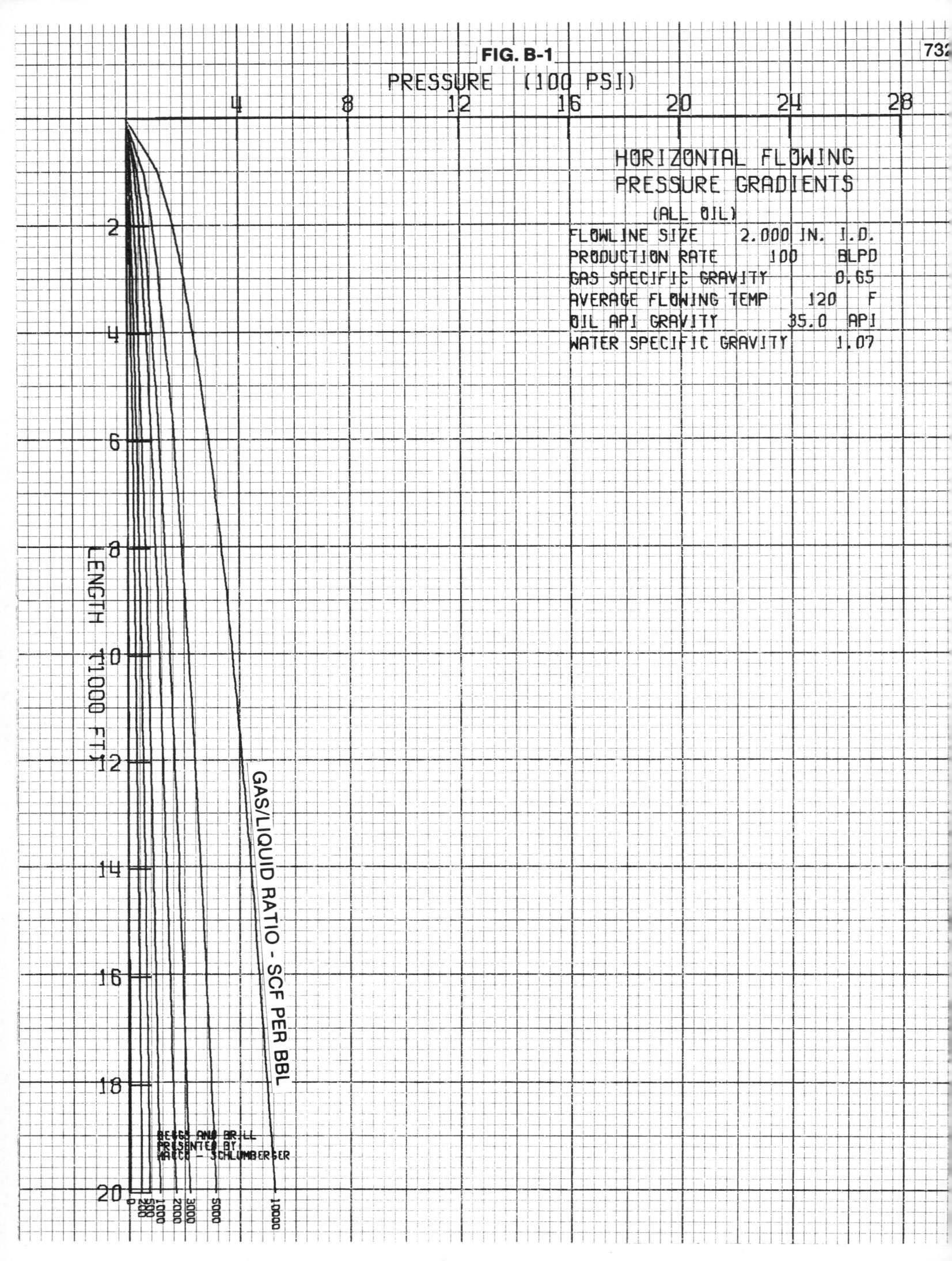

FIG. B-1

732

HORIZONTAL FLOWING
PRESSURE GRADIENTS
(ALL OIL)

FLOWLINE SIZE   2.000 IN. I.D.
PRODUCTION RATE   100   BLPD
GAS SPECIFIC GRAVITY   0.65
AVERAGE FLOWING TEMP   120   F
OIL API GRAVITY   35.0   API
WATER SPECIFIC GRAVITY   1.07

PRESSURE (100 PSI)

LENGTH (1000 FT)

GAS/LIQUID RATIO - SCF PER BBL

BEGGS AND BRILL
PRESENTED BY
MAYCO - SCHLUMBERGER

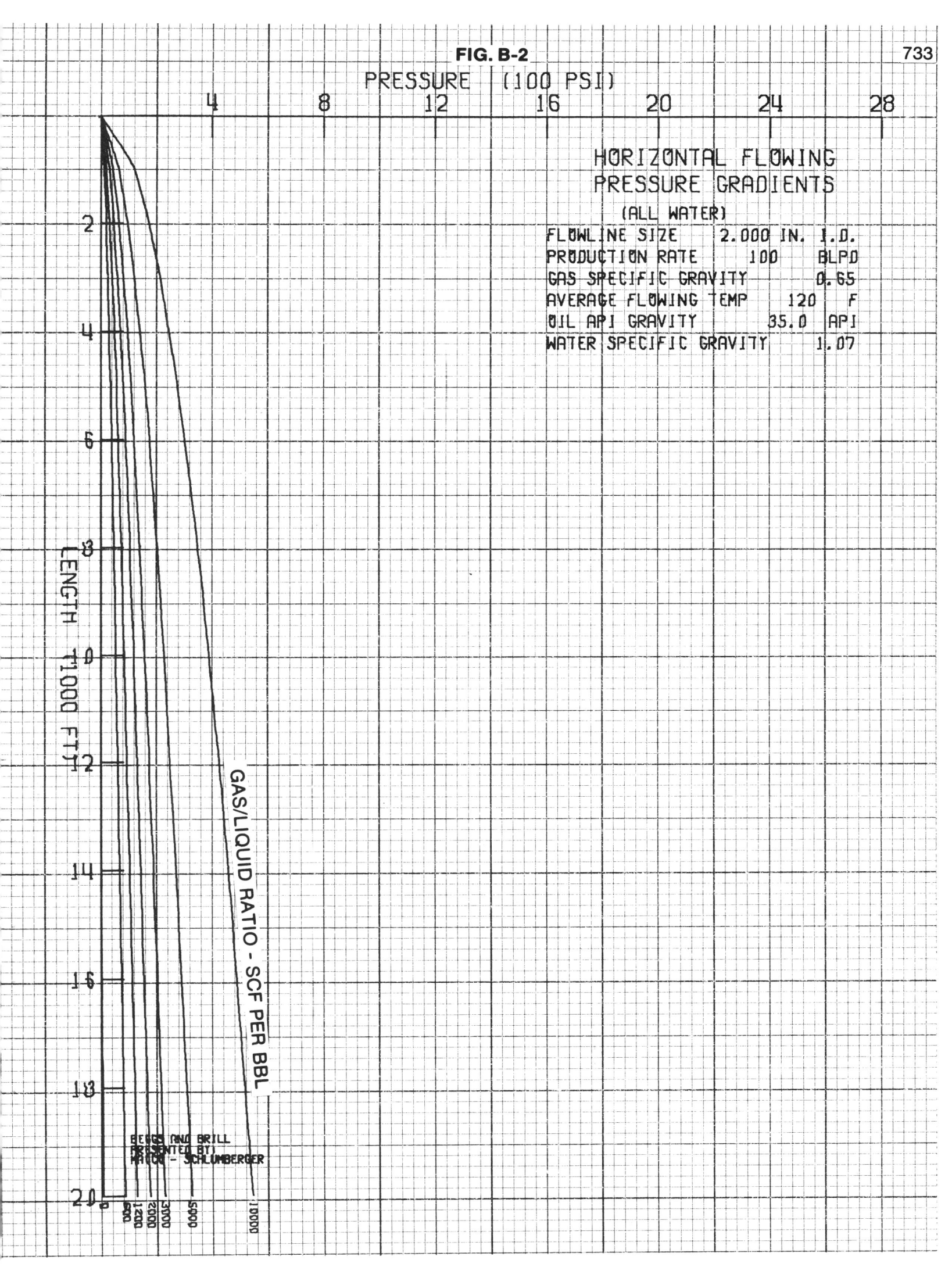

PRESSURE (100 PSI)

HORIZONTAL FLOWING
PRESSURE GRADIENTS

(ALL WATER)

FLOWLINE SIZE	2.000 IN.	I.D.
PRODUCTION RATE	100	BLPD
GAS SPECIFIC GRAVITY	0.65	
AVERAGE FLOWING TEMP	120	F
OIL API GRAVITY	35.0	API
WATER SPECIFIC GRAVITY	1.07	

LENGTH 1000 FT.

GAS/LIQUID RATIO - SCF PER BBL

BEGGS AND BRILL
PRESENTED BY:
MAICO - SCHLUMBERGER

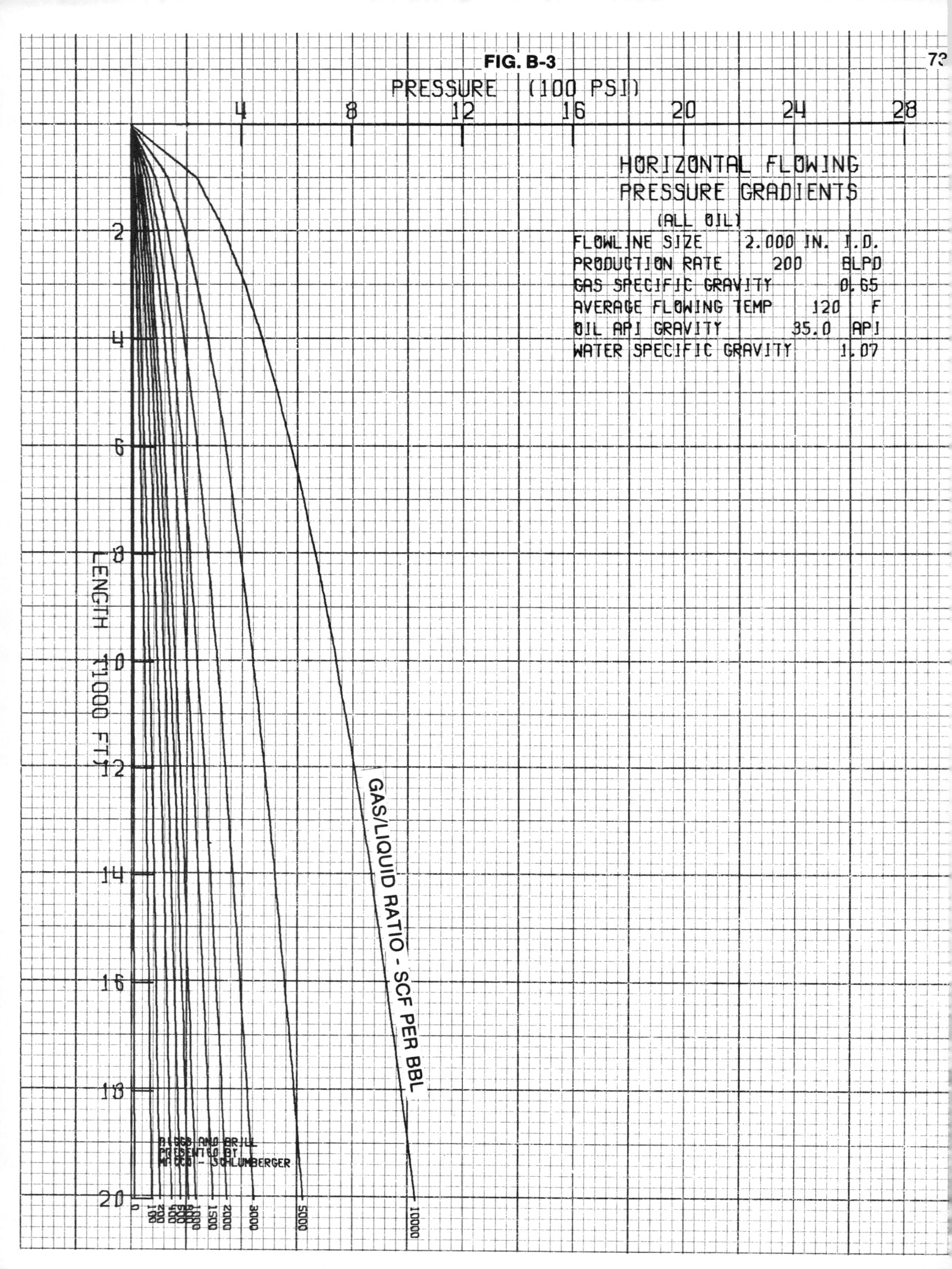

FIG. B-3

73

PRESSURE (100 PSI)

HORIZONTAL FLOWING
PRESSURE GRADIENTS
(ALL OIL)

FLOWLINE SIZE	2.000 IN. I.D.	
PRODUCTION RATE	200	BLPD
GAS SPECIFIC GRAVITY	0.65	
AVERAGE FLOWING TEMP	120	F
OIL API GRAVITY	35.0	API
WATER SPECIFIC GRAVITY	1.07	

LENGTH (1000 FT)

GAS/LIQUID RATIO - SCF PER BBL

BEGGS AND BRILL
PRESENTED BY
MICROI - SCHLUMBERGER

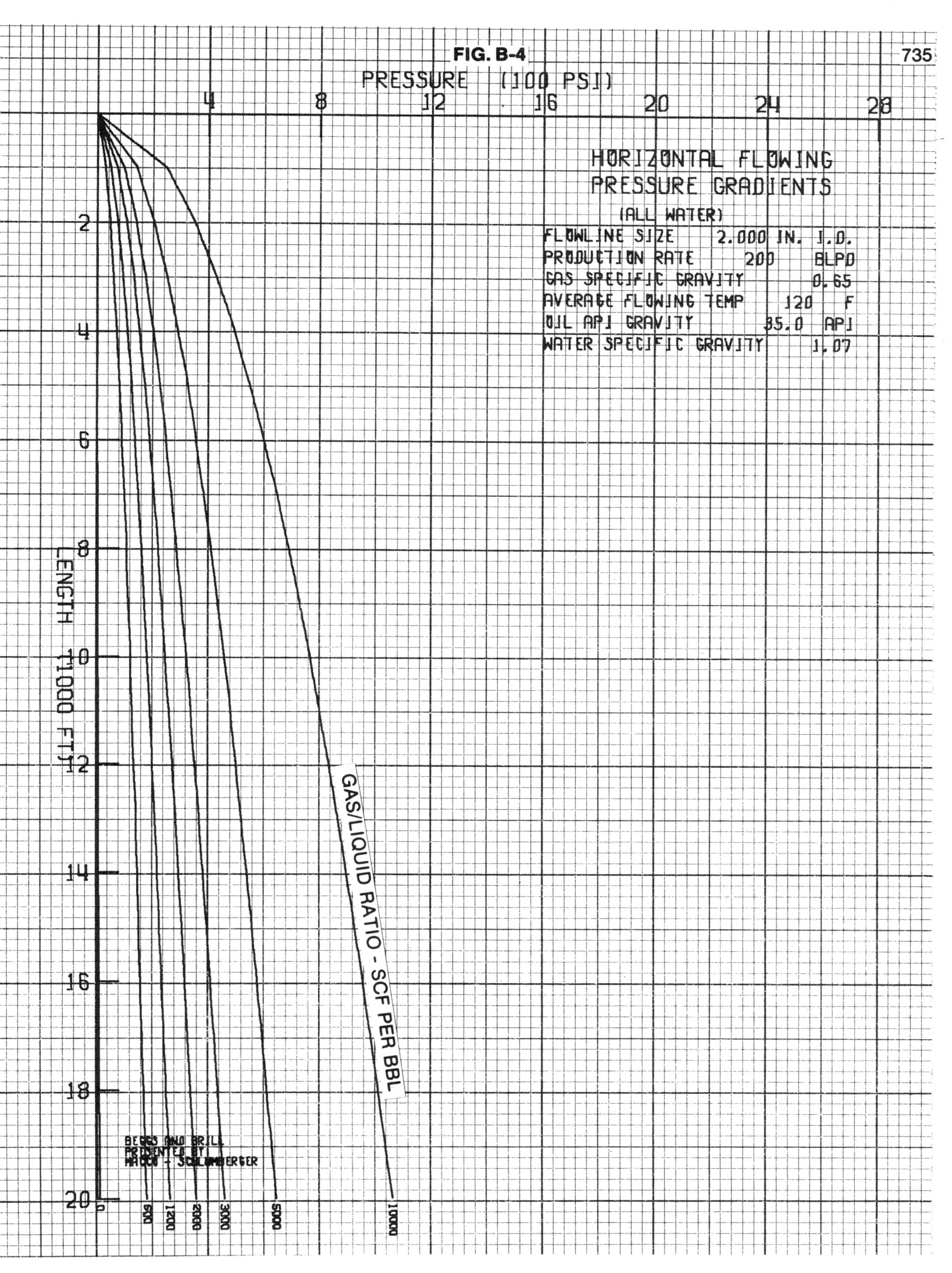

PRESSURE (100 PSI)

HORIZONTAL FLOWING
PRESSURE GRADIENTS
(ALL WATER)

FLOWLINE SIZE	2.000 IN. I.D.
PRODUCTION RATE	200 BLPD
GAS SPECIFIC GRAVITY	0.65
AVERAGE FLOWING TEMP	120 F
OIL API GRAVITY	35.0 API
WATER SPECIFIC GRAVITY	1.07

LENGTH (1000 FT)

GAS/LIQUID RATIO - SCF PER BBL

BEGGS AND BRILL
PRESENTED BY
MADCO - SCHLUMBERGER

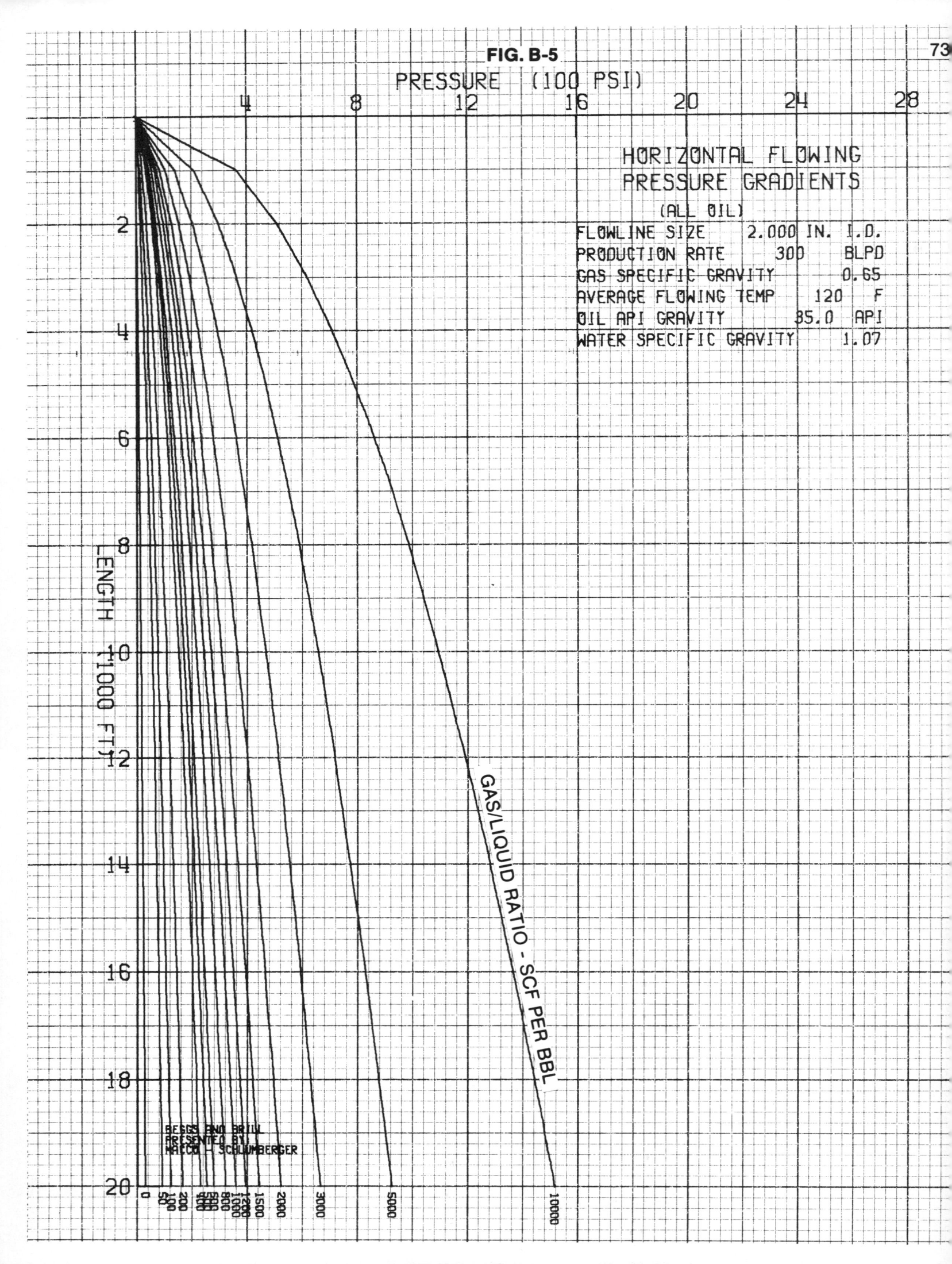

FIG. B-5

73

PRESSURE (100 PSI)

HORIZONTAL FLOWING
PRESSURE GRADIENTS
(ALL OIL)

FLOWLINE SIZE	2.000 IN.	I.D.
PRODUCTION RATE	300	BLPD
GAS SPECIFIC GRAVITY	0.65	
AVERAGE FLOWING TEMP	120	F
OIL API GRAVITY	85.0	API
WATER SPECIFIC GRAVITY	1.07	

LENGTH (1000 FT.)

GAS/LIQUID RATIO - SCF PER BBL

BEGGS AND BRILL
PRESENTED BY
HAECO H SCHLUMBERGER

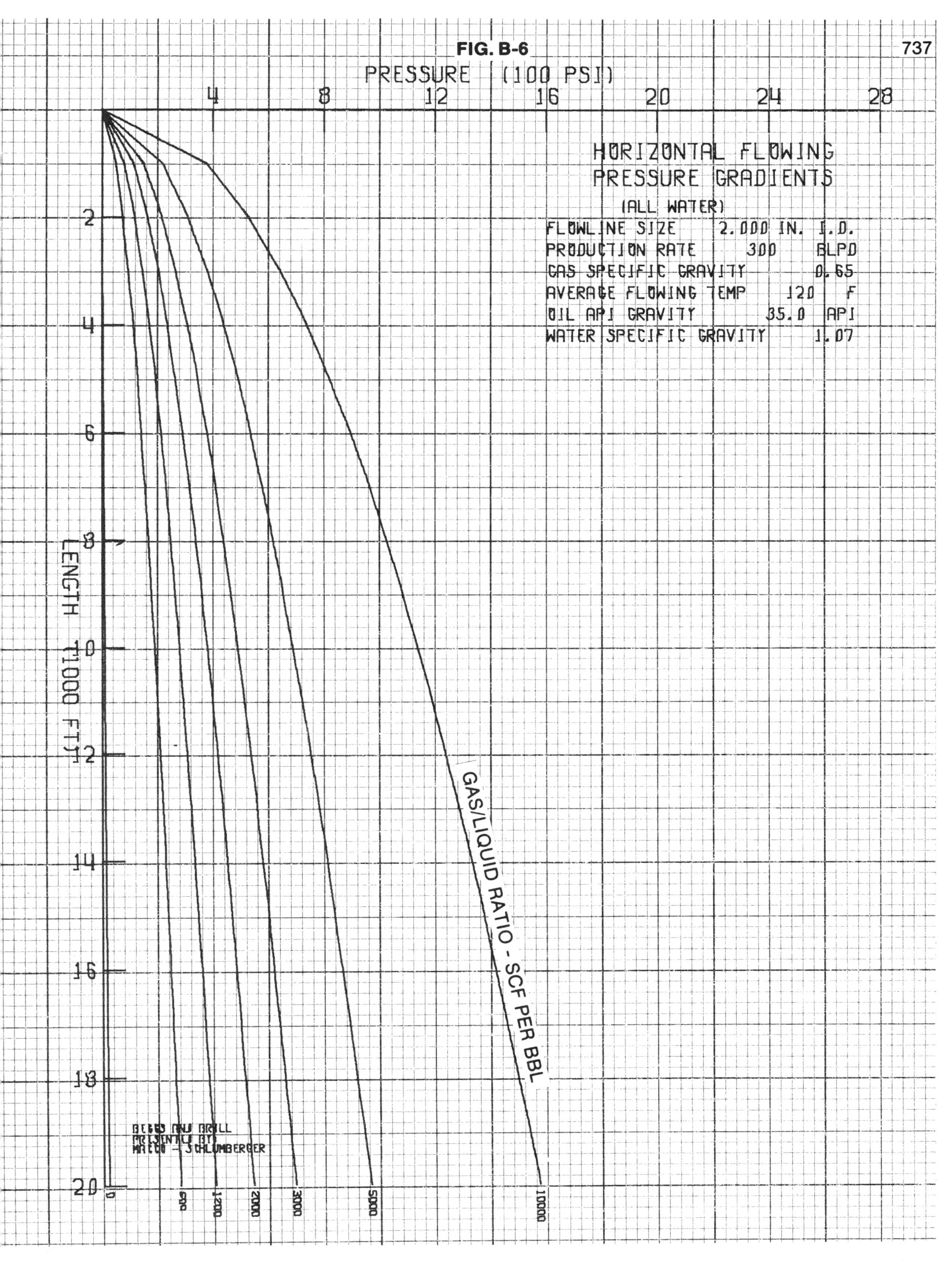

PRESSURE (100 PSI)

HORIZONTAL FLOWING
PRESSURE GRADIENTS
(ALL WATER)

FLOWLINE SIZE	2.000 IN. I.D.
PRODUCTION RATE	300 BLPD
GAS SPECIFIC GRAVITY	0.65
AVERAGE FLOWING TEMP	120 F
OIL API GRAVITY	35.0 API
WATER SPECIFIC GRAVITY	1.07

LENGTH (1000 FT)

GAS/LIQUID RATIO - SCF PER BBL

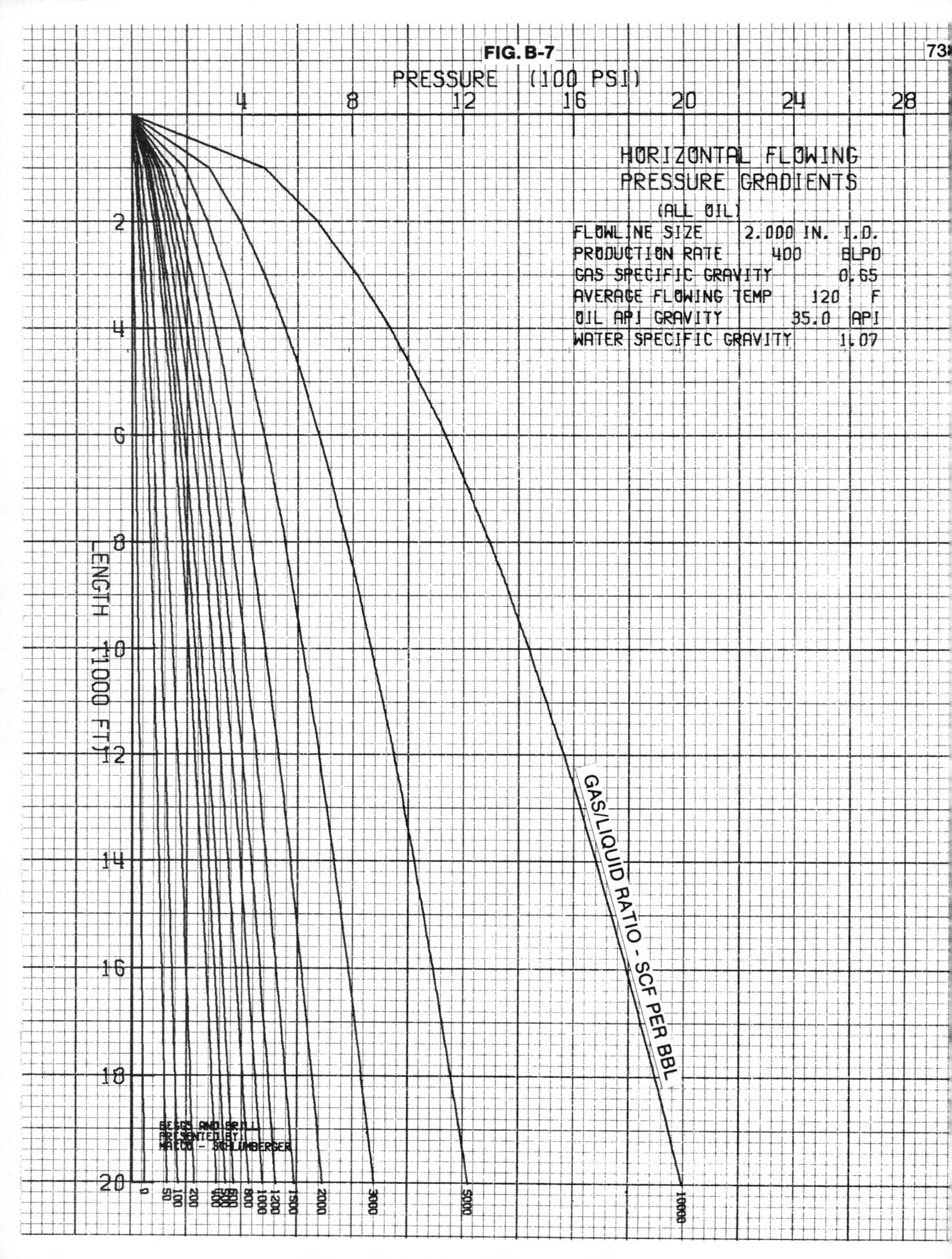

FIG. B-7

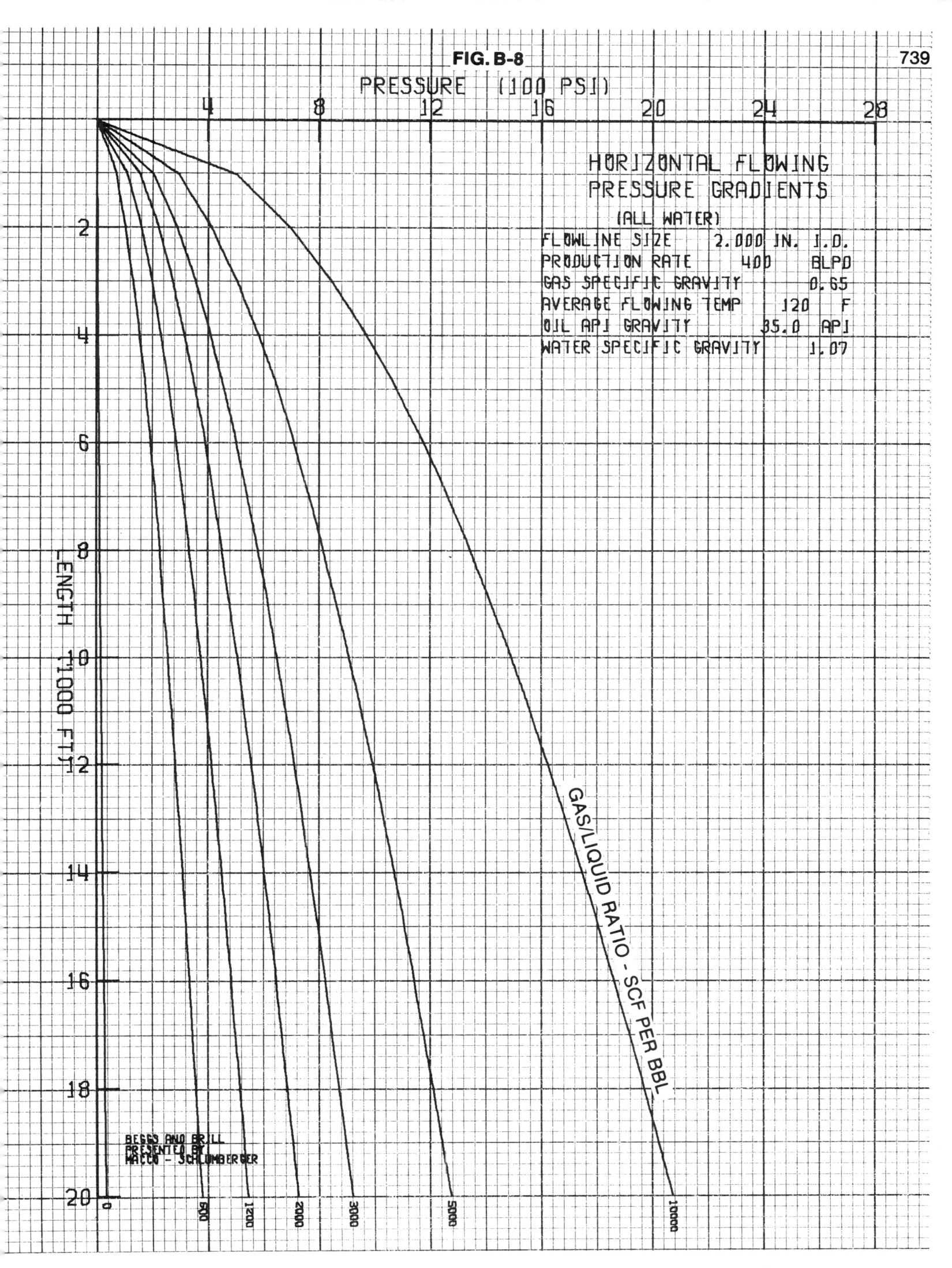

PRESSURE (100 PSI)

HORIZONTAL FLOWING
PRESSURE GRADIENTS
(ALL WATER)

FLOWLINE SIZE	2.000 IN. I.D.
PRODUCTION RATE	400 BLPD
GAS SPECIFIC GRAVITY	0.65
AVERAGE FLOWING TEMP	120 F
OIL API GRAVITY	35.0 API
WATER SPECIFIC GRAVITY	1.07

LENGTH (1000 FT)

GAS/LIQUID RATIO - SCF PER BBL

BEGGS AND BRILL
PRESENTED BY
MAECO - SCHLUMBERGER

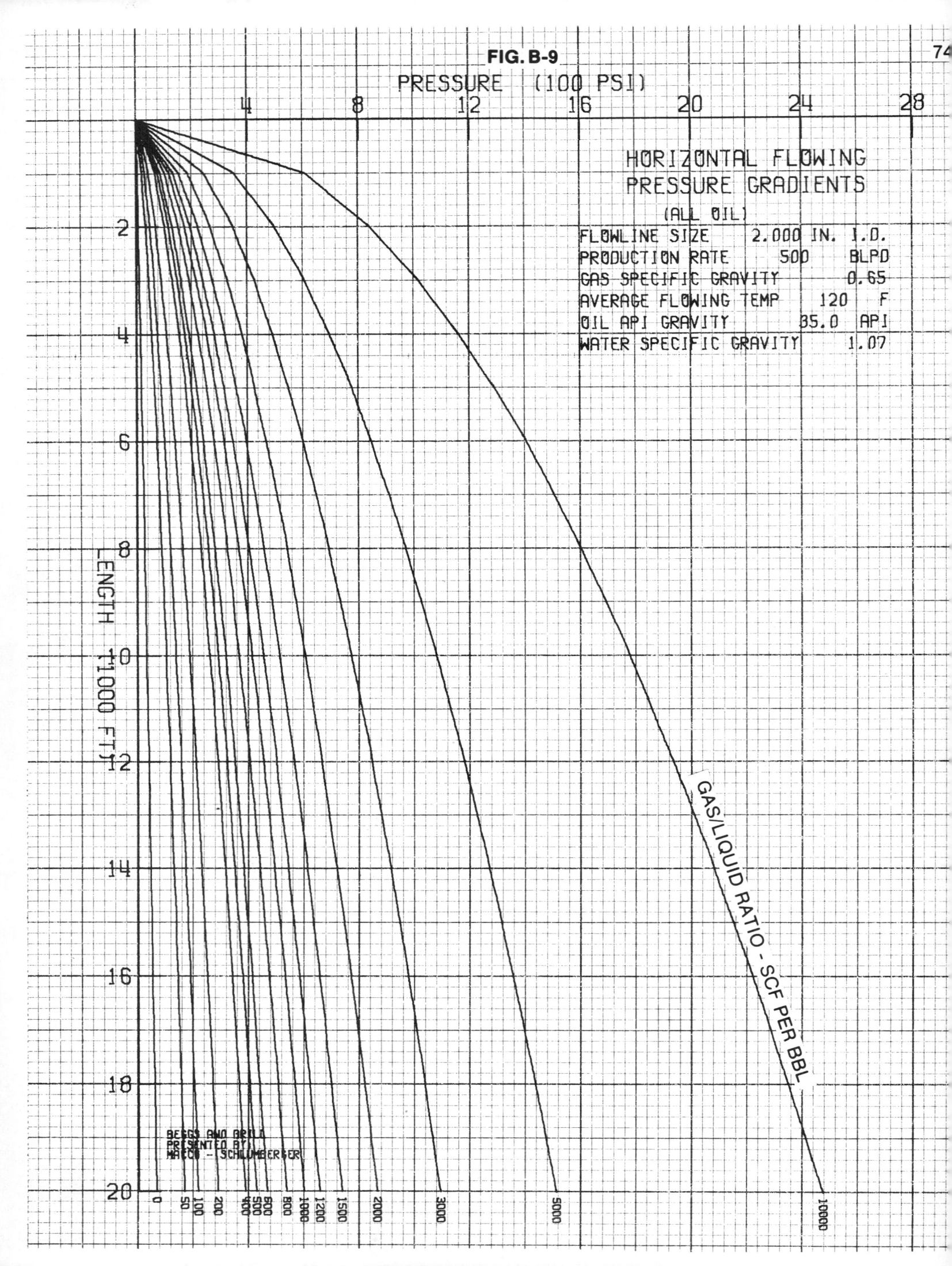

PRESSURE (100 PSI)

HORIZONTAL FLOWING
PRESSURE GRADIENTS
(ALL OIL)

FLOWLINE SIZE	2.000 IN. I.D.
PRODUCTION RATE	500 BLPD
GAS SPECIFIC GRAVITY	0.65
AVERAGE FLOWING TEMP	120 F
OIL API GRAVITY	35.0 API
WATER SPECIFIC GRAVITY	1.07

LENGTH (1000 FT.)

GAS/LIQUID RATIO - SCF PER BBL

BEGGS AND BRILL
PRESENTED BY
MAECO - SCHLUMBERGER

0 50 100 200 300 400 500 800 1000 1200 1500 2000 3000 5000 10000

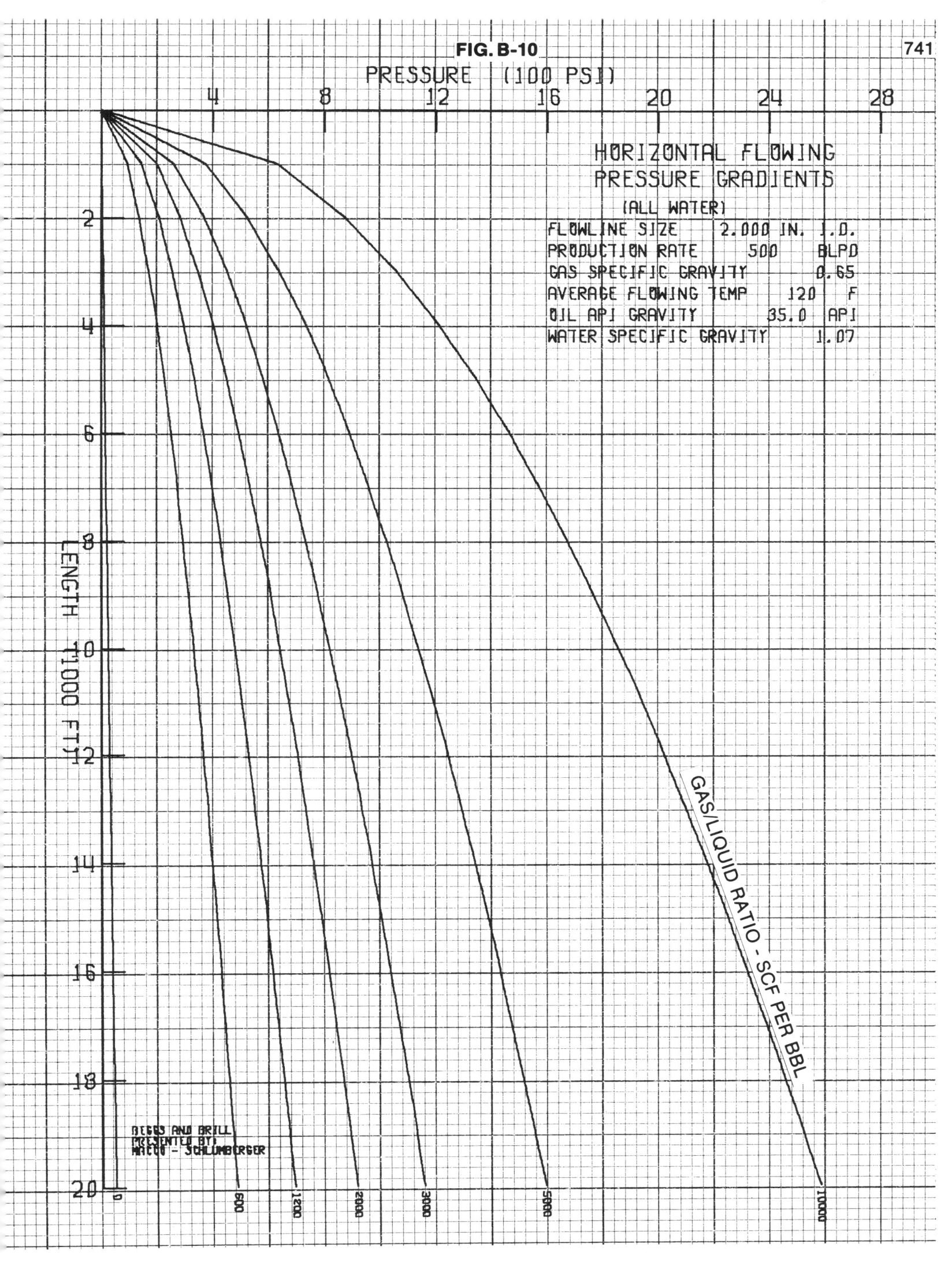

FIG. B-10   741

HORIZONTAL FLOWING
PRESSURE GRADIENTS
(ALL WATER)

FLOWLINE SIZE	2.000 IN. I.D.
PRODUCTION RATE	500 BLPD
GAS SPECIFIC GRAVITY	0.65
AVERAGE FLOWING TEMP	120 F
OIL API GRAVITY	35.0 API
WATER SPECIFIC GRAVITY	1.07

PRESSURE (100 PSI)

LENGTH (1000 FT)

GAS/LIQUID RATIO - SCF PER BBL

BEGGS AND BRILL
PRESENTED BY:
WAECO - SCHLUMBERGER

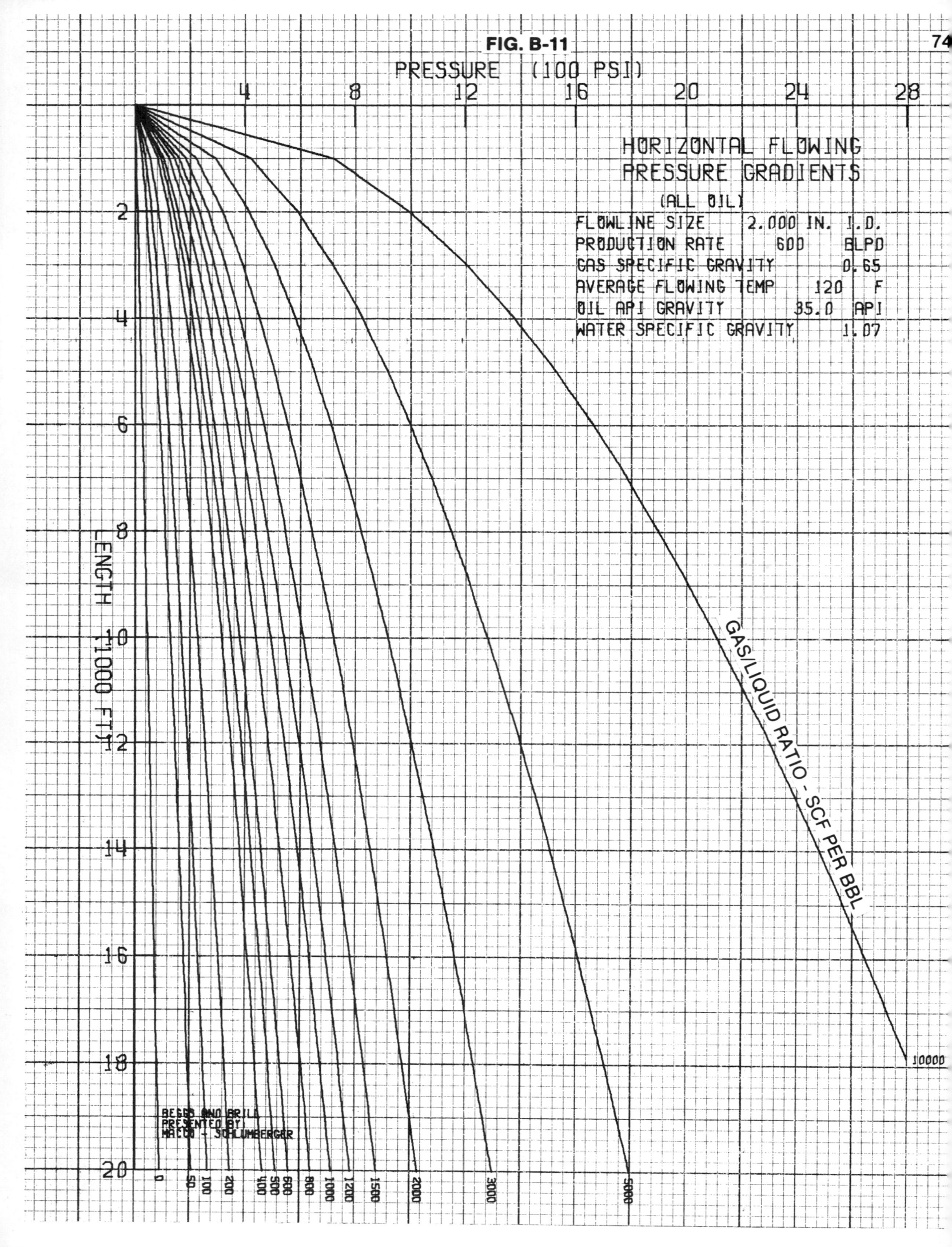

PRESSURE (100 PSI)

HORIZONTAL FLOWING
PRESSURE GRADIENTS
(ALL OIL)

FLOWLINE SIZE	2.000 IN.	I.D.
PRODUCTION RATE	600	BLPD
GAS SPECIFIC GRAVITY	0.65	
AVERAGE FLOWING TEMP	120	F
OIL API GRAVITY	35.0	API
WATER SPECIFIC GRAVITY	1.07	

LENGTH (1000 FT)

GAS/LIQUID RATIO - SCF PER BBL

BEGGS AND BRILL
PRESENTED BY
HALCO - SCHLUMBERGER

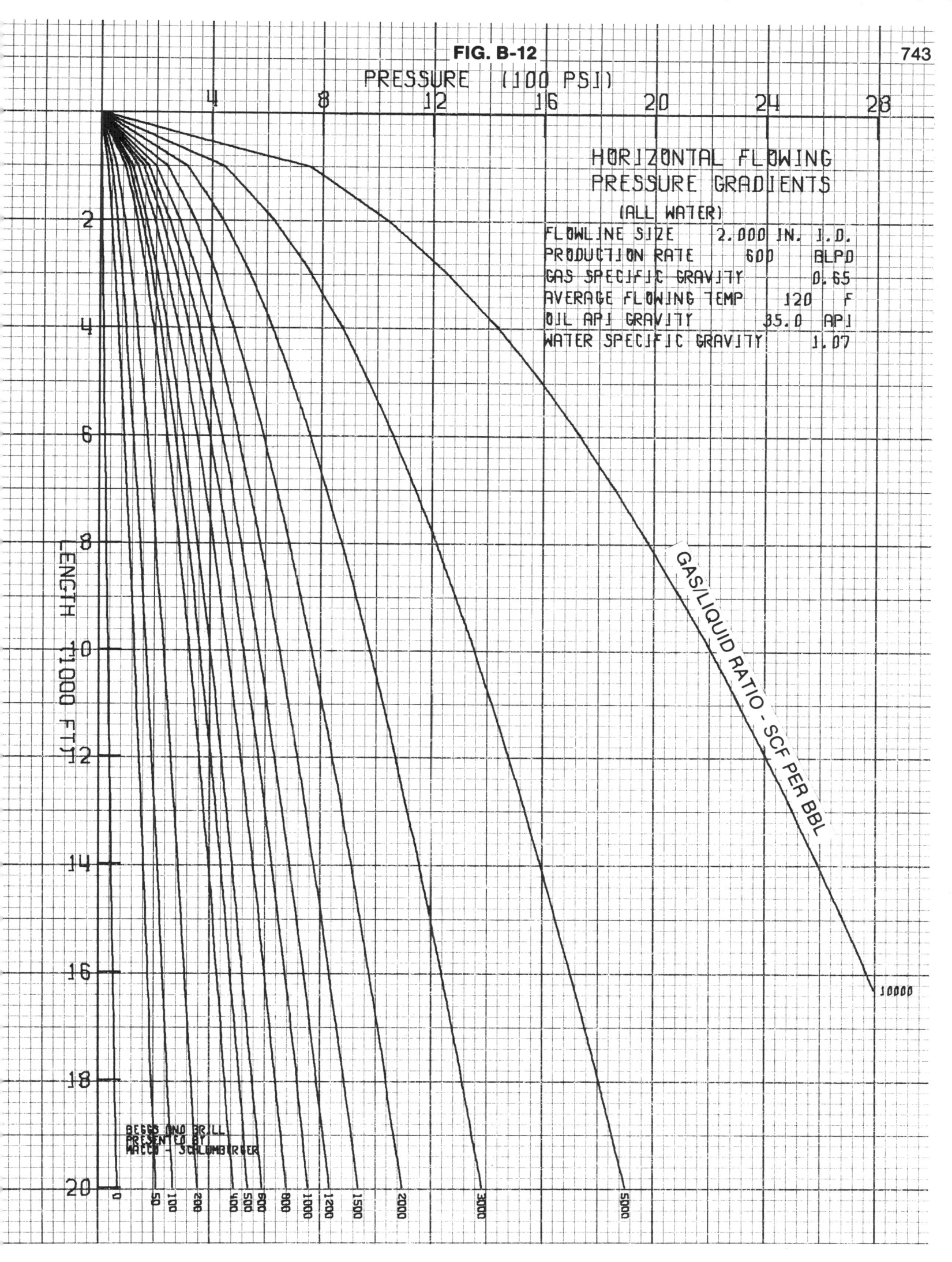

PRESSURE (100 PSI)

HORIZONTAL FLOWING
PRESSURE GRADIENTS
(ALL WATER)

FLOWLINE SIZE   2.000 IN. I.D.
PRODUCTION RATE   600   BLPD
GAS SPECIFIC GRAVITY   0.65
AVERAGE FLOWING TEMP   120   F
OIL API GRAVITY   35.0 API
WATER SPECIFIC GRAVITY   1.07

GAS/LIQUID RATIO - SCF PER BBL

10000

LENGTH (1000 FT)

BEGGS AND BRILL
PRESENTED BY
MAECO - SCHLUMBERGER

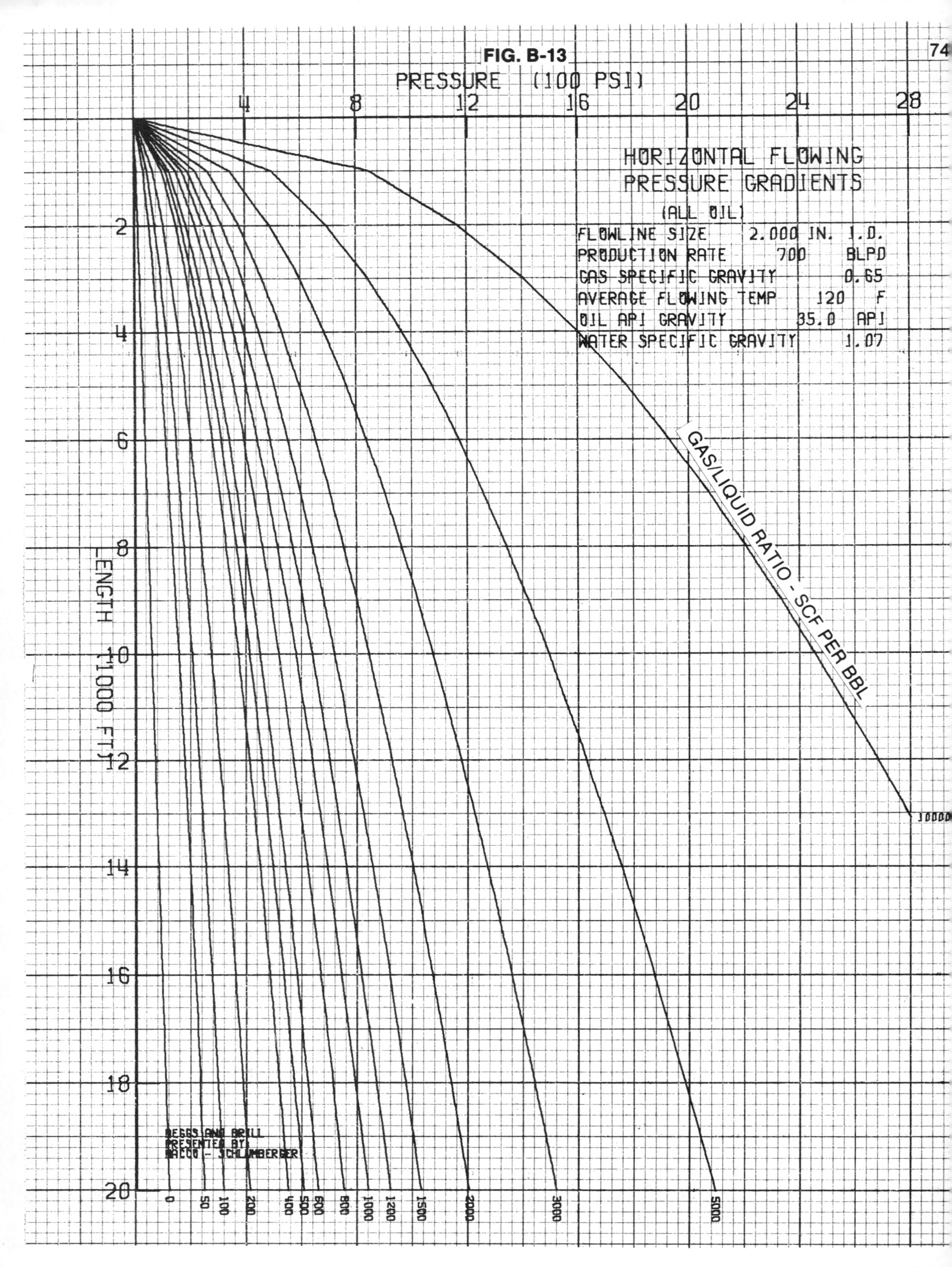

FIG. B-13

74

HORIZONTAL FLOWING
PRESSURE GRADIENTS
(ALL OIL)

FLOWLINE SIZE	2.000 IN.	I.D.
PRODUCTION RATE	700	BLPD
GAS SPECIFIC GRAVITY	0.65	
AVERAGE FLOWING TEMP	120	F
OIL API GRAVITY	35.0	API
WATER SPECIFIC GRAVITY	1.07	

GAS/LIQUID RATIO - SCF PER BBL

BEGGS AND BRILL
PRESENTED BY
NAECO - SCHLUMBERGER

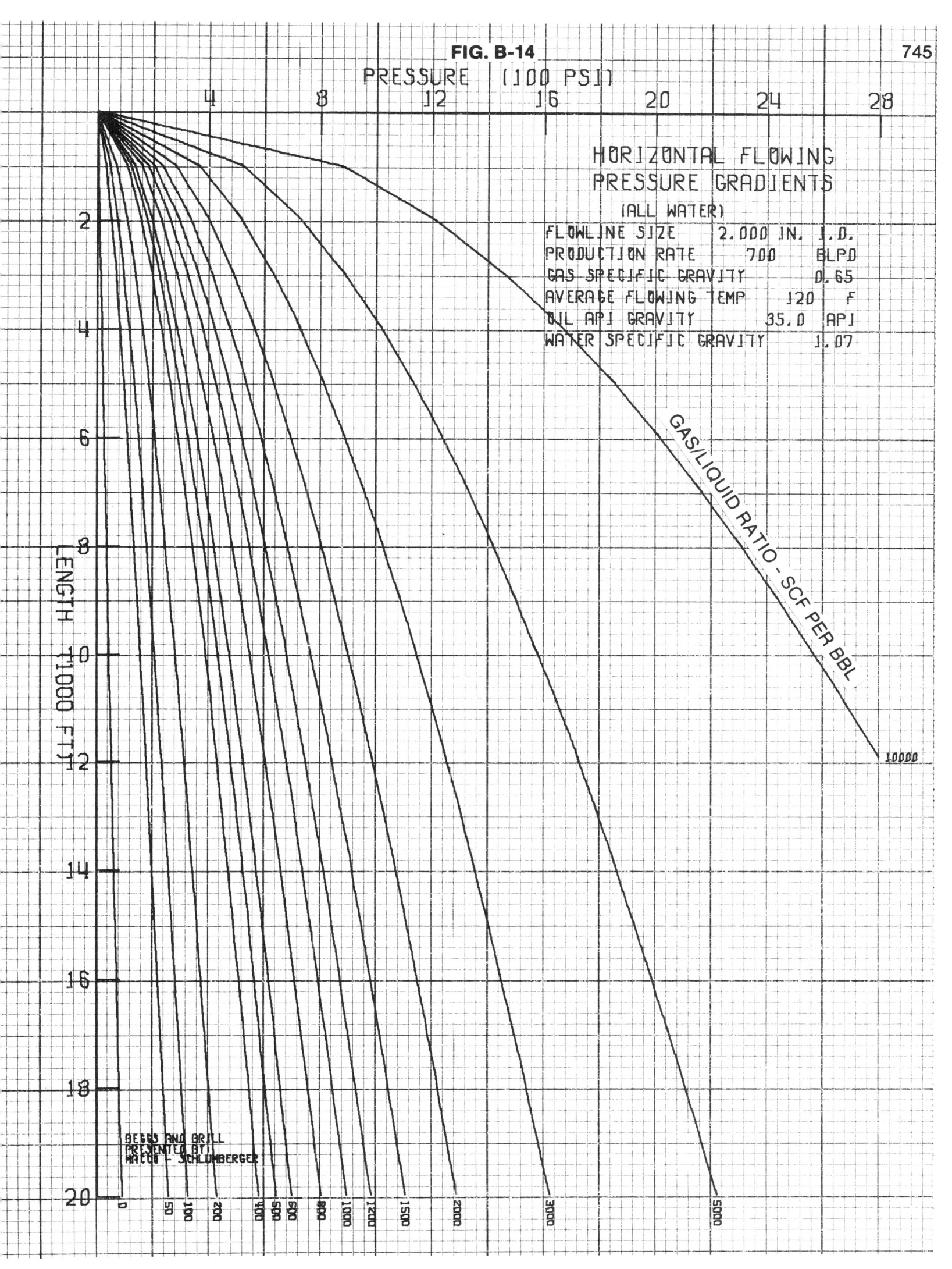

FIG. B-14

745

HORIZONTAL FLOWING
PRESSURE GRADIENTS
(ALL WATER)

FLOWLINE SIZE	2.000 IN. I.D.
PRODUCTION RATE	700 BLPD
GAS SPECIFIC GRAVITY	0.65
AVERAGE FLOWING TEMP	120 F
OIL API GRAVITY	35.0 API
WATER SPECIFIC GRAVITY	1.07

PRESSURE (100 PSI)

LENGTH (1000 FT)

GAS/LIQUID RATIO - SCF PER BBL

BEGGS AND BRILL
PRESENTED BY
WAECO - SCHLUMBERGER

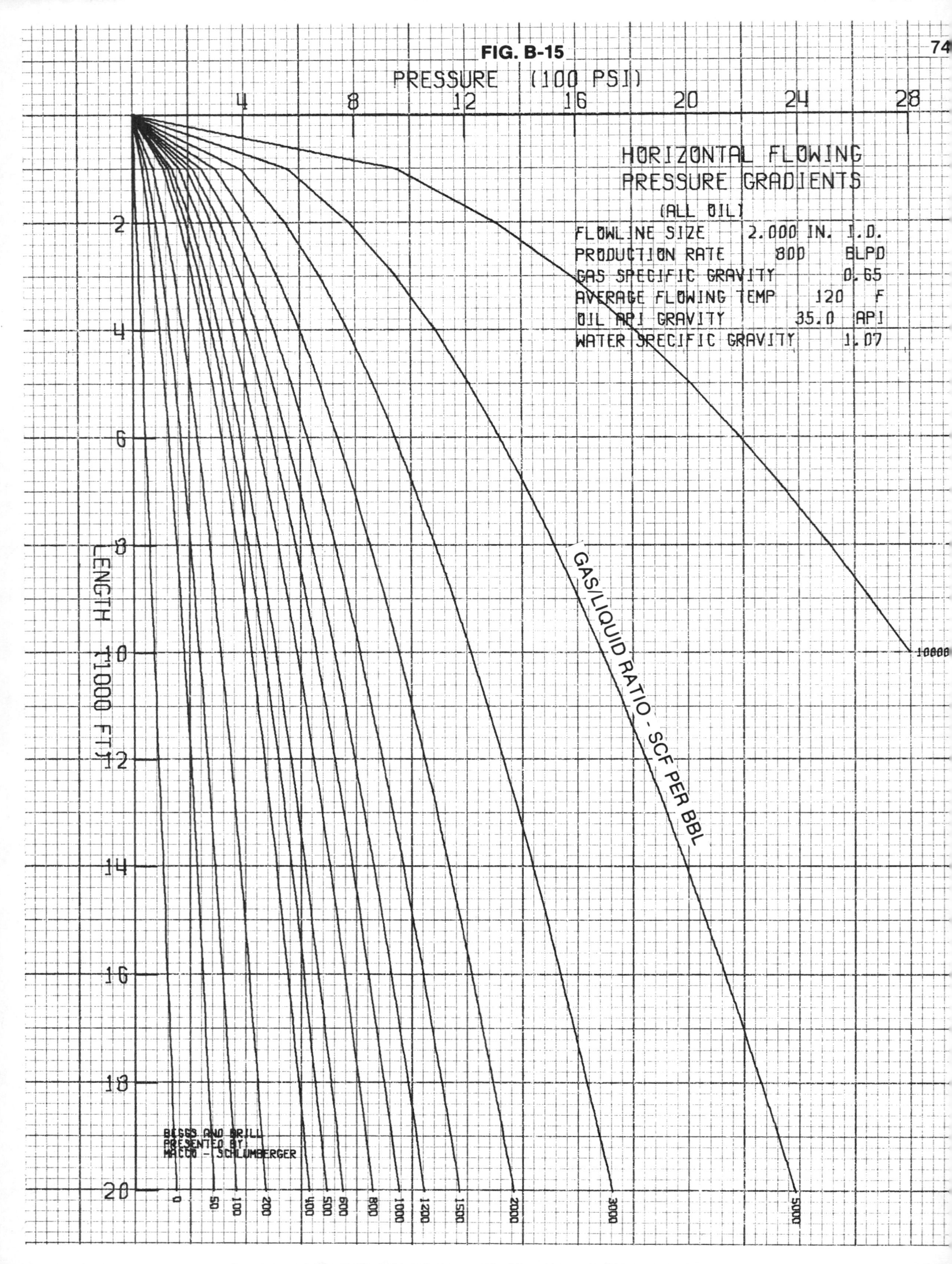

HORIZONTAL FLOWING
PRESSURE GRADIENTS

(ALL OIL)

FLOWLINE SIZE	2.000 IN.	I.D.
PRODUCTION RATE	800	BLPD
GAS SPECIFIC GRAVITY	0.65	
AVERAGE FLOWING TEMP	120	F
OIL API GRAVITY	35.0	API
WATER SPECIFIC GRAVITY	1.07	

PRESSURE (100 PSI)

LENGTH (1000 FT)

GAS/LIQUID RATIO - SCF PER BBL

BEGGS AND BRILL
PRESENTED BY
MACCO - SCHLUMBERGER

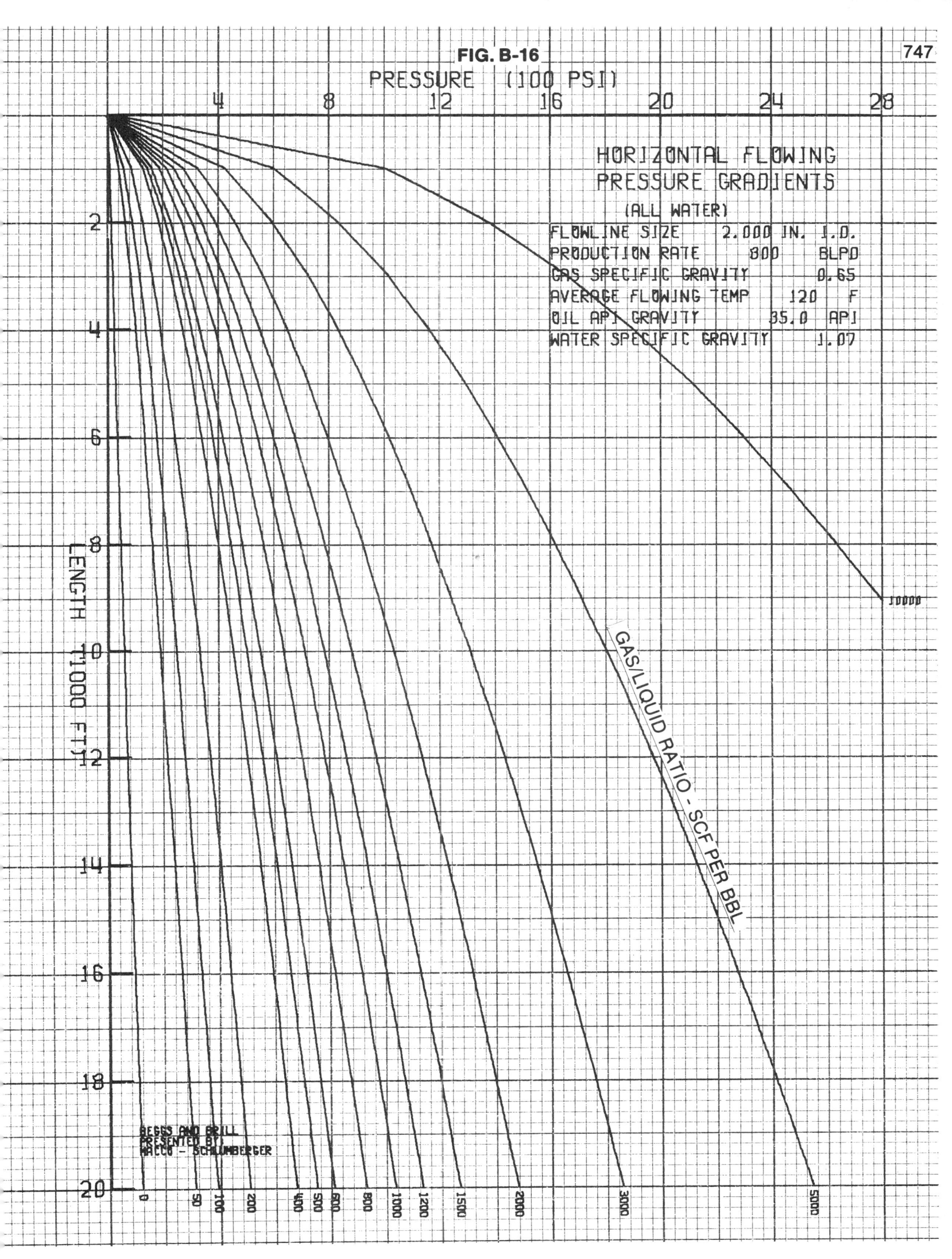

FIG. B-16

747

PRESSURE (100 PSI)

HORIZONTAL FLOWING
PRESSURE GRADIENTS
(ALL WATER)

FLOWLINE SIZE	2.000 IN. I.D.
PRODUCTION RATE	800 BLPD
GAS SPECIFIC GRAVITY	0.65
AVERAGE FLOWING TEMP	120 F
OIL API GRAVITY	35.0 API
WATER SPECIFIC GRAVITY	1.07

LENGTH (1000 FT)

GAS/LIQUID RATIO - SCF PER BBL

BEGGS AND BRILL
PRESENTED BY
HALCO - SCHLUMBERGER

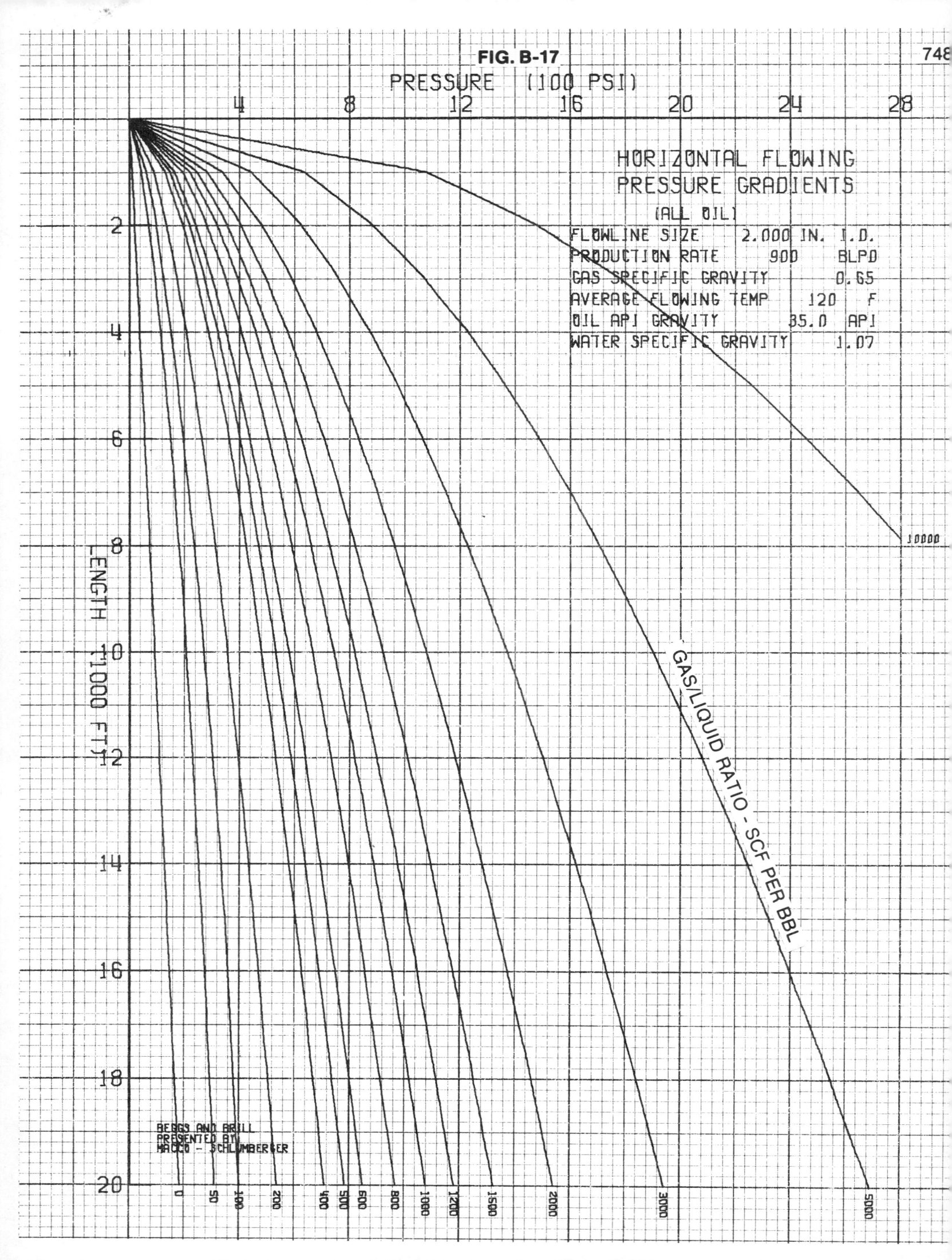

FIG. B-17

748

HORIZONTAL FLOWING
PRESSURE GRADIENTS
(ALL OIL)

FLOWLINE SIZE	2.000 IN. I.D.
PRODUCTION RATE	900 BLPD
GAS SPECIFIC GRAVITY	0.65
AVERAGE FLOWING TEMP	120 F
OIL API GRAVITY	35.0 API
WATER SPECIFIC GRAVITY	1.07

PRESSURE (100 PSI)

LENGTH (1000 FT)

GAS/LIQUID RATIO - SCF PER BBL

BEGGS AND BRILL
PRESENTED BY
MADCO - SCHLUMBERGER

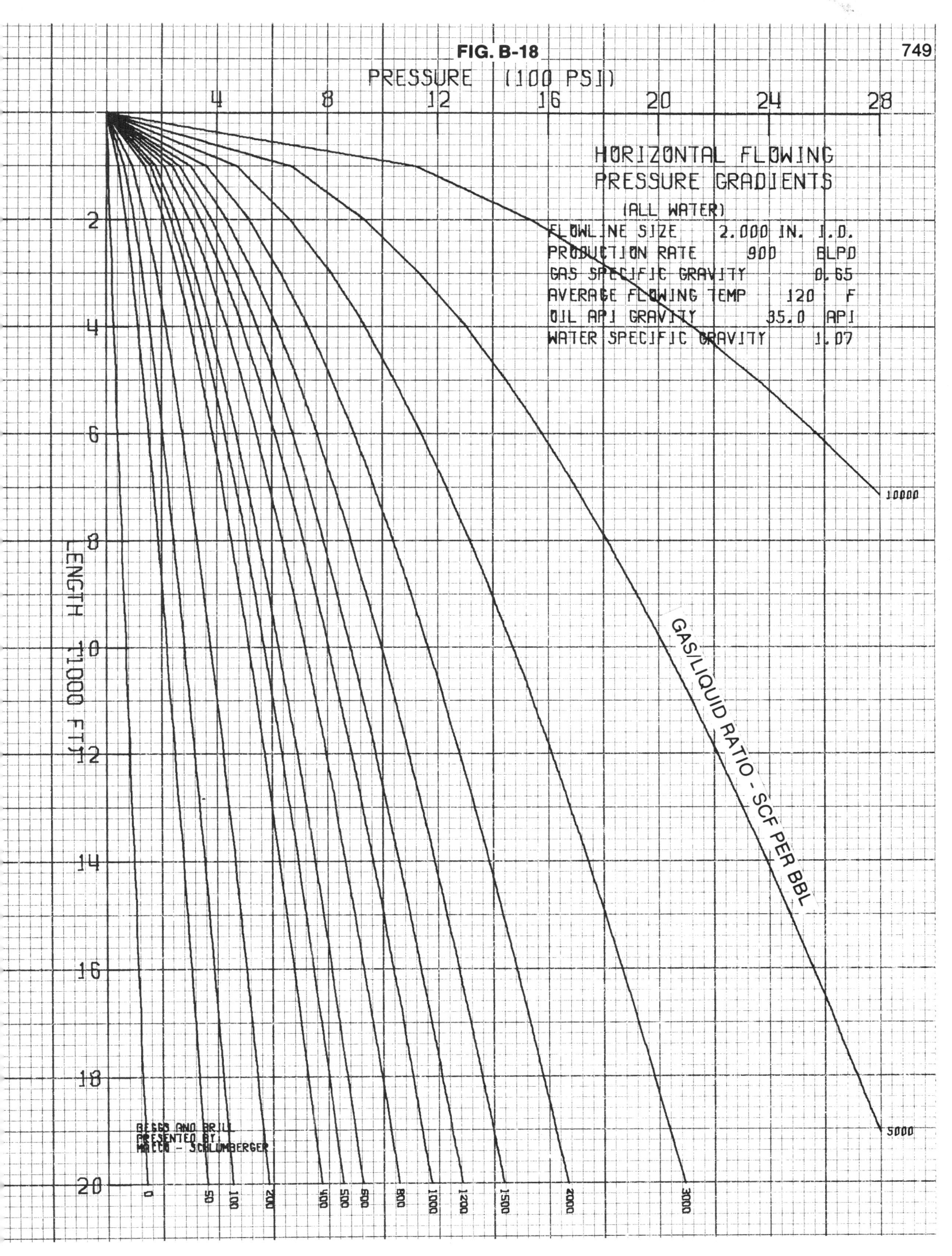

FIG. B-18

749

HORIZONTAL FLOWING
PRESSURE GRADIENTS
(ALL WATER)

FLOWLINE SIZE	2.000 IN. I.D.
PRODUCTION RATE	900 BLPD
GAS SPECIFIC GRAVITY	0.65
AVERAGE FLOWING TEMP	120 F
OIL API GRAVITY	35.0 API
WATER SPECIFIC GRAVITY	1.07

PRESSURE (100 PSI)

LENGTH (1000 FT)

GAS/LIQUID RATIO - SCF PER BBL

BEGGS AND BRILL
PRESENTED BY
MAECO - SCHLUMBERGER

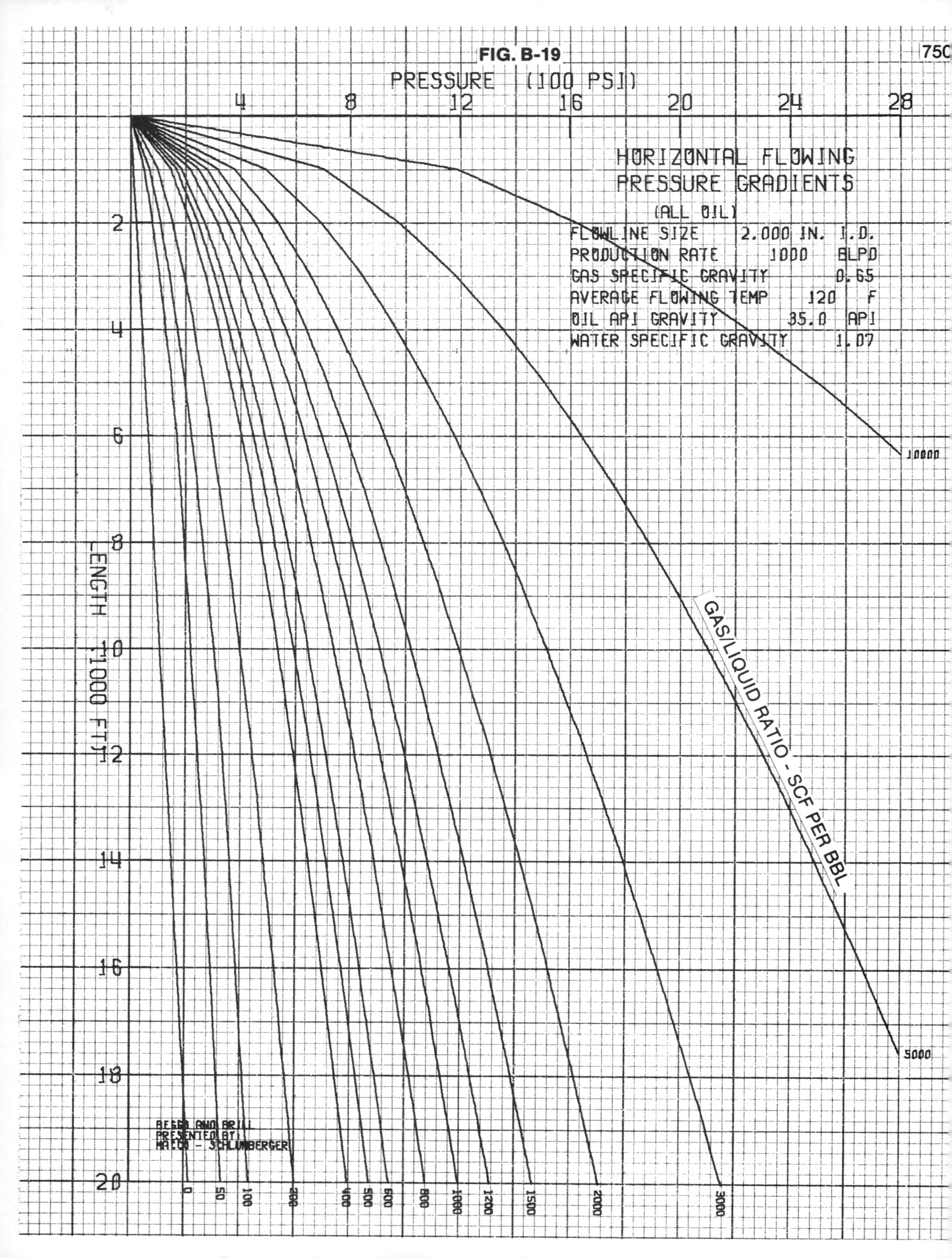

PRESSURE (100 PSI)

HORIZONTAL FLOWING
PRESSURE GRADIENTS
(ALL OIL)
FLOWLINE SIZE   2.000 IN. I.D.
PRODUCTION RATE   1000   BLPD
GAS SPECIFIC GRAVITY   0.65
AVERAGE FLOWING TEMP   120   F
OIL API GRAVITY   35.0 API
WATER SPECIFIC GRAVITY   1.07

LENGTH (1000 FT)

GAS/LIQUID RATIO - SCF PER BBL

REES AND BRILL
PRESENTED BY
MACCO - SCHLUMBERGER

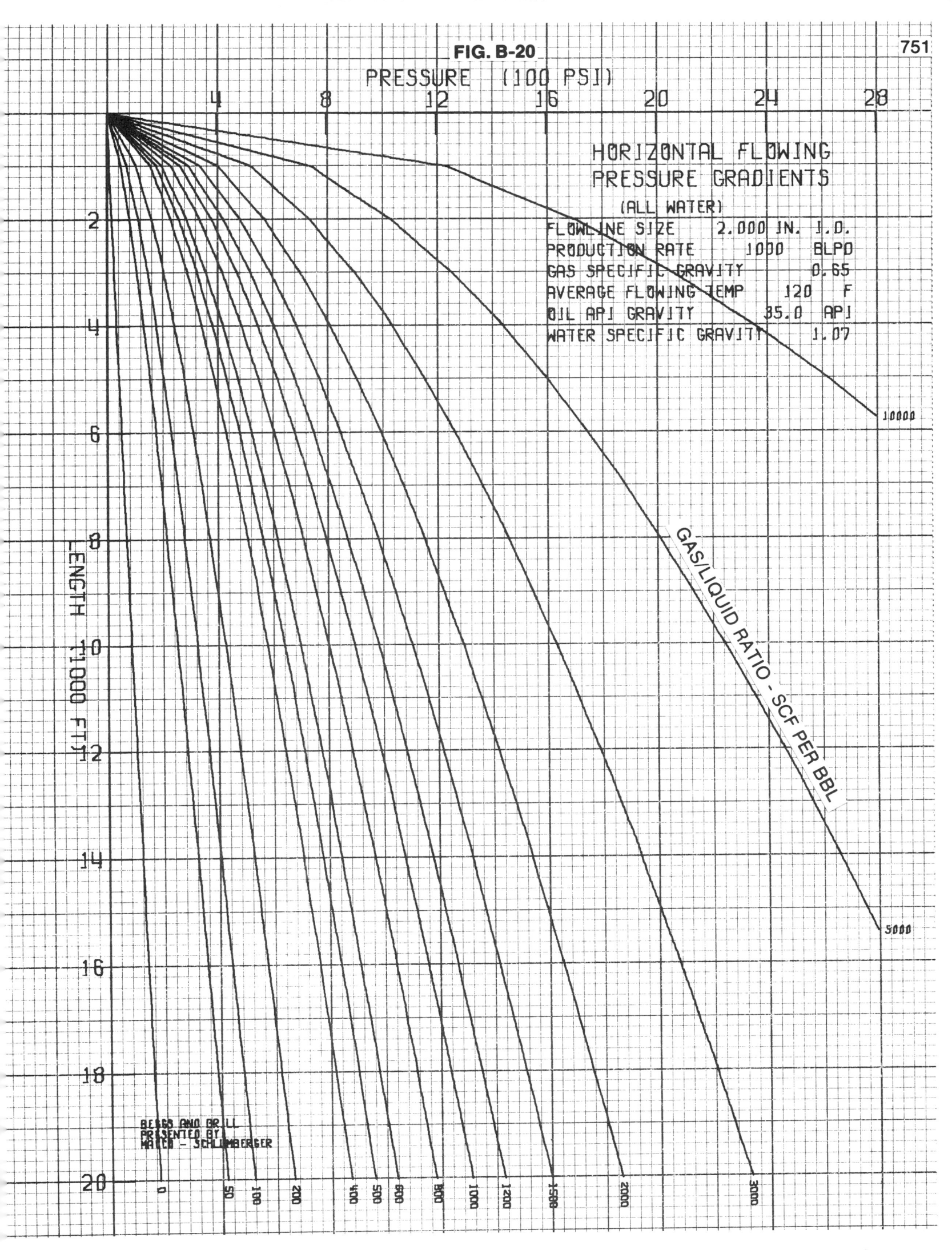

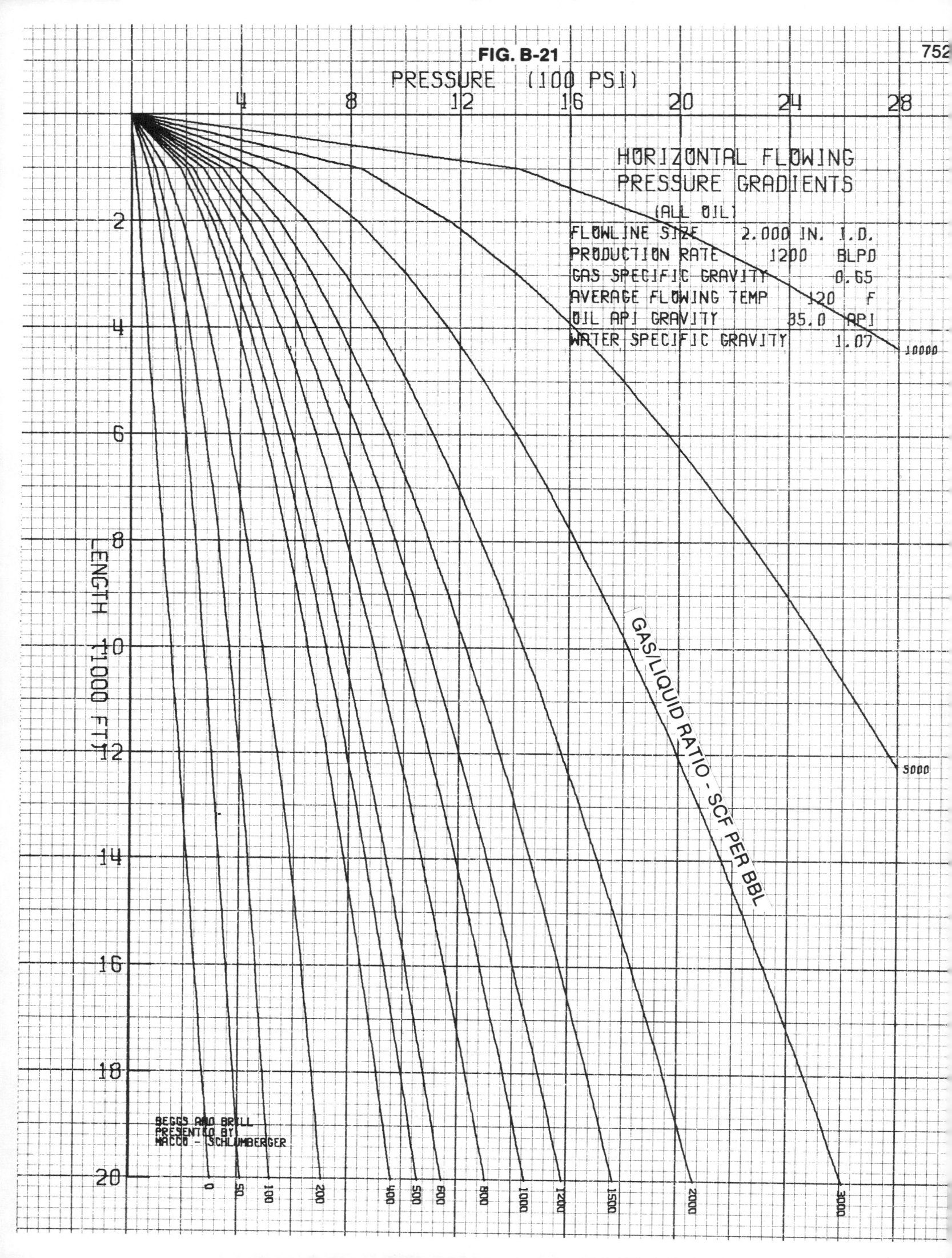

PRESSURE (100 PSI)

HORIZONTAL FLOWING
PRESSURE GRADIENTS
(ALL OIL)

FLOWLINE SIZE	2.000 IN. I.D.
PRODUCTION RATE	1200 BLPD
GAS SPECIFIC GRAVITY	0.65
AVERAGE FLOWING TEMP	120 F
OIL API GRAVITY	35.0 API
WATER SPECIFIC GRAVITY	1.07

LENGTH (1000 FT)

GAS/LIQUID RATIO - SCF PER BBL

BEGGS AND BRILL
PRESENTED BY
MACCO - SCHLUMBERGER

FIG. B-22

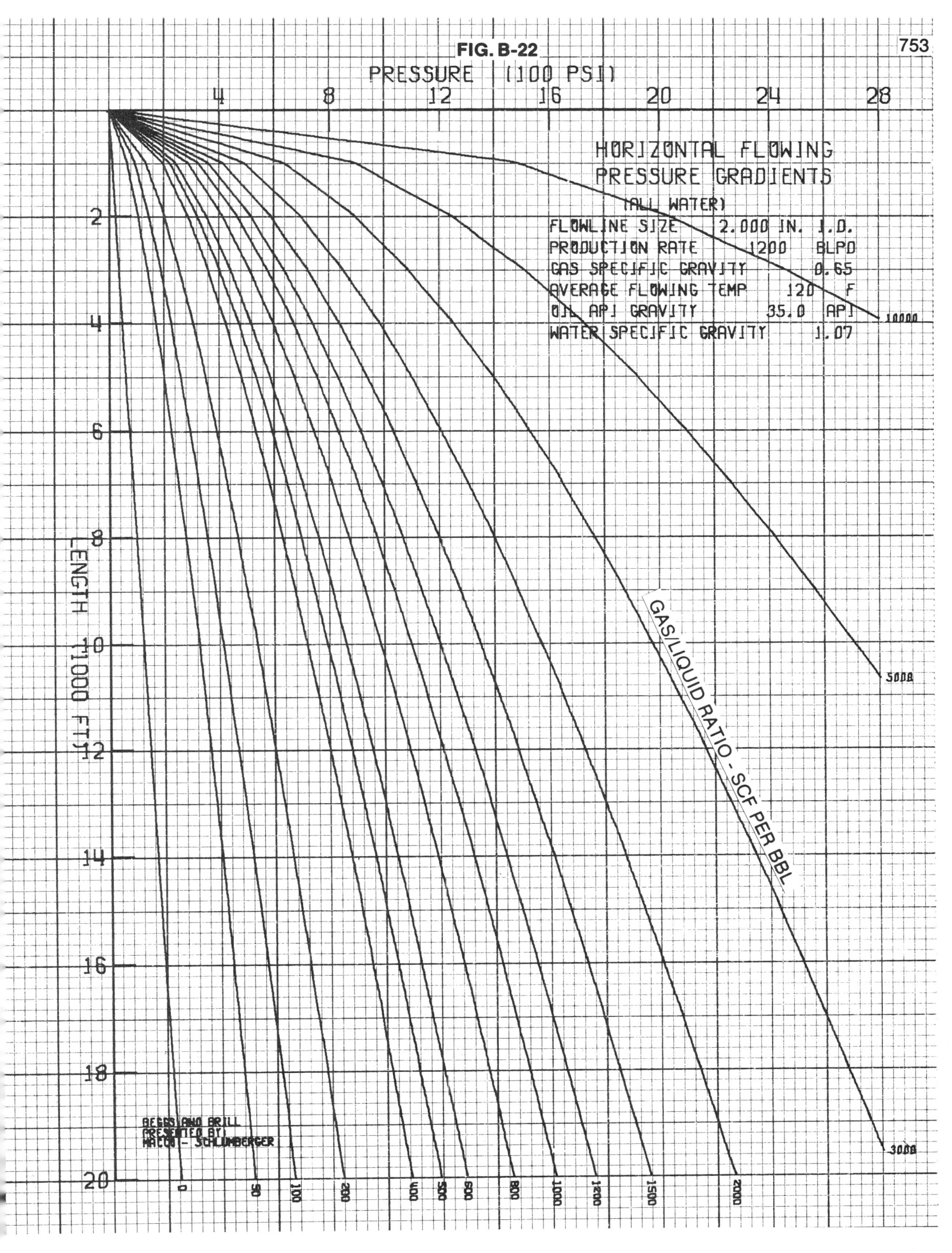

PRESSURE (100 PSI)

HORIZONTAL FLOWING
PRESSURE GRADIENTS
(ALL WATER)

FLOWLINE SIZE	2.000 IN. I.D.
PRODUCTION RATE	1200 BLPD
GAS SPECIFIC GRAVITY	0.65
AVERAGE FLOWING TEMP	120 F
OIL API GRAVITY	35.0 API
WATER SPECIFIC GRAVITY	1.07

LENGTH (1000 FT)

GAS/LIQUID RATIO - SCF PER BBL

BEGGS AND BRILL
PRESENTED BY:
MACCO - SCHLUMBERGER

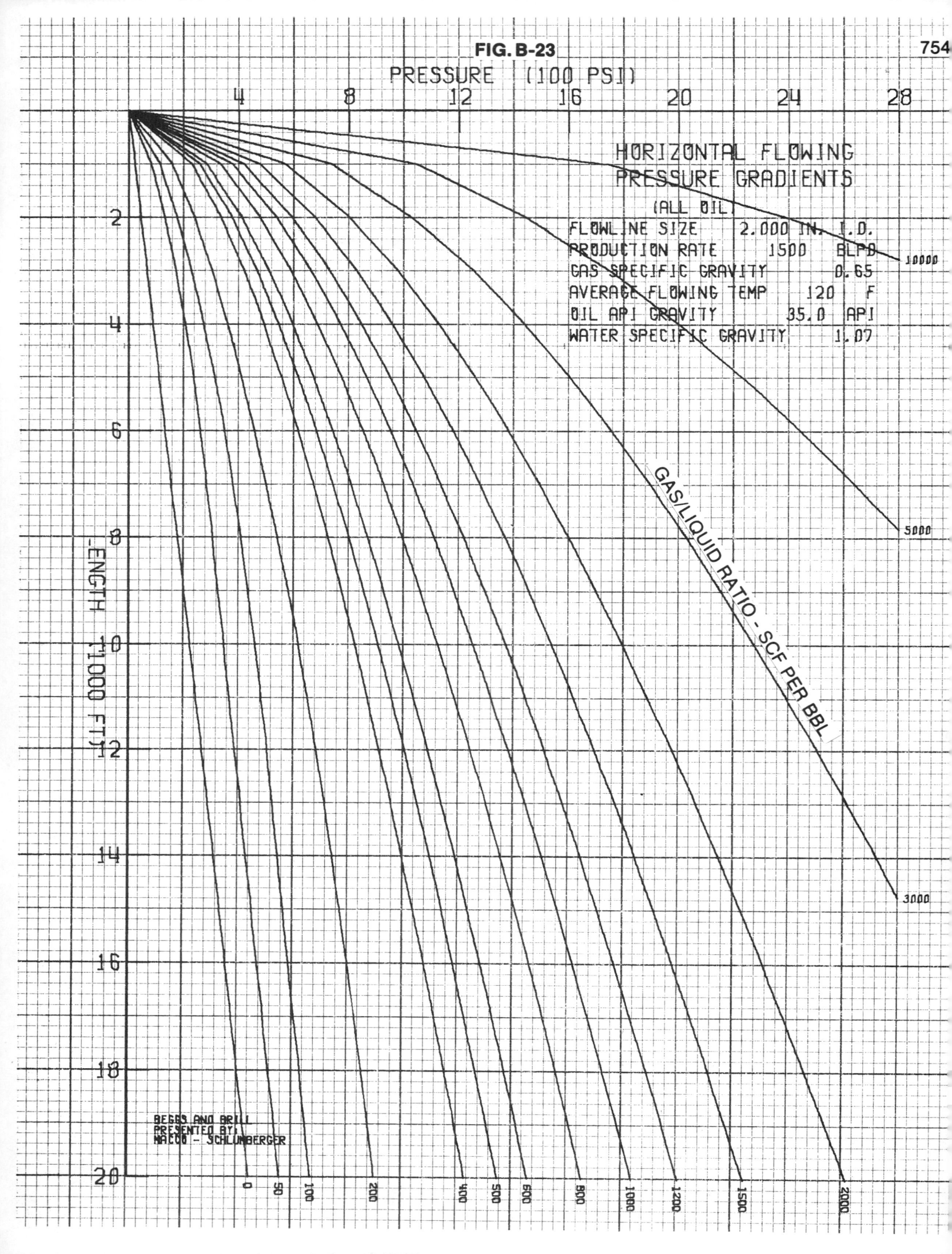

PRESSURE (100 PSI)

HORIZONTAL FLOWING
PRESSURE GRADIENTS
(ALL OIL)

FLOWLINE SIZE	2.000 IN. I.D.
PRODUCTION RATE	1500 BLPD
GAS SPECIFIC GRAVITY	0.65
AVERAGE FLOWING TEMP	120 F
OIL API GRAVITY	35.0 API
WATER SPECIFIC GRAVITY	1.07

GAS/LIQUID RATIO - SCF PER BBL

LENGTH (1000 FT)

BEGGS AND BRILL
PRESENTED BY:
MAECO - SCHLUMBERGER

FIG. B-24

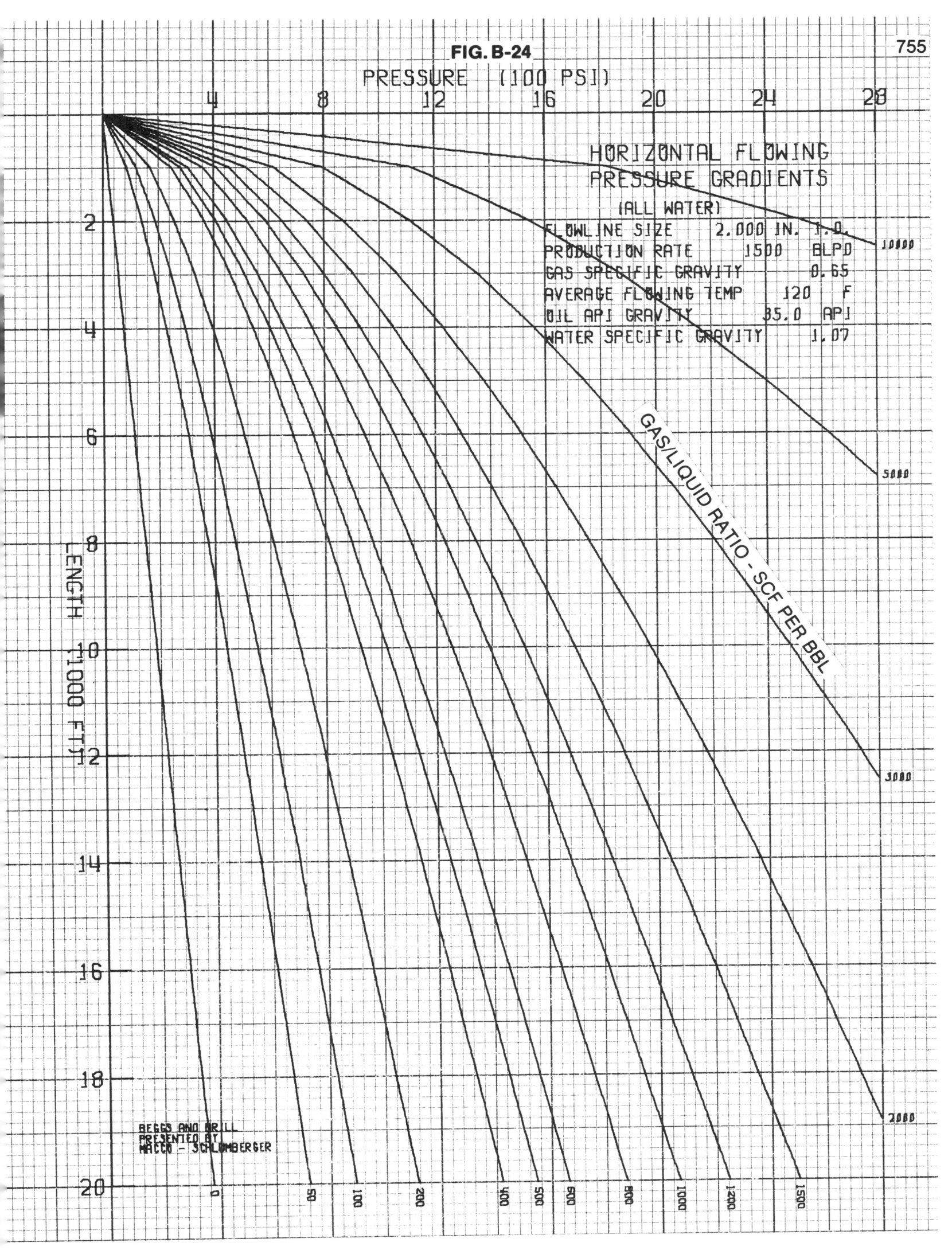

HORIZONTAL FLOWING
PRESSURE GRADIENTS
(ALL WATER)

FLOWLINE SIZE   2.000 IN. I.D.
PRODUCTION RATE   1500 BLPD
GAS SPECIFIC GRAVITY   0.65
AVERAGE FLOWING TEMP   120   F
OIL API GRAVITY   35.0 API
WATER SPECIFIC GRAVITY   1.07

GAS/LIQUID RATIO - SCF PER BBL

PRESSURE (100 PSI)

LENGTH (1000 FT.)

BEGGS AND BRILL
PRESENTED BY
HACCO - SCHLUMBERGER

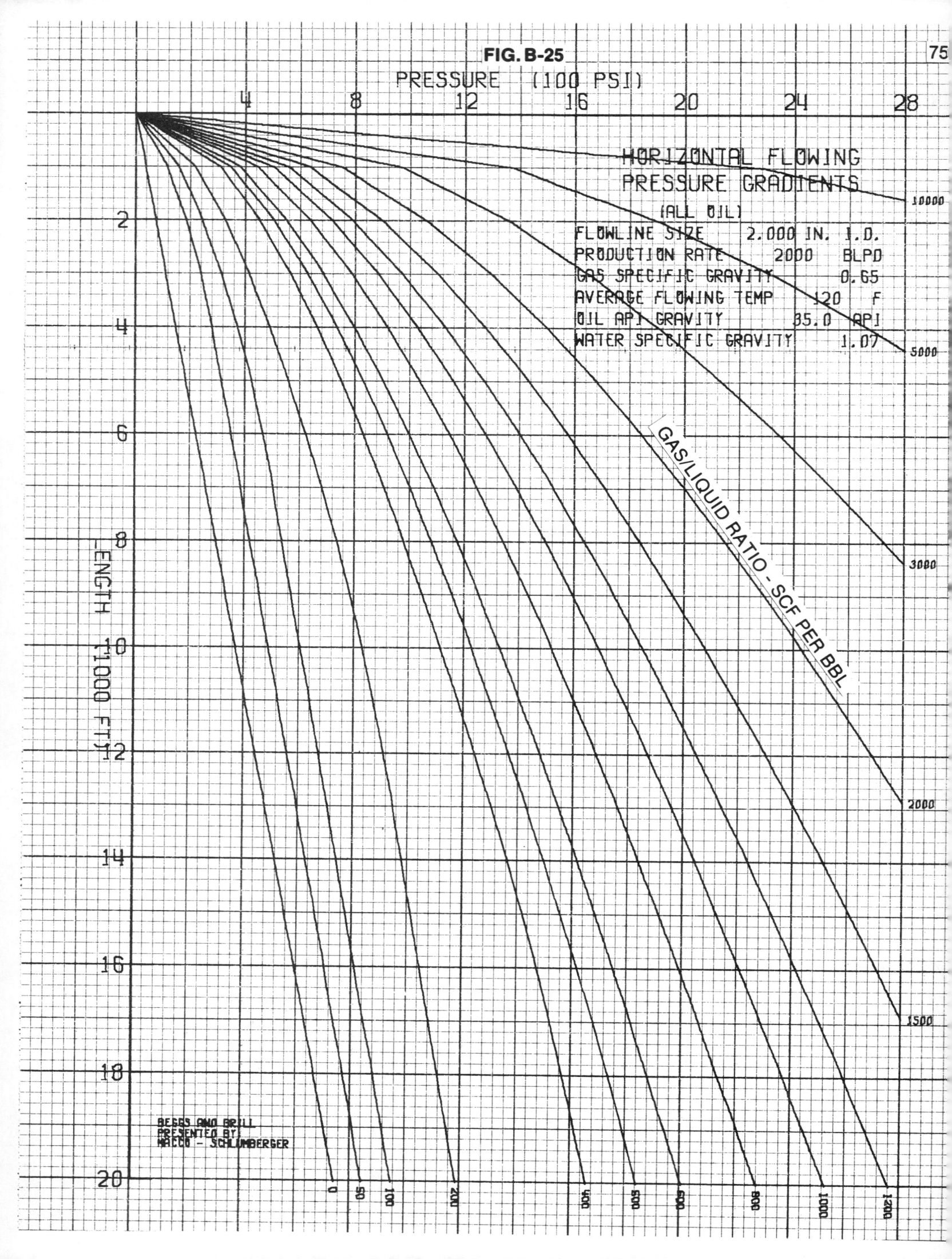

FIG. B-25

75

PRESSURE (100 PSI)

HORIZONTAL FLOWING
PRESSURE GRADIENTS
(ALL OIL)

FLOWLINE SIZE   2.000 IN. I.D.
PRODUCTION RATE   2000   BLPD
GAS SPECIFIC GRAVITY   0.65
AVERAGE FLOWING TEMP   120   F
OIL API GRAVITY   35.0 API
WATER SPECIFIC GRAVITY   1.07

GAS/LIQUID RATIO - SCF PER BBL

LENGTH (1000 FT.)

BEGGS AND BRILL
PRESENTED BY
HALCO - SCHLUMBERGER

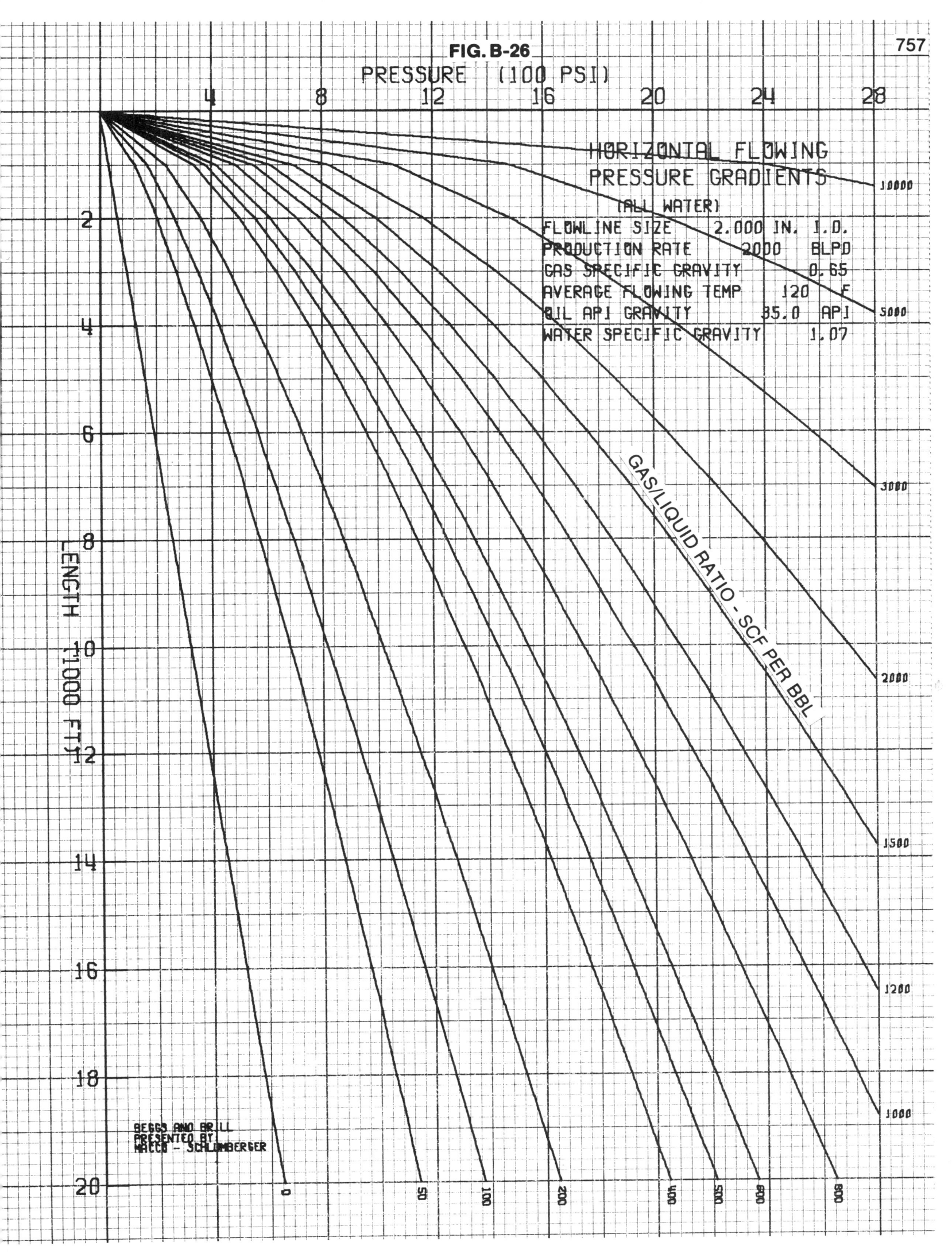

FIG. B-26

PRESSURE (100 PSI)

HORIZONTAL FLOWING
PRESSURE GRADIENTS
(ALL WATER)

FLOWLINE SIZE	2.000 IN. I.D.
PRODUCTION RATE	2000 BLPD
GAS SPECIFIC GRAVITY	0.65
AVERAGE FLOWING TEMP	120 F
OIL API GRAVITY	35.0 API
WATER SPECIFIC GRAVITY	1.07

GAS/LIQUID RATIO - SCF PER BBL

LENGTH (1000 FT)

BEGGS AND BRILL
PRESENTED BY
HALCO - SCHLUMBERGER

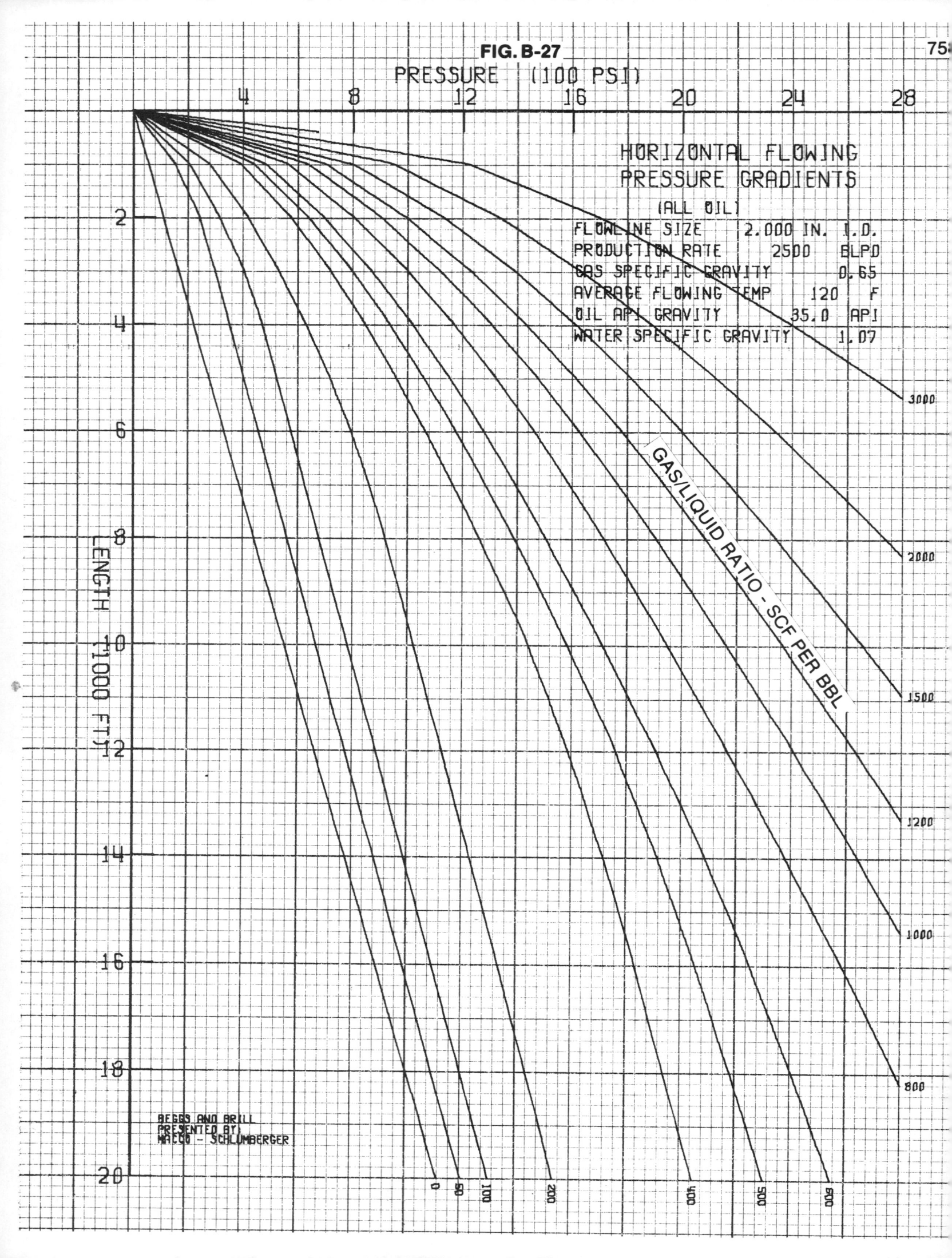

FIG. B-27

HORIZONTAL FLOWING
PRESSURE GRADIENTS
(ALL OIL)

FLOWLINE SIZE	2.000 IN. I.D.
PRODUCTION RATE	2500 BLPD
GAS SPECIFIC GRAVITY	0.65
AVERAGE FLOWING TEMP	120 F
OIL API GRAVITY	35.0 API
WATER SPECIFIC GRAVITY	1.07

PRESSURE (100 PSI)

LENGTH (1000 FT)

GAS/LIQUID RATIO - SCF PER BBL

BEGGS AND BRILL
PRESENTED BY
HALCO - SCHLUMBERGER

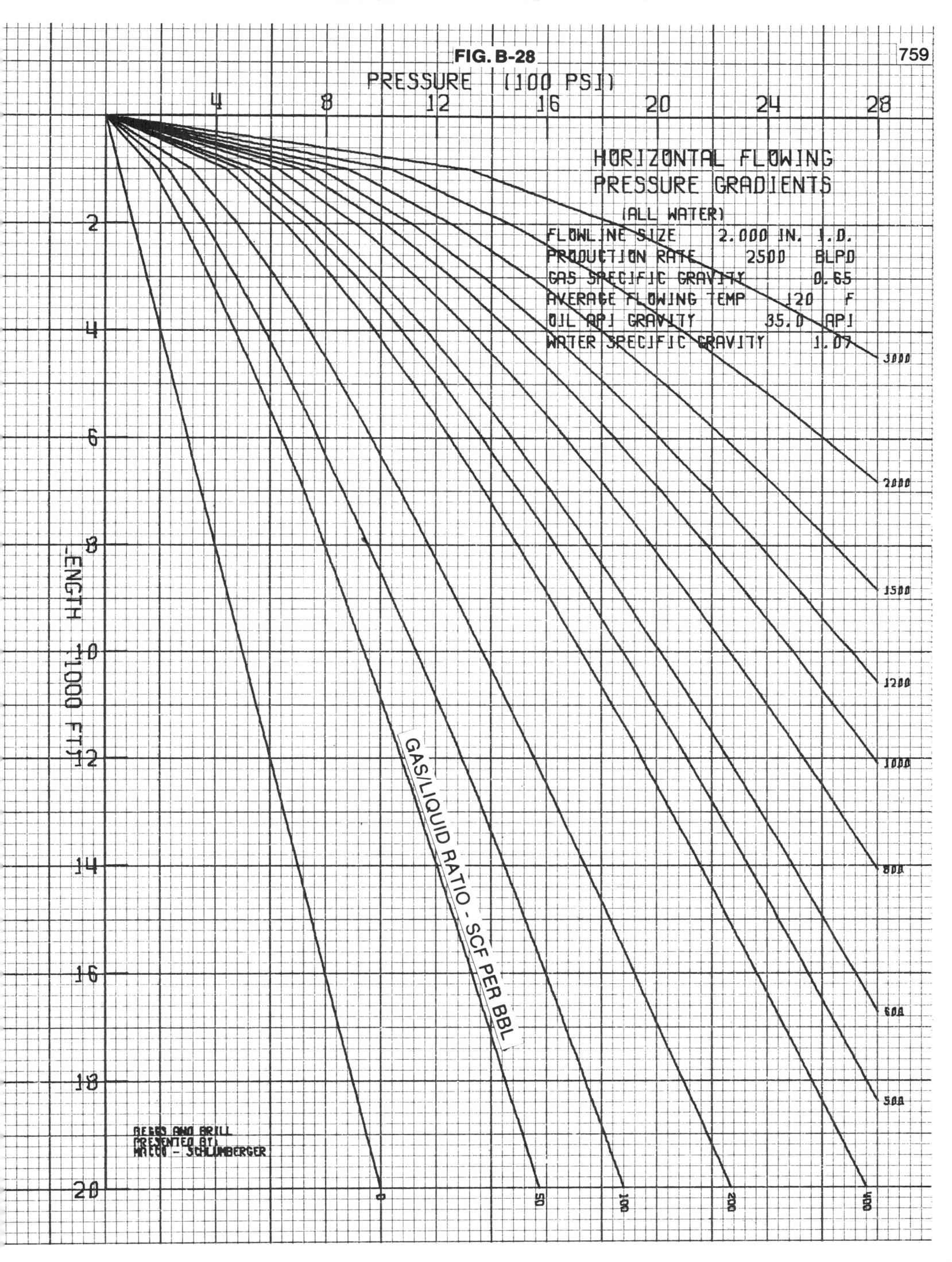

FIG. B-28

759

PRESSURE (100 PSI)

HORIZONTAL FLOWING
PRESSURE GRADIENTS
(ALL WATER)
FLOWLINE SIZE   2.000 IN. I.D.
PRODUCTION RATE   2500   BLPD
GAS SPECIFIC GRAVITY   0.65
AVERAGE FLOWING TEMP   120   F
OIL API GRAVITY   35.0   API
WATER SPECIFIC GRAVITY   1.07

LENGTH (1000 FT)

GAS/LIQUID RATIO - SCF PER BBL

BEGGS AND BRILL
PRESENTED BY:
MAECO - SCHLUMBERGER

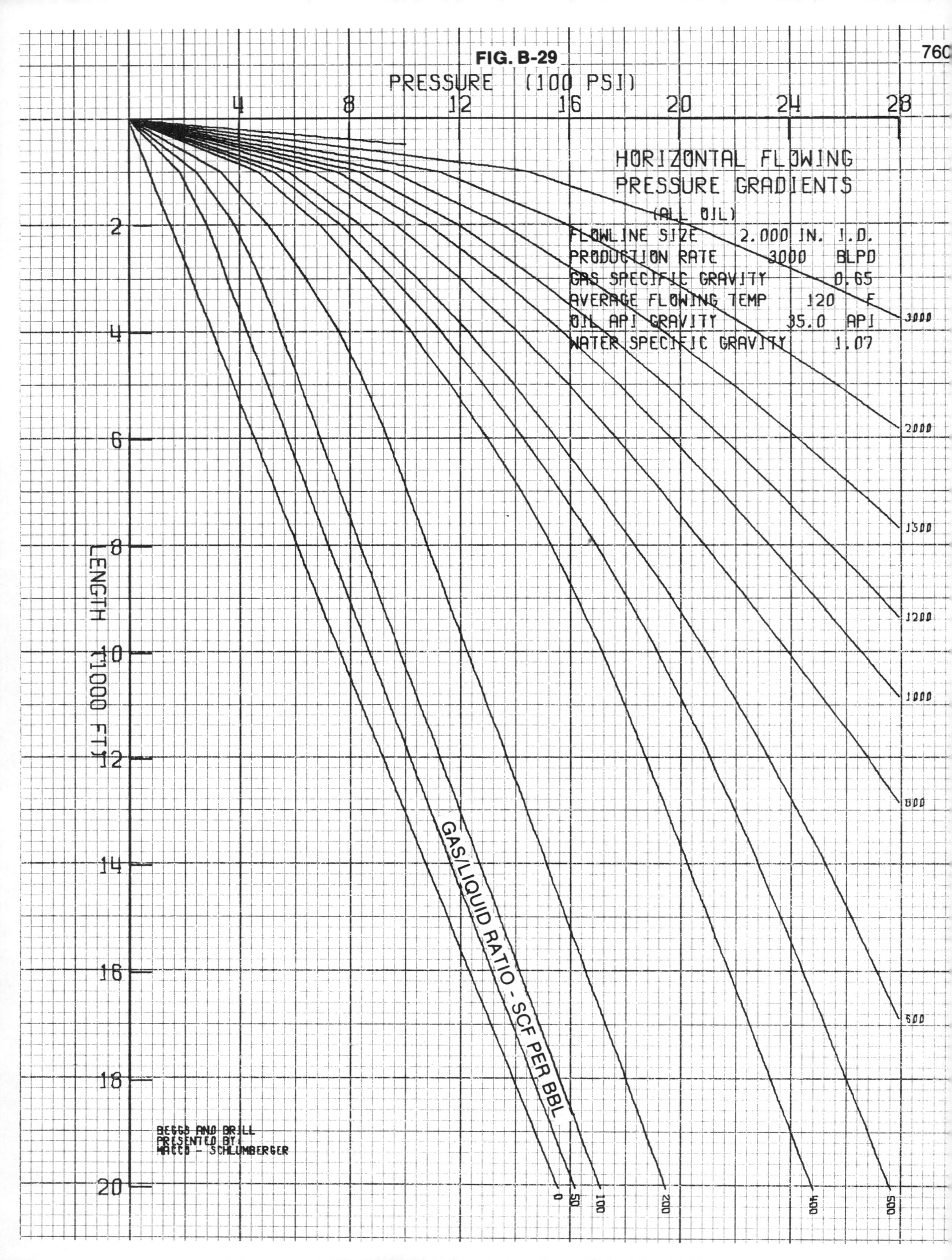

FIG. B-29

760

HORIZONTAL FLOWING
PRESSURE GRADIENTS
(ALL OIL)

FLOWLINE SIZE   2.000 IN. I.D.
PRODUCTION RATE   3000   BLPD
GAS SPECIFIC GRAVITY   0.65
AVERAGE FLOWING TEMP   120   F
OIL API GRAVITY   35.0   API
WATER SPECIFIC GRAVITY   1.07

PRESSURE (100 PSI)

LENGTH (1000 FT)

GAS/LIQUID RATIO - SCF PER BBL

BEGGS AND BRILL
PRESENTED BY
MACCO - SCHLUMBERGER

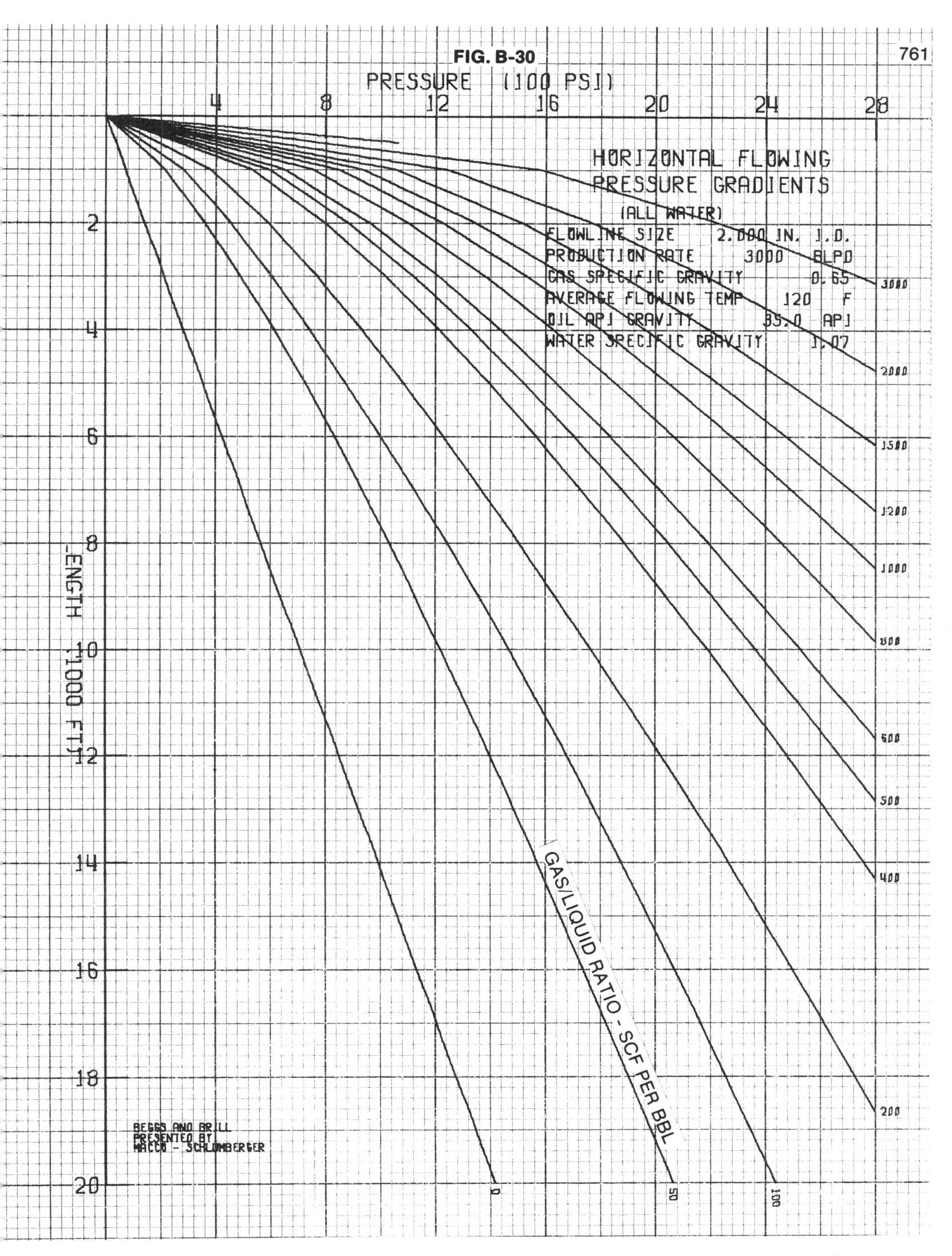

FIG. B-30

761

PRESSURE (100 PSI)

HORIZONTAL FLOWING
PRESSURE GRADIENTS
(ALL WATER)

FLOWLINE SIZE   2.000 IN. I.D.
PRODUCTION RATE   3000   BLPD
GAS SPECIFIC GRAVITY   0.65
AVERAGE FLOWING TEMP   120   F
OIL API GRAVITY   35.0 API
WATER SPECIFIC GRAVITY   1.07

LENGTH (1000 FT.)

GAS/LIQUID RATIO - SCF PER BBL

BEGGS AND BRILL
PRESENTED BY
MACCO - SCHLUMBERGER

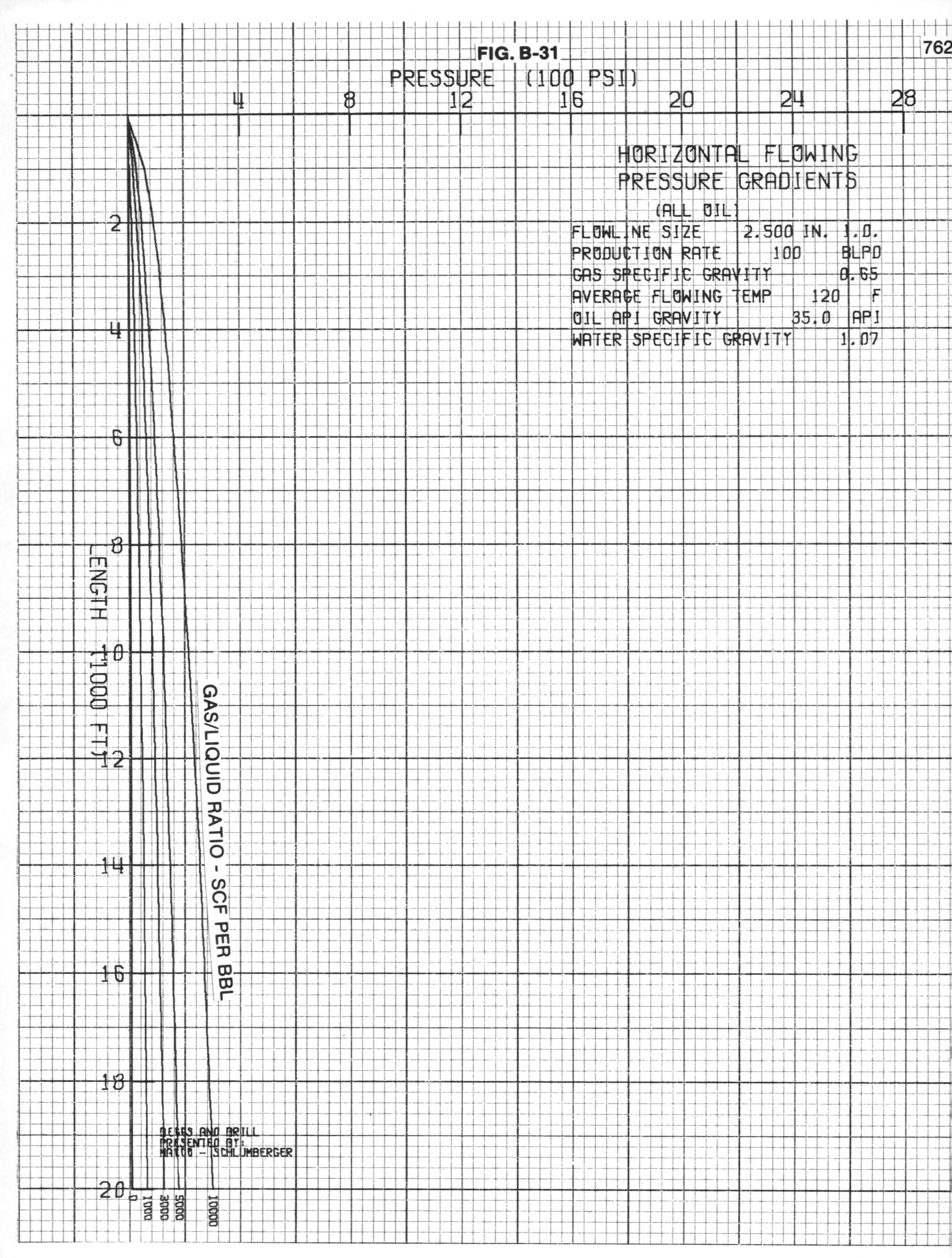

PRESSURE (100 PSI)

HORIZONTAL FLOWING
PRESSURE GRADIENTS
(ALL OIL)

FLOWLINE SIZE	2.500 IN.	I.D.
PRODUCTION RATE	100	BLPD
GAS SPECIFIC GRAVITY		0.65
AVERAGE FLOWING TEMP	120	F
OIL API GRAVITY	35.0	API
WATER SPECIFIC GRAVITY		1.07

LENGTH (1000 FT)

GAS/LIQUID RATIO - SCF PER BBL

BEGGS AND BRILL
PRESENTED BY:
NATCO - SCHLUMBERGER

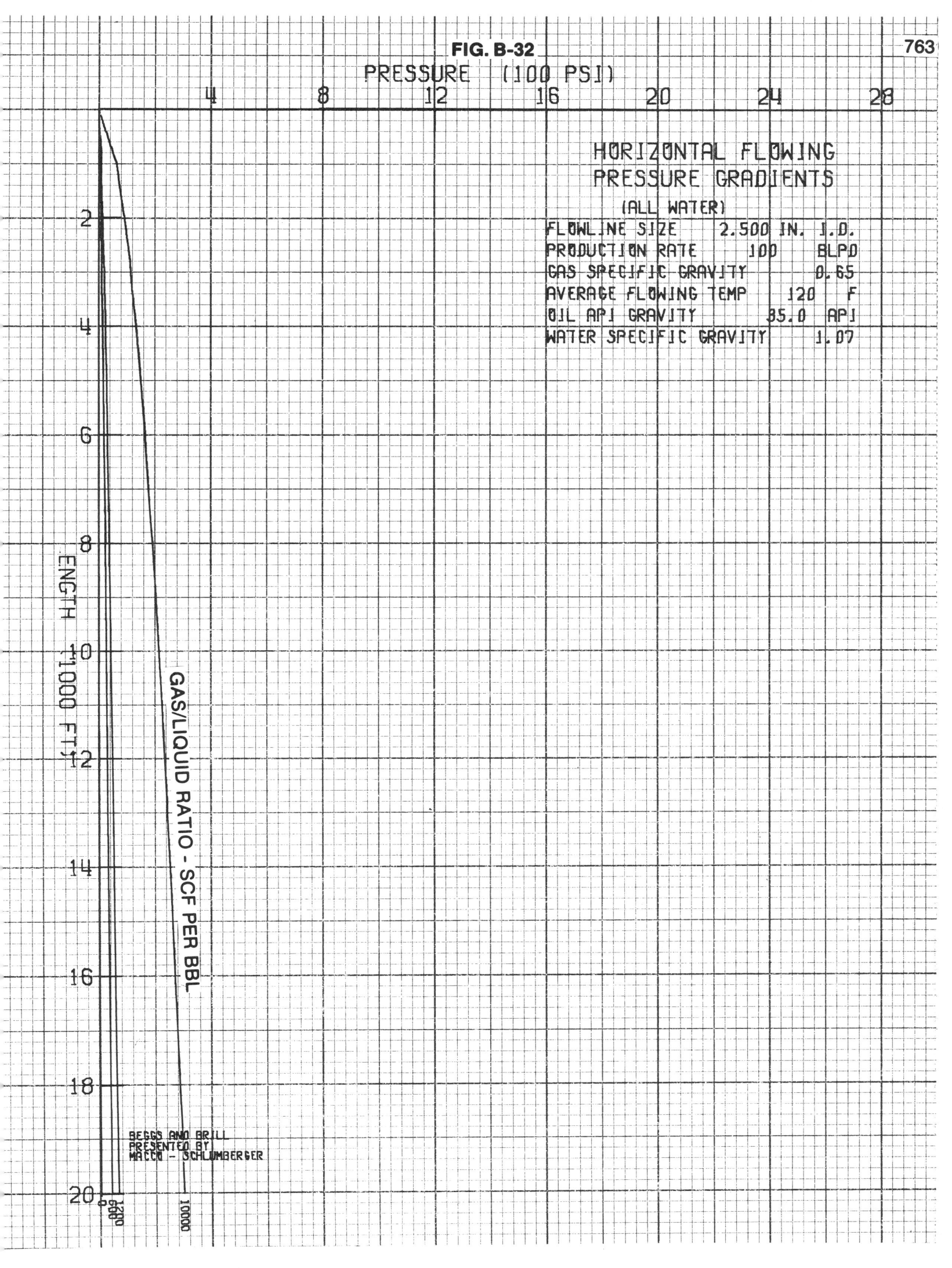

PRESSURE (100 PSI)

HORIZONTAL FLOWING
PRESSURE GRADIENTS
(ALL WATER)

FLOWLINE SIZE	2.500 IN. I.D.
PRODUCTION RATE	100 BLPD
GAS SPECIFIC GRAVITY	0.65
AVERAGE FLOWING TEMP	120 F
OIL API GRAVITY	35.0 API
WATER SPECIFIC GRAVITY	1.07

LENGTH (1000 FT)

GAS/LIQUID RATIO - SCF PER BBL

BEGGS AND BRILL
PRESENTED BY
MACCO - SCHLUMBERGER

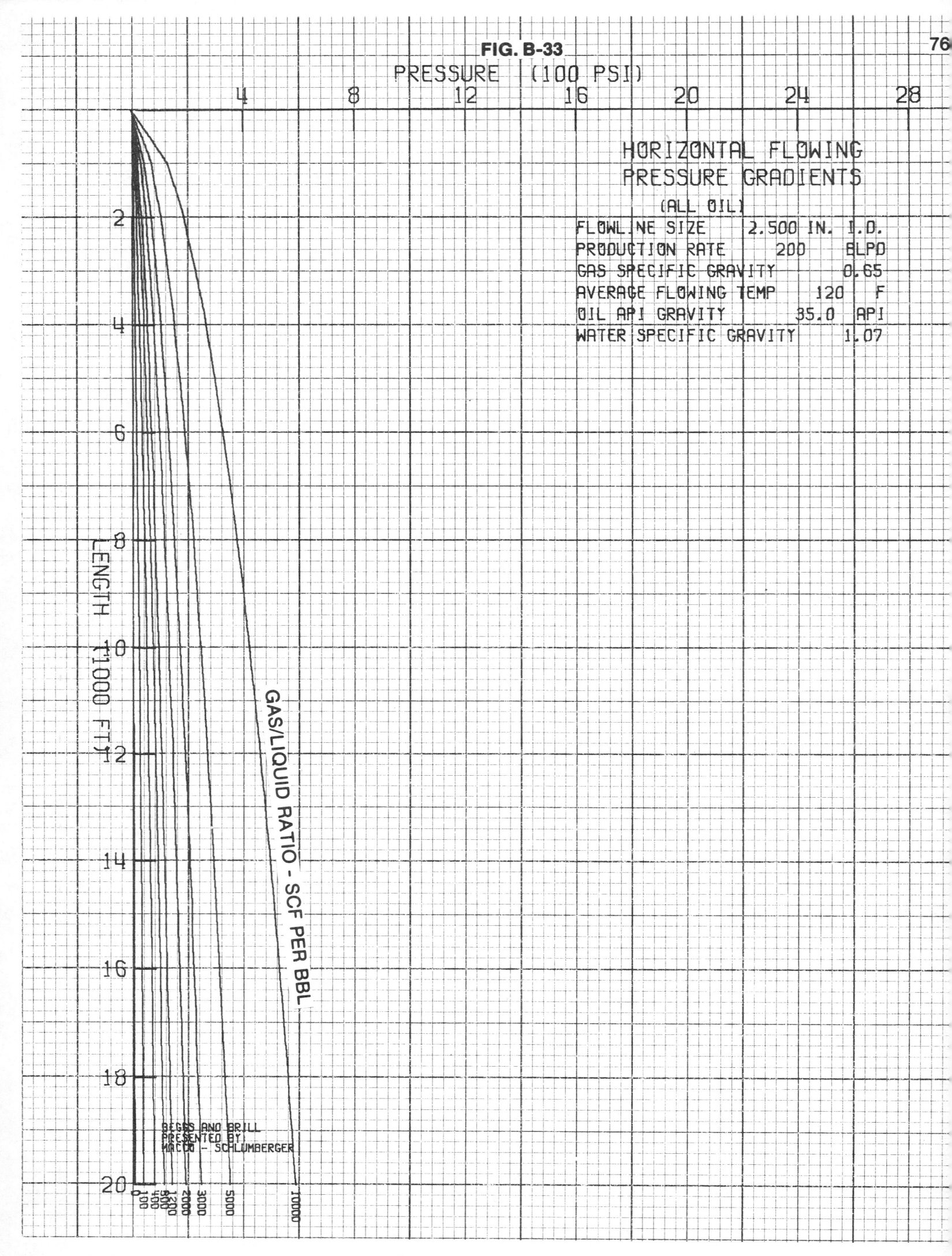

PRESSURE (100 PSI)

HORIZONTAL FLOWING
PRESSURE GRADIENTS
(ALL OIL)

FLOWLINE SIZE	2.500 IN. I.D.
PRODUCTION RATE	200 BLPD
GAS SPECIFIC GRAVITY	0.65
AVERAGE FLOWING TEMP	120 F
OIL API GRAVITY	35.0 API
WATER SPECIFIC GRAVITY	1.07

LENGTH (1000 FT.)

GAS/LIQUID RATIO - SCF PER BBL.

BEGGS AND BRILL
PRESENTED BY
KACOO - SCHLUMBERGER

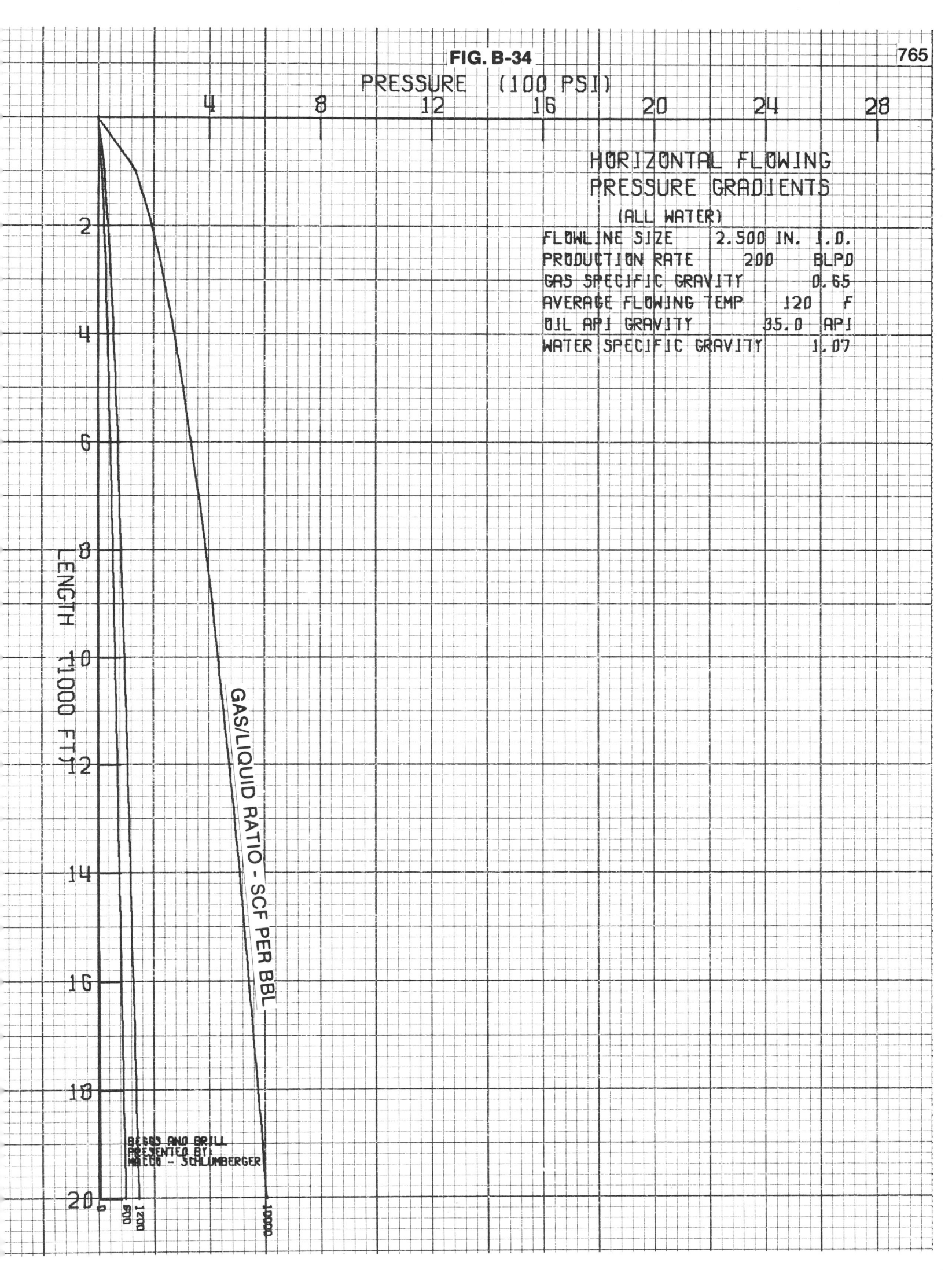

FIG. B-34

765

PRESSURE (100 PSI)

HORIZONTAL FLOWING
PRESSURE GRADIENTS
(ALL WATER)

FLOWLINE SIZE	2.500 IN.	I.D.
PRODUCTION RATE	200	BLPD
GAS SPECIFIC GRAVITY	0.65	
AVERAGE FLOWING TEMP	120	F
OIL API GRAVITY	35.0	API
WATER SPECIFIC GRAVITY	1.07	

LENGTH (1000 FT)

GAS/LIQUID RATIO - SCF PER BBL

BEGGS AND BRILL
PRESENTED BY:
HALCO - SCHLUMBERGER

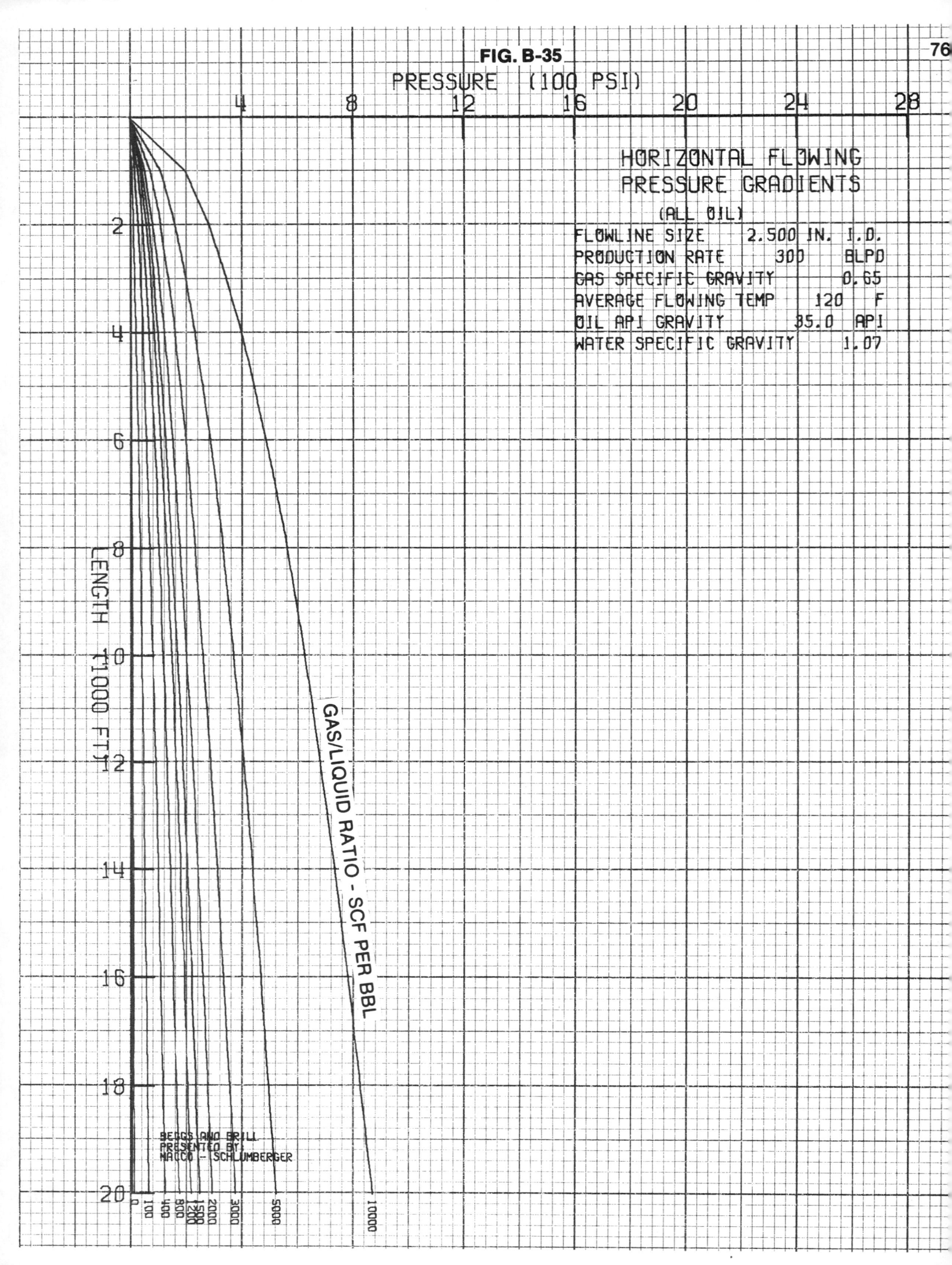

FIG. B-35

76

PRESSURE (100 PSI)

HORIZONTAL FLOWING
PRESSURE GRADIENTS
(ALL OIL)

FLOWLINE SIZE	2.500 IN. I.D.	
PRODUCTION RATE	300	BLPD
GAS SPECIFIC GRAVITY	0.65	
AVERAGE FLOWING TEMP	120	F
OIL API GRAVITY	85.0	API
WATER SPECIFIC GRAVITY	1.07	

LENGTH (1000 FT)

GAS/LIQUID RATIO - SCF PER BBL

BEGGS AND BRILL
PRESENTED BY:
NACCO - SCHLUMBERGER

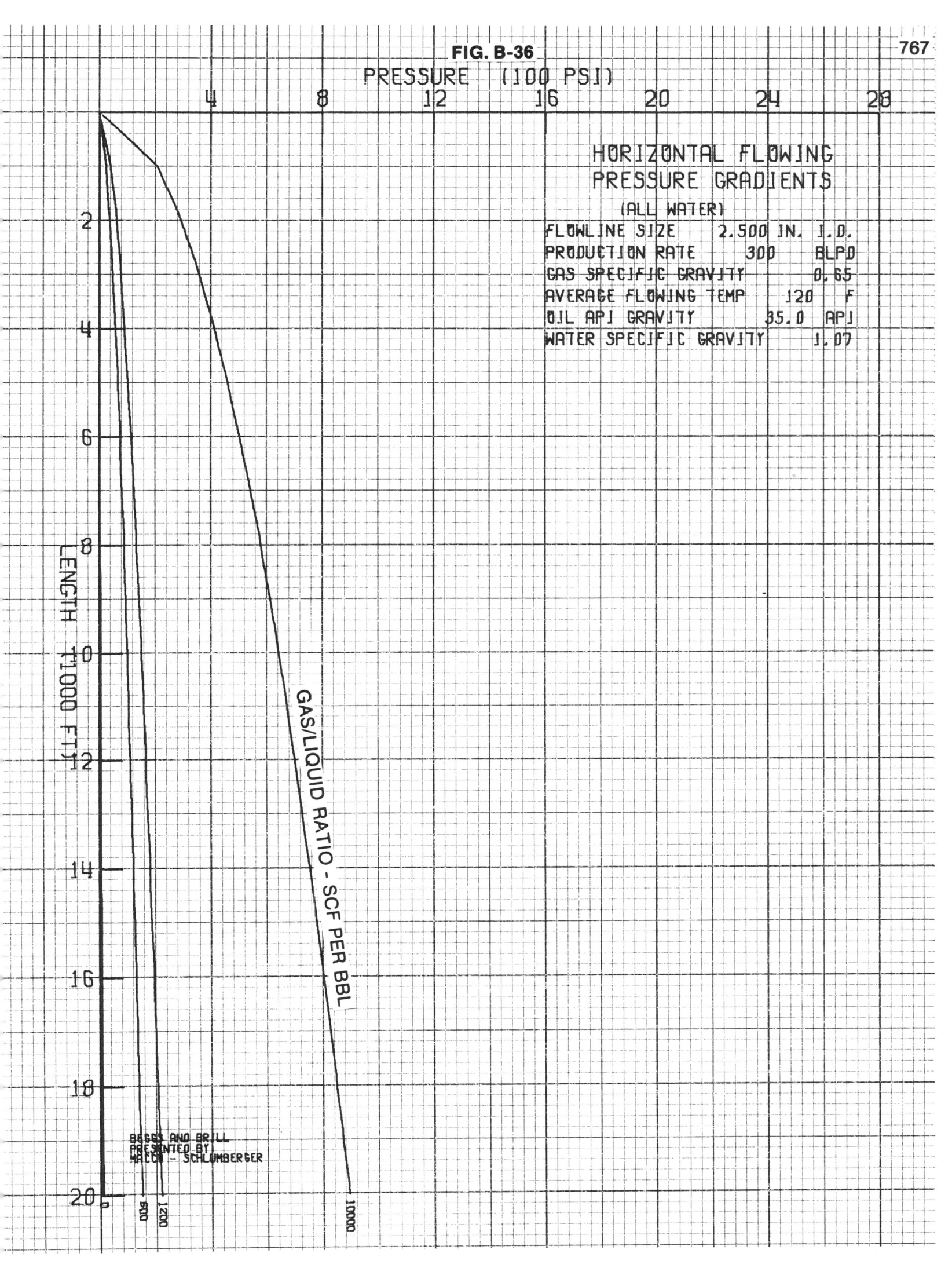

PRESSURE (100 PSI)

LENGTH (1000 FT)

HORIZONTAL FLOWING
PRESSURE GRADIENTS
(ALL WATER)

FLOWLINE SIZE   2.500 IN. I.D.
PRODUCTION RATE   300   BLPD
GAS SPECIFIC GRAVITY   0.65
AVERAGE FLOWING TEMP   120   F
OIL API GRAVITY   35.0   API
WATER SPECIFIC GRAVITY   1.07

GAS/LIQUID RATIO - SCF PER BBL

BEGGS AND BRILL
PRESENTED BY
MACCO - SCHLUMBERGER

500
1200
10000

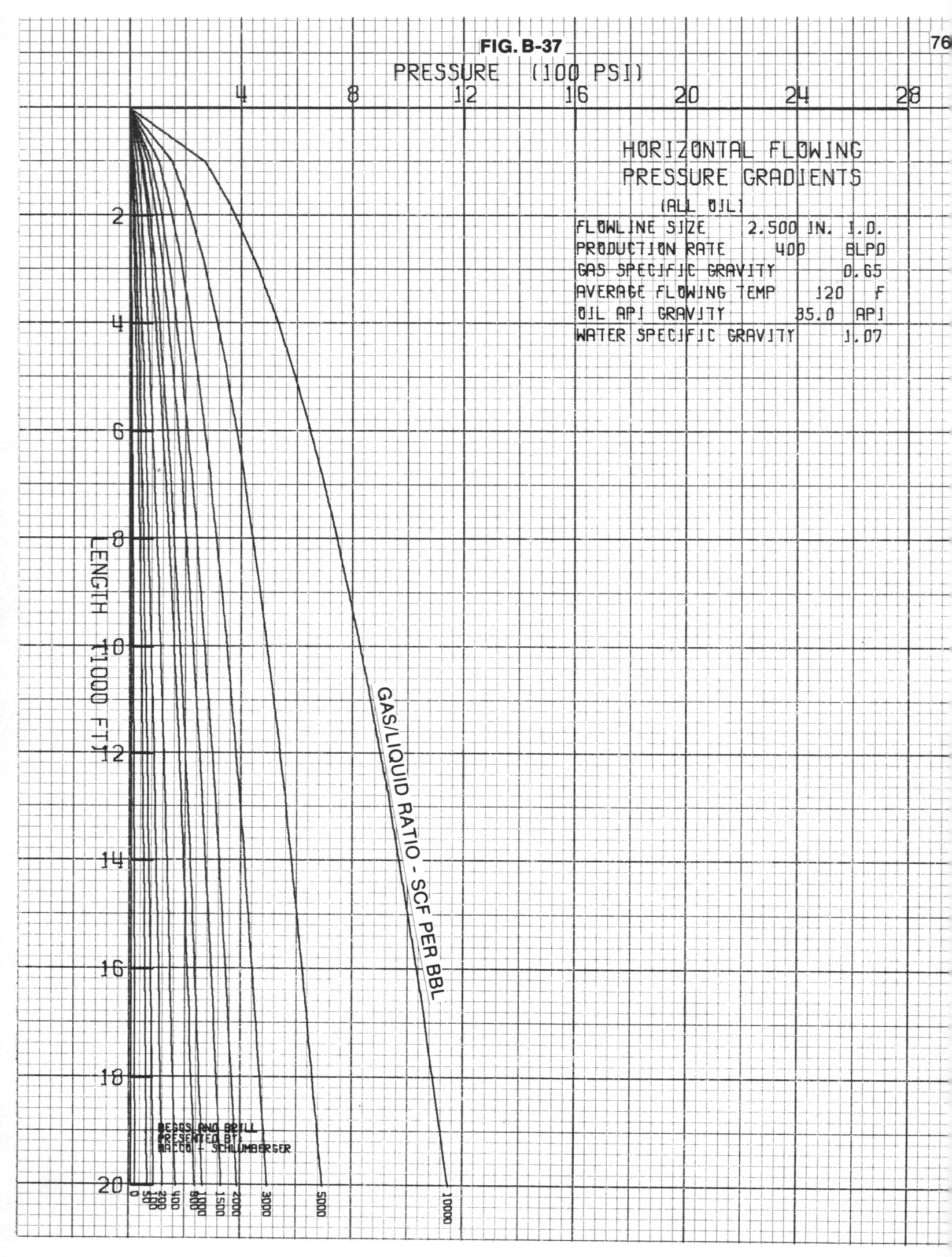

PRESSURE (100 PSI)

HORIZONTAL FLOWING
PRESSURE GRADIENTS
(ALL OIL)

FLOWLINE SIZE	2.500 IN. I.D.
PRODUCTION RATE	400 BLPD
GAS SPECIFIC GRAVITY	0.65
AVERAGE FLOWING TEMP	120 F
OIL API GRAVITY	35.0 API
WATER SPECIFIC GRAVITY	1.07

LENGTH (1000 FT)

GAS/LIQUID RATIO - SCF PER BBL

BEGGS AND BRILL
PRESENTED BY:
NAECO - SCHLUMBERGER

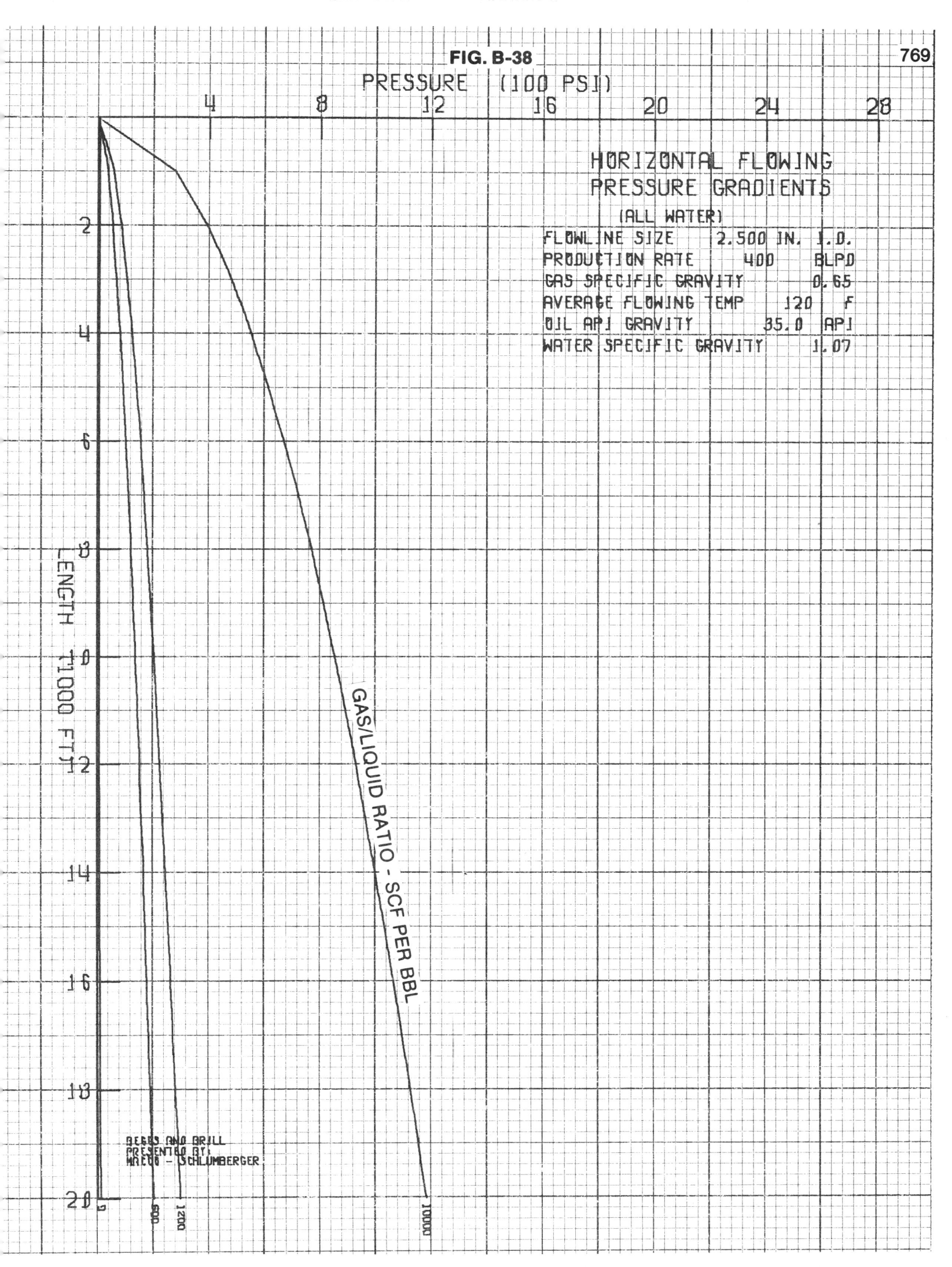

PRESSURE (100 PSI)

LENGTH (1000 FT)

HORIZONTAL FLOWING
PRESSURE GRADIENTS
(ALL WATER)

FLOWLINE SIZE	2.500 IN. I.D.
PRODUCTION RATE	400 BLPD
GAS SPECIFIC GRAVITY	0.65
AVERAGE FLOWING TEMP	120 F
OIL API GRAVITY	35.0 API
WATER SPECIFIC GRAVITY	1.07

GAS/LIQUID RATIO - SCF PER BBL

BEGGS AND BRILL
PRESENTED BY:
MAECO - SCHLUMBERGER

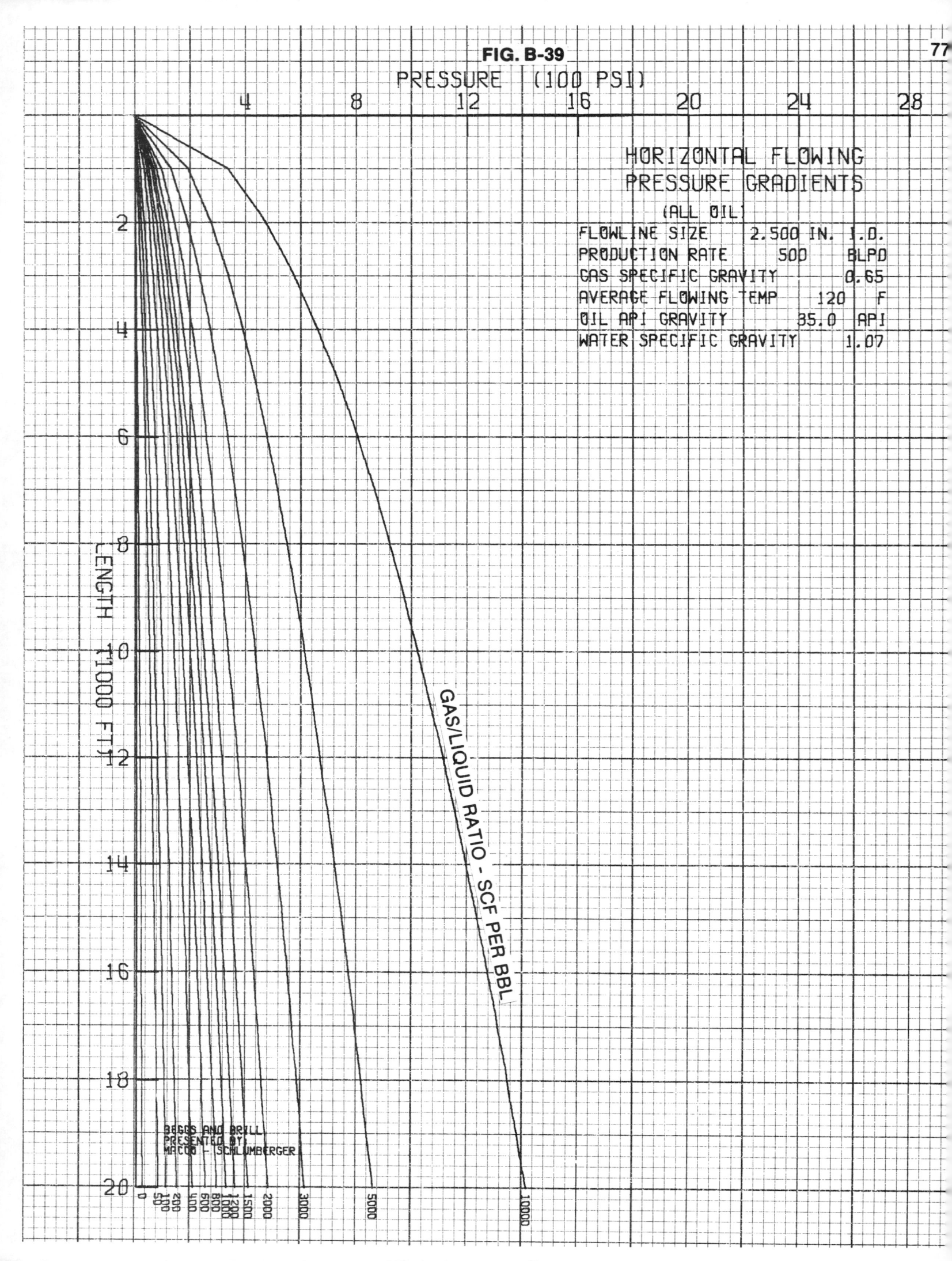

PRESSURE (100 PSI)

HORIZONTAL FLOWING
PRESSURE GRADIENTS
(ALL OIL)

FLOWLINE SIZE	2.500 IN. I.D.
PRODUCTION RATE	500 BLPD
GAS SPECIFIC GRAVITY	0.65
AVERAGE FLOWING TEMP	120 F
OIL API GRAVITY	35.0 API
WATER SPECIFIC GRAVITY	1.07

LENGTH (1000 FT.)

GAS/LIQUID RATIO - SCF PER BBL

BEGGS AND BRILL
PRESENTED BY
MACCO - SCHLUMBERGER

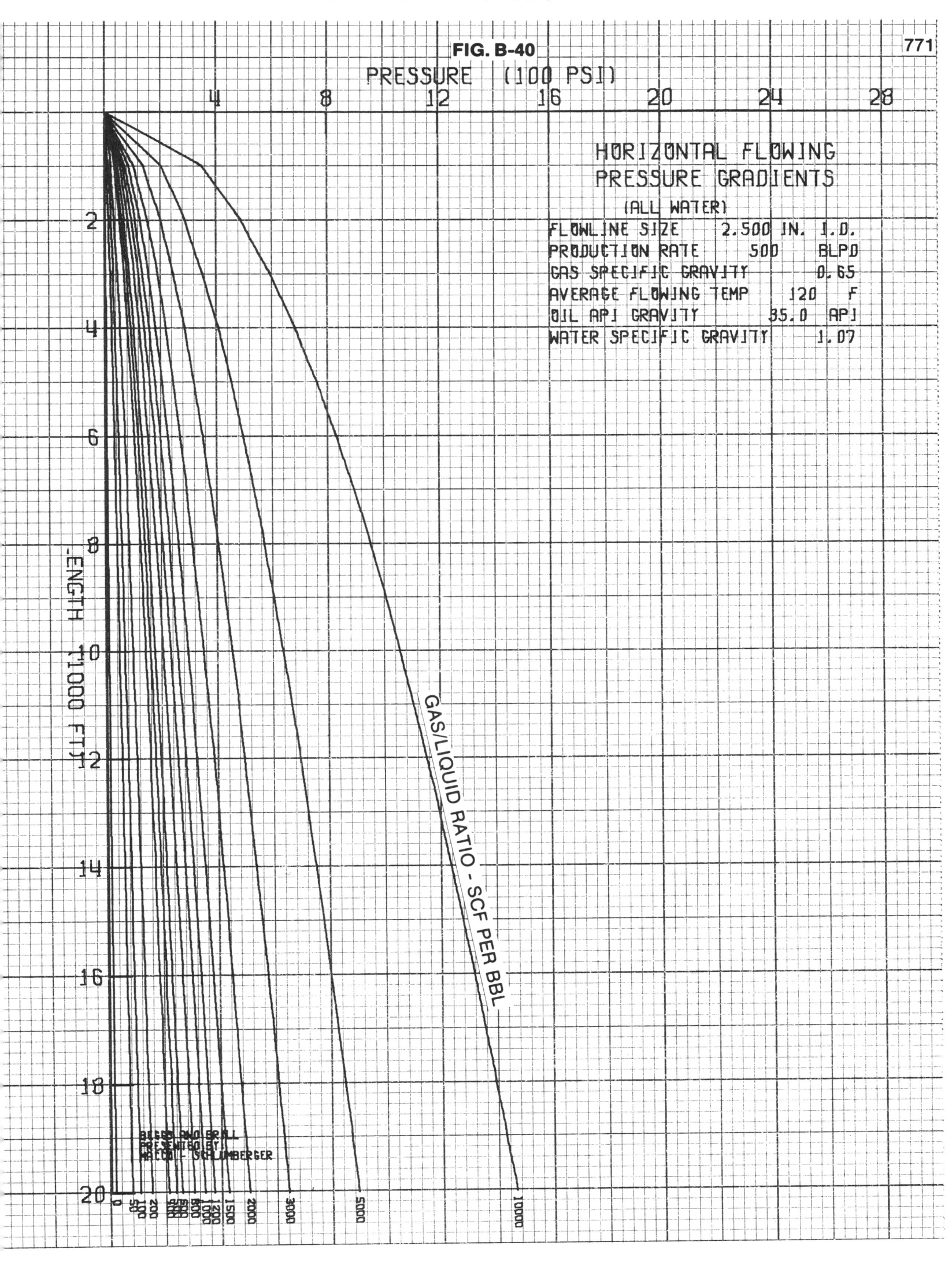

FIG. B-40

PRESSURE (100 PSI)

HORIZONTAL FLOWING
PRESSURE GRADIENTS
(ALL WATER)

FLOWLINE SIZE   2.500 IN. I.D.
PRODUCTION RATE   500   BLPD
GAS SPECIFIC GRAVITY   0.65
AVERAGE FLOWING TEMP   120   F
OIL API GRAVITY   35.0 API
WATER SPECIFIC GRAVITY   1.07

LENGTH (1000 FT.)

GAS/LIQUID RATIO - SCF PER BBL

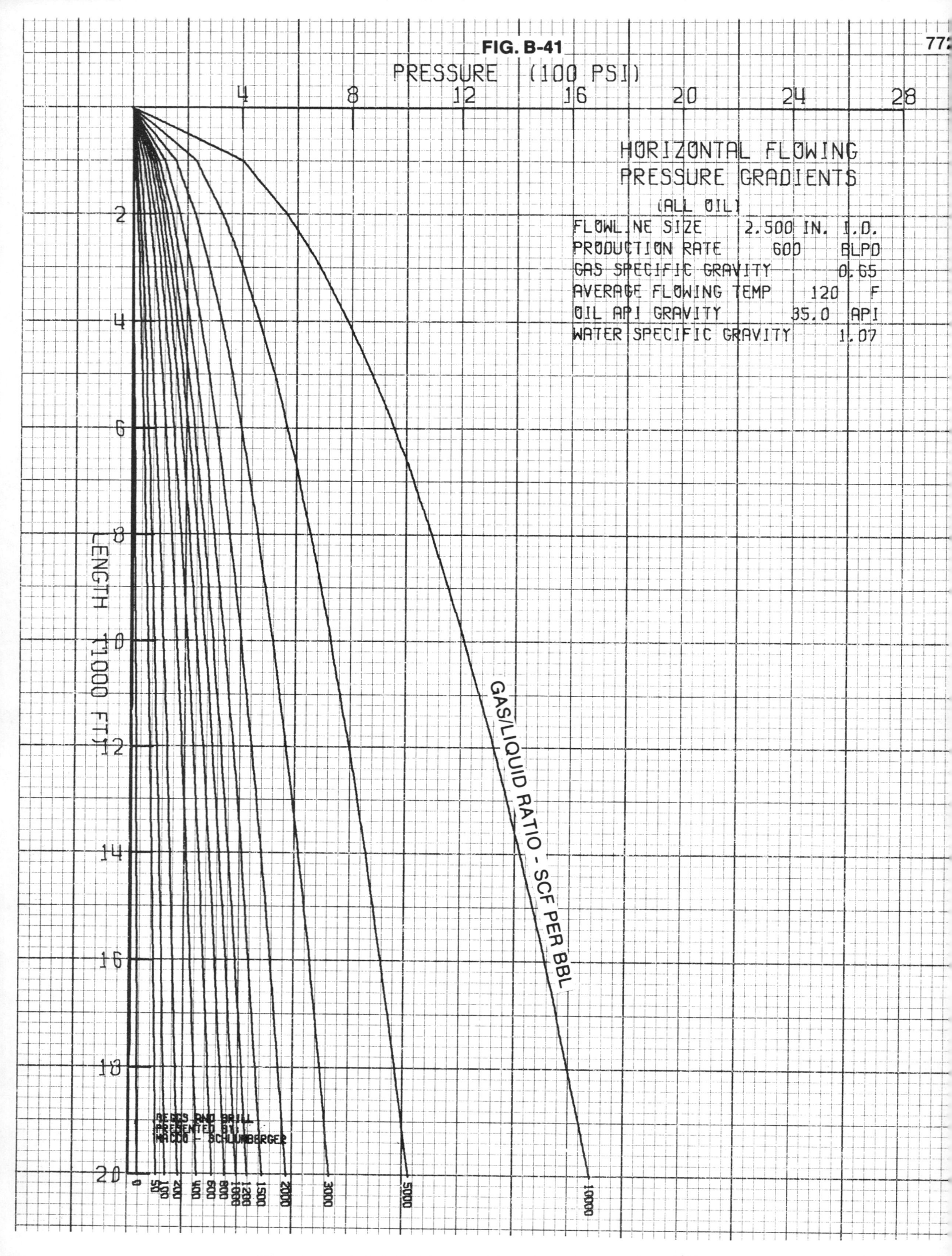

FIG. B-41

772

PRESSURE (100 PSI)

HORIZONTAL FLOWING
PRESSURE GRADIENTS
(ALL OIL)

FLOWLINE SIZE	2.500 IN. I.D.
PRODUCTION RATE	600 BLPD
GAS SPECIFIC GRAVITY	0.65
AVERAGE FLOWING TEMP	120 F
OIL API GRAVITY	35.0 API
WATER SPECIFIC GRAVITY	1.07

LENGTH (1000 FT)

GAS/LIQUID RATIO - SCF PER BBL

REGGS AND BRILL
PRESENTED BY
HACOO - SCHLUMBERGER

FIG. B-42

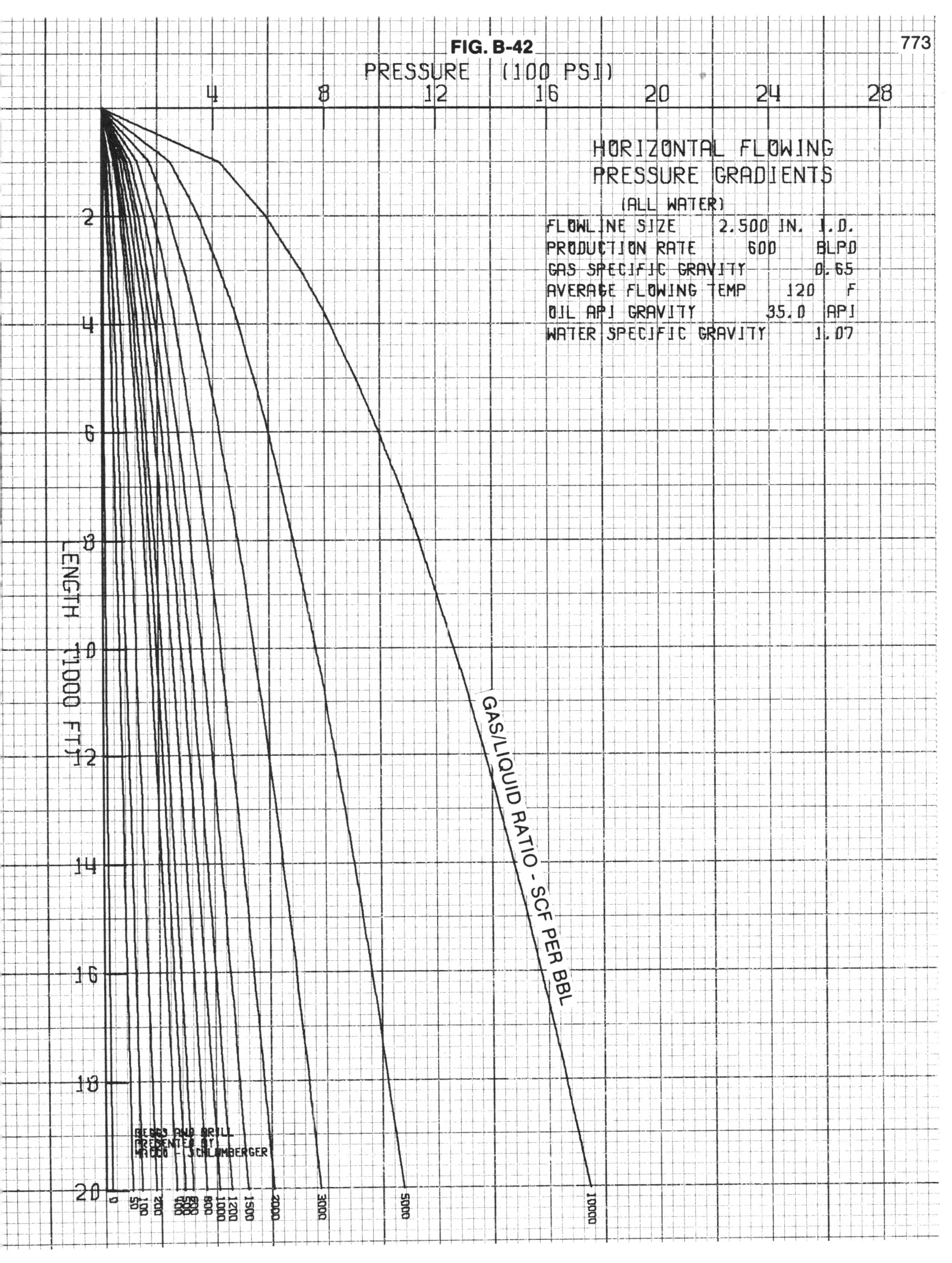

PRESSURE (100 PSI)

HORIZONTAL FLOWING
PRESSURE GRADIENTS
(ALL WATER)

FLOWLINE SIZE	2.500 IN. I.D.
PRODUCTION RATE	600 BLPD
GAS SPECIFIC GRAVITY	0.65
AVERAGE FLOWING TEMP	120 F
OIL API GRAVITY	35.0 API
WATER SPECIFIC GRAVITY	1.07

LENGTH (1000 FT)

GAS/LIQUID RATIO - SCF PER BBL

BEGGS AND BRILL
PRESENTED BY
MADCO - SCHLUMBERGER

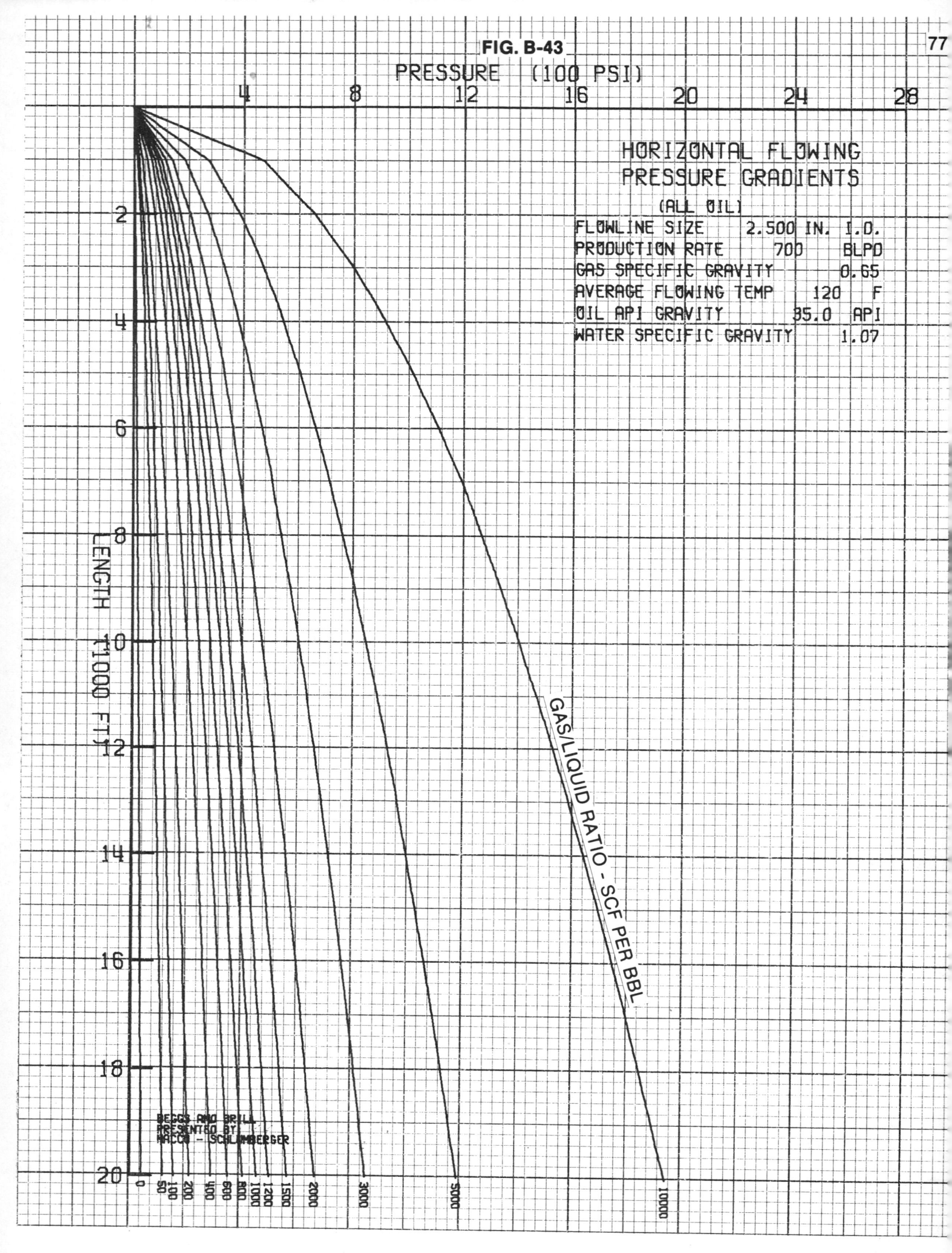

HORIZONTAL FLOWING
PRESSURE GRADIENTS
(ALL OIL)

FLOWLINE SIZE	2.500 IN.	I.D.
PRODUCTION RATE	700	BLPD
GAS SPECIFIC GRAVITY	0.65	
AVERAGE FLOWING TEMP	120	F
OIL API GRAVITY	35.0	API
WATER SPECIFIC GRAVITY	1.07	

PRESSURE (100 PSI)

LENGTH (1000 FT)

GAS/LIQUID RATIO - SCF PER BBL

BEGGS AND BRILL
PRESENTED BY
NACCO - SCHLUMBERGER

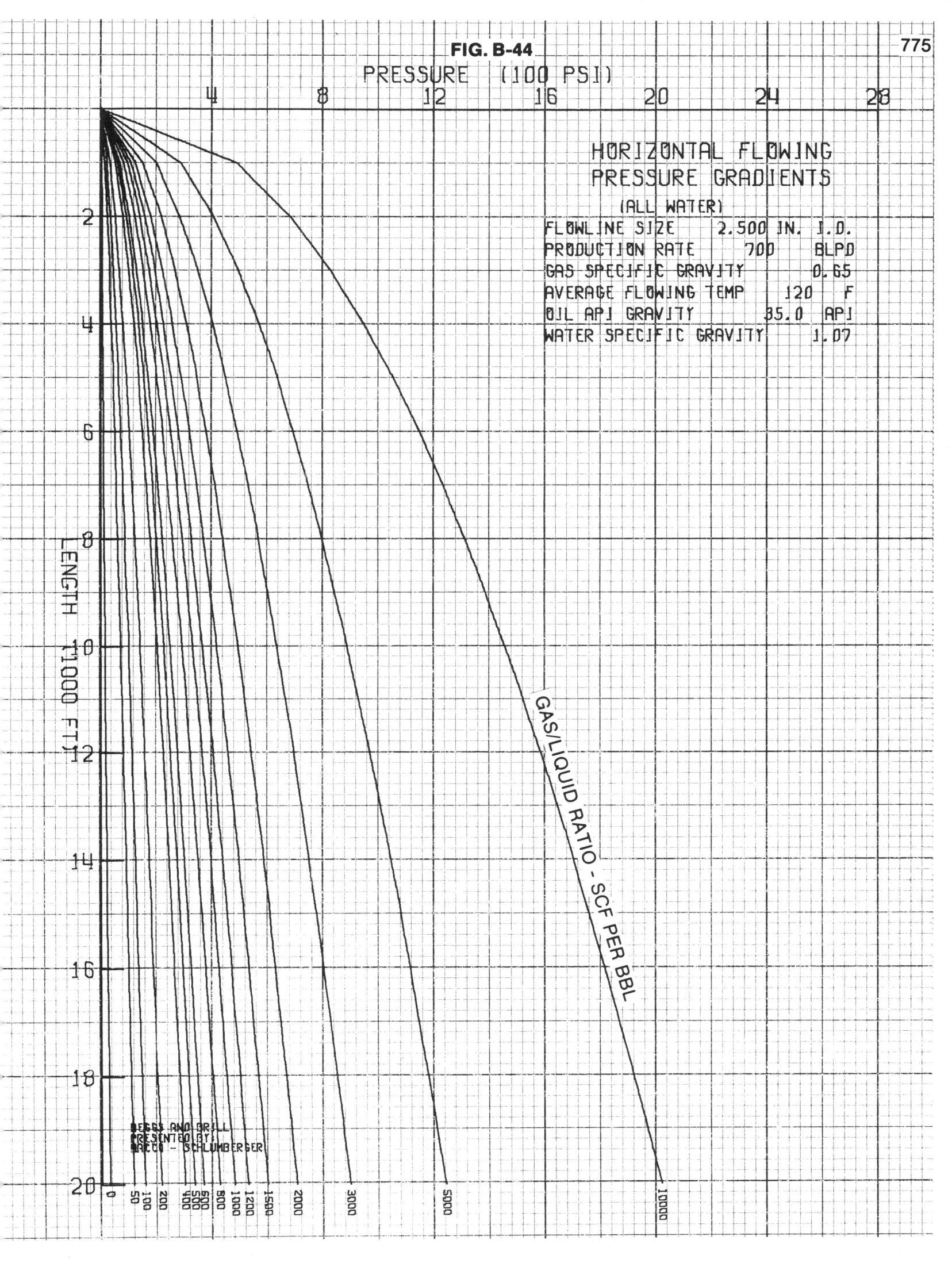

FIG. B-44

PRESSURE (100 PSI)

HORIZONTAL FLOWING
PRESSURE GRADIENTS
(ALL WATER)

FLOWLINE SIZE	2.500 IN. I.D.
PRODUCTION RATE	700 BLPD
GAS SPECIFIC GRAVITY	0.65
AVERAGE FLOWING TEMP	120 F
OIL API GRAVITY	35.0 API
WATER SPECIFIC GRAVITY	1.07

LENGTH (1000 FT)

GAS/LIQUID RATIO - SCF PER BBL

BEGGS AND BRILL
PRESENTED BY
NACCO - SCHLUMBERGER

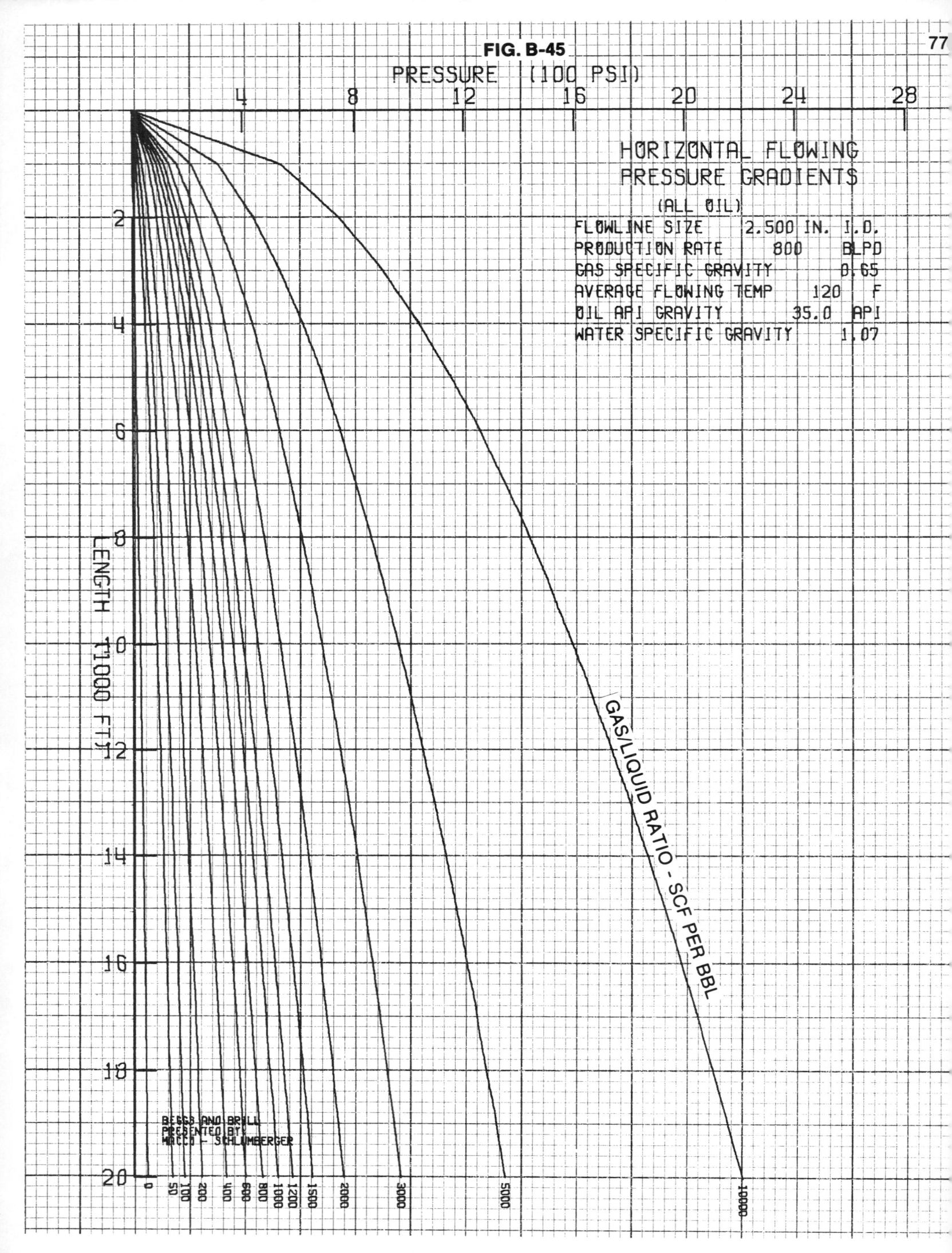

FIG. B-45

HORIZONTAL FLOWING
PRESSURE GRADIENTS
(ALL OIL)

FLOWLINE SIZE	2.500 IN.	I.D.
PRODUCTION RATE	800	BLPD
GAS SPECIFIC GRAVITY		0.65
AVERAGE FLOWING TEMP	120	F
OIL API GRAVITY	35.0	API
WATER SPECIFIC GRAVITY		1.07

PRESSURE (100 PSI)

LENGTH (1000 FT.)

GAS/LIQUID RATIO - SCF PER BBL

BEGGS AND BRILL
PRESENTED BY:
MACCO - SCHLUMBERGER

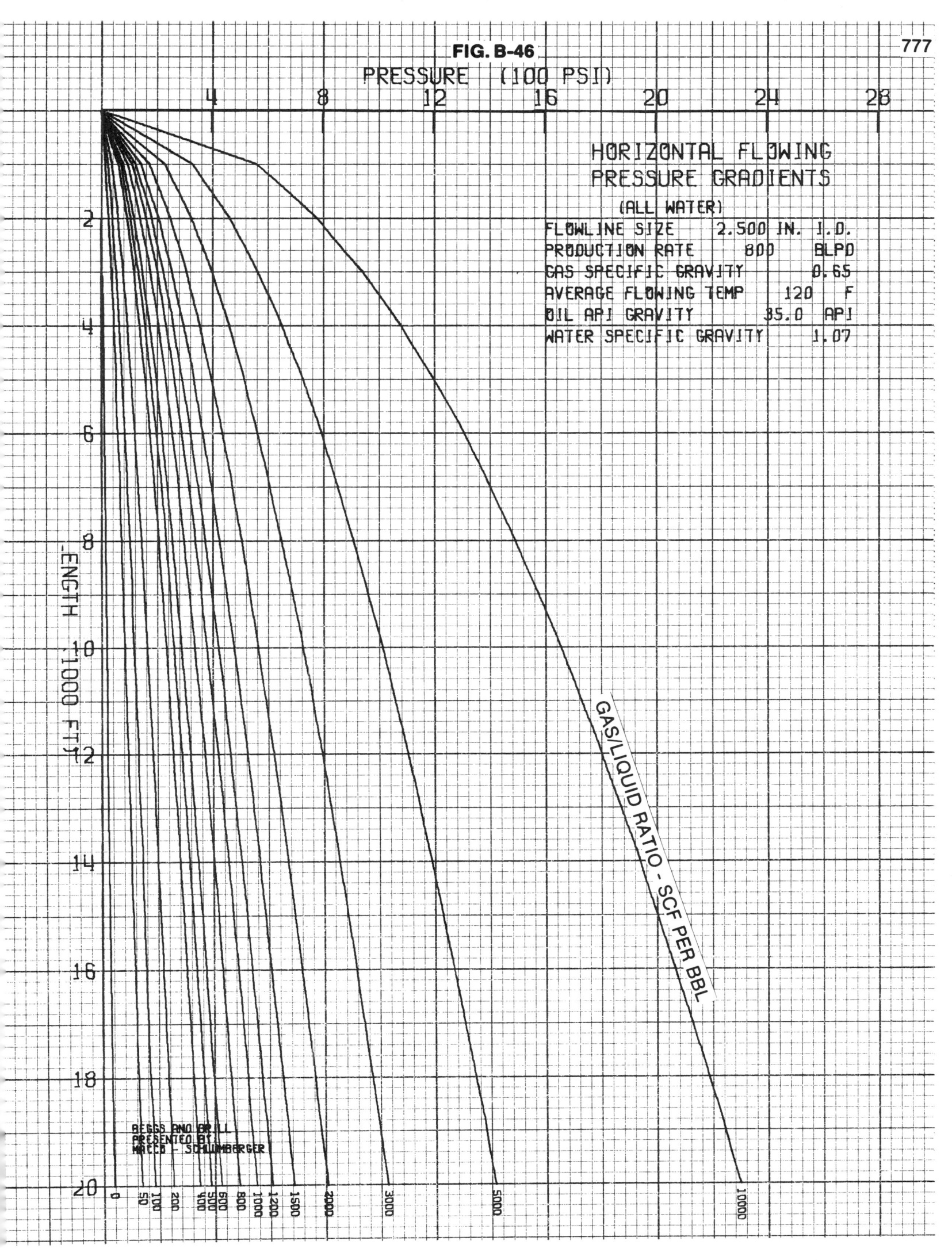

FIG. B-46

PRESSURE (100 PSI)

HORIZONTAL FLOWING
PRESSURE GRADIENTS
(ALL WATER)

FLOWLINE SIZE	2.500 IN.	I.D.
PRODUCTION RATE	800	BLPD
GAS SPECIFIC GRAVITY	0.65	
AVERAGE FLOWING TEMP	120	F
OIL API GRAVITY	35.0	API
WATER SPECIFIC GRAVITY	1.07	

LENGTH 1000 FT

GAS/LIQUID RATIO - SCF PER BBL

BEGGS AND BRILL
PRESENTED BY
MACCO - SCHLUMBERGER

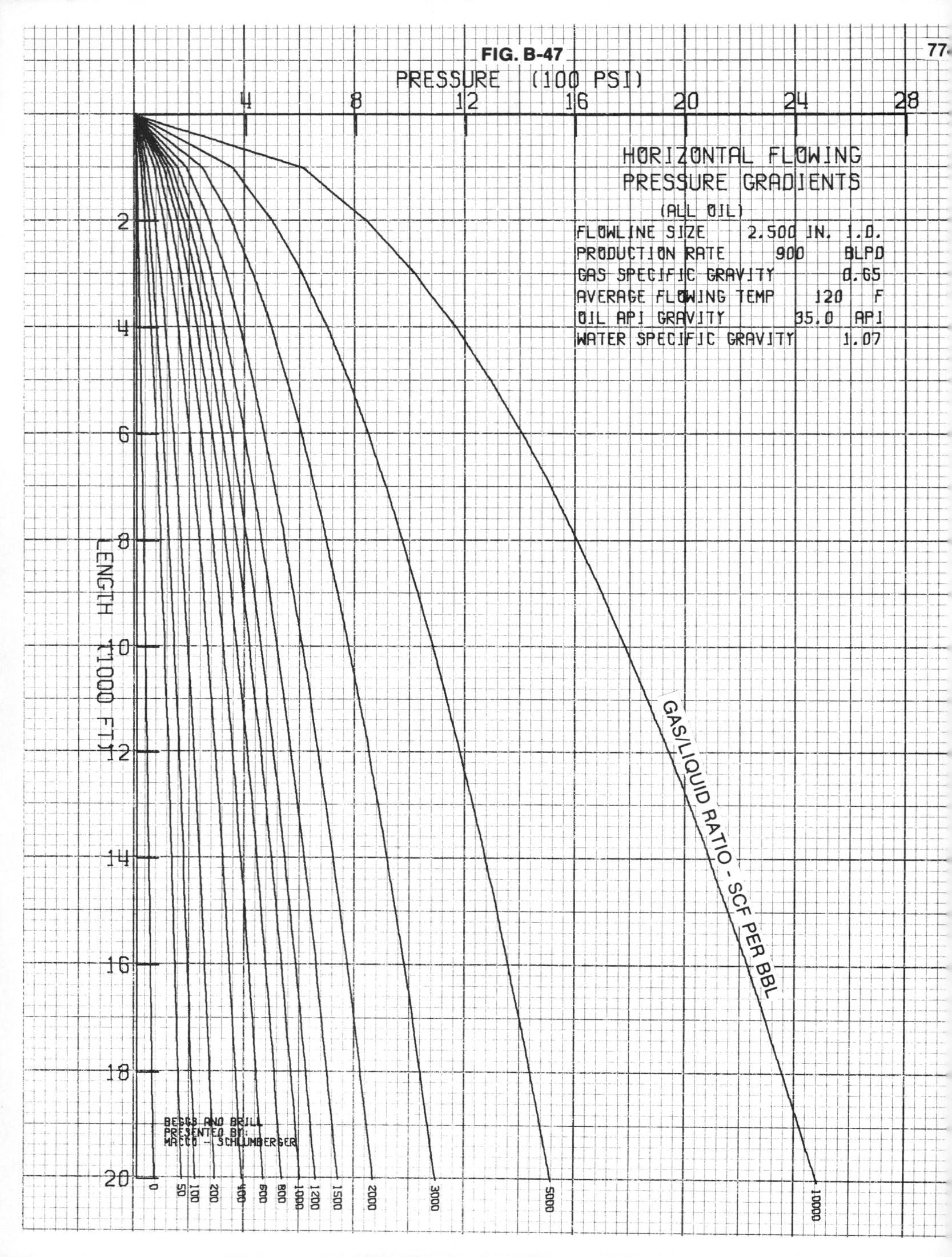

FIG. B-47

77.

PRESSURE (100 PSI)

HORIZONTAL FLOWING
PRESSURE GRADIENTS
(ALL OIL)

FLOWLINE SIZE	2.500 IN. I.D.
PRODUCTION RATE	900 BLPD
GAS SPECIFIC GRAVITY	0.65
AVERAGE FLOWING TEMP	120 F
OIL API GRAVITY	35.0 API
WATER SPECIFIC GRAVITY	1.07

LENGTH (1000 FT)

GAS/LIQUID RATIO - SCF PER BBL

BEGGS AND BRILL
PRESENTED BY:
MACCO - SCHLUMBERGER

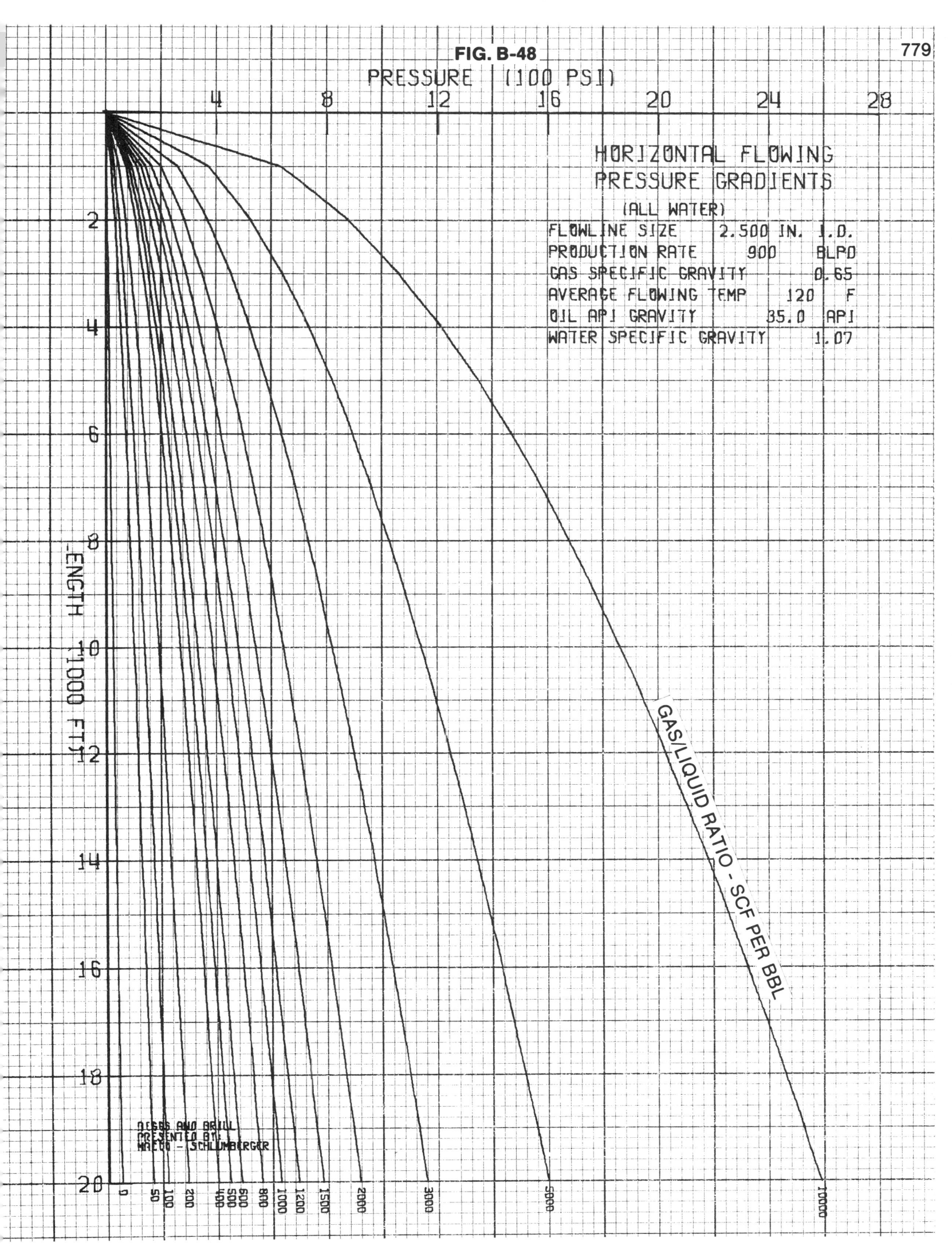

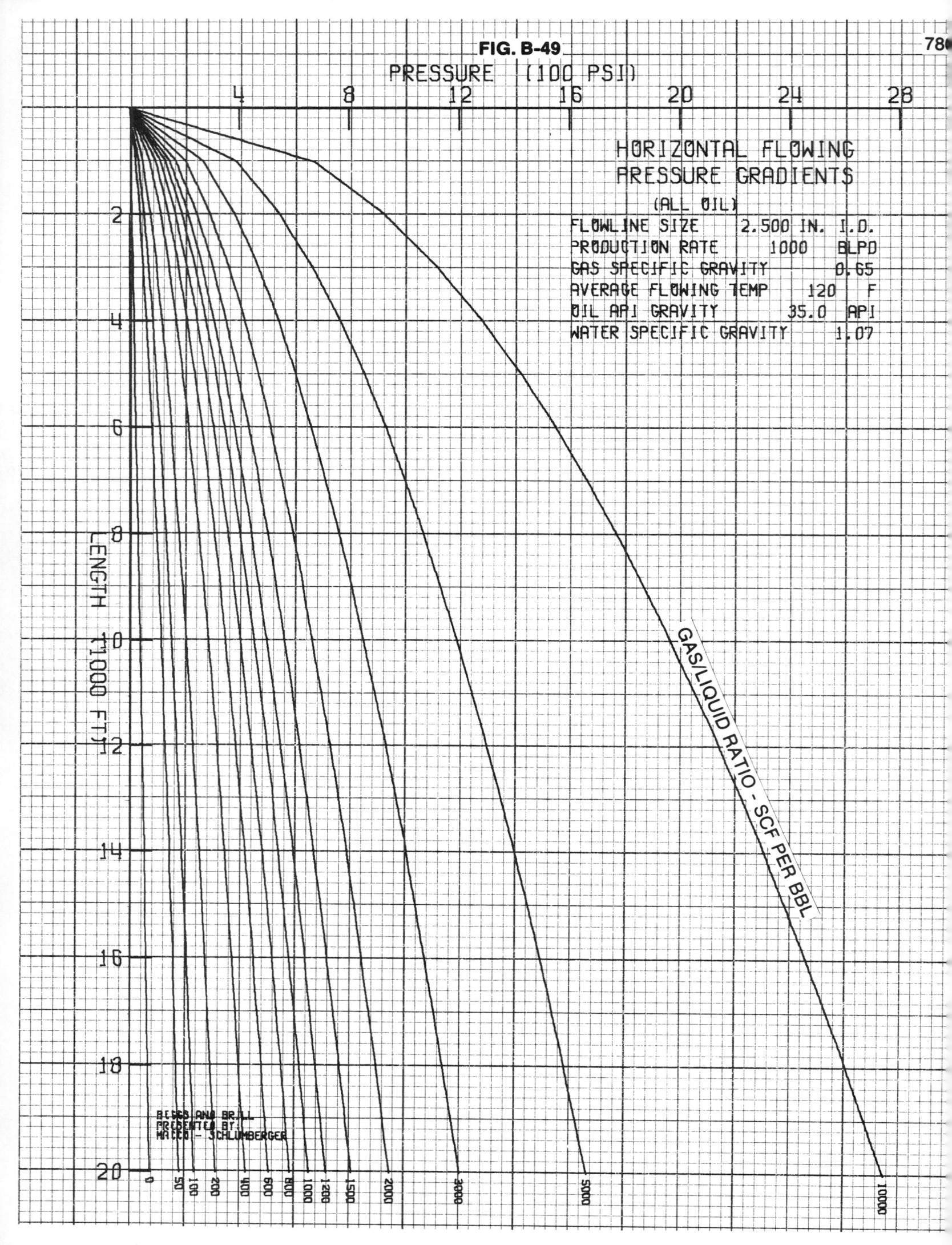

PRESSURE (100 PSI)

HORIZONTAL FLOWING
PRESSURE GRADIENTS
(ALL OIL)

FLOWLINE SIZE	2.500 IN. I.D.
PRODUCTION RATE	1000 BLPD
GAS SPECIFIC GRAVITY	0.65
AVERAGE FLOWING TEMP	120 F
OIL API GRAVITY	35.0 API
WATER SPECIFIC GRAVITY	1.07

LENGTH (1000 FT)

GAS/LIQUID RATIO - SCF PER BBL

BEGGS AND BRILL
PRESENTED BY:
WATCO - SCHLUMBERGER

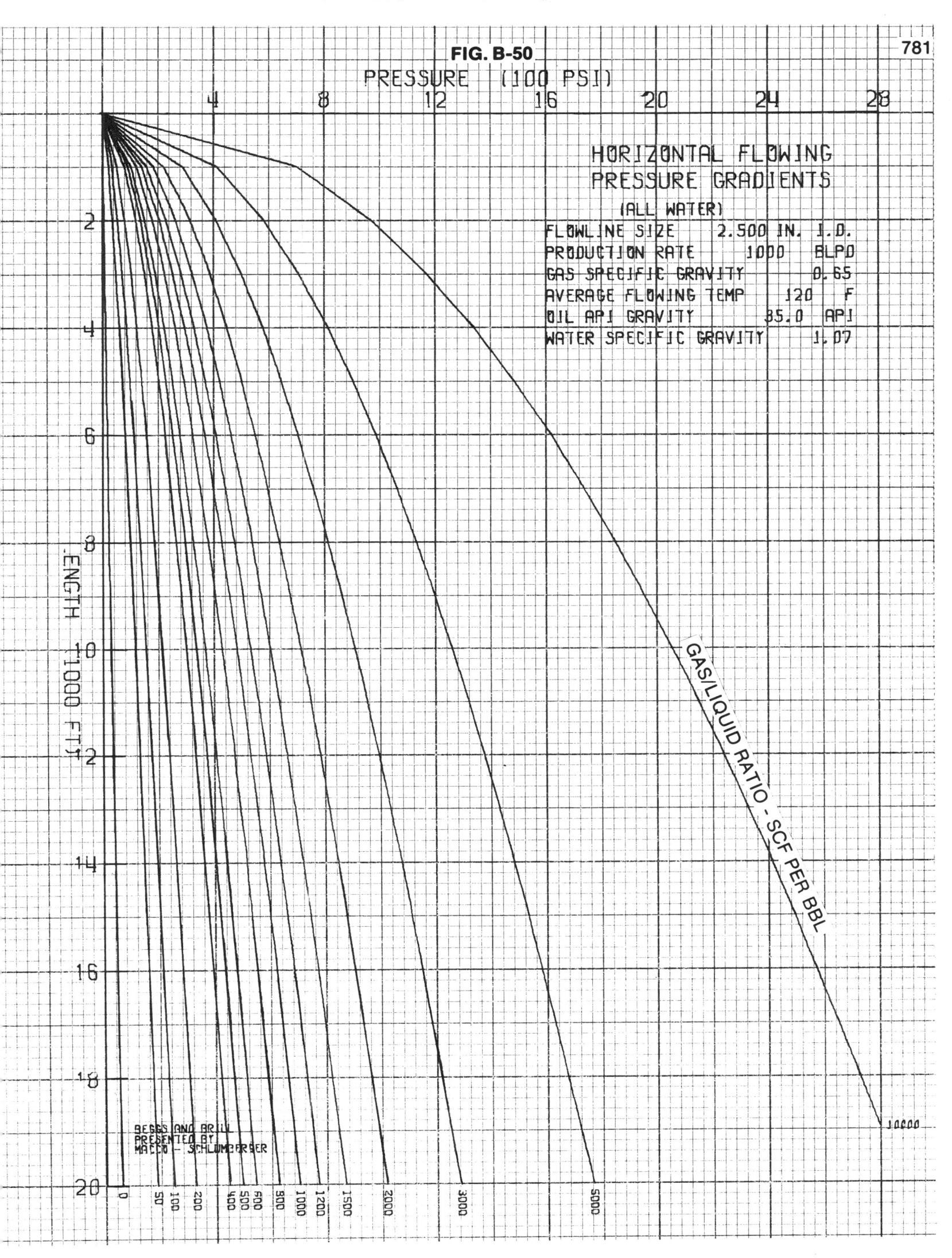

FIG. B-50

HORIZONTAL FLOWING
PRESSURE GRADIENTS
(ALL WATER)

FLOWLINE SIZE	2.500 IN. I.D.
PRODUCTION RATE	1000 BLPD
GAS SPECIFIC GRAVITY	0.65
AVERAGE FLOWING TEMP	120 F
OIL API GRAVITY	35.0 API
WATER SPECIFIC GRAVITY	1.07

PRESSURE (100 PSI)

LENGTH (1000 FT)

GAS/LIQUID RATIO - SCF PER BBL

BEGGS AND BRILL
PRESENTED BY
MAECO - SCHLUMBERGER

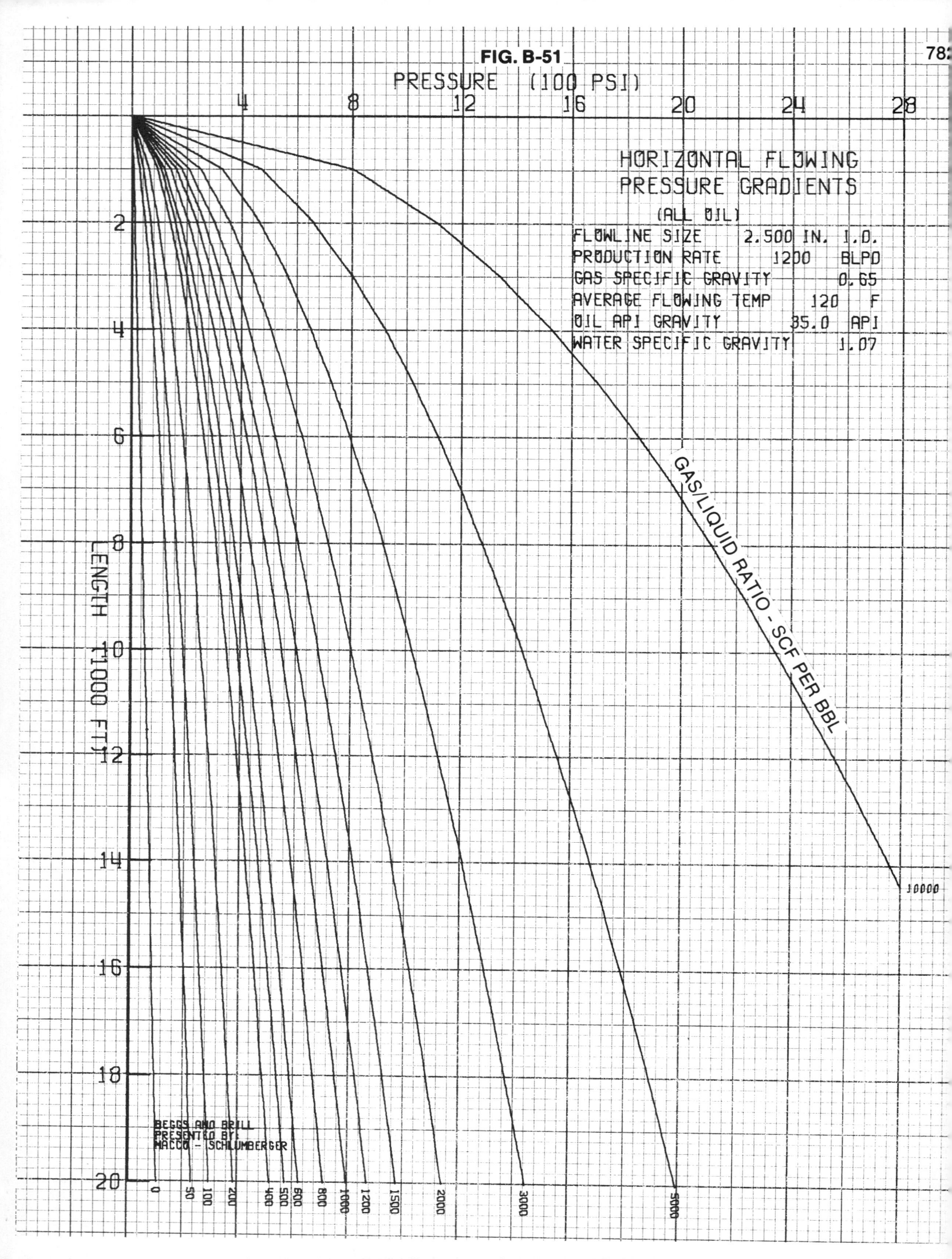

FIG. B-51

782

PRESSURE (100 PSI)

LENGTH (1000 FT)

HORIZONTAL FLOWING
PRESSURE GRADIENTS

(ALL OIL)

FLOWLINE SIZE	2.500 IN.	I.D.
PRODUCTION RATE	1200	BLPD
GAS SPECIFIC GRAVITY	0.65	
AVERAGE FLOWING TEMP	120	F
OIL API GRAVITY	35.0	API
WATER SPECIFIC GRAVITY	1.07	

GAS/LIQUID RATIO - SCF PER BBL

BEGGS AND BRILL
PRESENTED BY:
MACCO - SCHLUMBERGER

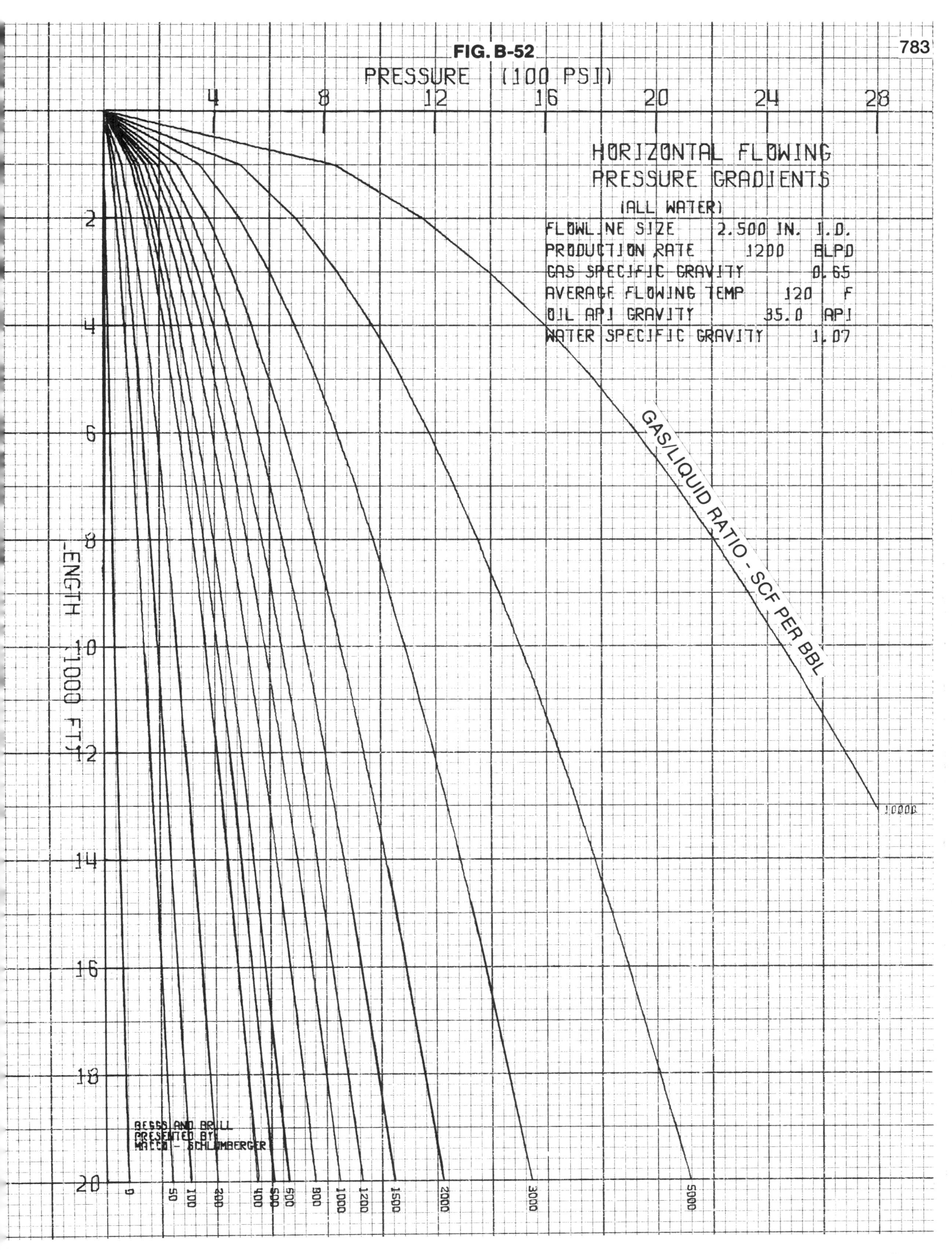

PRESSURE (100 PSI)

HORIZONTAL FLOWING
PRESSURE GRADIENTS
(ALL WATER)

FLOWLINE SIZE	2.500 IN. I.D.
PRODUCTION RATE	1200 BLPD
GAS SPECIFIC GRAVITY	0.65
AVERAGE FLOWING TEMP	120 F
OIL API GRAVITY	35.0 API
WATER SPECIFIC GRAVITY	1.07

GAS/LIQUID RATIO - SCF PER BBL

LENGTH (1000 FT)

BEGGS AND BRILL
PRESENTED BY:
MACCO - SCHLUMBERGER

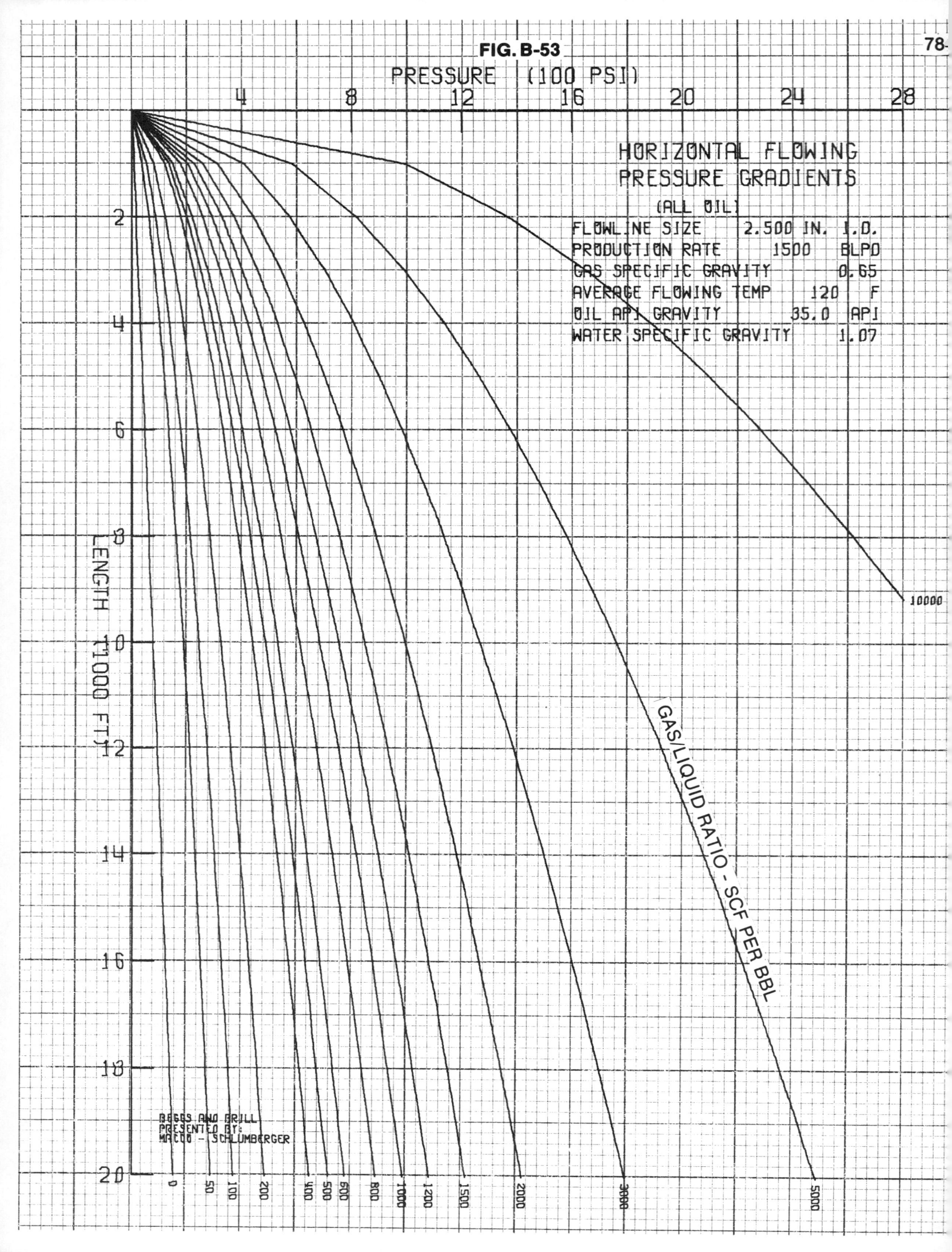

PRESSURE (100 PSI)

HORIZONTAL FLOWING
PRESSURE GRADIENTS
(ALL OIL)

FLOWLINE SIZE	2.500 IN. I.D.
PRODUCTION RATE	1500 BLPD
GAS SPECIFIC GRAVITY	0.65
AVERAGE FLOWING TEMP	120 F
OIL API GRAVITY	35.0 API
WATER SPECIFIC GRAVITY	1.07

LENGTH (1000 FT)

GAS/LIQUID RATIO - SCF PER BBL

BEGGS AND BRILL
PRESENTED BY:
MACCO - SCHLUMBERGER

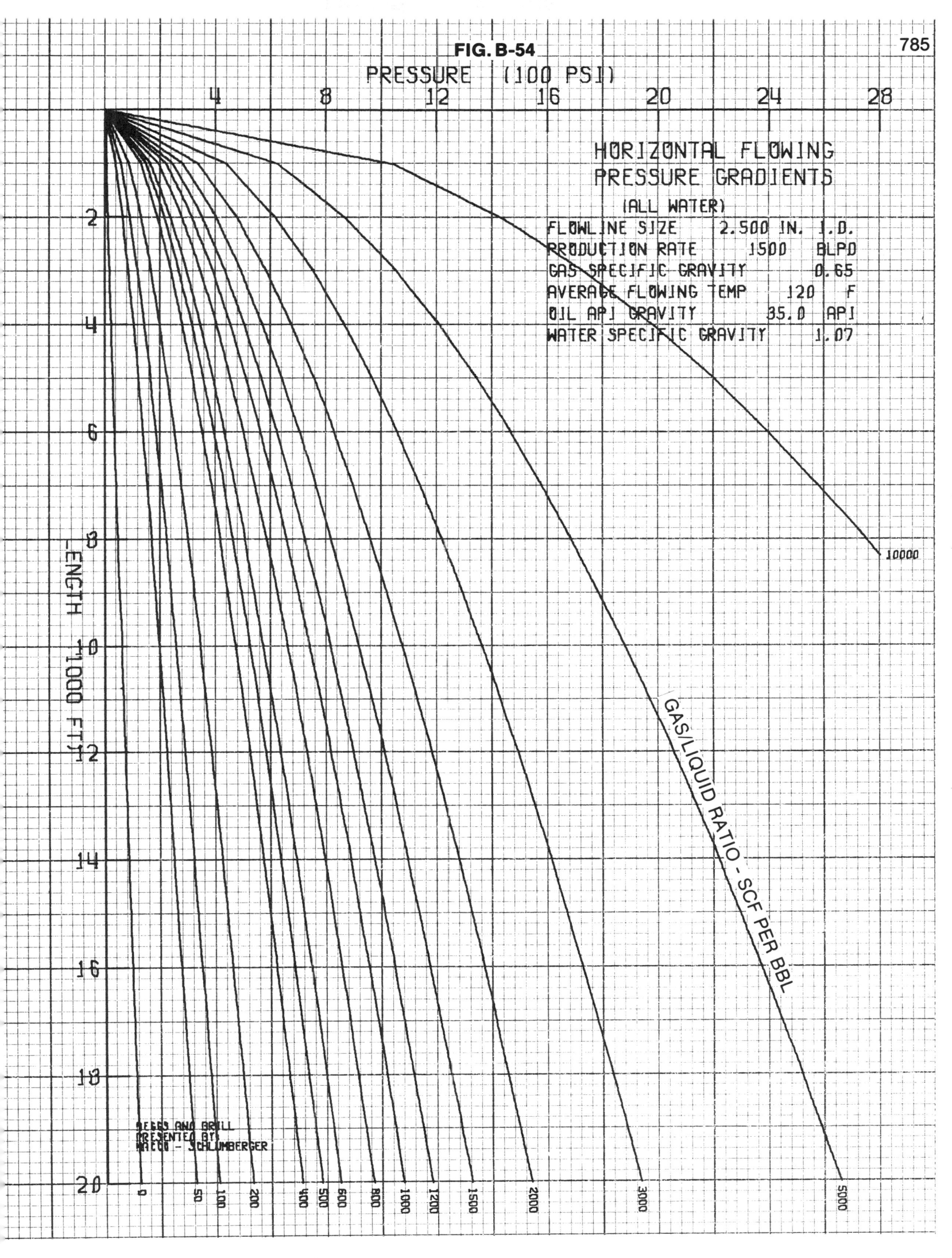

FIG. B-54

PRESSURE (100 PSI)

HORIZONTAL FLOWING
PRESSURE GRADIENTS
(ALL WATER)

FLOWLINE SIZE	2.500 IN. I.D.
PRODUCTION RATE	1500 BLPD
GAS SPECIFIC GRAVITY	0.65
AVERAGE FLOWING TEMP	120 F
OIL API GRAVITY	35.0 API
WATER SPECIFIC GRAVITY	1.07

LENGTH (1000 FT.)

GAS/LIQUID RATIO - SCF PER BBL

BEGGS AND BRILL
PRESENTED BY
MAECO - SCHLUMBERGER

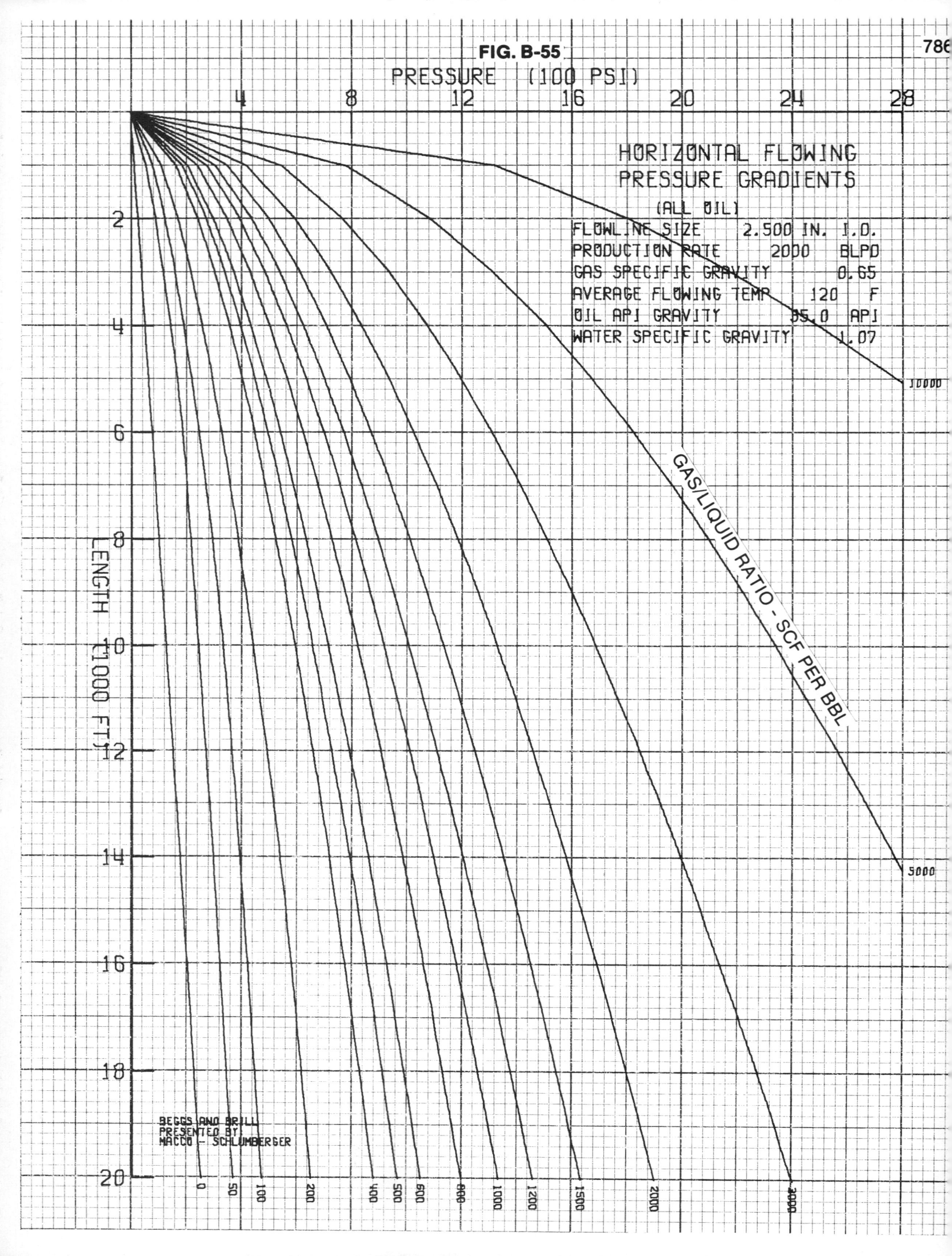

FIG. B-55

786

PRESSURE (100 PSI)

HORIZONTAL FLOWING
PRESSURE GRADIENTS
(ALL OIL)

FLOWLINE SIZE	2.500 IN. I.D.
PRODUCTION RATE	2000 BLPD
GAS SPECIFIC GRAVITY	0.65
AVERAGE FLOWING TEMP	120 F
OIL API GRAVITY	35.0 API
WATER SPECIFIC GRAVITY	1.07

GAS/LIQUID RATIO - SCF PER BBL

LENGTH (1000 FT)

BEGGS AND BRILL
PRESENTED BY
NACCO - SCHLUMBERGER

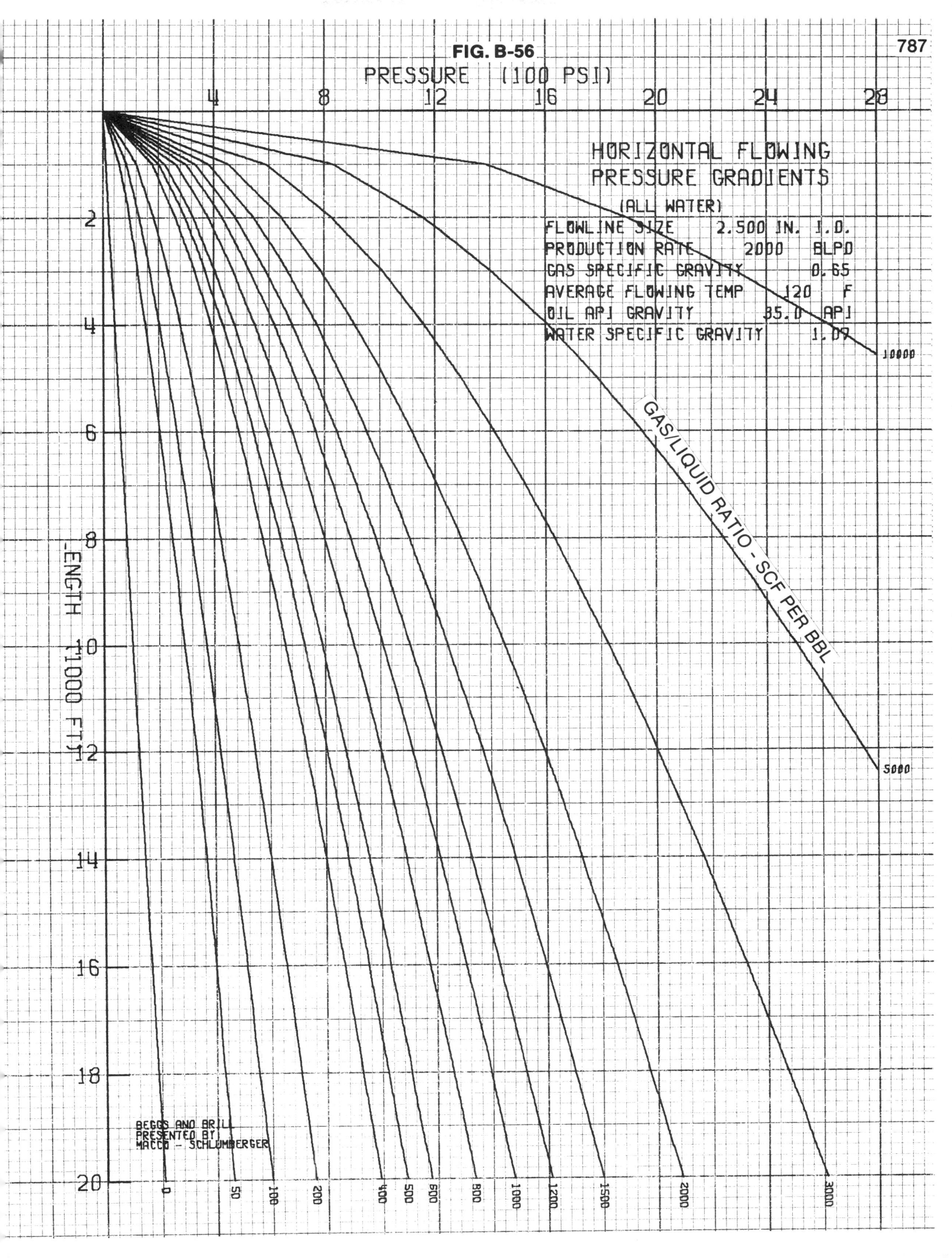

FIG. B-56

PRESSURE (100 PSI)

HORIZONTAL FLOWING
PRESSURE GRADIENTS
(ALL WATER)

FLOWLINE SIZE   2.500 IN. I.D.
PRODUCTION RATE   2000   BLPD
GAS SPECIFIC GRAVITY   0.65
AVERAGE FLOWING TEMP   120   F
OIL API GRAVITY   35.0   API
WATER SPECIFIC GRAVITY   1.07

GAS/LIQUID RATIO - SCF PER BBL

LENGTH (1000 FT)

BEGGS AND BRILL
PRESENTED BY
MACCO - SCHLUMBERGER

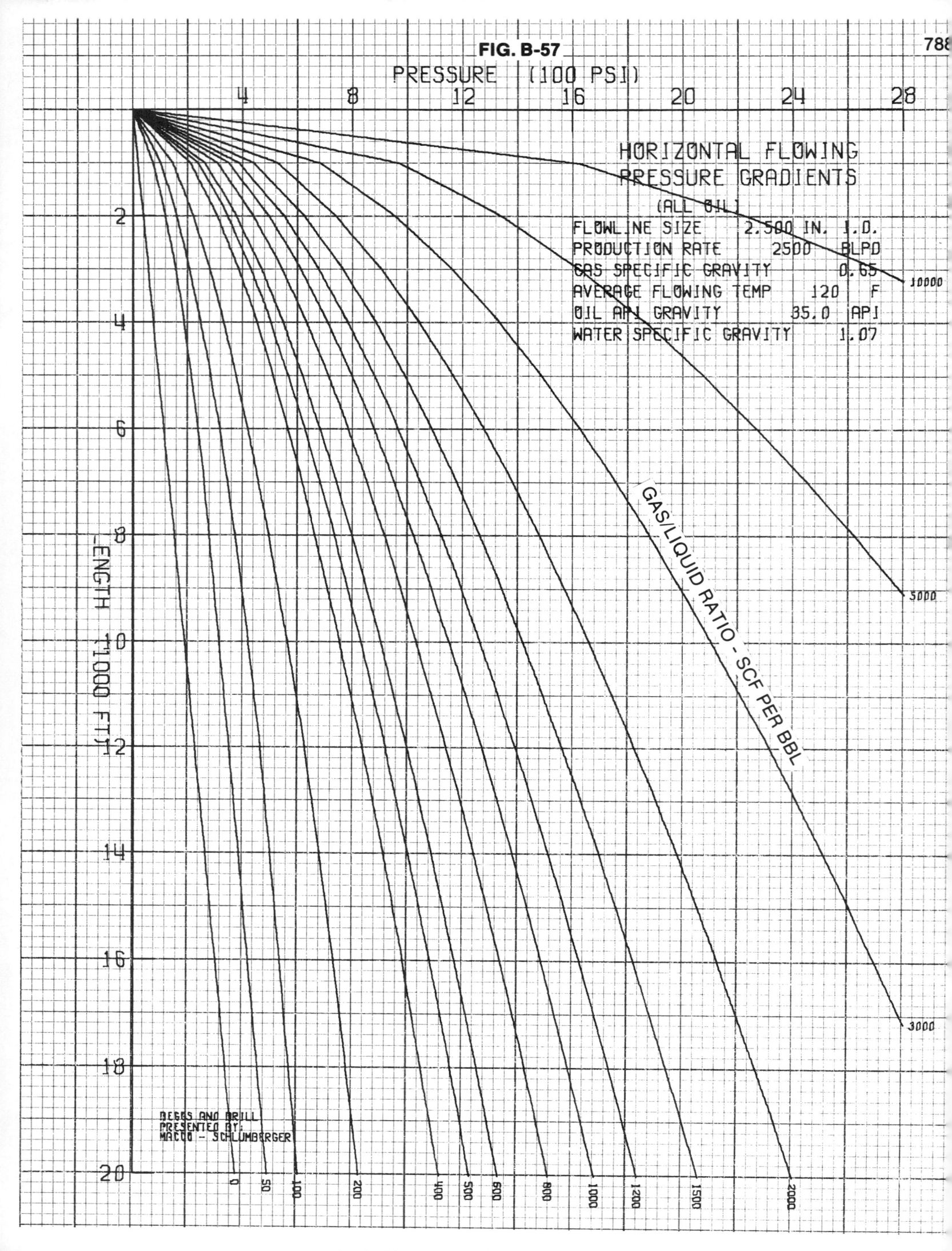

PRESSURE (100 PSI)

HORIZONTAL FLOWING
PRESSURE GRADIENTS
(ALL OIL)

FLOWLINE SIZE	2.500	IN. I.D.
PRODUCTION RATE	2500	BLPD
GAS SPECIFIC GRAVITY	0.65	
AVERAGE FLOWING TEMP	120	F
OIL API GRAVITY	35.0	API
WATER SPECIFIC GRAVITY	1.07	

GAS/LIQUID RATIO - SCF PER BBL

LENGTH (1000 FT)

BEGGS AND BRILL
PRESENTED BY:
MACCO - SCHLUMBERGER

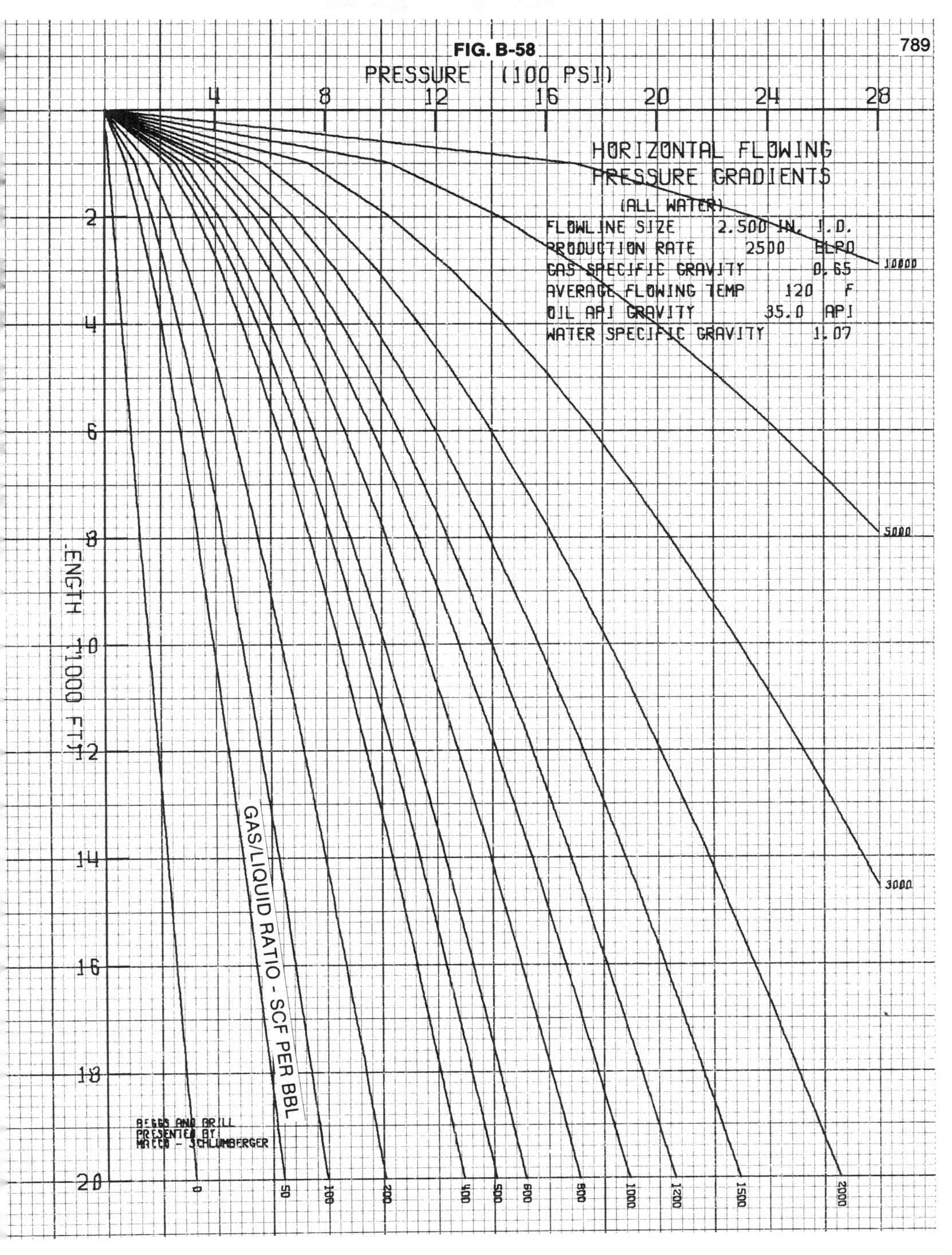

FIG. B-58

789

PRESSURE (100 PSI)

HORIZONTAL FLOWING
PRESSURE GRADIENTS
(ALL WATER)

FLOWLINE SIZE   2.500 IN. I.D.
PRODUCTION RATE   2500   BLPD
GAS SPECIFIC GRAVITY   0.65
AVERAGE FLOWING TEMP   120   F
OIL API GRAVITY   35.0   API
WATER SPECIFIC GRAVITY   1.07

LENGTH (1000 FT)

GAS/LIQUID RATIO - SCF PER BBL

BEGGS AND BRILL
PRESENTED BY
MACCO - SCHLUMBERGER

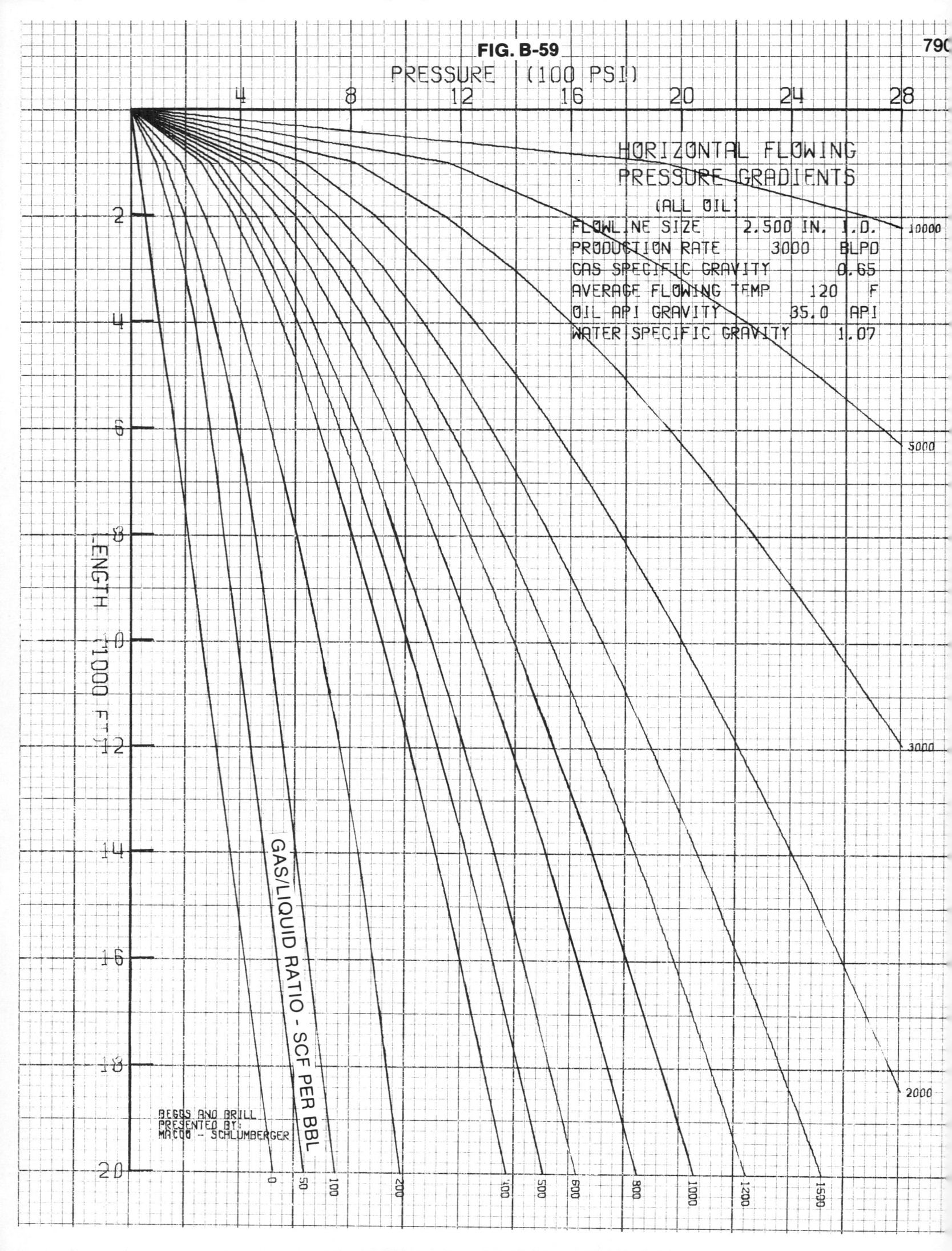

PRESSURE (100 PSI)

HORIZONTAL FLOWING
PRESSURE GRADIENTS
(ALL OIL)

FLOWLINE SIZE	2.500 IN. I.D.
PRODUCTION RATE	3000 BLPD
GAS SPECIFIC GRAVITY	0.65
AVERAGE FLOWING TEMP	120 F
OIL API GRAVITY	35.0 API
WATER SPECIFIC GRAVITY	1.07

LENGTH (1000 FT)

GAS/LIQUID RATIO - SCF PER BBL

BEGGS AND BRILL
PRESENTED BY:
MACOD -- SCHLUMBERGER

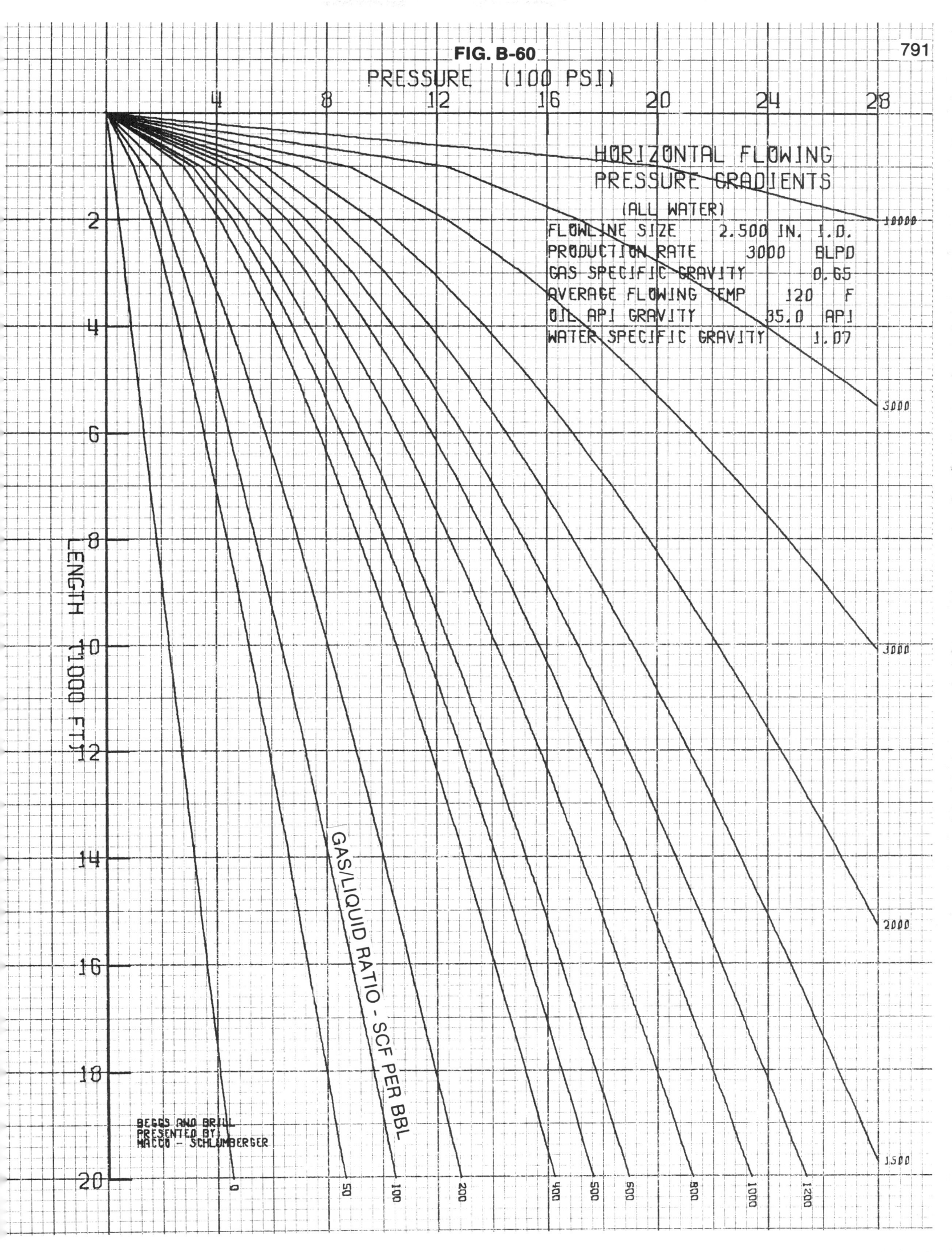

PRESSURE (100 PSI)

HORIZONTAL FLOWING
PRESSURE GRADIENTS
(ALL WATER)

FLOWLINE SIZE	2.500 IN.	I.D.
PRODUCTION RATE	3000	BLPD
GAS SPECIFIC GRAVITY	0.65	
AVERAGE FLOWING TEMP	120	F
OIL API GRAVITY	35.0	API
WATER SPECIFIC GRAVITY	1.07	

LENGTH (1000 FT.)

GAS/LIQUID RATIO - SCF PER BBL

BEGGS AND BRILL
PRESENTED BY
MACCO - SCHLUMBERGER

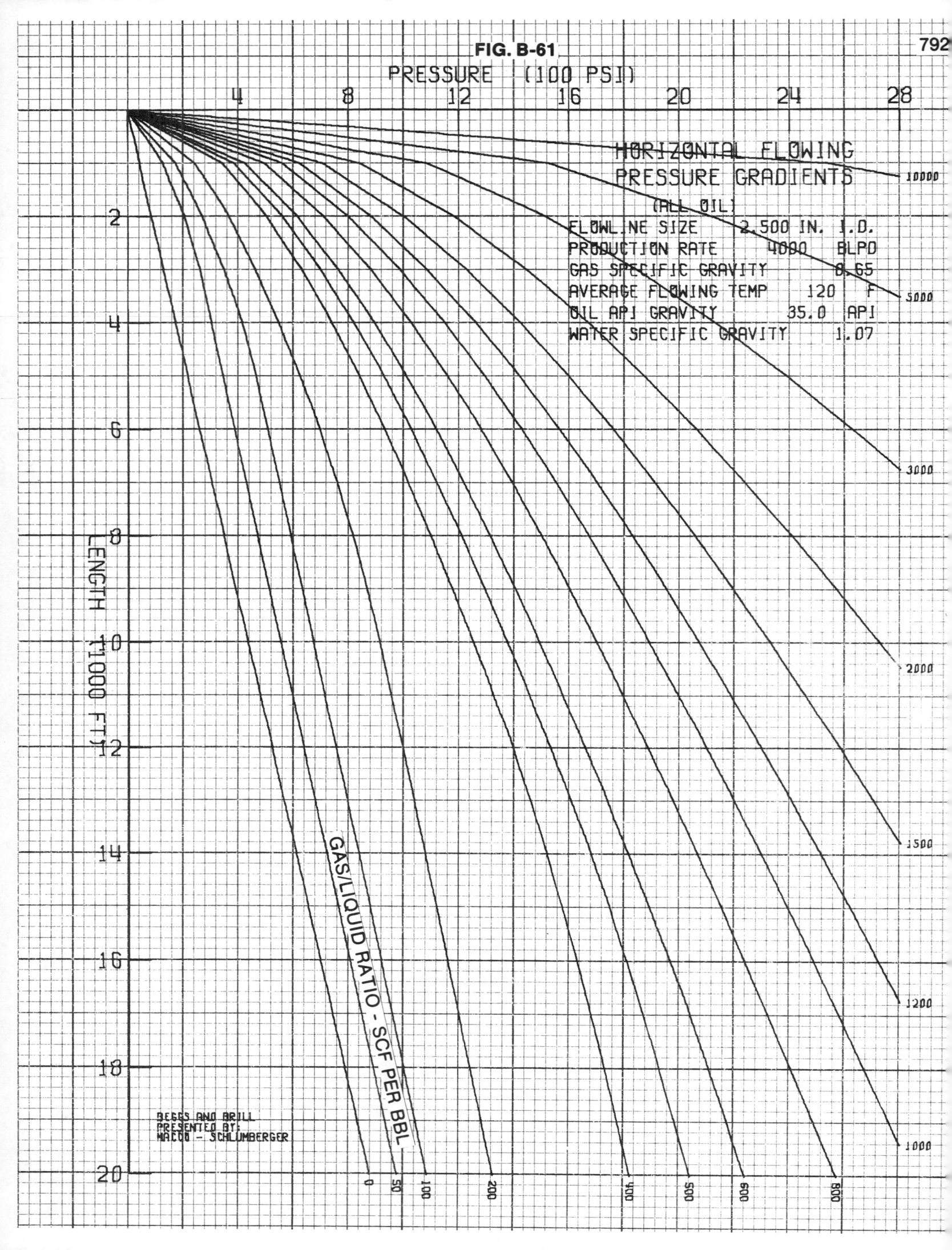

FIG. B-61

PRESSURE (100 PSI)

HORIZONTAL FLOWING
PRESSURE GRADIENTS
(ALL OIL)

FLOWLINE SIZE	2.500 IN. I.D.
PRODUCTION RATE	4000 BLPD
GAS SPECIFIC GRAVITY	0.65
AVERAGE FLOWING TEMP	120 F
OIL API GRAVITY	35.0 API
WATER SPECIFIC GRAVITY	1.07

LENGTH (1000 FT)

GAS/LIQUID RATIO - SCF PER BBL

BEGGS AND BRILL
PRESENTED BY:
NACCO - SCHLUMBERGER

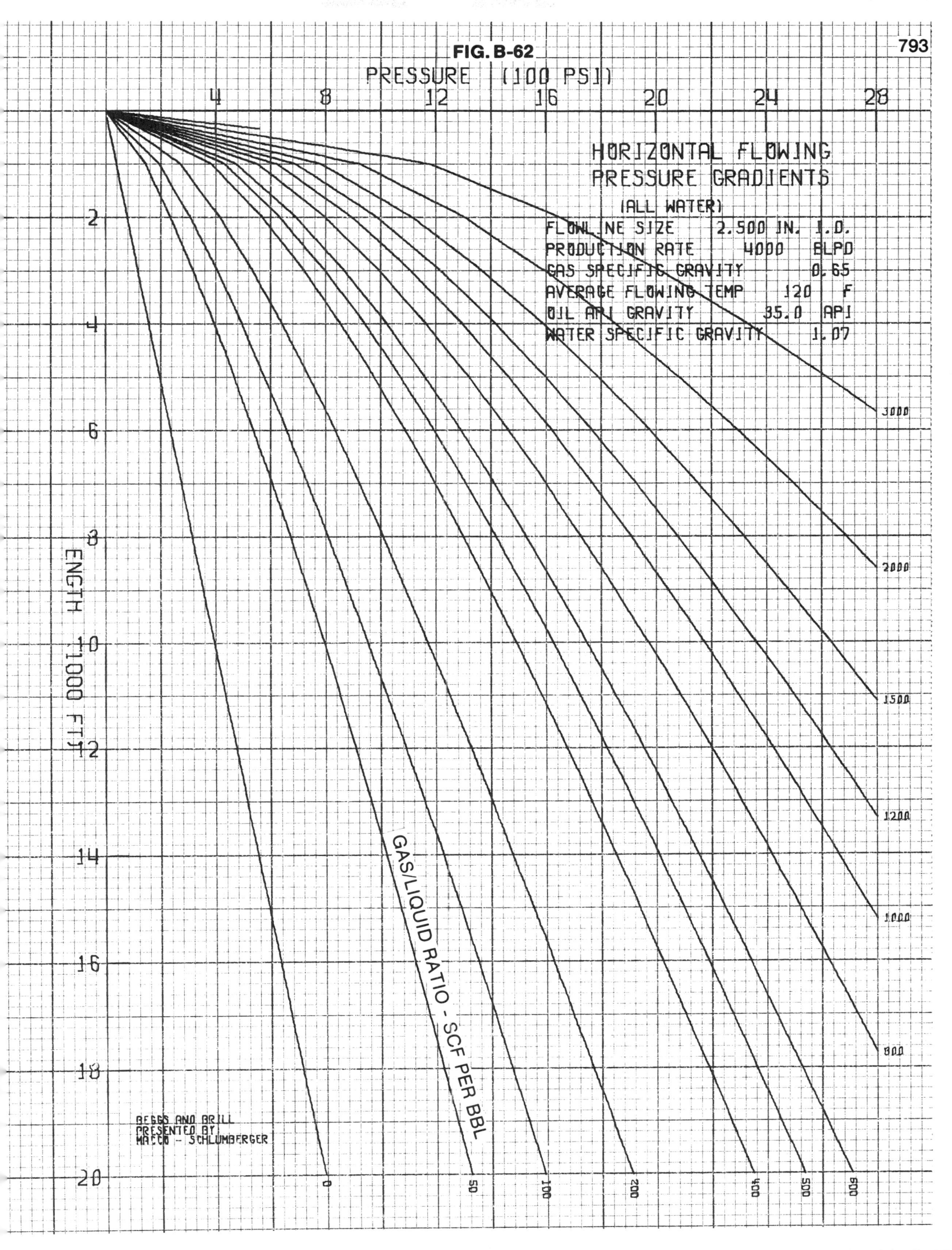

PRESSURE (100 PSI)

HORIZONTAL FLOWING
PRESSURE GRADIENTS
(ALL WATER)

FLOWLINE SIZE	2.500 IN. I.D.
PRODUCTION RATE	4000 BLPD
GAS SPECIFIC GRAVITY	0.65
AVERAGE FLOWING TEMP	120 F
OIL API GRAVITY	35.0 API
WATER SPECIFIC GRAVITY	1.07

LENGTH (1000 FT.)

GAS/LIQUID RATIO - SCF PER BBL

BEGGS AND BRILL
PRESENTED BY
MACCO - SCHLUMBERGER

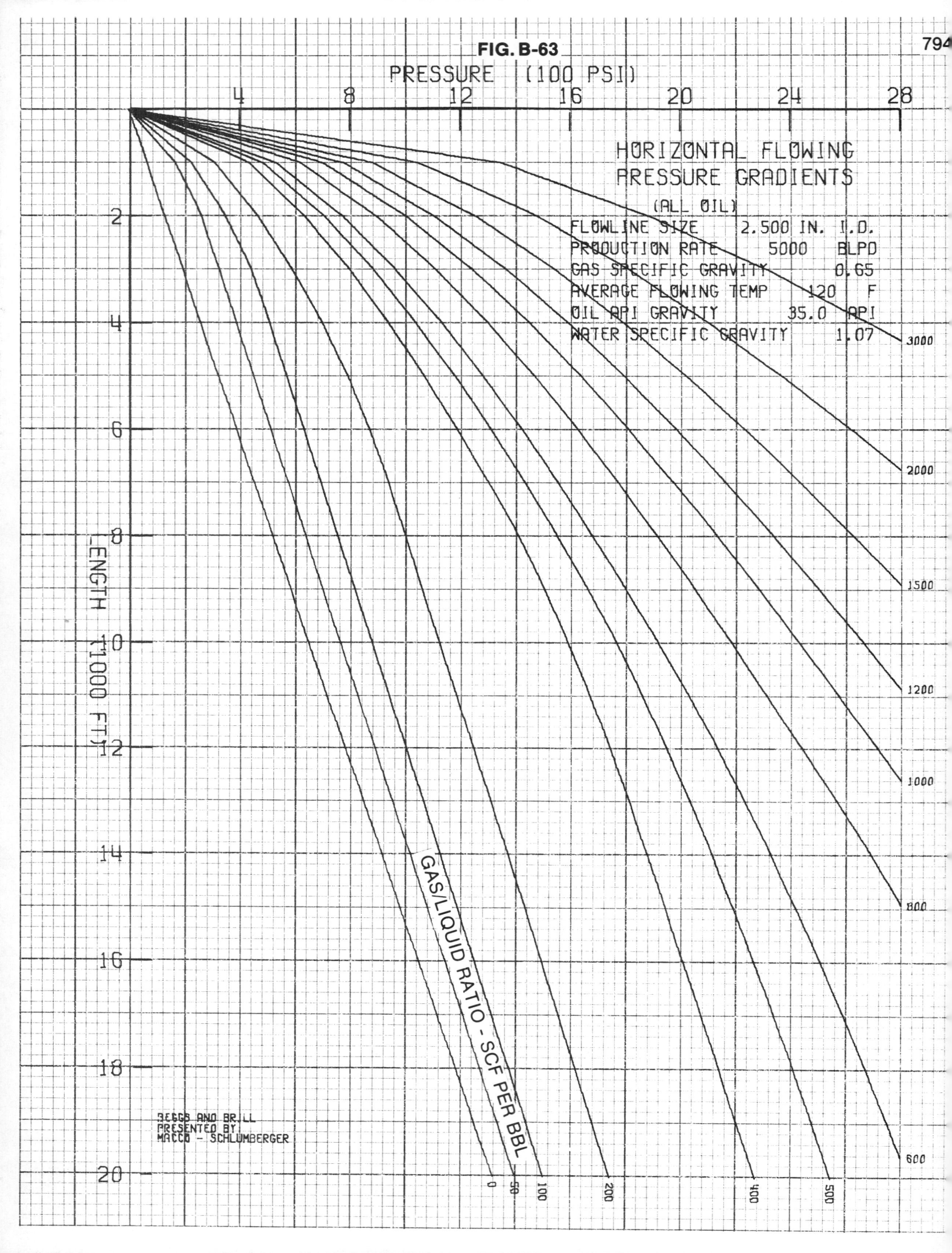

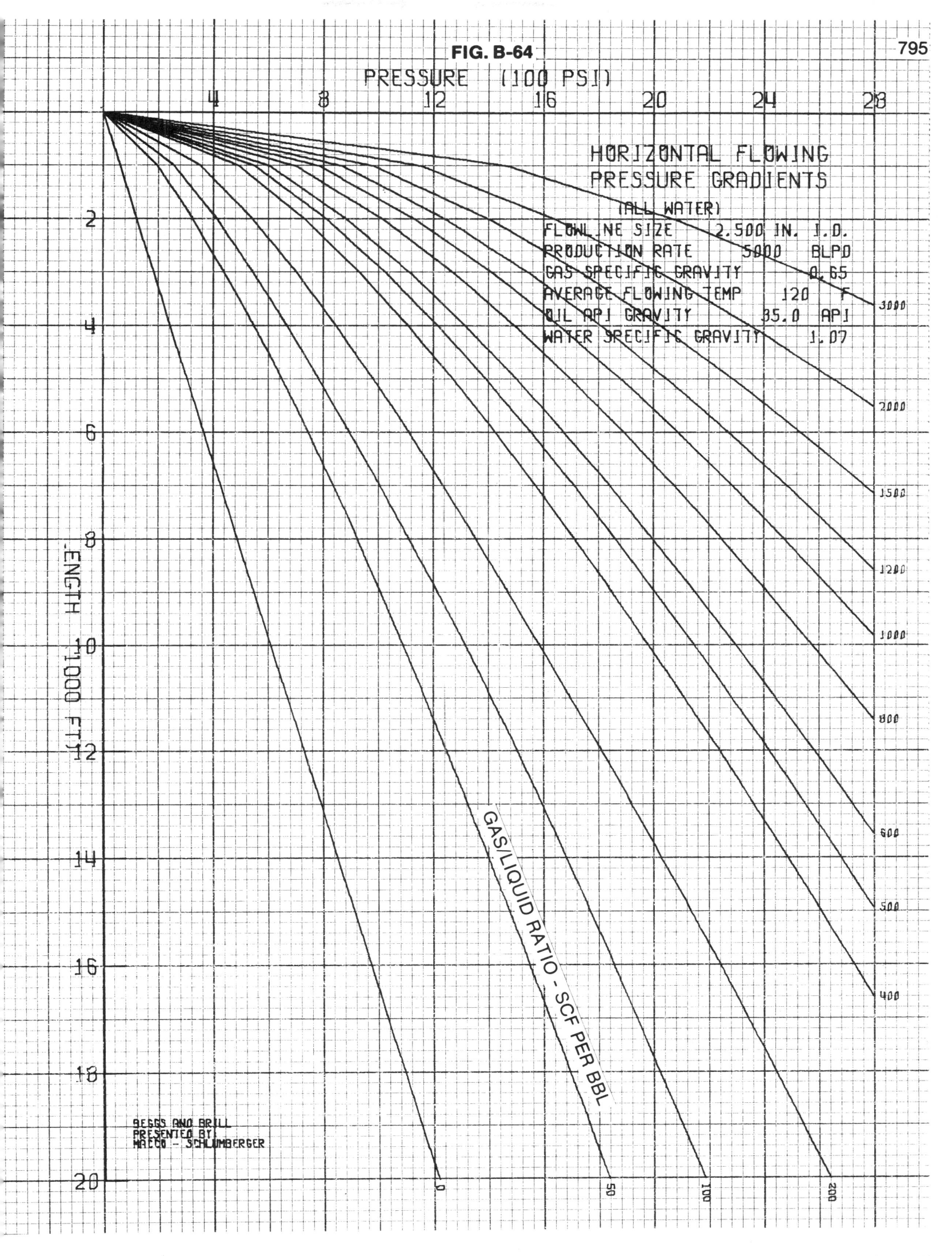

FIG. B-64

795

HORIZONTAL FLOWING
PRESSURE GRADIENTS
(ALL WATER)

FLOWLINE SIZE   2.500 IN. I.D.
PRODUCTION RATE   5000   BLPD
GAS SPECIFIC GRAVITY   0.65
AVERAGE FLOWING TEMP   120   F
OIL API GRAVITY   35.0 API
WATER SPECIFIC GRAVITY   1.07

PRESSURE (100 PSI)

LENGTH (1000 FT)

GAS/LIQUID RATIO - SCF PER BBL

BEGGS AND BRILL
PRESENTED BY
MAECO - SCHLUMBERGER

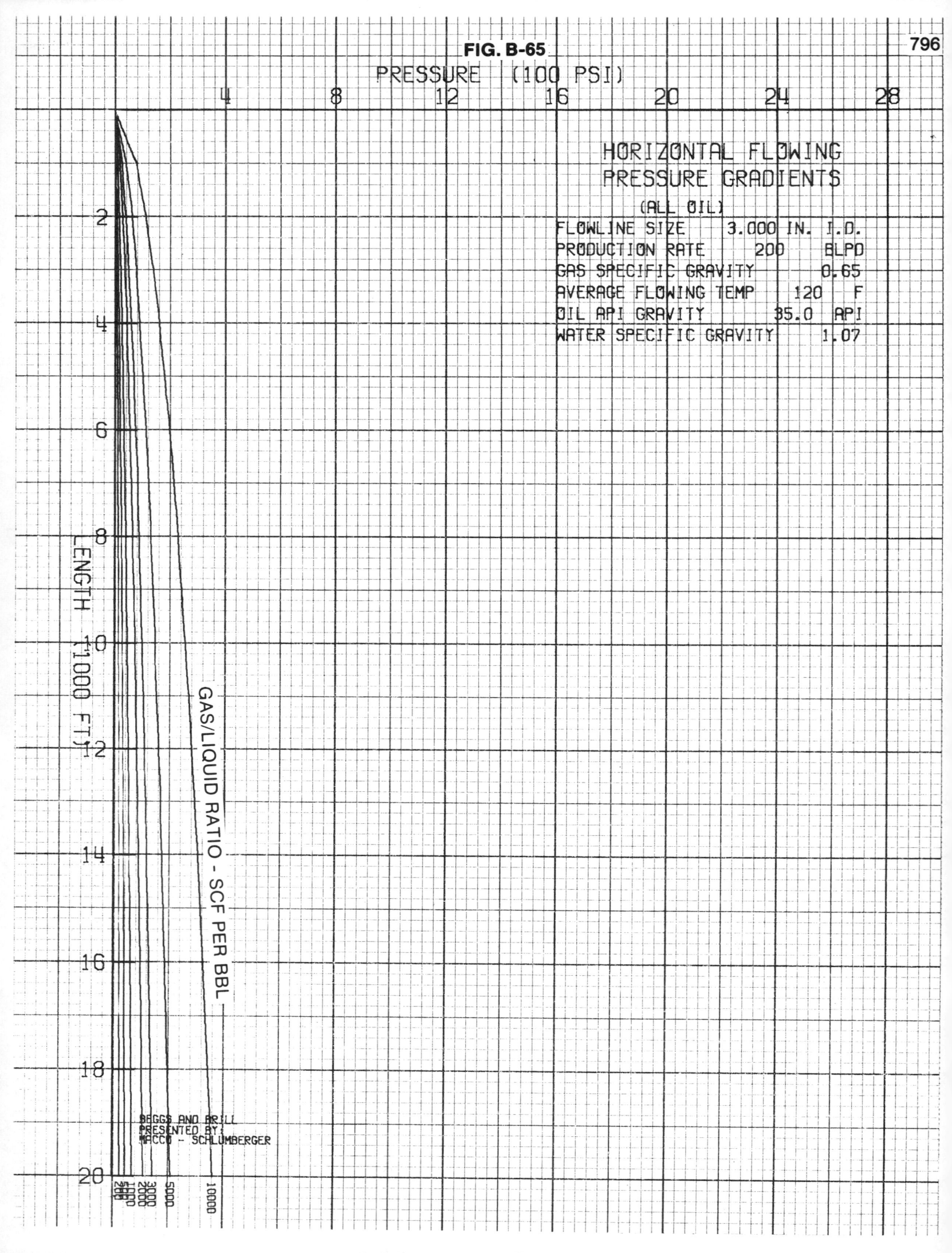

PRESSURE (100 PSI)

HORIZONTAL FLOWING
PRESSURE GRADIENTS
(ALL OIL)

FLOWLINE SIZE	3.000 IN.	I.D.
PRODUCTION RATE	200	BLPD
GAS SPECIFIC GRAVITY	0.65	
AVERAGE FLOWING TEMP	120	F
OIL API GRAVITY	35.0	API
WATER SPECIFIC GRAVITY	1.07	

LENGTH (1000 FT)

GAS/LIQUID RATIO - SCF PER BBL

BEGGS AND BRILL
PRESENTED BY:
MACCO - SCHLUMBERGER

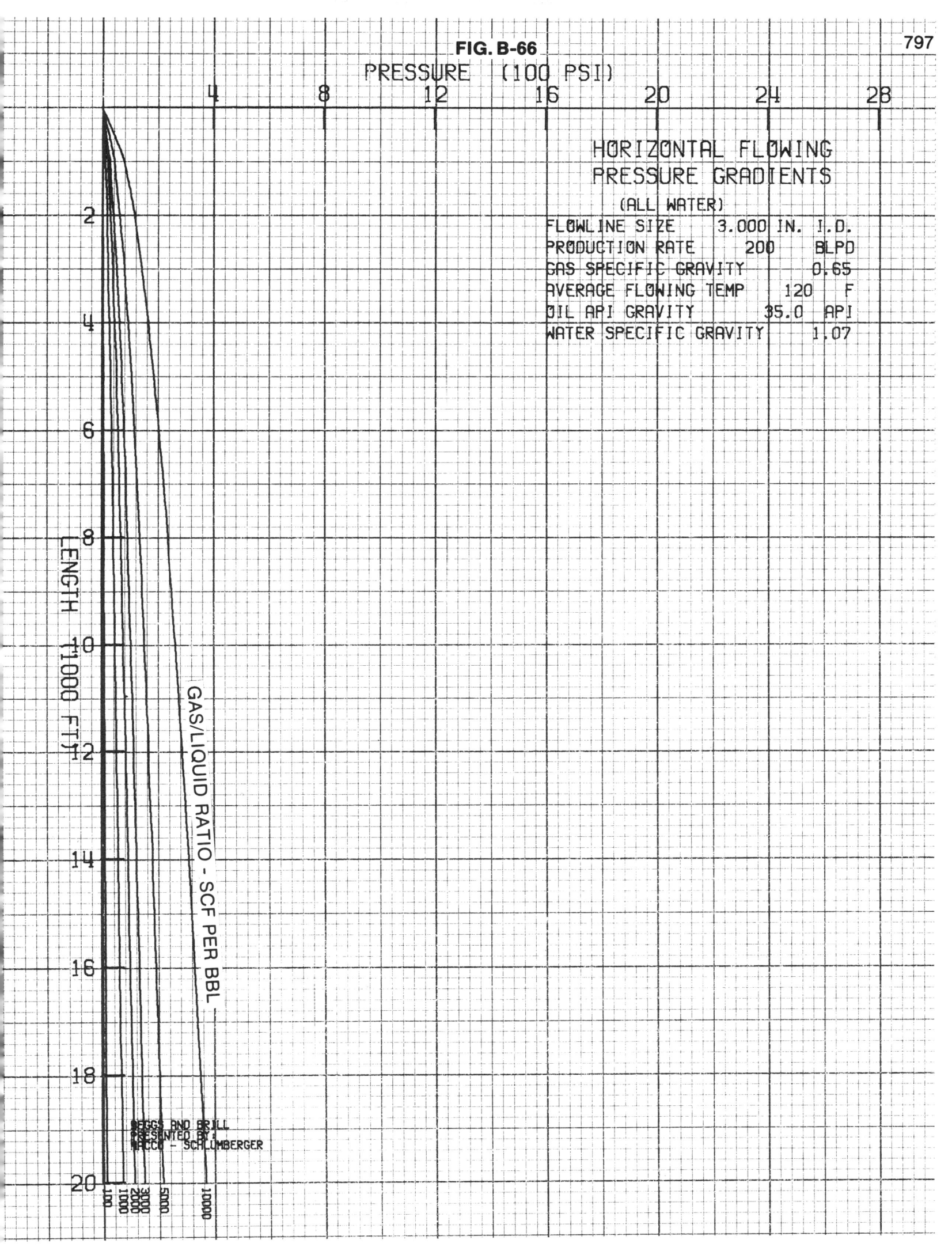

FIG. B-66

PRESSURE (100 PSI)

HORIZONTAL FLOWING
PRESSURE GRADIENTS
(ALL WATER)

FLOWLINE SIZE	3.000 IN. I.D.
PRODUCTION RATE	200 BLPD
GAS SPECIFIC GRAVITY	0.65
AVERAGE FLOWING TEMP	120 F
OIL API GRAVITY	35.0 API
WATER SPECIFIC GRAVITY	1.07

LENGTH (1000 FT)

GAS/LIQUID RATIO - SCF PER BBL

BRIGGS AND BRILL
PRESENTED BY:
NACCO - SCHLUMBERGER

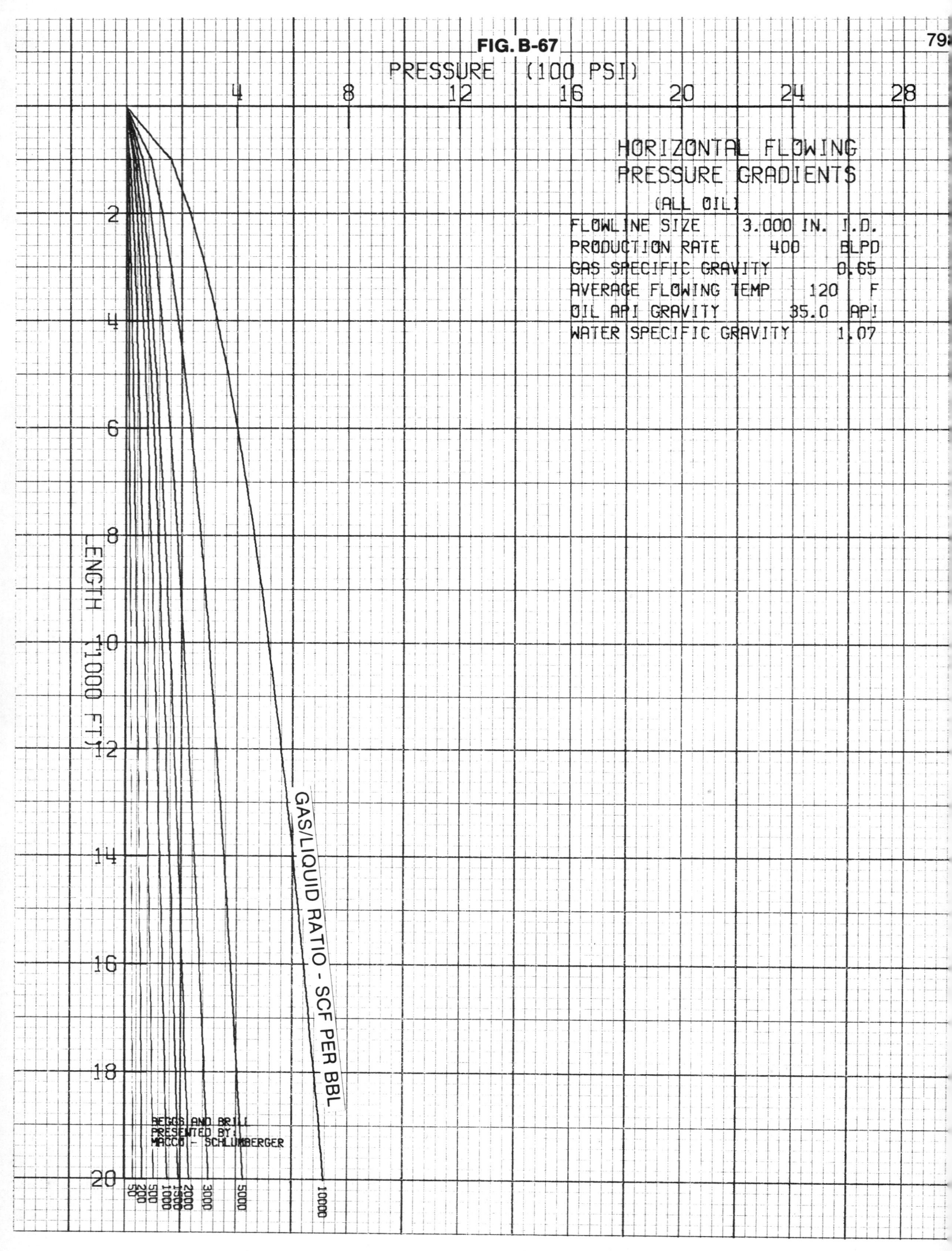

PRESSURE (100 PSI)

HORIZONTAL FLOWING
PRESSURE GRADIENTS
(ALL OIL)

FLOWLINE SIZE	3.000 IN.	I.D.
PRODUCTION RATE	400	BLPD
GAS SPECIFIC GRAVITY	0.65	
AVERAGE FLOWING TEMP	120	F
OIL API GRAVITY	35.0	API
WATER SPECIFIC GRAVITY	1.07	

LENGTH (1000 FT)

GAS/LIQUID RATIO - SCF PER BBL

BEGGS AND BRILL
PRESENTED BY:
MACCO - SCHLUMBERGER

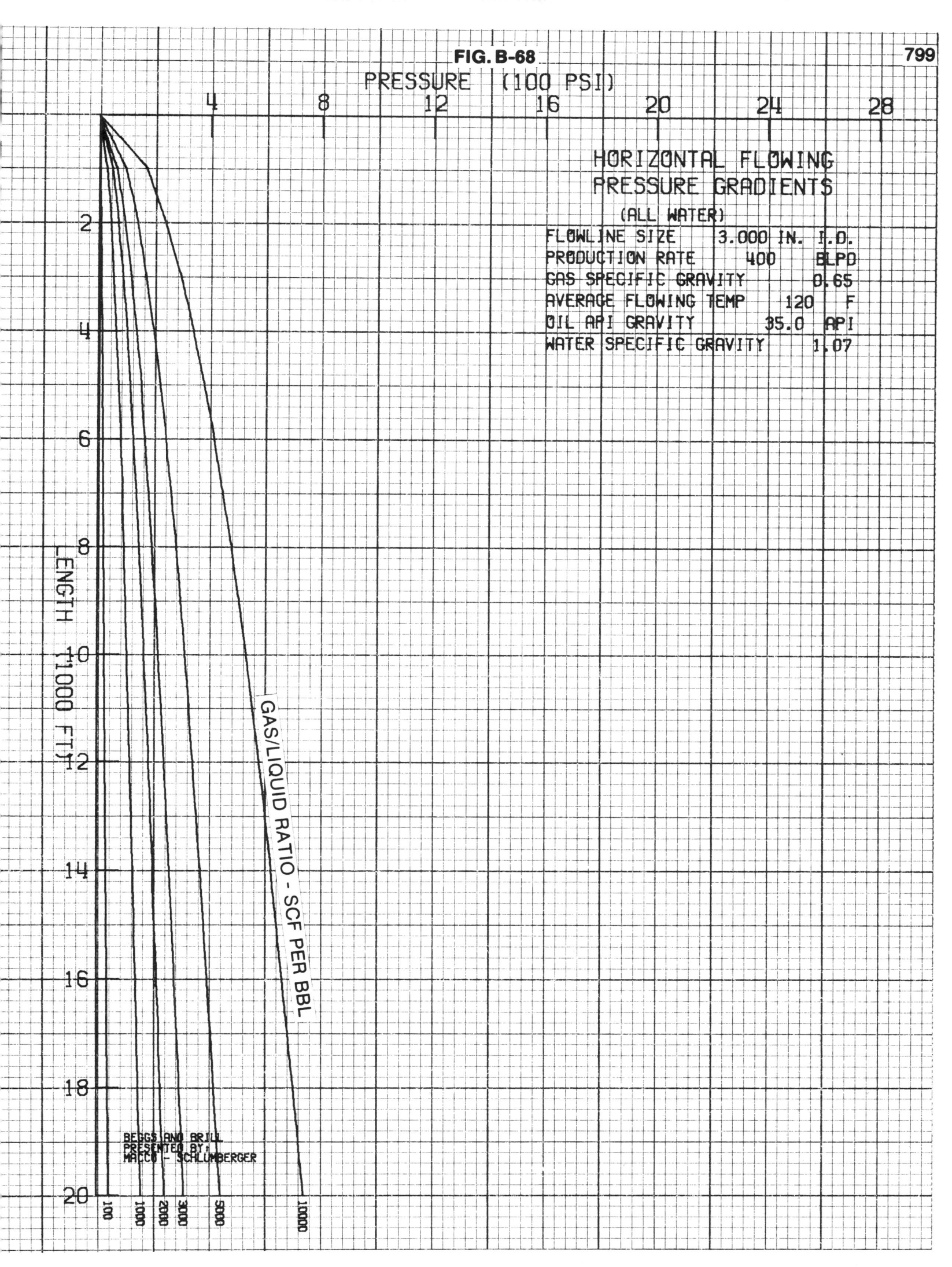

FIG. B-68

799

PRESSURE (100 PSI)

HORIZONTAL FLOWING
PRESSURE GRADIENTS
(ALL WATER)

FLOWLINE SIZE	3.000 IN.	I.D.
PRODUCTION RATE	400	BLPD
GAS SPECIFIC GRAVITY	0.65	
AVERAGE FLOWING TEMP	120	F
OIL API GRAVITY	35.0	API
WATER SPECIFIC GRAVITY	1.07	

LENGTH (1000 FT.)

GAS/LIQUID RATIO - SCF PER BBL

BEGGS AND BRILL
PRESENTED BY:
MACCO - SCHLUMBERGER

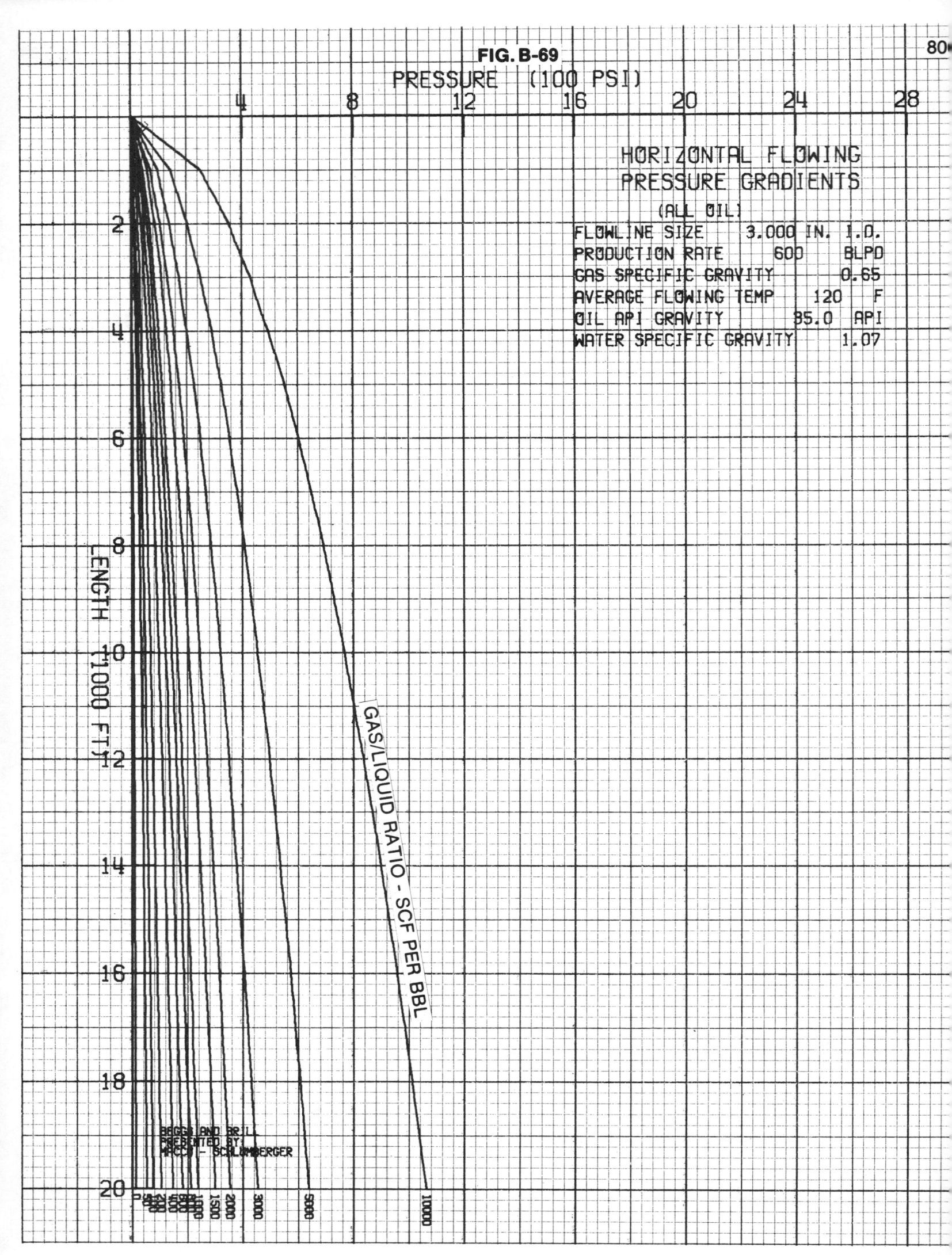

FIG. B-69

PRESSURE (100 PSI)

HORIZONTAL FLOWING
PRESSURE GRADIENTS
(ALL OIL)

FLOWLINE SIZE	3.000 IN. I.D.
PRODUCTION RATE	600 BLPD
GAS SPECIFIC GRAVITY	0.65
AVERAGE FLOWING TEMP	120 F
OIL API GRAVITY	35.0 API
WATER SPECIFIC GRAVITY	1.07

LENGTH (1000 FT)

GAS/LIQUID RATIO - SCF PER BBL

BAGGS AND BRILL
PRESENTED BY:
MACCO - SCHLUMBERGER

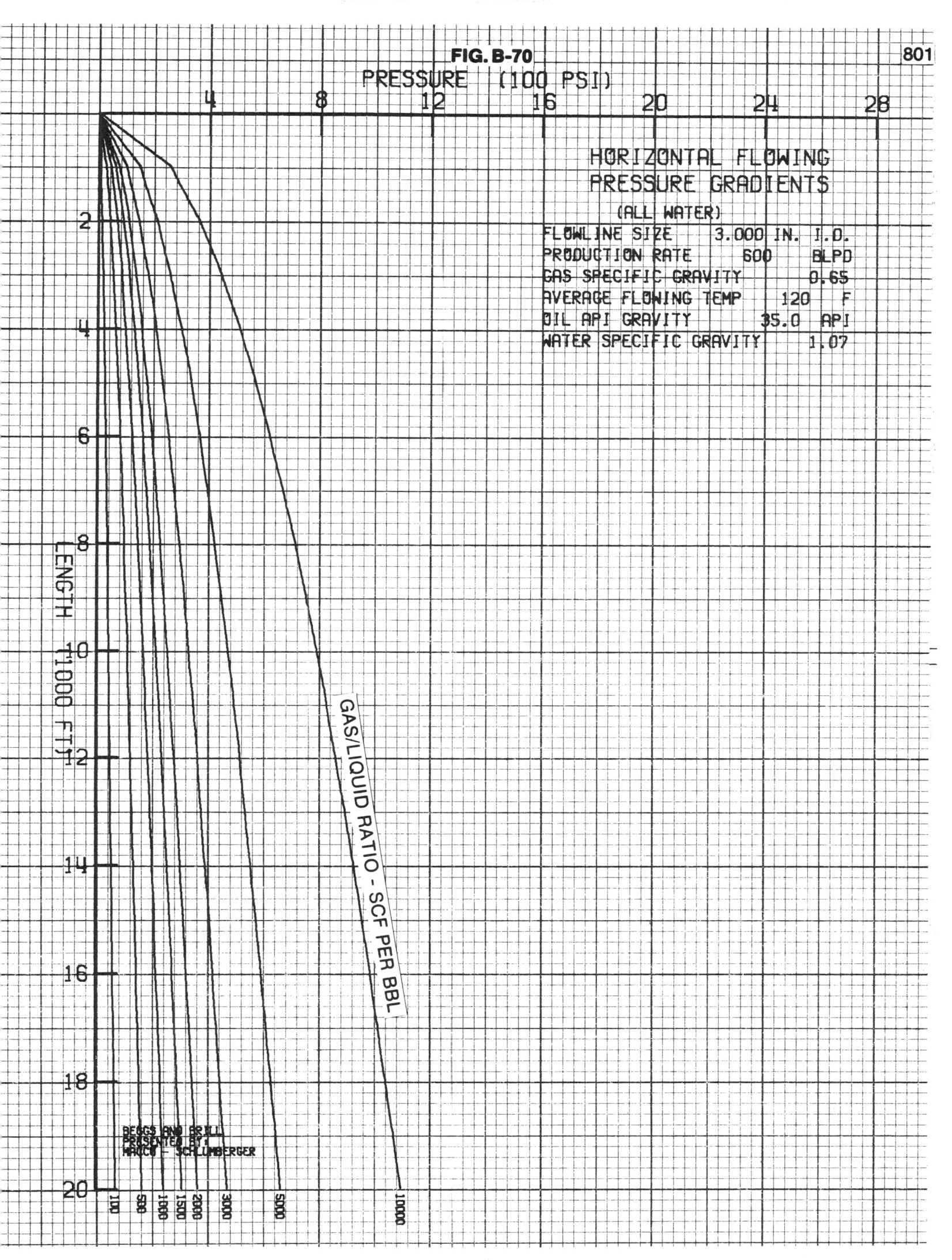

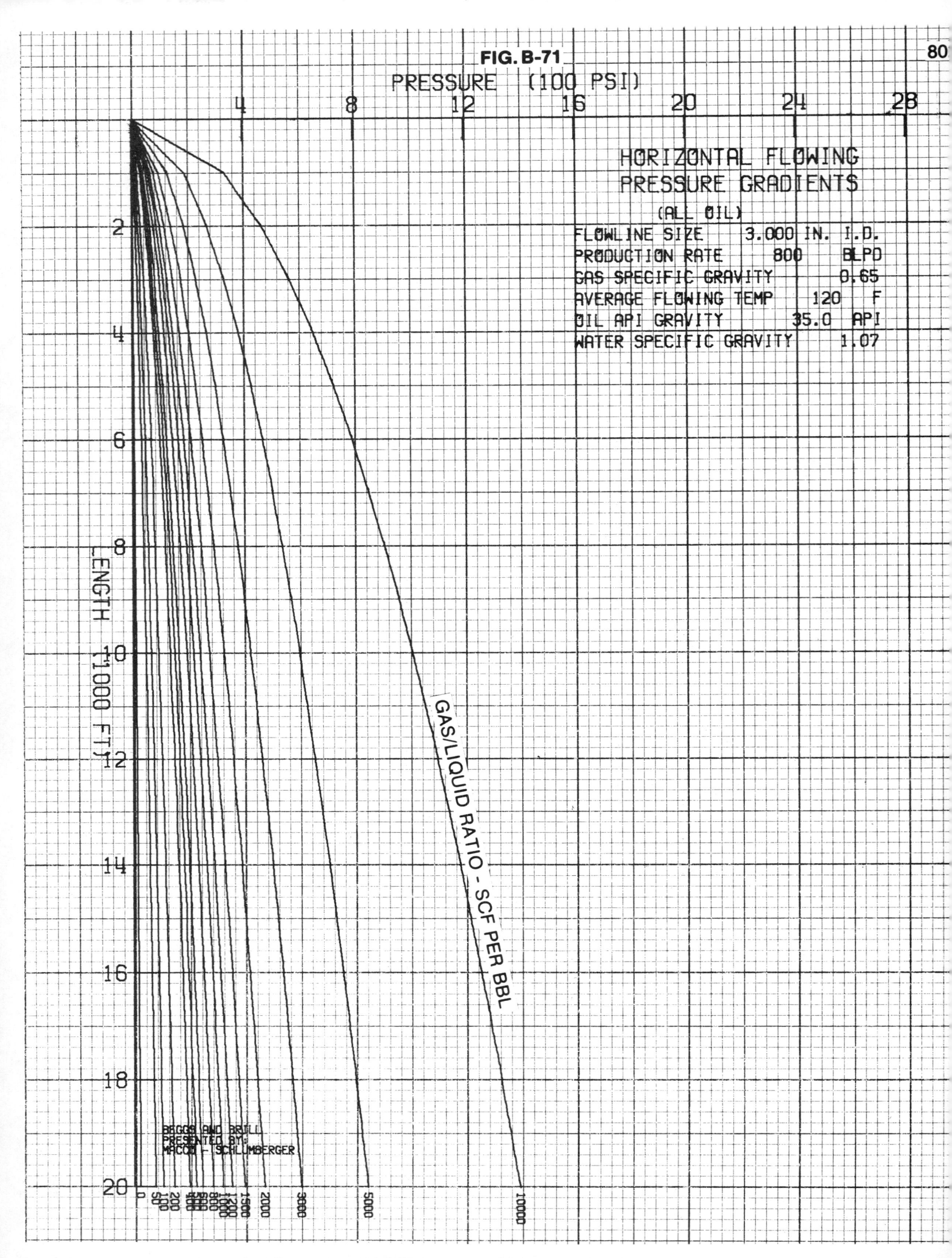

FIG. B-71

PRESSURE (100 PSI)

HORIZONTAL FLOWING
PRESSURE GRADIENTS
(ALL OIL)

FLOWLINE SIZE	3.000 IN. I.D.
PRODUCTION RATE	800 BLPD
GAS SPECIFIC GRAVITY	0.65
AVERAGE FLOWING TEMP	120 F
OIL API GRAVITY	35.0 API
WATER SPECIFIC GRAVITY	1.07

LENGTH (1000 FT)

GAS/LIQUID RATIO - SCF PER BBL

BEGGS AND BRILL
PRESENTED BY:
MACCO - SCHLUMBERGER

FIG. B-72

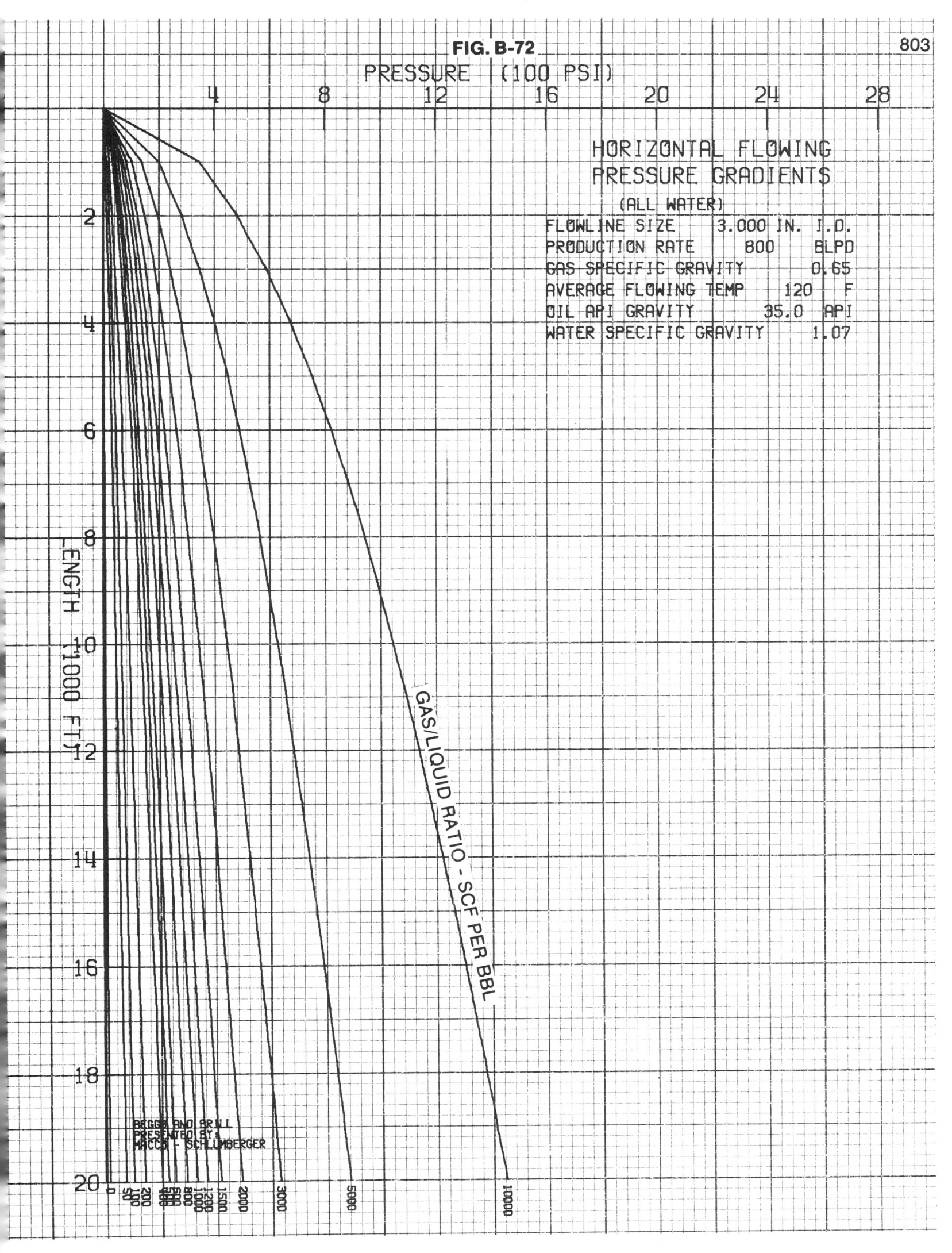

PRESSURE (100 PSI)

HORIZONTAL FLOWING
PRESSURE GRADIENTS
(ALL WATER)

FLOWLINE SIZE	3.000 IN.	I.D.
PRODUCTION RATE	800	BLPD
GAS SPECIFIC GRAVITY	0.65	
AVERAGE FLOWING TEMP	120	F
OIL API GRAVITY	35.0	API
WATER SPECIFIC GRAVITY	1.07	

LENGTH (1000 FT)

GAS/LIQUID RATIO - SCF PER BBL

BEGGS AND BRILL
PRESENTED BY
MACCO - SCHLUMBERGER

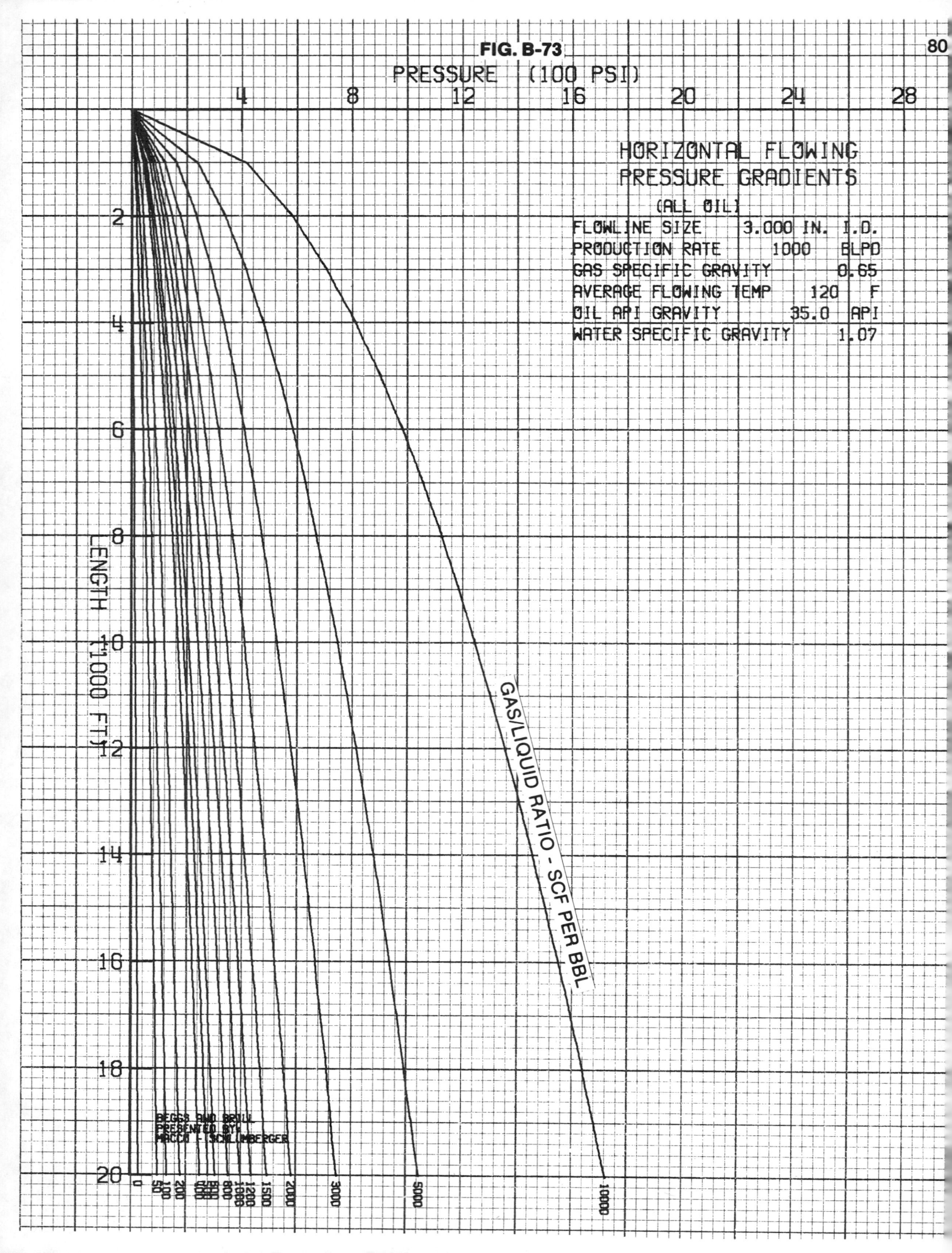

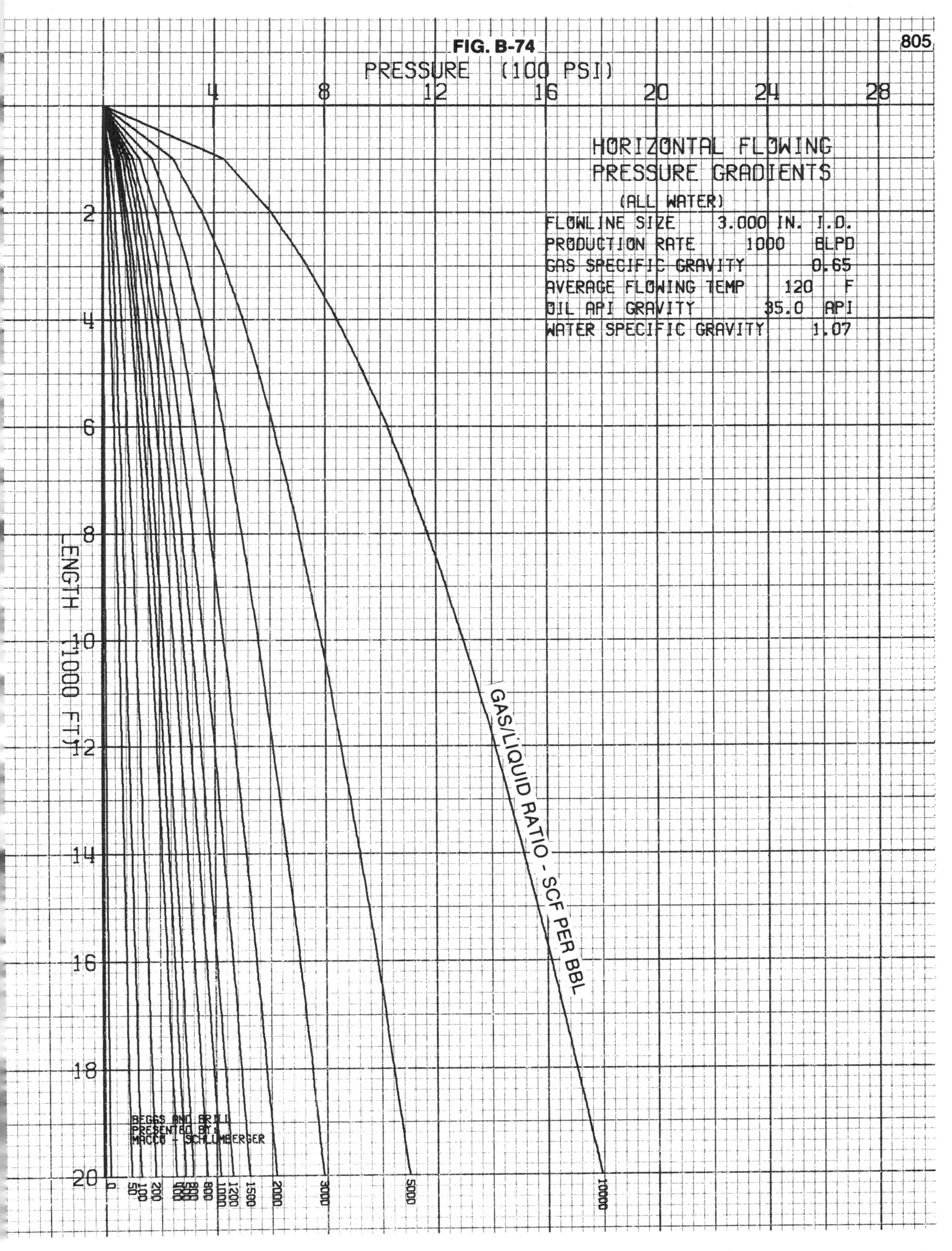

FIG. B-74

805

HORIZONTAL FLOWING
PRESSURE GRADIENTS
(ALL WATER)

FLOWLINE SIZE	3.000 IN.	I.D.
PRODUCTION RATE	1000	BLPD
GAS SPECIFIC GRAVITY	0.65	
AVERAGE FLOWING TEMP	120	F
OIL API GRAVITY	35.0	API
WATER SPECIFIC GRAVITY	1.07	

PRESSURE (100 PSI)

LENGTH (1000 FT)

GAS/LIQUID RATIO - SCF PER BBL

BEGGS AND BRILL
PRESENTED BY:
MACCO - SCHLUMBERGER

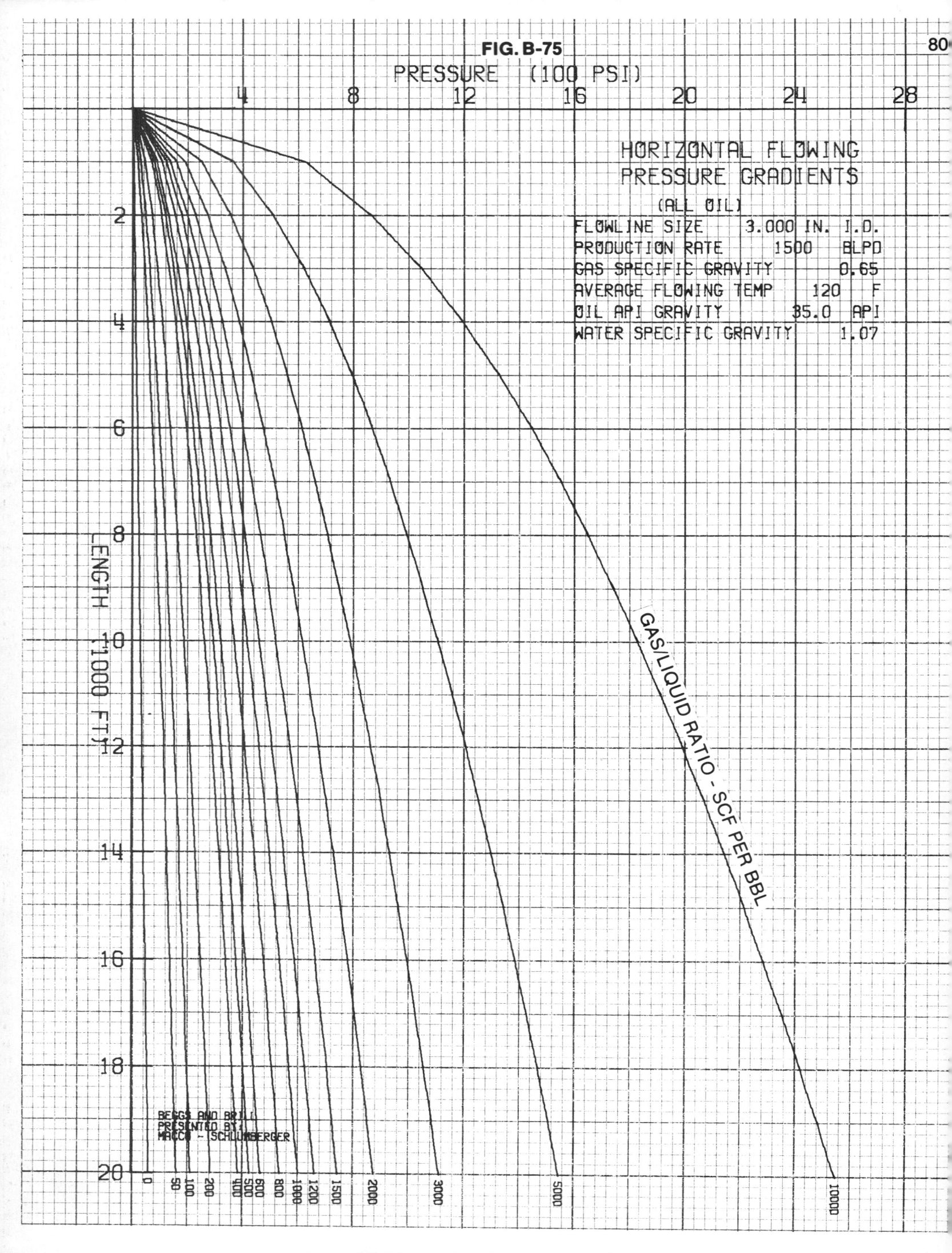

FIG. B-75

80●

PRESSURE (100 PSI)

HORIZONTAL FLOWING
PRESSURE GRADIENTS
(ALL OIL)

FLOWLINE SIZE	3.000 IN.	I.D.
PRODUCTION RATE	1500	BLPD
GAS SPECIFIC GRAVITY	0.65	
AVERAGE FLOWING TEMP	120	F
OIL API GRAVITY	35.0	API
WATER SPECIFIC GRAVITY	1.07	

LENGTH (1000 FT.)

GAS/LIQUID RATIO - SCF PER BBL

BEGGS AND BRILL
PRESENTED BY:
MACCO - SCHLUMBERGER

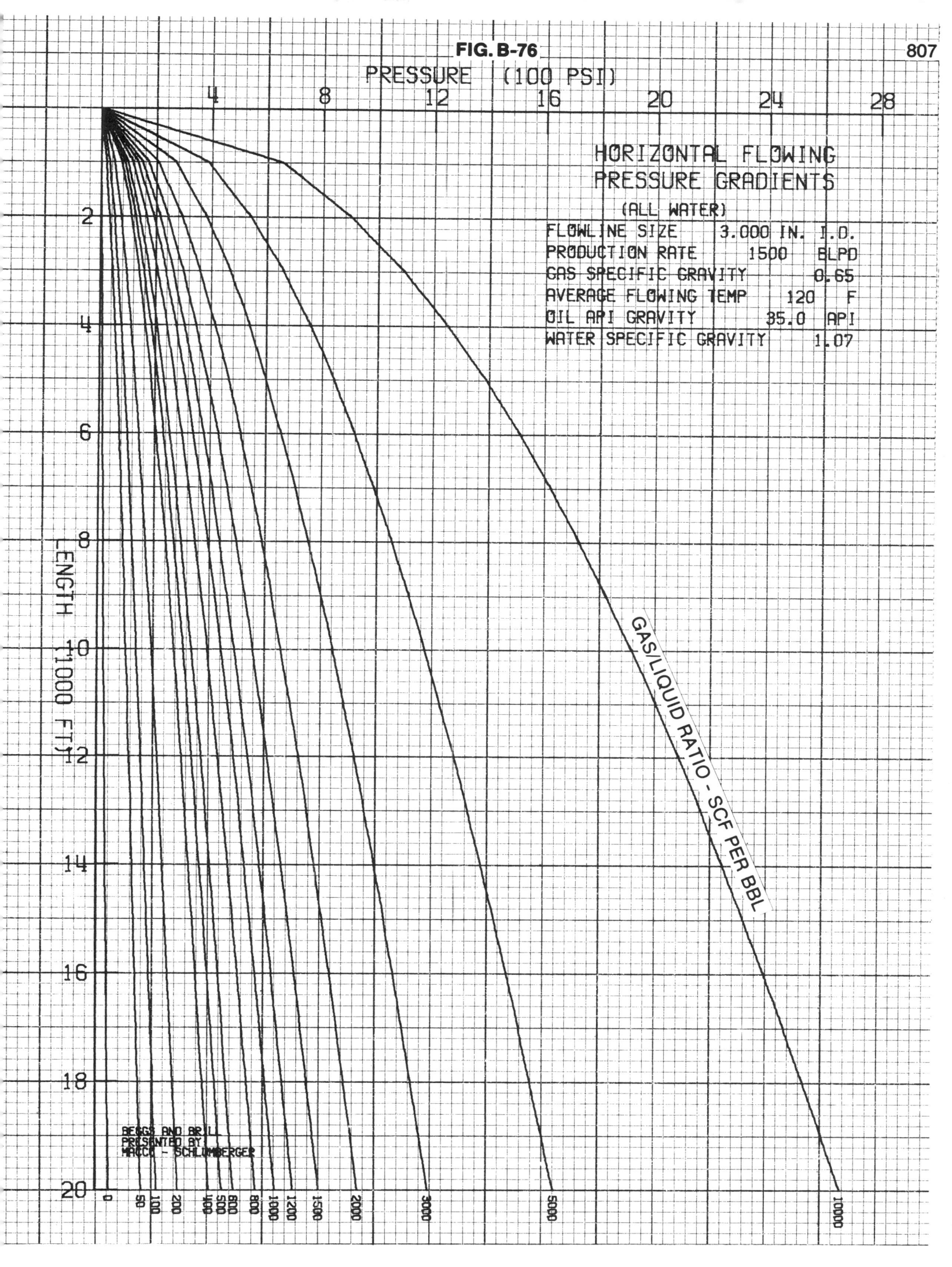

PRESSURE (100 PSI)

HORIZONTAL FLOWING
PRESSURE GRADIENTS
(ALL WATER)

FLOWLINE SIZE	3.000 IN. I.D.	
PRODUCTION RATE	1500	BLPD
GAS SPECIFIC GRAVITY		0.65
AVERAGE FLOWING TEMP	120	F
OIL API GRAVITY	35.0	API
WATER SPECIFIC GRAVITY		1.07

LENGTH (1000 FT)

GAS/LIQUID RATIO - SCF PER BBL

BEGGS AND BRILL
PRESENTED BY
MACCO - SCHLUMBERGER

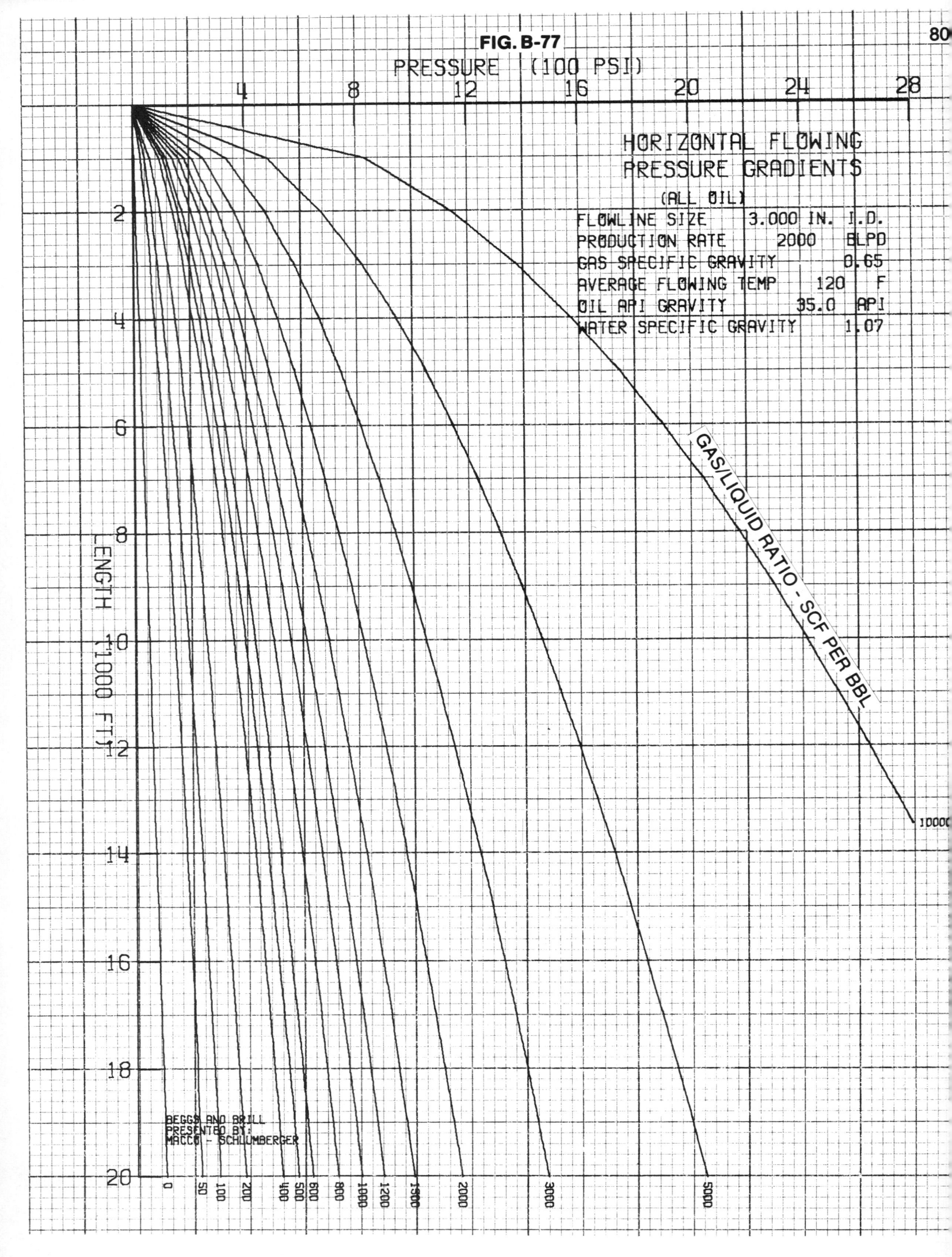

FIG. B-77

HORIZONTAL FLOWING
PRESSURE GRADIENTS
(ALL OIL)

FLOWLINE SIZE	3.000 IN.	I.D.
PRODUCTION RATE	2000	BLPD
GAS SPECIFIC GRAVITY	0.65	
AVERAGE FLOWING TEMP	120	F
OIL API GRAVITY	35.0	API
WATER SPECIFIC GRAVITY	1.07	

PRESSURE (100 PSI)

LENGTH (1000 FT)

GAS/LIQUID RATIO - SCF PER BBL

BEGGS AND BRILL
PRESENTED BY:
MACCO - SCHLUMBERGER

FIG. B-78

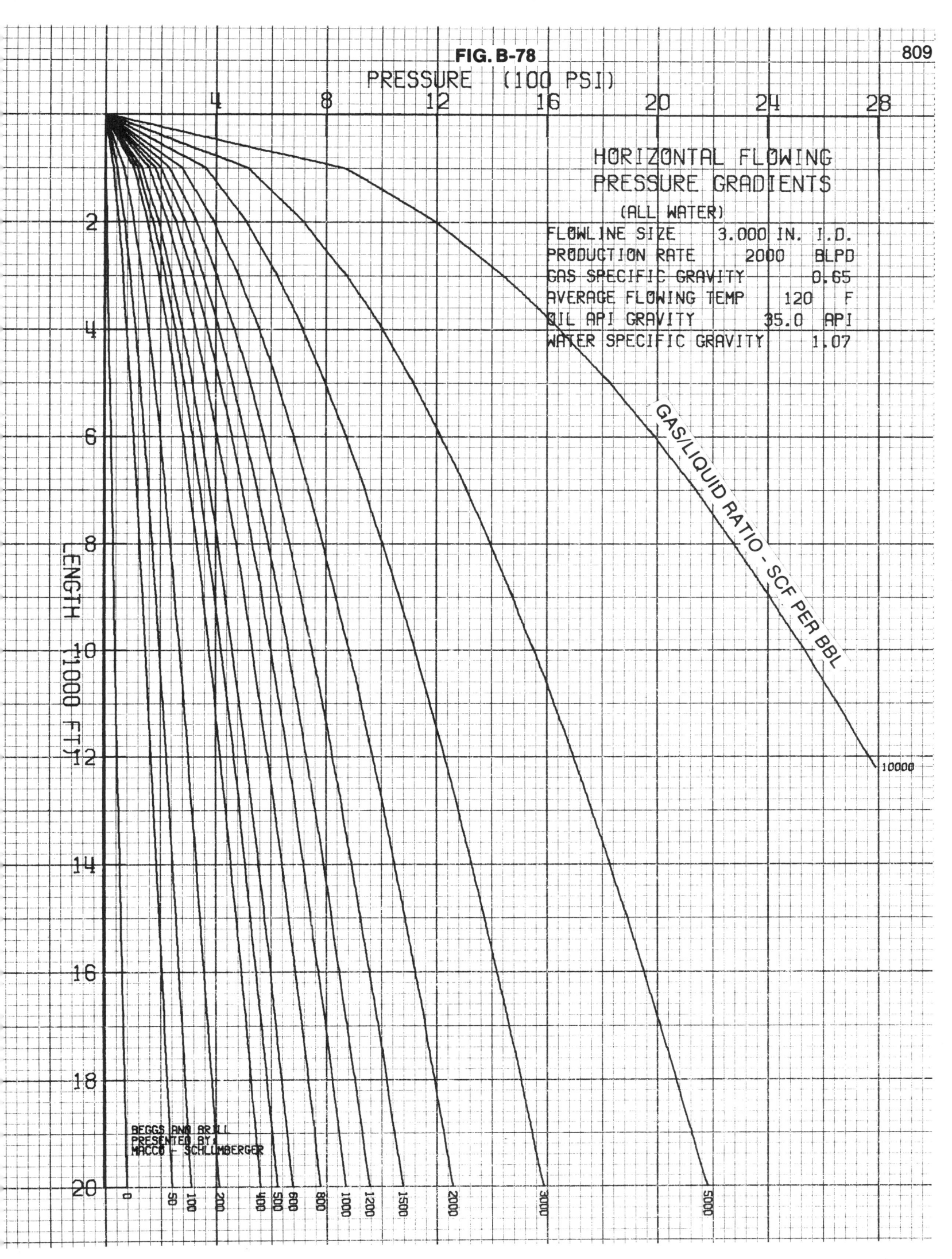

PRESSURE (100 PSI)

HORIZONTAL FLOWING
PRESSURE GRADIENTS

(ALL WATER)

FLOWLINE SIZE	3.000 IN.	I.D.
PRODUCTION RATE	2000	BLPD
GAS SPECIFIC GRAVITY	0.65	
AVERAGE FLOWING TEMP	120	F
OIL API GRAVITY	35.0	API
WATER SPECIFIC GRAVITY	1.07	

GAS/LIQUID RATIO - SCF PER BBL

LENGTH (1000 FT.)

BEGGS AND BRILL
PRESENTED BY
MACCO - SCHLUMBERGER

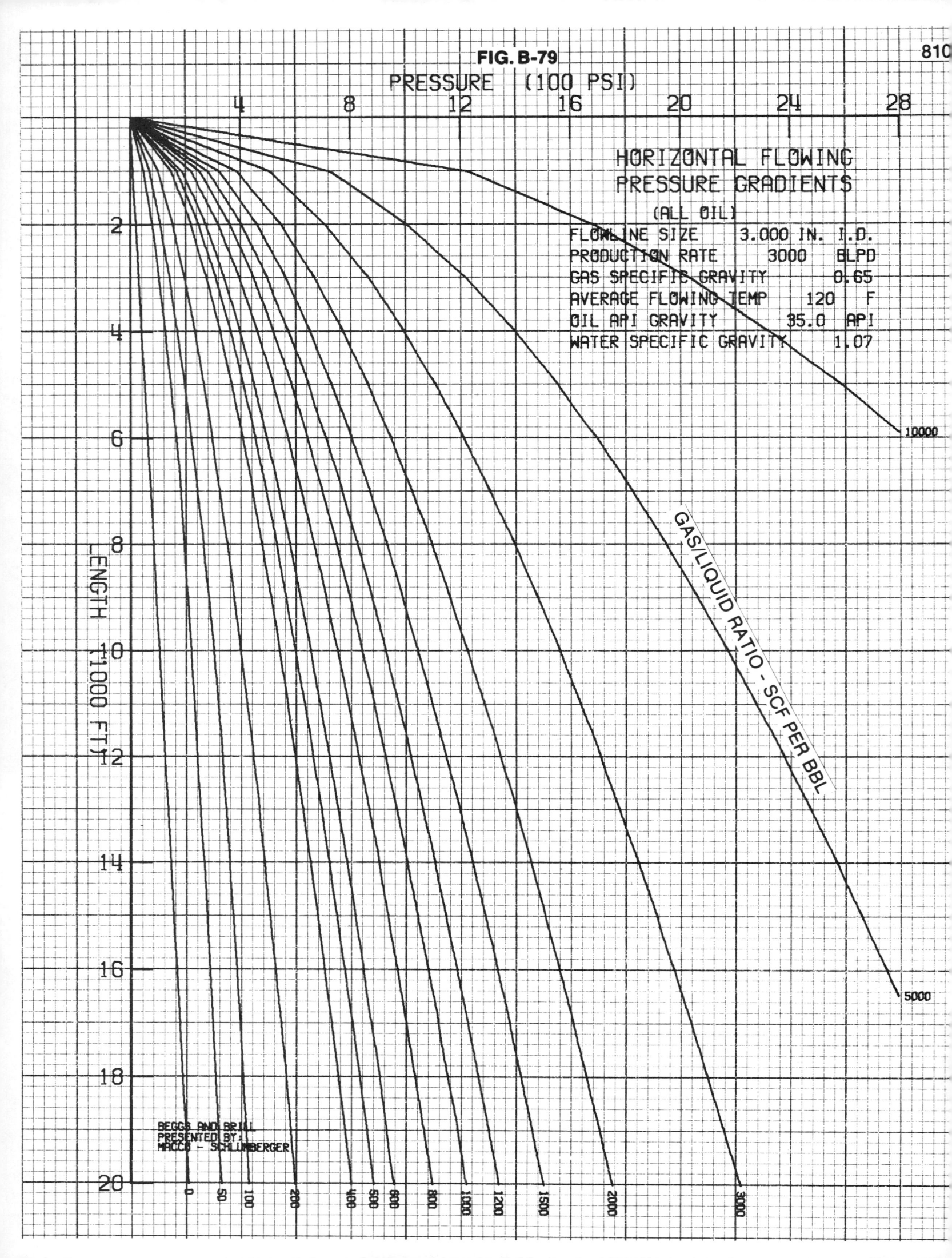

FIG. B-79

810

PRESSURE (100 PSI)

HORIZONTAL FLOWING
PRESSURE GRADIENTS
(ALL OIL)

FLOWLINE SIZE	3.000 IN. I.D.
PRODUCTION RATE	3000 BLPD
GAS SPECIFIC GRAVITY	0.65
AVERAGE FLOWING TEMP	120 F
OIL API GRAVITY	35.0 API
WATER SPECIFIC GRAVITY	1.07

LENGTH (1000 FT)

GAS/LIQUID RATIO - SCF PER BBL

BEGGS AND BRILL
PRESENTED BY:
MACCO - SCHLUMBERGER

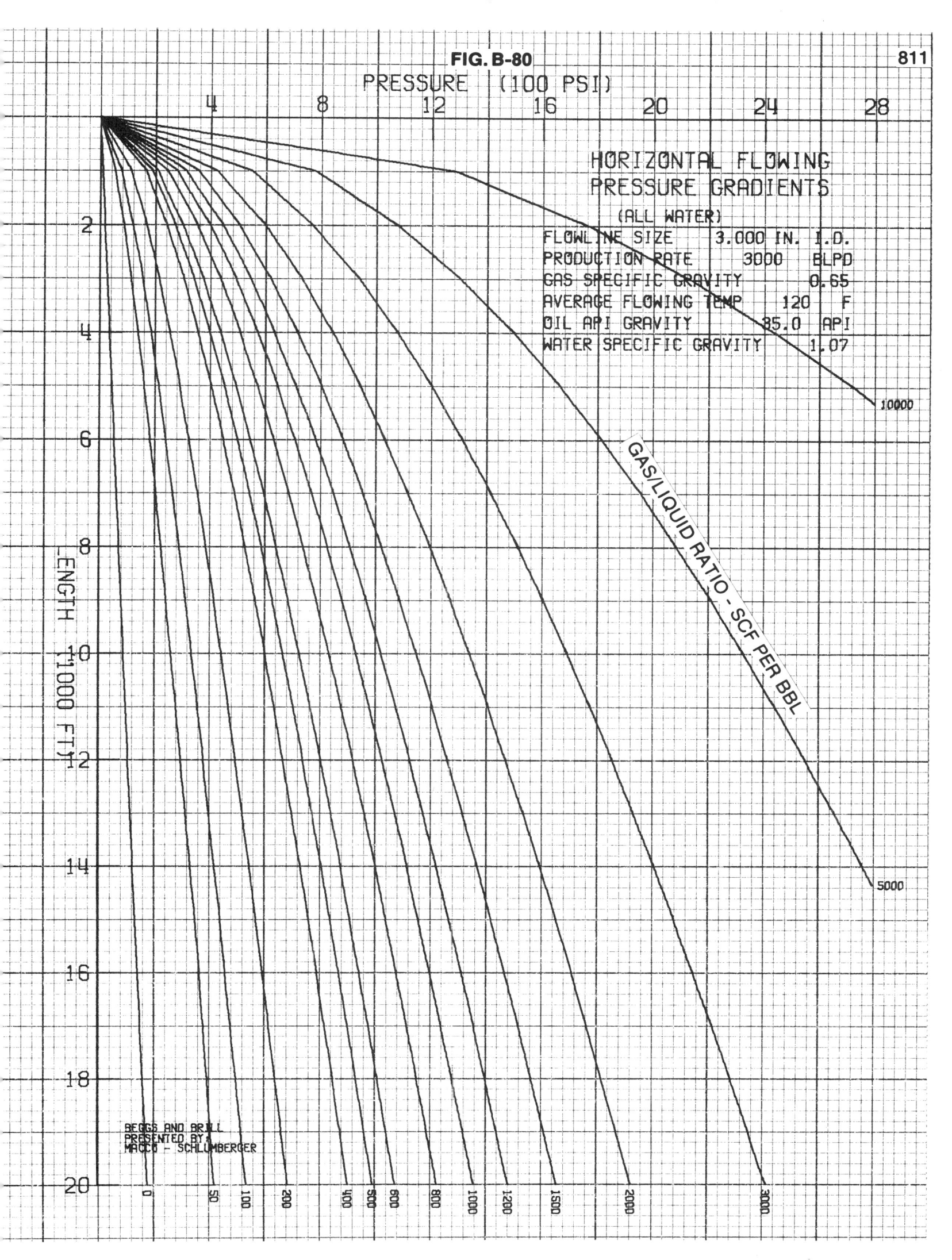

FIG. B-80   811

PRESSURE (100 PSI)

HORIZONTAL FLOWING
PRESSURE GRADIENTS
(ALL WATER)

FLOWLINE SIZE	3.000 IN. I.D.
PRODUCTION RATE	3000 BLPD
GAS SPECIFIC GRAVITY	0.65
AVERAGE FLOWING TEMP	120 F
OIL API GRAVITY	35.0 API
WATER SPECIFIC GRAVITY	1.07

GAS/LIQUID RATIO - SCF PER BBL

LENGTH (1000 FT.)

BEGGS AND BRILL
PRESENTED BY:
MACCO - SCHLUMBERGER

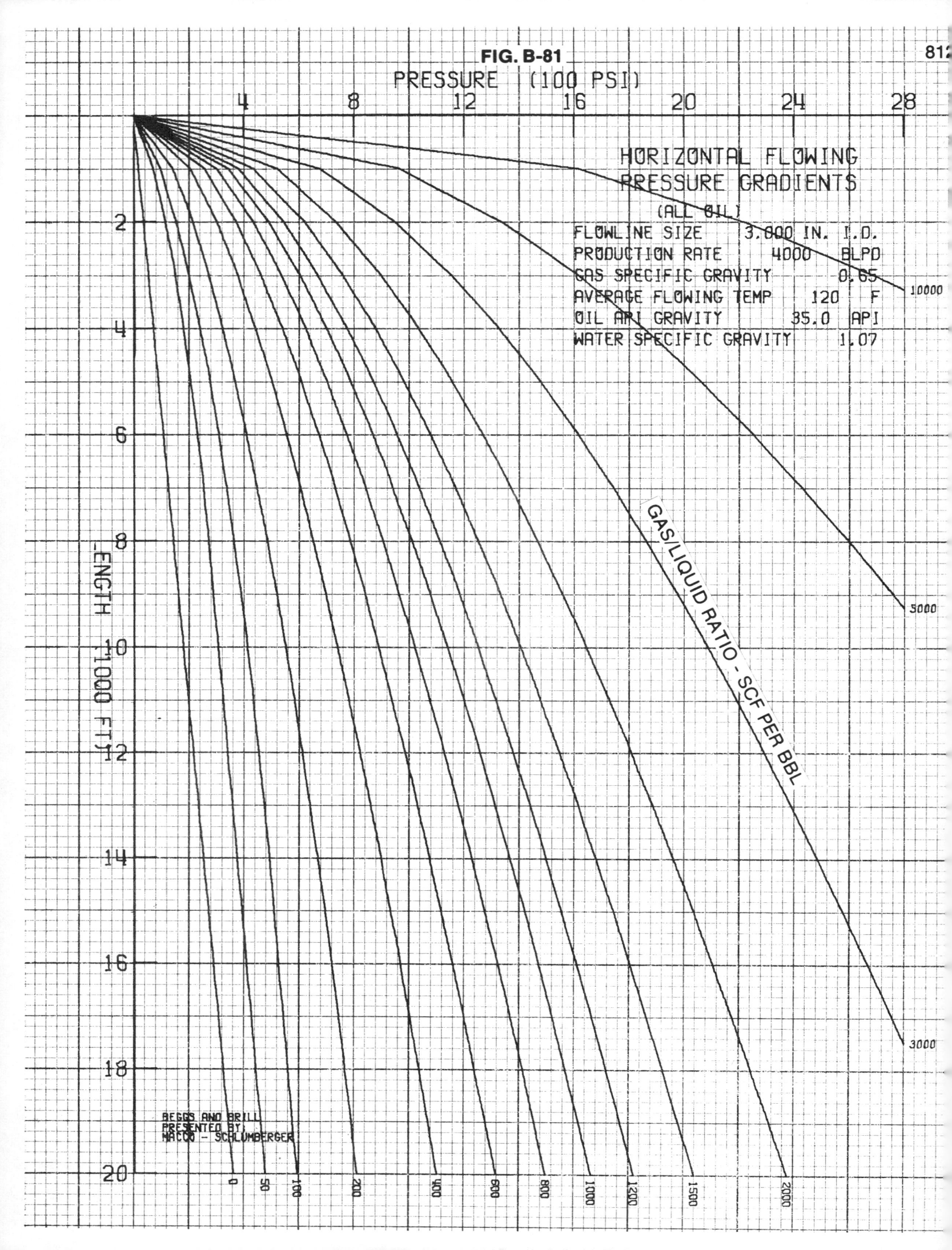

PRESSURE (100 PSI)

HORIZONTAL FLOWING
PRESSURE GRADIENTS
(ALL OIL)

FLOWLINE SIZE	3.000 IN. I.D.
PRODUCTION RATE	4000 BLPD
GAS SPECIFIC GRAVITY	0.65
AVERAGE FLOWING TEMP	120 F
OIL API GRAVITY	35.0 API
WATER SPECIFIC GRAVITY	1.07

GAS/LIQUID RATIO - SCF PER BBL

LENGTH (1000 FT.)

BEGGS AND BRILL
PRESENTED BY:
NACO - SCHLUMBERGER

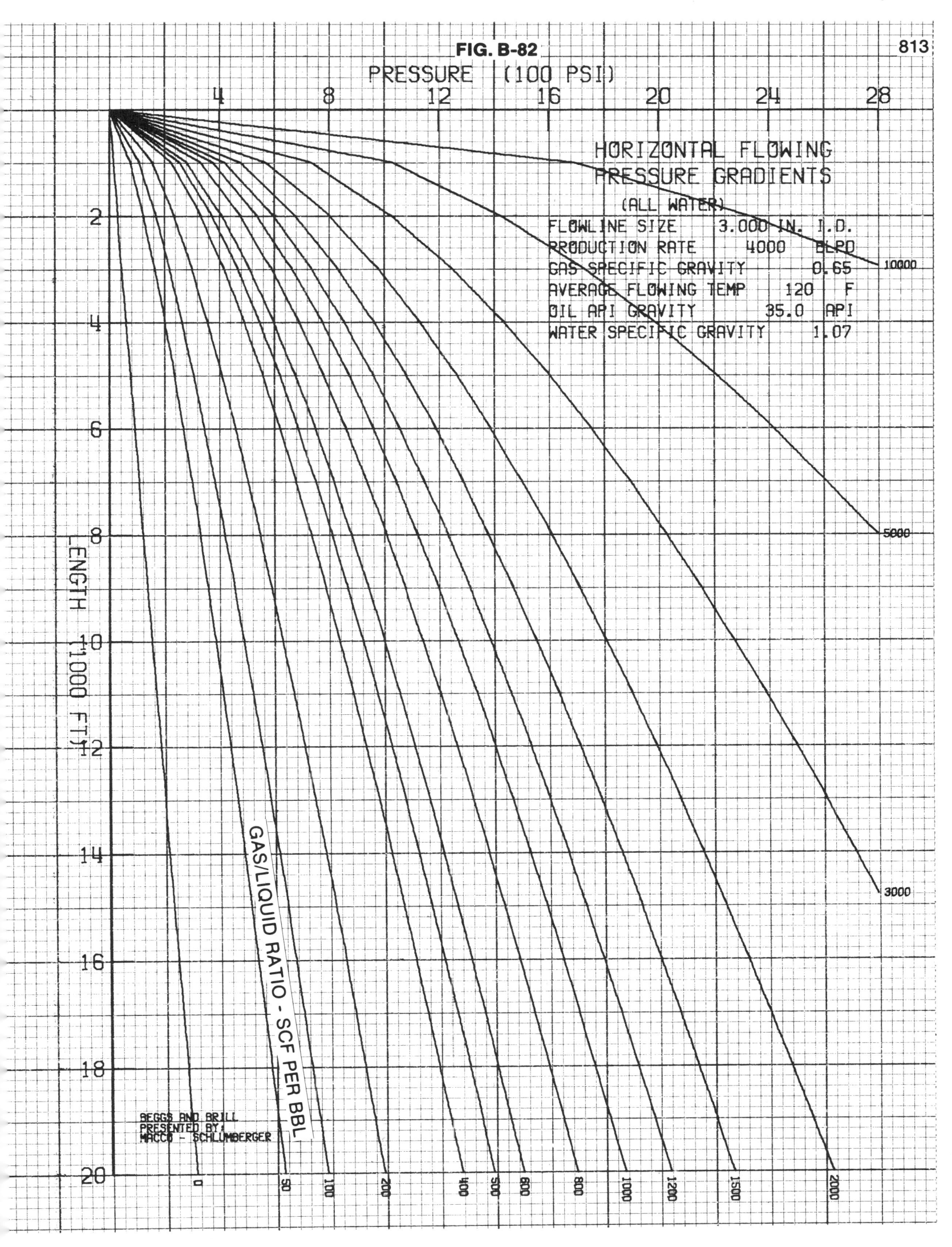

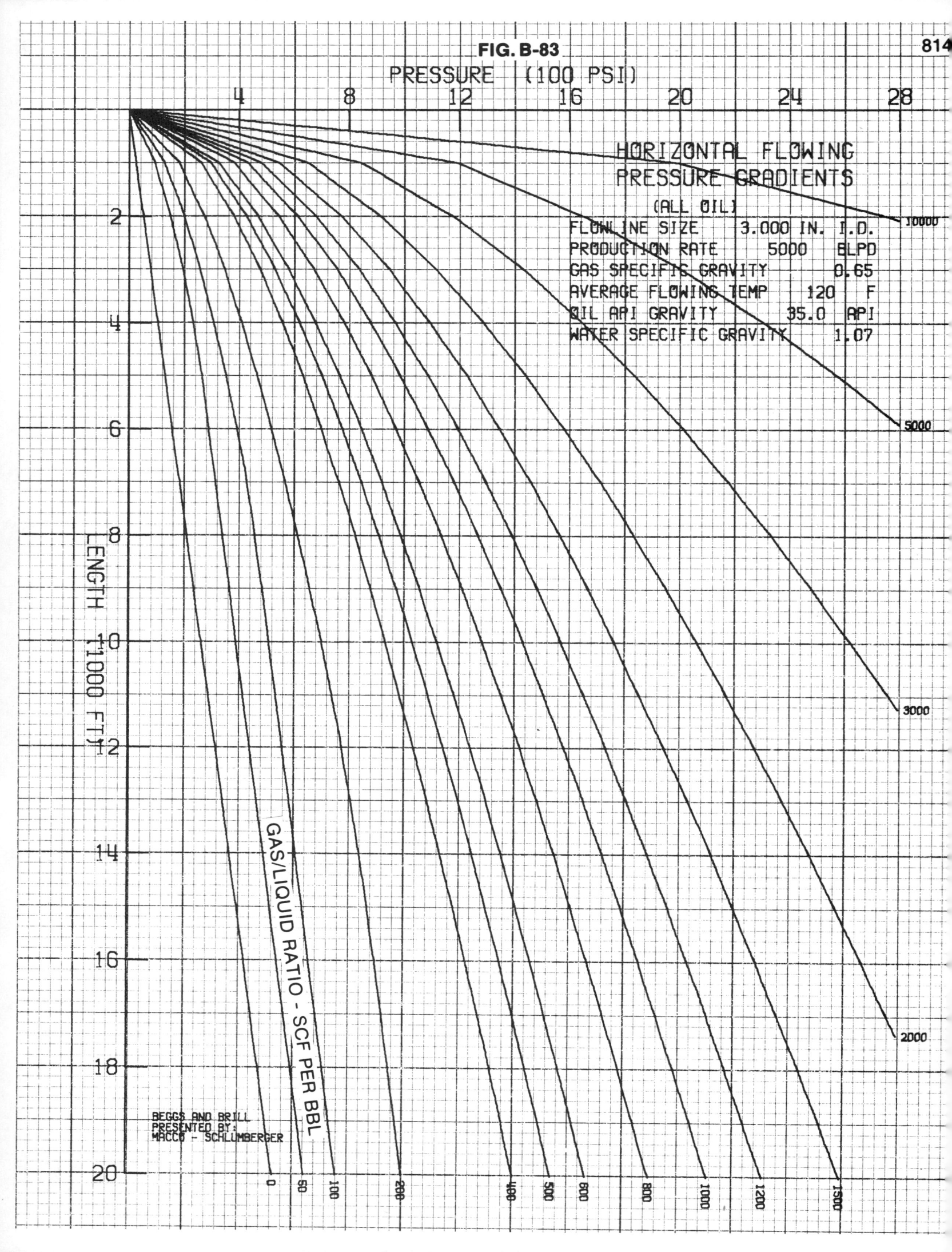

PRESSURE (100 PSI)

HORIZONTAL FLOWING
PRESSURE GRADIENTS
(ALL OIL)

FLOWLINE SIZE	3.000 IN. I.D.
PRODUCTION RATE	5000 BLPD
GAS SPECIFIC GRAVITY	0.65
AVERAGE FLOWING TEMP	120 F
OIL API GRAVITY	35.0 API
WATER SPECIFIC GRAVITY	1.07

LENGTH (1000 FT.)

GAS/LIQUID RATIO - SCF PER BBL

BEGGS AND BRILL
PRESENTED BY:
MACCO - SCHLUMBERGER

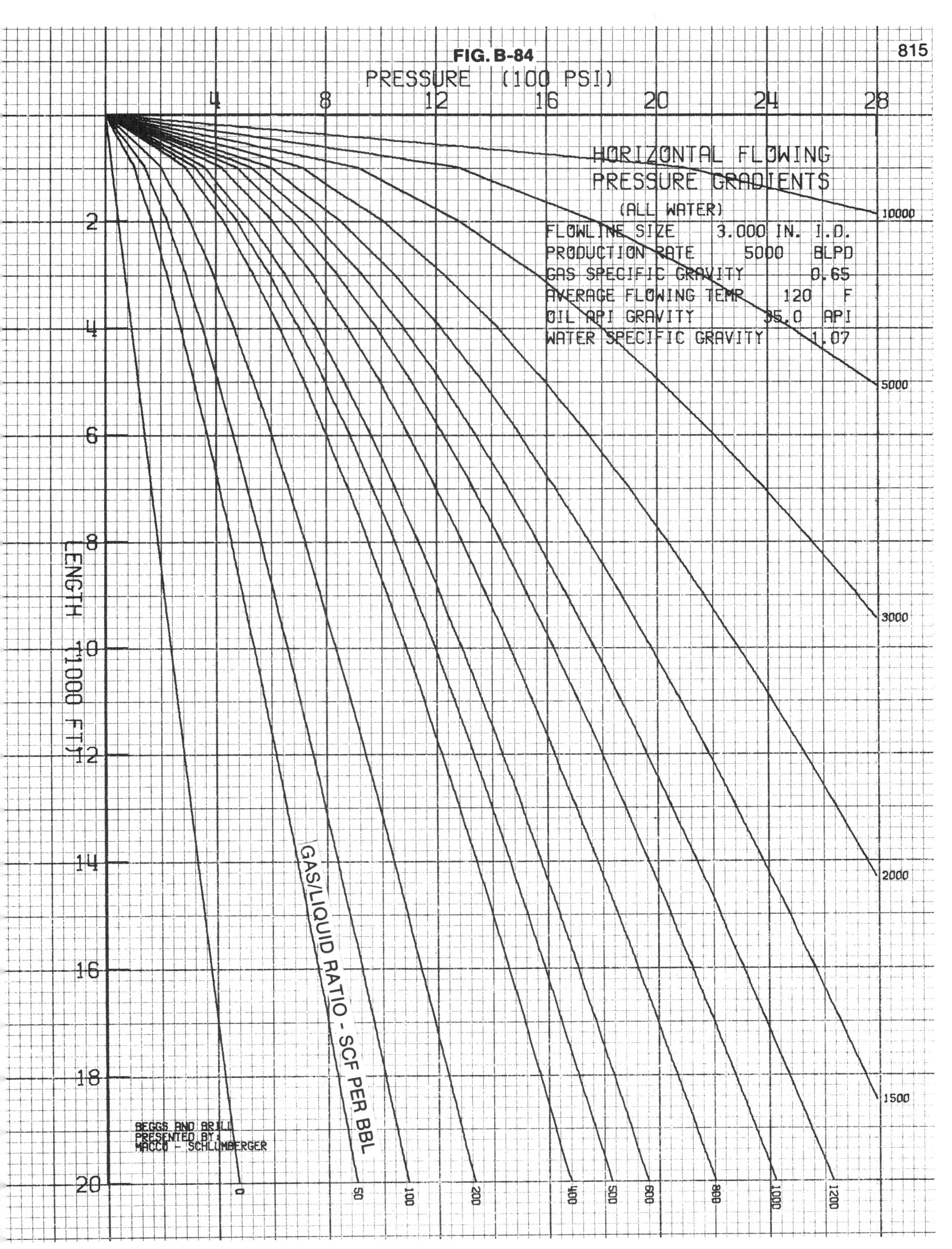

FIG. B-84

PRESSURE (100 PSI)

HORIZONTAL FLOWING
PRESSURE GRADIENTS
(ALL WATER)

FLOWLINE SIZE	3.000 IN. I.D.
PRODUCTION RATE	5000 BLPD
GAS SPECIFIC GRAVITY	0.65
AVERAGE FLOWING TEMP	120 F
OIL API GRAVITY	35.0 API
WATER SPECIFIC GRAVITY	1.07

LENGTH (1000 FT)

GAS/LIQUID RATIO - SCF PER BBL

BEGGS AND BRILL
PRESENTED BY :
MACCO - SCHLUMBERGER

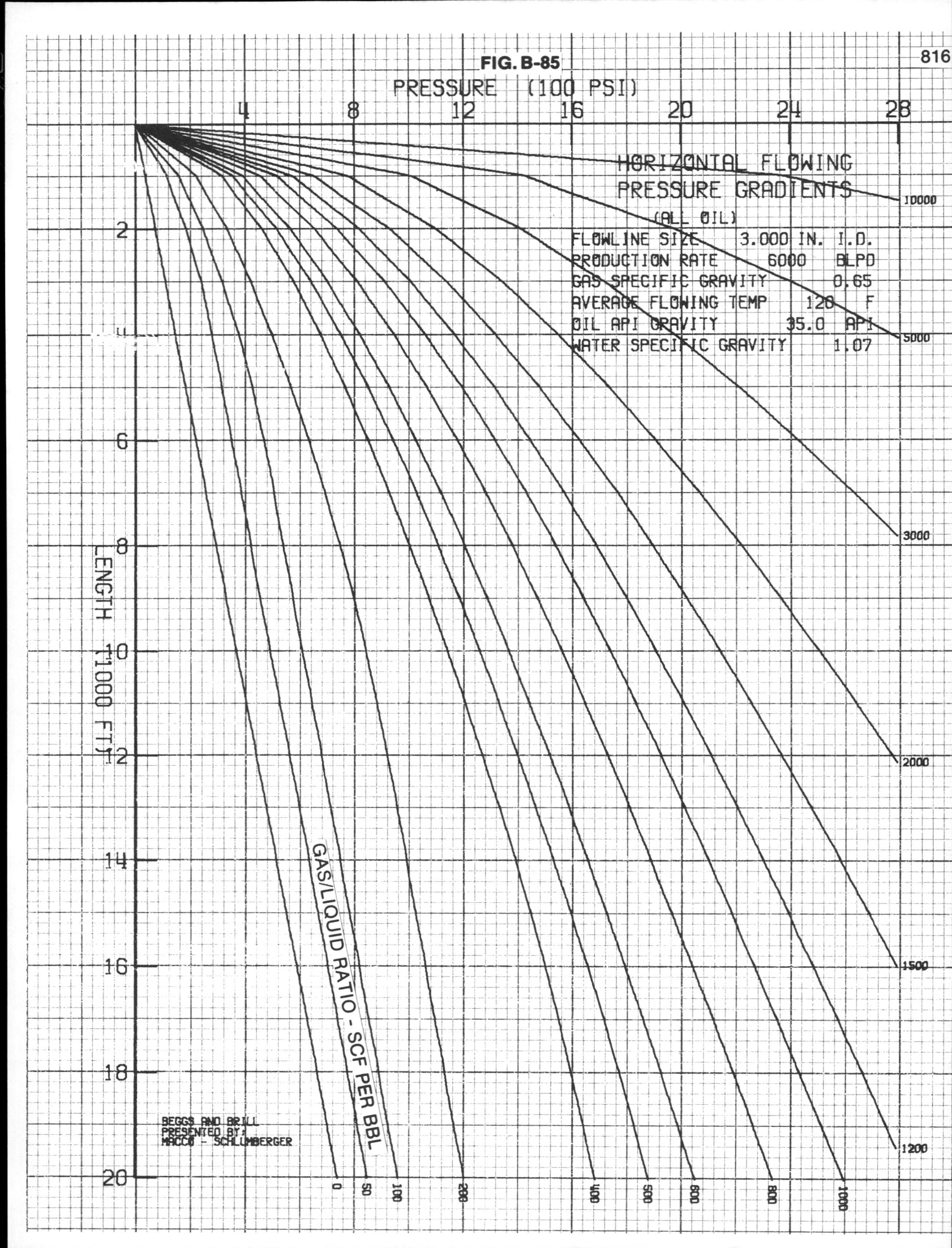

FIG. B-85

PRESSURE (100 PSI)

HORIZONTAL FLOWING
PRESSURE GRADIENTS
(ALL OIL)

FLOWLINE SIZE   3.000 IN. I.D.
PRODUCTION RATE   6000   BLPD
GAS SPECIFIC GRAVITY   0.65
AVERAGE FLOWING TEMP   120   F
OIL API GRAVITY   35.0 API
WATER SPECIFIC GRAVITY   1.07

LENGTH (1000 FT)

GAS/LIQUID RATIO - SCF PER BBL

BEGGS AND BRILL
PRESENTED BY:
MACCO - SCHLUMBERGER

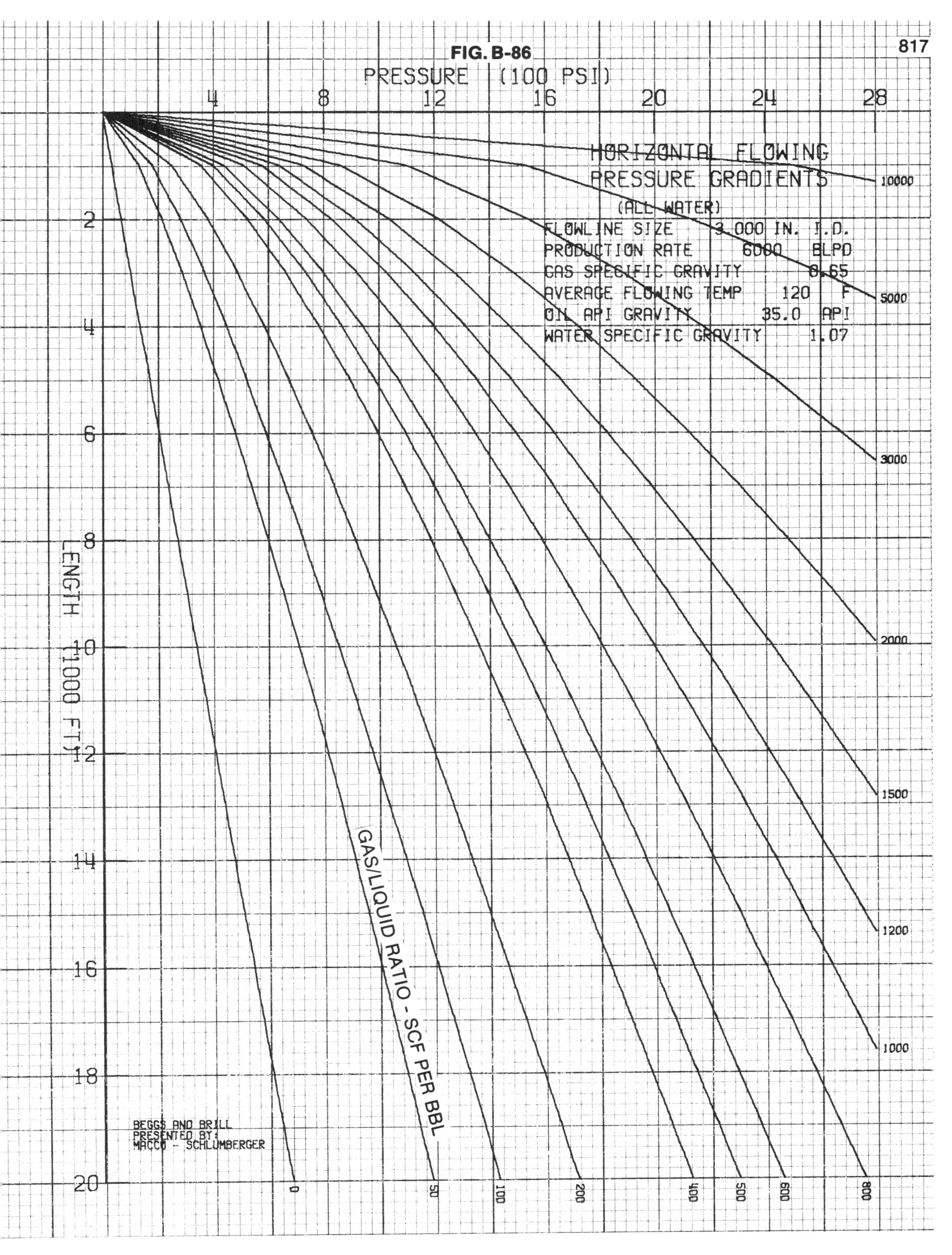

FIG. B-86

HORIZONTAL FLOWING PRESSURE GRADIENTS
(ALL WATER)

FLOWLINE SIZE	3.000 IN. I.D.
PRODUCTION RATE	6000 BLPD
GAS SPECIFIC GRAVITY	0.65
AVERAGE FLOWING TEMP	120 F
OIL API GRAVITY	35.0 API
WATER SPECIFIC GRAVITY	1.07

PRESSURE (100 PSI)

LENGTH (1000 FT.)

GAS/LIQUID RATIO - SCF PER BBL

BEGGS AND BRILL
PRESENTED BY:
MACCO - SCHLUMBERGER

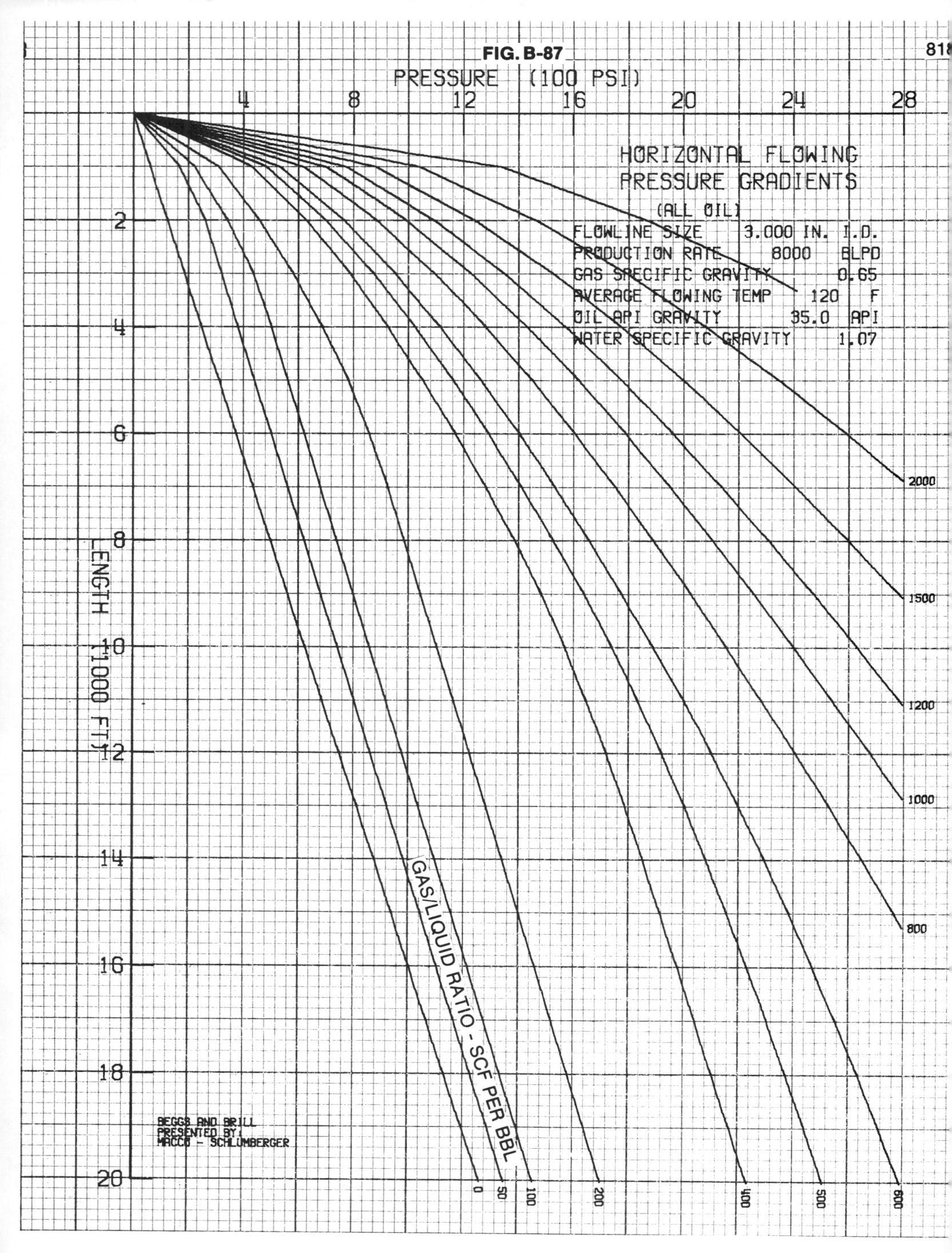

FIG. B-87

818

HORIZONTAL FLOWING
PRESSURE GRADIENTS
(ALL OIL)

FLOWLINE SIZE   3.000 IN. I.D.
PRODUCTION RATE   8000   BLPD
GAS SPECIFIC GRAVITY   0.65
AVERAGE FLOWING TEMP   120   F
OIL API GRAVITY   35.0 API
WATER SPECIFIC GRAVITY   1.07

PRESSURE (100 PSI)

LENGTH (1000 FT)

GAS/LIQUID RATIO - SCF PER BBL

BEGGS AND BRILL
PRESENTED BY
MACCO - SCHLUMBERGER

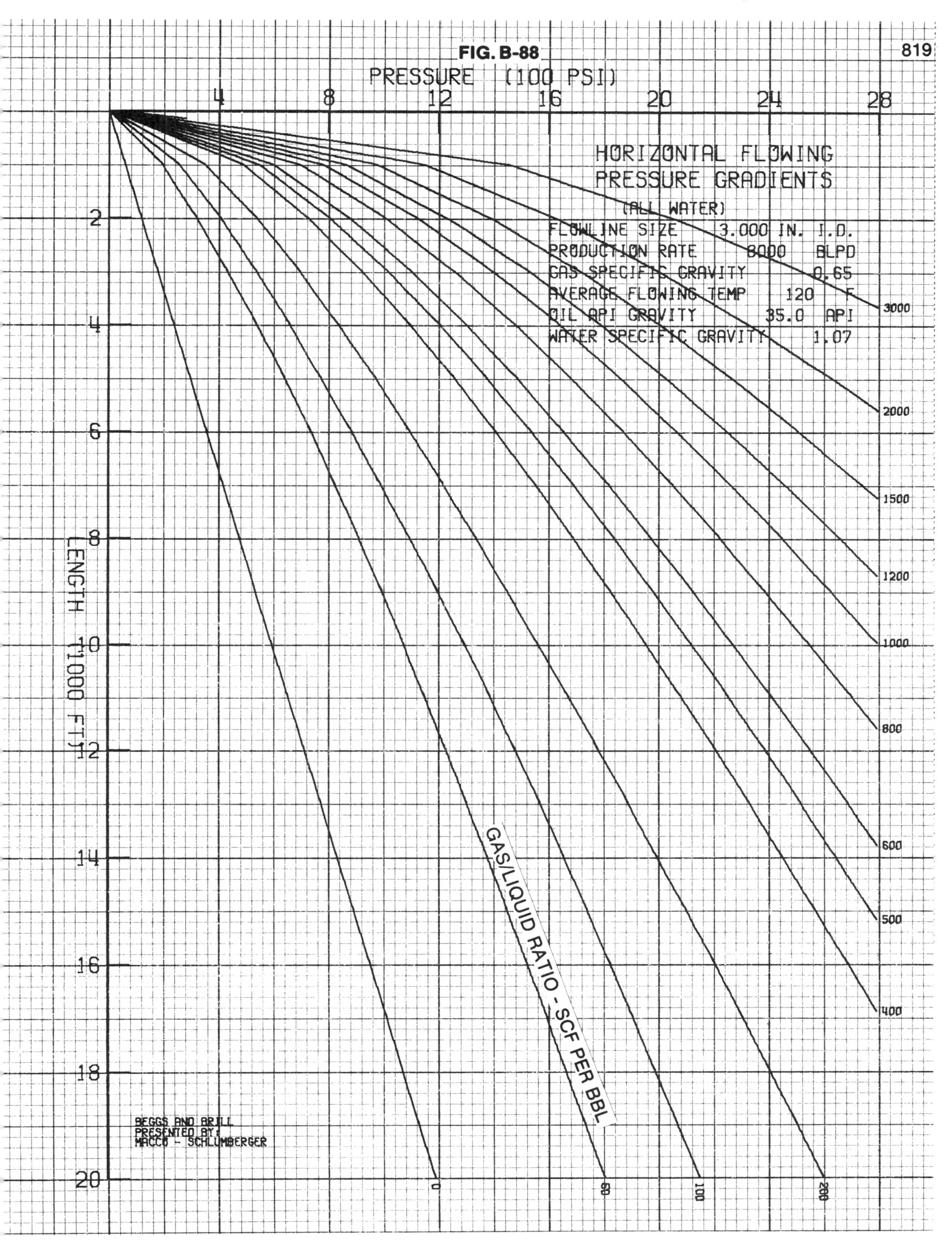

FIG. B-88

819

PRESSURE (100 PSI)

HORIZONTAL FLOWING
PRESSURE GRADIENTS

(ALL WATER)

FLOWLINE SIZE   3.000 IN. I.D.
PRODUCTION RATE   8000   BLPD
GAS SPECIFIC GRAVITY   0.65
AVERAGE FLOWING TEMP   120   F
OIL API GRAVITY   35.0   API
WATER SPECIFIC GRAVITY   1.07

LENGTH (1000 FT)

GAS/LIQUID RATIO - SCF PER BBL

BEGGS AND BRILL
PRESENTED BY:
MACCO - SCHLUMBERGER

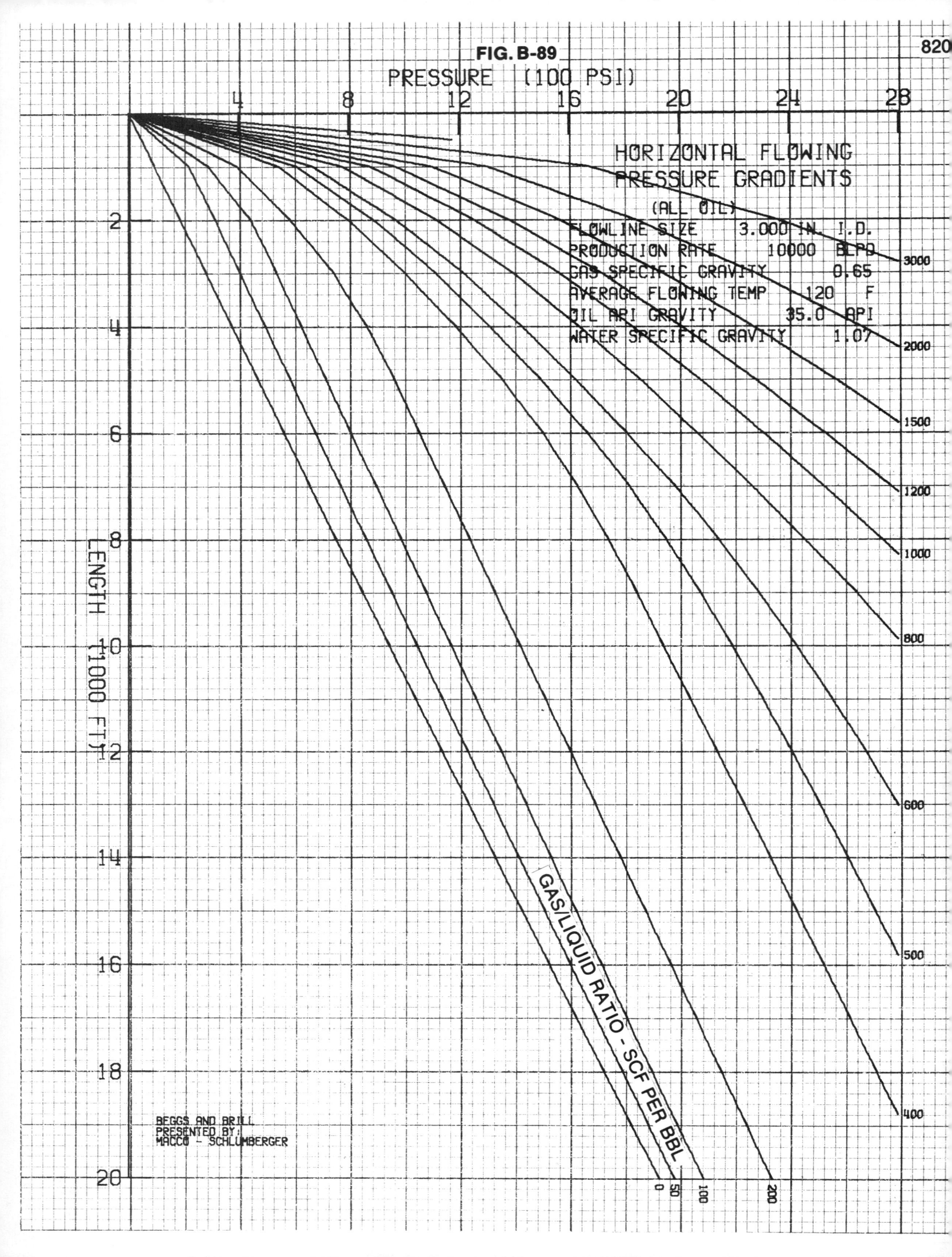

FIG. B-89

PRESSURE (100 PSI)

HORIZONTAL FLOWING
PRESSURE GRADIENTS
(ALL OIL)
FLOWLINE SIZE   3.000 IN. I.D.
PRODUCTION RATE   10000 BLPD
GAS SPECIFIC GRAVITY   0.65
AVERAGE FLOWING TEMP   120   F
OIL API GRAVITY   35.0   API
WATER SPECIFIC GRAVITY   1.07

LENGTH (1000 FT)

GAS/LIQUID RATIO - SCF PER BBL

BEGGS AND BRILL
PRESENTED BY:
MACCO - SCHLUMBERGER

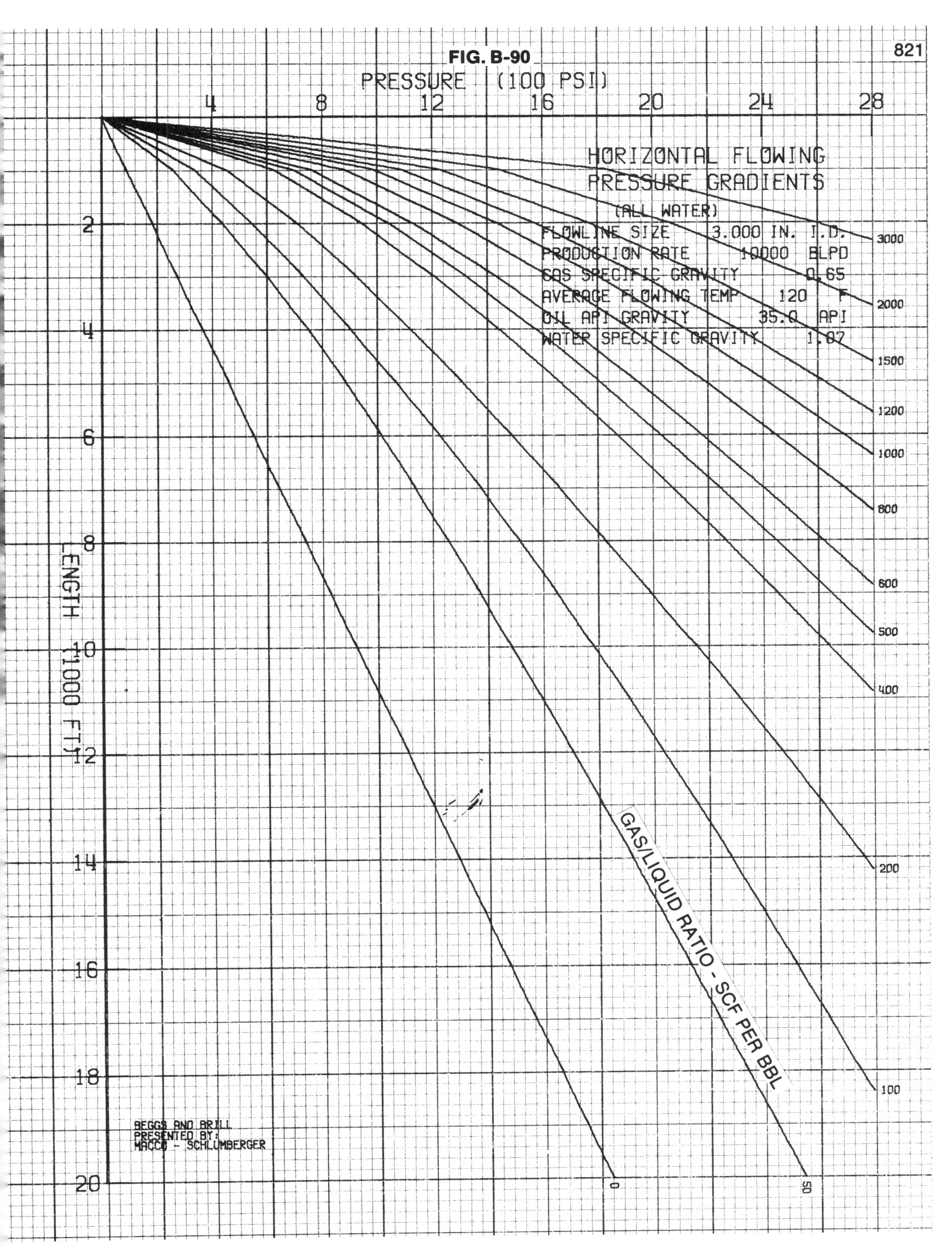

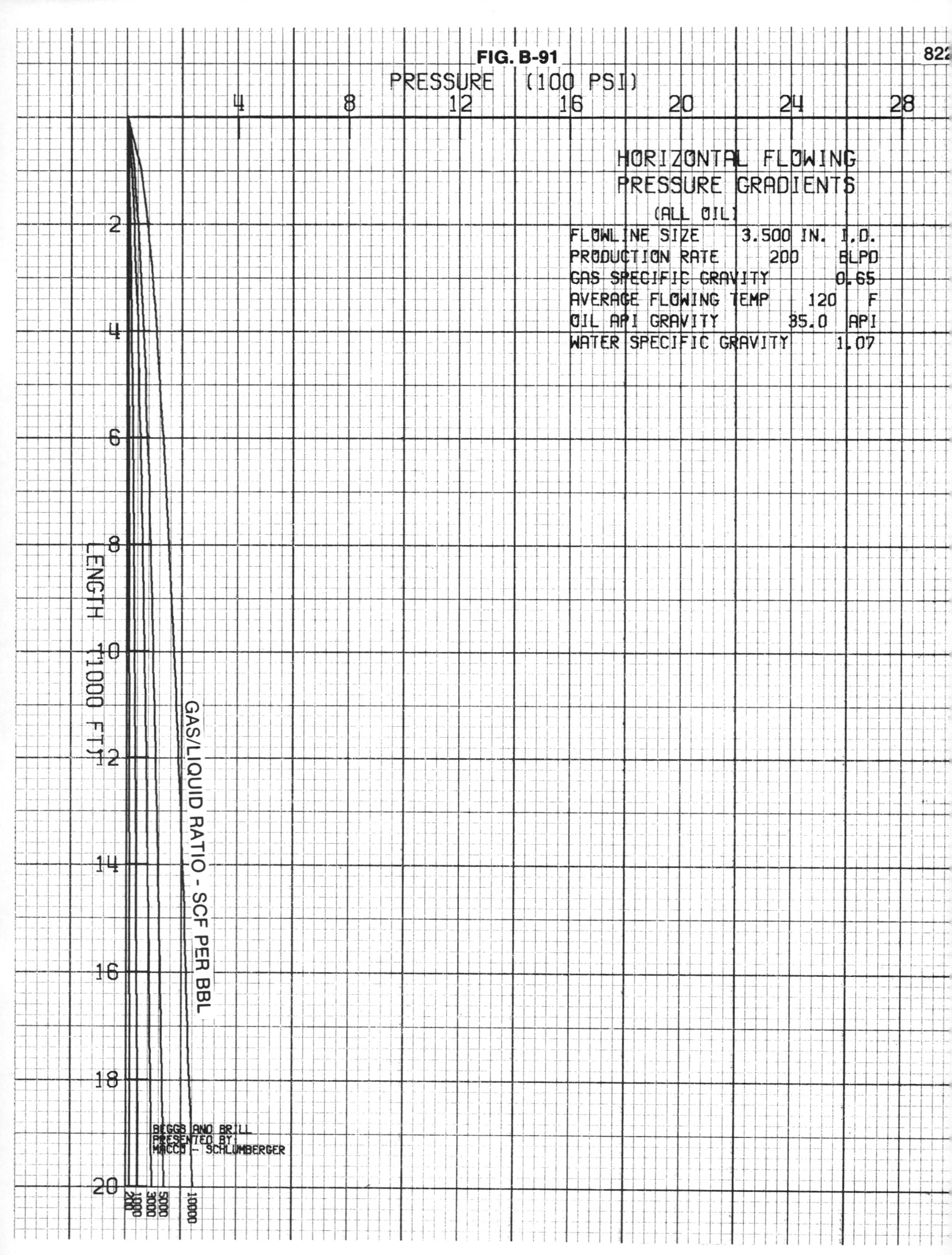

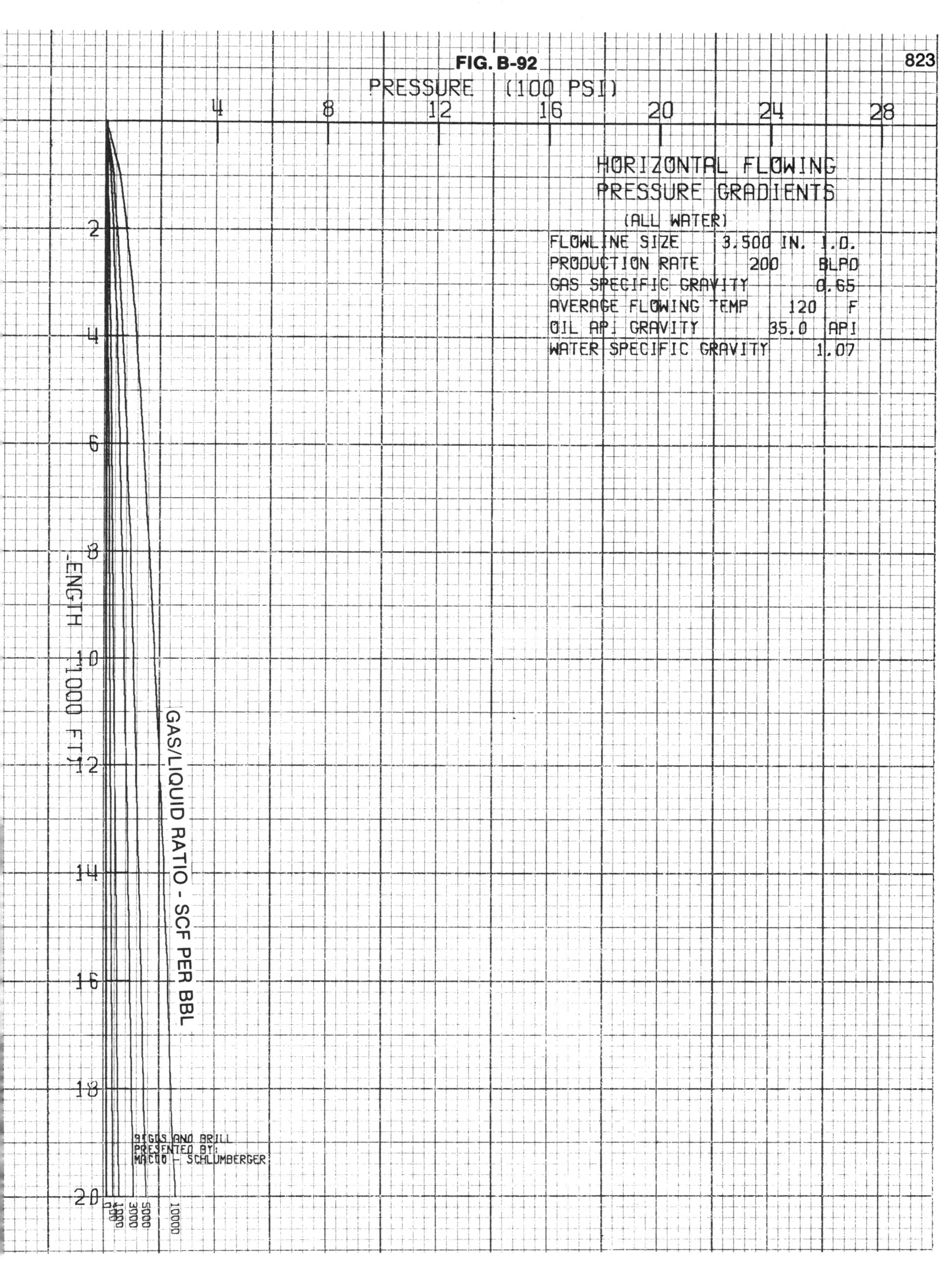

PRESSURE (100 PSI)

HORIZONTAL FLOWING
PRESSURE GRADIENTS
(ALL WATER)

FLOWLINE SIZE   3.500 IN. I.D.
PRODUCTION RATE   200   BLPD
GAS SPECIFIC GRAVITY   0.65
AVERAGE FLOWING TEMP   120   F
OIL API GRAVITY   35.0   API
WATER SPECIFIC GRAVITY   1.07

LENGTH (1000 FT.)

GAS/LIQUID RATIO - SCF PER BBL

BY: GGS AND BRILL
PRESENTED BY:
MACCO - SCHLUMBERGER

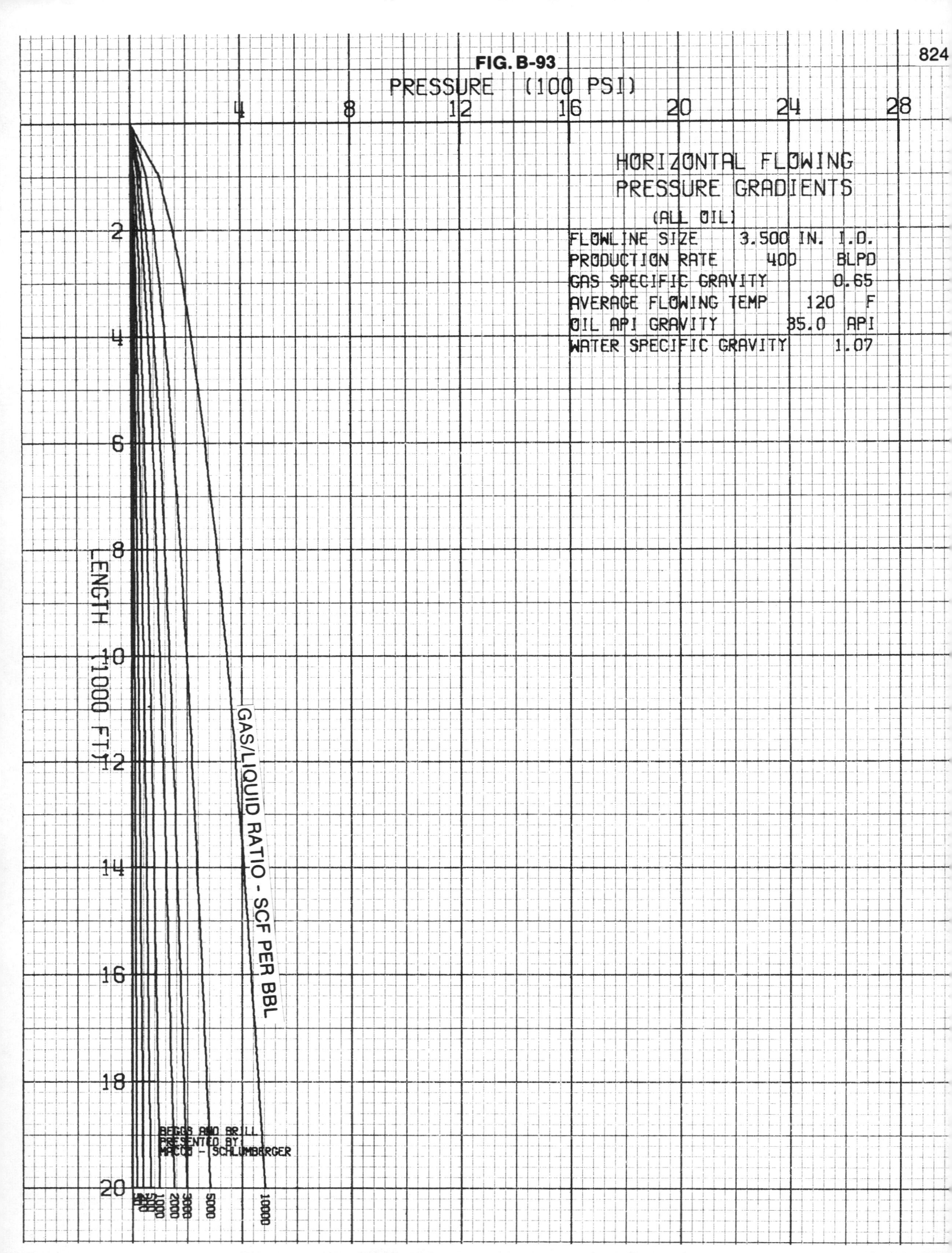

FIG. B-93

PRESSURE (100 PSI)

HORIZONTAL FLOWING
PRESSURE GRADIENTS
(ALL OIL)

FLOWLINE SIZE	3.500 IN. I.D.
PRODUCTION RATE	400 BLPD
GAS SPECIFIC GRAVITY	0.65
AVERAGE FLOWING TEMP	120 F
OIL API GRAVITY	35.0 API
WATER SPECIFIC GRAVITY	1.07

LENGTH (1000 FT)

GAS/LIQUID RATIO - SCF PER BBL

BEGGS AND BRILL
PRESENTED BY
MACCO - SCHLUMBERGER

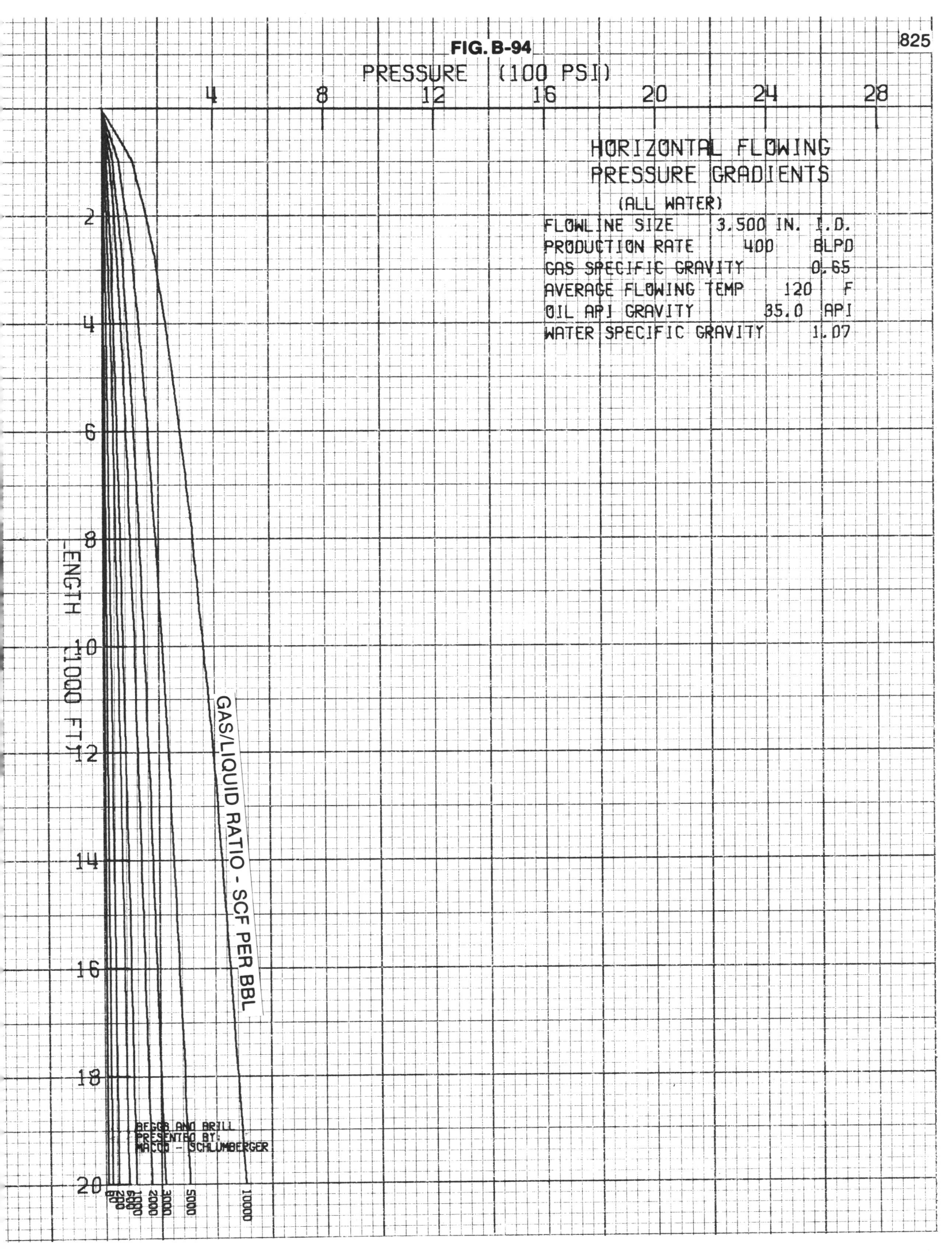

PRESSURE (100 PSI)

HORIZONTAL FLOWING
PRESSURE GRADIENTS

(ALL WATER)

FLOWLINE SIZE	3.500 IN. I.D.
PRODUCTION RATE	400 BLPD
GAS SPECIFIC GRAVITY	0.65
AVERAGE FLOWING TEMP	120 F
OIL API GRAVITY	35.0 API
WATER SPECIFIC GRAVITY	1.07

LENGTH (1000 FT.)

GAS/LIQUID RATIO - SCF PER BBL

BEGGS AND BRILL
PRESENTED BY:
MACCO - SCHLUMBERGER

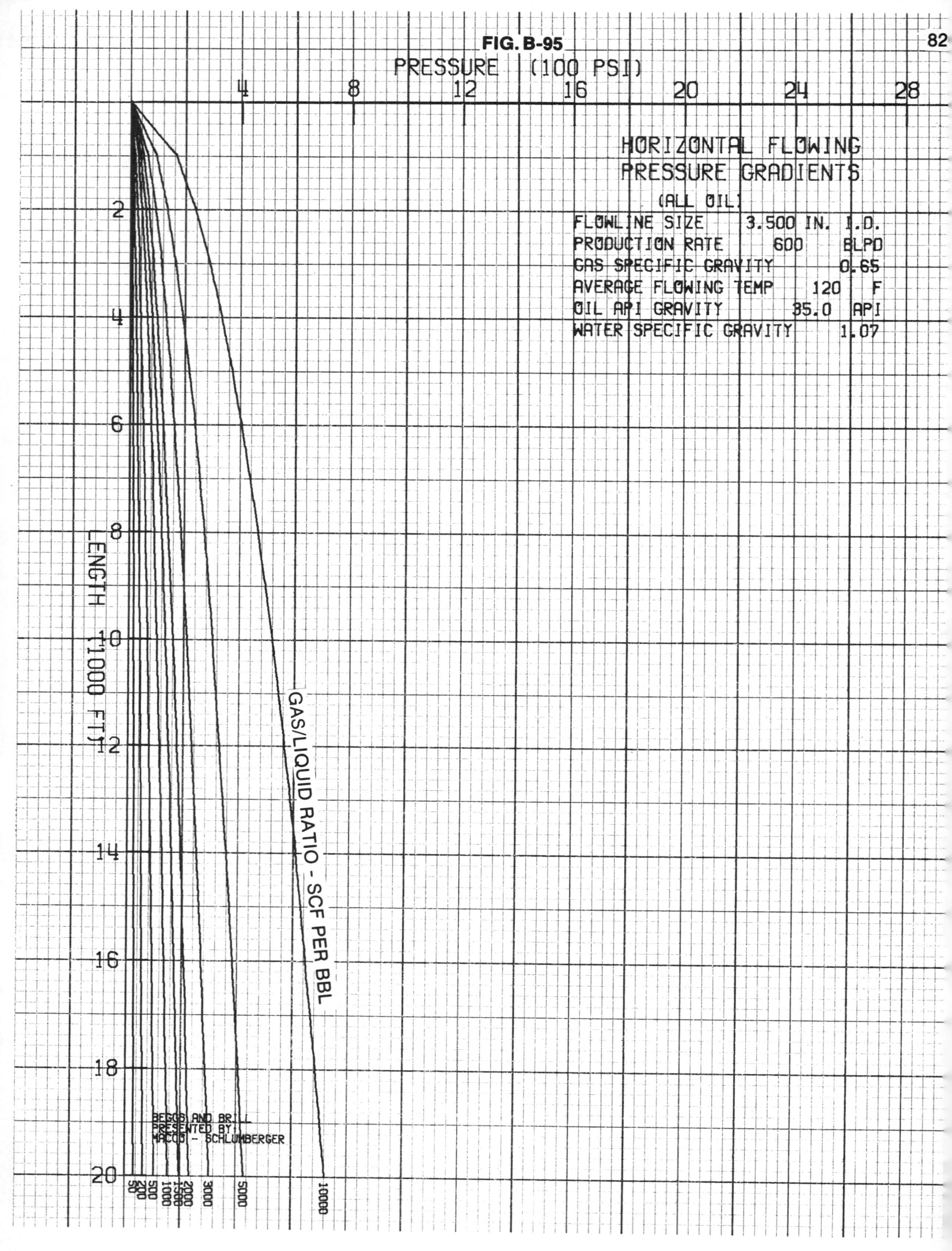

FIG. B-95

82

PRESSURE (100 PSI)

HORIZONTAL FLOWING
PRESSURE GRADIENTS
(ALL OIL)

FLOWLINE SIZE   3.500 IN. I.D.
PRODUCTION RATE   600   BLPD
GAS SPECIFIC GRAVITY   0.65
AVERAGE FLOWING TEMP   120   F
OIL API GRAVITY   35.0 API
WATER SPECIFIC GRAVITY   1.07

LENGTH (1000 FT)

GAS/LIQUID RATIO - SCF PER BBL

BEGGS AND BRILL
PRESENTED BY:
MACCO - SCHLUMBERGER

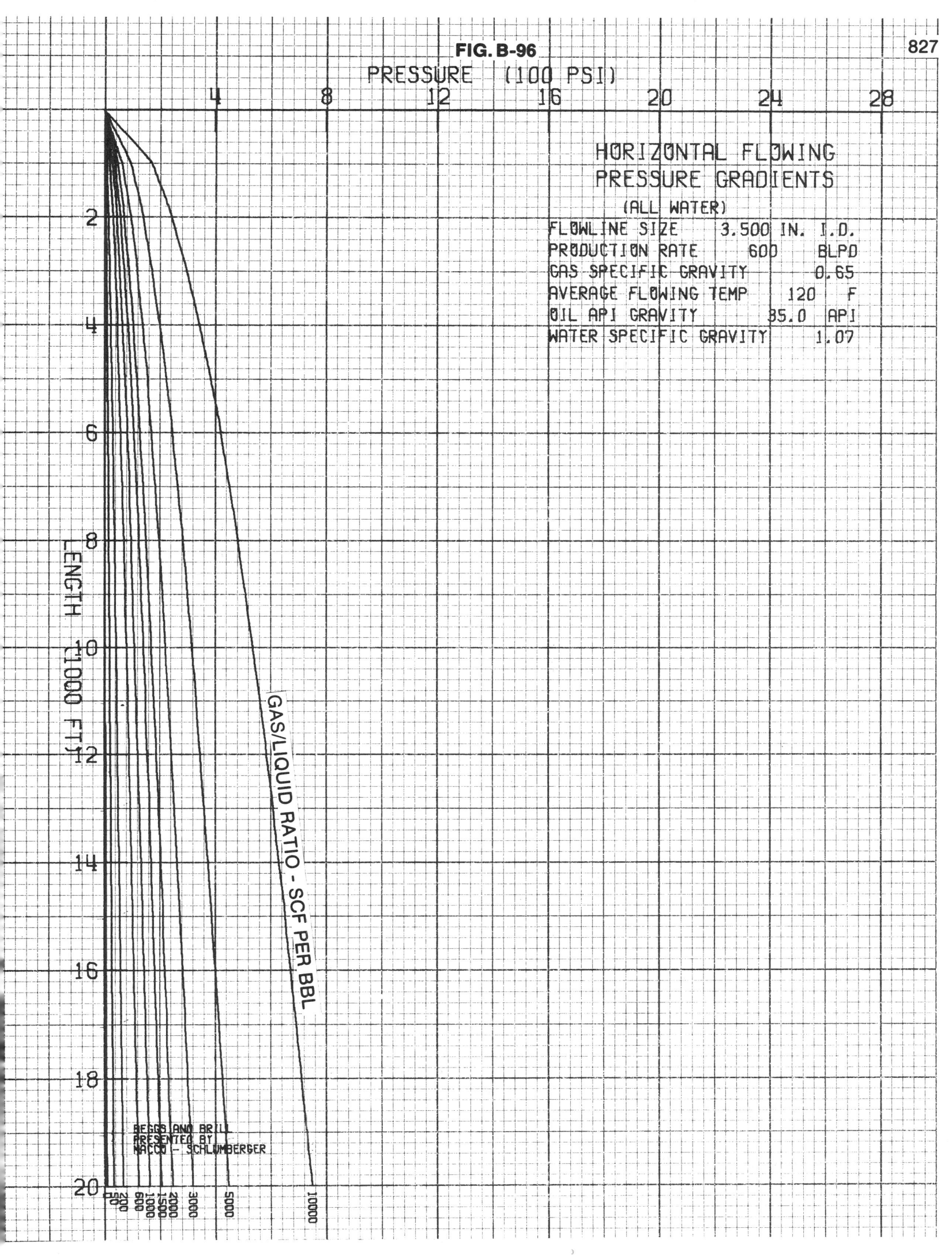

PRESSURE (100 PSI)

HORIZONTAL FLOWING
PRESSURE GRADIENTS
(ALL WATER)

FLOWLINE SIZE	3.500 IN. I.D.	
PRODUCTION RATE	600	BLPD
GAS SPECIFIC GRAVITY	0.65	
AVERAGE FLOWING TEMP	120	F
OIL API GRAVITY	35.0	API
WATER SPECIFIC GRAVITY	1.07	

LENGTH (1000 FT)

GAS/LIQUID RATIO - SCF PER BBL

BEGGS AND BRILL
PRESENTED BY
NACCO - SCHLUMBERGER

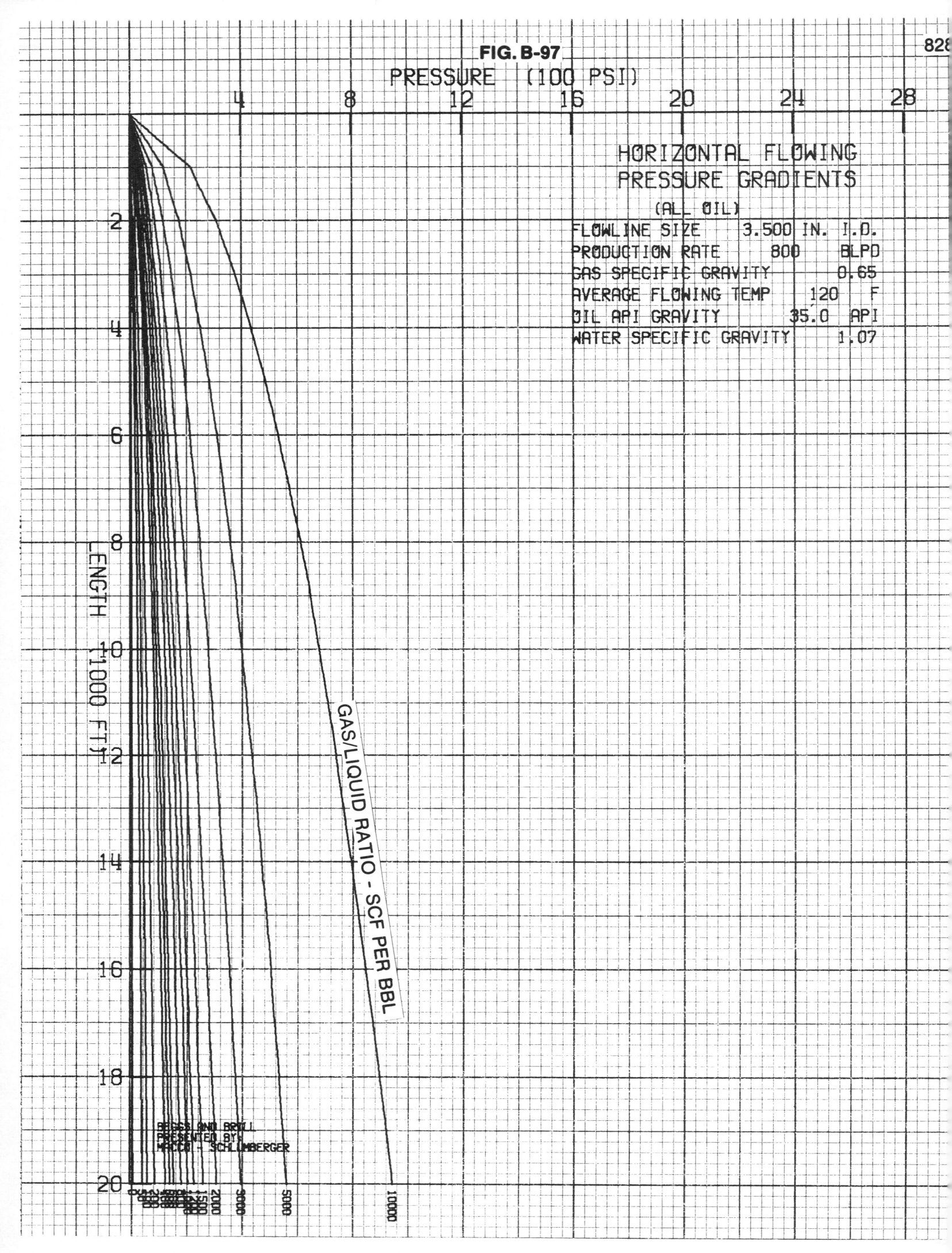

FIG. B-97

828

PRESSURE (100 PSI)

LENGTH (1000 FT)

GAS/LIQUID RATIO - SCF PER BBL

HORIZONTAL FLOWING
PRESSURE GRADIENTS
(ALL OIL)

FLOWLINE SIZE	3.500 IN.	I.D.
PRODUCTION RATE	800	BLPD
GAS SPECIFIC GRAVITY	0.65	
AVERAGE FLOWING TEMP	120	F
OIL API GRAVITY	35.0	API
WATER SPECIFIC GRAVITY	1.07	

BEGGS AND BRILL
PRESENTED BY
MACCO - SCHLUMBERGER

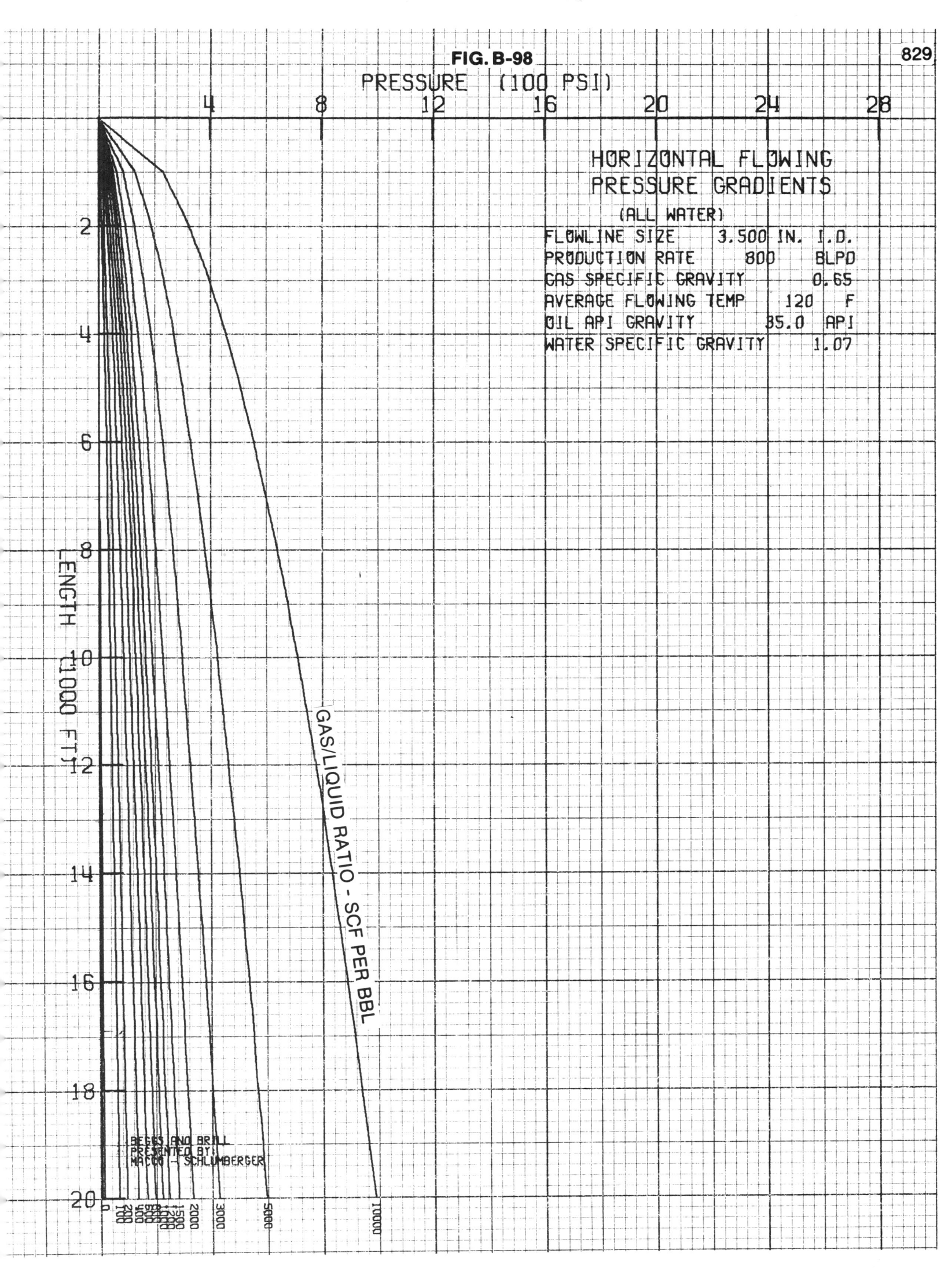

PRESSURE (100 PSI)

HORIZONTAL FLOWING
PRESSURE GRADIENTS
(ALL WATER)

FLOWLINE SIZE	3.500 IN. I.D.
PRODUCTION RATE	800 BLPD
GAS SPECIFIC GRAVITY	0.65
AVERAGE FLOWING TEMP	120 F
OIL API GRAVITY	35.0 API
WATER SPECIFIC GRAVITY	1.07

LENGTH (1000 FT)

GAS/LIQUID RATIO - SCF PER BBL

BEGGS AND BRILL
PRESENTED BY:
MACCO - SCHLUMBERGER

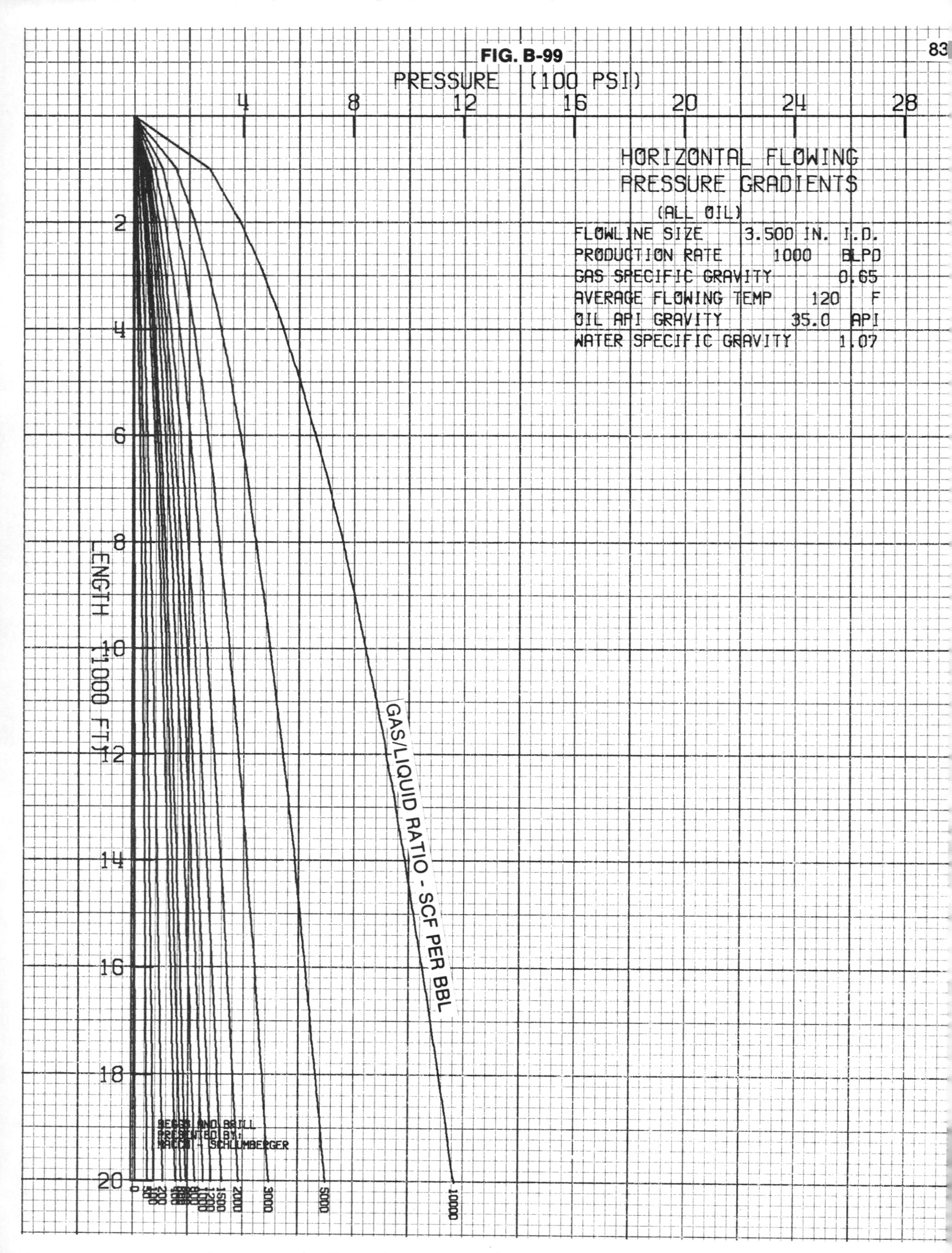

FIG. B-99

83

PRESSURE (100 PSI)

HORIZONTAL FLOWING
PRESSURE GRADIENTS
(ALL OIL)

FLOWLINE SIZE	3.500 IN.	I.D.
PRODUCTION RATE	1000	BLPD
GAS SPECIFIC GRAVITY	0.65	
AVERAGE FLOWING TEMP	120	F
OIL API GRAVITY	35.0	API
WATER SPECIFIC GRAVITY	1.07	

LENGTH (1000 FT)

GAS/LIQUID RATIO - SCF PER BBL

BEGGS AND BRILL
PRESENTED BY:
NACCO - SCHLUMBERGER

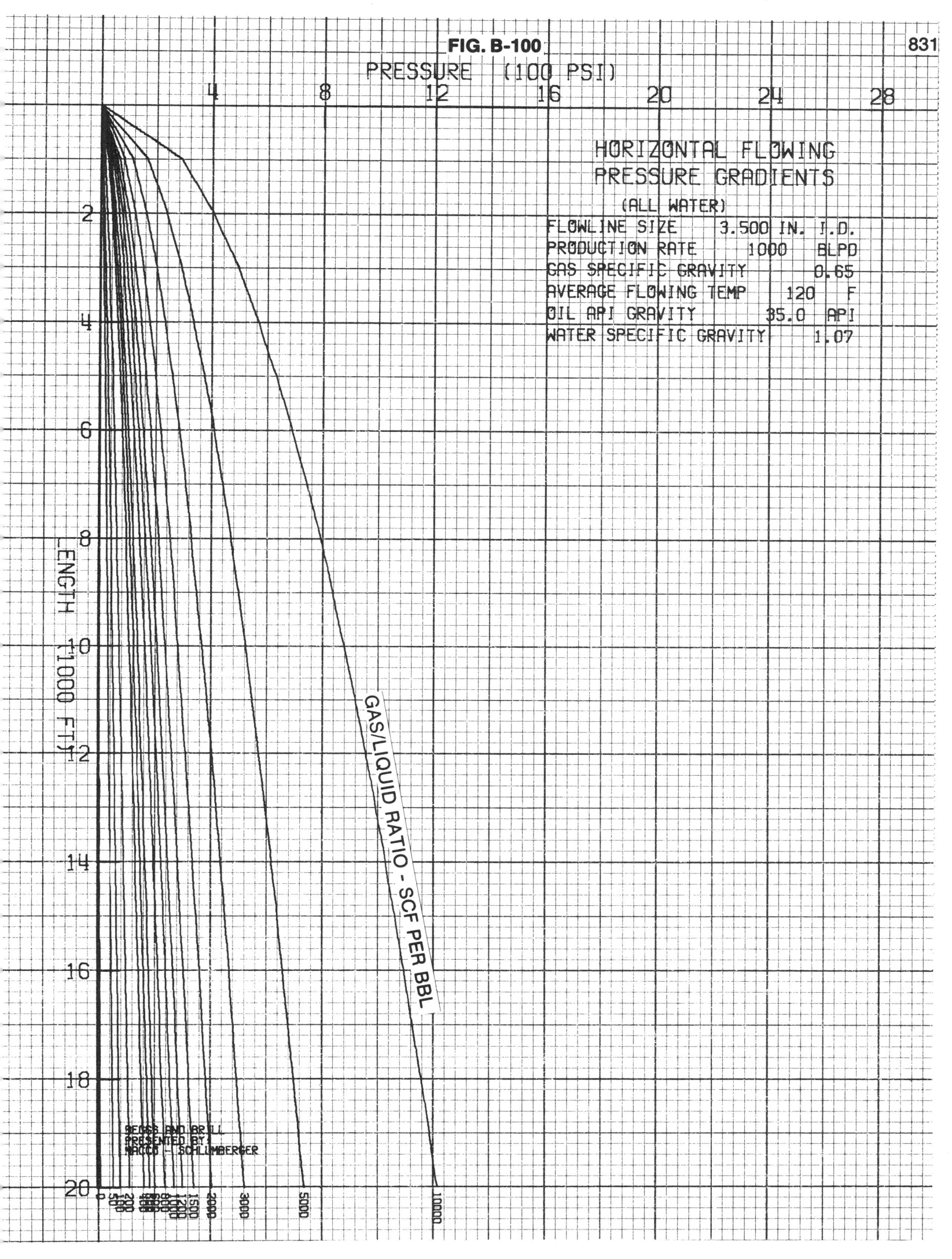

FIG. B-100

831

PRESSURE (100 PSI)

HORIZONTAL FLOWING
PRESSURE GRADIENTS
(ALL WATER)

FLOWLINE SIZE   3.500 IN. I.D.
PRODUCTION RATE   1000   BLPD
GAS SPECIFIC GRAVITY   0.65
AVERAGE FLOWING TEMP   120   F
OIL API GRAVITY   35.0 API
WATER SPECIFIC GRAVITY   1.07

LENGTH (1000 FT)

GAS/LIQUID RATIO - SCF PER BBL

BEGGS AND BRILL
PRESENTED BY:
NACCO - SCHLUMBERGER

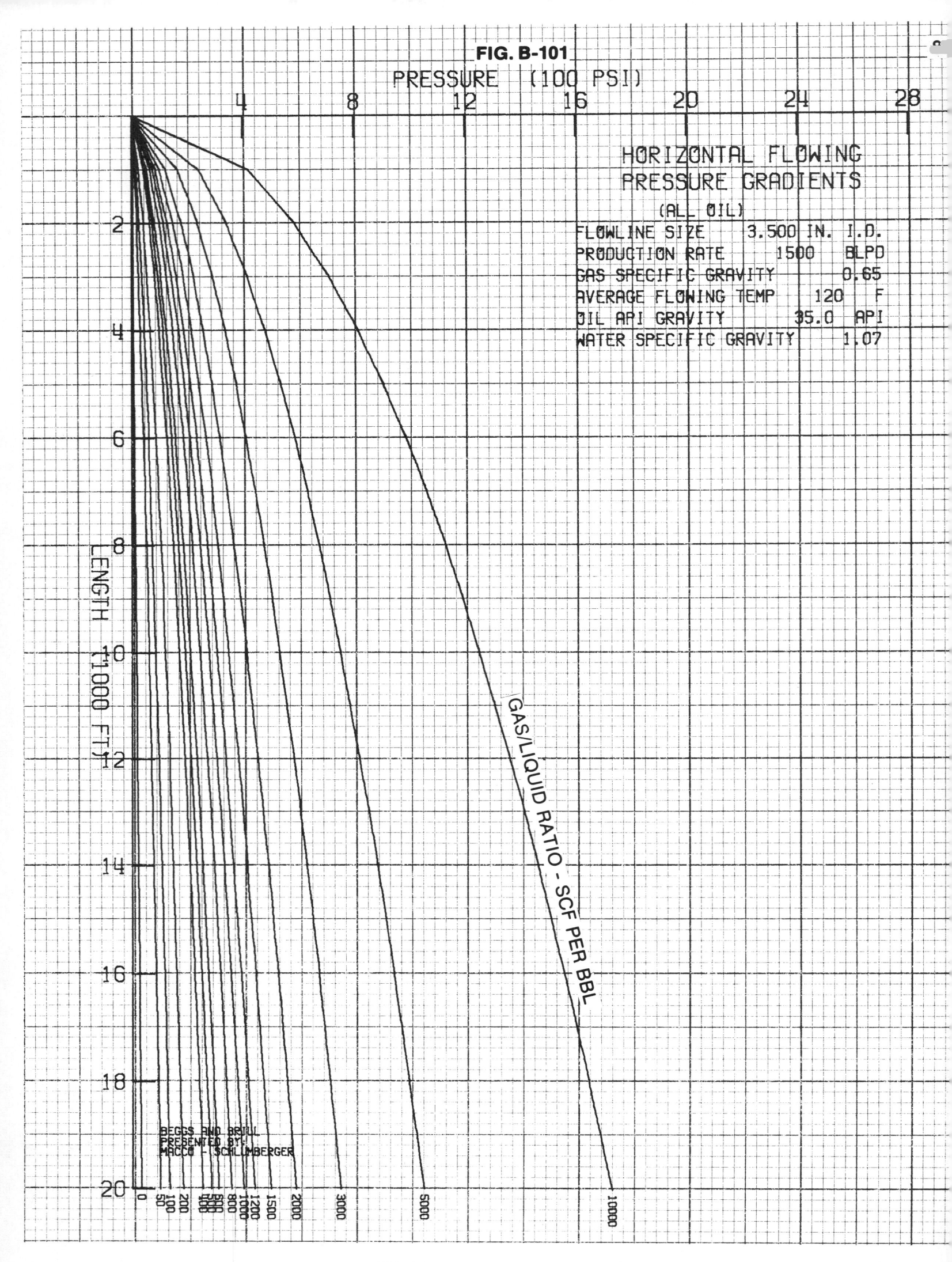

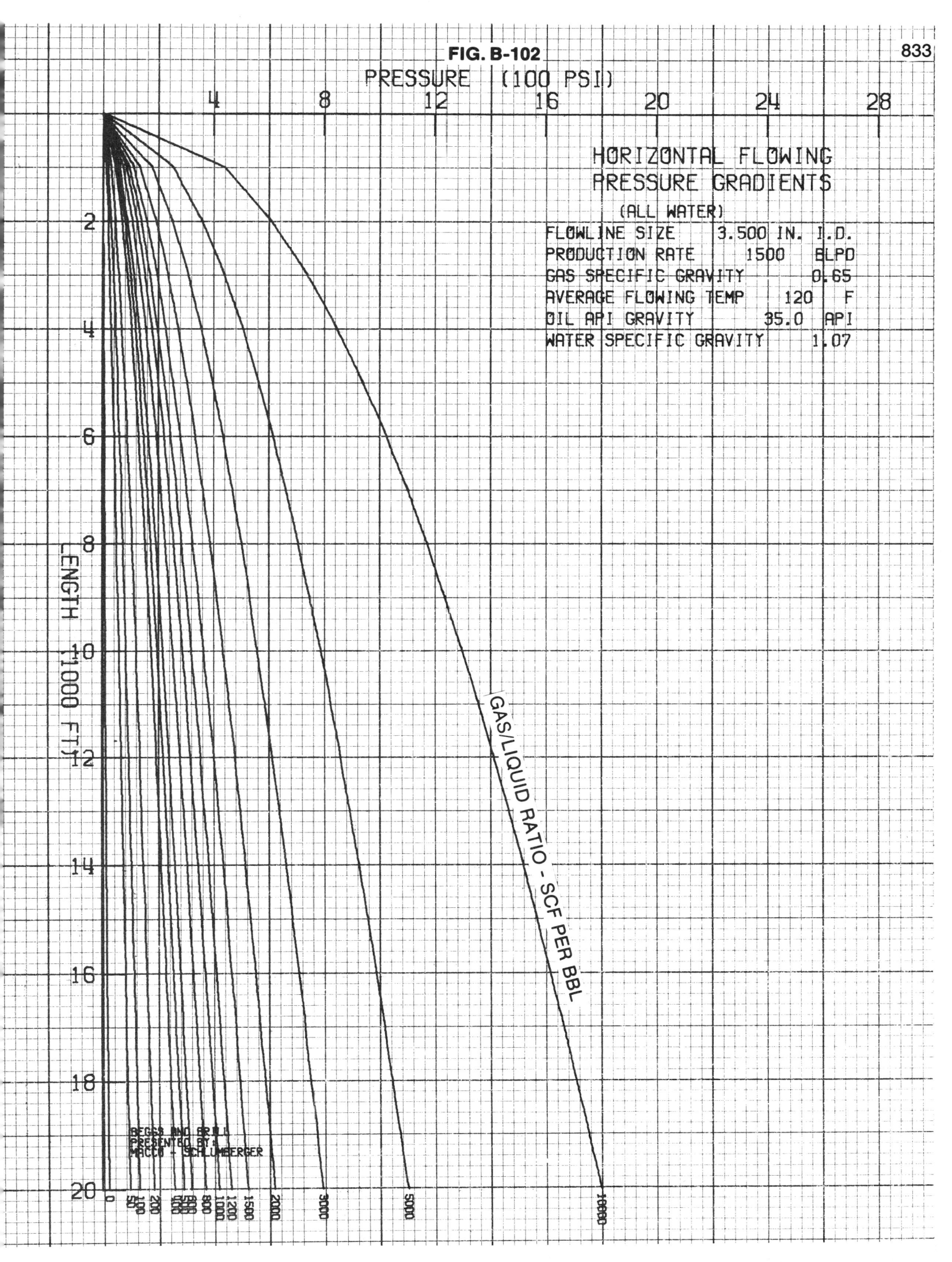

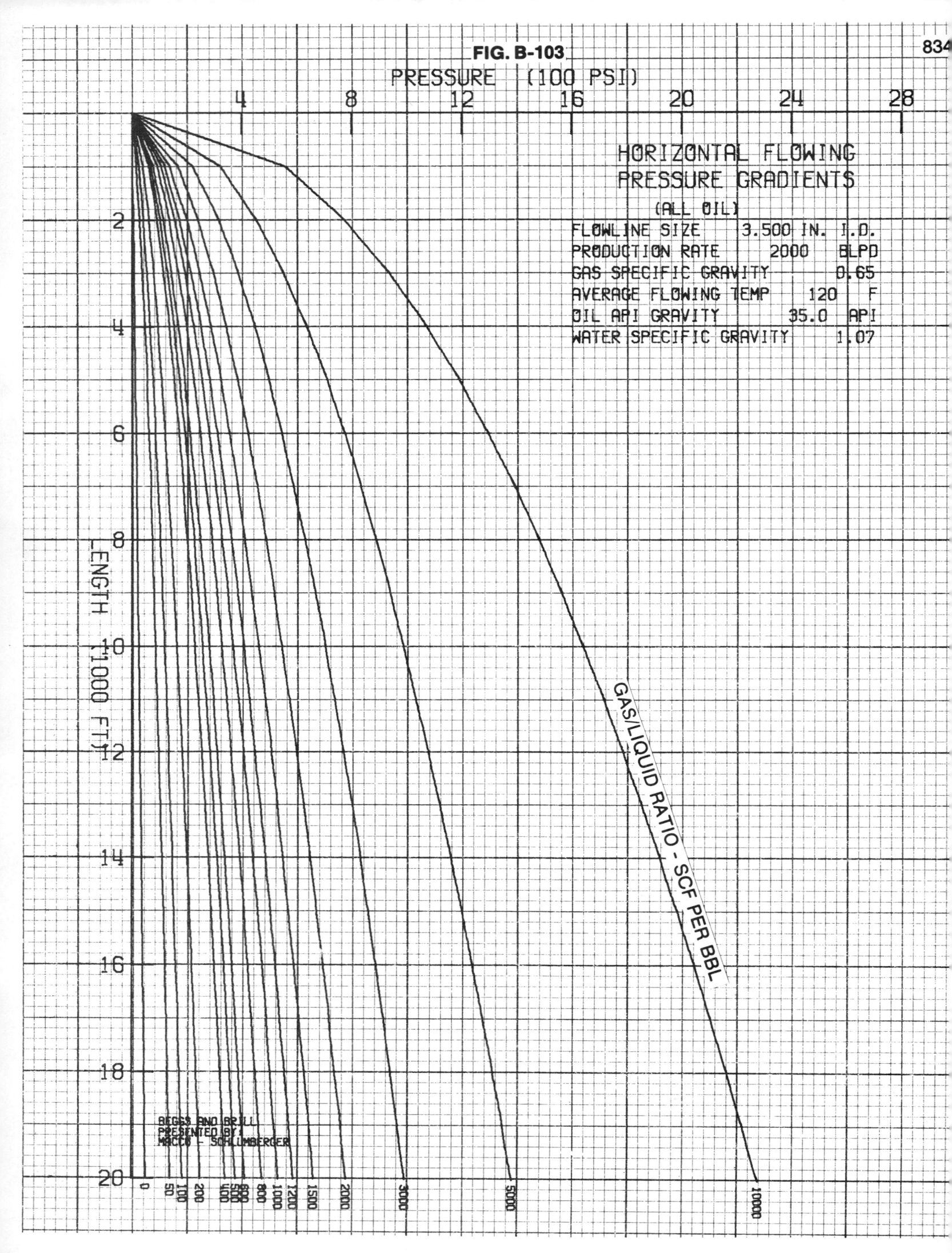

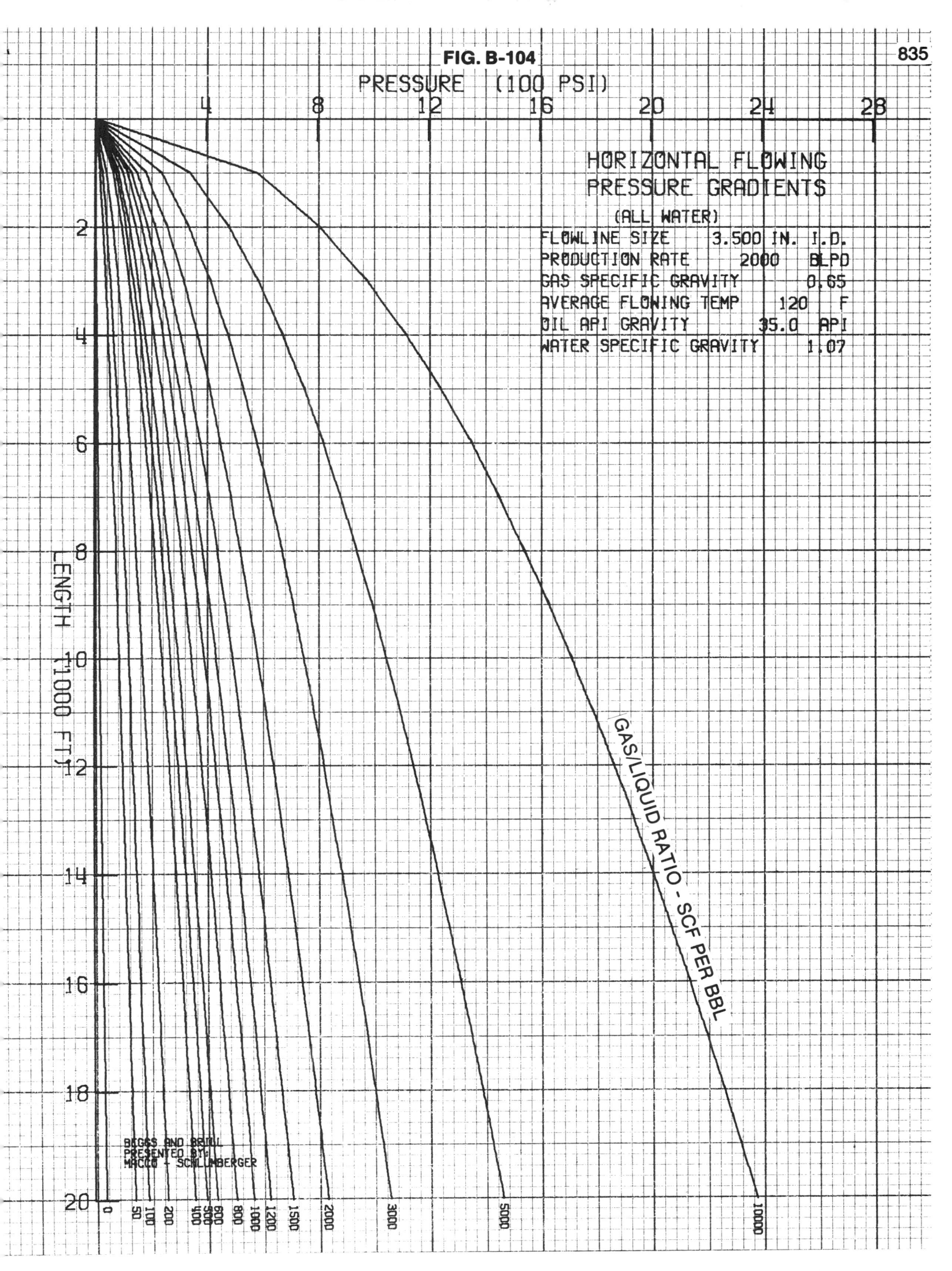

PRESSURE (100 PSI)

HORIZONTAL FLOWING
PRESSURE GRADIENTS
(ALL WATER)

FLOWLINE SIZE	3.500 IN.	I.D.
PRODUCTION RATE	2000	BLPD
GAS SPECIFIC GRAVITY	0.65	
AVERAGE FLOWING TEMP	120	F
OIL API GRAVITY	35.0	API
WATER SPECIFIC GRAVITY	1.07	

LENGTH (1000 FT)

GAS/LIQUID RATIO - SCF PER BBL

BEGGS AND BRILL
PRESENTED BY:
MACCO - SCHLUMBERGER

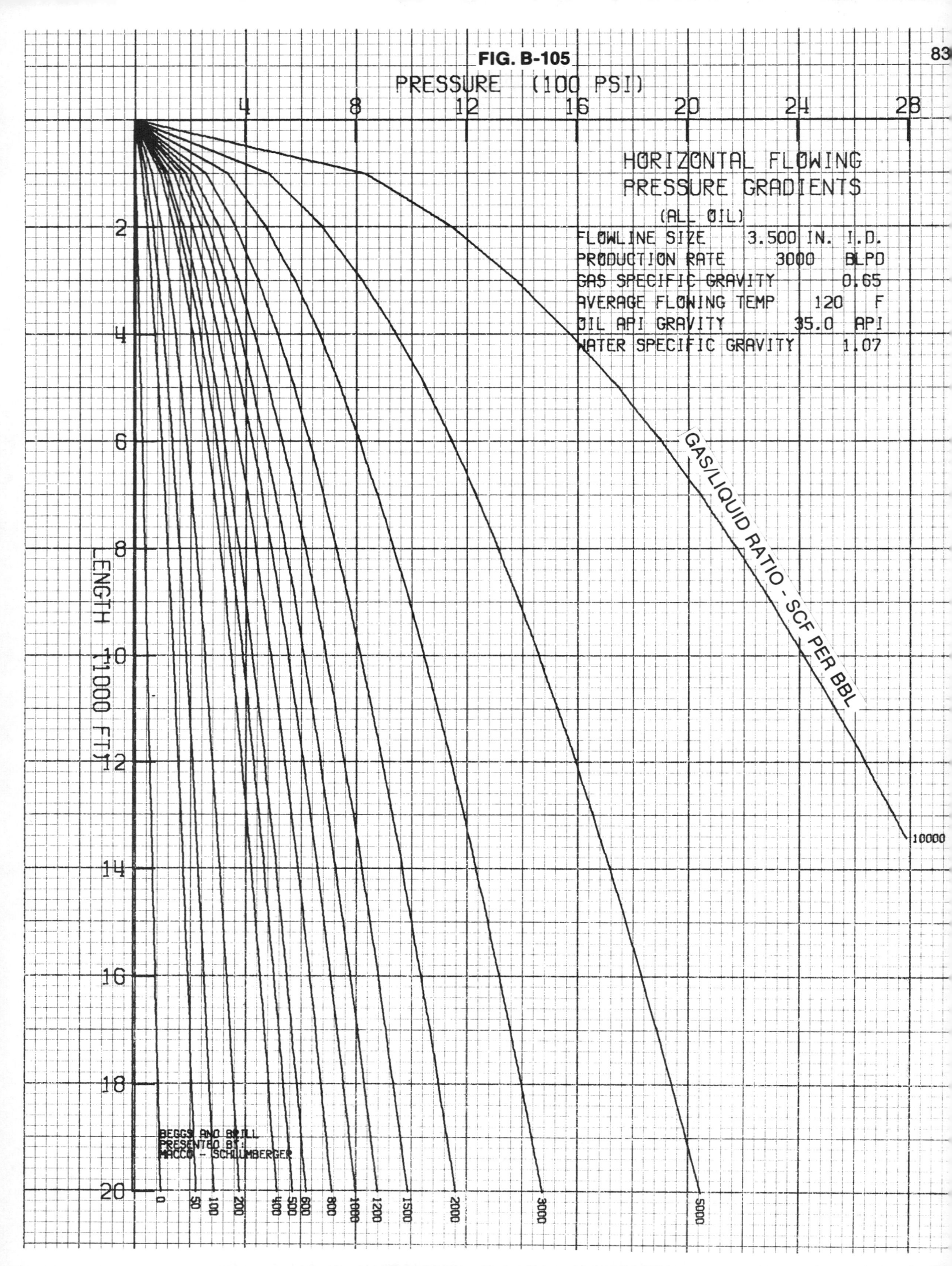

FIG. B-105

HORIZONTAL FLOWING
PRESSURE GRADIENTS
(ALL OIL)

FLOWLINE SIZE	3.500 IN. I.D.
PRODUCTION RATE	3000 BLPD
GAS SPECIFIC GRAVITY	0.65
AVERAGE FLOWING TEMP	120 F
OIL API GRAVITY	35.0 API
WATER SPECIFIC GRAVITY	1.07

GAS/LIQUID RATIO - SCF PER BBL

PRESSURE (100 PSI)

LENGTH (1000 FT)

BEGGS AND BRILL
PRESENTED BY:
MACCO - SCHLUMBERGER

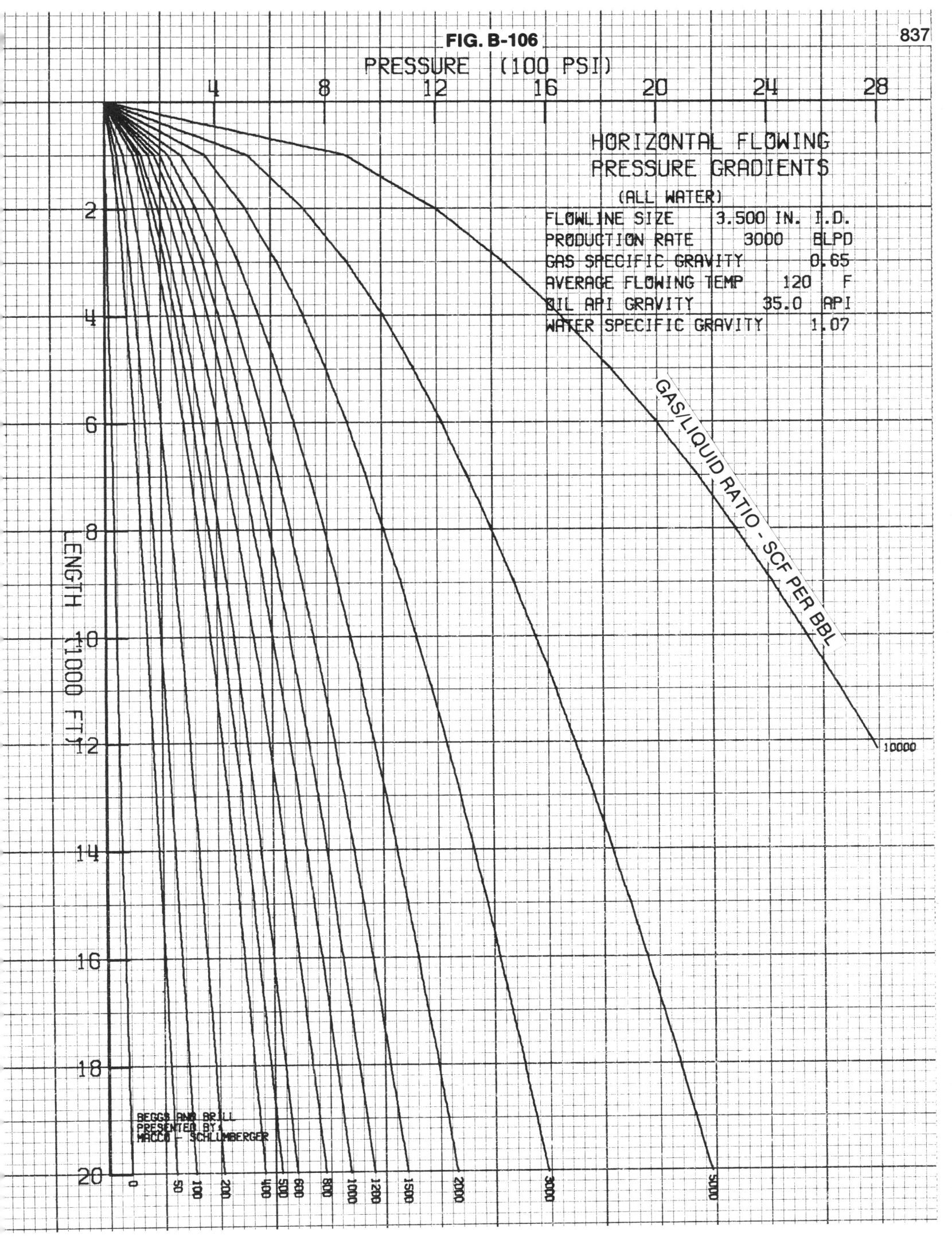

FIG. B-106

837

HORIZONTAL FLOWING
PRESSURE GRADIENTS
(ALL WATER)

FLOWLINE SIZE   3.500 IN. I.D.
PRODUCTION RATE   3000   BLPD
GAS SPECIFIC GRAVITY   0.65
AVERAGE FLOWING TEMP   120   F
OIL API GRAVITY   35.0   API
WATER SPECIFIC GRAVITY   1.07

GAS/LIQUID RATIO - SCF PER BBL

BEGGS AND BRILL
PRESENTED BY:
MACCO - SCHLUMBERGER

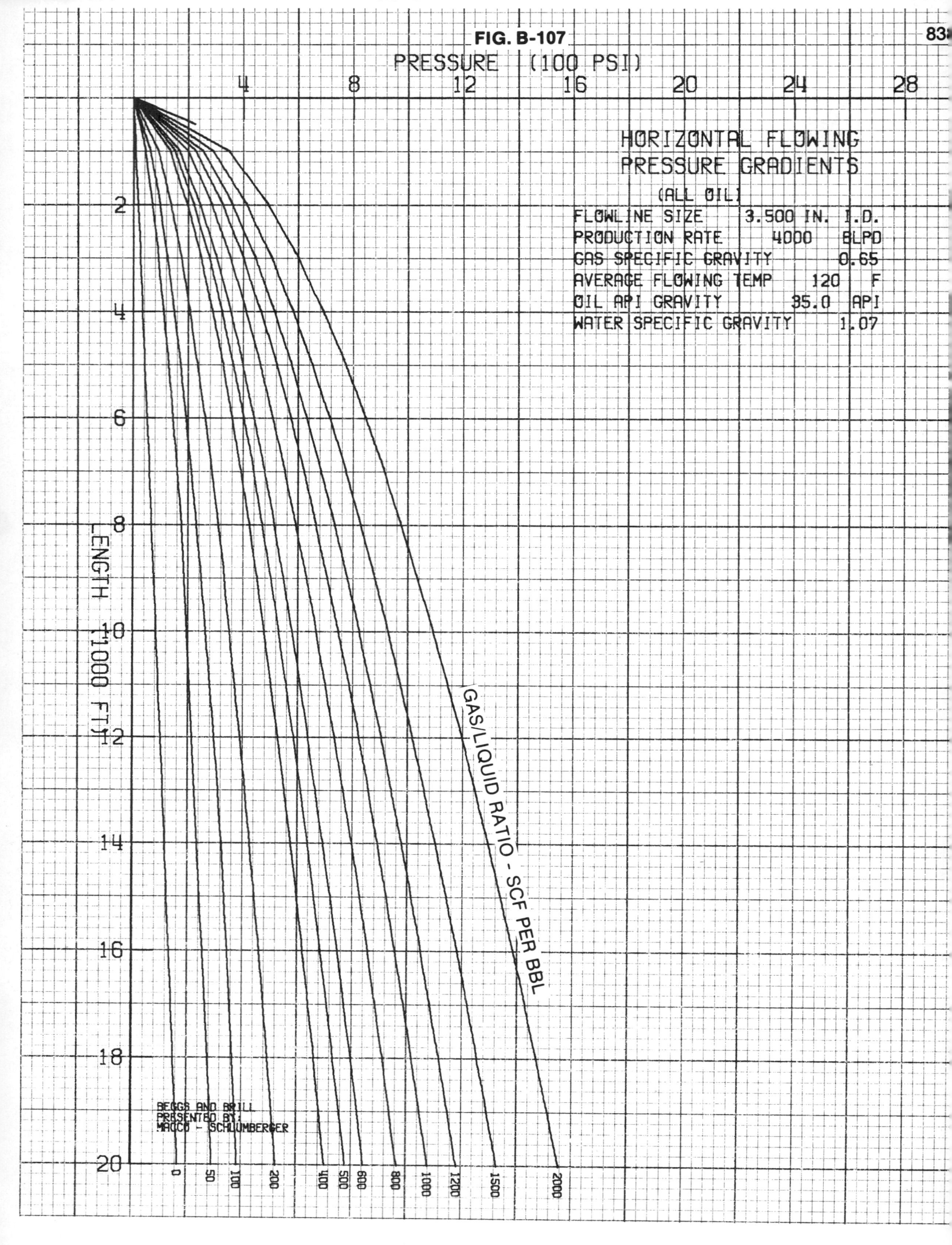

PRESSURE (100 PSI)

HORIZONTAL FLOWING
PRESSURE GRADIENTS

(ALL OIL)

FLOWLINE SIZE	3.500 IN. I.D.
PRODUCTION RATE	4000 BLPD
GAS SPECIFIC GRAVITY	0.65
AVERAGE FLOWING TEMP	120 F
OIL API GRAVITY	35.0 API
WATER SPECIFIC GRAVITY	1.07

LENGTH (1000 FT.)

GAS/LIQUID RATIO - SCF PER BBL

BEGGS AND BRILL
PRESENTED BY:
MAGCO - SCHLUMBERGER

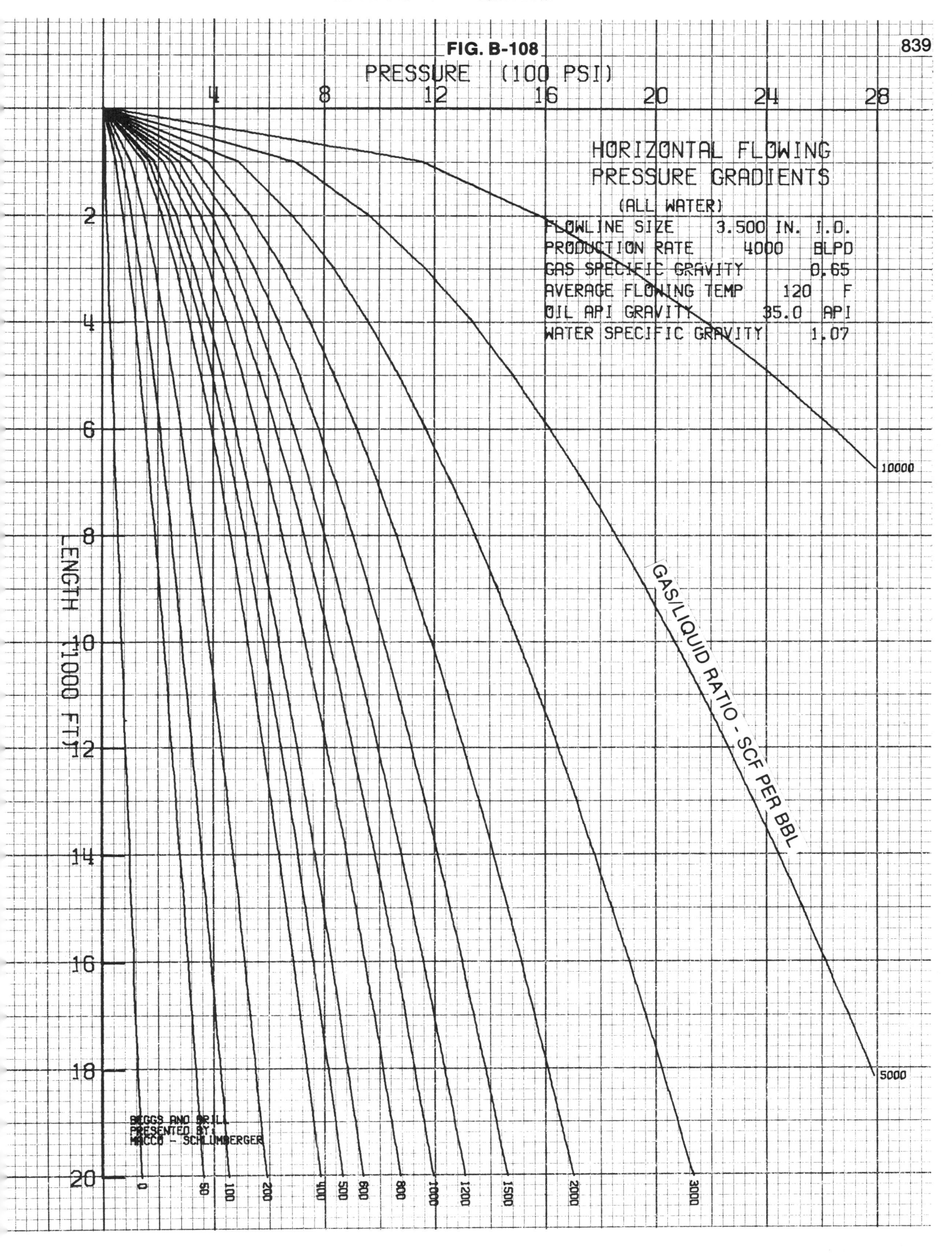

FIG. B-108

839

HORIZONTAL FLOWING
PRESSURE GRADIENTS
(ALL WATER)

FLOWLINE SIZE	3.500 IN. I.D.
PRODUCTION RATE	4000 BLPD
GAS SPECIFIC GRAVITY	0.65
AVERAGE FLOWING TEMP	120 F
OIL API GRAVITY	35.0 API
WATER SPECIFIC GRAVITY	1.07

PRESSURE (100 PSI)

LENGTH (1000 FT)

GAS/LIQUID RATIO - SCF PER BBL

BEGGS AND BRILL
PRESENTED BY:
MACCO - SCHLUMBERGER

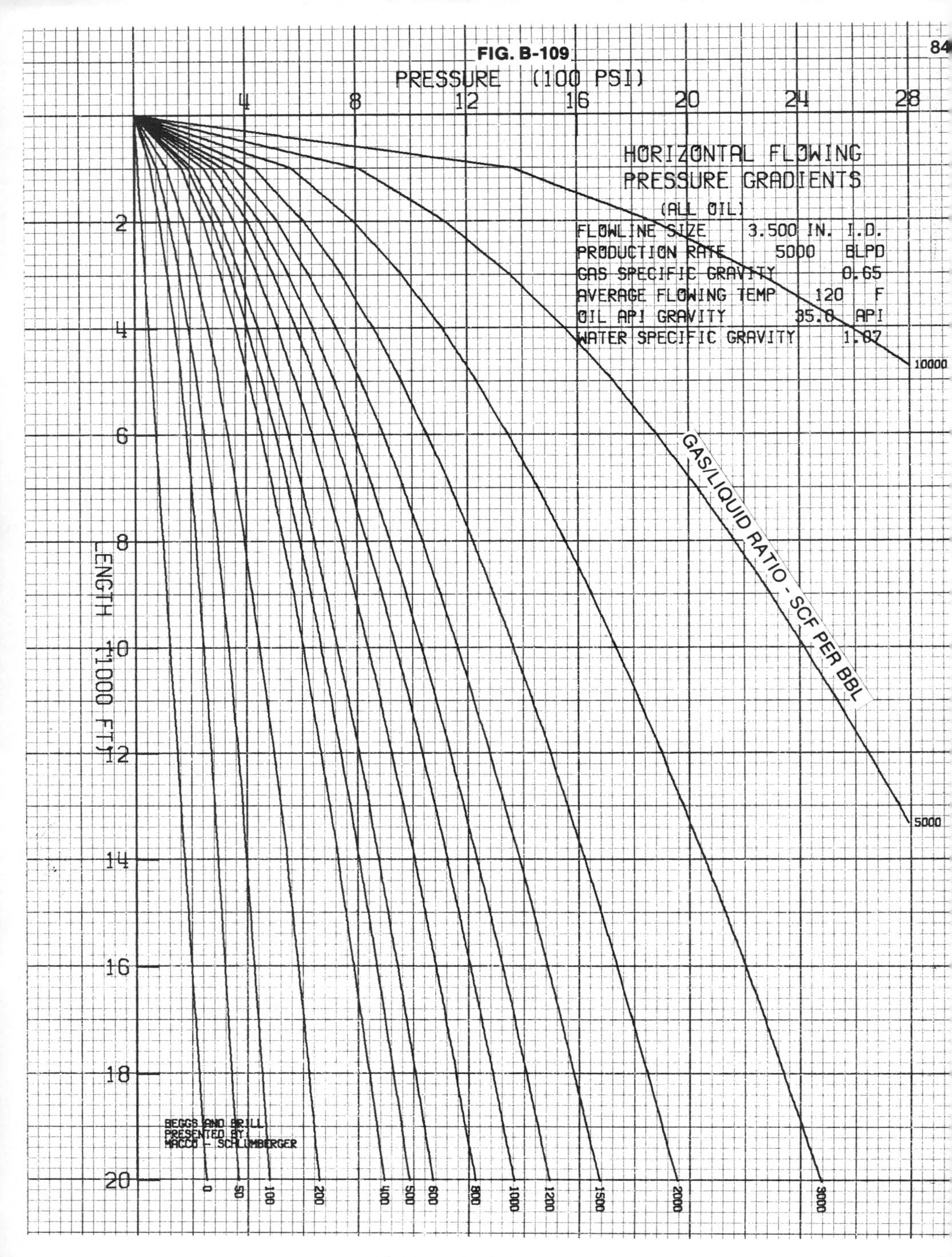

PRESSURE (100 PSI)

HORIZONTAL FLOWING
PRESSURE GRADIENTS
(ALL OIL)

FLOWLINE SIZE	3.500 IN. I.D.
PRODUCTION RATE	5000 BLPD
GAS SPECIFIC GRAVITY	0.65
AVERAGE FLOWING TEMP	120 F
OIL API GRAVITY	35.0 API
WATER SPECIFIC GRAVITY	1.07

GAS/LIQUID RATIO - SCF PER BBL

LENGTH (1000 FT.)

BEGGS AND BRILL
PRESENTED BY
MACCO - SCHLUMBERGER

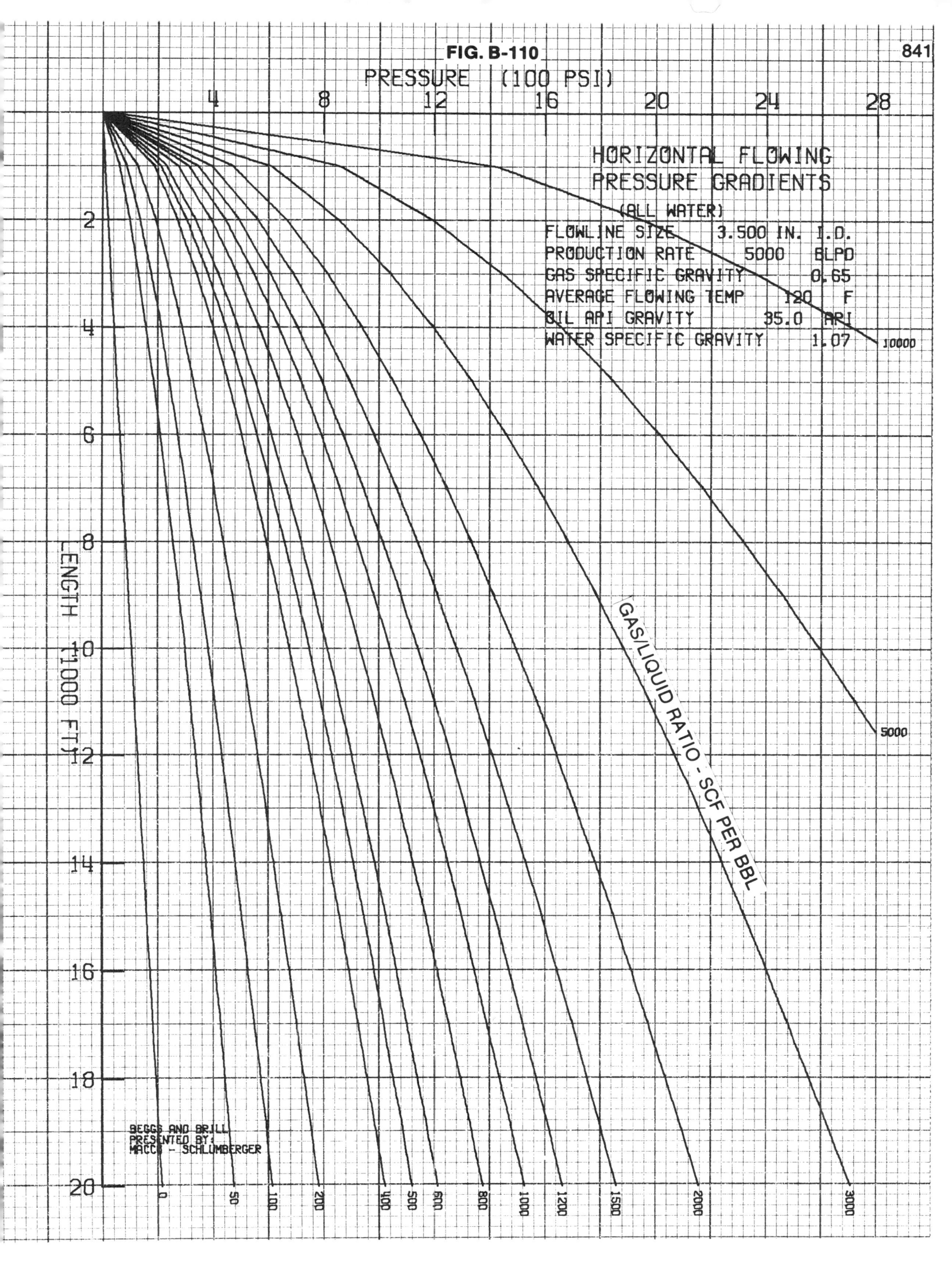

PRESSURE (100 PSI)

HORIZONTAL FLOWING
PRESSURE GRADIENTS
(ALL WATER)

FLOWLINE SIZE	3.500 IN. I.D.
PRODUCTION RATE	5000 BLPD
GAS SPECIFIC GRAVITY	0.65
AVERAGE FLOWING TEMP	120 F
OIL API GRAVITY	35.0 API
WATER SPECIFIC GRAVITY	1.07

LENGTH (1000 FT.)

GAS/LIQUID RATIO - SCF PER BBL

BEGGS AND BRILL
PRESENTED BY:
MACCO - SCHLUMBERGER

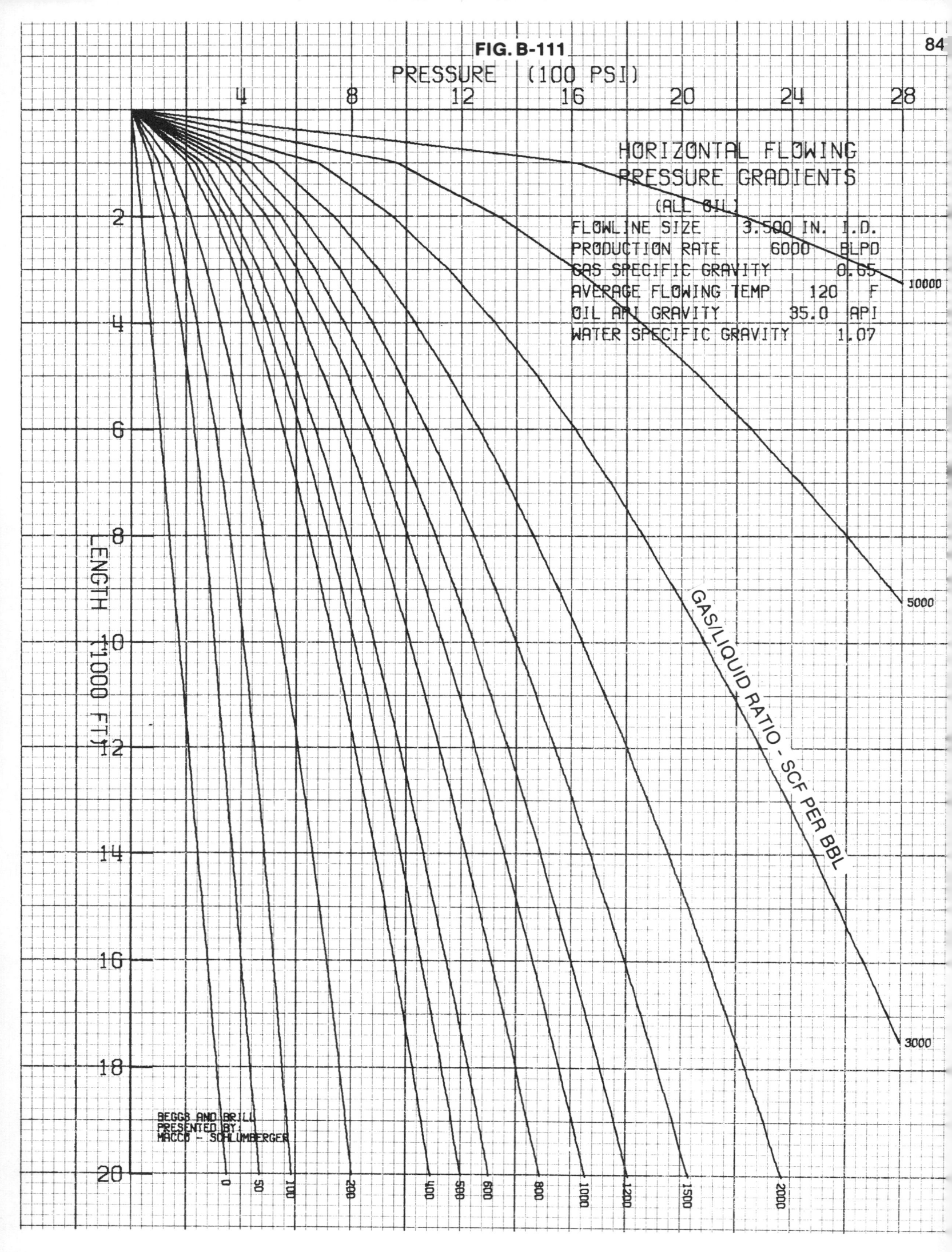

PRESSURE (100 PSI)

HORIZONTAL FLOWING
PRESSURE GRADIENTS
(ALL OIL)

FLOWLINE SIZE	3.500 IN. I.D.	
PRODUCTION RATE	6000	BLPD
GAS SPECIFIC GRAVITY	0.65	
AVERAGE FLOWING TEMP	120	F
OIL API GRAVITY	35.0	API
WATER SPECIFIC GRAVITY	1.07	

LENGTH (1000 FT.)

GAS/LIQUID RATIO - SCF PER BBL

BEGGS AND BRILL
PRESENTED BY:
MACCO - SCHLUMBERGER

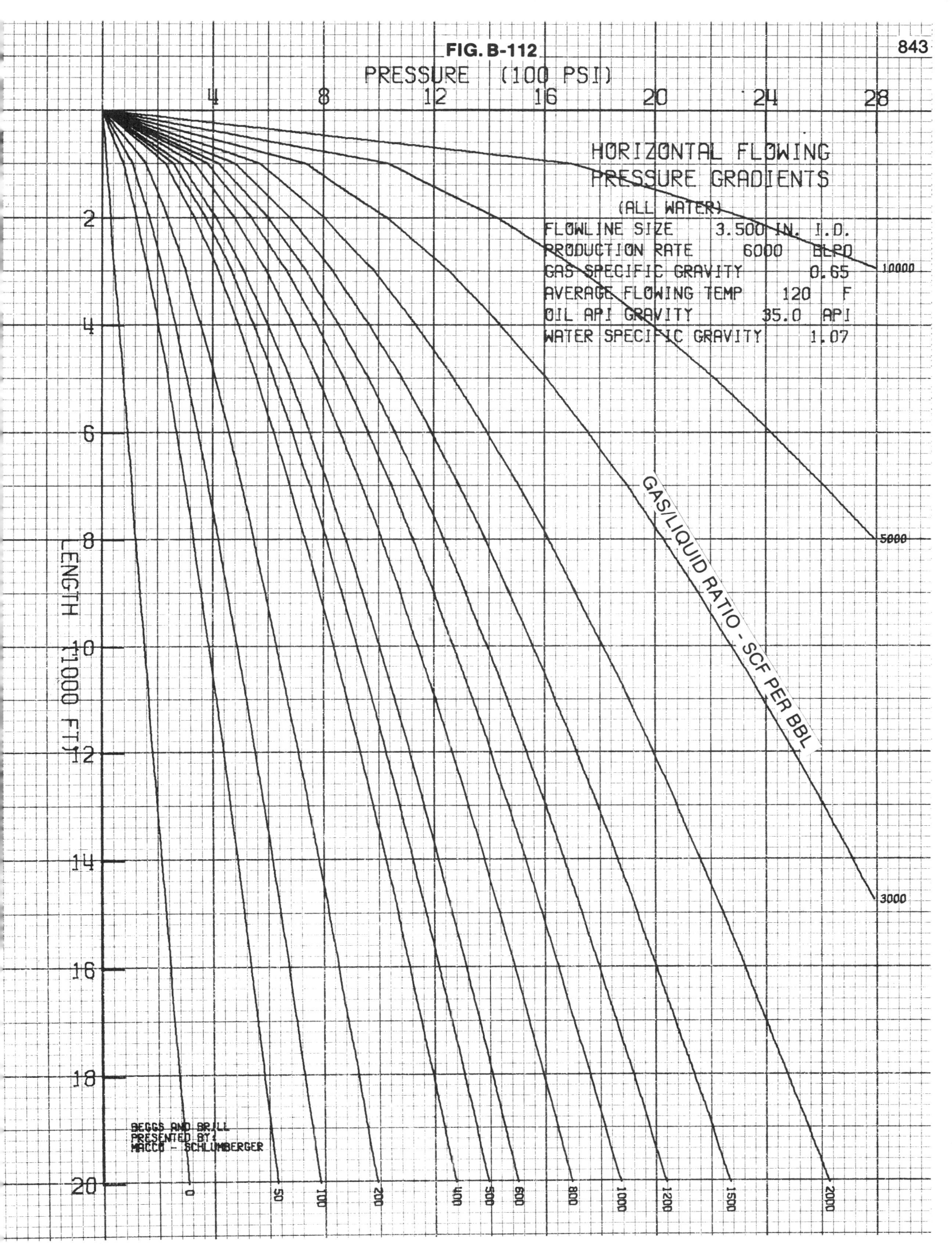

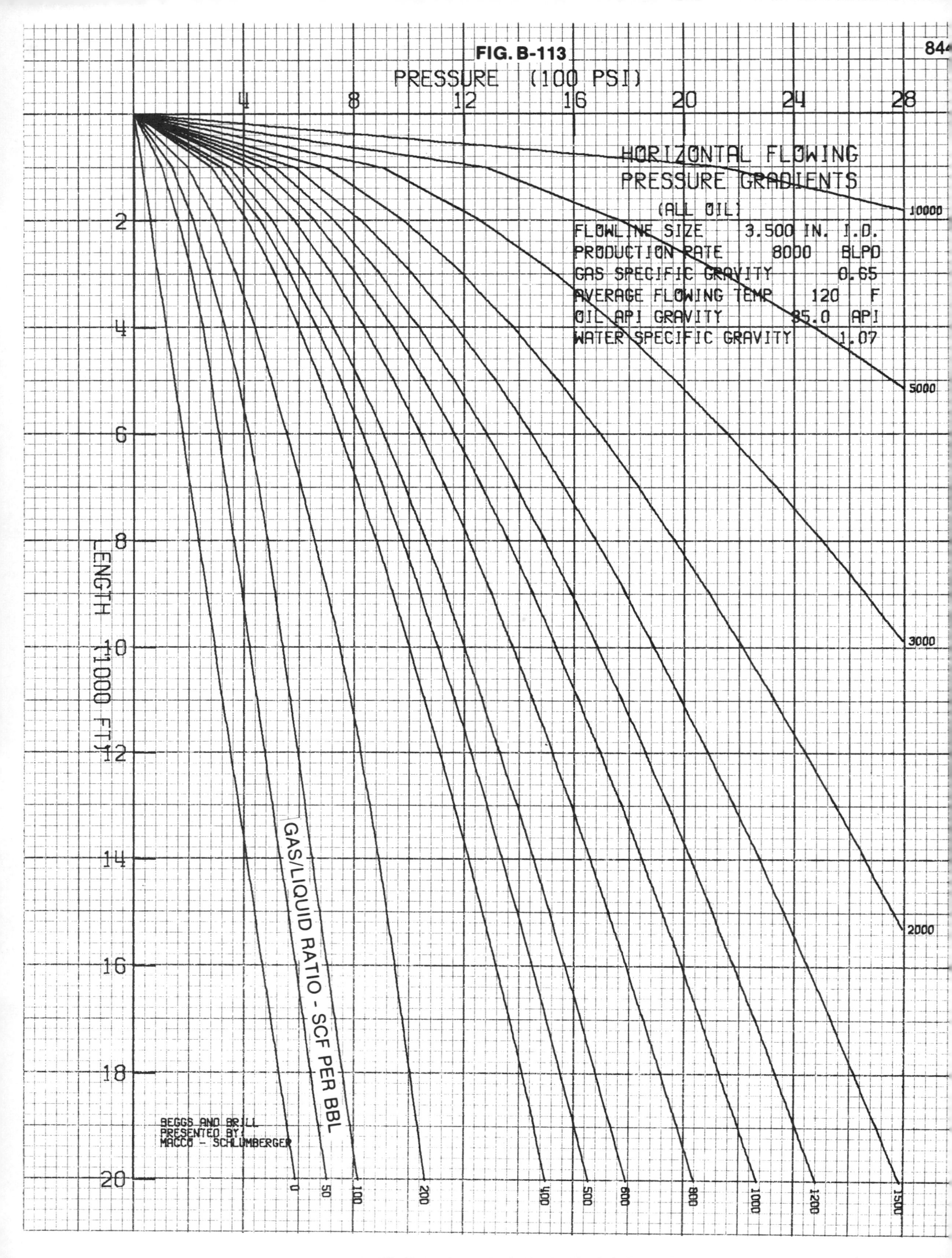

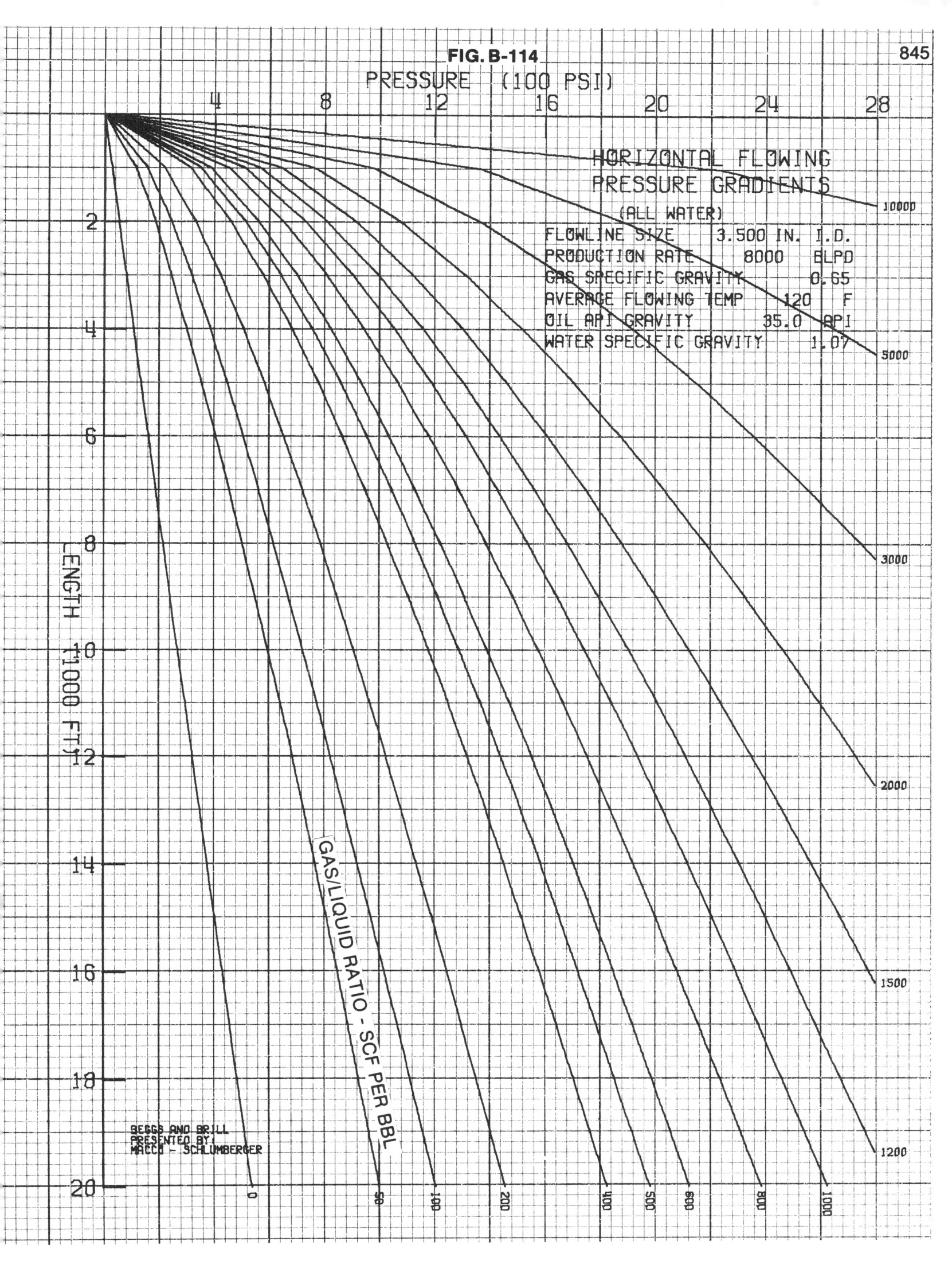

FIG. B-114

845

PRESSURE (100 PSI)

HORIZONTAL FLOWING
PRESSURE GRADIENTS
(ALL WATER)

FLOWLINE SIZE	3.500 IN. I.D.
PRODUCTION RATE	8000 BLPD
GAS SPECIFIC GRAVITY	0.65
AVERAGE FLOWING TEMP	120 F
OIL API GRAVITY	35.0 API
WATER SPECIFIC GRAVITY	1.07

LENGTH (1000 FT)

GAS/LIQUID RATIO - SCF PER BBL

BEGGS AND BRILL
PRESENTED BY:
MACCO - SCHLUMBERGER

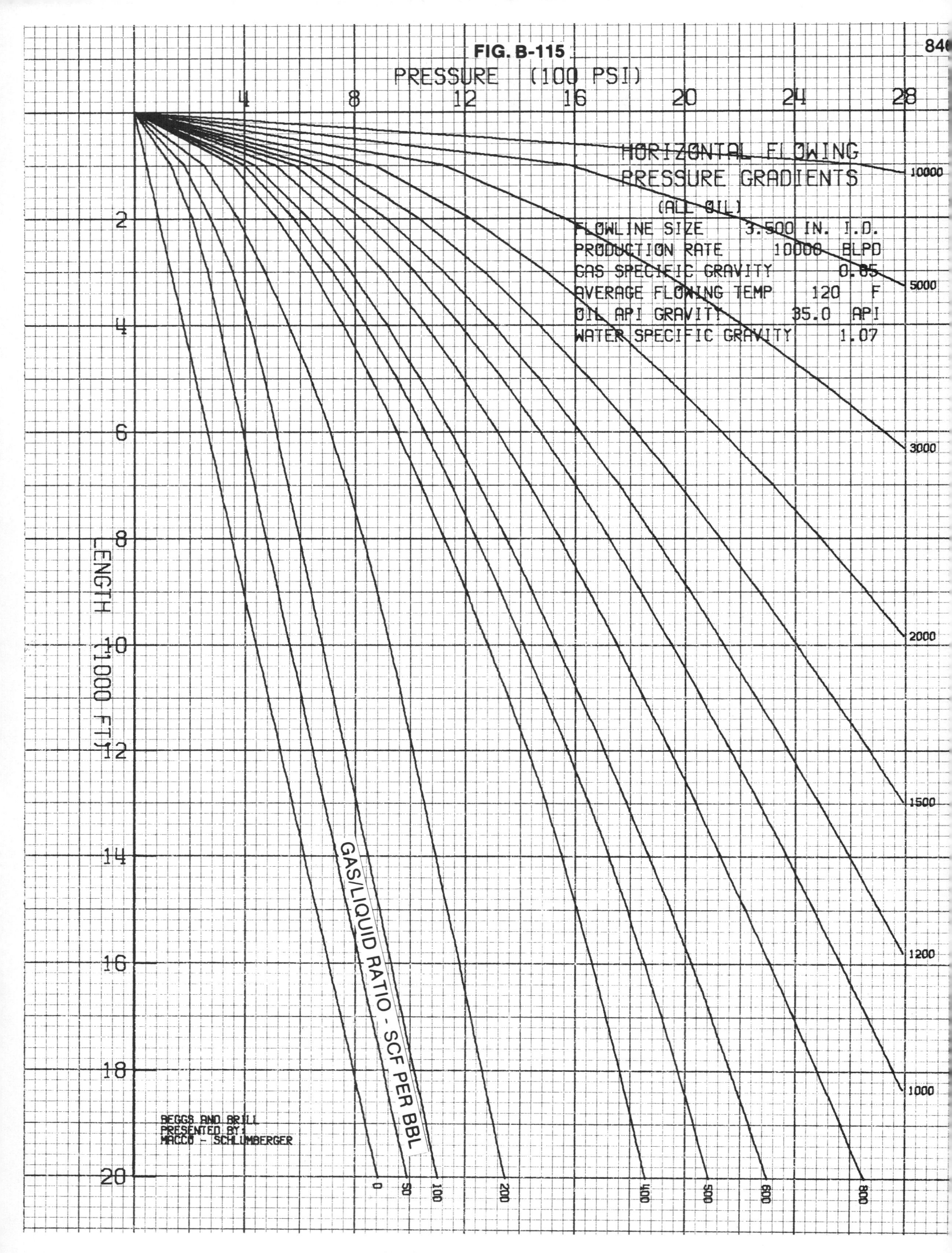

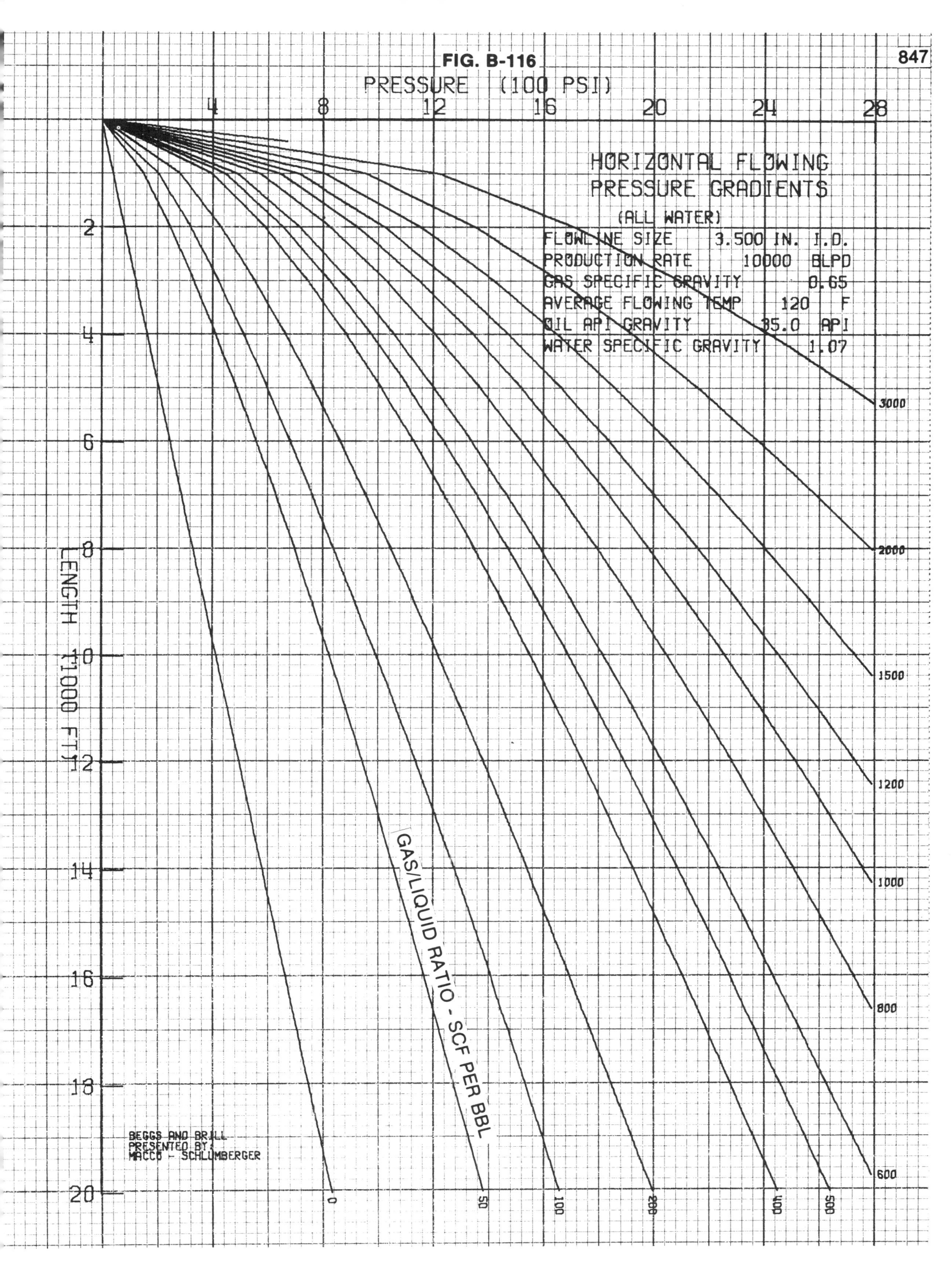

FIG. B-116

PRESSURE (100 PSI)

HORIZONTAL FLOWING
PRESSURE GRADIENTS
(ALL WATER)

FLOWLINE SIZE	3.500 IN. I.D.
PRODUCTION RATE	10000 BLPD
GAS SPECIFIC GRAVITY	0.65
AVERAGE FLOWING TEMP	120 F
OIL API GRAVITY	35.0 API
WATER SPECIFIC GRAVITY	1.07

LENGTH (1000 FT.)

GAS/LIQUID RATIO - SCF PER BBL

BEGGS AND BRILL
PRESENTED BY:
MACCO - SCHLUMBERGER

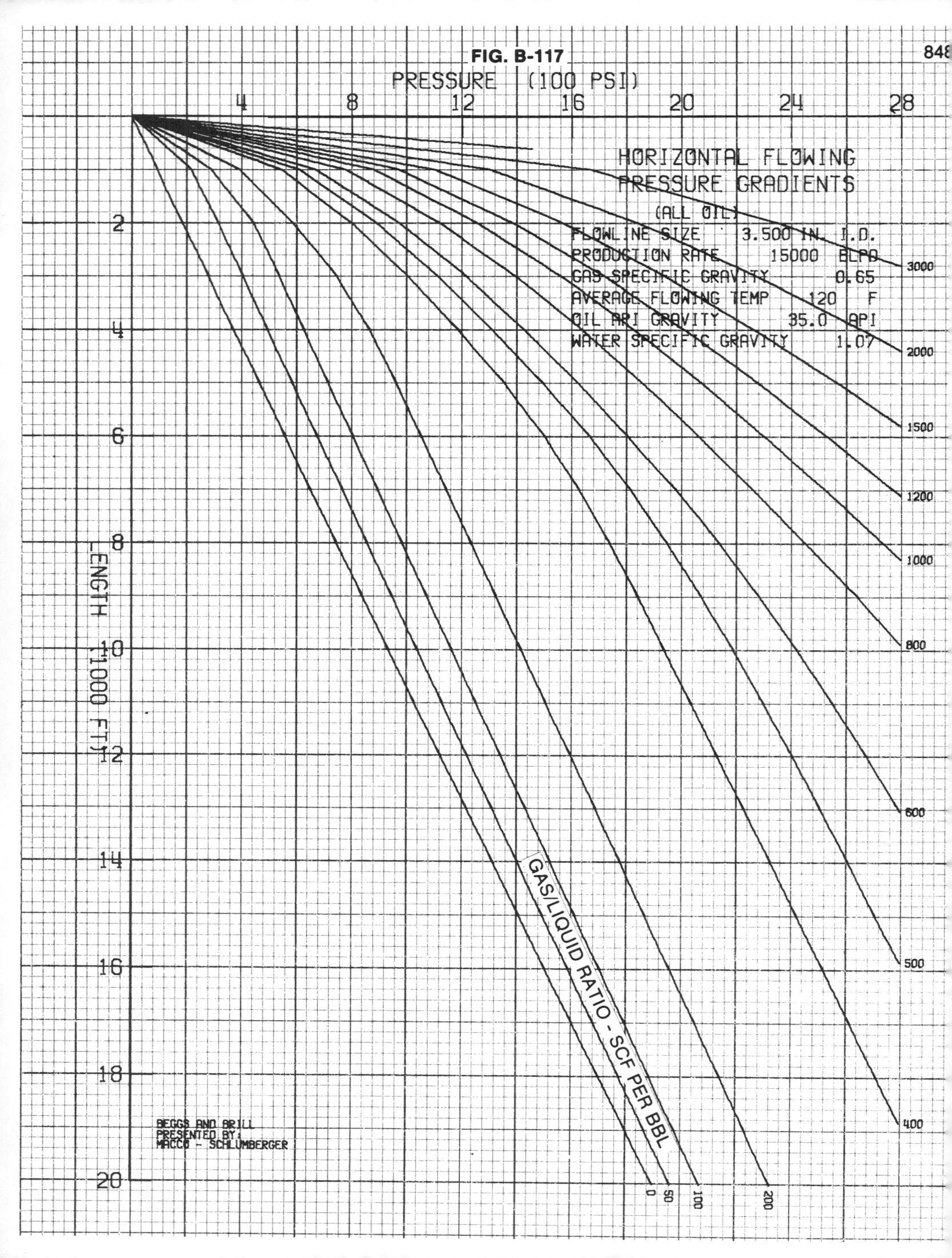

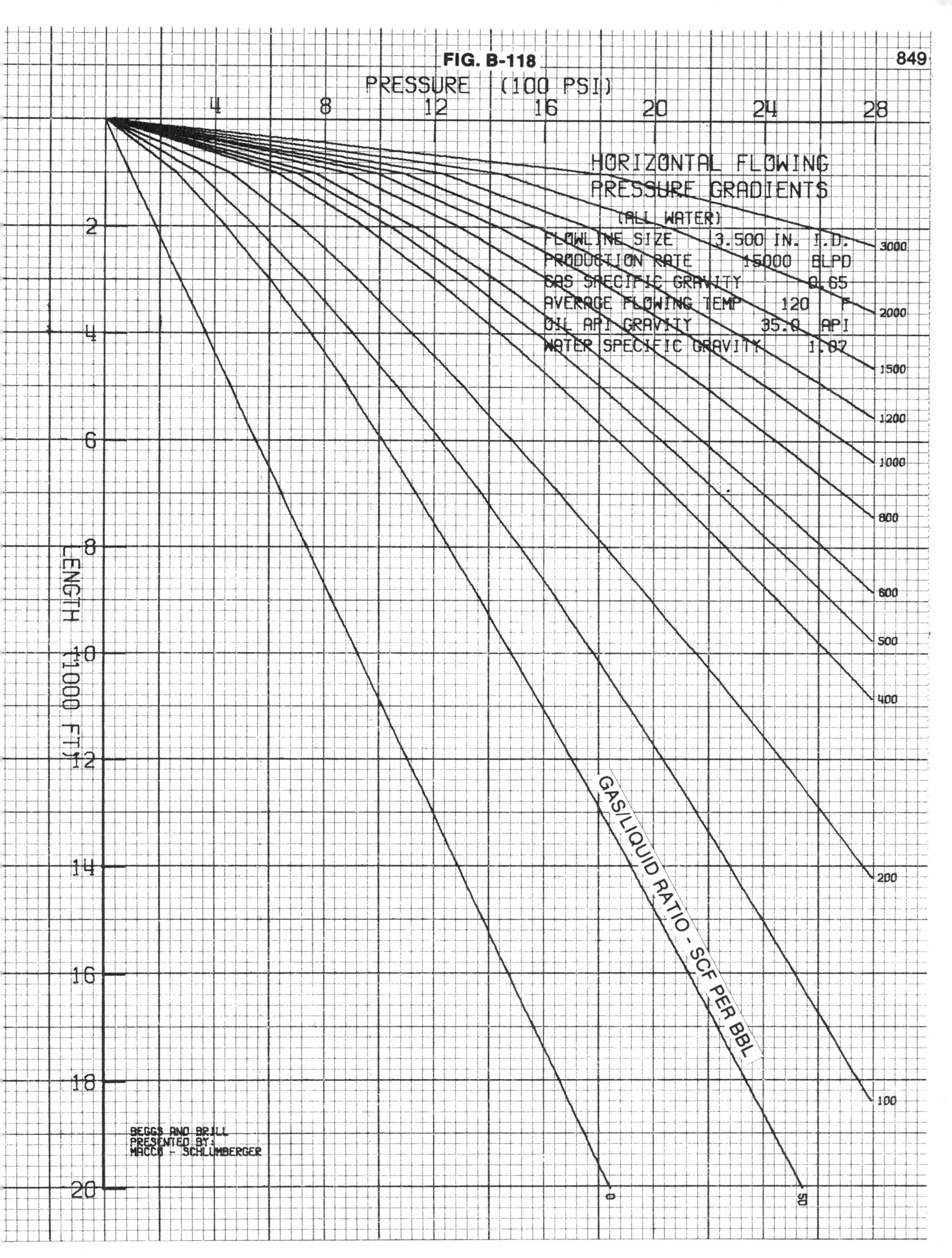

FIG. B-118

849

PRESSURE (100 PSI)

HORIZONTAL FLOWING
PRESSURE GRADIENTS
(ALL WATER)

FLOWLINE SIZE   3.500 IN. I.D.
PRODUCTION RATE   15000 BLPD
GAS SPECIFIC GRAVITY   0.65
AVERAGE FLOWING TEMP   120 F
OIL API GRAVITY   35.0 API
WATER SPECIFIC GRAVITY 1.07

LENGTH (1000 FT)

GAS/LIQUID RATIO - SCF PER BBL

BEGGS AND BRILL
PRESENTED BY:
MACCO - SCHLUMBERGER

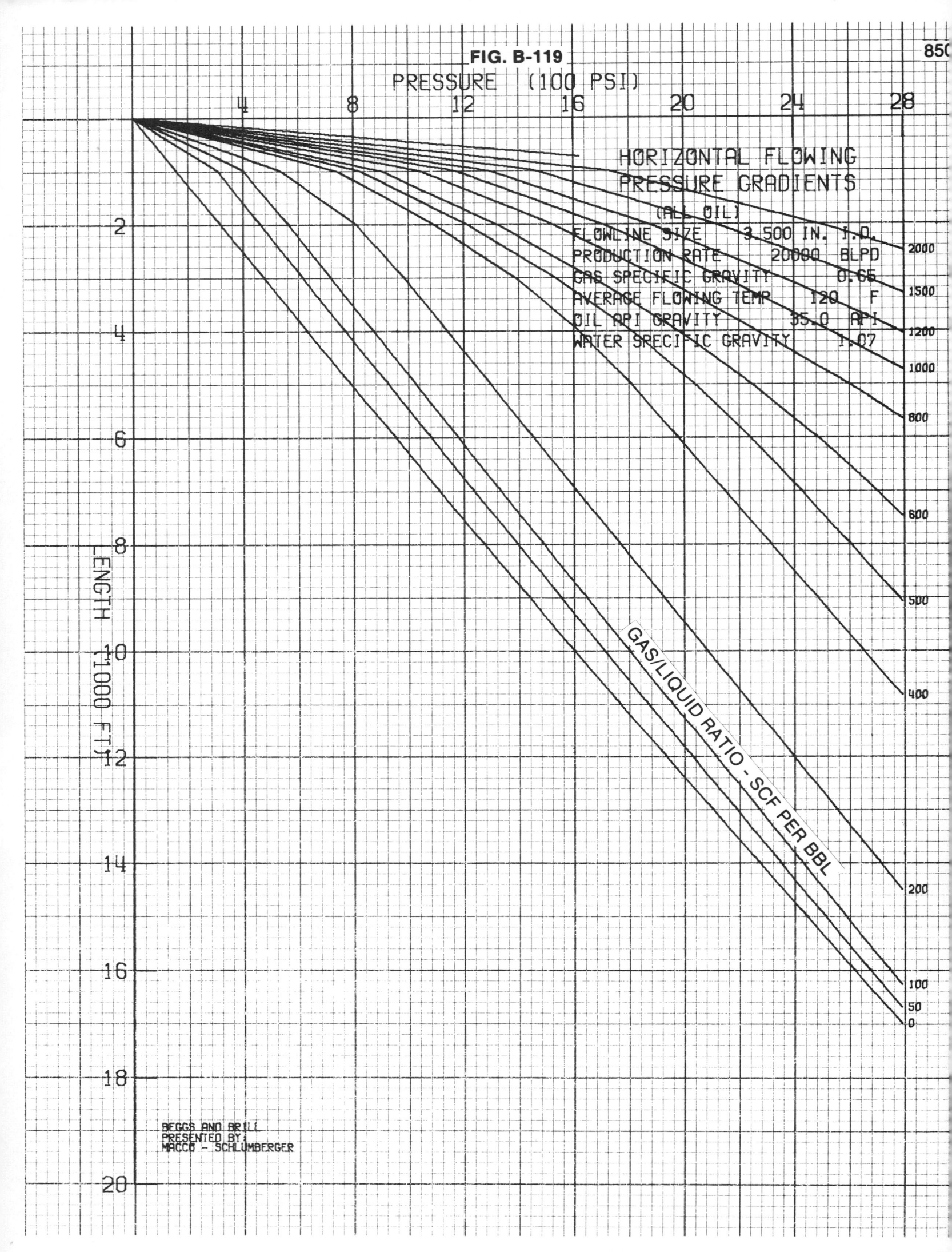

FIG. B-119

850

PRESSURE (100 PSI)

HORIZONTAL FLOWING
PRESSURE GRADIENTS
(ALL OIL)
FLOWLINE SIZE   3.500 IN. I.D.
PRODUCTION RATE   20000 BLPD
GAS SPECIFIC GRAVITY   0.65
AVERAGE FLOWING TEMP   120   F
OIL API GRAVITY   35.0 API
WATER SPECIFIC GRAVITY   1.07

LENGTH (1000 FT)

GAS/LIQUID RATIO - SCF PER BBL

BEGGS AND BRILL
PRESENTED BY:
MACCO - SCHLUMBERGER

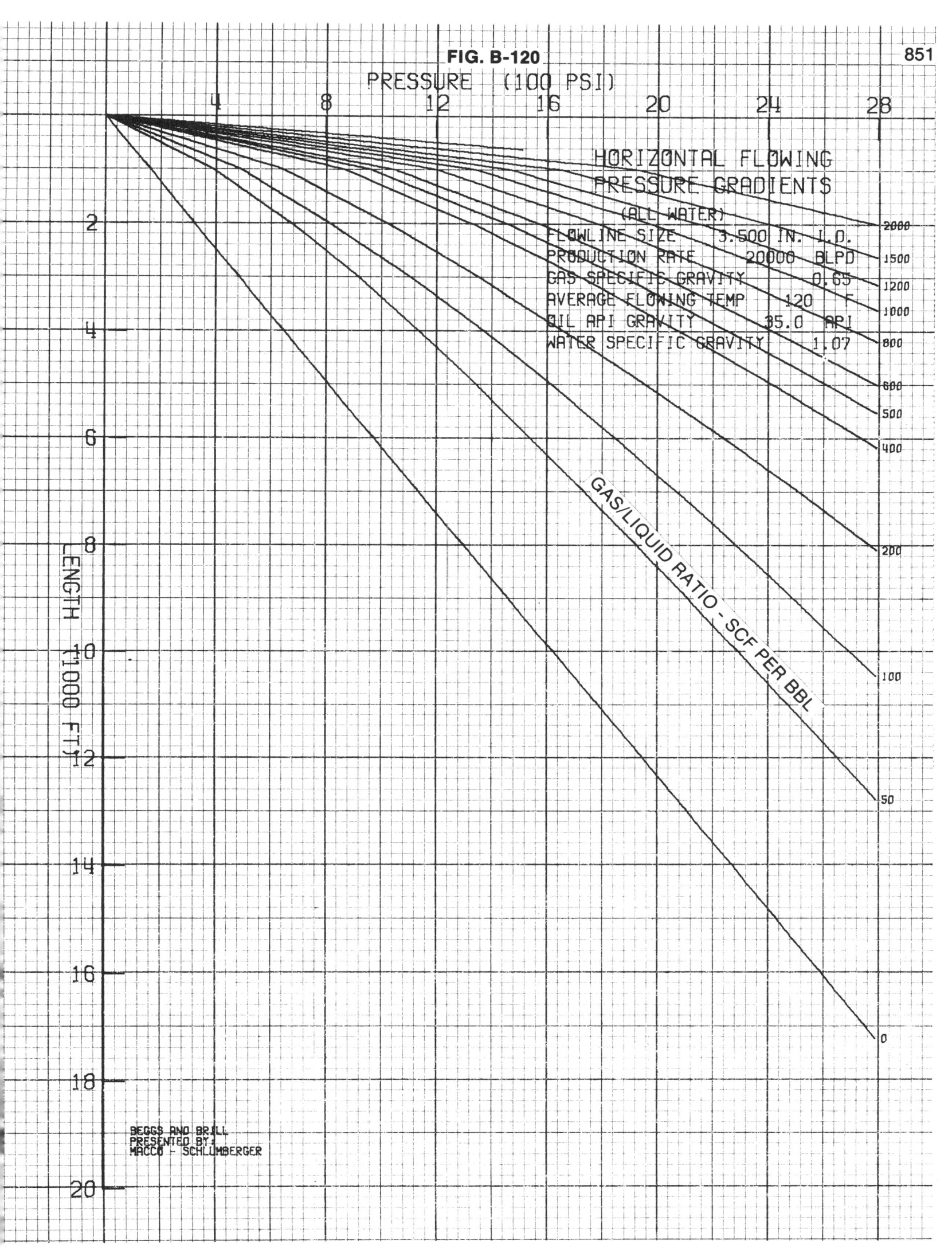

FIG. B-120

851

PRESSURE (100 PSI)

HORIZONTAL FLOWING
PRESSURE GRADIENTS
(ALL WATER)
FLOWLINE SIZE   3.500 IN. I.D.
PRODUCTION RATE   20000 BLPD
GAS SPECIFIC GRAVITY   0.65
AVERAGE FLOWING TEMP   120 F
OIL API GRAVITY   35.0 API
WATER SPECIFIC GRAVITY   1.07

GAS/LIQUID RATIO - SCF PER BBL

LENGTH (1000 FT)

BEGGS AND BRILL
PRESENTED BY:
MACCO - SCHLUMBERGER

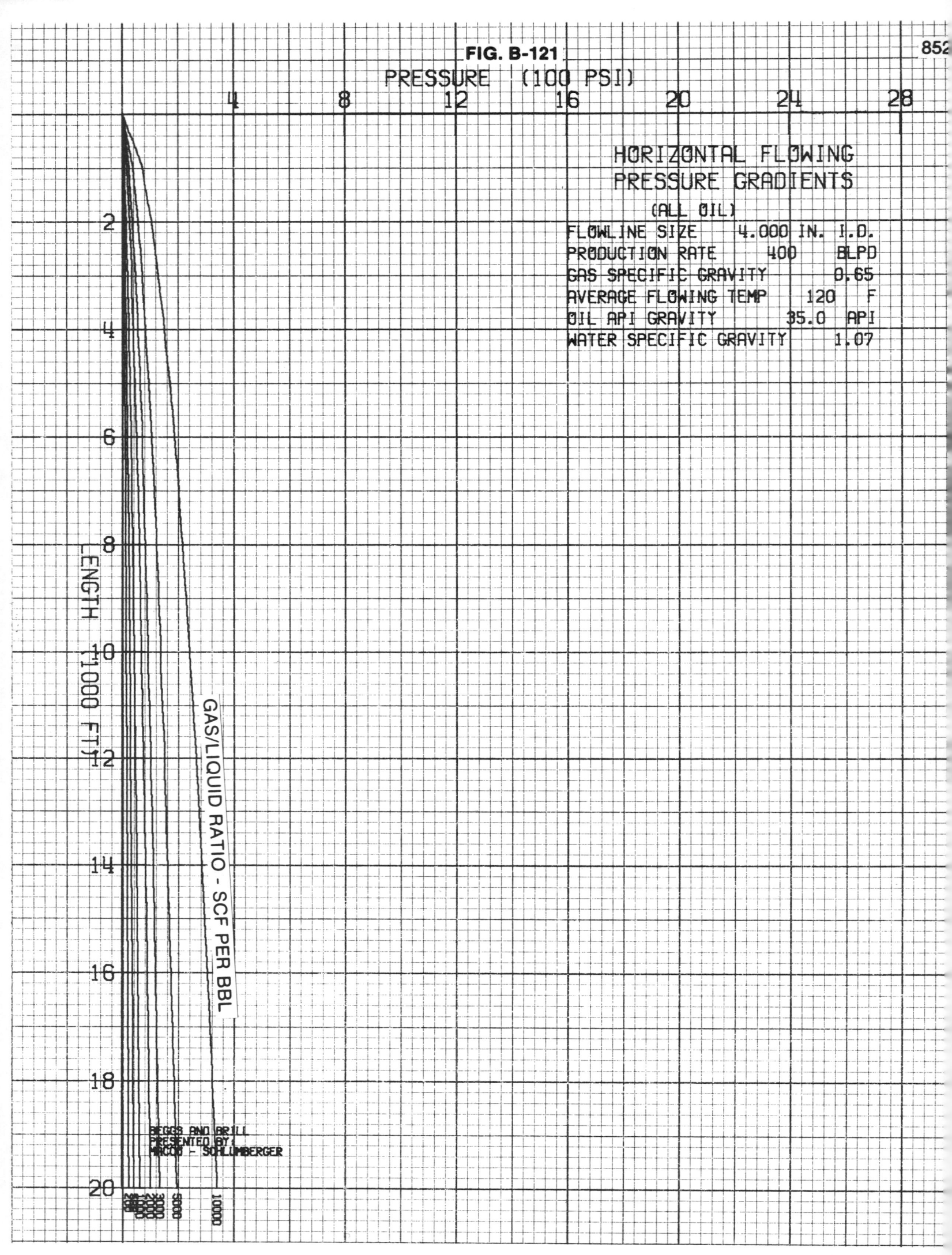

FIG. B-121

852

PRESSURE (100 PSI)

HORIZONTAL FLOWING
PRESSURE GRADIENTS
(ALL OIL)

FLOWLINE SIZE	4.000 IN. I.D.
PRODUCTION RATE	400 BLPD
GAS SPECIFIC GRAVITY	0.65
AVERAGE FLOWING TEMP	120 F
OIL API GRAVITY	35.0 API
WATER SPECIFIC GRAVITY	1.07

LENGTH (1000 FT)

GAS/LIQUID RATIO - SCF PER BBL

BEGGS AND BRILL
PRESENTED BY
MACOB - SCHLUMBERGER

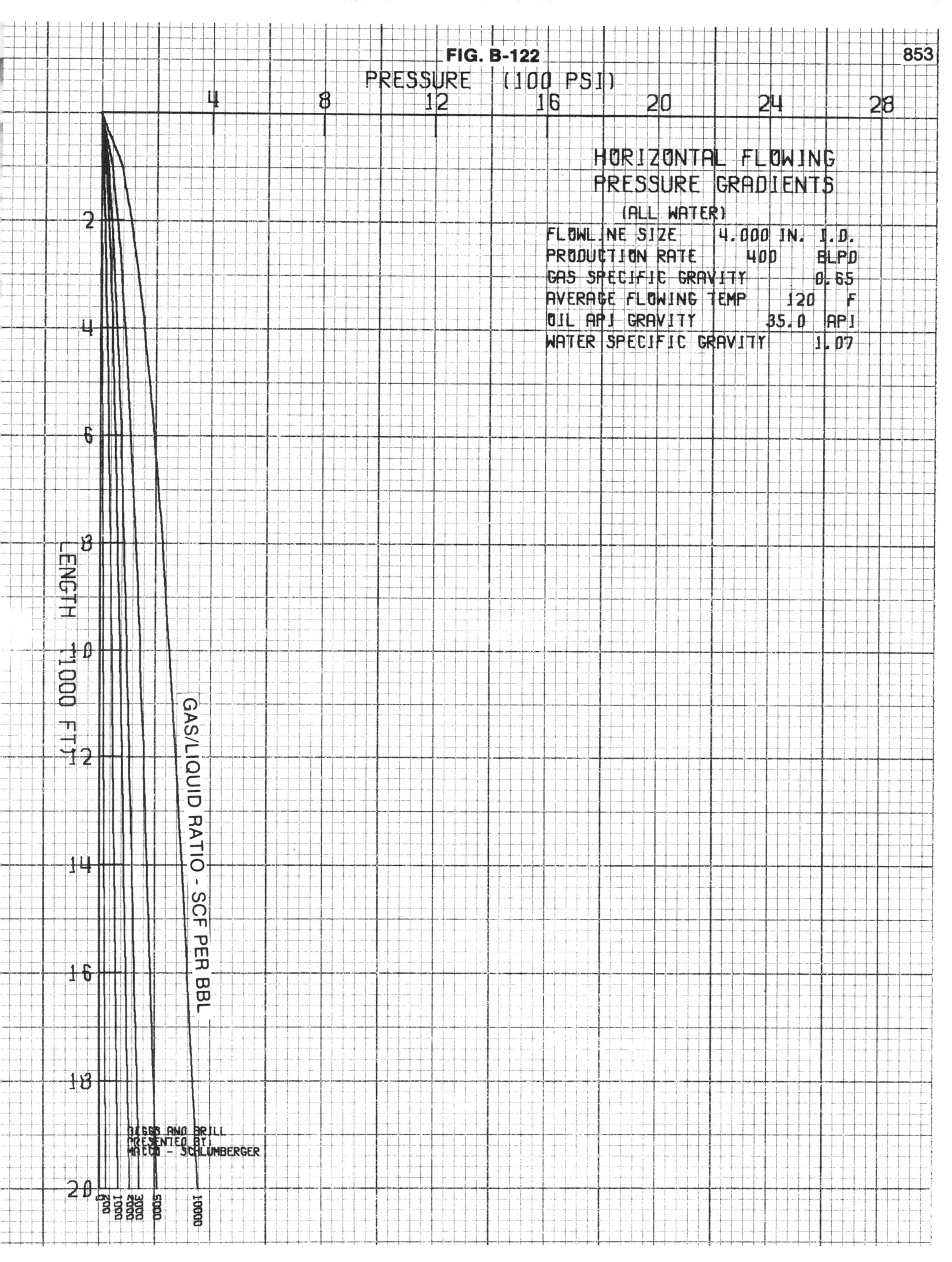

FIG. B-122

853

PRESSURE (100 PSI)

HORIZONTAL FLOWING
PRESSURE GRADIENTS
(ALL WATER)

FLOWLINE SIZE	4.000 IN.	I.D.
PRODUCTION RATE	400	BLPD
GAS SPECIFIC GRAVITY	0.65	
AVERAGE FLOWING TEMP	120	F
OIL API GRAVITY	35.0	API
WATER SPECIFIC GRAVITY	1.07	

LENGTH (1000 FT)

GAS/LIQUID RATIO - SCF PER BBL

BEGGS AND BRILL
PRESENTED BY
MAECO - SCHLUMBERGER

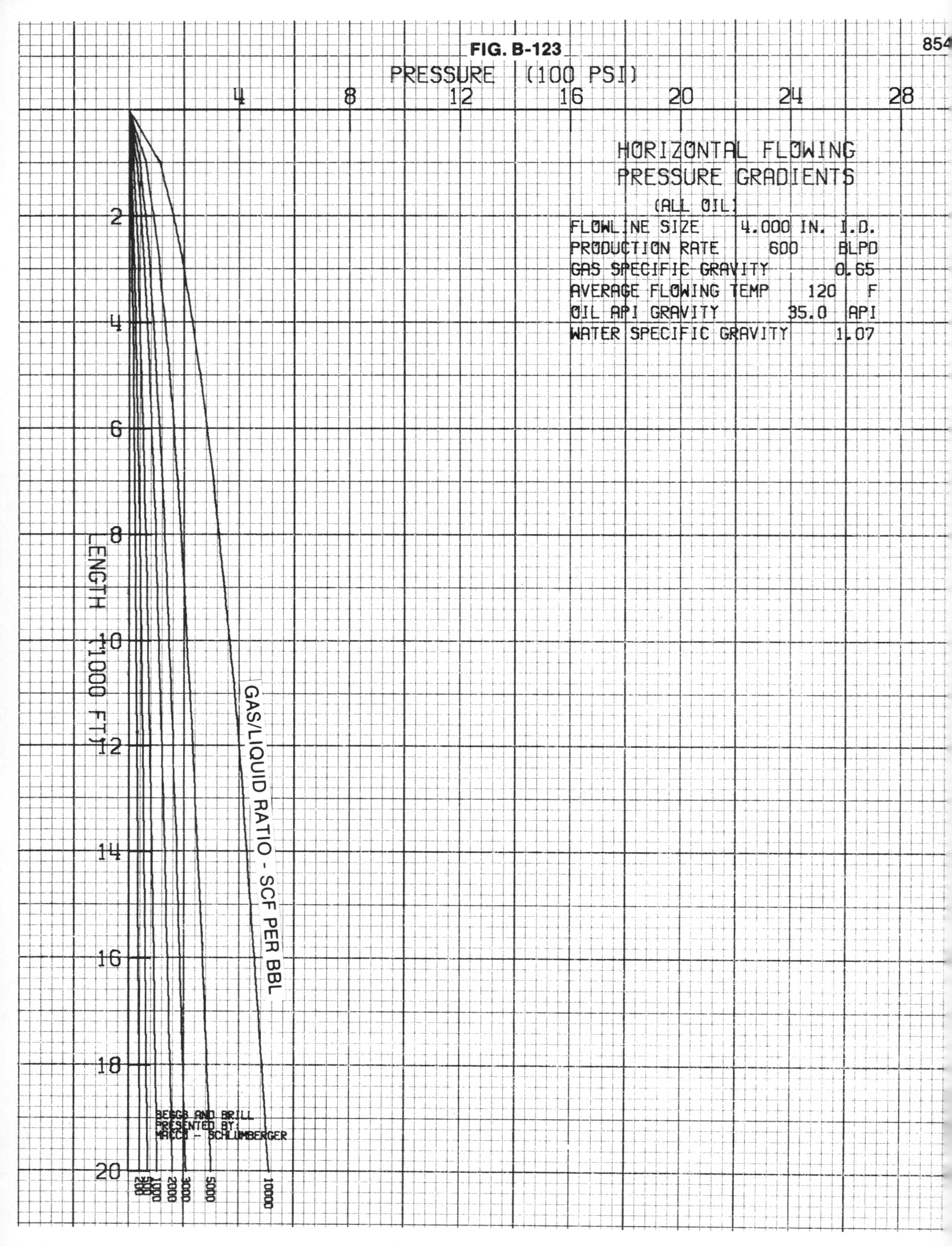

FIG. B-123

854

PRESSURE (100 PSI)

HORIZONTAL FLOWING
PRESSURE GRADIENTS
(ALL OIL)

FLOWLINE SIZE	4.000 IN.	I.D.
PRODUCTION RATE	600	BLPD
GAS SPECIFIC GRAVITY	0.65	
AVERAGE FLOWING TEMP	120	F
OIL API GRAVITY	35.0	API
WATER SPECIFIC GRAVITY	1.07	

LENGTH (1000 FT)

GAS/LIQUID RATIO - SCF PER BBL

BEGGS AND BRILL
PRESENTED BY
MACCO - SCHLUMBERGER

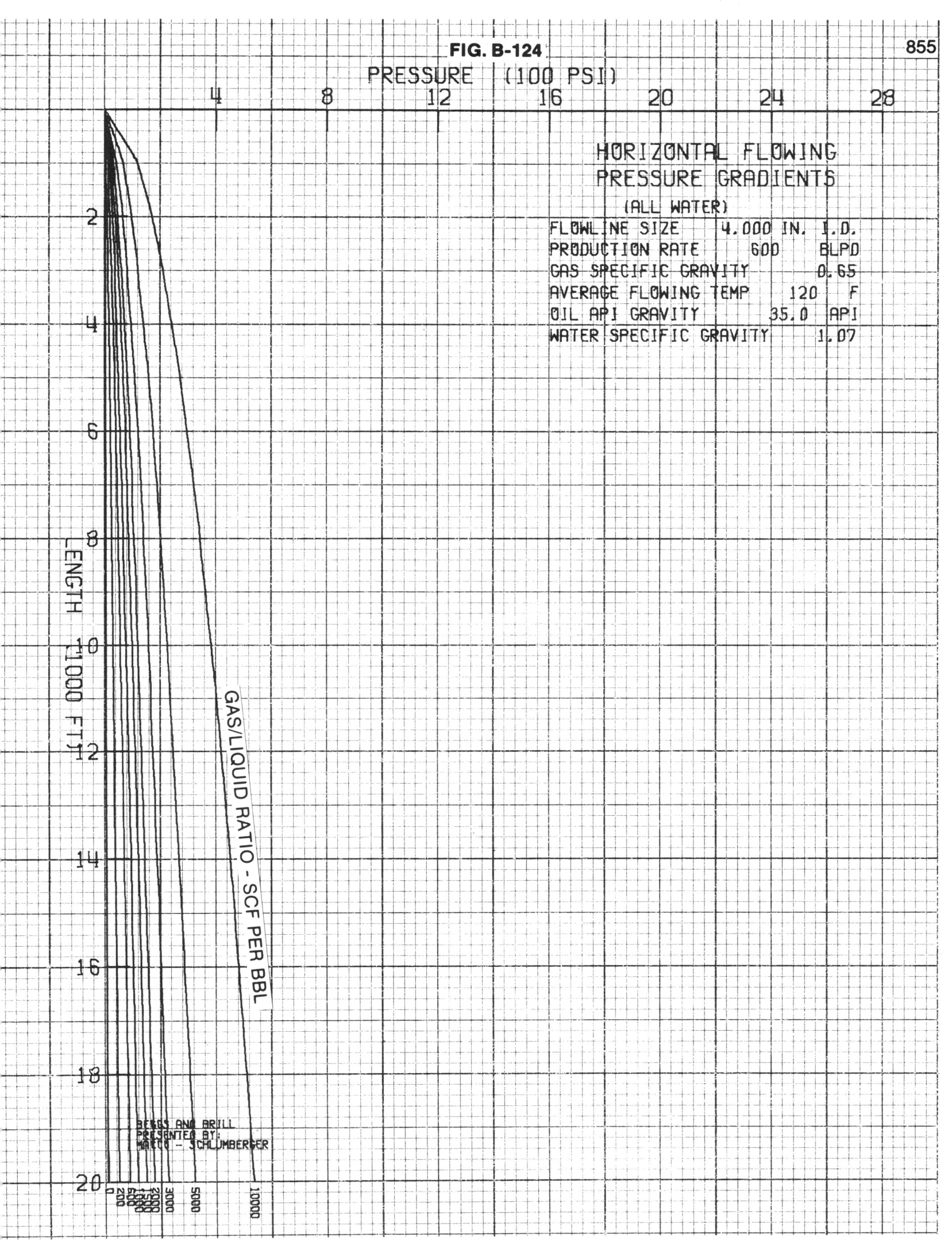

FIG. B-124

855

PRESSURE (100 PSI)

HORIZONTAL FLOWING
PRESSURE GRADIENTS
(ALL WATER)

FLOWLINE SIZE	4.000 IN. I.D.
PRODUCTION RATE	600 BLPD
GAS SPECIFIC GRAVITY	0.65
AVERAGE FLOWING TEMP	120 F
OIL API GRAVITY	35.0 API
WATER SPECIFIC GRAVITY	1.07

LENGTH (1000 FT.)

GAS/LIQUID RATIO - SCF PER BBL

BEGGS AND BRILL
PRESENTED BY:
WATCO - SCHLUMBERGER

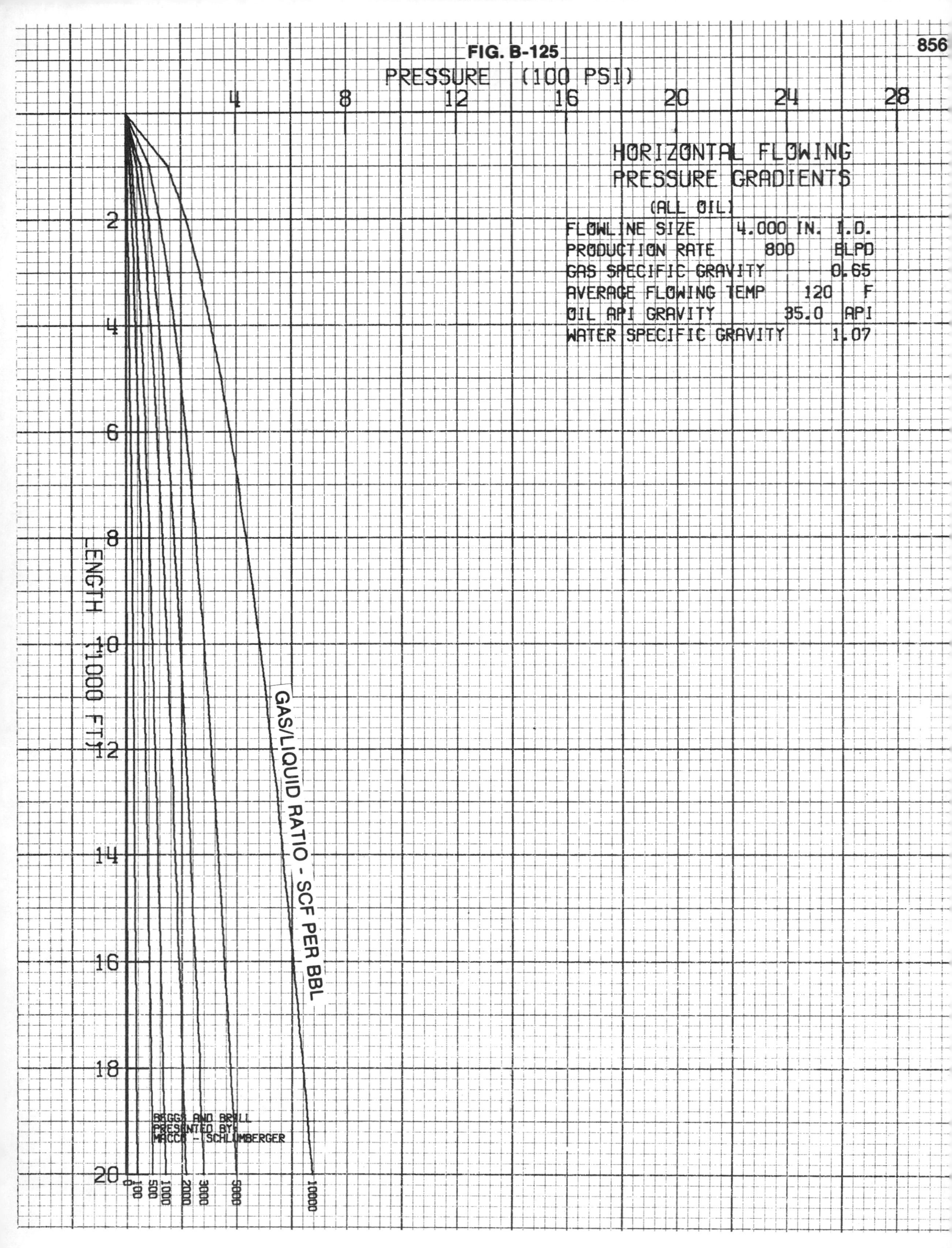

FIG. B-125

856

HORIZONTAL FLOWING
PRESSURE GRADIENTS
(ALL OIL)

FLOWLINE SIZE	4.000 IN. I.D.
PRODUCTION RATE	800 BLPD
GAS SPECIFIC GRAVITY	0.65
AVERAGE FLOWING TEMP	120 F
OIL API GRAVITY	35.0 API
WATER SPECIFIC GRAVITY	1.07

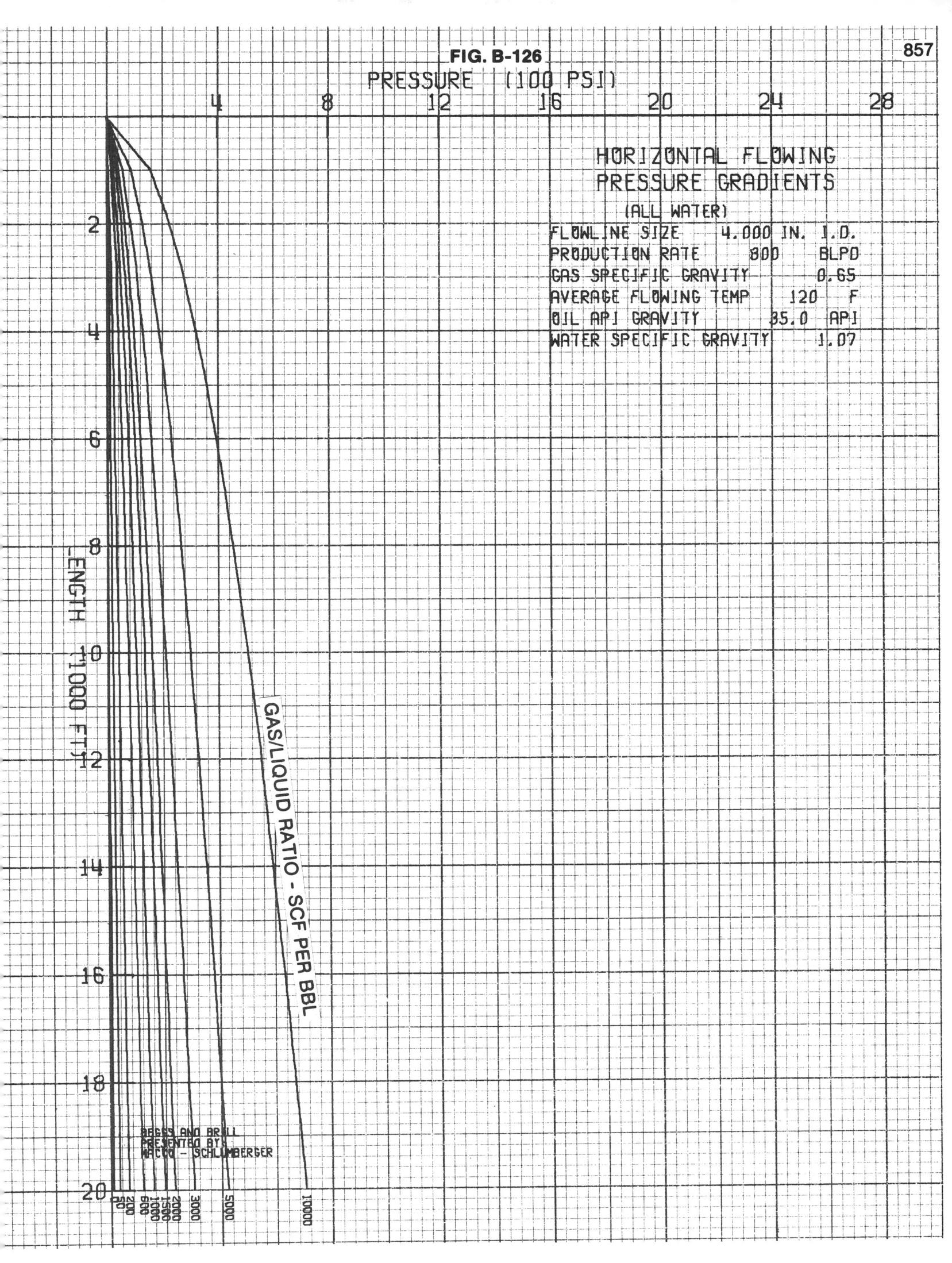

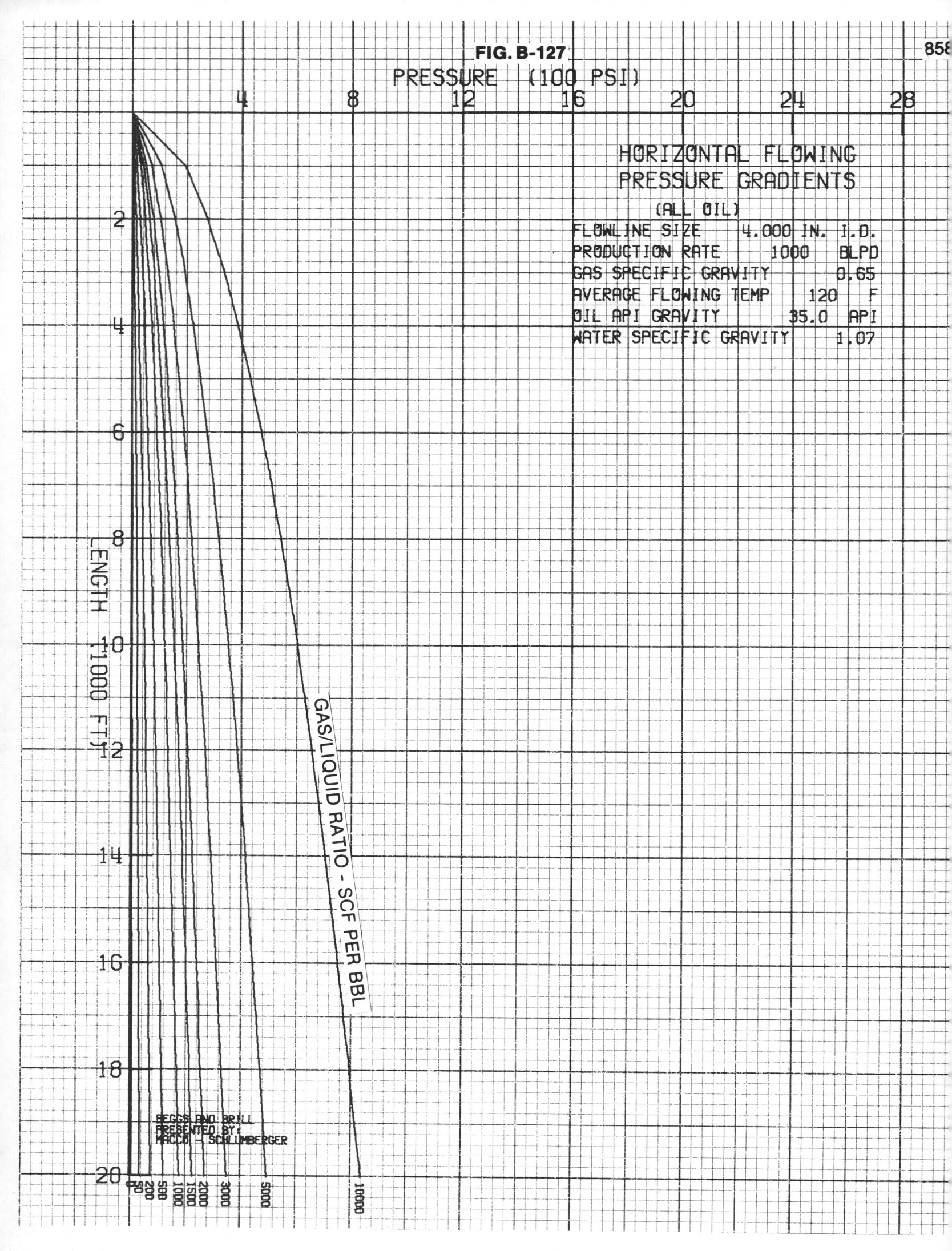

FIG. B-127

858

PRESSURE (100 PSI)

HORIZONTAL FLOWING
PRESSURE GRADIENTS
(ALL OIL)

FLOWLINE SIZE	4.000 IN. I.D.
PRODUCTION RATE	1000 BLPD
GAS SPECIFIC GRAVITY	0.65
AVERAGE FLOWING TEMP	120 F
OIL API GRAVITY	35.0 API
WATER SPECIFIC GRAVITY	1.07

LENGTH (1000 FT)

GAS/LIQUID RATIO - SCF PER BBL

BEGGS AND BRILL
PRESENTED BY:
MACCO - SCHLUMBERGER

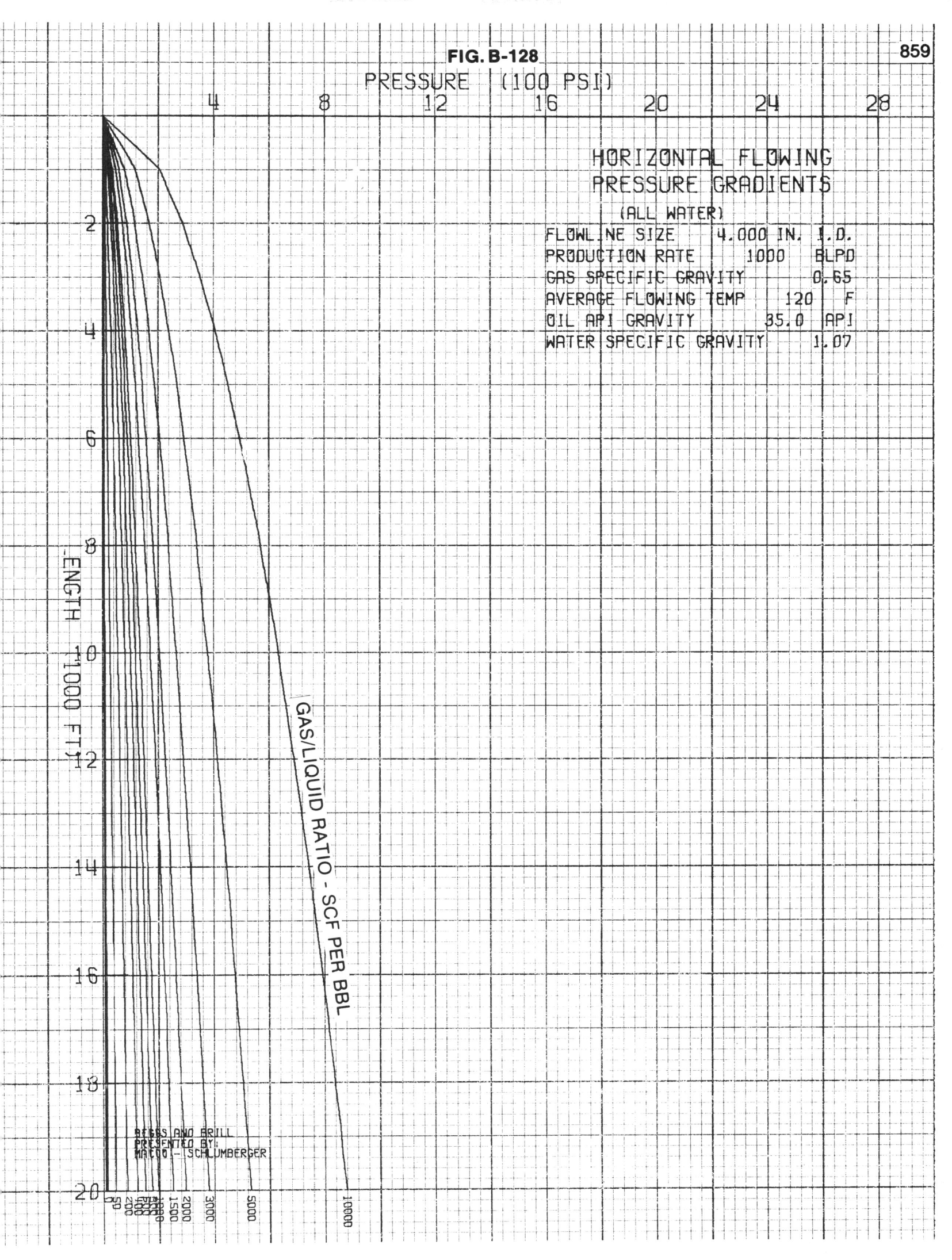

PRESSURE (100 PSI)

HORIZONTAL FLOWING
PRESSURE GRADIENTS
(ALL WATER)

FLOWLINE SIZE   4.000 IN. I.D.
PRODUCTION RATE   1000   BLPD
GAS SPECIFIC GRAVITY   0.65
AVERAGE FLOWING TEMP   120   F
OIL API GRAVITY   35.0   API
WATER SPECIFIC GRAVITY   1.07

LENGTH 1000 FT.

GAS/LIQUID RATIO - SCF PER BBL

BEGGS AND BRILL
PRESENTED BY:
MATCOL-SCHLUMBERGER

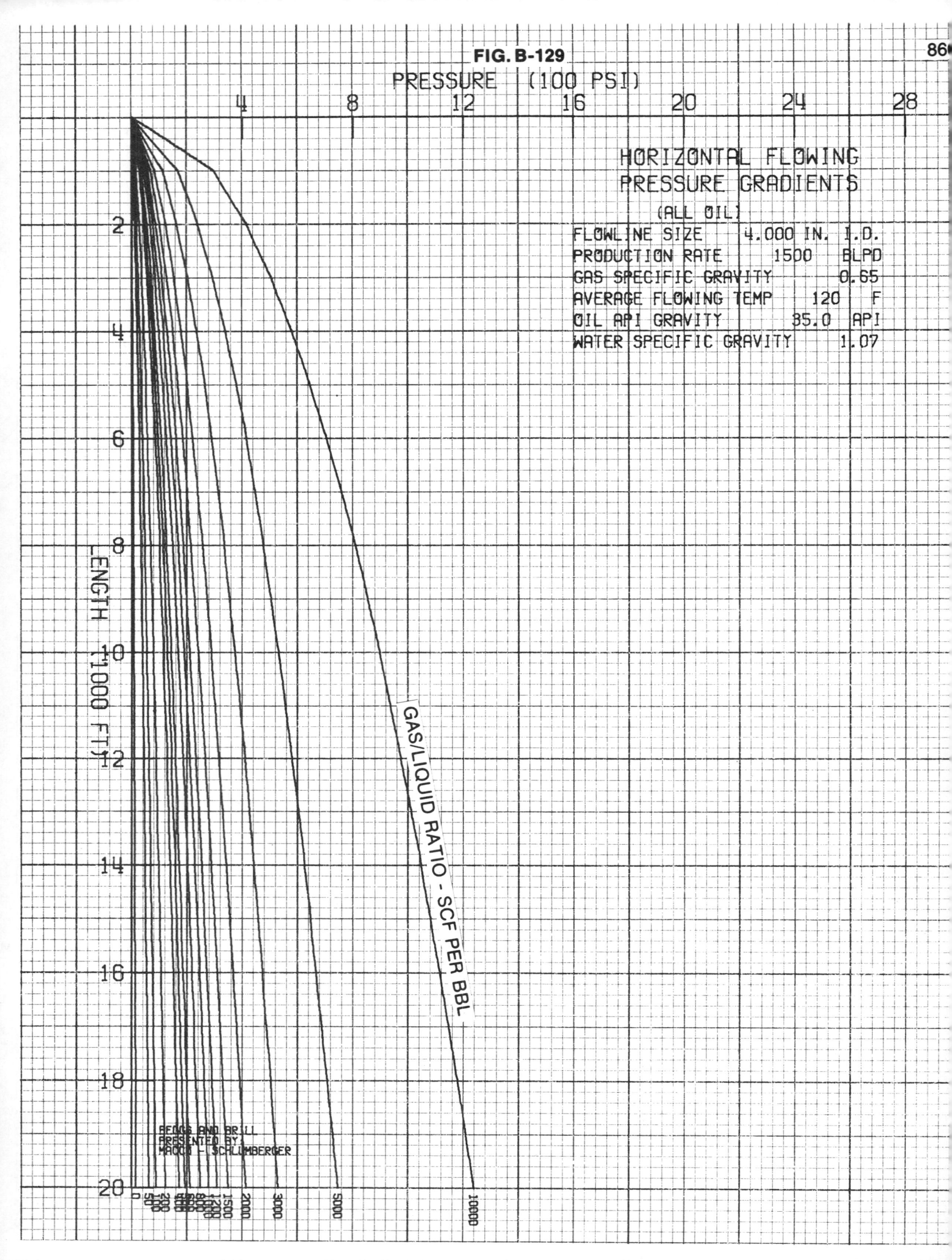

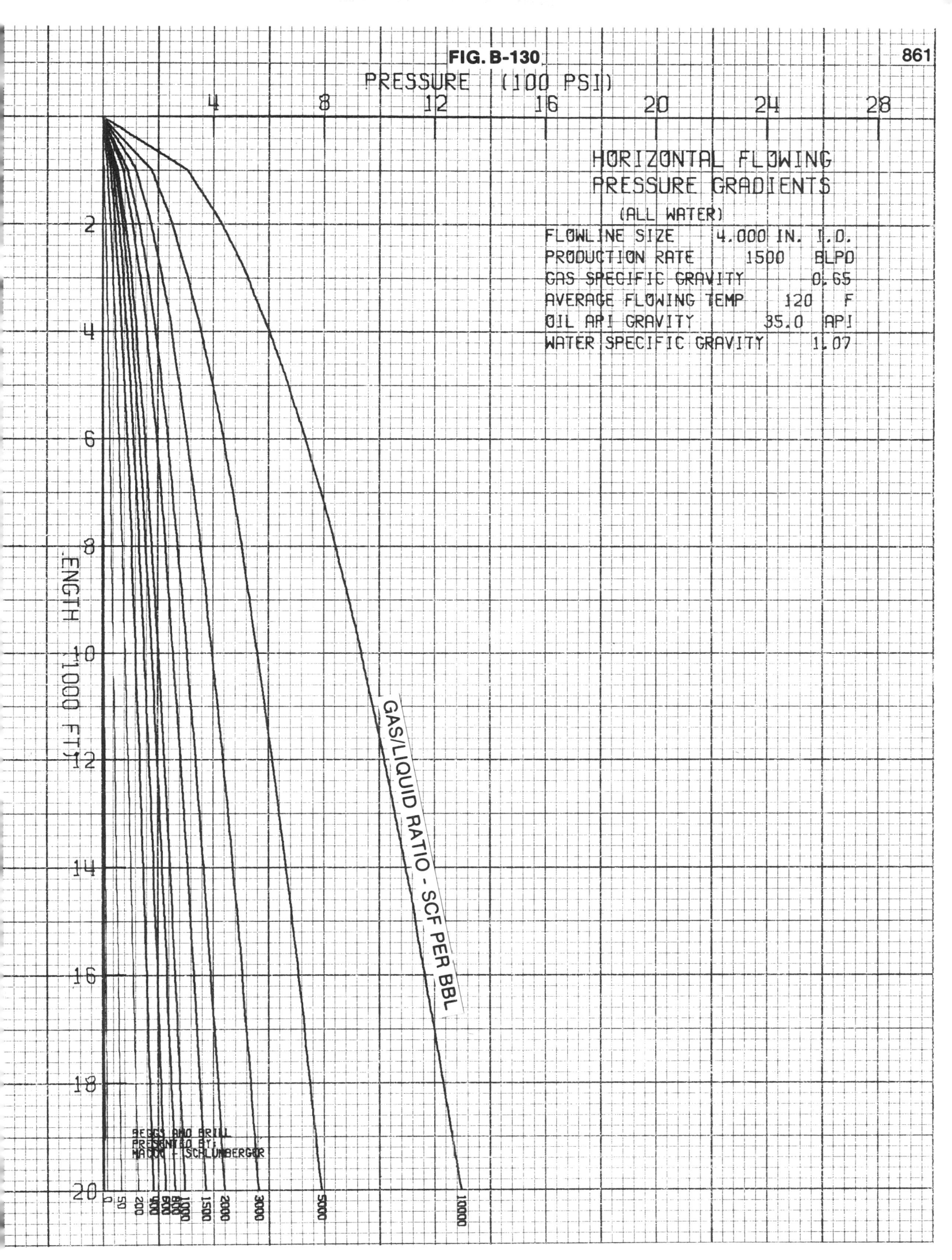

FIG. B-130

861

PRESSURE (100 PSI)

HORIZONTAL FLOWING
PRESSURE GRADIENTS
(ALL WATER)

FLOWLINE SIZE	4.000 IN.	I.D.
PRODUCTION RATE	1500	BLPD
GAS SPECIFIC GRAVITY	0.65	
AVERAGE FLOWING TEMP	120	F
OIL API GRAVITY	35.0	API
WATER SPECIFIC GRAVITY	1.07	

LENGTH (1000 FT)

GAS/LIQUID RATIO - SCF PER BBL

BEGGS AND BRILL
PRESENTED BY:
MADOG & SCHLUMBERGER

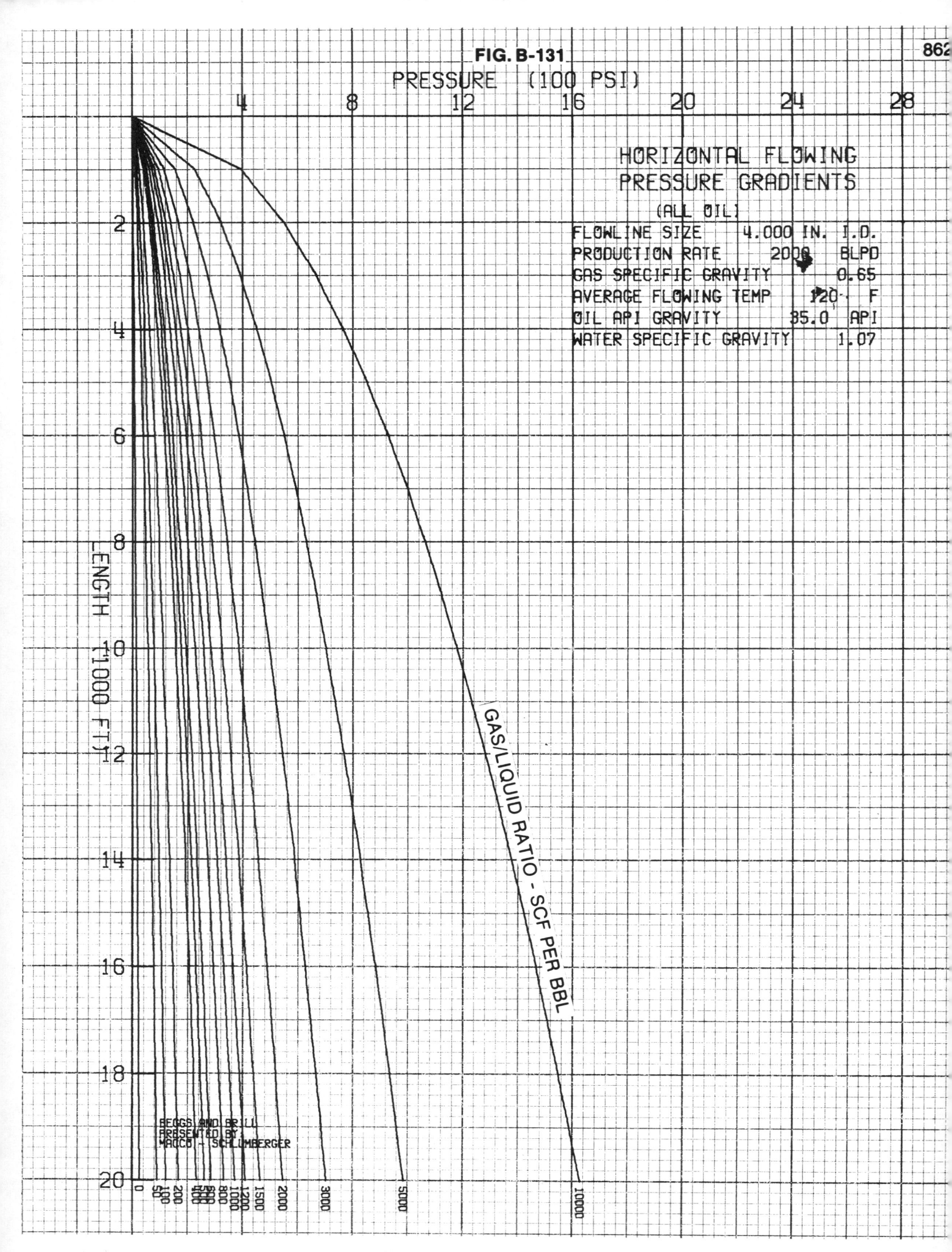

FIG. B-131

862

PRESSURE (100 PSI)

HORIZONTAL FLOWING
PRESSURE GRADIENTS
(ALL OIL)

FLOWLINE SIZE	4.000 IN. I.D.
PRODUCTION RATE	2000 BLPD
GAS SPECIFIC GRAVITY	0.65
AVERAGE FLOWING TEMP	120° F
OIL API GRAVITY	35.0 API
WATER SPECIFIC GRAVITY	1.07

LENGTH (1,000 FT)

GAS/LIQUID RATIO - SCF PER BBL

BEGGS AND BRILL
PRESENTED BY
MADCO - SCHLUMBERGER

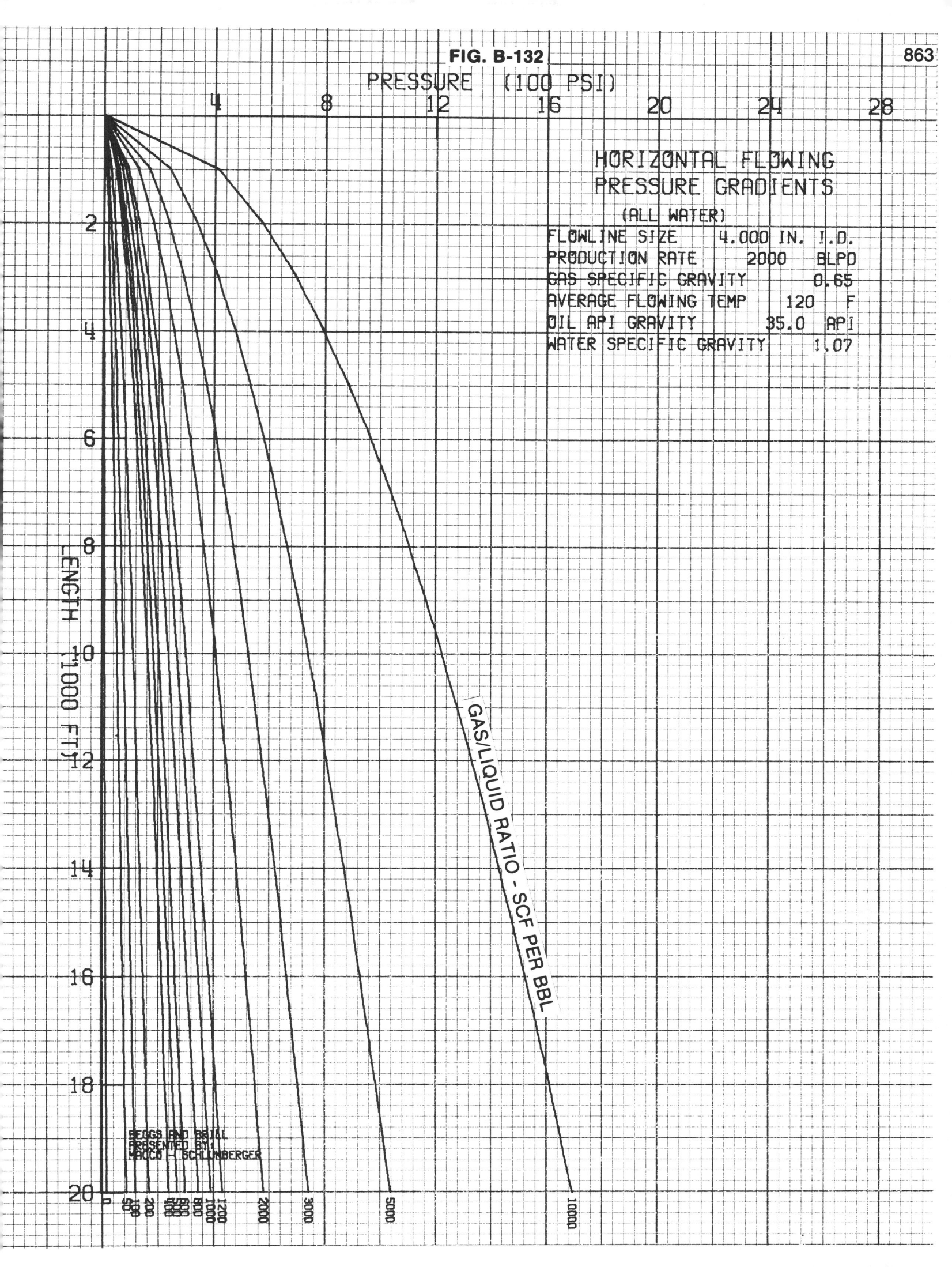

FIG. B-132

863

PRESSURE (100 PSI)

HORIZONTAL FLOWING
PRESSURE GRADIENTS
(ALL WATER)

FLOWLINE SIZE   4.000 IN. I.D.
PRODUCTION RATE   2000   BLPD
GAS SPECIFIC GRAVITY   0.65
AVERAGE FLOWING TEMP   120   F
OIL API GRAVITY   35.0   API
WATER SPECIFIC GRAVITY   1.07

LENGTH (1000 FT.)

GAS/LIQUID RATIO - SCF PER BBL

BEGGS AND BRILL
PRESENTED BY
MADCO - SCHLUMBERGER

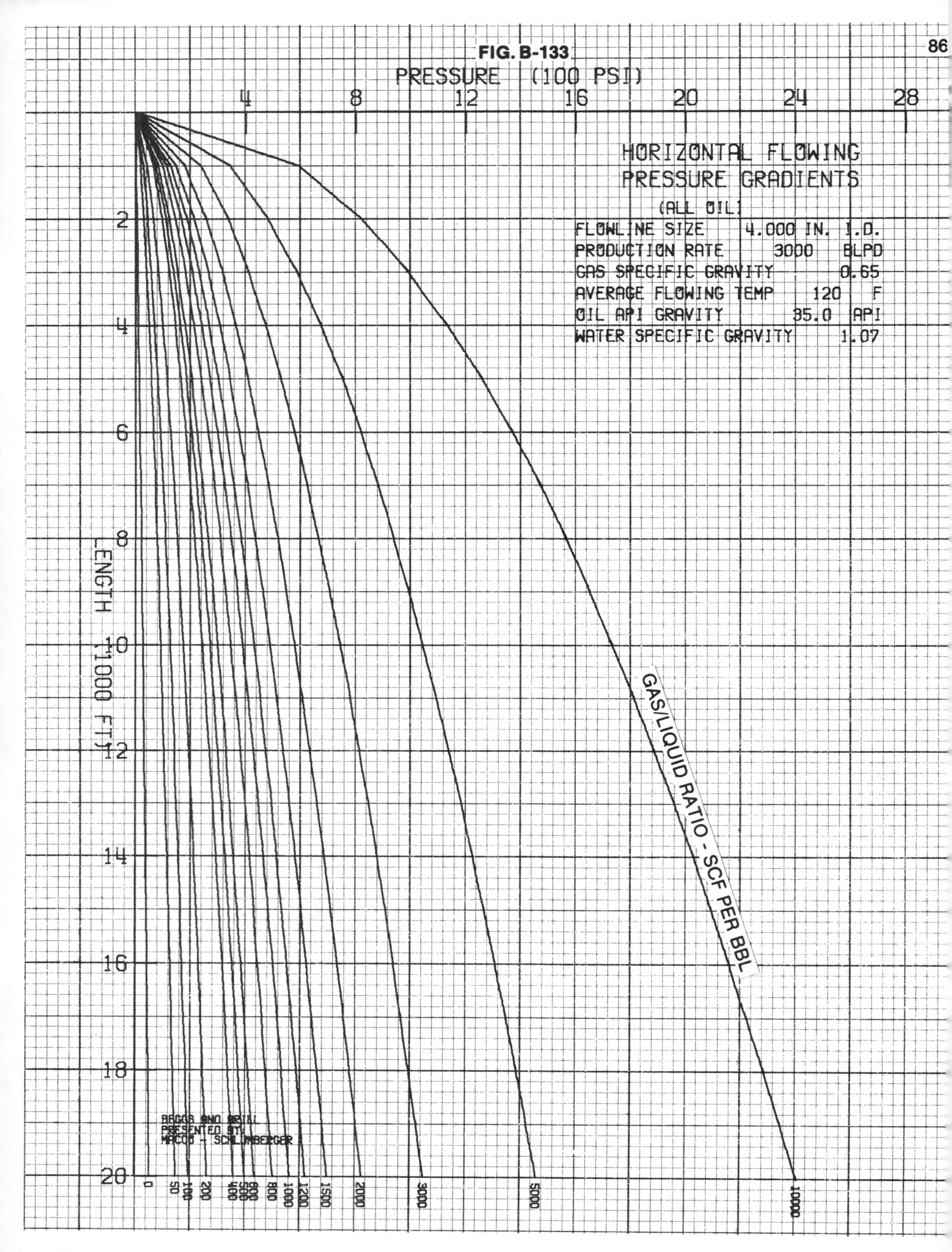

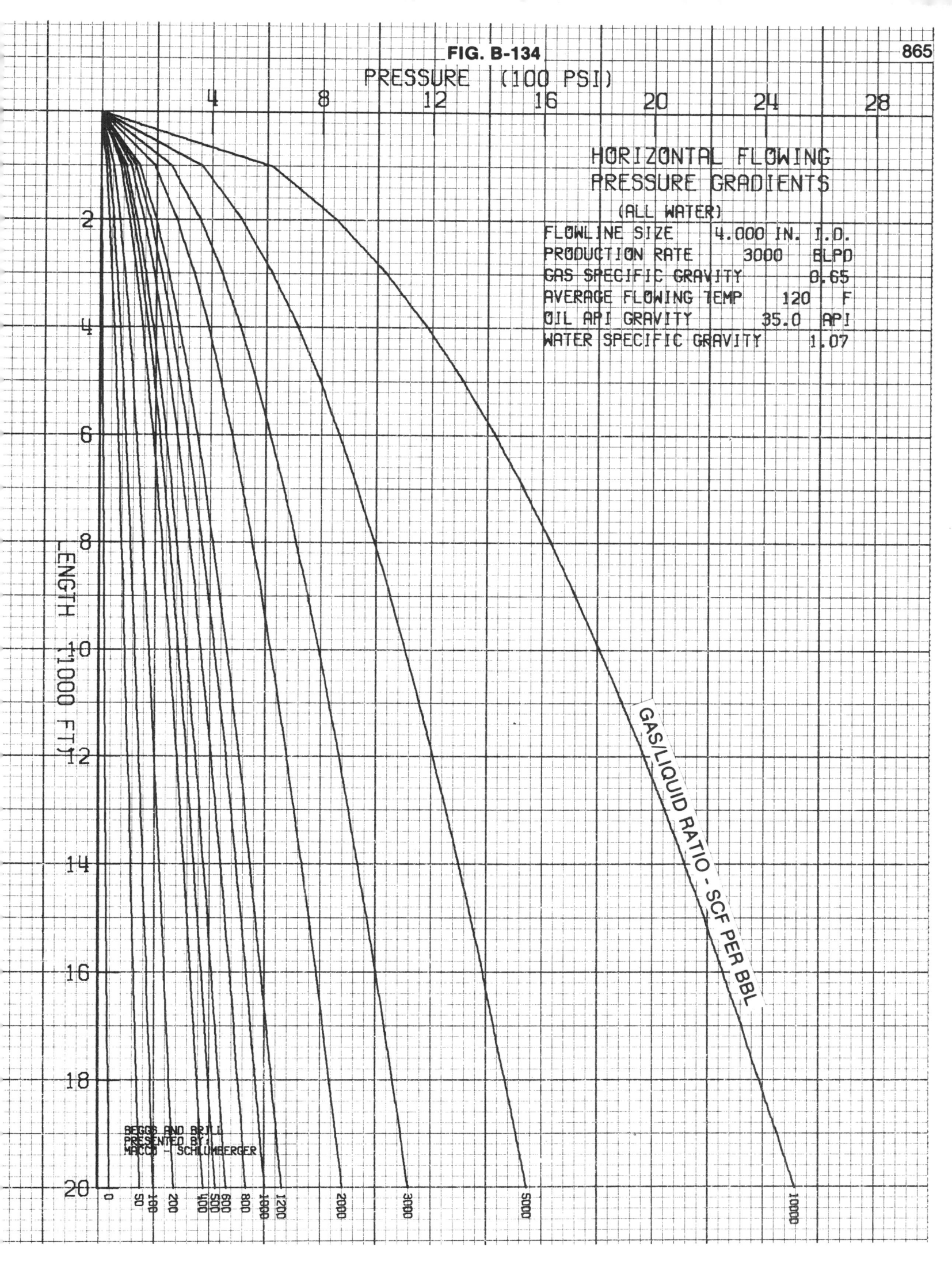

PRESSURE (100 PSI)

HORIZONTAL FLOWING
PRESSURE GRADIENTS
(ALL WATER)

FLOWLINE SIZE   4.000 IN. I.D.
PRODUCTION RATE   3000 BLPD
GAS SPECIFIC GRAVITY   0.65
AVERAGE FLOWING TEMP   120   F
OIL API GRAVITY   35.0 API
WATER SPECIFIC GRAVITY   1.07

LENGTH (1000 FT)

GAS/LIQUID RATIO - SCF PER BBL

BEGGS AND BRILL
PRESENTED BY:
MACCO - SCHLUMBERGER

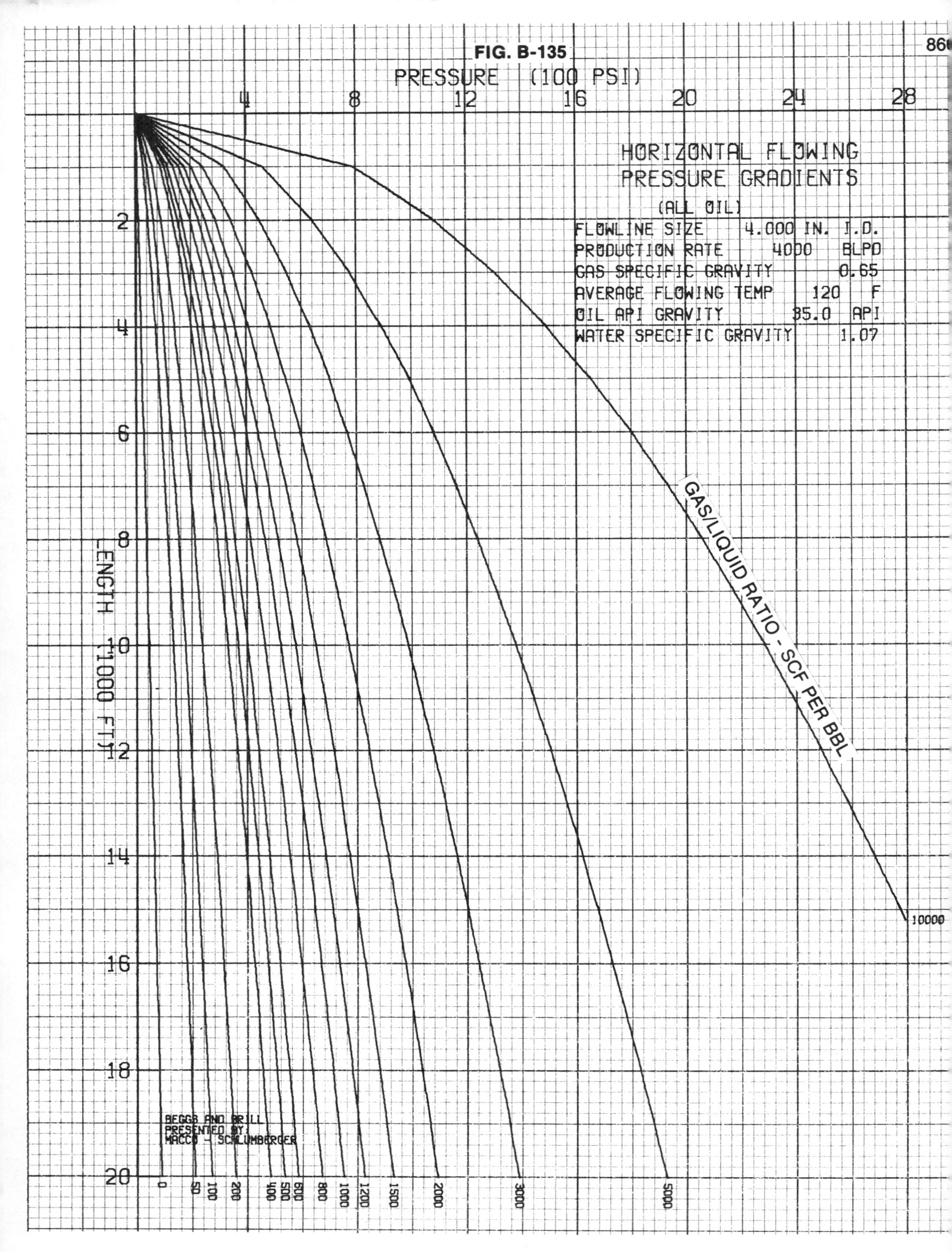

FIG. B-135

860

PRESSURE (100 PSI)

HORIZONTAL FLOWING
PRESSURE GRADIENTS
(ALL OIL)

FLOWLINE SIZE	4.000 IN. I.D.
PRODUCTION RATE	4000 BLPD
GAS SPECIFIC GRAVITY	0.65
AVERAGE FLOWING TEMP	120 F
OIL API GRAVITY	35.0 API
WATER SPECIFIC GRAVITY	1.07

GAS/LIQUID RATIO - SCF PER BBL

LENGTH (1000 FT)

BEGGS AND BRILL
PRESENTED BY:
MACCO - SCHLUMBERGER

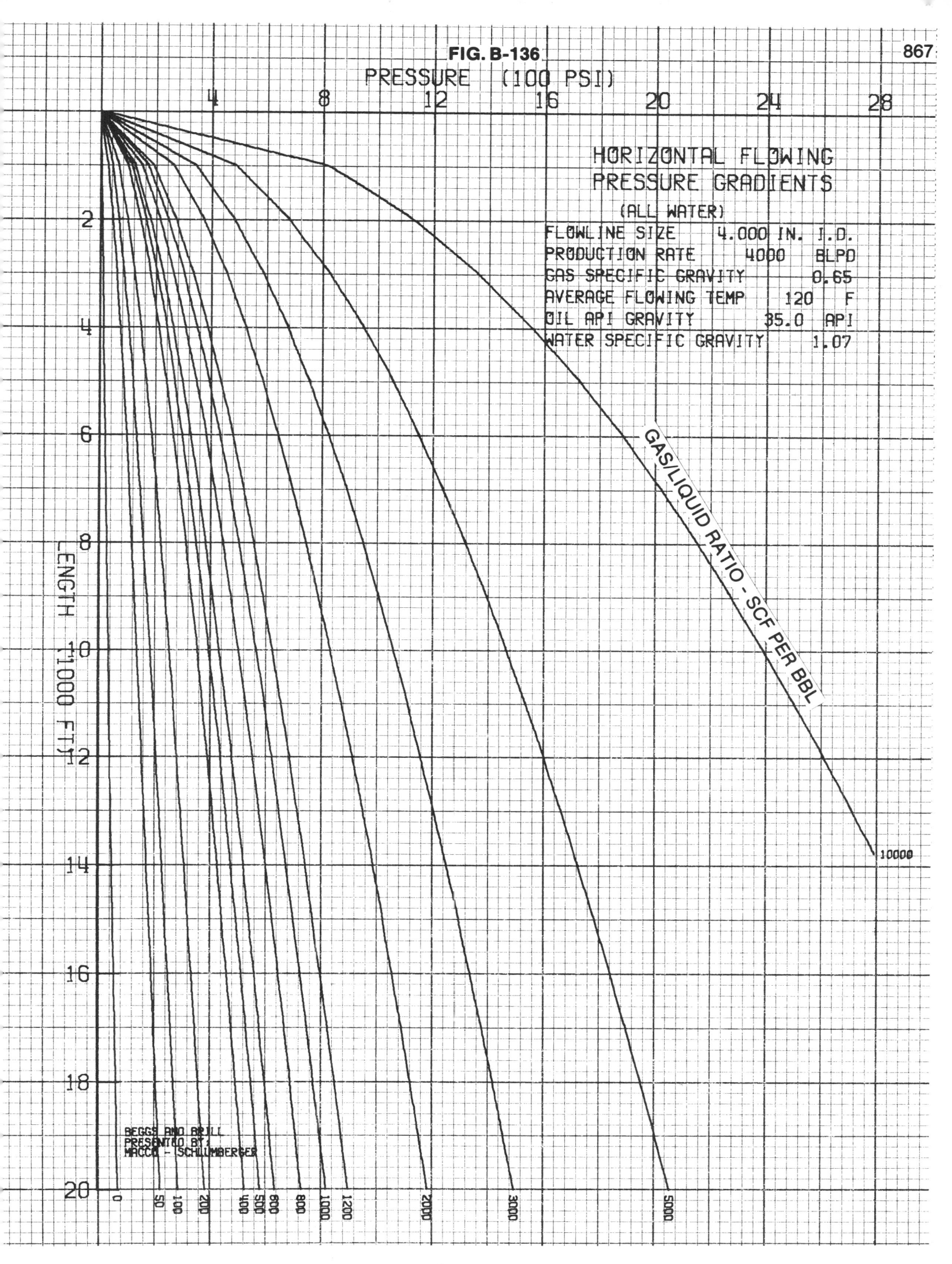

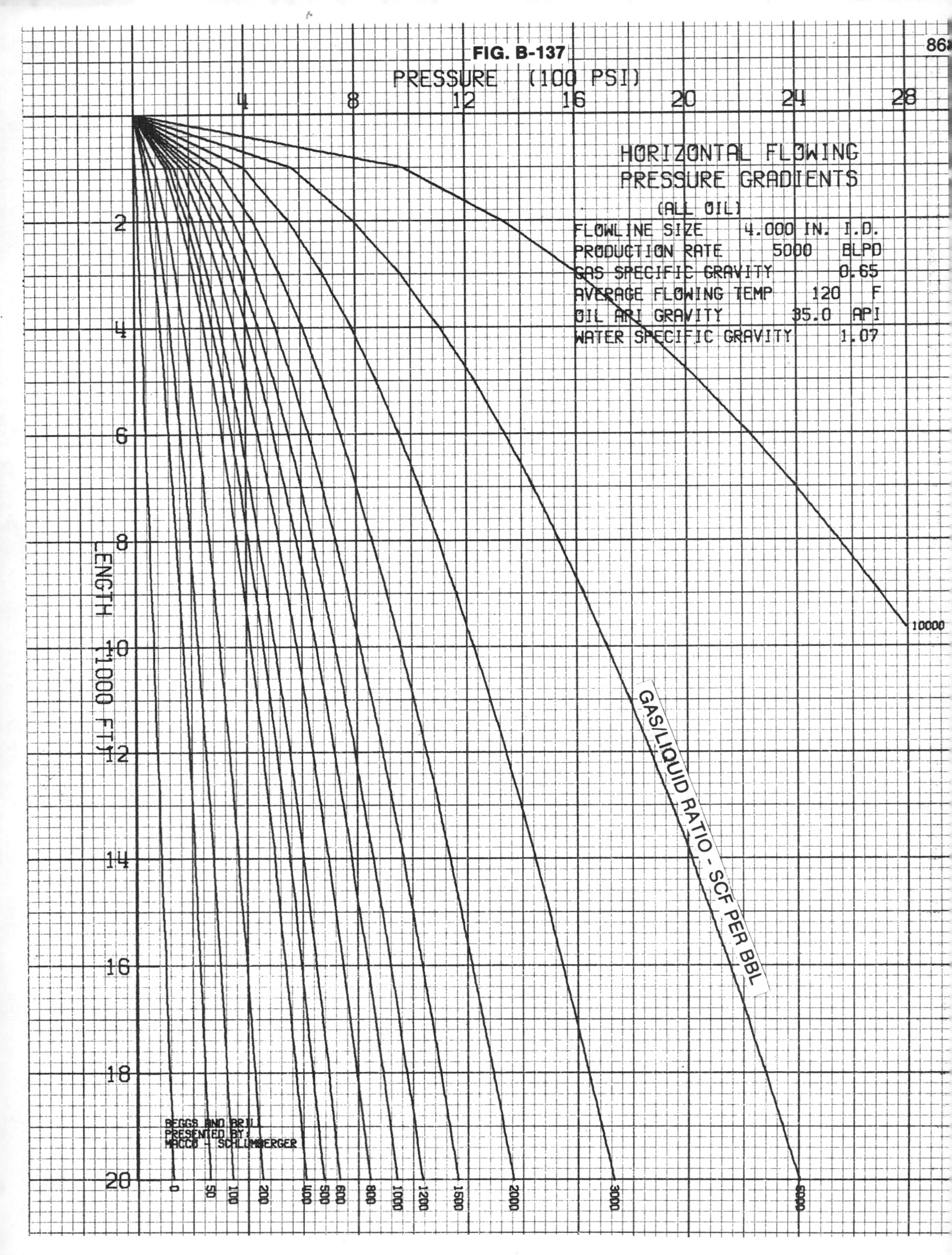

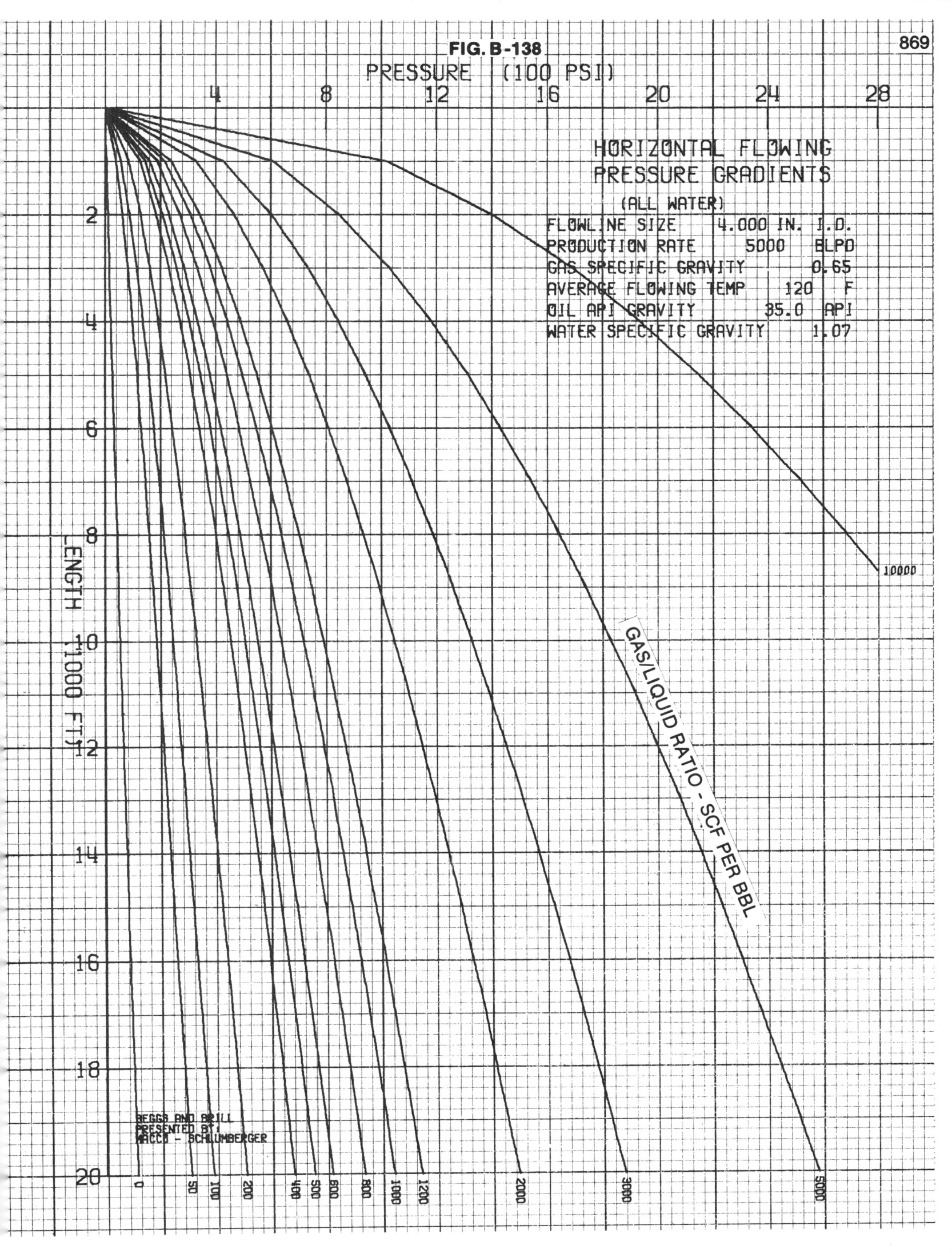

FIG. B-138

PRESSURE (100 PSI)

HORIZONTAL FLOWING
PRESSURE GRADIENTS
(ALL WATER)

FLOWLINE SIZE	4.000 IN. I.D.
PRODUCTION RATE	5000 BLPD
GAS SPECIFIC GRAVITY	0.65
AVERAGE FLOWING TEMP	120 F
OIL API GRAVITY	35.0 API
WATER SPECIFIC GRAVITY	1.07

LENGTH 1000 FT.

GAS/LIQUID RATIO - SCF PER BBL

10000

BEGGS AND BRILL
PRESENTED BY:
MACCO - SCHLUMBERGER

FIG. B-139

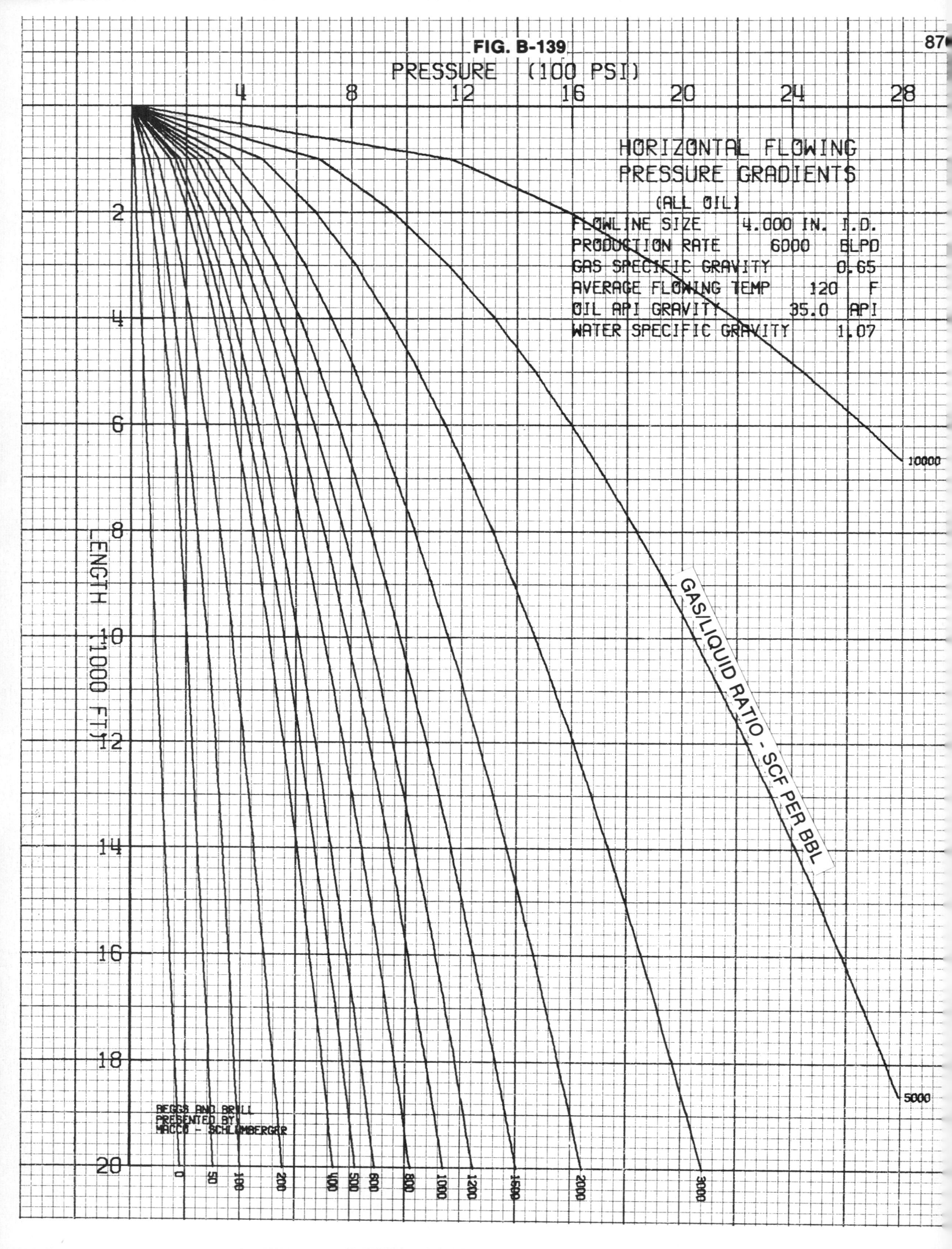

PRESSURE (100 PSI)

HORIZONTAL FLOWING
PRESSURE GRADIENTS
(ALL OIL)

FLOWLINE SIZE	4.000 IN. I.D.
PRODUCTION RATE	6000 BLPD
GAS SPECIFIC GRAVITY	0.65
AVERAGE FLOWING TEMP	120 F
OIL API GRAVITY	35.0 API
WATER SPECIFIC GRAVITY	1.07

LENGTH (1000 FT)

GAS/LIQUID RATIO - SCF PER BBL

10000

5000

BEGGS AND BRILL
PRESENTED BY
MACCO - SCHLUMBERGER

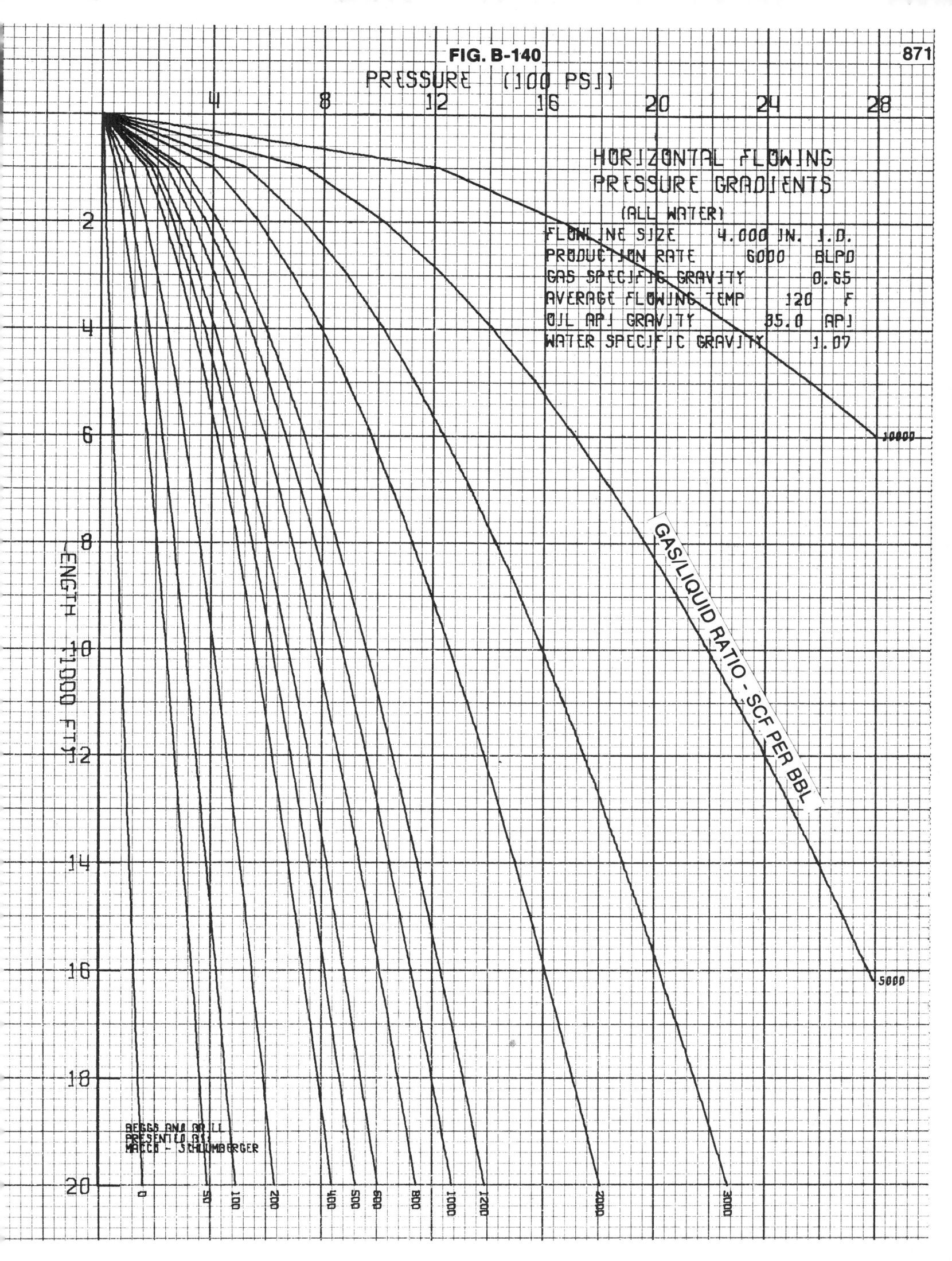

PRESSURE (100 PSI)

LENGTH (1000 FT)

HORIZONTAL FLOWING
PRESSURE GRADIENTS

(ALL WATER)

FLOWLINE SIZE	4.000 IN. I.D.
PRODUCTION RATE	6000 BLPD
GAS SPECIFIC GRAVITY	0.65
AVERAGE FLOWING TEMP	120 F
OIL API GRAVITY	35.0 API
WATER SPECIFIC GRAVITY	1.07

GAS/LIQUID RATIO - SCF PER BBL

BEGGS AND BRILL
PRESENTED AS
MACCO - SCHLUMBERGER

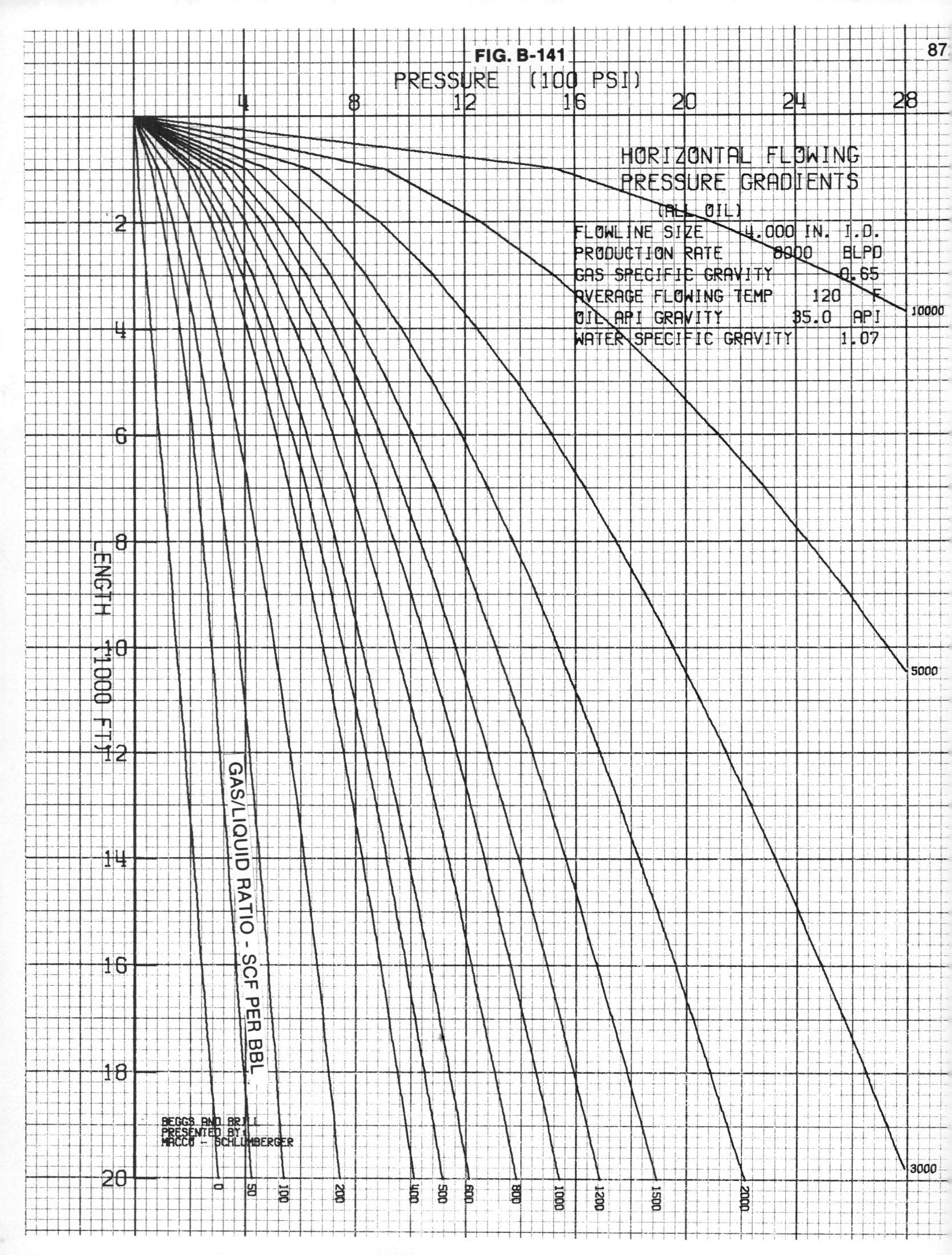

FIG. B-141

87

PRESSURE (100 PSI)

HORIZONTAL FLOWING
PRESSURE GRADIENTS
(ALL OIL)

FLOWLINE SIZE	4.000 IN. I.D.
PRODUCTION RATE	8000 BLPD
GAS SPECIFIC GRAVITY	0.65
AVERAGE FLOWING TEMP	120 F
OIL API GRAVITY	35.0 API
WATER SPECIFIC GRAVITY	1.07

LENGTH (1000 FT)

GAS/LIQUID RATIO - SCF PER BBL

BEGGS AND BRILL
PRESENTED BY:
MACCO - SCHLUMBERGER

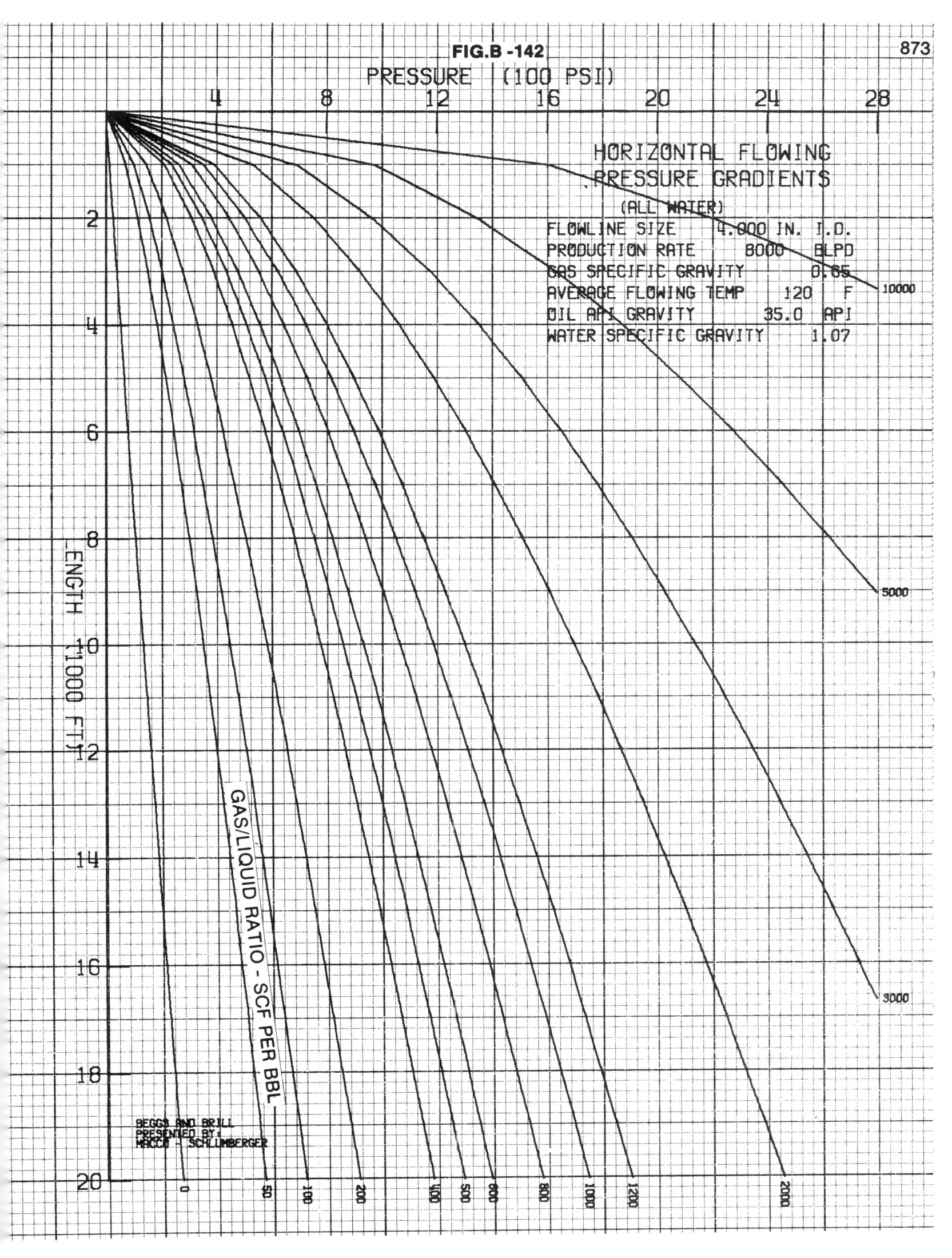

FIG. B-143

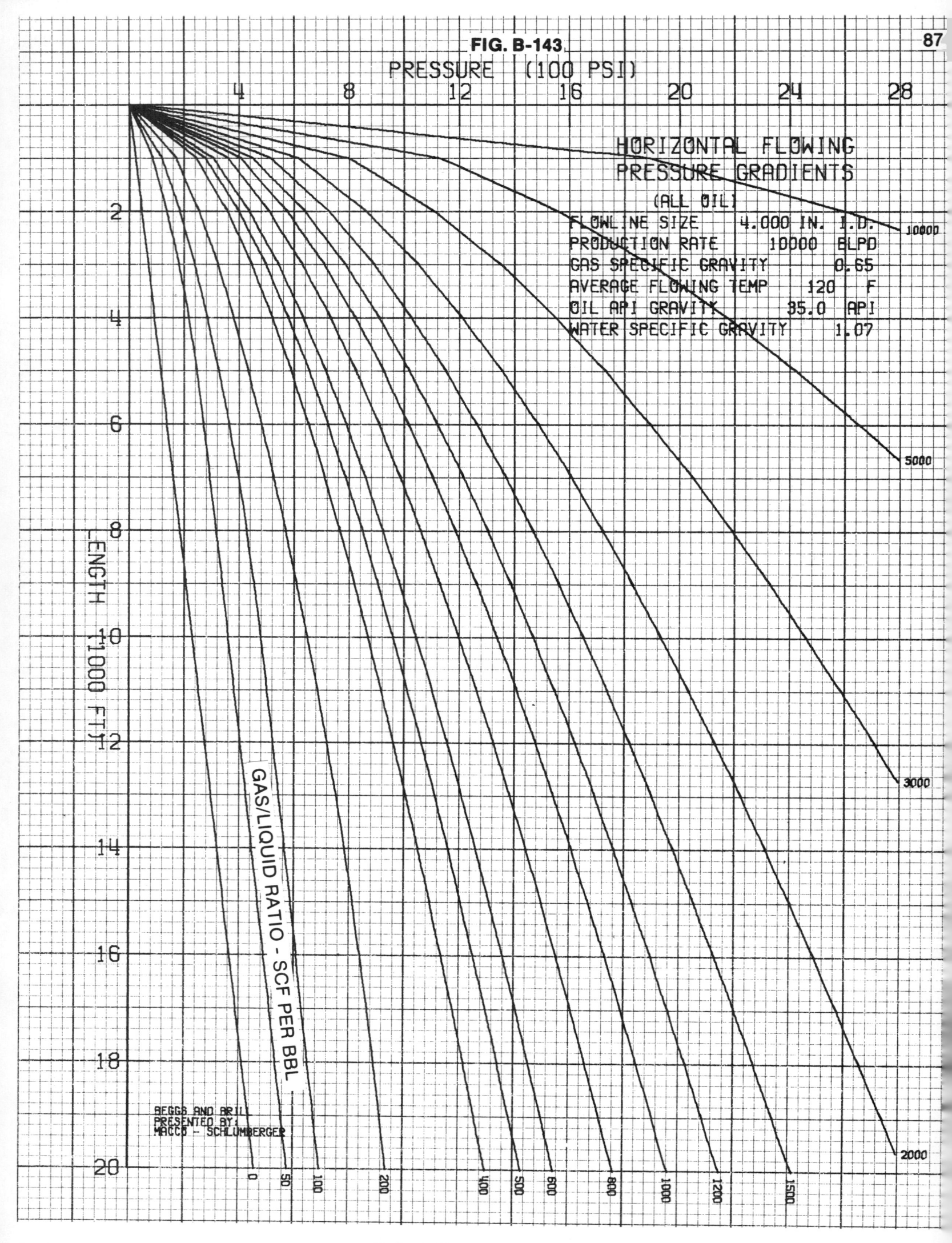

PRESSURE (100 PSI)

HORIZONTAL FLOWING
PRESSURE GRADIENTS
(ALL OIL)

FLOWLINE SIZE	4.000 IN. I.D.
PRODUCTION RATE	10000 BLPD
GAS SPECIFIC GRAVITY	0.65
AVERAGE FLOWING TEMP	120 F
OIL API GRAVITY	35.0 API
WATER SPECIFIC GRAVITY	1.07

LENGTH (1000 FT.)

GAS/LIQUID RATIO - SCF PER BBL

BEGGS AND BRILL
PRESENTED BY:
MACCO - SCHLUMBERGER

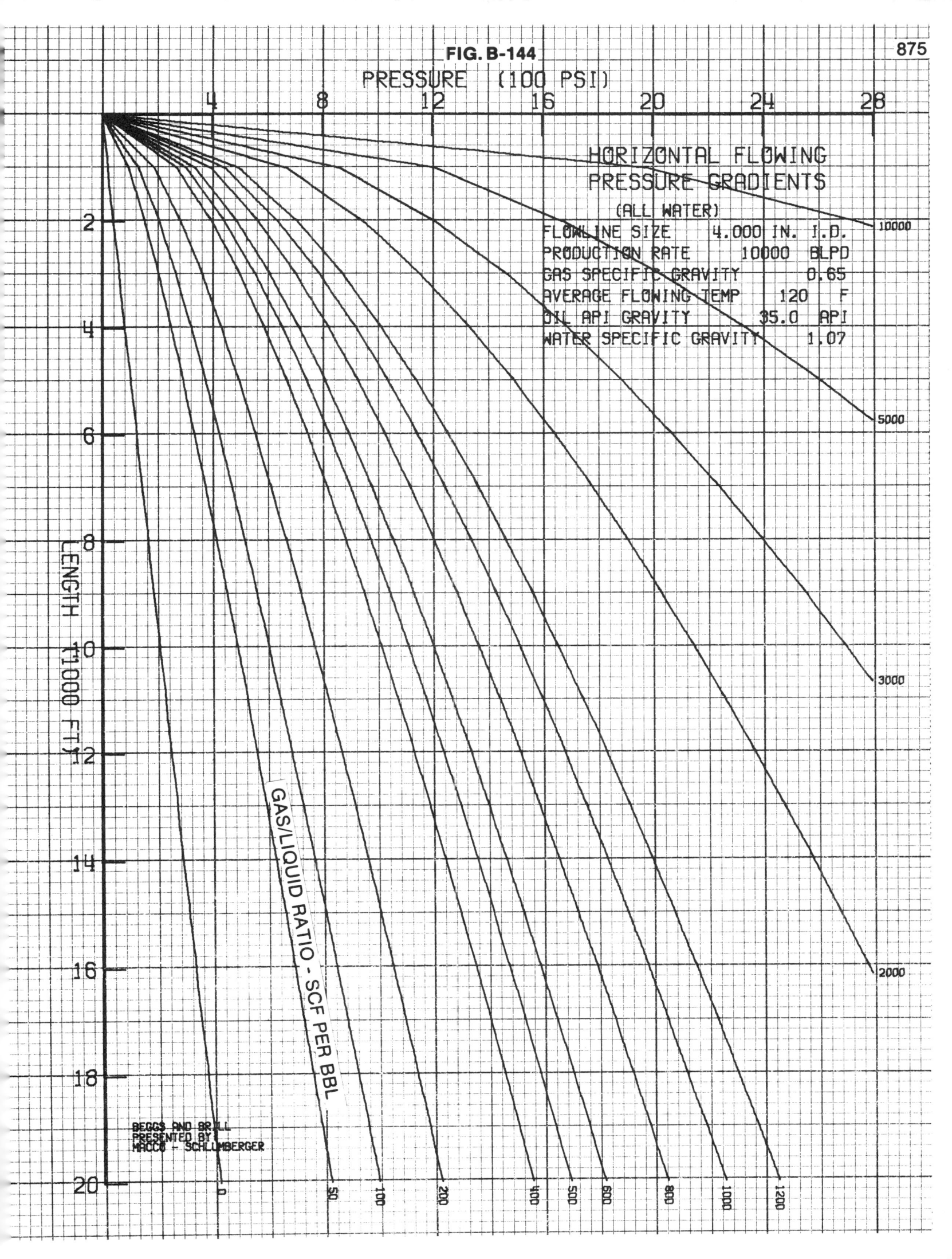

PRESSURE (100 PSI)

HORIZONTAL FLOWING
PRESSURE GRADIENTS
(ALL WATER)

FLOWLINE SIZE	4.000 IN. I.D.
PRODUCTION RATE	10000 BLPD
GAS SPECIFIC GRAVITY	0.65
AVERAGE FLOWING TEMP	120 F
OIL API GRAVITY	35.0 API
WATER SPECIFIC GRAVITY	1.07

LENGTH (1000 FT)

GAS/LIQUID RATIO - SCF PER BBL

BEGGS AND BRILL
PRESENTED BY
MACCO - SCHLUMBERGER

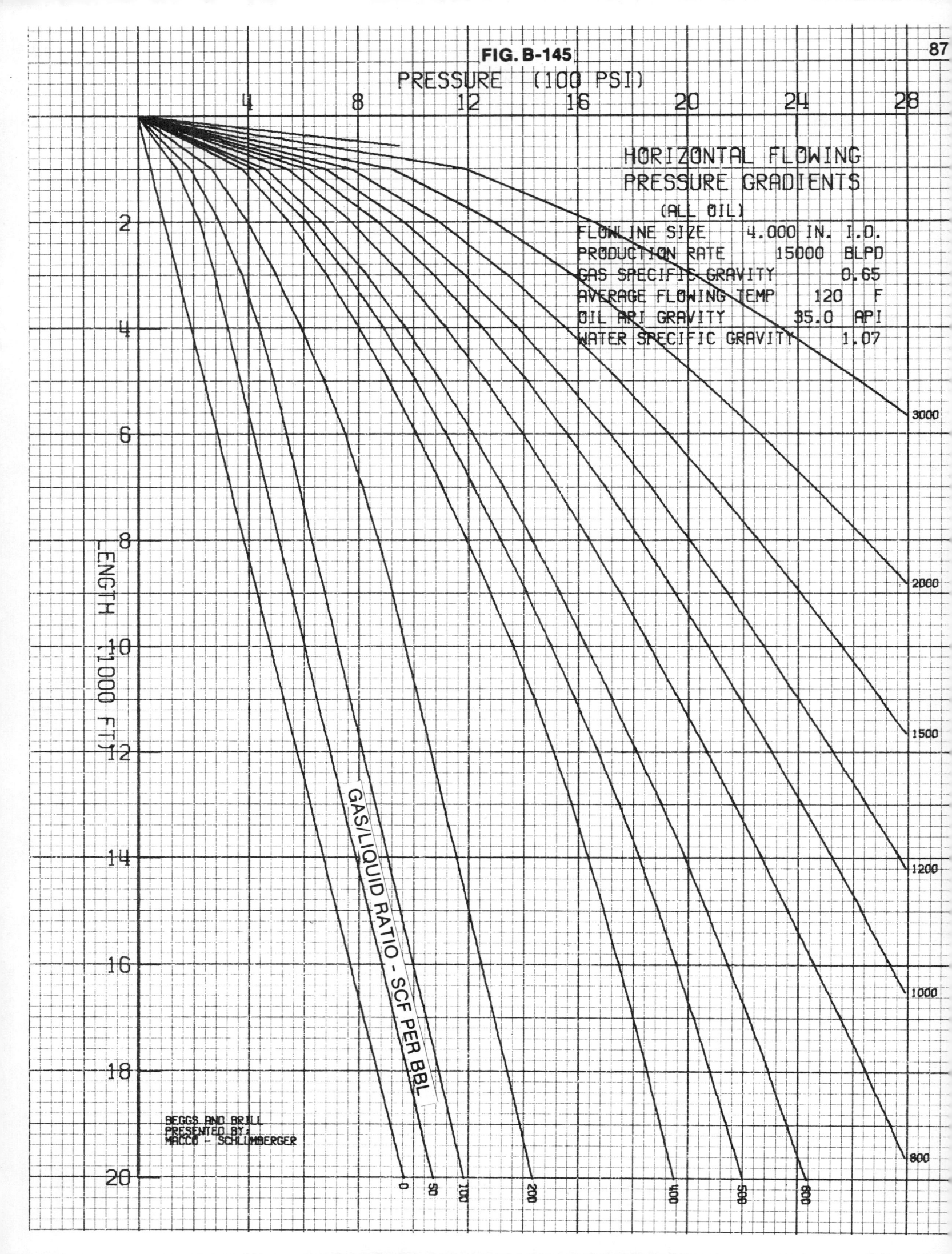

FIG. B-145

87

PRESSURE (100 PSI)

HORIZONTAL FLOWING
PRESSURE GRADIENTS
(ALL OIL)

FLOWLINE SIZE	4.000 IN. I.D.
PRODUCTION RATE	15000 BLPD
GAS SPECIFIC GRAVITY	0.65
AVERAGE FLOWING TEMP	120 F
OIL API GRAVITY	35.0 API
WATER SPECIFIC GRAVITY	1.07

LENGTH (1000 FT)

GAS/LIQUID RATIO - SCF PER BBL

BEGGS AND BRILL
PRESENTED BY:
MACCO - SCHLUMBERGER

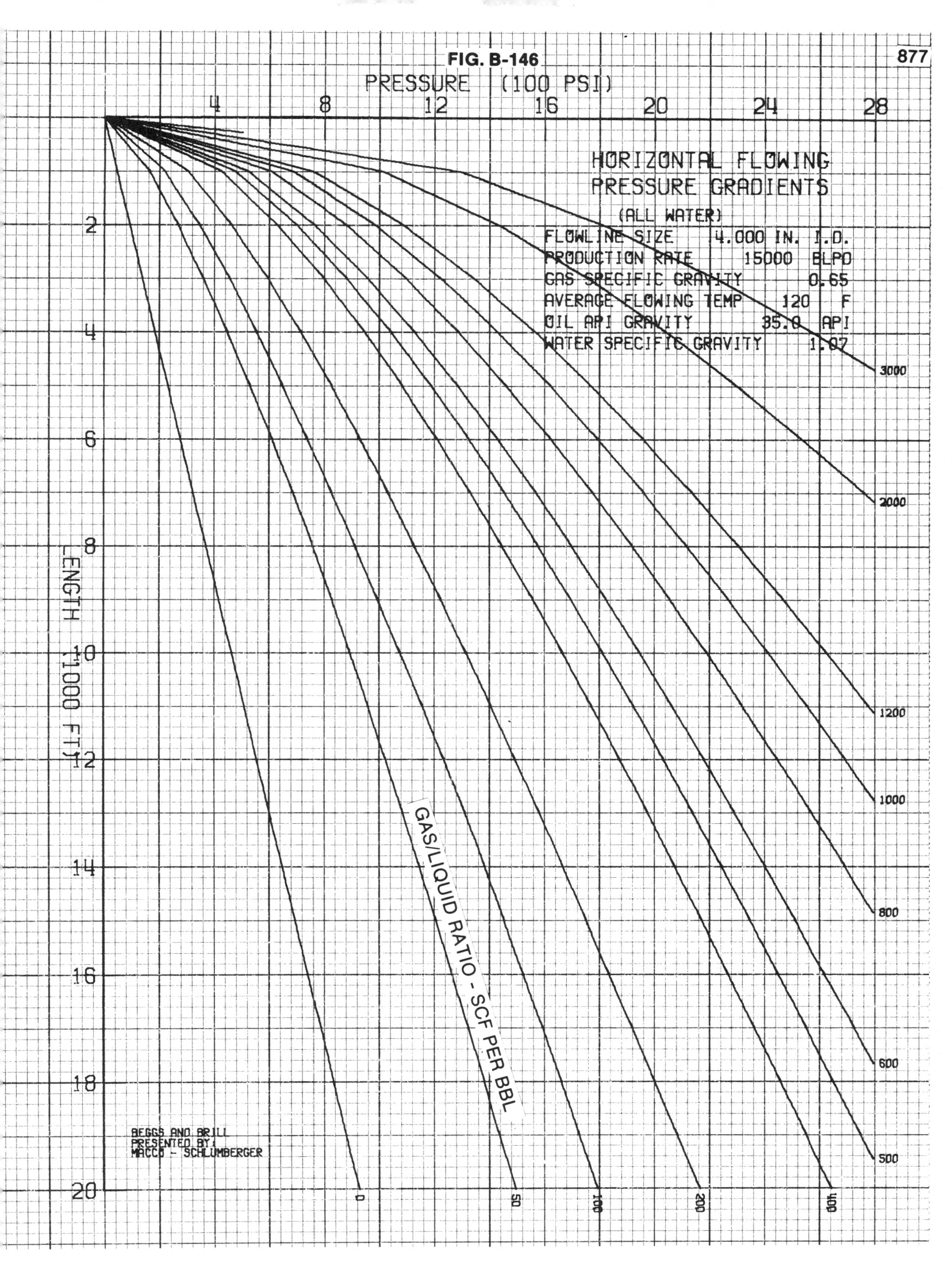

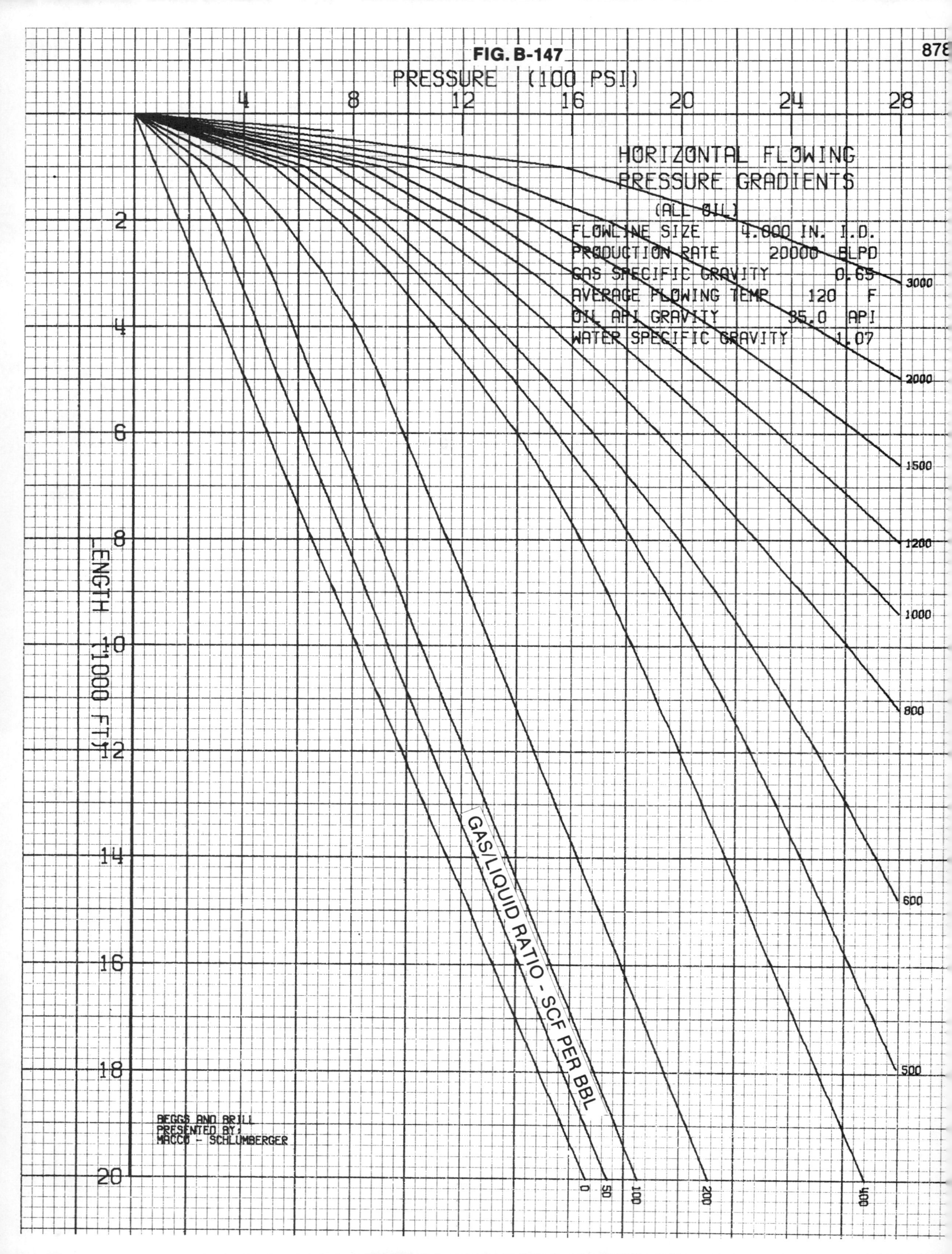

FIG. B-147

878

PRESSURE (100 PSI)

HORIZONTAL FLOWING
PRESSURE GRADIENTS
(ALL OIL)

FLOWLINE SIZE	4.000 IN. I.D.
PRODUCTION RATE	20000 BLPD
GAS SPECIFIC GRAVITY	0.65
AVERAGE FLOWING TEMP	120 F
OIL API GRAVITY	35.0 API
WATER SPECIFIC GRAVITY	1.07

LENGTH (1000 FT)

GAS/LIQUID RATIO - SCF PER BBL

BEGGS AND BRILL
PRESENTED BY:
MACCO - SCHLUMBERGER

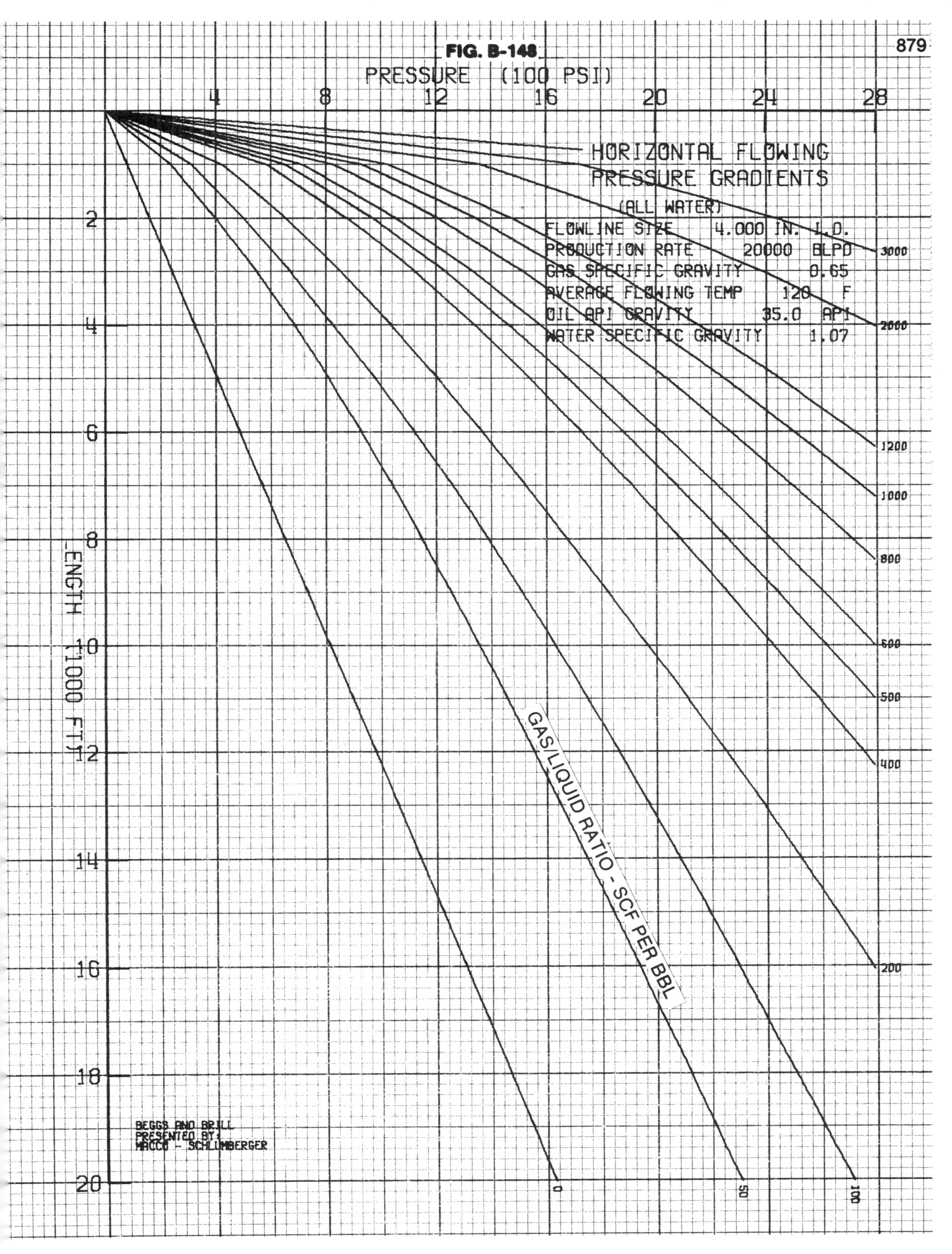

FIG. B-148

PRESSURE (100 PSI)

HORIZONTAL FLOWING
PRESSURE GRADIENTS
(ALL WATER)

FLOWLINE SIZE	4.000 IN. I.D.
PRODUCTION RATE	20000 BLPD
GAS SPECIFIC GRAVITY	0.65
AVERAGE FLOWING TEMP	120 F
OIL API GRAVITY	35.0 API
WATER SPECIFIC GRAVITY	1.07

LENGTH (1000 FT)

GAS/LIQUID RATIO - SCF PER BBL

BEGGS AND BRILL
PRESENTED BY:
MACCO - SCHLUMBERGER

879

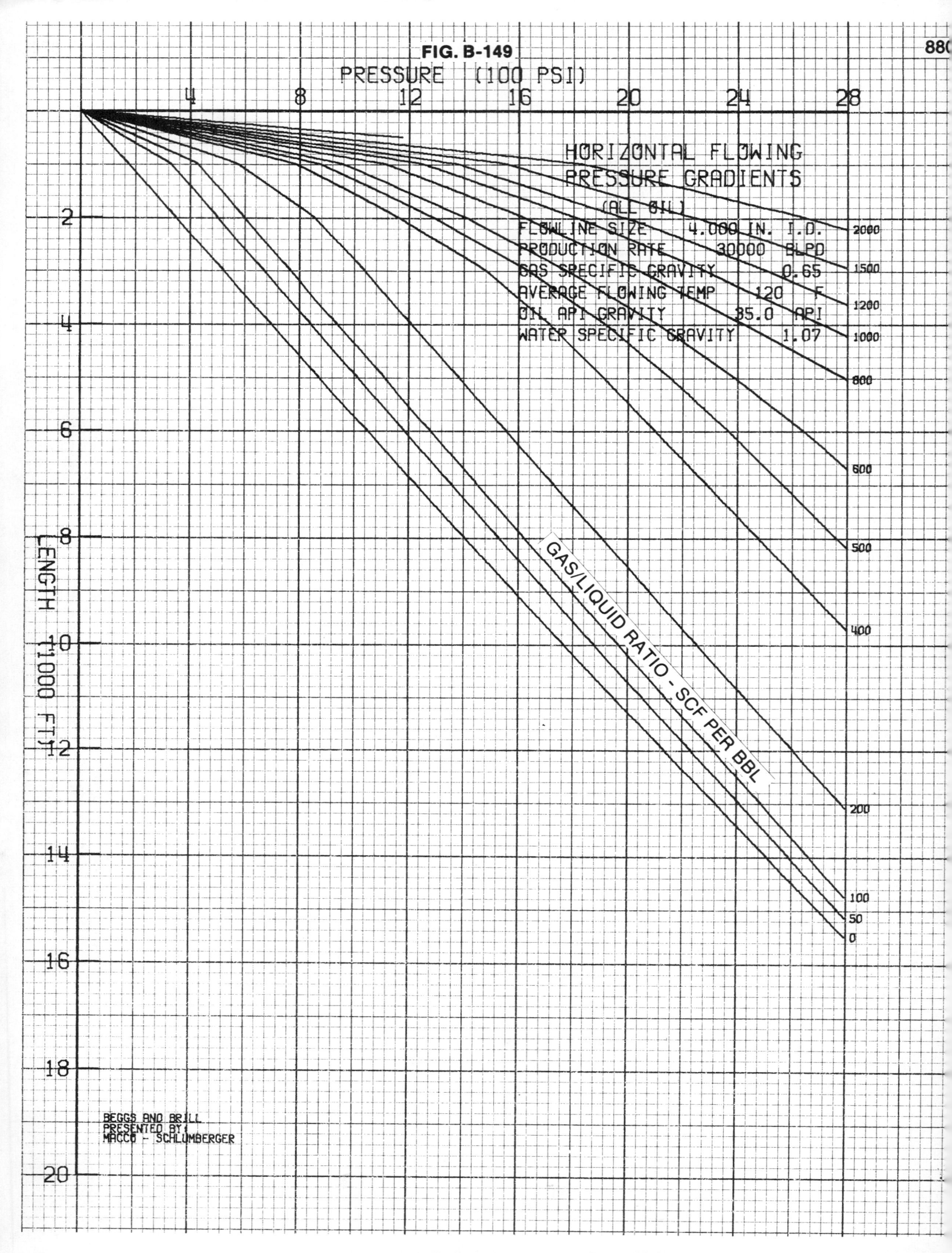

FIG. B-149

880

HORIZONTAL FLOWING
PRESSURE GRADIENTS
(ALL OIL)

FLOWLINE SIZE	4.000 IN. I.D.
PRODUCTION RATE	30000 BLPD
GAS SPECIFIC GRAVITY	0.65
AVERAGE FLOWING TEMP	120 F
OIL API GRAVITY	35.0 API
WATER SPECIFIC GRAVITY	1.07

GAS/LIQUID RATIO - SCF PER BBL

PRESSURE (100 PSI)

LENGTH (1000 FT.)

BEGGS AND BRILL
PRESENTED BY:
MACCO - SCHLUMBERGER

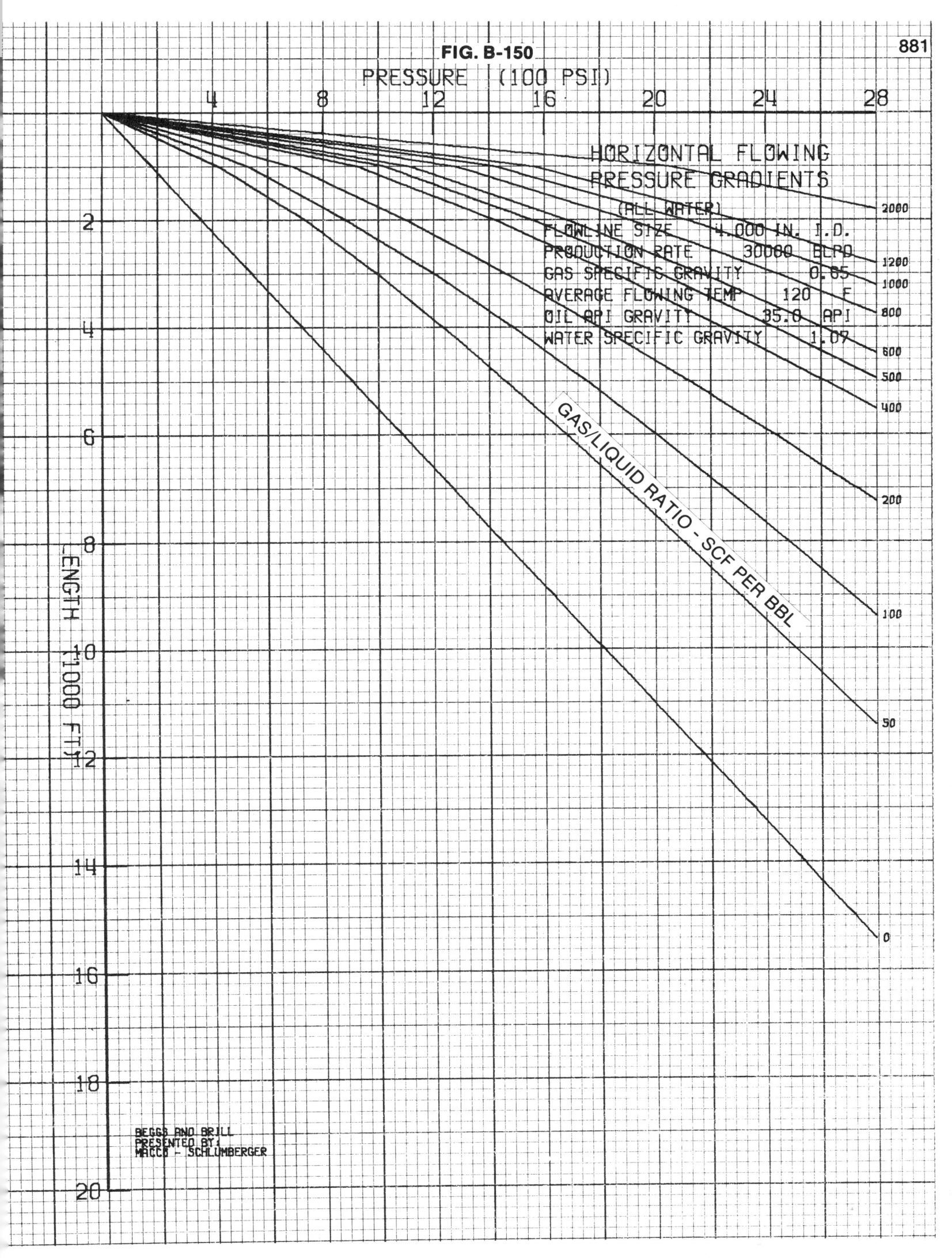

FIG. B-150

HORIZONTAL FLOWING
PRESSURE GRADIENTS
(ALL WATER)

FLOWLINE SIZE	4.000 IN. I.D.
PRODUCTION RATE	30000 BLPD
GAS SPECIFIC GRAVITY	0.65
AVERAGE FLOWING TEMP	120 F
OIL API GRAVITY	35.0 API
WATER SPECIFIC GRAVITY	1.07

GAS/LIQUID RATIO - SCF PER BBL

BEGGS AND BRILL
PRESENTED BY:
MACCO - SCHLUMBERGER

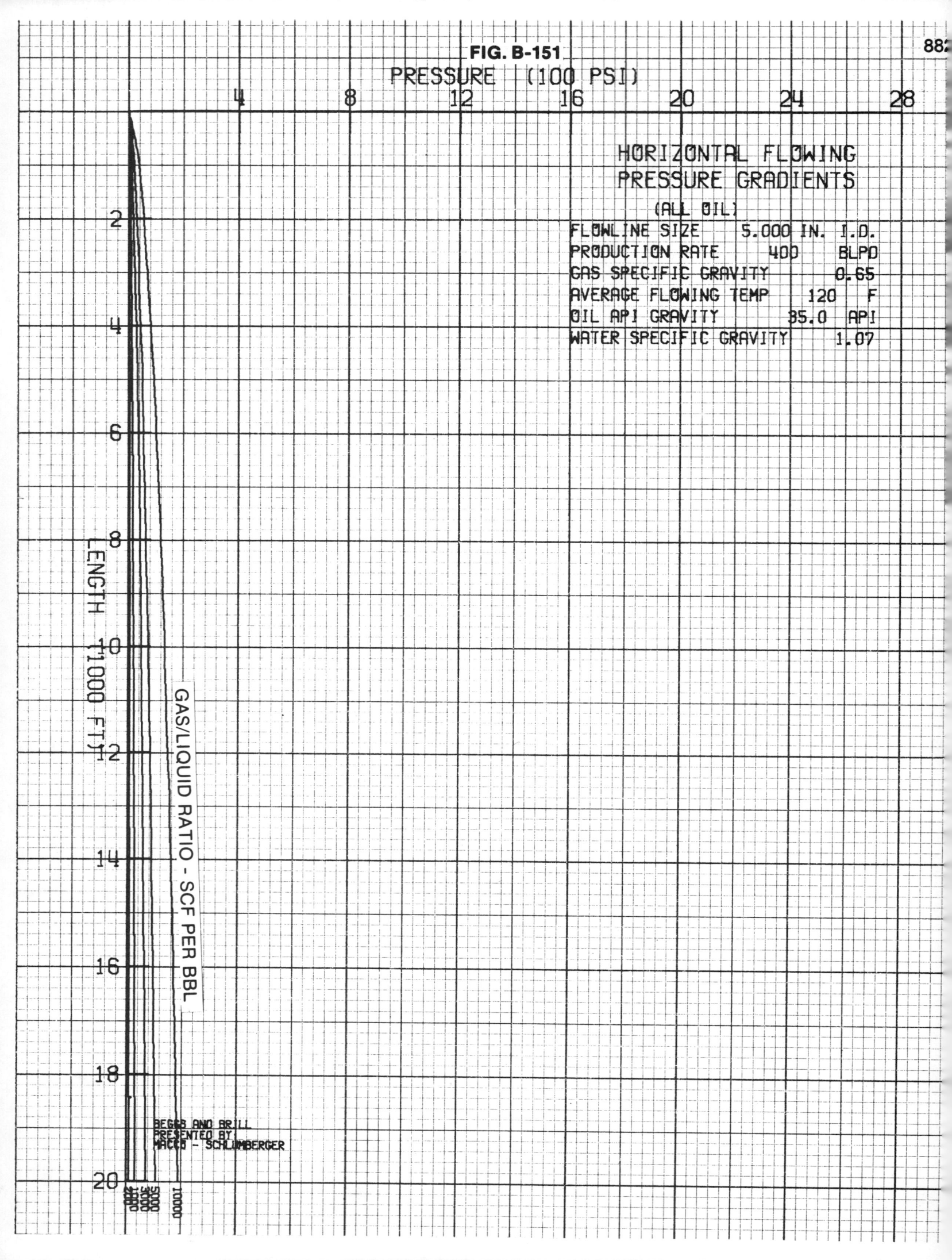

FIG. B-151

PRESSURE (100 PSI)

HORIZONTAL FLOWING
PRESSURE GRADIENTS
(ALL OIL)

FLOWLINE SIZE	5.000 IN.	I.D.
PRODUCTION RATE	400	BLPD
GAS SPECIFIC GRAVITY	0.65	
AVERAGE FLOWING TEMP	120	F
OIL API GRAVITY	35.0	API
WATER SPECIFIC GRAVITY	1.07	

LENGTH (1000 FT)

GAS/LIQUID RATIO - SCF PER BBL

BEGGS AND BRILL
PRESENTED BY
WACCO - SCHLUMBERGER

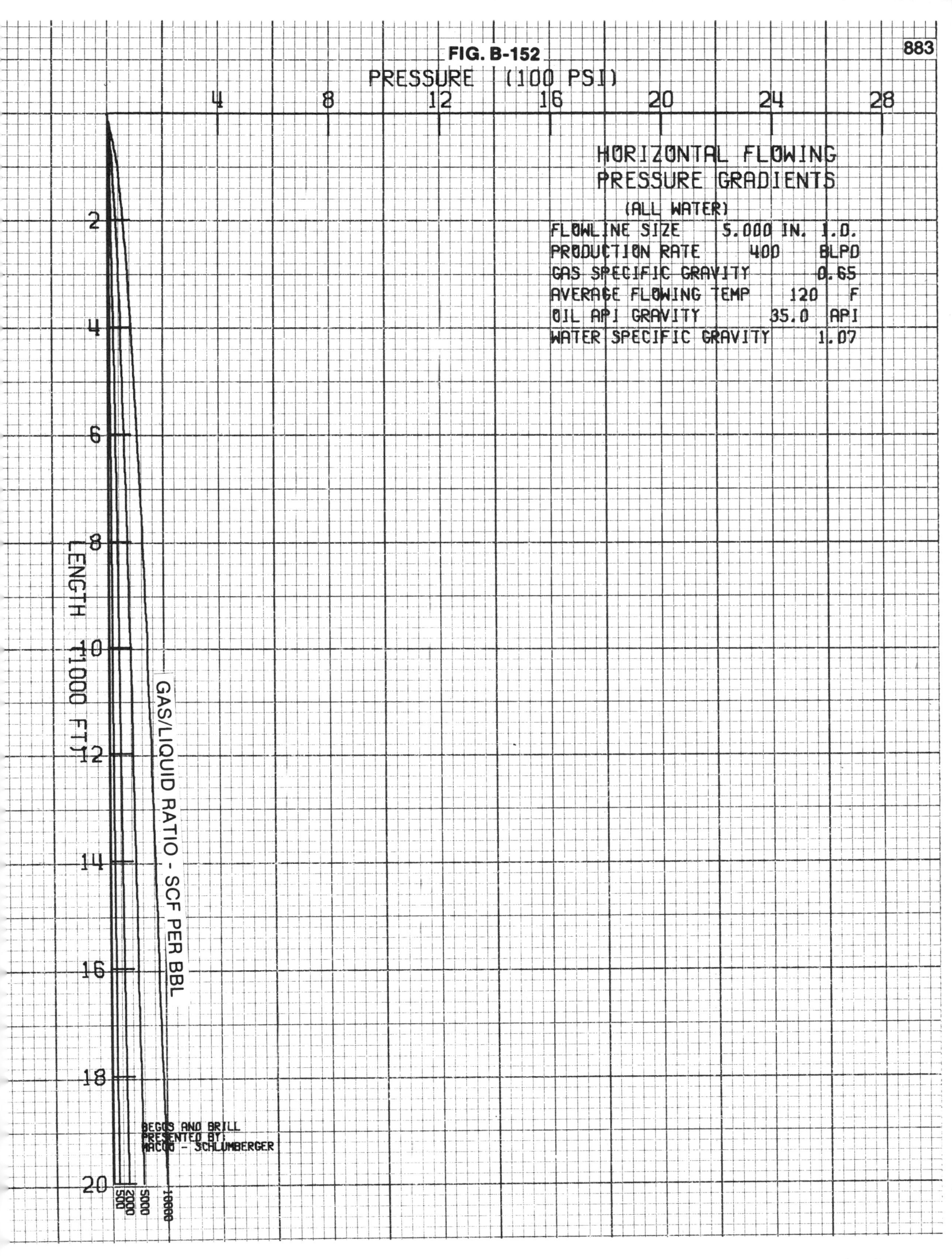

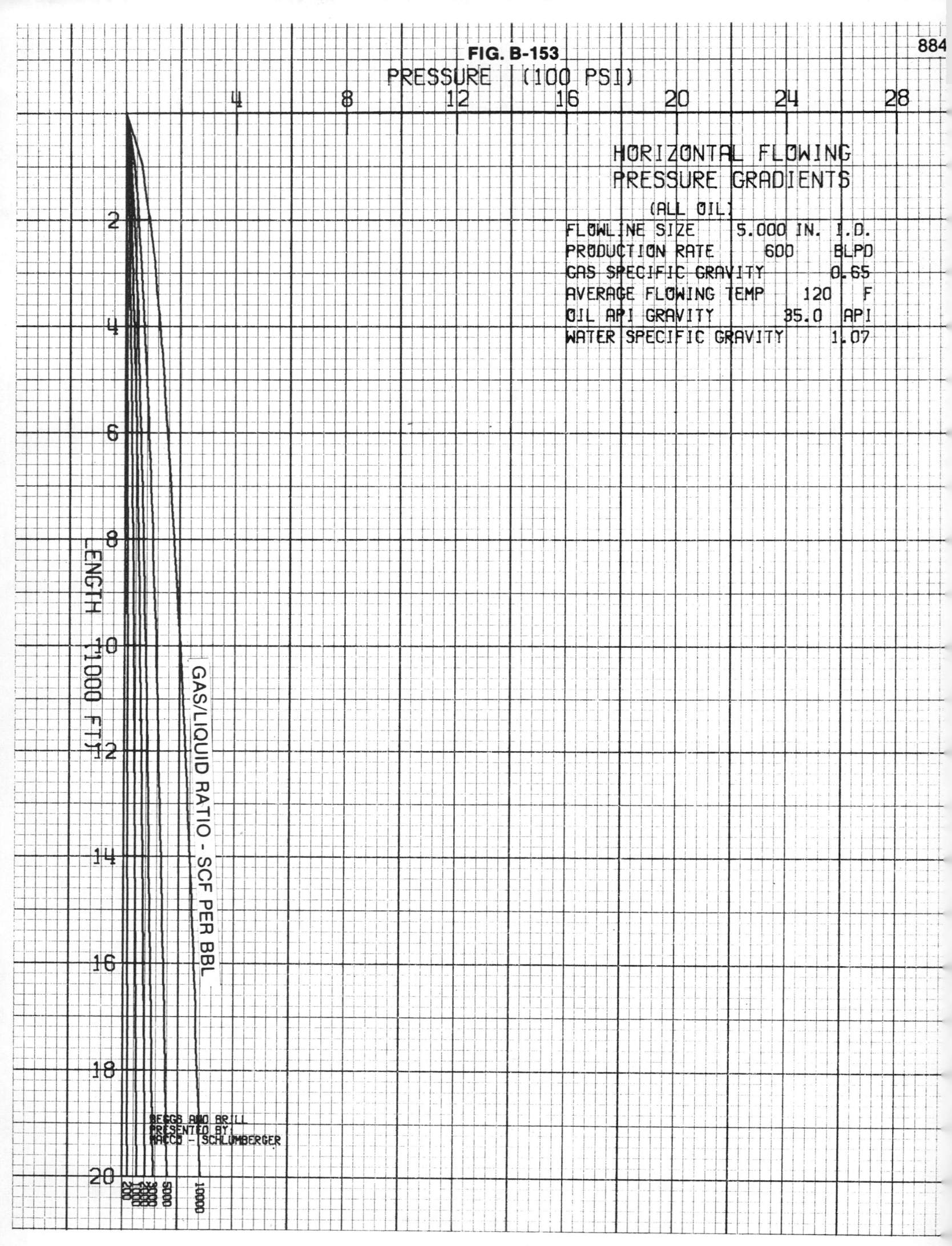

FIG. B-153

PRESSURE (100 PSI)

LENGTH (1000 FT)

GAS/LIQUID RATIO - SCF PER BBL

HORIZONTAL FLOWING
PRESSURE GRADIENTS
(ALL OIL)

FLOWLINE SIZE	5.000 IN. I.D.
PRODUCTION RATE	600 BLPD
GAS SPECIFIC GRAVITY	0.65
AVERAGE FLOWING TEMP	120 F
OIL API GRAVITY	35.0 API
WATER SPECIFIC GRAVITY	1.07

BEGGS AND BRILL
PRESENTED BY
MACCO - SCHLUMBERGER

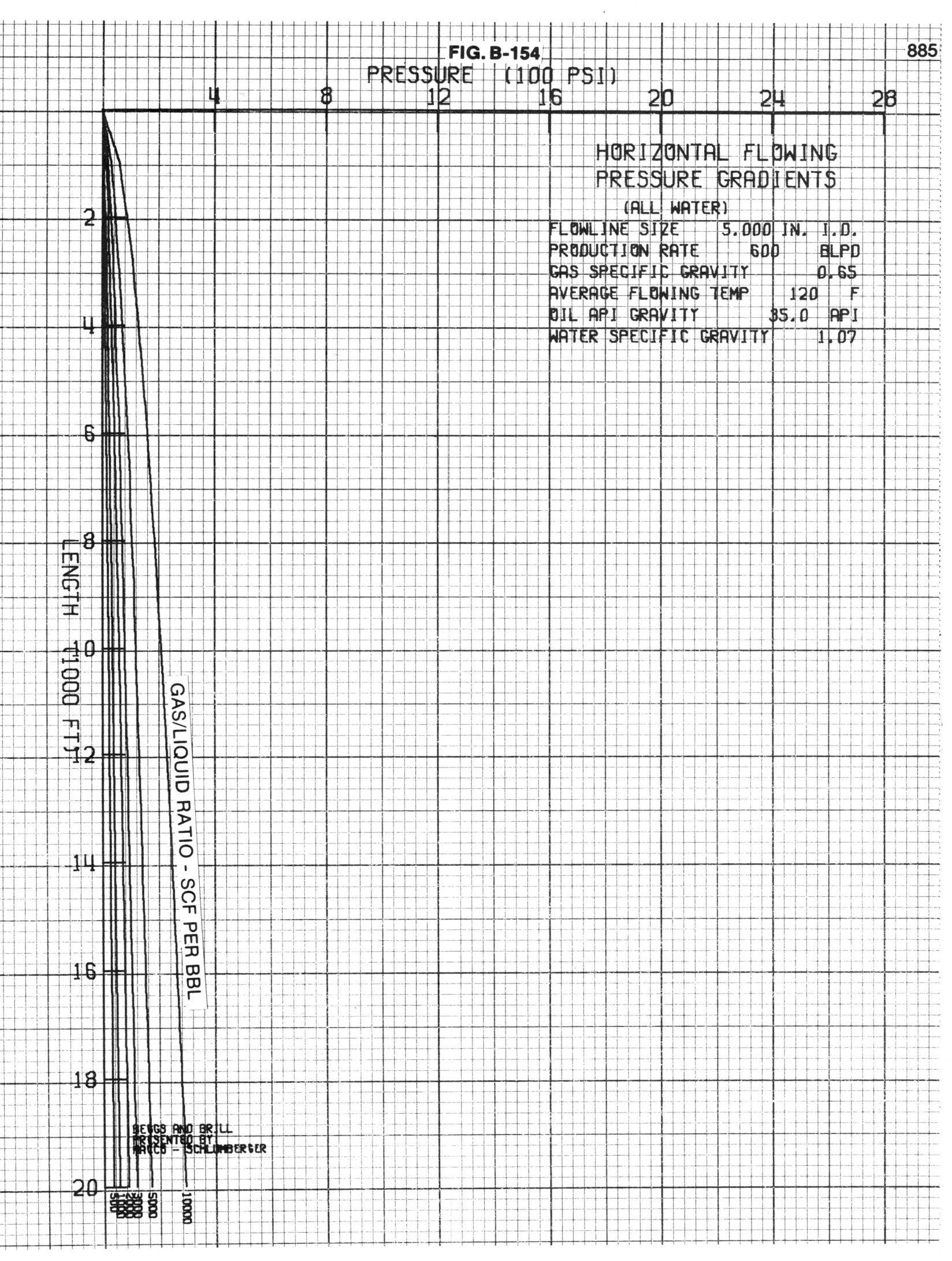

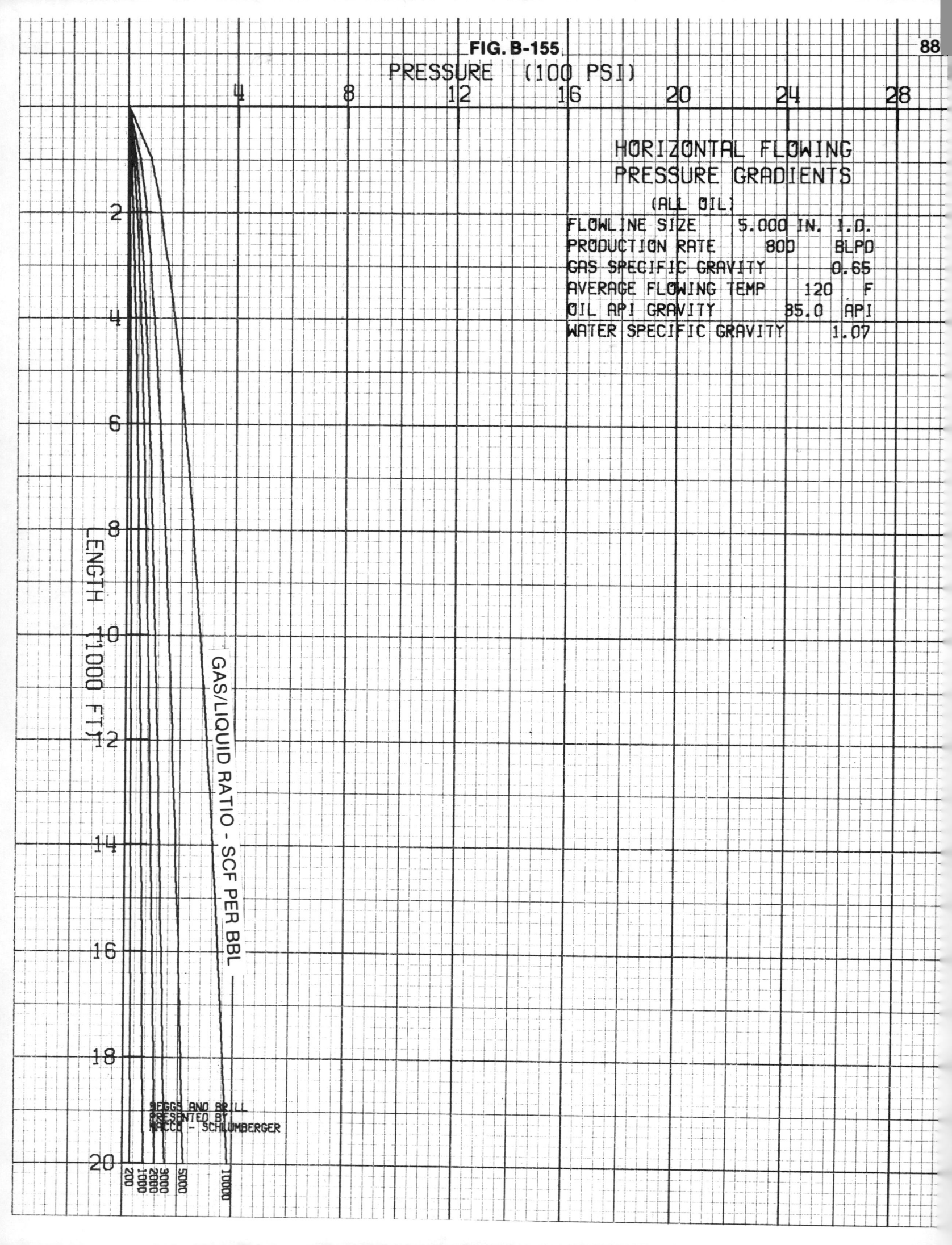

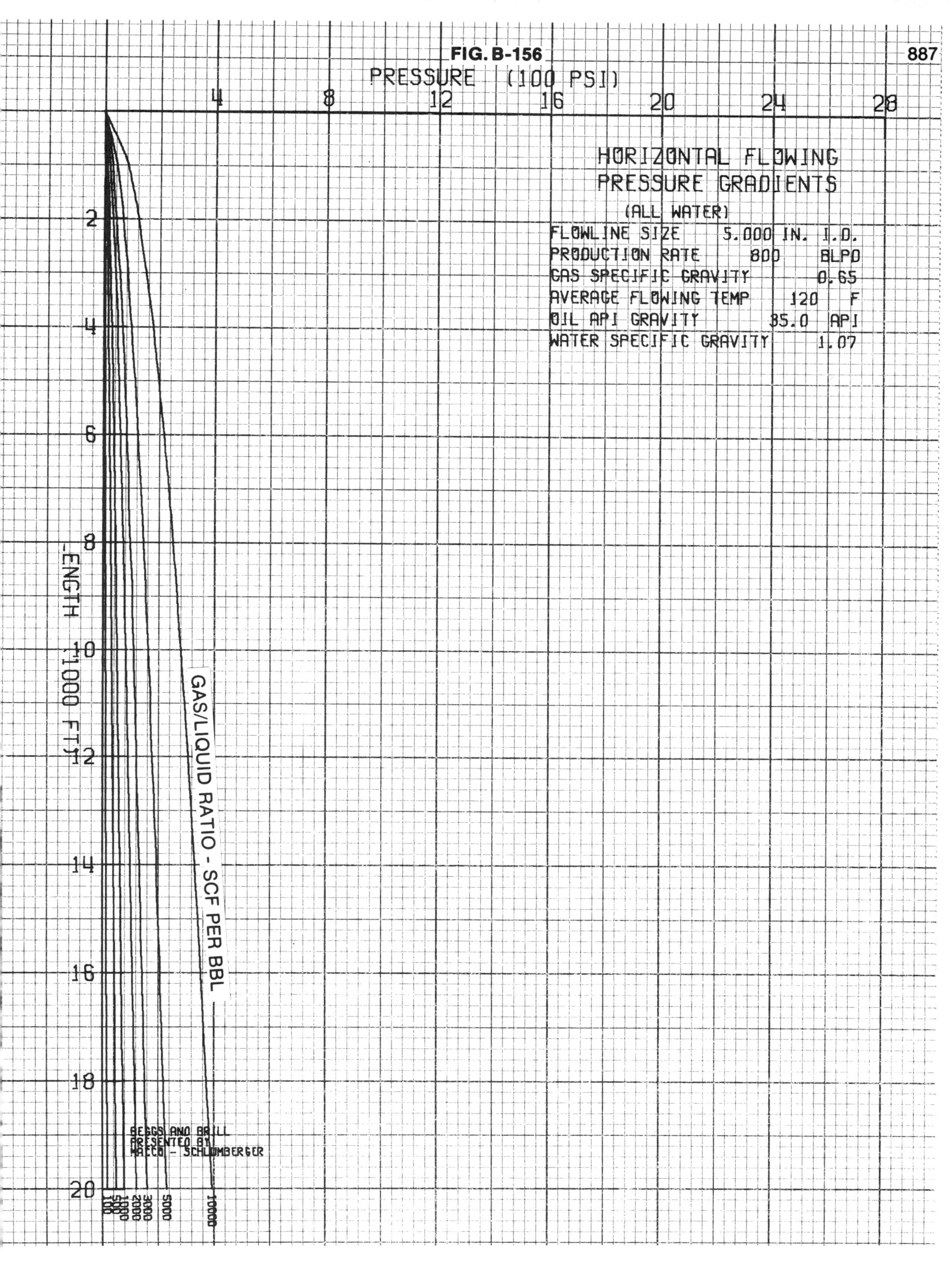

FIG. B-156

PRESSURE (100 PSI)

887

HORIZONTAL FLOWING
PRESSURE GRADIENTS
(ALL WATER)

FLOWLINE SIZE	5.000 IN. I.D.
PRODUCTION RATE	800 BLPD
GAS SPECIFIC GRAVITY	0.65
AVERAGE FLOWING TEMP	120 F
OIL API GRAVITY	35.0 API
WATER SPECIFIC GRAVITY	1.07

LENGTH (1000 FT.)

GAS/LIQUID RATIO - SCF PER BBL

BEGGS AND BRILL
PRESENTED BY:
MAECO - SCHLUMBERGER

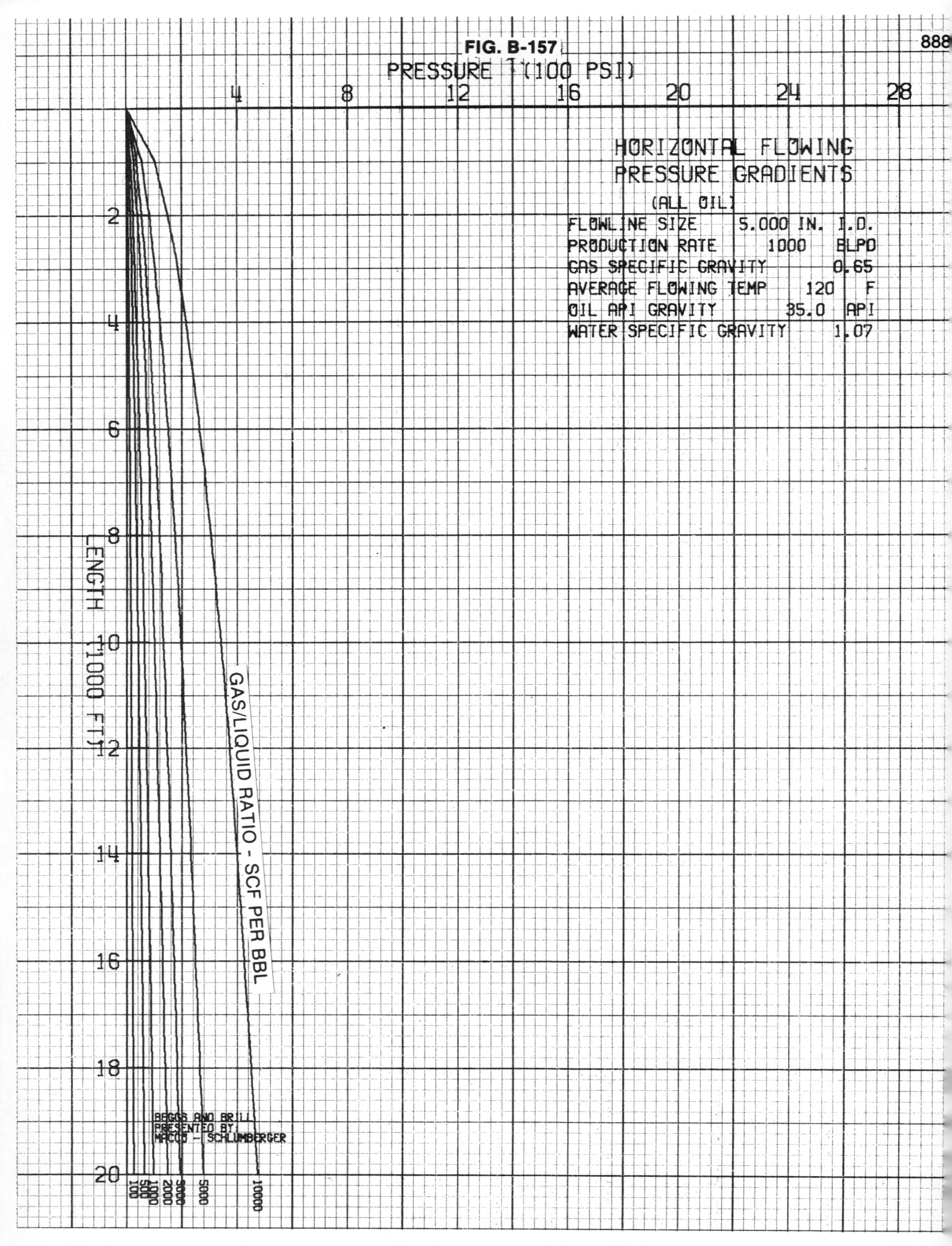

FIG. B-157

888

PRESSURE (100 PSI)

HORIZONTAL FLOWING
PRESSURE GRADIENTS
(ALL OIL)

FLOWLINE SIZE	5.000 IN. I.D.
PRODUCTION RATE	1000 BLPD
GAS SPECIFIC GRAVITY	0.65
AVERAGE FLOWING TEMP	120 F
OIL API GRAVITY	35.0 API
WATER SPECIFIC GRAVITY	1.07

LENGTH (1000 FT)

GAS/LIQUID RATIO - SCF PER BBL

BRIGGS AND BRILL
PRESENTED BY
MACCO - SCHLUMBERGER

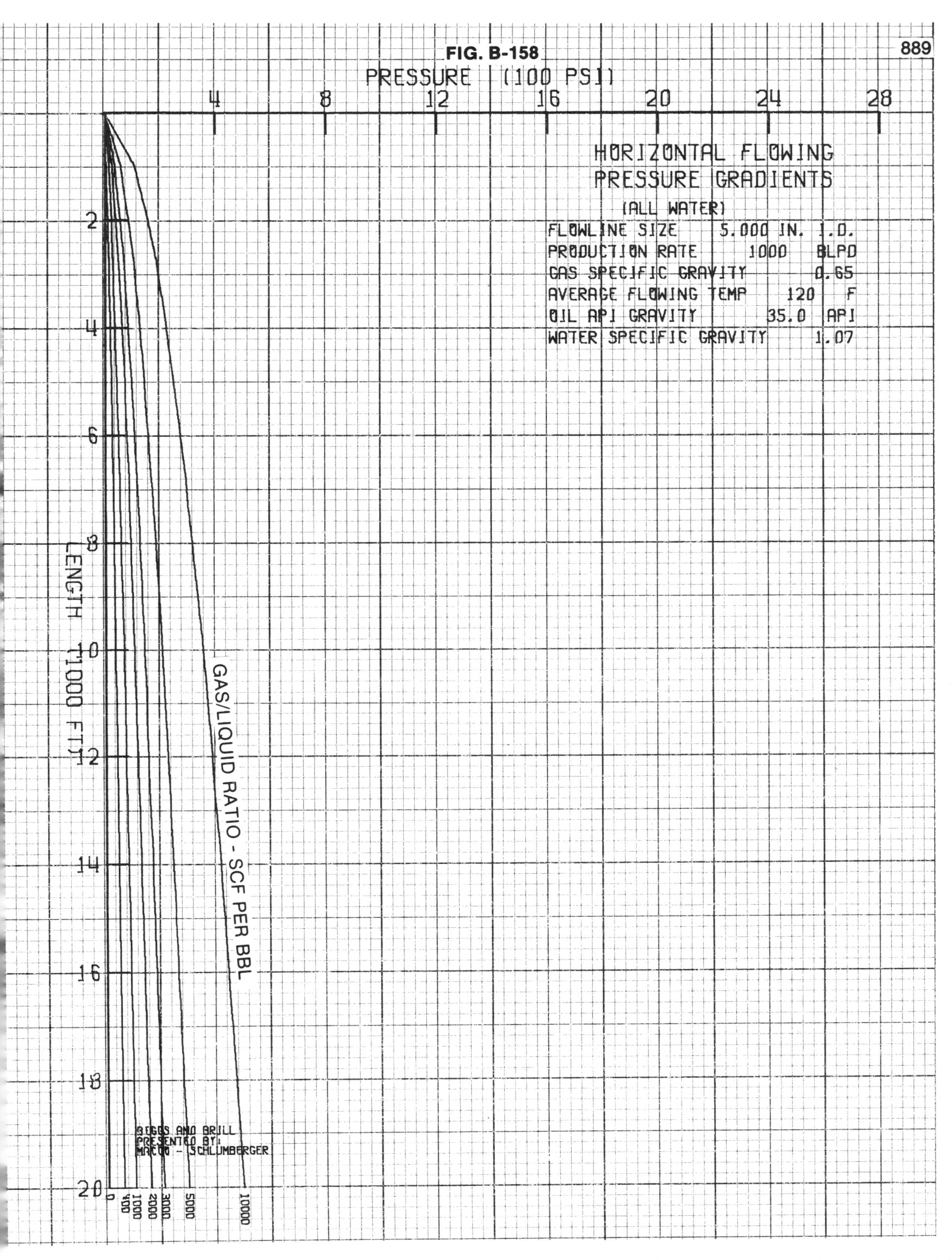

PRESSURE (100 PSI)

HORIZONTAL FLOWING
PRESSURE GRADIENTS
(ALL WATER)

FLOWLINE SIZE	5.000 IN.	I.D.
PRODUCTION RATE	1000	BLPD
GAS SPECIFIC GRAVITY	0.65	
AVERAGE FLOWING TEMP	120	F
OIL API GRAVITY	35.0	API
WATER SPECIFIC GRAVITY	1.07	

LENGTH (1000 FT)

GAS/LIQUID RATIO - SCF PER BBL

BEGGS AND BRILL
PRESENTED BY:
MACON - SCHLUMBERGER

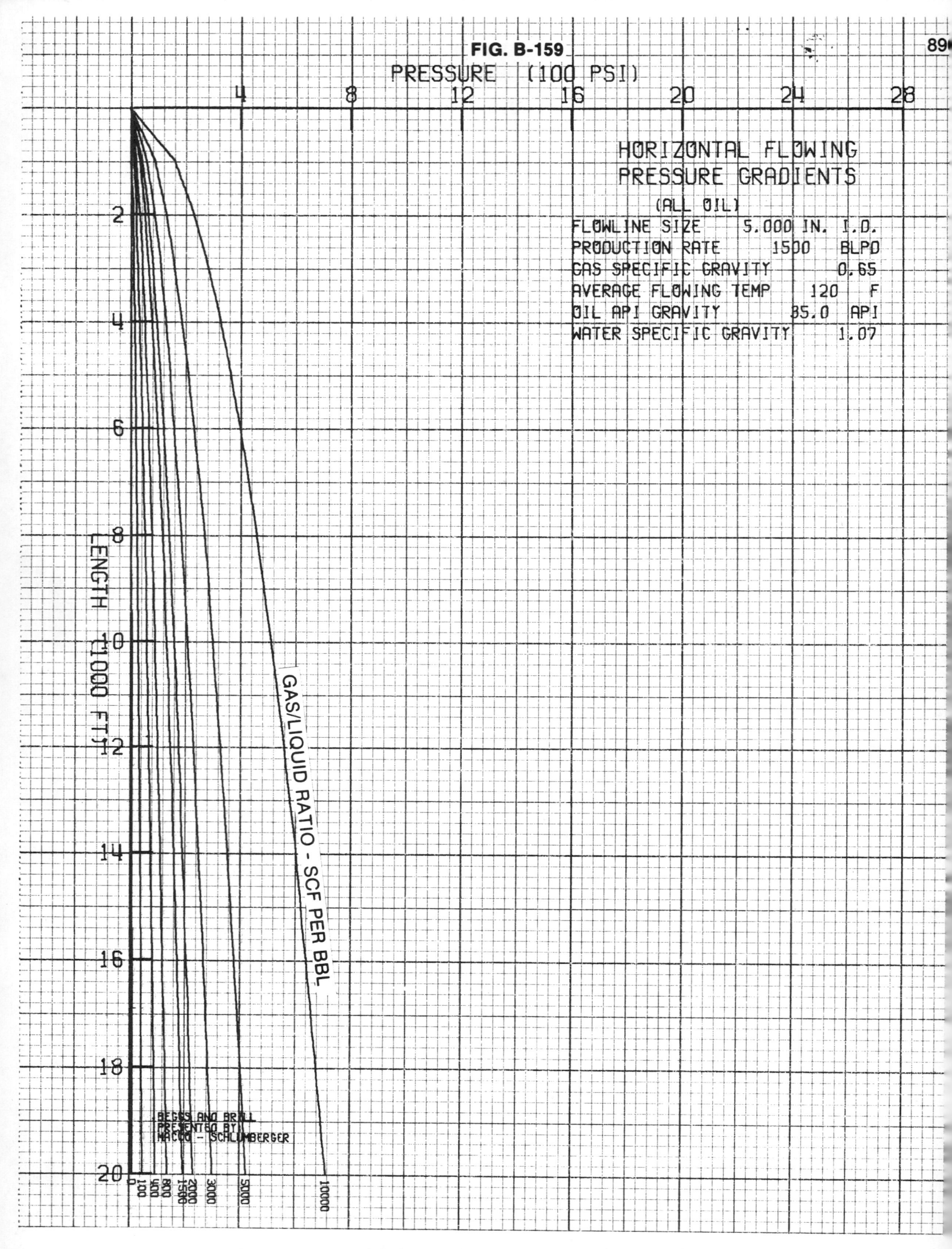

HORIZONTAL FLOWING
PRESSURE GRADIENTS
(ALL OIL)

FLOWLINE SIZE   5.000 IN. I.D.
PRODUCTION RATE   1500   BLPD
GAS SPECIFIC GRAVITY   0.65
AVERAGE FLOWING TEMP   120   F
OIL API GRAVITY   35.0 API
WATER SPECIFIC GRAVITY   1.07

PRESSURE (100 PSI)

LENGTH (1000 FT)

GAS/LIQUID RATIO - SCF PER BBL

BEGGS AND BRILL
PRESENTED BY
HACCO - SCHLUMBERGER

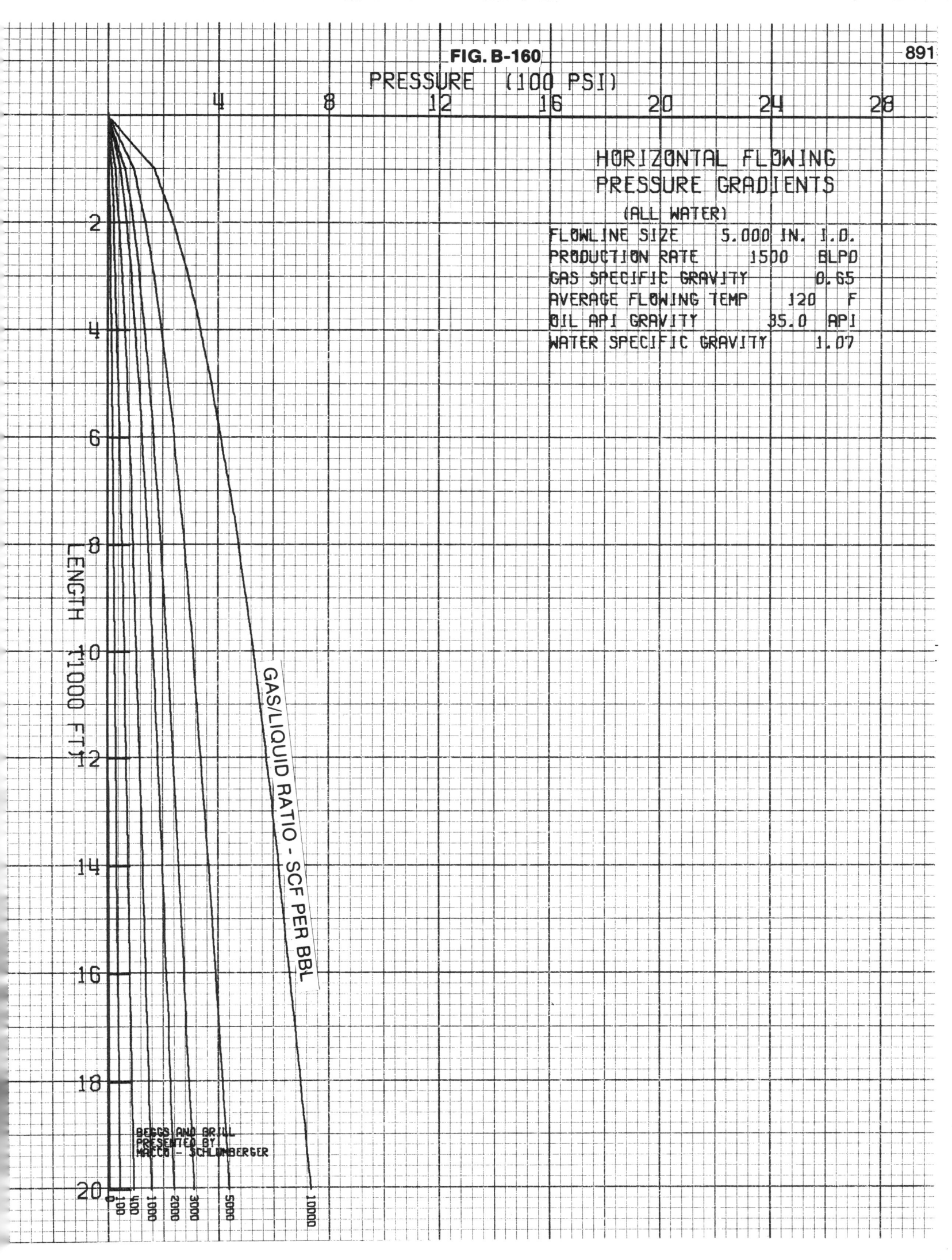

FIG. B-160

891

PRESSURE (100 PSI)

HORIZONTAL FLOWING
PRESSURE GRADIENTS
(ALL WATER)

FLOWLINE SIZE	5.000 IN.	I.D.
PRODUCTION RATE	1500	BLPD
GAS SPECIFIC GRAVITY	0.65	
AVERAGE FLOWING TEMP	120	F
OIL API GRAVITY	35.0	API
WATER SPECIFIC GRAVITY	1.07	

LENGTH (1000 FT)

GAS/LIQUID RATIO - SCF PER BBL

BEGGS AND BRILL
PRESENTED BY
MAECO1 - SCHLUMBERGER

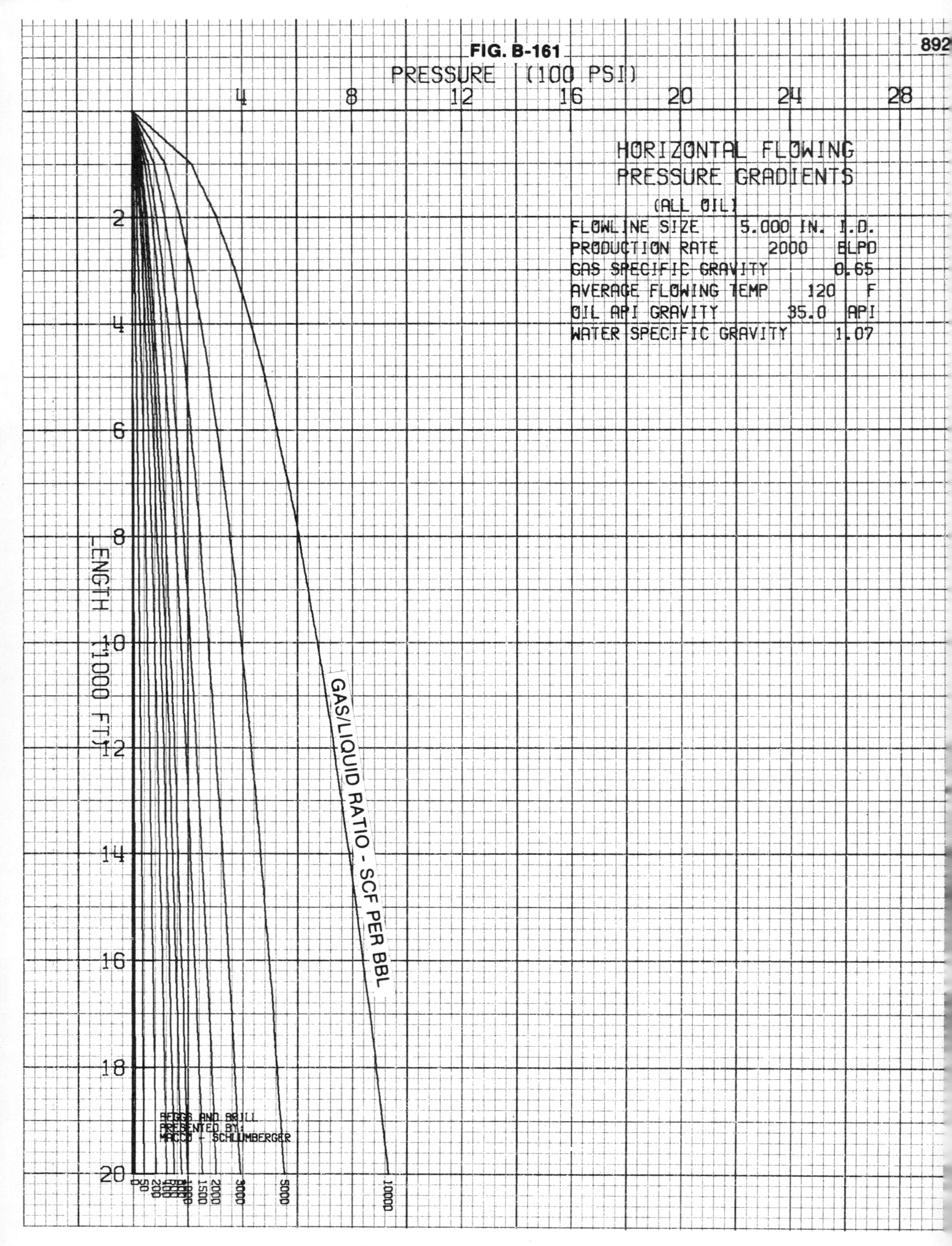

PRESSURE (100 PSI)

HORIZONTAL FLOWING
PRESSURE GRADIENTS
(ALL OIL)

FLOWLINE SIZE	5.000 IN. I.D.
PRODUCTION RATE	2000 BLPD
GAS SPECIFIC GRAVITY	0.65
AVERAGE FLOWING TEMP	120 F
OIL API GRAVITY	35.0 API
WATER SPECIFIC GRAVITY	1.07

LENGTH (1000 FT)

GAS/LIQUID RATIO - SCF PER BBL

BEGGS AND BRILL
PRESENTED BY
MACCO - SCHLUMBERGER

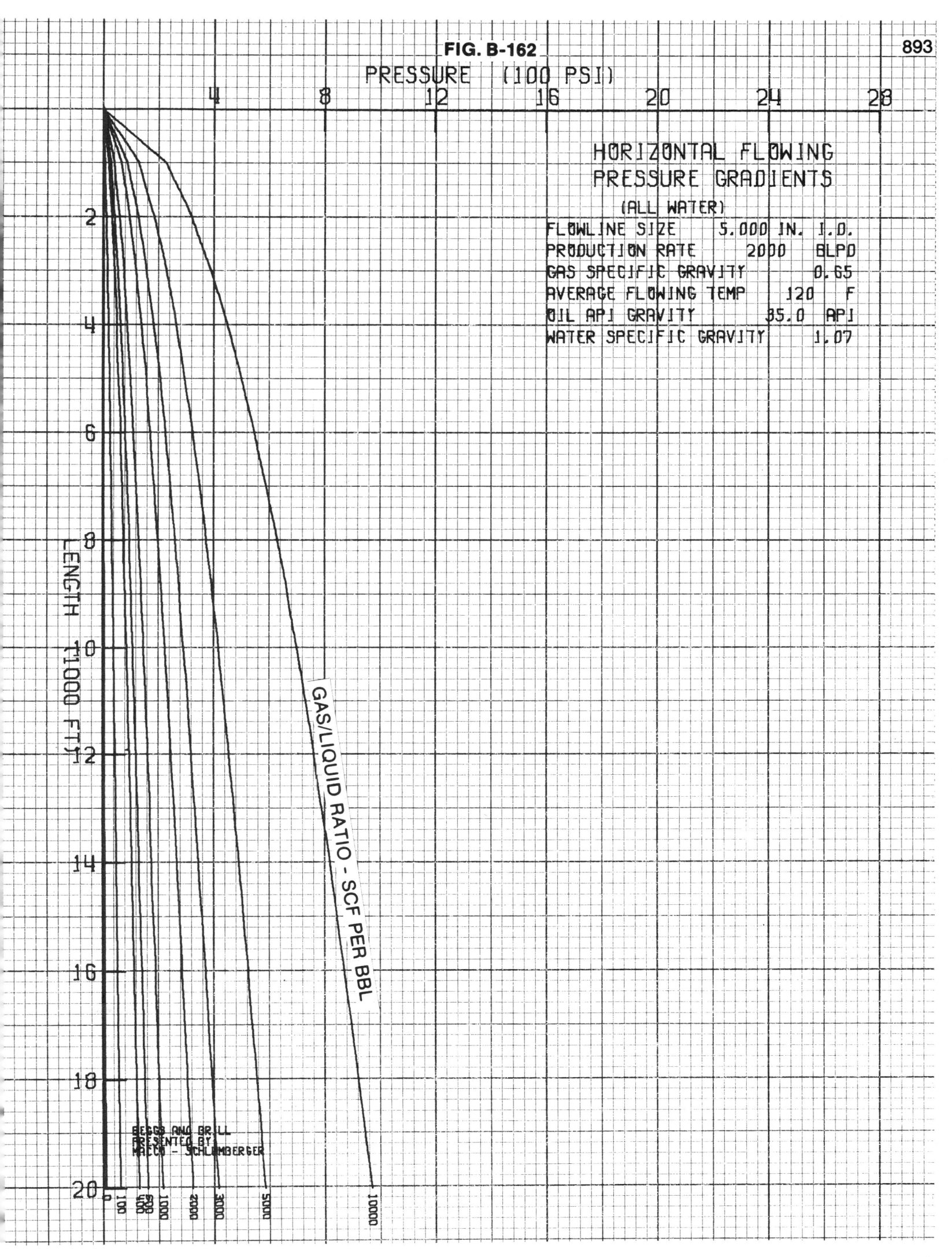

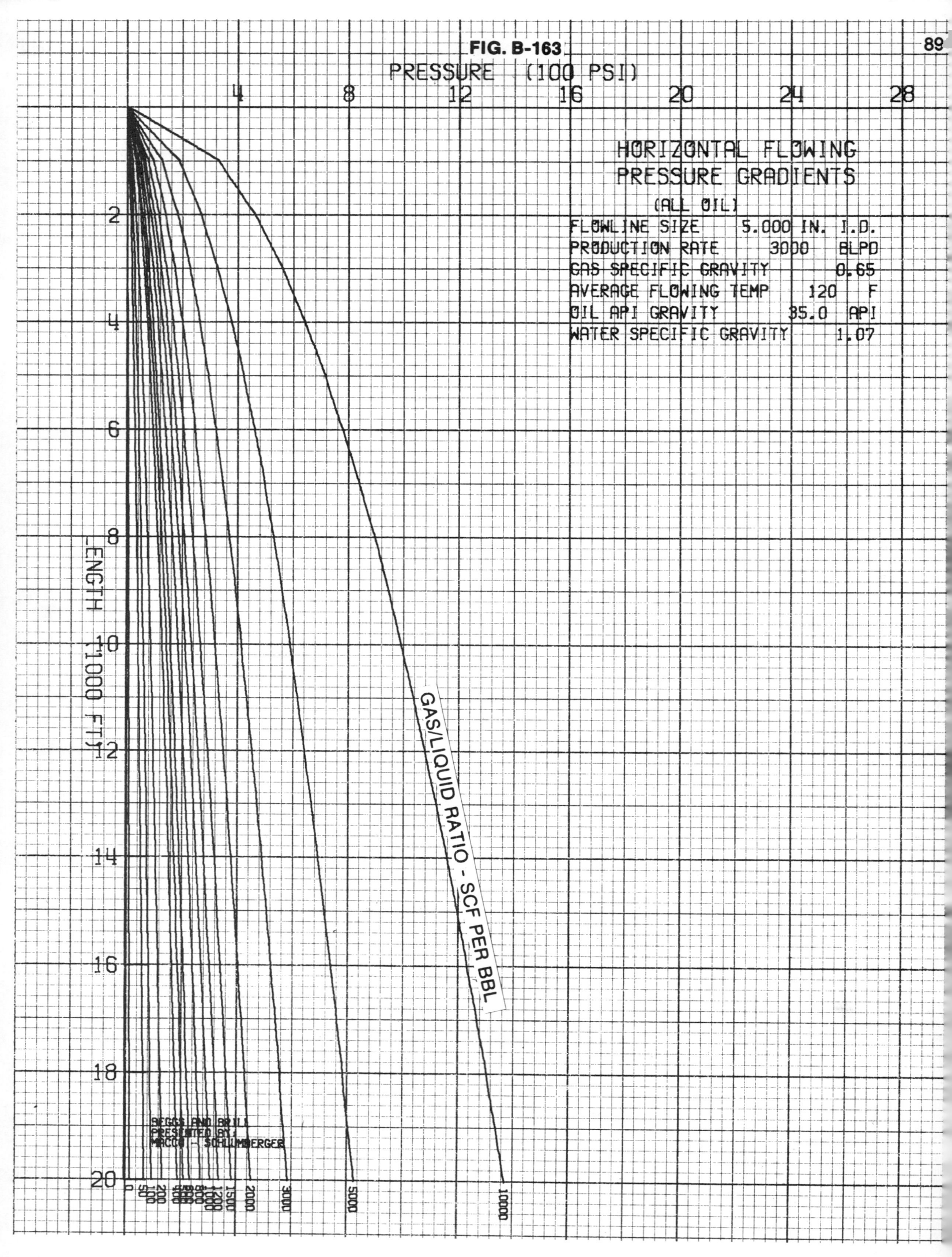

FIG. B-163

89

PRESSURE (100 PSI)

HORIZONTAL FLOWING
PRESSURE GRADIENTS
(ALL OIL)

FLOWLINE SIZE	5.000 IN. I.D.
PRODUCTION RATE	3000 BLPD
GAS SPECIFIC GRAVITY	0.65
AVERAGE FLOWING TEMP	120 F
OIL API GRAVITY	35.0 API
WATER SPECIFIC GRAVITY	1.07

LENGTH (1000 FT.)

GAS/LIQUID RATIO - SCF PER BBL

BEGGS AND BRILL
PRESENTED BY :
MACCO - SCHLUMBERGER

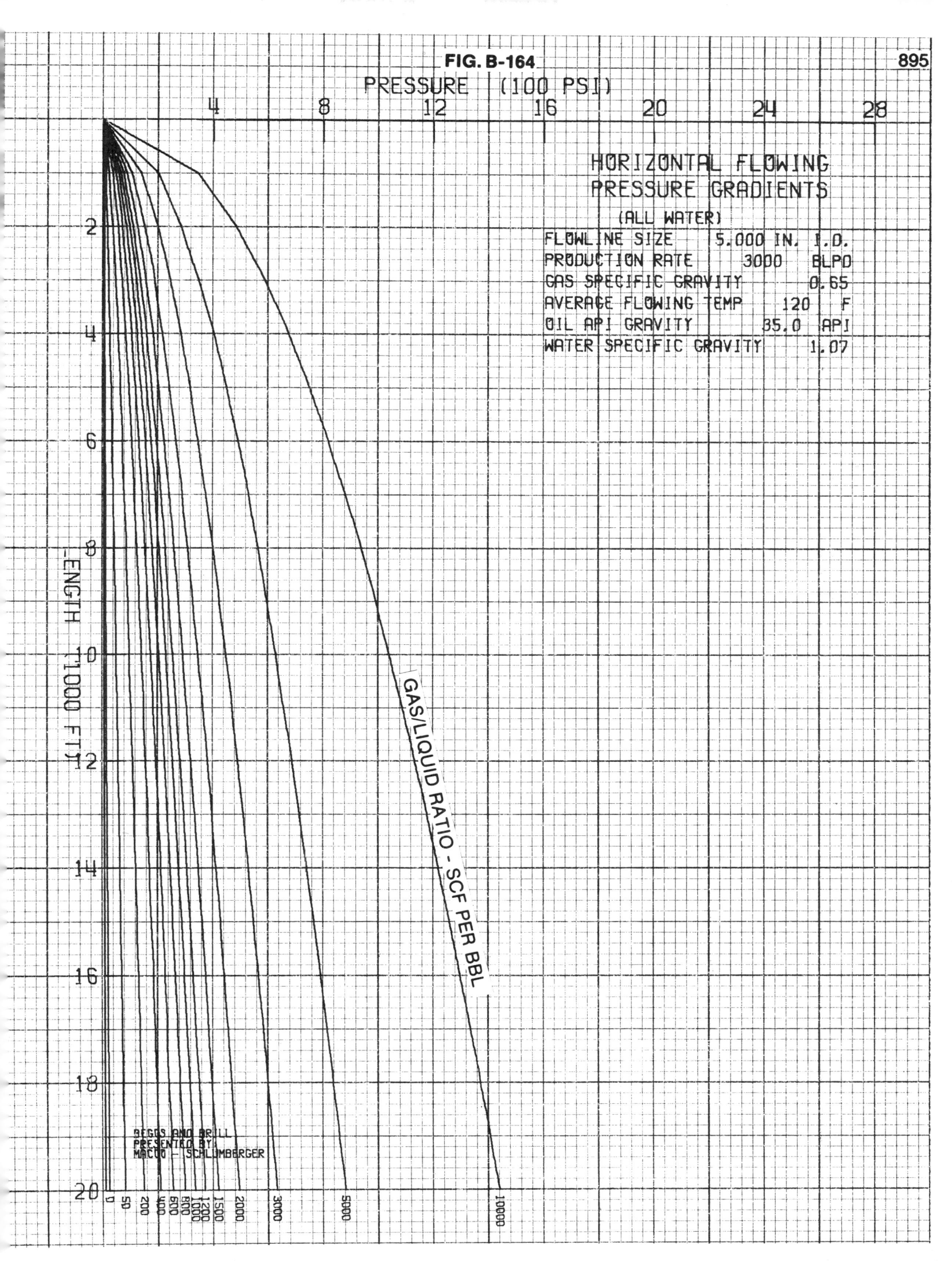

PRESSURE (100 PSI)

HORIZONTAL FLOWING
PRESSURE GRADIENTS
(ALL WATER)

FLOWLINE SIZE	5.000 IN. I.D.	
PRODUCTION RATE	3000	BLPD
GAS SPECIFIC GRAVITY	0.65	
AVERAGE FLOWING TEMP	120	F
OIL API GRAVITY	35.0	API
WATER SPECIFIC GRAVITY	1.07	

GAS/LIQUID RATIO - SCF PER BBL

LENGTH (1000 FT)

BEGGS AND BRILL
PRESENTED BY:
MACCO - SCHLUMBERGER

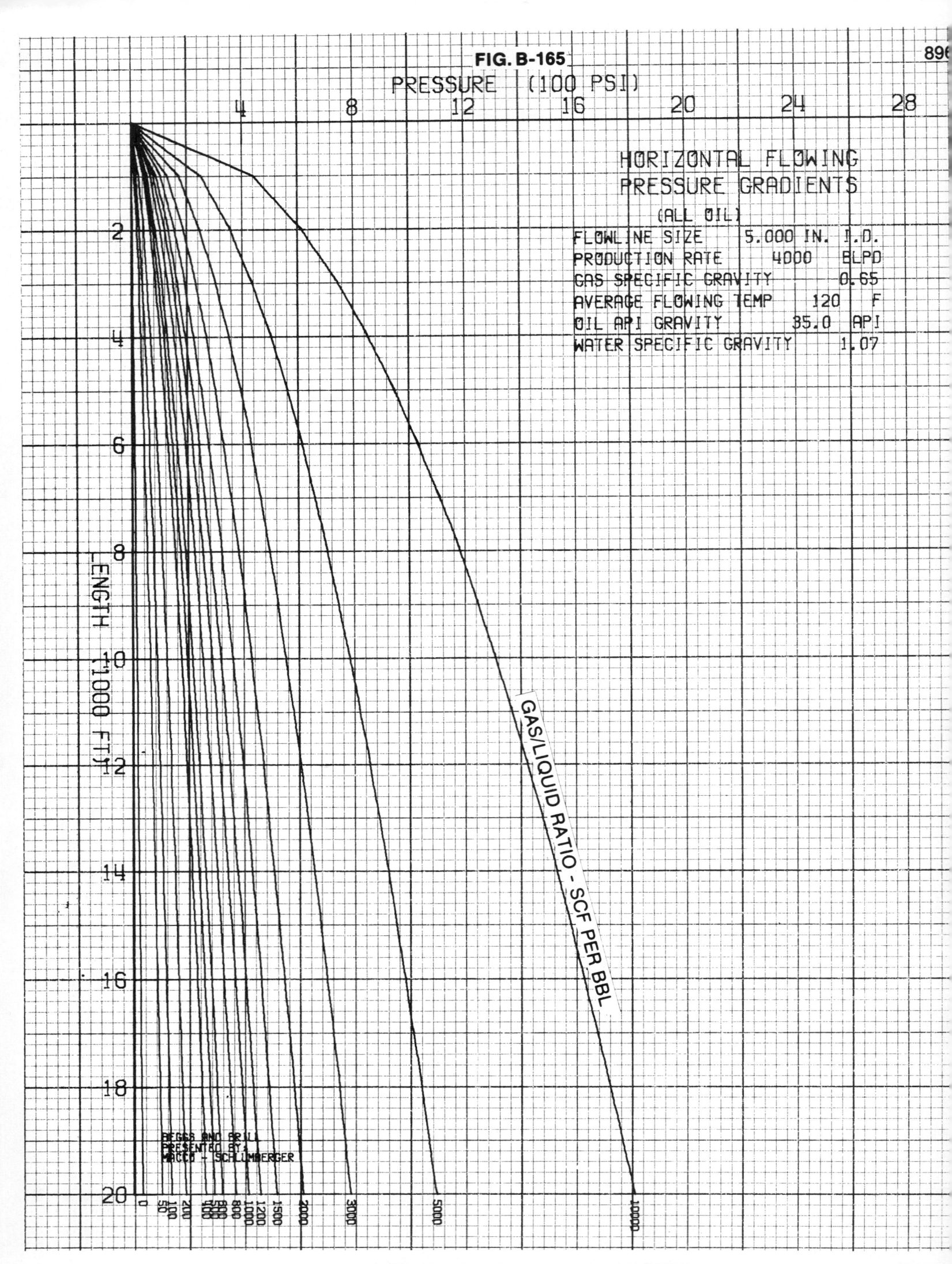

FIG. B-165

896

PRESSURE (100 PSI)

HORIZONTAL FLOWING
PRESSURE GRADIENTS
(ALL OIL)

FLOWLINE SIZE	5.000 IN. I.D.	
PRODUCTION RATE	4000	BLPD
GAS SPECIFIC GRAVITY		0.65
AVERAGE FLOWING TEMP	120	F
OIL API GRAVITY	35.0	API
WATER SPECIFIC GRAVITY		1.07

LENGTH (1000 FT)

GAS/LIQUID RATIO - SCF PER BBL

BEGGS AND BRILL
PRESENTED BY:
MACCO - SCHLUMBERGER

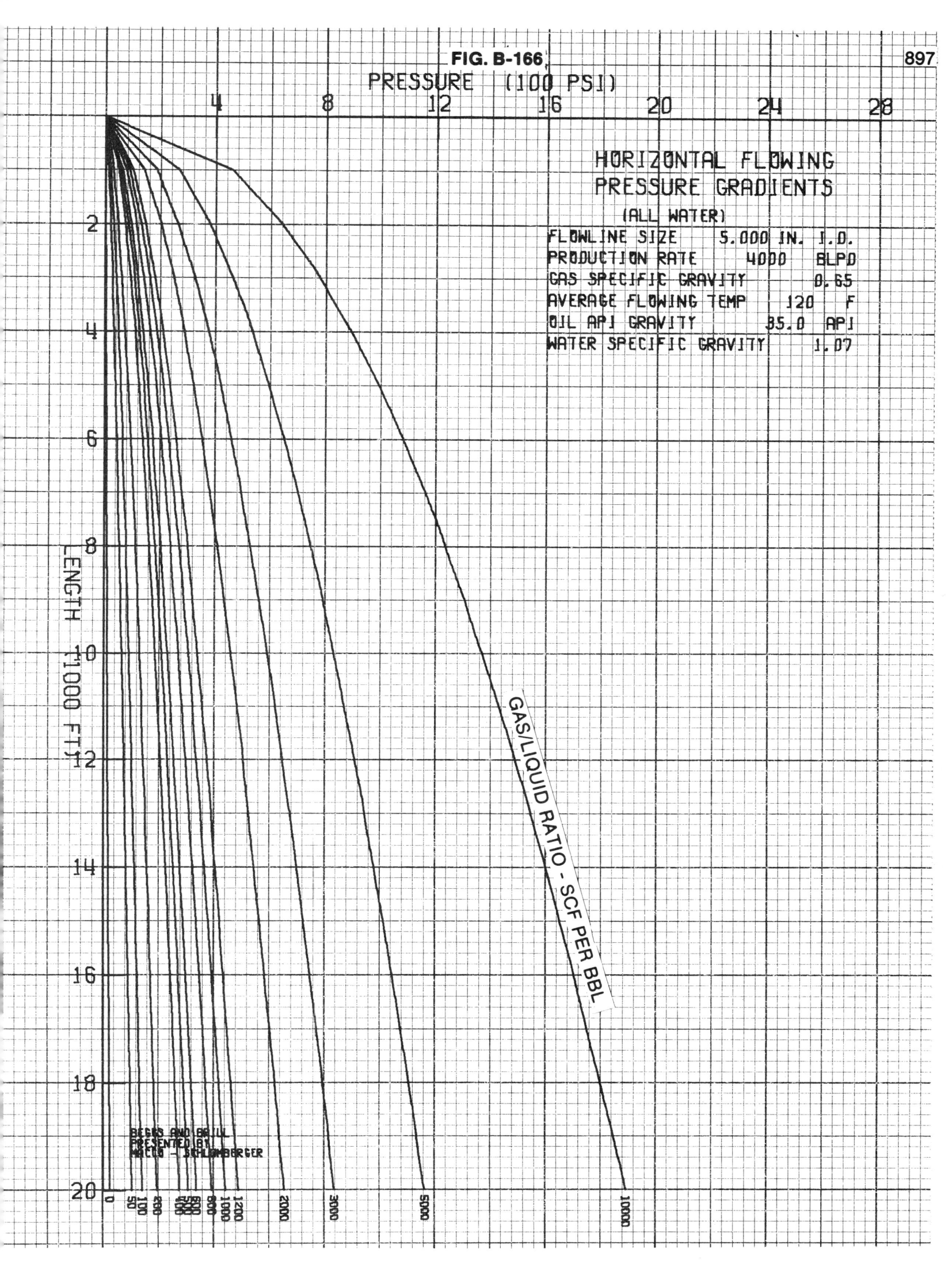

PRESSURE (100 PSI)

HORIZONTAL FLOWING
PRESSURE GRADIENTS
(ALL WATER)

FLOWLINE SIZE   5.000 IN. I.D.
PRODUCTION RATE   4000   BLPD
GAS SPECIFIC GRAVITY   0.65
AVERAGE FLOWING TEMP   120   F
OIL API GRAVITY   35.0   API
WATER SPECIFIC GRAVITY   1.07

LENGTH (1000 FT.)

GAS/LIQUID RATIO - SCF PER BBL

BEGGS AND BRILL
PRESENTED BY
MERLA - SCHLUMBERGER

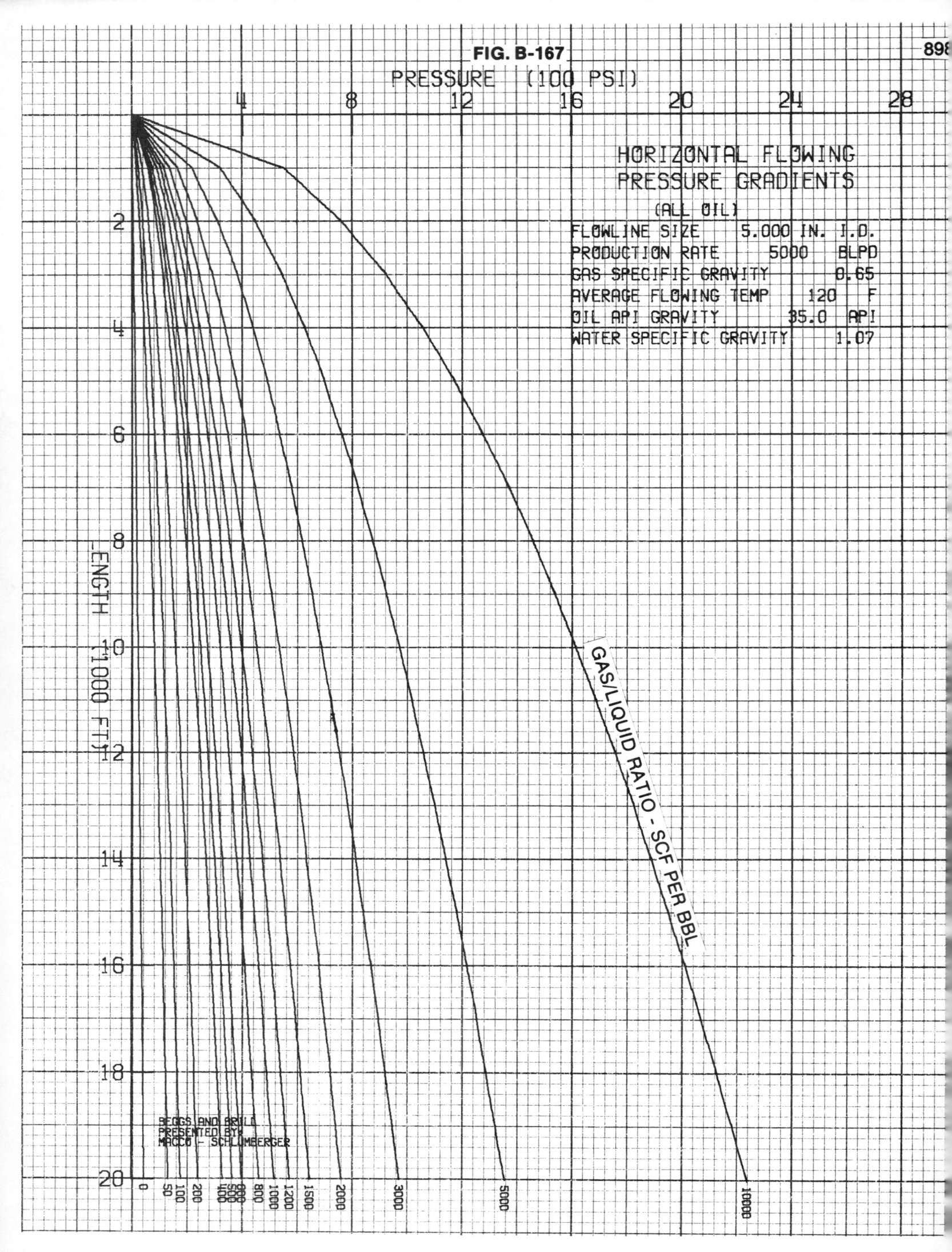

FIG. B-167

898

HORIZONTAL FLOWING
PRESSURE GRADIENTS
(ALL OIL)

FLOWLINE SIZE	5.000 IN. I.D.
PRODUCTION RATE	5000 BLPD
GAS SPECIFIC GRAVITY	0.65
AVERAGE FLOWING TEMP	120 F
OIL API GRAVITY	35.0 API
WATER SPECIFIC GRAVITY	1.07

PRESSURE (100 PSI)

LENGTH (1000 FT)

GAS/LIQUID RATIO - SCF PER BBL

BEGGS AND BRILL
PRESENTED BY
MACCO - SCHLUMBERGER

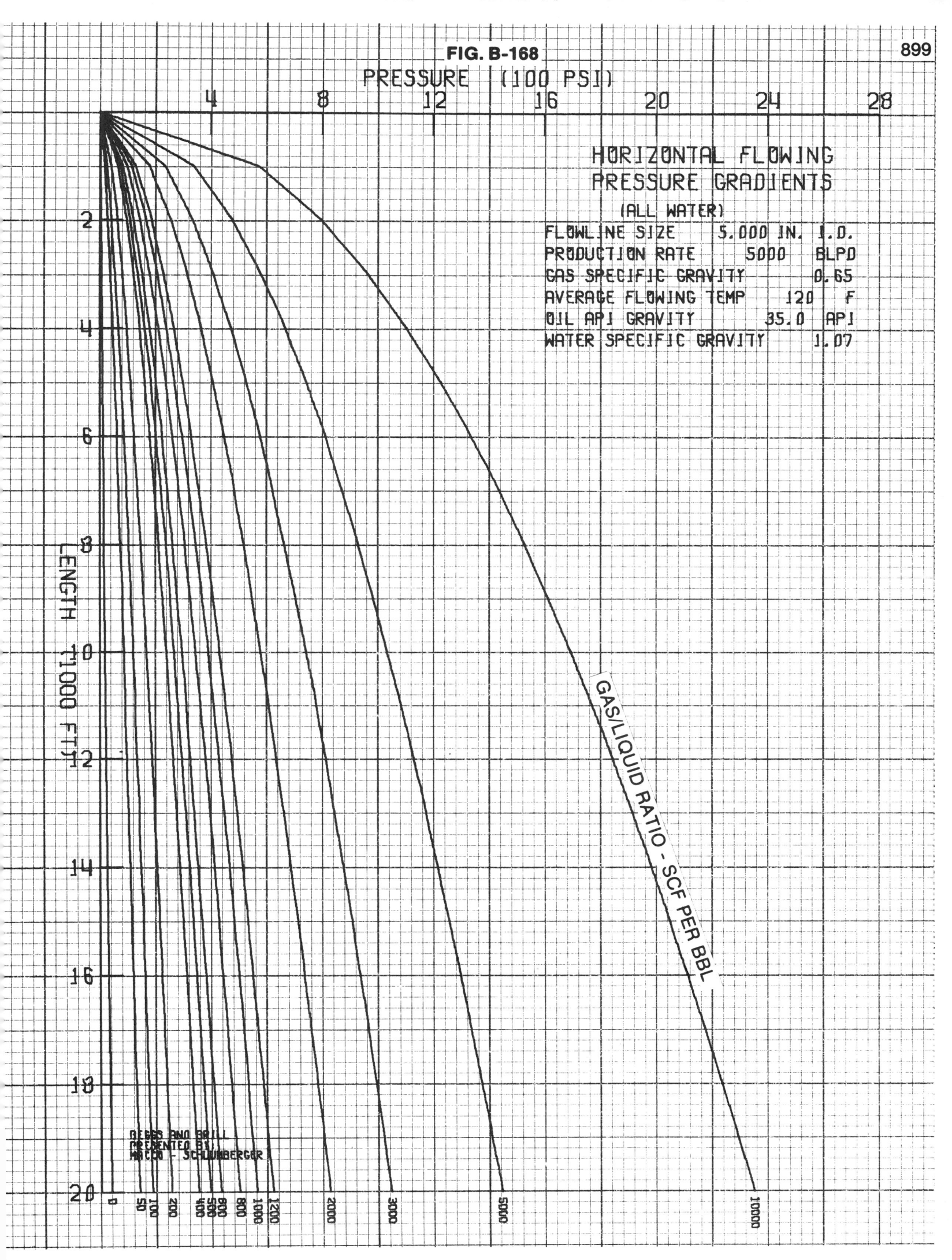

PRESSURE (100 PSI)

HORIZONTAL FLOWING
PRESSURE GRADIENTS
(ALL WATER)

FLOWLINE SIZE	5.000 IN.	I.D.
PRODUCTION RATE	5000	BLPD
GAS SPECIFIC GRAVITY	0.65	
AVERAGE FLOWING TEMP	120	F
OIL API GRAVITY	35.0	API
WATER SPECIFIC GRAVITY	1.07	

LENGTH (1000 FT)

GAS/LIQUID RATIO - SCF PER BBL

BEGGS AND BRILL
PRESENTED BY
HALCO - SCHLUMBERGER

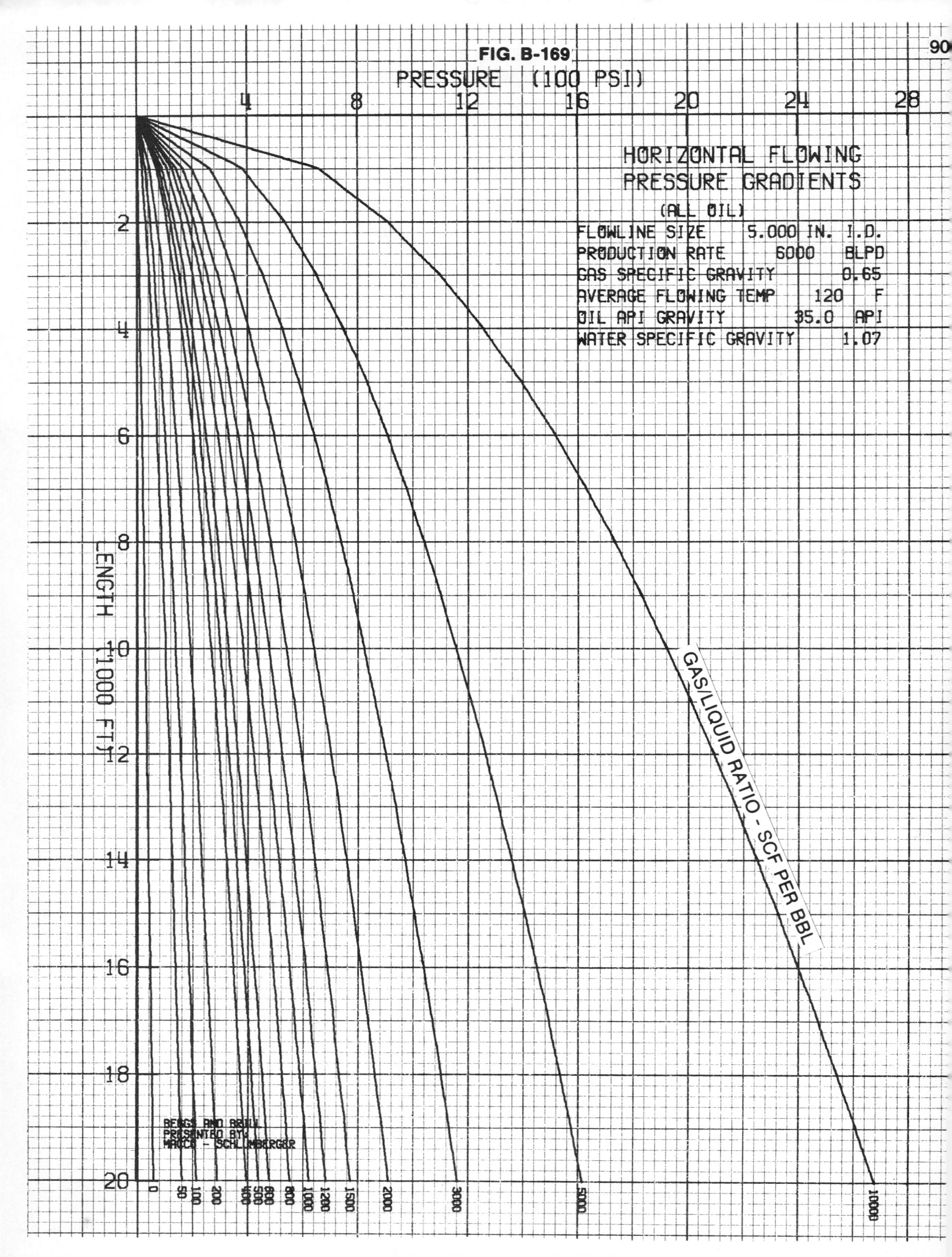

FIG. B-169

PRESSURE (100 PSI)

LENGTH (1000 FT)

HORIZONTAL FLOWING
PRESSURE GRADIENTS

(ALL OIL)

FLOWLINE SIZE	5.000 IN. I.D.
PRODUCTION RATE	6000 BLPD
GAS SPECIFIC GRAVITY	0.65
AVERAGE FLOWING TEMP	120 F
OIL API GRAVITY	35.0 API
WATER SPECIFIC GRAVITY	1.07

GAS/LIQUID RATIO - SCF PER BBL

BEGGS AND BRILL
PRESENTED BY:
MACCO - SCHLUMBERGER

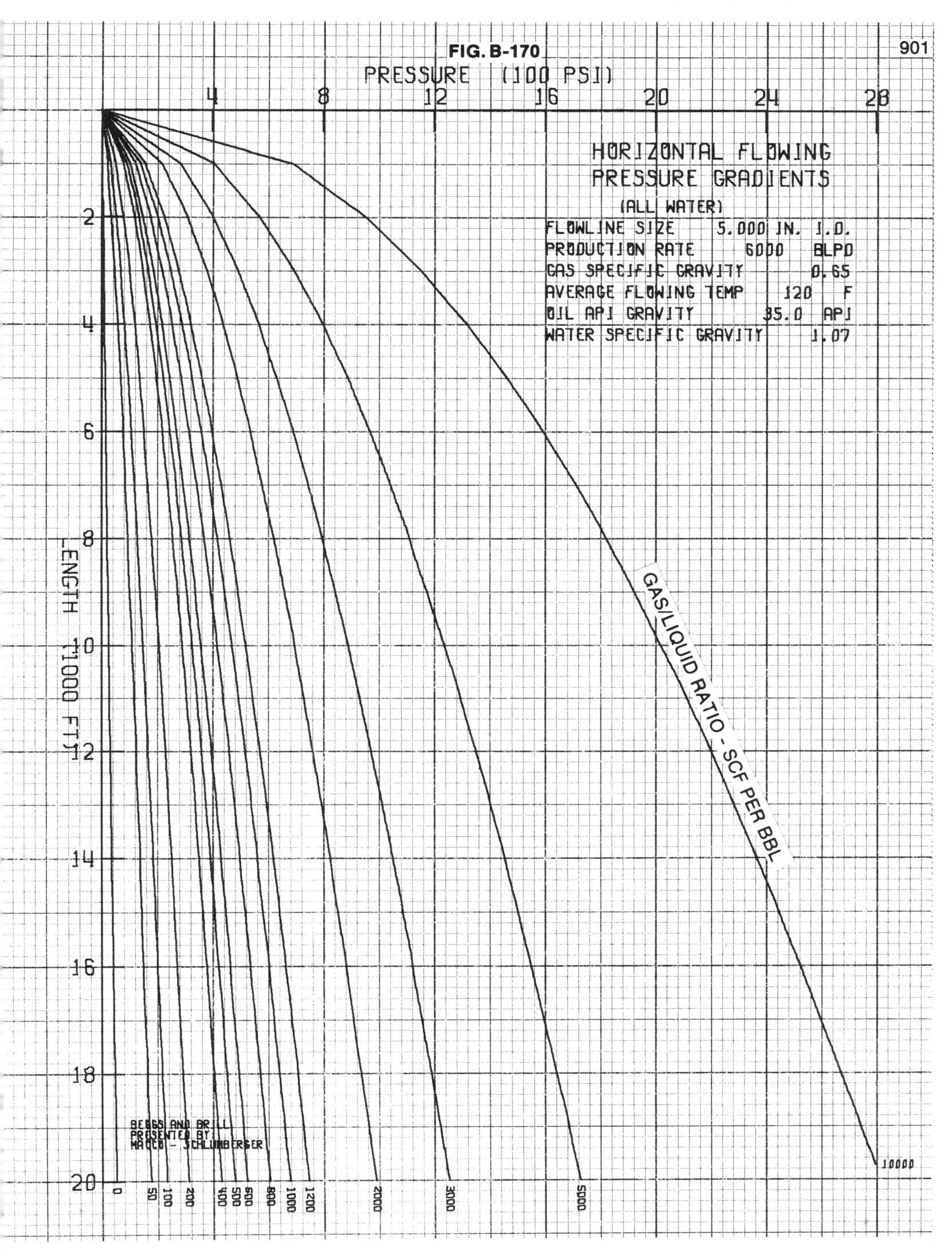

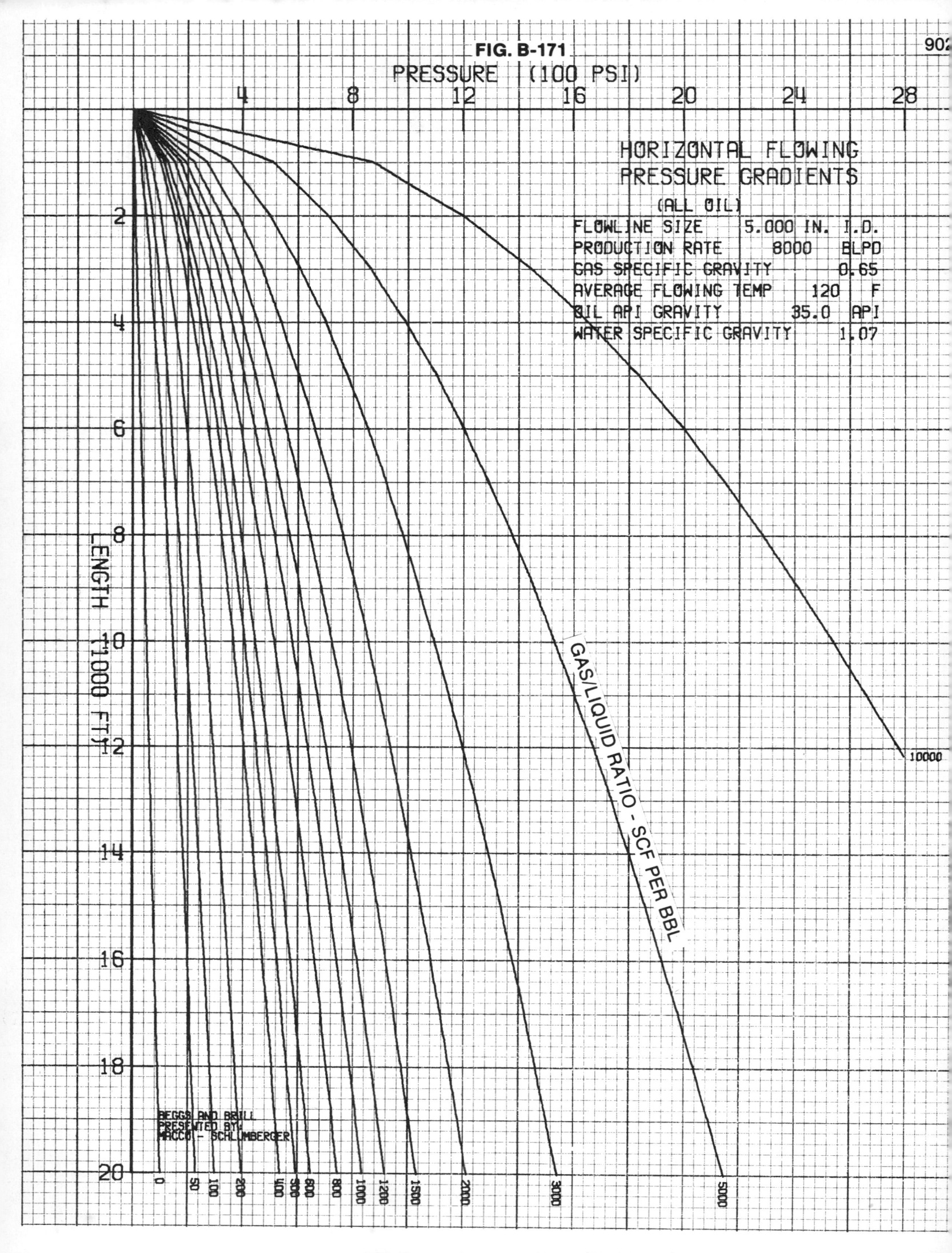

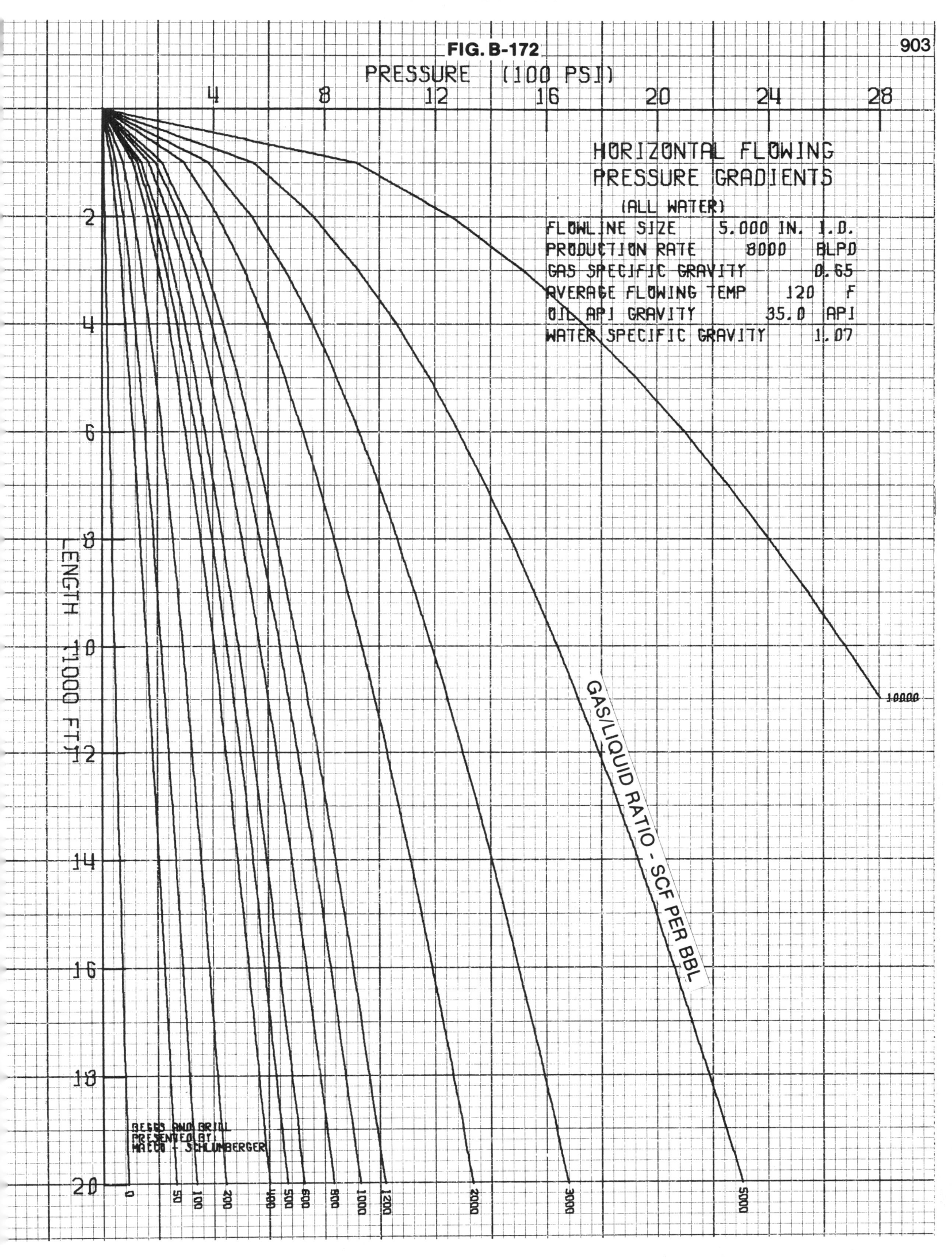

PRESSURE (100 PSI)

HORIZONTAL FLOWING
PRESSURE GRADIENTS
(ALL WATER)

FLOWLINE SIZE	5.000 IN. I.D.
PRODUCTION RATE	8000 BLPD
GAS SPECIFIC GRAVITY	0.65
AVERAGE FLOWING TEMP	120 F
OIL API GRAVITY	35.0 API
WATER SPECIFIC GRAVITY	1.07

LENGTH (1000 FT)

GAS/LIQUID RATIO - SCF PER BBL

10000

BEGGS AND BRILL
PRESENTED AT
MAECO + SCHLUMBERGER

0   50 100 200 400 500 600 800 1000 1200 2000 3000 5000

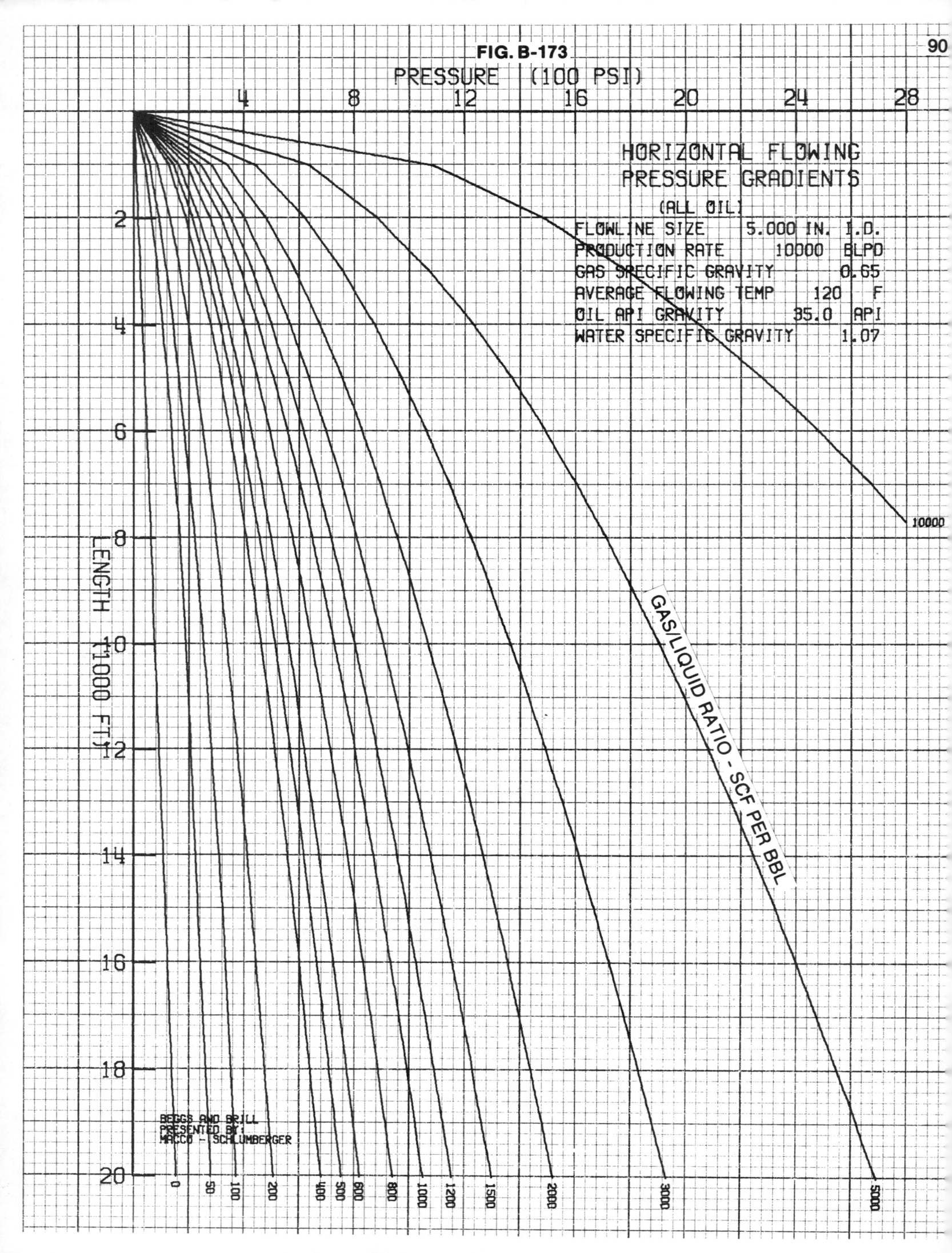

PRESSURE (100 PSI)

HORIZONTAL FLOWING
PRESSURE GRADIENTS
(ALL OIL)

FLOWLINE SIZE	5.000 IN. I.D.
PRODUCTION RATE	10000 BLPD
GAS SPECIFIC GRAVITY	0.65
AVERAGE FLOWING TEMP	120 F
OIL API GRAVITY	35.0 API
WATER SPECIFIC GRAVITY	1.07

LENGTH (1000 FT)

GAS/LIQUID RATIO - SCF PER BBL

BEGGS AND BRILL
PRESENTED BY:
MACCO - SCHLUMBERGER

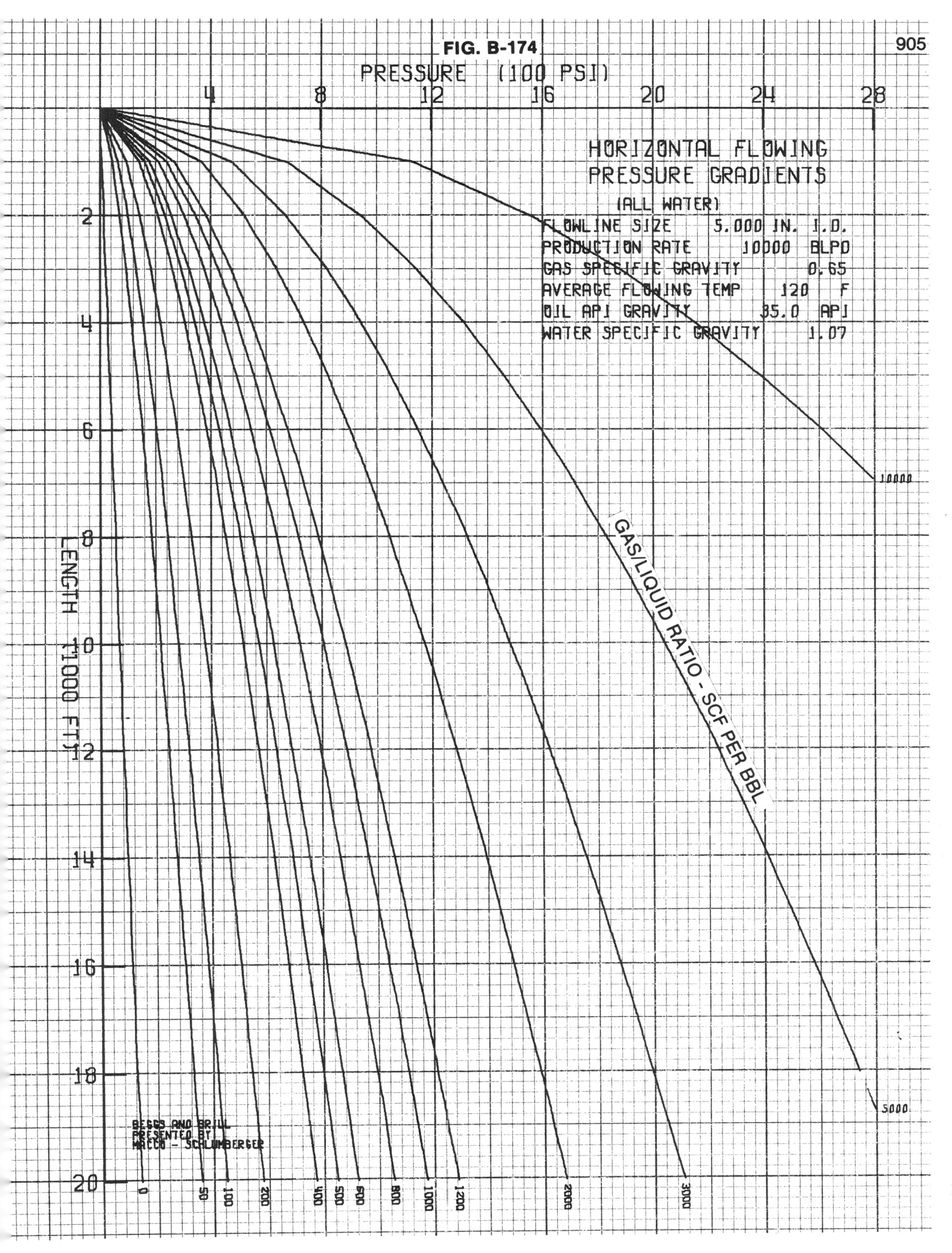

PRESSURE (100 PSI)

HORIZONTAL FLOWING
PRESSURE GRADIENTS
(ALL WATER)

FLOWLINE SIZE	5.000 IN. I.D.
PRODUCTION RATE	10000 BLPD
GAS SPECIFIC GRAVITY	0.65
AVERAGE FLOWING TEMP	120 F
OIL API GRAVITY	35.0 API
WATER SPECIFIC GRAVITY	1.07

LENGTH (1000 FT)

GAS/LIQUID RATIO - SCF PER BBL

BEGGS AND BRILL
PRESENTED BY
WACCO - SCHLUMBERGER

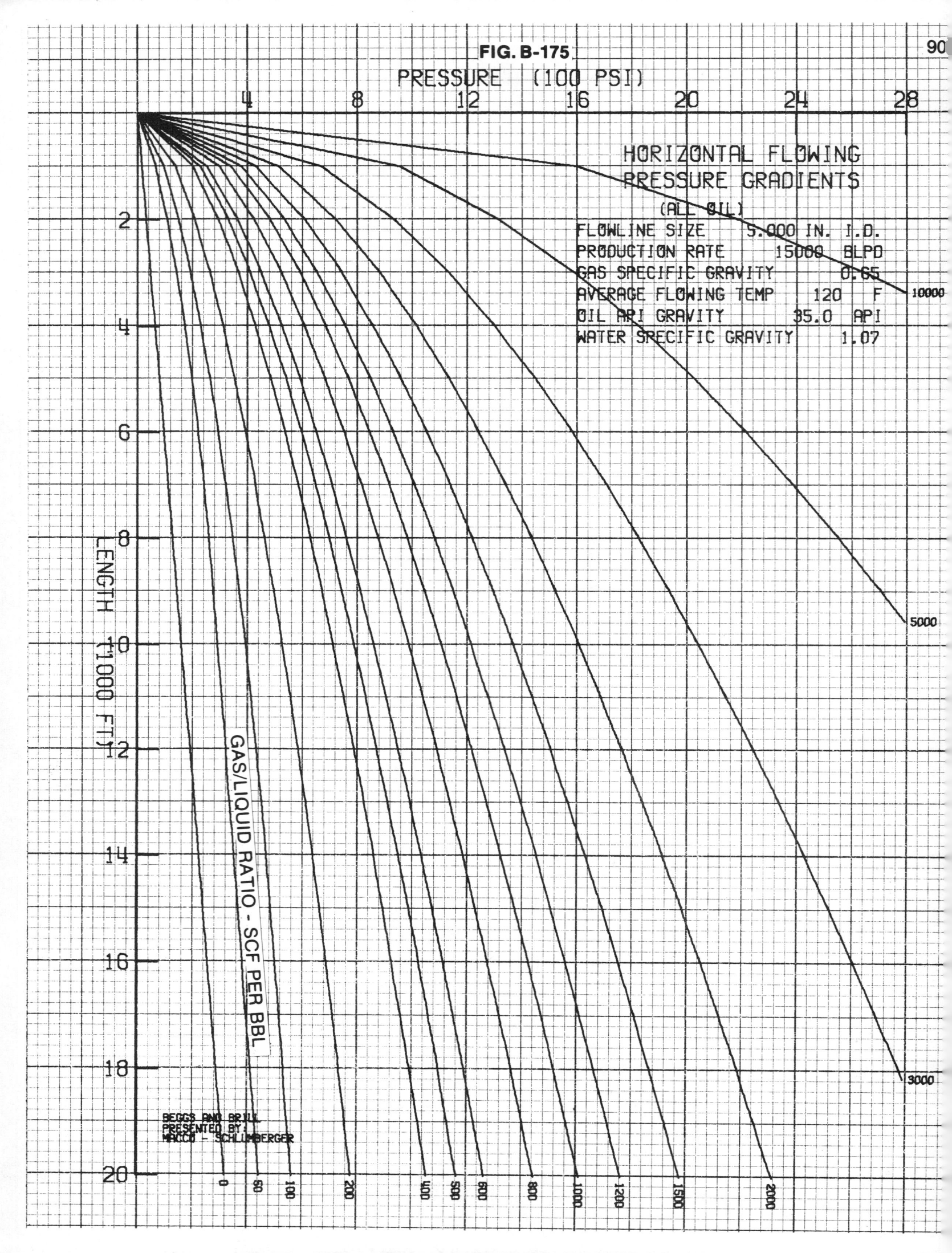

FIG. B-175

PRESSURE (100 PSI)

HORIZONTAL FLOWING
PRESSURE GRADIENTS
(ALL OIL)

FLOWLINE SIZE	5.000 IN. I.D.
PRODUCTION RATE	15000 BLPD
GAS SPECIFIC GRAVITY	0.65
AVERAGE FLOWING TEMP	120 F
OIL API GRAVITY	35.0 API
WATER SPECIFIC GRAVITY	1.07

LENGTH (1000 FT)

GAS/LIQUID RATIO - SCF PER BBL

BEGGS AND BRILL
PRESENTED BY:
MACCO - SCHLUMBERGER

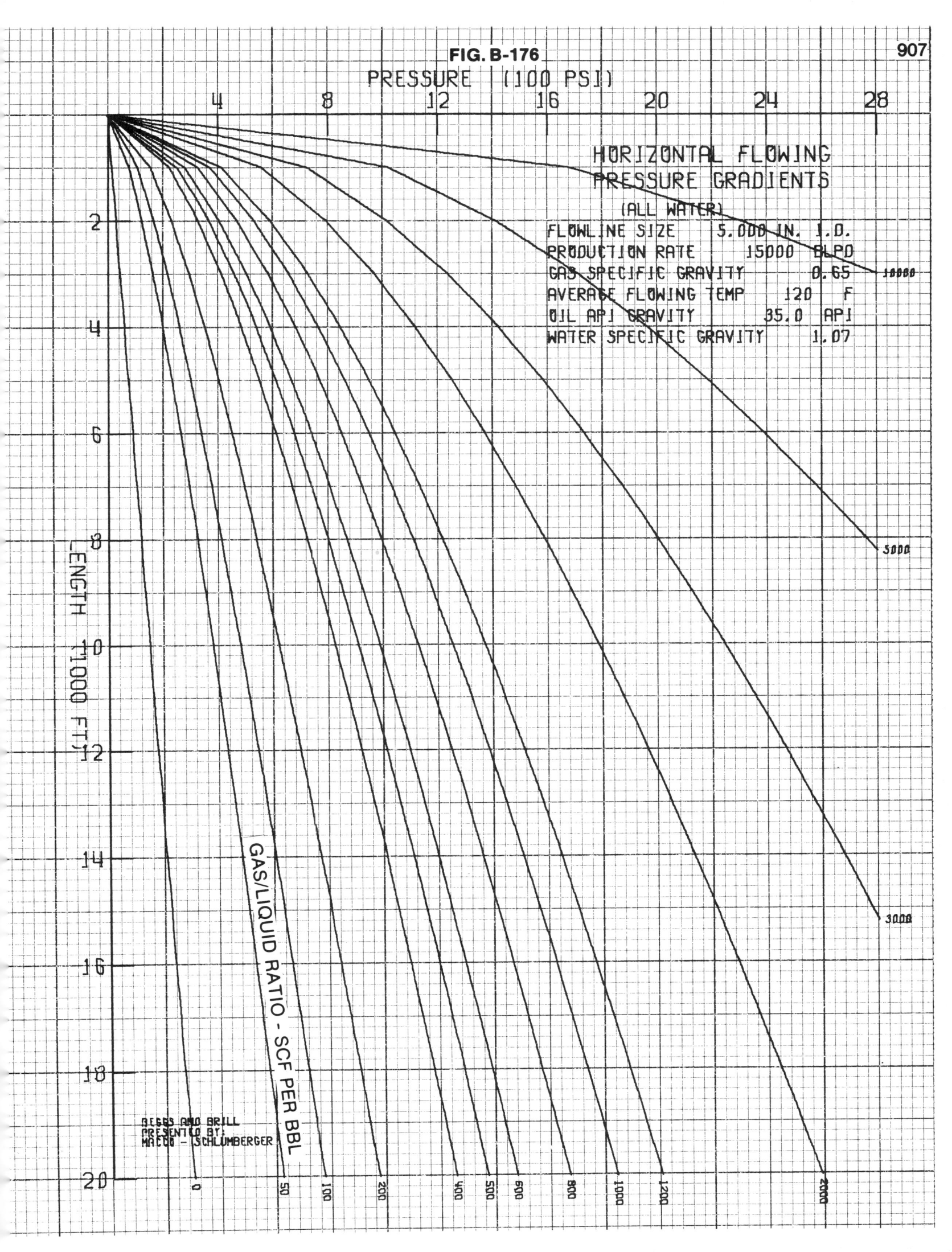

PRESSURE (100 PSI)

HORIZONTAL FLOWING
PRESSURE GRADIENTS
(ALL WATER)
FLOWLINE SIZE   5.000 IN. I.D.
PRODUCTION RATE   15000 BLPD
GAS SPECIFIC GRAVITY   0.65
AVERAGE FLOWING TEMP   120   F
OIL API GRAVITY   35.0 API
WATER SPECIFIC GRAVITY   1.07

LENGTH (1000 FT)

GAS/LIQUID RATIO - SCF PER BBL

BEGGS AND BRILL
PRESENTED BY
MAGCO - SCHLUMBERGER

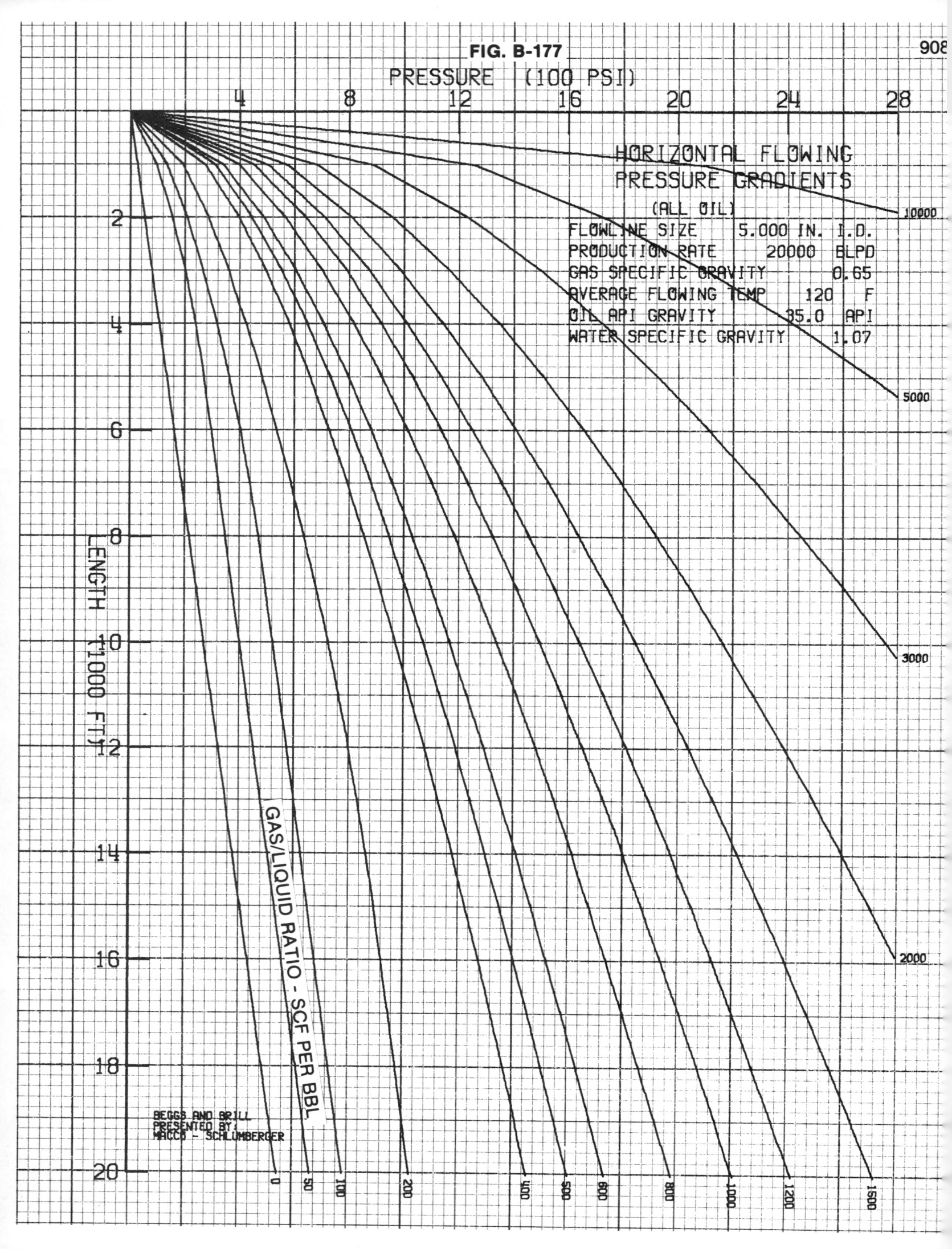

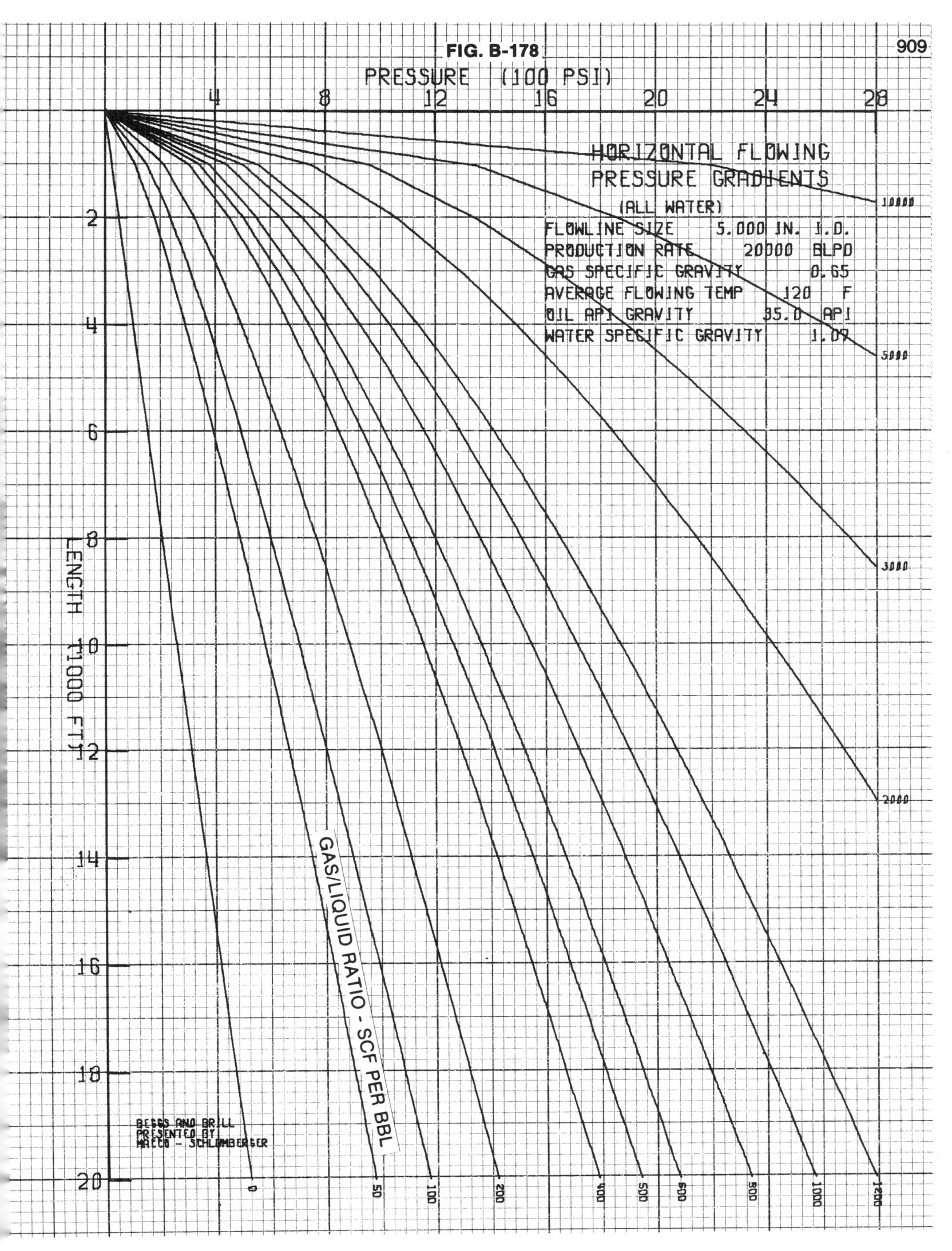

FIG. B-178

909

PRESSURE (100 PSI)

HORIZONTAL FLOWING
PRESSURE GRADIENTS
(ALL WATER)

FLOWLINE SIZE   5.000 IN. I.D.
PRODUCTION RATE   20000 BLPD
GAS SPECIFIC GRAVITY   0.65
AVERAGE FLOWING TEMP   120 F
OIL API GRAVITY   35.0 API
WATER SPECIFIC GRAVITY 1.07

LENGTH (1000 FT)

GAS/LIQUID RATIO - SCF PER BBL

BEGGS AND BRILL
PRESENTED BY
MAECO - SCHLUMBERGER

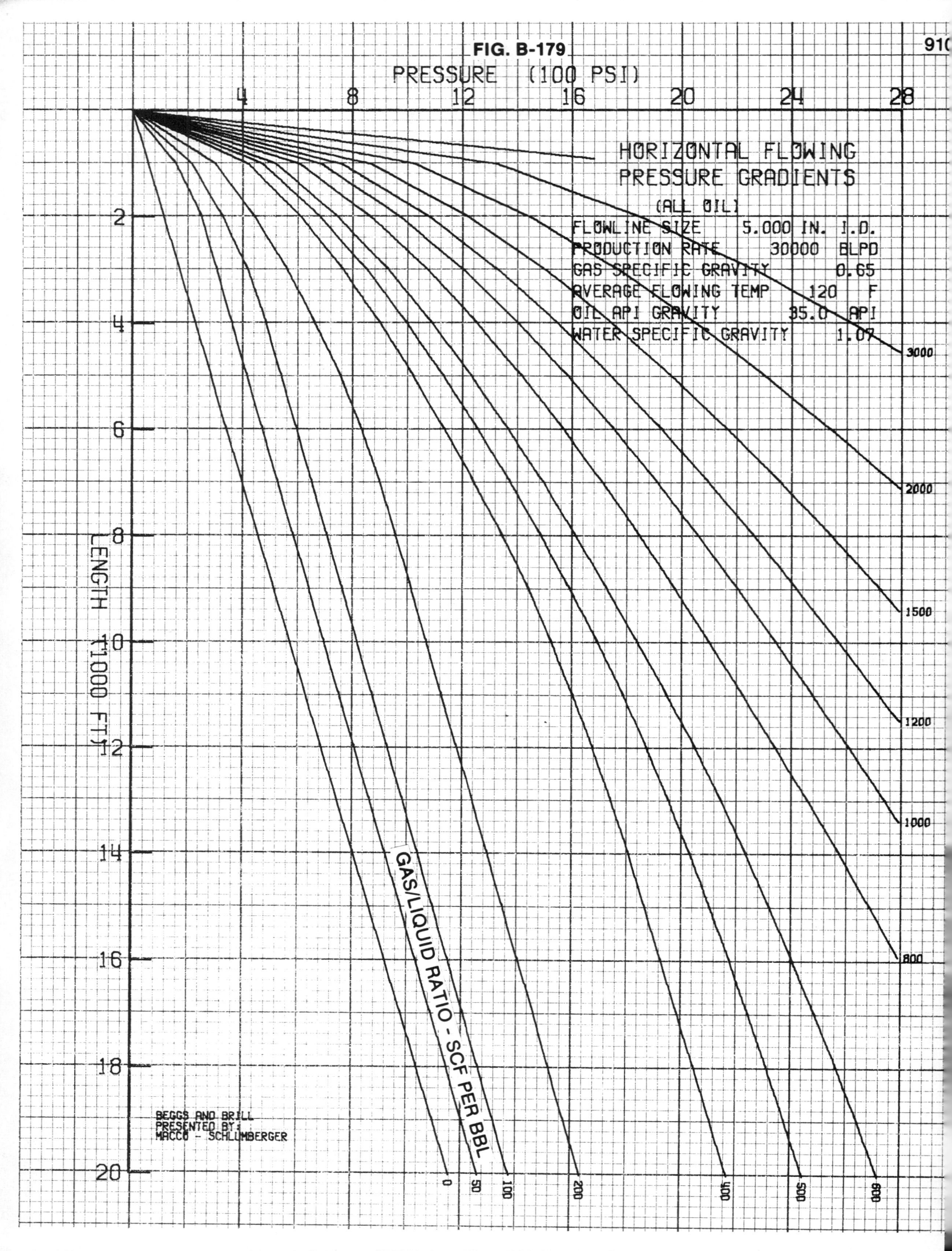

FIG. B-179

910

HORIZONTAL FLOWING
PRESSURE GRADIENTS
(ALL OIL)

FLOWLINE SIZE	5.000 IN. I.D.
PRODUCTION RATE	30000 BLPD
GAS SPECIFIC GRAVITY	0.65
AVERAGE FLOWING TEMP	120 F
OIL API GRAVITY	35.0 API
WATER SPECIFIC GRAVITY	1.07

PRESSURE (100 PSI)

LENGTH (1000 FT)

GAS/LIQUID RATIO - SCF PER BBL

BEGGS AND BRILL
PRESENTED BY:
MACCO - SCHLUMBERGER

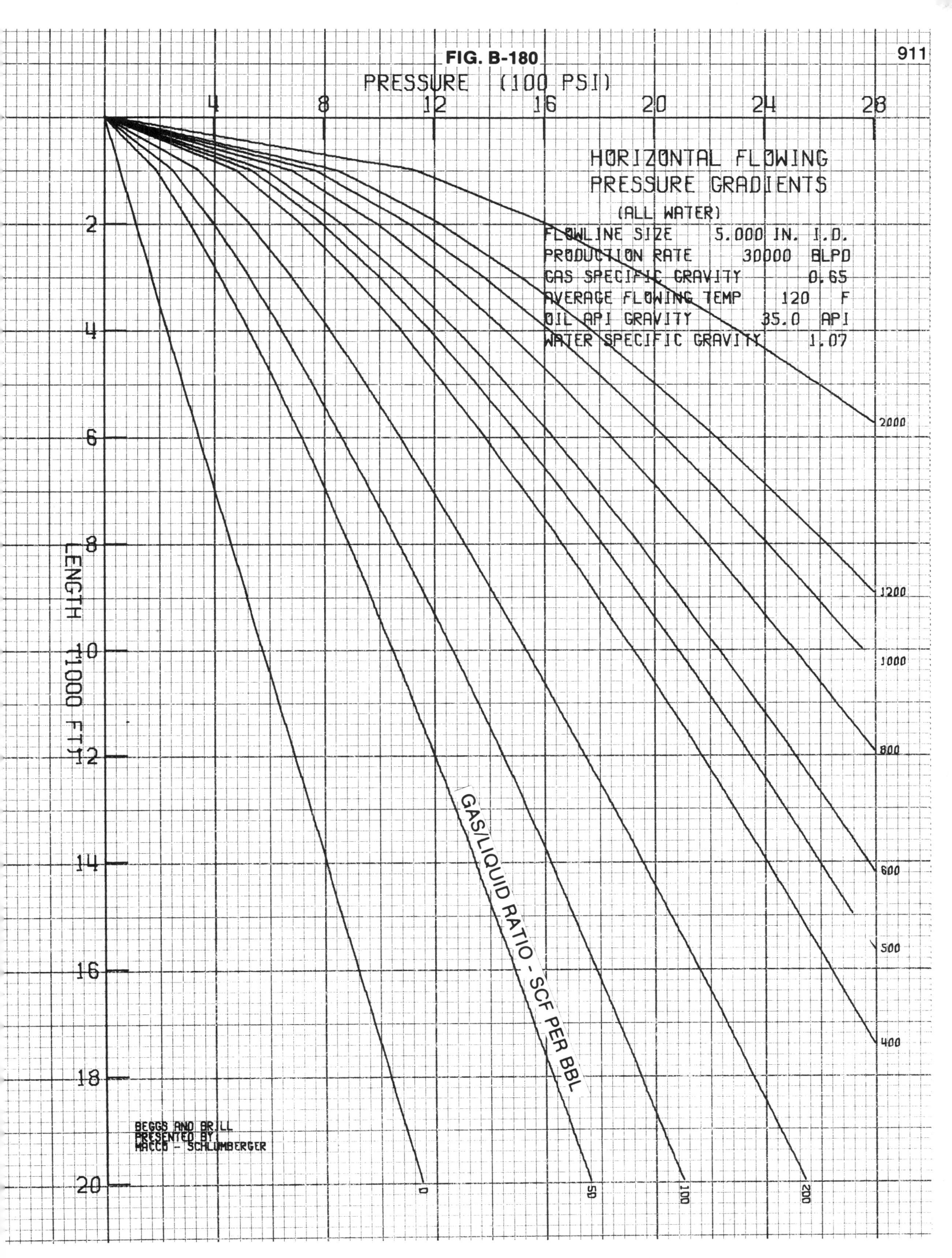

FIG. B-180

911

PRESSURE (100 PSI)

HORIZONTAL FLOWING
PRESSURE GRADIENTS
(ALL WATER)

FLOWLINE SIZE   5.000 IN. I.D.
PRODUCTION RATE   30000 BLPD
GAS SPECIFIC GRAVITY   0.65
AVERAGE FLOWING TEMP   120   F
OIL API GRAVITY   35.0 API
WATER SPECIFIC GRAVITY   1.07

LENGTH (1000 FT)

GAS/LIQUID RATIO - SCF PER BBL

BEGGS AND BRILL
PRESENTED BY
MACCO - SCHLUMBERGER

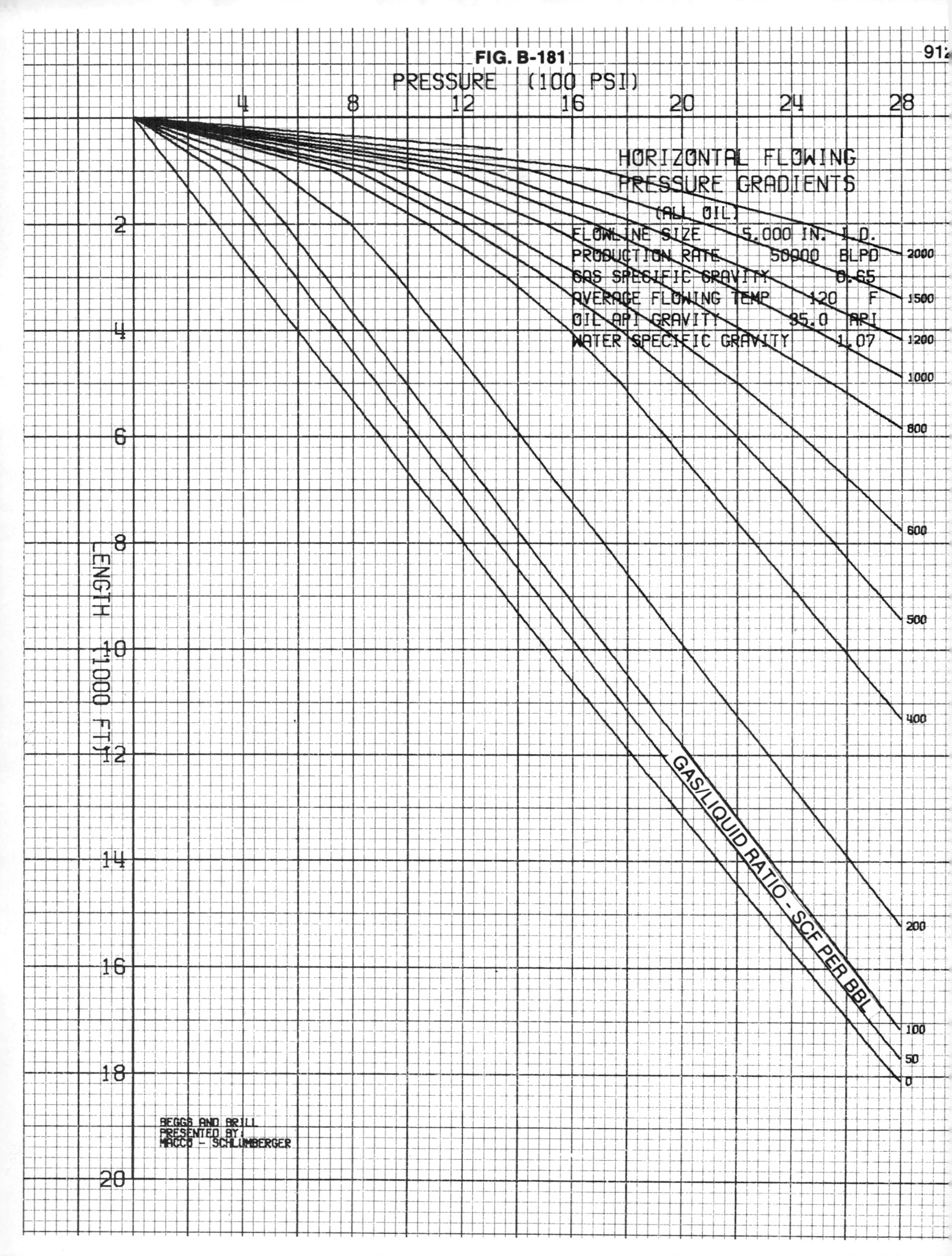

PRESSURE (100 PSI)

HORIZONTAL FLOWING
PRESSURE GRADIENTS
(ALL OIL)

FLOWLINE SIZE	5.000 IN. I.D.
PRODUCTION RATE	50000 BLPD
GAS SPECIFIC GRAVITY	0.65
AVERAGE FLOWING TEMP	120 F
OIL API GRAVITY	35.0 API
WATER SPECIFIC GRAVITY	1.07

LENGTH (1000 FT)

GAS/LIQUID RATIO - SCF PER BBL

BEGGS AND BRILL
PRESENTED BY:
MACCO - SCHLUMBERGER

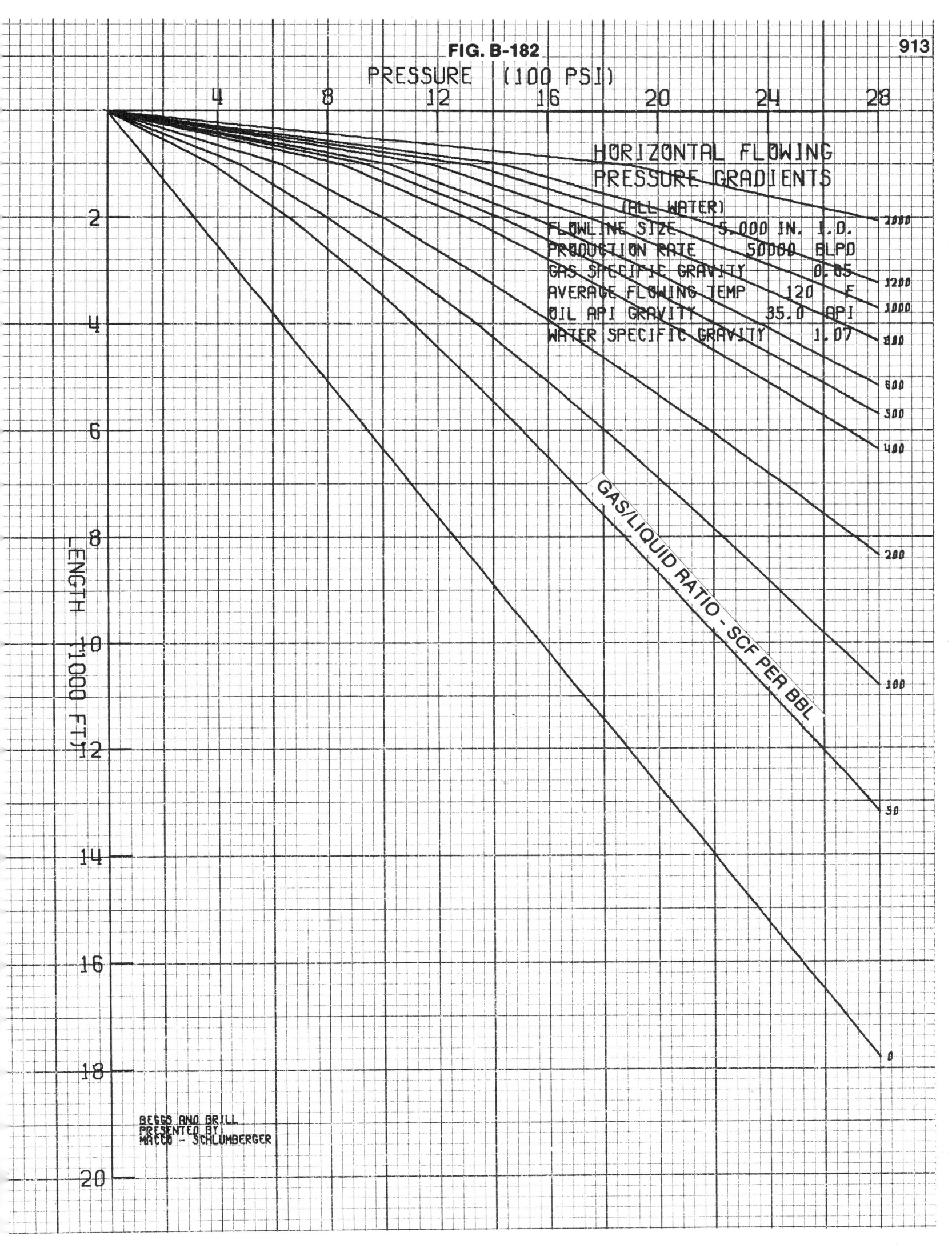

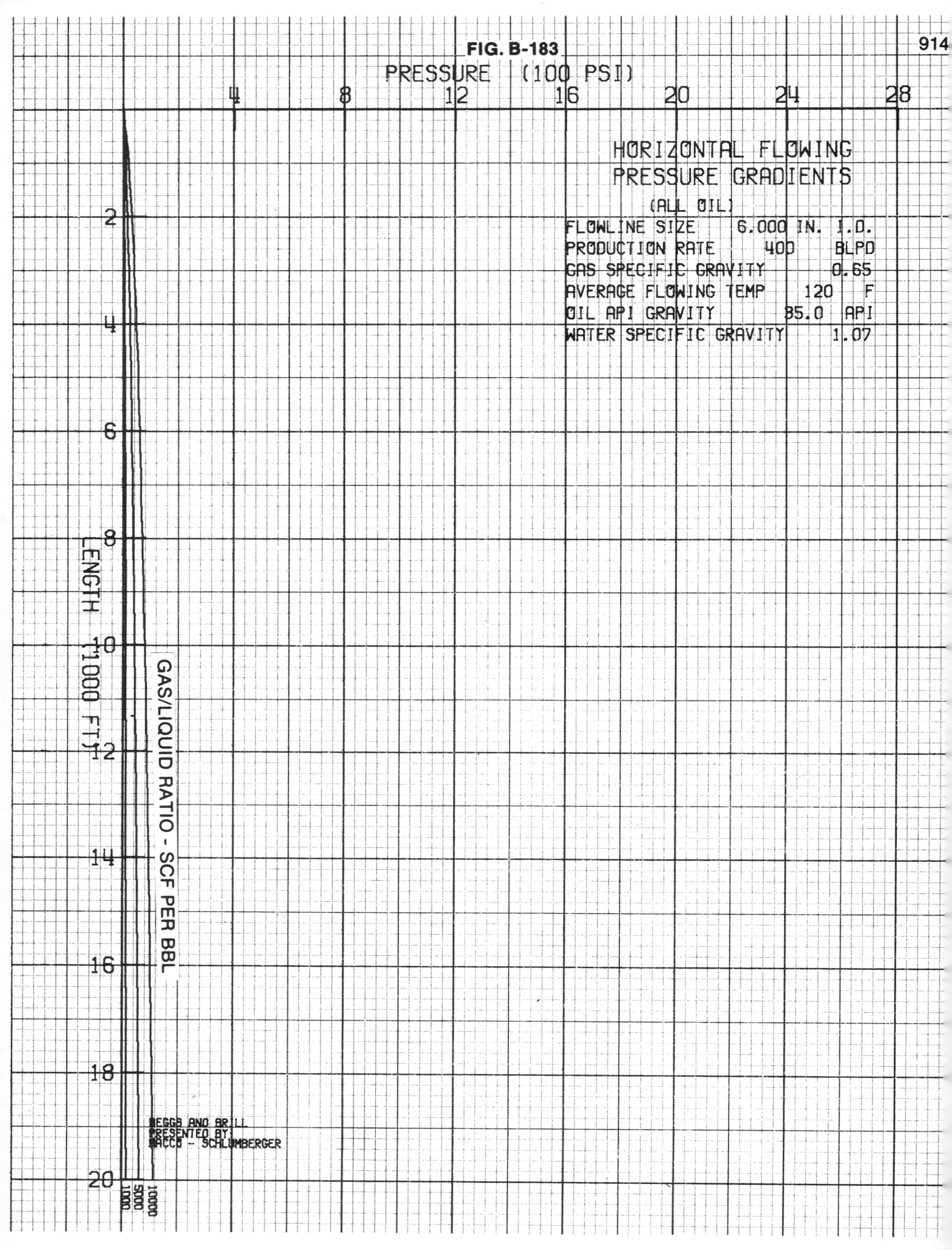

PRESSURE (100 PSI)

HORIZONTAL FLOWING
PRESSURE GRADIENTS
(ALL OIL)

FLOWLINE SIZE   6.000 IN. I.D.
PRODUCTION RATE   400   BLPD
GAS SPECIFIC GRAVITY   0.65
AVERAGE FLOWING TEMP   120   F
OIL API GRAVITY   35.0 API
WATER SPECIFIC GRAVITY   1.07

LENGTH (1000 FT)

GAS/LIQUID RATIO - SCF PER BBL

BEGGS AND BRILL
PRESENTED BY
MACCO – SCHLUMBERGER

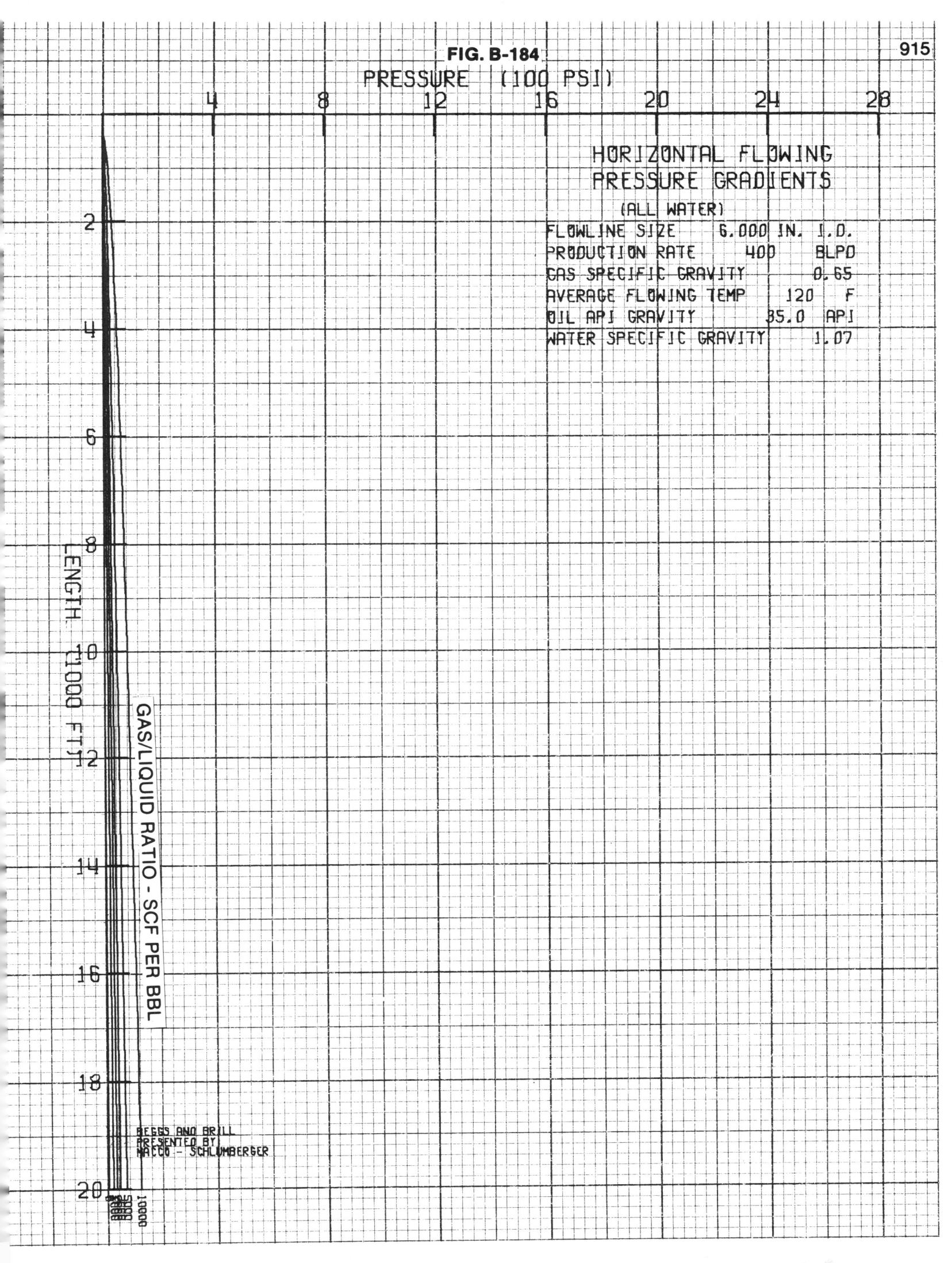

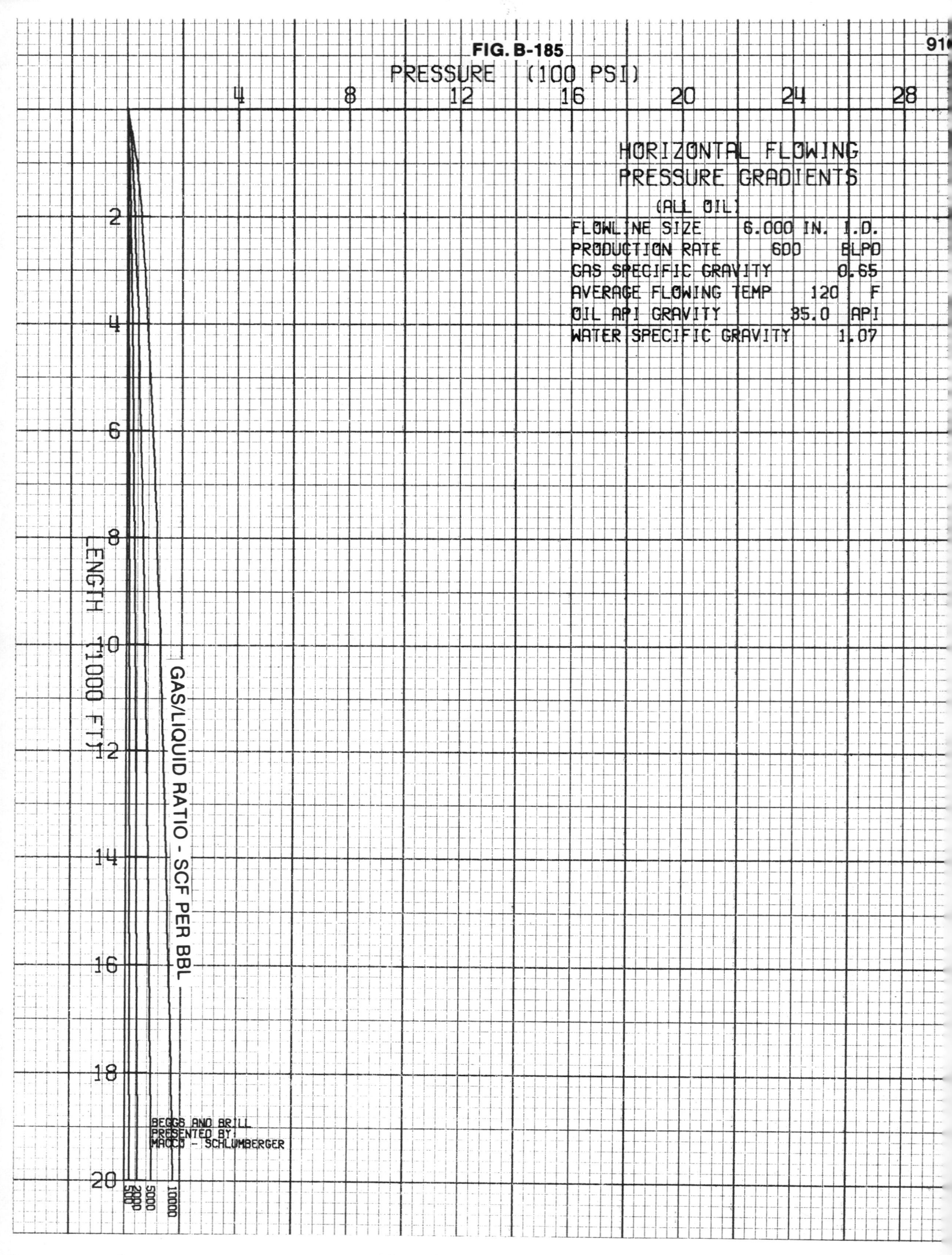

FIG. B-185

916

PRESSURE (100 PSI)

HORIZONTAL FLOWING
PRESSURE GRADIENTS
(ALL OIL)

FLOWLINE SIZE	6.000 IN. I.D.
PRODUCTION RATE	600 BLPD
GAS SPECIFIC GRAVITY	0.65
AVERAGE FLOWING TEMP	120 F
OIL API GRAVITY	35.0 API
WATER SPECIFIC GRAVITY	1.07

LENGTH (1000 FT.)

GAS/LIQUID RATIO - SCF PER BBL

BEGGS AND BRILL
PRESENTED BY:
MACCO - SCHLUMBERGER

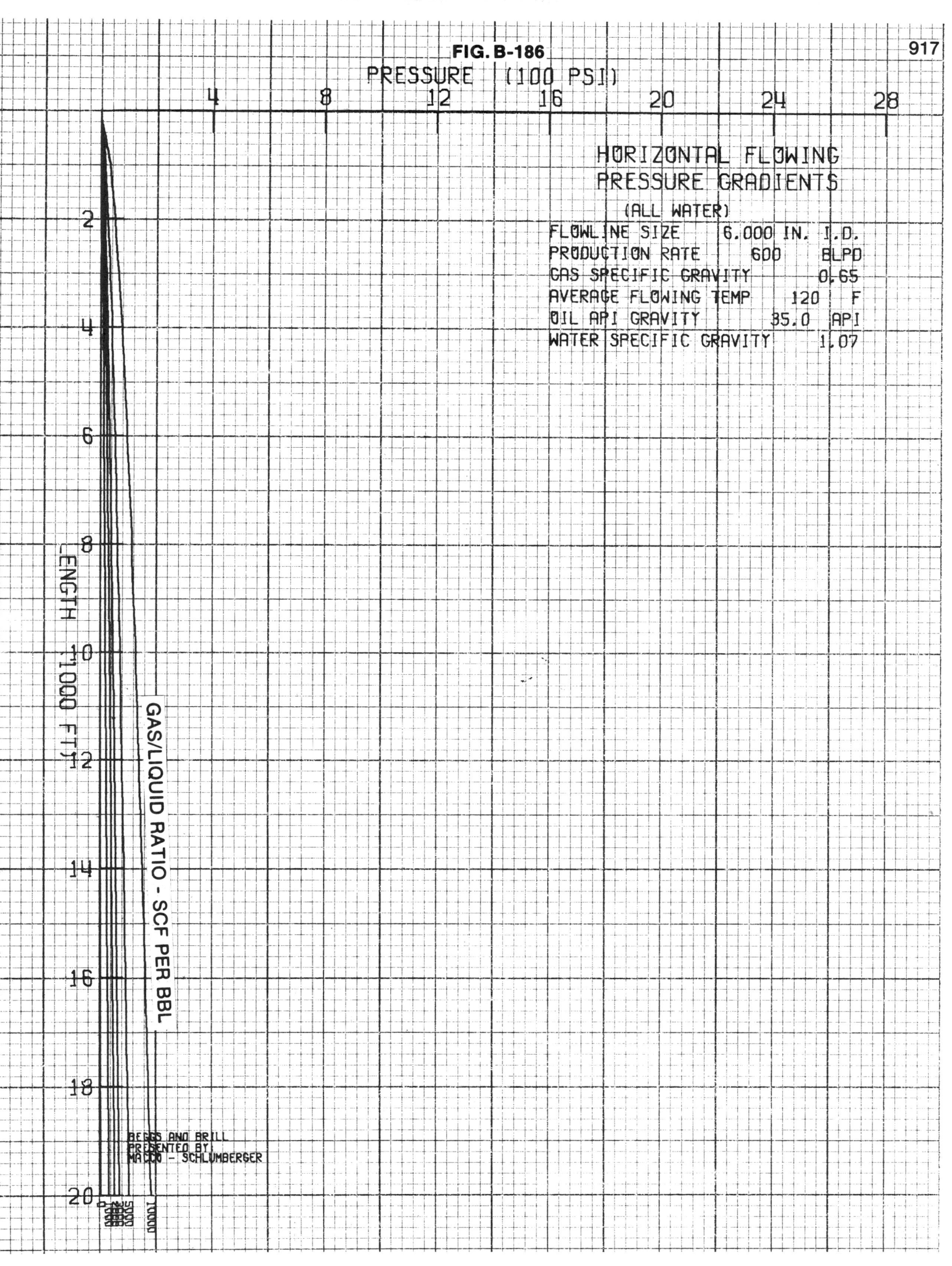

FIG. B-186

917

HORIZONTAL FLOWING
PRESSURE GRADIENTS
(ALL WATER)

FLOWLINE SIZE   6.000 IN. I.D.
PRODUCTION RATE   600   BLPD
GAS SPECIFIC GRAVITY   0.65
AVERAGE FLOWING TEMP   120   F
OIL API GRAVITY   35.0   API
WATER SPECIFIC GRAVITY   1.07

PRESSURE (100 PSI)

LENGTH (1000 FT.)

GAS/LIQUID RATIO - SCF PER BBL

BEGGS AND BRILL
PRESENTED BY
NATCO - SCHLUMBERGER

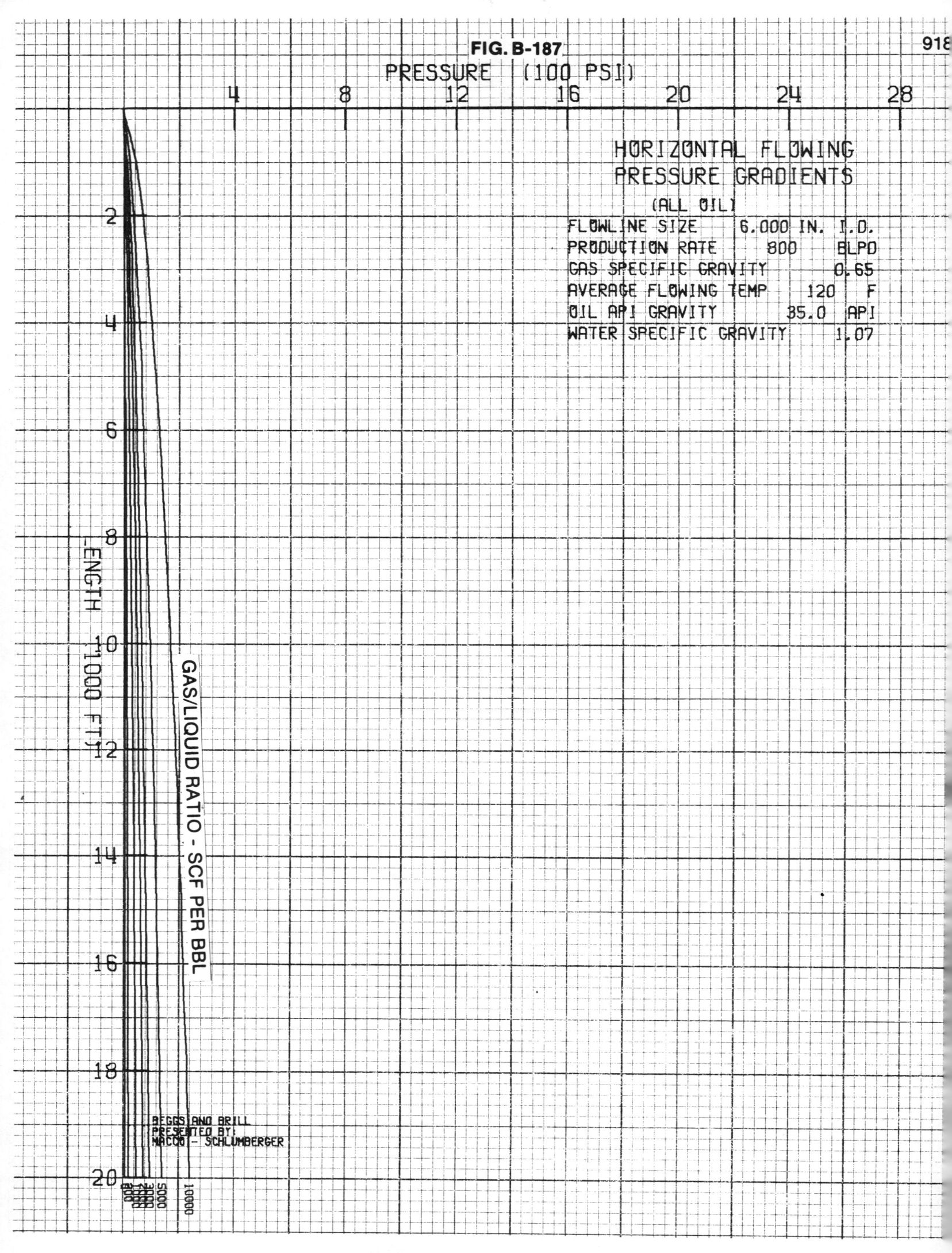

PRESSURE (100 PSI)

HORIZONTAL FLOWING
PRESSURE GRADIENTS
(ALL OIL)

FLOWLINE SIZE	6.000 IN. I.D.
PRODUCTION RATE	800 BLPD
GAS SPECIFIC GRAVITY	0.65
AVERAGE FLOWING TEMP	120 F
OIL API GRAVITY	35.0 API
WATER SPECIFIC GRAVITY	1.07

LENGTH (1000 FT)

GAS/LIQUID RATIO - SCF PER BBL

BEGGS AND BRILL
PRESENTED BY:
MACCO - SCHLUMBERGER

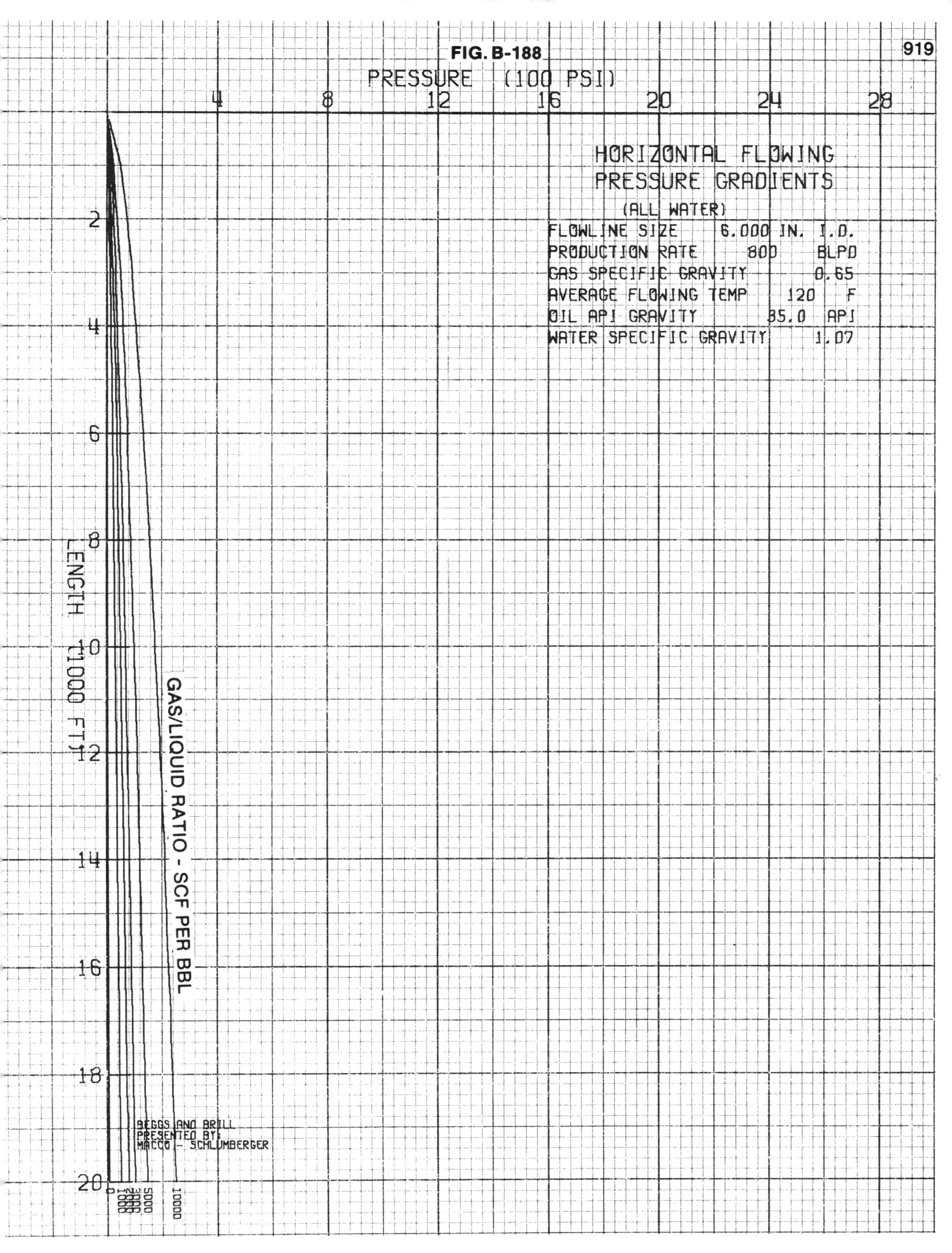

PRESSURE (100 PSI)

HORIZONTAL FLOWING
PRESSURE GRADIENTS
(ALL WATER)

FLOWLINE SIZE	6.000 IN.	I.D.
PRODUCTION RATE	800	BLPD
GAS SPECIFIC GRAVITY	0.65	
AVERAGE FLOWING TEMP	120	F
OIL API GRAVITY	35.0	API
WATER SPECIFIC GRAVITY	1.07	

LENGTH (1000 FT.)

GAS/LIQUID RATIO - SCF PER BBL

BEGGS AND BRILL
PRESENTED BY:
MACCO – SCHLUMBERGER

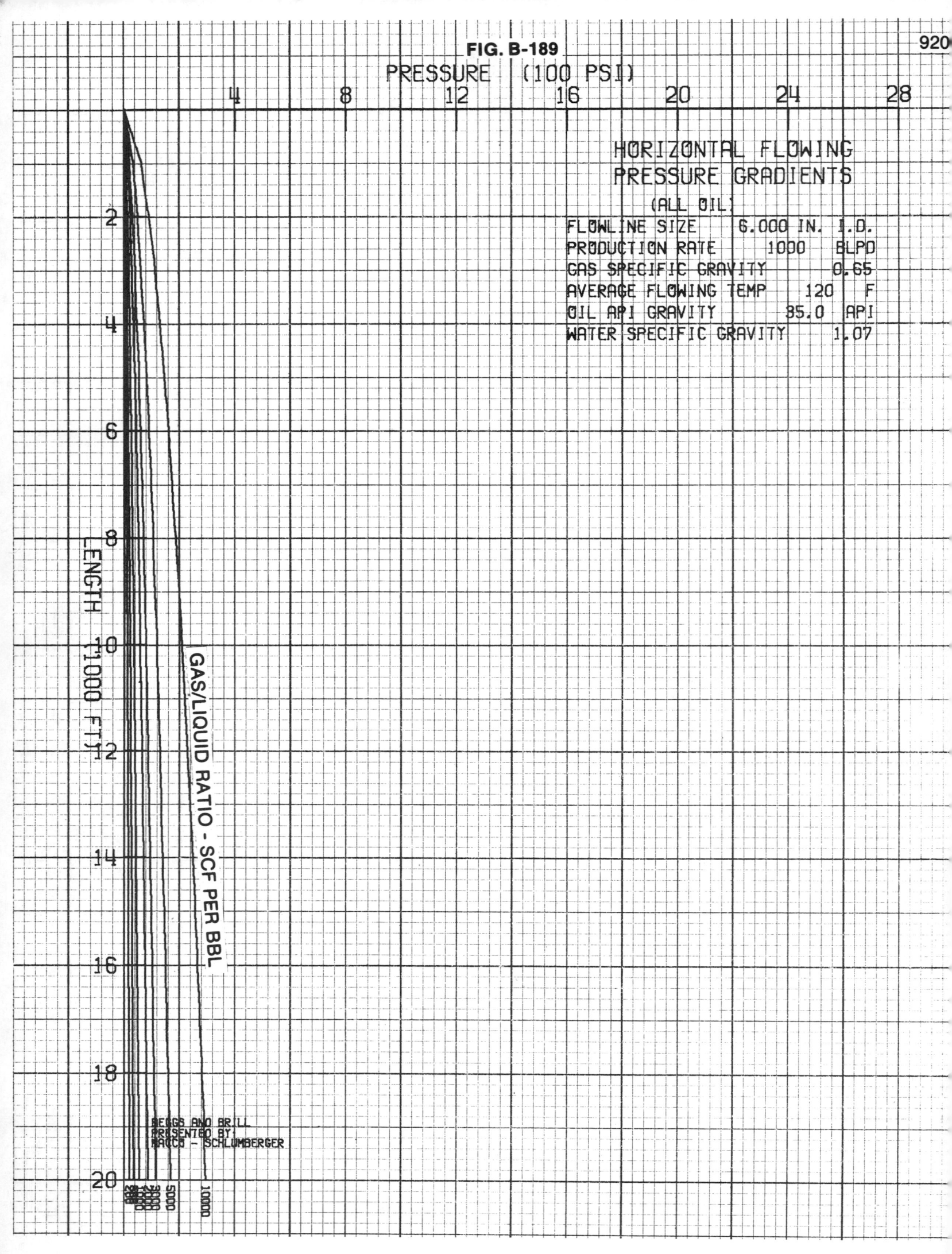

PRESSURE (100 PSI)

HORIZONTAL FLOWING
PRESSURE GRADIENTS
(ALL OIL)

FLOWLINE SIZE	6.000 IN.	I.D.
PRODUCTION RATE	1000	BLPD
GAS SPECIFIC GRAVITY	0.65	
AVERAGE FLOWING TEMP	120	F
OIL API GRAVITY	35.0	API
WATER SPECIFIC GRAVITY	1.07	

LENGTH (1000 FT)

GAS/LIQUID RATIO - SCF PER BBL

BEGGS AND BRILL
PRESENTED BY:
MACCO - SCHLUMBERGER

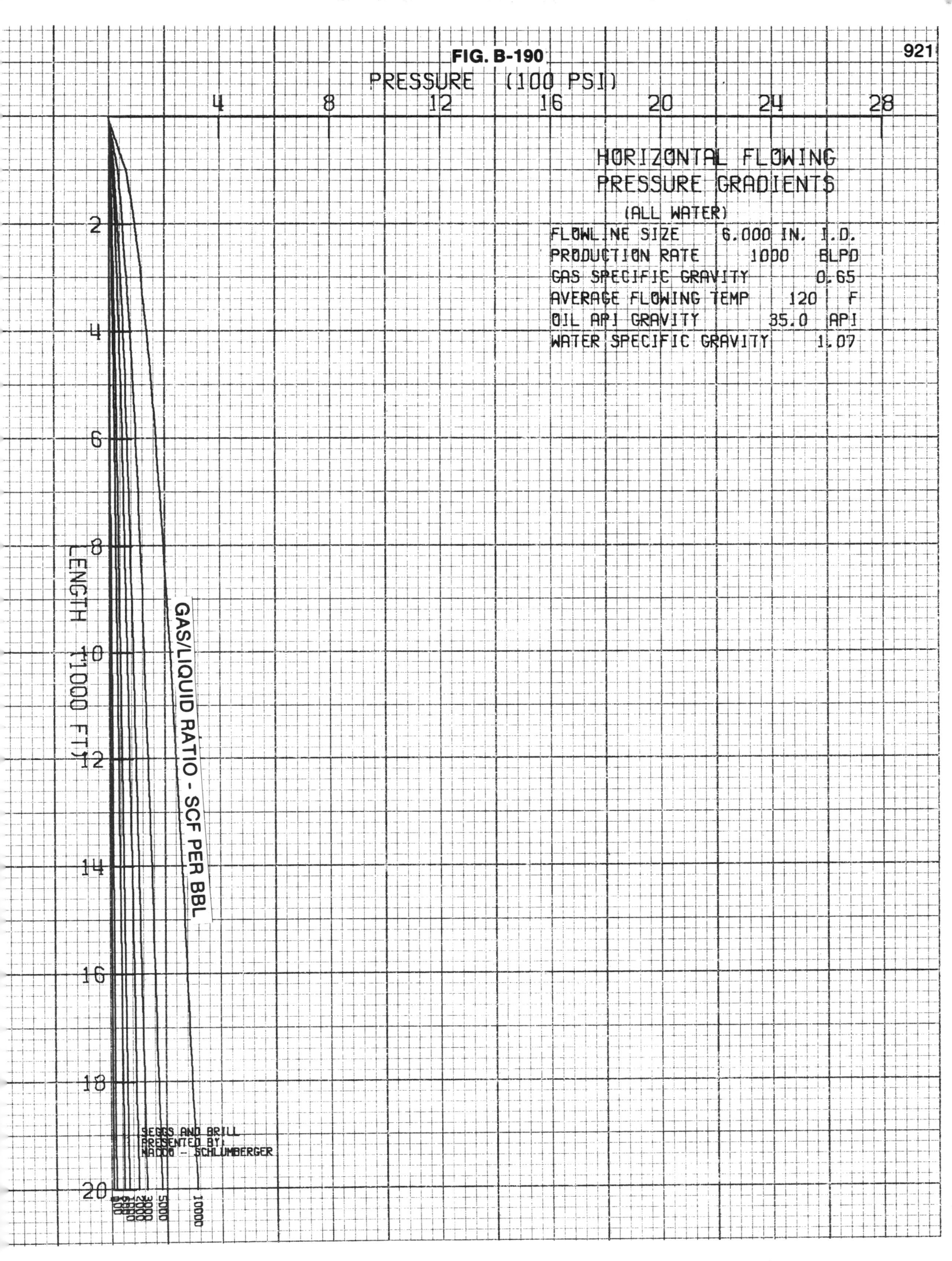

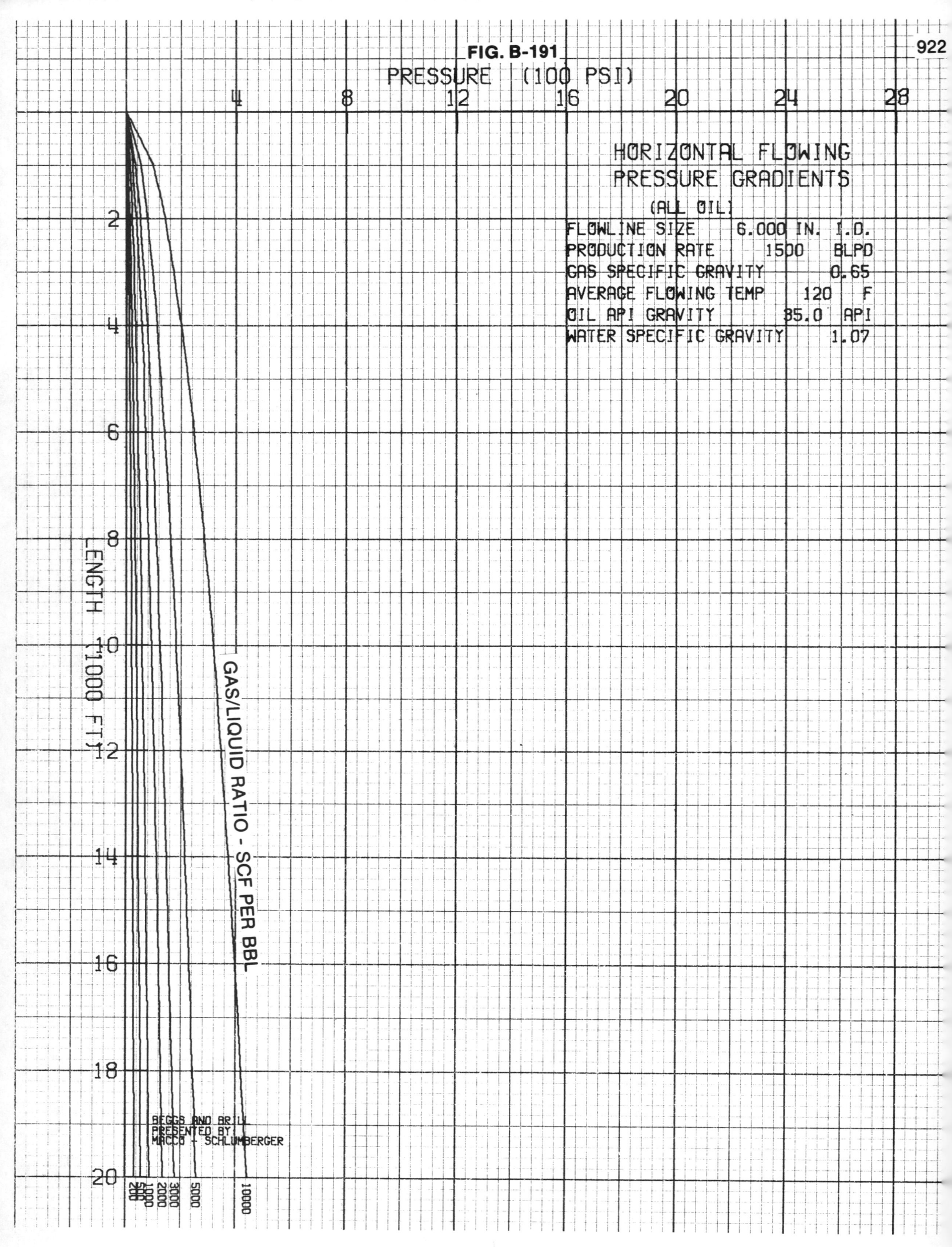

FIG. B-191

922

HORIZONTAL FLOWING
PRESSURE GRADIENTS
(ALL OIL)

FLOWLINE SIZE	6.000 IN. I.D.
PRODUCTION RATE	1500 BLPD
GAS SPECIFIC GRAVITY	0.65
AVERAGE FLOWING TEMP	120 F
OIL API GRAVITY	35.0 API
WATER SPECIFIC GRAVITY	1.07

PRESSURE (100 PSI)

LENGTH (1000 FT)

GAS/LIQUID RATIO - SCF PER BBL

BEGGS AND BRILL
PRESENTED BY
MACCO - SCHLUMBERGER

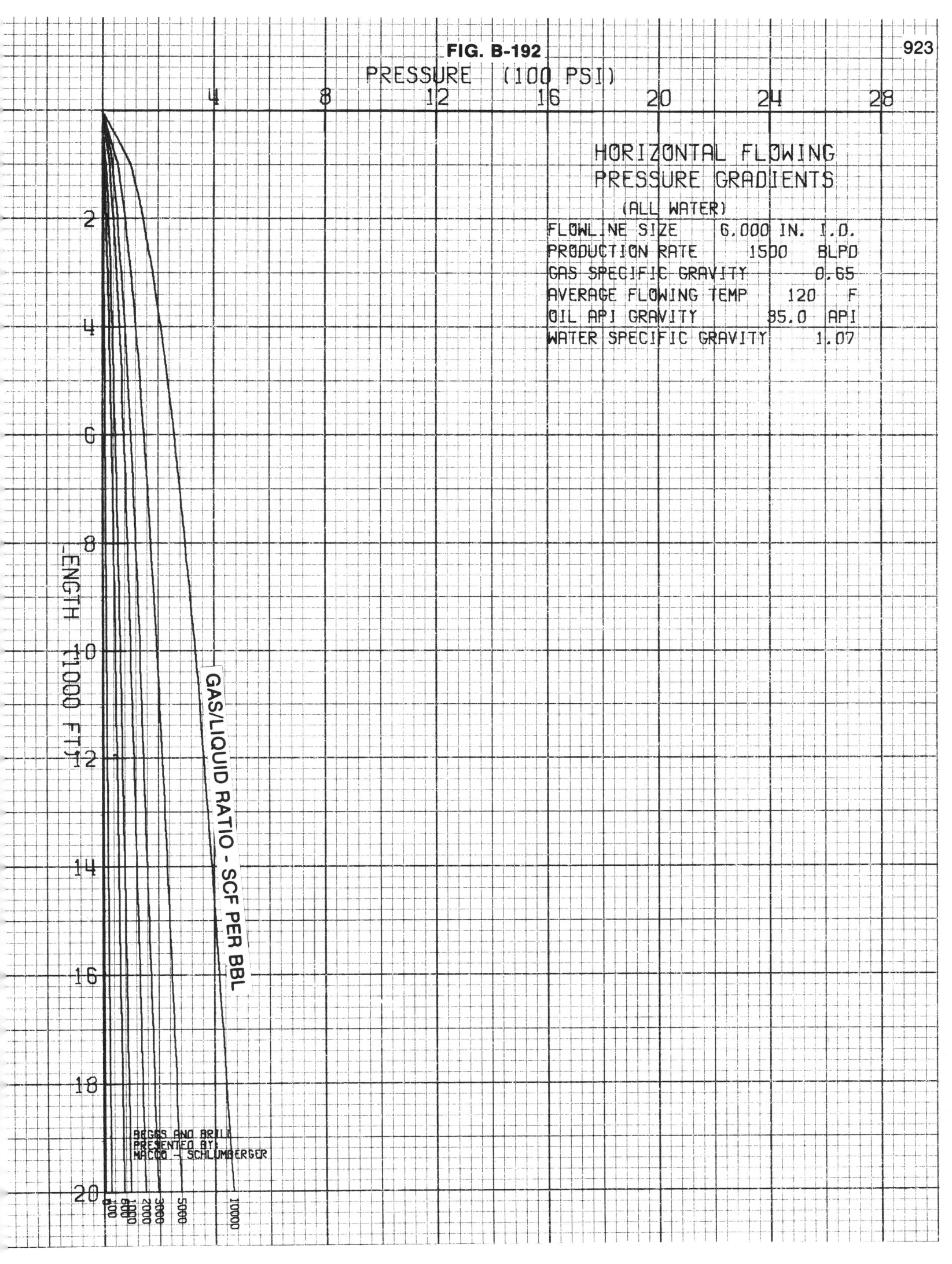

PRESSURE (100 PSI)

HORIZONTAL FLOWING
PRESSURE GRADIENTS

(ALL WATER)

FLOWLINE SIZE   6.000 IN. I.D.
PRODUCTION RATE   1500   BLPD
GAS SPECIFIC GRAVITY   0.65
AVERAGE FLOWING TEMP   120   F
OIL API GRAVITY   85.0   API
WATER SPECIFIC GRAVITY   1.07

LENGTH (1000 FT.)

GAS/LIQUID RATIO - SCF PER BBL

BEGGS AND BRILL
PRESENTED BY:
MACCO - SCHLUMBERGER

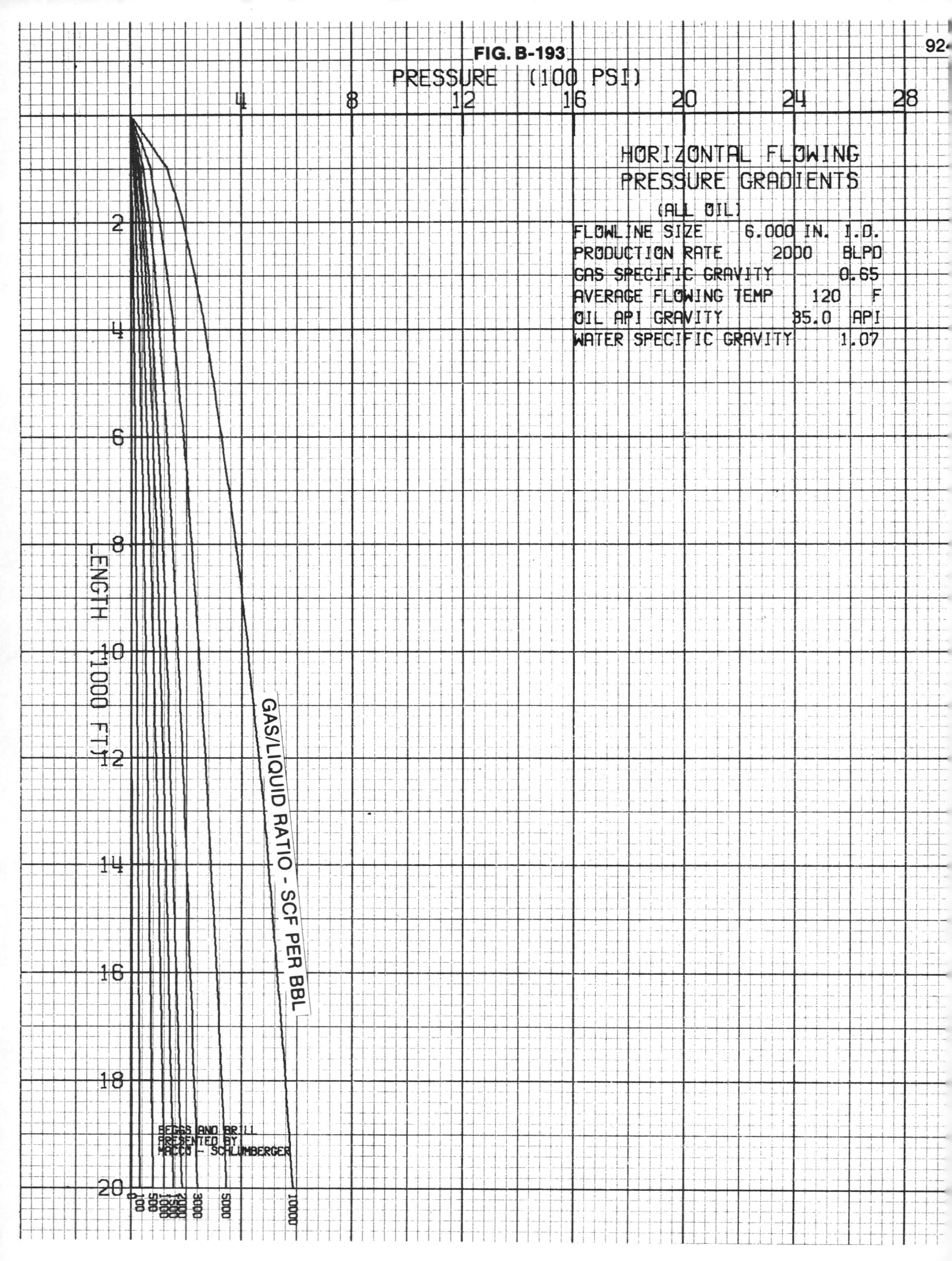

FIG. B-193

92

PRESSURE (100 PSI)

HORIZONTAL FLOWING
PRESSURE GRADIENTS
(ALL OIL)

FLOWLINE SIZE	6.000 IN.	I.D.
PRODUCTION RATE	2000	BLPD
GAS SPECIFIC GRAVITY		0.65
AVERAGE FLOWING TEMP	120	F
OIL API GRAVITY	35.0	API
WATER SPECIFIC GRAVITY		1.07

LENGTH (1000 FT)

GAS/LIQUID RATIO - SCF PER BBL

BEGGS AND BRILL
PRESENTED BY
MACCO - SCHLUMBERGER

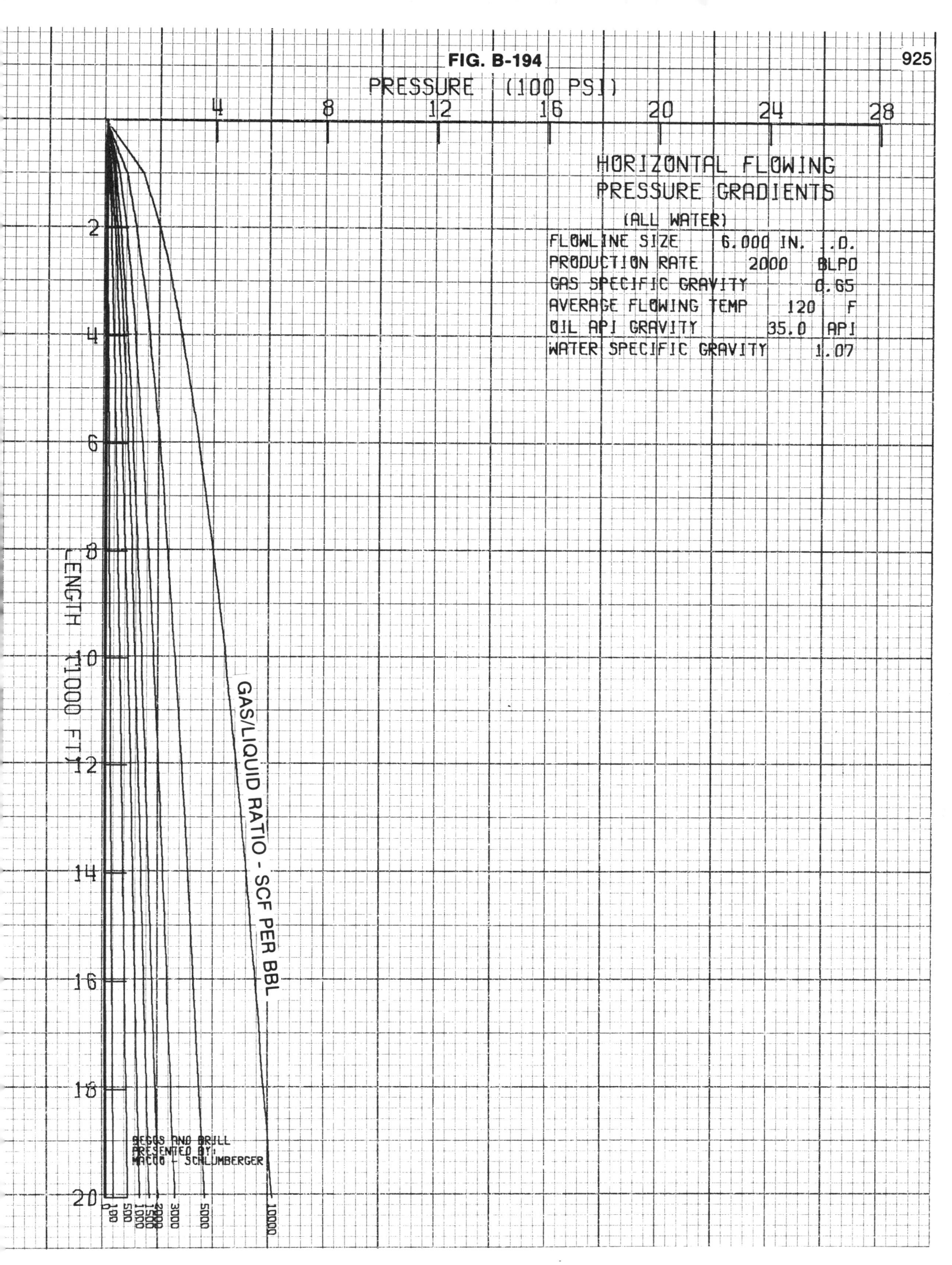

PRESSURE (100 PSI)

HORIZONTAL FLOWING
PRESSURE GRADIENTS
(ALL WATER)

FLOWLINE SIZE	6.000 IN.	I.D.
PRODUCTION RATE	2000	BLPD
GAS SPECIFIC GRAVITY	0.65	
AVERAGE FLOWING TEMP	120	F
OIL API GRAVITY	35.0	API
WATER SPECIFIC GRAVITY	1.07	

LENGTH (1000 FT)

GAS/LIQUID RATIO - SCF PER BBL

BEGGS AND BRILL
PRESENTED BY:
MACCO - SCHLUMBERGER

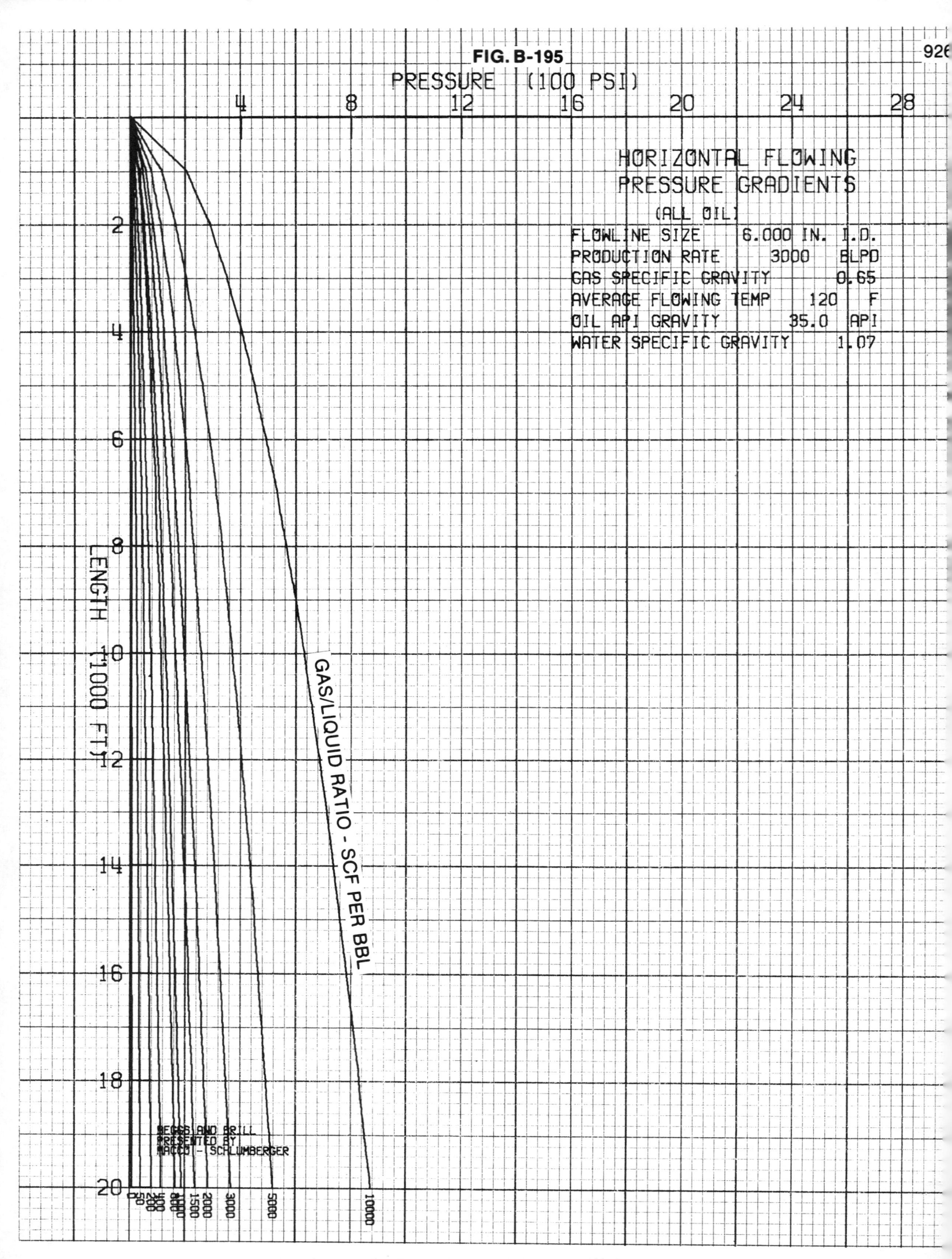

FIG. B-195

926

PRESSURE (100 PSI)

HORIZONTAL FLOWING
PRESSURE GRADIENTS
(ALL OIL)

FLOWLINE SIZE	6.000 IN.	I.D.
PRODUCTION RATE	3000	BLPD
GAS SPECIFIC GRAVITY	0.65	
AVERAGE FLOWING TEMP	120	F
OIL API GRAVITY	35.0	API
WATER SPECIFIC GRAVITY	1.07	

LENGTH (1000 FT)

GAS/LIQUID RATIO - SCF PER BBL

BEGGS AND BRILL
PRESENTED BY
MACCO - SCHLUMBERGER

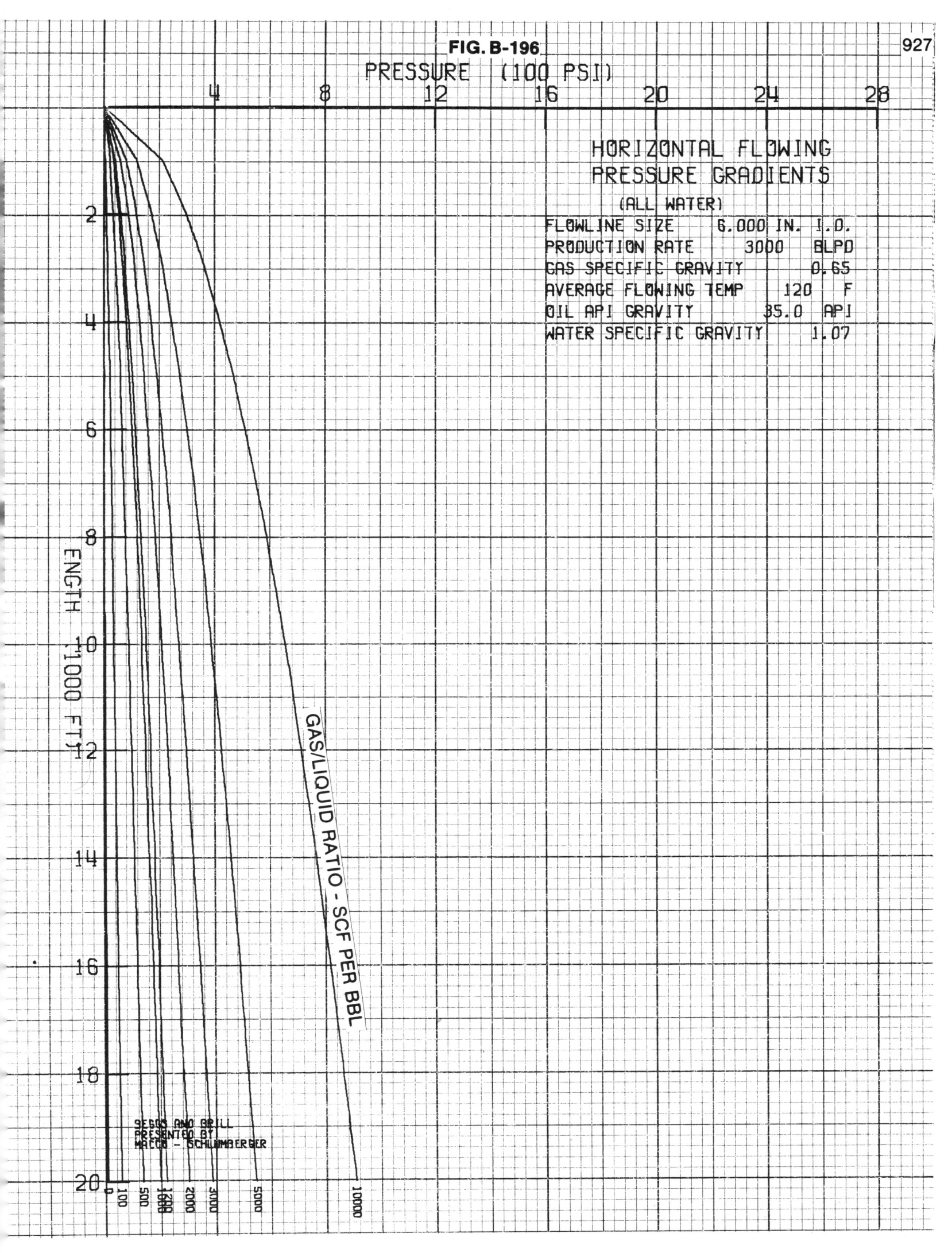

PRESSURE (100 PSI)

HORIZONTAL FLOWING
PRESSURE GRADIENTS
(ALL WATER)

FLOWLINE SIZE	6.000 IN.	I.D.
PRODUCTION RATE	3000	BLPD
GAS SPECIFIC GRAVITY		0.65
AVERAGE FLOWING TEMP	120	F
OIL API GRAVITY	35.0	API
WATER SPECIFIC GRAVITY		1.07

LENGTH (1000 FT)

GAS/LIQUID RATIO - SCF PER BBL

BEGGS AND BRILL
PRESENTED BY
MAECO - SCHLUMBERGER

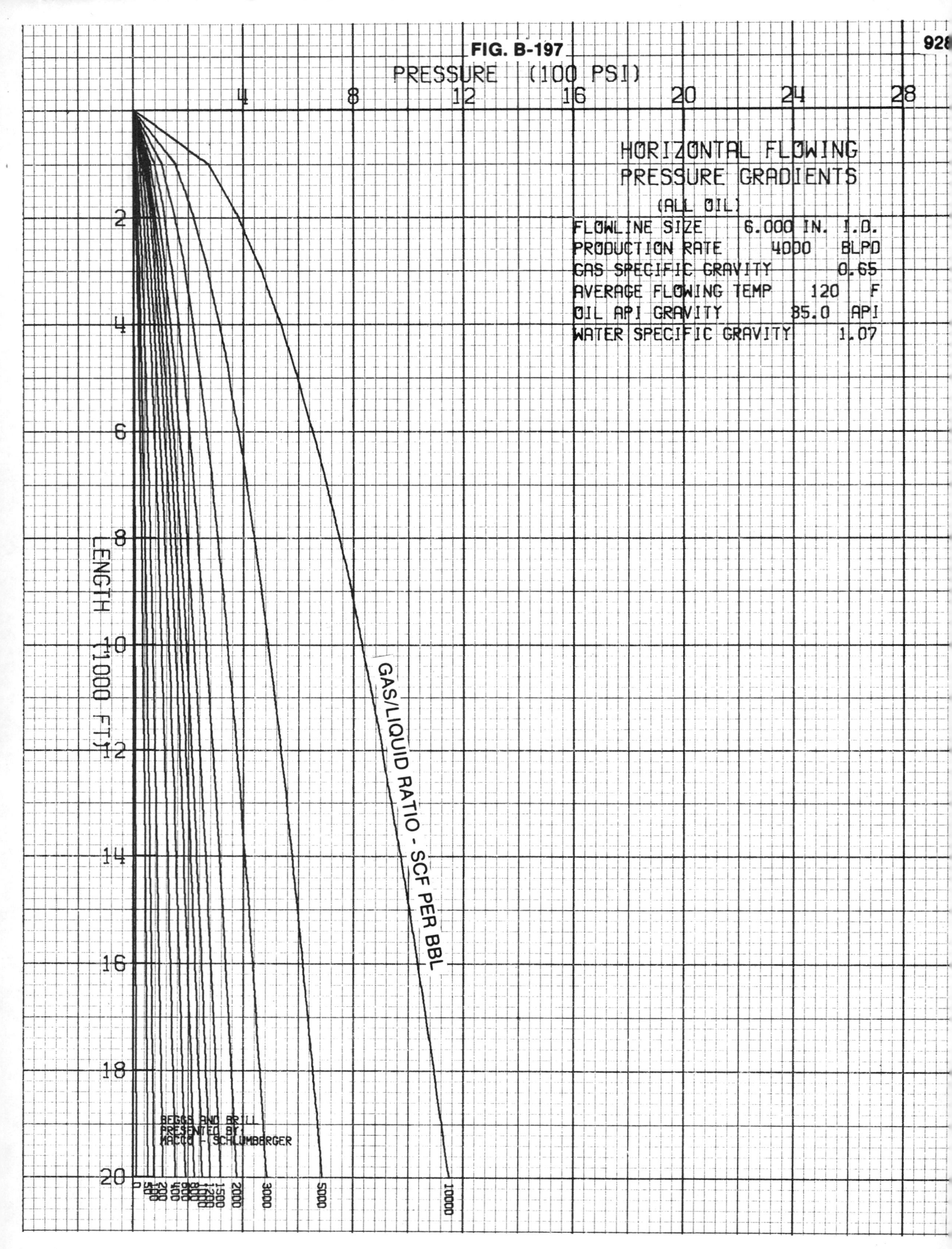

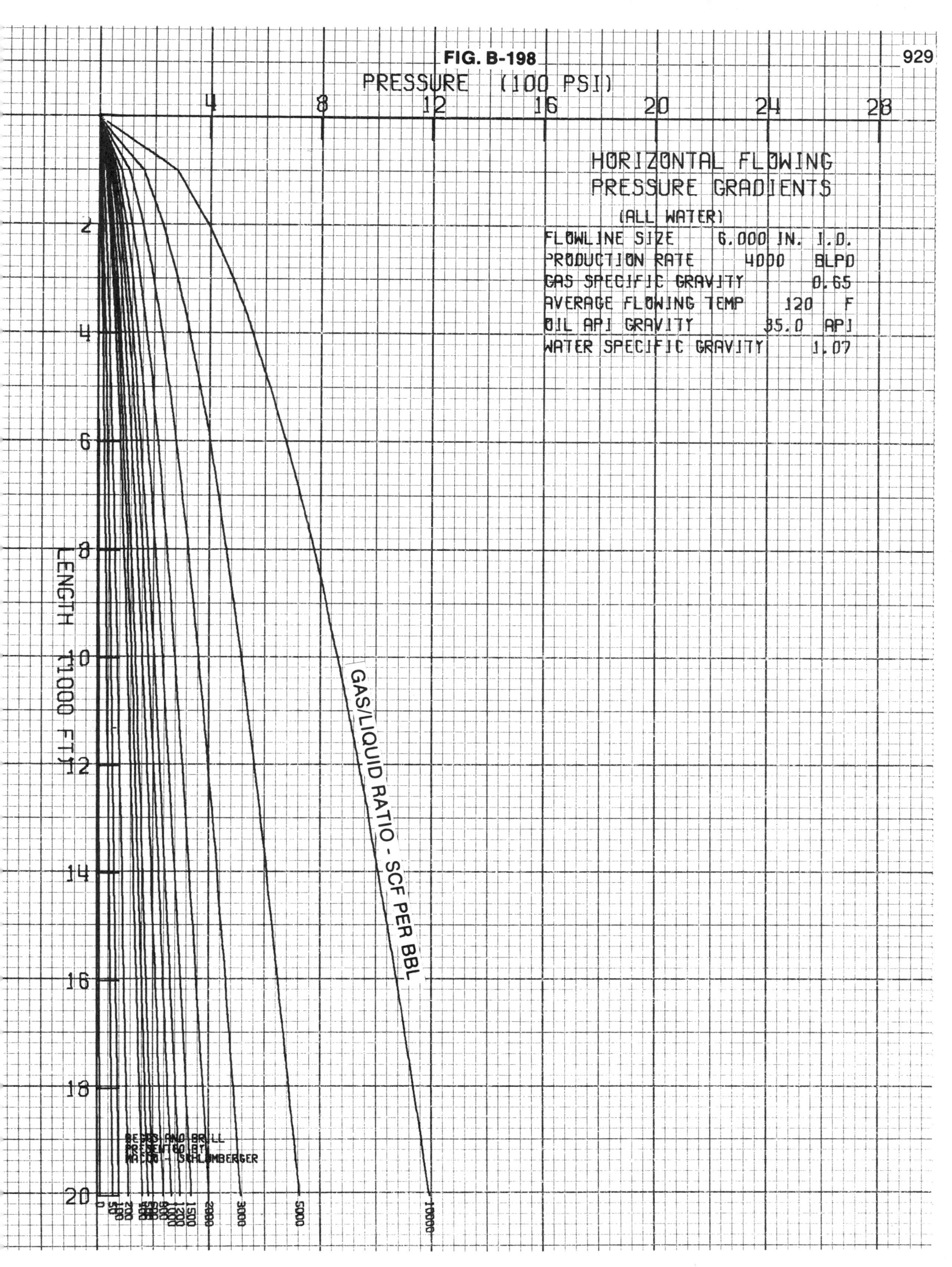

PRESSURE (100 PSI)

HORIZONTAL FLOWING
PRESSURE GRADIENTS
(ALL WATER)

FLOWLINE SIZE   6.000 IN. I.D.
PRODUCTION RATE   4000   BLPD
GAS SPECIFIC GRAVITY   0.65
AVERAGE FLOWING TEMP   120   F
OIL API GRAVITY   35.0 API
WATER SPECIFIC GRAVITY   1.07

LENGTH (1000 FT)

GAS/LIQUID RATIO - SCF PER BBL

BEGGS AND BRILL
PRESENTED BY
MACCO - SCHLUMBERGER

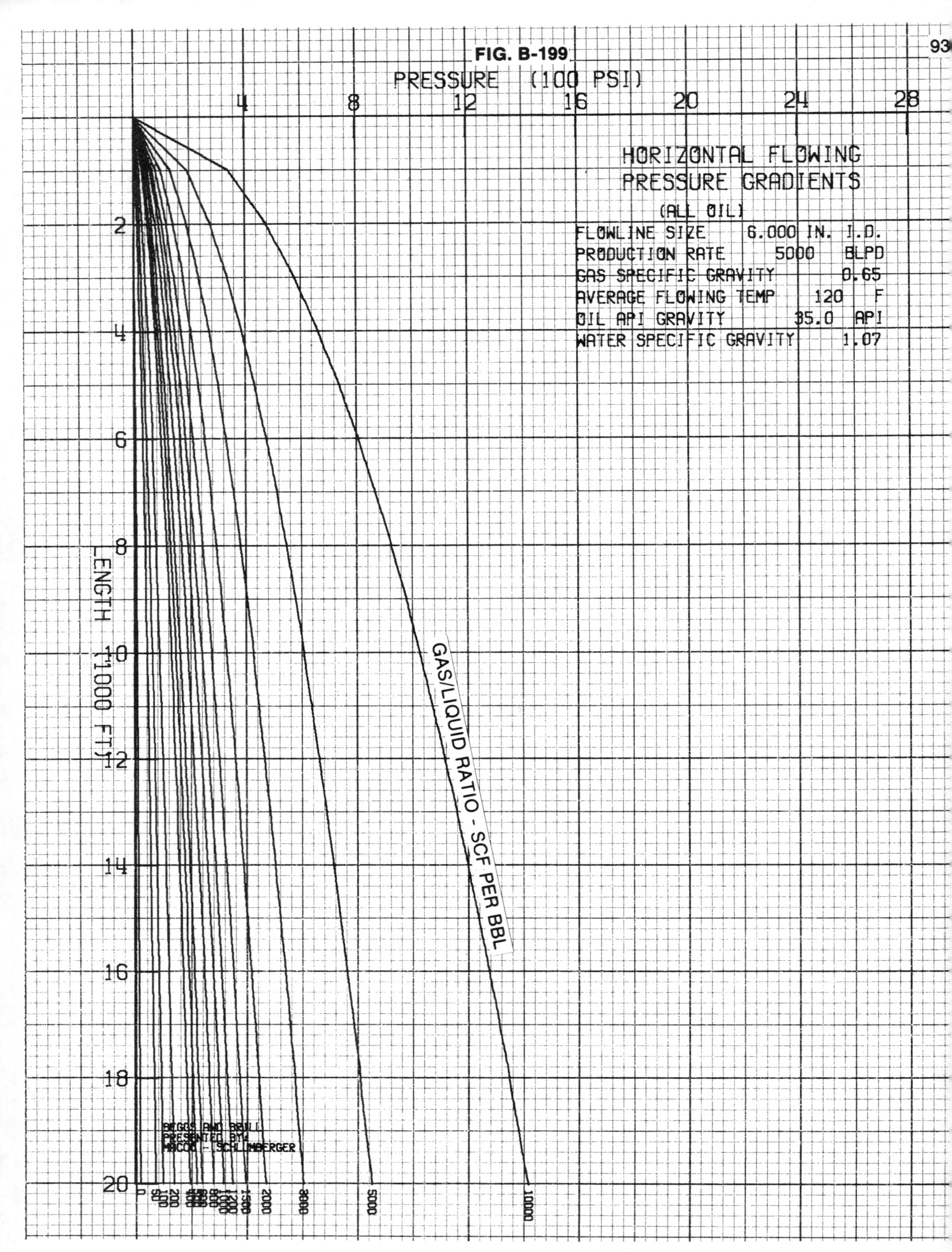

FIG. B-199

PRESSURE (100 PSI)

HORIZONTAL FLOWING
PRESSURE GRADIENTS
(ALL OIL)

FLOWLINE SIZE	6.000 IN. I.D.
PRODUCTION RATE	5000 BLPD
GAS SPECIFIC GRAVITY	0.65
AVERAGE FLOWING TEMP	120 F
OIL API GRAVITY	35.0 API
WATER SPECIFIC GRAVITY	1.07

LENGTH (1000 FT)

GAS/LIQUID RATIO - SCF PER BBL

BEGGS AND BRILL
PRESENTED BY:
MACCO - SCHLUMBERGER

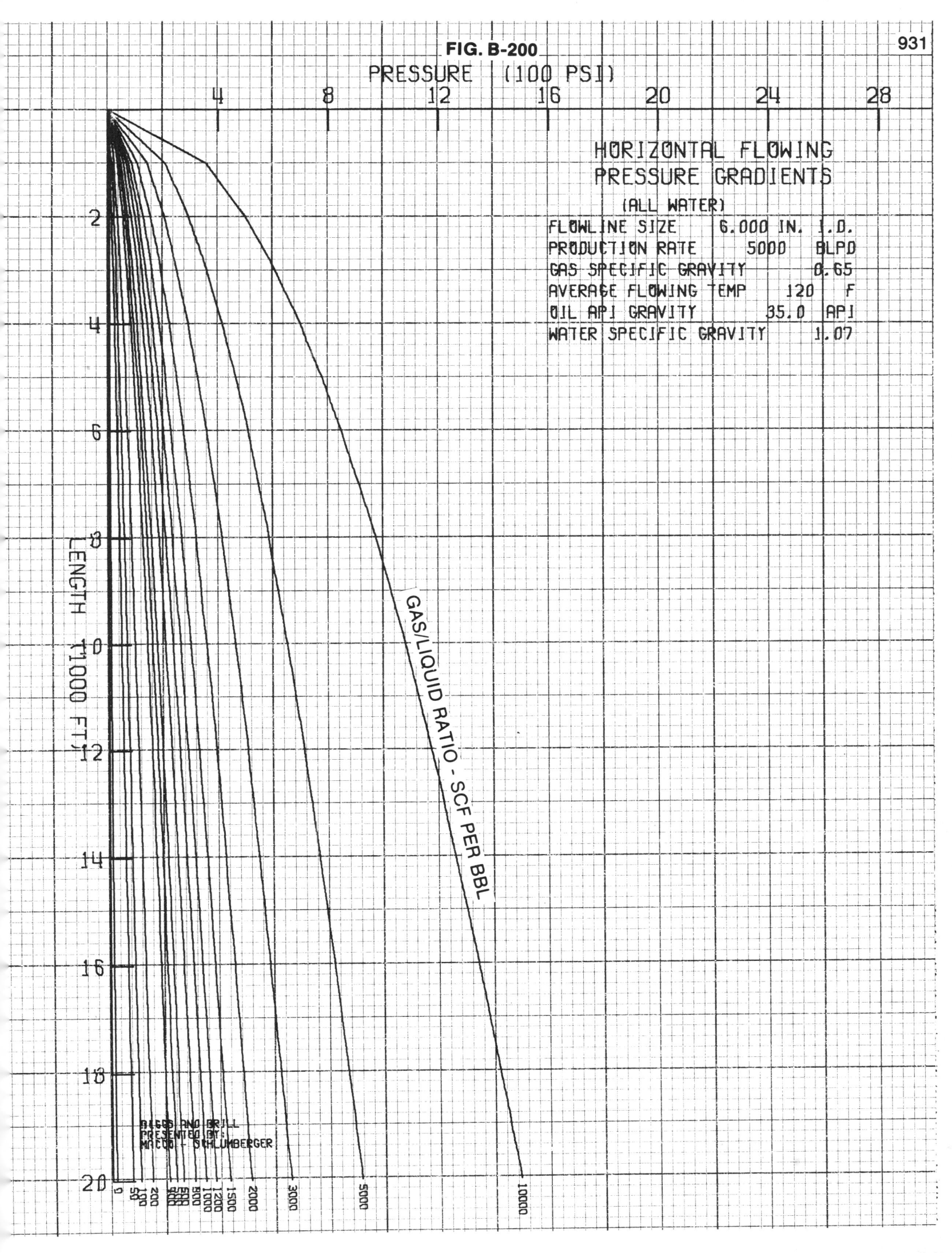

FIG. B-200

PRESSURE (100 PSI)

HORIZONTAL FLOWING
PRESSURE GRADIENTS
(ALL WATER)

FLOWLINE SIZE	6.000 IN.	I.D.
PRODUCTION RATE	5000	BLPD
GAS SPECIFIC GRAVITY	0.65	
AVERAGE FLOWING TEMP	120	F
OIL API GRAVITY	35.0	API
WATER SPECIFIC GRAVITY	1.07	

LENGTH (1000 FT.)

GAS/LIQUID RATIO - SCF PER BBL

BEGGS AND BRILL
PRESENTED BY:
MACCO - SCHLUMBERGER

FIG. B-201

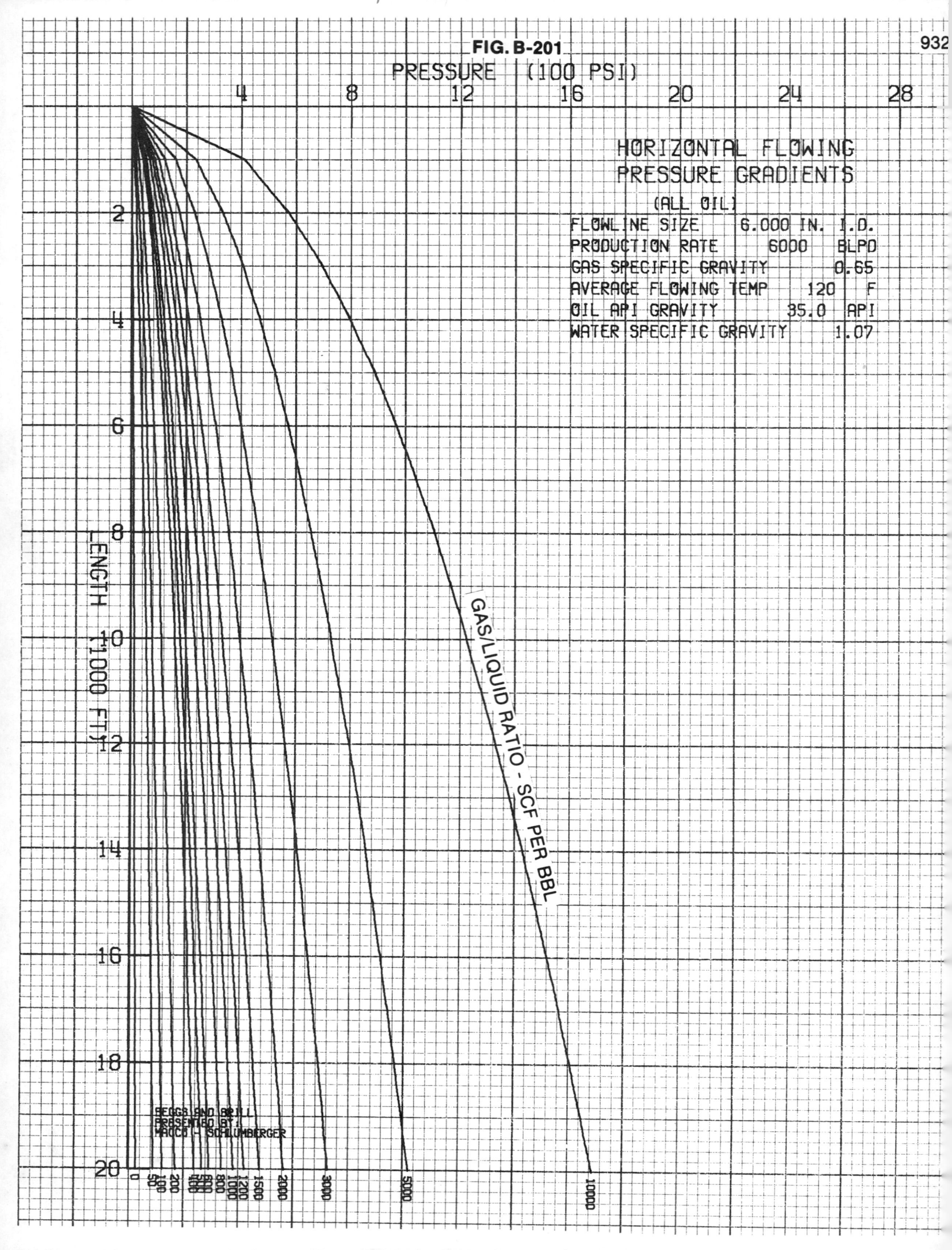

PRESSURE (100 PSI)

HORIZONTAL FLOWING
PRESSURE GRADIENTS
(ALL OIL)

FLOWLINE SIZE	6.000 IN. I.D.	
PRODUCTION RATE	6000	BLPD
GAS SPECIFIC GRAVITY		0.65
AVERAGE FLOWING TEMP	120	F
OIL API GRAVITY	35.0	API
WATER SPECIFIC GRAVITY		1.07

LENGTH (1,000 FT.)

GAS/LIQUID RATIO - SCF PER BBL

BEGGS AND BRILL
PRESENTED BY
MADCO - SCHLUMBERGER

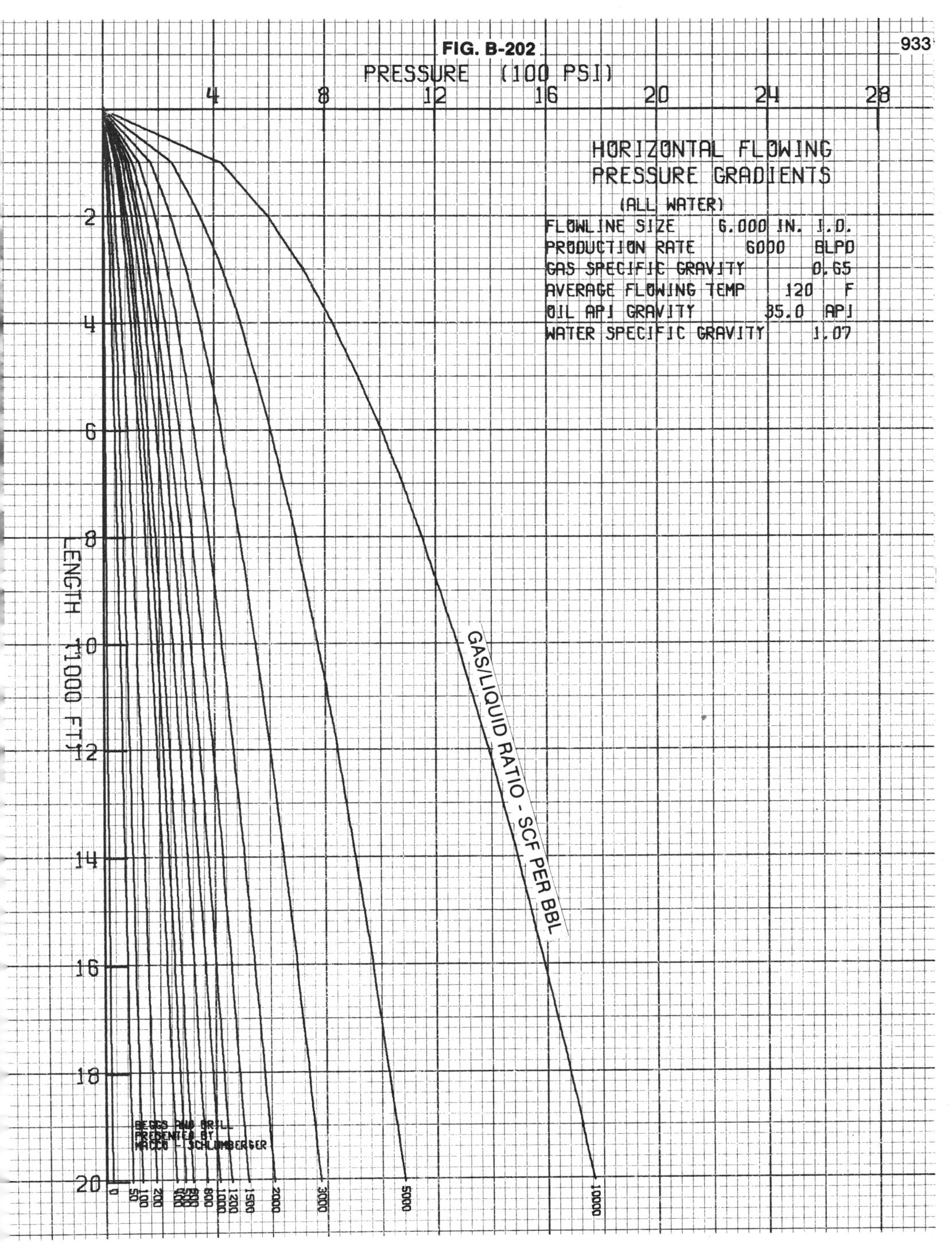

PRESSURE (100 PSI)

HORIZONTAL FLOWING
PRESSURE GRADIENTS
(ALL WATER)

FLOWLINE SIZE	6.000 IN. I.D.	
PRODUCTION RATE	6000	BLPD
GAS SPECIFIC GRAVITY	0.65	
AVERAGE FLOWING TEMP	120	F
OIL API GRAVITY	35.0	API
WATER SPECIFIC GRAVITY	1.07	

LENGTH (1000 FT)

GAS/LIQUID RATIO - SCF PER BBL

BEGGS AND BRILL
PRESENTED BY
MACCO - SCHLUMBERGER

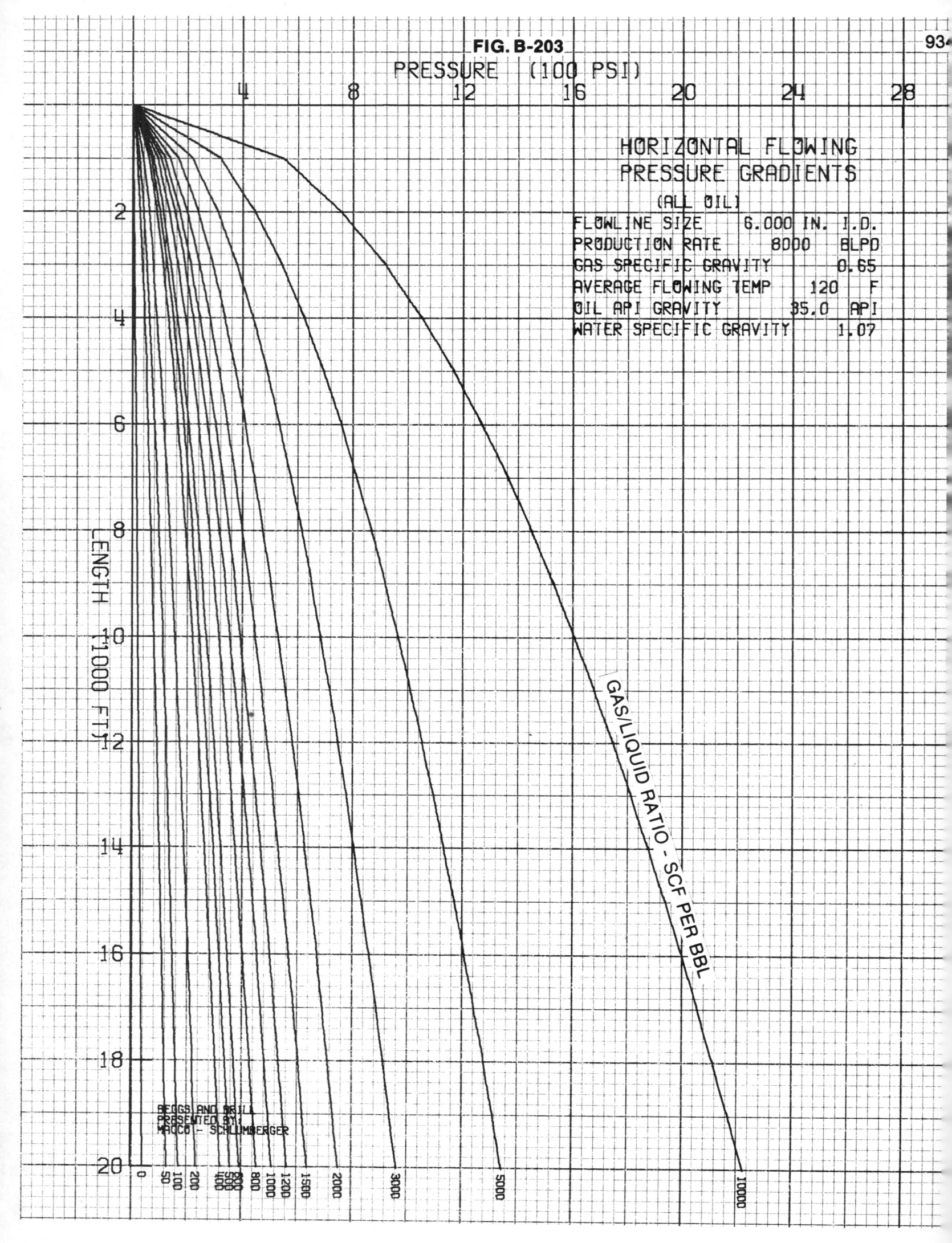

PRESSURE (100 PSI)

HORIZONTAL FLOWING
PRESSURE GRADIENTS
(ALL OIL)

FLOWLINE SIZE	6.000 IN. I.D.
PRODUCTION RATE	8000 BLPD
GAS SPECIFIC GRAVITY	0.65
AVERAGE FLOWING TEMP	120 F
OIL API GRAVITY	35.0 API
WATER SPECIFIC GRAVITY	1.07

GAS/LIQUID RATIO - SCF PER BBL

LENGTH (1000 FT)

BEGGS AND BRILL
PRESENTED BY:
MADCO - SCHLUMBERGER

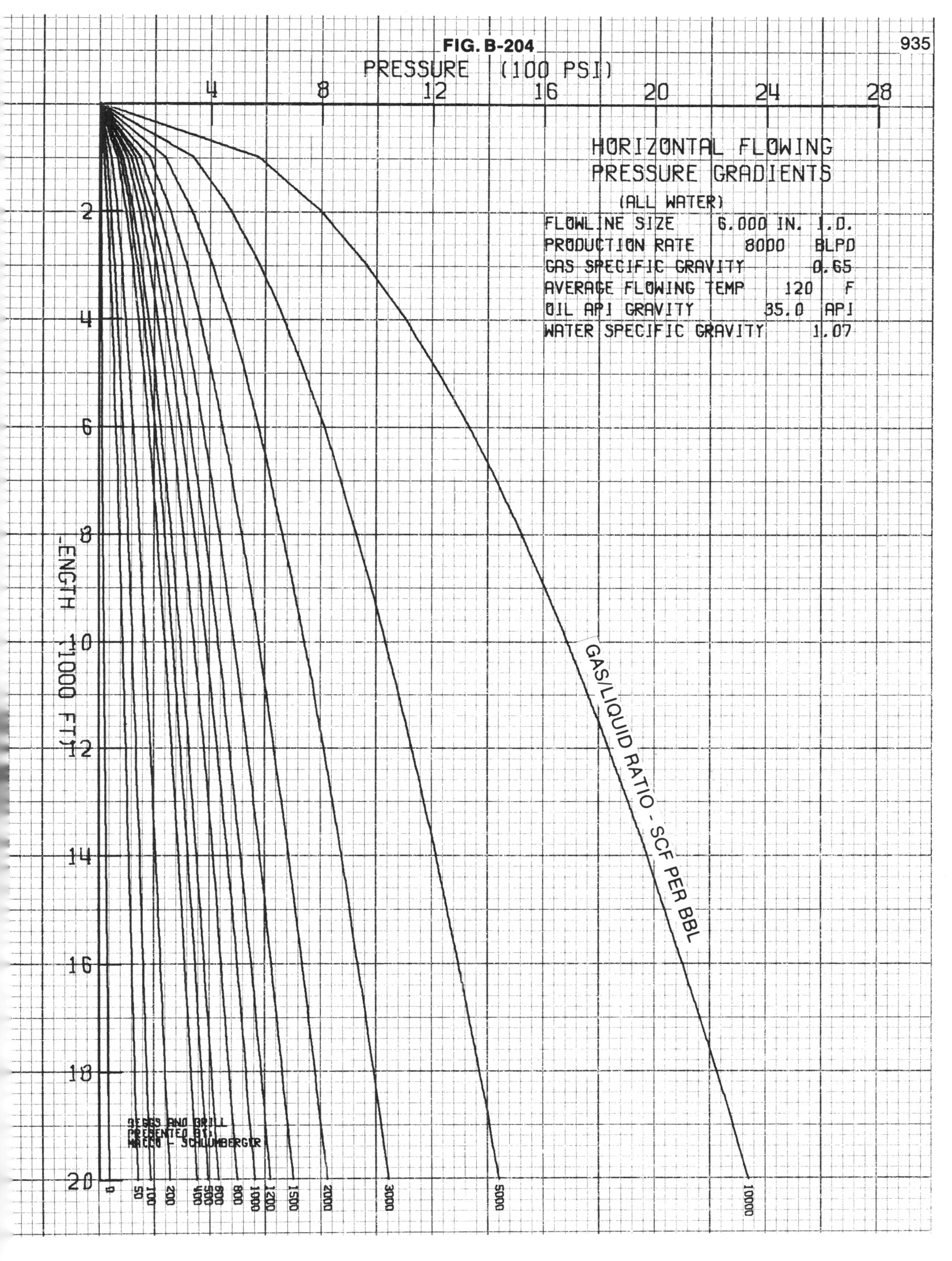

FIG. B-204

PRESSURE (100 PSI)

HORIZONTAL FLOWING
PRESSURE GRADIENTS
(ALL WATER)

FLOWLINE SIZE   6.000 IN. I.D.
PRODUCTION RATE   8000   BLPD
GAS SPECIFIC GRAVITY   0.65
AVERAGE FLOWING TEMP   120   F
OIL API GRAVITY   35.0 API
WATER SPECIFIC GRAVITY   1.07

LENGTH (1000 FT)

GAS/LIQUID RATIO - SCF PER BBL

BEGGS AND BRILL
PRESENTED BY:
WILCO - SCHLUMBERGER

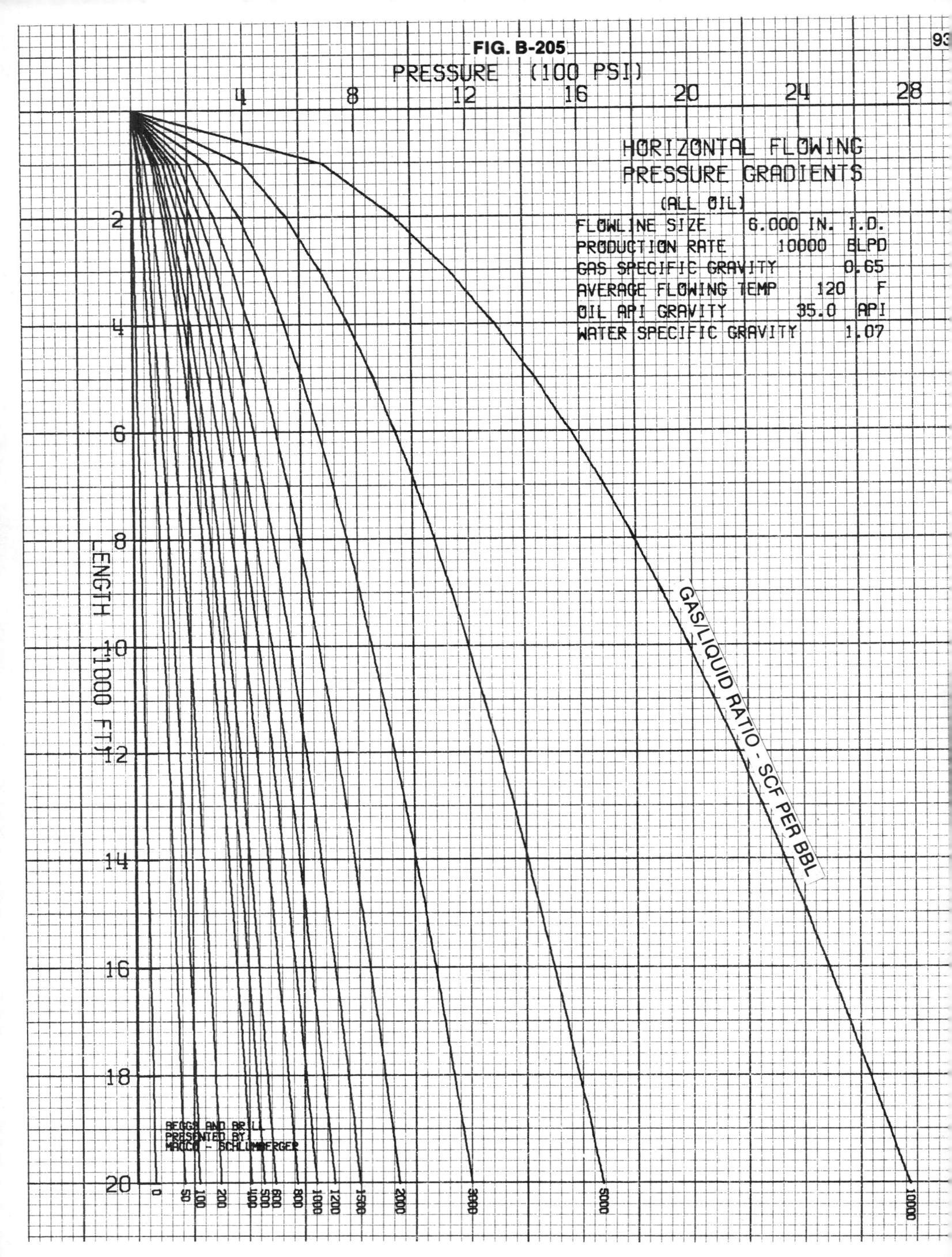

FIG. B-205

PRESSURE (100 PSI)

HORIZONTAL FLOWING
PRESSURE GRADIENTS
(ALL OIL)

FLOWLINE SIZE	6.000 IN. I.D.
PRODUCTION RATE	10000 BLPD
GAS SPECIFIC GRAVITY	0.65
AVERAGE FLOWING TEMP	120 F
OIL API GRAVITY	35.0 API
WATER SPECIFIC GRAVITY	1.07

GAS/LIQUID RATIO - SCF PER BBL

LENGTH (1000 FT.)

BEGGS AND BRILL
PRESENTED BY:
MADCO - SCHLUMBERGER

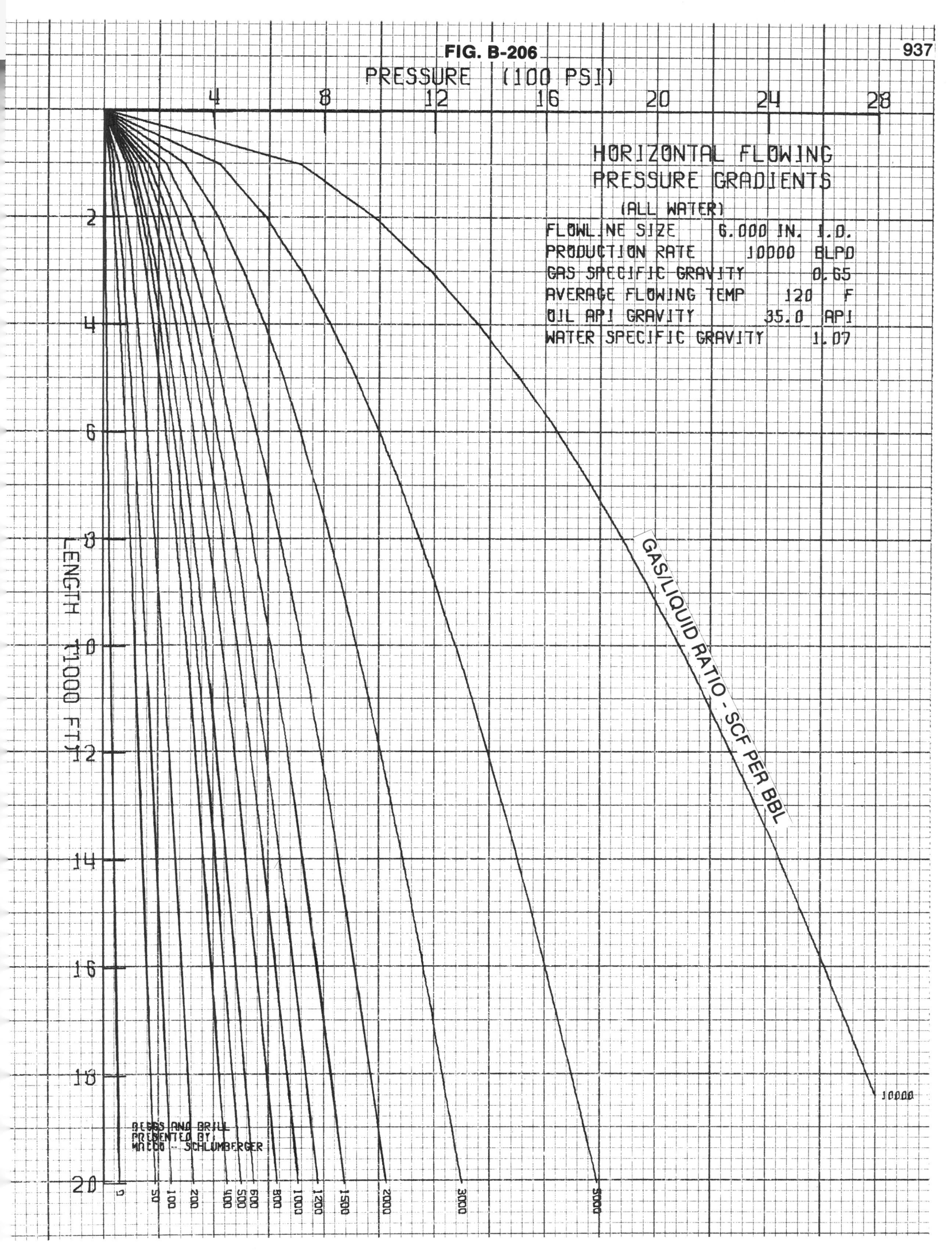

FIG. B-206   937

PRESSURE (100 PSI)

HORIZONTAL FLOWING
PRESSURE GRADIENTS
(ALL WATER)

FLOWLINE SIZE   6.000 IN. I.D.
PRODUCTION RATE   10000 BLPD
GAS SPECIFIC GRAVITY   0.65
AVERAGE FLOWING TEMP   120 F
OIL API GRAVITY   35.0 API
WATER SPECIFIC GRAVITY   1.07

GAS/LIQUID RATIO - SCF PER BBL

LENGTH (1000 FT)

BEGGS AND BRILL
PRESENTED BY:
MAECO - SCHLUMBERGER

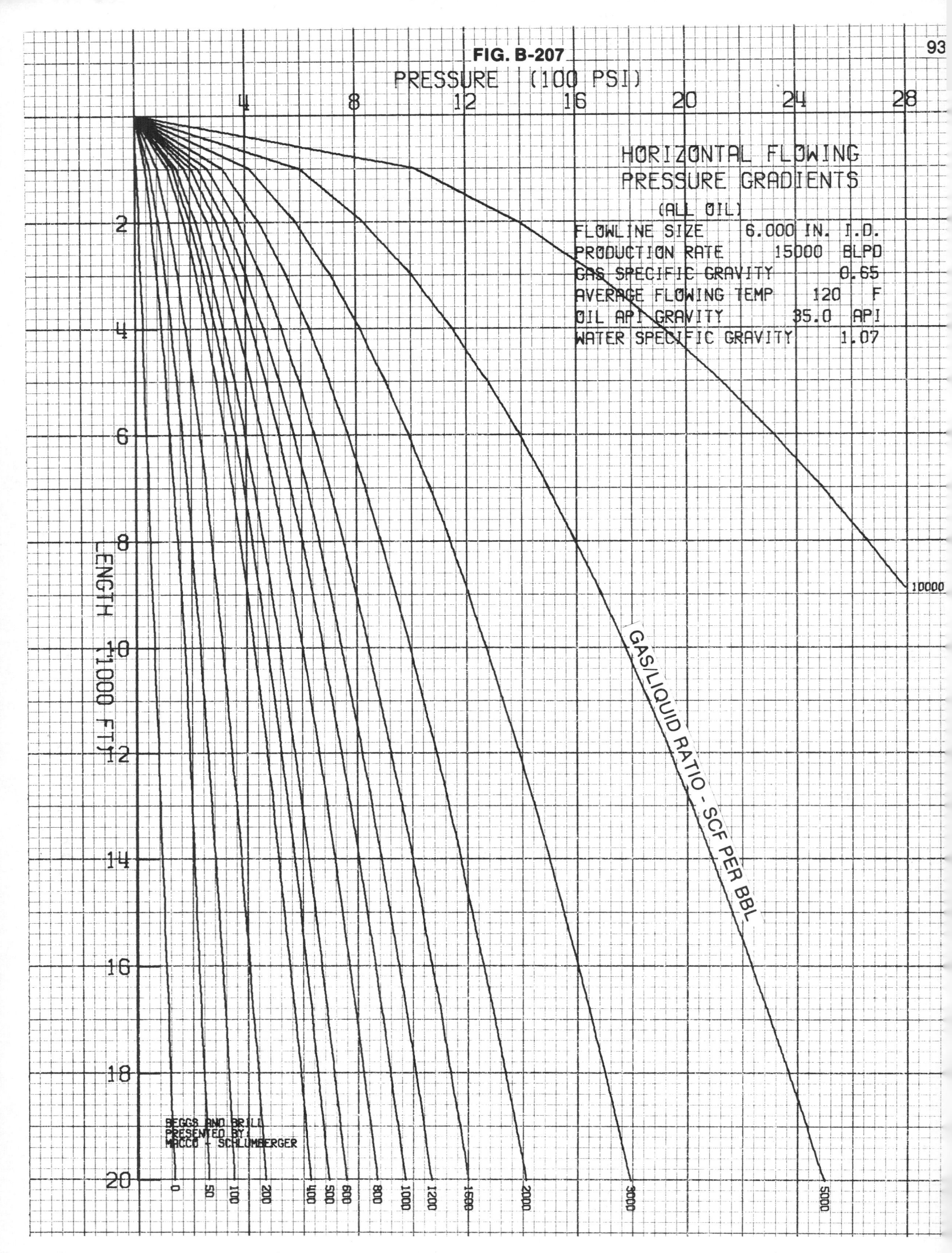

FIG. B-207

HORIZONTAL FLOWING
PRESSURE GRADIENTS
(ALL OIL)

FLOWLINE SIZE   6.000 IN. I.D.
PRODUCTION RATE   15000 BLPD
GAS SPECIFIC GRAVITY   0.65
AVERAGE FLOWING TEMP   120 F
OIL API GRAVITY   35.0 API
WATER SPECIFIC GRAVITY 1.07

PRESSURE (100 PSI)

LENGTH (1000 FT)

GAS/LIQUID RATIO - SCF PER BBL

BEGGS AND BRILL
PRESENTED BY:
MACCO - SCHLUMBERGER

93

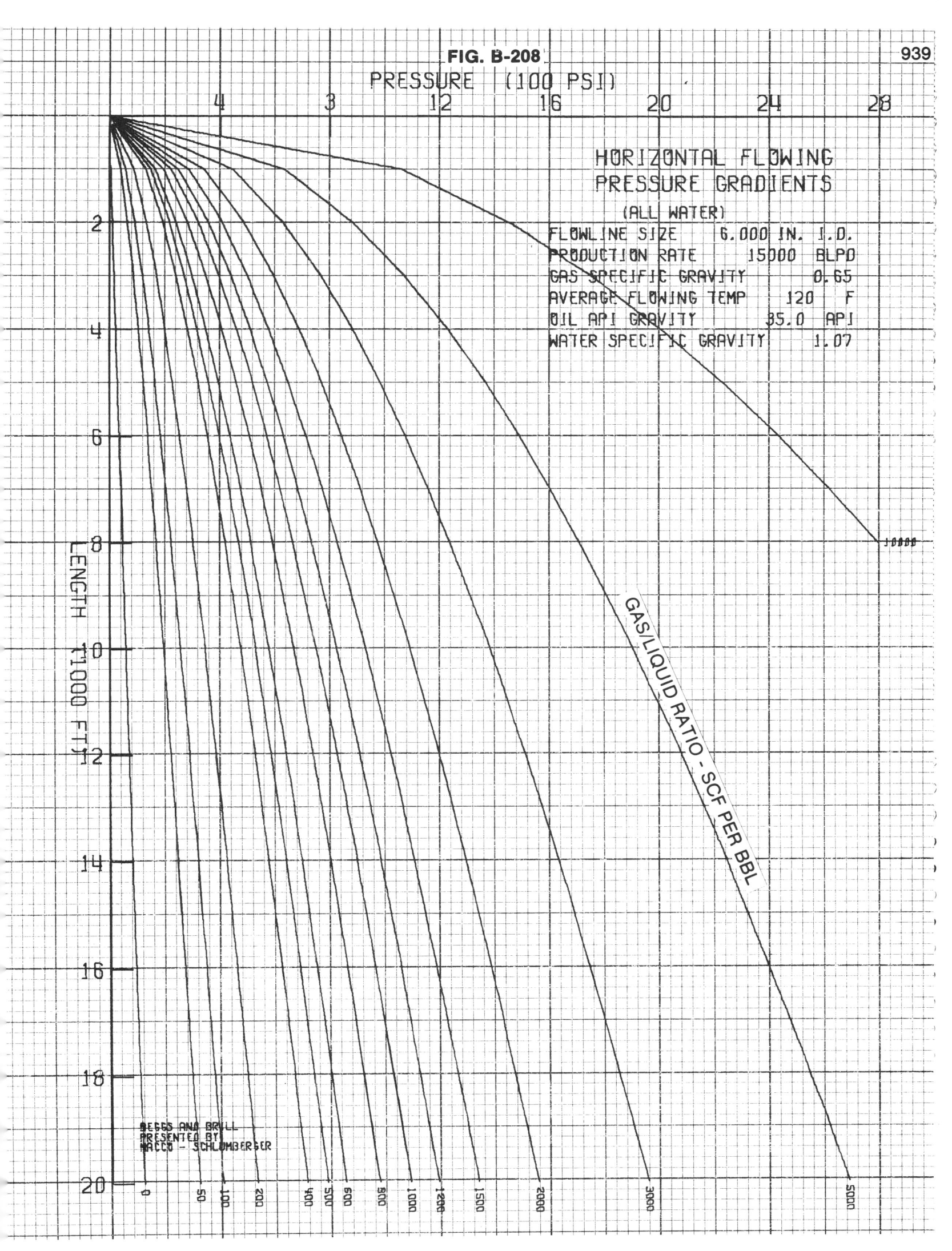

FIG. B-208

939

PRESSURE (100 PSI)

HORIZONTAL FLOWING
PRESSURE GRADIENTS
(ALL WATER)

FLOWLINE SIZE	6.000 IN.	I.D.
PRODUCTION RATE	15000	BLPD
GAS SPECIFIC GRAVITY	0.65	
AVERAGE FLOWING TEMP	120	F
OIL API GRAVITY	35.0	API
WATER SPECIFIC GRAVITY	1.07	

GAS/LIQUID RATIO - SCF PER BBL

LENGTH (1000 FT)

BEGGS AND BRILL
PRESENTED BY
NACCO - SCHLUMBERGER

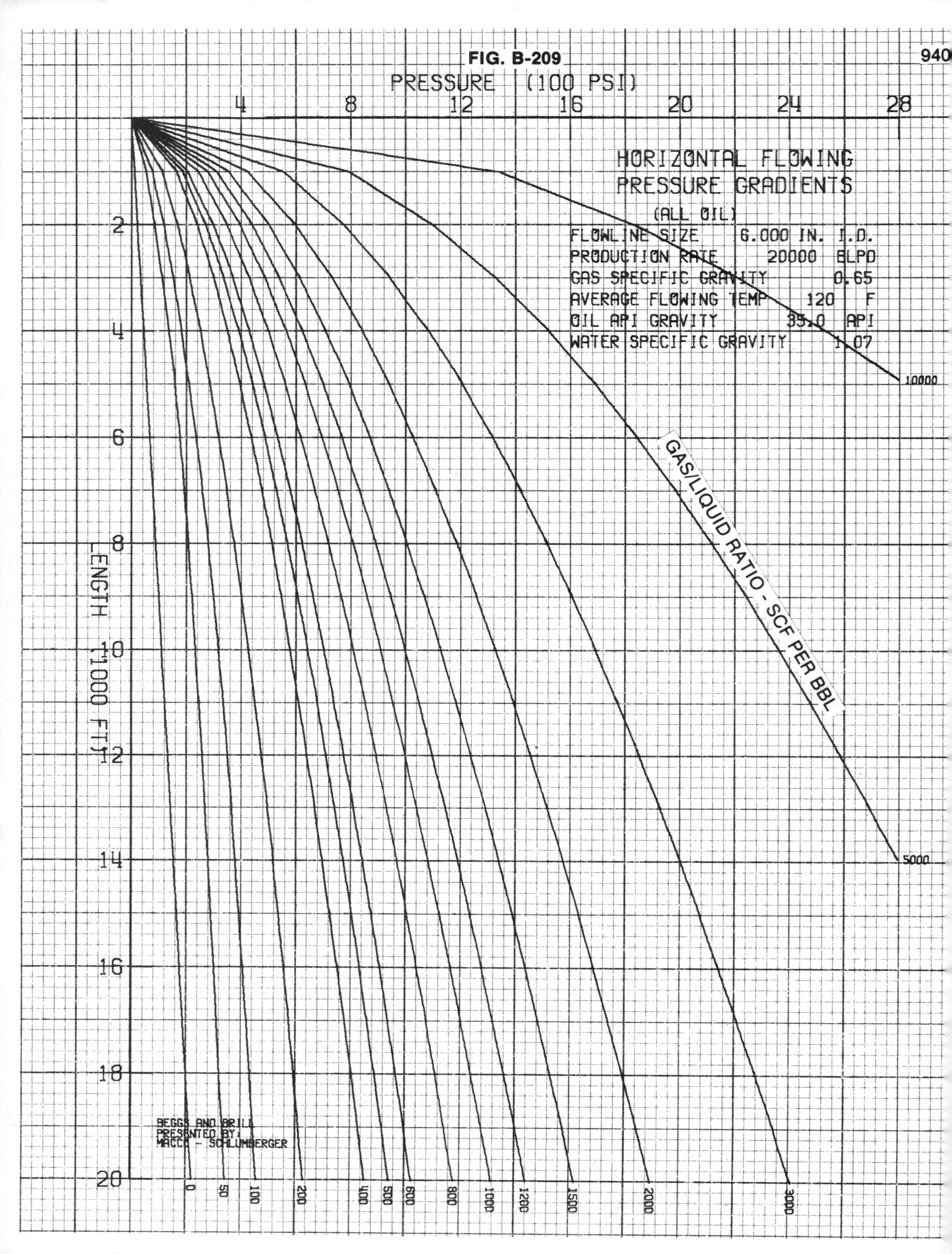

PRESSURE (100 PSI)

HORIZONTAL FLOWING
PRESSURE GRADIENTS
(ALL OIL)
FLOWLINE SIZE   6.000 IN. I.D.
PRODUCTION RATE   20000 BLPD
GAS SPECIFIC GRAVITY   0.65
AVERAGE FLOWING TEMP   120   F
OIL API GRAVITY   35.0   API
WATER SPECIFIC GRAVITY   1.07

LENGTH (1000 FT)

GAS/LIQUID RATIO - SCF PER BBL

BEGGS AND BRILL
PRESENTED BY:
MACCO - SCHLUMBERGER

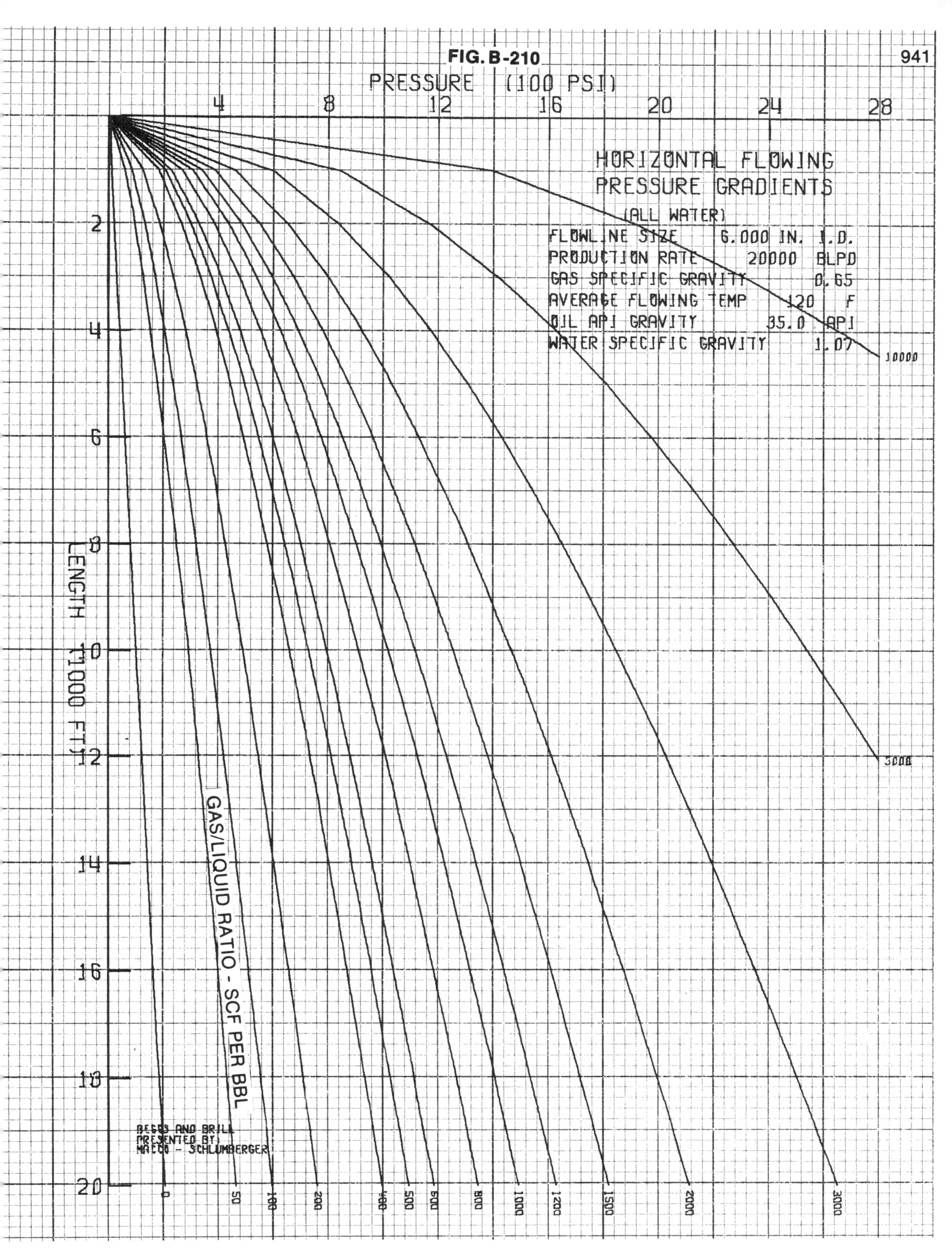

PRESSURE (100 PSI)

HORIZONTAL FLOWING
PRESSURE GRADIENTS
(ALL WATER)

FLOWLINE SIZE	6.000 IN. I.D.
PRODUCTION RATE	20000 BLPD
GAS SPECIFIC GRAVITY	0.65
AVERAGE FLOWING TEMP	120 F
OIL API GRAVITY	35.0 API
WATER SPECIFIC GRAVITY	1.07

LENGTH (1000 FT)

GAS/LIQUID RATIO - SCF PER BBL

BEGGS AND BRILL
PRESENTED BY:
MAECO - SCHLUMBERGER

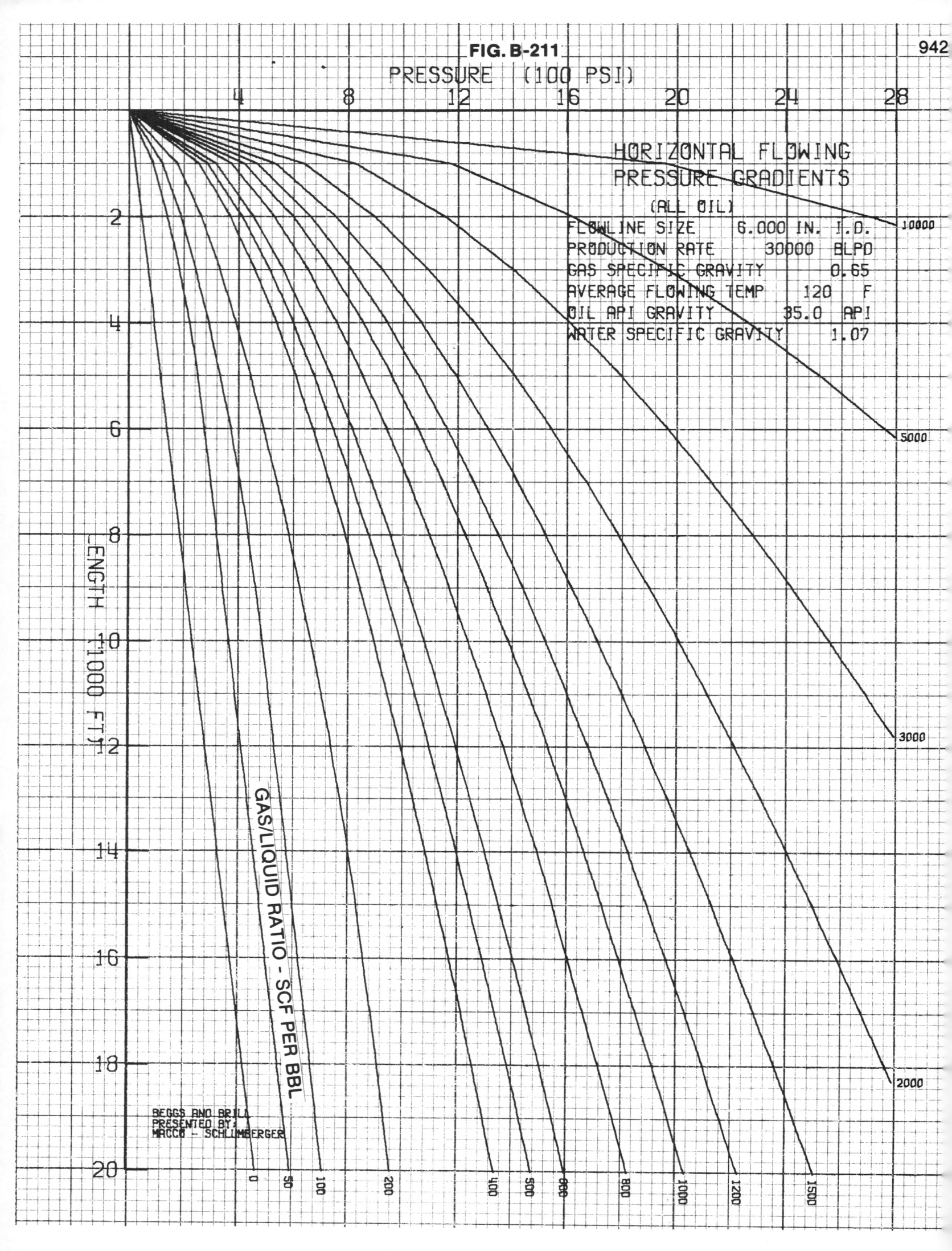

FIG. B-211

942

HORIZONTAL FLOWING
PRESSURE GRADIENTS
(ALL OIL)

FLOWLINE SIZE   6.000 IN. I.D.
PRODUCTION RATE   30000 BLPD
GAS SPECIFIC GRAVITY   0.65
AVERAGE FLOWING TEMP   120   F
OIL API GRAVITY   35.0 API
WATER SPECIFIC GRAVITY   1.07

PRESSURE (100 PSI)

LENGTH (1000 FT)

GAS/LIQUID RATIO - SCF PER BBL

BEGGS AND BRILL
PRESENTED BY:
MACCO - SCHLUMBERGER

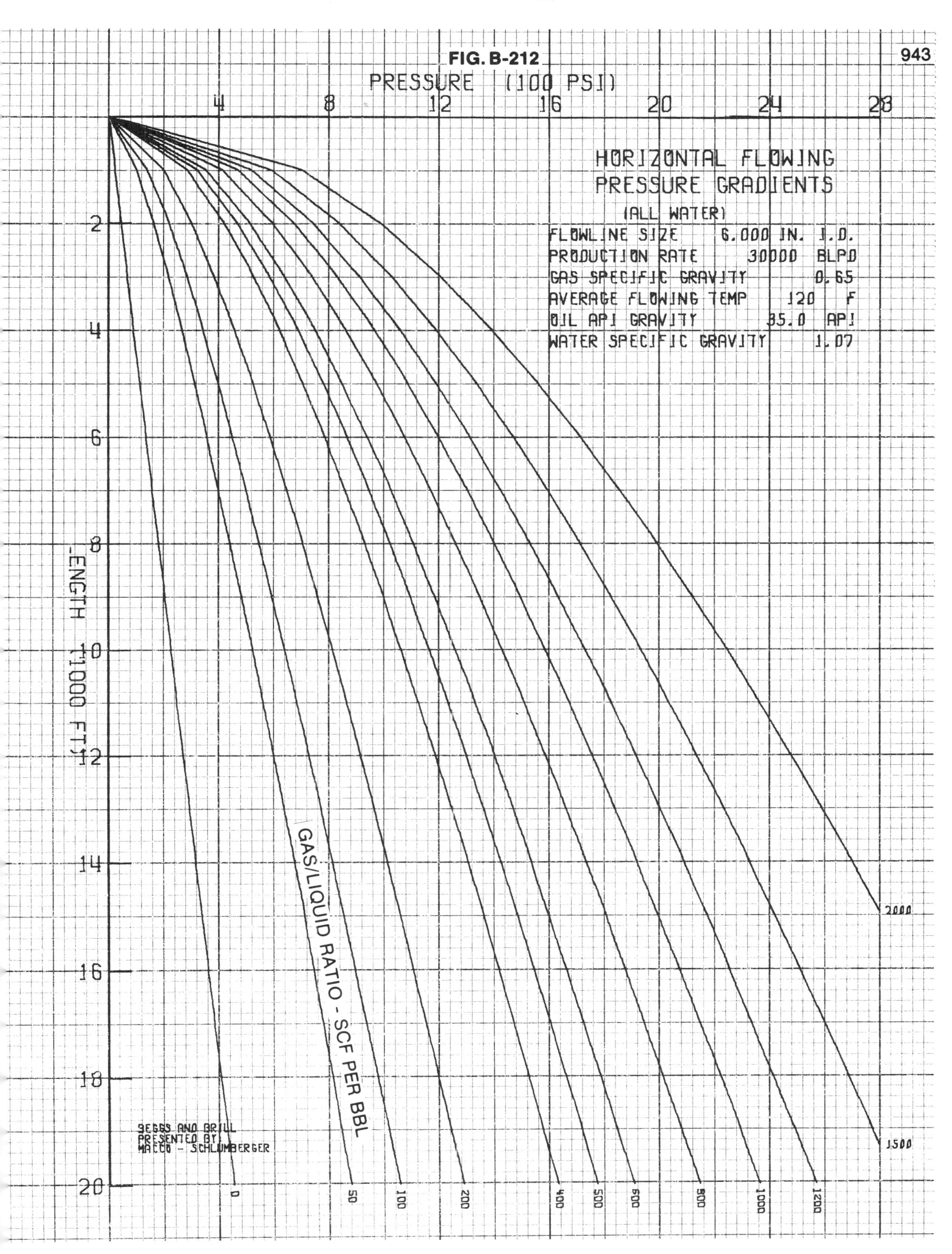

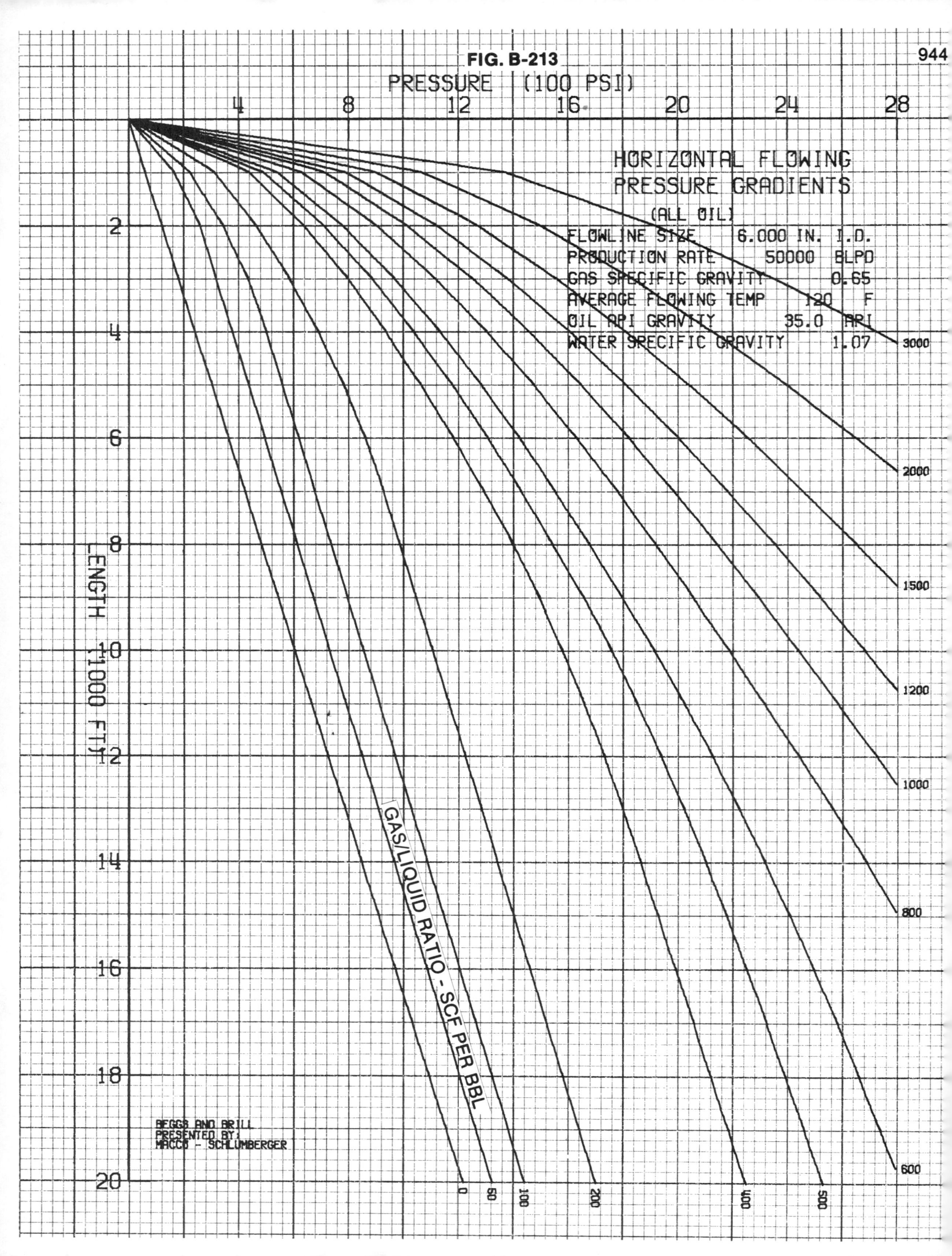

PRESSURE (100 PSI)

HORIZONTAL FLOWING
PRESSURE GRADIENTS
(ALL OIL)

FLOWLINE SIZE   6.000 IN. I.D.
PRODUCTION RATE   50000   BLPD
GAS SPECIFIC GRAVITY   0.65
AVERAGE FLOWING TEMP   120   F
OIL API GRAVITY   35.0   API
WATER SPECIFIC GRAVITY   1.07

LENGTH (1000 FT)

GAS/LIQUID RATIO - SCF PER BBL

BEGGS AND BRILL
PRESENTED BY:
MACCO - SCHLUMBERGER

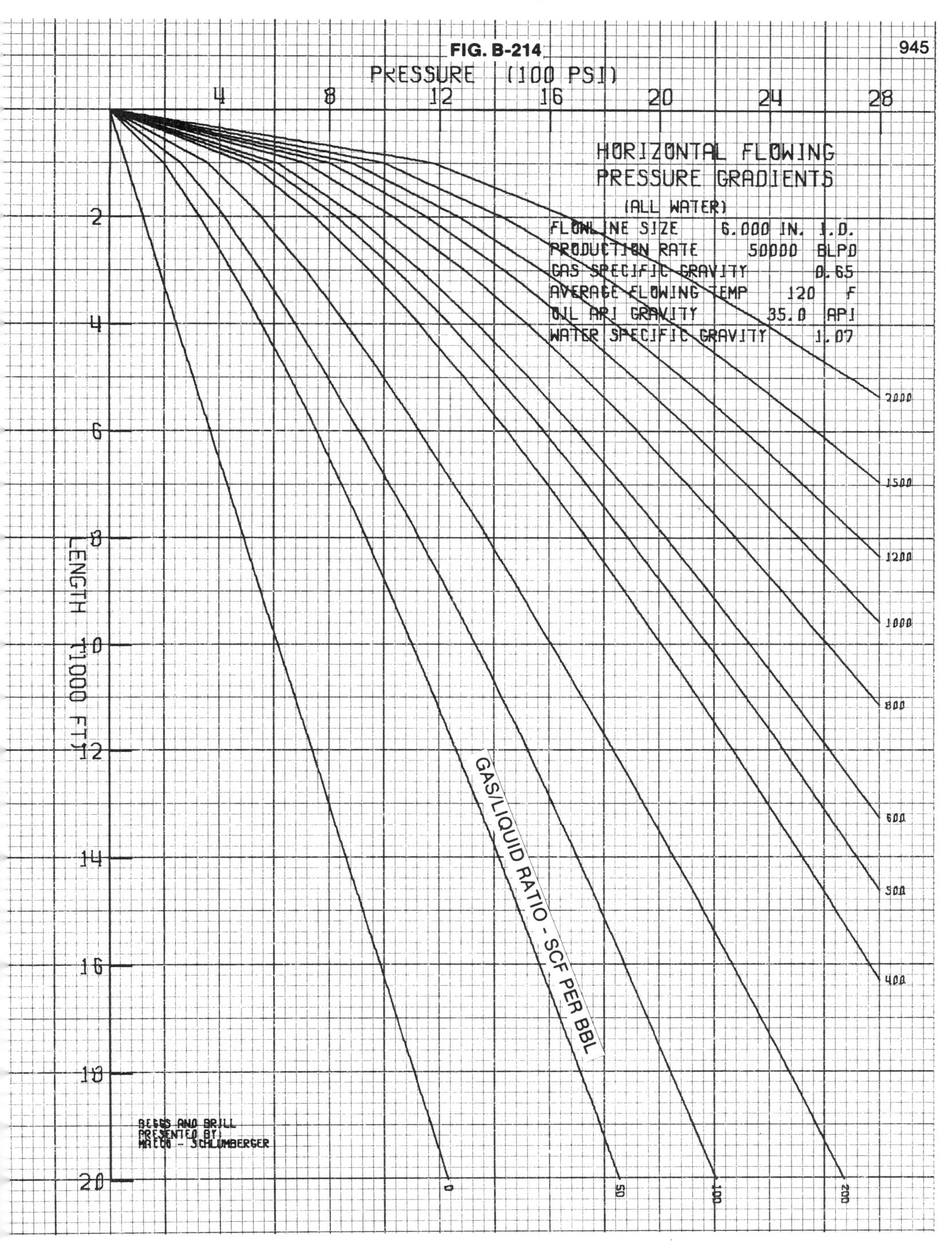

HORIZONTAL FLOWING
PRESSURE GRADIENTS
(ALL WATER)

FLOWLINE SIZE	6.000 IN. I.D.
PRODUCTION RATE	50000 BLPD
GAS SPECIFIC GRAVITY	0.65
AVERAGE FLOWING TEMP	120 F
OIL API GRAVITY	35.0 API
WATER SPECIFIC GRAVITY	1.07

PRESSURE (100 PSI)

LENGTH (1000 FT)

GAS/LIQUID RATIO - SCF PER BBL

BEGGS AND BRILL
PRESENTED BY:
MAECO - SCHLUMBERGER

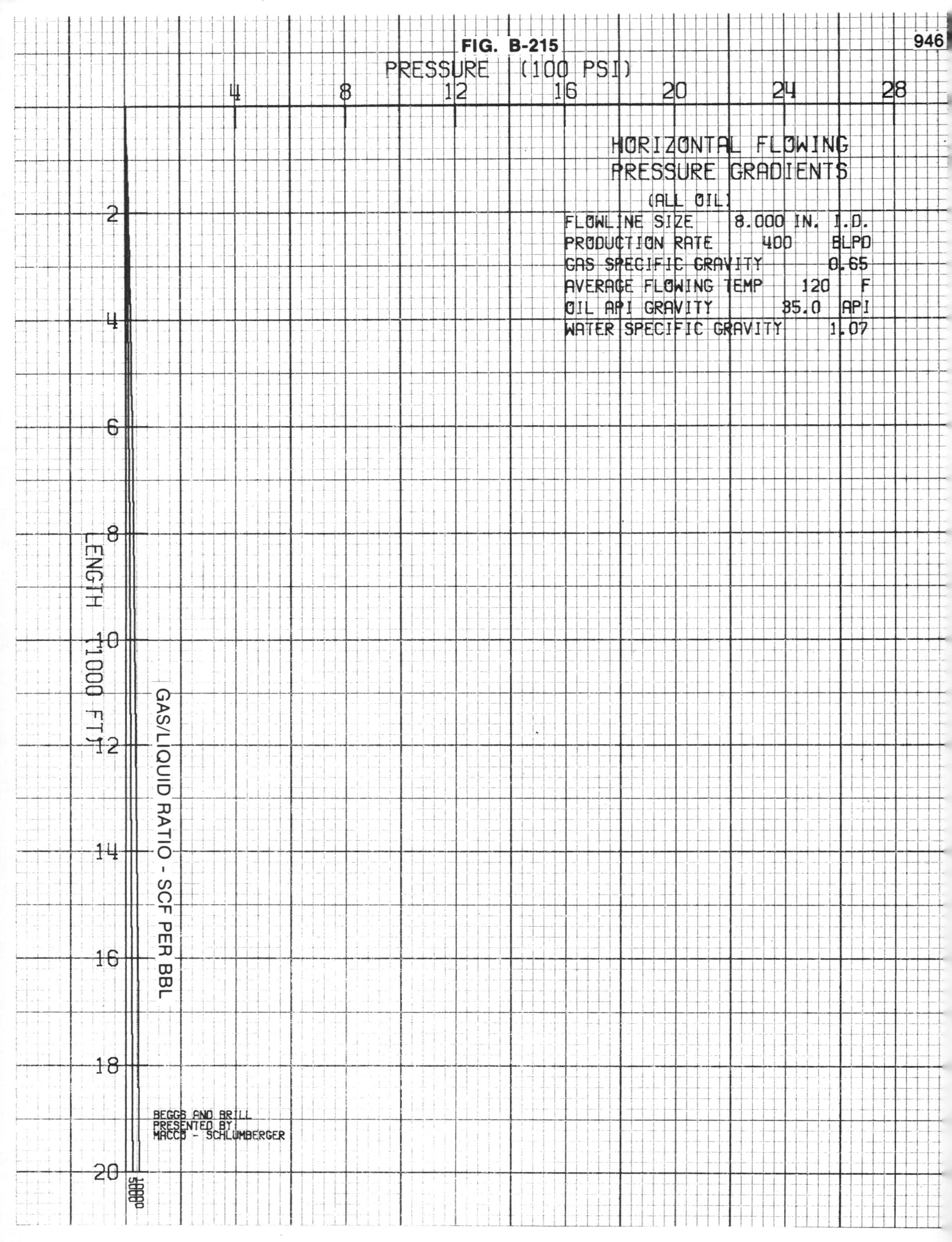

PRESSURE (100 PSI)

HORIZONTAL FLOWING
PRESSURE GRADIENTS
(ALL OIL)

FLOWLINE SIZE	8.000 IN. I.D.
PRODUCTION RATE	400 BLPD
GAS SPECIFIC GRAVITY	0.65
AVERAGE FLOWING TEMP	120 F
OIL API GRAVITY	35.0 API
WATER SPECIFIC GRAVITY	1.07

LENGTH (1000 FT)

GAS/LIQUID RATIO - SCF PER BBL

BEGGS AND BRILL
PRESENTED BY
MACCO - SCHLUMBERGER

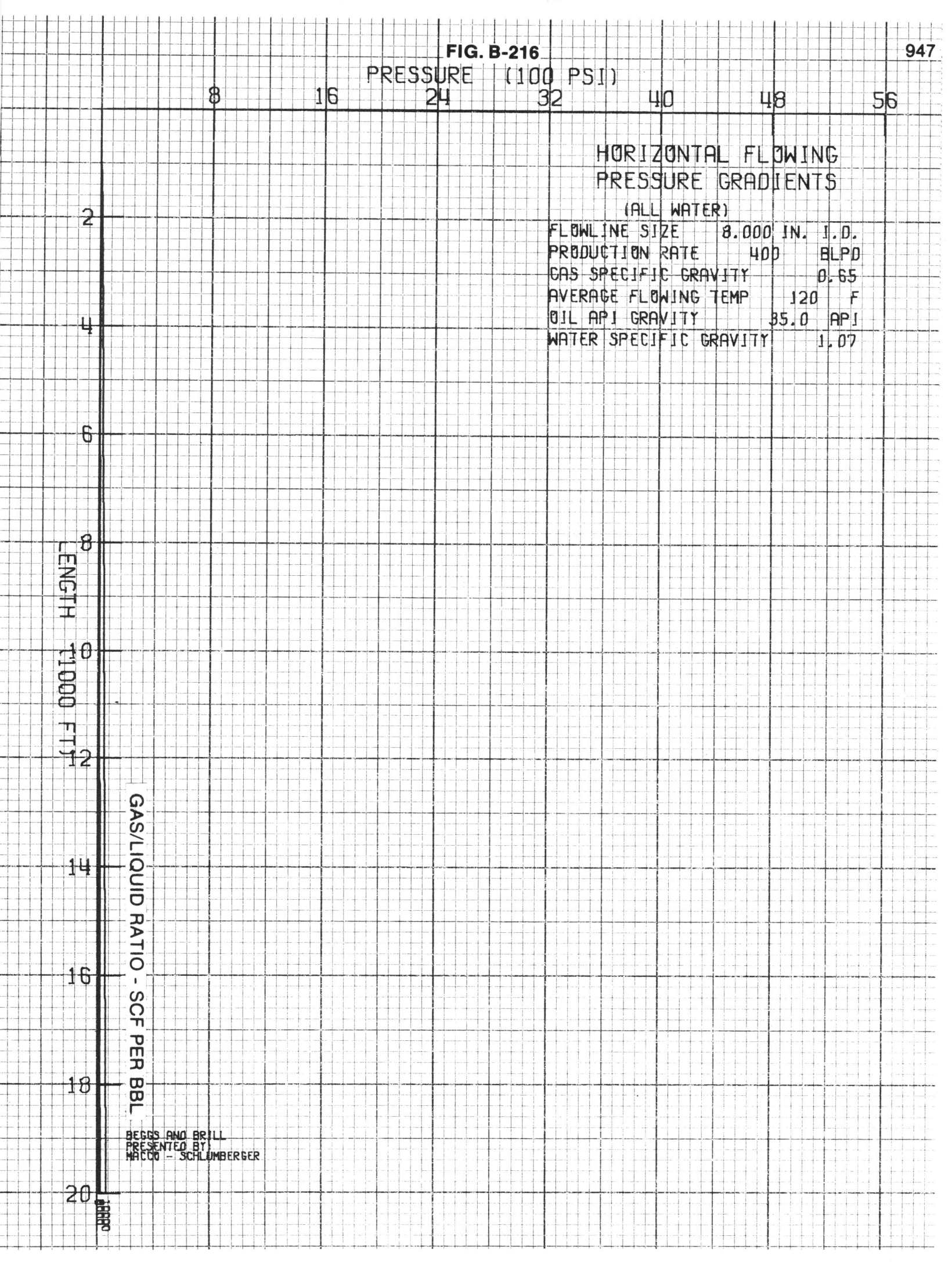

PRESSURE (100 PSI)

HORIZONTAL FLOWING
PRESSURE GRADIENTS
(ALL WATER)

FLOWLINE SIZE   8.000 IN. I.D.
PRODUCTION RATE   400   BLPD
GAS SPECIFIC GRAVITY   0.65
AVERAGE FLOWING TEMP   120   F
OIL API GRAVITY   35.0 API
WATER SPECIFIC GRAVITY   1.07

LENGTH (1000 FT)

GAS/LIQUID RATIO - SCF PER BBL

BEGGS AND BRILL
PRESENTED BY
MACCO - SCHLUMBERGER

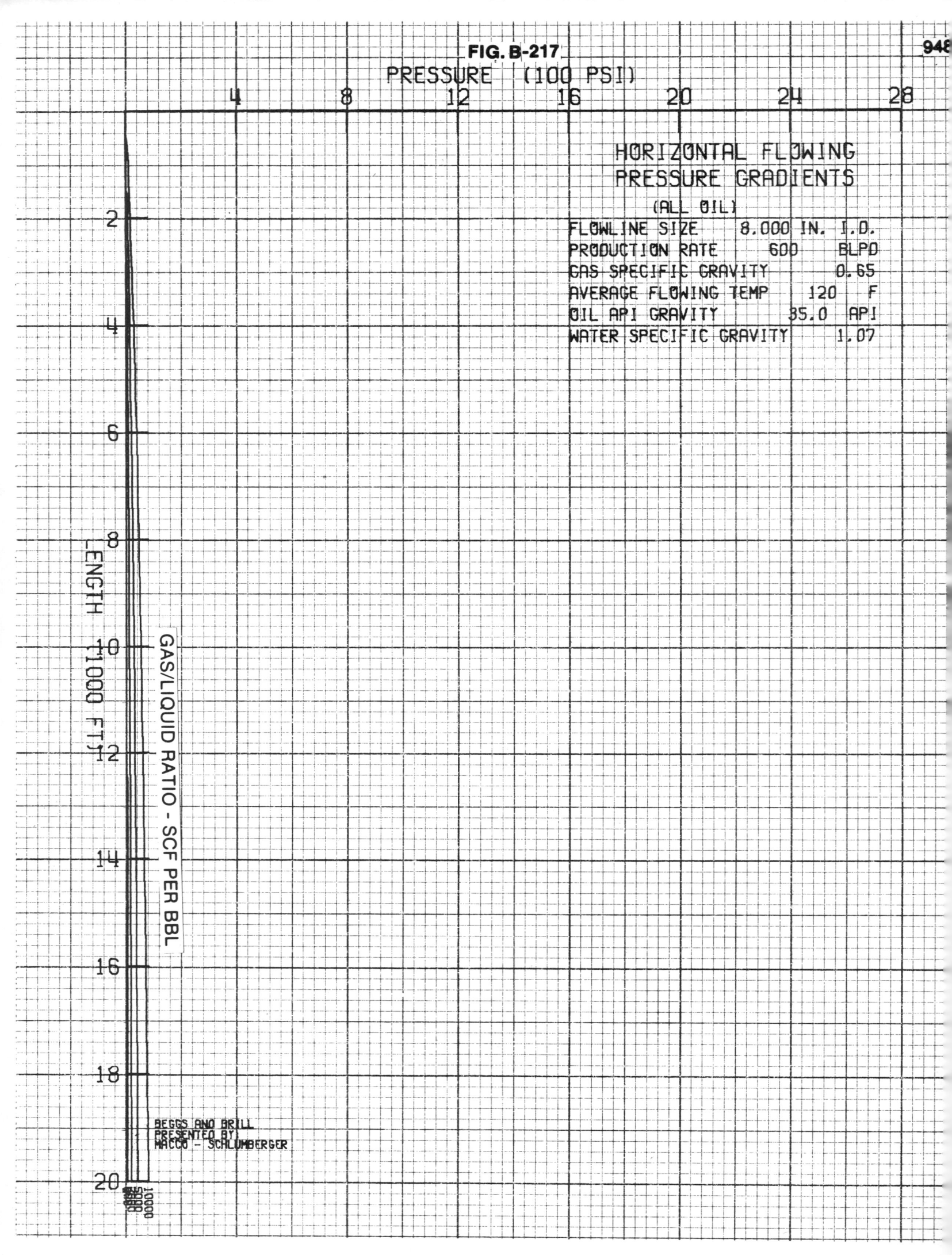

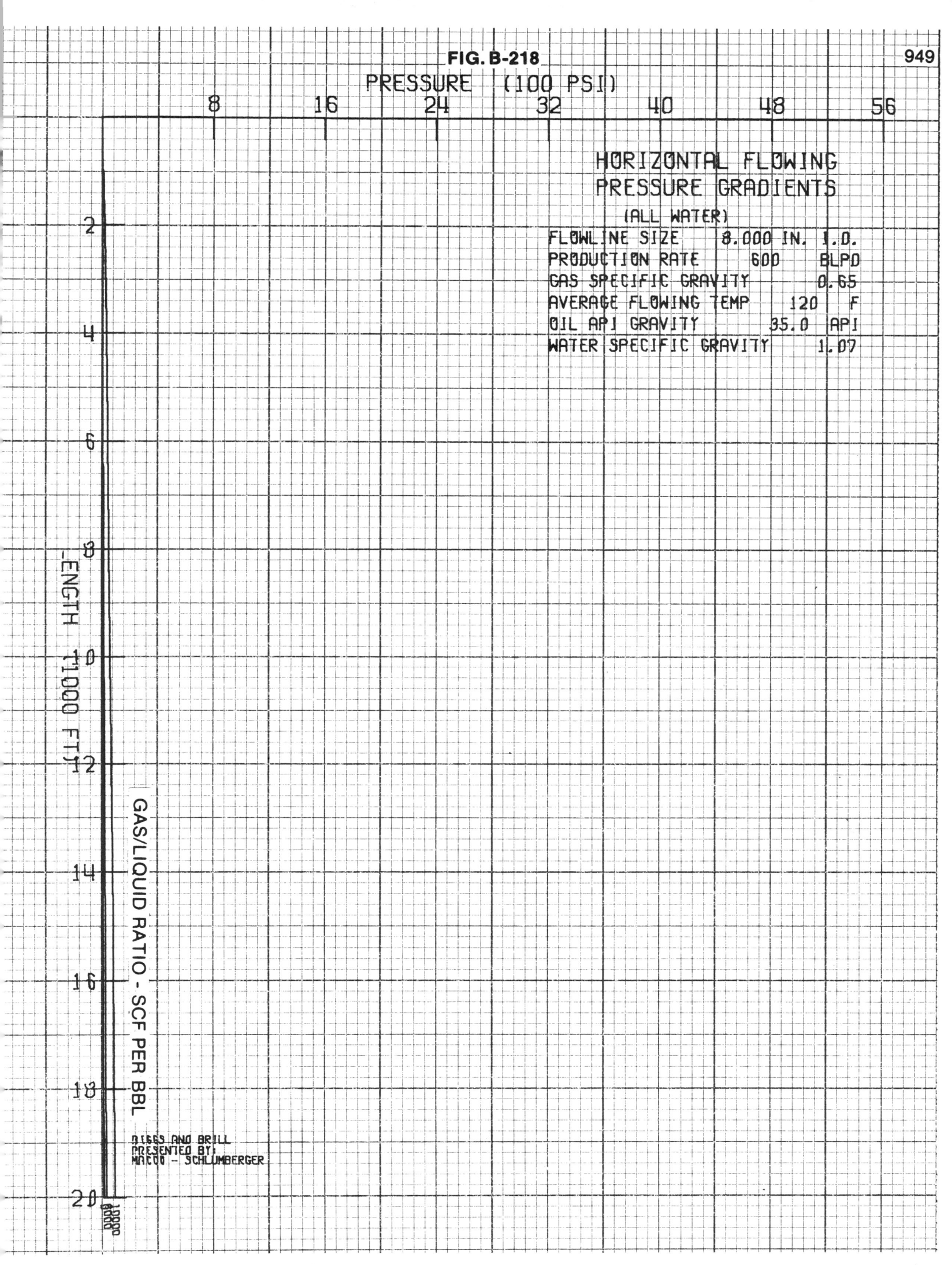

PRESSURE (100 PSI)

HORIZONTAL FLOWING
PRESSURE GRADIENTS
(ALL WATER)

FLOWLINE SIZE	8.000 IN.	I.D.
PRODUCTION RATE	600	BLPD
GAS SPECIFIC GRAVITY	0.65	
AVERAGE FLOWING TEMP	120	F
OIL API GRAVITY	35.0	API
WATER SPECIFIC GRAVITY	1.07	

LENGTH (1000 FT)

GAS/LIQUID RATIO - SCF PER BBL

DIGGS AND BRILL
PRESENTED BY:
MACCO - SCHLUMBERGER

PRESSURE (100 PSI)

4 8 12 16 20 24 28

HORIZONTAL FLOWING
PRESSURE GRADIENTS
(ALL OIL)

FLOWLINE SIZE	8.000 IN. I.D.
PRODUCTION RATE	800 BLPD
GAS SPECIFIC GRAVITY	0.65
AVERAGE FLOWING TEMP	120 F
OIL API GRAVITY	35.0 API
WATER SPECIFIC GRAVITY	1.07

LENGTH (1000 FT)

2

4

6

8

10

12

14

16

18

20

GAS/LIQUID RATIO - SCF PER BBL

BEGGS AND BRILL
PRESENTED BY:
MACCO - SCHLUMBERGER

5000
10000

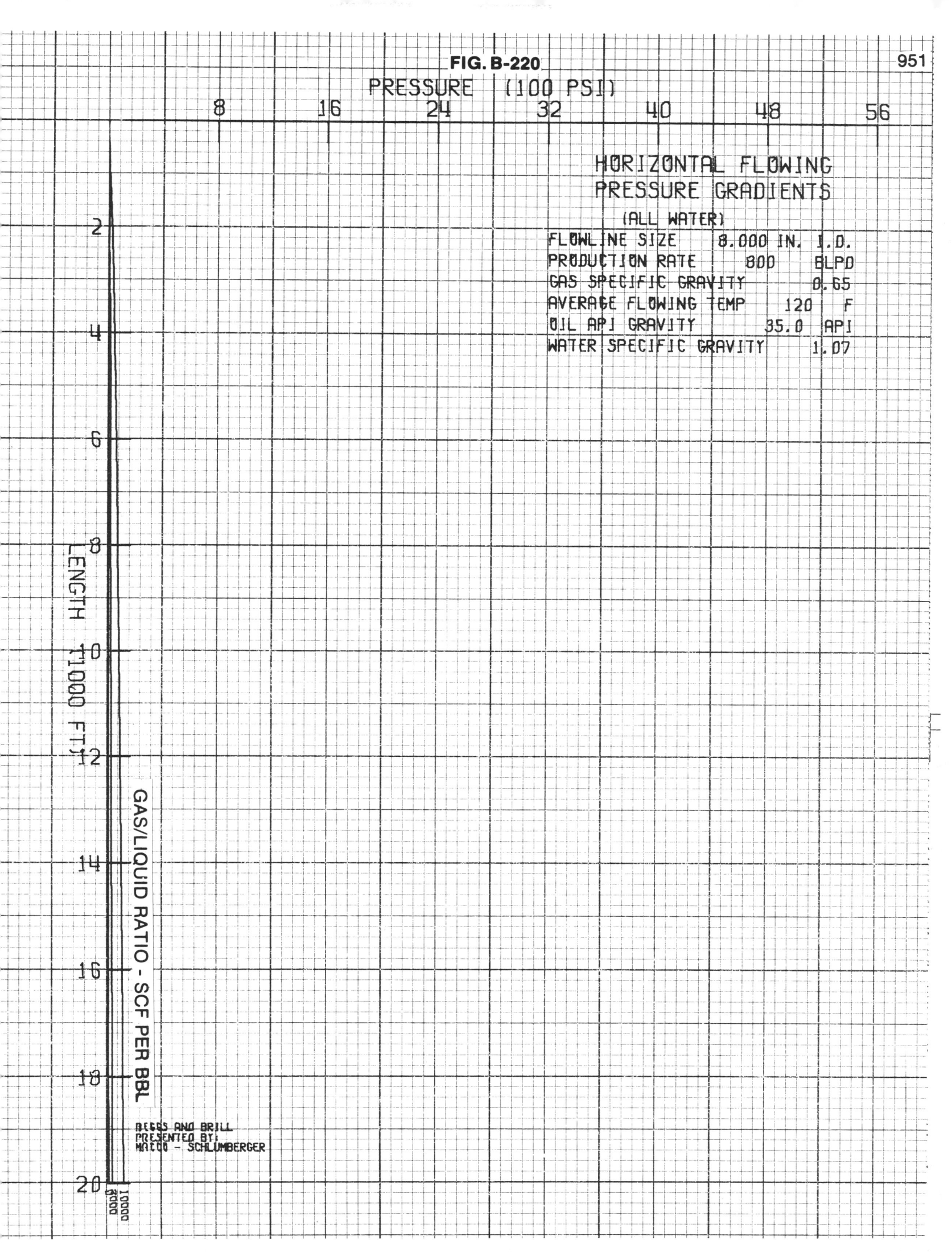

PRESSURE (100 PSI)

8   16   24   32   40   48   56

HORIZONTAL FLOWING
PRESSURE GRADIENTS
(ALL WATER)

FLOWLINE SIZE   8.000 IN. I.D.
PRODUCTION RATE   800   BLPD
GAS SPECIFIC GRAVITY   0.65
AVERAGE FLOWING TEMP   120   F
OIL API GRAVITY   35.0   API
WATER SPECIFIC GRAVITY   1.07

LENGTH (1000 FT.)

2   4   6   8   10   12   14   16   18   20

GAS/LIQUID RATIO - SCF PER BBL

10000
5000

BEGGS AND BRILL
PRESENTED BY:
WATCO - SCHLUMBERGER

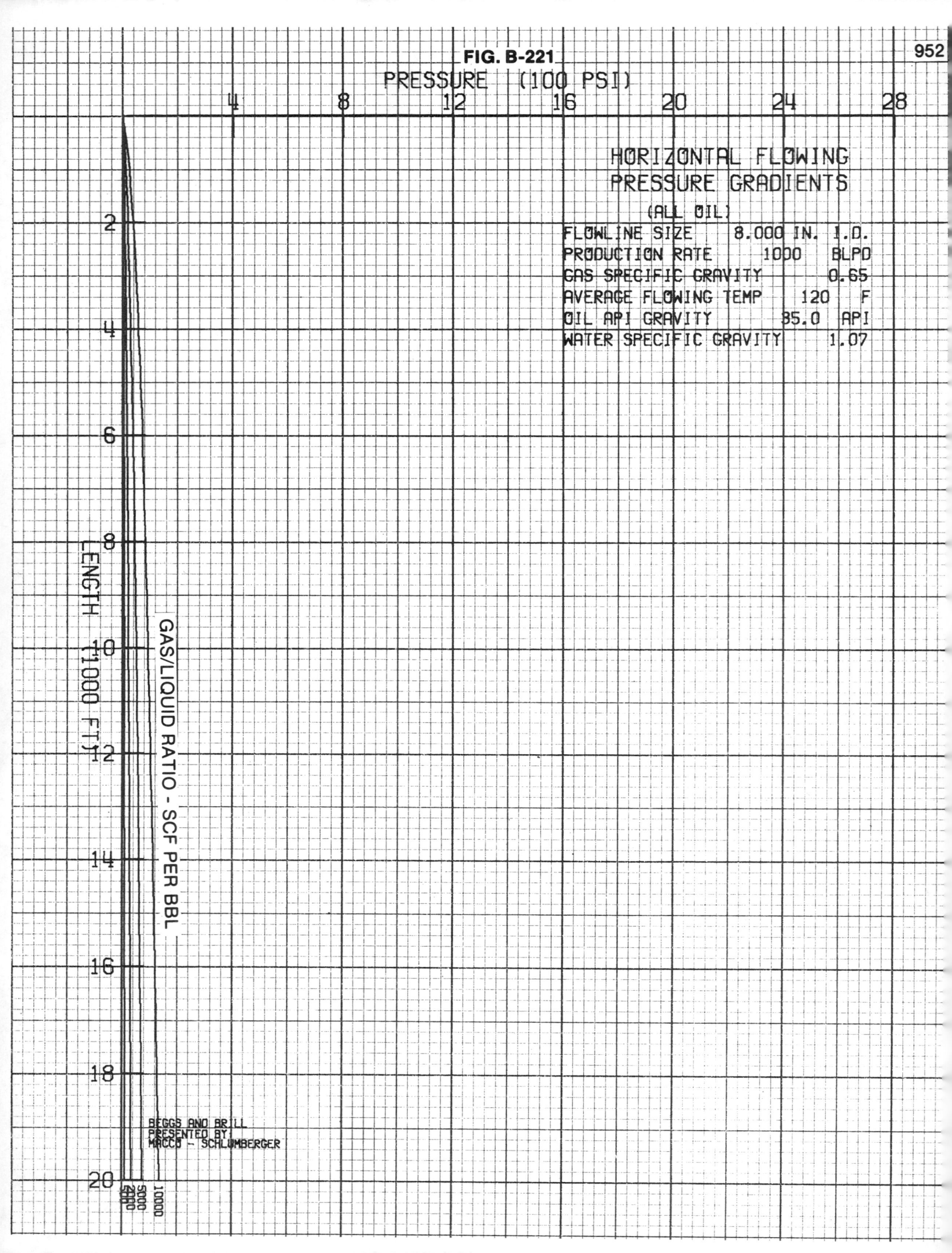

FIG. B-221

952

PRESSURE (100 PSI)

HORIZONTAL FLOWING
PRESSURE GRADIENTS
(ALL OIL)

FLOWLINE SIZE   8.000 IN. I.D.
PRODUCTION RATE   1000   BLPD
GAS SPECIFIC GRAVITY   0.65
AVERAGE FLOWING TEMP   120   F
OIL API GRAVITY   85.0   API
WATER SPECIFIC GRAVITY   1.07

LENGTH (1000 FT)

GAS/LIQUID RATIO - SCF PER BBL

BEGGS AND BRILL
PRESENTED BY:
MACCO - SCHLUMBERGER

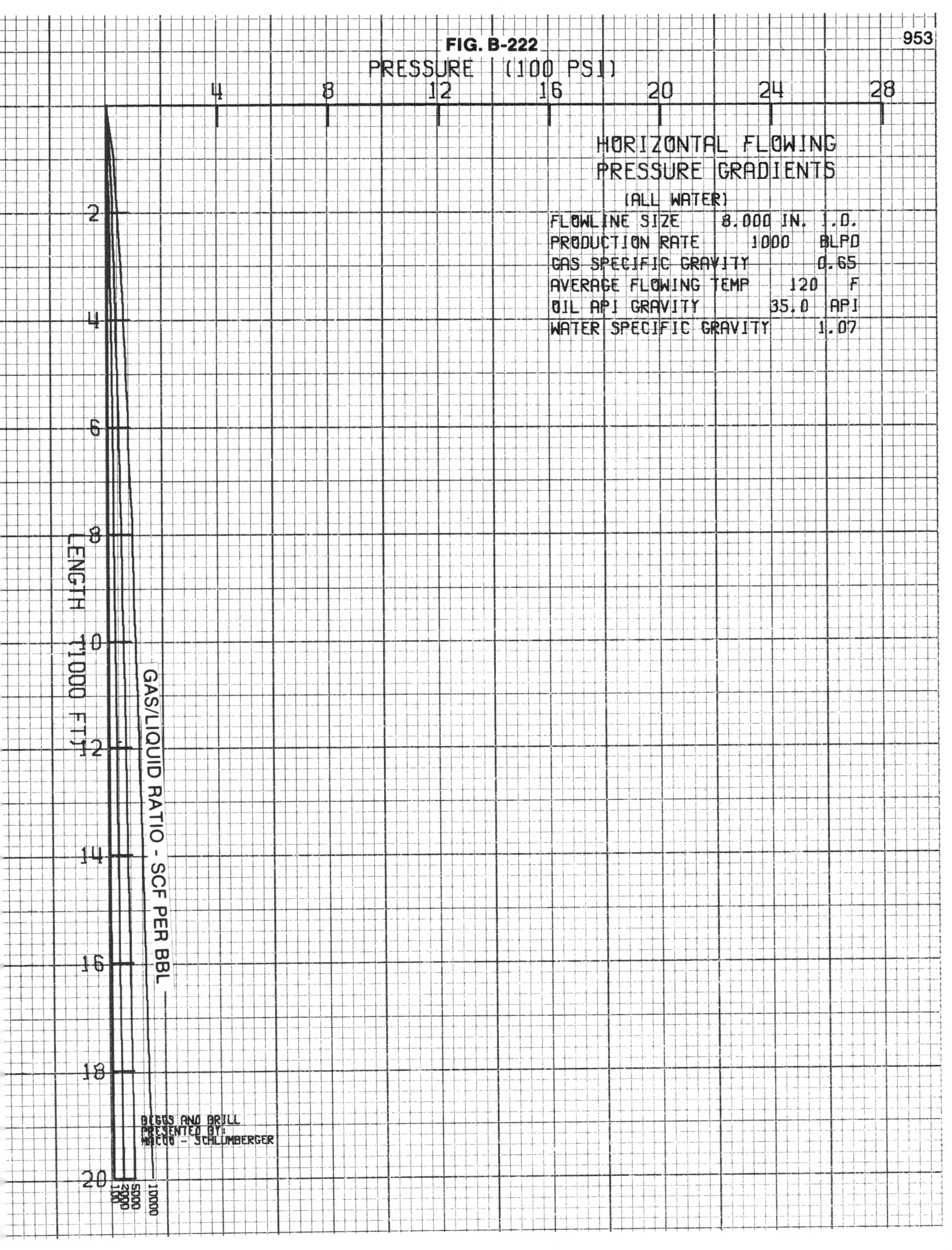

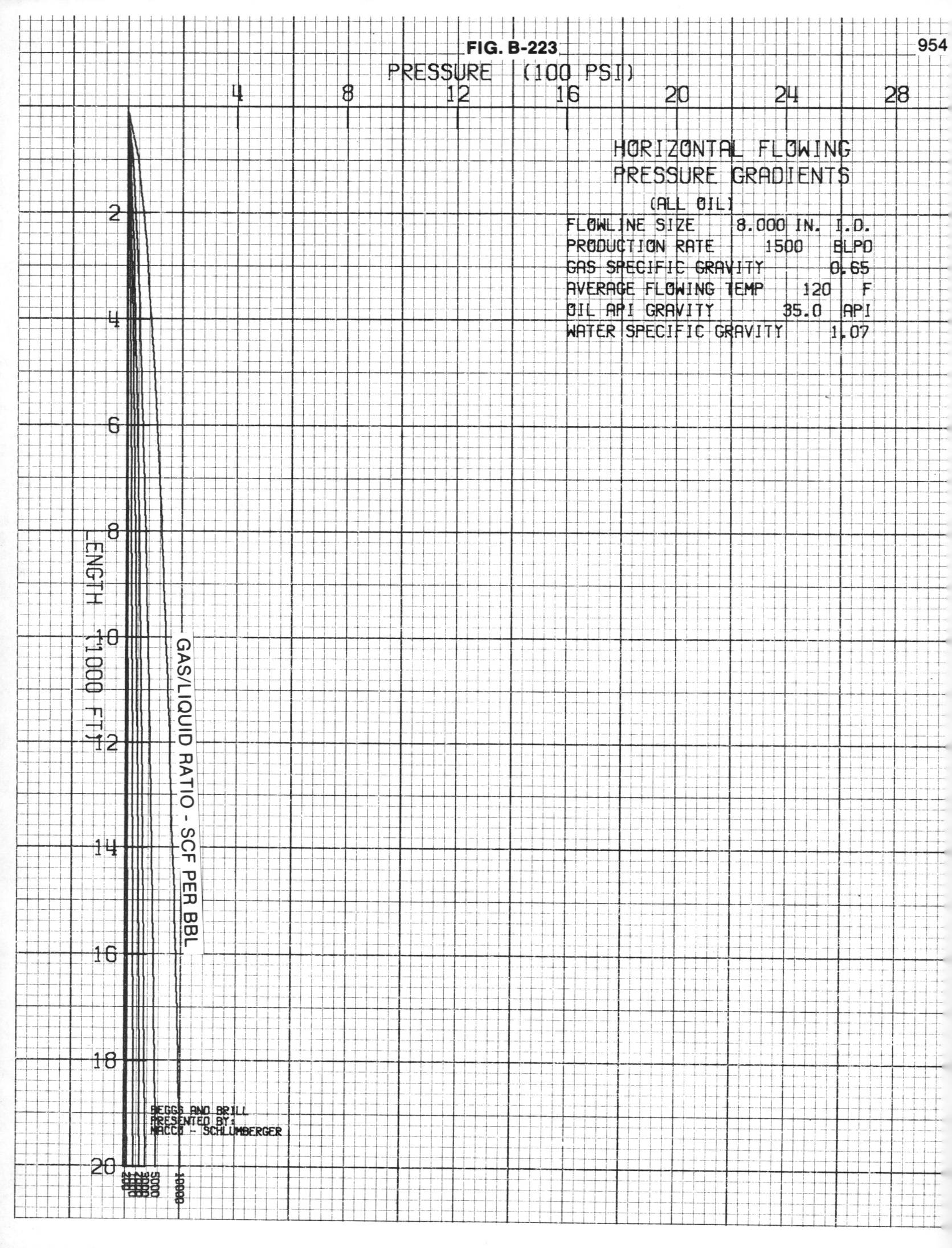

FIG. B-223

PRESSURE (100 PSI)

HORIZONTAL FLOWING
PRESSURE GRADIENTS
(ALL OIL)

FLOWLINE SIZE	8.000 IN. I.D.
PRODUCTION RATE	1500 BLPD
GAS SPECIFIC GRAVITY	0.65
AVERAGE FLOWING TEMP	120 F
OIL API GRAVITY	35.0 API
WATER SPECIFIC GRAVITY	1.07

LENGTH (1000 FT.)

GAS/LIQUID RATIO - SCF PER BBL

BEGGS AND BRILL
PRESENTED BY:
NACCO - SCHLUMBERGER

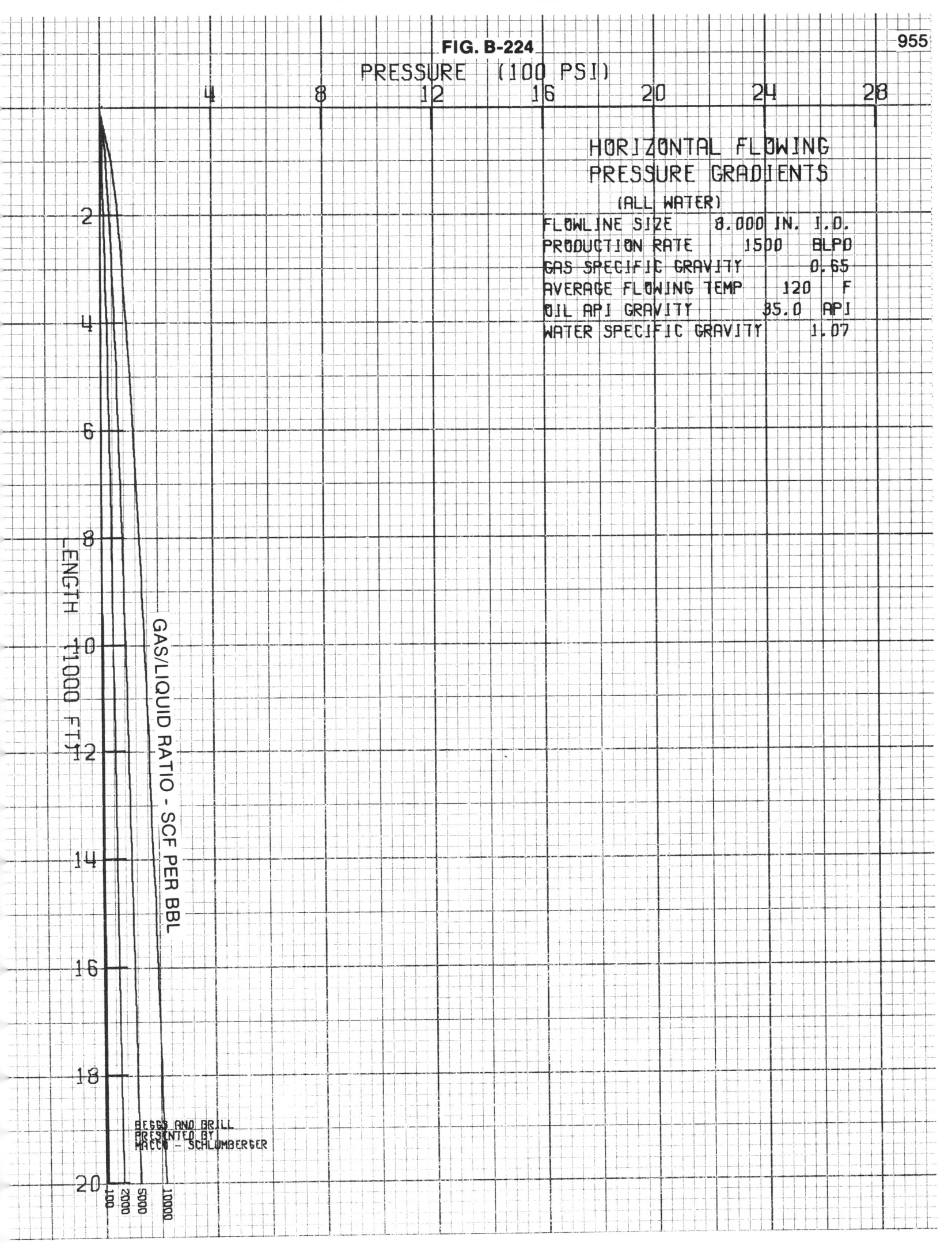

FIG. B-224

955

HORIZONTAL FLOWING
PRESSURE GRADIENTS
(ALL WATER)

FLOWLINE SIZE   8.000 IN. I.D.
PRODUCTION RATE   1500   BLPD
GAS SPECIFIC GRAVITY   0.65
AVERAGE FLOWING TEMP   120   F
OIL API GRAVITY   35.0 API
WATER SPECIFIC GRAVITY   1.07

PRESSURE (100 PSI)

LENGTH (1000 FT.)

GAS/LIQUID RATIO - SCF PER BBL

BEGGS AND BRILL
PRESENTED BY
MACCO - SCHLUMBERGER

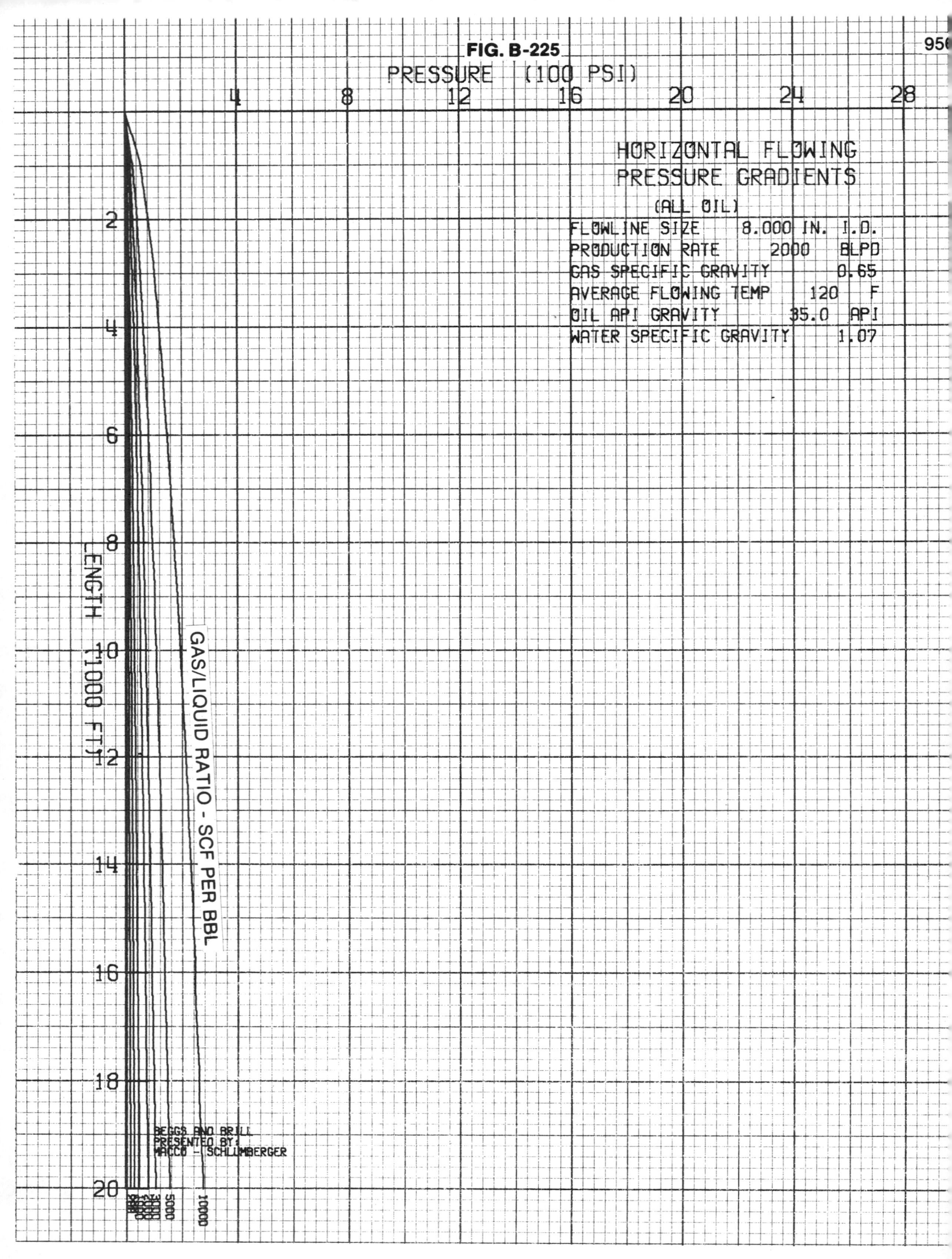

FIG. B-225

PRESSURE (100 PSI)

HORIZONTAL FLOWING
PRESSURE GRADIENTS
(ALL OIL)

FLOWLINE SIZE   8.000 IN. I.D.
PRODUCTION RATE   2000   BLPD
GAS SPECIFIC GRAVITY   0.65
AVERAGE FLOWING TEMP   120   F
OIL API GRAVITY   35.0 API
WATER SPECIFIC GRAVITY   1.07

LENGTH (1000 FT)

GAS/LIQUID RATIO - SCF PER BBL

BEGGS AND BRILL
PRESENTED BY:
MACCO - SCHLUMBERGER

950

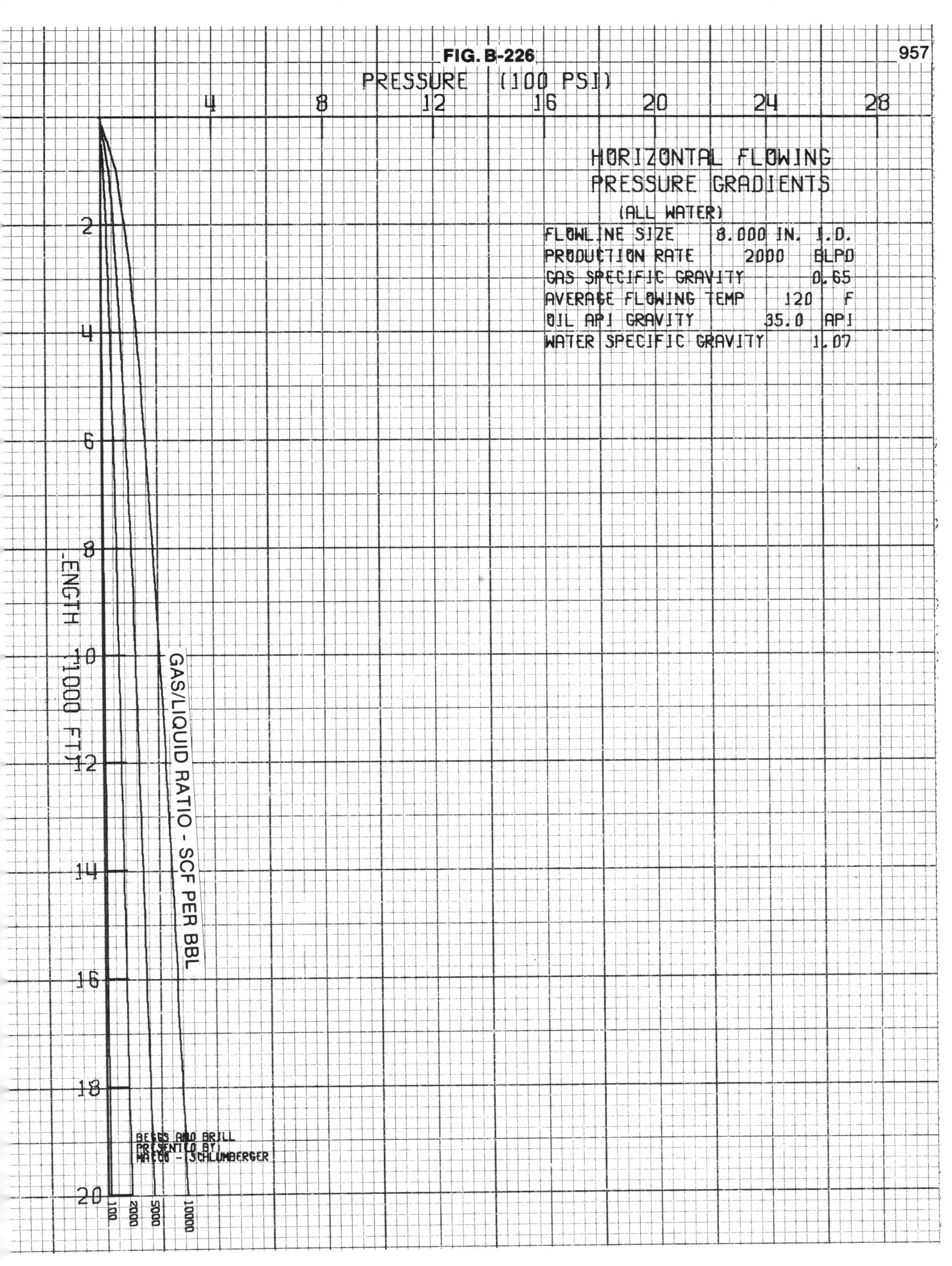

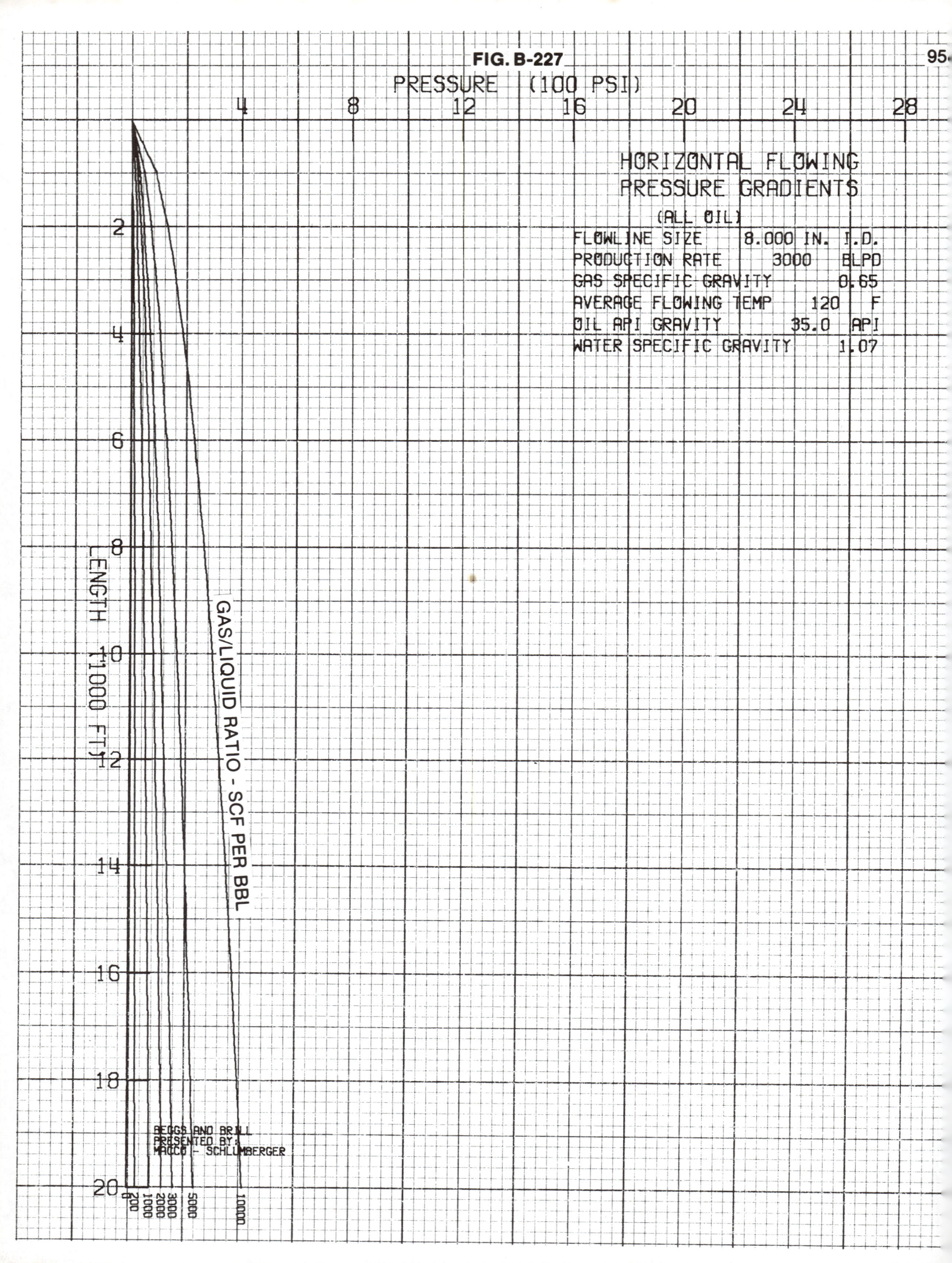

HORIZONTAL FLOWING
PRESSURE GRADIENTS
(ALL OIL)

FLOWLINE SIZE	8.000 IN.	I.D.
PRODUCTION RATE	3000	BLPD
GAS SPECIFIC GRAVITY	0.65	
AVERAGE FLOWING TEMP	120	F
OIL API GRAVITY	35.0	API
WATER SPECIFIC GRAVITY	1.07	

PRESSURE (100 PSI)

LENGTH (1000 FT.)

GAS/LIQUID RATIO - SCF PER BBL

BEGGS AND BRILL
PRESENTED BY:
MACCO - SCHLUMBERGER

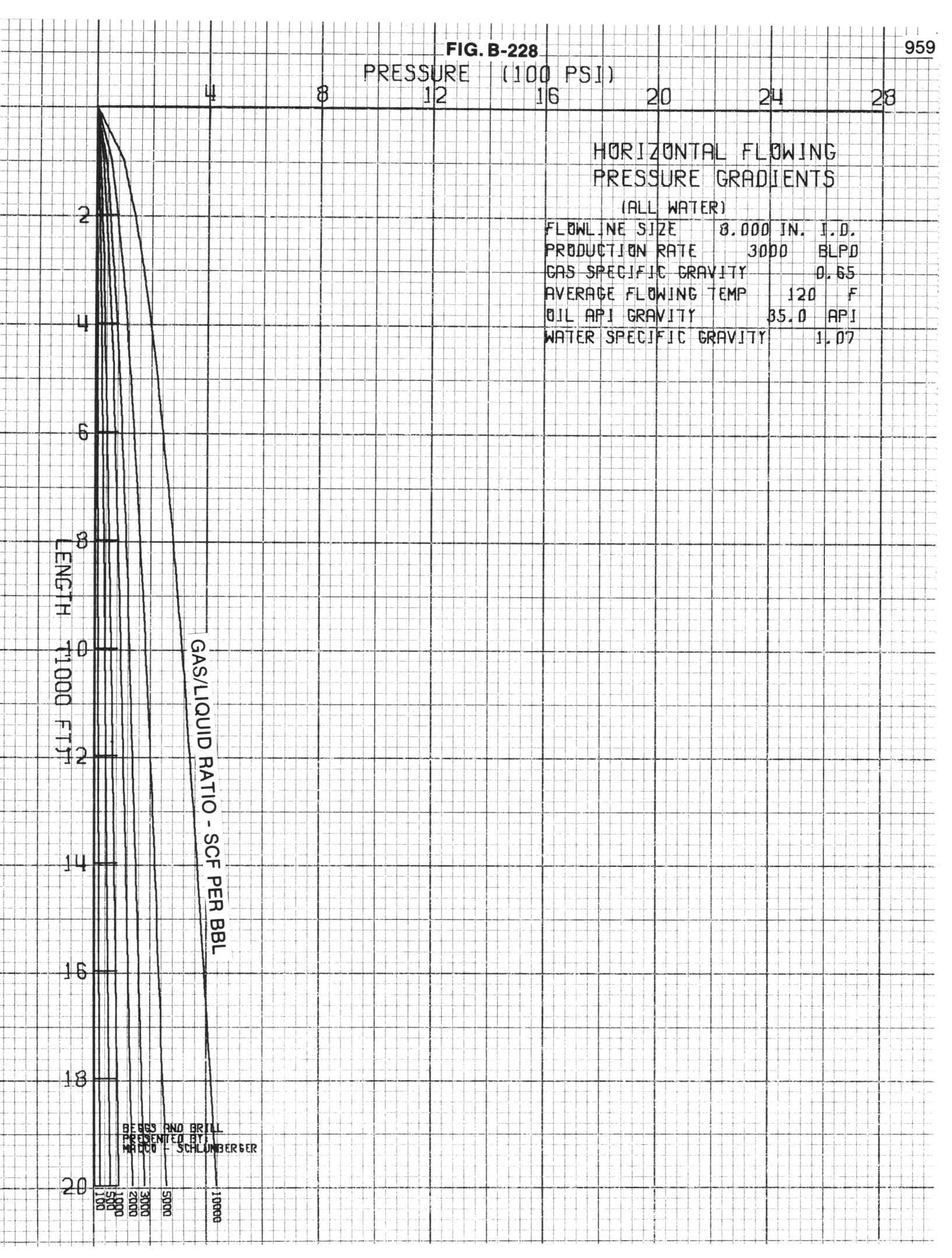

PRESSURE (100 PSI)

HORIZONTAL FLOWING
PRESSURE GRADIENTS
(ALL WATER)

FLOWLINE SIZE   8.000 IN. I.D.
PRODUCTION RATE   3000   BLPD
GAS SPECIFIC GRAVITY   0.65
AVERAGE FLOWING TEMP   120   F
OIL API GRAVITY   35.0 API
WATER SPECIFIC GRAVITY   1.07

LENGTH (1000 FT)

GAS/LIQUID RATIO - SCF PER BBL

BEGGS AND BRILL
PRESENTED BY
MADCO - SCHLUMBERGER

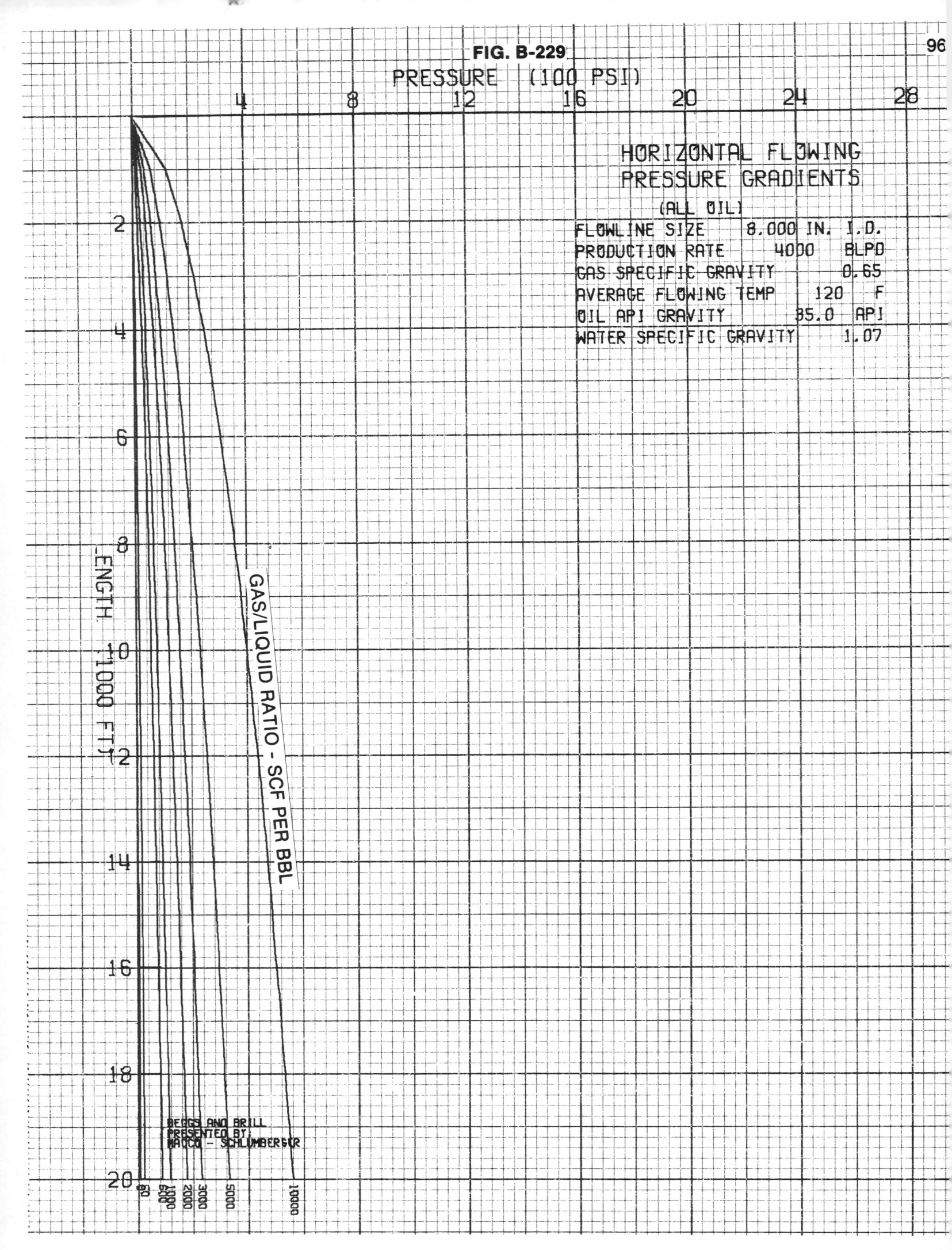

FIG. B-229

PRESSURE (100 PSI)

HORIZONTAL FLOWING
PRESSURE GRADIENTS
(ALL OIL)

FLOWLINE SIZE   8.000 IN. I.D.
PRODUCTION RATE   4000   BLPD
GAS SPECIFIC GRAVITY   0.65
AVERAGE FLOWING TEMP   120   F
OIL API GRAVITY   35.0 API
WATER SPECIFIC GRAVITY   1.07

LENGTH (1000 FT)

GAS/LIQUID RATIO - SCF PER BBL

BEGGS AND BRILL
PRESENTED BY
AMOCO - SCHLUMBERGER

96

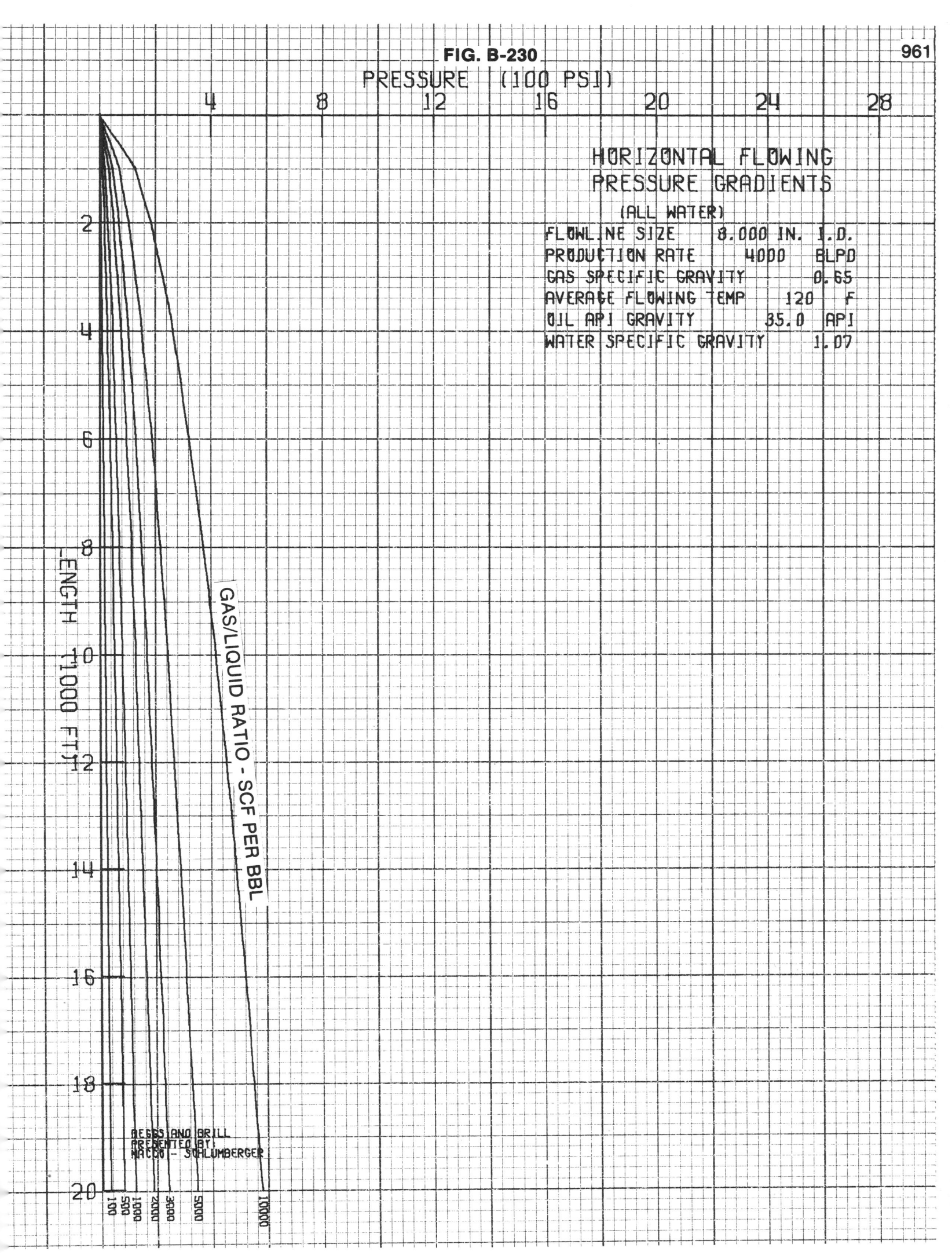

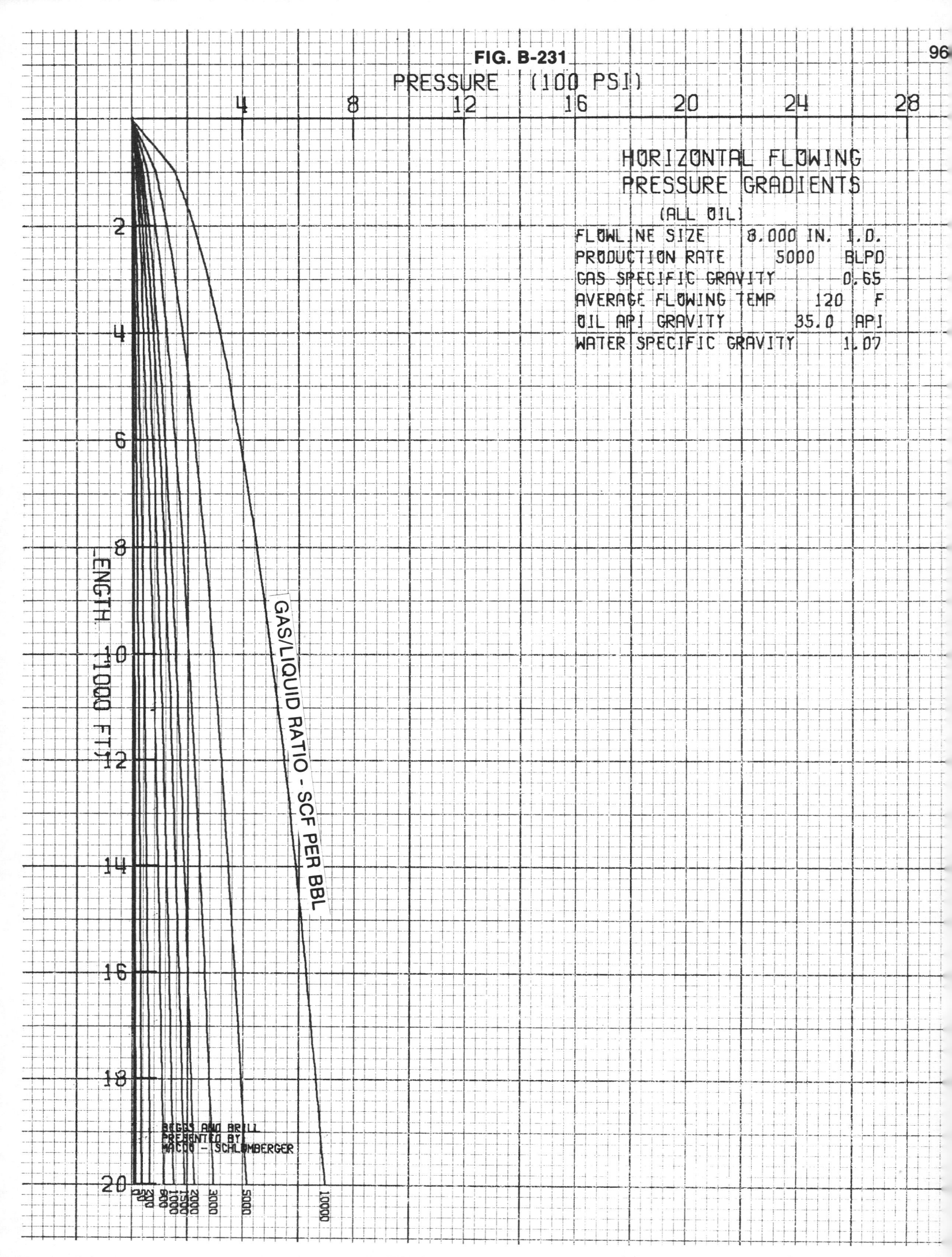

FIG. B-231

96

PRESSURE (100 PSI)

HORIZONTAL FLOWING
PRESSURE GRADIENTS
(ALL OIL)

FLOWLINE SIZE   8.000 IN. I.D.
PRODUCTION RATE   5000   BLPD
GAS SPECIFIC GRAVITY   0.65
AVERAGE FLOWING TEMP   120   F
OIL API GRAVITY   35.0 API
WATER SPECIFIC GRAVITY   1.07

LENGTH (1000 FT)

GAS/LIQUID RATIO - SCF PER BBL

BEGGS AND BRILL
PRESENTED BY
MACCO - SCHLUMBERGER

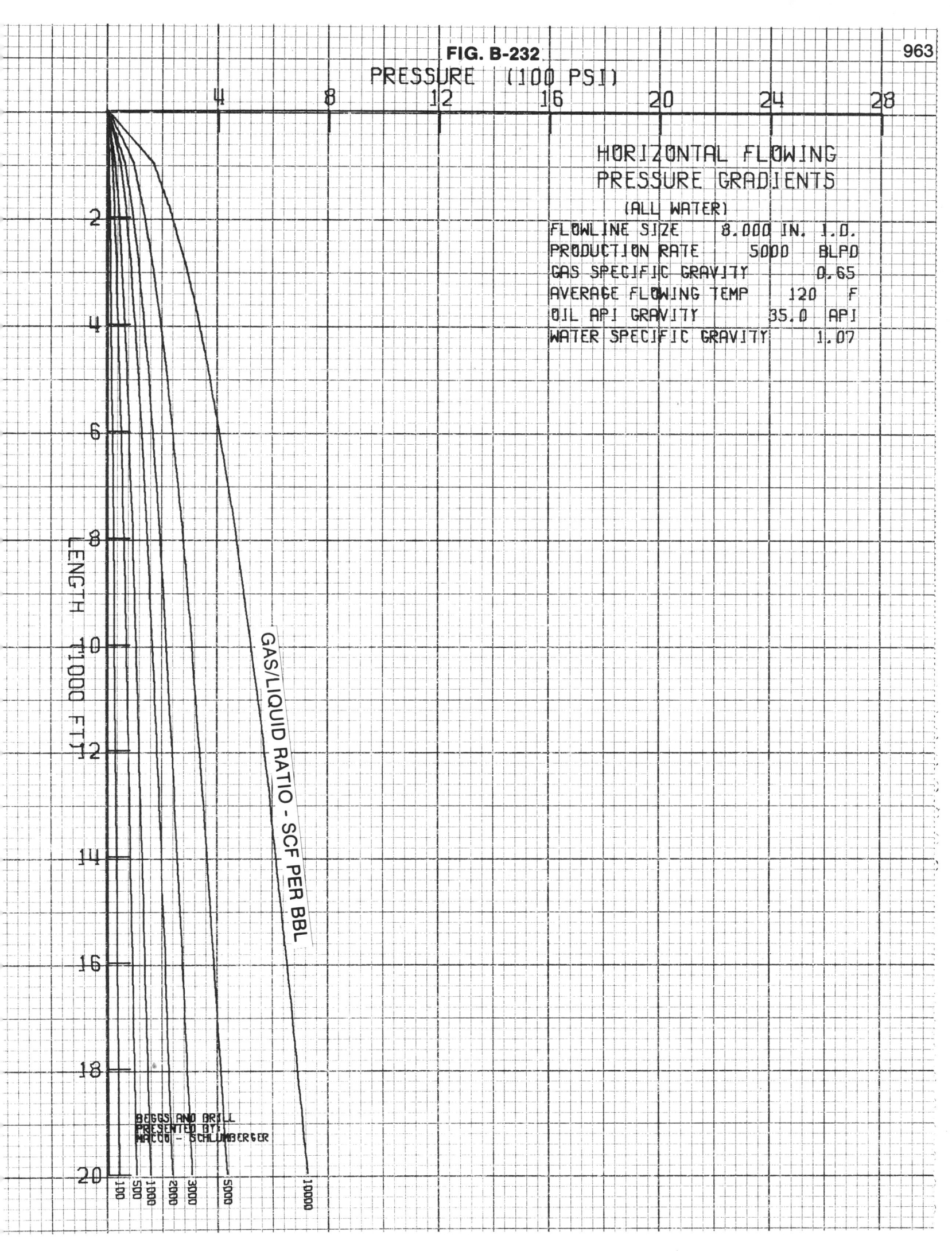

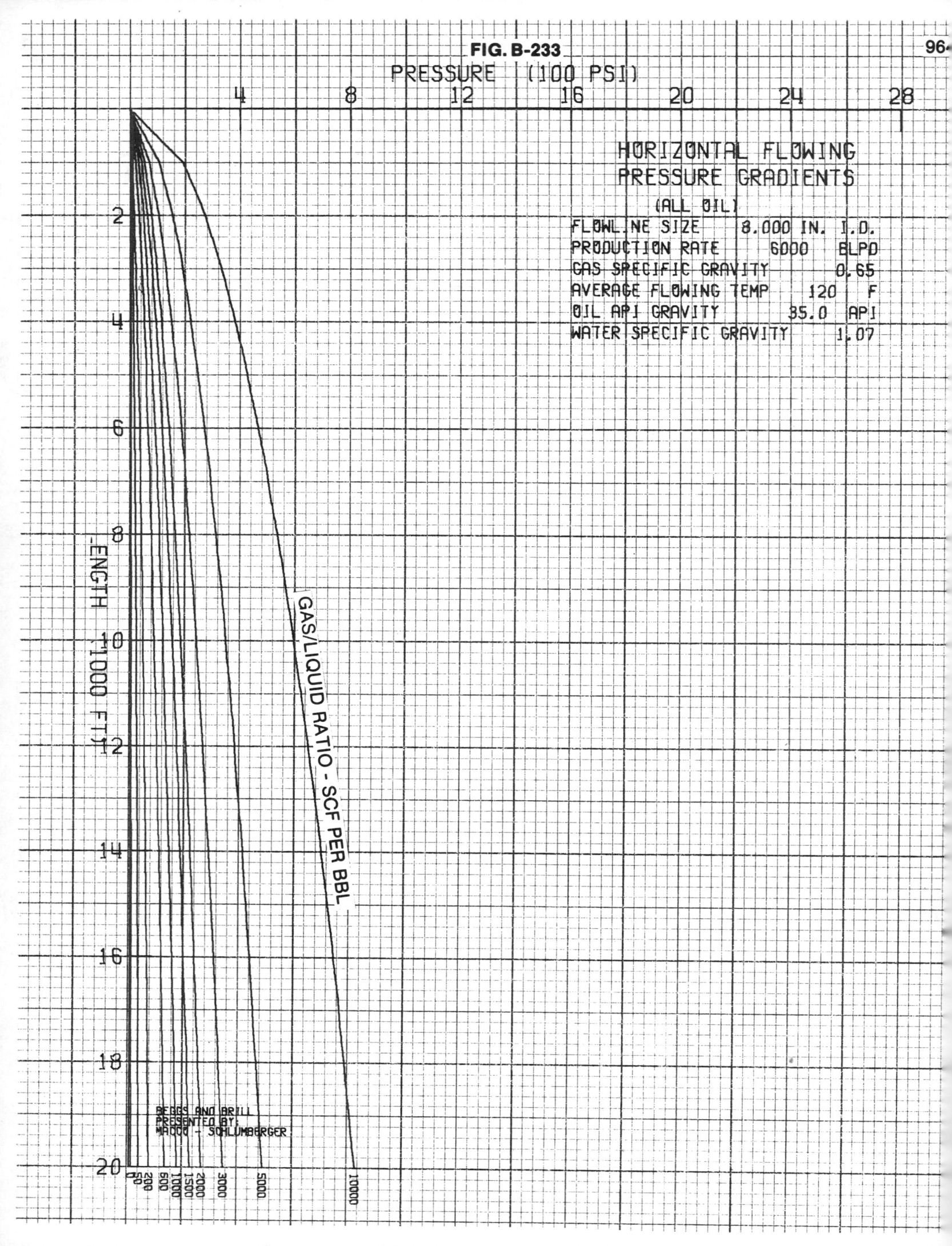

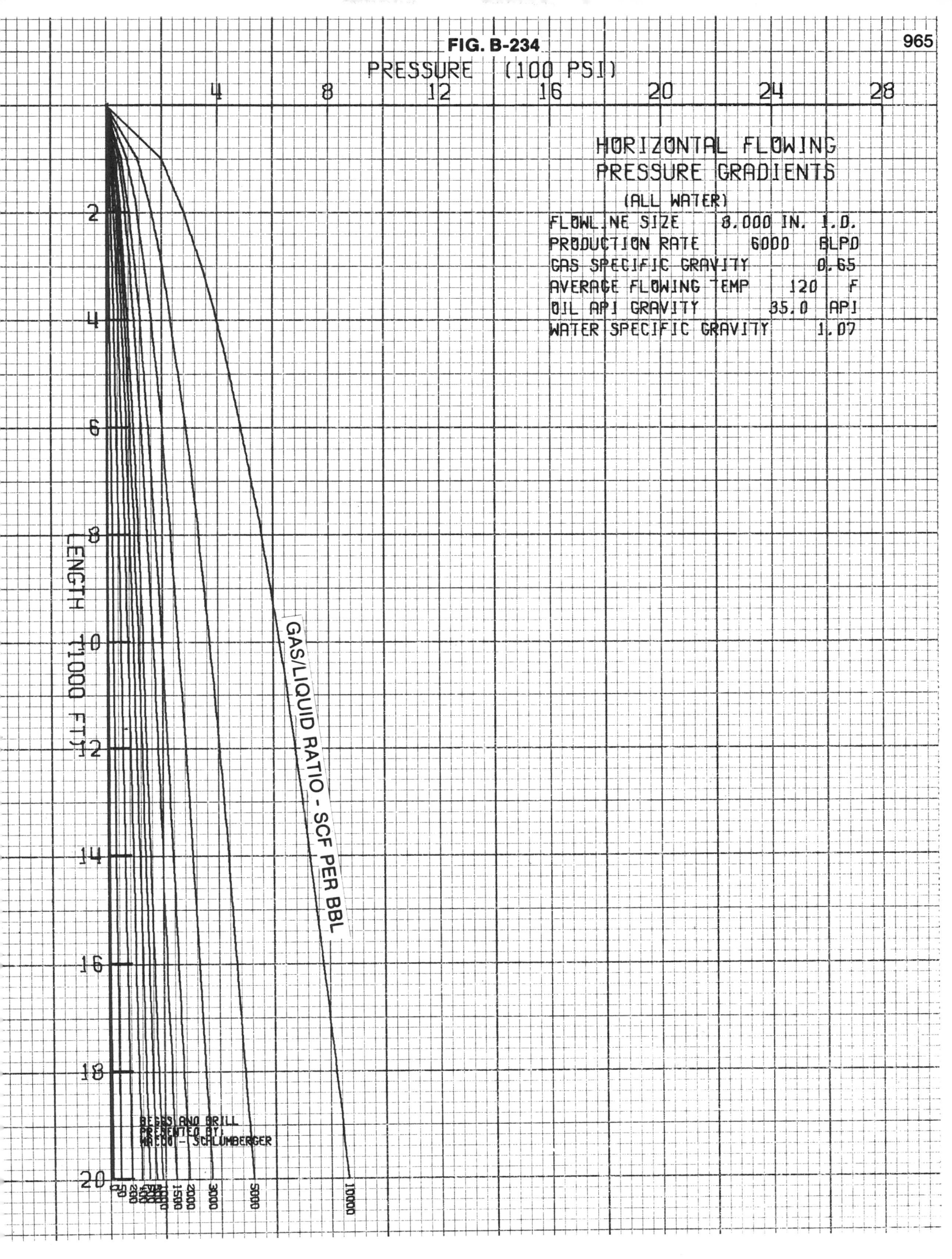

HORIZONTAL FLOWING
PRESSURE GRADIENTS
(ALL WATER)

FLOWLINE SIZE	8.000 IN.	I.D.
PRODUCTION RATE	6000	BLPD
GAS SPECIFIC GRAVITY	0.65	
AVERAGE FLOWING TEMP	120	F
OIL API GRAVITY	35.0	API
WATER SPECIFIC GRAVITY	1.07	

PRESSURE (100 PSI)

LENGTH (1000 FT)

GAS/LIQUID RATIO - SCF PER BBL

BEGGS AND BRILL
PRESENTED BY:
INTERA-SCHLUMBERGER

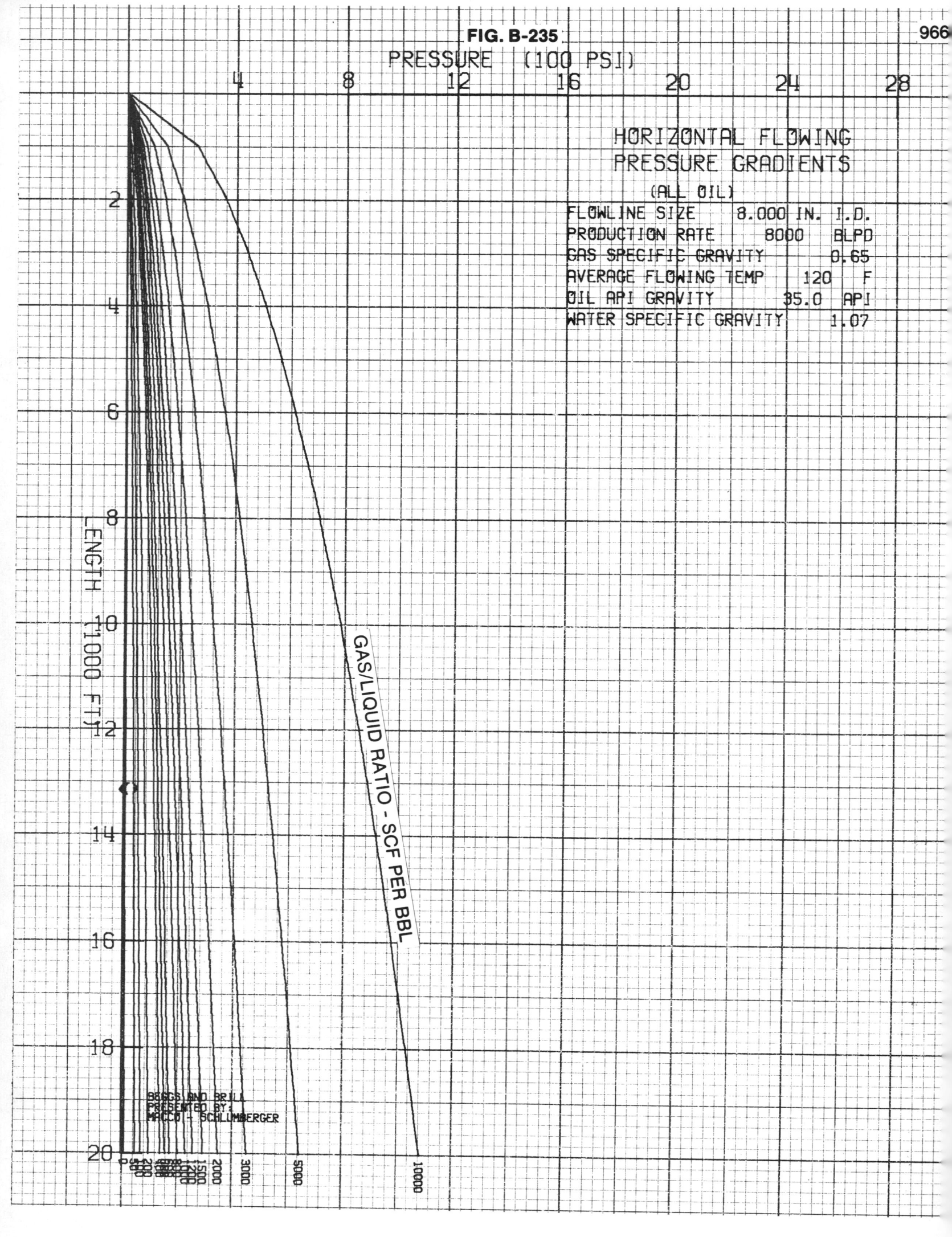

PRESSURE (100 PSI)

HORIZONTAL FLOWING
PRESSURE GRADIENTS
(ALL OIL)

FLOWLINE SIZE	8.000	IN.	I.D.
PRODUCTION RATE	8000		BLPD
GAS SPECIFIC GRAVITY		0.65	
AVERAGE FLOWING TEMP	120		F
OIL API GRAVITY	35.0		API
WATER SPECIFIC GRAVITY		1.07	

LENGTH (1000 FT)

GAS/LIQUID RATIO - SCF PER BBL

BEGGS AND BRILL
PRESENTED BY:
MACCO - SCHLUMBERGER

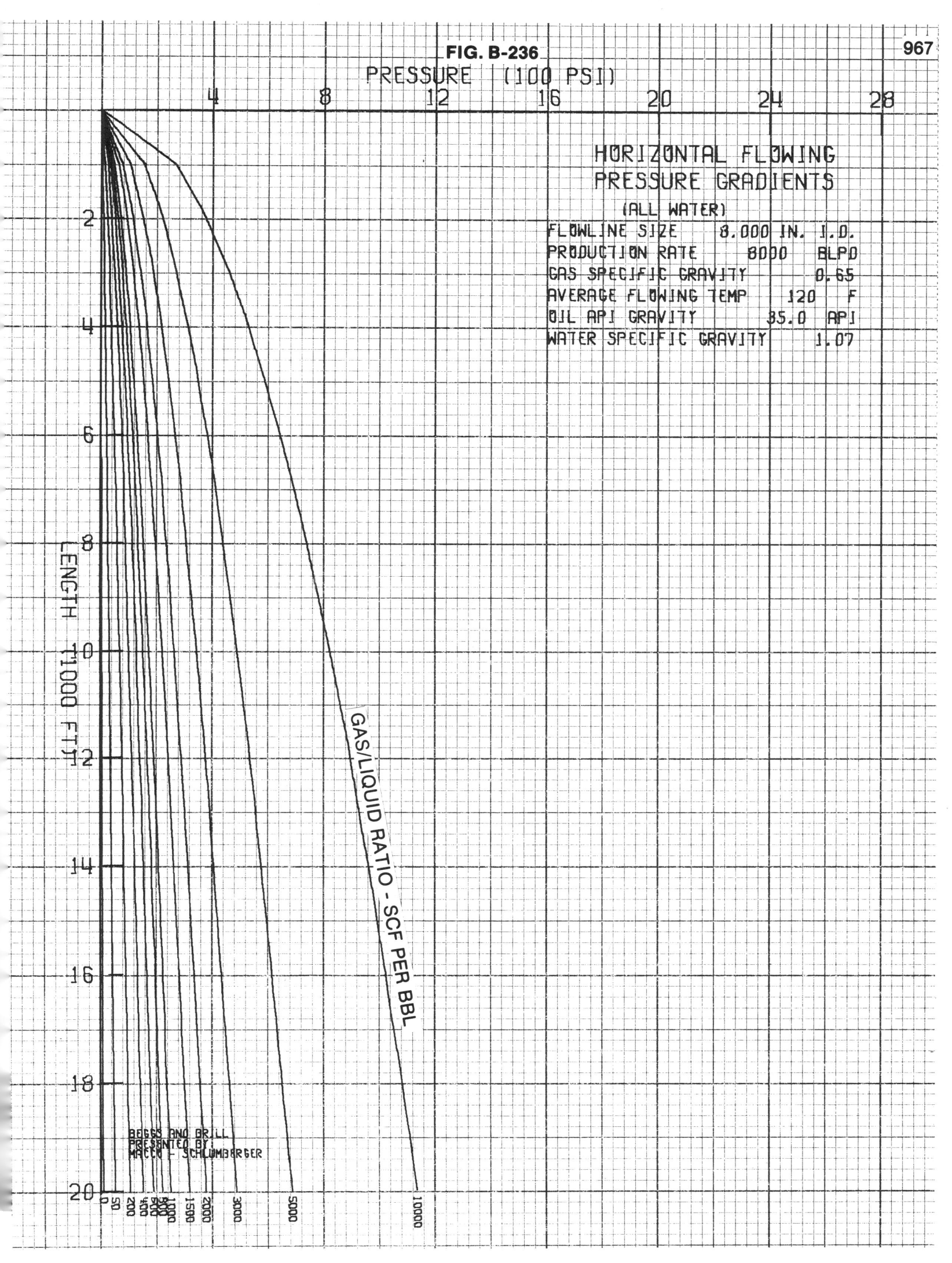

PRESSURE (100 PSI)

HORIZONTAL FLOWING
PRESSURE GRADIENTS
(ALL WATER)

FLOWLINE SIZE	8.000 IN. I.D.
PRODUCTION RATE	8000 BLPD
GAS SPECIFIC GRAVITY	0.65
AVERAGE FLOWING TEMP	120 F
OIL API GRAVITY	35.0 API
WATER SPECIFIC GRAVITY	1.07

LENGTH (1000 FT)

GAS/LIQUID RATIO - SCF PER BBL

BEGGS AND BRILL
PRESENTED BY
MAERSK - SCHLUMBERGER

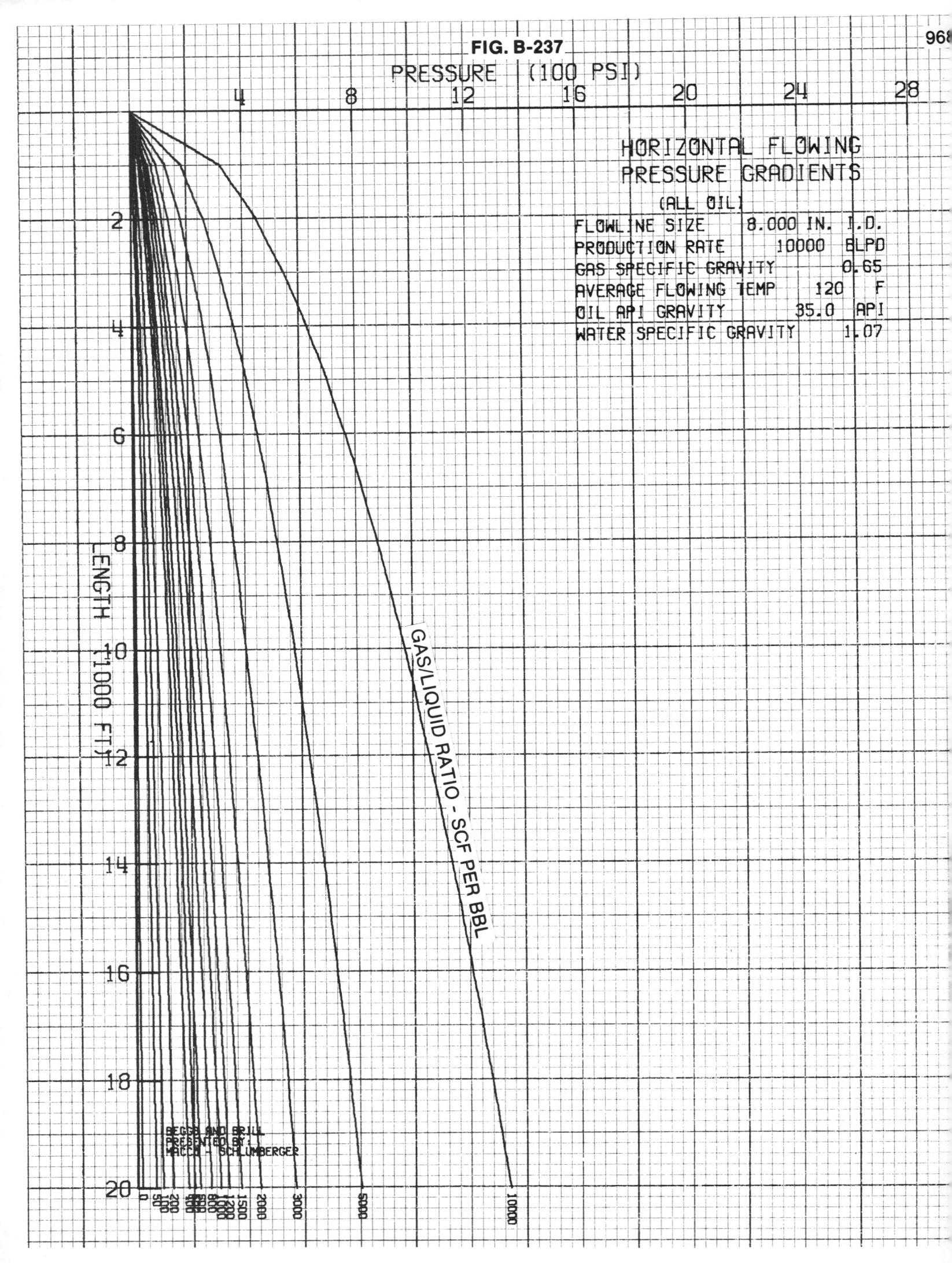

FIG. B-237

PRESSURE (100 PSI)

HORIZONTAL FLOWING
PRESSURE GRADIENTS
(ALL OIL)

FLOWLINE SIZE	8.000 IN. I.D.
PRODUCTION RATE	10000 BLPD
GAS SPECIFIC GRAVITY	0.65
AVERAGE FLOWING TEMP	120 F
OIL API GRAVITY	35.0 API
WATER SPECIFIC GRAVITY	1.07

LENGTH (1000 FT)

GAS/LIQUID RATIO - SCF PER BBL

BEGGS AND BRILL
PRESENTED BY:
MACCO - SCHLUMBERGER

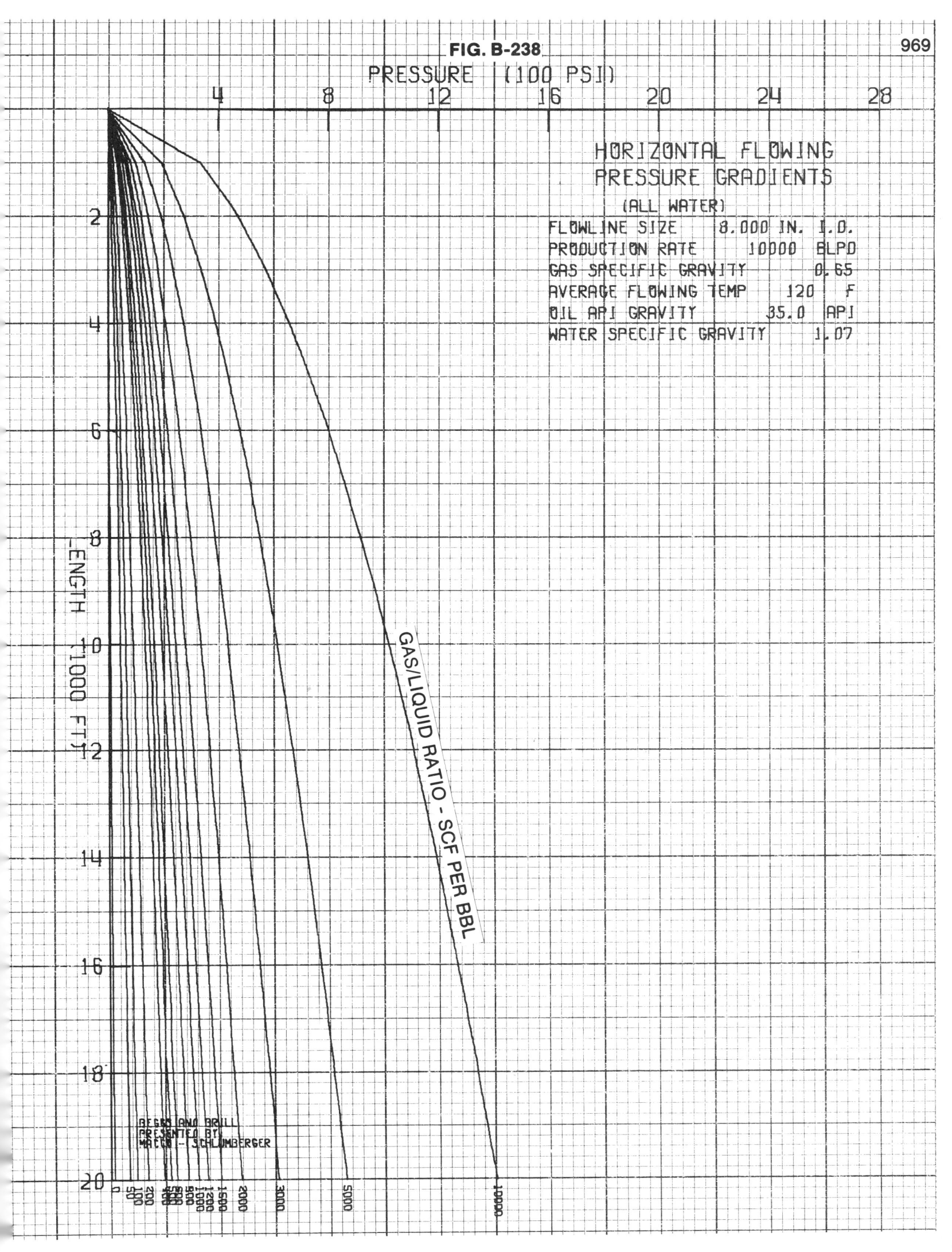

HORIZONTAL FLOWING
PRESSURE GRADIENTS
(ALL WATER)

FLOWLINE SIZE	8.000 IN.	I.D.
PRODUCTION RATE	10000	BLPD
GAS SPECIFIC GRAVITY	0.65	
AVERAGE FLOWING TEMP	120	F
OIL API GRAVITY	35.0	API
WATER SPECIFIC GRAVITY	1.07	

GAS/LIQUID RATIO - SCF PER BBL

BEGGS AND BRILL
PRESENTED BY
MACCO - SCHLUMBERGER

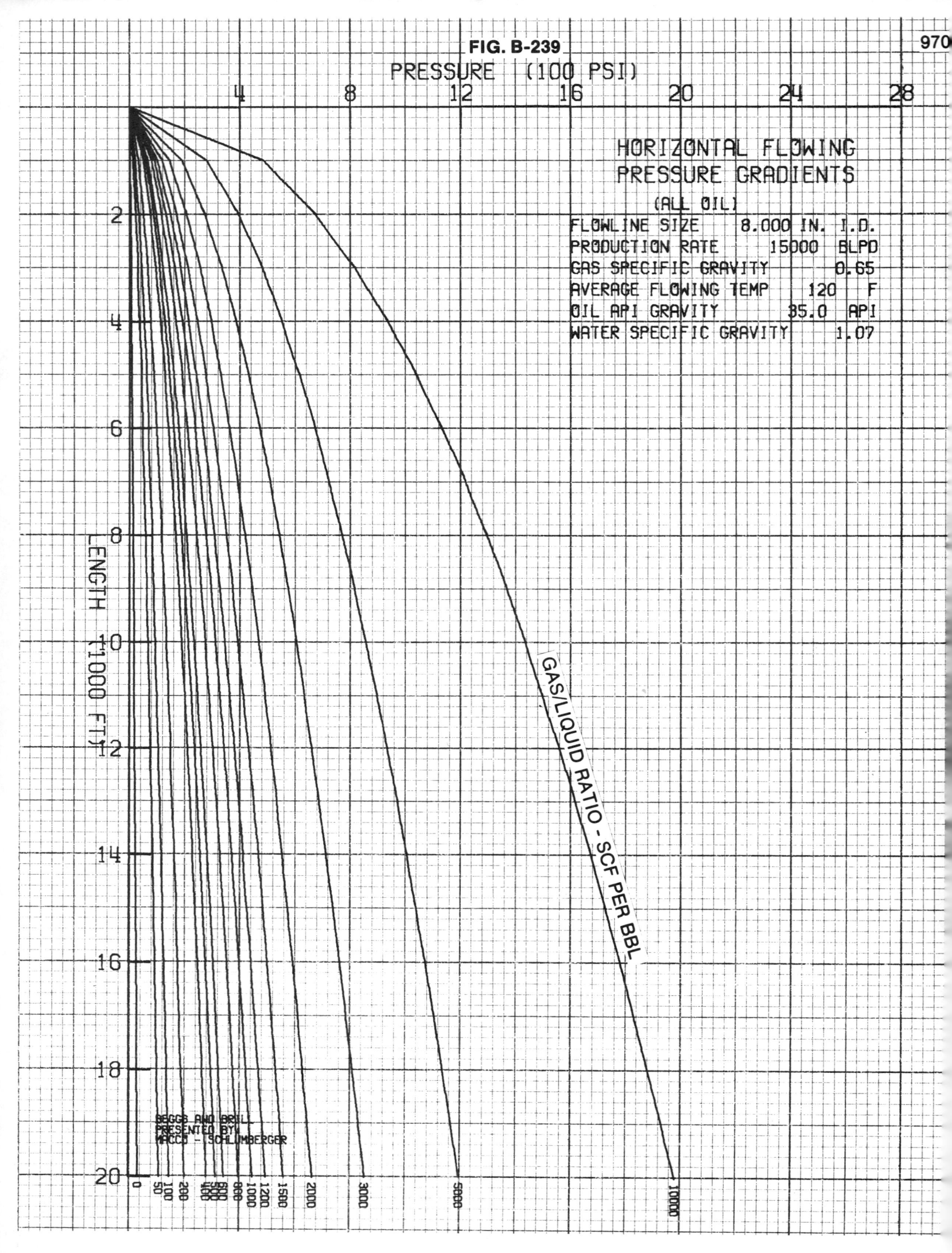

HORIZONTAL FLOWING
PRESSURE GRADIENTS
(ALL OIL)

FLOWLINE SIZE	8.000 IN. I.D.
PRODUCTION RATE	15000 BLPD
GAS SPECIFIC GRAVITY	0.65
AVERAGE FLOWING TEMP	120 F
OIL API GRAVITY	35.0 API
WATER SPECIFIC GRAVITY	1.07

PRESSURE (100 PSI)

LENGTH (1000 FT)

GAS/LIQUID RATIO - SCF PER BBL

BEGGS AND BRILL
PRESENTED BY
MACCO - SCHLUMBERGER

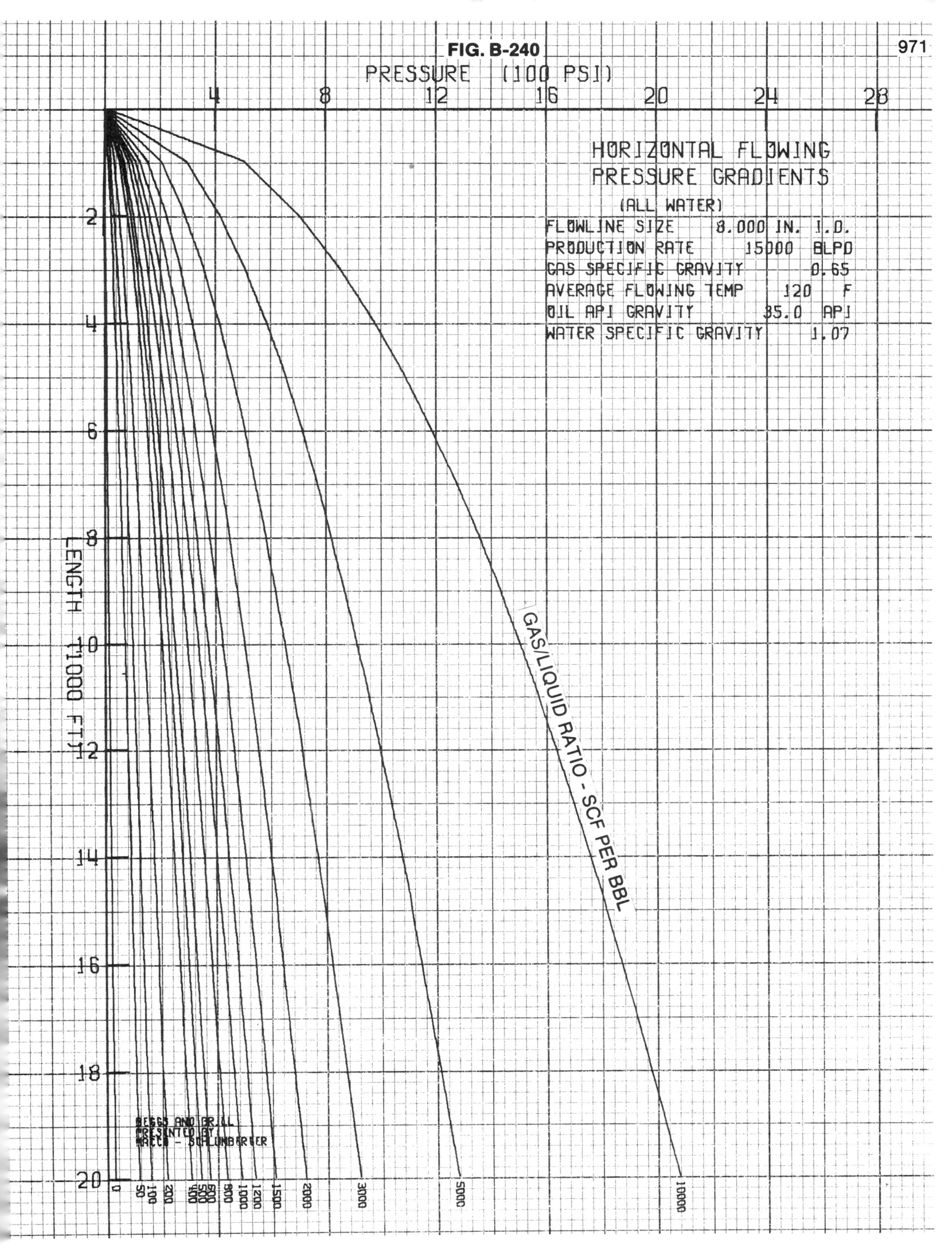

HORIZONTAL FLOWING
PRESSURE GRADIENTS
(ALL WATER)

FLOWLINE SIZE	8.000 IN. I.D.
PRODUCTION RATE	15000 BLPD
GAS SPECIFIC GRAVITY	0.65
AVERAGE FLOWING TEMP	120 F
OIL API GRAVITY	35.0 API
WATER SPECIFIC GRAVITY	1.07

PRESSURE (100 PSI)

LENGTH (1000 FT)

GAS/LIQUID RATIO - SCF PER BBL

BEGGS AND BRILL
PRESENTED BY
CAMCO - SCHLUMBERGER

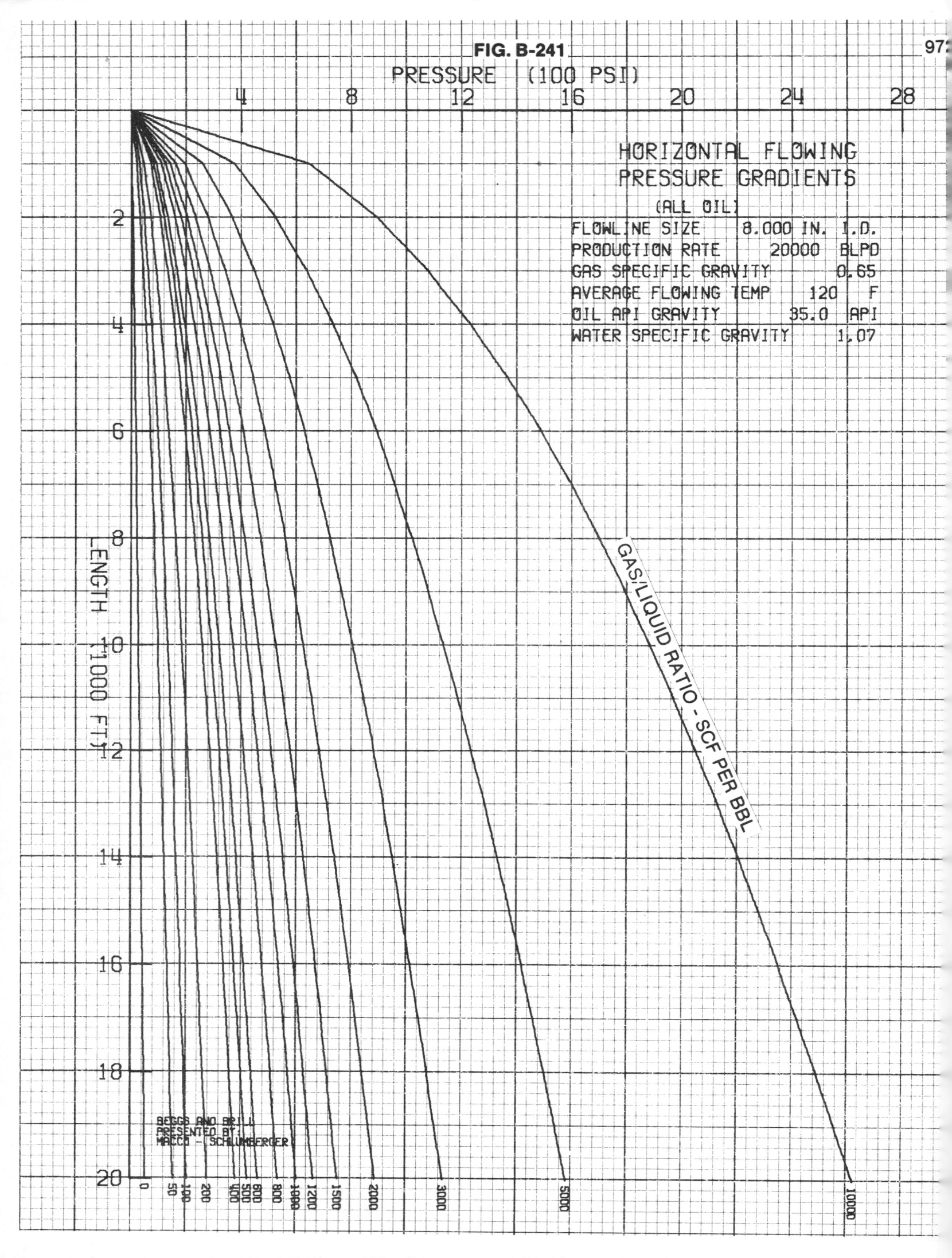

PRESSURE (100 PSI)

HORIZONTAL FLOWING
PRESSURE GRADIENTS
(ALL OIL)

FLOWLINE SIZE	8.000 IN. I.D.
PRODUCTION RATE	20000 BLPD
GAS SPECIFIC GRAVITY	0.65
AVERAGE FLOWING TEMP	120 F
OIL API GRAVITY	35.0 API
WATER SPECIFIC GRAVITY	1.07

LENGTH (1000 FT)

GAS/LIQUID RATIO - SCF PER BBL

BEGGS AND BRILL
PRESENTED BY
MACCO - SCHLUMBERGER

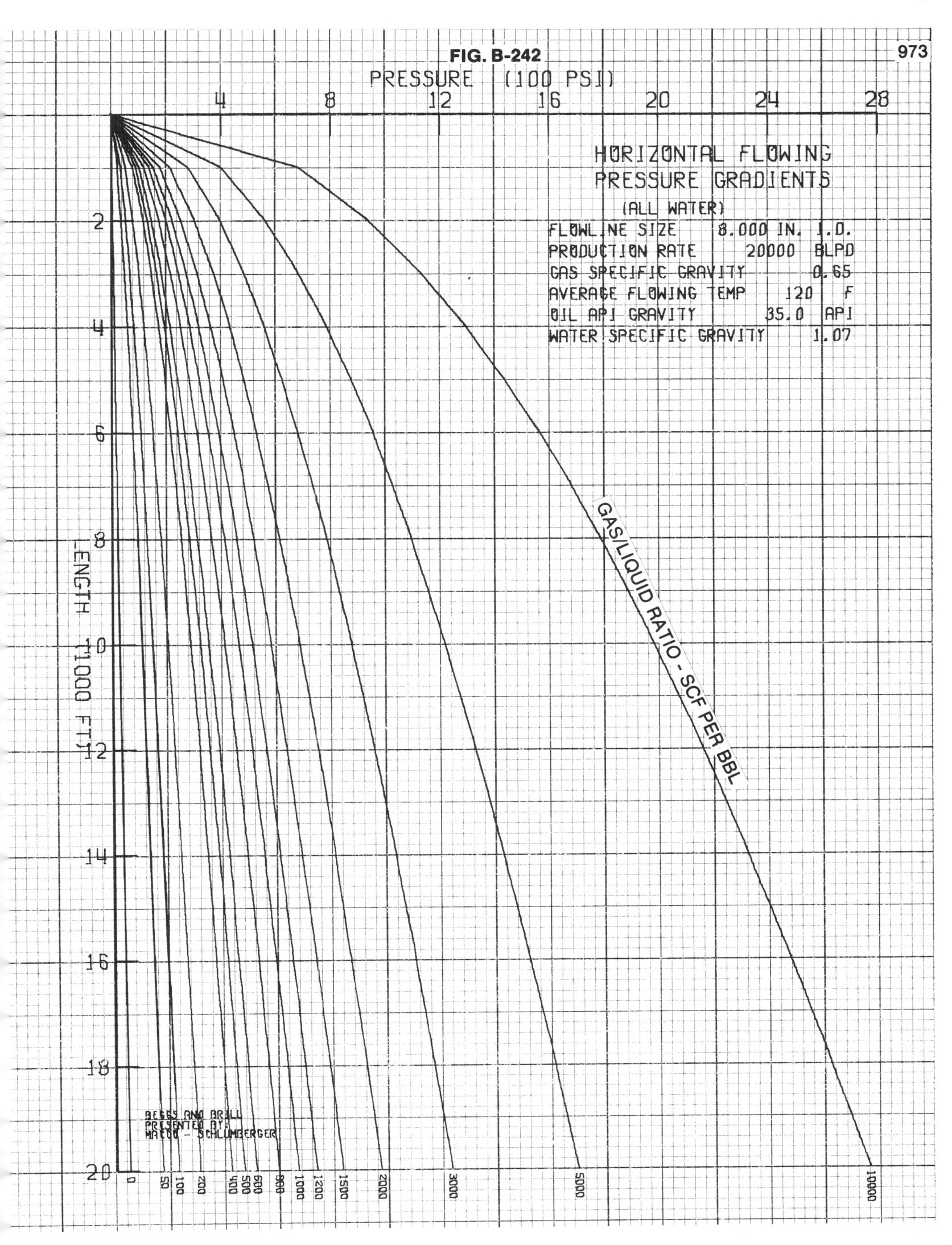

FIG. B-242

973

PRESSURE (100 PSI)

HORIZONTAL FLOWING
PRESSURE GRADIENTS
(ALL WATER)

FLOWLINE SIZE	8.000 IN. I.D.
PRODUCTION RATE	20000 BLPD
GAS SPECIFIC GRAVITY	0.65
AVERAGE FLOWING TEMP	120 F
OIL API GRAVITY	35.0 API
WATER SPECIFIC GRAVITY	1.07

LENGTH (1000 FT.)

GAS/LIQUID RATIO - SCF PER BBL

BEGGS AND BRILL
PRESENTED BY:
HALCO - SCHLUMBERGER

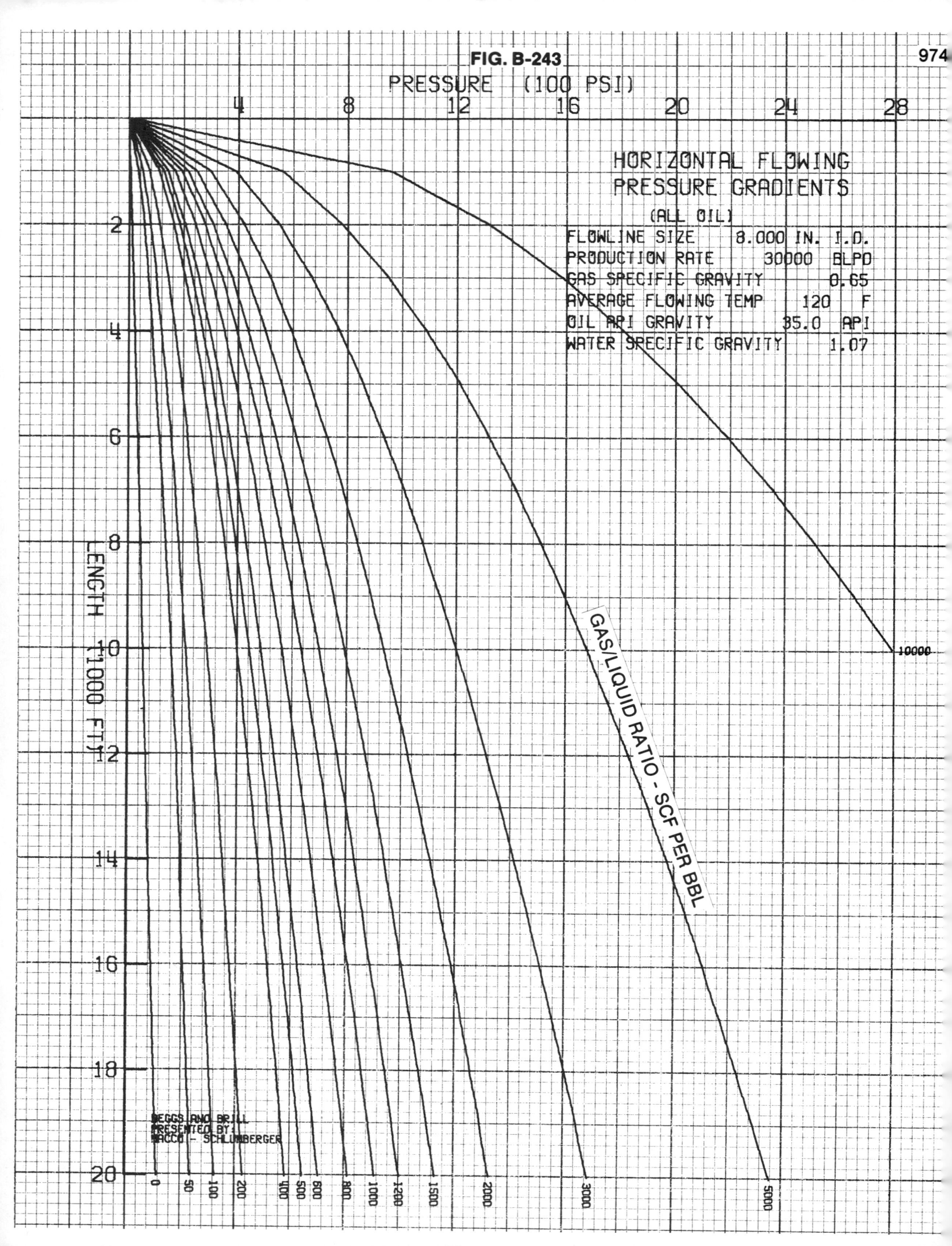

FIG. B-243

974

PRESSURE (100 PSI)

HORIZONTAL FLOWING
PRESSURE GRADIENTS
(ALL OIL)
FLOWLINE SIZE   8.000 IN. I.D.
PRODUCTION RATE   30000 BLPD
GAS SPECIFIC GRAVITY   0.65
AVERAGE FLOWING TEMP   120 F
OIL API GRAVITY   35.0 API
WATER SPECIFIC GRAVITY   1.07

LENGTH (1000 FT)

GAS/LIQUID RATIO - SCF PER BBL

BEGGS AND BRILL
PRESENTED BY
MACCO - SCHLUMBERGER

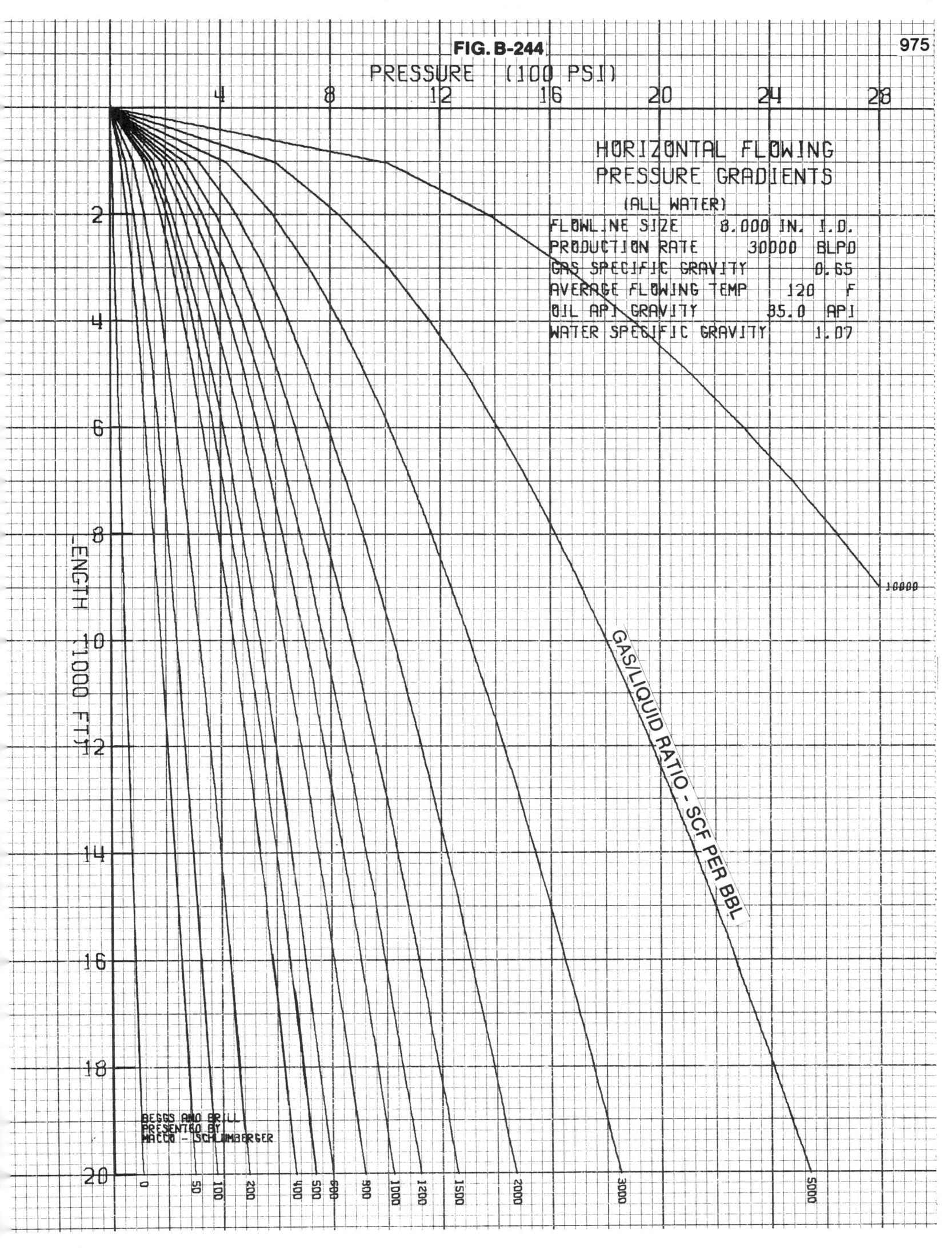

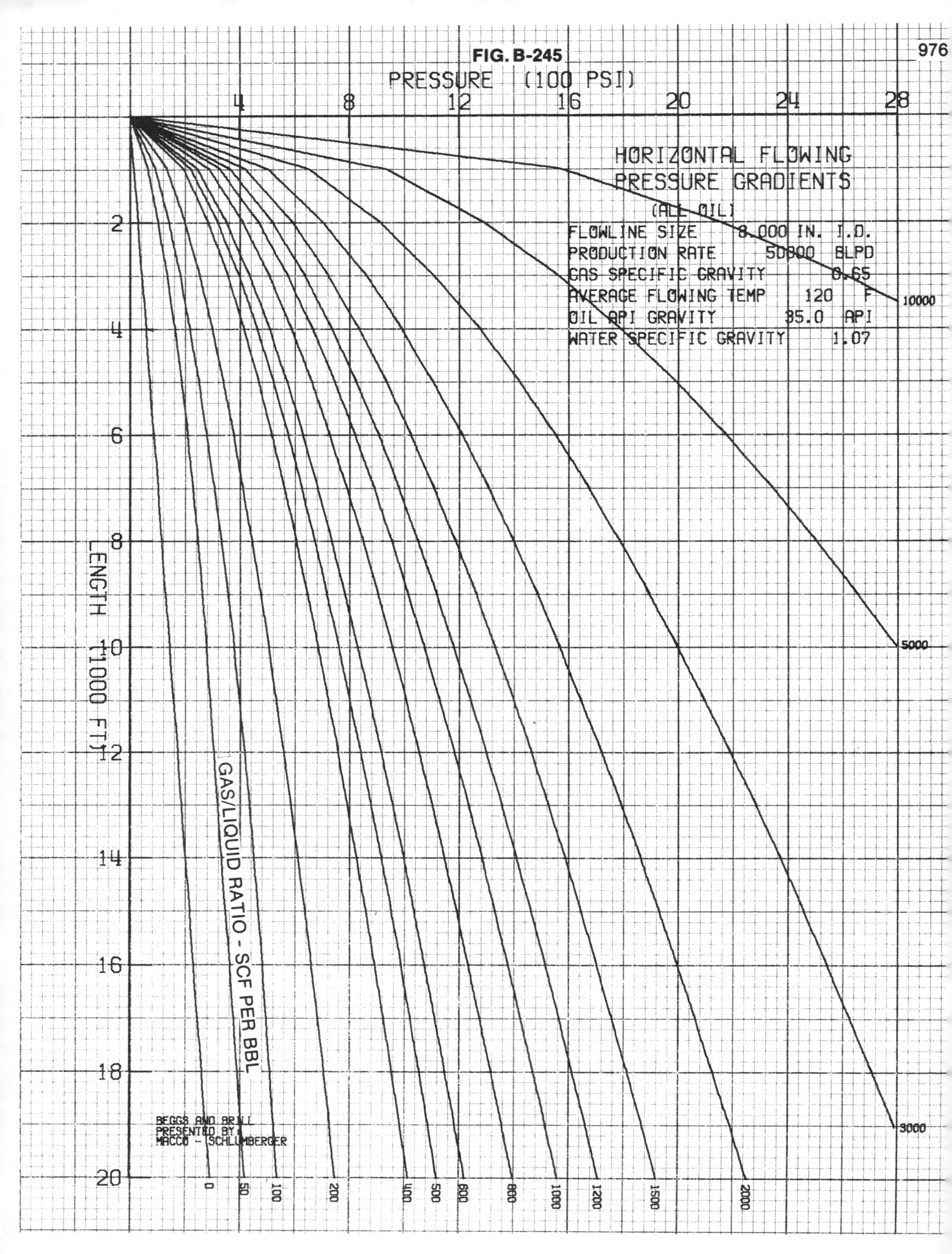

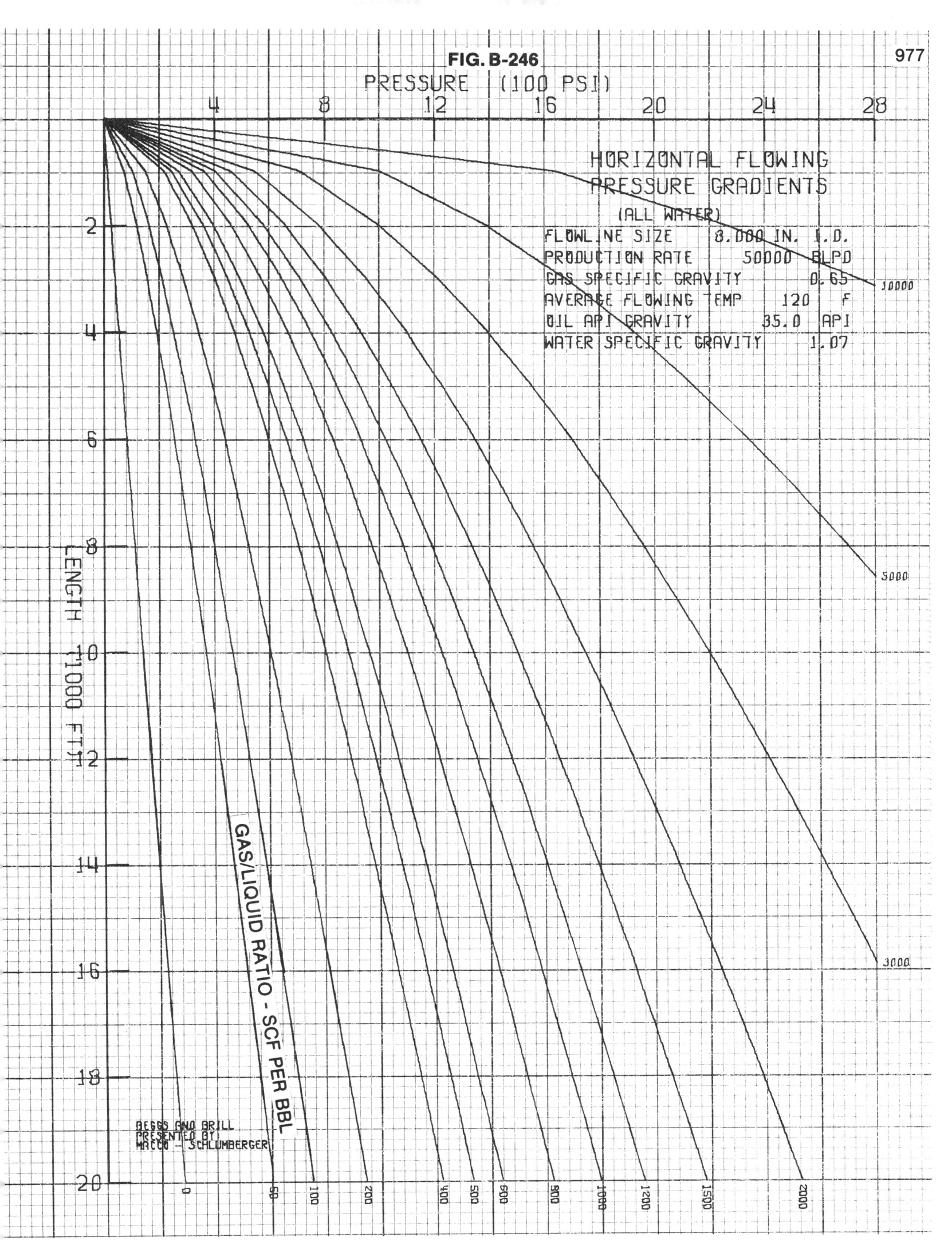

FIG. B-246

PRESSURE (100 PSI)

HORIZONTAL FLOWING
PRESSURE GRADIENTS
(ALL WATER)

FLOWLINE SIZE	8.000 IN. I.D.
PRODUCTION RATE	50000 BLPD
GAS SPECIFIC GRAVITY	0.65
AVERAGE FLOWING TEMP	120 F
OIL API GRAVITY	35.0 API
WATER SPECIFIC GRAVITY	1.07

LENGTH (1000 FT)

GAS/LIQUID RATIO - SCF PER BBL

BEGGS AND BRILL
PRESENTED BY
MACCO - SCHLUMBERGER

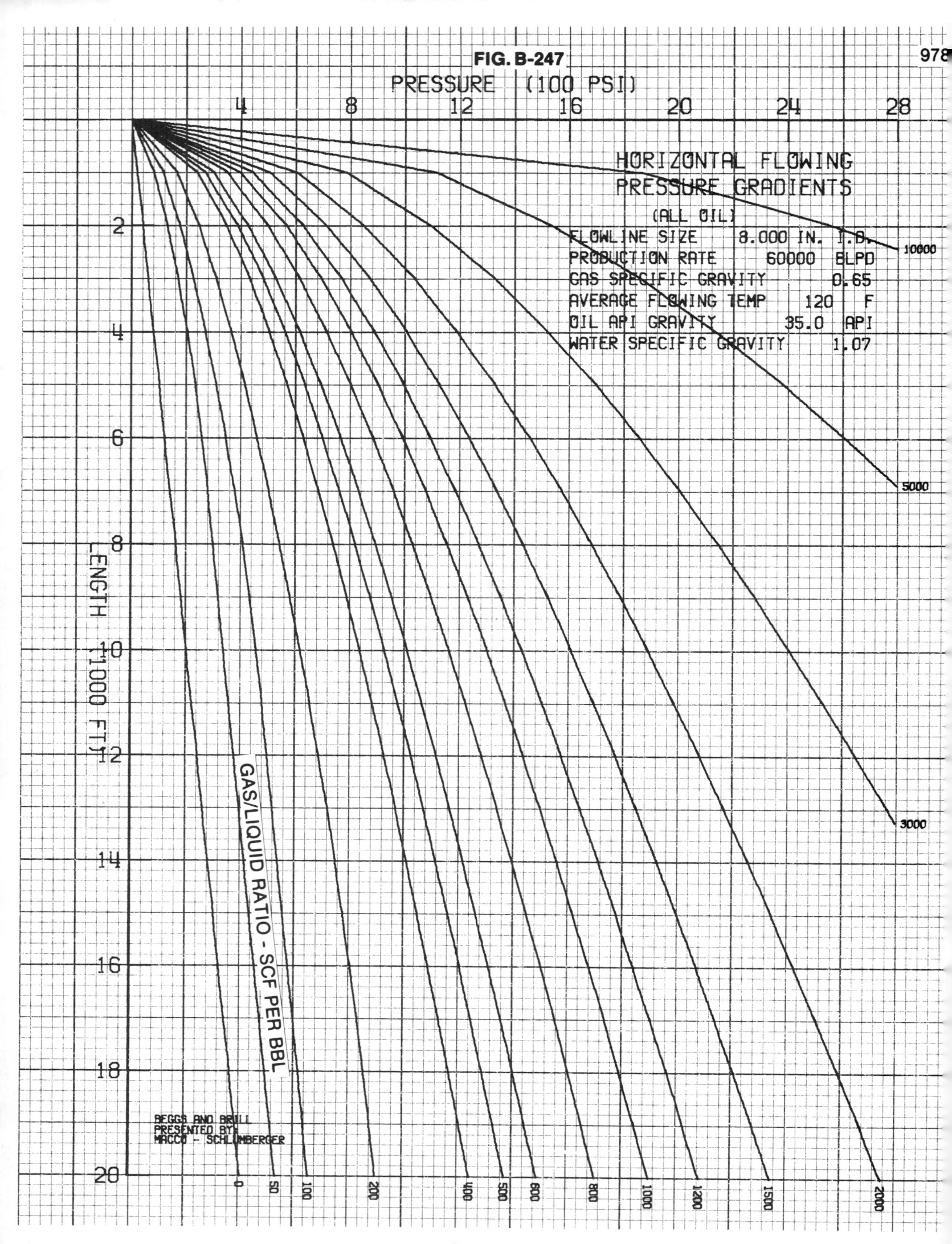

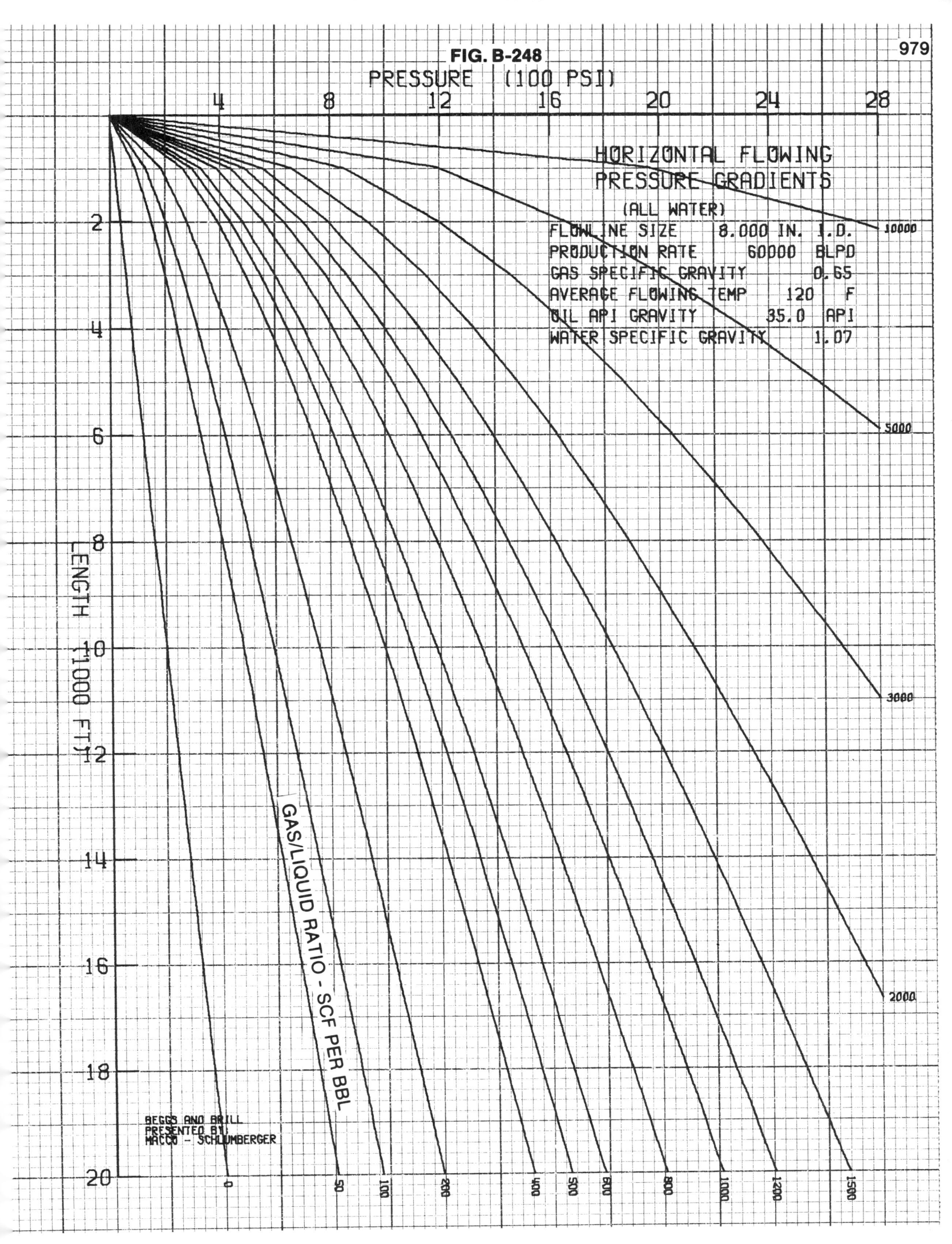

FIG. B-248

PRESSURE (100 PSI)

HORIZONTAL FLOWING
PRESSURE GRADIENTS

(ALL WATER)

FLOWLINE SIZE   8.000 IN. I.D.
PRODUCTION RATE   60000 BLPD
GAS SPECIFIC GRAVITY   0.65
AVERAGE FLOWING TEMP   120 F
OIL API GRAVITY   35.0 API
WATER SPECIFIC GRAVITY   1.07

LENGTH (1000 FT)

GAS/LIQUID RATIO - SCF PER BBL

BEGGS AND BRILL
PRESENTED BY
MACCO - SCHLUMBERGER

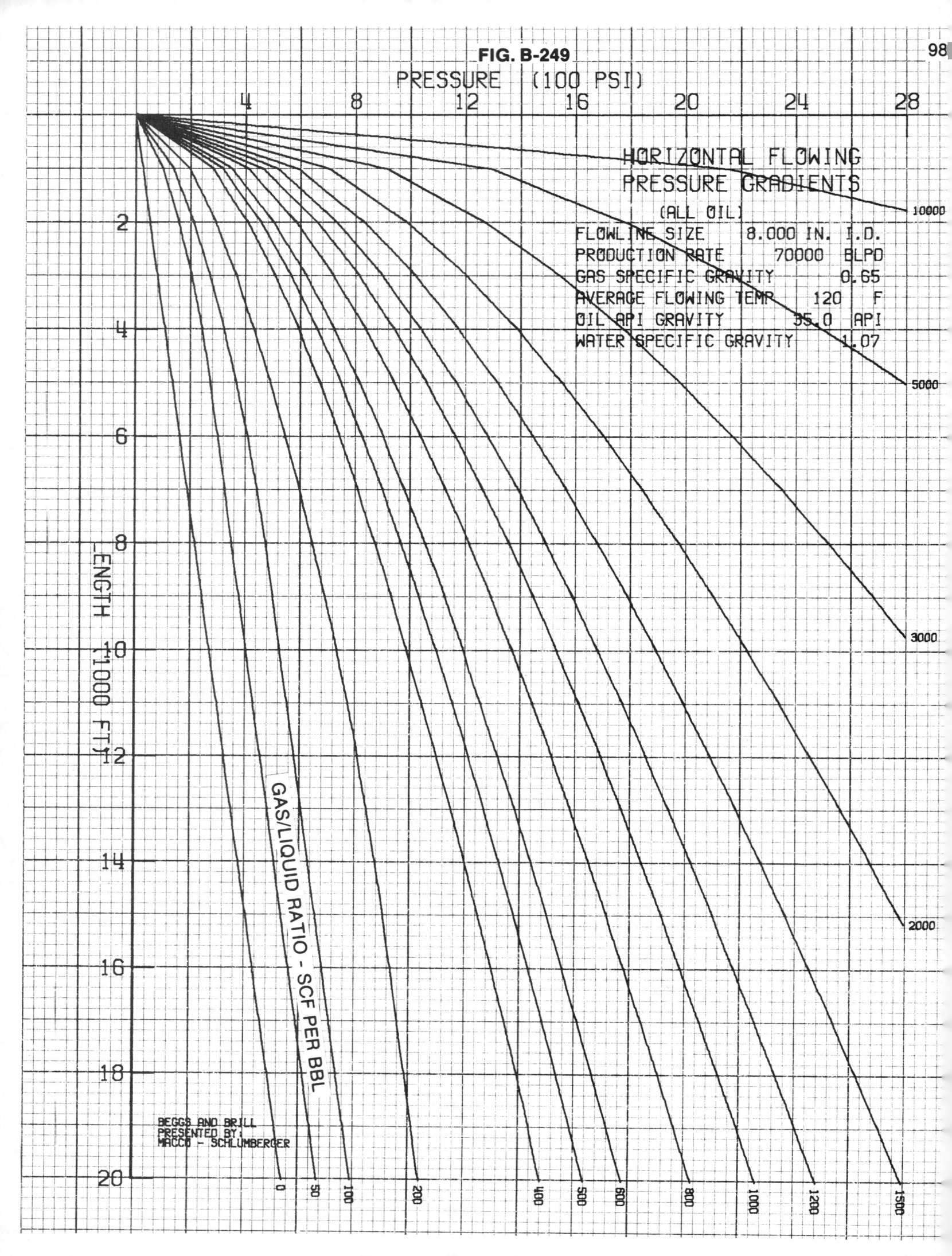

PRESSURE (100 PSI)

HORIZONTAL FLOWING
PRESSURE GRADIENTS
(ALL OIL)

FLOWLINE SIZE   8.000 IN. I.D.
PRODUCTION RATE   70000 BLPD
GAS SPECIFIC GRAVITY   0.65
AVERAGE FLOWING TEMP   120 F
OIL API GRAVITY   35.0 API
WATER SPECIFIC GRAVITY   1.07

LENGTH (1000 FT)

GAS/LIQUID RATIO - SCF PER BBL

BEGGS AND BRILL
PRESENTED BY:
MACCO - SCHLUMBERGER

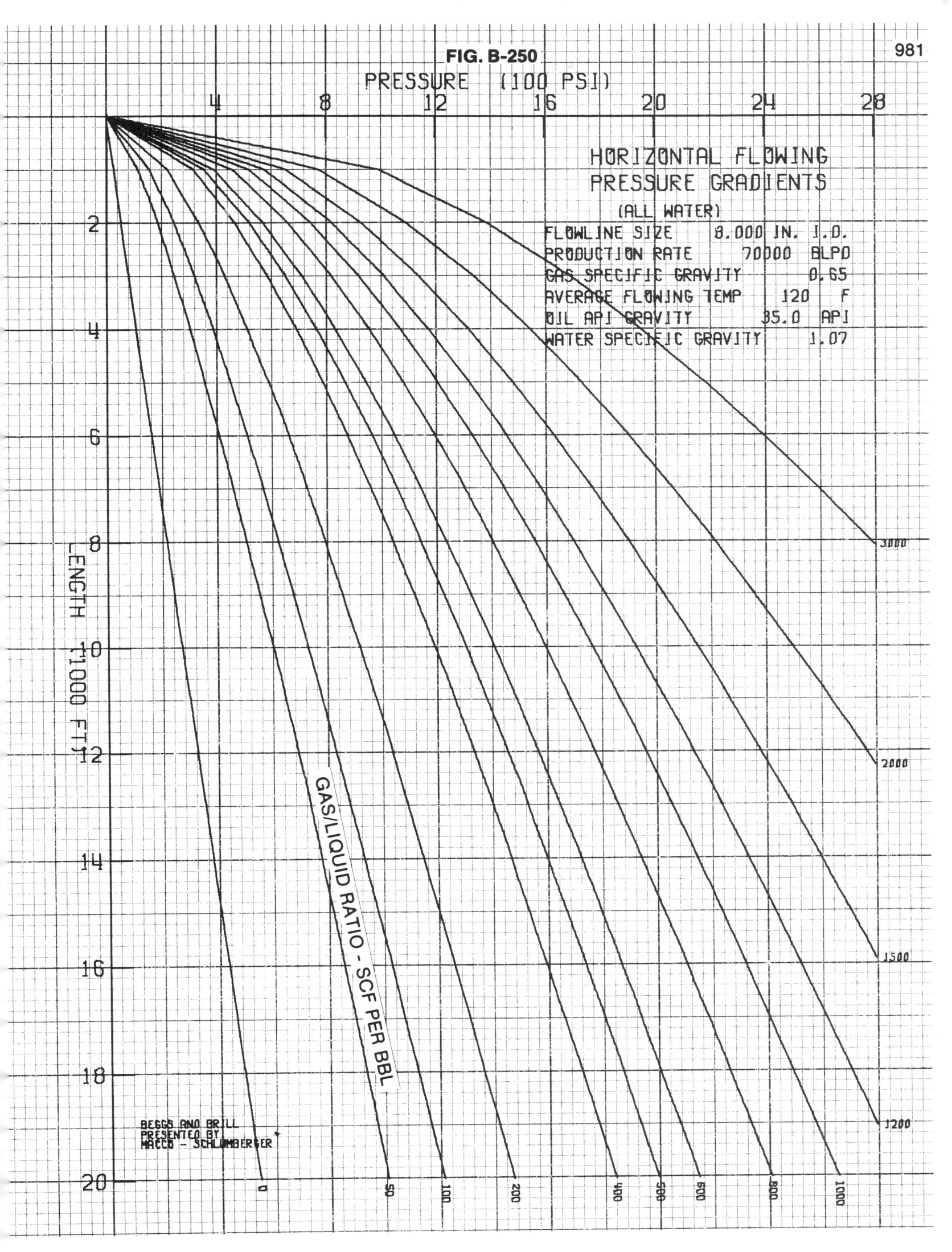

FIG. B-250

981

HORIZONTAL FLOWING
PRESSURE GRADIENTS
(ALL WATER)

FLOWLINE SIZE	8.000	IN. I.D.
PRODUCTION RATE	70000	BLPD
GAS SPECIFIC GRAVITY	0.65	
AVERAGE FLOWING TEMP	120	F
OIL API GRAVITY	35.0	API
WATER SPECIFIC GRAVITY	1.07	

PRESSURE (100 PSI)

LENGTH (1000 FT)

GAS/LIQUID RATIO - SCF PER BBL

BEGGS AND BRILL
PRESENTED BY
MACCO - SCHLUMBERGER

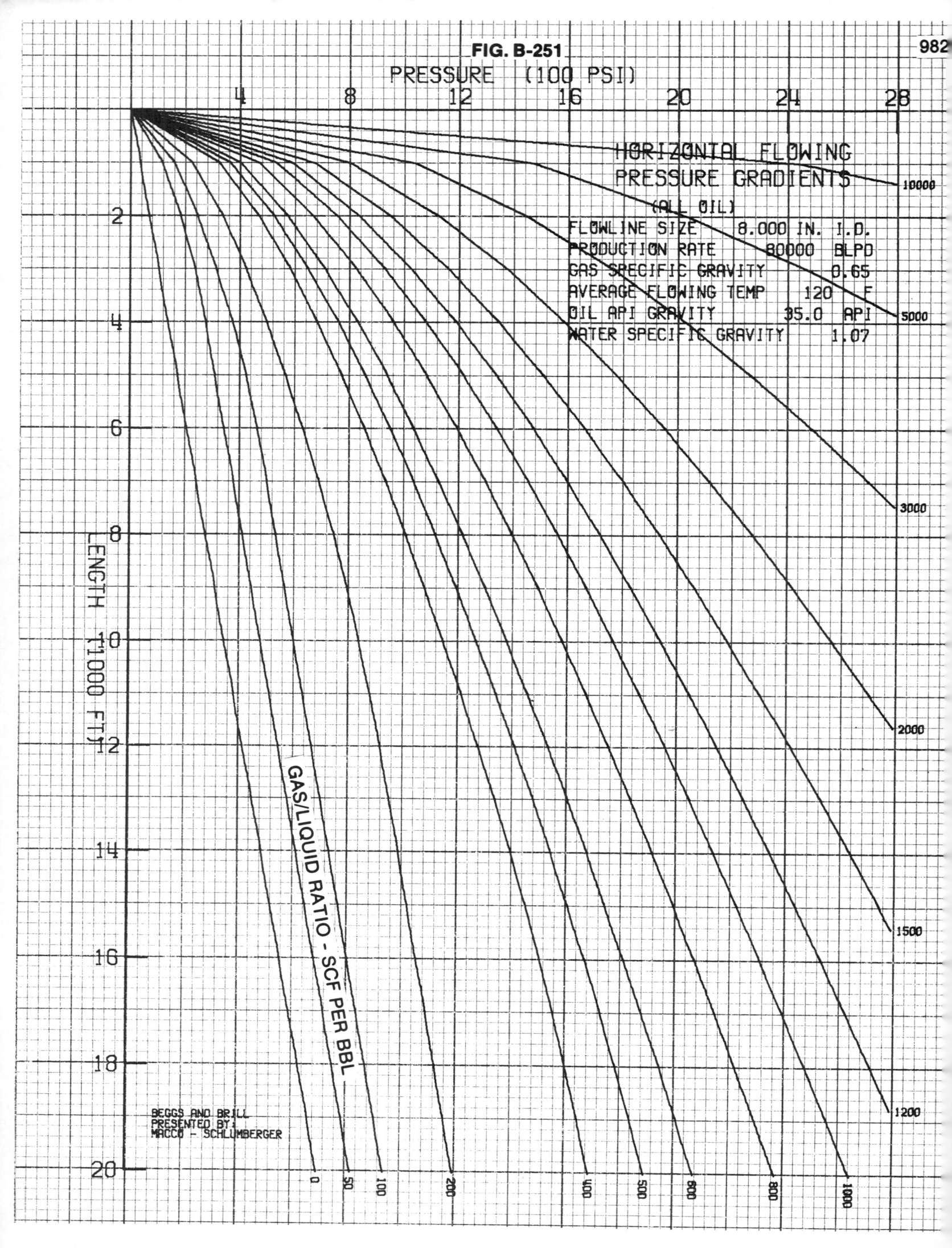

FIG. B-251

982

PRESSURE (100 PSI)

HORIZONTAL FLOWING
PRESSURE GRADIENTS
(ALL OIL)

FLOWLINE SIZE	8.000 IN. I.D.
PRODUCTION RATE	80000 BLPD
GAS SPECIFIC GRAVITY	0.65
AVERAGE FLOWING TEMP	120 F
OIL API GRAVITY	35.0 API
WATER SPECIFIC GRAVITY	1.07

LENGTH (1000 FT)

GAS/LIQUID RATIO - SCF PER BBL

BEGGS AND BRILL
PRESENTED BY:
MACCO - SCHLUMBERGER

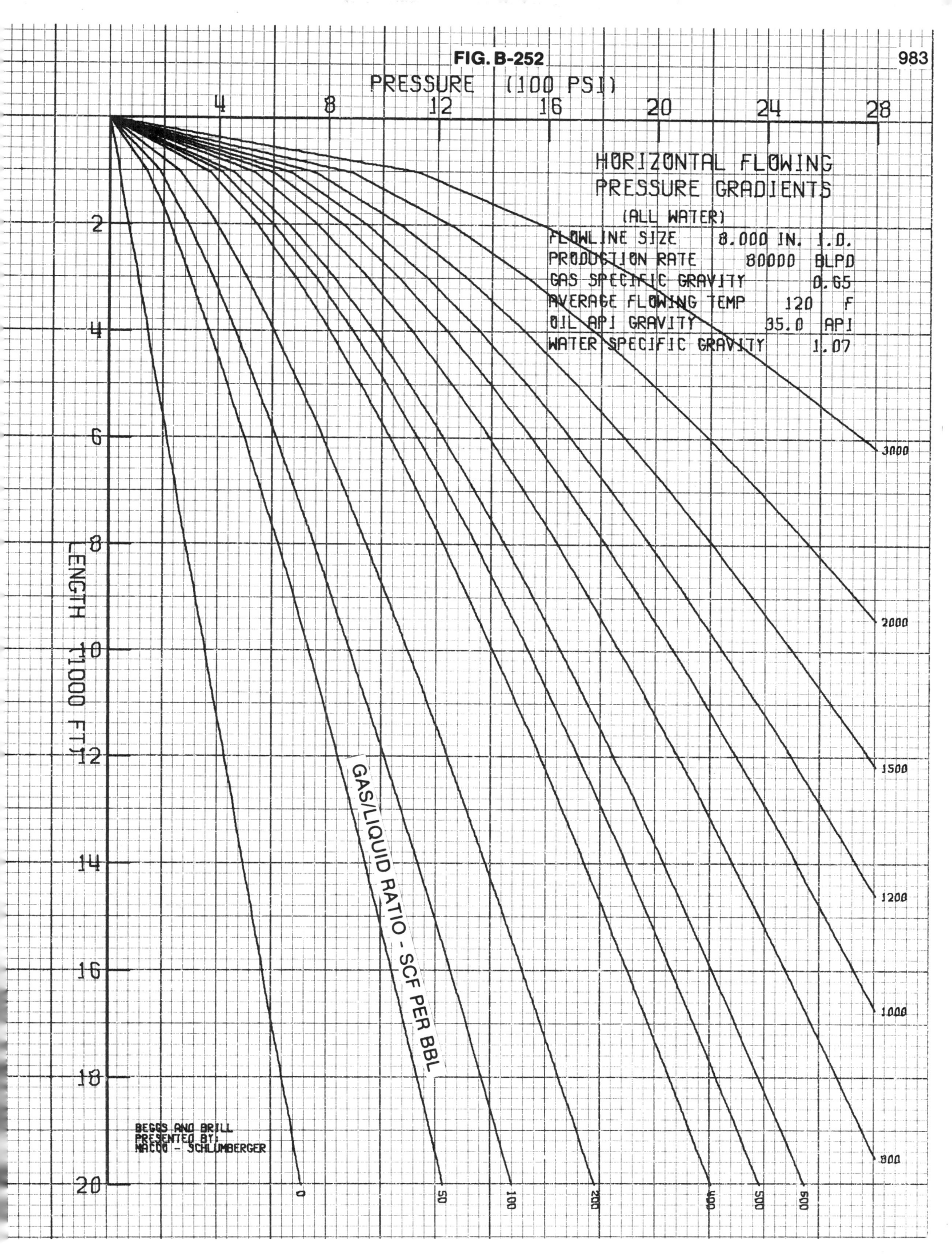

FIG. B-252

983

PRESSURE (100 PSI)

HORIZONTAL FLOWING
PRESSURE GRADIENTS
(ALL WATER)

FLOWLINE SIZE	8.000 IN. I.D.
PRODUCTION RATE	80000 BLPD
GAS SPECIFIC GRAVITY	0.65
AVERAGE FLOWING TEMP	120 F
OIL API GRAVITY	35.0 API
WATER SPECIFIC GRAVITY	1.07

LENGTH (1000 FT.)

GAS/LIQUID RATIO - SCF PER BBL.

BEGGS AND BRILL
PRESENTED BY:
MAEDO - SCHLUMBERGER

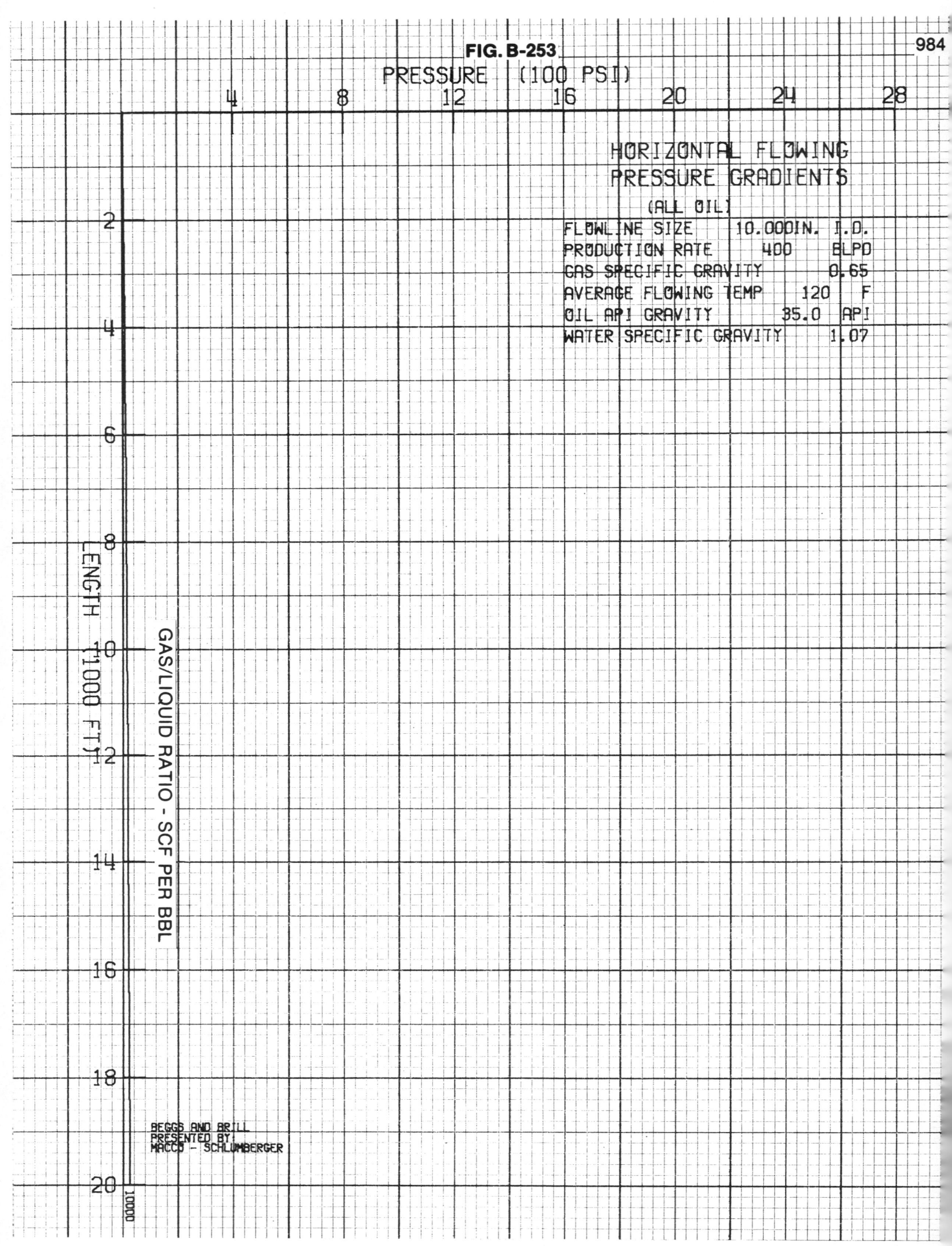

PRESSURE (100 PSI)

HORIZONTAL FLOWING
PRESSURE GRADIENTS
(ALL OIL)

FLOWLINE SIZE	10.000 IN. I.D.
PRODUCTION RATE	400 BLPD
GAS SPECIFIC GRAVITY	0.65
AVERAGE FLOWING TEMP	120 F
OIL API GRAVITY	35.0 API
WATER SPECIFIC GRAVITY	1.07

LENGTH (1000 FT)

GAS/LIQUID RATIO - SCF PER BBL

BEGGS AND BRILL
PRESENTED BY
MACCO - SCHLUMBERGER

FIG. B-254

PRESSURE (100 PSI)

8   16   24   32   40   48   56

HORIZONTAL FLOWING
PRESSURE GRADIENTS
(ALL WATER)

FLOWLINE SIZE	10.000 IN. I.D.
PRODUCTION RATE	400 BLPD
GAS SPECIFIC GRAVITY	0.65
AVERAGE FLOWING TEMP	120 F
OIL API GRAVITY	35.0 API
WATER SPECIFIC GRAVITY	1.07

LENGTH (1000 FT.)

2   4   6   8   10   12   14   16   18   20

GAS/LIQUID RATIO - SCF PER BBL

BEGGS AND BRILL
PRESENTED BY:
MACCO - SCHLUMBERGER

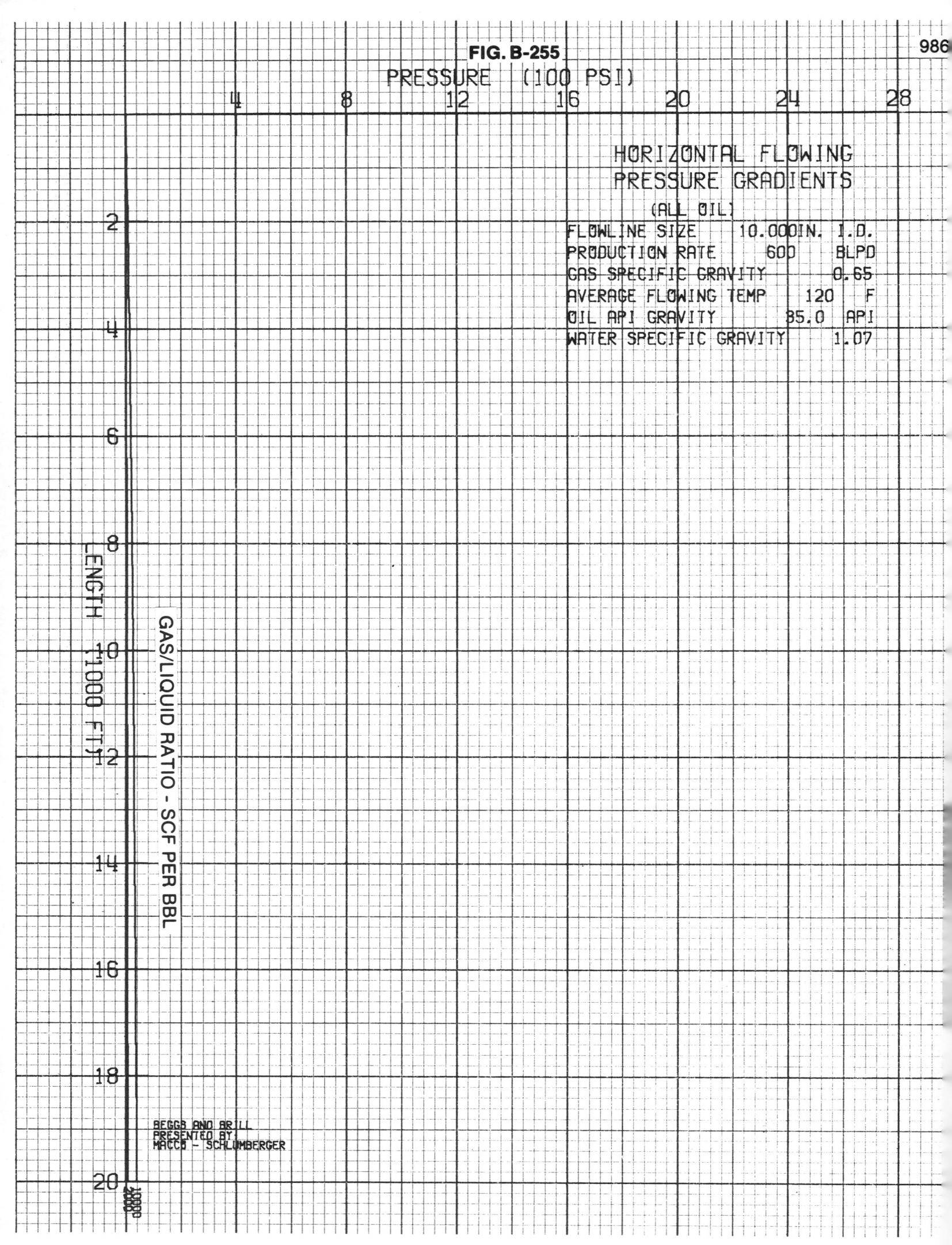

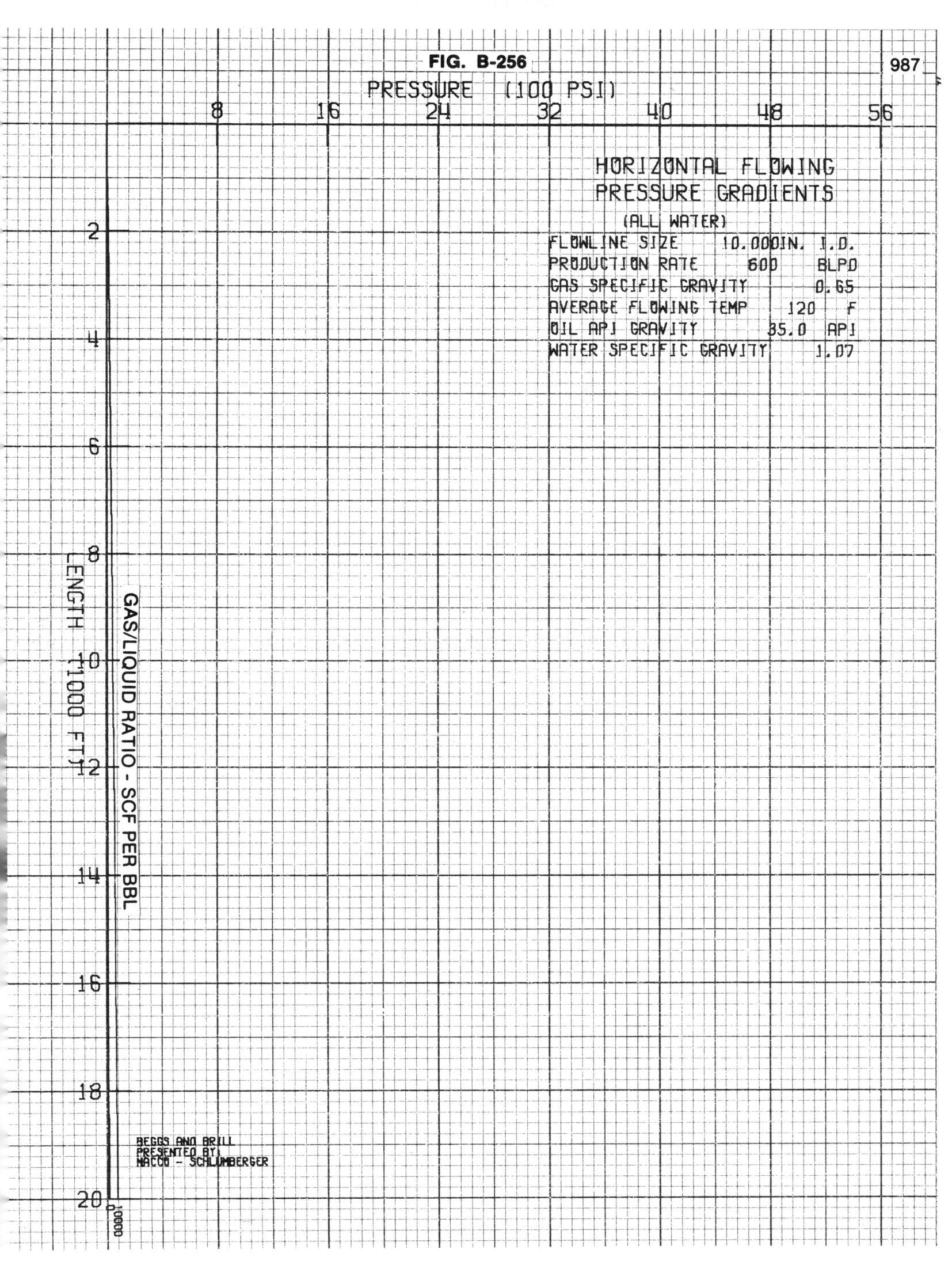

FIG. B-256

987

PRESSURE (100 PSI)

HORIZONTAL FLOWING
PRESSURE GRADIENTS
(ALL WATER)

FLOWLINE SIZE	10.000IN.	I.D.
PRODUCTION RATE	600	BLPD
GAS SPECIFIC GRAVITY	0.65	
AVERAGE FLOWING TEMP	120	F
OIL API GRAVITY	35.0	API
WATER SPECIFIC GRAVITY	1.07	

LENGTH (1000 FT)

GAS/LIQUID RATIO - SCF PER BBL

BEGGS AND BRILL
PRESENTED BY
NACOO - SCHLUMBERGER

PRESSURE (100 PSI)

HORIZONTAL FLOWING
PRESSURE GRADIENTS
(ALL OIL)

FLOWLINE SIZE	10.000IN. I.D.
PRODUCTION RATE	800 BLPD
GAS SPECIFIC GRAVITY	0.65
AVERAGE FLOWING TEMP	120 F
OIL API GRAVITY	35.0 API
WATER SPECIFIC GRAVITY	1.07

LENGTH (1000 FT.)

GAS/LIQUID RATIO - SCF PER BBL

BEGGS AND BRILL
PRESENTED BY
MACCO - SCHLUMBERGER

10000
3000

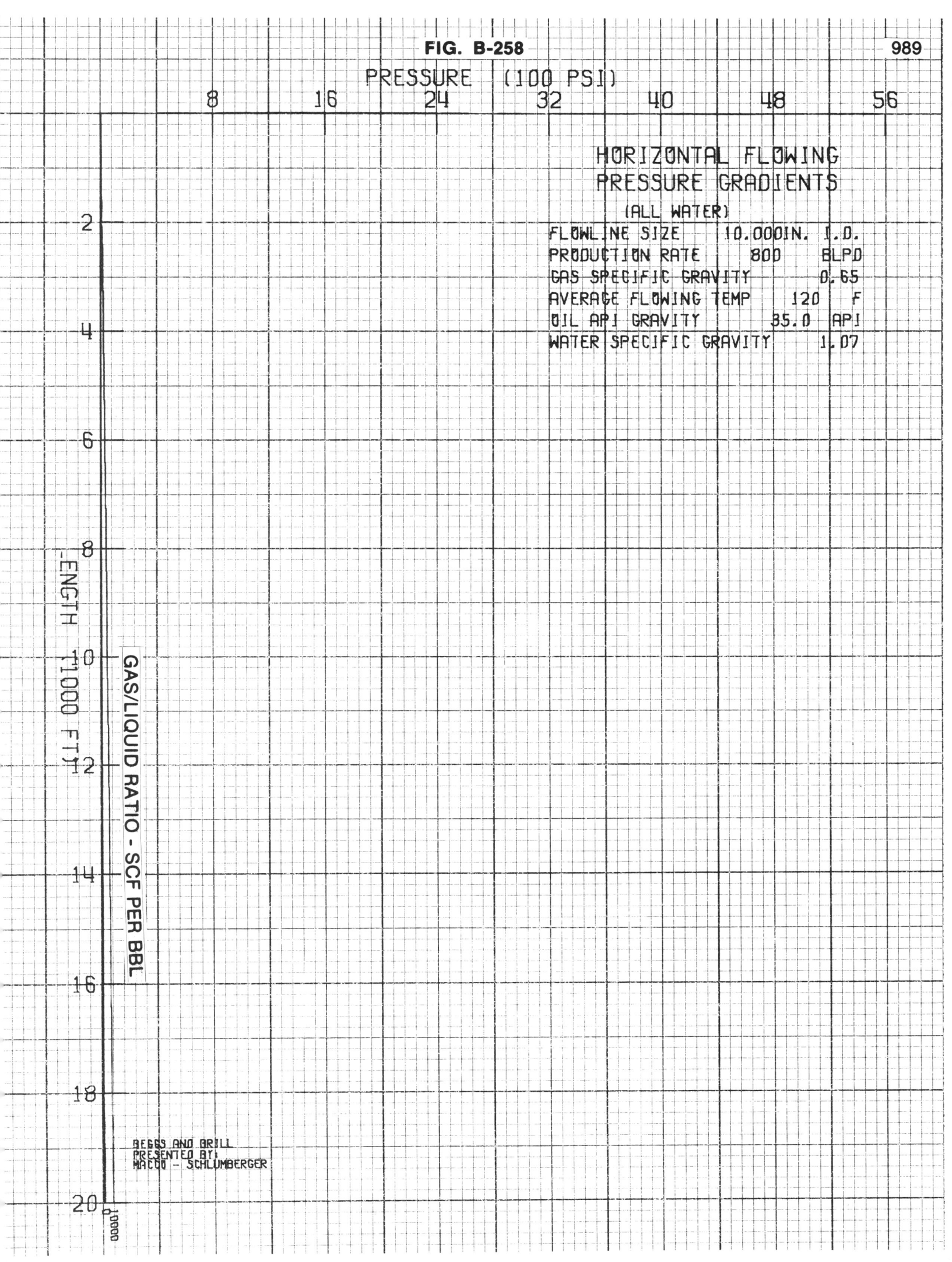

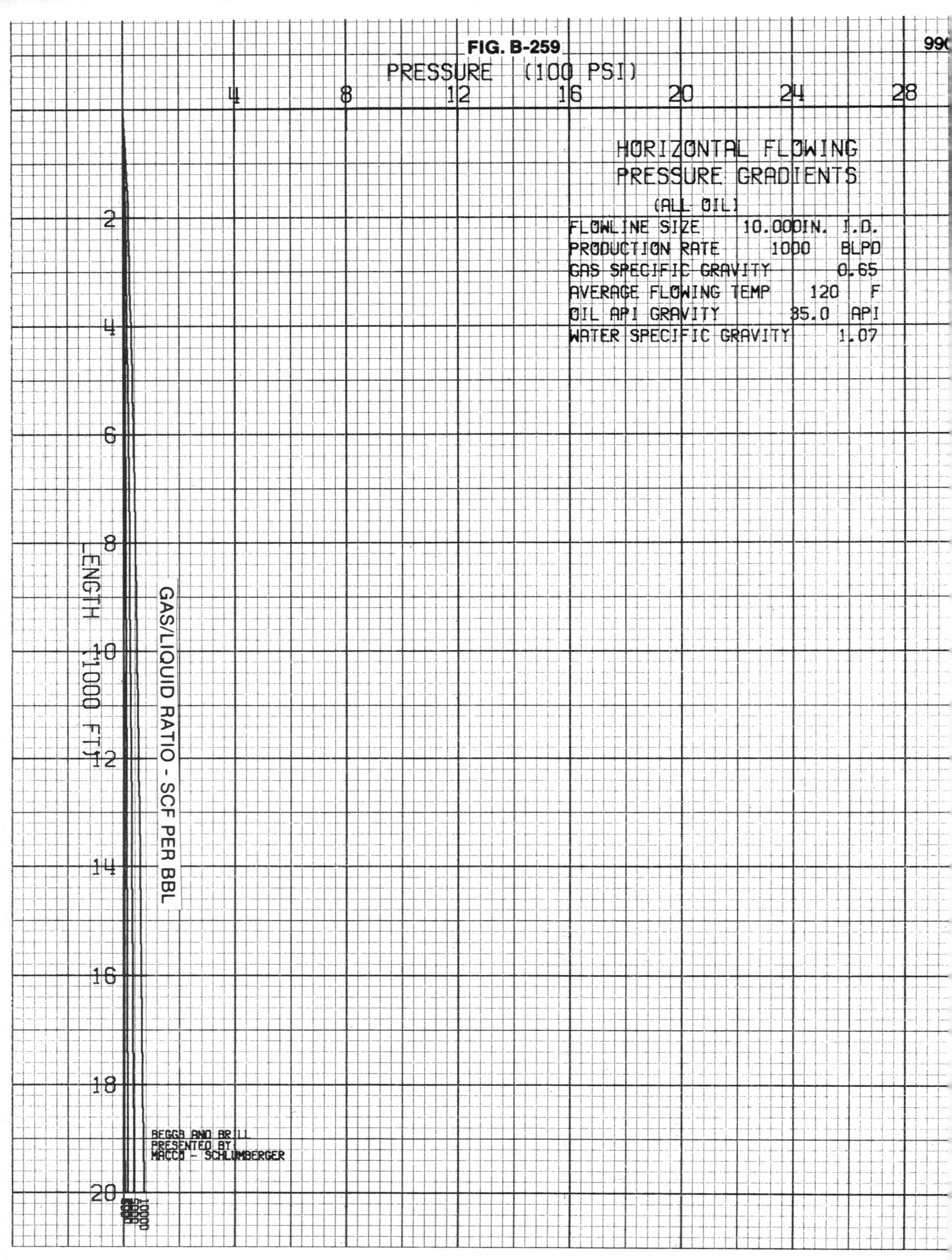

FIG. B-259

990

PRESSURE (100 PSI)

HORIZONTAL FLOWING
PRESSURE GRADIENTS
(ALL OIL)

FLOWLINE SIZE	10.000 IN. I.D.
PRODUCTION RATE	1000 BLPD
GAS SPECIFIC GRAVITY	0.65
AVERAGE FLOWING TEMP	120 F
OIL API GRAVITY	35.0 API
WATER SPECIFIC GRAVITY	1.07

LENGTH (1000 FT)

GAS/LIQUID RATIO - SCF PER BBL

BEGGS AND BRILL
PRESENTED BY
MACCO - SCHLUMBERGER

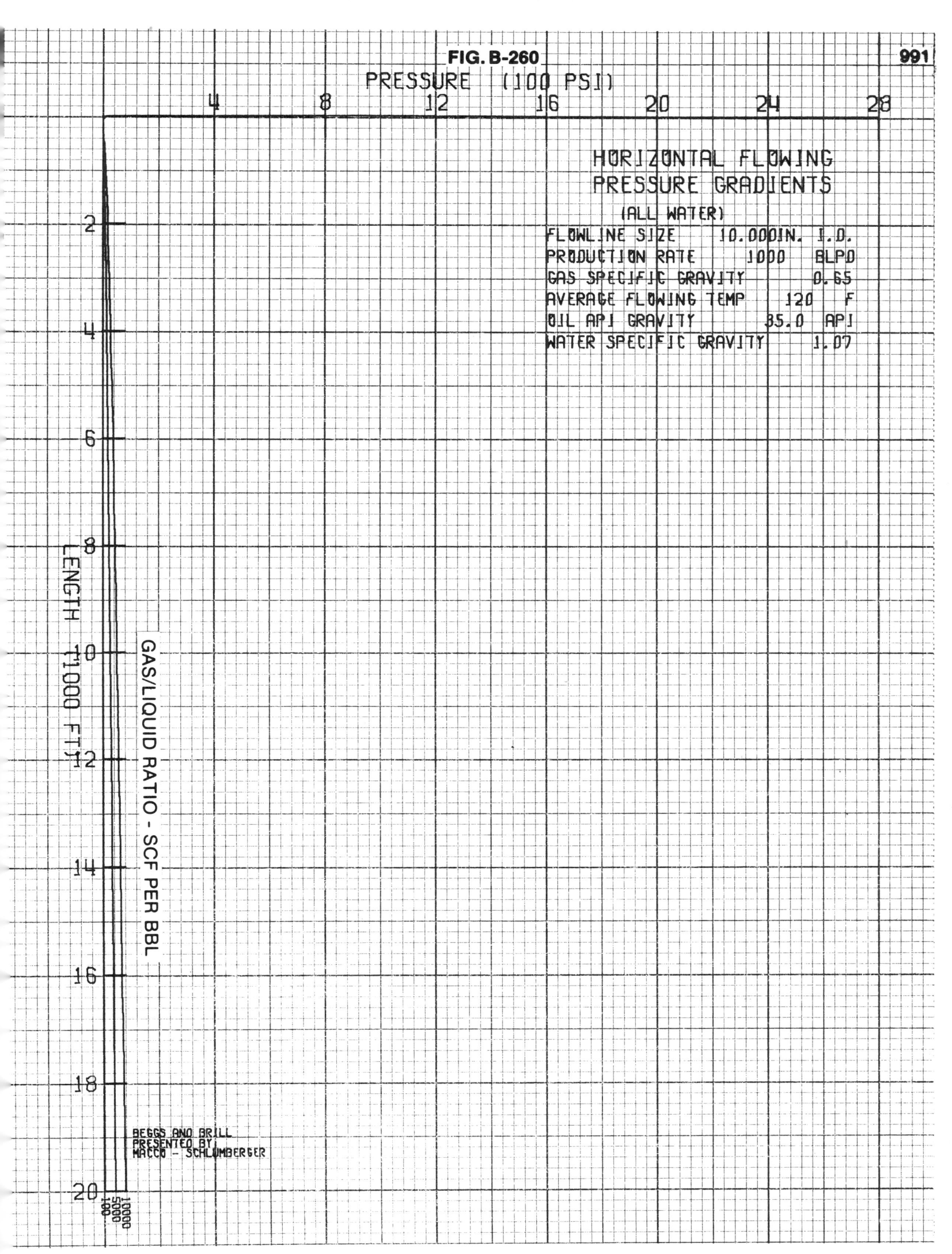

PRESSURE (100 PSI)

HORIZONTAL FLOWING
PRESSURE GRADIENTS
(ALL OIL)

FLOWLINE SIZE	10.000IN.	I.D.
PRODUCTION RATE	1500	BLPD
GAS SPECIFIC GRAVITY	0.65	
AVERAGE FLOWING TEMP	120	F
OIL API GRAVITY	35.0	API
WATER SPECIFIC GRAVITY	1.07	

LENGTH (1000 FT)

GAS/LIQUID RATIO - SCF PER BBL

BEGGS AND BRILL
PRESENTED BY
NACCO - SCHLUMBERGER

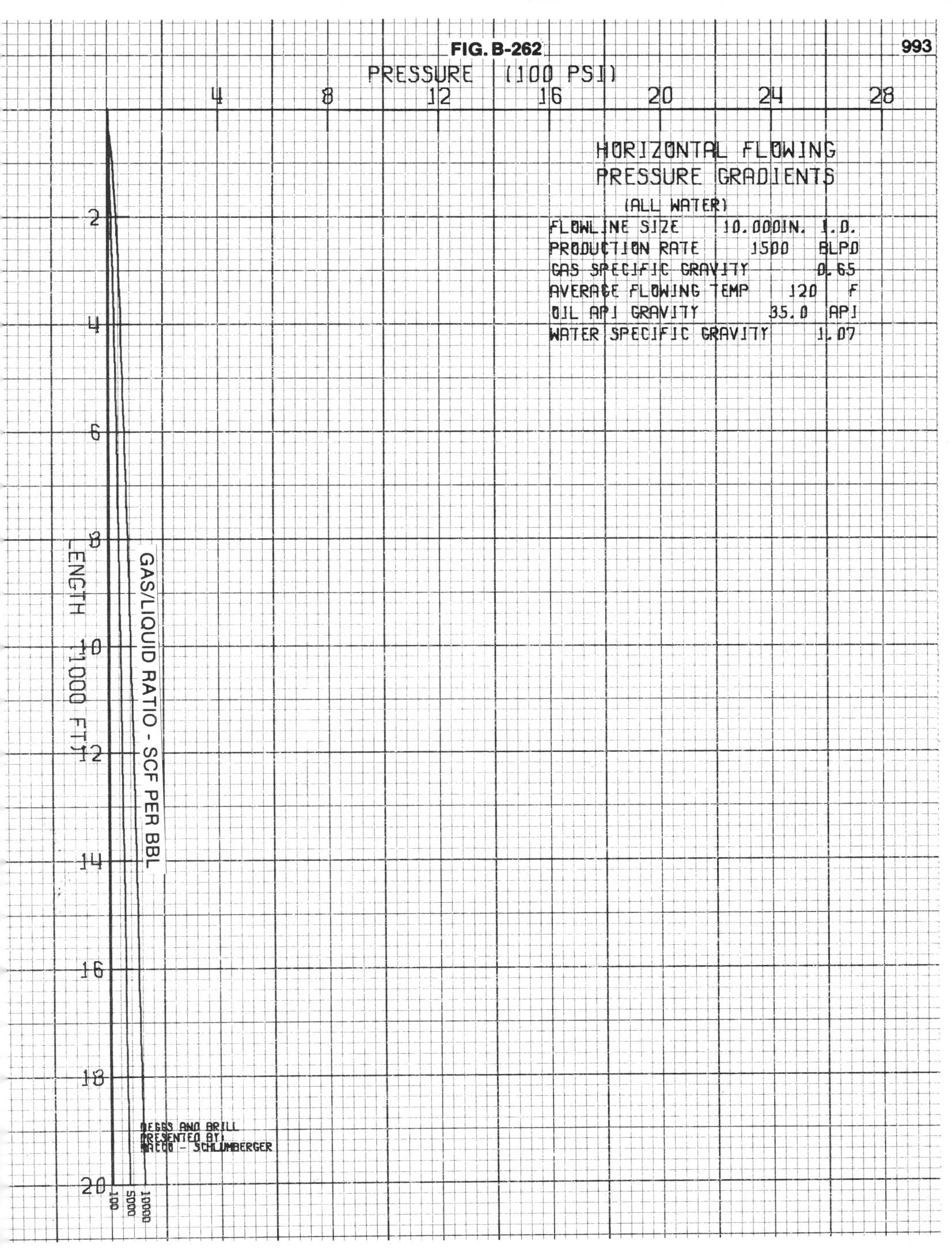

FIG. B-262

993

PRESSURE (100 PSI)

HORIZONTAL FLOWING
PRESSURE GRADIENTS
(ALL WATER)

FLOWLINE SIZE	10.000IN.	I.D.
PRODUCTION RATE	1500	BLPD
GAS SPECIFIC GRAVITY	0.65	
AVERAGE FLOWING TEMP	120	F
OIL API GRAVITY	35.0	API
WATER SPECIFIC GRAVITY	1.07	

LENGTH (1000 FT)

GAS/LIQUID RATIO - SCF PER BBL

BEGGS AND BRILL
PRESENTED BY:
MAECO - SCHLUMBERGER

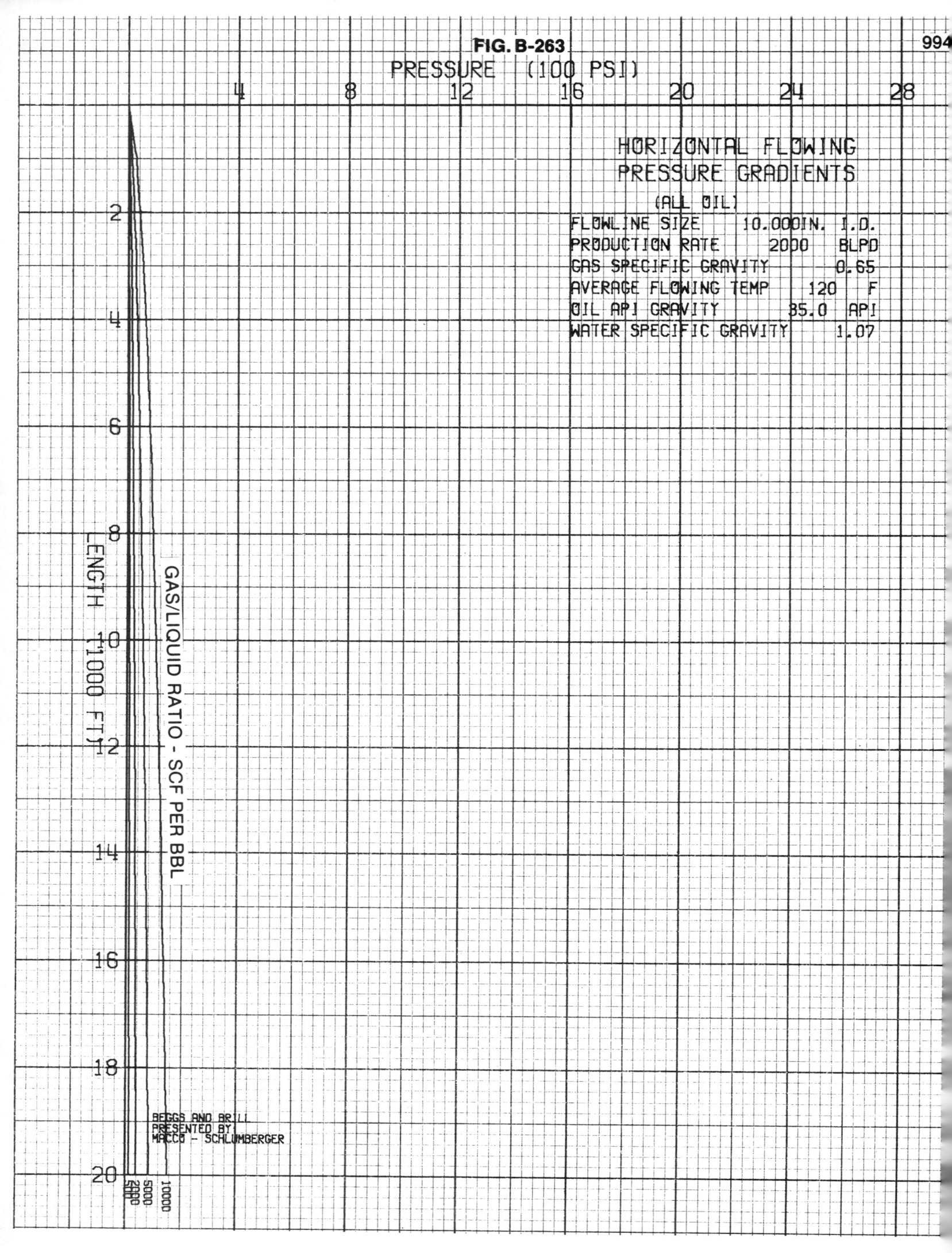

FIG. B-263

994

HORIZONTAL FLOWING
PRESSURE GRADIENTS
(ALL OIL)

FLOWLINE SIZE   10.000IN. I.D.
PRODUCTION RATE   2000   BLPD
GAS SPECIFIC GRAVITY   0.65
AVERAGE FLOWING TEMP   120   F
OIL API GRAVITY   35.0   API
WATER SPECIFIC GRAVITY   1.07

PRESSURE (100 PSI)

LENGTH (1000 FT)

GAS/LIQUID RATIO - SCF PER BBL

BEGGS AND BRILL
PRESENTED BY
MACCO - SCHLUMBERGER

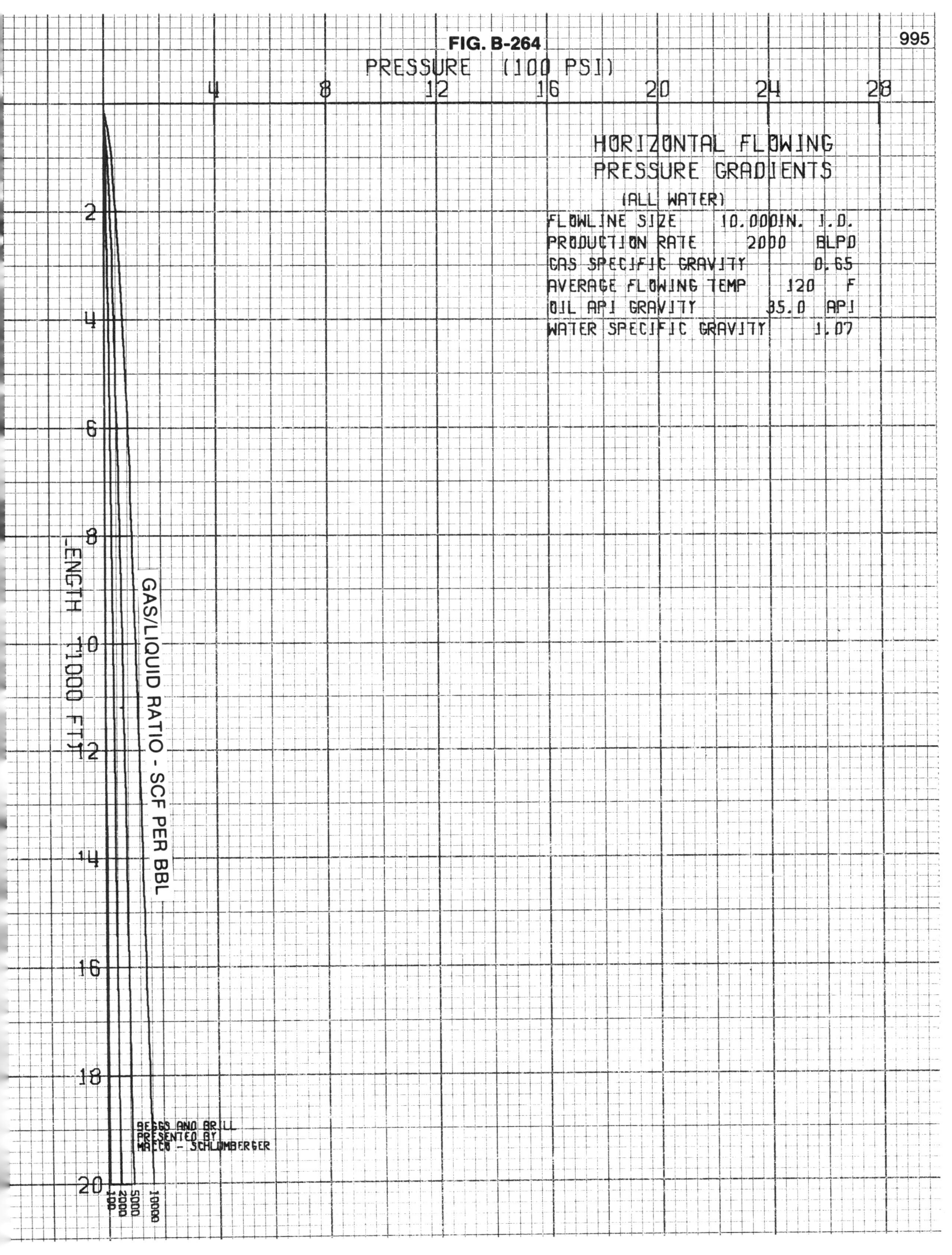

PRESSURE (100 PSI)

HORIZONTAL FLOWING
PRESSURE GRADIENTS
(ALL WATER)

FLOWLINE SIZE   10.000IN. I.D.
PRODUCTION RATE   2000 BLPD
GAS SPECIFIC GRAVITY   0.65
AVERAGE FLOWING TEMP   120   F
OIL API GRAVITY   35.0 API
WATER SPECIFIC GRAVITY   1.07

LENGTH (1000 FT.)

GAS/LIQUID RATIO - SCF PER BBL

BEGGS AND BRILL
PRESENTED BY
MACCO - SCHLUMBERGER

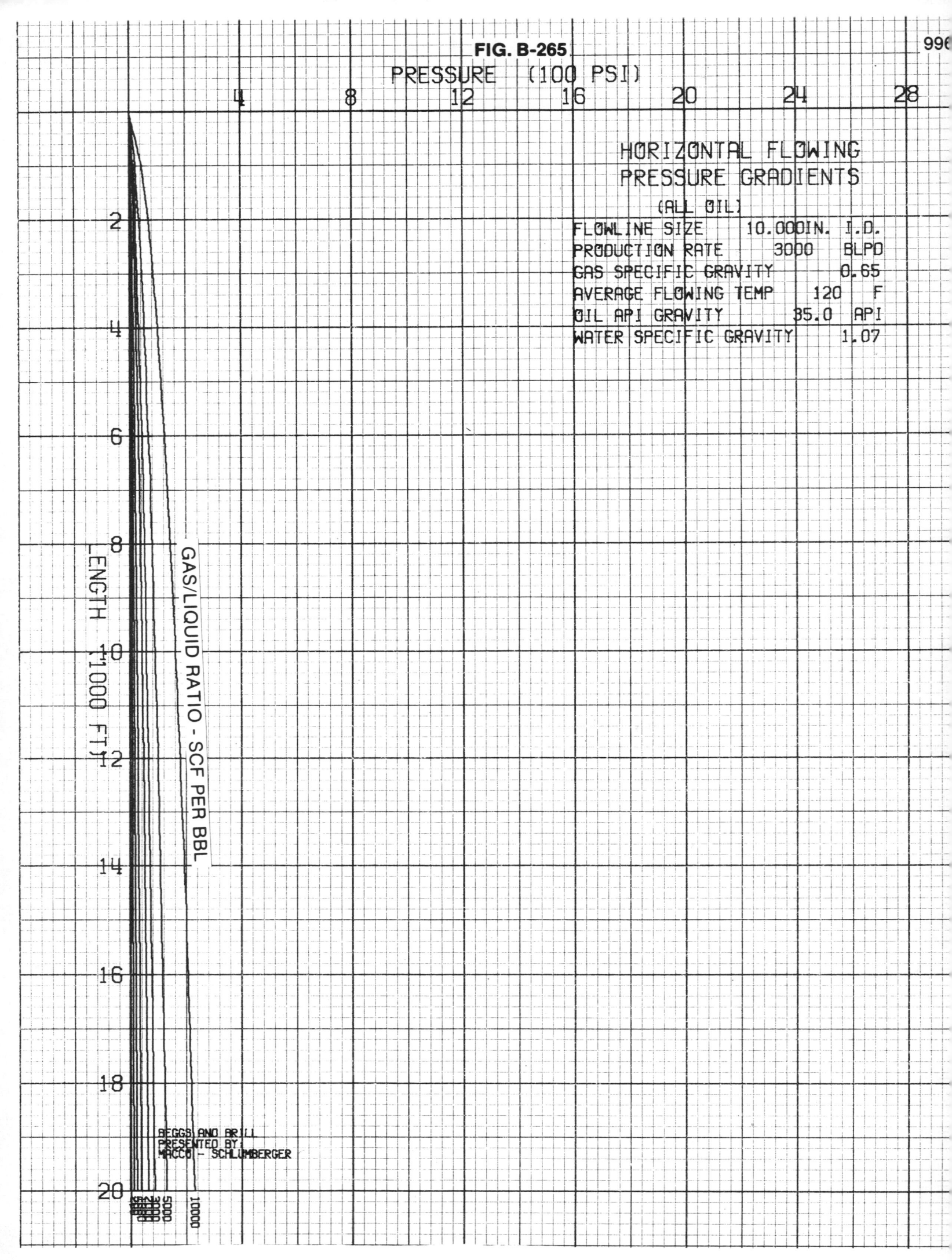

FIG. B-265

PRESSURE (100 PSI)

HORIZONTAL FLOWING
PRESSURE GRADIENTS
(ALL OIL)

FLOWLINE SIZE	10.000 IN. I.D.
PRODUCTION RATE	3000 BLPD
GAS SPECIFIC GRAVITY	0.65
AVERAGE FLOWING TEMP	120 F
OIL API GRAVITY	35.0 API
WATER SPECIFIC GRAVITY	1.07

GAS/LIQUID RATIO - SCF PER BBL

LENGTH (1000 FT)

BEGGS AND BRILL
PRESENTED BY:
MACCO - SCHLUMBERGER

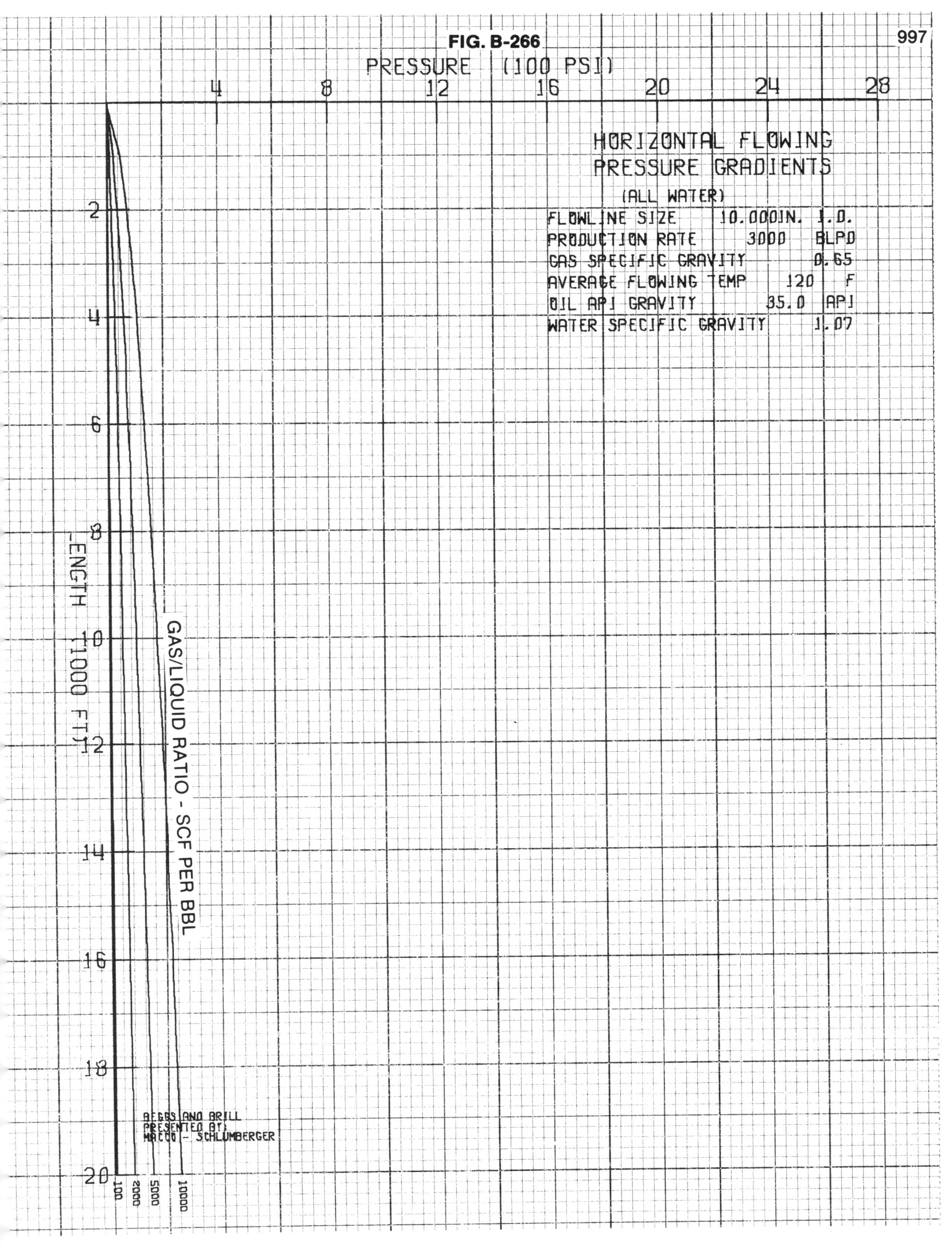

PRESSURE (100 PSI)

HORIZONTAL FLOWING
PRESSURE GRADIENTS
(ALL WATER)

FLOWLINE SIZE	10.000 IN. I.D.
PRODUCTION RATE	3000 BLPD
GAS SPECIFIC GRAVITY	0.65
AVERAGE FLOWING TEMP	120 F
OIL API GRAVITY	35.0 API
WATER SPECIFIC GRAVITY	1.07

LENGTH (1000 FT.)

GAS/LIQUID RATIO - SCF PER BBL

BEGGS AND BRILL
PRESENTED BY:
MACCO - SCHLUMBERGER

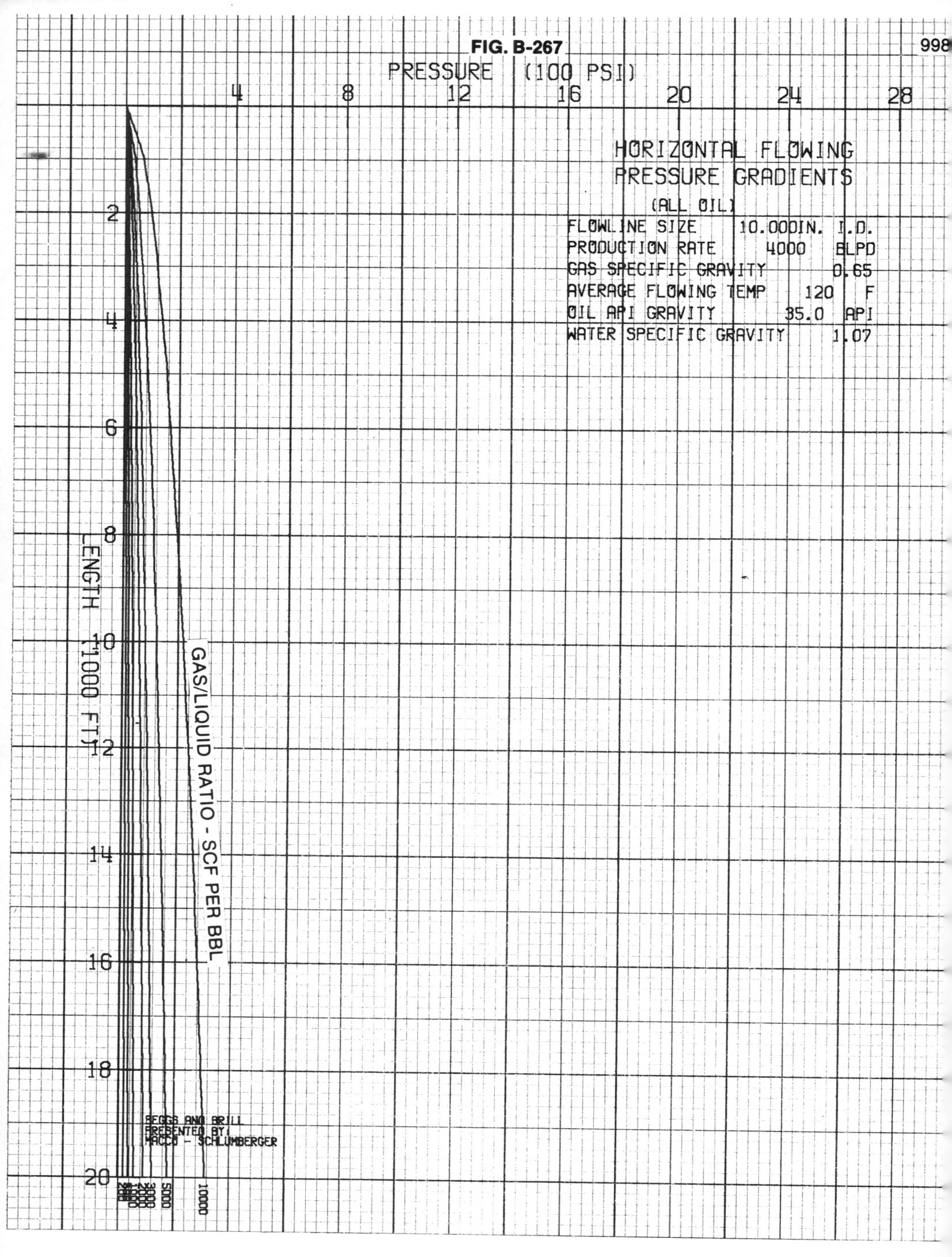

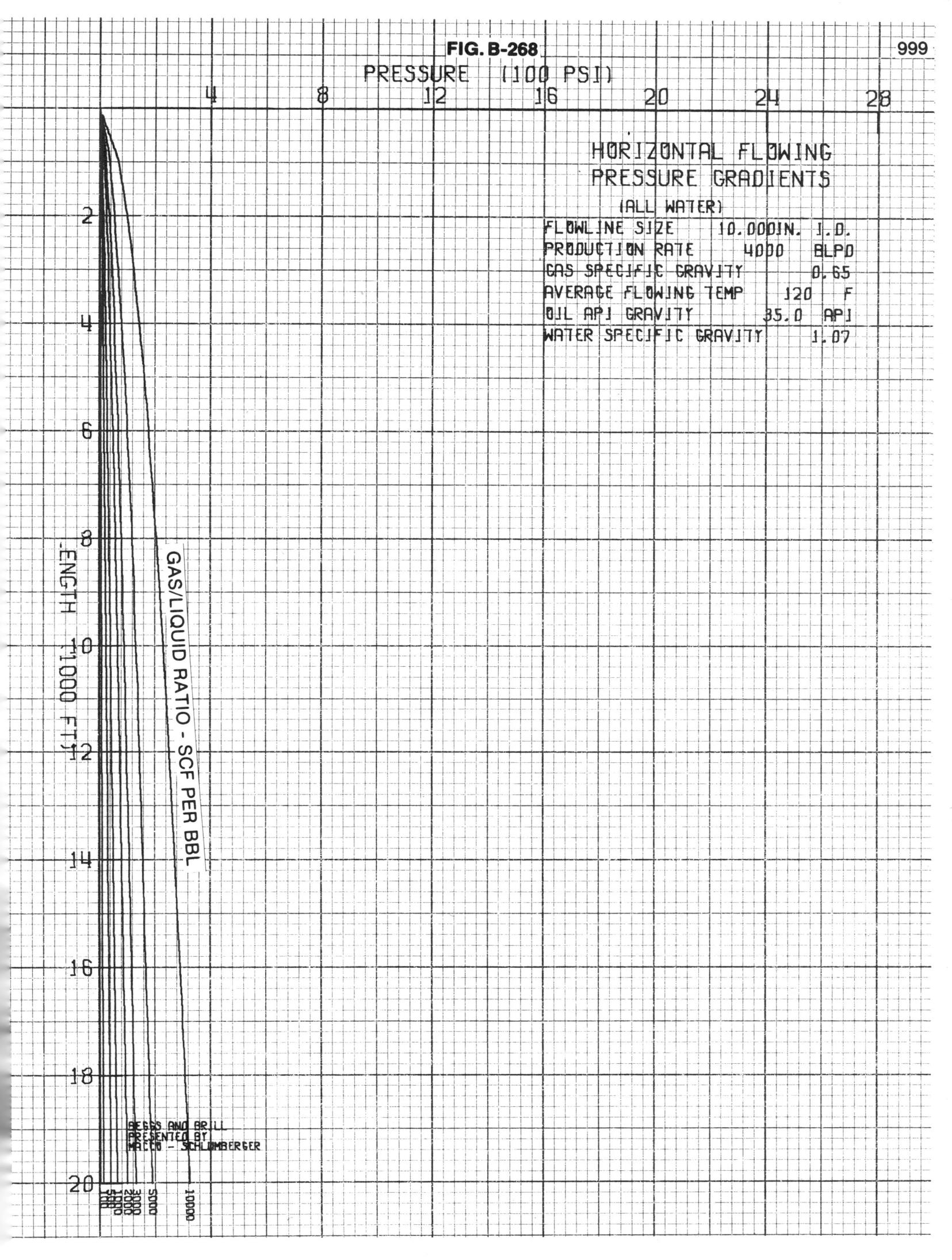

FIG. B-268

999

PRESSURE (100 PSI)

HORIZONTAL FLOWING
PRESSURE GRADIENTS
(ALL WATER)

FLOWLINE SIZE   10.000IN. I.D.
PRODUCTION RATE   4000   BLPD
GAS SPECIFIC GRAVITY   0.65
AVERAGE FLOWING TEMP   120   F
OIL API GRAVITY   35.0   API
WATER SPECIFIC GRAVITY   1.07

LENGTH (1000 FT)

GAS/LIQUID RATIO - SCF PER BBL

BEGGS AND BRILL
PRESENTED BY
MAECO - SCHLUMBERGER

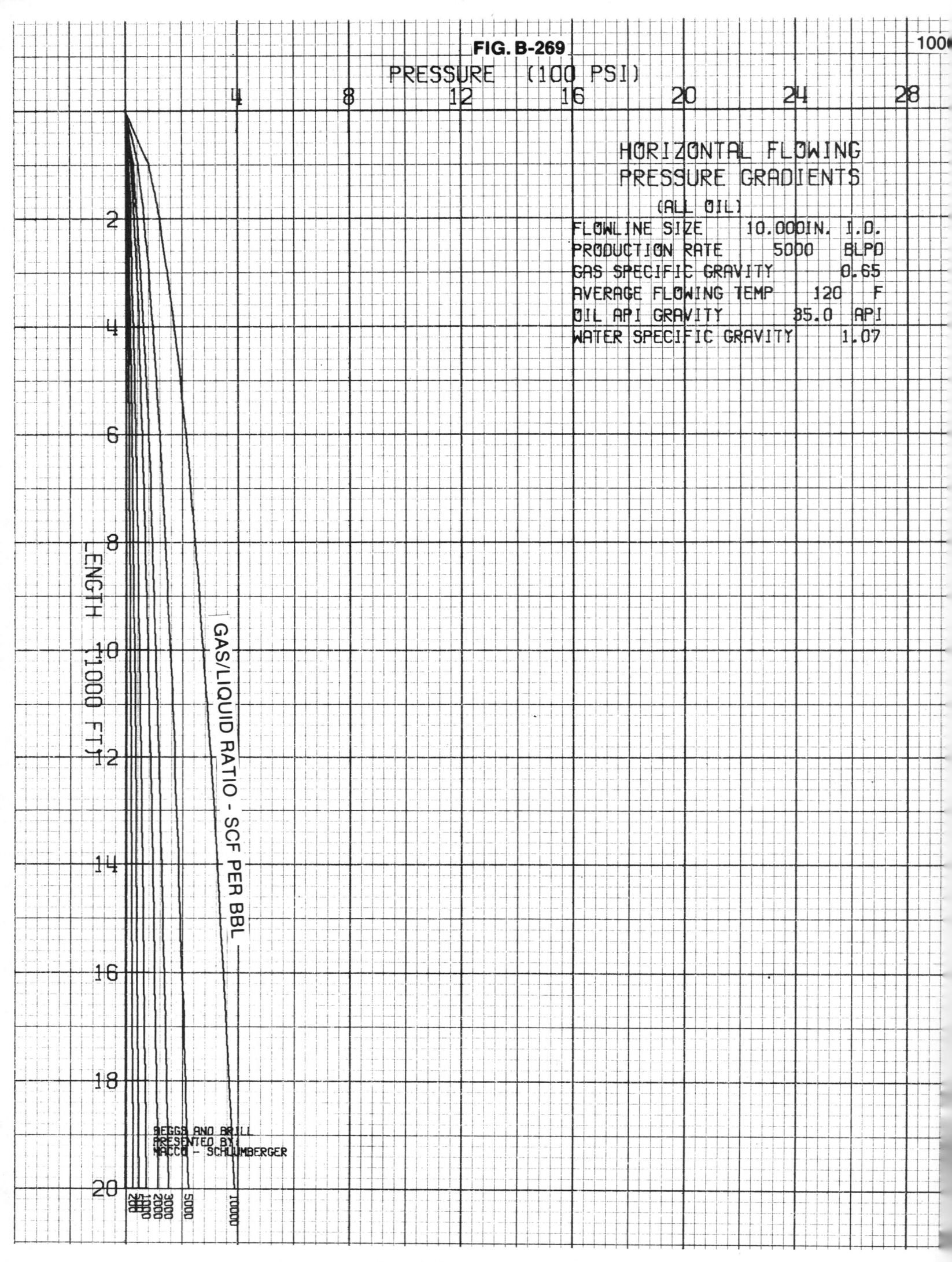

FIG. B-269

HORIZONTAL FLOWING
PRESSURE GRADIENTS
(ALL OIL)

FLOWLINE SIZE	10.000IN. I.D.
PRODUCTION RATE	5000 BLPD
GAS SPECIFIC GRAVITY	0.65
AVERAGE FLOWING TEMP	120 F
OIL API GRAVITY	35.0 API
WATER SPECIFIC GRAVITY	1.07

PRESSURE (100 PSI)

LENGTH (1000 FT.)

GAS/LIQUID RATIO - SCF PER BBL

BEGGS AND BRILL
PRESENTED BY
NACCO - SCHLUMBERGER

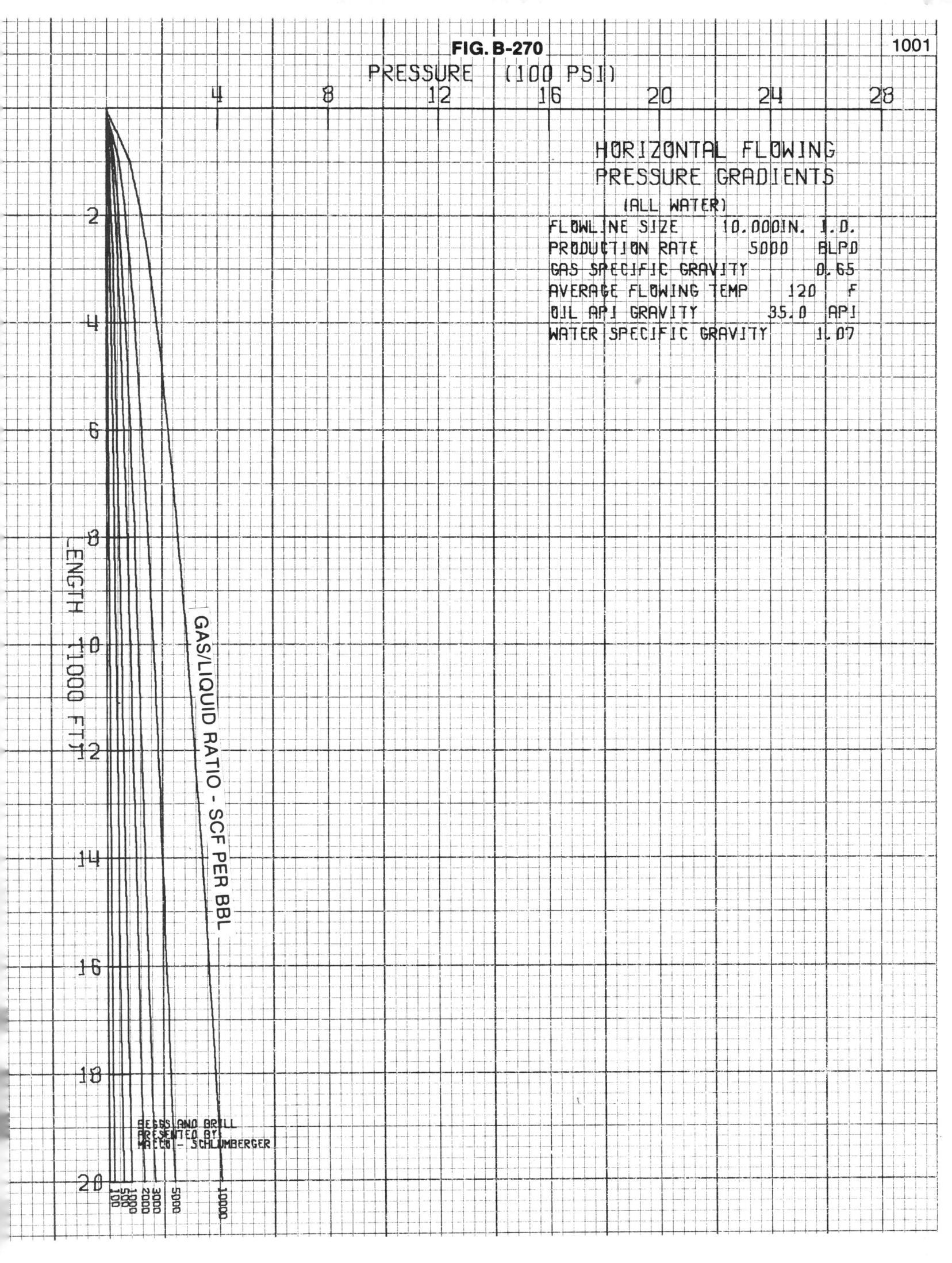

FIG. B-270

1001

PRESSURE (100 PSI)

HORIZONTAL FLOWING
PRESSURE GRADIENTS
(ALL WATER)

FLOWLINE SIZE	10.000IN. I.D.
PRODUCTION RATE	5000 BLPD
GAS SPECIFIC GRAVITY	0.65
AVERAGE FLOWING TEMP	120 F
OIL API GRAVITY	35.0 API
WATER SPECIFIC GRAVITY	1.07

LENGTH (1000 FT)

GAS/LIQUID RATIO - SCF PER BBL

BEGGS AND BRILL
PRESENTED BY
MAECO - SCHLUMBERGER

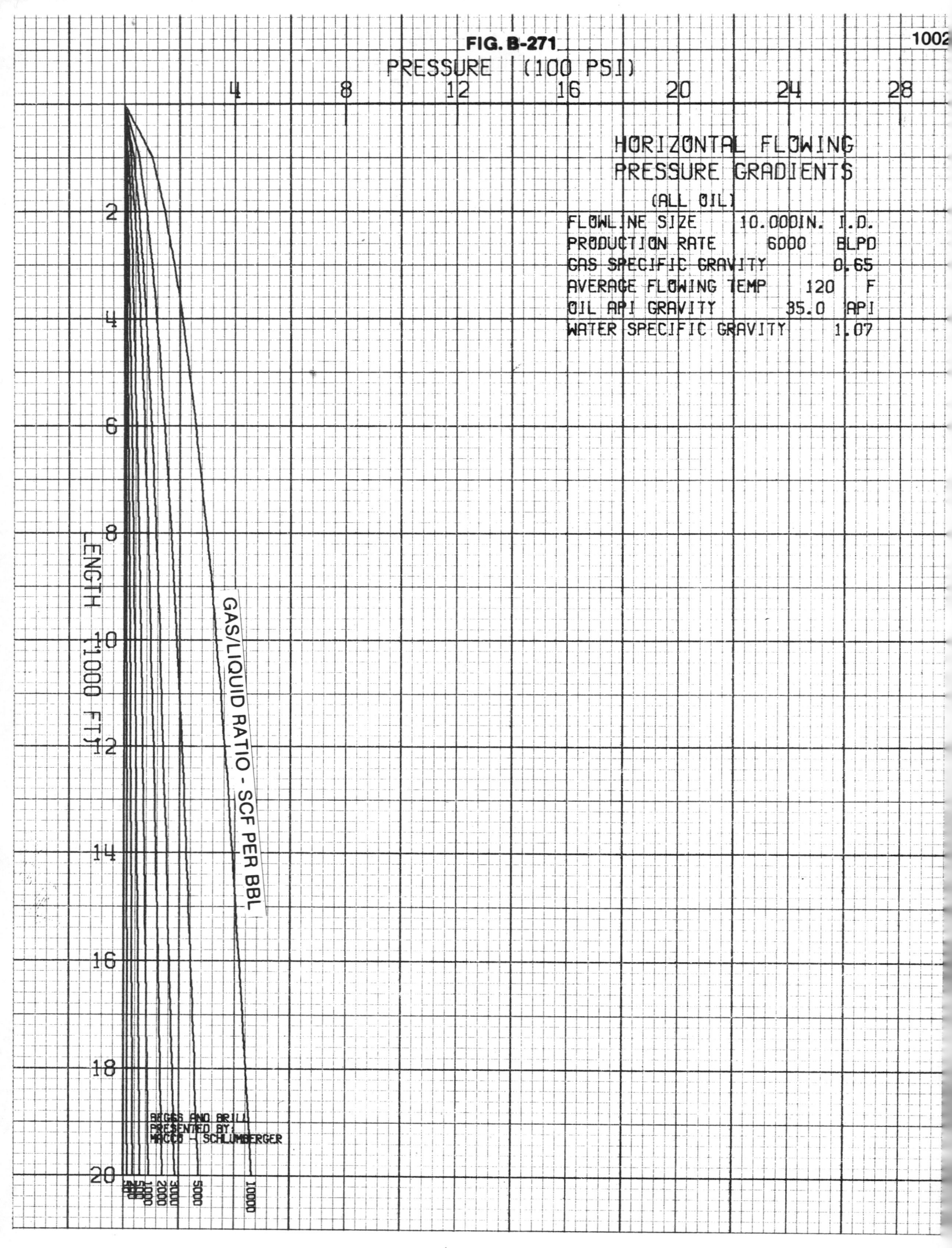

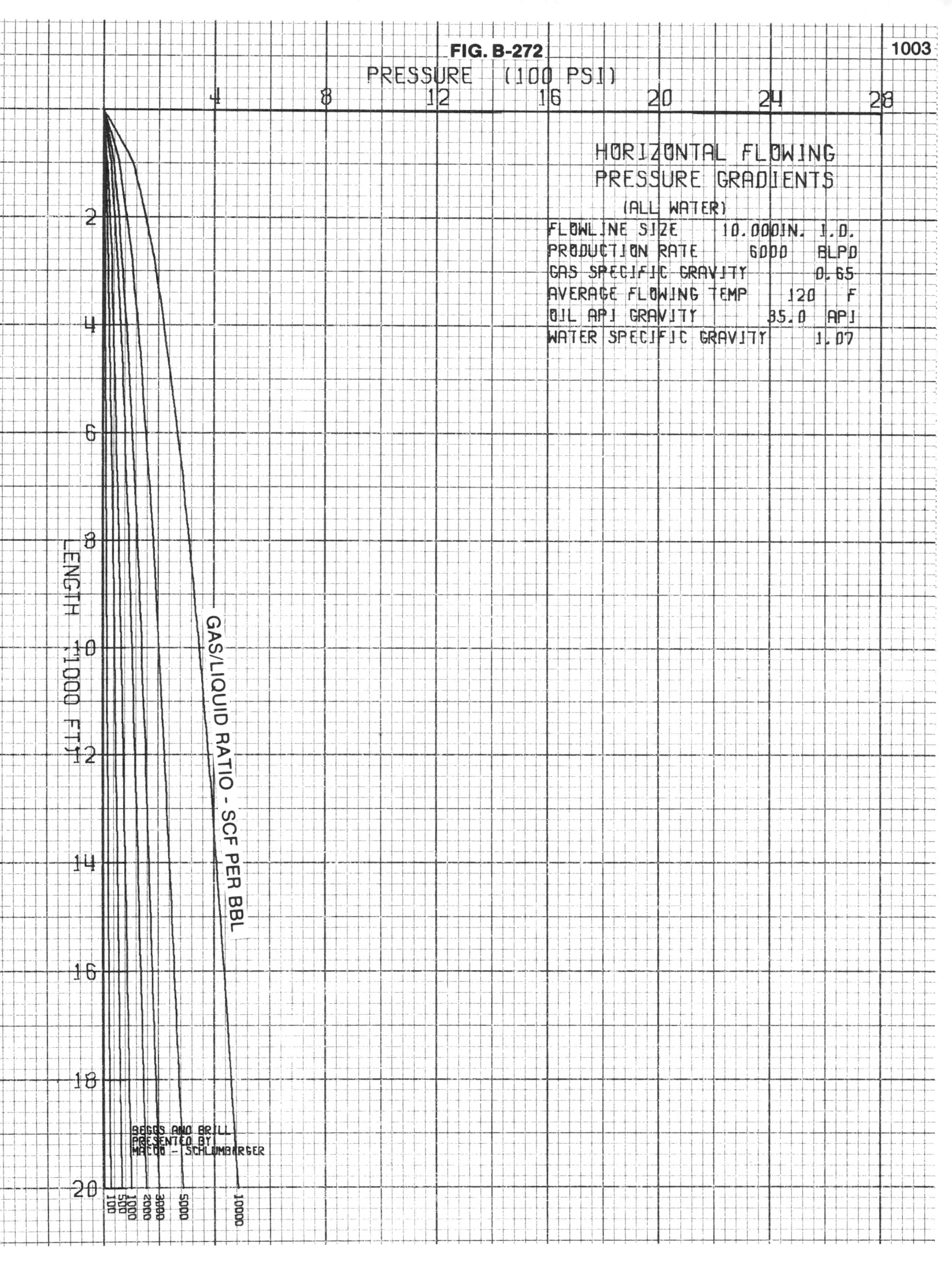

PRESSURE (100 PSI)

HORIZONTAL FLOWING
PRESSURE GRADIENTS
(ALL WATER)

FLOWLINE SIZE	10.000IN.	I.D.
PRODUCTION RATE	6000	BLPD
GAS SPECIFIC GRAVITY	0.65	
AVERAGE FLOWING TEMP	120	F
OIL API GRAVITY	35.0	API
WATER SPECIFIC GRAVITY	1.07	

LENGTH (1000 FT.)

GAS/LIQUID RATIO - SCF PER BBL

BEGGS AND BRILL
PRESENTED BY
MACCO - SCHLUMBERGER

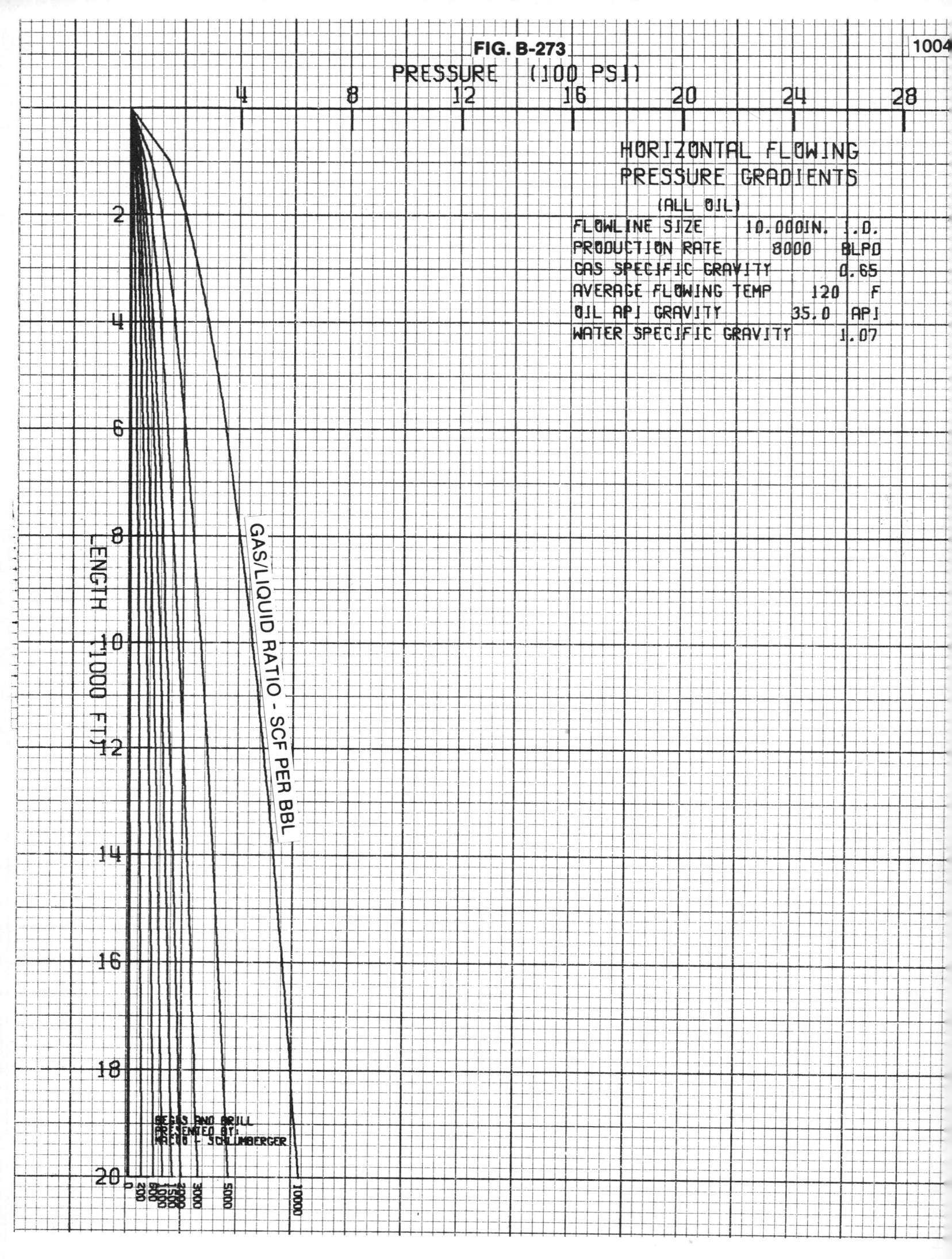

FIG. B-273

PRESSURE (100 PSI)

1004

HORIZONTAL FLOWING
PRESSURE GRADIENTS
(ALL OIL)

FLOWLINE SIZE   10.000IN. I.D.
PRODUCTION RATE   8000 BLPD
GAS SPECIFIC GRAVITY   0.65
AVERAGE FLOWING TEMP   120   F
OIL API GRAVITY   35.0 API
WATER SPECIFIC GRAVITY   1.07

GAS/LIQUID RATIO - SCF PER BBL

LENGTH (1000 FT)

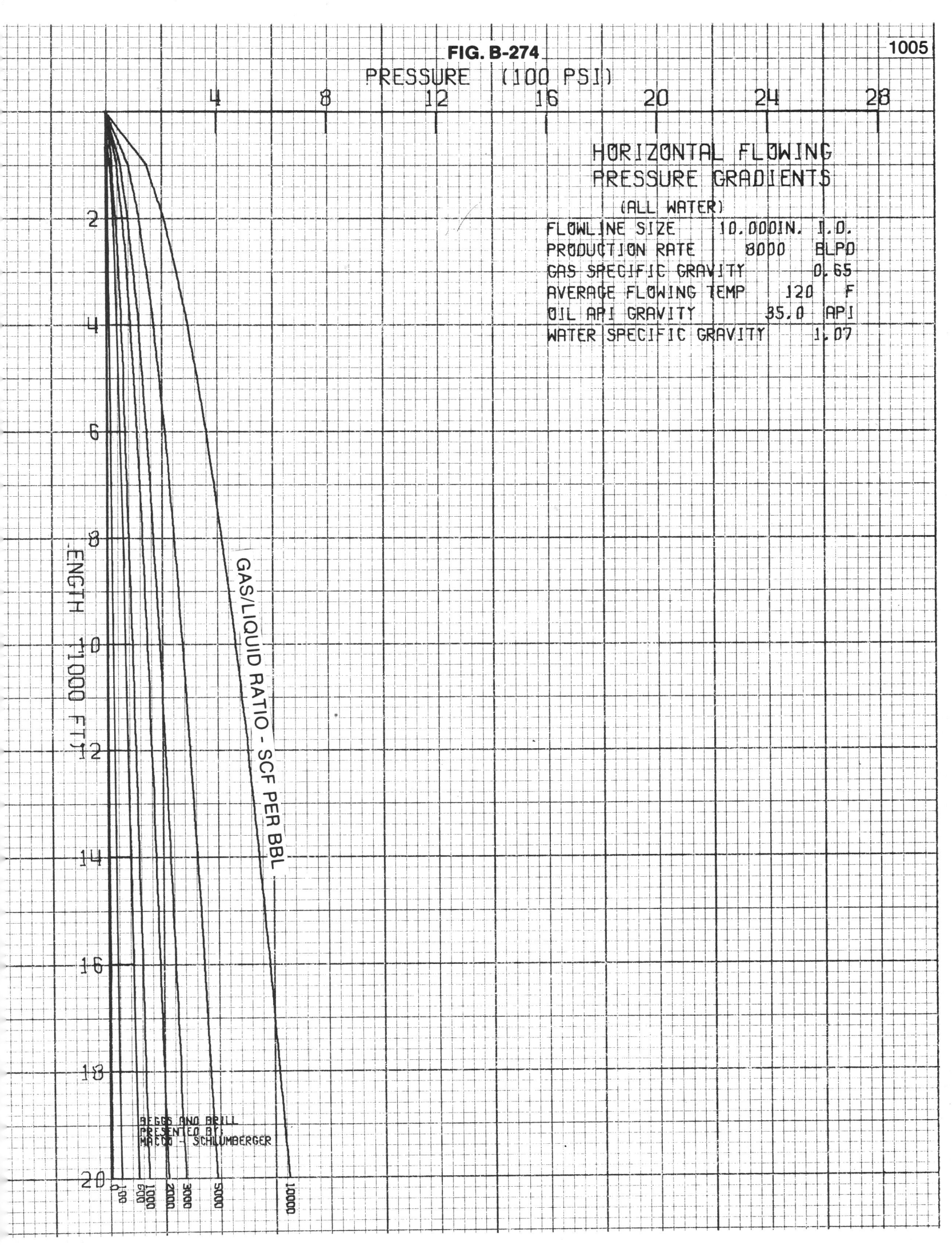

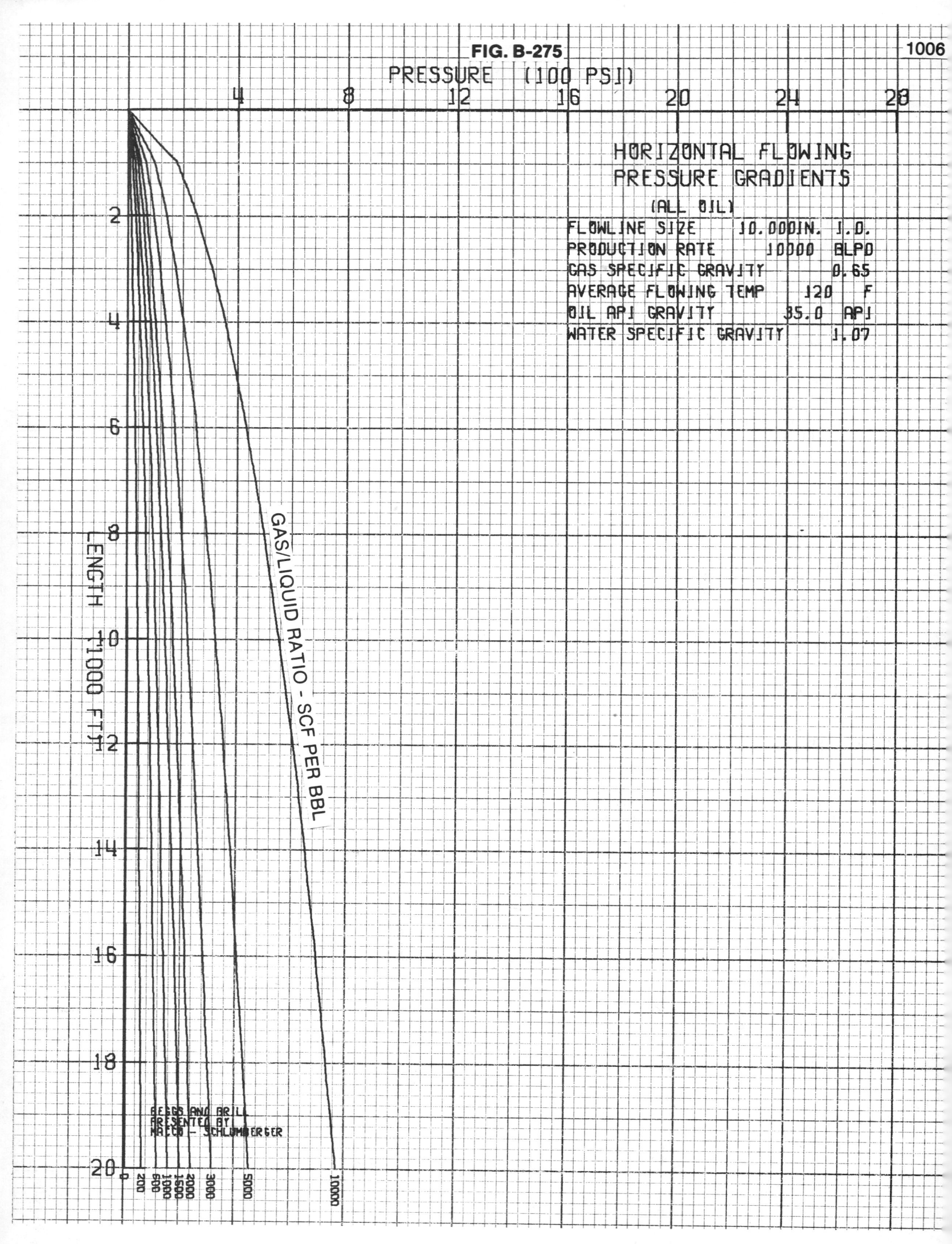

PRESSURE (100 PSI)

HORIZONTAL FLOWING
PRESSURE GRADIENTS
(ALL OIL)

FLOWLINE SIZE	10.000IN. I.D.
PRODUCTION RATE	10000 BLPD
GAS SPECIFIC GRAVITY	0.65
AVERAGE FLOWING TEMP	120 F
OIL API GRAVITY	35.0 API
WATER SPECIFIC GRAVITY	1.07

LENGTH 1000 FT.

GAS/LIQUID RATIO - SCF PER BBL

BEGGS AND BRILL
PRESENTED BY
HALCO - SCHLUMBERGER

200 600 1000 1500 2000 3000 5000 10000

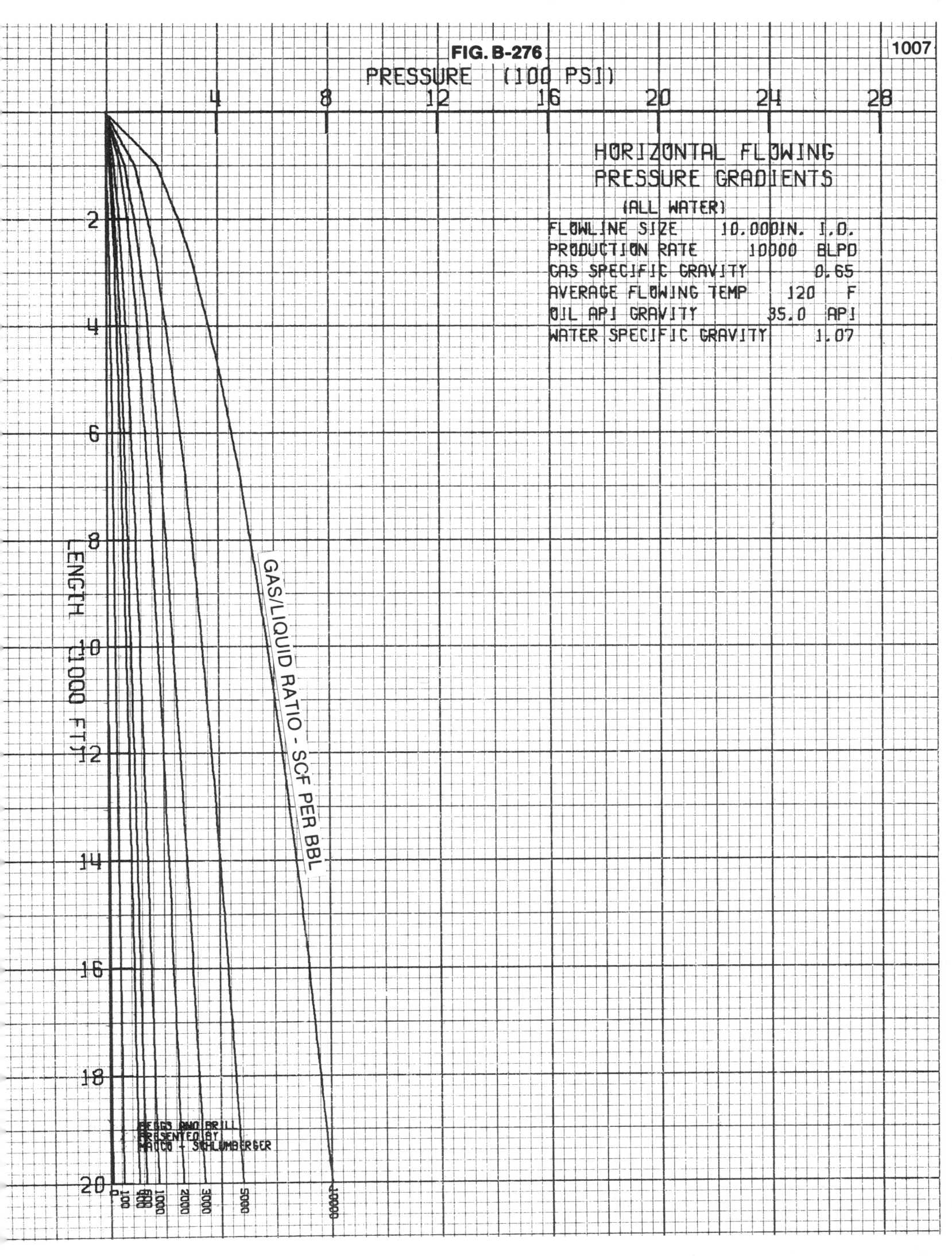

FIG. B-276

1007

PRESSURE (100 PSI)

HORIZONTAL FLOWING
PRESSURE GRADIENTS
(ALL WATER)

FLOWLINE SIZE   10.000IN. I.D.
PRODUCTION RATE   10000 BLPD
GAS SPECIFIC GRAVITY   0.65
AVERAGE FLOWING TEMP   120 F
OIL API GRAVITY   35.0 API
WATER SPECIFIC GRAVITY   1.07

LENGTH (1000 FT)

GAS/LIQUID RATIO - SCf PER BBL

BEGGS AND BRILL
PRESENTED BY
MATCO + SCHLUMBERGER

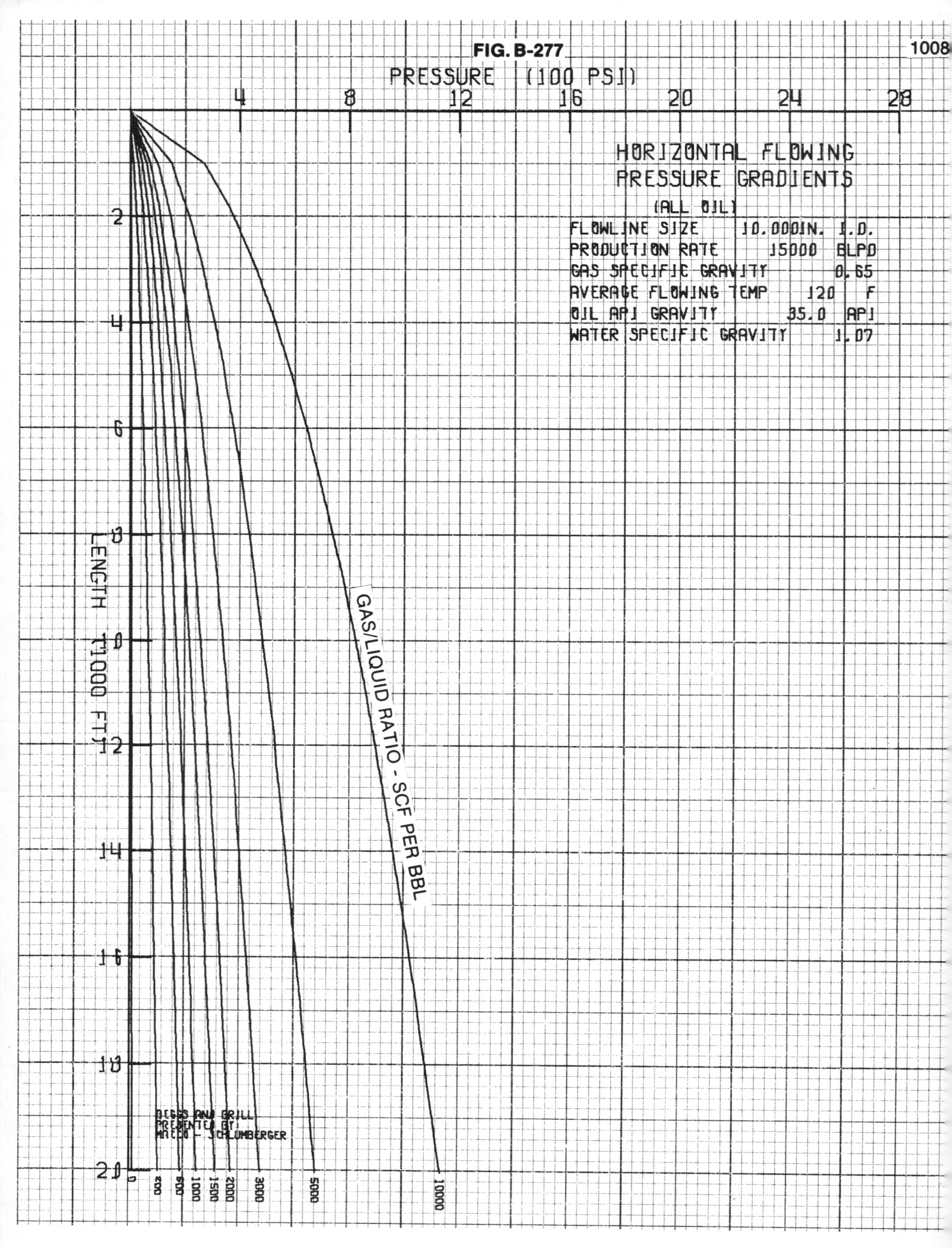

FIG. B-277

1008

PRESSURE (100 PSI)

HORIZONTAL FLOWING
PRESSURE GRADIENTS
(ALL OIL)

FLOWLINE SIZE   10.000IN. I.D.
PRODUCTION RATE   15000 BLPD
GAS SPECIFIC GRAVITY   0.65
AVERAGE FLOWING TEMP   120 F
OIL API GRAVITY   35.0 API
WATER SPECIFIC GRAVITY   1.07

LENGTH (1000 FT)

GAS/LIQUID RATIO - SCF PER BBL

BEGGS AND BRILL
PRESENTED BY
MADCO - SCHLUMBERGER

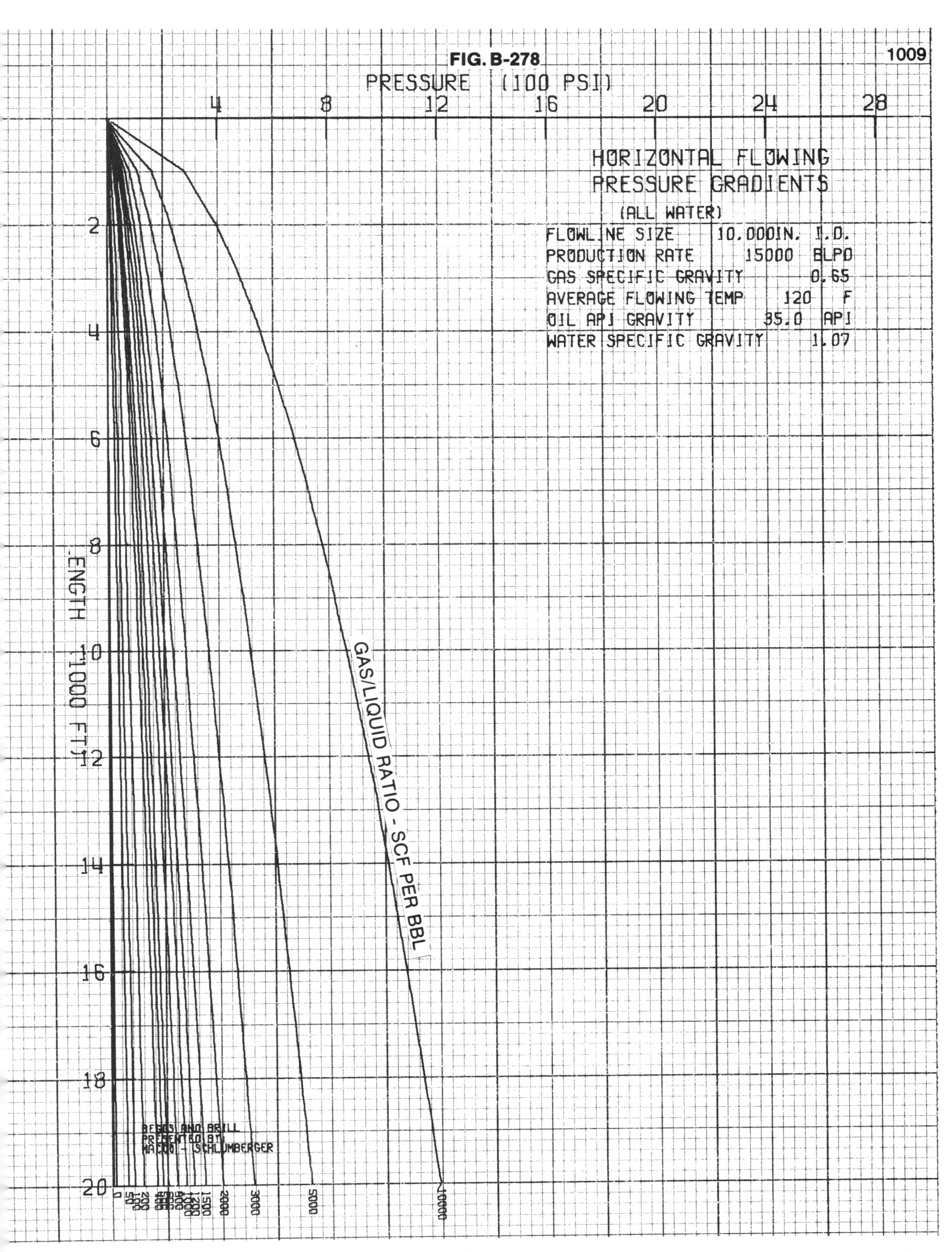

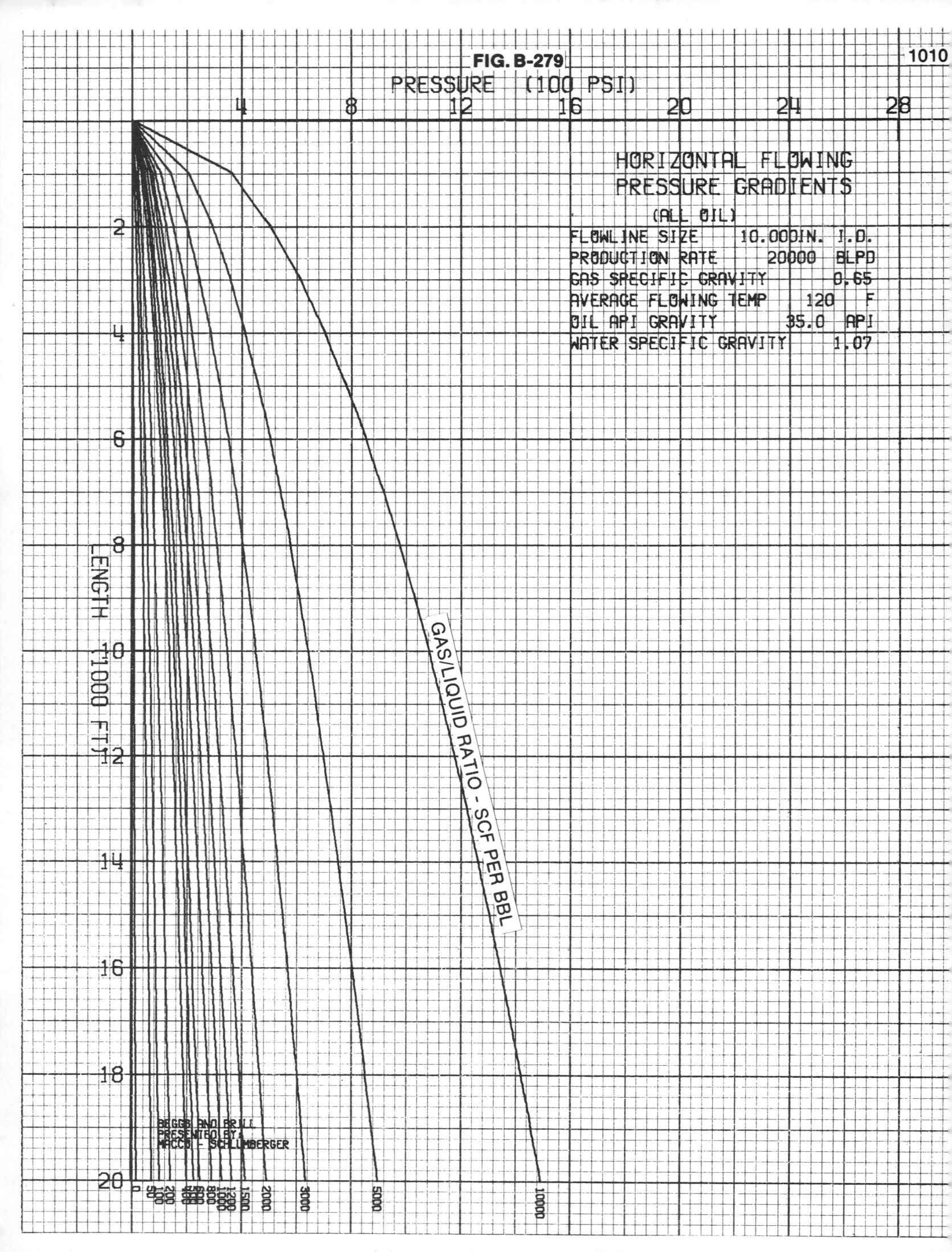

FIG. B-279

1010

PRESSURE (100 PSI)

HORIZONTAL FLOWING PRESSURE GRADIENTS
(ALL OIL)

FLOWLINE SIZE	10.000 IN. I.D.
PRODUCTION RATE	20000 BLPD
GAS SPECIFIC GRAVITY	0.65
AVERAGE FLOWING TEMP	120 F
OIL API GRAVITY	35.0 API
WATER SPECIFIC GRAVITY	1.07

LENGTH (1000 FT)

GAS/LIQUID RATIO - SCF PER BBL

BRIGGS AND BRILL
PRESENTED BY:
MACCO - SCHLUMBERGER

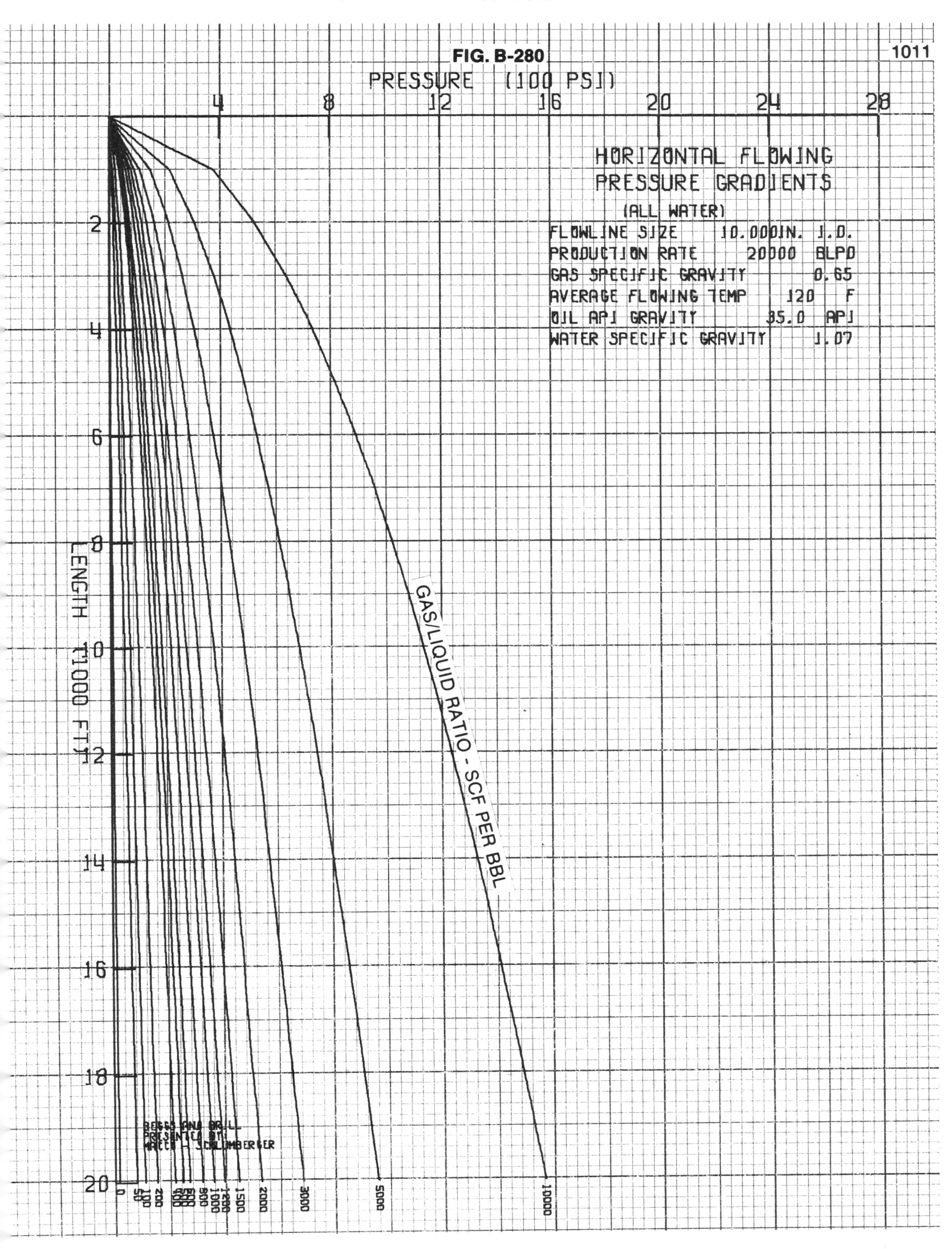

FIG. B-280

1011

PRESSURE (100 PSI)

HORIZONTAL FLOWING
PRESSURE GRADIENTS
(ALL WATER)

FLOWLINE SIZE   10.000IN. I.D.
PRODUCTION RATE   20000 BLPD
GAS SPECIFIC GRAVITY   0.65
AVERAGE FLOWING TEMP   120   F
OIL API GRAVITY   35.0 API
WATER SPECIFIC GRAVITY   1.07

LENGTH (1000 FT)

GAS/LIQUID RATIO - SCF PER BBL

BEGGS AND BRILL
PRESENTED BY
WATER - SCHLUMBERGER

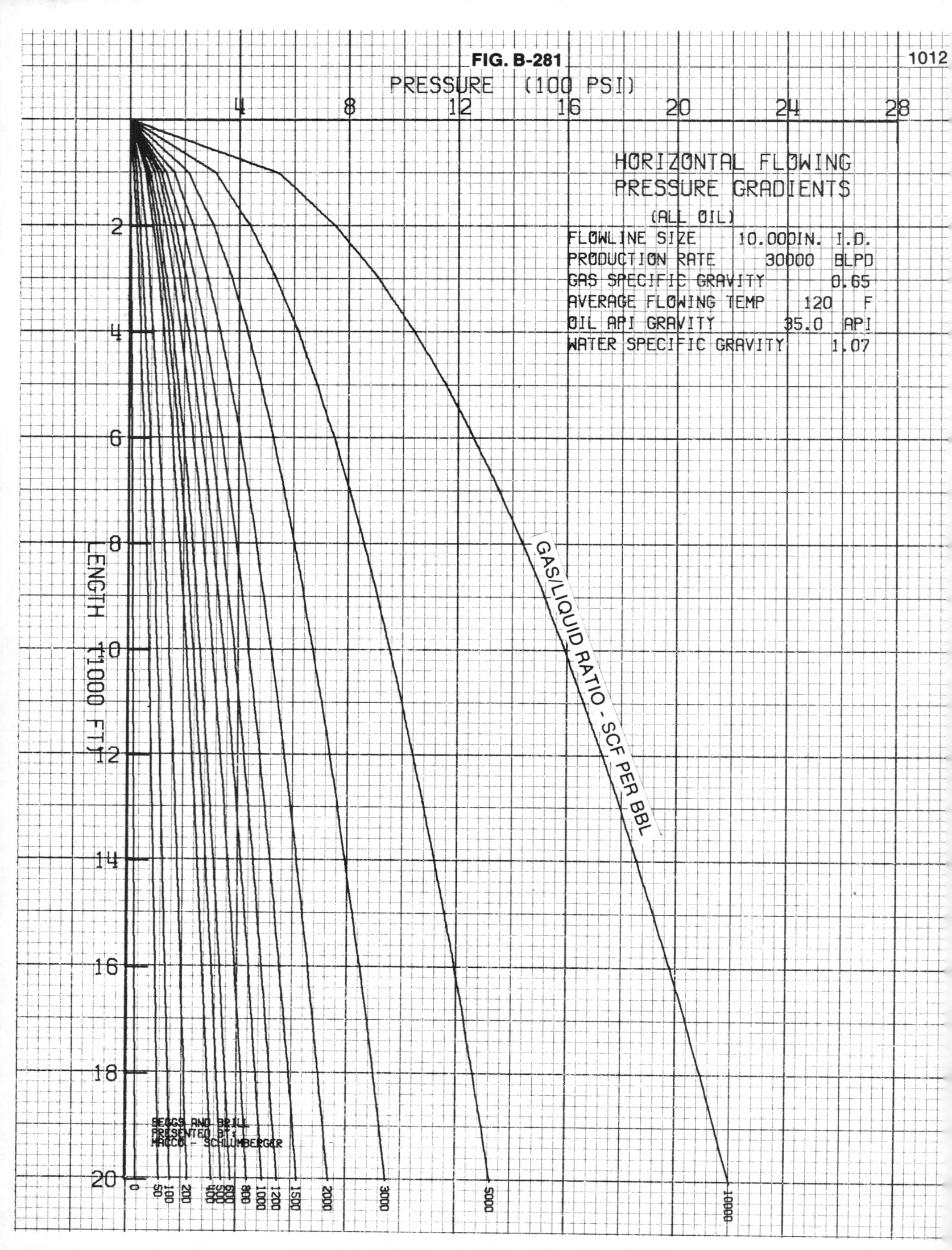

FIG. B-281

PRESSURE (100 PSI)

HORIZONTAL FLOWING
PRESSURE GRADIENTS
(ALL OIL)

FLOWLINE SIZE	10.000IN. I.D.
PRODUCTION RATE	30000 BLPD
GAS SPECIFIC GRAVITY	0.65
AVERAGE FLOWING TEMP	120 F
OIL API GRAVITY	35.0 API
WATER SPECIFIC GRAVITY	1.07

GAS/LIQUID RATIO - SCF PER BBL

LENGTH (1000 FT)

BEGGS AND BRILL
PRESENTED BY:
MACCO - SCHLUMBERGER

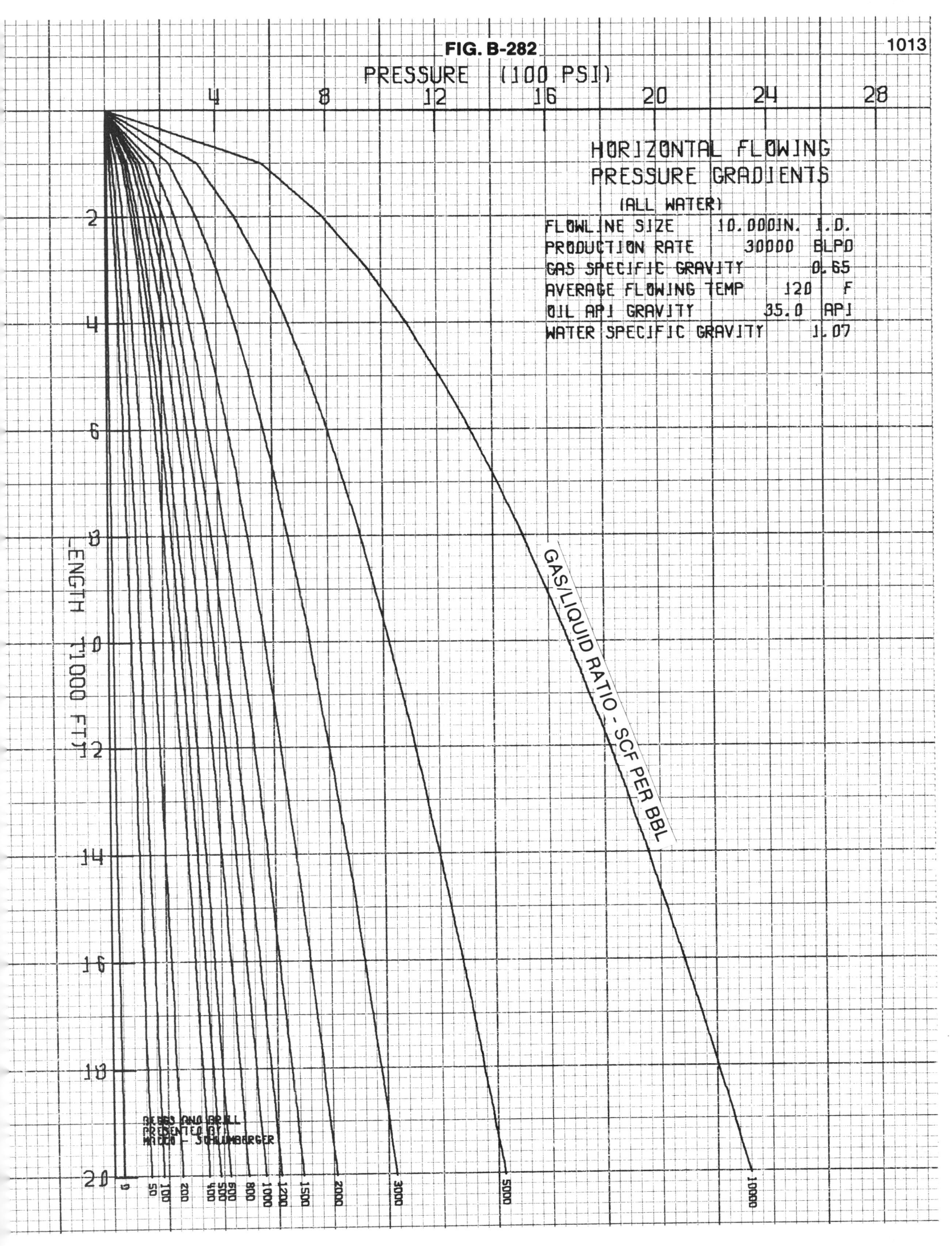

PRESSURE (100 PSI)

HORIZONTAL FLOWING
PRESSURE GRADIENTS
(ALL WATER)

FLOWLINE SIZE   10.000IN. I.D.
PRODUCTION RATE   30000 BLPD
GAS SPECIFIC GRAVITY   0.65
AVERAGE FLOWING TEMP   120   F
OIL API GRAVITY   35.0 API
WATER SPECIFIC GRAVITY   1.07

GAS/LIQUID RATIO - SCF PER BBL

LENGTH (1000 FT)

BEGGS AND BRILL
PRESENTED BY
MACEO - SCHLUMBERGER

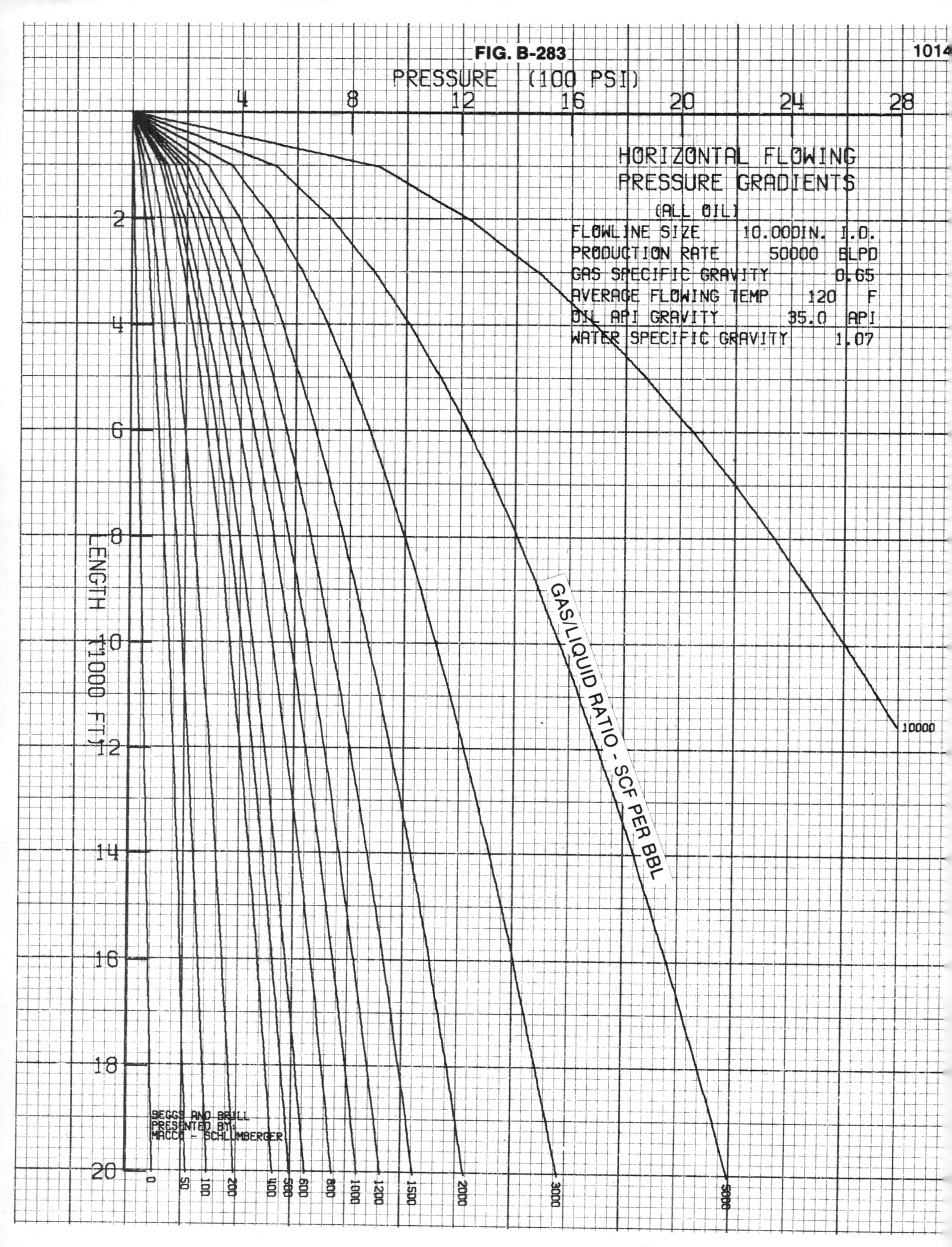

FIG. B-283

1014

PRESSURE (100 PSI)

HORIZONTAL FLOWING
PRESSURE GRADIENTS
(ALL OIL)

FLOWLINE SIZE	10.000 IN. I.D.
PRODUCTION RATE	50000 BLPD
GAS SPECIFIC GRAVITY	0.65
AVERAGE FLOWING TEMP	120 F
OIL API GRAVITY	35.0 API
WATER SPECIFIC GRAVITY	1.07

LENGTH (1000 FT)

GAS/LIQUID RATIO - SCF PER BBL

10000

BEGGS AND BRILL
PRESENTED BY:
MACCO - SCHLUMBERGER

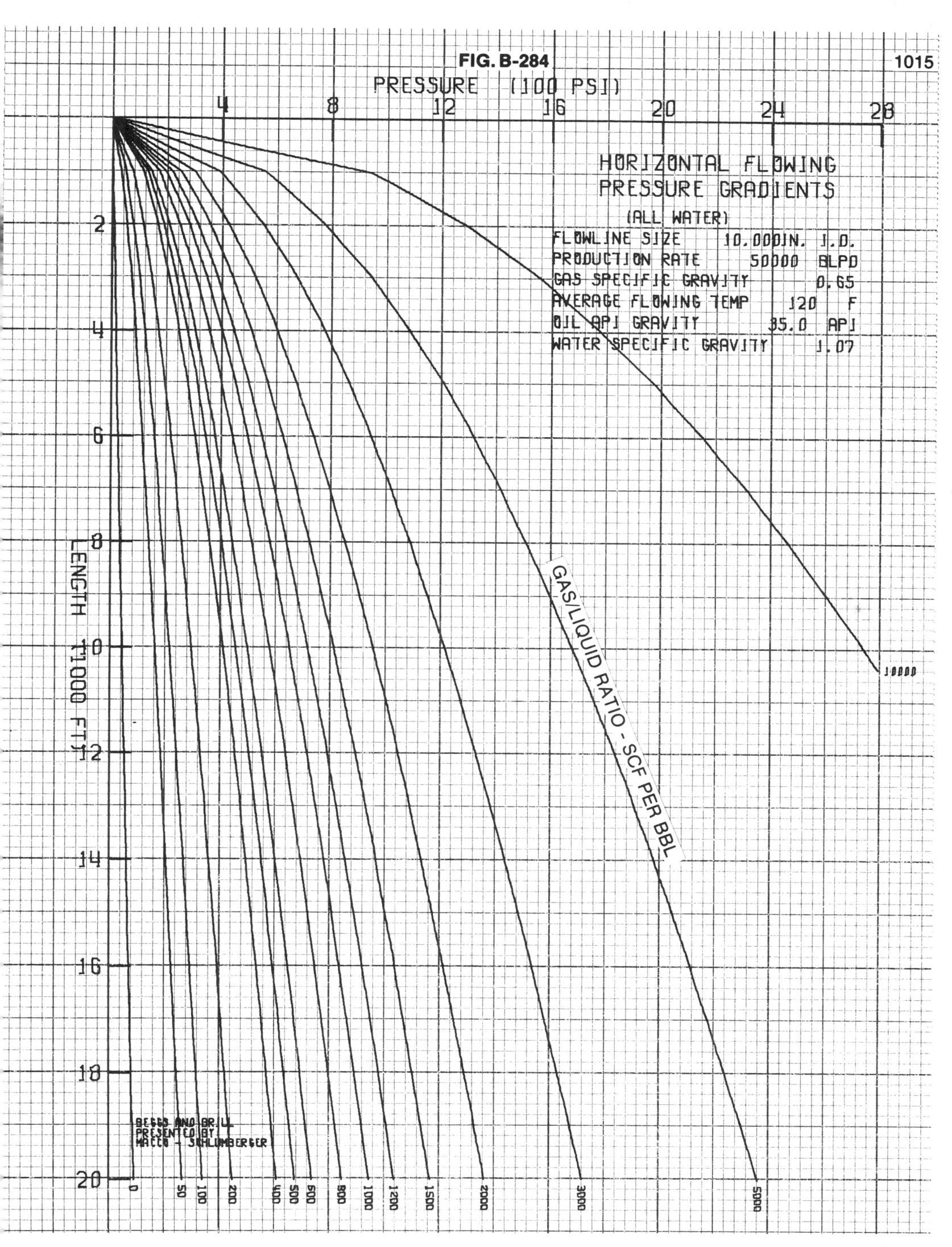

FIG. B-284

1015

PRESSURE (100 PSI)

HORIZONTAL FLOWING
PRESSURE GRADIENTS
(ALL WATER)

FLOWLINE SIZE   10.000 IN. I.D.
PRODUCTION RATE   50000 BLPD
GAS SPECIFIC GRAVITY   0.65
AVERAGE FLOWING TEMP   120 F
OIL API GRAVITY   35.0 API
WATER SPECIFIC GRAVITY   1.07

GAS/LIQUID RATIO - SCF PER BBL

LENGTH (1000 FT.)

BEGGS AND BRILL
PRESENTED BY
HALCO - SCHLUMBERGER

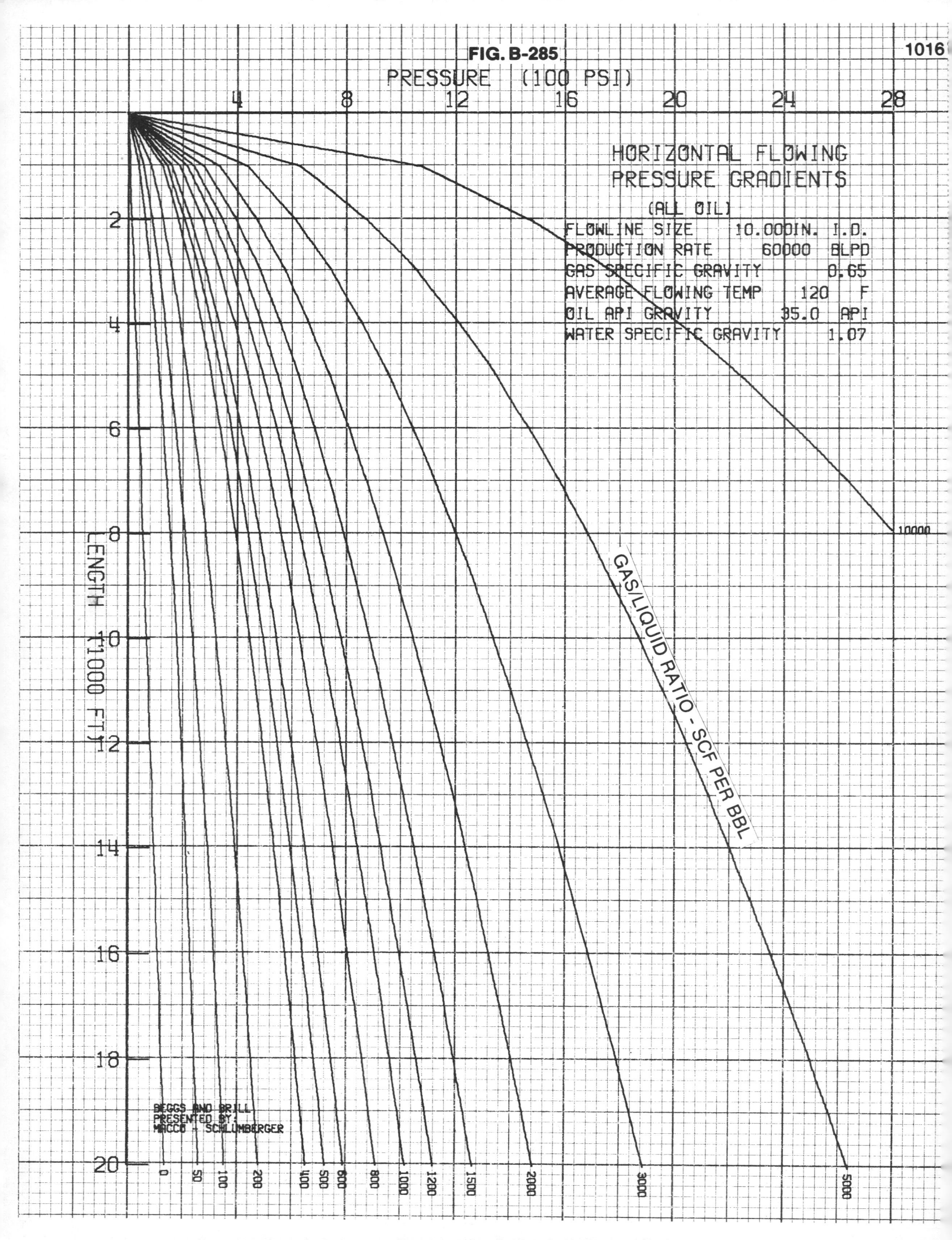

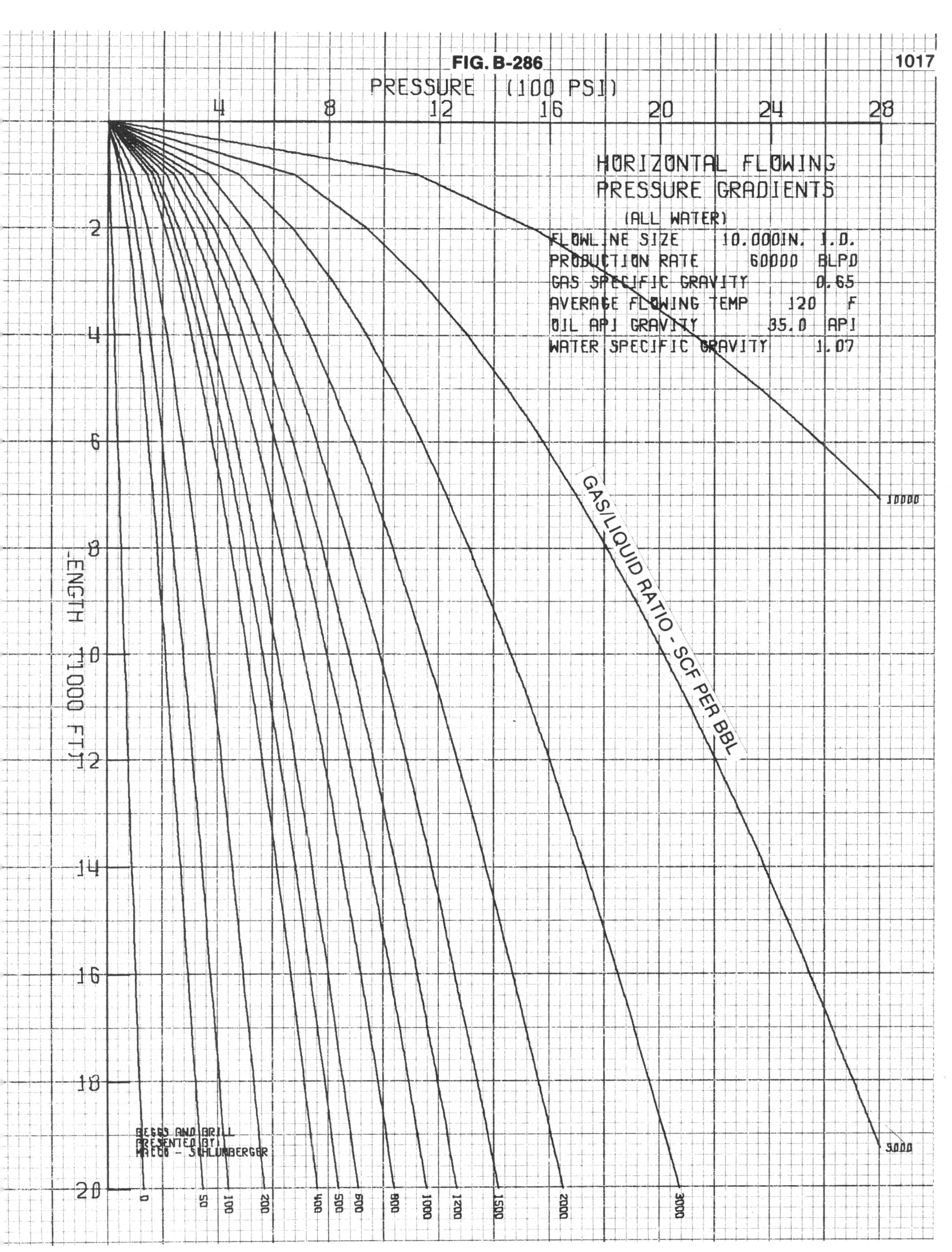

PRESSURE (100 PSI)

HORIZONTAL FLOWING
PRESSURE GRADIENTS
(ALL WATER)

FLOWLINE SIZE   10.000IN. I.D.
PRODUCTION RATE   60000 BLPD
GAS SPECIFIC GRAVITY   0.65
AVERAGE FLOWING TEMP   120   F
OIL API GRAVITY   35.0 API
WATER SPECIFIC GRAVITY   1.07

GAS/LIQUID RATIO - SCF PER BBL

LENGTH (1000 FT)

BEGGS AND BRILL
PRESENTED BY:
MAECO - SCHLUMBERGER

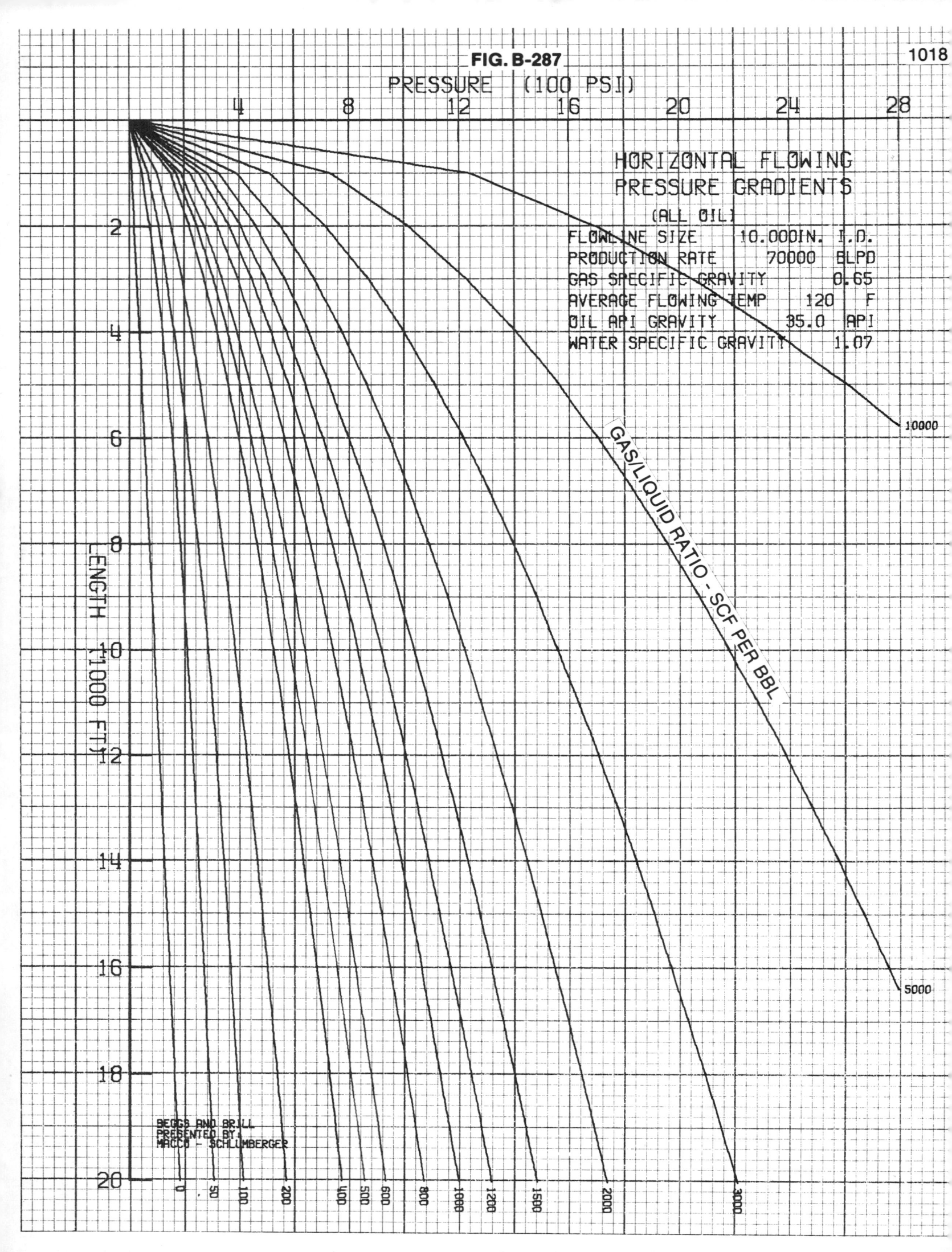

FIG. B-287

1018

PRESSURE (100 PSI)

HORIZONTAL FLOWING
PRESSURE GRADIENTS
(ALL OIL)

FLOWLINE SIZE	10.000IN. I.D.
PRODUCTION RATE	70000 BLPD
GAS SPECIFIC GRAVITY	0.65
AVERAGE FLOWING TEMP	120 F
OIL API GRAVITY	35.0 API
WATER SPECIFIC GRAVITY	1.07

GAS/LIQUID RATIO - SCF PER BBL

LENGTH (1000 FT)

BEGGS AND BRILL
PRESENTED BY:
MACCO - SCHLUMBERGER

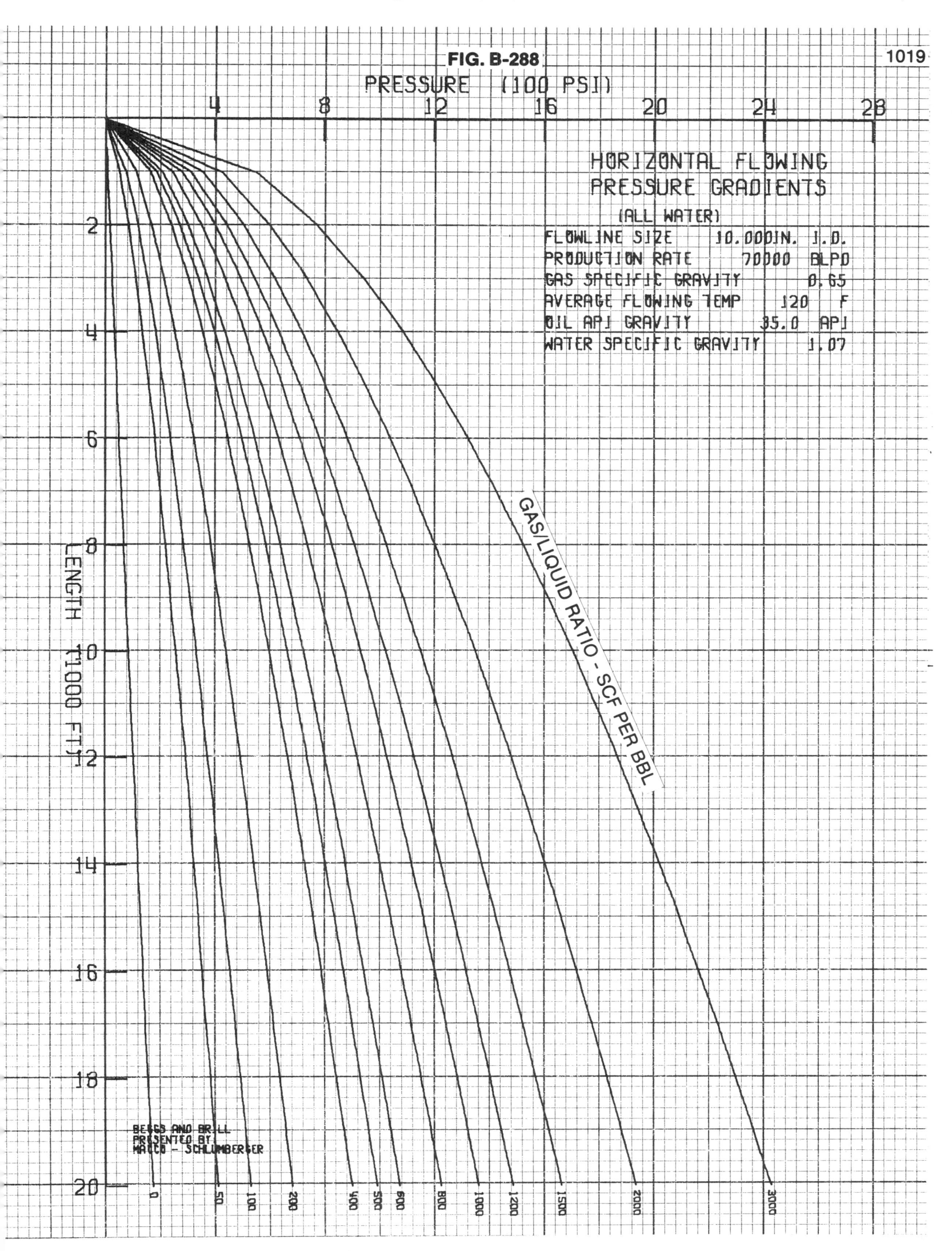

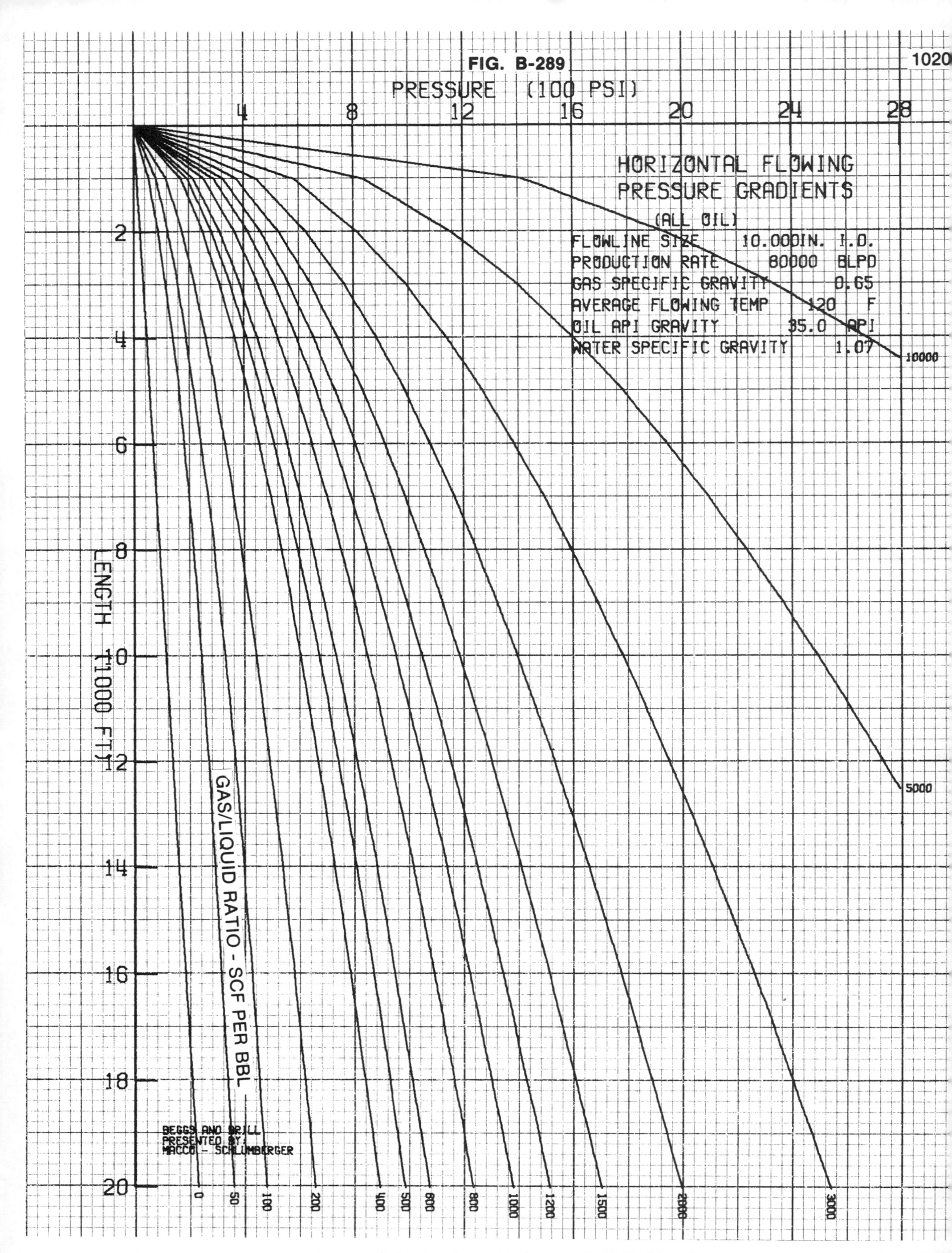

FIG. B-289

1020

PRESSURE (100 PSI)

HORIZONTAL FLOWING
PRESSURE GRADIENTS
(ALL OIL)

FLOWLINE SIZE	10.000IN. I.D.
PRODUCTION RATE	80000 BLPD
GAS SPECIFIC GRAVITY	0.65
AVERAGE FLOWING TEMP	120 F
OIL API GRAVITY	35.0 API
WATER SPECIFIC GRAVITY	1.07

LENGTH (1000 FT)

GAS/LIQUID RATIO - SCF PER BBL

BEGGS AND BRILL
PRESENTED BY
MACCO - SCHLUMBERGER

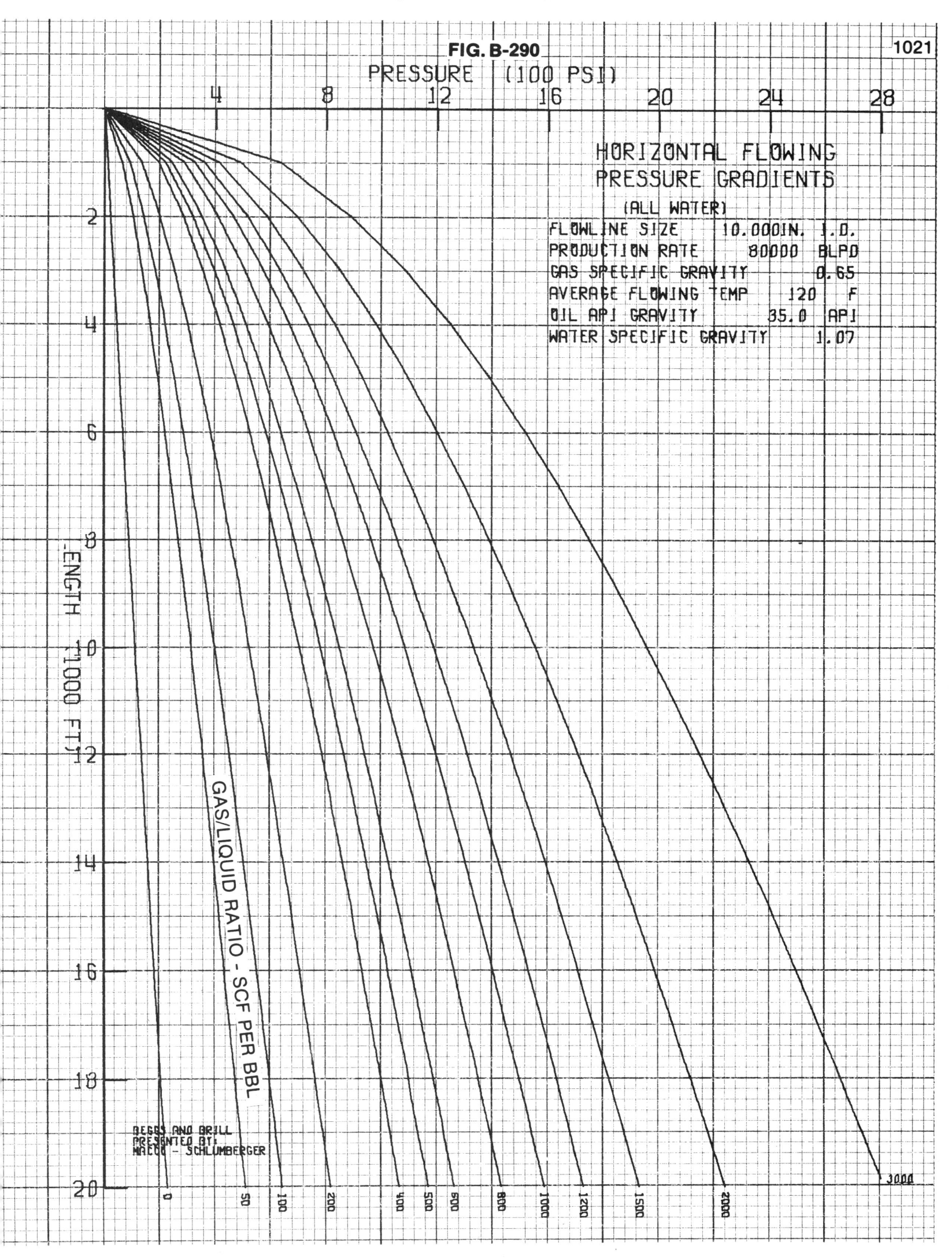

FIG. B-290

1021

PRESSURE (100 PSI)

HORIZONTAL FLOWING
PRESSURE GRADIENTS
(ALL WATER)

FLOWLINE SIZE	10.000IN. I.D.
PRODUCTION RATE	80000 BLPD
GAS SPECIFIC GRAVITY	0.65
AVERAGE FLOWING TEMP	120 F
OIL API GRAVITY	35.0 API
WATER SPECIFIC GRAVITY	1.07

LENGTH (1000 FT.)

GAS/LIQUID RATIO - SCF PER BBL

BEGGS AND BRILL
PRESENTED BY:
NAECO - SCHLUMBERGER

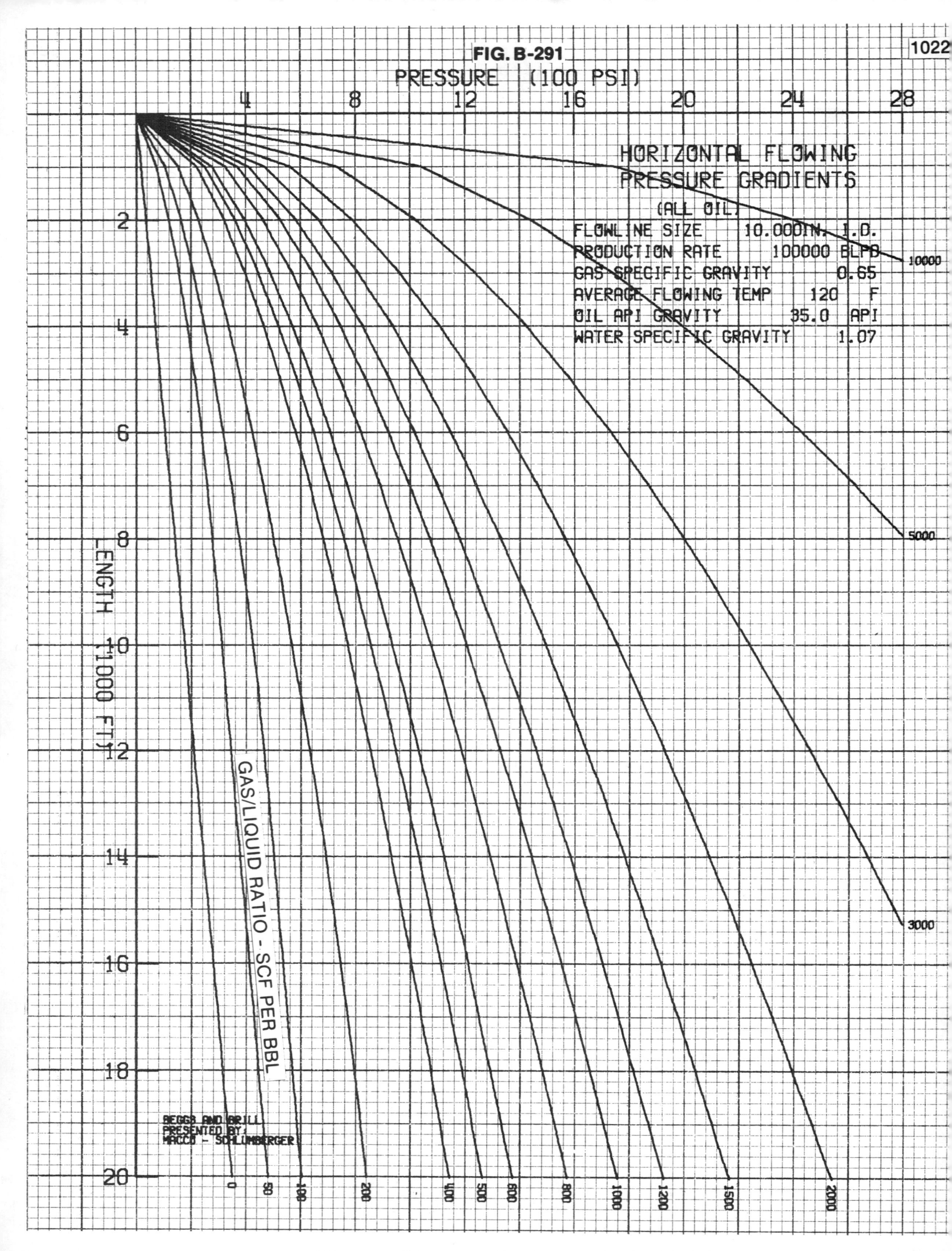

PRESSURE (100 PSI)

HORIZONTAL FLOWING
PRESSURE GRADIENTS
(ALL OIL)

FLOWLINE SIZE	10.000 IN. I.D.
PRODUCTION RATE	100000 BLPD
GAS SPECIFIC GRAVITY	0.65
AVERAGE FLOWING TEMP	120 F
OIL API GRAVITY	35.0 API
WATER SPECIFIC GRAVITY	1.07

LENGTH (1000 FT)

GAS/LIQUID RATIO - SCF PER BBL

BEGGS AND BRILL
PRESENTED BY
MACCO - SCHLUMBERGER

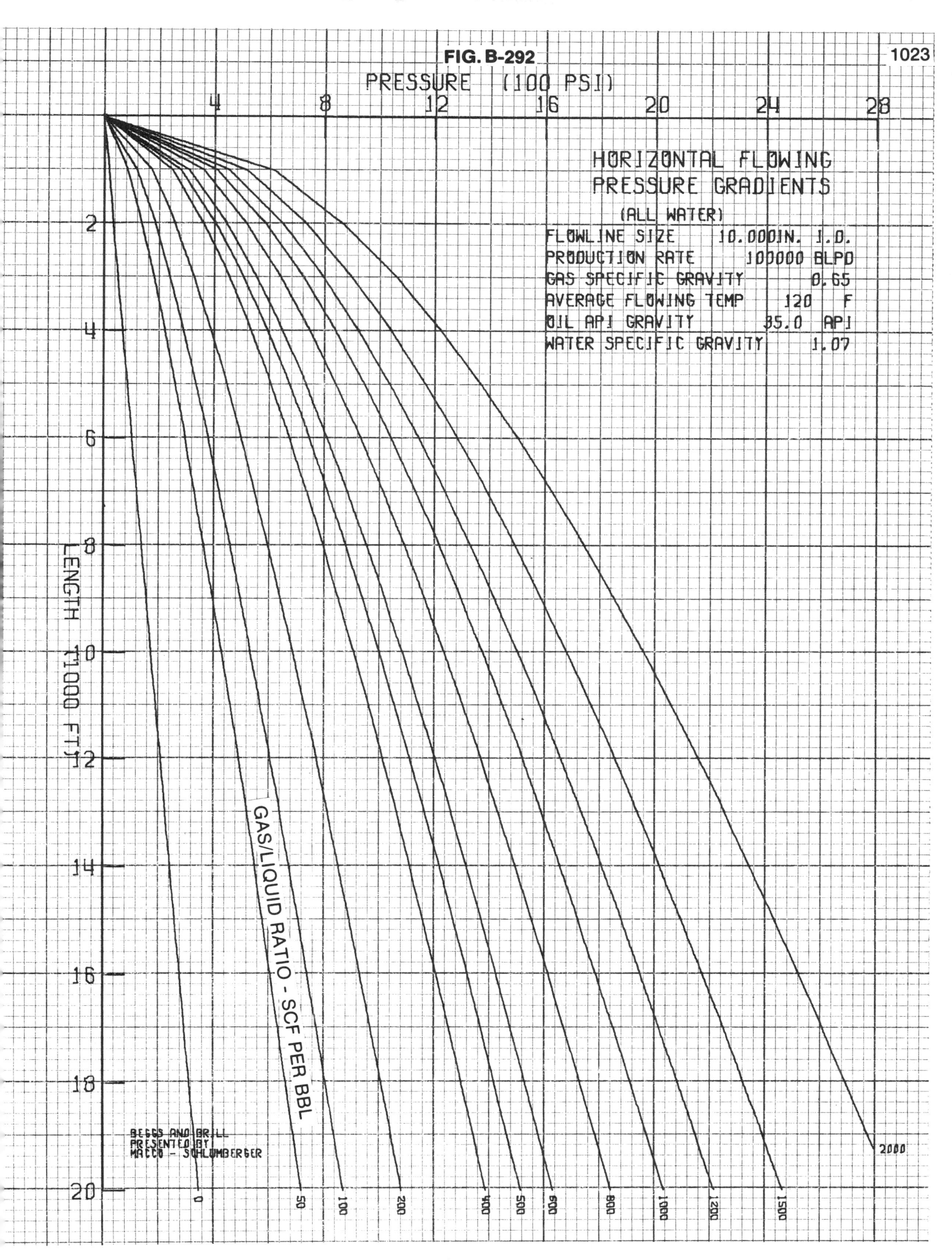

HORIZONTAL FLOWING
PRESSURE GRADIENTS
(ALL WATER)

FLOWLINE SIZE	10.000IN. I.D.
PRODUCTION RATE	100000 BLPD
GAS SPECIFIC GRAVITY	0.65
AVERAGE FLOWING TEMP	120 F
OIL API GRAVITY	35.0 API
WATER SPECIFIC GRAVITY	1.07

PRESSURE (100 PSI)

LENGTH (1000 FT.)

GAS/LIQUID RATIO - SCF PER BBL

BEGGS AND BRILL
PRESENTED BY
MACCO - SCHLUMBERGER

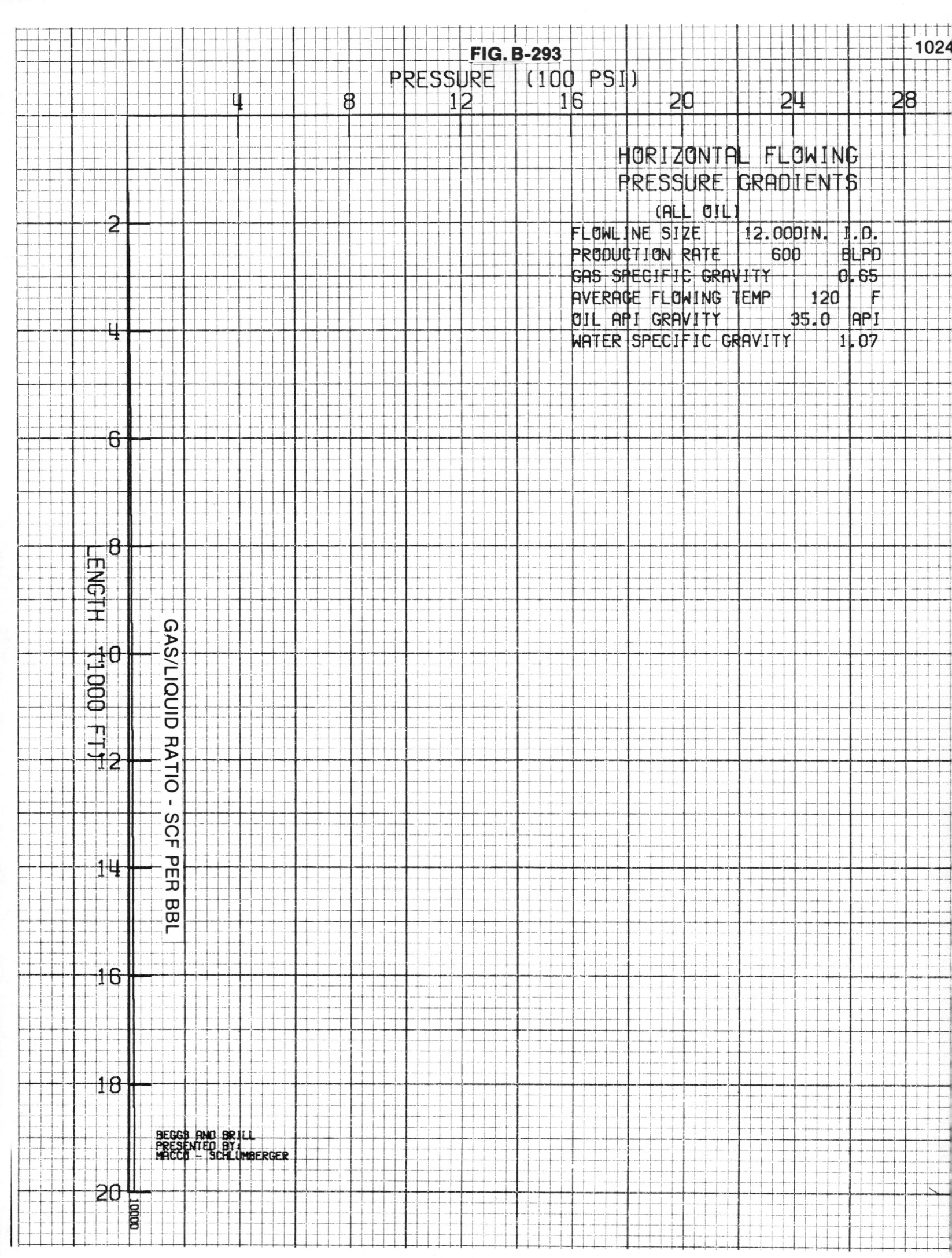

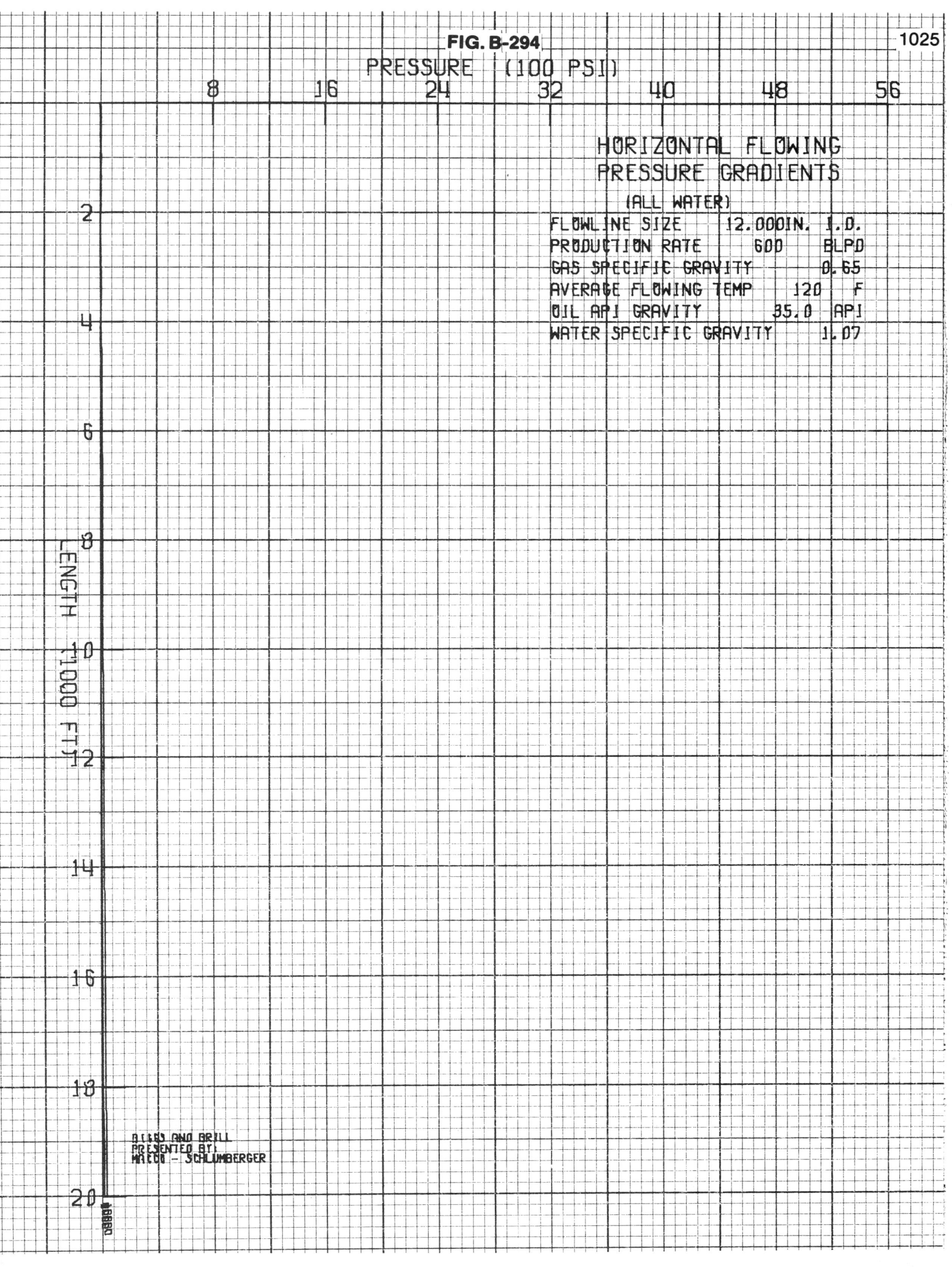

FIG. B-294

1025

PRESSURE (100 PSI)

HORIZONTAL FLOWING
PRESSURE GRADIENTS
(ALL WATER)

FLOWLINE SIZE	12.000IN. I.D.
PRODUCTION RATE	600 BLPD
GAS SPECIFIC GRAVITY	0.65
AVERAGE FLOWING TEMP	120 F
OIL API GRAVITY	35.0 API
WATER SPECIFIC GRAVITY	1.07

LENGTH (1000 FT)

BEGGS AND BRILL
PRESENTED BY:
MAECO - SCHLUMBERGER

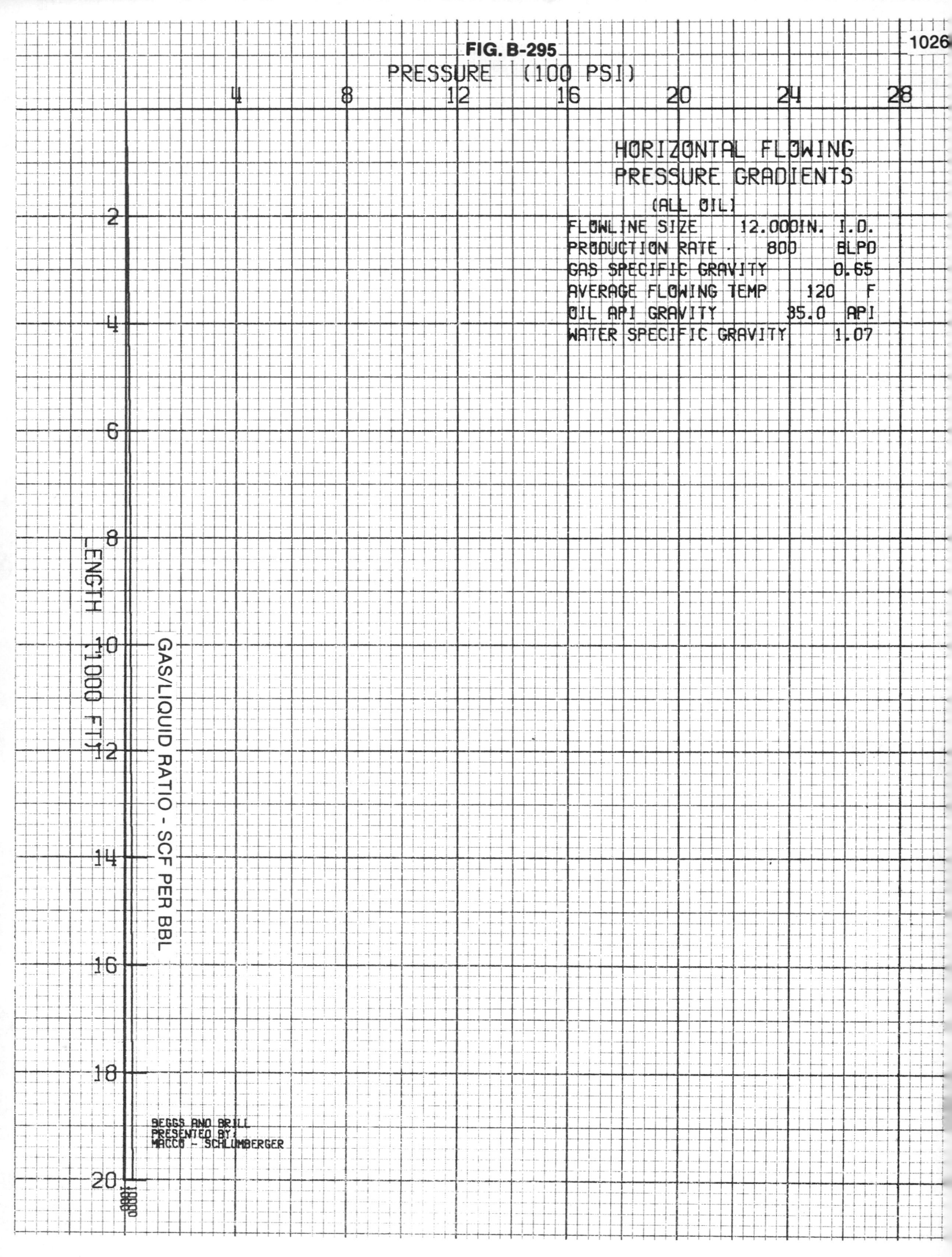

FIG. B-295

1026

PRESSURE (100 PSI)

HORIZONTAL FLOWING
PRESSURE GRADIENTS
(ALL OIL)

FLOWLINE SIZE   12.000IN. I.D.
PRODUCTION RATE ·   800   BLPD
GAS SPECIFIC GRAVITY   0.65
AVERAGE FLOWING TEMP   120   F
OIL API GRAVITY   35.0   API
WATER SPECIFIC GRAVITY   1.07

LENGTH (1000 FT)

GAS/LIQUID RATIO - SCF PER BBL

BEGGS AND BRILL
PRESENTED BY
MACCO - SCHLUMBERGER

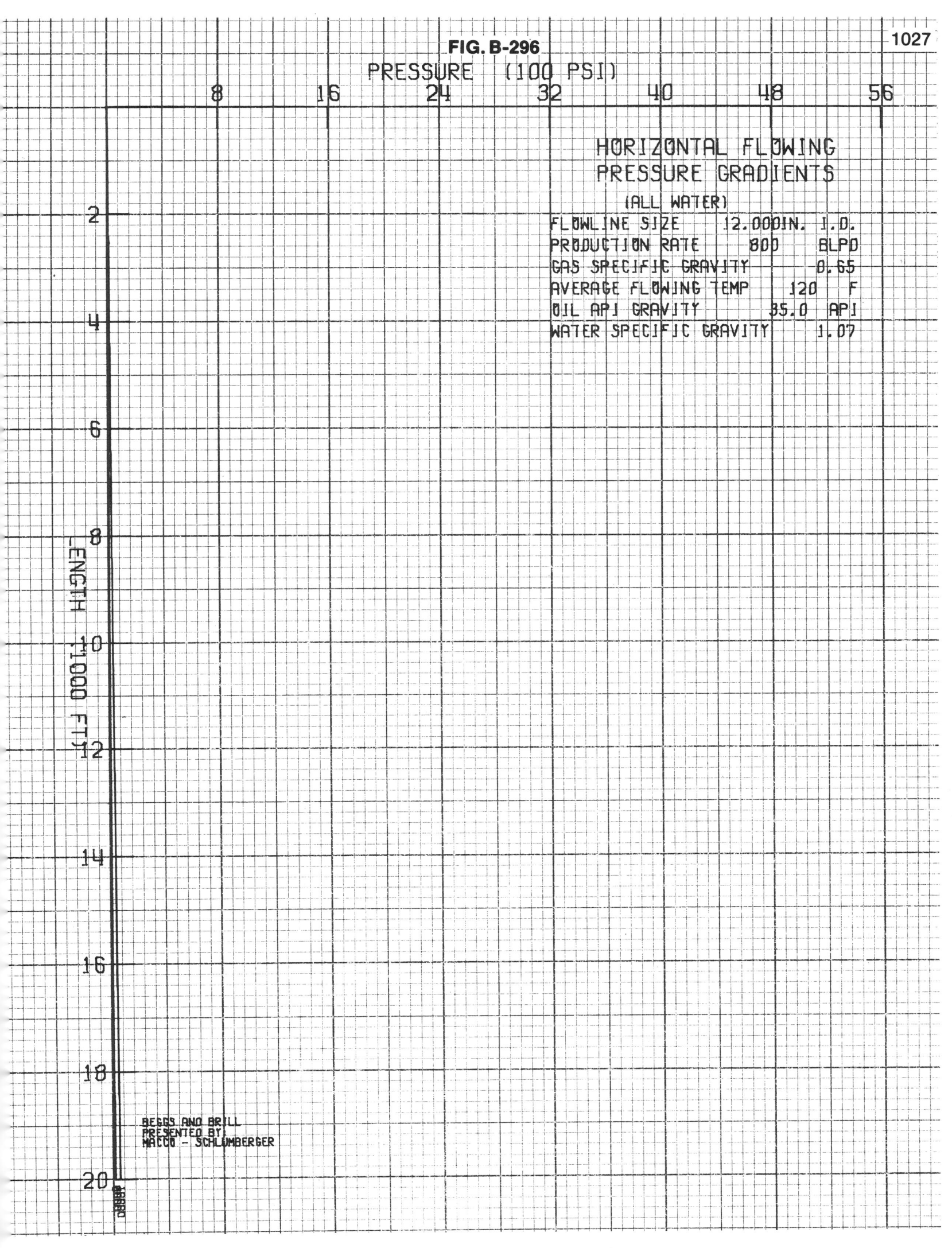

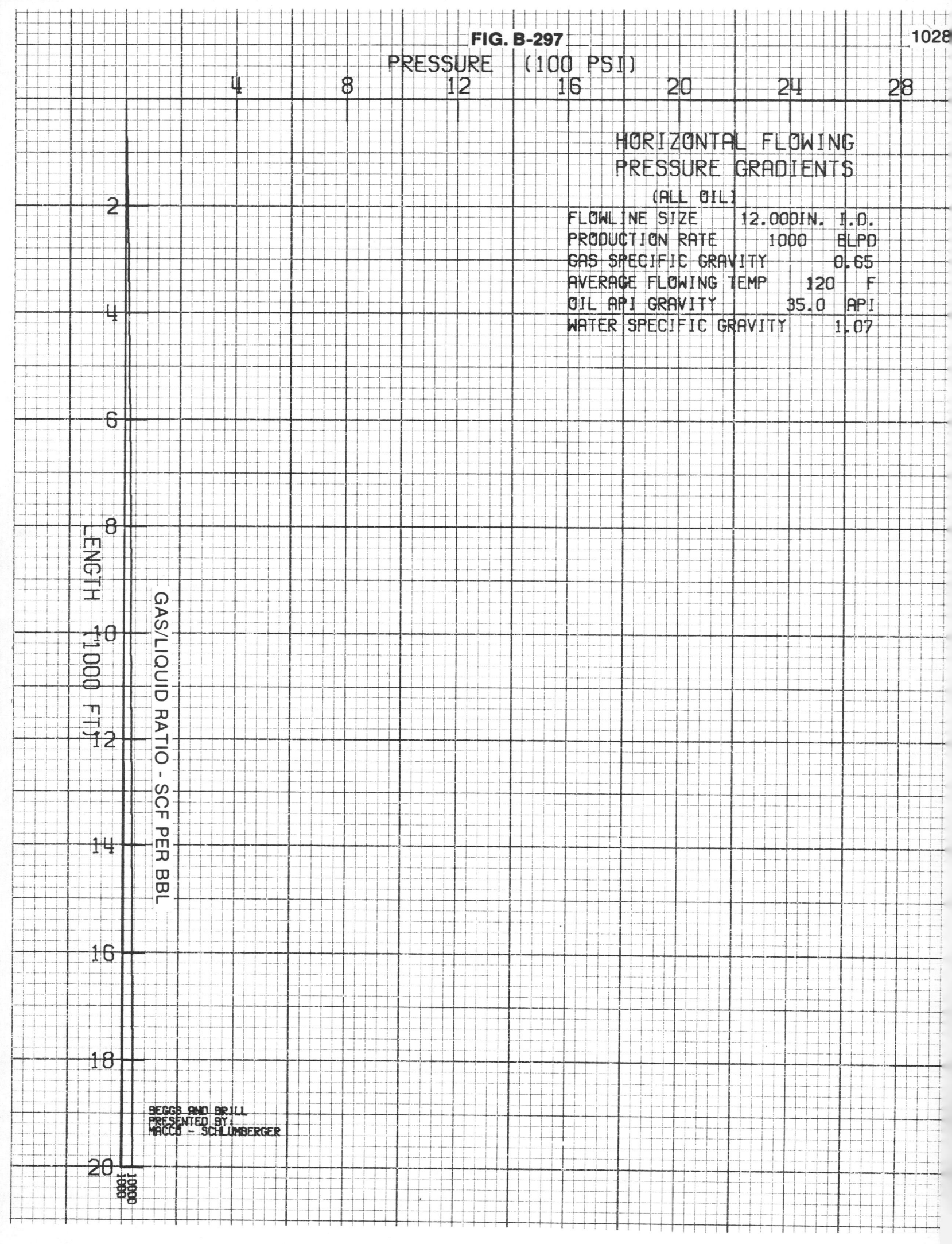

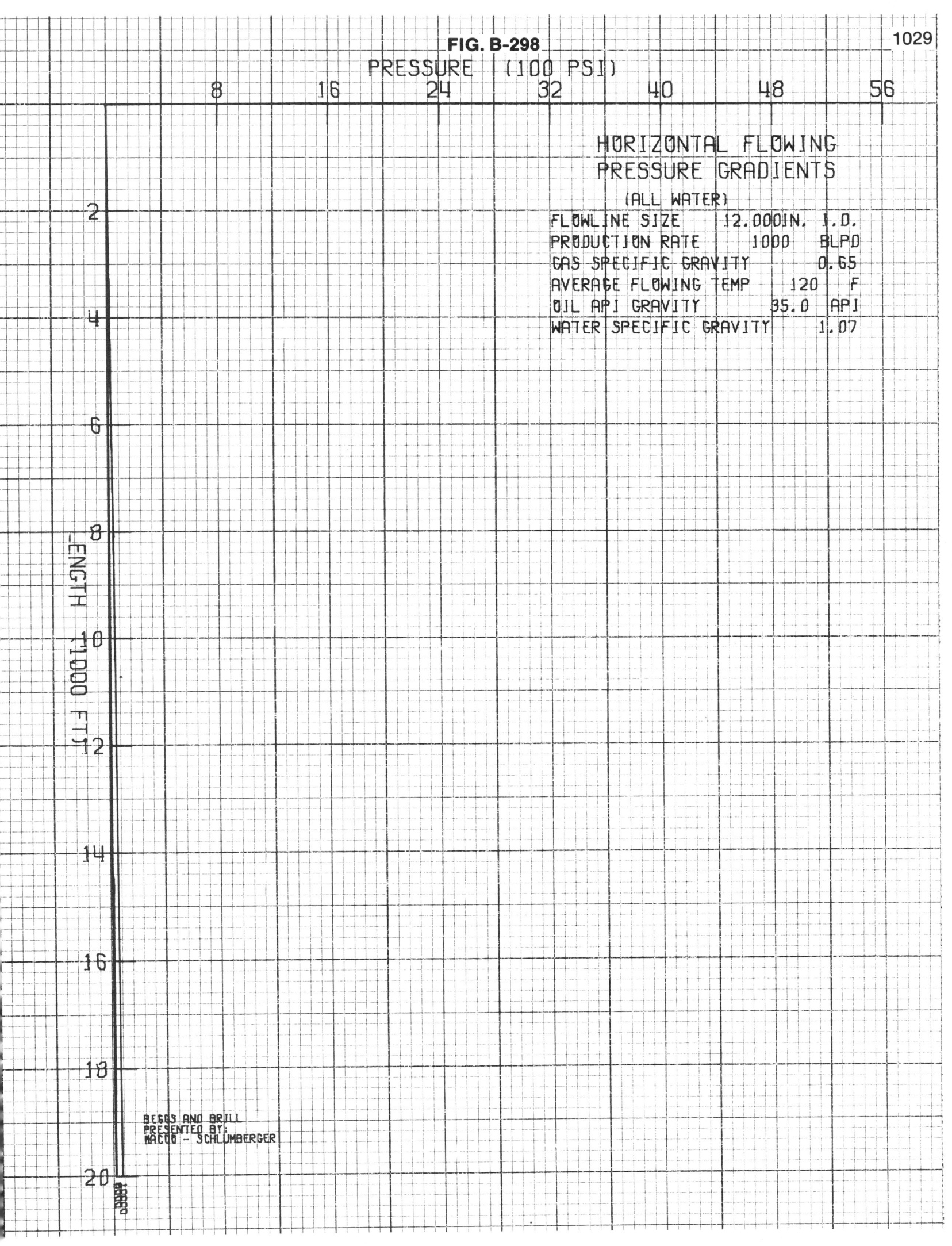

FIG. B-298

1029

PRESSURE (100 PSI)

HORIZONTAL FLOWING
PRESSURE GRADIENTS
(ALL WATER)

FLOWLINE SIZE	12.000IN. I.D.
PRODUCTION RATE	1000 BLPD
GAS SPECIFIC GRAVITY	0.65
AVERAGE FLOWING TEMP	120 F
OIL API GRAVITY	35.0 API
WATER SPECIFIC GRAVITY	1.07

LENGTH (1000 FT.)

BEGGS AND BRILL
PRESENTED BY:
MACCO – SCHLUMBERGER

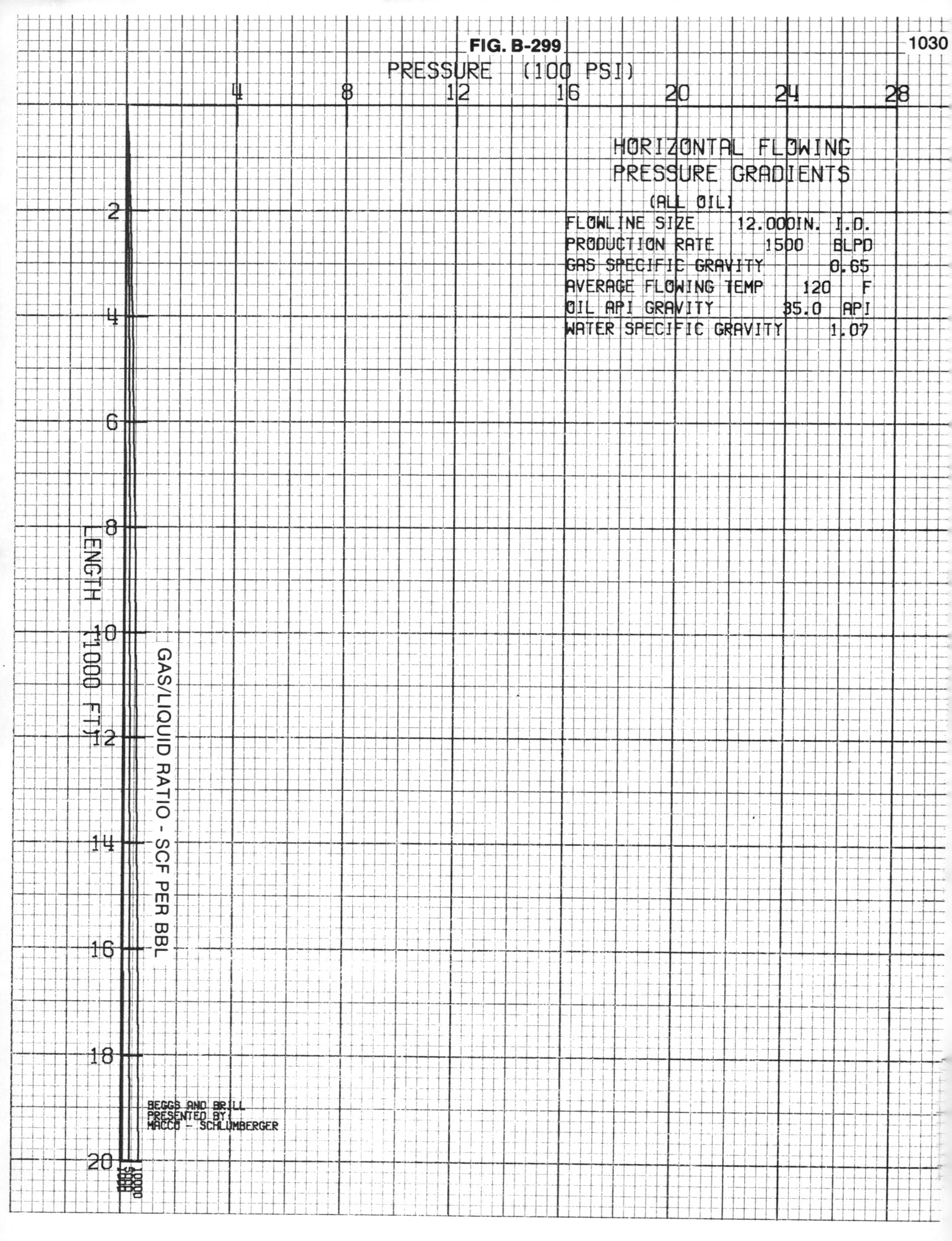

PRESSURE (100 PSI)

HORIZONTAL FLOWING
PRESSURE GRADIENTS
(ALL OIL)

FLOWLINE SIZE   12.000IN. I.D.
PRODUCTION RATE   1500   BLPD
GAS SPECIFIC GRAVITY   0.65
AVERAGE FLOWING TEMP   120   F
OIL API GRAVITY   35.0   API
WATER SPECIFIC GRAVITY   1.07

LENGTH (1000 FT)

GAS/LIQUID RATIO - SCF PER BBL

BEGGS AND BRILL
PRESENTED BY:
MACCO - SCHLUMBERGER

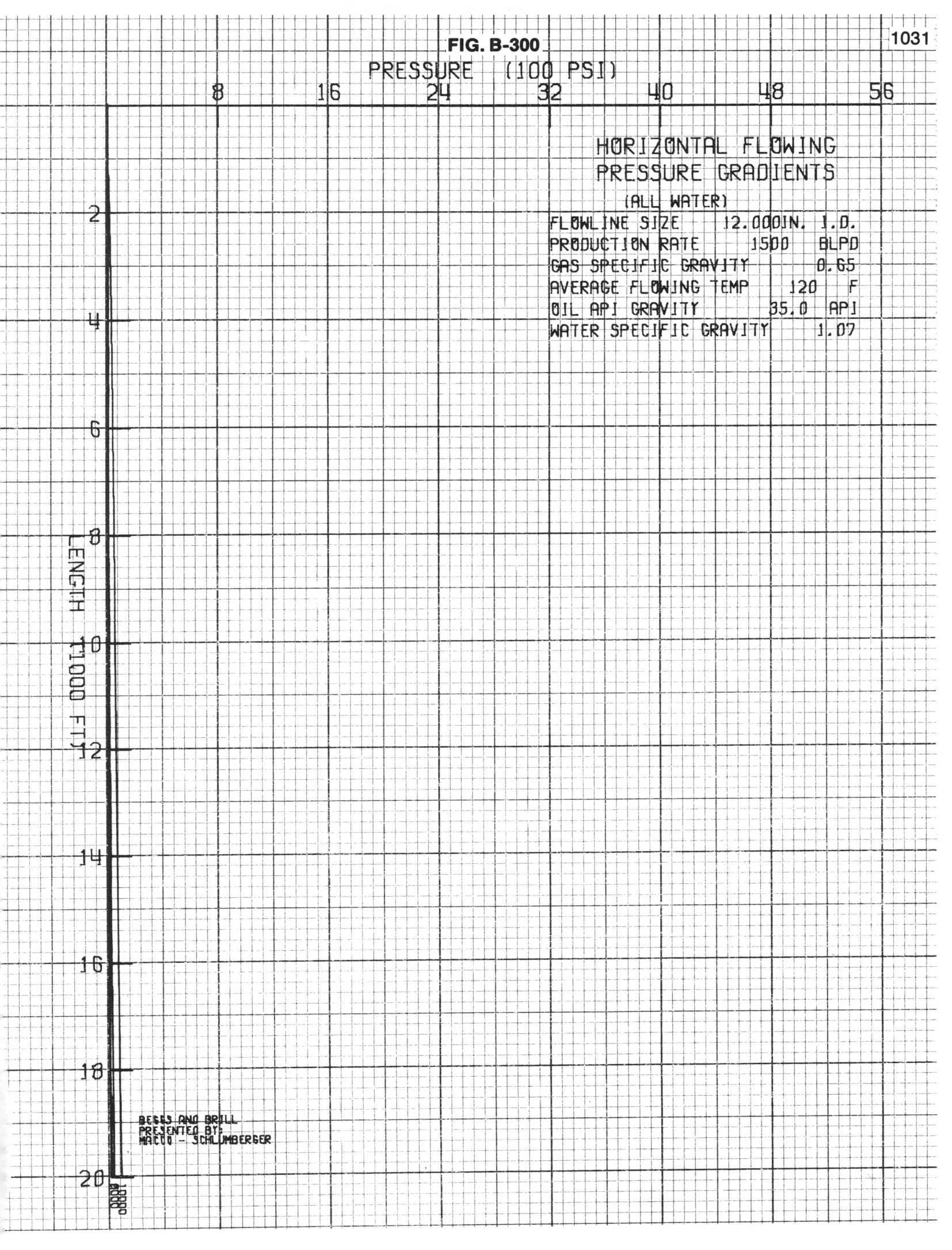

PRESSURE (100 PSI)

HORIZONTAL FLOWING
PRESSURE GRADIENTS
(ALL WATER)

FLOWLINE SIZE	12.000IN. I.D.
PRODUCTION RATE	1500 BLPD
GAS SPECIFIC GRAVITY	0.65
AVERAGE FLOWING TEMP	120 F
OIL API GRAVITY	35.0 API
WATER SPECIFIC GRAVITY	1.07

LENGTH (1000 FT.)

BEGGS AND BRILL
PRESENTED BY:
MAICO - SCHLUMBERGER

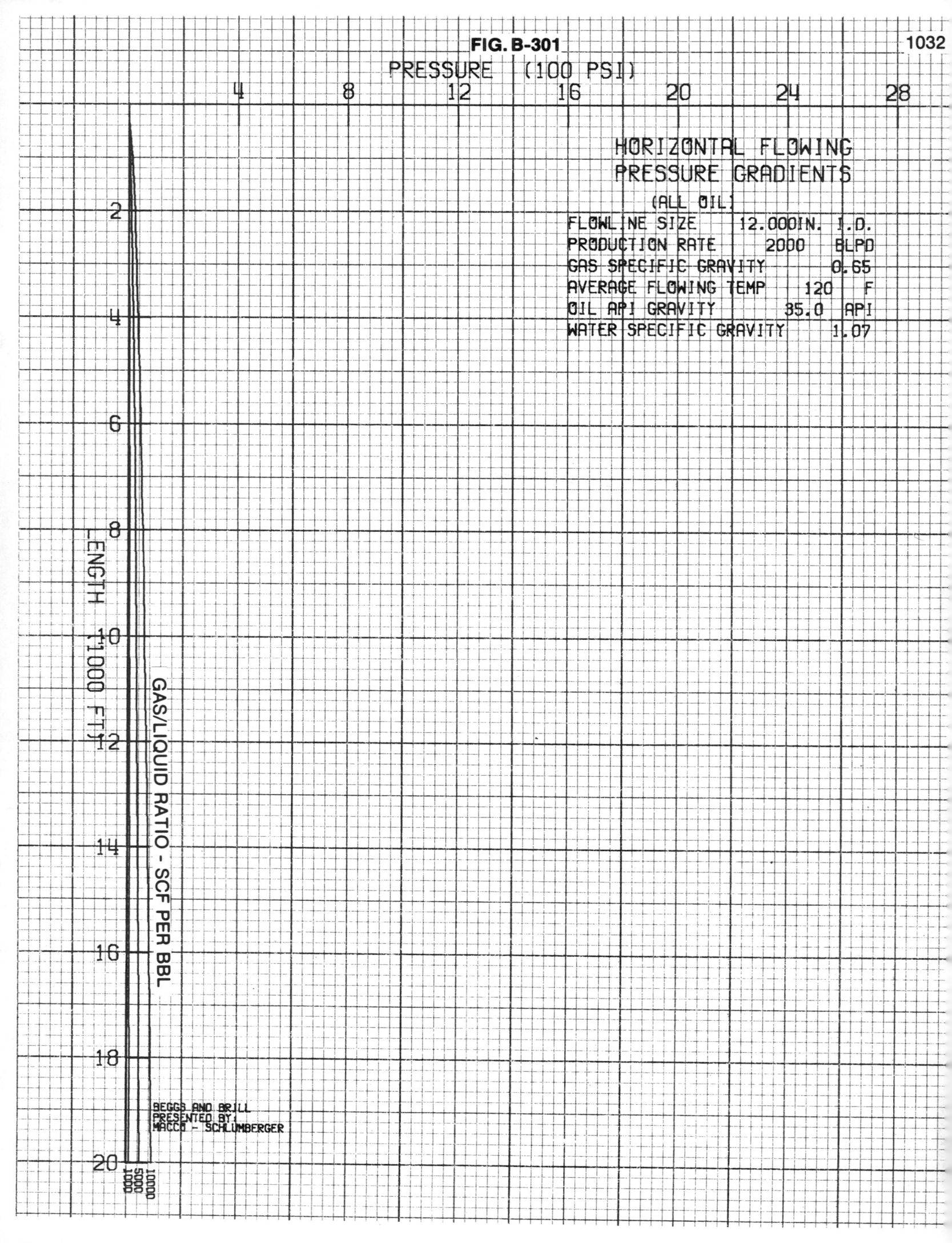

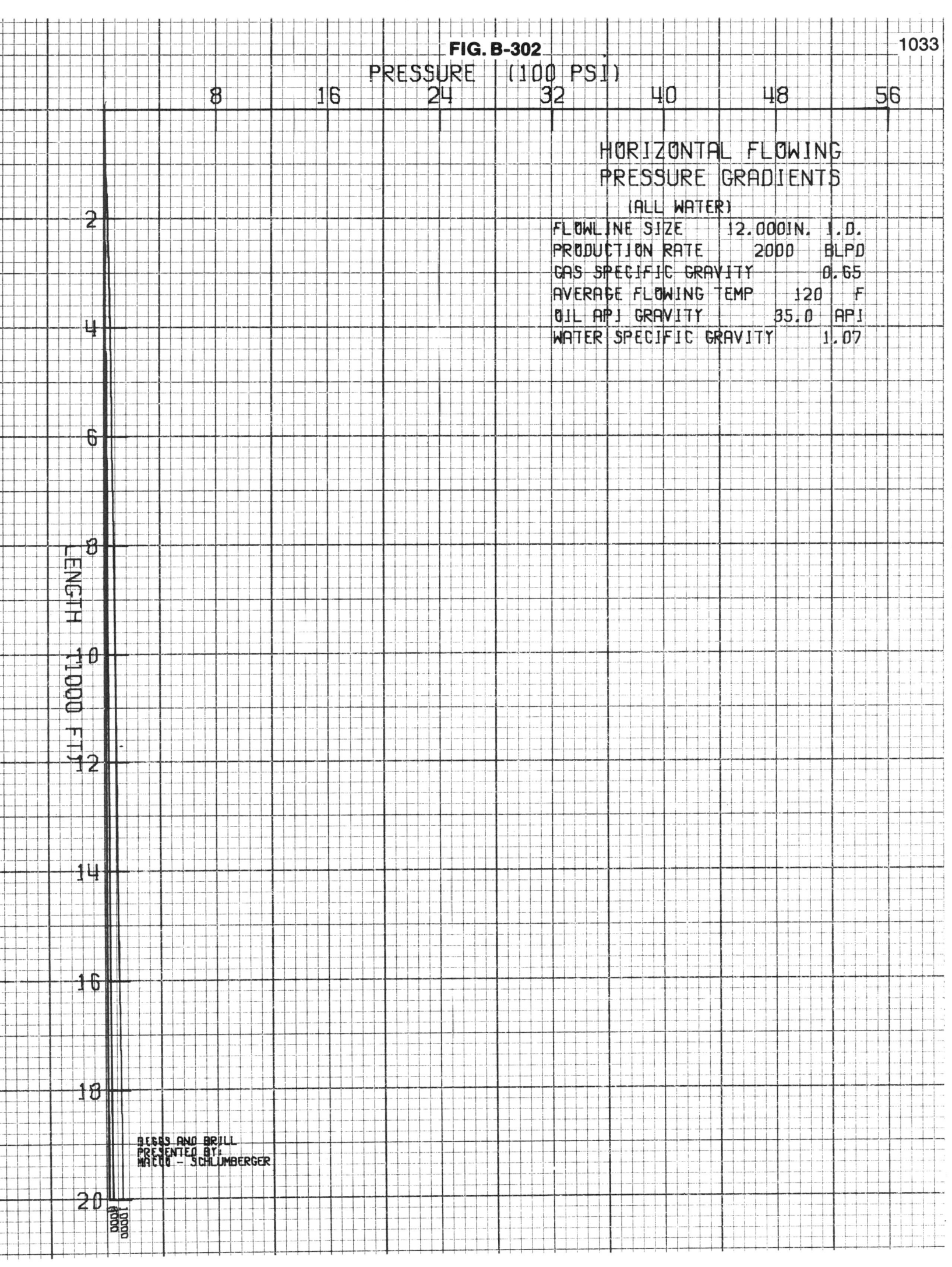

FIG. B-302

PRESSURE (100 PSI)

HORIZONTAL FLOWING
PRESSURE GRADIENTS
(ALL WATER)

FLOWLINE SIZE	12.000IN. I.D.
PRODUCTION RATE	2000 BLPD
GAS SPECIFIC GRAVITY	0.65
AVERAGE FLOWING TEMP	120 F
OIL API GRAVITY	35.0 API
WATER SPECIFIC GRAVITY	1.07

LENGTH (1000 FT.)

BEGGS AND BRILL
PRESENTED BY:
MACCO - SCHLUMBERGER

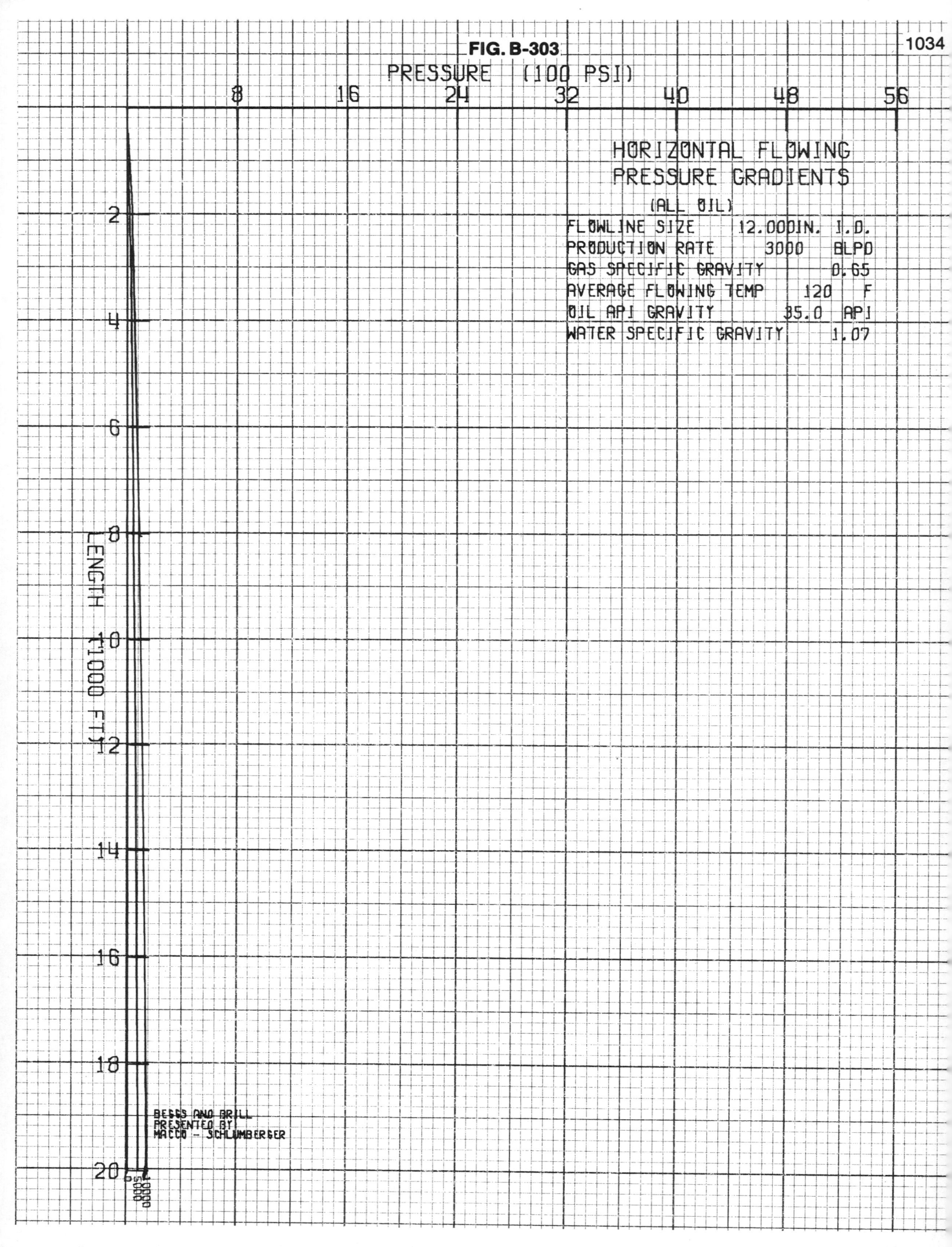

PRESSURE (100 PSI)

HORIZONTAL FLOWING
PRESSURE GRADIENTS

(ALL OIL)

FLOWLINE SIZE	12.000 IN. I.D.
PRODUCTION RATE	3000 BLPD
GAS SPECIFIC GRAVITY	0.65
AVERAGE FLOWING TEMP	120 F
OIL API GRAVITY	35.0 API
WATER SPECIFIC GRAVITY	1.07

LENGTH (1000 FT)

BEGGS AND BRILL
PRESENTED BY
MACCO - SCHLUMBERGER

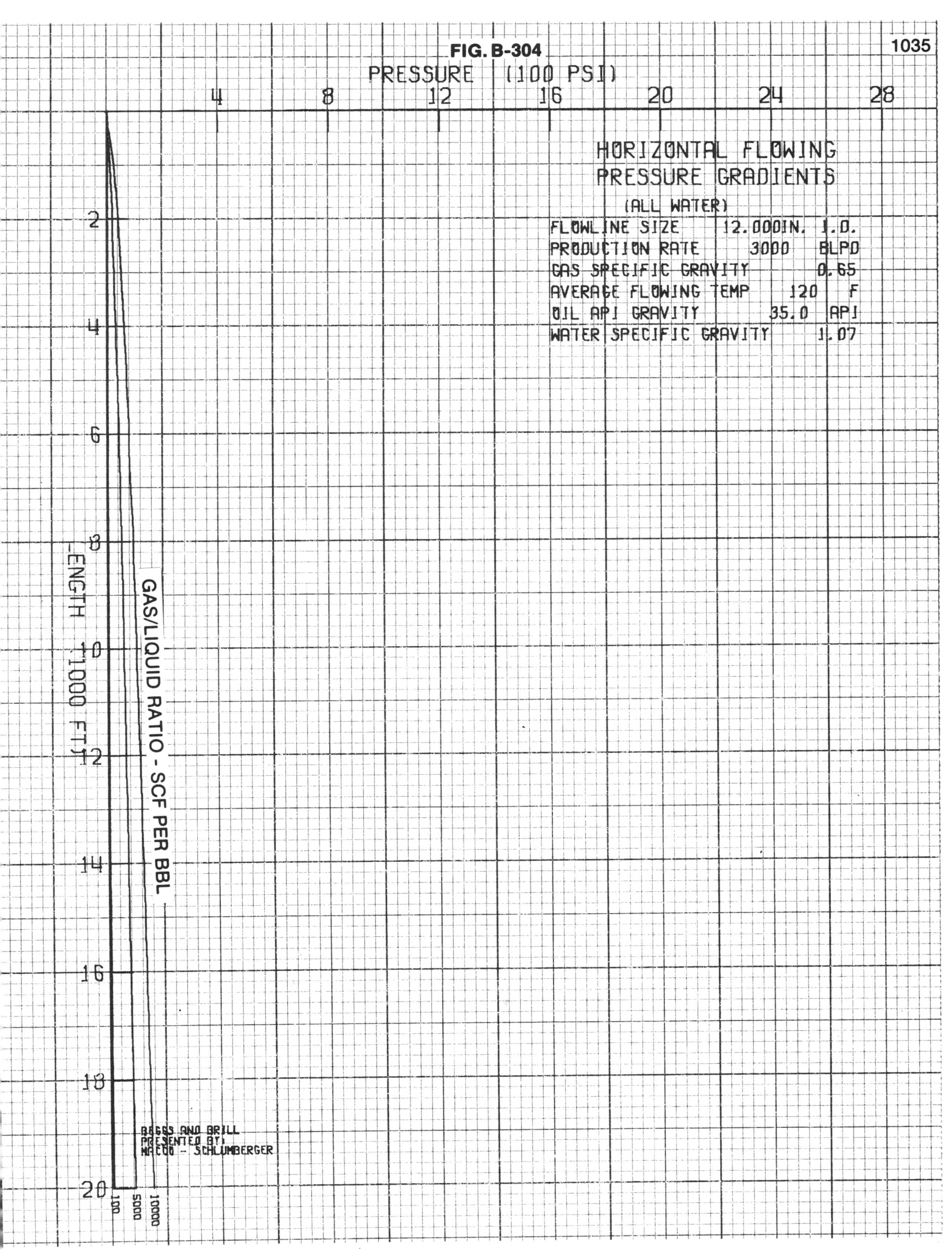

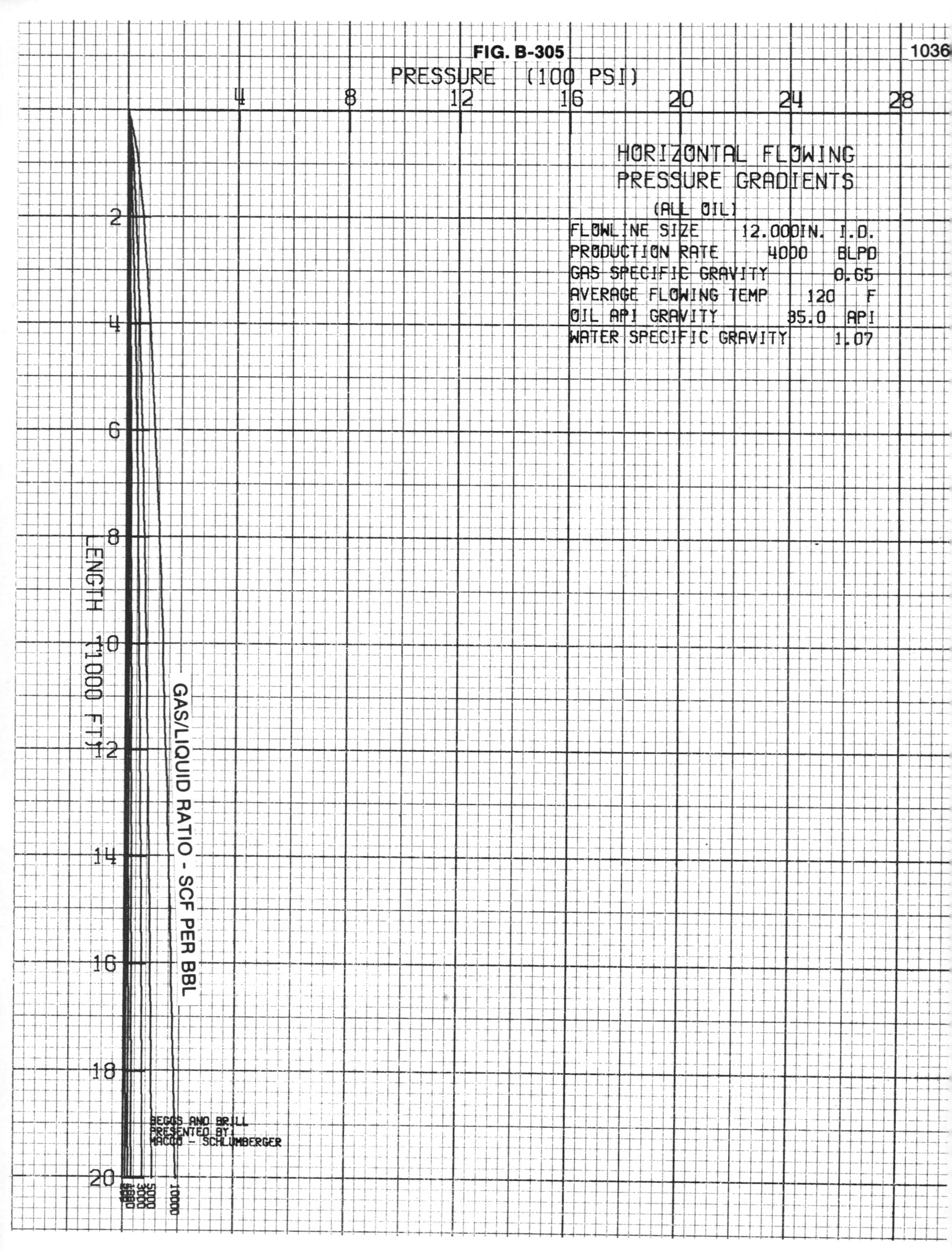

FIG. B-305

PRESSURE (100 PSI)

1036

HORIZONTAL FLOWING
PRESSURE GRADIENTS
(ALL OIL)

FLOWLINE SIZE	12.000 IN. I.D.
PRODUCTION RATE	4000 BLPD
GAS SPECIFIC GRAVITY	0.65
AVERAGE FLOWING TEMP	120 F
OIL API GRAVITY	35.0 API
WATER SPECIFIC GRAVITY	1.07

LENGTH (1000 FT)

GAS/LIQUID RATIO - SCF PER BBL

BEGGS AND BRILL
PRESENTED BY:
MACCO - SCHLUMBERGER

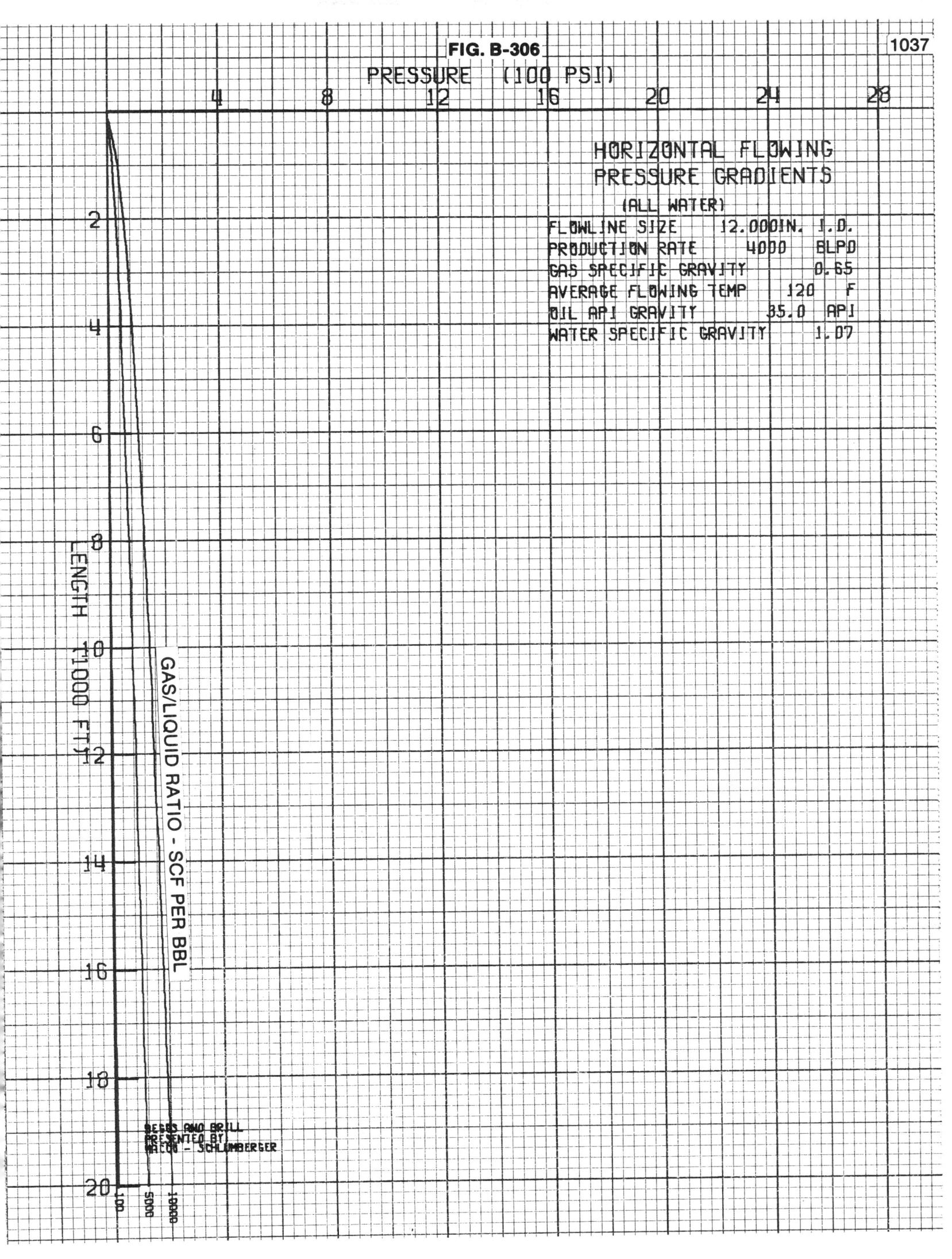

FIG. B-306

1037

PRESSURE (100 PSI)

HORIZONTAL FLOWING
PRESSURE GRADIENTS
(ALL WATER)

FLOWLINE SIZE   12.000IN. I.D.
PRODUCTION RATE   4000   BLPD
GAS SPECIFIC GRAVITY   0.65
AVERAGE FLOWING TEMP   120   F
OIL API GRAVITY   35.0 API
WATER SPECIFIC GRAVITY   1.07

LENGTH (1000 FT)

GAS/LIQUID RATIO - SCF PER BBL

BEGGS AND BRILL
PRESENTED BY
INTCOM - SCHLUMBERGER

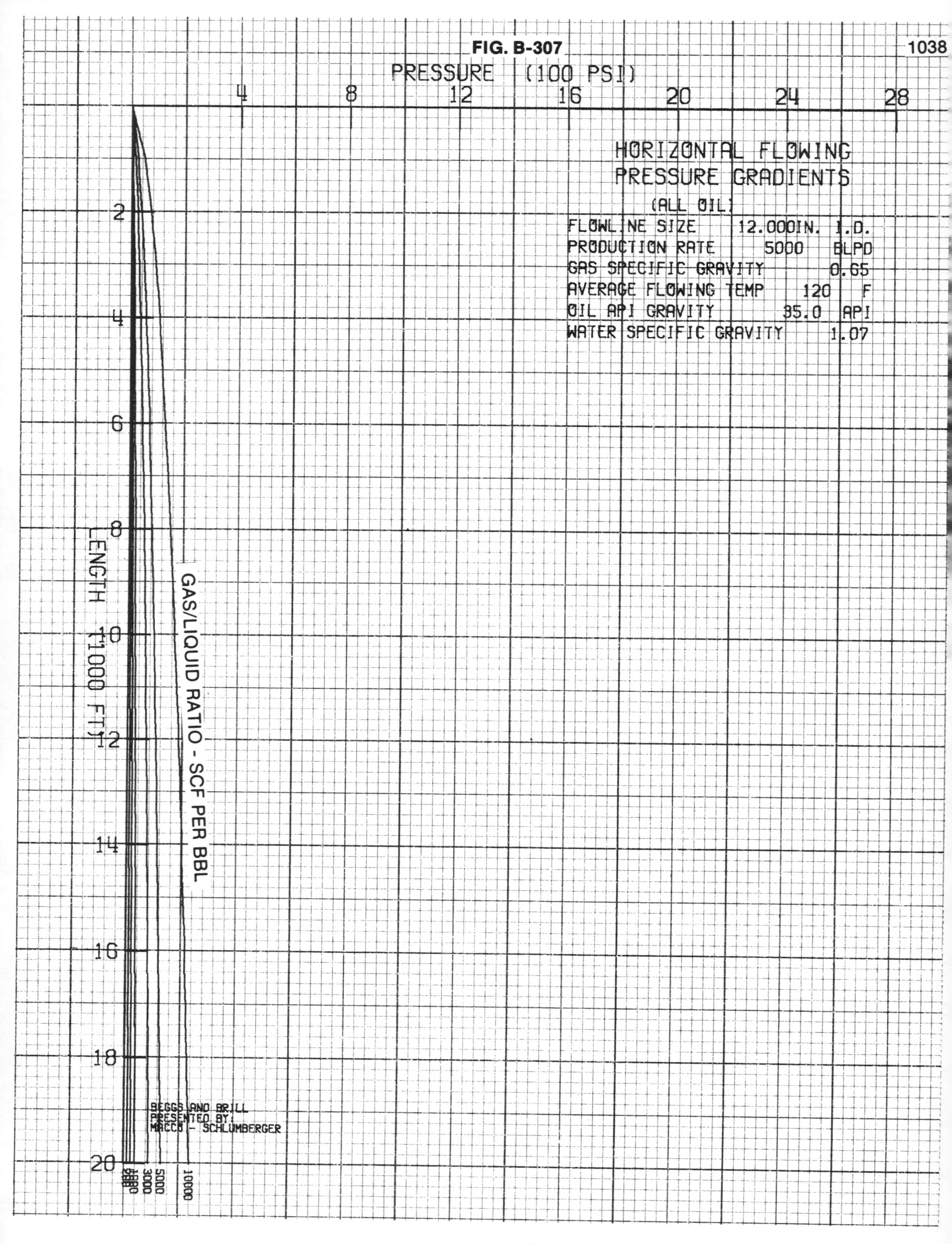

FIG. B-307

1038

HORIZONTAL FLOWING
PRESSURE GRADIENTS
(ALL OIL)

FLOWLINE SIZE	12.000IN. I.D.
PRODUCTION RATE	5000 BLPD
GAS SPECIFIC GRAVITY	0.65
AVERAGE FLOWING TEMP	120 F
OIL API GRAVITY	35.0 API
WATER SPECIFIC GRAVITY	1.07

PRESSURE (100 PSI)

LENGTH (1000 FT)

GAS/LIQUID RATIO - SCF PER BBL

BEGGS AND BRILL
PRESENTED BY:
MACCO - SCHLUMBERGER

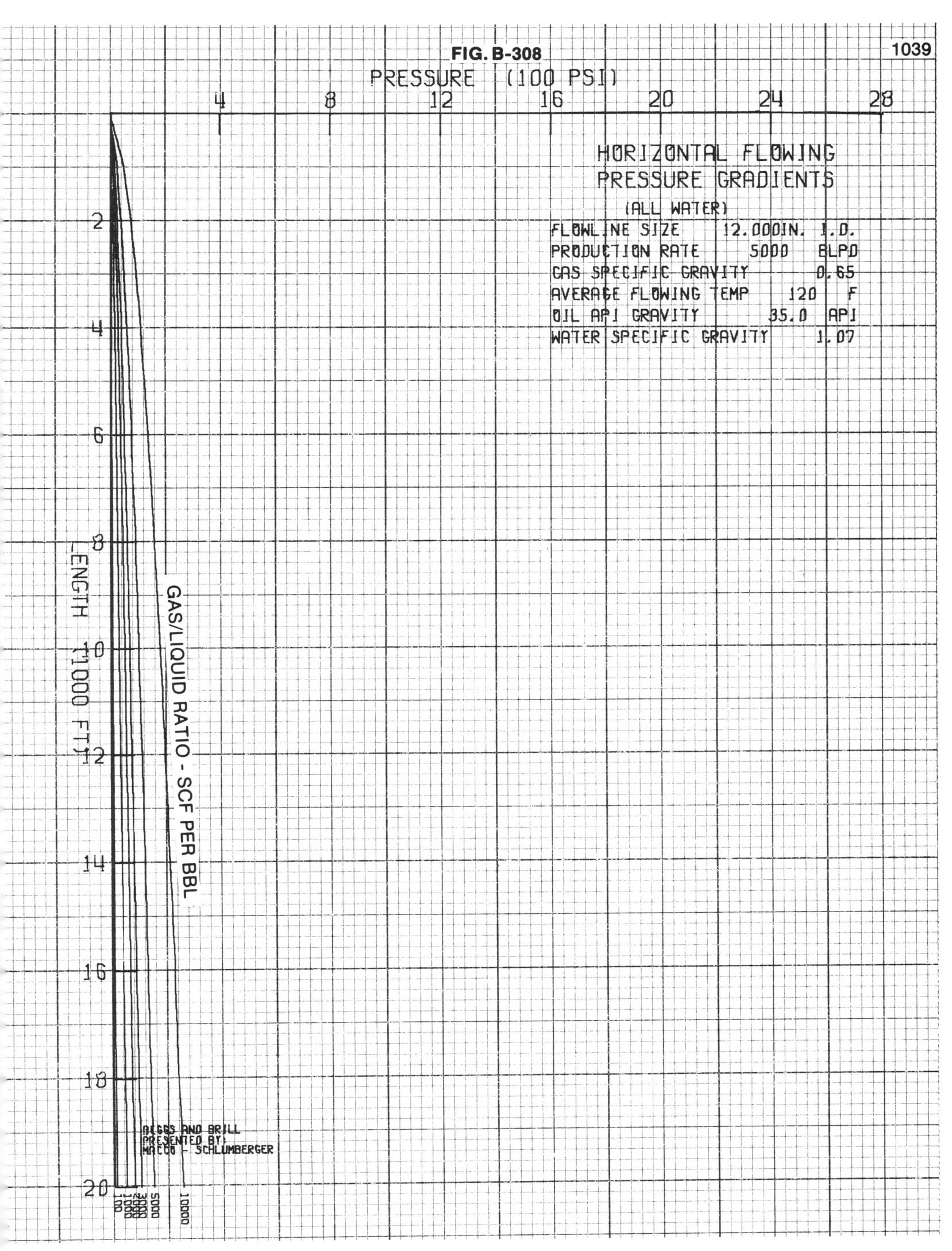

FIG. B-308

1039

PRESSURE (100 PSI)

HORIZONTAL FLOWING
PRESSURE GRADIENTS
(ALL WATER)

FLOWLINE SIZE   12.000IN. I.D.
PRODUCTION RATE   5000   BLPD
GAS SPECIFIC GRAVITY   0.65
AVERAGE FLOWING TEMP   120   F
OIL API GRAVITY   35.0   API
WATER SPECIFIC GRAVITY   1.07

LENGTH (1000 FT)

GAS/LIQUID RATIO - SCF PER BBL

BEGGS AND BRILL
PRESENTED BY:
MATCO - SCHLUMBERGER

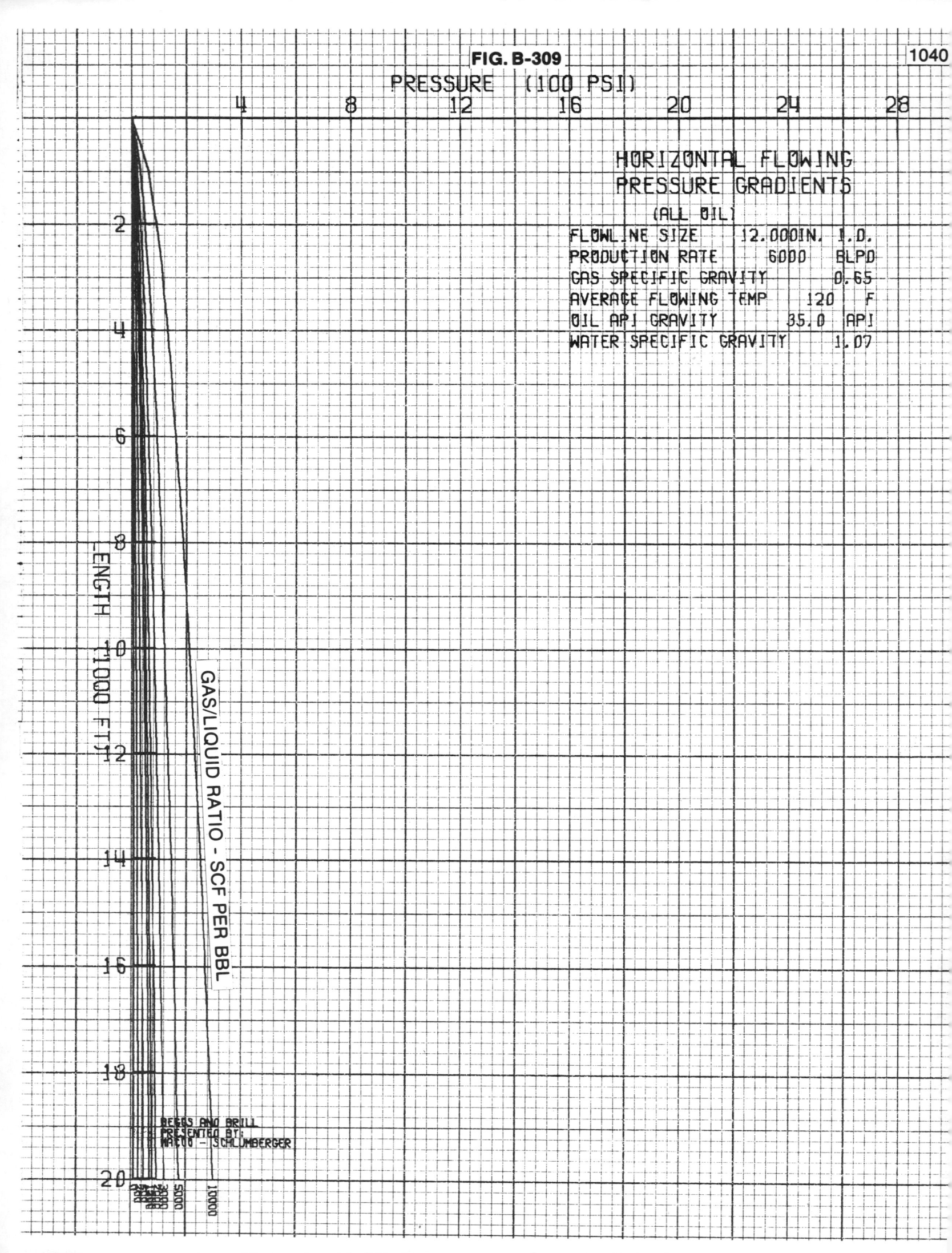

PRESSURE (100 PSI)

HORIZONTAL FLOWING
PRESSURE GRADIENTS
(ALL OIL)

FLOWLINE SIZE	12.000IN.	I.D.
PRODUCTION RATE	6000	BLPD
GAS SPECIFIC GRAVITY	0.65	
AVERAGE FLOWING TEMP	120	F
OIL API GRAVITY	35.0	API
WATER SPECIFIC GRAVITY	1.07	

LENGTH (1000 FT)

GAS/LIQUID RATIO - SCF PER BBL

BEGGS AND BRILL
PRESENTED BY:
WHITCO - SCHLUMBERGER

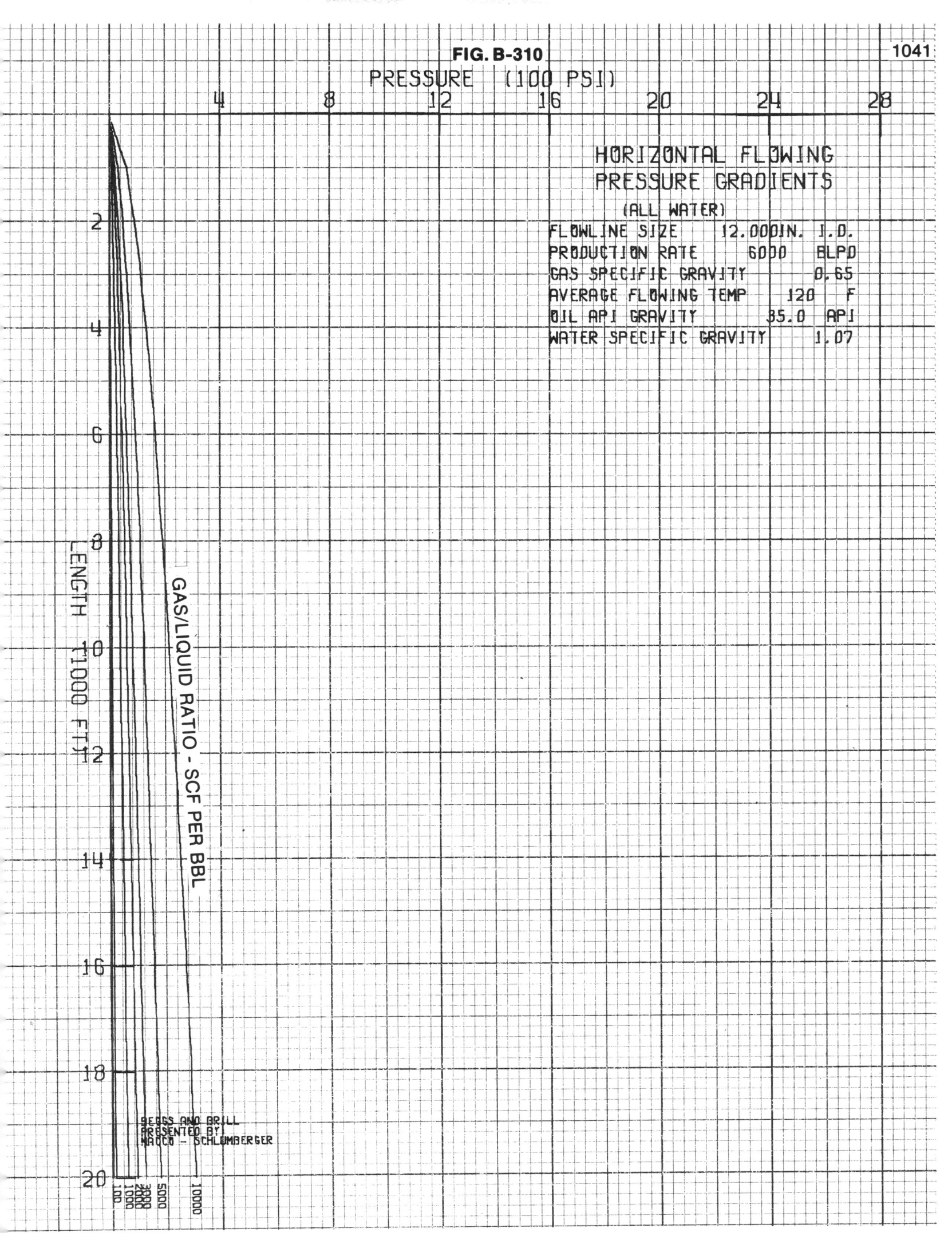

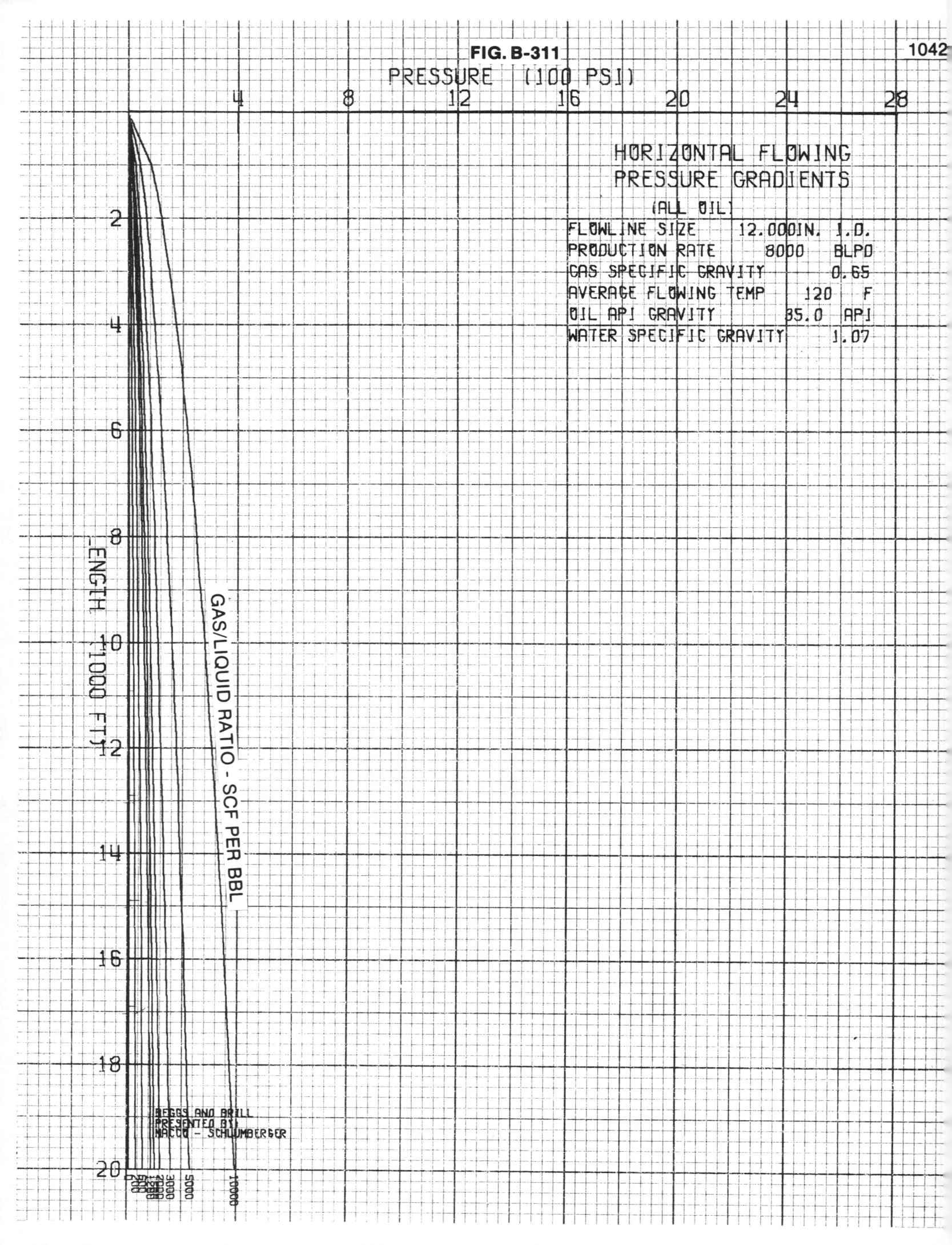

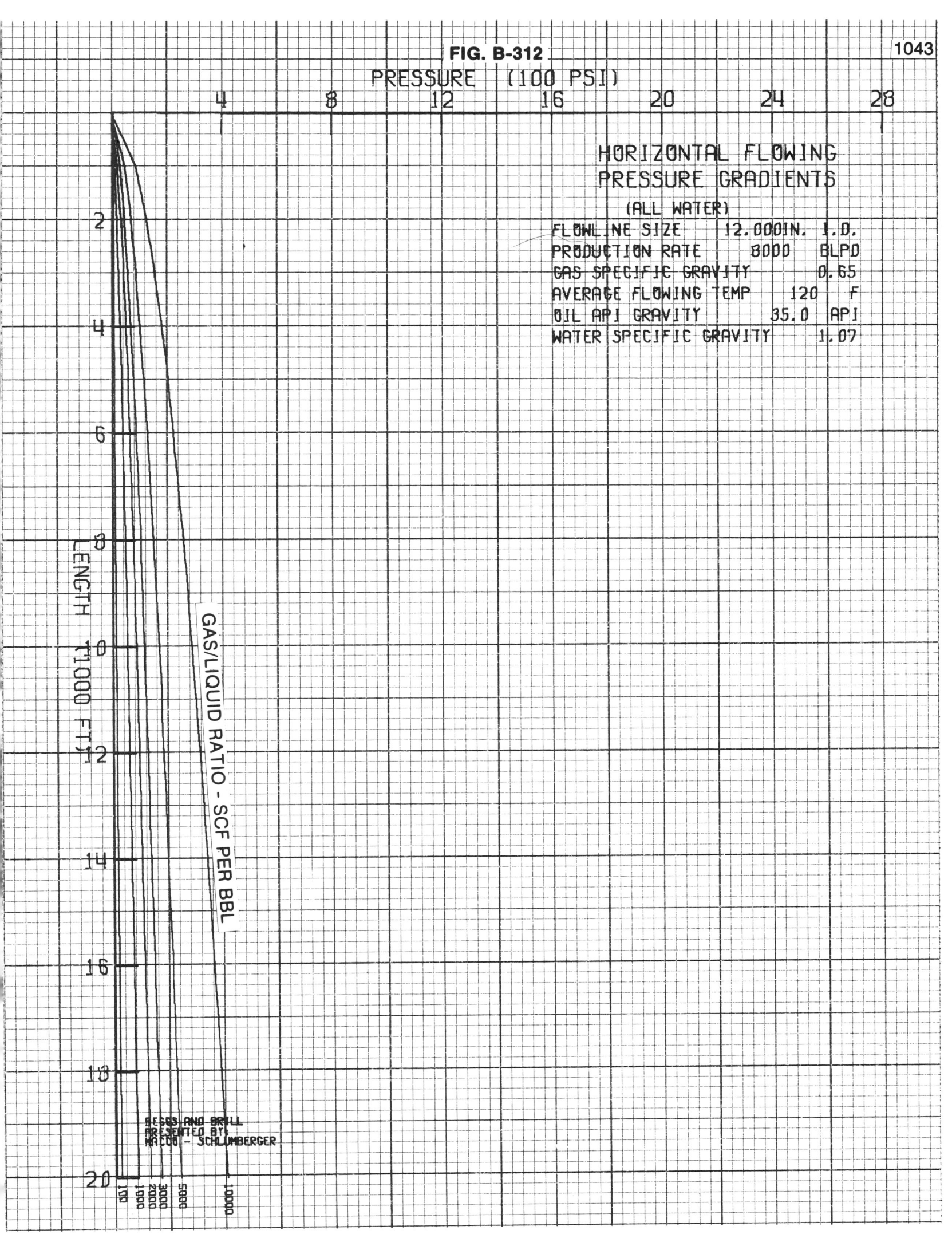

PRESSURE (100 PSI)

HORIZONTAL FLOWING
PRESSURE GRADIENTS
(ALL WATER)

FLOWLINE SIZE	12.000 IN. I.D.
PRODUCTION RATE	8000 BLPD
GAS SPECIFIC GRAVITY	0.65
AVERAGE FLOWING TEMP	120 F
OIL API GRAVITY	35.0 API
WATER SPECIFIC GRAVITY	1.07

LENGTH (1000 FT)

GAS/LIQUID RATIO - SCF PER BBL

BEGGS AND BRILL
PRESENTED BY
HACCO - SCHLUMBERGER

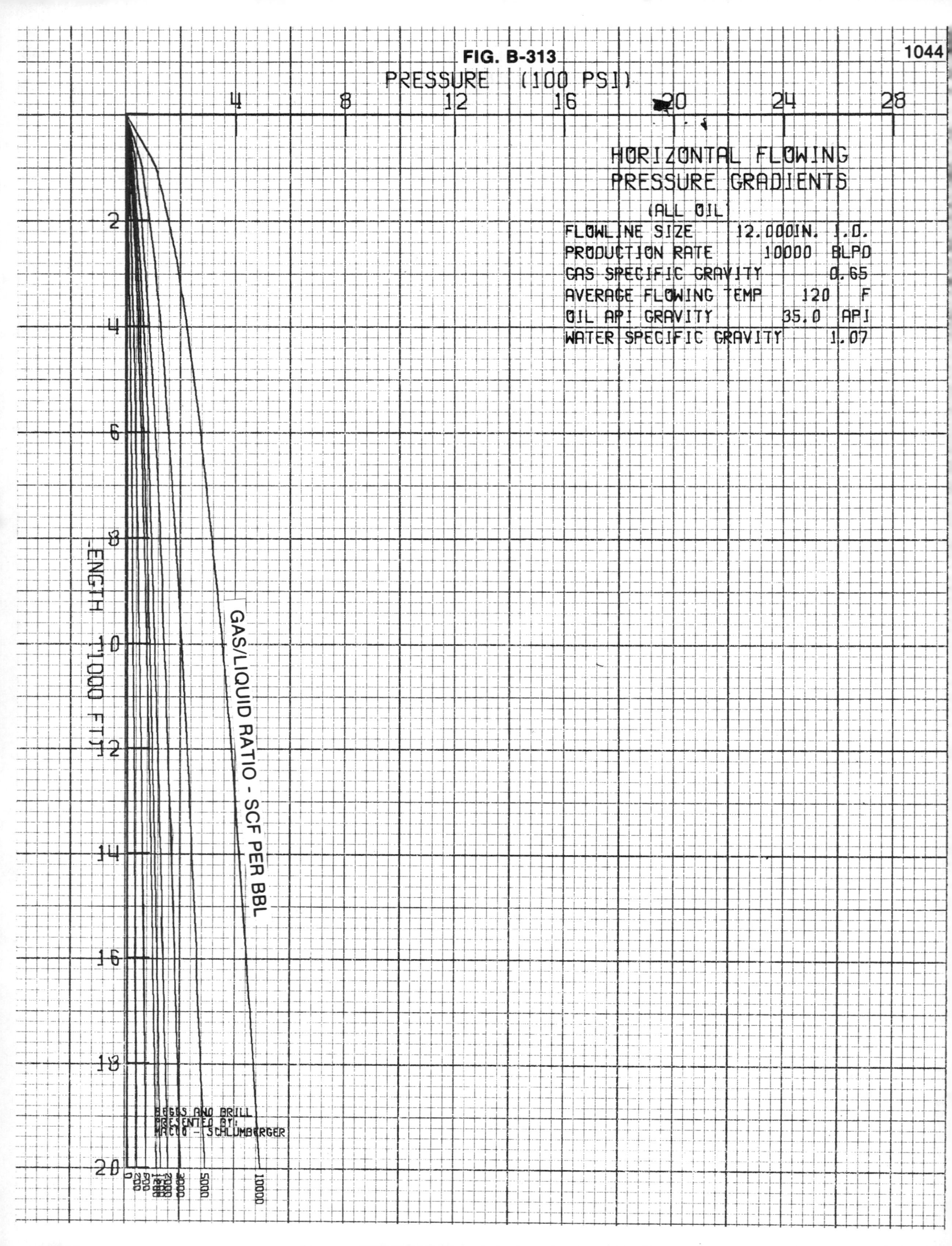

PRESSURE (100 PSI)

HORIZONTAL FLOWING
PRESSURE GRADIENTS
(ALL OIL)

FLOWLINE SIZE	12.000 IN. I.D.
PRODUCTION RATE	10000 BLPD
GAS SPECIFIC GRAVITY	0.65
AVERAGE FLOWING TEMP	120 F
OIL API GRAVITY	35.0 API
WATER SPECIFIC GRAVITY	1.07

LENGTH (1000 FT)

GAS/LIQUID RATIO - SCF PER BBL

BEGGS AND BRILL
PRESENTED BY:
MACIO - SCHLUMBERGER

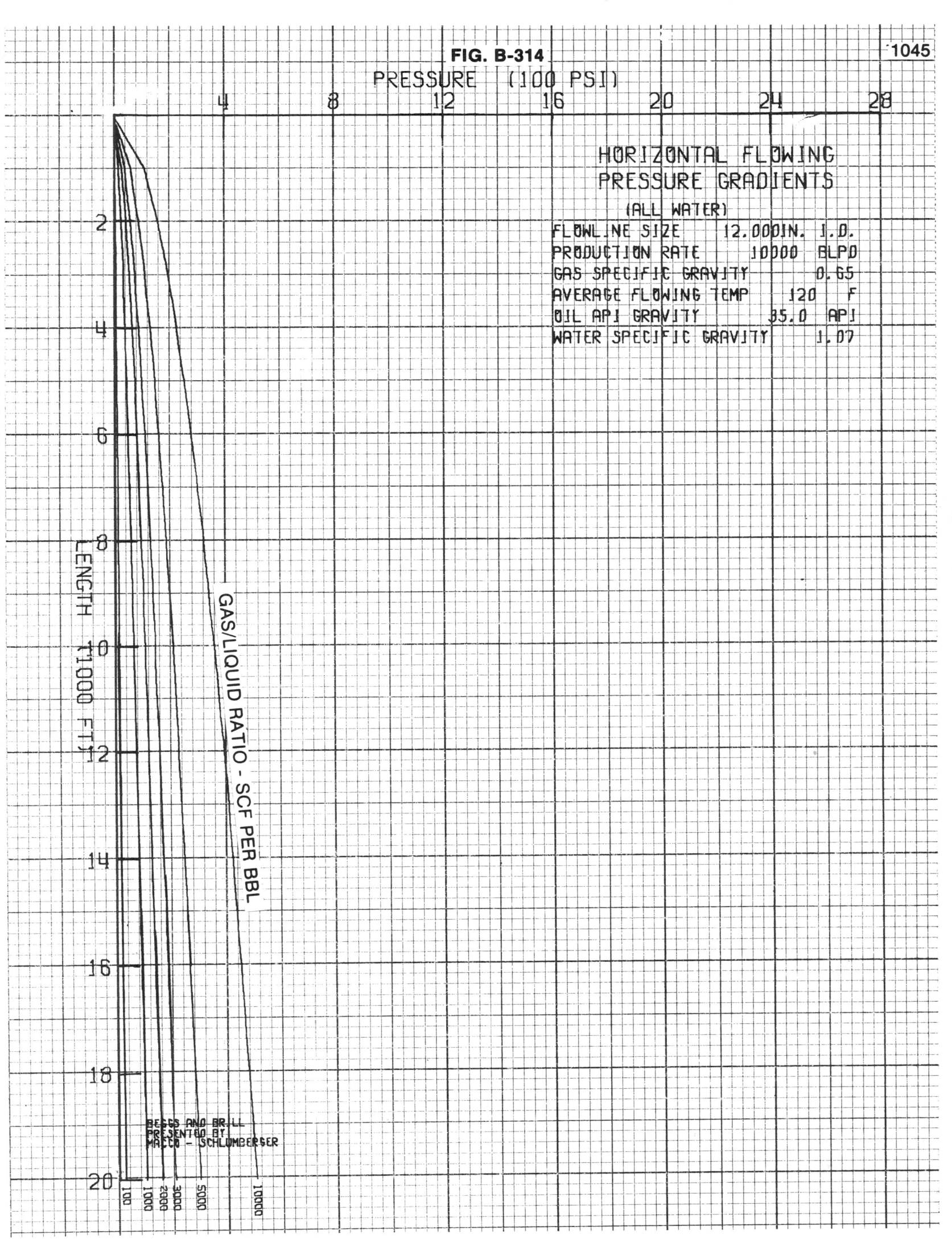

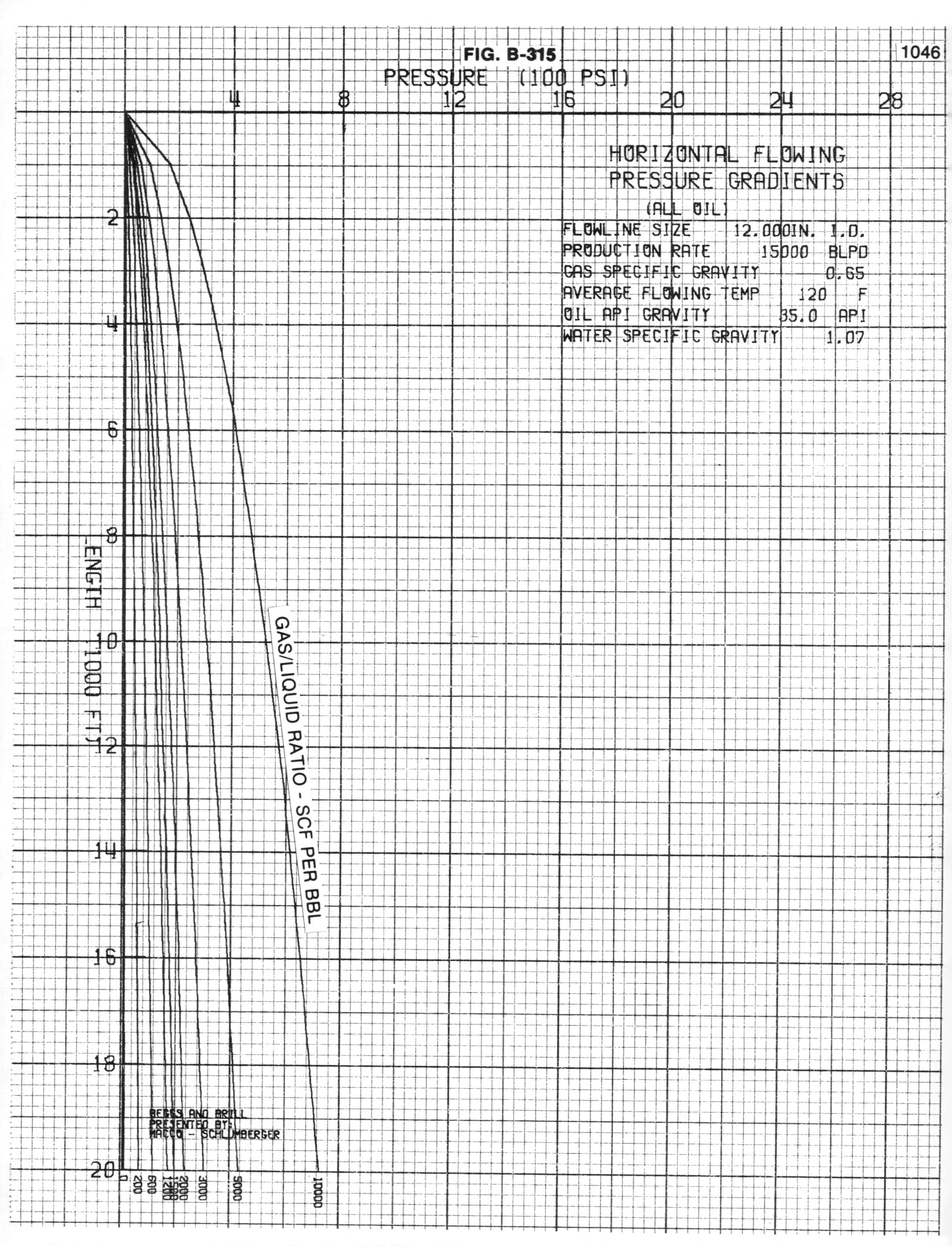

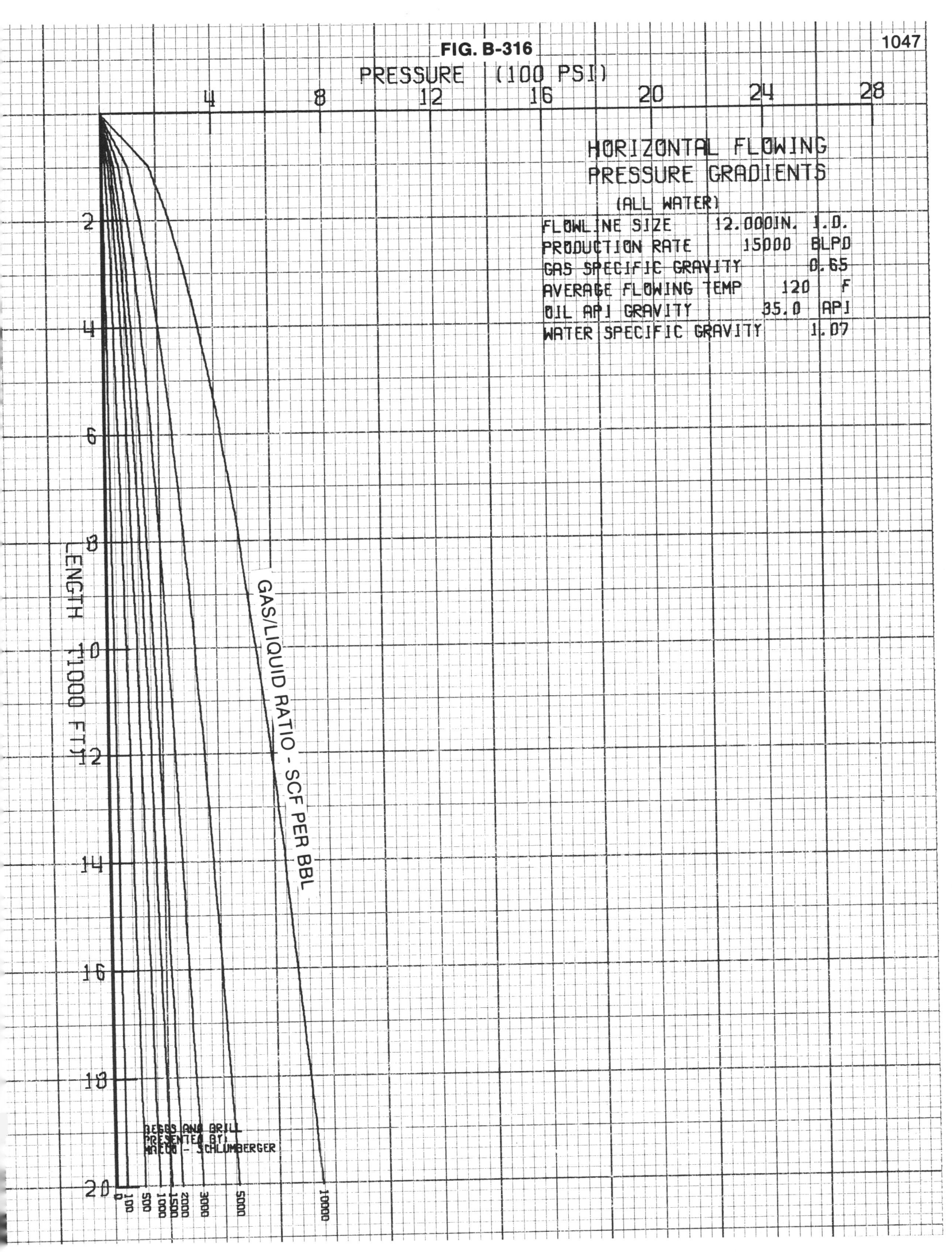

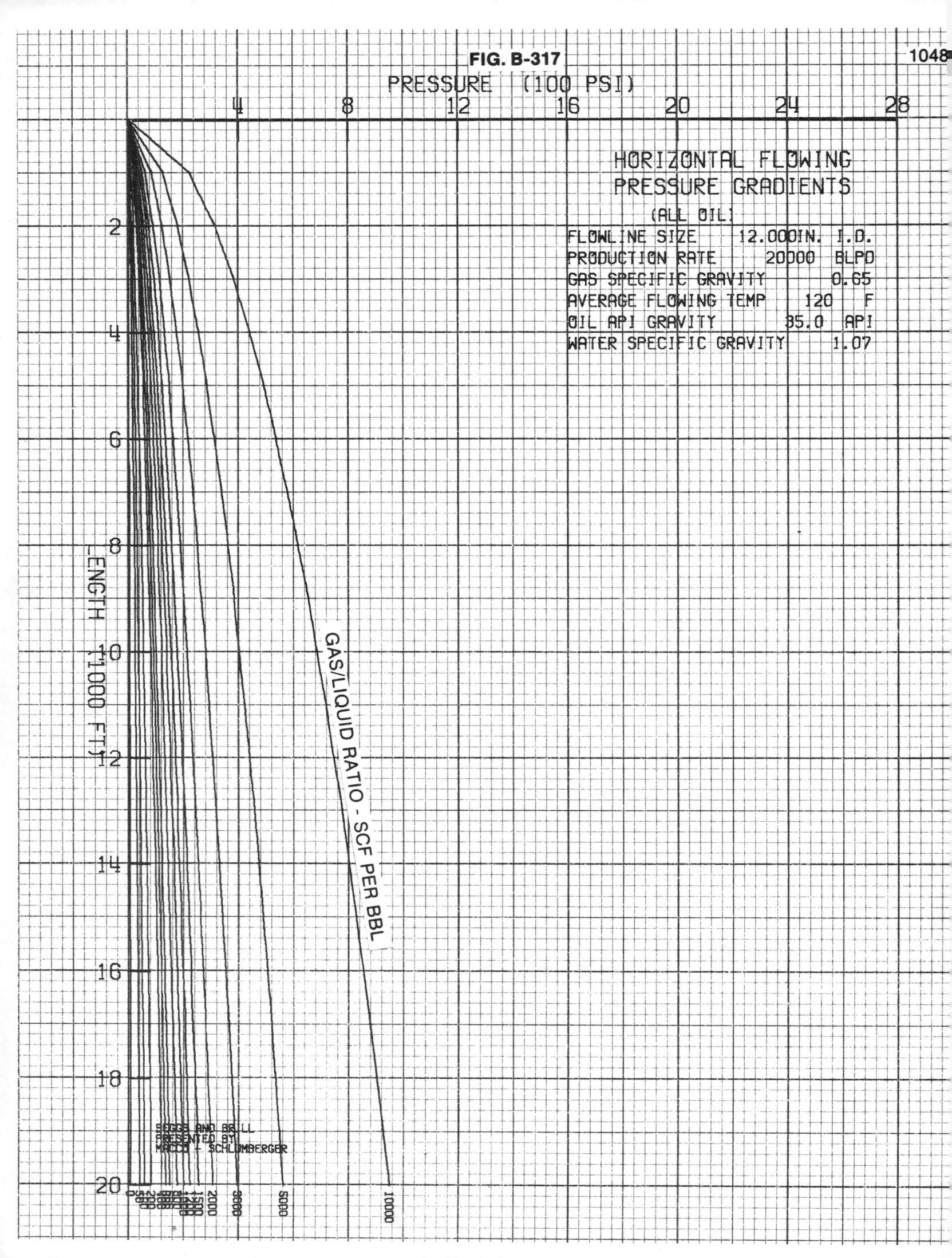

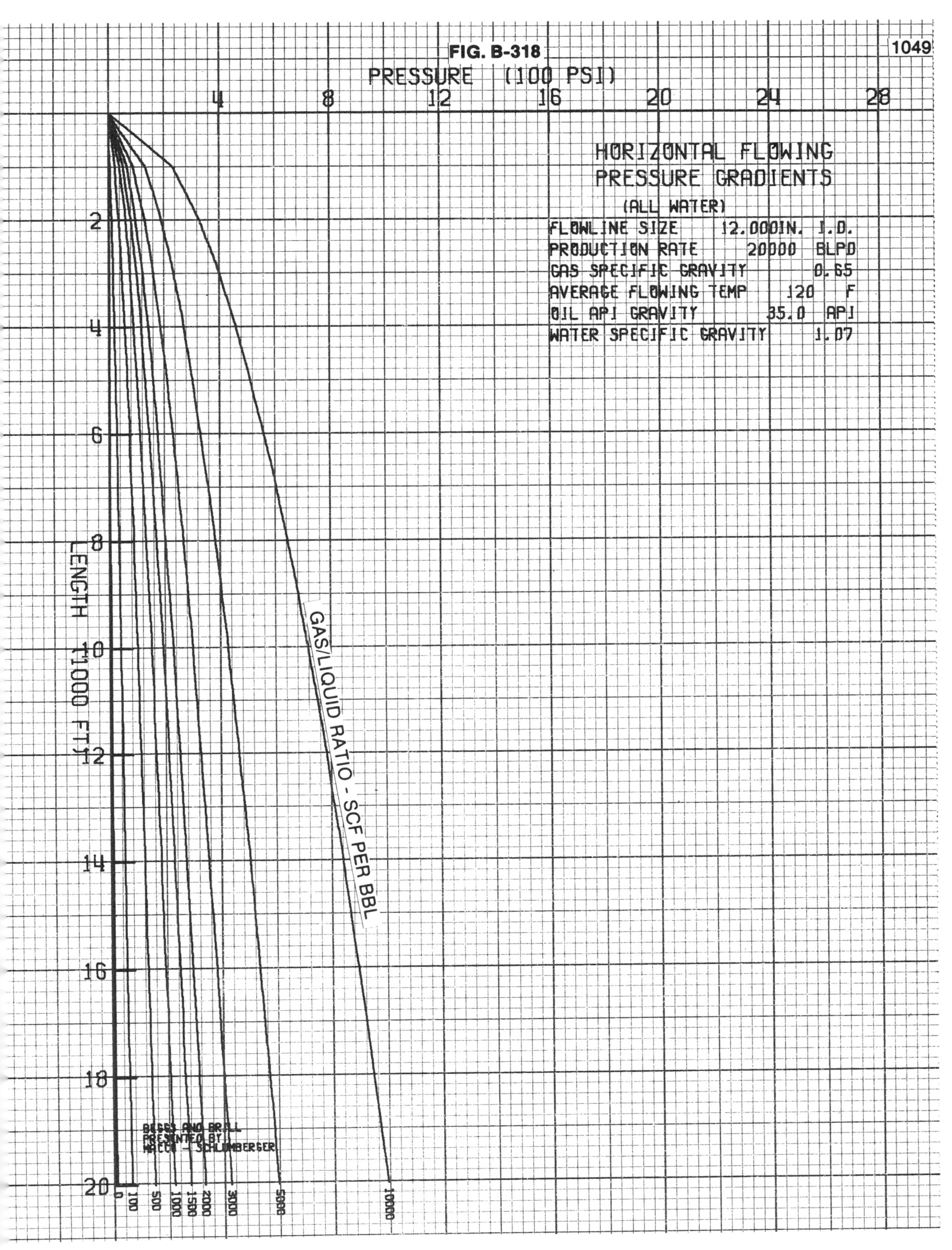

PRESSURE (100 PSI)

HORIZONTAL FLOWING
PRESSURE GRADIENTS
(ALL WATER)

FLOWLINE SIZE   12.000IN. I.D.
PRODUCTION RATE   20000 BLPD
GAS SPECIFIC GRAVITY   0.65
AVERAGE FLOWING TEMP   120   F
OIL API GRAVITY   35.0 API
WATER SPECIFIC GRAVITY   1.07

LENGTH (1000 FT.)

GAS/LIQUID RATIO - SCF PER BBL

BEGGS AND BRILL
PRESENTED BY
INTERA - SCHLUMBERGER

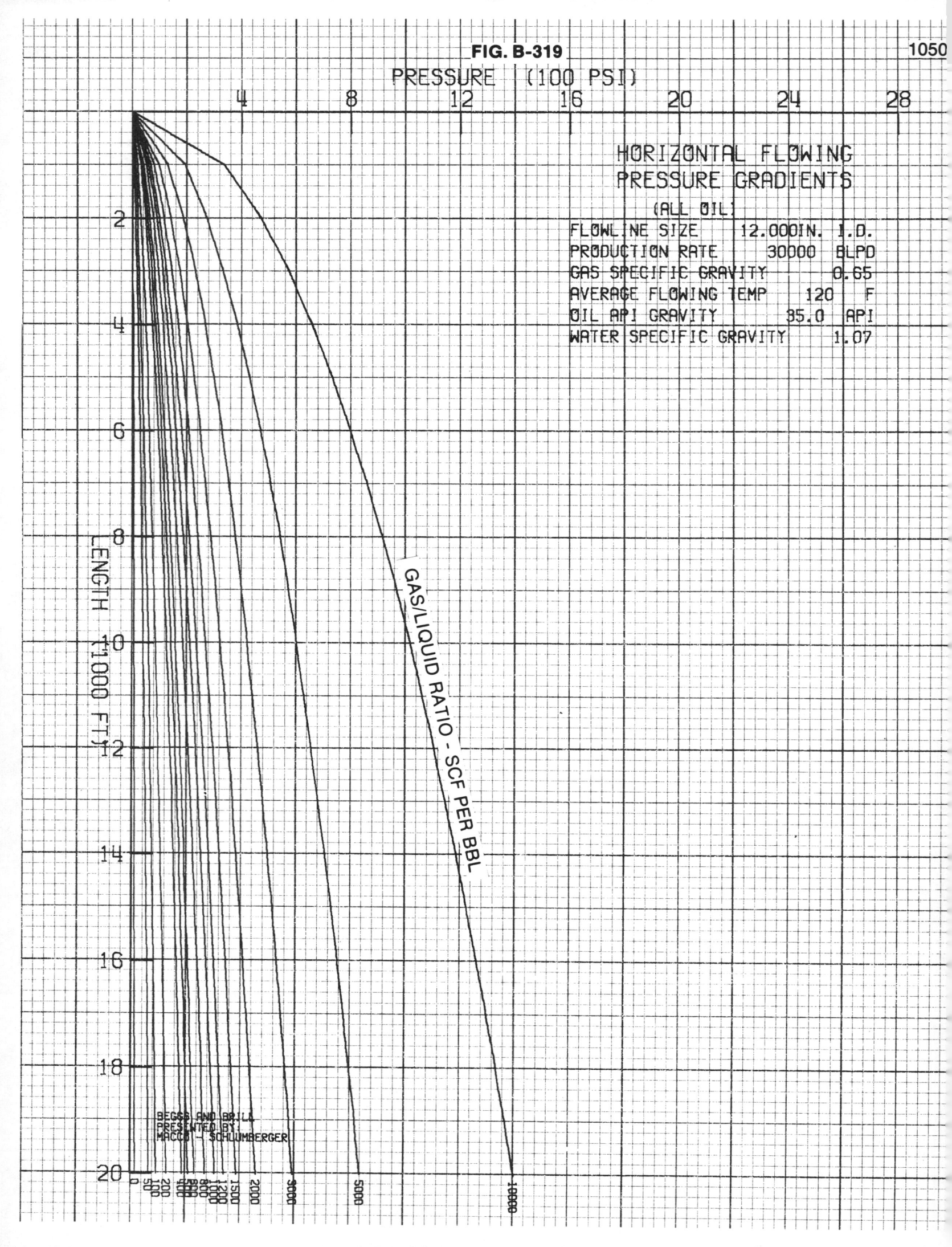

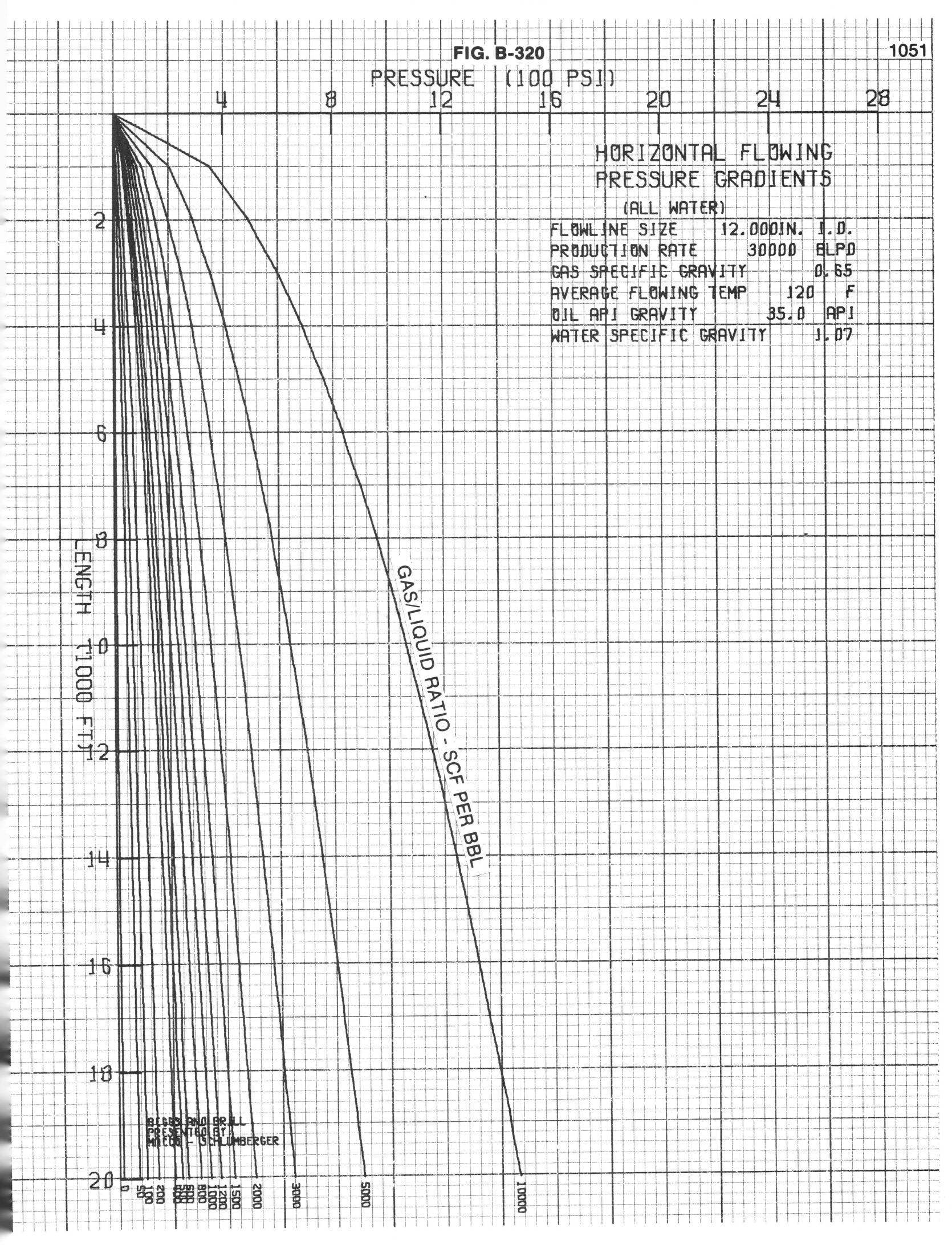

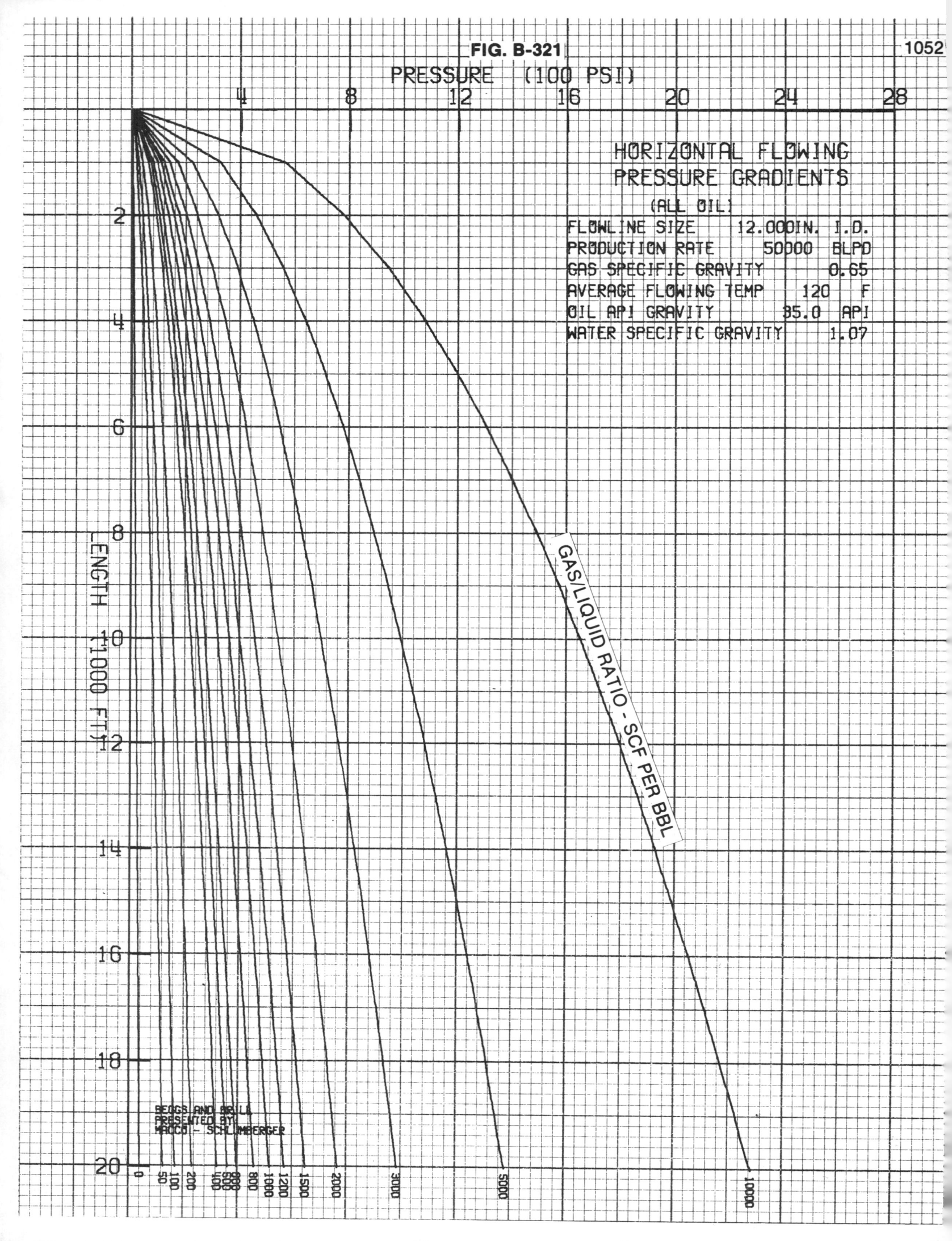

FIG. B-321

1052

PRESSURE (100 PSI)

HORIZONTAL FLOWING
PRESSURE GRADIENTS
(ALL OIL)

FLOWLINE SIZE	12.000IN. I.D.
PRODUCTION RATE	50000 BLPD
GAS SPECIFIC GRAVITY	0.65
AVERAGE FLOWING TEMP	120 F
OIL API GRAVITY	35.0 API
WATER SPECIFIC GRAVITY	1.07

GAS/LIQUID RATIO - SCF PER BBL

LENGTH (1000 FT)

BEGGS AND BRILL
PRESENTED BY
MADCO - SCHLUMBERGER

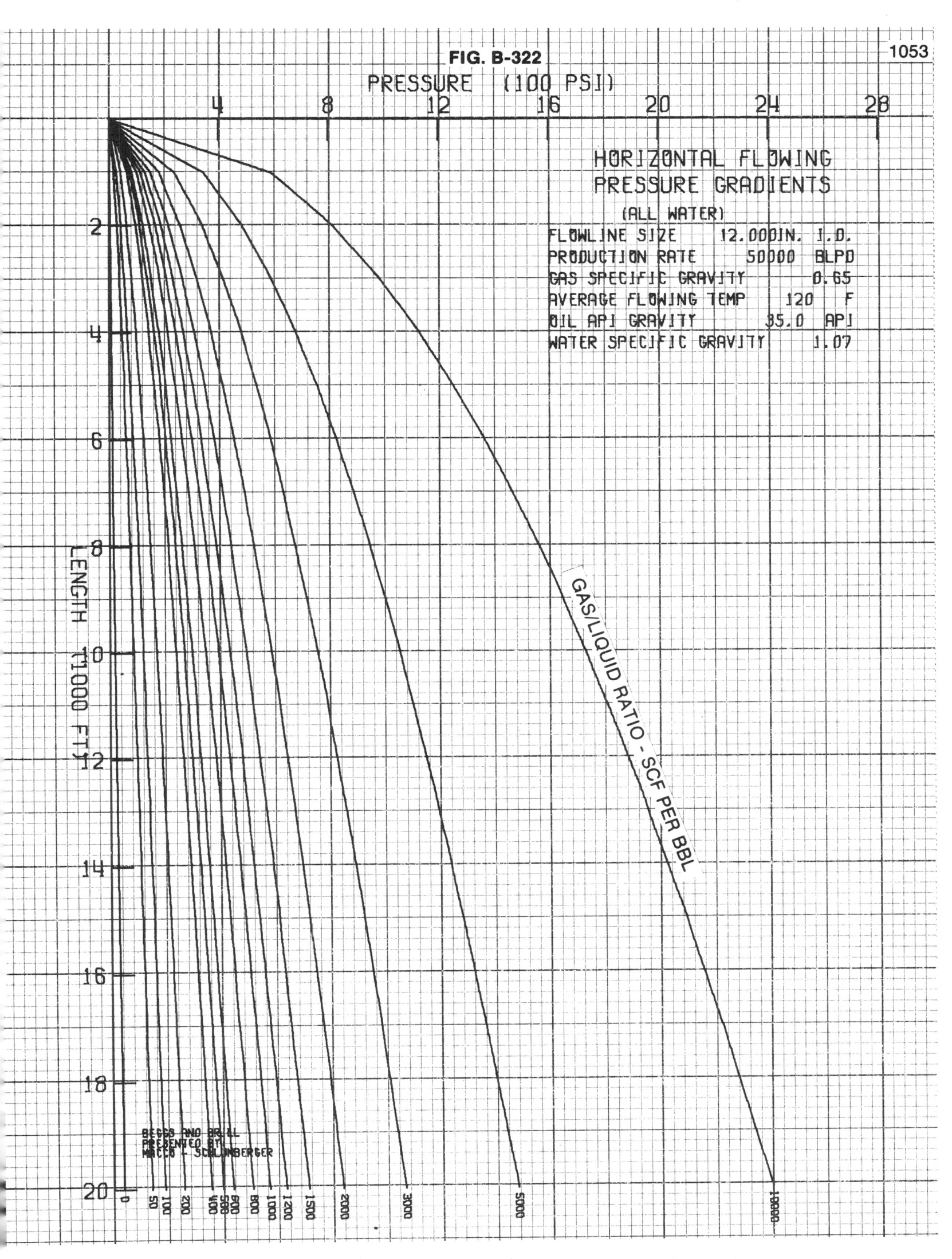

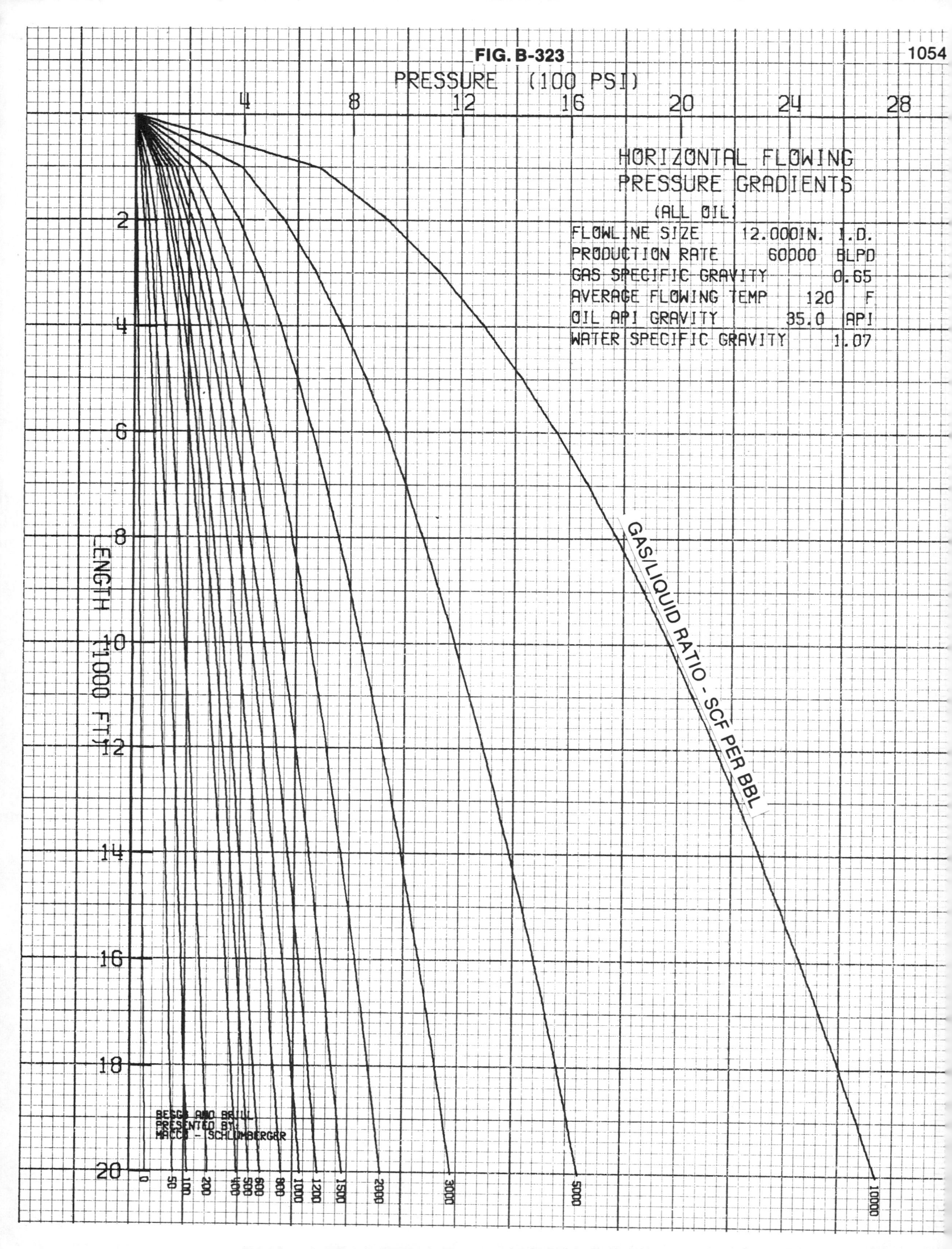

FIG. B-323

1054

PRESSURE (100 PSI)

HORIZONTAL FLOWING
PRESSURE GRADIENTS
(ALL OIL)

FLOWLINE SIZE	12.000IN. I.D.
PRODUCTION RATE	60000 BLPD
GAS SPECIFIC GRAVITY	0.65
AVERAGE FLOWING TEMP	120 F
OIL API GRAVITY	35.0 API
WATER SPECIFIC GRAVITY	1.07

GAS/LIQUID RATIO - SCF PER BBL

LENGTH (1000 FT)

BEGGS AND BRILL
PRESENTED BY:
MACCO - SCHLUMBERGER

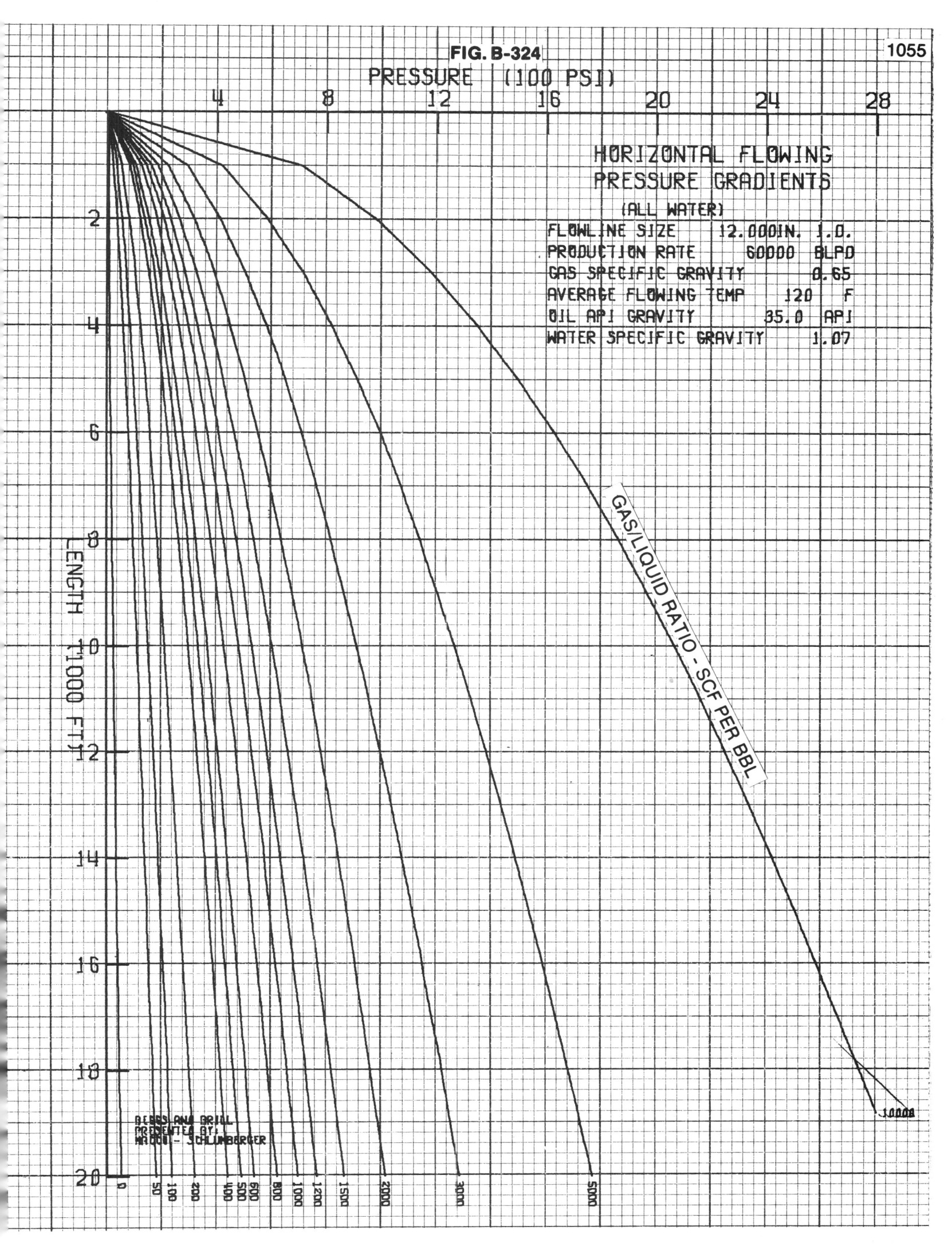

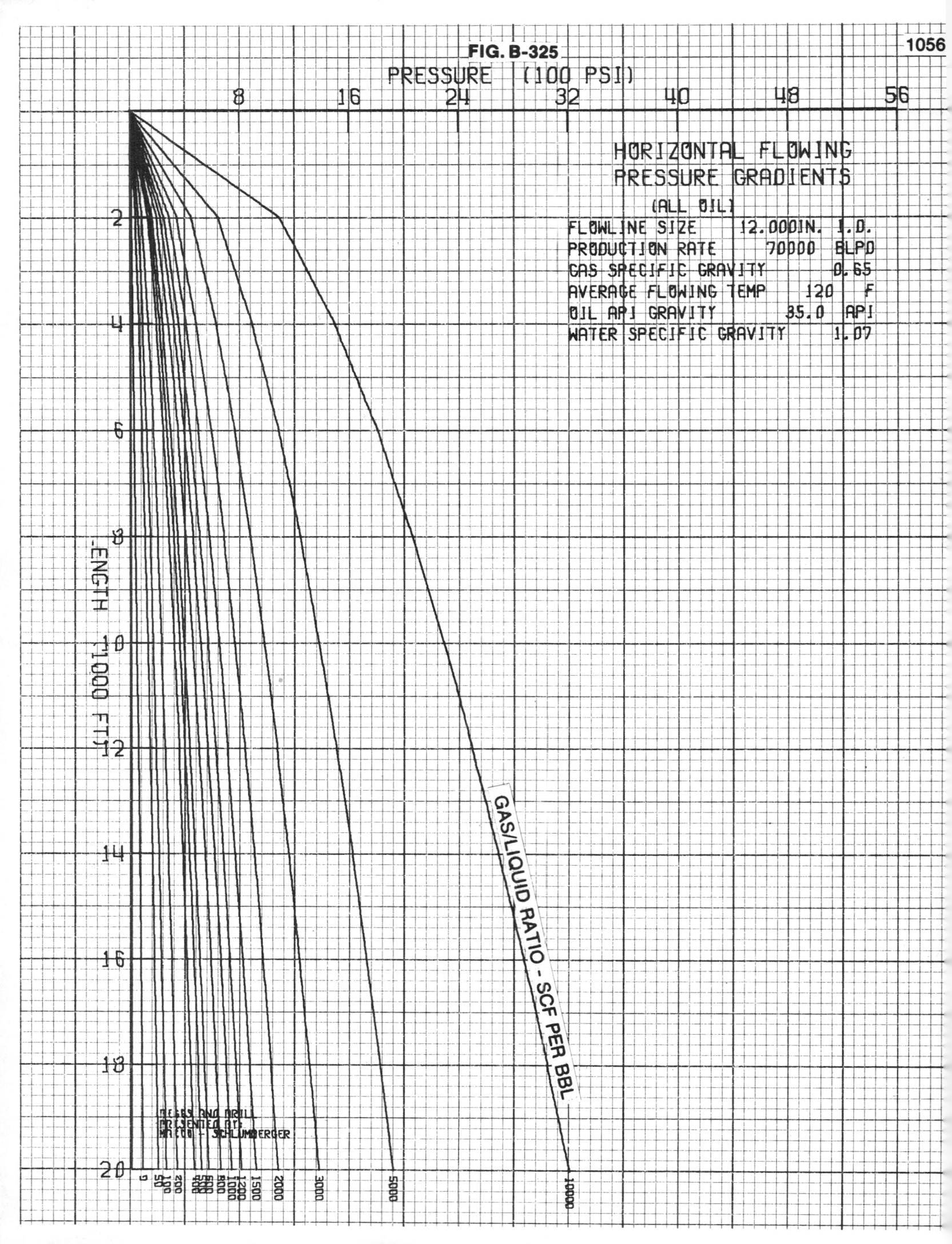

FIG. B-325

1056

PRESSURE (100 PSI)

HORIZONTAL FLOWING
PRESSURE GRADIENTS
(ALL OIL)

FLOWLINE SIZE	12.000 IN. I.D.
PRODUCTION RATE	70000 BLPD
GAS SPECIFIC GRAVITY	0.65
AVERAGE FLOWING TEMP	120 F
OIL API GRAVITY	35.0 API
WATER SPECIFIC GRAVITY	1.07

LENGTH (1000 FT.)

GAS/LIQUID RATIO - SCF PER BBL

TESES AND BRILL
PRESENTED AT:
HALCO - SCHLUMBERGER

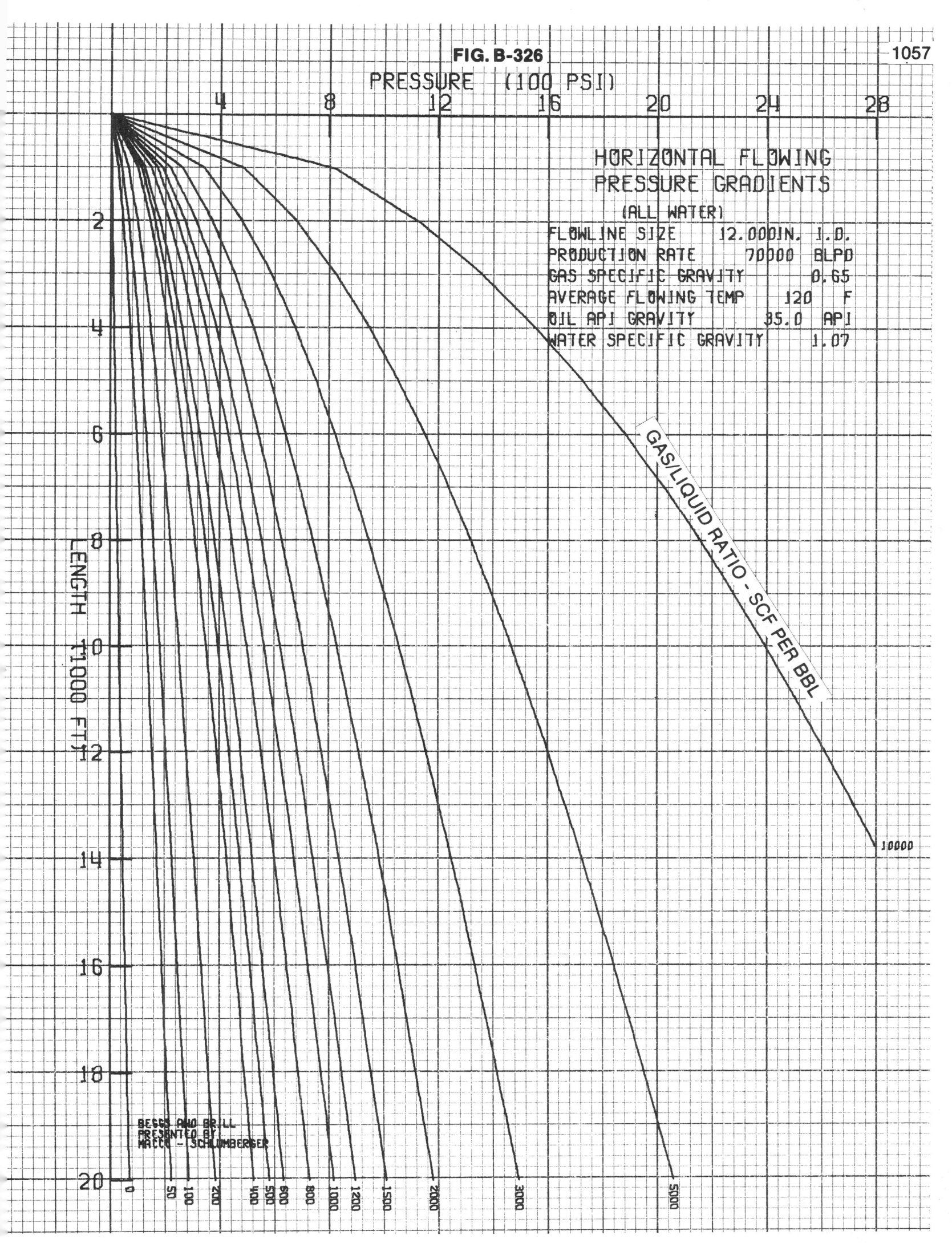

PRESSURE (100 PSI)

HORIZONTAL FLOWING
PRESSURE GRADIENTS
(ALL WATER)

FLOWLINE SIZE	12.000IN.	I.D.
PRODUCTION RATE	70000	BLPD
GAS SPECIFIC GRAVITY	0.65	
AVERAGE FLOWING TEMP	120	F
OIL API GRAVITY	35.0	API
WATER SPECIFIC GRAVITY	1.07	

GAS/LIQUID RATIO - SCF PER BBL

LENGTH (1000 FT)

BEGGS AND BRILL
PRESENTED BY
MACCO - SCHLUMBERGER

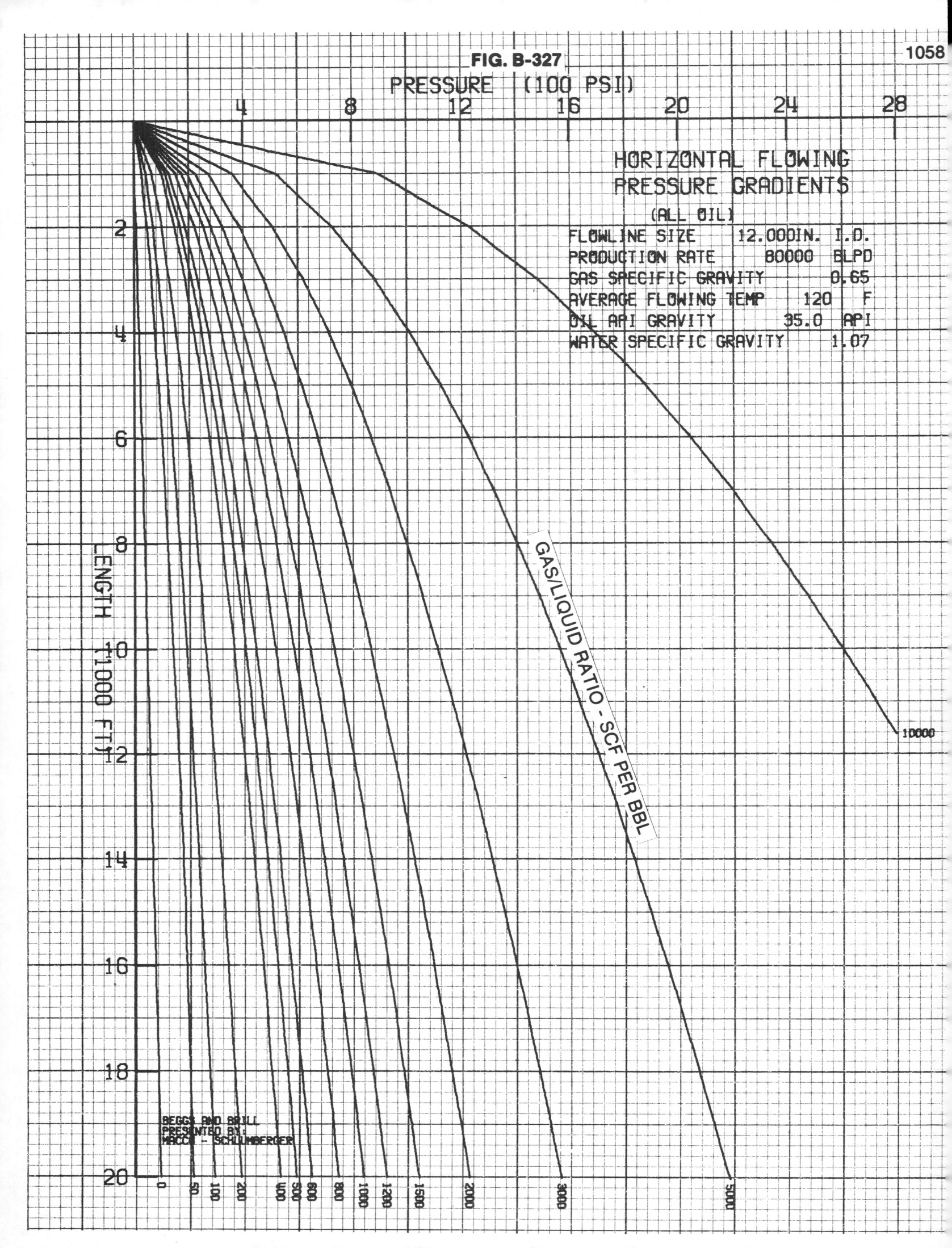

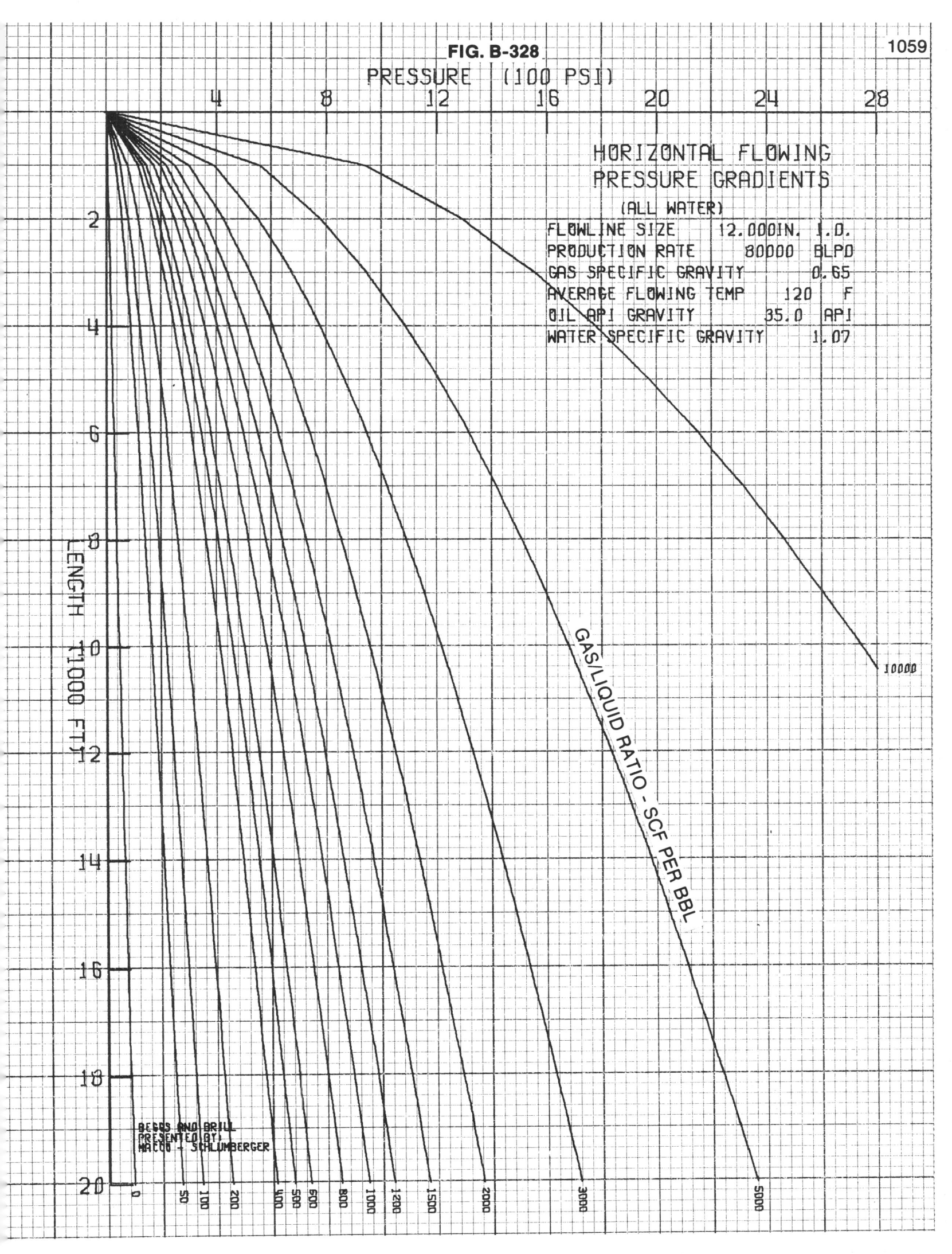

FIG. B-328

1059

PRESSURE (100 PSI)

HORIZONTAL FLOWING
PRESSURE GRADIENTS
(ALL WATER)

FLOWLINE SIZE	12.000IN. I.D.
PRODUCTION RATE	80000 BLPD
GAS SPECIFIC GRAVITY	0.65
AVERAGE FLOWING TEMP	120 F
OIL API GRAVITY	35.0 API
WATER SPECIFIC GRAVITY	1.07

LENGTH (1000 FT)

GAS/LIQUID RATIO - SCF PER BBL

BEGGS AND BRILL
PRESENTED BY:
HALCO - SCHLUMBERGER

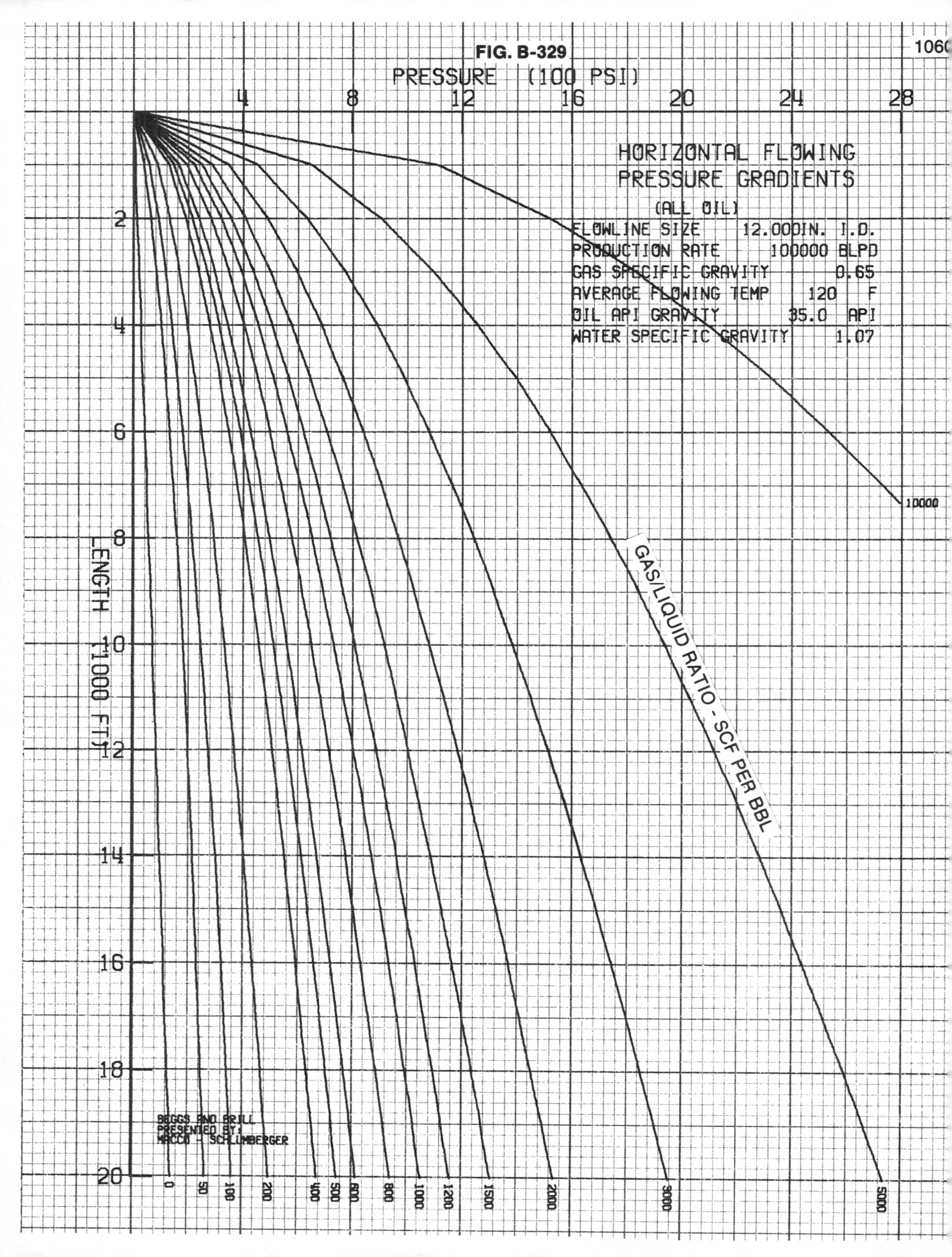

FIG. B-329

1060

PRESSURE (100 PSI)

HORIZONTAL FLOWING
PRESSURE GRADIENTS
(ALL OIL)

FLOWLINE SIZE   12.000IN. I.D.
PRODUCTION RATE   100000 BLPD
GAS SPECIFIC GRAVITY   0.65
AVERAGE FLOWING TEMP   120   F
OIL API GRAVITY   35.0 API
WATER SPECIFIC GRAVITY   1.07

LENGTH (1000 FT)

GAS/LIQUID RATIO - SCF PER BBL

BEGGS AND BRILL
PRESENTED BY:
KROCO - SCHLUMBERGER

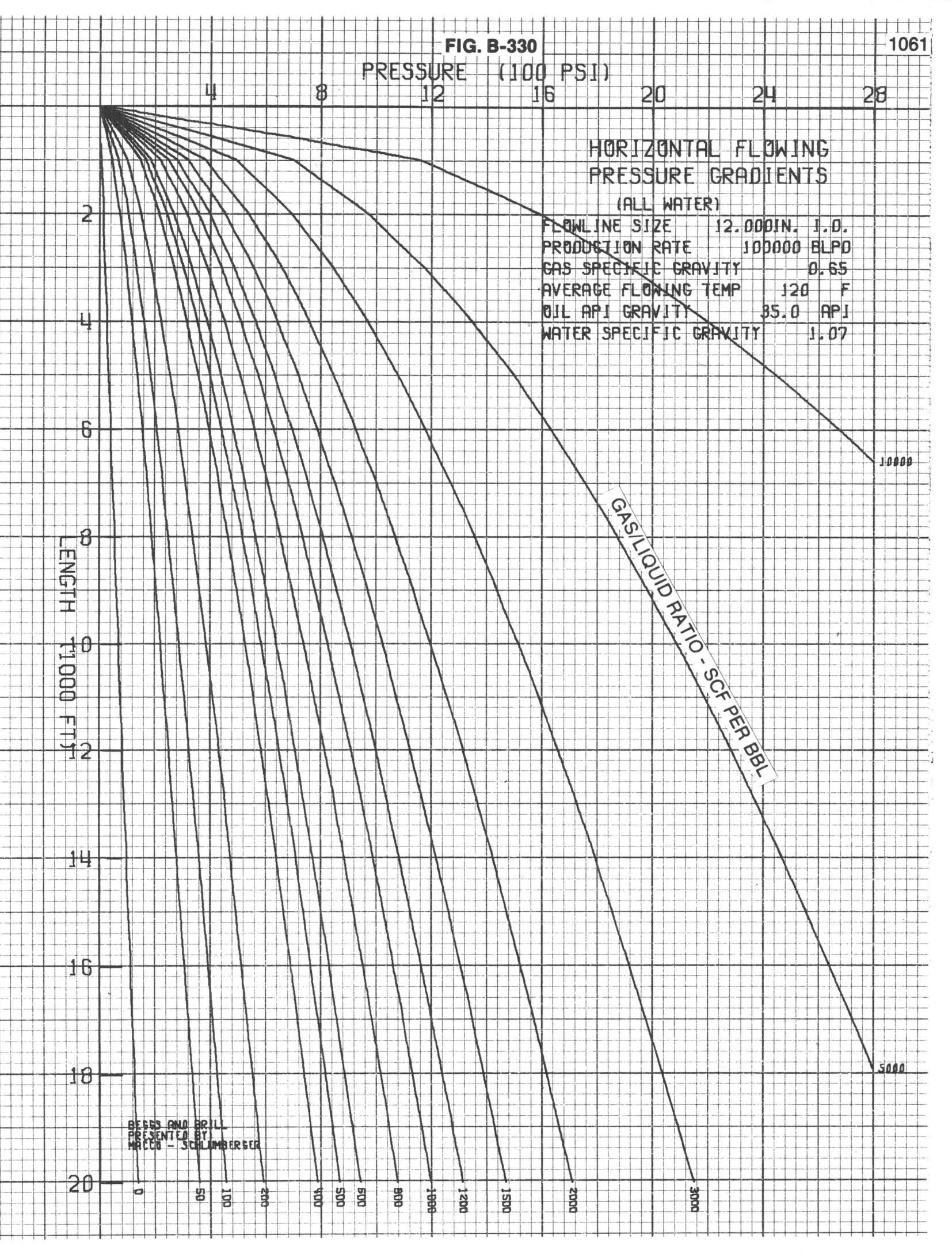

FIG. B-330

HORIZONTAL FLOWING
PRESSURE GRADIENTS
(ALL WATER)

FLOWLINE SIZE   12.000IN. I.D.
PRODUCTION RATE   100000 BLPD
GAS SPECIFIC GRAVITY   0.65
AVERAGE FLOWING TEMP   120   F
OIL API GRAVITY   35.0 API
WATER SPECIFIC GRAVITY   1.07

GAS/LIQUID RATIO - SCF PER BBL

BEGGS AND BRILL
PRESENTED BY
MAECO - SCHLUMBERGER